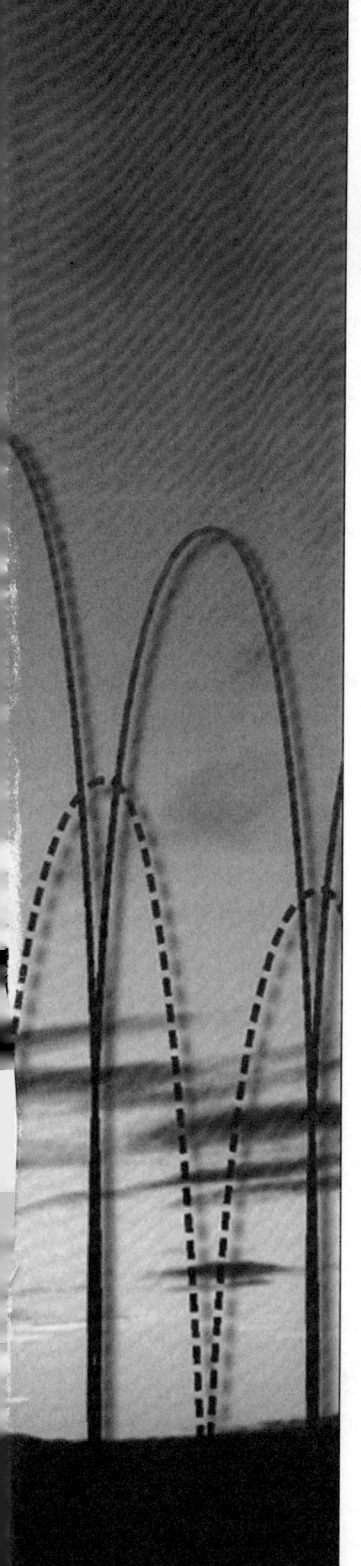

THE COMMUNICATIONS HANDBOOK

The Electrical Engineering Handbook Series

Series Editor
Richard C. Dorf
University of California, Davis

Titles Included in the Series
The Electrical Engineering Handbook, Richard C. Dorf
The Biomedical Engineering Handbook, Joseph D. Bronzino
The Circuits and Filters Handbook, Wai-Kai Chen
The Transforms and Applications Handbook, Alexander D. Poularikas
The Control Handbook, William S. Levine
The Electronics Handbook, Jerry C. Whitaker
The Industrial Electronics Handbook, J. David Irwin
The Communications Handbook, Jerry D. Gibson
The Mobile Communications Handbook, Jerry D. Gibson

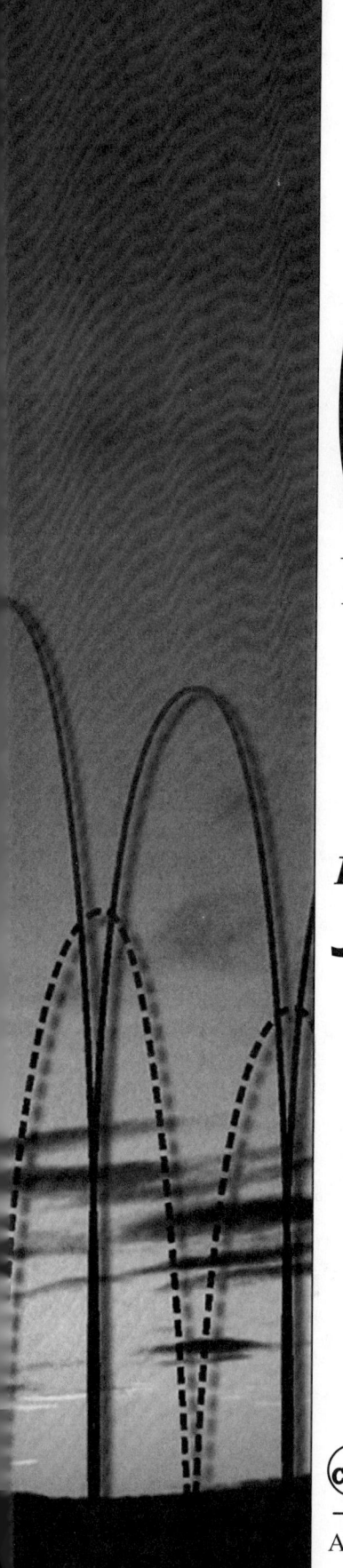

THE COMMUNICATIONS HANDBOOK

Editor-in-Chief

Jerry D. Gibson

CEDAR RAPIDS
INFORMATION CENTER
COLLINS DIVISIONS
ROCKWELL INTERNATIONAL

 CRC PRESS   

A CRC Handbook Published in Cooperation with IEEE Press

Library of Congress Cataloging-in-Publication Data

The communications handbook / editor-in-chief, Jerry D. Gibson.
      p.   cm. -- (Electrical engineering handbook series)
   Includes bibliographical references and index.
   ISBN 0-8493-8349-8 (alk. paper)
   1. Telecommunication--Handbooks, manuals, etc.   I. Gibson, Jerry
D. II. Series.
TK5101.C6583 1996
621.382--dc20   96-27623
                                                                                                                                               CIP

    This book contains information obtained from authentic and highly regarded sources. Reprinted material is quoted with permission, and sources are indicated. A wide variety of references are listed. Reasonable efforts have been made to publish reliable data and information, but the author and the publisher cannot assume responsibility for the validity of all materials or for the consequences of their use.
    Neither this book nor any part may be reproduced or transmitted in any form or by any means, electronic or mechanical, including photocopying, microfilming, and recording, or by any information storage or retrieval system, without prior permission in writing from the publisher.
    All rights reserved. Authorization to photocopy items for internal or personal use, or the personal or internal use of specific clients, may be granted by CRC Press, Inc., provided that $.50 per page photocopied is paid directly to Copyright Clearance Center, 27 Congress Street, Salem, MA 01970 USA. The fee code for users of the Transactional Reporting Service is ISBN 0-8493-8349-8/97/$0.00+$.50. The fee is subject to change without notice. For organizations that have been granted a photocopy license by the CCC, a separate system of payment has been arranged.
    The consent of CRC Press does not extend to copying for general distribution, for promotion, for creating new works, or for resale. Specific permission must be obtained in writing from CRC Press for such copying.
    Direct all inquiries to CRC Press, Inc., 2000 Corporate Blvd., N.W., Boca Raton, Florida 33431.

© 1997 by CRC Press, Inc.

No claim to original U.S. Government works
International Standard Book Number 0-8493-8349-8
Library of Congress Card Number 96-27623
Printed in the United States of America 1 2 3 4 5 6 7 8 9 0
Printed on acid-free paper

Preface

The Handbook series published by CRC Press represents a truly unique approach to disseminating technical information. Starting with the first edition of *The Electrical Engineering Handbook* edited by Richard Dorf and published in 1993, this series is dedicated to the idea that a reader should be able to pull one of these handbooks off the shelf and at least 80% of the time find what they need to know about a subject area. As handbooks, these books are also different in that they are more than just a dry listing of facts and data, filled mostly with tables. In fact, a hallmark of these handbooks is that the articles or chapters are designed to be relatively short, written as tutorials or overviews, so that once the reader locates the broader topic, it is easy to find an answer to a specific question.

At the inception of this project, I prepared an outline containing the main headings and chapter titles, about 135 chapters. With the counsel of the Advisory Board, this first table of contents was refined and I began to seek authors. My approach here was fairly simple—ask the best people I know. I then had the opportunity to talk to these very active people about their field. Sometimes as we spoke, prospective authors would indicate that a particular chapter topic or two included in the original outline could be eliminated or combined with another, or was being superceded by later developments. In response, the contents were modified accordingly.

Of course, the authors are the key to achieving the overall goal of the handbook, and having read all of the chapters personally, the results are impressive. The chapters are authoritative, to-the-point, and enjoyable to read. Answers to frequently asked questions, facts, and figures are available almost at a glance. Since the authors are experts in their field, it is understandable that the content is excellent. Additionally, the authors were encouraged to put some of their own interpretations and insights into the chapters, which greatly enhances the readability. However, I am most impressed by the ability of the authors to condense so much information into so few pages. These chapters are unlike any most people have ever written—they are not research journal articles, they are not textbooks, they are not long tutorial review articles for magazines. They really are a new format.

In reading drafts of the chapters, I applied two tests. If the chapter covered a topic that I felt familiar with, I checked to see if it contained what I thought were the essential facts and ideas. If the chapter was in an area of communications less familiar to me, I looked for definitions of terms that I had heard of, or for a discussion that brought forth to me why this area is important and what is happening in the field today. I was amazed at what I learned. You will be too.

Using *The Communications Handbook* is simple. Look up the topic of interest to you either in the Table of Contents or in the Index. Go directly to that chapter. As you are reading the chosen chapter, you may wish to refer to other chapters in the same main heading. If you need some background information in the general communications field, you need only consult the Basic Principles section. There is no need to read the *Handbook* beginning-to-end, start-to-finish. Look up what you need right now, read it, and go back to the task at hand.

The pleasure of such a project as this is in working with the authors, and I am gratified to have had this opportunity and to be associated with each of them. Additionally, Mary Kugler at CRC Press and the personnel at TechBooks have been extraordinarily helpful and patient as the handbook evolved and was finally brought to fruition. Elaine M. Gibson served as Managing Editor of the *Handbook*, working directly with the authors, the publisher, and the production staff, using her creative talents and enthusiasm to improve operations, solve problems, and make each step of the project as seamless as possible for everyone involved. Elaine and I also acknowledge our editors at CRC Press, first Joel Claypool, and now Kristen Maus, for their insights, guidance, and encouragement in making *The Communications Handbook* a reality.

Jerry D. Gibson
College Station, Texas

Editor-in-Chief

Jerry D. Gibson currently holds the J.W. Runyon, Jr. Professorship in the Department of Electrical Engineering at Texas A&M University. He has held positions at General Dynamics—Fort Worth (1969–72), the University of Notre Dame (1973–74), and the University of Nebraska—Lincoln (1974–76), and during the fall of 1991, Dr. Gibson was on sabbatical with the Information Systems Laboratory and the Telecommunications Program in the Department of Electrical Engineering at Stanford University.

He is coauthor of the book *Introduction to Nonparametric Detection with Applications* (Academic Press, 1975 and IEEE Press, 1995) and author of the textbook, *Principles of Digital and Analog Communications* (Prentice-Hall, second ed., 1993). He was Associate Editor for Speech Processing for the *IEEE Transactions on Communications* from 1981–1985 and Associate Editor for Communications for the *IEEE Transactions on Information Theory* from 1988–1991. He has served as a member of the Speech Technical Committee of the IEEE Signal Processing Society (1992–1995) and is currently a member of the IEEE Information Theory Society Board of Governors (1990–1996) and a member of the Editorial Board for the *Proceedings of the IEEE*. He is serving as President of the IEEE Information Theory Society in 1996.

Dr. Gibson is Editor-in-Chief of *The Mobile Communications Handbook* (CRC Press, 1995), Editor-in-Chief of *The Communications Handbook* (CRC Press, 1996), and Editor of the IEEE Press book series on Signal Processing.

In 1990, Dr. Gibson received The Frederick Emmons Terman Award from The American Society for Engineering Education, and in 1992, was elected Fellow of the IEEE "for contributions to the theory and practice of adaptive prediction and speech waveform coding." He was co-recipient of the 1993 IEEE Signal Processing Society Senior Paper Award for the Speech Processing area.

His research interests include data, speech, image, and video compression, multimedia over networks, wireless communications, information theory, and digital signal processing.

Advisory Board

Vijay K. Bhargava received the B.Sc., M.Sc., and Ph.D. degrees from Queen's University of Kingston, Canada in 1970, 1972 and 1974 respectively. He is a Professor of Electrical and Computer Engineering at the University of Victoria, co-author of the book *Digital Communications by Satellite* (New York: Wiley, 1981), and co-editor of the IEEE Press Book *Reed-Solomon Codes and Their Applications*. Dr. Bhargava is the founding President of Binary Communications, Inc. (est. 1983), a company which has successfully produced a programmable (31,k) Reed-Solomon CODEC using VLSI technology. He is very active in the IEEE and is currently a Director of the IEEE and a member of the Board of Governors of the IEEE Information Theory Society.

Chris D. Heegard received the B.S. and the M.S. degrees in electrical and computer engineering from the University of Massachusetts, Amherst, MA, in 1975 and 1976, respectively, and the Ph.D. degree in electrical engineering from Stanford University, Stanford, CA, in 1981.

From 1976–1978, he was an R&D Engineer at Linkabit Corp., San Diego, CA, and in 1981, he joined Cornell University, Ithaca, NY, as an Assistant Professor; he was appointed to Associate Professor in 1987.

In 1984 Dr. Heegard received the Presidential Young Investigator Award from the National Science Foundation and the IBM Faculty Development Award. In 1986, he was elected to the Board of Governors of the Information Theory Society of the IEEE, reelected in 1989, and he served as President in 1994.

Laurence B. Milstein received the B.E.E. degree from the City College of New York, New York, NY, in 1964, and the M.S. and Ph.D. degrees in electrical engineering from the Polytechnic Institute of Brooklyn, Brooklyn, NY, in 1966 and 1968, respectively.

From 1968 to 1974, he was employed by the Space and Communications Group of Hughes Aircraft Company, and from 1974 to 1976 he was a member of the Department of Electrical and Systems Engineering, Rensselaer Polytechnic Institute, Troy, NY. Since 1976, he has been with the Department of Electrical and Computer Engineering, University of California, San Diego, La Jolla, CA, where he is a Professor and former Department Chairman. He was an Associate Editor for Communication Theory for the *IEEE Transactions on Communications* and an Associate Technical Editor for the *IEEE Communications Magazine*, and he is currently a Senior Editor for the *IEEE Journal on Selected Areas in Communications*. He was the Vice-President for Technical Affairs in 1990 and 1991 of the IEEE Communications Society, and has served as a member of the Board of Governors of both the IEEE Communications Society and the IEEE Information Theory Society.

James W. Modestino received the B.S. degree from Northeastern University, Boston, MA, in 1962, and the M.S. degree from the University of Pennsylvania, PA in 1964, and the M.A. and Ph.D. degrees from Princeton University, Princeton, NJ, in 1968 and 1969, respectively.

He has held industrial positions with RCA Communication Systems Division, Camden, NJ; AVCO Systems Division, Wilmington, MA; GTE Laboratories, Inc., Waltham, MA and M.I.T. Lincoln Laboratory, Lexington, MA. From 1970 to 1972, he was an Assistant Professor in the Department of Electrical Engineering, Northeastern University. In 1972, he joined Rensselaer Polytechnic Institute, Troy, NY, where he is currently a Professor in the Department of Electrical, Computer, and Systems Engineering. Dr. Modestino is a past Associate Editor of the *IEEE Transactions on Information Theory*, and in 1984, he was co-recipient of the Stephen O. Rice Prize Paper Award from the IEEE Communications Society.

Joseph C. Palais has been a member of the electrical engineering faculty at Arizona State University since 1964, where is presently serving as a professor. He received the B.S.E.E. degree from the University of Arizona, Tucson, in 1959 and the M.S.E. and Ph.D. degrees from the University of Michigan, Ann Arbor, in 1962 and 1964 respectively.

Professor Palais has worked as a Microwave Engineer for Motorola and Stanford Research Institute and has been a consultant to several firms, including Sperry Flight Systems, Sylvania, and Polaroid. In 1967 he was awarded a Ford Foundation Residency in Engineering Practice, and in 1973 he served as a Visiting Associate Professor at the Technion-Israel Institute of Technology.

In 1993, he received the IEEE Educational Activities Board Meritorious Achievement Award in Continuing Education.

Dr. Palais is the author of the textbook *Fiber Optic Communications*, Prentice-Hall, 3rd edition 1992. Professor Palais has also coauthored the software manual, *Fiber Optic Systems and Devices with Microcomputer Software*, with D. Johnson, Kern International, Inc., 1987.

Jonathan S. Turner received the M.S. and Ph.D. degrees in computer science from Northwestern University in 1979 and 1981.

He is a Professor and Chairman of the Department of Computer Science at Washington University, where he has been since 1983. His primary research interest is the design and analysis of switching systems, with special interest in systems supporting multicast communication. He has been awarded more than a dozen patents for his work on switching systems and has a number of widely cited publications in this area. His research interests also include the study of algorithms and computational complexity, with particular interest in the probably performance of heuristic algorithms for NP-complete problems.

Dr. Turner is a member of the ACM and SIAM.

Managing Editor
Elaine M. Gibson

Contributors

Saf Asghar
Advanced Micro Devices, Inc.
Austin, Texas

Chris B. Autry
Georgia Institute of Technology
Atlanta, Georgia

Joseph Bannister
The Aerospace Corporation
Los Angeles, California

Melbourne Barton
Bell Communications Research
Morristown, New Jersey

Vijay K. Bhargava
University of Victoria
Victoria, B.C., Canada

Ezio Biglieri
Politecnico di Torino
Torino, Italia

Anders Bjarklev
Technical University of Denmark
Lyngby, Denmark

Daniel J. Blumenthal
Georgia Institute of Technology
Atlanta, Georgia

Paula M. Bordogna
Lucent Technologies
North Andover, Massachusetts

Madhukar Budagavi
Texas A&M University
College Station, Texas

Pierre Catala
Texas A&M University
College Station, Texas

Wai-Yip Chan
Illinois Institute of Technology
Chicago, Illinois

Biao Chen
University of Texas-Dallas
Dallas, Texas

Giovanni Cherubini
IBM Zurich Research Laboratory
Ruschlikon, Switzerland

Stanley Chia
BT Laboratories
Ipswich, United Kingdom

Youn Ho Choung
TRW Space & Defense
Redondo Beach, California

John M. Cioffi
Stanford University
Stanford, California

Leon W. Couch
University of Florida
Gainesville, Florida

Rene L. Cruz
University of California-San Diego
La Jolla, California

Marc Delprat
Alcatel Telecom
Mobile Communication Division
Colombes, France

Paul Diament
Columbia University
New York, New York

Spiros Dimolitsas
Lawrence Livermore National
 Laboratory
Livermore, California

Robert L. Douglas
University of Memphis
Memphis, Tennessee

Eric Dubois
University of Quebec
Verdun, Quebec, Canada

Niloy K. Dutta
Lucent Technologies
Murray Hill, New Jersey

Bruce Elbert
Hughes Space and Communications
 International, Inc.
El Segundo, California

Ahmed K. Elhakeem
Concordia University
Montreal, Quebec, Canada

Ivan J. Fair
Technical University of Nova Scotia
Halifax, Nova Scotia, Canada

John L. Fike
Texas A&M University
College Station, Texas

Michael D. Floyd
Motorola Semiconductor Products
Austin, Texas

Lew E. Franks
University of Massachusetts
Amherst, Massachusetts

Susan A.R. Garrod
Purdue University
West Lafayette, Indiana

Costas Georghiades
Texas A&M University
College Station, Texas

Ira A. Gerson
Motorola Semiconductor Products
Schaumburg, Illinois

Klein Gilhousen
Qualcomm, Inc.
San Diego, California

David Haccoun
Ecole Polytechnique de Montreal
Montreal, Quebec, Canada

Frederick Halsall
University of Wales
Swansea, Wales, United Kingdom

Jeff Hamilton
General Instrument Corporation
Hatboro, Pennsylvania

Lajos Hanzo
University of Southampton
Highfield, Southampton
United Kingdom

Roger Haskin
IBM Almaden Research Center
San Jose, California

Tor Helleseth
University of Bergen
Bergen, Norway

Garth D. Hillman
Motorola, Inc.
Austin, Texas

Stephen J. Hinterlong
Lucent Technologies
Naperville, Illinois

Michael L. Honig
Northwestern University
Evanston, Illinois

Hwei P. Hsu
Farleigh Dickinson University
Teaneck, New Jersey

Erwin Hudson
TRW Electronics Systems &
   Technology Division
Redondo Beach, California

Yeongming Hwang
City University of Hong Kong
Kowloon, Hong Kong

Louis J. Ippolito, Jr.
Stanford Telecom Inc.
Reston, Virginia

Bijan Jabbari
George Mason University
Fairfax, Virginia

Ravi Jain
Bell Communications Research
Red Bank, New Jersey

Amos E. Joel, Jr.
Executive Consultant
South Orange, New Jersey

Varun Kapoor
Virginia Polytechnic Institute and
   State University
Blacksburg, Virginia

B.L. Kasper
Lucent Technologies
Breiningsville, Pennsylvania

Kota Kinoshita
NTT Do Co Mo
Yokosuka Shi, Japan

Mark Kolber
General Instrument Corporation
Hatboro, Pennsylvania

Boneung Koo
Kyonggi University
Kyonggi-Do, Korea

Nitin C. Kothari
Georgia Institute of Technology
Atlanta, Georgia

Andrew D. Kucar
4U Comm. Research Inc.
Ottawa, Ontario, Canada

Vijay P. Kumar
University of Southern California
Los Angeles, California

Vinod Kumar
Alcatel Telecom
Mobile Communication Division
Colombes, France

B.P. Lathi
California State University
Sacramento, California

Allen H. Levesque
GTE Laboratories, Inc.
Needham, Massachusetts

Curtis A. Levis
Ohio State University
Columbus, Ohio

Chung Sheng Li
IBM T.J. Watson Research Center
Yorktown Heights, New Jersey

Yi-Bing Lin
National Chaio Tung University
Hsinchu, Taiwan, R.O.C.

Joseph L. LoCicero
Illinois Institute of Technology
Chicago, Illinois

John H. Lodge
Communications Research Centre
Ottawa, Ontario, Canada

Mari Maeda
Bell Communications Research
Red Bank, New Jersey

Nicholas Malcolm
Hewlett-Packard (Canada) Ltd.
Burnaby, BC, Canada

Masud Mansuripur
University of Arizona
Tucson, Arizona

Gerald A. Marin
IBM
Triangle Park, North Carolina

Nasir D. Memon
Northern Illinois University
DeKalb, Illinois

Paul Mermelstein
Bell Northern Research
Verdun, Quebec, Canada

Toshio Miki
Mobile Communication Network, Inc.
Yokosuka-shi, Kanagawa, Japan

Laurence B. Milstein
University of California at San Diego
La Jolla, California

Abdi R. Modarressi
Bell Laboratories
Columbus, Ohio

Seshadri Mohan
Bell Communications Research
Morristown, New Jersey

Michael Moher
Communications Research Centre
Ottawa, Ontario, Canada

Jaekyun Moon
University of Minnesota
Minneapolis, Minnesota

Rias Muhamed
Virginia Polytechnic Institute and
  State University
Blacksburg, Virginia

Masao Nakagawa
Keio University
Hiyoshi, Yokohama, Japan

Madihally J. Narasimha
Telecom Solutions
San Jose, California

A. Michael Noll
University of Southern California
Los Angeles, California

Peter Noll
Technische Universitaet Berlin,
Berlin, West Germany

Peter P. Nuspl
W.L. Pritchard & Co.
Bethesda, Maryland

Michael O'Flynn
San Jose State University
San Jose, California

Raif Onvural
Duke University
Durham, North Carolina

Geoffrey C. Orsak
George Mason University
Fairfax, Virginia

Henry W.L. Owen III
Georgia Institute of Technology
Atlanta, Georgia

Kaveh Pahlavan
Worcester Polytechnic Institute
Worcester, Massachusetts

Joseph C. Palais
Arizona State University
Tempe, Arizona

Bernd-Peter Paris
George Mason University
Fairfax, Virginia

Lance Parr
Texas A&M University
College Station, Texas

Bhasker P. Patel
Illinois Institute of Technology
Chicago, Illinois

Achille Pattavina
Politecnico di Milano
Milano, Italy

Arogyaswami Paulraj
Stanford University
Stanford, California

Ken D. Pedrotti
Rockwell International
Thousand Oaks, California

Eric Petajan
Lucent Technologies
Murray Hill, New Jersey

Roman Pichna
University of Victoria
Victoria, B.C., Canada

Samuel Pierre
Universite du Quebec
Montreal, Quebec, Canada

Alistair J. Price
Rockwell International
Thousand Oaks, California

David R. Pritchard
Norlight Telecommunications/
  Teleport Chicago
Skokie, Illinois

Wilbur L. Pritchard
W.L. Pritchard & Co.
Bethesda, Maryland

John G. Proakis
Northeastern University
Boston, Massachusetts

Ramesh R. Rao
University of California at San Diego
La Jolla, California

Theodore S. Rappaport
Virginia Polytechnic Institute and
  State University
Blacksburg, Virginia

Erwin P. Rathgeb
Siemens AG
Muenchen, Germany

Narasimha Reddy
Texas A&M University
College Station, Texas

Whitham D. Reeve
Reeve Engineers
Anchorage, Alaska

Bixio Rimoldi
Washington University
St. Louis, Missouri

Thomas G. Robertazzi
State University of New York
Stony Brook, New York

Martin S. Roden
California State University
Los Angeles, California

Arthur H.M. Ross
Qualcomm, Inc.
San Diego, California

Izhak Rubin
University of California at
  Los Angeles
Los Angeles, California

Khalid Sayood
University of Nebraska
Lincoln, Nebraska

Charles Schell
General Instrument Corporation
Hatboro, Pennsylvania

A. Udaya Shankar
University of Maryland
College Park, Maryland

Curtis A. Siller, Jr.
Lucent Technologies
North Andover, Massachusetts

Marvin K. Simon
Jet Propulsion Laboratory
Pasadena, California

Suresh P. Singh
University of South Carolina
Columbia, South Carolina

Bernard Sklar
Communications Engineering
　Services
Tarzana, California

R.A. Skoog
AT&T
Holmdel, New Jersey

David R. Smith
George Washington University
Ashburn, Virginia

Jonathan M. Smith
University of Pennsylvania
Philadelphia, Pennsylvania

Richard G. Smith
Center Valley, Pennsylvania

Raymond Steele
Multiple Access Communications
　Limited
Southampton, Hampshire
United Kingdom

Ferrel G. Stremler
University of Wisconsin
Madison, Wisconsin

Gordon Stuber
Georgia Institute of Technology
Atlanta, Georgia

Len Taupier
General Instrument Corporation
Hatboro, Pennsylvania

Chong Kwan Un
Korea Advanced Institute of Science
　and Technology
Taejon, Korea

Reudiger Urbanke
Lucent Technologies
Murray Hill, New Jersey

Harrell J. Van Norman
Dayton, Ohio

David Vlack
Bell Labs
St. Charles, Illinois

Qiang Wang
University of Victoria
Victoria, B.C., Canada

Richard H. Williams
University of New Mexico
Albuquerque, New Mexico

Ping Wah Wong
Hewlett Packard Laboratories
Palo Alto, California

Maynard A. Wright
Electrical Engineering Consultant
Citrus Heights, California

Michel D. Yacoub
The University of Campinas
San Paulo, Brazil

Shinj Yamashita
RCAST University of Tokyo
Tokyo, Japan

Chong Ho Yoon
University of Arizona
Tucson, Arizona

William C. Young
Bell Communications Research
Red Bank, New Jersey

Wei Zhao
Texas A&M University
College Station, Texas

Rodger E. Ziemer
University of Colorado at
　Colorado Springs
Colorado Springs, Colorado

Contents

SECTION I Basic Principles

1. Analog Modulation *Ferrel G. Stremler* 3
2. Sampling *Hwei P. Hsu* 13
3. Pulse Code Modulation *Leon W. Couch II* 23
4. Probabilities and Random Variables *Michael O'Flynn* 35
5. Random Processes, Autocorrelation, and Spectral Densities *Lew E. Franks* 62
6. Queuing *Richard H. Williams* 78
7. Multiplexing *Martin S. Roden* 87
8. Pseudonoise Sequences *Tor Helleseth and P. Vijay Kumar* 94
9. D/A and A/D Converters *Susan A.R. Garrod* 107
10. Signal Space *Rodger E. Ziemer* 118
11. Channel Models *David R. Smith* 131
12. Optimum Receivers *Geoffrey C. Orsak* 141
13. Standards Setting Bodies *Spiros Dimolitsas* 155
14. Forward Error Correction Coding *Vijay Bhargava and Ivan J. Fair* 166
15. Automatic Repeat Request *David Haccoun and Samuel Pierre* 181
16. Spread Spectrum Communications *Laurence B. Milstein and Marvin K. Simon* 199
17. Diversity Techniques *Arogyaswami Paulraj* 213
18. Information Theory *Bixio Rimoldi and Reudiger Urbanke* 224

19	Digital Communication System Performance *Bernard Sklar*	236
20	Synchronization *Costas N. Georghiades*	255
21	Digital Modulation Techniques *Ezio Biglieri*	273

SECTION II Telephony

22	POTS (Plain Old Telephone Service) *A. Michael Noll*	291
23	FDM Hierarchy *Pierre Catala*	301
24	Analog Telephone Channels and the Subscriber Loop *Whitham D. Reeve*	308
25	Baseband Signalling and Pulse Shaping *Michael L. Honig and Melbourne Barton*	318
26	Channel Equalization *John G. Proakis*	339
27	PCM Codec Filters *Michael D. Floyd and Garth D. Hillman*	364
28	Digital Hierarchy *B.P. Lathi and Maynard Wright*	377
29	Line Coding *Joseph L. LoCicero and Bhasker P. Patel*	386
30	Telecommunications Network Synchronization *Madihally J. Narasimha*	404
31	Echo Cancellation *Giovanni Cherubini*	414
32	Switching Fabrics *Amos E. Joel, Jr.*	425
33	Customer Premises Equipment *John L. Fike and M. Lance Parr*	433
34	Asymmetric Digital Subscriber Lines *John M. Cioffi*	450
35	Overview of Common Channel Signalling *Abdi R. Modarressi and R.A. Skoog*	480
36	Digital Cross-Connect Systems *Paula M. Bordogna and Curtis A. Siller, Jr.*	496
37	Building Future Networks by Using Photonics in Switching *Stephen J. Hinterlong and David Vlack*	513
38	Asynchronous Transfer Mode (ATM) *Raif O. Onvural*	529
39	Synchronous Optical Network (SONET) *Henry L. Owen and Chris B. Autry*	542
40	Synchronous Digital Hierarchy (SDH) *Henry L. Owen and Chris B. Autry*	554

SECTION III   Networks

41	The OSI Seven-Layer Model *Fred Halsall*	567
42	ISDN and B-ISDN *Erwin P. Rathgeb*	577
43	Ethernet Networks *Ramesh R. Rao*	591
44	FDDI *Biao Chen, Wei Zhao, and Nicholas Malcolm*	597
45	Broadband Local Area Networks *Joseph A. Bannister*	611
46	Multiple Access Methods *Izhak Rubin*	622
47	Routing and Flow Control *Rene L. Cruz*	650
48	Transport Layer *A. Udaya Shankar*	661
49	Gigabit Networks *Jonathan M. Smith*	672
50	Local Area Networks *Thomas G. Robertazzi*	681
51	Asynchronous Time Division Switching *Achille Pattavina*	686
52	Internetworking *Harrell J. Van Norman*	701
53	Architectural Framework for ATM Networks: Broadband Network Services *Gerald A. Marin and Raif O. Onvural*	717

SECTION IV   Optical

54	Fiber Optic Communications Systems *Joseph C. Palais*	731
55	Optical Fibers and Lightwave Propagation *Paul Diament*	740
56	Optical Sources *Niley K. Dutta*	751
57	Optical Transmitters *Alistair J. Price and Ken D. Pedrotti*	774
58	Optical Receivers *Richard G. Smith and B.L. Kasper*	789
59	Fiber Optic Connectors and Splices *William C. Young*	803
60	Passive Optical Components *Joseph C. Palais*	824
61	Semiconductor Optical Amplifiers *Daniel J. Blumenthal and Nitin C. Kothari*	832
62	Optical Amplifiers *Anders Bjarklev*	848
63	Coherent Systems *Shinji Yamashita*	862

| 64 | Fiber Optic Applications *Chung-Sheng Li* | 872 |
| 65 | Wavelength Division Multiplexed Systems and Applications *Mari W. Maeda* | 883 |

SECTION V Satellite

66	Geostationary Communications Satellites and Applications *Bruce R. Elbert*	893
67	Satellite Systems *Robert L. Douglas*	912
68	The Earth Station *David R. Pritchard*	922
69	Satellite Transmission Impairments *Louis J. Ippolito, Jr.*	935
70	Satellite Link Design *Peter P. Nuspl and Jahangir A. Tehrani*	949
71	The Calculation of System Temperature for a Microwave Receiver *Wilbur L. Pritchard*	966
72	Onboard Switching and Processing *Ahmed K. Elhakeem*	976
73	Path Diversity *Curt A. Levis*	996
74	Mobile Satellite Systems *John H. Lodge and Michael Moher*	1015
75	Satellite Antennas *Yeongming Hwang and Youn Ho Choung*	1032
76	Tracking and Data Relay Satellite System (TDRSS) *Erwin C. Hudson*	1054

SECTION VI Wireless

77	Mobile Radio: An Overview *Andrew D. Kucar*	1069
78	Base Station Subsystems *Chong Kwan Un and Chong Ho Yoon*	1090
79	Access Methods *Bernd-Peter Paris*	1104
80	Location Strategies for Personal Communications Services *Ravi Jain, Yi-Bing Lin, and Seshadri Mohan*	1116
81	Cell Design Principles *Michel Daoud Yacoub*	1146
82	Microcellular Radio Communications *Raymond Steele*	1160
83	Fixed and Dynamic Channel Assignment *Bijan Jabbari*	1175
84	Propagation Models *Theodore S. Rappaport, Rias Muhamed, and Varun Kapoor*	1182
85	Power Control *Roman Pichna and Qiang Wang*	1197
86	Second Generation Systems *Marc Delprat and Vinod Kumar*	1208

87	The Pan-European Cellular System *Lajos Hanzo* 1226
88	The IS-54 Digital Cellular Standard *Paul Mermelstein* 1246
89	CDMA Technology and the IS-95 North American Standard *Arthur H.M. Ross and Klein S. Gilhousen* 1257
90	Japanese Cellular Standard *Kota Kinoshita and Masao Nakagawa* 1276
91	The British Cordless Telephone Standard: CT-2 *Lajos Hanzo* 1289
92	DECT (Digital European Cordless Telephone) *Saf Asghar* 1305
93	The RACE Program *Stanley Chia* 1327
94	Half-Rate Standards *Wai-Yip Chan, Ira Gerson, and Toshio Miki* 1338
95	Modulation Methods *Gordon L. Stüber* 1353
96	Wireless LANs *Suresh P. Singh* 1367
97	Wireless Data *Allen H. Levesque and Kaveh Pahlavan* 1380

SECTION VII Source Compression

98	Lossless Compression *Khalid Sayood and Nasir D. Memon* 1397
99	Facsimile *Nasir D. Memon and Khalid Sayood* 1411
100	Speech *Boneung Koo* 1424
101	Still Image Compression and Halftoning *Ping Wah Wong* 1437
102	Video *Eric Dubois* 1449
103	The High-Definition Television Grand Alliance System *Eric Petajan* 1462
104	Audio Coding *Peter Noll* 1475
105	Cable *Jeff Hamilton, Mark Kolber, Charles Schell, and Len Taupier* 1488
106	Video Servers *Narasimha Reddy and Roger Haskin* 1502
107	Desktop Videoconferencing *Madhukar Budagavi* 1516

SECTION VIII Data Recording

108	Magnetic Storage *Jaekyun Moon* 1531
109	Magneto-Optical Disk Data Storage *Masud Mansuripur* 1546
Index	 1567

I

Basic Principles

1. **Analog Modulation** *Ferrel G. Stremler* 3
   Introduction • Amplitude Modulation • Angle Modulation

2. **Sampling** *Hwei P. Hsu* 13
   Introduction • Instantaneous Sampling • Sampling Theorem • Sampling of Sinusoidal Signals • Sampling of Bandpass Signals • Practical Sampling • Sampling Theorem in the Frequency Domain • Summary and Discussion

3. **Pulse Code Modulation** *Leon W. Couch II* 23
   Introduction • Generation of PCM • Percent Quantizing Noise • Practical PCM Circuits • Bandwidth of PCM • Effects of Noise • Nonuniform Quantizing: μ-Law and A-Law Companding • Example: Design of a PCM System

4. **Probabilities and Random Variables** *Michael O'Flynn* 35
   Introduction • Discrete Probability Theory • The Theory of One Random Variable • The Theory of Two Random Variables • Summary and Future Study

5. **Random Processes, Autocorrelation, and Spectral Densities** *L.E. Franks* 62
   Introduction • Basic definitions • Properties and Interpretation • Baseband Digital Data Signals • Coding for Power Spectrum Control • Bandpass Digital Data Signals • Appendix: The Poisson Sum Formula

6. **Queuing** *Richard H. Williams* 78
   Introduction • Little's Formula • The M/M/1 Queuing System: State Probabilities • The M/M/1 Queuing System: Averages and Variances • Averages for the Queue and the Server

7. **Multiplexing** *Martin S. Roden* 87
   Introduction • Frequency Multiplexing • Time Multiplexing • Space Multiplexing • Techniques for Multiplexing in Spread Spectrum • Concluding Remarks

8. **Pseudonoise Sequences** *Tor Helleseth and P. Vijay Kumar* 94
   Introduction • m Sequences • The q-ary Sequences with Low Autocorrelation • Families of Sequences with Low Crosscorrelation • Aperiodic Correlation • Other Correlation Measures

9. **D/A and A/D Converters** *Susan A.R. Garrod* 107
   D/A and A/D Circuits

10. **Signal Space** *Rodger E. Ziemer* 118
    Introduction • Fundamentals • Application of Signal Space Representation to Signal Detection • Application of Signal Space Representation to Parameter Estimation

11. **Channel Models** *David R. Smith* 131
    Introduction • Fading Dispersive Channel Model • Line-of-Sight Channel Models • Digital Channel Models

12. **Optimum Receivers** *Geoffrey C. Orsak* 141
    Introduction • Preliminaries • Karhunen–Loève Expansion • Detection Theory • Performance • Signal Space • Standard Binary Signalling Schemes • M-ary Optimal Receivers • More Realistic Channels • Dispersive Channels

13 **Standards Setting Bodies** *Spiros Dimolitsas* 155
Introduction • Global Standardization • Regional Standardization • National Standardization • Standards Coordination • Scientific • Standards Development Cycle

14 **Forward Error Correction Coding** *V.K. Bhargava and I.J. Fair* 166
Introduction • Fundamentals of Block Coding • Structure and Decoding of Block Codes • Important Classes of Block Codes • Principles of Convolutional Coding • Decoding of Convolutional Codes • Trellis-Coded Modulation • Additional Measures • Applications

15 **Automatic Repeat Request** *David Haccoun and Samuel Pierre* 181
Introduction • Fundamentals and Basic Automatic Repeat Request Schemes • Performance Analysis and Limitations • Variants of the Basic Automatic Repeat Request Schemes • Hybrid Forward Error Control/Automatic Repeat Request Schemes • Application Problem • Conclusion

16 **Spread Spectrum Communications** *Laurence B. Milstein and Marvin K. Simon* 199
A Brief History • Why Spread Spectrum? • Basic Concepts and Terminology • Spread Spectrum Techniques • Applications of Spread Spectrum

17 **Diversity Techniques** *A. Paulraj* 213
Introduction • Diversity Schemes • Diversity Combining Techniques • Effect of Diversity Combining on Bit-Error Rate (BER) • Conclusions

18 **Information Theory** *Bixio Rimoldi and Rüdiger Urbanke* 224
Introduction • The Communication Problem • Source Coding for Discrete-Alphabet Sources • Universal Source Coding • Rate Distortion Theory • Channel Coding • Simple Binary Codes

19 **Digital Communication System Performance** *Bernard Sklar* 236
Introduction • Bandwidth and Power Considerations • Example 19.1: Bandwidth-Limited Uncoded System • Example 19.2: Power-Limited Uncoded System • Example 19.3: Bandwidth-Limited and Power-Limited Coded System • Example 19.4: Direct-Sequence (DS) Spread-Spectrum Coded System • Conclusion • Appendix: Received E_b/N_0 Is Independent of the Code Parameters

20 **Synchronization** *Costas N. Georghiades* 255
Introduction • Carrier Synchronization • Symbol Synchronization • Frame Synchronization

21 **Digital Modulation Techniques** *Ezio Biglieri* 273
Introduction • The Challenge of Digital Modulation • One-Dimensional Modulation: Pulse-Amplitude Modulation (PAM) • Two-Dimensional Modulations • Multidimensional Modulations: Frequency-Shift Keying (FSK) • Multidimensional Modulations: Lattices • Modulations with Memory

1
Analog Modulation

1.1 Introduction 3
1.2 Amplitude Modulation 3
    Double-Sideband–Suppressed Carrier (DSB-SC) • Double-Sideband–Large Carrier (DSB-LC) • Quadrature Multiplexing (QM) • Single Sideband (SSB) • Vestigial-Sideband (VSB) Modulation
1.3 Angle Modulation 9
    FM • PM

Ferrel G. Stremler
University of Wisconsin

1.1 Introduction

Modulation is that process by which a property or a parameter of a given signal is varied in proportion to a second signal, which we term the input. Analog modulation generally refers to a modulation of the continuous complex exponential signal $a(t)\exp[j\theta(t)]$.[1] To make use of bandpass notation (and bandpass circuitry) we identify a carrier frequency ω_c about which we allow a bandwidth W (radians per second). Using the real-part notation, the modulated signal $\varphi(t)$ can be written as[2]

$$\varphi(t) = \text{Re}\{a(t)e^{j\theta(t)}\} = \text{Re}\{a(t)e^{j[\omega_c t + \gamma(t)]}\} \quad (1.1)$$

where $a(t)$ is the (time-varying) amplitude, ω_c the carrier frequency, and $\gamma(t)$ the (time-varying) angle with respect to the carrier phase $\omega_c t$. We assume that $a(t)$ and $\gamma(t)$ are slowly varying with respect to $\exp(j\omega_c t)$, i.e., we assume that $W \ll \omega_c$.

1.2 Amplitude Modulation

In **amplitude modulation** (AM), the angle term $\gamma(t)$ in Eq. (1.1) is a constant (often assumed to be zero) and $a(t)$ is made proportional to the input signal $f(t)$,

$$\varphi(t) = \text{Re}\{k_a f(t)e^{j\omega_c t}\} \quad (1.2)$$

where k_a is the proportionality constant of the modulator. The specific way in which this proportionality is achieved distinguishes several basic types of amplitude modulation.

---
[1] A relatively slow switching of a high-frequency sinusoidal signal on and off is sometimes referred to as continuous-wave (CW) modulation.
[2] The phasor notation is valid for the narrowband case $W \ll \omega_c$. Complex envelope notation and the concept of analytic signals can be used to extend this type of analysis to the wide bandwidth case; see, e.g., Proakis, J. G. 1989. *Digital Communications*, 2nd ed. McGraw-Hill, New York, Chap. 3.

Double-Sideband–Suppressed Carrier (DSB-SC)

If we assume that $f(t)$ is real valued, Eq. (1.2) simplifies to

$$\varphi(t) = k_a f(t) \cos \omega_c t \qquad (1.3)$$

An example of this type of modulation is shown in Fig. 1.1. Note that sign changes in the input $f(t)$ are conveyed as phase changes (of π radians) in carrier in the modulated waveform. Applying the frequency-shift property of the Fourier transform to Eq. (1.3), we find that the spectral density of $\varphi(t)$ is

$$\Phi(\omega) = \mathcal{F}\left\{\frac{1}{2}k_a f(t) e^{j\omega_c t} + \frac{1}{2}k_a f(t) e^{-j\omega_c t}\right\} = \frac{1}{2}F(\omega - \omega_c) + \frac{1}{2}F(\omega + \omega_c) \quad (1.4)$$

Thus, we see that this type of amplitude modulation translates the spectral density of $f(t)$ by $\pm\omega_c$ rad/s but leaves the spectral shape unaltered. This type of amplitude modulation is called **suppressed carrier** (SC) because the spectral density of $\varphi(t)$ has no (averaged) carrier in it, although the spectrum is centered at $\pm\omega_c$.

From the first term on the right-hand side of Eq. (1.4), we see that one-half of both the positive and the negative frequency content of $f(t)$ is shifted upward to positive frequencies in $\varphi(t)$, so that the

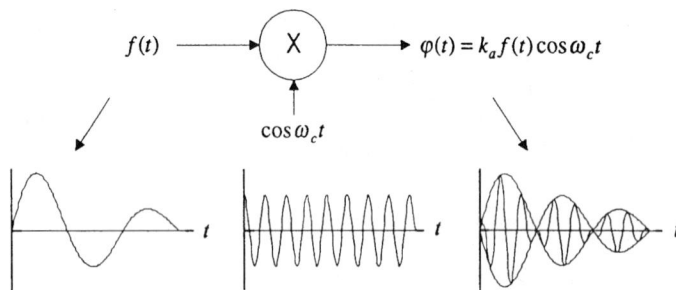

FIGURE 1.1 A suppressed-carrier amplitude modulation system.

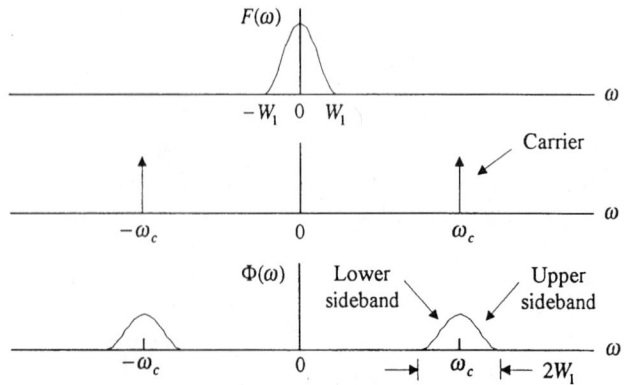

FIGURE 1.2 Spectral densities of signals shown in Fig. 1.1.

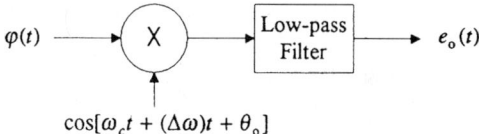

FIGURE 1.3 Demodulation of amplitude-modulated suppressed-carrier signals.

bandwidth is doubled.[3] This is illustrated in Fig. 1.2. The spectral content for positive frequencies above ω_c is called the *upper sideband* of $\varphi(t)$, and the spectral content for positive frequencies below ω_c is called the *lower sideband*. The fact that we have two sidebands and no identifiable averaged carrier present suggests the convenient designation: **double-sideband**, suppressed carrier (DSB-SC).

Demodulation of DSB-SC requires a multiplication operation to translate a portion of the spectral density down in frequency to baseband. A diagram of a basic receiver for DSB-SC is shown in Fig. 1.3. Introducing a frequency error $\Delta\omega$ and a phase error θ_0 in the locally generated reference in the receiver, the output signal $e_0(t)$ is

$$e_0(t) = \{f(t)\cos\omega_c t \cos[\omega_c t + (\Delta\omega)t + \theta_0]\}_{LP} = \frac{1}{2}f(t)\cos[(\Delta\omega)t + \theta_0] \quad (1.5)$$

The phase error θ_0 in the locally generated carrier causes a variable gain in the output signal; for small, fixed-phase errors this is tolerable. Any nonzero frequency error $\Delta\omega$, however, results in inacceptable multiplicative distortion. Thus, the use of *synchronous* detection of a DSB-SC modulated signal yields the original signal $f(t)$ with excellent fidelity.

Generation of a DSB-SC signal requires use of a nonlinear or a time-varying system. The former uses a nonlinear gain to generate the product-type term required; this nonlinearity may also introduce some carrier unless a signal balancing arrangement is provided to null out the carrier. The latter method generally uses a periodic switch-type multiplication that shifts a spectral replica of the input upward in frequency by the fundamental frequency of the switching waveform.

Double-Sideband–Large Carrier (DSB-LC)

Use of suppressed-carrier modulation requires provisions for acquiring and maintaining the necessary receiver synchronization. If we wish to use very inexpensive receivers at the expense of extra transmitter power, an alternative is to send a large enough carrier that no π jumps in carrier phase are necessary in the modulated waveform to convey sign changes of $f(t)$. To distinguish this case from the previous case, we designate this as double-sideband, **large carrier** (DSB-LC) modulation. Because commercial broadcast stations use this method of modulation, it is also commonly known as AM.

The DSB-LC modulated waveform can be described mathematically simply by adding a carrier term, $A\cos\omega_c t$, to the DSB-SC signal in Eq. (1.3),

$$\varphi_{AM}(t) = A\cos\omega_c t + k_a f(t)\cos\omega_c t = [A + k_a f(t)]\cos\omega_c t \quad (1.6)$$

If A is large enough, the envelope (magnitude) of the modulated waveform is proportional to $f(t)$. Demodulation in this case simply reduces to envelope detection (i.e., a diode and low-pass filter), plus deletion of an artificially generated average value caused by the added carrier term.

---

[3]By convention, the concepts of bandwidth, upper sideband, and lower sideband are defined for positive frequencies.

For the special case of a single-frequency sinusoid $f(t) = a \cos \omega_m t$, we define a dimensionless scale factor $m = ak_a/A$ to control the ratio of the peak sideband amplitude to the peak carrier amplitude. Using this sinusoidal input, Eq. (1.6) becomes

$$\varphi_{\text{AM}}(t) = A(1 + m \cos \omega_m t) \cos \omega_c t \tag{1.7}$$

The parameter m controls the relative proportions of peak sideband to peak carrier and is called the amplitude *modulation index*. A modulation index less than 100% on negative peaks of $f(t)$ does not result in a 180° phase change in the modulated waveform to convey a sign change in $f(t)$, and an inexpensive envelope detector can be used for demodulation. The economic advantages of use of an envelope detector for demodulation of DSB-LC instead of the synchronous detector necessary for DSB-SC come at the cost of a loss of low-frequency response (to provide for blocking the artificially generated nonzero average value) and a lowered power efficiency.

Generation of a DSB-LC signal can be accomplished by using gain nonlinearities to generate the product term required, or by a periodic switch-type time-varying system. The former is often used in low-power modulator applications, and the latter is used more in high-power modulator applications.

Commercial AM stations in the U.S. are assigned carrier frequencies at 10-kHz intervals from 520 to 1710 kHz. Required carrier stability is ±20 Hz, and the bandwidth of transmissions is nominally about 10 kHz. Stations in local proximity are usually assigned carrier frequencies that are separated by 30 kHz or more. Permissible transmitted (average) power levels range from 0.1 to 50 kW; licensed power output is for an unmodulated carrier. Transmission is primarily via ground-wave propagation during the day; sky-wave propagation may become more dominant at night in some circumstances. Interference between transmissions is controlled by a combination of frequency allocation, transmitter power, transmitting antenna pattern, and possible nighttime operating restrictions.

Quadrature Multiplexing (QM)

If we do not make the assumption that the input signal $f(t)$ is real valued in Eq. (1.2), we have

$$\varphi(t) = \text{Re}\{[x(t) - jy(t)]e^{j\omega_c t}\} \tag{1.8}$$

where $x(t)$ and $y(t)$ are the real and imaginary parts, respectively, of the input signal $f(t)$. Use of an identity for the real part of a product allows us to rewrite Eq. (1.8) as[4]

$$\varphi(t) = x(t) \cos \omega_c t + y(t) \sin \omega_c t = I(t) + Q(t) \tag{1.9}$$

where $I(t)$ is the in-phase component of $\varphi(t)$ and $Q(t)$ is the quadrature component. Because each is an example of DSB-SC modulation, the fact that the signals $\cos \omega_c t$ and $\sin \omega_c t$ are mutually orthogonal, and $x(t)$ and $y(t)$ are slowly varying with respect to $\exp(j\omega_c t)$, we can transmit (and recover) the two signals $x(t)$ and $y(t)$ simultaneously. This is referred to as **quadrature multiplexing** (QM). A quadrature multiplexed communication system is shown in Fig. 1.4.

Single Sideband (SSB)

The doubling of bandwidth in DSB modulation is a disadvantage if a given frequency band is crowded. Conversion to one sideband can be accomplished, in theory, by first generating DSB and then attenuating the undesired sideband. The **single-sideband** (SSB) filter requirements, however,

---

[4]Given two complex-valued quantities z_1, z_2, then $\text{Re}\{z_1 z_2\} = \text{Re}\{z_1\} \text{Re}\{z_2\} - \text{Im}\{z_1\} \text{Im}\{z_2\}$.

Analog Modulation

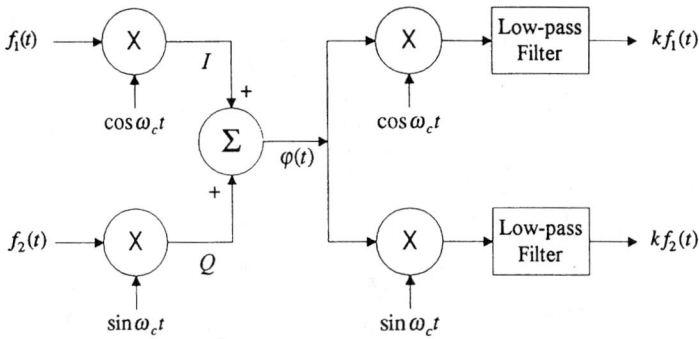

FIGURE 1.4 A quadrature-multiplexed communication system.

are stringent; ideally, the filter must pass all spectral components to one side of a carrier and attenuate all spectral components to the other side of the carrier. Using a complex-valued low-pass equivalent of the required bandpass filter, the frequency transfer function of the single-sideband filter is[5]

$$H(\omega) = \begin{cases} 2 & \omega > 0 \\ 0 & \omega < 0 \end{cases} \tag{1.10}$$

We designate the real and imaginary parts of this low-pass frequency transfer function in the following manner:

$$H(\omega) = H_r(\omega) + jH_i(\omega) \tag{1.11}$$

To maintain a phase characteristic that is an odd function of frequency and yet satisfies Eq. (1.10) requires[6]

$$H_r(\omega) = 1,$$
$$H_i(\omega) = \begin{cases} -j & \omega > 0 \\ j & \omega < 0 \end{cases} = -j\,\mathrm{sgn}(\omega) \tag{1.12}$$

Using the known Fourier transform pair $\mathcal{F}\{\mathrm{sgn}(\omega)\} = (-j\pi t)^{-1}$, we see that the required unit impulse response of the single-sideband filter is

$$h(t) = \delta(t) + j\frac{1}{\pi t} \tag{1.13}$$

Thus, for an input $f(t)$, the corresponding filter output $g(t)$ is

$$g(t) = f(t) \otimes h(t) = f(t) + jf(t) \otimes \frac{1}{\pi t} \tag{1.14}$$

where $\otimes$ is used to indicate the convolution operation.

The second term in the right-hand side of Eq. (1.14) is known as the *Hilbert transform* of $f(t)$, which we designate as $\hat{f}(t)$. Referring to Eq. (1.12) we see that each spectral component in $\hat{f}(t)$ is

---

[5] A gain factor of 2 is used here for convenience in the notation.
[6] Inequalities shown are for an upper-sideband filter; if the inequalities are interchanged, then we have a lower sideband filter.

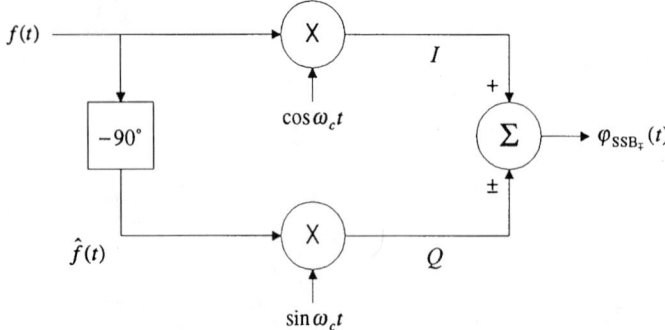

FIGURE 1.5 The phase-shift method of generating SSB modulation.

in phase quadrature ($-90°$) with that of $f(t)$. The Hilbert transform operation in time is not well defined at $t = 0$, and this causes peaks of undefined amplitude at points of finite discontinuity in $f(t)$. As a result, the use of SSB modulation is restricted to smooth waveforms to retain reasonable control of the overall signal envelope.

An interesting application of the preceding results and quadrature multiplexing leads to the phase shift method of generating SSB shown in Fig. 1.5. A major problem in the design of a phase-shift SSB system is the practical realization of the 90° phase-shift filter for $f(t)$, because all spectral components in the band must be shifted by exactly 90°. On the other hand, the SSB filter method must allow some finite attenuation rate at the band edge. Mathematically, the SSB filter method and the SSB phase-shift method are equivalent. In either case, the resultant SSB-SC signal can be written as

$$\varphi_{SSB_{\mp}}(t) = \text{Re}\{[f(t) \pm j\hat{f}(t)]e^{j\omega_c t}\} \qquad (1.15)$$

where SSB_+ designates upper sideband and SSB_- designates lower sideband. If the SSB-SC signal described by Eq. (1.15) is applied to the input of the synchronous detector shown in Fig. 1.3, the output can be written in the form

$$e_0(t) = \text{Re}\{[f(t) \pm j\hat{f}(t)]e^{j\omega_c t}e^{-j(\omega_c t + \Delta\omega t + \theta_0)}\} \qquad (1.16)$$

where $\Delta\omega$ and θ_0 are the frequency and phase errors, respectively, between transmitter and receiver. Phase distortion, i.e., a mix of $f(t)$ and $\hat{f}(t)$, results if $\theta_0 \neq 0$ and, in addition, a spectral shift results if $\Delta\omega \neq 0$. Although the fidelity is not very good, the SSB-SC receiver does not have the serious multiplicative distortion effect that occurs in DSB-SC reception using synchronous detection, and intelligible speech can be received using only half of the bandwidth of DSB systems, but at a cost of reduced fidelity. Single-sideband modulation is used in many of the amateur radio bands as a primary method for transmission of voice signals for communication.

Vestigial-Sideband (VSB) Modulation

To conserve spectral occupancy, a compromise can be made between the bandwidth requirements of SSB and DSB modulation by transmission of most of one sideband and only a portion or vestige of the second sideband. This **vestigial-sideband** (VSB) modulation is transmitted in such a way that synchronous detection of the modulated signal reproduces the original input signal. The VSB modulator can be represented by a DSB modulator followed by a VSB filter with frequency transfer function $H_V(\omega)$; the spectral density of the filter output is

$$\Phi_{\text{VSB}}(\omega) = \Phi_{\text{DSB}}(\omega) H_V(\omega) = \left[\frac{1}{2} F(\omega - \omega_c) + \frac{1}{2} F(\omega + \omega_c)\right] H_V(\omega) \quad (1.17)$$

The output of a synchronous detector with the input signal $\varphi_{\text{VSB}}(t)$ is

$$e_0(t) = [\varphi_{\text{VSB}}(t) \cos \omega_c t]_{LP}$$
$$E_0(\omega) = \frac{1}{4} F(\omega) H_V(\omega + \omega_c) + \frac{1}{4} F(\omega) H_V(\omega - \omega_c) \quad (1.18)$$

Thus, for a faithful reproduction of $f(t)$ we require that

$$[H_V(\omega - \omega_c) + H_V(\omega + \omega_c)]_{LP} = \text{const} \qquad (|\omega| - \omega_c) < \omega_m \quad (1.19)$$

Choosing the constant in Eq. (1.19) to be $2H_V(\omega_c)$, we find we require that $|H_V(\omega)|$ be antisymmetric about the carrier frequency ω_c. Although we have used synchronous detection in the preceding results, these principles also hold if a large carrier is present and if envelope detection is used.

Commercial television stations in the U.S. use VSB-LC modulation for transmission of the luminance (black and white) video signal, and quadrature-multiplexed DSB-SC and VSB-SC modulation for the color video signals. The color signal subcarrier is 3.579545 MHz above the video carrier, and is sent during a brief burst in the blanked portion of each horizontal trace line. The audio is sent using **frequency modulation** (FM) (cf. next section) with a carrier frequency 4.5 MHz above the frequency of the video carrier and with a peak frequency deviation of 25 kHz. The (audio) stereo is sent using DSB-SC/FM with a subcarrier frequency equal to twice the horizontal trace frequency of the video signal. Stations are assigned video carrier frequencies at 6-MHz intervals from 55.25 to 83.25 MHz and 175.25 to 211.25 MHz in the VHF band, and from 471.25 to 801.25 MHz in the UHF band. Maximum effective radiated (average) power is 100–316 kW in the VHF band and 5 MW in the UHF band. The audio carrier is transmitted 7–10 dB lower average power level than the video carrier.[7]

1.3 Angle Modulation

A general relation between the instantaneous angular rate $\omega_i(t)$ and the angle $\theta(t)$ in circular motion is

$$\theta(t) = \int_0^t \omega_i(\tau) \, d\tau + \theta_0 \quad (1.20)$$

Taking the derivative of both sides of Eq. (1.20), we have

$$\omega_i(t) = \frac{d\theta}{dt} \quad (1.21)$$

When applied to the phasor signal representation in Eq. (1.1), $\omega_i(t)$ is called the instantaneous frequency of the signal $\varphi(t)$. This definition of frequency differs, in general, from that associated with spectral content in Fourier analysis. The latter is time averaged and, thus, cannot be time dependent; in contrast, the instantaneous frequency can be time dependent.

---

[7]For more information on television systems, see, e.g., Benson, K. B. ed. 1986. *Television Engineering Handbook.* McGraw-Hill, New York.

In **angle modulation** we assume a constant amplitude in Eq. (1.1) and modulate the phase angle with respect to the carrier in proportion to the input signal $f(t)$. For a direct proportionality, we have

$$\theta(t) = \omega_c t + k_p f(t) + \theta_0 \qquad \text{(PM)} \tag{1.22}$$

and this type of angle modulation is called **phase modulation** (PM). Another possibility to make the instantaneous frequency $\omega_i(t)$ proportional to $f(t)$ is

$$\omega_i(t) = \omega_c + k_f f(t), \qquad \text{(FM)} \tag{1.23}$$

and this type is called **frequency modulation** (FM).

FM

Because the angle vs amplitude transformation is not linear—and thus superposition cannot be used—we choose a specific signal for the input,

$$f(t) = a \cos \omega_m t \tag{1.24}$$

The resulting instantaneous frequency [cf. Eqs. (1.23) and (1.24)] is

$$\omega_i(t) = \omega_c + a k_f \cos \omega_m t \tag{1.25}$$

where k_f is the frequency modulator constant; typical units are radians per second per volt.

Defining a new constant $\Delta \omega = a k_f$, called the *peak frequency deviation* (from carrier), we rewrite Eq. (1.25) as

$$\omega_i(t) = \omega_c + \Delta \omega \cos \omega_m t \tag{1.26}$$

Using Eq. (1.26) in Eq. (1.20), the phase angle of the FM signal with sinusoidal input is

$$\theta(t) = \omega_c t + \beta \sin \omega_m t + \theta_0 \tag{1.27}$$

where

$$\beta = \frac{\Delta \omega}{\omega_m} \tag{1.28}$$

is called the frequency *modulation index*. Combining Eqs. (1.1) and (1.27), we have[8]

$$\varphi_{\text{FM}}(t) = \text{Re}\{A e^{j\omega_c t} e^{j\beta \sin \omega_m t}\} \tag{1.29}$$

To obtain a result in terms of averaged spectral content, we note that the second exponential term in Eq. (1.29) is a periodic function (with a fundamental frequency of ω_m rad/s). Thus, it can be expanded in a Fourier series; doing this gives the result[9]

$$\varphi_{\text{FM}}(t) = \text{Re}\left\{ A e^{j\omega_c t} \sum_{n=-\infty}^{\infty} J_n(\beta) e^{jn\omega_m t} \right\} \tag{1.30}$$

---

[8] The constant phase θ_0 has been absorbed into A for convenience in this equation.
[9] Notation differs in describing Eq. (1.30). Some authors refer to the $J_n(\beta)$ as sidebands, and then we can say that $\varphi_{\text{FM}}(t)$ has an infinite number of sidebands. Others refer to only two sidebands, one above the carrier and one below, and then each sideband contains multiple side frequencies.

2
Sampling

Hwei P. Hsu
Fairleigh Dickinson University

2.1 Introduction 13
2.2 Instantaneous Sampling 13
      Ideal Sampled Signal • Band-Limited Signals
2.3 Sampling Theorem 15
2.4 Sampling of Sinusoidal Signals 16
2.5 Sampling of Bandpass Signals 16
2.6 Practical Sampling 18
      Natural Sampling • Flat-Top Sampling
2.7 Sampling Theorem in the Frequency Domain 21
2.8 Summary and Discussion 21

2.1 Introduction

To transmit analog message signals, such as speech signals or video signals, by digital means, the signal has to be converted into digital form. This process is known as analog-to-digital conversion. The sampling process is the first process performed in this conversion, and it converts a continuous-time signal into a discrete-time signal or a sequence of numbers. Digital transmission of analog signals is possible by virtue of the sampling theorem, and the sampling operation is performed in accordance with the sampling theorem.

In this chapter, using the Fourier transform technique, we present this remarkable sampling theorem and discuss the operation of sampling and practical aspects of sampling.

2.2 Instantaneous Sampling

Suppose we sample an arbitrary analog signal $m(t)$ shown in Fig. 2.1(a) instantaneously at a uniform rate, once every T_s seconds. As a result of this sampling process, we obtain an infinite sequence of samples $\{m(nT_s)\}$, where n takes on all possible integers. This form of sampling is called *instantaneous sampling*. We refer to T_s as the **sampling interval**, and its reciprocal $1/T_s = f_s$ as the **sampling rate**. Sampling rate (samples per second) is often cited in terms of sampling frequency expressed in hertz.

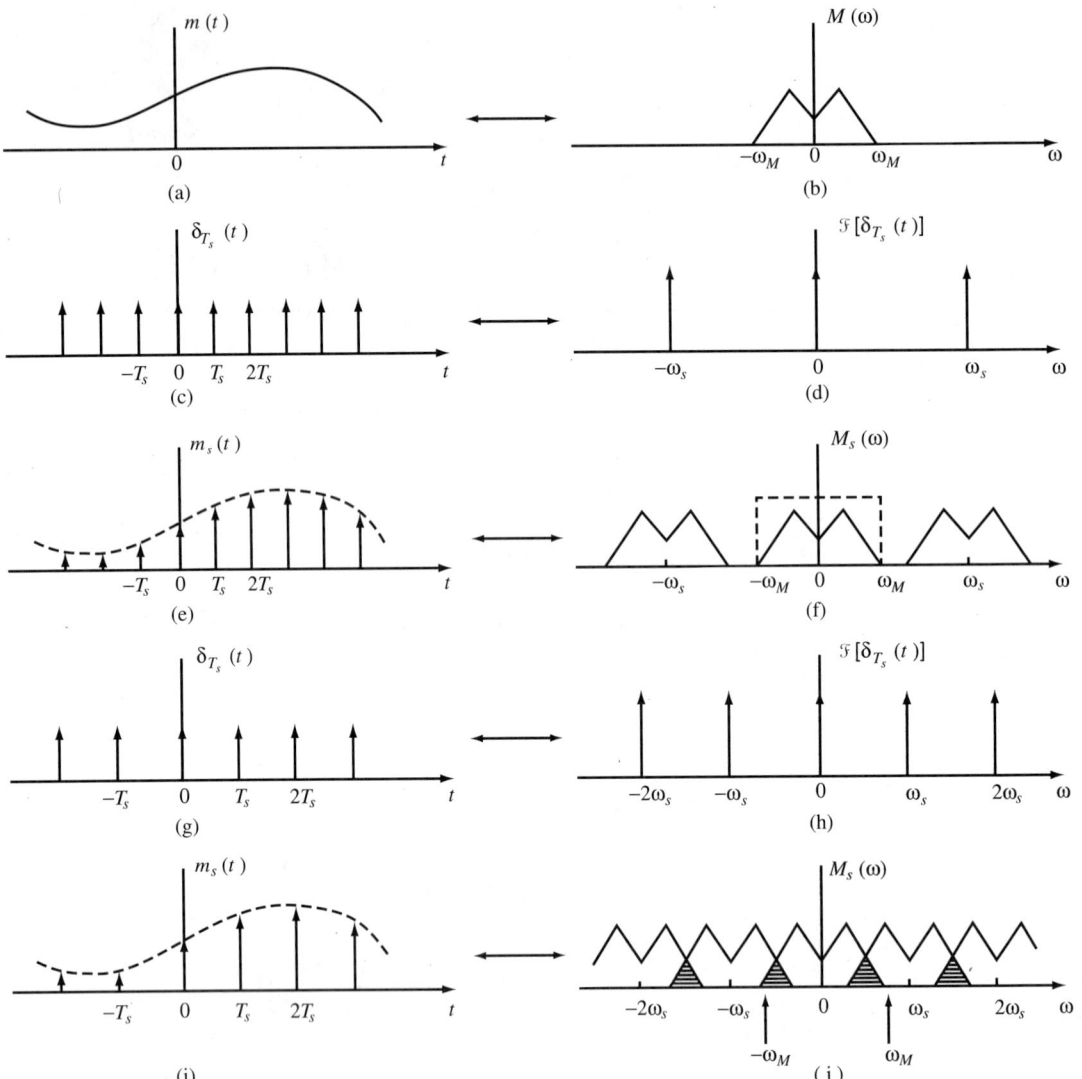

FIGURE 2.1 Illustration of instantaneous sampling and sampling theorem.

Ideal Sampled Signal

Let $m_s(t)$ be obtained by multiplication of $m(t)$ by the unit impulse train $\delta_T(t)$ with period T_s [Fig. 2.1(c)], that is,

$$m_s(t) = m(t)\delta_{T_s}(t) = m(t) \sum_{n=-\infty}^{\infty} \delta(t - nT_s)$$

$$= \sum_{n=-\infty}^{\infty} m(t)\delta(t - nT_s) = \sum_{n=-\infty}^{\infty} m(nT_s)\delta(t - nT_s) \qquad (2.1)$$

where we used the property of the δ function, $m(t)\delta(t - t_0) = m(t_0)\delta(t - t_0)$. The signal $m_s(t)$ [Fig. 2.1(e)] is referred to as the **ideal sampled signal**.

Sampling

Band-Limited Signals

A real-valued signal $m(t)$ is called a **band-limited signal** if its Fourier transform $M(\omega)$ satisfies the condition

$$M(\omega) = 0 \quad \text{for } |\omega| > \omega_M \tag{2.2}$$

where $\omega_M = 2\pi f_M$ [Fig. 2.1(b)]. A band-limited signal specified by Eq. (2.2) is often referred to as a *low-pass signal*.

2.3 Sampling Theorem

The sampling theorem states that a band-limited signal $m(t)$ specified by Eq. (2.2) can be uniquely determined from its values $m(nT_s)$ sampled at uniform interval T_s if $T_s \leq \pi/\omega_M = 1/(2f_M)$. In fact, when $T_s = \pi/\omega_M$, $m(t)$ is given by

$$m(t) = \sum_{n=-\infty}^{\infty} m(nT_s) \frac{\sin \omega_M(t - nT_s)}{\omega_M(t - nT_s)} \tag{2.3}$$

which is known as the **Nyquist–Shannon interpolation formula** and it is also sometimes called the *cardinal series*. The sampling interval $T_s = 1/(2f_M)$ is called the *Nyquist interval* and the minimum rate $f_s = 1/T_s = 2f_M$ is known as the **Nyquist rate**.

Illustration of the instantaneous sampling process and the sampling theorem is shown in Fig. 2.1. The Fourier transform of the unit impulse train is given by [Fig. 2.1(d)]

$$\mathcal{F}\{\delta_{T_s}(t)\} = \omega_s \sum_{n=-\infty}^{\infty} \delta(\omega - n\omega_s) \quad \omega_s = 2\pi/T_s \tag{2.4}$$

Then, by the convolution property of the Fourier transform, the Fourier transform $M_s(\omega)$ of the ideal sampled signal $m_s(t)$ is given by

$$M_s(\omega) = \frac{1}{2\pi}\left[M(\omega) * \omega_s \sum_{n=-\infty}^{\infty} \delta(\omega - n\omega_s)\right]$$

$$= \frac{1}{T_s} \sum_{n=-\infty}^{\infty} M(\omega - n\omega_s) \tag{2.5}$$

where $*$ denotes convolution and we used the convolution property of the δ-function $M(\omega) * \delta(\omega - \omega_0) = M(\omega - \omega_0)$. Thus, the sampling has produced images of $M(\omega)$ along the frequency axis. Note that $M_s(\omega)$ will repeat periodically without overlap as long as $\omega_s \geq 2\omega_M$ or $f_s \geq 2f_M$ [Fig. 2.1(f)]. It is clear from Fig. 2.1(f) that we can recover $M(\omega)$ and, hence, $m(t)$ by passing the sampled signal $m_s(t)$ through an ideal low-pass filter having frequency response

$$H(\omega) = \begin{cases} T_s, & |\omega| \leq \omega_M \\ 0, & \text{otherwise} \end{cases} \tag{2.6}$$

where $\omega_M = \pi/T_s$. Then

$$M(\omega) = M_s(\omega) H(\omega) \tag{2.7}$$

Taking the inverse Fourier transform of Eq. (2.6), we obtain the impulse response $h(t)$ of the ideal low-pass filter as

$$h(t) = \frac{\sin \omega_M t}{\omega_M t} \qquad (2.8)$$

Taking the inverse Fourier transform of Eq. (2.7), we obtain

$$\begin{aligned} m(t) &= m_s(t) * h(t) \\ &= \sum_{n=-\infty}^{\infty} m(nT_s)\delta(t-nT_s) * \frac{\sin \omega_M t}{\omega_M t} \\ &= \sum_{n=-\infty}^{\infty} m(nT_s) \frac{\sin \omega_M(t-nT_s)}{\omega_M(t-nT_s)} \end{aligned} \qquad (2.9)$$

which is Eq. (2.3).

The situation shown in Fig. 2.1(j) corresponds to the case where $f_s < 2f_M$. In this case there is an overlap between $M(\omega)$ and $M(\omega - \omega_M)$. This overlap of the spectra is known as *aliasing* or *foldover*. When this aliasing occurs, the signal is distorted and it is impossible to recover the original signal $m(t)$ from the sampled signal. To avoid aliasing, in practice, the signal is sampled at a rate slightly higher than the Nyquist rate. If $f_s > 2f_M$, then as shown in Fig. 2.1(f), there is a gap between the upper limit ω_M of $M(\omega)$ and the lower limit $\omega_s - \omega_M$ of $M(\omega - \omega_s)$. This range from ω_M to $\omega_s - \omega_M$ is called a *guard band*. As an example, speech transmitted via telephone is generally limited to $f_M = 3.3$ kHz (by passing the sampled signal through a low-pass filter). The Nyquist rate is, thus, 6.6 kHz. For digital transmission, the speech is normally sampled at the rate $f_s = 8$ kHz. The guard band is then $f_s - 2f_M = 1.4$ kHz. The use of a sampling rate higher than the Nyquist rate also has the desirable effect of making it somewhat easier to design the low-pass reconstruction filter so as to recover the original signal from the sampled signal.

2.4 Sampling of Sinusoidal Signals

A special case is the sampling of a sinusoidal signal having the frequency f_M. In this case we require that $f_s > 2f_M$ rather that $f_s \geq 2f_M$. To see that this condition is necessary, let $f_s = 2f_M$. Now, if an initial sample is taken at the instant the sinusoidal signal is zero, then all successive samples will also be zero. This situation is avoided by requiring $f_s > 2f_M$.

2.5 Sampling of Bandpass Signals

A real-valued signal $m(t)$ is called a **bandpass signal** if its Fourier transform $M(\omega)$ satisfies the condition

$$M(\omega) = 0 \quad \text{except for} \quad \begin{cases} \omega_1 < \omega < \omega_2 \\ -\omega_2 < \omega < -\omega_1 \end{cases} \qquad (2.10)$$

where $\omega_1 = 2\pi f_1$ and $\omega_2 = 2\pi f_2$ [Fig. 2.2(a)].

The sampling theorem for a band-limited signal has shown that a sampling rate of $2f_2$ or greater is adequate for a low-pass signal having the highest frequency f_2. Therefore, treating $m(t)$ specified by Eq. (2.10) as a special case of such a low-pass signal, we conclude that a sampling rate of $2f_2$ is adequate for the sampling of the bandpass signal $m(t)$. But it is not necessary to sample this fast. The minimum allowable sampling rate depends on f_1, f_2, and the bandwidth $f_B = f_2 - f_1$.

Let us consider the direct sampling of the bandpass signal specified by Eq. (2.10). The spectrum of the sampled signal is periodic with the period $\omega_s = 2\pi f_s$, where f_s is the sampling frequency,

Sampling

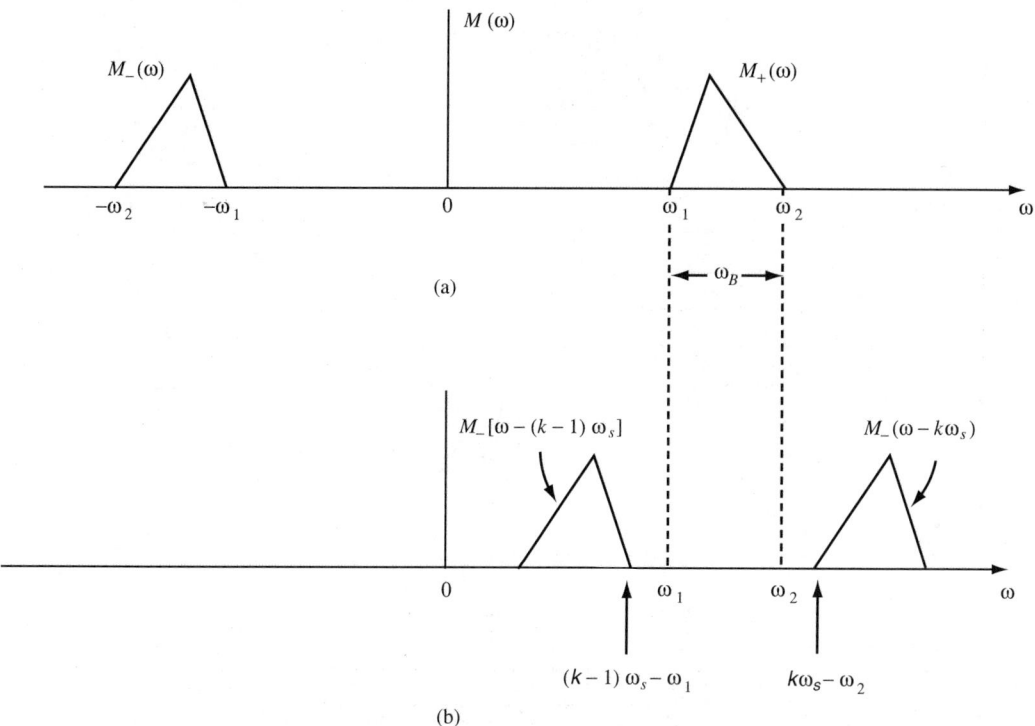

FIGURE 2.2 (a) Spectrum of a bandpass signal; (b) Shifted spectra of $M_-(\omega)$.

as in Eq. (2.4). Shown in Fig. 2.2(b) are the two right shifted spectra of the negative side spectrum $M_-(\omega)$. If the recovering of the bandpass signal is achieved by passing the sampled signal through an ideal bandpass filter covering the frequency bands $(-\omega_2, -\omega_1)$ and (ω_1, ω_2), it is necessary that there be no aliasing problem. From Fig. 2.2(b), it is clear that to avoid overlap it is necessary that

$$\omega_s \geq 2(\omega_2 - \omega_1) \tag{2.11}$$
$$(k-1)\omega_s - \omega_1 \leq \omega_1 \tag{2.12}$$

and

$$k\omega_s - \omega_2 \geq \omega_2 \tag{2.13}$$

where $\omega_1 = 2\pi f_1$, $\omega_2 = 2\pi f_2$, and k is an integer ($k = 1, 2, \ldots$). Since $f_1 = f_2 - f_B$, these constraints can be expressed as

$$1 \leq k \leq \frac{f_2}{f_B} \leq \frac{k}{2} \frac{f_s}{f_B} \tag{2.14}$$

and

$$\frac{k-1}{2} \frac{f_s}{f_B} \leq \frac{f_2}{f_B} - 1 \tag{2.15}$$

A graphical description of Eqs. (2.14) and (2.15) is illustrated in Fig. 2.3. The unshaded regions represent where the constraints are satisfied, whereas the shaded regions represent the regions where the constraints are not satisfied and overlap will occur. The solid line in Fig. 2.3 shows the locus of

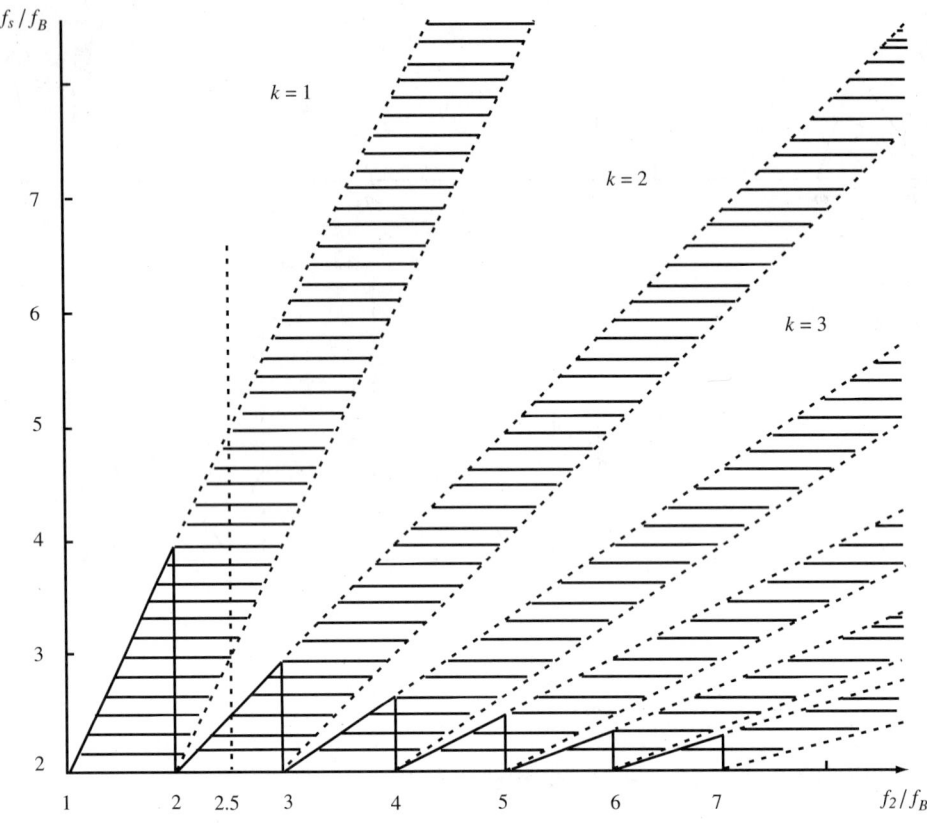

FIGURE 2.3 Minimum and permissible sampling rates for a bandpass signal.

the minimum sampling rate. The minimum sampling rate is given by

$$\min\{f_s\} = \frac{2f_2}{m} \qquad (2.16)$$

where m is the largest integer not exceeding f_2/f_B. Note that if the ratio f_2/f_B is an integer, then the minimum sampling rate is $2f_B$. As an example, consider a bandpass signal with $f_1 = 1.5$ kHz and $f_2 = 2.5$ kHz. Here $f_B = f_2 - f_1 = 1$ kHz, and $f_2/f_B = 2.5$. Then from Eq. (2.16) and Fig. 2.3 we see that the minimum sampling rate is $2f_2/2 = f_2 = 2.5$ kHz, and allowable ranges of sampling rate are 2.5 kHz $\leq f_s \leq$ 3 kHz and $f_s \geq 5$ kHz ($= 2f_2$).

2.6 Practical Sampling

In practice, the sampling of an analog signal is performed by means of high-speed switching circuits, and the sampling process takes the form of *natural sampling* or **flat-top sampling**.

Natural Sampling

Natural sampling of a band-limited signal $m(t)$ is shown in Fig. 2.4. The sampled signal $m_{\text{ns}}(t)$ can be expressed as

$$m_{\text{ns}}(t) = m(t)x_p(t) \qquad (2.17)$$

Sampling

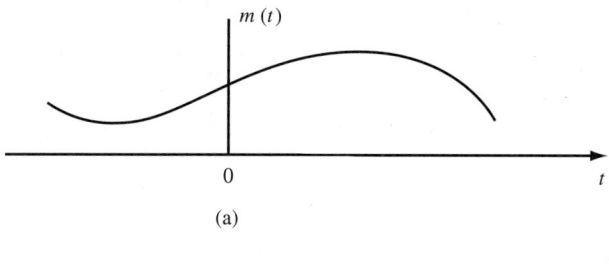

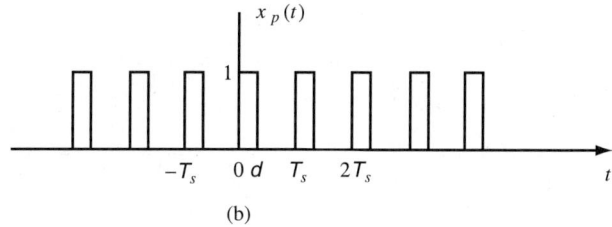

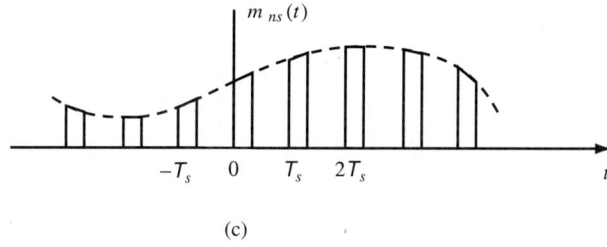

FIGURE 2.4 Natural sampling.

where $x_p(t)$ is the periodic train of rectangular pulses with fundamental period T_s, and each rectangular pulse in $x_p(t)$ has duration d and unit amplitude [Fig. 2.4(b)]. Observe that the sampled signal $m_{ns}(t)$ consists of a sequence of pulses of varying amplitude whose tops follow the waveform of the signal $m(t)$ [Fig. 2.4(c)].

The Fourier transform of $x_p(t)$ is

$$X_p(\omega) = \sum_{n=-\infty}^{\infty} c_n \delta(\omega - n\omega_s) \qquad \omega_s = 2\pi/T_s \tag{2.18}$$

where

$$c_n = \frac{d}{T_s} \frac{\sin(n\omega_s d/2)}{n\omega_s d/2} e^{-jn\omega_s d/2} \tag{2.19}$$

Then the Fourier transform of $m_{ns}(t)$ is given by

$$M_{ns}(\omega) = M(\omega) * X_p(\omega) = \sum_{n=-\infty}^{\infty} c_n M(\omega - n\omega_s) \tag{2.20}$$

from which we see that the effect of the natural sampling is to multiply the nth shifted spectrum $M(\omega - n\omega_s)$ by a constant c_n. Thus, the original signal $m(t)$ can be reconstructed from $m_{ns}(t)$ with no distortion by passing $m_{ns}(t)$ through an ideal low-pass filter if the sampling rate f_s is equal to or greater than the Nyquist rate $2f_M$.

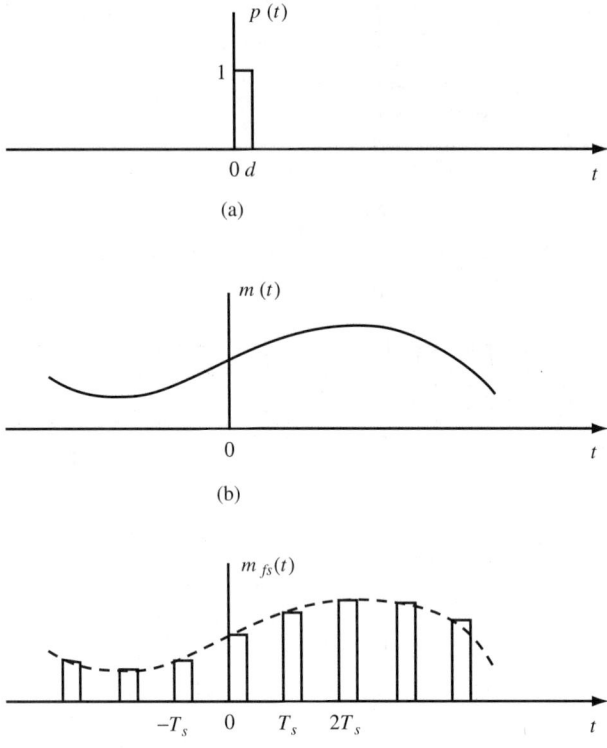

FIGURE 2.5 Flat-top sampling.

Flat-Top Sampling

The sampled waveform, produced by practical sampling devices that are the sample and hold types, has the form [Fig. 2.5(c)]

$$m_{\text{fs}}(t) = \sum_{n=-\infty}^{\infty} m(nT_s)p(t - nT_s) \qquad (2.21)$$

where $p(t)$ is a rectangular pulse of duration d with unit amplitude [Fig. 2.5(a)]. This type of sampling is known as **flat-top sampling**. Using the ideal sampled signal $m_s(t)$ of Eq. (2.1), $m_{\text{fs}}(t)$ can be expressed as

$$m_{\text{fs}}(t) = p(t) * \left[\sum_{n=-\infty}^{\infty} m(nT_s)\delta(t - nT_s) \right] = p(t) * m_s(t) \qquad (2.22)$$

Using the convolution property of the Fourier transform and Eq. (2.4), the Fourier transform of $m_{\text{fs}}(t)$ is given by

$$M_{\text{fs}}(\omega) = P(\omega)M_s(\omega) = \frac{1}{T_s} \sum_{n=-\infty}^{\infty} P(\omega)M(\omega - n\omega_s) \qquad (2.23)$$

where

$$P(\omega) = d\frac{\sin(\omega d/2)}{\omega d/2}e^{-j\omega d/2} \qquad (2.24)$$

From Eq. (2.23) we see that by using flat-top sampling we have introduced amplitude distortion and time delay, and the primary effect is an attenuation of high-frequency components. This effect is known as the *aperture effect*. The aperture effect can be compensated by an equalizing filter with a frequency response $H_{eq}(\omega) = 1/P(\omega)$. If the pulse duration d is chosen such that $d \ll T_s$, however, then $P(\omega)$ is essentially constant over the baseband and no equalization may be needed.

2.7 Sampling Theorem in the Frequency Domain

The sampling theorem expressed in Eq. (2.4) is the time-domain sampling theorem. There is a dual to this time-domain sampling theorem, i.e., the sampling theorem in the frequency domain.

Time-limited signals: A continuous-time signal $m(t)$ is called **time limited** if

$$m(t) = 0 \qquad \text{for} \quad |t| > |T_0| \qquad (2.25)$$

Frequency-domain sampling theorem: The frequency-domain sampling theorem states that the Fourier transform $M(\omega)$ of a time-limited signal $m(t)$ specified by Eq. (2.25) can be uniquely determined from its values $M(n\omega_s)$ sampled at a uniform rate ω_s if $\omega_s \leq \pi/T_0$. In fact, when $\omega_s = \pi/T_0$, then $M(\omega)$ is given by

$$M(\omega) = \sum_{n=-\infty}^{\infty} M(n\omega_s)\frac{\sin T_0(\omega - n\omega_s)}{T_0(\omega - n\omega_s)} \qquad (2.26)$$

2.8 Summary and Discussion

The sampling theorem is the fundamental principle of digital communications. We state the sampling theorem in two parts.

Theorem 2.1. If the signal contains no frequency higher than f_M Hz, it is completely described by specifying its samples taken at instants of time spaced $1/2f_M$ s.

Theorem 2.2. The signal can be completely recovered from its samples taken at the rate of $2f_M$ samples per second or higher.

The preceding sampling theorem assumes that the signal is strictly band limited. It is known that if a signal is band limited it cannot be time limited and vice versa. In many practical applications, the signal to be sampled is time limited and, consequently, it cannot be strictly band limited. Nevertheless, we know that the frequency components of physically occurring signals attenuate rapidly beyond some defined bandwidth, and for practical purposes we consider these signals are band limited. This approximation of real signals by band limited ones introduces no significant error in the application of the sampling theorem. When such a signal is sampled, we band limit the signal by filtering before sampling and sample at a rate slightly higher than the nominal Nyquist rate.

Defining Terms

Band-Limited Signal: A signal whose frequency content (Fourier transform) is equal to zero above some specified frequency.

Bandpass Signal: A signal whose frequency content (Fourier transform) is nonzero only in a band of frequencies not including the origin.

Flat-top Sampling: Sampling with finite width pulses that maintain a constant value for a time period less than or equal to the sampling interval. The constant value is the amplitude of the signal at the desired sampling instant.

Ideal Sampled Signal: A signal sampled using an ideal impulse train.

Nyquist Rate: The minimum allowable sampling rate of $2f_M$ samples per second, to reconstruct a signal band limited to f_M hertz.

Nyquist-Shannon Interpolation Formula: The infinite series representing a time domain waveform in terms of its ideal samples taken at uniform intervals.

Sampling Interval: The time between samples in uniform sampling.

Sampling Rate: The number of samples taken per second (expressed in Hertz and equal to the reciprocal of the sampling interval).

Time-limited: A signal that is zero outside of some specified time interval.

References

Brown, J.L. Jr. 1980. First order sampling of bandpass signals—A new approach. *IEEE Trans. Information Theory.* IT-26(5):613–615.

Byrne, C.L. and Fitzgerald, R. M. 1982. Time-limited sampling theorem for band-limited signals, *IEEE Trans. Information Theory.* IT-28(5):807–809.

Hsu, H.P. 1984. *Applied Fourier Analysis,* Harcourt Brace Jovanovich, San Diego, CA.

Hsu, H.P. 1993. *Analog and Digital Communications,* McGraw-Hill, New York.

Hulthén, R. 1983. Restoring causal signals by analytical continuation: A generalized sampling theorem for causal signals. *IEEE Trans. Acoustics, Speech, and Signal Processing.* ASSP-31(5):1294–1298.

Jerri, A.J. 1977. The Shannon sampling theorem—Its various extensions and applications: A tutorial review, *Proc. IEEE.* 65(11):1565–1596.

Further Information

For a tutorial review of the sampling theorem, historical notes, and earlier references see Jerri (1977).

3
Pulse Code Modulation[1]

Leon W. Couch II
University of Florida

    3.1 Introduction 23
    3.2 Generation of PCM 24
    3.3 Percent Quantizing Noise 25
    3.4 Practical PCM Circuits 26
    3.5 Bandwidth of PCM 27
    3.6 Effects of Noise 28
    3.7 Nonuniform Quantizing: μ-Law and A-Law Companding 30
    3.8 Example: Design of a PCM System 33

3.1 Introduction

Pulse code modulation (PCM) is analog-to-digital conversion of a special type where the information contained in the instantaneous samples of an analog signal is represented by digital words in a serial bit stream.

If we assume that each of the digital words has n binary digits, there are $M = 2^n$ unique code words that are possible, each code word corresponding to a certain amplitude level. Each sample value from the analog signal, however, can be any one of an infinite number of levels, so that the digital word that represents the amplitude closest to the actual sampled value is used. This is called **quantizing**. That is, instead of using the exact sample value of the analog waveform, the sample is replaced by the closest allowed value, where there are M allowed values, and each allowed value corresponds to one of the code words.

PCM is very popular because of the many advantages it offers. Some of these advantages are as follows.

- Relatively inexpensive digital circuitry may be used extensively in the system.
- PCM signals derived from all types of analog sources (audio, video, etc.) may be time-division multiplexed with data signals (e.g., from digital computers) and transmitted over a common high-speed digital communication system.
- In long-distance digital telephone systems requiring repeaters, a *clean* PCM waveform can be regenerated at the output of each repeater, where the input consists of a noisy PCM waveform. The noise at the input, however, may cause bit errors in the regenerated PCM output signal.

---

[1] *Source:* Leon W. Couch, II 1993. *Digital and Analog Communication Systems*, 4th ed., Macmillan Publishing Co., New York. With permission.

- The noise performance of a digital system can be superior to that of an analog system. In addition, the probability of error for the system output can be reduced even further by the use of appropriate coding techniques.

These advantages usually outweigh the main disadvantage of PCM: a much wider bandwidth than that of the corresponding analog signal.

3.2 Generation of PCM

The PCM signal is generated by carrying out three basic operations: sampling, quantizing, and encoding (see Fig. 3.1). The sampling operation generates an instantaneously-sampled flat-top **pulse-amplitude modulated** (PAM) signal.

The quantizing operation is illustrated in Fig. 3.2 for the $M = 8$ level case. This quantizer is said to be *uniform* since all of the steps are of equal size. Since we are approximating the analog sample values by using a finite number of levels ($M = 8$ in this illustration), *error* is introduced into the recovered output analog signal because of the quantizing effect. The error waveform is illustrated in Fig. 3.2c. The quantizing error consists of the difference between the analog signal at the sampler input and the output of the quantizer. Note that the peak value of the error (± 1) is one-half of the quantizer step size (2). If we sample at the Nyquist rate ($2B$, where B is the absolute bandwidth, in hertz, of the input analog signal) or faster and there is negligible channel noise, there will still be noise, called *quantizing noise*, on the recovered analog waveform due to this error. The quantizing noise can also be thought of as a round-off error. The quantizer output is a *quantized* (i.e., only M possible amplitude values) PAM signal.

TABLE 3.1 3-b Gray Code for $M = 8$ levels

Quantized Sample Voltage	Gray Code Word (PCM Output)
+7	110
+5	111
+3	101
+1	100
	Mirror image except for sign bit
−1	000
−3	001
−5	011
−7	010

Source: Couch, L.W. II 1993. *Digital and Analog Communication Systems*, 4th ed. Macmillan Publishing Co., New York, p. 145. With permission.

The PCM signal is obtained from the quantized PAM signal by encoding each quantized sample value into a digital word. It is up to the system designer to specify the exact code word that will represent a particular quantized level. If a Gray code of Table 3.1 is used, the resulting PCM signal is shown in Fig. 3.2d where the PCM word for each quantized sample is strobed out of the encoder by the next clock pulse. The Gray code was chosen because it has only 1-b change for each step change in the quantized level. Consequently, single errors in the received PCM code word will cause minimum errors in the recovered analog level, provided that the sign bit is not in error.

Here we have described PCM systems that represent the quantized analog sample values by *binary* code words. Of course, it is possible to represent the quantized analog samples by digital words using other than

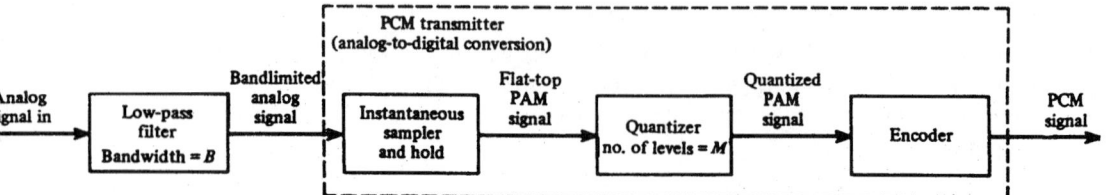

FIGURE 3.1 A PCM transmitter. *Source:* Couch, L.W. II 1993. *Digital and Analog Communication Systems*, 4th ed., Macmillan Publishing Co., New York, p. 143. With permission.

Pulse Code Modulation

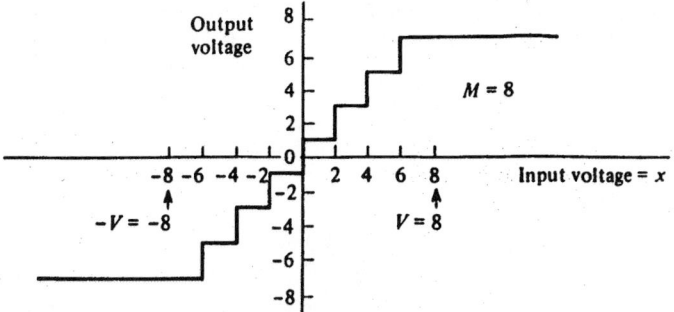

(a) Quantizer Output-Input Characteristics

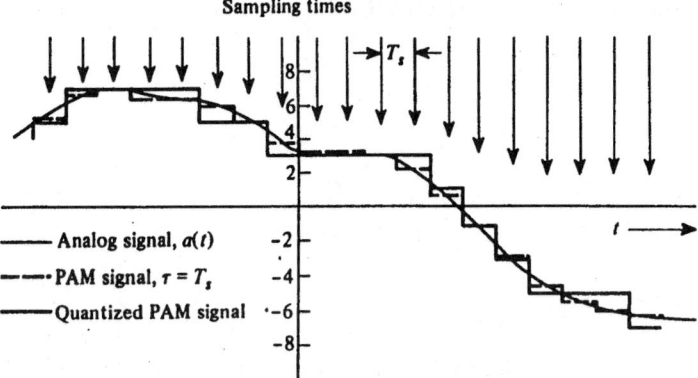

(b) Analog Signal, Flat-top PAM Signal, and Quantized PAM Signal

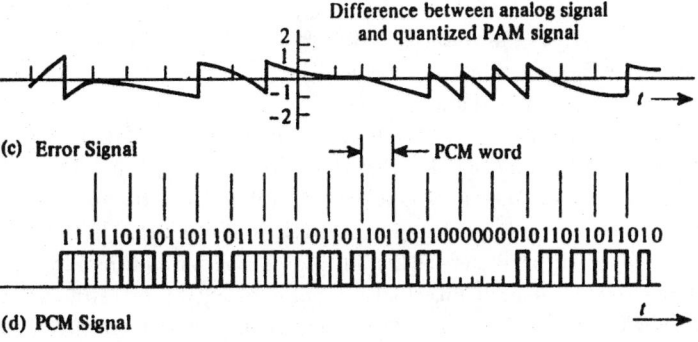

(c) Error Signal

(d) PCM Signal

FIGURE 3.2 Illustration of waveforms in a PCM system. *Source:* Couch, L.W. II 1993. *Digital and Analog Communication Systems*, 4th ed., Macmillan Publishing Co., New York, p. 144. With permission.

base 2. That is, for base q, the number of quantized levels allowed is $M = q^n$, where n is the number of q base digits in the code word. We will not pursue this topic since binary ($q = 2$) digital circuits are most commonly used.

3.3 Percent Quantizing Noise

The quantizer at the PCM encoder produces an error signal at the PCM decoder output as illustrated in Fig. 3.2c. The peak value of this error signal may be expressed as a percentage of the maximum possible analog signal amplitude. Referring to Fig. 3.2c, a peak error of 1 V occurs for a maximum analog signal amplitude of $M = 8$ V as shown in Fig. 3.1c. Thus, in general,

$$\frac{2P}{100} = \frac{1}{M} = \frac{1}{2^n}$$

or

$$2^n = \frac{50}{P} \tag{3.1}$$

where P is the peak percentage error for a PCM system that uses n bit code words. The design value of n needed in order to have less than P percent error is obtained by taking the base 2 logarithm of both sides of Eq. (3.1), where it is realized that $\log_2(x) = [\log_{10}(x)]/\log_{10}(2) = 3.32 \log_{10}(x)$. That is,

$$n \geq 3.32 \log_{10}\left(\frac{50}{P}\right) \tag{3.2}$$

where n is the number of bits needed in the PCM word in order to obtain less than P percent error in the recovered analog signal (i.e., decoded PCM signal).

3.4 Practical PCM Circuits

Three techniques are used to implement the analog-to-digital converter (ADC) encoding operation. These are the *counting* or *ramp*, *serial* or *successive approximation*, and *parallel* or *flash* encoders.

In the counting encoder, at the same time that the sample is taken, a ramp generator is energized and a binary counter is started. The output of the ramp generator is continuously compared to the sample value; when the value of the ramp becomes equal to the sample value, the binary value of the counter is read. This count is taken to be the PCM word. The binary counter and the ramp generator are then reset to zero and are ready to be reenergized at the next sampling time. This technique requires only a few components, but the speed of this type of ADC is usually limited by the speed of the counter. The Intersil ICL7126 CMOS ADC integrated circuit uses this technique.

The serial encoder compares the value of the sample with trial quantized values. Successive trials depend on whether the past comparator outputs are positive or negative. The trial values are chosen first in large steps and then in small steps so that the process will converge rapidly. The trial voltages are generated by a series of voltage dividers that are configured by (on-off) switches. These switches are controlled by digital logic. After the process converges, the value of the switch settings is read out as the PCM word. This technique requires more precision components (for the voltage dividers) than the ramp technique. The speed of the feedback ADC technique is determined by the speed of the switches. The National Semiconductor ADC0804 8-b ADC uses this technique.

The parallel encoder uses a set of parallel comparators with reference levels that are the permitted quantized values. The sample value is fed into all of the parallel comparators simultaneously. The high or low level of the comparator outputs determines the binary PCM word with the aid of some digital logic. This is a fast ADC technique but requires more hardware than the other two methods. The RCA CA3318 8-b ADC integrated circuit is an example of the technique.

All of the integrated circuits listed as examples have parallel digital outputs that correspond to the digital word that represents the analog sample value. For generation of PCM, the parallel output (digital word) needs to be converted to serial form for transmission over a two-wire channel. This is accomplished by using a parallel-to-serial converter integrated circuit, which is also known as a **serial-input-output** (SIO) chip. The SIO chip includes a shift register that is set to contain the parallel data (usually, from 8 or 16 input lines). Then the data are shifted out of the last stage of

Pulse Code Modulation

the shift register bit by bit onto a single output line to produce the serial format. Furthermore, the SIO chips are usually full duplex; that is, they have two sets of shift registers, one that functions for data flowing in each direction. One shift register converts parallel input data to serial output data for transmission over the channel, and, simultaneously, the other shift register converts received serial data from another input to parallel data that are available at another output. Three types of SIO chips are available: the *universal asynchronous receiver/transmitter* (UART), the *universal synchronous receiver/transmitter* (USRT), and the *universal synchronous/asynchronous receiver transmitter* (USART). The UART transmits and receives asynchronous serial data, the USRT transmits and receives synchronous serial data, and the USART combines both a UART and a USRT on one chip.

At the receiving end the PCM signal is decoded back into an analog signal by using a digital-to-analog converter (DAC) chip. If the DAC chip has a parallel data input, the received serial PCM data are first converted to a parallel form using a SIO chip as described in the preceding paragraph. The parallel data are then converted to an approximation of the analog sample value by the DAC chip. This conversion is usually accomplished by using the parallel digital word to set the configuration of electronic switches on a resistive current (or voltage) divider network so that the analog output is produced. This is called a *multiplying* DAC since the analog output voltage is directly proportional to the divider reference voltage multiplied by the value of the digital word. The Motorola MC1408 and the National Semiconductor DAC0808 8-b DAC chips are examples of this technique. The DAC chip outputs samples of the quantized analog signal that approximates the analog sample values. This may be smoothed by a low-pass reconstruction filter to produce the analog output.

The Electrical Engineering Handbook [Dorf, 1993, pp. 771–782] gives more details on ADC, DAC, and PCM circuits.

3.5 Bandwidth of PCM

A good question to ask is: What is the spectrum of a PCM signal? For the case of PAM signalling, the spectrum of the PAM signal could be obtained as a function of the spectrum of the input analog signal because the PAM signal is a linear function of the analog signal. This is not the case for PCM. As shown in Figs. 3.1 and 3.2, the PCM signal is a nonlinear function of the input signal. Consequently, the spectrum of the PCM signal is not directly related to the spectrum of the input analog signal. It can be shown that the spectrum of the PCM signal depends on the bit rate, the correlation of the PCM data, and on the PCM waveform pulse shape (usually rectangular) used to describe the bits [Couch, 1993; Couch, 1995]. From Fig. 3.2, the bit rate is

$$R = nf_s \tag{3.3}$$

where n is the number of bits in the PCM word ($M = 2^n$) and f_s is the sampling rate. For no aliasing we require $f_s \geq 2B$ where B is the bandwidth of the analog signal (that is to be converted to the PCM signal). The dimensionality theorem [Couch, 1993; Couch, 1995] shows that the bandwidth of the PCM waveform is bounded by

$$B_{\text{PCM}} \geq \frac{1}{2}R = \frac{1}{2}nf_s \tag{3.4}$$

where equality is obtained if a $(\sin x)/x$ type of pulse shape is used to generate the PCM waveform. The exact spectrum for the PCM waveform will depend on the pulse shape that is used as well as on the type of line encoding. For example, if one uses a rectangular pulse shape with polar nonreturn to zero (NRZ) line coding, the first null bandwidth is simply

$$B_{\text{PCM}} = R = nf_s \text{ Hz} \tag{3.5}$$

TABLE 3.2 Performance of a PCM System with Uniform Quantizing and No Channel Noise

Number of Quantizer Levels Used, M	Length of the PCM Word, n (bits)	Bandwidth of PCM Signal (First Null Bandwidth)[a]	Recovered Analog Signal Power-to-Quantizing Noise Power Ratios (dB) $(S/N)_{\text{out}}$
2	1	2B	6.0
4	2	4B	12.0
8	3	6B	18.1
16	4	8B	24.1
32	5	10B	30.1
64	6	12B	36.1
128	7	14B	42.1
256	8	16B	48.2
512	9	18B	54.2
1,024	10	20B	60.2
2,048	11	22B	66.2
4,096	12	24B	72.2
8,192	13	26B	78.3
16,384	14	28B	84.3
32,768	15	30B	90.3
65,536	16	32B	96.3

[a] B is the absolute bandwidth of the input analog signal. *Source:* Couch, L.W. II 1993. *Digital and Analog Communication Systems,* 4th ed. Macmillan Publishing Co., New York, p. 148. With permission.

Table 3.2 presents a tabulation of this result for the case of the minimum sampling rate, $f_s = 2B$. Note that Eq. (3.4) demonstrates that the bandwidth of the PCM signal has a lower bound given by

$$B_{\text{PCM}} \geq nB \tag{3.6}$$

where $f_s > 2B$ and B is the bandwidth of the corresponding analog signal. Thus, for reasonable values of n, the bandwidth of the PCM signal will be significantly larger than the bandwidth of the corresponding analog signal that it represents. For the example shown in Fig. 3.2 where $n = 3$, the PCM signal bandwidth will be at least three times wider than that of the corresponding analog signal. Furthermore, if the bandwidth of the PCM signal is reduced by improper filtering or by passing the PCM signal through a system that has a poor frequency response, the filtered pulses will be elongated (stretched in width) so that pulses corresponding to any one bit will smear into adjacent bit slots. If this condition becomes too serious, it will cause errors in the detected bits. This pulse smearing effect is called **intersymbol interference** (ISI).

3.6 Effects of Noise

The analog signal that is recovered at the PCM system output is corrupted by noise. Two main effects produce this noise or distortion: 1) quantizing noise that is caused by the M-step quantizer at the PCM transmitter and 2) bit errors in the recovered PCM signal. The bit errors are caused by *channel noise* as well as improper channel filtering, which causes ISI. In addition, if the input analog signal is not strictly band limited, there will be some aliasing noise on the recovered analog signal [Spilker, 1977]. Under certain assumptions, it can be shown that the recovered analog *average* signal

Pulse Code Modulation

power to the average noise power [Couch, 1993] is

$$\left(\frac{S}{N}\right)_{\text{out}} = \frac{M^2}{1 + 4(M^2 - 1)P_e} \tag{3.7}$$

where M is the number of uniformly spaced quantizer levels used in the PCM transmitter and P_e is the probability of bit error in the recovered binary PCM signal at the receiver DAC before it is converted back into an analog signal. Most practical systems are designed so that P_e is negligible. Consequently, if we assume that there are no bit errors due to channel noise (i.e., $P_e = 0$), the S/N due only to quantizing errors is

$$\left(\frac{S}{N}\right)_{\text{out}} = M^2 \tag{3.8}$$

Numerical values for these S/N ratios are given in Table 3.2.

To realize these S/N ratios, one critical assumption is that the peak-to-peak level of the analog waveform at the input to the PCM encoder is set to the design level of the quantizer. For example, referring to Fig. 3.2, this corresponds to the input traversing the range $-V$ to $+V$ volts where $V = 8$ V is the design level of the quantizer. Equation (3.7) was derived for waveforms with equally likely values, such as a triangle waveshape, that have a peak-to-peak value of $2V$ and an rms value of $V/\sqrt{3}$, where V is the design peak level of the quantizer.

From a practical viewpoint, the quantizing noise at the output of the PCM decoder can be categorized into four types depending on the operating conditions. The four types are overload noise, random noise, granular noise, and hunting noise. As discussed earlier, the level of the analog waveform at the input of the PCM encoder needs to be set so that its peak level does not exceed the design peak of V volts. If the peak input does exceed V, the recovered analog waveform at the output of the PCM system will have flat tops near the peak values. This produces *overload noise*. The flat tops are easily seen on an oscilloscope, and the recovered analog waveform sounds distorted since the flat topping produces unwanted harmonic components. For example, this type of distortion can be heard on PCM telephone systems when there are high levels such as dial tones, busy signals, or off-hook warning signals.

The second type of noise, *random noise*, is produced by the random quantization errors in the PCM system under normal operating conditions when the input level is properly set. This type of condition is assumed in Eq. (3.8). Random noise has a white hissing sound. If the input level is not sufficiently large, the S/N will deteriorate from that given by Eq. (3.8); the quantizing noise will still remain more or less random.

If the input level is reduced further to a relatively small value with respect to the design level, the error values are not equally likely from sample to sample, and the noise has a harsh sound resembling gravel being poured into a barrel. This is called *granular noise*. This type of noise can be randomized (noise power decreased) by increasing the number of quantization levels and, consequently, increasing the PCM bit rate. Alternatively, granular can be reduced by using a nonuniform quantizer, such as the μ-law or A-law quantizers that are described in the Sec. 3.7.

The fourth type of quantizing noise that may occur at the output of a PCM system is *hunting noise*. It can occur when the input analog waveform is nearly constant, including when there is no signal (i.e., zero level). For these conditions the sample values at the quantizer output (see Fig. 3.2) can oscillate between two adjacent quantization levels, causing an undesired sinusoidal type tone of frequency $1/2 f_s$ at the output of the PCM system. Hunting noise can be reduced by filtering out the tone or by designing the quantizer so that there is no vertical step at the constant value of the inputs, such as at 0-V input for the no signal case. For the no signal case, the hunting noise is also called *idle channel noise*. Idle channel noise can be reduced by using a horizontal step at the origin of the quantizer output–input characteristic instead of a vertical step as shown in Fig. 3.2.

Recalling that $M = 2^n$, we may express Eq. (3.8) in decibels by taking $10 \log_{10}(\cdot)$ of both sides of the equation,

$$\left(\frac{S}{N}\right)_{\text{dB}} = 6.02n + \alpha \qquad (3.9)$$

where n is the number of bits in the PCM word and $\alpha = 0$. This equation—called the 6-dB rule—points out the significant performance characteristic for PCM: an additional 6-dB improvement in S/N is obtained for each bit added to the PCM word. This is illustrated in Table 3.2. Equation (3.9) is valid for a wide variety of assumptions (such as various types of input waveshapes and quantification characteristics), although the value of α will depend on these assumptions [Jayant and Noll, 1984]. Of course, it is assumed that there are no bit errors and that the input signal level is large enough to range over a significant number of quantizing levels.

One may use Table 3.2 to examine the design requirements in a proposed PCM system. For example, high fidelity enthusiasts are turning to digital audio recording techniques. Here PCM signals are recorded instead of the analog audio signal to produce superb sound reproduction. For a dynamic range of 90 dB, it is seen that at least 15-b PCM words would be required. Furthermore, if the analog signal had a bandwidth of 20 kHz, the first null bandwidth for rectangular bit-shape PCM would be 2×20 kHz $\times 15 = 600$ kHz. Consequently, video-type tape recorders are needed to record and reproduce high-quality digital audio signals. Although this type of recording technique might seem ridiculous at first, it is realized that expensive high-quality analog recording devices are hard pressed to reproduce a dynamic range of 70 dB. Thus, digital audio is one way to achieve improved performance. This is being proven in the marketplace with the popularity of the digital compact disk (CD). The CD uses a 16-b PCM word and a sampling rate of 44.1 kHz on each stereo channel [Miyaoka, 1984; Peek, 1985]. Reed–Solomon coding with interleaving is used to correct burst errors that occur as a result of scratches and fingerprints on the compact disk.

3.7 Nonuniform Quantizing: μ-Law and A-Law Companding

Voice analog signals are more likely to have amplitude values near zero than at the extreme peak values allowed. For example, when digitizing voice signals, if the peak value allowed is 1 V, weak passages may have voltage levels on the order of 0.1 V (20 dB down). For signals such as these with nonuniform amplitude distribution, the granular quantizing noise will be a serious problem if the step size is not reduced for amplitude values near zero and increased for extremely large values. This is called nonuniform quantizing since a variable step size is used. An example of a nonuniform quantizing characteristic is shown in Fig. 3.3.

The effect of nonuniform quantizing can be obtained by first passing the analog signal through a compression (nonlinear) amplifier and then into the PCM circuit that uses a uniform quantizer. In the U.S., a μ-law type of compression characteristic is used. It is defined [Smith, 1957] by

$$|w_2(t)| = \frac{\ln(1 + \mu|w_1(t)|)}{\ln(1 + \mu)} \qquad (3.10)$$

where the allowed peak values of $w_1(t)$ are ± 1 (i.e., $|w_1(t)| \leq 1$), μ is a positive constant that is a parameter. This compression characteristic is shown in Fig. 3.3(b) for several values of μ, and it is noted that $\mu \to 0$ corresponds to linear amplification (uniform quantization overall). In the United States, Canada, and Japan, the telephone companies use a $\mu = 255$ compression characteristic in their PCM systems [Dammann, McDaniel, and Maddox, 1972].

Pulse Code Modulation

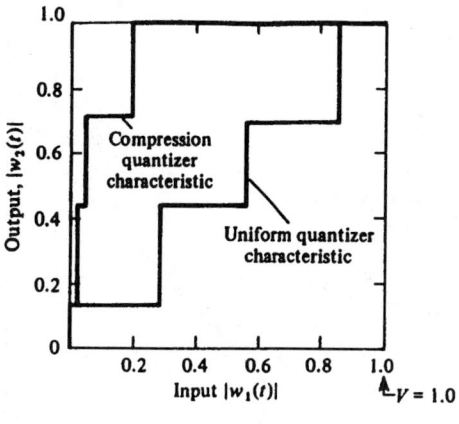

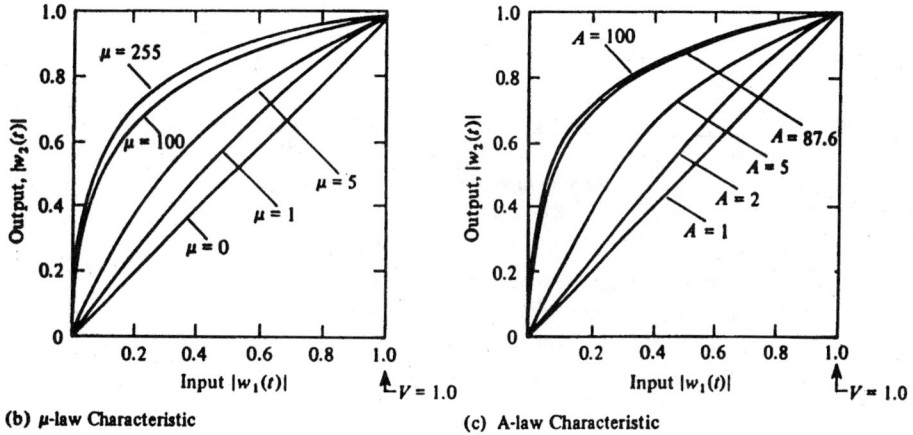

FIGURE 3.3 Compression characteristics (first quadrant shown). *Source:* Couch, L.W. II 1993. *Digital and Analog Communication Systems*, 4th ed., Macmillan Publishing Co., New York, p. 153. With permission.

Another compression law, used mainly in Europe, is the A-law characteristic. It is defined [Cattermole, 1969] by

$$|w_2(t)| = \begin{cases} \dfrac{A|w_1(t)|}{1 + \ln A}, & 0 \leq |w_1(t)| \leq \dfrac{1}{A} \\ \dfrac{1 + \ln(A|w_1(t)|)}{1 + \ln A}, & \dfrac{1}{A} \leq |w_1(t)| \leq 1 \end{cases} \quad (3.11)$$

where $|w_1(t)| < 1$ and A is a positive constant. The A-law compression characteristic is shown in Fig. 3.3(c). The typical value for A is 87.6.

When compression is used at the transmitter, *expansion* (i.e., decompression) must be used at the receiver output to restore signal levels to their correct relative values. The *expandor* characteristic is the inverse of the compression characteristic, and the combination of a compressor and an expandor is called a *compandor*.

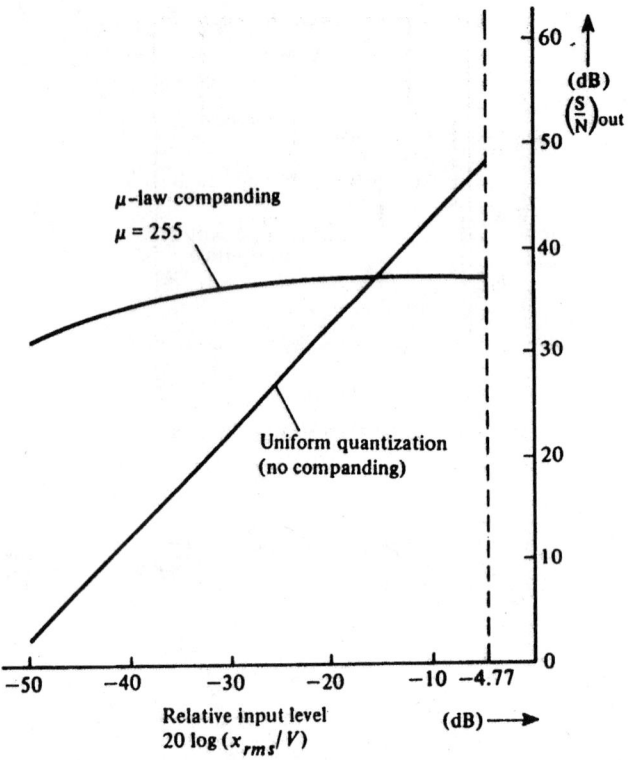

FIGURE 3.4 Output S/N of 8-b PCM systems with and without companding. *Source:* Couch, L.W. II 1993. *Digital and Analog Communication Systems*, 4th ed., Macmillan Publishing Co., New York, p. 155. With permission.

Once again, it can be shown that the output S/N follows the 6-dB law [Couch, 1993]

$$\left(\frac{S}{N}\right)_{dB} = 6.02 + \alpha \tag{3.12}$$

where for uniform quantizing

$$\alpha = 4.77 - 20\log(V/x_{rms}) \tag{3.13}$$

and for sufficiently large input levels[2] for μ-law companding

$$\alpha \approx 4.77 - 20\log[\ln(1+\mu)] \tag{3.14}$$

and for A-law companding [Jayant and Noll, 1984]

$$\alpha \approx 4.77 - 20\log[1+\ln A] \tag{3.15}$$

n is the number of bits used in the PCM word, V is the peak design level of the quantizer, and x_{rms} is the rms value of the input analog signal. Notice that the output S/N is a function of the

---

[2]See Couch, 1993 or Lathi, 1989 for a more complicated expression that is valid for any input level.

input level for the uniform quantizing (no companding) case but is relatively insensitive to input level for μ-law and A-law companding, as shown in Fig. 3.4. The ratio V/x_{rms} is called the *loading factor*. The input level is often set for a loading factor of 4 (12 dB) to ensure that the overload quantizing noise will be negligible. In practice this gives $\alpha = -7.3$ for the case of uniform encoding as compared to $\alpha = 0$, which was obtained for the ideal conditions associated with Eq. (3.8).

3.8 Example: Design of a PCM System

Assume that an analog voice-frequency signal, which occupies a band from 300 to 3400 Hz, is to be transmitted over a binary PCM system. The minimum sampling frequency would be $2 \times 3.4 = 6.8$ kHz. In practice the signal is oversampled, and in the U.S. a sampling frequency of 8 kHz is the standard used for voice-frequency signals in telephone communication systems. Assume that each sample value is represented by 8 b; then the bit rate of the PCM signal is

$$R = (f_s \text{ samples/s})(n \text{ b/s})$$
$$= (8\,k \text{ samples/s})(8 \text{ b/s}) = 64 \text{ kb/s} \qquad (3.16)$$

Referring to the dimensionality theorem [Eq. (3.4)], we realize that the theoretically minimum absolute bandwidth of the PCM signal is

$$B_{\min} = \frac{1}{2}D = 32 \text{ kHz} \qquad (3.17)$$

and this is realized if the PCM waveform consists of $(\sin x)/x$ pulse shapes. If rectangular pulse shaping is used, the absolute bandwidth is infinity, and the first null bandwidth [Eq. (3.5)] is

$$B_{\text{null}} = R = \frac{1}{T_b} = 64 \text{ kHz} \qquad (3.18)$$

That is, we require a bandwidth of 64 kHz to transmit this digital voice PCM signal where the bandwidth of the original analog voice signal was, at most, 4 kHz. Using $n = 8$ in Eq. (3.1), the error on the recovered analog signal is $\pm 0.2\%$. Using Eqs. (3.12) and (3.13) for the case of uniform quantizing with a loading factor, V/x_{rms}, of 10 (20 dB), we get for uniform quantizing

$$\left(\frac{S}{N}\right)_{\text{dB}} = 32.9 \text{ dB} \qquad (3.19)$$

Using Eqs. (3.12) and (3.14) for the case of $\mu = 255$ companding, we get

$$\left(\frac{S}{N}\right) = 38.05 \text{ dB} \qquad (3.20)$$

These results are illustrated in Fig. 3.4.

Defining Terms

Intersymbol interference: Filtering of a digital waveform so that a pulse corresponding to 1 b will smear (stretch in width) into adjacent bit slots.

Pulse amplitude modulation: An analog signal is represented by a train of pulses where the pulse amplitudes are proportional to the analog signal amplitude.

Pulse code modulation: A serial bit stream that consists of binary words which represent quantized sample values of an analog signal.

Quantizing: Replacing a sample value with the closest allowed value.

References

Cattermole, K.W. 1969. *Principles of Pulse-code Modulation*, American Elsevier, New York, NY.

Couch, L.W. 1993. *Digital and Analog Communication Systems*, 4th ed., Macmillan Publishing Co., New York, NY.

Couch, L.W. 1995. *Modern Communication Systems: Principles and Applications*, Macmillan Publishing Co., New York, NY.

Dammann, C.L., McDaniel, L.D., and Maddox, C.L. 1972. D2 Channel Bank—Multiplexing and Coding. *B. S. T. J.* 12(10):1675–1700.

Dorf, R.C. 1993. *The Electrical Engineering Handbook*, CRC Press, Inc., Boca Raton, FL.

Jayant, N.S. and Noll, P. 1984. *Digital Coding of Waveforms*, Prentice Hall, Englewood Cliffs, NJ.

Lathi, B.P. 1989. *Modern Digital and Analog Communication Systems*, 2nd ed, Holt, Rinehart and Winston, New York, NY.

Miyaoka, S. 1984. Digital Audio Is Compact and Rugged. *IEEE Spectrum.* 21(3):35–39.

Peek, J.B.H. 1985. Communication Aspects of the Compact Disk Digital Audio System. *IEEE Comm. Mag.* 23(2):7–15.

Smith, B. 1957. Instantaneous Companding of Quantized Signals. *B. S. T. J.* 36(5):653–709.

Spilker, J.J. 1977. *Digital Communications by Satellite*, Prentice Hall, Englewood Cliffs, NJ.

Further Information

Many practical design situations and applications of PCM transmission via twisted-pair T-1 telephone lines, fiber optic cable, microwave relay, and satellite systems are given in Couch, 1993 and Couch, 1995.

4
Probabilities and Random Variables

4.1	Introduction 35
4.2	Discrete Probability Theory .... 36 Counting Formulas • Axiomatic Formulas of Probability Theory • The Theorem of Total Probability and Bayes' Theorem
4.3	The Theory of One Random Variable............................... 41 Finding Probabilities from Density and Mass Functions • The Density or Mass Function of a Function of a Known Random Variable • Statistics of a Random Variable • Time Averages as Statistical Averages for Finite Power Waveforms
4.4	The Theory of Two Random Variables.............................. 51 Definitions of Joint Distribution, Density and Mass Functions • Finding Density and Mass Functions from Joint Functions • The Density (or Mass) Function of a Function of Two Random Variables • Statistics of Two Random Variables • Second-Order Time Averages for Finite Power Waveforms (FPW)
4.5	Summary and Future Study .... 59 Special Terminology for Random Finite Power Waveforms

Michael O'Flynn
San Jose State University

4.1 Introduction

Probability theory is presented here for those who use it in communication theory and signal processing of random waveforms. The concept of **sampling a finite power waveform** where a value of time t or n is uniformly chosen over a long time and $x(t)$ or $x(n)$ observed is introduced early. All **first-** and **second-order time averages** are visualized statistically. The topics covered with some highlights are as follows:

1. *Discrete probability theory.* The definitions of probability and conditional probability and their evaluation for outcomes of a **random phenomenon** using relative frequency and *axiomatic* formulas are given. The *theorem of total probability*, which elegantly subdivides compound problems into weighted subproblems, is highlighted along with Bayes' theorem.
2. *The theory of one random variable.* The use of **random variables** allows for the solution of probabilistic problems using integration and summations for sequences. The bridge from theory to applications is very short as in the relationship between finding the density or mass function of a function of a random variable and the communication problem of finding statistics of a system's random output given statistics of its random input.

3. *The theory of two random variables.* **Joint distribution, density, and mass functions** give complete probabilistic information about two random variables and are the key to solving most applications involving continuous and discrete random waveforms and later random processes. Again, the bridge from theory to application is short as in the relationship between finding the density or mass function of a function of two random variables and the communication problem of finding statistics of a system's random output when two random inputs are combined by operations such as summing or multiplying (modulation).

4.2 Discrete Probability Theory

Probability theory concerns itself with assumed random phenomena, which are characterized by outcomes which occur with statistical regularity. These statistical regularities define the probability $p(A)$ of any outcome A and the conditional probability of A given B, $p(A/B)$ by

$$p(A) = \lim_{N \to \infty} \frac{N_A}{N} \quad \text{and} \quad p(A/B) = \lim_{N \to \infty} \frac{N_{AB}}{N_B} \tag{4.1}$$

where N, N_A, N_{AB}, and N_B denote the total number of trials, those for which A occurs, those for which both A and B occur, and those for which B occurs, respectively. In order to use these definitions it is required to know *counting theory* or *permutations and combinations*.

Counting Formulas

A permutation of size k is an ordered array of k objects and a combination of size k is a group or set of size k where order of inclusion is immaterial. From four basic counting formulas, which are inductively developed, plus an understanding of counting factors, probabilities of complex outcomes may be found. These counting formulas are:

$$P_{n,k} = (n)_k = n(n-1) \times \cdots \times (n-k+1) \tag{4.2}$$

which is the number of permutations of size k from n different objects,

$$P_{n,n(n_1,n_2,\ldots,n_k)} = \frac{n!}{n_1! n_2! \times \cdots \times n_k!} \tag{4.3}$$

which is the number of permutations of size n from a total of n objects of which n_1 are identical of type 1 and so on until n_k are identical of type k, and

$$P_{\infty,k(a \text{ types})} = (a)^k \tag{4.4}$$

which is the number of permutations of size k from an infinite supply of a different types.

For example there are $(10)_3 = 720$ different possible results from a race with 10 competitors where only the first three positions are listed, there are 56 different 8 bit words that can be formed from five 1s and three 0s and there are $6^5 = 7776$ different results from rolling a die five times and stating the result of each roll.

The only formula used for combinations, which is the most versatile formula in counting, is

$$C_{n,k} = \binom{n}{k} = \frac{(n)_k}{k!} \tag{4.5}$$

for the number of combinations of size k from n different objects.

Probabilities and Random Variables

For example, given 52 balls which come in four colors with each color numbered from 1 to 13, we can form

$$\binom{13}{1}\binom{12}{1}\binom{4}{3}\binom{4}{2}$$

combinations of size 5 where each contains three of one number and two of another number and

$$\binom{13}{3}\binom{4}{2}\binom{4}{2}\binom{4}{2}$$

combinations of size 6 where each contains two each of three different numbers. Also the number of permutations of size 5 where each contains three of one number and two of another number is

$$\binom{13}{1}\binom{12}{1}\binom{4}{3}\binom{4}{2} 5!$$

A sample problem will be solved using the relative frequency definitions of probabilities and counting formulas.

Problem 1. Consider an urn with 50 balls of which 20 are red and 30 have other colors. If a sample of size 6 is drawn from the urn find the following probabilities:

1. Exactly four red balls are drawn.
2. Exactly four red balls given at least three are red.

Consider the case of drawing the sample with and without replacement.

Solution.

1. Case of without replacement between draws:

$$P[\text{exactly 4 red}] = \frac{\binom{20}{4}\binom{30}{2}}{\binom{50}{6}}$$

or

$$= \frac{(20)_4 (30)_2 \frac{6!}{4!\,2!}}{(50)_6} = 0.13$$

$$P[(\text{exactly 4 red})/(\text{at least 3 red})]$$

$$= \frac{\binom{20}{4}\binom{30}{2}}{\binom{50}{6} - \left[\binom{30}{6} + \binom{30}{5}\binom{20}{1} + \binom{30}{4}\binom{20}{2}\right]}$$

or

$$= \frac{(20)_4 (30)_2 \frac{6!}{4!\,2!}}{(50)_6 - \left[(30)_6 + (30)_5 (20)\frac{6!}{5!\,1!} + (30)_4 (20)_2 \frac{6!}{4!\,2!}\right]}$$

$$= 0.291$$

2. Case with replacement between draws:

$$P[\text{exactly 4 red}] = \frac{20^4 30^2 \frac{6!}{4!\,2!}}{50^6} = (0.4)^4(0.6)^2 \frac{6!}{4!\,2!} = 0.138$$

$P[(\text{exactly 4 red})/(\text{at least 3 red})]$

$$= \frac{(20)^4 (30)^2 \frac{6!}{4!\,2!}}{(50)^6 - \left[(30)^6 + (30)^5(20)\frac{6!}{5!\,2!} + (30)^4(20)^2 \frac{6!}{4!\,2!}\right]}$$

$$= 0.303$$

On reflection, much philosophy is involved in this problem. In order to use the relative frequency formulas we imagine all of the balls are different and from symmetry know the answer is correct. In the case of without replacement we can consider drawing a sample of six balls at once or drawing one at a time and noting the result.

Axiomatic Formulas of Probability Theory

The axioms of probability state that a probability is between zero and one and that the probability of the union of mutually exclusive outcomes is the sum of their probabilities. The axiomatic formulas are:

$$P(A/B) = \frac{P(AB)}{P(B)} \quad \text{if } P(B) \neq 0 \tag{4.6}$$

and

$$p(AB) = P(A)P(B/A) \tag{4.7}$$

Equation (4.7) may be extended for the intersection of n outcomes to give

$$P\left[\bigcap_{i=1}^{n} A_i\right] = p(A_1)p(A_2/A_1) \times \cdots \times p\left(A_n \bigg/ \bigcap_{i=1}^{n-1} A_i\right) \tag{4.8}$$

Also two outcomes are independent if

$$P(AB) = P(A)P(B) \quad \text{or} \quad P(A/B) = P(A) \tag{4.9}$$

and this may be extended to n outcomes. Let us now resolve part of Problem 1 using the axiomatic formulas.

Solution.

1. Without replacement:

$$P[(\text{exactly 4 red})] = \left(\frac{20}{50} \times \frac{19}{49} \times \frac{18}{48} \times \frac{17}{47} \times \frac{30}{46} \times \frac{29}{45}\right)\frac{6!}{4!\,2!}$$

The outcome of exactly 4 red is the union of $6!/4!\,2!$ mutually exclusive outcomes each with the same probability.

Probabilities and Random Variables

2. With replacement:

$$P[(\text{exactly 4 red})] = \left[(0.4)^4(0.6)^2\right]\frac{6!}{4!\,2!}$$

as before.

The Theorem of Total Probability and Bayes' Theorem

An historically elegant theorem which simplifies compound problems is *Bayes' Theorem*, and we will state it in a casual manner and note some applications.

Bayes' Theorem(s)

Consider a random phenomenon where one trial consists in performing a trial of one of m random phenomena $B_1, B_2, \ldots, B_m$ where the probability of performing a trial of B_1 is $P(B_1)$ and so on until $P(B_m)$ is the probability of performing a trial of B_m. Then for any outcome A,

$$P(A) = P(B_1)P(A/B_1) + P(B_2)P(A/B_2) + \cdots + P(B_m)P(A/B_m) \quad (4.10)$$

This is the *theorem of total probability*. Using the axiomatic formula, we obtain

$$P(B_k/A) = \frac{P(B_k)P(A/B_k)}{\sum_{i=1}^{m} P(B_i)P(A/B_i)} \quad (4.11)$$

This is Bayes' theorem.

Problem 2. Consider the random phenomenon of sampling the periodic waveform

$$x(t) = \sum_{n} g(t - 5n)$$

which is shown plotted in Fig. 4.1. A value of time is uniformly chosen over a long time (or over exactly one period) and $x(t)$ is observed. First find

$$P[-0.2 < x(t) \le 2.3]$$

then find

$$P\left[\left(x(t) \text{ from } \frac{1}{t} \text{ part}\right) \Big/ (-0.2 < x(t) \le 2.3)\right]$$

Solution. Let us define the Bayesian space (B_1, B_2, B_3) to denote sampling the $\frac{1}{t}$ portion, the $x(t) = -1$ portion or the $2t - 8$ portion respectively, with probabilities $p(B_1) = 0.4$, $p(B_2) = 0.2$, and $p(B_3) = 0.4$. If we let A be the event $(-0.2 < x(t) \le 2.3)$ then using the theorem of total probability

$$P(A) = P(B_1)P(A/B_1) + P(B_2)P(A/B_2) + P(B_3)P(A/B_3)$$
$$= 0.4\frac{(2 - 0.43)}{2} + 0.2(0) + 0.4\frac{5 - 3.9}{5 - 3} = 0.534$$

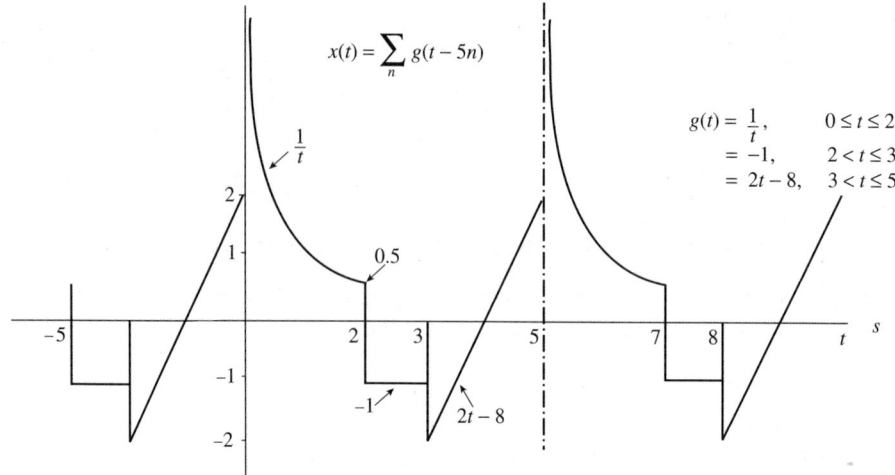

FIGURE 4.1 The periodic function for Problem 2.

and

$$p(B_1/A) = \frac{0.2(2 - 0.43)}{0.534} = 0.59$$

using Bayes' theorem.

Structured Problem Solving

Whenever many probabilistic questions are asked about a random phenomenon, a structured outline is given for the solution in three steps.

In step 1 an appropriate *event space* or the **sample description space** of the phenomenon is listed. The sample description space is the set of finest grain, mutually exclusive, collectively exhaustive outcomes, whereas an event space is a listing of mutually exclusive, collectively exhaustive outcomes. In step 2, a probability is assigned to each outcome of the chosen space using the relative frequency or axiomatic formulas. In step 3 any desired probabilities are found by using the axiomatic formulas.

For example, consider a seven game playoff series between two teams A and B, which terminates when a team wins four games. Assume that for any game, $P(A \text{ wins}) = 0.6$ and $P(B \text{ wins}) = 0.4$. If our interest is in how long the series lasts and who wins in how many games then an appropriate event space is

$$E = \{e_{A4}, e_{A5}, e_{A6}, e_{A7}, e_{B4}, e_{B5}, e_{B6}, e_{B7}\}$$

where, for example, e_{B6} is the outcome B wins in exactly six games and

$$p(e_{B6}) = (0.4)^4 (0.6)^2 \frac{5!}{3!\, 2!} = 0.092$$

Similarly, we can find the probabilities for the other seven outcomes. Some problems which now can be easily solved are

$$P[(\text{series lasts at least 6 games})] = p(e_{A6}) + p(e_{A7}) + p(e_{B6}) + p(e_{B7})$$

Probabilities and Random Variables 41

and

$$P[(A \text{ wins in at most 6 games})/(\text{series lasts at least 5 games})]$$
$$= (p[e_{A5}] + p[e_{A6}]) \div \left(\sum_{i=5}^{7} p(e_{Ai}) + \sum_{i=5}^{7} p(e_{Bi}) \right)$$

4.3 The Theory of One Random Variable

A random variable X assigns to every outcome of the sample description space of a random phenomenon a real number $X(s_i)$. Associated with X are three very important functions:

1. The **cumulative distribution function** $F_X(\alpha)$ defined by

$$F_X(\alpha) = P[X \le \alpha], \quad -\infty < \alpha < \infty \quad (4.12)$$

2. The density function $f_X(\alpha)$ defined by

$$f_X(\alpha) = \frac{d}{d\alpha} F_X(\alpha)$$

3. If appropriate, the probability mass function $p_X(\alpha_i)$ defined by $p_X(\alpha_i) = P[X = \alpha_i]$.

Given a random phenomenon, finding $F_X(\alpha)$, $f_X(\alpha)\Delta\alpha$, or $p_X(\alpha_i)$ are problems in discrete probability.

Problem 3. Given the periodic waveform

$$x(t) = \sum_{n} g(t - 5n)$$

of Fig. 4.1, let the **random variable** X describe sampling $x(t)$ with a one to one mapping. Find and plot $F_X(\alpha)$ and $f_X(\alpha)$.

Solution. Observing $x(t)$ in Fig. 4.1 we will define X_1 as sampling $x(t) = 1/t$, X_2 as sampling $x(t) = -1$, and X_3 as sampling $x(t) = 2t - 8$. Figure 4.2(a) shows the solution for $F_{X_i}(\alpha)$ and $f_{X_i}(\alpha)$ and Fig. 4.2(b) uses the theorem of total probability to find $F_X(\alpha)$ as $0.4F_{X_1}(\alpha) + 0.2F_{X_2}(\alpha) + 0.4F_{X_3}(\alpha)$ and, similarly, $f_X(\alpha)$ is found. In this case, X is said to be a mixed random variable as $F_X(\alpha)$ changes both continuously and in jumps. The delta function allows for the inclusion of a point taking on a specific probability in $f_X(\alpha)$.

In Fig. 4.3 are shown density or mass functions for four of the most commonly occurring random variables: the continuous uniform and Gaussian density functions and the discrete binomial and Poisson mass functions. Also included are some statistics to be encountered later.

Finding Probabilities from Density and Mass Functions

Density and mass functions allow for the use of integration and summations to answer probabilistic questions about a random variable. This is illustrated with two problems.

Problem 4. Consider the density function $f_X(\alpha) = e^{2\alpha}$, $-\infty < \alpha \le 0$; $f_X(\alpha) = 0.11(\alpha - 2)$, $2 \le \alpha \le 5$; and $f_X(\alpha) = 0$, otherwise. Find $P[(-1.0 < X < 4.0)/(-X^2 < -5)]$.

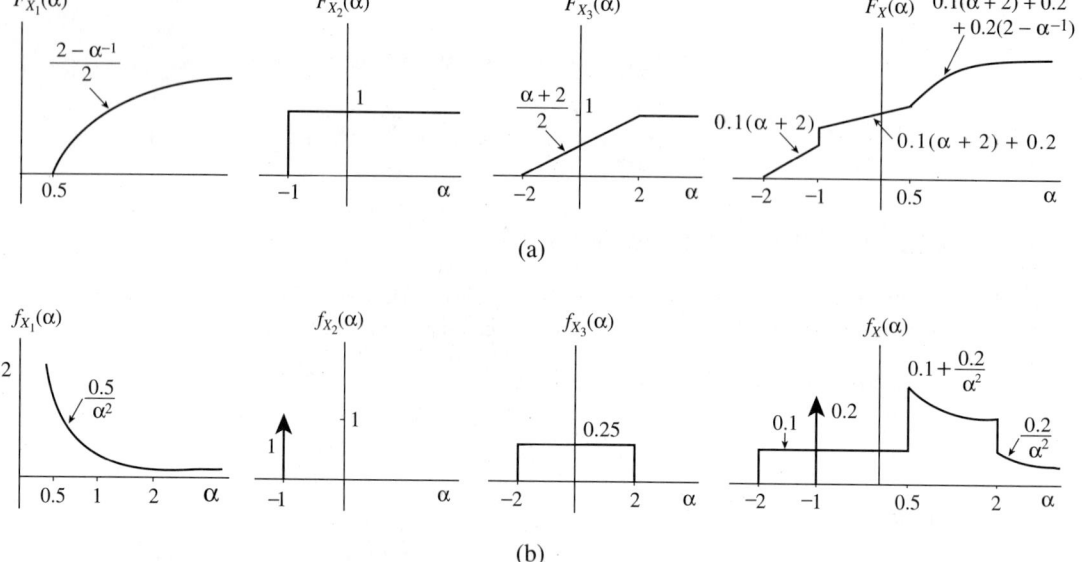

FIGURE 4.2 (a) $F_X(\alpha)$ for $x(t)$ in Fig. 4.1 for Problem 3 and (b) $f_X(\alpha)$ for $x(t)$ in Fig. 4.1.

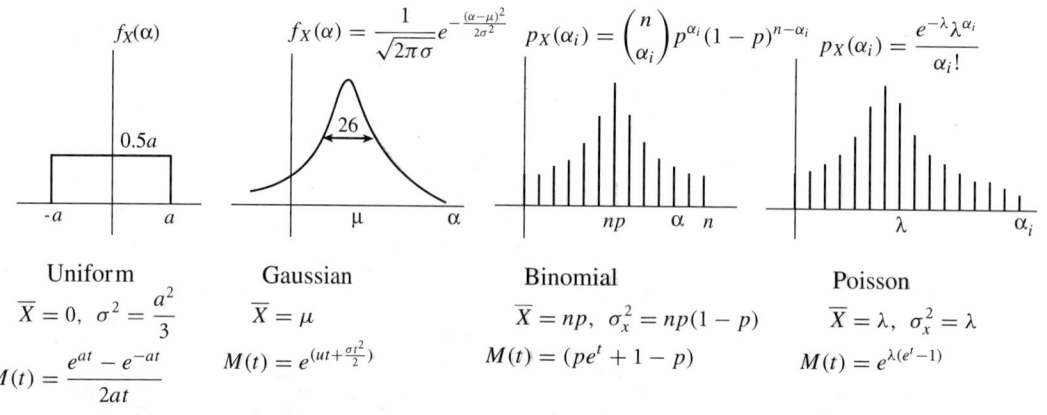

FIGURE 4.3 Important density and mass functions.

Solution. A sketch of $f_X(\alpha)$ is shown in Fig. 4.4. Using the axiomatic formula

$$P[(-1.0 < X < 4.0)/(-X^2 < -5)] = \frac{0 + \int_{2.24}^{4} 0.11(\alpha - 2)\, d\alpha}{\int_{-\infty}^{-2.24} 0.5 e^{\alpha}\, d\alpha + \int_{2.24}^{5} 0.11(\alpha - 2)\, d\alpha}$$

$$= 0.398$$

Problem 5. Given the probability mass function $p_X(\alpha_i) = 0.01$, $-40 \leq \alpha_1 \leq -10$, α_i integer; $p_X(\alpha_i) = 0.10(0.9)^{\alpha_i}$, $\alpha_i \geq 5$, α_i integer; and $p_X(\alpha_i) = 0$, otherwise. Find (1) $P[(|X| \leq 20)/(X^2 > 200)]$ and (2) $p_X[\alpha_i/(X^2 > 200)]$.

Probabilities and Random Variables 43

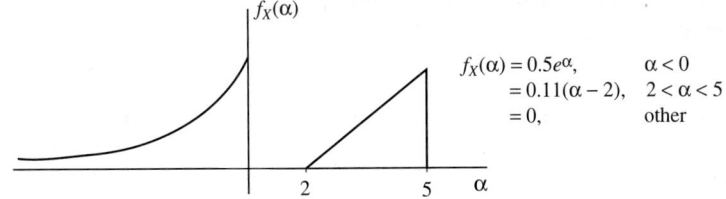

FIGURE 4.4 Density function $f_X(\alpha)$ for Problem 4.

Solution.

1. A sketch of $p_X(\alpha_i)$ is shown in Fig. 4.5(a). Using the axiomatic formula,

$$P[(|X| \leq 20)/(X^2 > 200)] = \frac{\sum_{-20}^{-15} 0.01 + \sum_{15}^{20} 0.10(0.9)^{\alpha_i}}{\sum_{-40}^{-15} 0.01 + \sum_{15}^{\infty} 0.10(0.9)^{\alpha_i}} = \frac{0.156}{0.467} = 0.34$$

The formulas for a finite geometric progression

$$\sum_{p=0}^{n} \alpha^p = (1 - \alpha^{n+1}) \div (1 - \alpha)$$

and for a geometric progression

$$\sum_{p=0}^{\infty} \alpha^p = (1 - \alpha)^{-1}, \quad |\alpha| < 1$$

were used in the calculation.

2. The second solution is

$$p_X(\alpha_i/(X^2 > 200)) = \frac{p_X(\alpha_i)}{P[X^2 > 200]}, \quad \alpha_i^2 > 200$$
$$= 0, \quad \text{otherwise}$$

and is shown plotted in Fig. 4.5(b).

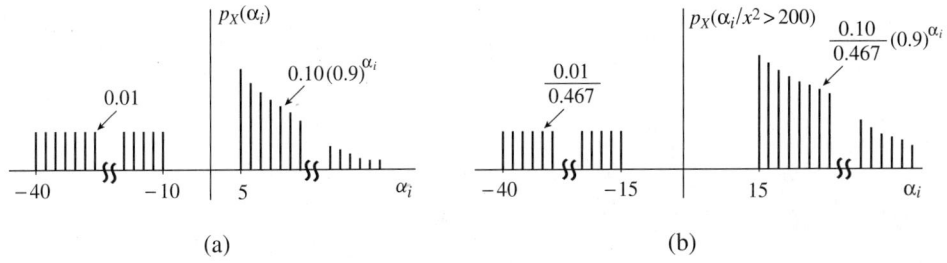

FIGURE 4.5 Probability mass function $p_X(\alpha_i)$ and $p_X(\alpha_i/X^2 > 200)$ for Problem 5.

The Density or Mass Function of a Function of a Known Random Variable

Continuous Case

Consider the problem: Given $Y = g(X)$ where $f_X(\alpha)$ is known, find $f_Y(\beta)$.

$$F_Y(\beta) = P[g(X) \leq \beta] = \int_\alpha f_X(\alpha)\, d\alpha \qquad \text{where } g(\alpha) \leq \beta$$

In order to find $f_Y(\beta)$ directly, Leibnitz' rule is used. Leibnitz' rule states

$$\frac{d}{d\beta} \int_{k_1(\beta)}^{k_2(\beta)} f_X(\alpha)\, d\alpha = f_X(k_2(\beta)) k_2'(\beta) - f_X(k_1(\beta)) k_1'(\beta) \qquad (4.13)$$

For example, if

$$f_X(\alpha) = \frac{1}{\sqrt{2\pi}\sigma} e^{-\frac{(\alpha-\mu)^2}{2\sigma^2}}$$

and $Y = aX + b$ where $a > 0$

$$F_Y(\beta) = P[aX + b \leq \beta] = \int_{-\infty}^{\frac{\beta-b}{a}} f_X(\alpha)\, d\alpha$$

Using Leibnitz' rule

$$f_Y(\beta) = \frac{1}{\sqrt{2\pi}\sigma} e^{-\frac{(\beta-b-a\mu)^2}{2a^2\sigma^2}} \frac{1}{a} = N(a\mu + b, (a\sigma)^2)$$

which means Y is also Gaussian with $\mu_Y = a\mu + b$ and $\sigma_Y^2 = (a\sigma)^2$.

Discrete Case

The discrete case is much more direct. If $Y = g(x)$ where $p_X(\alpha_i)$ is known, then $p_Y(\beta_j) = p_X(g^{-1}(\beta))$. For example, if $Y = X^2$ then $p_Y(\beta_j) = p_X(\sqrt{\beta_j}) + p_X(-\sqrt{\beta_j})$.

Problem 6. Given the probability mass function $p_X(\alpha_i) = 0.096(0.9)^{\alpha_i+5}$, $-5 \leq \alpha_i \leq 30$, α_i integer; $p_X(\alpha_i) = 0$ otherwise, and $Y = |2X|$, find $p_Y(\beta_j)$.

Solution.

$$Y = |2X| = 2X, \qquad X \geq 0$$
$$= -2X, \qquad X < 0$$
$$p_Y(0) = 0.096(0.9)^5$$

Range 1: $\beta_j = 2, 4, 6, 8, 10$ or $\beta_j = |2\alpha_i|$, $\alpha_i = \pm 1, \ldots, \pm 5$,

$$p_Y(\beta_j) = 0.096 \left[(0.9)^{\frac{\beta_j+10}{2}} + (0.9)^{\frac{-\beta_j+10}{2}} \right]$$

Range 2: $\beta_j = 12, 14, \ldots, 60,$ $\quad \beta_j = 2\alpha_i, \alpha_i = 6, 30$

$$p_Y(\beta_j) = 0.096 \left[(0.9)^{\frac{\beta_j+10}{2}} \right]$$

Statistics of a Random Variable

The three most widely used statistics of a random variable are

$$\overline{X} = \int_{-\infty}^{\infty} \alpha f_X(\alpha) \, d\alpha$$

or

$$= \sum_{\alpha_i} \alpha_i p_X(\alpha_i)$$

$$\overline{X^2} = \int_{-\infty}^{\infty} \alpha^2 f_X(\alpha) \, d\alpha$$

or

$$= \sum_{\alpha_i} \alpha_i^2 p_\alpha(\alpha_i)$$

and

$$\sigma_X^2 = \overline{(X - \overline{X})^2} = \int_{-\infty}^{\infty} (\alpha - \overline{X})^2 f_X(\alpha) \, d\alpha = \overline{X^2} - (\overline{X})^2$$

These are the mean or average value, the mean squared value, and the variance, respectively. The formulas follow from a relative frequency interpretation based on a large number of trials of a phenomenon. Higher order statistics are defined by

$$m_n = \overline{X^n} = \int_{-\infty}^{\infty} \alpha^n f_X(\alpha) \, d\alpha$$

or

$$= \sum_{\alpha_i} \alpha_i^n p_X(\alpha_i)$$

and

$$\overline{(X - \overline{X})^n} = \int_{-\infty}^{\infty} (\alpha - \overline{X})^n f_X(\alpha) \, d\alpha$$

or

$$= \sum_{\alpha_i} (\alpha - \overline{X})^n p_X(\alpha_i)$$

These are called moments and central moments, respectively. General expressions for higher order statistics may be found using the moment generation function $M(t)$ defined by

$$M(t) = \overline{e^{tX}} = \sum_{\alpha_i} e^{t\alpha_i} p_X(\alpha_i)$$

or

$$= \int_{-\infty}^{\infty} e^{t\alpha} f_X(\alpha) \, d\alpha \tag{4.14}$$

Time Averages as Statistical Averages for Finite Power Waveforms

Continuous Waveforms

A finite power waveform is defined as one for which

$$P_{av} = \lim_{T \to \infty} \frac{1}{2T} \int_{-T}^{T} x^2(t)\, dt$$

exists.

The time average definition for some first-order time averages are

$$\widetilde{x(t)} = \lim_{T \to \infty} \frac{1}{2T} \int_{-T}^{T} x(t)\, dt \quad \text{and} \quad \widetilde{x^2(t)} = \lim_{T \to \infty} \frac{1}{2T} \int_{-T}^{T} x^2(t)\, dt \quad (4.15)$$

called the average and mean squared value, respectively.

Applying these definitions to the periodic waveform

$$x(t) = \sum_n g(t - 3n)$$

shown in Fig. 4.6(a), it is easy to find the average and mean squared values,

$$\widetilde{x(t)} = \frac{1}{3} \int_0^3 x(t)\, dt = \frac{1}{3}\left[\int_0^2 (-1 + 2t)\, dt + \int_2^3 1\, dt\right] = 1.0$$

$$\widetilde{x^2(t)} = \frac{1}{3}\left[\int_0^2 (-1 + 2t)^2\, dt + \int_2^3 1\, dt\right] = 1.89 \quad \text{and} \quad \sigma^2_{x(t)} = 0.89$$

We will now interpret these averages statistically. Let the random variable X describe sampling t over a long time and observing $x(t)$. The density function $f_X(\alpha)$ is shown plotted in Fig. 4.6(a), and it is easy to show

$$\widetilde{x(t)} = \overline{X} = \left[\int_{-1}^{3} \alpha \frac{1}{6}\, d\alpha + \int_{-\infty}^{\infty} \alpha \frac{1}{3} \delta(\alpha - 1)\, d\alpha\right] = 1.0$$

and

$$\widetilde{x^2(t)} = \overline{X^2} = \left[\int_{-1}^{3} \alpha^2 \frac{1}{6}\, d\alpha + \int_{-\infty}^{\infty} \alpha^2 \frac{1}{3} \delta(\alpha - 1)\, d\alpha\right] = 1.89$$

The general formula for any first-order statistic is

$$\widetilde{g(x(t))} = \int_{-\infty}^{\infty} g(\alpha) f_X(\alpha)\, d\alpha \quad (4.16)$$

Probabilities and Random Variables

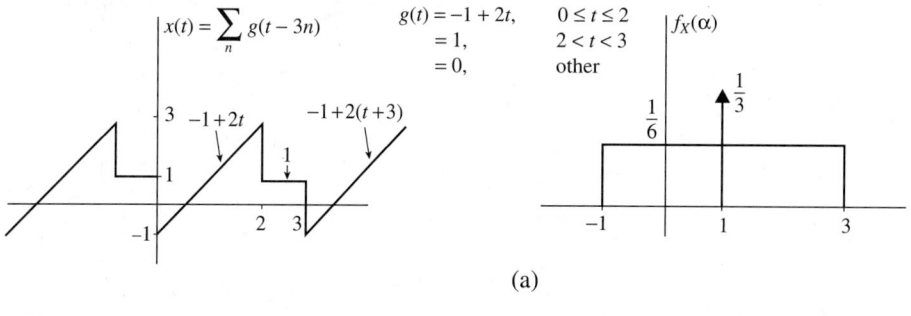

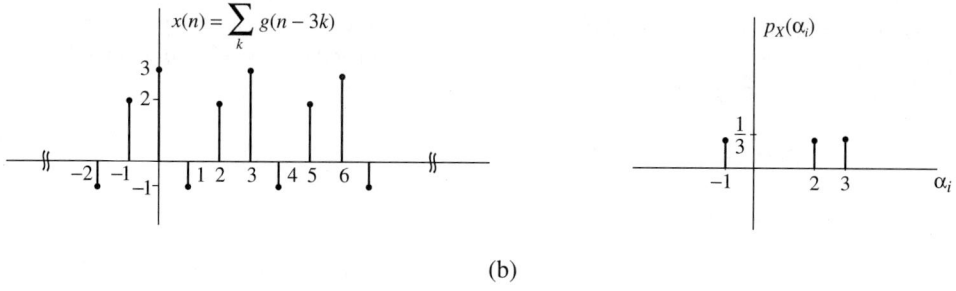

(b)

FIGURE 4.6 (a) Waveform $x(t) = \sum_n g(t - 3n)$ and $f_X(\alpha)$ for sampling $x(t)$ and (b) waveform $x(n) = \sum_k g(n - 3k)$ and $p_X(\alpha_i)$ for sampling $x(n)$.

Discrete Waveforms

A finite power discrete waveform is one for which

$$P_{av} = \lim_{N \to \infty} \frac{1}{2N+1} \sum_{n=-N}^{N} x^2(n)$$

exists.

The time average definitions for some first-order time averages are

$$\widetilde{x(n)} = \lim_{N \to \infty} \frac{1}{2N+1} \sum_{-N}^{N} x(n) \quad \text{and} \quad \widetilde{x^2(n)} = \lim_{N \to \infty} \frac{1}{2N+1} \sum_{-N}^{N} x^2(n) \quad (4.17)$$

called the average and mean square value, respectively. Applying these definitions to the periodic waveform

$$x(n) = \sum_{k=-\infty}^{\infty} g(n - 3k)$$

shown in Fig. 4.6(b), it is easy to find the average, mean square and variance of $x(n)$,

$$\widetilde{x(n)} = \frac{1}{3}[3 + (-1) + 2] = 1.33$$

$$\widetilde{x^2(n)} = \frac{1}{3}[9 + 1 + 4] = 4.67 \quad \text{and} \quad \sigma^2_{x(n)} = 4.67 - 1.78 = 2.89$$

If the random variable X describes uniformly choosing an integer n, $-N \leq n \leq N$ and observing $x(n)$, then the statistical definition for any first-order time average $\widetilde{g(x(n))}$ is

$$\widetilde{g[x(n)]} = \overline{g(X)} = \int g(\alpha) f_X(\alpha) \, d\alpha$$

or

$$= \sum g(\alpha_i) p_X(\alpha_i)$$

which is identical to Eq. (4.16). $P_X(\alpha_i)$ is shown in Fig. 4.6(b), and $\widetilde{x(n)} = \overline{X} = 3(\frac{1}{3}) + (-1)(\frac{1}{3}) + 2(\frac{1}{3}) = 1.33$, and $\widetilde{x^2(n)} = \overline{X^2} = 9(\frac{1}{3}) + 1(\frac{1}{3}) + 4(\frac{1}{3}) = 4.67$ as before.

Random Waveforms

A finite-power random or noise waveform is defined as being nondeterministic and such that statistics of a large section or of the whole waveform exist. Often $x(t)$ or $x(n)$ is represented by a formula involving time and random variables,

$$x(t) = f(t, \text{r. vs}) \quad \text{or} \quad x(n) = f(n, \text{r. vs})$$

where "r. vs" denotes "random variables." For example,

$$x(t) = \sum_{n=-\infty}^{\infty} A_n g(t - nT_0)$$

is a periodic function with each period weighted by a value statistically determined by the density or joint density functions for the As. Consider when $g(t) = 2$, $0 \leq t \leq 1$, and $T_0 = 1$ and the As are independent, each with $p_{A_i}(1) = 0.6$ and $p_{A_i}(0) = 0.4$.

A section of $x(t)$ may be as shown in Fig. 4.7(a). Consider

$$x(n) = \sum_k A_k g(n - 2k)$$

where $g(n) = 4\delta(n) - \delta(n-1)$ and the As are independent random variables each with $f_{A_i}(\alpha) = 0.5$, $0 \leq \alpha \leq 2$; $f_{A_i}(\alpha) = 0$, otherwise. A typical section of $x(n)$ may appear as in Fig. 4.7(b).

If X describes sampling $x(t)$ or $x(n)$ where t is uniformly chosen $-T < t < T$ and $x(t)$ observed or n is discretely uniformly chosen $-N \leq n \leq N$ and $x(n)$ observed, then as T or N approach infinity $f_X(\alpha)$ or $p_X(\alpha_i)$ may be found.

Any time average may be found by the statistical formula (4.16)

$$\widetilde{g(x(t))} \quad \text{or} \quad \widetilde{g[x(n)]} = \int_{-\infty}^{\infty} g(\alpha) f_X(\alpha) \, d\alpha$$

or

$$= \sum_{\alpha_i} g(\alpha_i) p_X(\alpha_i)$$

Two simple examples will be used for demonstration.

Problem 7. Given the random waveform

$$x(t) = \sum_n A_n g(t - 3n)$$

Probabilities and Random Variables

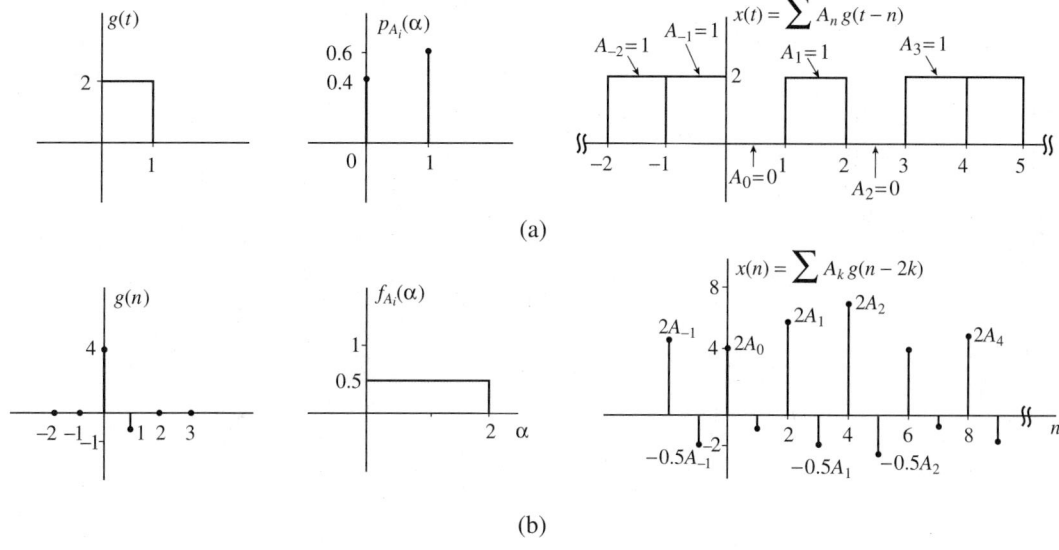

FIGURE 4.7 The random waveforms: (a) $x(t) = \sum_n A_n g(t-n)$ and (b) $x(n) = \sum_k A_k g(n-2k)$.

where $g(t) = 3, 0 \leq t \leq 2$; $g(t) = -4, 2 < t \leq 3$; $g(t) = 0$, otherwise and the A are independent random variables with $f_{A_i}(\alpha) = \frac{1}{3}, -1 \leq \alpha \leq 2$; $f_{A_i}(\alpha) = 0$, otherwise, find $f_X(\alpha), \widetilde{x(t)}$, and $\widetilde{x^2(t)}$.

Solution. Figure 4.8(a) shows a plot of a typical section of $x(t)$. We notice the front two units of a pulse takes on values uniformly from -3 to 6 and that the last unit takes on that value multiplied by -1.33. Using the theorem of total probability $f_X(\alpha) = \frac{2}{3} f_{X_1}(\alpha) + \frac{1}{3} f_{X_2}(\alpha)$ where X_1 and X_2 describe sampling the first two and last unit of a pulse, respectively. The density function $f_X(\alpha)$ is shown in Fig. 4.8(a) and it is easy to find; $\widetilde{x(t)} = \overline{X} = \frac{2}{3}(\overline{3A_i}) + \frac{1}{3}(\overline{-4A_i}) = 0.33$ and $\widetilde{x^2(t)} = \overline{X^2} = \frac{2}{3}(\overline{9A_i^2}) + \frac{1}{3}(\overline{16A_i^2}) = 11.33$.

Problem 8. Given the random waveform

$$x(n) = \sum_k A_k \delta(n-k)$$

where the A are independent random variables with $p_{A_i}(1) = 0.7$ and $p_{A_i}(0) = 0.3$, find $p_X(\alpha_i), \widetilde{x(t)}$ and $\widetilde{x^2(t)}$.

Solution. Figure 4.8(b) shows a plot for a typical section of $x(n)$ and $p_X(\alpha_i)$. $\widetilde{x(n)} = \overline{X} = (1)0.7$ and $\widetilde{x^2(t)} = \overline{X^2} = 0.7$.

Finally, let us consider an application for the density function of a known random variable.

The Square Wave Detector

Consider a noise waveform input to a system where the output and input are related by $y(t) = x^2(t)$. If the density function for sampling the input $x(t)$ is

$$f_X(\alpha) = \frac{1}{\sqrt{2\pi}\sigma} e^{-\frac{(\alpha-\mu)^2}{2\sigma^2}}$$

find $f_Y(\beta)$ for sampling the output $y(t)$.

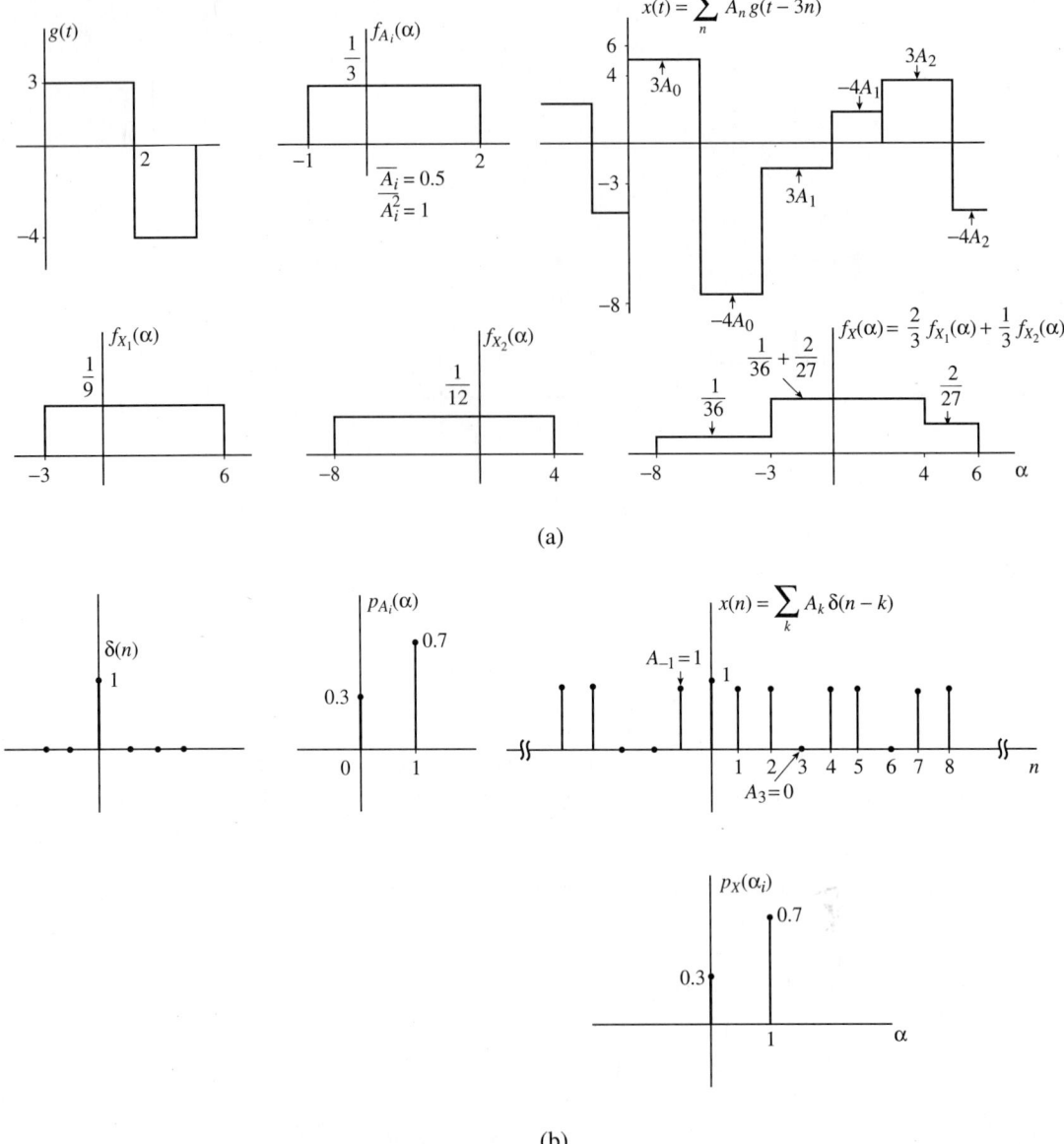

FIGURE 4.8 The random waveform of (a) Problem 7 and $f_X(\alpha)$ and (b) Problem 8 and $p_X(\alpha_i)$.

Solution.

$$F_Y(\beta) = 2\int_0^{\sqrt{\beta}} \frac{1}{\sqrt{2\pi}\sigma} e^{-\frac{(\alpha-\mu)^2}{2\sigma^2}}\, d\alpha$$

and by Leibnitz' rule

$$f_Y(\beta) = \frac{2}{\sqrt{2\pi}\sigma} e^{-\frac{(\sqrt{\beta}-\mu)^2}{2\sigma}} 0.5(\beta)^{-0.5}$$

which is the famous chi-squared density function.

Probabilities and Random Variables

4.4 The Theory of Two Random Variables

Definitions of Joint Distribution, Density and Mass Functions

Given a random phenomenon, two random variables X and Y map every point of the sample description space onto an α–β plane via $X(s_i)$ and $Y(s_i)$. The following important functions are associated with X and Y:

$$F_{XY}(\alpha, \beta) = P[(X \leq \alpha) \cap (Y \leq \beta)], \quad -\infty < \alpha \leq \infty, \quad -\infty < \beta < \infty \tag{4.18}$$

This is the joint cumulative distribution function.

$$f_{XY}(\alpha, \beta) = \frac{\partial^2}{\partial \alpha \partial \beta} F_{XY}(\alpha, \beta) \tag{4.19}$$

is the joint density function and

$$p_{XY}(\alpha_i, \beta_j) = P[(X = \alpha_i) \cap (Y = \beta_j)] \tag{4.20}$$

is the joint mass function when appropriate.

Finding $F_{XY}(\alpha, \beta)$ or $f_{XY}(\alpha, \beta)\Delta\alpha\Delta\beta$ or $p_{XY}(\alpha_i, \beta_j)$ are problems in discrete probability theory. Knowing $f_{XY}(\alpha, \beta)$ and $p_{XY}(\alpha_i, \beta_j)$ or $F_{XY}(\alpha, \beta)$ it is possible to answer any probabilistic question about X and Y, X given a condition on Y or Y given a condition on X. The following are some formulas which are easy to derive starting from a statement with probabilistic meaning:

$$f_X(\alpha) = \int_{-\infty}^{\infty} f_{XY}(\alpha, \beta) \, d\beta, \quad p_Y(\beta_j) = \sum_{\beta_j} p_{XY}(\alpha_i, \beta_j) \tag{4.21}$$

$$f_Y(\beta/g(X) < C_1) = \frac{\int f_{XY}(\alpha, \beta) \, d\alpha}{\int_{-\infty}^{\infty} \int f_{XY}(\alpha, \beta) \, d\alpha \, d\beta} \tag{4.22}$$

where the two missing limits of integration are: all α such that $g(\alpha) < C_1$,

$$p_X(\alpha_i/Y^2 \leq 25) = \frac{\sum_{\beta_j=-5}^{5} p_{XY}(\alpha_i, \beta_j)}{\sum_{\alpha_i} \sum_{\beta_j=-5}^{5} p_{XY}(\alpha_i, \beta_j)} \tag{4.23}$$

and

$$f_Y(\beta/X = \alpha) = \frac{f_{XY}(\alpha, \beta)}{f_X(\alpha)} \quad \text{all } \beta \text{ when } \alpha = \alpha \tag{4.24}$$

The use of these formulas involve detailed attention and care with different ranges.

Finding Density and Mass Functions from Joint Functions

Problem 9. Given $f_{XY}(\alpha, \beta) = -0.0010\beta$, $0.2\beta \leq \alpha \leq -0.2\beta$, $-20 \leq \beta \leq -8$; $f_{XY}(\alpha, \beta) = 0$, otherwise. Find and plot $f_Y(\beta)$ and $f_X(\alpha/(Y = \beta))$.

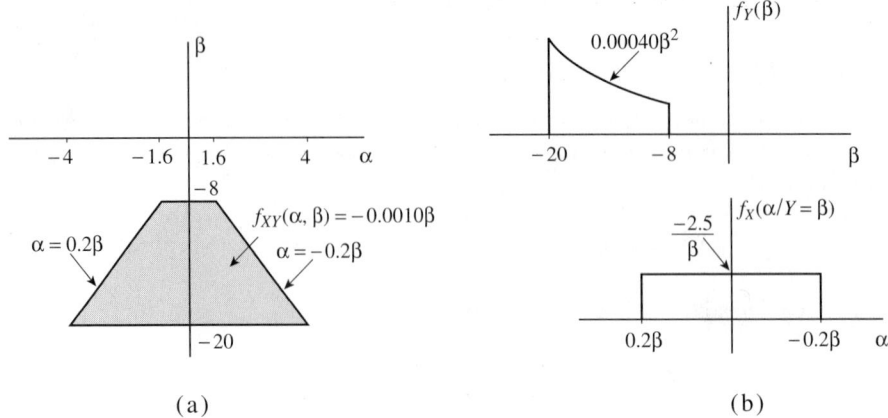

FIGURE 4.9 (a) Region for $f_{XY}(\alpha, \beta)$ for Problem 9 and (b) $f_Y(\beta)$ and $f_X(\alpha/(Y = \beta))$.

Solution. The region for which $f_{XY}(\alpha, \beta)$ is defined is shown in Fig. 4.9(a). Using

$$f_Y(\beta) = \int_{-\infty}^{\infty} f_{XY}(\alpha, \beta) \, d\alpha$$

we obtain the following.

Range 1: $-20 \leq \beta \leq -8$

$$f_Y(\beta) = \int_{0.2\beta}^{-0.2\beta} -0.0010\beta \, d\alpha = 0.0004\beta^2, \quad -20 \leq \beta \leq -8$$

$$= 0 \quad \text{otherwise,} \quad \text{and}$$

$$f_X(\alpha/Y = \beta) = \frac{f_{XY}(\alpha, \beta)}{f_Y(\beta)}$$

$$= \frac{-0.0010\beta}{0.0004\beta^2} = \frac{-2.5}{\beta}, \quad 0.2\beta < \alpha < -0.2\beta$$

for any β, $-20 \leq \beta \leq -8$

$$= 0 \quad \text{otherwise}$$

The density functions are plotted in Fig. 4.9(b).

Problem 10. Given $p_{XY}(\alpha_i, \beta_j) = 0.0033|\beta_j|$, $-\alpha_i \leq \beta_j \leq \alpha_i$, $6 \leq \alpha_i \leq 20$, both integer, $p_{XY}(\alpha_i, \beta_j) = 0$, otherwise, find $p_X(\alpha_i)$ and $p_X(\alpha_i/(Y = \beta_j))$ for $-20 \leq \beta_j \leq -6$. Are X and Y independent?

Solution. The region for which $p_{XY}(\alpha_i, \beta_j)$ is defined is shown in Fig. 4.10(a).
Range 1: $6 \leq \alpha_i \leq 20$

$$p_X(\alpha_i) = \sum_{\beta_j = -\alpha_i}^{\alpha_i} 0.0033|\beta_j| = 2 \sum_{1}^{\alpha_i} 0.0033\beta_j$$

$$= 0.0033(2) \frac{(\alpha_i + 1)\alpha_i}{2} = 0.0033(\alpha_i + 1)\alpha_i$$

Probabilities and Random Variables

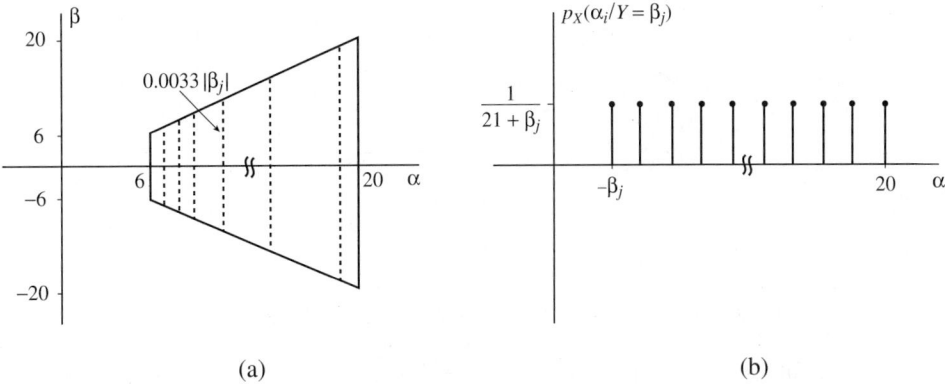

FIGURE 4.10 (a) Region for $p_{XY}(\alpha_i, \beta_j)$ for Problem 10 and (b) $p_X(\alpha_i/(Y=\beta_j))$.

using the formula for an arithmetic progression. Otherwise, $p_X(\alpha_i) = 0$.

$$p_X[\alpha_i/Y = \beta_j] = \frac{p_{XY}(\alpha_i, \beta_j)}{p_Y(\beta_j)}$$

Looking at Fig. 4.10(a) for the region when $-20 \leq \beta_j \leq -6$ the formula

$$p_Y(\beta_j) = \sum_{\alpha_i} p_{XY}(\alpha_i, \beta_j)$$

becomes

$$p_Y(\beta_j) = \sum_{\alpha_i=-\beta_j}^{20} p_{XY}(\alpha_i, \beta_j) = 0.0033(21 + \beta_j)(-\beta_j)$$

and

$$p_X(\alpha_i/Y = \beta_j) = \frac{0.0033(-\beta_j)}{0.0033(21 + \beta_j)(-\beta_j)} = \frac{1}{21 + \beta_j}, \quad -\beta_j \leq \alpha_i \leq 20$$
$$= 0 \quad \text{otherwise.}$$

This conditional mass function is plotted in Fig. 4.10(b). Since $p_X(\alpha_i/(Y = \beta_j))$ does not equal $p_X(\alpha_i)$ for all α_i, then X and Y are not independent.

The Density (or Mass) Function of a Function of Two Random Variables

Continuous Case

Given $Z = g(X, Y)$ where $f_{XY}(\alpha, \beta)$ is known, $f_Z(\gamma)$ is found as

$$f_Z(\gamma) = \frac{d}{d\gamma} F_Z(\gamma) = \frac{d}{d\gamma} \iint f_{XY}(\alpha, \beta) \, d\alpha \, d\beta \qquad (4.25a)$$

where the limits on the double integral satisfy $g(\alpha, \beta) \leq \gamma$.

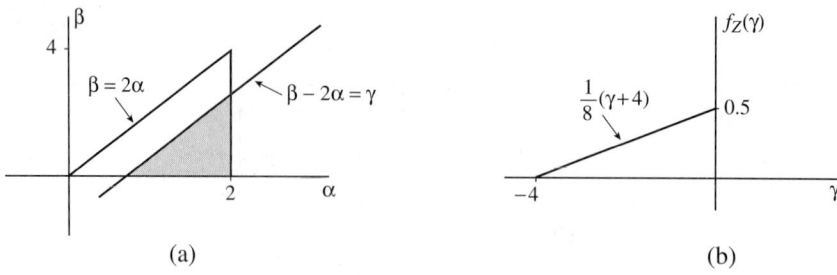

FIGURE 4.11 (a) Region of $f_{XY}(\alpha, \beta)$ for Problem 11 and (b) $f_Z(\gamma)$, $Z = Y - 2X$.

By Leibnitz' rule Eq. (4.25a) becomes for the case $Z = aX + bY$

$$f_Z(\gamma) = \int_{-\infty}^{\infty} C_1 f_{XY}(\alpha, g_1(\alpha, \gamma))\, d\alpha$$

or

$$= \int_{-\infty}^{\infty} C_2 f_{XY}(g_2(\beta, \gamma), \beta)\, d\beta \qquad (4.25b)$$

Problem 11. Given $f_{XY}(\alpha, \beta) = 0.25$; $0 \le \beta \le 2\alpha$, $0 < \alpha < 2$; $f_{XY}(\alpha, \beta) = 0$, otherwise, find $f_Z(\gamma)$ when $Z = Y - 2X$.

Solution. Figure 4.11 shows the region for which $f_{XY}(\alpha, \beta)$ is defined. If we visualize the family of curves $\beta - 2\alpha = \gamma$ as γ varies $-\infty < \gamma < \infty$ we note $F_Z(\gamma)$ is the weighed integral of all points to the right of $\beta - 2\alpha = \gamma$. $F_Z(\gamma) = 0$, $\gamma < -4$ and $F_Z(\gamma) = 1$, $\gamma > 0$.

In general $F_Z(\gamma) = \int_{-\infty}^{\infty} (\int_{-\infty}^{2\alpha+\gamma} f_{XY}(\alpha, \beta)\, d\beta)\, d\alpha$ and using Leibnitz' rule

$$f_Z(\gamma) = \int_{-\infty}^{\infty} f_{XY}(\alpha, 2\alpha + \gamma)\, d\alpha$$

Observing Fig. 4.11(a)

$$f_Z(\gamma) = \int_{-\frac{\gamma}{2}}^{2} 0.25\, d\alpha$$

$$= \frac{1}{8}(\gamma + 4), \quad -4 < \gamma < 0$$

which is shown in Fig. 4.11(b).

An important situation occurs when X and Y are independent random variables or when $f_{XY}(\alpha, \beta) = f_X(\alpha) f_Y(\beta)$ for all α, β. In this case if $Z = X + Y$, then

$$f_Z(\gamma) = f_X(\gamma) * f_Y(\gamma) = \int_{-\infty}^{\infty} f_X(\gamma - \beta) f_Y(\beta)\, d\beta$$

or

$$= \int_{-\infty}^{\infty} f_X(\alpha) f_Y(\gamma - \alpha)\, d\alpha \qquad (4.26)$$

which is the convolution of the two density functions. For example, if $f_{XY}(\alpha, \beta) = \frac{1}{32}$, $-3 \le \alpha \le 5$, $-1 \le \beta \le 3$, $f_{XY}(\alpha, \beta) = 0$, otherwise, then it is easy to find $f_X(\alpha) = \frac{1}{8}$, $-3 \le \alpha \le 5$, $f_X(\alpha) = 0$, otherwise and $f_Y(\beta) = \frac{1}{4}$, $-1 \le \beta \le 3$, $f_Y(\beta) = 0$, otherwise. Also X and Y are independent. Figure 4.12 shows $f_{XY}(\alpha, \beta)$, $f_X(\alpha)$, and $f_Y(\beta)$ and gives the convolution of $f_X(\gamma)$ with $f_Y(\gamma)$.

Probabilities and Random Variables

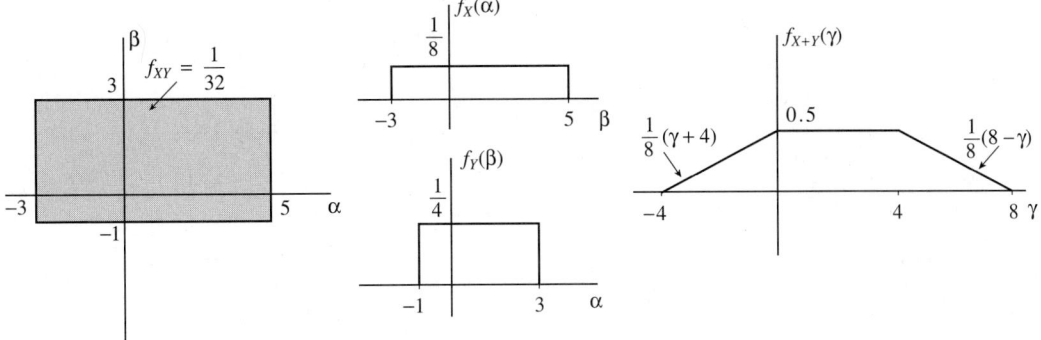

FIGURE 4.12 Function $f_Z(\gamma)$ by convolution when $f_{XY}(\alpha, \beta) = \frac{1}{32}$; $-3 \leq \alpha \leq 5$, $-1 \leq \beta \leq 3$ and $Z = X + Y$.

Discrete Case

Given $Z = g(X, Y)$, where $p_{XY}(\alpha_i, \beta_j)$ is known,

$$p_Z(\gamma_i) = \sum_{\beta_j} \sum_{\alpha_i} p_{XY}(\alpha_i, \beta_j)$$

for all α_i, β_j such that $g(\alpha_i, \beta_j) = \gamma_i$. For example, if $Z = XY$

$$p_Z(\gamma_i) = \sum_{\alpha_i} p_{XY}\left(\alpha_i, \frac{\gamma_i}{\alpha_i}\right) = \sum_{\beta_j} p_{XY}\left(\frac{\gamma_i}{\beta_j}, \beta_j\right)$$

Problem 12. Given $p_{XY}(\alpha_i, \beta_j) = \frac{1}{3}(0.5)^{\alpha_i}$, $1 \leq \alpha_i \leq \infty$, $-1 \leq \beta_j \leq 1$, both integers, $p_{XY}(\alpha_i, \beta_j) = 0$, otherwise and $Z = X + Y$, find (1) $p_Z(\gamma_i)$ directly and (2) if X and Y are independent find $p_Z(\gamma_i)$ using discrete convolution.

Solution.

1. A sketch of the region for which $p_{XY}(\alpha_i, \beta_j)$ exists is shown in Fig. 4.13. Since $Z = X + Y$, then when $Z = \gamma_i$,

$$p_Z(\gamma_i) = \sum_{\alpha_i} p_{XY}(\alpha_i, \gamma_i - \alpha_i) = \sum_{\beta_j} p_{XY}(\gamma_i - \beta_j, \beta_j)$$

We will use the latter formula as β_j has only three values:

$$p_Z(0) = p_{XY}(1, -1) = \frac{1}{6}$$

$$p_Z(1) = \frac{1}{3}\left[\left(\frac{1}{2}\right)^2 + \left(\frac{1}{2}\right)\right] = \frac{1}{4}$$

and for the range $\gamma_i \geq 2$,

$$p_Z(\gamma_i) = \sum_{\beta_j=-1}^{1} p_{XY}(\gamma_i - \beta_j, \beta_j) = \frac{1}{3}\left[\left(\frac{1}{2}\right)^{\gamma_i-1} + \left(\frac{1}{2}\right)^{\gamma_i} + \left(\frac{1}{2}\right)^{\gamma_i+1}\right]$$

$$= \frac{7}{6}\left(\frac{1}{2}\right)^{\gamma_i} = 1.17(0.5)^{\gamma_i}$$

Figure 4.13(b) shows $p_Z(\gamma_i)$ plotted.

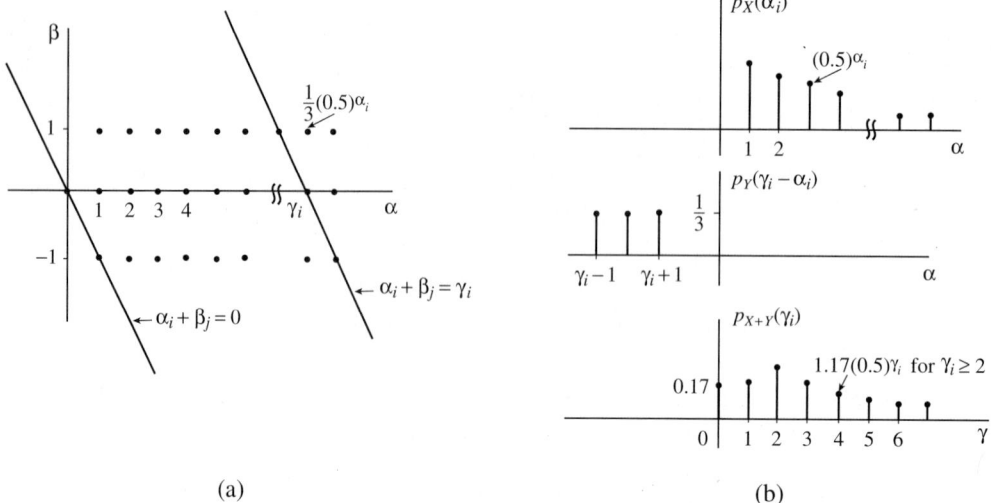

FIGURE 4.13 (a) Region of $p_{XY}(\alpha_i)$ for Problem 12 and (b) $p_X(\alpha_i)$, $p_Y(\gamma_i - \alpha_i)$, $p_{X+Y}(\gamma_i)$, $Z = X + Y$.

2. It is easy to show that $p_X(\alpha_i) = (0.5)^{\alpha_i}$, $\alpha_i \geq 1$ and is zero otherwise and

$$p_Y(\beta_j) = \frac{1}{3}\sum_1^\infty (0.5)^{\alpha_i} = \frac{1}{3}, \quad -1 \leq \beta_j \leq 1;$$
$$= 0, \text{ otherwise}$$

Since $p_{XY}(\alpha_i, \beta_j) = p_X(\alpha_i)p_Y(\beta_j)$ then X and Y are independent and

$$p_Z(\gamma_i) = \sum_{\beta_j} p_X(\gamma_i - \beta_j)p_Y(\beta_j) \quad \text{or} \quad \sum_{\alpha_i} p_X(\alpha_i)p_Y(\gamma_i - \alpha_i) \quad (4.27)$$

which is the discrete convolution of $p_X(\gamma_i)$ with $p_Y(\gamma_i)$. We will use

$$p_Z(\gamma_i) = \sum_{\alpha_i} p_X(\alpha_i)p_Y(\gamma_i - \alpha_i)$$

For the range $\gamma_i < 0$, $p_Z(\gamma_i) = 0$ as $p_Y(\gamma_i - \alpha_i)$ has not reached $\alpha_i = 1$. $p_Z(0) = \frac{1}{3}(\frac{1}{2}) = \frac{1}{6}$, $p_Z(1) = \frac{1}{3}(\frac{1}{4}) + \frac{1}{3}(\frac{1}{2}) = \frac{1}{4}$, and

$$p_Z(\gamma_i) = \sum_{\gamma_i - 1}^{\gamma_i + 1}(0.5)^{\alpha_i}\frac{1}{3} = \frac{7}{6}(0.5)^{\gamma_i} = 1.17(0.5)^{\gamma_i}, \quad \gamma_i \geq 2.$$

Statistics of Two Random Variables

Associated with two random variables X and Y the general second-order moment is

$$M_{p,q} = \overline{X^p Y^q} = \iint \alpha^p \beta^q f_{XY}(\alpha, \beta)\, d\alpha\, d\beta$$
or
$$= \sum\sum \alpha_i^p \beta_j^q\, p_{XY}(\alpha_i, \beta_j) \quad (4.28)$$

Probabilities and Random Variables

The general second-order central moment $L_{XY} = \overline{(X - \overline{X})^p (Y - \overline{Y})^q}$ is similarly defined. The two most widely used of these are the correlation $R_{XY} = \overline{XY}$ and the covariance $L_{XY} = \overline{(X - \overline{X})(Y - \overline{Y})}$. Associated with these statistics a two-dimensional moment generating function is defined,

$$M(s, t) = \overline{e^{sX+tY}} = \iint e^{sX+tY} f_{XY}(\alpha, \beta) \, d\alpha \, d\beta$$

and

$$\overline{X^p Y^q} = \frac{\delta^{p+q}}{(\delta s)^p (\delta l)^q} M(s, t)|_{s=t=0}$$

The moment generating function is particularly useful for obtaining results for jointly Gaussian random variables.

Second-Order Time Averages for Finite Power Waveforms (FPW)

The time autocorrelation function for a deterministic or random finite power waveform (FPW) is

$$\widetilde{R}_{xx}(\tau) = \widetilde{x(t)x(t+\tau)} = \lim_{T \to \infty} \frac{1}{2T} \int_{-T}^{T} x(t)x(t+\tau) \, dt \quad (4.29)$$

or

$$\widetilde{R}_{xx}(k) = \widetilde{x(n)x(n+k)} = \lim_{N \to \infty} \frac{1}{2N+1} \sum_{n=-N}^{N} x(n)x(n+k) \quad (4.30)$$

The $\sim$ symbol is used to distinguish them from the fundamental ensemble definition for $R_{xx}(\tau)$ or $R_{xx}(k)$ used for random processes.

These time-average definitions may be assigned an equivalent statistical interpretation. For a continuous FPW $x(t)$, X and Y describe uniformly sampling $x(t)$ at t and $t + \tau$ and observing $X = x(t)$ and $Y = x(t + \tau)$. For N trials as $N \to \infty$

$$\widetilde{R}_{xx}(\tau) = \lim_{N \to \infty} \frac{1}{N} \sum_{i=1}^{N} x(t_i)x(t_i + \tau) \quad (4.31)$$

where $x(t_i)$ is the value of $x(t)$ on the ith trial, or

$$\widetilde{R}_{xx}(\tau) = \widetilde{x(t)x(t+\tau)} = \overline{XY}$$

$$= \iint \alpha \beta f_{XY}(\alpha, \beta) \, d\alpha \, d\beta$$

or

$$= \sum \sum \alpha_i \beta_j p_{XY}(\alpha_i, \beta_j) \quad (4.32)$$

For a discrete FPW $x(n)$, X and Y describe uniformly sampling $x(n)$ at n and $n + k$ and observing $X = x(n)$ and $Y = x(n + k)$. For N trials as $N \to \infty$

$$\widetilde{R}_{xx}(k) = \lim_{N \to \infty} \frac{1}{N} \sum_{i=1}^{N} x(n_i)x(n_i + k) \quad (4.33)$$

where $x(n_i)$ is the value of $x(n)$ on the ith trial, or

$$\tilde{R}_{xx}(k) = \widetilde{x(n)x(n+k)} = \overline{XY}$$

$$= \iint \alpha \beta f_{xy}(\alpha, \beta) \, d\alpha \, d\beta$$

or

$$= \sum \sum \alpha_i \beta_j p_{XY}(\alpha_i, \beta_j) \qquad (4.34)$$

To clarify the relative frequency notation of Eq. (4.31) or (4.33), for example on the sixth trial t is chosen uniformly to obtain t_6 and $x(t_6)x(t_6 + \tau)$ and n is chosen uniformly to obtain n_6 and $x(n_6)x(n_6 + k)$. We also note Eqs. (4.32) and (4.34) are identical.

Two illustrative problems will be solved.

Problem 13. Consider the discrete random waveform

$$x(n) = \sum_k A_k g(n - 2k)$$

where $g(n) = 3\delta(n) - \delta(n-1)$ and the As are independent random variables each with mass function $p_{A_i}(1) = p_{A_i}(-1) = 0.5$. Find $\tilde{R}_{xx}(k)$.

Solution. A typical section of $x(n)$ is shown in Fig. 4.14(a). Using formula (4.34) for $\tilde{R}_{xx}(k)$, $\tilde{R}_{xx}(0) = [(3)^2 + (-3)^2 + (1)^2 + (-1)^2]\frac{1}{4} = 5.0$, $\tilde{R}_{xx}(\pm 1) = [3(-1) + (-3)(1)]\frac{1}{4} + [(1)(-3) + (1)(3)]\frac{1}{8}$ + other terms adding to zero $= -1.5$, and $\tilde{R}_{xx}(k) = 0$ for all $|k| > 1$. $\tilde{R}_{xx}(k)$ is shown plotted in Fig. 4.14(b).

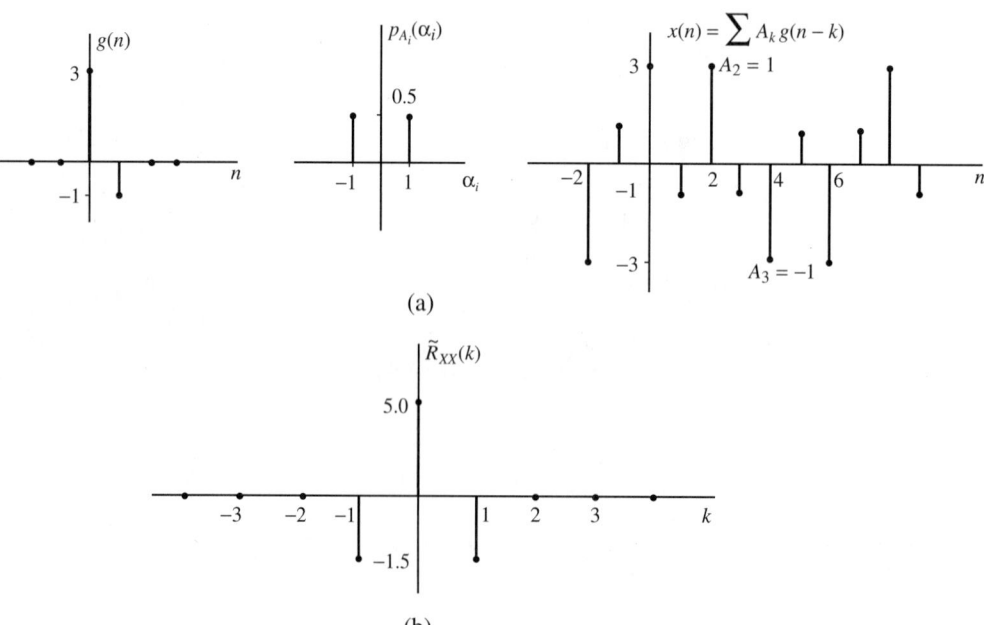

FIGURE 4.14 (a) A typical section of $x(n)$ for Problem 13 and (b) $\tilde{R}_{XX} = \widetilde{x(n)x(n+k)} = \overline{XY}$.

Probabilities and Random Variables

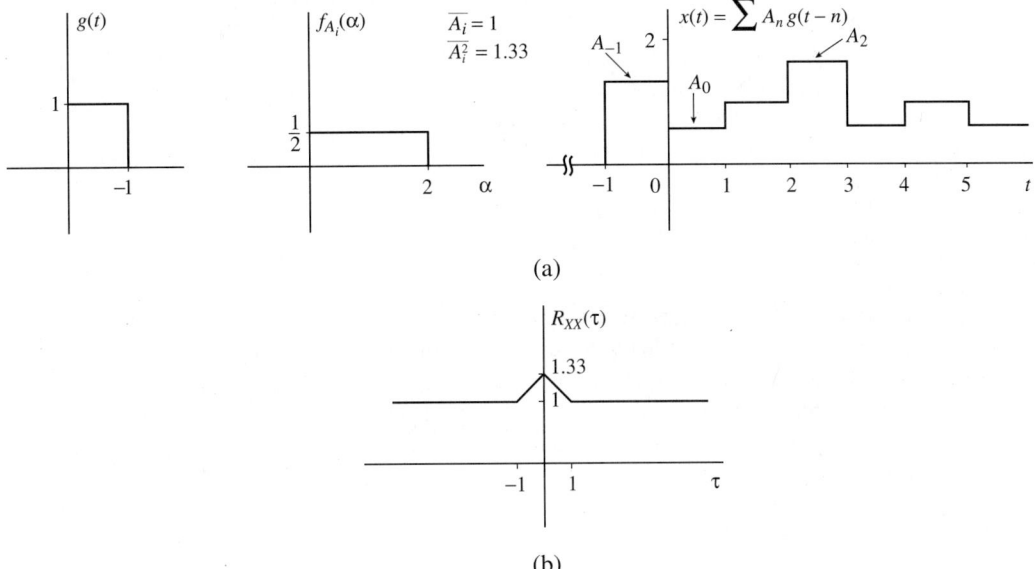

FIGURE 4.15 (a) A typical section of $x(t)$ for Problem 14 and (b) $\widetilde{R}_{XX}(\tau)$.

Problem 14. Consider the continuous random waveform

$$x(t) = \sum_n A_n g(t-n)$$

where $g(t) = 1, 0 \leq t \leq 1$; $g(t) = 0$, otherwise and the As are independent random variables with $f_{A_i}(\alpha) = \frac{1}{2}, 0 \leq \alpha \leq 2$, $f_{A_i}(\alpha) = 0$, otherwise. Find $\widetilde{R}_{xx}(\tau)$.

Solution. A typical section of $x(t)$ is shown in Fig. 4.15(a). Since the As are independent $\overline{A_i} = 1$, $\overline{A_i^2} = 1.33$, and $\overline{A_i A_j} = 1$ for all $i \neq j$. Observing Fig. 4.15(a) we obtain the following.
For $0 < \tau < 1$:

$$\widetilde{R}_{xx}(\tau) = \overline{A_i^2}(1-\tau) + \overline{A_i A_{i+1}}\,\tau$$
$$= 1.33 - 0.33\tau$$

For $\tau > 1$:

$$\widetilde{R}_{xx}(\tau) = \overline{A_i A_j} = 1$$

and $\widetilde{R}_{xx}(-\tau) = \widetilde{R}_{xx}(\tau)$ by symmetry. $\widetilde{R}_{xx}(\tau)$ is shown plotted in Fig. 4.15(b). We also note that the same result is obtained when the As are independent and $f_{A_i}(\alpha)$ or $p_{A_i}(\alpha)$ are such $\overline{A_i} = 1$ and $\overline{A_i^2} = 1.33$. Thus, many different waveforms have the same autocorrelation function.

4.5 Summary and Future Study

The elements of probability and random variable theory were presented as a prerequisite to the study of random processes and communication theory. Highlighted were the concepts of finding first- and second-order time averages statistically, by uniformly sampling a continuous random finite power waveform at t or at t and $t + \tau$ or discretely uniformly sampling a random finite power discrete waveform at n or at n and $n + k$. The illustrative problems chosen were simple. When ergodic random processes are later encountered the time averages for any one member of the ensemble will be equivalent to corresponding ensemble averages. A future task is to approximate time averages on a computer. Here from a section of a waveform T with its values sampled every τ seconds an estimate for the autocorrelation function and its Fourier transom $S_{XX}(f)$ called the power spectral density will be found.

Defining Terms

Cumulative distribution, density and mass function: $F_X(\alpha) = P[X \leq \alpha]$, $f_X(\alpha) = d/d\alpha\, F_X(\alpha)$, $p_X(\alpha_i) = P[X = \alpha_i]$. Other noted distribution or density functions are: total signal energy or total signal power vs frequency, corresponding to $F_X(\alpha)$; energy or power spectral density vs frequency, corresponding to $f_X(\alpha)$; total signal power vs frequency for a periodic waveform, corresponding to $F_X(\alpha)$ for a discrete random variable.

Joint distribution, density and mass function: $F_{XY}(\alpha, \beta) = P[(X \leq \alpha) \cap (Y \leq \beta)]$, $f_{XY}(\alpha, \beta) = \partial^2/\partial\alpha\,\partial\beta\, F_{XY}(\alpha, \beta)$, $p_{XY}(\alpha_i, \beta_j) = P[(X = \alpha_i) \cap (Y = \beta_j)]$.

Joint random variables X and Y: $X(s_i)$ and $Y(s_i)$ jointly map each outcome of the SDS onto an α–β plane.

Probability: $P(A)$ is the statistical regularity of A, $P(A/B)$ is the statistical regularity of A based only on trials when B occurs, axiomatically $P(A) = \sum_{i=1}^{n} p(s_i)$ if $A = \bigcup_{i=1}^{n} s_i$ and $s_i s_j = 0, i \neq j$. $P(A/B) = P(AB)/P(B)$ if $P(B) \neq 0$.

Random phenomenon: An experiment which when performed many times N, yields outcomes that occur with statistical regularity as $N \to \infty$.

Random variable X: $X(s_i)$ maps the outcomes of the SDS onto a real axis.

Sample description space (SDS): A listing of mutually exclusive, collectively exhaustive, finest grain outcomes.

Special Terminology for Random Finite Power Waveforms

Finite power waveform:

$$P_{\text{av}} = \lim_{T \to \infty} \frac{1}{2T} \int_{-T}^{T} x^2(t)\, dt$$

or

$$= \lim_{N \to \infty} \frac{1}{2N+1} \sum_{n=-N}^{N} x^2(n) \quad \text{exists.}$$

First-order time averages:

$$\widetilde{g(x(t))} = \lim_{T \to \infty} \frac{1}{2T} \int_{-T}^{T} g(x(t))\, dt = \int_{-\infty}^{\infty} g(\alpha) f_X(\alpha)\, d\alpha$$

and

$$\widetilde{g(x(n))} = \lim_{N \to \infty} \frac{1}{2N+1} \sum_{-N}^{N} g(x(n)) = \int_{-\infty}^{\infty} g(\alpha) f_X(\alpha)\, d\alpha$$

The most noted first-order statistics are $\widetilde{x(t)}$ or $\widetilde{x(n)} = \overline{X}$, $\widetilde{x^2(t)}$ or $\widetilde{x^2(n)} = \overline{X^2}$ and $\sigma_X^2 = \overline{(X - \overline{X})^2}$.

Probabilities and Random Variables

Jointly sampling a waveform: Uniformly choosing t and observing $X = x(t)$ and $Y = x(t+\tau)$ or discretely uniformly choosing n and observing $X = x(n)$ and $Y = x(n+k)$.

Sampling a waveform: For a continuous waveform uniformly choosing a value of time over a long time T and observing $X = x(t)$; for a discrete waveform discretely uniformly choosing $n, -N \leq n \leq N$ over a long time $2N+1$ and observing $x(n)$. The random variable X describes $x(t)$ or $x(n)$.

Second-order time averages:

$$\widetilde{g[x(t), x(t+\tau)]} = \overline{g(X,Y)} = \lim_{T \to \infty} \frac{1}{2T} \int_{-T}^{T} g[x(t), x(t+\tau)] \, dt$$

$$\widetilde{g[x(n), x(n+k)]} = \overline{g(X,Y)} = \lim_{N \to \infty} \frac{1}{2N+1} \sum_n g[x(n), x(n+k)]$$

and statistically

$$\overline{g(X,Y)} = \iint g(\alpha, \beta) f_{XY}(\alpha, \beta) \, d\alpha \, d\beta$$

or

$$= \sum \sum g(\alpha_i, \beta_j) p_{XY}(\alpha_i, \beta_j)$$

The most noted second-order statistics are $\widetilde{R}_{xx}(\tau)$ or $\widetilde{R}_{xx}(k) = \overline{XY}$, the time autocorrelation function, and/or $\widetilde{L}_{xx}(\tau)$ or $\widetilde{L}_{xx}(k) = \overline{(X-\overline{X})(Y-\overline{Y})} = \overline{XY} - \overline{X}\,\overline{Y}$, the covariance function.

References

Olkin, Gleser, and Derman. 1994. *Probability Models and Applications*, Macmillan, New York.
O'Flynn, M. 1982. *Probabilities, Random Variables and Random Processes*, Wiley, New York.
Papoulis, A. 1965. *Probability, Random Variables and Stochastic Processes*, McGraw-Hill, New York.
Parzen, E. 1960. *Modern Probability Theory and Its Applications*, Wiley, New York.
Ross, S. 1994. *A First Course in Probability*, Macmillan, New York.

Further Information

Some important topics were omitted or briefly mentioned here. The bivariate Gaussian or normal joint density function is given an excellent treatment in Olkin, Gleser, and Derman [1994, Chap. 12]. The topic of finding the density function of a function of two known random variables can be extended. The case of finding the joint density function for U and V each defined as a function of two known random variables X and Y is again well handled in Olkin, Gleser, and Derman [1994, Chap. 13]. Finally, the most general transformation problem of finding the joint density function of m random variables each defined as a linear combination of n known jointly Gaussly random variables is given in Chap. 9 of O'Flynn [1982].

5
Random Processes, Autocorrelation, and Spectral Densities

5.1	Introduction 62
5.2	Basic Definitions 62
5.3	Properties and Interpretation 65
5.4	Baseband Digital Data Signals................................ 67
5.5	Coding for Power Spectrum Control..................... 71
5.6	Bandpass Digital Data Signals................................. 73
5.7	Appendix: The Poisson Sum Formula 76

L.E. Franks
University of Massachusetts

5.1 Introduction

The modeling of signals and noise in communication systems as random processes is a well-established and proven tradition. Much of the system performance characterization and evaluation can be done in terms of a few low-order statistical moments of these processes. Hence a system designer often needs to be able to evaluate mean values and the **autocorrelation** of a process and the **cross correlation** of two processes. It is especially true in communication system applications that a frequency-domain-based evaluation is often easier to perform and interpret. Therefore, a knowledge of spectral properties of random processes is valuable.

More recently, engineers are finding that modeling of typical communication signals as **cyclostationary** (CS) processes, rather than stationary processes, more accurately captures their true nature and permits the design of more effective signal processing operations. In the book edited by Gardner [1994], several authors describe applications that have benefited from the exploitation of cyclostationarity. Although these processes are nonstationary, frequency-domain methods prove to be very effective, as in the stationary case.

Another notion that is becoming well accepted by communication engineers is the utility of complex-valued representations of signals. Accordingly, all of the results and relations presented here will be valid for complex-valued random processes.

5.2 Basic Definitions

We consider the random process, $x(t)$, defined as a set of jointly distributed random variables $\{x(t)\}$ indexed by a time parameter t. If the set of values of t is the real line, we say the process

Random Processes, Autocorrelation, and Spectral Densities

is a *continuous-time* process. If the set of values is the integers, we call the process a *discrete-time* process. The *mean* and *autocorrelation* of the process are given by the expected values $E[x(t)]$ and $E[x(t_1)x^*(t_2)]$, respectively. In dealing with cyclostationary processes, it is often more convenient to define an autocorrelation function in terms of a simple transformation on the t_1 and t_2 variables; namely, let $t = (t_1 + t_2)/2$ and $\tau = t_1 - t_2$ denote the mean value and difference of t_1 and t_2. Accordingly, we define the autocorrelation function for the process as

$$R_{xx}(t, \tau) = E[x(t + \tau/2) x^*(t - \tau/2)] \tag{5.1}$$

A process is *cyclostationary*, in the wide sense, with period T if both the mean and autocorrelation are periodic in t with period T. Hence, for a cyclostationary process CS(T) we can write[1]

$$m_x(t) = E[x(t)] = \sum_m M_x(m) e^{j2\pi mt/T} \tag{5.2}$$

$$R_{xx}(t, \tau) = \sum_k T c_{xx}(t - kT, \tau)$$

$$= \sum_m \tilde{c}_{xx}\left(\frac{m}{T}, \tau\right) e^{j2\pi mt/T} \tag{5.3}$$

Equation (5.3) results from an application of the **Poisson sum formula** (see Appendix), where $\tilde{c}_{xx}(\nu, \tau)$ is the Fourier transform of $c_{xx}(t, \tau)$ with respect to the t variable; that is,

$$\tilde{c}_{xx}(\nu, \tau) = \int c_{xx}(t, \tau) e^{-j2\pi \nu t} \, dt \tag{5.4}$$

hence,

$$\tilde{c}_{xx}(m/T, \tau) = \frac{1}{T} \int_0^T R_{xx}(t, \tau) e^{-j2\pi mt/T} \, dt \tag{5.5}$$

is the mth Fourier coefficient for the t-variation in $R_{xx}(t, \tau)$. It is called the **cyclic autocorrelation function** for $x(t)$ with **cycle frequency** m/T [Gardner, 1989]. Note that the $c_{xx}(t, \tau)$ function in Eq. (5.3) is not uniquely determined by the autocorrelation; only a discrete set of values of its Fourier transform are specified, as indicated in Eq. (5.3). Nevertheless, the notation is useful because the $c_{xx}(t, \tau)$ function arises naturally in modeling typical communication signal time-sequential formats that use sampling, scanning, or multiplexing. The **pulse amplitude modulation (PAM)** signal discussed in Sec. 5.4 is a good example of this.

A frequency-domain characterization for the CS(T) process is given by the double Fourier transform of $R_{xx}(t, \tau)$.

$$S_{xx}(\nu, f) = \int \int R_{xx}(t, \tau) e^{-j2\pi(\nu t + f\tau)} \, dt \, d\tau$$

$$= \sum_m C_{xx}\left(\frac{m}{T}, f\right) \delta\left(\nu - \frac{m}{T}\right) \tag{5.6}$$

where $C_{xx}(\nu, f)$ is the double Fourier transform of $c_{xx}(t, \tau)$; hence, it is the Fourier transform, with respect to the τ-variation, of the cyclic autocorrelation. It is called the *cyclic spectral density*

---

[1] All summation indices will run from $-\infty$ to ∞ unless otherwise indicated. The same is true for integration limits.

for the cycle frequency m/T. Notice that Eq. (5.6) is a set of *impulse fences* located at discrete values of ν in the dual-frequency (ν, f) plane. The impulse fence at $\nu = 0$ has a special significance as the (one-dimensional) **power spectral density** of the process. The other impulse fences characterize the correlation between signal components in different spectral regions. The physical significance of these concepts is discussed in the next section.

A process is *wide-sense stationary* (WSS) if the mean and autocorrelation are independent of the t variable; that is, if only the $m = 0$ terms in Eqs. (5.2) and (5.3) are nonzero. For the WSS process, we write the autocorrelation function as

$$R_{xx}(\tau) = R_{xx}(0, \tau) = \tilde{c}_{xx}(0, \tau) \tag{5.7}$$

since the cyclic autocorrelation is nonzero only for a cycle frequency of $\nu = 0$. The *power spectral density* of the process is simply the Fourier transform of $R_{xx}(\tau)$

$$S_{xx}(f) = \int R_{xx}(\tau) e^{-j2\pi f \tau} \, d\tau \tag{5.8}$$

A CS(T) process can be converted to a WSS process by phase randomization. Let $x'(t) = x(t - \theta)$ where the time shift θ is a random variable which is independent of the other parameters of the process. Then the mean and autocorrelation of the new process are obtained by convolution, with respect to the t variable, with the probability density function $p_\theta(\cdot)$ for the θ variable. Letting $P_\theta(\cdot)$ denote the Fourier transform of $p_\theta(\cdot)$, we have

$$m_{x'}(t) = \sum_m P_\theta\left(\frac{m}{T}\right) M_x(m) e^{j2\pi mt/T}; \qquad P_\theta(0) = 1 \tag{5.9a}$$

$$R_{x'x'}(t, \tau) = \sum_m P_\theta\left(\frac{m}{T}\right) \tilde{c}_{xx}\left(\frac{m}{T}, \tau\right) e^{j2\pi mt/T} \tag{5.9b}$$

which indicates that phase randomization tends to suppress the cyclic components with cycle frequencies different from zero. In fact, for a uniform distribution of θ over a T-second interval, only the $m = 0$ terms in Eq. (5.9) remain; hence $x'(t)$ is WSS. Using the Poisson sum formula, we see that any $p_\theta(\sigma)$ satisfying $\sum_k p_\theta(\sigma - kT) = 1/T$ makes $P_\theta(m/T) = 0$ for $m \neq 0$, hence making $x'(t)$ WSS.

It is important to note [from Eq. (5.5)] that the autocorrelation of the phase-randomized process is the same as the time average of the autocorrelation of the original CS(T) process [Franks, 1969].

For a WSS discrete-time process $a(k)$, where k is an integer, we write the mean and autocorrelation as

$$\bar{a} = E[a(k)] \tag{5.10a}$$

$$R_{aa}(m) = E[a(k+m)a^*(k)] \tag{5.10b}$$

The power spectral density of the discrete-time process is the *discrete-time Fourier transform* (DTFT) of the autocorrelation function; that is,

$$S_{aa}(f) = \sum_m R_{aa}(m) e^{-j2\pi mTf} \tag{5.11}$$

which is always a periodic function of f with a period of $1/T$. A case of frequent interest is when the $a(k)$ are uniformly spaced sample values of a continuous-time WSS random process. Suppose

Random Processes, Autocorrelation, and Spectral Densities

$a(k) = y(kT)$, then using the alternate form of the Poisson sum formula (Appendix), we have $R_{aa}(m) = R_{yy}(mT)$ and

$$S_{aa}(f) = \sum_m R_{yy}(mT)e^{-j2\pi mTf} = \frac{1}{T}\sum_n S_{yy}\left(f - \frac{n}{T}\right) \quad (5.12)$$

In communication system applications, the $a(k)$ sequence usually represents digital information to be transmitted, and the WSS assumption is reasonable. In some cases, such as when synchronizing pulses or other header information are added to the data stream, the $a(k)$ sequence is more appropriately modeled as a cyclostationary process also. The examples we consider here, however, will assume that $a(k)$ is a WSS process.

5.3 Properties and Interpretation

Many interesting features of a random process are revealed when the effects of a linear transformation on the first- and second-moment statistics are examined. For this purpose we use linear transformations which are a combination of frequency shifting and time-invariant filtering, as shown in Fig. 5.1. The exponential multipliers (modulators) provide the frequency shifts, and the filters with tranfer functions $H_1(f)$ and $H_2(f)$ provide frequency selective filtering. If $H_1(f)$ and $H_2(f)$ are narrowband low-pass filters, each path of the structure in Fig. 5.1 can be regarded as a sort of heterodyne technique for spectrum analysis. The cross correlation between the outputs of the two paths provides further useful characterization of the input process.

The cross-correlation function for a pair of jointly random complex processes is defined as

$$R_{yz}(t, \tau) = E[y(t + \tau/2)z^*(t - \tau/2)] \quad (5.13)$$

Substituting the indicated expressions for $y(t)$ and $z(t)$ in Fig. 5.1 and using Eq. (5.3) for the autocorrelation of the $x(t)$ process, we can derive the following general expression for the cross correlation:

$$R_{yz}(t, \tau) = \sum_m e^{j\pi(f_1+f_2)\tau} \cdot e^{j2\pi(f_1-f_2)t} \cdot e^{j2\pi mt/T} \cdot \int H_1\left(f + f_1 + \frac{m}{2T}\right)$$
$$\times H_2^*\left(f + f_2 - \frac{m}{2T}\right) C_{xx}\left(\frac{m}{T}, f\right) e^{j2\pi f\tau}\, df \quad (5.14)$$

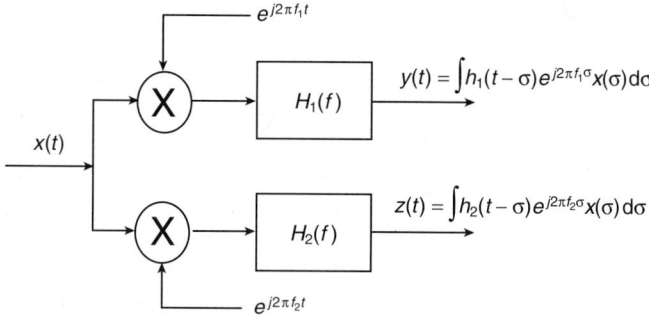

FIGURE 5.1 Conceptual apparatus for analysis of random processes.

First, we consider the case where $H_1(f) = H_2(f) = B(f)$ which is a very narrowband low-pass analysis filter. Let $B(0) = 1$ and let Δ denote the *noise bandwidth* of the filter given by

$$\Delta = \int \left|\frac{B(f)}{B(0)}\right|^2 df \tag{5.15}$$

Hence, an ideal low-pass filter with a rectangular gain shape and a cutoff frequency at $f = \Delta/2$ has a noise bandwidth of Δ. On the other hand, a low-pass filter resulting from averaging a signal over the last T_0 s, that is, one with an impulse response $b(t) = 1/T_0$ in $0 < t < T_0$ and zero elsewhere, has a noise bandwidth of $\Delta = 1/T_0$. Whatever kind of analysis filter we want to use, we shall consider the results as Δ approaches zero, so that we can assume that factors of the form $B(f + m/2T)B^*(f - m/2T)$ which will appear in Eq. (5.14) will vanish except for $m = 0$.

For example, let $f_1 = f_2 = -f_0$. Then we see that both paths in the figure are identical, so that $z(t) = y(t)$ and

$$R_{yy}(t, \tau) = e^{-j2\pi f_0 \tau} \int |B(f - f_0)|^2 C_{xx}(0, f) e^{j2\pi f \tau} df \tag{5.16}$$

because only the $m = 0$ term in Eq. (5.14) is nonzero under these conditions. Note also that $y(t)$ is WSS, and its mean-squared value (also called the power of the process) is given by

$$E[|y(t)|^2] = R_{yy}(0) = \int |B(f - f_0)|^2 C_{xx}(0, f) df$$
$$\cong \Delta C_{xx}(0, f_0) = \Delta S_{xx}(f_0) \qquad \text{as } \Delta \to 0 \tag{5.17}$$

This expression, valid for small Δ, gives the power in $x(t)$ due to components in a narrow band centered at a frequency of f_0. This provides the motivation for calling $S_{xx}(f) = C_{xx}(0, f)$ the power spectral density of the $x(t)$ process. Note that our spectrum analyzer gives the same results for a CS(T) process and its phase-randomized WSS version because of the time averaging inherent in the $B(f)$ analysis filter.

Now, keeping the same low-pass filters, $H_1(f) = H_2(f) = B(f)$, let $f_0 = -(f_1 + f_2)/2$ and $n/T = (f_2 - f_1)$, so that the mean value of the two analysis frequencies is f_0 and their difference is an integer multiple of $1/T$. Then only the $m = n$ term in Eq. (5.14) is nonzero and we have the following expression for the cross correlation of the $y(t)$ and $z(t)$ processes:

$$R_{yz}(t, 0) = \int |B(f - f_0)|^2 C_{xx}\left(\frac{n}{T}, f\right) df$$
$$\cong \Delta C_{xx}(n/T, f_0) \tag{5.18}$$

for small Δ. Thus, a physical significance of the cyclic spectral density, defined implicity in Eq. (5.6), is that it represents the correlation that exists between spectral components separated by the cycle frequency. Note that no such spectral correlation exisits for a WSS process. This spectral correlation is used to advantage in communication systems to extract timing or synchronization information or to eliminate certain kinds of interference by means of periodically time-varying filtering of signals [Gardner, 1994].

Equation (5.14) is useful to establish many other useful relationships. For example, let $x(t)$ be WSS and let $f_1 = f_2 = 0$, then $y(t)$ and $z(t)$ are jointly WSS and the cross-spectral density is

$$S_{yz}(f) = H_1(f) H_2^*(f) S_{xx}(f) \tag{5.19}$$

Random Processes, Autocorrelation, and Spectral Densities

and so if $H_1(f) = H_2(f)$, then $z(t) = y(t)$ and the power spectral density of a filtered WSS process is given by

$$S_{yy}(f) = |H_1(f)|^2 S_{xx}(f) \tag{5.20}$$

whereas letting $H_2(f) = 1$ makes $z(t) = x(t)$ so that the cross-spectral density of the input and output of a filtered WSS process is

$$S_{yx}(f) = H_1(f) S_{xx}(f) \tag{5.21}$$

If $H_1(f)H_2^*(f) = 0$ for all f, then $S_{yz}(f) = 0$, indicating that spectral components in nonoverlapping frequency bands are always uncorrelated. This is a fundamental property of all WSS processes.

5.4 Baseband Digital Data Signals

Perhaps the most basic form of mapping from a discrete-time data sequence $\{a(k)\}$ into a continuous-time signal $x(t)$ for transmission over a communication channel is baseband pulse amplitude modulation. The mapping uses a single carrier pulse shape, $g(t)$, and is a linear function of the data, which we will assume is a WSS discrete-time process [Eq. (5.10)].

$$x(t) = \sum_k a(k) g(t - kT) \tag{5.22}$$

The $x(t)$ process is CS(T) with

$$m_x(t) = \bar{a} \sum_k g(t - kT) = (\bar{a}/T) \sum_m G\left(\frac{m}{T}\right) e^{j2\pi m t/T} \tag{5.23a}$$

$$R_{xx}(t, \tau) = \sum_k \sum_m R_{aa}(m) g\left(t + \frac{\tau}{2} - kT - mT\right) g^*\left(t - \frac{\tau}{2} - kT\right) \tag{5.23b}$$

By inspection, the $c_{xx}(t, \tau)$ function in Eq. (5.3) is

$$c_{xx}(t, \tau) = \frac{1}{T} \sum_m R_{aa}(m) g\left(t + \frac{\tau}{2} - mT\right) g^*\left(t - \frac{\tau}{2}\right) \tag{5.24}$$

so that the cyclic autocorrelation is

$$\tilde{c}_{xx}(n/T, \tau) = \frac{1}{T} \int S_{aa}(f) G\left(\frac{n}{T} + f\right) G^*(f) e^{j2\pi f \tau} \, df \cdot e^{j\pi n \tau/T} \tag{5.25}$$

Note that the strength of the cyclic autocorrelations (equivalently, the amount of spectral correlation) depends strongly on the amount of overlap of $G(f)$ and its frequency-translated versions. Data carrier pulses are often sharply bandlimited, with **excess bandwidth** factors not exceeding 100%. A pulse with 100% excess bandwidth is bandlimited to the frequency interval $|f| < 1/T$. In this case, $x(t)$ has cycle frequencies of 0 and $\pm 1/T$ only. Note also that an excess bandwidth of less than 100% means that $G(n/T) = 0$ for $n \neq 0$ so that the mean of the PAM signal [Eq. (5.23a)] is constant.

Using time averaging to get the power spectrum of the PAM signal, we can write

$$R_{xx}(\tau) = \frac{1}{T} \sum_m R_{aa}(m) r_g(\tau - mT) \qquad (5.26)$$

where

$$r_g(\tau) = \int g(t+\tau) g^*(t)\, dt \qquad (5.27)$$

is the **time-ambiguity function** for the data pulse, $g(t)$ [Franks, 1969]. Its Fourier transform is $|G(f)|^2$, called the *energy spectral density* of the pulse. Hence, taking the Fourier transform of Eq. (5.26), the power spectral density of the PAM signal is

$$S_{xx}(f) = \frac{1}{T} S_{aa}(f) |G(f)|^2 \qquad (5.28)$$

which nicely separates the influence of the pulse shape from the correlation structure of the data.

Although it may not be apparent from Eq. (5.28), the power spectrum expression may contain δ-function terms, which represent nonzero concentrations of power at a discrete set of frequencies. In order to make evident the discrete part of the power spectrum, we express the continuous part of the spectrum as the Fourier transform of the autocovariance function, which is the autocorrelation of the zero-mean process, $x'(t) = x(t) - m_x(t)$. Then

$$R_{xx}(t,\tau) = R'_{xx}(t,\tau) + m_x(t+\tau/2) m_x^*(t-\tau/2) \qquad (5.29)$$

where

$$R'_{xx}(t,\tau) = R_{x'x'}(t,\tau) = \frac{1}{T} \sum_m R'_{aa}(m) r_g(\tau - mT) \qquad (5.30)$$

and

$$R'_{aa}(m) = R_{aa}(m) - |\bar{a}|^2 \qquad (5.31)$$

Now performing the time averaging, the power spectrum is expressed as

$$S_{xx}(f) = \frac{1}{T} S'_{aa}(f) |G(f)|^2 + \left|\frac{\bar{a}}{T}\right|^2 \sum_n \left|G\left(\frac{n}{T}\right)\right|^2 \delta\left(f - \frac{n}{T}\right) \qquad (5.32)$$

The second term in Eq. (5.32) is the discrete part of the spectrum. Notice that there are no discrete components for zero-mean data ($\bar{a} = 0$) or if the data pulse has less than 100% excess bandwidth ($G(n/T) = 0$ for $n \neq 0$). For zero-mean, independent data, pulse amplitudes are uncorrelated and the power spectrum is simply

$$S_{xx}(f) = \frac{\sigma_a^2}{T} |G(f)|^2 \qquad (5.33)$$

where σ_a^2 is the data variance.

A more general format for baseband data signals is to use a separate waveform for each data symbol from the symbol alphabet. Familiar examples are *frequency-shift keying* (FSK), *phase-shift keying* (PSK), and *pulse-position modulation* (PPM). For the binary alphabet case, $[a(k) \in \{0, 1\}]$, the format is basically PAM. Using waveforms $g_0(t)$ and $g_1(t)$ to transmit a zero or one, respectively, we have

$$x(t) = \sum_k [1 - a(k)]g_0(t - kT) + a(k)g_1(t - kT) \quad (5.34)$$

which can be rewritten, using $d(t) = g_1(t) - g_0(t)$, as

$$x(t) = \sum_k g_0(t - kT) + \sum_k a(k)d(t - kT) \quad (5.35)$$

which is a periodic fixed term plus a binary PAM signal. Let $p = Pr[a(k) = 1] = \bar{a}$ and let $P_{11}(m)$ denote the conditional probability, $Pr[a(k + m) = 1 \mid a(k) = 1]$, then with $a'(k) = a(k) - \bar{a}$,

$$\begin{aligned} R'_{aa}(m) &= E[a'(k+m)a'(k)] \\ &= R_{aa}(m) - p^2 = pP_{11}(m) - p^2 \end{aligned} \quad (5.36)$$

and carrying out the calculations, the power spectral density is

$$S_{xx}(f) = \frac{1}{T} s'_{aa}(f) |G_1(f) - G_0(f)|^2 \\ + \left(\frac{1}{T}\right)^2 \sum_n \left|(1-p)G_0\left(\frac{n}{T}\right) + pG_1\left(\frac{n}{T}\right)\right|^2 \delta\left(f - \frac{n}{T}\right) \quad (5.37)$$

where the second term gives the discrete components. Note that the strength of the discrete components depends only on the mean value of the data and not the correlation between data symbols.

To generalize the preceding result to an M-ary symbol alphabet for the data, we write

$$x(t) = \sum_k \sum_{m=0}^{M-1} a_m(k) g_m(t - kT) \quad (5.38)$$

with $a_m(k) \in \{0, 1\}$ and, for each k, only one of the $a_m(k)$ is nonzero for each realization of the process. Then calculating the autocorrelation function, we find that $c_{xx}(t, \tau)$ in Eq. (5.3) can be expressed as

$$c_{xx}(t, \tau) = \frac{1}{T} \sum_i \sum_{m,n=0}^{M-1} E[a_m(i+j)a_n(i)] g_m\left(t + \frac{\tau}{2} - iT\right) g_n^*\left(t - \frac{\tau}{2}\right) \quad (5.39)$$

We assume that the data process can be modeled as a *Markov chain* characterized by a probability transition matrix $P_{mn}(i)$ giving the probability that $a_m(i + j) = 1$ under the condition that $a_n(j) = 1$. We further assume the process is in a steady-state condition so that it is WSS. This means that the symbol probabilities, $p_n = Pr[a_n(j) = 1]$ must satisfy the homogeneous equation

$$p_m = \sum_{n=0}^{M-1} P_{mn}(1) p_n \quad (5.40)$$

and all of the transition matrices can be generated by recursion;

$$P_{mn}(i+1) = \sum_{k=0}^{M-1} P_{mk}(i)P_{kn}(1); \qquad i \geq 1 \qquad (5.41)$$

Now calculation of the autocorrelation can proceed using $E[a_m(i+j)a_n(j)] = q_{mn}(i)$ in Eq. (5.39), where

$$\begin{aligned} q_{mn}(i) &= Pr[a_m(i+j) = 1 \text{ and } a_n(j) = 1] \\ &= P_{mn}(i)p_n & \text{for } i > 0 \\ &= \delta_{mn} p_n & \text{for } i = 0 \\ &= P_{nm}(-i)p_m & \text{for } i < 0 \end{aligned} \qquad (5.42)$$

Taking the DTFT of the $q_{mn}(i)$ sequence

$$Q_{mn}(f) = \sum_i q_{mn}(i)e^{-j2\pi i Tf} \qquad (5.43)$$

we arrive at a remarkably compact expression for the power spectral density of this general form for the baseband data signal

$$S_{xx}(f) = \frac{1}{T} \sum_{m,n=0}^{M-1} Q_{mn}(f)G_m(f)G_n^*(f) \qquad (5.44)$$

The power spectrum may contain discrete components. The continuous part of the spectrum can be extracted by replacing the $q_{mn}(i)$ in Eqs. (5.42–5.44) by $q_{mn}(i) - p_m p_n$. Then the remaining part is the discrete spectrum given by

$$S_{xx}(f)_{\text{discrete}} = \left(\frac{1}{T}\right)^2 \sum_k \left|\sum_{m=0}^{M-1} p_m G_m\left(\frac{k}{T}\right)\right|^2 \delta\left(f - \frac{k}{T}\right) \qquad (5.45)$$

For the special case of independent data from symbol to symbol, the probability transition matrix takes on the special form $P_{mn}(i) = p_m$ for each m and each $i > 1$; that is, all columns are identical and consist of the set of steady-state symbol probabilities. In this case the continuous part of the power spectrum is

$$S_{xx}(f)_{\text{continuous}} = \frac{1}{T}\left[\sum_{m=0}^{M-1} p_m |G_m(f)|^2 - \left|\sum_{m=0}^{M-1} p_m G_m(f)\right|^2\right] \qquad (5.46)$$

and the discrete part is the same as in Eq. (5.45).

As an example of signalling with M different waveforms and with independent, equiprobable data, consider the case of pulse position modulation where

$$g_m(t) = g\left(t - m\frac{T}{M}\right); \qquad p_m = 1/M; \; m = 0, 1, 2, \ldots, M-1 \qquad (5.47)$$

Random Processes, Autocorrelation, and Spectral Densities

The resulting power spectrum is

$$S_{xx}(f) = \left(\frac{1}{T}\right) Q(f)|G(f)|^2 + \left(\frac{1}{T}\right)^2 \sum_k \left|G\left(\frac{kM}{T}\right)\right|^2 \delta\left(f - \frac{kM}{T}\right) \quad (5.48)$$

where

$$Q(f) = \left[1 - \left(\frac{1}{M}\right) \frac{\sin^2(\pi T f)}{\sin^2(\pi T f/M)}\right]$$

The continuous part of the spectrum will be broad because pulse widths will be narrow (on the order of T/M s duration) for a reliable determination of which symbol was transmitted. Note, however, that the spectrum has nulls at multiples of $1/T$ due to the shape of the $Q(f)$ factor in Eq. (5.48).

5.5 Coding for Power Spectrum Control

From Eq. (5.28) or (5.44), we see that that the shape of the continuous part of the power spectral density of data signals can be controlled either by choice of pulse shapes or by manipulating the correlation structure of the data sequence. Often the latter method is more convenient to implement. This can be accomplished by mapping the data sequence into a new sequence by a procedure called *line coding* [Lee and Messerschmitt, 1988]. The simpler types of line codes are easily handled by Markov chain models. As an example, we consider the *differential binary* (also called NRZI) coding scheme applied to a binary data sequence. Let $a(k) \in \{0, 1\}$ represent the data sequence and $b(k)$ the coded binary sequence. The coding rule is that $b(k)$ differs from $b(k-1)$ if $a(k) = 1$, otherwise $b(k) = b(k-1)$. This can also be expressed as $b(k) = [b(k-1) + a(k)] \mod 2$. Letting $q = Pr[a(k) = 1]$, the probability transition matrix can be expressed as

$$P(1) = \begin{bmatrix} 1-q & q \\ q & 1-q \end{bmatrix} \quad (5.49)$$

from which it follows that the steady-state probabilities for $b(k) = 1$ or 0 [Eq. (5.40)] are given by $p = 1 - p = 1/2$, regardless of the uncoded symbol probability q. Because the columns of a transition matrix sum to 1, the evolution of the transition probability [Eq. (5.41)] can be written as a single scalar equation which in this case is a first-order difference equation with an initial condition of $P_{11}(0) = 1$.

$$P_{11}(i+1) = (1 - 2q) P_{11}(i) + q; \quad i \geq 0 \quad (5.50)$$

Using $P_{11}(i) = P_{11}(-i)$, the solution is

$$P_{11}(i) = \frac{1}{2}\left[(1-2q)^{|i|} + 1\right] \quad (5.51)$$

Supposing that the transmitted signal is binary PAM using the $b(k)$ sequence, that is,

$$x(t) = \sum_k b(k) g(t - kT) \quad (5.52)$$

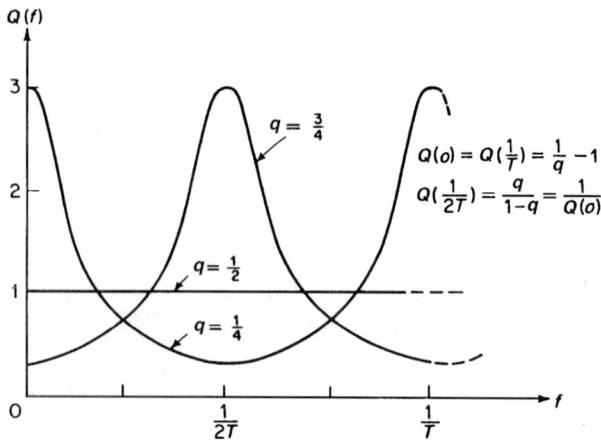

FIGURE 5.2 Spectral density shaping function for differential binary coding.

then, taking the DTFT of Eq. (5.51) and using Eq. (5.32), the power spectral density becomes [Franks, 1969]

$$S_{xx}(f) = \left(\frac{1}{4T}\right)Q(f)|G(f)|^2 + \left(\frac{1}{2T}\right)^2 \sum_n \left|G\left(\frac{n}{T}\right)\right|^2 \delta\left(f - \frac{n}{T}\right) \quad (5.53)$$

where the spectrum shaping function is

$$Q(f) = q(1-q)[q^2 + (1-2q)\sin^2(\pi T f)]^{-1} \quad (5.54)$$

It is sometimes stated that differential binary (NRZI) coding does not alter the shape of the power spectrum. We see from Eq. (5.54) that this is true only in the case that the original data are equiprobable ($q = 1/2$). As shown in Fig. 5.2, the shaping factor $Q(f)$ differs greatly from a constant as q approaches 0 or 1. Note, however, that the spectrum of the coded signal depends only on the symbol probability q, and not the specific correlation structure of the original data sequence.

A more practical line coding scheme is *bipolar coding*, also called *alternate mark inversion* (AMI). It is easily derived from the differential binary code previously discussed. The bipolar line signal is a ternary PAM signal

$$x(t) = \sum_k c(k)g(t - kT) \quad (5.55)$$

where $c(k) = 0$ if $a(k) = 0$ and $c(k) = +1$ or -1 if $a(k) = 1$, with the two pulse polarities forming an alternating sequence; that is, a positive pulse will always be followed by a negative pulse, and vice versa. Hence, the bipolar code can be expressed in terms of the differential binary code as

$$c(k) = b(k) - b(k-1) \quad (5.56)$$

which itself is a form of *partial response coding* [Lee and Messerschmitt, 1988; Proakis and Salehi, 1994]. The autocorrelation function for $c(k)$ is easily expressed in terms of the autocorrelation for $b(k)$,

$$R_{cc}(m) = 2R_{bb}(m) - R_{bb}(m-1) - R_{bb}(m+1) \quad (5.57)$$

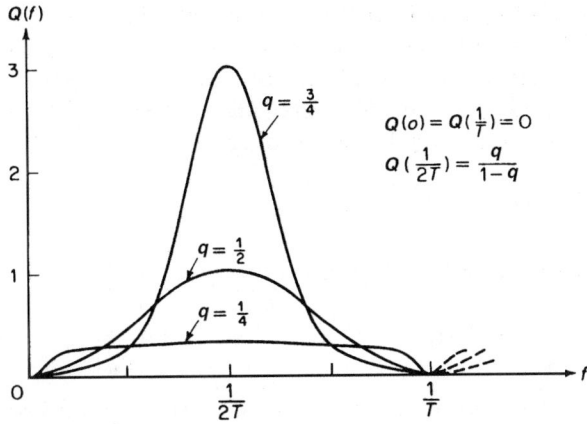

FIGURE 5.3 Spectral density shaping function for bipolar coding.

and since

$$R_{bb}(m) = \frac{1}{2}P_{11}(m) = \frac{1}{4}\left[(1-2q)^{|m|} + 1\right] \tag{5.58}$$

we get

$$R_{cc}(0) = q$$
$$R_{cc}(m) = -q^2(1-2q)^{|m-1|}; \qquad |m| \geq 1 \tag{5.59}$$

and taking the DTFT of $R_{cc}(m)$, the power spectral density for the bipolar signal is

$$S_{xx}(f) = \frac{1}{T}Q(f)|G(f)|^2 \tag{5.60}$$

where

$$Q(f) = [q(1-q)\sin^2(\pi T f)][q^2 + (1-2q)\sin^2(\pi T f)]^{-1}$$

Since $E[c(k)] = \bar{c} = \bar{b} - \bar{b} = 0$, the spectrum has no discrete components, which can be a substantial practical advantage. The spectral shaping factor $Q(f)$, shown in Fig. 5.3, produces spectral nulls at $f = 0$ and $f = 1/T$. This was used to advantage in early wire-cable PCM systems to mitigate the effects of no low-frequency transmission and the effects of increased crosstalk between wire pairs at the higher frequencies. Note that the spectrum shape is still dependent on data symbol probability, but not dependent on the correlation structure of the data.

Another simple line code, found useful in optical and magnetic recording systems, is the *Miller code* whose power spectrum can be determined with a four-state Markov chain [Proakis and Salehi, 1994].

5.6 Bandpass Digital Data Signals

In most communication systems, signals are positioned in a frequency band which excludes low frequencies. These are called bandpass signals. For digital data signals, we could simply consider the baseband formats discussed previously, but with data pulse shapes, $g(t)$, which have a bandpass nature. However, it is more common to produce bandpass signals by modulation of a sinusoidal carrier with low-pass (baseband) data signals. Hence, we examine here the correlation and spectral properties of **quadrature amplitude modulation (QAM)** carrier signals. Let $u(t)$ and $v(t)$ be real baseband signal processes, then the QAM signal with a carrier frequency of f_0 is

$$x(t) = u(t)\cos(2\pi f_0 t) - v(t)\sin(2\pi f_0 t) \tag{5.61}$$

which can be more conveniently handled in terms of complex signals

$$x(t) = \Re[w(t)e^{j2\pi f_0 t}]; \qquad w(t) = u(t) + jv(t) \tag{5.62}$$

The signals $u(t)$ and $v(t)$ are often called the *in-phase* and *quadrature* components, respectively, of the QAM signal. The autocorrelation function for $x(t)$ becomes

$$\begin{aligned} R_{xx}(t,\tau) &= \frac{1}{2}\Re[R_{ww}(t,\tau)e^{j2\pi f_0 \tau}] \\ &\quad + \frac{1}{2}\Re[R_{ww^*}(t,\tau)e^{j2\pi(2f_0)t}] \end{aligned} \tag{5.63}$$

First we consider the situation where $u(t)$ and $v(t)$ are jointly WSS processes, which might be a good model for some types of analog systems. In this case there are three cycle frequencies $(0, +2f_0, -2f_0)$. We obtain the cyclic autocorrelation functions from Eq. (5.5) by time averaging over a period of $T = 1/f_0$ with the result that

$$\tilde{c}_{xx}(0,\tau) = \frac{1}{2}\Re[R_{ww}(\tau)e^{j2\pi f_0 \tau}] \tag{5.64a}$$

$$\tilde{c}_{xx}(2f_0,\tau) = \frac{1}{4}R_{ww^*}(\tau) = \tilde{c}_{xx}^*(-2f_0,\tau) \tag{5.64b}$$

From the second term in Eq. (5.63), in the WSS case, it is evident that a period of $T = 1/2f_0$ would also work; however, we choose to regard the QAM signal as $CS(1/f_0)$ since the mean value could have a component at f_0. The autocorrelation result is the same in either case.

We note that $x(t)$ itself would be WSS under the condition that the cross correlation, $R_{ww^*}(\tau)$, vanishes, making the second term in Eq. (5.63) disappear. This condition is equivalent to $R_{uu}(\tau) = R_{vv}(\tau)$ and $R_{vu}(\tau) = -R_{uv}(\tau) = -R_{vu}(-\tau)$; in other words, $x(t)$ is WSS if the in-phase and quadrature components have the same autocorrelation and their cross-correlation function is odd. If $u(t)$ and $v(t)$ are independent, zero-mean processes, then we require only that their autocorrelations match. This is called *balanced* QAM and its power spectral density is

$$S_{xx}(f) = C_{xx}(0, f) = \frac{1}{2}S_{uu}(f - f_0) + \frac{1}{2}S_{uu}(f + f_0) \tag{5.65}$$

For the case of digital data signals, we examine the QAM/PAM signal where the complex baseband signal is a pair of PAM signals,

$$w(t) = \sum_k c(k)g(t - kT); \qquad c(k) = a(k) + jb(k) \tag{5.66}$$

Here we encounter a problem regarding the period of the cyclostationarity, because the PAM signal exhibits a period of T while the carrier modulation introduces a period of $1/f_0$, and these periods

Random Processes, Autocorrelation, and Spectral Densities

are not necessarily commensurate. The problem is easily handled by methods used for almost periodic functions, by extending the integration interval indefinitely,

$$R_{xx}(t, \tau) = \sum_i \tilde{c}_{xx}(\nu_i, \tau) e^{j 2\pi \nu_i t} \qquad (5.67a)$$

$$\tilde{c}_{xx}(\nu_i, \tau) = \lim_{T_0 \to \infty} \frac{1}{2T_0} \int_{-T_0}^{T_0} R_{xx}(t, \tau) e^{-j 2\pi \nu_i t} \, dt \qquad (5.67b)$$

The values of ν_i for which Eq. (5.67b) are nonzero are the cycle frequencies. The process is sometimes called *polycyclostationary*. Now using Eq. (5.25) for the cyclic autocorrelations of $w(t)$ and Eq. (5.63) for the effect of the modulation, then taking Fourier transforms, we get the following results for the cyclic spectral density functions for $x(t)$:

$$C_{xx}\left(\frac{m}{T}, f\right) = \left(\frac{1}{4T}\right) S_{cc}\left(f - f_0 - \frac{m}{2T}\right) K(f - f_0; m)$$

$$+ \left(\frac{1}{4T}\right) S_{cc}\left(-f - f_0 + \frac{m}{2T}\right) K(f + f_0; m) \qquad (5.68a)$$

$$C_{xx}\left(\frac{m}{T} + 2f_0, f\right) = \left(\frac{1}{4T}\right) S_{cc^*}\left(f - \frac{m}{2T}\right) K(f; m) \qquad (5.68b)$$

$$C_{xx}\left(\frac{m}{T} - 2f_0, f\right) = \left(\frac{1}{4T}\right) S_{cc^*}^*\left(-f + \frac{m}{2T}\right) K(f; m) \qquad (5.68c)$$

where, for convenience, we have defined the function

$$K(f; m) = G\left(f + \frac{m}{2T}\right) G^*\left(f - \frac{m}{2T}\right) \qquad (5.69)$$

The cycle frequencies are $\{m/T, m/T + 2f_0, m/T - 2f_0\}$. For example, if the data pulse $g(t)$ had less than 100% excess bandwidth, then $K(f; m)$ vanishes for $|m| > 2$ and there are a total of nine cycle frequencies. If the QAM is balanced, for example, by making the $a(k)$ and $b(k)$ identically distributed and independent, then the cross-spectral density $S_{cc^*}(f)$ vanishes and only three cycle frequencies remain.

As an illustration, let us assume that the $a(k)$ and $b(k)$ sequences are independent and each sequence has zero-mean, independent data, but with variances, σ_a^2 and σ_b^2, for in-phase and quadrature data, respectively. Then $S_{cc}(f) = \sigma_a^2 + \sigma_b^2$ and $S_{cc^*}(f) = \sigma_a^2 - \sigma_b^2$, giving the following simple expressions for the cyclic spectral densities:

$$C_{xx}\left(\frac{m}{T}, f\right) = \frac{\sigma_a^2 + \sigma_b^2}{4T} [K(f - f_0; m) + K(f + f_0; m)] \qquad (5.70a)$$

$$C_{xx}\left(\frac{m}{T} + 2f_0, f\right) = C_{xx}\left(\frac{m}{T} - 2f_0, f\right) = \frac{\sigma_a^2 - \sigma_b^2}{4T} K(f; m) \qquad (5.70b)$$

and the power spectral density is

$$S_{xx}(f) = C_{xx}(0, f) = \frac{\sigma_a^2 + \sigma_b^2}{4T} [|G(f - f_0)|^2 + |G(f + f_0)|^2] \qquad (5.71)$$

which is simply a pair of frequency-translated versions of the corresponding baseband PAM spectrum.

The case of *double-sideband amplitude modulation* (DSB-AM/PAM) is covered in the preceding equations simply by making all of the $b(k)$ equal to zero. This can be regarded as extremely unbalanced QAM; hence, the nonzero cyclic components will tend to be greater than in the QAM case. Other modulation formats can be handled by essentially the same methods. For example, *vestigial-sideband modulation* (VSB-AM/PAM) uses a single data stream and a complex data pulse; $w(t) = \sum a(k)[g(t - kT) + j\tilde{g}(t - kT)]$. A more general version of QAM uses a different pulse shape in the in-phase and quadrature channels. Staggered QAM (SQAM) uses time-shifted versions of the same pulse; for example, $w(t) = \sum a(k)g(t - kT) + jb(k)g(t - T/2 - kT)$. *Quaternary phase-shift keying* (QPSK) is a special case of QAM with $a(k), b(k) \in \{+1, -1\}$.

5.7 Appendix: The Poisson Sum Formula

For any time function, $g(t)$, having Fourier transform, $G(f)$, the following relation is often useful in system analysis:

$$\sum_k g(t - kT) = \frac{1}{T} \sum_m G\left(\frac{m}{T}\right) e^{j2\pi mt/T} \tag{A.5.1}$$

The relation is easily verified by making a Fourier series expansion of the periodic function on the left-hand side of Eq. (A.5.1). The mth Fourier coefficient is

$$c(m) = \frac{1}{T} \int_0^T \sum_k g(t - kT) e^{-j2\pi mt/T} \, dt \tag{A.5.2}$$

and by a change of integration variable,

$$c(m) = \frac{1}{T} \int_{-\infty}^{\infty} g(\sigma) e^{-j2\pi m\sigma/T} \, d\sigma = \frac{1}{T} G\left(\frac{m}{T}\right) \tag{A.5.3}$$

Another useful version of the Poisson sum formula is the time/frequency dual of Eq. (A.5.1).

$$\sum_k g(kT) e^{-j2\pi kTf} = \frac{1}{T} \sum_m G\left(f - \frac{m}{T}\right) \tag{A.5.4}$$

Defining Terms

Autocorrelation: The expected value of the product of two elements (time samples) of a single random process. The variables of the autocorrelation function are the two time values indexing the random variables in the product.

Cross correlation: The expected value of the product of elements from two distinct random processes.

Cycle frequency: The frequency parameter associated with a sinusoidal variation of a component of the autocorrelation of a cyclostationary process.

Cyclic autocorrelation: A parameter characterizing the amount of contribution to the overall autocorrelation at a particular cycle frequency.

Cyclostationary random process (wide-sense): A random process whose mean and autocorrelation functions vary periodically with time.

Excess bandwidth: The amount of bandwidth of a PAM signal that exceeds the minimum bandwidth for no intersymbol interference. It is usually expressed as a fraction of the Nyquist frequency of $1/2T$, where $1/T$ is the PAM symbol rate.

Poisson sum formula: A useful identify relating the Fourier series and sum-of-translates representations of a periodic function.

Power spectral density: A real, nonnegative function of frequency that characterizes the contribution to the overall mean-squared value (power) of a random process due to components in any specified frequency interval.

Pulse amplitude modulation (PAM): A signal format related to a discrete-time sequence by superposing time-translated versions of a single waveform, each scaled by the corresponding element of the sequence.

Quadrature amplitude modulation (QAM): A signal format obtained by separately modulating the amplitude of two components of a sinusoidal carrier differing in phase by 90°. In digital communications, the two components (referred to as in-phase and quadrature components) can each be regarded as a PAM signal.

Time-ambiguity function: A property of a waveform, similar to convolution of the waveform with itself, which characterizes the degree to which the waveform can be localized in time. The Fourier transform of the time-ambiguity function is the energy spectral density of the waveform.

References

Franks, L.E. 1969. *Signal Theory*, Prentice–Hall, Englewood Cliffs, NJ; rev. ed. 1981. Dowden and Culver.

Gardner, W.A. 1989. *Introduction to Random Processes with Applications to Signals and Systems*, 2nd ed., McGraw–Hill, New York.

Gardner, W.A., ed. 1994. *Cyclostationarity in Communications and Signal Processing*, IEEE Press.

Lee, E.A. and Messerschmitt, D.G. 1988. *Digital Communication*, Kluwer Academic.

Papoulis, A. 1991. *Probability, Random Variables, and Stochastic Processes*, 3rd ed., McGraw–Hill, New York.

Proakis, J.G. 1989. *Digital Communications*, McGraw–Hill, New York.

Proakis, J.G. and Salehi, M. 1994. *Communication Systems Engineering*, Prentice–Hall, Englewood Cliff, NJ.

Stark, H. and Woods, J.W. 1986. *Probability Random Processes, and Estimation Theory for Engineers*, 2nd ed., Prentice–Hall, Englewood Cliffs, NJ.

Further Information

The text by Papoulis [1991] is the third edition of a well known and referenced work on probability and random processes that first appeared in 1965. It is a concise and comprehensive treatment of the topic and is widely regarded as an ideal reference source. Gardner [1989] provides the most complete textbook treatment of cyclostationary processes. He presents a careful development of the theory of such processes as well as several examples of applications relevant to the communications field. Stark and Woods [1986] is a somewhat more advanced textbook. It uses modern notation and viewpoints and gives the reader a good introduction to the related topics of parameter estimation and decision theory.

6
Queuing[1]

Richard H. Williams
University of New Mexico

6.1 Introduction 78
6.2 Little's Formula 79
6.3 The M/M/1 Queuing System: State Probabilities 80
6.4 The M/M/1 Queuing System: Averages and Variances 83
6.5 Averages for the Queue and the Server 85

6.1 Introduction

Everyone has experienced being in a **queue**: waiting in line at a checkout counter, waiting in traffic as we enter a highway's tollgate, or waiting for a time-sharing computer to run our program. Examples of queues from everyday life and technical experiences are almost endless.

A study of queues is important for the following reasons:

1. From the user's point of view, a queue, that is, waiting in line, is a nuisance, and will be tolerated only at some minimal level of inconvenience.
2. From the point of view of a provider of some service, queues are necessary. It is extremely uneconomical to provide a large number of servers to handle the biggest possible rush, only to have many of them idle most of the time.

One motivation to study queues is to find some balance between these two conflicting viewpoints. Probability theory is the tool we use for a study of queues because arrivals and departures from a **queuing system** occur randomly. In Fig. 6.1, a **customer** arrives at the input to the queuing system,

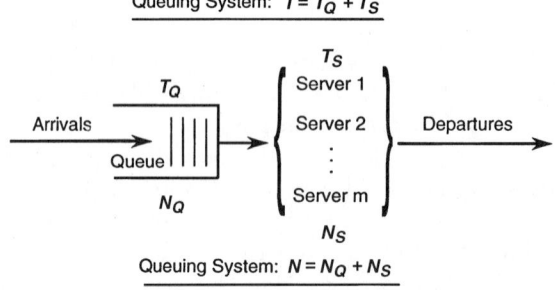

FIGURE 6.1 Illustration of queuing system: Customers arrive at a queue and, after some time, depart from a server that serves the queue.

---

[1] Portions are reprinted by permission. Williams, R.H. 1991. *Electrical Engineering Probability*, Copyright © by West, St. Paul, MN. All Rights Reserved.

Queuing

and if no **server** is free, the customer must wait in a queue until it is his, her, or its turn to be served. A **queue discipline** is the name given to a rule by which a customer is selected from a queue for service.

6.2 Little's Formula

Let the average rate at which customers arrive at a queuing system be the constant λ. Departures from the queuing system occur when a server finishes with a customer. In the steady state the average rate at which customers depart the queuing system is also λ. Arrivals to a queuing system may be described as a counting process $A(t)$ such as shown in Fig. 6.2. We estimate an average rate of arrivals with

$$\bar{\lambda} = \frac{A(t)}{t} \quad \text{(per unit time)} \tag{6.1}$$

where $A(t)$ is the number of arrivals in $[0, t]$. Departures from the system $D(t)$ are also described by a counting process, and an example is also shown in Fig. 6.2. Note that $A(t) \geq D(t)$ must be true at all times. The time that each customer spends in the queuing system is the time difference between the customer's arrival and subsequent departure from a server. The time spent in the queuing system for arrival i is denoted $t_i, i = 1, 2, 3, \ldots$.

The average time a customer spends waiting in a queuing system is

$$\bar{t} = \frac{1}{A(t)} \sum_{i=1}^{A(t)} t_i \quad \text{(units of time)} \tag{6.2}$$

The average number of customers in a queuing system $\bar{n}$ is, from Eqs. (6.1) and (6.2),

$$\bar{n} = \frac{1}{t} \sum_{i=1}^{A(t)} t_i = \bar{\lambda}\bar{t} \quad \text{(dimensionless)} \tag{6.3}$$

If we assume that, with enough data, sample averages approximate their expected values, then $\bar{\lambda} \to \lambda, \bar{t} \to E[T]$, and $\bar{n} \to E[N]$, where N and T are random variables. Also, Eq. (6.3) becomes

$$E[N] = \lambda E[T] \tag{6.4}$$

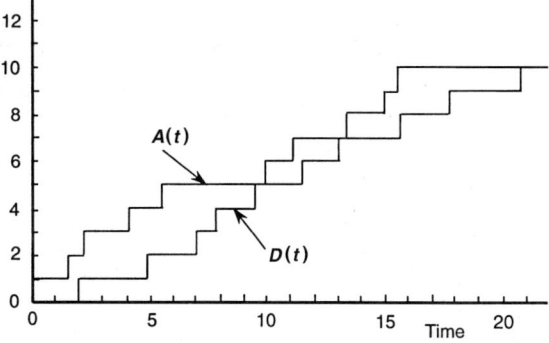

FIGURE 6.2 Illustration of the arrival and departure counting processes for a queuing system.

This relation is known as Little's formula [Williams, 1991]; it has a generality much beyond what one might suppose, given the heuristic demonstration leading to Eq. (6.4).

Example 1. The data in Table 6.1 are the data used to plot $A(t)$ and $D(t)$ in Fig. 6.1. In Table 6.1, the times, given in the units of minutes, are: arrival time t_A, time spent in the queue t_Q, time spent with the server t_S, total time in the queuing system for the ith arrival t_i, and the departure time t_D. For each i we see in Table 6.1 that $t_i = t_Q + t_S$, and that $t_D = t_A + t_i$. These data describe a first-in–first-out (FIFO) queuing discipline: If a customer arrives before the previous customer departs then the customer must wait his turn in the queue until the server is available. If the previous customer has departed prior to a customer's arrival, then the waiting time in the queue is $t_Q = 0$.

TABLE 6.1 Data from a FIFO Queue Used with Example 1

i	t_A	t_Q	t_S	t_i	t_D
1	0.0	0.0	2.0	2.0	2.0
2	1.5	0.5	2.9	3.4	4.9
3	2.2	2.7	2.1	4.8	7.0
4	4.1	2.9	0.8	3.7	7.8
5	5.5	2.3	1.7	4.0	9.5
6	10.0	0.0	1.5	1.5	11.5
7	11.3	0.2	1.6	1.8	13.1
8	13.4	0.0	2.3	2.3	15.7
9	15.0	0.7	2.0	2.7	17.7
10	15.5	2.2	3.0	5.2	20.7

According to the data in Table 6.1, when $t = 20.7$ then $A(t) = i = 10$. From Eq. (6.1),

$$\bar{\lambda} = \frac{10}{20.7} = 0.48 \quad (\text{min}^{-1})$$

The sum of all the t_i in Table 6.1 is 31.4 min. Therefore, from Eq. (6.2)

$$\bar{t} = \frac{31.4}{10} = 3.14 \quad (\text{min})$$

Finally, using Little's formula in Eq. (6.3), the sample average number of customers in the queuing system is

$$\bar{n} = \frac{10}{20.7} \times \frac{31.4}{10} = 1.52$$

6.3 The M/M/1 Queuing System: State Probabilities

A generally accepted notation used to name a queuing system is A/S/m: A is a description of the dynamics by which arrivals come to the queuing system; S is a description of the dynamics by which a server processes a customer taken from the queue; and m is the number of servers. The arrivals and the server are independent of each other except that a server cannot serve a customer who has not yet arrived into the queue.

We define the **state** of a queuing system as the total number of customers within it, that is, either within the queue or being served. The only queuing system that we discuss in this chapter is M/M/1: The arrival dynamics are Markov (with rate constant λ), the server dynamics are Markov (with rate constant λ_S), and there is only one server.

There can be at most only *one* arrival, and at most *one* departure, in a time interval Δt when $\Delta t \to 0$. As far as arrivals are concerned, when Δt is a small increment of time, the probability of an arrival into the queuing system in the interval Δt is

$$P(A) = \lambda \Delta t \tag{6.5}$$

Correspondingly, the probability of no arrival in Δt is

$$P(\bar{A}) = 1 - \lambda \Delta t \tag{6.6}$$

Queuing

The probability of serving (discharging) a customer $P(D)$ is actually conditioned by the event that there is at least one customer in the queuing system. Thus, *given* that the state is at least one,

$$P(D) = \lambda_S \Delta t \tag{6.7}$$

The probability of continuing to serve a customer who is already being served is

$$P(\overline{D}) = 1 - \lambda_S \Delta t \tag{6.8}$$

Suppose that the state of a queuing system is zero at time $t + \Delta t$. This can occur in only one of three mutually exclusive ways. If the state at time t is zero, then either there is no arrival in the increment Δt, or if there is an arrival in Δt then there is also a departure within the same Δt. If the state at t is one, then there must be one departure and no arrival within Δt. Let $P_n(t)$ denote the probability of a queuing system being in the state n at time t. Then, with this notation, a summary of the preceding statement is

$$P_0(t + \Delta t) = P_0(t)\{P(\overline{A}) + P(A)P(D)\} + P_1(t)P(\overline{A})P(D)$$

Using Eqs.(6.5–6.8), this may be arranged into

$$\frac{P_0(t + \Delta t) - P_0(t)}{\Delta t} = -\lambda P_0(t)\{1 - \lambda_S \Delta t\} + \lambda_S P_1(t)\{1 - \lambda \Delta t\}$$

Then, taking the limit as $\Delta t \to 0$, we have the derivative

$$\frac{dP_0(t)}{dt} = \lambda_S P_1(t) - \lambda P_0(t)$$

When a queuing system is in the steady state, the rate of change of the probability $P_0(t)$ with respect to time is zero, and we have

$$\lambda P_0(t) = \lambda_S P_1(t) \tag{6.9}$$

If at time $t + \Delta t$ the state of the queuing system is greater than zero, we have

$$P_n(t + \Delta t) = P_n(t)\{P(\overline{A})P(\overline{D}) + P(A)P(D)\} \\ + P_{n+1}(t)P(\overline{A})P(D) + P_{n-1}(t)P(A)P(\overline{D})$$

The first term on the right of this equation describes two ways for the state to remain the same when time increases from t to $t + \Delta t$: either no arrival and no departure occur, or exactly one arrival occurs and exactly one departure occurs. The second term describes the state $n + 1$ changing to n by exactly one departure with no arrival. The third term describes the change from state $n - 1$ to n by exactly one arrival and no departure. No other combinations of arrivals and departures can happen using our assumptions when $n > 0$. Then, as before, finding the derivative of $P_n(t)$ and setting it equal to zero in the steady-state condition, we find

$$(\lambda + \lambda_S)P_n(t) = \lambda_S P_{n+1}(t) + \lambda P_{n-1}(t) \tag{6.10}$$

We introduce the ratio

$$\rho = \frac{\lambda}{\lambda_S} \tag{6.11}$$

Then, when $n = 0$, we have from Eq. (6.9),

$$P_1(t) = \rho P_0(t)$$

When $n = 1$, Eq. (6.10) becomes

$$(\lambda + \lambda_S) P_1(t) = \lambda_S P_2(t) + \lambda P_0(t)$$

Combining this with $P_1(t)$ we have an expression, which when simplified becomes

$$P_2(t) = \rho^2 P_0(t)$$

Continuing in this way, it follows that

$$P_n(t) = \rho^n P_0(t) \tag{6.12}$$

Since the queuing system can theoretically be in any state $n = 0, 1, 2, \ldots$ (that is, there can be any nonnegative number of customers in the queuing system), we must require that

$$\sum_{n=0}^{\infty} P_n(t) = 1 \tag{6.13}$$

Combining Eqs. (6.12) and (6.13) we have

$$P_0(t) \sum_{n=0}^{\infty} \rho^n = 1$$

By stipulating that $\rho < 1$ we can sum the geometric progression [Beyer, 1987, p. 8],

$$\sum_{n=0}^{\infty} \rho^n = \frac{1}{1 - \rho} \tag{6.14}$$

Thus, when $n = 0$ we have

$$P_0(t) = 1 - \rho \tag{6.15}$$

Combining Eq. (6.15) with Eq. (6.12) gives the probability of any state n for the M/M/1 queuing system:

$$P_n(t) = (1 - \rho) \rho^n \tag{6.16}$$

As long as the state of a queuing system is greater than zero, then one of the customers must be with the server. Therefore, the probability that the server is busy is $P(N > 0)$. Since $P(N > 0) + P_0(t) = 1$, it follows, using Eq. (6.15), that

$$P(N > 0) = \rho \tag{6.17}$$

The result in Eq. (6.17) is why the ratio ρ is called the **utilization** of a M/M/1 queuing system.

Queuing

Example 2. A manager of a queuing system wants to know the probability that k or more customers are in the queue waiting to be served by the server; $k = 1, 2, 3, \ldots$. If even one customer is in the queue, then there must also be one customer with the server. Therefore, what the manager wants to know is the probability that the state of the queuing system is $k + 1$ or more. Then, with Eq. (6.16),

$$P(N_Q \geq k) = P(N \geq k + 1) = \sum_{n=k+1}^{\infty} (1 - \rho)\rho^n$$

If we use the change of variables $m = n - (k + 1)$, and Eq. (6.14), we then answer the manager's question:

$$P(N_Q \geq k) = (1 - \rho)\rho^{k+1} \sum_{m=0}^{\infty} \rho^m = \rho^{k+1}$$

6.4 The M/M/1 Queuing System: Averages and Variances

The purpose of this section is to find the statistical parameters of mean and variance for both random variables N and T shown in Fig. 6.1 for an M/M/1 queue. In the following, the lower case n and t stand for realizations of the random variables N and T.

If the geometric progression in Eq. (6.14) is differentiated once with respect to ρ we have

$$\sum_{n=0}^{\infty} n\rho^n = \frac{\rho}{(1-\rho)^2}$$

Then, differentiating this once again,

$$\sum_{n=0}^{\infty} n^2 \rho^n = \frac{\rho(1+\rho)}{(1-\rho)^3}$$

The first of these two derivatives can be used along with Eq. (6.16) to show that the mean of N is

$$E[N] = \sum_{n=0}^{\infty} n P_n(t) = \sum_{n=0}^{\infty} n(1-\rho)\rho^n = \frac{\rho}{1-\rho} \tag{6.18}$$

The second of these two derivatives is used to express the mean of the square of N,

$$E[N^2] = \sum_{n=0}^{\infty} n^2 P_n(t) = \sum_{n=0}^{\infty} n^2 (1-\rho)\rho^n = \frac{\rho(1+\rho)}{(1-\rho)^2} \tag{6.19}$$

Equations (6.18) and (6.19) are used to find the variance of N,

$$\text{Var}[N] = E[N^2] - (E[N])^2 = \frac{\rho}{(1-\rho)^2} \tag{6.20}$$

One notes that as $\lambda_S \to \lambda$, the utilization approaches unity, and both the mean and the variance of N for the M/M/1 queue become unbounded.

Assume that a customer arrives at an M/M/1 queuing system at time $t = 0$. The customer finds that the queuing system is in state n where n could be $0, 1, 2, \ldots$. We use the notation T_n to denote the random variable for the total time in the queuing system *given* that the state found by a customer is n. The probability of that customer leaving the queuing system in the interval between t and $t + \Delta t$ is

$$P(t < T_n \leq t + \Delta t) = F_{T_n}(t + \Delta t) - F_{T_n}(t)$$

where $F_{T_n}(t)$ is the probability cumulative distribution function (cdf) for T_n. Another way of expressing this probability is

$$P(t < T_n \leq t + \Delta t) = P(\text{exactly } n \text{ departures from the server between 0 and } t)$$
$$\times P(\text{the customer departs from the server in } \Delta t)$$

The first of the probabilities is Poisson [Williams, 1991] with rate parameter λ_S and time interval t. The second probability is obtained from Eq. (6.7). Thus,

$$P(t < T_n \leq t + \Delta t) = \left\{\frac{(\lambda_S t)^n}{n!} e^{-\lambda_S t}\right\} \lambda_S \Delta t$$

Combining these two expressions for $P(t < T_n \leq t + \Delta t)$, we have the following:

$$F_{T_n}(t + \Delta t) - F_{T_n}(t) = \left\{\frac{(\lambda_S t)^n}{n!} e^{-\lambda_S t}\right\} \lambda_S \Delta t$$

We can use this to construct the following derivative:

$$\lim_{\Delta t \to 0} \frac{F_{T_n}(t + \Delta t) - F_{T_n}(t)}{\Delta t} = \frac{d}{dt} F_{T_n}(t) = f_{T_n}(t)$$

where $f_{T_n}(t)$ is the probability density function (pdf) for T_n,

$$f_{T_n}(t) = \frac{\lambda_S^{n+1} t^n}{n!} e^{-\lambda_S t}, \qquad t \geq 0 \qquad (6.21)$$

The random variable T is the time in the queuing system considering all possible states. Then, using Eqs. (6.16), (6.21), and the series expansion for e^x [Beyer, 1987, p. 299]

$$f_T(t) = \sum_{n=0}^{\infty} f_{T_n}(t) P_n(0)$$

$$= \sum_{n=0}^{\infty} \frac{\lambda_S^{n+1} t^n}{n!} e^{-\lambda_S t} (1 - \rho) \rho^n$$

$$= (1 - \rho) \lambda_S e^{-\lambda_S t} \sum_{n=0}^{\infty} \frac{\lambda_S^n t^n \rho^n}{n!}$$

$$= (1 - \rho) \lambda_S e^{-\lambda_S t} e^{\rho \lambda_S t}$$

Using Eq. (6.11), we finally have the result

$$f_T(t) = (\lambda_S - \lambda) e^{(\lambda_S - \lambda) t}, \qquad t \geq 0 \qquad (6.22)$$

Queuing

This is recognized as the pdf for an exponential random variable. It is therefore known that [Williams, 1991]

$$E[T] = \frac{1}{\lambda_S - \lambda} = \frac{1/\lambda_S}{1 - \rho} \tag{6.23}$$

$$\text{Var}[T] = \frac{1}{(\lambda_S - \lambda)^2} = \frac{1/\lambda_S^2}{(1 - \rho)^2} \tag{6.24}$$

Example 3. Assume that an M/M/1 queuing system has a utilization of $\rho = 0.6$; this means that the server is idle 40% of the time. The average number of customers in the queuing system is, from Eq. (6.18),

$$E[N] = \frac{\rho}{1 - \rho} = 1.50$$

The standard deviation for this value is, from Eq. (6.20),

$$\sqrt{\text{Var}[N]} = \frac{\sqrt{\rho}}{1 - \rho} = 1.94$$

From Eqs. (6.23) and (6.24),

$$E[T] = \sqrt{\text{Var}[T]} = \left(\frac{1}{1 - \rho}\right)\frac{1}{\lambda_S} = \frac{2.5}{\lambda_S}$$

Thus, the mean and the standard deviation of the time a customer spends in the M/M/1 queuing system are the same.

Realizations of either N and T in an M/M/1 queuing system with $\rho = 0.6$ should be expected to fluctuate markedly about their means because of their relatively large standard deviations. A similar behavior may occur with other values of the utilization.

6.5 Averages for the Queue and the Server

The average number of customers in the server can be calculated using Eq. (6.17),

$$E[N_S] = (1)P(N > 0) + (0)P_0(t) = P(N > 0) = \rho \tag{6.25}$$

Combining Eq. (6.18) with Eq. (6.25), we find the average number of customers in the queue,

$$E[N_Q] = E[N] - E[N_S]$$
$$= \frac{\rho}{1 - \rho} - \rho$$
$$= \frac{\rho^2}{1 - \rho} \tag{6.26}$$

The average time a customer spends with the server is the reciprocal of the server's average rate

$$E[T_S] = 1/\lambda_S \tag{6.27}$$

Therefore, using Eqs. (6.23) and (6.27), the average time spent waiting in the queue, before being served, is

$$E[T_Q] = E[T] - E[T_S]$$
$$= \frac{1/\lambda_S}{1-\rho} - \frac{1}{\lambda_S}$$
$$= \frac{\rho/\lambda_S}{1-\rho} \qquad (6.28)$$

Example 4. An M/M/1 queuing system has a utilization of $\rho = 0.6$ and a rate at which the server can work of $\lambda_S = \frac{1}{45}(\text{s}^{-1})$. Using Eq. (6.28), customers should expect to spend the following time in the queue before being served:

$$E[T_Q] = \frac{45\rho}{1-\rho} = 1.13 \qquad (\text{min})$$

At any time, the expected number of customers in the queue waiting to be served is, using Eq. (6.26),

$$E[N_Q] = \frac{\rho^2}{1-\rho} = 0.90$$

And, the expected number of customers with the server is its utilization, Eq.(6.25),

$$E[N_S] = \rho = 0.6$$

Defining Terms

Customer: A person or object that is directed to a server.
Queue: A line of people or objects waiting to be served by a server.
Queue discipline: A rule by which a customer is selected from a queue for service by a server.
Queuing system: A combination of a queue, a server, and a queue discipline.
Server: A person, process, or a subsystem designed to perform some specific operation on or for a customer.
State: The total number of customers within a queuing system.
Utilization: The percentage of time that the server is busy.

References

Beyer, W.H. 1987. *CRC Standard Mathematical Tables*, 28th ed., CRC Press, Boca Raton, FL.
Williams, R.H. 1991. *Electrical Engineering Probability*, West, St. Paul, MN.

Further Information

Asmussen, S. 1987. *Applied Probability and Queues*, Wiley, Chichester, UK.
Kleinrock, L. 1975. *Queuing Systems*, Vol. 1, Wiley, New York.
Leon-Garcia, A. 1994. *Probability and Random Processes for Electrical Engineering*, 2nd ed., Addison–Wesley, Reading, MA.
Payne, J.A. 1982. *Introduction to Simulation*, McGraw–Hill, New York.

7
Multiplexing

Martin S. Roden
California State University, Los Angeles

7.1 Introduction 87
7.2 Frequency Multiplexing 87
7.3 Time Multiplexing 89
   Need for Frame Synchronization • Dissimilar Channels
7.4 Space Multiplexing 91
7.5 Techniques for Multiplexing in Spread Spectrum 91
7.6 Concluding Remarks 92

7.1 Introduction

A *channel* is the bridge between a source and a receiver (sink). In the early days of electrical wire communications, each channel was used to transmit only a single signal. As an example, if you were to view an early photograph of the streets of New York, you would note that the telephone poles dominate the scene. Literally hundreds of wires formed a web above the city streets. Some consider this beautiful, particularly on an icy winter morning. Of course, aside from aesthetic considerations, the proliferation of separate channels could not be permitted to continue as communications expanded.

In the majority of applications, a variety of signals must be transmitted on a single channel. The channel must, therefore, be shared among the various users. The process of sending multiple signals on a single channel is called **multiplexing**.

If multiple signals are to be sent on the same channel, the various signals must not overlap; they must be separable. In a mathematical sense, the various signals must be *orthogonal*.

The form of multiplexing used in everyday speech requires time separation. In a conversation, the participants try not to speak at the same time. There are many other ways that signals can be nonoverlapping. In this section, we examine time multiplexing, frequency multiplexing, space multiplexing, and several *multiple access* schemes.

7.2 Frequency Multiplexing

Frequency-division multiplexing (**FDM**) is the technique used in standard analog broadcast systems such as AM radio, FM radio, and television. It takes advantage of the observation that all frequencies of a particular message waveform can be easily shifted by a constant amount. The shifting is performed using a carrier signal. The original message signal is multiplied by a sinusoid. This shifts the frequencies of the message signal by the frequency of the sinusoid. If, for example, the original signal is baseband (composed of low frequencies around DC), multiplication of the signal

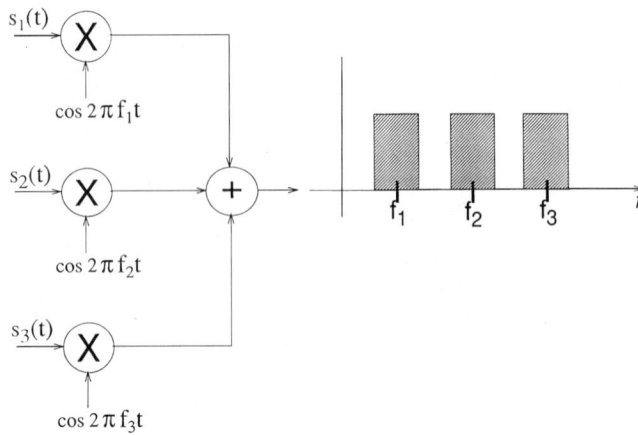

FIGURE 7.1   Frequency-division multiplexing.

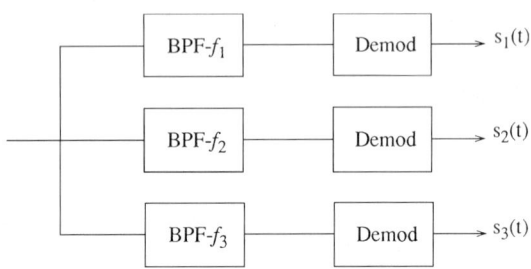

FIGURE 7.2   Demultiplexing of FDM.

by a sinusoid shifts all frequencies to a range centered about the sinusoidal frequency. By using multiple sinusoids of differing frequencies, the signals can be stacked in frequency in a nonoverlapping manner. This is illustrated in Fig. 7.1.

The right portion of the figure illustrates typical frequency spectra (Fourier transforms). To separate the signals in frequency, the various carriers must be separated by at least twice the highest frequency of the message signals.

The multiplexed signals can be separated using frequency gates (bandpass filters). The frequencies can then be shifted back to baseband using demodulators. This is shown in Fig. 7.2.

The demodulator shown in Fig. 7.2 takes various forms. The coherent form simply shifts the signal back down to baseband using a replica of the original sinusoid. If the original shifted signal is modified to contain the carrier sinusoid, incoherent demodulators (e.g., the envelope detector) may obviate the need for reproducing the original sinusoid.

In designing FDM systems, attention must be given to the minimum separation between carriers. Although the minimum spacing is twice the highest signal frequency, using this exact value would place unrealistic constraints on the bandpass filters of Fig. 7.2. For this reason, a *guard slot* is usually included between multiplexed frequency components. The larger the guard band, the easier it is to design the filters. However, the price being paid is a reduction in the number of channels that can be multiplexed within a given bandwidth.

Frequency-division multiplexing is sometimes used to create a composite baseband signal. For example, in FM stereo, the two audio signals are frequency multiplexed to produce a new baseband signal. One of the audio signals occupies the band of frequencies between DC and 15 kHz, whereas the second audio signal is shifted by 38 kHz. It then occupies a band between 23 and 53 kHz. The composite signal can then be considered as a new baseband waveform with frequencies between DC and 53 kHz. This baseband waveform frequency modulates a carrier, and is frequency multiplexed a second time: this time with other modulated composite signals.

7.3 Time Multiplexing

There is a complete duality between the time and frequency domain. Therefore, the discussion of FDM of the preceding section can be extended to **time-division multiplexing (TDM)**. We begin by considering a pulse modulated waveform.

An analog signal can be transmitted by first sampling the waveform. In accordance with the *sampling theorem*, the number of samples required each second is at least twice the highest frequency of the waveform. For example, a speech waveform, which can usually be thought of as limited to 4 kHz in frequency, can be fully represented by samples taken at a rate of 8000 samples/s. If each sample value modulates the height of a pulse, the signal can be sent using 8000 varying-height pulses/s. [This is pulse-amplitude modulation (PAM).] If each pulse occupies only a fraction of the time spacing between samples, the time axis can be shared with other sampled signals. Figure 7.3 shows an example for three signals. The notation used designates s_{ij} as the ith time sample of the jth signal. Thus, for example, s_{23} is the second sample of the third signal. The three signals are said to be time-division multiplexed. If the pulses are made narrower, additional signals can be multiplexed. Of course, the narrower the pulses, the wider the bandwidth is. Once you are given the bandwidth of the channel, the minimum pulse width is set. In using TDM, one must also be sensitive to channel distortion, which widens the pulses before they enter the receiver. Pulses that do not overlap at the transmitter may, therefore, overlap at the receiver, thus causing *intersymbol interference*.

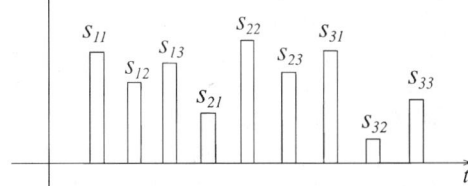

FIGURE 7.3 Time-division multiplexing of PAM signals.

If the transmission is digital rather than analog, each sample is coded into a binary number. Instead of sending a single pulse, multiple binary pulses are sent for each sample. [This is *pulse code modulation*, or (PCM).] TDM can still be used, but the number of pulses in each sampling interval increases.

The **T-1 carrier** transmission system is an example of a PCM system with TDM. It was developed by the Bell System (before divestiture) in the early 1960s. It is important to us, both for historical significance (it truly started the transition to digital voice transmission) and because it is still regularly used as the first level in a hierarchy that multiplexes ever increasing numbers of channels.

The T-1 carrier system develops a 1.544-Mb/s pulsed digital signal by multiplexing 24 voice channels. Each channel is sampled at a rate of 8000 samples/s. Each sample is then converted into 8 bits using companded PCM. Although all 8 bits corresponding to a particular sample can be available for sending that sample, one of the bits is sometimes devoted to *signaling*, which includes bookkeeping needed to set up the call and keep track of billing. The 24 channels are then time-division multiplexed, as shown in Fig. 7.4.

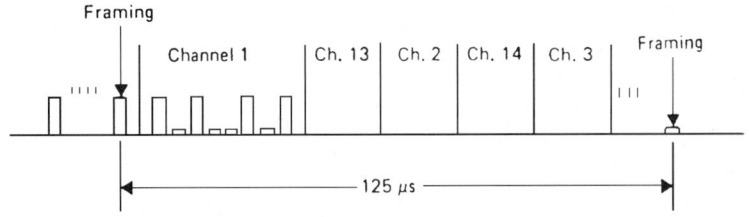

FIGURE 7.4 Multiplexing structure of T-1 carrier.

Each frame, which corresponds to one sampling period, contains 192 bits (8 times 24), plus a 193rd bit for **frame synchronization**. Since frames occur 8000 times/s, the system transmits 1.544 Mb/s.

Need for Frame Synchronization

When a receiver obtains and demodulates a TDM signal, the result is a continuous sequence of numbers (usually binary). These numbers must be sorted to associate the correct numbers with the correct signal. This sorting process is known as demultiplexing. To perform this function, the receiver needs to know when each frame begins. If nothing is known about the value of the signals, frame synchronization must be performed using overhead. A known synchronizing signal is transmitted along with the signal. This synchronizing signal can be sent as one of the multiplexed signals, or it can be a framing bit as in the case of T-1 carrier. The receiver searches for the known pattern and locks on to this. Of course, there is some probability that one of the actual signals will resemble the synchronizing signal, and false sync lock can occur. The probability of this decreases as the synchronizing signal is made longer.

Dissimilar Channels

The examples given so far in this section assume that the signals to be multiplexed are of the same form. For example, in the T-1 carrier system, each of the 24 signals is sampled at the same rate (8 kHz), and converted to the same number of bits.

Dissimilar signals must often be multiplexed. In the general case, we refer to this as **asynchronous multiplexing**. We present two examples. In the first example, the data rates of the various channels are rationally related.

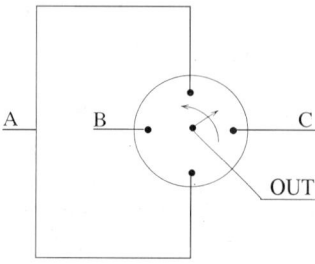

Suppose that we had to multiplex three channels, A, B, and C. Assume further that the data rates for both channels B and C are one-half that of A. That is, A is producing symbols at

FIGURE 7.5 Supercommutation.

twice the rate of B or C. A simple multiplexing scheme would include two symbols of signal A for each symbol of B or C. Thus, the transmitted sequence could be

A B A C A B A C A B A C A B A C ...

This is known as *supercommutation*. It can be visualized as a rotating switch with one contact each for signal B and C and two contacts for signal A. This is shown in Fig. 7.5.

Supercommutation can be combined with *subcommutation* to yield greater flexibility. Suppose, in the preceding example, the signal C were replaced by four signals, C1, C2, C3, and C4, each of which produces symbols at 1/4 the rate of C. Thus, every time the **commutation** of Fig. 7.5 reaches the C contact, we wish to feed a sample of one of the lower rate signals. This is accomplished using sub-commutation, as shown in Fig. 7.6.

If the rational relationships needed to do sub- and supercommutation are not present, we need to use a completely asynchronous multiplexing approach. Symbols from the various sources are stored in a buffer, and the buffer sorts things out and transmits the interleaved symbols at the appropriate fixed rate. This is a variation of the **statistical multiplexer** (**stat-MUX**) commonly used in packet switching and

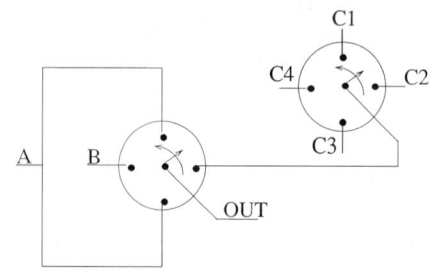

FIGURE 7.6 Combination of sub- and supercommutation.

other asynchronous schemes. If the individual channels are not synchronous, the symbols arrive at varying rates and times. Consider, for example, the output of a workstation connected to a network. The human operator certainly does not create symbols at a synchronous rate. The buffer in the multiplexer, therefore, must be carefully sized. If it is too small, it could become overloaded and lose information. If symbols exit the MUX too rapidly, then the buffer could become empty and stuffing symbols must be added.

7.4 Space Multiplexing

Time and frequency are the traditional domains used for multiplexing. However, there are other forms of multiplexing that force the signals to be orthogonal to each other. Among the most important of these are space and polarization multiplexing.

When terrestrial multipoint communication became popular, the general approach was to maximize the distance over which signals could be transmitted. In broadcast commercial systems, the advertising revenue is related to the number of households the signal reaches. Obviously, higher power and greater distances translate into more revenue. However, with the proliferation of high-power signal sources, frequency and time separation became impractical. Mobile radio was the driving force for a reversal of philosophy. The **cellular radio** concept is based on intentionally reducing the signal coverage area. With signal power low enough to limit range, signals that overlap in all other domains (e.g., time and frequency) can be transmitted simultaneously. The practicality of this approach had to wait until developments in electronics that permitted reliable handing off among antennas as the location of the transmitter and/or receiver changes.

Space-division multiplexing can also be accomplished with highly directional antennas. Spot beam antennas can focus a signal along a particular direction, and multiple signals can be transmitted using slightly different directions. Some satellite systems (e.g., INTELSAT) divide the Earth into regions and do simultaneous transmission of different signals to these regions using directional antennas.

Polarization multiplexing is yet another way to simultaneously send multiple signals on the same channel. If two signals occupy the same frequency band on the same channel, yet are polarized in different planes, they can be separated at the receiver. The receiving antenna must be polarized in the same direction as the desired signal.

7.5 Techniques for Multiplexing in Spread Spectrum

We have examined three areas of nonoverlapping signals: time, frequency, and space. There are hybrid variations of these in which the signals appear to overlap in all three domains, yet are still separable. The proliferation of spread spectrum systems has opened up the domain of **code-division multiple access** (**CDMA**). Spread spectrum is a transmission method that intentionally widens the signal bandwidth by modulating the bandlimited message with a wideband noiselike signal. Since the spreading function is pseudorandom (i.e., completely known although noiselike in properties), different widening functions can be made orthogonal to each other.

One way to visualize this is by considering the **frequency hopping** form of spread spectrum. This is a form of spread spectrum where the bandlimited message modulates a carrier whose frequency jumps around over a very wide range. The form of this jumping follows a noiselike sequence. However, since the jumping is completely specified, a second signal could occupy the same wideband, but jump across carrier frequencies that are always different from the first. The signals can be demodulated using the same jumped carrier that was used in the transmitter. This is illustrated in Fig. 7.7. A pseudonoise sequence generator (labeled PN in the figure) produces a noiselike sequence. This sequence controls the frequency of an oscillator (known as a frequency hopper). The two pseudonoise sequences, PN1 and PN2, result from different initiating sequences, so their resulting frequency patterns appear unrelated.

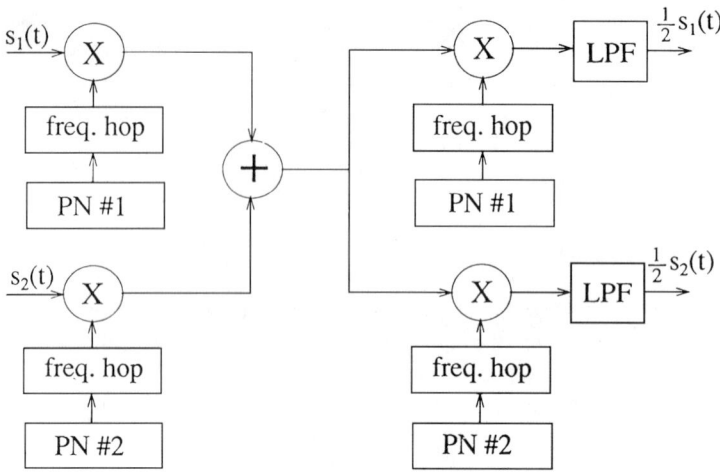

FIGURE 7.7 Code-division multiple access.

Our previous examples of multiplexing assume that the sources produce simultaneous messages. An alternate way of sharing a channel is to first process the messages so that they are nonoverlapping prior to transmission. The channel is then shared by assigning slots to each user.

In **time-division multiple access (TDMA)**, the channel is shared among multiple users by assigning time slots to each user. Each source therefore must transmit its signal in bursts that occur only during the time allocated to that signal.

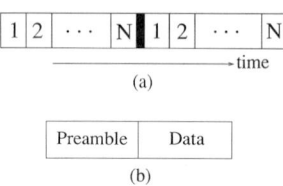

FIGURE 7.8 Typical frame for fixed-assignment TDMA.

In *fixed-assignment TDMA*, the time axis is divided into frames. Each frame is composed of time slots assigned to various users. Figure 7.8(a) shows a typical frame structure for N users. Two consecutive frames are illustrated. If a particular user generates a signal with a high symbol rate, that user could be assigned more than one time slot within the frame.

Figure 7.8(b) shows a representative time slot. Note that the transmission begins with a *preamble*, which contains addressing information. Each slot must be self-synchronizing (i.e., an individual receiver needs to be able to establish symbol synchronization), and so the preamble might also contain a unique waveform to help establish synchronization. In some systems, the preamble also contains error control data.

Sources sharing the channel need not be located in the same place. Therefore, as the signals line up in the frame, they experience varying time delays depending on the length of the transmission path. For this reason, *guard time slots* are included in the frame structure to assure that the delayed signals do not overlap in time.

7.6 Concluding Remarks

We have seen that signals can be nonoverlapping in a variety of domains. As it becomes possible to implement increasingly complex hardware and/or software systems, we can expect additional forms of multiplexing to be developed. The information superhighway will significantly increase the volume of signals being transmitted. As competition for channel use expands at an accelerating pace, the ingenuity and imagination of engineers will be needed.

Defining Terms

Asynchronous multiplexing: A technique for combining the output of various sources when those sources are producing information at unrelated rates.

Cellular radio: Each transmitting antenna uses a sufficiently low power so that the transmission range is relatively short (within a cell).

Code-division multiple access (CDMA): A technique for simultaneous transmission of wideband signals (spread spectrum) that occupy the same band of frequencies.

Commutation: Interspersing of signal pulses using the equivalent of a rotating multicontact switch.

Frame synchronization: In time-division multiplexing, frame synchronization allows the receiver to associate pulses with the correct original sources.

Frequency-division multiplexing (FDM): Multiplexing technique that requires that signals be confined to assigned, nonoverlapping frequency bands.

Frequency hopping: Spread spectrum technique that uses carriers whose frequencies hop around in a prescribed way to create a wideband modulated signal.

Multiplexing: The process of combining signals for transmission on a single channel.

Polarization multiplexing: More than one signal can share a channel if the signals are polarized in different planes.

Space-division multiplexing: Sharing of a channel by concentrating individual signals in nonoverlapping narrow beams.

Statistical multiplexer (Stat-MUX): Multiplexer that can accept asynchronous inputs and produce a synchronous output signal.

T-1 carrier: Frequency-division multiplexing technique for combining 24 voice channels.

Time-division multiple access (TDMA): A channel is shared among various users by assigning time slots to each user. Transmissions are bursts within the assigned time slots.

Time-division multiplexing (TDM): Multiplexing technique which requires that signals be confined to assigned, nonoverlapping portions of the time axis.

References

Couch, L.W. II. 1993. *Digital and Analog Communication Systems*, 4th ed., Macmillan, New York.
Gibson, J.D. 1993. *Principles of Digital and Analog Communications*, 2nd ed., Macmillan, New York.
Haykin, S. 1994. *Communication Systems*, 3rd ed., Wiley, New York.
Proakis, J.G. and Salehi, M. 1994. *Communication Systems Engineering*, Prentice–Hall, Englewood Cliffs, NJ.
Roden, M.S. 1996. *Analog and Digital Communication Systems*, 4th ed., Prentice–Hall, Englewood Cliffs, NJ.
Roden, M.S. 1988. *Digital Communication Systems Design*, Prentice–Hall, Englewood Cliffs, NJ.
Schwartz, M. 1990. *Information Transmission, Modulation, and Noise*, 4th ed., McGraw–Hill, New York.
Sklar, B. 1988. *Digital Communications*, Prentice–Hall, Englewood Cliffs, NJ.

Further Information

The basics of multiplexing are covered in any general communications textbook. The more recent state-of-the-art information can be found in the technical literature, primarily that of the IEEE. In particular, the *IEEE Transactions on Communications* contain articles on advances in research areas. Other IEEE Transactions often contain related articles (e.g., the *IEEE Transactions on Vehicular Technology* contain articles related to multiplexing in mobile radio systems). The *IEEE Communications Magazine* is a highly readable periodical that covers the broad spectrum of communications in the form of tutorial papers. *IEEE Spectrum Magazine* occasionally contains articles of interest related to communications. An annual issue of this publication is devoted to developments in technology, and contains sections related to communications.

8
Pseudonoise Sequences

8.1	Introduction	94
8.2	m Sequences	95
8.3	The q-ary Sequences with Low Autocorrelation	97
8.4	Families of Sequences with Low Crosscorrelation	98
	Gold and Kasami Sequences • Quaternary Sequences with Low Crosscorrelation • Kerdock Sequences	
8.5	Aperiodic Correlation	101
	Barker Sequences • Sequences with High Merit Factor • Sequences with Low Aperiodic Crosscorrelation	
8.6	Other Correlation Measures	103
	Partial-Period Correlation • Mean Square Correlation • Optical Orthogonal Codes	

Tor Helleseth
University of Bergen

P. Vijay Kumar
University of Southern California

8.1 Introduction

Pseudonoise sequences (PN sequences), also referred to as pseudorandom sequences, are sequences that are deterministically generated and yet possess some properties that one would expect to find in randomly generated sequences. Applications of PN sequences include signal synchronization, navigation, radar ranging, random number generation, spread-spectrum communications, multipath resolution, cryptography, and signal identification in multiple-access communication systems. The *correlation* between two sequences $\{x(t)\}$ and $\{y(t)\}$ is the complex inner product of the first sequence with a shifted version of the second sequence. The correlation is called 1) an autocorrelation if the two sequences are the same, 2) a crosscorrelation if they are distinct, 3) a periodic correlation if the shift is a cyclic shift, 4) an aperiodic correlation if the shift is not cyclic, and 5) a partial-period correlation if the inner product involves only a partial segment of the two sequences. More precise definitions are given subsequently.

Binary m **sequences**, defined in the next section, are perhaps the best-known family of PN sequences. The balance, run-distribution, and autocorrelation properties of these sequences mimic those of random sequences. It is perhaps the random-like correlation properties of PN sequences that makes them most attractive in a communications system, and it is common to refer to any collection of low-correlation sequences as a family of PN sequences.

Section 8.2 begins by discussing m sequences. Thereafter, the discussion continues with a description of sequences satisfying various correlation constraints along the lines of the accompanying self-explanatory figure, Fig. 8.1. Expanded tutorial discussions on pseudorandom sequences may be found in Sarwate and Pursley, 1980, in Chapter 5 of Simon et al., 1994, and in Helleseth and Kumar, 1996.

Pseudonoise Sequences 95

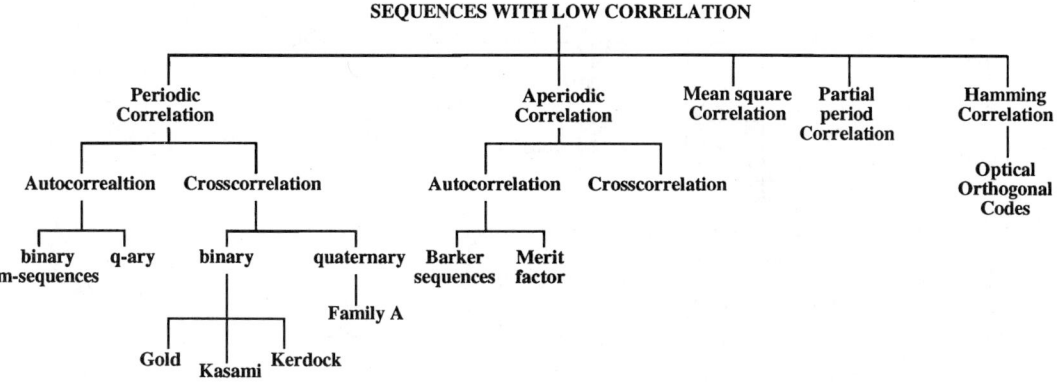

FIGURE 8.1 Overview of pseudonoise sequences.

8.2 m Sequences

A binary $\{0, 1\}$ **shift-register sequence** $\{s(t)\}$ is a sequence that satisfies a linear recurrence relation of the form

$$\sum_{i=0}^{r} f_i s(t+i) = 0, \qquad \text{for all } t \geq 0 \tag{8.1}$$

where $r \geq 1$ is the *degree* of the recursion; the coefficients f_i belong to the finite field $GF(2) = \{0, 1\}$ where the leading coefficient $f_r = 1$. Thus, both sequences $\{a(t)\}$ and $\{b(t)\}$ appearing in Fig. 8.2 are shift-register sequences. A sequence satisfying a recursion of the form in Eq. (8.1) is said to have *characteristic polynomial* $f(x) = \sum_{i=0}^{r} f_i x^i$. Thus, $\{a(t)\}$ and $\{b(t)\}$ have characteristic polynomials given by $f(x) = x^3 + x + 1$ and $f(x) = x^3 + x^2 + 1$, respectively.

Since an r-bit binary shift register can assume a maximum of 2^r different states, it follows that every shift-register sequence $\{s(t)\}$ is eventually periodic with period $n \leq 2^r$, i.e.,

$$s(t) = s(t+n), \qquad \text{for all } t \geq N$$

for some integer N. In fact, the maximum period of a shift-register sequence is $2^r - 1$, since a shift register that enters the all-zero state will remain forever in that state. The upper shift register in Fig. 8.2 when initialized with starting state 0 0 1 generates the periodic sequence $\{a(t)\}$ given by

$$0010111 \quad 0010111 \quad 0010111 \quad \cdots \tag{8.2}$$

of period $n = 7$. It follows then that this shift register generates sequences of maximal period starting from any nonzero initial state.

An m sequence is simply a binary shift-register sequence having maximal period. For every $r \geq 1$, m sequences are known to exist. The periodic **autocorrelation** function θ_s of a binary $\{0, 1\}$ sequence $\{s(t)\}$ of period n is defined by

$$\theta_s(\tau) = \sum_{t=0}^{n-1} (-1)^{s(t+\tau)-s(t)}, \qquad 0 \leq \tau \leq n-1$$

An m sequence of length $2^r - 1$ has the following attributes. 1) *Balance property:* in each period of the m sequence there are 2^{r-1} ones and $2^{r-1} - 1$ zeros. 2) *Run property:* every nonzero binary s-tuple,

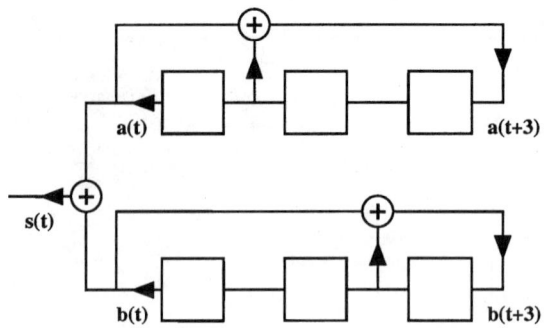

FIGURE 8.2 An example Gold sequence generator. Here $\{a(t)\}$ and $\{b(t)\}$ are m sequences of length 7.

$s \leq r$ occurs 2^{r-s} times, the all-zero s-tuple occurs $2^{r-s} - 1$ times. 3) *Two-level autocorrelation function:*

$$\theta_s(\tau) = \begin{cases} n & \text{if } \tau = 0 \\ -1 & \text{if } \tau \neq 0 \end{cases} \quad (8.3)$$

The first two properties follow immediately from the observation that every nonzero r-tuple occurs precisely once in each period of the m sequence. For the third property, consider the difference sequence $\{s(t+\tau) - s(t)\}$ for $\tau \neq 0$. This sequence satisfies the same recursion as the m sequence $\{s(t)\}$ and is clearly not the all-zero sequence. It follows, therefore, that $\{s(t+\tau) - s(t)\} \equiv \{s(t+\tau')\}$ for some τ', $0 \leq \tau' \leq n-1$, i.e., is a different cyclic shift of the m sequence $\{s(t)\}$. The balance property of the sequence $\{s(t+\tau')\}$ then gives us attribute 3. The m sequence $\{a(t)\}$ in Eq. (8.2) can be seen to have the three listed properties.

If $\{s(t)\}$ is any sequence of period n and d is an integer, $1 \leq d \leq n$, then the mapping $\{s(t)\} \to \{s(dt)\}$ is referred to as a *decimation* of $\{s(t)\}$ by the integer d. If $\{s(t)\}$ is an m sequence of period $n = 2^r - 1$ and d is an integer relatively prime to $2^r - 1$, then the decimated sequence $\{s(dt)\}$ clearly also has period n. Interestingly, it turns out that the sequence $\{s(dt)\}$ is always also an m sequence of the same period. For example, when $\{a(t)\}$ is the sequence in Eq. (8.2), then

$$a(3t) = 0011101 \quad 0011101 \quad 0011101 \quad \cdots \quad (8.4)$$

and

$$a(2t) = 0111001 \quad 0111001 \quad 0111001 \quad \cdots \quad (8.5)$$

The sequence $\{a(3t)\}$ is also an m sequence of period 7, since it satisfies the recursion

$$s(t+3) + s(t+2) + s(t) = 0 \quad \text{for all } t$$

of degree $r = 3$. In fact $\{a(3t)\}$ is precisely the sequence labeled $\{b(t)\}$ in Fig. 8.2. The sequence $\{a(2t)\}$ is simply a cyclically shifted version of $\{a(t)\}$ itself; this property holds in general. If $\{s(t)\}$ is any m sequence of period $2^r - 1$, then $\{s(2t)\}$ will always be a shifted version of the same m sequence. Clearly, the same is true for decimations by any power of 2.

Starting from an m sequence of period $2^r - 1$, it turns out that one can generate all m sequences of the same period through decimations by integers d relatively prime to $2^r - 1$. The set of integers d, $1 \leq d \leq 2^r - 1$ satisfying $(d, 2^r - 1) = 1$ forms a group under multiplication modulo $2^r - 1$, with

Pseudonoise Sequences

the powers $\{2^i \mid 0 \leq i \leq r-1\}$ of 2 forming a subgroup of order r. Since decimation by a power of 2 yields a shifted version of the same m sequence, it follows that the number of distinct m sequences of period $2^r - 1$ is $[\phi(2^r - 1)/r]$ where $\phi(n)$ denotes the number of integers d, $1 \leq d \leq n$, relatively prime to n. For example, when $r = 3$, there are just two cyclically distinct m sequences of period 7, and these are precisely the sequences $\{a(t)\}$ and $\{b(t)\}$ discussed in the preceding paragraph. Tables provided in Peterson and Weldon, 1972 can be used to determine the characteristic polynomial of the various m sequences obtainable through the decimation of a single given m sequence. The classical reference on m sequences is Golomb, 1982.

If one obtains a sequence of some large length n by repeatedly tossing an unbiased coin, then such a sequence will very likely satisfy the balance, run, and autocorrelation properties of an m sequence of comparable length. For this reason, it is customary to regard the extent to which a given sequence possesses these properties as a measure of randomness of the sequence. Quite apart from this, in many applications such as signal synchronization and radar ranging, it is desirable to have sequences $\{s(t)\}$ with low autocorrelation sidelobes i.e., $|\theta_s(\tau)|$ is small for $\tau \neq 0$. Whereas m sequences are a prime example, there exist other methods of constructing binary sequences with low out-of-phase autocorrelation.

Sequences $\{s(t)\}$ of period n having an autocorrelation function identical to that of an m sequence, i.e., having θ_s satisfying Eq. (8.3) correspond to well-studied combinatorial objects known as *cyclic Hadamard difference sets*. Known infinite families fall into three classes 1) Singer and Gordon, Mills and Welch, 2) quadratic residue, and 3) twin-prime difference sets. These correspond, respectively, to sequences of period n of the form $n = 2^r - 1, r \geq 1$; n prime; and $n = p(p + 2)$ with both p and $p + 2$ being prime in the last case. For a detailed treatment of cyclic difference sets, see Baumert, 1971.

8.3 The q-ary Sequences with Low Autocorrelation

As defined earlier, the autocorrelation of a binary $\{0, 1\}$ sequence $\{s(t)\}$ leads to the computation of the inner product of an $\{-1, +1\}$ sequence $\{(-1)^{s(t)}\}$ with a cyclically shifted version $\{(-1)^{s(t+\tau)}\}$ of itself. The $\{-1, +1\}$ sequence is transmitted as a phase shift by either $0°$ and $180°$ of a radio-frequency carrier, i.e., using binary phase-shift keying (PSK) modulation. If the modulation is q-ary PSK, then one is led to consider sequences $\{s(t)\}$ with symbols in the set Z_q, i.e., the set of integers modulo q. The relevant autocorrelation function $\theta_s(\tau)$ is now defined by

$$\theta_s(\tau) = \sum_{t=0}^{n-1} \omega^{s(t+\tau)-s(t)}$$

where n is the period of $\{s(t)\}$ and ω is a complex primitive qth root of unity. It is possible to construct sequences $\{s(t)\}$ over Z_q whose autocorrelation function satisfies

$$\theta_s(\tau) = \begin{cases} n & \text{if } \tau = 0 \\ 0 & \text{if } \tau \neq 0 \end{cases}$$

For obvious reasons, such sequences are said to have an *ideal autocorrelation function*.

We provide without proof two sample constructions. The sequences in the first construction are given by

$$s(t) = \begin{cases} t^2/2 \pmod{n} & \text{when } n \text{ is even} \\ t(t+1)/2 \pmod{n} & \text{when } n \text{ is odd} \end{cases}$$

Thus, this construction provides sequences with ideal autocorrelation for any period n. Note that the size q of the sequence symbol alphabet equals n when n is odd and $2n$ when n is even.

The second construction also provides sequences over Z_q of period n but requires that n be a perfect square. Let $n = r^2$ and let π be an arbitrary permutation of the elements in the subset $\{0, 1, 2, \ldots, (r-1)\}$ of Z_n: Let g be an arbitrary function defined on the subset $\{0, 1, 2, \ldots, r-1\}$ of Z_n. Then any sequence of the form

$$s(t) = rt_1\pi(t_2) + g(t_2) \pmod{n}$$

where $t = rt_1 + t_2$ with $0 \le t_1, t_2 \le r - 1$ is the base-r decomposition of t, has an ideal autocorrelation function. When the alphabet size q equals or divides the period n of the sequence, ideal-autocorrelation sequences also go by the name *generalized bent functions*. For details, see Helleseth and Kumar, 1996.

8.4 Families of Sequences with Low Crosscorrelation

Given two sequences $\{s_1(t)\}$ and $\{s_2(t)\}$ over Z_q of period n, their **crosscorrelation** function $\theta_{1,2}(\tau)$ is defined by

$$\theta_{1,2}(\tau) = \sum_{t=0}^{n-1} \omega^{s_1(t+\tau)-s_2(t)}$$

where ω is a primitive qth root of unity. The crosscorrelation function is important in code-division multiple-access (CDMA) communication systems. Here, each user is assigned a distinct signature sequence and to minimize interference due to the other users, it is desirable that the signature sequences have pairwise, low values of crosscorrelation function. To provide the system in addition with a self-synchronizing capability, it is desirable that the signature sequences have low values of the autocorrelation function as well.

Let $\mathcal{F} = \{\{s_i(t)\} \mid 1 \le i \le M\}$ be a family of M sequences $\{s_i(t)\}$ over Z_q each of period n. Let $\theta_{i,j}(\tau)$ denote the crosscorrelation between the ith and jth sequence at shift τ, i.e.

$$\theta_{i,j}(\tau) = \sum_{t=0}^{n-1} \omega^{s_i(t+\tau)-s_j(t)}, \qquad 0 \le \tau \le n-1$$

The classical goal in sequence design for CDMA systems has been minimization of the parameter

$$\theta_{\max} = \max\{|\theta_{i,j}(\tau)| \mid \text{either } i \ne j \text{ or } \tau \ne 0\}$$

for fixed n and M. It should be noted though that, in practice, because of data modulation the correlations that one runs into are typically of an aperiodic rather than a periodic nature (see Sec. 8.5). The problem of designing for low aperiodic correlation, however, is a more difficult one. A typical approach, therefore, has been to design based on periodic correlation, and then to analyze the resulting design for its aperiodic correlation properties. Again, in many practical systems, the mean square correlation properties are of greater interest than the worst-case correlation represented by a parameter such as $\theta_{\max}$. The mean square correlation is discussed in Sec. 8.6.

Bounds on the minimum possible value of $\theta_{\max}$ for given period n, family size M, and alphabet size q are available that can be used to judge the merits of a particular sequence design. The most efficient bounds are those due to Welch, Sidelnikov, and Levenshtein, see Helleseth and Kumar, 1996. In CDMA systems, there is greatest interest in designs in which the parameter $\theta_{\max}$ is in the range $\sqrt{n} \le \theta_{\max} \le 2\sqrt{n}$. Accordingly, Table 8.1 uses the Welch, Sidelnikov, and Levenshtein

Pseudonoise Sequences

TABLE 8.1 Bounds on Family Size M for Given n, θ_{max}

θ_{max}	Upper bound on M $q = 2$	Upper Bound on M $q > 2$
$\sqrt{n}$	$n/2$	n
$\sqrt{2n}$	n	$n^2/2$
$2\sqrt{n}$	$3n^2/10$	$n^3/2$

bounds to provide an order-of-magnitude upper bound on the family size M for certain θ_{max} in the cited range.

Practical considerations dictate that q be small. The bit-oriented nature of electronic hardware makes it preferable to have q a power of 2. With this in mind, a description of some efficient sequence families having low auto- and crosscorrelation values and alphabet sizes $q = 2$ and $q = 4$ are described next.

Gold and Kasami Sequences

Given the low autocorrelation sidelobes of an m sequence, it is natural to attempt to construct families of low correlation sequences starting from m sequences. Two of the better known constructions of this type are the families of Gold and Kasami sequences.

Let r be odd and $d = 2^k + 1$ where k, $1 \leq k \leq r - 1$, is an integer satisfying $(k, r) = 1$. Let $\{s(t)\}$ be a cyclic shift of an m sequence of period $n = 2^r - 1$ that signifies $S(dt) \neq 0$ and let $\mathcal{G}$ be the *Gold* family of $2^r + 1$ sequences given by

$$\mathcal{G} = \{s(t)\} \cup \{s(dt)\} \cup \{\{s(t) + s(d[t + \tau])\} \mid 0 \leq \tau \leq n - 1\}$$

Then each sequence in $\mathcal{G}$ has period $2^r - 1$ and the maximum-correlation parameter θ_{max} of $\mathcal{G}$ satisfies

$$\theta_{max} \leq \sqrt{2^{r+1}} + 1$$

An application of the Sidelnikov bound coupled with the information that θ_{max} must be an odd integer yields that for the family $\mathcal{G}$, θ_{max} is as small as it can possibly be. In this sense the family $\mathcal{G}$ is an optimal family. We remark that these comments remain true even when d is replaced by the integer $d = 2^{2k} - 2^k + 1$ with the conditions on k remaining unchanged.

The Gold family remains the best-known family of m sequences having low crosscorrelation. Applications include the Navstar Global Positioning System whose signals are based on Gold sequences.

The family of Kasami sequences has a similar description. Let $r = 2v$ and $d = 2^v + 1$. Let $\{s(t)\}$ be a cyclic shift of an m sequence of period $n = 2^r - 1$ that satisfies $s(dt) \neq 0$, and consider the family of Kasami sequences given by

$$\mathcal{K} = \{s(t)\} \cup \{\{s(t) + s(d[t + \tau])\} \mid 0 \leq \tau \leq 2^v - 2\}$$

Then the Kasami family $\mathcal{K}$ contains 2^v sequences of period $2^r - 1$. It can be shown that in this case

$$\theta_{max} = 1 + 2^v$$

This time an application of the Welch bound and the fact that θ_{max} is an integer shows that the Kasami family is optimal in terms of having the smallest possible value of θ_{max} for given n and M.

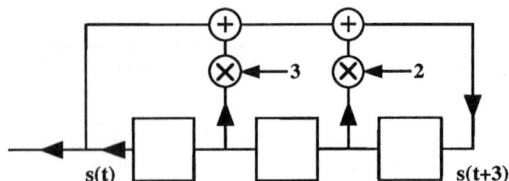

FIGURE 8.3 Shift register that generates family $\mathcal{A}$ quaternary sequences $\{s(t)\}$ of period 7.

Quaternary Sequences with Low Crosscorrelation

The entries in Table 8.1 suggest that nonbinary (i.e., $q > 2$) designs may be used for improved performance. A family of quaternary sequences that outperform the Gold and Kasami sequences is now discussed below.

Let $f(x)$ be the characteristic polynomial of a binary m sequence of length $2^r - 1$ for some integer r. The coefficients of $f(x)$ are either 0 or 1. Now, regard $f(x)$ as a polynomial over Z_4 and form the product $(-1)^r f(x) f(-x)$. This can be seen to be a polynomial in x^2. Define the polynomial $g(x)$ of degree r by setting $g(x^2) = (-1)^r f(x) f(-x)$. Let $g(x) = \sum_{i=0}^{r} g_i x^i$ and consider the set of all quaternary sequences $\{a(t)\}$ satisfying the recursion $\sum_{i=0}^{r} g_i a(t+i) = 0$ for all t.

It turns out that with the exception of the all-zero sequence, all of the sequences generated in this way have period $2^r - 1$. Thus, the recursion generates a family $\mathcal{A}$ of $2^r + 1$ cyclically distinct quaternary sequences. Closer study reveals that the maximum correlation parameter $\theta_{\max}$ of this family satisfies $\theta_{\max} \leq 1 + \sqrt{2^r}$. Thus, in comparison to the family of Gold sequences, the family $\mathcal{A}$ offers a lower value of $\theta_{\max}$ (by a factor of $\sqrt{2}$) for the same family size. In comparison to the set of Kasami sequences, it offers a much larger family size for the same bound on $\theta_{\max}$.

We illustrate with an example. Let $f(x) = x^3 + x + 1$ be the characteristic polynomial of the m sequence $\{a(t)\}$ in Eq. (8.1). Then over Z_4

$$g(x^2) = (-1)^3 f(x) f(-x) = x^6 + 2x^4 + x^2 + 3$$

so that $g(x) = x^3 + 2x^2 + x + 3$. Thus, the sequences in family $\mathcal{A}$ are generated by the recursion $s(t+3) + 2s(t+2) + s(t+1) + 3s(t) = 0 \mod 4$. The corresponding shift register is shown in Fig. 8.3. By varying initial conditions, this shift register can be made to generate nine cyclically distinct sequences, each of length 7. In this case $\theta_{\max} \leq 1 + \sqrt{8}$.

Kerdock Sequences

The Gold and Kasami families of sequences are closely related to binary linear cyclic codes. It is well known in coding theory that there exists nonlinear binary codes whose performance exceeds that of the best possible linear code. Surprisingly, some of these examples come from binary codes, which are images of *linear quaternary* ($q = 4$) *codes* under the Gray map: $0 \to 00$, $1 \to 01$, $2 \to 11$, $3 \to 10$. A prime example of this is the Kerdock code, which recently has been shown to be the Gray image of a quaternary linear code. Thus, it is not surprising that the Kerdock code yields binary sequences that significantly outperform the family of Kasami sequences.

The Kerdock sequences may be constructed as follows: let $f(x)$ be the characteristic polynomial of an m sequence of period $2^r - 1$, r odd. As before, regarding $f(x)$ as a polynomial over Z_4 (which happens to have $\{0, 1\}$ coefficients), let the polynomial $g(x)$ over Z_4 be defined via $g(x^2) = -f(x) f(-x)$. [Thus, $g(x)$ is the characteristic polynomial of a family $\mathcal{A}$ sequence set of period

Pseudonoise Sequences

$2^r - 1$.] Set $h(x) = -g(-x) = \sum_{i=0}^{r} h_i x^i$, and let S be the set of all Z_4 sequences satisfying the recursion $\sum_{i=0}^{r} h_i s(t+i) = 0$. Then S contain 4^r-distinct sequences corresponding to all possible distinct initializations of the shift register.

Let T denote the subset S of size 2^r-consisting of those sequences corresponding to initializations of the shift register only using the symbols 0 and 2 in Z_4. Then the set $S - T$ of size $4^r - 2^r$ contains a set $\mathcal{U}$ of 2^{r-1} cyclically distinct sequences each of period $2(2^r - 1)$. Given $x = a + 2b \in Z_4$ with $a, b \in \{0, 1\}$, let μ denote the most significant bit (MSB) map $\mu(x) = b$. Let $\mathcal{K}_E$ denote the family of 2^{r-1} binary sequences obtained by applying the map μ to each sequence in $\mathcal{U}$. It turns out that each sequence in $\mathcal{U}$ also has period $2(2^r - 1)$ and that, furthermore, for the family $\mathcal{K}_E$, $\theta_{max} \leq 2 + \sqrt{2^{r+1}}$. Thus, $\mathcal{K}_E$ is a much larger family than the Kasami family, while having almost exactly the same value of θ_{max}.

For example, taking $r = 3$ and $f(x) = x^3 + x + 1$, we have from the previous family $\mathcal{A}$ example that $g(x) = x^3 + 2x^2 + x + 3$, so that $h(x) = -g(-x) = x^3 + 2x^2 + x + 1$. Applying the MSB map to the head of the shift register, and discarding initializations of the shift register involving only 0's and 2's yields a family of four cyclically distinct binary sequences of period 14.

8.5 Aperiodic Correlation

Let $\{x(t)\}$ and $\{y(t)\}$ be complex-valued sequences of length (or period) n, not necessarily distinct. Their *aperiodic correlation* values $\{\rho_{x,y}(\tau)| -(n-1) \leq \tau \leq n-1\}$ are given by

$$\rho_{x,y}(\tau) = \sum_{t=\max\{0,-\tau\}}^{\min\{n-1, n-1-\tau\}} x(t+\tau) y^*(t)$$

where $y^*(t)$ denotes the complex conjugate of $y(t)$. When $x \equiv y$, we will abbreviate and write ρ_x in place of $\rho_{x,y}$. The sequences described next are perhaps the most famous example of sequences with low-aperiodic autocorrelation values.

Barker Sequences

A binary $\{-1, +1\}$ sequence $\{s(t)\}$ of length n is said to be a *Barker sequence* if the aperiodic autocorrelation values $\rho_s(\tau)$ satisfy $|\rho_s(\tau)| \leq 1$ for all τ, $-(n-1) \leq \tau \leq n-1$. The Barker property is preserved under the following transformations:

$$s(t) \to -s(t), \qquad s(t) \to (-1)^t s(t) \qquad \text{and} \qquad s(t) \to s(n-1-t)$$

as well as under compositions of the preceding transformations. Only the following Barker sequences are known:

$$
\begin{align}
n &= 2 \quad ++ \\
n &= 3 \quad ++- \\
n &= 4 \quad +++- \\
n &= 5 \quad +++-+ \\
n &= 7 \quad +++--+- \\
n &= 11 \quad +++----+--+- \\
n &= 13 \quad +++++--++-+-+
\end{align}
$$

where + denotes +1 and − denotes −1 and sequences are generated from these via the transformations already discussed. It is known that if any other Barker sequence exists, it must have length $n > 1,898,884$, that is a multiple of 4.

For an upper bound to the maximum out-of-phase aperiodic autocorrelation of an m sequence, see Sarwate, 1984.

Sequences with High Merit Factor

The *merit factor F* of a $\{-1, +1\}$ sequence $\{s(t)\}$ is defined by

$$F = \frac{n^2}{2 \sum_{\tau=1}^{n-1} \rho_s^2(\tau)}$$

Since $\rho_s(\tau) = \rho_s(-\tau)$ for $1 \leq |\tau| \leq n-1$ and $\rho_s(0) = n$, factor F may be regarded as the ratio of the square of the in-phase autocorrelation, to the sum of the squares of the out-of-phase aperiodic autocorrelation values. Thus, the merit factor is one measure of the aperiodic autocorrelation properties of a binary $\{-1, +1\}$ sequence. It is also closely connected with the signal to self-generated noise ratio of a communication system in which coded pulses are transmitted and received.

Let F_n denote the largest merit factor of any binary $\{-1, +1\}$ sequence of length n. For example, at length $n = 13$, the Barker sequence of length 13 has a merit factor $F = F_{13} = 14.08$. Assuming a certain ergodicity postulate it was established by Golay that $\lim_{n \to \infty} F_n = 12.32$. Exhaustive computer searches carried out for $n \leq 40$ have revealed the following.

1. For $1 \leq n \leq 40, n \neq 11, 13$,

$$3.3 \leq F_n \leq 9.85,$$

2. $F_{11} = 12.1$, $F_{13} = 14.08$.

The value F_{11} is also achieved by a Barker sequence. From partial searches, for lengths up to 117, the highest known merit factor is between 8 and 9.56; for lengths from 118 to 200, the best-known factor is close to 6. For lengths > 200, statistical search methods have failed to yield a sequence having merit factor exceeding 5.

An *offset sequence* is one in which a fraction θ of the elements of a sequence of length n are chopped off at one end and appended to the other end, i.e., an offset sequence is a cyclic shift of the original sequence by $n\theta$ symbols. It turns out that the asymptotic merit factor of m sequences is equal to 3 and is independent of the particular offset of the m sequence. There exist offsets of sequences associated with quadratic-residue and twin-prime difference sets that achieve a larger merit factor of 6. Details may be found in Jensen, Jenson, and Høholdt, 1991.

Sequences with Low Aperiodic Crosscorrelation

If $\{u(t)\}$ and $\{v(t)\}$ are sequences of length $2n - 1$ defined by

$$u(t) = \begin{cases} x(t) & \text{if } 0 \leq t \leq n-1 \\ 0 & \text{if } n \leq t \leq 2n-2 \end{cases}$$

and

$$v(t) = \begin{cases} y(t) & \text{if } 0 \leq t \leq n-1 \\ 0 & \text{if } n \leq t \leq 2n-2 \end{cases}$$

then

$$\{\rho_{x,y}(\tau) \mid -(n-1) \leq \tau \leq n-1\} = \{\theta_{u,v}(\tau) \mid 0 \leq \tau \leq 2n-2\} \quad (8.6)$$

Given a collection

$$U = \{\{x_i(t)\} \mid 1 \leq i \leq M\}$$

of sequences of length n over Z_q, let us define

$$\rho_{max} = \max\{|\rho_{a,b}(\tau)| \mid a, b \in U, \text{ either } a \neq b \text{ or } \tau \neq 0\}$$

It is clear from Eq. (8.6) how bounds on the *periodic* correlation parameter θ_{max} can be adapted to give bounds on ρ_{max}. Translation of the Welch bound gives that for every integer $k \geq 1$,

$$\rho_{max}^{2k} \geq \left(\frac{n^{2k}}{M(2n-1)-1}\right)\left\{\frac{M(2n-1)}{\binom{2n+k-2}{k}} - 1\right\}$$

Setting $k = 1$ in the preceding bound gives

$$\rho_{max} \geq n\sqrt{\frac{M-1}{M(2n-1)-1}}$$

Thus, for fixed M and large n, Welch's bound gives

$$\rho_{max} \geq \mathcal{O}(n^{1/2})$$

There exist sequence families which asymptotically achieve $\rho_{max} \approx \mathcal{O}(n^{1/2})$, [Mow, 1994].

8.6 Other Correlation Measures

Partial-Period Correlation

The *partial-period (p-p) correlation* between the sequences $\{u(t)\}$ and $\{v(t)\}$ is the collection $\{\Delta_{u,v}(l, \tau, t_0) \mid 1 \leq l \leq n, 0 \leq \tau \leq n-1, 0 \leq t_0 \leq n-1\}$ of inner products

$$\Delta_{u,v}(l, \tau, t_0) = \sum_{t=t_0}^{t=t_0+l-1} u(t+\tau)v^*(t)$$

where l is the length of the partial period and the sum $t + \tau$ is again computed modulo n.

In direct-sequence CDMA systems, the pseudorandom signature sequences used by the various users are often very long for reasons of data security. In such situations, to minimize receiver hardware complexity, correlation over a partial period of the signature sequence is often used to demodulate data, as well as to achieve synchronization. For this reason, the p-p correlation properties of a sequence are of interest.

Researchers have attempted to determine the moments of the p-p correlation. Here the main tool is the application of the Pless power-moment identities of coding theory [MacWilliams and Sloane, 1977]. The identities often allow the first and second p-p correlation moments to be completely determined. For example, this is true in the case of m sequences (the remaining moments turn out to depend upon the specific characteristic polynomial of the m sequence). Further details may be found in Simon et al., 1994.

Mean Square Correlation

Frequently in practice, there is a greater interest in the mean-square correlation distribution of a sequence family than in the parameter $\theta_{\max}$. Quite often in sequence design, the sequence family is derived from a linear, binary cyclic code of length n by picking a set of cyclically distinct sequences of period n. The families of Gold and Kasami sequences are so constructed. In this case, as pointed out by Massey, the mean square correlation of the family can be shown to be either optimum or close to optimum, under certain easily satisfied conditions, imposed on the minimum distance of the dual code. A similar situation holds even when the sequence family does not come from a linear cyclic code. In this sense, mean square correlation is not a very discriminating measure of the correlation properties of a family of sequences. An expanded discussion of this issue may be found in Hammons and Kumar, 1993.

Optical Orthogonal Codes

Given a pair of $\{0, 1\}$ sequences $\{s_1(t)\}$ and $\{s_2(t)\}$ each having period n, we define the *Hamming correlation* function $\theta_{12}(\tau)$, $0 \leq \tau \leq n - 1$, by

$$\theta_{12}(\tau) = \sum_{t=0}^{n-1} s_1(t + \tau) s_2(t)$$

Such correlations are of interest, for instance, in optical communication systems where the 1's and 0's in a sequence correspond to the presence or absence of pulses of transmitted light.

An (n, w, λ) optical orthogonal code (OOC) is a family $\mathcal{F} = \{\{s_i(t)\} \mid i = 1, 2, \ldots, M\}$, of M $\{0, 1\}$ sequences of period n, constant Hamming weight w, where w is an integer lying between 1 and $n - 1$ satisfying $\theta_{ij}(\tau) \leq \lambda$ whenever either $i \neq j$ or $\tau \neq 0$.

Note that the Hamming distance $d_{a,b}$ between a period of the corresponding codewords $\{a(t)\}$, $\{b(t)\}$, $0 \leq t \leq n - 1$ in an (n, w, λ) OOC having Hamming correlation ρ, $0 \leq \rho \leq \lambda$, is given by $d_{a,b} = 2(w - \rho)$, and, thus, OOCs are closely related to constant-weight error correcting codes. Given an (n, w, λ) OOC, by enlarging the OOC to include every cyclic shift of each sequence in the code, one obtains a constant-weight, minimum distance $d_{\min} \geq 2(w - \lambda)$ code. Conversely, given a constant-weight cyclic code of length n, weight w and minimum distance $d_{\min}$, one can derive an (n, w, λ) OOC code with $\lambda \leq w - d_{\min}/2$ by partitioning the code into cyclic equivalence classes and then picking precisely one representative from each equivalence class of size n.

By making use of this connection, one can derive bounds on the size of an OOC from known bounds on the size of constant-weight codes. The bound given next follows directly from the Johnson bound for constant weight codes [MacWilliams and Sloane, 1977]. The number $M(n, w, \lambda)$ of codewords in a (n, w, λ) OOC satisfies

$$M(n, w, \lambda) \leq \frac{1}{w} \left\lfloor \frac{n-1}{w-1} \cdots \left\lfloor \frac{n-\lambda+1}{w-\lambda+1} \left\lfloor \frac{n-\lambda}{w-\lambda} \right\rfloor \right\rfloor \cdots \right\rfloor$$

An OOC code that achieves the Johnson bound is said to be optimal. A family $\{\mathcal{F}_n\}$ of OOCs indexed by the parameter n and arising from a common construction is said to be asymptotically optimum if

$$\lim_{n \to \infty} \frac{|\mathcal{F}_n|}{M(n, w, \lambda)} = 1$$

Constructions for optical orthogonal codes are available for the cases when $\lambda = 1$ and $\lambda = 2$. For larger values of λ, there exist constructions which are asymptotically optimum. Further details may be found in Helleseth and Kumar, 1996.

Defining Terms

Autocorrelation of a sequence: The complex inner product of the sequence with a shifted version itself.

Crosscorrelation of two sequences: The complex inner product of the first sequence with a shifted version of the second sequence.

***m* Sequence:** A periodic binary {0, 1} sequence that is generated by a shift register with linear feedback and which has maximal possible period given the number of stages in the shift register.

Pseudonoise sequences: Also referred to as pseudorandom sequences (PN), these are sequences that are deterministically generated and yet possess some properties that one would expect to find in randomly generated sequences.

Shift-register sequence: A sequence with symbols drawn from a field, which satisfies a linear-recurrence relation and which can be implemented using a shift register.

References

Baumert, L.D. 1971. *Cyclic Difference Sets*, Lecture Notes in Mathematics 182, Springer–Verlag, New York.

Golomb, S.W. 1982. *Shift Register Sequences*, Aegean Park Press, San Francisco, CA.

Hammons, A.R., Jr. and Kumar, P.V. 1993. On a recent 4-phase sequence design for CDMA. *IEICE Trans. Commun.* E76-B(8).

Helleseth, T. and Kumar, P.V. 1996. (planned). Sequences with low correlation. In *Handbook of Coding Theory*, ed. R. Brualdi, C. Huffman, and V. Pless, Elsevier Science Publishers, Amsterdam.

Jensen, J.M., Jensen, H.E., and Høholdt, T. 1991. The merit factor of binary sequences related to difference sets. *IEEE Trans. Inform. Theory*. IT-37(May):617–626.

MacWilliams, F.J. and Sloane, N.J.A. 1977. *The Theory of Error-Correcting Codes*, North-Holland, Amsterdam.

Mow, W.H. 1994. On McEliece's open problem on minimax aperiodic correlation. In *Proc. IEEE Intern. Symp. Inform. Theory*, p. 75.

Peterson, W.W. and Weldon, E.J., Jr. 1972. *Error-Correcting Codes*, 2nd ed. MIT Press, Cambridge, MA.

Sarwate, D.V. 1984. An upper bound on the aperiodic autocorrelation function for a maximal-length sequence. *IEEE Trans. Inform. Theory*. IT-30(July):685–687.

Sarwate, D.V. and Pursley, M.B. 1980. Crosscorrelation properties of pseudorandom and related sequences. *Proc. IEEE*, 68(May):593–619.

Simon, M.K., Omura, J.K., Scholtz, R.A., and Levitt, B.K. 1994. *Spread Spectrum Communications Handbook*, revised ed. McGraw Hill, New York.

Further Information

A more in-depth treatment of pseudonoise sequences, may be found in the following.

Golomb, S.W., 1982, *Shift Register Sequences*, Aegean Park Press, San Francisco.

Helleseth, T., and Kumar, P.V., 1996 (planned), "Sequences with Low Correlation," in *Handbook of Coding Theory*, edited by R. Brualdi, C. Huffman and V. Pless, Elsevier Science Publishers, Amsterdam.

Sarwate, D.V., and Pursley, M.B., 1980, "Crosscorrelation Properties of Pseudorandom and Related Sequences," *Proc. IEEE*, Vol. 68, May, pp. 593–619.

Simon, M.K., Omura, J.K., Scholtz, R.A., and Levitt, B.K., 1994, *Spread Spectrum Communications Handbook*, revised ed. McGraw Hill, New York.

9
D/A and A/D Converters

Susan A.R. Garrod
Purdue University

9.1 D/A and A/D Circuits 107
D/A and A/D Converter Performance Criteria • D/A Conversion Processes • D/A Converter ICs • A/D Conversion Processes • A/D Converter ICs • Grounding and Bypassing on D/A and A/D ICs • Selection Criteria for D/A and A/D Converter ICs

Digital-to-analog (D/A) conversion is the process of converting digital codes into a continuous range of analog signals. *Analog-to-digital (A/D) conversion* is the complementary process of converting a continuous range of analog signals into digital codes. Such conversion processes are necessary to interface real-world systems, which typically monitor continuously varying analog signals, with digital systems that process, store, interpret, and manipulate the analog values.

D/A and A/D applications have evolved from predominately military-driven applications to consumer-oriented applications. Up to the mid-1980s, the military applications determined the design of many D/A and A/D devices. The military applications required very high performance coupled with hermetic packaging, radiation hardening, shock and vibration testing, and military specification and record keeping. Cost was of little concern, and "low power" applications required approximately 2.8 W. The major applications up the mid-1980s included military radar warning and guidance systems, digital oscilloscopes, medical imaging, infrared systems, and professional video.

The applications requiring D/A and A/D circuits in the 1990s have different performance criteria from those of earlier years. In particular, low power and high speed applications are driving the development of D/A and A/D circuits, as the devices are used extensively in battery-operated consumer products. The predominant applications include cellular telephones, hand-held camcorders, portable computers, and set-top cable TV boxes. These applications generally have low power and long battery life requirements, or they may have high speed and high resolution requirements, as is the case with the set-top cable TV boxes.

9.1 D/A and A/D Circuits

D/A and A/D conversion circuits are available as integrated circuits (ICs) from many manufacturers. A huge array of ICs exists, consisting of not only the D/A or A/D conversion circuits, but also closely

TABLE 9.1 D/A and A/D Integrated Circuits

D/A Converter ICs	Resolution, b	Multiplying vs Fixed Reference	Settling Time, μs	Input Data Format
Analog devices AD558	8	Fixed reference	3	Parallel
Analog devices AD7524	8	Multiplying	0.400	Parallel
Analog devices AD390	Quad, 12	Fixed reference	8	Parallel
Analog devices AD1856	16	Fixed reference	1.5	Serial
Burr–Brown DAC729	18	Fixed reference	8	Parallel
DATEL DACHF8	8	Multiplying	0.025	Parallel
National DAC0800	8	Multiplying	0.1	Parallel

A/D Converter ICs	Resolution, b	Signal Inputs	Conversion Speed, μs	Output Data Format
Analog devices AD572	12	1	25	Serial and Parallel
Burr–Brown ADC803	12	1	1.5	Parallel
Burr–Brown ADC701	16	1	1.5	Parallel
National ADC1005B	10	1	50	Parallel
TI, National ADC0808	8	8	100	Parallel
TI, National ADC0834	8	4	32	Serial
TI TLC0820	8	1	1	Parallel
TI TLC1540	10	11	21	Serial

Interface ICs A/D and D/A	Resolution, b	Onboard Filters	Sampling Rate, kHz	Data Format
TI TLC32040	14	Yes	19.2 (programmable)	Serial
TI 2914 PCM codec and filter	8	Yes	8	Serial

related circuits such as sample-and-hold amplifiers, analog multiplexers, voltage-to-frequency and frequency-to-voltage converters, voltage references, calibrators, operation amplifiers, isolation amplifiers, instrumentation amplifiers, active filters, DC-to-DC converters, analog interfaces to digital signal processing systems, and data acquisition subsystems. Data books from the IC manufacturers contain an enormous amount of information about these devices and their applications to assist the design engineer.

The ICs discussed in this chapter will be strictly the D/A and A/D conversion circuits. Table 9.1 lists a small sample of the variety of the D/A and A/D converters currently available. The ICs usually perform either D/A or A/D conversion. There are serial interface ICs, however, typically for high-performance audio and digital signal processing applications, that perform both A/D and D/A processes.

D/A and A/D Converter Performance Criteria

The major factors that determine the quality of performance of D/A and A/D converters are *resolution, sampling rate, speed,* and *linearity*.

The *resolution* of a D/A circuit is the smallest change in the output analog signal. In an A/D system, the resolution is the smallest change in voltage that can be detected by the system and that can produce a change in the digital code. The resolution determines the total number of digital codes, or *quantization levels*, that will be recognized or produced by the circuit.

The *resolution* of a D/A or A/D IC is usually specified in terms of the bits in the digital code or in terms of the least significant bit (LSB) of the system. An n-bit code allows for 2^n quantization levels, or $2^n - 1$ steps between quantization levels. As the number of bits increases, the step size between quantization levels decreases, therefore increasing the accuracy of the system when a conversion is made between an analog and digital signal. The system resolution can be specified also as the voltage step size between quantization levels. For A/D circuits, the resolution is the smallest input voltage that is detected by the system.

The *speed* of a D/A or A/D converter is determined by the time it takes to perform the conversion process. For D/A converters, the speed is specified as the *settling time*. For A/D converters, the speed is specified as the *conversion time*. The settling time for D/A converters will vary with supply voltage and transition in the digital code; thus, it is specified in the data sheet with the appropriate conditions stated.

A/D converters have a maximum *sampling rate* that limits the speed at which they can perform continuous conversions. The sampling rate is the number of times per second that the analog signal can be sampled and converted into a digital code. For proper A/D conversion, the minimum sampling rate must be at least two times the highest frequency of the analog signal being sampled to satisfy the Nyquist sampling criterion. The conversion speed and other timing factors must be taken into consideration to determine the maximum sampling rate of an A/D converter. **Nyquist A/D converters** use a sampling rate that is slightly more than twice the highest frequency in the analog signal. **Oversampling A/D converters** use sampling rates of N times rate, where N typically ranges from 2 to 64.

Both D/A and A/D converters require a voltage reference in order to achieve absolute conversion accuracy. Some conversion ICs have internal voltage references, whereas others accept external voltage references. For high-performance systems, an external precision reference is needed to ensure long-term stability, load regulation, and control over temperature fluctuations. External precision voltage reference ICs can be found in manufacturer's data books.

Measurement accuracy is specified by the converter's *linearity*. *Integral linearity* is a measure of linearity over the entire conversion range. It is often defined as the deviation from a straight line drawn between the endpoints and through zero (or the offset value) of the conversion range. Integral linearity is also referred to as *relative accuracy*. The *offset* value is the reference level required to establish the zero or midpoint of the conversion range. *Differential linearity* is the linearity between code transitions. Differential linearity is a measure of the *monotonicity* of the converter. A converter is said to be monotonic if increasing input values result in increasing output values.

The accuracy and linearity values of a converter are specified in the data sheet in units of the LSB of the code. The linearity can vary with temperature, and so the values are often specified at $+25°C$ as well as over the entire temperature range of the device.

D/A Conversion Processes

Digital codes are typically converted to analog voltages by assigning a voltage weight to each bit in the digital code and then summing the voltage weights of the entire code. A general D/A converter consists of a network of precision resistors, input switches, and level shifters to activate the switches to convert a digital code to an analog current or voltage. D/A ICs that produce an analog current output usually have a faster settling time and better linearity than those that produce a voltage output. When the output current is available, the designer can convert this to a voltage through the selection of an appropriate output amplifier to achieve the necessary response speed for the given application.

D/A converters commonly have a fixed or variable reference level. The reference level determines the switching threshold of the precision switches that form a controlled impedance network, which in turn controls the value of the output signal. **Fixed reference D/A converters** produce an output signal that is proportional to the digital input. **Multiplying D/A** converters produce an output signal that is proportional to the product of a varying reference level times a digital code.

D/A converters can produce bipolar, positive, or negative polarity signals. A four-quadrant multiplying D/A converter allows both the reference signal and the value of the binary code to have a positive or negative polarity. The four-quadrant multiplying D/A converter produces bipolar output signals.

D/A Converter ICs

Most D/A converters are designed for general-purpose control applications. Some D/A converters, however, are designed for special applications, such as video or graphic outputs, high-definition video displays, ultra high-speed signal processing, digital video tape recording, digital attenuators, or high-speed function generators.

D/A converter ICs often include special features that enable them to be interfaced easily to microprocessors or other systems. Microprocessor control inputs, input latches, buffers, input registers, and compatibility to standard logic families are features that are readily available in D/A ICs. In addition, the ICs usually have laser-trimmed precision resistors to eliminate the need for user trimming to achieve full-scale performance.

A/D Conversion Processes

Analog signals can be converted to digital codes by many methods, including integration, **successive approximation**, parallel (flash) conversion, **delta modulation, pulse code modulation**, and **sigma–delta conversion**. Two of the most common A/D conversion processes are successive approximation A/D conversion and parallel or **flash A/D** conversion. Very high-resolution digital audio or video systems require specialized A/D techniques that often incorporate one of these general techniques as well as specialized A/D conversion processes. Examples of specialized A/D conversion techniques are pulse code modulation (PCM), and sigma–delta conversion. PCM is a common voice encoding scheme used not only by the audio industry in digital audio recordings but also by the telecommunications industry for voice encoding and multiplexing. Sigma–delta conversion is an oversampling A/D conversion where signals are sampled at very high frequencies. It has very high resolution and low distortion and is being used in the digital audio recording industry.

Successive approximation A/D conversion is a technique that is commonly used in medium- to high-speed data acquisition applications. It is one of the fastest A/D conversion techniques that requires a minimum amount of circuitry. The conversion times for successive approximation A/D conversion typically range from 10 to 300 μs for 8-b systems.

The successive approximation A/D converter can approximate the analog signal to form an n-bit digital code in n steps. The successive approximation register (SAR) individually compares an analog input voltage to the midpoint of one of n ranges to determine the value of 1 b. This process is repeated a total of n times, using n ranges, to determine the n bits in the code. The comparison is accomplished as follows. The SAR determines if the analog input is above or below the midpoint and sets the bit of the digital code accordingly. The SAR assigns the bits beginning with the most significant bit. The bit is set to a 1 if the analog input is greater than the midpoint voltage, or it is set to a 0 if it is less than the midpoint voltage. The SAR then moves to the next bit and sets it to a 1 or a 0 based on the results of comparing the analog input with the midpoint of the next allowed range. Because the SAR must perform one approximation for each bit in the digital code, an n-bit code requires n approximations.

A successive approximation A/D converter consists of four functional blocks, as shown in Fig. 9.1: the SAR, the analog comparator, a D/A converter, and a clock.

Parallel or flash A/D conversion is used in high-speed applications such as video signal processing, medical imaging, and radar detection systems. A flash A/D converter simultaneously compares the input analog voltage to $2^n - 1$ threshold voltages to produce an n-bit digital code representing the analog voltage. Typical flash A/D converters with 8-b resolution operate at 20–100 MHz.

The functional blocks of a flash A/D converter are shown in Fig. 9.2. The circuitry consists of a precision resistor ladder network, $2^n - 1$ analog comparators, and a digital priority encoder. The resistor network establishes threshold voltages for each allowed quantization level. The analog comparators indicate whether or not the input analog voltage is above or below the threshold at each level. The output of the analog comparators is input to the digital priority encoder. The priority encoder produces the final digital output code that is stored in an output latch.

An 8-b flash A/D converter requires 255 comparators. The cost of high-resolution A/D comparators escalates as the circuit complexity increases and as the number of analog converters rises by $2^n - 1$. As a low-cost alternative, some manufacturers produce modified flash A/D converters that perform the A/D conversion in two steps to reduce the amount of circuitry required. These modified flash A/D converters are also referred to as *half-flash* A/D converters, since they perform only half of the conversion simultaneously.

A/D Converter ICs

A/D converter ICs can be classified as general-purpose, high-speed, flash, and sampling A/D converters. The *general-purpose A/D converters* are typically low speed and low cost, with conversion times ranging from 2 μs to 33 ms. A/D conversion techniques used by these devices typically include successive approximation, tracking, and integrating. The general-purpose A/D converters often have control signals for simplified microprocessor interfacing. These ICs are appropriate for many process control, industrial, and instrumentation applications, as well as for environmental monitoring such as seismology, oceanography, meteorology, and pollution monitoring.

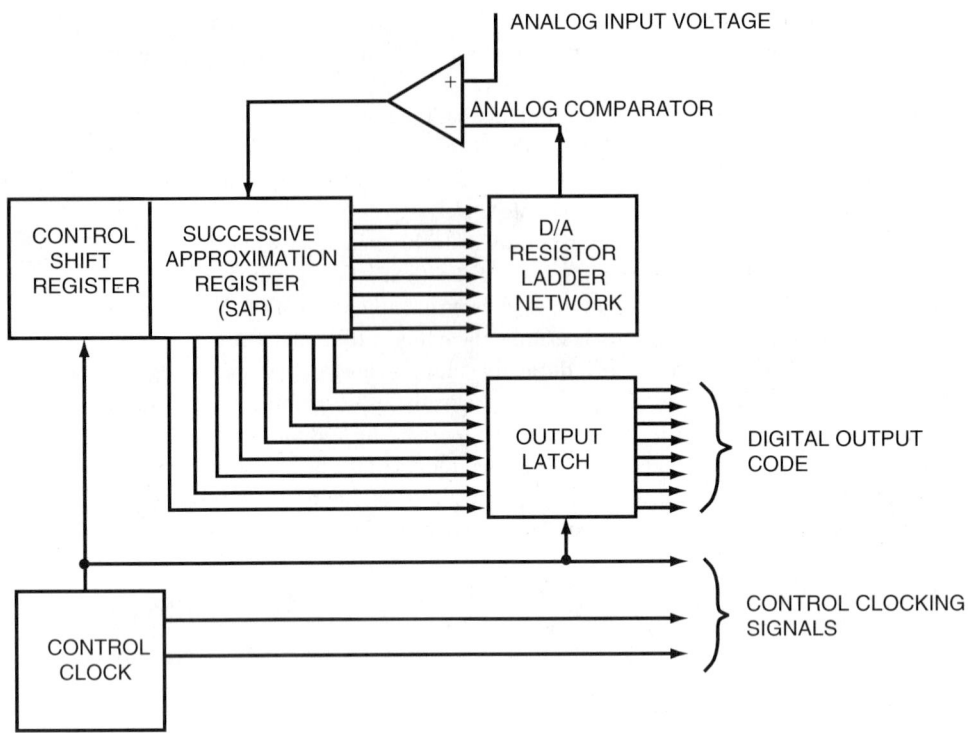

FIGURE 9.1 Successive approximation A/D converter block diagram. (*Source:* Garrod, S. and Borns, R. 1991. *Digital Logic: Analysis, Application, and Design*, p. 919. Copyright ©1991 by Saunders College Publishing, Philadelphia, PA. Reprinted by permission of the publisher.)

High-speed A/D converters have conversion times typically ranging from 400 ns to 3 μs. The higher speed performance of these devices is achieved by using the successive approximation technique, modified flash techniques, and statistically derived A/D conversion techniques. Applications appropriate for these A/D ICs include fast Fourier transform (FFT) analysis, radar digitization, medical instrumentation, and multiplexed data acquisition. Some ICs have been manufactured with an extremely high degree of linearity, to be appropriate for specialized applications in digital spectrum analysis, vibration analysis, geological research, sonar digitizing, and medical imaging.

Flash A/D converters have conversion times ranging typically from 10 to 50 ns. Flash A/D conversion techniques enable these ICs to be used in many specialized high-speed data acquisition applications such as TV video digitizing (encoding), radar analysis, transient analysis, high-speed digital oscilloscopes, medical ultrasound imaging, high-energy physics, and robotic vision applications.

Sampling A/D converters have a sample-and-hold amplifier circuit built into the IC. This eliminates the need for an external sample-and-hold circuit. The throughput of these A/D converter ICs ranges typically from 35 kHz to 100 MHz. The speed of the system is dependent on the A/D technique used by the sampling A/D converter.

A/D converter ICs produce digital codes in a serial or parallel format, and some ICs offer the designer both formats. The digital outputs are compatible with standard logic families to facilitate

D/A and A/D Converters

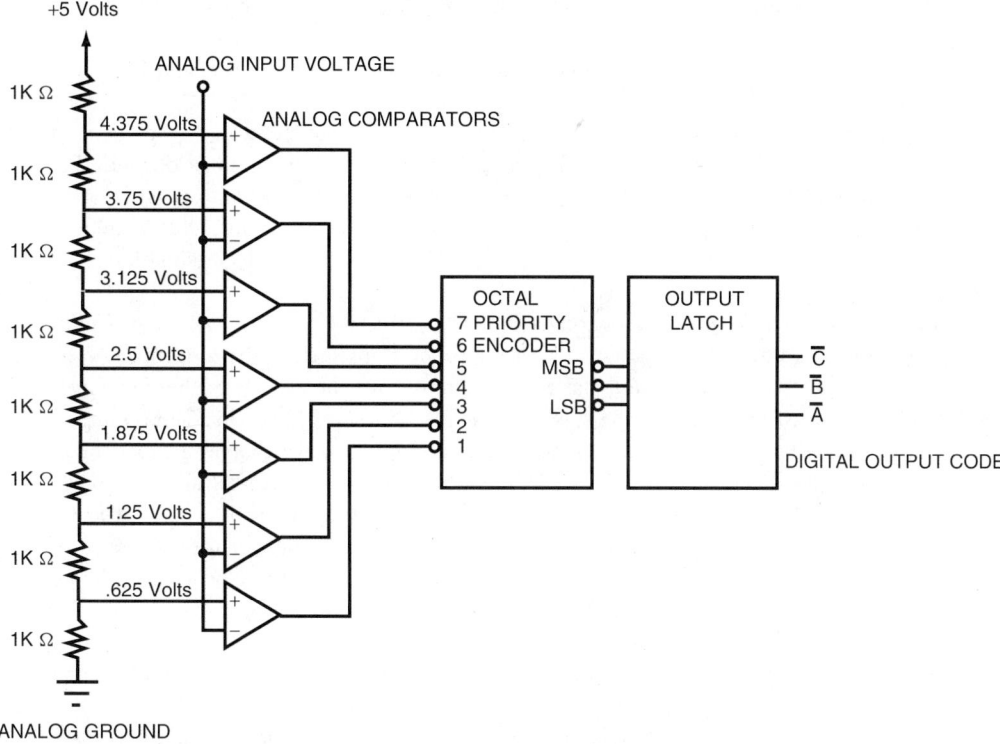

FIGURE 9.2 Flash A/D converter block diagram. (*Source*: Garrod, S. and Borns, R. *Digital Logic: Analysis, Application, and Design*, p. 928. Copyright ©1991 by Saunders College Publishing, Philadelphia, PA. Reprinted by permission of the publisher.)

interfacing to other digital systems. In addition, some A/D converter ICs have a built-in analog multiplexer and therefore can accept more than one analog input signal.

Pulse code modulation (PCM) ICs are high-precision A/D converters. The PCM IC is often refered to as a PCM *codec* with both encoder and decoder functions. The encoder portion of the codec performs the A/D conversion, and the decoder portion of the codec performs the D/A conversion. The digital code is usually formatted as a serial data stream for ease of interfacing to digital transmission and multiplexing systems.

PCM is a technique where an analog signal is sampled, quantized, and then encoded as a digital word. The PCM IC can include successive approximation techniques or other techniques to accomplish the PCM encoding. In addition, the PCM codec may employ nonlinear data compression techniques, such as companding, if it is necessary to minimize the number of bits in the output digital code. Companding is a logarithmic technique used to compress a code to fewer bits before transmission. The inverse logarithmic function is then used to expand the code to its original number of bits before converting it to the analog signal. Companding is typically used in telecommunications transmission systems to minimize data transmission rates without degrading the resolution of low-amplitude signals. Two standardized companding techniques are used extensively: A-law and μ-law. The A-law companding is used in Europe, whereas the μ-law is used predominantly in the United States and Japan. Linear PCM conversion is used in high-fidelity audio systems to preserve the integrity of the audio signal throughout the entire analog range.

Digital signal processing (DSP) techniques provide another type of A/D conversion ICs. Specialized A/D conversion such as *adaptive differential pulse code modulation* (ADPCM), sigma–delta modulation, *speech subband encoding, adaptive predictive speech encoding,* and *speech recognition* can be accomplished through the use of DSP systems. Some DSP systems require analog front ends that employ traditional PCM codec ICs or DSP interface ICs. These ICs can interface to a digital signal processor for advanced A/D applications. Some manufacturers have incorporated DSP techniques on board the single-chip A/D IC, as in the case of the DSP56ACD16 sigma–delta modulation IC by Motorola.

Integrating A/D converters are used for conversions that must take place over a long period of time, such as digital voltmeter applications or sensor applications such as thermocouples. The integrating A/D converter produces a digital code that represents the average of the signal over time. Noise is reduced by means of the signal averaging, or integration. Dual-slope integration is accomplished by a counter that advances while an input voltage charges a capacitor in a specified time interval, T. This is compared to another count sequence that advances while a reference voltage is discharging across the same capacitor in a time interval, *delta t*. The ratio of the charging count value to the discharging count value is proportional to the ratio of the input voltage to the reference voltage. Hence, the integrating converter provides a digital code that is a measure of the input voltage averaged over time. The conversion accuracy is independent of the capacitor and the clock frequency since they affect both the charging and discharging operations. The charging period, T, is selected to be the period of the fundamental frequency to be rejected. The maximum conversion rate is slightly less than $1/(2\,T)$ conversions per second. While this limits the conversion rate to be too slow for high-speed data acquisition applications, it is appropriate for long-duration applications of slowly varying input signals.

Grounding and Bypassing on D/A and A/D ICs

D/A and A/D converter ICs require correct grounding and capacitive bypassing in order to operate according to performance specifications. The digital signals can severely impair analog signals. To combat the electromagnetic interference induced by the digital signals, the analog and digital grounds should be kept separate and should have only one common point on the circuit board. If possible, this common point should be the connection to the power supply.

Bypass capacitors are required at the power connections to the IC, the reference signal inputs, and the analog inputs to minimize noise that is induced by the digital signals. Each manufacturer specifies the recommended bypass capacitor locations and values in the data sheet. The 1-μF tantalum capacitors are commonly recommended, with additional high-frequency power supply decoupling sometimes being recommended through the use of ceramic disc shunt capacitors. The manufacturers' recommendations should be followed to ensure proper performance.

Selection Criteria for D/A and A/D Converter ICs

Hundreds of D/A and A/D converter ICs are available, with prices ranging from a few dollars to several hundred dollars each. The selection of the appropriate type of converter is based on the application requirements of the system, the performance requirements, and cost. The following issues should be considered in order to select the appropriate converter.

1. What are the input and output requirements of the system? Specify all signal current and voltage ranges, logic levels, input and output impedances, digital codes, data rates, and data formats.
2. What level of accuracy is required? Determine the resolution needed throughout the analog voltage range, the dynamic response, the degree of linearity, and the number of bits encoding.
3. What speed is required? Determine the maximum analog input frequency for sampling in an A/D system, the number of bits for encoding each analog signal, and the rate of change of input digital codes in a D/A system.
4. What is the operating environment of the system? Obtain information on the temperature range and power supply to select a converter that is accurate over the operating range.

Final selection of D/A and A/D converter ICs should be made by consulting manufacturers to obtain their technical specifications of the devices. Major manufacturers of D/A and A/D converters include Analog Devices, Burr–Brown, DATEL, Maxim, National, Phillips Components, Precision Monolithics, Signetics, Sony, Texas Instruments, Ultra Analog, and Yamaha. Information on contacting these manufacturers and others can be found in an IC *Master Catalog*.

Defining Terms

Companding: A process designed to minimize the transmission bit rate of a signal by compressing it prior to transmission and expanding it upon reception. It is a rudimentary "data compression" technique that requires minimal processing.

Delta modulation: An A/D conversion process where the digital output code represents the change, or slope, of the analog input signal, rather than the absolute value of the analog input signal. A 1 indicates a rising slope of the input signal. A 0 indicates a falling slope of the input signal. The sampling rate is dependent on the derivative of the signal, since a rapidly changing signal would require a rapid sampling rate for acceptable performance.

Fixed reference D/A converter: The analog output is proportional to a fixed (nonvarying) reference signal.

Flash A/D: The fastest A/D conversion process available to date, also referred to as parallel A/D conversion. The analog signal is simultaneously evaluated by $2^n - 1$ comparators to produce an n-bit digital code in one step. Because of the large number of comparators required, the circuitry for flash A/D converters can be very expensive. This technique is commonly used in digital video systems.

Integrating A/D: The analog input signal is integrated over time to produce a digital signal that represents the area under the curve, or the integral.

Multiplying D/A: A D/A conversion process where the output signal is the product of a digital code multiplied times an analog input reference signal. This allows the analog reference signal to be scaled by a digital code.

Nyquist A/D converters: A/D converters that sample analog signals that have a maximum frequency that is less than the Nyquist frequency. The Nyquist frequency is defined as one-half of the sampling frequency. If a signal has frequencies above the Nyquist frequency, a distortion called *aliasing* occurs. To prevent aliasing, an *antialiasing filter* with a flat passband and very sharp rolloff is required.

Oversampling converters: A/D converters that sample frequencies at a rate much higher than the Nyquist frequency. Typical oversampling rates are 32 and 64 times the sampling rate that would be required with the Nyquist converters.

Pulse code modulation (PCM): An A/D conversion process requiring three steps: the analog signal is sampled, quantized, and encoded into a fixed length digital code. This technique is used in many digital voice and audio systems. The reverse process reconstructs an analog signal from the PCM code. The operation is very similar to other A/D techniques, but specific PCM circuits are optimized for the particular voice or audio application.

Sigma–delta A/D conversion: An *oversampling* A/D conversion process where the analog signal is sampled at rates much higher (typically 64 times) than the sampling rates that would be required with a Nyquist converter. Sigma–delta modulators integrate the analog signal before performing the delta modulation. The integral of the analog signal is encoded rather than the change in the analog signal, as is the case for traditional delta modulation. A digital sample rate reduction filter (also called a digital decimation filter) is used to provide an output sampling rate at twice the Nyquist frequency of the signal. The overall result of oversampling and digital sample rate reduction is greater resolution and less distortion compared to a Nyquist converter process.

Successive approximation: An A/D conversion process that systematically evaluates the analog signal in n steps to produce an n-bit digital code. The analog signal is successively compared to determine the digital code, beginning with the determination of the most significant bit of the code.

References

Analog Devices. 1989. *Analog Devices Data Conversion Products Data Book.* Analog Devices, Inc., Norwood, MA.
Burr–Brown. 1989. *Burr-Brown Integrated Circuits Data Book.* Burr–Brown, Tucson, AZ.
DATEL. 1988. *DATEL Data Conversion Catalog.* DATEL, Inc., Mansfield, MA.
Drachler, Will, and Bill Murphy. 1995. New High-Speed, Low-Power Data-Acquisition ICs. *Analog Dialogue* 29(2):3–6. Analog Devices, Inc., Norwood, MA.
Garrod, S. and Borns, R. 1991. *Digital Logic: Analysis, Application and Design,* Chap. 16. Saunders College Publishing, Philadelphia, PA.
Jacob, J.M. 1989. *Industrial Control Electronics,* Chap. 6. Prentice–Hall, Englewood Cliffs, NJ.
Keiser, B. and Strange, E. 1995. *Digital Telephony and Network Integration,* 2nd ed. Van Nostrand Reinhold, New York.
Motorola. 1989. *Motorola Telecommunications Data Book.* Motorola, Inc., Phoenix, AZ.
National Semiconductor. 1989. *National Semiconductor Data Acquisition Linear Devices Data Book.* National Semiconductor Corp., Santa Clara, CA.
Park, S. 1990. *Principles of Sigma–Delta Modulation for Analog-to-Digital Converters.* Motorola, Inc., Phoenix, AZ.
Texas Instruments. 1986. *Texas Instruments Digital Signal Processing Applications with the TMS320 Family.* Texas Instruments, Dallas, TX.
Texas Instruments. 1989. *Texas Instruments Linear Circuits Data Acquisition and Conversion Data Book.* Texas Instruments, Dallas, TX.

Further Information

Analog Devices, Inc. has edited or published several technical handbooks to assist design engineers with their data acquisition system requirements. These references should be consulted for extensive technical information and depth. The publications include *Analog-Digital Conversion Handbook,* by the engineering staff of Analog Devices, published by Prentice–Hall, Englewood Cliffs, NJ, 1986; *Nonlinear Circuits Handbook, Transducer Interfacing Hand-*

book, and *Synchro and Resolver Conversion,* all published by Analog Devices Inc., Norwood, MA.

Engineering trade journals and design publications often have articles describing recent A/D and D/A circuits and their applications. These publications include *EDN Magazine, EE Times,* and *IEEE Spectrum.* Research-related topics are covered in *IEEE Transactions on Circuits and Systems* and also *IEEE Transactions on Instrumentation and Measurement.*

10
Signal Space

10.1 Introduction 118
10.2 Fundamentals 118
10.3 Application of Signal Space Representation to Signal Detection 125
10.4 Application of Signal Space Representation to Parameter Estimation 126
   Wavelet Transforms • Mean Square Estimation—the Orthogonality Principle

Rodger E. Ziemer
University of Colorado at Colorado Springs

10.1 Introduction

Signal space concepts have their roots in the mathematical theory of inner product spaces known as **Hilbert spaces** [Stakgold, 1967]. Many books on linear systems touch on the subject of signal spaces in the context of Fourier series and transforms [Ziemer, Tranter, and Fannin, 1993; Frederick and Carlson, 1971]. The applications of signal space concepts in communication theory find their power in the representation of signal detection and estimation problems in geometrical terms, which provides much insight into signalling techniques and communication system design. The first person to have apparently exploited the power of signal space concepts in communication theory was the Russian Kotel'nikov [1968] who presented his doctoral dissertation in January, 1947. Wozencraft and Jacobs [1965] expanded on this approach and is still today widely referenced. Arthurs and Dym [1962] made use of signal space concepts in the performance analysis of several digital modulation schemes. A one-chapter summary of the use of signal space methods in signal detection and estimation is provided in Ziemer and Tranter [1995]. Another application of signal space concepts is in signal and image compression. Wavelet theory [Rioul and Vetterli, 1991] is currently finding use in these application areas. In the next section, the fundamentals of generalized vector spaces are summarized, followed by an overview of several applications to signal representations.

10.2 Fundamentals

A linear space or **vector space** (signal space) [Stakgold, 1967] is a collection of elements (called **vectors**) $x, y, z, \ldots$, for which the following axioms are satisfied:

1. To every pair of vectors x and y there corresponds a vector $x + y$, with the properties:
   a. $x + y = y + x$
   b. $x + (y + z) = (x + y) + z$

Signal Space

c. There exists a unique element 0 such that $x + 0 = x$ for every x

d. To every x, there exists a unique vector labeled $-x$ such that $x + (-x) = 0$

2. To all vectors x and y, and all numbers α and β (in general, complex), the following commutative and associative rules hold:

   a. $\alpha(\beta x) = (\alpha\beta)x$
   b. $(\alpha + \beta)x = \alpha x + \beta y$
   c. $\alpha(x + y) = \alpha x + \alpha y$
   d. $1x = x$, where 1 is the **identity element**

A vector is said to be a **linear combination** of the vectors $x_1, x_2, \ldots, x_k$ in a vector space if there exist numbers (in general, complex) $\alpha_1, \alpha_2, \ldots, \alpha_k$ such that

$$x = \sum_{i=1}^{k} \alpha_i x_i \tag{10.1}$$

The vectors $x_1, x_2, \ldots, x_k$ are said to be **linearly dependent** (or form a dependent set) if there exist complex numbers $\alpha_1, \alpha_2, \ldots, \alpha_k$, not all zero, such that

$$\alpha_1 x_1 + \alpha_2 x_2 + \cdots + \alpha_k x_k = 0 \tag{10.2}$$

If Eq. (10.2) can be satisfied only for $\alpha_1 = \alpha_2 = \cdots = \alpha_k = 0$, the vectors are **linearly independent**.

One is tempted to use the infinite-sum version of Eq. (10.2) in defining the notion of independence for an infinite set of vectors. This is not true in general; one needs the notion of convergence, which is based on the concept of distance between vectors.

With the idea of linear independence firmly in mind, the concept of **dimension** of a vector space readily follows. A vector space is n *dimensional* if it possesses a set of n **independent vectors**, but every set of $n + 1$ vectors is a dependent set. If for every positive integer k, a set of k independent vectors in the space can be found, the space is said to be infinite dimensional. By a **basis** for a vector space is meant a finite set of vectors $e_1, e_2, \ldots, e_k$ with the following attributes:

1. They are linearly independent.
2. Every vector x in the space can be written as a linear combination of the basis vectors; that is

$$x = \sum_{i=1}^{k} \xi_i e_i \tag{10.3}$$

It can be proved that the representation (10.3) is unique and that if the space is n dimensional, any set of n independent vectors $e_1, e_2, \ldots, e_n$ forms a basis.

The next concept to be developed is that of a **metric space**. In addition to the addition of vectors and multiplication of vectors by scalars, as is true of ordinary three-dimensional vectors, it is important to have the notions of length and direction of a vector imposed. In other words, a metric structure must be added to the algebraic structure already defined. A collection of elements $x, y, z, \ldots$ in a space will be called a metric space if to each pair of elements x, y there corresponds a real number $d(x, y)$ satisfying the properties:

1. $d(x, y) = d(y, x)$
2. $d(x, y) \geq 0$ with equality if and only if $x = y$
3. $d(x, z) \leq d(x, y) + d(x, z)$ (called the **triangle inequality**)

The function $d(x, y)$ is called a **metric** (or distance function). Note that the elements in the space *need not be vectors*; there may not be any way of adding elements or multiplying them by scalars as required for a vector space.

With the definition of a metric, one can now discuss the idea of convergence of a sequence $\{x_k\}$ of elements in the space. Note that $d(x_k, x)$ is a sequence of real numbers. Therefore, it is sensible to write that

$$\lim_{k \to \infty} x_k = x \tag{10.4}$$

if the sequence of numbers $d(x_k, x)$ converges to 0 in the ordinary sense of convergence of sequences of real numbers. If

$$\lim_{m, p \to \infty} d(x_m, x_p) = 0 \tag{10.5}$$

the sequence $\{x_k\}$ is said to be a **Cauchy sequence**. It can be shown that if a sequence $\{x_k\}$ converges, it is a Cauchy sequence. The converse is not necessarily true, for the limit may have carelessly been excluded from the space. If the converse is true, then the metric space is said to be **complete**.

The next vector space concept to be defined is that of length or norm of a vector. A **normed vector space** (or linear space) is a vector space in which a real-valued function $\|x\|$ (known as the *norm* of x) is defined, with the properties

1. $\|x\| \geq 0$ with equality if and only if $x = 0$
2. $\|\alpha x\| = |\alpha| \|x\|$
3. $\|x_1 + x_2\| \leq \|x_1\| + \|x_2\|$

A normed vector space is automatically a metric space if the metric is defined as

$$d(x, y) = \|x - y\| = \langle x - y, x - y \rangle^{1/2} \tag{10.6}$$

(see definition below) which is called the natural metric for the space. A normed vector space may be viewed either as a linear space, a metric space, or both. Its elements may be interpreted as *vectors* or *points*.

The structure of a normed vector space will now be refined further with the definition of the notion of angle between two vectors. In particular, it will be possible to tell whether two vectors are perpendicular. The notion of angle between two vectors will be obtained by defining the **inner product** (also known as a scalar or dot product). In general, an inner product in a vector space is a complex-valued function of ordered pairs x, y with the properties

1. $\langle x, y \rangle = \langle y, x \rangle^*$ (the asterisk denotes complex conjugate)
2. $\langle \alpha x, y \rangle = \alpha \langle x, y \rangle$
3. $\langle x_1 + x_2, y \rangle = \langle x_1, y \rangle + \langle x_2, y \rangle$
4. $\langle x, x \rangle \geq 0$ with equality if and only if $x = 0$

From the first two properties, it follows that

$$\langle x, \alpha y \rangle = \alpha^* \langle x, y \rangle \tag{10.7}$$

Also, **Schwarz's inequality** can be proved and is given by

$$|\langle x, y \rangle| \leq \langle x, x \rangle^{1/2} \langle y, y \rangle^{1/2} \tag{10.8}$$

with equality if and only if $x = \alpha y$. The real, nonnegative quantity $\langle x, x \rangle^{1/2}$ satisfies all of the properties of a norm. Therefore, it is adopted as the definition of the norm, and Schwartz's inequality assumes the form

$$|\langle x, y \rangle| \leq \|x\| \|y\| \tag{10.9}$$

Signal Space

The natural metric in the space is given by Eq. (10.6). An inner product space, which is complete in its natural metric, is called a Hilbert space.

Example 10.1

Consider the space of all complex-valued functions $x(t)$ defined on $a \leq t \leq b$ for which the integral

$$E_x = \int_a^b |x(t)|^2 \, dt \qquad (10.10)$$

exists (i.e., the space of all finite-energy signals in the interval $[a, b]$). The inner product is defined as

$$\langle x, y \rangle = \int_a^b x(t) y^*(t) \, dt \qquad (10.11)$$

The natural norm is

$$\|x\| = \left[\int_a^b |x(t)|^2 \, dt \right]^{1/2} \qquad (10.12)$$

and the metric is

$$d(x, y) = \|x - y\| = \left[\int_a^b |x(t) - y(t)|^2 \, dt \right]^{1/2} \qquad (10.13)$$

respectively. Schwarz's inequality becomes

$$\left| \int_a^b x(t) y^*(t) \, dt \right| \leq \left[\int_a^b |x(t)|^2 \, dt \right]^{1/2} \left[\int_a^b |y(t)|^2 \, dt \right]^{1/2} \qquad (10.14)$$

It can be shown that this space is complete and, hence, is a Hilbert space.

An additional requirement that can be imposed on a Hilbert space is separability, which, roughly speaking, restricts the number of elements in the space. A Hilbert space $\mathcal{H}$ is **separable** if there exists a **countable** (i.e., can be put in one-to-one correspondence with the positive integers) set of elements $(f_1, f_2, \ldots, f_n, \ldots)$ whose finite linear combinations are such that for any element f in $\mathcal{H}$ there exist an index N and constants $\alpha_1, \alpha_2, \ldots, \alpha_N$ such that

$$\left\| f - \sum_{k=1}^N \alpha_k f_k \right\| < \epsilon \qquad (10.15)$$

The set $(f_1, f_2, \ldots, f_n, \ldots)$ is called a **spanning set**. The discussions from here on are limited to separable Hilbert spaces.

Any finite-dimensional Hilbert space E_n is separable. In fact, there exists a set of n vectors $(f_1, f_2, \ldots, f_n)$ such that each vector x in E_n has the representation

$$x = \sum_{k=1}^n \alpha_k f_k \qquad (10.16)$$

It can be shown that the spaces consisting of square-integrable functions on the intervals $[a, b]$, $(-\infty, b]$, $[a, \infty)$, and $(-\infty, \infty)$ are all separable, where a and b are finite, and spanning sets exist for each of these spaces. For example, a spanning set for the space of square-integrable functions on $[a, b]$ is the set $(1, t, t^2, \ldots)$, which is clearly countable.

The concepts of convergence and Cauchy sequences carry over to Hilbert spaces, with *convergence in the mean* defined as

$$\lim_{k \to \infty} \int_a^b |x_k(t) - x(t)|^2 = 0 \quad (10.17)$$

Similarly, the ideas of independence and basis sets apply to Hilbert spaces in infinite-dimensional form.

It is necessary to distinguish between the concepts of a *basis* and a *spanning* set consisting of independent vectors. As ε is reduced in Eq. (10.15), it is expected that N must be increased, and it may also be necessary to change the previously found coefficients $\alpha_1, \ldots, \alpha_N$. Hence, there might not exist a fixed sequence of constants $\xi_1, \xi_2, \ldots, \xi_n, \ldots$ with the property

$$x = \sum_{k=1}^{\infty} \xi_k f_k \quad (10.18)$$

as would be required if the set $\{f_k\}$ were a basis. For example, on the space of square-integrable functions on $[-1, 1]$, the independent set $f_0 = 1, f_1 = t, f_2 = t^2, \ldots$, is a spanning set, but not a basis, since there are many square-integrable functions on $[-1, 1]$ that cannot be expanded in a series, like Eq. (10.19) (an example is $|t|$). Odd as it may seem at first, it is possible if the powers of t in this spanning set are regrouped into the set of polynomials known as the Legendre polynomials, $P_k(t)$.

Two vectors x, y are **orthogonal** or perpendicular if $\langle x, y \rangle = 0$. A finite or countably infinite set of vectors $\{\phi_1, \phi_2, \ldots, \phi_k, \ldots\}$ is said to be an *orthogonal set* if $\langle \phi_i, \phi_j \rangle = 0, i \neq j$. A *proper orthogonal set* is an orthogonal set none of whose elements is the zero vector. A proper orthogonal set is an independent set. A set is **orthonormal** if

$$\langle \phi_i, \phi_j \rangle = \begin{cases} 0, & i \neq j \\ 1, & i = j \end{cases} \quad (10.19)$$

An important concept is that of a **linear manifold** in a Hilbert space. A set M is said to be a linear manifold if, for x and y belonging to M, so does $\alpha x + \beta y$ for arbitrary complex numbers α and β; thus, M is itself a linear space. If a linear manifold is a closed set, it is called a closed linear manifold (i.e., every Cauchy sequence has a limit in the space) and is itself a Hilbert space. In three-dimensional Euclidean space, linear manifolds are simply lines and planes containing the origin. In a finite-dimensional space, every linear manifold is necessarily closed.

Let M be a linear manifold, closed or not. Consider the set $M^\perp$ of all vectors which are orthogonal to every vector in M. It is a linear manifold, which can be shown to be closed. If M is closed, M and $M^\perp$ are known as **orthogonal complements**. Given a linear manifold M, each vector in the space can be decomposed in a unique manner, as a sum $x_p + z$, where x_p is in M and z is in $M^\perp$.

Given an infinite orthonormal set $\{\phi_1, \phi_2, \ldots, \phi_n, \ldots\}$ in the space of all square-integrable functions on the interval $[a, b]$, let $\{a_n\}$ be a sequence of complex numbers. The **Riesz–Fischer theorem** tells how to represent an element in the space and states:

1. If

$$\sum_{n=1}^{\infty} |a_n|^2 \quad (10.20)$$

Signal Space

diverges, then

$$\sum_{n=1}^{\infty} a_n \phi_n \tag{10.21}$$

diverges.

2. If Eq. (10.20) converges, then Eq. (10.21) also converges to some element g in the space and

$$a_n = \langle g, \phi_n \rangle \tag{10.22}$$

The next question that arises is how to construct an orthonormal set from an independent set $\{e_1, e_2, \ldots, e_k, \ldots\}$. A way to do this is known as the **Gram–Schmidt procedure**. The construction is as follows:

1. Pick a function from the set $\{e_k\}$, say e_1. Let

$$\phi_1 = \frac{e_1}{\langle e_1, e_1 \rangle^{\frac{1}{2}}} \tag{10.23}$$

2. Remove from a second function in the set $\{e_k\}$, say e_2, its projection on ϕ_1. This yields

$$g_2 = e_2 - \langle e_2, \phi_1 \rangle \phi_1 \tag{10.24}$$

The vector g_2 is a linear combination of e_1 and e_2, and is orthogonal to ϕ_2. To normalize it, form

$$\phi_2 = \frac{g_2}{\langle g_2, g_2 \rangle^{\frac{1}{2}}} \tag{10.25}$$

3. Pick another function from the set $\{e_k\}$, say e_3, and form

$$g_3 = e_3 - \langle e_3, \phi_2 \rangle \phi_2 - \langle e_3, \phi_1 \rangle \phi_1 \tag{10.26}$$

Normalize g_3 in a manner similar to that used for g_2.

4. Continue until all functions in the set $\{e_k\}$ have been used.

Note that the sets $\{e_k\}$, $\{g_k\}$, and $\{\phi_k\}$ all generate the same linear manifold.

A basis consisting of orthonormal vectors is known as an **orthonormal basis**. If the basis vectors are not normalized, it is simply an *orthogonal basis*.

Example 10.2

Consider the interval $[-1, 1]$ and the independent set $e_0 = 1$, $e_2 = t, \ldots, e_k = t^k, \ldots$. The Gram–Schmidt procedure applied to this set without normalization, but with the requirement that all orthogonal functions take on the value 1 at $t = 1$, gives the set of *Lengendre polynomials*, which is

$$\psi_0(t) = 1, \quad \psi_1(t) = t, \quad \psi_2(t) = \tfrac{1}{2}(3t^2 - 1), \quad \psi_3(t) = \tfrac{1}{2}(5t^3 - 3t), \ldots \tag{10.27}$$

It is next desired to approximate an arbitrary vector x in a Hilbert space in terms of a linear combination of the independent set $\{e_1, \ldots, e_k\}$, where $k \leq n$ if the space is an n-dimensional Euclidean space, and k is an arbitrary integer if the space is infinite dimensional. First, the orthonormal set $\{\phi_1, \ldots, \phi_k\}$ is constructed from $\{e_1, \ldots, e_k\}$. The unique, best approximation to x is the **Fourier sum**

$$\sum_{i=1}^{k} \langle x, \phi_i \rangle \phi_i \tag{10.28}$$

which is geometrically the projection of x onto the linear manifold generated by $\{\phi_1, \ldots, \phi_k\}$, or equivalently, the sum of the projections along the individual axes defined by $\phi_1, \phi_2, \ldots, \phi_k$. The square of the distance between x and its projection is

$$\left\| x - \sum_{i=1}^{k} \langle x, \phi_i \rangle \phi_i \right\|^2 = \|x\|^2 - \sum_{i=1}^{k} |\langle x, \phi_i \rangle|^2 \tag{10.29}$$

Since the left-hand side is nonnegative, Eq. (10.29) gives **Bessel's inequality**, which is

$$\|x\|^2 \geq \sum_{i=1}^{k} |\langle x, \phi_i \rangle|^2 \tag{10.30}$$

A convenient feature of the Fourier sum is the following: If another vector ϕ_{k+1} is added the orthonormal approximating set of vectors, the best approximation now becomes

$$\sum_{i=1}^{k+1} \langle x, \phi_i \rangle \phi_i \tag{10.31}$$

Thus, an additional term is added to the series expansion without changing previously computed coefficients, which makes the extension to a countably infinite orthonormal approximating set simple to envision. In the case of an infinite orthonormal approximating set, Bessel's inequality (10.30) now has an infinite limit on the sum. Does the approximating sum (10.31) converge to x? The answer is that convergence can be guaranteed only if the set $\{\phi_k\}$ is extensive enough, that is, if it is a *basis or a complete orthonormal set*. In such cases, Eq. (10.30) becomes an equality. In fact, a number of equivalent criteria can be stated to determine whether an orthonormal set $\{\phi_k\}$ is a basis or not [Stakgold, 1967]. These are:

1. In finite n-dimensional Euclidean space, $\{\phi_k\}$ has exactly n elements for completeness.
2. For every x in the space of square-integrable functions

$$x = \sum_{i} \langle x, \phi_i \rangle \phi_i \tag{10.32}$$

3. For every x in the space of square-integrable functions

$$\|x\|^2 = \sum_{i=1}^{k} |\langle x, \phi_i \rangle|^2 \tag{10.33}$$

(known as **Parseval's equality**).
4. The only x in the space of square-integrable functions for which all of the Fourier coefficients vanish is the 0 function.
5. There exists no function $\phi(t)$ in the space of square-integrable functions such that $\{\phi, \phi_1, \phi_2, \ldots, \phi_k, \ldots\}$ is an orthonormal set.

Signal Space

Examples of complete orthonormal sets of trigonometric functions over the interval $[0, T]$ are as follows: (1) The complex exponentials with frequencies equal to the harmonics of the fundamental frequency $\omega_0 = 2\pi/T$, or

$$\frac{1}{\sqrt{2\pi}}, \quad \frac{e^{j\omega_0 t}}{\sqrt{2\pi}}, \quad \frac{e^{-j\omega_0 t}}{\sqrt{2\pi}}, \quad \frac{e^{j2\omega_0 t}}{\sqrt{2\pi}}, \quad \frac{e^{-j2\omega_0 t}}{\sqrt{2\pi}}, \ldots \tag{10.34}$$

The factor $(2\pi)^{1/2}$ in the denominator is necessary to normalize the functions and is often not included in the definition of the complex exponential Fourier series. (2) The sines and cosines with frequencies equal to harmonics of the fundamental frequency $\omega_0 = 2\pi/T$, or

$$\frac{1}{\sqrt{2\pi}}, \quad \frac{(\cos \omega_0 t)}{\sqrt{\pi}}, \quad \frac{(\sin \omega_0 t)}{\sqrt{\pi}}, \quad \frac{(\cos 2\omega_0 t)}{\sqrt{2\pi}}, \quad \frac{(\sin 2\omega_0 t)}{\sqrt{2\pi}}, \ldots \tag{10.35}$$

Note that if any function is left out of these sets, the basis is incomplete.

10.3 Application of Signal Space Representation to Signal Detection

The M-ary signal detection problem is as follows: given M signals, $s_0(t), s_1(t), \ldots, s_M(t)$, defined over $0 \leq t \leq T$. One is chosen at random and sent each T-second interval through a channel that adds white, Gaussian noise of power spectral density $N_0/2$ to it. The challenge is to design a receiver that will decide which signal was sent through the channel during each T-second interval with minimum probability of making an error.

An approach to this problem, as expanded upon in greater detail by Wozencraft and Jacobs [1965, Chap. 4] and Ziemer and Tranter [1995, Chap. 9], is to construct a linear manifold, called the signal space, using the Gram–Schmidt procedure on the M signals. Suppose that this results in the orthonormal basis set $\{\phi_1, \phi_2, \ldots, \phi_K\}$ where $K \leq M$. The received signal plus noise is represented in this signal space as vectors with coordinates (note that they depend on the signal transmitted)

$$Z_{ij} = A_{ij} + N_j, \quad \begin{matrix} j = 1, 2, \ldots, K, \\ i = 1, 2, \ldots, M \end{matrix} \tag{10.36}$$

where

$$A_{ij} = \int_0^T s_i(t)\phi_j(t)\,dt \tag{10.37}$$

The numbers Z_{ij} are components of vectors referred to as the **signal vectors**, and the space of all signal vectors is called the **observation space**. An apparent problem with this approach is that not all possible noise waveforms added to the signal can be represented as vectors in this K-dimensional observation space. The part of the noise that is represented is

$$n_|(t) = \sum_{j=1}^{K} N_j \phi_j(t) \tag{10.38}$$

where

$$N_j = \int_0^T n(t)\phi_j(t)\,dt \tag{10.39}$$

In terms of Hilbert space terminology, Eq. (10.38) is the projection of the noise waveform onto the

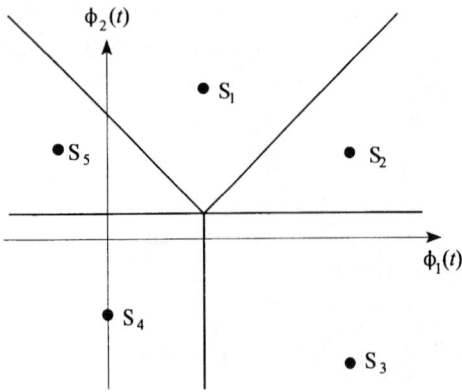

FIGURE 10.1 Observation space in two dimensions, showing five signal points with boundaries for decision regions providing the minimum distance decision in each case.

observation space (i.e., a linear manifold). The unrepresented part of the noise is

$$n_\perp(t) = n(t) - n_|(t) \qquad (10.40)$$

and is the part of the noise that must be represented in the orthogonal complement of the observation space. The question is, will the decision process be harmed by ignoring this part of the noise? It can be shown that $n_\perp(t)$ is uncorrelated with $n_|(t)$. Thus, they are statistically independent and $n_\perp(t)$ has no bearing on the decision process; nothing is lost by ignoring $n_\perp(t)$.

The decision process can be shown to reduce to choosing the signal $s_\ell(t)$, which minimizes the distance to the data vector; that is,

$$d(z, s_\ell) = \|z - s_\ell\| = \sum_{j=1}^{K}(Z_{ij} - A_{\ell j})^2 = \text{minimum}, \qquad \ell = 1, 2, \ldots, M \quad (10.41)$$

where

$$z(t) = \sum_{j=1}^{K} Z_{ij}\phi_j(t) \qquad (10.42)$$

and

$$s_\ell(t) = \sum_{j=1}^{K} A_{\ell j}\phi_j(t) \qquad (10.43)$$

Thus, the signal detection problem is reduced to a geometrical one, where the observation space is subdivided into decision regions in order to make a decision as shown in Fig. 10.1.

10.4 Application of Signal Space Representation to Parameter Estimation

The procedure used in applying signal space concepts to estimation is similar to that used for signal detection. Consider the observed waveform consisting of additive signal and noise of the form

$$y(t) = s(t, A) + n(t), \qquad 0 \le t \le T \qquad (10.44)$$

where A is a parameter to be estimated and the noise is white as before. Let $\{\phi_k(t)\}, k = 1, 2, \ldots$, be a complete orthonormal basis set. The observed waveform can be represented as

$$y(t) = \sum_{k=1}^{\infty} S_k(A)\phi_k(t) + \sum_{k=1}^{\infty} N_k\phi_k(t) \qquad (10.45)$$

where

$$S_k(A) = \langle s, \phi_k \rangle = \int_0^T s(t, A)\phi_k(t)\,dt \qquad (10.46)$$

Signal Space

and N_j is defined by Eq. (10.39). Hence, an estimate can be made on the basis of the set of coefficients

$$Z_k = S_k(A) + N_k \tag{10.47}$$

or on the basis of a vector in the signal space with these coordinates. A reasonable criterion for estimating A is to maximize the likelihood ratio, or a monotonic function thereof. Its logarithm can be shown to reduce to

$$\ell(A) = \lim_{K \to \infty} L_K(A) = \lim_{k \to \infty} \left[\frac{2}{N_0} \sum_{k=1}^{K} Z_k S_k(A) - \frac{1}{N_0} \sum_{k=1}^{K} S_k^2(A) \right] \tag{10.48}$$

In the limit as $k \to \infty$, this becomes

$$\ell(A) = \frac{2}{N_0} \int_0^T z(t) s(t, A) \, dt - \frac{1}{N_0} \int_0^T s^2(t, A) \, dt \tag{10.49}$$

A necessary condition for the value of A that maximizes Eq. (10.49) is

$$\frac{\partial \ell(A)}{\partial A} = \frac{2}{N_0} \int_0^T [z(t) - s(t, A)] \frac{\partial s(t, A)}{\partial A} \, dt \bigg|_{A=\hat{A}} = 0 \tag{10.50}$$

The value of A that maximizes Eq. (10.49), denoted $\hat{A}$, is called the **maximum likelihood estimate**.

Wavelet Transforms

Wavelet transforms can be continuous time or discrete time. They find applications in speech and image compression, signal and image classification, and pattern recognition. Wavelet representations have recently been adopted by the U.S. Federal Bureau of Investigation for fingerprint compression.

The continuous-time **wavelet transform** of a signal $x(t)$ takes the form [Rioul and Vetterli, 1991]

$$W_x(\tau, a) = \int_{-\infty}^{\infty} x(t) h_{a,\tau}^*(t) \, dt \tag{10.51}$$

where

$$h_{a,\tau}(t) = \frac{1}{\sqrt{a}} h\left(\frac{t - \tau}{a}\right) \tag{10.52}$$

are basis functions called **wavelets**. Thus, the wavelets defined in Eq. (10.52) are scaled and translated versions of the basic wavelet prototype $h(t)$ (also known as the *mother wavelet*), and the wavelet transform is seen to be a convolution of the conjugate of a wavelet with the signal $x(t)$. Substitution of Eq. (10.52) into Eq. (10.51) yields

$$W_x(\tau, a) = \frac{1}{\sqrt{a}} \int_{-\infty}^{\infty} x(t) h^*\left(\frac{t - \tau}{a}\right) dt \tag{10.53}$$

Note that $h(t/a)$ is contracted if $a < 1$ and expanded if $a > 1$. Thus, an interpretation of Eq. (10.53) is that as a increases, the function $h(t/a)$ becomes spread out over time and takes the long-term behavior of $x(t)$ into account; as a decreases the short-time behavior of $x(t)$ is taken into account. A change of variables in Eq. (10.53) gives

$$W_x(\tau, a) = \sqrt{a} \int_{-\infty}^{\infty} x(at) h^* \left(t - \frac{\tau}{a} \right) dt \tag{10.54}$$

Now the interpretation of Eq. (10.54) is as the scale increases ($a < 1$) an increasingly contracted version of the signal is seen through a constant-length sifting function, $h(t)$. This is only the barest of introductions to wavelets, and the reader is urged to consult the references to learn more about wavelets and their applications, particularly their discrete-time implementation.

Mean Square Estimation—the Orthogonality Principle

Given n random variables, $X_1, X_2, \ldots, X_n$, it is desired to find n constants $a_1, a_2, \ldots, a_n$ such that when another random variable S is estimated by the sum

$$\hat{S} = \sum_{i=1}^{n} a_i X_i \tag{10.55}$$

then the mean square value

$$\text{MSE} = E \left\{ \left| S - \sum_{i=1}^{n} a_i X_i \right|^2 \right\} \tag{10.56}$$

is a minimum, where $E\{\}$ denotes expectation or statistical average. It is shown in [Papoulis, 1984] that the **mean square estimate** (MSE) is minimized when the error is orthogonal to the data, or when

$$E \left\{ \left[S - \sum_{i=1}^{n} a_i X_i \right] X_j^* \right\} = 0, \qquad j = 1, 2, \ldots, n \tag{10.57}$$

This is known as the **orthogonality principle** or **projection theorem**, and can be interpreted as stating that the MSE is minimized when the error vector $S - \hat{S}$ is orthogonal to the subspace (linear manifold) spanned by the vectors $X_1, X_2, \ldots, X_n$ as shown in Fig. 10.2. The projection theorem has many applications including MSE filtering of noisy signals, known as Wiener filtering.

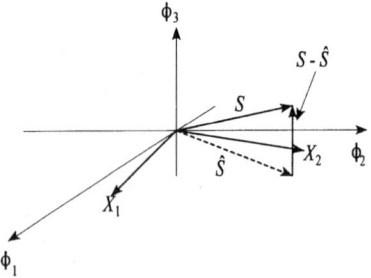

FIGURE 10.2 Schematic representation of mean square estimation in three dimensions with the estimate expressed in terms of a linear combination of two random variables, X_1 and X_2, which lie in the ϕ_1–ϕ_2 plane.

Defining Terms

Basis (basis functions): An independent set of vectors in a linear space in terms of which any vector in the space can be represented as a linear combination.

Bessel's inequality: The statement that the norm squared of any vector in a vector space is greater or equal to the sum of the squares of the projections onto a set of orthonormal basis vectors.

Cauchy sequence: A convergent sequence whose members become progressively closer as the limit is approached.

Signal Space

Complete: The idea that a basis set is extensive enough to represent any vector in the space as a Fourier sum.

Countable: A set that can be put in one-to-one correspondence with the positive integers.

Dimension: A vector space is n dimensional if it possesses a set of n independent vectors, but every set of $n + 1$ vectors is linearly dependent.

Fourier sum: An approximation of a vector in a Hilbert space as a linear combination of orthonormal basis functions where the coefficient of each basis function is the projection of the vector onto the respective basis function.

Gram–Schmidt procedure: An algorithm that produces a set of orthonormal vectors from an arbitrary set of vectors.

Hilbert space: A normed linear space complete in its natural metric.

Identity element: An element in a vector space that when multiplied times any vector reproduces that vector.

Independent vectors: A set of vectors is independent if any one of them cannot be expressed as a linear combination of the others.

Inner product: A function of ordered pairs of vectors in a vector space, which is analogous to the dot product in ordinary Euclidean vector space.

Linear combination: A linear sum of a set of vectors in a vector space with each member in the sum, in general, multiplied by a different scalar.

Linear manifold: A subspace of a Hilbert space that contains the origin.

Linearly dependent (dependent): A set of vectors that is not linearly independent.

Linearly independent (independent): A set of vectors no one of which can be expressed as a linear combination of the others.

Maximum likelihood estimate: An estimate for a parameter that maximizes a likelihood function.

Mean square estimation: Estimation of a random variable by a linear combination of n other random variables that minimizes the expectation of the squared difference between the random variable to be estimated and the approximating linear combination.

Metric: A real-valued function of two vectors in a vector space that is analogous to the distance between them in the ordinary Euclidean sense.

Metric space: A vector space in which a metric is defined.

Norm: A function of a single vector in a vector space that is analogous to length of a vector in ordinary Euclidean vector space.

Normed vector space: A vector space in which a norm has been defined.

Observation space: The space of all possible received data vectors in a signal detection or estimation problem.

Orthogonal: The property of two vectors expressed by their inner product being zero.

Orthogonal complement: The space of vectors that are orthogonal to a linear manifold or subspace.

Orthogonality principle: The theorem that states that the minimum mean-square estimate of a random variable in terms of a linear combination of n other random variables requires the difference between the random variable and linear combination, or error, to be statistically orthogonal to each random variable in the linear combination (i.e., the expectation of the product of the error and each random variable is zero). Also called the *projection theorem*.

Orthonormal: Orthogonal vectors that have unit norm.

Orthonormal basis: A basis set for which the basis vectors are orthonormal.

Parseval's equality: The statement that the norm squared of a vector in a complete Hilbert space equals the sum of the squares of the projections onto an orthonormal basis set in the space.

Projection theorem: See **orthogonality principle**.

Riesz–Fischer theorem: A theorem that states the conditions under which an element of a space of square-integrable functions can be represented in terms of an infinite orthonormal set.

Schwarz's inequality: An inequality expressing the fact that the absolute value of the inner product of any pair of vectors in a Hilbert space is less than or equal to the product of their respective norms.

Separable: A Hilbert space in which a countable set of elements exists that can be used to represent any element in the space to any degree of accuracy desired as a linear combination of the members of the set.

Signal vector: A vector representing a received signal in a signal detection or estimation problem.

Spanning set: The set of elements of a separable Hilbert space used to represent an arbitrary element with any degree of accuracy desired.

Triangle inequality: An inequality of a normed linear space that is analogous to the fact that length of the sum of any two sides of a triangle is less than or equal to the sum of their respective lengths.

Vector: An element of a linear, or vector, space.

Vector space: A space of elements, called vectors, which obey certain laws of associativity and commutativity, and have identity elements for scalar multiplication and element addition defined.

Wavelet: See wavelet transform.

Wavelet transform: Resolution of a signal into a set of basis functions called wavelets. As a parameter of the wavelet is changed, the behavior of the signal over progressively shorter time intervals is resolved.

References

Arthurs, E. and Dym, H. 1962. On the optimum detection of digital signals in the presence of white gaussian noise—A geometric approach and a study of three basic data transmission systems. *IRE Trans. Commun. Syst.*, CS-10 (Dec.):336–372.

Frederick, D.K. and Carlson, A.B. 1971. *Linear Systems in Communications and Control*, Wiley, New York.

Kotel'nikov, V.A. 1968. *The Theory of Optimum Noise Immunity* (trans. R.A. Silverman), Dover, New York.

Papoulis, A. 1984. *Probability, Random Variables, and Stochastic Processes*, 2nd ed., McGraw–Hill, New York.

Rioul, O. and Vetterli, M. 1991. Wavelets in signal processing. *IEEE Signal Proc. Mag.*, 8(Oct.):14–38.

Stakgold, I. 1967. *Boundary Value Problems of Mathematical Physics*, Vol. 1, Macmillan, Collier–Macmillan Ltd., London.

Wozencraft, J.M. and Jacobs, I.M. 1965. *Principles of Communication Engineering*, Wiley, New York.

Ziemer, R.E., Tranter, W.H., and Fannin, D.R. 1993. *Signals and Systems: Continuous and Discrete*, 3rd ed., Macmillan, New York.

Ziemer, R.E. and Tranter, W.H. 1995. *Principles of Communications: Systems, Modulation, and Noise*, 4th ed., Houghton Mifflin, Boston, MA.

Further Information

Very readable expositions of signal space concepts as applied to signal detection and estimation are found in Chapter 4 of the, by now, classic book by Wozencraft and Jacobs cited in the references. The paper by Arthurs and Dym listed in the references is also very readable. A treatment of signal space concepts and applications to signal detection and parameter estimation is given in Chapter 9 of the book by Ziemer and Tranter listed in the references. For the mathematical theory behind signal space concepts, the book by Stakgold, also listed in the references, is recommended. For those interested in wavelet transforms and their relationship to signal spaces, the April 1992 issue of the *IEEE Signal Processing Magazine* provides a tutorial article on wavelet transforms.

11
Channel Models

David R. Smith
George Washington University

11.1 Introduction 131
11.2 Fading Dispersive Channel Model 132
    Frequency-Selective Channel • Time-Selective Channel • Time- and Frequency-Selective Channel • Nonselective Channel
11.3 Line-of-Sight Channel Models 136
11.4 Digital Channel Models 138

11.1 Introduction

Early channel models used in conjunction with analysis of communication systems assumed a binary symmetric channel (BSC) [Wozencraft and Jacobs, 1967]. In this case, the probability of bit error is assumed constant. This channel is memoryless, that is, the presence or absence of an error for a particular symbol has no influence on past or future symbols. Few transmission paths can be characterized as a BSC; most channels of interest are subject to some degree of fading and dispersion which introduce memory (correlation) between symbol errors. Therefore, it is necessary that a fading dispersive channel model be described to allow complete characterization of communication systems for typical applications.

Modeling of the fading dispersive channel is based on the use of mathematics to describe the physical or observed properties of the channel. Stein, Schwartz, and Bennett [1966] and Kennedy [1969] first used a deterministic characterization of the channel, and then introduced dynamics into the model to account for the time-varying nature of a fading channel. Bello [1963] described a channel model using a tapped delay line representation, which is based on knowledge of the correlation properties of the channel.

The fading dispersive channel is described in Sec. 11.2 as a linear time-varying filter. Mobile and scatter channels produce a received signal consisting of the sum of multiple time-variant paths. Such channels are assumed to impart zero mean Gaussian statistics on a transmitted signal. Characterization of the channel then reduces to the specification of a correlation function or power spectral density. Many radio channels are adequately modeled by making these assumptions, including tropospheric scatter, high-frequency (HF) skywave, VHF/UHF mobile channels, and lunar reflection channels. Line-of-sight (LOS) channels, however, consist of a direct ray plus multiple indirect rays. The LOS channel, therefore, is assumed to impart nonzero mean Gaussian statistics on the transmitted signal. Modeling of the LOS channel can be somewhat simplified as compared to the general fading dispersive channel model, as shown in Sec. 11.3.

Digital communication channels introduce errors in bursts, which invalidates the use of the BSC. To characterize error clustering effects, various error models have been proposed, which can be divided into two basic categories, *descriptive* and *generative*. Descriptive models attempt to characterize channels by use of statistics that describe the general behavior of the channel. The Neyman type A distribution [Neyman, 1939] is an example of a descriptive model; here channel errors are modeled as a compound Poisson distribution in which error clusters have a Poisson distribution and errors within a cluster also have a Poisson distribution. Such models have been shown to apply to channels that have relatively few error-causing mechanisms, such as cable transmission [Becam et al., 1984], but have not been effectively used with more complex channels such as digital radio. Generative models, however, are able to match channel data by iteratively expanding the size of the model to fit the data. The generative model allows error statistics to be generated from a model made up of a finite-state Markov chain, using the states and transitions to represent the error behavior in the communications channel. These error models are usually developed to represent empirical results, but extensions are often made to more general cases.

11.2 Fading Dispersive Channel Model

Fading arises from destructive interference among multiple propagation paths, hence the term *multipath fading*. These paths arise due to reflection, refraction, or diffraction in the channel. The amplitude and phase of each path vary in time due to changes in the structure of the medium. The resultant signal at a receiver will experience fading, defined as changes in the received signal level in time, where resulting destructive or constructive interference depends on the relative amplitudes and phases of the multiple paths. The received signal may also experience dispersion, defined as spreading of the signal in time or frequency. Channels may be characterized as having multiple, randomly behaving paths or as having a dominant path in addition to multiple secondary paths. Forward scatter and mobile channels tend to have a large number of indirect paths, whereas line-of-sight channels tend to have a direct path that predominates over a few indirect paths.

The fading dispersive channel model may be best presented by using complex representation of the transmitted and received waveforms. Let us define $s(t)$ to be the transmitted signal and $y(t)$ to be the received signal. Assuming both $s(t)$ and $y(t)$ are narrowband signals, we can represent them as

$$s(t) = \text{Re}[z(t)e^{j2\pi f_c t}] \tag{11.1}$$

$$y(t) = \text{Re}[w(t)e^{j2\pi f_c t}] \tag{11.2}$$

where f_c is the carrier frequency and $z(t)$ and $w(t)$ are the complex envelopes of $s(t)$ and $y(t)$, respectively. The complex envelope notation carries information about the amplitude and phase of the signal waveforms. The magnitude of the complex envelope is the conventional envelope of the signal, and the angle of the complex envelope is the phase of the signal with respect to the carrier frequency f_c.

The first assumption made in characterizing the fading dispersive channel is that a large number of paths exist, and so the central limit theorem can be applied. We then can represent the channel impulse response as a complex Gaussian process, $g(t, \xi)$, which displays the time-varying nature of the channel. Assuming a linear channel

$$w(t) = \int_{-\infty}^{\infty} z(t - \xi) g(t, \xi) \, d\xi \tag{11.3}$$

We also need to define the time-varying transfer function of the channel, which is the Fourier transform dual of the channel impulse response

$$G(f, t) = \int_{-\infty}^{\infty} g(t, \xi) e^{-j2\pi f \xi} \, d\xi \tag{11.4}$$

Channel Models

In this section, we will further assume that no direct or dominant path is present in the channel and that the scatterers behave as a random process, so that the statistics are Gaussian with zero mean. The received envelope then displays Rayleigh law statistics. Mobile and scatter communication systems are examples of such channels.

The final assumption required for characterization of the channel model involves the stationarity of the channel impulse response. Although the fluctuations in the channel are due to non-stationary statistical phenomena, on a short enough time scale and for small enough bandwidth, the fluctuations in time and frequency can be characterized as approximately stationary. This approximate stationarity is often called quasistationarity. Since our interest here is in short-term fading, it is reasonable to assume that $g(t, \xi)$ is stationary in a time sense. Channels displaying stationarity in time are called *wide-sense stationary* (WSS). For most scatter and mobile channels, the channel may be modeled as a continuum of uncorrelated scatters. If the channel impulse response $g(t, \xi)$ is independent for different values of delay (ξ), the channel is said to exhibit *uncorrelated scattering* (US). When the time-varying impulse response is assumed to have stationary fluctuations in time and frequency, the channel is said to be wide-sense stationary with uncorrelated scattering (WSSUS).

A WSSUS complex Gaussian process is completely determined statistically by specifying its autocorrelation function or its power spectral density. Because of the dual (time and frequency) nature of the channel, there exist several correlation functions and power spectral densities that are used to characterize the channel, and the most applicable of these are defined and briefly described as follows:

1. The *time-frequency correlation function* (sometimes called the two-frequency correlation function) is defined as

$$R(\Omega, \tau) = E[G^*(f, t)G(f + \Omega, t + \tau)] \qquad (11.5)$$

   Note that the complex character of $G(f, t)$ requires a slightly altered definition of autocorrelation, with the asterisk denoting the complex conjugate operation. $R(\Omega, \tau)$ represents the cross-correlation function between the complex envelopes of received carriers transmitted Ω Hz apart. Since the transfer function is assumed stationary with uncorrelated scattering, $R(\Omega, \tau)$ is dependent only on the frequency separation Ω and time delay τ

2. The *tap gain correlation function* (sometimes called the path gain correlation function) is defined as

$$Q(\tau, \xi)\delta(\Delta\xi) = E[g^*(t, \xi)g(t + \tau, \xi + \Delta\xi)] \qquad (11.6)$$

   where $\delta(\bullet)$ is the unit impulse function and $Q(\tau, \xi)$ is the Fourier transform of $R(\Omega, \tau)$ on the Ω variable. $Q(\tau, \xi)$ represents the autocorrelation function of the gain fluctuations for paths providing delays in the interval $(\xi, \xi + \Delta\xi)$. The form of Eq. (11.6) implies that the fluctuations of the complex gains at different delays are uncorrelated, which is a result of the uncorrelated scattering assumption.

3. The *scattering function* is defined as the power spectrum of the complex gain fluctuations at delay ξ and is obtained from $Q(\tau, \xi)$ by applying the Fourier transform on the τ variable,

$$S(\xi, \nu) = \int_{-\infty}^{\infty} Q(\tau, \xi)e^{-j2\pi\xi\nu}\, d\tau \qquad (11.7)$$

   The scattering function directly exhibits the delay and Doppler spreading characteristics of the dispersive channel. The power spectral density associated with values of time delay ξ and Doppler frequency ν in this interval ξ to $\xi + \Delta\xi$ and ν to $\nu + \Delta\nu$, respectively, is then $S(\xi, \nu)\Delta\xi\Delta\nu$. The scattering function is also seen to be the double Fourier transform of the time-frequency correlation function.

Special cases of the described channel model that are of interest include the frequency-selective channel, time-selective channel, time- and frequency-selective channel, and nonselective channel. These cases are described by first showing how the general model is simplified and then showing the physical conditions for which these special cases arise.

Frequency-Selective Channel

The frequency-selective channel is characterized by a transfer function that varies in a more or less random fashion with frequency. An approximately constant amplitude and phase can only be observed over a frequency interval sufficiently small. The fading, which occurs at each frequency, will be very nearly the same, that is, highly correlated, for closely spaced frequencies and essentially independent, that is, no correlation for frequencies sufficiently far apart. The lack of correlation in fading of spaced frequencies is thus called **frequency-selective fading**.

A fading channel which is selective only in frequency is then implicitly also a nontime selective or *time-flat* channel. This time-flat channel is characterized by a channel impulse response that is time invariant. When the fading rate is so slow that little change in the channel impulse response occurs over the duration of a transmitted pulse, that channel may be approximated as a time-flat channel. Assuming the channel impulse response to be time invariant, the input–output relationship (11.3) may be simplified to the usual convolution integral

$$w(t) = \int_{-\infty}^{\infty} z(t - \xi) g(\xi) \, d\xi \qquad (11.8)$$

in which $g(\xi)$ is now the time-invariant channel impulse response. With the time-invariance assumption, the complex correlation function of $g(\xi)$ becomes

$$E[g^*(\xi) g(\xi + \Delta\xi)] = Q(\xi) \delta(\Delta\xi) \qquad (11.9)$$

where the function

$$Q(\xi) = Q(O, \xi) \qquad (11.10)$$

is called the *delay power spectrum*. The *frequency correlation function*, which is the Fourier transform of the delay power spectrum, is given by

$$q(\Omega) = R(\Omega, \xi) = E[G^*(f) G(f + \Omega)] \qquad (11.11)$$

where $G(f)$ is the time-invariant channel transfer function. A sufficient characterization of this channel is provided by either the delay power spectrum $Q(\xi)$ or the frequency correlation function $q(\Omega)$.

The frequency correlation function $q(\Omega)$ would be determined experimentally by transmitting two sine waves of different frequency and measuring the correlation between the complex envelopes as a function of frequency separation of the sine waves. When the frequency separation Ω is such that the correlation function $q(\Omega)$ is very near the maximum value $q(O)$ for all $\Omega < B_c$, all transmitted frequencies less than B_c will be received fading in a highly correlated fashion. For this reason B_c is called the **coherence bandwidth**, where B_c has been defined as the frequency separation for which $q(\Omega)$ equals $1/e$, or alternatively, $1/2$.

The delay power spectrum $Q(\xi)$ is the average power at time delay ξ in the multipath structure of the channel. One may define a **multipath** or **delay spread** parameter L as the width of $Q(\xi)$. Two measures of L that occur frequently are *total* and 2σ. The total delay spread L_T is meant to define the spread of $Q(\xi)$ for values of ξ where $Q(\xi)$ is significantly different from zero. The 2σ

Channel Models

delay spread $L_{2\sigma}$ is defined as twice the standard deviation of $Q(\xi)$ when it has been normalized to unit area and regarded as a probability density function. Note that the coherence bandwidth and delay spread are reciprocals of one another, due to the Fourier transform relationship between $q(\Omega)$ and $Q(\xi)$.

If we assume that the received signal consists of scattered components ranging over a delay interval L, then the received signal corresponding to a transmitted symbol of duration T will be spread over $T + L$. As long as $L > T$, the received signal is spread in time, and the corresponding channel is said to be *time dispersive*. In a digital communications system, the effect of time dispersion is to cause intersymbol interference and a corresponding degradation in probability of symbol error.

A channel exhibits frequency selectivity when the transmitted bandwidth of the signal is sufficiently larger than the coherence bandwidth of the channel. Stated another way, frequency selectivity is observed if the delay spread L of the channel is sufficiently larger than the transmitted signal duration T. This channel is simultaneously time flat since the duration of a transmitted pulse is sufficiently small compared to the fading correlation time. For example, a high-speed troposcatter channel (in the range of megabits per second) often exhibits frequency selectivity since the transmitted bandwidth is larger than the channel coherence bandwidth.

Time-Selective Channel

The time-selective channel has the property that the impulse response is time variant. An approximately constant amplitude and phase can only be observed over a time interval sufficiently small. The fading which occurs at some frequency will be very nearly the same, that is, highly correlated, for closely spaced times, and essentially independent for times sufficiently far apart. The lack of correlation in fading of a given frequency over time intervals is thus called **time-selective fading**.

A fading channel that is selective only in time is then implicitly also a nonfrequency selective or *frequency-flat* channel. Such a channel has the property that the amplitude and phase fluctuations observed on the channel response to a given frequency are the same for any frequency. The frequency-flat fading channel is observed when the signal bandwidth is smaller than the correlation bandwidth of the channel. Low-speed data transmission over an HF channel (in the range of kilobits per second) is an example of time selectivity wherein the transmitted pulse width is greater than the average fade duration.

For the frequency-flat fading channel the channel impulse response is independent of the frequency variable, so that the input–output relationship (11.3) becomes

$$w(t) = z(t)g(t) \tag{11.12}$$

in which $g(t)$ is the time-varying, frequency-invariant, impulse response. The complex correlation function of $g(t)$ then becomes

$$E[g^*(t)g(t+\tau)] = p(\tau) = R(0, \tau) \tag{11.13}$$

where the function $p(\tau)$ is the *time (fading) correlation function*. A sufficient characterization of this channel is provided by either the time correlation function $p(\tau)$, or its Fourier transform $P(\nu)$, the *Doppler power spectrum*.

To make an experimental determination of $p(\tau)$, one could simply transmit a sine wave and evaluate the autocorrelation function of the received process. Here one may define a **coherence time** parameter (also called *fading time constant*) T_c in terms of $p(\tau)$ in the same way as B_c was defined in terms of $q(\Omega)$. The channel parameter T_c is then a measure of the average fade duration and is particularly useful in predicting the occurrence of time-selective fading in the channel, just as B_c is useful in predicting the occurrence of frequency-selective fading.

The Fourier transform of the time-correlation function is the Doppler power spectrum $P(v)$, which yields signal power as a function of Doppler shift. The total **Doppler spread** B_T and 2σ Doppler spread $B_{2\sigma}$ parameters are defined analogous to delay spread L_T and $L_{2\sigma}$, respectively, and have analogous utilities. Note that the coherence time and Doppler spread are reciprocals of one another, due to the Fourier transform relationship between $p(\tau)$ and $P(v)$. If the Doppler spread B is greater than the transmitted signal bandwidth W, then the received signal will be spread over $W + B$. For the case $B > W$, the received signal is spread in frequency and the corresponding channel is said to be *frequency dispersive*.

Observation of Doppler effects is evidence that the channel is behaving as a time-varying filter. These time variations in the channel can be related directly to motion in the medium or motion by the communications device. In the case of a troposcatter channel, fluctuations caused by motion in the scatterers within the medium cause Doppler effects. Motion of the user or vehicle in a mobile communications channel likewise gives rise to Doppler effects.

Time- and Frequency-Selective Channel

Channels that exhibit both time and frequency selectivity are called doubly dispersive. Such channels are neither time flat nor frequency flat. In general, real channels do not exhibit time and frequency selectivity simultaneously. To simultaneously exhibit time and frequency selectivity, the coherence bandwidth must simultaneously be both larger and smaller than the transmitted bandwidth, or alternatively, the multipath spread must simultaneously be both larger and smaller than the signal duration. In view of this contradiction, we assume that either time or frequency selectivity effects can occur with this channel but not simultaneously. For this channel we then can make the often used approximation

$$S(\xi, v) = P(v)Q(\xi) \tag{11.14}$$

which thus separates the frequency and time selectivity effects on the channel. Equation (11.14) points out the fact that time and frequency selectivity are independent phenomena.

Nonselective Channel

The nonselective fading channel exhibits neither frequency nor time selectivity. This channel is also called *time flat, frequency flat* or just flat-flat. This channel is then the random-phase Rayleigh fading channel. For a transmitted signal given by Eq. (11.1), the received signal is

$$y(t) = A \operatorname{Re}\left[z(t)e^{j(2\pi f_c t - \theta)}\right] \tag{11.15}$$

where A and θ are statistically independent random variables that are time invariant, the former being Rayleigh distributed and the latter being uniformly distributed in the interval $-\pi$ to π.

The flat-flat fading model can closely approximate the behavior of a channel if the bandwidth of the transmitted signal is much less than the correlation bandwidth of the channel and if the transmitted pulse width is very much less than the correlation time of the fading. However, these two requirements are usually incompatible since the time-bandwidth product of the transmitted signal has a lower bound.

11.3 Line-of-Sight Channel Models

Multipath channel models for line-of-sight (LOS) radio systems assume a direct path in addition to one or more secondary paths observed at the receiver. The channel impulse response is modeled as a nonzero mean, complex Gaussian process, where the envelope has a Ricean distribution and the

Channel Models

channel is known as a *Ricean fading channel.* Both multipath and polynomial transfer functions have been used to model the LOS channel.

Several multipath transfer function models have been developed, usually based on the presence of two [Jakes, 1978] or three [Rummler, 1979] rays. In general, the multipath channel transfer function can be written as

$$H(\omega) = 1 + \sum_{i=1}^{n} \beta_i e^{j\omega\tau_i} \quad (11.16)$$

where the direct ray has been normalized to unity and the β_i and τ_i are amplitude and delay of the interfering rays relative to the direct ray. The two-ray model can thus be characterized by two parameters, β and τ. In this case, the amplitude of the resultant signal is

$$R = (1 + \beta^2 + 2\beta \cos \omega\tau)^{1/2} \quad (11.17)$$

and the phase of the resultant is

$$\phi = \arctan\left(\frac{\beta \sin \omega\tau}{1 + \beta \cos \omega\tau}\right) \quad (11.18)$$

The group delay is then

$$T(\omega) = \frac{d\phi}{d\omega} = \beta\tau \left(\frac{\beta + \cos \omega\tau}{1 + 2\beta \cos \omega\tau + \beta^2}\right) \quad (11.19)$$

The deepest fade occurs with

$$\omega_d \tau = \pi(2n - 1) \quad (n = 1, 2, 3, \ldots) \quad (11.20)$$

where both R and T are at a minimum, with

$$R_{\min} = 1 - \beta \quad (11.21)$$

$$T_{\min} = \frac{\beta\tau}{1 - \beta} \quad (11.22)$$

The frequency defined by ω_d is known as the notch frequency and is related to the carrier frequency ω_c by

$$\omega_d = \omega_c + \omega_0 \quad (11.23)$$

where ω_0 is referred to as the offset frequency.

Although the two-ray model described here is easy to understand and apply, most multipath propagation research points toward the presence of three (or more) rays during fading conditions. Out of this research, Rummler's three-ray model is the most widely accepted [Rummler, 1979].

Experiments conducted to develop or verify multipath channel models are limited by the narrowband channels available in existing radio systems. These limited channel bandwidths are generally insufficient to resolve parameters of a two- or three-ray multipath model. One alternative is to use polynomial functions of frequency to analytically fit measured data. Such polynomials have been used to describe amplitude and group delay distortion. Curve fits can be made to individual records of amplitude and group delay distortion using an Mth-order polynomial,

$$p(\omega) = C_0 + C_1\omega + C_2\omega^2 + \cdots + C_M\omega^M \quad (11.24)$$

To obtain the coefficients $\{C_1, C_2, \ldots, C_M\}$, a least squares fit has been used with polynomials typically of order $M = 2, 4$, and 6. For fades that are frequency selective, the most suitable order has been found to be $M = 4$. During periods of nonfading or flat fading, polynomials of order $M = 0$ or 2 have provided acceptable accuracy [Smith and Cormack, 1982]. As shown by Greenstein and Czekaj [1980] these polynomial coefficients can be related to a power series model of the channel transfer function

$$H(\omega) = A_0 + \sum_{k=1}^{n}(A_k + jB_k)(j\omega)^k \qquad (11.25)$$

Studies have shown that at least a first-order polynomial, with three coefficients $\{A_0, A_1,$ and $B_1\}$, is required for LOS channels, and that a second-order polynomial with five coefficients may be required for highly dispersive channels [Greenstein and Czekaj, 1980].

11.4 Digital Channel Models

For digital communications channels, one must be able to relate probability of (symbol, bit) error to the behavior of real channels, which tend to introduce errors in bursts. Two approaches may be taken. The classical method relates signal level variation to error rate statistics. Here, an equation for probability of error for a particular modulation technique is used to develop error statistics such as outage probability, outage rate, and outage duration for a given probability density function of signal level representing a certain channel. The drawback to this approach is the limitation of the results to only classical channels, such as Gaussian or Rayleigh, for which the signal statistics are known. As shown earlier, real channels tend to exhibit complex and time-varying behavior not easily represented by known statistics. For example, radio channels may experience time-selective fading, frequency-selective fading, rain attenuation, interference, and other forms of degradation at one time or another.

The second approach uses error models for channels with memory, that is, for channels where errors occur in bursts. If based on actual channel data, error models have the advantage of exactly reflecting the vagaries of the channel and, depending on the model, allowing various error statistics to be represented as a function of the model's parameters. The chief problem with error modeling is the selection of the model and its parameters to provide a good fit to the data and to the channel(s) in general. The Gilbert model focused on the use of a Markov chain composed of a good state G, which is error free, and a bad state B, with a certain probability of error [Gilbert, 1960]. Subsequent applications used a larger number of states in the Markov chain to better simulate error distributions in real channels. Fritchman [1967] first introduced the simple partitioned Markov chain with k error-free states and $N - k$ error states, which has shown good agreement with experimental results [Knowles and Drukarev, 1988; Semmar et al., 1991].

The single-error partitioned Fritchman model, in which the N states are divided into $N - 1$ error-free states and one error state, with transition probabilities $P_{ij}, i, j = 1$ to N, has been popularly used to characterize sequences of error-free and errored events (e.g., bits or symbols), known as gaps and clusters, respectively. For a binary error sequence where the presence or absence of a bit error is represented by 1 and 0, the error gap distribution $P(0^m \mid 1)$ is the probability that at least m error-free bits will follow given that an errored bit has occurred. Similarly, the error cluster distribution $P(1^m \mid 0)$ is the probability that at least m errored bits will follow given that an error-free bit has occurred. For the single-error Fritchman model [1967],

$$P(0^m \mid 1) = \sum_{i=1}^{k} \frac{P_{N,i}}{P_{i,i}}(P_{i,i})^m \qquad (11.26)$$

or

$$P(0^m \mid 1) = \sum_{i=1}^{k} \alpha_i \beta_i^m \qquad (11.27)$$

where the values α_i or β_i are the Fritchman model parameters. For $i = 1$ to k and $N - k = 1$, the transition probabilities are related to Fritchman's parameters by

$$P_{i,i} = \beta_i \qquad (11.28)$$

$$P_{i,N} = 1 - \beta_i \qquad (11.29)$$

$$P_{N,i} = \alpha_i \beta_i \qquad (11.30)$$

$$P_{N,N} = 1 - \sum_{i=1}^{k} \alpha_i \beta_i \qquad (11.31)$$

After obtaining $P(0^m \mid 1)$ empirically for a given channel, curve fitting techniques can be employed to represent $P(0^m \mid 1)$ as the sum of $(N - 1)$ exponentials, with coefficients α_i and β_i. Even simpler, for the single error state model, the error cluster distribution can be described with a single exponential with suitable coefficients α_N and β_N. Fritchman [1967] and others have found that a model with $k = 2$ or 3 exponential functions is generally sufficient. For high error rate sequences, a model with up to four exponential functions is required [Semmar et al., 1991].

Defining Terms

Coherence bandwidth: Bandwidth of a received signal for which fading is highly correlated.
Coherence time: Time interval of a received signal for which fading is highly correlated.
Doppler spread: Range of frequency over which the Doppler power spectrum is significantly different from zero.
Frequency-selective fading: Fading with high correlation for closely spaced frequencies and no correlation for widely spaced frequencies.
Multipath spread or delay spread: Range of time over which the delay power spectrum is significantly different from zero.
Time-selective fading: Fading with high correlation for closely spaced times and no correlation for widely spaced times.

References

Becam, D., Brigant, P., Cohen, R., and Szpirglas, J. 1984. Testing Neyman's model for error performance of 2 and 140 Mb/s line sections. In *1984 International Conference on Communications*, pp. 1362–1365.
Bello, P.A. 1963. Characterization of randomly time-variant linear channels. *IEEE Trans. Commun.*, CS-11(Dec.):360–393.
Fritchman, B.D. 1967. A binary channel characterization using partitioned Markov chains. *IEEE Trans. Commun.*, IT-13(April):221–227.
Gilbert, E.N. 1960. Capacity of a burst noise channel. *Bell Syst. Tech. J.*, 39(Sept.):1253–1265.
Greenstein, L.C. and Czekaj, B.A. 1980. A polynomial model for multipath fading channel responses. *Bell Syst. Tech. J.*, 59(Sept.):1197–1205.
Jakes, W.C. 1978. An approximate method to estimate an upper bound on the effect of multipath delay distortion on digital transmission. In *1978 International Conference on Communications*, pp. 47.1.1–47.1.5.

Kennedy, R.S. 1969. *Fading Dispersive Communication Channels*, Wiley, New York.

Knowles, M.D. and Drukarev, A.I. 1988. Bit error rate estimation for channels with memory. *IEEE Trans. Commun.*, COM-36(June):767–769.

Neyman, J. 1939. On a new class of contagious distribution, applicable in entomology and bacteriology. *Ann. Math. Stat.*, 10:35–57.

Rummler, W.D. 1979. A new selective fading model: Application to propagation data. *Bell Syst. Tech. J.*, 58(May/June):1037–1071.

Semmar, A., Lecours, M., Chouinard, J., and Ahern, J. 1991. Characterization of error sequences in UHF digital mobile radio channels. *IEEE Trans. Veh. Tech.*, 40(Nov.):769–776.

Smith, D.R. and Cormack, J.J. 1982. Measurement and characterization of a multipath fading channel. In *1982 International Conference on Communications*, pp. 7B.4.1–7B.4.6.

Stein, S., Schwartz, M., and Bennett, W. 1966. *Communication Systems and Techniques*, McGraw–Hill, New York.

Wozencraft, J.M. and Jacobs, I.M. 1967. *Principles of Communication Engineering*, Wiley, New York.

Further Information

A comprehensive treatment of channel models is given in *Fading Dispersive Communication Channels* by R.S. Kennedy. *Data Communications via Fading Channels*, an IEEE volume edited by K. Brayer, provides an excellent collection of papers on the subject.

12
Optimum Receivers

12.1	Introduction	141
12.2	Preliminaries	142
12.3	Karhunen–Loève Expansion	142
12.4	Detection Theory	144
12.5	Performance	145
12.6	Signal Space	146
12.7	Standard Binary Signalling Schemes	146
12.8	M-ary Optimal Receivers	147
12.9	More Realistic Channels	149
	Random Phase Channels • Rayleigh Channel	
12.10	Dispersive Channels	152

Geoffrey C. Orsak
George Mason University

12.1 Introduction

Every engineer strives for optimality in design. This is particularly true for communications engineers since in many cases implementing suboptimal receivers and sources can result in dramatic losses in performance. As such, this chapter focuses on design principles leading to the implementation of optimum receivers for the most common communication environments.

The main objective in digital communications is to transmit a sequence of bits to a remote location with the highest degree of accuracy. This is accomplished by first representing bits (or more generally short bit sequences) by distinct waveforms of finite time duration. These time-limited waveforms are then transmitted (broadcasted) to the remote sites in accordance with the data sequence.

Unfortunately, because of the nature of the **communication channel**, the remote location receives a corrupted version of the concatenated signal waveforms. The most widely accepted model for the communication channel is the so-called **additive white Gaussian noise**[1] **channel (AWGN channel)**. Mathematical arguments based upon the central limit theorem [Shiryayev, 1984], together with supporting empirical evidence, demonstrate that many common communication channels are accurately modeled by this abstraction. Moreover, from the design perspective, this is quite fortuitous since design and analysis with respect to this channel model is relatively straightforward.

---

[1] For those unfamiliar with AWGN, a random process (waveform) is formally said to be white Gaussian noise if all collections of instantaneous observations of the process are jointly Gaussian and mutually independent. An important consequence of this property is that the power spectral density of the process is a constant with respect to frequency variation (spectrally flat). For more on AWGN, see Papoulis, 1991.

12.2 Preliminaries

To better describe the digital communications process, we shall first elaborate on so-called binary communications. In this case, when the source wishes to transmit a bit value of 0, the transmitter broadcasts a specified waveform $s_0(t)$ over the **bit interval** $t \in [0, T]$. Conversely, if the source seeks to transmit the bit value of 1, the transmitter alternatively broadcasts the signal $s_1(t)$ over the same bit interval. The received waveform $R(t)$ corresponding to the first bit is then appropriately described by the following hypotheses testing problem:

$$\begin{aligned} H_0 &: R(t) = s_0(t) + \eta(t) \\ H_1 &: R(t) = s_1(t) + \eta(t) \end{aligned} \qquad 0 \le t \le T \qquad (12.1)$$

where, as stated previously, $\eta(t)$ corresponds to AWGN with spectral height nominally given by $N_0/2$. It is the objective of the receiver to determine the bit value, i.e., the most accurate hypothesis from the received waveform $R(t)$.

The optimality criterion of choice in digital communication applications is the **total probability of error** normally denoted as P_e. This scalar quantity is expressed as

$$\begin{aligned} P_e = &\; Pr(\text{declaring } 1 \mid 0 \text{ transmitted}) Pr(0 \text{ transmitted}) \\ &+ Pr(\text{declaring } 0 \mid 1 \text{ transmitted}) Pr(1 \text{ transmitted}) \end{aligned} \qquad (12.2)$$

The problem of determining the optimal binary receiver with respect to the probability of error is solved by applying stochastic representation theory [Wong and Hajek, 1985] to detection theory [Poor, 1988; Van Trees, 1968]. The specific waveform representation of relevance in this application is the **Karhunen–Loève (KL) expansion**.

12.3 Karhunen–Loève Expansion

The Karhunen–Loève expansion is a generalization of the Fourier series designed to represent a random process in terms of deterministic basis functions and uncorrelated random variables derived from the process. Whereas the Fourier series allows one to model or represent deterministic time-limited energy signals in terms of linear combinations of complex exponential waveforms, the Karhunen–Loève expansion allows us to represent a second-order random process in terms of a set of **orthonormal** basis functions scaled by a sequence of random variables. The objective in this representation is to choose the basis of time functions so that the coefficients in the expansion are mutually uncorrelated random variables.

To be more precise, if $R(t)$ is a zero mean second-order random process defined over $[0, T]$ with covariance function $K_R(t, s)$, then so long as the basis of deterministic functions satisfy certain integral constraints [Van Trees, 1968], one may write $R(t)$ as

$$R(t) = \sum_{i=1}^{\infty} R_i \phi_i(t) \qquad 0 \le t \le T \qquad (12.3)$$

where

$$R_i = \int_0^T R(t) \phi_i(t) \, dt$$

In this case the R_i will be mutually uncorrelated random variables with the ϕ_i being deterministic basis functions that are complete in the space of square integrable time functions over $[0, T]$.

Optimum Receivers

Importantly, in this case, equality is to be interpreted as **mean-square equivalence**, i.e.,

$$\lim_{N \to \infty} E\left[\left(R(t) - \sum_{i=1}^{N} R_i \phi_i(t)\right)^2\right] = 0$$

for all $0 \leq t \leq T$.

Fact 12.1. If $R(t)$ is AWGN, then any basis of the vector space of square integrable signals over $[0, T]$ results in uncorrelated and therefore independent Gaussian random variables.

The use of fact 12.1 allows for a conversion of a continuous time detection problem into a finite-dimensional detection problem. Proceeding, to derive the optimal binary receiver, we first construct our set of basis functions as the set of functions defined over $t \in [0, T]$ beginning with the signals of interest $s_0(t)$ and $s_1(t)$. That is,

$\{s_0(t), s_1(t),$ plus a countable number of functions which complete the basis$\}$

In order to insure that the basis is orthonormal, we must apply the Gramm–Schmidt procedure[2] [Proakis, 1989] to the full set of functions beginning with $s_0(t)$ and $s_1(t)$ to arrive at our final choice of basis $\{\phi_i(t)\}$.

Fact 12.2. Let $\{\phi_i(t)\}$ be the resultant set of basis functions.
Then for all $i > 2$, the $\phi_i(t)$ are orthogonal to $s_0(t)$ and $s_1(t)$. That is,

$$\int_0^T \phi_i(t) s_j(t) \, dt = 0$$

for all $i > 2$ and $j = 0, 1$.

Using this fact in conjunction with Eq. (12.3), one may recognize that only the coefficients R_1 and R_2 are functions of our signals of interest. Moreover, since the R_i are mutually independent, the optimal receiver will, therefore, only be a function of these two values.

Thus, through the application of the KL expansion, we arrive at an equivalent hypothesis testing problem to that given in Eq. (12.1),

$$H_0 : R = \begin{bmatrix} \int_0^T \phi_1(t) s_0(t) \, dt \\ \int_0^T \phi_2(t) s_0(t) \, dt \end{bmatrix} + \begin{bmatrix} \eta_1 \\ \eta_2 \end{bmatrix}$$

$$H_1 : R = \begin{bmatrix} \int_0^T \phi_1(t) s_1(t) \, dt \\ \int_0^T \phi_2(t) s_1(t) \, dt \end{bmatrix} + \begin{bmatrix} \eta_1 \\ \eta_2 \end{bmatrix} \qquad (12.4)$$

where it is easily shown that η_1 and η_2 are mutually independent, zero-mean, Gaussian random variables with variance given by $N_0/2$, and where ϕ_1 and ϕ_2 are the first two functions from our orthonormal set of basis functions. Thus, the design of the optimal binary receiver reduces to a simple two-dimensional detection problem that is readily solved through the application of detection theory.

---

[2]The Gramm-Schmidt procedure is a deterministic algorithm that simply converts an arbitrary set of basis functions (vectors) into an equivalent set of orthonormal basis functions (vectors).

12.4 Detection Theory

It is well known from detection theory [Poor, 1988] that under the minimum P_e criterion, the optimal detector is given by the *maximum a posteriori rule* (*MAP*),

$$\text{choose}_i \text{ largest } p_{H_i|R}(H_i \mid R = r) \qquad (12.5)$$

i.e., determine the hypothesis that is most likely, given that our observation vector is r. By a simple application of Bayes theorem [Papoulis, 1991], we immediately arrive at the central result in detection theory: the optimal binary detector is given by the likelihood ratio test (LRT),

$$L(R) = \frac{p_{R|H_1}(R)}{p_{R|H_0}(R)} \overset{H_1}{\underset{H_0}{\gtrless}} \frac{\pi_0}{\pi_1} \qquad (12.6)$$

where the π_i are the a priori probabilities of the hypotheses H_i being true. Since in this case we have assumed that the noise is white and Gaussian, the LRT can be written as

$$L(R) = \frac{\prod_1^2 \frac{1}{\sqrt{\pi N_0}} \exp\left(-\frac{1}{2} \frac{(R_i - s_{1,i})^2}{N_0/2}\right)}{\prod_1^2 \frac{1}{\sqrt{\pi N_0}} \exp\left(-\frac{1}{2} \frac{(R_i - s_{0,i})^2}{N_0/2}\right)} \overset{H_1}{\underset{H_0}{\gtrless}} \frac{\pi_0}{\pi_1} \qquad (12.7)$$

where

$$s_{j,i} = \int_0^T \phi_i(t) s_j(t)\, dt$$

By taking the logarithm and cancelling common terms, it is easily shown that the optimum binary receiver can be written as

$$\frac{2}{N_0} \sum_1^2 R_i (s_{1,i} - s_{0,i}) - \frac{1}{N_0} \sum_1^2 \left(s_{1,i}^2 - s_{0,i}^2\right) \overset{H_1}{\underset{H_0}{\gtrless}} \ln \frac{\pi_0}{\pi_1} \qquad (12.8)$$

This finite-dimensional version of the optimal receiver can be converted back into a continuous time receiver by the direct application of Parseval's theorem [Papoulis, 1991] where it is easily shown that

$$\begin{aligned}\sum_{i=1}^2 R_i s_{k,i} &= \int_0^T R(t) s_k(t)\, dt \\ \sum_{i=1}^2 s_{k,i}^2 &= \int_0^T s_k^2(t)\, dt\end{aligned} \qquad (12.9)$$

By applying Eq. (12.9) to Eq. (12.8) the final receiver structure is then given by

$$\int_0^T R(t)[s_1(t) - s_0(t)]\, dt - \frac{1}{2}(E_1 - E_0) \overset{H_1}{\underset{H_0}{\gtrless}} \frac{N_0}{2} \ln \frac{\pi_0}{\pi_1} \qquad (12.10)$$

Optimum Receivers

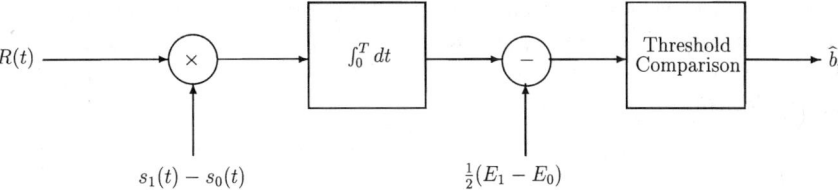

FIGURE 12.1 Optimal correlation receiver structure for binary communications.

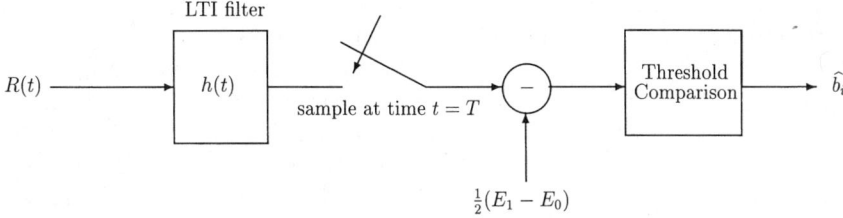

FIGURE 12.2 Optimal matched filter receiver structure for binary communications. In this case $h(t) = s_1(T-t) - s_0(t-t)$.

where E_1 and E_0 are the energies of signals $s_1(t)$ and $s_0(t)$, respectively. (See Fig. 12.1 for a block diagram.) Importantly, if the signals are equally likely ($\pi_0 = \pi_1$), the the optimal receiver is independent of the typically unknown spectral height of the background noise.

One can readily observe that the optimal binary communication receiver correlates the received waveform with the difference signal $s_1(t) - s_0(t)$ and then compares the statistic to a threshold. This operation can be interpreted as identifying the signal waveform $s_i(t)$ that best correlates with the received signal $R(t)$. Based on this interpretation, the receiver is often referred to as the **correlation receiver**.

As an alternate means of implementing the correlation receiver, we may reformulate the computation of the left-hand side of Eq. (12.10) in terms of standard concepts in filtering. Let $h(t)$ be the impulse response of a linear, time-invariant (LTI) system. By letting $h(t) = s_1(T-t) - s_0(T-t)$, then it is easily verified that the output of $R(t)$ to a LTI system with impulse response given by $h(t)$ and then sampled at time $t = T$ gives the desired result. (See Fig. 12.2 for a block diagram.) Since the impulse response is matched to the signal waveforms, this implementation is often referred to as the **matched filter receiver**.

12.5 Performance

Because of the nature of the statistics of the channel and the relative simplicity of the receiver, performance analysis of the optimal binary receiver in AWGN is a straightforward task. Since the conditional statistics of the log likelihood ratio are Gaussian random variables, the probability of error can be computed directly in terms of Marcum Q functions[3] as

$$P_e = Q\left(\frac{\|s_0 - s_1\|}{\sqrt{2N_0}}\right)$$

where the s_i are the two-dimensional signal vectors obtained from Eq. (12.4), and where $\|x\|$ denotes the Euclidean length of the vector x. Thus, $\|s_0 - s_1\|$ is best interpreted as the distance

---

[3]The Q function is the probability that a standard normal random variable exceeds a specified constant, i.e., $Q(x) = \int_x^\infty 1/\sqrt{2\pi} \exp(-z^2/2)\, dz$.

between the respective signal representations. Since the Q function is monotonically decreasing with an increasing argument, one may recognize that the probability of error for the optimal receiver decreases with an increasing separation between the signal representations, i.e., the more dissimilar the signals, the lower the P_e.

12.6 Signal Space

The concept of a **signal space** allows one to view the signal classification problem (receiver design) within a geometrical framework. This offers two primary benefits: first it supplies an often more intuitive perspective on the receiver characteristics (e.g., performance) and second it allows for a straightforward generalization to standard M-ary signalling schemes.

To demonstrate this, in Fig. 12.3, we have plotted an arbitrary signal space for the binary signal classification problem. The axes are given in terms of the basis functions $\phi_1(t)$ and $\phi_2(t)$. Thus, every point in the signal space is a time function constructed as a linear combination of the two basis functions. By fact 12.2, we recall that both signals $s_0(t)$ and $s_1(t)$ can be constructed as a linear combination of $\phi_1(t)$ and $\phi_2(t)$ and as such we may identify these two signals in this figure as two points.

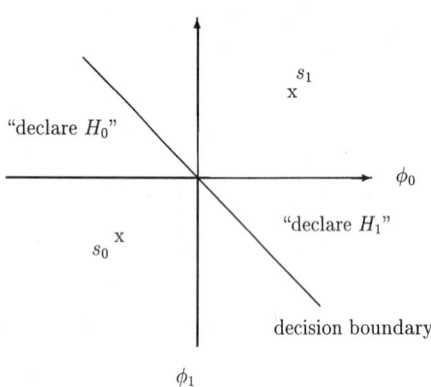

FIGURE 12.3 Signal space and decision boundary for optimal binary receiver.

Since the decision statistic given in Eq. (12.8) is a linear function of the observed vector $\boldsymbol{R}$ which is also located in the signal space, it is easily shown that the set of vectors under which the receiver declares hypothesis H_i is bounded by a line in the signal space. This so-called **decision boundary** is obtained by solving the equation $\ln[L(\boldsymbol{R})] = 0$. (Here again we have assumed equally likely hypotheses.) In the case under current discussion, this decision boundary is simply the hyperplane separating the two signals in signal space. Because of the generality of this formulation, many problems in communication system design are best cast in terms of the signal space, that is, signal locations and decision boundaries.

12.7 Standard Binary Signalling Schemes

The framework just described allows us to readily analyze the most popular signalling schemes in binary communications: amplitude-shift keying (ASK), frequency-shift keying (FSK), and phase-shift keying (PSK). Each of these examples simply constitute a different selection for signals $s_0(t)$ and $s_1(t)$.

In the case of ASK, $s_0(t) = 0$, while $s_1(t) = \sqrt{2E/T}\sin(2\pi f_c t)$, where E denotes the energy of the waveform and f_c denotes the frequency of the carrier wave with $f_c T$ being an integer. Because $s_0(t)$ is the null signal, the signal space is a one-dimensional vector space with $\phi_1(t) = \sqrt{2/T}\sin(2\pi f_c t)$. This, in turn, implies that $\|s_0 - s_1\| = \sqrt{E}$. Thus, the corresponding probability of error for ASK is

$$P_e(\text{ASK}) = Q\left(\sqrt{\frac{E}{2N_0}}\right)$$

For FSK, the signals are given by equal amplitude sinusoids with distinct center frequencies, that is, $s_i(t) = \sqrt{2E/T}\sin(2\pi f_i t)$ with $f_i T$ being two distinct integers. In this case, it is easily verified that the signal space is a two-dimensional vector space with $\phi_i(t) = \sqrt{2/T}\sin(2\pi f_i t)$ resulting in

Optimum Receivers

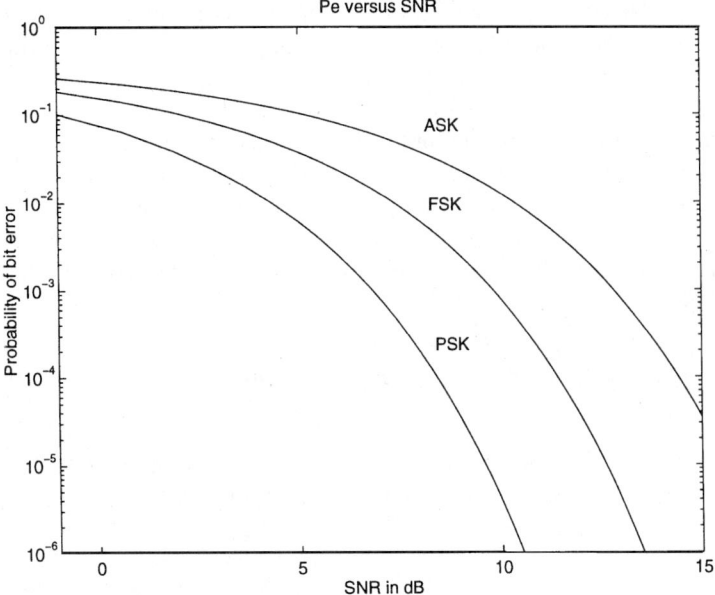

FIGURE 12.4 P_e vs the signal to noise ratio in decibels [dB = $10\log(E/N_0)$] for amplitude-shift keying, frequency-shift keying, and phase-shift keying; note that there is a 3-dB difference in performance from ASK to FSK to PSK.

$\|s_0 - s_1\| = \sqrt{2E}$. The corresponding error rate is given to be

$$P_e(\text{FSK}) = Q\left(\sqrt{\frac{E}{N_0}}\right)$$

Finally, with regard to PSK signalling, the most frequently utilized binary PSK signal set is an example of an antipodal signal set. Specifically, the antipodal signal set results in the greatest separation between the signals in the signal space subject to an energy constraint on both signals. This, in turn, translates into the energy constrained signal set with the minimum P_e. In this case, the $s_i(t)$ are typically given by $\sqrt{2E/T}\sin[2\pi f_c t + \theta(i)]$, where $\theta(0) = 0$ and $\theta(1) = \pi$. As in the ASK case, this results in a one-dimensional signal space, however, in this case $\|s_0 - s_1\| = 2\sqrt{E}$ resulting in probability of error given by

$$P_e(\text{PSK}) = Q\left(\sqrt{\frac{2E}{N_0}}\right)$$

In all three of the described cases, one can readily observe that the resulting performance is a function of only the signal-to-noise ratio E/N_0. In the more general case, the performance will be a function of the intersignal energy to noise ratio. To gauge the relative difference in performance of the three signalling schemes, in Fig. 12.4, we have plotted the P_e as a function of the SNR. Please note the large variation in performance between the three schemes for even moderate values of SNR.

12.8 *M*-ary Optimal Receivers

In binary signalling schemes, one seeks to transmit a single bit over the bit interval $[0, T]$. This is to be contrasted with M-ary signalling schemes where one transmits multiple bits simultaneously over the so-called symbol interval $[0, T]$. For example, using a signal set with 16 separate waveforms will allow one to transmit a length four-bit sequence per symbol (waveform). Examples of M-ary waveforms are quadrature phase-shift keying (QPSK) and quadrature amplitude modulation (QAM).

The derivation of the optimum receiver structure for M-ary signalling requires the straightforward application of fundamental results in detection theory. As with binary signalling, the Karhunen–Loève expansion is the mechanism utilized to convert a hypotheses testing problem based on continuous waveforms into a vector classification problem. Depending on the complexity of the M waveforms, the signal space can be as large as an M-dimensional vector space.

By extending results from the binary signalling case, it is easily shown that the optimum M-ary receiver computes

$$\xi_i[R(t)] = \int_0^T s_i(t) R(t)\, dt - \frac{E_i}{2} + \frac{N_0}{2} \ln \pi_i \qquad i = 1, \ldots, M$$

where, as before, the $s_i(t)$ constitute the signal set with the π_i being the corresponding a priori probabilities. After computing M separate values of ξ_i, the minimum probability of error receiver simply chooses the largest amongst this set. Thus, the M-ary receiver is implemented with a bank of correlation or matched filters followed by choose-largest decision logic.

In many cases of practical importance, the signal sets are selected so that the resulting signal space is a two-dimensional vector space irrespective of the number of signals. This simplifies the receiver structure in that the sufficient statistics are obtained by implementing only two matched filters. Both QPSK and QAM signal sets fit into this category. As an example, in Fig. 12.5, we have depicted

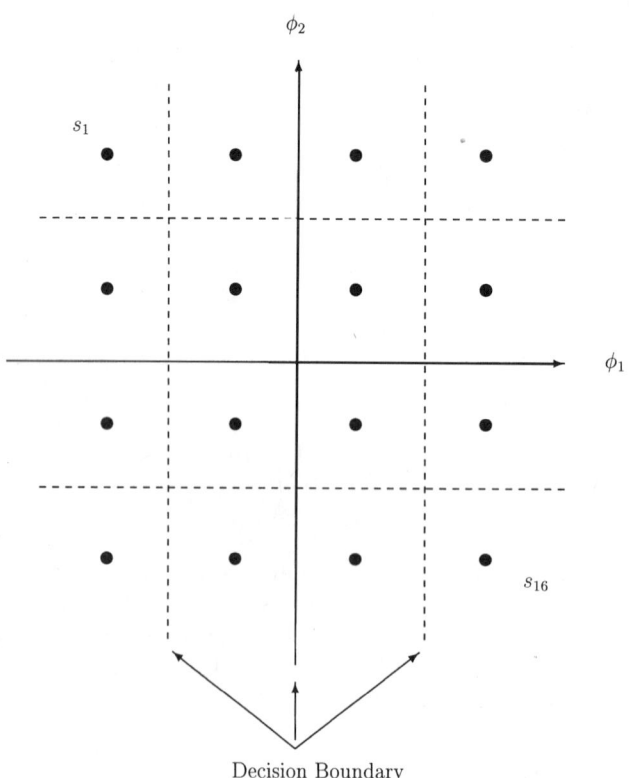

FIGURE 12.5 Signal space representation of 16-QAM signal set. Optimal decision regions for equally likely signals are also noted.

Optimum Receivers

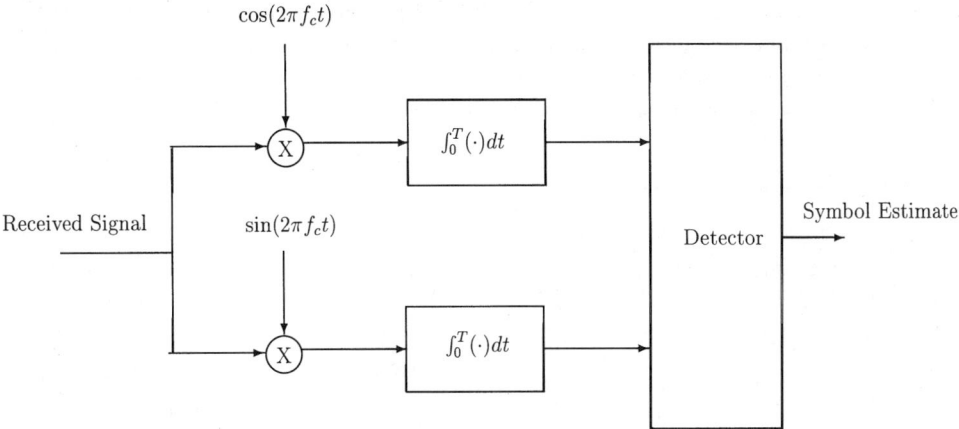

FIGURE 12.6 Optimum receiver structure for noncoherent (random or unknown phase) ASK demodulation.

the signal locations for standard 16-QAM signalling with the associated decision boundaries. In this case we have assumed an equally likely signal set. As can be seen, the optimal decision rule selects the signal representation that is closest to the received signal representation in this two-dimensional signal space.

12.9 More Realistic Channels

As is unfortunately often the case, many channels of practical interest are not accurately modeled as simply an AWGN channel. It is often that these channels impose nonlinear effects on the transmitted signals. The best example of this are channels that impose a random phase and random amplitude onto the signal. This typically occurs in applications such as in mobile communications, where one often experiences rapidly changing path lengths from source to receiver.

Fortunately, by the judicious choice of signal waveforms, it can be shown that the selection of the ϕ_i in the Karhunen–Loève transformation is often independent of these unwanted parameters. In these situations, the random amplitude serves only to scale the signals in signal space, whereas the random phase simply imposes a rotation on the signals in signal space.

Since the Karhunen–Loève basis functions typically do not depend on the unknown parameters, we may again convert the continuous time classification problem to a vector channel problem where the received vector $\mathbf{R}$ is computed as in Eq. (12.3). Since this vector is a function of both the unknown parameters (i.e., in this case amplitude A and phase ν), to obtain a likelihood ratio test independent of A and ν, we simply apply Bayes theorem to obtain the following form for the LRT:

$$L(\mathbf{R}) = \frac{E[p_{\mathbf{R}|H_1,A,\nu}(\mathbf{R} \mid H_1, A, \nu)]}{E[p_{\mathbf{R}|H_0,A,\nu}(\mathbf{R} \mid H_0, A, \nu)]} \overset{H_1}{\underset{H_0}{\gtrless}} \frac{\pi_0}{\pi_1}$$

where the expectations are taken with respect to A and ν, and where $p_{\mathbf{R}|H_i,A,\nu}$ are the conditional probability density functions of the signal representations. Assuming that the background noise is AWGN, it can be shown that the LRT simplifies to choosing the largest amongst

$$\xi_i[R(t)] = \pi_i \int_{A,\nu} \exp\left\{\frac{2}{N_0} \int_0^T R(t) s_i(t \mid A, \nu) \, dt - \frac{E_i(A,\nu)}{N_0}\right\} p_{A,\nu}(A, \nu) \, dA \, d\nu$$

$$i = 1, \ldots, M \quad (12.11)$$

It should be noted that in the Eq. (12.11) we have explicitly shown the dependence of the transmitted signals s_i on the parameters A and ν. The final receiver structures, together with their corresponding performance are, thus, a function of both the choice of signal sets and the probability density functions of the random amplitude and random phase.

Random Phase Channels

If we consider first the special case where the channel simply imposes a uniform random phase on the signal, then it can be easily shown that the so-called in-phase and quadrature statistics obtained from the received signal $R(t)$ (denoted by R_I and R_Q, respectively) are sufficient statistics for the signal classification problem. These quantities are computed as

$$R_I(i) = \int_0^T R(t) \cos[2\pi f_c(i)t]\, dt$$

and

$$R_Q(i) = \int_0^T R(t) \sin[2\pi f_c(i)t]\, dt$$

where in this case the index i corresponds to the center frequencies of hypotheses H_i, (e.g., FSK signalling). The optimum binary receiver selects the largest from amongst

$$\xi_i[R(t)] = \pi_i \exp\left(-\frac{E_i}{N_0}\right) I_0\left[\frac{2}{N_0}\sqrt{R_I^2(i) + R_Q^2(i)}\right] \quad i = 1, \ldots, M$$

where I_0 is a zeroth-order, modified Bessel function of the first kind. If the signals have equal energy and are equally likely (e.g., FSK signalling), then the optimum receiver is given by

$$R_I^2(1) + R_Q^2(1) \underset{H_0}{\overset{H_1}{\gtrless}} R_I^2(0) + R_Q^2(0)$$

One may readily observe that the optimum receiver bases its decision on the values of the two envelopes of the received signal $\sqrt{R_I^2(i) + R_Q^2(i)}$ and, as a consequence, is often referred to as an envelope or square-law detector. Moreover, it should be observed that the computation of the envelope is independent of the underlying phase of the signal and is as such known as a noncoherent receiver.

The computation of the error rate for this detector is a relatively straightforward exercise resulting in

$$P_e(\text{noncoherent}) = \frac{1}{2} \exp\left(-\frac{E}{2N_0}\right)$$

As before, note that the error rate for the noncoherent receiver is simply a function of the SNR.

Rayleigh Channel

As an important generalization of the described random phase channel, many communication systems are designed under the assumption that the channel introduces both a random amplitude

Optimum Receivers

and a random phase on the signal. Specifically, if the original signal sets are of the form $s_i(t) = m_i(t)\cos(2\pi f_c t)$ where $m_i(t)$ is the baseband version of the message (i.e., what distinguishes one signal from another), then the so-called **Rayleigh channel** introduces random distortion in the received signal of the following form:

$$s_i(t) = A m_i(t)\cos(2\pi f_c t + \nu)$$

where the amplitude A is a Rayleigh random variable[4] and where the random phase ν is a uniformly distributed between zero and 2π.

To determine the optimal receiver under this distortion, we must first construct an alternate statistical model for $s_i(t)$. To begin, it can be shown from the theory of random variables [Papoulis, 1991] that if X_I and X_Q are statistically independent, zero mean, Gaussian random variables with variance given by σ^2, then

$$A m_i(t)\cos(2\pi f_c t + \nu) = m_i(t) X_I \cos(2\pi f_c t) + m_i(t) X_Q \sin(2\pi f_c t)$$

Equality here is to be interpreted as implying that both A and ν will be the appropriate random variables. From this, we deduce that the combined uncertainty in the amplitude and phase of the signal is incorporated into the Gaussian random variables X_I and X_Q. The in-phase and quadrature components of the signal $s_i(t)$ are given by $s_{Ii}(t) = m_i(t)\cos(2\pi f_c t)$ and $s_{Qi}(t) = m_i(t)\sin(2\pi f_c t)$, respectively. By appealing to Eq. (12.11), it can be shown that the optimum receiver selects the largest from

$$\xi_i[R(t)] = \frac{\pi_i}{1 + \frac{2 E_i}{N_0}\sigma^2} \exp\left[\frac{\sigma^2}{\frac{1}{2} + \frac{E_i}{N_0}\sigma^2}\left(\langle R(t), s_{Ii}(t)\rangle^2 + \langle R(t), s_{Qi}(t)\rangle^2\right)\right]$$

where the inner product

$$\langle R(t), S_i(t)\rangle = \int_0^T R(t) s_i(t)\, dt$$

Further, if we impose the conditions that the signals be equally likely with equal energy over the symbol interval, then optimum receiver selects the largest amongst

$$\xi_i[R(t)] = \sqrt{\langle R(t), s_{Ii}(t)\rangle^2 + \langle R(t), s_{Qi}(t)\rangle^2}$$

Thus, much like for the random phase channel, the optimum receiver for the Rayleigh channel computes the projection of the received waveform onto the in-phase and quadrature components of the hypothetical signals. From a signal space perspective, this is akin to computing the length of the received vector in the subspace spanned by the hypothetical signal. The optimum receiver then chooses the largest amongst these lengths.

---

[4]The density of a Rayleigh random variable is given by $p_A(a) = a/\sigma^2 \exp(-a^2/2\sigma^2)$ for $a \geq 0$.

As with the random phase channel, computing the performance is a straightforward task resulting in (for the equally likely, equal energy case)

$$P_e(\text{Rayleigh}) = \frac{\frac{1}{2}}{\left(1 + \frac{E\sigma^2}{N_0}\right)}$$

Interestingly, in this case the performance depends not only on the SNR, but also on the variance (spread) of the Rayleigh amplitude A. Thus, if the amplitude spread is large, we expect to often experience what is known as deep fades in the amplitude of the received waveform and as such expect a commensurate loss in performance.

12.10 Dispersive Channels

The **dispersive channel** model assumes that the channel not only introduces AWGN but also distorts the signal through a filtering process. This model incorporates physical realities such as multipath effects and frequency selective fading. In particular, the standard model adopted is depicted in the block diagram given in Fig. 12.7. As can be seen, the receiver observes a filtered version of the signal plus AWGN. If the impulse response of the channel is known, then we arrive at the optimum receiver design by applying the previously presented theory. Unfortunately, the duration of the filtered signal can be a complicating factor. More often than not, the channel will increase the duration of the transmitted signals, hence leading to the description, dispersive channel.

However, if the designers take this into account by shortening the duration of $s_i(t)$ so that the duration of $s_i^*(t)$ is less than T, then the optimum receiver chooses the largest amongst

$$\xi_i(R(t)) = \frac{N_0}{2} \ln \pi_i + \langle R(t), s_i^*(t) \rangle - \frac{1}{2} E_i^*$$

If we limit our consideration to equally likely binary signal sets, then the minimum P_e matches the received waveform to the filtered versions of the signal waveforms. The resulting error rate is given by

$$P_e(\text{dispersive}) = Q\left(\frac{\|s_0^* - s_0^*\|}{\sqrt{2N_0}}\right)$$

Thus, in this case the minimum P_e is a function of the separation of the filtered version of the signals in the signal space.

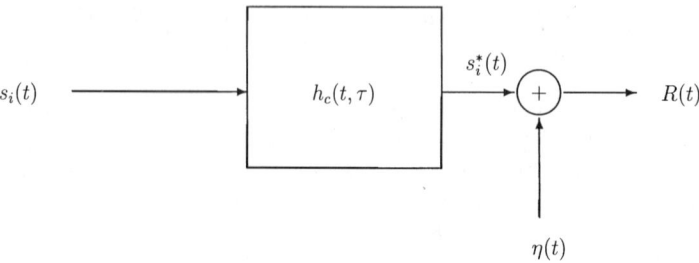

FIGURE 12.7 Standard model for dispersive channel. The time varying impulse response of the channel is denoted by $h_c(t, \tau)$.

The problem becomes substantially more complex if we cannot insure that the filtered signal durations are less than the symbol lengths. In this case we experience what is known as **intersymbol interference (ISI)**. That is, observations over one symbol interval contain not only the symbol information of interest but also information from previous symbols. In this case we must appeal to optimum sequence estimation [Poor, 1988] to take full advantage of the information in the waveform. The basis for this procedure is the maximization of the joint likelihood function conditioned on the sequence of symbols. This procedure not only defines the structure of the optimum receiver under ISI but also is critical in the decoding of convolutional codes and coded modulation. Alternate adaptive techniques to solve this problem involve the use of channel equalization.

Defining Terms

Additive white Gaussian noise (AWGN) channel: The channel whose model is that of corrupting a transmitted waveform by the addition of white (i.e., spectrally flat) Gaussian noise.

Bit (symbol) interval: The period of time over which a single symbol is transmitted.

Communication channel: The medium over which communication signals are transmitted. Examples are fiber optic cables, free space, or telephone lines.

Correlation or matched filter receiver: The optimal receiver structure for digital communications in AWGN.

Decision boundary: The boundary in signal space between the various regions where the receiver declares H_i. Typically a hyperplane when dealing with AWGN channels.

Dispersive channel: A channel that elongates and distorts the transmitted signal. Normally modeled as a time-varying linear system.

Intersymbol interference: The ill-effect of one symbol smearing into adjacent symbols thus interfering with the detection process. This is a consequence of the channel filtering the transmitted signals and therefore elongating their duration, see dispersive channel.

Karhunen–Loève expansion: A representation for second-order random processes. Allows one to express a random process in terms of a superposition of deterministic waveforms. The scale values are uncorrelated random variables obtained from the waveform.

Mean-square equivalence: Two random vectors or time-limited waveforms are mean-square equivalent if and only if the expected value of their mean-square error is zero.

Orthonormal: The property of two or more vectors or time-limited waveforms being mutually orthogonal and individually having unit length. Orthogonality and length are typically measured by the standard Euclidean inner product.

Rayleigh channel: A channel that randomly scales the transmitted waveform by a Rayleigh random variable while adding an independent uniform phase to the carrier.

Signal space: An abstraction for representing a time limited waveform in a low-dimensional vector space. Usually arrived at through the application of the Karhunen–Loève transformation.

Total probability of error: The probability of classifying the received waveform into any of the symbols that were not transmitted over a particular bit interval.

References

Gibson, J.D. 1993. *Principles of Digital and Analog Communications*, 2nd ed. MacMillan, New York.

Haykin, S. 1994. *Communication Systems*, 3rd ed. Wiley, New York.

Lee, E.A. and Messerschmitt, D.G. 1988. *Digital Communication*, Kluwer Academic Publishers, Norwell, MA.

Papoulis, A. 1991. *Probability, Random Variables, and Stochastic Processes*, 3rd ed. McGraw-Hill, New York.

Poor, H.V. 1988. *An Introduction to Signal Detection and Estimation*, Springer–Verlag, New York.

Proakis, J.G. 1989. *Digital Communications*, 2nd ed. McGraw-Hill, New York.

Shiryayev, A.N. 1984. *Probability*, Springer–Verlag, New York.
Sklar, B. 1988. *Digital Communications, Fundamentals and Applications*, Prentice Hall, Englewood Cliffs, NJ.
Van Trees, H.L. 1968. *Detection, Estimation, and Modulation Theory, Part I*, Wiley, New York.
Wong, E. and Hajek, B. 1985. *Stochastic Processes in Engineering Systems*, Springer–Verlag, New York.
Wozencraft, J.M. and Jacobs, I. 1990. *Principles of Communication Engineering*, reissue, Waveland Press, Inc, Prospect Heights, Illinois.
Ziemer, R.E. and Peterson, R.L. 1992. *Introduction to Digital Communication*, Macmillan, New York.

Further Information

The fundamentals of receiver design were put in place by Wozencraft and Jacobs in their seminal book. Since that time, there have been many outstanding textbooks in this area. For a sampling see Haykin, 1994; Sklar, 1988; Lee and Messerschmitt, 1988; Gibson, 1993; and Ziemer and Peterson, 1992. For a complete treatment on the use and application of detection theory in communications see van Trees, 1968 and Poor, 1988. For deeper insights into the Karhunen–Loève expansion and its use in communications and signal processing see Wong and Hajek, 1985.

13
Standards Setting Bodies

Spiros Dimolitsas
Lawrence Livermore National Laboratory

13.1	Introduction	155
13.2	Global Standardization	156
	ITU-T • ITU-R • BDT • ISO/IEC JTC 1	
13.3	Regional Standardization	159
13.4	National Standardization	161
	ANSI T1 • TIA • TTC	
13.5	Standards Coordination	163
13.6	Scientific	164
13.7	Standards Development Cycle	165

13.1 Introduction

National economies are increasingly becoming information based, where networking and information transport provide a foundation for productivity and economic growth. Concurrently, many countries are rapidly adopting deregulation policies that are resulting in a telecommunications industry that is increasingly multicarrier and multivendor based, and where interconnectivity and compatibility between different networks is emerging as key to the success of this technological and regulatory transition. The communications industry has, consequently, become more interested in standardization; standards give manufacturers, service providers, and users freedom of choice at reasonable cost.

In this chapter, a review is provided of the primary telecommunications standards setting bodies. As will be seen, these bodies are often driven by slightly different underlying philosophies, but the output of their activities, i.e., the standards, possess essentially the same characteristics. An all-encompassing review of standardization bodies is not attempted here; this would clearly take many volumes to describe. Furthermore, as country after country increasingly deregulates its telecommunication industry, new standards setting bodies emerge to fill in the void of the de-facto (but no longer existing) standards setting bodies: the national telecommunications administration.

The principal communications standards bodies that will be covered are the following: the International Telecommunications Union (ITU); the United States ANSI Committee T1 on Telecommunications and the Telecommunications Industry Association (TIA); the European Telecommunications Standards Institute (ETSI); the Japanese Telecommunications Technology Committee (TTC); and the Institute of Electrical and Electronics Engineers (IEEE). Not addressed explicitly are other standards setting bodies that are either national and regional in character; even though it is recognized that sometimes there is overlap in scope with the bodies explicitly covered here.

Most notably, standards setting bodies that are not covered, but that are worth noting, include: the United States ANSI Committee X3, the Inter-American Telecommunications Commission (CITEL), the International Standards Organization (ISO), the International Electrotechnical Commission (IEC) [except ISO/IEC joint technical committee (JTC) 1], the Telecommunications Standards Advisory Council of Canada (TSACC), the Australian Telecommunications Standardization Committee (ATSC), the Telecommunication Technology Association (TTA) in Korea, and several forums (whose scope is, in principle, somewhat different) such as the asynchronous transfer mode (ATM) forum, the frame relay forum, the integrated digital services network (ISDN) users' forum, and telocator. As will be described later, many of these bodies operate in a coherent fashion through a mechanism developed by the interregional telecommunications standards conference (ITSC) and its successor, the global standards collaboration (GSC).

13.2 Global Standardization

When it comes to setting global communications standards, the ITU comes to the forefront. The ITU is an intergovernmental organization, whereby each sovereign state that is a member of the United Nations may become a member of the ITU. Member governments (in most cases represented by their telecommunications administrations) are constitutional members with a right to vote. Other organizations, such as network and service providers, manufacturers, and scientific and industrial organizations also participate in ITU activities but with a lower legal status.

ITU traces its history back to 1865 in the era of telegraphy. The supreme organ of the ITU is the plenipotentiary conference, which is held not less than every five years and plays a major role in the management of ITU. In 1993 the ITU as a U.N.-specialized agency was reorganized into three sectors (see Fig. 13.1): The *telecommunications standardization* sector (ITU-T), the *radiocommunications* sector (ITU-R), and the *development* sector (BDT). These sectors' activities are, respectively, standardization of telecommunications, including radio communications (although during a transition period considerable role in this field will be played by the ITU-R); regulation of telecommunications (mainly for radio communications); and promotion of telecommunications in developing countries.

It should be noted that, in general, the ITU-T is the successor of the international telephone and telegraph consultative committee (CCITT) of the ITU with additional responsibilities for standardization of network-related radio communications. Similarly, the ITU-R is the successor of the international radio consultative committee (CCIR) and the international frequency registration bureau (IFRB) of the ITU (after transferring some of its standardization activities to the ITU-T). The BDT is a new sector, which became operational in 1989.

ITU-T

Within the ITU structure, standardization work is undertaken by a number of study groups (SG) dealing with specific areas of communications. There are currently 15 study groups, as shown in Table 13.1.

Study groups develop standards for their respective work areas, which then have to be agreed upon by consensus—a process that for the time being is reserved to administrations only. The standards so developed are called recommendations to indicate their legal nonbinding nature. Technically, however, there is no distinction between recommendations developed by the ITU and standards developed by other standards setting bodies.

The study groups' work is undertaken by members, or delegates sent or sponsored by their national administrations. Because an ITU-T study group can have anywhere from 100 to more than 500 participating members and deal with 20–50 project standards, the work of each study group is often divided among working parties (WP). Such working parties are usually split further into experts' groups led by a chair or rapporteur with responsibility for a single standards' topic.

Standards Setting Bodies

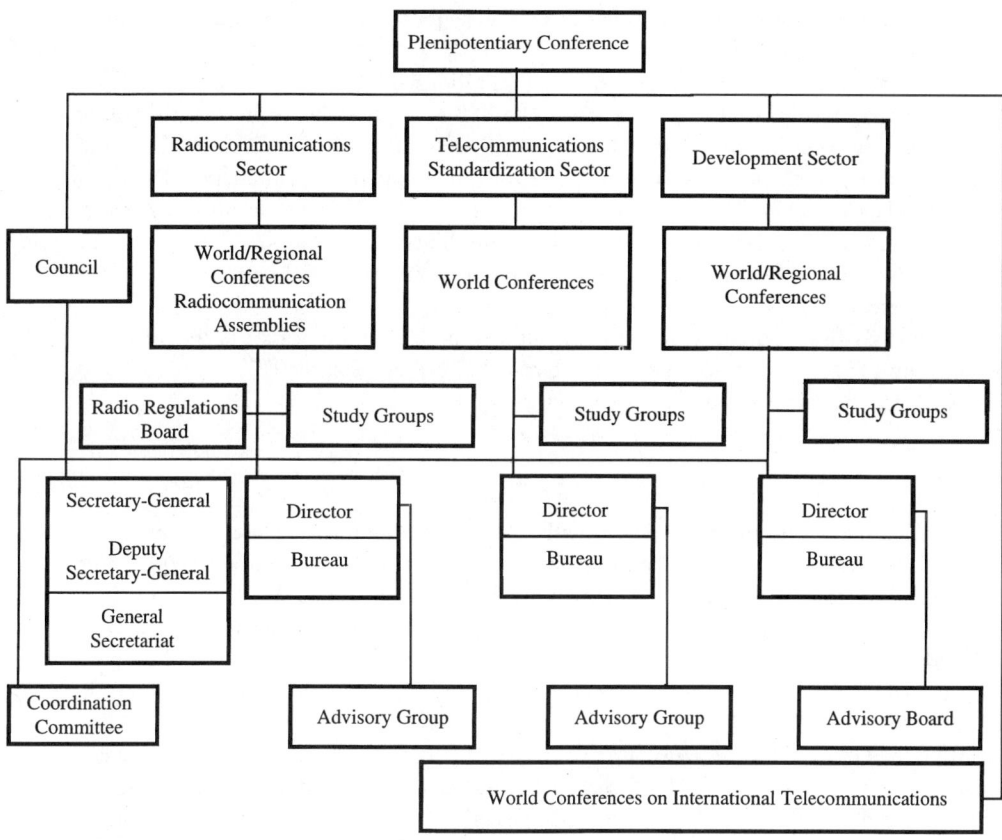

FIGURE 13.1 The (new) ITU structure.

TABLE 13.1 ITU-T Study Group Structure

SG 1	Service definition
SG 2	Network operation
SG 3	Tariff and accounting principles
SG 4	Network maintenance
SG 5	Protection against electromagnetic environmental effects
SG 6	Outside plant
SG 7	Data networks and open systems communications
SG 8	Terminals for telematic services
SG 9	Television and sound transmission (former CMTT)
SG 10	Languages for telecommunications applications
SG 11	Switching and signalling
SG 12	End-to-end transmission performance of networks and terminals
SG 13	General network aspects
SG 14	Modems and transmission techniques for data, telegraph, and telematic services
SG 15	Transmission systems and equipment

To coordinate standardization work that spans several study groups, a number of joint coordination groups (JCG) have also been established (not shown in Fig. 13.1). Presently, there are five such groups dealing with universal personal communications (UPT), transmission management network (TMN), audiovisual and multimedia services (AVMMS), quality of service and network performance (NP), and broadband ISDN (ATM/B-ISDN).

Such groups do not have executive powers but are merely there to coordinate work of pervasive interest within the ITU-T sector.

Also part of the ITU-T structure is the telecommunications standardization bureau (TSB) or, as it was formerly called, the CCITT secretariat. The TSB is responsible for the organization of numerous meetings held by the sector each year as well as all other support services required to ensure the smooth and efficient operation of the sector (including, but not limited to, document production and distribution). The TSB is headed by a director, who holds the executive power and, in collaboration with the study groups, bears full responsibility for the ITU-T activities. In this structure, unlike other U.N. organizations, the secretary general is the legal representative of the ITU, with the executive powers being vested in the director.

Finally, the ITU-T is supported by an advisory group, i.e. the telecommunications standardization advisory group (TSAG), which together with interested ITU members, the ITU-T Director, and ITU-T SG chairmen, guides standardization activities.

ITU-R

As noted earlier, standardization related work within the radiocommunications sector is gradually being transferred to the ITU-T, thus minimizing this sector's role in standardization and emphasizing its duties on regulatory and pure radio-interface aspects. The functional structure of the ITU-R includes a number (currently nine) of study groups, a radiocommunications bureau, and an advisory board. The role of the latter two elements is very similar to the ITU-T and, thus, need not be repeated here. The partitioning of the ITU-R sector into its nine study groups are shown in Table 13.2.

As within the ITU-T, there are areas of pervasive interest, and so areas of common interest can be found between the ITU-T and ITU-R where activities need to be coordinated. To achieve this objective, two intersector coordination groups (ICG) have been established (not shown in Fig. 13.1) dealing with future public land mobile telecommunications systems (FPLMTS) and ISDN and satellite matters.

BDT

Unlike the ITU-T (and to some extent ITU-R), which deals with standardization, the BDT deals with aspects that promote the integration and deployment of communications in developing countries. Typical outputs from this sector include implementation guides that expand the utility of ITU recommendations and ensure their expeditious implementation.

ISO/IEC JTC 1

Two global organizations are active in the information processing systems area, the ISO and the IEC, particularly through JTC 1.

TABLE 13.2 ITU-R Study Group Structure

SG 1	Spectrum management techniques
SG 2	Interservice sharing and compatibility
SG 3	Radio wave propagation
SG 4	Fixed satellite service
SG 7	Science services
SG 8	Mobile, radio determination, amateur and related services
SG 9	Fixed service
SG 10	Broadcasting services: sound
SG 11	Broadcasting services: television

Standards Setting Bodies

TABLE 13.3 ISO/IEC/JTC1 Subcommittees

SC 1	Vocabulary
SC 2	Character sets and information coding
SC 6	Telecommunications information exchange between systems
SC 7	Software systems
SC 11	Flexible magnetic media for digital data interchange
SC 14	Representation of data elements
SC 15	Labeling and file structure
SC 17	Identification cards and related devices
SC 18	Document processing and related communication
SC 21	Information retrieval, transfer, and management for open systems interconnection
SC 22	Languages
SC 23	Optical disk cartridges for information interchange
SC 24	Computer graphics and image processing
SC 25	Interconnection information technology management
SC 26	Microprocessor systems
SC 27	Security techniques
SC 28	Office equipment
SC 29	Coded representation of picture, audio and multimedia/hypermedia information.

The ISO comprises national standards bodies, which have the responsibility for promoting and distributing ISO standards within their own countries. ISO technical work is carried out by some 170 technical committees (TC). Technical committees are established by the ISO council and their work program is approved by the technical board on behalf of the council.

The IEC comprises national committees (one from each country) and deals with almost all spheres of electrotechnology, including power, electronics, telecommunications, and nuclear energy. IEC technical work is performed by some 80 technical committees set up by its council. In 1987 a joint technical committee was established incorporating ISO TC97, IEC TC83, and subcommittee 47B to deal with generic information technology. The international standards developed by JTC1 are published under the ISO and IEC logos. The activities of ISO/IEC/JTC 1 are listed in Table 13.3 expressed in terms of its subcommittees (SC).

The ISO and IEC jointly issue directions for the work of the technical committees. The first step toward an international standard is the committee draft (CD). When agreement is reached within the relevant TC, the CD is sent to the central secretariat for registration as a draft international standard (DIS), which is subsequently circulated for voting.

13.3 Regional Standardization

Today the ETSI comes closest to being a true regional standards setting body, although CITEL, once fully developed, will most likely present a second regional (Latin-American) standardization body.

ETSI is the result of the Single act of the European community and the EC commission green paper in 1987 that analyzed the consequences of the Single act and recommended that a European telecommunications standards body be created to develop common standards for telecommunications equipment and networks. Out of this recommendation, the committee for harmonization (CCH) and the European conference for post and telecommunications (CEPT) evolved into ETSI, which formally came into being in March 1988. It should be noted, however, that even though ETSI attributes at least part of its existence to the European Community, its membership is wider than just the European Union Nations.

Because of the way ETSI came into being, ETSI is characterized by a unique aspect, namely, it is often called upon by the European commission to develop standards that are necessary to implement legislation. Such standards, which are referred to as technical basis reports (TBR) and whose application is usually mandatory, are often needed in public procurements, as well as in provisioning

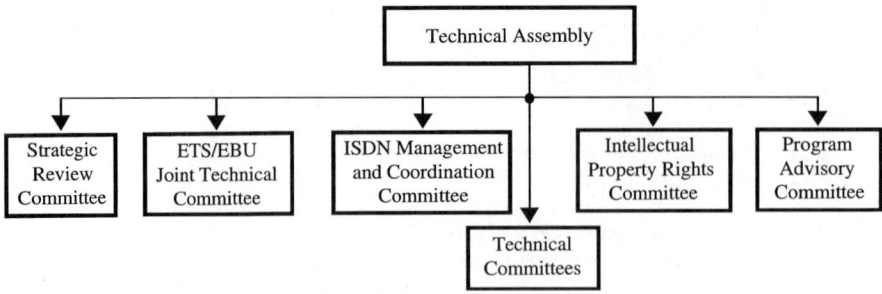

FIGURE 13.2 ETSI technical organization.

for open network interconnection as national telecommunications administrations are being deregulated. Like ITU, however, ETSI also develops voluntary standards in accordance with common international understanding against which industry is not obliged to produce conforming products. These standards fall into either the European technical standard (ETS) class when fully approved, or into the inter im-ETS class, when not fully stable or proven.

ETSI standards are typically sought when either the subject matter is not studied at the global level (such as when it may be required to support some piece of legislation), or the development of the standard is justified by market needs that exist in Europe and not in other parts of the world. In some cases, it may be necessary to adapt ITU standards for the European continent, although a simple endorsement of an ITU standard as a European standard is also possible. A more delicate case arises when both the ITU and ETSI are pursuing parallel standards activities, in which case close coordination with the ITU is sought either through member countries that may input ETSI standards to the ITU for consideration or through the global standards collaboration process.

TABLE 13.4 ETSI Technical Committees

NA	Network aspects
BT	Business telecommunications
SPS	Signaling protocols and switching
EE	Equipment engineering
RES	Radio equipment and systems
SES	Satellite earth stations
PS	Paging systems
TE	Terminal equipment
TM	Transmission & multiplexing
HF	Human factors
ATM	Advanced testing methods
SMG	Special mobile group

The highest authority of ETSI is the general assembly, which determines ETSI's policy, appoints its director and deputy, adopts the budget, and approves the audited accounts. The more technical issues are addressed by the technical assembly, which approves technical standards, advises on the work to be undertaken, and sets priorities. The structure of the technical part of ETSI is shown in Fig. 13.2, and the ETSI technical committees are listed in Table 13.4.

It can be seen that ETSI currently comprises 11 technical committees reporting to the technical assembly. These committees are responsible for the development of technical standards. In addition, these committees are responsible for prestandardization activities, that is, activities leading to ETSI technical reports (ETR) that eventually become the basis for future standards.

In addition to the technical assembly, a strategic review committee (SRC) is responsible for prospective examination of a single technical domain, whereas an intellectual property rights committee defines ETSI's policy in the area of intellectual property. Although by no means unique to ETSI, the rapid pace of technological progress has resulted in standards being adopted that embrace technologies that are still under patent protection. This creates a fundamental conflict between the private, exclusive nature of industrial property rights, and the open, public nature of standards. Harmonizing those conflicting claims has emerged as a thorny issue in all standards organizations; ETSI has established a formal function for this purpose. Finally, the ETS/EBU technical committee coordinates activities with the European broadcasting union (EBU), whereas the ISDN committee is in charge of managing and coordinating the standardization process for narrowband ISDN.

13.4 National Standardization

As standardization moves from global to regional and then to national levels, the number of actual participating entities rapidly grows. Here, the function of two national standards bodies are reviewed, primarily because these have been in existence the longest and secondarily because they also represent major markets for commercial communications.

ANSI T1

Unlike the ETSI, which came into being partly as a consequence of legislative recommendations, the ANSI Committee T1 on telecommunications came into being as a result of the realization that with the breakup of the Bell System, de-facto standards could no longer be expected. In fact, T1 came into being the very same year (1984) that the breakup of the Bell System came into effect.

The T1 membership comprises four types of interest groups: users and general interest groups, manufacturers, interexchange carriers, and exchange carriers. This rather broad membership is reflected, to some extent, by the scope to which T1 standards are being applied; this means that nontraditional telecommunications service providers are utilizing the technologies standardized by committee T1. This situation is the result of the rapid evolution and convergence of the telecommunications, computer, and cable television industries in the United States, and advances in wireless technology.

Committee T1 currently addresses approximately approved 150 projects, which led to the establishment of six, primarily functionally oriented, technical subcommittees (TSC), as shown in Table 13.5 and Fig. 13.3 [although not evident from Table 13.3, subcommittee T1P1 has primary responsibility for management of activities on personal communications systems (PCS)]. In turn, each of these six subcommittees is divided into a number of subtending working groups, and subworking groups.

Committee T1 also has an advisory group (T1AG) made up of representatives from each of the four interest groups to carry out committee T1 directives and to develop proposals for consideration by the T1 membership.

In parallel to serving as the forum that establishes ANSI telecommunications network standards, committee T1 technical subcommittees draft candidate U.S. technical contributions to the ITU. These contributions are submitted to the U.S. Department of State National Committee for the ITU,

TABLE 13.5 T1 Subcommittee Structure

TSC: T1A1	Performance and signal processing
TSC: T1E1	Network interfaces and environmental considerations
TSC: T1M1	Interwork operations, administration, maintenance, and provisioning
TSC: T1P1	Systems engineering, standards planning, and program management
TSC: T1S1	Services, architecture, and signalling
TSC: T1X1	Digital hierarchy and synchronization

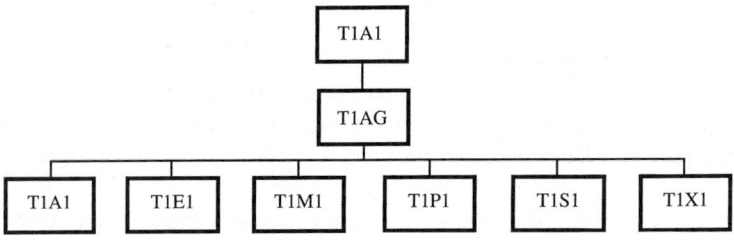

FIGURE 13.3 T1 committee structure.

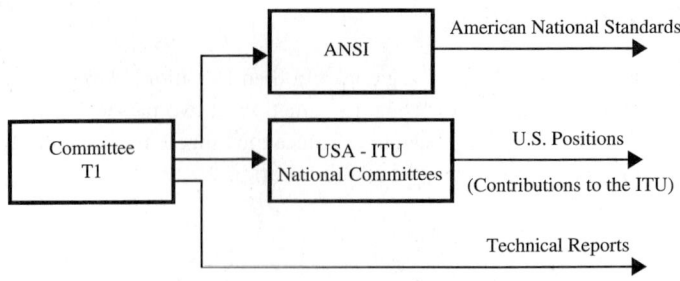

FIGURE 13.4 Committee T1 output.

which administers U.S. participation and contributions to the ITU (see Fig. 13.4). In this manner, activities within T1 are coordinated with those of the ITU. This coordination with other standards setting bodies is also reflected in T1's involvement with Latin-American standards, through the formation of an ad hoc group with CITEL's permanent technical committee 1 (PTC 1/T1). Further coordination with ETSI and other standards setting bodies is accomplished through the global standards collaboration process.

TIA

The TIA is a full-service trade organization that provides its members with numerous services including government relations, market support activities, educational programs, and standards setting activities.

TIA is a member-driven organization. Policy is formulated by 25 board members selected from member companies, and is carried out by a permanent professional staff located in Washington D.C. TIA comprises six issue-oriented standing committees, each of which is chaired by a board member. The six committees are membership scope and development, international, marketing and trade shows, public policy and government relations, and technical. It is this last committee that in 1992 was accredited by ANSI in the United States to standardize telecommunications products. Technology standardization activities are reflected by TIA's four product-oriented divisions, namely, user premises equipment, network equipment, mobile and personal communications equipment, and fiber optics.

In these divisions the legislative and regulatory concerns of product manufacturers and the preparation of standards dealing with performance testing and compatibility are addressed. For example, modem and telematic standards, as well as much of the cellular standards technology, has been standardized in the United States under the mandate of TIA.

TTC

The second national committee to be addressed is the TTC in Japan. TTC was established in October 1985 to develop and disseminate Japanese domestic standards for deregulated technical items and protocols. It is a nongovernmental, nonprofit standards setting organization established to ensure fair and transparent standardization procedures.

TTC's primary emphasis is to develop, conduct studies and research, and disseminate protocols and standards for the connection of telecommunications networks. TTC is organized along six technical subcommittees that report to a board of directors through a technical assembly (see Fig. 13.5).

The TTC organization comprises a general assembly, which is in charge of matters such as business plans and budgets. The councilors meeting examines standards development procedures in order to assure impartiality and clarity. The secretariat provides overall support to the organization;

Standards Setting Bodies

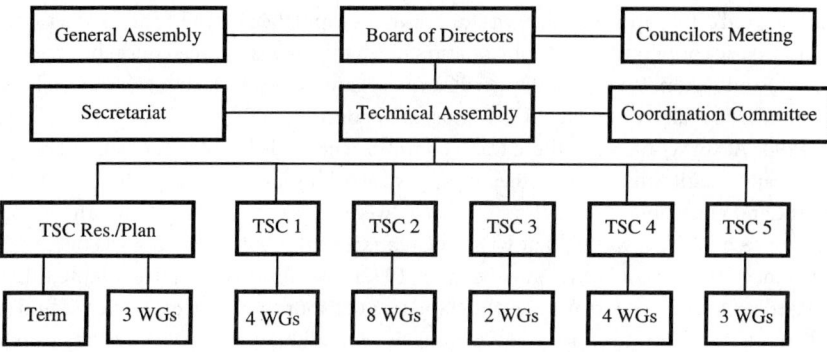

FIGURE 13.5 Organization of TTC.

the technical assembly develops standards and handles technical matters including surveys and research. Each technical subcommittee is partitioned into two or more working groups (WG). The coordination committee handles all issues in or between the TSCs and WGs, and it assures the smooth running of all technical committee meetings.

Under the coordination committee, a subcommittee examines users' requests and studies their applicability to the five-year standardization-project plan. This subcommittee also conducts user-request surveys.

TTC membership is divided into four categories. Type I telecommunications carriers, that is, those carriers that own telecommunications circuits and facilities; type II telecommunications carriers, that is, those with telecommunications circuits leased from type I carriers; related equipment manufacturers; and others, including users.

Underlying objectives that guide TTC's approach to standards development are 1) to conform to international recommendations or standards; 2) standardize items, where either international recommendations or standards are not clear, or where national standards need to be set, and where a consensus is achieved; and 3) to conduct further studies into any of the items just mentioned whenever the technical assembly is unable to arrive at a consensus.

These objectives, which give highest priority in developing standards that are compatible with international recommendations or standards, have often driven TTC to adapt international standards for national use through the use of supplements that:

- Give guidelines on users of TTC standards on how to apply them
- Help clarify the contents of standards
- Help with the implementation of standards in terminal equipment and adaptors
- Assure interconnection between terminal equipment and adaptors
- Provide background information regarding the content of standards
- Assure interconnection.

These supplements also include questions and answers that help in implementing the standards, including encoding examples of various parameters and explanation of the practical meaning of a standard.

13.5 Standards Coordination

Despite the growth of global telecommunications standards activities (between 1989 and 1992 the ITU has produced in three years 19,000 pages of standards, which is almost equal to the number produced in the prior 20 years), at the same time the industry is also witnessing an opposite trend towards regionalization.

In order to avoid duplication of work, waste of resources, and the possibility of arriving at conflicting standards, basic cooperation and coordination mechanisms were agreed upon between the ITU (then CCITT) and regional/national standardization organizations at the interregional telecommunications standardization conference (ITSC) hosted by committee T1 in Fredericksburg, Virginia in February 1990. At this conference there was a commitment made by all organizations represented to achieving and maintaining these objectives. Two working parties, the global standardization management and electronic information exchange, were set up to further pursue these objectives in detail. The second ITSC was held at ETSI's headquarters in Sophia/Antipolis (France), and the third conference was held in Tokyo in November 1992. The third conference adopted the Tokyo plan, a revised structure that streamlines the standards collaboration process while recognizing the rapidly changing telecommunications environment. Accordingly, it approved terms of reference that merged all necessary collaborative activities under a new global standards collaboration (GSC) group. The group will oversee the collaborative process, including work on electronic document handling (EDH) and five high-interest subjects:

- Broadband integrated services digital network (B-ISDN)
- Intelligent networks (IN)
- Transmission management network (TMN)
- Universal personal telecommunications (UPT)
- Synchronous digital hierarchy/synchronous optical network (SDH/SONET).

13.6 Scientific

Another global, scientifically based organization that has been particularly active in standards development (more recently emphasizing information processing) is the IEEE. Responsibility for standards adoption within the IEEE lies with the IEEE standards board. The board is supported by nine standing committees (see Fig. 13.6).

Proposed standards are normally developed in the technical committees of the IEEE Societies. There are occasions, however, when the scope of activity is too broad to be encompassed by a single society or where the societies are not able to do so for other reasons. In this case the standards board establishes its own standards developing committees, namely, the standards coordinating committees (SCC), to perform this function.

The adoption of IEEE standards is based on projects that have been approved by the IEEE standards board, while each project is the responsibility of a sponsor. Sponsors need not be an SCC but can also include technical committees of IEEE Societies; a standards, or standards coordinating committee

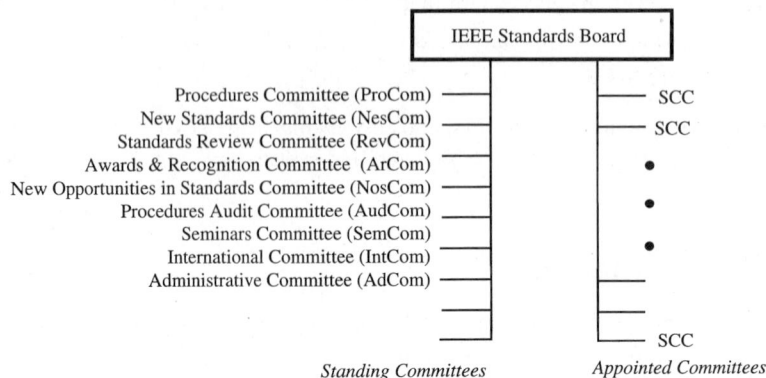

FIGURE 13.6 IEEE standards board organization.

Standards Setting Bodies

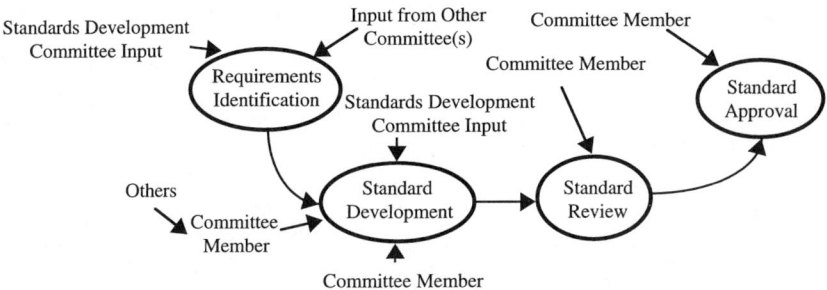

FIGURE 13.7 Typical standards development and approval process.

of an IEEE Society; an accredited standards committee; or another organization approved by the IEEE standards board.

13.7 Standards Development Cycle

Although the manner in which standards are developed and approved somewhat varies between standards organizations, there are common characteristics to be found.

For most standards, first a set of requirements is defined. This may be done either by the standards committee actually developing the standard or by another entity in collaboration with such a committee. Subsequently, the technical details of a standard are developed. The actual entity developing a standard may be a member of the standards committee, or the actual standards committee itself. Outsiders may also contribute to standards development but, typically, only if sponsored by a committee member. Membership in the standards committee and the right to contribute technical information towards the development of the standard differs among the various standards' organizations, as indicated. This process is illustrated in Fig. 13.7.

Finally, once the standard has been fully developed, it is placed under an approval cycle. Each standards setting body typically has precisely defined and often complex procedures for reviewing and then approving proposed standards, which although different in detail, are typically consensus driven.

Further Information

Irmer, T. 1994. Shaping future telecommunications: the challenge of global standardization, *IEEE Comm. Mag.* 32(1):20–28.

Matute, M.A. 1994. CITEL: formulating telecommunications in the Americas. *IEEE Comm. Mag.* 32(1):38–39.

Robin, G. 1994. The European perspective for telecommunications standards. *IEEE Comm. Mag.* 32(1):40–50.

Reilly, A.K. 1994. A US perspective on standards development. *IEEE Comm. Mag.* 32(1):30–36.

Iida, T. 1994. Domestic standards in a changing world. *IEEE Comm. Mag.* 32(1):46–50.

Habara, K. 1994. Cooperation in standardization. *IEEE Comm. Mag.* 32(1):78–84.

IEEE Standards Board Bylaws. 1993. Institute of Electrical and Electronic Engineers. Dec.

Chiarottino, W. and Pirani, G. 1993. International telecommunications standards organizations, *CSELT Tech. Repts.* XXI(2):207–236.

ITU, 1993. Book No. 1. Resolutions; Recommendations on the organization of the Work of ITU-T (series A); study groups and other groups; list of study questions (1993-1996). World Standardization Conf. Helsinki, 1–12, March.

Standards Committee T1. 1992. *Telecommunications.* Procedures Manual. 7th Iss. June.

14
Forward Error Correction Coding

14.1 Introduction 166
14.2 Fundamentals of Block Coding 167
14.3 Structure and Decoding of Block Codes 168
14.4 Important Classes of Block Codes 170
14.5 Principles of Convolutional Coding 171
14.6 Decoding of Convolutional Codes 174
14.7 Trellis-Coded Modulation 176
14.8 Additional Measures 177
14.9 Applications 177

V.K. Bhargava
University of Victoria

I.J. Fair
University of Victoria

14.1 Introduction

In 1948, Claude Shannon issued a challenge to communications engineers by proving that communication systems could be made arbitrarily reliable as long as a fixed percentage of the transmitted signal was redundant [Shannon, 1948]. He did not, however, indicate how this could be achieved. Subsequent research has led to a number of techniques that introduce redundancy to allow for correction of errors without retransmission. These techniques, collectively known as forward error correction (FEC) coding techniques, are used in systems where a reverse channel is not available for requesting retransmission, the delay with retransmission would be excessive, the expected number of errors would require a large number of retransmissions, or retransmission would be awkward to implement [Sklar, 1988].

A simplified model of a digital communication system which incorporates FEC coding is shown in Fig. 14.1. The FEC code acts on a **discrete data channel** comprising all system elements between the encoder output and decoder input. The encoder maps the source data to q-ary code symbols which are modulated and transmitted. During transmission, this signal can be corrupted, causing errors to arise in the demodulated symbol sequence. The FEC decoder attempts to correct these errors and restore the original source data.

A demodulator which outputs only a value for the q-ary symbol received during each symbol interval is said to make **hard decisions**. In the **binary symmetric channel** (BSC), hard decisions are made on binary symbols and the probability of error is independent of the value of the symbol. One example of a BSC is the coherently demodulated binary phase-shift-keyed (BPSK) signal corrupted by additive white Gaussian noise (AWGN). The conditional probability density functions which

Forward Error Correction Coding

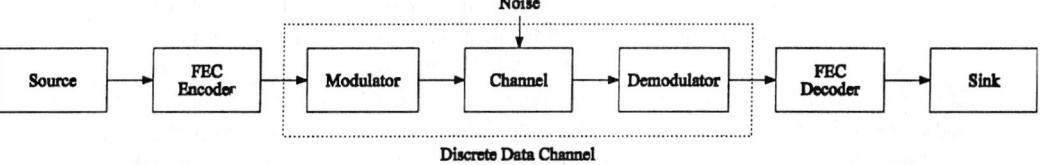

FIGURE 14.1 Block diagram of a digital communication system with forward error correction.

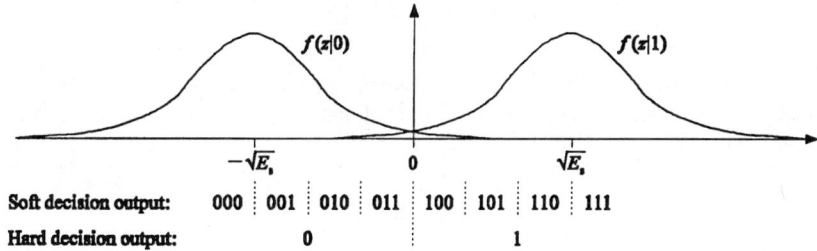

FIGURE 14.2 Hard and soft decision demodulation of a coherently demodulated BPSK signal corrupted by AWGN. $f(z \mid 1)$ and $f(z \mid 0)$ are the Gaussianly distributed conditional probability density functions at the threshold device.

result with this system are depicted in Fig. 14.2. The probability of error is given by the area under the density functions that lies across the decision threshold, and is a function of the symbol energy E_s and the one-sided noise power spectral density N_0.

Alternatively, the demodulator can make **soft decisions** or output an estimate of the symbol value along with an indication of its confidence in this estimate. For example, if the BPSK demodulator uses three-bit quantization, the two least significant bits can be taken as a confidence measure. Possible soft-decision thresholds for the BPSK signal are depicted in Fig. 14.2. In practice, there is little to be gained by using many soft-decision quantization levels.

Block and convolutional codes introduce redundancy by adding parity symbols to the message data. They map k source symbols to n code symbols and are said to have **code rate** $R = k/n$. With fixed information rates, this redundancy results in increased bandwidth and lower energy per transmitted symbol. At low signal-to-noise ratios, these codes cannot compensate for these impairments, and performance is degraded. At higher ratios of information symbol energy E_b to noise spectral density N_0, however, there is **coding gain** since the performance improvement offered by coding more than compensates for these impairments. Coding gain is usually defined as the reduction in required E_b/N_0 to achieve a specific error rate in an error-control coded system over one without coding. In contrast to block and convolutional codes, trellis-coded modulation introduces redundancy by expanding the size of the signal set rather than increasing the number of symbols transmitted, and so offers the advantages of coding to band-limited systems.

Each of these coding techniques is considered in turn. Following a discussion of interleaving and concatenated coding, this chapter concludes with a brief overview of FEC applications.

14.2 Fundamentals of Block Coding

In block codes there is a one-to-one mapping between k-symbol source words and n-symbol codewords. With q-ary signalling, q^k out of the q^n possible n-tuples are valid code vectors. The set of all n-tuples forms a **vector space** in which the q^k code vectors are distributed. The **Hamming** distance between any two code vectors is the number of symbols in which they differ; the **minimum distance** $d_{\min}$ of the code is the smallest Hamming distance between any two codewords.

There are two contradictory objectives of block codes. The first is to distribute the code vectors in the vector space such that the distance between them is maximized. Then, if the decoder receives a corrupted vector, by evaluating the nearest valid code vector it will decode the correct word with high probability. The second is to pack the vector space with as many code vectors as possible to reduce the redundancy in transmission.

When code vectors differ in at least $d_{\min}$ positions, a decoder which evaluates the nearest code vector to each received word is guaranteed to correct up to t random symbol errors per word if

$$d_{\min} \geq 2t + 1 \qquad (14.1)$$

Alternatively, all $q^n - q^k$ illegal words can be detected, including all error patterns with $d_{\min} - 1$ or fewer errors. In general, a block code can correct all patterns of t or fewer errors and detect all patterns of u or fewer errors provided that $u \geq t$ and

$$d_{\min} \geq t + u + 1 \qquad (14.2)$$

If $q = 2$, knowledge of the positions of the errors is sufficient for their correction; if $q > 2$, the decoder must determine both the positions and values of the errors. If the demodulator indicates positions in which the symbol values are unreliable, the decoder can assume their value unknown and has only to solve for the value of these symbols. These positions are called **erasures**. A block code can correct up to t errors and v erasures in each word if

$$d_{\min} \geq 2t + v + 1 \qquad (14.3)$$

14.3 Structure and Decoding of Block Codes

Shannon showed that the performance limit of codes with fixed code rate improves as the block length increases. As n and k increase, however, practical implementation requires that the mapping from message to code vector not be arbitrary but that an underlying structure to the code exist. The structures developed to date limit the error correcting capability of these codes to below what Shannon proved possible, on average, for a code with random codeword assignments. The search for good constructive codes continues.

A property which simplifies implementation of the coding operations is that of code linearity. A code is **linear** if the addition of any two code vectors forms another code vector, which implies that the code vectors form a subspace of the vector space of n-tuples. This subspace, which contains the all-zero vector, is spanned by any set of k linearly independent code vectors. Encoding can be described as the multiplication of the information k-tuple by a **generator matrix** G, of dimension $k \times n$, which contains these basis vectors as rows. That is, a message vector m_i is mapped to a code vector c_i according to

$$c_i = m_i G, \qquad i = 0, 1, \ldots, q^k - 1 \qquad (14.4)$$

where elementwise arithmetic is defined in the **finite field** $GF(q)$. In general, this encoding procedure results in code vectors with nonsystematic form in that the values of the message symbols cannot be determined by inspection of the code vector. However, if G has the form $[I_k, P]$ where I_k is the $k \times k$ identity matrix and P is a $k \times (n-k)$ matrix of parity checks, then the k most significant symbols of each code vector are identical to the message vector and the code has **systematic** form. This notation assumes that vectors are written with their most significant or first symbols in time on the left, a convention used throughout this chapter.

Forward Error Correction Coding

For each generator matrix there is an $(n-k) \times k$ **parity check matrix H** whose rows are orthogonal to the rows in G, i.e., $GH^T = 0$. If the code is systematic, $H = [-P^T, I_{n-k}]$. Since all codewords are linear sums of the rows in G, it follows that $c_i H^T = 0$ for all $i, i = 0, 1, \ldots, q^k - 1$, and that the validity of the demodulated vectors can be checked by performing this multiplication. If a codeword c is corrupted during transmission so that the hard-decision demodulator outputs the vector $\hat{c} = c + e$, where e is a nonzero error pattern, the result of this multiplication is an $(n-k)$-tuple that is indicative of the validity of the sequence. This result, called the **syndrome s**, is dependent only on the error pattern since

$$s = \hat{c}H^T = (c+e)H^T = cH^T + eH^T = eH^T \qquad (14.5)$$

If the error pattern is a code vector, the errors go undetected. For all other error patterns, however, the syndrome is nonzero. Since there are $q^{n-k} - 1$ nonzero syndromes, $q^{n-k} - 1$ error patterns can be corrected. When these patterns include all those with t or fewer errors and no others, the code is said to be a **perfect code**. Few codes are perfect; most codes are capable of correcting some patterns with more than t errors. **Standard array decoders** use lookup tables to associate each syndrome with an error pattern but become impractical as the block length and number of parity symbols increases. Algebraic decoding algorithms have been developed for codes with stronger structure. These algorithms are simplified with imperfect codes if the patterns corrected are limited to those with t or fewer errors, a simplification called **bounded distance decoding**.

Cyclic codes are a subclass of linear block codes with an algebraic structure that enables encoding to be implemented with a linear feedback shift register and decoding to be implemented without a lookup table. As a result, most block codes in use today are cyclic or are closely related to cyclic codes. These codes are best described if vectors are interpreted as polynomials and the arithmetic follows the rules for polynomials where the elementwise operations are defined in GF(q). In a cyclic code, all codeword polynomials are multiples of a **generator polynomial** $g(x)$ of degree $n - k$. This polynomial is chosen to be a divisor of $x^n - 1$ so that a cyclic shift of a code vector yields another code vector, giving this class of codes its name. A message polynomial $m_i(x)$ can be mapped to a codeword polynomial $c_i(x)$ in nonsystematic form as

$$c_i(x) = m_i(x)g(x), \qquad i = 0, 1, \ldots, q^k - 1 \qquad (14.6)$$

In systematic form, codeword polynomials have the form

$$c_i(x) = m_i(x)x^{n-k} - r_i(x), \qquad i = 0, 1, \ldots, q^k - 1 \qquad (14.7)$$

where $r_i(x)$ is the remainder of $m_i(x)x^{n-k}$ divided by $g(x)$. Polynomial multiplication and division can be easily implemented with shift registers [Blahut, 1983].

The first step in decoding the demodulated word is to determine if the word is a multiple of $g(x)$. This is done by dividing it by $g(x)$ and examining the remainder. Since polynomial division is a linear operation, the resulting syndrome $s(x)$ depends only on the error pattern. If $s(x)$ is the all-zero polynomial, transmission is errorless or an undetectable error pattern has occurred. If $s(x)$ is nonzero, at least one error has occurred. This is the principle of the **cyclic redundancy check** (CRC). It remains to determine the most likely error pattern that could have generated this syndrome.

Single error correcting binary codes can use the syndrome to immediately locate the bit in error. More powerful codes use this information to determine the locations and values of multiple errors. The most prominent approach of doing so is with the iterative technique developed by Berlekamp. This technique, which involves computing an error-locator polynomial and solving for its roots, was subsequently interpreted by Massey in terms of the design of a minimum-length shift register. Once the location and values of the errors are known, Chien's search algorithm efficiently corrects them. The implementation complexity of these decoders increases only as the square of the number of errors to be corrected [Bhargava, 1983] but does not generalize easily to accomodate soft-decision information. Other decoding techniques, including Chase's algorithm and threshold decoding, are easier to implement with soft-decision input [Clark and Cain, 1981]. Berlekamp's algorithm can be used in conjunction with transform-domain decoding, which involves transforming the received block with a finite field Fourier-like transform and solving for errors in the transform domain. Since the implementation complexity of these decoders depends on the block length rather than the number of symbols corrected, this approach results in simpler circuitry for codes with high redundancy [Wu et al., 1987].

Other block codes have also been constructed, including product codes that extend the ideas to two dimensions, codes that are based on transform-domain spectral properties, codes that are designed specifically for correction of burst errors, and codes that are decodable with straightforward threshold or majority logic decoders [Blahut, 1983; Clark and Cain, 1981; Lin and Costello, 1983].

14.4 Important Classes of Block Codes

When errors occur independently, Bose–Chaudhuri–Hocquenghem (BCH) codes provide one of the best performances of known codes for a given block length and code rate. They are cyclic codes with $n = q^m - 1$, where m is any integer greater than 2. They are designed to correct up to t errors per word and so have **designed distance** $d = 2t + 1$; the minimum distance may be greater. Generator polynomials for these codes are listed in many texts, including [Clark and Cain, 1981]. These polynomials are of degree less than or equal to mt, and so $k \geq n - mt$. BCH codes can be shortened to accomodate system requirements by deleting positions for information symbols.

Some subclasses of these codes are of special interest. Hamming codes are perfect single error correcting binary BCH codes. Full length codes have $n = 2^m - 1$ and $k = n - m$ for any m greater than 2. The duals of these codes are maximal-length codes, with $n = 2^m - 1$, $k = m$, and $d_{\min} = 2^{m-1}$. All $2^m - 1$ nonzero code vectors in these codes are cyclic shifts of a single nonzero code vector. Reed–Solomon (RS) codes are nonbinary BCH codes defined over GF(q), where q is often taken as a power of two so that symbols can be represented by a sequence of bits. In these cases, correction of even a single symbol allows for correction of a burst of bit errors. The block length is $n = q - 1$, and the minimum distance $d_{\min} = 2t + 1$ is achieved using only $2t$ parity symbols. Since RS codes meet the Singleton bound of $d_{\min} \leq n - k + 1$, they have the largest possible minimum distance for these values of n and k and are called **maximum distance separable** codes.

The Golay codes are the only nontrivial perfect codes that can correct more than one error. The (11, 6) ternary Golay code has minimum distance 5. The (23, 12) binary code is a triple error correcting BCH code with $d_{\min} = 7$. To simplify implementation, it is often extended to a (24, 12) code through the addition of an extra parity bit. The extended code has $d_{\min} = 8$.

The (23, 12) Golay code is also a binary quadratic residue code. These cyclic codes have prime length of the form $n = 8m \pm 1$, with $k = (n + 1)/2$ and $d_{\min} \geq \sqrt{n}$. Some of these codes are as good as the best codes known with these values of n and k, but it is unknown if there are good quadratic residue codes with large n [Blahut, 1983].

Reed–Muller codes are equivalent to binary cyclic codes with an additional overall parity bit. For any m, the rth-order Reed–Muller code has $n = 2^m$, $k = \sum_{i=0}^{r} \binom{m}{i}$, and $d_{\min} = 2^{m-r}$. The rth-order and $(m - r - 1)$th-order codes are duals, and the first-order codes are similar to maximal-length codes. These codes, and the closely related Euclidean geometry and projective geometry codes, can be decoded with threshold decoding.

The performance of several of these block codes is shown in Fig. 14.3 in terms of decoded bit error probability vs E_b/N_0 for systems using coherent, hard-decision demodulated BPSK signalling. Many other block codes have also been developed, including Goppa codes, quasicyclic codes, burst error correcting Fire codes, and other lesser known codes.

Forward Error Correction Coding

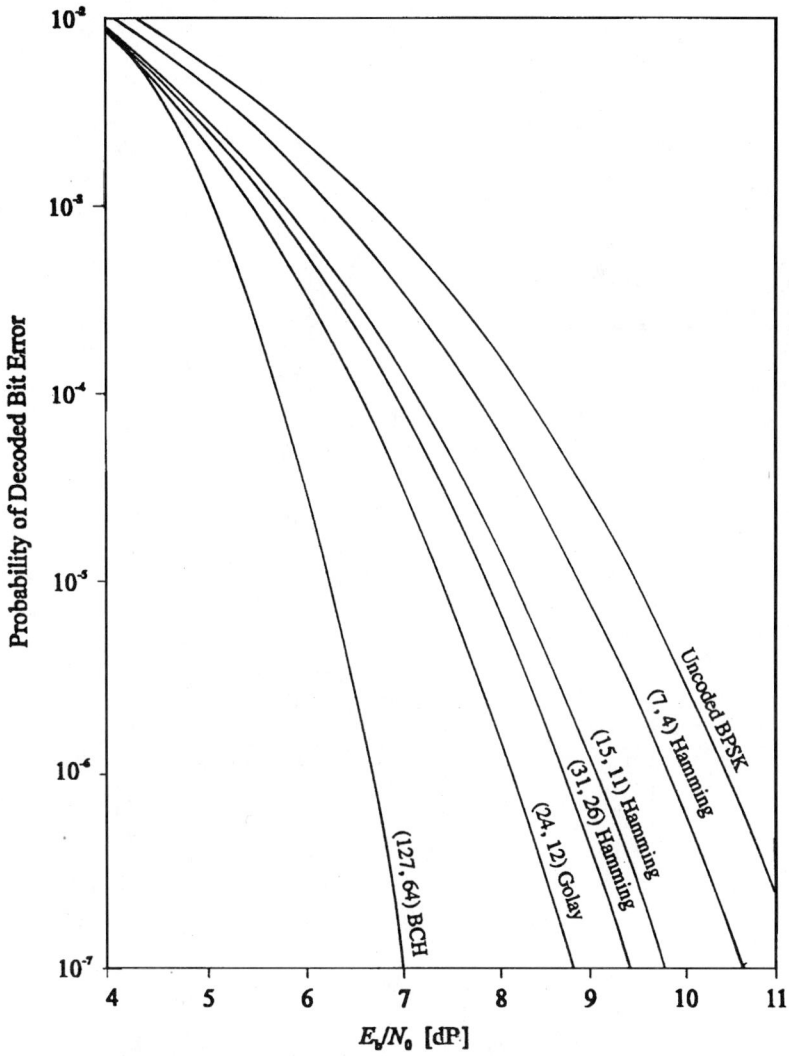

FIGURE 14.3 Block code performance. *Source*: Sklar, B., 1988, *Digital Communications: Fundamentals and Applications*, © 1988, p. 300. Reprinted by permission of Prentice-Hall, Inc., Englewood Cliffs, NJ.

14.5 Principles of Convolutional Coding

Convolutional codes map successive information k-tuples to a series of n-tuples such that the sequence of n-tuples has distance properties that allow for detection and correction of errors. Although these codes can be defined over any alphabet, their implementation has largely been restricted to binary signals, and only binary convolutional codes are considered here.

In addition to the code rate $R = k/n$, the **constraint length** K is an important parameter for these codes. Definitions vary; we will use the definition that K equals the number of k-tuples that affect formation of each n-tuple during encoding. That is, the value of an n-tuple depends on the k-tuple that arrives at the encoder during that encoding interval as well as the $K - 1$ previous information k-tuples.

Binary convolutional encoders can be implemented with kK-stage shift registers and n modulo-2 adders, an example of which is given in Fig. 14.4(a) for a rate 1/2, constraint length 3 code. The

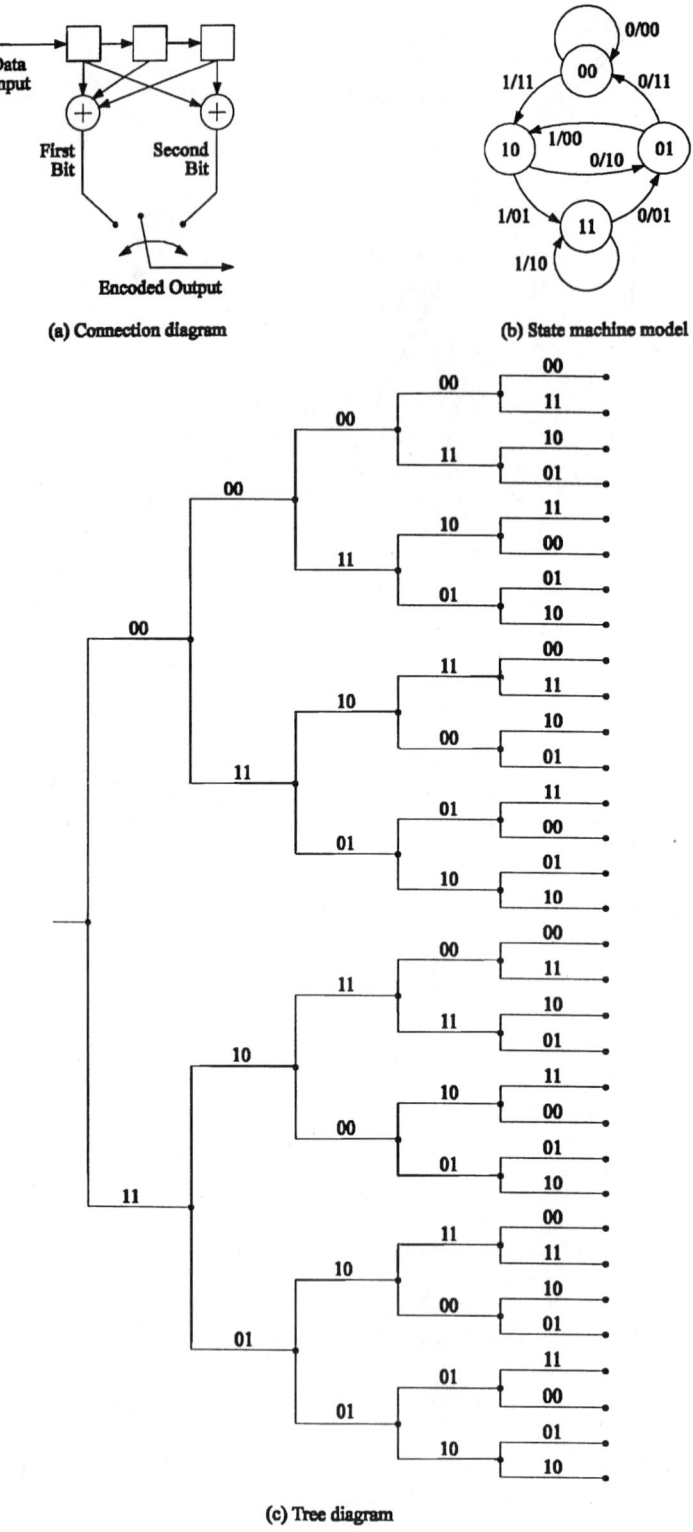

FIGURE 14.4 A rate 1/2, constraint length 3 convolutional code.
(*Continues*)

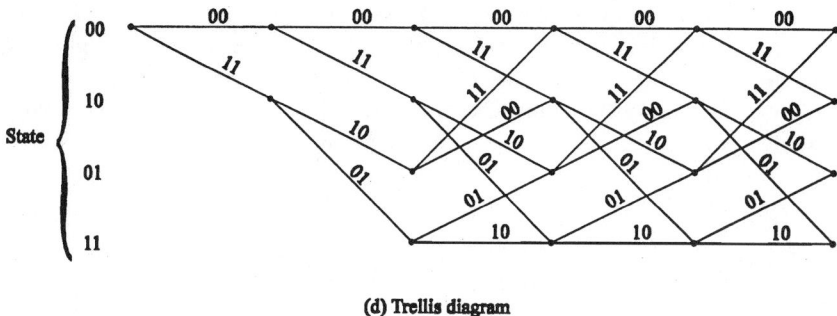

(d) Trellis diagram

FIGURE 14.4 (*Continued*)

encoder shifts in a new k-tuple during each encoding interval and samples the outputs of the adders sequentially to form the coded output.

Although connection diagrams similar to that of Fig. 14.4(a) completely describe the code, a more concise description can be given by stating the values of n, k, and K and giving the adder connections in the form of vectors or polynomials. For instance, the rate 1/2 code has the generator vectors $g_1 = 111$ and $g_2 = 101$, or equivalently, the generator polynomials $g_1(x) = x^2 + x + 1$ and $g_2(x) = x^2 + 1$. Alternatively, a convolutional code can be characterized by its impulse response, the coded sequence generated due to input of a single logic-1. It is straightforward to verify that the circuit in Fig. 14.4(a) has the impulse response 111011. Since modulo-2 addition is a linear operation, convolutional codes are linear, and the coded output can be viewed as the convolution of the input sequence with the impulse response, hence the name of this coding technique. Shifted versions of the impulse response or generator vectors can be combined to form an infinite-order generator matrix which also describes the code.

Shift register circuits can be modeled as finite state machines. A Mealy machine description of a convolutional encoder requires $2^{k(K-1)}$ states, each describing a different value of the $K - 1$ k-tuples which have most recently entered the shift register. Each state has 2^k exit paths which correspond to the value of the incoming k-tuple. A state machine description for the rate 1/2 encoder depicted in Fig. 14.4(a) is given in Fig. 14.4(b). States are labeled with the contents of the two leftmost register stages; edges are labeled with information bit values and their corresponding coded output.

The dimension of time is added to the description of the encoder with tree and trellis diagrams. The tree diagram for the rate 1/2 convolutional code is given in Fig. 14.4(c), assuming the shift register is initially clear. Each node represents an encoding interval, from which the upper branch is taken if the input bit is a 0 and the lower branch is taken if the input bit is a 1. Each branch is labeled with the corresponding output bit sequence. A drawback of the tree representation is that it grows without bound as the length of the input sequence increases. This is overcome with the trellis diagram depicted in Fig. 14.4(d). Again, encoding results in left-to-right movement, where the upper of the two branches is taken whenever the input is a 0, the lower branch is taken when the input is a 1, and the output is the bit sequence which weights the branch taken. Each level of nodes corresponds to a state of the encoder as shown on the left-hand side of the diagram.

If the received sequence contains errors, it may no longer depict a valid path through the tree or trellis. It is the job of the decoder to determine the original path. In doing so, the decoder does not so much correct errors as find the closest valid path to the received sequence. As a result, the error correcting capability of a convolutional code is more difficult to quantify than that of a block code; it depends on how valid paths differ. One measure of this difference is the **column distance** $d_c(i)$, the minimum Hamming distance between all coded sequences generated over i encoding intervals which differ in the first interval. The nondecreasing sequence of column distance values is the **distance profile** of the code. The column distance after K intervals is the minimum distance of the code and is important for evaluating the performance of a code that uses threshold decoding. As i increases, $d_c(i)$ approaches the **free distance** of the code, d_{free}, which is the minimum Hamming distance in the set of arbitrarily long paths that diverge and then remerge in the trellis.

With maximum likelihood decoding, convolutional codes can generally correct up to t errors within three to five constraint lengths, depending on how the errors are distributed, where

$$d_{\text{free}} \geq 2t + 1 \tag{14.8}$$

The free distance can be calculated by exhaustively searching for the minimum-weight path that returns to the all-zero state, or evaluating the term of lowest degree in the generating function of the code.

The objective of a convolutional code is to maximize these distance properties. They generally improve as the constraint length of the code increases, and nonsystematic codes generally have better properties than systematic ones. Good codes have been found by computer search and are tabulated in many texts, including [Clark and Cain, 1981]. Convolutional codes with high code rate can be constructed by **puncturing** or periodically deleting coded symbols from a low rate code. A list of low rate codes and perforation matrices that result in good high rate codes can be found in many sources, including [Wu et al., 1987]. The performance of good punctured codes approaches that of the best convolutional codes known with similar rate, and decoder implementation is significantly less complex.

Convolutional codes can be **catastrophic**, having the potential to generate an unlimited number of decoded bit errors in response to a finite number of errors in the demodulated bit sequence. Catastrophic error propagation is avoided if the code has generator polynomials with a greatest common divisor of the form x^a for any a or, equivalently, if there are no closed-loop paths in the state diagram with all-zero output other than the one taken with all-zero input. Systematic codes are not catastrophic.

14.6 Decoding of Convolutional Codes

In 1967, Viterbi developed a maximum likelihood decoding algorithm that takes advantage of the trellis structure to reduce the complexity of the evaluation. This algorithm has become known as the **Viterbi algorithm**. With each received n-tuple, the decoder computes a **metric** or measure of likelihood for all paths that could have been taken during that interval and discards all but the most likely to terminate on each node. An arbitrary decision is made if path metrics are equal. The metrics can be formed using either hard or soft decision information with little difference in implementation complexity.

If the message has finite length and the encoder is subsequently flushed with zeros, a single decoded path remains. With a BSC, this path corresponds to the valid code sequence with minimum Hamming distance from the demodulated sequence. Full-length decoding becomes impractical as the length of the message sequence increases. The most likely paths tend to have a common stem, however, and selecting the trace value four or five times the constraint length prior to the present decoding depth results in near-optimum performance. Since the number of paths examined during each interval increases exponentially with the constraint length, the Viterbi algorithm also becomes impractical for codes with large constraint length. To date, Viterbi decoding has been implemented for codes with constraint lengths up to ten. Other decoding techniques, such as sequential and threshold decoding, can be used with larger constraint lengths.

Sequential decoding was proposed by Wozencraft, and the most widely used algorithm was developed by Fano. Rather than tracking multiple paths through the trellis, the sequential decoder operates on a single path while searching the code tree for a path with high probability. It makes tentative decisions regarding the transmitted sequence, computes a metric between its proposed path and the demodulated sequence, and moves forward through the tree as long as the metric indicates

Forward Error Correction Coding

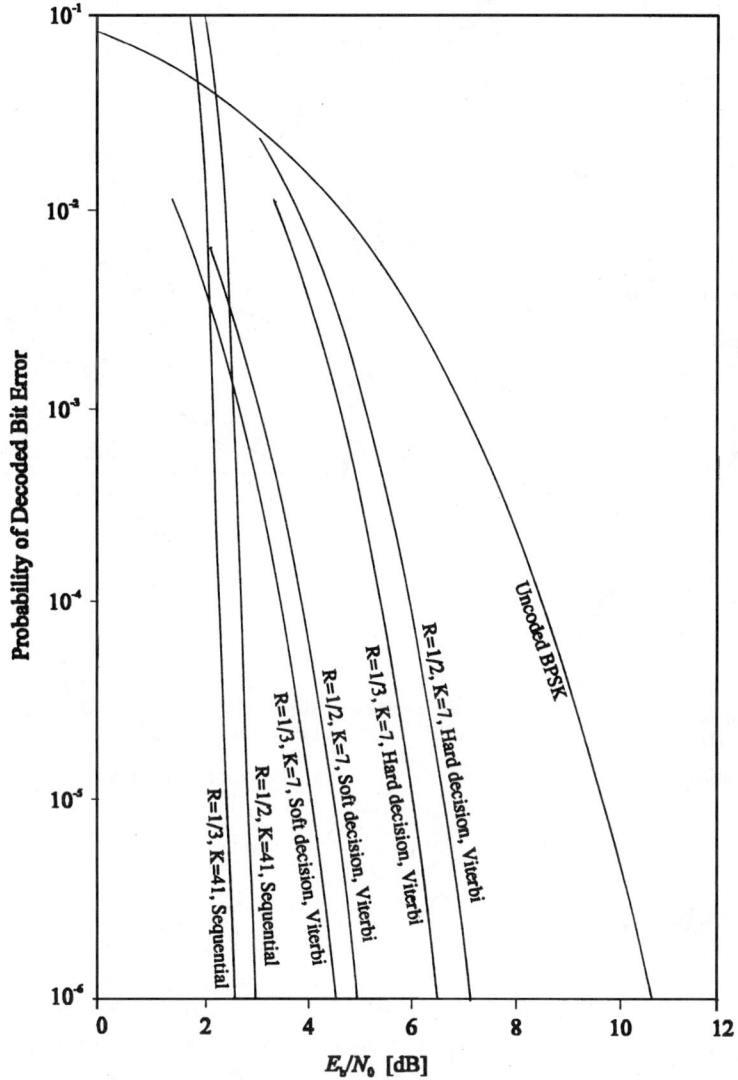

FIGURE 14.5 Convolutional code performance. *Source*: Omura, J.K. and Levitt, B.K., © 1982 IEEE, "Coded Error Probability Evaluation for Antijam Communication Systems," *IEEE Trans. Commun.*, vol. COM-30, no. 5, pp. 896–903. Reprinted by permission of IEEE.

that the path is likely. If the likelihood of the path becomes low, the decoder moves backward, searching other paths until it finds one with high probability. The number of computations involved in this procedure is almost independent of the constraint length and is typically quite small, but it can be highly variable, depending on the channel. Buffers must be provided to store incoming sequences as the decoder searches the tree. Their overflow is a significant limiting factor in the performance of these decoders.

Figure 14.5 compares the performance of the Viterbi and sequential decoding algorithms for several convolutional codes operating on coherently demodulated BPSK signals corrupted by AWGN. Other decoding algorithms have also been developed, including syndrome decoding methods such as table look-up feedback decoding and threshold decoding [Clark and Cain, 1981]. These algorithms are easily implemented but offer suboptimal performance.

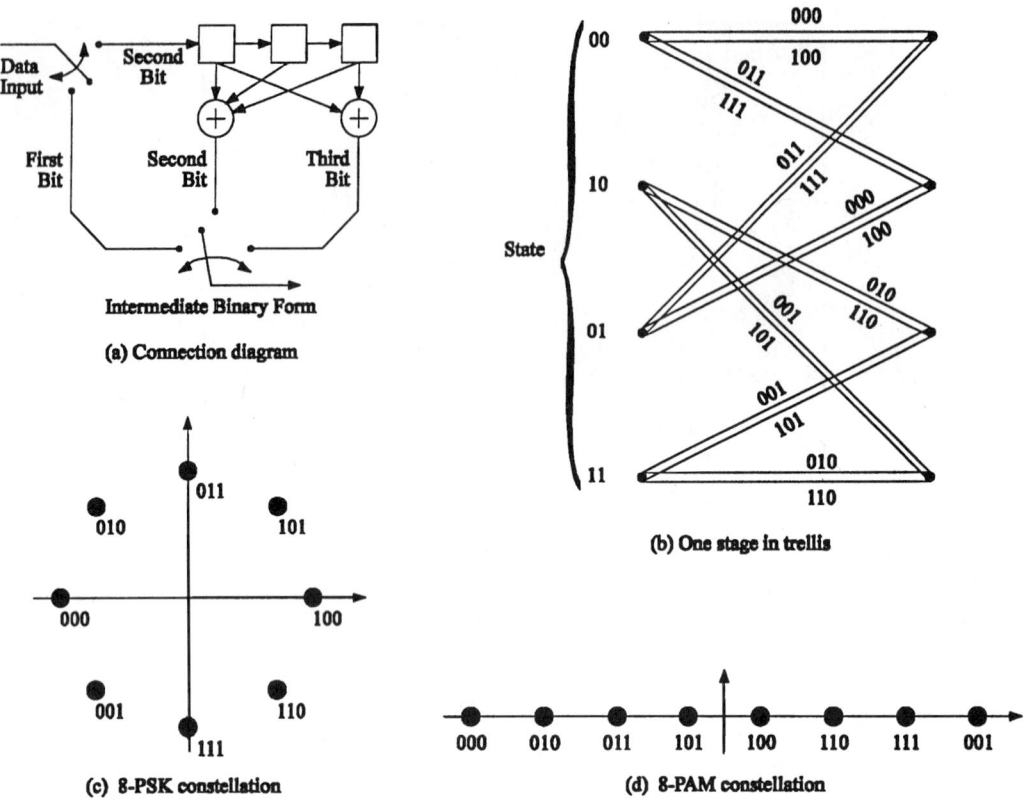

FIGURE 14.6 A rate 2/3 TCM code.

14.7 Trellis-Coded Modulation

Trellis-coded modulation (TCM) has received considerable attention since its development by Ungerboeck in the late 1970s [Ungerboeck, 1987]. Unlike block and convolutional codes, TCM schemes achieve coding gain by increasing the size of the signal alphabet and using multilevel/phase signalling. Like convolutional codes, sequences of coded symbols are restricted to certain valid patterns. In TCM, these patterns are chosen to have large Euclidean distance from one another so that a large number of corrupted sequences can be corrected. The Viterbi algorithm is often used to decode these sequences. Since the symbol transmission rate does not increase, coded and uncoded signals require the same transmission bandwidth. If transmission power is held constant, the signal constellation of the coded signal is denser. The loss in symbol separation, however, is more than overcome by the error correction capability of the code.

Ungerboeck investigated the increase in channel capacity that can be obtained by increasing the size of the signal set and restricting the pattern of transmitted symbols, and concluded that almost all of the additional capacity can be gained by doubling the number of points in the signal constellation. This is accomplished by encoding the binary data with a rate $R = k/(k+1)$ code and mapping sequences of $k+1$ coded bits to points in a constellation of 2^{k+1} symbols. For example, the rate 2/3 encoder of Fig. 14.6(a) encodes pairs of source bits to three coded bits. Figure 14.6(b) depicts one stage in the trellis of the coded output where, as with the convolutional code, the state of the encoder is defined by the values of the two most recent bits to enter the shift register. Note that unlike the trellis for the convolutional code, this trellis contains parallel paths between nodes.

The key to improving performance with TCM is to map the coded bits to points in the signal space such that the Euclidean distance between transmitted sequences is maximized. A method that ensures improved Euclidean distance is the method of **set partitioning**. This involves separating all parallel paths on the trellis with maximum distance and assigning the next greatest distance to paths that diverge from or merge onto the same node. Figures 14.6(c) and 14.6(d) give examples of mappings for the rate 2/3 code with 8-PSK and 8-PAM signal constellations respectively.

As with convolutional codes, the free distance of a TCM code is defined as the minimum distance between paths through the trellis, where the distance of concern is now Euclidean distance rather than Hamming distance. The free distance of an uncoded signal is defined as the distance between the closest signal points. When coded and uncoded signals have the same average power, the coding gain of the TCM system is defined as

$$\text{coding gain} = 20 \log_{10}\left(\frac{d_{\text{free, coded}}}{d_{\text{free, uncoded}}}\right) \quad (14.9)$$

It can be shown that the simple, rate 2/3 8 phase-shift keying (PSK) and 8 pulse-amplitude modulation (PAM) TCM systems provide gains of 3 dB and 3.3 dB, respectively [Clark and Cain, 1981]. More complex TCM systems yield gains up to 6 dB. Tables of good codes are given in [Ungerboeck, 1987].

14.8 Additional Measures

When the demodulated sequence contains bursts of errors, the performance of codes designed to correct independent errors improves if coded sequences are **interleaved** prior to transmission and deinterleaved prior to decoding. Deinterleaving separates the burst errors, making them appear more random and increasing the likelihood of accurate decoding. It is generally sufficient to interleave several block lengths of a block coded signal or several constraint lengths of a convolutionally encoded signal. Block interleaving is the most straightforward approach, but delay and memory requirements are halved with convolutional and helical interleaving techniques. Periodicity in the way sequences are combined is avoided with pseudorandom interleaving.

Concatenated codes, first investigated by Forney, use two levels of coding to achieve a level of performance with less complexity than a single coding stage would require. The inner code interfaces with the modulator and demodulator and corrects the majority of the errors; the outer code corrects errors that appear at the output of the inner-code decoder. A convolutional code with Viterbi decoding is usually chosen as the inner code, and an RS code is often chosen as the outer code due to its ability to correct the bursts of bit errors which can result with incorrect decoding of trellis-coded sequences. Interleaving and deinterleaving outer-code symbols between coding stages offers further protection against the burst error output of the inner code.

14.9 Applications

FEC coding remained of theoretical interest until advances in digital technology and improvements in decoding algorithms made their implementation possible. It has since become an attractive alternative to improving other system components or boosting transmission power. FEC codes are commonly used in digital storage systems, deep-space and satellite communication systems, terrestrial radio and band limited wireline systems, and have also been proposed for fiber optic transmission. Accordingly, the theory and practice of error correcting codes now occupies a prominent position in the field of communications engineering.

Deep-space systems began using forward error correction in the early 1970s to reduce transmission power requirements, and used multiple error correcting RS codes for the first time in 1977 to protect against corruption of compressed image data in the Voyager missions [Wicker and Bhargava, 1994]. The Consultative Committee for Space Data Systems (CCSDS) has since recommended use of a concatenated coding system which uses a rate 1/2, constraint length 7 convolutional inner code and a (255, 223) RS outer code.

Coding is now commonly used in satellite systems to reduce power requirements and overall hardware costs and to allow closer orbital spacing of geosynchronous satellites [Berlekamp et al., 1987]. FEC codes play integral roles in the VSAT, MSAT, INTELSAT, and INMARSAT systems [Wu et al., 1987]. Further, a (31, 15) RS code is used in the joint tactical information distribution system (JTIDS), a (7, 2) RS code is used in the air force satellite communication system (AFSATCOM), and a (204, 192) RS code has been designed specifically for satellite time division multiple access (TDMA) systems. Another code designed for military applications involves concatenation of a Golay and RS code with interleaving to ensure an imbalance of 1's and 0's in the transmitted symbol sequence and enhance signal recovery under severe noise and interference [Berlekamp et al., 1987].

TCM has become commonplace in transmission of data over voiceband telephone channels. Modems developed since 1984 use trellis coded QAM modulation to provide robust communication at rates above 9.6 kb/s.

FEC codes have also been widely used in digital recording systems, most prominently in the compact disc digital audio system. This system uses two levels of coding and interleaving in the cross-interleaved RS coding (CIRC) system to correct errors that result from disc imperfections and dirt and scratches which accumulate during use. Steps are also taken to mute uncorrectable sequences [Wicker and Bhargava, 1994].

Defining Terms

Binary symmetric channel: A memoryless discrete data channel with binary signalling, hard-decision demodulation, and channel impairments that do not depend on the value of the symbol transmitted.

Bounded distance decoding: Limiting the error patterns which are corrected in an imperfect code to those with t or fewer errors.

Catastrophic code: A convolutional code in which a finite number of code symbol errors can cause an unlimited number of decoded bit errors.

Code rate: The ratio of source word length to codeword length, indicative of the amount of information transmitted per encoded symbol.

Coding gain: The reduction in signal-to-noise ratio required for specified error performance in a block or convolutional coded system over an uncoded system with the same information rate, channel impairments, and modulation and demodulation techniques. In TCM, the ratio of the squared free distance in the coded system to that of the uncoded system.

Column distance: The minimum Hamming distance between convolutionally encoded sequences of a specified length with different leading n-tuples.

Concatenated codes: Two levels of codes that achieve a level of performance with less complexity than a single coding stage would require. The inner code is often a convolutional code, and the outer code is usually an RS code.

Cyclic code: A block code in which cyclic shifts of code vectors are also code vectors.

Cyclic redundancy check: When the syndrome of a cyclic block code is used to detect errors.

Designed distance: The guaranteed minimum distance of a BCH code designed to correct up to t errors.

Discrete data channel: The concatenation of all system elements between FEC encoder output and decoder input.

Distance profile: The minimum Hamming distance after each encoding interval of convolutionally encoded sequences which differ in the first interval.

Erasure: A position in the demodulated sequence where the symbol value is unknown.

Finite field: A finite set of elements and operations of addition and multiplication that satisfy specific properties. Often called Galois fields and denoted $GF(q)$, where q is the number of elements in the field. Finite fields exist for all q which are prime or the power of a prime.

Forward Error Correction Coding 179

Free distance: The minimum Hamming weight of convolutionally encoded sequences that diverge and remerge in the trellis. Equals the maximum column distance and the limiting value of the distance profile.

Generator matrix: A matrix used to describe a linear code. Code vectors equal the information vectors multiplied by this matrix.

Generator polynomial: The polynomial that is a divisor of all codeword polynomials in a cyclic block code; a polynomial that describes circuit connections in a convolutional encoder.

Hamming distance: The number of symbols in which codewords differ.

Hard decision: Demodulation that outputs only a value for each received symbol.

Interleaving: Shuffling the coded bit sequence prior to modulation and reversing this operation following demodulation. Used to separate and redistribute burst errors over several codewords (block codes) or constraint lengths (trellis codes) for higher probability of correct decoding by codes designed to correct random errors.

Linear code: A code whose code vectors form a vector space. Equivalently, a code where the addition of any two code vectors forms another code vector.

Maximum distance separable: A code with the largest possible minimum distance given the block length and code rate. These codes meet the Singleton bound of $d_{min} \leq n - k + 1$.

Metric: A measure of goodness against which items are judged. In the Viterbi algorithm, an indication of the probability of a path being taken given the demodulated symbol sequence.

Minimum distance: In a block code, the smallest Hamming distance between any two codewords. In a convolutional code, the column distance after K intervals.

Parity check matrix: A matrix whose rows are orthogonal to the rows in the generator matrix of a linear code. Errors can be detected by multiplying the received vector by this matrix.

Perfect code: A t error correcting (n, k) block code in which $q^{n-k} - 1 = \sum_{i=1}^{t} \binom{n}{i}$.

Puncturing: Periodic deletion of code symbols from the sequence generated by a convolutional encoder for purposes of constructing a higher rate code. Also, deletion of parity bits in a block code.

Set partitioning: Rules for mapping coded sequences to points in the signal constellation that always result in a larger Euclidean distance for a TCM system than an uncoded system, given appropriate construction of the trellis.

Soft decision: Demodulation that outputs an estimate of the received symbol value along with an indication of the reliability of this value. Usually implemented by quantizing the received signal to more levels than there are symbol values.

Standard array decoding: Association of an error pattern with each syndrome by way of a lookup table.

Syndrome: An indication of whether or not errors are present in the demodulated symbol sequence.

Systematic code: A code in which the values of the message symbols can be identified by inspection of the code vector.

Vector space: An algebraic structure comprised of a set of elements in which operations of vector addition and scalar multiplication are defined. For our purposes, a set of n-tuples consisting of symbols from GF(q) with addition and multiplication defined in terms of elementwise operations from this finite field.

Viterbi algorithm: A maximum-likelihood decoding algorithm for trellis codes that discards low-probability paths at each stage of the trellis, thereby reducing the total number of paths that must be considered.

References

Berlekamp, E.R., Peile, R.E., and Pope, S.P. 1987. The application of error control to communications. *IEEE Commun. Mag.* 25(4):44–57.

Bhargava, V.K. 1983. Forward error correction schemes for digital communications. *IEEE Commun. Mag.* 21(1):11–19.

Blahut, R.E. 1983. *Theory and Practice of Error Control Codes*, Addison-Wesley, Reading, MA.

Clark, G.C., Jr. and Cain, J.B. 1981. *Error Correction Coding for Digital Communications*, Plenum Press, New York.

Lin, S. and Costello, D.J., Jr. 1983. *Error Control Coding: Fundamentals and Applications*, Prentice-Hall, Englewood Cliffs, NJ.

Shannon, C.E. 1948. A mathematical theory of communication. *Bell Syst. Tech. J.* 27(3):379–423 and 623–656.

Sklar, B. 1988. *Digital Communications: Fundamentals and Applications*, Prentice-Hall, Englewood Cliffs, NJ.

Ungerboeck, G. 1987. Trellis-coded modulation with redundant signal sets. *IEEE Commun. Mag.* 25(2):5–11 and 12–21.

Wicker, S.B. and Bhargava, V.K. 1994. *Reed-Solomon Codes and Their Applications*, IEEE Press, NJ.

Wu, W.W., Haccoun, D., Peile, R., and Hirata, Y. 1987. Coding for satellite communication. *IEEE J. Selected Areas in Commun.* SAC-5(4):724–748.

Further Information

There is now a large amount of literature on the subject of FEC coding. An introduction to the philosophy and limitations of these codes can be found in the second chapter of Lucky's book *Silicon Dreams: Information, Man, and Machine*, St. Martin's Press, New York, 1989. More practical introductions can be found in overview chapters of many communications texts. The number of texts devoted entirely to this subject also continues to grow. Although these texts summarize the algebra underlying block codes, more in-depth treatments can be found in mathematical texts. Survey papers appear occasionally in the literature, but the interested reader is directed to the seminal papers by Shannon, Hamming, Reed and Solomon, Bose and Chaudhuri, Hocquenghem, Wozencraft, Fano, Forney, Berlekamp, Massey, Viterbi, and Ungerboeck, among others. The most recent advances in the theory and implementation of error control codes are published in *IEEE Transactions on Information Theory* and *IEEE Transactions on Communications*.

15
Automatic Repeat Request

David Haccoun
École Polytechnique de Montréal

Samuel Pierre
Université du Québec

15.1 Introduction 181
15.2 Fundamentals and Basic Automatic Repeat Request Schemes . 182
   Basic Principles • Stop-and-Wait Automatic Repeat Request • Sliding Window Protocols
15.3 Performance Analysis and Limitations 187
   Stop-and-Wait Automatic Repeat Request • Continuous Automatic Repeat Request: Sliding Window Protocols
15.4 Variants of the Basic Automatic Repeat Request Schemes 193
15.5 Hybrid Forward Error Control/Automatic Repeat Request Schemes 194
15.6 Application Problem 195
   Solution
15.7 Conclusion 197

15.1 Introduction

In most digital communication systems, whenever error events occur in the transmitted messages, some action must be taken to correct these events. This action may take the form of an **error correction** procedure that attempts to correct these errors. In some applications where a two-way communication link exists between the sender and the receiver, the receiver may inform the sender that a message has been received in error and, hence, request a repeat of that message. In principle, the procedure may be repeated as many times as necessary until that message is received error free. An error control system in which the erroneously received messages are simply retransmitted is called *automatic-repeat-request* (ARQ).

In ARQ systems, the receiver must perform only an **error detection** procedure on the received messages without attempting to correct the errors. Hence, an error detecting code, in the form of specific redundant or parity-check symbols, must be added to the information-bearing sequence. In general, as the error detecting capability of the code increases, the number of added redundant symbols must also be increased. Clearly, with such a system, an erroneously received message is delivered to the user only if the receiver fails to detect the presence of errors. Since error detection coding is simple, powerful, and quite robust, ARQ systems constitute a simple and efficient method for providing highly reliable transfer of messages from the source to the user over a variety of transmission channels. ARQ systems are therefore widely used in data communication systems that are highly sensitive to errors, such as in computer to computer communications.

This chapter presents and discusses principles, performances, limitations, and variants of basic ARQ strategies. The fundamentals of the basic ARQ schemes are presented in Sec. 15.2, whereas the performance analysis and the limitations are carried out in Sec. 15.3. Section 15.4 presents some of the most common variants of the basic ARQ schemes, and hybrid forward error control (FEC)/ARQ techniques are outlined in Sec. 15.5. Finally, an application problem is provided in Sec. 15.6.

15.2 Fundamentals and Basic Automatic Repeat Request Schemes

Messages transmitted on a communication channel are subjected to errors, which must be detected and/or corrected at the receiver. This section presents the basic principles underlying the concept of error control, using retransmission schemes.

Basic Principles

In the seven-layer open system interconnection (OSI) model, error control refers to some basic functions that may be performed at several layers, especially at the *transport layer* (level 4) and at the *data link layer* (level 2). The transport layer is responsible for the end-to-end transport of information processed as *protocol data unit* (PDU) through a network infrastructure. An intermediate layer, called *network layer* (level 3), ensures that messages, which are broken down into smaller blocks called *packets*, can successfully be switched from one node to another of the network.

The error control on each individual link of the network is performed by the data link layer, which transforms packets received from the network layer into *frames* by adding address, control, and error check fields prior to their transmission into the channel. Figure 15.1 shows an example of the frame structure, whose length, which depends on the particular data link protocol used, usually varies approximately between 50 and 200 bytes. The length of each field also depends on the particular data link control protocol employed. The data link layer must ensure a correct and orderly delivery of packets between switching nodes in the network. The objective of the error control is to maintain the integrity of the data in the frames as they transit through the links, overcoming the loss, duplication, desequencing, and damage of data and control information bits at these various levels. Most ARQ error control techniques integrate error detection using standard *cyclic redundancy check* (CRC), as well as requests for retransmission using PDU.

Error control coding consists essentially of adding to the information bits to be transmitted some extra or redundant bits, called *parity-check* bits, on a regular and controlled manner. These parity-check bits, which are removed at the receiver, do not convey information but allow the receiver to detect and possibly correct errors in the received frames.

The single parity-check bit used in ASCII is an example of a redundant bit for error detection purposes. It is an eighth bit added to the seven information bits in the frame representing the alphanumeric character to be transmitted. This bit takes the value 1 if the number of 1s in the first

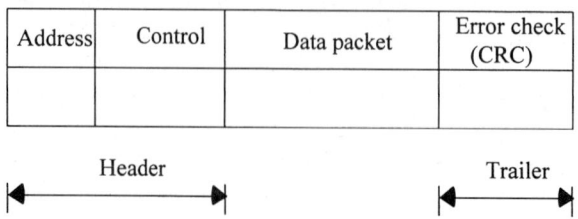

FIGURE 15.1 Structure of a frame.

seven bits is odd (odd parity), and 0 otherwise. At the receiver, the parity check bit of the received frame is computed again in order to verify whether or not it agrees with the received parity check bit. A mismatch of these parity check bits indicates that an odd number of errors has occurred during transmission, leading the receiver to request a retransmission of the erroneous frame.

ARQ refers to retransmission techniques, which basically operate as follows. Erroneously received frames are retransmitted until they are received or detected as being error free; errors are detected using a simple detection code. Positive (ACK) or negative (NAK) acknowledgments are sent back by the receiver to the sender over a reliable **feedback channel**, in order to report whether a previously transmitted frame has been received error free or with errors. In principle, a positive acknowledgment signals the transmitter to send the next data frame, whereas a negative acknowledgment is a request for frame retransmission. *stop-and-wait, go-back-N,* and *selective-repeat* are the three basic versions of ARQ, the last two schemes being usually referred to as *sliding window protocols*.

Stop-and-Wait Automatic Repeat Request

Stop-and-wait ARQ is a very simple acknowledgment scheme, which works as follows. Sender and receiver communicate through a **half-duplex** point-to-point **link**. Following transmission of a single frame to the receiver, the sender waits for an acknowledgment from the receiver before sending the next data frame or repeating the same frame. Upon receiving a frame, the receiver sends back to the sender a positive acknowledgment (ACK) if the frame is correct; otherwise it transmits a negative acknowledgment (NAK).

Figure 15.2 illustrates the frame flow for stop-and-wait ARQ. Frames $F1, F2, F3, \ldots$ are in queue for transmission. At time t_0, the sender transmits frame $F1$ to the receiver and stays idle until either an ACK or NAK is received for that frame. At time t_1, $F1$ is correctly received, hence an ACK is sent back. At time t_2, the sender receives this ACK and transmits the next frame $F2$, which is received at time t_3 and detected in error. Frame $F2$ is then discarded at the receiver and a NAK is therefore returned to the sender. At time t_4, the sender receiving a NAK for $F2$ retransmits a copy of that frame and waits again, until it receives at time t_6 either an ACK or NAK for $F2$. Assuming it is an ACK, as illustrated in Fig. 15.2, the sender then proceeds to transmit the next frame $F3$, and so on. Clearly, such a technique requires the sender to store in a buffer a copy of a transmitted frame until a positive acknowledgment has been received for that frame.

Obviously, this technique does not take into account all contingencies. Indeed, it is possible that some transmitted frames damaged by transmission impairments do not even reach the receiver, in which case the receiver cannot even acknowledge them. To overcome this situation, a timer is integrated in the sender for implementing a *time-out* mechanism. A time out is some pre-established

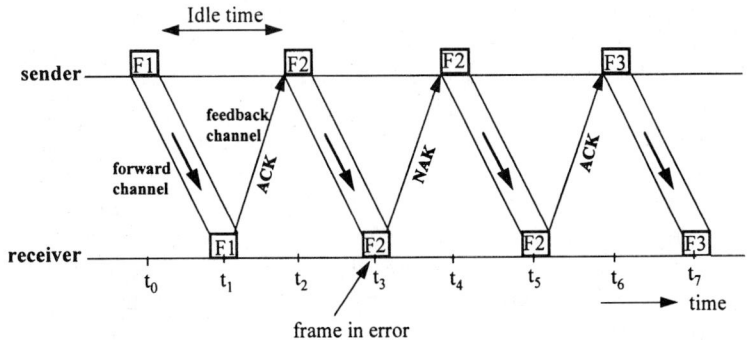

FIGURE 15.2 Stop-and-wait ARQ.

time interval beyond which a transmitted frame is considered lost if its acknowledgment is not received; a timed-out frame is usually simply retransmitted.

Another undesirable situation can be described as follows. The sender transmits a copy of some frame F, which is received correctly by the receiver. A positive acknowledgment (ACK) is then returned to the sender. Unfortunately, this ACK may be corrupted during its transmission in the feedback channel so that it becomes unrecognizable by the sender which therefore transmits another copy of the same frame. As a result, there is a possibility of frame duplication at the receiver leading to an ambiguity between a frame and its immediate predecessor or successor. This ambiguity can be easily resolved by attaching a sequence number in each transmitted frame. In practice, a single-bit (0 or 1) inserted in the header of each frame is sufficient to distinguish a frame from its duplicate at the receiver.

Stop-and-wait ARQ schemes offer the advantage of a great simplicity. However, they are not very suitable for high-speed modern communication systems since data is transmitted in one direction only. Protocols in which the ACK for a frame must be received by the sender prior to the transmission of the next frame are referred to as *positive acknowledgment with retransmission* (PAR) protocols. The principal weakness of PAR schemes is an inefficient link utilization related to the time lost awaiting for acknowledgments before sending another frame, especially if the propagation delay is significantly longer than the packet transmission time, such as in satellite links. To overcome this shortcoming, PAR schemes have been adapted to more practical situations, which require transmitting frames in both directions, using **full-duplex links** between sender and receiver. Clearly then, each direction of transmission acts as the **forward channel** for the frames and as the return channel for the acknowledgments. Now in order to improve further the efficiency of each channel, ACK/NAK control packets and information frames need not be sent separately. Instead, the acknowledgment about each previously received frame is simply appended to the next transmitted frame. In this technique called *piggybacking*, clearly, transmission of the acknowledgment of a received frame is delayed by the transmission of the next data frame. A fixed maximum waiting time for the arrival of a new frame is thus set. If no new frame has arrived by the end of that waiting time, then the acknowledgment frame is sent separately. Even with piggybacking, PAR schemes still require the sender to wait for an acknowledgment before transmitting another frame. An obvious improvement of the efficiency may be achieved by relaxing this condition and allow the sender to transmit up to W, $W > 1$ frames without waiting for an acknowledgment. A further improvement still is to allow a continuous transmission of the frames without any waiting for acknowledgments. These protocols are called *Continuous ARQ* and are associated with the concept of sliding window protocols.

Sliding Window Protocols

To overcome the inefficiency of PAR schemes, an obvious solution is not to stay idle between frame transmissions. One approach is to use a sliding window protocol where a window refers to a subset of consecutive frames.

Consider a window that contains N frames numbered $\omega, \omega + 1, \ldots, \omega + N - 1$, as shown in Fig. 15.3. Every frame using a number smaller than ω has been sent and acknowledged as identified in Fig. 15.3 as *frames before the window*, whereas no frame with a number larger than or equal to $\omega + N$, identified as *frames beyond the window*, has yet been sent. Frames in the window that have been sent but that have not yet been acknowledged are said to be *outstanding frames*. As outstanding frames are being acknowledged, the window shifts to the right to exclude acknowledged frames and, subsequently, to include the new frames that have to be sent. Since the window must always contain frames that are numbered consecutively, the frames are excluded from the window in the same order in which they were included.

To limit the number of outstanding frames, a limit on the window size may be determined. When this limit is reached, the sender accepts no more frames. A window size $N = 1$ corresponds to the

Automatic Repeat Request

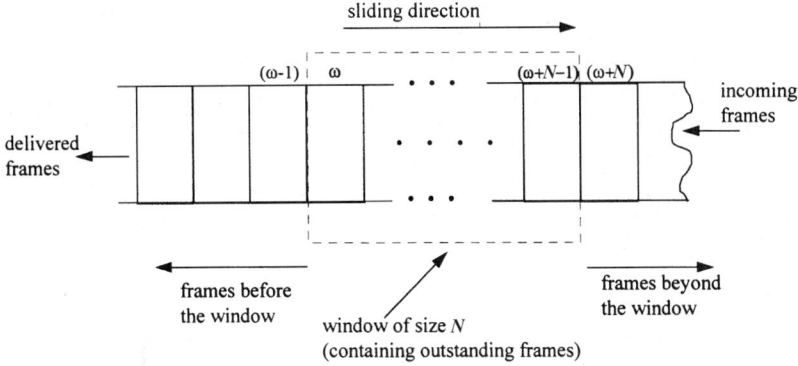

FIGURE 15.3 Frame sequencing in a sliding window protocol (sender side).

stop-and-wait protocol, whereas a window size greater than the total number of frames that can be represented in the header field of the frame refers to an *unrestricted protocol*. Clearly, the window size has some influence on the network traffic and on buffering requirements. Go-back-N and selective-repeat may be considered as two common implementations of a sliding window protocol as will be described next.

Go-Back-N Automatic Repeat Request

Go-back-N ARQ schemes use continuous transmission without waiting for ACK between frames. Clearly, a full-duplex link between the sender and the receiver is required, allowing a number of consecutive frames to be sent without receiving an acknowledgment. In fact, ACKs may not even be transmitted. Upon detecting a frame in error, the receiver sends a NAK for that frame and discards that frame and all succeeding frames, until that erroneous frame has been correctly received. Following reception of a NAK, the sender completes the transmission of the current frame and retransmits the erroneous frame and all subsequent ones.

In accordance with the sliding window protocol, each frame is numbered by an integer $\ell = 0, 1, 2, \ldots, 2^k - 1$, where k denotes the number of bits in the number field of the frame. In general, in order to facilitate their interpretation, frames are all numbered consecutively modulo-2^k. For example, for $k = 3$, frames are numbered 0–7 repeatedly.

It is important to fix the maximum window size, that is, the maximum number of outstanding frames waiting to be transmitted at any one time. To avoid two different outstanding frames having the same number, as well as to prevent pathological protocol failures, the maximum window size is $N = 2^k - 1$. It follows that the number of discarded frames is at most equal to N; however, the actual exact number of discarded frames depends also on the propagation delay. For most usual applications, a window of size $N = 7$ (i.e., $k = 3$) is adequate, whereas for some satellite links the window size N is usually equal to 127.

Figure 15.4 illustrates the frame flow for a go-back-N ARQ scheme, with $k = 3$ and, hence, $N = 7$. Thus the sender can transmit sequences of frames numbered $F0, F1, F2, \ldots, F7$. As shown in Fig. 15.4, since the receiver detects an error on frame $F3$, it returns NAK3 to the sender and discards both $F3$ and the succeeding frames $F4$, $F5$, and $F6$, which because of propagation delays have already been transmitted by the sender before it received the negative acknowledgment NAK3. Then, the sender retransmits $F3$ and the sequence of frames $F4$, $F5$, and $F6$, and proceeds to transmit $F7$ (using the current sequencing) as well as frames $F0, F1, \ldots$, using the next sequencing. Obviously should any of the preceding frames be received in error, then the same repeat procedure is implemented, starting from that frame in error.

Go-back-N procedures have been implemented in various data link protocols such as high-level data link control (HDLC), which refers to layer 2 of the seven-layer OSI model, and transmission

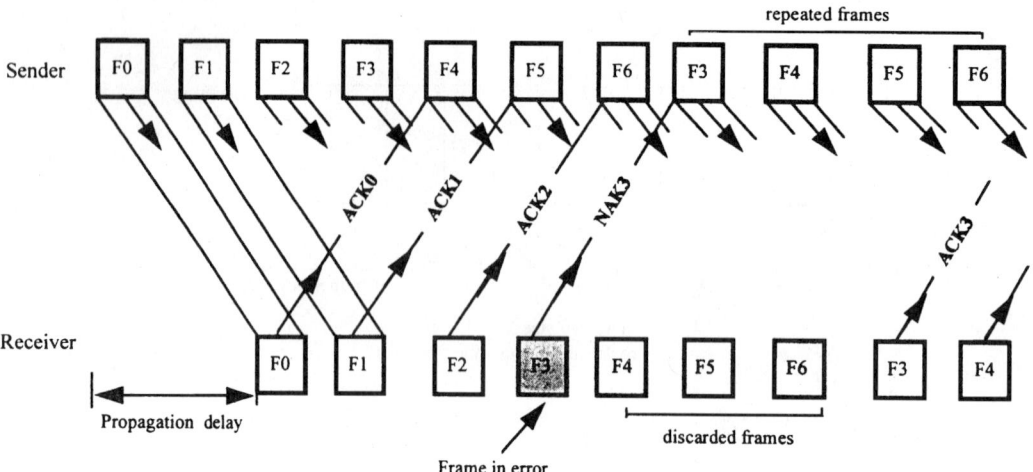

FIGURE 15.4 Go-back-N ARQ.

control protocol/internet protocol (TCP/IP), which supports the Internet, the world's largest interconnections of computer networks. Indeed, to ensure reliable end-to-end connections and improve the performance, TCP uses a sliding window protocol with timers. In accordance with this protocol, the sender can transmit several TCP messages or frames before an acknowledgment is received for any of the frames. The number of frames allowed in transit is negotiated dynamically using the window field in the TCP header.

As an implementation of the sliding window protocol, the go-back-N scheme is more efficient and more general than the stop-and-wait scheme, which is clearly equivalent to go-back-N, with $N = 1$. However, go-back-N presents the disadvantage of requiring a prerequisite numbering of frames as well as the buffering in the window of N frames awaiting positive acknowledgments. Furthermore, in the absence of a controlled source, all incoming frames not yet accepted by the window must be buffered while they wait to be inserted in the window. Inefficiency is also due to the discarding of all frames that follow the frame received in error, even though these frames may be error free. On the other hand, such a procedure presents the advantage of maintaining the proper sequencing of the frames accepted at the receiver.

Selective-Repeat Automatic Repeat Request

To overcome the inefficiency resulting from unnecessarily retransmitting the error-free frames, retransmission can be restricted to only those frames that have been detected in error. Repeated frames correspond to erroneous or damaged frames that are negatively acknowledged as well as frames for which the time out has expired. Such a procedure is referred to as selective-repeat ARQ, which obviously should improve the performance over both stop-and-wait and go-back-N schemes. Naturally, a full-duplex link is again assumed.

Figure 15.5 illustrates the frame flow for selective-repeat ARQ. The frames appear to be disjointed but there is no idle time between consecutive frames. The sender transmits a sequence of frames $F1, F2, F3, \ldots$. As illustrated in Fig. 15.5, the receiver detects an error on the third frame $F3$ and thus returns a negative acknowledgment denoted NAK3 to the sender. However, due to the continuous nature of the protocol, the succeeding frames $F4, F5$, and $F6$ have already been sent and are in the pipeline by the time the sender receives NAK3. Upon receiving NAK3, the sender completes the transmission of the current frame ($F6$) and then retransmits $F3$ before sending the next frame $F7$ in the sequence. The correctly received but incomplete sequence $F4, F5, F6$ must

Automatic Repeat Request

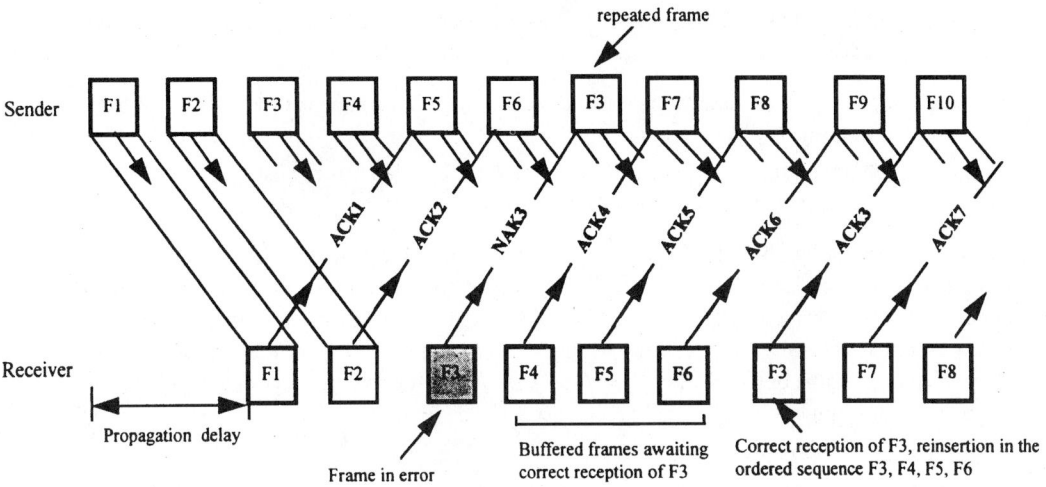

FIGURE 15.5 Selective-repeat ARQ.

be buffered at the receiver, until $F3$ has been correctly received and inserted in its proper place to complete the sequence $F3, F4, F5, F6, \ldots$, which is then delivered.

If $F3$ is correctly received at the first retransmission, the buffered sequence is $F4, F5$, and $F6$. However, if $F3$ is received in error during its retransmission and hence must be retransmitted again, then clearly the buffered sequence is $F4, F5, F6, F7, F8, F9$. Therefore, multiple retransmissions of a given frame leads to an ever larger buffering, which may be theoretically unbounded. To circumvent such a possibility and the ensuing buffer overflow, a time out mechanism is usually incorporated in all practical selective-repeat ARQ systems.

In practice, each correctly received frame prior to the appearance of an erroneous frame denoted F is delivered to the user (i.e., the physical layer). Thereafter, all error free successive frames are buffered at the receiver, until frame F is correctly received. It is then inserted at the beginning of the buffered sequence, which can thus be safely delivered to the user up to the next frame received in error. However, the reordering becomes somewhat more complex if other frames are received in error before frame F is finally received error free. As a consequence, proper buffering must be provided by the sender, which must also integrate the appropriate logic in order to send frames that are out of sequence. The receiver must also be provided with adequate buffering to store the out-of-order frames within the window until the frame in error is correctly received, in addition to be provided with the appropriate logic for inserting the accepted frames in their proper place in the sequence. The required buffers may be rather large for satellite transmission with long propagation delays and long pipelines corresponding to the many frames in transit between sender and receiver. As a result, selective-repeat ARQ procedure tends to be less commonly used than the go-back-N ARQ scheme, even though it may be more efficient from a throughput point of view.

15.3 Performance Analysis and Limitations

In ARQ systems, the performance is usually measured using two parameters: throughput efficiency and undetected error probability on the data bits. The throughput efficiency is defined as the ratio of the average number of information bits per second delivered to the user, to the average number of bits per second that have been transmitted in the system. This throughput efficiency is obviously smaller than 100%. For example, using an error-detecting scheme with a coding rate $R = 0.98$, an error-free transmission would then correspond to a throughput efficiency of 98%, and clearly, any frame retransmission yields a further decrease of the throughput efficiency.

The reliability of an ARQ system is measured by its frame error probability, which may take one of the following forms:

- P = Probability of error detection. It is the probability of an erroneously received frame detected in error by the error detecting code.
- P_u = Probability of undetected error. It is the probability of an erroneously received frame *not* detected in error by the error-detecting code.

Let P_c denote the probability of receiving an error-free frame. We have then $P_c = 1 - P - P_u$. All of these probabilities depend on both the channel error statistics and the error detection code implemented by CRC. By a proper selection of the CRC, these error probabilities can be made very small. Assuming white noise on a channel having a bit error probability p, the probabilities of correctly receiving a frame of length L is $P_c = (1 - p)^L$. As for the undetectable error probability P_u, it is usually vanishingly small, typically $P_u < 1 \times 10^{-10}$.

Stop-and-Wait Automatic Repeat Request

For error-free operation using stop-and-wait ARQ technique, *the utilization rate U_R of a link is* defined as

$$U_R = \frac{T_s}{T_T} \qquad (15.1)$$

where T_s denotes the time required to transmit a single frame, and T_T the overall time between transmission of two consecutive frames, including processing and ACK/NAK transmissions. As shown in Fig. 15.6, the total time T_T can be expressed as

$$T_T = T_p + T_s + T_c + T_p + T_a + T_c \qquad (15.2)$$

In this expression, T_p denotes the propagation delay, that is, the time needed for a transmitted bit to reach the receiver; T_c is the processing delay, that is, the time required for either the sender or the receiver to perform the necessary processing and error checking, whereas T_s and T_a denote the transmission duration of a data frame and of an ACK/NAK frame, respectively.

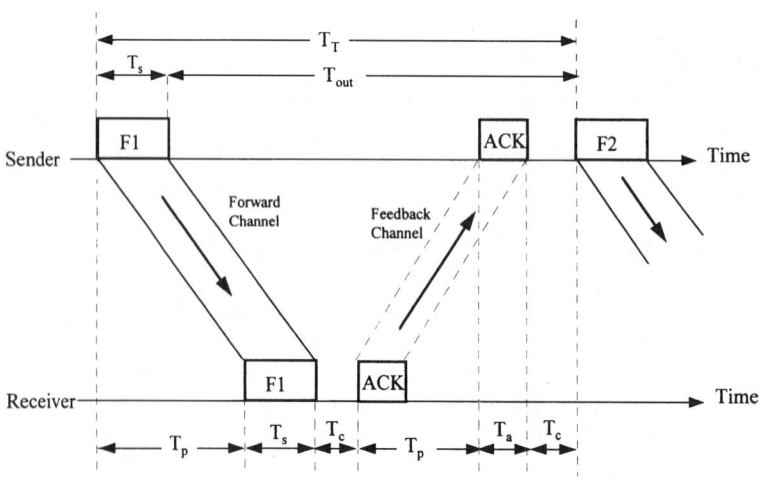

FIGURE 15.6 Timing of successive frames in stop-and-wait ARQ scheme.

Automatic Repeat Request

Assuming the processing time T_c is negligible with respect to T_p and that the sizes of the ACK/NAK frames are very small, leading to negligible value for T_a, then Eq. (15.2) becomes

$$T_T \approx T_s + 2T_p \tag{15.3}$$

Defining the propagation delay ratio $\alpha = T_p/T_s$, then Eq. (15.1) can be written as

$$U_R = \frac{1}{1 + 2\alpha} \tag{15.4}$$

Clearly Eq. (15.4) may be seen as a lower bound on the utilization rate or throughput efficiency that the communication link can achieve over all of the frames.

The utilization rate expressed by Eq. (15.4) may be used to evaluate the throughput efficiency on a per frame basis. In the remainder of the chapter, unless specified otherwise, the throughput efficiency is considered on a per frame basis. Because of repetitions caused by transmission errors, using Eq. (15.1), the *average utilization rate*, or *throughput efficiency*, is defined as

$$\eta = \frac{U_R}{N_t} = \frac{T_s}{N_t T_T} \tag{15.5}$$

where N_t is the expected number of transmissions per frame.

In this definition of the throughput efficiency, the coding rate of the error detecting code, as well as the other overhead bits, is not taken into account. That is, definition (15.5) represents the throughput efficiency on a per frame performance basis. Some other definitions of the throughput efficiency may be related to the number of information bits actually delivered to the destination. In such a case, then the new throughput efficiency η^* is simply equal to

$$\eta^* = \eta(1 - \rho)$$

where ρ represents the fraction of all redundant and overhead bits in the frame.

Assuming error-free transmissions of ACK and NAK frames, and assuming independence for each frame transmission, the probability of requiring exactly k attempts to receive successfully a given frame is $P_k = P^{k-1}(1 - P)$, $k = 1, 2, 3, \ldots$, where P is the frame error probability. The average number of transmissions N_t that are required before a frame is accepted by the receiver is then

$$N_t = \sum_{k=1}^{\infty} k P_k = \sum_{k=1}^{\infty} k P^{k-1}(1 - P) = \frac{1}{1 - P} \tag{15.6}$$

Using Eqs. (15.4–15.6) the throughput efficiency can be thus written as

$$\eta = \frac{1 - P}{1 + 2\alpha} \tag{15.7}$$

The average overall transmission time T_v, for an accepted frame, is then given by

$$T_v = \sum_{k=1}^{\infty} k P_k T_T = \frac{T_T}{1 - P} \tag{15.8}$$

As expected, Eq. (15.8) indicates that if the channel is error free and thus $P = 0$, then $T_v = T_T$, whereas if the channel is very noisy with $P \to 1$, then the average transmission time T_v may become theoretically unbounded.

Continuous Automatic Repeat Request: Sliding Window Protocols

For continuous ARQ schemes, the basic expression of the throughput efficiency Eq. (15.5) must be used with the following assumptions: the transmission duration of a data frame is normalized to $T_s = 1$; the transmission time of an ACK/NAK frame, T_a, as well as the processing delay of any frame, T_c, are negligible. Since $\alpha = T_p/T_s$ and $T_s = 1$, then the propagation delay T_p is equal to α.

Let the sender start transmitting the first frame $F1$ at time t_0. Since $T_p = \alpha$, the beginning of that frame will reach the receiver at time $t_0 + \alpha$. Therefore, given that $T_s = 1$, frame $F1$ will be entirely received at time $(t_0 + \alpha + 1)$. The processing delay being negligible, the receiver can immediately acknowledge frame $F1$, by returning ACK1. As a result, since the transmission duration T_a of an ACK/NAK frame is negligible, ACK1 will be delivered to the sender at time $(t_0 + 2\alpha + 1)$. Let N be the window size, then two cases may be considered: $N \geq 2\alpha + 1$ and $N < 2\alpha + 1$.

If $N \geq 2\alpha + 1$, the acknowledgment ACK1 for frame $F1$ will reach the sender before it has exhausted (e.g., emptied) its own window, and hence the sender can transmit continuously without stopping. In this case, the utilization rate of the transmission link is 100%, that is, $U_R = 1$.

If $N < 2\alpha + 1$, the sender will have exhausted its window at time $t_0 + N$, and thus cannot transmit additional frames until time $(t_0 + 2\alpha + 1)$. In this case, the utilization rate is obviously smaller than 100% and can be expressed as $U_R = N/(2\alpha + 1)$. Furthermore, the protocol is no longer a continuous protocol since there is a break in the sender transmission.

Assuming there are no transmission errors, it follows that the utilization rate of the transmission link for a continuous ARQ using a sliding window protocol of size N is given by

$$U_R = \begin{cases} 1, & N \geq 2\alpha + 1 \\ \dfrac{N}{1 + 2\alpha}, & N < 2\alpha + 1 \end{cases} \quad (15.9)$$

Selective-Repeat Automatic Repeat Request

For selective-repeat ARQ scheme, the expression for the throughput efficiency can be obtained by dividing Eq. (15.9) by $N_t = 1/(1 - P)$ yielding

$$\eta_{SR} = \begin{cases} 1 - P, & N \geq 2\alpha + 1 \\ \dfrac{N(1 - P)}{1 + 2\alpha}, & N < 2\alpha + 1 \end{cases} \quad (15.10)$$

Go-Back-N Automatic Repeat Request

For the go-back-N ARQ scheme, each frame in error necessitates the retransmission of M frames ($M \leq N$) rather than just one frame, where M depends on the roundtrip propagation delay and the frame size. Let $g(k)$ denote the total number of transmitted frames corresponding to a particular frame being transmitted k times. Since each repetition involves M frames, we can write [Stallings, 1994]

$$g(k) = 1 + (k - 1)M = (1 - M) + kM$$

Using the same approach as in Eq. (15.6), the average number of transmitted frames N_t to successfully transmit one frame can be expressed as

$$N_t = \sum_{k=1}^{\infty} g(k) P^{k-1}(1 - P) = \frac{1 - P + PM}{1 - P} \quad (15.11)$$

Automatic Repeat Request

If $N \geq 2\alpha + 1$, the sender transmits continuously and, consequently, M is approximately equal to $2\alpha + 1$. In this case, in accordance with Eq. (15.11), $N_t = (1 + 2\alpha P)/(1-P)$. Dividing Eq. (15.9) by N_t, the throughput efficiency becomes $\eta = (1-P)/(1+2\alpha P)$.

If $N < 2\alpha + 1$, then $M = N$, and hence $N_t = (1 - P + PN)/(1-P)$. Dividing Eq. (15.9) again by N_t, we obtain $U = N(1-P)/(1+2\alpha)(1-P+PN)$.

Summarizing, the throughput efficiency of go-back-N ARQ is given by

$$\eta_{GB} = \begin{cases} \dfrac{1-P}{1+2\alpha p}, & N \geq 2\alpha + 1 \\ \dfrac{N(1-P)}{(1+2\alpha)(1-P+PN)}, & N < 2\alpha + 1 \end{cases} \quad (15.12)$$

Assuming negligible processing times, the average transmission time of a frame as seen by the sender is given by

$$T_v = T_s + N_t T_T = T_s + \sum_{k=1}^{\infty} k P^k (1-P) T_T \quad (15.13)$$

where T_s is the actual transmission duration of a frame and where again $T_T = T_s + 2T_p$. Since $\alpha = T_p/T_s$, then the total time T_T can be written as: $T_T = T_s + 2\alpha T_s = T_s(1 + 2\alpha)$. As a result, using Eq. (15.13) the average transmission time of a frame becomes

$$T_v = T_s + \frac{PT_T}{1-P} = T_s\left(\frac{1+2\alpha P}{1-P}\right) \quad (15.14)$$

Figures 15.7–15.10 show the throughput efficiency η of the three basic ARQ schemes, as a function of the propagation delay ratio α, for frame error probabilities $P = 1 \times 10^{-1}, 1 \times 10^{-2}, 1 \times 10^{-3}$, and 1×10^{-4}, respectively.

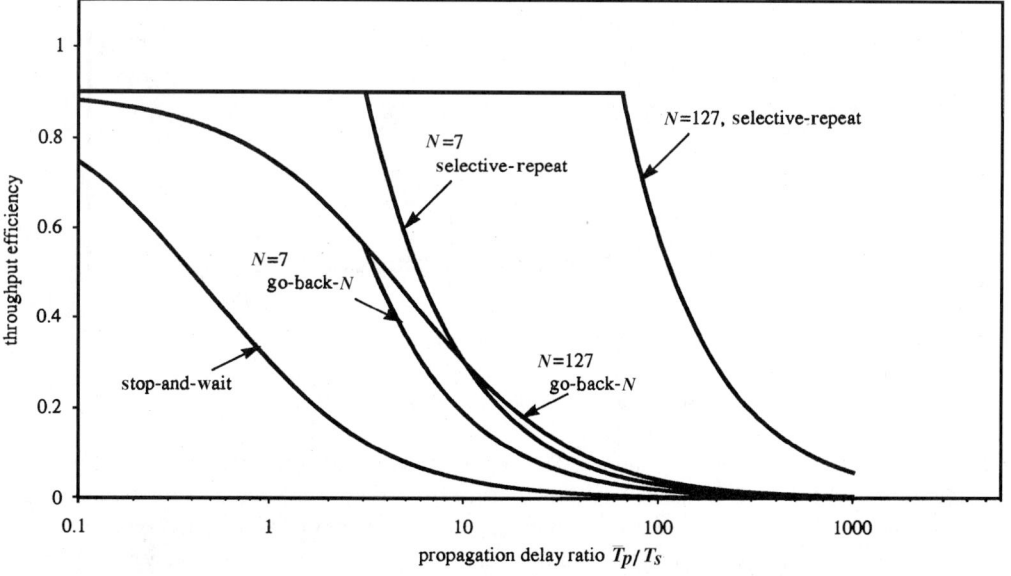

FIGURE 15.7 Throughput efficiency for various ARQ protocols ($P = 0.1$).

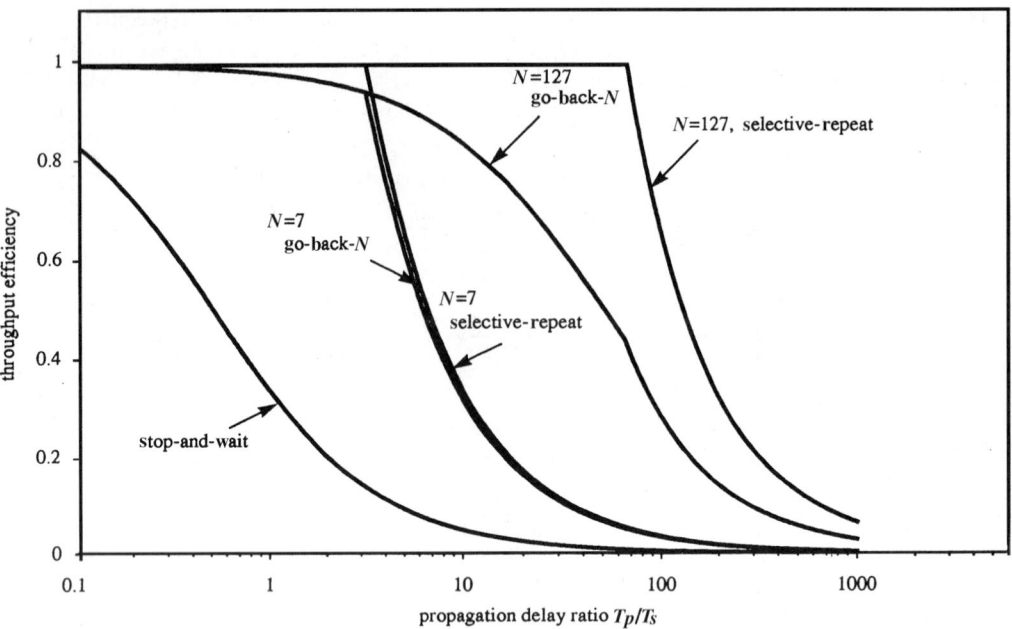

FIGURE 15.8 Throughput efficiency for various ARQ protocols ($P = 0.01$).

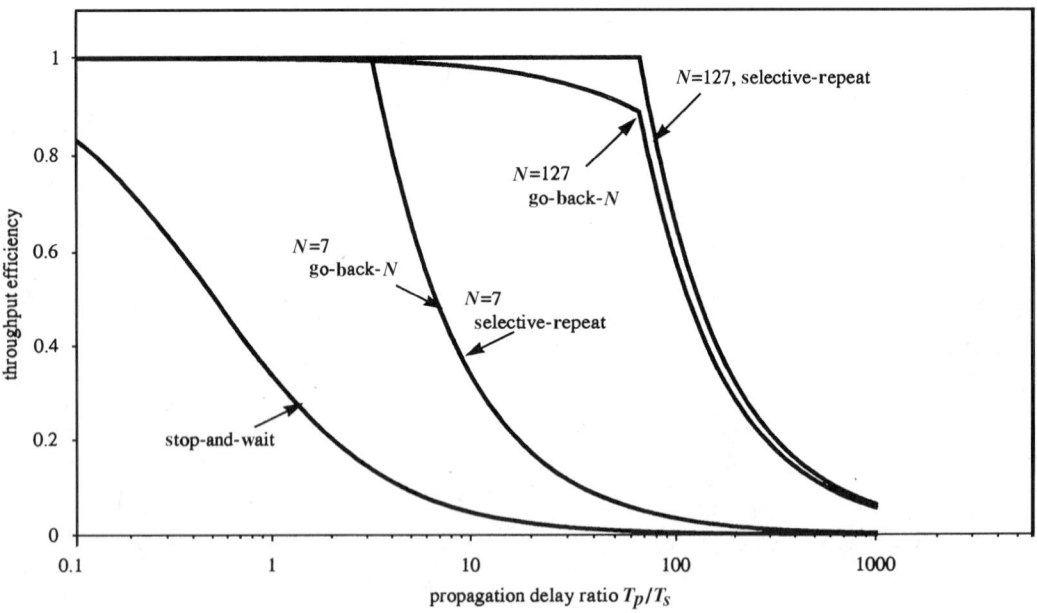

FIGURE 15.9 Throughput efficiency for various ARQ protocols ($P = 0.001$).

Basic ARQ schemes present some limitations. The stop-and-wait scheme suffers from inefficiency due to the fact that the channel is idle between the transmission of the frame and the reception of the acknowledgment from the receiver. As shown in Figs. 15.7–15.10, this inefficiency is especially severe for large values of α, that is, when the roundtrip delay between the sender and the receiver is long compared to the transmission duration of a frame. In systems where the channel roundtrip delay is large and/or when the frame error probability P is relatively high, such as in satellite

Automatic Repeat Request

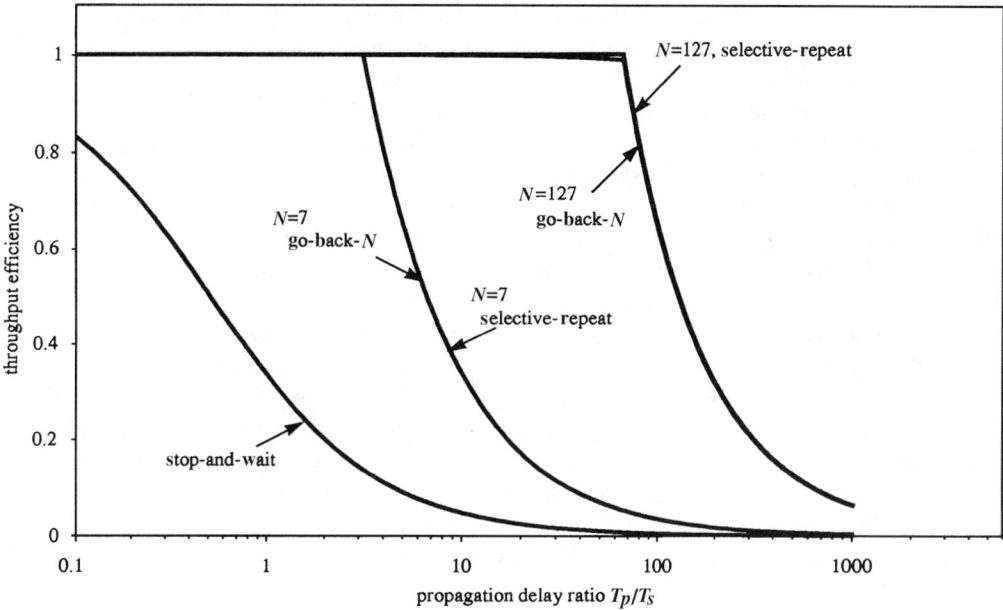

FIGURE 15.10 Throughput efficiency for various ARQ protocols ($P = 0.0001$).

broadcast channels, the throughput efficiency of a stop-and-wait ARQ scheme becomes too poor to be acceptable. On the other hand, the selective-repeat ARQ scheme offers the best performance in terms of throughput efficiency and as shown in Figs. 15.7–15.10, it is insensitive to the propagation delay ratio. To overcome some of these limitations, hybrid FEC/ARQ schemes have been proposed.

15.4 Variants of the Basic Automatic Repeat Request Schemes

Several variants of the basic ARQ schemes with the objective of increasing the throughput efficiency and decreasing both buffer size and average transmission time have been proposed in the literature [Sastry, 1995; Despins and Haccoun, 1993]. Some variants have been proposed with the objective of alleviating particular drawbacks of specific ARQ schemes or satisfying special constraints for specific applications. However, these objectives may not be all satisfied, and they may even be conflicting.

The main idea of these variants is to send multiple copies of a retransmitted frame rather than just sending a single copy. Variants of both stop-and-wait and continuous ARQ schemes have been proposed by Sastry [1995] for improving the throughput for channels having high error rates and long transmission delays such as satellite links. The modification consists of sending consecutively n copies, $n > 1$, of an erroneous frame before waiting for an ACK. This procedure substantially improves the probability of correctly receiving at least one error-free copy of that frame, hence, decreases multiple requests for retransmission of a given frame. For channels with long transmission delays, the increased transmission time for the n copies is more than compensated for by the fewer retransmission requests, leading to improved throughput effficiency over the usual stop-and-wait scheme. For continuous ARQ, whenever a NAK for a frame is received by the sender, that same frame is repeatedly transmitted until reception by the sender of its corresponding ACK. Here again the throughput may be substantially improved, especially for high error rate channels having a transmission delay much larger than the transmission duration of a frame.

A variation of the usual selective-repeat ARQ protocol has been recently proposed and analyzed by Despins and Haccoun [1993] for full-duplex channels suffering from a high error rate in both the forward and reverse directions. It consists essentially of sending a frame composed entirely of NAKs instead of piggybacking a single NAK on a data frame over the feedback channel. The idea is to provide a virtual error-free feedback channel for the NAKs, thus eliminating the need for implementing a time out and buffer control procedure. For very severely degraded channels, the procedure is further refined by sending a number m ($m > 1$) of NAK frames for each incorrectly received data frame. The number of NAK frames m is calculated to insure a practically error free feedback channel. It is shown that the elimination of needless retransmissions outweighs the additional overhead of the NAK frames, leading to higher throughput efficiency and smaller buffer requirements over conventional selective-repeat ARQ, especially over fading channels.

For point-to-multipoint communication, Wang and Silvester [1993] have proposed adaptive multireceiver ARQ schemes. These variants can be applied to all three basic ARQ schemes. They differ from the classic schemes in the way the sender utilizes the outcome of a previous transmission to alter its next transmission. Basically, upon receiving a NAK for an erroneous frame from some given receivers, instead of repeating the transmission of that frame to all of the receivers in the system, the sender transmits a number of copies of that frame to only those receivers that previously received that frame in error. It has been shown that the optimal number of copies that the sender should transmit depends on the transmission error probability, the roundtrip propagation delay, and the number of receivers that have incorrectly received the data frame.

15.5 Hybrid Forward Error Control/Automatic Repeat Request Schemes

In systems where the channel roundtrip delay is large and/or where there is a large number of receivers, such as in satellite broadcast channels, the throughput efficiencies of the basic ARQ schemes, especially stop-and-wait, become unacceptable. Thus hybrid ARQ schemes, consisting of an FEC subsystem integrated into an ARQ system, can overcome the drawbacks of ARQ and FEC schemes considered separately. Such combined schemes are commonly named *hybrid FEC/ARQ schemes*.

The principle of hybrid FEC/ARQ schemes is to reduce the number of retransmissions by improving the channel using an FEC procedure. As shown in Fig. 15.11, an hybrid FEC/ARQ system is a concatenated system consisting of an inner FEC system and an outer ARQ system. The function of the inner FEC system is to improve the quality of the channel as seen by the ARQ system, by correcting as many frames in error as possible, within its error correcting capability. Since only those erroneous frames that have not been corrected by the FEC will be retransmitted, clearly, under a poor channel condition where frame error rates are high and retransmission requests numerous,

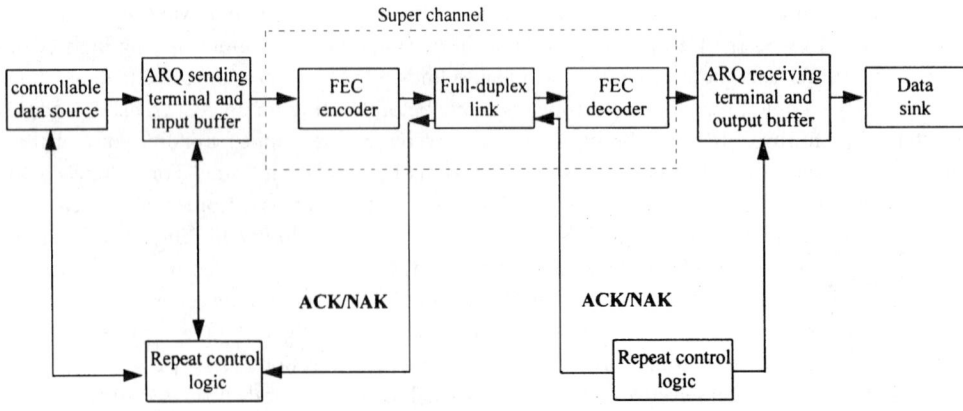

FIGURE 15.11 Principle of hybrid FEC/ARQ schemes.

hybrid FEC/ARQ schemes can be especially useful for improving the efficiency of conventional ARQ schemes.

Hybrid FEC/ARQ schemes in data point-to-point communications over noisy channels are retransmission techniques that employ both error correction and error detection coding in order to achieve high throughput and low undetected error probabilities. There are two basic types of hybrid FEC/ARQ schemes: *type I hybrid* ARQ scheme, which includes parity bits for both error detection and error correction in each transmitted frame, and *type II hybrid* ARQ scheme where parity bits for error correction only are sent on the first transmission.

In type I hybrid FEC/ARQ scheme, when a frame is detected in error, the receiver first attempts to correct these errors. If the error-correcting capability allows the receiver to correct all of the errors detected in the frame, those errors are then corrected and the decoded frame is considered ready to be delivered to the user. In the case where a frame detected in error is revealed uncorrectable, the receiver rejects that frame and requests its retransmission, until that frame is successfully received or decoded. Clearly, a code designed to perform simultaneous error detection and correction is required. As a result, type I hybrid FEC/ARQ schemes require more parity-check bits than a code used only for error detection in a basic ARQ scheme. These extra redundant bits obviously increase the overhead of each transmission.

For low channel error probabilities, type I hybrid FEC/ARQ schemes provide lower throughput efficiencies than any basic ARQ scheme. However, for high channel error probabilities, since the number of retransmissions is reduced by the error-correcting capability of the system, type I hybrid FEC/ARQ schemes provide higher throughput efficiencies than any basic ARQ scheme. These schemes require that a fixed number of parity-check bits are sent with every frame for error correction purposes. Consequently, they are not adaptive to changing channel conditions and are best suited for channels with a fairly constant noise or interference levels. For varying channel conditions, other adaptive error control techniques called type II hybrid FEC/ARQ scheme have been proposed [Rice and Wicker, 1994].

Type II hybrid FEC/ARQ schemes basically operate as follows. A frame F to be sent is coded, for its first transmission, with parity-check bits which only allow error detection. If that frame is detected in error in the form F^*, the receiver stores F^* in its buffer and then requests a retransmission. Instead of retransmitting F in its original form, the sender retransmits a parity-check frame F_1 consisting of the original frame F and an error-correcting code. The receiver, which receives this parity-check frame F_1, uses it to correct the erroneous frame F^* currently stored in its buffer. In the case of error correction failure, a second retransmission of the original frame F is requested by the receiver. Depending on the retransmission strategy and the type of error-correcting codes that are used, the second retransmission may be either a repetition of the original frame F or the transmission of another parity-check frame denoted F_2 [Lin, Costello, and Miller, 1984].

Both type I and type II hybrid FEC/ARQ schemes may use block or convolutional error-correcting codes. It is shown, by means of analysis as well as computer simulations, that both hybrid schemes are capable of providing high throughput efficiencies over a wide range of signal-to-noise ratios. More precisely, performance analysis of type II hybrid FEC/ARQ scheme shows that it is fairly robust against nonstationary channel noise due to the use of parity retransmission. In summary, hybrid FEC/ARQ schemes are suitable for large file transfers in applications where high throughput and high reliability are required over noisy channels, such as in satellite communications.

15.6 Application Problem

In this section, some numerical calculations on the performance are presented for illustration purposes.

A satellite channel with a data rate of 2 Mb/s transmits 1000-b frames. The roundtrip propagation time is 250 ms and the frame error probability is $P = 1 \times 10^{-3}$. Assuming that ACK and NAK frames are error-free, and assuming negligible processing and acknowledgment times:

1. Calculate the throughput efficiency for stop-and-wait ARQ.
2. Determine the average time for correct reception of a frame.
3. Find the average transmission time and the throughput efficiency for error-free operation using go-back-N ARQ, with a window size $N = 127$.
4. Compare the throughput efficiency for go-back-N ARQ and for selective-repeat ARQ, with the same window size $N = 127$.
5. Repeat step 4 using a window size $N = 2047$. Discuss.

Solution

The propagation delay $T_p = 1/2 \times 250 = 125$ ms. For a channel transmitting 1000-b frames, at 2 Mb/s, the transmission duration of a single frame is

$$T_s = 1000/(2 \times 10^6) = 500 \times 10^{-6} \text{ s} = 0.5 \text{ ms}$$

The propagation delay ratio is then

$$\alpha = T_p/T_s = 125/0.5 = 250$$

and, hence,

$$(2\alpha + 1) = 501$$

1. Given that ACK and NAK frames are error-free, and assuming the processing and acknowledgment times are negligible, the throughput efficiency for stop-and-wait ARQ can be calculated using Eq. (15.7)

$$\eta_{SW} = (1 - P)/(1 + 2\alpha) = 0.999/501 = 2 \times 10^{-3}$$

The throughput efficiency of stop-and-wait ARQ scheme over such a channel is vanishingly low as also observed in Fig. 15.9. Clearly, stop-and-wait ARQ on satellite channels is not suitable.

2. Using Eq. (15.8), the average time for correct reception of a frame is $T_v = T_T/(1 - P)$, where T_T can be computed using Eq. (15.3)

$$T_T = T_s + 2T_p = 0.5 + 2 \times 125 = 250.5 \text{ ms}$$

It follows that

$$T_v = 250.5/0.999 = 250.75 \text{ ms}$$

Hence, due to the low value of P, the average time for the correct retransmission of a frame is nearly equal to the roundtrip propagation time.

3. The average transmission time for go-back-N ARQ is given by Eq. (15.14)

$$T_v = 0.5[(1 + 500 \times 10^{-3})/0.999] = 751 \text{ ms}$$

The throughput efficiency for go-back-N ARQ can be computed using Eq. (15.12). Since the window size is $N = 127$ and $(2\alpha + 1) = 501$, it follows that

$$\eta_{GB} = 127(0.999)/[(501)(0.999 + 0.127)] = 0.225$$

This value for the throughput efficiency may also be read directly from Fig. 15.9. The improvement in throughput efficiency of the go-back-N ARQ over the stop-and-wait scheme is then substantial.

4. The throughput efficiency for selective-repeat ARQ is given by Eq. (15.10). With a window size $N = 127$ and $(2\alpha + 1) = 501$, then

$$\eta_{SR} = 127(0.999)/501 = 0.253$$

Again, this value can be read directly from Fig. 15.9, which shows that for these values of N and α, the throughput efficiencies of go-back-N and selective-repeat ARQ are approximately equal.

5. Since $N = 2047$ and $(2\alpha + 1) = 501$, using Eqs. (15.12) and (15.10), respectively, it follows that for go-back-N the throughput efficiency is

$$\eta_{GB} = (0.999)/[1 + (2 \times 250 \times 0.001)] = 0.667$$

For selective-repeat ARQ we obtain

$$\eta_{SR} = (1 - 0.001) = 0.999$$

Therefore the selective-repeat ARQ scheme provides nearly the maximum possible throughput efficiency, and is substantially more efficient than go-back-N.

15.7 Conclusion

Fundamentals, performances, limitations, and variants of the basic ARQ schemes have been presented. These schemes offer reliability and robustness at buffering cost and reduced throughput. In systems such as packet radio and satellite networks, which are characterized by relatively large frame lengths and high noise or interference levels, the use of error correction coding may not provide the desired reliability. Combination of error correction and error detection coding may be therefore especially attractive to overcome this shortcoming. Furthermore, over real channel systems characterized by great variations in the quality of the channel, ARQ systems tend to provide the robustness of the required error performance. New trends in digital communications systems using ARQ tend toward increasing the throughput efficiency without increasing buffering. These tendencies are being integrated into many applications involving point-to-multipoint communications over broadcast links used for file distribution, videotex systems, and teleconferencing. Such applications are becoming more and more popular, especially with the deployment of *Integrated Services Digital Networks* (ISDN) linked worldwide by satellite communication systems. Similarly, for wireless communication channels, which are corrupted by noise, fading, interference and shadowing, ARQ schemes can be used in order to provide adequate robustness and performance.

Defining Terms

Error correction: Procedure by which additional redundant symbols are added to the messages in order to allow the correction of errors in those messages.

Error detection: Procedure by which additional redundant symbols are added to the messages in order to allow the detection of errors in those messages.

Feedback channel: Return channel used by the receiver in order to inform the sender that erroneous frames have been received.

Forward channel: One-way channel used by the sender to transmit data frames to the receiver.

Full-duplex (or duplex) link: Link used for exchanging data or ACK/NAK frames between two connected devices in both directions simultaneously.

Half-duplex link: Link used for exchanging data or ACK/NAK frames between two connected devices in both directions but alternatively. Thus, the two devices must be able to switch between send and receive modes after each transmission.

References

Deng, R.H. 1994. Hybrid ARQ schemes employing coded modulation and sequence combining. *IEEE Trans. Commun.*, 42(June):2239–2245.

Despins, C. and Haccoun, D. 1993. A new selective-repeat ARQ protocol and its application to high rate indoor wireless cellular data links. Proceedings of the 4th International Symposium on Personal, Indoor and Mobile Radio Communications, Yokohama, Japan, Sept.

Lin, S., Costello, D.J., and Miller, M.J. 1984. Automatic repeat request error-control schemes. *IEEE Commun. Mag.*, 22(12):5–16.

Rice, M. and Wicker, S.B. 1994. Adaptive error control for slowly varying channels. *IEEE Trans. Commun.*, 42(Feb./March/April):917–926.

Sastry, A.R.K. 1975. Improving ARQ performance on satellite channels under high error rate conditions. *IEEE Trans. Commun.*, COM-23(April):436–439.

Stallings, W. 1994. *Data and Computer Communications*, 4th ed., Macmillan, New York.

Wang, J.L. and Silvester, J.A. 1993. Optimal adaptive multireceiver ARQ protocols. *IEEE Trans. Commun.*, COM-41(Dec.):1816–1829.

Further Information

Halsall, F. 1995. *Data Communications, Computer Networks, and Open Systems*. 4th ed., Chap. 4, pp. 168–214. Addison–Wesley, Reading, MA.

Bertsekas, D. and Gallager, R.G. 1992. *Data Networks*, 2nd ed. Chap. 2, pp. 37–127. Prentice–Hall, Englewood Cliffs, NJ.

Tanenbaum, A.S. *Computer Networks*, 2nd ed., Chap. 4, pp. 196–264. Prentice–Hall, Englewood Cliffs, NJ.

Stallings, W. 1994. *Data and Computer Communications*, 4th ed., Chaps. 4 and 5, pp. 133–197. Macmillan, New York.

16

Spread Spectrum Communications

Laurence B. Milstein
University of California

Marvin K. Simon
Jet Propulsion Laboratory

16.1 A Brief History 199
16.2 Why Spread Spectrum? 199
16.3 Basic Concepts and Terminology 200
16.4 Spread Spectrum Techniques 200
   Direct Sequence Modulation • Frequency Hopping Modulation • Time Hopping Modulation • Hybrid Modulations
16.5 Applications of Spread Spectrum 206
   Military • Commercial

16.1 A Brief History

Spread spectrum (SS) has its origin in the military arena where the friendly communicator is 1) susceptible to detection/interception by the enemy and 2) vulnerable to intentionally introduced unfriendly interference (jamming). Communication systems that employ spread spectrum to reduce the communicator's detectability and combat the enemy-introduced interference are respectively referred to as **low probability of intercept (LPI)** and **antijam (AJ) communication systems**. With the change in the current world political situation wherein the U.S. Department of Defense (DOD) has reduced its emphasis on the development and acquisition of new communication systems for the original purposes, a host of new commercial applications for SS has evolved, particularly in the area of cellular mobile communications. This shift from military to commercial applications of SS has demonstrated that the basic concepts that make SS techniques so useful in the military can also be put to practical peacetime use. In the next section, we give a simple description of these basic concepts using the original military application as the basis of explanation. The extension of these concepts to the mentioned commercial applications will be treated later on in the chapter.

16.2 Why Spread Spectrum?

Spread spectrum is a communication technique wherein the transmitted modulation is *spread* (increased) in bandwidth prior to transmission over the channel and then *despread* (decreased) in bandwidth by the same amount at the receiver. If it were not for the fact that the communication channel introduces some form of narrowband (relative to the spread bandwidth) interference, the receiver performance would be transparent to the spreading and despreading operations (assuming that they are identical inverses of each other). That is, after **despreading** the received signal would be identical to the transmitted signal prior to **spreading**. In the presence of narrowband interference, however, there is a significant advantage to employing the spreading/despreading procedure described. The reason for this is as follows. Since the interference is introduced after the transmitted signal is spread, then, whereas the despreading operation at the receiver shrinks the desired signal back to its original bandwidth, at the same time it spreads the undesired signal (interference) in bandwidth by the same amount, thus reducing its power spectral density. This, in turn, serves to diminish the effect of the interference on the receiver performance, which depends on the amount of interference power in the despread bandwidth. It is indeed this very simple explanation, which is at the heart of all spread spectrum techniques.

16.3 Basic Concepts and Terminology

To describe this process analytically and at the same time introduce some terminology that is common in spread spectrum parlance, we proceed as follows. Consider a communicator that desires to send a message using a transmitted power S Watts (W) at an information rate R_b bits/s (bps). By introducing a SS modulation, the bandwidth of the transmitted signal is increased from R_b Hz to W_{ss} Hz where $W_{ss} \gg R_b$ denotes the **spread spectrum bandwidth**. Assume that the channel introduces, in addition to the usual thermal noise (assumed to have a single-sided power spectral density (PSD) equal to N_0 W/Hz), an additive interference (jamming) having power J distributed over some bandwidth W_J. After despreading, the desired signal bandwidth is once again now equal to R_b Hz and the interference PSD is now $N_J = J/W_{ss}$. Note that since the thermal noise is assumed to be white, i.e., it is uniformly distributed over all frequencies, its PSD is unchanged by the despreading operation and, thus, remains equal to N_0. Regardless of the signal and interferer waveforms, the equivalent bit energy-to-total noise spectral density ratio is, in terms of the given parameters,

$$\frac{E_b}{N_t} = \frac{E_b}{N_0 + N_J} = \frac{S/R_b}{N_0 + J/W_{ss}} \qquad (16.1)$$

For most practical scenarios, the jammer limits performance and, thus, the effects of receiver noise in the channel can be ignored. Thus, assuming $N_J \gg N_0$, we can rewrite Eq. (16.1) as

$$\frac{E_b}{N_t} \cong \frac{E_b}{N_J} = \frac{S/R_b}{J/W_{ss}} = \frac{S}{J}\frac{W_{ss}}{R_b} \qquad (16.2)$$

where the ratio J/S is the *jammer-to-signal power ratio* and the ratio W_{ss}/R_b is the **spreading ratio** and is defined as the **processing gain** of the system. Since the ultimate error probability performance of the communication receiver depends on the ratio E_b/N_J, we see that from the communicator's viewpoint his goal should be to minimize J/S (by choice of S) and maximize the processing gain (by choice of W_{ss} for a given desired information rate). The possible strategies for the jammer will be discussed in the section on military applications dealing with AJ communications.

16.4 Spread Spectrum Techniques

By far the two most popular spreading techniques are **direct sequence (DS) modulation** and **frequency hopping (FH) modulation**. In the following subsections, we present a brief description of each.

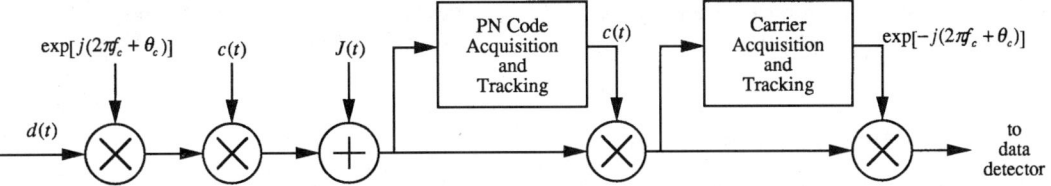

FIGURE 16.1 A DS-BPSK system (complex form).

Direct Sequence Modulation

A direct sequence modulation $c(t)$ is formed by linearly modulating the output sequence $\{c_n\}$ of a pseudorandom number generator onto a train of pulses, each having a duration T_c called the **chip time**. In mathematical form,

$$c(t) = \sum_{n=-\infty}^{\infty} c_n p(t - nT_c) \qquad (16.3)$$

where $p(t)$ is the basic pulse shape and is assumed to be of rectangular form. This type of modulation is usually used with binary phase-shift-keyed (BPSK) information signals, which have the complex form $d(t) \exp\{j(2\pi f_c t + \theta_c)\}$, where $d(t)$ is a binary-valued data waveform of rate $1/T_b$ bits/s and f_c and θ_c are the frequency and phase of the data-modulated carrier, respectively. As such, a DS/BPSK signal is formed by multiplying the BPSK signal by $c(t)$ (see Fig. 16.1), resulting in the real transmitted signal

$$x(t) = \mathrm{Re}\{c(t)d(t) \exp[j(2\pi f_c t + \theta_c)]\} \qquad (16.4)$$

Since T_c is chosen so that $T_b \gg T_c$, then relative to the bandwidth of the BPSK information signal, the bandwidth of the DS/BPSK signal[1] is effectively increased by the ratio $T_b/T_c = W_{ss}/2R_b$, which is one-half the spreading factor or processing gain of the system. At the receiver, the sum of the transmitted DS/BPSK signal and the channel interference $I(t)$ (as discussed before, we ignore the presence of the additive thermal noise) are ideally multiplied by the identical DS modulation (this operation is known as despreading), which returns the DS/BPSK signal to its original BPSK form whereas the real interference signal is now the real wideband signal $\mathrm{Re}\{I(t)c(t)\}$. In the previous sentence, we used the word ideally, which implies that the PN waveform used for despreading at the receiver is identical to that used for spreading at the transmitter. This simple implication covers up a multitude of tasks that a practical DS receiver must perform. In particular, the receiver must first acquire the PN waveform. That is, the local PN random generator that generates the PN waveform at the receiver used for despreading must be aligned (synchronized) to within one chip of the PN waveform of the received DS/BPSK signal. This is accomplished by employing some sort of **search algorithm** which typically steps the local PN waveform sequentially in time by a fraction of a chip (e.g., half a chip) and at each position searches for a high degree of correlation between the received and local PN reference waveforms. The search terminates when the correlation exceeds a given threshold, which is an indication that the alignment has been achieved. After bringing the two PN waveforms into **coarse alignment**, a **tracking algorithm** is employed to maintain **fine alignment**. The most popular forms of tracking loops are the continuous time **delay-locked loop** and its time-multiplexed version the **tau–dither loop**. It is the difficulty in synchronizing the receiver PN

---

[1]For the usual case of a rectangular spreading pulse $p(t)$, the PSD of the DS/BPSK modulation will have $(\sin x/x)^2$ form with first zero crossing at $1/T_c$, which is nominally taken as one-half the spread spectrum bandwidth W_{ss}.

generator to subnanosecond accuracy that limits PN chip rates to values on the order of hundreds of Mchips/s, which implies the same limitation on the DS spread spectrum bandwidth W_{ss}.

Frequency Hopping Modulation

A **frequency hopping (FH) modulation** $c(t)$ is formed by nonlinearly modulating a train of pulses with a sequence of pseudorandomly generated frequency shifts $\{f_n\}$. In mathematical terms, $c(t)$ has the complex form

$$c(t) = \sum_{n=-\infty}^{\infty} \exp\{j(2\pi f_n + \phi_n)\} p(t - nT_h) \qquad (16.5)$$

where $p(t)$ is again the basic pulse shape having a duration T_h, called the **hop time** and $\{\phi_n\}$ is a sequence of random phases associated with the generation of the hops. FH modulation is traditionally used with multiple-frequency-shift-keyed (MFSK) information signals, which have the complex form $\exp\{j[2\pi(f_c + d(t))t]\}$, where $d(t)$ is an M-level digital waveform (M denotes the symbol alphabet size) representing the information frequency modulation at a rate $1/T_s$ symbols/s (sps). As such, an FH/MFSK signal is formed by complex multiplying the MFSK signal by $c(t)$ resulting in the real transmitted signal

$$x(t) = \text{Re}\{c(t) \exp\{j[2\pi(f_c + d(t))t]\}\} \qquad (16.6)$$

In reality, $c(t)$ is never generated in the transmitter. Rather, $x(t)$ is obtained by applying the sequence of pseudorandom frequency shifts $\{f_n\}$ directly to the frequency synthesizer that generates the carrier frequency f_c (see Fig. 16.2). In terms of the actual implementation, successive (not necessarily disjoint) k-chip segments of a PN sequence drive a frequency synthesizer, which hops the carrier over 2^k frequencies. In view of the large bandwidths over which the frequency synthesizer must operate, it is difficult to maintain phase coherence from hop to hop, which explains the inclusion of the sequence $\{\phi_n\}$ in the Eq. (16.5) model for $c(t)$. On a short term basis, e.g., within a given hop, the signal bandwidth is identical to that of the MFSK information modulation, which is typically much smaller than W_{ss}. On the other hand, when averaged over many hops, the signal bandwidth is equal to W_{ss}, which can be on the order of several GHz, i.e., an order of magnitude larger than

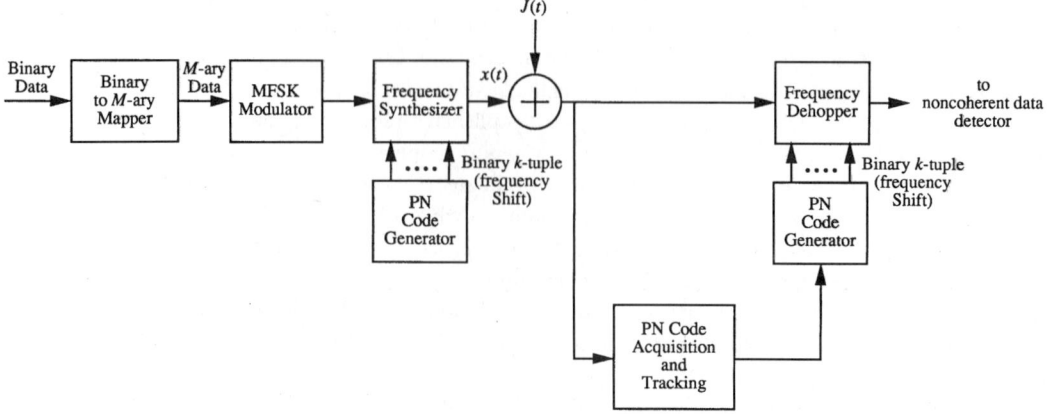

FIGURE 16.2 An FH-MFSK system.

that of implementable DS bandwidths. The exact relation between W_{ss}, T_h, T_s and the number of frequency shifts in the set $\{f_n\}$ will be discussed shortly.

At the receiver, the sum of the transmitted FH/MFSK signal and the channel interference $I(t)$ is ideally complex multiplied by the identical FH modulation (this operation is known as **dehopping**), which returns the FH/MFSK signal to its original MFSK form, whereas the real interference signal is now the wideband (in the average sense) signal $\text{Re}\{I(t)c(t)\}$. Analogous to the DS case, the receiver must acquire and track the FH signal so that the dehopping waveform is as close to the hopping waveform $c(t)$ as possible.

FH systems are traditionally classified in accordance with the relationship between T_h and T_s. **Fast frequency-hopped (FFH)** systems are ones in which there exists one or more hops per data symbol, that is, $T_s = NT_h$ (N an integer) whereas **slow frequency-hopped (SFH)** systems are ones in which there exists more than one symbol per hop, that is, $T_h = NT_s$. It is customary in SS parlance to refer to the FH/MFSK tone of shortest duration as a "chip", despite the same usage for the PN chips associated with the code generator that drives the frequency synthesizer. Keeping this distinction in mind, in an FFH system where, as already stated, there are multiple hops per data symbol, a chip is equal to a hop. For SFH, where there are multiple data symbols per hop, a chip is equal to an MFSK symbol. Combining these two statements, the chip rate R_c in an FH system is given by the larger of $R_h = 1/T_h$ and $R_s = 1/T_s$ and, as such, is the highest system clock rate.

The frequency spacing between the FH/MFSK tones is governed by the chip rate R_c and is, thus, dependent on whether the FH modulation is FFH or SFH. In particular, for SFH where $R_c = R_s$, the spacing between FH/MFSK tones is equal to the spacing between the MFSK tones themselves. For noncoherent detection (the most commonly encountered in FH/MFSK systems), the separation of the MFSK symbols necessary to provide orthogonality[2] is an integer multiple of R_s. Assuming the minimum spacing, i.e., R_s, the entire spread spectrum band is then partitioned into a total of $N_t = W_{ss}/R_s = W_{ss}/R_c$ equally spaced FH tones. One arrangement, which is by far the most common, is to group these N_t tones into $N_b = N_t/M$ contiguous, nonoverlapping bands, each with bandwidth $MR_s = MR_c$; see Fig. 16.3(a). Assuming symmetric MFSK modulation around the carrier frequency, then the center frequencies of the $N_b = 2^k$ bands represent the set of hop carriers, each of which is assigned to a given k-tuple of the PN code generator. In this fixed arrangement, each of the N_t FH/MFSK tones corresponds to the combination of a unique hop carrier (PN code k-tuple) and a unique MFSK symbol. Another arrangement, which provides more protection against the sophisticated interferer (jammer), is to overlap adjacent M-ary bands by an amount equal to R_c; see Fig. 16.3(b). Assuming again that the center frequency of each band corresponds to a possible hop carrier, then since all but $M-1$ of the N_t tones are available as center frequencies, the number of hop carriers has been increased from N_t/M to $N_t - (M-1)$, which for $N_t \gg M$ is approximately an increase in randomness by a factor of M.

For FFH, where $R_c = R_h$, the spacing between FH/MFSK tones is equal to the hop rate. Thus, the entire spread spectrum band is partitioned into a total of $N_t = W_{ss}/R_h = W_{ss}/R_c$ equally spaced FH tones, each of which is assigned to a unique k-tuple of the PN code generator that drives the frequency synthesizer. Since for FFH there are R_h/R_s hops per symbol, then the metric used to make a noncoherent decision on a particular symbol is obtained by summing up R_h/R_s detected chip (hop) energies, resulting in a so-called *noncoherent combining loss*.

Time Hopping Modulation

Time hopping (TH) is to spread spectrum modulation what pulse position modulation (PPM) is to information modulation. In particular, consider segmenting time into intervals of T_f seconds

---

[2] An optimum noncoherent MFSK detector consists of a bank of energy detectors each matched to one of the M frequencies in the MFSK set. In terms of this structure, the notion of *orthogonality* implies that for a given transmitted frequency there will be no crosstalk (energy spillover) in any of the other $M-1$ energy detectors.

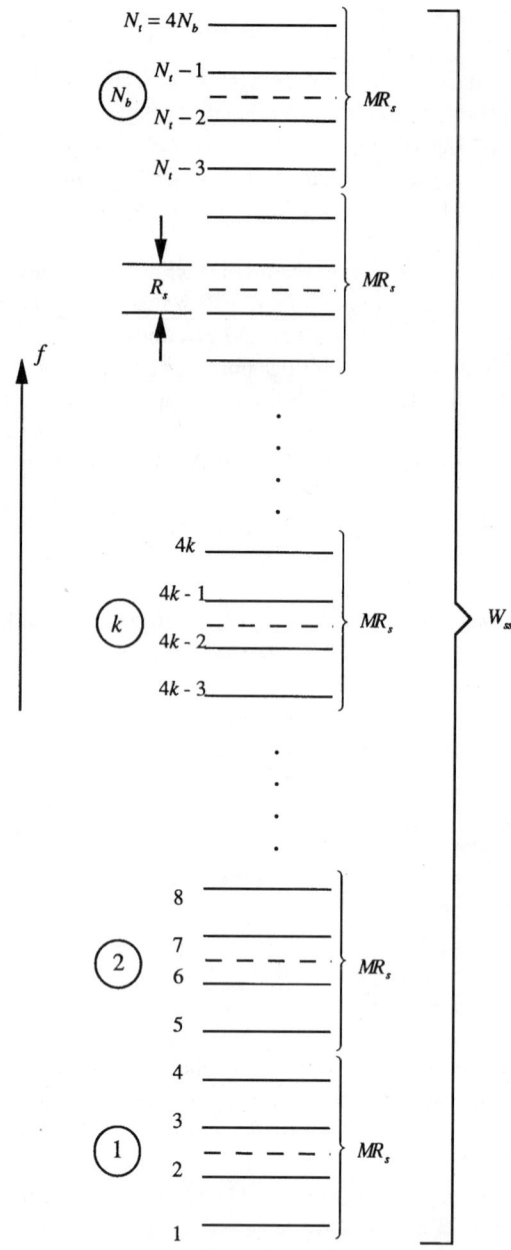

FIGURE 16.3(a) Frequency distribution for FH-4FSK —nonoverlapping bands. Dashed lines indicate location of hop frequencies.

and further segment each T_f interval into M_T increments of width T_f/M_T. Assuming a pulse of maximum duration equal to T_f/M_T, then a **time hopping spread spectrum** modulation would take the form

$$c(t) = \sum_{n=-\infty}^{\infty} p\left[t - \left(n + \frac{a_n}{M_T}\right)T_f\right] \qquad (16.7)$$

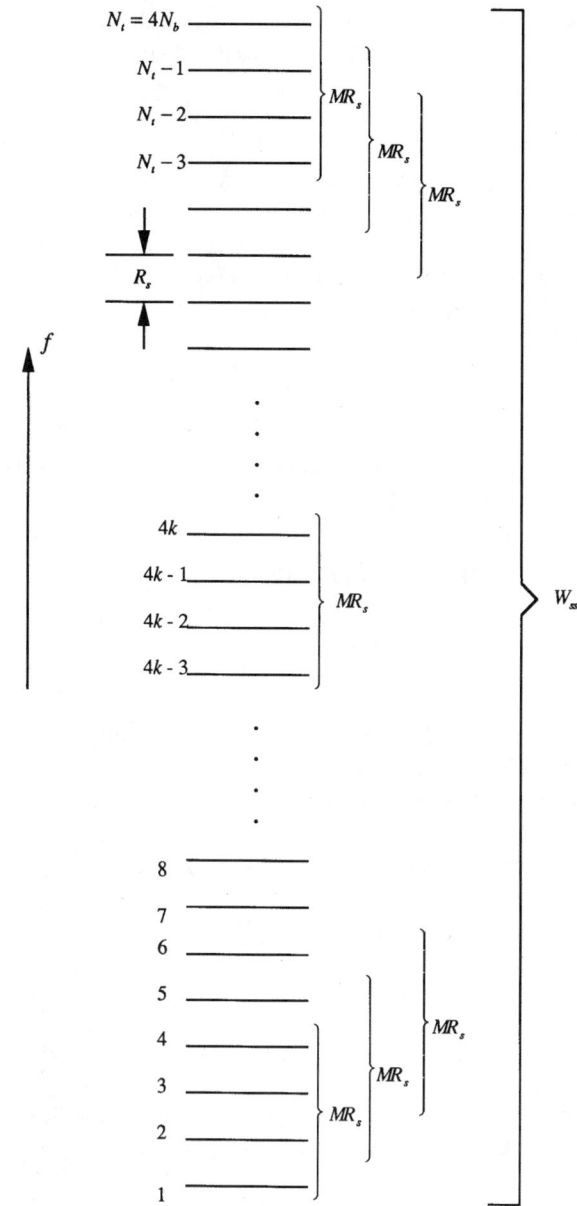

FIGURE 16.3(b) Frequency distribution for FH-4FSK—overlapping bands.

where a_n denotes the pseudorandom position (one of M_T uniformly spaced locations) of the pulse within the T_f-second interval.

For DS and FH, we saw that *multiplicative* modulation, that is the transmitted signal is the product of the SS and information signals, was the natural choice. For TH, *delay* modulation is the natural choice. In particular, a TH-SS modulation takes the form

$$x(t) = \text{Re}\{c(t - d(t)) \exp[j(2\pi f_c + \phi_T)]\} \quad (16.8)$$

where $d(t)$ is a digital information modulation at a rate $1/T_s$ sps. Finally, the dehopping procedure at the receiver consists of removing the sequence of delays introduced by $c(t)$, which restores the information signal back to its original form and spreads the interferer.

Hybrid Modulations

By blending together several of the previous types of SS modulation, one can form **hybrid** modulations that, depending on the system design objectives, can achieve a better performance against the interferer than can any of the SS modulations acting alone. One possibility is to multiply several of the $c(t)$ wideband waveforms [now denoted by $c^{(i)}(t)$ to distinguish them from one another] resulting in a SS modulation of the form

$$c(t) = \prod_i c^{(i)}(t) \tag{16.9}$$

Such a modulation may embrace the advantages of the various $c^{(i)}(t)$, while at the same time mitigating their individual disadvantages.

16.5 Applications of Spread Spectrum

Military

Antijam (AJ) Communications

As already noted, one of the key applications of spread spectrum is for antijam communications in a hostile environment. The basic mechanism by which a **direct sequence spread spectrum** receiver attenuates a noise jammer was illustrated in Sec. 16.3. Therefore, in this section, we will concentrate on tone jamming.

Assume the received signal, denoted $r(t)$, is given by

$$r(t) = Ax(t) + I(t) + n_w(t) \tag{16.10}$$

where $x(t)$ is given in Eq. (16.4), A is a constant amplitude,

$$I(t) = \alpha \cos(2\pi f_c t + \theta) \tag{16.11}$$

and $n_w(t)$ is additive white Gaussian noise (AWGN) having two-sided spectral density $N_0/2$. In Eq. (16.11), α is the amplitude of the tone jammer and θ is a random phase uniformly distributed in $[0, 2\pi]$.

If we employ the standard correlation receiver of Fig. 16.4, it is straightforward to show that the final test statistic out of the receiver is given by

$$g(T_b) = AT_b + \alpha \cos\theta \int_0^{T_b} c(t)\, dt + N(T_b) \tag{16.12}$$

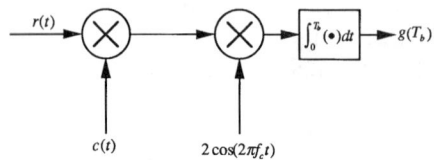

FIGURE 16.4

Spread Spectrum Communications

where $N(T_b)$ is the contribution to the test statistic due to the AWGN. Noting that, for rectangular chips, we can express

$$\int_0^{T_b} c(t)\,dt = T_c \sum_{i=1}^{M} c_i \qquad (16.13)$$

where

$$M \triangleq \frac{T_b}{T_c} \qquad (16.14)$$

is one-half of the processing gain, it is straightforward to show that, for a given value of θ, the signal-to-noise-plus-interference ratio, denoted by S/N_{total}, is given by

$$\frac{S}{N_{total}} = \frac{1}{\frac{N_0}{2E_b} + \left(\frac{J}{MS}\right)\cos^2\theta} \qquad (16.15)$$

In Eq. (16.15), the jammer power is

$$J \triangleq \frac{\alpha^2}{2} \qquad (16.16)$$

and the signal power is

$$S \triangleq \frac{A^2}{2} \qquad (16.17)$$

If we look at the second term in the denominator of Eq. (16.15), we see that the ratio J/S is divided by M. Realizing that J/S is the ratio of the jammer power to the signal power before despreading, and J/MS is the ratio of the same quantity after despreading, we see that, as was the case for noise jamming, the benefit of employing direct sequence spread spectrum signalling in the presence of tone jamming is to reduce the effect of the jammer by an amount on the order of the processing gain.

Finally, one can show that an estimate of the average probability of error of a system of this type is given by

$$P_e = \frac{1}{2\pi} \int_0^{2\pi} \phi\left(-\sqrt{\frac{S}{N_{total}}}\right) d\theta \qquad (16.18)$$

where

$$\phi(x) \triangleq \frac{1}{\sqrt{2\pi}} \int_{-\infty}^{x} e^{-y^2/2}\,dy \qquad (16.19)$$

If Eq. (16.18) is evaluated numerically and plotted, the results are as shown in Fig. 16.5. It is clear from this figure that a large initial power advantage of the jammer can be overcome by a sufficiently large value of the processing gain.

Low-Probability of Intercept (LPI)

The opposite side of the AJ problem is that of LPI, that is, the desire to hide your signal from detection by an intelligent adversary so that your transmissions will remain unnoticed and, thus,

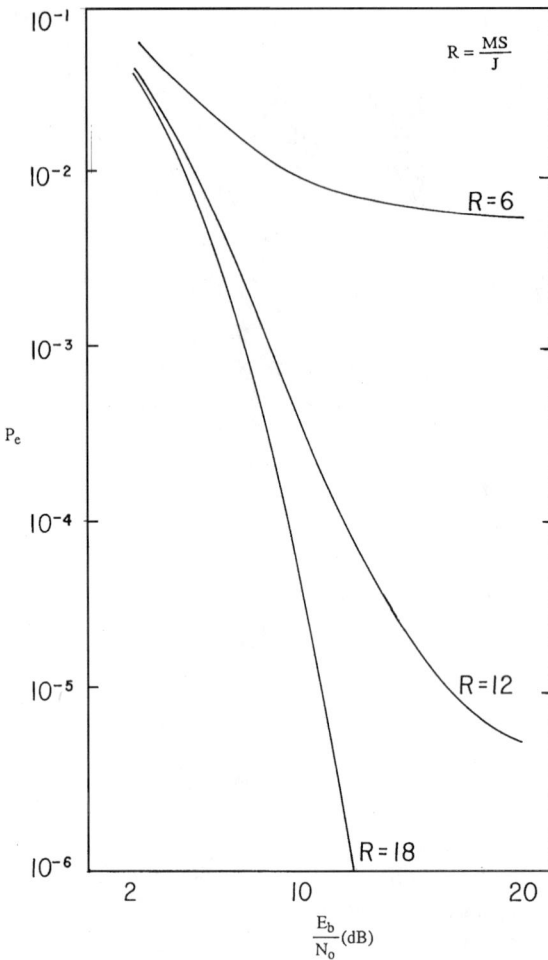

FIGURE 16.5

neither jammed nor exploited in any manner. This idea of designing an LPI system is achieved in a variety of ways, including transmitting at the smallest possible power level, and limiting the transmission time to as short an interval in time as is possible. The choice of signal design is also important, however, and it is here that spread spectrum techniques become relevant.

The basic mechanism is reasonably straightforward; if we start with a conventional narrowband signal, say a BPSK waveform having a spectrum as shown in Fig. 16.6(a), and then spread it so that its new spectrum is as shown in Fig. 16.6(b), the peak amplitude of the spectrum after spreading has been reduced by an amount on the order of the processing gain relative to what it was before spreading. Indeed, a sufficiently large processing gain will result in the spectrum of the signal after spreading falling below the ambient thermal noise level. Thus, there is no easy way for an unintended listener to determine that a transmission is taking place.

That is not to say the spread signal cannot be detected, however, merely that it is more difficult for an adversary to learn of the transmission. Indeed, there are many forms of so-called intercept receivers that are specifically designed to accomplish this very task. By way of example, probably the best known and simplest to implement is a **radiometer**, which is just a device that measures the total power present in the received signal. In the case of our intercept problem, even though we have lowered the power spectral density of the transmitted signal so that it falls below the noise

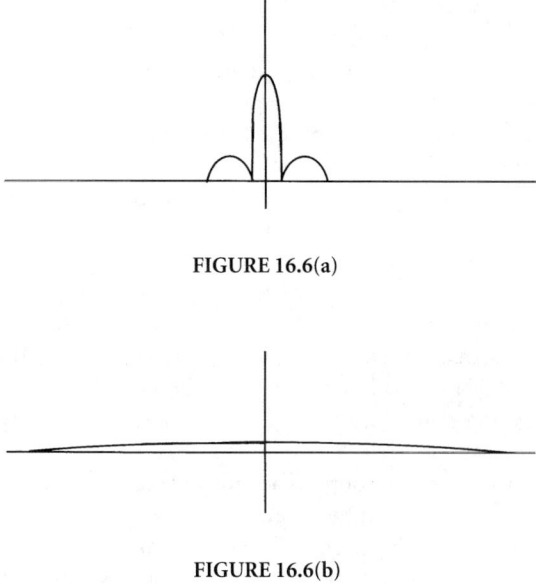

FIGURE 16.6(a)

FIGURE 16.6(b)

floor, we have not lowered its power (i.e., we have merely spread its power over a wider frequency range). Thus, if the radiometer integrates over a sufficiently long period of time, it will eventually determine the presence of the transmitted signal buried in the noise. The key point, of course, is that the use of the spreading makes the interceptor's task much more difficult, since he has no knowledge of the spreading code and, thus, cannot despread the signal.

Commercial

Multiple Access Communications

From the perspective of commercial applications, probably the most important use of spread spectrum communications is as a multiple accessing technique. When used in this manner, it becomes an alternative to either frequency division multiple access (FDMA) or time division multiple access (TDMA) and is typically referred to as either code division multiple access (CDMA) or spread spectrum multiple access (SSMA). When using CDMA, each signal in the set is given its own spreading sequence. As opposed to either FDMA, wherein all users occupy disjoint frequency bands but are transmitted simultaneously in time, or TDMA, whereby all users occupy the same bandwidth but transmit in disjoint intervals of time, in CDMA, all signals occupy the same bandwidth and are transmitted simultaneously in time; the different waveforms in CDMA are distinguished from one another at the receiver by the specific spreading codes they employ.

Since most CDMA detectors are correlation receivers, it is important when deploying such a system to have a set of spreading sequences that have relatively low-pairwise cross-correlation between any two sequences in the set. Further, there are two fundamental types of operation in CDMA, synchronous and asynchronous. In the former case, the symbol transition times of all of the users are aligned; this allows for orthogonal sequences to be used as the spreading sequences and, thus, eliminates interference from one user to another. Alternately, if no effort is made to align the sequences, the system operates asynchronously; in this latter mode, multiple access interference limits the ultimate channel capacity, but the system design exhibits much more flexibility.

CDMA has been of particular interest recently for applications in wireless communications. These applications include cellular communications, personal communications services (PCS), and wireless local area networks. The reason for this popularity is primarily due to the performance that spread spectrum waveforms display when transmitted over a multipath fading channel.

To illustrate this idea, consider DS signalling. As long as the duration of a single chip of the spreading sequence is less than the multipath delay spread, the use of DS waveforms provides the system designer with one of two options. First, the multipath can be treated as a form of interference, which means the receiver should attempt to attenuate it as much as possible. Indeed, under this condition, all of the multipath returns that arrive at the receiver with a time delay greater than a chip duration from the multipath return to which the receiver is synchronized (usually the first return) will be attenuated because of the processing gain of the system.

Alternately, the multipath returns that are separated by more than a chip duration from the main path represent independent "looks" at the received signal and can be used constructively to enhance the overall performance of the receiver. That is, because all of the multipath returns contain information regarding the data that is being sent, that information can be extracted by an appropriately designed receiver. Such a receiver, typically referred to as a RAKE receiver, attempts to resolve as many individual multipath returns as possible and then to sum them coherently. This results in an *implicit* diversity gain, comparable to the use of *explicit* diversity, such as receiving the signal with multiple antennas.

The condition under which the two options are available can be stated in an alternate manner. If one envisions what is taking place in the frequency domain, it is straightforward to show that the condition of the chip duration being smaller than the multipath delay spread is equivalent to requiring that the spread bandwidth of the transmitted waveform exceed what is called the coherence bandwidth of the channel. This latter quantity is simply the inverse of the multipath delay spread and is a measure of the range of frequencies that fade in a highly correlated manner. Indeed, anytime the coherence bandwidth of the channel is less than the spread bandwidth of the signal, the channel is said to be *frequency selective* with respect to the signal. Thus, we see that to take advantage of DS signalling when used over a multipath fading channel, that signal should be designed such that it makes the channel appear frequency selective.

In addition to the desirable properties that spread spectrum signals display over multipath channels, there are two other reasons why such signals are of interest in cellular-type applications. The first has to do with a concept known as the reuse factor. In conventional cellular systems, either analog or digital, in order to avoid excessive interference from one cell to its neighbor cells, the frequencies used by a given cell are not used by its immediate neighbors (i.e., the system is designed so that there is a certain spatial separation between cells that use the same carrier frequencies). For CDMA, however, such spatial isolation is typically not needed, so that so-called *universal reuse* is possible.

Further, because CDMA systems tend to be interference limited, for those applications involving voice transmission, an additional gain in the capacity of the system can be achieved by the use of *voice activity detection*. That is, in any given two-way telephone conversation, each user is typically talking only about 50% of the time. During the time when a user is quiet, he is not contributing to the instantaneous interference. Thus, if a sufficiently large number of users can be supported by the system, statistically only about one-half of them will be active simultaneously, and the effective capacity can be doubled.

Interference Rejection

In addition to providing multiple accessing capability, spread spectrum techniques are of interest in the commercial sector for basically the same reasons they are in the military community, namely their AJ and LPI characteristics. However, the motivations for such interest differ. For example, whereas the military is interested in ensuring that systems they deploy are robust to interference generated by an intelligent adversary (i.e., exhibit jamming resistance), the interference of concern in commercial applications is unintentional. It is sometimes referred to as co-channel interference (CCI) and arises naturally as the result of many services using the same frequency band at the same time. And while such scenarios almost always allow for some type of spatial isolation between the interfering waveforms, such as the use of narrow-beam antenna patterns, at times the use of the inherent interference suppression property of a spread spectrum signal is also desired. Similarly,

Spread Spectrum Communications

whereas the military is very much interested in the LPI property of a spread spectrum waveform, as indicated in Sec. 16.3, there are applications in the commercial segment where the same characteristic can be used to advantage.

To illustrate these two ideas, consider a scenario whereby a given band of frequencies is somewhat sparsely occupied by a set of conventional (i.e., nonspread) signals. To increase the overall spectral efficiency of the band, a set of spread spectrum waveforms can be overlaid on the same frequency band, thus forcing the two sets of users to share common spectrum. Clearly, this scheme is feasible only if the mutual interference that one set of users imposes on the other is within tolerable limits. Because of the interference suppression properties of spread spectrum waveforms, the despreading process at each spread spectrum receiver will attenuate the components of the final test statistic due to the overlaid narrowband signals. Similarly, because of the LPI characteristics of spread spectrum waveforms, the increase in the overall noise level as seen by any of the conventional signals, due to the overlay, can be kept relatively small.

Defining Terms

Antijam communication system: A communication system designed to resist intentional jamming by the enemy.

Chip time (interval): The duration of a single pulse in a direct sequence modulation; typically much smaller than the information symbol interval.

Coarse alignment: The process whereby the received signal and the despreading signal are aligned to within a single chip interval.

Dehopping: Despreading using a frequency-hopping modulation.

Delay-locked loop: A particular implementation of a closed-loop technique for maintaining fine alignment.

Despreading: The notion of decreasing the bandwidth of the received (spread) signal back to its information bandwidth.

Direct sequence modulation: A signal formed by linearly modulating the output sequence of a pseudorandom number generator onto a train of pulses.

Direct sequence spread spectrum: A spreading technique achieved by multiplying the information signal by a direct sequence modulation.

Fast frequency-hopping: A spread spectrum technique wherein the hop time is less than or equal to the information symbol interval, i.e., there exist one or more hops per data symbol.

Fine alignment: The state of the system wherein the received signal and the despreading signal are aligned to within a small fraction of a single chip interval.

Frequency-hopping modulation: A signal formed by nonlinearly modulating a train of pulses with a sequence of pseudorandomly generated frequency shifts.

Hop time (interval): The duration of a single pulse in a frequency-hopping modulation.

Hybrid spread spectrum: A spreading technique formed by blending together several spread spectrum techniques, e.g., direct sequence, frequency-hopping, etc.

Low-probability-of-intercept communication system: A communication system designed to operate in a hostile environment wherein the enemy tries to detect the presence and perhaps characteristics of the friendly communicator's transmission.

Processing gain (spreading ratio): The ratio of the spread spectrum bandwidth to the information data rate.

Radiometer: A device used to measure the total energy in the received signal.

Search algorithm: A means for coarse aligning (synchronizing) the despreading signal with the received spread spectrum signal.

Slow frequency-hopping: A spread spectrum technique wherein the hop time is greater than the information symbol interval, i.e., there exists more than one data symbol per hop.

Spread spectrum bandwidth: The bandwidth of the transmitted signal after spreading.

Spreading: The notion of increasing the bandwidth of the transmitted signal by a factor far in excess of its information bandwidth.

Tau–dither loop: A particular implementation of a closed-loop technique for maintaining fine alignment.

Time-hopping spread spectrum: A spreading technique that is analogous to pulse position modulation.

Tracking algorithm: An algorithm (typically closed loop) for maintaining fine alignment.

Further Information

M.K. Simon, J.K. Omura, R.A. Scholtz, and B.K. Levitt, *Spread Spectrum Communications Handbook*, McGraw Hill, 1994 (previously published as *Spread Spectrum Communications*, Computer Science Press, 1985).

R.E. Ziemer and R.L. Peterson, *Digital Communications and Spread Spectrum Techniques*, Macmillan, 1985.

J.K. Holmes, *Coherent Spread Spectrum Systems*, John Wiley and Sons, Inc. 1982.

R.C. Dixon, *Spread Spectrum Systems*, 3rd ed., John Wiley and Sons, Inc. 1994.

C.F. Cook, F.W. Ellersick, L.B. Milstein, and D.L. Schilling, *Spread Spectrum Communications*, IEEE Press, 1983.

17
Diversity Techniques

17.1 Introduction 213
17.2 Diversity Schemes 215
   Space Diversity • Polarization Diversity • Angle Diversity • Frequency Diversity • Path Diversity • Time Diversity
17.3 Diversity Combining Techniques 217
   Selection Combining • Maximal Ratio Combining • Equal Gain Combining • Loss of Diversity Gain Due to Branch Correlation and Unequal Branch Powers
17.4 Effect of Diversity Combining on Bit-Error Rate (BER) 221
17.5 Conclusions 222

A. Paulraj
Stanford University

17.1 Introduction

Diversity is a commonly used technique in mobile radio systems to combat signal fading. The basic principle of diversity is as follows. If several replicas of the same information carrying signal are received over multiple channels with comparable strengths and that exhibit independent fading, then there is a good likelihood that at least one or more of these of the received signals will not be in a fade at any given instant in time, thus making it possible to deliver adequate signal level to the receiver. Without diversity techniques, in noise limited conditions, the transmitter will have to deliver a much higher power level to protect the link during the short intervals when the channel is severely faded. In mobile radio, the power available on the reverse link is severely limited by the battery capacity in handheld subscriber units. Diversity methods play a crucial role in reducing transmit power needs. Also, cellular communication networks are mostly interference limited and once again mitigation of channel fading through use of diversity can translate into improved interference tolerance, which in turn means greater ability to support additional users and therefore higher system capacity.

The basic principles of diversity have been known since 1927 when the first experiments in space diversity were reported. There are many techniques for obtaining independently fading branches, and these can be subdivided into two main classes. The first called explicit techniques, uses explicit redundant signal transmission to exploit diversity channels. Use of dual polarized signal transmission and reception in many point-to-point radios is an example of explicit diversity. Clearly, such redundant signal transmission involves a penalty in frequency spectrum or additional power. On the other hand, in the second main class called implicit techniques, the signal is transmitted only once, but the decorrelating effects in the propagation medium, such as multipaths, are exploited to receive signals over multiple diversity channels. A good example of implicit diversity is the **RAKE receiver** in code division multiple access (CDMA) systems that use independent fading of

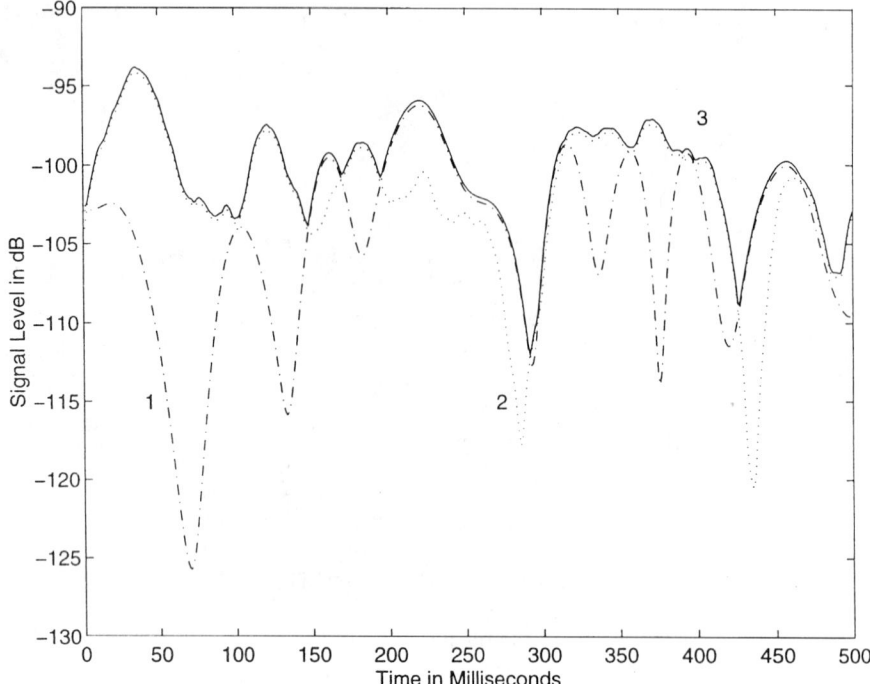

FIGURE 17.1 Example of diversity combining; two independently fading signals 1 and 2; the signal 3 is the result of selecting the strongest signal.

resolvable multipaths to achieve diversity gain. Figure 17.1 illustrates the principle of diversity where two independently fading signals are shown along with selection diversity output which selects the stronger signal. The fades in the resulting signal have been substantially smoothed out while also yielding higher average power.

Exploiting diversity needs careful design of the communication link. In explicit diversity, multiple copies of the same signal will have to be transmitted in channels using either frequency, or time, or polarization dimension. At the receiver end we need arrangements to receive the different diversity branches (this is true for both explicit and implicit diversity). The different diversity branches are then combined to reduce signal **outage probability** or bit-error rate.

In practice, the signals in the diversity branches may not show completely independent fading. The envelope cross correlation ρ between these signals is a measure of their independence.

$$\rho = \frac{E[[r_1 - \bar{r}_1][r_2 - \bar{r}_2]]}{\sqrt{E|r_1 - \bar{r}_1|^2 E|r_2 - \bar{r}_2|^2}}$$

where r_1 and r_2 represent the instantaneous envelope levels of the normalized signals at the two receivers and $\bar{r}_1$ and $\bar{r}_2$ are their respective means. It has been shown [Jakes, 1974] that a cross correlation of 0.7 between signal envelopes is sufficient to provide a reasonable degree of diversity gain. Depending on the type of the diversity employed, these diversity channels must be sufficiently *separated* along the appropriate diversity dimension. For spatial diversity, the antennas should be separated larger than the *coherence distance* to ensure a cross correlation less than 0.7. Likewise, in frequency diversity, the frequency separation must be larger than the *coherence bandwidth*, and in time diversity the separation between channel reuse in time should larger than the *coherence time*. These coherence factors, in turn, depend on the channel characteristics. The coherence distance,

Diversity Techniques 215

coherence bandwidth, and coherence time vary inversely as the angle spread, delay spread, and Doppler spread, respectively.

Once the receiver has a number of diversity branches, it has to combine these branches to maximize the signal level. Several techniques have been studied for diversity combining. We will describe three main techniques: selection combining, equal gain combining, and maximal ratio combining.

Finally, we should note that diversity is primarily used to combat fading, and if the signal does not have significant fading in the first place, as, for example, when there is a direct path component, diversity combining will not provide the gains normally expected.

17.2 Diversity Schemes

There are several techniques for obtaining diversity branches or, as they are sometimes also known, diversity dimensions. The most important of these are described in the following.

Space Diversity

This has historically been the most common form of diversity in mobile radio base stations. It is easy to implement and does not require additional frequency spectrum resources. Space diversity is exploited on the reverse link at the base station receiver by spacing antennas apart so as to obtain sufficient decorrelation. The key for obtaining uncorrelated fading of antenna outputs is adequate spacing of the antennas. The required spacing depends on the degree of multipath angle spread. For example, if the multipath signals arrive from all directions in the azimuth, as is usually the case at the mobile, antenna spacing of the order of 0.5λ–0.8λ is quite adequate [Lee, 1982]. On the other hand, if the multipath angle spread is small, as in the case of base stations, the coherence distance is much larger. Also, empirical measurements show a strong coupling between antenna height and spatial correlation. Larger antenna heights imply larger coherence distances. Typically, 10λ–20λ separation is adequate to achieve $\rho = 0.7$ at base stations in suburban settings when the signals arrive from the broadside direction. The coherence distance can be 3–4 times larger for endfire arrivals. The endfire problem is averted in base stations with trisectored antennas as each sector needs to handle only signals arriving ± 60 deg off the broadside. The coherence distance depends strongly on the terrain. Also, base stations normally use space diversity in the horizontal plane only. Separation in the vertical plane can also be used, and the necessary spacing depends on vertical multipath angle spread. This can be small for distant mobiles, making vertical plane diversity less attractive in most applications.

Polarization Diversity

In mobile radio environments, signals transmitted on orthogonal polarizations exhibit decorrelated fading and, therefore, offer potential for diversity combining. Polarization diversity can be obtained either by explicit or implicit techniques. Note that with polarization only two diversity branches are available as opposed to space diversity where several branches can be obtained using multiple antennas. In explicit polarization diversity, the signal is both transmitted and received in two orthogonal polarizations. For a fixed total transmit power, the power in each branch will be 3 dB lower than if single polarization was used. In the implicit polarization technique, the signal is launched in a single polarization but is received with cross-polarized antennas. The propagation medium couples some energy into the cross-polarization plane. The observed cross-polarization coupling factor has been observed to be 10–12 dB in mobile ratio frequencies [Vaughan, 1990; Adachi et al., 1986]. Also, the cross-polarization envelope correlation has been found to be better than 0.7 and, therefore, provides significant diversity gain.

In recent years, with the increasing use of pocket telephones, the handset can be held at random orientations during a call. This results in energy being launched with varying polarization angles ranging from vertical to horizontal. This further increases the advantage of cross-polarized antennas at the base station since at least one of the two antennas will be well matched to the signal launch polarization. Recent work [Jefford, 1995] has shown that with variable launch polarization, a cross-polarized antenna can give comparable performance to a vertically polarized space diversity antenna.

Finally, we should note that cross-polarized antennas can be deployed in a compact antenna assembly and do not need the potentially large physical separation needed in space diversity antennas. This is an important advantage in the personal communication service (PCS) base stations, where low profile antennas are needed.

Angle Diversity

In situations where the angle spread is very high, as in the case of indoors or at the mobile unit in urban locations, signals collected from multiple nonoverlapping beams offer low-fade correlation with balanced power in the diversity branches. Clearly, since directional beams imply use of antenna aperture, angle diversity is closely related to the space diversity. Angle diversity has been utilized in indoor wireless local area networks (LANs), where its use allows substantial increase in LAN throughputs [Freeburg, 1991].

Frequency Diversity

Another technique to obtain decorrelated diversity branches is to transmit the same signal over different frequencies. The frequency separation between carriers should be larger than the coherence bandwidth. The coherence bandwidth, of course, depends on the multipath delay spread of the channel. The larger the delay spread is, the smaller the coherence bandwidth and the more closely the frequency diversity channels can be spaced. Clearly, frequency diversity is an explicit diversity technique and needs additional frequency spectrum.

A common form of frequency diversity is multicarrier (also known as multitone) modulation. This technique involves sending redundant data over a number of closely spaced carriers to benefit from frequency diversity, which is then exploited by applying **forward error correction** (**FEC**) across the carriers.

Path Diversity

A sophisticated form of diversity is based on using a signal bandwidth much larger than the channel coherence bandwidth, as is used in the so-called direct sequence spread spectrum modulation techniques. This modulation scheme is used in the CDMA mobile networks, one example of which is the IS-95 standard for 800-MHz cellular band. Spread spectrum signals can resolve multipath arrivals as long as the path delays are separated by at least one *chip* period. If the signal in each path shows low-fade correlation, as is usually the case, these paths offer a valuable source of diversity. A receiver that resolves the multipaths via code correlation and then combines them is referred to as a RAKE receiver [Viterbi, 1995]; see also Fig. 17.2.

In CDMA, diversity gain provided by the multiple paths (and other diversity branches, if any) not only reduces transmit power needs but also increases the number of users that can be supported per cell for a given bandwidth.

Time Diversity

In mobile communications channels, the mobile motion together with scattering in the vicinity of the mobile causes time selective fading of the signal with Rayleigh fading statistics for the signal envelope. Signal fade levels separated by the coherence time show low correlation and can be used as diversity branches if the same signal can be transmitted at multiple instants separated by the

Diversity Techniques

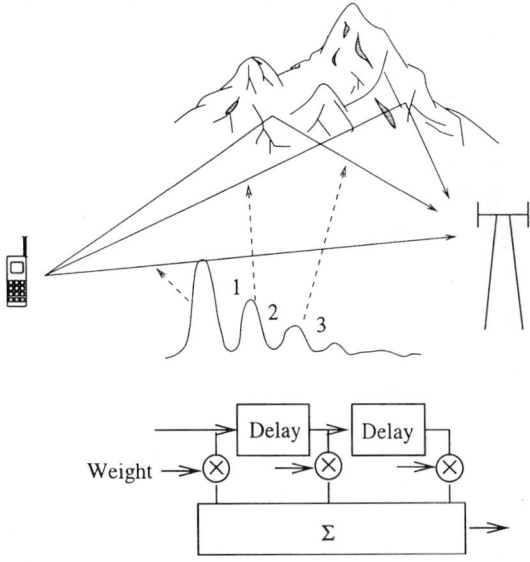

FIGURE 17.2 Multipath and RAKE receivers: three paths with independent fading are combined to provide path diversity.

coherence time. The coherence time depends on the Doppler spread of the signal, which in turn is a function of the mobile speed and the carrier frequency.

Time diversity is usually exploited via **interleaving**, forward-error correction coding, and **automatic request for repeat** (**ARQ**). These are sophisticated techniques to exploit channel coding and time diversity. One fundamental drawback with time diversity approaches is the delay needed to collect the repeated or interleaved transmissions. If the coherence time is large, as, for example, when the vehicle is slow moving, the required delay becomes too large to be acceptable for interactive voice conversation.

The statistical properties of a fading signal depend on the field component used by the antenna, the vehicular speed, and the carrier frequency. For an idealized case of a mobile surrounded by scatterers in all directions, the auto-correlation function of the received signal $x(t)$ [note this is not the envelope $r(t)$] can be shown to be

$$E[x(t)x(t+\tau)] = J_0(2\pi\tau v/\lambda)$$

where J_0 is a Bessel function of zeroth order and v is the mobile velocity.

17.3 Diversity Combining Techniques

Several diversity combining methods are known. We describe three main techniques: selection, maximal ratio, and equal gain. They can be used with each of the diversity schemes just discussed.

Selection Combining

This is the simplest and perhaps the most frequently used form of diversity combining. In this technique, one of the two diversity branches with the highest carrier to noise ratio (C/N) is connected to the output; see Fig. 17.3(a).

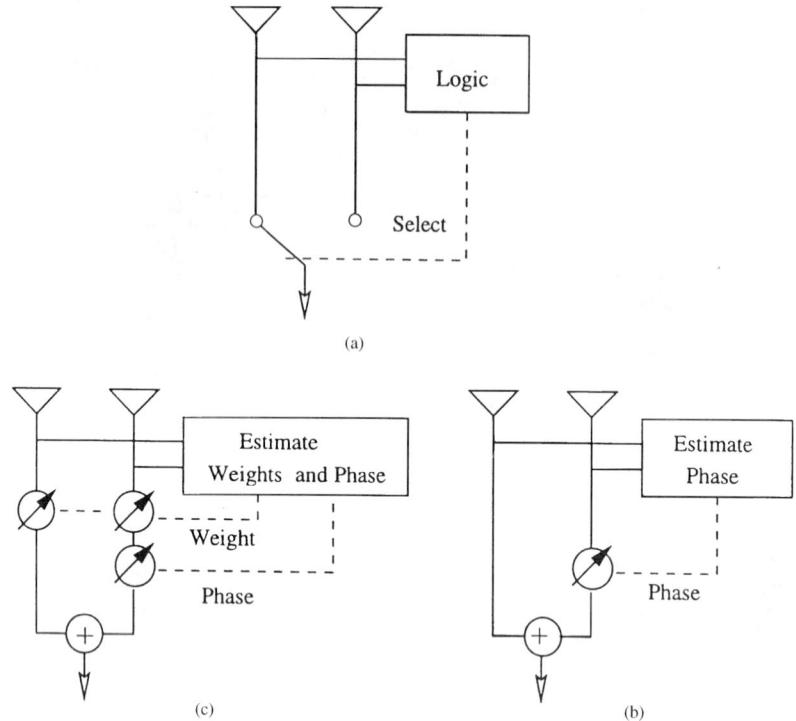

FIGURE 17.3 Diversity combining methods for two diversity branches. (a) Selection combining; (b) Maximal ratio combining; (c) Equal gain combining.

The performance improvement due to selection diversity can be seen as follows. Let the signal in each branch exhibit Rayleigh fading with mean power σ^2. The density function of the envelope is given by

$$p(r_i) = \frac{r_i}{\sigma^2} e^{\frac{-r_i^2}{2\sigma^2}} \tag{17.1}$$

where r_i is the signal envelope in each branch. If we define two new variables

$$\gamma_i = \frac{\text{instantaneous signal power in each branch}}{\text{mean noise power}}$$

$$\Gamma = \frac{\text{mean signal power in each branch}}{\text{mean noise power}}$$

then the probability that the C/N is less than or equal to some specified value γ_s is

$$\text{prob}[\gamma_i \leq \gamma_s] = 1 - e^{-\gamma_s/\Gamma} \tag{17.2}$$

The probability that γ_i in all branches with independent fading will be simultaneously less than or equal to γ_s is then

$$\text{prob}[\gamma_1, \gamma_2, \ldots, \gamma_M \leq \gamma_s] = (1 - e^{-\gamma_s/\Gamma})^M \tag{17.3}$$

This is the distribution of the best signal envelope from the two diversity branches. Figure 17.4 shows

Diversity Techniques

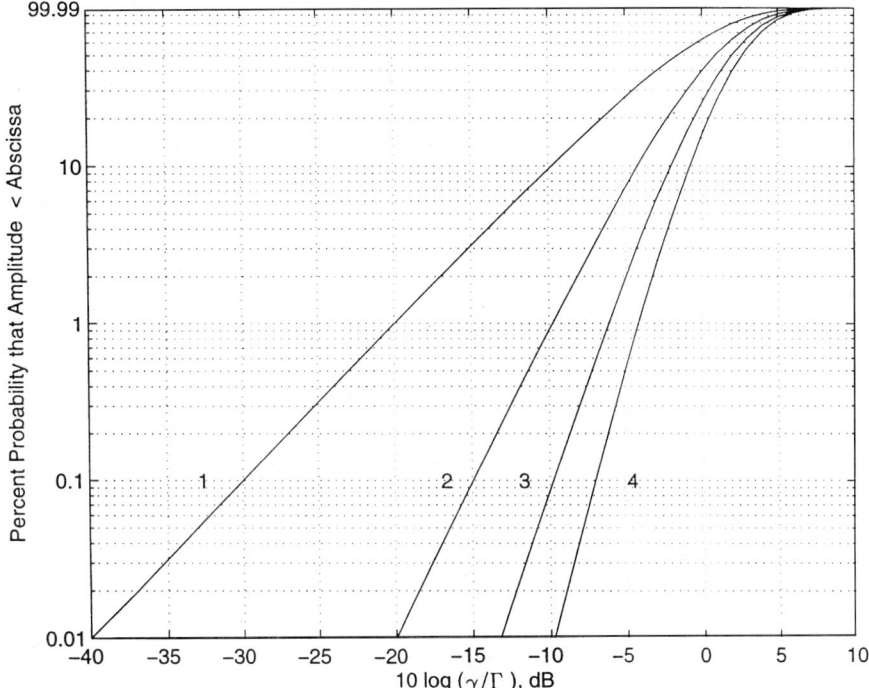

FIGURE 17.4 Probability distribution for signal envelope for selection combining.

the distribution of the combiner output C/N for $M = 1, 2, 3$, and 4 branches. The improvement in signal quality is significant. For example, at 99% reliability level, the improvement in C/N is 10 dB for 2 branches and 16 dB for 4 branches.

Selection combining also increases the mean C/N of the combiner output and can be shown [Jakes, 1974] to be

$$\text{mean}(\gamma_s) = \Gamma \sum_{k=1}^{M} \frac{1}{k} \qquad (17.4)$$

This indicates that with 4 branches, for example, the mean C/N of the selected branch is 2.03 better than the mean C/N in any one branch.

Maximal Ratio Combining

In this technique the M diversity branches are first cophased and then weighted proportional to their signal level before summing; see Fig. 17.3(b). The distribution of maximal ratio combining has been shown [Lee, 1982] to be

$$\text{prob}[\gamma \leq \gamma_m] = 1 - e^{(-\gamma_m/\Gamma)} \sum_{k=1}^{M} \frac{(\gamma_m/\Gamma)^{k-1}}{(k-1)!} \qquad (17.5)$$

The distribution of the output of a maximal ratio combiner is shown in Fig. 17.5. Maximal ratio combining is known to be optimum in the sense that it yields the best statistical reduction of fading of any linear diversity combiner. In comparison to the selection combiner, at 99% reliability level,

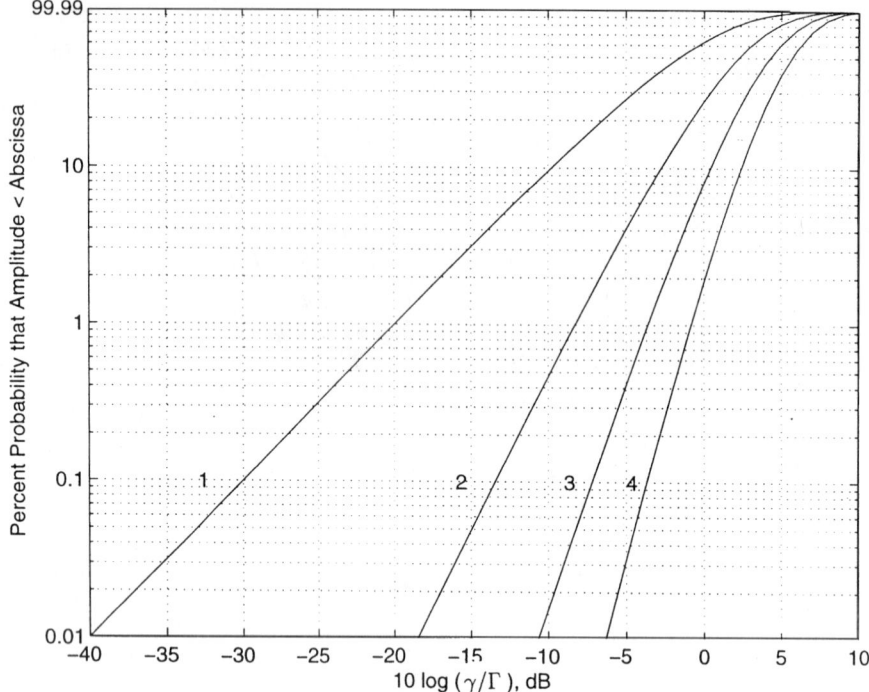

FIGURE 17.5 Probability distribution for signal envelope for maximal ratio combining.

the maximal ratio combiner provides a 11.5-dB gain for 2 branches and a 19-dB gain for 4 branches. An improvement of 1.5 and 3 dB, respectively, over the selection diversity combiner.

The mean C/N of the combined signal may be easily shown to be

$$\text{mean}(\gamma_m) = M\Gamma \tag{17.6}$$

Therefore, combiner output mean varies linearly with M. This confirms the intuitive result that the output C/N averaged over fades should provide gain proportional to the number of diversity branches. This is a situation similar to conventional beam forming.

Equal Gain Combining

In some applications, it may be difficult to estimate the amplitude accurately, and the combining gains may all be set to unity, and the diversity branches merely summed after cophasing; see Fig. 17.3(c).

The distribution of the equal gain combiner does not have a neat expression and has been computed by numerical evaluation. Its performance has been shown to be very close (less than a decibel) to maximal ratio combining. The mean C/N can be shown [Jakes, 1974] to be

$$\text{mean}(\gamma_e) = \Gamma\left[1 + \frac{\pi}{4}(M-1)\right] \tag{17.7}$$

Like maximal ratio combining, the mean C/N for equal gain combining grows almost linearly with M and is approximately only 1 dB poorer than the maximal ratio combiner even with infinite number of branches.

Diversity Techniques

Loss of Diversity Gain Due to Branch Correlation and Unequal Branch Powers

The preceding analysis assumed that the fading signals in the diversity branches were all uncorrelated and of equal power. In practice, this may be difficult to achieve and, as we saw earlier, the branch cross-correlation coefficient $\rho = 0.7$ is considered to be acceptable. Also, equal mean power in diversity branches is rarely available. In such cases, we can expect a certain loss of diversity gain. Since most of the damage in fading is due to deep fades, however, and also since the chance of coincidental deep fades is small even for moderate branch correlation, one can expect reasonable tolerance to branch correlation.

The distribution of the output signal envelope of the maximal ratio combiner has been shown [Pahlavan and Levesque, 1995] to be

$$\text{prob}[\gamma_m] = \sum_{n=1}^{M} \frac{A_n}{2\lambda_n} e^{-\gamma_m/2\lambda_n} \tag{17.8}$$

where λ_n are the eigenvalues of the $M \times M$ branch envelope covariance matrix whose elements are defined by

$$\mathbf{R}_{ij} = E\left[r_i r_j^*\right] \tag{17.9}$$

and A_n is defined by

$$A_n = \prod_{\substack{k=1 \\ k \neq n}}^{M} \frac{1}{1 - \lambda_k/\lambda_n} \tag{17.10}$$

Recently extensive field measurements were carried out in the 1800-MHz band, and compact empirical results have been obtained for diversity gain at 90% signal reliability level as a function of branch correlation ρ and mean signal level difference Δ; see Jefford, 1995, for a full discussion.

Selection:

$$G = 5.71\, e^{(-0.87\rho - 0.16\Delta)} \tag{17.11}$$

Equal gain:

$$G = -8.98 + 15.22\, e^{(-0.20\rho - 0.04\Delta)} \tag{17.12}$$

Maximal ratio:

$$G = 7.14\, e^{(-0.59\rho - 0.11\Delta)} \tag{17.13}$$

where G is in decibels.

17.4 Effect of Diversity Combining on Bit-Error Rate (BER)

So far we have studied the distribution of the instantaneous envelope or C/N after diversity combining. We will now briefly survey how diversity combining affects BER performance in digital radio links; we assume maximal ratio combining.

To begin let us first examine the effect of Rayleigh fading on BER performance of digital transmission links. This has been studied by several authors and is summarized in Proakis, 1989. Table 17.1

TABLE 17.1 Comparison of BER Performance for Unfaded and Rayleigh Faded Signals

Modulation	Unfaded BER	Faded BER
Coh BPSK	$\frac{1}{2}\text{erfc}(\sqrt{E_b/N_0})$	$\frac{1}{4(\overline{E}_b/N_0)}$
Coh FSK	$\frac{1}{2}\text{erfc}\left(\sqrt{\frac{1}{2}E_b/N_0}\right)$	$\frac{1}{2(\overline{E}_b/N_0)}$

TABLE 17.2 BER Performance for Coherent BPSK and FSK with Diversity

Modulation	Postdiversity BER
Coherent BPSK	$\left(\frac{1}{4\overline{E}_b/N_0}\right)^L \binom{2L-1}{L}$
Coherent FSK	$\left(\frac{1}{2\overline{E}_b/N_0}\right)^L \binom{2L-1}{L}$

gives the BER expressions in the large E_b/N_0 case for coherent binary phase-shift keying (PSK) and coherent binary orthogonal frequency-shift keying (FSK) for unfaded and Rayleigh faded additive white Gaussian noise channels (AWGN) channels. $\overline{E}_b/N_0$ represents the average E_b/N_0 for the fading channel.

Observe that error rates decrease only inversely with SNR as against exponential decrease for the unfaded channel. Also note that for fading channels, coherent binary PSK is 3 dB better than coherent binary FSK, exactly the same advantage as in the unfaded case. Even for modest target BER of 10^{-2} that is usually needed in mobile communications, the loss due to fading can be significant.

To obtain the BER with maximal ratio diversity combining we have to average the BER expression for the unfaded BER with the distribution obtained for the maximal ratio combiner given in Fig. 17.5. Analytical expressions have been derived for these in Proakis, 1989. For branch SNR greater than 10 dB, the BER after maximal ratio diversity combining is given in Table 17.2.

We observe that the probability of error varies as $1/(\overline{E}_b/N_0)$ raised to the Lth power. Thus, diversity reduces the error rate exponentially as the number of independent branches increases.

17.5 Conclusions

Diversity provides a powerful technique for combating fading in mobile communication systems. Diversity techniques seek to generate and exploit multiple branches over which the signal shows low-fade correlation. To obtain the best diversity performance, the multiple access, modulation, coding, and antenna design of the wireless link must all be carefully chosen so as to provide a rich and reliable level of well-balanced, low-correlation diversity branches in the target propagation environment. Successful diversity exploitation can impact a mobile network in several ways. Reduced power requirements can result in increased coverage or improved battery life. Low-signal outage improves voice quality and handoff performance. Finally, reduced fade margins directly translate to increased system capacity, particularly in CDMA networks.

Defining Terms

Automatic request for repeat: An error control mechanism in which received packets that cannot be corrected are retransmitted.

Fading: Fluctuation in the signal level due to shadowing and multipath effects.

Forward error correction (FEC): A technique that inserts redundant bits during transmission to help detect and correct bit errors during reception.

Interleaving: A form of data scrambling that spreads burst of bit errors evenly over the received data allowing efficient forward error correction.

Outage probability: The probability that the signal level falls below a specified minimum level.

RAKE receiver: A receiver used in direct sequence spread spectrum signals. The receiver extracts energy in each path and then adds them together with appropriate weighting and delay.

References

Adachi, F., Feeney, M.T., Williamson, A.G., and Parsons, J.D. 1986. Crosscorrelation between the envelopes of 900 MHz signals received at a mobile radio base station site. *Proc. IEE*, 133(6): 506–512.

Freeburg, T.A. 1991. Enabling technologies for in-building network communications—four technical challenges and four solutions. *IEEE Trans. Veh. Tech.* 29(4):58–64.

Jakes, W.C. 1974. *Microwave Mobile Communications*, John Wiley, New York.

Jefford, P.A., Turkmani, A.M.D., Arowojolu, A.A., and Kellett, C.J. 1995. An experimental evaluation of the performance of the two branch space and polarization schemes at 1800 MHz. *IEEE Trans. Veh. Tech.* VT-44(2):318–326.

Lee, W.C.Y. 1982. *Mobile Communications Engineering*, McGraw-Hill, New York.

Pahlavan, K. and Levesque, A. H. 1995. *Wireless Information Networks*, John Wiley, New York.

Proakis, J.G. 1989. *Digital Communications*, McGraw-Hill, New York.

Vaughan, R.G. 1990. Polarization diversity system in mobile communications. *IEEE Trans. Veh. Tech.* VT-39(3):177–86.

Viterbi, A.J. 1995. *CDMA: Principle of Spread Spectrum Communications*, Addison–Wesley, Reading, MA.

18
Information Theory

18.1	Introduction 224
18.2	The Communication Problem 224
18.3	Source Coding for Discrete-Alphabet Sources 226
18.4	Universal Source Coding 228
18.5	Rate Distortion Theory 229
18.6	Channel Coding 231
18.7	Simple Binary Codes 233
	Repetition Codes • Single-Parity-Check Codes • Hamming Codes

Bixio Rimoldi
Washington University

Rüdiger Urbanke
Washington University

18.1 Introduction

The field of information theory has its origin in Claude Shannon's 1948 paper "A mathematical theory of communication." Shannon's motivation was to study "[The problem] of reproducing at one point either exactly or approximately a message selected at another point." Whereas in this section we will be concerned only with Shannon's original problem, one should keep in mind that information theory is a growing field of research whose profound impact has reached various areas such as statistical physics, computer science, statistical inference, and probability theory. For an excellent treatment of information theory that extends beyond the area of communication we recommend Cover and Thomas [1991]. For the reader who is strictly interested in communication problems we also recommend Gallager [1968], Blahut [1987], and McEliece [1977].

18.2 The Communication Problem

A rather general diagram for a (point-to-point) communication system is shown in Fig. 18.1.

The source might be digital (e.g., a data file) or analog (e.g., a video signal). Since an analog source can be sampled without loss of information (see chapter on sampling, this volume), without loss of generality we consider only discrete-time sources and model them as discrete-time stochastic processes.

The channel could be a pair of wires, an optical fiber, a radio link, etc. The channel model specifies the set of possible channel inputs and, for each input, specifies the output process. Most real-world channels are waveform channels, meaning that the input and output sets are sets of waveforms. It is often the case that the communication engineer is given a waveform channel with a modulator and a demodulator. In a satellite communication system, the modulator might be a phase-shift keying modulator whose input alphabet V is the set $\{0, 1\}$; the channel might be modeled by the

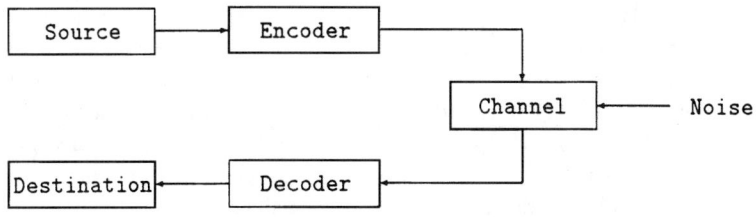

FIGURE 18.1 Block diagram of a communications system.

additive white Gaussian noise channel, and the demodulator attempts to output a guess (perhaps incorrect) of the modulator input. In this case the modulator output alphabet W equals V. In more sophisticated cases, where $|W| > |V|$ and $|\cdot|$ denotes the number of elements in the enclosed set, the digital data demodulator can also furnish information about the reliability of the decision. Either way, one would consider the modulator and the demodulator as part of the channel. If the statistics of the channel output at a given time depend only on the value of the corresponding position of the input sequence, then the channel is a discrete memoryless channel. This is an important class of channel models, which will receive particular attention in this chapter. Even if the modulator and the demodulator are not given, assuming that the bandwidth of the waveform channel is limited, one can always use the sampling theorem to convert a waveform channel into a channel with discrete-time input and output.

The destination is merely a place holder to remind us that the user has some expectation concerning the quality of the reproduced signal. If the source output symbols are elements of a finite set, then the destination may specify a maximum value for the probability of error. Otherwise, it may specify a maximum distortion computed according to some specified criterion (e.g., mean square error).

It is generally assumed that the source, the channel, and the destination are given and fixed. On the other hand, the communication engineer usually is completely free to design the encoder and the decoder to meet the desired performance specifications.

An important result of information theory is that, without loss of optimality, the encoder in Fig. 18.1 can be decomposed into two parts, a source encoder that produces a sequence of binary data and a channel encoder as shown in Fig. 18.2. Similarly, the decoder may be split into a channel decoder and a source decoder.

The objective of the source encoder is to compress the source, that is, to minimize the average number of bits necessary to represent a source symbol. The most important questions concerning the source encoder are the following: What is the minimum number of bits required (on average) to represent a source output symbol? How do we design a source encoder that achieves this minimum? These questions will be considered here. It turns out that the output of an ideal source encoder is a sequence of independent and uniformly distributed binary symbols. The purpose of the channel

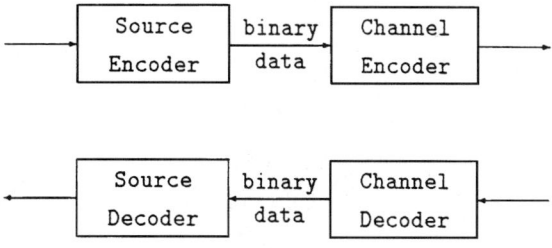

FIGURE 18.2 Block diagram of an encoder and a decoder, each split into two parts.

encoder/decoder pair is to create a reliable bit pipe for the binary sequence produced by the source encoder. Here the most important question is whether or not it is possible to design such an encoder/decoder pair. This question will also be considered in some detail. The fact that we can split the encoder into two parts as described is of fundamental importance. First of all, it allows us to study the source and the channel separately and to assume that they are connected by a binary interface. Second, it tells us that for a given channel we may design the channel encoder and the channel decoder to maximize the rate of the resulting (virtually error-free) bit pipe without having to know the nature of the source that will use it.

18.3 Source Coding for Discrete-Alphabet Sources

We start by considering the simplest possible source, namely, a discrete-time discrete-alphabet memoryless information source modeled by a random variable X taking values on a finite alphabet $\mathcal{X}$. A (binary source) **code C** for X is a mapping from $\mathcal{X}$ into the set of finite length binary sequences called codewords. Such a code can be represented by a binary tree as shown in the following example.

Example 1. Let X take on values in $\{1, 2, 3, 4, 5, 6, 7\}$. A possible binary code for X is given by the (ordered) set of binary sequences $\{1, 010, 011, 0000, 0001, 0010, 0011\}$. The corresponding tree is shown in Fig. 18.3. Notice that the codeword corresponding to a given source output symbol is the label sequence from the root to the node corresponding to that symbol.

In Example 1 source output symbols correspond to leaves in the tree. Such a code is called *prefix free* since no codeword is the prefix of another codeword. Given a concatenation of codewords of a prefix-free code we can parse it in an unique way into codewords. Codes with this property are called *uniquely decodable*. Although there are uniquely decodable codes which are *not* prefix free we will restrict our attention to prefix-free codes, as it can be shown that the performance of general uniquely decodable codes is no better than that of prefix-free codes (see Cover and Thomas [1991]).

For each i in $\mathcal{X}$ let $C(i)$ be the codeword associated to the symbol i, with l_i its length, $L \equiv \max_i l_i$, and let p_i be the probability that $X = i$. A complete binary tree of depth L has 2^L leaves. To each leaf at depth l, $l \leq L$, of the binary code tree (which is not necessarily complete) there correspond 2^{L-l} leaves at depth L of the corresponding complete tree (obtained by extending the binary code tree). Further, any two distinct leaves in the code tree have distinct associated leaves at depth L in the complete binary tree. Hence, summing up over all leaves of the code tree (all codewords) we get

$$2^L \geq \sum_i 2^{L-l_i} \Leftrightarrow \sum_i 2^{-l_i} \leq 1$$

The right-hand side is called the *Kraft inequality*, and is a necessary condition on the codeword lengths of any prefix-free code. Conversely, for any set of codeword lengths satisfying the displayed

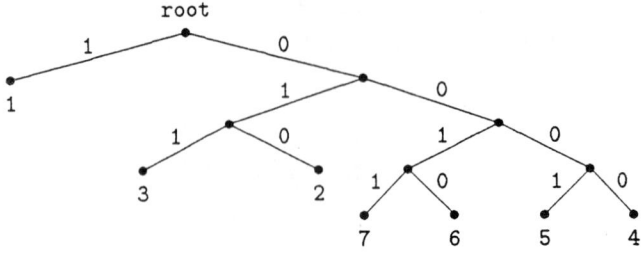

FIGURE 18.3 Example of a code tree.

Information Theory

inequality we can construct a prefix-free code having codeword lengths l_i (start with a complete binary tree of sufficient length; to each l_i associate a node in this tree at level l_i and make this node a leaf by deleting all of its descendants). The problem of finding the best source code then reduces to that of finding a set of lengths that satisfies the Kraft inequality and minimizes the average codeword lengths $L^* = \sum_i l_i p_i$. Such a code is called *optimal*.

To any source X with probabilities p_i we associate the quantity $H(X)$, which we call **entropy** and which is defined as

$$H(X) = -\sum_i p_i \log p_i$$

Usually we take the base of the logarithm to be 2 in which case the units of entropy are called *bits*.

It is straightforward to show by means of Lagrange multipliers that if we neglect the integer constraint on l_i then the minimization of L^* subject to the Kraft inequality yields $l_i^* = -\log p_i$. This noninteger choice of codeword lengths yields

$$L^* = \sum_i p_i l_i^* = -\sum_i p_i \log p_i = H(X)$$

On the other hand, the (not necessarily optimal) integer choice $l_i = \lceil -\log p_i \rceil$ satisfies the Kraft inequality and yields an expected codeword length

$$L(\mathbf{C}) = \sum_i p_i \left\lceil \log \frac{1}{p_i} \right\rceil \leq -\sum_i p_i \log p_i + \sum_i p_i = H(X) + 1$$

Hence, we have the following theorem.

Theorem 1. *The average length L^* of the optimal prefix-free code for the random variable X satisfies $H(X) \leq L^* \leq H(X) + 1$.*

Example 2. Let X be a random variable that takes on the values 1 and 0 with probabilities θ ($0 \leq \theta \leq 1$) and $1 - \theta$, respectively. Then $H(X) = h(\theta)$, where $h(\theta) = -\theta \log \theta - (1 - \theta \log(1 - \theta))$. The function $h(\theta)$ is called the *binary entropy function*. In particular, it can be shown that $h(\theta) \leq 1$ with equality if and only if $\theta = \frac{1}{2}$. More generally, if X takes on $|\mathcal{X}|$ values then $H(X) \leq \log |\mathcal{X}|$ with equality if and only if X is uniformly distributed.

A simple algorithm that generates optimal prefix-free codes is the Huffman algorithm, which can be described as follows:

Step 1. Arrange the probabilities in decreasing order so that $p_1 \geq p_2 \geq \cdots \geq p_m$.

Step 2. Form a subtree by combining the last two probabilities p_{m-1} and p_m into a single node of weight $p'_m = p_{m-1} + p_m$.

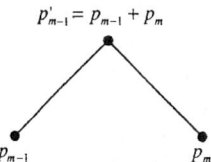

Step 3. Recursively execute steps 1 and 2, decreasing the number of nodes each time, until a single node is obtained.

Step 4. Use the tree constructed above to assign codewords.

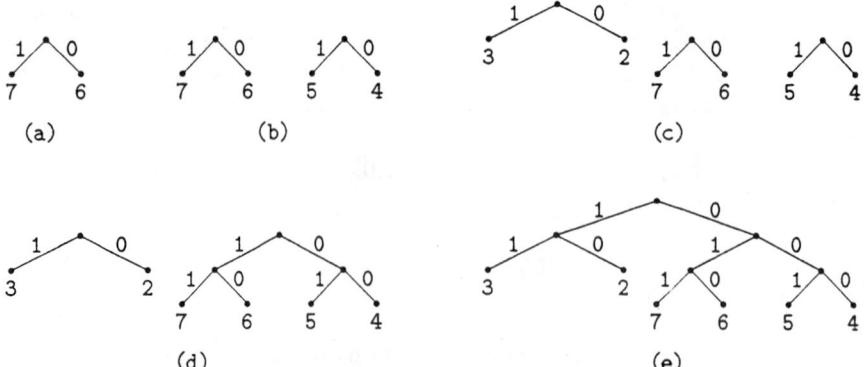

FIGURE 18.4 Construction of a Huffman code for X as given in Example 1.

Example 3. As in Example 1, let X take on the values $\{1, 2, 3, 4, 5, 6, 7\}$. Assume their corresponding probabilities are $\{0.4, 0.1, 0.1, 0.1, 0.1, 0.1, 0.1\}$. Figure 18.4 shows the steps in the construction of the **Huffman code**. For simplicity, probabilities are not shown. The final tree is given in Example 1.

Given a block of n random variables $X_1, \ldots, X_n$, with joint distribution $p(X_1, \ldots, X_n)$, define their entropy as

$$H(X_1, \ldots, X_n) = -\sum_{x_1}\sum_{x_2}\cdots\sum_{x_n} p(X_1, \ldots, X_n) \log p(X_1, \ldots, X_n)$$

Note that for the special case of n independent random variables we have $H(X_1, \ldots, X_n) = nH(X)$. This agrees with the intuitive notion that if it takes $H(X)$ bits to encode one source output symbol, then it should take $nH(X)$ bits to encode n independent output symbols. For a sequence of random variables $X_1, X_2, \ldots, X_n$ we define $H_n = \frac{1}{n}H(X_1, \ldots, X_n)$. If $X_1, X_2, \ldots, X_n$ is a stationary random process, then as n goes to infinity H_n converges to a limit of H_∞, which is called the *entropy rate* of the process.

Applying Theorem 1 to blocks of random variables we obtain

$$H_n \leq \frac{L^*}{n} < H_n + \frac{1}{n}$$

In the special case that the process $X_1, X_2, \ldots, X_n$ is stationary, the left and the right side will both tend to H_∞ as $n \to \infty$. This shows that by encoding a sufficient number of source output symbols at a time, one can get arbitrarily close to the fundamental limit of H_∞ bits per symbol.

18.4 Universal Source Coding

The Huffman algorithm produces optimal codes, but it requires the statistics of the source. It is a surprising fact that there are source coding algorithms that are asymptotically optimal without requiring prior knowledge of the source statistics. Such algorithms are called *universal* source coding algorithms. The best known universal source coding algorithm is the **Lempel–Ziv** algorithm, which has been implemented in various forms on personal computers (PCs) (as PKZIP command and as part of Microsoft's MS-DOS-6) and on UNIX-based systems (as *compress* command). There are two versions of the Lempel–Ziv algorithm: the LZ77, see Ziv and Lempel [1977], and the LZ78, see Lempel and Ziv [1978]. We will describe the basic idea of the LZ77.

Consider the sequence $X_1, X_2, \ldots, X_k$ taking values in $\{a, b, c\}$, shown in Fig. 18.5, and assume that the first w letters ($w = 6$ in our example) are passed to the decoder with no attempt at

Information Theory

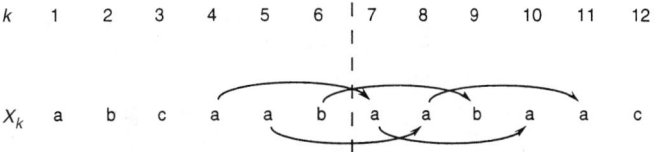

FIGURE 18.5 Encoding procedure of the Lempel–Ziv algorithm.

compression. At this point the encoder identifies the longest string $X_{w+1}, X_{w+2}, \ldots, X_{w+L}$ such that a copy, $X_{w-p+1}, \ldots, X_{w-p+L}$, begins (but not necessarily ends) in the portion of data already available to the decoder. In our example this string is *aabaa*, $L = 5$, and $p = 3$, see Fig. 18.5. Next the encoder transmits a binary representation of the pair (L, p) as this is sufficient for the decoder to reconstruct $X_{w+1}, \ldots, X_{w+L}$. At this point the procedure can be repeated with w replaced by $w + L$. There are two exceptions: The first occurs when X_{w+l} is a new letter that does not appear in $X_1, \ldots, X_w$; the second exception occurs when it is more efficient to encode $X_{w+1}, \ldots, X_{w+L}$ directly than to encode (L, p). In order to handle such special cases, the algorithm has the option of encoding a substring directly, without attempt at compression. It is surprising that the Lempel–Ziv algorithm is asymptotically optimal for all finite alphabet stationary ergodic sources. This has been recently shown in Wyner and Ziv [1994].

18.5 Rate Distortion Theory

Thus far we have assumed that $\mathcal{X}$ is a discrete alphabet and that the decoder must be able to perfectly reconstruct the source output. Now we drop both assumptions. This allows, for example, for sources that output real numbers.

Let $X_1, X_2, \ldots$, be a sequence of independent identically distributed random variables. Let $\mathcal{X}$ denote the source alphabet and $\mathcal{Y}$ be a suitably chosen representation alphabet (where $\mathcal{Y}$ is not necessarily equal to $\mathcal{X}$). A *distortion measure* d is a function $d : \mathcal{X} \times \mathcal{Y} \to \mathbb{R}^+$. The most common distortion measures are the *Hamming* distortion d_H and the *squared error* distortion d_E defined as follows:

$$d_H(x, y) = \begin{cases} 0 & \text{if } x = y \\ 1 & \text{if } x \neq y \end{cases}$$

$$d_E(x, y) = (x - y)^2$$

Such a single-letter distortion measure can be extended to a distortion measure on n-tuples by defining

$$d(x^n, y^n) = \frac{1}{n} \sum_{i=1}^{n} d(x_i, y_i)$$

The *encoder* maps an n-length source output X^n to an index $U \in \{1, 2, \ldots, 2^{\lfloor nR \rfloor}\}$, where R denotes the number of bits per source symbol that we are allowed or willing to use. The *decoder* maps U into an n-tuple Y^n, called the representation of X^n. We will call U together with the associated encoding and decoding function a *rate distortion code*. Let f be the function that maps X^n to its representation Y^n, that is, f is the mapping describing the concatenation of the encoding and the decoding function. The expected distortion D is then given by

$$D = \sum_{x^n} p(x^n) d(x^n, f(x^n))$$

The objective is to minimize the average number of bits per symbol, denoted by R, for a given average distortion D. What is the minimum R?

Definition. The rate distortion pair (R, D) is said to be *achievable* if there exists a rate distortion code of rate R with expected distortion D. The **rate distortion function** $R(D)$ is the infimum of rates R such that (R, D) is achievable for a given D.

Entropy played a key role in describing the limits of lossless coding. When distortion is allowed, a similar role is played by a quantity called mutual information.

Definition. Given the random variables X and Y with their respective distributions $p(x)$, $p(y)$, and $p(x, y)$, the **mutual information** $I(X; Y)$ between X and Y is defined as

$$I(X; Y) = \sum_{x \in X} \sum_{y \in Y} p(x, y) \log \frac{p(x, y)}{p(x)p(y)}$$

This definition can be extended to blocks of random variables X^n and Y^n by replacing $p(x)$, $p(y)$, and $p(x, y)$ with their higher dimensional counterparts. We can now state the fundamental theorem of rate distortion theory. For its proof see Cover and Thomas [1991].

Theorem 2. The rate distortion function for an independent identically distributed source X with distribution $p(x)$ and distortion function $d(x, y)$ is

$$R(D) = \min_{p(y|x): \sum_{x,y} p(x)p(y|x)d(x,y) \leq D} I(X; Y)$$

The rate distortion functions for a *binary source* that outputs 1 with probability θ and Hamming distortion and for a *Gaussian source* of variance σ^2 and squared error distortion are as follows:

$$R_B(D) = \begin{cases} h(\theta) - h(D), & 0 \leq D \leq \min\{\theta, 1 - \theta\} \\ 0, & D > \min\{\theta, 1 - \theta\} \end{cases}$$

$$R_G(D) = \begin{cases} \frac{1}{2} \log \frac{\sigma^2}{D}, & 0 \leq D \leq \sigma^2 \\ 0, & D > \sigma^2 \end{cases}$$

These two functions are plotted in Fig. 18.6.

Some insight can be gained by considering the extreme values of both $R_B(D)$ and $R_G(D)$. Assume that $X_1, X_2, \ldots$ is a sequence of independent and identically distributed binary random variables and let θ be the probability that $X_i = 1$. Without loss of generality, we may assume that $\theta \geq 1 - \theta$. If we let $Y^n = (0, \ldots, 0)$ regardless of X^n, then the expected Hamming distortion equals θ. Hence it must be true that $R_B(D) = 0$ for $D \geq \theta$. On the other hand $D = 0$ means that the reproduction Y^n must be a perfect copy of X^n. From Theorem 1 and Example 2 we know that

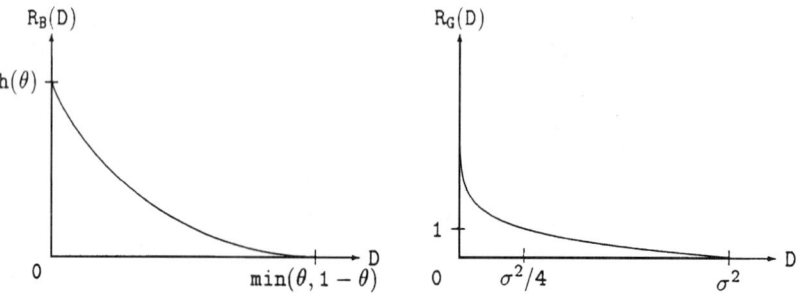

FIGURE 18.6 Examples of rate distortion functions.

Information Theory

this means $R_B(0) = H(X) = h(q)$. Similar considerations hold for $R_G(D)$. It is interesting to observe that $R_G(D)$ tells us that if we are allowed to use R bits per symbol to describe a sequence of independent Gaussian random variables with variance σ^2, then we need to accept a mean squared error of $\sigma^2 2^{-2R}$.

18.6 Channel Coding

Now we consider the fundamental limit for the amount of information that can be transmitted through a noisy channel and discuss practical methods to approach it. For simplicity we assume that the channel can be modeled as a discrete memoryless channel, defined as follows:

A **discrete memoryless channel** (**DMC**) is a system consisting of an input alphabet $\mathcal{X}$, an output alphabet $\mathcal{Y}$, and a collection of probability mass functions $p(y \mid x)$, one for each $x \in \mathcal{X}$. A specific DMC will be denoted by $(\mathcal{X}, p(y \mid x), \mathcal{Y})$.

A (M, n) channel code for the DMC $(\mathcal{X}, p(y \mid x), \mathcal{Y})$ consists of the following:

1. An index set $\{1, 2, \ldots, M\}$
2. An encoding function $X^n : \{1, 2, \ldots, M\} \to \mathcal{X}^n$, yielding codewords $X^n(1), X^n(2), \ldots, X^n(M)$ (The set of codewords is called the *code book*.)
3. A decoding function $g: \mathcal{Y}^n \to \{1, 2, \ldots, M\}$, which is a deterministic rule that assigns a guess to each possible received vector

The most important figures of merit for a channel code for a DMC are the *maximal probability of error* defined by

$$\lambda^{(n)} \equiv \max_{i \in \{1,2,\ldots,M\}} \Pr\{g(Y^n) \neq i \mid X^n = X^n(i)\}$$

and the *transmission rate*

$$R = \frac{\log M}{n} \qquad \text{bits/channel use}$$

It makes sense to define the rate in this way since the M codewords can be labeled with $\log M$ bits. Hence, every time we transmit a codeword we actually transmit $\log M$ bits of information. Since it takes n uses of the channel to transmit one codeword, the resulting information rate is $(\log M/n)$ bits per channel use.

For a given DMC, a rate R is said to be *achievable* if for any desired probability of error P_e and sufficiently large block length n there exists a $(\lceil 2^{nR} \rceil, n)$ code for that DMC with maximal probability of error $\lambda^{(n)} \leq P_e$.

The *operational capacity* of a discrete memoryless channel is the supremum of all achievable rates. One would expect that the determination of the operational capacity is a formidable task. One of the most remarkable results of information theory is an easy-to-compute way to determine the operational capacity.

We define the **information channel capacity** of a discrete memoryless channel $(\mathcal{X}, p(y \mid x), \mathcal{Y})$ as

$$C \equiv \max_{p(x)} I(X; Y)$$

where the maximum is taken over all possible input distributions $p(x)$. As we will show in the next example, it is straightforward to compute the information channel capacity for most discrete memoryless channels of interest. Shannon's channel coding theorem establishes that the information channel capacity equals the operational channel capacity.

Theorem (Shannon's channel coding theorem). For a DMC the *information* channel capacity equals the *operational* channel capacity, that is, for every rate $R < C$, there exists a sequence of $(\lceil 2^{nR} \rceil, n)$ codes with maximum probability of error $\lambda^{(n)} \to 0$. Conversely, any sequence of $(\lceil 2^{nR} \rceil, n)$ codes with $\lambda^{(n)} \to 0$ must have $R \leq C$.

This is perhaps the most important result of information theory. In order to shed some light on the interpretation of mutual information and to provide an efficient way to determine C, it is convenient to rewrite $I(X;Y)$ in terms of conditional entropy. Given two random variables X and Y with a joint probability mass function $p(x, y)$, marginal probability mass function $p(x) = \sum_y p(x, y)$, and conditional probability mass function $p(x \mid y) = p(x, y)/p(y)$, we define the conditional entropy of X given that $Y = y$ as

$$H(X \mid Y = y) = -\sum_x p(x \mid y) \log p(x \mid y)$$

Its average is the conditional entropy of X given Y, denoted by $H(X \mid Y)$ and defined as

$$H(X \mid Y) = \sum_y p(y) H(X \mid Y = y)$$
$$= -\sum_{x,y} p(x, y) \log p(x \mid y)$$
$$= E[-\log p(x \mid y)]$$

Recall that for the discrete variable X with probability mass function $p(x)$, $H(X)$ is the average number of binary symbols necessary to describe X. Now let X be the random variable at the input of a DMC and Y be the output. The knowledge that $Y = y$ changes the probability mass function of the channel input from $p(x)$ to $p(x \mid y)$ and its entropy from $H(X)$ to $H(X \mid Y = y)$. According to our previous result, $H(X \mid Y = y)$ is the average number of bits necessary to describe X after the observation that $Y = y$. Since $H(X)$ is the average number of bits needed to describe X (without knowledge of Y), $H(X) - H(X \mid Y = y)$ is the average amount of information about X acquired by the observation that $Y = y$, and

$$H(X) - H(X \mid Y)$$

is the average amount of information about X acquired by the observation of Y. One can easily verify that the latter expression is another way to write $I(X; Y)$. Moreover,

$$I(X; Y) = H(X) - H(X \mid Y)$$
$$= H(Y) - H(Y \mid X)$$
$$= I(Y; X)$$

Hence, the amount of information that Y gives about X is the same as the amount of information that X gives about Y. For this reason, $I(X; Y)$ is called the *mutual* information between X and Y.

Example (Binary Symmetric Channel). Let $\mathcal{X} = \{0, 1\}$, $\mathcal{Y} = \{0, 1\}$, and $p(y \mid x) = p$ when $x \neq y$ and $1 - p$ otherwise. This DMC can be conveniently depicted as in Fig. 18.7. For obvious reasons it is called the *binary symmetric channel* (BSC).

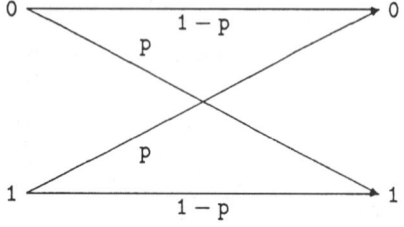

FIGURE 18.7 Discrete memoryless channel with cross over probability p.

Information Theory

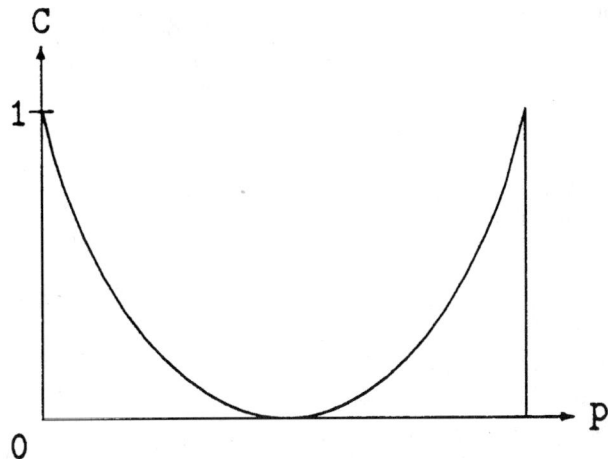

FIGURE 18.8 Capacity of the BSC as a function of p.

For the BSC we bound the mutual information by

$$\begin{aligned} I(X;Y) &= H(Y) - H(Y\mid X) \\ &= H(Y) - \sum_x p(x) H(Y \mid X = x) \\ &= H(Y) - h(p) \\ &\leq 1 - h(p) \end{aligned}$$

where the last inequality follows because Y is a binary random variable (see Example 2). Equality is achieved when the input distribution is uniform since in this case the output distribution is also uniform. Hence, the information capacity of a binary symmetric channel with parameter p is

$$C = 1 - h(p) \qquad \text{[bits/channel use]}$$

This function is plotted in Fig. 18.8.

18.7 Simple Binary Codes

Shannon's channel coding theorem promises the existence of block codes that allow us to transmit reliably through an unreliable channel at any rate below the channel capacity. Unfortunately, the codes constructed in all known proofs of this fundamental result are impractical in that they do not exhibit any structure which can be employed for an efficient implementation of the encoder and the decoder. This is an important issue, as efficient codes must have long block lengths and, hence, a large number of codewords. For these codes simple table lookup encoding and decoding procedures are not feasible. The search for good and practical codes has led to the development of coding theory. Although powerful block codes are in general described by sophisticated techniques, the main idea is simple. We collect a number k of information symbols which we wish to transmit, we append r check symbols, and transmit the entire block of $n = k + r$ channel symbols. Assuming that the channel changes a sufficiently small number of symbols within an n-length block, the r check symbols may provide the receiver with sufficient information to detect and/or correct the errors. We now show the main idea via three families of codes that are simple to describe. We assume the binary symmetric channel described earlier.

Repetition Codes

This is a class of codes with $k = 1$, r arbitrary, and $n = k + r = r + 1$. Any repetition code contains two codewords, the all-zero codeword and the all-one codeword. The decoder decides that the information symbol is one (zero) if the number of ones (zeros) within a codeword exceeds the number of zeros (ones). Typically, n is odd so that there is no possibility of ties. This code can correct any error pattern, provided that less than $n/2$ errors occur within a codeword. If n is sufficiently large, the probability of decoding error is very small. However, the rate of a repetition code is $R = \log M/n = 1/n$, which goes to zero as n goes to infinity.

Single-Parity-Check Codes

This is a class of codes with k arbitrary, $r = 1$, and $n = k + r = k + 1$. The check symbol is chosen in such a way that the codeword has an even number of ones. If the received word has an even number of ones, then the decoder assumes that no errors occurred during transmission. If the number of ones in the received word is odd, then the decoder announces a transmission error (since at least one transmission error must have occurred). Hence, the decoder will decode correctly only if no channel errors occur in the transmitted block, it will announce a transmission error if an odd number of channel errors occurred, and it will make a decoding error if there is an even number of channel errors. The code has $M = 2^k$ codewords and its rate is $R = \log M/n = k/(k+1)$, which goes to 1 as n goes to infinity. The repetition codes and the single-parity-check codes represent two extremes in the spectrum of possible code performances. The former have high error detection/correction capabilities and very poor rates, whereas the opposite is true for the latter.

Hamming Codes

For any positive integer m, there is a Hamming code with parameters $k = 2^m - m - 1$ and $n = 2^m - 1$. To illustrate, we consider the $k = 4$, $n = 7$ Hamming code. All operations will be done modulo 2. Consider the set of all nonzero binary vectors of length $m = 3$. Arrange them in columns to form a matrix

$$H = \begin{pmatrix} 1 & 0 & 1 & 0 & 1 & 0 & 1 \\ 0 & 1 & 1 & 0 & 0 & 1 & 1 \\ 0 & 0 & 0 & 1 & 1 & 1 & 1 \end{pmatrix}$$

The row space of H is a vector space of binary 7-tuples of dimension $m = 3$. The $k = 4$, $n = 7$ Hamming code is the null space of H, that is, the vector space of dimension $n - m = 4$ consisting of all binary 7-tuples c such that $cH^T = 0$. Hence it contains the $M = 2^k = 16$ codewords shown in Table 18.1 and its rate is $R = \log M/n = k/n = 4/7$. If c is transmitted and u is received, then the error is the unique binary n-tuple e such that $u = c + e$, where addition is componentwise modulo 2. The decoder performs the operation

$$uH^T = (c+e)H^T = cH^T + eH^T = eH^T$$

If there is a single error at position i, that is, if e is zero except for the ith component (which is one), then uH^T is the ith row of H^T. This tells us the error location. If e contains 2 ones (2 channel errors), then uH^T is the sum of two rows of H^T. Since this cannot be zero, two channel errors are detectable. They are not correctable, however, since as one can easily verify when two channel errors occur there is always a codeword that agrees with the received word in $n - 1$ positions. This is the codeword that would be selected by the decoder if it were forced to make a decision since the transmitted codeword agrees with the received word in only $n - 2$ positions. A simple encoding procedure can be specified as follows: Let G be a $k \times n$ matrix whose rows span

TABLE 18.1 Codewords of the (7, 4) Hamming Code

0 0 0 0 0 0 0	1 0 0 0 1 0 1
0 0 0 1 0 1 1	1 0 0 1 1 1 0
0 0 1 0 1 1 0	1 0 1 0 0 1 1
0 0 1 1 1 0 1	1 0 1 1 0 0 0
0 1 0 0 1 1 1	1 1 0 0 0 1 0
0 1 0 1 1 0 0	1 1 0 1 0 0 1
0 1 1 0 0 0 1	1 1 1 0 1 0 0
0 1 1 1 0 1 0	1 1 1 1 1 1 1

the null space of H. Such a matrix is

$$G = \begin{pmatrix} 1 & 0 & 0 & 0 & 1 & 1 & 0 \\ 0 & 1 & 0 & 0 & 1 & 0 & 1 \\ 0 & 0 & 1 & 0 & 0 & 1 & 1 \\ 0 & 0 & 0 & 1 & 1 & 1 & 1 \end{pmatrix}$$

Then an information k-tuple a can be mapped into a codeword $c = aG$ by a simple matrix multiplication.

Defining Terms

Binary source code: A mapping from a set of messages into binary strings.

Channel capacity: The highest rate at which information can be transmitted reliably across a channel.

Discrete memoryless channel: A channel model characterized by discrete input and output alphabets and a probability mass function on the output conditioned on the input.

Entropy: A measure of the average uncertainty of a random variable. For a random variable with distribution $p(x)$, the entropy $H(X)$ is defined as $-\sum_x p(x) \log p(x)$.

Huffman coding: A procedure that constructs a code of minimum average length for a random variable.

Lempel–Ziv coding: A procedure for coding that does not use the probability distribution of the source but nevertheless is asymptotically optimal.

Mutual information: A measure for the amount of information that a random variable gives about another.

Rate distortion function: The minimum rate at which a source can be described to the given average distortion.

References

Blahut, R. 1987. *Principles and Practice of Information Theory*, Addison–Wesley, Reading, MA.

Csiszár, I. and Körner, J. 1981. *Information Theory: Coding Theorems for Discrete Memoryless Systems*, Academic Press, New York.

Cover, T.M. and Thomas, J.A. 1991. *Elements of Information Theory*, Wiley, New York.

Gallager, R.G. 1968. *Information Theory and Reliable Communication*, Wiley, New York.

McEliece, R.J. 1977. *The Theory of Information and Coding*, Addison–Wesley, Reading, MA.

Wyner, A.D. and Ziv, J. 1994. The sliding-window Lempel–Ziv algorithm is asymptotically optimal. *Communications and Cryptography*, edited by Blahut, R.E., Costello, D.J., Jr., Maurer, V., and Mittelholzer, T., Kluwer Academic Publishers, Boston, MA.

Ziv, J. and Lempel, A. 1977. A universal algorithm for sequential data compression. *IEEE Trans. Inf. The.*, IT-23(May):337–343.

Ziv, J. and Lempel, A. 1978. Compression of individual sequences by variable rate coding. *IEEE Trans. Inf. The.*, IT-24(Sept.):530–536.

Further Information

For a lucid and up-to-date treatment of information theory extending beyond the area of communication, a most recommended reading is Cover and Thomas [1991]. For readers interested in continuous-time channels and sources we recommend Gallager [1968]. For mathematically inclined readers who do not require much physical motivation, strong results for discrete memoryless channels and sources may be found in Csiszár and Körner [1981]. Other excellent readings are Blahut [1987] and McEliece [1977]. Most results in information theory are published in *IEEE Transactions on Information Theory*.

19

Digital Communication System Performance[1]

19.1 Introduction 236
   The Channel • The Link
19.2 Bandwidth and Power Considerations 238
   The Bandwidth Efficiency Plane • M-ary Signalling • Bandwidth-Limited Systems • Power-Limited Systems • Minimum Bandwidth Requirements for MPSK and MFSK Signalling
19.3 Example 19.1: Bandwidth-Limited Uncoded System 242
   Solution to Example 19.1
19.4 Example 19.2: Power-Limited Uncoded System 244
   Solution to Example 19.2
19.5 Example 19.3: Bandwidth-Limited and Power-Limited Coded System 245
   Solution to Example 19.3 • Calculating Coding Gain
19.6 Example 19.4: Direct-Sequence (DS) Spread-Spectrum Coded System 250
   Processing Gain • Channel Parameters for Example 19.4 • Solution to Example 19.4
19.7 Conclusion 253
   Appendix: Received E_b/N_0 Is Independent of the Code Parameters 253

Bernard Sklar
Communications Engineering Services

19.1 Introduction

In this section we examine some fundamental tradeoffs among bandwidth, power, and error performance of digital communication systems. The criteria for choosing modulation and coding schemes, based on whether a system is bandwidth limited or power limited, are reviewed for several system examples. Emphasis is placed on the subtle but straightforward relationships we encounter when transforming from data-bits to channel-bits to symbols to chips.

The design or definition of any digital communication system begins with a description of the communication link. The *link* is the name given to the communication transmission path from the modulator and transmitter, through the channel, and up to and including the receiver and

[1]A version of this chapter has appeared as a paper in the *IEEE Communications Magazine*, November 1993, under the title "Defining, Designing, and Evaluating Digital Communication Systems."

Digital Communication System Performance

demodulator. The *channel* is the name given to the propagating medium between the transmitter and receiver. A link description quantifies the average signal power that is received, the available bandwidth, the noise statistics, and other impairments, such as fading. Also needed to define the system are basic requirements, such as the data rate to be supported and the error performance.

The Channel

For radio communications, the concept of *free space* assumes a channel region free of all objects that might affect radio frequency (RF) propagation by absorption, reflection, or refraction. It further assumes that the atmosphere in the channel is perfectly uniform and nonabsorbing, and that the earth is infinitely far away or its reflection coefficient is negligible. The RF energy arriving at the receiver is assumed to be a function of distance from the transmitter (simply following the inverse-square law of optics). In practice, of course, propagation in the atmosphere and near the ground results in refraction, reflection, and absorption, which modify the free space transmission.

The Link

A radio transmitter is characterized by its average output signal power P_t and the gain of its transmitting antenna G_t. The name given to the product $P_t G_t$, with reference to an isotropic antenna is *effective radiated power* (*EIRP*) in watts (or dBW). The predetection average signal power S arriving at the output of the receiver antenna can be described as a function of the *EIRP*, the gain of the receiving antenna G_r, the path loss (or space loss) L_s, and other losses, L_o, as follows [Sklar, 1988; 1979]:

$$S = \frac{EIRP\, G_r}{L_s L_o} \qquad (19.1)$$

The path loss L_s can be written as follows [Sklar, 1988]:

$$L_s = \left(\frac{4\pi d}{\lambda}\right)^2 \qquad (19.2)$$

where d is the distance between the transmitter and receiver and λ is the wavelength.

We restrict our discussion to those links distorted by the mechanism of additive white Gaussian noise (AWGN) only. Such a noise assumption is a very useful model for a large class of communication systems. A valid approximation for average received noise power N that this model introduces is written as follows [Johnson, 1928; Nyquist, 1928]:

$$N \cong kT^\circ W \qquad (19.3)$$

where k is Boltzmann's constant (1.38×10^{-23} joule/K), T° is effective temperature in kelvin, and W is bandwidth in hertz. Dividing Eq. (19.3) by bandwidth, enables us to write the received noise-power spectral density N_0 as follows:

$$N_0 = \frac{N}{W} = kT^\circ \qquad (19.4)$$

Dividing Eq. (19.1) by N_0 yields the received average signal-power to noise-power spectral density S/N_0 as

$$\frac{S}{N_0} = \frac{EIRP\, G_r/T^\circ}{k L_s L_o} \qquad (19.5)$$

where $G_r/T°$ is often referred to as the receiver figure of merit. A link budget analysis is a compilation of the power gains and losses throughout the link; it is generally computed in decibels, and thus takes on the bookkeeping appearance of a business enterprise, highlighting the assets and liabilities of the link. Once the value of S/N_0 is specified or calculated from the link parameters, we then shift our attention to optimizing the choice of signalling types for meeting system bandwidth and error performance requirements.

Given the received S/N_0, we can write the received bit-energy to noise-power spectral density E_b/N_0, for any desired data rate R, as follows:

$$\frac{E_b}{N_0} = \frac{ST_b}{N_0} = \frac{S}{N_0}\left(\frac{1}{R}\right) \qquad (19.6)$$

Equation (19.6) follows from the basic definitions that received bit energy is equal to received average signal power times the bit duration and that bit rate is the reciprocal of bit duration. Received E_b/N_0 is a key parameter in defining a digital communication system. Its value indicates the apportionment of the received waveform energy among the bits that the waveform represents. At first glance, one might think that a system specification should entail the symbol-energy to noise-power spectral density E_s/N_0 associated with the arriving waveforms. We will show, however, that for a given S/N_0 the value of E_s/N_0 is a function of the modulation and coding. The reason for defining systems in terms of E_b/N_0 stems from the fact that E_b/N_0 depends only on S/N_0 and R and is unaffected by any system design choices, such as modulation and coding.

19.2 Bandwidth and Power Considerations

Two primary communications resources are the received power and the available transmission bandwidth. In many communication systems, one of these resources may be more precious than the other and, hence, most systems can be classified as either bandwidth limited or power limited. In bandwidth-limited systems, spectrally efficient modulation techniques can be used to save bandwidth at the expense of power; in power-limited systems, power efficient modulation techniques can be used to save power at the expense of bandwidth. In both bandwidth- and power-limited systems, error-correction coding (often called channel coding) can be used to save power or to improve error performance at the expense of bandwidth. Recently, trellis-coded modulation (TCM) schemes have been used to improve the error performance of bandwidth-limited channels without any increase in bandwidth [Ungerboeck, 1987], but these methods are beyond the scope of this chapter.

The Bandwidth Efficiency Plane

Figure 19.1 shows the abscissa as the ratio of bit-energy to noise-power spectral density E_b/N_0 (in decibels) and the ordinate as the ratio of throughput, R (in bits per second), that can be transmitted per hertz in a given bandwidth W. The ratio R/W is called bandwidth efficiency, since it reflects how efficiently the bandwidth resource is utilized. The plot stems from the Shannon–Hartley capacity theorem [Sklar, 1988; Shannon, 1948; 1949], which can be stated as

$$C = W \log_2\left(1 + \frac{S}{N}\right) \qquad (19.7)$$

where S/N is the ratio of received average signal power to noise power. When the logarithm is taken to the base 2, the capacity C, is given in bits per second. The capacity of a channel defines the maximum number of bits that can be reliably sent per second over the channel. For the case where the data (information) rate R is equal to C, the curve separates a region of practical communication systems from a region where such communication systems cannot operate reliably [Sklar, 1988; Shannon, 1948].

Digital Communication System Performance 239

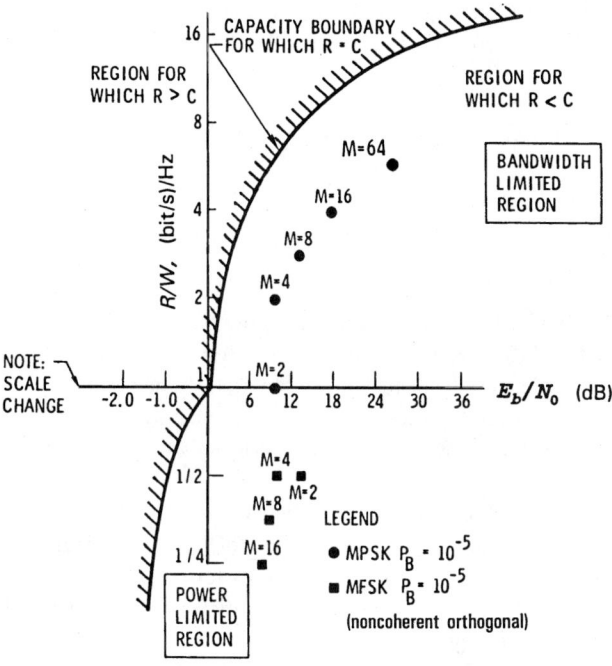

FIGURE 19.1 Bandwidth-efficiency plane.

M-ary Signalling

Each symbol in an *M*-ary alphabet can be related to a unique sequence of *m* bits, expressed as

$$M = 2^m \quad \text{or} \quad m = \log_2 M \tag{19.8}$$

where *M* is the size of the alphabet. In the case of digital transmission, the term symbol refers to the member of the *M*-ary alphabet that is transmitted during each symbol duration T_s. To transmit the symbol, it must be mapped onto an electrical voltage or current waveform. Because the waveform represents the symbol, the terms symbol and waveform are sometimes used interchangeably. Since one of *M* symbols or waveforms is transmitted during each symbol duration T_s, the data rate *R* in bits per second can be expressed as

$$R = \frac{m}{T_s} = \frac{\log_2 M}{T_s} \tag{19.9}$$

Data-bit-time duration is the reciprocal of data rate. Similarly, symbol-time duration is the reciprocal of symbol rate. Therefore, from Eq. (19.9), we write that the effective time duration T_b of each bit in terms of the symbol duration T_s or the symbol rate R_s is

$$T_b = \frac{1}{R} = \frac{T_s}{m} = \frac{1}{mR_s} \tag{19.10}$$

Then, using Eqs. (19.8) and (19.10) we can express the symbol rate R_s in terms of the bit rate *R* as follows:

$$R_s = \frac{R}{\log_2 M} \tag{19.11}$$

From Eqs. (19.9) and (19.10), any digital scheme that transmits $m = \log_2 M$ bits in T_s seconds, using a bandwidth of W hertz, operates at a bandwidth efficiency of

$$\frac{R}{W} = \frac{\log_2 M}{WT_s} = \frac{1}{WT_b} \quad \text{(b/s)/Hz} \quad (19.12)$$

where T_b is the effective time duration of each data bit.

Bandwidth-Limited Systems

From Eq. (19.12), the smaller the WT_b product, the more bandwidth efficient will be any digital communication system. Thus, signals with small WT_b products are often used with bandwidth-limited systems. For example, the European digital mobile telephone system known as Global System for Mobile Communications (GSM) uses Gaussian minimum shift keying (GMSK) modulation having a WT_b product equal to 0.3 Hz/(b/s), where W is the 3-dB bandwidth of a Gaussian filter [Hodges, 1990].

For uncoded bandwidth-limited systems, the objective is to maximize the transmitted information rate within the allowable bandwidth, at the expense of E_b/N_0 (while maintaining a specified value of bit-error probability P_B). The operating points for coherent M-ary phase-shift keying (MPSK) at $P_B = 10^{-5}$ are plotted on the bandwidth-efficiency plane of Fig. 19.1. We assume Nyquist (ideal rectangular) filtering at baseband [Nyquist, April 1928]. Thus, for MPSK, the required double-sideband (DSB) bandwidth at an intermediate frequency (IF) is related to the symbol rate as follows:

$$W = \frac{1}{T_s} = R_s \quad (19.13)$$

where T_s is the symbol duration and R_s is the symbol rate. The use of Nyquist filtering results in the minimum required transmission bandwidth that yields zero intersymbol interference; such ideal filtering gives rise to the name Nyquist minimum bandwidth.

From Eqs. (19.12) and (19.13), the bandwidth efficiency of MPSK modulated signals using Nyquist filtering can be expressed as

$$R/W = \log_2 M \quad \text{(b/s)/Hz} \quad (19.14)$$

The MPSK points in Fig. 19.1 confirm the relationship shown in Eq. (19.14). Note that MPSK modulation is a bandwidth-efficient scheme. As M increases in value, R/W also increases. MPSK modulation can be used for realizing an improvement in bandwidth efficiency at the cost of increased E_b/N_0. Although beyond the scope of this chapter, many highly bandwidth-efficient modulation schemes are under investigation [Anderson and Sundberg, 1991].

Power-Limited Systems

Operating points for noncoherent orthogonal M-ary FSK (MFSK) modulation at $P_B = 10^{-5}$ are also plotted on Fig. 19.1. For MFSK, the IF minimum bandwidth is as follows [Sklar, 1988]

$$W = \frac{M}{T_s} = MR_s \quad (19.15)$$

where T_s is the symbol duration and R_s is the symbol rate. With MFSK, the required transmission bandwidth is expanded M-fold over binary FSK since there are M different orthogonal waveforms, each requiring a bandwidth of $1/T_s$. Thus, from Eqs. (19.12) and (19.15), the bandwidth efficiency of noncoherent orthogonal MFSK signals can be expressed as

$$\frac{R}{W} = \frac{\log_2 M}{M} \quad \text{(b/s)/Hz} \qquad (19.16)$$

The MFSK points plotted in Fig. 19.1 confirm the relationship shown in Eq. (19.16). Note that MFSK modulation is a bandwidth-expansive scheme. As M increases, R/W decreases. MFSK modulation can be used for realizing a reduction in required E_b/N_0 at the cost of increased bandwidth.

In Eqs. (19.13) and (19.14) for MPSK, and Eqs. (19.15) and (19.16) for MFSK, and for all the points plotted in Fig. 19.1, ideal filtering has been assumed. Such filters are not realizable! For realistic channels and waveforms, the required transmission bandwidth must be increased in order to account for realizable filters.

In the examples that follow, we will consider radio channels that are disturbed only by additive white Gaussian noise (AWGN) and have no other impairments, and for simplicity, we will limit the modulation choice to constant-envelope types, i.e., either MPSK or noncoherent orthogonal MFSK. For an uncoded system, MPSK is selected if the channel is bandwidth limited, and MFSK is selected if the channel is power limited. When error-correction coding is considered, modulation selection is not as simple, because coding techniques can provide power-bandwidth tradeoffs more effectively than would be possible through the use of any M-ary modulation scheme considered in this chapter [Clark and Cain, 1981].

In the most general sense, M-ary signalling can be regarded as a waveform-coding procedure, i.e., when we select an M-ary modulation technique instead of a binary one, we in effect have replaced the binary waveforms with better waveforms—either better for bandwidth performance (MPSK) or better for power performance (MFSK). Even though orthogonal MFSK signalling can be thought of as being a coded system, i.e., a first-order Reed-Muller code [Lindsey and Simon, 1973], we restrict our use of the term coded system to those traditional error-correction codes using redundancies, e.g., block codes or convolutional codes.

Minimum Bandwidth Requirements for MPSK and MFSK Signalling

The basic relationship between the symbol (or waveform) transmission rate R_s and the data rate R was shown in Eq. (19.11). Using this relationship together with Eqs. (19.13–19.16) and $R = 9600$ b/s, a summary of symbol rate, minimum bandwidth, and bandwidth efficiency for MPSK and noncoherent orthogonal MFSK was compiled for $M = 2, 4, 8, 16$, and 32 (Table 19.1). Values of E_b/N_0 required to achieve a bit-error probability of 10^{-5} for MPSK and MFSK are also given for each value of M. These entries (which were computed using relationships that are presented later in this chapter) corroborate the tradeoffs shown in Fig. 19.1. As M increases, MPSK signalling provides more bandwidth efficiency at the cost of increased E_b/N_0, whereas MFSK signalling allows for a reduction in E_b/N_0 at the cost of increased bandwidth.

TABLE 19.1 Symbol Rate, Minimum Bandwidth, Bandwidth Efficiency, and Required E_b/N_0 for MPSK and Noncoherent Orthogonal MFSK Signalling at 9600 bit/s

M	m	R (b/s)	R_s (symb/s)	MPSK Minimum Bandwidth (Hz)	MPSK R/W	MPSK E_b/N_0 (dB) $P_B = 10^{-5}$	Noncoherent Orthog MFSK Min Bandwidth (Hz)	MFSK R/W	MFSK E_b/N_0 (dB) $P_B = 10^{-5}$
2	1	9600	9600	9600	1	9.6	19,200	1/2	13.4
4	2	9600	4800	4800	2	9.6	19,200	1/2	10.6
8	3	9600	3200	3200	3	13.0	25,600	3/8	9.1
16	4	9600	2400	2400	4	17.5	38,400	1/4	8.1
32	5	9600	1920	1920	5	22.4	61,440	5/32	7.4

19.3 Example 19.1: Bandwidth-Limited Uncoded System

Suppose we are given a bandwidth-limited AWGN radio channel with an available bandwidth of $W = 4000$ Hz. Also, suppose that the link constraints (transmitter power, antenna gains, path loss, etc.) result in the ratio of received average signal-power to noise-power spectral density S/N_0 being equal to 53 dB-Hz. Let the required data rate R be equal to 9600 b/s, and let the required bit-error performance P_B be at most 10^{-5}. The goal is to choose a modulation scheme that meets the required performance. In general, an error-correction coding scheme may be needed if none of the allowable modulation schemes can meet the requirements. In this example, however, we shall find that the use of error-correction coding is not necessary.

Solution to Example 19.1

For any digital communication system, the relationship between received S/N_0 and received bit-energy to noise-power spectral density, E_b/N_0 was given in Eq. (19.6) and is briefly rewritten as

$$\frac{S}{N_0} = \frac{E_b}{N_0} R \qquad (19.17)$$

Solving for E_b/N_0 in decibels, we obtain

$$\frac{E_b}{N_0} \text{(dB)} = \frac{S}{N_0} \text{(dB-Hz)} - R \text{(dB-b/s)}$$
$$= 53 \text{ dB-Hz} - (10 \times \log_{10} 9600) \text{ dB-b/s}$$
$$= 13.2 \text{ dB (or 20.89)} \qquad (19.18)$$

Since the required data rate of 9600 b/s is much larger than the available bandwidth of 4000 Hz, the channel is bandwidth limited. We therefore select MPSK as our modulation scheme. We have confined the possible modulation choices to be constant-envelope types; without such a restriction, we would be able to select a modulation type with greater bandwidth efficiency. To conserve power, we compute the *smallest possible* value of M such that the MPSK minimum bandwidth does not exceed the available bandwidth of 4000 Hz. Table 19.1 shows that the smallest value of M meeting this requirement is $M = 8$. Next we determine whether the required bit-error performance of $P_B \leq 10^{-5}$ can be met by using 8-PSK modulation alone or whether it is necessary to use an error-correction coding scheme. Table 19.1 shows that 8-PSK alone will meet the requirements, since the required E_b/N_0 listed for 8-PSK is less than the received E_b/N_0 derived in Eq. (19.18). Let us imagine that we do not have Table 19.1, however, and evaluate whether or not error-correction coding is necessary.

Figure 19.2 shows the basic modulator/demodulator (MODEM) block diagram summarizing the functional details of this design. At the modulator, the transformation from data bits to symbols yields an output symbol rate R_s, that is, a factor $\log_2 M$ smaller than the input data-bit rate R, as is seen in Eq. (19.11). Similarly, at the input to the demodulator, the symbol-energy to noise-power spectral density E_S/N_0 is a factor $\log_2 M$ larger than E_b/N_0, since each symbol is made up of $\log_2 M$ bits. Because E_S/N_0 is larger than E_b/N_0 by the same factor that R_s is smaller than R, we can expand Eq. (19.17), as follows:

$$\frac{S}{N_0} = \frac{E_b}{N_0} R = \frac{E_s}{N_0} R_s \qquad (19.19)$$

The demodulator receives a waveform (in this example, one of $M = 8$ possible phase shifts) during each time interval T_s. The probability that the demodulator makes a symbol error $P_E(M)$ is well

Digital Communication System Performance

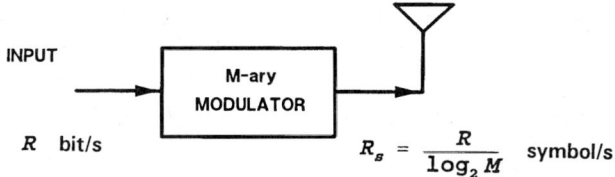

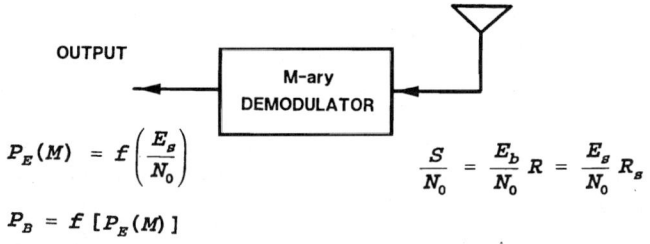

FIGURE 19.2 Basic modulator/demodulator (MODEM) without channel coding.

approximated by the following equation for $M > 2$ [Korn, 1985]:

$$P_E(M) \cong 2Q\left[\sqrt{\frac{2E_s}{N_0}} \sin\left(\frac{\pi}{M}\right)\right] \qquad (19.20)$$

where $Q(x)$, sometimes called the complementary error function, represents the probability under the tail of a zero-mean unit-variance Gaussian density function. It is defined as follows [Van Trees, 1968]:

$$Q(x) = \frac{1}{\sqrt{2\pi}} \int_x^\infty \exp\left(-\frac{u^2}{2}\right) du \qquad (19.21)$$

A good approximation for $Q(x)$, valid for $x > 3$, is given by the following equation [Borjesson and Sundberg, 1979]:

$$Q(x) \cong \frac{1}{x\sqrt{2\pi}} \exp\left(-\frac{x^2}{2}\right) \qquad (19.22)$$

In Fig. 19.2 and all of the figures that follow, rather than show explicit probability relationships, the generalized notation $f(x)$ has been used to indicate some functional dependence on x.

A traditional way of characterizing communication efficiency in digital systems is in terms of the received E_b/N_0 in decibels. This E_b/N_0 description has become standard practice, but recall that there are no bits at the input to the demodulator; there are only waveforms that have been assigned bit meanings. The received E_b/N_0 represents a bit-apportionment of the arriving waveform energy.

To solve for $P_E(M)$ in Eq. (19.20), we first need to compute the ratio of received symbol-energy to noise-power spectral density E_s/N_0. Since from Eq. (19.18)

$$\frac{E_b}{N_0} = 13.2 \text{ dB (or 20.89)}$$

and because each symbol is made up of $\log_2 M$ bits, we compute the following using $M = 8$.

$$\frac{E_s}{N_0} = (\log_2 M)\frac{E_b}{N_0} = 3 \times 20.89 = 62.67 \tag{19.23}$$

Using the results of Eq. (19.23) in Eq. (19.20), yields the symbol-error probability $P_E = 2.2 \times 10^{-5}$. To transform this to bit-error probability, we use the relationship between bit-error probability P_B and symbol-error probability P_E, for multiple-phase signalling [Lindsey and Simon, 1973] for $P_E \ll 1$ as follows:

$$P_B \cong \frac{P_E}{\log_2 M} = \frac{P_E}{m} \tag{19.24}$$

which is a good approximation when Gray coding is used for the bit-to-symbol assignment [Korn, 1985]. This last computation yields $P_B = 7.3 \times 10^{-6}$, which meets the required bit-error performance. No error-correction coding is necessary, and 8-PSK modulation represents the design choice to meet the requirements of the bandwidth-limited channel, which we had predicted by examining the required E_b/N_0 values in Table 19.1.

19.4 Example 19.2: Power-Limited Uncoded System

Now, suppose that we have exactly the same data rate and bit-error probability requirements as in Example 19.1, but let the available bandwidth W be equal to 45 kHz, and the available S/N_0 be equal to 48 dB-Hz. The goal is to choose a modulation or modulation/coding scheme that yields the required performance. We shall again find that error-correction coding is not required.

Solution to Example 19.2

The channel is clearly not bandwidth limited since the available bandwidth of 45 kHz is more than adequate for supporting the required data rate of 9600 bit/s. We find the received E_b/N_0 from Eq. (19.18), as follows:

$$\frac{E_b}{N_0} \text{ (dB)} = 48 \text{ dB-Hz} - (10 \times \log_{10} 9600) \text{ dB-b/s} = 8.2 \text{ dB (or 6.61)} \tag{19.25}$$

Since there is abundant bandwidth but a relatively small E_b/N_0 for the required bit-error probability, we consider that this channel is power limited and choose MFSK as the modulation scheme. To conserve power, we search for the *largest possible M* such that the MFSK minimum bandwidth is not expanded beyond our available bandwidth of 45 kHz. A search results in the choice of $M = 16$ (Table 19.1). Next, we determine whether the required error performance of $P_B \leq 10^{-5}$ can be met by using 16-FSK alone, i.e., without error-correction coding. Table 19.1 shows that 16-FSK alone meets the requirements, since the required E_b/N_0 listed for 16-FSK is less than the received E_b/N_0 derived in Eq. (19.25). Let us imagine again that we do not have Table 19.1, and evaluate whether or not error-correction coding is necessary.

The block diagram in Fig. 19.2 summarizes the relationships between symbol rate R_s, and bit rate R, and between E_s/N_0 and E_b/N_0, which is identical to each of the respective relationships in Example 19.1. The 16-FSK demodulator receives a waveform (one of 16 possible frequencies) during each symbol time interval T_s. For noncoherent orthogonal MFSK, the probability that the demodulator makes a symbol error $P_E(M)$ is approximated by the following upper bound [Viterbi, 1979]:

$$P_E(M) \leq \frac{M-1}{2}\exp\left(-\frac{E_s}{2N_0}\right) \tag{19.26}$$

Digital Communication System Performance

To solve for $P_E(M)$ in Eq. (19.26), we compute E_S/N_0, as in Example 19.1. Using the results of Eq. (19.25) in Eq. (19.23), with $M = 16$, we get

$$\frac{E_s}{N_0} = (\log_2 M)\frac{E_b}{N_0} = 4 \times 6.61 = 26.44 \qquad (19.27)$$

Next, using the results of Eq. (19.27) in Eq. (19.26), yields the symbol-error probability $P_E = 1.4 \times 10^{-5}$. To transform this to bit-error probability, P_B, we use the relationship between P_B and P_E for orthogonal signalling [Viterbi, 1979], given by

$$P_B = \frac{2^{m-1}}{(2^m - 1)}P_E \qquad (19.28)$$

This last computation yields $P_B = 7.3 \times 10^{-6}$, which meets the required bit-error performance. Thus, we can meet the given specifications for this power-limited channel by using 16-FSK modulation, without any need for error-correction coding, as we had predicted by examining the required E_b/N_0 values in Table 19.1.

19.5 Example 19.3: Bandwidth-Limited and Power-Limited Coded System

We start with the same channel parameters as in Example 19.1 ($W = 4000$ Hz, $S/N_0 = 53$ dB-Hz, and $R = 9600$ b/s), with one exception. In this example, we specify that P_B must be at most 10^{-9}. Table 19.1 shows that the system is both bandwidth limited and power limited, based on the available bandwidth of 4000 Hz and the available E_b/N_0 of 20.2 dB, from Eq. (19.18); 8-PSK is the only possible choice to meet the bandwidth constraint; however, the available E_b/N_0 of 20.2 dB is certainly insufficient to meet the required P_B of 10^{-9}. For this small value of P_B, we need to consider the performance improvement that error-correction coding can provide within the available bandwidth. In general, one can use convolutional codes or block codes.

The Bose–Chaudhuri–Hocquenghem (BCH) codes form a large class of powerful error-correcting cyclic (block) codes [Lin and Costello, 1983]. To simplify the explanation, we shall choose a block code from the BCH family. Table 19.2 presents a partial catalog of the available BCH codes in terms of n, k, and t, where k represents the number of information (or data) bits that the code transforms into a longer block of n coded bits (or channel bits), and t represents the largest number of incorrect channel bits that the code can correct within each n-sized block. The rate of a code is defined as the ratio k/n; its inverse represents a measure of the code's redundancy [Lin and Costello, 1983].

TABLE 19.2 BCH Codes (Partial Catalog)

n	k	t
7	4	1
15	11	1
	7	2
	5	3
31	26	1
	21	2
	16	3
	11	5
63	57	1
	51	2
	45	3
	39	4
	36	5
	30	6
127	120	1
	113	2
	106	3
	99	4
	92	5
	85	6
	78	7
	71	9
	64	10

Solution to Example 19.3

Since this example has the same bandwidth-limited parameters given in Example 19.1, we start with the same 8-PSK modulation used to meet the stated bandwidth constraint. We now employ error-correction coding, however, so that the bit-error probability can be lowered to $P_B \leq 10^{-9}$.

To make the optimum code selection from Table 19.2, we are guided by the following goals.

1. The output bit-error probability of the combined modulation/coding system must meet the system error requirement.
2. The rate of the code must not expand the required transmission bandwidth beyond the available channel bandwidth.
3. The code should be as simple as possible. Generally, the shorter the code, the simpler will be its implementation.

The uncoded 8-PSK minimum bandwidth requirement is 3200 Hz (Table 19.1) and the allowable channel bandwidth is 4000 Hz, and so the uncoded signal bandwidth can be increased by no more than a factor of 1.25 (i.e., an expansion of 25%). The very first step in this (simplified) code selection example is to eliminate the candidates in Table 19.2 that would expand the bandwidth by more than 25%. The remaining entries form a much reduced set of bandwidth-compatible codes (Table 19.3).

TABLE 19.3 Bandwidth-Compatible BCH Codes

n	k	t	Coding Gain, G (dB) MPSK, $P_B = 10^{-9}$
31	26	1	2.0
63	57	1	2.2
	51	2	3.1
127	120	1	2.2
	113	2	3.3
	106	3	3.9

In Table 19.3, a column designated Coding Gain G (for MPSK at $P_B = 10^{-9}$) has been added. Coding gain in decibels is defined as follows:

$$G = \left(\frac{E_b}{N_0}\right)_{\text{uncoded}} - \left(\frac{E_b}{N_0}\right)_{\text{coded}} \quad (19.29)$$

G can be described as the reduction in the required E_b/N_0 (in decibels) that is needed due to the error-performance properties of the channel coding. G is a function of the modulation type and bit-error probability, and it has been computed for MPSK at $P_B = 10^{-9}$ (Table 19.3). For MPSK modulation, G is relatively independent of the value of M. Thus, for a particular bit-error probability, a given code will provide about the same coding gain when used with any of the MPSK modulation schemes. Coding gains were calculated using a procedure outlined in the subsequent Calculating Coding Gain section.

A block diagram summarizes this system, which contains both modulation and coding (Fig. 19.3). The introduction of encoder/decoder blocks brings about additional transformations. The relationships that exist when transforming from R b/s to R_c channel-b/s to R_s symbol/s are shown at the encoder/modulator. Regarding the channel-bit rate R_c, some authors prefer to use the units of channel-symbol/s (or code-symbol/s). The benefit is that error-correction coding is often described more efficiently with nonbinary digits. We reserve the term symbol for that group of bits mapped onto an electrical waveform for transmission, and we designate the units of R_c to be channel-b/s (or coded-b/s).

We assume that our communication system cannot tolerate any message delay, so that the channel-bit rate R_c must exceed the data-bit rate R by the factor n/k. Further, each symbol is made up of $\log_2 M$ channel bits, and so the symbol rate R_s is less than R_c by the factor $\log_2 M$. For a system containing both modulation and coding, we summarize the rate transformations as follows:

$$R_c = \left(\frac{n}{k}\right) R \quad (19.30)$$

$$R_s = \frac{R_c}{\log_2 M} \quad (19.31)$$

Digital Communication System Performance   247

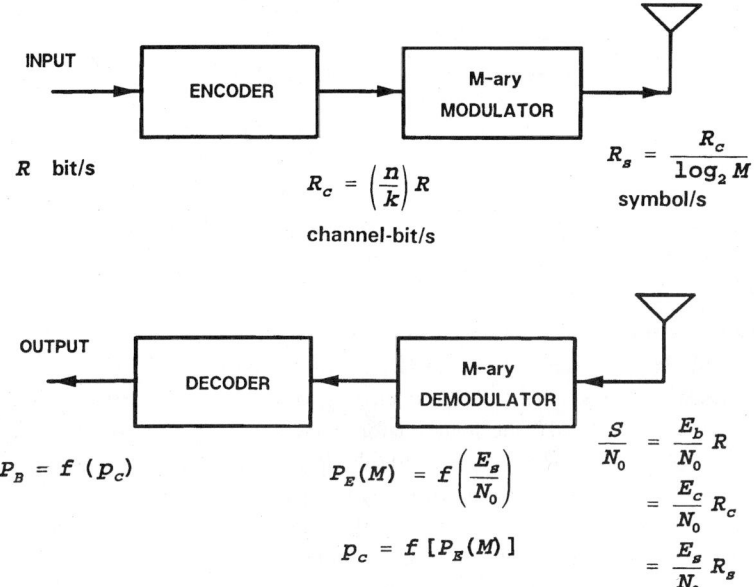

FIGURE 19.3 MODEM with channel coding.

At the demodulator/decoder in Fig. 19.3, the transformations among data-bit energy, channel-bit energy, and symbol energy are related (in a reciprocal fashion) by the same factors as shown among the rate transformations in Eqs. (19.30) and (19.31). Since the encoding transformation has replaced k data bits with n channel bits, then the ratio of channel-bit energy to noise-power spectral density E_c/N_0 is computed by decrementing the value of E_b/N_0 by the factor k/n. Also, since each transmission symbol is made up of $\log_2 M$ channel bits, then E_S/N_0, which is needed in Eq. (19.20) to solve for P_E, is computed by incrementing E_c/N_0 by the factor $\log_2 M$. For a system containing both modulation and coding, we summarize the energy to noise-power spectral density transformations as follows:

$$\frac{E_c}{N_0} = \left(\frac{k}{n}\right)\frac{E_b}{N_0} \qquad (19.32)$$

$$\frac{E_s}{N_0} = (\log_2 M)\frac{E_c}{N_0} \qquad (19.33)$$

Using Eqs. (19.30) and (19.31), we can now expand the expression for S/N_0 in Eq. (19.19), as follows(Appendix).

$$\frac{S}{N_0} = \frac{E_b}{N_0}R = \frac{E_c}{N_0}R_c = \frac{E_s}{N_0}R_s \qquad (19.34)$$

As before, a standard way of describing the link is in terms of the received E_b/N_0 in decibels. However, there are no data bits at the input to the demodulator, and there are no channel bits; there are only waveforms that have bit meanings and, thus, the waveforms can be described in terms of bit-energy apportionments.

Since S/N_0 and R were given as 53 dB-Hz and 9600 b/s, respectively, we find as before, from Eq. (19.18), that the received $E_b/N_0 = 13.2$ dB. The received E_b/N_0 is fixed and independent of n, k, and t (Appendix). As we search, in Table 19.3 for the ideal code to meet the specifications, we can iteratively repeat the computations suggested in Fig. 19.3. It might be useful to program on a personal computer (or calculator) the following four steps as a function of n, k, and t. Step 1 starts by combining Eqs. (19.32) and (19.33), as follows.

Step 1:

$$\frac{E_s}{N_0} = (\log_2 M)\frac{E_c}{N_0} = (\log_2 M)\left(\frac{k}{n}\right)\frac{E_b}{N_0} \tag{19.35}$$

Step 2:

$$P_E(M) \cong 2Q\left[\sqrt{\frac{2E_s}{N_0}}\sin\left(\frac{\pi}{M}\right)\right] \tag{19.36}$$

which is the approximation for symbol-error probability P_E rewritten from Eq. (19.20). At each symbol-time interval, the demodulator makes a symbol decision, but it delivers a channel-bit sequence representing that symbol to the decoder. When the channel-bit output of the demodulator is quantized to two levels, 1 and 0, the demodulator is said to make hard decisions. When the output is quantized to more than two levels, the demodulator is said to make soft decisions [Sklar, 1988]. Throughout this paper, we shall assume hard-decision demodulation.

Now that we have a decoder block in the system, we designate the channel-bit-error probability out of the demodulator and into the decoder as p_c, and we reserve the notation P_B for the bit-error probability out of the decoder. We rewrite Eq. (19.24) in terms of p_c for $P_E \ll 1$ as follows.

Step 3:

$$p_c \cong \frac{P_E}{\log_2 M} = \frac{P_E}{m} \tag{19.37}$$

relating the channel-bit-error probability to the symbol-error probability out of the demodulator, assuming Gray coding, as referenced in Eq. (19.24).

For traditional channel-coding schemes and a given value of received S/N_0, the value of E_s/N_0 with coding will always be less than the value of E_s/N_0 without coding. Since the demodulator with coding receives less E_s/N_0, it makes more errors! When coding is used, however, the system error-performance does not only depend on the performance of the demodulator, it also depends on the performance of the decoder. For error-performance improvement due to coding, the decoder must provide enough error correction to more than compensate for the poor performance of the demodulator.

The final output decoded bit-error probability P_B depends on the particular code, the decoder, and the channel-bit-error probability p_c. It can be expressed by the following approximation [Odenwalder, 1976].

Step 4:

$$P_B \cong \frac{1}{n}\sum_{j=t+1}^{n} j\binom{n}{j}p_c^j(1-p_c)^{n-j} \tag{19.38}$$

where t is the largest number of channel bits that the code can correct within each block of n bits. Using Eqs. (19.35–19.38) in the four steps, we can compute the decoded bit-error probability P_B as a function of n, k, and t for each of the codes listed in Table 19.3. The entry that meets the stated error requirement with the largest possible code rate and the smallest value of n is the double-error correcting (63, 51) code. The computations are as follows.

Step 1:

$$\frac{E_s}{N_0} = 3\left(\frac{51}{63}\right)20.89 = 50.73$$

where $M = 8$, and the received $E_b/N_0 = 13.2$ dB (or 20.89).

Digital Communication System Performance

Step 2:

$$P_E \cong 2Q\left[\sqrt{101.5} \times \sin\left(\frac{\pi}{8}\right)\right] = 2Q(3.86) = 1.2 \times 10^{-4}$$

Step 3:

$$p_c \cong \frac{1.2 \times 10^{-4}}{3} = 4 \times 10^{-5}$$

Step 4:

$$P_B \cong \frac{3}{63}\binom{63}{3}(4 \times 10^{-5})^3(1 - 4 \times 10^{-5})^{60}$$

$$+ \frac{4}{63}\binom{63}{4}(4 \times 10^{-5})^4(1 - 4 \times 10^{-5})^{59} + \cdots$$

$$= 1.2 \times 10^{-10}$$

where the bit-error-correcting capability of the code is $t = 2$. For the computation of P_B in step 4, we need only consider the first two terms in the summation of Eq. (19.38) since the other terms have a vanishingly small effect on the result. Now that we have selected the (63, 51) code, we can compute the values of channel-bit rate R_c and symbol rate R_s using Eqs. (19.30) and (19.31), with $M = 8$,

$$R_c = \left(\frac{n}{k}\right)R = \left(\frac{63}{51}\right)9600 \approx 11{,}859 \text{ channel-b/s}$$

$$R_s = \frac{R_c}{\log_2 M} = \frac{11859}{3} = 3953 \text{ symbol/s}$$

Calculating Coding Gain

Perhaps a more direct way of finding the simplest code that meets the specified error performance is to first compute how much coding gain G is required in order to yield $P_B = 10^{-9}$ when using 8-PSK modulation alone; then, from Table 19.3, we can simply choose the code that provides this performance improvement. First, we find the uncoded E_s/N_0 that yields an error probability of $P_B = 10^{-9}$, by writing from Eqs. (19.24) and (19.36), the following:

$$P_B \cong \frac{P_E}{\log_2 M} \cong \frac{2Q\left[\sqrt{\frac{2E_s}{N_0}}\sin\left(\frac{\pi}{M}\right)\right]}{\log_2 M} = 10^{-9} \qquad (19.39)$$

At this low value of bit-error probability, it is valid to use Eq. (19.22) to approximate $Q(x)$ in Eq. (19.39). By trial and error (on a programmable calculator), we find that the uncoded $E_s/N_0 = 120.67 = 20.8$ dB, and since each symbol is made up of $\log_2 8 = 3$ bits, the required $(E_b/N_0)_{\text{uncoded}} = 120.67/3 = 40.22 = 16$ dB. From the given parameters and Eq. (19.18), we know that the received $(E_b/N_0)_{\text{coded}} = 13.2$ dB. Using Eq. (19.29), the required coding gain to meet the bit-error performance of $P_B = 10^{-9}$ in decibels is

$$G = \left(\frac{E_b}{N_0}\right)_{\text{uncoded}} - \left(\frac{E_b}{N_0}\right)_{\text{coded}} = 16 - 13.2 = 2.8$$

To be precise, each of the E_b/N_0 values in the preceding computation must correspond to exactly the same value of bit-error probability (which they do not). They correspond to $P_B = 10^{-9}$

and $P_B = 1.2 \times 10^{-10}$, respectively. At these low probability values, however, even with such a discrepancy, this computation still provides a good approximation of the required coding gain. In searching Table 19.3 for the simplest code that will yield a coding gain of at least 2.8 dB, we see that the choice is the (63, 51) code, which corresponds to the same code choice that we made earlier.

19.6 Example 19.4: Direct-Sequence (DS) Spread-Spectrum Coded System

Spread-spectrum systems are not usually classified as being bandwidth- or power-limited. They are generally perceived to be power-limited systems, however, because the bandwidth occupancy of the information is much larger than the bandwidth that is intrinsically needed for the information transmission. In a direct-sequence spread-spectrum (DS/SS) system, spreading the signal bandwidth by some factor permits lowering the signal-power spectral density by the same factor (the total average signal power is the same as before spreading). The bandwidth spreading is typically accomplished by multiplying a relatively narrowband data signal by a wideband spreading signal. The spreading signal or spreading code is often referred to as a pseudorandom code or PN code.

Processing Gain

A typical DS/SS radio system is often described as a two-step BPSK modulation process. In the first step, the carrier wave is modulated by a bipolar data waveform having a value +1 or −1 during each data-bit duration; in the second step, the output of the first step is multiplied (modulated) by a bipolar PN-code waveform having a value +1 or −1 during each PN-code-bit duration. In reality, DS/SS systems are usually implemented by first multiplying the data waveform by the PN-code waveform and then making a single pass through a BPSK modulator. For this example, however, it is useful to characterize the modulation process in two separate steps—the outer modulator/demodulator for the data, and the inner modulator/demodulator for the PN code (Fig. 19.4).

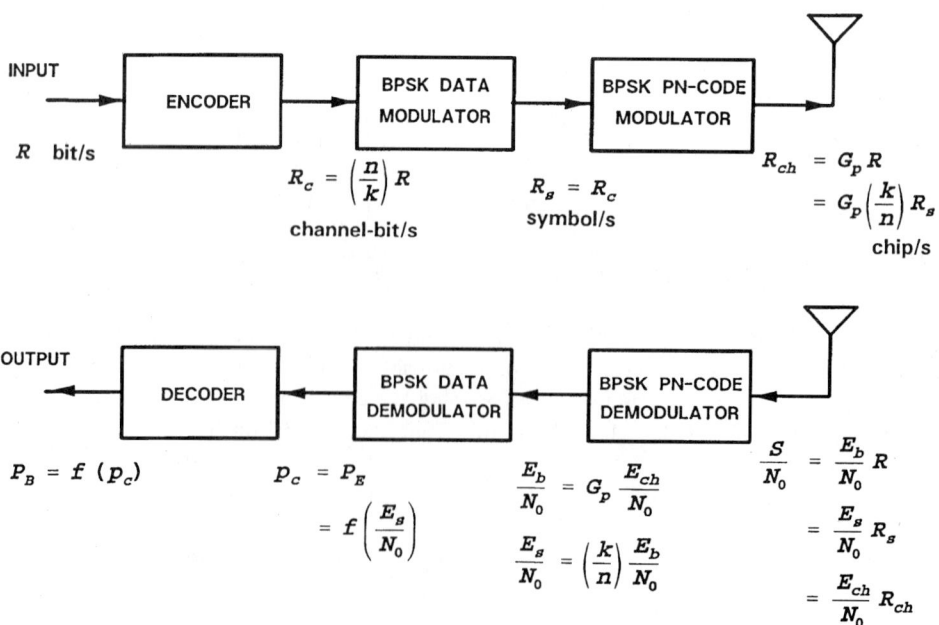

FIGURE 19.4 Direct-sequence spread-spectrum MODEM with channel coding.

Digital Communication System Performance

A spread-spectrum system is characterized by a processing gain G_p, that is defined in terms of the spread-spectrum bandwidth W_{ss} and the data rate R as follows [Viterbi, 1979]:

$$G_p = \frac{W_{ss}}{R} \qquad (19.40)$$

For a DS/SS system, the PN-code bit has been given the name chip, and the spread-spectrum signal bandwidth can be shown to be about equal to the chip rate R_{ch} as follows:

$$G_p = \frac{R_{ch}}{R} \qquad (19.41)$$

Some authors define processing gain to be the ratio of the spread-spectrum bandwidth to the symbol rate. This definition separates the system performance that is due to bandwidth spreading from the performance that is due to error-correction coding. Since we ultimately want to relate all of the coding mechanisms relative to the information source, we shall conform to the most usually accepted definition for processing gain, as expressed in Eqs. (19.40) and (19.41).

A spread-spectrum system can be used for interference rejection and for multiple access (allowing multiple users to access a communications resource simultaneously). The benefits of DS/SS signals are best achieved when the processing gain is very large; in other words, the chip rate of the spreading (or PN) code is much larger than the data rate. In such systems, the large value of G_p allows the signalling chips to be transmitted at a power level well below that of the thermal noise. We will use a value of $G_p = 1000$. At the receiver, the despreading operation correlates the incoming signal with a synchronized copy of the PN code and, thus, accumulates the energy from multiple (G_p) chips to yield the energy per data bit. The value of G_p has a major influence on the performance of the spread-spectrum system application. We shall see, however, that the value of G_p has no effect on the received E_b/N_0. In other words, spread spectrum techniques offer no error-performance advantage over thermal noise. For DS/SS systems, there is no disadvantage either! Sometimes such spread-spectrum radio systems are employed only to enable the transmission of very small power-spectral densities and thus avoid the need for FCC licensing [Title, 47].

Channel Parameters for Example 19.4

Consider a DS/SS radio system that uses the same (63, 51) code as in the previous example. Instead of using MPSK for the data modulation, we shall use BPSK. Also, we shall use BPSK for modulating the PN-code chips. Let the received $S/N_0 = 48$ dB-Hz, the data rate $R = 9600$ b/s, and the required $P_B \leq 10^{-6}$. For simplicity, assume that there are no bandwidth constraints. Our task is simply to determine whether or not the required error performance can be achieved using the given system architecture and design parameters. In evaluating the system, we will use the same type of transformations used in the previous examples.

Solution to Example 19.4

A typical DS/SS system can be implemented more simply than the one shown in Fig. 19.4. The data and the PN code would be combined at baseband, followed by a single pass through a BPSK modulator. We will, however, assume the existence of the individual blocks in Fig. 19.4 because they enhance our understanding of the transformation process. The relationships in transforming from data bits, to channel bits, to symbols, and to chips (Fig. 19.4) have the same pattern of subtle but straightforward transformations in rates and energies as previous relationships (Figs. 19.2 and 19.3). The values of R_c, R_s, and R_{ch} can now be calculated immediately since the (63, 51) BCH code has already been selected. From Eq. (19.30) we write

$$R_c = \left(\frac{n}{k}\right) R = \left(\frac{63}{51}\right) 9600 \approx 11{,}859 \text{ channel-b/s}$$

Since the data modulation considered here is BPSK, then from Eq. (19.31) we write

$$R_s = R_c \approx 11{,}859 \text{ symbol/s}$$

and from Eq. (19.41), with an assumed value of $G_p = 1000$

$$R_{ch} = G_p R = 1000 \times 9600 = 9.6 \times 10^6 \text{ chip/s}$$

Since we have been given the same S/N_0 and the same data rate as in Example 19.2, we find the value of received E_b/N_0 from Eq. (19.25) to be 8.2 dB (or 6.61). At the demodulator, we can now expand the expression for S/N_0 in Eq. (19.34) and the Appendix as follows:

$$\frac{S}{N_0} = \frac{E_b}{N_0} R = \frac{E_c}{N_0} R_c = \frac{E_s}{N_0} R_s = \frac{E_{ch}}{N_0} R_{ch} \qquad (19.42)$$

Corresponding to each transformed entity (data bit, channel bit, symbol, or chip) there is a change in rate and, similarly, a reciprocal change in energy-to-noise spectral density for that received entity. Equation (19.42) is valid for any such transformation when the rate and energy are modified in a reciprocal way. There is a kind of *conservation of power* (or energy) phenomenon that exists in the transformations. The total received average power (or total received energy per symbol duration) is fixed regardless of how it is computed, on the basis of data bits, channel bits, symbols, or chips.

The ratio E_{ch}/N_0 is much lower in value than E_b/N_0. This can be seen from Eqs. (19.42) and (19.41), as follows:

$$\frac{E_{ch}}{N_0} = \frac{S}{N_0}\left(\frac{1}{R_{ch}}\right) = \frac{S}{N_0}\left(\frac{1}{G_p R}\right) = \left(\frac{1}{G_p}\right)\frac{E_b}{N_0} \qquad (19.43)$$

But, even so, the despreading function (when properly synchronized) accumulates the energy contained in a quantity G_p of the chips, yielding the same value $E_b/N_0 = 8.2$ dB, as was computed earlier from Eq. (19.25). Thus, the DS spreading transformation has no effect on the error performance of an AWGN channel [Sklar, 1988], and the value of G_p has no bearing on the value of P_B in this example.

From Eq. (19.43), we can compute, in decibels,

$$\frac{E_{ch}}{N_0} = E_b/N_0 - G_p$$
$$= 8.2 - (10 \times \log_{10} 1000)$$
$$= -21.8 \qquad (19.44)$$

The chosen value of processing gain ($G_p = 1000$) enables the DS/SS system to operate at a value of chip energy well below the thermal noise, with the same error performance as without spreading.

Since BPSK is the data modulation selected in this example, each message symbol therefore corresponds to a single channel bit, and we can write

$$\frac{E_s}{N_0} = \frac{E_c}{N_0} = \left(\frac{k}{n}\right)\frac{E_b}{N_0} = \left(\frac{51}{63}\right) \times 6.61 = 5.35 \qquad (19.45)$$

where the received $E_b/N_0 = 8.2$ dB (or 6.61). Out of the BPSK data demodulator, the symbol-error probability P_E (and the channel-bit error probability p_c) is computed as follows [Sklar, 1988]:

$$p_c = P_E = Q\left(\sqrt{\frac{2E_c}{N_0}}\right) \qquad (19.46)$$

Digital Communication System Performance

Using the results of Eq. (19.45) in Eq. (19.46) yields

$$p_c = Q(3.27) = 5.8 \times 10^{-4}$$

Finally, using this value of p_c in Eq. (19.38) for the (63, 51) double-error correcting code yields the output bit-error probability of $P_B = 3.6 \times 10^{-7}$. We can, therefore, verify that for the given architecture and design parameters of this example the system does, in fact, achieve the required error performance.

19.7 Conclusion

The goal of this section has been to review fundamental relationships used in evaluating the performance of digital communication systems. First, we described the concept of a link and a channel and examined a radio system from its transmitting segment up through the output of the receiving antenna. We then examined the concept of bandwidth-limited and power-limited systems and how such conditions influence the system design when the choices are confined to MPSK and MFSK modulation. Most important, we focused on the definitions and computations involved in transforming from data bits to channel bits to symbols to chips. In general, most digital communication systems share these concepts; thus, understanding them should enable one to evaluate other such systems in a similar way.

Appendix: Received E_b/N_0 Is Independent of the Code Parameters

Starting with the basic concept that the received average signal power S is equal to the received symbol or waveform energy, E_s, divided by the symbol-time duration, T_s (or multiplied by the symbol rate, R_s), we write

$$\frac{S}{N_0} = \frac{E_s/T_s}{N_0} = \frac{E_s}{N_0} R_s \qquad (A19.1)$$

where N_0 is noise-power spectral density.

Using Eqs. (19.27) and (19.25), rewritten as

$$\frac{E_s}{N_0} = (\log_2 M) \frac{E_c}{N_0} \quad \text{and} \quad R_s = \frac{R_c}{\log_2 M}$$

let us make substitutions into Eq. (A19.1), which yields

$$\frac{S}{N_0} = \frac{E_c}{N_0} R_c \qquad (A19.2)$$

Next, using Eqs. (19.26) and (19.24), rewritten as

$$\frac{E_c}{N_0} = \left(\frac{k}{n}\right) \frac{E_b}{N_0} \quad \text{and} \quad R_c = \left(\frac{n}{k}\right) R$$

let us now make substitutions into Eq. (A19.2), which yields the relationship expressed in Eq. (19.11)

$$\frac{S}{N_0} = \frac{E_b}{N_0} R \qquad (A19.3)$$

Hence, the received E_b/N_0 is only a function of the received S/N_0 and the data rate R. It is independent of the code parameters, n, k, and t. These results are summarized in Fig. 19.3.

References

Anderson, J.B. and Sundberg, C.-E.W. 1991. Advances in constant envelope coded modulation, *IEEE Commun., Mag.*, 29(12):36–45.

Borjesson, P.O. and Sundberg, C.E. 1979. Simple approximations of the error function $Q(x)$ for communications applications, *IEEE Trans. Comm.*, COM-27(March):639–642.

Clark, G.C., Jr. and Cain, J.B. 1981. *Error-Correction Coding for Digital Communications*, Plenum Press, New York.

Hodges, M.R.L. 1990. The GSM radio interface, *British Telecom Technol. J.* 8(1):31–43.

Johnson, J.B. 1928. Thermal agitation of electricity in conductors, *Phys. Rev.* 32(July):97–109.

Korn, I. 1985. *Digital Communications*, Van Nostrand Reinhold Co., New York.

Lin, S. and Costello, D.J., Jr. 1983. *Error Control Coding: Fundamentals and Applications*, Prentice-Hall, Englewood Cliffs, NJ.

Lindsey, W.C. and Simon, M.K. 1973. *Telecommunication Systems Engineering*, Prentice-Hall, Englewood Cliffs, NJ.

Nyquist, H. 1928. Thermal agitation of electric charge in conductors, *Phys. Rev.* 32(July):110–113.

Nyquist, H. 1928. Certain topics on telegraph transmission theory, *Trans. AIEE.* 47(April):617–644.

Odenwalder, J.P. 1976. *Error Control Coding Handbook*. Linkabit Corp., San Diego, CA, July 15.

Shannon, C.E. 1948. A mathematical theory of communication, *BSTJ.* 27:379–423, 623–657.

Shannon, C.E. 1949. Communication in the presence of noise, *Proc. IRE.* 37(1):10–21.

Sklar, B. 1988. *Digital Communications: Fundamentals and Applications*, Prentice-Hall Inc., Englewood Cliffs, N.J.

Sklar, B. 1979. What the system link budget tells the system engineer or how I learned to count in decibels, *Proc. of the Int'l. Telemetering Conf.*, San Diego, CA, Nov.

Title 47, *Code of Federal Regulations*, Part 15 Radio Frequency Devices.

Ungerboeck, G. 1987. Trellis-coded modulation with redundant signal sets, Pt. I and II, *IEEE Comm. Mag.*, 25(Feb.):5–21.

Van Trees, H.L. 1968. *Detection, Estimation, and Modulation Theory*, Pt. I, John Wiley and Sons, Inc., New York.

Viterbi, A.J. 1966. *Principles of Coherent Communication*, McGraw-Hill Book Co., New York.

Viterbi, A.J. 1979. Spread spectrum communications—myths and realities, *IEEE Comm. Mag.*, (May):11–18.

Further Information

A useful compilation of selected papers can be found in: *Cellular Radio & Personal Communications–A Book of Selected Readings*, edited by Theodore S. Rappaport, Institute of Electrical and Electronics Engineers, Inc., Piscataway, New Jersey, 1995. Fundamental design issues, such as propagation, modulation , channel coding, speech coding, multiple-accessing and networking, are well represented in this volume.

Another useful sourcebook that covers the fundamentals of mobile communications in great detail is: *Mobile Radio Communications*, edited by Raymond Steele, Pentech Press, London 1992. This volume is also available through the Institute of Electrical and Electronics Engineers, Inc., Piscataway, New Jersey.

For spread spectrum systems, an excellent reference is: *Spread Spectrum Communications Handbook*, by Marvin K. Simon, Jim K Omura, Robert A. Scholtz, and Barry K. Levitt, McGraw-Hill Inc., New York, 1994.

20
Synchronization

20.1	Introduction	255
20.2	Carrier Synchronization	256
	Unmodulated Carrier • Synchronization from a Modulated Carrier • To Explore Further: Carrier Acquisition for Quadrature Amplitude Modulation Constellations	
20.3	Symbol Synchronization	264
	To Browse Further	
20.4	Frame Synchronization	268
	Performance	

Costas N. Georghiades
Texas A&M University

20.1 Introduction

Etymologically, the word synchronization refers to the process of making two or more events occur at the same time, or, by duality, at the same frequency. In a digital communication context, various levels of synchronization must be established before data decoding can take place, including *carrier synchronization, symbol synchronization*, and *frame synchronization*.

In radio frequency communications, carrier synchronization refers to the process of generating a sinusoidal signal that closely tracks the *phase* and *frequency* of a received noisy carrier, transmitted by a possibly distant transmitter. Thus, carrier synchronization in general refers to both frequency and phase acquisition and tracking. Of the two, in many cases the more difficult problem is extracting carrier phase, which is often a much faster varying process compared to frequency offset. This is especially so since the advent of highly stable crystal oscillators operating in the ultra-high-frequency (UHF) or lower frequency bands, although the problem is still prevalent at the higher microwave frequencies where crystal oscillators are not available. Frequency acquisition is a also a problem in mobile radio applications where, due to the Doppler effect, there is an offset in the frequency of the received carrier.

Communication systems that have available or somehow extract and make use of good (theoretically perfect) carrier frequency and phase information are known as *coherent* systems, in contrast to *incoherent* systems that neglect the carrier phase. Systems that attempt to acquire phase information, but do not do a perfect job, are known as *partially coherent* systems. Coherent systems are known to perform better than incoherent ones, at the price, however, of more complexity required for carrier synchronization. In this chapter we will look at some of the classical techniques for carrier acquisition, as well as some of the modern techniques, which often involve operating on sampled instead of analog data.

Symbol synchronization is the process of deriving at the receiver timing signals that indicate where in time the transmitted symbols are located. The decision part of the receiver subsequently uses this information in order to decide what the symbols are. As with carrier synchronization, the data available to the receiver for making timing estimates is noisy. Thus, perfect timing information cannot be obtained in practice, although practical systems come close.

Once symbol synchronization is achieved, the next highest synchronization level is frame synchronization. Frame synchronization is necessary in systems for which the unit of information is not a symbol, but rather a sequence of symbols. Such systems are, for example, coded systems where the unit of information is a code word which consists of a number of symbols. In this case, it is clear that knowing where the symbols are is not enough, and further knowledge of where the code words are is needed. It is easily seen that the existence of frame synchronization automatically implies symbol synchronization, but the converse is not true. Thus, one might be tempted to attempt frame synchronization before symbol synchronization is achieved, thus achieving both at once. Such an approach, although in theory resulting in better performance, has the disadvantage of requiring more complex processing than the approach of first achieving lower level synchronization before higher ones are attempted. In practice, almost invariably the latter approach is followed.

A rather standard approach to achieving frame synchronization is for the transmitter to insert at the start of every frame a special synchronization pattern, whose detection at the receiver locates frame boundaries. We will look at the optimal frame synchronization processing, as well as some of the desirable characteristics of synchronization sequences later.

There are two general methodologies for achieving the various synchronization levels needed by the receiver. One way is to provide at the receiver side an unmodulated carrier, or a carrier modulated by a known sequence, which can be used solely for the purpose of synchronization. This approach has the advantage of decoupling the problem of data detection and synchronization, and makes the synchronization system design easier. On the other hand, the overall communication efficiency suffers since signal energy and time is used but no information is sent. A second approach, which is often preferred, is to derive synchronization from the data-modulated carrier, the same signal used for symbol decisions. In this way, no efficiency is sacrificed, but the processing becomes somewhat more involved. In the sequel, we will only consider algorithms that utilize a modulated received signal to derive synchronization. We first start with a look at carrier synchronization algorithms. Excellent general treatments of this and the other synchronization problems studied later can be found in the classic texts of Stiffler [1971] and Lindsey and Simon [1973], and more recently in Meyr [1990].

20.2 Carrier Synchronization

In this section we will formulate the carrier synchronization problem in mathematical terms as an estimation problem and then see how the optimal equations derived can be practically implemented through appropriate approximations. Our optimality criterion is in the sense of maximum likelihood (ML), under which estimates maximize with respect to the parameter to be estimated the conditional probability of the data given the parameter (see, for example, Van Trees [1968]). As pointed out in the Introduction, carrier synchronization may be achieved rather easily by tracking the phase of an unmodulated carrier that is frequency multiplexed with the modulated carrier. In this case, we need not worry about the noise (uncertainty) introduced by the random modulation. On the other hand, the resulting system is inefficient in that part of the transmitter power carries no information and is used solely for carrier phase estimation. Although we will look at such synchronizers, our emphasis will be on the more efficient carrier synchronizers that derive their carrier phase estimates from suppressed carrier signals.

Unmodulated Carrier

The received unmodulated signal $r(t)$ is

$$r(t) = s(t; \phi) + n(t), \qquad t \in T_0 \tag{20.1}$$

Synchronization

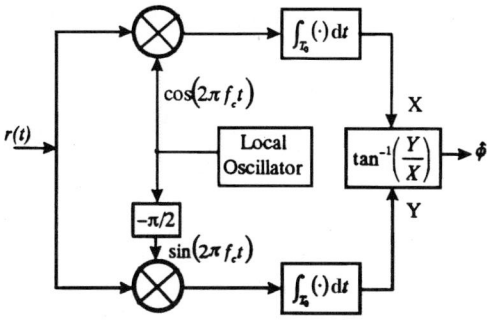

FIGURE 20.1 ML synchronizer: open-loop.

where $s(t; \phi) = A \cos(2\pi f_c t - \phi)$ is the transmitted (unmodulated) signal, ϕ is the carrier phase, and $n(t)$ is a zero-mean, white Gaussian noise process with spectral density $N_0/2$.

The maximum-likelihood estimate of the received noisy carrier phase is the value of ϕ that maximizes the likelihood function, given by (see, for example, Van Trees [1968])

$$L(\phi) = \exp\left[\frac{2}{N_0} \int_{T_0} r(t) s(t; \phi)\, dt - \frac{1}{N_0} \int_{T_0} s^2(t; \phi)\, dt\right] \qquad (20.2)$$

Since the second integral is not a function of ϕ, it can be dropped. Taking the logarithm of the resulting expression, we can equivalently maximize the log-likelihood function given by

$$\ell(\phi) = \frac{2}{N_0} \int_{T_0} r(t) s(t; \phi)\, dt \qquad (20.3)$$

Differentiating with respect to ϕ and setting to zero we obtain the following necessary condition for the maximum-likelihood estimate $\hat{\phi}$:

$$\int_{T_0} r(t) \sin(2\pi f_c t - \hat{\phi})\, dt = 0 \qquad (20.4)$$

Solving for $\hat{\phi}$ we get

$$\hat{\phi} = \tan^{-1}\left[\frac{\int_{T_0} r(t) \sin(2\pi f_c t)\, dt}{\int_{T_0} r(t) \cos(2\pi f_c t)\, dt}\right]$$

Figure 20.1 shows how this estimator can be implemented in block-diagram form in what is referred to as on *open-loop* realization.

A *closed-loop* or tracking synchronizer that uses the optimality condition in Eq. (20.4) in a tracking loop referred to as a *phase-locked loop* (PLL) is shown in Fig. 20.2. In this figure, VCO is the voltage controlled oscillator and is a device that produces a sinusoid at the carrier frequency f_c with an instantaneous frequency which is proportional to its input (or equivalently, a phase that is the integral of its input). The integrator in Fig. 20.2 over the interval T_0 is a linear filter that, in general, can be modeled by an impulse response $g(t)$ or a transfer function $G(s)$ and is referred to as the *loop filter*.

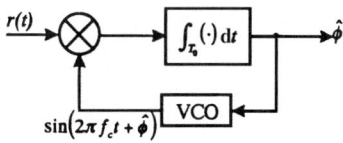

FIGURE 20.2 Phase-locked loop estimator.

Let us now investigate the performance of the PLL synchronizer. We have the following equation describing the PLL in Fig. 20.2:

$$\frac{d\hat{\phi}(t)}{dt} = \{r(t)\sin[2\pi f_c t + \hat{\phi}(t)]\} * g(t) \tag{20.5}$$

where the asterisk represents convolution. Substituting for $r(t)$ from Eq. (20.1) the product term becomes

$$r(t)\sin[2\pi f_c t + \hat{\phi}(t)] = \frac{A}{2}\{\sin[4\pi f_c t + \hat{\phi}(t) + \phi] + \sin[\hat{\phi}(t) - \phi]\} + n'(t) \tag{20.6}$$

where $n'(t)$ can be argued to be white and Gaussian with zero mean and spectral density $N_0/4$. Dropping the double-frequency term in Eq. (20.6) since it will be filtered out by the much lower bandwidth of the loop, we obtain the following model for the PLL:

$$\frac{d\hat{\phi}(t)}{dt} = \left[\frac{A}{2}\sin[\hat{\phi}(t) - \phi] + n'(t)\right] * g(t) \tag{20.7}$$

Equation (20.7) is a nonlinear, stochastic differential equation that describes the evolution of the phase estimate and is modeled in Fig. 20.3. If we let $e(t) = [\hat{\phi}(t) - \phi]$ be the phase error, then it is easily seen (since ϕ is not a function of time) that the phase error is described by

$$\frac{de(t)}{dt} = \left[\frac{A}{2}\sin[e(t)] + n'(t)\right] * g(t) \tag{20.8}$$

An analytical solution for the density of the error $e(t)$ in steady state has been derived (see Viterbi, [1966]) for the special case when $G(s) = 1$ and is given by the *Tichonov density function*

$$p(e) = \frac{\exp[\alpha \cos(e)]}{2\pi I_0(\alpha)}, \quad -\pi \le e \le \pi \tag{20.9}$$

where $\alpha = 4A/N_0$ and $I_0(\cdot)$ is the zero-order, modified Bessel function. For large α, that is, large signal-to-noise ratios (SNRs), the variance of the phase error computed from Eq. (20.9) can be approximated by $\sigma_e^2 \cong 1/\alpha$. Further, for large SNRs, the error will be small on the average, and so in Eq. (20.8) $\sin(e) \cong e$. Under this approximation, Eq. (20.8) [and consequently Eq. (20.7)] become linear. Fig. 20.4 shows the linearized model for a PLL.

For the linear model in Fig. 20.4, an expression for the variance of the estimation error in steady state can be derived easily by computing the variance of the output $\hat{\phi}$ when the input is just the noise $n'(t)$. The following expression for the variance of the estimation error can be derived (which

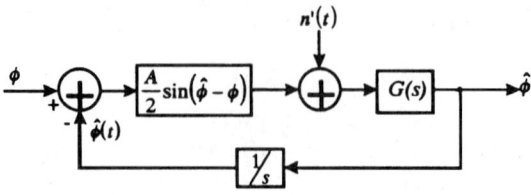

FIGURE 20.3 Equivalent PLL model for performance analysis.

Synchronization

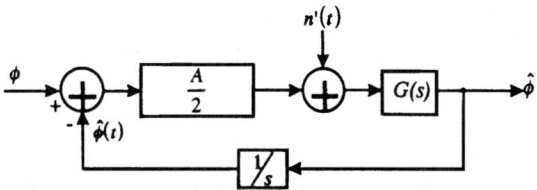

FIGURE 20.4 Linearized PLL model for performance computation.

is a good approximation to the actual variance at high SNRs when the linearized model is valid):

$$\sigma_c^2 = \frac{2N_0 B_L}{A^2} \qquad (20.10)$$

where the parameter B_L is known as the (one-sided) *loop noise equivalent bandwidth* and is given by

$$B_L = \frac{1}{\max |H(f)|^2} \int_0^\infty |H(f)|^2 \, df \qquad (20.11)$$

In Eq. (20.11), $H(s)$ is the closed-loop transfer function of the (linearized) loop, given by

$$H(s) = \frac{AG(s)}{2s + AG(s)} \qquad (20.12)$$

A simple computation of the variance given by Eq. (20.9) for the special case when $G(s) = 1$ yields $1/\alpha$, as expected. The advantage of Eq. (20.10) is that it can be used for general $G(s)$, provided the signal-to-noise ratio is large enough for the linearity approximation to hold. Note also that the estimation error becomes smaller with decreasing bandwidth B_L. Thus, one might be tempted to make B_L close to zero. The problem in this case is that the transient performance of the loop degrades to the extent that it may take a long time before a small error is achieved. Also, in practice, the input phase ϕ is time varying, in which case the loop bandwidth should be large enough so that the loop can track the changes in the input. We now turn our attention to finding the performance of the ML carrier synchronizer in Fig. 20.1.

We make use of the Cramer–Rao lower bound [Van Trees, 1968] that states that under some conditions the variance of the estimation error for *any* estimate of a parameter ϕ is lower bounded by

$$\sigma_c^2 = \frac{-1}{E\left[\dfrac{\partial^2 \ln p(r \mid \phi)}{\partial \phi^2}\right]} \qquad (20.13)$$

where $\ln p(r \mid \phi)$ is the log-likelihood function, given by Eq. (20.3) for the phase estimation problem. Using Eq. (20.13), we obtain

$$\sigma_c^2 \geq \left\{ \frac{2A}{N_0} \int_{T_0} E[r(t)] \cos(2\pi f_c t + \phi) \, dt \right\} = \frac{2N_0}{A^2} \frac{1}{2T_0} \qquad (20.14)$$

A brief computation shows that $1/T_0$ corresponds to the equivalent noise bandwidth of the integrator. Thus, if the equivalent noise bandwidths in Eq. (20.10) and (20.14) are the same, the performance of the ML and the PLL synchronizers are the same [assuming that the bound in Eq. (20.14) is achieved closely].

Next, we address carrier extraction from a modulated carrier, which, as pointed out earlier, is in practice the preferred approach for efficiency reasons.

Synchronization from a Modulated Carrier

A suboptimal approach, often used in practice, to extract carrier synchronization from a modulated signal is to first nonlinearly preprocess the signal to wipe off the modulation, and then follow that by, for example, a PLL, as previously described. For M-ary phase-shift keying (PSK), the nonlinear preprocessing involves taking the Mth power of the signal. This has the effect of multiplying the carrier phase by M, and thus a subsequent division by M is needed to match the original carrier phase. For PSK signalling, a side effect of the power-law nonlinearity is to introduce a phase ambiguity, which must be resolved either by using pilot symbols or (which is the case in practice) through the use of *differential encoding* of the data (information is conveyed by the change in the phase relative to the previous baud interval, rather than absolute phase).

Besides the suboptimal power-law technique, ML estimation can be used to suggest the optimal processing and possible approximations, which we study next. Let the modulated data be

$$r(t) = s(t; d, \phi) + n(t) \tag{20.15}$$

where d is a sequence of N modulation symbols. For simplicity we will assume binary antipodal signals, in which case each component, d_k of d is either 1 or -1. If we let the baud rate be $1/T$ symbols/s, then the signal part in Eq. (20.15) can be expressed as

$$s(t; d, \phi) = A \sum_{k=0}^{N-1} d_k \cos(2\pi f_c t + \phi) p(t - kT) \tag{20.16}$$

where $p(t)$ is a baseband pulse, which determines to a large extent the spectral content of the transmitted signal, and we are assuming a data window of length N. For simplicity, we will assume a unit height rectangular pulse next, but the results can be generalized. Assuming for the moment that the modulation sequence d is known, we have the following conditional likelihood function:

$$L(\phi) = \exp\left[\frac{2}{N_0} \int_0^{NT} r(t) s(t; d, \phi) \, dt\right]$$

$$= \exp\left[\frac{2A}{N_0} \sum_{k=0}^{N-1} d_k \int_{kT}^{(k+1)T} r(t) \cos(2\pi f_c t + \phi) \, dt\right] \tag{20.17}$$

All we need to do now is take the expectation of the conditional likelihood function in Eq. (20.17) with respect to the random modulation sequence. Assuming that each binary symbol occurs with probability 1/2 and that symbols are independent, we obtain the following log-likelihood function after dropping terms that are not functions of ϕ and taking the logarithm of the resulting expression:

$$\Lambda(\phi) = \sum_{k=0}^{N-1} \ln \cosh\left[\frac{2A}{N_0} \int_{kT}^{(k+1)T} r(t) \cos(2\pi f_c t + \phi) dt\right] \tag{20.18}$$

A maximum likelihood estimator maximizes this expression with respect to the phase ϕ. In practice, to reduce complexity we use the following substitutions; $\ln \cosh(x) \propto x^2$, and $\ln \cosh(x) \propto |x|$,

Synchronization

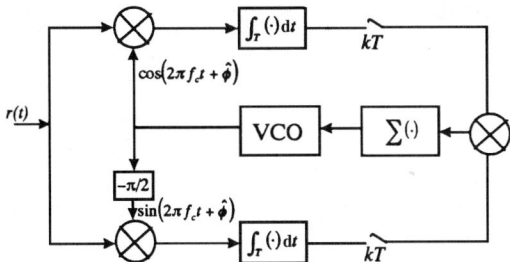

FIGURE 20.5 Carrier synchronization from modulated data.

valid for small and large signal-to-noise ratios, respectively. In the case of small SNRs, the log-likelihood simplifies to

$$\ell(\phi) = \sum \left[\int_{kT}^{(k+1)T} r(t) \cos(2\pi f_c t + \phi) \, dt \right]^2 \quad (20.19)$$

Taking the derivative of Eq. (20.19) with respect to ϕ we obtain the following necessary condition for the ML estimate $\hat{\phi}$ of ϕ:

$$\sum_{k=0}^{N-1} \int_{kT}^{(k+1)T} r(t) \cos(2\pi f_c t + \hat{\phi}) \, dt \times \int_{kT}^{(k+1)T} r(t) \sin(2\pi f_c t + \hat{\phi}) \, dt = 0 \quad (20.20)$$

A tracking loop that dynamically forces the condition in Eq. (20.20) is shown in Fig. 20.5. Note that the product of the two integrals effectively removes the modulation. In practice, the summer may be replaced by a digital filter that applies different weights to past and present data in order to improve response.

To Explore Further: Carrier Acquisition for Quadrature Amplitude Modulation Constellations

The need for high throughputs required by several high-speed applications (such as digital TV) has pushed system designers toward more throughput-efficient modulation schemes. Because of their relatively good performance, large quadrature amplitude modulation (QAM) constellations are being used in many of these applications. One of the problems associated with their use is that of carrier acquisition, which for efficiency reasons must often be done without the use of a preamble. The problem is further complicated for cross QAM constellations, for which the high SNR corner points used by some simple carrier phase estimators are not available. Clearly, due to the phase symmetry of QAM constellations, only phase offsets modulo $\pi/2$ are detectable and differential encoding is used to resolve the ambiguity.

The phase synchronization problem is invariably divided into an acquisition and a tracking part. In many practical systems, tracking is done simply and efficiently in a decision directed (DD) mode after acquisition has been established, and it is the acquisition problem that is the most problematic, especially in applications where no preamble is allowed. For square QAM constellations a simple technique for phase acquisition is based on detecting the signals at the four corners and using them to produce an estimate of the phase offset, which can be averaged in time to converge to a reliable estimate. The problem is more complicated for cross constellations, which do not have the corner points. We first look at the ML carrier phase estimator. Let

$$r_k = d_k e^{j\theta} + n_k$$

be the baud-rate samples of the output of a matched filter, where d_k is a complex number denoting the transmitted QAM symbol at time KT ($1/T$ is the signalling rate), and θ denotes the unknown phase offset to be estimated; the n_k are complex, independent identically distributed (i.i.d.), zero-mean, Gaussian random variables with independent real and imaginary parts of variance σ^2, modeling the effects of noise in the system and channel. Without loss of generality, we assume that $E[|d_k|^2] = 1$ (i.e., a unit average energy constellation), in which case the signal-to-noise ratio per symbol is SNR $= 1/2\sigma^2$. Then the ML phase estimate of θ from data over a window of length N is easily obtained as the value of ϕ that maximizes the log-likelihood function

$$L(\phi) = \sum_{k=1}^{N} \ln\left[\sum_{d} \exp\left(-\frac{1}{2\sigma^2}|r_k - d \cdot e^{j\theta}|^2\right)\right]$$

where the inner summation is over all data d in the constellation. The complexity of the ML algorithm is due in part to this inner summation, which even for small constellations will require more computations than possible, especially for high-speed systems. Another complication in implementing the ML estimator is the need to solve a nonlinear maximization problem in order to find the ML estimate of ϕ. Some simplification of the ML estimator can be obtained for square constellations, but it is not enough to make it practical.

For square constellations, a simple algorithm can be used to extract carrier phase by detecting the presence of one of the four corner points in the constellation. These points can be detected by setting an amplitude threshold that is between the peak amplitude of the corner points and the second largest amplitude. The angle of these four points can be expressed as

$$\phi^i = \frac{\pi}{4} + i \cdot \frac{\pi}{2} \qquad i = 0, 1, 2, 3$$

Thus,

$$\phi^i = \frac{\pi}{4} \operatorname{mod}\left(\frac{\pi}{2}\right)$$

Since only phase rotations modulo $\pi/2$ are required, a simple estimator looks at the angle of the received sample r_k modulo $\pi/2$ and subtracts it from $\pi/4$. The result is the required estimate, which can be refined in time as more data are observed.

Another often used algorithm, which has been shown in Moeneclaey and de Jonghe [1994] to be asymptotically ML in the limit of small SNRs, is the Mth power-law estimator, where $M = 4$ for QAM constellations, and it equals the size of the constellation for PSK signalling. For QAM signalling, the fourth power estimator extracts a phase estimate according to

$$\hat{\varphi} = \frac{1}{4} \arg\left[E[d^{*4}] \cdot \sum_{k=1}^{N} r_k^4\right]$$

The approximate mean-square error performance of this estimator was also obtained in Moeneclaey and de Jonghe [1994] and is given by

$$\text{MSE} \cong \frac{A}{2N \cdot \text{SNR}} + \frac{B}{N}$$

Synchronization

where the constants A and B are functions of the QAM constellation given by

$$A = \frac{E[|d_k^2|] \cdot E[|d_k^2|^3]}{|E[d_k^4]|^2}$$

$$B = \frac{2|E[d_k^4]|^2 \cdot E[|d_k^4|^4] - E^2[d_k^4] \cdot E[d_k^{*8}] - E^2[d_k^{*4}] \cdot E[d_k^8]}{64|E[d_k^4]|^4}$$

More recently another algorithm that seems to work well for both square and cross constellations is described in Georghiades [1996]. The algorithm referred to as the histogram algorithm (HA) does the following: (1) For each received sample r_k find the set of signals whose magnitude is closest to $|r_k|$. (2) Then compute the angle of the subset of signals from step 1 belonging to the first quadrant. (3) Subtract each of the angles computed at step 2 from the angle of r_k. (4) Uniformly quantize the angle interval from 0 to 90° into L bins, and associate a counter with each; increment each counter whose associated angles were computed at step 3. (5) Repeat this process for new data r_k as they arrive. (6) When enough data is received, find the bin that has the largest counter value. The angle corresponding to this bin is produced as the phase estimate.

Figure 20.6 compares the performance of the ML and HA algorithms obtained through simulation to the Cramer–Rao bound for the 128-QAM (cross) constellation. Results for the fourth power estimator are shown separately in Fig. 20.7 since the performance of this estimator is about two orders of magnitude worse than the ML and HA algorithms. The reason behind the bad performance of the fourth power estimator is the existence of large self-noise, partly due to the absence of the corner points. These results indicate that the fourth power estimator is not an option for cross constellations, at least not for sizes greater than or equal to 128.

Figure 20.8 compares the HA and the fourth power estimators for the 256-QAM (square) constellation. As can be seen, the fourth power estimator performs much better with square constellations, and in fact it outperforms the HA for a range of data sequence lengths at 25-dB SNR. As the SNR increases, however, the self-noise dominates and the performance of the fourth power estimator degrades.

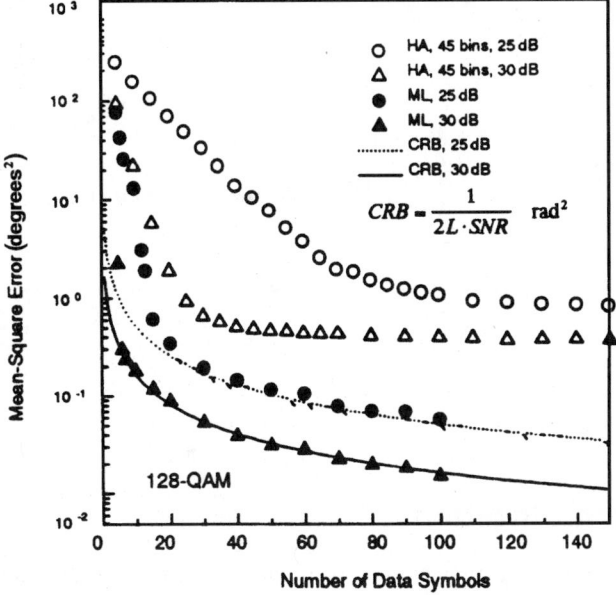

FIGURE 20.6 Mean square error for the various algorithms and 128QAM.

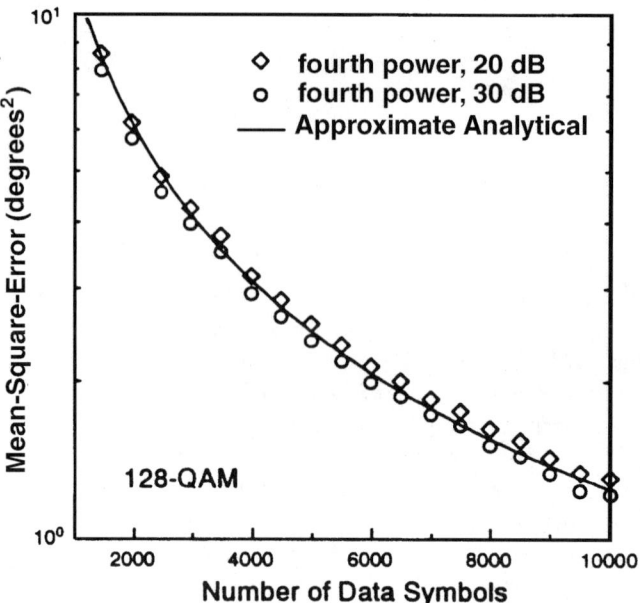

FIGURE 20.7 The performance of the fourth power estimator for 128QAM.

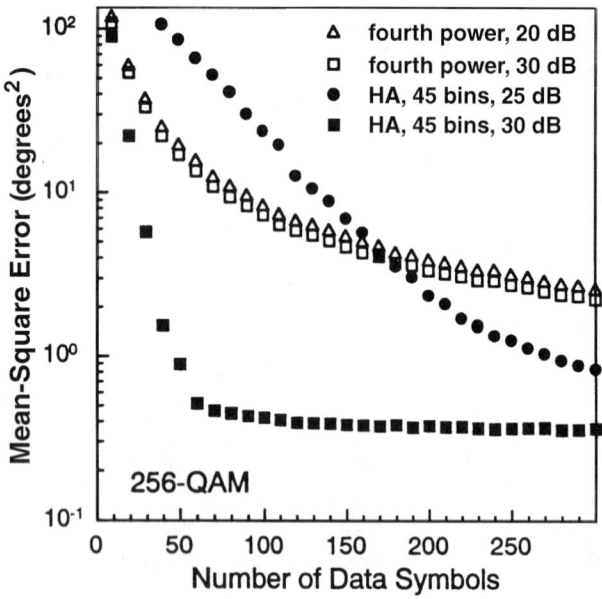

FIGURE 20.8 The HA the fourth power algorithms for 256QAM.

20.3 Symbol Synchronization

We first investigate symbol synchronizers that are optimal in a maximum-likelihood sense. The ML symbol synchronizer can be used to suggest suboptimal but more easily implementable algorithms, and provides a benchmark against which the performance of other synchronizers can be compared.

Synchronization

Let T be the symbol duration and $r(t)$ be the received data. We assume an additive, white Gaussian noise channel, so that $r(t)$ is modeled by

$$r(t) = s(t; d, \tau) + n(t) \tag{20.21}$$

where $n(t)$ is zero-mean white Gaussian noise having spectral density $N_0/2$, d is a sequence of modulation symbols d_k, $k = \ldots, -1, 0, 1, \ldots$, and τ is the timing error. Assuming pulse-amplitude modulation (PAM), the signal can be described explicitly by

$$s(t; d, \tau) = \sum_k d_k p(t - kT - \tau) \tag{20.22}$$

where $p(t)$ is a baseband pulse. Our problem is to process the received signal $r(t)$ in order to obtain an estimate of the timing error τ. To avoid loss in communication efficiency, we will do this in the presence of modulation symbols. For simplicity, we will assume a binary system with antipodal signals (one signal is just the negative of the other), in which case $d_k \in \{1, -1\}$.

As previously for carrier phase estimation, ML synchronization requires the probability density function of the data given the timing error τ. Conditioned on knowing the data sequence d, the likelihood-function is given by Eq. (20.2) [with the signal part now given by Eq. (20.22)]. Again, the quadratic term can be dropped as it is not a function of τ, resulting in

$$L(\tau, d) = \exp\left[\frac{2}{N_0} \sum_k d_k \int_{-\infty}^{\infty} r(t) p(t - kT - \tau)\, dt\right] \tag{20.23}$$

All we need to do now is take the expectation of this conditional likelihood function with respect to the data sequence to obtain the likelihood function. Performing the expectation, assuming independent and equiprobable data, we obtain (after taking the logarithm of the resulting expression) the required log-likelihood function

$$\ell(\tau) = \sum \ln \cosh\left[\frac{2q_k(\tau)}{N_0}\right] \tag{20.24}$$

where

$$q_k(\tau) = \int_{-\infty}^{\infty} r(t) p(t - kT - \tau)\, dt \tag{20.25}$$

A maximum-likelihood synchronizer finds the value $\hat{\tau}$ of τ, which maximizes the log likelihood function in Eq. (20.24) for the received data. There are several problems in implementing the optimal synchronizer: (1) obtaining a ML estimate requires maximizing in real time Eq. (20.24), an impossible task in most cases; (2) implementation of Eq. (20.24) requires knowledge of the signal-to-noise ratio, which is not readily available and must in practice be estimated; and (3) there is no obvious, simple way to use Eq. (20.24) to extract timing estimates in time as more data arrives.

The first problem can be partly alleviated by approximating the $\ln \cosh(\cdot)$ function for large and small SNRs as was done before, which results in

$$\begin{aligned}\ell(\tau) &\cong \sum_k q_k^2(\tau), \quad &\text{low SNR} \\ \ell(\tau) &\cong \sum_k |q_k(\tau)|, \quad &\text{high SNR}\end{aligned} \tag{20.26}$$

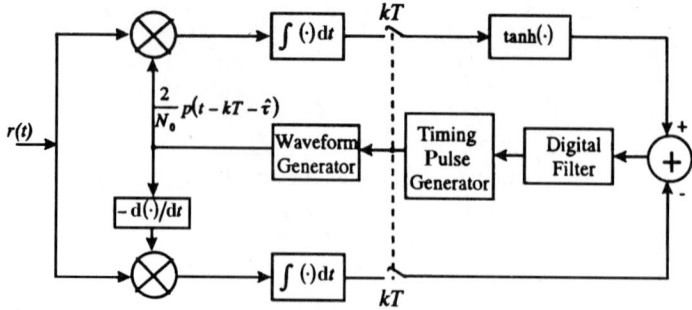

FIGURE 20.9 Closed-loop timing synchronizer.

In addition to simplifying the log-likelihood function, the approximations also obviate the need for knowing the SNR. The need, however, to maximize a nonlinear function in real time still exists.

As for the carrier synchronization problem, a number of open-loop realizations of the ML timing estimator exist. Perhaps the more interesting practically, however, are closed-loop algorithms, two of which we motivate next. Taking the derivative of Eq. (20.24) with respect to τ and equating to zero results in an equation whose solution is the ML timing estimate,

$$\left.\frac{\partial \ell(\tau)}{\partial \tau}\right|_{\tau=\hat{\tau}} = \sum_k \left[\frac{2}{N_0} \int_{-\infty}^{\infty} r(t) \frac{\partial p(t - kT - \tau)}{\partial \tau} \, dt \right.$$
$$\left. \times \tanh\left[\frac{2}{N_0} \int_{-\infty}^{\infty} r(t) p(t - kT - \tau) \, dt\right]\right] = 0 \qquad (20.27)$$

where we have assumed that $p(-\infty) = p(\infty) = 0$. Note that if the timing τ is other than the ML estimate $\hat{\tau}$ that makes the left-hand side of Eq. (20.27) zero, the derivative will be either positive or negative depending on the sign of the error $(\tau - \hat{\tau})$. Thus, the derivative can be used in a tracking loop to provide a correcting signal in a system that dynamically produces the ML timing estimate. Such a system is shown in Fig. 20.9, where the timing pulse generator adjusts the phase of the timing depending on the output of the accumulator once every T seconds. In practice the accumulator may be replaced by a digital filter whose response is such that it puts more emphasis on recent data and less on past data. Clearly, when the timing jitter is fast changing, better results may be obtained by having a short memory filter. On the other hand, if the timing jitter is slowly varying, a filter with a long memory (low bandwidth) will result in better results.

Further simplifications to the previous tracking synchronizer can be made under the assumptions of low or high signal-to-noise ratios, in which case Eq. (20.26) instead of Eq. (20.24) may be used. Further, if the derivative of the likelihood function with respect to τ is approximated by a difference

$$\frac{\partial \ell(\tau)}{\partial \tau} \cong \frac{\ell(\tau + \delta/2) - \ell(\tau - \delta/2)}{\delta} \qquad (20.28)$$

then, under the high SNR approximation (for example), Eq. (20.27) can be replaced by

$$\frac{\partial \ell(\tau)}{\partial \tau} \cong \sum_k \left[\left| \int_{-\infty}^{\infty} r(t) p\left(t - kT - \tau - \frac{\delta}{2}\right) dt \right| \right.$$
$$\left. - \left| \int_{-\infty}^{\infty} r(t) p(t - kT - \tau + \delta/2) \, dt \right| \right] \qquad (20.29)$$

Synchronization

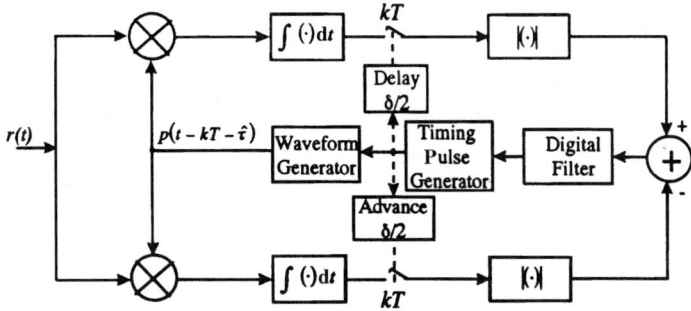

FIGURE 20.10 The early–late gate synchronizer.

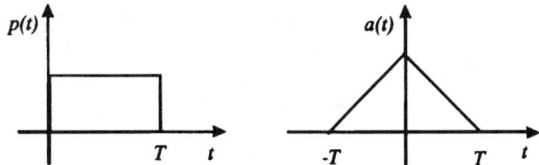

FIGURE 20.11 Example for NRZ pulses.

A tracking-loop synchronizer, known as an *early–late gate symbol synchronizer* that implements Eq. (20.29) is shown in Fig. 20.10. A similar synchronizer for a low SNR approximation can be implemented similarly. The intuitive explanation of how early–late gate symbol synchronizers work is simple and can be easily illustrated for the case of non-return to zero (NRZ) pulses whose shape $p(t)$ and autocorrelation function $a(t)$ are shown in Fig. 20.11.

In the absence of timing error, the receiver samples the output of the matched filter at the times corresponding to the peak of the autocorrelation function of the NRZ pulse (which results in the largest SNR). When a timing error exists, the samples occur at either side of the peak depending on whether the error is positive or negative. In either case, because of the symmetry of the autocorrelation function, the samples are of the same value (on the average). In an early–late gate synchronizer, two samples are taken, separated by δ seconds and centered around the current timing estimate. Depending on whether the error is positive or negative, the difference between the absolute values of these samples will be positive or negative, thus providing a control signal to increase or decrease $\hat{\tau}$ in the desired direction to reduce the error. Notice that on the average, due to the even symmetry of the autocorrelation function, the difference between the absolute values of the two samples is zero at the actual symbol timing phase, that is, when the timing error is zero. Thus, at least intuitively, the system in Fig. 20.10 is a stable loop for tracking the symbol timing phase.

We end this section by noting that a number of practical implementations based on some further simplification of the synchronizers just discussed are used in practice.

To Browse Further

In modern receivers, more and more of the processing is done in the discrete domain, which allows for more complicated algorithms to be accurately implemented compared to analog implementations. For timing recovery, the preferred technique is to process samples taken at the output of a matched (or other suitable) filter at rates as low as the baud rate to a few samples per baud. The advantage of baud-rate sampling is that it uses the same samples that the detector uses to make symbol decisions, and it is the lowest rate possible, making the sampler less costly and the processing of samples faster. The disadvantage is that baud-rate sampling is below the Nyquist rate and thus acquisition performance tends to suffer somewhat. For bandlimited signalling, two or more samples per baud are at or above the Nyquist rate, and thus all information contained in the original analog signal is preserved by the sampling process. This means that the sampler can be free running without the need to adjust its sampling phase since that can be done in the discrete domain through interpolation.

Perhaps the most known paper on timing-recovery from baud-rate samples is that by Mueller and Muller [1976]. The timing algorithms studied therein are decision directed (make use of tentative symbol decisions) and use the baud-rate samples in order to estimate the timing error. The timing error information is then used to adjust the phase of the sampler toward reducing the timing error. Other work on two or more samples per baud has been reported in Agazzi et al. [1985], Gardner [1926], and Georghiades and Moeneclaey [1991].

We close this section by giving a lower-bound on the performance of any timing estimator operating on samples taken at the output of a matched filter at the rate of $1/T_s$ samples/s. It can be shown [Georghiades and Moeneclaey, 1991]

$$E[(\tau - \hat{\tau})^2] \geq \left\{ 2N \cdot \text{SNR} \cdot \int_{-\infty}^{\infty} 4\pi^2 f^2 R(f) \frac{R(f)}{\sum_k R(f - k/T_s)} df \right\}^{-1}$$

$$\geq \left\{ 2N \cdot \text{SNR} \cdot \int_{-\infty}^{\infty} 4\pi^2 f^2 R(f) \, df \right\}^{-1}$$

where N is the number of baud intervals observed,

$$R(f) = |P(f)|^2$$

and $P(f)$ is the Fourier transform of the (baseband) signalling pulse $p(t)$. The second inequality is achieved when the sampling rate $1/T_s$ is greater than or equal to twice the bandwidth of $p(t)$.

20.4 Frame Synchronization

As noted in the Introduction, frame synchronization is obtained in practice by locating at the receiver the position of a frame synchronization pattern (referred to also as a *marker*), periodically inserted in the data stream by the transmitter. In most systems, partly for simplicity and partly because the periodicity of the marker insertion makes it easily identifiable when enough frames are processed, the marker is not prevented from appearing in the random data stream. This means that it should be long enough compared to the frame length to make the probability of it appearing in the data small. If we let the synchronization pattern be of length L and the frame size (including the marker) be of length N, then the efficiency of such a system, measured by the number of data symbols per channel symbol, is

$$e = 1 - \frac{L}{N} \tag{20.30}$$

The efficiency can be made arbitrarily close to one by increasing N for a fixed L or by decreasing L for a fixed N. In both cases, however, the probability of correctly detecting the position of the marker is reduced. In practice, good first pass acquisition probabilities can be achieved with efficiencies of about 97%. Figure 20.12 shows the contents of a frame. As for the symbol synchronization case, we

FIGURE 20.12   The composition of a frame.

Synchronization

will first introduce the optimum (ML) frame synchronizer and then investigate some suboptimum synchronizers. For simplicity, we will only look at binary antipodal baseband signalling, although in qualitative terms similar results hold for nonbinary systems. In the sequel, we assume that perfect symbol synchronization is present.

As usual, we assume a white Gaussian noise channel, in which case the sufficient statistic (loosely, the simplest function of the data required by the optimum synchronizer) is the baud-rate samples of the output of a matched filter. Let $\mathbf{r} = (\mathbf{r}_1, \mathbf{r}_2, \ldots, \mathbf{r}_N)$ be the vector of observed data obtained by sampling the output of a matched filter (matched to the baseband pulse) at the correct symbol rate and phase (no symbol timing error), but not necessarily at the correct frame phase. Under a Gaussian noise assumption, the discrete matched filter samples can be modeled by

$$\mathbf{r}_k = \sqrt{E} d_k + n_k \quad (20.31)$$

where E is the signal energy, $d_k \in \{1, -1\}$ is the kth modulation symbol and the n_k constitute a sequence of i.i.d. Gaussian random variables with zero mean and variance σ^2.

It is clear that since the frame length is N, there is exactly one frame marker within the observation window. Our problem is to locate the position $m \in (0, 1, 2, \ldots, N-1)$ of the marker from the observed data $\mathbf{r}$. If m is the actual position of the marker, then the data vector corresponding to the observed vector $\mathbf{r}$ is

$$\mathbf{d} = (\mathbf{d}_1, \mathbf{d}_2, \ldots, \mathbf{d}_{m-1}, \mathbf{d}_m, \ldots, \mathbf{d}_{m+L-1}, \ldots, \mathbf{d}_N)$$

where the L-symbol sequence starting at position m is the marker. If we denote the marker by S, this means

$$S = (s_1, s_2, \ldots, s_L) = (\mathbf{d}_m, \mathbf{d}_{m+1}, \ldots, \mathbf{d}_{m+L-1})$$

For maximum-likelihood estimation of m, we need to maximize the following conditional density

$$p(r \mid m) = \sum_{d'} p[r \mid m, d'] \Pr(d') \quad (20.32)$$

where d' is the $(N-L)$-symbol data sequence that surrounds the marker. Assuming equiprobable symbols, Massey [1972] derived the following log-likelihood function

$$L(m) = \sum_{k=1}^{L} s_k r_{k+m} - \frac{\sigma^2}{\sqrt{E}} \sum \ln \cosh\left(\frac{\sqrt{E}}{\sigma^2} r_{k+m}\right) \quad (20.33)$$

To account for the periodicity of the marker, indices in Eq. (20.33) are interpreted modulo N. An optimal frame synchronizer computes the preceding expression for all values of m and chooses as its best estimate of the marker position the value that maximizes $L(m)$.

A few observations are in order regarding the preceding likelihood function. First, we note that $L(m)$ is the sum of two terms: a linear term and a nonlinear term. The first term can be recognized as the correlation between the received data $\mathbf{r}$ and the known marker; the second term can be interpreted as an energy correction term that accounts for the random data surrounding the marker.

For practical implementation, some approximations to the optimal rule can be obtained easily by approximating $\ln \cosh(\cdot)$. Thus, for high SNRs, replacing $\ln \cosh(x)$ by $|x|$ we obtain

$$L(m) \cong \sum_{k=1}^{L} s_k \mathbf{r}_{k+m} - \sum_{k=1}^{L} |r_{k+m}| \quad (20.34)$$

For low SNRs, replacing $\ln \cosh(x)$ by $x^2/2$, the optimal rule becomes

$$L(m) \cong \sum_{k=1}^{L} s_k r_{k+m} - \frac{\sqrt{E}}{2\sigma^2} \sum_{k=1}^{L} r_{k+m}^2 \qquad (20.35)$$

A further approximation that is quite often used is to drop the second nonlinear term altogether. The resulting rule then becomes

$$L(m) \cong \sum_{k=1}^{L} s_k r_{k+m} \qquad (20.36)$$

and is known as the *simple correlation rule* for obvious reasons. The high SNR approximation and the simple correlation rule have the added advantage that no knowledge of the SNR is needed for implementation, compared to the optimum and the low SNR approximation rules.

Practical frame synchronizers use the periodicity of the marker in order to improve performance in time, and usually include algorithms for detecting loss of synchronization in which reacquisition is initiated. The preceding algorithms can be used as the basis for these practical synchronizers in estimating the marker position from a frame's worth of data. Their performance in correctly identifying the marker position affects significantly the overall performance of the frame synchronizer, as measured not only by the probability of correct acquisition, but also by the acquisition time for the algorithm.

Other techniques for marker acquisition (besides those based on the maximum-likelihood principle) can be used as well. For example, in some practical implementations of the simple correlation rule, often sequential detection of the marker is implemented: the correlation of the marker with the data is computed sequentially for each frame position and the result is compared to some threshold. When for some value of m the computed correlation exceeds the threshold, the frame synchronizer declares a marker presence. Otherwise the search continues. The value of the threshold is critical for performance and it is usually chosen to minimize the time to marker acquisition.

Another important aspect of frame synchronization design is the design of good marker sequences. Although the preceding algorithms work with any chosen sequence, the resulting performance of the synchronizer depends critically on the sequence used. In general, sequences that have good autocorrelation properties perform well as frame markers. These sequences have the property that their autocorrelation function is uniformly small for all shifts other than the zero shift. Examples of such sequences include the *Barker sequences* [Barker, 1953] and the *Neuman–Hofman sequences* [Neuman and Hofman, 1971]. Barker sequences are binary sequences whose largest side lobe (nonzero shift correlation) is at most 1. Unfortunately, the largest known Barker sequence is of length 13, and there is proof that no Barker sequences of length between 14 and 6084 exist. In many cases, however, there is a need for larger sequences to improve performance. Neuman–Hofman sequences were specifically designed to maximize performance when a simple correlation rule is used. Thus, these sequences perform somewhat better than Barker sequences when a correlation rule is used. What is more important though is that Neuman–Hofman sequences of large length exist. Examples of Barker and Neuman–Hofman sequences of length 7 and 13 are

$(1, -1, 1, 1, -1, -1, -1)$,	Barker, $L = 7$
$(1, 1, 1, 1, 1, -1, -1, 1, 1, -1, 1, -1, 1)$	Barker, $L = 13$
$(-1, -1, -1, -1, -1, -1, 1, 1, -1, -1, 1, -1, 1)$,	Neuman–Hofman, $L = 13$

Performance

We address briefly next the performance of the ML and two suboptimal synchronizers given, previously, as measured by the probability of erroneous synchronization. First, we look at the question

Synchronization

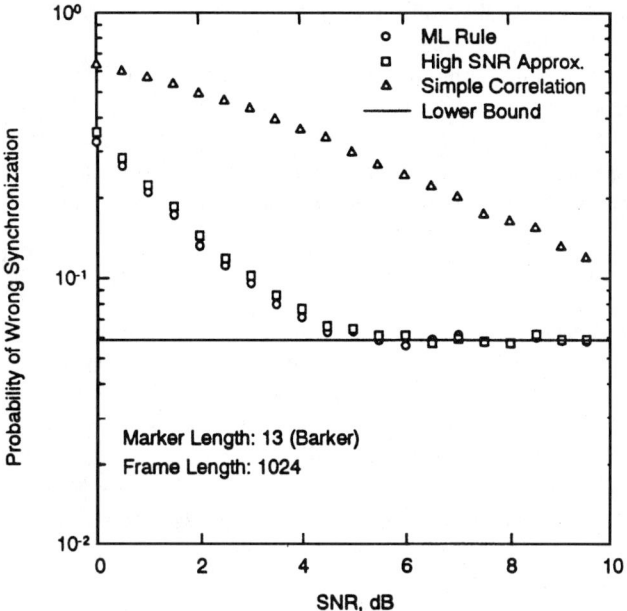

FIGURE 20.13 The performance of the ML and other rules.

of how well *any* frame synchronizer can perform, as a function of the frame length N, marker length L, and SNR. Clearly, in the limit of infinite SNR we obtain the best performance (smallest probability of erroneous marker detection). In this case, an error can be made when the marker appears randomly in one or more positions in the random data part. For *bifix-free* sequences (i.e., sequences for which no prefix is also a suffix) Nielsen [1973] has obtained the following expression for probability of erroneous synchronization in this case:

$$P_{LB} = \sum_{k=1}^{R} \frac{(-1)^{k+1}}{k+1} \cdot \binom{N - L - k(L-1)}{k} \cdot M^{-kL} \quad (20.37)$$

where

$$R = \left\lfloor \frac{N-L}{L} \right\rfloor \quad (20.38)$$

and M is the size of the modulation ($M = 2$ for binary signalling). The bifix-free condition guarantees that no partial overlap of the marker with itself results in a perfect match for the overlapped parts. Figure 20.13 shows simulation results for the performance of the ML, high SNR approximation, and simple-correlation rules of Eqs. (20.33), (20.34), and (20.36), respectively. Shown also is the lower-bound in Eq. (20.37), which is achieved by the ML rule and its high SNR approximation. On the other hand, the simple-correlation rule performs significantly worse.

References

Agazzi, O., Tzeng, C.-P.J., Messerschmitt, D.G., and Hodges, D.A. 1985. Timing recovery in digital subscriber loops. *IEEE Trans. Commun.*, COM-33 (June):558–569.

Barker, R.H. 1953. Group synchronization of binary systems. *Communication Theory*, ed. W. Jackson, pp. 273–287. London.

Gardner, F.M. 1986. A BPSK/QPSK timing-error detector for sampled receivers. *IEEE Trans. Commun.*, COM-34 (May):423–429.

Georghiades, C.N. 1996. A simple blind carrier phase acquisition algorithm for QAM constellations. *Proceedings of the Conference on Information Sciences and Systems*, Princeton, NJ, March.

Georghiades, C.N. and Moeneclaey, M. 1991. Sequence estimation and synchronization from non-synchronized samples. *IEEE Trans. Inf. Theory*, 37(Nov.):1649–1657.

Lindsey, W. and Simon, M. 1973. *Telecommunication Systems Engineering*, Prentice–Hall, Englewood Cliffs, NJ.

Massey, J.L. 1972. Optimum frame synchronization. *IEEE Trans. Commun.*, COM-20 (April):115–119.

Meyr, H. 1990. *Synchronization in Digital Communications*, Wiley, New York.

Moeneclaey, M. and de Jonghe, G. 1994. ML-oriented NDA carrier synchronization for general rotationally symmetric signal constellations. *IEEE Trans. Commun.*, 42(Aug.):2531–2533.

Mueller, K.H. and Muller, M. 1976. Timing recovery in digital synchronous data receivers. *IEEE Trans. Commun.*, COM-24(May):516–531.

Neuman, F. and Hofman, L. 1971. New pulses sequences with desirable correlation properties. *Proceedings of the National Telemetry Conference*, pp. 272–282. Washington, DC, April.

Nielsen, P.T. 1973. On the expected duration of a search for a fixed pattern in random data. *IEEE Trans. Inf. Theory*, (Sept):702–704.

Stiffler, J.J. 1970. *Theory of Synchronous Communications*, Prentice–Hall, Englewood Cliffs, NJ.

Van Trees, H.L. 1968. *Detection, Estimation, and Modulation Theory*, Wiley, New York.

Viterbi, A. 1966. *Principles of Coherent Communication*, McGraw–Hill, New York.

Further Information

The Proceedings of the International Communications Conference (ICC) and Global Telecommunications Conference (GLOBECOM) are good sources of current information on synchronization work. Other sources of archival value are the *IEEE Transactions on Communications* and the *IEEE Transactions on Information Theory*.

21

Digital Modulation Techniques

21.1	Introduction	273
21.2	The Challenge of Digital Modulation	274
	Bandwidth • Signal-to-Noise Ratio • Error Probability	
21.3	One-Dimensional Modulation: Pulse-Amplitude Modulation (PAM)	278
21.4	Two-Dimensional Modulations	279
	Phase-Shift Keying (PSK) • Quadrature Amplitude Modulation (QAM)	
21.5	Multidimensional Modulations: Frequency-Shift Keying (FSK)	281
21.6	Multidimensional Modulations: Lattices	282
	Examples of Lattices • The Coding Gain of a Lattice • Carving a Signal Constellation out of a Lattice	
21.7	Modulations with Memory	284

Ezio Biglieri
Politecnico di Torino

21.1 Introduction

The goal of a digital communication system is to deliver information represented by a sequence of binary symbols, through a physical channel, to a user. The mapping of these symbols into signals, selected to match the features of the physical channel, is called **digital modulation**.

The digital modulator is the device used to achieve this mapping. The simplest type of modulator has no memory, that is, the mapping of blocks of binary digits into signals is performed independently of the blocks transmitted before or after. If the modulator maps the binary digits 0 and 1 into a set of two different waveforms, then the modulation is called *binary*. Alternatively, the modulator may map symbols formed by h binary digits at a time onto $M = 2^h$ different waveforms. This modulation is called *multilevel*.

The general expression of a signal modulated by a modulator with memory is

$$v(t) = \sum_{n=-\infty}^{\infty} s(t - nT; \xi_n, \sigma_n)$$

where $\{s(t; i, j)\}$ is a set of waveforms, ξ_n is the symbol emitted by the source at time nT, and σ_n is the state of the modulator at time n. If the modulator has no memory, then there is no dependence on σ_n. A *linear* memoryless modulation scheme is one such that $s(t - nT; \xi_n; \sigma_n) = \xi_n s(t - nT)$.

Given a signal set $\{s_i(t)\}_{i=1}^{M}$ used for modulation, its compact characterization may be given in terms of a **geometric representation**. From the set of M waveforms we first construct a set of $N \leq M$ orthonormal waveforms $\{\psi_i(t)\}_{i=1}^{N}$ (the Gram–Schmidt procedure is the appropriate algorithm to do this). Next we express each signal $s_k(t)$ as a linear combination of these waveforms,

$$s_k(t) = \sum_{i=1}^{N} s_{ki} \psi_i(t)$$

The coefficients of this expansion may be interpreted as the components of M vectors that geometrically represent the original signal set, or, equivalently, as the coordinates in a Euclidean N-dimensional space of a set of points called the **signal constellation**.

21.2 The Challenge of Digital Modulation

The selection of a digital modulation scheme should be done by making the best possible use of the resources available for transmission, namely, bandwidth, power, and complexity, in order to achieve the reliability required.

Bandwidth

There is no unique definition of signal bandwidth. Actually, any signal $s(t)$ strictly limited to a time interval T would have an infinite bandwidth if the latter were defined as the support of the Fourier transform of $s(t)$. For example, consider the bandpass linearly modulated signal

$$v(t) = \Re\left[\sum_{n=-\infty}^{\infty} \xi_k s(t - nT) e^{j2\pi f_0 t}\right]$$

where $\Re$ denotes real part, f_0 is the carrier frequency, $s(t)$ is a rectangular pulse with duration T and amplitude 1, and (ξ_k) is a stationary sequence of complex uncorrelated random variables with $\mathbf{E}(\xi_n) = 0$ and $\mathbf{E}(|\xi_n|^2) = 1$. Then the power density spectrum of $v(t)$ is given by

$$\mathcal{G}(f) = \frac{1}{4}[G(-f - f_0) + G(f - f_0)]$$

where

$$G(f) = T\left[\frac{\sin \pi f T}{\pi f T}\right]^2 \tag{21.1}$$

The function $G(f)$ is shown in Fig. 21.1.

The following are possible definitions of the bandwidth:

- *Half-power bandwidth.* This is the interval between the two frequencies at which the power spectrum is 3 dB below its peak value.
- *Equivalent noise bandwidth.* This is given by

$$B_{\text{eq}} = \frac{1}{2} \frac{\int_{-\infty}^{\infty} \mathcal{G}(f) \, df}{\max_f \mathcal{G}(f)}$$

This measures the basis of a rectangle whose height is $\max_f \mathcal{G}(f)$ and whose area is one-half of the power of the modulated signal.

Digital Modulation Techniques

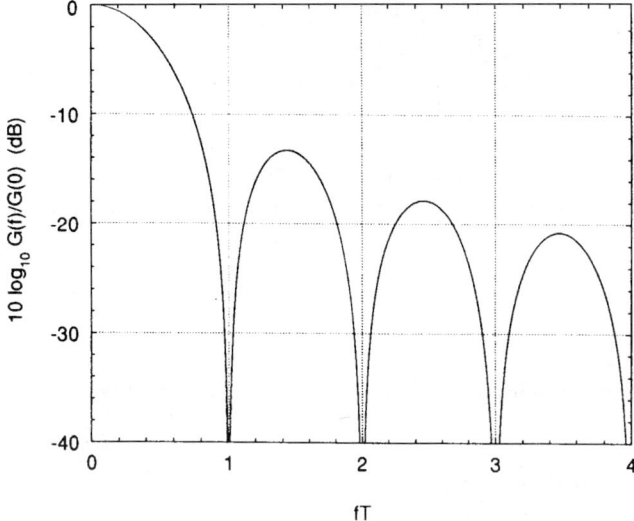

FIGURE 21.1 Power density spectrum of a linearly modulated signal with rectangular waveforms.

- *Null-to-null bandwidth.* This represents the width of the main spectral lobe.
- *Fractional power containment bandwidth.* This bandwidth definition states that the occupied bandwidth is the band that contains $(1 - \varepsilon)$ of the total signal power.
- *Bounded-power spectral density bandwidth.* This states that everywhere outside this bandwidth the power spectral density must fall at least a certain level (e.g., 35 or 50 dB) below its maximum value.

Although the actual value of the signal bandwidth depends on the definition that has been accepted for the specific application, in general, we can say that

$$B = \frac{\alpha}{T}$$

where T is the duration of one of the waveforms used by the modulator, and α reflects the definition of bandwidth and the selection of waveforms. For example, for (Eq. 21.1) the null-to-null bandwidth provides $B = 2/T$, that is, $\alpha = 2$. For 3-dB bandwidth, $\alpha = 0.88$. For equivalent-noise bandwidth, we have $\alpha = 1$.

Shannon Bandwidth

To make it possible to compare different modulation schemes in terms of their bandwidth efficiency, it is useful to consider the following definition of bandwidth. Consider a signal set, and its geometric representation based on the orthonormal set of signals $\{\psi_i(t)\}_{i=1}^{N}$ defined over a time interval with duration T. The value of N is called the *dimensionality* of the signal set. We say that a real signal $x(t)$ with Fourier transform $X(f)$ is time limited to the interval $-T/2 < t < T/2$ at level ϵ if

$$\int_{|t|>T/2} x^2(t)\,dt < \epsilon$$

and is bandlimited with bandwidth B at level ϵ if

$$\int_{|f|>B} |X(f)|^2 \, df < \epsilon$$

Then for large BT the space of signals that are time limited and bandlimited at level ϵ has dimensionality $N = 2BT$ [Slepian, 1976]. Consequently, the Shannon bandwidth [Massey, 1995] of the signal set is defined as

$$B = \frac{N}{2T}$$

and is measured in dimensions per second.

Signal-to-Noise Ratio

Assume from now on that the information source emits independent, identically distributed binary digits with rate R_s digits per second, and that the transmission channel adds to the signal a realization of a white Gaussian noise process with power spectral density $N_0/2$.

The rate, in bits per second, that can be accepted by the modulator is

$$R_s = \frac{\log_2 M}{T}$$

where M is the number of signals of duration T available at the modulator, and $1/T$ is the signaling rate. The average signal power is

$$\mathcal{P} = \frac{\mathcal{E}}{T} = \mathcal{E}_b R_s$$

where $\mathcal{E}$ is the average signal energy and $\mathcal{E}_b = \mathcal{E}/\log_2 M$ is the energy required to transmit one binary digit. As a consequence, if B denotes the bandwidth of the modulated signal, the ratio between signal power and noise power is

$$\frac{\mathcal{P}}{N_0 B} = \frac{\mathcal{E}_b}{N_0} \frac{R_s}{B}$$

This shows that the **signal-to-noise ratio** is the product of two quantities, namely, the ratio $\mathcal{E}_b/N_0$, the energy per transmitted bit divided by twice the noise spectral density, and the ratio R_s/B, representing the *bandwidth efficiency* of the modulation scheme.

In some instances the *peak energy* $\mathcal{E}_p$ is of importance. This is the energy of the signal with the maximum amplitude level.

Error Probability

The performance of a modulation scheme is measured by its **symbol error probability** $P(e)$, which is the probability that a waveform is detected incorrectly, and by its *bit error probability*, or *bit error rate* (BER) $P_b(e)$, the probability that a bit sent is received incorrectly. A simple relationship between the two quantities can be obtained by observing that, since each symbol carries $\log_2 M$ bits, one symbol error causes at least one and at most $\log_2 M$ bits to be in error,

$$\frac{P(e)}{\log_2 M} \leq P_b(e) \leq P(e)$$

When the transmission takes place over a channel affected by additive white Gaussian noise, and the modulation scheme is memoryless, the symbol error probability is upper bounded as follows:

$$P(e) \leq \frac{1}{2M} \sum_{i=1}^{M} \sum_{\substack{j=1 \\ j \neq i}}^{M} \text{erfc}\left(\frac{d_{ij}}{2\sqrt{N_0}}\right)$$

where d_{ij} is the *Euclidean distance* between signals $s_i(t)$ and $s_j(t)$,

$$d_{ij}^2 = \int_0^T [s_i(t) - s_j(t)]^2 \, dt$$

and erfc($\cdot$) denotes the Gaussian integral function

$$\text{erfc}(x) = \frac{2}{\sqrt{\pi}} \int_x^\infty e^{-z^2} \, dz.$$

Another function, denoted $Q(x)$, is often used in lieu of erfc($\cdot$). This is defined as

$$Q(x) = \frac{1}{2} \text{erfc}\left(\frac{x}{\sqrt{2}}\right)$$

A simpler upper bound on error probability is given by

$$P(e) \leq \frac{M-1}{2} \text{erfc}\left(\frac{d_{\min}}{2\sqrt{N_0}}\right)$$

where $d_{\min} = \min_{i \neq j} d_{ij}$.

A simple lower bound on symbol error probability is given by

$$P(e) \geq \frac{1}{M} \text{erfc}\left(\frac{d_{\min}}{2\sqrt{N_0}}\right)$$

By comparing the upper and the lower bound we can see that the symbol error probability depends exponentially on the term $d_{\min}$, the minimum Euclidean distance among signals of the constellation. In fact, upper and lower bounds coalesce asymptotically as the signal-to-noise ratio increases. For intermediate signal-to-noise ratios a fair comparison among constellations should take into account the *error coefficient* as well as the minimum distance. This is the average number ν of nearest neighbors [i.e., the average number of signals at distance $d_{\min}$ from a signal in the constellation; for example, this is equal to 2 for M-ary **phase-shift keying** (PSK), $M > 2$]. A good approximation to $P(e)$ is given by

$$P(e) \approx \frac{\nu}{2} \text{erfc}\left(\frac{d_{\min}}{2\sqrt{N_0}}\right)$$

Roughly, at $P(e) = 10^{-6}$ doubling ν accounts for a loss of 0.2 dB in the signal-to-noise ratio.

21.3 One-Dimensional Modulation: Pulse-Amplitude Modulation (PAM)

Pulse-amplitude modulation (PAM) is a linear modulation scheme in which a signal $s(t)$ is modulated by random variables ξ_n taking on values in the set of amplitudes $\{a_i\}_{i=1}^{M}$, where

$$a_i = (2i - 1 - M)\frac{d_{\min}}{2}, \qquad i = 1, 2, \ldots, M$$

The transmitter uses the set of waveforms $\{a_i s(t)\}_{i=1}^{M}$, where $s(t)$ is a unit-energy pulse. The geometric representation of this signal set is shown in Fig. 21.2 for $M = 4$.

The symbol-error probability is given by

$$P(e) = \left(1 - \frac{1}{M}\right) \text{erfc}\left(\sqrt{\frac{3 \log_2 M}{M^2 - 1} \frac{\mathcal{E}_b}{N_0}}\right)$$

The ratio between peak energy and average energy is

$$\frac{\mathcal{E}_p}{\mathcal{E}} = 3 \frac{M - 1}{M + 1}$$

The bandwidth efficiency of this modulation is

$$\frac{R_s}{B} = \frac{\log_2 M}{T} \cdot 2T = 2 \log_2 M$$

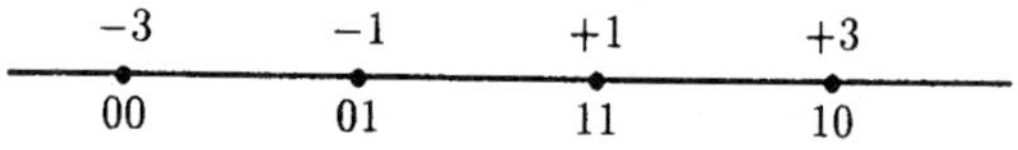

```
    -3      -1      +1      +3
    ●       ●       ●       ●
    00      01      11      10
```

FIGURE 21.2 Quaternary PAM constellation.

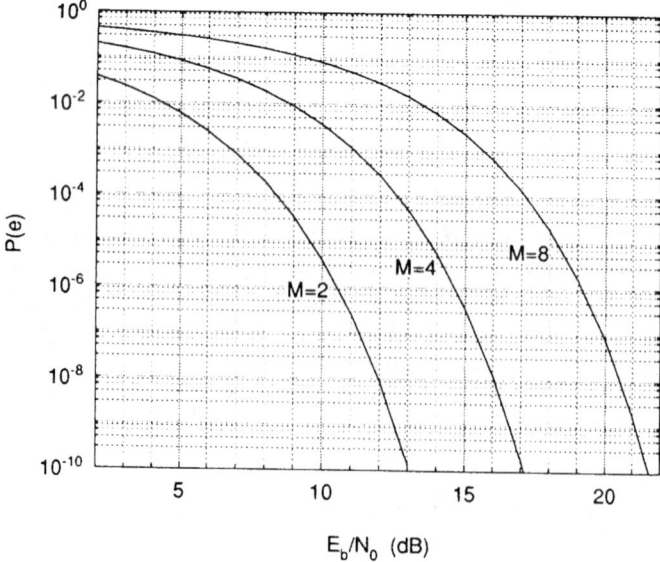

FIGURE 21.3 Symbol error probability of coherently demodulated PAM.

21.4 Two-Dimensional Modulations

Phase-Shift Keying (PSK)

This is a linear modulation scheme generating the signal

$$v(t) = \Re\left\{\sum_{n=-\infty}^{\infty} \xi_n s(t-nT)e^{j2\pi f_0 t}\right\}$$

where $\xi_n = e^{j\phi_n}$ takes values in the set

$$\left\{\frac{2\pi}{M}(i-1) + \Phi\right\}_{i=1}^{M}$$

where Φ is an arbitrary phase. The signal constellation is shown, for $M = 8$, in Fig. 21.4.

The symbol-error probability of M-ary PSK is closely approximated by

$$P(e) \approx \mathrm{erfc}\left(\sqrt{\frac{\mathcal{E}_b}{N_0}\log_2 M}\sin\frac{\pi}{M}\right)$$

for high signal-to-noise ratios: (See Fig. 21.5). The bandwidth efficiency of PSK is

$$\frac{R_s}{B} = \frac{\log_2 M}{T}\cdot T = \log_2 M$$

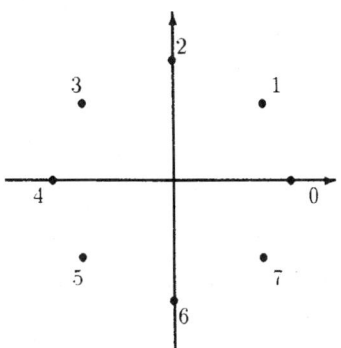

FIGURE 21.4 An 8-PSK constellation.

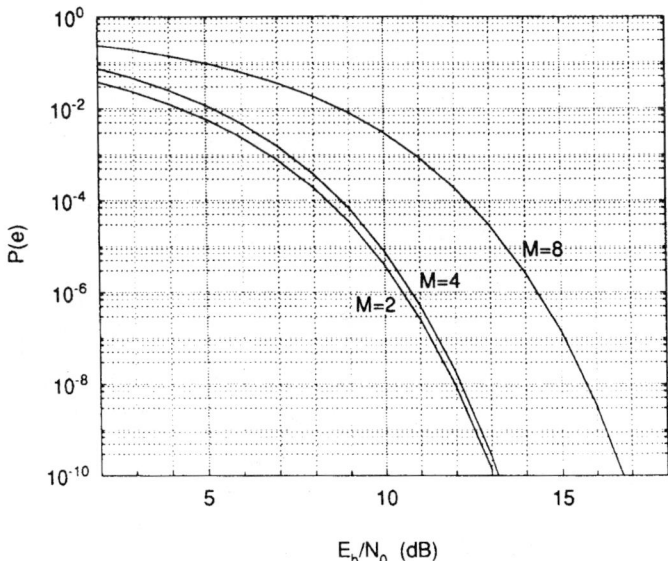

FIGURE 21.5 Symbol error probability of coherently demodulated PSK.

Quadrature Amplitude Modulation (QAM)

Quadrature amplitude modulation (QAM) is a linear modulation scheme for which the modulated signal takes the form

$$v(t) = \Re\left\{\sum_{n=-\infty}^{\infty} \xi_n s(t-nT) e^{j2\pi f_0 t}\right\}$$

where

$$\xi_n = \alpha_n + j\beta_n$$

and α_n, β_n take on equally spaced values. Figure 21.6 shows a QAM constellation with 16 points.

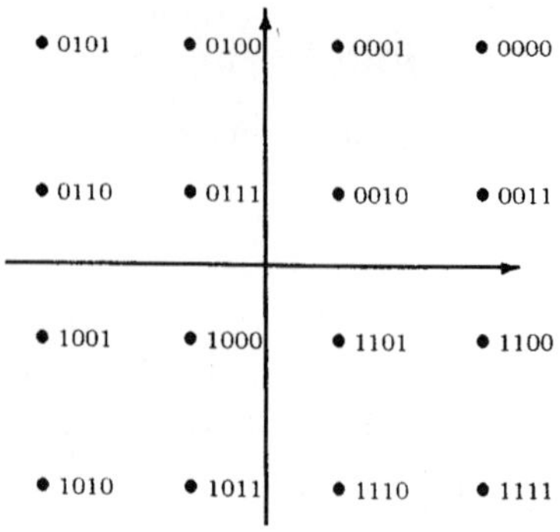

FIGURE 21.6 A 16-QAM constellation.

When $\log_2 M$ is an even integer, we have

$$P(e) = 1 - (1-p)^2$$

with

$$p = \left(1 - \frac{1}{\sqrt{M}}\right) \text{erfc}\left(\sqrt{\frac{3 \log_2 M}{2(M-1)} \frac{\mathcal{E}_b}{N_0}}\right)$$

When $\log_2 M$ is odd, the following upper bound holds:

$$P(e) < 2\, \text{erfc}\left(\sqrt{\frac{3 \log_2 M}{2(M-1)} \frac{\mathcal{E}_b}{N_0}}\right)$$

(See Fig. 21.7.)

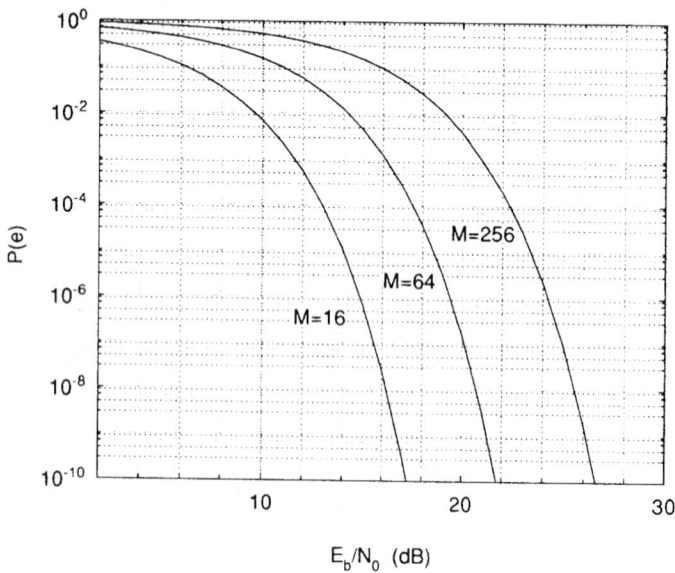

FIGURE 21.7 Symbol error probability of coherently demodulated QAM.

21.5 Multidimensional Modulations: Frequency-Shift Keying (FSK)

In the modulation technique called **frequency-shift keying (FSK)** the transmitter uses the waveforms

$$s_i(t) = A \cos 2\pi f_i t, \qquad 0 \le t < T$$

where $f_i = f_0 + (2i - 1 - M) f_d$, for $i = 1, \ldots, M$, and f_0 denotes the carrier frequency. The waveforms have constant energy $\mathcal{E} = A^2 T/2$, and the modulated signal has constant envelope A. The distance properties of these waveforms for a given energy depend on the value of their normalized frequency separation $2 f_d T$.

Generation of these waveforms may be accomplished with a set of M separate oscillators, each tuned to the frequency f_i. In this case, the phase discontinuities at frequency-transition times cause large sidelobes in the power density spectrum of the modulated signal. A narrower spectrum can be obtained by using a single oscillator whose frequency is modulated by the source bits. This results in a modulation scheme with memory called continuous-phase FSK (discussed in Sec. 21.7).

The signals $s_i(t)$ are orthogonal, that is,

$$\int_0^T \cos(2\pi f_i t) \cos(2\pi f_j t) \, dt = 0 \qquad i \ne j$$

if the frequency separation between adjacent signals is such that $2 f_d T = m/2$, where m is any nonzero integer. Thus, the minimum frequency separation for orthogonality if $2 f_d = 1/2T$.

The symbol error probability for M orthogonal signals with energy $\mathcal{E}$ is upper and lower bounded by

$$\frac{1}{2} \operatorname{erfc}\left(\sqrt{\frac{\mathcal{E}_b \log_2 M}{2 N_0}}\right) \le P(e) \le \frac{M-1}{2} \operatorname{erfc}\left(\sqrt{\frac{\mathcal{E}_b \log_2 M}{2 N_0}}\right).$$

For $M = 2$ the bounds merge into the exact value. The upper bound to $P(e)$ is plotted in Fig. 21.8.

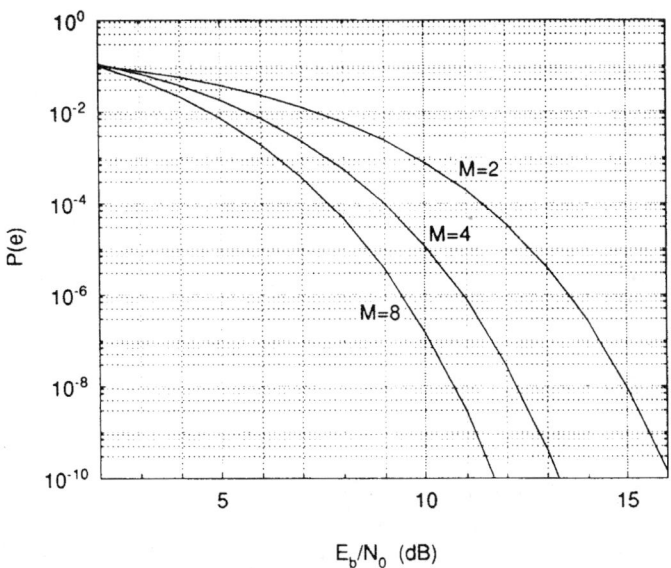

FIGURE 21.8 Symbol error probability of coherently demodulated orthogonal FSK.

It can be observed from Fig. 21.8 that the error probability for a given value of signal-to-noise ratio decreases as M increases, contrary to what is observed for PAM, PSK, and QAM. On the other hand, the bandwidth efficiency *decreases* as M increases, its value being given by

$$\frac{R_s}{B} = \frac{2 \log_2 M}{M}$$

21.6 Multidimensional Modulations: Lattices

The QAM constellations may be considered as carved from an infinite constellation of regularly spaced points in the plane, called the **lattice** Z^2. This is represented in Fig. 21.9. This concept can be generalized to other lattices, both two dimensional and multi dimensional. In this framework, the design of a signal constellation consists of choosing a lattice, then carving a finite constellation of signal points from the lattice.

In general, a lattice Λ is defined as a set closed under ordinary addition and multiplication by integers. As such, it forms an additive group as well as a vector space. Specifically, an N-dimensional lattice Λ is an infinite set of N vectors.

If $d_{\min}$ is the minimum distance between any two points in the lattice, the number v of adjacent lattice points located at distance $d_{\min}$, that is, the number of nearest neighbors of any lattice point, is called its *kissing number*.

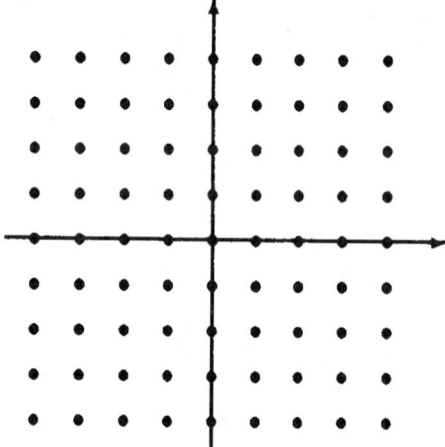

FIGURE 21.9 The lattice Z^2.

Examples of Lattices

The Lattice Z^N

The set Z^N of all N-tuples with integer coordinates is called the *cubic lattice*, or *integer lattice*. The minimum distance is $d_{\min} = 1$, and the kissing number is $v = 2N$.

The N-Dimensional Lattices A_N and A_N^*

A_N is the set of all vectors with $(N + 1)$ integer coordinates whose sum is zero. This lattice may be viewed as the intersection of Z^{N+1} and a hyperplane cutting the origin. The minimum distance is $d_{\min} = \sqrt{2}$, and the kissing number is $v = N(N + 1)$.

The N-Dimensional Lattice D_N

D_N is the set of all N-dimensional points whose integer coordinates have an even sum. It may be viewed as a punctured version of Z^N, in which the points are colored alternately black and white with a checkerboard coloring, and the white points (those with odd sums) are removed. We have $d_{\min} = \sqrt{2}$, and $v = 2N(N - 1)$. D_4 represents the densest lattice packing in R^4. This means that if unit-radius, four-dimensional spheres with centers in the lattice points are used to pack R^4, then D_4 is the lattice with the largest number of spheres per unit volume.

Digital Modulation Techniques

The Gosset Lattice E_8

In the *even coordinate system* E_8 consists of the points

$$\left\{ (x_1, \ldots, x_8) : \quad \forall x_i \in Z \text{ or } \forall x_i \in Z + \tfrac{1}{2}, \quad \sum_{i=1}^{8} x_i \equiv 0 \mod 2 \right\}$$

In words, E_8 consists of the eight-vectors whose components are all integers, or all halves of odd integers, and whose sum is even.

The *odd* coordinate system is obtained by changing the sign of any coordinate: the points are

$$\left\{ (x_1, \ldots, x_8) : \quad \forall x_i \in Z \text{ or } \forall x_i \in Z + \tfrac{1}{2}, \quad \sum_{i=1}^{8} x_i \equiv 2x_8 \mod 2 \right\}$$

This lattice has $d_{min} = \sqrt{2}$ and $\nu = 240$.

Other Lattices

The description and the properties of other important lattices, such as the 16-dimensional Barnes–Wall lattice Λ_{16} and the 24-dimensional Leech lattice Λ_{24}, can be found in Chapter 4 of Conway and Sloane [1988].

The Coding Gain of a Lattice

The *coding gain* $\gamma_c(\Lambda)$ of the lattice Λ is defined as

$$\gamma_c(\Lambda) = \frac{d_{min}^2(\Lambda)}{V(\Lambda)^{2/N}} \tag{21.2}$$

where $V(\Lambda)$ is the *fundamental lattice volume*, that is, the reciprocal of the number of lattice points per unit volume [for example, $V(Z^N) = 1$]. Table 21.1 lists the coding gains of some of the most popular lattices.

Carving a Signal Constellation out of a Lattice

We denote with $C(\Lambda, \mathcal{R})$ a constellation obtained from Λ (or from its translate $\Lambda + a$) by retaining only the points that fall in the region $\mathcal{R}$. The resulting constellation has a number of signals approximately equal to the ratio between the volume of $\mathcal{R}$ and the volume of Λ,

$$M \approx \frac{V(\mathcal{R})}{V(\Lambda)}$$

provided that $V(\mathcal{R}) \gg V(\Lambda)$, that is, that M is large enough.

The *figure of merit* (CFM) of $C = C(\Lambda, \mathcal{R})$ is the ratio between d_{min}^2 and the average energy of the constellation *per dimension pair*,

$$\text{CFM}(C) = \frac{d_{min}^2}{\mathcal{E}/(N/2)}$$

To express the figure of merit of the constellation $C(\Lambda, \mathcal{R})$ in terms of parameters related to the lattice Λ and to the region $\mathcal{R}$, we introduce the definition of the

TABLE 21.1 Coding Gains of Lattices

Name	Λ	N	$\gamma_c(\Lambda)$, (dB)
Integer lattice	Z	1	0.00
Hexagonal lattice	A_2	2	0.62
Schläfli	D_4	4	1.51
Gosset	E_8	8	3.01
Barnes-Wall	Λ_{16}	16	4.52
Leech	Λ_{24}	24	6.02

shape gain $\gamma_s(\mathcal{R})$ of the region $\mathcal{R}$ [Forney and Wei, 1990]. This is the reduction in average energy (per dimension pair) required by a constellation bounded by $\mathcal{R}$ compared to that which would be required by a constellation bounded by an N-dimensional cube of the same volume $V(\mathcal{R})$. In formulas, the shape gain is the ratio between the *normalized second moment* of any N-dimensional cube (which is equal to $1/12$) and the normalized second moment of $\mathcal{R}$,

$$\gamma_s(\mathcal{R}) = \frac{1/12}{G(\mathcal{R})} \tag{21.3}$$

where

$$G(\mathcal{R}) = \frac{\int_\mathcal{R} \|\mathbf{r}\|^2 \, d\mathbf{r}}{N V(\mathcal{R})^{1+2/N}} \tag{21.4}$$

The following result holds [Forney and Wei, 1990]: The figure of merit of the constellation $C(\Lambda, \mathcal{R})$ is given by

$$\text{CFM}(C) \approx \gamma_0 \cdot \gamma_c(\Lambda) \cdot \gamma_s(\mathcal{R}) \tag{21.5}$$

where γ_0 is the figure of merit of the one-dimensional PAM constellation with the same bit rate (chosen as the baseline), $\gamma_c(\Lambda)$ is the coding gain of the lattice Λ [see Eq. (21.2)], and $\gamma_s(\mathcal{R})$ is the shaping gain of the region $\mathcal{R}$. The approximation holds for large constellations.

This result shows that at least for large constellations the gain from shaping by the region $\mathcal{R}$ is almost completely decoupled from the coding gain due to Λ. Thus, for a good design it makes sense to optimize separately $\gamma_c(\Lambda)$ (i.e., the choice of the lattice) and $\gamma_s(\mathcal{R})$ (i.e., the choice of the region).

Spherical Constellations

The maximum shape gain achieved by an N-dimensional region $\mathcal{R}$ is that of a sphere, for which

$$\gamma_s = \frac{\pi(n+1)}{6(n!)^{1/n}}$$

where $n = N/2$ and N is even. As $N \to \infty$, the γ_s approaches $\pi e/6$, or 1.53 dB. The last figure is the maximum achievable shaping gain. A problem with spherical constellations is that the complexity of the encoding procedure (mapping input symbols to signals) may be too high. The main goal of N-dimensional lattice-constellation design is to obtain a shape gain as close to that of the N sphere as possible, while maintaining a reasonable implementation complexity and other desirable constellation characteristics.

21.7 Modulations with Memory

The modulation schemes considered so far are *memoryless*, in the sense that the waveform transmitted in one symbol interval depends only on the symbol emitted by the source in that interval. In contrast, there are modulation schemes with memory.

One example of a modulation scheme with memory is given by *continuous-phase frequency-shift keying* (CPFSK), which in turn is a special case of **continuous-phase modulation** (CPM). To describe CPFSK, consider standard FSK modulation, whose signals are generated by separate oscillators. As a consequence of this generation technique, at the end of each symbol interval the phase of the carrier changes abruptly whenever the frequency changes, because the oscillators are not phase synchronized. Since spectral occupancy increases by decreasing the smoothness of a signal,

Digital Modulation Techniques

these sudden phase changes cause spectrum broadening. To reduce the spectral occupancy of a frequency-modulated signal one option is CPFSK, in which the frequency is changed continuously.

To describe CPFSK, consider the PAM signal

$$v(t) = \sum_{n=-\infty}^{\infty} \xi_n \, q(t - nT)$$

where (ξ_n) is the sequence of source symbols, taking on values $\pm 1, \pm 3, \ldots, \pm(M-1)$, and $q(t)$ is a rectangular pulse of duration T and area $1/2$. If the signal $v(t)$ is used to modulate the frequency of a sinusoidal carrier, we obtain the signal

$$u(t) = A \cos\left[2\pi f_0 t + 2\pi h \int_{-\infty}^{t} v(\tau) \, d\tau + \varphi_0\right]$$

where f_0 is the unmodulated-carrier frequency, φ_0 is the initial phase, and h is a constant called the *modulation index*. The phase shift induced on the carrier, that is,

$$\theta(t) = 2\pi h \int_{-\infty}^{t} v(\tau) \, d\tau$$

turns out to be a continuous function of time t, so that a continuous-phase signal is generated. The trajectories followed by the phase, as reduced mod 2π, form the *phase trellis* of the modulated signal. Figure 21.10 shows a segment (for 7 symbol intervals) of the phase trellis of binary CPFSK with $h = 1/2$ [this is called *minimum-shift keying* or (MSK)].

With MSK, the carrier-phase shift induced by the modulation in the time interval $nT \leq t \leq (n+1)T$ is given by

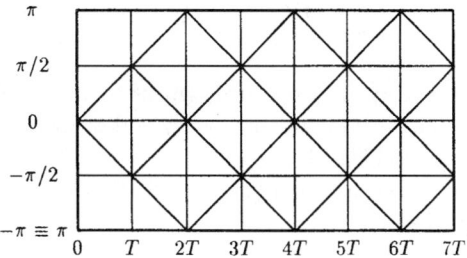

FIGURE 21.10 Phase trellis of binary CPFSK with $h = 1/2$ (MSK).

$$\theta(t) = \theta_n + \frac{\pi}{2}\left(\frac{t - nT}{T}\right)\xi_n \qquad nT \leq t \leq (n+1)T$$

where

$$\theta_n = \frac{\pi}{2} \sum_{k=-\infty}^{n-1} \xi_k$$

The corresponding transmitted signal is

$$u(t) = A \cos\left[2\pi\left(f_0 + \frac{1}{4T}a_n\right)t - \frac{n\pi}{2}a_n + \theta_n\right]$$

which shows that MSK is an FSK using the two frequencies

$$f_1 = f_0 - \frac{1}{4T} \qquad f_2 = f_0 + \frac{1}{4T}$$

The frequency separation $f_2 - f_1 = 1/2T$ is the minimum separation for orthogonality of two sinusoids, which explains the name given to this modulation scheme.

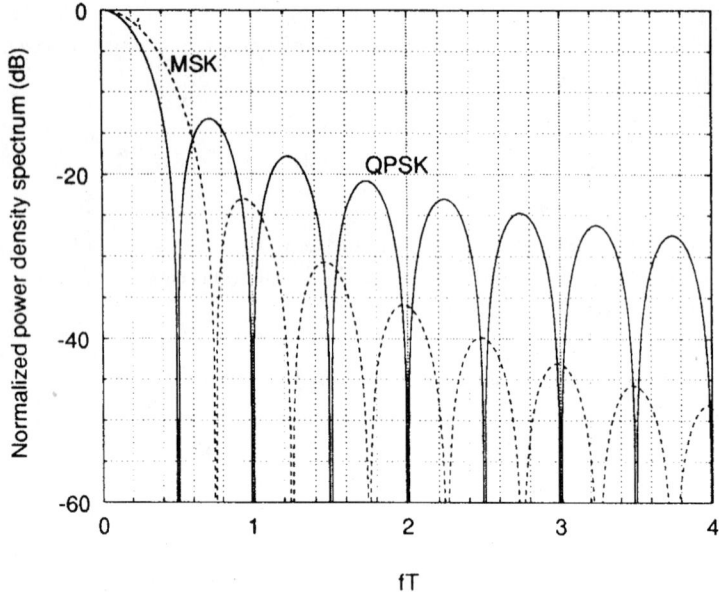

FIGURE 21.11 Power density spectrum of MSK and of quaternary PSK.

The spectrum of CPFSK depends of the value of h. For $h < 1$, as the modulation index decreases so does the spectral occupancy of the modulated signal. For MSK the power density spectrum is shown in Fig. 21.11.

CPM is a generalization of CPFSK. Here the carrier-phase shift in the interval $nT \leq t \leq (n+1)T$ is given by

$$\theta(t) = 2\pi \sum_{k=-\infty}^{n} a_k h_k q(t - nT)$$

where the waveform $q(t)$ is the integral of a frequency pulse $g(t)$ of arbitrary shape

$$q(t) = \int_{-\infty}^{t} g(\tau)\, d\tau$$

subject to the constraints $g(t) = 0$ for $t < 0$ and

$$\int_{0}^{\infty} g(\tau)\, d\tau = \frac{1}{2}$$

When the modulation index h_k varies (usually with a periodic law) from one symbol to another, the modulation is called *multi-h*. If $g(t) \neq 0$ for $t > T$, then the modulated signal is called *partial-response* CPM. Otherwise, it is called *full-response* CPM.

An example of partial-response CPM is given by Gaussian MSK (GMSK), whose frequency pulse is obtained by passing a rectangular waveform into a low-pass filter whose impulse response $h(t)$ has a Gaussian shape,

$$h(t) = \sqrt{\frac{2\pi}{\ln 2}} B \exp\left\{ -\frac{2\pi^2 B^2}{\ln 2} t^2 \right\}$$

where B denotes the filter bandwidth. By decreasing B the shape of the frequency pulse becomes smoother and the spectrum occupancy of the modulated signal is reduced. This modulation scheme is employed in the Global System for Mobile Communications (GSM) standard for cellular mobile radio with the choice $BT = 0.3$.

Defining Terms

Bandwidth: The frequency interval in which the density spectrum of a signal is significantly different from zero.

Continuous-phase modulation: A digital modulation scheme derived from phase-shift keying in which the carrier has no phase jumps, that is, the phase transitions occur continuously.

Digital modulation: The mapping of information-source symbols into signals, performed to carry information through the transmission channel.

Error probability: Probability that a symbol emitted by the information source will be received incorrectly by the end user.

Frequency-shift keying: A digital modulation scheme in which the source information is carried by the frequency of a sinusoidal waveform.

Geometric signal representation: Representation of a finite set of signals as a set of vector.

Lattice: An infinite signal constellation of points regularly located in space.

Phase-shift keying: A digital modulation scheme in which the source information is carried by the phase of a sinusoidal waveform, called the carrier.

Pulse-amplitude modulation: A digital modulation scheme in which the source information is carried by the amplitude of a waveform.

Quadrature amplitude modulation: A digital modulation scheme in which the source information is carried by the amplitude and by the phase of a sinusoidal waveform.

Signal constellation: A set of signals geometrically represented in the form of a set of vectors.

Signal-to-noise ratio: The ratio of the signal power and noise power. It is an index of channel quality.

References

Anderson, J.B., Aulin, T., and Sundberg, C.-E.W. 1986. *Digital Phase Modulation*, Plenum, New York.

Benedetto, S., Biglieri, E., and Castellani, V. 1987. *Digital Transmission Theory*, Prentice–Hall, Englewood Cliffs, NJ.

Conway, J.H. and Sloane, N.J.A. 1988. *Sphere Packings, Lattices and Groups*, Springer–Verlag, New York.

Forney, G.D. Jr. and Wei, L.-F. 1990. Multidimensional constellations—Part I: Introduction, figures of merit, and generalized cross constellations. *IEEE J. Selected Areas Comm.*, 7(6):877–892.

Massey, J.L. 1995. Towards an information theory of spread-spectrum systems. In: *Code Division Multiple Access Communications*, eds. S.G. Glisic and P.A. Leppänen, Kluwer Academic, Boston.

Proakis, J.G. 1995. *Digital Communications*, 3rd ed., McGraw–Hill, New York.

Simon, M.K., Hinedi, S.M., and Lindsey, W.C. 1995. *Digital Communication Techniques. Signal Design and Detection*, Prentice–Hall, Englewood Cliffs, NJ.

Slepian, D. 1976. On bandwidth. *IEEE Proc.*, 64(2):292–300.

Further Information

The monthly journal *IEEE Transactions on Communications* reports advances in digital modulation techniques.

The books by Benedetto, Biglieri, and Castellani 1987, Proakis 1995, and Simon, Hinedi, and Lindsey 1995 are good introductions to the theory of digital modulation.

II

Wireless

22 Plain Old Telephone Service (POTS) *A. Michael Noll* 291
Introduction • The Network • Station Apparatus • Transmission • Switching • Signalling • Functionality • The Future

23 FDM Hierarchy *Pierre Catala* 301
Introduction • Background Information • Frequency-Division Multiplexing (FDM) • The Hierarchy • Pilots • Direct to Line (DTL) • Summary

24 Analog Telephone Channels and the Subscriber Loop *Whitham D. Reeve* 308
Telephone Band • Noise • Crosstalk • Circuit Noise • Impulse Noise • Attenuation Distortion • Envelope Delay Distortion • Line Conditioning

25 Baseband Signalling and Pulse Shaping *Michael L. Honig and Melbourne Barton* 318
Communications System Model • Intersymbol Interference and the Nyquist Criterion • Nyquist Criterion with Matched Filtering • Eye Diagrams • Partial-Response Signalling • Additional Considerations • Examples

26 Channel Equalization *John G. Proakis* 339
Characterization of Channel Distortion • Characterization of Intersymbol Interference • Linear Equalizers • Decision-Feedback Equalizer • Maximum-Likelihood Sequence Detection • Conclusions

27 Pulse-Code Modulation Codec Filters *Michael D. Floyd and Garth D. Hillman* 364
Introduction and General Description of a Pulse-Code Modulation (PCM) Codec Filter • Where PCM Codec Filters Are Used in the Telephone Network • Design of Voice PCM Codec Filters: Analog Transmission Performance and Voice Quality for Intelligibility • Linear PCM Codec Filter for High-Speed Modem Applications • Concluding Remarks

28 Digital Hierarchy *B.P. Lathi and Maynard Wright* 377
Introduction • North American Asynchronous Digital Hierarchy

29 Line Coding *Joseph L. LoCicero and Bhasker P. Patel* 386
Introduction • Common Line Coding Formats • Alternate Line Codes • Multilevel Signalling, Partial Response Signalling, and Duobinary Coding • Bandwidth Comparison • Concluding Remarks

30 Telecommunications Network Synchronization *M.J. Narasimha* 404
Introduction • Synchronization Distribution Networks • Effect of Synchronization Impairments • Characterization of Synchronization Impairments • Synchronization Standards • Summary and Conclusions

31 Echo Cancellation *Giovanni Cherubini* 414
Introduction • Baseband Transmission • Passband Transmission • Summary and Conclusions

32 Switching Fabrics *Amos E. Joel, Jr.* 425
Introduction • Virtual Circuit Switching • Packet Switching • Crosspoints • Multiplexing Versus Divisions • Switching System Functions • Topology of Switch Fabrics • Concentration,

Expansion and Distribution • Multistage Fabrics • Fabric Control • Topology • Fabrics for Asynchronous Transfer Mode (ATM)

33 **Customer Premises Equipment** *John L. Fike and M. Lance Parr* 433
Introduction • Customer Interface • Standards • Restrictions • A Functional Definition of CPE • Private Branch Exchanges

34 **Asymmetric Digital Subscriber Lines** *John M. Cioffi* 450
Introduction • ADSL System Description • Discrete Multitone: The ADSL Standard Line Code • ADSL Data Rates, Ranges, and Performance • The Future

35 **Overview of Common Channel Signal** *A.R. Modarressi and R.A. Skoog* 480
Introduction • Historical Background • Signalling Transport in Signalling System 7 Networks • Signalling User Parts in SS7 Networks

36 **Digital Cross-Connect System** *Paula Bordogna and Curtis A. Siller, Jr.* 496
Introduction • Functional Description • Digital Cross-Connect Application: Public Networks • Digital Cross-Connection Applications: Private and Hybrid Networks • Operations Support Systems • Digital Cross Connects in Asynchronous Transfer Mode Networks • Conclusion

37 **Building Future Networks by Using Photonics in Switching** *Stephen J. Hinterlong and David Vlack* 513
Introduction • Bit Sensing Switching Versus Bit Rate Transparent • Bit-Sensing Photonic Devices in the Network • Multistage, Call-by-Call Switching Fabric Demonstration • Bit Transparent Photonic Devices in the Network • Lithium Niobate 16 × 16 Switching System Demonstration

38 **Asynchronous Transfer Mode (ATM) Networking Standards: A Review**
Raif O. Onvural 529
Introduction • ATM Architecture • Asynchronous Transfer Mode Interfaces • Conclusions

39 **Synchronous Optical Network (SONET)** *Chris B. Autry and Henry L. Owen* 542
Introduction • SONET Frame • Optical Issues • SONET Equipment • SONET Standards

40 **Synchronous Digital Hierarchy (SDH)** *Chris B. Autry and Henry L. Owen* 554
Introduction • Frame Structure • Mapping of Asynchronous Transfer Mode (ATM) Cells • Standards

22
Plain Old Telephone Service (POTS)

A. Michael Noll
University of Southern California, Los Angeles

22.1	Introduction 291
	Bell's Vision
22.2	The Network 292
22.3	Station Apparatus 293
22.4	Transmission 295
22.5	Switching 296
22.6	Signalling 297
22.7	Functionality 299
22.8	The Future 299

22.1 Introduction

The acronym POTS stands for plain old telephone service. Over the years POTS acquired an undeserved reputation for connoting *old fashioned* and even *obsolete*. The telephone and the switched public network enables us to reach anyone anywhere on this planet at anytime and to speak to them in our natural voices. The continued growth in minutes of telephone usage, in the number of access lines, and in the revenue of telephone companies attests to the central importance of telephone service in today's information-age, global economy.

The *old* in the acronym implies familiarity and ease of use, a major reason for the continued popularity of telephone service. The *telephone* and the nuances expressed by natural human speech is what it is all about. *Service* used to mean responsiveness to the public, and it is sad that this dimension of the acronym has become so threatened by emphases on short-term profits, particularly in the new world of competition that characterizes telephone service on all levels. The term *plain* is indeed obsolete, and the provision of today's telephone service utilizes some of the most sophisticated and advanced transmission and switching technology. In fact, telephone service today with all its intelligent and functional new features, advanced technology, and improved quality is truly fantastic. Allowing for a little misspelling, the acronym POTS can still be used, but with the "P" standing for phantastic! So, POTS it was, and POTS is still where the real action is.

Part of the excitement about the telephone network is that it can be used with a wide variety of devices to create exciting and useful services, such as facsimile for graphical communication and modems for data communication. And cellular telephony, pagers, and phones in airplanes extend telephony wherever we travel. Increased functionality in the network brings us voice mail, call forwarding, call waiting, caller ID, and a host of such services. The future of an ever evolving POTS, indeed, looks bright and exciting.

Bell's Vision

The telegraph is older than the telephone, but special knowledge of Morse code was necessary to use the telegraph thereby relegating it to use by specialists. The telephone uses normal human speech and therefore can be used by anyone and can convey all the nuances of the inflections of human speech. Dr. Alexander Graham Bell was very prophetic in his vision of a world wired for two-way telecommunication using natural human speech. Bell and his assistant Thomas A. Watson demonstrated the first working model of a telephone on March 10, 1876, but Bell had applied for a patent a month earlier on February 14, 1876. Elisha Grey had submitted a disclosure of invention for a telephone on that same day in February, but ultimately the U.S. Supreme Court upheld Bell's invention, although in a split decision.

22.2 The Network

The telephone has come to signify a public switched network capable of reaching any other telephone on Earth. This switched network interconnects not only telephones but also facsimile machines, cellular telephones, and personal computers—anything that is connected to the network. As shown in Fig. 22.1 the telephones and other station apparatus in homes are all connected by pairs of copper wire to a single point, called the protector block, which offers simple protection to the network from overvoltages. A twisted pair of wires then connects the protector block all the way back to the central office. Many twisted pairs are all carried together in a cable that can be buried underground, placed in conduit, or strung between telephone poles. The twisted pair of wires connecting station apparatus to the central office is called the local loop.

The very first stage of switching occurs at the central office. From there, telephone calls may be connected to other central offices, over interoffice trunks. Calls may also be carried over much greater distances by connection to the long-distance networks operated by a number of **interexchange carriers (IXC)**, such as AT&T, MCI, and Sprint. The point where the connection is made from the local service provider to the interexchange carrier is known as the **point of presence (POP)**.

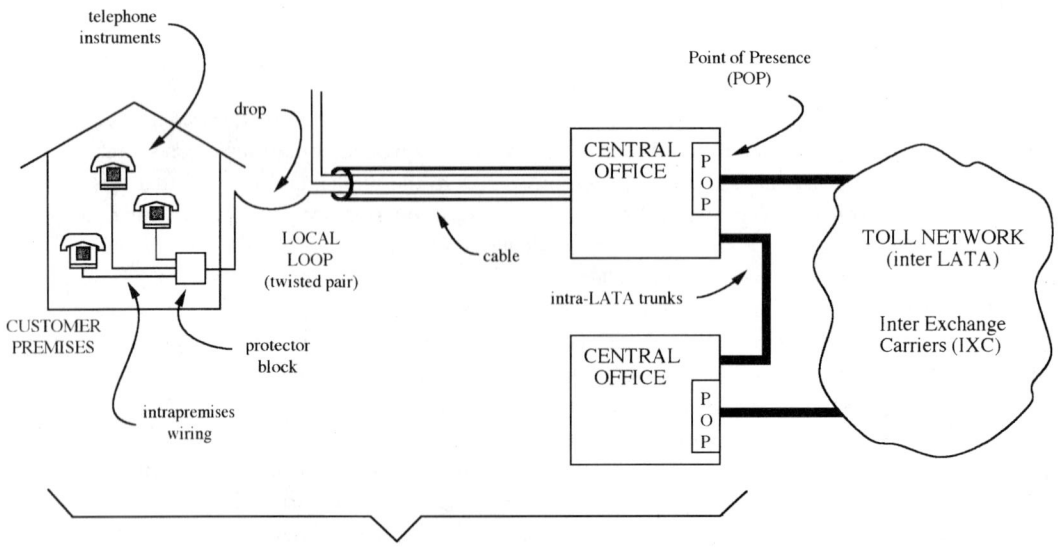

FIGURE 22.1   Network.

Plain Old Telephone Service (POTS)

The local portion of the telephone network is today known as a **local access and transport area** (**LATA**). The local Bell telephone companies (commonly called the Baby Bells) were restricted by the Bell breakup of 1984 solely to the provision of intra-LATA service and were forbidden from providing inter-LATA service. Although the technology of telephony has progressed impressively over the last 100 years, policy and regulation have also had great impact on the telephone industry.

In the past, telephone service in the U.S. was mostly under the control of AT&T and its Bell System and a number of smaller independent telephone companies. Today, a number of competing companies own and operate their own long-distance networks, and some futurists believe that local competition will occur soon. The various long-distance networks and local network all interconnect and offer what has been called *a network of networks*.

22.3 Station Apparatus

Telephones of the past were black with a rotary dial with limited functionality. Today's telephones come is many colors and sizes and offer push-button dialing along with a variety of intelligent features such as repertory dialing and display of the dialed number. However, the basic functions of the telephone instrument have not changed, and are shown in Fig. 22.2. The telephone needs to signal the central office when service is desired. This is accomplished by lifting the handset, which then closes contacts in the telephone—the switch hook—so that the telephone draws DC over the local loop from the central office. The switching machine at the central office senses this flow of DC and thus knows that the customer desires service. The common battery at the central office has an electromotive force (EMF) of 48 V, and the telephone draws at least about 20 mA of current over the loop. The maximum loop resistance can not exceed about 1300 Ω.

The user needs to specify the telephone number of the called party. This is accomplished by a process called dialing. Older telephones accomplished dialing with a rotary dial that interrupted the flow of DC with short pulses at a rate of about 10 dial pulses per second. Most telephones today

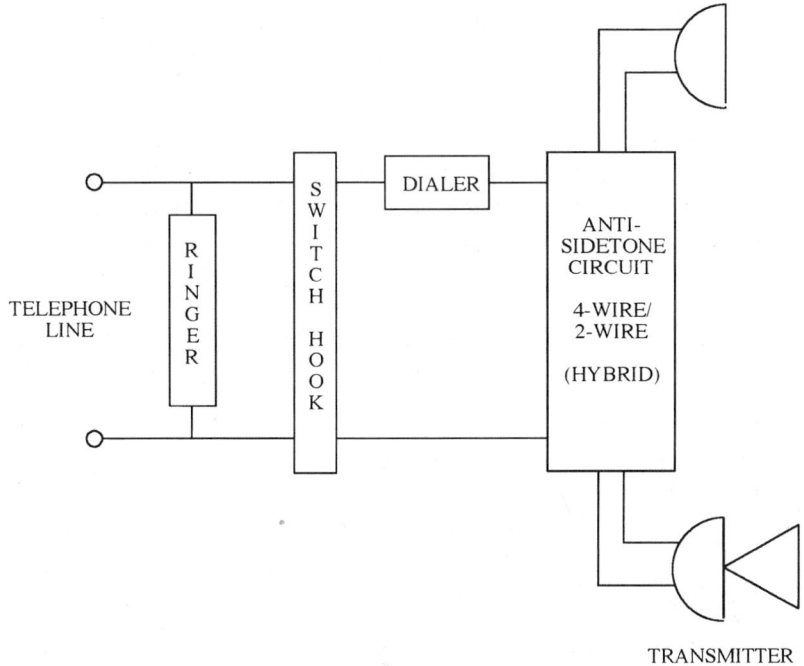

FIGURE 22.2 Telephone.

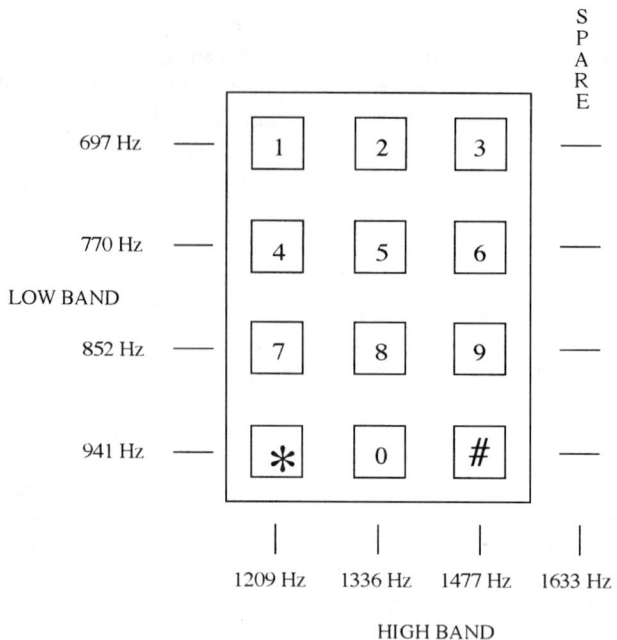

FIGURE 22.3 Touch Tone.

use touch-tone dialing and push buttons, as shown in Fig. 22.3. When a specific digit is pushed, a unique combination of two sinusoidal tones is transmitted over the line to the central office. For example, an 8 is indicated by the combination of sine waves at 852 Hz and 1336 Hz. Filters are used at the switching machine at the central office to detect the frequencies of the tones and thus decode the dialed digits. Touchtone is also known as **dual-tone multifrequency** (**DTMF**) dialing.

Bell's first telephone microphone, or transmitter, used the principle of variable resistance to create a large varying current. It consisted of a wire that moved in and out of a small cup of acid in response to the acoustic speech signal. Clearly, such a transmitter was not very practical, and it was soon replaced by the use of loosely packed carbon, invented in 1878 by the Britisher Henry Hummings. In 1886, Thomas Alva Edison improved on the carbon transmitted by using roasted granules of anthracite coal. Today's telephone transmitters use high-quality, small, variable-capacitance, electret microphones. The telephone receiver is a small loudspeaker using a permanent magnet, coil of wire, and metal diaphragm. It was invented in 1876 by Thomas Watson, Bell's assistant, and the basic principles have changed little since then.

The pair of wires going to the telephone transmitter and receiver constitutes a four-wire circuit. The transmitter sends a speech signal down the telephone line, and the receiver receives the signal from the central office. However, a two-wire local loop connects the telephone instrument to the central office, and hence two-wire to four-wire conversion is needed within the telephone instrument. A center-tapped transformer, called a **hybrid**, accomplishes this conversion. The leakage current in the secondary receiver circuit depends on how well a balance network exactly matches the impedance of the telephone line. Since this balance network can never match the line perfectly, a small amount of the received transmitted signal leaks into the receiver circuit, and the user hears one's own speech, an effect known as sidetone. Actually, a small amount of sidetone is desirable because it makes the telephone seem live and natural, and thus the balance network is designed to allow an optimum amount of sidetone. Too much sidetone results in the user pulling the handset away from the head, which reduces the transmitted speech signal—an undesirable effect. The use of an induction coil to balance the electrical sidetone was patented in 1918 by G.A. Campbell, an AT&T research engineer. The induction coil has been replaced in modern telephones by a speech network that electronically cancels the sidetone leakage and performs the two-wire to four-wire conversion.

The telephone ringer is connected in parallel across the telephone line before the switch hook's contacts. Thomas Watson applied for the first ringer patent in 1878, and today's electromechanical ringers have changed little since then. A hammer attached to an armature with a magnetic field strengthened by a permanent magnet moves in response to the ringer current loudly striking two bells. The high-impedance ringer was invented in 1890 by John J. Carty, a Bell engineer who had invented the two-wire local loop in 1881. A capacitor is placed in series with the ringer to prevent DC from flowing through it. The ringer signal consists of a short 2-s burst of a 75-V (rms), 20-Hz sine wave followed by 4 s of silence. Piezoelectric transducers and small loudspeakers are replacing electromechanical ringers in today's telephones.

Telephone instruments have progressed greatly in their functionality from the basic-black, rotary-dial phones of the past. Today's phones frequently include repertory dialers, speakerphones, and liquid crystal displays (LCDs). Tomorrow's phones will most likely build up this functionality and extend it to control central office features, to perform e-mail, and to integrate voice and data. Although some people still believe that the telephone of the future will also include a two-way video capability, the videophone, most consumers do not want to be seen while speaking on the phone, and thus the videophone will most probably remain an element of science fiction.

The public switched network can be used to transmit and switch any signal that remains within its baseband, namely, about 4 kHz. Thus, devices other than just a telephone can be used on the telephone network. The recent success of facsimile is one example; modems operating at speeds of 28.8 kb/s are another.

22.4 Transmission

A wide variety of transmission media have been and are used in providing telephone service. At the local level, twisted pairs of copper wire today connect most customers to the central office, although open copper wire was used in the distant past and in rural areas. Many pairs of wire are placed together in a cable, which is then either placed underground or strung between telephone poles. Coaxial cable carried telephone calls across the country. Microwave radio carried telephone calls terrestrially from microwave tower to tower across the country, with each tower located about 26 mi from the next. Microwave radio also carries telephone calls across oceans and continents by communication satellites located in geosynchronous orbits 22,300 mi above the surface of the earth. Today's transmission medium of choice for carrying telephone calls over long distances and between central offices is optical fiber.

Multiplexing is the means by which a number of communication signals are combined together to share a single communication medium. With analog multiplexing, signals are combined by frequency-division multiplexing; with digital multiplexing, signals are combined by time-division multiplexing.

Analog multiplexing is today obsolete in telephony. AT&T replaced all its analog multiplexing with digital multiplexing in the late 1980s; MCI followed suit in the early 1990s. Analog multiplexing was accomplished by A-type channel banks. Each baseband telephone channel was shifted in frequency to its own unique 4-kHz channel. The frequency-division multiplexing was accomplished in hierarchial stages. A hierarchy of multiplexing was created with 12 baseband channels forming a group, 5 groups forming a supergroup, 10 supergroups forming a mastergroup, and 6 mastergroups forming a jumbo group. A jumbo multiplex group contained 10,800 telephone channels and occupied a frequency range from 654 to 17,548 kHz.

With digital multiplexing, each baseband analog voice signal is converted to digital using a sampling rate of 8,000 samples/s with 8-b nonlinear quantization for an overall bit rate of 64,000 b/s. A hierarchy of digital multiplexing has evolved with 24 digital telephone signals forming a DS1 signal requiring 1.544 Mb/s, 4 DS1 signals forming a DS2 signal, 7 DS2 signals forming a DS3 signal, 6 DS3 signals forming a DS4 signal. A single digital telephone signal at 64 kb/s is called a DS0 signal. A DS4 signal multiplexes 4032 DS0 signals and requires an overall bit rate of about 274 Mb/s.

The transmission media and systems used for long-distance telephone service have progressed over the decades. The L1-carrier system, first installed in 1946, utilized three working pairs of coax in a buried cable to carry 1800 telephone circuits across the country using analog, frequency-division multiplexing. The L5E-carrier system, installed in 1978, carried 132,000 telephone circuits in 10 coax pairs. Terrestrial microwave radio has been used to carry telephone signals from towers located about every 26 mi across the country. The first system, TD-2, became available in 1950 and carried 2400 voice circuits. The use of polarized radio waves to reduce channel interference, the horn antenna to allow simultaneous operation in both the 6-GHz and 4-GHz bands, solid-state technology, and single-sideband suppressed-carrier amplitude modulation resulted in a total system capacity in 1981 of 61,800 voice circuits.

Communication satellites located in a geosynchronous orbit 22,300 mi above the Earth's equator have been used to carry telephone signals. Here, too, the technology has progressed offering ever increasing capacities. But geosynchronous communication satellites suffer a serious shortcoming. The time required for the radio signals to travel back and forth between and satellite and the Earth stations creates a round-trip delay of about 0.5 s, which is quite annoying to most people.

Today's transmission medium of choice is optical fiber utilizing digital, time-division multiplexing of the voice circuits. The basic idea of guiding light through thin glass fibers is quite old and was described by the British physicist Charles Vernon Boys in 1887. Today's optical fiber utilizes ultrapure silica. Optical fiber itself has progressed from multimode stepped index and graded index fibers to today's single-mode fiber. Solid-state lasers have also progressed in their use as light sources, and detector technology is also an area of much technological advancement. Today's fiber strands each carry a few gigabits per second. Technological advances include color mutliplexing in which a number of light signals at different frequencies are carried on the same fiber strand and erbium-doped fiber amplifiers that increase the strength of the light signal without the need to convert signals back to an electrical form for regeneration. Usually, a number of fiber strands are placed together in a single cable, but the capacity of each fiber strand is so great that many of the strands are not used; what is called *dark fiber*.

Long-distance transmission systems and local carrier systems all utilize separate paths for each direction of transmission thereby creating four-wire circuits. These four-wire circuits need to be connected to the two-wire local loop. The hybrids that accomplish this connection and conversion can not match perfectly the transmission characteristics of each and every local loop. The result is that a small portion of the signal leaks through the hybrid and is heard by the speaking party as a very annoying echo. The echo suppressor senses which party is speaking and then introduces loss in the return path to prevent the echo, but this solution also prevents simultaneous talking. Today, echo elimination is accomplished by an **echo canceler**, which uses an adaptive filter to create a synthetic echo, which is then subtracted from the return signal thereby eliminating the echo entirely but also allowing simultaneous double talking. An echo canceler is required at each end of the transmission circuit.

22.5 Switching

In the old days, one telephone was connected to another telephone at switchboards operated by humans. The human operators used cords with a plugs at each end to make the connections. Each plug had a tip and a ring about the tip to create the electric circuit to carry the signals. A sleeve was used for signalling purposes to indicate whether a circuit was in use. Each human operator could reach as many as 10,000 jacks. The automation of the switchboard came early in the history of telephony with Almon B. Strowger's invention in 1892 of an electromechanical automatic switch and the creation of his Automatic Electric Company to manufacture and sell his switching systems, mostly to non-Bell telephone companies. The Strowger switching was ignored by the Bell System until 1919 when it was finally adopted. Now electromechanical switching is totally obsolete in

Plain Old Telephone Service (POTS)

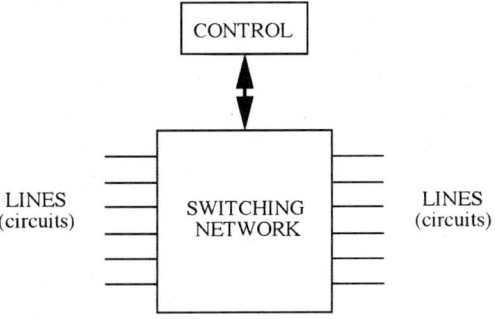

FIGURE 22.4 Switching System.

the U.S., and today's telephone system utilizes electronic switching systems. However, the Strowger system, known as step-by-step in the Bell System, was a thing of great mechanical ingenuity that gave an intuitive grasp of switching with turning and stepping switch contacts that is not possible with today's computerized electronic systems. But electromechanical switching was subject to much wear and tear and required considerable space and costly maintenance. Furthermore, electromechanical switching was inflexible and could not be reprogrammed.

In general, a switching system consists of two major functional parts, as shown in Fig. 22.4: (1) the switching network itself where one telephone call is connected to another and (2) the means of control that determines the specific connections. Calls can be connected by physically connecting wires to create an electrical path, a technique called **space switching**. With space switching, individual telephone circuits are connected physically to each other by some form of electromechanical or electronic switch. Calls can also be connected by reordering the time sequence of digitized samples, a technique **called time switching**. Modern digital switching systems frequently utilize both techniques in the switching network.

In the past, the switching network utilized electromechanical technology to accomplish space switching. This technology progressed over time from the automated Strowger switch to the Bell System's crossbar switch. The first crossbar switching system was installed in 1938, and crossbar switching systems were still in use in the U.S. in the early 1990s.

The switching network in today's switching systems is completely digital. Telephone signals either arrive in digital or are converted to digital. The digital signals are then switched, usually using a combination of electronic space switching along with time switching of the sequence of digitized samples. The space switches are shared by a number of digital calls connecting each of them for short durations while a small number of bits in each sample are transferred.

Yesterday's switching systems were controlled by hard-wired electromechanical relays. Today's switching systems are controlled by programmable digital computers thereby offering great flexibility. The use of a digital computer to control the operation of a switching network is called electronic switching or stored-program control. The intelligence of stored-program control, coupled with the capabilities of modern signalling systems, enables a wide variety of functional services tailored to the needs of individual users.

22.6 Signalling

A variety of signals are sent over the telephone network to control its operation, an aspect of POTS known as signalling. The familiar dial tone, busy signal, and ring-back tone are signals presented to the calling party. Ringing is accomplished by a 20-Hz signal that is on for 2 s and off for 4 s. In addition to these more audible signals that tell us when to dial, whether lines are busy, and when to answer the telephone, other signals are sent over the telephone network itself to control its operation.

The telephone network needed to know whether a trunk is idle or not, and the presence or absence of DC indicated whether a local trunk was in use or idle. Long-distance trunks used in-band and out-of-band tones to indicate whether a circuit was idle or not. A single-frequency tone of 2600 Hz, which is within the voice band, was placed on an idle circuit to indicate its availability. The telephone number was sent as a sequence of two tones at a rate of 10 combinations per second, a technique known as multifrequency key pulsing (MFKP). Signalling today is accomplished mostly by common channel interoffice signalling (CCIS).

With **common-channel signalling**, a separate dedicated data channel is used solely to carry signalling information in the form of short packets of data. Common-channel signalling is known as signalling system 7 (SS7) in the U.S. It offers advanced 800-service such as time of day routing, identification of the calling party, and various software defined network features. In addition to new features and services, common-channel signalling offers more efficient assignment of telephone circuits and operation of the telephone network. Although first used for long-distance networks, common-channel signalling is being quickly installed at the local level.

Signalling has been integrated into a modern telecommunications network, depicted in Fig. 22.5, and when coupled with centralized data-bases, provides many of the features associated with today's intelligent network. The database, known as a service control point (SCP), contains the information needed to translate 800-numbers to the appropriate telephone location, among other items. The signalling information is sent over its own signalling links from one signalling processor to another, located at nodes called signal transfer points (STP). The signalling processors determine the actual switching of the customer circuits, performed by switching systems at service switching points. The bulk traffic carried over transmission media can be switched in times of service failures or to balance loads by digital cross-connect systems (DCS). The signalling links connect to the local network at a signalling point of interface (SPI), and the customer circuits connect at a point of presence. Today's signalling systems add much functionality to the network.

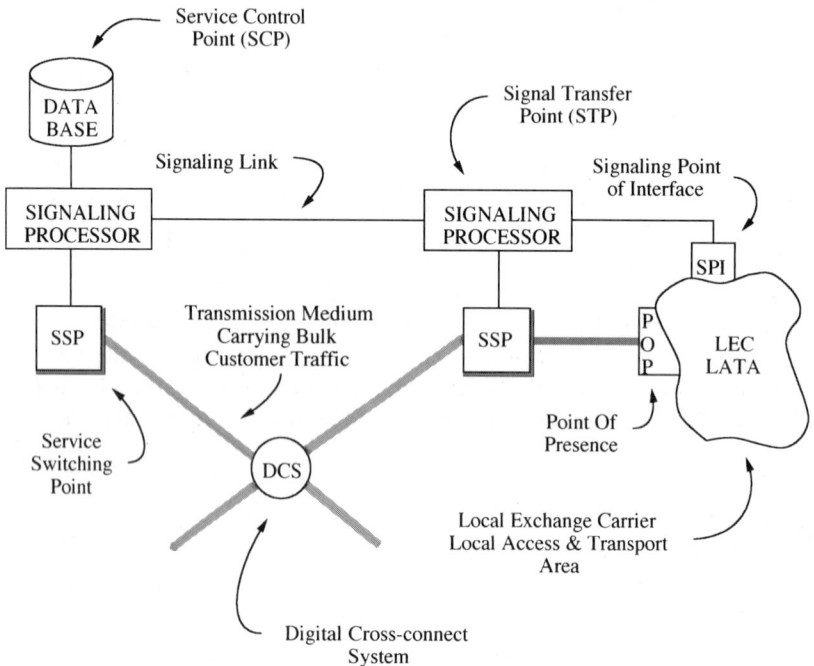

FIGURE 22.5 Intelligent Network.

22.7 Functionality

A wide variety of intelligent features was easily available when humans operated the telephone network and its switchboards. The operator could announce the name of the calling party, hold calls if you were busy, transfer calls to other phones and locations, and interrupt a call if another more important one arrived. However, human operators were far too costly and were replaced by automated electromechanical switching systems. This made telephone service more affordable to more people, but lost the functionality of the human intelligence of the operators. Today's telephone switching systems are controlled by programmable computers and once again intelligence has returned to the telephone network so that the functional services of the past can again be offered using today's computer-controlled technology. Call waiting, call forwarding, and caller-ID are examples of some of these functional services. But not all of these services are wanted by all telephone users.

Caller-ID transmits the telephone number of the calling party over the local loop to the called party where the number is displayed on a small visual display. The number is encoded as digital data and is sent in a short burst, using phase shift keying, during the interval between the first and second ringing signal. Newer caller-ID systems also transmit the name associated with the directory listing for the calling number. With caller-ID it is possible to know who is calling before answering the telephone. However, some people consider their telephone number to be very private and personal and do not want it transmitted to others. This privacy issue has stalled the availability of caller-ID in some states and is a good example of the importance of need for understanding the social impacts of the telephone.

22.8 The Future

Based on all of the false promises of such new products and services as picturephones, videotex, and high-definition television (HDTV), skepticism is warranted toward most new ideas, particularly when so many of them are really reincarnations of past failures. But accepting these words of warning, some last thoughts about the future of POTS will nevertheless be opined.

The telephone system invented by Alexander Graham Bell simply enabled people to convey their speech over distance. Bell's network has evolved to a system that enables people to stay in contact wherever they may be through the use of paging and radio-based cellular telephone services, and even telephones in commercial airplanes. The telephone network carries not only voice signals but also facsimile and data signals.

Bell's vision of a wired world communicating by human speech has been mostly achieved, although there are still many places on this planet for which a simple telephone call is a luxury. A global system of low earth orbit satellites could solve this problem, but would be far too costly to offer telephone service to the poorer inhabitants of those places without conventional telephone service. Bell was very wise in emphasizing human speech over the Morse code of telegraphy. However, alphanumeric keyboards negate the need for knowledge of Morse code or American Standard Code for Information Interchange (ASCII) bits, and now everyone can communicate by text, or e-mail as such communication is generally called. Textual communication can have long-holding times but very low average data transmission. Today's packet-switched networks are most appropriate and efficient for this form of communication. A public packet-switched network akin to and as seamless as the public telephone network is clearly needed, perhaps some form of local area data network to our homes and offices that would hasten the arrival of an all-digital end-to-end network. And perhaps such an integrated network would bring a reality to the hopes of an integrated services digital network (ISDN).

Much is said about the convergence of telephony and television, of telecommunication and entertainment, of telephony and community antenna television (CATV). Yet the purpose of passive entertainment seems much different than the interactivity and two-way nature of telephony and most telecommunication. The entertainment center is quite different than the communication centers in most homes. Yet the myth of convergence continues, although the past would tell us that convergence really is a blurring of boundaries between technologies and industry segments.

A trip to a central office will show that although much progress has been made in substituting electronic switching for electromechanical switching, there are still tens of thousands of physical wires for the provision of service over the local loops. The solution is the use of time-division multiplexing so that thousands of circuits are carried over a few physical optical fibers. However, there are still many engineering and cost challenges that must be solved before copper local loops are eliminated, but some day they clearly will be, resulting in lower costs and further increases in productivity.

Although the progress of the technology of telephony has been most impressive over the last century of the provision of POTS, many other factors are equally important in shaping the future of telecommunication. Policy, regulatory, and competitive factors caused the breakup of the Bell System in 1984, and this breakup has had tremendous impact on the provision of telecommunication in the U.S. The entry of long-distance companies into local service and the entry of local telephone companies into long distance will likewise have considerable impact on the balance of power within the industry. Consumer reactions have halted picturephones and delayed caller-ID. The financial risks of the provision of CATV and entertainment have had sobering impact on the plans of telephone companies to expand telephone service to these other businesses.

Defining Terms

Common channel signalling: Uses a separate dedicated path to carry signalling information in the form of short packets of data.

Dual tone multifrequency (DTMF) dialing: Touchtone dialing where pairs of tones denote a dialed number.

Echo canceler: A device that uses an adaptive filter to create synthetic echo to subtract from the return signal to eliminate the echo.

Hybrid: A center tapped transformer that accomplishes two-wire to four-wire conversion.

Interexchange carriers: Companies that operate long-distance networks, such as AT&T, MCI, and Sprint.

Local access and transport area (LATA):   The local portion of the telephone network.

Multiplexing: Method by which a number of communication channels are combined together to share a single to eliminate the echo.

Point of presence (POP):   The point where the connection is made from the local service provider to the interexchange carrier.

Space switching: Physically connecting wires to create an electrical path.

Time switching: Connecting calls by reordering the time sequence of digitized samples.

Further Information

Much of the material in this chapter is based on:

Noll, A.M. 1991. *Introduction to Telephones and Telephone Systems*, 2nd ed., Artech House, Norwood, MA.

Pierce, J.R. and Noll, A.M. 1990. *Signals: The Science of Telecommunications*, Scientific American Library, New York.

Other references:

Elbert, B.R. 1987. *Introduction to Satellite Communication*, Artech House, Norwood, MA.

Hills, M.T. 1979. *Telecommunications Switching Principles*, MIT Press, Cambridge, MA.

Noll, A.M. 1996. *Highway of Dreams: A Critical View Along the Information Superhighway*, Lawrence Erlbaum Associates, Mahwah, NJ.

Parker, S.P. ed. 1987. *Communications Source Book*, McGraw–Hill Company, New York.

23
FDM Hierarchy

Pierre Catala
Texas A&M University

23.1 Introduction 301
23.2 Background Information 301
    Voice-Channel Bandwidth
23.3 Frequency-Division Multiplexing (FDM) 302
    Implementation Considerations
23.4 The Hierarchy 303
    Group • Supergroup • Mastergroup • Higher Levels • Jumbo-group
23.5 Pilots 306
23.6 Direct to Line (DTL) 306
23.7 Summary 307

23.1 Introduction

Circuits linking telephone central offices nationwide carry from dozens to thousands of voice channels, all operating simultaneously. It would be very inefficient and prohibitively expensive to let each pair of copper wires carry a single voice communication. Therefore, very early in the history of the telephone network, telecommunications engineers searched for ways to combine telephone conversations so that several of them could be simultaneously transmitted over a single copper pair. At the receive end, the combined channels would be separated back into individual channels.

Such a combining method is called *multiplexing*. There are two main methods of multiplexing: time-division multiplexing (TDM), which can be used only with digital signals and is a more recent (1960s) technology, and frequency-division multiplexing (FDM), which was first implemented in 1918 by "Ma Bell" between Baltimore and Pittsburgh and could carry four simultaneous conversations per pair of wires. By the early 1970s, FDM microwave links commonly supported close to 2000 (and sometimes 2700) voice channels on a single radio carrier. In 1981 AT&T introduced its AR6A single sideband (SSB) microwave radio, which carried 6000 channels [Rey, 1987]. Coaxial cable carrier systems, such as the AT&T L5E, could carry in excess of 13,000 channels using frequency-division multiplexing.

This paper reviews the principle of FDM then describes the details of how FDM channels are combined in a hierarchy sometimes referred to as the *analog hierarchy*.

23.2 Background Information

Voice-Channel Bandwidth

The voice-spectrum bandwidth in telephone circuits is internationally limited to frequencies in the range from 0.3 to 3.4 kHz. Trunk signaling for that voice channel is generally done *out of band* at 3825 Hz. Furthermore, since bandpass filters are not perfect (the flanks of the filters are not perfectly vertical) it is necessary to provide a guard band between channels. To take all of these factors into account, the overall voice channel is standardized as a 4-kHz channel.

Although most of the time the voice channels will, indeed, be carrying analog voice signals, they can also be used to transmit *analog data* signals, that is, signals generated by data modems. This consideration is important in the calculation of the loading created by FDM **baseband signals**.

23.3 Frequency-Division Multiplexing (FDM)

Frequency division multiplexing combines the different voice channels by stacking them one above the other in the frequency domain, as shown in Fig. 23.1, before transmitting them. Therefore, each 4-kHz voice channel is shifted up to a frequency 4 kHz above the previous channel.

The FDM concept is not specific to telephony transmission. The TV channels in cable TV (or on-the-air broadcasting for that matter) are also stacked in frequency (channels 1–90, etc.) and are, therefore, frequency-division multiplexed. The TV receiver is the demultiplexer in that case. The difference here is that the TV receiver only needs to receive one channel at a time, whereas the telephone FDM demultiplexer must receive all channels simultaneously.

Implementation Considerations

Practical reasons preclude the assembly (*stacking*) of the different channels, one above the other, continuously from DC on. Problems such as AC power supply hum make the use of low frequencies unwise, thus the first voice channel is usually *transposed* to a frequency of 60 kHz (although for small capacity systems, the first channel can be as low as 12 kHz).

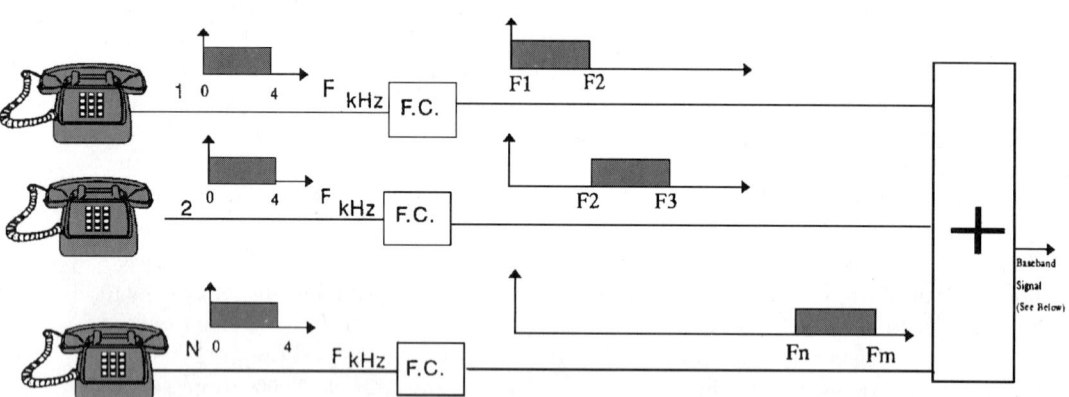

F.C.: Frequency Converter (also called "Channel Translation Equipment")

a) Frequency Translation

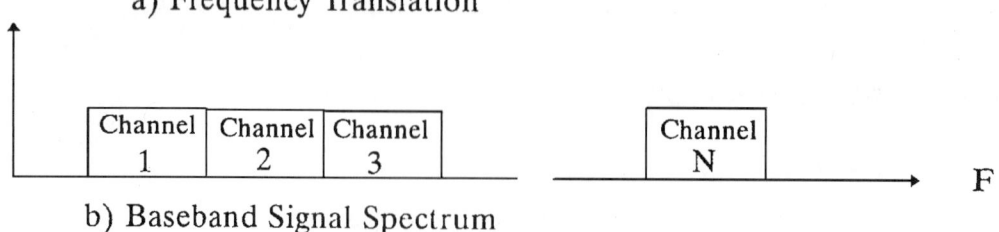

b) Baseband Signal Spectrum

FIGURE 23.1 FDM Principle.

FDM Hierarchy

The necessity of having compatible equipment so that FDM links could be established between equipment of different vendors and even between different countries, dictated the establishment of standards that specified how channels should be grouped for multiplexing. This also allowed the manufacturing of modular equipment, which can be combined to increase the capacity of the links.

All of these considerations led to the development of FDM baseband signals that follow a specific organization and occupy given sets of frequencies as described in the following sections.

23.4 The Hierarchy

Group

The first step in FDM is to combine 12 voice channels together. This first level of the analog multiplexing hierarchy is called a *group* [also called *basic group* by the Consultative Committee on International Telephony and Telegraphy (CCITT) and *primary group* by several countries]. The FDM group is, therefore, 48 kHz wide and composed of 12 voice channels stacked from 60 to 108 kHz, as shown in Fig. 23.2. For *thin route*, that is, small capacity systems, the group can instead be translated to a range of 12–60 kHz, so as to reduce the overall baseband bandwidth required.

The equipment performing the frequency translation of each channel uses a combination of mixers and SSB modulation with appropriate bandpass filters to create this baseband group. That equipment is called *channel bank* in North America and *channel translation equipment* (CTE) in other English-speaking countries.

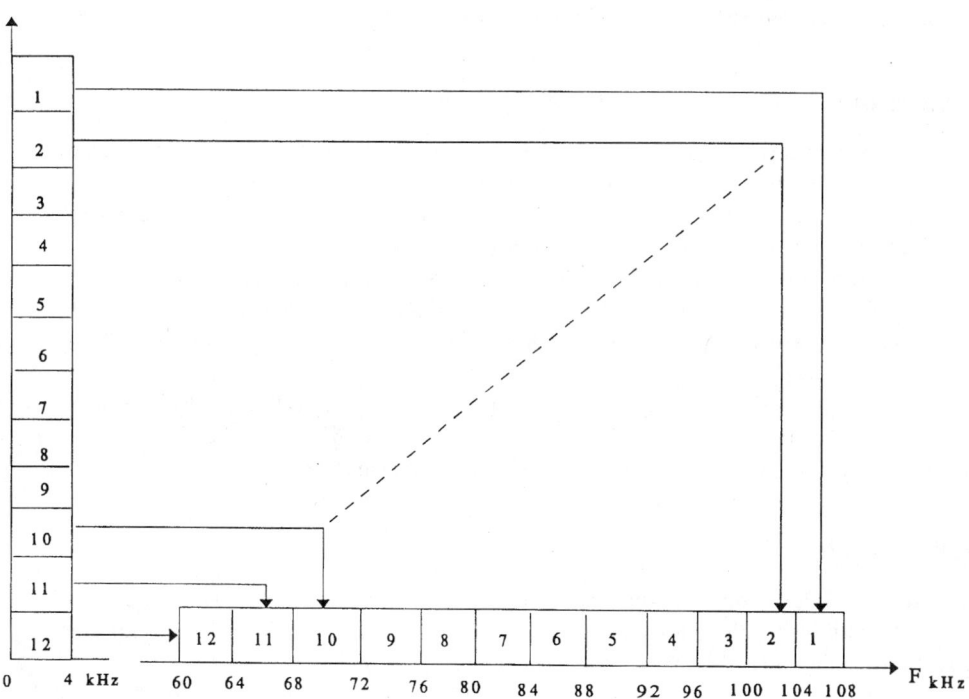

The group of 12 channels occupies a bandwidth of 48 kHz from 60 to 108 kHz.

FIGURE 23.2 Basic FDM group.

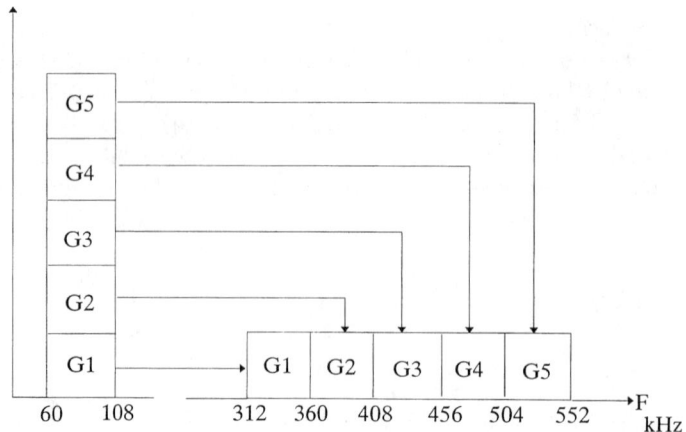

The supergroup of 60 channels (5 groups) occupies a bandwidth of 240 kHz between 312 and 552 kHz.

FIGURE 23.3   FDM supergroup.

Supergroup

If more than 12 channels are needed, the next level in the hierarchy is created by combining 5 groups together, thus providing a capacity of 60 voice channels. This second level is called a *basic supergroup*, usually abbreviated as supergroup, and has a bandwidth of 240 kHz going from 312 to 552 kHz, as shown in Fig. 23.3. Some countries call this 60-channel assembly a *secondary group*.

The related frequency translation equipment is called *group bank* in the U.S. and Canada but *group translation equipment* (GTE) in most other countries.

Mastergroup

The international agreements concerning FDM standards unfortunately stop at the supergroup level. Although both the CCITT [now ITU-T] and the U.S./Canadian standards include a third level called *mastergroup*, the number of voice channels differ and the two are, therefore, not compatible! The most common mastergroup in North America was the Western Electric U600 scheme, which combines 10 supergroups, thus creating a 600-channel system as shown in Fig. 23.4.

There are several variations of the Western Electric U600 mastergroup such as the AT&T L1 coax-based carrier system, which also carries 600 channels but translated between 60 and 2788 kHz rather than 564–3084 kHz.

The frequency translation equipment creating mastergroups is called *supergroup bank* or *supergroup translation equipment* (STE). The CCITT mastergroup on the other hand combines only five supergroups to create a 300-channel baseband as shown in Fig. 23.5.

Higher Levels

There are many different multiplexing schemes for the higher density FDM systems. They become quite complex and will only be listed here. The reference section of this paper lists several sources where details of these higher level multiplexed signals can be found.

- *CCITT supermastergroup.* This combines three CCITT mastergroups for a total of 900 channels occupying a 3.9 MHz baseband spectrum from 8.516 to 12.388 MHz.
- *CCITT 15-Supergroup Assembly.* This also provides 900 channels but bypasses the mastergroup level by directly combining 15 supergroups (15 × 60 = 900) occupying a 3.7-MHz

FDM Hierarchy

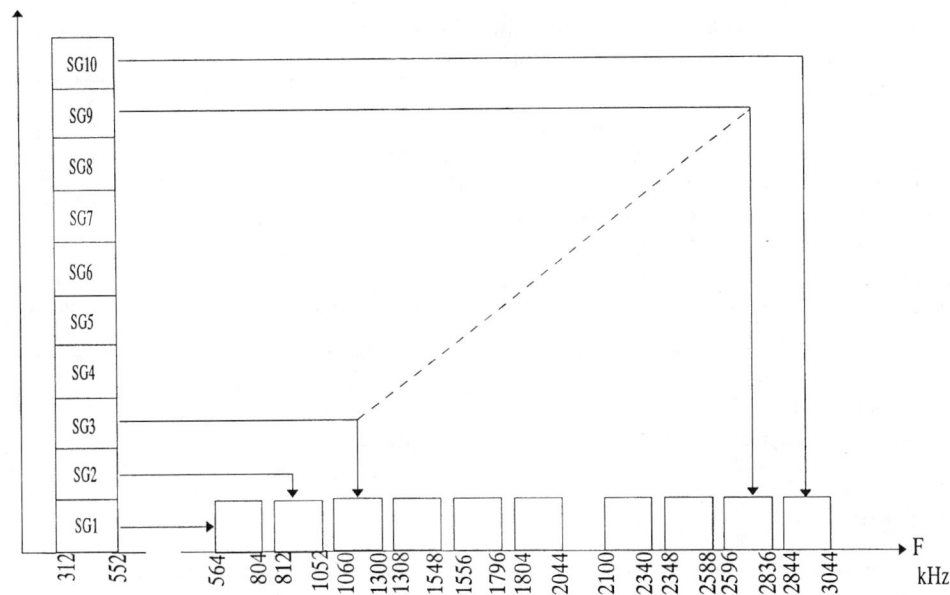

The US Mastergroup of 600 channels (10 supergroups) occupies a bandwidth of 2.52 MHz from .564 to 3.084 MHz.

FIGURE 23.4 U.S. (AT&T) basic mastergroup.

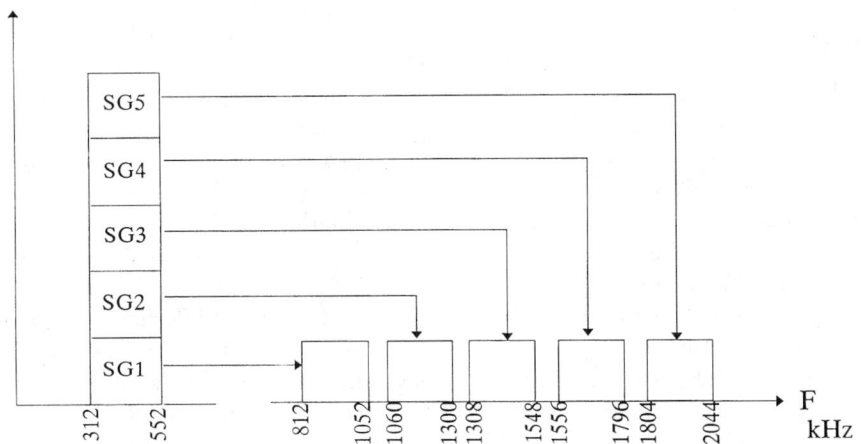

The CCITT Mastergroup of 300 channels (5 supergroups) occupies a bandwidth of 1232 kHz from 812 to 2044 kHz.

FIGURE 23.5 CCITT basic mastergroup.

spectrum from 312 to 4028 kHz. This assembly is sometimes called a *hypergroup*. There are variations of the 15-supergroup assembly such as a 16-supergroup assembly (960 channels).

For the very-high capacity FDM systems, grouping of supermastergroups or of 15-supergroup assemblies are used, with the preference on the latter method.

The AT&T high-level FDM hierarchy is, of course, based on the 600-channel mastergroup. The AT&T *mastergroup multiplex* (MMX) combines these mastergroups to a variety of levels, shown in

Table 23.1, by multiplexing from two to eight mastergroups (except for the four-mastergroup level, which is not used).

Two of the 3000-channel intermediate levels are further combined to form the baseband signal for the 6000-channel AR6A microwave radio. At 6000 channels per radio, the AR6A radio system can transmit 42,000 voice channels on a single antenna, using seven different RF carriers [Rey, 1987].

Two of the 4200-channel intermediate levels are combined with one 4800-channel intermediate level to form the line signal for the L5E, 13,200-channel, coaxial cable system, which occupies a frequency spectrum of 61.592 MHz starting at 3.252 MHz.

TABLE 23.1 AT&T Mastergroup Multiplex Hierarchy

No. of Multiplexed Mastergroups	Resulting No. of Channels	AT&T Application
2	1200	TD microwave radio
3	1800	TH microwave radio
5	3000	Intermediate level
6	3600	Jumbogroup
7	4200	Intermediate level
8	4800	Intermediate level

Jumbogroup

As was seen in Table 23.1, the basic jumbogroup is composed of six mastergroups (3600 channels). It was the basis for the L4 coaxial-cable system and has a spectrum ranging from 0.564 to 17.548 MHz.

When the L5 coaxial system was implemented in 1974, it used a line signal created by multiplexing three jumbogroups, thus providing a capacity of 10,800 channels occupying frequencies from 3.124 to 60.566 MHz.

23.5 Pilots

In addition to the voice channels (and related signalling tone), special tones called pilots are introduced to provide the receive end (demultiplexer) with frequency and level references.

Each level of the FDM hierarchy (group, supergroup, mastergroup, etc.) will have its own *reference* pilot tone. Each of the respective demultiplexers monitors the appropriate tone to detect interruptions (faults) in which case it generates an alarm indicating which group, supergroup, etc., is defective. Alarms are also generated if the level of the tone is 4 dB below (or above) its normal value.

Other pilots, called *line regulating*, used to be found in gaps between the FDM building groups (e.g., supergroups). Their name came from the fact that they were used for automatic gain control (AGC) of the miscellaneous amplifiers/repeaters. As FDM equipment became better, the regulation function was transferred to the reference pilots just mentioned and line regulating pilots were no longer implemented. A regulation of 0.5 dB is a typical objective [Freeman, 1989].

Microwave radios use an additional type of pilot tone, the *continuity pilot*, which is generally inserted at the top of the baseband spectrum and controls the receiver AGC and also serves as an overall continuity pilot.

In older FDM equipment there were also *frequency synchronization* pilots used to ensure that the demultiplexing carriers were within a few hertz of the multiplexing ones. Otherwise, audio distortion would occur. Newer FDM equipment have sufficiently precise and stable oscillators so that synchronization pilots are no longer needed.

23.6 Direct to Line (DTL)

If interfacing with existing FDM equipment is not needed on a particular link, then a cheaper method of multiplexing can be done with modern multiplexing equipment, at least for the low-capacity systems. The direct-to-line (DTL) method of forming the baseband signal moves the voice channels directly to their respective line frequencies, thus bypassing the group and supergroup building blocks. Such equipment is frequently used on thin-line microwave links carrying 60 channels or less but can also be found in multiplexers providing basebands of up to 600 channels.

The DTL method decreases the number of steps in the multiplexing process and, consequently, the number of circuit boards and filters. Therefore, it is cheaper, more reliable, and more flexible. DTL equipment, however, is not compatible with the standard FDM equipment.

23.7 Summary

Multiplexing in the frequency domain (FDM) was the mainstay of carrier systems, on twisted pairs, coaxial cable and microwave links, until digital carrier systems became prevalent in the early 1980s. FDM is done by stacking voice channels, one above the other, in the frequency domain. In other words, each voice channel is translated from its original 0–4-kHz spectrum to a frequency above the previous channel. This is generally done in building blocks called groups (12 channels), supergroups (5 groups), mastergroups (5 or 10 supergroups), etc., which follow international (ITU/CCITT) standards or AT&T standards. FDM equipment with fewer than 600 channels can also be implemented using a simpler and cheaper technique called direct to line, where the building blocks are bypassed and the voice channels are transposed directly to the appropriate line (baseband) frequency. However, DTL is not a standard, and, therefore, direct interconnection with the public telephone network is not possible.

Defining Terms

Baseband signal: Modulating signal at the transmitter (i.e., signal found before modulation of the transmitter). At the receiver it would be the demodulated signal, that is, the signal found after the demodulator. In the case of FDM, the baseband signal (also called line frequency) is therefore the combined signal of all the voice channels being transmitted.

References

CCITT (ITU-T). 1989. *International Analogue Carrier Systems.* Blue Book, Vol. III, Fascicle III. 2, Consultative Committee on International Telephony and Telegraphy/International Telecommunications Union, Geneva, Switzerland.

Freeman, R. 1989. *Telecommunication System Engineering,* Wiley-Interscience, New York.

Rey, R.F., tech. ed. 1987. Transmission systems. In *Engineering and Operations in the Bell System,* 2nd ed., pp. 345–356, AT&T Bell Labs., Murray Hill, NJ.

Tomasi, W. 1992. *Advanced Electronics Communications Systems,* Prentice Hall, Englewood Cliffs, NJ.

Further Information

General concepts of multiplexing can be found in Chapter 7 of this handbook.

Principles of mixers, frequency converters (frequency translation), and single sideband modulation can be found in any electronic communications or communication systems type of book.

24

Analog Telephone Channels and the Subscriber Loop

Whitham D. Reeve

24.1 Telephone Band 308
24.2 Noise 309
24.3 Crosstalk 310
24.4 Circuit Noise 310
24.5 Impulse Noise 312
24.6 Attenuation Distortion 312
24.7 Envelope Delay Distortion 312
24.8 Line Conditioning 313

24.1 Telephone Band

A voice channel is considered to require a nominal bandwidth of 4000 Hz. For all practical purposes, however, the usable bandwidth of an end-to-end call in the **Public Switched Telephone Network (PSTN)**, including the loop, is considered to fall between approximately 300 and 3400 Hz (this is called the *voice band*) giving a bandwidth of 3100 Hz. This bandwidth is entirely acceptable from a voice transmission point of view, giving subscriber satisfaction level above 90% [IEEE, 1984].

Although it is not unduly difficult to provide a 3100-Hz bandwidth in the subscriber loop, the loop and the associated terminal equipment have evolved in such a way that this bandwidth is essentially fixed at this value and will continue to be for analog transmission. The bandwidth is somewhat restrictive for data transmission when using modems. However, the problem of bandwidth restriction has encouraged a number of innovative solutions in modem design.

Obviously, on a loop derived entirely from copper cable, the frequency response of the loop itself would extend down to DC (zero frequency). The lower response is lost, however, once the loop is switched or connected to other transmission and signalling equipment, all of which are AC coupled. Where DC continuity is not available or not practical, special tone signalling equipment is used to replace the DC signals. When voice signals or other signals with frequency content approaching zero frequency are placed on the loop, the transmission is considered to be in the baseband.

Similarly, the upper frequency limit is not exactly 3400 Hz. Depending on how it is specified or the type of cable, the limit may be much higher. In practice, the loop does not generally set the upper limit of a voice channel; the upper limit is mostly due to the design of the equipment that interfaces with the loop.

Analog Telephone Channels and the Subscriber Loop 309

For voice-band transmission, the bandwidth (or frequency response) of a telecommunication channel is defined as the limiting frequencies where loop loss is down by 10 dB from its 1000 Hz value [IEEE, 1984]. The field measurement of bandwidth usually does not proceed with measurement of the 10-dB points. Instead, simple slope tests are made; these provide an indirect, but reliable, indicator of the transmission channel bandwidth. If the slope, as defined subsequently, is within predetermined limits, the bandwidth of the channel can be assumed to be acceptable.

Slope tests are loss measurements at 404, 1004 and 2804 Hz (this is also called the *three-tone slope*). The loss at 1004 Hz is subtracted from the loss at 404 Hz to give the low-frequency slope, and from 2804 Hz to give the high-frequency slope.

24.2 Noise

Noise is any interfering signal on the telecommunication channel. There must be a noise source, a coupling mechanism and a receptor. The relationship among the three is illustrated in Fig. 24.1. Noise sources are either manmade or natural. Practically any piece of electrical or electronic equipment can be a manmade noise source, and power lines are perhaps the most pervasive of all of these. Natural noise comes from lightning and other atmospherics and random thermal motion of electrons and galactic sources, as well as electrostatic discharges.

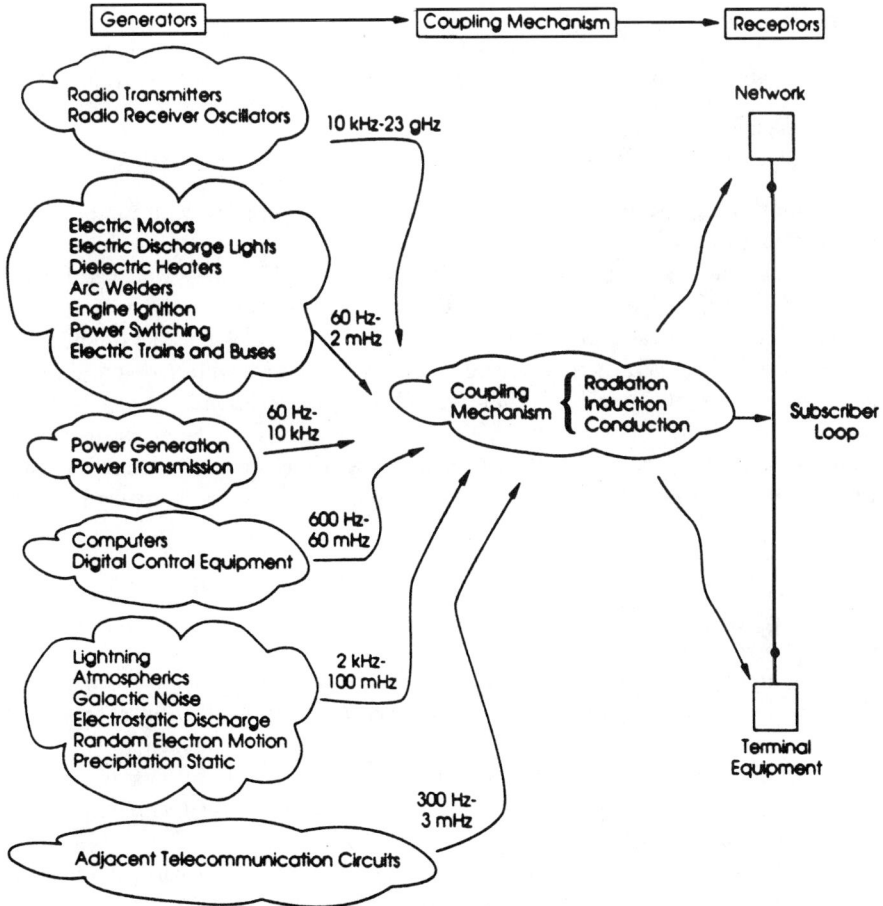

FIGURE 24.1 Noise diagram.

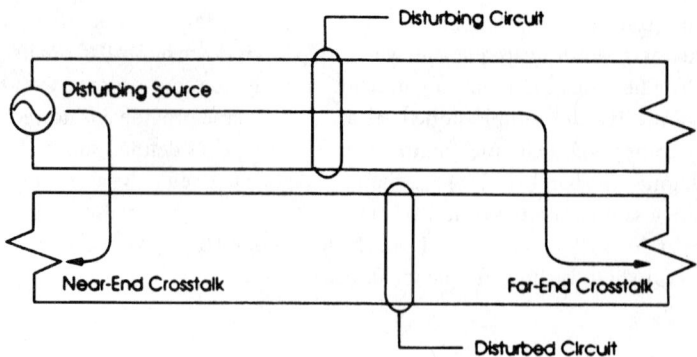

FIGURE 24.2 Crosstalk.

Noise is coupled by radiation, induction, and conduction. The predominant coupling mode in subscriber loops is by induction from nearby power lines. The other coupling modes exist to some extent, too, depending on the situation. For example, noise can be conducted into the loop through insulation faults or poor or faulty grounding methods.

24.3 Crosstalk

Crosstalk falls in two categories:

1. Unintelligible crosstalk (babble)
2. Intelligible crosstalk

The latter is most disturbing because it removes any impression of privacy. It can be caused by a single disturbing channel with enough coupling to spill into adjacent circuits. Unintelligible crosstalk is usually caused by a large number of disturbing channels, none of which is of sufficient magnitude to be understood, or extraneous modulation products in carrier systems.

Crosstalk can be further categorized as near end and far end. As these names imply, near-end crosstalk is caused by crosstalk interference at the near end of a circuit with respect to the listener; far-end crosstalk is crosstalk interference at the far end as shown in Fig. 24.2.

Crosstalk of any kind is caused by insufficient shielding, excessively large disparity between signal levels in adjacent circuits, unbalanced lines, or overloaded carrier systems. Crosstalk is a statistical quantity because the number of sources and coupling paths is usually too large to quantify.

24.4 Circuit Noise

The noise that appears across the two conductors (tip and ring) of a loop, and is heard by the subscriber, is called **circuit noise** (also called *message circuit noise, noise metallic,* or *differential noise*). The noise can be due to random thermal motion of electrons (known as *white noise* or *Gaussian noise*) or static from lightning storms, but on subscriber loops its most likely source is interference from power line induction. For the purposes of this discussion, then, circuit noise and interference are assumed to be the same.

The total noise power on a loop is related to the noise bandwidth. Since particular frequencies differently affect the various services (for example, voice, data, and radio studio material), weighting curves have been designed to restrict the frequency response of the noise measuring sets with which objective tests are made. This frequency response restriction is called *weighting*.

Analog Telephone Channels and the Subscriber Loop

Noise, in voice applications, is described in terms of decibels above a noise reference when measured with a noise meter containing a special weighting filter. There are four common filters that are used to provide the necessary weighting:

- C-message
- 3-kHz flat
- Program
- 15-kHz flat

The most common filter is called a C-message filter and measurements are based on decibels with respect to reference noise, C-message weighted (dBrnC). The noise reference is 1 pW(−90 dBrn); therefore, a properly calibrated meter will read 0 dBrnC when measuring a 1000-Hz tone having a power of −90 dBrn.

C-message weighting is primarily used to measure noise that affects voice transmission when common telephone instruments are used, but it also is used to evaluate the effects of noise on data circuits. It weights the various frequencies according to their perceived annoyance such that frequencies below 600 or 700 Hz and above 3000 Hz have less importance.

The 3-kHz-flat weighting curve is used on voice circuits, too, but all frequencies within the 3000-Hz bandwidth carry equal importance. This filter rolls off above 3000 Hz and approximates the response of common modems. It generally is used to investigate problems caused by power induction at the lower power harmonic frequencies or by higher interfering frequencies. Frequencies in these ranges can affect data transmission as well as voice frequency signalling equipment.

The *Program* weighting curve is used with voice circuits that require bandwidth in the order of 8 kHz or more (such as in the distribution of radio or television program audio material). This curve emphasizes frequencies between 1 and 8 kHz. The 15-kHz-flat weighting curve is also used on these types of circuits, but it actually carries little weighting (it is essentially flat from 20 Hz to 15 kHz).

A comparison of the various weighting curves is shown in Fig. 24.3.

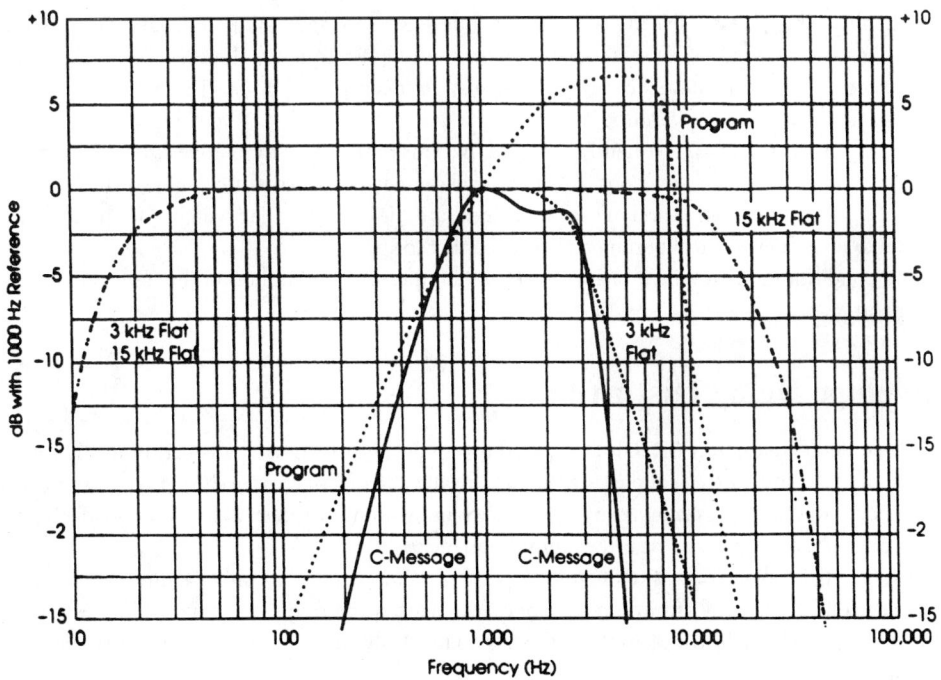

FIGURE 24.3 Weighting curve comparison.

24.5 Impulse Noise

Data circuits are particularly sensitive to **impulse noise**. Impulse noise, heard as clicks, is usually defined as a voltage increase of 12 dB or more above the background [root mean square (rms)] noise lasting 10 ms or less. Its main source is from switching transients in electromechanical switching systems, maintenance activity on an adjacent circuit, or electrical system switching transients. It is less of a problem with properly grounded electronic switching systems.

Impulse noise objectives vary with the type and makeup of the circuit. Usually, a threshold is established and counts are made of any impulse noise that exceeds that threshold in a given time period. When impulse noise tests are made on a single circuit, the usual specification is 15 counts in 15 min. When a group of circuits is being tested, shorter time intervals are used.

On **subscriber loops**, the background noise threshold is 59 dBrnC when measured at the central office [BELLCORE, 1986]. On long-haul circuits, the threshold can vary from 54 to 71 dBrnC0. The latter figure (71 dBrnC0) implies a 6-dB signal-to-impulse noise ratio with a -13-dBm0 signal. The 0 (zero) suffix in these parameters indicates the parameter is referred to the 0 Transmission Level Point (0TLP).

24.6 Attenuation Distortion

Attenuation distortion is the change in circuit loss with frequency. It is also known as *frequency response*. Ideally, attenuation should be constant throughout the frequency band of interest. Unfortunately, this is not usually the case. Unless it is excessive, however, attenuation distortion is not noticeable to the human ear. Attenuation distortion manifests itself on voice calls by changing the sound of the talker's voice as it is heard by the listener. The change may be dramatic enough to render the voice unrecognizable. On a data circuit, attenuation distortion can manifest itself in the form of errors through loss of signal energy at critical frequencies.

The inherent attenuation distortion in subscriber loops used in regular switched service is not objectionable except in extreme cases. On voice transmissions, excessive low-frequency slope degrades voice quality, whereas excessive high-frequency slope degrades intelligibility. On data transmissions using *phase-shift keying* (PSK) methods, both low- and high-frequency slope affects performance, whereas only high-frequency slope affects *frequency-shift keying* (FSK) [CCITT, 1989]. Some slope is considered necessary for stability.

Attenuation distortion is frequently specified in terms of the three tone slope as discussed previously. The slope objectives for loops used in special switched services [for example, private branch exchange (PBX) trunks and foreign exchange lines] are similar to regular switched loop objectives. It is not an extremely critical parameter on subscriber loops.

Line conditioning (explained later) is used to compensate for the loop's inherent attenuation distortion.

24.7 Envelope Delay Distortion

Envelope delay distortion (EDD), also called group delay distortion, is distortion in the rate of change of phase shift with frequency of a signal. Ideally, the rate of change should be constant with frequency, and it is approximately so in the voice band with nonloaded cables. With loaded cables this is not the case, especially near the cutoff frequency.

Envelope delay distortion is defined as the difference, in time units such as microseconds, between the maximum and minimum envelope delay within the frequency band of interest. If the difference is zero then, by this definition, there is no EDD, but this is hardly ever the case in practical systems.

Voice-band signals (not just voice signals but all signals in the voice band, including data signals) are made up of many frequencies. Each particular frequency propagates at a different velocity (called *phase velocity*) due to the facility's inherent transmission characteristics. This causes phase delay. If the relationship between the resulting phase shift and frequency is nonlinear, the facility will cause delay distortion. EDD at the upper edge of the voice band can cause *ringing* in the conversation. Excessive EDD at the lower edge can cause *speech blurring* [CCITT, 1989].

Not all EDD results from the loop; in fact, very little does. EDD results from any part of the circuit where a nonlinear relationship exists between phase shift and frequency. Usually this occurs at the terminal and intermediate equipment. Some data transmission modulation techniques are more susceptible to intersymbol interference caused by EDD than others, which can explain why some modems of a given speed give better performance than others.

24.8 Line Conditioning

When high-speed data signals are applied to the loop, the various transmission parameters, especially attenuation and envelope delay distortion, must be held within well-defined limits. These parameters are controlled by amplifiers and equalizers, which usually are installed at the ends of the cable facility.

If required, equalization is almost always applied to the receive side of a four-wire circuit. Sometimes, however, predistortion equalization is provided on the transmit side. (Note that present high-speed data circuits are almost always point-to-point arrangements made up from a four-wire circuit with separate transmit and receive paths.)

The inherent loss of a loop used in private line applications can be reduced by adding gain with amplifiers. This has to be done carefully to prevent the loop from singing if two-wire conversion is used at any point in the circuit.

Equalizers are used to offset a loop's high-frequency rolloff characteristic, which affects both attenuation and envelope delay distortion. The application of equalizers and other devices to control transmission characteristics is called *line conditioning*.

TABLE 24.1 C Conditioning Specifications

Freq. Band, Hz	Atten. Dist.,[a] dB	Freq. Band, Hz	EDD, μs
Basic (no conditioning)			
500–2500	−2 to +8	800–2600	1750
300–3000	−3 to +12		
C1 Conditioning			
1000–2400	−1 to +3	1000–2400	1000
300–2700	−2 to +6	800–2600	1750
2700–3000	−3 to +12		
C2 Conditioning			
500–2800	−1 to +3	1000–2600	500
300–3200	−2 to +6	600–2600	1500
		500–2800	3000
C4 Conditioning			
500–3000	−2 to +3	1000–2600	300
300–3200	−2 to +6	800–2800	500
		600–3000	1500
		500–3000	3000
C5 Conditioning			
500–2800	−0.5 to +1.5	1000–2600	100
300–3000	−3 to +3	600–2600	300
		500–2800	600

[a] With respect to 1004 Hz.

TABLE 24.2 BELLCORE Voice Grade Types Acceptance Limits

Freq. Band, Hz	Atten. Dist.,[a] dB	Freq. Band, Hz	EDD, μs
Voice grade 1			
504–2504	−1.5 to +7.5		None specified
404–2804	−1.5 to +9.5		
304–3004	−2.5 to +11.5		
Voice grade 2			
404–2804	−0.5 to +3.5		None specified
304–3004	−0.5 to +4.5		
Voice grade 3			
404–2804	−0.5 to +2.5		None specified
304–3004	−0.5 to +4.5		
Voice grade 4			
304–504	−0.5 to +3.0		None specified
504–2504	−0.5 to +1.5		
2504–2804	−0.5 to +2.5		
2804–3004	−0.5 to +3.5		
Voice grade 5			
404–2804	−0.5 to +4.5		None specified
Voice grade 6			
504–2504	−0.5 to +2.5	804–2604	650
404–2804	−0.5 to +3.5		
2804–3004	−0.5 to +4.5		
Voice grade 7			
404–2804	−0.5 to +1.5	804–2604	650
304–3004	−0.5 to +4.5		
Voice grade 8			
404–2804	−0.5 to +1.5	804–2604	650
304–3004	−0.5 to +4.5		
Voice grade 9			
404–2804	−0.5 to +1.5	804–2604	650
304–3004	−0.5 to +4.5		
Voice grade 10			
504–2504	−1.5 to +7.5	804–2604	1700
404–2804	−1.5 to +9.5		
304–3004	−2.5 to +11.5		
Voice grade 11			
304–3004	−0.5 to +4.5	804–2604	650
1204–2604	−0.5 to +0.5		
Voice grade 12			
504–2804	−0.5 to +0.5	804–2604	650
304–3004	−0.5 to +2.0		

[a] With respect to 1004 Hz.

On switched, two-wire loops, loop treatment is straightforward and does not require special techniques if the proper equipment is used. With four-wire dedicated circuits, however, line conditioning is used to tailor a loop to particular transmission characteristics. Modern transmission equipment allows this to be done on a cookbook or prescription basis.

Most companies (telephone companies and circuit vendors) have standardized the various types of line conditioning for four-wire circuits. Three tables are provided that describe the basic transmission characteristics of leased lines available from many companies. In each of these tables, the attenuation distortion is shown with respect to the attenuation at 1004 Hz. The envelope delay

TABLE 24.3 AT&T Service types Acceptance Limits

Freq. Band, Hz	Atten. Dist.,[a] dB	Freq. Band, Hz	EDD, μs
Service type 1			
304–404	−3 to +12		None specified
404–2804	−2 to +9		
2804–3004	−3 to +12		
Service type 2			
304–404	−3 to +12		None specified
404–2804	−2 to +6		
2804–3004	−2 to +12		
Service type 3			
304–504	−2 to +9		None specified
504–2504	−2 to +6		
2504–2804	−2 to +8		
2804–3004	−2 to +11		
Service type 4			
404–2804	−4 to +12		None specified
Service type 5			
304–404	−3 to +12	804–2604	1750
404–504	−2 to +10		
504–2504	−2 to +8		
2504–2804	−2 to +10		
2804–3004	−3 to +12		
Service type 6			
304–404	−3 to +12	804–2604	1250
404–2804	−2 to +6		
2804–3004	−3 to +12		
Service type 7			
304–404	−3 to +12	804–2604	550
404–2804	−2 to +5		
2804–3004	−3 to +12		
Service type 8			
304–404	−3 to +12	804–2604	400
404–2804	−1 to +4		
2804–3004	−3 to +12		
Service type 9			
304–1204	−3 to +12	804–2604	600
1204–2604	−3 to +3		
2604–3004	−3 to +12		
Service type 10			
304–504	−2 to +6	804–2604	2000
504–2804	−1 to +3		
2804–3004	−2 to +6		

[a]With respect to 1004 Hz.

distortion is given in terms of the difference between the maximum and minimum envelope delay (in microseconds) within the frequency band shown. All values are for circuits from vendor demarcation point to vendor demarcation point.

Table 24.1 compares the characteristics of lines with various types of C conditioning [AT&T, 1975]. The nomenclature used (C1, C2, etc.) originated with the Bell System (AT&T). It was adopted by many non-Bell operating companies as well. This list is not exhaustive. Additional types of conditioning may be available under this nomenclature from a particular telephone company. The conditioning specifications listed in Table 24.1 are considered to be end-to-end, so they will include the characteristics of the end links (loops) as well as interoffice transmission facilities. The circuit noise depends on the circuit mileage. The circuit noise limit for all conditioning types is 31 dBrnC0 for a circuit less than 50 mi long.

Most modern high-speed modems (up to around 20 kb/s) will function properly over a single-link C2 conditioned line. Higher grade lines (such as C4 and C5) are specified to provide improved performance, higher speeds, or to ensure that the overall circuit composed of cascaded links meets the performance criteria of C2 conditioning. Other conditioning specifications exist, such as D conditioning, which controls signal-to-noise ratio and nonlinear distortion.

Although some companies still use the C conditioning nomenclature, it is becoming obsolete. With the divestiture of AT&T, the specification of end-to-end transmission facilities has become more difficult because the public network now has multivendor and multidimensional characteristics. A given link vendor will provide lines with predetermined characteristics according to the facilities owned and operated by them and them only. As a result, BELLCORE has developed standardized *Voice Grade* (VG) circuit types with the characteristics shown in Table 24.2 [BELLCORE, 1989].

Table 24.3 shows some of the characteristics of the most commonly used AT&T service types [FAA, 1976]. Not all characteristics are shown in this table; as with VG types, many service types also will include specifications for impulse noise, and some include phase jitter and distortion limits. For most service types, the circuit noise is measure in absolute terms (dBrnC0) or as a signal-to-noise ratio with a holding tone, or both. Virtually all interoffice facility links provided by AT&T use some type of compandored carrier rather than voice grade metallic loops, so signal-to-noise ratio is an appropriate parameter for this situation.

Defining Terms

Circuit noise: The noise heard by the subscriber that appears across the pair of conductors in the loop.

Crosstalk: Interference coupled into the present telephone connection from adjacent channels or connections.

Envelope delay distortion (EDD): Distortion in the rate of change of phase shift as a function of frequency.

Impulse noise: Short voltage increases of 12 dB or more usually caused by electromechanical switching transients, maintenance activity, or electrical system switching transients.

Line conditioning: The use of amplifiers and equalizers to control channel transmission characteristics.

Public Switched Telephone Network (PSTN): Traditionally, the network developed to provide dial-up voice communications.

Subscriber loop: The subscriber loop is the transmission and signalling channel between a telephone subscriber's terminal equipment and the network.

Voice band: The usable bandwidth of a telephone voice channel, often taken to be 300–3400 Hz.

References

AT&T. 1975. *Telecommunications Transmission Engineering*, Vols. 1–3. Tech. Pub. Western Electric Company, Inc. Winston-Salem, NC (Available from AT&T Customer Information Center).

BELLCORE. 1986. *Notes on the BOC Intra-LATA Networks*. Tech. Ref. TR-NPL-000275. BELLCORE Customer Service. Morristown, NJ.

BELLCORE. 1989. *Voice Grade Special Access Service Transmission Parameter Limits and Interface Combinations*, Tech. Advisory RA-TSY-000335, Dec. BELLCORE Document Registrar. Morristown, NJ.

CCITT. 1989. *Telephone Transmission Quality, Series P Recommendations*, CCITT Blue Book, Vol. V, Consultative Committee on International Telephony and Telegraphy. Geneva, Switz.

FAA. 1976. *Maintenance of Two-point Private Lines.* FAA Order 6000.22, Aug. 9. Dept. of Transportation, Federal Aviation Administration, Washington, DC.

IEEE. 1984. *IEEE Standard Telephone Loop Performance Characteristics.* IEEE Standard 820-1984, Institute of Electrical and Electronics Engineers, New York.

Further Information

See *Subscriber Loop Signaling and Transmission Handbook: Analog* by Whitham D. Reeve, IEEE Press, 1992, and the three volume set prepared by AT&T entitled *Telecommunications Transmission Engineering*, 1977, Vols. 1–3.

25
Baseband Signalling and Pulse Shaping

25.1	Communications System Model	318
25.2	Intersymbol Interference and the Nyquist Criterion	321
	Raised Cosine Pulse	
25.3	Nyquist Criterion with Matched Filtering	325
25.4	Eye Diagrams	327
	Vertical Eye Opening • Horizontal Eye Opening • Slope of the Inner Eye	
25.5	Partial-Response Signalling	329
	Precoding	
25.6	Additional Considerations	334
	Average Transmitted Power and Spectral Constraints • Peak-to-Average Power • Channel and Receiver Characteristics • Complexity • Tolerance to Interference • Probability of Intercept and Detection	
25.7	Examples	336
	Global System for Mobile Communications (GSM) • U.S. Digital Cellular (IS-54) • Interim Standard-95 • Personal Access Communications System (PACS)	

Michael L. Honig
Northwestern University

Melbourne Barton
Bellcore

Many physical communications channels, such as radio channels, accept a continuous-time waveform as input. Consequently, a sequence of source bits, representing data or a digitized analog signal, must be converted to a continuous-time waveform at the transmitter. In general, each successive group of bits taken from this sequence is mapped to a particular continuous-time pulse. In this chapter we discuss the basic principles involved in selecting such a pulse for channels that can be characterized as linear and time invariant with finite bandwidth.

25.1 Communications System Model

Figure 25.1(a) shows a simple block diagram of a communications system. The sequence of source bits $\{b_i\}$ are grouped into sequential blocks (vectors) of m bits $\{\bm{b}_i\}$, and each binary vector $\bm{b}_i$ is mapped to one of 2^m pulses, $p(\bm{b}_i; t)$, which is transmitted over the channel. The transmitted signal as a function of time can be written as

$$s(t) = \sum_i p(\bm{b}_i; t - iT) \qquad (25.1)$$

Baseband Signalling and Pulse Shaping

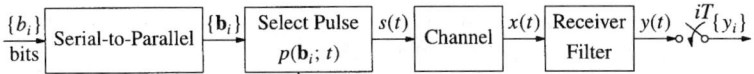

FIGURE 25.1(a) Communication system model. The source bits are grouped into binary vectors, which are mapped to a sequence of pulse shapes.

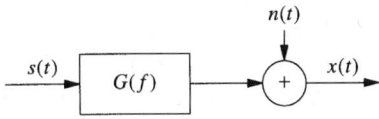

FIGURE 25.1(b) Channel model consisting of a linear, time-invariant system (transfer function) followed by additive noise.

where $1/T$ is the rate at which each group of m bits, or pulses, is introduced to the channel. The information (bit) rate is therefore m/T.

The channel in Fig. 25.1(a) can be a radio link, which may distort the input signal $s(t)$ in a variety of ways. For example, it may introduce pulse dispersion (due to finite bandwidth) and multipath, as well as additive background noise. The output of the channel is denoted as $x(t)$, which is processed by the receiver to determine estimates of the source bits. The receiver can be quite complicated; however, for the purpose of this discussion, it is sufficient to assume only that it contains a front-end filter and a sampler, as shown in Fig. 25.1(a). This assumption is valid for a wide variety of detection strategies. The purpose of the receiver filter is to remove noise outside of the transmitted frequency band and to compensate for the channel frequency response.

A commonly used channel model is shown in Fig. 25.1(b) and consists of a linear, time-invariant filter, denoted as $G(f)$, followed by additive noise $n(t)$. The channel output is, therefore,

$$x(t) = [g(t) * s(t)] + n(t) \tag{25.2}$$

where $g(t)$ is the channel impulse response associated with $G(f)$, and the asterisk denotes convolution,

$$g(t) * s(t) = \int_{-\infty}^{\infty} g(t - \tau) s(\tau) \, d\tau$$

This channel model accounts for all linear, time-invariant channel impairments, such as finite bandwidth and time-invariant multipath. It does not account for time-varying impairments, such as rapid fading due to time-varying multipath. Nevertheless, this model can be considered valid over short time periods during which the multipath parameters remain constant.

In Fig. 25.1 it is assumed that all signals are **baseband signals**, which means that the frequency content is centered around $f = 0$ (DC). The channel passband, therefore, partially coincides with the transmitted spectrum. In general, this condition requires that the transmitted signal be modulated by an appropriate carrier frequency and demodulated at the receiver. In that case, the model in Fig. 25.1 still applies; however, *baseband-equivalent* signals must be derived from their modulated (passband) counterparts. *Baseband signalling* and *pulse shaping* refers to the way in which a group of source bits is mapped to a baseband transmitted pulse.

As a simple example of baseband signalling, we can take $m = 1$ (map each source bit to a pulse), assign a 0 bit to a pulse $p(t)$, and a 1 bit to the pulse $-p(t)$. Perhaps the simplest example of a baseband pulse is the *rectangular* pulse given by $p(t) = 1, 0 < t \leq T$, and $p(t) = 0$ elsewhere. In this case, we can write the transmitted signal as

$$s(t) = \sum_i A_i p(t - iT) \tag{25.3}$$

where each symbol A_i takes on a value of $+1$ or -1, depending on the value of the ith bit, and $1/T$ is the *symbol rate*, namely, the rate at which the symbols A_i are introduced to the channel.

The preceding example is called *binary* **pulse amplitude modulation (PAM)**, since the data symbols A_i are binary valued, and they amplitude modulate the transmitted pulse $p(t)$. The information rate (bits per second) in this case is the same as the symbol rate $1/T$. As a simple extension of this signalling technique, we can increase m and choose A_i from one of $M = 2^m$ values to transmit at bit rate m/T. This is known as M-ary PAM. For example, letting $m = 2$, each pair of bits can be mapped to a pulse in the set $\{p(t), -p(t), 3p(t), -3p(t)\}$.

In general, the transmitted symbols $\{A_i\}$, the baseband pulse $p(t)$, and channel impulse response $g(t)$ can be *complex valued*. For example, each successive pair of bits might select a symbol from the set $\{1, -1, j, -j\}$, where $j = \sqrt{-1}$. This is a consequence of considering the baseband equivalent of passband modulation. (That is, generating a transmitted spectrum which is centered around a carrier frequency f_c.) Here we are not concerned with the relation between the passband and baseband equivalent models and simply point out that the discussion and results in this chapter apply to complex-valued symbols and pulse shapes.

As an example of a signalling technique which is not PAM, let $m = 1$ and

$$p(0; t) = \begin{cases} \sqrt{2}\sin(2\pi f_1 t) & 0 < t < T \\ 0 & \text{elsewhere} \end{cases}$$
$$p(1; t) = \begin{cases} \sqrt{2}\sin(2\pi f_2 t) & 0 < t < T \\ 0 & \text{elsewhere} \end{cases} \quad (25.4)$$

where f_1 and $f_2 \neq f_1$ are fixed frequencies selected so that $f_1 T$ and $f_2 T$ (number of cycles for each bit) are multiples of $1/2$. These pulses are *orthogonal*, namely,

$$\int_0^T p(1; t) p(0; t) \, dt = 0$$

This choice of pulse shapes is called binary **frequency-shift keying (FSK)**.

Another example of a set of orthogonal pulse shapes for $m = 2$ bits/T is shown in Fig. 25.2. Because these pulses may have as many as three transitions within a symbol period, the transmitted spectrum occupies roughly four times the transmitted spectrum of binary PAM with a rectangular pulse shape. The spectrum is, therefore, spread across a much larger band than the smallest required for reliable transmission, assuming a data rate of $2/T$. This type of signalling is referred

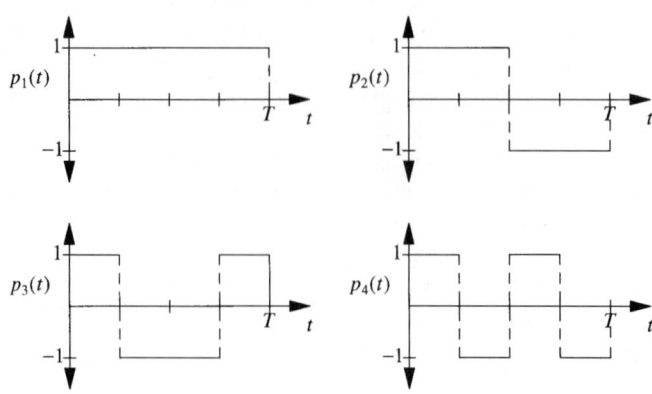

FIGURE 25.2 Four orthogonal spread-spectrum pulse shapes.

Baseband Signalling and Pulse Shaping

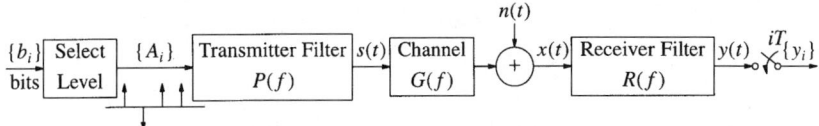

FIGURE 25.3 Baseband model of a pulse amplitude modulation system.

to as **spread-spectrum**. Spread-spectrum signals are more robust with respect to interference from other transmitted signals than are narrowband signals.[1]

25.2 Intersymbol Interference and the Nyquist Criterion

Consider the transmission of a PAM signal illustrated in Fig. 25.3. The source bits $\{b_i\}$ are mapped to a sequence of levels $\{A_i\}$, which modulate the transmitter pulse $p(t)$. The channel input is, therefore, given by Eq. (25.3) where $p(t)$ is the impulse response of the transmitter *pulse-shaping filter* $P(f)$ shown in Fig. 25.3. The input to the transmitter filter $P(f)$ is the modulated sequence of delta functions $\sum_i A_i \delta(t - iT)$. The channel is represented by the transfer function $G(f)$ (plus noise), which has impulse response $g(t)$, and the receiver filter has transfer function $R(f)$ with associated impulse response $r(t)$.

Let $h(t)$ be the overall impulse response of the combined transmitter, channel, and receiver, which has transfer function $H(f) = P(f)G(f)R(f)$. We can write $h(t) = p(t) * g(t) * r(t)$. The output of the receiver filter is then

$$y(t) = \sum_i A_i h(t - iT) + \tilde{n}(t) \qquad (25.5)$$

where $\tilde{n}(t) = r(t) * n(t)$ is the output of the filter $R(f)$ with input $n(t)$. Assuming that samples are collected at the output of the filter $R(f)$ at the symbol rate $1/T$, we can write the kth sample of $y(t)$ as

$$\begin{aligned} y(kT) &= \sum_i A_i h(kT - iT) + \tilde{n}(kT) \\ &= A_k h(0) + \sum_{i \neq k} A_i h(kT - iT) + \tilde{n}(kT) \end{aligned} \qquad (25.6)$$

The first term on the right-hand side of Eq. (25.6) is the kth transmitted symbol scaled by the system impulse response at $t = 0$. If this were the only term on the right side of Eq. (25.6), we could obtain the source bits without error by scaling the received samples by $1/h(0)$. The second term on the right-hand side of Eq. (25.6) is called **intersymbol interference**, which reflects the view that neighboring symbols interfere with the detection of each desired symbol.

One possible criterion for choosing the transmitter and receiver filters is to minimize intersymbol interference. Specifically, if we choose $p(t)$ and $r(t)$ so that

$$h(kT) = \begin{cases} 1 & k = 0 \\ 0 & k \neq 0 \end{cases} \qquad (25.7)$$

---

[1]This example can also be viewed as coded binary PAM. Namely, each pair of two source bits are mapped to 4 coded bits, which are transmitted via binary PAM with a rectangular pulse. The current IS-95 air interface uses an extension of this signalling method in which groups of 6 b are mapped to 64 orthogonal pulse shapes with as many as 63 transitions during a symbol.

then the kth received sample is

$$y(kT) = A_k + \tilde{n}(kT) \tag{25.8}$$

In this case, the intersymbol interference has been eliminated. This choice of $p(t)$ and $r(t)$ is called a **zero-forcing** solution, since it forces the intersymbol interference to zero. Depending on the type of detection scheme used, a zero-forcing solution may not be desirable. This is because the probability of error also depends on the noise intensity, which generally increases when intersymbol interference is suppressed. It is instructive, however, to examine the properties of the zero-forcing solution.

We now view Eq. (25.7) in the frequency domain. Since $h(t)$ has Fourier transform

$$H(f) = P(f)G(f)R(f) \tag{25.9}$$

where $P(f)$ is the Fourier transform of $p(t)$, the bandwidth of $H(f)$ is limited by the bandwidth of the channel $G(f)$. We will assume that $G(f) = 0$, $|f| > W$. The sampled impulse response $h(kT)$ can, therefore, be written as the inverse Fourier transform

$$h(kT) = \int_{-W}^{W} H(f) e^{j2\pi f kT} \, df$$

Through a series of manipulations, this integral can be rewritten as an inverse discrete Fourier transform,

$$h(kT) = T \int_{-1/(2T)}^{1/(2T)} H_{eq}(e^{j2\pi fT}) e^{j2\pi f kT} \, df \tag{25.10a}$$

where

$$H_{eq}(e^{j2\pi fT}) = \frac{1}{T} \sum_k H\left(f + \frac{k}{T}\right)$$

$$= \frac{1}{T} \sum_k P\left(f + \frac{k}{T}\right) G\left(f + \frac{k}{T}\right) R\left(f + \frac{k}{T}\right) \tag{25.10b}$$

This relation states that $H_{eq}(z)$, $z = e^{j2\pi fT}$, is the discrete Fourier transform of the sequence $\{h_k\}$, where $h_k = h(kT)$. Sampling the impulse response $h(t)$ therefore changes the transfer function $H(f)$ to the *aliased* frequency response $H_{eq}(e^{j2\pi fT})$. From Eqs. (25.10) and (25.6) we conclude that $H_{eq}(z)$ is the transfer function that relates the sequence of input data symbols $\{A_i\}$ to the sequence of received samples $\{y_i\}$, where $y_i = y(iT)$, in the absence of noise. This is illustrated in Fig. 25.4. For this reason, $H_{eq}(z)$ is called the **equivalent discrete-time transfer function** for the overall system transfer function $H(f)$.

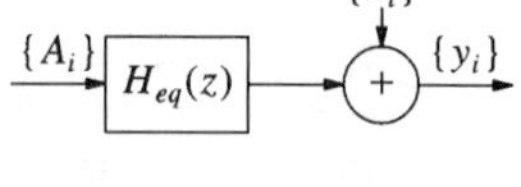

FIGURE 25.4 Equivalent discrete-time channel for the PAM system shown in Fig. 25.3 $[y_i = y(iT), \tilde{n}_i = \tilde{n}(iT)]$

Since $H_{eq}(e^{j2\pi fT})$ is the discrete Fourier transform of the sequence $\{h_k\}$, the time-domain, or sequence condition (4.7) is equivalent to the frequency-domain condition

$$H_{eq}(e^{j2\pi fT}) = 1 \tag{25.11}$$

This relation is called the **Nyquist criterion**. From Eqs. (25.10b) and (25.11) we make the following observations.

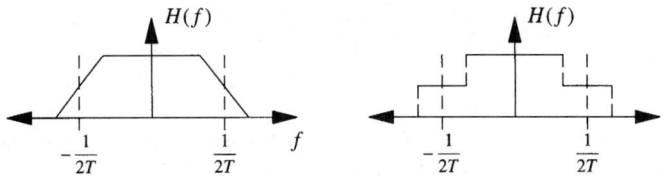

FIGURE 25.5 Two examples of frequency responses that satisfy the Nyquist criterion.

1. To satisfy the Nyquist criterion, the channel bandwidth W must be at least $1/(2T)$. Otherwise, $G(f + n/T) = 0$ for f in some interval of positive length for all n, which implies that $H_{eq}(e^{j2\pi fT}) = 0$ for f in the same interval.
2. For the minimum bandwidth $W = 1/(2T)$, Eqs. (25.10b) and (25.11) imply that $H(f) = T$ for $|f| < 1/(2T)$ and $H(f) = 0$ elsewhere. This implies that the system impulse response is given by

$$h(t) = \frac{\sin(\pi t/T)}{\pi t/T} \qquad (25.12)$$

(Since $\int_{-\infty}^{\infty} h^2(t)\,dt = T$, the transmitted signal $s(t) = \sum_i A_i h(t - iT)$ has power equal to the symbol variance $E[|A_i|^2]$.) The impulse response in Eq. (25.12) is called a *minimum bandwidth* or Nyquist pulse. The frequency band $[-1/(2T), 1/(2T)]$ [i.e., the passband of $H(f)$] is called the **Nyquist band**.

3. Suppose that the channel is bandlimited to twice the Nyquist bandwidth. That is, $G(f) = 0$ for $|f| > 1/T$. The condition (25.11) then becomes

$$H(f) + H\left(f - \frac{1}{T}\right) + H\left(f + \frac{1}{T}\right) = T \qquad (25.13)$$

Assume for the moment that $H(f)$ and $h(t)$ are both real valued, so that $H(f)$ is an even function of $f[H(f) = H(-f)]$. This is the case when the receiver filter is the matched filter (see Sec. 25.3). We can then rewrite Eq. (25.13) as

$$H(f) + H\left(\frac{1}{T} - f\right) = T, \qquad 0 < f < \frac{1}{2T} \qquad (25.14)$$

which states that $H(f)$ must have odd symmetry about $f = 1/(2T)$. This is illustrated in Fig. 25.5, which shows two different transfer functions $H(f)$ that satisfy the Nyquist criterion.
4. The pulse shape $p(t)$ enters into Eq. (25.11) only through the product $P(f)R(f)$. Consequently, either $P(f)$ or $R(f)$ can be fixed, and the other filter can be adjusted or adapted to the particular channel. Typically, the pulse shape $p(t)$ is fixed, and the receiver filter is adapted to the (possibly time-varying) channel.

Raised Cosine Pulse

Suppose that the channel is ideal with transfer function

$$G(f) = \begin{cases} 1, & |f| < W \\ 0, & |f| > W \end{cases} \qquad (25.15)$$

To maximize bandwidth efficiency, Nyquist pulses given by Eq. (25.12) should be used where $W = 1/(2T)$. This type of signalling, however, has two major drawbacks. First, Nyquist pulses are noncausal and of infinite duration. They can be approximated in practice by introducing an appropriate delay, and truncating the pulse. The pulse, however, decays very slowly, namely, as $1/t$, so that the truncation window must be wide. This is equivalent to observing that the ideal bandlimited frequency response given by Eq. (25.15) is difficult to approximate closely. The second drawback, which is more important, is the fact that this type of signalling is not robust with respect to sampling jitter. Namely, a small sampling offset ε produces the output sample

$$y(kT + \varepsilon) = \sum_i A_i \frac{\sin[\pi(k - i + \varepsilon/T)]}{\pi(k - i + \varepsilon/T)} \tag{25.16}$$

Since the Nyquist pulse decays as $1/t$, this sum is not guaranteed to converge. A particular choice of symbols $\{A_i\}$ can, therefore, lead to very large intersymbol interference, no matter how small the offset. Minimum bandwidth signalling is therefore impractical.

The preceding problem is generally solved in one of two ways in practice:

1. The pulse bandwidth is increased to provide a faster pulse decay than $1/t$.
2. A *controlled* amount of intersymbol interference is introduced at the transmitter, which can be subtracted out at the receiver.

The former approach sacrifices bandwidth efficiency, whereas the latter approach sacrifices power efficiency. We will examine the latter approach in Sec. 25.5. The most common example of a pulse, which illustrates the first technique, is the **raised cosine pulse**, given by

$$h(t) = \left[\frac{\sin(\pi t/T)}{\pi t/T}\right]\left[\frac{\cos(\alpha \pi t/T)}{1 - (2\alpha t/T)^2}\right] \tag{25.17}$$

which has Fourier transform

$$H(f) = \begin{cases} T & 0 \leq |f| \leq \frac{1-\alpha}{2T} \\ \frac{T}{2}\left\{1 + \cos\left[\frac{\pi T}{\alpha}\left(|f| - \frac{1-\alpha}{2T}\right)\right]\right\} & \frac{1-\alpha}{2T} \leq |f| \leq \frac{1+\alpha}{2T} \\ 0 & |f| > \frac{1+\alpha}{2T} \end{cases} \tag{25.18}$$

where $0 \leq \alpha \leq 1$.

Plots of $p(t)$ and $P(f)$ are shown in Fig. 25.6 for different values of α. It is easily verified that $h(t)$ satisfies the Nyquist criterion (25.7) and, consequently, $H(f)$ satisfies Eq. (25.11). When $\alpha = 0$, $H(f)$ is the Nyquist pulse with minimum bandwidth $1/(2T)$, and when $\alpha > 0$, $H(f)$ has bandwidth $(1 + \alpha)/(2T)$ with a raised cosine rolloff. The parameter α, therefore, represents the additional, or **excess bandwidth** as a fraction of the minimum bandwidth $1/(2T)$. For example, when $\alpha = 1$, we say that that the pulse is a raised cosine pulse with 100% excess bandwidth. This is because the pulse bandwidth $1/T$ is twice the minimum bandwidth. Because the raised cosine pulse decays as $1/t^3$, performance is robust with respect to sampling offsets.

The raised cosine frequency response (25.18) applies to the combination of transmitter, channel, and receiver. If the transmitted pulse shape $p(t)$ is a raised cosine pulse, then $h(t)$ is a raised cosine pulse only if the combined receiver and channel frequency response is constant. Even with an ideal (transparent) channel, however, the optimum (matched) receiver filter response is generally not constant in the presence of additive Gaussian noise. An alternative is to transmit the *square-root raised cosine* pulse shape, which has frequency response $P(f)$ given by the square-root of the raised cosine frequency response in Eq. (25.18). Assuming an ideal channel, setting the receiver frequency response $R(f) = P(f)$ then results in an overall raised cosine system response $H(f)$.

Baseband Signalling and Pulse Shaping

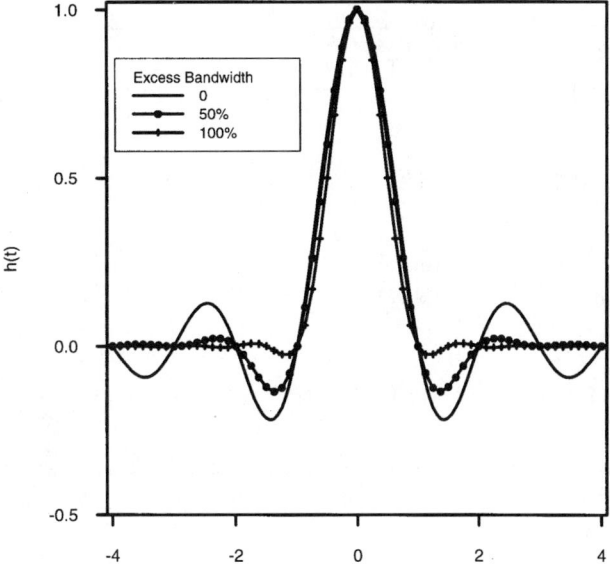

FIGURE 25.6(a) Raised cosine pulse.

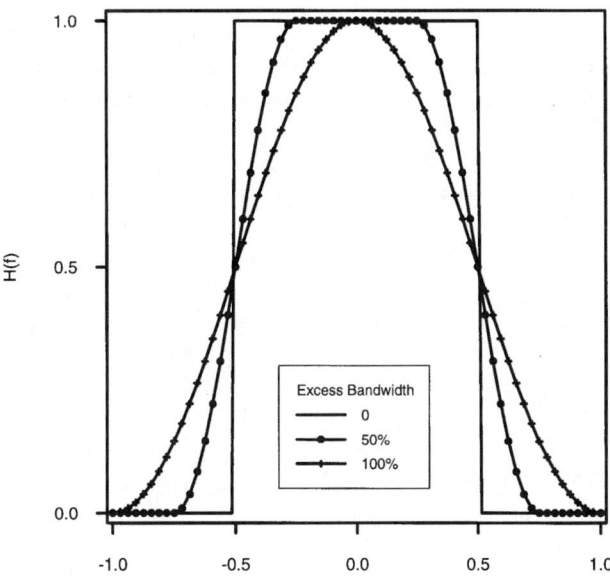

FIGURE 25.6(b) Raised cosine spectrum.

25.3 Nyquist Criterion with Matched Filtering

Consider the transmission of an isolated pulse $A_0\delta(t)$. In this case the input to the receiver in Fig. 25.3 is

$$x(t) = A_0\tilde{g}(t) + n(t) \qquad (25.19)$$

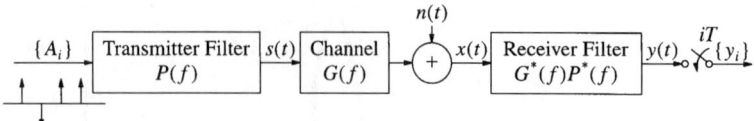

FIGURE 25.7 Baseband PAM model with a matched filter at the receiver.

where $\tilde{g}(t)$ is the inverse Fourier transform of the combined transmitter-channel transfer function $\tilde{G}(f) = P(f)G(f)$. We will assume that the noise $n(t)$ is white with spectrum $N_0/2$. The output of the receiver filter is then

$$y(t) = r(t) * x(t) = A_0[r(t) * \tilde{g}(t)] + [r(t) * n(t)] \tag{25.20}$$

The first term on the right-hand side is the desired signal, and the second term is noise. Assuming that $y(t)$ is sampled at $t = 0$, the ratio of signal energy to noise energy, or signal-to-noise ratio (SNR) at the sampling instant, is

$$\text{SNR} = \frac{E[|A_0|^2]\left|\int_{-\infty}^{\infty} r(-t)\tilde{g}(t)\,dt\right|^2}{\dfrac{N_0}{2}\int_{-\infty}^{\infty} |r(t)|^2\,dt} \tag{25.21}$$

The receiver impulse response that maximizes this expression is $r(t) = \tilde{g}^*(-t)$ [complex conjugate of $\tilde{g}(-t)$], which is known as the **matched filter** impulse response. The associated transfer function is $R(f) = \tilde{G}^*(f)$.

Choosing the receiver filter to be the matched filter is optimal in more general situations, such as when detecting a sequence of channel symbols with intersymbol interference (assuming the additive noise is Gaussian). We, therefore, reconsider the Nyquist criterion when the receiver filter is the matched filter. In this case, the baseband model is shown in Fig. 25.7, and the output of the receiver filter is given by

$$y(t) = \sum_i A_i h(t - iT) + \tilde{n}(t) \tag{25.22}$$

where the baseband pulse $h(t)$ is now the impulse response of the filter with transfer function $|\tilde{G}(f)|^2 = |P(f)G(f)|^2$. This impulse response is the *autocorrelation* of the impulse response of the combined transmitter-channel filter $\tilde{G}(f)$,

$$h(t) = \int_{-\infty}^{\infty} \tilde{g}^*(s)\tilde{g}(s+t)\,ds \tag{25.23}$$

With a matched filter at the receiver, the equivalent discrete-time transfer function is

$$H_{\text{eq}}(e^{j2\pi fT}) = \frac{1}{T}\sum_k \left|\tilde{G}\left(f - \frac{k}{T}\right)\right|^2$$

$$= \frac{1}{T}\sum_k \left|P\left(f - \frac{k}{T}\right)G\left(f - \frac{k}{T}\right)\right|^2 \tag{25.24}$$

Baseband Signalling and Pulse Shaping

which relates the sequence of transmitted symbols $\{A_k\}$ to the sequence of received samples $\{y_k\}$ in the absence of noise. Note that $H_{eq}(e^{j2\pi fT})$ is positive, real valued, and an even function of f. If the channel is bandlimited to twice the Nyquist bandwidth, then $H(f) = 0$ for $|f| > 1/T$, and the Nyquist condition is given by Eq. (25.14) where $H(f) = |G(f)P(f)|^2$. The aliasing sum in Eq. (25.10b) can therefore be described as a folding operation in which the channel response $|H(f)|^2$ is folded around the Nyquist frequency $1/(2T)$. For this reason, $H_{eq}(e^{j2\pi fT})$ with a matched receiver filter is often referred to as the folded channel spectrum.

25.4 Eye Diagrams

One way to assess the severity of distortion due to intersymbol interference in a digital communications system is to examine the **eye diagram**. The eye diagram is illustrated in Fig. 25.8 for a raised cosine pulse shape with 25% excess bandwidth and an ideal bandlimited channel. Figure 25.8(a) shows the data signal at the receiver

$$y(t) = \sum_i A_i h(t - iT) + n(t) \qquad (25.25)$$

where $h(t)$ is given by Eq. (25.17), $\alpha = 1/4$, each symbol A_i is independently chosen from the set $\{\pm 1, \pm 3\}$, where each symbol is equally likely, and $n(t)$ is bandlimited white Gaussian noise. (The received SNR is 30 dB.) The eye diagram is constructed from the time-domain data signal $y(t)$ as follows (assuming nominal sampling times at $kT, k = 0, 1, 2, \ldots$):

1. Partition the waveform $y(t)$ into successive segments of length T starting from $t = T/2$.
2. Translate each of these waveform segments $[y(t), (k + 1/2)T \leq t \leq (k + 3/2)T, k = 0, 1, 2, \ldots]$ to the interval $[-T/2, T/2]$, and superimpose.

The resulting picture is shown in Fig. 25.8(b) for the $y(t)$ shown in Fig. 25.8(a). (Partitioning $y(t)$ into successive segments of length iT, $i > 1$, is also possible. This would result in i successive eye diagrams.) The number of eye openings is one less than the number of transmitted signal levels. In practice, the eye diagram is easily viewed on an oscilloscope by applying the received waveform $y(t)$ to the vertical deflection plates of the oscilloscope and applying a sawtooth waveform at the

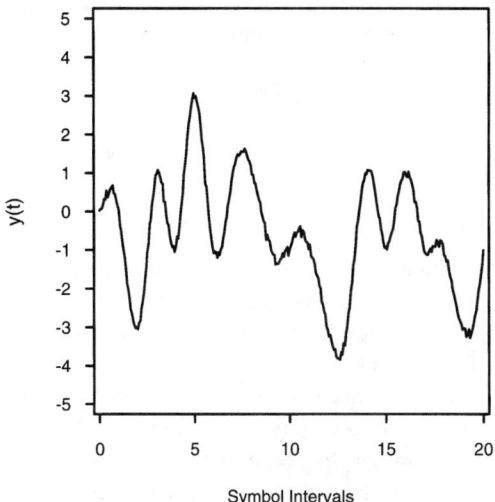

FIGURE 25.8(a) Received signal $y(t)$.

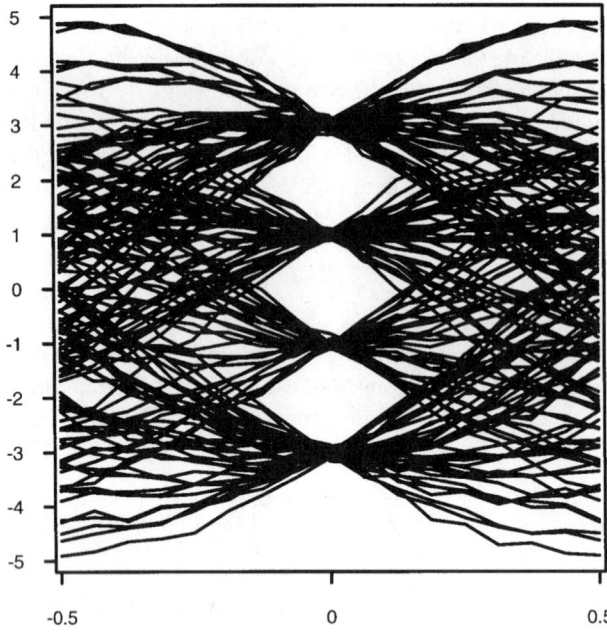

FIGURE 25.8(b) Eye diagram for received signal shown in Fig. 25.8(a).

symbol rate $1/T$ to the horizontal deflection plates. This causes successive symbol intervals to be translated into one interval on the oscilloscope display.

Each waveform segment $y(t)$, $(k+1/2)T \leq t \leq (k+3/2)T$, depends on the particular sequence of channel symbols surrounding A_k. The number of channel symbols that affects a particular waveform segment depends on the extent of the intersymbol interference, shown in Eq. (25.6). This, in turn, depends on the duration of the impulse response $h(t)$. For example, if $h(t)$ has most of its energy in the interval $0 < t < mT$, then each waveform segment depends on approximately m symbols. Assuming binary transmission, this implies that there are a total of 2^m waveform segments that can be superimposed in the eye diagram. (It is possible that only one sequence of channel symbols causes significant intersymbol interference, and this sequence occurs with very low probability.) In current digital wireless applications the impulse response typically spans only a few symbols.

The eye diagram has the following important features which measure the performance of a digital communications system.

Vertical Eye Opening

The vertical openings at any time t_0, $-T/2 \leq t_0 \leq T/2$, represent the separation between signal levels with worst-case intersymbol interference, assuming that $y(t)$ is sampled at times $t = kT + t_0$, $k = 0, 1, 2, \ldots$. It is possible for the intersymbol interference to be large enough so that this vertical opening between some, or all, signal levels disappears altogether. In that case, the eye is said to be closed. Otherwise, the eye is said to be open. A closed eye implies that if the estimated bits are obtained by thresholding the samples $y(kT)$, then the decisions will depend primarily on the intersymbol interference rather than on the desired symbol. The probability of error will, therefore, be close to $1/2$. Conversely, wide vertical spacings between signal levels imply a large degree of immunity to additive noise. In general, $y(t)$ should be sampled at the times $kT + t_0$, $k = 0, 1, 2, \ldots$, where t_0 is chosen to maximize the vertical eye opening.

Baseband Signalling and Pulse Shaping

Horizontal Eye Opening

The width of each opening indicates the sensitivity to timing offset. Specifically, a very narrow eye opening indicates that a small timing offset will result in sampling where the eye is closed. Conversely, a wide horizontal opening indicates that a large timing offset can be tolerated, although the error probability will depend on the vertical opening.

Slope of the Inner Eye

The slope of the inner eye indicates sensitivity to timing jitter or variance in the timing offset. Specifically, a very steep slope means that the eye closes rapidly as the timing offset increases. In this case, a significant amount of jitter in the sampling times significantly increases the probability of error.

The shape of the eye diagram is determined by the pulse shape. In general, the faster the baseband pulse decays, the wider the eye opening. For example, a rectangular pulse produces a box-shaped eye diagram (assuming binary signalling). The minimum bandwidth pulse shape Eq. (25.12) produces an eye diagram which is closed for all t except for $t = 0$. This is because, as shown earlier, an arbitrarily small timing offset can lead to an intersymbol interference term that is arbitrarily large, depending on the data sequence.

25.5 Partial-Response Signalling

To avoid the problems associated with Nyquist signalling over an ideal bandlimited channel, bandwidth and/or power efficiency must be compromised. Raised cosine pulses compromise bandwidth efficiency to gain robustness with respect to timing errors. Another possibility is to introduce a controlled amount of intersymbol at the transmitter, which can be removed at the receiver. This approach is called **partial-response (PR) signalling**. The terminology reflects the fact that the sampled system impulse response does not have the full response given by the Nyquist condition Eq. (25.7).

To illustrate PR signalling, suppose that the Nyquist condition Eq. (25.7) is replaced by the condition

$$h_k = \begin{cases} 1 & k = 0, 1 \\ 0 & \text{all other } k \end{cases} \qquad (25.26)$$

The kth received sample is then

$$y_k = A_k + A_{k-1} + \tilde{n}_k \qquad (25.27)$$

so that there is intersymbol interference from one neighboring transmitted symbol. For now we focus on the spectral characteristics of PR signalling and defer discussion of how to detect the transmitted sequence $\{A_k\}$ in the presence of intersymbol interference. The equivalent discrete-time transfer function in this case is the discrete Fourier transform of the sequence in Eq. (25.26),

$$H_{eq}(e^{j2\pi fT}) = \frac{1}{T} \sum_k H\left(f + \frac{k}{T}\right)$$

$$= 1 + e^{-j2\pi fT} = 2e^{-j\pi fT} \cos(\pi fT) \qquad (25.28)$$

As in the full-response case, for Eq. (25.28) to be satisfied, the *minimum* bandwidth of the channel $G(f)$ and transmitter filter $P(f)$ is $W = 1/(2T)$. Assuming $P(f)$ has this minimum bandwidth implies

$$H(f) = \begin{cases} 2Te^{-j\pi fT} \cos(\pi fT) & |f| < 1/(2T) \\ 0 & |f| > 1/(2T) \end{cases} \quad (25.29a)$$

and

$$h(t) = T\{\text{sinc}(t/T) + \text{sinc}[(t-T)/T]\} \quad (25.29b)$$

where $\text{sinc}\, x = (\sin \pi x)/(\pi x)$. This pulse is called a *duobinary* pulse and is shown along with the associated $H(f)$ in Fig. 25.9. [Notice that $h(t)$ satisfies Eq. (25.26).] Unlike the ideal bandlimited frequency response, the transfer function $H(f)$ in Eq. (25.29a) is continuous and is, therefore, easily approximated by a physically realizable filter. Duobinary PR was first proposed by Lender, 1963, and later generalized by Kretzmer, 1966.

The main advantage of the duobinary pulse Eq. (25.29b), relative to the minimum bandwidth pulse Eq. (25.12), is that signalling at the Nyquist symbol rate is feasible with zero excess bandwidth. Because the pulse decays much more rapidly than a Nyquist pulse, it is robust with respect to timing errors. Selecting the transmitter and receiver filters so that the overall system response is duobinary is appropriate in situations where the channel frequency response $G(f)$ is near zero or has a rapid rolloff at the Nyquist band edge $f = 1/(2T)$.

As another example of PR signaling, consider the *modified* duobinary partial response

$$h_k = \begin{cases} 1 & k = -1 \\ -1 & k = 1 \\ 0 & \text{all other } k \end{cases} \quad (25.30)$$

which has equivalent discrete-time transfer function

$$H_{\text{eq}}(e^{j2\pi fT}) = e^{j2\pi fT} - e^{-j2\pi fT}$$
$$= j2\sin(2\pi fT) \quad (25.31)$$

With zero excess bandwidth, the overall system response is

$$H(f) = \begin{cases} j2T\sin(2\pi fT) & |f| < 1/(2T) \\ 0 & |f| > 1/(2T) \end{cases} \quad (25.32a)$$

and

$$h(t) = T\{\text{sinc}[(t+T)/T] - \text{sinc}[(t-T)/T]\} \quad (25.32b)$$

These functions are plotted in Fig. 25.10. This pulse shape is appropriate when the channel response $G(f)$ is near zero at both DC ($f = 0$) and at the Nyquist band edge. This is often the case for wire (twisted-pair) channels where the transmitted signal is coupled to the channel through a transformer. Like duobinary PR, modified duobinary allows minimum bandwidth signalling at the Nyquist rate.

A particular partial response is often identified by the polynomial

$$\sum_{k=0}^{K} h_k D^k$$

where D (for delay) takes the place of the usual z^{-1} in the z transform of the sequence $\{h_k\}$. For example, duobinary is also referred to as $1 + D$ partial response.

Baseband Signalling and Pulse Shaping

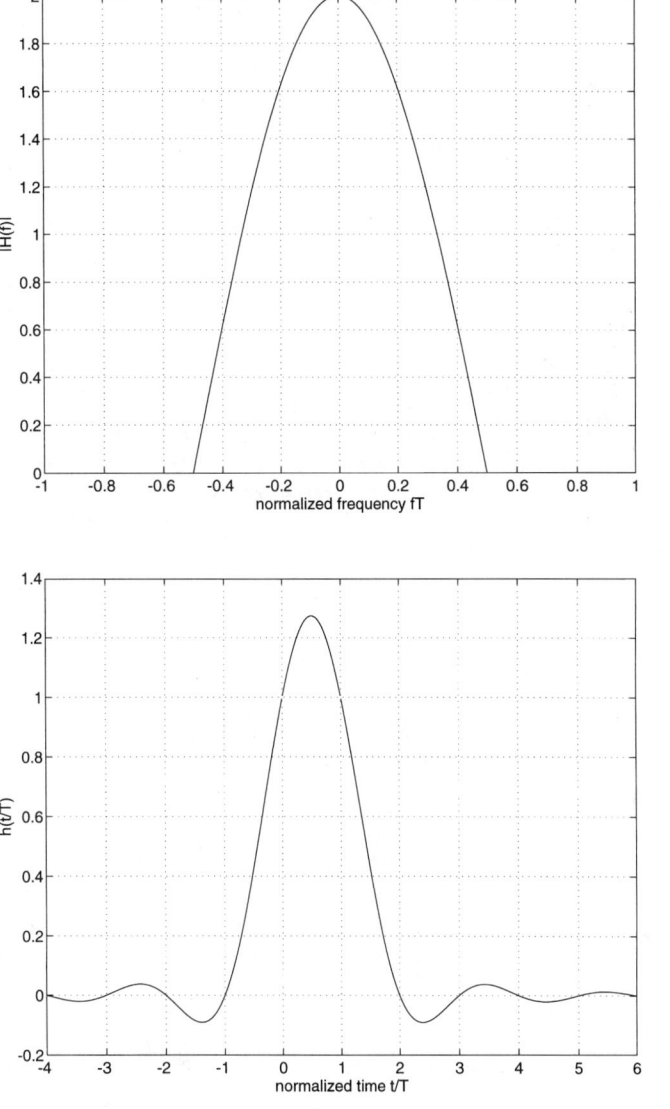

FIGURE 25.9 Duobinary frequency response and minimum bandwidth pulse.

In general, more complicated system responses than those shown in Figs. 25.9 and 25.10 can be generated by choosing more nonzero coefficients in the sequence $\{h_k\}$. This complicates detection, however, because of the additional intersymbol interference that is generated.

Rather than modulating a PR pulse $h(t)$, a PR signal can also be generated by filtering the sequence of transmitted levels $\{A_i\}$. This is shown in Fig. 25.11. Namely, the transmitted levels are first passed through a discrete-time (digital) filter with transfer function $P_d(e^{j2\pi fT})$ (where the subscript d indicates discrete). [Note that $P_d(e^{j2\pi fT})$ can be selected to be $H_{\text{eq}}(e^{j2\pi fT})$.] The outputs of this filter form the PAM signal, where the pulse shaping filter $P(f) = 1$, $|f| < 1/(2T)$ and is zero elsewhere. If the transmitted levels $\{A_k\}$ are selected independently and are identically distributed, then the transmitted spectrum is $\sigma_A^2 |P_d(e^{j2\pi fT})|^2$ for $|f| < 1/(2T)$ and is zero for $|f| > 1/(2T)$, where $\sigma_A^2 = E[|A_k|^2]$.

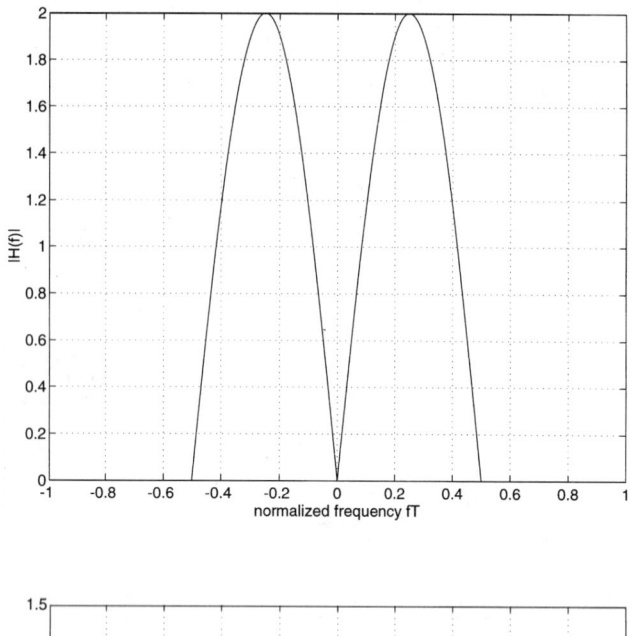

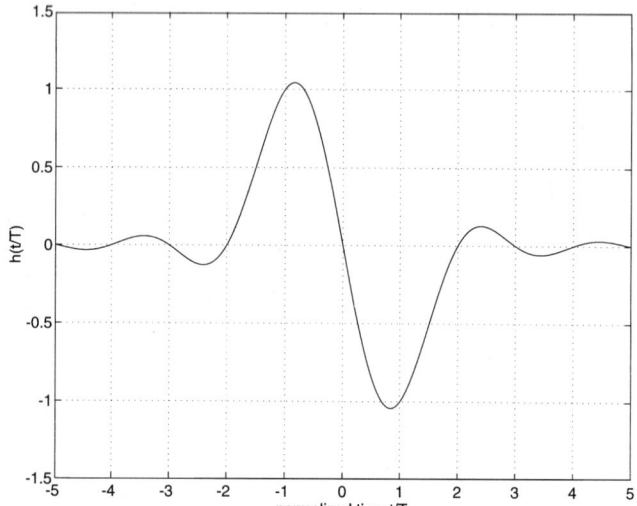

FIGURE 25.10 Modified duobinary frequency response and minimum bandwidth pulse.

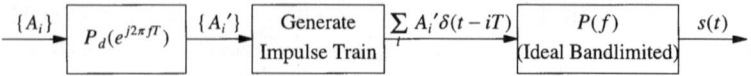

FIGURE 25.11 Generation of PR signal.

Shaping the transmitted spectrum to have nulls coincident with nulls in the channel response potentially offers significant performance advantages. By introducing intersymbol interference, however, PR signalling increases the number of received signal levels, which increases the complexity of the detector and may reduce immunity to noise. For example, the set of received signal levels for duobinary signalling is $\{0, \pm 2\}$ from which the transmitted levels $\{\pm 1\}$ must be estimated. The

Baseband Signalling and Pulse Shaping

FIGURE 25.12 Precoding for a PR channel.

TABLE 25.1 Example of Precoding for Duobinary PR.

$\{b_i\}$:		1	0	0	1	1	1	0	0	1	0
$\{b'_i\}$:	0	1	1	1	0	1	0	0	0	1	1
$\{A_i\}$:	-1	1	1	1	-1	1	-1	-1	-1	1	1
$\{y_i\}$:		0	2	2	0	0	0	-2	-2	0	2

performance of a particular PR scheme depends on the channel characteristics, as well as the type of detector used at the receiver. We now describe a simple suboptimal detection strategy.

Precoding

Consider the received signal sample Eq. (25.27) with duobinary signalling. If the receiver has correctly decoded the symbol A_{k-1}, then in the absence of noise A_k can be decoded by subtracting A_{k-1} from the received sample y_k. If an error occurs, however, then subtracting the preceding symbol estimate from the received sample will cause the error to propogate to successive detected symbols. To avoid this problem, the transmitted levels can be **precoded** in such a way as to compensate for the intersymbol interference introduced by the overall partial response.

We first illustrate precoding for duobinary PR. The sequence of operations is illustrated in Fig. 25.12. Let $\{b_k\}$ denote the sequence of source bits where $b_k \in \{0, 1\}$. This sequence is transformed to the sequence $\{b'_k\}$ by the operation

$$b'_k = b_k \oplus b'_{k-1} \qquad (25.33)$$

where $\oplus$ denotes modulo 2 addition (exclusive OR). The sequence $\{b'_k\}$ is mapped to the sequence of binary transmitted signal levels $\{A_k\}$ according to

$$A_k = 2b'_k - 1 \qquad (25.34)$$

That is, $b'_k = 0$ ($b_k = 1$) is mapped to the transmitted level $A_k = -1$ ($A_k = 1$). In the absence of noise, the received symbol is then

$$y_k = A_k + A_{k-1} = 2(b'_k + b'_{k-1} - 1) \qquad (25.35)$$

and combining Eqs. (25.33) and (25.35) gives

$$b_k = \left(\frac{1}{2}y_k + 1\right) \bmod 2 \qquad (25.36)$$

That is, if $y_k = \pm 2$, then $b_k = 0$, and if $y_k = 0$, then $b_k = 1$. Precoding, therefore, enables the detector to make *symbol-by-symbol* decisions that do not depend on previous decisions. Table 25.1 shows a sequence of transmitted bits $\{b_i\}$, precoded bits $\{b'_i\}$, transmitted signal levels $\{A_i\}$, and received samples $\{y_i\}$.

The preceding precoding technique can be extended to multilevel PAM and to other PR channels. Suppose that the PR is specified by

$$H_{eq}(D) = \sum_{k=0}^{K} h_k D^k$$

where the coefficients are integers and that the source symbols $\{b_k\}$ are selected from the set $\{0, 1, \ldots, M-1\}$. These symbols are transformed to the sequence $\{b'_k\}$ via the precoding operation

$$b'_k = \left(b_k - \sum_{i=1}^{K} h_i b'_{k-i}\right) \bmod M \qquad (25.37)$$

Because of the modulo operation, each symbol b'_k is also in the set $\{0, 1, \ldots, M-1\}$. The kth transmitted signal level is given by

$$A_k = 2b'_k - (M-1) \qquad (25.38)$$

so that the set of transmitted levels is $\{-(M-1), \ldots, (M-1)\}$ (i.e., a shifted version of the set of values assumed by b_k). In the absence of noise the received sample is

$$y_k = \sum_{i=0}^{K} h_i A_{k-i} \qquad (25.39)$$

and it can be shown that the kth source symbol is given by

$$b_k = \frac{1}{2}(y_k + (M-1) \cdot H_{eq}(1)) \bmod M \qquad (25.40)$$

Precoding the symbols $\{b_k\}$ in this manner, therefore, enables symbol-by-symbol decisions at the receiver. In the presence of noise, more sophisticated detection schemes (e.g., maximum likelihood) can be used with PR signalling to obtain improvements in performance.

25.6 Additional Considerations

In many applications, bandwidth and intersymbol interference are not the only important considerations for selecting baseband pulses. Here we give a brief discussion of additional practical constraints that may influence this selection.

Average Transmitted Power and Spectral Constraints

The constraint on average transmitted power varies according to the application. For example, low-average power is highly desirable for mobile wireless applications that use battery-powered transmitters. In many applications (e.g., digital subscriber loops, as well as digital radio), constraints are imposed to limit the amount of interference, or crosstalk, radiated into neighboring receivers and communications systems. Because this type of interference is frequency dependent, the constraint may take the form of a spectral mask that specifies the maximum allowable transmitted power as a function of frequency. For example, crosstalk in wireline channels is generally caused by capacitive coupling and increases as a function of frequency. Consequently, to reduce the amount of crosstalk generated at a particular transmitter, the pulse shaping filter generally attenuates high frequencies more than low frequencies.

Baseband Signalling and Pulse Shaping

In radio applications where signals are assigned different frequency bands, constraints on the transmitted spectrum are imposed to limit *adjacent-channel interference.* This interference is generated by transmitters assigned to adjacent frequency bands. Therefore, a constraint is needed to limit the amount of *out-of-band power* generated by each transmitter, in addition to an overall average power constraint. To meet this constraint, the transmitter filter in Fig. 25.3 must have a sufficiently steep rolloff at the edges of the assigned frequency band. (Conversely, if the transmitted signals are time multiplexed, then the duration of the system impulse response must be contained within the assigned time slot.)

Peak-to-Average Power

In addition to a constraint on average transmitted power, a *peak-power* constraint is often imposed as well. This constraint is important in practice for the following reasons:

1. The dynamic range of the transmitter is limited. In particular, saturation of the output amplifier will "clip" the transmitted waveform.
2. Rapid fades can severely distort signals with high peak-to-average power.
3. The transmitted signal may be subjected to nonlinearities. Saturation of the output amplifier is one example. Another example that pertains to wireline applications is the companding process in the voice telephone network [Kalet and Saltzberg, 1994]. Namely, the compander used to reduce quantization noise for pulse-code modulated voice signals introduces amplitude-dependent distortion in data signals.

The preceding impairments or constraints indicate that the transmitted waveform should have a low peak-to-average power ratio. The peak-to-average power ratio is minimized by using binary signalling with rectangular pulse shapes. However, this compromises bandwidth efficiency. In applications where peak-to-average ratio should be low, binary signalling with rounded pulses are often used.

Channel and Receiver Characteristics

The type of channel impairments encountered and the type of detection scheme used at the receiver can also influence the choice of a transmitted pulse shape. For example, a constant amplitude pulse is appropriate for a fast fading environment with noncoherent detection. The ability to track channel characteristics, such as phase, may allow more bandwidth efficient pulse shapes in addition to multilevel signalling.

High-speed data communications over time-varying channels requires that the transmitter and/or receiver adapt to the changing channel characteristics. Adapting the transmitter to compensate for a time-varying channel requires a feedback channel through which the receiver can notify the transmitter of changes in channel characteristics. Because of this extra complication, adapting the receiver is often preferred to adapting the transmitter pulse shape. However, the following examples are notable exceptions.

1. The current IS-95 air interface for direct-sequence code-division multiple access adapts the transmitter power to control the amount of interference generated and to compensate for channel fades. This can be viewed as a simple form of adaptive transmitter pulse shaping in which a single parameter associated with the pulse shape is varied.
2. Multitone modulation divides the channel bandwidth into small subbands, and the transmitted power and source bits are distributed among these subbands to maximize the information rate. The received signal-to-noise ratio for each subband must be transmitted back to the transmitter to guide the allocation of transmitted bits and power [Bingham, 1990].

In addition to multitone modulation, *adaptive precoding* (also known as Tomlinson–Harashima precoding [Tomlinson, 1971; Harashima and Miyakawa, 1972]) is another way in which the transmitter can adapt to the channel frequency response. Adaptive precoding is an extension of the technique described earlier for partial-response channels. Namely, the equivalent discrete-time channel impulse response is measured at the receiver and sent back to the transmitter, where it is used in a precoder. The precoder compensates for the intersymbol interference introduced by the channel, allowing the receiver to detect the data by a simple threshhold operation. Both multitone modulation and precoding have been used with wireline channels (voiceband modems and digital subscriber loops).

Complexity

Generation of a bandwidth-efficient signal requires a filter with a sharp cutoff. In addition, bandwidth-efficient pulse shapes can complicate other system functions, such as timing and carrier recovery. If sufficient bandwidth is available, the cost can be reduced by using a rectangular pulse shape with a simple detection strategy (low-pass filter and threshold).

Tolerance to Interference

Interference is one of the primary channel impairments associated with digital radio. In addition to adjacent-channel interference described earlier, *cochannel interference* may be generated by other transmitters assigned to the same frequency band as the desired signal. Co-channel interference can be controlled through frequency (and perhaps time slot) assignments and by pulse shaping. For example, assuming fixed average power, increasing the bandwidth occupied by the signal lowers the power spectral density and decreases the amount of interference into a narrowband system that occupies part of the available bandwidth. Sufficient bandwidth spreading, therefore, enables wideband signals to be overlaid on top of narrowband signals without disrupting either service.

Probability of Intercept and Detection

The broadcast nature of wireless channels generally makes eavesdropping easier than for wired channels. A requirement for most commercial, as well as military applications, is to guarantee the privacy of user conversations (low probability of intercept). An additional requirement, in some applications, is that determining whether or not communications is taking place must be difficult (low probability of detection). Spread-spectrum waveforms are attractive in these applications since spreading the pulse energy over a wide frequency band decreases the power spectral density and, hence, makes the signal less visible. Power-efficient modulation combined with coding enables a further reduction in transmitted power for a target error rate.

25.7 Examples

We conclude this chapter with a brief description of baseband pulse shapes used in existing and emerging standards for digital mobile cellular and Personal Communications Services (PCS).

Global System for Mobile Communications (GSM)

The European GSM standard for digital mobile cellular communications operates in the 900-MHz frequency band, and is based on time-division multiple access (TDMA) [Rahnema, 1993]. A special variant of binary FSK is used called *Gaussian minimum-shift keying (GMSK)*. The GMSK modulator is illustrated in Fig. 25.13. The input to the modulator is a binary PAM signal $s(t)$, given by Eq. (25.3), where the pulse $p(t)$ is a Gaussian function and $|s(t)| < 1$. This waveform frequency modulates the carrier f_c, so that the (passband) transmitted signal is

$$w(t) = K \cos\left[2\pi f_c t + 2\pi f_d \int_{-\infty}^{t} s(\tau)\, d\tau\right]$$

Baseband Signalling and Pulse Shaping

FIGURE 25.13 Generation of GMSK signal; LPF is low-pass filter.

The maximum frequency deviation from the carrier is $f_d = 1/(2T)$, which characterizes minimum-shift keying. This technique can be used with a noncoherent receiver that is easy to implement. Because the transmitted signal has a constant envelope, the data can be reliably detected in the presence of rapid fades that are characteristic of mobile radio channels.

U.S. Digital Cellular (IS-54)

The IS-54 air interface operates in the 800-MHz band and is based on TDMA [EIA/TIA, 1991]. The baseband signal is given by Eq. (25.3) where the symbols are complex-valued, corresponding to quadrature phase modulation. The pulse has a square-root raised cosine spectrum with 35% excess bandwidth.

Interim Standard-95

The IS-95 air interface for digital mobile cellular uses spread-spectrum signalling (CDMA) in the 800-MHz band [TIA, 1993]. The baseband transmitted pulse shapes are analogous to those shown in Fig. 25.2, where the number of square pulses (chips) per bit is 128. To improve spectral efficiency the (wideband) transmitted signal is filtered by an approximation to an ideal low-pass response with a small amount of excess bandwidth. This shapes the chips so that they resemble minimum bandwidth pulses.

Personal Access Communications System (PACS)

Both PACS and the Japanese personal handy phone (PHP) system are TDMA systems which have been proposed for personal communications systems (PCS), and operate near 2 Ghz [Cox, 1995]. The baseband signal is given by Eq. (25.3) with four complex symbols representing four-phase quadrature modulation. The baseband pulse has a square-root raised cosine spectrum with 50% excess bandwidth.

Defining Terms

Baseband signal: A signal with frequency content centered around DC.
Equivalent discrete-time transfer function: A discrete-time transfer function (z transform) that relates the transmitted amplitudes to received samples in the absence of noise.
Excess bandwidth: That part of the baseband transmitted spectrum which is not contained within the Nyquist band.
Eye diagram: Superposition of segments of a received PAM signal that indicates the amount of intersymbol interference present.
Frequency-shift keying: A digital modulation technique in which the transmitted pulse is sinusoidal, where the frequency is determined by the source bits.
Intersymbol interference: The additive contribution (interference) to a received sample from transmitted symbols other than the symbol to be detected.

Matched filter: The receiver filter with impulse response equal to the time-reversed, complex conjugate impulse response of the combined transmitter filter-channel impulse response.

Nyquist band: The narrowest frequency band that can support a PAM signal without intersymbol interference (the interval $[-1/(2T), 1/(2T)]$ where $1/T$ is the symbol rate).

Nyquist criterion: A condition on the overall frequency response of a PAM system that ensures the absence of intersymbol interference.

Partial-response signalling: A signalling technique in which a controlled amount of intersymbol interference is introduced at the transmitter in order to shape the transmitted spectrum.

Precoding: A transformation of source symbols at the transmitter that compensates for intersymbol interference introduced by the channel.

Pulse amplitude modulation (PAM): A digital modulation technique in which the source bits are mapped to a sequence of amplitudes that modulate a transmitted pulse.

Raised cosine pulse: A pulse shape with Fourier transform that decays to zero according to a raised cosine; see Eq. (25.18). The amount of excess bandwidth is conveniently determined by a single parameter (α).

Spread spectrum: A signalling technique in which the pulse bandwidth is many times wider than the Nyquist bandwidth.

Zero-forcing criterion: A design constraint which specifies that intersymbol interference be eliminated.

References

Bingham, J.A.C. 1990. Multicarrier modulation for data transmission: an idea whose time has come. *IEEE Commun. Mag.* 28(May):5–14.

Cox, D.C. 1995. Wireless personal communications: what is it? *IEEE Personal Comm.* 2(2):20–35.

Electronic Industries Association/Telecommunications Industry Association. 1991. Recommended minimum performance standards for 800 MHz dual-mode mobile stations. Incorp. EIA/TIA 19B, EIA/TIA Project No. 2216, March.

Harashima, H. and Miyakawa, H. 1972. Matched-transmission technique for channels with intersymbol interference. *IEEE Trans. on Commun.* COM-20(Aug.):774–780.

Kalet, I. and Saltzberg, B.R. 1994. QAM transmission through a companding channel—signal constellations and detection. *IEEE Trans. on Comm.* 42(2–4):417–429.

Kretzmer, E.R. 1966. Generalization of a technique for binary data communication. *IEEE Trans. Comm. Tech.* COM-14 (Feb.):67, 68.

Lender, A. 1963. The duobinary technique for high-speed data Transmission. *AIEE Trans. on Comm. Electronics*, 82 (March):214–218.

Rahnema, M. 1993. Overview of the GSM system and protocol architecture. *IEEE Commun. Mag.* (April):92–100.

Telecommunication Industry Association. 1993. Mobile station-base station compatibility standard for dual-mode wideband spread spectrum cellular system. TIA/EIA/IS-95. July.

Tomlinson, M. 1971. New automatic equalizer employing modulo arithmetic. *Electron. Lett.* 7(March):138, 139.

Further Information

Baseband signalling and pulse shaping is fundamental to the design of any digital communications system and is, therefore, covered in numerous texts on digital communications. For more advanced treatments see E.A. Lee and D.G. Messerschmitt, *Digital Communication*, Kluwer 1994, and J.G. Proakis, *Digital Communications*, McGraw-Hill 1995.

26
Channel Equalization

26.1	Characterization of Channel Distortion	339
26.2	Characterization of Intersymbol Interference	343
26.3	Linear Equalizers	348
	Adaptive Linear Equalizers	
26.4	Decision-Feedback Equalizer	356
26.5	Maximum-Likelihood Sequence Detection	359
26.6	Conclusions	362

John G. Proakis
Northeastern University

26.1 Characterization of Channel Distortion

Many communication channels, including telephone channels, and some radio channels, may be generally characterized as band-limited linear filters. Consequently, such channels are described by their frequency response $C(f)$, which may be expressed as

$$C(f) = A(f)e^{j\theta(f)} \tag{26.1}$$

where $A(f)$ is called the *amplitude response* and $\theta(f)$ is called the *phase response*. Another characteristic that is sometimes used in place of the phase response is the *envelope delay* or *group delay*, which is defined as

$$\tau(f) = -\frac{1}{2\pi}\frac{d\theta(f)}{df} \tag{26.2}$$

A channel is said to be nondistorting or ideal if, within the bandwidth W occupied by the transmitted signal, $A(f) = $ const and $\theta(f)$ is a linear function of frequency [or the envelope delay $\tau(f) = $ const]. On the other hand, if $A(f)$ and $\tau(f)$ are not constant within the bandwidth occupied by the transmitted signal, the channel distorts the signal. If $A(f)$ is not constant, the distortion is called *amplitude distortion* and if $\tau(f)$ is not constant, the distortion on the transmitted signal is called *delay distortion*.

As a result of the amplitude and delay distortion caused by the nonideal channel frequency response characteristic $C(f)$, a succession of pulses transmitted through the channel at rates comparable to the bandwidth W are smeared to the point that they are no longer distinguishable as well-defined pulses at the receiving terminal. Instead, they overlap and, thus, we have **intersymbol interference** (**ISI**). As an example of the effect of delay distortion on a transmitted pulse, Fig. 26.1(a) illustrates a band-limited pulse having zeros periodically spaced in time at points labeled $\pm T$, $\pm 2T$,

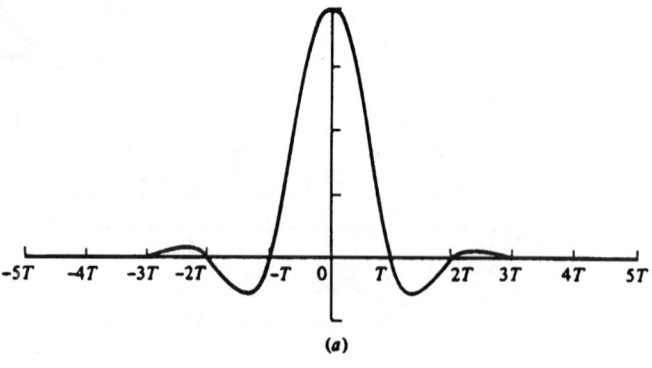

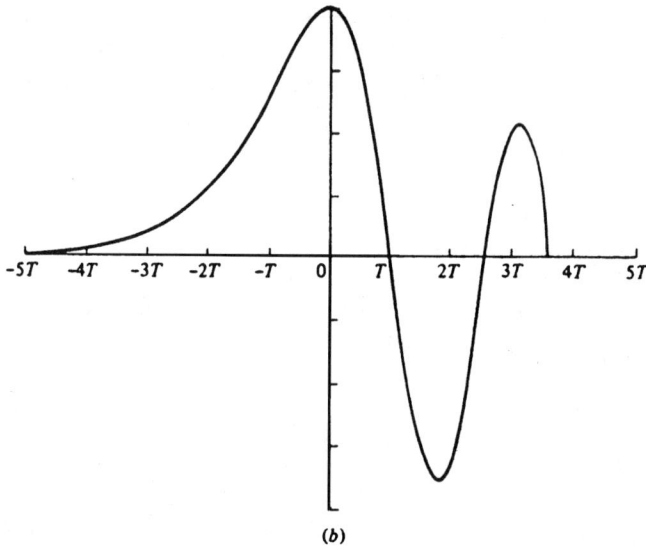

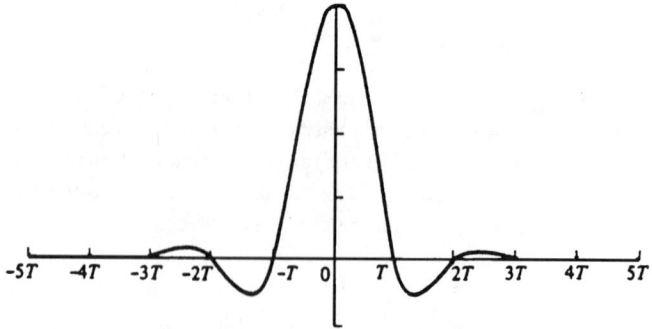

FIGURE 26.1 Effect of channel distortion: (a) channel input, (b) channel output, (c) equalizer output.

Channel Equalization

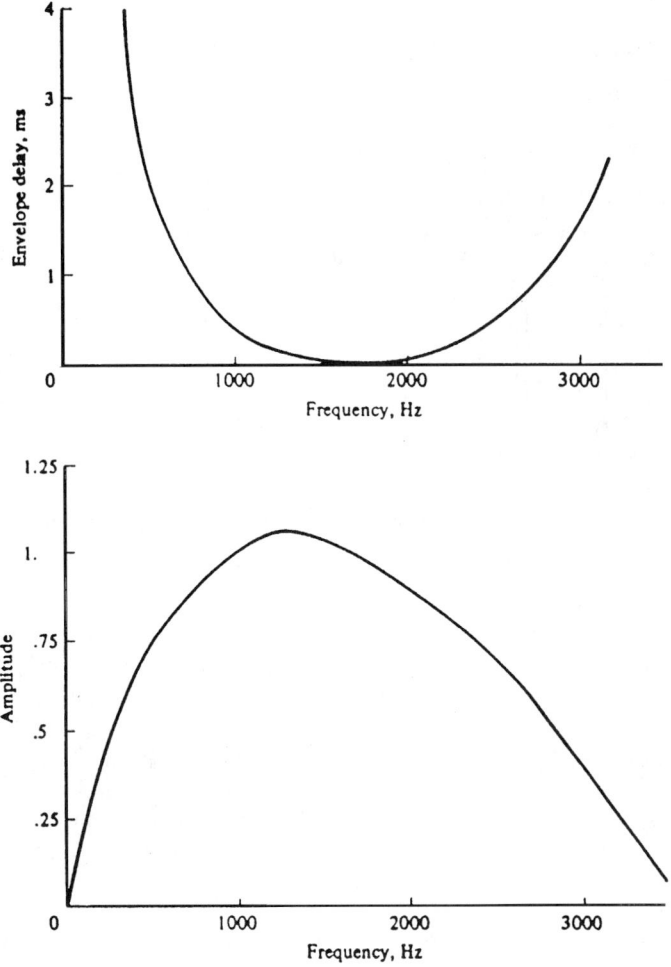

FIGURE 26.2 Average amplitude and delay characteristics of medium-range telephone channel.

etc. If information is conveyed by the pulse amplitude, as in pulse amplitude modulation (PAM), for example, then one can transmit a sequence of pulses, each of which has a peak at the periodic zeros of the other pulses. Transmission of the pulse through a channel modeled as having a linear envelope delay characteristic $\tau(f)$ [quadratic phase $\theta(f)$], however, results in the received pulse shown in Fig. 26.1(b) having zero crossings that are no longer periodically spaced. Consequently a sequence of successive pulses would be smeared into one another, and the peaks of the pulses would no longer be distinguishable. Thus, the channel delay distortion results in intersymbol interference. As will be discussed in this chapter, it is possible to compensate for the nonideal frequency response characteristic of the channel by use of a filter or equalizer at the demodulator. Figure 26.1(c) illustrates the output of a linear equalizer that compensates for the linear distortion in the channel.

The extent of the intersymbol interference on a telephone channel can be appreciated by observing a frequency response characteristic of the channel. Figure 26.2 illustrates the measured average amplitude and delay as a function of frequency for a medium-range (180–725 mi) telephone channel of the switched telecommunications network as given by Duffy and Tratcher, 1971. We observe that the usable band of the channel extends from about 300 Hz to about 3000 Hz. The corresponding impulse response of the average channel is shown in Fig. 26.3. Its duration is about 10 ms. In

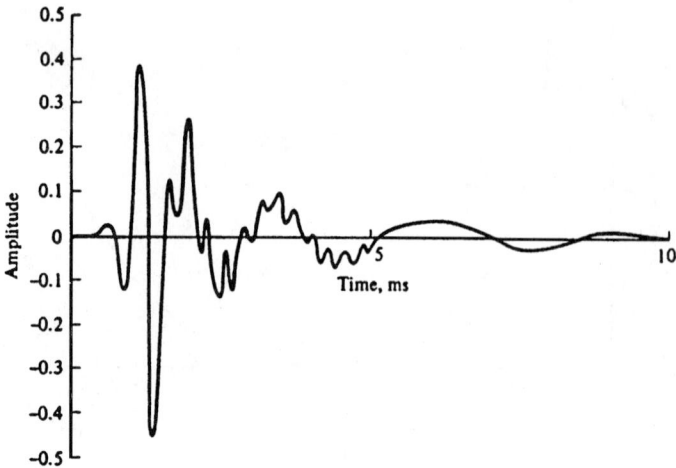

FIGURE 26.3 Impulse response of average channel with amplitude and delay shown in Fig. 26.2.

comparison, the transmitted symbol rates on such a channel may be of the order of 2500 pulses or symbols per second. Hence, intersymbol interference might extend over 20–30 symbols.

Besides telephone channels, there are other physical channels that exhibit some form of time dispersion and, thus, introduce intersymbol interference. Radio channels, such as short-wave ionospheric propagation (HF), tropospheric scatter, and mobile cellular radio are three examples of time-dispersive wireless channels. In these channels, time dispersion and, hence, intersymbol interference is the result of multiple propagation paths with different path delays. The number

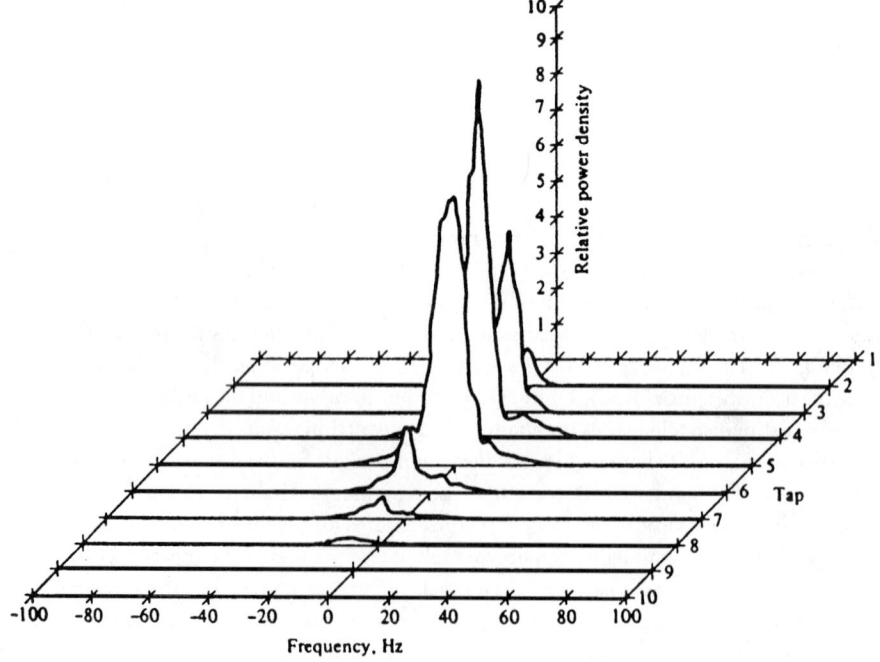

FIGURE 26.4 Scattering function of a medium-range tropospheric scatter channel.

Channel Equalization

of paths and the relative time delays among the paths vary with time and, for this reason, these radio channels are usually called time-variant multipath channels. The time-variant multipath conditions give rise to a wide variety of frequency response characteristics. Consequently, the frequency response characterization that is used for telephone channels is inappropriate for time-variant multipath channels. Instead, these radio channels are characterized statistically in terms of the scattering function, which, in brief, is a two-dimensional representation of the average received signal power as a function of relative time delay and Doppler frequency (see Proakis, 1995).

For illustrative purposes, a scattering function measured on a medium-range (150 mi) tropospheric scatter channel is shown in Fig. 26.4. The total time duration (multipath spread) of the channel response is approximately 0.7 μs on the average, and the spread between half-power points in Doppler frequency is a little less than 1 Hz on the strongest path and somewhat larger on the other paths. Typically, if one is transmitting at a rate of 10^7 symbols/s over such a channel, the multipath spread of 0.7 μs will result in intersymbol interference that spans about seven symbols.

26.2 Characterization of Intersymbol Interference

In a digital communication system, channel distortion causes intersymbol interference, as illustrated in the preceding section. In this section, we shall present a model that characterizes the ISI. The digital modulation methods to which this treatment applies are PAM, phase-shift keying (PSK) and quadrature amplitude modulation (QAM). The transmitted signal for these three types of modulation may be expressed as

$$s(t) = v_c(t) \cos 2\pi f_c t - v_s(t) \sin 2\pi f_c t$$
$$= \text{Re}[v(t) e^{j2\pi f_c t}] \qquad (26.3)$$

where $v(t) = v_c(t) + jv_s(t)$ is called the *equivalent low-pass signal*, f_c is the carrier frequency, and Re[ ] denotes the real part of the quantity in brackets.

In general, the equivalent low-pass signal is expressed as

$$v(t) = \sum_{n=0}^{\infty} I_n g_T(t - nT) \qquad (26.4)$$

where $g_T(t)$ is the basic pulse shape that is selected to control the spectral characteristics of the transmitted signal, $\{I_n\}$ the sequence of transmitted information symbols selected from a signal constellation consisting of M points, and T the signal interval ($1/T$ is the symbol rate). For PAM, PSK, and QAM, the values of I_n are points from M-ary signal constellations. Figure 26.5 illustrates the signal constellations for the case of $M = 8$ signal points. Note that for PAM, the signal constellation is one dimensional. Hence, the equivalent low-pass signal $v(t)$ is real valued, i.e., $v_s(t) = 0$ and $v_c(t) = v(t)$. For M-ary ($M > 2$) PSK and QAM, the signal constellations are two dimensional and, hence, $v(t)$ is complex valued.

The signal $s(t)$ is transmitted over a bandpass channel that may be characterized by an equivalent low-pass frequency response $C(f)$. Consequently, the equivalent low-pass received signal can be represented as

$$r(t) = \sum_{n=0}^{\infty} I_n h(t - nT) + w(t) \qquad (26.5)$$

where $h(t) = g_T(t) * c(t)$, and $c(t)$ is the impulse response of the equivalent low-pass channel, the asterisk denotes convolution, and $w(t)$ represents the additive noise in the channel.

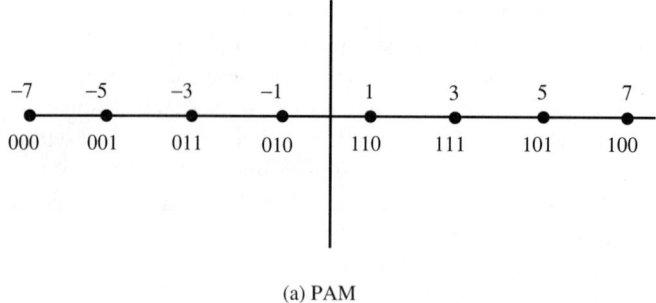

(a) PAM

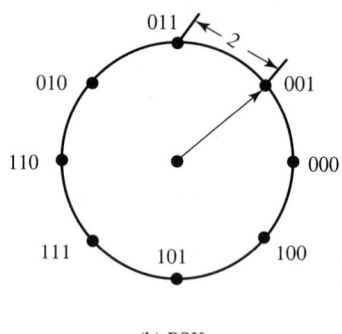

(b) PSK

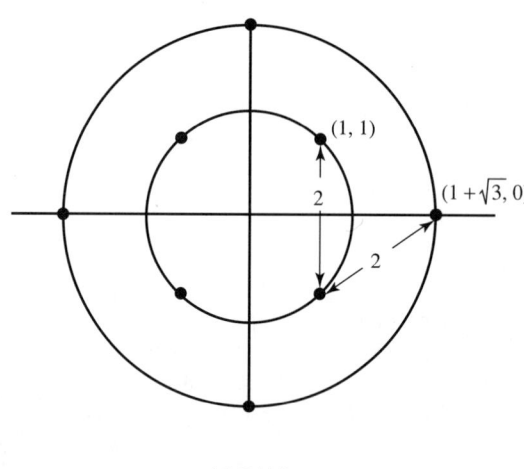

(c) QAM

FIGURE 26.5 $M = 8$ signal constellations for PAM, PSK, and QAM.

To characterize the ISI, suppose that the received signal is passed through a receiving filter and then sampled at the rate $1/T$ samples/s. In general, the optimum filter at the receiver is matched to the received signal pulse $h(t)$. Hence, the frequency response of this filter is $H^*(f)$. We denote its output as

$$y(t) = \sum_{n=0}^{\infty} I_n x(t - nT) + v(t) \qquad (26.6)$$

Channel Equalization

where $x(t)$ is the signal pulse response of the receiving filter, i.e., $X(f) = H(f)H^*(f) = |H(f)|^2$, and $v(t)$ is the response of the receiving filter to the noise $w(t)$. Now, if $y(t)$ is sampled at times $t = kT$, $k = 0, 1, 2, \ldots$, we have

$$y(kT) \equiv y_k = \sum_{n=0}^{\infty} I_n x(kT - nT) + v(kT)$$

$$= \sum_{n=0}^{\infty} I_n x_{k-n} + v_k, \qquad k = 0, 1, \ldots \qquad (26.7)$$

The sample values $\{y_k\}$ can be expressed as

$$y_k = x_0 \left(I_k + \frac{1}{x_0} \sum_{\substack{n=0 \\ n \neq k}}^{\infty} I_n x_{k-n} \right) + v_k, \qquad k = 0, 1, \ldots \qquad (26.8)$$

The term x_0 is an arbitrary scale factor, which we arbitrarily set equal to unity for convenience. Then

$$y_k = I_k + \sum_{\substack{n=0 \\ n \neq k}}^{\infty} I_n x_{k-n} + v_k \qquad (26.9)$$

The term I_k represents the desired information symbol at the kth sampling instant, the term

$$\sum_{\substack{n=0 \\ n \neq k}}^{\infty} I_n x_{k-n} \qquad (26.10)$$

represents the ISI, and v_k is the additive noise variable at the kth sampling instant.

The amount of ISI, and noise in a digital communications system can be viewed on an oscilloscope. For PAM signals, we can display the received signal $y(t)$ on the vertical input with the horizontal sweep rate set at $1/T$. The resulting oscilloscope display is called an *eye pattern* because of its resemblance to the human eye. For example, Fig. 26.6 illustrates the eye patterns for binary and four-level PAM modulation. The effect of ISI is to cause the eye to close, thereby reducing the margin

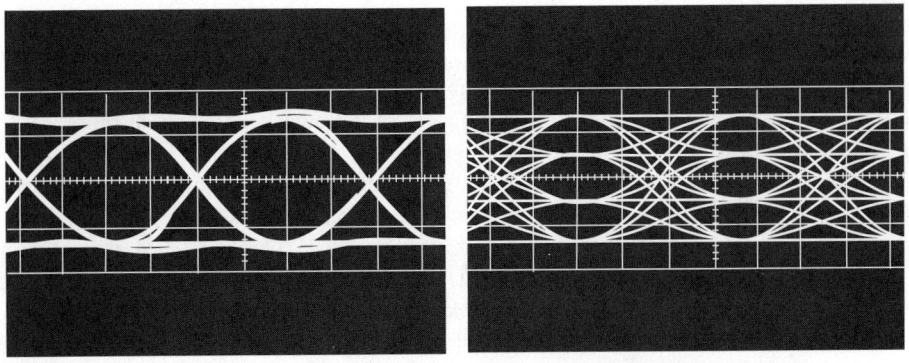

BINARY   QUATERNARY

FIGURE 26.6 Examples of eye patterns for binary and quaternary amplitude shift keying (or PAM).

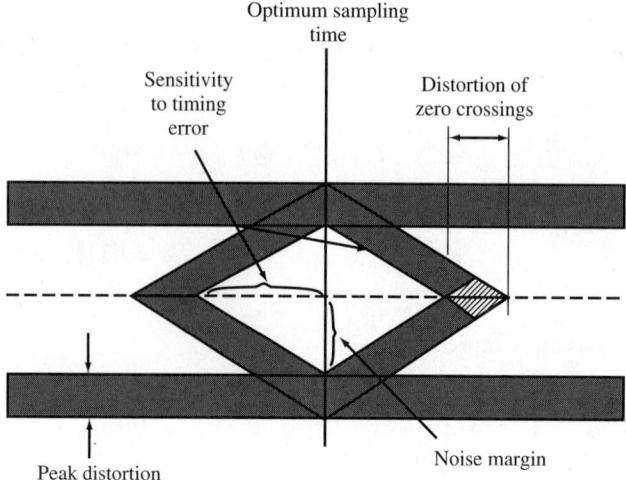

FIGURE 26.7 Effect of intersymbol interference on eye opening.

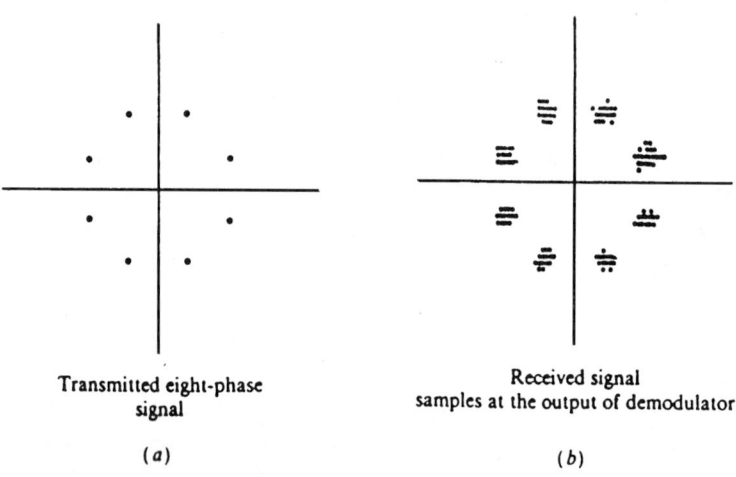

FIGURE 26.8 Two-dimensional digital eye patterns.

for additive noise to cause errors. Figure 26.7 graphically illustrates the effect of ISI in reducing the opening of a binary eye. Note that intersymbol interference distorts the position of the zero crossings and causes a reduction in the eye opening. Thus, it causes the system to be more sensitive to a synchronization error.

For PSK and QAM it is customary to display the eye pattern as a two-dimensional scatter diagram illustrating the sampled values $\{y_k\}$ that represent the decision variables at the sampling instants. Figure 26.8 illustrates such an eye pattern for an 8-PSK signal. In the absence of intersymbol interference and noise, the superimposed signals at the sampling instants would result in eight distinct points corresponding to the eight transmitted signal phases. Intersymbol interference and noise result in a deviation of the received samples $\{y_k\}$ from the desired 8-PSK signal. The larger the intersymbol interference and noise, the larger the scattering of the received signal samples relative to the transmitted signal points.

In practice, the transmitter and receiver filters are designed for zero ISI at the desired sampling times $t = kT$. Thus, if $G_T(f)$ is the frequency response of the transmitter filter and $G_R(f)$ is the

Channel Equalization

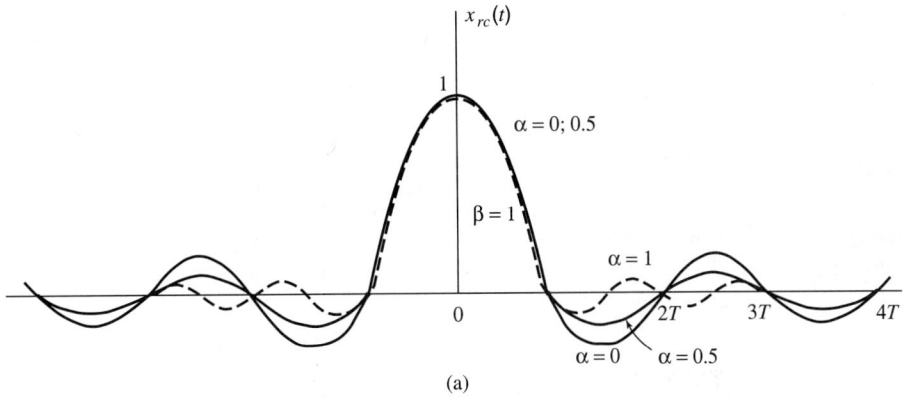

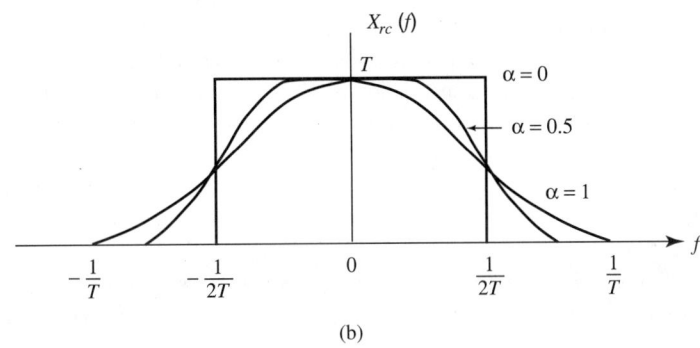

FIGURE 26.9 Pulses having a raised cosine spectrum.

frequency response of the receiver filter, then the product $G_T(f)\,G_R(f)$ is designed to yield zero ISI. For example, the product $G_T(f)\,G_R(f)$ may be selected as

$$G_T(f)G_R(f) = X_{rc}(f) \qquad (26.11)$$

where $X_{rc}(f)$ is the raised-cosine frequency response characteristic, defined as

$$X_{rc}(f) = \begin{cases} T, & 0 \le |f| \le (1-\alpha)/2T \\ \dfrac{T}{2}\left[1 + \cos\dfrac{\pi T}{\alpha}\left(|f| - \dfrac{1-\alpha}{2T}\right)\right], & \dfrac{1-\alpha}{2T} \le |f| \le \dfrac{1+\alpha}{2T} \\ 0, & |f| > \dfrac{1+\alpha}{2T} \end{cases} \qquad (26.12)$$

where α is called the *rolloff* factor, which takes values in the range $0 \le \alpha \le 1$, and $1/T$ is the symbol rate. The frequency response $X_{rc}(f)$ is illustrated in Fig. 26.9(a) for $\alpha = 0$, $1/2$, and 1. Note that when $\alpha = 0$, $X_{rc}(f)$ reduces to an ideal brick wall physically nonrealizable frequency response with bandwidth occupancy $1/2T$. The frequency $1/2T$ is called the *Nyquist frequency*. For $\alpha > 0$, the bandwidth occupied by the desired signal $X_{rc}(f)$ beyond the Nyquist frequency $1/2T$ is called the *excess bandwidth*, and is usually expressed as a percentage of the Nyquist frequency. For example, when $\alpha = 1/2$, the excess bandwidth is 50% and when $\alpha = 1$, the excess bandwidth is 100%. The signal pulse $x_{rc}(t)$ having the raised-cosine spectrum is

$$x_{rc}(t) = \frac{\sin \pi t/T}{\pi t/T} \frac{\cos(\pi \alpha t/T)}{1 - 4\alpha^2 t^2/T^2} \qquad (26.13)$$

Figure 26.9(b) illustrates $x_{rc}(t)$ for $\alpha = 0$, $1/2$, and 1. Note that $x_{rc}(t) = 1$ at $t = 0$ and $x_{rc}(t) = 0$ at $t = kT$, $k = \pm 1, \pm 2, \ldots$. Consequently, at the sampling instants $t = kT$, $k \neq 0$, there is no ISI from adjacent symbols when there is no channel distortion. In the presence of channel distortion, however, the ISI given by Eq. (26.10) is no longer zero, and a channel equalizer is needed to minimize its effect on system performance.

26.3 Linear Equalizers

The most common type of channel equalizer used in practice to reduce ISI is a linear transversal filter with adjustable coefficients $\{c_i\}$, as shown in Fig. 26.10.

On channels whose frequency response characteristics are unknown, but time invariant, we may measure the channel characteristics and adjust the parameters of the equalizer; once adjusted, the parameters remain fixed during the transmission of data. Such equalizers are called **preset equalizers**. On the other hand, **adaptive equalizers** update their parameters on a periodic basis during the transmission of data and, thus, they are capable of tracking a slowly time-varying channel response.

First, let us consider the design characteristics for a linear equalizer from a frequency domain viewpoint. Figure 26.11 shows a block diagram of a system that employs a linear filter as a **channel equalizer**.

The demodulator consists of a receiver filter with frequency response $G_R(f)$ in cascade with a channel equalizing filter that has a frequency response $G_E(f)$. As indicated in the preceding section, the receiver filter response $G_R(f)$ is matched to the transmitter response, i.e., $G_R(f) = G_T^*(f)$, and the product $G_R(f)G_T(f)$ is usually designed so that there is zero ISI at the sampling instants as, for example, when $G_R(t)G_T(f) = X_{rc}(f)$.

For the system shown in Fig. 26.11, in which the channel frequency response is not ideal, the desired condition for zero ISI is

$$G_T(f)C(f)G_R(f)G_E(f) = X_{rc}(f) \qquad (26.14)$$

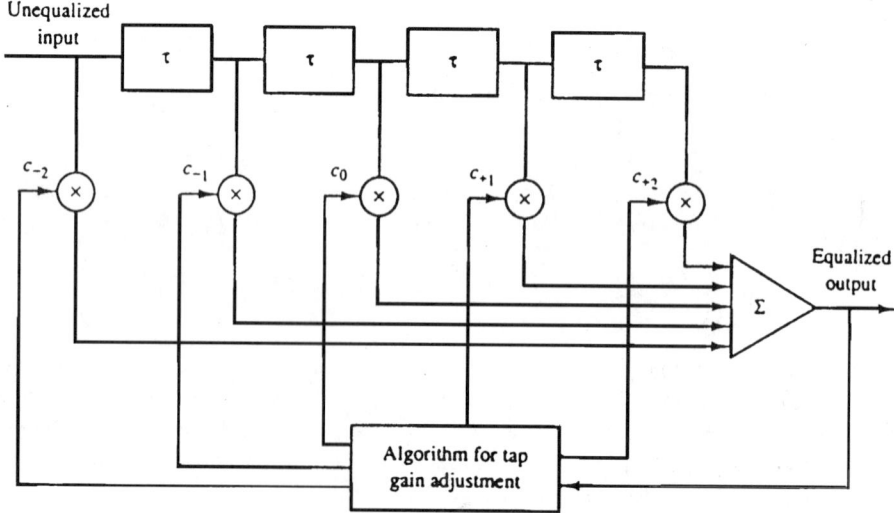

FIGURE 26.10 Linear transversal filter.

Channel Equalization

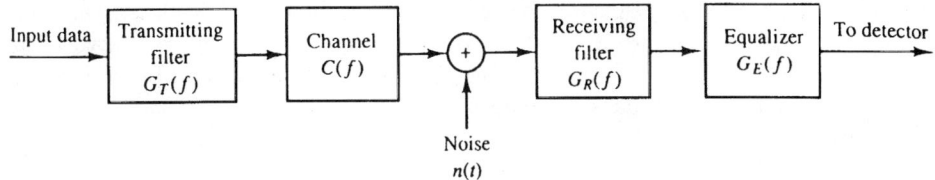

FIGURE 26.11 Block diagram of a system with an equalizer.

where $X_{rc}(f)$ is the desired raised-cosine spectral characteristic. Since $G_T(f)G_R(f) = X_{rc}(f)$ by design, the frequency response of the equalizer that compensates for the channel distortion is

$$G_E(f) = \frac{1}{C(f)} = \frac{1}{|C(f)|} e^{-j\theta_c(f)} \tag{26.15}$$

Thus, the amplitude response of the equalizer is $|G_E(f)| = 1/|C(f)|$ and its phase response is $\theta_E(f) = -\theta_c(f)$. In this case, the equalizer is said to be the *inverse channel filter* to the channel response.

We note that the inverse channel filter completely eliminates ISI caused by the channel. Since it forces the ISI to be zero at the sampling instants $t = kT$, $k = 0, 1, \ldots$, the equalizer is called a **zero-forcing equalizer**. Hence, the input to the detector is simply

$$z_k = I_k + \eta_k, \qquad k = 0, 1, \ldots \tag{26.16}$$

where η_k represents the additive noise and I_k is the desired symbol.

In practice, the ISI caused by channel distortion is usually limited to a finite number of symbols on either side of the desired symbol. Hence, the number of terms that constitute the ISI in the summation given by Eq. (26.10) is finite. As a consequence, in practice the channel equalizer is implemented as a finite duration impulse response (FIR) filter, or transversal filter, with adjustable tap coefficients $\{c_n\}$, as illustrated in Fig. 26.10. The time delay τ between adjacent taps may be selected as large as T, the symbol interval, in which case the FIR equalizer is called a **symbol-spaced equalizer**. In this case, the input to the equalizer is the sampled sequence given by Eq. (26.7). We note that when the symbol rate $1/T < 2W$, however, frequencies in the received signal above the folding frequency $1/T$ are aliased into frequencies below $1/T$. In this case, the equalizer compensates for the aliased channel-distorted signal.

On the other hand, when the time delay τ between adjacent taps is selected such that $1/\tau \geq 2W > 1/T$, no aliasing occurs and, hence, the inverse channel equalizer compensates for the true channel distortion. Since $\tau < T$, the channel equalizer is said to have *fractionally spaced taps* and it is called a **fractionally spaced equalizer**. In practice, τ is often selected as $\tau = T/2$. Notice that, in this case, the sampling rate at the output of the filter $G_R(f)$ is $2/T$.

The impulse response of the FIR equalizer is

$$g_E(t) = \sum_{n=-N}^{N} c_n \delta(t - n\tau) \tag{26.17}$$

and the corresponding frequency response is

$$G_E(f) = \sum_{n=-N}^{N} c_n e^{-j2\pi f n\tau} \tag{26.18}$$

where $\{c_n\}$ are the $(2N+1)$ equalizer coefficients and N is chosen sufficiently large so that the equalizer spans the length of the ISI, i.e., $2N+1 \geq L$, where L is the number of signal samples spanned by the ISI. Since $X(f) = G_T(f)C(f)G_R(f)$ and $x(t)$ is the signal pulse corresponding to $X(f)$, then the equalized output signal pulse is

$$q(t) = \sum_{n=-N}^{N} c_n x(t - n\tau) \quad (26.19)$$

The zero-forcing condition can now be applied to the samples of $q(t)$ taken at times $t = mT$. These samples are

$$q(mT) = \sum_{n=-N}^{N} c_n x(mT - n\tau), \quad m = 0, \pm 1, \ldots, \pm N \quad (26.20)$$

Since there are $2N+1$ equalizer coefficients, we can control only $2N+1$ sampled values of $q(t)$. Specifically, we may force the conditions

$$q(mT) = \sum_{n=-N}^{N} c_n x(mT - n\tau) = \begin{cases} 1, & m = 0 \\ 0, & m = \pm 1, \pm 2, \ldots, \pm N \end{cases} \quad (26.21)$$

which may be expressed in matrix form as $\mathbf{Xc} = \mathbf{q}$, where $\mathbf{X}$ is a $(2N+1) \times (2N+1)$ matrix with elements $\{x(mT - n\tau)\}$, $\mathbf{c}$ is the $(2N+1)$ coefficient vector and $\mathbf{q}$ is the $(2N+1)$ column vector with one nonzero element. Thus, we obtain a set of $2N+1$ linear equations for the coefficients of the zero-forcing equalizer.

We should emphasize that the FIR zero-forcing equalizer does not completely eliminate ISI because it has a finite length. As N is increased, however, the residual ISI can be reduced, and in the limit as $N \to \infty$, the ISI is completely eliminated.

Example 26.1 Consider a channel distorted pulse $x(t)$, at the input to the equalizer, given by the expression

$$x(t) = \frac{1}{1 + \left(\dfrac{2t}{T}\right)^2}$$

where $1/T$ is the symbol rate. The pulse is sampled at the rate $2/T$ and equalized by a zero-forcing equalizer. Determine the coefficients of a five-tap zero-forcing equalizer.

Solution 26.1 According to Eq. (26.21), the zero-forcing equalizer must satisfy the equations

$$q(mT) = \sum_{n=-2}^{2} c_n x(mT - nT/2) = \begin{cases} 1, & m = 0 \\ 0, & m = \pm 1, \pm 2 \end{cases}$$

Channel Equalization

The matrix X with elements $x(mT - nT/2)$ is given as

$$X = \begin{bmatrix} \frac{1}{5} & \frac{1}{10} & \frac{1}{17} & \frac{1}{26} & \frac{1}{37} \\ 1 & \frac{1}{2} & \frac{1}{5} & \frac{1}{10} & \frac{1}{17} \\ \frac{1}{5} & \frac{1}{2} & 1 & \frac{1}{2} & \frac{1}{5} \\ \frac{1}{17} & \frac{1}{10} & \frac{1}{5} & \frac{1}{2} & 1 \\ \frac{1}{37} & \frac{1}{26} & \frac{1}{17} & \frac{1}{10} & \frac{1}{5} \end{bmatrix} \qquad (26.22)$$

The coefficient vector c and the vector q are given as

$$c = \begin{bmatrix} c_{-2} \\ c_{-1} \\ c_0 \\ c_1 \\ c_2 \end{bmatrix} \qquad q = \begin{bmatrix} 0 \\ 0 \\ 1 \\ 0 \\ 0 \end{bmatrix} \qquad (26.23)$$

Then, the linear equations $Xc = q$ can be solved by inverting the matrix X. Thus, we obtain

$$c_{\text{opt}} = X^{-1} q = \begin{bmatrix} -2.2 \\ 4.9 \\ -3 \\ 4.9 \\ -2.2 \end{bmatrix} \qquad (26.24)$$

One drawback to the zero-forcing equalizer is that it ignores the presence of additive noise. As a consequence, its use may result in significant noise enhancement. This is easily seen by noting that in a frequency range where $C(f)$ is small, the channel equalizer $G_E(f) = 1/C(f)$ compensates by placing a large gain in that frequency range. Consequently, the noise in that frequency range is greatly enhanced. An alternative is to relax the zero ISI condition and select the channel equalizer characteristic such that the combined power in the residual ISI and the additive noise at the output of the equalizer is minimized. A channel equalizer that is optimized based on the minimum mean square error (MMSE) criterion accomplishes the desired goal.

To elaborate, let us consider the noise corrupted output of the FIR equalizer, which is

$$z(t) = \sum_{n=-N}^{N} c_n y(t - n\tau) \qquad (26.25)$$

where $y(t)$ is the input to the equalizer, given by Eq. (26.6). The equalizer output is sampled at times $t = mT$. Thus, we obtain

$$z(mT) = \sum_{n=-N}^{N} c_n y(mT - n\tau) \qquad (26.26)$$

The desired response at the output of the equalizer at $t = mT$ is the transmitted symbol I_m. The error is defined as the difference between I_m and $z(mT)$. Then, the mean square error (*MSE*) between the actual output sample $z(mT)$ and the desired values I_m is

$$\begin{aligned} MSE &= E|z(mT) - I_m|^2 \\ &= E\left[\left|\sum_{n=-N}^{N} c_n y(mT - n\tau) - I_m\right|^2\right] \\ &= \sum_{n=-N}^{N} \sum_{k=-N}^{N} c_n c_k R_Y(n-k) \\ &\quad - 2 \sum_{k=-N}^{N} c_k R_{IY}(k) + E(|I_m|^2) \end{aligned} \quad (26.27)$$

where the correlations are defined as

$$\begin{aligned} R_Y(n-k) &= E[y^*(mT - n\tau)y(mT - k\tau)] \\ R_{IY}(k) &= E[y(mT - k\tau)I_m^*] \end{aligned} \quad (26.28)$$

and the expectation is taken with respect to the random information sequence $\{I_m\}$ and the additive noise.

The minimum *MSE* solution is obtained by differentiating Eq. (26.27) with respect to the equalizer coefficients $\{c_n\}$. Thus, we obtain the necessary conditions for the minimum *MSE* as

$$\sum_{n=-N}^{N} c_n R_Y(n-k) = R_{IY}(k), \qquad k = 0, \pm 1, 2, \ldots, \pm N \quad (26.29)$$

These are the $(2N+1)$ linear equations for the equalizer coefficients. In contrast to the zero-forcing solution already described, these equations depend on the statistical properties (the autocorrelation) of the noise as well as the ISI through the autocorrelation $R_Y(n)$.

In practice, the autocorrelation matrix $R_Y(n)$ and the crosscorrelation vector $R_{IY}(n)$ are unknown a priori. These correlation sequences can be estimated, however, by transmitting a test signal over the channel and using the time-average estimates

$$\begin{aligned} \hat{R}_Y(n) &= \frac{1}{K} \sum_{k=1}^{K} y^*(kT - n\tau)y(kT) \\ \hat{R}_{IY}(n) &= \frac{1}{K} \sum_{k=1}^{K} y(kT - n\tau)I_k^* \end{aligned} \quad (26.30)$$

in place of the ensemble averages to solve for the equalizer coefficients given by Eq. (26.29).

Adaptive Linear Equalizers

We have shown that the tap coefficients of a linear equalizer can be determined by solving a set of linear equations. In the zero-forcing optimization criterion, the linear equations are given by Eq. (26.21). On the other hand, if the optimization criterion is based on minimizing the *MSE*,

Channel Equalization

the optimum equalizer coefficients are determined by solving the set of linear equations given by Eq. (26.29).

In both cases, we may express the set of linear equations in the general matrix form

$$Bc = d \tag{26.31}$$

where B is a $(2N + 1) \times (2N + 1)$ matrix, c is a column vector representing the $2N + 1$ equalizer coefficients, and d a $(2N + 1)$-dimensional column vector. The solution of Eq. (26.31) yields

$$c_{\text{opt}} = B^{-1} d \tag{26.32}$$

In practical implementations of equalizers, the solution of Eq. (26.31) for the optimum coefficient vector is usually obtained by an iterative procedure that avoids the explicit computation of the inverse of the matrix B. The simplest iterative procedure is the method of steepest descent, in which one begins by choosing arbitrarily the coefficient vector c, say c_0. This initial choice of coefficients corresponds to a point on the criterion function that is being optimized. For example, in the case of the *MSE* criterion, the initial guess c_0 corresponds to a point on the quadratic *MSE* surface in the $(2N + 1)$-dimensional space of coefficients. The gradient vector, defined as g_0, which is the derivative of the *MSE* with respect to the $2N + 1$ filter coefficients, is then computed at this point on the criterion surface, and each tap coefficient is changed in the direction opposite to its corresponding gradient component. The change in the jth tap coefficient is proportional to the size of the jth gradient component.

For example, the gradient vector denoted as g_k, for the MSE criterion, found by taking the derivatives of the MSE with respect to each of the $2N + 1$ coefficients, is

$$g_k = Bc_k - d, \qquad k = 0, 1, 2, \ldots \tag{26.33}$$

Then the coefficient vector c_k is updated according to the relation

$$c_{k+1} = c_k - \Delta g_k \tag{26.34}$$

where Δ is the *step-size parameter* for the iterative procedure. To ensure convergence of the iterative procedure, Δ is chosen to be a small positive number. In such a case, the gradient vector g_k converges toward zero, i.e., $g_k \to 0$ as $k \to \infty$, and the coefficient vector $c_k \to c_{\text{opt}}$ as illustrated in Fig. 26.12 based on two-dimensional optimization. In general, convergence of the equalizer tap coefficients to c_{opt} cannot be attained in a finite number of iterations with the steepest-descent method. The optimum solution c_{opt}, however, can be approached as closely as desired in a few hundred iterations. In digital communication systems that employ channel equalizers, each iteration corresponds to a time interval for sending one symbol and, hence, a few hundred iterations to achieve convergence to c_{opt} corresponds to a fraction of a second.

Adaptive channel equalization is required for channels whose characteristics change with time. In such a case, the ISI varies with time. The channel equalizer must track such time variations in the channel response and adapt its coefficients to reduce the ISI. In the context of the preceding discussion, the optimum coefficient vector c_{opt} varies with time due to time variations in the matrix B and, for the case of the *MSE* criterion, time variations in the vector d. Under these conditions, the iterative method described can be modified to use estimates of the gradient components. Thus, the algorithm for adjusting the equalizer tap coefficients may be expressed as

$$\hat{c}_{k+1} = \hat{c}_k - \Delta \hat{g}_k \tag{26.35}$$

where $\hat{g}_k$ denotes an estimate of the gradient vector g_k and $\hat{c}_k$ denotes the estimate of the tap coefficient vector.

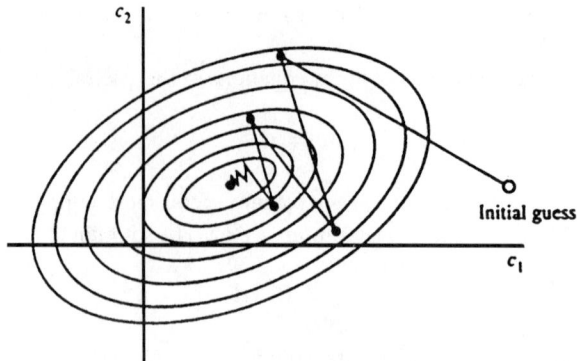

FIGURE 26.12 Examples of convergence characteristics of a gradient algorithm.

In the case of the *MSE* criterion, the gradient vector g_k given by Eq. (26.33) may also be expressed as

$$g_k = -E\left(e_k y_k^*\right)$$

An estimate $\hat{g}_k$ of the gradient vector at the kth iteration is computed as

$$\hat{g}_k = -e_k y_k^* \qquad (26.36)$$

where e_k denotes the difference between the desired output from the equalizer at the kth time instant and the actual output $z(kT)$, and y_k denotes the column vector of $2N + 1$ received signal values contained in the equalizer at time instant k. The *error signal* e_k is expressed as

$$e_k = I_k - z_k \qquad (26.37)$$

where $z_k = z(kT)$ is the equalizer output given by Eq. (26.26) and I_k is the desired symbol. Hence, by substituting Eq. (26.36) into Eq. (26.35), we obtain the adaptive algorithm for optimizing the tap coefficients (based on the *MSE* criterion) as

$$\hat{c}_{k+1} = \hat{c}_k + \Delta e_k y_k^* \qquad (26.38)$$

Since an estimate of the gradient vector is used in Eq. (26.38) the algorithm is called a **stochastic gradient algorithm**; it is also known as the **LMS algorithm**.

A block diagram of an adaptive equalizer that adapts its tap coefficients according to Eq. (26.38) is illustrated in Fig. 26.13. Note that the difference between the desired output I_k and the actual output z_k from the equalizer is used to form the error signal e_k. This error is scaled by the step-size parameter Δ, and the scaled error signal Δe_k multiplies the received signal values $\{y(kT - n\tau)\}$ at the $2N + 1$ taps. The products $\Delta e_k y^*(kT - n\tau)$ at the $(2N + 1)$ taps are then added to the previous values of the tap coefficients to obtain the updated tap coefficients, according to Eq. (26.38). This computation is repeated as each new symbol is received. Thus, the equalizer coefficients are updated at the symbol rate.

Initially, the adaptive equalizer is trained by the transmission of a known pseudo-random sequence $\{I_m\}$ over the channel. At the demodulator, the equalizer employs the known sequence to adjust its coefficients. Upon initial adjustment, the adaptive equalizer switches from a **training mode** to a **decision-directed mode**, in which case the decisions at the output of the detector are sufficiently reliable so that the error signal is formed by computing the difference between the

Channel Equalization

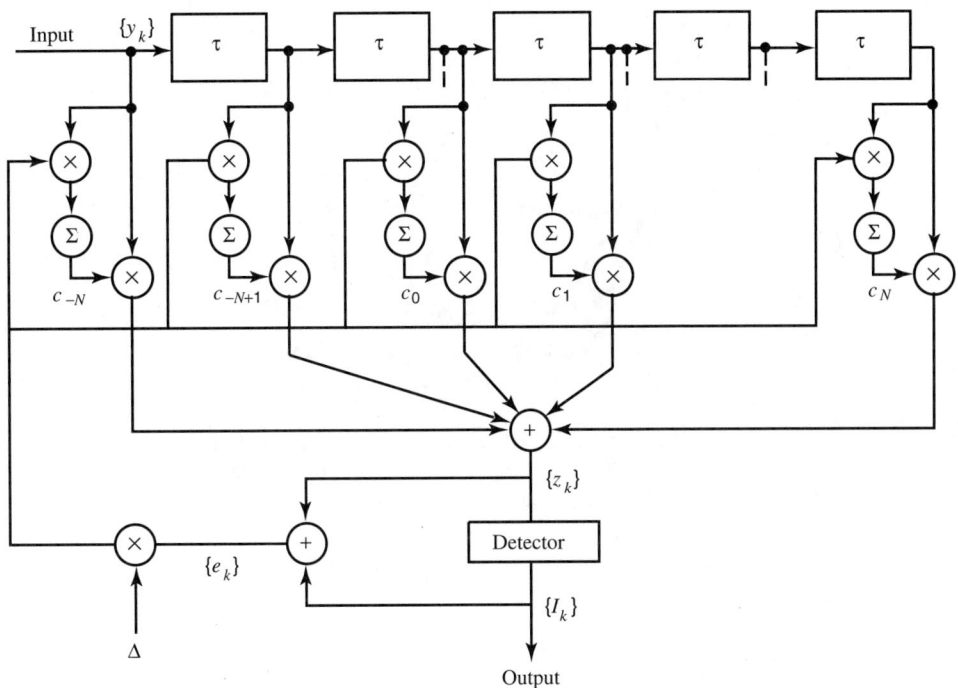

FIGURE 26.13 Linear adaptive equalizer based on the *MSE* criterion.

detector output and the equalizer output, i.e.,

$$e_k = \tilde{I}_k - z_k \quad (26.39)$$

where $\tilde{I}_k$ is the output of the detector. In general, decision errors at the output of the detector occur infrequently and, consequently, such errors have little effect on the performance of the tracking algorithm given by Eq. (26.38).

A rule of thumb for selecting the step-size parameter so as to ensure convergence and good tracking capabilities in slowly varying channels is

$$\Delta = \frac{1}{5(2N+1)P_R} \quad (26.40)$$

where P_R denotes the received signal-plus-noise power, which can be estimated from the received signal (see Proakis, 1995).

The convergence characteristic of the stochastic gradient algorithm in Eq. (26.38) is illustrated in Fig. 26.14. These graphs were obtained from a computer simulation of an 11-tap adaptive equalizer operating on a channel with a rather modest amount of ISI. The input signal-plus-noise power P_R was normalized to unity. The rule of thumb given in Eq. (26.40) for selecting the step size gives $\Delta = 0.018$. The effect of making Δ too large is illustrated by the large jumps in *MSE* as shown for $\Delta = 0.115$. As Δ is decreased, the convergence is slowed somewhat, but a lower MSE is achieved, indicating that the estimated coefficients are closer to c_{opt}.

Although we have described in some detail the operation of an adaptive equalizer that is optimized on the basis of the MSE criterion, the operation of an adaptive equalizer based on the zero-forcing method is very similar. The major difference lies in the method for estimating the gradient vectors g_k at each iteration. A block diagram of an adaptive zero-forcing equalizer is shown in Fig. 26.15.

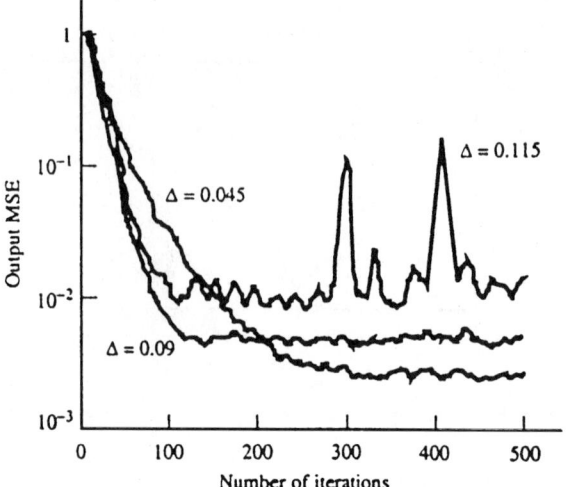

FIGURE 26.14 Initial convergence characteristics of the *LMS* algorithm with different step sizes.

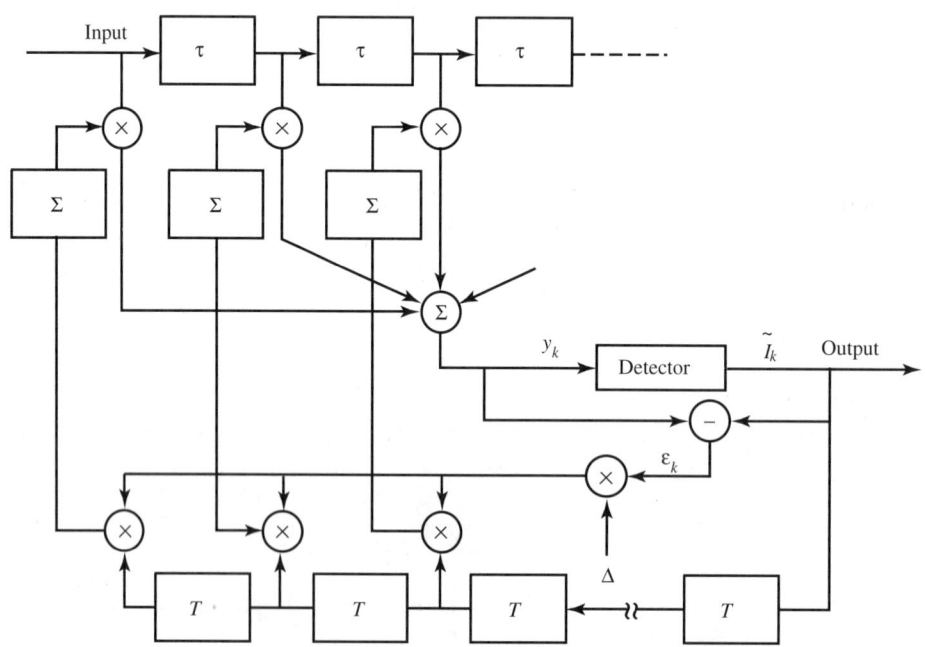

FIGURE 26.15 An adaptive zero-forcing equalizer.

For more details on the tap coefficient update method for a zero-forcing equalizer, the reader is referred to the papers by Lucky, 1965 and 1966, and the text by Proakis, 1995.

26.4 Decision-Feedback Equalizer

The linear filter equalizers described in the preceding section are very effective on channels, such as wire line telephone channels, where the ISI is not severe. The severity of the ISI is directly related

Channel Equalization

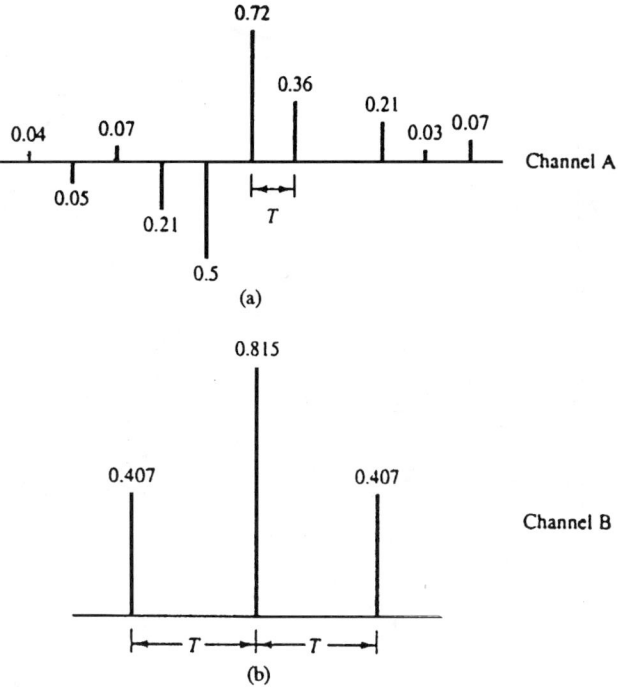

FIGURE 26.16 Two channels with ISI.

to the spectral characteristics and not necessarily to the time span of the ISI. For example, consider the ISI resulting from the two channels that are illustrated in Fig. 26.16. The time span for the ISI in channel A is 5 symbol intervals on each side of the desired signal component, which has a value of 0.72. On the other hand, the time span for the ISI in channel B is one symbol interval on each side of the desired signal component, which has a value of 0.815. The energy of the total response is normalized to unity for both channels.

In spite of the shorter ISI span, channel B results in more severe ISI. This is evidenced in the frequency response characteristics of these channels, which are shown in Fig. 26.17. We observe that channel B has a spectral null [the frequency response $C(f) = 0$ for some frequencies in the band $|f| \leq W$] at $f = 1/2T$, whereas this does not occur in the case of channel A. Consequently, a linear equalizer will introduce a large gain in its frequency response to compensate for the channel null. Thus, the noise in channel B will be enhanced much more than in channel A. This implies that the performance of the linear equalizer for channel B will be sufficiently poorer than that for channel A. This fact is borne out by the computer simulation results for the performance of the two linear equalizers shown in Fig. 26.18. Hence, the basic limitation of a linear equalizer is that it performs poorly on channels having spectral nulls. Such channels are often encountered in radio communications, such as ionospheric transmission at frequencies below 30 MHz and mobile radio channels, such as those used for cellular radio communications.

A **decision-feedback equalizer** (**DFE**) is a nonlinear equalizer that employs previous decisions to eliminate the ISI caused by previously detected symbols on the current symbol to be detected. A simple block diagram for a DFE is shown in Fig. 26.19. The DFE consists of two filters. The first filter is called a *feedforward filter* and it is generally a fractionally spaced FIR filter with adjustable tap coefficients. This filter is identical in form to the linear equalizer already described. Its input is the received filtered signal $y(t)$ sampled at some rate that is a multiple of the symbol rate, e.g., at rate $2/T$. The second filter is a *feedback filter*. It is implemented as an FIR filter with symbol-spaced taps having adjustable coefficients. Its input is the set of previously detected symbols. The output of the feedback filter is subtracted from the output of the feedforward filter to form the input to the

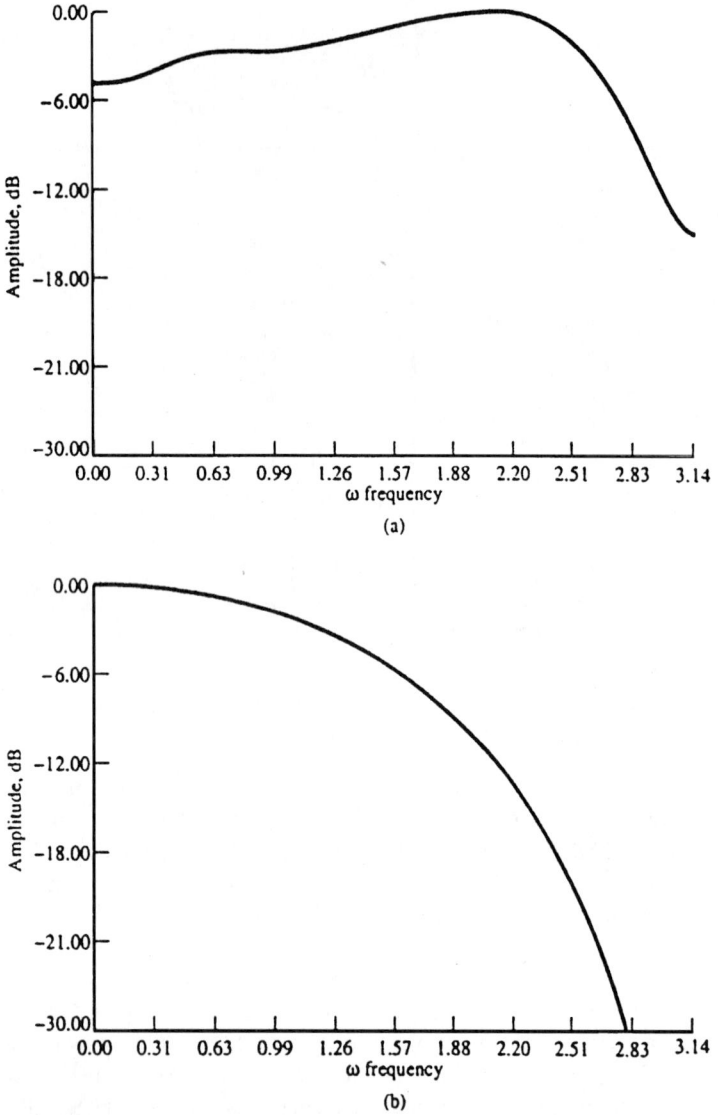

FIGURE 26.17 Amplitude spectra for (a) channel A shown in Fig. 26.16(a) and (b) channel B shown in Fig. 26.16(b).

detector. Thus, we have

$$z_m = \sum_{n=-N_1}^{0} c_n y(mT - n\tau) - \sum_{n=1}^{N_2} b_n \tilde{I}_{m-n} \qquad (26.41)$$

where $\{c_n\}$ and $\{b_n\}$ are the adjustable coefficients of the feedforward and feedback filters, respectively, $\tilde{I}_{m-n}, n = 1, 2, \ldots, N_2$ are the previously detected symbols, $N_1 + 1$ is the length of the feedforward filter, and N_2 is the length of the feedback filter. Based on the input z_m, the detector determines which of the possible transmitted symbols is closest in distance to the input signal I_m. Thus, it makes its decision and outputs $\tilde{I}_m$. What makes the DFE nonlinear is the nonlinear characteristic of the detector that provides the input to the feedback filter.

Channel Equalization

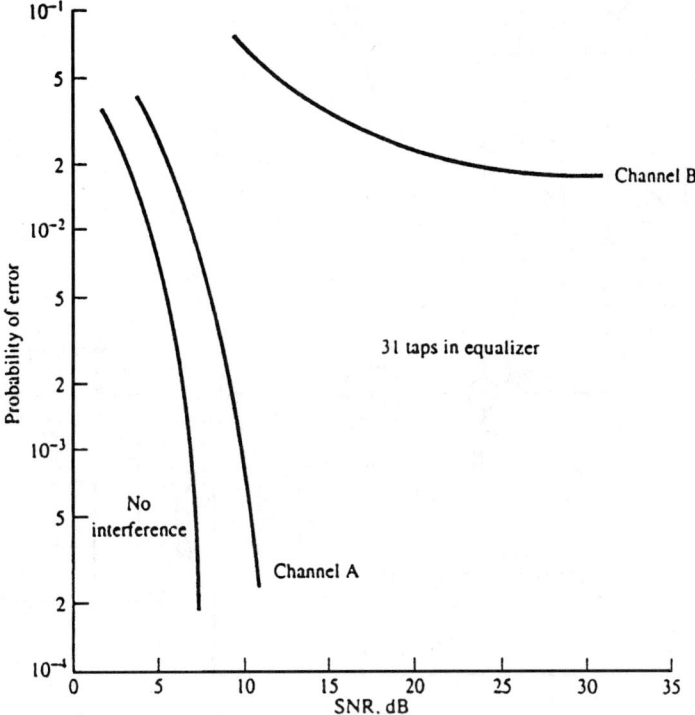

FIGURE 26.18 Error-rate performance of linear *MSE* equalizer.

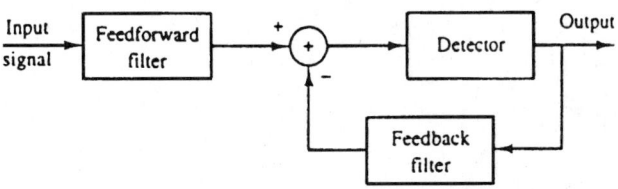

FIGURE 26.19 Block diagram of DFE.

The tap coefficients of the feedforward and feedback filters are selected to optimize some desired performance measure. For mathematical simplicity, the *MSE* criterion is usually applied, and a stochastic gradient algorithm is commonly used to implement an adaptive DFE. Figure 26.20 illustrates the block diagram of an adaptive DFE whose tap coefficients are adjusted by means of the LMS stochastic gradient algorithm. Figure 26.21 illustrates the probability of error performance of the DFE, obtained by computer simulation, for binary PAM transmission over channel B. The gain in performance relative to that of a linear equalizer is clearly evident.

We should mention that decision errors from the detector that are fed to the feedback filter have a small effect on the performance of the DFE. In general, a small loss in performance of one to two decibels is possible at error rates below 10^{-2}, as illustrated in Fig. 26.21, but the decision errors in the feedback filters are not catastrophic.

26.5 Maximum-Likelihood Sequence Detection

Although the DFE outperforms a linear equalizer, it is not the optimum equalizer from the viewpoint of minimizing the probability of error in the detection of the information sequence $\{I_k\}$ from the

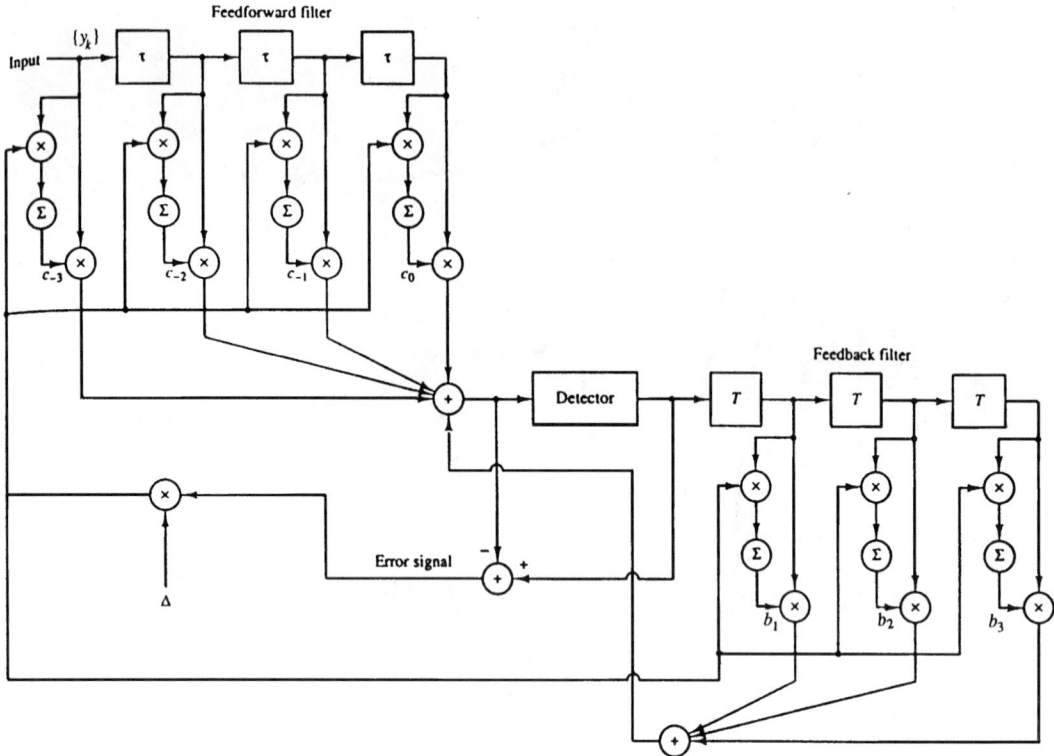

FIGURE 26.20 Adaptive DFE.

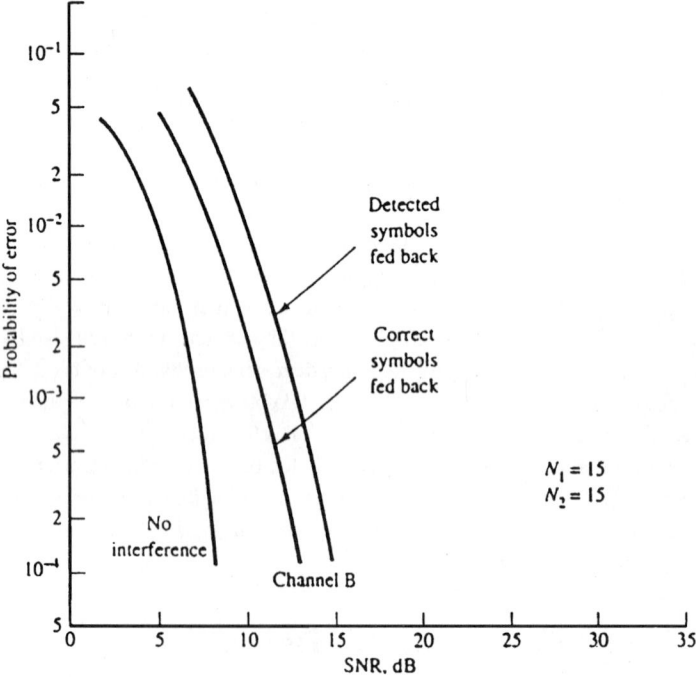

FIGURE 26.21 Performance of DFE with and without error propagation.

Channel Equalization

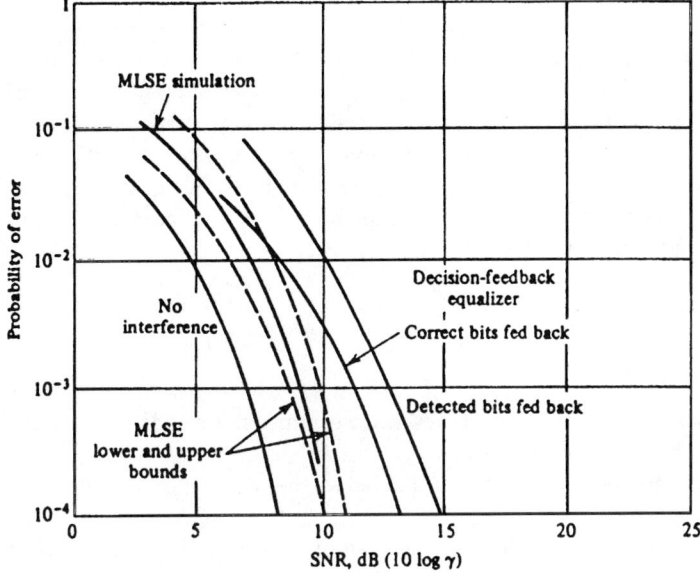

FIGURE 26.22 Comparison of performance between MLSE and decision-feedback equalization for channel B of Fig. 26.16.

received signal samples $\{y_k\}$ given in Eq. (5). In a digital communication system that transmits information over a channel that causes ISI, the optimum detector is a maximum-likelihood symbol sequence detector which produces at its output the most probable symbol sequence $\{\tilde{I}_k\}$ for the given received sampled sequence $\{y_k\}$. That is, the detector finds the sequence $\{\tilde{I}_k\}$ that maximizes the *likelihood function*

$$\Lambda(\{I_k\}) = \ln p(\{y_k\} \mid \{I_k\}) \tag{26.42}$$

where $p(\{y_k\} \mid \{I_k\})$ is the joint probability of the received sequence $\{y_k\}$ conditioned on $\{I_k\}$. The sequence of symbols $\{\tilde{I}_k\}$ that maximizes this joint conditional probability is called the **maximum-likelihood sequence detector.**

An algorithm that implements maximum-likelihood sequence detection (MLSD) is the Viterbi algorithm, which was originally devised for decoding convolutional codes. For a description of this algorithm in the context of sequence detection in the presence of ISI, the reader is referred to the paper by Forney, 1972 and the text by Proakis, 1995.

The major drawback of MLSD for channels with ISI is the exponential behavior in computational complexity as a function of the span of the ISI. Consequently, MLSD is practical only for channels where the ISI spans only a few symbols and the ISI is severe, in the sense that it causes a severe degradation in the performance of a linear equalizer or a decision-feedback equalizer. For example, Fig. 26.22 illustrates the error probability performance of the Viterbi algorithm for a binary PAM signal transmitted through channel B (see Fig. 26.16). For purposes of comparison, we also illustrate the probability of error for a DFE. Both results were obtained by computer simulation. We observe that the performance of the maximum likelihood sequence detector is about 4.5 dB better than that of the DFE at an error probability of 10^{-4}. Hence, this is one example where the ML sequence detector provides a significant performance gain on a channel with a relatively short ISI span.

26.6 Conclusions

Channel equalizers are widely used in digital communication systems to mitigate the effects of ISI cause by channel distortion. Linear equalizers are widely used for high-speed modems that transmit data over telephone channels. For wireless (radio) transmission, such as in mobile cellular communications and interoffice communications, the multipath propagation of the transmitted signal results in severe ISI. Such channels require more powerful equalizers to combat the severe ISI. The decision-feedback equalizer and the MLSD are two nonlinear channel equalizers that are suitable for radio channels with severe ISI.

Defining Terms

Adaptive equalizer: A channel equalizer whose parameters are updated automatically and adaptively during transmission of data.

Channel equalizer: A device that is used to reduce the effects of channel distortion in a received signal.

Decision-directed mode: Mode for adjustment of the equalizer coefficient adaptively based on the use of the detected symbols at the output of the detector.

Decision-feedback equalizer (DFE): An adaptive equalizer that consists of a feedforward filter and a feedback filter, where the latter is fed with previously detected symbols that are used to eliminate the intersymbol interference due to the tail in the channel impulse response.

Fractionally spaced equalizer: A tapped-delay line channel equalizer in which the delay between adjacent taps is less than the duration of a transmitted symbol.

Intersymbol interference: Interference in a received symbol from adjacent (nearby) transmitted symbols caused by channel distortion in data transmission.

LMS algorithm: See stochastic gradient algorithm.

Maximum-likelihood sequence detector: A detector for estimating the most probable sequence of data symbols by maximizing the likelihood function of the received signal.

Preset equalizer: A channel equalizer whose parameters are fixed (time-invariant) during transmission of data.

Stochastic gradient algorithm: An algorithm for adaptively adjusting the coefficients of an equalizer based on the use of (noise-corrupted) estimates of the gradients.

Symbol-spaced equalizer: A tapped-delay line channel equalizer in which the delay between adjacent taps is equal to the duration of a transmitted symbol.

Training mode: Mode for adjustment of the equalizer coefficients based on the transmission of a known sequence of transmitted symbols.

Zero-forcing equalizer: A channel equalizer whose parameters are adjusted to completely eliminate intersymbol interference in a sequence of transmitted data symbols.

References

Forney, G.D., Jr. 1972. Maximum-likelihood sequence estimation of digital sequences in the presence of intersymbol interference. *IEEE Trans. Inform. Theory*, IT-18(May):363–378.

Lucky, R.W. 1965. Automatic equalization for digital communications. *Bell Syst. Tech. J.*, 44(April): 547–588.

Lucky, R.W. 1966. Techniques for adaptive equalization of digital communication. *Bell Syst. Tech. J.* 45(Feb.):255–286.

Proakis, J.G. 1995. *Digital Communications*, 3rd ed. McGraw-Hill, New York.

Further Information

For a comprehensive treatment of adaptive equalization techniques and their performance characteristics, the reader may refer to the book by Proakis, 1995. The two papers by Lucky, 1965 and 1966, provide a treatment on linear equalizers based on the zero-forcing criterion. Additional information on decision-feedback equalizers may be found in the journal papers "An Adaptive Decision-Feedback Equalizer" by D.A. George, R.R. Bowen, and J.R. Storey, *IEEE Transactions on Communications Technology*, Vol. COM-19, pp. 281–293, June 1971, and "Feedback Equalization for Fading Dispersive Channels" by P. Monsen, *IEEE Transactions on Information Theory*, Vol. IT-17, pp. 56–64, January 1971. A through treatment of channel equalization based on maximum-likelihood sequence detection is given in the paper by Forney, 1972.

27
Pulse-Code Modulation Codec-Filters

Michael D. Floyd
Motorola Semiconductor Products

Garth D. Hillman
Motorola Semiconductor Products

27.1 Introduction and General Description of a Pulse-Code Modulation (PCM) Codec-Filter 364
27.2 Where PCM Codec-Filters Are Used in the Telephone Network 365
27.3 Design of Voice PCM Codec-Filters: Analog Transmission Performance and Voice Quality for Intelligibility 367
Filtering • Quantizing Distortion • Gain Calibration: Transmission Level Point • Idle Channel Noise • Gain Tracking: Gain Variations as a Function of Level
27.4 Linear PCM Codec-Filter for High-Speed Modem Applications 374
27.5 Concluding Remarks 376

27.1 Introduction and General Description of a Pulse-Code Modulation (PCM) Codec-Filter

This chapter introduces the reader to the pulse-code modulation (PCM) codec-filter function and where it is used in the telephone network. A PCM codec-filter was originally used in the switching offices and network for digitizing and reconstructing the human voice. Over the last five years linear PCM codec-filters have been used in high-speed modems at the subscriber's premise.

The name codec is an acronym from coder for the analog-to-digital converter (ADC) used to digitize voice, and decoder for the digital-to-analog converter (DAC) used for reconstructing voice. A codec is a single device that does both the ADC and DAC conversions. A PCM codec-filter includes the bandlimiting filter for the ADC and the reconstruction smoothing filter for the output of the DAC, in addition to the ADC and DAC functions. PCM codec-filter is often referred to as a PCM codec, Combo^{TM} (from National Semiconductor), Monocircuit, or Cofidec.

PCM codec-filters were developed to transmit telephone conversations over long distances with improved performance and at lower cost. Digitizing the voice channel is relatively economical compared to expensive low-noise analog transmission equipment.

Multiple digitized voice channels can be multiplexed into one higher data rate digital channel without concern for interference or crosstalk of the analog voice information. Digitized voice data can also be received and retransmitted without attenuation and noise degradation. This digitized data could be transmitted via T1, microwave, satellite, fiber optics, RF carrier, etc., without loss to the digitized voice channel.

27.2 Where PCM Codec-Filters Are Used in the Telephone Network

With the advancements in microelectronics, the PCM codec-filter function has evolved from the early 1960s technology of passive resistor–capacitor–inductor filters with discrete analog-to-digital converter and discrete digital-to-analog converter implementations to the fully integrated devices that were introduced in the late 1970s and early 1980s. As monolithic devices, the PCM codec-filter has experienced performance improvements and cost reductions by taking advantage of the improvements in integrated circuit technology developed within the semiconductor industry. Through this evolutionary progression, the cost reductions of the PCM codec-filter function has increased their applicability to include switching systems for telephone central offices (CO), private branch exchanges (PBX), and key systems. The transmission applications have increased also to include digital loop carriers, pair gain multiplexers, telephone loop extenders, integrated services digital network (ISDN) terminals, digital cellular telephones, and digital cordless telephones. New applications have developed, which include voice recognition equipment, voice storage, voice mail and digital tapeless answering machines.

Figure 27.1 is a simplified diagram showing some of the services that the local telephone service provider has to offer. The telephone network must be able to operate in extreme weather conditions

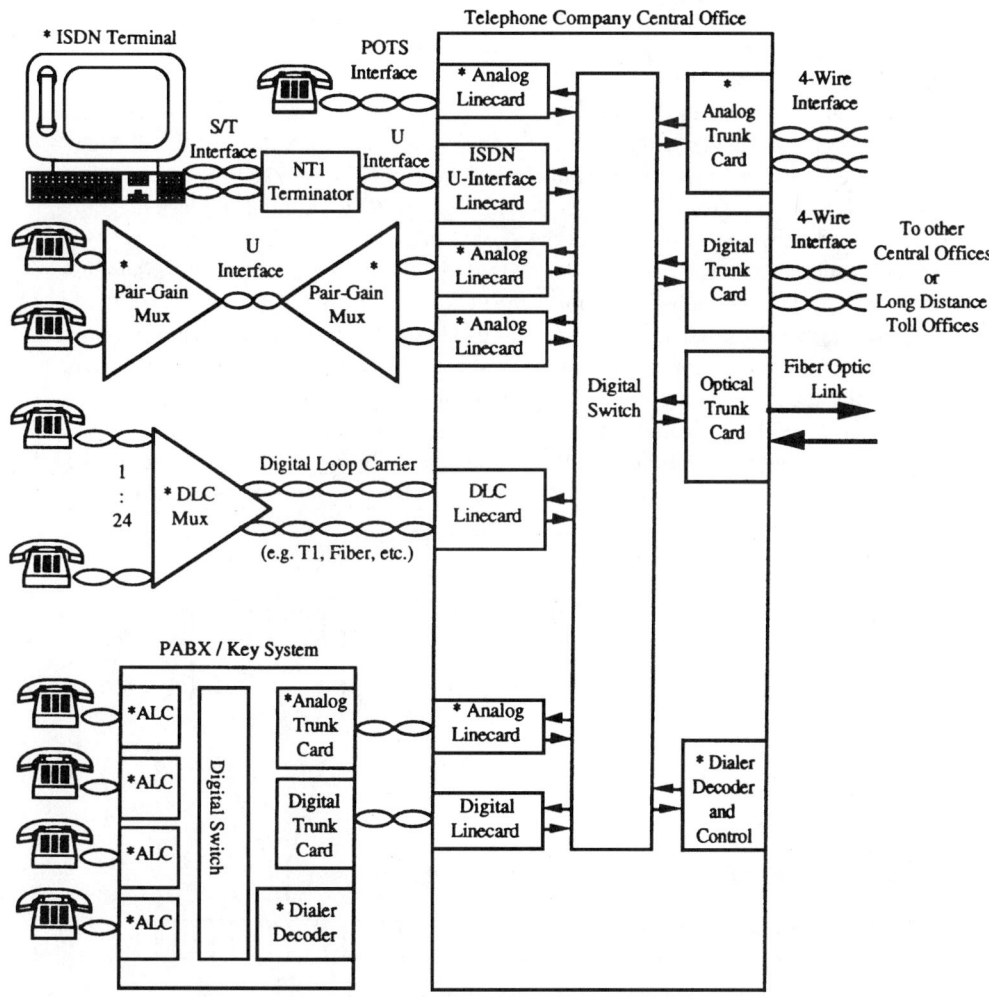

FIGURE 27.1 Public switching telephone network (PSTN).

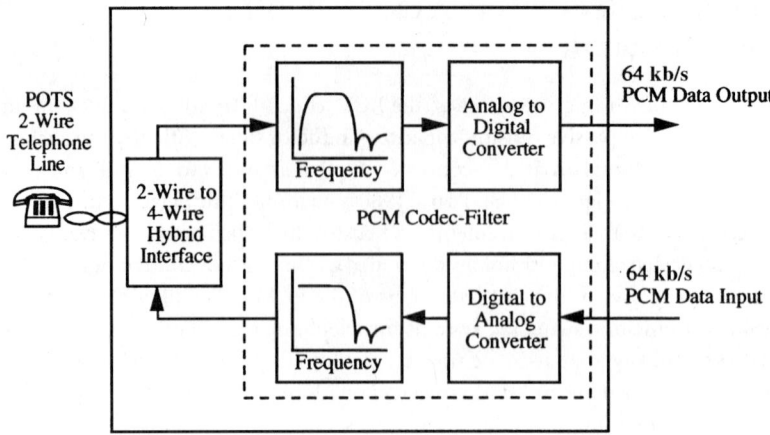

FIGURE 27.2 Analog telephone linecard.

and during the loss of AC power by emergency battery backup. This requires low-power dissipation of the PCM codec-filter. Complementary metal-oxide-semiconductor (CMOS) integrated circuit (IC) fabrication technology has proven to be a reliable low-power semiconductor solution for this type of complex analog and digital very large-scale integration (VLSI) circuit. A PCM codec-filter is used in each location with an asterisk (*).

The analog linecard is the interface between the switch and the telephone wires that reach from the central office or service provider to the subscriber's telephone. The analog linecard also provides the digital interface for the switching or transmission equipment. The accepted name for analog telephone service is plain old telephone service (POTS). Figure 27.2 is a simplified block diagram for an analog linecard showing the functions of the PCM codec-filter.

The PCM codec-filter converts voice into 8-b PCM words at a conversion rate of 8 kilosamples per second. This PCM data is serially shifted out of the PCM codec-filter in the form of a serial 8-b word. This results in a 64-kb/s digital bit stream. This 64 kb/s PCM data from an analog linecard is typically multiplexed onto a PCM data bus or highway for transfer to a digital switch. Where it will be routed to the appropriate output PCM highway, which routes the data to another linecard, completing the signal path for the conversation. (Refer to Fig. 27.1.)

Transmission equipment will have multiple PCM codec-filters shifting their PCM words onto a single conductor. This is referred to as a serial bus or PCM highway. Figure 27.3 shows the 8-b

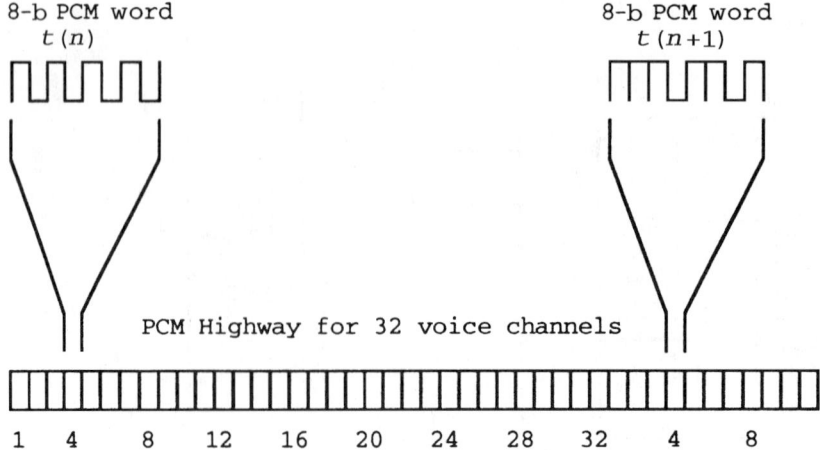

FIGURE 27.3 Multiplexed PCM highway.

PCM words from a PCM codec-filter for one voice channel being multiplexed into a time slot on a PCM highway that can accommodate 32 voice channels. For more information on PCM switching, see the chapters in this handbook on switching systems, and common channel signalling. For more information on hybrid interfaces, see the chapters in this handbook on POTS, analog hierarchy, and analog telephone channels and conditioning.

27.3 Design of Voice PCM Codec-Filters: Analog Transmission Performance and Voice Quality for Intelligibility

Filtering

The pass-band frequency response of the voice channel is roughly 300–3000 Hz. The 300–3000 Hz spectrum is where most of the energy in the human voice is located. Voice contains spectral energy below 300 Hz and above 3 kHz, but its absence is not detrimental to intelligibility. The frequency response within the passband of 300 Hz–3 kHz must be tightly controlled, because the public switching telephone network (PSTN) is allowed to have as many as seven ADC/DAC conversions end-to-end. The cumulative effects of the network plus private equipment such as PBX or key systems could audibly distort the frequency response of the voice channel with less stringent requirements. Figure 27.4 shows the typical half-channel pass-band frequency response requirement of $+/-$ 0.25 dB.

The 3-kHz bandwidth for voice signals determines the sample rate for the ADC and DAC. In a sampling environment, Nyquist theory states that to properly sample a continuous signal, it must be sampled at a frequency higher than twice the signal's highest frequency component. Minimizing the sample rate for digitizing voice is a priority since it determines the system data rates which are directly proportional to cost. The voice bandwidth of 3 kHz plus the filter transition band sampled at 8 kHz represents the best compromise that meets the Nyquist criterion.

The amount of attenuation required for frequencies of 4 kHz and higher is dictated by the required signal-to-distortion ratio of about 30 dB for acceptable voice communication as determined by the telephone industry. Frequencies in the filter transition band of 3.4–4.6 kHz must be attenuated enough such that their reconstructed alias frequencies will have a combined attenuation of about 30 dB. This permits typical filter transition band attenuation of 15 dB at 4 kHz for both the input filter for the ADC and the output reconstruction filter for the DAC. All frequencies above 4.6 kHz must be satisfactorily attenuated before the ADC conversion to prevent aliasing in-band. The requirement for out-of-band attenuation is specified by the country of interest, but is typically from 25 to 32 dB for frequencies of 4600 Hz and higher.

The requirement for limiting frequencies below 300 Hz is also regionally dependent and application dependent, ranging from flat response to 20 dB or more attenuation at 50 and 60 Hz. The telephone line is susceptible to 50/60-Hz power line coupling, which must be attenuated from the signal by a high-pass filter before the ADC. Attenuation of power line frequencies is desirable to prevent 60 Hz noise during voice applications. Figure 27.4 shows a typical frequency response requirement for North American telephone equipment.

The digital-to-analog conversion process reconstructs a pulse-amplitude modulated (PAM) staircase version of the desired in-band signal, which has spectral images of the in-band signal modulated about the sample frequency and its harmonics. These high-frequency spectral images are called aliasing components which need to be attenuated to meet performance specifications. The low-pass filter used to attenuate these aliasing components is typically called a reconstruction or smoothing filter. The low-pass filter characteristics of the reconstruction filter are similar to the low-pass filter characteristics of the input antialiasing filter for the ADC.

The accuracy of these filters requires op amps with both high gain and low noise on the same monolithic substrate with precision matched capacitors which are used in the switched-capacitor

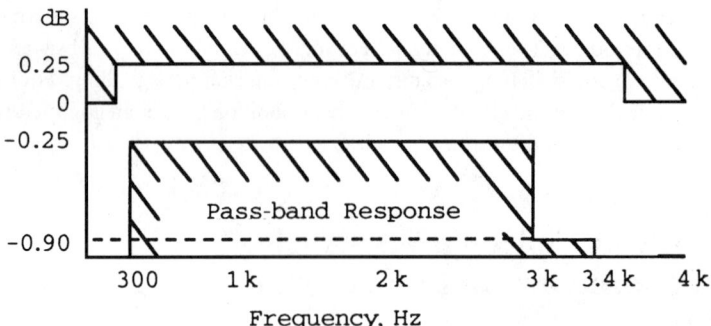
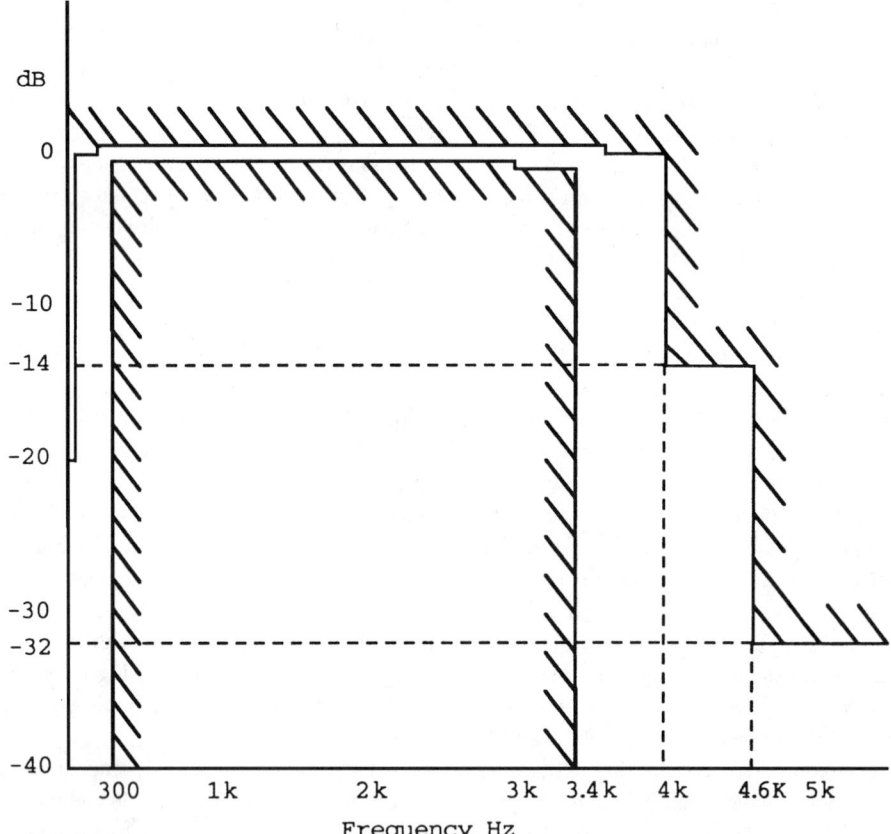

FIGURE 27.4 Typical frequency response requirement for the input filter: pass-band and stop band.

filter structures. For more information on filter requirements and sampling theory, refer to the chapters in this handbook on analog modulation, sampling, and pulse-code modulation.

Quantizing Distortion

To digitize voice intelligibly requires a signal-to-distortion ratio of about 30 dB over a dynamic range of about 40 dB. This may be accomplished with a linear 13-b ADC and DAC, but will far exceed the required signal-to-distortion ratio for voice at amplitudes greater than 40 dB below the peak overload amplitude.

Pulse-Code Modulation Codec-Filters

This excess performance comes at the expense of increased bits per sample which directly translates into system cost, as stated earlier. Figure 27.5 shows the signal-to-quantization distortion ratio performance of a 13-b linear DAC compared to a companded mu-law DAC. A high-speed modem application is discussed later in this chapter, which requires linear 13 b or better performance.

Two methods of data reduction are implemented both of which use compressing the 13-b linear codes during the analog-to-digital conversion process and expanding the codes back during the digital-to-analog conversion process. These compression and expansion schemes are referred to as companding. The two companding schemes are: mu-255 law, primarily used in North America and Japan, and A-law, primarily used in Europe. Companding effectively trades the signal to distortion performance at larger amplitudes for fewer bits for each sample. These companding schemes map the 13-b linear codes into pseudologarithmic 8-b codes. These companding schemes follow a segmented or piecewise-linear input–output transfer curve formatted as sign bit, three chord bits, and four step bits. For a given chord, all 16 of the steps have the same voltage weighting. As the voltage of the analog input increases, the four step bits increment and carry over to the three chord

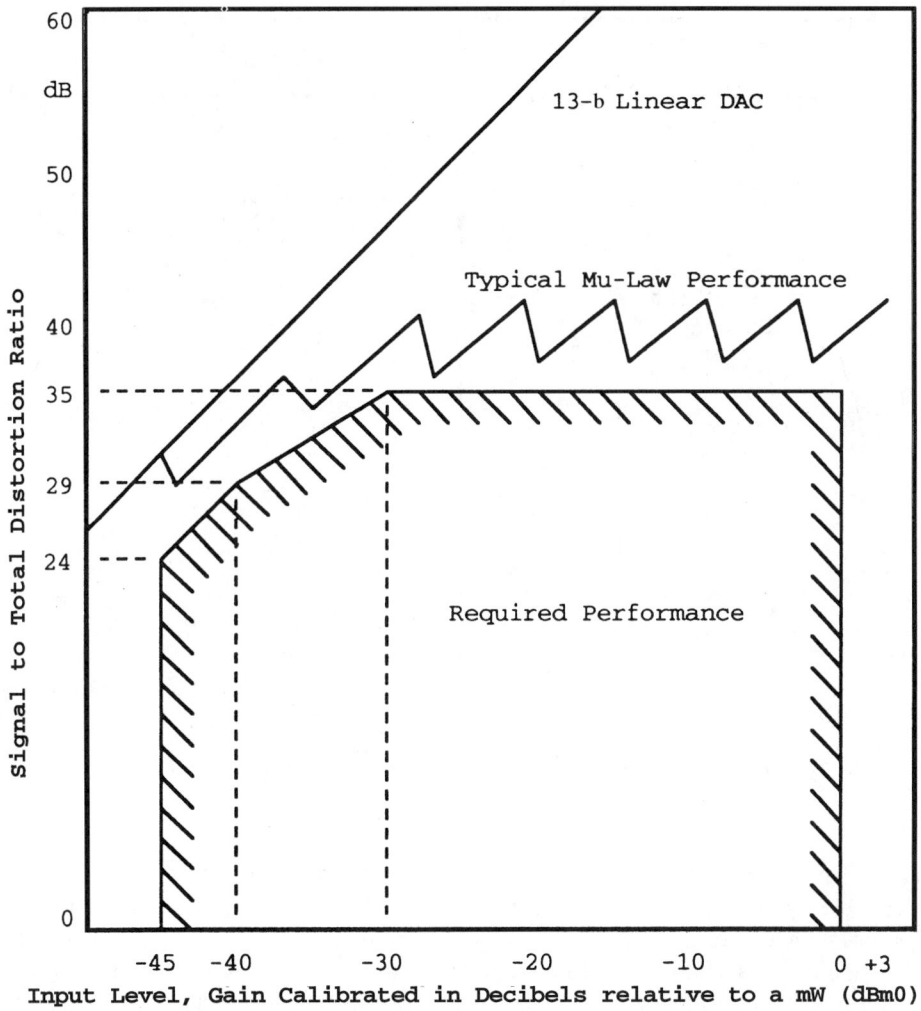

FIGURE 27.5 Signal-to-quantization distortion performance for mu-law and 13-b linear DACs compared to the telephone network requirements.

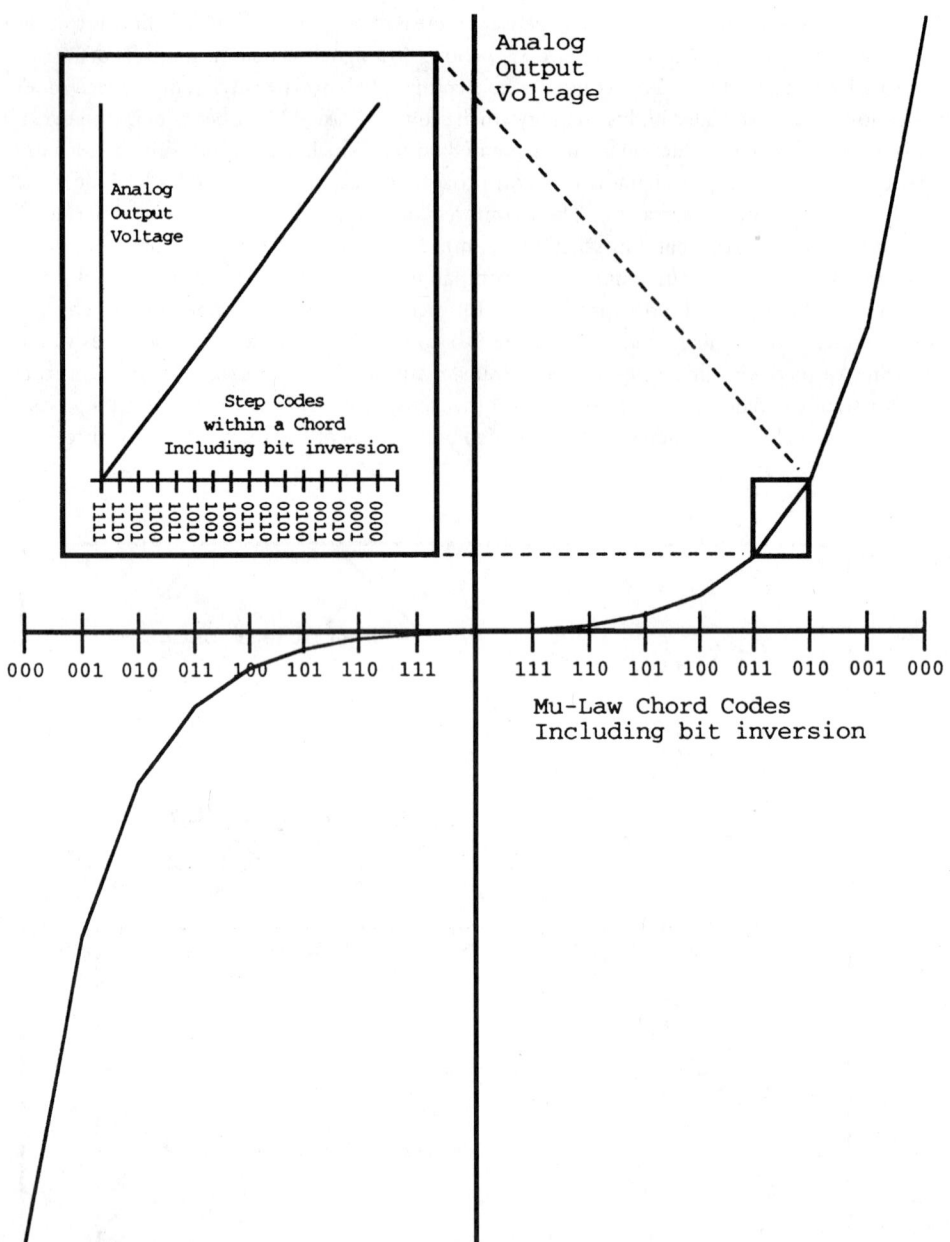

FIGURE 27.6 Mu-law companding showing piecewise-linear approximation of a logarithmic curve.

bits, which increment. When the chord bits increment, the step bits double their voltage weighting. This results in an effective resolution of six bits (1 sign bit + 1 effective chord bit + 4 step bits) with a sinusoidal stimulus across a 42-dB dynamic range (seven chords above zero, by 6 dB per chord, i.e., effectively 1 b). This satisfies the original requirement of 30-dB signal-to-distortion over a 40-dB dynamic range. Figure 27.6 is a graphical explanation of mu-law companding.

Note that to minimize die size area, which further cost reduces the PCM codec-filter, compressing ADC designs and expanding DAC designs are used instead of linear ADC/DAC designs with read-only memory (ROM) conversion lookup tables.

	Mu-Law			A-Law		
Level	Sign Bit	Chord Bits	Step Bits	Sign Bit	Chord Bits	Step Bits
+Full scale	1	0 0 0	0 0 0 0	1	0 1 0	1 0 1 0
+Zero	1	1 1 1	1 1 1 1	1	1 0 1	0 1 0 1
−Zero	0	1 1 1	1 1 1 1	0	1 0 1	0 1 0 1
−Full scale	0	0 0 0	0 0 0 0	0	0 1 0	1 0 1 0

FIGURE 27.7 Full scale and zero codes for mu-law and A-law.

A-law companding is very similar to mu-law with three differences. The first difference is that A-law has equal voltage weighting per step for the two positive or negative chords near zero volts. In mu-law the step weighting voltage for chord zero is one-half the step weighting voltage for A-law chord zero. This reduces the A-law resolution for small signals near the voltage origin. The second difference is that the smallest voltage of the DAC curve does not include zero volts, and instead produces a positive half-step for positive zero and a negative half-step for negative zero. This is in contrast to mu-law companding, which produces a zero voltage output for both positive and negative zero PCM codes. Both mu-law and A-law have symmetric transfer curves about the zero volts origin, but Mu-law has redundancy at zero volts and, therefore, effectively an unused code. The third difference between the two companding schemes deals with the data bit format. A-law inverts the least significant bit (LSB) and every other bit, leaving the sign bit unchanged relative to a conventional binary code. For example, applying the A-law data format formula to binary zero, s0000000, results in s1010101, the A-law code for +/− zero. Mu-law maintains a positive polarity for the sign bit with all of the magnitude bits inverted. Figure 27.7 shows the zero codes and full-scale codes for both mu-law and A-law companding schemes.

Mu-law and A-law companding schemes are recognized around the world. Many equipment manufacturers build products for distribution worldwide. This distribution versatility is facilitated by the PCM codec-filter being programmable for either mu-law or A-law. With the deployment of ISDN, the PCM codec-filter is located at the subscriber's location. The need to place international telephone calls dictates the need for a PCM codec-filter that is both mu-law and A-law compatible.

Gain Calibration: Transmission Level Point

The test tone amplitude for analog channels is 3 dB down from the sinusoidal clip level. This test level for a POTS line is generally 1 mW, which is 0 dBm. The telephone line historically was referred to as having a 0 dB transmission level point (TLP). (With the improvements in transmission quality and fewer losses, telephone lines often have attenuation pads to maintain talker/listener appropriate levels.) Amplitudes that have been TLP calibrated are given unit abbreviations that have a zero added to them, for example, dBm0. The test signal may experience gain or attenuation as it is transmitted through the network. As an example the absolute magnitude of one electrical node for the voice channel could result in a measured value of +6 dBm when a 0 dBm0 test signal is applied. This node would be gain calibrated as having a +6 dB TLP. This would result in absolute measurements for noise having this same +6 dB of gain. To properly reference the noise, all measurements must be gain calibrated against the 0 dBm0 level, which in this example would result in 6 dB being subtracted from all measurements. All signals in this channel, including distortion and noise, would be attenuated by 6 dB, by the time they get to the subscriber's telephone.

For PCM channels the 0 dBm0 calibration level is nominally 3 dB down from maximum sinusoidal clip level; specifically, it is 3.17 dB down for mu-law and 3.14 dB down for A-law. These minor deviations from an ideal 3 dB result from using a simple code series, called the digital milliwatt, for generating the digital calibration signal. The digital milliwatt is a series of eight PCM codes that are

	Mu-Law			A-Law		
Phase	Sign Bit	Chord Bits	Step Bits	Sign Bit	Chord Bits	Step Bits
$\pi/8$	0	0 0 1	1 1 1 0	0	0 1 1	0 1 0 0
$3\pi/8$	0	0 0 0	1 0 1 1	0	0 1 0	0 0 0 1
$5\pi/8$	0	0 0 0	1 0 1 1	0	0 1 0	0 0 0 1
$7\pi/8$	0	0 0 1	1 1 1 0	0	0 1 1	0 1 0 0
$9\pi/8$	1	0 0 1	1 1 1 0	1	0 1 1	0 1 0 0
$11\pi/8$	1	0 0 0	1 0 1 1	1	0 1 0	0 0 0 1
$13\pi/8$	1	0 0 0	1 0 1 1	1	0 1 0	0 0 0 1
$15\pi/8$	1	0 0 1	1 1 1 0	1	0 1 1	0 1 0 0

FIGURE 27.8 PCM codes for digital milliwatt.

repeated to generate the 1 kHz calibration tone. Figure 27.8 shows the eight PCM codes which are repeated to generate a digital milliwatt for both mu-law and A-law in the DAC.

The digital milliwatt is used to calibrate the gains of telephone networks using voice grade PCM codec-filters.

Idle Channel Noise

Idle channel noise is the noise when there is no signal applied to the voice channel, which is different from the quantization distortion caused by the ADC/DAC process. This noise can be from any circuit element in the channel, in addition to the PCM codec-filter. Noise may be coupled in from other circuits including, but not limited to, hybrid transformer interfaces, power supply induced noise, digital circuitry, radio frequency radiation, and resistive noise sources.

The PCM codec-filter itself has many opportunities to contribute to the idle channel noise level. The input high-pass filter often is third order, whereas both the input and the output low-pass filters are typically fifth-order elliptic designs. The potential for noise generation is proportional to the order of the filter. The ADC and DAC arrays have many components that are controlled by sequencing digital circuits. The power supply rejection capability of PCM codec-filters once dominated the noise level of the channel. The power supply rejection ratio of these devices has been improved to a level where the device is virtually immune to power supply noise within the allowable power supply voltage range. This performance was attained by differential analog circuit designs in combination with tightly controlled matching between on-chip components.

Noise measurements require a different decibel unit as they usually involve some bandwidth or filter conditioning. One such unit commonly used (especially in North America) is decibels above reference noise (dBrn). The reference noise level is defined as 1 pW or −90 dBm. Telephone measurements typically refer to dBrnC, which is the noise level measured through a C-message weighting filter (a filter that simulates the frequency response of the human ear's noise sensitivity). European systems use a related term called dBmp, which is the dBm level noise measured through a psophometric filter. Noise measurements made by either dBrnC or dBmp filter weightings have units to show that the noise has been gain referenced to 0 dBm0 by adding a zero, hence dBrnC0 and dBm0p. Two examples are shown to illustrate the use of these units:

1. Mu-law: If 0 dB TLP = +6 dB, then a noise measurement of 20 dBrnC equals 14 dBrnC0.
2. A-law: If 0 dB TLP = +4 dB, then a noise measurement of −70 dBmp equals −74 dBm0p.

The examples are representative of typical idle channel noise measurements for a full-duplex digital channel. Idle channel noise measuring 14 dBrnc0 at a 0 dB TLP output is about 123-μV root mean square (rms). This low-noise level is very susceptible to degradation by the noise sources mentioned earlier.

The idle channel noise of the voice channel directly impacts the dynamic range, which is the ratio of maximum power the channel can handle to this noise level. Typical dynamic range for a mu-law PCM codec-filter is about 78 dB. The term dynamic range is similar to signal-to-noise ratio for linear circuits. The signal-to-noise plus distortion ratio for a companded channel is generally limited by the nonlinear companding except at very low-signal amplitudes.

Gain Tracking: Gain Variations as a Function of Level

The quantization curves, as discussed earlier, are not linear and may cause additional nonlinear distortions. The concern is that the gain through the channel may be different if the amplitude of the signal is changed. Gain variations as a function of level or how well the gain tracks with level is also called gain tracking. This is a type of distortion, but could easily be missed by a tone stimulus signal-to-distortion test alone. Gain tracking errors can cause ringing in the 2-wire–4-wire hybrid circuits if the gain at any level gets too large. Gain tracking performance is dominated by the IC fabrication technology, more specifically, the matching consistency of capacitors and resistors on a monolithic substrate. Figure 27.9 shows the half-channel gain tracking performance recommended for a PCM digital interface.

Figure 27.10 is a block diagram for a monolithic MC145480 5 V PCM Codec-Filter manufactured by Motorola Inc. The MC145480 uses CMOS for reliable low-power operation. The operating temperature range is from $-40°$ to $+85°C$, with a typical power dissipation of 23 mW. All analog signal processing circuitry is differential utilizing Motorola's six-sigma quality IC fabrication process. This device includes both antialiasing and reconstruction switched capacitor filters, compressing ADC, expanding DAC, a precision voltage reference, shift registers, and additional features to facilitate interfacing to both the analog and digital circuitry on a linecard.

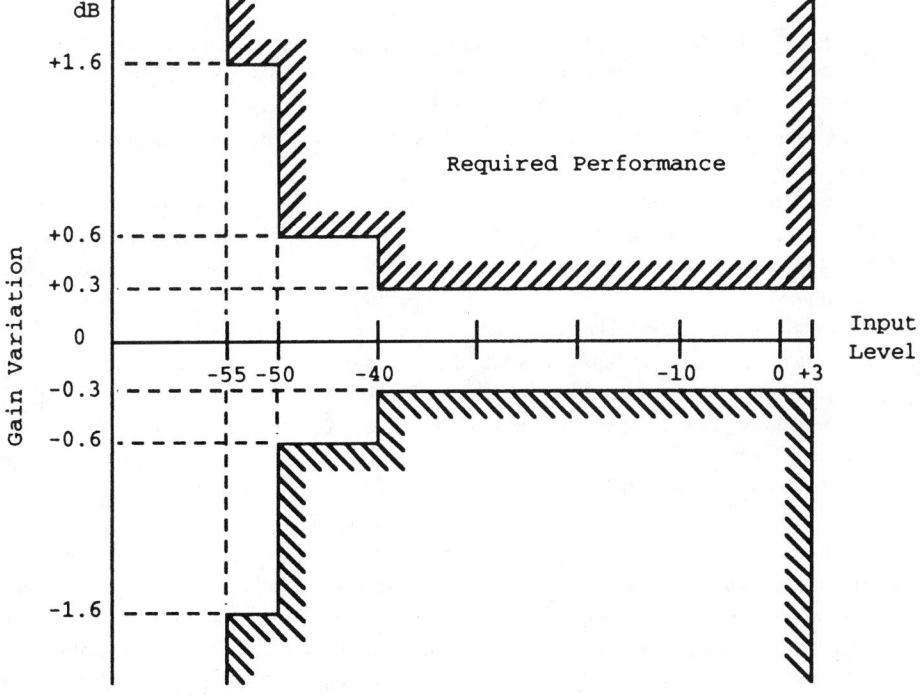

FIGURE 27.9 Variation of gain with input level using a tone stimulus.

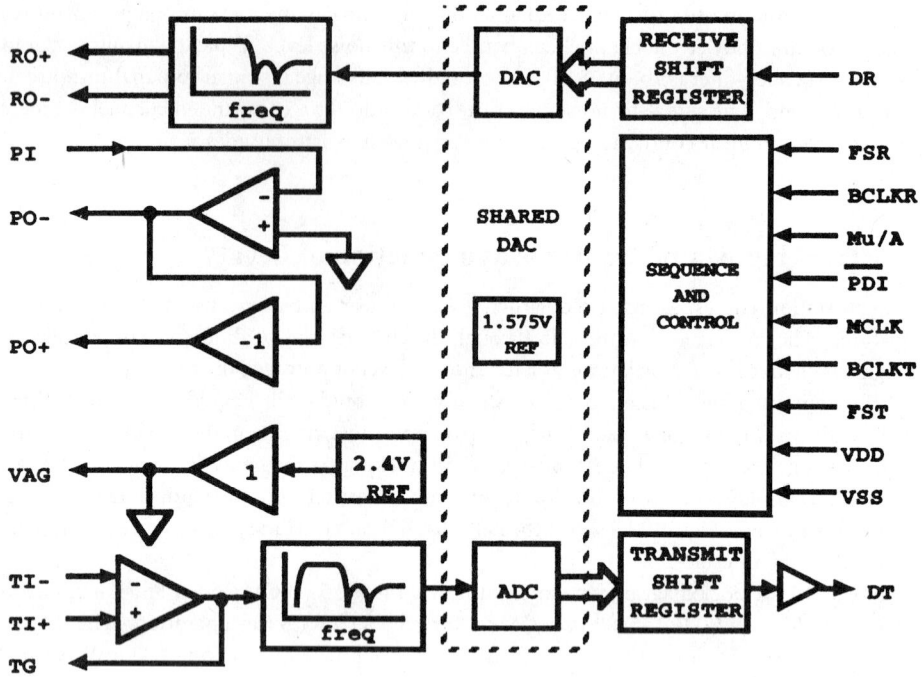

FIGURE 27.10 MC145480 5 V PCM Codec-Filter.

27.4 Linear PCM Codec-Filter for High-Speed Modem Applications

As discussed in previous sections of this chapter the original PCM codec-filters were developed to facilitate the digital switching of voice signals in the central office of the PSTN. The signal processing in the digital domain consisted of simply moving digitized samples of voice from one time slot to another to accomplish the switching function. As such the A/D and D/A conversion processes could be nonlinear, and companding was employed.

With the advent of the personal computer, the facsimile machine, and most recently the Internet, the need to transmit data originating at the subscriber over the PSTN at the highest possible rates has grown to the point where the theoretical limits as defined by Shannon's law [Shannon, 1948] are being approached. For data transmission the signal processing is in a sense the inverse of that which spawned the original PCM codecs in that the initial signal is data that must be modulated to look like a voice signal in order to be transmitted over the PSTN. At data rates above 2400 b/s the modulation is done using sophisticated digital signal processing techniques, which have resulted in a whole new class of linear PCM codec-filters being developed.

Codec-filters for V.32 (9.6 kb/s), V.32bis (14.4 kb/s) and V.34 (28.8 kb/s) modems must be highly linear with a signal to distortion much greater than 30 dB. This linearity and signal to distortion is required to implement the echo cancellation function with sufficient precision to resolve the voltages corresponding to the data points in the dense constellations of high-speed data modems. Please refer to the chapters in this handbook on signal space and echo cancellation for more information. The critical signal-to-noise plus distortion plots for two popular commercially available codec-filters used in V.32bis (T7525) and V.34bis (STLC 7545) modems are shown in Fig. 27.11. Both of these codec-filters are implemented using linear high ratio oversampling sigma–delta ($\Sigma\Delta$) A/D and D/A conversion technology and digital filters. This can be contrasted with the nonlinear companding converters and switched capacitor filters of voice grade PCM codec-filters.

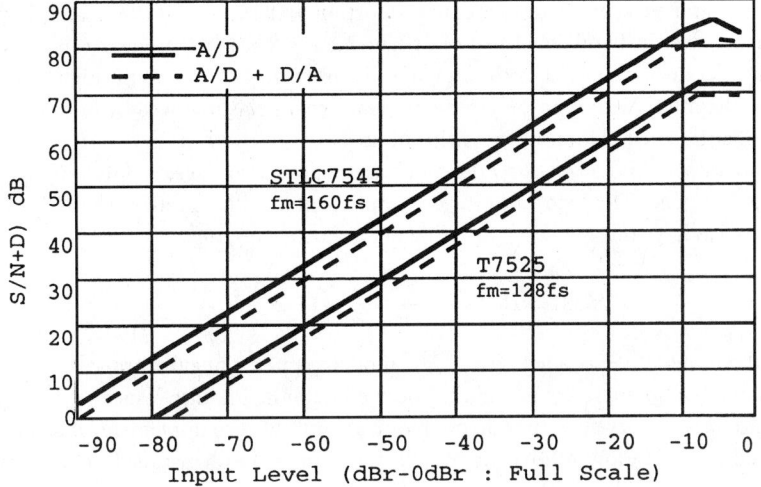

FIGURE 27.11 Linear codec dynamic range, where fm is the modulator frequency and fs is the sample frequency.

Sigma–delta conversion technology is based on coarsely sampling a signal at a high rate and filtering the resulting noise [Candy and Temes, 1992; Park, 1990]. A generalized block diagram of sigma–delta PCM codec-filters for modem applications is shown in Fig. 27.12. The transmit and receive signal paths are the inverse of each other. The second-order modulators basically consist of two integrators and a comparator with two feedback loops. The action of the modulators is to low-pass filter the signal and high-pass filter the noise so that greater than 12 b of resolution (70 dB) can be achieved in-band. If it is assumed that the input signal is sufficiently active to make the quantization noise random then it can be shown that [Candy and Oconnell, 1981] for a simple first-order modulator

$$No^2 = (1/fo)^3$$

Where:

No = net rms noise in band
fo = oversampling frequency

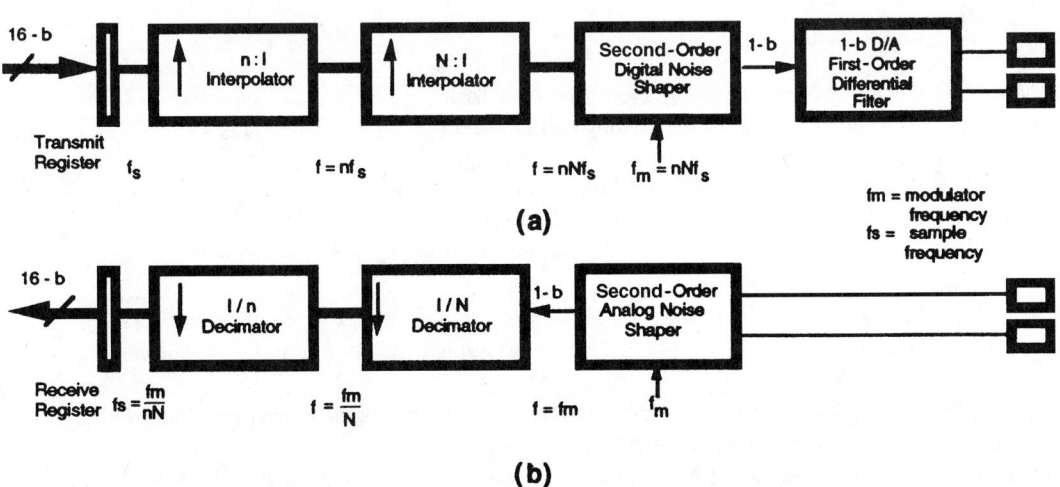

FIGURE 27.12 Generalized $\Sigma\Delta$ PCM modem codec-filter: (a) transmit channel and (b) receive channel.

That is, the in-band noise power decreases in proportion to the cube of the oversampling frequency resulting in improved in-band signal-to-noise performance. The high-frequency quantization noise in the receive channel is removed by the digital decimation filters, whereas the high-frequency alias noise due to the interpolation process in the transmit channel is removed by interpolation filters. The transmit and receive channels are synchronous.

The decimation and interpolation operations are done in two stages: a low-ratio (n) stage and a high-ratio (N) stage. The high-ratio stage is implemented with cascaded comb filters which are simply moving average filters having the following frequency response:

$$H(z) = 1/N((1 - z^{-N})/(1 - z^{-1}))^i$$

where i is the number of cascaded filters. For modem applications the order i of the cascade is generally three. Comb filters are essentially linear phase finite impulse response (FIR) filters with unity coefficients and, therefore, they do not require a multiplier to implement, which makes them silicon area efficient. The low-ratio stage is usually implemented with cascaded second-order infinite impulse response (IIR) filters. These filters do require a multiplier to implement. The frequency response of the IIR filters compensate for the $\sin x/x$ droop of the comb filters and shape the pass-band to meet telephone specifications. The total oversampling product, $nN = fo$, is typically greater than or equal to 128 with $N = 32$ or 64. For V.32bis modems fs is fixed at 9.6 kHz, whereas for V.34bis modems fs varies from 9.6 to $\sim$14 kHz.

27.5 Concluding Remarks

The PCM voice codec-filter has continuously evolved since the late 1950s. With each improvement in performance and each reduction in cost, the number of applications has increased. The analog IC technology incorporated in today's PCM voice codec-filters represents one of the true bargains in the semiconductor industry.

Sigma–delta-based PCM codec-filters have become the standard for implementing high-speed modems because they can offer sufficient resolution at the lowest cost. Sigma–delta codecs are cost/performance effective because they are 90% digital and digital circuitry scales with integrated circuit (IC) process density improvements and associated speed enhancements. It is interesting to note that although the analog portion of the sigma–delta codec-filter is small and does not require precision or matched components like other converter technologies, it (the D/A output differential filter) limits the ultimate performance of the codec-filter.

References

Bellamy, J. 1991. *Digital Telephony*, Wiley, New York, New York.
Bellcore, 1987. Functional criteria for digital loop carrier systems. Tech. Ref. Doc., TR-TSY-000057, Issue 1, April.
Candy, J.C. and Oconnell, J.B. 1981. The Structure of quantization noise from sigma-delta modulation. *IEEE Trans. Comm.*, COM-29 (Sept.):1316–1323.
Candy, J.C. and Temes, G.C. 1992. *Oversampling Delta-Sigma Data Converters Theory, Design and Simulation*, IEEE Press, New York.
ITU. 1984. ITU-T digital networks, transmission systems and multiplexing equipment. CCITT Red Book, G.711, G.712, G.713, G.714, International Telecommunications Union—Telecommunications .
Park, S. 1990. Principles of sigma-delta modulation for analog-to-digital converters. Motorola Digital Signal Processing Division Application Note, Austin, TX.
Shannon, C.E. 1948. A mathematical theory of communication. *Bell Syst. Tech. J.*, 27 (July):379–423 and 27 (Oct.):623–656.

28
Digital Hierarchy

B.P. Lathi
California State University, Sacramento

Maynard A. Wright
Electrical Engineering Consultant

28.1 Introduction 377
28.2 North American Asynchronous Digital Hierarchy 378
　　　Digital Signal Level 0 (DS0) • Digital Signal Level 1 (DS1) • Higher Rate Formats

28.1 Introduction

Following the introduction of digital encoding and transmission of voice signals in the early 1960s, multiplexing schemes for increasing the number of channels that could be transported were developed by the Consultative Committee on International Telephony and Telegraphy (CCITT) in Europe and by the Bell System in North America. Initial development of such systems was aimed at coaxial cable and millimeter waveguide transmission systems. The advent of low-loss optical fiber and efficient modulation schemes for digital radio greatly increased the need to transport digital channels at rates higher than the primary multiplexing rate.

Separate multiplexing schemes, called digital hierarchies, were developed by CCITT and by the Bell System. During the period in which the two hierarchies were conceived, there was no reliable way to distribute highly accurate clock signals to each central office which might serve as a multiplexing point. Local clock signals were therefore provided by relatively inaccurate crystal oscillators, which were the only economical signal sources available. Therefore, each hierarchy was designed with the expectation that the clocks controlling the various stages of the multiplexing and demultiplexing processes would not be accurate and that means would be required for compensating for the asynchronicities among the various signals.

The mechanism chosen to accomplish synchronization of the multiplexing process was positive bit stuffing as described in Chapter 7. The CCITT development was termed the *plesiochronous digital hierarchy* (PDH) and, in North America, the scheme was designated the *asynchronous digital hierarchy*. The two differ significantly. The North American digital hierarchy will be described in the sections which follow.

The advent of accurate and stable clock sources has made possible the deployment of multiplexing schemes which anticipate that most of the signals involved will be synchronous. The North American synchronous multiplexing scheme is the Synchronous Optical Network and is described in Chapter 39. The International Telecommunications Union–Telecommunications Standardization Sector (ITU-T) version is termed the synchronous digital hierarchy (SDH). In spite of the development of these synchronous multiplexing hierarchies, the asynchronous hierarchies remain important because there are substantial numbers of asynchronous legacy systems and routes, which must be maintained and which must support further growth.

28.2 North American Asynchronous Digital Hierarchy

Following dissolution of the Bell System in 1984, the Exchange Carrier Standards Association (**ECSA**), an American National Standards Institute (ANSI) chartered standards development body, became responsible for the maintenance and development of the North American telecommunications standards. The role played by ECSA has since been assumed by **Alliance for Telecommunications Industry Solutions (ATIS)**, which continues to sponsor Committee T1 (telecommunications). Committee T1 maintains ANSI Standard T1.107 [ANSI, 1995a], which describes the North American asynchronous digital hierarchy.

The capacities of the bit streams of the various multiplexing levels in the North American digital hierarchy are divided into **overhead** and **payload**. Overhead functions include framing, error checking, and maintenance functions, which will be described separately for each level. Payload capacity is used to carry signals delivered from the next lower level or delivered directly to the multiplexer by a customer.

The various rates in the digital hierarchy are termed **digital signal n (DS-n)**, where n is the specified level in the hierarchy. The various rates are summarized in Table 28.1. In addition, ANSI T1.107 [ANSI, 1995a] provides standards for multiplexing various sub-DS0 digital data signals into a 64 kb/s DS0.

The relationships among the various signals are illustrated in Fig. 28.1. Note that DS1C is a dead-end rate. It is useful for channelizing paired cable plant but it cannot be multiplexed into higher level signals. DM in Fig. 28.1 represents digital multiplexer.

Digital Signal Level 0 (DS0)

The DS0 signal is a 64-kb/s signal, which is usually organized in 8-b bytes and which may be developed by encoding analog signals as described elsewhere in this handbook. Alternatively, the DS0 may be built from digital signals which are synchronous with the DS1 clock. There is no provision for handling asynchronous signals at the DS0 rate.

Digital Signal Level 1 (DS1)

The DS1 signal is built by interleaving 1.536 Mb/s of payload capacity with 8 kb/s of overhead. Following each overhead bit, 8-b bytes from each of the 24 DS0 channels to be multiplexed are byte interleaved. The resulting structure appears in Fig. 28.2. The DS0 bytes (channels) are numbered sequentially from 1 to 24 in Fig. 28.2 and that scheme is the standard method for interleaving DS0 bytes. Other schemes, which use a nonsequential ordering of DS0 channels, have been used in the past.

TABLE 28.1 Rates in the North American Digital Hierarchy

DS0	64-kb/s signal which may contain digital data or $\mu = 255$ encoded analog signals
DS1	1.544-Mb/s signal which byte-interleaves 24 DS0s
DS1C	3.152-Mb/s signal which bit-interleaves 2 DS1s
DS2	6.312-Mb/s signal which bit-interleaves 4 DS1s
DS3	44.736-Mb/s signal which bit-interleaves 7 DS2s
DS4NA	139.264-Mb/s signal which bit-interleaves 3 DS3s

Digital Hierarchy

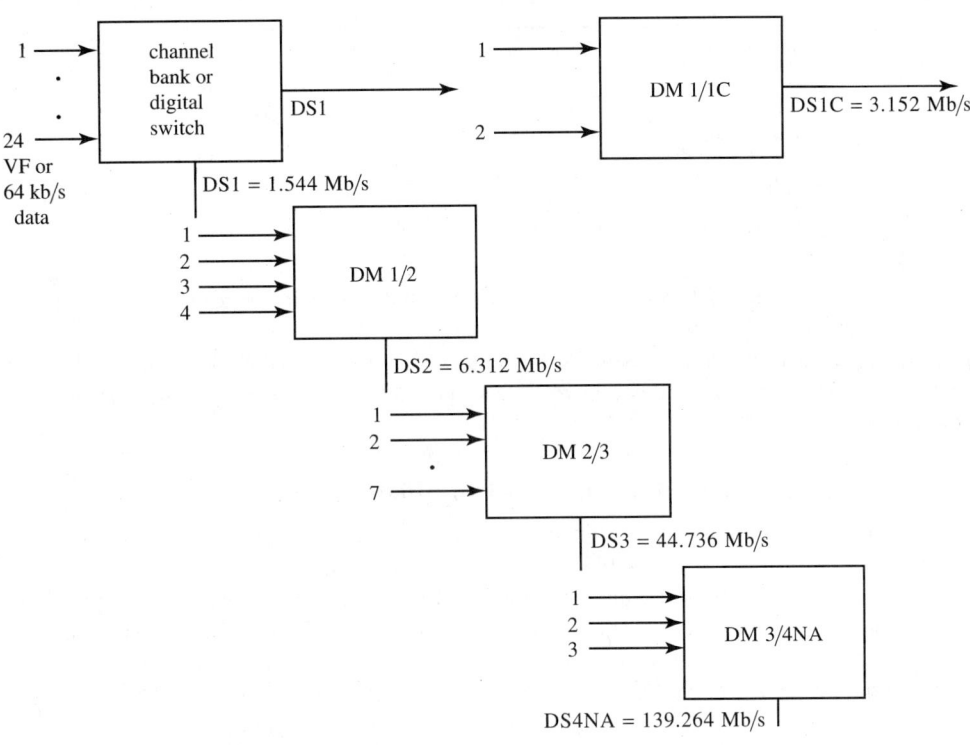

FIGURE 28.1 Multiplexing in the North American Digital Hierarchy.

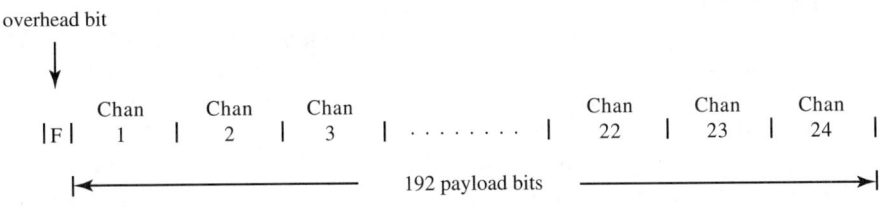

FIGURE 28.2 DS1 frame structure.

This 193-b pattern is called a *frame* and is repeated 8000 times per second to coincide with the sampling rate for encoding analog signals. Each 8-b encoded sample of a particular channel may be transmitted in the appropriate channel position of a single frame.

Digital Signal Level 1 Overhead: Superframe Format

In the earliest 1.544 Mb/s systems, the framing pattern was simply used to identify the beginning of each frame so that the channels could be properly demultiplexed by the receiver. With the advent of robbed-bit signalling, it became necessary to identify the particular frames in which robbed-bit signalling occurs. A superframe SF of 12 frames was developed in which robbed-bit signalling occurs in the 6th and 12th frames. The overhead capacity in the superframe format is organized into two streams. The pattern carried by the overhead bits (Ft bits-terminal framing) in odd-numbered frames identifies the beginning of the frame structure and serves to locate the overhead stream. The pattern in the even-numbered frames (Fs bits-signal framing) locates the beginning of each superframe so that the robbed-bit signalling can be properly demultiplexed. The SF framing

Frame Number:	1	2	3	4	5	6	7	8	9	10	11	12
Ft bits	1	–	0	–	1	–	0	–	1	–	0	–
Fs bits	–	0	–	0	–	1	–	1	–	1	–	0
Composite pattern	1	0	0	0	1	1	0	1	1	1	0	0

FIGURE 28.3 DS1 SF overhead bit assignments.

pattern is shown in Fig. 28.3. Note that each frame in which the logical state of the Fs bit changes is a signalling frame which may carry robbed-bit signalling.

There are 2 b per superframe devoted to signalling. They are denoted A and B and may be used to carry four-state signalling although two-state signalling (on- and off-hook indication) is more common.

Digital Signal Level 1 Overhead: Extended Superframe Format

Improvements in electronics and framing algorithms have made it possible for receivers to frame rapidly and efficiently on patterns with less information than is contained in the 8-kb/s superframe overhead stream. The **extended superframe format** (**ESF**) uses a 2-kb/s framing pattern to locate a 24-frame extended superframe. The frame structure remains as for the SF format, and the ESF format therefore provides 6 kb/s of overhead capacity which is not devoted to framing.

A communications channel [**data link** (DL)] consumes 4 kb/s of overhead. This channel is used to convey performance monitoring information from one DS1 terminal location to the other (scheduled messages) and to send alarm messages and loopback requests (unscheduled messages). The use of this channel is detailed in ANSI T1.403 [1995b]. When no message is being sent in the DL, it is usually filled with unconcatenated **high-level data link control** (**HDLC**) flags (01111110), although some older equipment may fill with an all ones pattern.

The remaining 2 kb/s of overhead in the ESF format carries a **cyclic redundancy code with a 6-b remainder** (CRC-6) channel to provide error checking. The divisor polynomial is $X^6 + X + 1$. Division is carried out over all 4632 b of an extended superframe with all of the overhead bits set to logical ones. The 6-b remainder resulting from the division is written into the CRC-6 bits of the following extended superframe.

Figure 28.4 contains a summary of the overhead bit assignments in the ESF format where the **frame alignment signal** (**FAS**) is given. The logical values of the bits written into the FAS are shown in the figure. The bits forming the 4 kb/s DL are represented by "D" and the six individual BUS positions for the CRC-6 remainder from the previous extended superframe are shown as "C."

Digital System Level 1C Format

Although, as mentioned previously, DS1C cannot be multiplexed to higher levels in the hierarchy, it has proved useful for channelizing inter-office cable pairs with more channels than can be carried by a comparable DS1 system. A transmission facility for DS1C, the T1C line, was introduced and deployed during the 1970s. Although lightwave interoffice transmission has diminished the importance of DS1C, there are still a number of T1C lines in service in the North American network.

Frame Number:	1	2	3	4	5	6	7	8	9	10	11	12	13	14	15	16	17	18	19	20	21	22	23	24
FAS				0				0				1				0				1				1
DL	D		D		D		D		D		D		D		D		D		D		D		D	
CRC		C				C				C				C				C				C		

FIGURE 28.4 DS1 extended superframe overhead bit assignments.

Digital Hierarchy

M-frame bit number	M-subframe number	Overhead bit assignment	Logical value of bit
0	1	M1	0
53	1	C1	first DS1 stuff control
106	1	F1	0
159	1	C2	first DS1 stuff control
212	1	C3	first DS1 stuff control
265	1	F2	1
318	2	M2	1
371	2	C1	second DS1 stuff control
424	2	F1	0
477	2	C2	second DS1 stuff control
530	2	C3	second DS1 stuff control
583	2	F2	1
636	3	M3	1
689	3	C1	first DS1 stuff control
742	3	F1	0
795	3	C2	first DS1 stuff control
848	3	C3	first DS1 stuff control
901	3	F2	1
954	4	X	X
1007	4	C1	second DS1 stuff control
1060	4	F1	0
1113	4	C2	second DS1 stuff control
1166	4	C3	second DS1 stuff control
1219	4	F2	1

FIGURE 28.5 DS1C overhead bit assignments.

M-subframe 1	M-subframe 2	M-subframe 3	M-subframe 4	Overhead bit assignment
0	318	636	954	M1/M2/M3/X
53	371	689	1007	C1
106	424	742	1060	F1
159	477	795	1113	C2
212	530	848	1166	C3
265	583	901	1219	F2

FIGURE 28.6 DS1C M-subframe structure.

The DS1C format multiplexes two 1.544-Mb/s DS1 signals into a single 3.152-Mb/s stream. Unlike the other levels in the digital hierarchy, DS1C signals cannot be multiplexed to higher levels. Two DS1Cs together exceed the capacity of a single DS2, making such multiplexing impossible and making DS1C an orphan rate.

As is the case for all of the hierarchical rates above DS1, the DS1C format is organized into M-frames and M-subframes. The length of a single M-frame is 1272 b. Overhead bits occur every 53 b, with 52 payload bits interspersed between adjacent overhead bits. Each M-frame is divided into 4 M-subframes of 318 b each. Overhead assignments are shown in Fig. 28.5.

Digital System Level 1C Frame Alignment

The information of Fig. 28.5 is reorganized into columns by M-subframe in Fig. 28.6. Note that certain of the overhead bits, including the F1 and F2 bits, recur in the same position in every M-subframe. Typical framing practice, therefore, is to frame on the F bits to locate the boundaries of the M-subframes and to then frame on the first bits of the M-subframes to locate the boundaries of the M-frame structure.

Digital Signal Level 1C X Bit

The X bit provides a communications channel between DS1C terminals which runs at just under 2500 b/s. The usual use of the X-bit channel is to provide a **remote alarm indication** (**RAI**) to the distant terminal. When no alarm condition exists, the X-bit is set to logical 1.

Digital Signal Level 1C Bit Stuffing and Stuffing Control Bits

Of the 1272 b in a DS1C M-frame, 24 are overhead bits. The bandwidth available for transporting payload is, therefore,

$$PCmax = (Mfbits - OHbits)/Mfbits * rate$$
$$= (1272 - 24)/1272 * 3.152 \, \text{Mb/s}$$
$$= 3.092 \, \text{Mb/s}$$

Where:

$PCmax$ = maximum payload capacity of the DS1C signal
$Mfbits$ = number of bits per M-frame, 1272
$OHbits$ = number of overhead bits per M-frame, 24
$rate$ = DS1C bit rate, 3.152 Mb/s

Note that this is more than the payload required to transport two DS1s running at their nominal rates, which is

$$PCreq = 2 * 1.544 \, \text{Mb/s}$$
$$= 3.088 \, \text{Mb/s}$$

where $PCreq$ is the rate required to transport two nominal DS1s.

Each M-subframe, however, contains a payload bit which is designated as a stuff bit. It can be used to carry a payload bit or it may be passed over and left unused. If it is unused in every M-subframe the bandwidth made available for the payload will be

$$PCmin = (Mfbits - OHbits - Sbits)/Mfbits * rate$$
$$= (1272 - 24 - 4)/1272 * 3.152 \, \text{Mb/s}$$
$$= 3.083 \, \text{Mb/s}$$

where:

$PCmin$ = minimum payload capacity of the DS1C signal
$Sbits$ = number of stuff bits (opportunities) per M-frame

If all of the stuff bits are skipped, the capacity of the DS1C channel is less than the amount required by two DS1s. Note that, by either using or skipping the stuff bits, $PCmin$ and $PCmax$ may be varied enough to handle the full range of DS1 signals rates allowed by ANSI [T1.102-1993].

Stuff bits for the first DS1 occur in the first and third M-subframes and for the second DS1 in the second and fourth M-subframes. The stuff bit for a particular M-subframe is always the third time slot allocated to the DS1 involved following overhead bit C3. For DS1 number 1, this is the fifth bit after C3 and for DS1 number 2, it is the sixth bit after C3. For a particular M-subframe, stuffing will be performed if the C bits (C1, C2 and C2) for that M-subframe are all set to logical ones. If the C bits are set to logical zeros, no stuffing will occur in that M-subframe. The use of three C bits

Digital Hierarchy

allows for majority voting by the receiver where one of the C bits may have been corrupted by a line error. This makes the process much more robust to such errors.

Digital Signal Level 1C Payload

The two DS1s which are to be multiplexed are bit interleaved together to form the DS1C payload. Prior to bit interleaving, DS1 number 2 is logically inverted. Prior to inserting the interleaved payload into the DS1C overhead structure, the payload is scrambled in a single-stage scrambler. The output of the scrambler is the modulo-2 sum of the current input bit and the previous output bit.

Higher Rate Formats

As does DS1C, the DS2, DS3, and DS4NA formats all use positive bit stuffing to reconcile the rates of the signals being multiplexed. All use the same M-frame and M-subframe structure with overhead bits assigned to the same tasks as for DS1C. Each of the rates uses bit interleaving to insert subsidiary bit streams into their payload bits. A synopsis of the characteristics of the various rates appears in Table 28.2. Each rate will be discussed in the sections which follow.

The Digital Signal Level 2 Rate

The DS2 rate is summarized in Table 28.2. It operates in a manner very similar to DS1C except that the rate is higher and four DS1s may be carried by a single DS2. A transmission system for carrying DS2 signals over paired copper cable called T2 was once available (1970s) but was never widely deployed. DS2 serves today primarily as a bridge between DS1 and DS3 and is only rarely found outside the confines of a single unit of equipment.

The Digital Signal Level 3 Rate

DS3 is heavily used as an interface to lightwave and digital radio systems. DS3 operates much as do DS1C and DS2. An additional feature of DS3 is the pair of parity bits carried by each M-frame. They are used to transmit a parity error indication for the preceding M-frame. If the modulo-2

TABLE 28.2 Characteristics of Levels in the Digital Hierarchy

	DS1C	DS2	DS3	DS4NA
Transmission rate, Mb/s	3.152	6.312	44.736	139.264
Subsidiary format	DS1	DS1	DS2	DS3
Number of subsidiaries multiplexed	2	4	7	3
Ratio of payload bits to overhead bits	52:1	48:1	679:1	954:33
M-frame length, b	1272	1176	4760	954
Number of M-sub-frames per M-frame	4	4	7	6
M-subframe length, b	318	294	680	159
Number of X bits per M-frame	1	1	2	1
Number of C bits per M-frame	12	12	21	15
Number of M bits per M-frame	3	3	3	12
Number of F bits per M-subframe	2	2	4	0
Number of P bits per M-frame	0	0	2	1

C bit	Application	C bit	Application
1	C-bit parity identifier	12	FEBE
2	N = 1 (future network use)	13	DL (data link)
3	FEAC	14	DL (data link)
4	application specific	15	DL (data link)
5	application specific	16	application specific
6	application specific	17	application specific
7	CP (path parity)	18	application specific
8	CP (path parity)	19	application specific
9	CP (path parity)	20	application specific
10	FEBE	21	application specific
11	FEBE		

FIGURE 28.7 Assignment of C bits in C-bit parity application.

sum of all the information bits in the preceding M-frame is one, then both P bits are set to one. If the modulo-2 sum of the information bits is zero, then the P bits are set to zero. The two P bits of an M-frame are always set to the same value.

The same is true of the two X-bits in an M-frame. They are used as an alarm channel as with DS1C but are always set to the same value.

C-Bit Parity Digital Signal Level 3

A DS3 which operates using positive bit stuffing to multiplex its constituent DS2s is known as an M23 application. Another DS3 application, C-bit parity, is also defined in ANSI T1.107 [1995a].

Since the DS2 rate is almost never used except internal to a multiplexer as a stepping stone from DS1 to DS3, it is possible to lock its rate to a particular value, which is slaved to the DS3 signal generator in the multiplexer. If the rate chosen for the DS2s provides for either no bit stuffing or for stuffing at every opportunity, the receiver can be made to *know* that and will be able to demultiplex the DS2s without reading the stuffing information, which is normally carried by the C-bits.

The C-bit parity format operates the DS2 with stuffing at every opportunity and so frees up the control bits for other uses. There are 21 C bits per M-frame, which provides a channel running at approximately 197 kb/s. The 21 C bits per M-frame are assigned as shown in Fig. 28.7.

The C-bit parity identifier is always set to logical one and is used as a tag to identify the DS3 signal as C-bit parity formatted. Note that the C-bit parity identifier is necessary but not sufficient for this purpose because it may be counterfeited by a DS2 in DS3 time-slot number one running at minimum rate, which therefore requires stuffing at every opportunity.

The **far end alarm and control channel** (**FEAC**) carries alarm and status information from one DS3 terminal to another and may be used as a channel to initiate DS1 and DS3 maintenance loopbacks at the distant DS3 terminal.

The path parity (CP) bits are set to the same value as are the P bits at the terminal which generates them. The CP bits are not to be changed by intermediate network elements and provide a more reliable end-to-end parity indication than do the DS3 P bits.

Defining Terms

Alliance for Telecommunications Industry Solutions (ATIS): An ANSI chartered group that develops telecommunications standards. This organization has taken over sponsorship of Committee T1, Telecommunications, which is responsible for the working groups that actually develop the standards. ATIS has replaced ECSA in this task. Note that the T1 in the committee name has no connection to the T1 line which operates at the DS1 rate.

Cyclic redundancy code (CRC) with an *n*-bit remainder (CRC-*n*): CRC codes provide highly reliable error checking of blocks of transmitted information.

Data link (DL): Used for transporting messages across a digital path using certain of the overhead bits as a data channel. The DL is sometimes called facility data link (FDL).

Digital signal level n (DS-n): Refers to one of the levels (rates) in the North American digital hierarchy that are discussed in this chapter.

ECSA: Exchange Carrier Standards Association.

Extended Superframe (ESF): A DS1 superframe that is 24 frames in length and makes more efficient use of the overhead bits than does the older SF format.

Far end alarm and control channel (FEAC): Used in C-bit parity DS3. The FEAC uses repeated 16-b code words to send status messages alarm signals and requests for loopback.

Far end block error indicator (FEBE): Bits which carry an indication that the terminal at the far end has detected an error.

Frame alignment signal (FAS): Serves to allow a receiver to locate significant repetitive points within the received bit stream so that the information may be extracted.

High-level data link control (HDLC): A format used to implement layer 2 (the data link layer) of the International Standards Organization (ISO) seven-layer model.

Overhead: Bits in a digital signal that do not carry the information signals the signal is intended to transport but that perform housekeeping functions such as framing, error detection, and the transport of maintenance data from one digital terminal to the other.

Payload: The aggregate of the information bits the digital signal is intended to transport.

Remote alarm indication (RAI): A signal that indicates to the terminal at one end of a digital path that the terminal at the other end has detected a failure in the incoming signal.

Superframe format (SF): A format in which the superframe is 12 frames in length and in which all of the overhead bits are used for framing.

References

ANSI. 1993. American National Standard for telecommunications-digital hierarchy—electrical interfaces. ANSI T1. 102-1993, American National Standards Institute.

ANSI. 1995a. American National Standard for telecommunications-digital hierarchy—formats specifications. ANSI T1.107-1995, American National Standards Institute.

ANSI. 1995b. American National Standard for telecommunications-network-to-customer installation—DS1 metallic interface. ANSI T1.403-1995, American National Standards Institute.

Further Information

For further information on the North American digital hierarchy, see B.P. Lathi, *Modern Digital and Analog Communication Systems*, 2nd ed., Holt, Rinehart & Winston, 1989. Information supplementing the standards may be found in various technical requirements of Bell Communications Research, Inc. including GR-499 and TR-TSY-000009.

Address for Bell Communications Research (Bellcore):

Information Exchange Management
Bellcore
445 South st., Room 2J-125
P.O. Box 1910
Morristown, NJ 07962-1910
201-829-4785

29
Line Coding

29.1	Introduction	386
29.2	Common Line Coding Formats	387

Unipolar NRZ (Binary On-Off Keying) • Unipolar RZ • Polar NRZ • Polar RZ [Bipolar, Alternate Mark Inversion (AMI), or Pseudoternary] • Manchester Coding (Split Phase or Digital Biphase)

29.3	Alternate Line Codes	395

Delay Modulation (Miller Code) • Split Phase (Mark) • Biphase (Mark) • Code Mark Inversion (CMI) • NRZ (I) • Binary N Zero Substitution (BNZS) • High-Density Bipolar N (HDBN) • Ternary Coding

29.4	Multilevel Signalling, Partial Response Signalling, and Duobinary Coding	400

Multilevel Signalling • Partial Response Signalling and Duobinary Coding

29.5	Bandwidth Comparison	401
29.6	Concluding Remarks	402

Joseph L. LoCicero
Illinois Institute of Technology

Bhasker P. Patel
Illinois Institute of Technology

29.1 Introduction

The terminology **line coding** originated in telephony with the need to transmit digital information across a copper telephone *line;* more specifically, binary data over a digital repeatered line. The concept of line coding, however, readily applies to any transmission line or channel. In a digital communication system, there exists a known set of symbols to be transmitted. These can be designated as $\{m_i\}$, $i = 1, 2, \ldots, N$, with a probability of occurrence $\{p_i\}$, $i = 1, 2, \ldots, N$, where the sequentially transmitted symbols are generally assumed to be statistically independent. The conversion or *coding* of these abstract symbols into real, temporal waveforms to be transmitted in baseband is the process of line coding. Since the most common type of line coding is for binary data, such a waveform can be succinctly termed a direct format for serial bits. The concentration in this section will be line coding for binary data.

Different channel characteristics, as well as different applications and performance requirements, have provided the impetus for the development and study of various types of line coding [Bellamy, 1991; Bell Telephone Laboratories (BTL), 1970]. For example, the channel might be ac coupled and, thus, could not support a line code with a dc component or large dc content. Synchronization or timing recovery requirements might necessitate a discrete component at the data rate. The channel bandwidth and **crosstalk** limitations might dictate the type of line coding employed. Even such factors as the complexity of the encoder and the economy of the decoder could determine the line code chosen. Each line code has its own distinct properties. Depending on the application, one property may be more important than the other. In what follows, we describe, in general, the most desirable features that are considered when choosing a line code.

It is commonly accepted [Bellamy, 1991; BTL, 1970; Couch, 1994; Lathi, 1989] that the dominant considerations effecting the choice of a line code are: 1) timing, 2) dc content, 3) power spectrum, 4) performance monitoring, 5) probability of error, and 6) transparency. Each of these are detailed in the following paragraphs.

1) *Timing:* The waveform produced by a line code should contain enough timing information such that the receiver can synchronize with the transmitter and decode the received signal properly. The timing content should be relatively independent of source statistics, i.e., a long string of 1s or 0s should not result in loss of timing or jitter at the receiver.

2) *DC content:* Since the repeaters used in telephony are ac coupled, it is desirable to have zero dc in the waveform produced by a given line code. If a signal with significant dc content is used in ac coupled lines, it will cause **dc wander** in the received waveform. That is, the received signal baseline will vary with time. Telephone lines do not pass dc due to ac coupling with transformers and capacitors to eliminate dc ground loops. Because of this, the telephone channel causes a droop in constant signals. This causes dc wander. It can be eliminated by dc restoration circuits, feedback systems, or with specially designed line codes.

3) *Power spectrum:* The power spectrum and bandwidth of the transmitted signal should be matched to the frequency response of the channel to avoid significant distortion. Also, the power spectrum should be such that most of the energy is contained in as small bandwidth as possible. The smaller is the bandwidth, the higher is the transmission efficiency.

4) *Performance monitoring:* It is very desirable to detect errors caused by a noisy transmission channel. The error detection capability in turn allows performance monitoring while the channel is in use (i.e., without elaborate testing procedures that require suspending use of the channel).

5) *Probability of error:* The average error probability should be as small as possible for a given transmitter power. This reflects the reliability of the line code.

6) *Transparency:* A line code should allow all the possible patterns of 1s and 0s. If a certain pattern is undesirable due to other considerations, it should be mapped to a unique alternative pattern.

29.2 Common Line Coding Formats

A line coding format consists of a formal definition of the line code that specifies how a string of binary digits are converted to a line code waveform. There are two major classes of binary line codes: **level codes** and **transition codes**. Level codes carry information in their voltage level, which may be high or low for a full bit period or part of the bit period. Level codes are usually instantaneous since they typically encode a binary digit into a distinct waveform, independent of any past binary data. However, some level codes do exhibit memory. Transition codes carry information in the change in level appearing in the line code waveform. Transition codes may be instantaneous, but they generally have memory, using past binary data to dictate the present waveform. There are two common forms of level line codes: one is called **return to zero (RZ)** and the other is called **nonreturn to zero (NRZ)**. In RZ coding, the level of the pulse returns to zero for a portion of the bit interval. In NRZ coding, the level of the pulse is maintained during the entire bit interval.

Line coding formats are further classified according to the polarity of the voltage levels used to represent the data. If only one polarity of voltage level is used, i.e., positive or negative (in addition to the zero level) then it is called **unipolar** signalling. If both positive and negative voltage levels are being used, with or without a zero voltage level, then it is called **polar** signalling. The term **bipolar** signalling is used by some authors to designate a specific line coding scheme with positive, negative, and zero voltage levels. This will be described in detail later in this section. The formal definition of five common line codes is given in the following along with a representative waveform, the *power spectral density* (PSD), the probability of error, and a discussion of advantages and disadvantages. In some cases specific applications are noted.

Unipolar NRZ (Binary On-Off Keying)

In this line code, a binary **1** is represented by a non- zero voltage level and a binary **0** is represented by a zero voltage level as shown in Fig. 29.1(a). This is an instantaneous level code. The PSD of this code with equally likely **1**s and **0**s is given by [Couch, 1994; Lathi, 1989]

$$S_1(f) = \frac{V^2 T}{4} \left(\frac{\sin \pi f T}{\pi f T} \right)^2 + \frac{V^2}{4} \delta(f) \qquad (29.1)$$

where V is the binary **1** voltage level, $T = 1/R$ is the bit duration, and R is the bit rate in bits per second. The spectrum of unipolar NRZ is plotted in Fig. 29.2(a). This PSD is a two-sided even spectrum, although only half of the plot is shown for efficiency of presentation. If the probability of a binary **1** is p, and the probability of a binary **0** is $(1-p)$, then the PSD of this code, in the most general case, is $4p(1-p) S_1(f)$. Considering the frequency of the first spectral null as the bandwidth of the waveform, the bandwidth of unipolar NRZ is R in hertz. The error rate performance of this code, for equally likely data, with additive white Gaussian noise (AWGN) and optimum, i.e., matched filter, detection is given by [Bellamy, 1991; Couch, 1994]

$$P_e = \frac{1}{2} \text{erfc} \left(\sqrt{\frac{E_b}{2 N_0}} \right) \qquad (29.2)$$

where E_b/N_0 is a measure of the signal-to-noise ratio (SNR) of the received signal. In general, E_b is the energy per bit and $N_0/2$ is the two-sided PSD of the AWGN. More specifically, for unipolar NRZ, E_b is the energy in a binary **1**, which is $V^2 T$. The performance of the unipolar NRZ code is plotted in Fig. 29.3.

The principle advantages of unipolar NRZ are ease of generation, since it requires only a single power supply, and a relatively low bandwidth of R Hz. There are quite a few disadvantages of this line code. A loss of synchronization and timing jitter can result with a long sequence of **1**s or **0**s because no pulse transition is present. The code has no error detection capability and, hence, performance cannot be monitored. There is a significant dc component as well as a dc content. The error rate performance is not as good as that of polar line codes.

Unipolar RZ

In this line code, a binary **1** is represented by a nonzero voltage level during a portion of the bit duration, usually for half of the bit period, and a zero voltage level for rest of the bit duration. A binary **0** is represented by a zero voltage level during the entire bit duration. Thus, this is an instantaneous level code. Figure 29.1(b) illustrates a unipolar RZ waveform in which the **1** is represented by a nonzero voltage level for half the bit period. The PSD of this line code, with equally likely binary digits, is given by [Couch, 1994; Feher, 1977; Lathi, 1989]

$$S_2(f) = \frac{V^2 T}{16} \left(\frac{\sin \pi f T/2}{\pi f T/2} \right)^2 \\ + \frac{V^2}{4\pi^2} \left[\frac{\pi^2}{4} \delta(f) + \sum_{n=-\infty}^{\infty} \frac{1}{(2n+1)^2} \delta(f - (2n+1)R) \right] \qquad (29.3)$$

where again V is the binary **1** voltage level, and $T = 1/R$ is the bit period. The spectrum of this code is drawn in Fig. 29.2(a). In the most general case, when the probability of a **1** is p, the continuous portion of the PSD in Eq. (29.3) is scaled by the factor $4p(1-p)$ and the discrete portion is scaled by the factor $4p^2$. The first null bandwidth of unipolar RZ is $2R$ Hz. The error rate performance

Line Coding

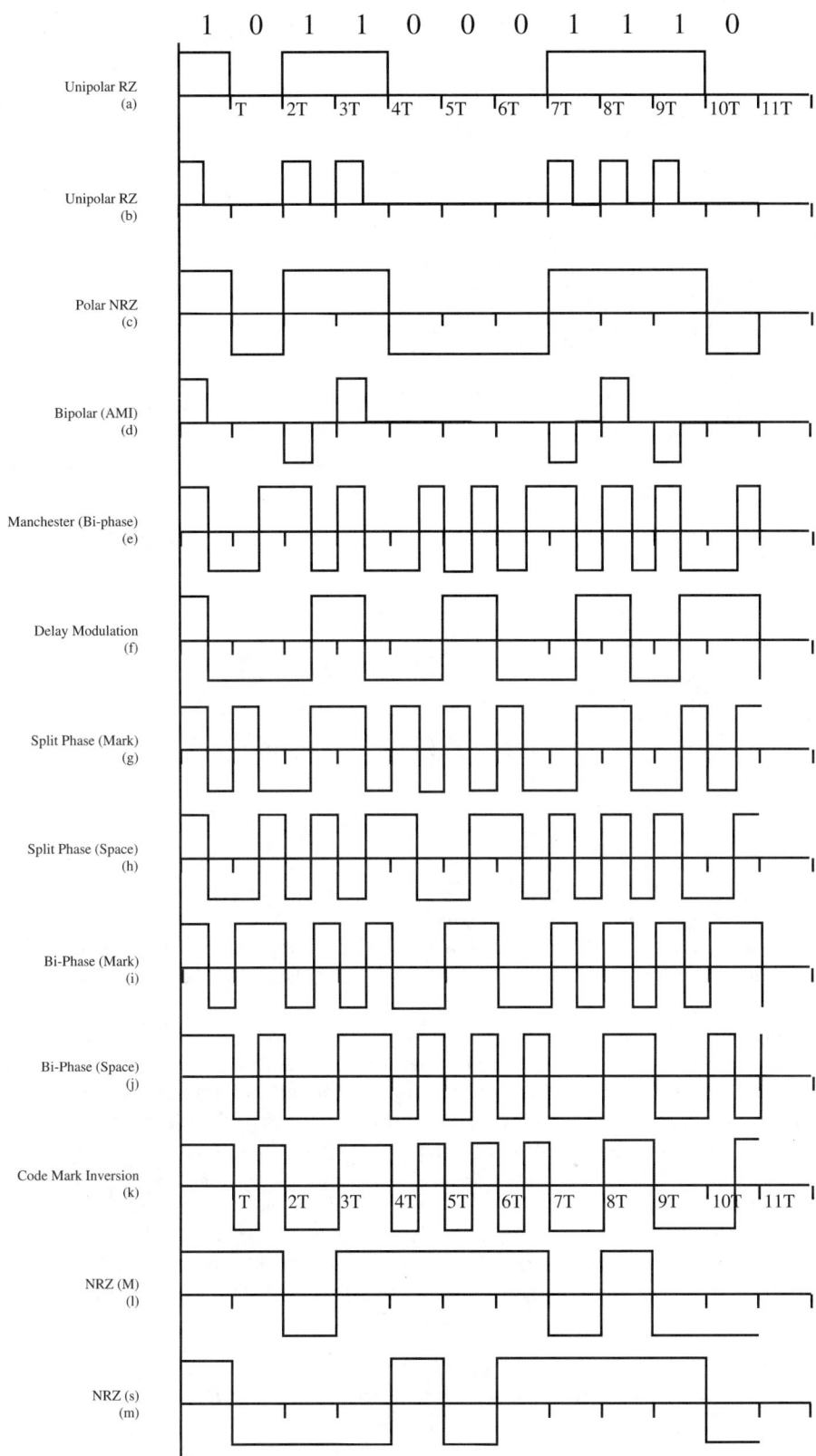

FIGURE 29.1 Waveforms for different line codes.

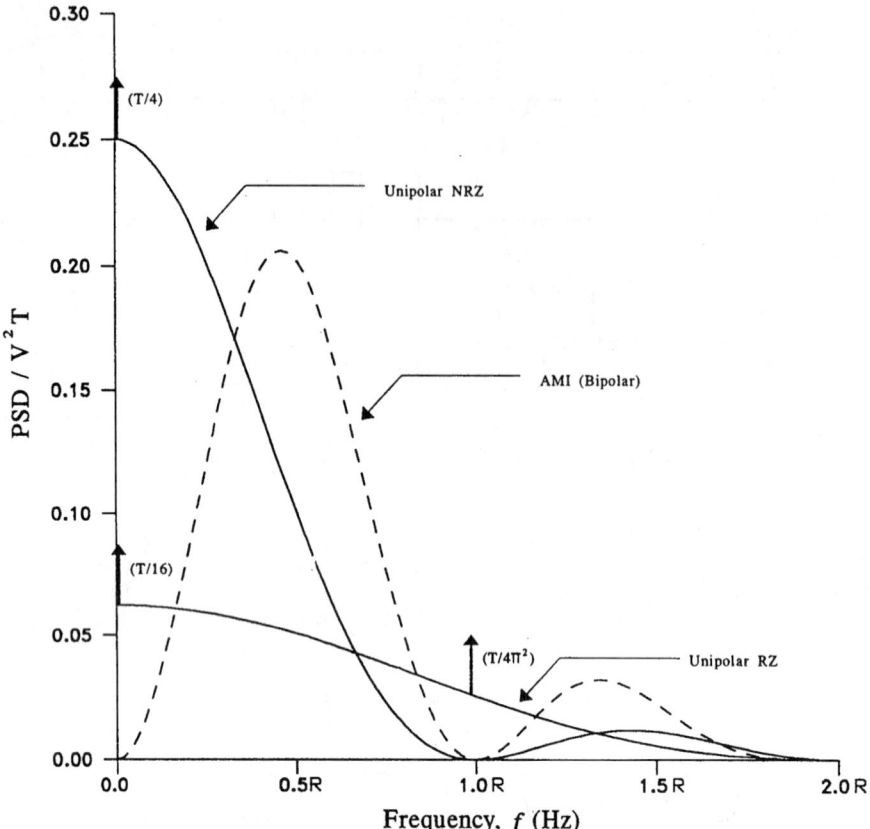

FIGURE 29.2(a) Power spectral density of different line codes, where $R = 1/T$ is the bit rate.

of this line code is the same as that of the unipolar NRZ provided we increase the voltage level of this code such that the energy in binary **1**, E_b, is the same for both codes. The probability of error is given by Eq. (29.2) and identified in Fig. 29.3. If the voltage level and bit period are the same for unipolar NRZ and unipolar RZ, then the energy in a binary **1** for unipolar RZ will be $V^2T/2$ and the probability of error is worse by 3 dB.

The main advantages of unipolar RZ are, again, ease of generation since it requires a single power supply and the presence of a discrete spectral component at the symbol rate, which allows simple timing recovery. A number of disadvantages exist for this line code. It has a nonzero dc component and nonzero dc content, which can lead to dc wander. A long string of **0**s will lack pulse transitions and could lead to loss of synchronization. There is no error detection capability and, hence, performance monitoring is not possible. The bandwidth requirement ($2R$ Hz) is higher than that of NRZ signals. The error rate performance is worse than that of polar line codes.

Unipolar NRZ as well as unipolar RZ are examples of pulse/no-pulse type of signalling. In this type of signalling, the pulse for a binary **0**, $g_2(t)$, is zero and the pulse for a binary **1** is specified generically as $g_1(t) = g(t)$. Using $G(f)$ as the Fourier transform of $g(t)$, the PSD of pulse/no-pulse signalling is given as [Feher, 1977; Gibson, 1993; Lindsey, 1973]

$$S_{\text{PNP}}(f) = p(1-p)R|G(f)|^2 + p^2 R^2 \sum_{n=-\infty}^{\infty} |G(nR)|^2 \delta(f - nR) \qquad (29.4)$$

where p is the probability of a binary **1**, and R is the bit rate.

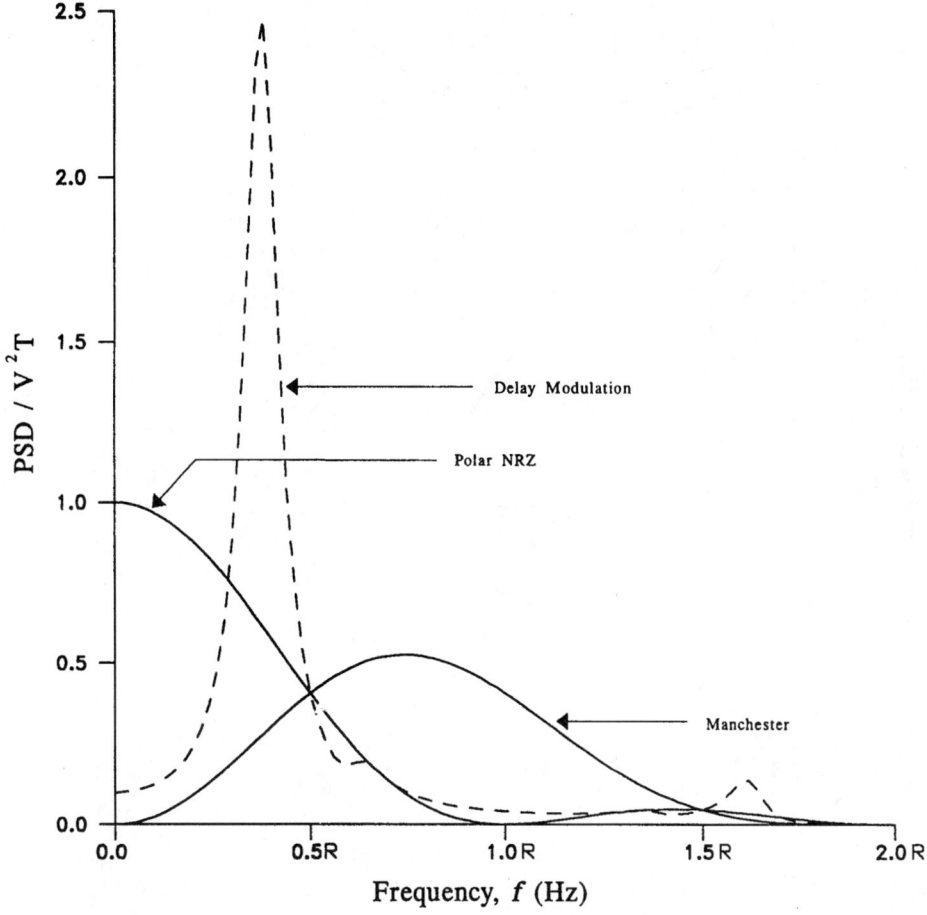

FIGURE 29.2(b) Power spectral density of different line codes, where $R = 1/T$ is the bit rate.

Polar NRZ

In this line code, a binary **1** is represented by a positive voltage $+V$ and a binary **0** is represented by a negative voltage $-V$ over the full bit period. This code is also referred to as NRZ (L), since a bit is represented by maintaining a level (L) during its entire period. A polar NRZ waveform is shown in Fig. 29.1(c). This is again an instantaneous level code. Alternatively, a **1** may be represented by a $-V$ voltage level and a **0** by a $+V$ voltage level, without changing the spectral characteristics and performance of the line code. The PSD of this line code with equally likely bits is given by [Couch, 1994; Lathi, 1989]

$$S_3(f) = V^2 T \left(\frac{\sin \pi f T}{\pi f T} \right)^2 \tag{29.5}$$

This is plotted in Fig. 29.2(b). When the probability of a **1** is p, and p is not 0.5, a dc component exists, and the PSD becomes [Lindsey, 1973]

$$S_{3p}(f) = 4V^2 T p(1-p) \left(\frac{\sin \pi f T}{\pi f T} \right)^2 + V^2(1-2p)^2 \delta(f) \tag{29.6}$$

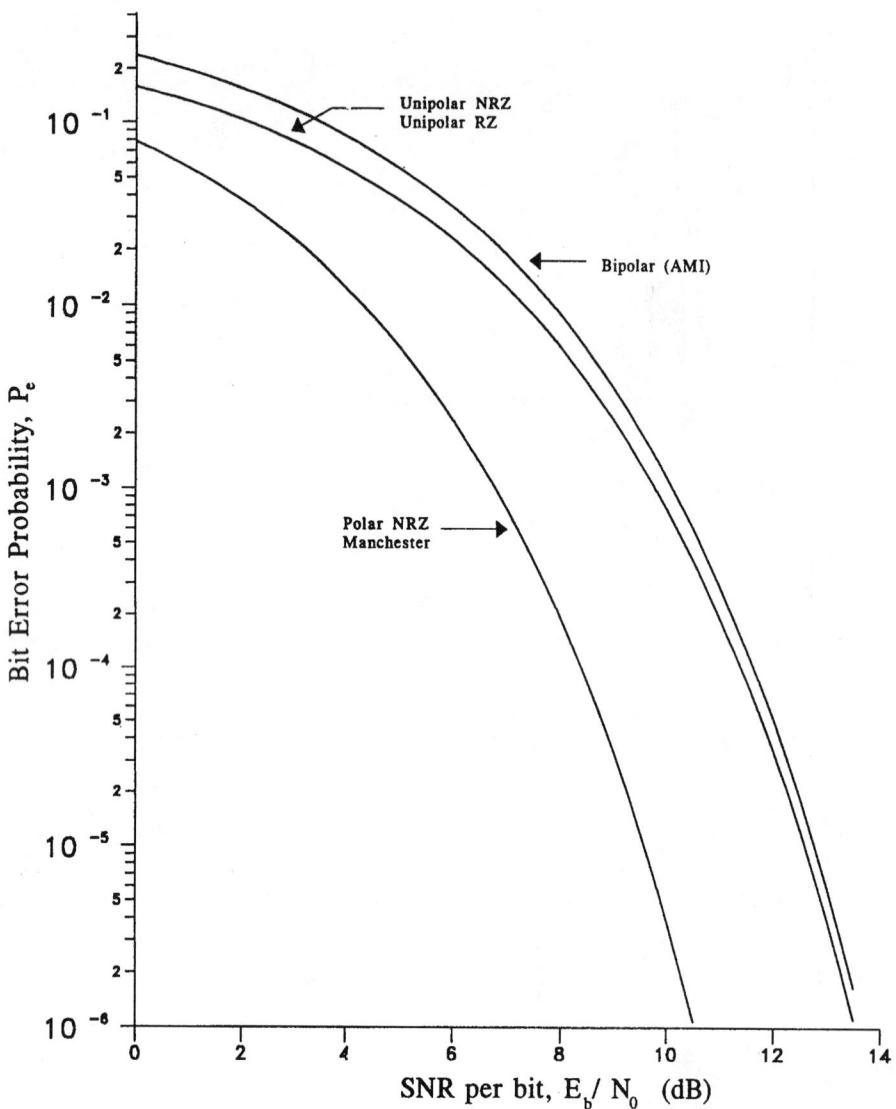

FIGURE 29.3 Bit error probability for different line codes.

The first null bandwidth for this line code is again R Hz, independent of p. The probability of error of this line code when $p = 0.5$ is given by [Bellamy, 1991; Couch, 1994]

$$P_e = \frac{1}{2}\text{erfc}\left(\sqrt{\frac{E_b}{N_0}}\right) \qquad (29.7)$$

The performance of polar NRZ is plotted in Fig. 29.3. This is better than the error performance of the unipolar codes by 3 dB.

The advantages of polar NRZ include a low-bandwidth requirement, R Hz, comparable to unipolar NRZ, very good error probability, and greatly reduced dc because the waveform has a zero dc component when $p = 0.5$ even though the dc content is never zero. A few notable disadvantages are that there is no error detection capability, and that a long string of **1**s or **0**s could result in loss of synchronization, since there are no transitions during the string duration. Two power supplies are required to generate this code.

Polar RZ [Bipolar, Alternate Mark Inversion (AMI), or Pseudoternary]

In this scheme, a binary **1** is represented by alternating the positive and negative voltage levels, which return to zero for a portion of the bit duration, generally half the bit period. A binary **0** is represented by a zero voltage level during the entire bit duration. This line coding scheme is often called **alternate mark inversion** (AMI) since 1s (marks) are represented by alternating positive and negative pulses. It is also called *pseudoternary* since three different voltage levels are used to represent binary data. Some authors designate this line code as bipolar RZ (BRZ). An AMI waveform is shown in Fig. 29.1(d). Note that this is a level code with memory. The AMI code is well known for its use in telephony. The PSD of this line code with memory is given by [Bellamy, 1991; BTL, 1970; Gibson, 1993]

$$S_{4p}(f) = 2p(1-p)R|G(f)|^2 \left(\frac{1 - \cos 2\pi fT}{1 + (2p-1)^2 + 2(2p-1)\cos 2\pi fT} \right) \quad (29.8)$$

where $G(f)$ is the Fourier transform of the pulse used to represent a binary **1**, and p is the probability of a binary **1**. When $p = 0.5$ and square pulses with amplitude $\pm V$ and duration $T/2$ are used to represent binary 1s, the PSD becomes

$$S_4(f) = \frac{V^2 T}{4} \left(\frac{\sin \pi fT/2}{\pi fT/2} \right)^2 \sin^2(\pi fT) \quad (29.9)$$

This PSD is plotted in Fig. 29.2(a). The first null bandwidth of this waveform is R Hz. This is true for RZ rectangular pulses, independent of the value of p in Eq. (29.8). The error rate performance of this line code for equally likely binary data is given by [Couch, 1994]

$$P_e \approx \frac{3}{4}\text{erfc}\left(\sqrt{\frac{E_b}{2N_0}}\right), \qquad E_b/N_0 > 2 \quad (29.10)$$

This curve is plotted in Fig. 29.3 and is seen to be no more than 0.5 dB worse than the unipolar codes.

The advantages of polar RZ (or AMI, as it is most commonly called) outweigh the disadvantages. This code has no dc component and zero dc content, completely avoiding the dc wander problem. Timing recovery is rather easy since squaring, or full-wave rectifying, this type of signal yields a unipolar RZ waveform with a discrete component at the bit rate, R Hz. Because of the alternating polarity pulses for binary 1s, this code has error detection and, hence, performance monitoring capability. It has a low-bandwidth requirement, R Hz, comparable to unipolar NRZ. The obvious disadvantage is that the error rate performance is worse than that of the unipolar and polar waveforms. A long string of **0**s could result in loss of synchronization, and two power supplies are required for this code.

Manchester Coding (Split Phase or Digital Biphase)

In this coding, a binary **1** is represented by a pulse that has positive voltage during the first-half of the bit duration and negative voltage during second-half of the bit duration. A binary **0** is represented by a pulse that is negative during the first-half of the bit duration and positive during the second-half of the bit duration. The negative or positive midbit transition indicates a binary **1** or binary **0**, respectively. Thus, a Manchester code is classified as an instantaneous transition code;

it has no memory. The code is also called diphase because a square wave with a 0° phase is used to represent a binary **1** and a square wave with a phase of 180° used to represent a binary **0**; or vice versa. This line code is used in Ethernet local area networks (LANs). The waveform for Manchester coding is shown in Fig. 29.1(e). The PSD of a Manchester waveform with equally likely bits is given by [Couch, 1994; Lathi, 1989]

$$S_5(f) = V^2 T \left(\frac{\sin \pi f T/2}{\pi f T/2}\right)^2 \sin^2(\pi f T/2) \qquad (29.11)$$

where $\pm V$ are used as the positive/negative voltage levels for this code. Its spectrum is plotted in Fig. 29.2(b). When the probability p of a binary **1**, is not equal to one-half, the continuous portion of the PSD is reduced in amplitude and discrete components appear at integer multiples of the bit rate, $R = 1/T$. The resulting PSD is [Feher, 1977; Lindsey, 1973]

$$S_{5p}(f) = V^2 T 4p(1-p)\left(\frac{\sin \pi f T/2}{\pi f T/2}\right)^2 \sin^2 \frac{\pi f T}{2}$$
$$+ V^2(1-2p)^2 \sum_{n=-\infty, n\neq 0}^{\infty} \left(\frac{2}{n\pi}\right)^2 \delta(f - nR) \qquad (29.12)$$

The first null bandwidth of the waveform generated by a Manchester code is $2R$ Hz. The error rate performance of this waveform when $p = 0.5$ is the same as that of polar NRZ, given by Eq. (29.9), and plotted in Fig. 29.3.

The advantages of this code include a zero dc content on an individual pulse basis, so no pattern of bits can cause dc buildup; midbit transitions are always present making it is easy to extract timing information; and it has good error rate performance, identical to polar NRZ. The main disadvantage of this code is a larger bandwidth than any of the other common codes. Also, it has no error detection capability and, hence, performance monitoring is not possible.

Polar NRZ and Manchester coding are examples of the use of pure polar signalling where the pulse for a binary **0**, $g_2(t)$ is the negative of the pulse for a binary **1**, i.e., $g_2(t) = -g_1(t)$. This is also referred to as an antipodal signal set. For this broad type of polar binary line code, the PSD is given by [Lindsey, 1973]

$$S_{BP}(f) = 4p(1-p)R|G(f)|^2 + (2p-1)^2 R^2 \sum_{n=-\infty}^{\infty} |G(nR)|^2 \delta(f - nR) \qquad (29.13)$$

where $|G(f)|$ is the magnitude of the Fourier transform of either $g_1(t)$ or $g_2(t)$.

A further generalization of the PSD of binary line codes can be given, wherein a continuous spectrum and a discrete spectrum is evident. Let a binary **1**, with probability p, be represented by $g_1(t)$ over the $T = 1/R$ second bit interval; and let a binary **0**, with probability $1-p$, be represented by $g_2(t)$ over the same T second bit interval. The two-sided PSD for this general binary line code is [Lindsey, 1973]

$$S_{GB}(f) = p(1-p)R|G_1(f) - G_2(f)|^2$$
$$+ R^2 \sum_{n=-\infty}^{\infty} |pG_1(nR) + (1-p)G_2(nR)|^2 \delta(f - nR) \qquad (29.14)$$

where the Fourier transform of $g_1(t)$ and $g_2(t)$ are given by $G_1(f)$ and $G_2(f)$, respectively.

29.3 Alternate Line Codes

Most of the line codes discussed thus far were instantaneous level codes. Only AMI had memory, and Manchester was an instantaneous transition code. The alternate line codes presented in this section all have memory. The first four are transition codes, where binary data is represented as the presence or absence of a transition, or by the direction of transition, i.e., positive to negative or vice versa. The last four codes described in this section are level line codes with memory.

Delay Modulation (Miller Code)

In this line code, a binary **1** is represented by a transition at the midbit position, and a binary **0** is represented by no transition at the midbit position. If a **0** is followed by another **0**, however, the signal transition also occurs at the end of the bit interval, that is, between the two **0**s. An example of delay modulation is shown in Fig. 29.1(f). It is clear that delay modulation is a transition code with memory. This code achieves the goal of providing good timing content without sacrificing bandwidth. The PSD of the Miller code for equally likely data is given by [Lindsey, 1973]

$$S_6(f) = \frac{V^2 T}{2(\pi f T)^2 (17 + 8\cos 2\pi f T)}$$
$$\times (23 - 2\cos \pi f T - 22\cos 2\pi f T$$
$$- 12\cos 3\pi f T + 5\cos 4\pi f T + 12\cos 5\pi f T$$
$$+ 2\cos 6\pi f T - 8\cos 7\pi f T + 2\cos 8\pi f T) \qquad (29.15)$$

This spectrum is plotted in Fig. 29.2(b). The advantages of this code are that it requires relatively low bandwidth, most of the energy is contained in less than $0.5R$. However, there is no distinct spectral null within the $2R$-Hz band. It has low dc content and no dc component. It has very good timing content, and carrier tracking is easier than Manchester coding. Error rate performance is comparable to that of the common line codes. One important disadvantage is that it has no error detection capability and, hence, performance cannot be monitored.

Split Phase (Mark)

This code is similar to Manchester in the sense that there are always midbit transitions. Hence, this code is relatively easy to synchronize and has no dc. Unlike Manchester, however, split phase (mark) encodes a binary digit into a midbit transition dependent on the midbit transition in the previous bit period [Stremler, 1980]. Specifically, a binary **1** produces a reversal of midbit transition relative to the previous midbit transition. A binary **0** produces no reversal of the midbit transition. Certainly this is a transition code with memory. An example of a split phase (mark) coded waveform is shown in Fig. 29.1(g), where the waveform in the first bit period is chosen arbitrarily. Since this method encodes bits differentially, there is no $180°$-phase ambiguity associated with some line codes. This phase ambiguity may not be an issue in most baseband links but is important if the line code is modulated. Split phase (space) is very similar to split phase (mark), where the role of the binary **1** and binary **0** are interchanged. An example of a split phase (space) coded waveform is given in Fig. 29.1(h); again, the first bit waveform is arbitrary.

Biphase (Mark)

This code, designated as Bi ϕ-M, is similar to a Miller code in that a binary **1** is represented by a midbit transition, and a binary **0** has no midbit transition. However, this code always has a transition at the beginning of a bit period [Lindsey, 1973]. Thus, the code is easy to synchronize and has no dc. An example of Bi ϕ-M is given in Fig. 29.1(i), where the direction of the transition at $t = 0$ is arbitrarily chosen. Biphase (space) or Bi ϕ-S is similar to Bi ϕ-M, except the role of the binary data is reversed. Here a binary **0** (space) produces a midbit transition, and a binary **1** does not have a midbit transition. A waveform example of Bi ϕ-S is shown in Fig. 29.1(j). Both Bi ϕ-S and Bi ϕ-M are transition codes with memory.

Code Mark Inversion (CMI)

This line code is used as the interface to a Consultative Committee on International Telegraphy and Telephony (CCITT) multiplexer and is very similar to Bi ϕ-S. A binary **1** is encoded as an NRZ pulse with alternate polarity, $+V$ or $-V$. A binary **0** is encoded with a definitive midbit transition (or square wave phase) [Bellamy, 1991]. An example of this waveform is shown in Fig. 29.1(k) where a negative to positive transition (or 180° phase) is used for a binary **0**. The voltage level of the first binary **1** in this example is chosen arbitrarily. This example waveform is identical to Bi ϕ-S shown in Fig. 29.1(j), except for the last bit. CMI has good synchronization properties and has no dc.

NRZ (I)

This type of line code uses an inversion (I) to designate binary digits, specifically, a change in level or no change in level. There are two variants of this code, NRZ mark (M) and NRZ space (S) [Couch, 1994, Stremler, 1980]. In NRZ (M), a change of level is used to indicate a binary **1**, and no change of level is used to indicate a binary **0**. In NRZ (S) a change of level is used to indicate a binary **0**, and no change of level is used to indicate a binary **1**. Waveforms for NRZ (M) and NRZ (S) are depicted in Fig. 29.1(l) and Fig. 29.1(m), respectively, where the voltage level of the first binary **1** in the example is chosen arbitrarily. These codes are level codes with memory. In general, line codes that use differential encoding, like NRZ (I), are insensitive to 180° phase ambiguity. Clock recovery with NRZ (I) is not particularly good, and dc wander is a problem as well. Its bandwidth is comparable to polar NRZ.

Binary N Zero Substitution (BNZS)

The common bipolar code AMI has many desirable properties of a line code. Its major limitation, however, is that a long string of zeros can lead to loss of synchronization and timing jitter because there are no pulses in the waveform for relatively long periods of time. **Binary N zero substitution (BNZS)** attempts to improve AMI by substituting a special code of length N for all strings of N zeros. This special code contains pulses that look like binary 1s but purposely produce violations of the AMI pulse convention. Two consecutive pulses of the same polarity violate the AMI pulse convention, independent of the number of zeros between the two consecutive pulses. These violations can be detected at the receiver, and the special code replaced by N zeros. The special code contains pulses facilitating synchronization even when the original data has long string of zeros. The special code is chosen such that the desirable properties of AMI coding are retained despite the AMI pulse convention violations, i.e., dc balance and error detection capability. The only disadvantage of BNZS compared to AMI is a slight increase in crosstalk due to the increased number of pulses and, hence, an increase in the average energy in the code.

Choosing different values of N yields different BNZS codes. The value of N is chosen to meet the timing requirements of the application. In telephony, there are three commonly used BNZS codes: B6ZS, B3ZS, and B8ZS. All BNZS codes are level codes with memory.

In a B6ZS code, a string of six consecutive zeros is replaced by one of two the special codes according to the rule:

If the last pulse was positive (+), the special code is: $0\ +\ -\ 0\ -\ +$.
If the last pulse was negative (−), the special code is: $0\ -\ +\ 0\ +\ -$.

Line Coding 397

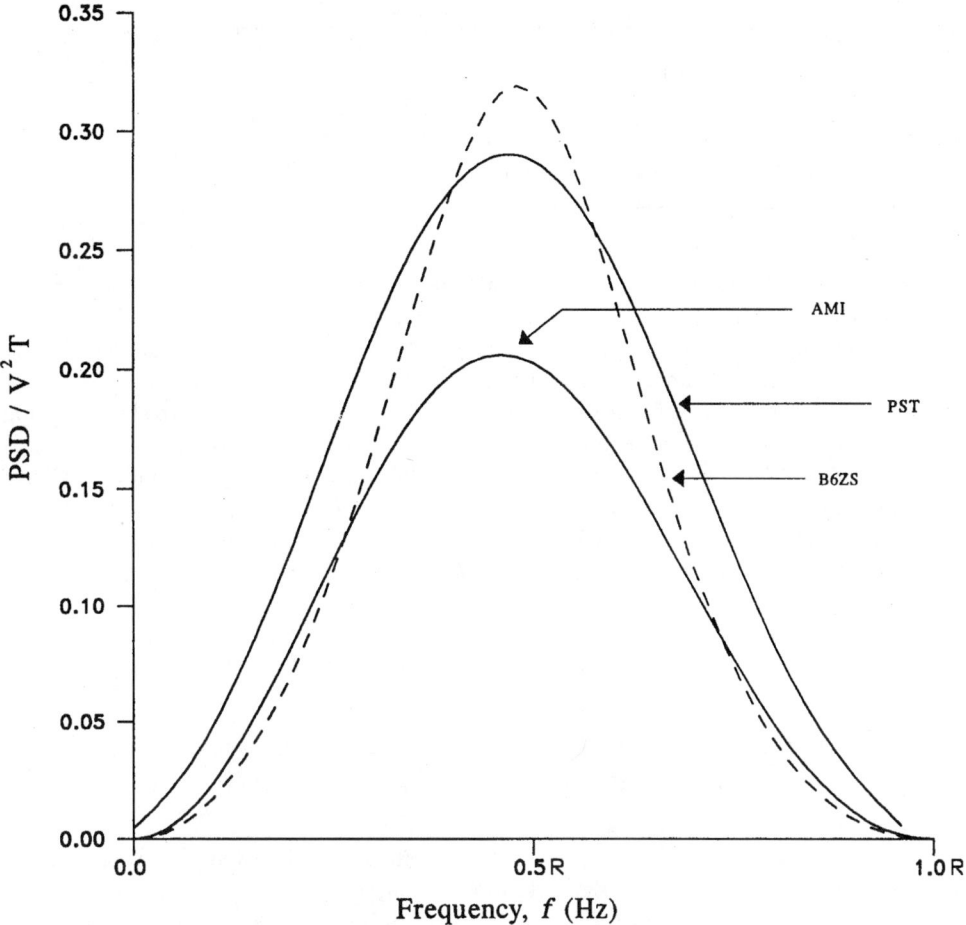

FIGURE 29.4 Power spectral density of different line codes, where $R = 1/T$ is the bit rate.

Here a zero indicates a zero voltage level for the bit period; a plus designates a positive pulse; and a minus indicates a negative pulse.

This special code causes two AMI pulse violations: in its second bit position and in its fifth bit position. These violations are easily detected at the receiver and zeros resubstituted. If the number of consecutive zeros is 12, 18, 24, ..., the substitution is repeated 2, 3, 4, ... times. Since the number of violations is even, the B6ZS waveform is the same as the AMI waveform outside the special code, i.e., between special code sequences.

There are four pulses introduced by the special code that facilitates timing recovery. Also, note that the special code is dc balanced. An example of the B6ZS code is given as follows, where the special code is indicated by the bold characters.

Original data: 0 1 0 0 0 0 0 0 1 1 0 1 0 0 0 0 0 0 1 1
B6ZS format: 0 + **0 + − 0 − +** − + 0 **− 0 − + 0 + −** + −

The computation of the PSD of a B6ZS code is tedious. Its shape is given in Fig. 29.4, for comparison purposes with AMI, for the case of equally likely data.

In a B3ZS code, a string of three consecutive zeros is replaced by either *B0V* or **00V**, where *B* denotes a pulse obeying the AMI (bipolar) convention and *V* denotes a pulse violating the

TABLE 29.1 B3ZS Substitution Rules

Number of B Pulses Since Last Violation	Polarity of Last B Pulse	Substitution Code	Substitution Code Form
Odd	Negative (−)	0 0 −	00V
Odd	Positive (+)	0 0 +	00V
Even	Negative (−)	+ 0 +	B0V
Even	Positive (+)	− 0 −	B0V

AMI convention. $B0V$ or $00V$ is chosen such that the number of bipolar (B) pulses between the violations is odd. The B3ZS rules are summarized in Table 29.1.

Observe that the violation always occurs in the third bit position of the substitution code, and so it can be easily detected and zero replacement made at the receiver. Also, the substitution code selection maintains dc balance. There is either one or two pulses in the substitution code, facilitating synchronization. The error detection capability of AMI is retained in B3ZS because a single channel error would make the number of bipolar pulses between violations even instead of being odd. Unlike B6ZS, the B3ZS waveform between violations may not be the same as the AMI waveform. B3ZS is used in the digital signal-3 (DS-3) signal interface in North America and also in the long distance-4 (LD-4) coaxial transmission system in Canada. Next is an example of a B3ZS code, using the same symbol meaning as in the B6ZS code.

Original data: 1 0 0 1 0 0 0 1 1 0 0 0 0 1 0 0 0 1

B3ZS format:

Even No. of B pulses: + 0 0 − + 0 + − + − 0 − 0 + 0 0 + −

Odd No. of B pulses: + 0 0 − 0 0 − + − + 0 + 0 − 0 0 − +

The last BNZS code considered here uses $N = 8$. A B8ZS code is used to provide transparent channels for the Integrated Services Digital Network (ISDN) on T1 lines and is similar to the B6ZS code. Here a string of eight consecutive zeros is replaced by one of two special codes according to the following rule:

If the last pulse was positive (+), the special code is: 0 0 0 + − 0 − +.

If the last pulse was negative (−), the special code is: 0 0 0 − + 0 + −.

There are two bipolar violations in the special codes, at the fourth and seventh bit positions. The code is dc balanced, and the error detection capability of AMI is retained. The waveform between substitutions is the same as that of AMI. If the number of consecutive zeros is 16, 24, ..., then the substitution is repeated 2, 3, ..., times.

High-Density Bipolar N (HDBN)

This coding algorithm is a CCITT standard recommended by the Conference of European Posts and Telecommunications Administrations (CEPT), a European standards body. It is quite similar to BNZS coding. It is thus a level code with memory. Whenever there is a string of $N + 1$ consecutive zeros, they are replaced by a special code of length $N + 1$ containing AMI violations. Specific codes can be constructed for different values of N. A specific **high-density bipolar N (HDBN)** code, HDB3, is implemented as a CEPT primary digital signal. It is very similar to the B3ZS code. In this code, a string of four consecutive zeros is replaced by either $B00V$ or $000V$. $B00V$ or $000V$ is chosen such that the number of bipolar (B) pulses between violations is odd. The HDB3 rules are summarized in Table 29.2.

Line Coding 399

TABLE 29.2 HDB3 Substitution Rules

Number of B Pulses Since Last Violation	Polarity of Last B Pulse	Substitution Code	Substitution Code Form
Odd	Negative (−)	0 0 0 −	0 0 0 V
Odd	Positive (+)	0 0 0 +	0 0 0 V
Even	Negative (−)	+ 0 0 +	B 0 0 V
Even	Positive (+)	− 0 0 −	B 0 0 V

Here the violation always occurs in the fourth bit position of the substitution code, so that it can be easily detected and zero replacement made at the receiver. Also, the substitution code selection maintains dc balance. There is either one or two pulses in the substitution code facilitating synchronization. The error detection capability of AMI is retained in HDB3 because a single channel error would make the number of bipolar pulses between violations even instead of being odd.

Ternary Coding

Many line coding schemes employ three symbols or levels to represent only one bit of information, like AMI. Theoretically, it should be possible to transmit information more efficiently with three symbols, specifically the maximum efficiency is $\log_2 3 = 1.58$ bits per symbol. Alternatively, the redundancy in the code signal space can be used to provide better error control. Two examples of ternary coding are described next [Bellamy, 1991; BTL, 1970]: **pair selected ternary (PST)** and **4 binary 3 ternary (4B3T)**. The PST code has many of the desirable properties of line codes, but its transmission efficiency is still 1 bit per symbol. The 4B3T code also has many of the desirable properties of line codes, and it has increased transmission efficiency.

In the PST code, two consecutive bits, termed a binary pair, are grouped together to form a word. These binary pairs are assigned codewords consisting of two ternary symbols, where each ternary symbol can be +, −, or 0, just as in AMI. There are nine possible ternary codewords. Ternary codewords with identical elements, however, are avoided, i.e., ++, −−, and 00. The remaining six codewords are transmitted using two modes called + mode and − mode. The modes are switched whenever a codeword with a single pulse is transmitted. The PST code and mode switching rules are summarized in Table 29.3.

PST is designed to maintain dc balance and include a strong timing component. One drawback of this code is that the bits must be framed into pairs. At the receiver, an *out-of-frame* condition is signalled when unused ternary codewords (++, −−, and 00) are detected. The mode switching property of PST provides error detection capability. PST can be classified as a level code with memory.

If the original data for PST coding contains only **1**s or **0**s, an alternating sequence of +− +− ⋯ is transmitted. As a result, an out-of-frame condition can not be detected. This problem can be minimized by using the modified PST code as shown in Table 29.4.

It is tedious to derive the PSD of a PST coded waveform. Again, Fig. 29.4 shows the PSD of the PST code along with the PSD of AMI and B6ZS for comparison purposes, all for equally likely binary data. Observe that PST has more power than AMI and, thus, a larger amount of energy per bit, which translates into slightly increased crosstalk.

In 4B3T coding, words consisting of four binary digits are mapped into three ternary symbols. Four bits imply $2^4 = 16$ possible binary words, whereas three ternary symbols allow $3^3 = 27$ possible ternary codewords. The binary-to-ternary conversion in 4B3T in-

TABLE 29.3 PST Codeword Assignment and Mode Switching Rules

Binary Pair	Ternary Codewords + Mode	− Mode	Mode Switching
11	+ −	+ −	No
10	+ 0	− 0	Yes
01	0 +	0 −	Yes
00	− +	− +	No

TABLE 29.4 Modified PST Codeword Assignment and Mode Switching Rules

Binary Pair	Ternary Codewords + Mode	− Mode	Mode Switching
11	+ 0	0 −	Yes
10	+ −	+ −	No
01	− +	− +	No
00	0 +	− 0	Yes

TABLE 29.5 4B3T Codeword Assignment

Binary Words	Ternary Codewords Column 1	Column 2	Column 3
0000	$---$		$+++$
0001	$--0$		$++0$
0010	$-0-$		$+0+$
0011	$0--$		$0++$
0100	$--+$		$++-$
0101	$-+-$		$+-+$
0110	$+--$		$-++$
0111	-00		$+00$
1000	$0-0$		$0+0$
1001	$00-$		$00+$
1010		$0+-$	
1011		$0-+$	
1100		$+0-$	
1101		$-0+$	
1110		$+-0$	
1111		$-+0$	

sures dc balance and a strong timing component. The specific codeword assignment is as shown in Table 29.5.

There are three types of codewords in Table 29.5, organized into three columns. The codewords in the first column have negative dc, codewords in the second column have zero dc, and those in the third column have positive dc. The encoder monitors the integer variable

$$I = N_p - N_n, \qquad (29.16)$$

where N_p is the number of positive pulses transmitted and N_n are the number of negative pulses transmitted. Codewords are chosen according to following rule:

If $I < 0$, choose the ternary codeword from columns 1 and 2.

If $I > 0$, choose the ternary codeword from columns 2 and 3.

If $I = 0$, choose the ternary word from column 2, and from column 1 if the previous $I > 0$ or from column 3 if the previous $I < 0$.

Note that the ternary codeword 000 is not used, but the remaining 26 codewords are used in a complementary manner. For example, the column 1 codeword for 0001 is $--0$, whereas the column 3 codeword is $++0$. The maximum transmission efficiency for the 4B3T code is 1.33 bits per symbol compared to 1 bit per symbol for the other line codes. The disadvantages of 4B3T are that framing is required and that performance monitoring is complicated. The 4B3T code is used in the T148 span line developed by ITT Telecommunications. This code allows transmission of 48 channels using only 50% more bandwidth than required by T1 lines, instead of 100% more bandwidth.

29.4 Multilevel Signalling, Partial Response Signalling, and Duobinary Coding

Ternary coding, such as 4B3T, is an example of the use of more than two levels to improve the transmission efficiency. To increase the transmission efficiency further, more levels and/or more signal processing is needed. Multilevel signalling allows an improvement in the transmission efficiency at the expense of an increase in the error rate, i.e., more transmitter power will be required to maintain a given probability of error. In partial response signalling, intersymbol interference is deliberately introduced by using pulses that are wider and, hence, require less bandwidth. The

Line Coding

controlled amount of interference from each pulse can be removed at the receiver. This improves the transmission efficiency, at the expense of increased complexity. **Duobinary coding**, a special case of partial response signalling, requires only the minimum theoretical bandwidth of $0.5R$ Hz. In what follows these techniques are discussed in slightly more detail.

Multilevel Signalling

The number of levels that can be used for a line code is not restricted to two or three. Since more levels or symbols allow higher transmission efficiency, multilevel signalling can be considered in bandwidth-limited applications. Specifically, if the signalling rate or baud rate is R_s and the number of levels used is L, the equivalent transmission bit rate R_b is given by

$$R_b = R_s \log_2[L]. \qquad (29.17)$$

Alternatively, multilevel signalling can be used to reduce the baud rate, which in turn can reduce crosstalk for the same equivalent bit rate. The penalty, however, is that the SNR must increase to achieve the same error rate. The T1G carrier system of AT&T uses multilevel signalling with $L = 4$ and a baud rate of 3.152 mega-symbols/s to double the capacity of the T1C system from 48 channels to 96 channels. Also, a four level signalling scheme at 80-kB is used to achieve 160 kb/s as a basic rate in a digital subscriber loop (DSL) for ISDN.

Partial Response Signalling and Duobinary Coding

This class of signalling is also called *correlative* coding because it purposely introduces a controlled or correlated amount of intersymbol interference in each symbol. At the receiver, the known amount of interference is effectively removed from each symbol. The advantage of this signalling is that wider pulses can be used requiring less bandwidth, but the SNR must be increased to realize a given error rate. Also, errors can propagate unless *precoding* is used.

There are many commonly used partial response signalling schemes, often described in terms of the delay operator D, which represents one signalling interval delay. For example, in $(1 + D)$ signalling the current pulse and the previous pulse are added. The T1D system of AT&T uses $(1+D)$ signalling with precoding, referred to as duobinary signalling, to convert binary (two level) data into ternary (three level) data at the same rate. This requires the minimum theoretical channel bandwidth without the deleterious effects of intersymbol interference and avoids error propagation. Complete details regarding duobinary coding are found in Lender, 1963 and Schwartz, 1980. Some partial response signalling schemes, such as $(1 - D)$, are used to shape the bandwidth rather than control it. Another interesting example of duobinary coding is a $(1 - D^2)$, which can be analyzed as the product $(1-D)(1+D)$. It is used by GTE in its modified T carrier system. AT&T also uses $(1-D^2)$ with four input levels to achieve an equivalent data rate of 1.544 Mb/s in only a 0.5-MHz bandwidth.

29.5 Bandwidth Comparison

We have provided the PSD expressions for most of the commonly used line codes. The actual bandwidth requirement, however, depends on the pulse shape used and the definition of bandwidth itself. There are many ways to define bandwidth, for example, as a percentage of the total power or the sidelobe suppression relative to the main lobe. Using the first null of the PSD of the code as the definition of bandwidth, Table 29.6 provides a useful bandwidth comparison.

TABLE 29.6 First Null Bandwidth Comparison

Bandwidth	Codes	
R	Unipolar NRZ	BNZS
	Polar NRZ	HDBN
	Polar RZ (AMI)	PST
$2R$	Unipolar RZ	Split Phase
	Manchester	CMI

The notable omission in Table 29.6 is delay modulation (Miller code). It does not have a first null in the $2R$-Hz band, but most of its power is contained in less than $0.5R$ Hz.

29.6 Concluding Remarks

An in-depth presentation of line coding, particularly applicable to telephony, has been included in this chapter. The most desirable characteristics of line codes were discussed. We introduced five common line codes and eight alternate line codes. Each line code was illustrated by an example waveform. In most cases expressions for the PSD and the probability of error were given and plotted. Advantages and disadvantages of all codes were included in the discussion, and some specific applications were noted. Line codes for optical fiber channels and networks built around them, such as fiber distributed data interface (FDDI) were not included in this section. A discussion of line codes for optical fiber channels, and other new developments in this topic area can be found in Bellamy, 1991, Bic, Duponteil, and Imbeaux, 1991, and Bylanski, 1976.

Defining Terms

Alternate mark inversion (AMI): A popular name for bipolar line coding using three levels: zero, positive, and negative.

Binary N zero substitution (BNZS): A class of coding schemes that attempts to improve AMI line coding.

Bipolar: A particular line coding scheme using three levels: zero, positive, and negative.

Crosstalk: An unwanted signal from an adjacent channel.

DC Wander: The dc level variation in the received signal due to a channel that cannot support dc.

Duobinary coding: A coding scheme with binary input and ternary output requiring the minimum theoretical channel bandwidth.

4 Binary 3 Ternary (4B3T): A line coding scheme that maps four binary digits into three ternary symbols.

High-density bipolar N (HDBN): A class of coding schemes that attempts to improve AMI.

Level codes: Line codes carrying information in their voltage levels.

Line coding: The process of converting abstract symbols into real, temporal waveforms to be transmitted through a baseband channel.

Nonreturn to zero (NRZ): A signal that stays at a nonzero level for the entire bit duration.

Pair selected ternary (PST): A coding scheme based on selecting a pair of three level symbols.

Polar: A line coding scheme using both polarity of voltages, with or without a zero level.

Return to zero (RZ): A signal that returns to zero for a portion of the bit duration.

Transition codes: Line codes carrying information in voltage level transitions.

Unipolar: A line coding scheme using only one polarity of voltage, in addition to a zero level.

References

Bellamy, J. 1991. *Digital Telephony*, John Wiley & Sons, Inc., New York, NY.

Bell Telephone Laboratories Technical Staff Members. 1970. *Transmission Systems for Communications*, 4th ed. Western Electric Company, Inc., Technical Publications, Winston-Salem, NC.

Bic, J.C., Duponteil, D., and Imbeaux, J.C. 1991. *Elements of Digital Communication*, John Wiley & Sons, Inc., New York, NY.

Bylanski, P. 1976. *Digital Transmission Systems*, Peter Peregrinus Ltd., Herts, England.

Couch, L.W. 1994. *Modern Communication Systems: Principles and Applications*, Prentice-Hall, Inc., Englewood Cliffs, NJ.

Feher, K. 1977. *Digital Modulation Techniques in an Interference Environment*, EMC Encyclopedia Series, Vol. IX. Don White Consultants, Inc., Germantown, MD.

Gibson, J.D. 1993. *Principles of Analog and Digital Communications*, MacMillan Publishing, Inc., New York, NY.

Lathi, B.P. 1989. *Modern Digital and Analog Communication Systems*, Holt, Rinehart and Winston, Inc., Philadelphia, PA.

Lender, A. 1963. Duobinary Techniques for High Speed Data Transmission, *IEEE Trans. Commun. Electron.*, CE-82(May):214–218.

Lindsey, W.C. and Simon, M.K. 1973. *Telecommunication Systems Engineering*, Prentice-Hall, Inc., Englewood Cliffs, NJ.

Schwartz, M. 1980. *Information Transmission, Modulation, and Noise*, McGraw-Hill Book Co., Inc., New York, NY.

Stremler, F.G. 1990. *Introduction to Communication Systems*, Addison-Wesley Publishing, Co., Reading, MA.

30
Telecommunications Network Synchronization

M.J. Narasimha
Telecom Solutions

30.1 Introduction 404
30.2 Synchronization Distribution Networks 405
30.3 Effect of Synchronization Impairments 406
30.4 Characterization of Synchronization Impairments 408
30.5 Synchronization Standards 410
30.6 Summary and Conclusions 412

30.1 Introduction

Today's telecommunications network comprises customer premises equipment and telephone central offices interconnected by suitable transmission facilities. Although analog technology still exists in the customer loop, digital time-division multiplex (TDM) technology is more prevalent in the central office switching and transmission systems.

A digital switch located within the interoffice network terminates TDM signals originating from many other offices. It performs a combination of time-slot interchange and space switching to accomplish the interconnect function amongst the individual channels of the multiplexed signals. In order to accomplish this function without impairments, the average rates of all of the TDM signals terminating on the switch have to be synchronized to within some achievable bound. Furthermore, the reference clock of the switch itself must also be synchronized to the common rate of the incoming signals. These synchronization requirements are also applicable to a digital cross-connect system since it realizes the channel interconnection function in a similar manner.

Synchronous multiplexers, such as synchronous optical network (SONET) and synchronous digital hierarchy (SDH) terminals, used in fiber optic systems, generate the high-speed output signal by interleaving the time slots of the lower speed input signals. Again, to accomplish this function properly, the rates of the incoming lines and that of the multiplexer clock have to be synchronized. Primary rate multiplexers (known as channel banks) also employ synchronous time interleaving of the 64-kb/s tributary signals, often generated by other network elements (digital switches and signalling transfer points, for example) within the same office. For unimpaired information transfer in this case, the network elements that terminate these 64-kb/s signals have to be synchronized in both frequency and phase (bit and byte synchronization!).

Network synchronization is the background technology that enables the operating clocks of the various network elements throughout the network to be synchronized. Robust and accurate synchronization networks are essential to the proper functioning of these elements and the reliable transfer of information between them. The growth of data services and the deployment of SONET and SDH transmission equipment has further emphasized the need for such networks.

30.2 Synchronization Distribution Networks

The goal of the synchronization distribution network is to provide reference clocks, traceable to a highly accurate clock called a primary reference source (PRS), to all of the network elements. Since the transport of timing signals over long distances incurs many impairments, the *interoffice* distribution of synchronization references is generally more difficult compared to the task of distributing these reference clocks to the network elements within an office (*intraoffice* distribution).

One method of achieving synchronization in the interoffice network is to designate a single PRS as the master clock for the entire network, and transport this clock to every office using a **master–slave** hierarchical discipline. This method is impractical for large networks because of noise accumulation through many levels of cascaded slave clocks, delay variations caused by the rerouting of clock distribution paths under failure conditions, and geopolitical problems. At the other extreme is the **plesiochronous** mode of operation where each office has its own PRS, and no interoffice clock distribution is necessary. This strategy is expensive to implement now, although it can be envisioned in the future because of the projected availability of affordable PRSs. Therefore, a combination of the two techniques is typically employed in practice. As shown in Fig. 30.1, the network is divided into small synchronization regions, each of which has a duplicated set of PRSs as the master clock. Within each region, the timing distribution is accomplished by following a master–slave hierarchy of **stratum clocks**. A stratum 1 clock, normally implemented with cesium beam technology, is required to have a long term accuracy of better than 1×10^{-11} completely autonomous of other references. The PRS is also required to have the same long-term accuracy as a stratum 1 clock. However, it can be realized either as an autonomous cesium clock or as a nonautonomous clock disciplined by precision global positioning system (GPS) or LORAN-C radio-navigational signals. Stratum 2, 3,

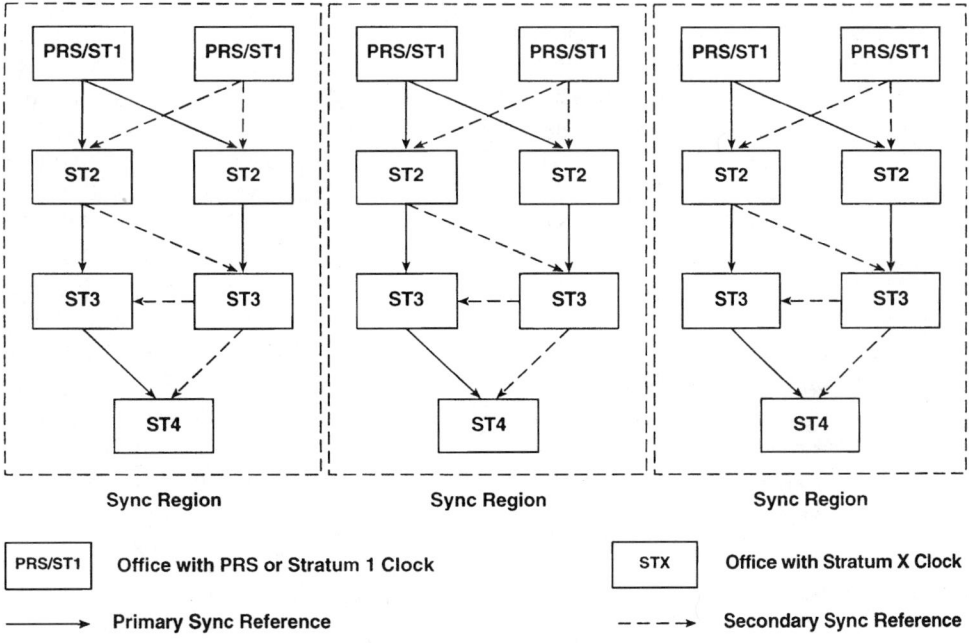

FIGURE 30.1 Network synchronization distribution plan.

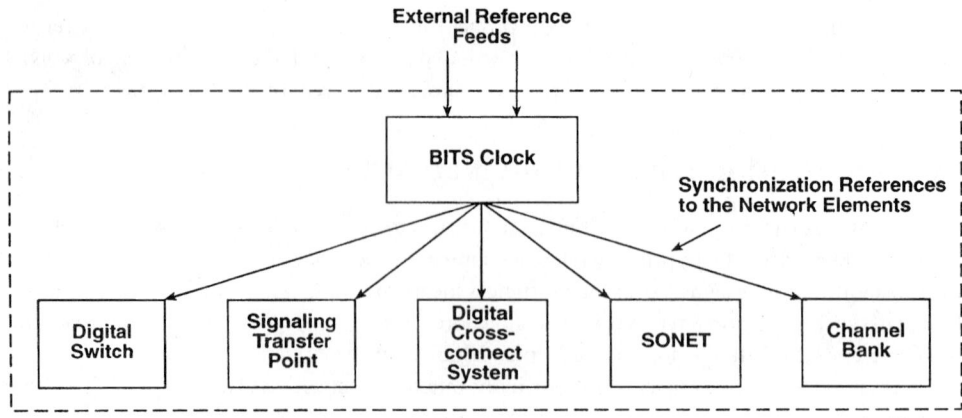

FIGURE 30.2 Intraoffice synchronization distribution plan.

and 4 clocks have progressively lower accuracy and performance requirements. Synchronization references are passed from higher performance master clocks to equivalent or lower performance slave clocks. Path redundancy is achieved by providing primary and secondary reference sources at the slave clocks.

Whereas the interoffice plan for distributing synchronization follows the regionalized master–slave hierarchical strategy previously described, the distribution of reference clocks within an office is realized with a discipline known as the building integrated timing supply (BITS) plan. As shown in Fig. 30.2, this plan calls for the deployment of a properly stratified slave clock called the BITS in each office, which distributes reference timing to all of the network elements. It is the only clock in the office that is directly synchronized to an external reference traceable to a PRS. For robustness, it accepts two external reference feeds and contains duplicated stratum oscillators. If both the external reference feeds are disrupted, the BITS enters the holdover mode of operation where the output reference timing signals are generated using the data acquired during the normal (synchronized) mode.

The master–slave synchronization distribution scheme functions well within a (small) region if the underlying network topology is of the star or tree type. However, self-healing ring topologies are popular with fiber optic networks using the SONET and SDH technology, especially in the loop environment. Feeding multiple synchronization references to slave clocks, or directly to the network elements at the nodes where the BITS plan is not implemented, in such networks can lead to **timing loops** under failure conditions [Bellcore, 1992]. Timing loops are undesirable since they lead to isolation of the clocks within the loop from the PRS, and can also cause frequency instabilities due to reference feedback. Providing only a single-synchronization reference source at the nodes is one method of avoiding the possibility of timing loops [Bellcore, 1992]. However, this compromises the path redundancy provided by dual feeds. An alternate procedure is the use of synchronization messages embedded in the overhead channels of the reference signals. Here the nodes indicate the synchronization status (e.g., reference traceable to stratum 1, traceability unknown, etc.) of the transmitted signals to neighboring nodes. The slave clocks (or the network elements) at the other nodes can then avoid the inadvertent creation of timing loops by deciding intelligently, based on the embedded messages, whether or not to switch to the secondary synchronization reference source upon failure of the primary.

30.3 Effect of Synchronization Impairments

The severity of impact on network traffic due to disruptions in the synchronization distribution system or accumulation of phase noise depends on many factors. These include buffering techniques

Telecommunications Network Synchronization

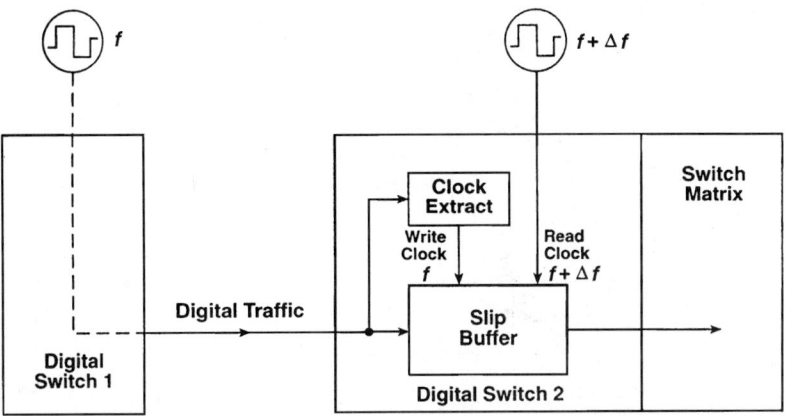

FIGURE 30.3 Slips in the digital network.

TABLE 30.1 Impact of Slips on Network Service

Service	Observed Effect
PCM voice	No noticeable degradation for moderate slip rates
Voice-band data	Serious degradation with some modems needing up to 6 s to recover from a slip
Group III fax	Each slip can wipe out 0.08 in of vertical space
Encrypted voice	Serious degradation requiring retransmission of the encryption key
Compressed video	Serious degradation resulting in freeze frames or missing lines
Digital data (e.g., SS7)	Loss of messages resulting in requests for retransmission and degraded throughput

employed to manage the incoming phase variations at the network elements, the type of information transported through the traffic network, and the architecture of the synchronization distribution system itself.

Figure 30.3 shows the situation where the reference clocks of the two interconnected digital switches (or two digital cross-connect systems) are operating at different frequencies due to disruptions in the synchronization distribution network, for example. The receiver at switch 2 has a 125-μs slip buffer to absorb the phase variations of the incoming line with respect to its clock. Since the rates of the write and read clocks at the slip buffer are different, however, an eventual overflow or underflow of the buffer is inevitable. This results in a frame of data either being repeated or deleted at the output of the buffer. Such an event is called a controlled slip. The duration between slips is given by $(125/\varepsilon)\mu$s, where ε is the fractional frequency deviation $|\Delta f|/f$. As an example, the duration between slips for plesiochronous operation ($\varepsilon = \pm 10^{-11}$) is about 72 days. The impact of slips on various network services is delineated in Table 30.1.

Slips occur infrequently when there are no synchronization disruptions since the 125-μs slip buffers can readily absorb the phase noise caused by the existing transmission systems. However, excessive phase noise either in the incoming lines or in the reference clock at a synchronous multiplexer can cause impairments to the transported data. To understand this, consider the operation of a SONET multiplexer, shown in Fig. 30.4, that combines three 51.84-Mb/s synchronous transport system level 1 (STS-1) signals to form a 155.52-Mb/s STS-3 signal. Data is written into the small-capacity (8–16 bytes) receive buffer in the pointer processor at the rate of the incoming line. If the buffer is near its normal half-full capacity, it is read out at the rate of the local reference clock. However, if the buffer shows signs of filling up or emptying (determined by the upper and lower

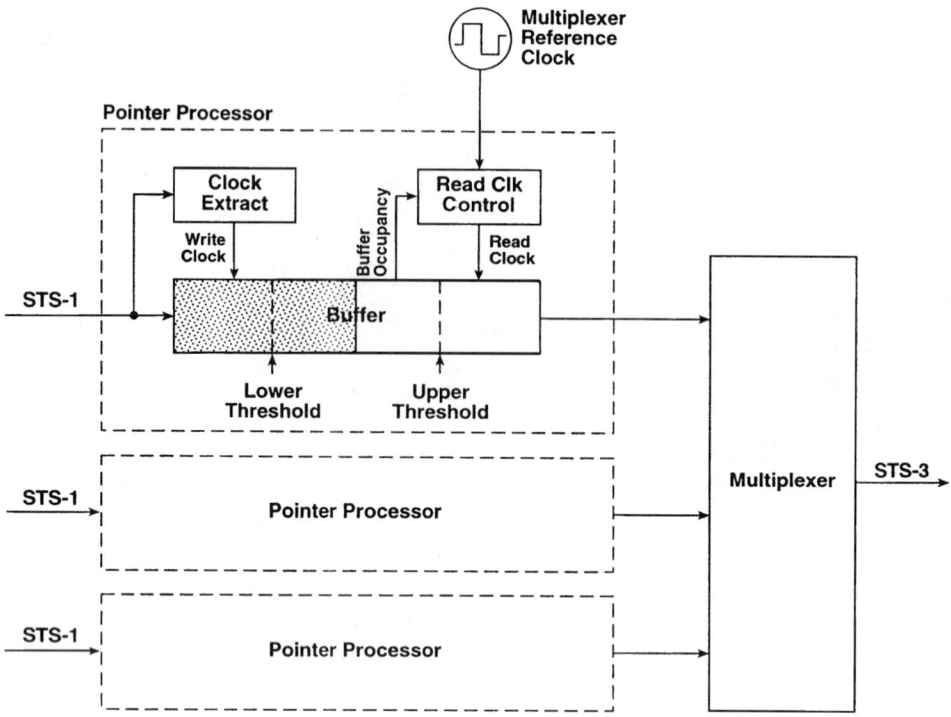

FIGURE 30.4 SONET multiplexer.

threshold settings), its read clock is modified by either augmenting or deleting eight pulses from the reference clock. This event, known as a *pointer adjustment*, results in a phase jump of eight unit intervals for the data read out of the buffer and multiplexed onto the STS-3 signal. Occasional pointer adjustments occur even in a synchronized network because of short-term phase variations of the incoming signals, and they cause no harm since data is not lost. However, existing asynchronous equipment may not function properly when they receive severely jittered signals, caused by frequent pointer adjustments, at gateway points between the synchronous and the asynchronous networks. Furthermore, the rate of pointer adjustments is not allowed to exceed certain limits in SONET and SDH networks. Severe phase noise may cause these limits to be exceeded resulting in a buffer overflow or underflow situation, which leads to loss or repetition of data as in the case of a slip.

30.4 Characterization of Synchronization Impairments

The synchronization performance of a telecommunications signal is typically determined from measured values of the deviations in time between when the signal transitions actually occur and when they are ideally expected. The master clock that provides the ideal time positions should be significantly more accurate than the signal under test. The measurement yields the raw phase error signal, also known as the *time delay* or simply *phase*, in units of time as a function of elapsed time. Postprocessing of this raw data is, however, necessary to extract useful parameters that help define the underlying synchronization impairment model.

The current synchronization impairment model used for the telecommunications signal is based mainly on the source of the impairments. It refers to the higher frequency components of the phase (error) oscillations as **jitter** and to the lower frequency components as **wander**, with 10 Hz being the demarcation frequency. Jitter is produced mainly by regenerative repeaters and asynchronous multiplexers, and normal levels of it can readily be buffered and filtered out. However, excessive jitter is a potential source of bit errors in the digital network. Wander, on the other hand, has many causes. These include temperature cycling effects in cables, waiting time effects in asynchronous multiplexers, and the effects of frequency and phase quantization in slave clocks which employ narrowband filters in their servocontrol loops. Since it contains very low-frequency components, wander cannot be completely filtered out, and hence is passed on. Excessive wander adversely affects the slip performance of the network and can also lead to an unacceptable rate of pointer adjustments in SONET and SDH multiplexers. Furthermore, wander on an input reference signal compromises the holdover performance of a slave clock during reference interruptions.

Although jitter and wander are adequate descriptors of synchronization impairments in the existing digital network, a more comprehensive model is necessary to characterize the phase variations caused by the newly introduced SONET and SDH equipment, and to appropriately specify the limits on timing noise at network interfaces. Recently, there is renewed interest to adapt the traditional noise model used with precision oscillators and clocks in time and frequency metrology to the telecommunications applications. This model permits the characterization of the complex timing impairments with a handful of parameters classified into systematic components (phase offset, frequency offset, and frequency drift) and stochastic power-law noise components (white PM, flicker PM, white FM, flicker FM, and random walk FM).

Maximum time interval error (MTIE) and time variance (TVAR) are two parameters [ANSI, 1994] for specifying the synchronization performance of telecommunications signals that are defined to capture the essential features of the traditional model. MTIE is effective in characterizing peak-to-peak phase movements, which are primarily due to systematic components, whereas TVAR is useful in characterizing the stochastic power-law noise components of the phase noise.

The algorithm for computing the MTIE from N time delay (phase error) samples $x_i, i = 1, 2, \ldots, N$, measured at time intervals τ_0, is illustrated in Fig. 30.5. For a given observation interval S spanning n samples, the peak-to-peak phase excursion values are noted for all possible positions of the observation window that can be accommodated by the collected data. The maximum of all such values yields MTIE(S). This computation can be expressed as

$$\text{MTIE}(S) = \max_{j=1,\ldots,N-n+1} x_{ppj}(S)$$

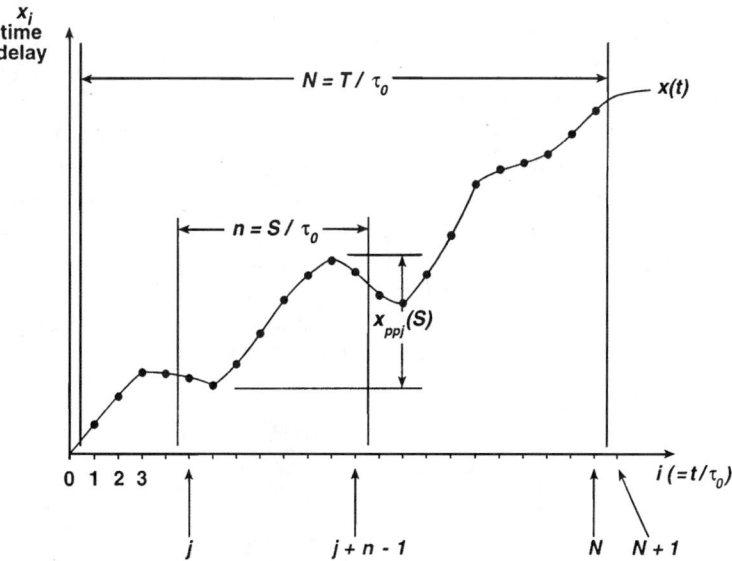

FIGURE 30.5 Computation of the MTIE from measured time delay samples.

where

$$x_{ppj}(S) = \max x_i - \min x_i, \quad i = j, \ldots, j+n-1$$

The calculation of TVAR for the same set of time delay samples is expressed by the equation:

$$\text{TVAR}(\tau) = \frac{1}{6n^2(N-3n+1)} \sum_{j=1}^{N-3n+1} \left[\sum_{k=0}^{n-1}(x_{j+2n+k} - 2x_{j+n+k} + x_{j+k})\right]^2$$

where the independent variable $\tau = n\tau_0$ spanning n samples is known as the *integration time*. This calculation may be viewed as a spectral estimation process with a nonuniform bandpass filter bank. The magnitude response of the bandpass filter corresponding to the integration time τ at frequency f is given by

$$\left|\sqrt{\frac{8}{3}} \frac{\sin^3(\pi\tau f)}{n \sin(\pi\tau_0 f)}\right|$$

A standard variance estimate of the filtered samples then yields TVAR(τ). The square root of TVAR is denoted by time deviation (TDEV).

The graphs of MTIE plotted as a function of the observation interval S and TDEV plotted as a function of the integration time τ are primarily used to characterize the synchronization performance of a telecommunications signal.

30.5 Synchronization Standards

Standards for synchronization in telecommunications systems are set by the American National Standards Institute (ANSI) in the U.S., and by the Consultative Committee on International Telephony and Telegraphy (CCITT), now known as the International Telecommunications Union (ITU), for international applications. In addition, Bellcore issues technical references and technical advisories that specify the requirements from the viewpoint of the regional Bell operating companies (RBOCs). These documents are listed in the References. Some highlights of these standards are reviewed here.

The various standards deal with two categories of specifications: the characteristics of the clocks used in the synchronization distribution network, and the synchronization performance of the reference signals at network interfaces.

The **free-run accuracy**, the **holdover stability**, and the **pull-in range** requirements of the stratum clocks employed in the synchronized network are delineated in Table 30.2. (The ITU uses a different terminology for the stratum levels. Refer to Annex D of the revised ANSI standard T1.101 for the differences.) In addition to the pull-in range requirements, the slave clocks should be able to tolerate certain amounts of jitter, wander, and **phase transients** at their reference inputs. Moreover, there are other performance constraints, besides accuracy and stability, on the output signal. These include wander generation, wander transfer, jitter generation, and phase transients. These are detailed in the Bellcore [1993b] Technical Reference TR-NWT-001244.

TABLE 30.2 Performance Requirements of Stratum Clocks

Stratum	Free-Run Accuracy	Holdover Stability	Pull-In Range
1	$\pm 1 \times 10^{-11}$	—	—
2	$\pm 1.6 \times 10^{-8}$	$\pm 1 \times 10^{-10}$/day	$\pm 1.6 \times 10^{-8}$
3E	$\pm 4.6 \times 10^{-6}$	$\pm 1 \times 10^{-8}$/day	$\pm 4.6 \times 10^{-6}$
3	$\pm 4.6 \times 10^{-6}$	$\pm 3.7 \times 10^{-7}$/day	$\pm 4.6 \times 10^{-6}$
4/4E	$\pm 32 \times 10^{-6}$	—	$\pm 32 \times 10^{-6}$

Telecommunications Network Synchronization

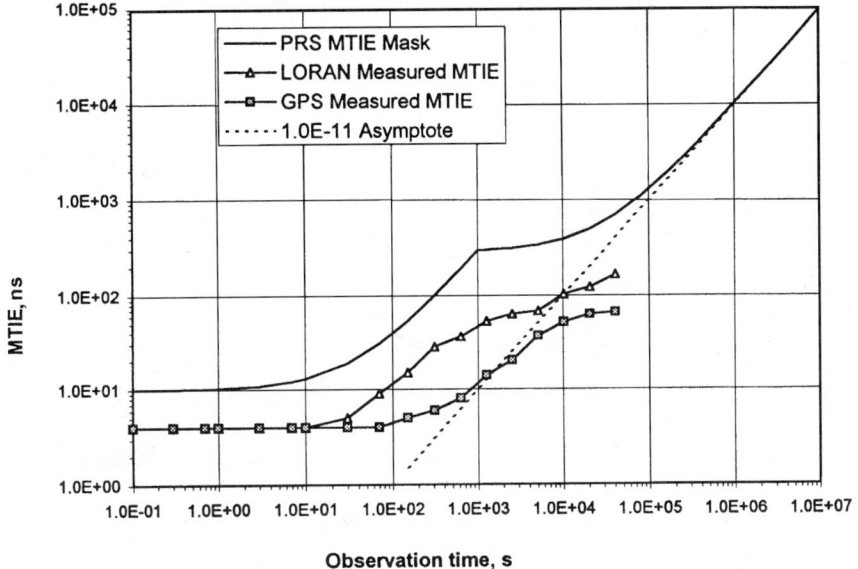

FIGURE 30.6 MTIE specification mask and measured performance of GPS and LORAN PRSs.

As an illustration, Fig. 30.6 shows the MTIE specification mask for a PRS. Also shown are the measured performance curves of typical PRS clocks based on GPS and LORAN-C receiver technology.

The performance requirements of synchronization references at network interfaces depend on whether optical rate signals [i.e., optical carrier level N (OC-N)] or primary rate electrical signals [e.g., digital signal level 1 (DS1)] are being considered. The specifications for optical interfaces are tighter because SONET and SDH pointer adjustments are sensitive to short-term phase noise. The MTIE and TDEV specification masks for DS1 and OC-N signals are shown in Figs. 30.7 and 30.8.

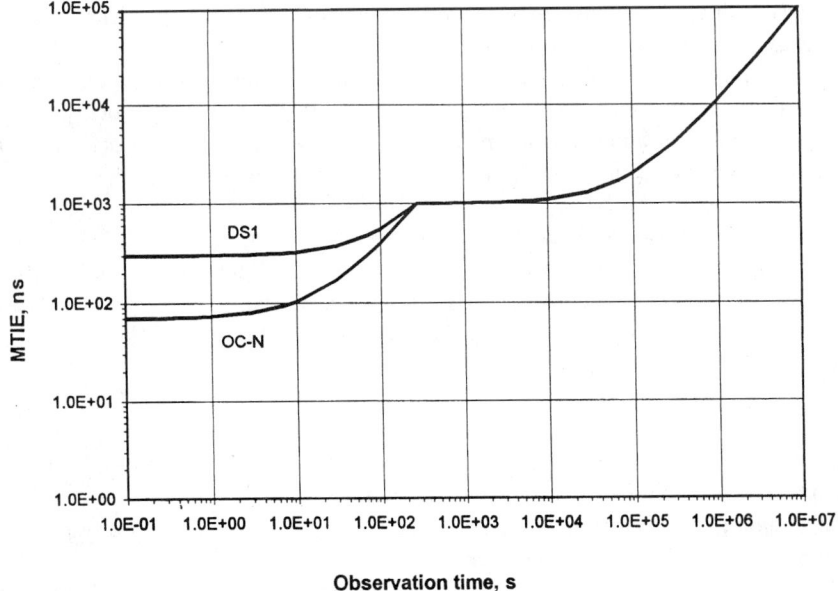

FIGURE 30.7 MTIE specification for synchronization reference signals at network interfaces.

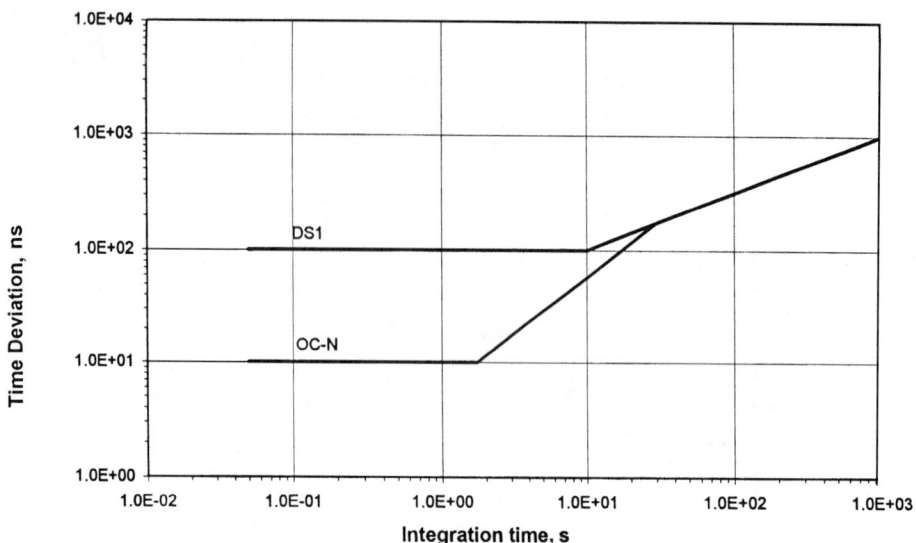

FIGURE 30.8 TDEV specification for synchronization reference signals at network interfaces.

These reference signals also have to satisfy certain constraints on the magnitude of phase transients. The revised ANSI Standard T1.101-1987 [ANSI, 1994] specifies these details.

30.6 Summary and Conclusions

The deployment of digital switching exchanges and cross-connect equipment in telecommunications systems created the necessity for robust synchronization distribution networks. These networks were originally designed to guarantee satisfactory slip performance for an end-to-end connection. However, recently introduced high-speed synchronous multiplexing and transmission systems based on SONET and SDH technology have emphasized the need for enhancing their accuracy and reliability. A disciplined approach to the design of such networks is delineated in many of the synchronization standards issued so far.

Our current understanding of the sources of synchronization impairments in telecommunications networks has not yet reached a mature point. The adaption of traditional clock noise models, such as those used in time and frequency metrology, to describe these impairments is a step in the right direction. However, much work is necessary to first evaluate reliably the parameters of this model from the phase error measurements, and then to identify the actual sources of impairments from them. As newer transport technologies such as asynchronous transfer mode (ATM) are introduced, the distribution of accurate synchronization references will be more complicated, and the impact of synchronization disruptions on network services will be harder to predict. These and other issues are currently being investigated by the various standards organizations.

Defining Terms

Free-run accuracy: The maximum long-term (20 years) deviation limit of a clock from the nominal frequency with no external frequency reference.

Holdover stability: The maximum rate of change of the clock frequency with respect to time upon loss of all input frequency references.

Jitter: The short-term variations of the significant instants (e.g., zero level crossings) of a digital signal from their ideal positions in time, where short term implies phase variations of frequency greater than or equal to 10 Hz.

Master–slave hierarchy: The hierarchical method where synchronization references are distributed from offices with higher performance master stratum clocks to offices with the same or lower performance slave stratum clocks.

Phase transient: Perturbations in phase of limited duration (typically several time constants of the slave clock which produces it) seen at synchronization interfaces.

Plesiochronous: Two signals are plesiochronous if their corresponding significant instants (e.g., zero level crossings) occur at nominally the same rate, any variation in rate being constrained within specified limits.

Pull-in range: Measure of the maximum reference frequency deviation from the nominal rate that can be overcome by a slave clock to pull itself into synchronization.

Stratum clocks: A classification of clocks in the synchronization network based on performance. Stratum 1 is the highest and stratum 4 is the lowest level of performance.

Timing loop: The situation where a slaved clock receives input reference timing from itself via a chain of other slaved clocks.

Wander: The long-term variations of the significant instants (e.g., zero level crossings) of a digital signal from their ideal positions in time, where long term implies phase variations of low frequency (less than 10 Hz).

References

ANSI. 1987. Synchronization interface standards for digital networks. ANSI Standard T1.101-1987, American National Standards Institute.

ANSI. 1994. Revision of ANSI Standard T1.101-1987. ANSI T1.101-1994, American National Standards Institute.

Bellcore. 1992. SONET synchronization planning guidelines. Special Rep. SR-NWT-002224, Issue 1, Feb.

Bellcore. 1993a. Digital network synchronization plan. Tech. Advisory TA-NWT-000436, Issue 2, June.

Bellcore. 1993b. Clocks for the synchronized network: Common generic criteria. Tech. Ref. TR-NWT-001244, Issue 1, June.

CCITT. 1988a. Timing requirements at the outputs of primary reference clocks suitable for plesiochronous operation of international digital links. Recommendation G.811, Consultative Committee on International Telephony and Telegraphy, Blue Book, Melbourne, Nov. 1988.

CCITT. 1988b. Timing requirements at the outputs of slave clocks suitable for plesiochronous operation of international digital links. Recommendation G.812, Consultative Committee on International Telephony and Telegraphy, Blue Book, Melbourne, Nov. 1988.

CCITT. 1992. Timing characteristics of slave clocks suitable for operation of SDH equipment. Draft Recommendation G.81s, Consultative Committee on International Telephony and Telegraphy, Geneva, June.

Zampetti, G. 1992. Synopsis of timing measurement techniques used in telecommunications. In *Proceedings of the 24th Annual Precise Time and Time Interval (PTTI) Applications and Planning Meeting*, McLean, VA, pp. 313–324, Dec.

Further Information

The ANSI, Bellcore, and ITU (CCITT) documents listed under the references are the most suitable sources for the synchronization standards. Information regarding the revisions to these standards and newer standards being contemplated is available from the recent contributions to the T1X1.3 working group of the ANSI and to the SG XIII/WP6 of the ITU. Zampetti's paper [Zampetti, 1992] provides a synopsis of the performance parameters MTIE and TVAR for characterizing synchronization impairments in the network, and also proposes an additional parameter to gain further insights.

31

Echo Cancellation

Giovanni Cherubini
IBM Zurich Research Laboratory

31.1 Introduction...414
31.2 Baseband Transmission..415
31.3 Passband Transmission 421
31.4 Summary and Conclusions 423

31.1 Introduction

Full-duplex data transmission over a single twisted-pair cable permits the simultaneous flow of information in two directions when the same frequency band is used. Examples of this technique are digital communication systems that operate over the telephone network. In a digital subscriber loop, at each end of the full-duplex link, a circuit known as a hybrid separates the two directions of transmission. To avoid signal reflections at the near- and far-end hybrid, a precise knowledge of the line impedance would be required. Since the line impedance depends on line parameters that, in general, are not exactly known, however, an attenuated and distorted replica of the transmit signal leaks to the receiver input as an echo signal. Data-driven adaptive echo cancellation mitigates the effects of impedance mismatch.

A similar problem is caused by crosstalk in transmission systems over voice-grade unshielded twisted-pair cables for local-area network applications, where multipair cables are used to physically separate the two directions of transmission. Crosstalk is a statistical phenomenon due to randomly varying differential capacitive and inductive coupling between adjacent two-wire transmission lines. At the rates of several megabits per second that are usually considered for local-area network applications, near-end crosstalk represents the dominant disturbance; hence near-end crosstalk cancellation must be performed to ensure reliable communication.

In voiceband data modems, the model for the echo channel is considerably different from the echo model adopted in baseband transmission. In fact, since the transmitted passband signal is obtained by modulating a complex-valued baseband signal, the far-end echo signal may experience significant jitter and frequency shift, which are caused by signal processing at intermediate points in the telephone network. Therefore, a digital adaptive echo canceller for passband transmission needs to embody algorithms that account for the presence of such additional impairments.

In this chapter, we describe the echo channel models and the structure of digital echo cancellers for baseband and passband transmission and address the tradeoffs between complexity, speed of adaptation, and accuracy of cancellation in adaptive echo cancellers.

31.2 Baseband Transmission

The model of a full-duplex data transmission system with adaptive echo cancellation is shown in Fig. 31.1. To describe system operations, we consider one end of the full-duplex link. The configuration for a baseband channel digital echo canceller is shown in Fig. 31.2. The transmitted data consist of a sequence $\{a_n\}$ of independent and identically distributed (i.i.d.) real-valued symbols from the M-ary alphabet $\mathcal{A} = \{\pm 1, \ldots, \pm(M-1)\}$. The sequence $\{a_n\}$ is converted into an analog signal by a digital-to-analog (D/A) converter. The conversion to a staircase signal by a zero-order hold D/A converter is described by the frequency response $H_{D/A}(f) = T \sin(\pi f T)/(\pi f T)$, where T is the modulation interval. The D/A converter output is filtered by the analog transmit filter and is input to the channel through the hybrid.

The signal $x(t)$ at the output of the low-pass analog receive filter has three components, namely, the signal from the far-end transmitter $r(t)$, the echo $u(t)$, and additive Gaussian noise $w(t)$. The signal $x(t)$ is given by

$$x(t) = r(t) + u(t) + w(t)$$
$$= \sum_{n=-\infty}^{\infty} a_n^{FE} h(t - nT) + \sum_{n=-\infty}^{\infty} a_n h_E(t - nT) + w(t) \quad (31.1)$$

where $\{a_n^{FE}\}$ is the sequence of symbols from the far-end transmitter, and $h(t)$ and $h_E(t) = \{h_{D/A} \otimes g_E\}(t)$ are the impulse responses of the overall channel and the echo channel, respectively. In the expression of $h_E(t)$, the function $h_{D/A}(t)$ is the inverse Fourier transform of $H_{D/A}(f)$, and the operator $\otimes$ denotes convolution. The signal obtained after echo cancellation is processed by a detector that outputs the sequence of estimated symbols $\{\hat{a}_n^{FE}\}$. In the case of data transmission for local-area network applications, where near-end crosstalk represents the main disturbance, the configuration of a digital near-end crosstalk canceller is obtained from Fig. 31.2, with the echo channel replaced by the crosstalk channel.

In general, we consider baseband signalling techniques such that the signal at the output of the overall channel has nonnegligible excess bandwidth, i.e., nonnegligible spectral components at frequencies larger than half of the modulation rate, $|f| \geq 1/2T$. Therefore, to avoid aliasing, the signal $x(t)$ is sampled at twice the modulation rate or at a higher sampling rate. Assuming a

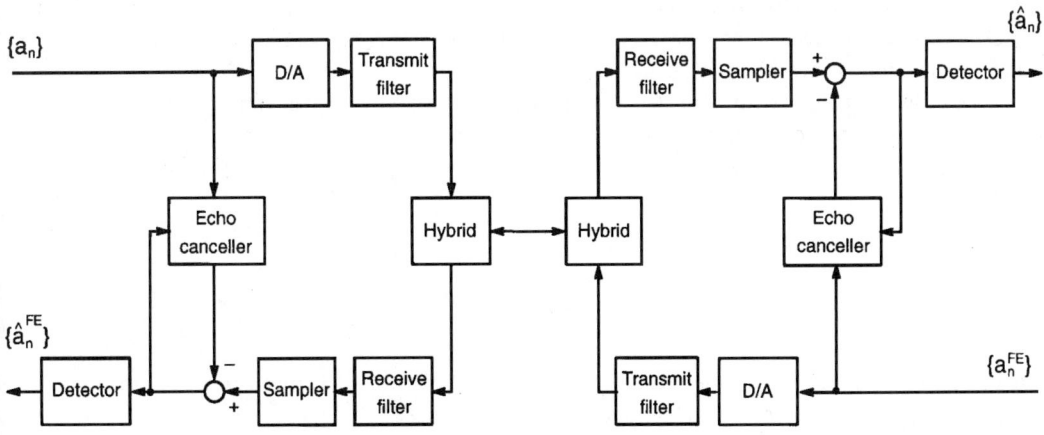

FIGURE 31.1 Model of a full-duplex transmission system.

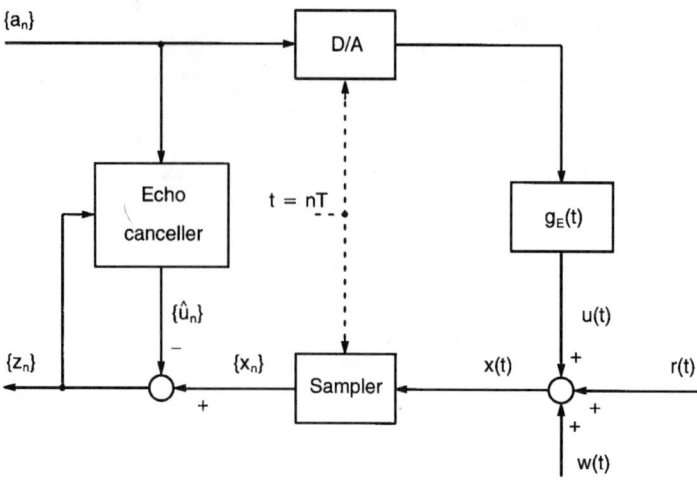

FIGURE 31.2 Configuration for a baseband channel echo canceller.

sampling rate equal to m/T, $m > 1$, the ith sample during the nth modulation interval is given by

$$x\left[(nm+i)\frac{T}{m}\right] = x_{nm+i} = r_{nm+i} + u_{nm+i} + w_{nm+i}, \quad i = 0, \ldots, m-1$$

$$= \sum_{k=-\infty}^{\infty} h_{km+i} a_{n-k}^{\text{FE}} + \sum_{k=-\infty}^{\infty} h_{km+i}^{E} a_{n-k} + w_{nm+i} \quad (31.2)$$

where $\{h_{nm+i}, i = 0, \ldots, m-1\}$ and $\{h_{nm+i}^{E}, i = 0, \ldots, m-1\}$ are the discrete-time impulse responses of the overall channel and the echo channel, respectively, and $\{w_{nm+i}, i = 0, \ldots, m-1\}$ is a sequence of Gaussian noise samples with zero mean and variance σ_w^2. Equation (31.2) suggests that the sequence of samples $\{x_{nm+i}, i = 0, \ldots, m-1\}$ be regarded as a set of m interleaved sequences, each with a sampling rate equal to the modulation rate. Similarly, the sequence of echo samples $\{u_{nm+i}, i = 0, \ldots, m-1\}$ can be regarded as a set of m interleaved sequences that are output by m independent echo channels with discrete-time impulse responses $\{h_{nm+i}^{E}\}, i = 0, \ldots, m-1$, and an identical sequence $\{a_n\}$ of input symbols [Lee and Messerschmitt, 1994]. Hence, echo cancellation can be performed by m interleaved echo cancellers, as shown in Fig. 31.3. Since the performance of each canceller is independent of the other $m-1$ units, in the remaining part of this section we will consider the operations of a single echo canceller.

The echo canceller generates an estimate $\hat{u}_n$ of the echo signal. If we consider a transversal filter realization, $\hat{u}_n$ is obtained as the inner product of the vector of filter coefficients at time $t = nT$, $c'_n = (c_{n,0}, \ldots, c_{n,N-1})$ and the vector of signals stored in the echo canceller delay line at the same instant, $a'_n = (a_n, \ldots, a_{n-N+1})$

$$\hat{u}_n = c'_n a_n = \sum_{k=0}^{N-1} c_{n,k} a_{n-k} \quad (31.3)$$

where c'_n denotes the transpose of the vector c_n. The estimate of the echo is subtracted from the received signal. The result is defined as the cancellation error signal

$$z_n = x_n - \hat{u}_n = x_n - c'_n a_n \quad (31.4)$$

Echo Cancellation

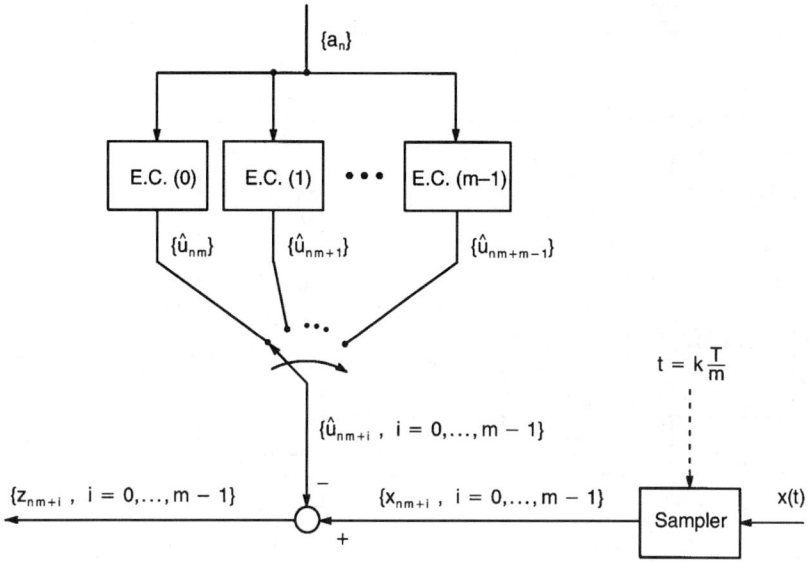

FIGURE 31.3 A set of m interleaved echo cancellers.

The echo attenuation that must be provided by the echo canceller to achieve proper system operation depends on the application. For example, for the Integrated Services Digital Network (ISDN) U-Interface transceiver, the echo attenuation must be larger than 55 dB [Messerschmitt, 1986]. It is then required that the echo signals outside of the time span of the echo canceller delay line be negligible, i.e., $h_n^E \approx 0$ for $n < 0$ and $n > N - 1$. As a measure of system performance, we consider the mean square error ε_n^2 at the output of the echo canceller at time $t = nT$, defined by

$$\varepsilon_n^2 = E\{z_n^2\} \qquad (31.5)$$

where $\{z_n\}$ is the error sequence and $E\{\cdot\}$ denotes the expectation operator. For a particular coefficient vector c_n, substitution of Eq. (31.4) into Eq. (31.5) yields

$$\varepsilon_n^2 = E\{x_n^2\} - 2c_n'q + c_n'R c_n \qquad (31.6)$$

where $q = E\{x_n a_n\}$ and $R = E\{a_n a_n'\}$. With the assumption of i.i.d. transmitted symbols, the correlation matrix R is diagonal. The elements on the diagonal are equal to the variance of the transmitted symbols, $\sigma_a^2 = (M^2 - 1)/3$. The minimum mean square error is given by

$$\varepsilon_{\min}^2 = E\{x_n^2\} - c_{\mathrm{opt}}' R c_{\mathrm{opt}} \qquad (31.7)$$

where the optimum coefficient vector is $c_{\mathrm{opt}} = R^{-1}q$. We note that proper system operation is achieved only if the transmitted symbols are uncorrelated with the symbols from the far-end transmitter. If this condition is satisfied, the optimum filter coefficients are given by the values of the discrete-time echo channel impulse response, i.e., $c_{n,k} = h_k^E$, $k = 0, \ldots, N - 1$.

By the decision-directed stochastic gradient algorithm, also known as the least mean square algorithm, the coefficients of the echo canceller converge in the mean to c_{opt}. The stochastic gradient algorithm for an N-tap adaptive linear transversal filter is formulated as follows:

$$c_{n+1} = c_n - \frac{1}{2}\alpha \nabla_c\{z_n^2\} = c_n + \alpha z_n a_n \qquad (31.8)$$

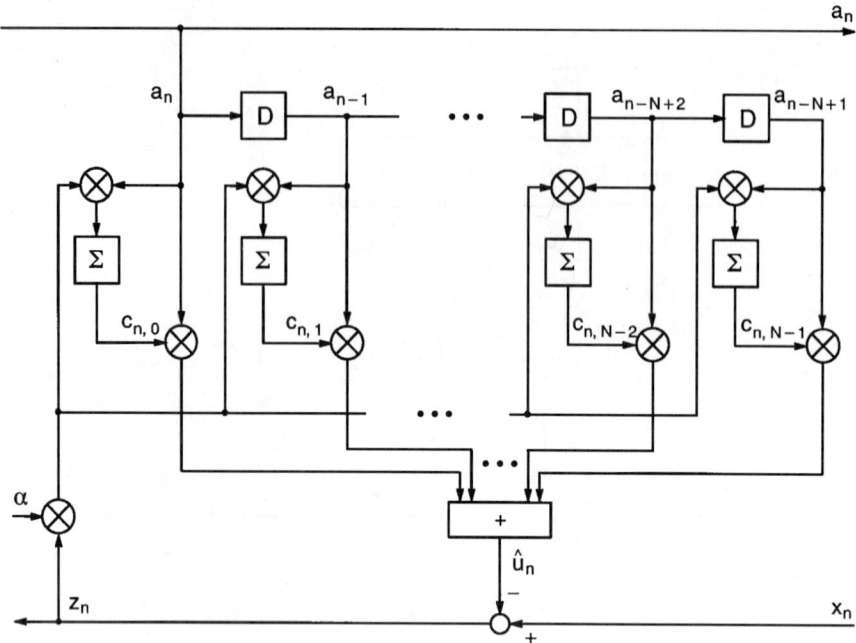

FIGURE 31.4 Block diagram of an adaptive transversal filter echo canceller.

where α is the adaptation gain and

$$\nabla_c\{z_n^2\} = \left(\frac{\partial z_n^2}{\partial c_{n,0}}, \ldots, \frac{\partial z_n^2}{\partial c_{n,N-1}}\right) = -2z_n \boldsymbol{a}_n$$

is the gradient of the squared error with respect to the vector of coefficients. The block diagram of an adaptive transversal filter echo canceller is shown in Fig. 31.4.

If we define the vector $\boldsymbol{p}_n = \boldsymbol{c}_{\text{opt}} - \boldsymbol{c}_n$, the mean square error can be expressed as

$$\varepsilon_n^2 = \varepsilon_{\min}^2 + \boldsymbol{p}_n' \boldsymbol{R} \boldsymbol{p}_n \tag{31.9}$$

where the term $\boldsymbol{p}_n' \boldsymbol{R} \boldsymbol{p}_n$ represents an excess mean square distortion due to the misadjustment of the filter settings. The analysis of the convergence behavior of the excess mean square distortion was first proposed for adaptive equalizers [Ungerboeck, 1972] and later extended to adaptive echo cancellers [Messerschmitt, 1984]. Under the assumption that the vectors $\boldsymbol{p}_n$ and $\boldsymbol{a}_n$ are statistically independent, the dynamics of the mean square error are given by

$$E\{\varepsilon_n^2\} = \varepsilon_0^2[1 - \alpha\sigma_a^2(2 - \alpha N\sigma_a^2)]^n + \frac{2\varepsilon_{\min}^2}{2 - \alpha N\sigma_a^2} \tag{31.10}$$

where ε_0^2 is determined by the initial conditions. The mean square error converges to a finite steady-state value ε_∞^2 if the stability condition $0 < \alpha < 2/(N\sigma_a^2)$ is satisfied. The optimum adaptation gain that yields fastest convergence at the beginning of the adaptation process is $\alpha_{\text{opt}} = 1/(N\sigma_a^2)$. The corresponding time constant and asymptotic mean square error are $\tau_{\text{opt}} = N$ and $\varepsilon_\infty^2 = 2\varepsilon_{\min}^2$, respectively.

We note that a fixed adaptation gain equal to α_{opt} could not be adopted in practice, since after echo cancellation the signal from the far-end transmitter would be embedded in a residual echo having approximately the same power. If the time constant of the convergence mode is not a critical system parameter, an adaptation gain smaller than α_{opt} will be adopted to achieve an asymptotic mean square error close to $\varepsilon_{\min}^2$. On the other hand, if fast convergence is required, a variable gain will be chosen.

Echo Cancellation

Several techniques have been proposed to increase the speed of convergence of the stochastic gradient algorithm. In particular, for echo cancellation in data transmission, the speed of adaptation is reduced by the presence of the signal from the far-end transmitter in the cancellation error. To mitigate this problem, the data signal can be adaptively removed from the cancellation error by a decision-directed algorithm [Falconer, 1982].

Modified versions of the stochastic gradient algorithm have been also proposed to reduce system complexity. For example, the sign algorithm suggests that only the sign of the error signal be used to compute an approximation of the stochastic gradient [Duttweiler, 1982]. An alternative means to reduce the implementation complexity of an adaptive echo canceller consists in the choice of a filter structure with a lower computational complexity than the transversal filter.

At high data rates, very large scale integration (VLSI) technology is needed for the implementation of transceivers for full-duplex data transmission. High-speed echo cancellers and near-end crosstalk cancellers that do not require multiplications represent an attractive solution because of their low complexity. As an example of an architecture suitable for VLSI implementation, we consider echo cancellation by a distributed-arithmetic filter, where multiplications are replaced by table lookup and shift-and-add operations [Smith, Cowan, and Adams, 1988]. By segmenting the echo canceller into filter sections of shorter lengths, various tradeoffs concerning the number of operations per modulation interval and the number of memory locations needed to store the lookup tables are possible. Adaptivity is achieved by updating the values stored in the lookup tables by the stochastic gradient algorithm.

To describe the principles of operations of a distributed-arithmetic echo canceller, we assume that the number of elements in the alphabet of input symbols is a power of two, $M = 2^W$. Therefore, each symbol is represented by the vector $(a_n^{(0)}, \ldots, a_n^{(W-1)})$, where $a_n^{(i)}, i = 0, \ldots, W-1$, are independent binary random variables, i.e.,

$$a_n = \sum_{w=0}^{W-1} (2a_n^{(w)} - 1)2^w = \sum_{w=0}^{W-1} b_n^{(w)} 2^w \qquad (31.11)$$

where $b_n^{(w)} = (2a_n^{(w)} - 1) \in \{-1, +1\}$. If we recall the expression (31.1) of the output of a transversal filter, by substituting Eq. (31.11) into Eq. (31.1) and segmenting the delay line of the echo canceller into L sections with $K = N/L$ delay elements each, we obtain

$$\hat{u}_n = \sum_{\ell=0}^{L-1} \sum_{w=0}^{W-1} 2^w \left[\sum_{k=0}^{K-1} b_{n-\ell K-k}^{(w)} c_{n,\ell K+k} \right] \qquad (31.12)$$

Equation (31.12) suggests that the filter output can be computed using a set of $L2^K$ lookup values that are stored in L lookup tables with 2^K memory locations each. The binary vectors $\mathbf{a}_{n,\ell}^{(w)} = (a_{n-(\ell+1)K+1}^{(w)}, \ldots, a_{n-\ell K}^{(w)}), w = 0, \ldots, W-1, \ell = 0, \ldots, L-1$, determine the addresses of the memory locations where the lookup values that are needed to compute the filter output are stored. The filter output is obtained by WL table lookup and shift-and-add operations.

We observe that $\mathbf{a}_{n,\ell}^{(w)}$ and its binary complement $\bar{\mathbf{a}}_{n,\ell}^{(w)}$ select two values that differ only in their sign. This symmetry is exploited to halve the number of lookup values to be stored. To determine the output of a distributed-arithmetic filter with reduced memory size, we use the identity

$$\hat{u}_n = \sum_{\ell=0}^{L-1} \sum_{w=0}^{W-1} 2^w b_{n-\ell K-k_0}^{(w)} \left[c_{\ell K+k_0} + b_{n-\ell K-k_0}^{(w)} \sum_{\substack{k=0 \\ k \neq k_0}}^{K-1} b_{n-\ell K-k}^{(w)} c_{n,\ell K+k} \right] \qquad (31.13)$$

where k_0 can be any element of the set $\{0, \ldots, K-1\}$. In the following, we take $k_0 = 0$. Then the binary symbols $b_{n-\ell K}^{(w)}$ determine whether the selected lookup values are to be added or subtracted. Each lookup table has now 2^{K-1} memory locations, and the filter output is given by

$$\hat{u}_n = \sum_{\ell=0}^{L-1} \sum_{w=0}^{W-1} 2^w b_{n-\ell K}^{(w)} d_n\left(i_{n,\ell}^{(w)}, \ell\right) \qquad (31.14)$$

where $d_n(k, \ell)$, $k = 0, \ldots, 2^{K-1} - 1$, $\ell = 0, \ldots, L - 1$, are the lookup values, and $i_{n,\ell}^{(w)}$, $w = 0, \ldots, W - 1$, $\ell = 0, \ldots, L - 1$, are the lookup indices computed as follows:

$$i_{n,\ell}^{(w)} = \begin{cases} \sum_{k=1}^{K-1} a_{n-\ell K-k}^{(w)} 2^{k-1} & \text{if } a_{n-\ell K}^{(w)} = 1 \\ \sum_{k=1}^{K-1} \bar{a}_{n-\ell K-k}^{(w)} 2^{k-1} & \text{if } a_{n-\ell K}^{(w)} = 0 \end{cases} \qquad (31.15)$$

We note that, as long as Eqs. (31.12) and (31.13) hold for some coefficient vector $(c_{n,0}, \ldots, c_{n,N-1})$, the distributed-arithmetic filter emulates the operation of a linear transversal filter. For arbitrary values $d_n(k, \ell)$, however, a nonlinear filtering operation results.

The expression of the stochastic gradient algorithm to update the lookup values of a distributed-arithmetic echo canceller takes the form

$$\mathbf{d}_{n+1} = \mathbf{d}_n - \frac{1}{2}\alpha \nabla_d\{z_n^2\} = \mathbf{d}_n + \alpha z_n \mathbf{y}_n \qquad (31.16)$$

where $\mathbf{d}'_n = [\mathbf{d}'_n(0), \ldots, \mathbf{d}'_n(L-1)]$, with $\mathbf{d}'_n(\ell) = [d_n(0, \ell), \ldots, d_n(2^{K-1} - 1, \ell)]$, and $\mathbf{y}'_n = [\mathbf{y}'_n(0), \ldots, \mathbf{y}'_n(L-1)]$, with

$$\mathbf{y}'_n(\ell) = \sum_{w=0}^{W-1} 2^w b_{n-\ell K}^{(w)} \left(\delta_{0,i_{n,\ell}^{(w)}}, \ldots, \delta_{2^{K-1}-1,i_{n,\ell}^{(w)}}\right)$$

are $L2^{K-1} \times 1$ vectors and where $\delta_{i,j}$ is the Kronecker delta. We note that at each iteration only those lookup values that are selected to generate the filter output are updated. The block diagram of an adaptive distributed-arithmetic echo canceller is shown in Fig. 31.5.

The analysis of the mean square error convergence behavior and steady-state performance has been extended to adaptive distributed-arithmetic echo cancellers [Cherubini, 1993]. The dynamics of the mean square error are given by

$$E\{\varepsilon_n^2\} = \varepsilon_0^2 \left[1 - \frac{\alpha \sigma_a^2}{2^{K-1}}(2 - \alpha L \sigma_a^2)\right]^n + \frac{2\varepsilon_{\min}^2}{2 - \alpha L \sigma_a^2} \qquad (31.17)$$

The stability condition for the echo canceller is $0 < \alpha < 2/(L\sigma_a^2)$. For a given adaptation gain, echo canceller stability depends on the number of lookup tables and on the variance of the transmitted symbols. Therefore, the time span of the echo canceller can be increased without affecting system stability, provided that the number L of lookup tables is kept constant. In that case, however, mean square error convergence will be slower. From Eq. (31.17), one finds that the optimum adaptation gain that permits the fastest mean square error convergence at the beginning of the adaptation process is $\alpha_{\text{opt}} = 1/(L\sigma_a^2)$. The time constant of the convergence mode is $\tau_{\text{opt}} = L2^{K-1}$. The smallest achievable time constant is thus proportional to the total number of lookup values. The realization of a distributed-arithmetic echo canceller can be further simplified by updating at each iteration only the lookup values that are addressed by the most significant bits of the symbols stored in the delay line. If the number of signal levels is large, the complexity required

Echo Cancellation

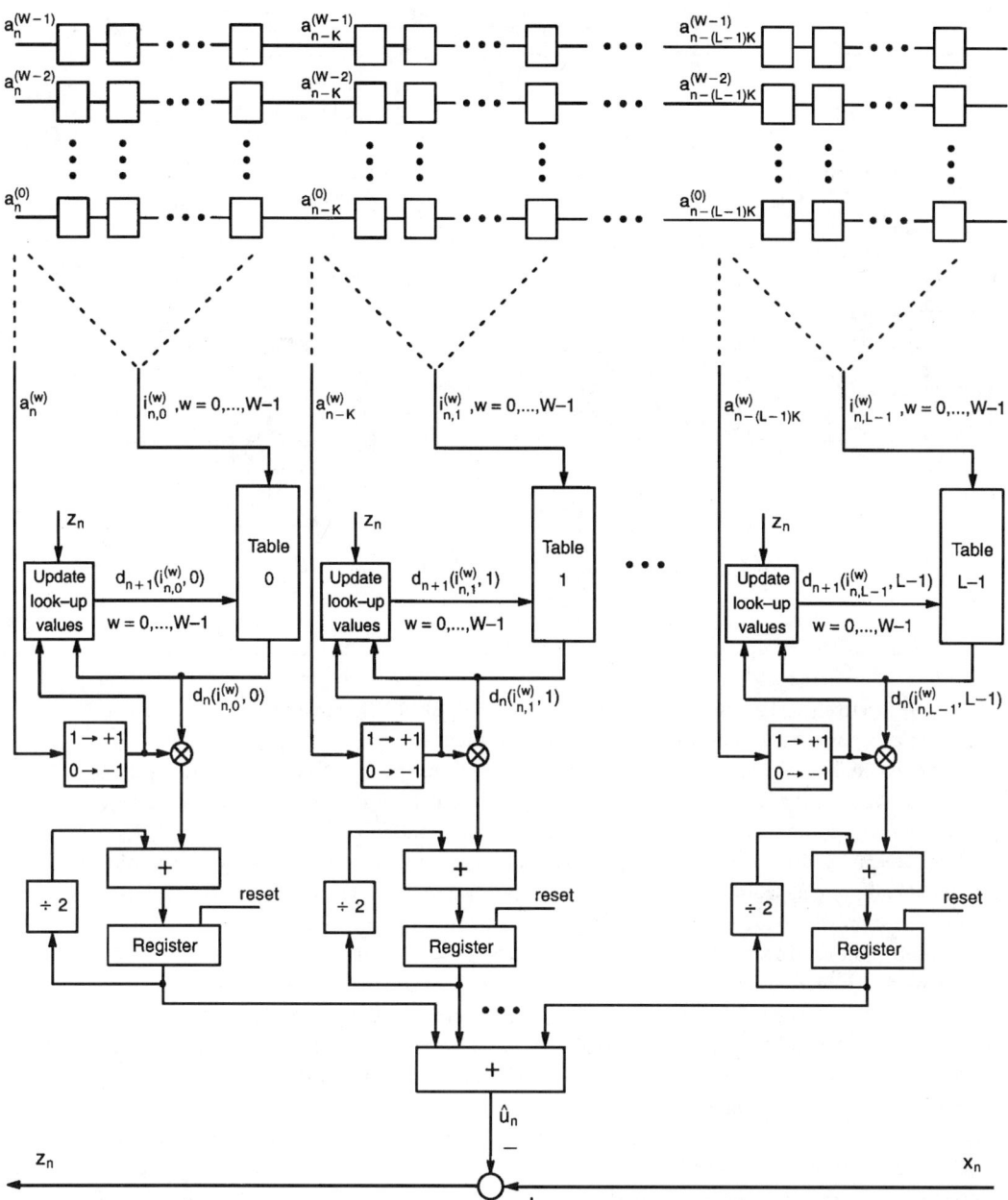

FIGURE 31.5 Block diagram of an adaptive distributed-arithmetic echo canceller.

for adaptation can be considerably reduced at the price of a small increase of the time constant of the convergence mode.

31.3 Passband Transmission

Although most of the concepts presented in the preceding sections can be readily extended to echo cancellation for passband transmission, the case of full-duplex transmission over a voiceband data channel requires a specific discussion. We consider the passband channel model shown in

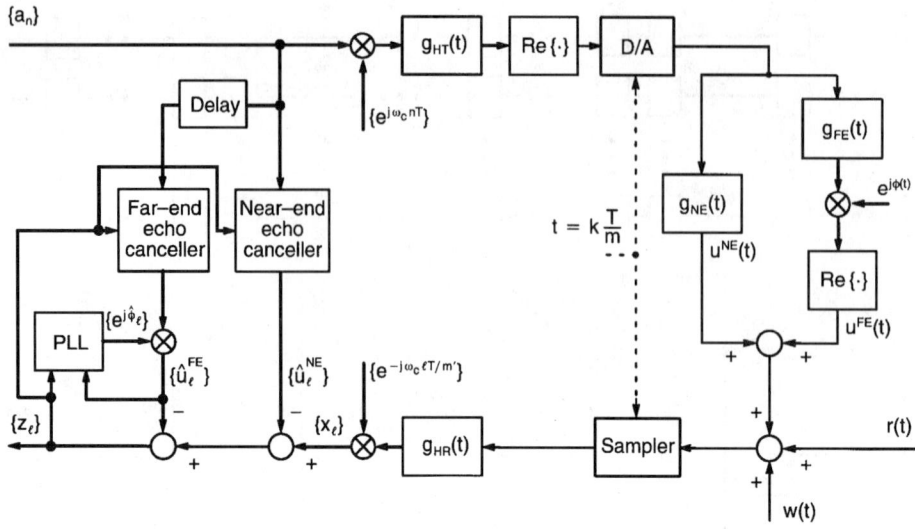

FIGURE 31.6 Configuration for a passband channel echo canceller.

Fig. 31.6. The transmitter generates a sequence $\{a_n\}$ of i.i.d. complex-valued symbols, which are modulated by the carrier $e^{j\omega_c nT}$, where T and ω_c denote the modulation interval and the carrier frequency, respectively. The discrete-time signal at the output of the transmit Hilbert filter may be regarded as an analytic signal, which is generated at the rate of m/T samples/s, $m > 1$. The real part of the analytic signal is converted into an analog signal by a D/A converter and input to the channel. We note that by transmitting the real part of a complex-valued signal positive- and negative-frequency components become folded. The image band attenuation of the transmit Hilbert filter thus determines the achievable echo suppression. In fact, the receiver cannot extract aliasing image-band components from desired passband frequency components, and the echo canceller is able to suppress only echo arising from transmitted passband components.

The output of the echo channel is represented as the sum of two contributions. The near-end echo $u^{\text{NE}}(t)$ arises from the impedance mismatch between the hybrid and the transmission line, as in the case of baseband transmission. The far-end echo $u^{\text{FE}}(t)$ represents the contribution due to echos that are generated at intermediate points in the telephone network. These echos are characterized by additional impairments, such as jitter and frequency shift, which are accounted for by introducing a carrier-phase rotation of an angle $\phi(t)$ in the model of the far-end echo.

At the receiver, samples of the signal at the channel output are obtained synchronously with the transmitter timing, at the sampling rate of m/T samples/s. The discrete-time received signal is converted to a complex-valued baseband signal $\{x_{nm'+i}, i = 0, \ldots, m'-1\}$, at the rate of m'/T samples/s, $1 < m' < m$, through filtering by the receive Hilbert filter, decimation, and demodulation. From delayed transmit symbols, estimates of the near- and far-end echo signals after demodulation, $\{\hat{u}^{\text{NE}}_{nm'+i}, i = 0, \ldots, m'-1\}$ and $\{\hat{u}^{\text{FE}}_{nm'+i}, i = 0, \ldots, m'-1\}$, respectively, are generated using m' interleaved near- and far-end echo cancellers. The cancellation error is given by

$$z_\ell = x_\ell - \hat{u}^{\text{NE}}_\ell - \hat{u}^{\text{FE}}_\ell \qquad (31.18)$$

A different model is obtained if echo cancellation is accomplished before demodulation. In this case, two equivalent configurations for the echo canceller may be considered. In one configuration, the modulated symbols are input to the transversal filter, which approximates the passband echo response. Alternatively, the modulator can be placed after the transversal filter, which is then called a baseband transversal filter [Weinstein, 1977].

Echo Cancellation

In the considered realization, the estimates of the echo signals after demodulation are given by

$$\hat{u}^{\text{NE}}_{nm'+i} = \sum_{k=0}^{N_{\text{NE}}-1} c^{\text{NE}}_{n,km'+i} a_{n-k}, \qquad i = 0, \ldots, m'-1 \qquad (31.19)$$

and

$$\hat{u}^{\text{FE}}_{nm'+i} = \left[\sum_{k=0}^{N_{\text{FE}}-1} c^{\text{FE}}_{n,km'+i} a_{n-k-D_{\text{FE}}}\right] e^{j\hat{\phi}_{nm'+i}}, \qquad i = 0, \ldots, m'-1 \qquad (31.20)$$

where $(c^{\text{NE}}_{n,0}, \ldots, c^{\text{NE}}_{n,m'N_{\text{NE}}-1})$ and $(c^{\text{FE}}_{n,0}, \ldots, c^{\text{FE}}_{n,m'N_{\text{FE}}-1})$ are the coefficients of the m' interleaved near- and far-end echo cancellers, respectively, $\{\hat{\phi}_{nm'+i}, i = 0, \ldots, m'-1\}$ is the sequence of far-end echo phase estimates, and D_{FE} denotes the bulk delay accounting for the round-trip delay from the transmitter to the point of echo generation. To prevent overlap of the time span of the near-end echo canceller with the time span of the far-end echo canceller, the condition $D_{\text{FE}} > N_{\text{NE}}$ must be satisfied. We also note that, because of the different nature of near- and far-end echo generation, the time span of the far-end echo canceller needs to be larger than the time span of the near-end echo canceller, i.e, $N_{\text{FE}} > N_{\text{NE}}$.

Adaptation of the filter coefficients in the near- and far-end echo cancellers by the stochastic gradient algorithm leads to

$$c^{\text{NE}}_{n+1,km'+i} = c^{\text{NE}}_{n,km'+i} + \alpha z_{nm'+i}(a_{n-k})^*$$
$$k = 0, \ldots, N_{\text{NE}} - 1, \qquad i = 0, \ldots, m'-1 \qquad (31.21)$$

and

$$c^{\text{FE}}_{n+1,km'+i} = c^{\text{FE}}_{n,km'+i} + \alpha z_{nm'+i}(a_{n-k-D_{\text{FE}}})^* e^{-j\hat{\phi}_{nm'+i}}$$
$$k = 0, \ldots, N_{\text{FE}} - 1, \qquad i = 0, \ldots, m'-1 \qquad (31.22)$$

respectively, where the asterisk denotes complex conjugation.

The far-end echo phase estimate is computed by a second-order phase-lock loop algorithm, where the following stochastic gradient approach is adopted:

$$\begin{cases} \hat{\phi}_{\ell+1} = \hat{\phi}_\ell - \frac{1}{2}\gamma_{\text{FE}} \nabla_{\hat{\phi}} |z_\ell|^2 + \Delta\phi_\ell & \pmod{2\pi} \\ \Delta\phi_{\ell+1} = \Delta\phi_\ell - \frac{1}{2}\zeta_{\text{FE}} \nabla_{\hat{\phi}} |z_\ell|^2 \end{cases} \qquad (31.23)$$

where $\ell = nm' + i$, $i = 0, \ldots, m'-1$, γ_{FE} and ζ_{FE} are step-size parameters, and

$$\nabla_{\hat{\phi}} |z_\ell|^2 = \frac{\partial |z_\ell|^2}{\partial \hat{\phi}_\ell} = -2\,\text{Im}\{z_\ell(\hat{u}^{\text{FE}}_\ell)^*\} \qquad (31.24)$$

We note that algorithm (31.23) requires m' iterations per modulation interval, i.e., we cannot resort to interleaving to reduce the complexity of the computation of the far-end echo phase estimate.

31.4 Summary and Conclusions

Digital signal processing techniques for echo cancellation provide large echo attenuation and eliminate the need for additional line interfaces and digital-to-analog and analog-to-digital converters that are required by echo cancellation in the analog signal domain.

The realization of digital echo cancellers in transceivers for high-speed full-duplex data transmission is today possible at a low cost thanks to the advances in VLSI technology. Digital techniques for echo cancellation are also appropriate for near-end crosstalk cancellation in transceivers for transmission over voice-grade cables at rates of several megabit per second for local-area network applications.

In voiceband modems for data transmission over the telephone network, digital techniques for echo cancellation also permit precise tracking of the carrier phase and frequency shift of far-end echos.

References

Cherubini, G. 1993. Analysis of the convergence behavior of adaptive distributed-arithmetic echo cancellers. *IEEE Trans. Commun.* 41(11):1703–1714.

Duttweiler, D.L. 1982. Adaptive filter performance with nonlinearities in the correlation multiplier. *IEEE Trans. Acoust., Speech, Signal Processing*, 30(8):578–586.

Falconer, D.D. 1982. Adaptive reference echo-cancellation. *IEEE Trans. Commun.* 30(9):2083–2094.

Lee, E.A. and Messerschmitt, D.G. 1994. *Digital Communication*, 2nd ed. Kluwer Academic Publishers, Boston MA.

Messerschmitt, D.G. 1984. Echo cancellation in speech and data transmission. *IEEE J. Sel. Areas Commun.* 2(2):283–297.

Messerschmitt, D.G. 1986. Design issues for the ISDN U-Interface transceiver. *IEEE J. Sel. Areas Commun.* 4(8):1281–1293.

Smith, M.J., Cowan, C.F.N., and Adams, P.F. 1988. Adaptive nonlinear digital filters using distributed arithmetic. *IEEE Trans. Circuits and Systems* 35(1):6–18.

Ungerboeck, G. 1972. Theory on the speed of convergence in adaptive equalizers for digital communication. *IBM J. Res. Develop.* 16(6):546–555.

Weinstein, S.B. 1977. A passband data-driven echo-canceller for full-duplex transmission on two-wire circuits. *IEEE Trans. Commun.* 25(7):654–666.

Further Information

For further information on adaptive transversal filters with application to echo cancellation, see *Adaptive Filters: Structures, Algorithms, and Applications*, M.L. Honig and D.G. Messerschmitt, Kluwer, 1984.

32

Switching Fabrics

32.1	Introduction	425
32.2	Virtual Circuit Switching	426
32.3	Packet Switching	426
32.4	Crosspoints	426
32.5	Multiplexing Versus Divisions	427
	Types of Divisions	
32.6	Switching System Functions	428
32.7	Topology of Switch Fabrics	428
32.8	Concentration, Expansion and Distribution	428
32.9	Multistage Fabrics	429
32.10	Fabric Control	429
32.11	Topology	430
	Time-Division Switching Elements • Time Multiplex Switching (TMS) • Time-Slot Interchange (TSI) • Time-Division Switch Fabrics	
32.12	Fabrics for Asynchronous Transfer Mode (ATM)	432

Amos E. Joel Jr.

32.1 Introduction

There are two distinct principles employed in electronic switching systems. They are known as *circuit switching* and *packet switching*.

In circuit switching a dedicated path is selected, established, and utilized for the entire duration a message session between a calling terminal and at least one selected called terminal. It usually has the capability of simultaneous transmission in both directions, known as duplex. Establishment of a channel through one or more switches in sequence is known as a *connection*. The process of requesting and establishing a connection is known as a *call*. The bandwidth associated with each service of a call is reserved for the duration of the service and is dedicated to that call.

In telecommunication systems the most important element is the message. If there are only two terminals, one to originate and another to terminate messages, no switching is required. When there are more than two terminals, unless messages are broadcast, switching is required to provide *selective* communication among a number of terminals. Generally privacy is required so that a message will be received by only the terminal for which the message is intended. Communication privacy requires that *contention* be included in the selection process to insure privacy and that a message reaches only the selected terminal over the selected path. *Selection* and *contention* are basic to switching.

Contention for the selection of channels occurs only at the time when a connection is being established. If a complete path is not possible when requested, the calling terminal is usually requested or scheduled to try to establish the connection again later.

The function within a switch that provides paths for connections is known as the *switch fabric*. This term is relatively new. In the past it has been known as the switching network or the switching center network.

In circuit switches, a message may be sent only one way, directed only from one terminal to another, or there may be interactive or two-way messages. For two-way messages that occur in real time, circuit switching systems establish a path between two or more terminals exchanging messages.[1]

32.2 Virtual Circuit Switching

Circuit switches are generally transparent to message content. However, to obtain efficiencies, circuit switching systems may remove or disconnect an established path in either or both directions during silent periods in the message and re-establish the circuit when the message information resumes. This technique is sometimes referred to as *virtual circuit switching*. Virtual circuit switching assumes that there is no perceptible degradation of the message. This perception depends on the service quality expectations of users.

The combination of transmission, switching, and terminals form a telecommunications *network*. Where the nature of messages may not require real-time interaction, other forms of networks with different economic criteria for service quality and message delay may be employed. They may also be used where the network serves more than one type of telecommunications, for example, voice, data, and video.

32.3 Packet Switching

In packet switching, messages are divided into segments or packets. These packets may be transmitted over different selected channels in a network. Each packet contains address information as well as other information regarding message treatment. Contention occurs for each packet at each switch. Most applications using packet switching are, traditionally at least, one way and not necessarily in real time; packets may be stored and delayed until the contention is resolved or the switch runs out of storage capacity. The memory used to store the packets is, for packet switching, the equivalent of the switching fabrics.[2]

32.4 Crosspoints

There are two basic technology elements used in switch fabrics. One is the memory previously referred to and the other is simply a switch often referred to as a *crosspoint*. Any device that has at least two stable states, such as one that opens and closes a circuit or changes from a high to low electrical impedance may be used as a crosspoint. Crosspoints are assembled in various ways to the form switch fabrics. The following are some technologies that have been used:

- Manual devices, for example, plugs and jacks. The plugs may be associated with flexible cords, several of which operate over a field of jacks that together constitute a fabric.
- Gross (large) motion electromechanical devices, for example, two motion (up and around) step-by-step switches and those that move in only one direction, such as rotary selectors and panel selectors that move over a vertical plane. In gross motion switches, wipers or brushes move over fixed banks of contacts. The combination of wiper and bank constitutes crosspoints.

---

[1] Circuit switching channels may also be established on a schedule.

[2] Asynchronous transfer mode (ATM) is a form of packet transmission with provision for real-time interactive (voice) messaging. The packets are of uniform length called *cells*. Switch fabrics for ATM (discussed subsequently) may employ crosspoints as well as memory.

- Fine motion electromechanical devices, for example, matrix arrays of contact devices where sets of contacts are operated at matrix intersections. The contact sets are crosspoints that may be operated individually, such as electromagnetic relays, or operated by cooperatively moving actuators, such as crossbar switches.

- Electronic, for example, *pnpn* or *cmos* bistable transistor devices or any other electronic devices that are inherently bistable, or are used to create a bistable condition, including such devices as gas and vacuum tubes. Also any gating device, such as an AND gate may be used as a crosspoint. Where electronic crosspoints are used to pass analog signals, such as speech, they should possess linear characteristics over the used range, 4 kHz in this case. Crosspoints that pass digital signals may be nonlinear. The electrical characteristics of the crosspoints and their interconnections determine the bandwidth capability of the fabric.

- Photonic, for example, arrays of *directional couplers* that are organized in a manner similar to crosspoints or optical beam switches operating in free space that are the equivalent of complete switch fabrics.

32.5 Multiplexing Versus Divisions

The design of switch fabrics relates to the organization of memory and crosspoints with the requirements for size, speed, interface, and control. Switch fabrics are organized into *divisions* that are analogous to transmission *multiplexing* techniques. The best known multiplexing techniques are in the frequency and time domains. When looked at broadly, transmission channels packaged into multipair cables might be considered to be multiplexed in the space domain. These are the corresponding switching divisions. Multiplexing is deterministic with a fix ordering of channels or time slots. Division usually involves selection and contention to produce what appears to be a random association between input and output channels.

Types of Divisions

Space Division

Here crosspoints are assembled into matrix arrays such that each call or connection uses a distinct crosspoint or set of crosspoints in series to establish a separate path, in space. The topology of a switch fabric determines its size and traffic capabilities. The topology includes the size(s) of the fabric and matrices, the number of stages, and the way links are interconnected between them.

Time Division

As in transmission, connection paths are represented by synchronized channels established periodically, called *time slots*. Time slots contain periodic samples of the message in a channel. The sample rate relates to the bandwidth of the message channel. A common value is 8000 samples/s to faithfully transmit 4-Hz speech signals. Unlike transmission multiplexing, switching time slots may carry analog or digital signals. For digital, each time slot is assigned a fixed number of bits. This establishes limits on the bandwidth of a connection in a time-division circuit switch. Greater bandwidth can be attained by *inverse multiplexing* several time slots.

Frequency Division

Frequency division uses a single transmission medium switch fabric to carry a plurality of calls simultaneously, each assigned to a different frequency. As in transmission, the capacity of a frequency-division switch fabric is limited by the bandwidth of the transmission medium divided by the bandwidth of each connection. Frequency division, although important for transmission multiplexing, is not practical in switching since a space division fabric is generally required to set the frequencies of the input modulators and output demodulators.

32.6 Switching System Functions

There are three functions used in most switches. These are signalling, control, and the switching fabric. There are two different control functions. One is for call or packet processing and the other is for the switch fabric. Signalling may be over the message channel or a separate signalling network. The processing control interprets the address of the called terminal, determines a routing for the call or packet, and establishes a path within the switch fabric,[3] as well as participates in the operation, maintenance, and administration of the switch. The selection of a path toward the called terminal is the principal function of the switch.

32.7 Topology of Switch Fabrics

The following covers the technology and techniques of space- and time-division circuit switching. When a switch fabric enables any desired connection to be established between idle ports, independent of traffic being carried, it is said to be *nonblocking*. Each input of a switch fabric must have at least one path to reach each of the output ports. This is known as *availability*. However, when due to congestion a desired connection cannot be established, a fabric is said to *block*. Blocking is a measure that denotes the quality of service rendered. With some services an alternative to blocking is *delay*. This is practical when the input signals are digital. When a service can tolerate time delay, storing all or part of messages may be introduced into the fabric. This *buffering* is limited inherently in capacity and therefore implies the possible loss of message content.

Two way ports or inputs may have a single appearance on a switch fabric. This fabric is called *single sided*. A more common type of fabric is *two sided* with ports on each side. Each path has an input and an output. With respect to the flow of traffic, one-sided fabrics are inherently bidirectional whereas two-sided fabrics may be unidirectional or bidirectional.

Traffic in most practical networks, where humans control the demand, is at *random*. In circuit switching, provision is made to serve the expected peak traffic. When all serviceable paths are busy, a request for service is rejected. In noncircuit switching where delay is allowed, messages or parts thereof may be delayed or, under peak traffic, lost.

32.8 Concentration, Expansion and Distribution

Taking advantage of the flow of traffic, networks are made economic by introducing the concept of *concentration*. With concentration a circuit switch[4] provides for fewer simultaneous connections than would be required than if each terminal were busy on a connection.

For two-sided unidirectional fabrics if the number of inputs is greater than the number of outputs, the switch fabric functions as a *concentrator* of traffic. Conversely, if the number of inputs is less than the number of outputs, the fabric is used for *expansion*. For bidirectional fabrics the two attributes can be combined, concentrating in one direction and expanding in the other. Generally, the single term concentration is used. These fabrics are always blocking.

When the number of inputs on each side of a two-sided fabric is the same, then the fabric is called a *distribution* switch fabric. This fabric may be blocking or nonblocking.

In concentration the selection taking place is to an idle link, channel, or path. With expansion the selection is for either the addressed terminal or provided for paths in different network directions,

---

[3] There are switch fabrics that utilize all or part of the message address to control directly the elements of the switch fabric. There are known as *self-routing* switch fabrics.

[4] For noncircuit switching, different names are used for the concentration function, for example, statistical multiplexing for packet switching.

which are expected at different times. Distribution is to select paths between concentration and expansion, which may be in different switch fabrics.

32.9 Multistage Fabrics

All space-division and some parts of time-division switch fabrics are implemented with crosspoint arrays. A nonblocking distribution fabric with n inputs is a square of $n \times n$ crosspoints. Similarly, for concentration or expansion, the number of crosspoints is $n \times m$, where n does not equal m.

When n and m are large, the number of crosspoints required in a single matrix becomes large. The number of crosspoints per input from m or n to a smaller number may be reduced by introducing additional stages of matrices resulting in *multistage* switch fabrics. The paths between matrices are known as *links*.

One interesting innovation in switch fabrics was the invention of Clos [1953]. He postulated multistage nonblocking fabrics that, for large numbers of inputs, have fewer crosspoints than the obvious single rectangular matrix. This capability was obtained by dividing the inputs into a number of smaller switch matrices. Each of these matrices introduced expansion into the first stage and concentration into the third-stage of a three-stage fabric. The expansion and concentration ratios are almost two-to-one. The center stage consists of a greater number of square matrices. With this configuration, if all inputs of a given matrix were to be connected to the corresponding outputs of a third stage concentration matrix, then there would be sufficient middle-stage matrices through which connections could be made between any other input and output.

Three stage nonblocking switching may be substituted for each of the middle-stage square switches. Therefore, any size nonblocking switch fabric may be achieved by employing multistage fabrics with an odd number of stages. The same principles may also be applied to one-sided fabrics. Clos networks are known as strictly nonblocking.

By gradually reducing the expansion and concentration ratios of the first-stage switches, thereby reducing the total number of crosspoints, blocking may be introduced into these odd-stage fabrics. As expected, blocking fabrics require less crosspoints than nonblocking fabrics. Also, a blocking network derived in this manner, until the center stage is the same size as the end stages, may assume the characteristics of nonblocking by rearranging connections already established.

Blocking fabrics may also use an even number of stages. Generally, these fabrics are composed of many small square matrices of crosspoints in each stage. Each matrix in one stage is connected by one or more links with each matrix in the next stage known as *grids*. With a single link between each switch in two adjacent stages, there is full availability; that is, each input can reach every output, but blocking is very high. By adding pairs of stages, parallel paths are introduced between end switch matrices. In this way, blocking is gradually reduced as more pairs of stages are employed.

Multimatrix multistage techniques are thus used to obtain fabrics of large capacity with minimum crosspoints at a desired blocking levels.

Another principle used to reduce crosspoints introduces blocking within the grids. This is known as *graded multipling*. Here access is reduced to less than the required availability; that is, each input cannot reach all outputs of a given matrix.

32.10 Fabric Control

Historically, the control of the switching fabrics has been identified with the technology of the switches or crosspoints employed in them together with the type of signalling. The remote control of the switching fabrics is basic to locating the switch at a central network point or node.

Gross motion electromechanical switches of the past, when wired together, form fabrics. For most applications these switches were organized in stages and controlled progressively. In many cases the switches were operated directly by the calling device, usually a dial, located at the calling station or terminal. In some respects self-routing fabrics (see footnote 3) used for ATM employ the same progressive control principle.

For switches of a type not compatible with the usually decimal calling device, the control signals would be recorded in a *register* that in turn would provide the nondecimal code to control the switches. This type of fabric control is known as *indirect progressive.*

As described earlier, stages of crosspoint matrices can be controlled progressively. By matching idle links between matrices of each stage, a *common control* can establish a required path concurrently. The links are accessed by the control. If more than one idle path is possible, the control makes a choice among them. Another path-seeking form of common control is known as *end marking*. Here, for example, opposing electrical polarities are placed on the input and selected output(s) of a two-sided switching fabric. If paths meet, one is selected and the crosspoint devices are actuated.

In space-division fabrics, each crosspoint includes some form of memory. It may be inherent as in a pnpn crosspoint. It may be a locking contact as part of a relaylike set of contacts, as in a crossbar switch. Or it may be a hard magnet that holds the crosspoints operated, as in sealed reed contact matrices.

In modern circuit switches, crosspoint memory is frequently found in a common random access memory (RAM). This memory may be for fabric control alone or shared as part of the general RAM requirements of the system. In either case this application of the RAM is known as the *network or fabric map*. They may indicate crosspoint or, more efficiently, busy/idle condition if interstage links.

32.11 Topology

Topologies for space and time division differ in technology. Space-division fabrics require a choice of technology. Time-division fabrics always use the electronic technology and the principles for these fabrics are given next.

Time-Division Switching Elements

Signals entering a time-division fabric are either analog or digital. Analog signals are sampled at a switch as part of the fabric. High-speed gates take analog pulse samples of input sources and connect them with output sinks. This is known as *pulse amplitude switching* (PAS). In a technique known as *resonant transfer* there is no loss of signal power in this process. There is a limit as to how far the pulse samples may be faithfully transmitted within a switch. Therefore, the PAS technique is used only in small fabrics where the common medium may serve only a few hundred gates. Another PAS technique uses two common media, one for transmitting and one for receiving, with a pulse amplifier between them.

Better known are time-division switching fabrics employing digital techniques. Conversion of signals from analog to digital may occur in the switch or they may reach the switch digital time multiplexed.

Switching digital signals permits the circuit switching of digitized voice or video message signals as well as signals from inherently digital sources such as computers. These systems are known as *time-division digital switching systems*, but for many the misappelation *digital switching* is utilized widely.

There are two functions in time-division switching, time slots carrying the message samples may be moved in time and/or assigned to different transmission lines. The movement in time is known as *time-slot interchange* and the multiplexing between different lines is known as *time multiplex switching* (TMS). To move the contents between time slots implies storage and a delay in transmission of at least one frame of time slots. When used for interactive speech, these delays must be held to levels that are imperceptible to the users.

Time Multiplex Switching (TMS)

Digital transmission is buffered at the input and output of the fabric so that complete frames of time slots of all transmission lines entering the switch are synchronized. Frames of synchronized digital transmission lines are connected so that information can pass in the same time slot from one line to another. An input time slot may be associated with the same time slot in the same or different output lines. Switching occurs at the time-slot rate; a new permutation of input to output lines is established for each time slot.

The TMS function is performed by high-speed electronic gates acting as crosspoints. For the duration of each time slot, one multiplexed transmission line is connected with another. The gates are arranged topologically as two-sided space-division fabrics. The simplest fabric is a single, square, ungraded, or nonblocking, matrix stage.

As the number of served time-division transmission lines increases, it becomes more economical to employ multistage fabrics. These fabrics are usually nonblocking or nearly nonblocking. Since TMS fabrics use the same topology as space-division fabrics, the TMS stage(s) are known as the *space* or *S* stages of a time-division switching system. The time multiplexed switch fabric acts in the time domain thereby representing many space division fabrics, and realizing the power of time-division switching.

Time-Slot Interchange (TSI)

Traffic capacity of a fabric is limited when switching is confined to within the same time slot. Time-slot interchange (TSI) enables a fabric to associate different time slots. This can be achieved by storing and delaying the message sample in one time slot so that it may be used in the desired time slot of the next frame or later in the same frame. Samples are stored in a RAM memory as received and read out in the next frame in a different order.

Since TSI involves the use of time delay, stages where this function is performed are known as *time* or T stages.

Time-Division Switch Fabrics

Time-division switch fabrics are made up of combinations of T and S stages. The order of the stages is a designer's choice. In general, it has been found that smaller systems may better employ S-T-S fabrics, where large systems use T-S-T.

Integrated circuit chips are available that provide an entire T-S-T switch fabric. Typically, eight incoming 24 or 32 channel lines may be connected to eight outgoing lines, thereby switching as many as 256 time slots including the line buffering on a single chip.

In transmission lines the digital signals travel only serially along the line. However, within a switch fabric the bits representing the samples may be sent serially and/or in parallel. For example, sending 8 bit samples through the switch on a parallel basis means that the 1.544-Mb/s technology can serve 192 time slots rather than only 24.

Also within a switch fabric, additional bits may be added to the sample for parity checking, signalling, or other functions. The digital line interface circuits between transmission and switching systems not only provide the buffering required for frame synchronization but also for the conversion from serial to parallel and the insertion of additional bits.

For most time-division digital switches, two fabrics are required but they are controlled usually by a single (memory) circuit. The fabrics are two sided with incoming transmission lines with their buffers appearing on one side whereas outgoing lines appear on the other. Interconnecting the receiving end of one line with the transmitting end of another line is only half of the connection required for a circuit switched connection. For the other half, a reciprocal relationship is established within the same fabric so that the inverse connection may be established.

Since time-division fabrics employ many active elements, it is not unusual to include redundancy or further duplication for this purpose. This redundancy is in addition to that required for the reciprocal transmission relationship.

Within the switch fabric the number of time slots per frame may exceed the standard number supported by the connected lines. This expansion, sometimes called *decorrelation*, is used to compensate for blocking that may occur in the S stages of the fabric. Therefore, the time slots used for connections within the switch fabric are not necessarily the same time slots used for the messages in the transmission media.

32.12 Fabrics for Asynchronous Transfer Mode (ATM)

In general, switch fabrics are designed to interface the signals that are received over the connected transmission media. Two types of switch fabrics are emerging for switching cell or ATM signals.

One is similar to those used for packet switching with memory not only at the ports but also at a central memory. The other is a form of space-division fabric that, as indicated in footnote 3, uses self-routing control. Because of the asynchronous nature of the signals, buffer memory at ports not only synchronize transport through the fabric but also reduces blocking by employing variable delay. Switching for ATM is generally not reciprocal.

Space-division fabrics are used to provide paths for the transport of cells from input to output. Linkage between stages are given the names of Banyan for routing and Batcher for sorting. Links may contain memory. Different arrangements of links sort out or trap the cells coincidentally destined for the same output.

References

Clos, C. 1953. A study of non-blocking switching networks. *Bell Syst. Tech. J.*, (March):406–424.

Hui, J.Y. 1990. *Switching and Traffic Theory For Integrated Broadband Networks*, Kluwer Academic, p. 347.

Marcus, M.J. 1977. The theory of connecting networks and their complexity: A review. *IEEE Proc.*, (Sept.):1263–1271.

Masson, G.M., Gingher, G.C., and Nakamura, S. 1979. A sampler of circuit switching networks. *Comput. Mag.*, (June):32–48.

Paull, M.C. 1962. Reswitching of connection networks. *Bell Syst. Tech. J.*, (May):833.

33
Customer Premises Equipment

John L. Fike
Texas A&M University

M. Lance Parr
Texas A&M University

33.1 Introduction... 433
33.2 Customer Interface 434
33.3 Standards 435
33.4 Restrictions.. 435
   Regulations Governing Interconnection • U.S. Interconnection Rules • Other Restrictions
33.5 A Functional Definition of CPE 438
   Analog Telephone Instruments • Electronic Telephones • Network Features • Key Telephones
33.6 Private Branch Exchanges 440
   Centrex • Voice Processing • Automatic Call Distributors • Modems • Facsimile • Channel Service Unit (CSU)/Data Service Unit (DSU) • Integrated Services Digital Network Arrangements • Integrated Services Digital Network Customer Premises Equipment • Premises Wiring and Connectors • Computer–Telephony Integration

33.1 Introduction

From the user's perspective, telecommunications is largely a business of interfaces, and no interface is more important than that between the customers application and the network. This is the purpose of Customer Premises Equipment (CPE): converting the customers application to information that can be transmitted and switched through the network, then reconverted to usable form at the destination.

Customer Premises Equipment may be defined as any device connected by a customer to the transmission facilities of the telephone company. Such devices are ordinarily located on the customer's property (premises), hence the term. More detailed definitions are provided by the U.S. Federal Communications Commission (FCC) in Part 68 of the FCC Rules, while the U.S. Telecommunication Act of 1996 defines CPE as "equipment employed on the premises of a person (other than a carrier) to originate, route, or terminate telecommunications." Note that the FCC's definition is for *terminal equipment*, which is the technical definition of customer premises equipment in North American practice (another term, just coming into vogue, is User Premises Equipment). It is worth noting that actual ownership of the equipment has little to do with the definition; a telephone set is referred to as CPE, even though it may be owned by the customer, the telephone company, or a third party.

CPE can be hierarchical: for example, the private branch exchange (PBX) is certainly CPE (in perhaps its largest embodiment), yet the telephones, modems, and fax machines that connect to the

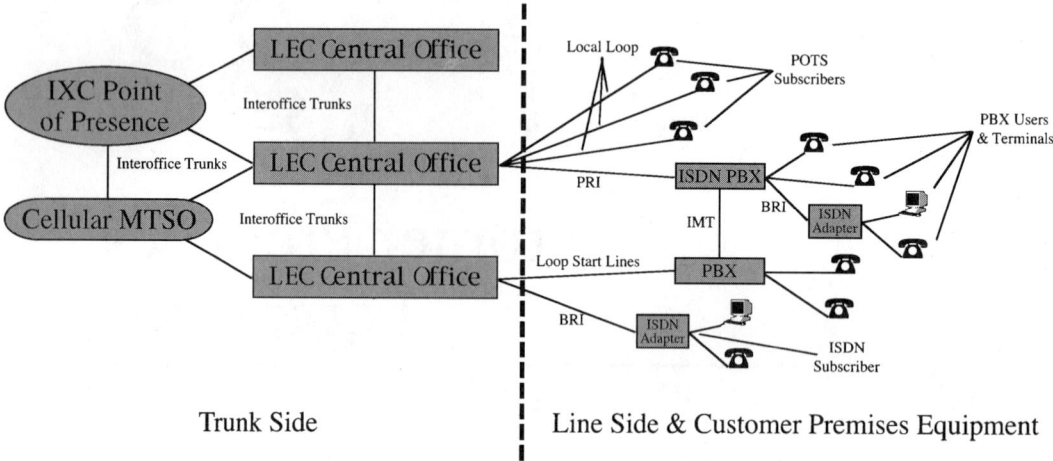

FIGURE 33.1 Location of customer premises equipment.

PBX are CPE also. This points out that CPE, in many instances, provides a networklike interface on the station side for connection of other CPE devices. Figure 33.1 illustrates this conception.

CPE may be viewed from three perspectives, each equally valid. The telephone service provider views CPE as the network terminator, and requires that the equipment operate correctly while not harming the network. The manufacturer is equally concerned with correct operation, and in addition must conform to the network's interface requirements. Finally, the user has little knowledge of the network interface, but considerable interest in the functionality of the equipment. It is likely that the user will not distinguish between features provided by the equipment and those furnished by the network itself.

33.2 Customer Interface

It is important that CPE, as well as the network itself, fulfill what might be termed the *cultural expectation* of users regarding their interface with telephone service. Customers expect a telephone to operate in certain ways, to enable access to various services at costs within a given range, to provide certain audible cues, and so forth. Designers of equipment and systems must satisfy these expectations to avoid embarrassing and expensive mistakes. The following examples illustrate problems that arose from misjudging the customers expectations:

- Some of the earliest operators of private pay phones refused to provide no-charge access to operator and emergency services such as 911 because no revenue was derived from such calls. Regulatory authorities and Congress eventually outlawed the practice; in the meantime, the private payphone industry received a large amount of bad publicity.
- PBX and key system manufacturers have occasionally offered equipment where the caller did not receive ringback and other call progress tones. The manufacturers were surprised to learn that the customers disliked this unexpected feature and told their friends, with the result that sales were poor.
- Some low-end telephones are equipped with * and # keys that are either nonfunctional, or that operate in unexpected ways. Once bitten, customers are disinclined to purchase additional units of this brand.
- In a recent manifestation of corporate ignorance, certain independent long-distance companies and payphone providers have teamed up to delete alphabetic characters from their keypads, giving an entirely new meaning to the term all-number dialing. The goal is to force

customers to use only the presubscribed long-distance company by discouraging the use of easily remembered alphabetic numbers such as 1-800-COLLECT. Predictably, state and national consumer groups have joined forces to oppose the action; an FCC rulemaking to disallow the practice is likely.

33.3 Standards

As CPE features and capabilities increase, so does the complexity of the interface with the network and with other CPE. Many of these issues are dealt with by international and national standards organizations, which devise standards in an effort to promote compatibility. The recommendations of the International Telecommunications Union (ITU) are worldwide in scope, although occasionally criticized as being self-serving to the national telecommunications administrations who control the ITU. Other regional or national standards organizations include the American National Standards Institute (ANSI), the Electronic Industries Association (EIA), the Telecommunications Industries Association (TIA), the European Telecom Standards Institute (ETSI), and many more. Some national standards organizations are also government entities that have regulatory powers and that certify or register CPE in their respective countries.

33.4 Restrictions

There are significant and complex restrictions governing the interconnection of customer premises equipment to the telephone network. This is because the improper design or use of such equipment can endanger the technical integrity of the network, or even the safety of telephone company personnel or other customers. Examples would be transmitting incorrect signalling information, or the connection of a high-voltage or high-current source into the network. In order to prevent these and other types of harm to the network, all telephone companies and administrations have regulations governing the attachment of customer premises equipment to the network. The incidental fact that such regulations may be used to protect the telephone company's equipment business from competition, or to shelter a domestic equipment manufacturer, is often an additional benefit.

Regulations Governing Interconnection

In the U.S., for many years the *Foreign Attachment Rule* in telco tariffs forbade the "... connection of any device to the telephone network not provided by the telephone company...." The word foreign here is telephone company parlance for outside, other, or similar usage. This provision was overturned by the FCC's Carterfone decision of 1968, which overnight created the *interconnect* (competitive CPE) industry.

Many, if not most, telephone administrations and operating companies throughout the world have similar regulations. The penalties for violating them can range from discontinuance of service (the usual practice in the U.S.), through confiscation of the offending equipment, to criminal charges, presumably on the basis of theft of service or endangering government property.

Obviously, the Foreign Attachment Rule or its equivalent simplifies the issue of responsibility. If the telephone provider is also the provider of CPE, then the customer can reasonably expect the CPE to operate correctly with the telephone network, to not endanger the network or any person, and to generally function in the manner intended. Unfortunately, these clear advantages of a single provider for terminal equipment carry with them a considerable set of disadvantages: the lack (or at least reduction) of innovation in the equipment itself as well as its application, little if any improvement in price-to-performance ratios, and the absence of other benefits that competition brings.

The problem that telephone administrations face is to balance the advantages of competition against the problems it brings. The integrity and safety of the network are paramount; this implies that, if the customer is allowed to supply their own CPE (which presumably is purchased from sources other than the telephone company), some sort of design validation and/or review is necessary to ensure its safe and correct operation when attached to the network.

The generic term for the process of determining a device's acceptability and conformity to local regulations is *homologation*. It may be as complex as the testing of actual equipment in a government laboratory, or as simple as type acceptance of another government's (or a manufacturer's) certification. The standards applied may be global [e.g., ITU-T or International Standards Organization (ISO)], regional (e.g., ETSI), or national (e.g., FCC). The process may take a few minutes if the proposed equipment is already on the approved list, or years if the proposed equipment competes with a domestic manufacturer or appears likely to reduce telephone company revenue. For example, the European Community requires homologation before a product can be used with a direct connect European Post Telephone and Telegraph (PTT) service. This includes testing and certification of the signalling, safety, and electromagnetic compatibility (EMC) aspects of the product. In any event, it should be assumed that the use of non-telco-provided CPE will slow deployment of private corporate networks, particularly in non-industrialized nations. Vendors and other users are the best sources for advice regarding the situation in a particular country.

U.S. Interconnection Rules

In the U.S., terminal equipment for direct connection to the network must be registered with the FCC. Part 68 of the FCC Rules and Regulations outlines the technical standards that must be met by terminal equipment in order to qualify. These requirements (which cover 131 pages) are designed to protect the network from electrical harm, and do not deal with performance characteristics of the equipment. Thus, registration is not a guarantee of satisfactory operation. Part 68 is the repository for a broad range of rules regulating equipment, and as such, changes fairly frequently. The portions which are discussed here have been in place, with few changes, for some 20 years.

There are a number of exceptions to the FCC's broad definition of Terminal Equipment, two of which are important here. First, coin-operated telephones using Central Office (CO) control are regarded as part of the Local Exchange Carrier's (LEC) network, and thus are not regulated (insofar as interconnection standards are concerned) by the FCC. Second, premise wiring does not lend itself to the registration process, and thus is not registered. Note, however, that there may be other requirements, such as electrical codes, that apply. Also, equipment directly connected to the network as of June 1, 1978 is grandfathered, and may remain connected without registration unless modified.

Part 68 specifications require that equipment withstand minimum environmental conditions, such as temperature, shock, vibration, and humidity. Limitations on signal power and longitudinal balance are also included. Equipment not meeting these requirements cannot be registered, and can only be connected to the network through protective circuitry that ensures compliance with Part 68. In other words, noncompliant equipment may be connected to the network via registered equipment. Figure 33.2 illustrates these two concepts; it is obvious that the Terminal Equipment Registration Program effectively moves the protective arrangement inside the CPE.

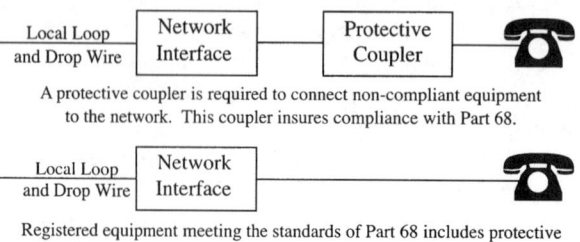

FIGURE 33.2 Protective circuitry.

Customer Premises Equipment

The U.S. rules are briefly outlined as follows:

1. Covered equipment: The rules cover all terminal equipment directly connected to the telephone company's network, with the exception of party-line and CO-controlled coin equipment. Coin-operated telephones that do not use CO-based coin service are considered Customer-Owned Coin-Operated Telephones (COCOTs) and must be registered. Any company wishing to sell CPE for interconnection with the U.S. Public-Switched Telephone Network (PSTN) must obtain registration for that equipment, regardless of origin.
2. Equipment not covered: CPE that is not directly connected to the telephone network requires no registration in the U.S. Data terminal equipment (DTE) for direct connection to another computing device, for instance, would not require FCC registration. An example of this is a personal computer connected to a modem; the modem must be registered, but the PC is not. Other equipment, such as handheld Personal Digital Assistants (PDAs) or portable computers with cellular modems, may be indirectly connected to the network through a host and require no registration in the U.S. CPE using any type of radio frequency (RF) technology is normally registered with the appropriate government body responsible for governing the use of spectrum (such as the FCC in the U.S.).
3. Demarc: The point of demarcation is the point of interconnection between the telephone company's facilities and the customer's; in other words, the point at which the telco's responsibility ends and the customer's begins. For ordinary one- or two-line installations (called simple premises wiring), this point is within 30 cm (12 in) of the lightning protector, or if there is no protector, within 30 cm of the point where the wire enters the customer's premises. For multiline installations (complex wiring) where there is no single protector, the demarc is the closest practicable point to where the wiring crosses the property line or enters the first building. It is important to note that the customer is also responsible for the wiring on his side of the demarc, even though wiring cannot be registered.
4. Application process: The appropriate application forms, information, and fees are submitted to the FCC. Information required includes the identification, technical description, purpose, means of limiting signal power, description of all circuitry employed to assure compliance, circuit diagrams, description of test methods and test results, photographs, and related data sufficient for the FCC staff to determine that the proposed equipment meets the requirements of Part 68. Note that the equipment itself is not submitted; the U.S. process is a design review and certification, not a laboratory test.
5. Registration number: Upon granting of the application, the FCC will assign a unique 14-character registration number, which must be affixed to the equipment and also noted in the user's manual. Figure 33.3 illustrates a typical registration number. A ringer equivalence number (REN) is also determined by the applicant and affixed as well; this informs the telco as to the type of ringing and the ringing current requirements of the device. Figure 33.4 illustrates these two numbers.
6. Telco notification: At the request of the telephone company, a customer who has connected equipment to the network (or who contemplates doing so) must provide the registration

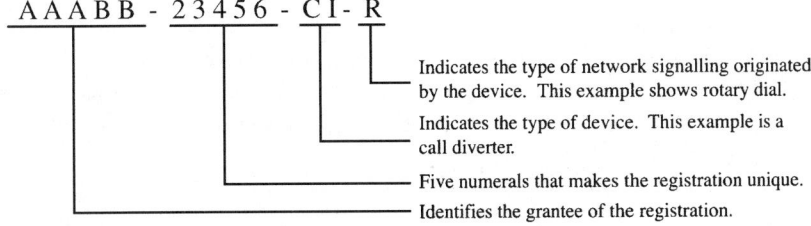

FIGURE 33.3 FCC registration number.

number and the ringer equivalence number. The telco will note their records accordingly. If service problems are found to be caused by the CPE, the telco may bill the customer on a time and materials basis for diagnosis. This rule replaced an earlier version that required prior notification; it was rarely observed, particularly by residential customers.

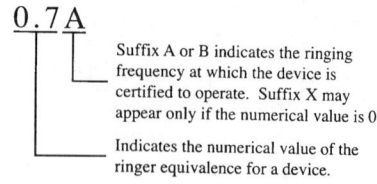

FIGURE 33.4 REN.

7. Service and alterations: Repairs and modifications to registered equipment are also covered by Part 68. In general, no modifications may be made, and all repairs must be performed by certified technicians using original equipment parts.

The registration process, although cumbersome, is not as arduous as it might sound. The required protective circuitry is often provided by readily available integrated circuit devices, which are placed inside the equipment and connected directly behind the network interface. The testing, certification, and documentation requirements may be met through the use of commercial laboratories that specialize in such services. Those unfamiliar with the process are advised to investigate such firms early in the design cycle; thus, decisions might not have to be changed to satisfy registration requirements.

Other Restrictions

In addition to rules and standards governing telecommunications equipment and the interconnection thereof, broader regulations also apply. For example, in the U.S. the National Electrical Code (as adopted by various government authorities) specifies many aspects of communications cabling, powering, grounding, and the like. Similar regulations exist in almost all countries. For regulatory purposes, it is useful to view CPE as a telecommunications appliance, which thus must conform to general requirements for safety as well as specific rules for interconnection.

33.5 A Functional Definition of CPE

With such a broad definition, i.e., anything connected to telephone company facilities (which might be wire, fiber, or radio link) by any means (which could include direct electrical connection or inductive coupling), it is helpful to define CPE in functional terms. By this approach, CPE includes any device that enables the customer to obtain a useful service from the telephone network. *Service* would typically include information transmission and switching, but as far as the customer is concerned, there is no interest in utilizing transmission. Rather, the customer wishes to speak with someone in another location (voice), communicate between the computer and a distant server (data), transmit an image to another location (facsimile), and so forth. Thus, it is seen that customer premises equipment may be classified by function as well as by type.

Analog Telephone Instruments

The most common CPE found in any country is the telephone. Equipped with a rotary dial or dual tone multifrequency (DTMF) keypad for signalling, a transmitter, and receiver, it is present in almost every home or office in most countries. It is used for placing voice calls over a single line to any other telephone in most of the world. The analog telephone set is the embodiment of plain old telephone service (POTS).

Although terms such as standard telephone, standard desk set, and single-line instrument appear frequently in the trade press, there is really no standard telephone, as such. In North America, such terms generally refer to analog sets similar to the Western Electric Type 500 rotary dial desk telephone set, or the equivalent Type 2500 DTMF set (later manufactured by AT&T). A relevant standard (for line interconnection, if not for appearance and operation) is EIA-470-A, "Telephone Instruments with Loop Signalling." The terms station, station set, and station equipment all refer to a single telephone instrument.

Customers would define the functionality of such a unit in terms of how it operated in the normal scenario of placing or receiving a telephone call. The telephone has a handset (usually with integrated speaker and microphone), an alerting device (ringer), a signalling device (rotary dial or keypad), and the switch hook (contained in the handset cradle in this instance). Other capabilities include automatic loop compensation, hybrid functions, and the ever present line powering.

The ringer is bridged across the loop to the CO when the handset is in the cradle; picking up the handset disconnects the ringer and connects the hybrid across the line. The hybrid divides the two-way conversation signals on the loop into two one-way paths: one from the CO to the receiver, the other from the transmitter to the CO. A small portion of the transmitted energy is fed back into the receiver; this function is called *side tone*, and allows the speaker to hear what they are saying. Studies have shown that this is necessary for most individuals to speak normally into the telephone.

Alerting devices include the traditional electromechanical ringer (mechanically and electrically tuned to the ringing frequency) or loudspeakers, which sound a ringing tone. Loop compensation is a circuit that automatically adjusts to the loop resistance; this replaces the older manual compensation, which had to be adjusted at installation.

Line powering means that the telephone set is powered by battery from the CO over the copper local loop. It is a feature that has existed since the telephone was invented, and is taken for granted by most customers. People count on their telephone to operate, even when utility power fails. But newer loop technologies (such as fiber-to-the-curb) may not furnish CO battery. This is a potentially difficult problem that may involve regulatory authorities.

Because of the pervasive effects of regulation, seemingly insignificant details can be expensive. For example, DTMF signalling is commonplace in many nations, yet regulators often make it an extra-cost service feature ($0.75 to $1.50 per line per month in the U.S.). In some locations, DTMF is unavailable. Thus, many subscribers will not have tone signalling to the CO, even though they may require it to interact with equipment and services (such as automated attendants) at the distant end of the call. The CPE designer must be aware of the customer's environment.

Electronic Telephones

Electronic telephones are generally defined as telephone sets where some or all of the functions of a standard analog set are implemented through microprocessors or microcontrollers. Because of the programmability of these devices, electronic sets normally contain an array of features. Manufacturing costs are usually far less than for older technology; consequently, electronic sets dominate the market, including those that are designed to resemble more traditional styles. Typical features include memory dialing, last number recall, volume controls, microphone muting, and number displays. Higher-end sets, or those that are part of a key system or a PBX may also have user-programmable buttons for speed dialing or to provide one-touch access to system features.

It is important to distinguish between electronic sets and digital sets, particularly since marketing people tend to equate the two. A digital telephone is an instrument where the information transmitted between the telephone and the switch [CO, PBX, or Key Service Unit (KSU) discussed subsequently] is in digital form. Voice is digitally encoded before it leaves the set. Digital telephones are electronic, but most electronic telephones are not digital; they are analog.

Network Features

New network services interact with CPE for greater functionality. Probably the most important of these is Calling Party Identification (CPI), better know as caller ID. Devices are now available built into telephones and as stand-alone units that identify the phone number (and in some cases the name) of the calling party before the call is answered. Although the discussion of caller ID is more properly covered in the chapter on network signalling, the impact of the sharing of network information on CPE is substantial.

Key Telephones

If the user requires access to more than one telephone line, some sort of on-premise switching is required (the alternative being that the user has more than one telephone instrument, each connected to a different line). The simplest method to accomplish this is known as a key telephone system (KTS). These systems operate by routing all lines to each telephone instrument, where the user selects the line desired by pressing a button.

For many years, the most ubiquitous type of key telephone equipment was the Western Electric type 1A2, commonly described as a six-button telephone. Depression of a line selection button connected the handset to that line and the button lights; similarly, if a call rang in on one of these outside lines, a ringing sound was provided, and the line button flashed. The cadence of both the ringing tone and the flash were used to differentiate an internal call (perhaps an intercom call from another station) from an outside call.

Most key systems are controlled from a central cabinet called a Key Service Unit (KSU). It is usually wall mounted in a closet, and contains the system control, memory, switching, and power supply. The CO lines terminate in one part of this cabinet; the cables from the key telephones terminate in another. The 1A2 system used relay technology and one wire per function; consequently, it required a 25-pair cable from the KSU to each station set. The cost of cabling, together with the inflexibility and maintenance problems inherent with electromechanical technology, created a market opportunity that was soon filled by electronic key systems. Such systems offer the basic functionality of the 1A2, along with many more features (which are relatively inexpensive in a computer-controlled device).

Hybrid key systems are systems that in some ways resemble a PBX; for example, not all lines may appear at each station, or the user dials 9 to get an available outside line. These systems typically serve larger installations (more than 20 stations), and have more features than smaller systems. An occasional difficulty is that the telco will classify hybrid systems as PBXs, with the result that the CO access lines are charged at trunk rates rather than as single lines.

Turrets are very large key sets, allowing direct selection from dozens of lines. They are primarily used in the securities industry, particularly on trading floors.

33.6 Private Branch Exchanges

The Private Branch Exchange (PBX) is a smaller version of the switching system (or *switch*) found at the telephone company's Central Office. International terminology differentiates between attendant-operated switches and automatic systems, where switching of internal calls (at least) is accomplished automatically in response to the dialed digits. The former is a Private Manual Branch Exchange (PMBX), the latter a Private Automatic Branch Exchange (PABX). Since virtually all systems sold today are PABXs, the term PBX is equivalent.

The PBX is located on the premises of the customer, and serves anywhere from 50 to thousands of lines. It mimics the functionality of the central office switch in that it provides dialtone to users, and connects calls to other users located on the customer's premises, or to the public-switched telephone network. Calls being placed to subscribers outside the PBX must be proceeded by a special access code, traditionally 9, for a nontoll outside line.

A PBX has its own numbering plan, normally with three- to five-digit station numbers. Each telephone served by the PBX has an individual extension number, and can be directly dialed from any other phone served by the PBX. Outside calls coming into the PBX from the local exchange are answered by a PBX operator and transferred (extended) to the desired station (extension). Many recent PBXs substitute an automated attendant for the traditional operator; these are not yet well received by the public.

Direct Inward Dialing (DID) allows calls coming in from the local exchange over special DID trunks to be connected directly to an extension without intervention by an operator or auto-attendant. Callers dial a standard seven-digit phone number corresponding to the particular PBX and extension number. Obviously, DID requires a signalling arrangement between the CO and the PBX for transferring the inbound number. It also requires that all of the numbers reachable by DID be unique in the exchange numbering scheme. Thus, a company with 1000 PBX stations would only require one telephone number from the outside, provided the incoming calls were extended by an attendant (automated or manual). With DID, the same company would require 1000 telephone numbers. The convenience of DID comes at a price; available numbers are used up at a faster rate, resulting in more frequent need for new area codes.

A modern PBX is similar to its larger cousin at the telephone company central office in that it is a digital switch complete with central control, program and call store, line and trunk interfaces, and many of the same features. Various options, such as dual-central control, backup power, and custom calling features, are available. Diagnostic software continuously monitors the PBX and reports problems to the attendant or service personnel. The PBX differs from a CO switch primarily in its capacity to serve lines and process calls. A PBX typically serves fewer than a thousand stations and can process between 50,000 and 100,000 busy hour call attempts, whereas a modern digital central office switch will serve between 20,000 and 80,000 lines with over a million busy hour call attempts. It is important to note that some very large PBX installations that serve tens of thousands of stations are actually CO switches. Vendors such as Lucent, Nortel, Ericsson, and Siemens furnish CO switching platforms which can also be configured as PBXs, depending on the line cards, software, and other details.

A PBX is viewed by the telephone company as CPE even though it connects to the telco's network through a trunk interface on the lineside of the CO. Connection may be by analog twisted pair copper or direct digital carrier interface. On the station side of the PBX, there may be three types of interfaces: analog, analog/digital, and Integrated Services Digital Network (ISDN). The analog interface is identical to that provided by the CO; it allows connection of any analog set that will work directly with the CO. The digital interface (which may include an analog voice channel) allows use of feature-rich electronic sets; unfortunately, these interfaces are almost always proprietary, limiting the choices of the customer. ISDN offers a way around this restriction; however, unless the PBX itself has ISDN connectivity to the CO (and through the CO to the rest of the ISDN world), the PBX and its users constitute an *ISDN island*. Data terminal support may be provided by digital telephone sets or data terminals; the former provides a data connector (usually RS-232) on the station set. ISDN interfaces allow an ISDN-ready PBX to connect to an ISDN Primary Rate Interface (PRI) on the trunk side and provide Basic Rate Interface (BRI) service to stations on the line side.

PBX station sets range from simple single-line telephones to multiline instruments with displays, attendant consoles, feature sets with built-in organizers, and telephones that combine voice and data functions. A single-line set connected to a PBX works exactly the same as its counterpart connected to a CO. It allows the users to access a single line, answer, and place calls. A multiline set provides functionality similar to that of a 1A2 keyset in that it allows access to many different lines from the same set through push-button selection. These instruments interoperate with the PBX through use of a command channel; the switching is actually accomplished at the PBX, and only one voice channel actually is present at the set. In this way the system operates much the same as a modern electronic key system, except that there is no need for a separate KSU. Attendant consoles are very large multiple-line sets that include extended functionality beyond that of individual multiline sets; they allow an attendant to extend (transfer) calls, monitor calls in progress, and determine the status of calls from an annunciator panel or screen. Feature telephones are multiple key sets that have programmable buttons for accessing features such as paging, call transfer, speed dialing, automatic redial of busy numbers, and hands-free calling.

PBX station equipment provides another example of the importance of user expectations. Because the first computer-controlled PBXs were substantially more expensive than the electromechanical systems they replaced, the telecommunications managers of user firms had to be rather creative in their financial justification for the new PBX. One way this was done was to show a credit for the discontinuance of all of the rented 1A2 key equipment which operated behind the PBX, replacing it with standard single-line sets that could do everything that the 1A2 could (selecting multiple lines, placing calls on hold, extending calls, etc.) through use of "easily remembered" sequences of switchhook depressions, dialed digits, #, *, etc. What invariably happened was that a few months after cutover of the new system, key equipment began to reappear, first on executive row, then throughout the company. The users had built their workday routines around the audible and visual cues that the key telephones provided, and they were not about to give them up. A secretary did not need to dial a magic code to determine if the boss was using the telephone; a glance at the boss' line button sufficed. So the heavy telephone users with influence in the organization (the executive secretaries and administrative assistants) got their key telephones back, regardless of the objections of the telecom manager. The solution to this problem was the development of electronic simulated key systems that operated in conjunction with the PBX software to provide keylike features, such as line indications and single-button feature access.

Major enhancements to digital PBXs include automatic call distribution, voice mail, and automated attendant features. These capabilities may require use of an auxiliary processors.

Most current PBX systems provide the capability for remote diagnosis and setup. This allows advanced support or administration from a remote site without having to send a technician to the PBX location. It also poses a substantial security risk since crackers can access the PBX via the maintenance port, break the password protection and reprogram the machine, or go into the long-distance business with the charges paid by the unwitting owner of the PBX.

Centrex

Centrex is a service provided by the local exchange carrier that mimics the functionality of a key system or PBX with station lines provisioned from the telephone company's central office switch. This is done through software that, in essence, partitions the CO switch to give each Centrex customer their own virtual PBX or key system. With a Centrex arrangement each extension is connected directly to the CO switch. The Centrex customer has a numbering plan similar to that of a PBX, and places calls to other extensions within the Centrex group exactly as a PBX user would. Calls from outside the Centrex group are handled as calls coming into a DID-equipped PBX in that a caller dials a seven-digit number to reach a Centrex group subscriber. Even though the users are served directly by the central office, an access code (such as 9) must be dialed for an outside line, further mimicking the functionality of a PBX. Thus, the Centrex user sees a system that provides service that is indistinguishable from that provided by a PBX.

CPE for Centrex service is virtually identical in functionality and appearance to that for PBX applications, although the signalling is often quite different. Centrex CPE can range from simple single-line analog telephones to ISDN sets. Such CPE is in some cases proprietary to the vendor of the central office switch that is providing the service, and is normally leased to the customer by the local exchange carrier as a part of the Centrex service contract.

Advantages of Centrex include reliability that is virtually the same as that of the central office, (which is normally less than 2 hours of total unavailability over 40 years), software that is always up to date (assuming that the CO software is up to date), and virtually unlimited call-handling capacity. Additionally, the customer does not have to provide floor space, power, or cooling for KSUs or PBX cabinets, nor is maintenance of any of this equipment required. Since it is provisioned from the CO, a Centrex system can span multiple locations within an exchange. Disadvantages include high cost and lack of control. Centrex is priced on a per-line, per-month basis based under contract with the local exchange carrier, whereas a PBX that is purchased has a fixed cost based on the purchase

Customer Premises Equipment

price of the hardware and installation. Also, some electronic PBX sets can only be used within a few thousand cable feet of the CO. An additional disadvantage is reliance on the telephone company for additions or changes to the service.

Voice Processing

CPE that combines computing and communications to manage voice calls automatically are known as *voice processing* systems. These include three classes of equipment comprising voice processing:

- *Voice mail*: A voice mail system stores messages in digital form on some type of computer memory. These may be stand-alone systems (often utilizing a personal computer for processing and message storage) which can be connected directly to the central office. Integrated voice mail systems are combined with a PBX for greater functionality, provided the PBX is designed for integration of external peripherals. Voice mail systems are sized by number of simultaneous callers (ports) and hours of storage.
- *Automated attendant*: The automated attendant is the most misused and maligned of the voice processing technologies. The system answers the telephone and offers the caller a menu of choices such as: "dial 1 for sales, 2 for service, 3 for engineering." A properly designed system can save time for callers, but systems with lengthy prompts and endless labyrinthine menus can be a source of great frustration to callers and may cause a loss of business. An automated attendant usually offers voice mail as an option, perhaps to callers who have arrived at a dead-end.
- *Interactive voice response*: Interactive Voice Response (IVR) or *audiotex* systems prompt the user to enter information such as an account number via DTMF dialing. This information is passed to a host computer for processing, and a voice synthesizer reads the information back to the caller. Banks and credit unions use IVR technology to enable customers to obtain their account balances without waiting for an agent.

Recent voice processing systems combine all three technologies. For example, a customer-service system may offer a menu of choices from an automated attendant, ask the caller to enter his or her customer number using IVR, and allow the caller to leave a message for a support specialist on voice mail.

Automatic Call Distributors

An Automatic Call Distributor (ACD) is a generic term for CPE that distributes incoming calls in a predetermined manner to a group of terminals, manned by agents who work in call centers. Call centers are utilized for such diverse applications as reservation systems, technical customer support, telemarketing, and many others. The object of ACD equipment is to improve call center performance and record keeping.

The acronym ACD is often used generically, to describe three types of systems that perform similar functions:

- A *call sequencer* is a device that routes incoming calls to a group of agents in sequence. Call sequencers may work with a PBX or key telephone system, or they may be connected directly to incoming lines. These devices are very simple; there may not be provision for record keeping, off-line work time, placing calls on hold, etc.
- A *uniform call distributor* (UCD) significantly improves call handling, and is an optional feature of many PBXs. The UCD routes incoming calls to the first available agent, using the PBX station set as the agent's telephone. If no agent is available, calls are routed to an appropriate announcement and/or held in queue until an agent is available. Agent performance and other management information is usually available on UCDs.

- A true automatic call distributor is the most sophisticated of the three devices. These machines, which often serve hundreds of agents, are either stand-alone or integrated with a PBX. An ACD routes calls to the least busy agent, equalizing the work load. In contrast to a PBX, which has more stations than trunks, an ACD has more trunks than agent positions. ACDs provide an array of management reports and controls, including the ability to reroute calls to other centers, routing calls by type of customer (using Automatic Number Identification (ANI)), and grouping agents into identifiable units called splits, queues, or groups.

Modems

What the telephone set is to the voice customer (a device that converts acoustic energy to electrical energy for transmission, and reverses the process at the receiving end), a modem is for data applications. A modem (modulator/demodulator) converts data (in binary form) into a modulated analog signal, which can be transmitted over facilities designed for voice. Again, conversion back to binary form is performed by a similar device at the receiving end. Similar to the telephone set itself, modems can be equipped with numerous features, such as the ability to dial outbound and answer inbound calls, to recognize call progress tones, to compress data where possible, and operate more efficiently and transparently in other ways. As modems have moved into the consumer market, sales volumes have resulted in more innovation and lower prices. Fax modems combine facsimile transmission and reception with data transmission; data-over-voice modems allow near-simultaneous conversation and data transmission over the same channel.

Facsimile

Facsimile (fax) machines are CPE that convert text or graphical information, originated on a sheet of paper or in a computer, into signals that can be interpreted at the receiving end by a sister machine, the result being the same or a very similar image or display.

Although fax systems resemble data communications, they are very different. The fax system transmits images, not characters; hence, there is no way for the fax system to operate on the content of the information transmitted. On the other hand, any image, no matter what the source, can be transmitted with equal efficiency.

The ITU-T divides facsimile standards into four groups:

Group 1. Frequency modulated analog encoding, 6-minutes per page transmission time, 100-lines-per-inch (lpi) resolution, 1500-Hz white frequency and 2400-Hz black frequency (in the U.S.); 1300-Hz white and 3100 Hz black in the rest of the world. Group 1 fax is so slow that it is virtually extinct in the U.S.

Group 2. Amplitude modulated analog encoding, 2–3 minutes per page transmission time, 100-lpi resolution, 2100-Hz carrier frequency using amplitude, phase, or vestigial sideband modulation. Group 2 is obsolete in the U.S.

Group 3. Compressed digital encoding using analog modems for transmission, 1 minute or less per page transmission time, 200-lpi resolution, 4800–9600 b/s data rate. Group 3 is the current standard product.

Group 4. Compressed digital using ISDN or other digital transmission, less than 1 minute per page transmission speed, 200–400 lpi resolution, data rates up to 64 kb/s.

The major changes in the facsimile market come from its integration into computers and word processing systems, and its coupling with scanners. Fax modems accept input directly from the word processing software, by appearing to be a printer. This allows transmission of computer-generated documents without the necessity of an intermediate paper copy, providing faster output and better image quality. On the receiving side, direct fax input to the computer allows storage of the image in the memory, permitting scanning for conversion into character and graphic form, editing,

Customer Premises Equipment

retransmission, and other functions. The overall picture (no pun intended) is one of increasing seamlessness between data and image communications.

Channel Service Unit (CSU)/Data Service Unit (DSU)

A channel service unit (CSU) is CPE that terminates a digital carrier circuit and provides line conditioning, diagnostic, and testing functions. A data service unit (DSU) is CPE that provides an interface to a digital carrier such as a T1.

The functionality of the DSU and CSU are normally combined into the same unit. Originally supplied by the LEC as part of end-to-end digital service, the FCC's 1980 decision in the Second Computer Inquiry made it clear that the CSU was considered CPE and must be supplied by the customer. In the U.S., the CSU is powered by the customer rather than by the LEC via the line.

Integrated Services Digital Network Arrangements

There are five functional groups of ISDN CPE in the U.S.:

Network Termination 1 (NT1): Performs signal conversion between the customer's equipment and the local loop, ensuring that the local loop signal conforms to the physical layer protocol for ISDN.

Network Termination 2 (NT2): Can switch, multiplex, or concentrate subscriber's lines, enabling a number of devices to share the ISDN service. Examples of NT2 include terminal controllers and ISDN PBXs.

Terminal Equipment 1 (TE1): Any device meeting the ISDN physical, data link and network layer protocols. These devices can be directly connected through an NT1 (such as an ISDN PBX) to an ISDN local loop.

Terminal Equipment 2 (TE2): Any device not meeting the ISDN physical, data link and network layer protocols, which must use a terminal adapter (TA) to adapt their output to ISDN protocols. A desktop computer is an example of a TE2, which requires an adapter card to connect to an ISDN network.

Terminal Adapter: Used to connect a non-ISDN device (such as a dumb terminal or a non-TA equipped personal computer) to an NT1.

ISDN is provided in two different channel arrangements, known as Primary Rate Interface (PRI) and Basic Rate Interface (BRI). PRI has 23 Bearer (B) channels, each of which provide 64-kb/s bandwidth, and one Delta (D) channel with 64-kb/s bandwidth that carries signalling information to the central office. BRI has 2 B channels and one D channel, that provides the same signalling function as the PRI D channel but with only 16 kb/s of bandwidth. PRI is formatted identically to the North American T1 carrier, and is normally terminated on an ISDN PBX (which is considered TE1 in this context). BRI provides much less bandwidth, and is normally terminated on a Terminal Adapter.

As an example of the versatility of systems using ISDN CPE, consider an insurance company utilizing telecommuting to allow some customer service representatives to work at home. A BRI connection is installed at the employee's home. Customers call the 800/888 number of the company, and are forwarded to the service representatives by the company's ACD equipped ISDN PBX to the next available agent. The PBX has PRI access to the CO, as well as to the computer on which the customer's records reside. An incoming call is answered by an automatic attendant on the PBX; the Calling Number Identification sent over the D channel to the PBX arrives with the call, and is used to match the caller's home telephone number with their customer information file. When an available work at home agent is identified, both the call and the customer information are forwarded to the agent over the BRI line to the agent's home, where the agent can simultaneously answer the call (using one B channels) and review the customer information (using the other B channel and their personal computer). Additional resources, such as FAX servers or order entry systems, can be accessed as needed.

BRI may be used for providing Centrex service. In this application, a specialized version of an ISDN telephone, customized with appropriate keys, may be used as a station set or attendant's console. These sets may share an ISDN line, allowing concentration of station lines. This multipoint capability of BRI allows the sharing of a line among several devices, such as telephones, fax machines, and computers, making ISDN particularly attractive for this type of application. A wide variety of CPE is coming onto the market that utilizes this capability, and may even obsolete separate modems, FAX machines, key systems, and small PBXs with integrated products.

Integrated Services Digital Network Customer Premises Equipment

ISDN CPE can be divided into three broad categories.

Integrated Services Digital Network Telephones

ISDN telephones are voice terminals, which are equipped with an S0 interface, and do not require an NT-1 interface. These voice terminals have a standard telephone handset, keypad, and hookswitch and mimic the appearance and functionality of a standard telephone; they also include interfaces for both digital and analog devices. ISDN telephones typically allow connection of other ISDN devices, such as Group 4 FAX machines.

An important difference between an ISDN telephone and a standard desk telephone is the method in which the set is powered. Whereas a standard analog telephone is powered from the central office battery, an ISDN telephone (voice terminal) must be powered at the customer premise. This means that if commercial power is lost the ISDN telephone will not operate unless alternate power from a battery or another uninterruptable power supply is available.

Integrated Services Digital Network Modems and Interface Cards

The term *ISDN modem* is a misnomer. Because no modulation or demodulation takes place, an ISDN modem is not actually a modem at all. Sometimes called a digital modem, these devices are Terminal Adapters with a built in NT-1. Such modems also include a POTS telephone jack so an analog telephone may be used to place an call using one of the B channels. It is a bidirectional interface that takes asynchronous data from a computer or terminal through a standard serial interface and performs the appropriate conversion necessary to produce a properly formatted synchronous ISDN channel. It is called a modem because of the appearance and operational similarity it has with the familiar computer modem. Personal computers equipped with these cards are ISDN compatible and do not require an external Terminal Adapter.

It is important to note that although conventional modems are available for use at a variety of different speeds over analog lines of various quality, an BRI ISDN modem operates at 128 kb/s, which is the data rate for 2 ISDN B channels, or 64 kb/s if a voice call is being placed at the same time on the same ISDN line. The term ISDN modem has been applied to this device by the vendors in a effort to avoid confusion on the part of ISDN users who may not understand or care about the technology, but simply want a device that performs the same function as a modem, that of connecting their computer to another computer or to an information service. It is important to note that the use of an ISDN modem may show only marginal improvement (if any) over its high-speed analog counterpart due to the built-in limitation of the serial port in some personal computers. Many of these are limited to 38.8 or 57.6 kb/s.

Integrated Services Digital Network Devices

ISDN bridges and routers provide local-area network (LAN) to LAN communication over ISDN using Transmission Control Protocol/Internet Protocol (TCP/IP) and Internetwork Packet Exchange (IPX). These devices may also include data compression and encryption as an option. An ISDN gateway allows connection of a LAN to the Internet, either directly or through an Internet service provider.

Premises Wiring and Connectors

Customer premise wiring is not considered CPE, and is not subject to the requirements for registration. This wiring is classified as either nonsystem premises wiring, or system premises wiring. Nonsystem premises wiring (often referred to as *simple inside wire*) is used with one- or two-line residential or business telephone service, and connects the CPE to the network at the telephone company demarcation point (demarc). Nonsystem premises wiring may be installed by parties other than the phone company, and must meet the requirements of Part 68 (such as insulation standards). The telephone company is not responsible for nonsystem premises wiring, although many LECs offer maintenance agreements for premises wiring.

System premises wiring connects separately housed equipment entities or system components to one another, or connects an equipment entity or system component with the telephone network interface located at the customer's premises. Installation, configuration, or modification of this wiring must be carried out in accordance with Part 68 under the supervision of a responsible supervisor. The telephone company is responsible for system premises wiring. See Fig. 33.5. Although simple premises wiring may consist of no more than a wire pair extended to several outlets in the home, it may also involve special cabling and highly engineered connectors to provide high-speed connectivity for data.

Prior the breakup of the Bell System there was little or no competition in the premises wiring market in the U.S. All premises wiring was provided and installed by the local telephone company, which also owned all of the CPE connected to it. For a time after the breakup the market for premise wiring systems was dominated by AT&T, Digital Equipment Corporation, IBM, and Northern Telecom. Each company used a proprietary and incompatible architecture. Demands on premises wiring have grown considerably during this time, with typical systems being called upon to support every type of communications need from a simple telephone to a mainframe computer and video. With the advent of the EIA/TIA-568 standard, a set of precise guidelines for premises wiring, cable type, connectors, topology, and distance was standardized. The standard is simple, yet rigid. It divides premises cabling down into several distinct subsystems for which detailed standards are published. Calling for a star topology, each node in the network is connected to a central wiring closet using one of four cable types: 100-Ω unshielded twisted pair (UTP), 150-Ω shielded twisted pair (STP), 50-Ω coaxial cable, or optical fiber of either 62.5 or 125 μm core diameter. Compliant installations will support equipment from many vendors without reconfiguration or replacement, making it possible for cabling of a building to proceed without knowledge of the exact brand and type of equipment to be supported by the network, and allowing users to choose any computer or vendor without concern for existing cable plant compatibility.

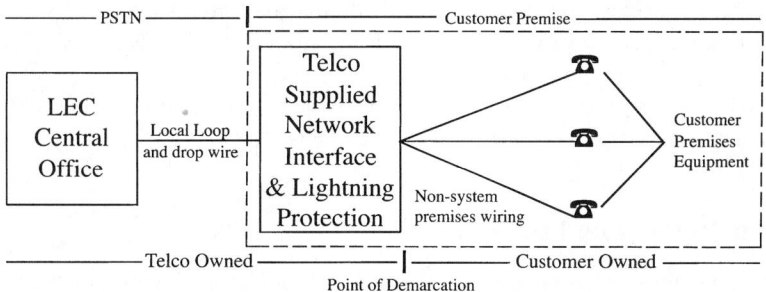

FIGURE 33.5 Customer premises.

Increased competition in the premises wiring market has resulted from opening the market to smaller vendors. Although EIA/TIA-568 is a North American standard, it serves as the foundation for premises wiring standards on a global basis. Cabling products meeting EIA/TIA-568 may be certified by independent testing labs such at ETL Testing Laboratories or Underwriters Laboratories, Inc.

Implementation of these standards for high-speed cabling can be challenging. The *component level* approach to standards proves particularly problematic for wideband (100 Mb/s or faster) wiring. A new systems-based approach based on the international standard ISO/IEC 11801 promises to solve many of these issues by taking the performance of the entire wiring plant (wiring, connectors, patch cables, etc.) into account. This new standard defines a number of link classes culminating with 100-MHz systems demanding EIA/TIA-568 Category Five system performance.

Special care must be taken when installing premise wiring for use with CPE that requires high bandwidth. For example, building a fully compliant Category Five system requires the use of only approved Type Five cable, connectors, jacks, plugs, and installation techniques. At the higher speeds (100 Mb/s) one must pay careful attention to avoiding kinks or sharp bends in cables, untwisting an excessive amount of wire where connectors are installed, and other seemingly insignificant details. Carelessness can result in premise wiring that is noncompliant and useless for the application for which it was installed, as well as being extremely difficult to troubleshoot.

A number of EIA/TIA standards deal with other aspects of premises wiring. The EIA/TIA-569 Commercial Building Standard for Telecommunications Pathways and Spaces provides architects, building designers, and engineers with a standard for minimum requirements for building space and pathways needed for installation of wiring for voice, image, data, etc. This includes space requirements for telecommunications closets and equipment rooms, facility entrances, and work areas. The 569 standard also provides specifications for ceiling heights, door sizes, floor loading, and lighting. A related standard is PN-2290, Telecommunications Administration Standard for Commercial Buildings, which standardizes gathering and documenting information needed for administration of building wiring. Aspects covered include pathways, cables, cross-connects, and terminating hardware. It outlines both required and optional information associated with each entity identified.

The EIA/TIA-570 Residential and Light Commercial Telecommunications Wiring Standard does not cover backbone wiring, but otherwise provides guidelines for a simple star topology wiring for homes or small businesses. It calls for four pair 24 AWG UTP for all station wiring, with outlets spaced every 12 ft along a wall and requires that no point in a room be more than 25 ft from an outlet. PN-2416 Backbone Wiring for Residential Buildings covers backbone wiring for multitenant buildings between the demarc and the leased space of a tenant. The wiring within the leased space would be covered by EIA/TIA-570.

Connection of CPE to the network at the demarc is generally simplified through the use of standardized telephone-company-provided jacks that provide either a series or bridging type connection to the tip and ring leads of the line. Standards for these connectors for the U.S. are found in part 68.500 of the FCC Rules. The popular RJ series of jacks and plugs are an example of this type of standardized connector. Grandfathered CPE may be connected to the network through nonstandard means (such as adapters or hard wiring) if a suitable jack is not available from the carrier.

The Building Industry Consulting Service International (BICSI) is an excellent source for information on premises wiring.

Computer–Telephony Integration

Computer–Telephony Integration (CTI) is the process of combining computer and telephone functionality, where the result is greater capability than either of the parts. The previous example of telecommuting for customer service representatives of the insurance company is an example of

CTI; the result in terms of customer satisfaction is greater, and the cost in terms of employee time is less, than for any combination of computing and telephony separately. This example also serves to illustrate that the greatest current area of CTI applications is in call centers.

Although call centers are a very important and growing aspect of CTI, the promise is much more. The concept of direct customer control of unbundled network functions (such as signalling, routing, and switching), accessed through CPE, is a compelling one. CPE hardware and software will, of necessity, be carefully designed to take maximum advantage of network capabilities while, at the same time, ensuring that no interference occurs with either network operation or with service to other customers. One can imagine extensions to FCC Part 68 to cover such contingencies.

There are many more aspects of CPE than this chapter can cover. One of the most exciting is integration of CPE with the Internet. Internet telephony is just one example; the availability of information on-line and worldwide may be more important in the long run. In its broad sense, computer-telephony integration is perhaps the ultimate embodiment of customer premises equipment: seamless integration of programmable terminal equipment and the advanced intelligent network to provide new levels of functionality and service to the customer.

34
Asymmetric Digital Subscriber Lines

34.1	Introduction	450
34.2	ADSL System Description	451
	ADSL Line Impairments • Compliance Testing	
34.3	Discrete Multitone: The ADSL Standard Line Code	454
	Basic DMT Description • DMT Implementation Basics • Performance Analysis • ADSL Data Rates, Framing, and Coding • ADSL Timing Methods	
34.4	ADSL Data Rates, Ranges, and Performance	471
	Range Projections and Test Results • ADSL Looking Forward: Very High-Speed Digital Subscriber Lines (VDSL)	
34.5	The Future	477
	Acknowledgment	

John M. Cioffi
Stanford University

34.1 Introduction

On March 10, 1876 at 4:10 p.m. in Boston, Massachusetts, Alexander Graham Bell spoke the words "Watson, Come here, I need you." Those words initiated the eventual worldwide deployment of the nearly one billion telephone lines that enable billions of people to speak to one another just as Bell spoke to his assistant Watson using the world's very first telephone. On March 10, 1993 at 4:10 p.m. in Miami, Florida, the American National Standard's Institute's T1E1.4 committee unanimously selected **discrete multitone** (**DMT**), a powerful and flexible transmission method, to enable the use of Bell's billion twisted-pair phone lines to carry a plethora of interactive multimedia digital signals. *Asymmetric digital subscriber lines* (ADSL) systems employ this DMT technology to implement digital TV broadcast and on-demand video, high-speed video-based Internet access, work-at-home digital file transfer and teleconferencing, interactive distance learning, home shopping; and information services, while still maintaining Bell's plain old telephone service (POTS) on the same simple existing phone line.

ADSL and its derivatives allow a feasible migration using Bell's billions of miles of twisted pair to a broadband network of the future that possibly replaces twisted pair with fiber as economically justified. At the time of writing, over 30 major phone companies worldwide are just beginning to evaluate and install ADSL services with the hope that ADSL will become a major new revenue-producing service.

Technically, the reliable transmission of the high-digital-data rates that support such desirable applications presents a challenging engineering design. Designs that achieve the ADSL standardized rate of 6 Mb/s over 2 mi of twisted pair under difficult environmental constraints press fundamental theoretical limits, but have been successful and have been and continue to be further integrated into low-cost integrated circuits. ADSL's recent offspring is very high-speed DSL (VDSL), looking at data rates from 25 to 50 (and later to 600) Mb/s over shorter distances of twisted pair with fiber used in part of the transmission path to a customer where appropriate. This chapter describes the standardized DMT technology as it is used in ADSL as well as aspects of ADSL that exploit the DMT technology. A complete system-level overview of ADSL is beyond the scope of this chapter, but a brief high-level overview begins Sec. 34.2 before progressing to a tutorial description of the phone-line environment and testing and qualification of ADSL.

Section 34.3 is the majority of this chapter and describes the basic DMT implementation, in particular the dynamic adaptive transmitter fundamentals. A method for simple analysis of the performance of DMT ADSL enables projection of data rates and twisted-pair loop lengths in Sec. 34.3. Section 34.3 also describes ADSL's multiplexing of digital bit streams with the DMT ADSL engine. Section 34.4 is a collection of publicly reported measurements by several independent groups of DMT modems, and also reviews the large improvements of this technique over other methods that led to DMTs standardization.

34.2 ADSL System Description

Figure 34.1 succinctly illustrates the system architecture for ADSL service. The ADSL modem at the telephone company sits in an access module in a telephone-company central office. The access module interfaces the modem to a variety of digital-service signals that ADSL can support. Internet connection allows high-speed support of video-based applications providing advertising/sales, information-on-demand, telecommuting access of computing systems, video games, education, and electronic mail. The telephone-company network can connect to similar packet-based services, but also readily accommodates movies/video on demand (VOD), video conferencing, and generally remote access of other networks and servers. Broadcast feeds to the access module provide real-time

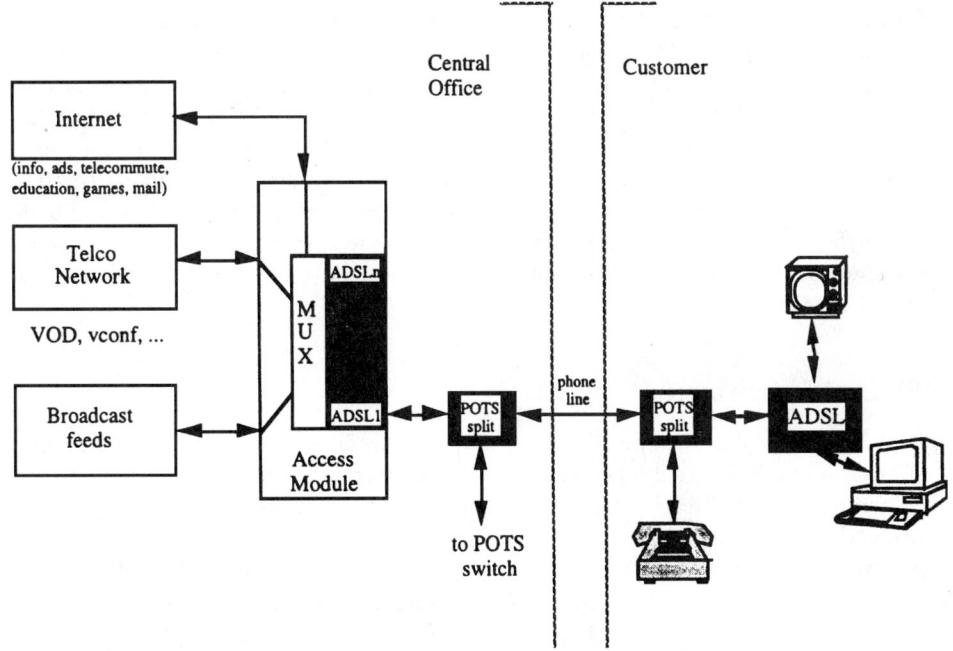

FIGURE 34.1 ADSL system applications architecture.

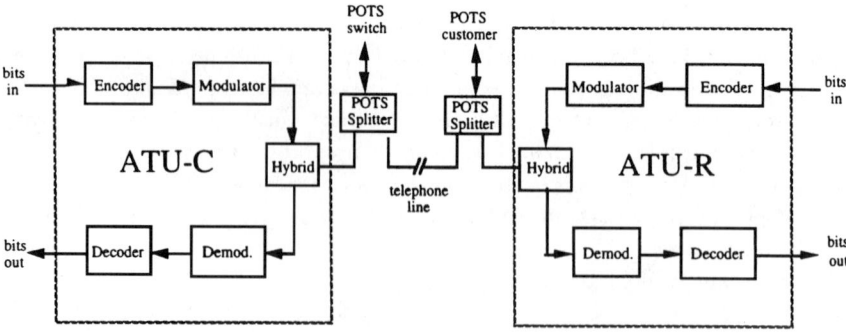

FIGURE 34.2 Basic ADSL modem function.

live video including television broadcasts, special closed-circuit programs, and wireless/satellite-based video provided by media companies/equipment.

The portion of ADSL of interest in this chapter is the shaded equipment consisting of the **ADSL transmission unit**-central office (**ATU-C**) in the access module, the line and splitter circuits, and the customer's **ADSL transmission unit-remote** (**ATU-R**), which may be connected to a television, computer, or other applications device. POTS is not altered and remains as is with ADSL. The POTS splitter circuits combine/separate the ADSL signals and POTS signal for common carriage on the length of phone line between central office and the remote site.

ADSL should be distinguished from a voice-band modem, which modulates digital data directly into the POTS band, preventing the use of the phone while the modem is on and also limiting the data rate to the bandwidth of a voice signal. Such data rates in voice-band modems are fundamentally limited (not including compression) to less than 30 kb/s. ADSL is 200 times faster and on the same twisted pair, and is fundamentally different in application and use.

The basic ADSL modem function is illustrated in Fig. 34.2. The input data bits are not suitable for transmission on the twisted pair and so therefore must be converted into equivalent analog waveforms that can be transmitted. This process is typically called *modulation*. These bits are first redundantly encoded for protection against errors. The reverse process of recovering the transmitted bits from the received analog waveforms is called *demodulation*, and a contraction for modulator–demodulator forms the name modem. With the usual noisy channel, as in the case of ADSL, the demodulation process is followed by a decoder, which decides which input bit(s) are most likely to have been transmitted. In ADSL, both binary coding (adding redundant parity bits to the transmitted set to facilitate distinction of noise-corrupted received bit sequences) and trellis coding (careful selection of transmitted waveform sequences to represent input bit sequences) are used. The demodulator/detector then becomes a Viterbi detector (for the trellis code) followed by a decoder for the binary code.

ADSL Line Impairments

Figure 34.3(a) simply illustrates the ADSL transmission environment. (An excellent description of the ADSL environment in detail appears in the reference paper by Werner [1991].)

The transmission line itself has significant attenuation distortion over the frequency band of approximately 0–1 MHz used by ADSL. The quantity known as insertion loss (ratio of line output voltage with transmission line present to voltage with null loop pent into a balanced load) illustrates the increasing attenuation with frequency, as plotted in Fig. 34.3(b). Well-designed ADSL systems can squeeze useful information through the channel at insertion losses as low as −85 dB to −90 dB. This means that the dynamic range of signals over the used bandwidth on the channels illustrated can be as much as 70 dB—this wide range is orders of magnitude more disparate with frequency

Asymmetric Digital Subscriber Lines

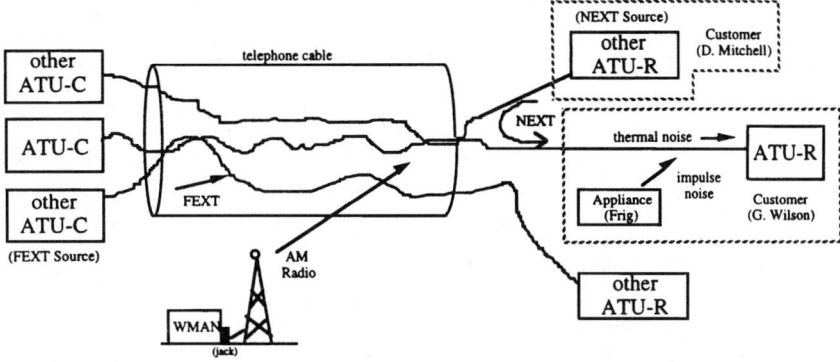

FIGURE 34.3(a) ADSL transmission environment.

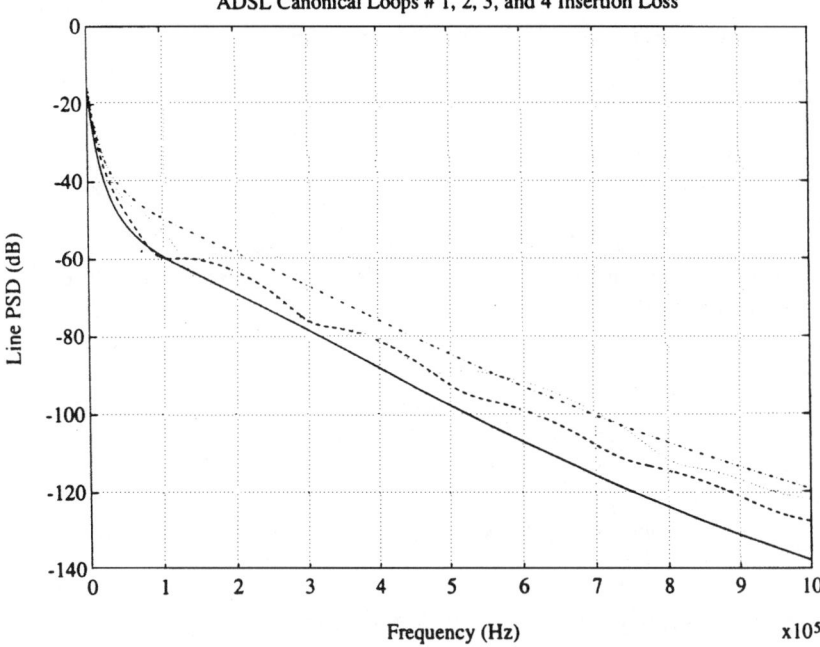

FIGURE 34.3(b) Insertion loss for some worst-case standardized phone lines.

than had ever been attempted in any transmission design prior to ADSL. Furthermore, short loops can show almost no attenuation, meaning the total receiver sensitivity can be nearly 100 dB. Analog components in the transmission path often must show over 100-dB linearity. Line attenuation distortion is the most challenging impairment of ADSL.

Near-end crosstalk (NEXT) corrupts ADSL waveforms, thus complicating the demodulation/decoding process. NEXT arises at the ATU-R when another ATU-R's upstream transmitted signals couple electromagnetically into the first ATU-R's signal, reverse direction, and thereby enter the first ATU-R's receiver. NEXT can also occur at the ATU-C end. NEXT coupling can be highly dependent on loop geometry and cabling practices, but a 1% worst-case approximation that is used almost universally is that the NEXT noise has a power spectrum given by

$$S_{\text{NEXT}}(f) = K_{\text{NEXT}} \cdot f^{1.5} \cdot S_x(f)$$

where f is the frequency variable, $K_{\text{NEXT}} = 10^{-13}$, and $S_x(f)$ is the power spectrum of the crosstalker's transmitted signal.[1] **Far-end crosstalk (FEXT)** is also shown in Fig. 34.3. FEXT arises when other ATU-C's transmitted signals electromagnetically couple into the line and then traverse the line along with the intended signals to the ATU-R. The noise power spectrum of FEXT is given by

$$S_{\text{FEXT}}(f) = K_{\text{FEXT}} \cdot f^2 \cdot d \cdot S_x(f)$$

where $K_{\text{FEXT}} = 10^{-19}$, d is the distance in feet of the loops, and $S_x(f)$ is now the ATU-C transmitted spectrum. At high frequencies, FEXT dominates NEXT (and vice versa). The data rate will not increase with transmit power when either NEXT or FEXT dominates transmission if the crosstalkers are also ADSL signals. AM radio signals can couple into phone lines, depending on the orientation of the line. AM radio signals are typically 10 kHz wide and can couple into phone lines leaving differential peak voltages of a few hundred millivolts [Hogue, 1995]. The AM signal can be a significant noise with ADSL transmission. Impulses are temporary disturbances, typically measuring between microvolts and tens of millivolts, that can last as long a millisecond; see the characterization in Lawrence [1993]. Impulses can be caused by everything from telephone company switch transients to refrigerators and light switches in the home.

Compliance Testing

ADSL modems are tested by transmitting ADSL signals through a twisted-pair phone line and then injecting additive noise, typically 1% worst-case NEXT, FEXT, and/or RF noises, into the output of that line. The bit error rate is measured (acceptable values vary from 10^{-9} to 10^{-3} depending on the application) while the noise is increased until the bit error rate exceeds the desired value. The amount by which the noise is increased in decibels with respect to the initial level is called the **ADSL performance margin** or just **margin**. Negative margin means the noise had to be decreased for acceptable performance. Telephone companies will typically not deploy products unless they can be assured that the measured margin is at least 6 dB. Margin is used for crosstalk, background noise, and AM-radio noise tests. For impulses, a more complicated measure is known as the error-free-second (EFS) percentage, which is difficult to describe. Instead, it suffices to say that few, if any, measured impulses should cause the ADSL system to make bit errors if the system is to be of use. Acceptable EFS percentages typically are in excess of 99%, roughly meaning that impulse-caused error bursts are rare. Section 34.4 briefly lists some compliance test results on various ADSL systems.

34.3 Discrete Multitone: The ADSL Standard Line Code

The American National Standards Institute selected discrete multitone over two more traditional line coding methods, quadrature amplitude modulation (QAM) and carrierless amplitude/phase modulation (CAP), for the ADSL standard for three primary reasons:

1. best transmission performance
2. lower implementation complexity than the QAM/CAP methods
3. bandwidth-on-demand flexibility.

The first subsection generically describes the basic joint transmitter/receiver adaptive optimization that distinguishes DMT from other line codes. The subjects of channel quality, bit loading, and channel identification are all introduced with the help of some examples. The second subsection illustrates the basic architecture of a DMT system, and describes how this powerful adaptive line

---

[1] When NEXT is caused by an existing T1 transmission system, K_{NEXT} should be reduced by 5.5 dB to account for the fact that the T1 transmitter can be in many different positions and is not generally collocated with the ADSL ATU-R (thus meaning more attenuation of T1 NEXT before it gets to the ATU-R).

code can be simply implemented. The third subsection analyzes the fundamental limits approached by DMT. That subsection also shows simple computational procedures that can be used to project data rates and/or performance margins. As the telecommunications network evolves towards asynchronous transfer mode (ATM) switching, multimedia applications of differing data rates and requirements need to be accommodated. The DMT line code of ADSL features a procedure called *rate negotiation* that is used to accommodate the evolving ATM network and the variety of multimedia applications that can be served, as the fourth subsection illustrates. Synchronization is necessary in any transmission system and the methods used for DMT-based ADSL are described in the last subsection of Sec. 34.3.

Basic DMT Description

DMT inherently transmits an optimized time-variable spectrum. This spectrum is adjusted according to the desired data rate and the transmission characteristics (transfer function and noise spectrum) on each and every channel. The basic concept of DMT is illustrated by the three examples in Fig. 34.4. In each case, DMT initializes by the ATU-C transmitting 256 4-kHz-wide tones downstream. The ATU-R measures the quality of each of these tones and then decides whether a tone has sufficient quality to be used for further transmission and, if so, how much data this tone should carry relative to the other tones that are used. This procedure maximizes transmission performance (minimizes the probability of bit errors in transmission).

For instance, the first channel in Fig. 34.4 illustrates a segment of single-gauge twisted-pair phone line. Transformer coupling eliminates low frequencies. Higher frequencies are increasingly attenuated. The degree of high-frequency attenuation depends on the length of the phone line: longer lines have greater attenuation. (For a tutorial on computing the transmission characteristics of a phone line, see Werner [1991].) An equal number of bits per tone are initially used to measure

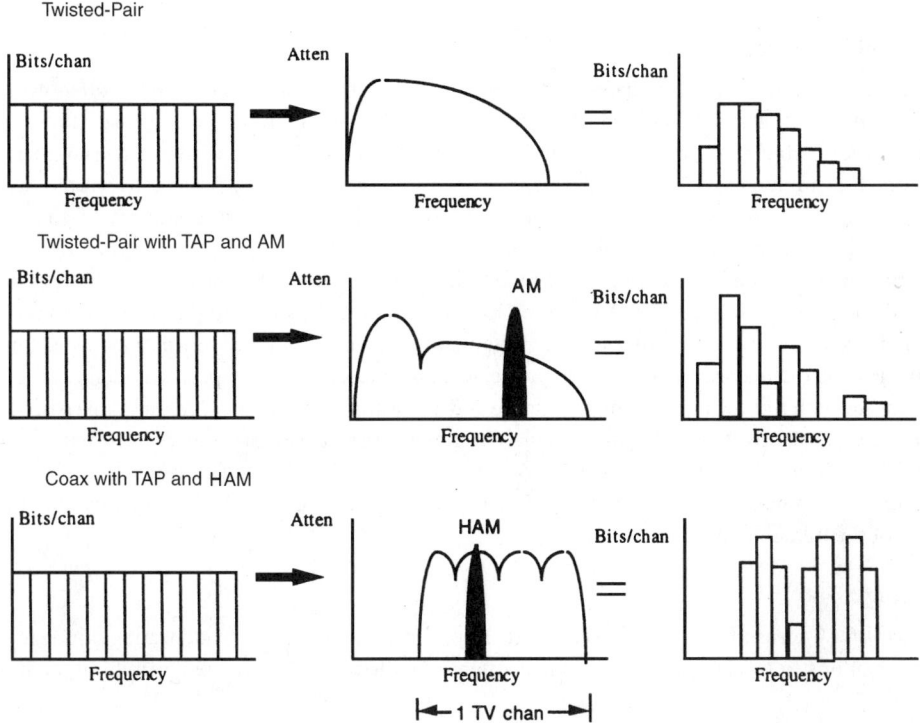

FIGURE 34.4 DMT examples.

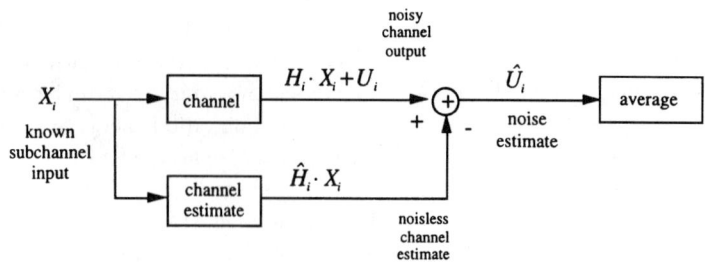

FIGURE 34.5 Noise estimation.

the transmission line. After processing the received signals, the ATU-R tells the ATU-C (by secure low-speed reverse transmission over the same phone line) to use the optimized bit distribution shown on the right in the first example. Note that higher frequencies transmit fewer bits because they are more attenuated. Essentially, the bit distribution follows the transmission characteristic.

The second example is more realistic and includes the notch in spectrum that is illustrative of bridge taps[2] and also the interference of an AM radio station. NEXT and FEXT noise also increase with frequency, emphasizing that lower frequencies see less distortion and so should carry more bits. Note that the resultant bit spectrum still reduces with frequency, but also note that fewer bits are used on tones in the vicinity of the notch caused by the bridge tap. No bits occupy tones that are corrupted by the AM radio interference. Given 400 million phone lines worldwide, there can correspondingly be a large variation in characteristics. No compromise design can handle this variation (as we will see in Sec. 34.4 where the independent phone company test laboratory reports compare the performance of DMT vs QAM/CAP methods on real phone lines).

A third example in Fig. 34.4 illustrates that DMT can be a good match to other transmission channels, such as coaxial cable-TV transmission networks which illustrate rippling of the frequency characteristic (bouncing of signals off the various neighborhood TVs) and HAM radio interference.

Channel Identification

The process of measuring the quality of the various tones as they are transmitted is often called *channel identification*, and consists of two basic measurements: (1) the channel transfer function and (2) the channel noise power spectral density. It is possible to combine these measurements into one measurement of essentially frequency-dependent signal-to-noise ratio SNR, but the two-part measurement is easier to describe for purposes of this introductory treatment of channel identification. Channel identification conceptually averages the channel response on each tone to a known transmitted sequence on that tone. The averaging mitigates the effect of the channel noise disturbances (NEXT, FEXT, AM, etc.). One can compute the gain and phase translation of a transmitted tone by averaging the quotient of the channel response and the known transmitted tone input. Having established the channel response, Fig. 34.5 illustrates noise spectrum identification. Continuing with a known transmitted sequence and a, presumably accurate, channel gain estimate, the receiver can subtract any transmitted signal component and then average the resulting remaining noise. Only the power spectrum (not phase) of the noise is necessary. Having measured both channel gain and noise power, the SNR can be computed for each tone, thus establishing a quality measure for all of the transmitted tones.

Loading

The use of the measured (or computed) SNRs to optimize performance by allocating a judicious number of bits to be carried on each tone is often called **loading**. There are a variety of loading

---

[2]A bridge tap is an open-circuited branched telephone line off the main used line that may correspond to another extension line that is not being used. Bridge taps are common in telephone company wiring practices.

algorithms (a good summary appears in Chow [1993]), most of which are patented, that select transmit energy levels and the number of bits to be transmitted on each tone. Each may have its relative merits and drawbacks. The optimum solution for a system operating with infinite complexity coding was determined by Shannon [1948] and is described in a later subsection. A method to analyze any energy/bit distribution is also given in a later subsection. The best methods will maximize margin at any given data rate.

Example

As an example, consider the magnitude of the transfer function of the loop channel that is described by the pole-zero polynomial

$$H(D) = \sum_k h_k D^k = \frac{0.1(1 - D^2)}{(1 - 0.9D)(1 - 0.6D)}$$

at a sampling rate of $1/T = 2.208$ MHz. Four DMT tones are used in this example. While 4 tones in DMT is not sufficient for the independent subchannels assumption to hold in this example, it simplifies this example for illustrative purposes. Once the example is understood, the procedure illustrated easily extends to larger numbers of subchannels with the assistance of a computer. The four center frequencies for this example are

$$f_1 = 276 \text{ kHz} \qquad f_2 = 552 \text{ kHz}$$
$$f_3 = 828 \text{ kHz} \qquad f_4 = 1.104 \text{ MHz}$$

One of the channels, $i = 4$, cannot be used because $H(0) = H(1.104 \text{ MHz}) = 0$. The other three subchannels have squared gains

$$|H_1|^2 = 0.073 \qquad |H_2|^2 = 0.016 \qquad |H_3|^2 = 0.0029$$

The three remaining subchannels can be independently modulated and demodulated. This example will be analyzed further subsequently. In this example, the portion of the channel characteristic with the greatest transfer magnitude (from 0 to 138 kHz) is ignored by this simple DMT system. However, by increasing the number of subchannels, an increasing fraction of the total *good* bandwidth (away from DC and 1.104 MHz) can be used by DMT. With the TEQ of the next subsection, 256 subchannels is more than sufficient on this channel.

For ADSL, the upstream transmission path from ATU-R to ATU-C uses only the lower 32 4-kHz-wide tones (instead of 256 downstream), thus occupying only 1/8 the bandwidth and typically transmitting about 1/8 the data rate (often with lower upstream transmit power than downstream because the lower tones see less attenuation). The ATU-R and ATU-C can share bandwidth through the use of an echo canceler as described in the next subsection.

It is important to distinguish DMT from coded orthogonal frequency division multiplexing (COFDM). COFDM methods have been standardized for broadcast digital radio and television transmission in Europe. In the broadcast application, the bit distribution cannot be adaptively optimized because there is no feedback path to the transmitter from the receiver. Instead, a fixed bandwidth and bit allocation are used. Various codes and interleaving techniques can be used to make such COFDM transmission systems very robust to time-varying multipath distortion, but the adaptive bit distribution of DMT, unfortunately, cannot be used. Thus, fundamentally, a DMT system, when it can be used, will outperform COFDM.

DMT Implementation Basics

The basic DMT implementation of the ATU-C transmitter for ADSL is given in Fig. 34.6. The incoming bits are coded in the ADSL standard through the use of forward-error-correcting (FEC)

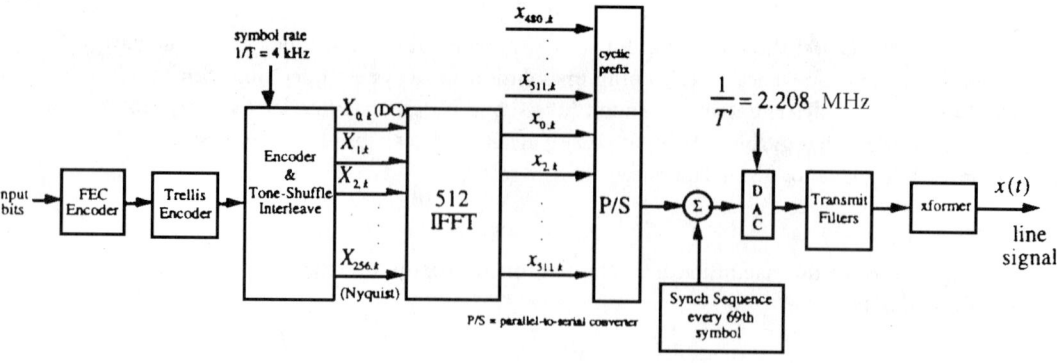

FIGURE 34.6 ATU-C transmitter block diagram.

codes, which appends (about 5% increase) extra bits to the original bit stream, and optionally the use of trellis coding. The resultant bits parse into groups of b_i, $i = 0, \ldots, 255$ b for each tone at the output of the encoder, which optionally uses a four-dimensional 16-state trellis code. The result of the coding is a two-dimensional quantity, X_i, that consists of a real part and imaginary part. There are 2^{b_i} such complex values for the ith tone, one to represent each of the 2^{b_i} input bit patterns for that tone. The process of encoding repeats for each of these 256 subchannels 4000 times per second. The amount of energy applied to each tone (effectively scaling the amplitude of X_i) is also determined by the loading process of the next subsection. An IFFT of size 512 (complex to real) converts the complex samples ($X_i = X^*_{512-i}$, $i = 0, \ldots, 255$) into a real sequence of 512 samples x_k. The sequence represents the sum of modulated sinusoids. The formula for the IFFT is

$$x_k = \frac{1}{\sqrt{512}} \sum_{n=0}^{511} X_n W^{kn}$$

where $W = \exp(j \frac{\pi}{256})$. The last 32 samples of each transmit symbol block repeat in a prefix that begins each block (or symbol) of $512 + 32 = 544$ samples.

Each such prefixed symbol is transmitted 4000 times per second. For reliability, a dummy known synchronization symbol of 544 samples is inserted every 69th symbol, so that the aggregate sampling rate of the ATU-C transmitter is thus $(544)(4000)(69/68) = 2.208$ MHz. The upstream path is identical, except that the 512 IFFT is replaced by a 64 IFFT, with prepending of $32/8 = 4$ samples, for an upstream sampling rate of 276 kHz. The upstream and downstream clocks can be, and often are, synchronized in what is called *loop-timing*.

Figure 34.7 illustrates the encoding of b bits, with (v_{b-1}, $v_{b-2}, \ldots, 0$) interpreted as an integer label for the constellation.

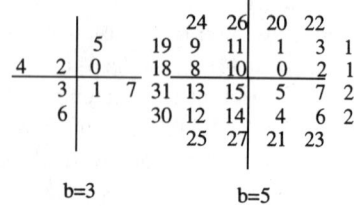

FIGURE 34.7 ADSL constellation examples.

The algorithm for generating the next size constellation for both even and odd b replaces any label n in the previous even (odd) constellation by the group of four labels

$$4n + 1 \qquad 4n + 3$$
$$4n \qquad 4n + 2$$

Asymmetric Digital Subscriber Lines

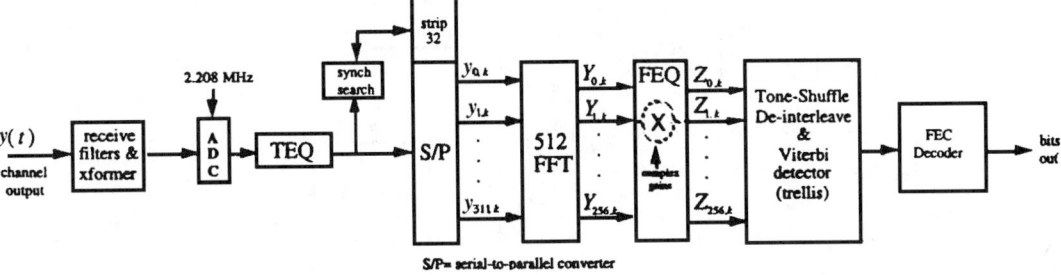

FIGURE 34.8 ATU-R receiver block diagram.

The bits undergo the encoding of forward error correction, and possibly trellis coding, prior to translation into constellation points. The horizontal axis in Fig. 34.7 corresponds to the real part of X_i, whereas the vertical component corresponds to the imaginary part of X_i. The subchannels X_i are not encoded in the order $i = 0, \ldots, 255$, but rather in a channel-dependent manner that first encodes those subchannels with indices corresponding to the smallest b_i and then progressing to the largest b_i in order of the sizes of the b_i for that channel. This is called *tone-shuffle interleaving* in Fig. 34.6. The channels with smallest b_i are encoded into the so-called fast path (discussed subsequently) with less protection against impulse noise, but lower latency. Impulse noise is typically concentrated in the spectrum on a transmission line below 300 kHz where the largest b_i usually occur in ADSL. Thus, the stronger longer-latency error correction is applied where needed most, to those subchannels that are most likely to contract impulse noise.

The receiver block diagram appears in Fig. 34.8. Ignoring the box labeled TEQ momentarily, the ADC output samples are processed by fast Fourier transform (FFT) after deletion of the samples corresponding to the cyclic prefix. The FFT undoes the IFFT of the ATU-C transmitter. Each subchannel output of the FFT will be scaled in magnitude and offset in phase by the values of the channel transfer function at the corresponding center frequency for the tone. A little more mathematical description appears in the next subsection. The frequency-domain equalizer (FEQ) simplifies decoder implementation and has no effect on performance. The FEQ adaptively scales each subchannel by the inverse of the channel gain and phase so that a common decision boundary can be used in decoding the constellations of Fig. 34.7. The FEQ is a one-tap complex adaptive filter for each subchannel in Fig. 34.8, typically using a simple gradient algorithm to update the complex coefficient. The resultant FEQ outputs are then decoded, possibly by Viterbi decoding for any trellis coding, and then by the forward-error correction decoder.

The FFT subchannel outputs do not interfere with one another if the cyclic prefix is longer in length than the impulse response of the channel, which is equivalent to saying that the channel input looks periodic so that circular convolution results apply (that is the FFT of the channel output is the product of the FFT of the channel and the FFT of the input). In practice, the twisted-pair channel impulse response is infinite in length so that some approximation is necessary in deciding the length of the cyclic prefix. In ADSL, the patented TEQ technique (see Chow [1993]) is a filter that adaptively alters the channel so that the impulse response is reduced to the length of the cyclic prefix or less in a minimum-mean-square-distortion sense. Although this filter is not necessary and not part of the standard, it can lead to enormous gains in performance and reduction in overall system complexity for a given performance level.

The 4-wire-to-2-wire interfaces of Fig. 34.2 introduce another problem for ADSL. Well-designed hybrid circuits prevent the leakage of transmitted ADSL signals into the received ADSL signals at the same end of the transmission line. Well-designed hybrids require knowledge of the transmission line, which, of course, varies widely as discussed previously. Thus, some leakage or echo inevitably occurs. An echo canceler is nominally a device in parallel with the hybrid, which accepts transmitted ADSL signals and reconstructs the echo so that it may be subtracted from the received ADSL signals.

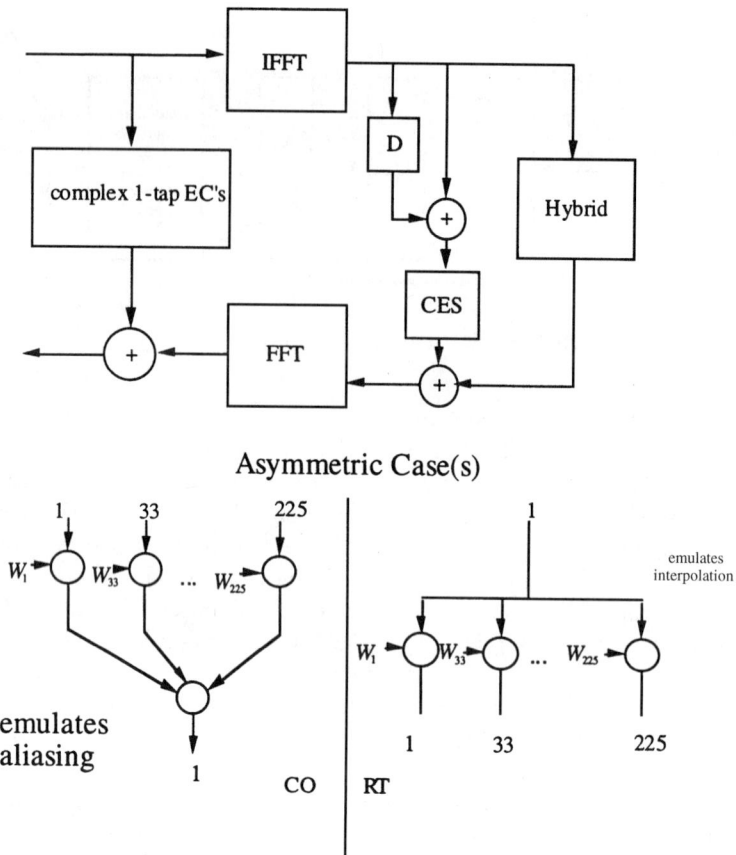

FIGURE 34.9 ADSL echo cancellation.

At the high sampling rates of ADSL, a nominal echo canceler (for instance, see Lee and Messerschmitt [1988]) implementation requires too much computation to be practical. A less costly approach (see Ho and Cioffi [in press]), shown in Fig. 34.9, uses the transmitted subchannel symbol values X_i to reconstruct the echo in each subchannel. For the echo in subchannel i to depend only on X_i (and not any other X_j), the transmitted signal would need to be periodic. The cyclic prefix is not long enough for the periodic convolution trick used in the transmission path to work for the echo cancellation path. Thus, a cyclic echo synthesizer (CES) uses the estimated echo response to subtract echo components from the last symbol and add in components in the corresponding previous-symbol time positions that make the echo appear periodic. This operation is far less computation intensive than a full echo canceler. Simple tone-by-tone echo cancellation is then very effective. Figure 34.9 also shows that the echo canceler easily decimates from the high-speed transmitter to the 1/8th speed receiver in the ATU-C by simply adding the output of every 8th tone in echo cancellation. A dual interpolation operation interpolates from the low-speed transmitter to the high-speed receiver in the ATU-R.

The central limit theorem of statistics roughly applies to the transmitted DMT signal in that the sum of a large number of randomly modulated sinusoids should tend toward a Gaussian distribution. Indeed, the DMT signal that enters the transmission line looks like *noise* on an oscilloscope in the time domain. This can lead to a very high peak-to-average ratio. There are a number of proprietary solutions to this problem that take advantage of the discrete frequency-domain nature of the DMT signal. One patent-pending technique (see Bingham [1990]) essentially looks ahead at the transmit signal, anticipates a clip, and alters the symbol to avert the clip. The clip-aversion mechanism is

Asymmetric Digital Subscriber Lines

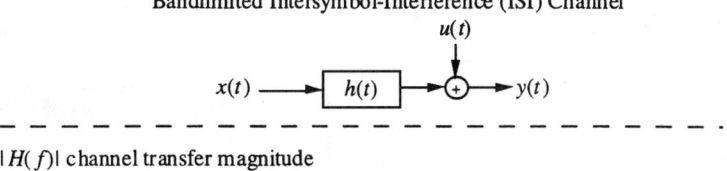
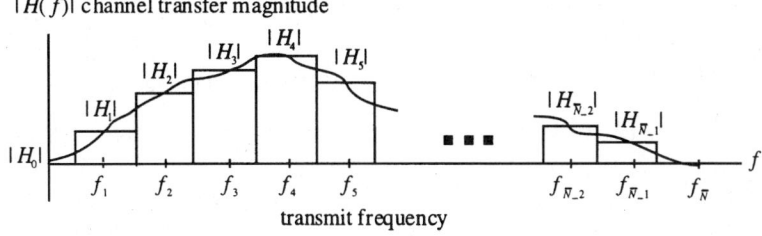

FIGURE 34.10 DMT decomposition of a transmission channel.

encoded into a low-rate signalling channel that is the first to be decoded by the receiver, which then reverses the alteration. The small overhead penalty in data rate can save an expensive bit of precision that might otherwise be necessary. Typically, 12-b conversion devices are sufficient for ADSL.

Performance Analysis

The DMT approach to a transmission line can conceptually be viewed as in Fig. 34.10. The transmission characteristic is viewed as a set of independent frequency indexed subchannels, with center frequencies, f_i, and complex channel gains $H_i = |H_i|e^j < H_i$. Each subchannel is approximately *flat* in that no transmission distortion is evident other than the multiplicative H_i and the addition of noise U_i. Clearly, as the number of subchannels N increases, this approximation becomes very accurate. ADSL uses 256 subchannels, which with a good TEQ design leads to excellent accuracy of this approximation on a twisted pair. The equivalent set of channels appears in Fig. 34.11. Each of the subchannels shown in Fig. 34.10 has an SNR_i, which is useful in the performance analysis.

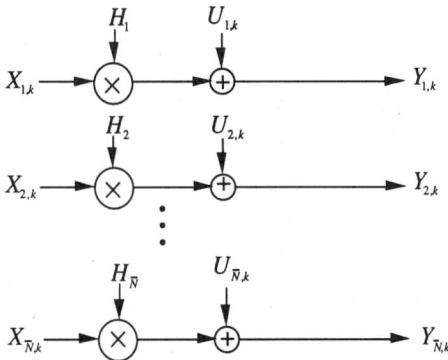

FIGURE 34.11 Equivalent set of parallel channels.

The maximum data rate, or capacity C_i, that can be transmitted over a single flat transmission channel was derived by Shannon in 1948 (see Gallager text), and is readily computed according to the formula

$$C_i = \log_2(1 + SNR_i)$$

Denoting the average squared value of the complex input X_i, as E_i (sometimes called the energy per complex sample) and the average square noise sample as σ_i^2, the FFT-output signal-to-noise ratio for any flat subchannel is

$$SNR_i = \frac{E_i |H_i|^2}{\sigma_i^2}$$

A well-known result from information theory is that the capacity of a set of independent channels is the sum of the capacities of the individual subchannels. The DMT system should allocate an amount of energy to each of the subchannels such that the overall capacity $C = \sum_i C_i$ is maximized subject to a total energy constraint $E = \sum_i E_i$. In 1948, Shannon solved this problem with the so-called *water-filling* solution that can be easily visualized as in Fig. 34.12 for a large, essentially infinite number of sub-

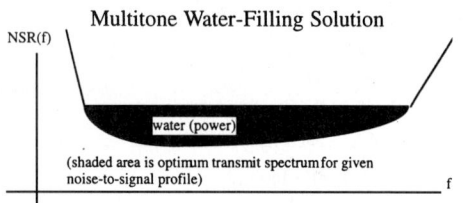

FIGURE 34.12 Illustration of water filling.

channels. The energy is viewed as water that can be poured into a bowl (curve) that represents essentially the inverse SNR of the transmission medium [ratio of $\sigma^2(f)/|H(f)|^2$] until no more water (energy) is left. Clearly, the subchannels with the highest transmission gain vs noise get the most energy and some subchannels may not get any energy. The optimum energy is just the amount occurring at each frequency. The case of a finite number of channels is the obvious extension with the discrete index i replacing the continuous frequency f. The number of bits optimally carried by each subchannel is then $C_i = \log_2(1 + SNR_i)$. Although conceptually simple, the water-fill distribution can be difficult to compute in actual use and only applies to the maximum theoretical data rate. Loading algorithms compute a good bit distribution and energy allocation for a DMT system at any data rate below the capacity.

The SNR Gap

The *SNR gap* is a useful concept introduced by Forney and Eyuboglu [1991] that characterizes how close a given code on a subchannel is to the theoretical maximum for that particular subchannel and its given transmit energy. An uncoded constellation that transmits b_i bits per successive value of X_i on subchannel i is characterized by the quantity d_i^2, the distance between any two closest points. Example constellations appear in Fig. 34.7. The energy for a square constellation is

$$E_i = \frac{2^{b_i} - 1}{6} d_i^2$$

This energy computation is exact when b_i is an even integer. However, it can be used for any value of b_i accurately when coding is used. Any small deviations are incorporated into a second coding parameter called the coding gain, which is to be subsequently described. Coding improves the distance d_i^2 over sequences of transmissions with a resultant improvement in d_i^2 with respect to uncoded transmission on the square constellation (or something like it for odd b). The improvement is called the coding gain γ_c. The probability of error of this code on a channel with gain H_i and $d_{\min,i}^2 = |H_i|^2 \cdot d_i^2 \cdot \gamma_c$ is

$$P_{e,i} \approx 4Q\left(\frac{d_{\min,i}}{2\sigma_i}\right)$$

The SNR gap Γ_i is proportional to the square of the Q-function[3] argument, namely

$$3\Gamma_i = \frac{d_{\min,i}^2}{4\sigma_i^2}$$

---

[3]The Q-function in digital communications is defined by

$$Q(x) = \int_x^\infty \frac{1}{\sqrt{2\pi}} \exp\left(-\frac{x^2}{2}\right) dx$$

Asymmetric Digital Subscriber Lines

With algebraic manipulation, one determines that the number of bits that are being transmitted with this code is

$$b_i = \log_2\left(1 + \frac{SNR_i}{\Gamma_i}\right) \quad \text{or} \quad \Gamma_i = \frac{SNR_i}{2^{b_i} - 1}$$

where one notes the reduction with respect to theoretical capacity by the factor of Γ_i in SNR, whence the name of SNR gap. For uncoded transmission and a probability of error of 10^{-7}, one finds that the SNR gap is 9.8 dB, meaning that 9.8 dB is lost with respect to an optimally coded system. The gap is reduced by the amount of coding gain. Coding gains with measured gains as high as 7.3 dB can be realized by DMT systems (see Zogakis [1995]), thus reducing the gap to about 2.5 dB. Although better codes exist, nearly infinite complexity is required for realization, so it is often stated that "well-designed DMT systems can come within 3 dB of capacity." What is meant is that codes augment the DMT design so that the gap is 3 dB.

ADSL compliance testing measures the margin that was described in Sec. 34.2. Requiring a margin can be construed as artificially increasing the gap so that the gap becomes, at error rate 10^{-7},

$$\Gamma = 9.8 - \gamma_c + \gamma_m \text{ dB}$$

or perhaps more usefully, the margin can be computed according to

$$\gamma_m = 10\log_{10}\left(\frac{SNR}{2^b - 1}\right) + \gamma_c - 9.8 \text{ dB}$$

Designing for the Weak Link

The probability of error for a multitone system is the average of the probabilities of error for each of the subchannels. The subchannels with largest probabilities of error dominate this average error probability. Well-designed multitone systems usually impose an equal probability of error on all subchannels so that no one subchannel is any better than the others.[4]

ADSL selects the probability of subsymbol error on all subchannels to be $P_e = 10^{-7}$. Then, the analysis of the preceding subsection directly applies to each subchannel. All subchannels have a constant SNR gap, $\Gamma = T_i$, $i = 0, \ldots, 255$. Thus,

$$b_i = \log_2\left(1 + \frac{SNR_i}{\Gamma}\right)$$

is the number of bits per symbol carried by subchannel i. This description imposes the additional restriction that $E_i = E$, that is, the energy is constant on those subchannels that are used and zero on subchannels not used. For first-order analyses, the loss between an on/off energy distribution, and one that is the water-fill is often small, as long as the best set of subchannels is used to carry data in both cases.

Computing Rate or Margin

The total number of bits that DMT transports in one symbol is the sum of the number of bits on each of the subchannels, and so

$$b = \sum_i b_i = \sum_i \log_2\left(1 + \frac{SNR_i}{\Gamma_i}\right)$$

---

[4] DMT ADSL also provides a mechanism that can assign more critical information to those subchannels for which the design ensures a lower probability of error than on other subchannels. The next subsection elaborates a little more on this mechanism.

Then the data rate is $R = b/T$, where $T = (512 + 32)/2.208 \times 10^6$ for ADSL. An alternative relationship is

$$b = \log_2 \left[\prod_i \left(1 + \frac{SNR_i}{\Gamma_i}\right) \right]$$

By defining an average SNR where $\overline{N}$ is the number of subchannels with nonzero transmit energy

$$1 + \frac{\overline{SNR}}{\Gamma} = \left[\prod_i \left(1 + \frac{SNR_i}{\Gamma_i}\right) \right]^{1/\overline{N}}$$

or

$$\overline{SNR} = \Gamma \left\{ \left[\prod_i \left(1 + \frac{SNR_i}{\Gamma_i}\right) \right]^{1/\overline{N}} - 1 \right\}$$

one obtains the convenient expression

$$b = \overline{N} \cdot \log_2 \left[\left(1 + \frac{\overline{SNR}}{\Gamma}\right) \right]$$

The form of the preceding expression shows direct computation of a margin for a multitone system with fixed data rate and probability of error. The $1+$ and -1 terms are often negligible so that the average SNR becomes the geometric average

$$\overline{SNR} = \left[\prod_{i=1}^{\overline{N}} SNR_i \right]^{1/\overline{N}}$$

which does not involve the gap (which is unknown if we are trying to compute the margin). One must take care in dropping the $1+$ and -1 terms to alter $\overline{N}$ to the number of used subchannels (that is, do not count channels with zero input energy) in computing the margin. Then, the margin becomes

$$\gamma_m = 10 \log_{10} \left(\frac{\overline{SNR}}{2^{b/\overline{N}}} \right) + \gamma_c - 9.8 \text{ dB}$$

where $\overline{N}$ is again the number of used subchannels.

Example

Returning to the example earlier in this section, let us use $E = 1$ for $i = 1, 2, 3$. Further, let us assume a crosstalking noise that increases in power with frequency as $f^{1.5}$ appears on this channel. The corresponding subchannel noise variances are $\sigma_1^2 = 5 \times 10^{-6}$, $\sigma_2^2 = 1.41 \times 10^{-5}$, and $\sigma_3^2 = 2.6 \times 10^{-5}$. The subchannel signal to noise ratios are

$$SNR_1 = \frac{1 \cdot (.073)}{10^{-5}} = 7300 \quad (38.6 \text{ dB})$$

$$SNR_2 = \frac{1 \cdot (.016)}{2.82 \times 10^{-5}} = 567 \quad (27.5 \text{ dB})$$

$$SNR_3 = \frac{1 \cdot (.0029)}{5.2 \times 10^{-5}} = 56 \quad (17.4 \text{ dB})$$

Asymmetric Digital Subscriber Lines

With no margin and no code, the achievable data rate is

$$R = b/T = 128 \text{ kHz} \left[\log_2 \left(1 + \frac{7300}{10^{.98}} \right) + \log_2 \left(1 + \frac{567}{10^{.98}} \right) + \log_2 \left(1 + \frac{56}{10^{.98}} \right) \right]$$
$$= 2.339 \text{ Mb/s}$$

With a powerful trellis code of gain 5 dB, we would like to know the margin at 3.864 Mb/s. First, we compute the geometric average SNR for the three channels as

$$\overline{SNR} = [(7300) \cdot (567) \cdot (56)]^{1/3} = 614$$

(The number of used subchannels is $\overline{N} = 3$.) The number of bits per symbol required to achieve 3.864 Mb/s is 14. Then, the margin is

$$\gamma_m = 10 \log_{10} \left(\frac{614}{2^{14/3} - 1} \right) + 5.0 - 9.8 \text{ dB} = 9.2 \text{ dB}$$

With this small of an FFT size, the subchannels are not really independent (and we have ignored the cyclic prefix), so only the method of analysis is illustrated with this example. If N is increased to 512, this analysis accurately projects the achievable data rate or margin for DMT on subscriber loops. The TEQ simply ensures that the analysis for block length $N = 512$ is accurate by forcing the 256 derived subchannels to be independent, and thus forcing the H_i slightly from $H(f = i/T)$.

Review of Performance Calculation

The procedure to analyze the maximum data rate for the multitone system can be summarized in the following five steps:

1. From the power budget, compute a preliminary subsymbol energy allocation according to

$$E = E_i = \frac{PT}{N}.$$

2. Compute the subchannel SNRs according to

$$SNR_i = \frac{\mathcal{E}|H_i|^2}{\sigma_i^2}$$

3. Compute the number of bits that can be transmitted on each subchannel with a given margin and given trellis code (thus determining $\Gamma = 9.8 + \gamma_m - \gamma_c$ dB) as

$$b_i = \log_2 \left(1 + \frac{SNR}{\Gamma} \right)_i$$

4. For those subchannels with $b_i < 0.5$, reset $E_i = 0$ and reallocate their energy to the other subchannels equally. Then, recompute b_i.
5. Compute b by summing the b_i, and then compute the maximum data rate $R = b/T$.

A margin can be computed using any number of used subchannels. For a fixed data rate considerably below theoretical optimums, the number of used subchannels often decreases with respect to the bandwidth used for the maximum data rate. The geometric average $\overline{SNR}$ can be computed recursively by ordering the $|H_i|^2/\sigma_i^2$ and incorporating the largest SNR_i first, then the next largest, etc. Each of these $\overline{SNR_i}$ can in turn be used to compute a margin. The bandwidth with the best margin is then used for this target (lower than maximum) data rate.

ADSL Data Rates, Framing, and Coding

ADSL exploits the flexible bandwidth-on-demand nature of DMT to permit simultaneous transmission of the signals corresponding to several asynchronous multimedia applications. The first subsection illustrates the transmitter structure and the ADSL options for multiple applications. The transmission path can be made more reliable by external coding. ADSL has multilayer coding possible, with both trellis coding and forward error correction. The next subsection lists the information about these codes. The last subsection describes ADSL data organization or frame structure.

Multiple Applications

A transmitter block diagram for an ATU-C is shown in Fig. 34.13. Digital signals enter the ATU-C, logically, on one or more of seven possible input paths: AS0, AS1, AS2, AS3, LS0, LS1, LS2. The AS stands for *asymmetric signal* and corresponds to a downstream simplex data path through the ATU-C and ATU-R to the customer. Up to four such ASX channels, possibly asynchronous with one another, may simultaneously be transmitted. The data rates on each of these ASX channels are determined through the rate negotiation described subsequently. The maximum ASX data rate in practice for ADSL (as in the standard) is about 8–9 Mb/s, meaning that the sum of the data rates on the ASX channels must be below this number. The LSX channels are similar, except that they

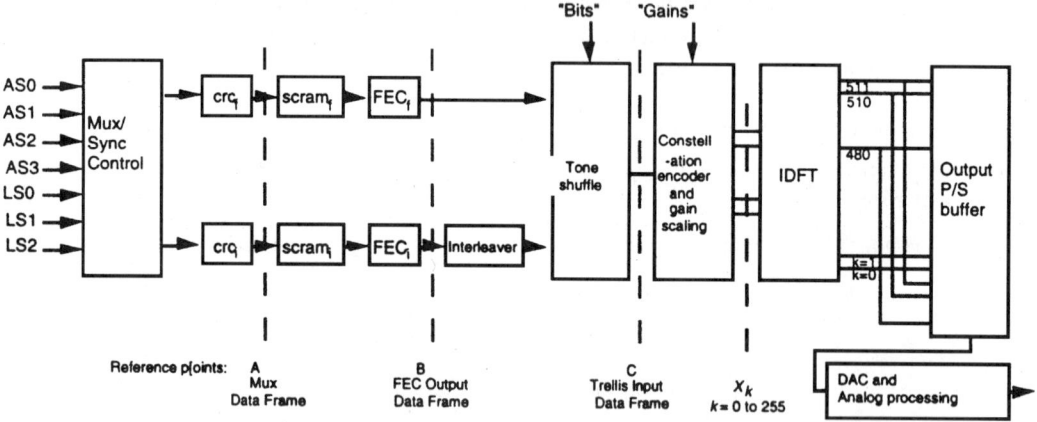

FIGURE 34.13(a) Detailed ATU-C data generation block diagram.

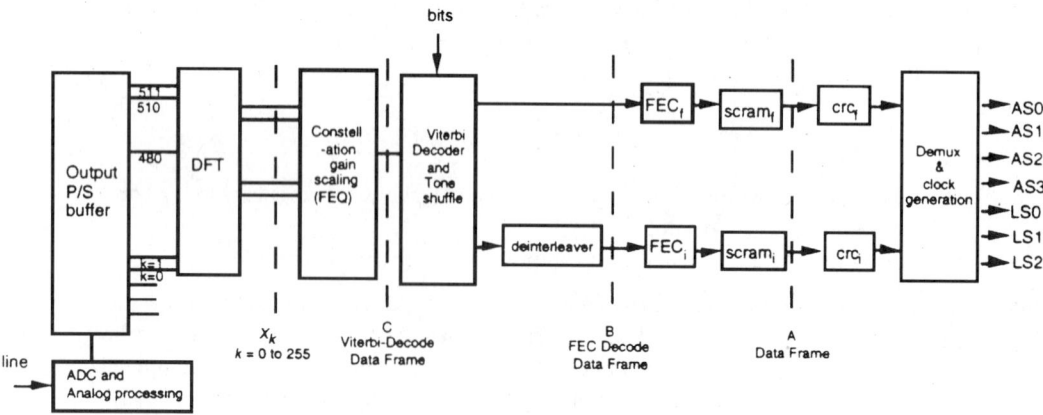

FIGURE 34.13(b) Detailed ATU-R data generation block diagram.

Asymmetric Digital Subscriber Lines

are bidirectional and typically lower speed, the sum data rate practically being limited to about 800 kb/s. Different ASX and LSX signals can correspond to different applications. The LSX channels can be used for videophone, digital telephony, data transfer, and signalling (LS0). LS0 usually carries signalling information for all applications.[5] The ASX channels typically are used for high-quality digital video, but can also be used for high-speed data transfer.

One of ADSL's most novel features, rate negotiation, is a facility for accommodating various applications and application sets within a single standard-compliant modem. During initialization of the ADSL link, the ATU-C transmits a set of data rates that it would prefer to use in a standardized manner. The ATU-R can implement these data rates, or signal that the requested data rate cannot be implemented. The number of bytes per symbol and the selected coding options are exchanged in rate negotiation for each of ASX and LSX channels. Rate negotiation can exploit the capability in some proprietary loading algorithms that permits software optioning of the number of bits to be carried per symbol. One way that a rate request might be denied by the ATU-R is if the receiver performance is determined during loading to be of insufficient quality to accommodate the requested data rate. Such a feature has many operational and diagnostic capabilities.

Another feature of the separate data paths through the ADSL link for different applications is the fast/interleave concept. Data signals requiring low latency transfer through the ADSL link can be assigned (again through the rate negotiation facility) to a fast data path with minimum latency. The fast path is the upper path crossing the A and B vertical dashed lines in Fig. 34.13(a). Data signals (like video) that can tolerate longer delay are usually assigned to an interleaved path that has much better immunity to impulse noise. Data bits in the interleave path are time interleaved over a long time interval to disperse any errors caused by temporary impulsive disturbances. The interleave path is the lower path illustrated in Fig. 34.13(a). A diagram of the corresponding ATU-R appears in Fig. 34.13(b). Fast path data in the ATU-R need not be deinterleaved and is decoded by the FEC decoder prior to descrambling and redundancy checking. Interleave path bits in the ATU-R are deinterleaved prior to decoding and descrambling. A demultiplexer in the ATU-R then separates the ASX and LSX signals for supply to applications modules in the home. The upstream fast and interleave paths are identical, except in the opposite direction from ATU-R to ATU-C, and only accept and process the upstream direction of the LSX signals.

Coding

The fast and interleave paths through the ADSL connection incur (up to) three successive stages of encoding (and the corresponding decoding). The outer layer of coding is simple scrambling, shown as scram$_f$ and scram$_i$ in Fig. 34.13(a). If data bits are denoted d_k coming into the scrambler, then the scrambled output bits $\overline{d}_k$ are computed in binary arithmetic as

$$\overline{d}_k = d_k \oplus \overline{d}_{k-18} \oplus \overline{d}_{k-23}$$

Descrambling in the ATU-R is then accomplished according to the inverse relation

$$d_k = \overline{d}_k \oplus \overline{d}_{k-18} \oplus \overline{d}_{k-23}$$

The next step in the process, independently applied to both the fast path signals and the interleave path signals as shown in Fig. 34.13(a), is forward error correction. Organizing data into K bytes of information with message polynomial description $M(D) = m_0 \cdot D^{k-1} \oplus m_1 \cdot D^{k-2} \oplus \cdots \oplus m_{k-1}$ where m_{k-1} is the last byte in a string of K bytes, and defining the R parity bytes appended to a

---

[5] Signalling information includes signals such as off-hook service requests, on-hook service termination, as well as more sophisticated signalling requests, menubrowsing, and service connection. Although carried by an ADSL modem, signalling information is usually not interpreted by the ADSL modem, but rather by application devices and the network.

Reed–Solomon code as $C(D)$, these check bytes are found through

$$C(D) = M(D) \cdot D^R \text{ modulo } G(D)$$

where arithmetic is performed in GF(256) and

$$G(D) = \prod_{i=0}^{R-1}[D \oplus \alpha^i]$$

where α is a root of the equation $x^8 \oplus x^4 \oplus x^3 \oplus x^2 \oplus 1 = 0$. Data bytes are identified with the element $d_7\alpha^7 \oplus d_6\alpha^6 \oplus \cdots \oplus d_0$. The FEC encoder output bytes are convolutionally interleaved with byte i delayed by $(L-1)i$ bytes, where L is the interleave depth, $i = 0, \ldots, K+R-1$. A good example of convolutional interleaving appears in ANSI [1994].

The corresponding FEC decoders for fast and interleave paths appear in Fig. 34.13(b). These decoders find the closest codeword to the received set of bits corresponding to the codeword. For the Reed–Solomon codeword, if $N-R$ parity bytes are appended, then $(N-R)/2$ byte errors can be corrected (assuming $N-R$ is an even number). Interleaving allows a group of successive byte errors to be distributed evenly in time. On the average, a burst of M byte errors is reduced to M/L byte errors with interleaving, but only if successive bursts occur more than ML bytes apart.

ADSL also allows for the optional use of a four-dimensional trellis code. Trellis codes translate a sequence of bits into a sequence of tone inputs that is more robust than the simple encoding described earlier in this section. A 16-state trellis diagram (see ANSI [1994]) describes the possible sequences of output bits. Trellis codes, as originally conceived by Ungerboeck [1989], allow for any number of bits per symbol to be used with essentially the same code. A larger number of bits simply corresponds to a larger number of parallel transitions between states. This feature of trellis codes naturally and trivially fits with the variable bit tables of DMT ADSL. A corresponding sequence or *Viterbi* decoder in the receiver traces all possible sequence possibilities and compares to the received sequence of tone outputs. The closest trellis code sequence is then selected and the corresponding bits are output to the FEC section of the ATU-R in Fig. 34.13(b). This code provides an additional 2–3 dB of coding gain over that already obtained by the use of FEC. Well-designed ADSL modems exhibit about 5.5 dB of measurable coding gain. Trellis codes are well documented in the tutorial papers by Ungerboeck [1989] and the detailed expose by Forney [1990].

ADSL Frame Format

The ADSL superframe appears in Fig. 34.14, where 68 data-carrying bytes are followed by a 69th *synch symbol*. The synch symbol is a known fixed repeated pattern, essentially a DMT tone where most of the tones are transmitted and contain 2 b per tone from a pseudorandom pattern. This symbol can be used for quick recovery of frame alignment if lost because of a substantial interruption, like a few-second line outage. This symbol can also be used to derive timing phase. Shorter interruptions typically do not result in loss of frame with well-designed DMT receivers.

Each of the 68 symbols within the superframe has a length that is determined by the data rate and fast/interleaving rate negotiation of ADSL. Within each symbol fast-path data occurs first with a control byte known as the *fast byte* followed by the bytes of fast data and the corresponding parity bytes. Fast path codewords are never longer than one DMT symbol to keep latency to a minimum. The fast data is followed by the interleaved data. The interleave data is also preceded by a control byte known as the *synch* byte in each symbol. Because many DMT symbols may correspond to a codeword in the interleave path, the parity bytes are deferred until all information bytes have occurred. Within the data portions of either the fast or interleave sections of a symbol, the data corresponding to AS0, AS1, AS2, AS3, LS0, LS1, and LS2 occur in order. Any one of these must be in either the fast buffer or the interleave buffer. If unused, then there are zero bytes for the unused ASX or LSX channel. Control bits in the fast and synch bytes allow the addition of a byte (AEX or LEX) to the information stream, or deletion of a byte. This add/delete mechanism allows

Asymmetric Digital Subscriber Lines

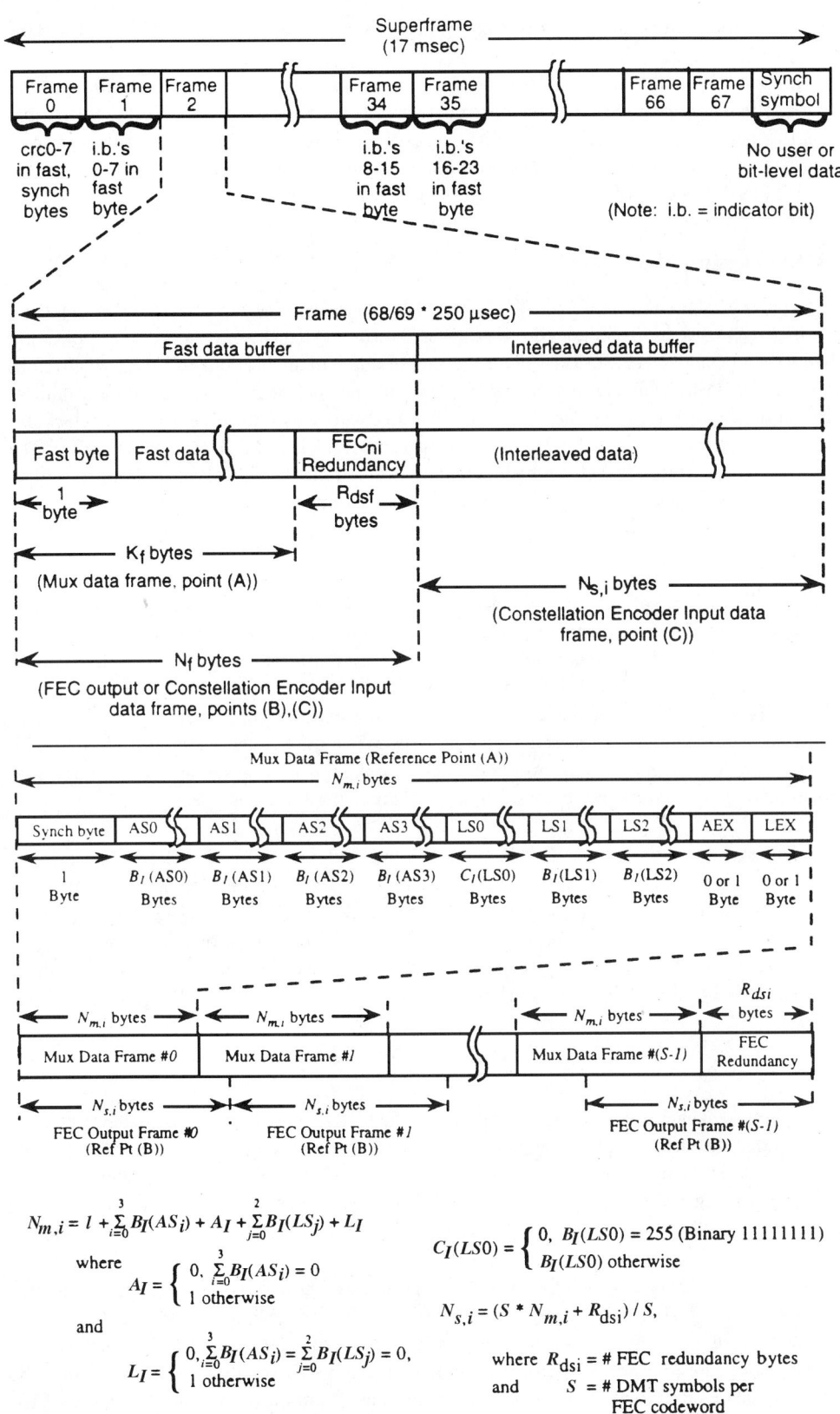

FIGURE 34.14 ADSL superframe data organization.

the various application signals to be asynchronous, as is commonly encountered in ADSL serving multiple applications.

Whereas the frame structure of ADSL is conceptually complicated, the implementation in very large-size integration (VLSI) is negligible in gate count. Further, this highly flexible format has begun to prove itself as anticipating the varied uses of ADSL for multiple signals that are now beginning to occur. (In other words, the initial superframe structure had to be flexible because the applications that would drive ADSL use could not be accurately described.)

ADSL Timing Methods

The synchronization of ATU-C and ATU-R in ADSL is necessary at several levels. The frame structure of given earlier allows for add/delete stuffing of incoming data bytes on any of the application channels (ASX and/or LSX). This allows the data of different applications to have asynchronous independent clocks, while sharing a common ADSL phone line. At the receive side of any of these application channels the queue depth of bytes delivered can be used to adjust an output clock frequency. The synchronization of transmit sampling/symbol clocks in ADSL requires more care. The accuracy requirements of DMT can be understood through the following basic Fourier transform relationship for a signal offset in time from some original time index

$$v(t + \varepsilon) \leftrightarrow V(f) \cdot e^{j2\pi f \varepsilon}$$

The Fourier transform of the original waveform is rotated by a linearly increasing phase offset with slope proportional to the amount of constant time offset to get the Fourier transform of the offset signal. Thus, for a constant phase offset, a received signal that is offset in time (sampled with an offset) can be adjusted in phase by a constant at each frequency so that it appears at the correct phase. This adjustment is automatic in the FEQ structure of earlier subsections. (The offset must be within reason, say, within one sample either way.) When the frequency (not just phase) of the receiver sampling clock is offset from the frequency of the transmit sampling clock, a number of synchronizing mechanisms and phase-error generation methods are possible.

Pilot Timing. In ADSL, the ATU-C transmits a pilot at 276 kHz (tone number 64) that has constant phase. Any offset at the output of the ATU-R FFT on this tone from the known nominal phase is a phase error that can be used to drive an analog phase lock loop that controls the sampling clock. The accuracy of this clock must be such that any residual phase error noise is so low that 15-b transmission may still occur. Using the previous phase error equation, one can easily compute a phase-error-induced SNR limit (for small phase error) of

$$SNR_{\text{phase}} = \frac{1}{4\pi^2 f^2 \varepsilon^2}$$

The highest frequency at which 15 b could practically be transmitted is approximately 100 kHz, meaning that the sampling phase deviation should be 500 ps or less for a SNR in excess of 50 dB. Such small deviation usually imposes a need for a crystal-based, analog, voltage-controlled oscillator (VCXO) if analog phase locking is used. Coincidentally, a similar jitter tolerance can be found necessary for sufficient accuracy of echo cancelers when used.

ROTORS. Another form of timing recovery that makes use of the basic preceding Fourier relationship is known as ROTORS. With ROTORS, a crystal (with center frequency near the 2.208-MHz sampling rate of ADSL, or some multiple of it) is used to sample the incoming DMT signal at the receiver. The phase error generated from the pilot can then be used to rotate each tone by the linearly increasing phase offset (with frequency) that is determined from the estimate of the timing-offset error. This method assumes that frequency offset within any symbol period is sufficiently close to constant for the basic relationship to hold. In practice, this requires a transmit crystal clock with accuracy at or below 5 ppm (15-b constellations require even yet more accurate crystals). Then, the offset is compensated digitally. Because a frequency offset will led to gradual growth with time of this phase offset, a sample will need to be added or deleted from the receiver's deletion of the cyclic prefix (and a corresponding adjustment of the internally digital phase). This technique avoids the need for a VCXO; the VCXO may be difficult to integrate in lowest possible cost implementations. However, the need for an accurate match of transmit and receive crystals, ensuring that timing offset is roughly constant within a single symbol of 250 μs, can cause a compensating increase in cost.

When the ATU-R does not loop time, then a second pilot is necessary upstream and this is placed at 69 kHz in the ADSL standard (tone number 16). A similar recovery mechanism can then be used upstream, although frequency tolerances are typically relaxed with respect to the downstream direction.

34.4 ADSL Data Rates, Ranges, and Performance

This section reviews measured ADSL performance to assess data rate vs range performance, which is of crucial importance for telephone company deployment practices. The analysis methods of Sec. 34.3 are found to project the measured performance of DMT-based ADSL modems accurately, and thus extrapolations can guide deployment practices. The first subsection compares projections and measured performance and also presents some of the test results on ADSL DMT modems that were publically released by Bell Communications Research and (independently) by GTE Research Laboratories. Some of these Bellcore results compare ADSL modems vs nonstandard QAM and CAP modems. The typical performance improvements of DMT refute claims that QAM/CAP methods can achieve the same performance levels in a implementable ADSL modem. The second subsection extrapolates ADSL performance, using the same accurate methods of Sec. 34.3, to data rates as high as 50 Mb/s.

Range Projections and Test Results

DMT ADSL modems have been measured at a variety of data rates, with the publically reported measurements occurring at 1.536 Mb/s (information rate) and 6.176 Mb/s downstream. The ADSL test-lab objectives were near-ubiquitous loop coverage at the lower data rate and **carrier service area (CSA) range** at 6 Mb/s (approximately 2-mi range). Test results show that these objectives were achieved. There are essentially two sets of agreed test loops used for ADSL [as well as other digital twisted-pair transmission systems like **Integrated Services Digital Network (ISDN)** basic rate service and **high-bit-rate digital subscriber lines (HDSL)**]. The first set is called the *18 ANSI loops*, and is shown in Fig. 34.15. The gauge and length of each segment of these loops is shown in Fig. 34.15, for instance, 16,500'/26 means 16,500 ft of 26-gauge twisted pair. The black dots in Fig. 34.15 indicate splice or tap junctions of different wire segments. These loops were originally defined for ISDN testing and have also been adopted for 1.5-Mb/s ADSL testing. The 18 ANSI loops represent the worst 10% of loops in the U.S. The second set of loops are the so-called *CSA loops*, and are shown in Fig. 34.16. The CSA loop deployment guidelines have been used for 25 years and about 80% of the nonloaded loops in the U.S. conform to these guidelines. The CSA loops are 1.75–2.5 mi loops, whereas the ANSI loops are 3–4 mi loops. Testing 6 Mb/s for ADSL is performed on CSA loops. Also shown in Fig. 34.16 are the so-called mid-CSA loops that were specifically determined by Bellcore for ADSL testing.

Prior to the selection of a transmission method for the ADSL standard, Bellcore and GTE engineers reported results (see Blake [1993a, 1993b] and Veeneman and Gross [1993] for some of the reported results) to ANSI concerning the performance of three line code alternatives for ADSL. Those results are summarized in Table 34.1.

A DMT ADSL modem using the same standardized parameters was tested also by Bellcore and GTE at 6.176-Mb/s data rate. The system had approximately 8% FEC overhead and trellis coding in a 16-ms convolutionally interleaved path through the pair of modems.

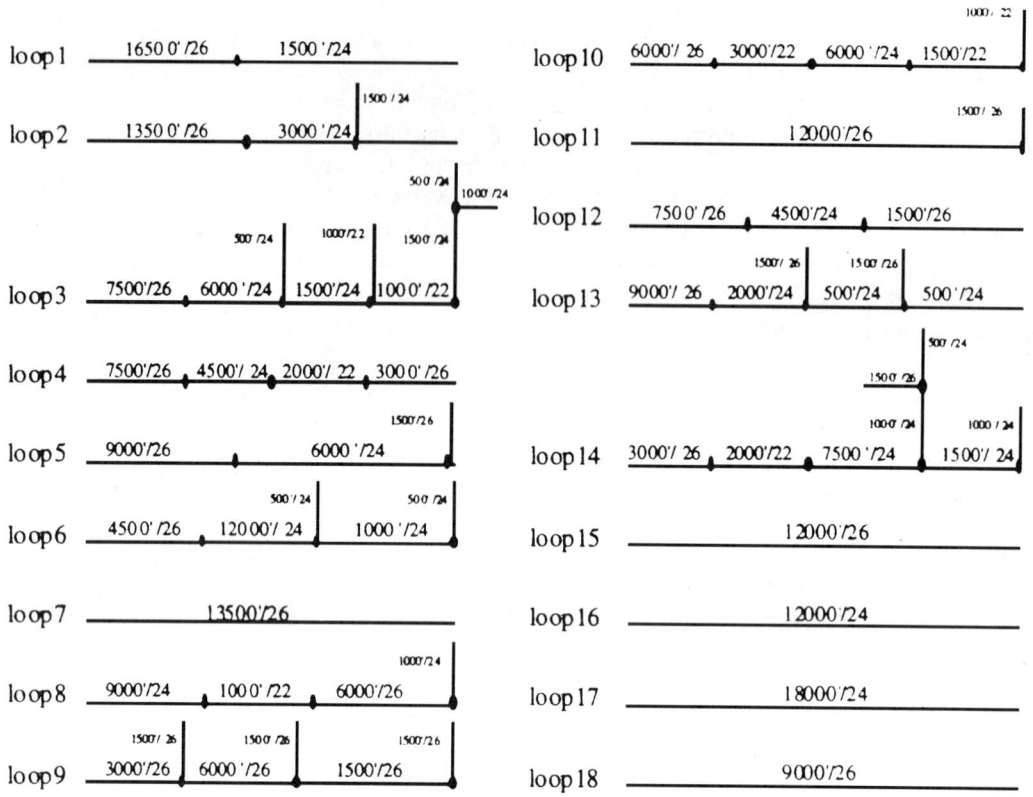

FIGURE 34.15 The ANSI test loops.

No echo cancellation was used, and the upstream data speed was essentially 16 kb/s through a very narrow upstream filter at low frequencies. Tones below 60 kHz were therefore blocked in the downstream path. The maximum number of bits per tone transmitted was 11 (well below the 15 allowed in the standard). Further, a transmit power mask of −40 dB(1 mW)/Hz [not −34 dB(1 mW)/Hz as is allowed in difficult cases by the standard]. Thus, all margins could be increased by as much as 10–15 dB if the improvements of 15-b cap, echo cancellation, and −34 dB(1 mW)/Hz were used. Nevertheless, CSA range performance was achieved unless **T1** crosstalk (NEXT) is present. The extra 10–15 dB is not quite enough to improve performance to CSA when T1 crosstalk is used, but is close.

Figure 34.17 compares margins computed using the methods of Sec. 34.3 with those measured. The agreement is close. Typically, the current DMT modems are 2–4 dB below optimum performance. Most of the performance loss is caused by FDM filters, which unduly distort the channel and make the TEQ's job difficult. An echo canceled system eliminates these costly FDM filters while simultaneously simplifying the TEQ.

Impulse noise has many sources, but there are two basic impulses that have been used for testing. The first is the so-called Cook pulse, after John Cook of BT Labs who created it: the Cook pulse is injected at various amplitude levels and frequency rates into ADSL test loops. The other impulses are a set of 12 introduced by NYNEX to test various digital subscriber services; they represent 12 of the worst of several thousand impulses that were recorded by NYNEX. Early testing at 1.5 Mb/s of DMT ADSL with the standardized amount of error correction showed essentially no errors. Some early 6-Mb/s testing by GTE showed excellent results also, except in the case when the worst-case T1 crosstalk was also present; since margins are tight on current systems, this is expected. Further interleaving (with latencies of 40 ms or 80 ms) can be used to eliminate impulse noise effects for all practical purposes.

Asymmetric Digital Subscriber Lines

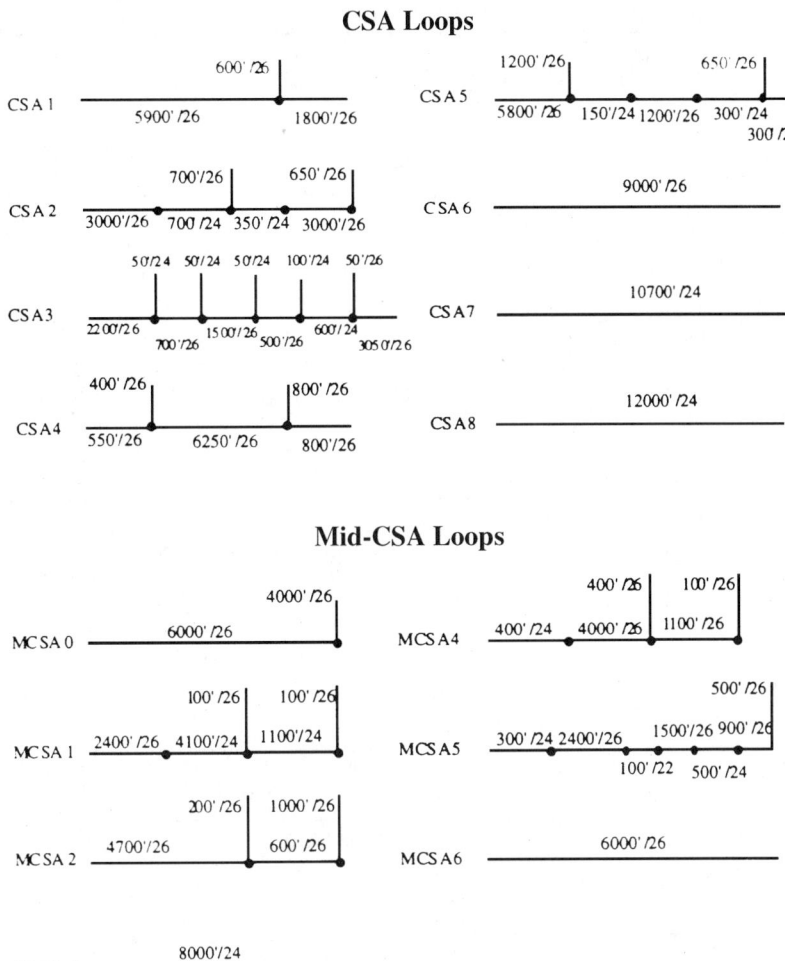

FIGURE 34.16 The CSA test loops.

ADSL Looking Forward: Very High-Speed Digital Subscriber Lines (VDSL)

The insertion loss of a telephone line increases (at all frequencies) with length. DMT uses all the subchannels on short loops because none are attenuated sufficiently to be silenced (unless AM radio interference is present in large amounts). On such loops, it is desirable to have the subchannels span a wider transmission bandwidth, which can be achieved by increasing the width of each tone. The wider subchannels can then lead to higher data rates on shorter loops. Figure 34.19 plots achievable data rate vs loop length for 26-gauge twisted pair for a system with transmit power spectral density -60 dB (1 mW)/Hz (the much lower transmit power is used for the practical reasons of realization and also to avoid emissions into radio bands at too high a level). With this transmit power, 52-Mb/s speeds achieve almost 2000-ft range and 26 Mb/s speeds achieve over 3000 ft.

For **very high-speed digital subscriber lines (VDSL)**, especially in business applications within a building on twisted pair at shorter ranges, there is also significant interest in symmetric transmission. Figure 34.20 plots the range of a DMT system with symmetric VDSL transmission. Because of the effects of increased crosstalk, range is reduced. However, 26 Mb/s symmetric service has range of approximately 2000 ft. Further extrapolation shows that 620-Mb/s transmission is theoretically possible on shorter distances of about 100 m.

TABLE 34.1 ADSL 1.5 Mb/s Bellcore Test Results[a]

Test Loop Number	Amati DMT −40(−45) dB(1 mW)/Hz		RTEC QAM −45 dB(1 mW)/Hz	AT&T CAP −45 dB(1 mW)/Hz
		24 ISDN NEXT:		
ANSI4	8	4	−6	no-connect(FAIL)
ANSI6	7	3	<−8	no-connect(FAIL)
ANSI9	>16	>16	7	no-connect(FAIL)
ANSI17	15	10	1	no-connect(FAIL)
		24 HDSL NEXT:		
CSA2	>11	>11	>11	<−8
CSA6	>11	>11	5	−8
CSA7	>11	>11	>11	−1
		1 T1 NEXT:		
CSA2	>10	>10	7	<0
CSA6	>10	8	2	<0
CSA7	>10	>10	8	−2

[a] A > sign means that the margin exceeds the measurement equipment's largest margin. Similarly, a < means that the noise could not be attenuated sufficiently for the modems to work. A negative margin means that the noise had to be reduced with respect to minimum acceptable levels. The Amati system was tested at two transmit power spectrum levels because the RTEC and AT&T designs could not achieve the higher spectrum level of −40 dB(1mW)/Hz. However, the higher spectrum level was recommended by and adopted by the ADSL committee even before the testing. The results at the higher level are now those used by the ANSI committee for low-speed 1.5 Mb/s ADSL.

TABLE 34.2 The 6-Mb/s ADSL Test Results[a]

Test Loop Number	Bellcore −40 dB(1 mW)/Hz	GTE[b] −40 dB(1 mW)/Hz	Manufacturer[b] −40 dB(1 mW)/Hz
	24 ISDN NEXT + 24 T1 NEXT:		
MID-CSA3	2		
MID-CSA4	6		
MID-CSA6	4	4	5
	10 ADSL + 10 HDSL NEXT:		
CSA4	5	6	
CSA6	4	7	9
CSA8	4	9	
	49 HDSL NEXT:		
CSA2	0	3	
CSA6	2	4	5
CSA7	2	5	
	49 DSL NEXT:		
CSA2	11	12	
CSA6	11	13	16
CSA7	11	13	

[a] All were performed on Amati DMT ADSL modems. FEC and trellis included, but no Echo canceler, no −34 dB(1 mW) power boost, and bit cap was 11.

[b] The GTE and manufacturer tests in the second group were actually done with 24 crosstalkers on both ADSL and HDSL, rather than 10; the measured results were boosted by 2 dB to conform to Bellcore's use of 10.

A feature of DMT is that modems that have different numbers of the same-width subchannels can interoperate on the common subchannels. An enhancement of DMT is known as DMT information bus (DIB), which allows a central-office modem to talk to several remote modems simultaneously (on different tones) without need for analog separation. In this case, all remote modems must

Asymmetric Digital Subscriber Lines

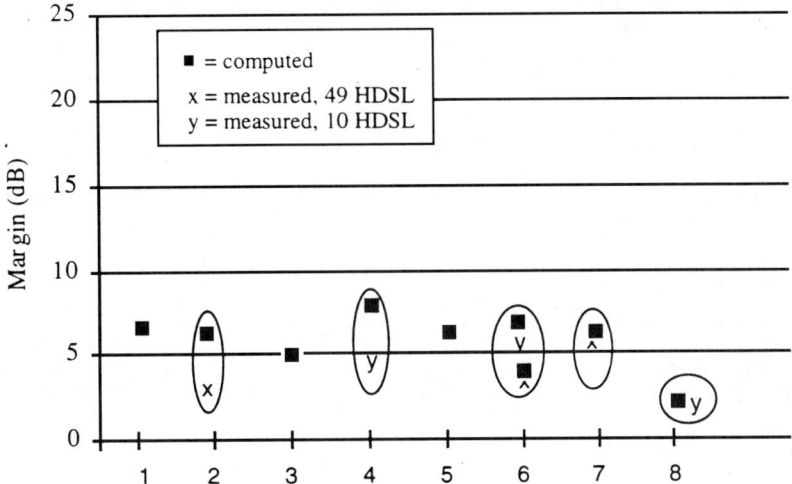

FIGURE 34.17 Comparison of computed and measured results.

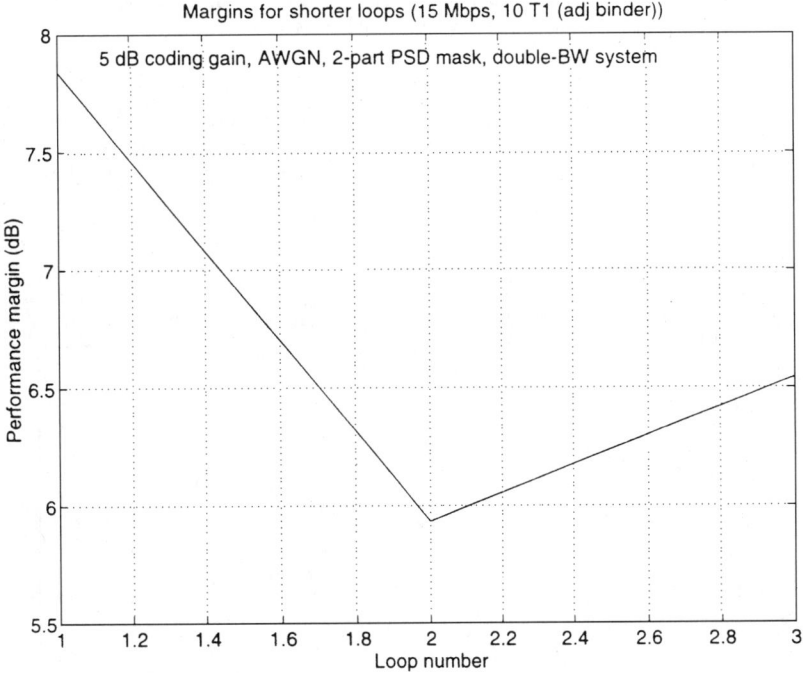

FIGURE 34.18 CSA margins for ADSL (with 512 tones) and T1 noise.

be loop timed in sample and symbol clock when their signals arrive at the central office. In this way, multidrop of ADSL applications to any point within the reach of the existing twisted pair can occur without rewiring the home or small business. This improved form of DMT ADSL can be used to simplify wiring also in apartment buildings where several users might best share a common access and it is difficult to rewire from this access to all tenants. An example of DIB for the home distribution appears in Fig. 34.21. In this case, the DMT ADSL electronics are incorporated in or near the *set-top* or *applications module*, eliminating the need for rewiring.

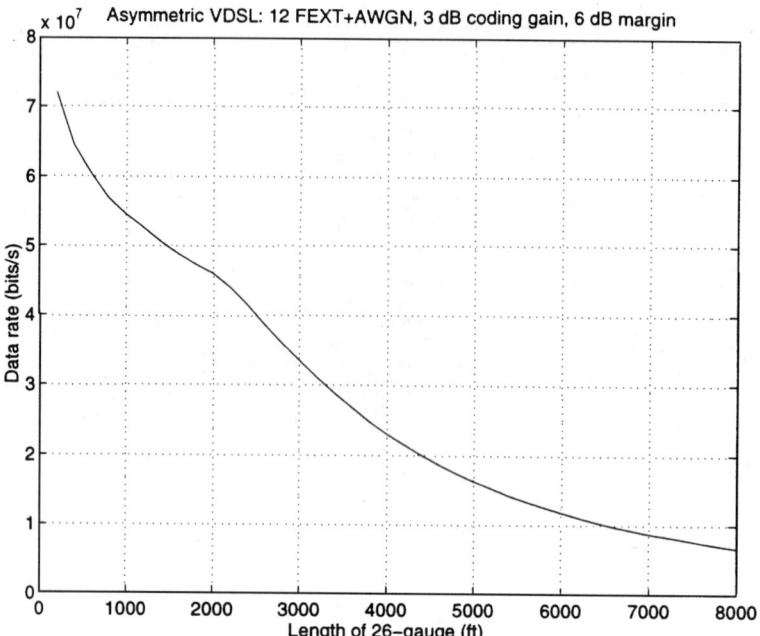

FIGURE 34.19 DMT VDSL asymmetric data rate vs range.

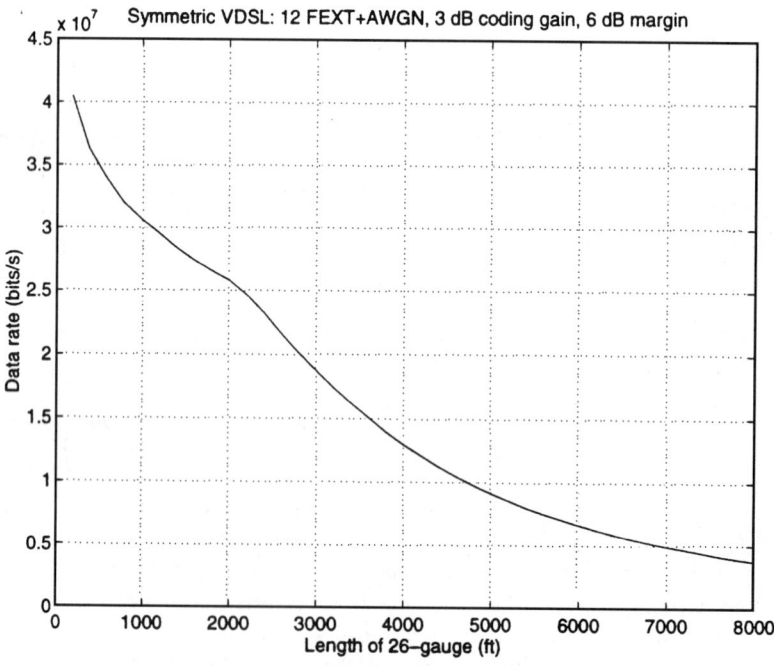

FIGURE 34.20 DMT VDSL symmetric data rate vs range.

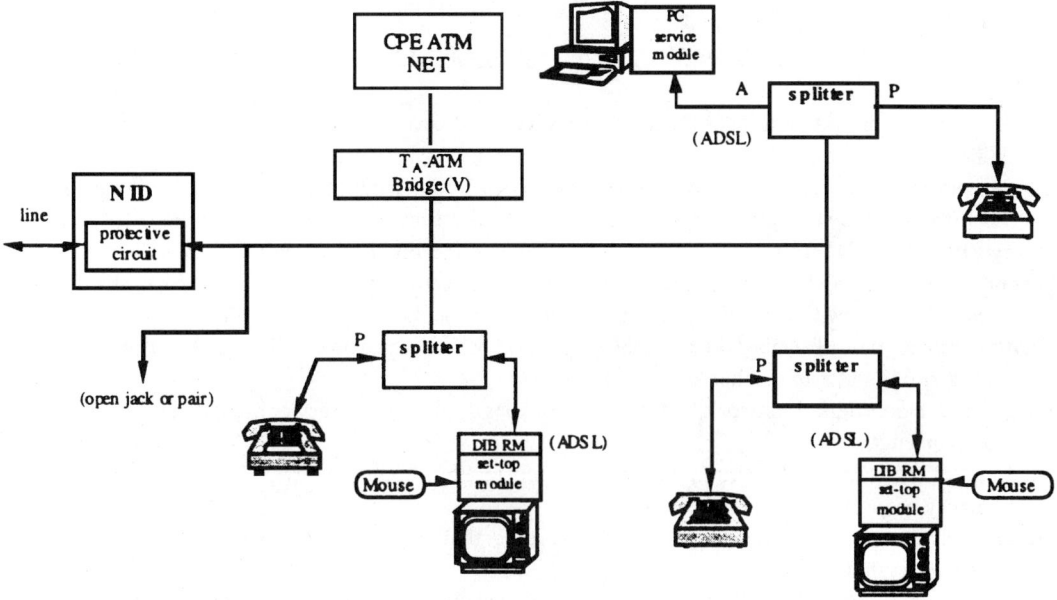

FIGURE 34.21 Multidrop extension of ADSL on existing wiring.

34.5 The Future

ADSL and the VDSL extension to 50-Mb/s speeds enable a cost-effective incremental and prudent deployment of a plethora of interactive digital multimedia services through the billions of miles of telephone lines initiated by Bell's first conversation. The bottlenecks to such service deployment are the applications equipment and the availability of reasonable cost ATM switches (or other digital switches), not the availability of transmission bandwidth on the media to the customer. Trials of ADSL systems are still in early stages with a few dozen scattered around the world at the time of writing. Successful application development is key to ADSL's success. The reuse of Bell's patented (1883) twisted pairs, while this applications development and assessment is ongoing, makes the economics of an information superhighway just barely tangible; anything else is cost prohibitive in reality.

It is perhaps too much for coincidence that the anniversary of working telephony is also that of ADSL's enabling: March 10, 4:10 p.m. Bell, a professor and teacher of the deaf, married one of his deaf students, Mabel, who coincidentally owned all but 10 shares of the early Bell System stock, earning the name MA BELL. Bell's sadness that she could not fully appreciate the use of the telephone was legendary; now, finally, as his history smiles upon ADSL through the legendary date, Mabel can finally use the phone.

Acknowledgment

The author owes a great debt to the many Stanford students who have helped to make ADSL and DMT a success and have contributed greatly to the many topics contained within, Jim Aslanis, Jacky Chow, Peter Chow, and Nick Zogakis (all now Amati employees); also, A. Ruiz, K. Jacobsen, M. Ho, H. Lou, S. K. Wilson, N. Al-Dhahir, and J. Tu. A number of Amati employees/consultants also deserve at least as much credit, including John Bingham, M. Mallory, M. Flowers, V. Whitis, R. Hunt, P. Tong, A. Gooch, R. Adolf, S. Macica, J. Gibbons, M. Agah, A. Granzotti, L. Rennick, L. Valdez, J. Ottinger, D. Pal, T. Ha, A. Hagedorn, and J. Grady, and especially J. Lennon and L. Kehoe.

Defining Terms

ADSL transmission unit-central office (ATU-C): The ADSL modem at the phone-company side of the phone line.

ADSL transmission unit-remote (ATU-R): The ADSL modem at the customer side of the phone line.

Carrier service area (CSA): A set of loops less than about 2 mi in length that are known to phone companies in the U.S. and represent a large fraction of the existing phone lines worldwide.

Discrete multitone (DMT): The standardized transmission technology of ADSL.

Far end crosstalk (FEXT): Electromagnetically coupled noise from radiation of signals on near-by telephone lines with transmitters at the far end of the telephone line.

High-bit-rate digital subscriber lines (HDSL): A 1.544 Mb/s service on CSA loops on two lines. Modernizes and extends the range of T1 circuits.

Integrated services digital network (ISDN): A 144 kb/s duplex digital service on phone lines to 3 mi in length.

Loading: The process of determining how much information should be carried in different frequency bands by ADSL DMT.

Margin: The amount of received signal power in excess of that required to achieve the ADSL target bit error rate of 10^{-7}.

Near end crosstalk (NEXT): Electromagnetically coupled noise from radiation of signals on near-by telephone lines with transmitters close by.

Range: The length of twisted pair over which a digital service can operate acceptably.

T1: An older 1.544 Mb/s duplex service to 1 mi on phone lines.

Very high-speed digital subscriber lines (VDSL): Higher speed extension of ADSL to data rates of 25–50 Mb/s.

References

ANSI. 1994. American National Standard for Telecommunications: Asymmetric digital subscriber line (ADSL) metallic interface specification. *Secretariat, Alliance for Telecommunications Industry Solutions—Subcommittee T1E1.4*, Sept. Draft 94-007R6. American National Standards Institute.

Bingham, J.A.C. 1990. Multicarrier modulation for data transmission: An idea whose time has come. *IEEE Commun. Mag.* 28(5):5–14.

Blake, B.A. 1993a. Results of crosstalk margin tests on the Amati DMT ADSL transceivers performed at Bellcore. *ANSI T1E1.4 Contribution 93-292*, Nov. Bellcore, Denver, CO.

Blake, B.A. 1993b. Results of transmission tests on ADSL transceivers. *ANSI T1E1.4 Contributions 93-030, -031 and -032*, March. Bell Communications Research, Miami.

Chow, J.S. and Cioffi, J.M. 1992. A cost effective maximum-likelihood receiver for multicarrier systems. *IEEE International Conference on Communications*, Paper No. 333.7 Chicago, IL, June.

Chow, P.S. 1993. Bandwidth optimized digital transmission techniques for spectrally shaped channels with impulse noise. Ph.D. Dissertation, Stanford Univ. May, Stanford, CA.

Forney, G.D. 1988. Coset codes I and II. *IEEE Trans. Inf. Theory*, (Oct.).

Forney, G.D. and Eyuboglu, V. 1991. Combined equalization and coding using precoding. *IEEE Commun. Mag.*, 29(12).

Foster, K. and Cook, J. 1992. A symbolic pulse for impulsive noise testing, *ANSI T1E1.4/92-143*, Aug. British Telecom, Portland, OR.

Ho, M., Bingham, J.A.C., and Cioffi, J. in press. Method and apparatus for echo cancellation with DMT modulation. *IEEE Trans. Commun.*

Ho, M. and Cioffi, J. Method and apparatus for echo cancellation with DMT modulation. *U.S. Patent No. 5,317,596.*

Hoque, M. 1995. Radio frequency interference aspects of VDSL. *ANSI T1E1.4/95-132*, Dec. Orlando, FL.

Lawrence, R.W. 1993. Impulse noise test results for 3 line-code technologies proposed for ADSL testing—preliminary report. *ANSI T1E1.4/93-078* March 10. NYNEX, Miami, FL.

Lechleider, J.W. 1991. High bit rate subscriber lines: A review of HDSL progress. *IEEE J. Selec. Areas Commun.*, 9(6):769–784.

Lee, E.A. and Messerschmitt, D.G. 1988. *Digital Communication*, Kluwer, Boston, MA.

Shannon, C.E. 1948. A mathematical theory of communication. *BSTJ*, 27:379–423, 623–656.

Ungerboeck, G. 1989. Trellis coding—Parts I and II. *IEEE Commun. Mag.*, 27(2).

Veeneman, D. and Gross, R. 1993. Results of objective transmission tests performed on an ADSL system from Amati communications. *ANSI T1E1.4 Contribution 93-295*, Nov. GTE, Denver, CO.

Werner, J.J. 1991. The HDSL environment. *IEEE J. Selec. Areas Commun.*, 9(6):785–800

Zogakis, T.N. 1995. A coded and shaped discrete multitone system. *IEEE Trans. Commun.*, 43(12): 2941–2949.

Further Information

See the references and "ADSL: Twisted-Pair Access to the Information Superhighway," by the ADSL Forum (web page http://www-adsl.com/adsl) and other references on that web page. Also, "VDSL System Requirements Report," American National Standards Institute document T1E1.4/95-117R4 (or ask for most current version from ANSI Secretariat, New York, NY).

35
Overview of Common Channel Signal

A.R. Modarressi
BellSouth Telecommunication

R.A. Skoog
AT&T

35.1	Introduction	480
35.2	Historical Background	481
35.3	Signalling Transport in Signalling System 7 Networks	482
	The Message Transfer Part • The Signalling Connection Control Part • Signalling Modes and Signalling Network Topologies	
35.4	Signalling User Parts in SS7 Networks	488
	The Integrated Services Digital Network User Part • The Transaction Capabilities Application Part	

35.1 Introduction

In the context of modern telecommunication, signalling can be defined as the *system* that enables stored program control exchanges, network databases, and other intelligent nodes of the network to exchange (1) messages related to call setup, supervision, and tear down (call/connection control); (2) information needed for distributed application processing in support of various services (service control); and (3) network management information. As such, signalling constitutes the control infrastructure of the modern telecommunication network.

Modern common channel signalling systems are essentially data communication systems using layered protocols. What distinguishes them from ordinary data communication systems are basically two things: their real-time performance and their reliability. No matter how complex the set of network interactions needed to set up a call, the call setup time (which is closely related to post-dialing delay) should not exceed a couple of seconds. This imposes a stringent end-to-end delay requirement on the signalling systems. On the other hand, because of the total reliance of the telecommunication network on its signalling system, requirements for signalling network reliability (message integrity, end-to-end availability, network robustness, recovery from failure, etc.) are quite demanding. For example, current objectives require the total cumulative downtime between an arbitrary pair of communicating nodes in the signalling network not to exceed 10 min per year. This is at least two orders of magnitude smaller than the corresponding requirement in a general-purpose data network. Requirements on real-time performance and reliability of signalling systems are likely to become even more stringent with advances in technology and new application needs.

Although signalling has evolved with the technology of telephony for more than a century, the pace of this evolution has never been faster than in the last two decades, a period characterized by the marriage of computer and switching technologies. The advent of the integrated services digital network (ISDN), and the emergence of an increasing number of intelligent network (IN) services have further accelerated the pace of development and deployment of common channel signalling systems. When viewed as an end-to-end capability, signalling in ISDN has two distinct components: signalling between a user and the network (access or UNI signalling), and signalling within or between networks (network or NNI signalling). The current set of international protocol standards for common channel *access signalling* is known as the digital subscriber signalling system no. 1 (DSS1). The current set of international protocol standards for common channel *network* signalling is known as the signalling system no. 7 (SS7).

From a functional viewpoint, any common channel signalling system has two distinct parts. The lower part has to do with providing a transport capability to carry messages between signalling nodes in a way that ensures high reliability even in the face of network failures. Once this reliable exchange capability is realized, each signalling node has to be able to process the transported messages in support of a useful service such as call setup. For an increasing number of services, call setup has to be preceded by invocation of distributed remote processes the outcome of which determine the nature and attributes of the subsequent call. These nodal capabilities of call/connection control and distributed process invocation and management constitute the upper parts, known as application or user parts, of a common channel signalling system. All application parts use the services of the transport part [Modarressi and Skoog, 1990].

35.2 Historical Background

Signalling has traditionally consisted of supervisory functions (e.g., on-hook/off-hook to indicate idle or busy status), addressing (e.g., called number), and providing call information (e.g., dial tone, busy signals, etc.). Early methods of interoffice signalling involved dial pulse techniques, as with step-by-step machines. Common control switching systems introduced single frequency (SF) and multifrequency (MF) signalling techniques. The SF method indicated the busy/idle state of a trunk (circuit) by putting a single-frequency tone (usually 2600 Hz) on the trunk when it was in the idle state. Removal of the tone would signal an off-hook condition. Multifrequency tones were used to pass addressing information. Some key attributes of these signalling methods are that they are in-band (i.e., signalling information is conveyed over the same channel that is used for speech), call setup times are long (10–20 s), limited information can be transferred resulting, among other things, in restrictive network routing capabilities (mostly hierarchical routing), calls in progress cannot be modified, and networks are subject to fraud (black box/blue box tone generators).

With the introduction of electronic processors in switching systems came the possibility of providing common channel signalling. This is an out-of-band signalling method in which a common data channel is used to convey signalling information related to a number of trunks. In 1976, AT&T introduced a common channel signalling system into its toll network called common channel interoffice signalling (CCIS). This signalling system (that we will refer to as CCS6) was based on the International Telegraph and Telephone Consultative Committee (CCITT) signalling system no. 6 recommendations [CCITT n.d.]. Initially CCS6 provided for trunk signalling only, and routing was on the basis of a permanent virtual circuit network. Messages were routed by a band and label scheme where 512 bands of 16 trunks could be assigned to a signalling link. Routing was done by translating the incoming band and terminal to an outgoing band and terminal. The CCS6 protocol structure was not layered; it was one monolithic structure. A major consideration in the design of CCS6 was efficiency. Since the links were originally planned to work at 2.4 kb/s and processing speeds were low, message lengths had to be kept as small as possible; thus, a layered structure, with its associated overhead, would not have met these needs. Furthermore, protocol layering was not a common practice in the early 1970s.

In 1980, a *direct signalling* capability (based on a datagram mode of operation) was added to AT&T's CCS6 network. This capability allowed messages in the form of queries to be sent to network databases from switches, and thus opened the way for a new network architecture. Now some of the call processing service logic would be placed in centralized databases. The first services to use this new architecture were the 800 (free-phone) service and calling card service. Direct signalling routing was done on a destination point basis rather than on the banded virtual circuit basis as for trunk signalling.

Although CCS6 provided major advancement over SF/MF signalling, it had some significant drawbacks: large administrative efforts for banded routing, limited message lengths, and low-speed links. In the mid-1970s, CCITT [now International Telecommunications Union (ITU)] began work on signalling system no. 7 to provide a signalling system for digital networks. In this time frame, the layered approach to designing protocols was being developed for open systems interconnection (OSI) data transport, and its value was recognized for signalling applications. Furthermore, technology had advanced to the point that the inefficiencies associated with layered protocols were far outweighed by their flexibility in realization and management of complex functions. Also, bit-oriented protocols such as higher-level data link control (HDLC) were known, and these had an influence on the development of signalling system no. 7. The first CCITT recommendations on this new signalling protocol were published in 1980, followed in 1984 by an expanded set of specifications, and again in 1988 and 1992 by even more expansions. Although the development of the signalling system no. 7 protocol has in many respects been tailored to the special needs of telephone signalling (in contrast with general data communication needs), its evolution has been influenced by the need to assume a broader role and become somewhat more aligned with the OSI reference model.

35.3 Signalling Transport in Signalling System 7 Networks

In this section, signalling system no. 7 transport protocols, which roughly correspond to the first three layers (physical, data link, and network) of the OSI reference model, are described. This component of the SS7 protocol is called the network service part (NSP), and it consists of the message transfer part (MTP) and the signalling connection control part (SCCP). Figure 35.1 shows how these relate to each other and to the other parts of the protocol. MTP consists of three levels: signalling data link (level 1), signalling link (level 2), and signalling network (level 3). MTP provides a connectionless message transfer system that enables signalling information to be transferred across the network to its desired destination. Functions are included in MTP that maintain transport reliability in the face of failures in the network. SCCP uses the services of MTP to provide additional functions for both connectionless and connection-oriented network services.

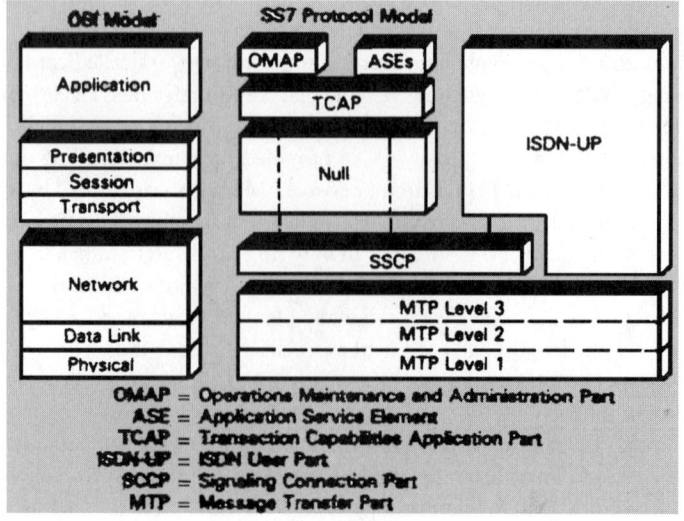

FIGURE 35.1 Signalling system no. 7 protocol architecture.

MTP was developed before SCCP and it was tailored to the real-time needs of telephony applications. Thus, a connectionless (datagram) capability was called for which avoids the administration and overhead of virtual circuit networks (one of the disadvantages of CCS6). Later, it became clear that there were other applications that would need additional network services such as an expanded addressing capability and connection-oriented message transfer. SCCP was developed to satisfy this need. The resulting structure, and specifically the splitting of the OSI network layer functions into MTP level 3 and SCCP, has certain advantages in the sense that the higher overhead SCCP services can be used only when needed, allowing the more efficient MTP to serve the needs of those applications that can use a connectionless message transfer with limited addressing capability.

The Message Transfer Part

The overall purpose of the MTP is to provide a reliable transfer and delivery of signalling information across the signalling network, and to take necessary actions in response to system and network failures to ensure that reliable transfer is maintained.

Signalling Data Link Functions (MTP Level 1)

A *signalling data link* is a bidirectional transmission path for signalling, consisting of two data channels operating together in opposite directions at the same data rate. It complies with the OSI's definition of the physical layer (layer 1). Transmission channels can be either digital or analog, terrestrial or satellite. For digital signalling data links, the recommended bit rate in the North American [American National Standards Institute (ANSI)] standard is 56 kb/s, and in the international (ITU) standard is 64 kb/s. Higher link speeds, and specifically 1.544 Mb/s in the U.S., are beginning to emerge as processing speeds and traffic loads increase.

Signalling Link Functions (MTP Level 2)

The *signalling link* functions correspond to the OSI's data link layer (layer 2). Using the services of the signalling data link, the signalling link functions provide a dedicated *reliable* channel for transfer of signalling messages between two *directly connected* signalling points. Signalling messages are transferred over the signalling link in variable length messages called *signal units*. There are three types of signal units, differentiated by the length indicator (LI) field contained in each, and their formats are shown in Fig. 35.2. The signalling information field (SIF) in a message signal unit (MSU) must have a length less than or equal to 272 octets. This limitation is imposed to control the delay a message can impose on other messages due to its emission time. This number may go up if link speed of 1.544 Mb/s is standardized.

The standard flag (01111110) is used to open and close signal units, and the standard 16-b CRC checksum is used for error detection. However, when there is no message traffic, fill-in signal units (FISUs) are sent rather than flags, in contrast to other data link protocols. The reason for this is to allow for a consistent error monitoring method (described subsequently) so that faulty links can be quickly detected and removed from service even when traffic is low.

Error Correction. Two forms of error correction are specified in the signalling link procedures. They are the *basic method* and the *preventive cyclic retransmission* (PCR) method. In both methods, only errored MSUs and link status signal units (LSSUs) are corrected, whereas errors in FISUs are detected but not corrected. Both methods are also designed to avoid out-of-sequence and duplicated messages. The PCR method (not described here) is used when the propagation delay is large (e.g., with satellite transmission).

The basic method of error correction is a noncompelled positive/negative acknowledgment retransmission error correction system. It uses the *go-back-N* technique of retransmission used in many other protocols. If a negative acknowledgment is received, the transmitting terminal stops sending new MSUs, rolls back to the MSU received in error, and retransmits everything from that

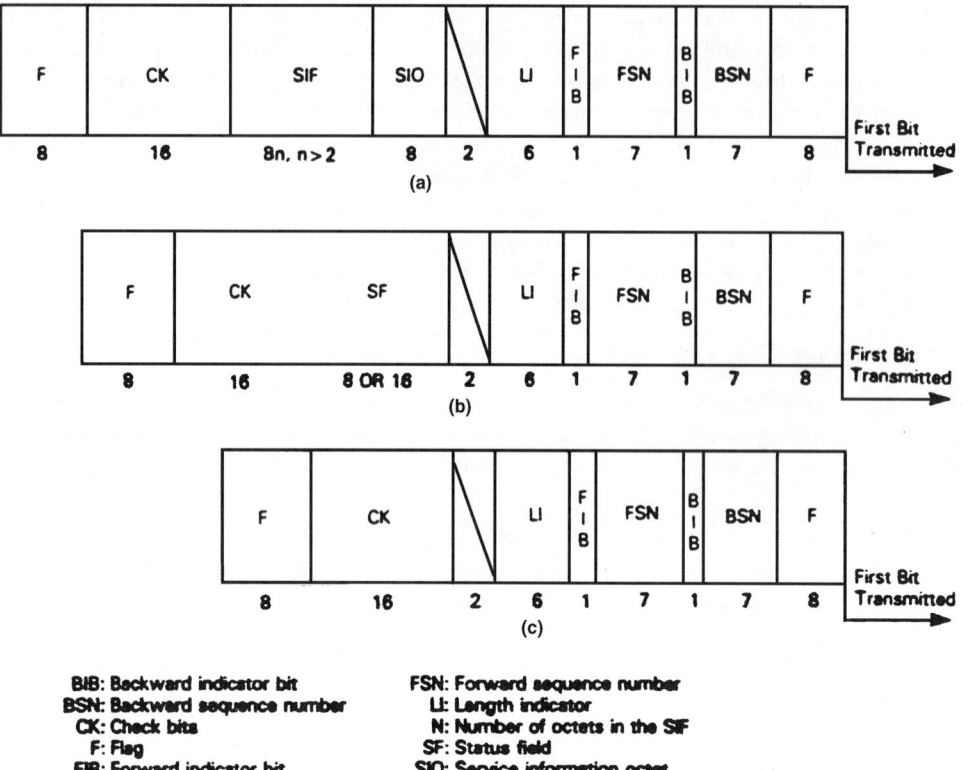

FIGURE 35.2 Signal unit format (MTP level 2 headers): (a) basic format of an MSU, (b) format of a link status signal unit (LSSU), and (c) format of FISU.

point before resuming transmission of new MSUs. Positive acknowledgments are used to indicate correct reception of MSUs, and as an indication that the positively acknowledged buffered MSUs can be discarded at the transmitting end. For sequence control, each signal unit is assigned forward and backward sequence numbers and forward and backward indicator bits (see Fig. 35.2). The sequence numbers are 7 b long, which means at most 127 messages can be transmitted without receiving a positive acknowledgment.

Error Monitoring. Two mechanisms for error rate monitoring are provided. A *signal unit error rate monitor* is used while a signalling link is in service, and it provides the criteria for taking a signalling link out of service due to an excessively high error rate. An *alignment error rate monitor* is used while a signalling link is in the proving state of the initial alignment procedure, and it provides the criteria for rejecting a signalling link for service during the initial alignment due to too high an error rate. Only the first error rate monitoring will be briefly described.

The signal unit error rate monitor is based on a signal unit (including FISU) error count, incremented and decremented using a *leaky bucket* algorithm. For each errored signal unit the count is increased by one, and for each 256 signal units received (errored or not), a positive count is decremented by one (a zero count is left at zero). When the count reaches 64, an excessive error rate indication is sent to MTP level 3, and the signalling link is taken out of service.

Flow Control. The flow control procedure is initiated when congestion is detected at the receiving end of the signalling link (e.g., when a certain threshold is reached in the receive buffer). The congested receiving end notifies the transmitting end of its congestion with a link status signal unit indicating busy, and withholds acknowledgments of all incoming signal units. This action stops the transmitting end from failing the link due to a time-out of acknowledgments. However, if the congestion condition lasts too long (3–6 s), the transmitting end will fail the link.

Another flow control indication called signalling indication processor outage (SIPO) is sent on the link by MTP level 2 when it recognizes that it cannot transfer messages to MTP level 3 (due to failure or congestion). The level 2 entity receiving a SIPO responds by sending only FISUs, and informing its own level 3 of the SIPO condition at the remote end of the link. The level 3 entity at the transmitting end in turn will reroute traffic in accordance with the signalling network management procedures to be described.

Signalling Network Functions (MTP Level 3)

The signalling network functions correspond to the lower-half of the OSI network layer, and they provide the functions and procedures for the transfer of messages between signalling points, which are the nodes of the signalling network. The signalling network functions can be divided into two basic categories: *signalling message handling* and *signalling network management*.

Signalling Message Handling. Signalling message handling consists of message routing, discrimination, and distribution functions. These functions are performed at each signalling point in a signalling network, and they are based on the part of the message called the *routing label*, as shown in Fig. 35.3, and the service information octet (SIO), as shown in Fig. 35.2. The routing label consists of the destination point code (DPC), the origination point code (OPC), and the signalling link selection (SLS) field. In the international standard, the DPC and OPC are 14 b each, whereas the SLS field is 4 b long. In North American standards, the OPC and DPC are each 24 b (to accommodate larger networks), whereas the SLS field has 5 b, with 3 spare bits in the routing label. The routing label is placed at the beginning of the signalling information field (SIF), and it is the common part of the label that is defined for each MTP user.

When a message comes from a user of level 3, or originates at level 3, the choice of the particular signalling link on which it is to be sent is made by the message routing function. When a message is received from level 2, the discrimination function is activated, which determines if the message is addressed to another signalling point or to itself based on the DPC in the message. If the received message is addressed to another signalling point, and the receiving signalling point has the transfer capability, that is, the signalling transfer point (STP) function, the message is sent to the message routing function. If the received message is addressed to the receiving signalling point, the message distribution function is activated, which delivers the message to the appropriate MTP user or MTP level 3 function based on the value of the service indicator, a subfield of the SIO field.

Generally, more than one outgoing signalling link can be used to route a message to a particular DPC. The selection of the particular link to use is made by use of the SLS field. This is called load sharing. A set of links between two signalling points is called a *link set*, and load sharing can be done over links in the same link set or over links not belonging to the same link set. A load sharing collection of one or more link sets is called a *combined link set*.

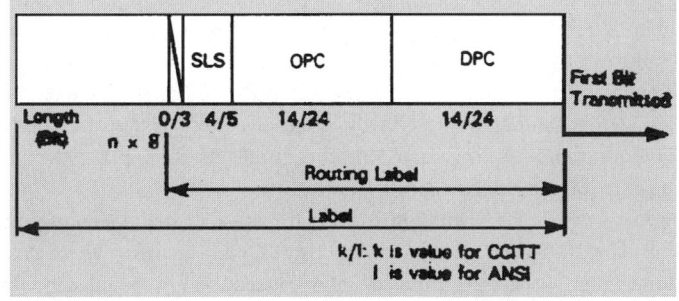

FIGURE 35.3 Routing label structure (MTP level 3 headers).

The objective of load sharing is to keep the load as evenly balanced as possible on the signalling links within a combined link set. For messages that should be kept in sequence, the same SLS code is specified by the MTP user so that such messages all follow the same path, arriving at the destination in the correct sequence.

Signalling Network Management. The purpose of the signalling network management is to provide reconfiguration of the signalling network in the case of signalling link or signalling point failures, and to control traffic in the case of congestion or blockage. In all cases, messages are not to be lost, duplicated, or put out of sequence, nor should message delays become excessive. Signalling network management consists of three functions: signalling traffic management, signalling route management, and signalling link management. Whenever a change in the status of a signalling link, signalling route, or signalling point occurs, these three functions are activated as summarized.

The *signalling traffic management* procedures are used to divert signalling traffic, without causing message loss, missequencing, or duplication, from unavailable signalling links or routes to one or more alternative signalling links or routes, and to reduce traffic in the case of congestion. When a signalling link becomes unavailable, a *changeover* procedure is used to divert signalling traffic to one or more alternative signalling links, as well as to retrieve for retransmission messages that have not been positively acknowledged. When a signalling link becomes available, a *changeback* procedure is used to re-establish signalling traffic on the signalling link made available. When signalling routes (succession of links from the origination to the destination signalling point) become unavailable or available, *forced rerouting* and *controlled rerouting* procedures are used, respectively, to divert the traffic to alternative routes or to the route made available. Controlled rerouting is also used to divert traffic to an alternate (more efficient) route when the original route becomes restricted (i.e., less efficient because of additional transfer points in the path).

The *signalling route management* procedures are used to distribute information about the signalling network status in order to block or unblock signalling routes. The following procedures are defined to take care of different situations. The *transfer-controlled* procedure is performed at a signalling transfer point in the case of signalling link congestion. In this procedure, for every message received having a congestion priority less than the congestion level of the signalling link, a control message is sent to the OPC of the message asking it to stop sending traffic (to the DPC of the message) that has a congestion priority less than the congestion level of the signalling link. In North American standards, four congestion priorities are used; in international networks only one is used. The *transfer-prohibited* procedure is performed at a signal transfer point to inform adjacent signalling points that they must no longer route to a particular DPC via that STP. This procedure would be invoked, for example, if the STP had no available routes to a particular destination. The *transfer-restricted* procedure is performed at a signalling transfer point to inform adjacent signalling points that, if possible, they should no longer route messages to a particular DPC via that STP. The *transfer-allowed* procedure is used to inform adjacent signalling points that routing to a DPC through that STP is now normal. In the ANSI standards, the previous procedures are also specified on a cluster basis (a cluster being a collection of signalling points), which significantly reduces the number of network management messages and related processing required when there is a cluster-level failure or recovery event. The *signalling-route-set-test* procedure is used by the signalling points receiving transfer prohibited and transfer restricted messages to obtain the signalling route availability information that may not have been received due to some failure. Finally, in ANSI standards the *signalling-route-set-congestion-test* procedure is used to update the congestion status associated with a route toward a particular destination.

The *signalling link management* function is used to restore failed signalling links, to activate new signalling links, and to deactivate aligned (working) signalling links. There exists a basic set of signalling link management procedures, and this set of procedures are provided for any international or national signalling system. The basic set of procedures are: *signalling link activation* (used for signalling links that have never been put into service, or that have been taken out of service), *signalling* link restoration (used for restoring otherwise active signalling links that have failed), *signalling link deactivation*, and *signalling link set activation*.

The Signalling Connection Control Part

SCCP enhances the services of the MTP level 3 to provide the functional equivalent of OSI's network layer (layer 3). The addressing capability of MTP is limited to delivering a message to a node and using a 4-b service indicator (a subfield of the SIO) to distribute messages to MTP users within the node. SCCP supplements this capability by providing an addressing capability that uses DPCs plus subsystem numbers (SSNs). The SSN is local addressing information used by SCCP to identify each of the SCCP users at a node. Another addressing enhancement to MTP provided by SCCP is the ability to address messages with global titles, addresses (such as dialed 800 or free-phone numbers) that are not directly usable for routing by MTP. For global titles, a translation capability is required in SCCP to translate the global title to a DPC plus SSN. This translation function can be performed at the originating point of the message, or at another signalling point in the network (e.g., at an STP).

In addition to enhanced addressing capability, SCCP provides four classes of service, two connectionless and two connection-oriented. The four classes are:

- Class 0:   Basic connectionless class
- Class 1:   Sequenced (MTP) connectionless class
- Class 2:   Basic connection-oriented class
- Class 3:   Flow control connection-oriented class

In class 0 service, a user-to-user information block, called a network service data unit (NSDU), is passed by higher layers to SCCP in the node of origin; it is transported to the SCCP function in the destination node in the user field of a *unitdata* message; at the destination node it is delivered by SCCP to higher layers. The NSDUs are transported independently and may be delivered out of sequence, so this class of service is pure connectionless.

In class 1, the features of class 0 are provided with an additional feature that allows the higher layer to indicate to SCCP that a particular stream of NSDUs should be delivered in sequence. SCCP does this by associating the stream members with a sequence control parameter and giving all messages in the stream the same SLS code.

In class 2, a bidirectional transfer of NSDUs is performed by setting up a temporary or permanent signalling connection (a virtual channel through the signalling network). Messages belonging to the same signalling connection are given the same SLS code to ensure sequencing. In addition, this service class provides a segmentation and reassembly capability. With this capability, if an NSDU is longer than 255 octets, it is split into multiple segments at the originating node, each segment is transported to the destination node in the user field of a *data* message. At the destination node, SCCP reassembles the original NSDU.

In class 3, the capabilities of class 2 are provided with the addition of flow control. Also, the detection of message loss and missequencing is provided. In the event of lost or missequenced messages, the signalling connection is reset and notification is given to the higher layers.

Signalling Modes and Signalling Network Topologies

Signalling networks consist of signalling points and signalling links connecting them. A signalling point that can only originate or terminate signalling messages is known as a signalling endpoint (SEP). A signalling point that can transfer messages from one signalling link to another at MTP level 3 is said to be a signal transfer point (STP). STPs can also provide higher layer functions, such as SCCP. STPs can be stand-alone (providing MTP level 3 switching and possibly SCCP capabilities), or integrated (into the same unit that also provides call/service processing capabilities).

In the signalling system no. 7 terminology, when two nodes are capable of exchanging signalling messages through the signalling network, a *signalling relation* is said to exist between them. Signalling

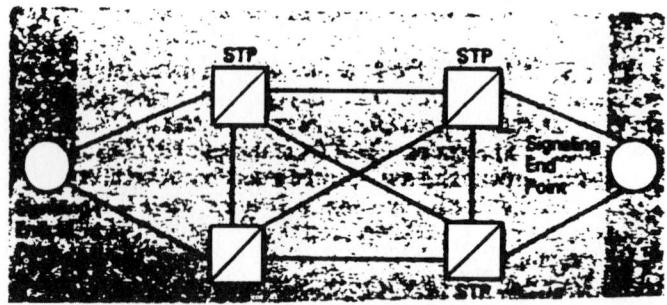

FIGURE 35.4 Signalling network topology (North American networks).

networks can use three different signalling *modes*, where mode refers to the association between the path taken by the signalling message and its corresponding signalling relation. In the *associated mode* of signalling, the messages corresponding to a signalling relation between two points are conveyed over a link set directly connecting them. In the *non-associated mode* of signalling, the messages corresponding to a signalling relation between two points are conveyed over two or more link sets in tandem, passing through one or more STPs. The *quasi-associated mode* of signalling is a non-associated signalling mode where the path taken by the message through the signalling network is predetermined and fixed, except for the rerouting caused by failure and recovery events.

The signalling system no. 7 NSP protocol is specified independent of the underlying signalling network topology and signalling mode. In the interest of keeping the cost down, however, the quasi-associated mode of signalling is used most of the time since the associated mode would require an excessive number of links and terminations in a network of realistic size. Thus, typical signalling networks generally include STPs. To meet stringent reliability requirements, any network topology must provide redundancies in signalling routes (which translates into redundancies in signalling links and STPs). A familiar signalling network topology is the *mesh* or *quad* structure illustrated in Fig. 35.4. The STPs in this structure are mated on a pairwise basis, and facility diversity rules enforced in the provisioning of signalling links ensure that the end-to-end downtime objective of 10 min per year between any pair of signalling end points can be realized. This is the type of topology used in the North American networks.

35.4 Signalling User Parts in SS7 Networks

In this section we briefly describe two major signalling system no. 7 user parts that use the transport services provided by MTP and SCCP: the ISDN user part (ISUP), and the transaction capabilities application part (TCAP). Other user parts such as the operations, maintenance, and administration part (OMAP), the telephone user part (TUP), and the data user part (DUP) will not be covered for the sake of brevity. The latter two user part's functionalities, however, are provided in the ISUP protocol.

The Integrated Services Digital Network User Part

The ISDN user part of the signalling system no. 7 protocol provides the signalling functions that are needed to support the basic bearer service, as well as ISDN supplementary services, for switched voice and non-voice applications. The following summary is based on the 1988 Blue Book version of ISUP [CCITT, 1989c].

ISUP uses the services of MTP for reliable in-sequence transport of signalling messages between exchanges. It can also use some services of SCCP as a method of end-to-end signalling. Figure 35.1 shows the relationship of ISUP with the other parts of the SS7 protocol. The ISDN user part message structure is shown in Fig. 35.5. As seen in this figure, ISUP messages have variable lengths

Overview of Common Channel Signal

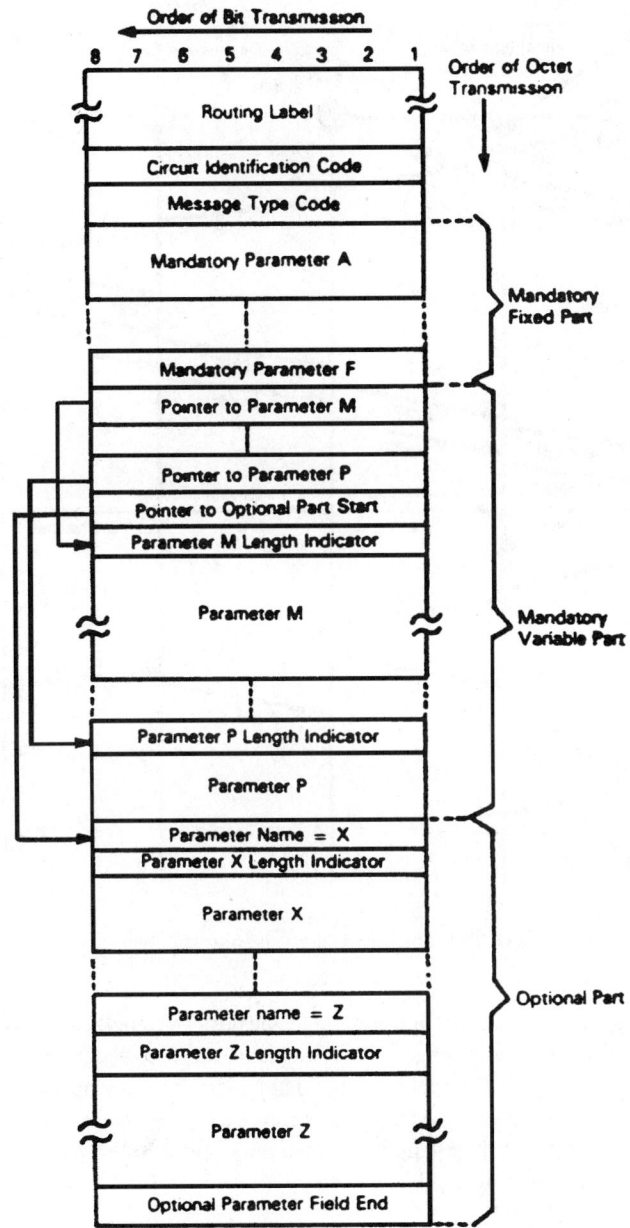

FIGURE 35.5 ISUP message structure.

(an ISUP message can consist of up to 272 octets including MTP level 3 headers). All ISUP messages include a routing label identifying the origin and destination of the message,[1] a circuit identification code (CIC), and a message-type code that uniquely defines the function and format of each ISUP

---

[1] As discussed in the MTP subsection, the routing label is actually an MTP level 3 header (see Fig. 35.3), and not an ISUP header. It is shown in Fig. 35.5 primarily to highlight the fact that (1) the ISUP fields are preceded by the routing label and (2) for each individual circuit connection, the same routing label must be used in all messages associated with that connection.

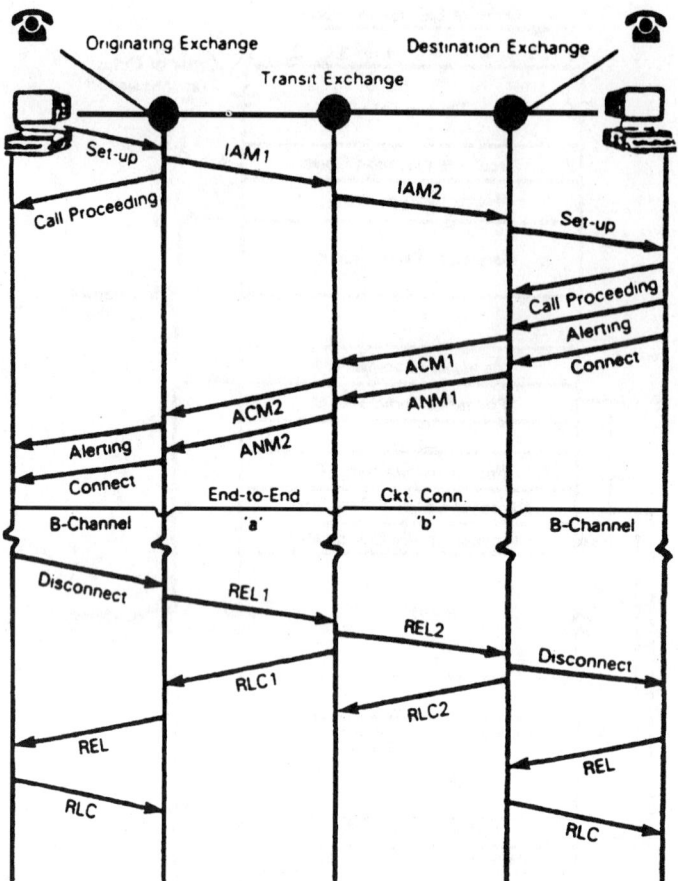

FIGURE 35.6 ISUP basic call setup example.

message.[2] In the mandatory fixed part of the message, the position, length, and order of parameters are uniquely determined by the message type. The mandatory fixed part of an ISUP message is followed by a series of pointers that point to mandatory (and possibly optional) variable-length parameters. Mandatory variable-length parameters are specified by a length field and a parameter-value field, whereas optional variable-length parameters are specified by a parameter name field, a length field, and a parameter-value field.

Basic Bearer Service

The basic service offered by ISUP is the control of circuit-switched network connections between exchange terminations. Figure 35.6 shows a typical basic call setup and release procedure between an originating exchange and a destination exchange, through an intermediate (or transit) exchange, using ISUP messages. The user-to-network or access signalling in this figure is performed by use of the DSS1 protocol (Q.931) on the D-channel. Thus, in response to a Q.931 setup message, the originating exchange launches an initial address message (IAM1) toward the transit exchange for the purpose of setting up trunk a. The transit exchange processes IAM1, sets up trunk a, and launches another initial address message (IAM2) toward the destination exchange requesting use of trunk b. As the destination exchange sets up trunk b, an ISUP-Q.931 interworking takes place in this exchange, resulting in transmission of a Q.931 setup message on the D-channel toward

[2] It is the CIC that provides the identification which relates all ISUP messages for a given *call*.

the subscriber. After the subscriber has been alerted, an address complete message (ACM) is sent by the destination exchange to the transit exchange, which processes it and launches another ACM toward the originating exchange. A Q.931 alerting message is then generated by the originating exchange and sent to the calling station. When the called party answers, a Q.931 connect message received on the D-channel at the destination exchange causes an ISUP answer message (ANM) to be sent to the originating exchange, which sends a Q.931 connect message to the calling station. The circuit-switched path between the calling and the called stations now consists of the cascade of a B-channel on the originating side, trunk a, trunk b, and a B-channel on the terminating side. Call tear down is effected by use of ISUP messages release (REL) and release complete (RLC) as shown in Fig. 35.6. Many other ISUP messages have been defined in support of the basic service for all contingencies that can arise on call setup, during the call, or on call tear down, and for maintenance of associated circuits [CCITT, 1989c].

Supplementary Services

Supplementary services supported by ISUP include calling line identification, call forwarding, user-to-user signalling, closed user group, etc. Short descriptions of the first two of these services which also happen to be the most popular, follow.

Calling Line Identification. The calling line identity presentation (CLIP) service is used to present the calling party's number to the called party, possibly, with additional subaddress information. The calling party may have the option of activating the calling line identity restriction (CLIR) facility, which would prevent the calling party's number from being presented to the called party. The transmission of the calling party's number to the destination exchange can be effected by either the originating exchange including it in the IAM, or by the destination exchange requesting it from the originating exchange through an ISUP information request message. If CLIR is activated by the calling party, the originating exchange will provide the destination exchange with an indication in the IAM that the calling party's number is not to be presented to the called station.

Call Forwarding. The call forwarding service provides for the redirection of a call from the destination originally intended to a different destination. Three types of call forwarding service have been defined: call forwarding unconditional, call forwarding busy, and call forwarding no reply. By requesting the call forwarding unconditional service, a subscriber is able to have the network redirect all calls, or just calls associated with a basic service, originally intended for a user's number to another number. The call forwarding busy service allows the user to do the same but only if the original destination is busy. The call forwarding no reply works in a similar way to the call forwarding unconditional but only after allowing the original destination to be alerted for a specified length of time before redirecting the unanswered call.

Upon receipt of an IAM with a called party number for which call forwarding is in effect, the destination exchange determines if the redirection number is in the same exchange. If so, it alerts that station and sends back an address complete message containing the redirection number to the originating exchange. If the redirection number is in another exchange, an IAM that contains the redirection number as the called party number is sent from the original destination exchange to the exchange with the redirection number. The latter exchange then sends an address complete message containing the redirection number to the originating exchange.

The Transaction Capabilities Application Part

Transaction capabilities (TC) refer to the set of protocols and functions used by a set of widely distributed applications in a network to communicate with one another. In the SS7 terminology, TC refers to the application-layer protocols, called transaction capabilities application part, plus any transport, session, and presentation layer services and protocols that support it. For all SS7

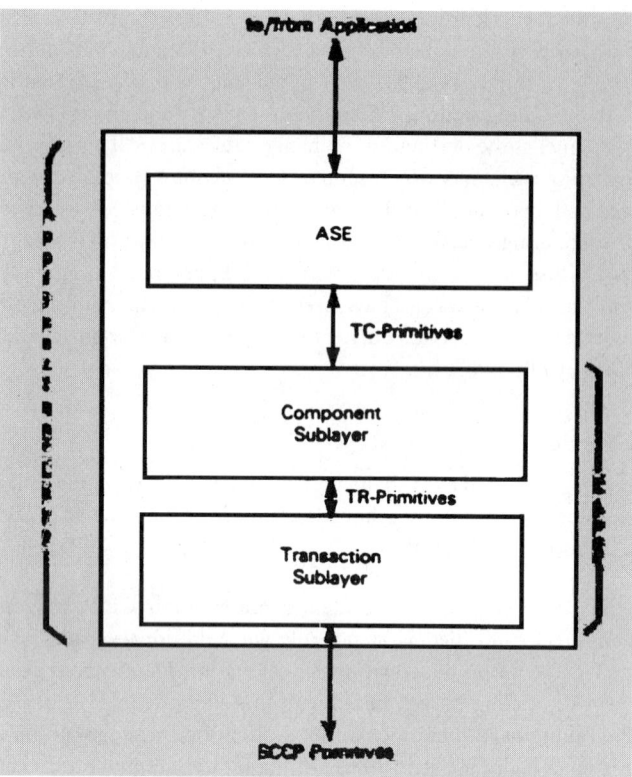

FIGURE 35.7 Application layer structure.

applications that have been designed thus far, TCAP directly uses the services of SCCP, which in turn uses the services of MTP, with transport, session, and presentation layers being null layers. In this context, then, the terms TC and TCAP are synonymous (see Fig. 35.1).

Essentially, TCAP provides a set of tools in a connectionless environment that can be used by an application at one node to invoke execution of a procedure at another node, and exchange the results of such invocation. As such, it includes protocols and services to perform remote operations. It is closely related and aligned (except for one extension [3]) with the OSI remote operation service element (ROSE) protocol specified in Recommendations X.219 and X.229 [CCITT, 1989a]. In the telecommunications network, the distributed applications that use TCAP can reside in exchanges or in network databases. The primary use of TCAP in these networks is for invoking remote procedures in support of intelligent network services such as 800-service (free phone). The application layer structure in TCAP is shown in Fig. 35.7. A TC-user application service element (ASE) provides the *specific* information that a particular application needs (e.g., information for querying a remote database to convert an 800 number into a network-routable telephone number). TCAP provides the tools needed by all applications that require remote operation. TCAP itself is divided into two sublayers: the component sublayer and the transaction sublayer. The component sublayer involves exchange of *components* [the equivalent of protocol data units (PDUs) in ROSE] between TC users. These components contain either requests for action at the remote end (e.g., invoking a process), or data indicating the response to the requested operation. The transaction sublayer deals with exchange of messages that contain such components. This involves establishment and management

---

[3]This extension is the return result–not last (RR-NL) component whose purpose is to carry the segments of a result that would otherwise be longer than the maximum allowed message size.

of a dialogue (transaction) between the TC users. A simplified discussion of the two TCAP sublayers and an example are now given.

The Transaction Sublayer

A transaction (or dialogue)[4] defines the *context* within which a complete remote operation involving, for example, exchange of queries and responses between two TC users, is executed. The transaction sublayer is responsible for management of such a dialogue.[5] Two kinds of dialogues can take place between peer transaction sublayers: unstructured dialogue and structured dialogue. In the unstructured dialogue service, the transaction sublayer provides a means for a TC user to send to its remote peer one or more components that do not require any responses. These components are received by the transaction sublayer from the TC user (through the intervening component sublayer), and are packaged and sent to the remote transaction sublayer in a unidirectional message. There is no explicit association established between peer transaction sublayers for this service.

The second kind of dialogue is the structured dialogue. Here, the TC user issues a TC-BEGIN primitive containing a unique dialogue identification (ID) to the component sublayer. All of the components that the TC user sends within this dialogue would contain the same dialogue ID. The component sublayer maps this TC-BEGIN primitive into a TR-BEGIN primitive containing a *transaction* ID and issues it to the underlying transaction sublayer. There is a one-to-one correspondence between transaction IDs and dialogue IDs. Multiple components received by the component sublayer from the TC user with the same dialogue ID can be grouped into a single TR-BEGIN message containing the appropriate transaction ID. The transaction sublayer manages each transaction (identified by its unique transaction ID), groups components belonging to the same transaction into appropriate BEGIN, CONTINUE, END, and ABORT messages, and transmits them to its peer at the remote end (see Fig. 35.7 and 35.8).

In short, the overall purpose of the transaction sublayer is to provide an end-to-end connection between two TC users over which they can exchange components related to one particular invocation of a distributed processing application. It avoids the OSI connection establishment and release overheads by packaging components in the connection setup and release messages. To reduce the number of signalling messages, it also supports a prearranged release facility where the peer transaction sublayers release their transaction resources related to a dialogue after a fixed period of time without the exchange of an END message.

The Component Sublayer

As alluded to earlier, a component consists of either a request to perform a remote operation, or a reply. Only one response may be sent to an operation request (which, however, could be segmented). The originating TC user may send several components to the component sublayer before the component sublayer transmits them in a single message to its peer at the remote end. Components is a message are delivered individually to the TC user at the remote end, and in the same order in which they were provided at the originating interface. Successive components exchanged between two TC users for the purpose of executing an application constitute a dialogue. The component sublayer allows several dialogues to be run concurrently between two TC users. Such dialogues can be unstructured or structured.

---

[4]A dialogue refers to an explicit association between two TC users, whereas a transaction refers to an explicit association between two peer transaction sublayers. In TCAP, there is a one-to-one correspondence between dialogues and transactions, except in the case of an unstructured dialogue, which does not use a transaction identification.

[5]The transaction sublayer is designed to provide an efficient *end-to-end* connection for exchange of components using the connectionless services of SCCP. In this role, it may be said to provide a very skinny and hybrid substitute for the missing layers (and specifically for the OSI transport layer).

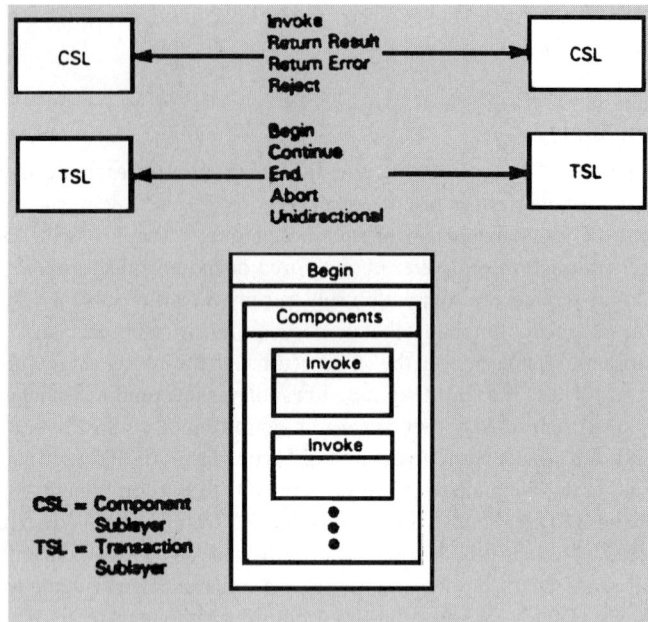

FIGURE 35.8 TCAP sublayer messages.

In the context of a structured dialogue, the component sublayer provides the function of associating replies with operations as well as handling abnormal situations. Associated with any invocation of an operation is a unique component ID. This allows several invocations of the same remote operation to be active simultaneously. The value of the invoke ID identifies an invocation of an operation unambiguously, and is returned in any reply to that operation. The component sublayer allows for four classes of remote operations. In class 1, both success and failure in performing the remote operation are reported. In class 2, only failure is reported, and in class 3 only success is reported. In class 4, neither failure nor success is reported. The replies to an operation could consist of one of the following components: return result (last), return error, or reject depending, respectively, on whether a result, error, or notification of syntax error in performing the remote operation is being provided (Fig. 35.8). Also, due to the signalling message size limitation, the segmentation of a successful result can be provided by the non-ROSE component return result–not last. In addition, any number of linked operations may be invoked prior to transmission of the reply to the original operation.

The reader is referred to CCITT Blue Book Recommendations Q.771–Q.774 for more details [CCITT, 1989d]. The example in the next subsection can help clarify the procedures involved.

A Simplified Transaction Capabilities Application Part Example

By way of a simplified example, consider an interactive 800 (free-phone) service and one way that it may be implemented using TCAP. Figure 35.9 shows the flow of TCAP messages between the originating exchange that has received the 800 call and the network database that contains the information for routing the call. In order for the database to provide the routing number, it is necessary for it to ask the exchange to play a certain announcement to the calling party, collect some more digits, and pass it on to the database.

The first TCAP message sent by the exchange is a BEGIN message that establishes a structured dialogue (and its associated transaction) between the exchange and the database in order to execute this application. Within the BEGIN message, a process that provides the routing number for the 800-service is invoked and given an invocation ID #1. The dialed 800 number is included as a

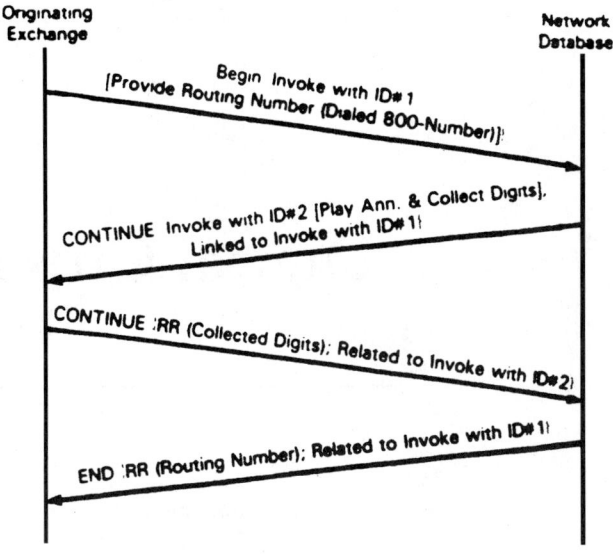

FIGURE 35.9 Example of an 800 service using TCAP.

parameter in the invoke component. The database sends back a CONTINUE message as part of the structured dialogue in which it invokes a linked operation that requires the exchange to play a certain announcement to the calling party and collects some digits. This invocation has invoke ID #2 and is linked (related) to the invocation with invoke ID #1. The exchange performs the required action and sends a CONTINUE message to the database. Within this CONTINUE, a return result component with invocation ID #2 is included containing the collected digits. Upon receipt and processing of this message, the database sends an END message to the originating exchange terminating the dialogue. Within the END message, however, a return result component is included with invocation ID #1, with the final routing number contained in it as a parameter.

References

CCITT. 1981. Specifications of signalling system no. 7. CCITT Study Group XI, CCITT Yellow Books, Geneva.
CCITT. 1985. Specifications of signalling system no. 7. CCITT Study Group XI, CCITT Red Books, Geneva.
CCITT. 1989a. Data communication networks. CCITT Study Group VI, CCITT Blue Books, Vol. VIII, Fascicles VIII.4 and VIII.5, Geneva.
CCITT. 1989b. Specifications of signalling system no. 7. CCITT Study Group XI, CCITT Red Books, Geneva (1992 versions published in White Books).
CCITT. 1989c. Specifications of signalling system no. 7. CCITT Study Group XI, Blue Book, Vol. VI, Fascicle VI.8, Geneva.
CCITT. 1989d. Specifications of signalling system no. 7. CCITT Study Group XI, Blue Book, Vol. VI, Fascicle VI.9, Geneva.
CCITT. n.d. CCITT Recommendations on signalling system no. 6. CCITT Orange Books, Geneva.
Modarressi, A.R and Skoog, R.A. 1990. Signalling system no. 7: A tutotial. *IEEE Commun. Mag.* 28(July):9–35.

36
Digital Cross-Connect System

36.1	Introduction	496
36.2	Functional Description	497
	Hierarchical Multiplex Control • Cross-Connection Nomenclature • Architectural Description • Central Office Architecture	
36.3	Digital Cross-Connect Application: Public Networks	500
	DCS 3/1/0 Applications • Wideband and Broadband DCS Applications • Automated Electronic Digital Signal Cross Connection (EDSX)	
36.4	Digital Cross-Connection Applications: Private and Hybrid Networks	506
	Digital Cross Connection in Private Networks • Digital Cross Connection in Hybrid Networks	
36.5	Operations Support Systems	508
36.6	Digital Cross Connects in Asynchronous Transfer Mode Networks	509
	Introduction • ATM Digital Cross-Connect Applications • Asynchronous Transfer Mode Digital Cross-Connect Fabric Considerations	
36.7	Conclusion	511

Paula Bordogna
Lucent Technologies

Curtis A. Siller, Jr.
Lucent Technologies

36.1 Introduction

Digital cross-connect systems (DCSs) are the almost invisible work horses of today's public telecommunication networks, where they support such disparate applications as facility reconfiguration, performance monitoring, automatic restoration, and remote test access. Since their introduction just over a decade ago, they continue to evolve in terms of technology and application support. With the recent advent of synchronous optical network (SONET) and the incipient market appearance of asynchronous transfer mode (ATM), the future promises to be as exciting as the past.

Lucent Technologies (formerly AT&T) Bell Laboratories commenced development of the first DCS in 1978, followed by equipment deployment in 1981. Known as a digital access and cross-connect system (DACS), the term subsequently came to appear as a generic name (and is still occasionally seen), although it refers to a family of products designed and manufactured by Lucent Technologies Network Systems. Today, DCSs are sold into a truly global market by a comparatively large number of international equipment vendors. They are extensively used in public, private, and hybrid networks, the latter referring to end-user control of DCSs sited in public carrier (e.g., local or interexchange) central offices.

Digital Cross-Connect System

The first DCSs were largely used to replace manual facility cross connection with electronic facility rearrangement at main distribution frames (MDFs), or to groom narrowband special-service circuits (e.g., digital data) from voice traffic, a function previously provided using back-to-back digital channel (D-channel) banks (essentially 64-kb/s channelized multiplexers interfacing with T-carrier networks). Since those early days, cross connects are today used to manage wideband and broadband facilities, plus offering a host of ancillary functions [including support of operations, administration, maintenance, and provisioning (OAM&P)] crucial to modern wide-area telecommunication networks.

The sections that follow provide a comprehensive overview of DCSs. We begin with a brief functional description that not only describes intrinsic capabilities, but also introduces important nomenclature and provides a necessary foundation to understand equipment applications. As previously noted, those applications address public, private, and hybrid network needs; subsequent subsections respectively consider public and then private/hybrid network roles for electronic cross connects. This chapter closes with remarks on ATM and its profound implications for future architectures and applications of cell-based DCSs.

36.2 Functional Description

Hierarchical Multiplex Control

Belying their importance in modern digital networks, DCSs actually perform relatively few intrinsic *functions* that serve as the basis for an impressive array of *applications*. Understanding those functions—**electronic cross connection**, consolidation and segregation (**grooming**), and **multiplexing**—requires, however, a brief review of the digital multiplex hierarchy (DMH).

The DMH is a structured approach for organizing tributary and primary time-division multiplexed digital signals. Without loss of generality, we reference the North American hierarchy (which differs, for example, from the DMH in Asia and Europe), where the rates significant to contemporary DCSs are: DS-0 (0th-level digital signal, frequently referred to as a 64-kb/s *time slot*), DS-1 (1.544 Mb/s), and DS-3 (44.736 Mb/s). Rates and formats below 64-kb/s, denoted DS-0A and DS-0B, are described later; most non-SONET rates above the DS-3 level are vendor specific. The DS-1 and DS-3 can be thought of as primary rates, insofar as they are used for wide-area multiplexed transport, with 24 DS-0s multiplexed into a DS-1 (to 1.536 Mb/s, with an additional 8 kb/s of framing overhead) and 28 DS-1s comprising a DS-3 (43.008 Mb/s, plus 1.728 Mb/s of overhead).

The newer SONET and corresponding international synchronous digital hierarchy (SDH) standards are much more flexible. They currently permit service rates from as low as a 64-kb/s DS-0, to 622 Mb/s ATM channels. SONET and SDH have no format limits; new rates can be created as services require them. In its most widespread implementations today, DS-1s are mapped into virtual tributaries (VT) (1.728 Mb/s, including VT overhead), which are in turn multiplexed into a synchronous transport system-1 (STS-1) synchronous payload envelope (the STS-n path rate is $n \cdot 50.112$ Mb/s), and are then multiplexed into an optical carrier-n (OC-n) optical line rate. The values of n most widely used are 1, 3, 12, and 48; the corresponding line rates are $n \cdot 51.840$ Mb/s (e.g., OC-48 is 2.48832 Gb/s).

As will be described, DCSs manipulate these tributary digital signals and associated time-division multiplexed rates in a provisioned manner. For that reason they are sometimes referred to as slow switches, channel switches, or nailed-up switches. Although the DCS cross-connect fabric itself may be similar, if not identical, to a true switch, use of the term *switch* to characterize this network element is a common misrepresentation of their capabilities and associated applications: DCSs do not handle signalling messages or perform call processing; they rely upon a locally stored cross-connect map that can be reprovisioned locally or remotely via software.

The three principal DCS functions mentioned earlier are conveniently understood in relation to Fig. 36.1, which shows a hypothetical cross-connect terminating seven facilities. For illustrative purposes, we further assume that trunks 1–6 are at the DS-1 rate and carry multiplexed DS-0s (by

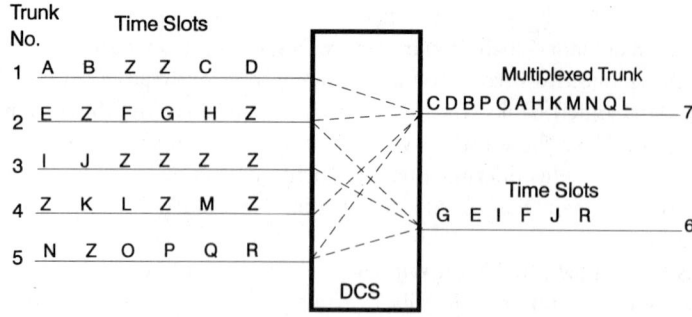

FIGURE 36.1 Digital cross-connect system illustrating cross connection, grooming, and multiplexing.

way of interest, DS-1s are often treated as clear channels, in which case it is immaterial whether or not multiplexed signals are within). Trunk 7 is assumed to be a DS-3. The letters A, B, C, ..., R denote individual DS-0 channels, and Z designates an idle (unused) time slot. For simplicity, framing bits within the DS-1s and DS-3 are not shown.

As the figure infers, select time slots on trunks 2, 3, and 5 are electronically cross connected through to trunk 6, and multiple time slots on trunks 1, 2, 4, and 5 are cross connected and multiplexed onto trunk 7. The simple redirection of channels from one facility interface to another is called *cross connection*; the fact that individual subrated signals can be rearranged (segregated) and directed onto another single trunk is referred to as *grooming*, and the placement of multiple lower speed signals onto a higher speed trunk is *multiplexing*. Additionally, the removal of idle time slots, Z, before live traffic is placed onto other trunks, is an example of *concentration*, and clearly is of significant economic import to service providers, to the extent it provides improved facility fill. Finally, the flexible management of tributary and primary rate signals permits the *adding, dropping,* and *passing through* of channels; this is a function obviously ascribed to *add/drop multiplexers* but often overlooked in the context of DCSs.

DCS cross-connect fabrics that permit completely unrestricted internal rearrangement of channelized bandwidth are said to be *nonblocking*. However, even nonblocking DCSs may corrupt very low-speed signals, carried via DS-1 or DS-3 overhead, that provide communication among intelligent customer premises equipment (CPE) used in private networks.

Cross-Connection Nomenclature

The preceding description is useful in describing a widely used nomenclature for classifying DCSs. Using our current example of a DCS that grooms at the DS-0 level, with cross connection of DS-1 and DS-3 facility rate interfaces, we have what is commonly referred to as a *DCS 3/1/0*. Another extremely important group of cross connects are labeled *DCS 3/1*, in that they offer facility interfaces at DS-1 and DS-3 rates, groom DS-1s within DS-3s, and can multiplex selected DS-1s to the DS-3 rate. Just as a DCS 1/0 obviates the need for back-to-back channel banks, a DCS 3/1 electronically administers DS-1s within DS-3s and, in so doing, eliminates the need for back-to-back M13 multiplexers, multiplexers that aggregate 28 DS-1s to the DS-3 rate. (Such a cross connect obviously offers no signal visibility at the DS-0 level.) As a final example, consider the *DCS 3/3*: This cross connect supports interfaces at only the DS-3 rate and offers simple cross connection. There is no processing of DS-1s, much less DS-0s. Yet each of these cross connects finds prominence in the modern digital network and affords distinct operational advantages or service opportunities to service providers, as described later in this chapter.

Unfortunately, the taxonomy mentioned has not been extended to SONET/SDH-compliant DCSs. Several reasons might be offered, but perhaps the most reasonable explanation is that the synchronous nature of this transmission format invites facile channel rearrangement of VTs, and STS-*n*s

Digital Cross-Connect System

within the higher-level multiplex hierarchy. (Indeed, this synchronous multiplexing format that permits simple add/drop is one of the three cardinal attributes of SONET/SDH, the other two being optical midspan meet for multivendor interworking and comprehensive network management.) It should also be noted that in the context of the DCS nomenclature introduced, some vendors' DCS 3/1/0 equipment may not offer SONET/SDH capabilities. Since their major applications support narrowband voice and special services, they barely reach high enough into the multiplex hierarchy to make SONET/SDH germane. On the other hand, modern DCS 3/1 and DCS 3/3 equipment are very much SONET/SDH network elements. A SONET-compliant DCS 3/1 would provide cross connection of a DS-1 signal from a DS-1 or DS-3 interface to an STS-1 interface, at which point the DS-1 is mapped into a VT within the STS-1, and SONET overhead is added. Further, both STS-1 and VT signals can be cross connected while maintaining continuity of SONET path overhead. A SONET/SDH-compliant DCS 3/3 maps a DS-3 to an STS-1, and also supports STS-1 clear-channel synchronous payload envelope (SPE) cross connection while preserving path overhead.

Architectural Description

As shown in Fig. 36.2, DCSs are architecturally made up of three principal components: a *cross-connect fabric*, *input/output (I/O) interfaces* to external facilities, and a *controller*. With regard to the first item, fabric functionality has been discussed (and will be revisited later for ATM DCSs).

The network external to DCSs is an exceptionally heterogeneous environment. In spite of the relatively simple DMH mentioned earlier, DCSs must deal with specific facility formats, not just rates. Though several of these interfaces are discussed subsequently with regard to applications, it is important that the following be highlighted: CEPT interfaces, as a gateway to international networks; SDH/SONET interfaces for synchronous networks conforming to international and North American standards; D4, extended superframe, and T1DM 1.544-Mb/s interfaces; digital loop carrier interfaces for switched-service access; binary 8-zero substitution (B8ZS), zero-byte time slot interchange (ZBTSI), and 64-kb/s clear channel formats; asynchronous DS-3; and *clear* DS-1 interfaces.

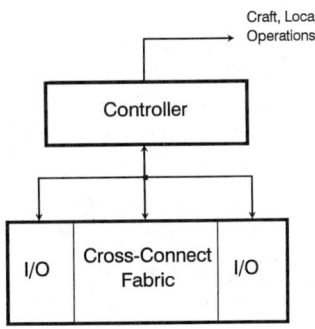

FIGURE 36.2

DCS controllers not only afford direct craft access, support local office alarms, and interface with service providers' operations systems (OSs), they also redundantly store the local DCS cross-connect maps. Backup can be in the form of bubble memory, tape drives, or disk storage.

Central Office Architecture

No central office (CO) is quite like any other. Nevertheless, it is useful to examine their architecture at a high level to appreciate the position of DCSs in relation to other carrier network equipment. As an example, consider the large SONET-compliant CO depicted in Fig. 36.3. To the left, an access network (not shown) provides service interfaces to customers; the right side of the diagram connects this CO to an interoffice fiber network.

Five types of equipment populate this office. They are: digital loop carrier (DLC) or network multiplex (NM) equipment, which terminate the loop feeder network out to carrier serving areas (CSAs); lightwave terminal equipment (LTE) optically interfacing with trunk networks; narrowband/wideband (DS-0 and $n \cdot$ DS-0, respectively) switches for POTS and integrated services digital network (ISDN) telephony, and an assemblage of DCSs.

DLC systems principally handle narrowband special service circuits and basic telephony applications, and are deployed in point-to-point, linear add/drop, or ring arrangements. NMs either

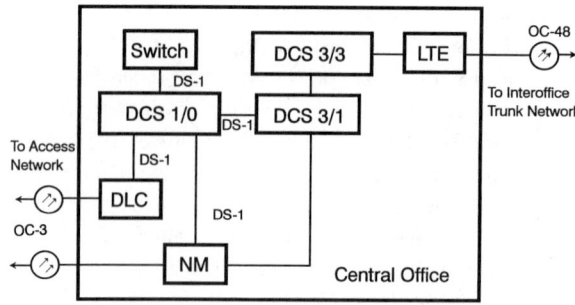

FIGURE 36.3 Representative central office.

interface to DLCs in the loop plant, or bring private line (generally DS-1, infrequently DS-3) traffic into the CO. In either case, in a SONET environment, they both will typically bring traffic into the office over OC-3 optical fiber, with intermediate to long-reach optics. The DCS 1/0 grooms locally switched traffic to the collocated switch, with traffic destined to the interoffice network sent to a DCS 3/1. After cross connection, grooming, and multiplexing, the DCS 3/1 feeds a DCS 3/3, which itself interfaces with an LTE system. LTEs may serve point-to-point, add/drop, or ring topologies. They provide line-level protection switching and performance monitoring above the SONET path level, generate threshold crossing alerts, and support interoffice optical regenerators. Today's SDH/SONET-compliant LTEs typically provide OC-48 facility interfaces to the interoffice network. Depending on the vintage, size, and sophistication of the CO, manual cross-connect panels (DSX-1, DSX-3) may or may not be present. Additionally, note that intraoffice connections among the NM, DCS 3/1, DCS 3/3, and LTE are not specified in this figure, since they can differ depending on equipment supplier and product generation.

36.3 Digital Cross-Connect Application: Public Networks

Digital cross-connect systems were initially introduced in the public network as a means to electronically groom or segregate message service channels from special service channels. It was soon recognized that the electronic nature of the DCS was readily extendible to higher bandwidth channels and to many applications where manual cross connects could be automated using DCSs. Because of operations savings attributed to automated reconfigurations and remote, centralized operations, DCSs were able to "pay for themselves," leading to wide deployment. This section addresses several of these DCS applications.

DCS 3/1/0 Applications

DCS 3/1/0 digital cross-connect systems provide automation and centralization for efficient management of DS-0 and subrate channels. This section describes several DCS 3/1/0 applications.

Grooming and Consolidation

Interoffice facilities typically contain a mix of message and special services, the former terminating on the trunk side of a local switch, and the latter part of a nonswitched or private line network. DS-0-level switched service channels, for example, can be groomed from special service channels using the 1/0 cross-connect capability of the DCS 3/1/0. In addition, consolidation capability of the DCS allows for the collection of channels such that the fill on the facilities is optimized. DCSs perform all activities digitally, retaining the integrity of the signal. Before the advent of the DCS, separate networks with suboptimum fill would need to be maintained for message and special services, or back-to-back channel banks would be deployed to allow for manual grooming and consolidation

Digital Cross-Connect System

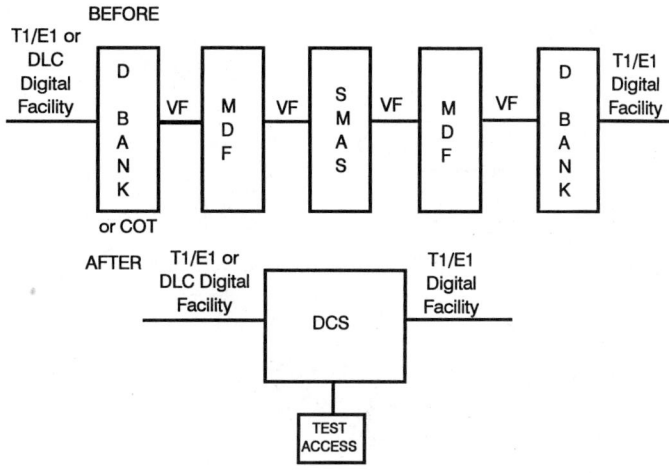

FIGURE 36.4 Digital cross-connect system back-to-back channel bank replacement.

in the central office. Even within a purely segregated special services network, back-to-back D-banks, which drop the signal within the digital facility back to analog voice frequency (VF) for manual manipulation, would have been required to groom and consolidate services for transport to the appropriate end office. Referring to Fig. 36.4, back-to-back D-banks, MDFs, and switched maintenance access systems (SMAS), used for manually switched test access, can all be replaced by the DCS. Not only is the requirement for manual channel manipulation eliminated by the DCS, but also the need for digital-to-analog-to-digital signal conversion and the associated signal degradation.

DCSs also eliminate the need for back-to-back DLC central office terminals (COTS)-to-D-bank arrangements often found in end offices. The DCS can be used to groom and consolidate message and special services traffic that is served via digital loop carrier technology. The DCS passes the message traffic through to the interface of a digital switch, leaving the enhanced facility intact for the operational capabilities required for direct digital switch capabilities. The DCS consolidates the special services and converts them to standard format for interoffice facility transport. This DCS application allows for optimal digital loop carrier facility fill by eliminating the inefficiencies of maintaining separate message and special service access networks.

Replacement of back-to-back D-banks provides ease of operation and increased signal quality. DCSs have been found to economically prove-in over back-to-back D-banks through hardware and operations savings for approximately eight special services digital facilities within an intermediate office, resulting in their popularity and wide deployment.

Hubbing Networks

Hubbing networks using digital cross-connect systems help network planners manage special services circuit demand efficiently. Special services circuits exhibit a large amount of unexpected requests for new or disconnect of services, known as churn, making forecasting difficult. Hubbing helps planners manage churn by providing flexible use of network capacity. As a result, customer service requests can be responded to quickly, and capital construction of networks can be managed effectively.

Hubbing networks using DCSs are often described using the airline analogy of central points or traffic hubs, used to maximize airplane facilities, vs dedicated point-to-point routes. DCSs serve as an excellent vehicle for hubbing, as illustrated in Fig. 36.5. Point-to-point routes may be justified where sufficient traffic between two points exist; however, in areas where little traffic exists between two points, unfilled capacity becomes costly. Some improvement in fill can result by manually routing traffic between interoffice routes to direct a circuit from one point to another, but such

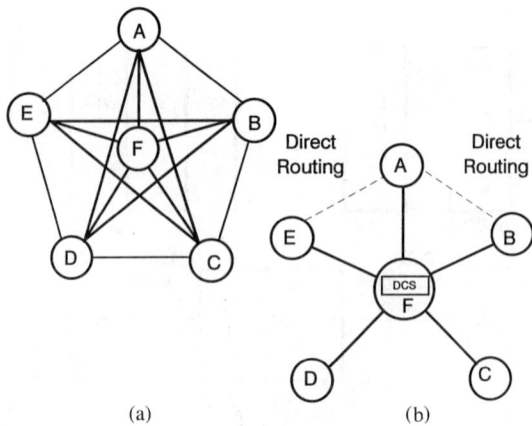

FIGURE 36.5 Digital cross-connect system hubbing application: (a) direct routing and (b) hubbing.

a scenario requires a high level of forecasting and facility engineering. Facility efficiency can be greatly improved using the DCS in a hubbing arrangement. The electronic capability of the DCS allows for a flexible network that is easily administered, highly responsive to unexpected demand, and is less expensive overall.

Where dedicated routes are not economically justified, DCSs provide the flexibility to use routes that optimize facility fill. As the network grows, the increase in traffic may justify a dedicated route. The hub DCS allows electronic rollover to a direct point-to-point facility, eliminating costly manual rearrangement, and freeing circuit capacity on the hub route.

Multipoint Voiceband Circuits

DCSs can be used to connect voice and voiceband data circuits into multipoint circuits or bridges without conversion to analog signals. Analog bridges, manual arrangement at MDFs, and SMAS test points are eliminated by the digital bridging capability built into DCSs. Symmetrical voice conference circuits, broadcast circuits, and polling data circuits are all digitally implemented within today's DCS systems.

Subrate and Multipoint Digital Data Circuits

DCS 3/1/0 systems support grooming, consolidation, and hubbing applications of subrate (2.4, 4.8, 9.6, and 19.2 kb/s) and 56-kb/s digital data services. Multiplexing a subrate circuit from a single customer into a DS-0 is referred to as a DS-0A format; multiple subrate circuits of differing speeds result in a DS-0B formatted time slot. The DCS rearranges the digital services on interoffice facilities, and adds and drops local service channels as required. Multipoint bridging of these services is also provided electronically within the DCS. Performing subrate capabilities in the DCS eliminates the needs for out-of-date subrate multiplexer and multijunction unit systems, which require separate, manual maintenance procedures. By supporting digital data services on the DCS, they can be provisioned, tested, and maintained in an integrated fashion with other DCS-supported services. The DCS also supports error correction features, allowing for the high quality of service demanded by data users.

DS-3 Signal Applications

DCS 3/1/0 systems allow electronic arrangement of signals on a DS-3 facility at the DS-1, DS-0, or subrate signal rates, for true 3/1/0/subrate cross-connect capability, without any external DS-3 facility equipment. This capability allows for reduced equipment expense, allowing direct connection to an electronic DCS 3/3 or a lightwave system for interoffice transport. For DS-3 signals containing DS-0 or subrate-level signals that require rearranging in the central office, the

Digital Cross-Connect System 503

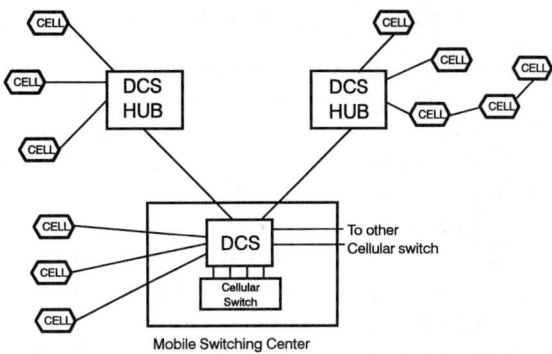

FIGURE 36.6 Digital cross-connect system cellular applications.

DS-3 termination on the DCS 3/1/0 eliminates additional 3/1 multiplexer equipment and offers increased integration of central office operations.

The DS-3 termination feature on the DCS 3/1/0 system also can perform the 3/1 cross-connect applications typically offered on a separate DCS 3/1 system. In small office applications or private networks, where a limited amount of 3/1 applications are needed, the 3/1 cross-connect capability integrated within the DCS 3/1/0 system can offer a more economical solution over a separate DCS 3/1 system. Furthermore, DCS 3/1/0 systems support standard DS-3 performance monitoring and transmittal of DS-3 performance data to far-end DS-3 equipment.

DCS 3/1/0 in Cellular Applications

The DCS 3/1/0 cross-connect capabilities can be extended into the wireless network to achieve more efficient use of facilities. The DCS 3/1/0 can be used to groom and consolidate services coming from various cell sites into the mobile switching center (MSC) to achieve better fill of the ports on a cellular switch. Similarly, a DCS 3/1/0 can be placed at a cellular hub location to cross-connect signals on the incoming DS-1s to ensure better utilization of the facilities into the MSC. Figure 36.6 illustrates this application.

International Gateway Applications

DCS 3/1/0 systems can serve as a **gateway** between international E1 (2.048 Mb/s) and North American DS-1-formatted T1 (1.544 Mb/s) for international applications [Reichard, 1988]. The gateway functionality provides A-law to μ-law processing conversion and signalling conversion between the E1 and T1 standards, and vice versa. Not only does this application allow for smooth interoperability between international and North American networks, but it also allows for other DCS 3/1/0 applications, for example, grooming, consolidation, hubbing, and remote operations, to interwork across global networks.

Wideband and Broadband DCS Applications

Wideband digital cross-connect systems provide automation and centralization for efficient management of DS-1, VT1.5, and STS-1 signals.[1] Wideband DCSs are also called DCS 3/1s, but because of their ability to manage SONET/SDH signals, the former terminology is preferable. Similarly, *broadband* DCSs are used to manage DS-3 and STS-1 signals. This section describes several SONET wideband and broadband DCS applications [Fabricius, 1991].

---

[1] Similar applications apply at the consultative committee on international telephony and telegraphy (CCITT) and SDH rates in international networks.

Automated Electronic Digital Signal Cross Connection (EDSX)

Without wideband and broadband DCSs, high-bandwidth services could only be managed manually at digital system cross connection (DSX) frames for connectivity among central office equipment. Also, external central office equipment was required for testing and performance monitoring. With wideband and broadband DCS systems, the network becomes *automated*, that is, it can be provisioned, maintained, and administered from remote, centralized operations centers.

Wideband and broadband DCSs electronically provide all of the functions traditionally performed at manual DSX-1 and DSX-3 cross-connect frames, respectively, as well as the functions required for SONET VT1.5 and STS-1 signals. These functions include cross connection, facility rolling, patching, loopbacks, and test access. Furthermore, wideband and broadband DCSs eliminate other manual central office equipment by providing full performance monitoring data for through and terminating facilities.

The operations cost savings provided by allowing the network to be intelligently managed justify replacement of the manual cross-connect frames with electronic DCSs. Eliminating manual wiring and operations significantly reduces human errors, such as pulling the wrong jumper cables. Via continuous monitoring of performance data at a centralized, remote site, service degradation can often be discovered and repaired before the customer notices, allowing for proactive, instead of reactive, network management. Service provisioning and turn-up can also be performed faster, by eliminating the need to dispatch craft to the central office.

Service Flexibility via Automated Grooming and Consolidation

SONET wideband and broadband DCSs' electronic capability increases the flexibility of facility and equipment rearrangement through centralized, remote administration and control. Figure 36.7(a) illustrates a central office arrangement without SONET DCSs. Fiber access facilities are terminated on back-to-back multiplexers, which are hard wired to DSX frames, where grooming and consolidation is manually performed. In addition, test access and performance monitoring are manual operations performed at the DSX and require external test equipment. In Fig. 36.7(b), the wideband DCS offers interoffice grooming for both SONET and asynchronous facilities. VT1.5 or DS-1 tributaries can be electronically rearranged on terminated STS-1 or DS-3 facilities. Without the wideband DCS VT1.5 cross-connect capability, grooming below the STS-1 signal level would require terminating the SONET signal to groom at the DS-1 level, hence breaking the SONET end-to-end path continuity.

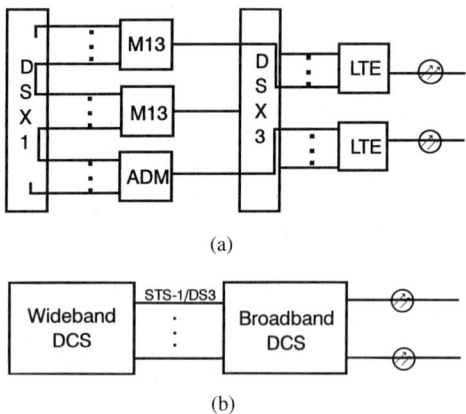

FIGURE 36.7 (a) Grooming and consolidation with back-to-back multiplexers. (b) Automated grooming and consolidation.

Wideband DCSs eliminate the need for the multiplexer equipment, the DSX frames, and all of the cabling associated with that equipment, resulting in reduced capital costs. Furthermore, operating costs are reduced significantly by eliminating the need of manual wiring and by automating OAM&P activities. Performance monitoring and test access capabilities are built into the DCS, allowing the network to be maintained from a central operations system. Provisioning and new service turn-up can beperformed remotely as well, significantly reducing the time to satisfy a customer's service request for a T1 service, for example.

The automated, electronic functionality in the wideband DCS lends itself to several other applications such as automated facility roll, analog-to-digital switch conversion, automated service restoration, and VT1.5/DS-1 network hubbing. For example, when an analog switch is upgraded to a digital switch, the wideband DCS can be used to assist in rolling the facilities from the old switch to the new, significantly reducing the amount of manual roll and the possibility of manual errors. Broadband DCSs offer similar functionality at the DS-3/STS-1 rates.

SONET Network Applications

Wideband and broadband DCSs offer flexibility and operational benefits to SONET networks as well as to traditional asynchronous networks [Couture and Steinhorn, 1990; Von Buttlar, 1993]. An important application that these DCSs have to offer is the asynchronous/SONET gateway function. SONET access and transport systems are being deployed rapidly in public networks; however, there will continue to be a period during which both SONET-based and asynchronous systems coexist. Wideband and broadband DCS systems can be used to maintain SONET path continuity within SONET, and can serve as the bridge between the asynchronous network and SONET.

Wideband DCSs maintain SONET paths while allowing grooming and consolidation at the VT1.5 and STS-1 signal rates. Also, the wideband DCS cross-connect matrix maintains the VT1.5 path while grooming/consolidating VT1.5 signals within STS-1 signals. Without the SONET networking capability built into the wideband DCS, the signal would need to be demultiplexed down to the asynchronous DS-1 level, and the SONET path would be terminated.

Wideband DCSs can be used to provide the gateway between asynchronous and SONET subnetworks. Asynchronous DS-1 signals can be crossconnected and mapped into VT1.5 signals within an STS-1, with the appropriate SONET overhead added, for transport into the SONET subnetwork. VT1.5 path performance monitoring is also provided by the wideband DCS. Conversely, the wideband DCS can terminate the SONET path and output DS-1s, or DS-1s within DS-3s, into the asynchronous network.

Similarly, broadband DCSs can cross connect the SONET STS-1 clear-channel SPE while supporting STS-1 path continuity, and can provide the gateway between asynchronous DS-3 signals and SONET STS-1 signals.

Network Restoration

Service outages can affect millions of people and can take hours or even days to restore. Service outages caused by human error, equipment failures, fiber cuts, or even complete central office disasters can be restored within minutes via the flexibility of a DCS-based network. Wideband and broadband DCSs can be used for monitoring and rerouting DS-1/DS-3-based services in response to an outage. Furthermore, broadband DCSs are instrumental in managing total network survivability by offering interoffice flexibility at the DS-3/STS-1 signal rates [Ponder, 1992]. Figure 36.8 illustrates a survivable network controlled by broadband DCSs. Following an outage in central office D, traffic between CO A and CO E is restored via DCS rerouting in CO A and CO B.

The broadband DCS is capable of reconnecting and rerouting traffic signals around the troubled area almost instantaneously. Through the centralized monitoring capability of the DCS-based network, the troubled areas can be detected. Upon notification of the failure, a centralized controller can then determine the network's free capacity and can automatically reroute traffic affected by the failure. Free capacity may be available by virtue of that set aside for future service growth, or can be made available by using capacity freed up by shedding low-priority traffic during the outage.

Network restoration can also be performed through the use of survivable SONET rings. DCSs play a role in interconnecting SONET rings within a network, and managing the ring capacity to optimize ring utilization [Liese, 1991]. Figure 36.9 illustrates the role of wideband and broadband DCSs in SONET ring interworking. In the figure, traffic is groomed and managed between access-to-access rings and access-to-interoffice rings at the V1.5 signal level within the wideband DCS. The broadband DCS manages the interoffice STS-1 signals.

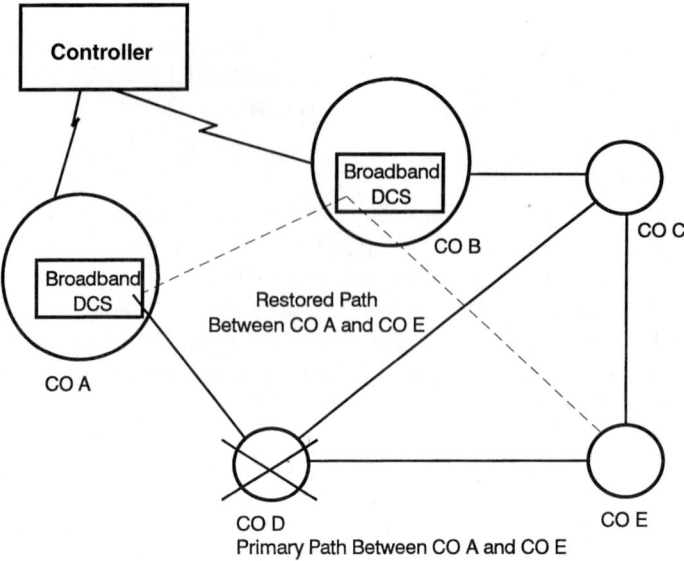

FIGURE 36.8 Digital cross-connect system network restoration.

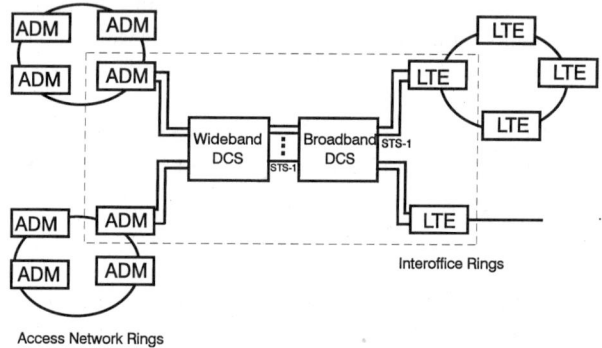

FIGURE 36.9 Digital cross-connect system traffic management between rings.

36.4 Digital Cross-Connection Applications: Private and Hybrid Networks

Digital Cross Connection in Private Networks

Digital cross connection, either explicitly as stand-alone equipment or implicitly as a networking function, plays an extremely important role in private networks. (The significance of private networks can be appreciated by noting that domestic equipment sales into private networks is estimated to be roughly that of communications equipment purchased by the combined regional Bell operating companies.)

Leased-line private networks are usually referenced to 1982, when AT&T filed Tariff No. 270, High Capacity Terrestrial Digital Service, making 1.544-Mb/s (T1) digital transport affordable between customer premises locations. The availability of T1 carrier facilities came at an opportune time: Interest was already increasing in automated offices and distributed data processing, both of which required economical and efficient transport of large amounts of data. Additionally, whether stimulated by new service offerings or developing independently, the concomitant trends of convenient data access with multiplexing at the network periphery and placing network management in customers hands were emerging. These factors, taken collectively, propelled the annual T1 multiplexer market from almost nothing in 1982 to over $1 billion today.

T1 multiplexers, which commonly rely on an internal DCS function or external DCS, are customer-located terminal equipment that multiplex a wide variety of voice, data, and video inputs to the DS-1 rate. They offer comprehensive network management from any network node, including remote I/O (soft) port provisioning, maintenance, and diagnostics. Further, these multiplexers can reconfigure a complex mesh network which exceeds 100 nodes, with reconfiguration initiated by craft or automatically (as to time-of-day scheduling or alarm activity). Typical low-speed inputs are analog voice, digital voice, asynchronous and synchronous data, and video up to the trunk speed. The multiplexers can be used to terminate only a few T1s at small remote customer sites (access multiplexers) or manage bandwidth at large private network nodes that may see 100 or more T1s converge at a single point (nodal processors). The need to add/drop and pass through lower speed time-division multiplexed (TDM) or virtual channels is an essential aspect of private T1 networks and a critical feature of T1 multiplexers. This identical function is, as described earlier, a principal aspect of DCSs.

The role of digital cross connection in private networks is not substantially different than the needs they meet in carrier networks, albeit private network DCSs are generally smaller than their public network counterparts, and are owned and solely dedicated to use in a single customer network. In private networks, DCSs are customarily used to: dynamically route around link or nodal failures; reallocate bandwidth on a time-of-day basis (e.g., shift available bandwidth from voice to data for bulk file transfers during evening hours); shed bandwidth, using priority classification, in the event of facility overload or failure; load balance private network traffic to minimize the impact of network outage; and optimize network configuration to take advantage of changing service tariffs. Management of T1 multiplexers and DCSs can be either centralized or distributed; cross-connect maps to accommodate the needs just mentioned can be either stored centrally, stored locally, or dynamically established using algorithmic methods.

Internodal communication among the smart CPE was at one time carried via low-speed data communication channels transported via the DS-1 superframe. For the most part, this approach has given way to in-band administrative channels, inasmuch as communication via the DS-1 superframe was observed to be blocked as private network DS-1 traffic passed through public network cross connects. This limitation in carrier-based DCSs has generally been resolved.

Finally, today's private network TDM DCSs function at the 1/0 level and commonly include cross connection and grooming of sub-DS-0 channels implemented using proprietary multiplexing formats. Some networks use frame relay packet transport, in which case the DCSs are more akin to ATM DCSs, as described later.

Digital Cross Connection in Hybrid Networks

Service providers recognized that central-office-based DCSs can be used to support the communication needs of private networks. In particular, such DCSs, either under direct end user control or administered by a carrier on behalf of a business customer, offer the functionality of end user owned cross connects, but at less cost by virtue of shared usage. Again, these are customarily DCS 1/0 equipment. And where end user control is offered, either in-band or out-of-band control messages communicate with one or more DCSs. At the same time, at least two limitations proscribe the utility of these hybrid arrangements. First, there exist no standards for carrier-wide DCS control; implementations are generally limited to a single service provider. Second, regulatory constraints in the U.S. limit Bell operating companies to geographically limited regions, their service offerings typically are not uniform with other operating companies, and the need for nationwide coverage will often require coordination with an interexchange carrier.

Placing market issues aside, it is important to note that such hybrid services are available and extend to DCS 1/1, DCS 1/0, DCS 3/1, and DCS 3/3 equipment. Hybrid networking places certain requirements on carrier-based DCSs, including the need to provide: security with user identifiers, passwords and partitioned databases; centralized operation to ensure resource synchronization; a software base, which addresses private network needs; high-reliability, fault-tolerant operation; and customer friendly user interfaces. As noted, these DCS-based services have experienced limited success in the market to date, though they could be the forerunner of more successful virtual networks based on ATM DCSs.

36.5 Operations Support Systems

Since digital cross-connect systems are software based, they fit well into the public network philosophy of remote, centralized OAM&P work centers [Doherty, 1990]. In fact, it is the DCSs' ability to be controlled from a central location, coupled with the electronic flexibility DCSs add to the network, that leads to overall reduced network costs. Figure 36.10 illustrates the network operations architecture for DCSs. The addition of a DCS controller provides an integrated environment for OAM&P functions.

The provisioning process without the DCS requires the generation of a work order and dispatch to the COs where the new service or facility traverses, for manual implementation of the order. With DCSs, the service circuit and facility provisioning process is now automated. Work orders can electronically flow from the provisioning center to the appropriate central offices, where the DCS electronically interconnects the central office equipment.

When the cross connections are complete, the DCS controller can notify the centralized test center, which then requests preservice test setups. Via its cross-connect capability, the DCSs provide various test access arrangements for circuit or facility testing. The automation of the provisioning process results in faster service turn-up and reduces the potential for manual error, thus allowing network providers to respond quickly to their customers' service requests.

DCSs also allow for more sophisticated network monitoring and faster trouble isolation. The DCS provides the centralized maintenance center with detailed performance monitoring and alarm information. DCSs enhance the maintenance center operations by allowing for potential problems to be isolated and corrected before it impacts the end-customer's service. These capabilities added to the network restoration applications provided by DCSs allow network providers the ability to offer enhanced service quality to their end customers.

Owing to the automated and remote operations capabilities inherent in a DCS system, DCSs have also become a crucial part of end-customer control service applications. Network providers can offer business customers the control of their own virtual network by providing these customers

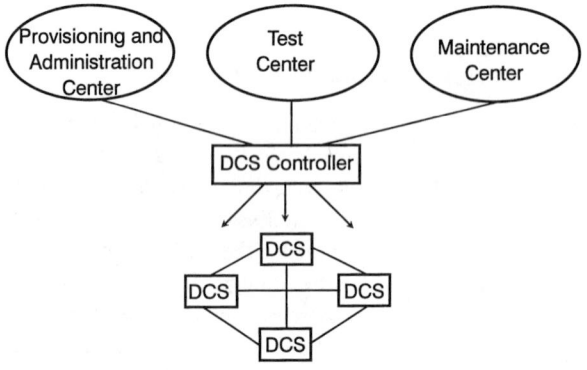

FIGURE 36.10 Operation support of a digital cross-connect system network.

access for various electronic rearrangement applications such as time-of-day circuit administrations and service restoration. This gives network providers new revenue-generating opportunities, while giving business customers greater service flexibility. DCS 3/1/0s, wideband DCSs, and broadband DCSs offer similar operational and end-customer capabilities at their respective cross-connect signal rates.

36.6 Digital Cross Connects in Asynchronous Transfer Mode Networks

Introduction

Prior discussion has largely focused on the role of DCSs in synchronous (DS-0, DS-1, SDH/SONET) and asynchronous (DS-3) circuit-oriented networks. Further, it was noted that in public networks broadband DCSs provide a facility cross connection function, and, in so doing, support network management via service provisioning, performance monitoring, fault detection and isolation, and topology reconfiguration. The latter is the basis for network resilience in response to link or nodal failures. Insofar as carrier-provided, wide-area ATM networks will be logically built on top of a facility network, many of the applications already cited remain pertinent, though effected by the underlying ATM transmission technology.

In the mid-1980s, standards bodies were studying three transport modes for the broadband Integrated Services Digital Network (B-ISDN): synchronous transfer mode (STM), packet transfer mode (PTM), and ATM. STM is the customary transfer mode for digital voice, with information transfer at fixed time intervals; PTM assumes variable-size units of information, each bearing a destination address, and introduced into the network at variable intervals; whereas ATM is similar to PTM, though based on fixed information blocks, customarily referred to as *cells*.

After much debate ATM was selected as the target transport modality, predicated upon asynchronous time-division multiplexing with SDH/SONET as the transport layer. (SDH/SONET offers two particular advantages for ATM: easy concatenation of bandwidth within the SPE, thereby forming very large channels, and a common transport and management platform that is independent of payload format.)

In 1989 agreement was reached that the ATM cell size be 53 bytes (often equivalently referred to as octets, either term referring to 8 b), where it will be noted that short cell size favors voice, a service that is very delay sensitive, whereas longer cell sizes were favored by the data communications communities. By 1992, the cell format and 5-byte header details had been agreed upon. Though outside the scope of this particular chapter (and treated elsewhere in this book), it is sufficient to note that the header provides a labeled routing field, which associates a particular channel with the cell. The routing field has only local significance. Just as the position of a time slot circuit within a time-division multiplexed channel can change on a node-by-node basis (discussed previously for circuit DCSs), so, too, the cell routing address will generally change as it traverses from input to output in an ATM DCS or switching element. The address field itself is divided into two components: a virtual path identifier (VPI) and virtual channel identifier (VCI). VCIs can be grouped within a VPI and routed as an aggregated entity, as in interconnection of ATM routers or private branch exchanges (PBXs).

Two final points are offered regarding the ATM transmission format. First, the 48-byte (53 total, minus the 5-byte header) payload field carries user information, though even within this field overhead bytes must be set aside to adapt certain traffic types to ATM; these overhead types are called the ATM adaptation layer (AAL). Second, most wide-area networking of ATM cells, especially in the early stages of this market, assume a connection-oriented (as opposed to connectionless) connection mode. That is, a connection—in this case a virtual circuit—is established among communicating parties before information exchange occurs, and that connection is maintained over a single path through the network for the duration of the session.

With this prolegomenon aside, we now ask: What are the implications for DCSs? By way of an answer we note ATM DCSs can be used to administer the existing hierarchically multiplexed facility network, ATM DCSs can be the basis for virtual network services, and the internal architecture of ATM DCSs differs fundamentally from circuit-oriented DCSs (e.g., DCS 3/1 and 3/3).

ATM Digital Cross-Connect Applications

ATM DCSs can mimic the functions of circuit-oriented DCSs. ATM AAL-1 provides for mapping synchronous, continuous bit rate (CBR) traffic into the ATM payload. (This is sometimes called synchronous-to-asynchronous conversion, or SACing.) Consequently, carriers can prepare their networks for ATM traffic by first seeding them with ATM DCSs that externally appear to function as circuit cross-connect nodes. Indeed, though these ATM DCSs have inherently different fabrics, the I/O ports can potentially draw upon existing, cost-reduced technology, and many carriers see great potential in simplified OAM&P using ATM. On the other hand, differences will arise in two areas, namely, facility management and the integration of LTE functionality into the ATM DCS. Today carriers typically manage their networks at the DS-3 level, reflecting extensive deployment of DCS 3/3s. With management at the DS-3 level, LTE equipment provides the necessary interfaces to the optical fiber interoffice trunk network. With extensive carrier deployment of ATM on the horizon, many expect the physical fiber network to remain in place, facility management to move to a higher, SDH/SONET-level in the DMH (e.g., STS-12), and integration of LTE functionality into the DCS via direct optical termination.

In time networks will evolve so that a greater fraction of traffic is native ATM and what SACing is necessary will move to the network boundary. In this environment, ATM DCSs provide fiber interfaces, with ATM cells multiplexed into the SPE. Concurrently, it may be necessary to adopt whole new approaches to survivable restoration. Where restoration is currently based on circuit levels in the DMH, reconfiguration for robust networking will likely be based on VPI/VCI readdressing, possibly in conjunction with higher level facility management. These are areas of intense investigation, and consensus has not been reached on the optimum approach(es).

In a more mature ATM network environment, ATM DCSs become service elements directly accommodating end user needs, in addition to their more traditional role in facilitating network management on behalf of the service provider (i.e., carrier). It will be remembered that many of today's business services are based on leased private lines. This situation is not expected to change in the future, and given a trend toward increased wide-area transport of bursty data, native ATM traffic can be expected to increase dramatically. At the same time, the virtual ATM network (sharing and controlling common transmission resources among many users) accrues economies that can be passed on to the end user via appropriate tariffs.

Asynchronous Transfer Mode Digital Cross-Connect Fabric Considerations

DCS 3/1/0, DCS 3/1 and DCS 3/3 equipment host cross-connect fabrics that can require product-specific design to optimally support facility interfaces and provide visibility to their respective groomed tributaries. Clearly, DCS realization based on channelized TDM technology impacts development resources, equipment price, and time to market. Though different technologies are available in ATM DCS fabric realization, conceptually, at least, any particular fabric allows cross connection of any tributary signal. Consider that in an ATM stream, regardless of the multiplexed stream speed, the principal distinction between applications with signals of differing rates are (1) their routing address and (2) the comparative periodicity of cell appearance. For example, in an ATM stream carrying CBR traffic, cells corresponding to a DS-3 application are expected to appear 28 times more often than an embedded CBR DS-1 application. From the perspective of a cross connect, since one cell is pretty much like any other, the predominant distinguishing feature is the

VPI/VCI cell address, not its temporal position in the multiplexed input. Even more significant are variable bit rate applications. For this case, one cannot easily anticipate the periodic appearance of individual cells. Rather, the DCS will likely see cell bursts, in which clusters of cells (not necessarily contiguous in time) with the same VPI/VCI address must be mapped to and then multiplexed onto a specific output facility interface.

Basic ATM cross-connect fabric design and the related challenge of designing interconnection stages to create ATM DCSs of varying throughputs is still evolving. This reflects not so much the status of international standards, but rather the newness of the market and uncertainties as to the important applications and how they affect fabric choice. Still, three general approaches currently seem to garner most attention: *shared memory, shared medium* and *self-routing* [Mollenauer, 1993]. With regard to shared-memory fabrics, incoming cells are basically placed in memory, with pointers identifying where cells destined for a specific output are to be found. Output port logic essentially steps through a pointer chain, retrieving stored cells for delivery to the multiplexed output. With parallel bit processing this fabric technology can be pushed to relatively high output speeds (tens of Gb/s) and the approach is straightforward provided the number of input/output ports is limited.

Shared-medium architectures see I/O controllers, processors, and memory attached to a common bus. Arriving cells on the bus are examined as to their distinctive (output) address and forwarded to the associated port. In this case, cells are processed one at a time, overhead for internal processing is minimal, internal delay is small, but the aggregate output is probably only several gigabits per second.

In the self-routing fabric, headers are used to self-route a cell through a series of gates until the destination port is finally reached. An especially common form of self-routing occurs in a Banyan fabric. Much like the tree and branch structure of the Banyan tree, each cell is examined at every node in the switch, sequential elements of the bit address are used to route the cell, a process which, though simple, ultimately finds the cell at the prescribed destination. This architecture obviously requires buffers to resolve internal cell collisions (the simultaneous arrival of cells at a decision point). Additionally, this fabric architecture is not amenable to broadcast or multicast applications that require cell replication.

Each of the fabric designs described has been implemented by major vendors in the development of ATM equipment that performs cross connection or switching. Each approach has pros and cons that suit their use to specific applications, and other architectural alternatives undoubtedly exist. Regardless of any specific implementation, fabrics must deal with buffering (to accommodate cell bursts and collisions), aggregate throughput, cell loss, delay, replication, and architectural considerations, which impact modularity.

36.7 Conclusion

DCSs play an important, if not always conspicuous, role in modern digital networks. They are not only found in carrier COs, where they support the administration and integrity of wide-area transmission, but also in private and hybrid network settings. Today these cross connects are predominantly used to manage DS-1 and DS-3 facilities. The introduction of synchronous networks is occurring rapidly, stimulated by international SDH/SONET standards, and the fact that future fiber networks will be the basis for ATM-based B-ISDN.

Defining Terms

Electronic Cross Connection: The most basic of digital cross-connect functions, in which traffic on one terminated facility is internally redirected via the cross-connect fabric to one or more other terminations.

Gateway: A digital cross-connect application in which the network element serves as a demarcation between two dissimilar networks, for example, between plesiochronous and synchronous (SONET) networks or North American and international networks (with differing standards for voice coding and channelization).

Grooming: One of the basic functions of a digital cross-connect system, wherein subrate channels on a terminated facility are segregated and electronically cross connected onto another facility. Hence, grooming is the basis for consolidating traffic.

Hubbing: The application of bringing transmission facilities to a single point for centralized digital cross connection and rerouting, vs a mesh network with extensive facility interconnection.

Multiplexing: The function of aggregating traffic on multiple low-speed facility terminations to a higher speed termination. As such, some digital cross-connect systems provide this capability in addition to electronic cross connection and grooming.

References

Couture, R.W. and Steinhorn, D. 1990. Broadband and wideband DCS applications in the asynchronous and synchronous environment. In *IEEE Global Telecommunications Conference Record*, Vol. 2, pp. 1251–1256, Dec. 2–5.

Doherty, D.K. et al. 1990. High capacity digital network management and control. In *IEEE Global Telecommunications Conference Record*, Vol. 1, pp. 60–64, Dec. 2–5.

Fabricius, W. 1991. DACS means world-class networking. *Telephony*, 221(8):22–26.

Liese, S.T. 1991. The network planner's dilemma. *Telephony*, 221(18):24–30.

Mollenauer, J.F. 1993. The impact of ATM on local and wide area networks. *Telecommun.*, 27(3):35–43.

Ponder, D. 1992. A network survival guide. *Telephony*, 222(8):36–41.

Reichard, D. W. 1988. Digital cross-connect systems CEPT gateway application. In *IEEE International Conference on Communications Conference Record*, Vol. 1, pp. 331–335, June 12–15.

Von Buttlar, R. 1993. Technology in transition the cross-connect revolution. *Telephony*, 224(8):27–33.

Further Information

Few articles provide a detailed view of cross-connect configuration, features, and internal subsystems. An especially good example is found in "DACS II: Core of the Flexible Network," by F. H. van Unen and published in *Trends in Communications*, Vol. 6, No. 3, pp. 119–134, December 1990.

For a particularly comprehensive article describing cross connects from an operational perspective, see "Digital Cross Connect Systems—A System Family for the Transport Network," by J.-O. Andersson, *Ericsson Review*, Vol. 67, No. 2, pp. 72–83, 1990.

A comprehensive, very readable overview of ATM in relationship to switch attributes, architectures, and market needs is presented in "Finding the Right ATM Switch for the Market," by R. Rooholamini et al., *Computer*, Vol. 27, No. 4, pp. 16–28, April 1994.

37
Building Future Networks by Using Photonics in Switching

Stephen J. Hinterlong
Lucent Technologies

David Vlack
Lucent Technologies

37.1 Introduction..513
Protection Switching • Call-by-Call Switching • The Role for Photonics in Switching
37.2 Bit Sensing Switching Versus Bit Rate Transparent 515
37.3 Bit-Sensing Photonic Devices in the Network....................516
Digital Regeneration of the Data • Large Pinout Capacity Reduces System Complexity • Repartitioning of Heat Loads • Improved Signal Simultaneity • Chip-to-Chip Registration
37.4 Multistage, Call-by-Call Switching Fabric Demonstration 520
Optical System • System Operation • Fast Reconfigurable Circuit Switch
37.5 Bit Transparent Photonic Devices in the Network...............525
Small Integration Density
37.6 Lithium Niobate 16 × 16 Switching System Demonstration... 526

37.1 Introduction

Migrating photonics into switching is occurring in a variety of ways. This is happening because of the large variety of switching tasks to be performed in a national communications network. The different switching tasks range from ensuring that major trunks have the ability to be switched from one route to another when a catastrophic accident destroys a route, to the real-time switching of high-speed calls on a per call basis. Different applications of photonics in switching are best served by different approaches. Two generic switching applications that are in use in the present network are **protection switching** and **call-by-call switching**. Both of these applications appear to be candidates for optical switching techniques.

Protection Switching

One application that demands extremely high bit rate communications and does not use the information bit stream as it flows through the switching element is that of protection switching high-bandwidth highways of information. Normally this switching or reconfiguration occurs at a relatively slow time frame (minutes to days) and is accomplished through human intervention.

At the request of a customer, or due to unusual circumstances, a complete fiber transmission facility may need to be switched. Consider the case where a fiber system is damaged in a construction accident. Due to prior arrangement, switches will be located to reroute the data away from the damaged fiber and onto another fiber. This type of switching is referred to as protection switching. Protection switching is only interested in restoring the entire fiber data channel so that the terminals of the fiber link can again send and receive information on the fiber conduit. The terminals of such high-bandwidth links are usually switching systems that aggregate many user's calls. Thus, if a fiber is cut, the result is hundreds to thousands of customers without service. The ability to quickly restore networks via protection switching is an important aspect of network operation.

Call-by-Call Switching

Another type of switching involves the call-by-call establishments of connections between endpoints across a network. In order for any connection to occur smoothly, a collection of network activities must be initiated and coordinated to successfully connect a user to the desired destination. Normally the user will initiate a connection by sending what is referred to as **signalling information** into the network. This signalling information must be correctly received by the first network entity, normally the local switching office, and interpreted to assess what must happen next. If the local switch determines that the call will be local, its activities are entirely local. It makes the connection locally and sends signalling information to the destination party, usually a ringing telephone. If the connection is not a local connection, then in addition to reading the incoming signalling information, the local switch relays signalling information to a switch remotely located on the network. The remote switch may initiate signalling to the destination or pass along the signalling to another switch in the path to the destination user. In any case all switches on the network must recognize the signalling information and react. As the number of calls increases or the signalling paradigm changes [e.g., Integrated Service Digital Network (ISDN)] to become more complicated, the computing power of the switch must increase to ensure the same call completion performance. An equivalent view is that new services and message style signalling will demand increased computing performance at a switch to terminate signalling information.

A new method of multiplexing is called **asynchronous transfer mode (ATM)**. ATM is one way of providing very high-bandwidth availability to the user of a network. In order not to encumber an entire high-bandwidth pipe from the user into the network, ATM specifies small packets of information be sent into the network instead of a continuous stream of data. If the user desires a high bandwidth, then many packets per second would be sent into the network. If the user requires less bandwidth, then fewer packets per second are sent into the network. This provides the user a choice of available bandwidths, which are selectable by the user.

To accomplish ATM switching, the switching elements that carry the users connection must recognize the individual packets and route them to the desired destination. Accompanying the packet is routing information for that particular packet. The standard for ATM packets (or cells as they are sometimes called) is that the data part of the packet be composed of 48 bytes and the signalling information or header be composed of 5 bytes. Roughly 10% of the information of the packet is header information for that packet. When the user wants a high-bandwidth connection, the amount of header information can be quite large. For example, if the user would send packets into the network at an average rate of 150 Mb/s the average signalling data rate would be approximately 15 Mb/s. If the switch were to terminate a thousand users fibers the potential average header bandwidth could be 15 Gb/s at the switch. Since each header must be processed to route the cell, extremely powerful computing machinery and algorithms are needed to route the packets with minimal delay. Fortunately, ATM cell header information is arranged in a way that routing can be accomplished with minimal examination of the header data in most cases.

The Role for Photonics in Switching

Both protection and call-by-call switching require specialized equipment suited especially for each application. In the case of call-by-call switching, the system must be capable of assessing the incoming signalling information, reconfiguring the local switching fabric and originating signalling for the outgoing call information. In the case of protection switching, the connections are maintained for a long duration until the network needs to reconfigure, yet the bandwidth available for the reconfigured transmission paths may be extremely high. Thus, protection switching represents a challenge because of the bandwidth that must flow through the system. Call-by-call, or *real-time switching* must be capable of reacting autonomously to information received from the users of the network. The challenge for a real-time switching system is to make connections as a response to customer input information. Real-time switching and protection switching are the extremes of the switching operation. Switches in a modern telecommunications network must have attributes of both kinds of switching. Efficient operation requires autonomous actions taken by the switching equipment as a consequence of user input and supplying potentially high bandwidths to satisfy individual and aggregated user needs. The following sections review different aspects of the devices that are used in photonic systems. By analyzing the different attributes of these technologies, a perspective on their appropriate use can be gained. Two representative experiments will be discussed to highlight two alternative approaches to building photonic switching systems.

37.2 Bit Sensing Switching Versus Bit Rate Transparent

Just as there are different needs (call-by-call and protection) in network switching, there are different photonic technologies available for the different applications. In a broad sense, two categories of photonic switching devices are **bit rate transparent** and **bit sensing**. Bit rate transparent photonic hardware is insensitive to the data that are flowing through the device. Thus, the specific bit rate is of little consequence in transporting the input data to the output of a photonic switch employing these devices. Bit sensing devices consume the data stream and register the change as individual bits. By sensing the data, it may be retimed and regenerated for further, downstream use. A comparison of some of the distinguishing features of the two types of devices is given in Table 37.1.

Generally, the bit transparent devices are useful in a limited number of input, high-bandwidth switching systems. Examples of bit transparent devices are lithium niobate [Alferness, Economou, and Buhl, 1981], indium phosphide laser amplifier arrays [Gustavsson et al., 1992], or indium gallium arsenide waveguide modulators [Zucker et al., 1992]. Bit sensing devices are most useful in more complex, large-number-of-input systems that do not require extremely high per channel

TABLE 37.1 Comparison of Bit Sensing and Bit Rate Transparent Devices

Bit Sensing Devices	Bit Rate Transparent Devices
Limited bit rate because signal energy consumed in changing the state of the digital system.	Analog transmission path allows unlimited bit rate through the device.
Complex systems are possible because of the data regeneration capability of the device.	Only simple systems are allowed due to the effects of analog data impairments such as loss and crosstalk.
Synchronization between the data and the fabric control possible because both are in digital format.	Synchronization between fabric control information and the data is difficult because they are in different representations.
Possible to integrate control and data in device because both are digital.	Difficult to integrate control and data circuitry on device because of mixed digital and analog representation.
Binary digital interface is standard.	Analog interface is variable. Defined by system.
Usually single wavelength operation.	Allows multiple, simultaneous data streams with different wavelengths.

data rates. Examples of the bit rate sensitive device would be symmetric self electro-optic effect devices (S-SEEDs) [Lentine et al., 1989], exciton absorption reflection switches (EARS) [Yamaguchi et al., 1994], double heterostructure optoelectronic switches (DOES) [Simmons and Taylor, 1987], or vertical to surface transmission electrophotonic (VSTEP) [Nishio et al., 1992] devices. In each case (transparent and sensing) there are applications which demand the unique advantages offered by both classes of device.

37.3 Bit-Sensing Photonic Devices in the Network

Sensing the bit stream in the optical domain is accomplished by various optically sensitive devices (S-SEEDs, EARS, and DOES devices). These devices are characterized by the ability to detect an optical signal's binary state, modulate a binary optical output, and in some cases perform logic on the data stream as it flows through the node. An example of a device that performs logic on data received optically is a field effect transistor-self-electro-optic effect device (FET-SEED) [Lentine et al., 1992]. This device can detect the optical signals incident on its inputs, interpret those signals as 0s and 1s, and provide the binary data to the logic section of the chip. When the logical operation is completed, the data is read from the FET-SEED chip by modulating beams of light. The ability to receive data optically, process the data electronically, and then resend the data optically relieves some of the problems with high-speed electronic circuits. For example, high-speed electronic chips receive the data via terminated transmission lines and send the data via terminated transmission lines. Whereas transmission lines provide an acceptable solution for small numbers of inputs and outputs, the space required for hundreds or thousands of transmission line inputs and outputs exceeds the space available. The only solution for the extremely high density of I/O per chip is optical imaging. In the optical domain the number of signals per area that may be accessed on the chip is larger than any other technology.

There are advantages to maintaining the digital format via digital optical inputs, digital optical outputs, and digital electronics for processing. A digital system has the advantage over an analog system of being much less sensitive to noise. The signal, represented in the digital domain, is more immune to interference. Thus, the digitally represented signal received will better represent the signal sent.

The heat generated on these devices is less than the heat generated on equivalent throughput electronic devices. This is true for devices that have many inputs–outputs (I/Os) at high-speed because of the communications energy needed. As mentioned earlier, current high-speed chips communicate to their surroundings via transmission lines. The circuits to drive large currents to quickly charge the transmission lines are not 100% efficient. The cost of the interface circuit inefficiency is in the heat generated. If the heat generated by a transmission line driver were only 10 mW, for a system with 1000 outputs the total heat generated by the interface circuitry would be 10 W. Avoiding the generation of heat in a complex system by minimizing the heat generated in communicating has to be a priority for system engineers. Optical communication avoids the heat generation because the output from the individual chips is achieved by illuminating a modulator. The power to change the state of a modulator in a short time is more like the power required to switch a transistor in the processing electronics rather than the power to drive a transmission line. Typical switching powers are between 100 μW and 1 mW for bit rates of between 100 Mb/s and 1 Gb/s.

All of these attributes allow system designs, with unique and beneficial partitioning. There are restrictions that optical interconnections impose on system designs, however. One particular restriction that must be addressed is that of maintaining the registration of the signals as they move from chip to chip. To achieve the large number of I/Os per chip while maintaining small chip size implies large optical I/O density. The spacing between adjacent I/Os is typically on the order of 100 μm. The optical detector and modulator active areas are on the order of 10 μm. The power required to change the state of the detector or modulator, which is proportional to the devices area,

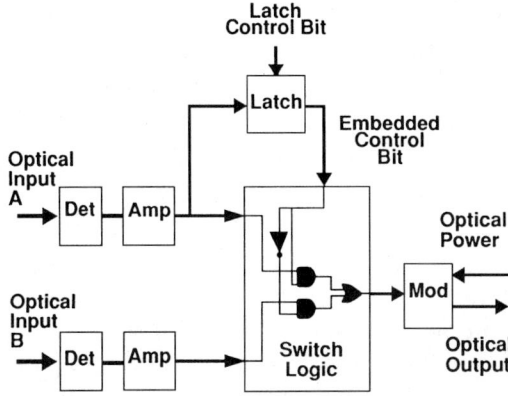

FIGURE 37.1 Schematic view of single FET-SEED broadband (2, 1, 1) switching node.

can remain small. Small-sized detectors and modulators require precise alignment in order for a system to operate successfully.

A concrete example of an optical device that satisfies the criteria of digital performance, large optical pinout, and minimized on-chip power dissipation is the FET-SEED device used at Bell Laboratories. The current FET-SEED node design is composed of two differential optical inputs, a differential optical output, and processing electronics. See Fig. 37.1. These fundamental circuits are replicated in a regular square array to form a complete chip. One version of the chip has an array of 16 nodes arranged as 4 rows by 4 columns. The goal of combining multiple nodes in an array was to build a complete switching fabric by optically combining multiple chips. Chips for the next switching experiment have more than 4000 optical I/Os.

Functionally, the nodes act as a router of information from either input A or input B to the output. Controlling the routing function is performed by inserting a control bit in the data stream. This control bit will be stored in the node and, if set to logical 0, will route the data from input B to the output, and, if set to logical 1, will route input A to the output. Electronic very large-scale integration (VLSI) performs functions that are difficult, in a small volume, to perform with optics. Electronics is used to receive and amplify the data and store the state of a routing bit. The routing information is received optically via input A and latched into the electronic memory with an electrical clock pulse that simultaneously supplies all of the nodes in the array. Once the data is received and stored in the FET-SEED memory, the output of the memory is used to determine which of the two optical inputs is relayed to the optical output. In this way the routing information for the network may be embedded within the optical data stream and latched as the data flows through the network. Once latched, the nodes are configured to route data from any of the FET-SEED inputs in the array to the outputs. This in itself is insufficient to build a complete switching fabric but by interconnecting the arrays of FET-SEEDs *optically* a complete multistage fabric may be constructed.

The FET-SEED device array shown is the first of many possible designs. Subsequent arrays may be built entirely differently with the use of more logic or more optical inputs and outputs. In any case the ability to recognize and act on the data stream as it flows through the nodes in a network switch is possible and desirable. The combination of the optical I/O with electronic processing plus the advantages described subsequently allow economical, high throughput switch designs.

Digital Regeneration of the Data

Many optical switching systems of the past have not had the advantage of being all digital. They suffered from the problems of analog systems, namely, loss, crosstalk, and noise impairments. Although engineering an analog system is possible, if the information is in the digital domain it is better to remain entirely in the digital domain for switching. The FET-SEED devices allow the switching of digital data at the expense of requiring enough energy to change the state of the electronics in the switching node. The advantages are repeatability, better noise immunity, and minimal crosstalk. They achieve this performance by building on the principles of digital electronics while simultaneously allowing the advantages of optical I/O.

Large Pinout Capacity Reduces System Complexity

An advantage of optics is its ability to make many, high-speed data connections to an integrated circuit. This ability is important if a system is to be smaller than a comparable electronics system. In comparison, an entire broadband two-input–one-output switching node may be constructed to be about the same size as the transistors used to drive a single output pin of a high-speed electronic circuit. Figure 37.2 shows one FET-SEED, broadband photonic switching node adjacent to the cascade of transistors required to drive a single high-speed transmission line. The two are at the

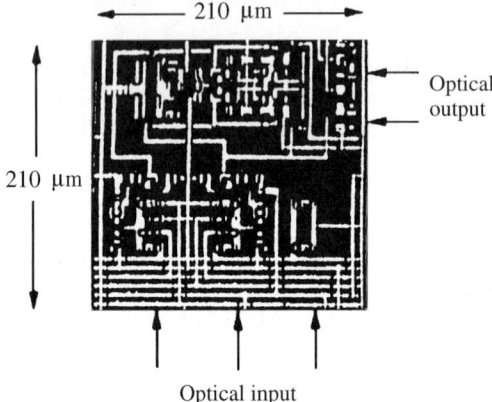

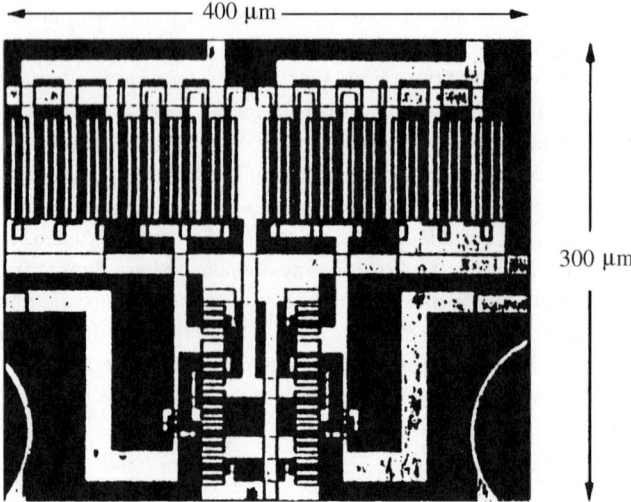

FIGURE 37.2 Comparision of the size of a broadband switching node to the size of a transistor cascade to drive a high-speed transmission line.

same scale. The switching node has two, dual rail optical inputs, processing electronics, and one dual rail optical output, as in Fig. 37.1. Everything that is required to select one or the other high-speed input and route that input to the output is contained within the area of 200 × 200 μm. In contrast, the transistor cascade begins with the connection of a smaller transistor to ever-increasing sized transistors so that the current supplied at the output of the largest transistor is capable of charging a transmission line in the time allowed for high-speed communication. All of the transistors are required because the *processing* transistors are made incapable of sinking or supplying large amounts of current. In order for the data from the processing transistors to communicate outside of the chip, a string of transistors with greater and greater current sinking or sourcing must be built. Larger transistors that supply larger currents require larger drive currents to function. Thus the cascade grows until the final transistor may ultimately drive a transmission line.

The capability to have many pinouts greatly enhances the alternatives that an engineer may use to build connection intensive systems. A comparison is that of a microprocessor system that, in the conventional way, has time-multiplexed data and address busses of a fixed size, for example, 32 b. The number of pins available on the package has defined the size and time-multiplexing scheme used for data and address busses. If the limitation of pinouts were eliminated, the possibility of having nonmultiplexed data and address busses exists. Thus, a microprocessor architecture could be built in which the address and data busses could be separate and distinct for all of the internal execution elements, memory, low-speed I/O, high-speed I/O, and cache memory. With separate busses the processor could be executing in cache with the cache dedicated busses while simultaneously updating memory from disk using the high-speed I/O dedicated busses and the memory dedicated busses. Photonic interconnections offer a single chip solution that will perform all of these functions.

Repartitioning of Heat Loads

Another important factor in the construction of high-speed integrated circuits that they generate more heat than equivalent circuits that run at lower speeds. Part of the difference is in the amount of power used to drive a transmission line. Shown in Fig. 37.3 is a comparison of the power used to drive a transmission line vs the power used to change the state of an optical modulator used for photonic connections. For distances more than about a centimeter and data rates of greater than 100 Mb/s, less power is needed for photonics than for electronics. More power means hotter systems. Heat generated in a system is another constraint on system design limiting construction of large, complex switches or computers. The reason for this is that the distributed capacitance of

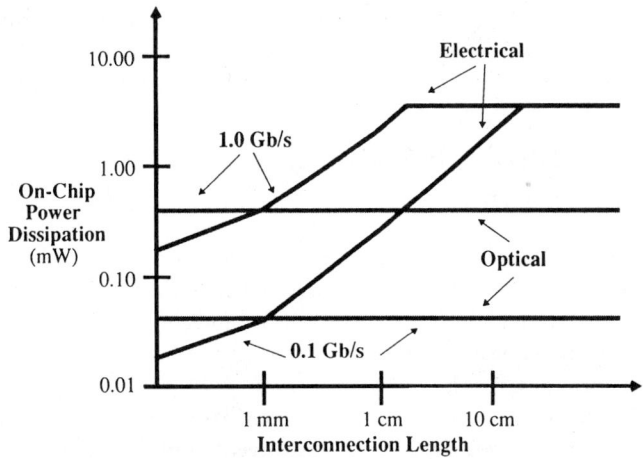

FIGURE 37.3 Optical versus electrical power dissipation on chip.

a transmission line needs to be charged entirely in order for the signal to propagate to the receiver. Thus, for distances of 1 cm and more, it is more energy advantageous to send the data optically for bandwidths of 100 Mb/s or more.

Improved Signal Simultaneity

Another advantage of optics over electronics is in minimizing arrival time differences for signals sent simultaneously on a bus but received at different times. In the electronics case this happens because of various and nonconstant reactive loading on the electrical bus. As the electrical signals propagate, some encounter the load of receivers on circuit boards plugged into the bus. Others, while boards are plugged into the bus, do not use certain signals; thus, all of the signals may not encounter the reactive loading of receivers. This is due to the needs of the architecture to provide connections on some signal lines and no connections to other signals. Because of the differences in the load presented to the bus, some of the signals will propagate faster than others. This leads to the signals arriving nonsimultaneously. If the signals are all to be used simultaneously, the result of this loading is that the wrong data will be used along the bus. This situation, through careful engineering, can be minimized, but as the signal rates increase, timing becomes more critical. The optical approach to the problem is to propagate all of the signals through the same medium, that is, through the glass lenses, beam splitters and *free-space* between the optical components. In this way all of the beams will incur the same delay; no subset of the beams will be delayed more than any others. Here, again, optical systems benefit the designer at high data rates.

Chip-to-Chip Registration

One of the challenges of using this technology is establishing and maintaining the connections between the logically adjacent devices in the multistage fabric. Because the optical I/O is so small many connections can be made between chips. Of course, this also means that the stability of the mounts for these devices must be very good. A way to look at the optical case is that the connections to the chip do not provide the mechanical mounting that an equivalent electrical chip would have. This is one of the drawbacks of the optical interconnection system. Mounting of the chips to ensure a sufficiently stable registration is one of the research topics that will continue to be worked on.

Another consideration of the mechanical mount being separate and distinct from the data connection is that the chips may be tested optically. Any electrical power connection could be made to the chip with the conventional probe technique. Once the power connections are made, the logical functions of the chip could be tested without physically contacting the chip. One could imagine that optical I/O chips could be quickly and thoroughly tested in the wafer form, diced from the wafer, and packaged. Being able to quickly access thousands of test points on a chip would facilitate the packaging process by ensuring that only chips that passed the test would be packaged.

37.4 Multistage, Call-by-Call Switching Fabric Demonstration

When arrays of FET-SEEDs are connected by beams of light it is possible to build a switch capable of call-by-call connections. The system that is described here is the fifth in a series of systems built to evaluate this technology for broadband switching systems, hence it has been called system 5. System 5 employed many new features that has allowed it to function without realignment for eight months at the time this text was written.

The FET-SEED technology operated at a rate of 155 Mb/s in the system. The FET-SEED chips sometimes are called smart pixels because of their organization into a regular array of optical devices that are capable of processing data locally to any element in the array. The shorthand naming convention for these kind of nodes is (2, 1, 1), that is, two inputs, one output from the node, and one path through the node. Despite its apparent simplicity, an array of this type of node can be used to build complex switching fabrics. The designer trades a more complicated switching node design for a more complex interconnection. Fortunately, the interconnection complexity problem is conveniently solved as a consequence of the advantages of free-space photonics.

With the control memory integrated in the switching node, it may be changed at the native rate of the data that is flowing through the node. Because the node can be reconfigured in 1-b time the entire multistage network can be reconfigured in twice the number of stages times the bit period. For example, in system 5 the native bit rate in the fabric is 155 Mb/s (bit period of 6.5 ns) and the fabric has five stages; thus, the reconfiguration time is 65 ns.

Because system 5 was a demonstration fabric it has fewer stages than a large fully interconnected fabric. The number of stages for a large fabric would be no more than three times the number of stages in system 5 and, therefore, the reconfiguration rate would be ~200 ns for that system. This very fast reconfiguration rate is possible because all of the nodes on a chip are simultaneously being reconfigured.

Shown in Fig. 37.4 is a photograph of the FET-SEED (2, 1, 1) switching node chip. On the chip one entire switching node is evident (with power supply and control lines defining the node). Surrounding this node are multiple identical switching nodes. In the center of the picture, to the left side of the single node, are the two receivers for the node data input. On the right side of the node is the latch circuitry and the modulator output. Both the receiver inputs and the modulator output are dual rail signals and, thus, they have two optical widows each.

FIGURE 37.4 Microscope picture of the surface of the FEED-SEED chip used in system 5.

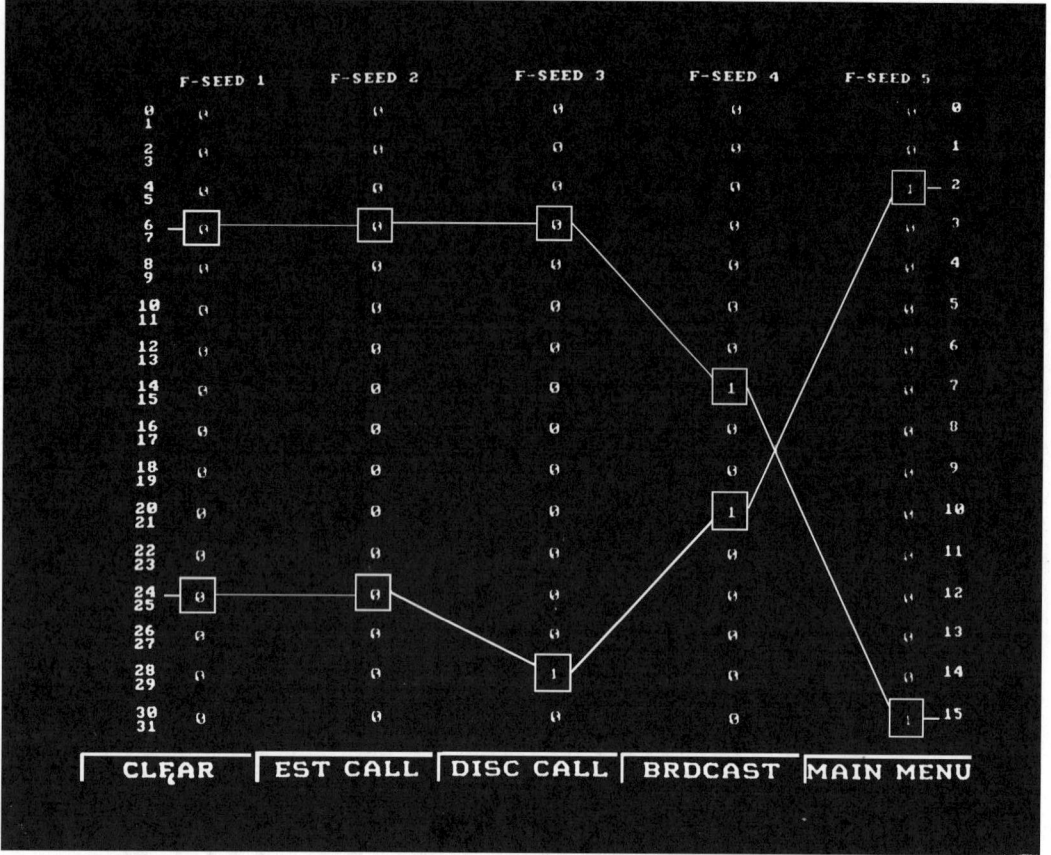

FIGURE 37.5 The interconnection pattern for a banyan network with 80 nodes and five stages.

In operation, when the signals from the receivers have propagated through the electronics to the modulator output it may be illuminated to extract the output signal in the optical domain. Output from the chip requires the modulator windows to be uniformly illuminated with optical power beams, the state of the bit at the modulator will be encoded as beams of light and sent to receivers on a succeeding array.

Optoelectronic broadband switching nodes such as those in Fig. 37.4 are integrated on a single gallium arsenide (GaAs) chip in arrays of 16 elements arranged as a 4 × 4 array. Thus, each chip can receive 32 input beam pairs and route 16 output beam pairs. The beams from the chips are routed in a Banyan interconnection scheme such as that in Fig. 37.5. With 32 input beam pairs and five stages of these FET-SEED devices, any input may be routed to any of the 16 outputs. Once the routing data has been stored in the latch, user data may then flow from either optical input A or optical input B through the switch until the network must again be reconfigured.

Optical System

Because the chips are configured in arrays, the data that emanates from the surface of the chip are themselves an array of spots. This array of spots is identically reproduced, shifted, and imaged with the original array of spots onto the subsequent FET-SEED array. Thus, with an array of 16 beam pairs from the modulators on the previous FET-SEED array, 32 beam pairs are produced to send to the 32 inputs of the subsequent FET-SEED array. It is this fan out by two that makes the routing of all inputs to all possible outputs in the five-stage network.

As the 32 beam pairs from the previous array arrive at the subsequent array they are imaged to the appropriate receivers via polarization optics. The polarization optics are used to provide two input ports to send images of arrays onto the FET-SEED and two output ports to received images of arrays of beams. Shown in Fig. 37.6 are the optics needed to provide these two input are imaging ports and the two output imaging ports. In operation, data from the previous FET-SEED or the input are imaged onto the surface of the FET-SEED devices. These 32 beam pairs propagate the input data, which will be routed to the output port. The 16 output beam pairs exit via the output port to be used by the next FET-SEED array as input data. The data are moved from the chip surface by illuminating the output modulators with an array of 16 equal intensity beam pairs sent via the power port. The light reflected from the modulators carries the information from the chip, encoded as low- and high-intensity beams. These beams emerge from the output port ready to be imaged onto the next array.

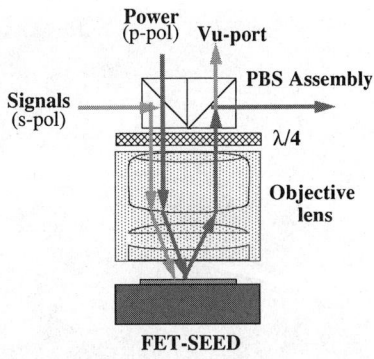

FIGURE 37.6 System 5 showing the input fiber bundle at the bottom, output fiber bundle at the top, power supply lasers under the cover at the left, and the FET-SEED devices at the right. In the center of the picture are the polarization optics and the objective lens used to image the spots onto the FET-SEED windows. The entire system is mounted on a rigid steel baseplate, which provides mechanical registration and temperature stability for the components.

The view port is used in constructing the system. By using infrared sensing cameras the surface of the FET-SEED chip can be observed. Registered on the surface of the array are two sets of spots, one from the signals port and one from the power port. The array of spots from the signals port carries the input data to the FET-SEED. These signal beams must be properly registered on the receiver windows. The second array of spots is from the power port. These are the spots that provide the power to illuminate the modulators. In a fashion similar to the input signals, the power signals must be aligned to the modulator windows.

System Operation

The complete system is shown in Fig. 37.7. Data flow is from the fiber bundle at the bottom of the picture to the fiber bundle at the top of the picture. Because the array of bits from the users is presented as an image at the first-stage FET-SEED, it is routed according to the configuration established in the setup phase of the fabric operation. Between the FET-SEED arrays are the optical interconnections provided by the binary phase gratings and imaging optics. For every output the signals are passively, optically split and directed to two receivers on the downstream array (in Fig. 37.7 this would be the array that is higher in the figure). At the receiving array the nodes route one signal and deny the routing to the other signal. The data routing at every stage proceeds in the same way, always based on the configuration established in the setup phase. By selecting the segments of the path from node outputs to node inputs, a complete path can be built through the fabric.

Just as in conventionally connected digital electronics, the data from the previous stage is regenerated. By regenerating the data at every stage the data integrity at the output is preserved without the problems associated with an analog system.

The system operated continuously for eight months before being decommissioned. Although it has run with 155-Mb/s data rates for every channel, the demonstration bit rate in the fabric is 105 Mb/s. This bit rate was selected based on the native bit rate for the video codecs, which feed the video system demonstration in which video sources are switched to alternate video displays. In the demonstration, switching between two video sources is shown.

FIGURE 37.7 Optical data flow through a FET-SEED showing two input ports and two output ports.

Fast Reconfigurable Circuit Switch

The demonstration described is intended to show the ability of an optical switching fabric to quickly reconfigure based on the input of an external controller. Because of the fast reconfigurability this switching fabric could be applied in a variety of different switching applications. Two that will be considered here are synchronous switching and asynchronous switching.

An attribute of the synchronous switching case is that the switch reconfigures at a constant fixed rate based on the multiplexing format of the data. For example, suppose that there are multiple inputs to this switching fabric and each of the inputs has fixed time slots. Each time slot on each input represents a unique channel of user data. The challenge is to route the data from a channel on one input to a time slot on an output. With buffering at the input and the output, a fast reconfigurable switch could route any input data to any output time slot. Once the timing relationships for all of the inputs to all of the outputs has been established, it could be repeated at the interval at which time slots recycle. In current systems the time slots recycle at a 125-μs rate. For 100 time slots to be routed in 125 μs the reconfiguration and data flow through the switch for each time slot must be completed in 1.25 μs.

Another switching technique is that of asynchronous switching in which cells from users arrive at the periphery of the switch randomly. Each user may send packets at any time into the switch; the more packets per second that are sent into the network, the greater the bandwidth. A user may fill the total bandwidth available on the facility to the switch. In this paradigm it is conceivable that multiple simultaneous logical connection can be made on one physical connection by associating different packets sent with different logical connections. In order to maintain which packets are to be associated with which logical connection, each packet has a unique routing tag enclosed in a header followed by the user data. Based on the header in each packet, the switch calculates onto which output to route the packet and reconfigures. The packets available at all of the inputs are routed, and the next echelon of packets will have their routes calculated in the fabric and eventually be routed. Although this sounds like a straightforward process, the details of calculating the routing from all of the inputs to a switch during the interpacket arrival time are not trivial. As an example, consider a switch with 1000 inputs and interpacket arrival time of approximately 700 ns (622 Mb/s with 424 b per packet, asynchronous transfer mode standard). Each routing calculation must take, on the average, less than 700 ps. Further research is needed to achieve ATM switching in large optical network switches. The basic optical switching fabric previously described could be used in ATM switches.

37.5 Bit Transparent Photonic Devices in the Network

An alternative for photonics protection switching networks is to provide a fundamental multiwavelength optical layer supplying information to a higher electronics layer [Johansson et al., 1992]. In this scheme, the optical layer is capable of wavelength switching and will provide protection switching for the high-layer electronics cross connect system. Consider the network shown in Fig. 37.8.

The optical nodes in this network are wavelength sensitive. That is, any signal of a particular wavelength may be routed to any particular destination within the network. In this case the signals may be routed to other optical nodes within the network via the optical fibers or these signals may be routed to the electronic cross connects that form the higher layer in the network. With optical

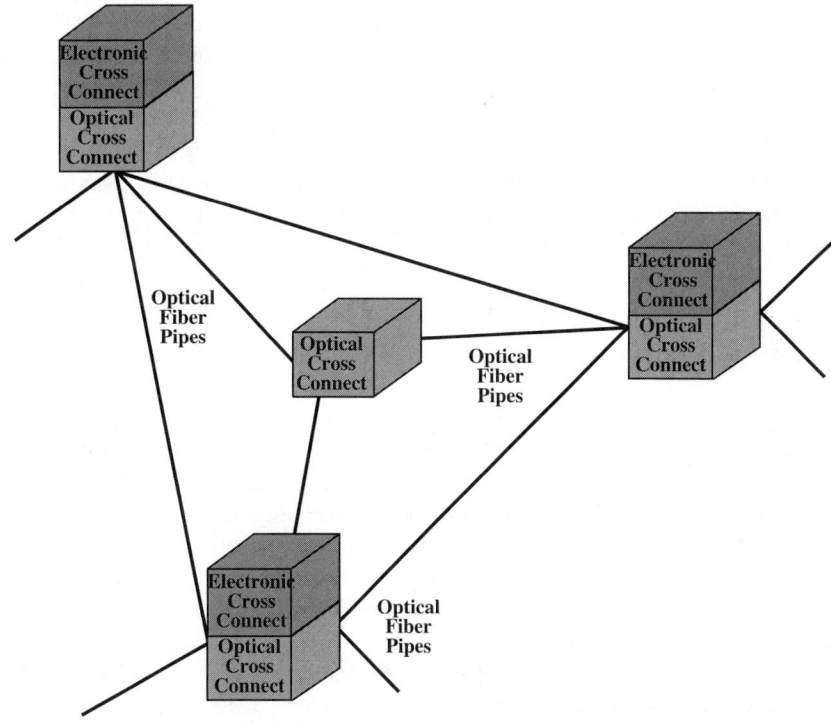

FIGURE 37.8 Optical and electronic cross connect.

cross connects, switching occurs based on the wavelength of the signals arriving at the optical cross-connect nodes. The sense of the bits that are carried on that wavelength for the optical node is unimportant. The signals that are delivered to the electronic cross connect, once decoupled from the other wavelength signals by the optical cross connect, can be switched based on the information contained within the bit stream. This is a case where the data information in the bit stream is unimportant to the optical network, and thus bit rate transparent devices may be used in the system.

This type of system is critical to the successful operation of a national network. The reliability of the network depends on the ability of the optical cross-connect system to reconfigure. Thus, the reliability requirements of the optical cross-connect system is of paramount importance. A failure of critical switching capabilities after a network failure could be catastrophic. Because of the reliability requirements, any protection switching system must be robust and immune to the changes experienced in its environment.

Small Integration Density

Current state-of-the-art guided wave switching elements suffer from not being able to integrate hundreds into a single device. With the lithium niobate technology, an 8×8 multistage switching element has been fabricated. Depending on the needs of the application this size element may be sufficient. If the throughput of the system is less than the product of the number of inputs and the maximum per channel data rate, then a single element will work. For example, if the 8×8 element were used with 1-Gb/s channel data rates, then 8 Gb/s throughput could theoretically be achieved in this switch. The maximum practical throughput for this switch may not be 8 Gb/s because of the guard time required while reconfiguring the switching fabric.

If the maximum throughput of the switch needed to be more than the 8 Gb/s, multiple 8×8 fabrics could be connected together to form a larger switching fabric. The technology to connect multiple fabrics together requires the alignment of multiple fibers to the edges of the substrates containing the lithium niobate fabrics. Single connections of single-mode fiber are difficult. Multiplying the number of fibers that are needed to achieve single-mode coupling is extremely difficult. Although it is possible to connect multiple substrates together, the transition to another material system such as InP seems better. The devices in InP are smaller, and thus many more devices can be placed on a substrate. Unfortunately, the InP material system brings increased index of refraction, and thus coupling standard optical fiber to InP is difficult. More work needs to be done in increasing the integration density of guided wave devices and improving the coupling efficiency of fibers to the InP or lithium niobate substrates.

37.6 Lithium Niobate 16 × 16 Switching System Demonstration

A system demonstration at AT&T Bell Laboratories was conducted to evaluate guided wave devices for applications in synchronous switching systems. The experiment was called distributed switching with centralized optics (DISCO) [Berglund, 1993]. It was designed to switch broadband, voice, data, and video calls over a time-multiplexed network connections through a guided wave photonic network.

The 8×8 photonic element forms the time-multiplexed space switch. The switch configuration changes every 488 ns while recycling to the original configuration after 125 μs. In this way there are 8000 frames/s just as in the standard synchronous hierarchy, and 256 time slots in every frame. Each input to the 8×8 element is connected to a time slot interchanger (TSI), which can demultiplex the time-multiplexed data received at every input. Likewise there is a TSI at every output of the 8×8 switch. Thus, at every time slot, one each of the eight inputs are routed to an output. At the output the receiving TSIs put the received data into their respective time slots on the outgoing channel.

Because the 8 × 8 switching fabric is bit rate insensitive, each time slot may be used for low-speed voice, or high-speed data. In either case the channel through the 8 × 8 element is available for 488 ns. At 622 Mb/s approximately 300 b can be sent between input and output. By concatenating multiple channels, multiples of 300 b can be sent. Alhough this represents a real switching system, a great deal of work remains. Work must be done both on the devices and architectures before practical guided wave switching systems can be realized.

Acknowledgments

This work is partially sponsored by ARPA under Air Force Rome Laboratories Contract F 30602-93-C-0166.

Defining Terms

Asynchronous transfer mode (ATM): A packet-oriented communication protocol that has the capability to make many logical connection on one physical link.

Bit rate transparent: Any devices that are capable of providing a connection between two points that does not detect the data in the stream; an example is a relay for electrical systems.

Bit sensing: A device that detects the logical data in a communications path; an example would be a logic gate in which the energy delivered to the gate is consumed and the logical result of flowing through the gate is reconstituted at the gate output.

Call-by-call switching: Switching that occurs as a result of real-time input from customers who want a connection made between themselves and other endpoints.

Protection switching: Switching in a national network that occurs to restore connections after catastrophic damage in the fiber or cable part of the network connecting switches.

Signalling information: Information that is sent from switch to switch in a national network indicating which connections are to mode through the switch, an example would be signalling system 7 messages (SS7).

References

Alferness, R.C, Economou, N.P., and Buhl, L.L. 1981. Fast compact optical waveguide switch modulator. *Appl. Physics Lett.*, 38:214.

Berglund, G.D. 1993. A photonic time-division switching experiment. In *Photonics in Switching Technical Digest*, p. 200, Optical Society of America, Washington, DC.

Gustavsson, M., Lagerstrom, B., Thylen, L., Janson, M., Lundgren, L., Morner, A.-C., Rask, M., and Stoltz, B. 1992. Monolithically integrated 4 × 4 In GaAsP/InP laser amplifier gate switch arrays. *Electron. Lett.*, 24:2223.

Johansson, S., Lindblom, M., Buhrgard, M., Granestrand, P., Lagerstrom, B., Thylen, L., and Wosinska, L. 1992. Photonic switching in high capacity transport networks. In *Proceedings of the XIV International Switching Symposium*, Yokohama, Japan.

Lentine, A.L., Cloonan, T.J., Hinton, H.S., Chirovsky, L.M.F., D'Asaro, L.A., Laskowski, E.J., Pei, S.S., Focht, M.W., Freund, J.M., Guth, G.D., Leibenguth, R.E., Smith, L.E., Boyd, G.D., and Woodward, T.K. 1992. 4 × 4 Arrays of FET-SEED embedded control 2 × 1 switching nodes. *IEEE Topical Meeting on Smart Pixels*, postdeadline paper #1.

Lentine, A.L., Hinton, H.S., Miller, D.A.B., Henry, J.E., Cunningham, J.E., and Chirovsky, L.M.F. 1989. Self electro-optic effect device: Optical set-reset latch, differential logic gate and differential modulator/detector. *IEEE J. Quantum Electron.*, QE-25:1928.

Nishio, M., Suzuki, S., Takagi, K., Ogura, I., Numai, T., Kasahara, K., and Kaede, K., 1992. Photonic ATM switch using vertical to surface transmission electro-optic devices (VSTEP). In *Proceedings of the XIV International Switching Symposium*, Yokohama, Japan.

Simmons, J.G. and Taylor, G.W. 1987. Theory of electron conduction in the double heterostructure optoelectronic switch (DOES). *IEEE Trans. on Electron Dev.*, ED-34(5).

Yamaguchi, M., Yamamoto, T., Yukimatsu, K., Matsuo, S., Amano, C., Nakano Y., and Kurokawa, T. 1994. Experimental investigation of a digital free-space photonic switch that uses exciton absorption reflection switch arrays. *App. Optics*, 33(8):1337.

Zucker, J.E., Marshall, J.L., Chang, T.Y., Sauer, N.J., Burrus, C.A., and Centanni, J.C. 1992. 15 GHz bandwidth quantum well electron transfer intensity modulator at 1.55 mm. *Electron. Lett.*, 24:2206.

Further Information

Transmission Networking: SONET and the Synchronous Digital Hierarchy, Mike Sexton and Andy Reid, Artech House.

ATM: Foundation For Broadband Networks, Uyless Black, Prentice–Hall.

38
Asynchronous Transfer Mode (ATM) Networking Standards: A Review

Raif O. Onvural
IBM

38.1 Introduction 529
38.2 ATM Architecture 531
   Physical Layer • ATM Layer • ATM Adaptation Layer (AAL)
38.3 Asynchronous Transfer Mode Interfaces 536
   User–Network Interface 3.0 • Data Exchange Interface (DXI) • Broadband Interchange Interface (B-ICI) • Network Node Interface (NNI)
38.4 Conclusions 541

38.1 Introduction

Broadband integrated services digital network (B-ISDN) is conceived as an all purpose digital network envisioned to facilitate worldwide information exchange between any two subscribers without limitations that can be imposed by the communication medium or the media. At least conceptually, B-ISDNs will not only support all types of networking applications that we know of today but also provide the framework to support future applications that we do not fully understand, or even know of, today.

B-ISDN standards are being developed in a number of national standards bodies around the world and by the International Telecommunications Union–Telecommunications Standards Sector (ITU-TS) [former Consultative Committee on International Telephony and Telecommunications (CCITT)]. Another major contributor to solving interoperability problems among asynchronous transfer mode (ATM) equipment is the ATM Forum, a consortium of more than 500 companies world wide. Its main mission is to speed up the development and deployment of ATM products through interoperability specifications. This chapter reviews various ATM standards and specifications. The basic ATM framework is introduced next. This is followed by a brief introduction of different ATM interfaces and a review of their features.

The B-ISDN standards and protocol layers are being developed around the B-ISDN protocol reference model illustrated in Fig. 38.1.

ATM is the transport mode of choice for B-ISDN. It is a connection-oriented packet-switching technique that uses 53-byte fixed size cells to transfer information in the network. The short packet size of ATM, at high-transmission rates, is expected to offer full bandwidth flexibility, provide the basic framework to support a wide range of services required by different applications, and achieve high-resource utilization through statistical multiplexing. With *statistical multiplexing*, the sum of maximum bit rates of connections multiplexed on a link may exceed the link transmission rate if their average bit rates are (much) less than their maximum. The term *asynchronous* states that the cells generated by a source may appear at irregular intervals in the network. The connection-oriented nature of ATM arises out of the need to reserve resources in the network to meet the quality of service requirements of applications.

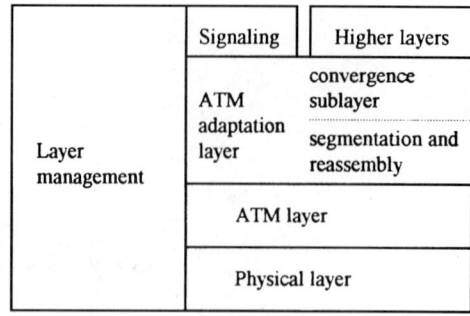

FIGURE 38.1 B-ISDN protocol reference model.

ITU-TS defines the transfer mode as a technique used for transmission, multiplexing, and switching aspects of communication networks. ATM, as the transfer mode of choice for B-ISDN, is envisioned to have the following properties:

- Support all existing services and those with yet unknown characteristics that would emerge in the future in an integrated manner, including voice, video, image, audio, and data
- Minimize switching complexity
- Minimize the processing time at intermediate nodes to support very high-transmission speeds
- Minimize the number of buffers required at the intermediate nodes to bound the delay and minimize buffer management complexity

It has been argued for a long time how well ATM can meet such challenges and at what expense. Independent of whether or not the arguments made against it are valid, ATM is the transfer mode of choice, and, most likely, we all have to learn how to live with it for several years to come.

An ATM cell consists of a 5-byte cell header and a 48-byte payload, as illustrated in Fig. 38.2. The cell header includes the following fields: *generic flow control* (GFC), *virtual path identifier* (VPI), *virtual channel identifier* (VCI), *payload type* (PT), *cell loss priority* (CLP), and *header error control* (HEC). GFC is a 4-b field providing flow control and fairness at a user–network interface. It is not used to control the traffic in the other direction, that is, network to user traffic flow. The GFC field has no use within the network and is meant to be used only by access mechanisms that implement different access levels and priorities.

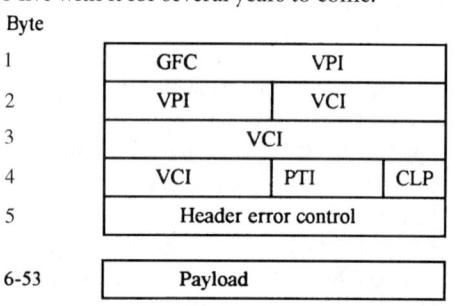

FIGURE 38.2 ATM cell format at user-network interface.

Accordingly, this field is part of the VPI field within an ATM network.

ATM requires connections to be established prior to data flow. It uses routing tables at each node along the path of a connection that maps the connection identifiers from the incoming links to the outgoing links. The two levels of routing hierarchies, *virtual paths* (VPs) and *virtual channels* (VCs), are defined as follows:

- VC is a concept used to describe unidirectional transport of ATM cells associated by a common unique identifier value, referred to as *virtual channel identifier* (VCI).
- VP is a concept used to describe the unidirectional transport of cells belonging to VCs that are associated by a common identifier value, referred to as *virtual path identifier* (VPI).

TABLE 38.1 Payload Type Indicator Field Values

PTI coding	Meaning
000	User data cell, congestion not experienced, SDU-type = 0
001	User data cell, congestion not experienced, SDU-type = 1
010	User data cell, congestion experienced, SDU-type = 0
011	User data cell, congestion experienced, SDU-type = 1
100	Segment OAM flow related cell
101	End to end OAM flow related cell
110	Resource management cell
111	Reserved

A VP is a collection of one or more VCs between two nodes. Each VP has a bandwidth associated with it limiting the number of VCs that can be multiplexed on a VP. VPIs are used to route packets between two nodes that originate, remove, or terminate the VPs whereas VCIs are used at these endnodes to distinguish between individual connections. It is noted that there is no difference between a VP and a VC when a VP is defined over a single physical link. When a VP is defined over two or more physical links, it reduces the size of the routing tables by allowing a number of VCs to be switched based on a single identifier, that is, VPI.

Payload type indicator (PTI) defines what is transmitted in the cell payload as illustrated in Table 38.1. The CLP field of the ATM cell header is a 1-b field that provides a limited flexibility by defining two cell loss priorities. The HEC field is used mainly for two purposes: to discard cells with corrupted headers and to delineate cells. The 8-b field provides single bit error correction and a low probability corrupted cell delivery capabilities. Cell delineation is determining the cell boundaries from the received bit stream.

38.2 ATM Architecture

The ATM architecture illustrated in Fig. 38.1 includes the physical, ATM, ATM adaptation, user, and signalling layers.

Physical Layer

The physical layer transports ATM cells between two ATM entities. It guarantees, within a certain probability, the cell header integrity and merges user cells to generate a continuous bit stream across the physical medium. The functions of the physical layer are grouped into two sublayers:

- The physical media (PM) dependent sublayer provides bit transmission capabilities, that is, the generation and reception of waveforms suitable for the medium, insertion and extraction of timing information, and electrical to optical and optical to electrical transformations whenever appropriate.
- The transmission convergence (TC) sublayer performs HEC generation and verification, frame and cell delineation, and line coding. On the transmit side, the TC sublayer receives cells from the ATM layer and packs them into the PM format used. At the link, a continuous stream of cells (bytes) is required. This necessitates the insertion of *idle cells* into the medium when no cell is passed from the ATM layer. Idle cells are identified with a specific VPI/VCI value and are not passed to the ATM layer at the receiver.

Various physical layer interfaces developed for ATM are given in Table 38.2. Developed originally by Bellcore, the synchronous optical network (SONET) hierarchy defines a set of interfaces with speeds of multiples of 51.84 Mb/s. Two SONET interfaces are currently defined for ATM at line rates of 155 Mb/s and 622 Mb/s. For early availability of ATM by taking advantage of existing infrastructure as much as possible, the ATM Forum defined additional ATM physical layer interfaces

TABLE 38.2 ATM Physical Layer Interfaces

Transmission medium	Total Transmission Rate, Mb/s	Information Transfer Rate, Mb/s	Coding
Multimode fiber	125	100	4B/5B
	155	149.76	SONET STS3-C
	194.4	149.76	8B/10B
Single mode fiber	622	599.04	SONET STS12-C
	155	149.76	SONET STS3-C
Shielded twisted pair	194.4	149.76	8B/10B
Coax	44.21	40.704	DS-3

that include DS-3 and 100-Mb/s multimode fiber. Others being defined include 51 Mb/s on UTP-3, 155 Mb/s UTP-5, DS-1, E1, E3, and E4.

ATM Layer

The ATM layer transfers fixed size ATM cells between the users of the ATM layer by preserving the cell sequence integrity. The ATM layer specification includes the following:

- ATM cell structure and encoding
- Services provided to AAL
- Traffic and congestion control
- Services expected from the physical layer
- ATM layer management specification

The ATM layer exchanges any 48-byte pattern [ATM service data units (SDU)] with the ATM layer user. Two primitives with associated parameters used for this exchange are defined as follows:

1. Data transmission from ATM layer user to the ATM layer

    ATM-DATA.request(ATM-SDU, SDU-type, submitted loss priority)

2. Data transmission from ATM layer to ATM layer user

    ATM-DATA.indication(ATM-SDU, SDU-type, received cell loss priority, congestion experienced)

Similarly, the ATM layer expects the physical layer to provide for the transport of 53-byte ATM cells between two (or more) communicating ATM entities. Two primitives defined between the ATM and physical layers are PHY-UNIT-DATA.request and PHY-UNIT-DATA.indication. With the former primitive, the ATM layer passes one cell and accepts one cell with the latter primitive.

The ATM layer management functions include alarm surveillance for VPs and connectivity verification for VPs and VCs. These two functions are carried out by using special management cells. Furthermore, ATM cells with invalid VPI/VCI values are discarded and layer management is informed.

ATM networks are expected to provide services to a wide variety of applications with different quality of service (QoS) requirements. The QoS provided to a connection may be agreed upon at connection establishment time for connections established dynamically or at subscription time if permanent connections are used. In either case, a traffic contract is negotiated between the user and the network for each ATM connection. Details of a traffic contract across a user–network interface (UNI) are discussed later. There are four service classes defined to support applications with different requirements:

- Class A: corresponds to constant bit rate (CBR) connection oriented services with timing relation between source and destination. Two examples of this class are 64-kb/s pulse code modulation (PCM) voice and CBR video.
- Class B: corresponds to variable bit rate (VBR) connection-oriented services with timing relation between source and destination. VBR encoded video is a typical example of this service class.
- Class C: corresponds to VBR connection-oriented services with no timing relation between source and destination. A typical service of this class is connection-oriented data transfer.
- Class D: corresponds to VBR connectionless services with no timing relation between source and destination. Connectionless data transfer between two local area networks over a wide area network is a typical example of this type of service.

In addition, a user-defined class of service is defined as class X.

- Class X: This is a raw cell service to allow a proprietary, vendor-defined ATM adaption layer (AAL).

A user, at the setup phase, requests an ATM layer quality of service from the QoS classes the network provides for ATM layer connections. Upon agreement, the network commits to meet the requested QoS as long as the user complies with the traffic contract. A QoS class has either specified performance parameters (referred to as specified QoS class) or unspecified performance parameters (referred to as unspecified QoS class). The former provides a QoS to an ATM connection in terms of a subset of performance parameters defined next. The specified QoS class is one of the classes 1, 2, 3, or 4, defined based on the service class of the application, that is, class A, B, C, or D. Unspecified QoS class, referred to as best effort service, is intended for connectionless traffic in which there is no explicitly specified QoS commitment on the cell flow.

The QoS requirement of an application is defined through a subset of parameters from the following list:

- Cell error ratio
- Severely errored cell block ratio
- Cell delay variation
- Cell loss ratio
- Cell misinsertion rate
- Cell transfer delay
- Mean cell transfer delay

Given this framework, a traffic contract between an ATM user and the network across a UNI include:

- Requested QoS class
- Connection traffic descriptor
- Conformance definition
- Compliant connection definition

A connection traffic descriptor includes a subset of source traffic descriptor, cell delay variation tolerance, conformance definition, and experimental parameters. A source traffic descriptor is a subset of traffic parameters used to capture the traffic characteristics of the connection requested by a source, which currently includes peak cell rate, sustainable cell rate, burst tolerance, and source type. Experimental parameters allow vendors to define and use additional metrics.

Cell delay variation tolerance is the difference between a cell's expected reference arrival time and its actual arrival time at the user–network interface. A generic cell rate algorithm (GCRA) is specified as the formal definition of traffic conformance. GCRA has two parameters: increment per unit time (I) and a capacity (L). In the leaky bucket version of the algorithm, the conformance of

TABLE 38.3 Allowable Combinations of Traffic Parameters in Signalling Messages

Combination	Traffic Parameters
1	PCR for CLP = 0 and PCR for CLP = 0 + 1
2	PCR for CLP = 0 and PCR for CLP = 0 + 1 with tagging requested
3	PCR for CLP = 0 + 1 and SCR for CLP = 0; BT for CLP = 0
4	PCR for CLP = 0 + 1, SCR for CLP = 0 and BT for CLP = 0 with tagging requested
5	PCR with CLP = 0 + 1
6	PCR for CLP = 0 + 1, SCR for CLP = 0 + 1; BT for CLP = 0 + 1
Best effort service	PCR for CLP = 0 + 1

a cell is determined by using a finite capacity bucket whose contents leak out at a continuous rate of one per time unit and increase by I units per cell arrival (as long as the capacity L is not exceeded). A cell that would cause the bucket to overflow is classified as nonconforming. A connection is said to be compliant as long as the number of nonconforming cells does not exceed a threshold that is defined by the network provider and specified in the traffic contract.

When a cell is detected to be nonconforming then there are two choices: either drop the cell at the interface or allow the cell to enter the network, hoping that there might be enough resources to deliver the cell to its destination. In the latter case, it is necessary to make sure that such nonconforming cells do not cause degradation to the service provided to connections that stay within their negotiated parameters. Based on this framework, *tagging* is defined as allowing nonconforming cells with CLP = 0 enter the network after their CLP bits are changed to one. These cells are discarded first when a congestion occurs at a network node so that the services provided to conforming sources are not affected. The allowable combinations of traffic parameters are defined in Table 38.3.

ATM Adaptation Layer (AAL)

The ATM layer deals with the functions of the cell header independent of the type of information carried in the payload. This simplicity and the flexibility are achieved by leaving out various services required by applications. In particular, at the ATM layer there is:

- No information on the frequency of the service clock
- No detection for misinserted cells
- No detection for lost cells
- No means to determine and handle cell delay variation
- No awareness on the content of user information

The main reason for not providing these functions at the ATM layer is that not all of these services are required by every application. For example, data traffic does not require any information on the frequency of the service clock, whereas voice may not require any awareness on the contents of the information. Accordingly, considering the commonality among existing and emerging applications in ATM networks, the functionality required by various services are grouped into a small number of classes and are defined as different AALs.

The AAL functions, for presentation purposes, can be classified into two categories: (1) CBR services adaptation functions and (2) VBR services adaptation functions. Examples of AAL functions for CBR services are cell assembly/disassembly, variable delay compensation, mapping control signals into ATM cell stream, clock recovery, and loss cell handling. Although CBR services can be used to support most VBR applications, doing so does not take into account the bursty nature of VBR services. In particular, CBR services do not take advantage of idle periods between information transmission and cause resource inefficiency for VBR traffic. VBR services are designed to provide bandwidth savings by taking the bursty nature of such applications into consideration.

AAL functions for VBR services include:

- Segmentation of information units into cells
- Handling partially filled cells
- Reassembling cells to form information units
- Action on lost cells

AAL consists of two sublayers: segmentation and reassembly (SAR) and convergence sublayer (CS). Transmit side SAR layer receives CS-protocol data units (CS-PDUs) and segments them (or collects in case of CBR services) so that when SAR header/trailer is added to the PDU, the final payload fits into one ATM cell, that is, 48 bytes. On the receiving end, the SAR layer reconstructs the CS-PDUs from received cells and passes them to the CS. The CS sublayer is subdivided into common part CS (CPCS) and service specific CS (SSCS). The CPCS performs functions common to particular AAL users such as multiplexing and lost cell detection. The service specific requirements of different classes of users, for example, timing recovery for real-time applications, are implemented in the SSCS. For services that do not require any specific function, the SSCP may be null.

CBR services in ATM framework are assumed to require timing relationship between the communicating entities. VBR services, on the other hand, are further classified into two subcategories depending on whether an end-to-end timing relationship is required or not. The former case is intended for applications such as VBR video and audio. The latter classification is mainly for data services. Finally, VBR services that do not require end-to-end timing relationship are further classified into two subcategories based on connection oriented vs connectionless application service.

Next, we briefly introduce various AAL types defined by ITU-TS.

AAL Type 1

AAL type 1 is used for CBR services that require a timing relation between the endpoints of connections. Of the 48-byte ATM cell payload, 4 b are used for *sequence numbering* (SN) and another 4 b for *SN field protection* (SNP), giving a net payload of 47 bytes per cell for user traffic. The SNP field provides error detection and correction capabilities on the 1-byte AAL type 1 header. The most significant bit of the SN field is used by the CS layer, whereas the remaining 3 b provide a cyclic counting capability from 0 to 7. This permits the detection of lost cells except eight consecutive cell losses. It also helps detecting misinserted cells. These cells, if detected, are discarded with no further action.

The explicit timing between the communicating entities may be provided using a time stamp transmitted by the source and used at the receiver to recover its local clock. Another method is to use implicit timing in which case the receiver monitors the buffer filling level and makes the required adjustments in its local clock. If explicit timing is used, then the most significant SN bit indicates to the receiver that the cell payload includes time stamp information. This bit is not used in implicit timing.

Next, we review AAL types 2, 3/4, and 5 defined for VBR services. Two optional modes of CS services are defined for VBR traffic: message mode and streaming mode. The message mode provides the transport of fixed size or variable length PDUs. The PDUs at the source are passed to the SSCP as a single unit. For fixed size PDUs, the SSCP may include the blocking function at the transmitter to collect a number of them together before passing the collection to the SAR layer as a single unit. At the receiver, the corresponding functionality is to deblock received data units into fixed size PDUs. In case of variable length PDUs, the SSCP may include the segmentation function to divide PDUs into smaller data units before passing them to the CPCS. Data units are reassembled at the receiving end to form the original PDU.

The stream mode provides the transport of variable length PDUs. A single PDU is passed to the CS layer as one or more data units possibly separated in time. Similar to the message mode, segmentation and reassembly functions are included in SSCP. In addition, an internal pipelining function may be applied providing means by which the sending AAL entity initiates the transfer to the receiving entity before the complete PDU is received.

Furthermore, both service modes may offer one of the two operational procedures: assured operations and nonassured operations. Assured mode provides reliable service by retransmitting missing or corrupted data units. Flow control is mandatory in assured operations. In case of nonassured operation, lost or corrupted PDUs are not corrected by retransmission. Both flow control and delivery of corrupted PDUs to the receiver are optional functions.

In VBR services, the CS-PDU size is not necessarily an integer multiple of the cell payload. Accordingly, the last segment of or a single-segment CS-PDU may not fill the entire cell payload. This necessitates the use of a *length indicator* (LI) which indicates the number of useful bytes in the cell payload. It is also necessary to indicate, either explicitly or implicitly, the information type (IT), that is, whether a cell is the first, middle, or last cell of a PDU. Furthermore, VBR services are more sensitive to errors. Cyclic redundancy check (CRC) may be included at the CS layer to detect bit errors.

AAL Type 2

AAL type 2 is intended for connection-oriented, variable bit rate service with timing relationship among the endnodes. The standardization of AAL type 2 has not yet been completed. SAR structure of type 2 AAL may include SN, SNP, IT, LI, and CRC fields. Various convergence sublayer functions expected may include the handling of lost or misdelivered cells, forward error correction, and clock recovery.

AAL Type 3/4

This type is defined for both connectionless and connection-oriented VBR services that do not require timing relationship between the source and the destination. The SAR structure of the protocol consists of LI, CRC, IT, and SN fields. Furthermore, a 10-b multiplexer identifier (MID) field is defined. In connectionless services, a number of CS-PDUs originally belonging to different sources may be transported with the same VPI/VCI. MID is used to differentiate among these packets when they are multiplexed on to a single connection by assigning a unique MID value to each CS-PDU.

AAL Type 5

AAL type 5 is defined for variable bit rate services that do not require timing relation among end stations. Unlike other AAL types, there is no per cell overhead in AAL type 5 payload. Instead, the data frame received from AAL user is encapsulated into an AAL type 5 CS-PDU. This PDU contains the user data field, a pad field to align the resulting PDU to fill an integral number of ATM cell payload (i.e., 48 bytes), a control field, a length field, and a CRC field. The total overhead excluding the pad field is 8 bytes for each CS-PDU. PTI at the cell header is used to distinguish the last cell of the PDU from the others. Multiplexing of different PDUs, unlike AAL type 3/4, is not allowed in AAL type 5 on a single connection. Compared with the AAL type 3/4, type 5 has the same effective payload usage for CS-PDU sizes of 88 bytes or less and smaller overhead for PDUs larger than 88 bytes, which decreases as the PDU size increases.

38.3 Asynchronous Transfer Mode Interfaces

In this section, we review the most recently approved ATM Forum specifications:

- User–network interface is the interface between an ATM network and an external device that communicates by means of ATM cells.
- Data exchange interface (DXI) enables routers to work with ATM networks without the use of special hardware.
- Broadband intercarrier interface (B-ICI) is a carrier-to-carrier interface that supports the multiplexing of a number of different services.

A high-level review of each interface is presented next.

User–Network Interface 3.0

The user-network interface is the interface between an ATM user or an endpoint equipment and the network. Depending on whether the network is private or public, the interface is, respectively, referred to as a private UNI or a public UNI. An ATM user in this context refers to a device that transmits ATM cells to the network. Accordingly, an ATM user may be an interworking unit that encapsulates data into ATM cells, a private ATM switch, or an ATM workstation. Transportation of user data across an ATM network is defined in the standards as the ATM bearer service. Implementation of an ATM bearer service may be based on virtual paths (VP service) or virtual circuits (VC service), or combined virtual path and virtual circuit service (VP/VC service). Connections in ATM networks are established either dynamically, referred to as switched virtual connections (SVC) or pre-configured, referred to as permanent virtual connections (PVC).

UNI 3.0 includes the specifications of the following:

- Physical layer interfaces
- ATM layer
- Interim local management interface
- UNI signalling

Physical and ATM layer specifications were discussed previously. In this section, we will only review the interim local management interface and UNI signalling. UNI 3.0 interim local management interface (ILMI) provides an ATM user with the status and configuration information concerning (both VP and VC) connections available at the UNI. The ILMI communication protocol is based on the SNMP network management standard. The term interim refers to the usage of this interface until related standards are completed by the standards organizations.

The main functions provided by the ILMI include the following: status, configuration, and control information about the link and physical layer parameters at the UNI; and address registration across UNI.

Various types of information available in the ATM UNI management information base (MIB) include: physical layer, ATM layer and its statistics, VP connections, VC connections, and address registration information. ILMI supports all physical layer interfaces defined by the ATM Forum. It provides a set of attributes and information associated with a particular physical layer interface and status information on the state of the physical link connecting two adjacent UNI management entities (UME) at each side of the UNI. Configuration information at the ATM layer provides information on the size of the VPI and VCI address fields that can be used by an ATM user, number of configured virtual path connections (VPCs) and virtual channel connections (VCCs), and the maximum number of connections allowed at the UNI.

VPC ILMI MIB (VCC ILMI MIB) status information indicates a UME's knowledge of the VPC (VCC) status (i.e., end-to-end, local, or unknown). Configuration information on the other hand relates to the QoS parameters for the VPC (VCC) local endpoint. Address registration identifies the mechanism for the exchange of identifier and address information between an ATM user and an ATM switch port across a UNI. ATM network addresses are manually configured by the network operator into a switch port. An ATM user has its own separate identifier information [i.e., its medium access control (MAC) address]. The end station attached to a particular switch port through the UNI would then exchange its identifier information for the ATM address information configured at the switch port through the address registration mechanism. As a result of this exchange, the endnode automatically acquires the ATM network address as configured by the network operator without any requirement for the same address to have been manually provisioned into the ATM user equipment. The ATM user then appends its own identifier forming its full ATM address. Similarly, the ATM end user identifier is registered at the network and it is associated with the respective network part of the address. With this scheme, several ATM addresses with the same network-defined part can be registered in the network.

Signalling in a communication network is the collection of procedures used to dynamically establish, maintain, and terminate connections. For each function performed, the corresponding signalling procedures define the sequence and the format of messages exchanged, which are specific to the network interface across which the exchange takes place. UNI 3.0 is built on the currently under development Q.2931 broadband signalling standard. Three major areas of extensions in UNI 3.0 are point-to-multipoint signalling, private ATM address formats, and additional traffic management capabilities.

The length of an ATM address is 20 bytes. The three private ATM addresses are defined based on the OSI network service access point (NSAP) format. An address format consists of two subfields: an initial domain part (IDP) and a domain specific part (DSP). IDP specifies a subdomain of the global address space and identifies the network addressing authority responsible for assigning ATM addresses in the specified subdomain. Accordingly, IDP is further subdivided into two fields: the authority and format identifier (AFI) and the initial domain identifier (IDI). AFI specifies the format of the IDI, the network addressing authority responsible for allocating values of the IDI, and the abstract syntax of the DSP. Three AFI's specified are data country code (DCC), international code designator (ICD), and E164. An IDI, on the other hand, specifies the network addressing domain from which values of the DSP are allocated and the network addressing authority responsible for allocating values of the DSP from that domain. DSP includes a number of fields that identify administrative authority, routing domain, area within a routing domain, and end system identifier.

UNI signalling architecture is shown in Fig. 38.3. It is a layer 3 protocol and runs on top of signalling AAL (SAAL), which defines the mechanisms used to transfer the signalling information for call/connection control reliably using the ATM layer on signalling virtual channels. SAAL consists of service specific and common parts. The service specific part further consists of service specific coordination function (SSCF) and service specific connection-oriented protocol (SSCOP). The SAAL common part is the same as AAL type 5 common part (AAL-5 CPCS) and segmentation and reassembly (AAL-5 SAR) sublayers.

UNI signaling
SSCF
SSCOP
AAL-5 CPCS
AAL-5 SAR
ATM layer
Physical

FIGURE 38.3 B-ISDN signalling structure.

As ATM requires connections to be established before any information can be sent, it is necessary to establish signalling channels before signalling messages can be exchanged between the two sides, that is, the user and the network, of the UNI interface. Currently, a dedicated point-to-point signalling virtual channel with VCI = 5 and VPCI = 0 is used for all UNI signalling. Each signalling message contains a number of information elements, which are the parameters describing some aspects of the interaction. The inclusion of information elements may be mandatory or optional. When an element is optional, notes explaining the circumstances under which such elements are included are specified in the standards documents.

A point-to-point connection is a collection of VC or VP links that connect two endpoints that wish to communicate to each other. As discussed previously, signalling between user and the network takes place at the edges of the network. The signalling messages, in this context, provide the network enough information to characterize the source and to locate the destination. Three groups of point-to-point call processing messages are defined as listed in Table 38.4.

A point-to-multipoint connection is a collection of VC or VP links with associated end nodes and has the following properties:

- The traffic on the connection is generated by a single node, referred to as the root, and received by all other endnodes of the connection, referred to as leaf nodes (or leaves).
- There is no bandwidth reserved from the leaf nodes toward the root and leaf nodes can not communicate with the root through the point-to-multipoint connection.
- Leaf nodes cannot communicate with each other directly through the point-to-multipoint connection.

TABLE 38.4 Point-to-Point Signalling Messages

Message Type	Message Name	Definition
Call establishment messages	Call proceeding	The requested call establishment has been initiated and no more call establishment information will be accepted
	Connect	Call acceptance
	Connect acknowledge	User has been awarded the call
	Set up	Initiate call establishment
Call clearing messages	Release	Equipment sending the message has disconnected the virtual connection and intends to release the virtual channel
	Release complete	Indicates the virtual channel and related information are released and the virtual channel is ready for reuse
Miscellaneous messages	Restart	Request to release all resources associated with the indicated VC(s) controlled by the signalling channel
	Restart acknowledge	Response to a start message to indicate requested restart is completed
	Status	Response to a status inquiry message or sent any time to report error conditions
	Status inquiry	Solicit a status message

TABLE 38.5 Point-to-Multipoint Signalling Messages

Message Name	Definition
ADD PARTY	Add a party to an existing connection
ADD PARTY ACKNOWLEDGE	Response to ADD PARTY message to acknowledge that the add party request was successful
ADD PARTY REJECT	Response to ADD PARTY message to acknowledge that the add party request was not successful
DROP PARTY	Clear a party from an existing connection point to multipoint connection
DROP PARTY ACKNOWLEDGE	Response to DROP PARTY message to indicate that the party was dropped from the connection

The same VPCI/VCI values are used to reach all leaf nodes and the traffic to the leaf nodes has all of the same QoS, bearer capability, and ATM cell rate.

A point-to-multipoint connection is set up by first establishing a point-to-point connection between the root and a leaf node. The first SET UP message is sent to a leaf node and has the endpoint reference value of zero. Furthermore, the SET UP message contains a so-called broadband bearer capability information element which indicates a point-to-multipoint connection in its user plane connection configuration field. Additional leaves are added to this connection one at a time or simultaneously by the use of ADD PARTY messages. The set of messages used for point-to-multipoint signalling is given in Table 38.5.

Data Exchange Interface (DXI)

DXI was developed to allow existing routers to interwork with ATM networks without requiring special hardware. In this specification, a DTE (a router) and a DCE (an ATM DSU) cooperate toward providing a user–network interface. DXI defines the data link protocol and physical layers that handle data transfer between a DTE and a DCE, as well as the ATM DXI local management interface.

Two physical interfaces defined for DTE/DCE are V.35 and HSSI. DSU/ATM-switch physical layer (i.e., across UNI) can be any one of the physical layers specified previously in Sec. 38.2. The data link layer defines the protocol used to transport DXI frames over the physical layer between the DTE and DCE. Across a DXI, three operational modes are defined.

In mode 1a, the transport of DTE-SDU is based on AAL type 5 CPCS and SAR sublayers. At the origination node, data link control layer receives a DTE-SDU and encapsulates it into a DXI data link control frame (DXI-PDU). The resulting PDU is transmitted to DCE across the DXI. The DCE strips off the DXI encapsulation and obtains the values of data frame address and cell loss priority. DCE then encapsulates the DTE-SDU into AAL type 5 CPCS PDU and segments the resulting PDU into 48 byte AAL type 5 SAR-SDUs. DCE also maps the frame address to the appropriate VPI/VCI of each cell. Cell loss priority bit value at the DXI header is also copied to the CLP bits of the transmitted cells. For data transmission from the ATM network to the destination DTE, the reverse process is followed. The only exception is the use of a congestion notification bit at the DXI header, which is set to one if one or more cells of the DTE-DU experience congestion in the network.

Mode 1b consists of mode 1a plus transport of DTE-SDU service based on AAL type 3/4. When AAL type 3/4 is used, DTE first encapsulates the DTE-SDU into AAL type 3/4 CPCS PDU by appending the corresponding CPCS header and trailers to the DTE-SDU. The CPCS-PDU is then encapsulated into DXI data link control frame. The exchanges take place similar to mode 1a.

In mode 2, DTE operations are the same as in mode 1 b with AAL type 3/4. At the DCE, both AAL type 5 and AAL type 3/4 connections are allowed. If AAL type 5 is used, then DCE strips off both the DXI and AAL type 3/4 CPCS encapsulations. The remaining PDU (i.e., DXI-DSU) is then encapsulated into AAL type 5 CPCS PDU, segmented into AAL type 5 SAR-SDUs and transmitted to the ATM network using the services of ATM and physical layers, similar to the mode 1a (and 1b with AAL type 5). If AAL type 3/4 connection is used, then DCE strips off only the DXI encapsulation and segments the AAL type 3/4 CPCS PDU into AAL type 3/4 SAR-SDUs.

DXI LMI defines the protocol for exchanging management information across DXI. It is designed to support a management station running SNMP and/or switch running UNI 3.0 ILMI protocol. LMI supports the exchange of various types of management information that include DXI specific, AAL specific, and ATM UNI specific information.

Broadband Interchange Interface (B-ICI)

B-ICI is a carrier-to-carrier public interface that supports the multiplexing of different services that include SMDS, frame relay (FRS), circuit emulation (CES), and cell relay (CRS). It includes physical layer, ATM layer, and service specific functions above the ATM layer required to transport, operate, and manage a variety of intercarrier services across a B-ICI. The document also includes traffic management, network performance, and operations and maintenance specifications.

The physical layer of the interface is based on ITU-TS defined network node interface (NNI), which includes SONET/synchronous digital hierarchy (SDH) physical and ATM layers with the addition of DS-3 physical layer. A service specific non-ATM network is connected to an ATM network via an interworking unit (IWU), which essentially defines an interface between the two networks. The main function performed at an IWU for each service supported across a B-ICI is to accept incoming frames and when necessary encapsulate them into AAL PDUs and convert them into ATM cells for transmission to the ATM network. These ATM cells are multiplexed together and passed across B-ICI over one or more VPC and/or VCC that are preconfigured at subscription time. That is, B-ICI version 1.0 supports only permanent virtual connections for each source–destination pair and the routing process provides a fixed path between two endnodes. It is noted that this multiplexing is done at the interface level. In particular, for each service supported, there is at least one connection and cells belonging to different services are not multiplexed onto the same connection.

CRS is a cell-based information transfer service that offers its users direct access to the ATM layer at rates up to the access link rate. Both VPC and VCC are supported. In CRS, a point-to-point PVC denotes an ATM layer VPC or VCC from a source service access point (ATM-SAP) to a destination ATM-SAP. CES supports the transport of CBR traffic. B-ICI supports the transport of DS-n signals across two public networks connecting users with UNIs at both ends, users connected to an ATM network via UNI with a DS-*n* interface used at the other end, and both users connected to the network with DS-*n* interfaces through interworking functions at both ends.

FRS is a connection-oriented data transport service. B-ICI defines an encapsulation of FRS frames and functions into AAL type 5 PDUs and AAL type 5 SAR is used to form AAL type 5 SDUs. Cells are transmitted over PVCs. Similarly, SMDS is a public packet-switched service that provides for the transport of data packets without the need for call establishment procedures, that is, connectionless service. Customer access to the SMDS will be over the SMDS subscriber network interface (SNI). B-ICI uses L3 PDUs encapsulation in intercarrier service protocol connectionless service (ICIP-CLS) PDUs and AAL type 3/4 is used for transmission across an ATM network

Network Node Interface (NNI)

NNI is the interface between either two private networks or two public networks, respectively, referred to as private NNI and public NNI. Private NNI is being defined by the ATM Forum between switching systems where a switching system can be a single switch or a subnetwork. Two current activities in the forum are the definitions of private-NNI routing and signalling. Public NNI is being defined by ITU-TS.

38.4 Conclusions

B-ISDNs are envisioned not only to support current services but also new services that are limited only by the human imagination. Standardization is necessary and will provide the basic framework for emerging ATM networks. Despite the tremendous amount of work completed in standards organizations and the ATM forum, much remains to be satisfactorily addressed beyond what can and will be standardized before ATM networks can be widely deployed. Some areas of interest include:

- Congestion control framework
- User protocols on top of AAL, particularly for multimedia applications
- Path selection framework
- Group management and multicast
- Other network services that include security, non-disruptive path switching, etc.

References

ATM. 1993. ATM User-Network Interface Specification, Version 3, ATM Forum, Aug.
ATM B-ICI. 1993. Version 1.0, ATM Forum, Aug.
ATM DXI. 1993. Version 1.0, ATM Forum, July.
ATM. 1993–1994. ATM Forum newsletters: 53 bytes.
CCITT. 1990. Broadband aspects of ISDN. CCITT Rec. I.121.
CCITT. 1990. B-ISDN service aspects. CCITT Rec. I.211.
CCITT. 1990. B-ISDN user network interface specification. CCITT Rec. I.432, SG XVIII, Rep. R-34.
CCITT. 1991. B-ISDN ATM adaptation layer (AAL) specification. CCITT Rec. I.363.
CCITT. 1991. Broadband aspects of ISDN. CCITT Rec. I.413.
De Prycker, M. 1993. *Asynchronous Transfer Mode: Solution for B-ISDN*, 2nd ed. Ellis Horwood, New York.
ITU. 1993. B-ISDN user-network interface layer 3 specification for basic call/bearer control, ITU-TS Rec. Q.93B, Dec.
ITU. 1993. ITU-TS Rec. SSCF and SSCOP.
Onvural, R.O. 1993. *ATM Networks: Performance Issues*, Artech House, Boston, MA.

39
Synchronous Optical Network (SONET)

Chris B. Autry
Georgia Institute of Technology

Henry L. Owen
Georgia Institute of Technology

39.1 Introduction...542
39.2 SONET Frame 543
   SONET Data Rates • Physical, Path, Line, and Section Layers • Overhead Byte Definitions • Pointers • Virtual Tributaries • Scrambling • Synchronous Transport Signal Level-*N* (STS-*N*) • STS Concatenation • Transmission of ATM in SONET
39.3 Optical Issues 550
39.4 SONET Equipment..551
39.5 SONET Standards 551

39.1 Introduction

Synchronous optical network (SONET) is a physical transmission vehicle that is capable of transmissions in the gigabit range. SONET is defined by a set of electrical as well as optical standards. SONET is intended to be the transmission means over the next several decades in the same manner that T1 technology has been the basic transport mechanism. Since SONET can use the capability of already installed fiber optic cable, eliminates complexity, and reduces equipment functionality requirements, local and interexchange carriers have incentive to install SONET over T3. Immediate savings in operational cost, as well as preparing for higher bandwidth applications, justify this installation. The technological step forward provided by SONET allows the realization of a new generation of high-bandwidth services in a more economical manner than has previously existed [Davidson and Muller, 1991].

There are at least two problems with the **plesiochronous** digital hierarchy (PDH) transmission system which motivate SONET. The first is the complexity involved in signal access. In accessing a given 1.544-Mb/s T1 from a 44.736-Mb/s T3, the entire T3 signal must be demultiplexed in order to access the given T1 of interest. The second problem with the existing PDH system is the complexity of network management. It has proven not to be a simple task to measure network performance, respond to network failures, or manage remote network equipment from control centers. SONET was created to solve these as well as other problems by [Davidson and Muller, 1991; Minoli, 1993; Sexton and Reid, 1992]:

- Unifying North American and International Standards
- Including intelligence in the multiplexers for protection switching, administration, operations, and maintenance
- Making multivendor networks possible

- Using a base rate that is able to accommodate existing T1 and T3 rates
- Using a **synchronous** network to simplify multiplexing and demultiplexing for easy signal access
- Supporting continually increasing optical bit rates
- Including sufficient overhead channels and functions to support facility maintenance

39.2 SONET Frame

The basic building block in SONET is the synchronous transport signal level-1 (STS-1). It is transported as a 51.840-Mb/s serial transmission using an optical carrier level-1 (OC-1) optical signal. Even though SONET is physically serially transmitted, it is conceptually simpler to think of it in terms of bytes. In fact, hardware implementations of SONET systems do the majority of the processing in terms of bytes not bits. The grouping of a specified set of bytes in the STS-1 is called a frame. An STS-1 frame consists of 6480 bits, which is 810 bytes. The 810 bytes are typically represented as 90 columns and 9 rows, as shown in Fig. 39.1. The bytes are numbered from left to right across the top row until 90 is reached, then continue on the second row again left to right starting with byte 91. In a given byte, the most significant bit is transmitted first. This group of bytes (the STS-1 frame) is transmitted in 125 μs so that 8000 frames occur per second.

SONET Data Rates

Higher data rates are transported in SONET by synchronously multiplexing N lower level modules together. Table 39.1 lists valid values of N and the corresponding data rates. SONET standards define both optical and electrical signals. **Optical carrier level-N (OC-N)** and **synchronous transport signal level-N (STS-N)** correspond to the optical and electrical transmissions, respectively, of the same data rate. The maximum value of N is limited by the requirement that each individual STS-1 is allocated only one 8-bit identification value and this value must be unique.

TABLE 39.1 SONET Optical Carrier and Line Rates [Bellcore, 1994]

Optical carrier OC-N/Electrical Signal	Line Rate Mb/s
OC-1/STS-1	51.84
OC-3/STS-3	155.52
OC-12	622.08
OC-24	1244.16
OC-48	2488.32

Physical, Path, Line, and Section Layers

As shown in Fig. 39.1, the STS-1 frame structure has two parts, the transport overhead and the **synchronous payload envelope (SPE)**. Starting with a payload (for example, DS1, DS3, etc.), which

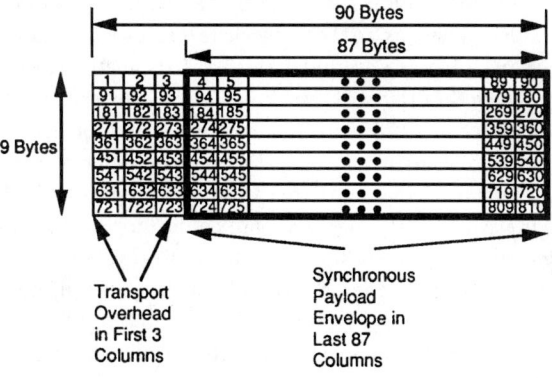

FIGURE 39.1 STS-1 frame.

is to be transported by SONET, the payload is first mapped into the SPE. This operation is defined to be the path layer and is accomplished using path terminating equipment. Associated with the path layer are some additional bytes named the **path overhead (POH)** bytes, which are also placed into the SPE. After the formation of the SPE, the SPE is placed into the frame along with some additional overhead bytes, which are named the line overhead LOH bytes. The LOH bytes are used to provide information for line protection and maintenance purposes. This LOH is created and used by line terminating equipment such as OC-N to OC-M multiplexers. The next layer is defined as the section layer. It is used to transport the STS-N frame over a physical medium. Associated with this layer are the section overhead (SOH) bytes. These bytes are used for framing, section error monitoring, and section level equipment communications. The physical layer is the final layer and transports bits serially as either optical or electrical entities. There is no defined overhead at this layer. Figure 39.2 [Bellcore, 1994] shows the relationship between physical, section, line, and path as defined in the SONET standards.

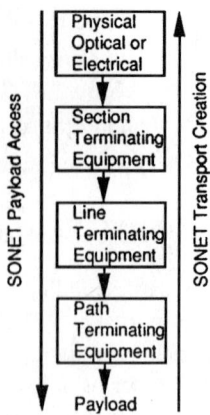

FIGURE 39.2 Relationship of physical, section, line, and path layers [Bellcore, 1994].

Overhead Byte Definitions

There are four types of overhead bytes: section, line, path, and **virtual tributary (VT)**. The first three columns in the SONET frame are where the section and the line overhead bytes are always located. The SONET overhead bytes are defined in Table 39.2 using the byte numbers from Fig. 39.1. A brief summary of each of the overhead bytes in an STS-1 frame is given in Table 39.2.

The framing bytes A1 and A2 are used by SONET hardware to identify the start of the STS-1 frame. Bytes H1 and H2 indicate the start of the SPE, and H3 is the negative justification opportunity byte. Bit interleaved parity (BIP) codes are used to detect errors. Bytes D1–D3 form a 192-kb/s data communication channel (DCC), and bytes D4–D12 comprise a 576-kb/s DCC. DCCs are used for alarms, maintenance, control, monitoring, and administration needs. An orderwire provides voice communications between maintenance personnel. Each identification byte is used to identify a connection at the specified layer. The growth bytes are available to support services and technologies of the future. Additional details on the overhead bytes may be found in ANSI [1991].

Pointers

The second portion of the STS-1 frame is the SPE. The SPE consists of 783 bytes and may be thought of as 87 columns by 9 rows. Two columns (30 and 59) are not used for carrying payload, instead they contain *fixed stuff* bytes. Another column contains STS-1 POH. The remaining 756 bytes are used for carrying payload. The SPE may have its first byte anywhere inside the 87 columns by 9 rows of the SPE area in the STS-1 and, in fact, can move around in this area. The method by which the starting location is determined is by the overhead bytes H1 and H2. These two bytes contain a 10-bit binary value, which contains a value between 0 and 782 corresponding to the location in the STS-1 frame in which the first payload byte is located. The contents of the H1 and H2 pointer bytes are shown in Table 39.3. Bits 1–4 (new data flag) contain bits that signal that a brand new **pointer** value is to be used, allowing for a complete change in payload. Bits 6 and 7 are not defined for use. Bits 7, 9, 11, 13, and 15 are defined to be *increment* bits, whereas bits 8, 10, 12, 14, and 16 are defined to be *decrement* bits.

In a SONET network element, there is a receive side clock and a transmit side clock, which are not always at the exact same frequency. In SONET it is an objective to recover and to distribute highly accurate clocks; however, this is not always possible in all operating scenarios. For example, in a failure mode a clock may be derived from a less accurate source to allow network operation to continue. In such situations it is possible that the STS-1 frame rate is faster or slower than the

Synchronous Optical Network (SONET)

TABLE 39.2 SONET Section, Line, Path, and VT Overhead Byte Allocations [ANSI, 1991; Bellcore, 1994]

Byte No.	Name	Type	Description of Usage
1	A1	Section	Framing
2	A2	Section	Framing
3	C1	Section	An 8-bit STS-1 identifier
91	B1	Section	Bit interleaved parity-8 (BIP-8) code over section
92	E1	Section	Orderwire
93	F1	Section	User channel
181	D1	Section	One byte of 192-kb/s data communications channel
182	D2	Section	One byte of 192-kb/s data communications channel
183	D3	Section	One byte of 192-kb/s data communications channel
271	H1	Line	Pointer most significant byte
272	H2	Line	Pointer least significant byte
273	H3	Line	Negative justification opportunity byte
361	B2	Line	Bit interleaved parity-8 (BIP-8) code over line
362	K1	Line	Automatic protection switching most significant byte
363	K2	Line	Automatic protection switching least significant byte
451	D4	Line	One byte of 576-kb/s data communications channel
452	D5	Line	One byte of 576-kb/s data communications channel
453	D6	Line	One byte of 576-kb/s data communications channel
541	D7	Line	One byte of 576-kb/s data communications channel
542	D8	Line	One byte of 576-kb/s data communications channel
543	D9	Line	One byte of 576-kb/s data communications channel
631	D10	Line	One byte of 576-kb/s data communications channel
632	D11	Line	One byte of 576-kb/s data communications channel
633	D12	Line	One byte of 576-kb/s data communications channel
721	Z1	Line	Growth
722	Z2	Line	Growth
723	E2	Line	Orderwire
In SPE	J1	Path	Repeating 64-byte identification
In SPE	B3	Path	Bit interleaved parity-8 (BIP-8) code over path
In SPE	C2	Path	Payload type identification
In SPE	G1	Path	Path status including far end block error (FEBE)
In SPE	F2	Path	User channel
In SPE	H4	Path	Multiframe frame count
In SPE	Z3	Path	Growth
In SPE	Z4	Path	Growth
In SPE	Z5	Path	Growth
In SPE	V5	VT	Error checking, signal label, and path status
In SPE	J2	VT	Path trace
In SPE	Z6	VT	Path growth
In SPE	Z7	VT	Path growth

TABLE 39.3 H1 and H2 Byte Contents [Bellcore, 1994]

Bit Number and Usage in Byte H1								Bit Number and Usage in Byte H2							
1	2	3	4	5	6	7	8	9	10	11	12	13	14	15	16
N	N	N	N	*	*	I	D	I	D	I	D	I	D	I	D

SPE rate. In this situation it is necessary to transmit one extra byte in the negative justification opportunity byte or one less byte (known as a positive stuff byte) in a given STS-1 frame so as to accommodate the SPE. In a positive justification action the normally used byte immediately after the H3 overhead byte is not used to carry payload. In a negative justification action the H3 overhead byte that normally does not carry payload is used to carry a payload byte.

The corresponding 10-bit pointer value is used to indicate which one of these two cases is occurring in the same STS-1 frame by the increment or decrement indication inside the 10-bit pointer value.

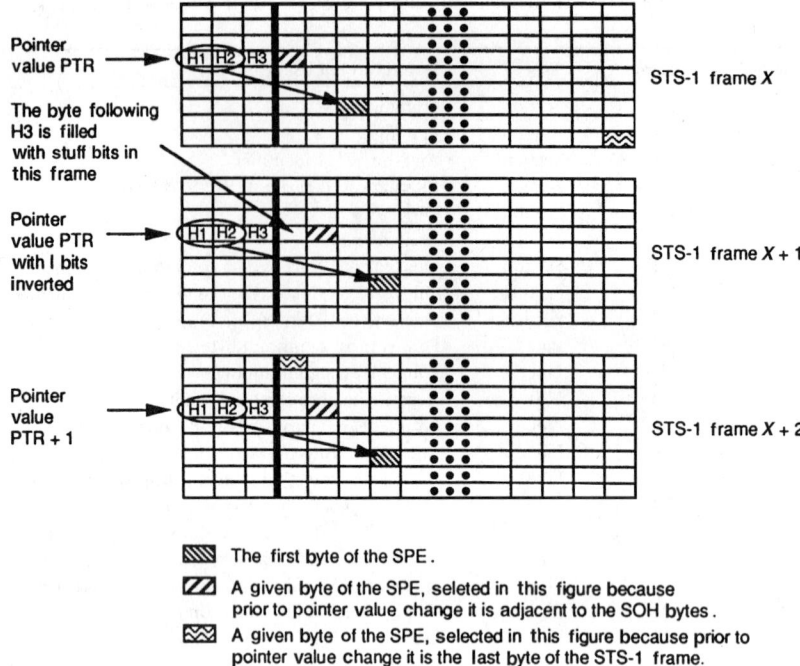

FIGURE 39.3 Positive pointer activity.

A positive stuff byte usage in this frame is indicated by inverting the bits 7, 9, 11, 13, and 15 (the I bits), which denotes that the positive stuff opportunity byte does not contain a payload byte and that all bytes of the payload may now be found one byte later in the SPE area of the STS frame. The pointer value must be incremented by one in this case. An example of this is shown in Fig. 39.3 [ANSI, 1991]. Conversely, if the negative justification byte opportunity is used the bits 8, 10, 12, 14, and 16 (the D bits) are inverted, which denotes that the negative stuff opportunity byte does contain payload data and that all bytes in the SPE may now be found one byte earlier. The pointer value is decremented by one. An example of this is shown in Fig. 39.4 [ANSI, 1991].

The physical byte locations and the corresponding pointer values, which range from 0 to 782, are shown in Fig. 39.5 [ANSI, 1991]. Note that when the pointer value is 522, the payload fits into the SPE area of a single frame. Conceptually this is the most simple situation to visualize. When the pointer value is 523, the first byte of the payload is found one byte later, and consequently the last byte of the payload is not contained in this same frame but instead is found in the following frame. This is conceptually more difficult to visualize. As shown in Fig. 39.5, if the pointer value is 0, the first byte of the payload is found in the byte immediately after the H3 byte. The last byte of the payload is found in the byte before the H1 byte in the following frame.

Virtual Tributaries

Four different sizes of payloads, which are named virtual tributaries (VT), fit into the SPE. They are the VT1.5, which is 1.728 Mb/s; VT2, which is 2.304 Mb/s; VT3, which is 3.456 Mb/s; and last the VT6, which is 6.912 Mb/s. A VT requires a 500-μs structure (four STS-1 frames) for transmission. These required four frames are named a **superframe**. Inside the bytes, which make up four frames of a given VT, are the special bytes named V1, V2, V3, and V4. The V1 and V2 bytes make up a VT pointer, which indicates the alignment of the VT within the allocated VT bytes independent of the other VTs in the same STS-1. The pointer bits contained within the two VT pointer bytes (named

Synchronous Optical Network (SONET)

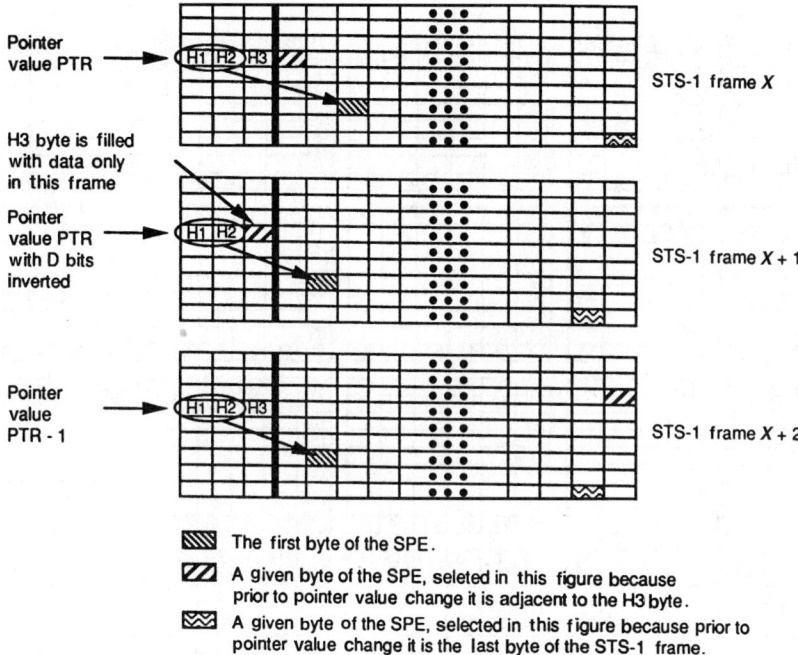

FIGURE 39.4 Negative pointer activity.

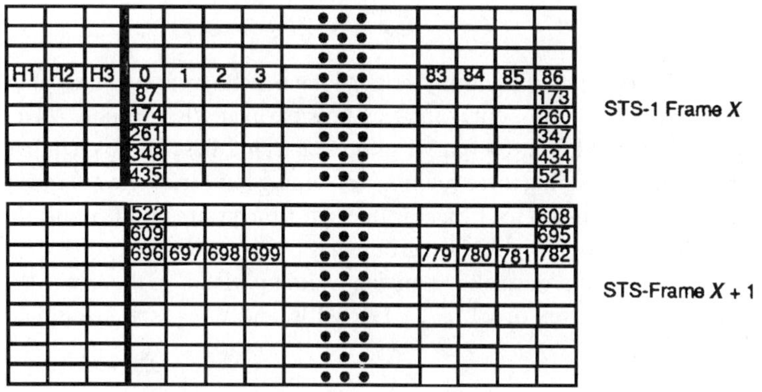

FIGURE 39.5 Pointer value locations in the STS-1 frame.

the V1 and V2 bytes) are defined similarly to the STS-1 pointer bytes shown in Table 39.3. The definitions of the locations for the VT pointer values are shown in Fig. 39.6 [ANSI, 1991]. These VT pointer bytes (V1 and V2) point to the first byte of the VT payload, a byte named the V5 byte. The V5 byte contains overhead information about the payload contained within the VT. The V5 byte includes error checking in the form of a 2-bit BIP-2, a **far end block error** (**FEBE**) bit, path signal identification, and a downstream equipment error indication. The V3 byte is allocated as the negative justification opportunity byte for the VT. The byte immediately after the V3 byte is the positive justification byte opportunity. The V4 byte is an undefined byte. The remaining bytes in a VT are used for the various mappings of DS1, DS2, E1, and so forth into the various VTs.

In a given STS-1 frame, a single VT1.5 occupies 3 columns out of 87. It is possible to have 28 VT1.5s located in an STS-1 frame. A single VT2 occupies 4 columns; it is possible to have 21 VT2s in

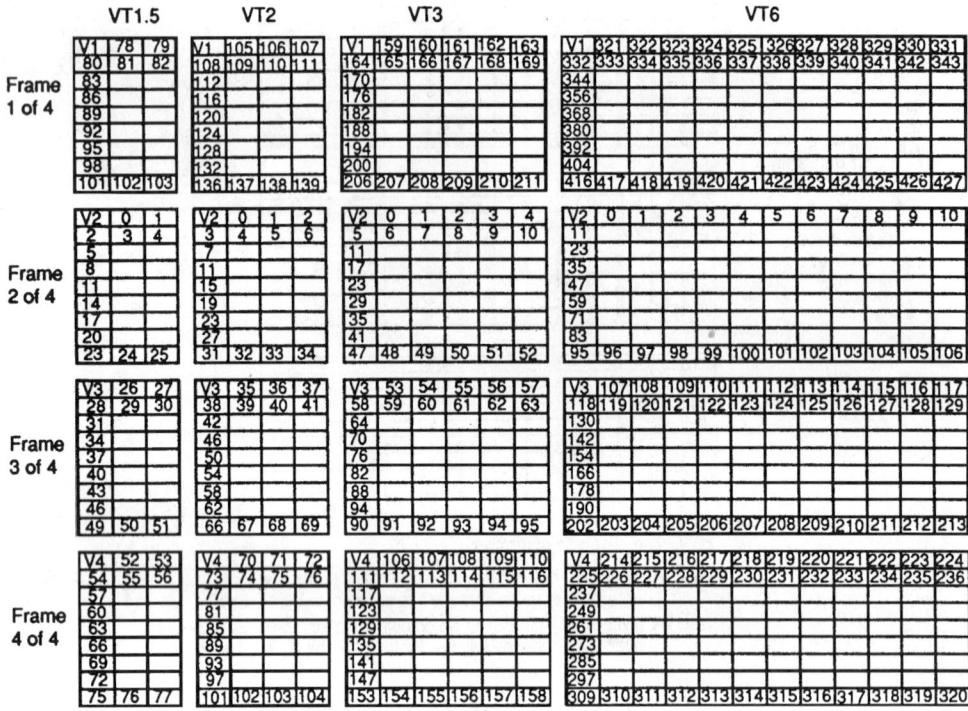

FIGURE 39.6 VT pointer values: V1 and V2 contain the VT pointer, V3 is the negative justification byte opportunity, and V4 is unused.

an STS-1 frame. A VT3 occupies 6 columns; thus 14 total may be accommodated in a single STS-1 frame. A VT6 occupies 12 columns; thus 7 total are possible in an STS-1 frame. Mixes of these VTs are allowed by dividing the SPE into seven VT groups. Each of these groups requires 12 of the 87 columns with two of the remaining three columns containing fixed stuff and the remaining column contains POH. The SPE is constructed by taking one byte from group number one, the second byte from group number two, and so on. A given group may be constructed out of 4 VT1.5s, 3 VT2s, 2 VT3s, or a single VT6. This approach allows mixing various payloads in the same SPE. An example SPE constructed out of a mixture of VT1.5s, VT2s, VT3s, and VT6s is shown in Fig. 39.7 [ANSI, 1991].

Scrambling

A scrambling function is used on all bytes of the STS-1 frame except for the A1, A2, and C1 bytes. This scrambling function is used to prevent long strings of ones or zeros from occurring. A generating polynomial of $1 + X^6 + X^7$ with a length of 127 is used in each frame. The scrambler is reset to all ones on the most significant bit of the byte following the C1 byte. A functional diagram of a scrambler is shown in Fig. 39.8 [ANSI, 1991].

Synchronous Transport Signal Level-*N* (STS-*N*)

An STS-*N* is formed by byte interleaving the STS-1 signals that comprise the STS-*N* signal. It may be thought of as $N \times 810$ bytes or as an $N \times 90$ column $\times$ 9 row structure. The transport overhead of the individual signals that make up the STS-*N* is frame aligned before interleaving. The SPEs that are contained are not required to be aligned because each STS-1 will have a unique payload pointer to indicate the position of each of the SPEs.

Synchronous Optical Network (SONET)

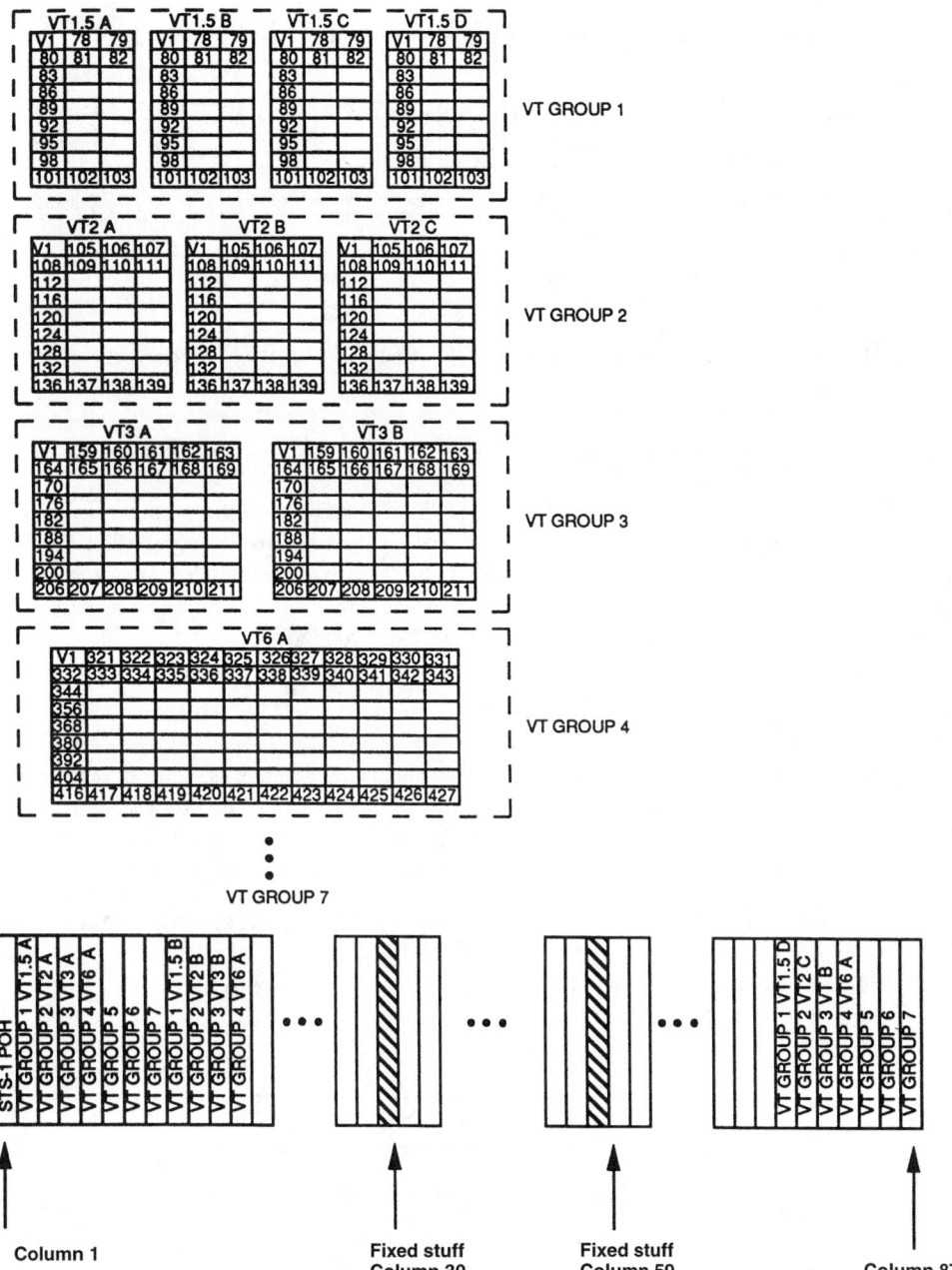

FIGURE 39.7 Example of an STS-1 SPE constructed from seven VT groups assuming first frame of the multiframe.

STS Concatenation

A concatenated STS (**STS-Nc**) is a number of STS-1s that are kept together. Certain services such as asynchronous transfer mode (ATM) payloads may use this superrate payload capability. The multiples of the STS-1 rate are mapped into an STS-Nc SPE. The STS-Nc is multiplexed, switched, and transported as a single unit. The SPE of an STS-Nc consists of $N \times 783$ bytes, which may be thought of as an $N \times 87$ column $\times$ 9 row structure. Only one set of STS POH is used in the STS-Nc;

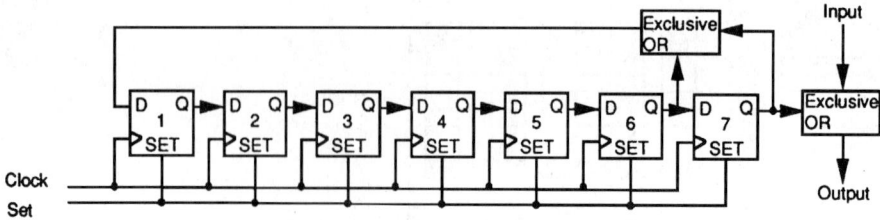

FIGURE 39.8 Scramble functionality.

the POH always appears in the first of the N STS-1s that make up the STS-Nc. At present only mappings for STS-3c and STS-12c have been defined [Bellcore, 1994].

Transmission of ATM in SONET

The STS-3c is used to transport ATM cells. The H4 byte contains a number in the lower order 6 bits that indicates the number of bytes between the H4 byte and the first byte of the first ATM cell that is contained in the SPE [ANSI, 1991]. The ATM cell payload is scrambled using a generator polynomial of $1 + X^{43}$. This is done to prevent the payload information from providing false A1A2 framing values. Figure 39.9 [ANSI, 1991] shows the ATM mapping.

39.3 Optical Issues

Signals using SONET can be transported by either optical or electrical means. Optical transmission allows for a wide range of applications and are limited by, among many other things, the choice of application involved, the transmitting and receiving devices, and the type of fiber used. SONET optical systems are conjoint and are split into the following broad application categories:

1. Long reach (LR) high-power interfaces are used typically for long haul telecommunication systems; system losses range between 10 and 28 dB.
2. Intermediate reach (IR) low-power systems have system losses of 0–12 dB.
3. Short reach (SR) systems have losses from 0–7 dB.

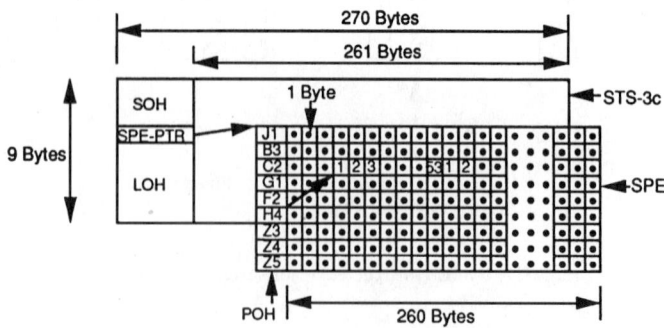

FIGURE 39.9 ATM mapping in STS-3c [ANSI, 1991].

Synchronous Optical Network (SONET)

Within each category it is possible to use either dispersion-unshifted single-mode fiber or dispersion-shifted single-mode fiber. LR, SR, and IR optical systems operate with a nominal 1310-nm source on dispersion-unshifted single-mode fiber. In addition, LR and IR optical systems are also capable of operating with a nominal 1550-nm source on both dispersion-unshifted and dispersion-shifted single-mode fibers. Possible choices for transmitter devices include light-emitting diodes (LEDs), single-longitudinal mode (SLM) lasers, or multilongitudinal mode (MLM) lasers.

39.4 SONET Equipment

A **terminal multiplexer (TM)** is a SONET multiplexer that terminates multiple DS1 signals and assembles them into STS-1 payloads. As an example, a TM that performs the termination of 28 DS1 signals is able to replace an existing M13 multiplexer from an existing PDH network. The output of this type of TM would be an OC-1 optical signal. A second example is the termination of 84 DS1 signals, which would be converted into an OC-3. A TM has only one bidirectional optical port.

An **add/drop multiplexer (ADM)** is a SONET multiplexer that allows DS1 signals to be added into or dropped from an STS-1 signal. ADMs have two bidirectional ports, commonly referred to as an east and a west port. ADMs may be used in SONET self-healing **ring** (SHR) network architectures. A SHR uses a collection of nodes equipped with ADMs in a physical closed loop so that each node is connected to two adjacent nodes with a duplex connection. Any loss of connection due to a single failure of a node or a connection between nodes may be automatically restored in this topology. (Note: The traffic sourced/sunk at a failed node is lost). There are two types of SHRs, unidirectional (USHR) and bidirectional (BSHR) as defined by the traffic flow in normal conditions. Bidirectional rings have a capacity carrying advantage over unidirectional rings because of the ability to share protection capacity among the links between nodes, as opposed to unidirectional rings, which dedicate capacity all the way around the ring [Ching and Say, 1993; Wu, 1992].

A **digital cross-connect system (DCS)** is used to terminate digital signals and cross connect them. A DCS integrates multiple functionalities such as signal adding and dropping, cross connect, and multiplexing and demultiplexing. A DCS provides an advantage over existing PDH networks because it eliminates back-to-back multiplexing and reduces the need for labor intensive jumpers. Operational cost savings are realized by a DCS through electronically controlling cross connections, test access and loopbacks, and maintenance.

There are two types of cross connects: wideband and broadband. Wideband DCS (W-DCS) terminates full duplex OC-Ns and DS3s, has VT cross-connection capability, and provides DS1 interfaces. A broadband DCS (B-DCS) terminates full-duplex OC-N signals and provides DS3 interfaces. The B-DCS makes two-way cross connections at the DS3, STS-1, and concatenated STS-Nc levels. An example use of STS-Nc is in broadband services such as high-definition television (HDTV) where STS-3c cross connection may be used to cross connect a single high-capacity STS-N channel [Bellcore, 1993].

39.5 SONET Standards

In 1985, Bellcore proposed to the T1X1 subcommittee of the ANSI accredited committee T1 to standardize the interconnection of multiowner, multimanufacturer fiber optic transmission terminals as well as the interconnection of fiber optic elements of varying functionality. This Bellcore proposal included a hierarchical family of digital signals whose rates were integer multiples of a basic module signal. A synchronous interleaving and multiplexing technique was proposed that would allow simple implementations. The term SONET was a result of this [Ballart and Ching, 1989]. Since several manufacturers had independently investigated this concept, it gained widespread support. In the beginning the early discussions were about the basic module rate and the idea of the VT as the basis for transporting DS1 services. In 1986 the international standards body, the Consultative

TABLE 39.4 A few SONET Standards Document Numbers and Titles

Standard	Title
ANSI T1.101	Synchronization interface standards for digital networks
ANSI T1.102	Digital hierarchy: electrical interfaces
ANSI T1.105	Digital hierarchy: optical interface rates and formats specifications (SONET)
ANSI T1.107	Digital hierarchy: formats specifications
ANSI T1.210	Operations, administration, maintenance, and provisioning (OAM&P): Principles of functions, architectures, and protocols for telecommunications management network (TMN) interfaces

Committee on International Telephony and Telegraphy (CCITT), now named the International Telecommunications Union (ITU), became involved. As a result, the ITU study group XVIII began a study of a synchronous digital hierarchy (SDH). Since the European signal hierarchy had no level near the proposed T1X1 50 Mb/s rate, ITU proposed a change to a 150 Mb/s rate. After some changes to the T1X1 proposals, it became possible to make the two proposals compatible. In 1988 ANSI T1.105-1988 and ANSI T1.106-1988, which are known as the SONET phase 1 North American specifications, were published. ITU also recommended an international standard based on OC-3, which allowed European multiplexing of 34 Mb/s signals as easily as North America could multiplex a DS3 signal. Some of the relevant standards documents for SONET are listed in Table 39.4.

Defining Terms

Add/drop multiplexer (ADM): Equipment that allows access to all or a portion of SONET payloads. The payload may be added to or dropped from the OC-N signal as it passes through the equipment. OC-N signals interface to this equipment through east as well as west ports.

Digital cross-connect system (DCS): Equipment that allows the switching or rearrangement of where payloads are located in SONET frames. DCSs also may allow adding and dropping of payloads as well, thus incorporating the functionality of both switching as well as ADM functionality.

Far end block error (FEBE): A signal that is returned to the source indicating that a parity error has been received.

Optical carrier level-N (OC-N): An optical signal that is derived from an electrical STS-N signal.

Path overhead (POH): Bytes that are added to the transport of a payload, which are used to implement functions necessary in the transport of the payload.

Plesiochronous: Corresponding signals are plesiochronous if their significant instances occur at nominally the same rate, any variation of rate being constrained within specified limits.

Pointer: The H1 and H2 overhead bytes contain a 10-bit value, which indicates where the first byte of the payload is located in the SONET frame.

Ring: The topology that allows for clockwise and counterclockwise transmission of SONET signals such that in the event of a fiber cut, source to destination connections may be automatically achieved through an alternative route. ADMs are typically used in this topology.

STS-Nc: A combination of STS-N signals into a single entity that must be transported together in contrast to several STS-N signals that may be carried as separate signals.

Synchronous: Two signals are synchronous if their significant instances occur at precisely the same average rate.

Synchronous optical network (SONET): A standardized set of rates and formats.

Synchronous payload envelope (SPE): The remainder of the SONET frame after the section and line overheads are removed. The pointer contained in the H1 and H2 overhead bytes point to the beginning of the SPE.

Synchronous transport signal level-N (STS-N): An electrical signal that is a byte interleaving of N STS-1 signals.

Terminal multiplexer (TM):   Equipment that has only one port (in contrast to an ADM that has two) and allows access to all or a portion of the payloads. A TM is used to add and drop payloads at the end of a single transmission path and is a less expensive and less capable equipment than an ADM, which is typically used in a ring topology.

Virtual tributary (VT):   A collection of bytes that is used for the transport of payloads such as DS1, DS2, and so forth. Four different VT sizes with four different data rates are defined in SONET.

Virtual tributary superframe: The group of four SONET frames required to hold the structure that contains the VT overhead bytes such as the V1, V2, V3, and V4 bytes. These overhead bytes repeat every four frames.

References

ANSI. 1991. Digital hierarchy optical interface rates and formats specifications (SONET). ANSI T1.105-1991, American National Standards Institute, New York.

Ballart, R. and Ching, Y. 1989. SONET: Now it's the standard optical network. *IEEE Commun. Mag.* 27(3):8–15.

Bellcore. 1993. Digital cross-connect systems in transport network survivability. Issue 1, SR-NWT-002514, Bell Communications Research, Piscataway, NJ.

Bellcore. 1994. Synchronous optical network (SONET) transport systems: Common generic criterion, GR-253-CORE, Bell Communications Research, Piscataway, NJ.

Ching, Y. and Say, S. 1993. SONET implementation. *IEEE Commun. Mag.* 31(9):30–73.

Davidson, R. and Muller, N. 1991. The *Guide to SONET*, Telecom Library, New York.

Minoli, D. 1993. *Enterprise Networking*, Artech House, Norwood, MA.

Sexton, M. and Reid, A. 1992. *Transmission Networking: SONET and the Synchronous Digital Hierarchy*, Artech House, Norwood, MA.

Wu, T. 1992. *Fiber Network Service Survivability*, Artech House, Norwood, MA.

Further Information

A good very high-level introduction to SONET may be found in *The Guide to SONET* [Davidson and Muller, 1991]. A more advanced treatment of SONET is *Transmission Networking: SONET and the Synchronous Digital Hierarchy* [Sexton and Reid, 1992]. Survivability of SONET networks is examined in *Fiber Network Service Survivability* [Wu, 1992]. An excellent chapter on SONET, an explanation of clock distribution in synchronous networks, and how SONET fits into the big picture of networks in general is contained in *Enterprise Networking* [Minoli, 1993]. A single technical document that contains a majority of the details of SONET is "Synchronous Optical Network (SONET) Transport Systems: Common Generic Criterion" GR-253-CORE, which may be obtained from Bellcore.

40
Synchronous Digital Hierarchy (SDH)

Chris B. Autry
Georgia Institute of Technology

Henry L. Owen
Georgia Institute of Technology

40.1 Introduction 554
40.2 Frame Structure 554
      Synchronous Transfer Module-*N* (STM-*N*) • Administration Unit (AU), Tributary Unit (TU), and Virtual Container (VC) Definitions • Administrative Unit Pointer Mechanism • Tributary Unit-3 Pointer Mechanism • Tributary Unit-1/Tributary Unit-2 (TU-1/TU-2) Pointer Mechanism • Section Overhead • Mapping of Tributaries • Synchronous Digital Hierarchy (SDH) Network Block Diagram
40.3 Mapping of Asynchronous Transfer Mode (ATM) Cells 562
40.4 Standards 562

40.1 Introduction

The **synchronous digital hierarchy (SDH)** is a set of international, digital transmission standards. SDH is the international version of the synchronous optical network (SONET), which is used in North America. The fundamental principles of SONET apply directly to SDH. The two major differences between SDH and SONET are the terminology and the basic line rates used. SONET uses a basic line rate of 51.84 Mb/s, and SDH uses a basic line rate of 155.52 Mb/s, which is exactly three times SONET's basic rate [Omidayr and Aldridge, 1993; Sexton, Roverano, and DeCremiers, 1993]. The compatibility between SDH and SONET allows for internetworking at the Administrative Unit-4 (AU-4) level [ITU, 1993a].

In contrast to the existing plesiochronous digital hierarchy (PDH), SDH allows direct access to tributary signals without demultiplexing the composite signal. As a result, network node costs are reduced because direct multiplexing is cheaper than step-by-step multiplexing. Furthermore, SDH supports advanced operations, administration, and maintenance (OA&M) by dedicating several embedded channels for this purpose. To prevent SDH from quickly becoming outdated, some overhead bytes have been reserved for growth to support services and technologies of the future. In addition, SDH has a **concatenation** mechanism so that lower rate payloads can be combined to form higher rate payloads. Therefore, SDH can support broadband services such as broadband integrated services digital network (B-ISDN) [Asatani, Harrison, and Ballart, 1990; Ballart and Ching, 1989].

40.2 Frame Structure

The **synchronous transport module-1 (STM-1) frame** consists of 2430 bytes (19,440 bits) and is represented as a 9 row by 270 column structure as depicted in Fig. 40.1 [ITU, 1993c]. The frame is

Synchronous Digital Hierarchy (SDH)

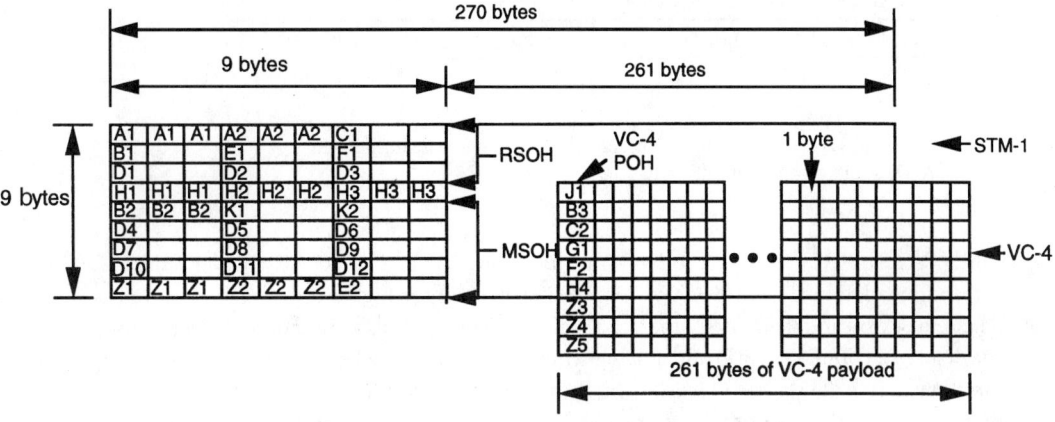

FIGURE 40.1 STM-1 frame structure with a VC-4 payload.

transmitted serially a bit at a time from left to right and from top to bottom. The frame consists of section overhead, administration unit (AU) pointers, and payload. An STM-1 frame is transmitted every 125 μs and thus has a bit rate of 155.52 Mb/s. This rate is defined as the basic rate in SDH [ITU, 1993a].

Synchronous Transfer Module-*N* (STM-*N*)

An STM-*N* signal is formed by byte interleaving the individual STM-1 signals. The rate of an STM-*N* is *N* times the rate of an STM-1. Currently, *N* is only defined for the values of 1, 4, and 16, but the International Telecommunications Union (ITU) is considering higher values for *N*. Table 40.1 shows the bit rates for the serial transmission of the corresponding STM.

TABLE 40.1 SDH Bit Rates [ITU, 1993b]

Synchronous Transport Module	Line Rate (Mb/s)
STM-1	155.52
STM-4	622.08
STM-16	2488.32
STM-*N*	$N \times 155.52$

Administration Unit (AU), Tributary Unit (TU), and Virtual Container (VC) Definitions

A **container-*n*** (C-*n*: *n* = 11, 12, 2, 3, or 4) as shown in Fig. 40.2 [ITU, 1993c] contains the client information to be transported and forms the payload of the corresponding **virtual container-*n*** (**VC-*n***). A VC-*n* (*n* = 11, 12, 3, or 4) is a structure consisting of payload from a container plus the addition of some path overhead information. A given VC is either classified as a lower order VC or a

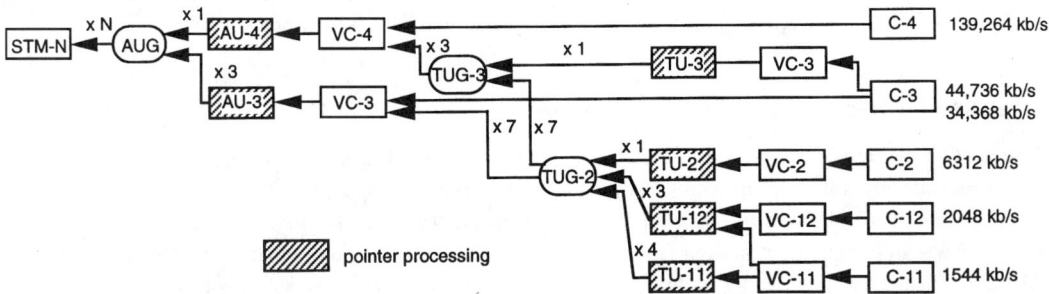

FIGURE 40.2 STM payload hierarchy.

TABLE 40.2 H1 and H2 Byte Contents [ITU, 1993c]

Bit Number and Usage in Byte H1								Bit Number and Usage in Byte H2							
1	2	3	4	5	6	7	8	9	10	11	12	13	14	15	16
N	N	N	N	a	a	I	D	I	D	I	D	I	D	I	D

[a] These bits are set to 10 when AU-4, AU-3, or TU-3, otherwise unspecified.
N These bits comprise the new data flag.

higher order VC. The VC-11, VC-12, and VC-2 are lower order VCs. The VC-4 is a higher order VC. In ITU recommendation [ITU, 1993a], VC-3s are classified only as higher order VCs. However, as pointed out in Sexton and Reid [1992] the VC-3 really has a dual role as a lower order VC when used as the payload of a tributary unit-3 and as a higher order VC when used as the payload of an AU-3.

A **tributary unit-n** (TU-n: $n = 11, 12, 2,$ or 3) contains a lower order VC and a TU pointer. The TU pointer indicates the start of the lower order VC relative to the start of the supporting higher order VC. An **administrative unit-n** (AU-n: $n = 3, 4$) consists of a higher order VC and a AU pointer. The AU pointer indicates the start of the higher order VC relative to the start of the STM-N frame. A **tributary unit group-n** (TUG-n: $n = 2, 3$) is a collection of one or more tributary units. A TUG-2 consists of one TU-2, three TU-12s, or four TU-11s. A TUG-3 contains one TU-3 or seven TUG-2s. Furthermore, an **administrative unit group-n** (AUG-n) is a collection of one or more administrative units. An AUG consists of either one AU-4 or three AU-3s [ITU, 1993a; Asatani, Harrison, and Ballart, 1990].

Administrative Unit Pointer Mechanism

The AU **pointer** indicates the start of the higher order VC relative to the start of the STM-N frame. Thus, the pointer mechanism allows the VC to float inside the frame. When the incoming clock is running faster than the outgoing clock at a node in the SDH network, a finite-size buffer inside the equipment at that SDH node will begin to fill up. Thus, the fill of the buffer will eventually reach an upper threshold and cause a negative justification pointer action. A negative justification pointer action prevents overflow by transmitting a data byte instead of a stuff byte in the negative justification opportunity, the H3 byte. Therefore, the buffer fill is reduced. A negative justification pointer action inverts the decrement (D) bits in the AU pointer (H1 and H2 as shown in Fig. 40.1 and Table 40.2) and reads a data byte from the buffer into the H3 byte. Normally, the negative justification opportunity carries a stuff (dummy) byte.

When the outgoing clock is running faster than the incoming clock at a node, the buffer will start emptying. To prevent underflow, the buffer fill will eventually reach a lower threshold, which results in a positive justification pointer action. A positive justification pointer action prevents underflow by sending a stuff byte instead of a data byte in the positive justification opportunity, byte 0 of the VC. Thus, the buffer fill will rise. To perform a positive justification pointer action, the increment (I) bits in the AU pointer (H1 and H2 as shown in Table 40.2) are inverted, and a stuff byte is transmitted in the positive justification opportunity. Normally, the positive justification opportunity contains data from the buffer. Consecutive pointer actions must be at least four frames apart (500 μs).

Tributary Unit-3 Pointer Mechanism

When a VC-4 contains three VC-3s, the TU-3 pointer (H1 and H2 as shown in Fig. 40.3 [ITU, 1993c]) indicates the offset of the VC-3 relative to the start of the supporting VC-4. The pointer mechanism works similarly to the AU pointer mechanism described in the preceding subsection. Each TU-3 has an independent pointer (H1 and H2), negative justification opportunity (H3), and positive justification opportunity (VC byte 0). When TUG-2s are carried inside a VC-4, the TU-3 pointer is not used and has a null pointer indication (NPI) value [ITU, 1993c].

Synchronous Digital Hierarchy (SDH)

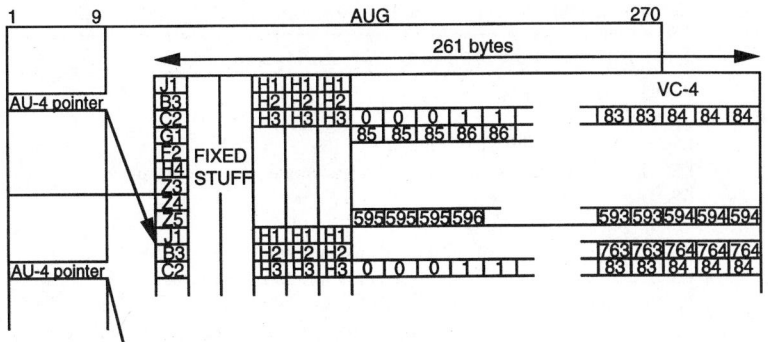

FIGURE 40.3 TU-3 pointer offset numbering.

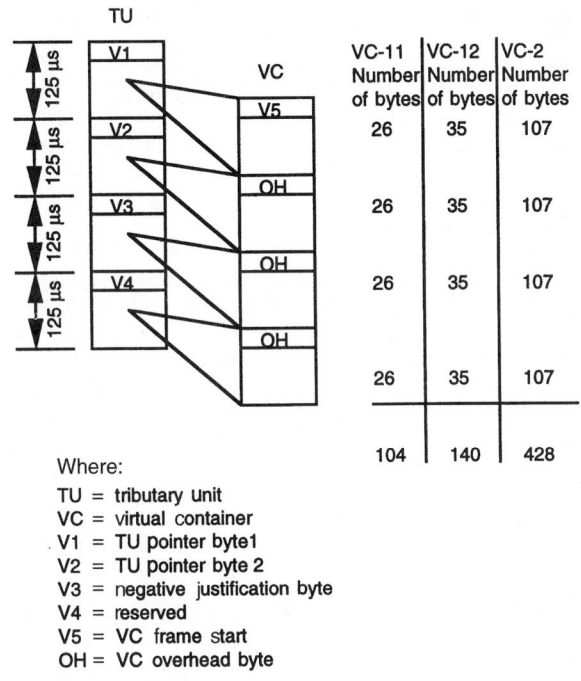

Where:
TU = tributary unit
VC = virtual container
V1 = TU pointer byte 1
V2 = TU pointer byte 2
V3 = negative justification byte
V4 = reserved
V5 = VC frame start
OH = VC overhead byte

FIGURE 40.4 Virtual container mapping in multiframe tributary unit.

Tributary Unit-1/Tributary Unit-2 (TU-1/TU-2) Pointer Mechanism

The TU pointer indicates the start of the lower order VC relative to the start of the associated higher order VC. The TU pointer mechanism works essentially the same way as the AU pointer mechanism and is only used in the floating mode. The negative justification opportunity is the V3 byte, and the positive justification opportunity is the VC byte after V3 in the TU multiframe. Bytes V1 and V2 as shown in Fig. 40.4 [ITU, 1993c] comprise the TU pointer, which points to the first byte in the VC, the V5 byte. Consecutive pointer actions must be separated by at least four multiframes (2 ms) [ITU, 1993c].

Section Overhead

Figure 40.1 [ITU, 1993c] shows the section overhead (SOH) of an STM-1 signal. As shown, the SOH is divided into the regenerator section overhead (RSOH) and the multiplex section overhead

TABLE 40.3 SDH RSOH, MSOH, and POH Byte Allocations [ITU, 1993a,1993c]

Name	Type	Description of Usage
A1	RSOH	Framing
A2	RSOH	Framing
C1	RSOH	An 8-bit STM-1 identifier.
B1	RSOH	Bit interleaved parity-8 (BIP-8) code over regenerator section
E1	RSOH	Orderwire
F1	RSOH	User channel
D1	RSOH	One byte of 192-kb/s data communications channel
D2	RSOH	One byte of 192-kb/s data communications channel
D3	RSOH	One byte of 192-kb/s data communications channel
B2	MSOH	BIP-$N \times 24$: Bit interleaved parity-$N \times 24$ code over multiplex section
K1	MSOH	Automatic protection switching communications
K2	MSOH	Automatic protection switching communications
D4	MSOH	One byte of 576-kb/s data communications channel
D5	MSOH	One Byte of 576-kb/s data communications channel
D6	MSOH	One Byte of 576-kb/s data communications channel
D7	MSOH	One Byte of 576-kb/s data communications channel
D8	MSOH	One Byte of 576-kb/s data communications channel
D9	MSOH	One Byte of 576-kb/s data communications channel
D10	MSOH	One Byte of 576-kb/s data communications channel
D11	MSOH	One Byte of 576-kb/s data communications channel
D12	MSOH	One Byte of 576-kb/s data communications channel
Z1	MSOH	Bits 1–4: growth; bits 5–8: synchronization status comm.
Z2	MSOH	Growth
E2	MSOH	Orderwire
J1	HO POH	Repeating 16-byte or 64-byte HO path access point identification
B3	HO POH	Bit interleaved parity-8 (BIP-8) code over HO path
C2	HO POH	Payload type identification
G1	HO POH	HO path status and performance monitoring
F2	HO POH	User channel
H4	HO POH	Payload position indicator
Z3	HO POH	User channel
Z4	HO POH	Growth
Z5	HO POH	Network operator communications channel
V5	LO POH	LO path status, signal label, and error checking
J2	LO POH	Repeating 16-byte LO path access point identifier
Z6	LO POH	Tandem connection monitoring
Z7	LO POH	Growth

(MSOH). Equipment that performs regeneration terminates the RSOH. Regenerators pass the MSOH through unmodified. Multiplexing equipment terminates the MSOH [ITU, 1993a]. In SONET, the RSOH is called the section overhead, and the MSOH is termed the line overhead. Table 40.3 indicates the basic function of each overhead byte. Most of the overhead bytes have the same function in both SDH and SONET.

The A1 and A2 bytes are used for framing, C1 is an STM identifier byte, and E1 is an orderwire for voice communications. Bytes D1, D2, and D3 form a 192-kb/s data communications channel (DCC) for the regenerator section. B1 provides error monitoring for the regenerator section by utilizing a bit interleaved parity 8 (BIP-8) code using even parity. B1 for the current STM-N frame before scrambling is derived from the entire previous STM-N frame after scrambling. F1 is a user channel. K1 and K2 provide for automatic protection switching (APS) communications. Bytes D4–D12 form a 576-kb/s DCC, and E2 is an orderwire for voice communications. Bits 5–8 of Z1 indicate the synchronization status, whereas bits 1–4 of Z1 and byte Z2 are reserved for future use. B2 is a BIP-24 for an STM-1 signal, a BIP-96 for an STM-4 signal, and a BIP-384 for an STM-16 signal using even parity for all cases. B2 for the current STM-N frame before scrambling is computed from the entire previous STM-N frame except for the RSOH bytes. The bytes J1, B3, C2, G1, F2,

Synchronous Digital Hierarchy (SDH)

H4, Z3, Z4, and Z5 are higher order (HO) POH and are shown in Fig. 40.1 in a VC-4. Byte J1 is a repeating 16-byte or 64-byte path access point identification, and B3 provides for error monitoring using an even parity BIP-8 code. C2 indicates the payload type, H4 gives the payload position, and G1 provides for path status and performance monitoring. F2 and Z3 are user communications channels. Z4 is reserved for future use, and Z5 is the network operator byte to be used for specific management purposes. The bytes V5, J2, Z6, and Z7 are provisionally allocated for lower order (LO) POH and are shown later in Fig. 40.7 and 40.8. Byte V5 indicates the LO path status and signal label. Also, V5 is used for error checking. J2 is a repeating 16-byte LO path access point identifier, and Z6 is used for tandem connection monitoring. Z7 is reserved for future use. More details on the overhead bytes may be found in [ITU, 1993a] and [ITU, 1993c].

Mapping of Tributaries

Both synchronous and asynchronous tributaries are allowed in SDH. These mappings are fully defined in reference [ITU, 1993c]. Asynchronous mappings of payloads into TUs involve the use of justification bits. Synchronous tributary mappings do not use justification bits in the mapping process. The asynchronous mapping of a 139,264-kb/s signal is accomplished by dividing each of the nine rows in the STM-1 frame into blocks of 13-bytes. Each of these 13-byte blocks contains a W, X, Y, or Z byte followed by 12 bytes that contain bits from the 139,264-kb/s signal. One of the nine rows from the STM-1 frame is shown in Fig. 40.5 [ITU, 1993c] (excluding the first nine overhead bytes of the STM-1 frame). The W byte is defined to contain 8 data bits from the 139,264-kb/s signal. The Z byte contains 6 or 7 data bits from the signal depending on the usage of the S-bit negative justification opportunity. The S-bit negative justification opportunity, which is used based on the fill of an incoming 139,264-kb/s bit-wide buffer, has a nominal justification ratio of 2/9. Hence, since each VC-4 row contains an independent negative bit justification opportunity, each STM-1 frame nominally contains two used S bits. The X and Y overhead bytes contain 0 data bits from the signal. The Y byte contains fixed stuff, whereas the X byte encodes the S-bit usage in each VC-4 row.

The asynchronous mapping of a 34,368-kb/s signal into a VC-3 is shown in Fig. 40.6 [ITU, 1993c]. The STM-1 frame (excluding the overhead) is broken up into three subframes T1, T2, and T3. One of these subframes is shown in detail in Fig. 40.6. Each VC-3 subframe spans three rows in the STM-1 frame. There are five types of bytes in the subframe. The first type is an I byte, which contains 8 bits from the 34,368-kb/s signal. In Fig. 40.6, the letter I, with a box around it, represents a group of three I bytes. The R overhead bytes, represented by a box with an X inside, each contain eight fixed stuff R bits and hence no signal data bits. The A byte contains 0 or 1 signal data bits

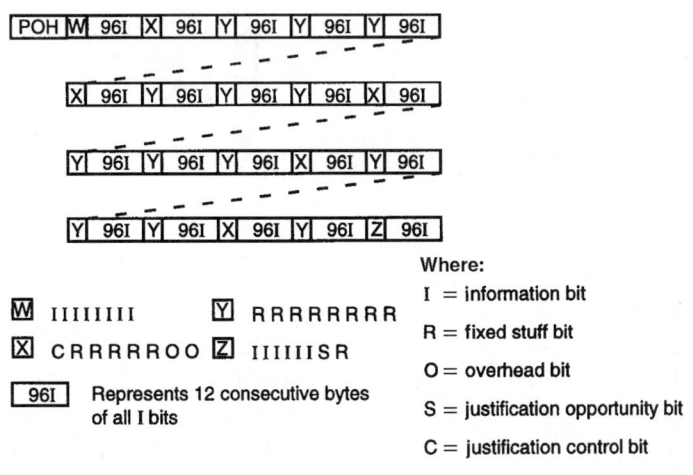

FIGURE 40.5 Asynchronous mapping of 139,264-kb/s tributary into VC-4. (Note only one row of the nine row VC-4 container structure is shown.)

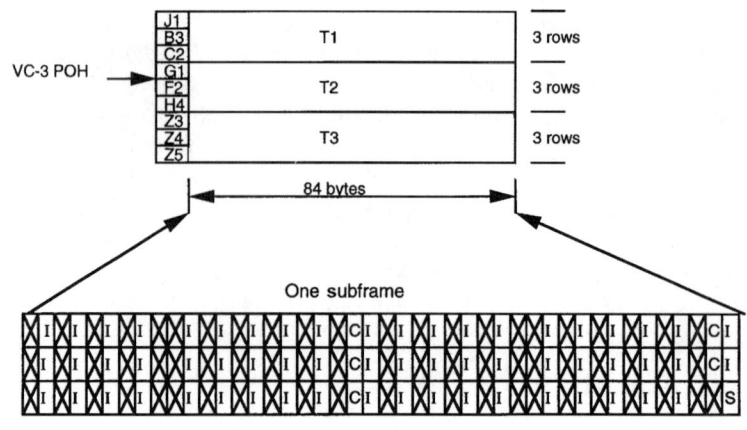

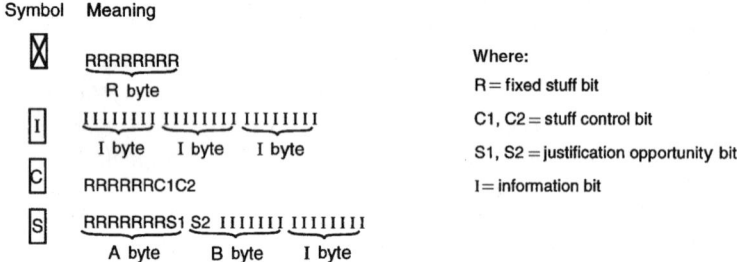

FIGURE 40.6 Asynchronous mapping of 34,368-kb/s tributary into a VC-3.

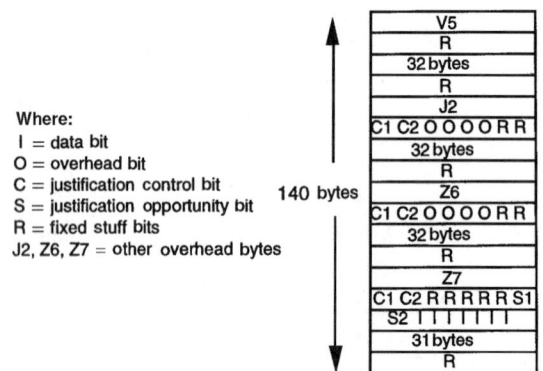

FIGURE 40.7 Mapping of asynchronous 2048-kb/s tributary into VC-12.

depending on the usage of the S1 negative bit justification opportunity. The B byte contains 7 or 8 signal data bits depending on the usage of the S2 positive bit justification opportunity. In Fig. 40.6, the letter S, with a box around it, represents a group of one A byte, one B byte, and one I byte. The S1 and S2 bit justification opportunities are either used or not used based on the fill in the incoming 34,368-kb/s bit-wide buffer. Finally, the C overhead bytes, which contain no signal data bits, encode the S1 and S2 bit usage in each subframe. Since each STM-1 frame contains three VC-3 subframes, each VC-3 has three independent positive/negative bit justification opportunities in each STM-1 frame.

The VC-12 structure shown in Fig. 40.7 [ITU, 1993c] consists of 1023 data bits, six justification control bits, two justification bits S1 and S2, and eight overhead communications channel bits. The

Synchronous Digital Hierarchy (SDH)

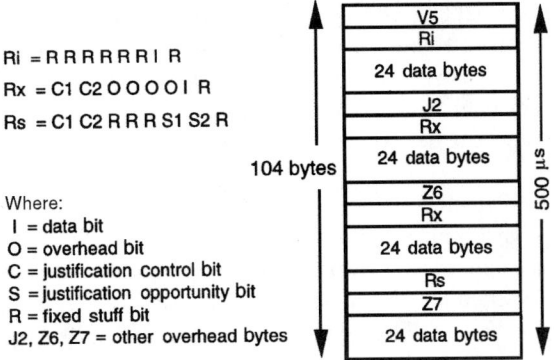

FIGURE 40.8 Asynchronous mapping of 1544-kb/s tributary into VC-11.

remaining bits are fixed stuff (R) bits. The O bits are reserved for future overhead communications purposes. For a nominal 2048-kb/s signal, the S2 justification bit is always used, and the S1 justification bit is not used. In the event that the synchronizer node's reference clock is too fast, S2 bit justification opportunities will occasionally not be used. Conversely, a slow synchronizer clock will result in both S2 and also occasionally S1 justification bit usage.

There are four types of bytes in the 1544-kb/s signal mapping shown in Fig. 40.8 [ITU, 1993c]. The first type is a normal 8-bit data byte. The second type is a Rx or Ri overhead byte, which contains one signal data bit. The third type is a purely overhead byte, which contains no signal data bits. The final type is a Rs overhead byte containing zero (an unused S2 positive bit justification opportunity), one (a normally used S2 bit and a normally unused S1 negative bit justification opportunity), or two signal data bits (both S1 and S2 bits are used). The S1 and S2 bit justification opportunities are either used or not used based on the fill of the synchronizer's bit-wide buffer.

Synchronous Digital Hierarchy (SDH) Network Block Diagram

A block diagram of an SDH network is shown in Fig. 40.9. The input signal is a serial bit stream. The first portion of an SDH network is the **synchronizer** (mapper) node. During each cycle of the mapper node byte clock, a byte of the mapping structure (shown in Figs. 40.5–40.8, for example)

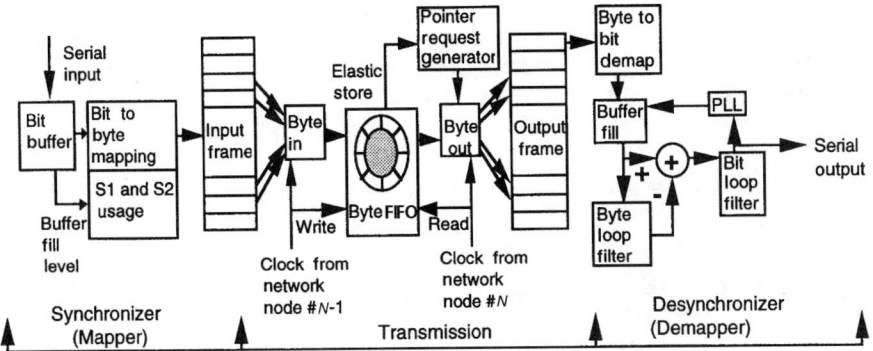

FIGURE 40.9 SDH network block diagram.

is examined to determine what type of byte is to be placed into the SDH network. In a 2048-kb/s signal, for example, the possible bytes are a normal 8-bit data byte, a byte containing seven data bits and a S2 justification bit, a byte containing only one data bit (the byte with the S1 justification bit opportunity), or an overhead byte, which contains zero data bits. Based on the buffer fill, the justification opportunities are either used or not used so as to keep this bit buffer within acceptable levels.

The next portion of a SDH network is the transmission portion. This portion of the network receives bytes from the synchronizer and places these bytes into the byte locations in the STM-1 frame. Each intermediate node in the network contains a byte-wide first in first out (FIFO) model as well as an incoming STM-1 frame and an outgoing STM-1 frame. When the byte FIFO in a given SDH network node becomes too full or too empty based on threshold values, a **pointer processor** located at that node will initiate a pointer action. The effect of these pointer actions is that one extra or one less byte will be taken from the byte FIFO. When an extra data byte is used, the data is transmitted in the negative justification opportunity byte. When one less data byte is used, stuff is transmitted in the positive justification opportunity byte.

The terminating point of a SDH network is a **desynchronizer** (demapper). A desynchronizer either speeds up or slows down the output of the serial data based on the fill of the desynchronizer bit FIFO.

40.3 Mapping of Asynchronous Transfer Mode (ATM) Cells

The 53-byte ATM cell mapping into SDH is accomplished by aligning the byte structure of every cell with the byte structure of the VC. Legal VC structures for transport of ATM cells include, but are not limited to, VC-4 (as shown in Fig. 40.10 [ITU, 1993c]) and VC-4-Xc. The ATM cell contains a header error control field, which is used as a cell alignment word to determine cell delineation. This method uses the correlation between the 32 header bits to be protected by the header error control field and the control bit of the 8-bit header error control field. A shortened cyclic code with the generator polynomial $g(x) = x^8 + x^2 + x + 1$ is used where the remainder is then added to the fixed pattern 01010101 in order to improve the cell delineation performance. This is similar to conventional frame alignment recovery where the alignment word is not fixed but varies from cell to cell. The H4 byte is not used to indicate cell offset [ITU, 1993c].

40.4 Standards

The Consultative Committee on International Telephony and Telegraphy (CCITT) now named the International Telecommunication Union Standardization Sector (ITU-T) approved in 1988 recommendations G.707, G.708, and G.709. These recommendations specify the bit rates, formats, multiplexing structure, and payload mappings for SDH. Some of the standards that are relevant to SDH are shown in Table 40.4 [Balcer et al., 1990].

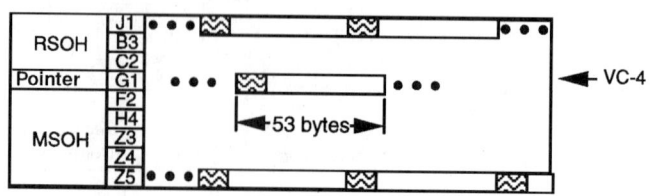

FIGURE 40.10 ATM cell mapping into a VC-4.

Synchronous Digital Hierarchy (SDH)

TABLE 40.4 Some of the Relevant SDH Standards

Standard	Title
ITU-T G.707	Synchronous digital hierarchy bit rates
ITU-T G.708	Network node interface for the SDH
ITU-T G.709	Synchronous multiplexing structure
ITU-T G.781	Structure of recommendations on equipment for the SDH equipment
ITU-T G.782	Types and general characteristics of SDH equipment
ITU-T G.823	The control of jitter and wander within digital networks which are based on the 2048-kb/s hierarchy
ITU-T G.823	The control of jitter and wander within digital networks which are based on the 1544-kb/s hierarchy
ITU-T G.823	The control of jitter and wander within digital networks which are based on the SDH
ITU-T G.783	Characteristics of SDH equipment functional blocks
ITU-T G.784	SDH management
ITU-T G.911	Parameters and calculation methodologies for reliability and availability of fiber optic systems
ITU-T G.955	Digital line systems based on the 1544-kb/s and the 2048-kb/s hierarchy on optical fiber cables
ITU-T G.957	Optical interfaces for equipments and systems relating to the SDH
ITU-T G.958	Digital line systems based on the SDH for use on optical fiber cables

Defining Terms

Administrative unit-*n* (AU-*n*): An information structure that provides compatibility between path and section layers. It is made up of a virtual container and an administrative unit pointer that indicates the phase of the VC within the AU.

Administrative unit group-*n* (AUG-*n*): A collection of one or more administrative units.

Concatenation: Several virtual containers combined together so that it can be used as a single higher data rate transport entity.

Container-*n*: The signal being transported in its application generated form is considered to be a container.

Desynchronizer: A demultiplexer that converts virtual containers back into payloads.

Pointer: A 10-bit binary value that is used to indicated the phase offset between the virtual container and the entity being transported. The pointer is contained within the H1 and H2 bytes or the V1 and V2 bytes depending on the virtual container type.

Pointer processor: The functionality that interprets and/or creates the values that indicate the phase of the virtual containers within the transport mechanisms.

Synchronizer: A multiplexer that converts (maps) payloads into virtual containers.

Synchronous digital hierarchy (SDH): A hierarchical set of digital transport structures that standardizes the transmission of payloads over physical transmission networks.

Synchronous transport module (STM) frame: A structure consisting of information payload and section overhead fields, which repeats every 125 μs. The basic STM-1 is defined to be 155,520 kb/s. The higher capacity STM-4 and STM-16 signals are formed by multiplexing 4 and 16 STM-1s, respectively. Higher rate STMs are possible.

Tributary unit-*n* (TU-*n*): An information structure that provides compatibility between lower order and higher order path layers. It is made up of a lower order virtual container and a tributary unit pointer that indicates the phase of the payload frame with respect to the VC that contains the TU.

Tributary unit group-*n* (TUG-*n*): One or more tributary units that are placed in a higher order VC. TUGs are defined to allow mixing and matching of various payloads into the STM.

Virtual container-*n* (VC-*n*): A structure consisting of information payload and path overhead used to carry payloads (digital signals). VCs are building blocks that are placed inside of STMs. VC-11, VC-12, and VC-2 structures repeat every 500 μs, whereas the VC-3 and VC-4 structures repeat every 125 μs.

References

Asatani, K., Harrison, K., and Ballart, R. 1990. CCITT standardization of network node interface of synchronous digital hierarchy. *IEEE Commun. Mag.*, 28(8):15–20.

Balcer, R., Eaves, J., Legras, J., McLintock, R., and Wright, T. 1990. An overview of emerging CCITT recommendations for the synchronous digital hierarchy: multiplexers, line systems, management, and network aspects. *IEEE Commun. Mag.*, 28(8):21–25.

Ballart, R. and Ching, Y. 1989. SONET: Now it's the standard optical network. *IEEE Commun. Mag.*, 27(3):8–15.

ITU. 1993a. Network node interface for the synchronous digital hierarchy. ITU-T Recommendation G.708, International Telecommunications Union, Geneva, Switzerland.

ITU. 1993b. Synchronous digital hierarchy bit rates. ITU-T Recommendation G.707, International Telecommunications Union, Geneva, Switzerland.

ITU. 1993c. Synchronous multiplexing structure. ITU-T Recommendation G.709, International Telecommunications Union, Geneva, Switzerland.

Omidayr, C. and Aldridge, A. 1993. Introduction to SONET/SDH. *IEEE Commun. Mag.*, 31(9):30–33.

Sexton, M. and Reid, A. 1992. *Transmission Networking: SONET and the Synchronous Digital Hierarchy*, Artech House, Norwood, MA.

Sexton, M., Roverano, F., and DeCremiers, F. 1993. SDH architecture and standards. *ALCATEL Electrical Commun.*, (4th Qtr.):299–312.

Further Information

An introductory source on SDH is the September 1993 *IEEE Communications Magazine* [Omidayr and Aldridge, 1993], which features SDH/SONET. For more details the book *Transmission Networking: SONET and the Synchronous Digital Hierarchy* Sexton and Reid [1992] is an excellent source of information. Finally, the details of SDH are best obtained directly from the standards. Many of the SDH standards are available on the Internet from ITU(gopher://info.itu.ch/).

Networks

41 The Open Systems Interconnections (OSI) Seven-Layer Model *Fred Halsall* 567
Computer Communications Requirements • Standards Evolution • International Standards Organization Reference Model • Open System Standards • Summary

42 Integrated Services Digital Network (ISDN) and Broadband (B-ISDN)
Erwin P. Rathgeb 577
Introduction • The ISDN Communication Services • The Generic Architecture of an ISDN • The ISDN Protocol Reference Model • Layers 1–3 of the ISDN User–Network Interface • Broadband ISDN (B-ISDN) • Structure of the B-ISDN • The B-ISDN Protocol Reference Model • Functions of the ATM Specific Layers • Signalling in the B-ISDN • Summary and Conclusion

43 Ethernet Networks *Ramesh R. Rao* 591
Overview • Historical Development • Standards • Operation

44 Fiber Distributed Data Interface and Its Use for Time-Critical Applications
Biao Chen, Nicholas Malcolm, and Wei Zhao 597
Introduction • Architecture and Fault Management • The Protocol and Its Timing Properties • Parameter Selection for Real-Time Applications • Final Remarks

45 Broadband Local Area Networks *Joseph A. Bannister* 611
Introduction • User Requirements • BLAN Technologies • ATM BLANs • Other BLANs • New Applications • Conclusion

46 Multiple Access Methods for Communications Networks *Izhak Rubin* 622
Introduction • Features of Medium Access Control Systems • Categorization of Medium Access Control Procedures • Polling-Based Multiple Access Networks • Random-Access Protocols • Multiple-Access Schemes for Wireless Networks • Multiple Access Methods for Spatial-Reuse Ultra-High-Speed Optical Communications Networks

47 Routing and Flow Control *Rene L. Cruz* 650
Introduction • Connection-Oriented and Connectionless Protocols, Services, and Networks • Routing in Datagram Networks • Routing in Virtual Circuit Switched Networks • Hierarchical Routing • Flow Control in Datagram Networks • Flow Control in Virtual Circuit Switched Networks

48 Transport Layer *A. Udaya Shankar* 661
Introduction • Transport Service • Data-Transfer Protocol • Connection-Management Protocol • Transport Protocols • Conclusions

49 Gigabit Networks *Jonathan M. Smith* 672
Introduction • Asynchronous Transfer Mode Host Interfacing • Multimedia Architectures • Distributed Shared Memory Communications • Conclusions

50 Local Area Networks *Thomas G. Robertazzi* 681
Introduction • Local Area Networks • Metropolitan Area Networks: The LAN Interconnect • The Future

565

51 Asynchronous Time Division Switching *Achille Pattavina* 686
Introduction • The ATM Standard • Switch Model • ATM Switch with Blocking Multistage IN and Minimum Depth • ATM Switch with Blocking Multistage IN and Arbitrary Depth • ATM Switch with Nonblocking IN • Conclusions

52 Internetworking *Harrell J. Van Norman* 701
Introduction • Internetworking Protocols • The Total Network Engineering Process • Internetwork Simulation • Internetwork Optimization • Summary

53 Architectural Framework for Asynchronous Transfer Mode Networks: Broadband Network Services *Gerald A. Marin and Raif O. Onvural* 717
Introduction • Broadband Integrated Services Digital Network Framework • Architectural Drivers • Review of Broadband Network Services • How Does It All Fit Together? • Broadband Network Services • Conclusions

41
The Open Systems Interconnections (OSI) Seven-Layer Model

Fred Halsall
University of Wales, Swansea

41.1 Computer Communications Requirements 567
41.2 Standards Evolution 568
41.3 International Standards Organization Reference Model 569
    The Application-Oriented Layers • The Presentation Layer • The Network-Dependent Layers
41.4 Open System Standards 574
41.5 Summary 576

41.1 Computer Communications Requirements

Although in many instances computers are used to perform their intended role in a stand-alone mode, in others there is a need to interwork and exchange data with other computers. In financial applications, for example, to carry out funds transfers from one institution computer to another, in travel applications to access the reservation systems belonging to various airlines, and so on. The general requirement in all of these applications is for application programs running in different computers to cooperate to achieve a specific distributed application function. To achieve this, three basic issues must be considered. These are shown in diagrammatic form in Fig. 41.1.

The fundamental requirement in all applications that involve two or more computers is the provision of a suitable data communications facility. This may comprise a **local area network** (LAN), if the computers are distributed around a single site, a **wide area network** (WAN), if the computers are situated at different sites or an **internetwork** if multiple interconnected network types are involved. Associated with these different network types is a set of access protocols which enable a communications path between two computers to be established and for data to be transferred across this path. Typically, these protocols differ for the different network types. In addition to these access protocols, the communication subsystem in each computer must provide additional functionality. For example, if the communicating computers are of different types, possibly with different word sizes and character sets, then a means of ensuring the transferred data is interpreted in the same way in each computer must be incorporated. Also, the computers may use different file systems and hence functionality to enable application programs, normally referred to as **application processes (APs)**, to access these in a standardized way must

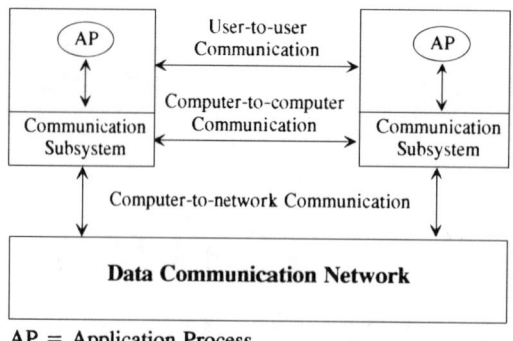

FIGURE 41.1 Computer communication schematic.

also be included. All of these issues must be considered when communicating data between two computers.

41.2 Standards Evolution

Until recently, the standards established for use in the computer industry by the various international bodies were concerned primarily with either the internal operation of a computer or the connection of a local peripheral device. The result was that early hardware and software communication subsystems offered by manufacturers only enables their own computers, and so-called *plug-compatible systems*, to exchange information. Such systems are known as *closed systems* since computers from other manufacturers cannot exchange information unless they adhere to the (proprietary) standards of a particular manufacturer.

In contrast, the various international bodies concerned with public-carrier telecommunication networks have for many years formulated internationally agreed standards for connecting devices to these networks. The *V-series recommendations*, for example, are concerned with the connection of equipment, normally referred to as a *data terminal equipment* (DTE), to a modem connected to the public switched telephone network (PSTN); the *X-series recommendations* for connecting a DTE to a public data network; and the *I-series recommendations* for connecting a DTE to the integrated services digital networks (ISDNs). The recommendations have resulted in compatibility between the equipment from different vendors, enabling a purchaser to select suitable equipment from a range of manufacturers.

Initially, the services provided by most public carriers were concerned primarily with data transmission, and hence the associated standards only related to the method of interfacing a device to these networks. More recently, however, the public carriers have started to provide more extensive distributed information services, such as the exchange of electronic messages (*teletex*) and access to public databases (*videotex*). To cater for such services, the standards bodies associated with the telecommunications industry have formulated standards not only for interfacing to such networks but also so-called higher level standards concerned with the format (syntax) and control of the exchange of information (data) between systems. Consequently, the equipment from one manufacturer that adheres to these standards can be interchangeable with equipment from any other manufacturer that complies with the standards. The resulting system is then known as an *open system* or, more completely, as an *open systems interconnection environment* (*OSIE*).

In the mid-1970s as different types of distributed systems (based on both public and private data networks) started to proliferate, the potential advantages of open systems were acknowledged by the computer industry. As a result, a range of standards started to be introduced. The first was concerned with the overall structure of the complete communication subsystem within each computer. This was produced by the *International Standards Organization* (*ISO*) and is known as the *ISO reference model* for **open systems interconnection** (**OSI**).

The aim of the ISO reference model is to provide a framework for the coordination of standards development and to allow existing and evolving standards activities to be set within a common framework. The aim is to allow an application process in any computer that supports a particular set of standards to communication freely with an application process in any other computer that supports the same standards, irrespective of its origin of manufacture.

Some examples of application processes that may wish to communicate in an open way are the following:

- A process (program) executing in a computer and accessing a remote file system
- A process acting as a central file service (server) to a distributed community of (client) processes
- A process in an office workstation (computer) accessing an electronic mail service
- A process acting as an electronic mail server to a distributed community of (client) processes
- A process in a supervisory computer controlling a distributed community of computer-based instruments or robot controllers associated with a process or automated manufacturing plant
- A process in an instrument or robot controller receiving commands and returning results to a supervisory system
- A process in a bank computer that initiates debit and credit operations on a remote system

Open systems interconnection is concerned with the exchange of information between such processes. The aim is to enable application processes to cooperate in carrying out a particular (distributed) information processing task irrespective of the computers on which they are running.

41.3 International Standards Organization Reference Model

A communication subsystem is a complex piece of hardware and software. Early attempts at implementing the software for such subsystems were often based on a single, complex, unstructured program (normally written in assembly language) with many interacting components. The resulting software was difficult to test and often very difficult to modify.

To overcome this problem, the ISO has adopted a layered approach for the reference model. The complete communication subsystem is broken down into a number of layers each of which performs a well-defined function. Conceptually, these layers can be considered as performing one of two generic functions, network-dependent functions and application-oriented functions. This in turn gives rise to three distinct operational environments:

1. The **network environment** is concerned with the protocols and standards relating to the different types of underlying data communication networks.
2. The **OSI environment** embraces the network environment and adds additional application-oriented protocols and standards to allow end systems (computers) to communicate with one another in an open way.
3. The **real systems environment** builds on the OSI environment and is concerned with a manufacturers own proprietary software and services which have been developed to perform a particular distributed information processing task.

This is shown in diagrammatic form in Fig. 41.2.

Both the network-dependent and application-oriented (network-independent) components of the OSI model are implemented as a number of layers. The boundaries between each layer and

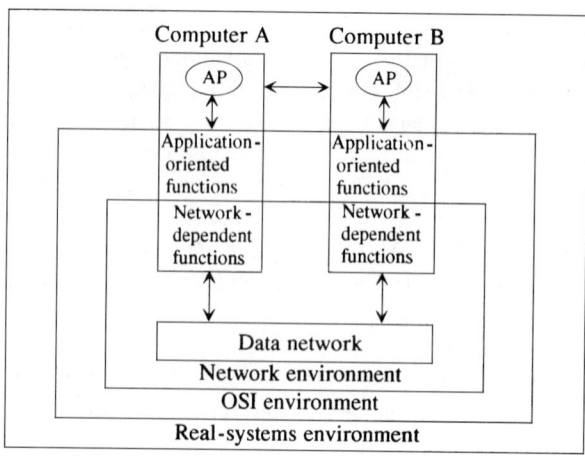

FIGURE 41.2 Operational environments.

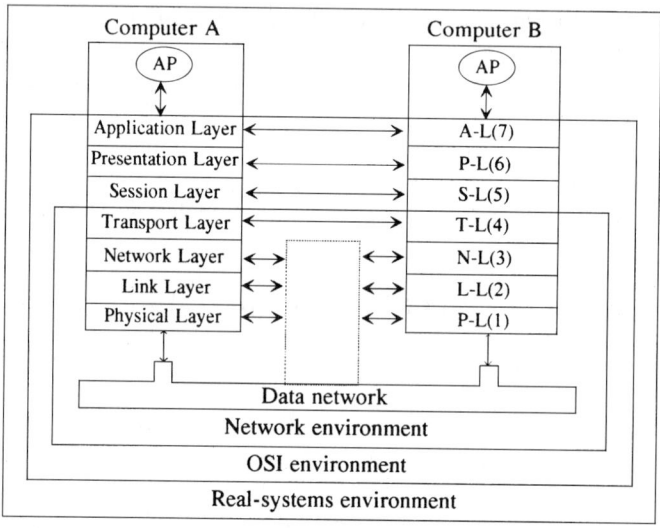

FIGURE 41.3 Overall structure of the ISO reference model.

the functions performed by each layer have been selected on the basis of experience gained during earlier standardization activity.

Each layer performs a well-defined function in the context of the overall communication subsystem. It operates according to a defined *protocol* by exchanging messages, both user data and additional control information, with a corresponding *peer* layer in a remote system. Each layer has a well-defined interface between itself and the layer immediately above and below. Consequently, the implementation of a particular protocol layer is independent of all other layers.

The logical structure of the ISO reference model is made up of seven protocol layers, as shown in Fig. 41.3. The three lowest layers (1–3) are network dependent and are concerned with the protocols associated with the data communication network being used to link the two communicating computers. In contrast, the three upper layers (5–7) are application oriented and are concerned with the protocols that allow two end user application processes to interact with each other, normally through a range of services offered by the local operating system. The intermediate transport layer (4) masks the upper application-oriented layers from the detailed operation of the lower network-dependent layers. Essentially, it builds on the services provided by the latter to provide the application-oriented layers with a network-independent message interchange service.

The function of each layer is specified formally as a protocol that defines the set of rules and conventions used by the layer to communicate with a similar peer layer in another (remote) system. Each layer provides a defined set of services to the layer immediately above. It also uses the services provided by the layer immediately below it to transport the message units associated with the protocol to the remote peer layer. For example, the transport layer provides a network-independent message transport service to the session layer above it and uses the service provided by the network layer below it to transfer the set of message units associated with the transport protocol to a peer transport layer in another system. Conceptually, therefore, each layer communicates with a similar peer layer in a remote system according to a defined protocol. However, in practice the resulting protocol message units of the layer are passed by means of the services provided by the next lower layer. The basic functions of each layer are summarized in Fig. 41.4.

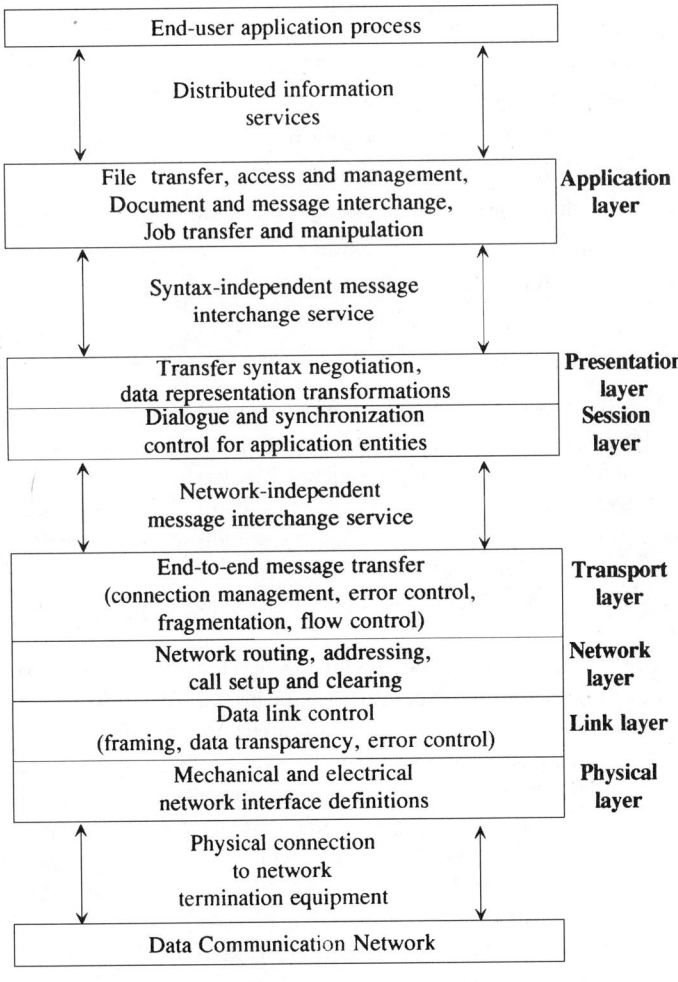

FIGURE 41.4 Protocol layer summary.

The Application-Oriented Layers

The Application Layer

The application layer provides the user interface, normally an application program/process, to a range of networkwide distributed information services. These include file transfer access and management, as well as general document and message interchange services such as electronic mail. A number of standard protocols are either available or are being developed for these and other types of service.

Access to application services is normally achieved through a defined set of primitives, each with associated parameters, which are supported by the local operating system. The access primitives are the same as other operating system calls (as used for access to, say, a local file system) and result in an appropriate operating system procedure (process) being activated. These operating system procedures use the communication subsystem (software and hardware) as if it is a local device, similar to a disk controller, for example. The detailed operation and implementation of the communication subsystem is thus transparent to the (user) application process. When the application process making the call is rescheduled (run), one (or more) status parameters are returned indicating the success (or otherwise) of the network transaction that has been attempted.

In addition to information transfer, the application layer provides such services as:

- Identification of the intended communication partner(s) by name or by address
- Determination of the current availability of an intended communication partner
- Establishment of authority to communicate
- Agreement on privacy (encryption) mechanisms
- Authentication of an intended communication partner
- Selection of the dialogue discipline, including the initiation and release procedures
- Agreement on responsibility for error recovery
- Identification of constraints on data syntax (character sets, data structures, etc.)

The Presentation Layer

The *presentation layer* is concerned with the representation (syntax) of data during transfer between two communicating application processes. To achieve true open systems interconnection, a number of common *abstract data syntax* forms have been defined for use by application processes together with associated *transfer* (or *concrete*) *syntaxes*. The presentation layer negotiates and selects the appropriate transfer syntax(es) to be used during a transaction so that the syntax (structure) of the messages being exchanged between two application entities is maintained. Then, if this form of representation is different from the internal abstract form, the presentation entity performs the necessary conversion.

To illustrate the services provided by the presentation layer, consider a telephone conversation between a French speaking person and a Spanish speaking person. Assume each uses an interpreter and that the only language understood by both interpreters is English. Each interpreter must translate from their local language to English, and vice versa. The two correspondents are thus analogous to two application processes with the two interpreters representing presentation layer entities. French and Spanish are the local syntaxes and English the transfer or concrete syntax. Note that there must be a universally understood language which must be defined to allow the agreed transfer language (syntax) to be negotiated. Also note that the interpreters do not necessarily understand the meaning (semantics) of the conversation.

Another function of the presentation layer is concerned with data security. In some applications, data sent by an application is first encrypted (enciphered) using a *key*, which is (hopefully) known only by the intended recipient presentation layer. The latter decrypts (deciphers) any received data using the corresponding key before passing it on to the intended recipient.

The Session Layer

The *session layer* provides the means that enables two application layer protocol entities to organize and synchronize their dialogue and manage their data exchange. It is thus responsible for setting up (and clearing) a communication (dialogue) channel between two communicating application layer protocol entities (presentation layer protocol entities in practice) for the duration of the complete network transaction. A number of optional services are provided, including the following:

- Interaction management: The data exchange associated with a dialogue may be duplex or half-duplex. In the latter case it provides facilities for controlling the exchange of data (dialogue units) in a synchronized way.
- Synchronization: For lengthy network transactions, the user (through the services provided by the session layer) may choose periodically to establish synchronization points associated with the transfer. Then, should a fault develop during a transaction, the dialogue may be restarted at an agreed (earlier) synchronization point.
- Exception reporting: Nonrecoverable exceptions arising during a transaction can be signaled to the application layer by the session layer.

The Transport Layer

The *transport layer* acts as the interface between the higher application-oriented layers and the underlying network-dependent protocol layers. It provides the session layer with a message transfer facility that is independent of the underlying network type. By providing the session layer with a defined set of message transfer facilities, the transport layer hides the detailed operation of the underlying network from the session layer.

The transport layer offers a number of **classes of service** which cater for the varying *quality of service* (*QOS*) provided by different types of network.

There are five classes of service ranging from class 0, which provides only the basic functions needed for connection establishment and data transfer, to class 4, which provides full error control and flow control procedures. As an example, class 0 may be selected for use with a packet-switched data network (PSDN), whereas class 4 may be used with a local area network (LAN) providing a best-try service; that is, if errors are detected in a frame then the frame is simply discarded.

The Network-Dependent Layers

As the lowest three layers of the ISO reference model are network dependent, their detailed operation varies from one network type to another. In general, however, the *network layer* is responsible for establishing and clearing a networkwide connection between two transport layer protocol entities. It includes such facilities as network routing (addressing) and, in some instances, flow control across the computer-to-network interface. In the case of internetworking it provides various harmonizing functions between the interconnected networks.

The *link layer* builds on the physical connection provided by the particular network to provide the network layer with a reliable information transfer facility. It is thus responsible for such functions as error detection and, in the event of transmission errors, the retransmission of messages. Normally, two types of service are provided:

1. *Connectionless* treats each information frame as a self-contained entity that is transferred using a best-try approach.
2. *Connection oriented* endeavors to provide an error-free information transfer facility.

Finally, the *physical layer* is concerned with the physical and electrical interfaces between the user equipment and the network terminating equipment. It provides the link layer with a means of transmitting a serial bit stream between the two equipments.

41.4 Open System Standards

The ISO reference model has been formulated simply as a template for the structure of a communication subsystem on which standards activities associated with each layer may be based. It is not intended that there should be a single standard protocol associated with each layer. Rather, a set of standards is associated with each layer, each offering different levels of functionality. Then, for a specific open systems interconnection environment, such as that linking numerous computer-based systems in a fully automated manufacturing plant, a selected set of standards is defined for use by all systems in that environment.

The three major international bodies actively producing standards for computer communications are the ISO, the American Institute of Electrical and Electronic Engineers (IEEE) and the International Telegraph and Telephone Consultative Committee (CCITT). Essentially, ISO and IEEE produce standards for use by computer manufacturers, whereas CCITT defines standards for connecting equipment to the different types of national and international public networks. As the degree of overlap between the computer and telecommunications industries increases, however, there is an increasing level of cooperation and commonality between the standards produced by these organizations.

In addition, prior to and concurrently with ISO standards activity, the U.S. Department of Defense has for many years funded research into computer communications and networking through its Defense Advanced Research Projects Agency (DARPA). As part of this research, the computer networks associated with a large number of universities and other research establishments were linked to those of DARPA. The resulting internetwork, known as ARPANET, has been extended to incorporate internets developed by other government agencies. The combined internet is now known simply as the *Internet.*

The protocol suite used with the Internet is known as **transmission control protocol/internet protocol (TCP/IP)**. It includes both network-oriented protocols and application support protocols. Because TCP/IP is in widespread use with an existing internet, many of the TCP/IP protocols have been used as the basis for ISO standards. Moreover, since all of the protocol specifications associated with TCP/IP are in the public domain, and hence no licence fees are payable, they have been used extensively by commercial and public authorities for creating open system networking environments. In practice, therefore, there are two major open system (vendor-independent) standards: the TCP/IP protocol suite and those based on the evolving ISO standards.

Figure 41.5 shows some of the standards associated with the TCP/IP protocol suite. As can be seen, since TCP/IP has developed concurrently with the ISO initiative, it does not contain specific protocols relating to all of the seven layers in the OSI model.

Moreover, the specification methodology used for the TCP/IP protocols differs from that used for the ISO standards. Nevertheless, most of the functionality associated with the ISO layers is embedded in the TCP/IP suite. A range of standards has been defined by the ISO/CCITT and a selection of these is shown in Fig. 41.5. Collectively they enable the administrative authority that is establishing the open system environment to select the most suitable set of standards for the application. The resulting protocol suite is known as the *open system interconnection profile.* A number of such profiles have now been defined, including: TOP, a protocol set for use in technical and office environments; MAP, for use in manufacturing automation; U.S. and U.K. GOSIP for use in U.S. and U.K. government projects, respectively; and a similar suite used in Europe known as the CEN functional standards. The latter has been defined by the Standards Promotion and Application Group (SPAG), a group of 12 European companies.

As Fig. 41.6 shows, the lower three layers vary for different network types. CCITT has defined V, X, and I series standards for use with public-carrier networks. The V series is for use with the existing switched telephone network (PSTN), the X series for use with existing switched data networks (PSDN), and the I series for use with the integrated services digital networks (ISDN). Those produced by ISO/IEEE for use with local area networks are known collectively as the 802 (IEEE) or 8802 (ISO) series.

The Open Systems Interconnections (OSI) Seven-Layer Model

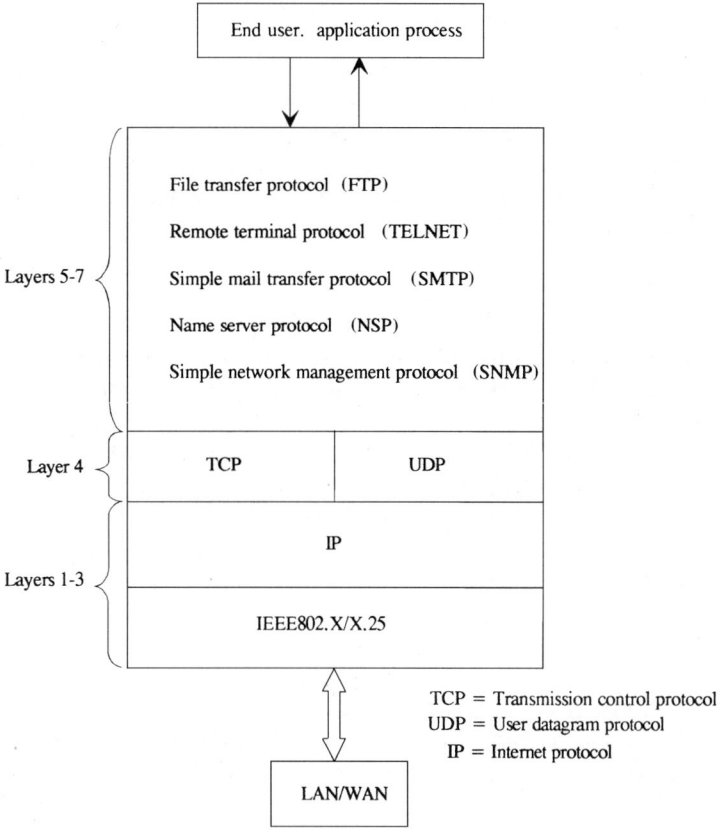

FIGURE 41.5 TCP/IP protocol suite.

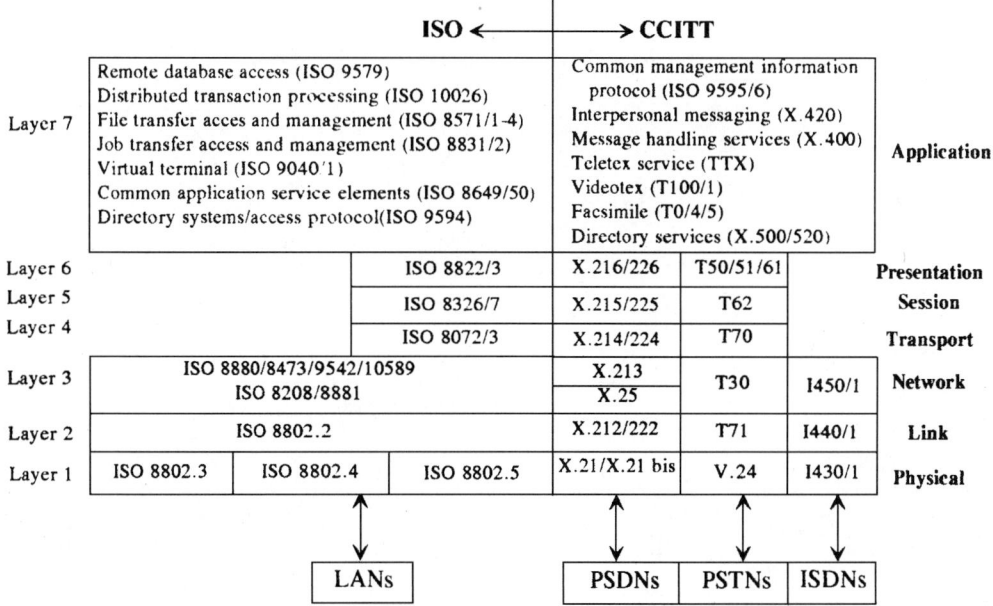

FIGURE 41.6 Standards summary.

41.5 Summary

This chapter has reviewed the requirements for the communications subsystem in each of a set of interconnected computers that enables them to communicate in an open way to perform various distributed application functions. The philosophy behind the structure of the ISO reference model for open systems interconnection has been presented and a description of the functionality of the seven layers that make up the reference model described. Finally, a selection of the ISO/CCITT standards that have been defined have been identified.

Defining Terms

Internetwork: A collection of different network types that are linked together.

Local area network (LAN): Network used to link computers that are distributed around a single site.

Open system interconnection (OSI): A standard set of communication protocols that are independent of a specific manufacturer.

Transmission control protocol/Internet protocol (TCP/IP): The standard set of communication protocols that are used by each computer that is connected to the Internet.

Wide area network (WAN): Network used to link computers that are situated at different sites.

Further Information

Further coverage of the material covered in this chapter can be found in:

Halsall, F. 1996. *Data Communications, Computer Networks and Open Systems*, 4th ed., Addison Wesley, Reading, MA.

42
Integrated Services Digital Network (ISDN) and Broadband (B-ISDN)

Erwin P. Rathgeb
Siemens AG

42.1 Introduction 577
42.2 The ISDN Communication Services 578
42.3 The Generic Architecture of an ISDN 579
42.4 The ISDN Protocol Reference Model 581
42.5 Layers 1–3 of the ISDN User–Network Interface 583
42.6 Broadband ISDN (B-ISDN) 584
42.7 Structure of the B-ISDN 586
42.8 The B-ISDN Protocol Reference Model 587
42.9 Functions of the ATM Specific Layers 587
42.10 Signalling in the B-ISDN 589
42.11 Summary and Conclusion 589

42.1 Introduction

Today, the public switched telephone network is the world's largest network providing interactive communications. In recent years an increasing percentage of the previously analog transmission and switching equipment has been replaced by digital systems for economic reasons. This digitization paves the way for the migration to the integrated services digital network (ISDN). The basic idea behind the ISDN is to provide one multipurpose communication network using a uniform technology to support a wide variety of telecommunication services[1] instead of having dedicated, service specific networks. This approach provides the opportunity for the user to conveniently cover most of his communication needs by subscribing to one single network. It also allows the network operator to optimize investment as well as operation costs. The inherent flexibility of this approach also stimulates the rapid introduction of new services.

This ISDN is internationally standardized in the I.-, Q.-, and G.-series of the Telecommunication Standardization Sector of the International Telecommunications Union (ITU-T), formerly called Consultative Committee on International Telephony and Telegraphy (CCITT), recommendations

---

[1] *Service* here is not used in the open systems interconnection (OSI) sense but describes the collection of all telecommunications means and facilities provided by the network operator to allow a certain type of communication over the network.

and can briefly be characterized as follows:

- It offers fully digital connectivity from subscriber to subscriber.
- It is essentially a circuit-switched network based on 64-kb/s basic rate channels. However, packet-switched services also can be supported by the ISDN.
- Signalling information is strictly separated from user data and is transported in dedicated signalling channels (out-of-band signalling). This signalling principle provides increased flexibility compared to the in-band signalling scheme used in the conventional telephone network because it also allows the exchange of signalling information during the information phase of a call and even without setting up a user information channel.
- The ISDN basic access, which is the primary interface between the network and the subscriber, provides two 64-kb/s basic channels (**B-channels**) for user data transfer plus a 16-kb/s signalling channel (**D-channel**). A primary rate access with 24 or 30 B-channels (depending on the transmission system used) and one 64-kb/s signalling channel is also offered, for examples to connect ISDN private branch exchanges to the public network.
- Both basic and primary rate access can be provided by using the copper twisted pair subscriber lines of the existing telephone network.
- The ISDN offers a wide variety of services ranging from transparent digital information transfer to fully standardized services including telephone, facsimile, text, data and even interactive video communication. The ISDN provides service discrimination, that is, it ensures that only compatible terminals will communicate.
- The standardized user–network interface not only provides a universal *communication socket* for all kinds of terminals but also a unified control of all services by using a common set of signalling procedures. Several service specific or multifunctional terminals in any combination can be connected to the user–network interface by using a passive bus or star structure.
- The subscriber can be reached under one single address (directory number) no matter what kind and mix of services being used.

The following chapter will provide an overview of the basic principles of ISDN together with an introduction to the respective protocol stacks.

42.2 The ISDN Communication Services

Since service integration is one of the main goals of the ISDN, principles for the flexible definition and description of a wide variety of communication services have been defined [I.210]. The ISDN services can be subdivided into two groups, namely, the **bearer services** and the **teleservices**. Bearer services [I.23x][2] provide only basic information transfer between the involved user–network interfaces of the ISDN and require only the functionality of the OSI layers 1–3. Thus, it is the responsibility of the users to ensure the compatibility of the higher protocol layers of the end systems (terminals). Circuit-mode as well as packet-mode bearer services have been defined. Typical categories for circuit-mode bearer services are: 64-kb/s unrestricted, speech, 2 × 64-kb/s unrestricted or 384-kb/s unrestricted, providing either clear digital channels only or basic low-level support for specific services as, for example, for the transfer of speech information. The packet-mode bearer services include categories for both connection-oriented and connectionless communication.

Teleservices [I.24x] on the other hand also define all communication functions implemented within the end systems, that is, all seven OSI layers, and therefore ensure full compatibility between the terminals for a specific teleservice. Apart from ISDN telephony, examples for teleservices are:

---

[2]This topic is covered by a combination of several recommendations, x denotes one digit.

ISDN telefax (Group 4), ISDN mixed mode (combination of text and facsimile in one document), and ISDN teleaction.

The services mentioned can be categorized as *conversational services* because they are dialog oriented and require bidirectional real-time end-to-end information transfer between individual users. In addition, there is the category of the messaging services offering communication among individual users by using storage devices located within the network. Examples here are the message handling systems based on the X.400 standards. There is also the so-called *retrieval service* category, where information stored in databases located in service centers within the network (or connected to it) can be accessed by individual users. One example for this service category is the ISDN videotex teleservice. In addition to these *interactive services* there is also a class of *distribution services* allowing an arbitrary number of users to access one information source.

Packet-oriented data services can be provided in various ways. One way is to provide a circuit switched B-channel connection to a [X.25] packet-switching function located within the ISDN [X.31] or to an interworking unit to the public X.25 data network. Alternatively, it is possible to transfer low-rate user data also via the D-channel, because its capacity of 16-kb/s is normally only used to a small extent for signalling. A new class of packet-switched services, *frame mode bearer services* (FMBS) has also been defined to provide enhanced data communication capabilities [I.233]. The most commonly used service of this class is the ISDN frame relaying bearer service.

To be able to describe and standardize the wide variety of services efficiently a combination of *service attributes* is used. These attributes describe, among others, all basic service characteristics as, for example, connection type (switched/permanent), transfer mode (circuit/packet), information transfer rate, and protocol stacks used. In addition to these basic communication functions, **supplementary services** [I.25x] are provided to enhance the services by providing features such as reverse charging, call forwarding, conference calls, closed user groups, call barring, or calling line identification. These supplementary services can be used in conjunction with arbitrary basic services (with some restrictions).

42.3 The Generic Architecture of an ISDN

The generic architecture of the ISDN is shown in Fig. 42.1 to the extent necessary for the understanding of the remainder of this chapter. A small set of standardized **user–network interfaces** (UNI) provides access to the ISDN. The primary UNI, the *ISDN basic access* offers to the subscriber two circuit-switched 64-kb/s B-channels for user information transfer and one packet-oriented 16-kb/s D-channel for the transfer of signalling information. Up to eight terminals can be connected per ISDN basic access via a passive bus called the *S-bus*. The subscriber lines are conventional copper twisted pair wires over which the B-channels and the D-channel are transmitted using a synchronous time division multiplexing principle.

Signalling between the terminals and the local exchange via the D-channel is done according to the *D-channel protocol*[3] [I.4xx]. Interexchange signalling within the ISDN is also done separately from the user information transfer. For this purpose a logically separate, packet-oriented signalling network based on the common channel signalling system 7 (CCS7) [Q.7xx] is used.

To describe network elements and interfaces in a unified, abstract way (independently from a physical implementation), a *reference configuration* has been developed [I.411]. The reference configuration consists of *functional groups* characterizing the functionality located in the various network elements such as terminals, public access branch exchanges (PABXs), or exchanges. *Reference points* have been defined in between these functional groups to identify possible locations of

---

[3]The signalling protocol standards are included in the I-series of recommendations defining the ISDN as well as in the Q-series defining signalling systems. The Q.9xx recommendations also describe the D-channel protocol, which is named Digital Subscriber Signalling System No. 1 (DSS1) in this context, the text of the corresponding recommendations is identical.

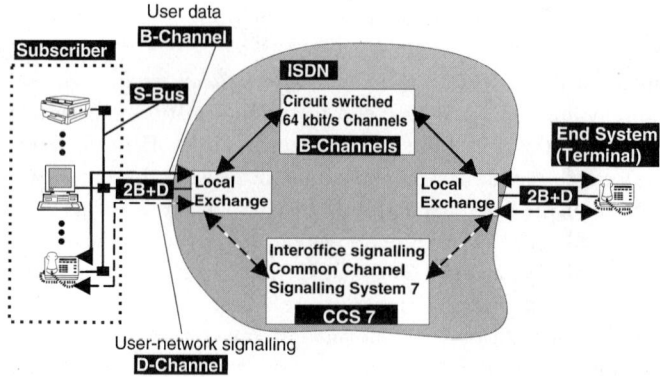

FIGURE 42.1 Generic architecture of the ISDN.

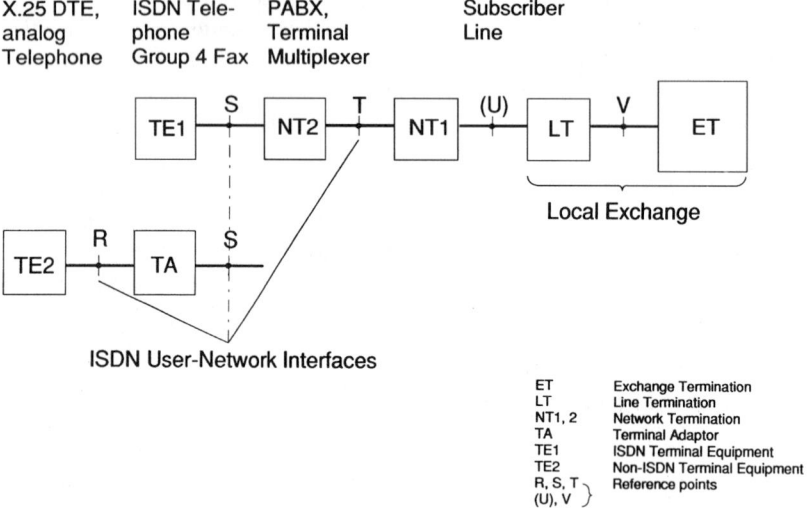

FIGURE 42.2 Reference configuration for the ISDN user–network interface.

physical interfaces. The functional blocks and reference points for the ISDN user–network interfaces are shown in Fig. 42.2.

The functional blocks can be characterized as follows:

- *Terminal equipment type 1 (TE1)*. End system (terminal) with standard ISDN interfaces implementing all seven layers of the OSI model as required. Examples are telephone sets, fax machines, or computers with ISDN interfaces.
- *Terminal equipment type 2 (TE2)*. End systems with non-ISDN interfaces. Examples are analog telephones, data terminal equipment (DTE) according to ITU-T recommendation X.25, or teletex terminals.
- *Terminal adaptor (TA)*. Adaptation device allowing connection of non-ISDN TEs to the ISDN. The TA has to adapt the user information as well as the signalling mechanisms. A TA for analog telephones, for example, has to perform the A/D conversion of the voice signal and has to convert the pulse or multifrequency code dialing information to ISDN signalling messages according to the D-channel protocol.
- *Network termination 2 (NT2)*. Network termination unit allowing the connection of several TEs to one UNI. The NT2 typically implements the functions of the OSI layers 1–3 including

multiplexing, concentration, or switching. Examples are PABXs, local-area networks (LANs) or terminal multiplexers. If this functionality is not required, the NT2 can degenerate to a null-NT2.

- *Network termination 1 (NT1).* Network termination unit to connect to the subscriber line. It implements the functions of OSI layer 1 necessary to provide a physical and electrical termination of the subscriber line as well as operation and maintenance capabilities (test loops). It also provides a conversion from the transmission method used on the subscriber side of the NT1 to the one used on the subscriber line.
- *Line termination (LT).* The LT is the counterpart of the NT1 and terminates the transmission functions on the exchange side of the subscriber line.
- *Exchange termination (ET).* The ET logically terminates the subscriber access and implements functions of the OSI layers 2–7, that is, it primarily terminates the user–network signalling protocols. The ET functions are located in the local exchange.

At the *R reference point*, standardized non-ISDN interfaces (X.25, V.24, a/b, etc.) can be used, whereas a connection of terminals at the *S reference point* requires standard ISDN interfaces. To allow the NT2 to degenerate to a null function, that is, to a physically nonexisting NT2, the interface specifications for the *S* and *T reference points* have been kept identical. Because of the big differences in the subscriber lines in various countries no international standards are available for the *U reference point* (which is, in fact, not even internationally standardized itself). The location where the domain of the network provider ends is also not standardized on an international level and may be the S, T, or U reference point, respectively.

At the *V reference point*, however, there exist internationally standardized interfaces. These interfaces are not necessarily accessible from outside the switch, because the LT and ET functions are often physically combined for cost and implementation reasons.

42.4 The ISDN Protocol Reference Model

To support communication across the ISDN three distinct but interacting protocol stacks are required, namely, the user–network signalling protocol on the D-channel, the interoffice signalling system CCS 7, and the user information protocol on the B-channel. The OSI model alone is not sufficient to describe this communication because it offers, for example, no means to model the out-of-band signalling scenario. Because the OSI model is generic, it also offers no adequate support for describing network specific properties as the information flows for control of supplementary services (reverse charging, call forwarding, change of service during a connection, etc.) or for modeling voice or video communication. Therefore, a specific *ISDN protocol reference model (PRM)* has been defined [I.320].

Some of the OSI concepts and related terminology are also used in the ISDN PRM. It exhibits a layered structure and is based on an OSI-like service concept using service primitives for the communication of adjacent layers. The idea of peer-to-peer protocols for the communication of entities within the same layer has also been preserved. To identify the various layers, only the numbers (1–7) are used, because the OSI names for them would be misleading in some instances. However, specific extensions have been made. The most obvious one is the introduction of several *protocol planes* needed to model the separation of user and signalling information.

As shown in Fig. 42.3, the generic protocol block used to model ISDN network elements (TE, NT, etc.) is structured into three of these planes:

- *Control plane (C).* Represents the protocols used for transfer of signalling information.
- *User plane (U).* Represents the protocols used for transfer of user information.
- *Management plane (M).* Represents the functions needed to coordinate the activities in the U- and C-plane. This includes the synchronization between signalling and user information transfer.

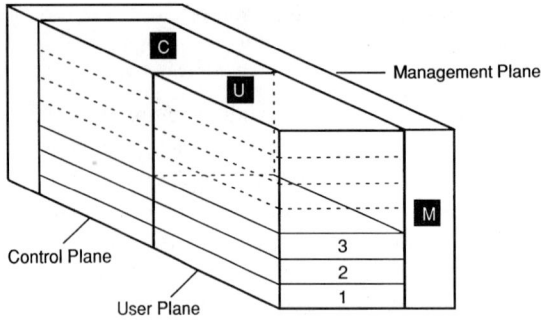

FIGURE 42.3 Generic protocol block of the ISDN protocol reference model.

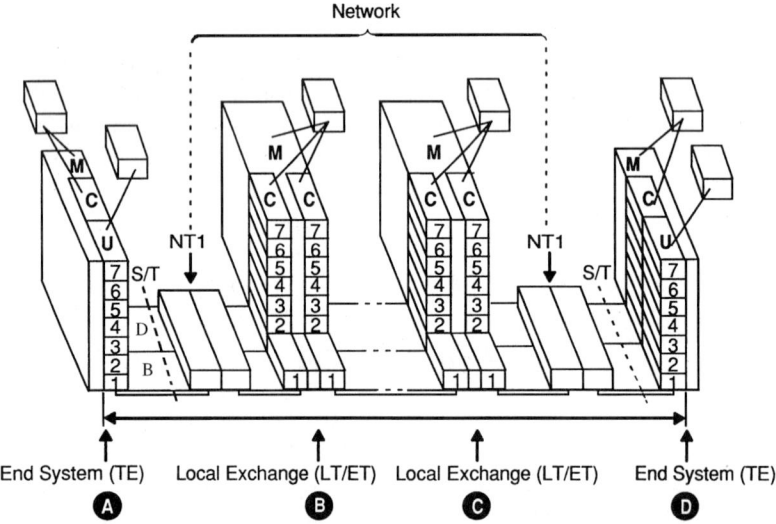

FIGURE 42.4 Modeling of a circuit-switched connection with the ISDN PRM.

C- and U-plane protocols are layered stacks, whereas the M-plane is usually represented as a monolithic block, because its functions can not be described properly by using the OSI concepts. However, the M-plane functions can be structured into plane related functions and functions which are related to specific layers in the other planes.

One example for the application of the ISDN PRM is shown in Fig. 42.4, where it is used to model a basic circuit-switched connection over an ISDN. Two terminals, A and D, are connected to the ISDN at the reference point S/T via a NT1. The NT1 is connected via the subscriber line to the local exchange implementing the LT/ET functionality. The intermediate transit exchanges are not shown. It can be seen that for the information transport in the U-plane only layer 1 functions are used to provide a transparent B-channel. The TEs implement all seven layers in the U-plane as required for the specific service. The control of the B-channel connection is performed in the C-plane. The TEs communicate to the corresponding local exchanges by using the user–network signalling protocol over the D-channel (the OSI layers 4–7 are normally empty in this case). The local exchanges convert this information to interexchange signalling information according to the CCS 7 protocol, which is transferred through the ISDN in separate signalling channels.

42.5 Layers 1–3 of the ISDN User–Network Interface

The *layer 1* for the ISDN basic access [I.430] allows a maximum of eight TEs to be connected to the UNI, where the typical configuration is a short passive bus with a length of at most 200 m. The limitation of the physical length of the bus is necessary due to the timing requirements of the layer 1 protocol.

The physical medium for this bus is usually the same unshielded twisted copper wire pair used in the conventional telephone system to allow reuse of the existing cabling (one wire pair for each direction of transmission).

To carry the two independent B-channels and the D-channel, a synchronous transmission frame with a length of 48 b and a duration of 250 μs, corresponding to a total transmission bit rate per direction of 192 kb/s, has been defined. The line code used is a pseudoternary (AMI-) code *alternate mark inversion* together with specific DC balancing bits in the frame, this code allows a transmission without a DC component, which would otherwise interfere with the transformer coupling method used.

Procedures and primitives are provided to activate and deactivate the layer 1 connection. Rapid and secure synchronization to the transmission frame is achieved by using two intentional AMI-code violations within the first 14 b of the frame. This mechanism allows all TEs to start sending their frame toward the NT at the same time.

The B-channels are exclusively allocated to specific connections during call setup. The D-channel, however, is operated in a packet-oriented mode and can be used by all connected TEs asynchronously. To solve the possible access collisions in direction from the TEs toward the NT, a *multiple access protocol* is being used. It takes advantage of the fact that a logical ZERO supersedes any logical ONEs sent simultaneously due to the definition of the electrical characteristics. In addition, the NT echoes back the received D-channel bits in a *D-echo channel* provided in the transmission frame sent toward the TEs. Since the D-channel bits are spaced out over the frame and, in addition, the frame toward the NT is delayed by 2-b durations with respect to the frame sent toward the TEs, the TEs can monitor which D-channel bit value has been received by the NT before they send the next D-channel bit. Before a TE can start transmitting, it has to encounter a specified number (at least 8) of consecutive logical ONEs on the D-echo channel. This pattern can only occur if the D-channel is idle, because it is guaranteed that during the transmission of a layer 2 frame there can be at most 6 consecutive logical ONEs.

If TEs start transmitting at exactly the same time, they will send an unambiguous TE specific address right after the start flag of the layer 2 frame. All TEs that are sending a logical ONE while at least one TE sends a logical ZERO will recognize this when receiving the logical ZERO in the D-echo channel. They will stop transmitting and the winning TE can transmit the frame without interference. By assigning different numbers of consecutive ONEs, a priority mechanism for different types of layer 2 frames as well as a fairness mechanism among the TEs is implemented.

Layer 2 of the ISDN UNI [I.440 and I.441[4]] uses the link access procedure on the D-channel (LAPD) protocol, which has been derived from the LAPB procedure of the high-level data link control (HDLC) protocol family. It provides the standard functions of *bit-oriented data link layer protocols*, namely, frame delimitation and transparency function, synchronization, error detection, error correction and window flow control [Tanenbaum, 1989]. Both the acknowledged and the unacknowledged information transfer service is available.

LAPD supports the point-to-multipoint configuration of the ISDN UNI. In such a configuration, the number and type of connected TEs may be unknown for the local exchange because they can be connected and disconnected anytime at the subscriber's discretion. To cope with this, a mechanism has been defined to enable the local exchange to assign, check, and deassign TE specific

---

[4]These recommendations are identical to recommendations Q.920 and Q.921 for the DSS1, the relevant text is only included in the Q-series recommendations.

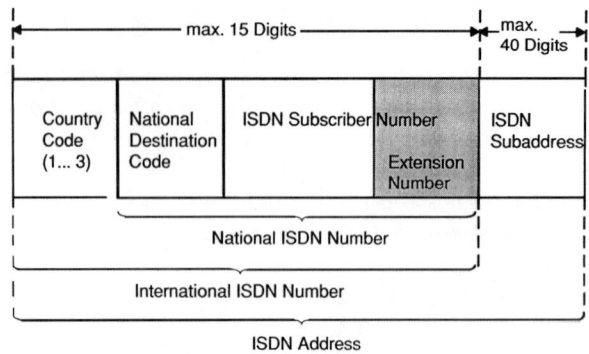

FIGURE 42.5 Structure of the ISDN address.

addresses: terminal endpoint identifiers (TEI). Together with service access point identifiers (SAPI) used to address specific service access points within a TE, the TEIs form the *data link connection identifier* (DLCI) used to identify the corresponding layer 2 connection. In addition to the signalling information, packet-oriented user information transfer also can be supported on the D-channel by using the link access procedure to frame mode bearer services (LAPF) protocol. LAPF [Q.922] is closely related to LAPD and is defined such that both protocols can be used simultaneously over one D-channel. In principle, however, the LAPF protocol can also be used on any other ISDN channel.

LAPD provides a broadcast capability for the indication of incoming calls to all TEs, whereas the other signalling activities are mainly supported via point-to-point connections. LAPD also supports multiplexing of several independent layer 3 connections onto one layer 2 connection.

Layer 3 of the UNI signalling protocol [I.45x/Q.93x] provides all functions required to control the U-plane connections of the various services in a uniform but flexible manner. Therefore, the layers above it are empty for the UNI signalling system and no formal service protocol has been defined for layer 3.

Recommendation I.451/Q.931 specifies the message formats and the procedures for the peer-to-peer protocol used to establish and clear down circuit-switched as well as packet-switched U-plane connections.

ISDN subscribers are identified by means of an ISDN number according to the *ISDN numbering plan* [E.164]. As shown in Fig. 42.5, an international ISDN number has a maximum length of 15 digits and contains a 1–3 digit country code, a national destination code (area code), and an individual subscriber number. There is also an extended addressing capability by means of an ISDN subaddress, which can carry, for example, an OSI-Network Service Access Point (OSI-NSAP) address for data communication services.

The layer 3 protocol also provides compatibility checks for TEs and for the service offered by the ISDN transport network. This guarantees that the transfer of user data through the ISDN is done in the way required by the telecommunication service and that only connections among compatible terminals can be set up. To support X.25 data services, specific messages and procedures have been defined to allow to control connections between TEs and the packet-switching unit of the ISDN. To control the ISDN supplementary services, generic procedures are specified in recommendation I.452/Q.932, the standardized supplementary services can be found in the Q.95x series of recommendations.

42.6 Broadband ISDN (B-ISDN)

Although offering significant advantages compared to the conventional public telephone network, the ISDN still has severe limitations with respect to the data transfer rates it offers, which is why it is often referred to as narrowband ISDN. Considering the data rates between 10 and more than 100 Mb/s nowadays common for data communication in local area networks, the 64-kb/s basic channels of the ISDN are by no means adequate to satisfy the increasing demand for LAN interconnections. Other examples are the field of medium to high quality video communication, where bit rates of 2 Mb/s and up are required, even when using the most advanced video coding techniques, or upcoming multimedia applications.

To be able to cope with the current and future demand for high bit rate communication in the wide area network, the basic idea of the ISDN has been consequently extended by adding appropriate broadband capabilities allowing an evolution toward a *broadband ISDN (B-ISDN)*. Thereby, the basic concepts described such as the generic architecture, the service concept [I.211], the definition of reference points and functional blocks [I.327], as well as the application of out-of-band signalling, have been preserved and extended where necessary. To distinguish the definitions from those for the narrowband ISDN, a B has been added to the names of reference points and functional blocks (B-TE, SB, etc.). While adhering to the basic ideas of the ISDN, the B-ISDN is based on a packet-oriented rather than on a circuit-oriented information transfer. This transfer principle offers more flexibility and has been internationally standardized in a new set of I.-series Recommendations under the name **asynchronous transfer mode** (ATM). It can be characterized as follows [I.150]:

- All user, control, and signalling information is transferred in small, fixed size packets called cells. Each cell consists of a 5-byte cell header containing the information needed to transfer the cell through the network and of a 48-byte information field carrying the actual information payload.
- The information transfer is connection oriented, that is, prior to the actual information transfer a route (virtual connection) through the network has to be established, which is then being followed by all cells of this specific connection. Because all cells of a connection take the same route through the network, the correct sequence of the cells within one connection is preserved.
- The asynchronous time-division multiplexing principle is applied. This means that network resources, such as time slots in a transmission frame, are not assigned to a connection for its whole duration (as in the synchronous time-division multiplexing case), but are only allocated when a cell actually needs them. To resolve the resulting access conflicts to the network resources without excessive information losses, limited buffering capabilities have to be foreseen in the network elements (multiplexers, switches, etc.).
- The primary[5] user–network interfaces for the B-ISDN are specified at a bit rate of 155.52 Mb/s and 622.08 Mb/s, respectively.

These characteristics result in some properties of the B-ISDN that are different from those of the narrowband ISDN. Because of the asynchronous nature of ATM no fixed channel structures are necessary and a wide range of peak bit rates with a fine granularity can be offered. Even connections with a bit rate that is variable over time can be supported in a natural way. The asynchronous nature of the traffic also allows one to effectively switch all connections irrespective of their bit rate within one common switch fabric. In addition, the ATM concept allows a flexible adaptation of the transfer characteristics to the properties and requirements of existing as well as new services.

However, the price that has to be paid for this flexibility is that coexisting connections influence each other due to the asynchronous sharing of resources. This results in cell delay variations among the cells within a connection and—in the extreme—in cell losses due to overflows of the finite buffers in the network elements. These unwanted effects of ATM have to be kept within specified bounds by a sophisticated set of interrelated traffic management procedures. Depending on the

---

[5] In practice, lower rate ATM based user–network interfaces are also required for economical reasons and have, therefore, been defined.

nature of the service transported over an ATM network, they also have to be leveled out by using appropriate, service specific mechanisms.

42.7 Structure of the B-ISDN

As shown in Fig. 42.6, the B-ISDN has a hierarchical structure where the functions of the higher layers are supported by the *ATM transport network*. The ATM transport network is responsible for transferring the ATM cells within the network and provides the functions of the physical and ATM layer. In the physical layer, which is subdivided into the regenerator section, digital section, and transmission path layer functions, bit or symbol streams are transmitted together with the corresponding clock information. To provide these functions all synchronous digital hierarch/synchronous optical network (SDH/SONET), most plesiochronous digital hierarchy (PDH), as well as newly defined ATM cell-based transmission systems can be used.

The ATM layer transfers cells by using the control information contained in the cell header. The information field of the cells is not processed in the ATM layer, which means that there is no error detection and correction for user information in the ATM transport network. The ATM layer transport functions are subdivided into two layers, namely, the **virtual channel** (**VC**) layer and the **virtual path** (**VP**) layer. A VC is defined as a unidirectional channel, which can be used to transport cells with a common identifier [VC identifier (VCI)]. The VCI values are assigned and removed at the connection endpoints of the virtual channel connections (VCCs), that is, at those points where the transport functionality of the VC layer is offered to the higher protocol layers. Normally this will be within the ATM terminals at the service access points of the ATM adaption layer (AAL) (discussed subsequently). VCCs crossing the B-ISDN will, in general, consist of several concatenated VC links, where one VC link extends between two adjacent points (VC connecting points) in the network where the VC is switched (multiplexers, switches, cross connects). At the VC connecting points the VCI values are translated.

A VP is defined as a collection of individual VCs that are logically combined by the use of a common identifier [VP identifier, (VPI)]. Virtual path connections (VPCs) across the B-ISDN consist of concatenated VP links, VPI values are assigned and removed at the VP endpoints and translated at the VP connecting points. At the VP endpoints, the individual VCs contained in a VP become visible and can, therefore, be switched individually. At the VP connecting points only the VP as a whole can be switched by translating the VPI value while transporting the VCI values transparently. VP endpoints can be located in various network elements. Therefore, VPCs can be used for various purposes, for example, to build logically separate private networks (end-to-end VPCs, similar to leased lines) or to support network internal functions such as rerouting or separation of VCs with different quality of service requirements (network-to-network VPCs). In general, VPCs are semipermanent or permanent connections controlled via the network management (M-plane), whereas VCCs are on-demand connections controlled via signalling protocols (C-plane).

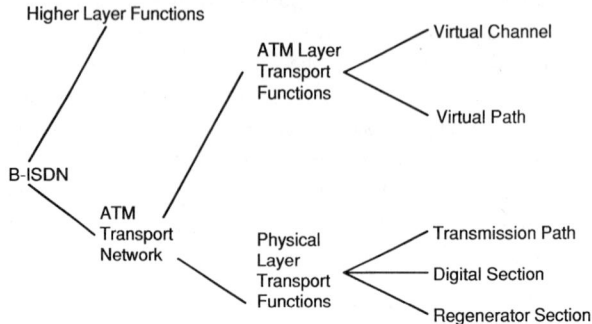

FIGURE 42.6 Hierarchical structure of the B-ISDN.

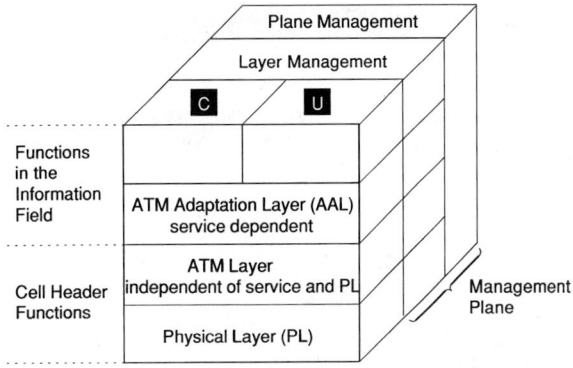

FIGURE 42.7 B-ISDN protocol reference model.

42.8 The B-ISDN Protocol Reference Model

The PRM for B-ISDN [I.321] shown in Fig. 42.7 has the same structure of U-, C-, and M-plane as the PRM for the narrowband ISDN described earlier. However, two specific layers, the ATM layer and the ATM adaptation layer have been introduced. The *physical layer* [I.432] depends on the physical transmission medium and on the transmission system used.[6] *The ATM layer* [I.361] has been kept as much as possible independent of the transmission system used and provides the functions to perform the link-by-link transport of ATM cells. Because of its asynchronous operation, the ATM layer introduces variable delays and, with a very low probability, also cell losses, whereas the correct sequence of cells within a connection is preserved. The *ATM adaptation layer* (AAL) [I.362, I.363] provides service specific functions to adapt the characteristics of the ATM transport network to the requirements of the higher layer protocols. The functions of the physical and ATM layers are supported by the cell header, whereas the AAL and higher layer functions have to be supported by appropriate fields in the information field. The layers up to the AAL, which can be viewed as belonging to the OSI layer 1, are used commonly by the U- and C-plane. The operation, administration, and maintenance (OAM) principles and functions (M-plane) for the B-ISDN are described in recommendation I.610.

42.9 Functions of the ATM Specific Layers

Figure 42.8 provides an overview of the functions of the ATM specific protocol layers. The leftmost and rightmost function stacks depict the TE functions in sending and receiving direction, respectively. The two stacks in the middle reflect the functions that have to be performed in the U-plane of an intermediate network element (switch, multiplexer, etc.). The *physical layer* (PL) is subdivided into functions depending on the physical medium (physical medium dependent sublayer, PM) and in functions specific for the transmission system used (transmission convergence sublayer, TC). The PM sublayer specifies electrical/optical characteristics of the transmission medium as well as line coding and bit synchronization mechanisms. In the TC sublayer, the specific frame structure of the transmission system is generated/recognized, the ATM cell mapping/demapping is performed, and the physical layer OAM functions are handled. In addition, the TC sublayer provides a header error check (HEC) mechanism for the cell header based on a cyclic code with 8 control bits. The HEC mechanism is also used for transmission system independent cell synchronization by computing the error syndrome for 32-b blocks and shifting bit by bit until the cell start is correctly recognized.

---

[6]In addition to the SDH and cell-based options defined in the recommendation, the transport of ATM cells has been standardized over most of the currently existing transmission systems.

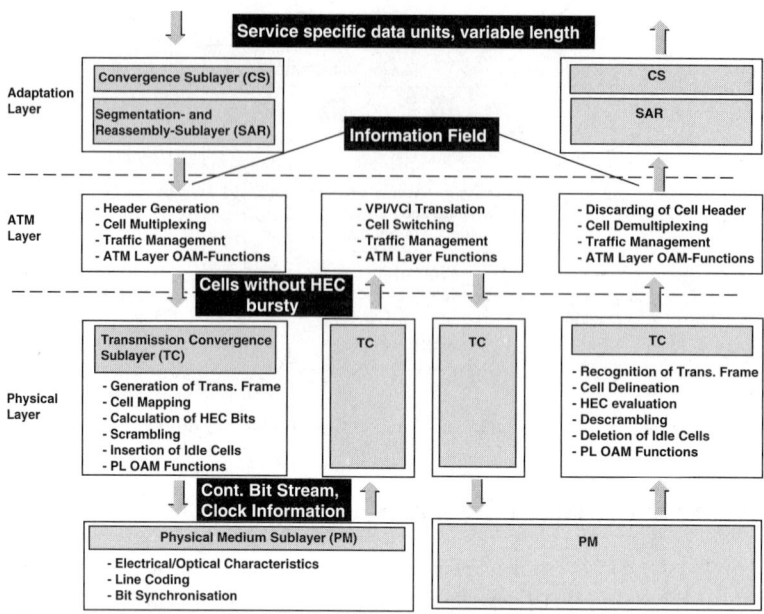

FIGURE 42.8 Functions of the ATM specific B-ISDN protocol layers.

Scrambling/descrambling of the information field is performed to increase the effectivity of the HEC mechanism. Another important task of the TC sublayer is to decouple the cell rate from the transmission bit rate. This is necessary because the cell stream provided by the ATM layer is not necessarily continuous as required by existing transmission systems. Therefore, fill cells have to be added/removed accordingly. The PM sublayer and the transmission frame structures are not ATM specific thus allowing the use of existing transmission systems to transport ATM cell streams.

The main functions of the ATM layer are as follows:

- *VPI/VCI translation.* Mapping of the VPI/VCI (or only the VPI in case of VP switching) valid for a connection on one transmission link into a new value valid on the adjacent link. The translation tables are loaded during call setup.
- *Cell multiplexing/demultiplexing.* Superposition of different cell streams by using the asynchronous time-division multiplexing scheme. This involves limited cell buffering and causes cell delay variations and possibly cell losses due to buffer overflows.
- *Cell header generation/extraction.* A valid cell header has to be added to the information fields received from the AAL in sending direction. Before information fields are forwarded to the AAL on the receiving side, the ATM cell header has to be stripped off.
- *ATM layer OAM functions.* OAM functions, specifically, fault management (AIS/RDI similar to SDH mechanisms), loopback, continuity check, and performance monitoring, have been standardized for the ATM layer [I.610]. ATM layer OAM information is transported in specific OAM cells. Separate OAM flows have been defined for the VP layer (F4 flow) and for the VC layer (F5 flow). These OAM flows can either span the complete connection (end-to-end OAM) or only specific connection segments (segment OAM).
- *Traffic management functions.* To be able to control the quality of service provided to the different connections asynchronously sharing the same resources (link capacity, buffers, etc.) in the network, several interrelated traffic management mechanisms have been defined [I.371]. These mechanisms include, among others, handling of two cell loss priority classes and rate-based flow control at the network boundaries [usage parameter control/network parameter

control (UPC/NPC)]. To be able to fully exploit the inherent capability of the asynchronous multiplexing principle to achieve higher link utilizations for variable bit rate and bursty traffic streams compared to the synchronous transfer mode (statistical multiplexing gain), highly sophisticated buffering and scheduling schemes are currently being developed and standardized. This also includes network-wide flow control mechanisms for specific traffic classes [e.g., available bit rate traffic (ABR)].

The AAL provides the interface to the higher layers. Since the higher layer data units can have different and possibly variable lengths depending on the service, their format has to be adapted to the 48-byte information field length of the ATM cells and vice versa (segmentation and reassembly). In addition, service-dependent functions have to be performed to adapt the characteristics of the ATM transport to the requirements of the service. Examples are the application of elastic buffering at the receiver to equalize variable delays among the cells of one connection for isochronous services or error protection and recovery mechanisms for the information field in case of data traffic. Different sets of AAL functions have been combined to define several AAL types. The most important AAL types are AAL 1 to support isochronous constant bit rate services (voice, $N \times 64$ kb/s, circuit emulation service) and AAL 5 to support data as well as video services. To support connectionless data services (SMDS, CBDS), the AAL 3/4 is used.

42.10 Signalling in the B-ISDN

Signalling in the B-ISDN is specified in the Q.2xxx series of ITU-T recommendations. Signalling between the user and the B-ISDN network is done based on a signalling protocol derived from the D-channel protocol used in the narrowband ISDN. This Digital Subscriber Signalling System No. 2 (DSS2) is defined in recommendation Q.2931. The signalling information will be transported over an ATM signalling connection by using a specific *signalling-AAL* (SAAL, Q.21xx). For interexchange signalling, the lower levels of CCS7 (up to MTP level 2) have been replaced by ATM transport and SAAL and a new *B-ISDN user part* (B-ISUP, Q.276x) has been defined. The new signalling protocols are being introduced in a phased approach by defining various *capability sets*.

Conceptually the main difference between the narrowband and the B-ISDN signalling systems is the clear call/connection separation. This implies that several (bearer) connections can be controlled by one call to support multipoint and multimedia communication.

Since the B-ISDN network will have to coexist for a long time with the current networks, interworking schemes between the B-ISDN and narrowband ISDN, frame relay and other networks have also been defined. This is a crucial aspect, because it allows the network providers as well as the subscribers to perform a smooth migration toward the B-ISDN.

42.11 Summary and Conclusion

The ever increasing demand for flexible, convenient and affordable telecommunication in the information age has led to the definition of an evolution concept from the conventional telephone network to the integrated services digital network and further on to a broadband ISDN offering virtually unlimited connectivity for a broad range of services. With the increasing deployment of ISDN and B-ISDN networks worldwide, the gap between the classical telephone and the data communication world will diminish and allow an integration of the network infrastructure that is beneficial both for the users and the network operators.

In addition, the fact that the basic transfer mode for the B-ISDN, ATM, also has been widely accepted as the basis for the next generation of local area and corporate network technology offers the unprecedented opportunity to create a truly universal communication infrastructure for both local and wide area networks. However, since the local and corporate networks are classically more data oriented whereas the wide area networks have been focused on voice services, some differences have already developed between the ATM Forum driven private and corporate ATM networks and the ITU-T driven B-ISDN, although they are based on the same principles. The challenge for the years to come will be to integrate both worlds to provide seamless high-performance communication irrespective of the distance covered.

Defining Terms

Asynchronous transfer mode (ATM): Basic packet-oriented transfer mode for the B-ISDN.
B-channel: Basic 64 kb/s circuit-switched channel for user information transfer.
Bearer service: Basic information transfer service with functions standardized up to OSI layer 3.
D-channel: 16 kb/s packet-switched channel for signalling information transfer.
Supplementary service: Additional feature used to provide enhanced control in connection with basic telecommunication services.
Teleservice: Fully standardized telecommunications service, functions of all seven OSI layers are standardized.
User–network interface (UNI): Standardized interface providing subscriber access to the ISDN.
Virtual channel (VC): Unidirectional channel allowing transport of cells with a common identifier (VCI).
Virtual path (VP): Logical combination of individual VCs by means of a common identifier (VPI).

References

The relevant B-ISDN standards cited in the text are published by ITU-T and revised as necessary. Therefore, no more detailed references are given.

Further Information

A good overview of concepts, methods and systems for ISDN is given in *ISDN—The Integrated Services Digital Network* by Peter Bocker (Springer-Verlag).

The book, *ATM Networks-Concepts, Protocols, Applications*, by Rainer Händel, Manfred N. Huber and Stefan Schröder (Addison-Wesley) provides an overview on the state of the art in B-ISDN with specific emphasis on concepts and standards.

A more implementation-oriented overview of ATM and B-ISDN is given in *Asynchronous Transfer Mode—Solution for B-ISDN* by Martin de Prycker (Ellis Horwood).

Contributions on ISDN and B-ISDN can be found regularly in many journals, for example, *IEEE Transactions on Communications*, *IEEE Journal on Selected Areas in Communications*, *International Journal of Communication Systems* (Wiley), and *Computer Networks and ISDN Systems* (North Holland) just to name a few.

An overview on the latest advances and trends in this area is given in the conference proceedings of the International Switching Symposium, which is held regularly every three years, and in the proceedings of many other conferences.

43
Ethernet Networks

Ramesh R. Rao
University of California, San Diego

43.1 Overview 591
43.2 Historical Development 592
43.3 Standards 592
   10BASE5 • 10BASE2 • 10BROAD36 • 1BASE5 • 10BASET • 100BASET
43.4 Operation 594

43.1 Overview

The term ethernet describes a collection of hardware, software, and control algorithms that together provide a technique for interconnecting dozens of nodes, spread over hundreds of meters, at aggregate rates ranging from 1 to 100 Mb/s. These nodes are typically computer work stations or peripheral devices that are part of a community of users that frequently exchange files, messages, and other types of data among each other. Shielded coaxial cables and unshielded twisted pairs have been used for the physical interconnection.

The primary hardware element used in an ethernet network is the *network interface card* (NIC). These cards are attached to a computer bus at one end and to a *media attachment unit* (MAU) at the other end via an attachment unit interface (AUI) cable. A MAU is an active device that includes transceivers and other elements matched to the nature of the physical transmission medium. Some vendors consolidate two or more of these elements into a single package thereby rendering certain elements, such as AUI cables, unnecessary. *Repeaters* are sometimes used to regenerate signals, to ensure reliable communication. The NICs perform a number of control functions in software, chief amongst these is the execution of a random access protocol for exchanging packets of information.

Nodes attached to an ethernet network exchange packets of information by broadcasting on a common communication channel. Simultaneous transmission by two or more nodes results in a detectable collision that triggers a collision resolution protocol. This decentralized protocol induces a stochastic rescheduling of the retransmission instants of the colliding nodes. If the overall traffic on the network is below a certain threshold, eventual delivery of all new and previously collided messages is guaranteed. The value of the threshold establishes the maximum aggregate information transfer rate that can be sustained. This threshold is influenced by the span of the network, the length of the packets, latency in collision detection, and the dynamics of the external traffic, as well as the rate at which bits can be transported over the physical medium. The aggregate information transfer rate can never exceed the bit transfer rate on the physical medium but understanding the impact of the other variables has been the subject of many studies.

Standards bodies stipulate the values of some of these variables. For example, both the 10BASE2 and 1BASE5 versions of the IEEE 802.3 standards restrict packet lengths to values between 512 and

12,144 b. The 10BASE2 version restricts the network span to 925 m and relies on a bit transfer rate of 10 Mb/s over 50-Ω coaxial cable with a diameter of 5 mm, whereas 1BASE5 restricts the span to 2.5 kms and relies on a bit transfer rate of 1 Mb/s over unshielded twisted pair. It is not uncommon to observe aggregate information transfer rates of 1.5 Mb/s or less in a 10BASE2 installation.

43.2 Historical Development

Ethernet evolved from a random access protocol that was developed at the University of Hawaii in the 1970s for use in the packet-switched Aloha broadcast radio network. The Aloha network was meant to serve many users that generated infrequent and irregularly spaced bursts of data. Techniques, such as synchronous time-division multiplexing, that require nodes to access a common channel only during precoordinated, nonoverlapping intervals of time, are ill suited to serving many bursty users. This drawback motivated the development of the Aloha random access protocol, in which users access a common channel freely and invoke a collision resolution process only when collisions occur.

Spurred by the success of Aloha, efforts were undertaken to make the protocol more efficient. Early refinements included the use of a common clock to avoid partial collisions (slotted Aloha) as well as the development of hybrid protocols that combined elements of random access with reservations (reservation Aloha). Another series of refinements became possible in the context of limited span local area networks (LANs) that use low-noise communication links, such as coaxial cables, for the transmission medium. In such environments, a node can accurately and quickly sense ongoing transmissions on the cable. This first led to the development of carrier sense multiple access (CSMA) in which nodes withhold transmissions whenever they sense activity on the channel, thereby reducing the occurrence of collisions and improving efficiency. The next refinement for the LAN environment was based on a node's ability to detect a collision in less time than is required for a complete packet transmission. Early collision detection allows the colliding users to abort the transmission of colliding packets, thereby reducing the duration of collisions and further improving efficiency.

The ethernet protocol uses carrier sense multiple access with collision detection (CSMA/CD). The topology of an Ethernet network may differ from the Aloha packet radio network. The Aloha packet radio network was configured with a controller node at the center of the network. The member nodes transmitted to the controller on a common inbound frequency band. On a different outbound frequency band, the controller broadcasts to the member nodes all of the packets it received, as well as information from which the occurrence of a collision could be inferred. In contrast, in some types of ethernet networks, nodes attach to a common communication channel and directly monitor it for carrier sensing and collision detection, as well as to transmit and receive packets. There is thus no central controller in some ethernet networks.

Ethernet was conceived at Xerox Palo Alto Research Center in 1973 and described in an 1976 Association for Computing Machinery (ACM) article by Robert M. Metcalfe and David R. Boggs. The first prototype, also developed at Xerox PARC, operated at a speed of 3 Mb/s. In 1980, the Digital-Intel-Xerox 10 Mb/s ethernet was announced. The non-proprietary nature of the technology contributed to its widespread deployment and use in local area networks. By the mid-1980s, the Institute of Electrical and Electronics Engineers/American National Standards Institute (IEEE/ANSI) standards effort resulted in a set of CSMA/CD standards that were all based on the original ethernet concept.

43.3 Standards

In the terminology of the open systems interconnections (OSI) protocol architecture, the 802.3 series of standards are media access control (MAC) protocols. There exists a separate logical link control (LLC) sublayer, called the 802.2, which is above the MAC layer and together with the MAC layer forms a data link layer.

Currently there are five IEEE/ANSI MAC standards in the 802.3 series. They are designated as 10BASE5, 10BASE2, 10BROAD36, 1BASE5, and 10BASET. A new version called 100BASET is currently under study. The numeric prefix in these designations refers to the speed in megabits per second. The term BASE refers to baseband transmissions and the term BROAD refers to transmission over a broad-band medium. The numeric suffix, when it appears, is related to limits on network span.

All 802.3 MAC standards share certain common features. Chief amongst them is a common MAC framing format for the exchange of information. This format includes: a fixed preamble for synchronization, destination and source addresses for managing frame exchanges, a start of frame delimiter and length of frame field to track the boundaries of the frames, a frame check sequence to detect errors, and a data field and a pad field, which have a role related to collision detection which is explained further in the section on Operation. In addition to a common framing format, all of the 802.3 MAC protocols constrain the frame lengths to be between 512 and 12,144 b. A 32-b jam pattern is used to enforce collisions. An interframe gap of 9.6 μs is used with the the 10-Mb/s standards and a gap of 96 μs is used with the 1 Mb/s standard. The parameters that characterize the retransmission backoff algorithm are also the same.

The primary difference between the various 802.3 standards is with regard to the physical communication medium used and consequences thereof.

10BASE5

This was the first of the IEEE/ANSI 802.3 standards to be finalized. A 10-mm coaxial cable with a characteristic impedance of 50 Ω is the physical medium. The length of any one segment cannot exceed 500 m, but a maximum of five segments can be attached together, through the use of four repeaters, to create a network span of 2.5 km. There can be no more than 100 attachments to a segment, and the distance between successive attachments to this cable must be at points that are multiples of 2.5 m to prevent reflections from adjacent taps from adding in phase. In view of the inflexibility of the 10-mm cable, a separate transceiver cable is used to connect the network interface cards to the coaxial cable. Transmitting nodes detect collisions when voltages in excess of the amount that can be attributed to their transmission are measured. Nontransmitting nodes can detect collisions when voltages in excess of the amount that can be produced by any one node is detected.

10BASE2

10BASE2 networks are based on a coaxial cable that has a characteristic impedance of 50 Ω and a diameter of 5 mm. The flexibility of the thinner cable makes it possible to dispense with the transceiver cable and directly connect the coaxial cable to the network interface card through the use of a t-connector. Furthermore, nodes can be spaced as close as 0.5 m. On the other hand, signals on the thinner cable are less immune to noise and suffer greater degradation. Consequently, segment lengths cannot exceed 185 m and no more than 30 nodes can be attached to a segment. As with 10BASE5, up to five segments can be attached together with repeaters, to create a maximum network span of 925 m. The 10BASE2 and 10BASE5 segments can be attached together as long as the less robust 10BASE2 segments are at the periphery of the network.

10BROAD36

The 10BROAD36 networks are based on standard cable TV coaxial cables with a characteristic impedance of 75 Ω. The 10BROAD36 networks operate in a manner that is closer to the original Aloha network than 10BASE2 or 10BASE5. Nodes on a 10BROAD36 generate differential phase-shift keyed (DPSK) RF signals. A head-end station receives these signals and repeats them on a different frequency band. Thus, sensing and transmissions are done on two different channels just as in the Aloha network. The distance between a node and the head-end station can be as much as 1.8 km for a total network span of 3.6 km. A separate band of frequencies is set aside for the transmission of a collision enforcement signal. Interestingly, in this environment, there is a possibility that when two signals collide, one of the two may be correctly captured by some or even all of the stations. Nonetheless, 10BROAD36 requires that transmitting stations compare their uplink transmissions with the downlink head-end transmission and broadcast a collision enforcement signal whenever differences are observed.

1BASE5

Also referred to as StarLAN, 1BASE5 networks operate at the lower rate of 1 Mb/s over unshielded twisted pair cables whose diameters range between 0.4 and 0.6 mm. The 1BASE5 networks are physically configured in the shape of a star. Each node is attached to a hub, via two separate pairs of unshielded twisted wires, one pair is for transmissions to the hub, and the other pair is for receptions from the hub. Neither of these pairs is shared with any other node. A hub has multiple nodes attached to it, and detects collisions by sensing activity on more than one incoming port. It is responsible for broadcasting all correctly received packets on all outgoing links or broadcasting a collision presence signal when one is detected. The 1BASE5 standard does not constrain the number of nodes that can be attached to a hub.

It is also possible to cascade hubs in a five-layer hierarchy with one header hub and a number of intermediate hubs. The node to hub distance and the inter-hub distances are limited to 250 m and, therefore the largest network span that can be created is limited to 2.5 kms. In view of the lower speed of 1BASE5 networks they can not be easily integrated with the other IEEE 802.3 networks.

10BASET

The 10BASET network is similar to the older 1BASE5 network in that both use unshielded twisted pairs, but 10BASET networks can sustain data rates of 10 Mb/s. This is accomplished partly by limiting individual segment lengths to 100 m. Instead of the hierarchical hubs used in 1BASE5, 10BASET uses multiport repeater sets, which are connected together in exactly the same way as nodes are connected to a repeater set. Two nodes on a 10BASET network can be separated by up to four repeater sets and up to five segments. Of the five segments, no more than three can be coaxial cable-based 10BASE2 or 10BASE5 segments, the remaining segments can be either unshielded twisted pair, point-to-point links of length less than 100 m each, or fiber optic point-to-point links of length less than 500 m each. The network span that can be supported depends on the mix of repeater sets and segments used.

100BASET

Currently, there is an effort to create a 100-Mb/s standard in the 802.3 series. The 100 BASET network will use multiple pairs of unshielded twisted pairs, a more efficient code for encoding the data stream and a slightly faster clock rate to generate the tenfold increase in bit transfer rates.

43.4 Operation

The core algorithm used in all of the 802.3 series of MAC standards and the original ethernet is essentially the same. The MAC layer receives LLC data and encapsulates into MAC frames. Of particular interest in the MAC frame is the pad field, whose role is related to collision detection and will be explained. Prior to transmitting frames, the MAC layer senses the physical channel. If transmission activity is detected, the node continues to monitor until a break is detected, at which time the node transmits its frame. In spite of the prior channel sensing performed, it is possible that the frame will collide with one or more other frames. Such collisions may occur either because two or more nodes concurrently initiate transmissions following the end of the previous transmission or because propagation delays prevent a node from sensing the transmission initiated by another node at an earlier instant.

Colliding users detect collisions through physical mechanisms that depend on the nature of the attachment to the underlying communication medium. As described in the section on standards, in 10BASE2 and 10BASE5 networks, transmitting nodes recognize electrical currents in excess of what they injected into the coaxial cable. In 1BASE5 and 10BASET networks, nodes recognize activity on both the transmitter and receiver ports, and in 10BROAD36 networks nodes spot differences in the transmitted and received waveforms. All of these collision sensing methods rely on a node's knowledge of its own transmission activity. Since this information is not available to all of the other nodes, the disposition of some partially received MAC frames may be in doubt. Therefore, to enforce a common view of the channel outcome, colliding users transmit a jam pattern as soon as they sense a collision.

Thus, the full duration of a collision is the sum of the maximum collision detect time and the jam pattern. The first of these two components, the time to detect a collision, is primarily determined by the time it takes for signals to propagate between any two nodes on the cable. By constraining the span of a network, standards such as the IEEE 802.3, limit the maximum collision detect time. The second component, the jam pattern length, is also constrained by standards. Thus, the longest period of collision activity, which can be thought of as an invalid frame, is limited. The pad field of the MAC frame can then be used to ensure that valid frames are larger than the longest invalid frame that may occur as a result of collisions. This provides a simple mechanism by which nodes that were not involved in a collision can also sense the collision.

In order to resolve collisions, ethernet nodes retransmit after a randomly chosen delay. Furthermore, the probability distribution of the retransmission delay of a given packet is updated each time the packet experiences a collision. The choice of the initial probability distribution of the retransmission delay and the dynamics of its updates determines whether or not the total population of new and retransmitted users can transmit successfully. Thus the retransmission algorithm lies at the heart of the ethernet protocol.

The IEEE 802.3 standard specifies the use of a truncated binary exponential backoff algorithm. In this algorithm, a packet that has collided n times is withheld for a duration that is proportional to a quantity that is uniformly distributed between 0 and $2^m - 1$, where $m = \min(10, n)$. The constant of proportionality is chosen to be large enough to ensure that nodes that chose to withhold for different durations do not collide, regardless of their location on the cable. Furthermore, when the number of times a packet has collided exceeds a user-defined threshold, which is commonly chosen to be 16, it is rejected by the MAC layer and has to be resubmitted at a later time by a higher layer.

As a consequence of the collision resolution process, the order in which ethernet serves packets depends on the traffic intensity. During periods of very light traffic, service is first come first serve. During periods of heavy contention, some colliding users will defer their transmissions into the future even as some latter arrivals transmit sooner. Because of this, packet delivery in ethernet may occur out of sequence and has to be rectified at the receiver.

The lack of first come first serve service and the stochastic nature of the rescheduling algorithm are sometimes considered to be weaknesses of the ethernet network. On the other hand, the ability to use unshielded twisted pair wires for 10-Mb/s communication, the large installed base of users, and the promise of even faster communication at rates of 100 Mb/s over multiple pairs of unshielded twisted pair wires makes ethernet networks very attractive.

References

Abramson, N. 1970. The Aloha system—Another alternative for computer communications. In Proceedings of the Fall Joint Computer Conference, AFIPS Conference, Vol. 37, pp. 281–285.

Metcalfe, R.M. and Boggs, D.R. 1976. Ethernet: Distributed packet switching for local computer networks. *Commun. ACM* (July). Reprinted in "The Ethernet Sourcebook." pp. 3–12.

Shotwell, R. ed. 1985. *The Ethernet Source Book*, North Holland, New York.

Stallings, W. 1989–1990. *Handbook of Computer-Communications Standards*, 2nd ed., H.W. Sams, Carmel, IN.

Walrand, J. 1991. *Communication Networks: A First Course*, Aksen Associates, Irwin, Homewood, IL.

Further Information

The Ethernet Source Book (1985) is a compilation of papers on various aspects of ethernet including the first published paper on ethernet entitled "Ethernet: Distributed Packet Switching for Local Computer Networks" by R.M. Metcalfe and D.R. Boggs. *The Ethernet Source Book* includes an introduction by Robert M. Metcalfe who, along with David Boggs, Chuck Thacker, and Tat Lampson, holds the patent on "Multipoint Data Communications System with Collision Detection." The *Handbook of Computer Communication Standards*, Vol. 2, by William Stallings describes most of the details of the IEEE 802.3 standards. N. Abramson described the ALOHA System in an article entitled "The ALOHA System—Another Alternative for Computer Communications." *Communication Networks—a First Course* by Jean Walrand is an accessible introductory book on all aspects of networks.

44
Fiber Distributed Data Interface and Its Use for Time-Critical Applications[1]

Biao Chen
University of Texas, Dallas

Nicholas Malcolm
Hewlett-Packard (Canada) Ltd.

Wei Zhao
Texas A&M University

44.1	Introduction.. 597
44.2	Architecture and Fault Management 598
44.3	The Protocol and Its Timing Properties........................... 600
	The Network and the Message Models • Constraints • Timing Properties
44.4	Parameter Selection for Real-Time Applications................. 603
	Synchronous Bandwidth Allocation • Selection of Target Token Rotation Time • Buffer Requirements
44.5	Final Remarks.. 608

44.1 Introduction

Fiber distributed data interface (FDDI) is an American National Standards Institute (ANSI) standard for a 100-Mb/s fiber optic token ring network. High-transmission speed and a bounded access time [Sevcik and Johnson, 1987] make FDDI very suitable for supporting real-time applications. Furthermore, FDDI employs a dual-loop architecture, which enhances its reliability; in the event of a fault on one loop, the other loop can be used for transmission. Several new civil and military networks have adopted FDDI as the backbone network.

In order to achieve efficiency, a high-speed network such as an FDDI network requires simple (i.e., low overhead) protocols. In a token ring network, the simplest protocol is the *token passing* protocol. In this protocol, a node transmits its message whenever it receives the token. After it completes its transmission, the node passes on the token to the next neighboring node. Although it

---

[1] The work described in this chapter was supported in part by the National Science Foundation under Grant NCR-9210583, by Office of Naval Research under Grant N00014-95-J-0238, and by Texas A&M University under an Engineering Excellence Grant. The views and conclusions contained in this document are those of the authors and should not be interpreted as representing official positions or policies of NSF, ONR, University of Texas at Dallas, Hewlett-Packard (Canada) Ltd., or Texas A&M University.

has the least overhead, the token passing protocol cannot bound the time between two consecutive visits of the token to a node (called the *token rotation time*), which makes it incapable of guaranteeing message delay requirements. The *timed token protocol*, proposed by Grow [1982], overcomes this problem. This protocol has been adopted as a standard for the FDDI networks. The idea behind the timed token protocol is to control the token rotation time. As a result, FDDI is able to support real-time applications.

In the rest of this chapter, we will first introduce the architecture of FDDI networks and discuss its fault management capabilities. We then address how to use FDDI to support time-critical applications.

44.2 Architecture and Fault Management

FDDI is a token ring network. The basic architecture of an FDDI network consists of nodes connected by two counter-rotating loops as illustrated in Fig. 44.1(a). A node in an FDDI network can be a single attachment station (SAS), a dual attachment station (DAS), a single attachment concentrator (SAC), or a dual attachment concentrator (DAC). Whereas stations constitute the sources and destinations for user frames, the concentrators provide attachments to other stations. The single attachment stations and concentrators are so called because they connect to only one of the two loops. The two fiber loops of an FDDI network are usually enclosed in a single cable. These two loops will be collectively referred to as the FDDI *trunk ring*.

The FDDI standards have been influenced by the need to provide a certain degree of built-in fault tolerance. The station management (SMT) layer of the FDDI protocol deals with initialization/ control, monitoring, and fault isolation/recovery in FDDI networks [Jain, 1994]. In ring networks such as FDDI, faults are broadly classified into two categories: node faults and link faults. The FDDI standard specifies explicit mechanisms and procedures for detection of and recovery from both kinds of faults. These fault management capabilities of FDDI provide a foundation for

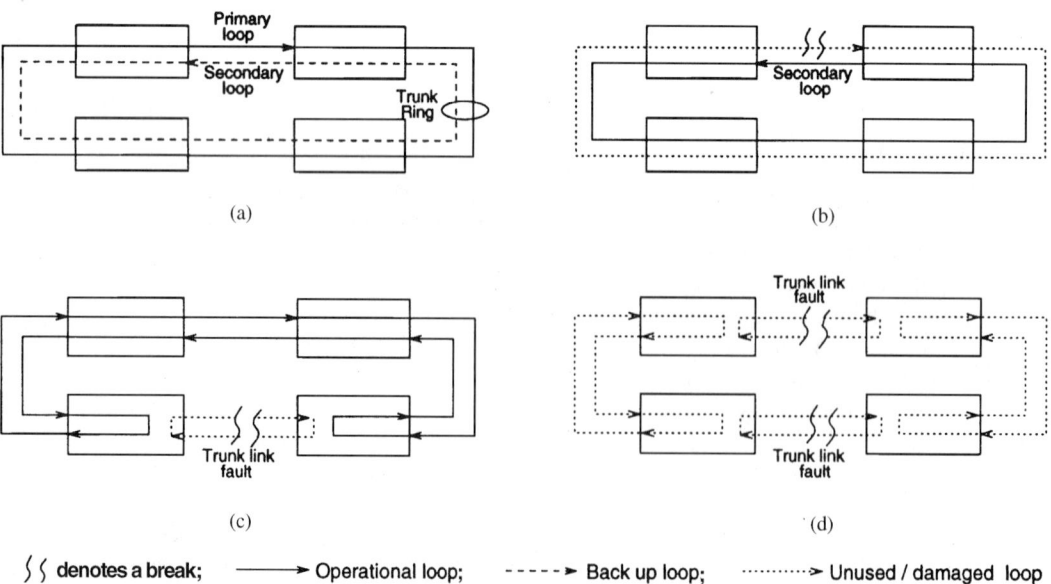

FIGURE 44.1 Architecture and fault management: (a) FDDI trunk ring consists of primary and secondary loops; (b) a fault on the primary loop, secondary loop used for transmission; (c) single trunk link fault, primary and secondary loops wrap up to form a new operational loop; and (d) two trunk link faults segment the network, no loop connects all the nodes.

capabilities of FDDI enable fault-tolerant operation. We present here a brief sketch of FDDI's fault management capabilities. For a comprehensive discussion, see ANSI [1992].

To deal with node faults, each node in FDDI is equipped with an optical bypass mechanism. Using this mechanism, a faulty node can be isolated from the ring, thus letting the network recover from a node fault. Link faults are handled by exploiting the dual-loop architecture of FDDI. Many FDDI networks use one loop as the primary loop and the other as a backup. In the event of a fault on the primary loop, the backup loop is used for transmission [Fig. 44.1(b)].

Because of the proximity of the two fiber loops, a link fault on one loop is quite likely to be accompanied by a fault on the second loop as well, at approximately the same physical location. This is particularly true of faults caused by destructive forces. Such link faults, with both loops damaged, may be termed as *trunk link faults*.

An FDDI network can recover from a single trunk link fault using the so called *wrap-up* operation. This operation is illustrated in Fig. 44.1(c). The wrap-up operation consists of connecting the primary loop to the secondary loop. Once a link fault is detected, the two nodes on either side of the fault perform the wrap-up operation. This process isolates the fault and restores a single closed loop.

The fault detection and the wrap-up operation is performed by a sequence of steps defined by FDDI's link fault management. Once the network is initialized, each node continuously executes the link fault management procedure. The flow chart of the link fault management procedure is shown in Fig. 44.2. Figure 44.2 only

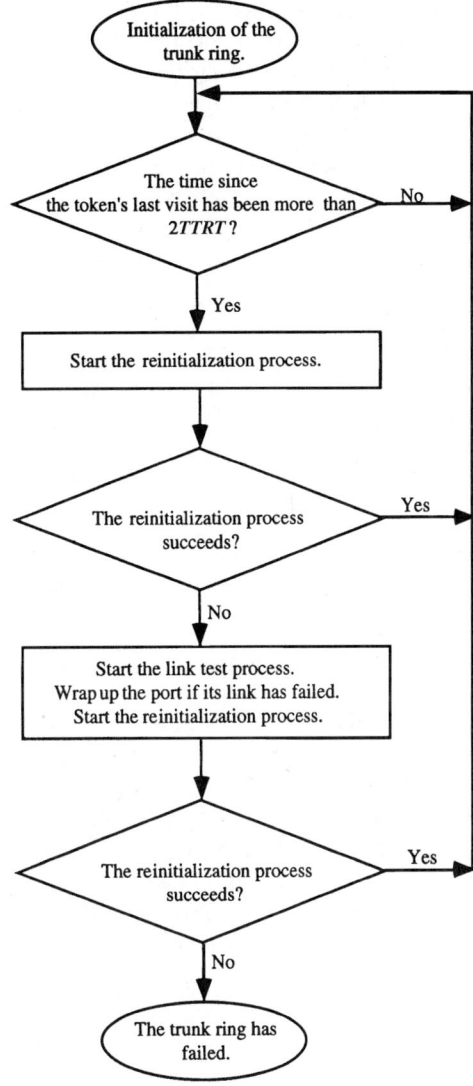

FIGURE 44.2 Flow chart of FDDI link fault management procedure.

presents the gist of the fault management procedure. For details, the reader is referred to ANSI [1992]. The major steps of this procedure are as follows:

- An FDDI node detects a fault by measuring the time elapsed since the previous token visit to that node. At network initialization time, a constant value is chosen for a protocol parameter called the target token rotation time (*TTRT*). The FDDI protocol ensures that under fault-free operating conditions, the duration between two consecutive token visits to any node is not more than 2*TTRT*. Hence, a node initiates the fault tracing procedure when the elapsed time since the previous token arrival at the node exceeds 2*TTRT*.

- The node first tries to reinitialize the network (using Claim frames). If the fault was temporary, the network will recover. Several nodes may initiate the reinitialization procedure asynchronously. However, the procedure converges within *TTRT* time (which is usually on the order of a few milliseconds). The details of this procedure are beyond the scope of this chapter. The reader is referred to ANSI [1992].

- If the reinitialization procedure times out, then the node suspects a permanent fault in the trunk ring. The nodes then execute a series of further steps in order to locate the ring's fault domain. The nodes within the fault domain perform link tests to check the status of the links on all their ports. If a link is faulty, the ring is wrapped up at the corresponding port. The node then tries to reinitialize the wrapped up ring. Once again, note that these operations are carried out concurrently and asynchronously by all the nodes in the trunk ring. That is, any of them may initiate the series of fault recovery steps but their actions will eventually converge to locate and recover from the fault(s).
- In the event that the last step has also failed, the network is considered unrecoverable. It is possible that operator intervention is required to solve the problem.

With the basic fault management mechanism just described, an FDDI network may not be able to tolerate two or more faults. This is illustrated in Fig. 44.1(d) where two trunk link faults leave the network disconnected. Hence, the basic FDDI architecture needs to be augmented to meet the more stringent fault tolerance requirements of mission-critical applications. In addition, the bandwidth of a single FDDI ring may not be sufficient for some high-bandwidth applications. FDDI-based reconfigurable networks (FBRNs) provide increased fault tolerance and transmission bandwidth as compared to standard FDDI networks. See Agrawal, Kamat, and Zhao [1994]; Chen, Kamart, and Zhao [1995]; and Kamat, Agrawal, and Zhao [1994] for a discussion on this subject.

44.3 The Protocol and Its Timing Properties

FDDI uses the timed token protocol [Grow, 1982]. The timed token protocol is a token passing protocol in which each node gets a guaranteed share of the network bandwidth. Messages in the timed token protocol are segregated into two separate classes: the *synchronous* class and the *asynchronous* class [Grow, 1982]. Synchronous messages are given a guaranteed share of the network bandwidth, and are used for real-time communication.

The idea behind the timed token protocol is to control the token rotation time. During network initialization, a protocol parameter called the target token rotation time (*TTRT*) is determined, which indicates the expected token rotation time. Each node is assigned a portion of the *TTRT*, known as its *synchronous bandwidth* (H_i), which is the maximum time a node is permitted to transmit synchronous messages *every time* it receives the token. When a node receives the token, it transmits its synchronous messages, if any, for a time no more than its allocated synchronous bandwidth. It can then transmit its asynchronous messages only if the time elapsed since the previous token departure from the same node is less than the value of *TTRT*, that is, only if the token arrives earlier than expected.

In order to use the timed token protocol for real-time communications, that is, to guarantee that the deadlines of synchronous messages are met, parameters such as the synchronous bandwidth, the target token rotation time, and the buffer size must be chosen carefully.

- The synchronous bandwidth is the most critical parameter in determining whether message deadlines will be met. If the synchronous bandwidth allocated to a node is too small, then the node may not have enough network access time to transmit messages before their deadlines. Conversely, large synchronous bandwidths can result in a long token rotation time, which can also cause message deadlines to be missed.
- Proper selection of *TTRT* is also important. Let τ be the token walk time around the network. The proportion of time taken due to token walking is given by $\tau/TTRT$. The maximum network utilization available to user applications is then $1 - \tau/TTRT$ [Ulm, 1982]. A smaller *TTRT* results in less available utilization and limits the network capacity. On the other hand, if *TTRT* is too large, the token may not arrive at a node soon enough to meet message deadlines.

- Each node has buffers for outgoing and incoming synchronous messages. The sizes of these buffers also affect the real-time performance of the network. A buffer that is too small can result in messages being lost due to buffer overflow. A buffer that is too large is wasteful of memory.

In Sec. 44.4, we will discuss parameter selection for real-time applications. Before that, we need to define network and message models and investigate protocol properties.

The Network and the Message Models

The network contains n nodes arranged in a ring. Message transmission is controlled by the timed token protocol, and the network is assumed to operate without any faults. Outgoing messages at a node are assumed to be queued in first in first out (FIFO) order.

Recall that the token walk time is denoted by τ, which includes the node-to-node delay and the token transmission time. Here τ is the portion of $TTRT$ that is not available for message transmission. Let α be the ratio of τ to $TTRT$, that is, $\alpha = \tau/TTRT$. Then α represents the proportion of time that is not available for transmitting messages.

There are n streams of synchronous messages, $S_1, \ldots, S_n$, with stream S_i originating at node i. Each synchronous message stream S_i may be characterized as $S_i = (C_i, P_i, D_i)$:

- C_i is the maximum amount of time required to transmit a message in the stream. This includes the time to transmit both the payload data and the message headers.
- P_i is the interarrival period between messages in the synchronous message stream. Let the first message in stream S_i arrive at node i at time $t_{i,1}$. The jth message in stream S_i will arrive at node i at time $t_{i,j} = t_{i,1} + (j-1)P_i$, where $j \geq 1$.
- D_i is the *relative deadline* of messages in the stream. The relative deadline is the maximum amount of time that may elapse between a message arrival and completion of its transmission. Thus, the transmission of the jth message in stream S_i, which arrives at $t_{i,j}$, must be completed by $t_{i,j} + D_i$, which is the *absolute deadline* of the message. To simplify the discussion, the terms relative and absolute will be omitted in the remainder of the chapter when the meaning is clear from the context.

Throughout this chapter, we make no assumptions regarding the destination of synchronous message streams. Several streams may be sent to one node. Alternatively, multicasting may occur in which messages from one stream are sent to several nodes.

If the parameters (C_i, P_i, D_i) are not completely deterministic, then their worst-case values must be used. For example, if the period varies between 80 and 100 ms, a period of 80 ms must be assumed. Asynchronous messages that have time constraints can have their deadlines guaranteed by using pseudosynchronous streams. For each source of time-constrained asynchronous messages, a corresponding synchronous message stream is created. The time-constrained asynchronous messages are then promoted to the synchronous class and sent as synchronous messages in the corresponding synchronous message stream.

Each synchronous message stream places a certain load on the system. We need a measure for the load. We define the *effective utilization* U_i of stream S_i as follows:

$$U_i = \frac{C_i}{\min(P_i, D_i)} \qquad (44.1)$$

Because a message of length C_i arrives every P_i time, $U_i = C_i/P_i$ is usually regarded as the load presented by stream S_i. An exception occurs when $D_i < P_i$. A message with such a small deadline must be sent relatively urgently, even if the period is very large. Thus, $U_i = C_i/D_i$ is used to reflect the load in this case.

The total effective utilization U of a synchronous message set can now be defined as

$$U = \sum_{i=1}^{n} U_i \qquad (44.2)$$

The total effective utilization is a measure of the demands placed on the system by the entire synchronous message set. In order to meet message deadlines, it is necessary that $U \leq 1$.

Constraints

The timed token protocol requires several parameters to be set. The synchronous bandwidths, the target token rotation time, and the buffer sizes parameters are crucial in guaranteeing the deadlines of synchronous messages. Any choice of these parameters must satisfy the following constraints.

The Protocol Constraint

This constraint states that the total bandwidth allocated to synchronous messages must be less than the available network bandwidth, that is,

$$\sum_{i=1}^{n} H_i \leq TTRT - \tau \qquad (44.3)$$

This constraint is necessary to ensure stable operation of the timed token protocol.

The Deadline Constraint

This constraint simply states that every synchronous message must be transmitted before its (absolute) deadline. Formally, let $s_{i,j}$ be the time that the transmission of the jth message in stream S_i is completed. The deadline constraint requires that for $i = 1, \ldots, n$ and $j = 1, 2, \ldots$,

$$s_{i,j} \leq t_{i,j} + D_i \qquad (44.4)$$

where $t_{i,j}$ is the arrival time and D_i is the (relative) deadline. Note that in this inequality, $t_{i,j}$ and D_i are given by the application, but $s_{i,j}$ depends on the synchronous bandwidth allocation and the choice of *TTRT*.

The Buffer Constraint

This constraint states that the size of the buffers at each node must be sufficient to hold the maximum number of outgoing or incoming synchronous messages that could be queued at the node. This constraint is necessary to ensure that messages are not lost due to buffer overflow.

Timing Properties

According to Eq. (44.4) the deadline constraint of message M is satisfied if and only if the minimum synchronous bandwidth available to M within its deadline is bigger than or equal to its message size. That is, for a given *TTRT*, whether or not deadline constraints can be satisfied is solely determined by bandwidth allocation. In order to allocate appropriate synchronous bandwidth to a message stream, we need to know the worst-case available synchronous bandwidth for a stream within any time period.

Johnson and Sevcik [1987] showed that the maximum amount of time that may pass between two consecutive token arrivals at a node can approach $2TTRT$. This bound holds regardless of the behavior of asynchronous messages in the network. To satisfy the deadline constraint, it is necessary for a node to have at least one opportunity to send each synchronous message before the message

Fiber Distributed Data Interface and Its Use for Time-Critical Applications

deadline expires. Therefore, in order for the deadline constraint to be satisfied, it is necessary that for $i = 1, \ldots, n$

$$D_i \geq 2TTRT \tag{44.5}$$

It is important to note that Eq. (44.5) is only a necessary but not a sufficient condition for the deadline constraint to be satisfied. For all message deadlines to be met it is also crucial to choose the synchronous bandwidths H_i appropriately.

Further studies on timing properties have been performed in Chen, Li, and Zhao [1994] and the results can be summarized by the following theorem.

Theorem 44.1. Let $X_i(t, t + I, \mathbf{H})$ be the minimum total transmission time available for node i to transmit its synchronous message within the time interval $(t, t + I)$ under bandwidth allocation $\mathbf{H} = (H_1, H_2, \ldots, H_n)^T$ then

$$X_i(t, t+I, \mathbf{H}) = \begin{cases} \left\lfloor \dfrac{I}{TTRT} - 1 \right\rfloor \cdot H_i + \max\left(0, \min\left(r_i - \tau - \sum_{j=1,\ldots,n, j \neq i} H_j, H_i\right)\right) & \text{if } I \geq TTRT \\ 0 & \text{otherwise} \end{cases} \tag{44.6}$$

where $r_i = I - \lfloor \frac{I}{TTRT} \rfloor \cdot TTRT$.

By Theorem 44.1, deadline constraint (44.4) is satisfied if and only if for $i = 1, \ldots, n$

$$X_i(t, t + D_i, \mathbf{H}) \geq C_i \tag{44.7}$$

44.4 Parameter Selection for Real-Time Applications

To support real-time applications on FDDI, we have to properly set the following three types of parameters: synchronous bandwidth, *TTRT*, and buffer size. We address this parameter selection problem in this section.

Synchronous Bandwidth Allocation

In FDDI synchronous bandwidths are assigned by a synchronous bandwidth allocation scheme. This subsection examines synchronous bandwidth allocation schemes, and discusses how to evaluate their effectiveness.

A Classification of Allocation Schemes

Synchronous bandwidth allocation schemes may be divided into two classes: local allocation schemes and global allocation schemes. These schemes differ in the type of information they may use. A *local* synchronous bandwidth allocation scheme can only use information available locally to node i in allocating H_i. Locally available information at node i includes the parameters of stream S_i (i.e., C_i, P_i, and D_i). *TTRT* and τ are also locally available at node i, because these values are known to all nodes. On the other hand, a *global* synchronous bandwidth allocation scheme can use global information in its allocation of synchronous bandwidth to a node. Global information includes both local information and information regarding the parameters of synchronous message streams originating at other nodes.

A local scheme is preferable from a network management perspective. If the parameters of stream S_i change, then only the synchronous bandwidth H_i of node i needs to be recalculated. The synchronous bandwidths at other nodes do not need to be changed because they were calculated independently of S_i. This makes a local scheme flexible and suited for use in dynamic environments where synchronous message streams are dynamically initiated or terminated.

In a global scheme, if the parameters of S_i change, it may be necessary to recompute the synchronous bandwidths for all nodes. Therefore, a global scheme is not well suited for a dynamic environment. In addition, the extra information employed by a global scheme may cause it to handle more traffic than a local scheme. However, it is known that local schemes can perform *very closely* to the optimal synchronous bandwidth allocation scheme when message deadlines are equal to message periods. Consequently, given the previously demonstrated good performance of local schemes and their desirable network management properties, we concentrate on local synchronous bandwidth allocation schemes in this chapter.

A Local Allocation Scheme

Several local synchronous bandwidth allocation schemes have been proposed, for both the case of $D_i = P_i$ [Agrawal, Chen, and Zhao, 1993] and the case of $D_i \neq P_i$ [Malcolm and Zhao, 1993; Zheng and Shin, 1993]. These schemes all have similar worst-case performance. Here we will consider the scheme proposed in Malcolm and Zhao [1993]. With this scheme, the synchronous bandwidth for node i is allocated according to the following formula:

$$H_i = \frac{U_i D_i}{\left\lfloor \dfrac{D_i}{TTRT} - 1 \right\rfloor} \tag{44.8}$$

Intuitively, this scheme follows the flow conservation principle. Between the arrival of a message and its absolute deadline, which is D_i time later, node i will have at least $\lfloor \frac{D_i}{TTRT} - 1 \rfloor H_i$ of transmission time available for synchronous messages by Eq. (44.6). This transmission time is available regardless of the number of asynchronous messages in the network. During the D_i time, $U_i D_i$ can loosely be regarded as the load on node i. Thus, the synchronous bandwidth in Eq. (44.8) is just sufficient to handle the load on node i between the arrival of a message and its deadline.

The scheme defined in Eq. (44.8) is a simplified version of those in Malcolm and Zhao [1993] and Zheng and Shin [1993]. In the rest of this chapter, we assume this scheme is used because it is simple, intuitive, and well understood. Another reason for concentrating on this scheme is that it has been adopted for use with the SAFENET standard, and thus will be used in distributed real-time systems in the future.

Schedulability Testing

We now consider schedulability testing. A message set is schedulable if the deadlines of its synchronous messages can be satisfied. This can be determined by referring to the tests that both the protocol and deadline constraints [Eqs. (44.3) and (44.4) or Eq. (44.7)] are satisfied.

Testing if the protocol constraint is satisfied is very straightforward. But the test of deadline constraints is more complicated and requires more information. This test can be greatly simplified if the bandwidths are allocated according to Eq. (44.8). It was shown that if this scheme is used, the protocol constraint condition defined in Eq. (44.3) implies the deadline constraint condition Eq. (44.4). Testing the protocol constraint alone is sufficient to ensure that both constraints are satisfied. This is a big advantage of using the allocation scheme defined in Eq. (44.8).

A second method of schedulability testing is to use the *worst-case achievable utilization* criteria. This criteria has been widely used in real-time systems. For a synchronous bandwidth allocation scheme, the worst-case achievable utilization U^* defines an upper bound on the effective utilization of a message set: if the effective utilization is no more than the upper bound, both the protocol

Fiber Distributed Data Interface and Its Use for Time-Critical Applications

and the deadline constraints are always satisfied. The worst-case achievable utilization U^* for the scheme defined in Eq. (44.8) is

$$U^* = \frac{\left\lfloor \dfrac{D_{\min}}{TTRT} \right\rfloor - 1}{\left\lfloor \dfrac{D_{\min}}{TTRT} \right\rfloor + 1}(1-\alpha) \qquad (44.9)$$

where $TTRT$ is the target token rotation time and $D_{\min}$ is the minimum deadline. For any message set, if its effective utilization Eq. (44.2) is less than U^*, both the protocol and deadline constraints are guaranteed to be satisfied.

We would like to emphasize that in practice, using this criteria can simplify network management considerably. The parameters of a synchronous message set can be freely modified while still maintaining schedulability, provided that the effective utilization remains less than U^*.

Let us examine the impact of $TTRT$ and $D_{\min}$ on the worst-case achievable utilization given in Eq. (44.9). Figure 44.3 shows the worst-case achievable utilization vs $D_{\min}$ for several different values of $TTRT$. This figure was obtained by plotting Eq. (44.9) with τ taken to be 1 ms (a typical value for an FDDI network). Several observations can be made from Fig. 44.3 and formula (44.8).

1. For a fixed value of $TTRT$, the worst-case achievable utilization increases as $D_{\min}$ increases. From Eq. (44.9) it can be shown that when $D_{\min}$ approaches infinity, U^* approaches $(1-\alpha) = (1 - \tau/TTRT)$. That is, as the deadlines become larger, the worst-case achievable utilization approaches the available utilization of the network.
2. In Agrawal, Chen, and Zhao [1993], it was shown that for a system in which all relative deadlines are equal to the corresponding message periods ($D_i = P_i$), a worst-case achievable utilization of $\frac{1}{3}(1-\alpha)$ can be achieved. That result can be seen as a special case of Eq. (44.9): if $D_{\min} = 2TTRT$, we have $\lfloor \frac{D_{\min}}{TTRT} \rfloor = 2$ and $U^* = \frac{1}{3}(1-\alpha)$.
3. $TTRT$ clearly has an impact on the worst-case achievable utilization. From Fig. 44.3, we see that when $D_{\min} = 50$ ms, the case of $TTRT = 5$ ms gives a higher worst-case achievable utilization than the other plotted values of $TTRT$. When $D_{\min} = 125$ ms, the case of $TTRT = 10$ ms gives a higher U^* than the other plotted values of $TTRT$. This observation provides motivation for maximizing U^* by properly selecting $TTRT$ once $D_{\min}$ is given.

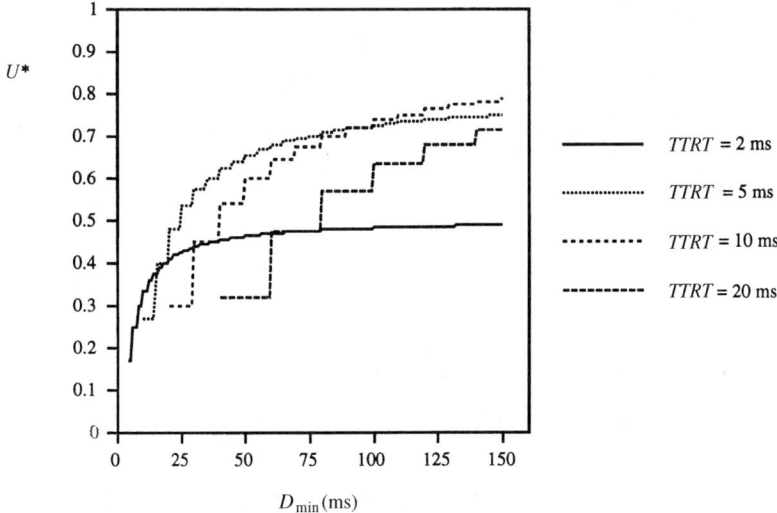

FIGURE 44.3 U^* vs $D_{\min}$.

Selection of Target Token Rotation Time

Recall that *TTRT*, the target token rotation time, determines the expected value of the token rotation time. In contrast to the synchronous bandwidths of individual nodes, *TTRT* is common to all of the nodes and should be kept constant during run time. As we observed in the last section, *TTRT* has an impact on the worst-case achievable utilization. Thus, we would like to choose *TTRT* in an optimal fashion, so that the worst-case achievable utilization U^* is maximized. This will increase the amount of real-time traffic that can be supported by the network.

The optimal value of *TTRT* has been derived in Malcolm and Zhao [1993] and is given by

$$TTRT = \frac{D_{min}}{\left[\frac{-3 + \sqrt{9 + \frac{8D_{min}}{\tau}}}{2}\right]} \quad (44.10)$$

The impact of an appropriate selection of *TTRT* is further evident from Fig. 44.4, which uses Eq. (44.10) to show U^* vs *TTRT* for several different values of D_{min}. As with Fig. 44.3, τ is taken to be 1 ms. From Fig. 44.4 the following observations can be made:

1. The curves in Fig. 44.4 verify the prediction of the optimal *TTRT* value given by Eq. (44.10). For example, consider the case of $D_{min} = 40$ ms. By Eq. (44.10), the optimal value of *TTRT* is 5 ms. The curve clearly indicates that at *TTRT* = 5 ms the worst-case achievable utilization is maximized. Similar observations can be made for the other cases.
2. As indicated in Eq. (44.10), the optimal *TTRT* is a function of D_{min}. This coincides with the expectations from the observations of Fig. 44.3. A general trend is that as D_{min} increases, the optimal *TTRT* increases. For example, the optimal values of *TTRT* are approximately 2.5 for $D_{min} = 10$ ms, 4 ms for $D_{min} = 20$ ms, 5 ms for $D_{min} = 40$ ms, and 6.67 ms for $D_{min} = 80$ ms.
3. The choice of *TTRT* has a large effect on the worst-case achievable utilization U^*. Consider the case of $D_{min} = 40$ ms shown in Fig. 44.4. If *TTRT* is too small (say, *TTRT* = 2 ms), U^* can be as low as 45%. If *TTRT* is too large (say, *TTRT* = 15 ms), U^* can be as low as 31%.

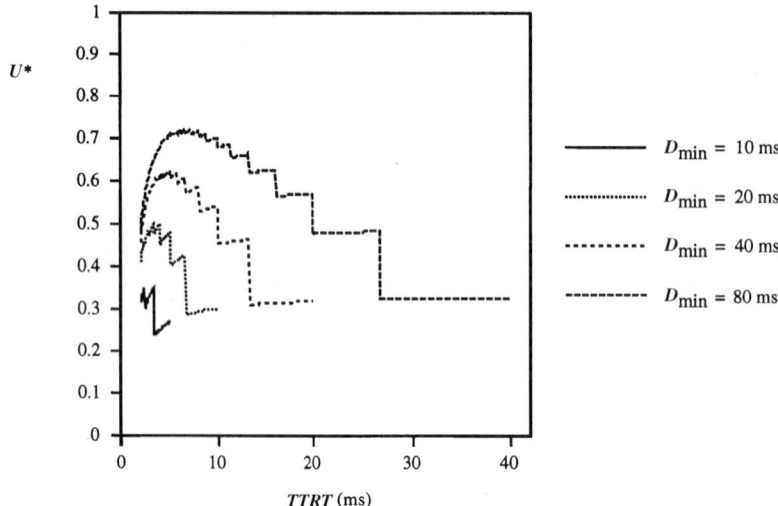

FIGURE 44.4 U^* vs *TTRT*.

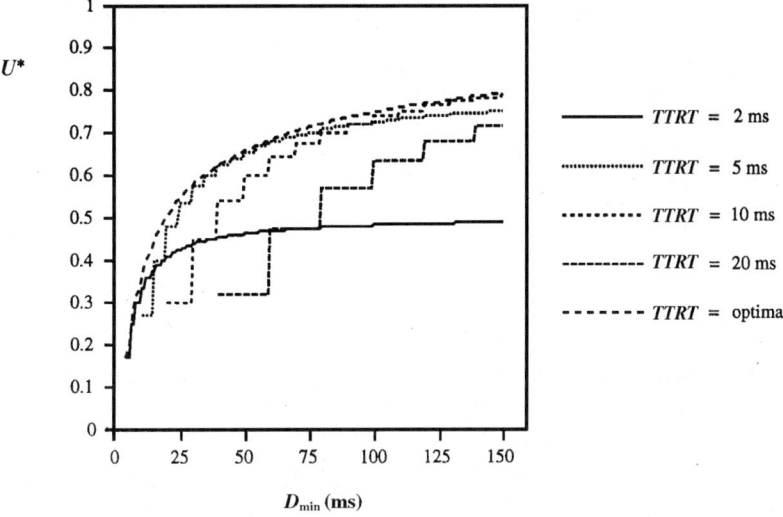

FIGURE 44.5 U^* vs D_{min} with optimal $TTRT$.

However, when the optimal value of $TTRT$ is used (i.e., $TTRT = 5$ ms), U^* is 62%. This is an improvement of 17% and 31%, respectively, over the previous two cases.

The effect of choosing an optimal value for $TTRT$ is also shown in Fig. 44.5. Figure 44.5 shows the worst-case achievable utilization vs D_{min} when an optimal value of $TTRT$ is used. For ease of comparison, the earlier results of Fig. 44.3 are also shown in Fig. 44.5.

Buffer Requirements

Messages in a network can be lost if there is insufficient buffer space at either the sending or the receiving node. To avoid such message loss, we need to study the buffer space requirements.

There are two buffers for synchronous messages on each node. One buffer is for messages waiting to be transmitted to other nodes and is called the *send buffer*. The other buffer is for messages that have been received from other nodes and are waiting to be processed by the host. This buffer is called the *receive buffer*.

We consider the send buffer first.

- Case of $D_i \leq P_i$. In this case, a message must be transmitted within the period it arrives. At most one message will be waiting. Hence, the send buffer need only accommodate one message.
- Case of $D_i > P_i$. In this case, a message may wait as long as D_i time without violating its deadline constraint. At node i, messages from stream S_i arrive every P_i time, requesting transmission. During the D_i time following the arrival of a message, there will be a further $\lfloor \frac{D_i}{P_i} \rfloor$ message arrivals. Thus, the send buffer may potentially have to accommodate $\lfloor \frac{D_i}{P_i} \rfloor + 1$ messages. When the deadline is very large (tens or hundreds times larger than the period, for example, which can occur in voice transmission), this might become a problem in practice due to excessive buffer requirements.

    However, when the synchronous bandwidths are allocated as in Eq. (44.8), the impact of D_i on required buffer size is limited: it can be shown that if Eq. (44.8) is used, the waiting time of a message from stream S_i is bounded by $\min(D_i, P_i + 2TTRT)$ [Malcolm, 1994]. Thus, the maximum number of messages from S_i that could be queued at node i is no more than $\lceil \frac{\min(D_i, P_i + 2TTRT)}{P_i} \rceil$. This allows us to reduce the required size of the send buffer. Let

BS_i denote the send buffer size at node i. The send buffer will never overflow if

$$BS_i \geq \left\lceil \frac{\min(D_i, P_i + 2TTRT)}{P_i} \right\rceil B_i \qquad (44.11)$$

where B_i is the number of bytes in a message of stream S_i.

An interesting observation is that if $D_i \geq P_i + 2TTRT$, then

$$BS_i \geq \left\lceil \frac{P_i + 2TTRT}{P_i} \right\rceil B_i \qquad (44.12)$$

That is, the send buffer requirements for stream S_i are independent of the deadline D_i. As mentioned earlier, one would expect that increasing the deadline of a stream could result in increased buffer requirements. When the scheme in Eq. (44.8) is used, however, the buffer requirements are not affected by increasing the deadline once the deadline reaches certain point.

Now let us consider the receive buffer size. Suppose that node j is the destination node of messages from nodes $j_1, \ldots, j_k$. This means that messages from streams $S_{j_1}, \ldots, S_{j_k}$ are being sent to node j. The size of the receive buffer at node j depends not only on the message traffic from these streams but also on the speed at which the host of node j is able to process incoming messages from other nodes. The host at node j has to be able to keep pace with the incoming messages, otherwise the receive buffer at node j can overflow. When a message from stream S_{j_i} arrives at node j, we assume that the host at node j can process the message within P_{j_i} time.

Assuming that the host is fast enough, then it can be shown that the number of messages from S_i that could be queued at the destination node is bounded by $\lceil \frac{\min(D_i, P_i + 2TTRT)}{P_i} \rceil + 1$. With this bound, we can derive the space requirements for the receive buffer at node j. Let BR_j denote the receive buffer size at node j. The receive buffer will never overflow if

$$BR_j \geq \sum_{i=1}^{k} \left\lceil \frac{\min(D_{j_i}, P_{j_i} + 2TTRT)}{P_{j_i}} + 1 \right\rceil B_{j_i} \qquad (44.13)$$

As in the case of the send buffer, we can observe from Eq. (44.13) that if the deadlines are large the required size of the receive buffer will not grow as the deadlines increase.

44.5 Final Remarks

In this chapter, we introduced the architecture of FDDI and discussed its fault-tolerant capability. We presented a methodology for the use of FDDI networks for real-time applications. In particular, we considered methods of selecting the network parameters and of schedulability testing. The parameter selection methods are compatible with current standards. The synchronous bandwidth allocation method has the advantage of only using information local to a node. This means that modifications in the characteristics of synchronous message streams, or the creation of new synchronous message streams, can be handled locally without reinitializing the network. Schedulability tests that determines whether the time constraints of messages will be met were presented. They are simple to implement and computationally efficient.

The materials presented in this chapter complement much of the published work on FDDI. Since the early 1980s, extensive research has been done on the timed token protocol and its use in FDDI networks [Albert and Jayasumana, 1994; Jain, 1994; Mills, 1995; Shah and Ramakrishnan, 1994]. The papers by Ross [1989], Iyer and Joshi [1985], and others [Mccool, 1988; Southard,

1988; Stallings, 1987] provided comprehensive discussions on the timed token protocol and its use in FDDI. Ulm [1982] discussed the protocol performance with respect to parameters such as the channel capacity, the network cable length, and the number of stations. Dykeman and Bux [1988] developed a procedure for estimating the maximum total throughput of asynchronous messages when using single and multiple asynchronous priority levels. The analysis done by Pang and Tobagi [1989] gives insight into the relationship between the bandwidth allocated to each class of traffic and the timing parameters. Valenzo, Montuschi, and Ciminiera [1990] concentrated on the asynchronous throughput and the average token rotation time when the asynchronous traffic is heavy. The performance of the timed token ring depends on both the network load and the system parameters. A study on FDDI by Jain [1991] suggests that a value of 8 ms for *TTRT* is desirable as it can achieve 80% utilization on all configurations and results in around 100 ms maximum access delay on typical rings. Further studies have been carried out by Sankar and Yang [1989] to consider the influence of the target token rotation time on the performance of varying FDDI ring configurations.

References

ANSI. 1992. FDDI station management protocol (SMT). ANSI Standard X3T9.5/84-89, X3T9/92-067, Aug.

Agrawal, G., Chen, B., and Zhao, W. 1993. Local synchronous capacity allocation schemes for guaranteeing message deadlines with the timed token protocol. In *Proc. IEEE Infocom'93*, pp. 186–193.

Agrawal, G., Kamat, S., and Zhao, W. 1994. Architectural support for FDDI-based reconfigurable networks. In *Workshop on Architectures for Real-Time Applications (WARTA)*.

Albert, B., and Jayasumana, A.P. 1994. *FDDI and FDDI-II Architecture, Protocols, and Performance*, Artech House.

Chen, B., Li, H., and Zhao, W. 1994. Some timing properties of timed token medium access control protocol. In *Proceedings of International Conference on Communication Technology*, pp. 1416–1419, June.

Chen, B., Kamart, S., and Zhao, W. 1995. Fault-tolerant real-time communication in fddi-based networks. *Proc. IEEE Real-Time Systems Symposium*, pp. 141–151, Dec.

Dykeman, D. and Bux, W. 1988. Analysis and tuning of the FDDI media access control protocol. *IEEE J. Selec. Areas Commun.*, 6(6).

Grow, R.M. 1982. A timed token protocol for local area networks. In *Proc. Electro/82, Token Access Protocols*, May.

Iyer, V. and Joshi, S.P. 1985. FDDI's 100 M-bps protocol improves on 802.5 Spec's 4-M-bps limit. *Elec. Design News*, (May 2):151–160.

Jain, R. 1991. Performance analysis of FDDI token ring networks: Effect of parameters and guidelines for setting TTRT. *IEEE Lett.* (May):16–22.

Jain, R. 1994. *FDDI Handbook—High-Speed Networking Using Fiber and Other Media*, Addison Wesley.

Kamat, S., Agrawal, G., and Zhao, W. 1994. On available bandwidth in FDDI-based reconfigurable networks. In *Proc. IEEE Infocom'94*.

Malcolm, N. and Zhao, W. 1993. Guaranteeing synchronous messages with arbitrary deadline constraints in an FDDI network. In *Proc. IEEE Conference on Local Computer Networks*, pp. 186–195, Sept.

Malcolm, N. 1994. *Hard real-time communication in high speed networks*. Ph.D. Thesis, Dept. of Computer Science, Texas A&M Univ., College Station.

Mccool, J. 1988. FDDI—getting to the inside of the ring. *Data Commun.*, (March):185–192.

Mills, A. 1995. *Understanding FDDI*, Prentice–Hall, Englewood Cliffs, NJ.

Pang, J. and Tobagi, F.A. 1989. Throughput analysis of a timer controlled token passing protocol under heavy load. *IEEE Trans. Commun.*, 37(7):694–702.

Ross, F.E. 1989. An overview of FDDI: The fiber distributed data interface. *IEEE J. Selec. Areas Commun.*, 7(Sept.):1043–1051.

Sankar, R. and Yang, Y.Y. 1989. Performance analysis of FDDI, In *Proc. IEEE Conference on Local Computer Networks*, pp. 328–332, Minneapolis MN, Oct. 10–12.

Sevcik, K.C. and Johnson, M.J. 1987. Cycle time properties of the FDDI token ring protocol. *IEEE Trans. Software Eng.*, SE-13(3):376–385.

Shah, A. and Ramakrishnan, G. 1994. *FDDI: A High Speed Network*, Prentice–Hall, Englewood Cliffs, NJ.

Southard, R. 1988. Fibre optics: A winning technology for LANs. *Electronics* (Feb.):111–114.

Stallings, W. 1987. *Computer Communication Standards, Vol 2: Local Area Network Standards*, Howard W. Sams.

Ulm, J.N. 1982. A timed token ring local area network and its performance characteristics. In *Proc. IEEE Conference on Local Computer Networks*, pp. 50–56, Feb.

Valenzano, A., Montuschi, P., and Ciminiera, L. 1990. Some properties of timed token medium access protocols. *IEEE Trans. Software Eng.*, 16(8).

Zheng, Q. and Shin, K.G. 1993. Synchronous bandwidth allocation in FDDI networks. In *Proc. ACM Multimedia'93*, pp. 31–38, Aug.

45
Broadband Local Area Networks

45.1 Introduction 611
45.2 User Requirements 612
45.3 BLAN Technologies 613
45.4 ATM BLANs 615
45.5 Other BLANs 617
45.6 New Applications 618
45.7 Conclusion 619

Joseph A. Bannister
The Aerospace Corporation

45.1 Introduction

The local area network (LAN) became a permanent fixture of the computer world during the 1980s, having achieved penetration primarily in the form of the Institute of Electrical and Electronics Engineers (IEEE) 802.3 ethernet[1] LAN. Thereafter, other types of LANs, such as the IEEE 802.5 token ring and the American National Standards Institute (ANSI) X3T9.5 fiber distributed data interface (FDDI), also established themselves as lesser—though potent—players. These LANs now form the dominant communications infrastructures that connect computers and related devices within individual enterprises. They provide the transmission substrate for common services such as file transfer, network file services, network window service, electronic mail, hypertext transfer, and others.

The installed base of ethernet, token ring, and FDDI LANs consists fundamentally of shared-media broadcast networks in which each node's transmission may be heard by all other nodes on the LAN. In any such **shared-media** LAN, a **media-access control protocol** must be provided to coordinate the different nodes' transmissions, lest they interfere adversely with one another. The most serious drawback of shared-media LANs is that, as more nodes are added, each node receives a proportionally smaller share of the media's bandwidth. The bandwidth of the media is, therefore, a limiting resource. This inability to scale up to meet increased demands for bandwidth makes shared-media LANs inherently unsuitable for new classes of communications services, such as the transmission of high-resolution image and video data. Furthermore, the nature of their media-access control protocols, which can introduce substantial and variable delays into transmissions,

---

[1] Originally, ethernet referred to a specific product offering, whereas the IEEE 802.3 LAN was formally known as the carrier sense multiple access with collision detection LAN. The two LANs differ minutely, and henceforth we consider them to be the same.

often makes shared-media LANs unable to meet the tight **latency** and **jitter** requirements of real-time video and audio transmission.

To address the problems of bandwidth scalability and real-time transmission, a new type of LAN, called the **broadband local area network** (**BLAN**), has been steadily taking root. BLANs are designed to accommodate increases in traffic demand by modular expansion. They are also capable of dedicating a set amount of bandwidth to a requesting application, guaranteeing that the latency of the application's transmissions is bounded and unvarying, if required. The BLAN will thus carry integrated traffic, comprising data, voice, and video. In this respect, the BLAN is a local manifestation of the **broadband integrated services digital network** (**BISDN**), which is intended for deployment in the global telecommunications system. The common ground of the BLAN and the BISDN is their reliance on **asynchronous transfer mode** (**ATM**). Special considerations, however, distinguish the BLAN from the BISDN, including different cost structures, the need to support different services, and different approaches to network management and security.

This chapter will describe the common approaches to BLANs, including discussion of the major requirements and the hardware and software technologies essential to their operation. The central role of ATM networks in BLANs, as well as other less well-known network technologies, will be covered. The protocols used in BLANs will also be discussed. Finally, some of the new applications being enabled by BLANs will be presented.

45.2 User Requirements

BLAN users expect to run network-based applications that require services not provided by conventional LANs. These requirements include bandwidth scalability, integrated support for different traffic classes, multicasting, error recovery, manageability, and security.

To meet the increased demand for bandwidth that accompanies the addition of new nodes and the replacement of old nodes by more-capable equipment, bandwidth scalability in shared-media LANs has been achieved by segmenting users into distinct **collision domains**, which are interconnected by bridges and routers so that intrasegment traffic is filtered out before leaving its originating domain, and intersegment traffic is forwarded to its destination. This approach is, however, not truly scalable, and the resulting web of LANs becomes fundamentally unmanageable. To overcome this difficulty, packet switches replace the collision domains, and each node is connected to its switch by a dedicated link. Growth of the network is then accomplished by adding switches or selectively replacing the interfaces between switches and nodes with higher speed interfaces.

Network traffic is traditionally divided into three classes: **isochronous, synchronous**, and **asynchronous** traffic. Isochronous traffic flows at a constant rate and has a strict timing relationship between any consecutively transmitted units of data. An example is periodically sampled fixed-size data blocks, such as one would encounter in the transmission of a video frame buffer every 1/30 of a second. Synchronous traffic flows at a variable rate and has a loose timing relationship between consecutively transmitted units of data. An example is the transmission of a compressed video frame buffer every 1/30 of a second. Asynchronous traffic flows at a variable rate but has no definite timing relationship among its data units. An example is file transfer. Although all conventional LANs provide support for asynchronous traffic, few support synchronous traffic. Support for isochronous traffic is not common in conventional LANs. Currently, only BLANs support all three classes of traffic.

Users who desire to run video applications will use a BLAN that supports **multicasting**, a generalization of broadcasting in which a sender may transmit the same message to a number of different receivers without sending multiple copies. Easily supported in shared-media LANs, this capability must be designed in to switch-based networks. Multicasting is not only used in disseminating video broadcasts, but has become entrenched into the way that shared-media LANs operate, for example in registering addresses. Thus, it must be supported in BLANs to accommodate legacy protocols.

As components of a complex system—such as a BLAN—fail, the system should continue to provide service with a minimum of interruption. Having assumed the responsibility for transporting nearly all of the traffic in an enterprise, the BLAN of necessity must be robust enough to provide continuous service even in the face of failures. Given that a BLAN can fail in many different ways, it is essential that error-recovery procedures are integrated into its design. The BLAN must detect transmission errors and notify the system administrator of them. Failures of resource (e.g., a switch or link) must also be recognized and bypassed. Since failures or oversubscription can cause congestion, there must be the means to recognize and clear congestion from the network.

Owned and operated by a single enterprise, the BLAN usually falls under the responsibility of a specialized department charged with its management. It is often necessary to track all resources within the BLAN so that the network manager knows the precise disposition of all equipment at any given time. Each resource also must be monitored continuously to assess its level of performance. Resources also are controlled and configured by the network manager. For the convenience of the network manager, these functions are normally concentrated in a network-management platform. The dynamically changing state of all resources within the BLAN is referred to as the **management information base** (MIB), which may be viewed as readable and writeable distributed database through which the network is monitored and controlled. User information may be extracted from the MIB in order to produce usage and billing information.

Although the BLAN resides within the relatively friendly confines of an enterprise, external and internal attacks on the BLAN's computers are an unignorable threat. External attacks on a BLAN connected to a larger public network are effectively controlled by use of a network firewall, which bars unauthorized outside access to the enterprise's computers. Because sensitive information might have to be denied to internal users, it is also necessary to provide for privacy within some BLANs. Techniques for encrypting stored or transmitted data and for providing access controls are then required. To enhance security further, it also could be necessary to provide for intrusion-detection software, which passively monitors BLAN resources for internal and external attacks.

45.3 BLAN Technologies

The spread of BLANs has been promoted by technological advances on several fronts. Fast packet switching has been one of the enabling technologies. So, too, has been high-speed transmission technology. Other important software technologies in BLANs are routing, signalling, and network management.

Packet switching networks are composed of switching nodes connected by links. The switches forward input packets to the appropriate output links after reading the packets' header information and possibly buffering packets temporarily to resolve contention for common output links. The move to packet switching LANs was motivated by the realization that shared-media LANs do not scale, because the total throughput of all nodes can never exceed the bandwidth of the media. Similarly, latency and jitter grow as nodes are added to a shared-media LAN, because the media-access control protocol tends to perform like a single-server queueing system, in which the mean and variance of delay increase as the server's workload increases. Switching plays a key role in BLANs, because a **mesh topology**, which consists of switches connected by links, is necessary to overcome the bandwidth bottleneck of the shared-media LAN, as illustrated in Fig. 45.1. In the shared-media LAN only one node at a time can transmit information, but in the mesh-topology LAN several nodes can transmit information simultaneously. The result is a dramatic increase in the throughput of the mesh-topology LAN, compared to the shared-media LAN. If one can build a packet switch with a sufficient number of input/output ports, then a mesh network may be easily expanded by adding new nodes, switches, and links without significantly reducing the amount of bandwidth that is available to other nodes. It is also true that latency and jitter in a mesh topology may be kept low, since the capacity of the network is an increasing function of the number of nodes.

Paced by advances in integrated circuits, switch technology has progressed rapidly. Packet switching fabrics based on high-speed backplane buses, shared memory, or multistage interconnection

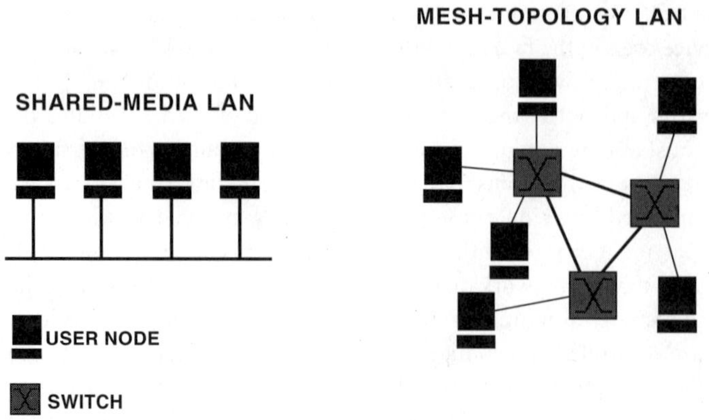

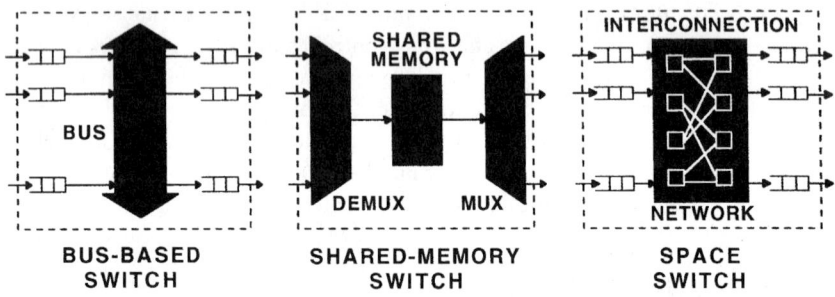

FIGURE 45.1 The mesh-topology LAN provides higher bandwidth and supports more nodes than the shared-media LAN.

FIGURE 45.2 Packet switch architectures.

networks (MINs), such as delta, omega, shuffle-exchange, or Clos networks [Tobagi, 1990], can be constructed to switch pre-established connections with a throughput of several gigabits per second. Examples of each of these switch architectures are shown in Fig. 45.2. To increase switch throughput, buses may be replicated and shared memories interleaved. Although backplane-bus fabrics cannot switch more than a few dozen input and output ports, because the electrical limits on bus length restrict the number of modules that may be attached, two-by-two switching elements can be combined in very deep configurations to implement MINs with arbitrary numbers of input and output ports. Switch latency can be reduced by the use of cut-through routing, which permits the head of a packet to emerge from the switch before its tail has even entered the switch. This is especially beneficial for long packets, but the switch design is more complex and special facilities must often be added to prevent deadlock. Bus-based switching fabrics support multicasting easily. MIN-based switching fabrics have also been implemented to support multicasting by forking copies of a packet at various stages or using a separate copy network [Turner, 1987]. Today's nonblocking, output-queued MIN-based fabrics can achieve very low latency and high throughput [Hluchyj and Karol, 1988].

High-performance microprocessors—increasingly with reduced instruction set computer architectures—often form an integral part of switch and interface architectures. These provide necessary decision-making and housekeeping functions such as routing-table maintenance and call processing. They also may be used for such functions as segmenting messages into smaller transmission units at the sender and then reassembling the message at the receiver. Such segmentation and reassembly functions are sometimes performed by special-purpose hardware, under the control of the microprocessor. High-speed random-access and first-in-first-out memories are also critical for storage and packet buffering.

Examples of chipsets that support high-speed switching are found in Denzel, Engbersen, and Iliadis [1995] and Collivignarelli et al. [1994]. Supporting chipsets are generally implemented as complementary metal-oxide-semiconductor devices. As higher switching and transmission speeds are needed, however, it is expected that supporting chipsets will be implemented as gallium arsenide or bipolar devices, which operate in a much faster regime.

Hand in hand with fast packet switching technology comes high-speed transmission technology. The enabling technology in this area has been optical fiber technology, which can achieve link speeds of several gigabits per second. The principal transmission technology for BLANs is multimode optical fibers (MOFs). Using infrared signals with a 1.3-μm wavelength, BLANs can achieve data rates of more than 100 Mb/s over distances of up to 2000 m. Transmitters are usually light-emitting diodes, receivers are photodiodes, and fibers are 125-μm cladding/62.5-μm core multimode silica fibers. To transmit over greater distances or at higher data rates, 100-μm cladding/9-μm core single-mode silica fibers are used with 1.3- or 1.5-μm wavelength laser-diode transmitters. Because LANs are of limited geographical reach, single-mode optical fibers can carry information at rates of several gigabits per second without the need for optical amplification or regeneration. Optical fiber BLAN links typically operate with direct direction and return-to-zero, 4B/5B, or 8B/10B encoding (nB/mB denotes a scheme in which n data bits are encoded as m transmission bits). Error-detecting or -correcting codes are used sparingly, because the transmission quality is normally very high.

The relatively high cost of optical systems and the installed base of metallic-wire systems make the use of inexpensive metallic-wire media very desirable. Most conventional LANs now use category-3 twisted-pair 24-gauge insulated wires, deployed in a manner similar to telephone wires. By upgrading to category-5 wires or using multiple category-3 wires, it is possible to transmit data at rates up to 155 Mb/s. BLANs are beginning to use category-5 twisted-pair wires for transmission at data rates greater than 100 Mb/s. This is especially attractive for linking office or laboratory equipment to telecommunications closets, which are designed to be situated no more than 100 m away from their users. Switches located inside distinct telecommunications closets are then usually connected to each other by means of optical fibers, which enjoy a greater geographical span and lower error rates than electronic links. To reduce electromagnetic interference, it is sometimes necessary to scramble high-speed data traveling over twisted-pair wires. Higher data rates can be achieved more easily by shielding wires to reduce crosstalk and electromagnetic emission. However, unshielded twisted-pair wires are desirable for BLANs, because they are compatible with the type of wire used to connect desktop telephones to telecommunications closet equipment.

LANs, being complex systems, require substantial software to operate properly. This is even more true of BLANs, which require more-sophisticated control to support all features of conventional LANs, as well as traffic integration. Common software technologies for BLANs include routing, signalling, and network management. Routing implies the ability to discover paths to a destination, signalling the ability to reserve resources along the best path, and network management the ability to monitor and control the state of those resources.

Wireless and mobile technologies are on the verge of being introduced into BLANs, but prototypes have yet to appear.

45.4 ATM BLANs

ATM occupies the dominant position among BLANs. Originally conceived as a method for transferring information over the public telecommunications network, ATM was quickly accepted as a way to provide high bandwidth within the premises of a private enterprise. Vendors arose to provide ATM products specialized for LAN use. This includes inexpensive switches, computer-interface cards, and software.

A factor that militates against the spread of ATM BLANs is the incomplete state of ATM standardization. ATM standards are officially under the aegis of the International Telecommunications Union–Telecommunications Standardization Sector (ITU-T). The overall framework of BISDN falls within the ITU's purview. The ATM Forum, a large consortium of organizations whose purpose is to accelerate the acceptance of ATM networking by producing agreements among its members on how to implement specific standards, augments the contributions of the ITU-T. The ATM Forum has defined many of the physical- and link-layer standards for ATM, including 100/140-Mb/s transparent asynchronous transceiver interface (TAXI) and 25-Mb/s Desktop 25 link protocols, which operate on MOFs and twisted pairs, respectively. The Internet Engineering Task Force (IETF) also plays a role in ATM BLANs by its consideration of how to integrate ATM into the Internet protocol suite. It produced the standard for running the classical internet protocol (IP) over ATM.

ATM is a **connection-oriented protocol**, in which a caller requests to exchange data with a recipient. The network is responsible for finding the route and reserving any resources needed to complete the exchange. Once the call has been established, each hop of the call is associated with a virtual path identifier/virtual circuit identifier (VPI/VCI). The VPI/VCI has only local significance at switches or interfaces and is used to map an incoming cell (the 53-octet [an octet equals 8 b] fixed-length ATM packet) to its output port. The connection guarantees a specific **quality of service** (QOS), which specifies a contract between the network and the user that the network will satisfy the bandwidth, delay, and jitter requirements of the user provided that the user's traffic meets the constraints of the contract. Several QOS classes may be specified:

- Constant bit rate (CBR), which supports isochronous traffic
- Variable bit rate/real time (VBR/RT), which supports synchronous traffic
- Variable bit rate/nonreal time (VBR/NRT), which supports asynchronous traffic that requires a minimum bandwidth guarantee
- Available bit rate (ABR), which supports asynchronous traffic with no QOS guarantees (best effort) but increases and reduces bandwidth according to competing demands of other connections
- Unspecified bit rate (UBR), which supports asynchronous traffic but possibly also drops cells

ABR and UBR correspond most closely to the QOS provided by existing LANs. CBR will be needed to carry uncompressed video, whereas VBRRT will be used to carry compressed video.

An important consideration is to be able to coexist with existing software packages. For IP-based networks this largely means supporting applications that have been written using the UNIX sockets application program interface. This type of BLAN uses the classical-IP-over-ATM approach defined by the IETF. For other networks, such as those based on Novell Netware or AppleTalk protocols, the answer is to use the LAN emulation (LANE) standard developed by the ATM forum. Both these approaches attempt to hide from the application the fact that its underlying network is a connection-oriented ATM BLAN rather than a connectionless ethernet or token ring shared-media LAN. One of the central difficulties in achieving this is that ATM must mimic the broadcasting that is used by the older LANs.

ATM BLANs may use a variety of media. Currently the most popular media are optical fibers, but unshielded twisted pair (UTP) is fast becoming the media of choice for ATM desktop connection. MOFs frequently use the synchronous digital hierarchy (SDH) to carry synchronous transport signal (STS) level-3 concatenated (STS-3c) frames, which consist of 2430 octets sent every 125 μs for an overall transmission rate of 155.5 Mb/s. In the U.S.A. SDH is known as synchronous optical network (SONET) and SDH-3c is known as optical carrier level 3 concatenated (OC-3c). The maximum reach of a segment must be less than 2000 m. STS-3c may also be used with UTP, but the maximum reach of a segment is only 100 m.

Many ATM BLAN devices were based on the TAXI standard, a 100-Mb/s MOF interface with 4B/5B encoding that was made popular by the FDDI standard. The appeal of TAXI for ATM reflected the fact that several TAXI chipsets were available at the time of initial ATM BLAN introduction. Also used were 140-Mb/s TAXI devices. The TAXI devices, however, have rapidly been supplanted by SDH devices, because of their more universal appeal.

A host of other physical interfaces has been defined for ATM BLANs. These include the Fibre Channel—compatible 155.5-Mb/s MOF and shielded twisted-pair interfaces with 8B/10B encoding. The STS-1/2 interface—also known as Desktop 25—is a 25.9-Mb/s category-3 UTP interface. An STS-1 UTP interface operates at 51.8 Mb/s. None of these interfaces is widely used in BLANs at this time.

Two principal types of ATM BLAN interfaces are the **private user–network interface** (PUNI) and the **private network–node interface** (PNNI), where private designates that the interfaces do not impinge on public networks. The PUNI is an interface between user equipment and BLAN switches, whereas the PNNI is an interface between BLAN switches. Each type of interface has a protocol associated with it to implement signalling, routing, connection admission control, traffic shaping, traffic policing, and congestion control. These protocols use reserved VCIs to communicate among peers.

Because of their popularity, a great deal of effort has gone into developing special protocols for ATM BLANs. As most networks use IP, the classical-IP-over-ATM protocol was designed to permit connectionless IP to establish a new ATM virtual connection or bind to an existing one by means of an address resolution protocol similar to that used in shared-media LANs. To support multicasting the ATM forum has also defined leaf-initiated–join protocols that allow point-to-multipoint multicast trees to be grown from the destination leaves to the source root; this is important for LANE operations, which rely heavily on multicasting. Security is also an important concern for BLAN administrators, and the U.S.A. government—sponsored Fastlane device allows key-agile end-to-end encryption and decryption of cells on a virtual-connection–by–virtual-connection basis [Stevenson, 1995].

45.5 Other BLANs

Although ATM offers the most capable and widely deployed approach to implementing BLANs, several other approaches are available. These approaches are IEEE 802.3 ethernet, IEEE 802.9 isochronous ethernet, and IEEE 802.12 100VG-AnyLAN.

Given ethernet's popularity, it is not surprising that there should be ethernet-based BLANs. Ethernet originally was designed to use the carriersense multiple access with collision detection protocol, which allows a node to share a LAN segment with other nodes by transmitting only when the medium is inactive and randomly deferring transmission in the event of a collision (simultaneous transmission by two or more nodes). The inefficiency and nondeterminism that accompany this protocol can be partially overcome by partitioning the LAN so that each node resides on its own dedicated segment. The simplest ethernet BLAN consists of 10-Mb/s IEEE 802.3 10BaseT dedicated links connecting nodes to switching hubs. Using ethernet in this way, one can implement a BLAN that delivers 10 Mb/s of bandwidth directly to each user, if the switch has adequate packet buffers and the user interface can keep up. Although they provide no mechanisms for supporting specific levels of QOS, such switched-ethernets are able to support high and scalable throughput. Nevertheless, the need of some applications for more than 10 Mb/s of bandwidth justifies the same switched-ethernet approach but with 100-Mb/s IEEE 802.3 links. The IEEE 802.3 100-Mb/s ethernet (sometimes called fast ethernet) is virtually identical to the 10-Mb/s standard, except that the data rate has been increased tenfold and almost all time constants have been decimated. User nodes are attached to switching or nonswitching hubs (the latter of which merely acts as a shared medium) either by UTPs or MOFs. The 100BaseTX option uses two category-5 UTPs with 4B/5B encoding, the 100BaseT4 option uses four category-3 UTPs with 8B/6T encoding (which encodes 8 b as six ternary digits), and the 100BaseFX option uses two MOFs with 4B/5B encoding. Moreover, all of these options (including 10BaseT) may be mixed in a single switching hub. Fast ethernet is rapidly gaining wide acceptance, and several vendors supply ethernet and fast ethernet BLAN equipment.

A drawback of using switched ethernet for a BLAN is that there is no explicit support for isochronous traffic. Isochronous ethernet (isoethernet), as defined by IEEE 802.9, overcomes those drawbacks. Isoethernet maintains compatibility with the 10BaseT wiring infrastructure and interoperates with 10BaseT at the hub level. Isoethernet multiplexes onto a category-3 UTP a 10-Mb/s ethernet and a 6.14-Mb/s ISDN stream. Using 4B/5B encoding, these streams are mapped into a 125-μs frame, which dedicates specific quartet (a quartet equals 4 b) positions of the frame to ethernet or ISDN traffic. The ethernet subframe is subject to the usual collision-backoff-retransmit

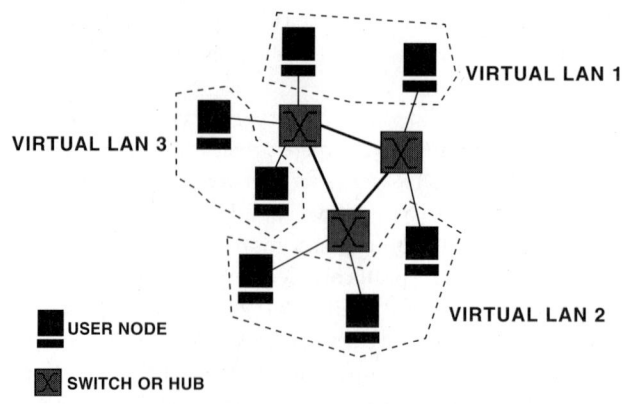

FIGURE 45.3 Virtual LANs.

cycle, but the ISDN subframe can be allocated to several CBR connections, which have a fundamental frequency of 8000 Hz. Its signalling protocol is similar to ATM's. The isoethernet BLAN has not yet been widely deployed.

The IEEE 802.12 100VG-AnyLAN standard is designed to be run over voice-grade (hence, VG) wiring infrastructure. It uses four category-3 bidirectional UTPs to connect user nodes to repeaters, which may be cascaded in a tree configuration. Repeaters poll user nodes for transmission requests, which are granted in a round-robin discipline. Transmissions are not restricted to any single protocol (hence, AnyLAN), and can therefore support traffic integration. Transmission requests have either a low or high priority, with high-priority requests pre-empting those of lower priority. The 100VG-AnyLAN BLAN has not been widely deployed.

The complete replacement of a legacy LAN being impractical, it is not uncommon to build a BLAN with a mix of technologies. A backbone ATM network is frequently the point of access for 802.3 LANs, which can provide from 10 to 100 Mb/s bandwidth to the desktop. User nodes that have higher bandwidth requirements or need a specific QOS may be attached directly to the ATM backbone. The attachment equipment in the backbone—called an edge device—provides ATM and other BLAN interfaces. Working together, these edge devices have the capability to form virtual LANs (VLANs) of nodes logically joined regardless of their physical locations, as shown in Fig. 45.3. Each VLAN forms a unique broadcast domain, in which broadcasts of a node in the VLAN are forwarded to all other members of the VLAN but prevented from propagating to nonmembers. In principle, VLANs should simplify BLAN administration and enhance security and performance. The edge devices and the configurations of VLANs that they support may be managed through MIB variables specially designed to manipulate VLANs. VLAN creation by address, by attachment port, or by several other attributes is currently supported.

45.6 New Applications

Conventional LANs are used principally for a narrow range of client–server applications. BLANs open the door to new computing applications.

Multimedia computing, which processes data, video, image, and audio, is the driving application area for BLANs. The local distribution of video to computer users requires the bandwidth and QOS of a BLAN. With the use of corporate video servers growing, it is essential to have a BLAN that can transport several simultaneous video streams. Such servers are commonly used in training and film production. Other applications include videoconferencing, with workstation-based frame grabbers, cameras, microphones, and electronic whiteboards.

The success of hypermedia browsers, especially those based on the hypertext transport protocol (HTTP), has given rise to the phenomenon of intranetworking, in which enterprises use HTTP to share data among members. As HTTP browsers embrace more functions such as audio, animation, video, and image transfer, the need for higher bandwidth LANs grows. Given the unparalleled traffic growth on the Internet that resulted from HTTP usage, it is reasonable to expect an intranet to experience similar growth. Deployment of BLANs within these intranets thus becomes imperative.

Today most enterprise LANs are collections of segmented collision domains interconnected by bridges and hubs. A cost-effective strategy for expanding a LAN is to use an ATM BLAN as the backbone. In this manner existing LANs can be attached to the ATM backbone and power users can also be attached as the need arises. The use of VLAN technology and edge devices just described makes it relatively simple to link existing LANs through an ATM BLAN.

Coupled with high-performance workstations, BLANs make it possible to approximate the power of special-purpose supercomputers for a small fraction of the cost. Needing high throughput and low latency between communicating processes, parallel–distributed computations of hundreds of processes can tax conventional LANs. The use of networks of workstations (NOWs) to perform tasks formerly delegated to expensive supercomputers demands the power of a BLAN. Experimental NOWs based on ATM BLANs are proving the effectiveness of this approach.

45.7 Conclusion

Although still in its infancy the BLAN is gaining wide acceptance for upgrading legacy LANs and enabling new applications that require a range of QOS. A wide variety of technologies is used to implement the BLAN, including high-speed integrated circuits, electronic and optical links, and high-level protocol software. By providing guaranteed bandwidth on demand, BLANs enable the integration of different classes of traffic in a single network. This in turn makes possible a host of new applications, such as video servers, image servers, hypermedia browsing, and parallel—distributed computing.

ATM has taken the early lead in the race to become the leading BLAN technology. Strong contenders, however, are fast ethernet, isoethernet, and 100BaseVG-AnyLAN. Moreover, these different technologies are often mixed together in edge switches to realize hybridized BLANs.

Although much of the hardware technology behind the BLAN is relatively mature, the complex software-based protocols needed to support broadband applications are only partially complete. Advances in signalling, QOS management, resource reservation, and internetworking must continue before BLANs become permanently established.

Defining Terms

Asynchronous traffic: Traffic with no particular timing relationship between successive transmission units.
Asynchronous transfer mode: A broadband network technology based on fast cell switching.
Broadband integrated services digital network: The cell-relay—based technology upon which the next-generation telecommunications infrastructure is to be based.
Broadband local area network: A high-speed LAN that carries integrated traffic.
Collision domains: A group of nodes that share a single broadcast medium.
Connection-oriented protocol: A protocol that must establish a logical association between two communicating peers before exchanging information and then release the association after the exchange is complete.
Desktop 25: A low-speed (25-Mb/s) link protocol for use in desktop ATM networks.
Isochronous traffic: Constant-rate traffic in which successive transmisson units have a strictly fixed timing relationship.
Jitter: The amount of variation in the time between successive arrivals of transmission units of a data stream.
Latency: The delay from transmission of a packet by the source until its reception by the destination.
Management information base: The collection of a network's parameters monitored and controlled by a network management system.

Media-access control protocol: The link-layer protocol in a shared-media LAN that guarantees fair and orderly access to the media by nodes with packets to transmit.
Mesh topology: An arrangement of packet switches connected by dedicated links.
Multicasting: Transmitting a packet within a network so that it will be received by a specified group of nodes.
Quality of service: The bandwidth, latency, and error characteristics provided by a network to a group of users.
Shared-media topology: Refers to a network in which any transmission is heard by all nodes.
Synchronous traffic: Variable-rate traffic in which successive transmission must be delivered within a specific deadline.
Transparent Asynchronous Transceiver Interface: A 100-Mb/s link protocol originally developed for FDDI and later adopted for use in ATM networks.
Virtual path/circuit identifier: A pair of designators in an ATM cell which uniquely identifies within a switch the logical connection to which the cell belongs.

References

Collivignarelli, M., Daniele, A., De Nicola, P., Licciardi, L., Turolla, M., and Zappalorto, A. 1994. A complete set of VLSI circuits for ATM switching. In *Proc. IEEE GLOBECOM'94*, pp. 134–138. Nov., Inst. of Electrical and Electronics Engineers, New York.

Denzel, W.E., Engbersen, A.P.J., and Iliadis, I. 1995. A flexible shared-buffer switch for ATM at Gb/s rates. *Comput. Networks ISDN Syst.*, 27(4):611–634.

Hluchyj, M.G. and Karol, M.J. 1988. Queueing in high-performance packet switching. *IEEE J. Select. Areas Commun.*, 6(9):1587–1597.

Stevenson, D., Hillery, N., and Byrd, G. 1995. Secure communications in ATM networks. *Commun. ACM*, 38(2):45–52.

Tobagi, F.A. 1990. Fast packet switch architectures for broadband integrated services digital networks. *Proc. IEEE*, 78(1):133–167.

Turner, J.S. 1987. Design of a broadcast packet switching network. *IEEE Trans. Commun.*, 36(6):734–743.

Further Information

Two leading textbooks on ATM networks and Fast Ethernet are

- M. Händel, N. Huber, and S. Schröder, *ATM Networks: Concepts, Protocols, Applications*, 2nd ed., Addison–Wesley, Reading, MA, 1994.
- W. Johnson, *Fast Ethernet: Dawn of a New Network*, Prentice–Hall, Englewood Cliffs, NJ, 1995.

Several conferences and workshops cover topics in BLANs, including

- IEEE INFOCOM (Conference on Computer Communications)
- ACM SIGCOMM (Conference on Applications, Technologies, Architectures, and Protocols for Computer Communications)
- IEEE Conference on Local Computer Networks
- IEEE LAN/MAN Workshop

The important journals that cover BLANs are

- *IEEE/ACM Transactions on Networking*
- *IEEE Network Magazine*
- *Computer Communications Review*

- *Computer Networks and ISDN Systems*
- *Internetworking Research and Experience*

Three prominent organizations that create and maintain BLAN standards are

- ATM Forum
- IETF
- IEEE Project 802

46
Multiple Access Methods for Communications Networks

Izhak Rubin
University of California, Los Angeles

46.1 Introduction 622
46.2 Features of Medium Access Control Systems 623
    Broadcast (Logical Bus) Topology • Mesh (Switching) Topology • Hybrid Logical Bus and Buffered Switching Topologies • Layered Protocols and the Medium Access Control Sublayer
46.3 Categorization of Medium Access Control Procedures 630
    Medium Access Control Dimensions • Medium Access Control Categories
46.4 Polling-Based Multiple Access Networks 634
    Token-Ring Local Area Network • The Fiber Data Distribution Interface Network • Implicit Polling Schemes • Positional-Priority and Collision Avoidance Schemes • Probing Schemes
46.5 Random-Access Protocols 639
    ALOHA Multiple Access • Carrier Sense Multiple-Access • CSMA/CD Local Area Networks
46.6 Multiple-Access Schemes for Wireless Networks 642
46.7 Multiple Access Methods for Spatial-Reuse Ultra-High-Speed Optical Communications Networks 644

46.1 Introduction

Modern computer communications networks, particularly local area networks (LANs) and metropolitan area networks (MANs), employ multiple access communications methods to share their communications resources. A multiple-access communications channel is a network system whose communications media are shared among distributed stations (terminals, computers, users). The stations are distributed in the sense that there exists no relatively low cost and low-delay mechanism for a single controlling station to gain access to the status of all stations. If such a mechanism exists, the resulting sharing mechanism is identified as a **multiplexing system**.

The procedure used to share a multiple access communications medium is the **multiple access algorithm**. The latter provides for the control, coordination, and supervision of the sharing of the system's communications resources among the distributed stations, which transport information across the underlying multiple access communications network system.

In the following, we present a categorization of the various medium access control (MAC) methods employed for the sharing of multiple access communications channel systems. We demonstrate these schemes by considering applications to many different classes of computer communication networks, including wired and wireless local and metropolitan area networks, satellite communications networks, and local area optical communications networks.

46.2 Features of Medium Access Control Systems

Typically employed topologies for shared-medium communications networks are shown in Fig. 46.1, and are characterized as follows [Rubin and Baker, 1990].

A *star* topology, under which each station can directly access a single central node, is shown in Fig. 46.1(a). A switching star network results when the star node provides store and forward buffering and switching functions, whereas a broadcast star network involves the employment of the star node as an unbuffered repeater, which reflects all of the incoming signals into all outgoing links. A wired star configuration is shown in Fig. 46.1(a1). In Fig. 46.1(a2), we show a wireless cell which employs radio channels for the mobile terminals to communicate with a central base station. The terminals use a multiple access algorithm to gain access to the shared (mobile to base-station,

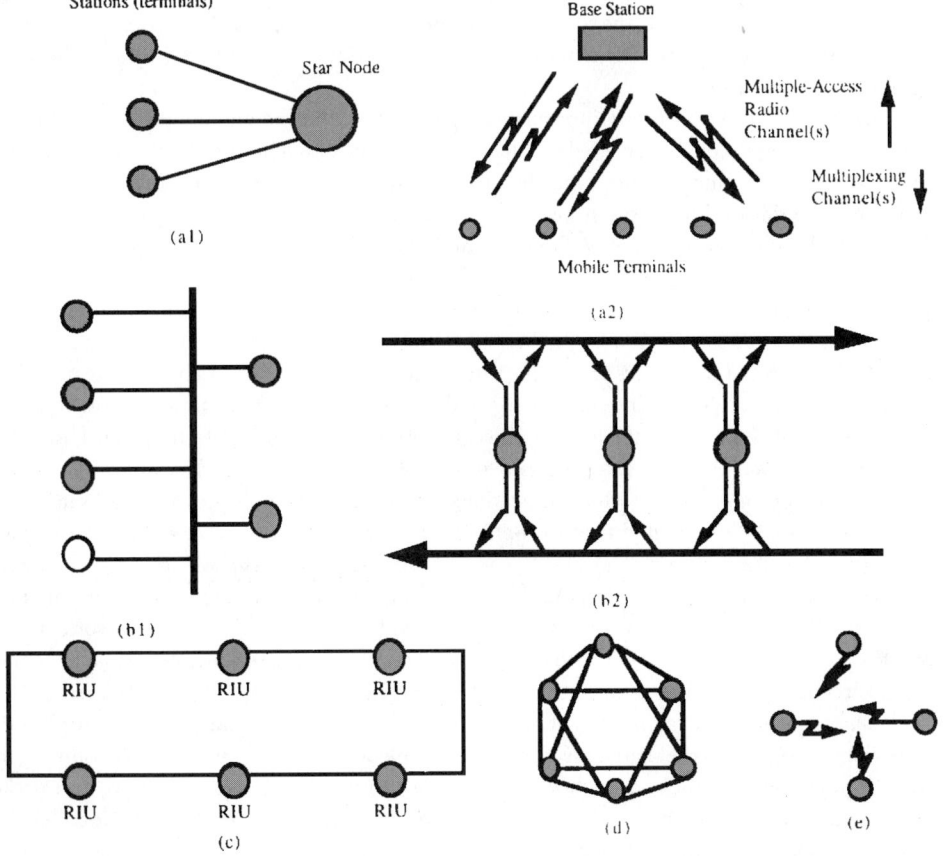

FIGURE 46.1 Multiple-access network topologies: (a) star, (b) bus, (c) ring, (d) mesh, and (e) broadcast radio net.

reverse access) radio channel(s), whereas the base-station employs a multiplexing scheme for the transmission of its messages to the mobile terminals across the (base-station to terminals, forward access) communications channels.

A *bus* topology, under which stations are connected to a bus backbone channel, is shown in Fig. 46.1(b). Under a logical bus topology, each station's message transmissions are broadcast to all network stations. Under a physical bus implementation, the station is passively connected to the bus, so that a station's failure does not interfere with the operation of the bus network. The station is then able to sense and receive all transmissions that cross its interface with the bus; however, it is not able to strip information off the bus or to properly overwrite information passed along the bus; it can transmit its own messages across the bus when assigned to do so by the employed MAC protocol.

It is noted that when fiber links are used, due to the unidirectional nature of the channel, to achieve full station-to-station connectivity, a bus network implementation necessitates the use of two buses. As depicted in Fig. 46.1(b2), a common approach is to employ two separate counterdirectional buses. Stations can then access each fiber bus through the use of a proper read (sense) tap followed by a write tap.

A *ring* topology, under which the stations are connected by point-to-point links in, typically, a closed-loop topology is shown in Fig. 46.1(c). In a physical ring network implementation, each station connects to the ring through an active interface, so that transmissions across the ring pass through and are delayed in the register of the ring interface units (RIUs) they traverse. To increase network reliability, such implementations include special circuits for ensuring a rapid elimination of a failed RIU, to prevent such a failure from leading to a prolonged overall network failure. Because of the use of an active interface, the station is now able to strip characters or messages it receives off the medium, as well as to overwrite onto symbols and messages transmitted across the medium, when the latter pass its interface.

An active interface into the medium further enables the station to amplify the signals it passes along, thus leading to considerable reduction of the insertion loss. This is of particular importance in interfacing a fiber optic medium, whereby a passive interface causes a distinct insertion loss, thus leading to a significant limitation in the number of stations that can be passively connected to a fiber bus, without the incorporation of optical amplifiers.

For fiber optic links, electrical amplification and MAC processing at the active interface involve double conversion: optical-to-electronic of the received signal and electronic-to-optical of the amplified transmitted signal. As a result, the station's access burst rate needs to be selected so that it is compatible with the electronics processing rate at the station interface, so that no rate mismatch exists at the electronics/optics interfaces between the potentially very high-transmission burst rate across the optical channel and the limited processing rate capacity of the electronically based very large-scale integrated (VLSI) access processor at the station's medium interface unit.

A *mesh* topology is one under which the stations are connected by point-to-point links in a more diverse and spatially distributed (mesh) topology [Fig. 46.1(d)]. To traverse a mesh topology, switching nodes are required. Through the use of cross-connect switches, multiple embedded multiple-access subnetworks are constructed. These subnetworks can be dynamically reconfigured (by adjusting the cross-connect matrices at the switches) to adapt to variations in network loading conditions and in interconnection patterns, and to allow the support of private virtual networks. Such an architecture, as demonstrated by the SMARTNet optical network, is presented in Sec. 46.7.

A *broadcast* multiaccess radio net, as shown in Fig. 46.1(e) provides, by the physical features of the radio propagation across the shared medium, a direct link between any pair of nodes (assuming the terrain topography to induce no blockages in the node-to-node line-of-sight clearances). A packet transmitted by a single station (node) will be received directly by all other stations. Note that this broadcast property is physically achieved also by the wireline two-way bus network systems shown in Figs. 46.1(b1) and 46.1(b2). In turn, the other network topologies shown in Fig. 46.1 require signal repeating or retransmission protocols to be used to endow them with the broadcast property.

In considering general high-speed local and metropolitan area networks, the following frame distribution topologies can be distinguished for basic implementations: broadcast (logical bus) topology and mesh (switching) topology.

Broadcast (Logical Bus) Topology

Under such a distribution method, station messages are routed through the use of a broadcasting method. Since each message frame (the MAC level protocol data unit) contains addressing information, it is copied automatically from the medium by the intended destination station or stations, so that MAC routing is automatically achieved. Typically bus, ring, or broadcast star topologies are used to simplify the characterization of the broadcasting path.

The communications link is set up as a bus (for a physical bus implementation), or as a point-to-point link (for a physical ring implementation). The corresponding station interface unit (SIU) acts as a passive or active MAC repeater. A passive MAC node does not interfere with on-going transmissions along the bus, while being able to copy the messages transmitted across the bus system.

An active MAC station interface unit operates in one out of two possible modes: (1) *repeat mode*, whereby it performs as a repeater, serving to repeat the frame it receives and (2) *nonrepeat mode*, under which the SIU is not repeating the information it receives from the medium. In the latter case, under a logical bus configuration, the SIU is also typically set to be in a *stripping mode*, stripping from the medium the information it receives. During this time, the SIU is able to transmit messages across the medium, provided it gains permission to do so from the underlying MAC protocol. Note that if the SIU is provided with a store-and-forward capability, it can store all of the information it receives while being in a nonrepeat mode, and retransmit this information, if so desired, at a subsequent time.

A logical-bus topology can also be associated with an active interface of the station onto the fiber bus/buses. This is, for example, the case for the Institute of Electrical and Electronics Engineers (IEEE) 802.6 distributed queue dual bus (DQDB) implementation [Rubin and Baker, 1990]. Under DQDB, the station can overwrite each passing bus bit through optical to electrical conversion, an electrical OR write operation, and an electrical to optical reconversion. The active interface of the station can provide it with the capacity to also strip bits, and thus message frames, off the bus.

MAC procedures for logical bus configurations can also be characterized by the following features relating to the method used for removing message frames and the constraints imposed upon the number of simultaneous transmissions carried along the medium. As to the latter feature, we differentiate between the following implementations.

Frame Transmission Initiations in Relation to the Number of Simultaneous Transmissions Along the Logical Bus

Single Message Frame Transmission Across the Medium. A single MAC-frame transmission is permitted across the medium at any time instant; thus, no more than one station can initiate a transmission onto the medium at any given time instant, and no other station can initiate a transmission until this later transmission is removed from the medium.

This is the technique employed by the IEEE 802.5 token ring LAN under a late token release mode, whereby a station holding the token does not release it until it receives its own frame (following the latter's full circulation around the ring). This simplifies the operation of the protocol and provides the station with the ability to review its acknowledged message frame prior to releasing the token, thus enabling it to immediately retransmit its MAC message frame if the latter is determined to have not been properly received by the destination.

The latter token ring LAN uses twisted-pair or coaxial media at transmission rates of 4–16 Mb/s. At 10 Mb/s, the transmission time of a 1000-b frame is equal to 100 μs, which is longer than the typical propagation time across the medium when the overall LAN length is shorter than 20 km, considering a propagation rate of about 5 μs/km.

In turn, when considering a 100-Mb/s LAN or MAN fiber-optic based system, such as the fiber distribution data interface (FDDI) token ring or the DQDB reservation bus systems, which can span longer distances of around 100 km, the corresponding frame transmission time and networkwide propagation delay are equal to 10 and 500 μs, respectively. For a corresponding MAN system, which operates at a channel rate of 1 Gb/s, the propagation delay of 500 μs is much larger than the frame transmission time of 1 μs. Thus, under such high-transmission rate conditions, each message transmission occupies only a small physical length of the logical bus network medium. Therefore, it is not efficient, from message delay and channel bandwidth MAC utilization considerations, to provide for only a single transmission at a time across the logical bus medium. The following mode of operation is thus preferred. [The following mode, single transmission initiation, can also be used for the token-ring network under the early token release option, which can lead to performance improvements at the higher data rate levels.]

Multiple Simultaneous Frame Transmissions Along the Logical Bus Medium. While permitting multiple frame transmissions across the medium, we can consider two different procedures as it relates to whether a single or multiple simultaneous message frame transmission initiations are allowed.

A *single transmission initiation* is allowed at any instant of time. When a fiber optic-based token ring MAC scheme, such as FDDI, is used, the station releases the token immediately following the transmission of the station's message, rather than waiting to fully receive its own message prior to releasing the token. In this manner, multiple simultaneous transmissions can take place across the bus, allowing for a better utilization of the bus spatial-bandwidth resources.

A slotted access scheme is often used for logical bus linear topologies. Under such a scheme, a bus controller is responsible for generating successive time slots within recurring time cycles. A slot propagates along the unidirectional fiber bus as an idle slot until captured by a busy station; the slot is then designated as busy and is used for carrying the inserted message segment. A busy station senses the medium and captures an available idle slot, which it then uses to transmit its own message segment. For fairness reasons, a busy station may be allowed to transmit only a single segment (or a limited maximum number of segments) during each time cycle. In this manner, such a MAC scheme strives to schedule busy station accesses onto the medium such that one will follow the other as soon as possible, so that a message train is generated efficiently utilizing the medium bandwidth and space (length) dimensions. However, note that no more than a single station is allowed at any time to initiate transmissions onto the unidirectional medium; transmission initiations occur in an order that matches the positional location of the stations along the fiber. As a result, the throughput capacity of such networks is limited by the shared medium's data rate.

Multiple simultaneous frame transmission initiations are included in the MAC protocol. In using such a MAC scheduling feature, multiple stations are permitted to initiate frame transmissions at the same time, accessing the medium at sufficiently distant physical locations, so that multiple transmissions can propagate simultaneously in time along the space dimension of the shared medium. The underlying MAC algorithm needs to ensure that these simultaneous transmissions do not cause any frame overlaps (collisions).

In a ring system, such a procedure can involve the proper use of multiple tokens or ring buffer insertions at each station's interface. In a logical bus operation, when a *slotted* channel structure is employed, such an operation can be implemented through the designation of slots for use by a proper station or group of stations, in accordance with various system requests and interconnectivity conditions. For example, in the DQDB MAN, stations indicate their requests for channel slots, so that propagating idle slots along each fiber bus can be identified with a proper station to which they are oriented; in this manner, multiple simultaneous frame transmission initiations and on-going message propagations can take place. Similarly, when time-division multiple access (TDMA) and demand-assigned circuit or packet-switched TDMA MAC schemes are used.

Further enhancements in bandwidth and space utilization can be achieved by incorporating in the MAC scheme an appropriate message frame removal method, as indicated in the following.

Frame Removal Method

When considering a logical bus network with active SIUs, the removal of frames from the logical bus system can be in accordance with the following methods.

Source Removal. Considering loop topologies, under a source removal method the source station is responsible for the removal of its own transmitted frames. This is, as previously noted, the scheme employed by the IEEE 802.5 and fiber-based FDDI token ring systems. Such a MAC feature permits the source station (following the transmission of a frame) to receive an immediate acknowledgment from its destination station (which is appended as a frame trailer), or to identify immediate no-response when the destination station is not operative.

Destination Removal. Under such a scheme, a station upon identifying a passing frame destined for itself removes this frame from the medium. Under such a removal policy, a frame is not broadcast across the overall length of medium, but just occupies a space segment of the medium that spans the distance between the source and destination stations. Such a method can lead to a more complex MAC protocol and management scheme. Improvement in delay and throughput levels can, however, be realized through spatial reuse, particularly when a noticeable fraction of the system traffic flows among stations that are closely located with respect to each other's position across the network medium.

To apply such a scheme to a token ring system, multiple tokens are allowed and are properly distributed across the ring. When two closely located stations are communicating, other tokens can be used by other stations, located away from the occupied segment(s) to initiate their own nonoverlapping communications paths. Concurrent transmissions are also attained by the use of a slotted access scheme or through the use of a buffer insertion ring architecture, as illustrated later.

When such a scheme is applied to a bus system with actively connected SIUs, which is controlled by a *slotted* access scheme, the destination station is responsible for stripping the information contained in the slot destined to itself from the bus, and for releasing the corresponding slot for potential use by a downstream station.

Removal by Supervisory Nodes. It can be beneficial to employ special supervisory nodes, located across the medium, to remove frames from the medium. In this manner, the frame destination removal process can be implemented in only supervisory nodes, relieving regular nodes of this task.

Using such frame removal supervisor nodes, the system interconnectivity patterns can be divided (statically or dynamically) into a number of modes. Under an extensively divisive mode, the supervisory stations allow only communications between stations that are located between two neighboring supervisory stations to take place. Under a less divisive connectivity mode, the system is divided into longer disjoint communications segments. Under a full broadcast mode, each frame is broadcast to all network stations. Depending on the network traffic pattern, time cycles can be defined such that a specific mode of operation is invoked during each time cycle.

If such supervisory stations (or any station) are operated as store-and-forward buffered switching units, then clearly the distribution processes across the medium segments can be isolated and intersegment message frames would then be delivered across the logical bus system in a *multiple-hop* (multiretransmission) fashion. This is the role played by buffer-insertion ring architectures for the specialized ring topology and, in general, by the mesh switching architectures discussed in the following.

Mesh (Switching) Topology

Under a mesh (switching) topology, the network topology can be configured as an arbitrary mesh graph. The network nodes provide buffering and switching services. Communications channels are set up as point-to-point links interconnecting the network nodes. Messages are routed through the network, through the use of specially developed routing algorithms.

Depending on whether the switching node performs store-and-forward or cross-connect switching functions, we distinguish between the following architectures:

Store-and-Forward Mesh Switching Architecture

Under a store-and-forward mesh switching architecture, the nodal switches operate in a store-and-forward fashion as packet switches. Thus, each packet received at an incoming port of the switch is examined, and based on its destination address it is queued, switched and forwarded on the appropriate outgoing port. Each point-to-point channel in such a mesh network needs to be efficiently shared among the multitude of messages and connections that are scheduled to traverse it. A statistical multiplexing scheme is selected (and implemented at each switch output module) for dynamically sharing the internodal links.

Cross-Connect Mesh Switching Architecture

In this case, each switching node operates as a cross-connect switch. The latter serves as a circuit (or virtual circuit) switch, which transfers the messages belonging to as established circuit (or virtual circuit) from their incoming line and time slot(s) [or logical connection groups for cross-connect virtual path switches used by asynchronous transfer mode (ATM) networks] to their outgoing line and time slot(s) (or logical group connections). The cross-connect matrix used by the node to implement this switching function is either preset and kept constant, or it can be readjusted periodically, as traffic characteristics vary, or even dynamically in response to the setup and establishment of end-to-end connections. (See also Sec. 46.7 for such an optical network identified as SMARTNet.)

Hybrid Mesh Switching Architecture

The nodal switch can integrate fixed assigned and statistical operations in multiplexing traffic across the mesh topology links. For example, in supporting an integrated circuit-switched and packet-switched implementation, a time cycle (time frame) is typically defined, during which a number of slots are allocated to the supported circuits that use this channel, while the remainder of the cycle slots are allocated for the transmission of the packets waiting in the buffers feeding this channel. Frequently, priority-based disciplines must be employed in statistically multiplexing the buffered packets across the packet-switched portion of the shared link, so that packets belonging to different service classes can be guaranteed their desired quality of service, as it relates to their characteristic delay and throughput requirements. For example, voice packets must subscribe to strict end-to-end time delay and delay jitter limits, whereas video packets induce high throughput support requirements. Asynchronous transfer mode network structures, which employ combined virtual path and virtual circuit switches, serve as another example.

Hybrid Logical Bus and Buffered Switching Topologies

High-speed multigigabit communications networking architectures that cover wider areas often need to combine broadcast (logical bus) and mesh (switching) architectures, to yield an efficient, reliable and responsive integrated-services fiber-based network.

In considering logical-bus topologies with active SIUs, we previously noted that certain nodes can be designated to act as store-and-forward processors and to also serve as frame removal supervisory nodes. Thus, such nodes actually operate as MAC bridge gateways. These gateways serve to isolate the segments interconnecting them, each segment operating as an independent logical bus network. These gateways act to filter MAC bridge interconnections between these segments. The individual segments can operate efficiently when they serve a local community of stations, noting that each segment spans a shorter distance, thus reducing the effect of the end-to-end propagation delay on message performance. To ensure their timely distribution, it can be effective to grant access priority, in each segment, to the intersegment packets that traverse this segment.

Layered Protocols and the Medium Access Control Sublayer

In relation to the open systems interconnect (OSI) reference model, the data link layer of multiple access (such as typical LAN and MAN) networks is subdivided into the MAC lower sublayer and the logical link control (LLC) upper sublayer. Services provided by the MAC layer allow the local protocol entity to exchange MAC message frames (which are the MAC sublayer protocol data units) with remote MAC entities.

In considering typical LAN and MAN systems, we note that the MAC sublayer provides the following services:

1. MAC sublayer services are provided to the higher layer (such as the LLC sublayer for LAN and MAN systems). LLC service data units (SDUs) are submitted to the MAC sublayer for transmission through proper multiple-access medium sharing. In turn, MAC protocol data units (PDUs) received from the medium and destined to the LLC are transferred to the LLC sublayer as proper LLC-SDUs. The underlying LLC-SDUs include source and destination addresses, the data itself and service class and quality of service parameters.
2. MAC sublayer services are similarly provided to directly connected isochronous and non-isochronous (connection-oriented circuit and packet-switched) channel users (CUs) allowing a local CU entity to exchange CU-data units with peer CU entities. These are connection-oriented services, whereby after an initial connection set-up, the channel user is able to directly access the communications channel through the proper mediation of the MAC sublayer. The CU generates and receives its data units through the MAC sublayer over an existing connection, on an isochronous or nonisochronous basis. Such CU-SDUs contain the data itself and possibly service quality parameters; no addressing information is needed, since an established connection is involved.

Corresponding services are provided for connectionless flows.

3. MAC sublayer services are provided to the local MAC station management entity, via the local MAC layer management interface.

Example of services include: the opening and closing of an isochronous or nonisochronous connection; its profile, features, and the physical medium it should be transmitted on; and the establishment and disestablishment of the binding between the channel user and the connection endpoint identifier.

The MAC sublayer requires services from the physical layer that provide for the physical transmission and reception of information bits. Thus, in submitting information (MAC frames) to the physical layer, the MAC sublayer implements the medium access control algorithm, which provides its clients (the LLC or other higher layer messages) access to the shared medium. In receiving information from the physical layer, the MAC sublayer uses its implemented access control algorithm to select the MAC frames destined to itself and then to provide them to the higher layer protocol entities. MAC layer addresses are used by the MAC layer entity to identify the destination(s) of a MAC frame.

In Fig. 46.2, we show the associated layers as implemented by various local and metropolitan area networks, as they relate to the OSI data link and physical layers.

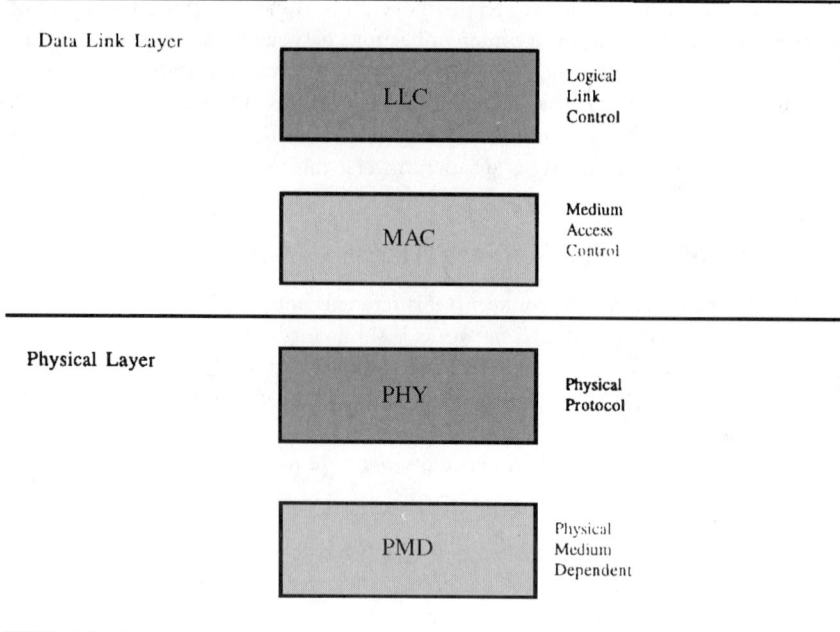

FIGURE 46.2 The MAC sublayer and its interfacing layers across the OSI data link and physical layers.

46.3 Categorization of Medium Access Control Procedures

A multitude of access control procedures are used by stations to share a multiaccess communications channel or network system. We provide in this section a classification of MAC schemes.

Medium Access Control Dimensions

The multiple-access communications medium resource can be shared among the network stations through the allocation of a number of key resource dimensions. The assignment space considered is the (T, F, C, S) = {time, frequency, code, space} space. The allocation of access to the shared medium is provided to stations' messages and sessions/calls through the assignment of slots, segments, and cells of the time, frequency, code, and space dimensions.

1. *Time-division scheduling.* Stations share a prescribed channel frequency band by having their transmissions scheduled to take place at different segments of time. Typically, only a single message can be transmitted successfully across the designated channel at any instant of time.
2. *Frequency- and wavelength-division allocation.* The bandwidth of the communications channel is divided into multiple disjoint frequency bands (or wavelength channels for an optical channel), so that a station can be allocated a group, consisting of one or more frequency/wavelength bands, for use in accessing the medium. Multiple time-simultaneous transmissions can take place across the channel, whereby each message transmission occupies a distinct frequency/wavelength channel.
3. *Code-division multiple access (CDMA).* Each station's message is properly encoded so that multiple messages can be transmitted simultaneously in time in a successful manner using a single-band of the shared communications channel, so that each message transmission is correctly received by its destined station. Typically, orthogonal (or nearly orthogonal) pseudonoise sequences are used to randomly spread segments of a message over a wide

frequency band (frequency hopping method), or to time correlate the message bit stream (direct sequencing method). A message can be encoded by an address-based key sequence that is associated with the identity of the source, the destination, the call/connection, or their proper combination. A wider frequency band is occupied by such code divided signals. In return, a common frequency band can be used by all network stations to successfully carry, simultaneously in time, multiple message transmissions.

4. *Space-division multiple access.* Communications channels are shared along their space dimension. For example, this involves the sharing of groups of physically distinct links, or multiple space segments located across a single high-speed logical-bus network.

Considering the structure of the access control procedure from the dimensional allocation point of view, note the similarity between the space-division and frequency/wavelength-division methods in that they induce a *channel selection* algorithm that provides an allocation of distinct frequency/wavelength or physical channels to a user or to a group of users. In turn, once a user has been assigned such a frequency/wavelength band, the sharing of this band can be controlled in accordance with a time-division and/or a code-division MAC method. Thus, under a combined use of these dimensions, the medium access control algorithm serves to schedule the transmission of a message by an active station by specifying the selected channel (in a frequency/wavelength-division or space-division manner) and subsequently specify the time slot(s) and/or multiple-access codes to be used, in accordance with the employed time-division and/or code-division methods.

Medium Access Control Categories

In Fig. 46.3, we show our categorization of medium access control procedures over the previously defined (T, F, C, S) assignment space. Three classes of access control policies are identified: **fixed assignment** (FA), **demand assignment** (DA), and **random access** (RA). Within each class, we identify the signalling (SIG) component and the information transmission (IT) method used. Note that within each class, circuit-switching as well as packet-switching mechanisms (under connectionless and/or connection-oriented modes) can be used, in isolation or in an integrated fashion.

Under a *fixed-assignment* scheme, a station is dedicated, over the (T, F, C, S) space, a communications channel resource, which it can permanently use for accessing the channel. Corresponding access control procedures thus include **time-division multiple access (TDMA), frequency-division**

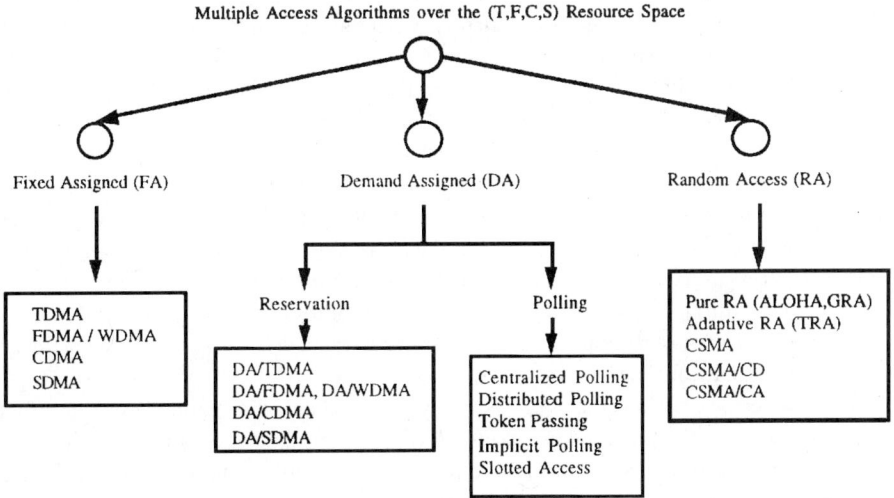

FIGURE 46.3 Classes of medium access control schemes with illustrative access procedures noted for each class.

multiple access (FDMA), **wavelength-division multiple access (WDMA)**, **code-division multiple access (CDMA)**, and **space-division multiple access (SDMA)** schemes. A corresponding signalling procedure is implemented to ensure that station transmission boundaries are well recognized by the participating stations, along the respective (T, F, C, S) dimensions.

Medium resources (along each dimension) can be assigned on a fixed basis to specified sessions/connections, to a station or to a group of stations. When allocated to a connection/call, such schemes provide the basis for the establishment of isochronous circuits and can lead to efficient sharing of the medium-bandwidth resources for the support of many voice, video, high-speed data, and real-time session connections, when a steady traffic stream of information is generated. Effective channel sharing can also result when such channel resources are dedicated for the exclusive use by a station or a group of stations, provided the latter generate steady traffic streams which can efficiently utilize the dedicated resources at an acceptable quality of service level.

For example, under a TDMA procedure, a station is allocated a fixed number of slots during each frame. Under a packet-switched TDMA (PS-TDMA) operation, the station is accessing the shared medium by multiplexing its packets into the allocated slots. In turn, under a circuit-switched TDMA (CS-TDMA) operation, the station uses its allocated time slots to establish circuits. A circuit can consist of a fixed number of the station's time slots allocated in each time frame (e.g., a single slot per frame). Connections are assigned by the stations available circuits at the requested rate. The messages generated by a connection are then transmitted in the time slots belonging to the circuit allocated to this connection.

Under a *random-access* scheme, stations contend for accessing the communications channel in accordance with an algorithm that can lead to time-simultaneous (overlapping, colliding) transmissions by several stations across the same frequency band, causing at times the retransmission of certain packets.

Under fixed random-access schemes, a station transmits its message frame across the channel at a random time, in a slotted or unslotted fashion, without coordinating its access with other stations. A station that detects (through channel sensing, or through nonreceipt of a positive acknowledgment from its destination) its transmission to collide with another transmission, will retransmit its message after a properly computed *random* retransmission delay (whose duration may depend on the estimated channel state of congestion). Under an adaptive channel-sensing random-access algorithm, a ready station first senses the channel to gain certain channel state information, and then uses this information to schedule its message for transmission in accordance with the underlying protocol, again without undertaking full access coordination with all other network stations. Various fixed (such as ALOHA [Abramson, 1973] and group random access [Rubin, 1978]) and adaptive channel-sensing random-access algorithms such as carrier sense multiple access (CSMA) [Kleinrock and Tobagi, 1975], carrier sense multiple access/collision detection (CSMA/CD), dynamic group random access (DGRA) [Rubin, 1983], tree-random-access collision resolution-based [Capetanakis, 1979] and others [Sachs, 1988; Maxemchuk, 1982; Stallings, 1993] can be invoked.

In general, random-access schemes are not well suited as information transmission methods for governing the access of information messages onto a very high-speed fiber communications channel, unless provisions are made to considerably limit the extent of message overlaps due to collisions, or to limit the use of this access technique to certain low-throughput messaging classes.

A random-access procedure is typically employed for supporting packet-switching services. At higher normalized throughput levels, high packet delay variances can occur. A proper flow control regulation mechanism must be employed to guarantee maximum delay limits for packets associated with isochronous and real-time based services.

Nonchannel-sensing random-access procedures yield low channel utilization levels. In turn, the efficiency of a channel-sensing random-access procedure critically depends upon the timeliness of the sensed channel occupancy information, and thus upon the overhead durations associated with sensing the channel. For high-speed logical-bus systems, when channel-sensing type random-access schemes are used, performance degradation is caused by the very high ratio of the channel propagation delay to the packet transmission time. An active station is required to obtain the channel state prior to initiating a transmission. However, since a station's transmission must propagate across the total logical-bus medium before its existence can be sensed by all network stations, and since stations make uncoordinated access decisions under a random-access scheme, these decisions can be based on stale channel state information; as a result, this can lead to the occurrence of excessive message collisions.

The degrading effects of the propagation delay to message transmission time ratio can be somewhat reduced by the following approaches.

Limit the overall propagation delay by dividing the overall network into multiple tiers or segments, as illustrated by the double-tier LAN and MAN architecture presented in Rubin and Tsai [1987]. The network stations are divided into groups. A shared backbone medium can be used to provide interconnection between groups (on a broadcast global basis using repeaters as gateways, or on a store-and-forward bridged/routing basis), using a noncontention- (nonrandom-access) based MAC algorithm. A random-access MAC strategy can be used for controlling the access of messages across the shared group (segment) medium, which now spans a much shorter distance and thus involves considerably reduced propagation delays. Similarly, such a network configuration is realized by interconnecting LAN (such as ethernet) segments by routers, bridges, or switches.

While the propagation delay depends on the conducting medium and is independent of the transmission speed, the packet transmission time is reduced as the transmission speed is increased. The MAC message frame transmission time can be increased by using FDMA/WDMA or SDMA methods to replace the single-band high-bandwidth medium with a group of shared multiple channels, each of lower bandwidth [Chlamtac and Ganz, 1988]. Clearly, this will provide an improvement only if the underlying services can be efficiently supported over such a reduced bandwidth channel when a random-access (or other MAC) scheme is used.

Random-access schemes can serve, as part of other MAC architectures, as effective procedures for transmitting signalling packets for demand-assigned schemes. Random-access schemes can also be used in conjunction with CDMA and spread spectrum multiple access (SSMA) procedures, over high-speed channels.

Random-access protocols are further presented and discussed in Sec. 46.5.

Under a *demand-assigned* scheme, a *signalling* procedure is implemented to allow certain network entities to be informed about the transmission and networking needs and demands of the network stations. Once these network entities are informed, a centralized or distributed algorithm is used to allocate to the demanding stations communications resource segments over the (T, F, C, S) assignment space.

The specific methods used in establishing the signalling channel and in controlling the access of individual demanding stations onto this channel characterize the features of the established demand-assignment scheme.

The signalling channel can be established as an out-of-band or in-band fixed channel to which communications resources are dedicated, as is the case for many DA/TDMA, DA/FDMA, or DA/WDMA schemes. In turn, a more dynamic and adaptive procedure can be used to announce the establishment of a signalling channel at certain appropriate times, as is the case for the many polling, implicit polling, slotted, and packet-train-type access methodologies.

The schemes used to provide for the access of stations onto the signalling channel, are divided by us into two categories: **polling** and **reservation** procedures.

Under a reservation scheme, it is up to the individual station to generate a reservation packet and transmit it across the signalling (reservation) channel to inform the network about its needs for communications resources. The station can identify its requirements as they occur or in advance, in accordance with the type of service required (isochronous vs asynchronous, for example) and the precedence level of its messages and sessions/connections. Reservation and assignment messages are often transmitted over an in-band or out-of-band multiple-access channel, so that a MAC procedure must be implemented to control the access of these MAC-signalling messages.

Under a polling procedure, it is the responsibility of the network system to query the network stations so that it can find out their transmission needs, currently or in advance. Polling and polling-response messages can be transmitted over an in-band or out-of-band multiple-access channel, so that a MAC procedure must be implemented to control the access of these MAC-signalling messages.

Under an implicit polling procedure, stations are granted access into the medium (one at a time, or in a concurrent noninterfering fashion) when certain network state conditions occur; such conditions can be deduced directly by each station without the need to require the station to capture an explicit polling message prior to access initiation. Such schemes are illustrated by slotted bus systems whereby the arrival of an idle slot at a station serves as an implicit polling message, and by buffer-insertion schemes whereby a station is permitted to access the ring if its ring buffer is empty. Additional constraints are typically imposed on the station in utilizing these implicit polling messages for access to the channel to ensure a fair allocation of communications network bandwidth resources. Such constraints often guarantee that a station can transmit a quota of packets within a properly defined cycle.

A multitude of access policies can be adopted in implementing integrated-services polling and reservation algorithms. Note that polling techniques are used in implementing the IEEE 802.4 token bus and IEEE 802.5 token ring LANs, the FDDI fiber LAN, and many other fiber bus and ring network systems. Reservation techniques are employed by the IEEE 802.6 DQDB metropolitan area network (whereby a slotted positional priority implicit polling access procedure is used to provide access to reservation bits [Newman and Hullett, 1986]), by demand-assigned TDMA, FDMA, and WDMA systems, and many other multiple-access network systems. Integrated reservation and polling schemes are also used. See the following section for further discussion of polling and implicit polling methods.

46.4 Polling-Based Multiple Access Networks

In this section, we describe a number of multiple-access networks whose medium access control architectures are based on polling procedures.

Token-Ring Local Area Network

The token ring network is a local area network whose physical, MAC and link layer protocol structures are based on the IEEE 802.5 Standard. The ring interface units are connected by point-to-point directional links, which operate at data rates of 4 Mb/s or 16 Mb/s. A distributed polling mechanism, known as a **token passing** protocol is used as the medium access control scheme. A polling packet, known as a token, is used to poll the stations (through their RIUs). When a station receives the token (buffering it for a certain number of bit times to allow the station to identify the packet as a token message and to react to its reception), it either passes the token along to its downstream neighboring station (when it has no message waiting for transmission across the ring network), or it seizes the token. A captured token is thus removed from the ring, so that only a single station can hold a token at a time. The station holding the token is then allowed to transmit its ready messages. The dwell time of this station on the ring is limited through the setting of the station's token holding timer. When the latter expires, or when the station finishes transmitting its frames, whichever occurs first, the station is required to generate a new token and pass it to its downstream neighboring station. The transmitted frames are broadcasted to all stations on the ring: they fully circulate the ring and are removed from the ring by their source station. Such a procedure is also known as a source removal mechanism. The token-ring protocol permits a message priority based access control operation. For this purpose, the token contains a priority field. A token of a certain priority level can be captured by only those stations which wish to transmit across the medium a message of a priority level equal to or higher than the token's priority level. Furthermore, stations can make a reservation for the issue of a token of a higher priority level by setting a priority request tag in a reservation field contained in circulating frames (such tags can also be marked in the reservation field of a busy token, which is a token that has been captured but not yet removed from the ring and tagged as busy by the station which has seized it). Upon its release of a new idle token, the releasing station marks the new token at a priority levels which is the highest of all of those levels included in the recently received reservations. At a later time, when no station is interested in using a token of such a high-priority level, this station is responsible for downgrading the priority of the token. In setting priority-based timeout levels for the token holding timer, different timeout values can be selected for different priority levels.

The *throughput capacity* attainable by a polling scheme, including a token passing distributed polling mechanism such as the token ring LAN, is dependent on the *walk time* parameter. The system *walk time* is equal to the total time it takes for the polling message (the token) to circulate around the network (the ring) when no station is active. For a token ring network, the walk time is thus calculated as the sum of the transmission times, propagation delay, and ring buffer delays incurred by the token in circulating across an idle ring. Clearly the throughput inefficiency of the network operation is proportional to the ratio between the walk time and the overall time occupied by message transmissions during a single rotation of the token around the ring. The network's maximum achievable throughput level, also known as the network's throughput capacity, is denoted as $L(C)$ (b/s); the normalized throughput capacity index is set to $s(C) = L(C)/R$, where R (b/s) denotes the data rate of each network link, noting that $0 \leq s(C) \leq 1$. Thus, if $S(C) = 0.8$, the network permits a maximum throughput level that is equal to 80% of the link's data rate. To assess the network's throughput capacity level, we assume the network to be highly loaded so that all stations have frames ready for transmission across the ring. In this case, each cycle (token circulation around the ring) has an average length of $E(C) = NK(F/R) + W$, where N denotes the number of ring stations (RIUs), K denotes the maximum number of frames that a station is allowed to transmit during a single visit of the token, F is the frame's average length (in bits), so that F/R represents the average frame transmission time across a ring's link, whereas W denotes the average duration (in seconds) of the token's walk time. The network's normalized throughput capacity index is thus given by $s(C) = NK(F/R)/E(C)$. Clearly, higher throughput levels are attained as the walk time durations (W) are reduced. Note that $W = R(p)L + (T/R) + N(M/R)$, where $R(p)$ denotes the propagation delay across the medium (typically equal to 5 μs/km for wired links); L (km) is the distance spanned by the ring network so that $R(p)L$ represents the overall propagation delay around the ring; T is the token's length, so that T/R represents the token's transmission time; M denotes the number of bit time delays incurred at the RIU's interface buffer, so that $N(M/R)$ expresses the overall delay incurred by a frame in being delayed at each interface around the ring.

The delay-throughput performance behavior exhibited by polling systems, such as the distributed polling token-passing scheme of the token-ring LAN, follows the behavior of a single server queueing system in which the server dynamically moves from one station to the next, staying (for a limited time) only at those stations that have messages requiring service (i.e., transmission across the shared medium). This is a stable operation for which message delays increase with the overall loading on the network, as long as the latter is lower than the throughput capacity level $L(C)$. As the loading approaches the latter capacity level, message delays rapidly increase and buffer overflows can be incurred.

The Fiber Data Distribution Interface Network

The fiber data distribution interface network also employs a token passing access control algorithm. It also uses a ring topology. Two counter-rotating fiberoptic rings are used (one of which is in a standby mode) so that upon the failure of a fiber segment the other ring is employed to provide for a closed-loop topology. The communications link operates at a data rate of 100 Mb/s. The network can span a looped distance of up to 200 km.

As previously described for the token ring network, on the receipt of a token, an idle station passes it to its downstream station after an interface delay. If the station is busy, it will capture the token and transmit its frames until its token holding timer timeouts. As for the token-ring network, a source removal mechanism is used, so that the transmitted frames are broadcasted to all stations on the ring; they fully circulate the ring and are removed from the ring by their source station. The timeout mechanism and the priority support procedure used by FDDI are different than those employed by the token ring network and are described subsequently. When a station terminates its dwell time on the medium, it immediately releases the token. This is identified as an *early token release* operation. For lower speed token-ring implementations, a late token release operation can be selected. Under the latter, the token is released by the station only after its has received all of its transmitted messages and removed them from the ring. Such an operation leads to throughput performance degradations when the network uses higher speed links since then the station holds the token for an extra time, which includes as a component the ring's propagation delay. The latter can be long relative to the frame's transmission time. As a result, higher speed token-ring networks are generally set to operate in the early token release mode.

The FDDI MAC scheme distinguishes between two key FDDI service types: synchronous and asynchronous. Up to eight priority levels can be selected for asynchronous services. Todate, most commonly employed FDDI adapter cards implement mostly only the asynchronous priority 1 service; some also provide a synchronous priority service.

The access of frames onto the medium is controlled by a *timed token rotation* (TTR) protocol. Each station continuously records the time elapsed since it has last received the token (denoted as the token rotation time). An initialization procedure is used to select a target token rotation time (TTRT) through a bidding process whereby each station bids for a token rotation time (TRT) and the minimum such time is selected. As previously noted, two classes of service are defined: synchronous, under which a station can capture the token whenever it has synchronous frames to transmit, and asynchronous, which permits a station to capture a token only if the current TRT is lower than the established TTRT. To support multiple priority levels for asynchronous frames, additional time thresholds are defined for each priority level. In this manner, a message of a certain priority level is allowed to be transmitted by its station, when the latter captures the token, only if the time difference between the time this station has already used (at this ring access) for transmitting higher priority messages, and the time since the token last visited this station is higher than the corresponding time threshold associated with the underlying message priority level. This priority-based access protocol is similar to the one used for the IEEE 802.4 token bus LAN system.

Using this procedure, stations can request and establish guaranteed bandwidth and response time for synchronous frames. A guaranteed maximum cycle latency-based response time is established for the ring, since the arrival time between two successive tokens at a station can be shown to not exceed the value of $2 \times$ TTRT.

As a polling scheme, the performance of the FDDI network is limited by the ring walk (W) time. The ring throughput is thus proportional to $1 - W/\text{TTRT}$. While lower TTRT values (such as 4–8 ms) yield lower guaranteed cycle response times (token intervisit times lower than 8–16 ms), higher TTRT values need to be selected to provide for better bandwidth utilization under higher load conditions. The ring latency varies from a small value of 0.081 ms for a 50 stations, 10-km LAN, to a value of 0.808 ms for a 500 stations, 100-km LAN. Using a TTRT value of 50 ms, for a LAN that supports 75 stations and 30 km of fiber, and having a ring latency $W = 0.25$ ms, a maximum utilization of 99.5% can be achieved [Ross, 1986].

To provide messages their desired delay performance behavior across the FDDI ring network, it is important to calibrate the FDDI network so that acceptable levels of queueing delays are incurred at the stations' access queues for each service class [Shah et al., 1992]. This can be achieved by the proper selection of the network's MAC parameters, such as the TTRT level, the timeout threshold levels when multipriority Asynchronous services are used, and the station synchronous bandwidth allocation level when a station's FDDI adapter card is set also to provide a synchronous FDDI service. The latter service is effective in guaranteeing quality of service support to real-time streams, such as voice, compressed video, sensor data, and high-priority critical message processes, which require strictly limited network delay jitter levels [Shah et al., 1992]. Note that when a token is received by a station that provides a FDDI synchronous service, the station is permitted to transmit its frames that receive such a service (for a limited time, which is equal to a guaranteed fraction of the TTRT) independently of the currently measured TRT. When no such messages are queued at the station at the arrival of the token, the token immediately starts to serve messages which receive asynchronous service, so that the network's bandwidth is dynamically shared among all classes of service.

The delay-throughput performance features of the FDDI networks follow the characteristics of a distributed polling scheme already discussed. The FDDI performance results reported in Shah et al. [1992] were obtained by using the PLANYST^{TM} tool developed by IRI Corporation. This tool has been also used to calibrate the parameters of FDDI network through the use of its expert based analytical routines.

Implicit Polling Schemes

Under an implicit polling multiple-access mechanism, the network stations monitor the shared medium and are then granted access to it by identifying proper status conditions or tags. To illustrate such structures, we consider the slotted channel, register-insertion, positional-priority, and collision-avoidance access protocols.

Under a *slotted channel access protocol*, the shared medium link(s) are time shared through the generation of time slots. Each time slot contains an header which identifies it as busy or idle. In addition, time slots may also be reserved to connections, so that a circuit-switched mode can be integrated with the packet-switched multiple-access mode described in the following. To regulate station access rates, to assign time circuits, and to achieve a fair allocation of channel resources, the time slots are normally grouped into recurring time frames (cycles). A ready station, with packets to transmit across the medium, monitors the medium. When this station identifies an idle slot, which it is allowed to capture, it marks it to be in a busy state and inserts a single segment into this slot. Clearly, in transmitting a packet, a station must break it into multiple segments, whereby the maximum length of a segment is selected such that it fits into a single slot. The packet segments must then be assembled into the original packet at the destination station.

To be able to insert packets into moving slots, the station must actively interface the medium by inserting an active buffer into the channel. A common configuration for such a network is the slotted ring topology. A slotted access protocol can also be used in sharing a linear logical bus topology, with active station interfaces. The later configuration is used by the distributed queue dual-bus MAN defined by the IEEE 802.6 standard. The latter uses a fiber optic-based dual-bus configuration so that each station is connected to two counterdirectional buses.

To regulate the maximum level of bandwidth allocated to each station, in accordance with the class of service provided to the station, and to control the fair allocation of network resources among stations that receive the same class of service, the access algorithm can limit (statically or dynamically) the number of slots that can be captured by a station during each frame. For the DQDB network, a reservation subchannel is established for stations to insert reservation tags requesting for slots to be used for the transmission of their packets. A station is allowed to capture an idle slot which passes its interface only if it has satisfied all of the earlier requests it has received signifying slot reservations made by other stations. The DQDB network also integrates a circuit-switched mode through the establishment of isochronous circuits as part of the call setup procedure. A frame header is used to identify the slots that belong to dedicated time circuits.

Under a *register-insertion* configuration, each station's medium interface card includes a register (buffer), which is actively inserted into the medium. Each packet is again broken down into segments. A station is permitted to insert its segment(s) into the medium when its register contains no in-transit segments (thus deferring the transmission of its own segments until no in-transit segments are passing by), or when the gap between in-transit segments resident in its register is sufficiently long. In-transit packets arriving at the station's interface when the station is in the process of transmitting its own segment are delayed in the register. To avoid register overflows, its size is set to be equal to at least the maximum segment length. The IBM Metaring/Orbit [Cidon and Ofek, 1989] network is an example of such a network system which employs a ring topology as well as the destination removal spatial reuse features to be presented.

At higher speeds, to further increase the throughput efficiency of the shared medium network, a *destination removal* mechanism is used. This leads to *spatial reuse* since different space segments of the network's medium can be used simultaneously in time by different source–destination pairs. For example, for a spatial-reuse slotted channel network (such as slotted ring or bus-based topologies), once a segment has reached its destination station, the latter marks the slot as idle allowing it to be reused by subsequent stations it visits. Similarly, the use of an actively inserted buffer (as performed by the register-insertion ring network), allows for operational isolation of the network links, providing spatial reuse features.

To assess the increase in throughput achieved by spatial reuse ring networks, assume the traffic matrix to be uniform (so that the same traffic loading level is assumed between any source–destination pair of stations). Also assume that the system employs two counter-rotating rings, so that a station transmits its segment across the ring that offers the shortest path to the destination. Clearly, the maximum path length is equal to half the ring length, while the average path length is equal to one-quater of the ring length. Hence, an average of four source–destination station pairs time simultaneously communicate across the dual ring network. As a result, the normalized throughput capacity achieved by the spatial reuse dual ring network is equal to 400% (across each one of the rings), as compared with a utilization capacity of 100% (per ring) realized when a destination removal mechanism is used. Hence, such spatial reuse methods lead to substantial throughput gains, particularly when the network links operate at high speeds. They are thus especially important when used in ultra-high-speed optical networks, as we note in Sec. 46.7.

Positional-Priority and Collision Avoidance Schemes

Under a hub-polling (centralized polling) positional-priority scheme, a central station (such as a computer controller) polls the network stations (such as terminals) in an order which is dictated by their physical position in the network (such as their location on a ring network with respect to the central station), or by following a service order table [Baker and Rubin, 1987]. Stations located in a higher position in the ordering list are granted higher opportunities for access to the network's shared medium. For example, considering a ring network, the polling cycle starts with the controller polling station 1 (the one allocated highest access priority). Subsequently station 2 is polled. If the latter is found to be idle, station 1 is polled again.

In a similar manner, a terminal priority-based distributed implicit polling system is implemented. Following a frame start tag, station 1 is allowed to transmit its packet across the shared medium. A ready station 2 must wait a single slot of duration T (which is sufficiently long to allow station 1 to start transmitting and for its transmission to propagate throughout the network so that all stations monitoring the shared channel can determine that this station is in the process of transmission) before it can determine whether it is allowed to transmit its packet (or segment). If this slot is monitored by station 2 to be idle, it can immediately transmit its packet (segment). Similarly, station i must wait for $(i-1)$ idle slots (giving a chance to the $i-1$ higher priority terminals to initiate their transmissions) following the end of a previous transmission (or following a frame start tag when the channel has been idle) before it can transmit its packet across the shared communications channel.

Such a multiple-access mechanism is also known as a collision avoidance scheme. It has been implemented as part of a MAC protocol for high-speed back-end LANs (which support a relatively small number of nodes) such as HyperChannel. It has also been implemented by wireless packet radio networks, such as the TACFIRE field artillery military nets. In assessing the delay-throughput behavior of such an implicit polling distributed control mechanism, we note that the throughput efficiency of the scheme depends critically on the monitoring slot duration T, whereas the message delay behavior depends on the terminal's priority level [Rubin and Baker, 1986]. At lower loading levels, when the number of network nodes is not too large, acceptable message delays may be incurred by all terminals. In turn, at higher loading levels, only higher priority terminals will manage to attain timely access to the network while lower priority ones will be effectively blocked from entering the shared medium.

Probing Schemes

When a network consists of a large number of terminals, each generating traffic in a low duty cycle manner, the polling process can become highly inefficient in that it will require relatively high bandwidth and will occupy the channel with unproductive polling message transmissions for long periods of time. This is induced by the need to poll a large number of stations when only a few of them will actually have a packet to transmit.

A *probing* [Hayes, 1984] scheme can then be employed to increase the efficiency of the polling process. For this purpose, rather then polling individual stations, groups of stations are polled. The responding individual stations are then identified through the use of a collision resolution algorithm. For example, the following *tree-random-access* ([Capetanakis, 1979; Hayes, 1984]) algorithm can be employed. Following an idle state, the first selected polling group consists of all of the net stations. A group polling message is then broadcasted to all stations. All stations belonging to this group which have a packet to transmit then respond. If multiple stations respond, a collision will occur. This will be recognized by the controller, which will subsequently subdivide the latest group into two equal subgroups. The process will proceed with the transmission of a subgroup polling message. Using such a binary search type algorithm, all currently active stations are eventually identified. At this point, the probing phase has been completed and the service phase is initiated. All stations which have been determined to be active are then allocated medium resources for the transmission of their messages and streams.

Note that this procedure is similar to a reservation scheme in that the channel use temporally alternates between a signalling period (which is used to identify user requests for network support) and a service period (during which the requesting stations are provided channel resources for the transmission of their messages). Under a *reservation* scheme, the stations themselves initiate the transmission of their requests during the signalling periods. For this purpose, the stations may use a random-access algorithm, or other access methods. When the number of network stations is not too large, dedicated minislots for the transmission of reservation tags can be used.

46.5 Random-Access Protocols

In this section, we describe a number of random-access protocols which are commonly used by many wireline and wireless networks. Random-access protocols are used for networks which require a distributed control multiple-access scheme, avoiding the need for a controller station which distributes (statically or dynamically) medium resources to active network stations. This results in a survivable operation, which avoids the need for investment of significant resources into the establishment and operation of a signalling subnetwork. This is of particular interest when the shared communications medium supports a relatively large number of terminals operating each at a low duty cycle. When a station has just one or a few packets it needs to transmit in a timely manner (at low delay levels) on infrequent occasions, it is not effective to allocate to the station a fixed resource of the network (as performed by a fixed assignment scheme). It is also not effective to go through a signalling procedure to identify the station's communications needs prior to allocating it a resource for the transport of its few packets (as performed by demand-assigned polling and reservation schemes). As a consequence, for many network systems, particularly for wireless networks, it is effective to use random-access techniques for the transport of infrequently generated station packets; or, when active connections must be sustained, to use random-access procedures for the multiaccess transport of signalling packets.

The key differences among the random-access methods described subsequently are reflected by the method used in performing shared medium status monitoring.

When stations use full-duplex radios and the network is characterized by a broadcast communications medium (so that every transmission propagates to all stations), each station receives the transmissions generated by the transmitting station, including the transmitting station itself. Stations can then rapidly assess whether their own transmission is successful (through data comparison or energy detection). This is the situation for many local area network implementations. In turn, when stations are equipped with half-duplex transceivers (so that they need to turn around their radios to transition between a reception mode and a transmission mode) and/or when a fully broadcast channel is not available (as is the case for mobile radio nets for which topographical conditions lead to the masking of certain stations, lacking line-of-sight connections between certain pairs of stations), the transmitting station cannot automatically determine the status of its transmission. The station must then rely on the receipt of a positive acknowledgment packet from the destination station.

ALOHA Multiple Access

Under the ALOHA random-access method, the network stations do not monitor the status of the shared communications channel. When a ready station receives a packet, it transmits it across the channel at any time (under an unslotted ALOHA scheme) or at the start of time slot (under a slotted ALOHA algorithm, where the length of a slot is set equal to the maximum MAC frame length). If two or more stations transmit packets (frames) at the same time (or at overlapping times), the corresponding receivers will not usually be able to correctly receive the involved packets, resulting in a destructive collision. (Under communications channel capture conditions, the stronger signal may capture the receiver and may be received correctly, while the weaker signals may be rejected.) When a station has determined that its transmitted packet has collided, it then schedules this packet for retransmission after a random retransmission delay. The latter delay can be selected at random from an interval whose length is dynamically determined based on the estimated level of congestion existing across the shared communications channel. Under a binary exponential backoff algorithm, each station adapts this retransmission interval on its own, by doubling its span each time its packet experiences an additional collision.

The throughput capacity of an unslotted ALOHA algorithm is equal to $s(C) = 1/(2e) = 18.4\%$, whereas that of a slotted ALOHA scheme is equal to $s(C) = 1/e = 36.8\%$. The remainder of the shared channel's used bandwidth is occupied by original and retransmitted colliding packets. In effect, to reduce the packet delay level and the delay variance (jitter), the loading on the medium must be reduced significantly below the throughput capacity level. Hence, the throughput efficiency of ALOHA channels is normally much lower than that attainable by fixed-assigned and demand-assigned methods. The random-access network system is, however, more robust to station failures, and is much simpler to implement, not requiring the use of complex signalling subnetworks.

The ALHOA shared communications channel exhibits a bistable system behavior. Two distinctly different local equilibrium points of operation are noted. Under sufficiently low-loading levels the system state resides at the first point, yielding acceptable delay-throughput behavior. In turn, under high-loading levels, the system can transition to operate around the second point. Loading fluctuations around this point can lead to very high packet delays and diminishing throughput levels. Thus, under high-loading bursts the system can experience a high level of collisions, which in turn lead to further retransmissions and collisions, causing the system to produce very few successful transmissions. To correct this unstable behavior of the random-access multiaccess channel, flow control mechanisms must be used. The latter regulate admission of new packets into the shared medium at times during which the network is congested. Of course, this in turn induces an increase in the packet blocking probability, or in the delay of the packet at its station buffer.

Carrier Sense Multiple-Access

Under the CSMA random-access method, the network stations monitor the status of the shared communications channel to determine if the channel is busy (carrying one or more transmissions) or is idle. A station must listen to the channel before it schedules its packet for transmission across the channel.

If a ready station senses the channel to be busy, it will avoid transmitting its packet. It then either (under the nonpersistent CSMA algorithm) takes a random delay, after which it will remonitor the channel, or (under the 1-persistent CSMA algorithm), it will keep persisting on monitoring the channel until it becomes idle.

Once the channel is sensed to be idle, the station proceeds to transmit its packet. If this station is the only one transmitting its packet at this time, a successful transmission results. Otherwise, the packet transmission normally results in a destructive collision. Once the station has determined that its packet has collided, it schedules its packet for retransmission after a random retransmission delay.

Many local and regional area packet-radio multiaccess networks supporting stationary and mobile stations, including those using half-duplex radio transceivers, have been designed to use a CSMA protocol. The performance efficiency of CSMA networks is determined by the *acquisition delay index* $a = t(a)/T(P)$, where $T(P)$ denotes the average packet transmission time and $t(a)$ denotes the system's acquisition time delay. The latter is defined as the time elapsed from the instant that the ready station initiates its transmission (following the termination of the last activity on the channel) to the instant that the packet's transmission has propagated to all network stations so that the latter can sense a transmitting station and thus avoid initiating their own transmissions. The acquisition delay $t(a)$ includes as components the network's end-to-end propagation delay, the radio turn-around time, radio detection times of channel busy-to-idle transitions, radio attack times (times to build up the radio's output power), various packet preamble times, and other components. As a result, for half-duplex radios, the network's acquisition delay $t(a)$ may assume a relatively large value. The efficiency of the operation is, however, determined by the factor a, which is given as the ratio of $t(a)$ and the packet transmission time $T(P)$, since once the station has acquired the channel and is the only one currently active on the channel [after a period of length $t(a)$], it can proceed with the uninterrupted transmission of its full packet [for a period of duration $T(P)$].

Clearly, this can be a more efficient mechanism for packet radio networks which operate at lower channel transmission rates. As the transmission rate increases, $T(P)$ decreases while the index a increases, so that the delay-throughput efficiency of the operation rapidly degrades. A CSMA network will attain good delay-throughput performance behavior for acquisition delay index levels lower than about 0.05. For index levels around or higher than 0.2, the CSMA network exhibits a throughput level that is lower than that obtained by a slotted (or even unslotted) ALOHA multiaccess net. Under such conditions, the channel sensing mechanism is relatively ineffective since the window of vulnerability for collisions [$t(a)$] is now relatively too long. It is thus highly inefficient to use a CSMA mechanism for higher data rate channels, as well as for channels which induce relatively long-propagation delays (such as a satellite communication network), or for systems which include other mechanisms that contribute to an increase in the value of the acquisition delay index.

As for the ALOHA scheme, a CSMA network exhibits a bistable behavior. Thus, under loading bursts the channel can enter a mode under which the number of collisions is excessive so that further loading of the channel results in diminishing throughput levels and higher packet delays. A flow control-based mechanism can be used to stabilize the behavior of the CSMA dynamic network system.

CSMA/CD Local Area Networks

A local area network that operates by using a CSMA/CD access control algorithm incorporates into the CSMA multiple-access scheme previously described the capability to perform collision detection (CD). The station's access module uses a full-duplex radio and appends to its CSMA-based sensing mechanism a CD operation. Once the ready station has determined the channel to be idle, it proceeds to transmit its packet across the shared channel while at the same time it is listening to the channel to determine whether its transmission has resulted in a collision. In the latter case, once the station has determined its packet to be involved in a process of collision, the station will immediately abort transmission. In this manner, colliding stations will occupy the channel with their colliding transmissions only for a limited period of time, the collision detection time $T(CD)$. Clearly, if $T(CD) < T(P)$, where $T(P)$ denotes the transmission time of a packet (MAC frame) across the medium, the CSMA/CD operation leads to improved delay-throughput performance over that of a CSMA operation.

The ethernet LAN developed by Xerox Corporation is a bus-based network operating at a channel data rate of 10 Mb/s and using a 1-persistent CSMA/CD medium access control algorithm. The physical, MAC and link layers of such a CSMA/CD network are defined by the IEEE 802.3 Standard (and a corresponding ISO standard). Ethernet nets can employ different media types: twisted pair, coaxial cable, fiber optic line, as well as radio links (in the case of an ethernet-based wireless LAN system). The configuration of the ethernet LAN is that of a logical bus so that a frame transmission by a station propagates across the bus medium to all other stations. In turn, the physical layout of the medium can assume a bus or a star topology. Under the latter configuration, all stations are connected to a central hub node at which point the access lines are connected by a reflecting repeater module so that a transmission received from an access line is repeated into all other lines. Typical ethernet layouts are limited in their geographical span to an overall distance of about 500–2000 m.

The delay-throughput efficiency of the CSMA/CD network operation is determined by the acquisition time delay $t(a)$ and index a, as for the CSMA network. In addition, the network efficiency also depends on the CD time $T(CD)$. It is noted that $T(CD)$ can be as long as twice the propagation delay across the overall span of the bus medium, plus the time required by the station to establish the occurrence of a collision. To ensure that a collision is reliably detected, the power levels of received packets must be sufficiently high. This also serves as a factor limiting the length of the bus and of the distance at which repeaters must be placed. As a result, the network stations are attached to several ethernet *segments*, each assuming a sufficiently short span. The different ethernet segments are interconnected by gateways or routers. The latter act as store-and-forward switches which serve to isolate the CSMA/CD multiaccess operation of each segment.

As for the random-access mechanisms previously discussed, burst loads applied to the CSMA/CD network can cause large delay-throughput degradations. In supporting application streams which require limited packet delay jitters, it is thus required to prevent the bus from being excessively loaded. Many implementations thus plan the net's offered traffic loading levels to be no higher than 40% (or 4 Mb/s). Also note that as for other random access schemes, large loading variations can induce an unstable behavior. A flow control mechanism must then be employed to regulate the maximum loading level of the network.

When shorter bus spans are used, or when shorter access link distances are employed in a star configuration, it is possible to operate this network at higher channel data rates. Under such conditions, an ethernet operation at a data rate of 100 Mb/s (or higher) has been implemented.

46.6 Multiple-Access Schemes for Wireless Networks

Under a cellular wireless network architecture, the geographical area is divided into cells. Each cell contains a central base station. The mobile terminals communicate with the base station controlling the cell in which they reside. The terminals use the reverse traffic and signalling channel(s) to transmit their messages to the base station. The base station multiplexes the messages it wishes to send to the cell's mobile terminals across the forward traffic and signalling channels.

First-generation cellular wireless networks are designed to carry voice connections employing a circuit switching method. Analog communications signals are used to transport the voice information. The underlying signalling subnetwork is used to carry the connection setup, termination, and handover signalling messages. The voice circuits are allocated through the use of a reservation based demand-assigned/FDMA scheme. A ready mobile uses an allocated signalling channel for its handset to transmit to its cell's base station a request message for the establishment of a voice circuit. If a frequency channel is available for the allocation of such a circuit (traffic channel), the base station will make this allocation by signalling the requesting handset.

Second-generation cellular wireless networks use digital communications channels. A circuit-switching method is still employed, with the primary service providing for the accommodation of voice connections. Circuits are formed across the shared radio medium in each cell by using either a TDMA-access control scheme (through the periodic allocation of time slots to form an established circuit, as performed by the European GSM and the U.S. IS-54 standards) or by employing a CDMA procedure (through the allocation of a code sequence to a connection's circuit, as carried out by the IS-95 standard). A signalling subnetwork is established. Reverse signalling channels (RSCs) are multiple-access channels which are used by the mobile terminals to transmit to their cell's base station their channel-request packets (for a mobile originating call), as well as for the transmission of paging response packets (by those mobiles which terminate calls). Forward signalling channels (FSCs) are multiplexing channels configured for the transmission of packets from the base station to the mobiles. Such packets include channel-allocation messages (which are sent in response to received channel request packets) and paging packets (which are broadcasted in the underlying location area in which the destination mobile may reside).

For the reverse signalling channels, a random access algorithm such as the ALOHA scheme is frequently employed. For TDMA systems, time circuit(s) are allocated for the random access transmission of signalling packets. For CDMA systems, codes are allocated for signalling channels; each code channel is time shared through the use of a random access scheme (employing, for example, a slotted or unslotted ALOHA multiple-access algorithm).

Paging and channel allocation packets are multiplexed (in a time-shared fashion) across the forward signalling channels. To reduce battery consumption at the mobile terminals, a slotted mode operation can be invoked. In this case, the cell mobiles are divided into groups, and paging messages destined to a mobile belonging to a certain group are transmitted within that group's allocated channels (time slots) [Rubin and Choi, 1996]. In this manner, an idle mobile handset needs to be activated for the purpose of listening to its FSC only during its group's allocated slots.

Under the IEEE 802.11 protocol, terminals access the base station by sharing the radio channel through the use of a CSMA/CA (collision avoidance) scheme.

Third-generation cellular wireless networks are planned to provide for the support of both voice and data services. These networks employ packet-switching principles. Many of the multiple-access schemes described in the previous sections can be employed to provide for the sharing of the mobile-to-base-station multiple-access reverse communications channels. In particular, random-access, polling (or implicit polling) and reservation protocols can be effectively implemented. For example, the following versions of a reservation method have been investigated. Under the packet reservation multiple access (PRMA) [Goodman et al., 1989] or random-access burst reservation procedure, a random-access mechanism is used for the mobile to reserve a time circuit for the transmission of a burst (which consists of a number of consecutively generated packets). For voice bursts, a random-access algorithm is used to govern the transmission of the first packet of the burst across the shared channel (by randomly selecting an unused slot, noting that, in each frame, the base station notifies all terminals as to which slots are unoccupied). If this packet's transmission is successful, its terminal keeps the captured slot's position in each subsequent frame until the termination of the burst's activity. Otherwise, the voice packet is discarded and the next voice packet is transmitted in the same manner. In selecting the parameters and capacity levels of such a multiaccess network, it is necessary to ensure that connections receive acceptable throughput levels and that the packet discard probability is not higher than a prescribed level ensuring an acceptable voice quality performance [Rubin and Shambayati, 1995].

For connection-oriented packet-switching network implementations which provide also data packet transport, including wireless ATM networks, it is necessary to avoid frequent occurrences of packet discards, to reduce the rate of packet retransmissions and to lower the chances for packet reordering procedures needed to guarantee the ordered delivery of packets. Reverse and forward signalling channels are established to set up the virtual-circuit connection. Channel resources can then be allocated to established connections in accordance with the statistical features of such connections. For example, real-time connections can be accommodated through the periodic allocation of time slots (or frequency/code resources) whereas burst sources can be supported by the allocation of such resources only for the limited duration of a burst activity. A mechanism must then be employed to identify the start and end times of burst activities. For example, the signalling channels can be used by the mobiles to signal the start of burst activity; whereas the in-band channel is used to detect the end of activity (directly and/or through the use of tagging marks).

46.7 Multiple Access Methods for Spatial-Reuse Ultra-High-Speed Optical Communications Networks

As we have seen above, token-ring and FDDI LAN's use token-passing methods to access a shared medium. Furthermore, these networks use a *source removal* procedure so that each station is responsible for removing its own transmitted frames from the ring. In this manner, each transmitted frame circulates the ring and is then removed by its source station. As a result, the throughput capacity of such networks is limited to a value which is not higher than the ring's channel data rate.

In turn, as observed in Sec. 46.4 (in connection with implicit polling-based multiple-access networks), the bandwidth utilization of shared medium networks, with particular applications to local and metropolitan area networks operating at high speed using generally fiber optic links, can be significantly upgraded through the employment of spatial-reuse methods. For example, consider a ring network which consists of two counter-rotating unidirectional fiber optic ring topologies. Each terminal (station) is connected to both rings through ring interface units. The bandwidth resources of the rings are shared among the active stations. Assume now that a *destination removal* method is employed. Thus, when a frame reaches its destination node (RIU), it is removed from the ring by the latter. The network communications resources occupied by this frame (such as a time slot) can then be made available to source nodes located downstream to the removing station (as well as to the removing station itself).

A source station which has a frame to transmit selects the ring that offers the shortest path to the destination node. In this manner, the length of a path will be no longer than 1/2 the ring's length. If we assume a uniform traffic matrix (so that traffic flows between the network terminal nodes are symmetrically distributed), the average path length will be equal to 1/4 the ring's length. As a result, using such a destination removal technique, we conclude that an average of four source-destination flows will be carried simultaneously in time by each one of the two rings. Therefore, the throughput capacity of such a spatial reuse network can reach a level which is equal to 400% the ring's channel data rate, for each one of the two rings.

Spatial reuse ring networks can be used to carry traffic on a circuit-switched or packet-switched (connectionless or connection-oriented) basis. For packet-switched network implementations, the following two distributed implicit polling multiple-access methods have been used. Under a slotted ring operation, time slots are generated and circulated across the ring. Each time slot contains a header which identifies it as either empty or busy. A station which senses an empty time slot can mark it as busy and insert a segment (a MAC frame, which for example can carry an ATM cell) into the slot. The destination station will remove the segment from its slot, and will designate the slot as idle so that it can be reused by itself or other stations. Under a buffer-insertion ring multiple-access scheme, each station inserts a buffer into the ring. A busy station will defer the transmission of its packet to packets currently being received from an upstream station. In turn, when its buffer is detected to be empty, the station will insert its segment into the ring. The buffer capacity is set equal to a maximum size packet so that a packet received by a station while it is in a process of transmission of its own packet can be stored in the station's inserted buffer. Each packet is removed from the ring by its destination node.

As previously noted, the DQDB MAN is a slotted dual-bus network which employs active station interfaces. To achieve spatial bandwidth reuse gains of the bus' bandwidth resources, frame removal stations can be positioned across the bus. These stations serve to remove frames which have already reached their destinations so that their slots can be reused by downstream stations. Note that such stations must use an inserted buffer of sufficient capacity to permit reading of the frame's destination node(s) so that they can determine whether a received frame has already reached its destination(s).

In designing optical communications networks, it is desirable to avoid the need for executing store-and-forward switching and operations at the network nodes. Such operations are undesirable due to the large differences existing between the optical communications rates across the fiber optic links and the electronic processing speeds at the nodes, as well as due to the difficulty involved in performing intelligent buffering and switching operations in the optical domain. Networks that avoid such operations (at least to a certain extent) are known as *all-optical networks.*

Systems commonly used for the implementation of a shared medium local area all-optical network employ a star topology. At the center of the star, an optical coupler is used. The coupler serves to repeat the frame transmissions received across any of its incoming fiber links to all of its outgoing fiber links. The coupler can be operated in a passive transmissive mode or in an active mode (using optical amplification to compensate for power losses incurred due to the distribution of the received optical signal across all output links). Each station is then connected by a fiber link to the star coupler, so that the stations share a multiple-access communications channel. A multiple-access algorithm must then be employed to control the sharing of this channel among all active stations. Typical schemes employed for this purpose use random-access, polling, and reservation methods. Furthermore, multiple wavelengths can be employed so that a WDMA component can be integrated into the multiple-access scheme. The wavelengths can be statically assigned or dynamically allocated. To illustrate the latter case, consider a reservation-based DA/WDMA procedure. Under a connection-oriented operation, wavelengths are assigned to source–destination nodes involved in a connection. Both transmitter and receiver based wavelength assignment configurations can be used: the transmitter's or receiver's (or both) operating wavelengths can be selected dynamically to accommodate a configured connection.

Because of its broadcast feature, the star-configured optical network does not offer any spatial or wavelength reuse advantages. In turn, the meshed-ring scalable multimedia adaptable meshed-ring terabit (SMARTNet) optical network introduced by Rubin [1994] capitalizes on the integrated use of spatial-reuse and wavelength-reuse methods. In the following, we illustrate the architecture of the SMARTNet configuration. Two counter-rotating fibers make up the ring periphery. The stations access the network through their RIUs, which are actively connected to the peripheral fiber rings. The fiber link resources are shared through the use of a wavelength-division scheme. Multiple wavelengths are used, and each station has access to all wavelengths (or to a limited number of them). The rings are divided into segments. Each segment provides access to multiple stations through their connected RIUs. Each segment is connected to its neighboring segments through the use of wavelength cross-connect routers. Such a router switches messages received at multiple wavelengths from its incoming ports to its outgoing ports. No store-and-forward switching operations are performed. The switching matrix is preset (on a static or adjustable basis) so that messages arriving across an incoming link at a certain wavelength are always immediately switched to a prescribed outgoing link (at the same wavelength; although wavelength translations can also take place). The wavelength routers are also connected (to certain other wavelength routers) by chord links (each consisting of a pair of counterdirectional fiber links). The routers and their chord links form the chord graph topology.

For each configuration of the chord graph, for each assignment of wavelengths, and for each switching table configuration of the routers, the SMARTNet topology can be divided into wavelength graphs. Each wavelength graph represents a subnetwork topology and is associated with a specific wavelength; all stations accessing this subnetwork (which is set here to be a ring or a bus topology) can communicate to each other across this subnetwork by using the associated wavelength. Within each subnetwork, destination-removal spatial-reuse techniques are used. Among the subnetworks, wavelengths can be reused by a number of subnetworks which do not share links.

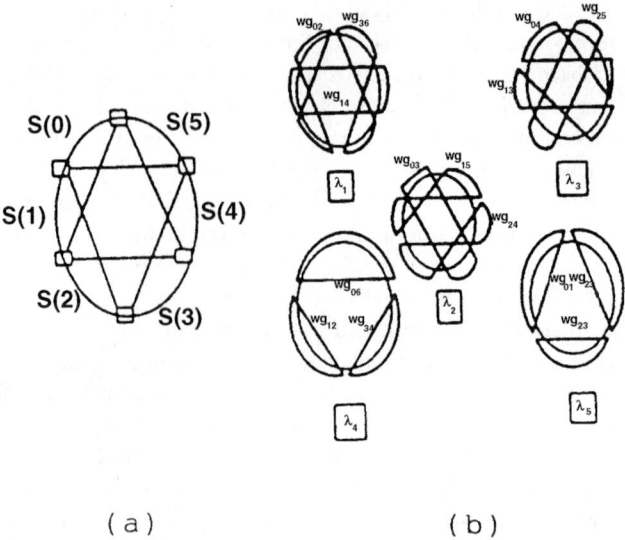

FIGURE 46.4 SMARTNet All-Optical Network: (a) the network layout; nodes represent wavelength routers; (b) group of wavelength graphs. Five groups are defined, and each group contains three wavelength graphs.

As previously discussed for the spatial-reuse ring networks, a SMARTNet system can be used to support connection-oriented circuit-switching and packet-switching (including ATM) architectures as well as connectionless packet-switching networking modes. For packet-switching networks, the links (with each wavelength channel) can be time shared by using slotted or buffer-insertion implicit polling schemes or through the use of other multiple-access methods.

As a network which internally employs cross-connect switches, no random delays or packet discards are incurred within the network. Such a performance behavior is advantageous for the effective support of multimedia services, including voice and video streams (which require low network delay jitter levels) and data flows (which prefer no network packet discards, avoiding the need for packet and segment retransmission and reordering operations). An integrated circuit and packet-switched operation is also readily implemented through the allocation of time, wavelength, and subnetwork channels. Such allocations can also be used to support and isolate virtual subnetworks and their associated user communities.

The SMARTNet configuration illustrated in Fig. 46.4 consists of six wavelength routers; each router has a nodal degree of 4, connected to four incoming and outgoing pairs of counterdirectional links. The chord topology is such that every router is connected to its second-neighboring router (rather than to its neighboring router). Analysis shows the network to require the use of a minimum of five wavelengths [Rubin and Hua, 1995]. Five different groups of wavelength graphs are noted. Each group includes three wavelength graph subnetworks (rings), which reuse the same wavelength. Hence, a wavelength reuse factor of 3 is attained. Each ring subnetwork offers a spatial reuse gain of 2.4 (rather than 4, due to the nonuniform distribution of the traffic across the wavelength graph). Hence, this SMARTNet configuration provides a throughput capacity level of $3 \times 2.4 = 720\%$ of the data rate across each wavelength channel, multiplied by the number of wavelengths used (equal to at least five, and can be selected to be any integral multiple of five to yield the same throughput efficiency factor) and by the number of parallel fiber links employed (two for the basic configuration previously discussed).

It can be shown that an upper bound on the throughput gain achieved by a SMARTNet configuration, which employs wavelength routers of degree four, is equal to 800%. Hence, the illustrated configuration, which requires only six wavelength routers and a minimum of only five wavelengths, yields a throughput gain efficiency which reaches 90% of the theoretical upper bound. Clearly, higher throughput gains can be obtained when the router nodes are permitted to support higher nodal degrees (thus increasing the number of ports sustained by each router). Furthermore, it can be shown that such a network structure offers a throughput performance level which is effectively equal to that obtained when store-and-forward switching nodes are employed for such network configurations.

Defining Terms

Carrier sense multiple access (CSMA): A random access scheme that requires each station to listen to (sense) the shared medium prior to the initiation of a packet transmission; if a station determines the medium to be busy, it will not transmit its packet.

Carrier sense multiple access with collision avoidance (CSMA/CA): A CSMA procedure under which each station uses the sensed channel state and an apriori agreed upon scheduling policy to determine whether it is permitted, at the underling instant of time, to transmit its packet across the shared medium; the scheduling function can be selected so that collision events are reduced or avoided.

Carrier sense multiple access with collision detection (CSMA/CD): A CSMA procedure under which each station is also equipped with a mechanism that enables it to determine if its on-going packet transmission is in the process of colliding with other packets; when a station detects its transmission to collide, it immediately acts to stop the remainder of its transmission.

Code-division multiple access (CDMA): A multiple access procedure under which different messages are encoded by using different (orthogonal) code words while sharing the same time and frequency resources.

Demand assignment multiple access (DAMA): A class of multiple access schemes that allocate dynamically (rather than statically) a resource of the shared medium to an active station; polling and reservation policies are protocols used by DAMA systems; reservation-based TDMA and FDMA systems are also known as demand assigned (DA)/TDMA and DA/FDMA systems.

Fixed assignment multiple access scheme: A multiple access procedure that allocates to each station a resource of the communications medium on a permanent dedicated basis.

Frequency-division multiple access (FDMA): A multiple access method under which different stations are allocated different frequency bands for the transmission of their messages.

Multiple access algorithm: A procedure for sharing a communications resource (such as a communications medium) among distributed end users (stations).

Multiplexing scheme: A system that allows multiple colocated stations (or message streams) to share a communications medium.

Polling scheme: A multiple access method under which the system controller polls the network stations to determine their information transmission (or medium resource) requirements; active stations are then allowed to use the shared medium for a prescribed period of time.

Random access algorithm: A multiple access procedure that allows active stations to transmit their packets across the shared medium without first coordinating their selected access time (or other selected medium's resource) among themselves; can result in multiple stations contending for access to the medium at overlapping time periods (or other overlapping resource sets), which may lead to destructive collisions.

Reservation scheme: A multiple access method under which a station uses a signalling channel to request a resource of the shared medium when it becomes active; also known as demand assigned scheme.

Time-division multiple access (TDMA): A multiple access scheme under which a station is assigned dedicated distinct time slots within (specified or recurring) time frames for the purpose of transmitting its messages; different stations are allocated different time periods for the transmission of their messages.

Token passing scheme: A polling multiple access procedure that employs a distributed control access mechanism; a token (query) message is passed among the network stations; a station that captures the token is permitted to transmit its messages across the medium for a specified period of time.

References

Abramson, N. 1973. The ALOHA system-another alternative for computer communications. In *Computer Communications Networks*, eds. N. Abramson and F. Kuo, Prentice–Hall, Englewood Cliffs, N.J.

Baker, J.E. and Rubin, I. 1987. Polling with a general service order table. *IEEE Trans. Commun.*, 35(3):283–288.

Capetanakis, J.I. 1979. Tree algorithm for packet broadcast channels. *IEEE Trans. Inf. The.*, 25(5):505–515.

Chlamtac, I. and Ganz, A. 1988. Frequency-time controlled multichannel networks for high-speed communication. *IEEE Trans. Commun.*, 36(April):430–440.

Cidon, I. and Ofek, Y. 1989. METARING—A ring with fairness and spatial reuse. *Research Report*, IBM T.J. Watson Research Center.

Goodman, D.J., Valenzuela, R.A., Gayliard, K.T., and Ramamurthi, B. 1989. Packet reservation multiple access for local wireless communications. *IEEE Trans. Commun.*, 37:885–890.

Hayes, J.F. 1984. *Modeling and Analysis of Computer Communications Networks*, Plenum Press, New York.

Kleinrock, L. and Tobagi, F.A. 1975. Packet switching in radio channels Part I—Carrier sense multiple access models and their throughput-delay characteristics. *IEEE Trans. Commun.*, 23(Dec.):1417–1433.

Maxemchuk, N.F. 1982. A variation on CSMA/CD that yields movable TDM slots in integrated voice/data local networks. *Bell Syst. Tech. J.*, 61(7).

Newman, R.M. and Hullett, J.L. 1986. Distributed queueing: A fast and efficient packet access protocol for QPSX. In *Proc. ICCC'86*, Munich.

Ross, F.E. 1986. FDDI-A tutorial. *IEEE Commun. Mag.*, 24(5):10–17.

Rubin, I. 1978. Group random-access disciplines for multi-access broadcast channels. *IEEE Trans. Inf. Th.*, Sept.

Rubin, I. 1983. Synchronous and carrier sense asynchronous dynamic group random access schemes for multiple-access communications. *IEEE Trans. Commun.*, 31(9):1063–1077.

Rubin, I. 1994. SMARTNet: A scalable multi-channel adaptable ring terabit network. Gigabit Network Workshop. *IEEE INFOCOM'94*, Toronto, Canada, June.

Rubin, I. and Baker, J.E. 1986. Performance analysis for a terminal priority contentionless access algorithm for multiple access communications. *IEEE Trans. Commun.*, 34(6):569–575.

Rubin, I. and Baker, J.E. 1990. Medium access control for high speed local and metropolitan area networks. *Proc. IEEE, Spec. Issue on High Speed Networks*, 78(1):168–203; also in 1994 *Advances in Local and Metropolitan Area Networks*, ed. W. Stallings, Chap. 2, IEEE Computer Society, pp. 96–131.

Rubin, I. and Choi, C.W. 1996. Delay analysis for forward signalling channels in wireless cellular networks. In *Proc. IEEE INFOCOM'96 Conf.*, San Francisco, CA, March.

Rubin, I. and Hua, H.K. 1995. SMARTNet: An all-optical wavelength-division meshed-ring packet switching network. In *Proc. IEEE GLOBECOM'95 Conf.*, Singapore, Nov.

Rubin, I. and Shambayati, S. 1995. Performance evaluation of a reservation access scheme for packetized wireless systems with call control and hand-off loading. *Wireless Networks J.*, 1(II):147–160.

Rubin, I. and Tsai, Z. 1987. Performance of double-tier access control schemes using a polling backbone for metropolitan and interconnected communication networks. *IEEE Trans. Spec. Topics Commun.*, Dec.

Sachs, S.R. 1988. Alternative local area network access protocols. *IEEE Commun. Mag.*, 26(3):25–45.

Shah, A., Staddon, D., Rubin, I., and Ratkovic, A. 1992. Multimedia over FDDI. In *Proc. IEEE Local Comput. Networks Conf.*, Minneapolis, MN; also in 1995. *Multimedia Networking Handbook*, ed. J.P. Cavanagh, Chap. 6.4, Auerbach.

Stallings, W. 1993. *Networking Standards: A Guide to OSI, ISDN, LAN, and MAN Standards*, Addison-Wesley.

Further Information

For performance analysis of multiple access local area networks, see *Performance Analysis of Local Computer Networks*, by J.J. Hammond and P.J.P. O'Reilly, Addison-Wesley, 1986. For further information on related networking standards for local and metropolitan area networks, consult *Networking Standards: A Guide to OSI, ISDN, LAN and MAN Standards*, W. Stallings, Addison-Wesley, 1993.

47

Routing and Flow Control

47.1	Introduction 650
47.2	Connection-Oriented and Connectionless Protocols, Services, and Networks 651
47.3	Routing in Datagram Networks 652
47.4	Routing in Virtual Circuit Switched Networks 655
47.5	Hierarchical Routing 655
47.6	Flow Control in Datagram Networks 656
47.7	Flow Control in Virtual Circuit Switched Networks 657

Rene L. Cruz
University of California, San Diego

47.1 Introduction

There are a large number of issues relating to routing and flow control, and we cannot hope to discuss the subject comprehensively in this short chapter. Instead, we outline generic concepts and highlight some important issues from a high-level perspective. We limit our discussion to the context of **packet switched networks.**

Routing and flow control are concerned with how information packets are transported from their source to their destination. Specifically, routing is concerned with *where* packets travel, that is, the paths traveled by packets through the network. Flow control is concerned with *when* packets travel, that is, the timing and rate of packet transmissions.

The behavior of routing and flow control can be characterized by a number of interrelated performance measures. These performance measures can be categorized into two groups: user metrics and network metrics. User metrics characterize performance from the perspective of a single user. Common user metrics are throughput (rate of information transfer), average delay, maximum delay, delay jitter (variations in delay), and packet loss probability. Network metrics characterize performance from the perspective of the network as a whole, and all users of it. Common network metrics are network throughput, number of users that can be supported, average delay (averaged over all users), and buffer requirements. Fairness issues may also arise from the perspective of the network.

The goals of routing and flow control are to balance these different performance measures. Optimization of user metrics are often at the expense of network metrics, and vice versa. For example, if the number of users allowed to access the network is too large, the quality of service seen by some users might fall below acceptable levels.

In reality, it is often even difficult to optimize a single performance measure, due to the uncertainty of demand placed on the network. Statistical models can be used, but the appropriate parameters of the models are often unknown. Even with an accurate model of demand, the problem of predicting performance for a given routing or flow control strategy is often intractable. For this reason, routing and flow control algorithms are typically based on several simplifying assumptions, approximations, and heuristics.

Before we discuss routing and flow control in more detail, we begin by classifying different modes of operation common in packet switched networks.

47.2 Connection-Oriented and Connectionless Protocols, Services, and Networks

In a connection-oriented protocol or service between two network terminals, the two terminals interact with one another prior to the transport of data, to set up a *connection* between them. The interaction prior to the exchange of data is required, for example, to identify the terminals to each other and to define the rules governing the exchange of data. For example, users of the telephone network engage in a connection-oriented protocol in the sense that a phone number must first be dialed, and typically the parties involved must identify themselves to each other before the essential conversation can begin. The transport control protocol (TCP) used on the Internet is an example of a connection-oriented protocol. Examples of connection-oriented services are rlogin and telnet, which can be built upon TCP. Connection-oriented protocols and services typically involve the exchange of data in both directions. In addition, connection-oriented protocols are frequently used to provide reliable transport of data packets between network endpoints, whereby the network endpoints continue to interact during the data exchange phase in order to manage retransmission of lost or errored packets.

In *connectionless* protocols or services between two network terminals, no interaction between the terminals takes place prior to the exchange of data. For example, when a letter is mailed over the postal network, a connectionless protocol is used in the sense that the recipient of the letter need not interact with the sender ahead of time. An example of a connectionless protocol is the current Internet protocol (IP) used to transport packets on the Internet. Electronic mail provides a connectionless service, in the same sense that the postal network provides a connectionless service. Because of the lack of interaction between terminals engaging in a connectionless protocol, retransmission of lost or errored packets is not possible, and hence connectionless protocols are not considered intrinsically reliable. On the other hand, connectionless protocols are often appropriate in environments where lost or errored packets are sufficiently rare.

A connection-oriented *network* is a network specifically designed to support connection-oriented protocols. When a connection request is made to a connection-oriented network, the network establishes a path between the source and destination terminal. The switching nodes are then configured to route packets from the connection along the established path. Each switching node is aware of the connections that pass through it and may reserve resources (e.g., buffers and bandwidth) on a per-connection basis. A packet switched network that is connection oriented is also referred to as a **virtual circuit switched network.** Asynchronous transfer mode (ATM) networks are an example of virtual circuit switched networks.

A connectionless *network* is a network designed to support connectionless protocols. Each node of a connectionless network typically knows how to route packets to each possible destination, since there is no prior coordination between terminals before they communicate. Connectionless networks are sometimes referred to as **datagram networks.**

The terminology as defined is not standard, and indeed there does not appear to be widespread agreement on terminology. We have made a distinction between a connection-oriented protocol and a connection-oriented network, and a distinction between a connectionless protocol and a connectionless network. The reason for these distinctions is because of the many ways in which protocols and networks can be layered. In particular, connection-oriented protocols can be built upon connectionless networks and protocols, and connectionless protocols can be built upon connection-oriented networks and protocols. For example, TCP is in widespread use on end hosts in the Internet, but the Internet itself is inherently a connectionless network which utilizes IP. Another example that has received attention recently is in using an ATM network (connection oriented) as a lower layer transport mechanism to support the IP protocol between Internet nodes. This may be considered somewhat odd, since IP will in turn often support the connection-oriented TCP.

Packet switched networks can be further classified as lossy or lossless according to whether or not network nodes will sometimes be forced to drop (destroy) packets without forwarding them to their intended destination. Connectionless networks are inherently lossy, since the lack of coordination of data sources creates the possibility that buffers inside the network will overflow. Connection-oriented networks can be designed to be lossless. On the other hand, connection-oriented networks can also be lossy, depending on if and how flow control is exercised.

47.3 Routing in Datagram Networks

We will first confine our discussion on routing to the context of connectionless networks. A common approach to routing in connectionless networks is for each node in the network to have enough intelligence to route packets to every possible destination in the network.

A common approach to route computation is *shortest path routing*. Each link in the network is assigned a weight, and the length of a given path is defined as the sum of the link weights along the path. The weight of a link can reflect the desirability of using that link and might be based on reliability or queueing delay estimates on the link, for example. However, another common choice for link weights is unity, and shortest paths in this case correspond to minimum hop paths.

We now briefly review some popular algorithms for computing shortest paths. Suppose there are n nodes in the network, labeled $1, 2, \ldots, n$. The label (i, j) is assigned to a link that connects node i to node j. The links are assumed to be oriented, that is, link (i, j) is distinct from (j, i). We assume that if there is a link (i, j) in the network, then there is also a link (j, i) in the network. We say that node i is a neighbor of node j if the link (i, j) exists. Let $N(i)$ be the set of nodes for which node i is a neighbor. Let w_{ij} denote the weight assigned to link (i, j), and define $w_{ij} = +\infty$ if node i is not a neighbor of node j. We assume that $w_{ij} > 0$ for all i and j. Formally, a *path* from node i_1 to node i_k is a sequence of nodes $(i_1, i_2, \ldots, i_k)$. The length of such a path is the sum of the weights of links between successive nodes, and the path is said to have $k - 1$ hops. We assume that the network is *strongly connected*, that is, there exists a path between all ordered pairs of nodes with finite length.

Consider the problem of finding shortest (i.e., minimum length) paths from each node to a given destination node. Since each node in the network is a potential destination, this problem will be solved separately for each destination. Without loss of generality, assume that the destination node is node 1. The distance from a node i to the destination node is defined to be the length of the shortest path from node i to the destination node, and is denoted as D_i. Define $D_1 = 0$.

Consider a shortest path from a node i to the destination: $(i, j, k, \ldots, 1)$. Note that the subpath that begins at node j, $(j, k, \ldots, 1)$, must be a shortest path from node j to the destination, for otherwise there exists a path from node i to the destination that is shorter than $(i, j, k, \ldots, 1)$. Thus, this subpath has length D_j, and $D_i = w_{ij} + D_j$. Furthermore, since D_i is the length of a shortest path, we have $D_i \leq w_{ik} + D_k$ for any k. Thus, we have $D_1 = 0$ and

$$D_i = \min\{w_{ij} + D_j : 1 \leq j \leq n\}, \qquad 1 < i \leq n$$

This is known as *Bellman's equation*, and forms the basis of an optimization technique known as dynamic programming. Let $D = (D_1, D_2, \ldots, D_n)$ denote the vector of shortest path lengths, and let f denote the vector valued mapping in Bellman's equation. With these definitions, Bellman's equation says that D is a fixed point of the mapping f, that is, $D = f(D)$. It can be shown that under our assumption of positive link weights, there is only one solution x to the equation $x = f(x)$, that is, D is the unique solution to Bellman's equation.

Routing and Flow Control

The *Bellman–Ford algorithm* for shortest paths finds the solution to Bellman's equation iteratively. Specifically, let $D^0 = (0, \infty, \infty, \ldots, \infty)$. In the first iteration of the Bellman–Ford algorithm, $D^1 = f(D^0)$ is computed. In the mth iteration, $D^m = f(D^{m-1})$ is computed. It can be shown by induction on m that D^m is the vector of shortest path lengths, where the paths are constrained to have at most m hops. Since link weights are positive, any shortest path visits a node at most once, and hence any shortest path has at most $n - 1$ hops. Thus, it follows that the Bellman–Ford algorithm converges within $n - 1$ iterations. In other words, we have $f(D^{n-1}) = D^{n-1} = D$. The shortest paths can be generated from D by examining the argument of the minimum in Bellman's equation. Specifically, for each i, let Next$(i) = j$, where j is an argument of the minimum in the equation for D_i, that is, $D_i = w_{ij} + D_j$. With this definition, Next(i) is clearly the first node along a shortest path from node i to the destination. The worst case time complexity of the Bellman–Ford algorithm is easily seen to be $\mathcal{O}(n^3)$.

The Bellman–Ford shortest path algorithm is sometimes called a *distance vector* routing algorithm. The algorithm lends itself easily to a distributed, asynchronous implementation that forms the basis for many commonly used routing algorithms. In such an implementation, each node i is responsible for computing the distance from itself to the destination D_i. Each node initially has an estimate of D_i. These initial estimates may be arbitrary, except the destination itself always knows the correct distance from itself to itself, that is, node 1 knows that $D_1 = 0$. Each node i sends, or advertises, to its neighbors the estimate of D_i that it currently has. This advertisement is either done periodically or whenever the estimate changes. Whenever a node receives an estimate from a neighbor, it updates its own estimate in accordance with Bellman's equation. Specifically, if $D_i(t)$ is the estimate of D_i at time t and if $D_j(t - \tau_j)$ is the estimate of D_j from node j that was most recently advertised to node i, then $D_i(t) = \min\{w_{ij} + D_j(t - \tau_j) : j \in N(i)\}$. Note that this minimum can be taken over only nodes j which are neighbors of node i, since $w_{ij} = \infty$ if node i is not a neighbor of node j. It can be shown that each node's estimate converges to the actual distance to the destination under essentially arbitrary assumptions regarding propagation delays of advertisements, and delays associated with computing estimates [Bertsekas and Gallager, 1992].

In the preceding discussion we assumed that the link weights were constant. In practice, however, the link weights may change with time. For example, this may be as a result of link failures, or changing traffic characteristics on the links. If the link weights change slowly enough, the distributed asynchronous Bellman–Ford algorithm will track the shortest paths correctly as the link weights change. However, if the link weights change fast enough so that the algorithm does not have a chance to converge, problems may arise. For example, if link weights change quickly in response to changing traffic characteristics, it may be possible for packets to travel around in loops indefinitely. In such a scenario, the link weights oscillate as the traffic pattern on the links change due to oscillating estimates of shortest paths [Bertsekas and Gallager, 1992].

Another popular approach to finding shortest paths is for each node to independently compute the shortest paths. Each node i broadcasts the link weight w_{ij} for all outgoing links (i, j) to all other nodes. Each node then has complete information about the network and can compute shortest paths independently. Networks that use this approach are sometimes said to use *link state* routing. Each node could use the Bellman–Ford algorithm as described to compute shortest paths. However, with link state routing it is common to use *Dijkstra's algorithm* for shortest paths, described next.

Dijkstra's algorithm for finding shortest paths first finds the closest node to the destination node (i.e., the node i other than the destination node with the smallest value of D_i), then the second closest node, then the third closest node, and so on. Let us assume again without loss of generality that node 1 is the destination node. Dijkstra's algorithm is summarized as follows:

Initialization: $m \leftarrow 1$, $P_m \leftarrow \{1\}$, $D_1 \leftarrow 0$.
For all i not in P_m:
$\quad X_i \leftarrow w_{i1} + D_1$
$\quad$Next$(i) \leftarrow 1$

Iteration:

$i^* \leftarrow \text{argmin}\{X_i : i \text{ not in } P_m\}$
$P_{m+1} \leftarrow P_m \cup \{i^*\}$
$D_{i^*} \leftarrow X_{i^*}$
For all k not in P_{m+1}:
    If $X_k > w_{ki^*} + D_{i^*}$ then
        $X_k \leftarrow w_{ki^*} + D_{i^*}$
        Next$(k) \leftarrow i^*$
$m \leftarrow m + 1$
If $m = n$, then terminate. Otherwise, execute iteration again.

The set P_m consists of the m closest nodes to the destination, including the destination. In other words, for each m we have $|P_m| = m$, and $D_i \geq D_j$ if j belongs to P_m and i does not belong to P_m. To see why Dijkstra's algorithm works, assume after executing the iteration m times that P_{m+1} is the set of $m + 1$ closest nodes, that $X_k = \min\{w_{kj} + D_j : j \in P_{m+1}\}$, and that D_j has been computed correctly for all $j \in P_{m+1}$. Clearly this is true for $m = 0$, and we now show by induction that it is true for all m. Consider the $(m + 1)$st iteration. Suppose that node i^* is a node that could be added to P_{m+1} to yield P_{m+2} with the desired property and that node k is an arbitrary node that cannot be added to P_{m+1} to yield P_{m+2} with the desired property. Note that $D_k > D_{i^*}$. By the induction hypothesis and the fact that all link weights are positive, it thus follows from Bellman's equation that $D_{i^*} = X_{i^*}$. Furthermore, we have $X_k \geq D_k > D_{i^*} = X_{i^*}$. Thus, in the $(m + 1)$st iteration, node i^* is chosen correctly to form P_{m+2} and $D_{i^*} = X_{i^*}$. Furthermore, we see that X_k is updated correctly to preserve the induction hypothesis.

It is seen that the time complexity of Dijkstra's algorithm is $\mathcal{O}(n^2)$, which is less than the worst case-time complexity of $\mathcal{O}(n^3)$ for the Bellman–Ford algorithm. The description of Dijkstra's algorithm can also easily be adapted to address the computation of shortest paths from a given source node to all possible destinations, which makes it well suited for link state routing.

Deflection routing [Maxemchuk, 1987] has been proposed for high-speed networks. With deflection routing, each node attempts to route packets along a shortest path. If a link becomes congested, however, a node will intentionally send a packet to a node which is *not* along a shortest path. Such a packet is said to be *deflected*. If the number of links incoming to a node is equal to the number of outgoing links, and the capacity of all links is the same, packet buffering can be essentially eliminated with deflection routing. It is possible for packets to circulate in loops with deflection routing. With a suitable priority structure, however, delay can be bounded. Deflection routing has also been proposed *within* high-speed packet switches [Cruz and Tsai, 1996].

In another approach to routing, sometimes called *optimal routing*, the rate of traffic flow between each possible source–destination pair is estimated. Suppose that $r(w)$ is the estimated rate (bits per second) of traffic flow for a given source–destination pair w. A number of possible paths for routing traffic between each source–destination pair are identified, say path 1 through P_w for source–destination pair w. The total flow $r(w)$ is allocated among all of the P_w paths according to a vector $(f_1, f_2, \ldots, f_{P_w})$ such that $r(w) = f_1 + f_2 + \cdots + f_{P_w}$, and such that f_p flows on path p. Typically, bifurcation is allowed, so that f_p may be positive for more than one path p. The path flow allocation for all source–destination pairs is determined simultaneously in order to minimize some cost function. Typically, the cost function is of the form $\Sigma C_e(f_e)$, where the sum is taken over all links e in the network and f_e denotes the total flow over all paths which cross link e. A common choice for $C_e(f_e)$ is $C_e(f_e) = f_e/(C - f_e)$, where C is the capacity in bits per second of link e. This would correspond to the average delay on link e if the output queue for link e were an M/M/1 queue. Nonlinear optimization techniques can be used to optimize the path flow allocations with respect to the cost function. The disadvantage of the optimal routing approach is that it is difficult to estimate the rates $r(w)$ a priori. In addition, the choice of the cost function to minimize is somewhat arbitrary.

Typically in connectionless networks, the network nodes will determine where to route a packet so that it gets forwarded to its destination. In such an environment, the source places the address of the destination in a **header field** of each packet. Upon receiving a packet, a network node examines the destination address, and uses this to index into a lookup table to determine the outgoing link that the packet should be forwarded on.

In contrast, *source routing* describes the situation where the route that a packet travels is determined by the source. In this case, the source explicitly specifies what route the packet should take within a header field of the packet. Typically, this specification consists of the list of node addresses along the path that the packet should travel. This eliminates the need for large lookup tables within each network node. For example, link state routing can be used, where the sources compute shortest paths.

47.4 Routing in Virtual Circuit Switched Networks

In connection-oriented networks, sources make connection requests to the network, transmit data to the destinations through the connection, and then terminate the connection. Since connection requests are distributed in time, a routing decision is typically made for each connection request as it is made. In other words, the sequence of connection requests that will be made is unknown, and a routing decision for a connection is not made until the connection request occurs.

Shortest path routing is still a viable approach in connection-oriented networks. The network nodes calculate shortest paths to all possible destinations. When a connection request is made, the connection request is forwarded along a shortest path to the destination, and the switching nodes along the path can determine if they have the resources (e.g., bandwidth, buffers) to support the connection request. If all network nodes determine that they can support the request, the network accepts the connection, link weights are updated to reflect the newly formed connection, and data for the connection is forwarded along the established path. If a network node along the shortest path is not able to support the connection request, the network would then reject the connection request.

It is possible for the network to attempt to route a connection request along another path if the first attempted path fails. However, this should be done carefully, since routing a connection on a path with many hops may use up bandwidth that could be better used to support other connection requests. **Trunk reservation** techniques may be used to advantage in this regard.

Alternatively, the sources might be responsible for specifying a desired path in a connection request, which would be a form of source routing. Obviously in this case, the sources would need knowledge of the network topology in order to make intelligent choices for the routes they request.

In connection-oriented networks, the source need not place the destination address into each packet. Rather, it could put a *virtual circuit identifier* (VCI) in each packet, which labels the connection that the packet belongs to. This can result in a savings of bandwidth if there are a large number of potential destinations, and a comparatively smaller number of connections passing through each link. A given connection can have a different VCI assigned on every link along the path assigned to it. When a connection is set up through a node, the node assigns a VCI to the connection to be used on the outgoing link, which may be different from the VCI used for the connection on the incoming link. The node then establishes a binding between the VCI used on the incoming link, the VCI assigned on the outgoing link, as well as the appropriate output link itself. Upon receiving a packet, the network node reads the VCI used on the incoming link and uses this as an index into a lookup table. In the lookup table, the node determines the appropriate outgoing link to forward the packet on, as well as the new VCI to use. This is sometimes referred to as VCI translation.

47.5 Hierarchical Routing

In a network with a very large number of nodes, lookup tables for routing can get very large. It is possible to reduce the size of the lookup tables, as well as the complexity of route computation, by using a hierarchical routing strategy. One approach to a hierarchical routing strategy is as follows. A subset of the network nodes are identified as *backbone* nodes. A link connecting two backbones is called a backbone link. The set of backbone nodes and backbone links is called the backbone network. The nonbackbone nodes are partitioned into clusters of nodes called domains. Each backbone node belongs to a distinct domain, and is called the *parent* of all nodes in the domain. The nodes within a domain are typically geographically close to one another. Routes between pairs of nodes in the same domain are constrained not to go outside the domain. Routes between pairs of nodes in different domains are constrained to be a concatenation of a path completely within the source domain, a path completely within the backbone network, and a path completely within the destination domain. Routes are computed at two levels, the backbone level and the domain level. Route computations at the backbone level need only information about the backbone network and can proceed independently of route computations at the domain level. Conversely, route computations at the domain level need only information about the specific domain for which routes are computed. Because of this decoupling, the problem of route computation and packet forwarding can be considerably simplified in large networks.

With a hierarchical routing approach it is often convenient to have a hierarchical addressing strategy as well. For example, for the hierarchical routing strategy just described, a node address could be grouped into two parts, the domain address and the address of the node within the domain. Packet forwarding mechanisms at the backbone level need only look at the domain part of the node address and packet forwarding mechanisms at the domain level can ignore the domain part of the node address. Of course, it is possible to have hierarchical routing strategies with more than two levels as described.

47.6 Flow Control in Datagram Networks

Since sources in a connectionless network do not interact with the network prior to the transmission of data, it is possible that the sources may overload the network, causing buffers to overflow and packets to be lost. The sources may become aware that their packets have been lost, and as a result retransmit the lost packets. In turn, it is possible that the retransmission of packets will cause more buffers to overflow, causing more retransmissions, and so on; the network may be caught in a traffic jam where the network throughput decreases to an unacceptable level. It is also possible that buffers may overflow at the destinations if the sources send data too fast. The objective of flow control in connectionless networks is to throttle the sources, but only as necessary, so that rate of packet loss due to buffer overflow is at an acceptably low level.

A number of schemes have been proposed for flow control in connectionless networks. One technique that has been proposed is that network nodes that experience congestion (e.g., buffer overflow) should send the network nodes causing the congestion special control packets, which in effect request that the offending nodes decrease the rate at which packets are injected into the network. This has been implemented in the Internet, and the control packets are called *source quench* packets. A practical problem with this approach is the lack of accepted standards that specify under what conditions source quench packets are sent, and that specify exactly how network nodes should react to receiving a source quench packet.

We now focus our discussion on a common flow control technique used on connectionless networks, whereby the network carries traffic predominately from connection-oriented protocols, and *window flow control* is exercised within the connection-oriented protocols.

Window flow control is a technique where a source may have to wait for acknowledgements of packets previously sent before sending new packets. Specifically, a *window size W* is assigned to a source, and the source can have at most W unacknowledged packets outstanding. Typically, window flow control is exercised end-to-end, and the acknowledgements are sent back to the source by the destination as soon as it is ready to receive another packet. By delaying acknowledgements, the destination can slow down the source to prevent buffer overflow within the destination.

Another motivation behind window flow control is that if the network becomes congested, packet delays will increase. This will slow down the delivery of acknowledgements back to the source, which will cause the source to decrease its rate of transmissions. The window size W is typically chosen large enough so that in the absence of congestion, the source will not be slowed down by waiting for acknowledgements to arrive.

Although it is possible to prevent buffer overflow by choosing the window sizes small enough and appropriately managing the creation of connections, this is typically not how datagram networks operate. In many cases, packet loss due to buffer overflow is possible, and window flow control is combined with an **automatic repeat request (ARQ) protocol** to manage the retransmission of lost packets. In particular, the source will wait only a limited time for an acknowledgement to arrive. After this amount of time passes without an acknowledgement arriving, the source will *timeout* and assume that the packet has been lost, causing one or more packets to be retransmitted. The amount of time that the source waits is called the *timeout value*.

As already mentioned, retransmissions should be carefully controlled to prevent the network throughput from becoming unacceptably low. We now provide a explanation of how this is commonly handled within the current Internet, specifically within the TCP protocol. The explanation is oversimplified, but should give a rough idea of how congestion control mechanisms have been built into the TCP protocol. This is the primary method of flow control within the current Internet.

The first idea [Jain, 1986] is that a timeout indicates congestion, and sources should wait for the congestion to clear and not add to it. In the TCP protocol [Jacobson, 1988], each timeout causes the timeout value to be doubled. This is called *exponential backoff*, and is similar to what happens on ethernet networks. In practice, so that the timeout values do not get unacceptably large, exponential backoff is stopped when the timeout value reaches a suitable threshold. After acknowledgements begin to arrive at the source again, the source can reset the timeout value appropriately.

In addition to exponential backoff, the TCP protocol dynamically adjusts the window size in response to congestion. In particular, a congestion window W' is defined. Initially, W' is set to the nominal window size W, which is appropriate for noncongestion conditions. After each timeout, W' is reduced by a factor of two, until $W' = 1$. This is called *multiplicative decrease*.

The window size used by the TCP protocol is at most W', and could be less. The idea is that after a period of congestion, a large number of sources could still retransmit a large number of packets and cause another period of congestion. To prevent this, each time a timeout occurs, the window size is reset to one. Hence, a source is then allowed to send one packet. If the source receives an acknowledgement of this packet, it increases the window size by one, and hence is then allowed to send two packets. In general, the source will increase the window size by one for each acknowledgement it receives, until the window size reaches the congestion window size W'. This is called *slow start* and is also used for new connections. After the window size reaches W', the window size is increased by one for every roundtrip delay, assuming no timeouts occur, until the original window size appropriate for noncongestion conditions is reached. This has been called *slower start*.

47.7 Flow Control in Virtual Circuit Switched Networks

In connection-oriented networks, flow control, in a broad sense, can be exercised at two different levels. At a coarse level, when a connection request is made, the network may assess whether it has the resources to support the requested connection. If the network decides that it cannot support the requested connection, the request is blocked (rejected). This is commonly called *admission control*. A connection request may carry information such as the bandwidth of the connection that is required, which the network can use to assess its ability to carry the connection. Some sources do not formally require any specified bandwidth from the network, but would like as high a bandwidth as possible. Such sources are not acutely sensitive to delay; as an example, a file transfer might be associated with such a source. These types of sources have been recently named *available bit rate* (ABR) sources. If all sources are ABR, it may be unnecessary for the network to employ admission control.

At a finer level, when a connection is made, flow control may be exercised at the packet level. In particular, the rate as well as the burstiness of sources may be controlled by the network. We focus our discussion on flow control entirely at the packet level. Proposals have also been made to exercise flow control at an intermediate level, on groups of packets called *bursts*, but we do not discuss that here.

We first discuss flow control for ABR sources. Given that the bandwidth allocated to an ABR source is variable, the issue of fairness arises. There are many possible ways to define fairness. One popular definition of fairness is *max–min fairness*, which is defined as follows. Each link in the network has a given maximum total bandwidth that is allocated to ABR sources using that link. Each ABR source has an associated path through the network that is used to deliver data from the source to the associated destination. An allocation of bandwidths to the ABR sources is called max–min fair if the minimum bandwidth that is allocated is as large as possible. Given that constraint, the next lowest bandwidth that is allocated is as large as possible, and so on.

A simple algorithm to compute the max–min fair allocation of bandwidths to the ABR sources is as follows. A link is said to be saturated if the sum of the bandwidths of all ABR sources using that link equals the total bandwidth available to ABR sources on that link. Each ABR source is initially assigned a bandwidth of zero. The bandwidth of all ABR sources is increased equally until a link becomes saturated. The bandwidth allocated to each ABR source that crosses a saturated link is then frozen. The bandwidth of all remaining ABR sources are then increased equally until another link becomes saturated, and the bandwidth of each ABR source that crosses a newly saturated link is then frozen. The algorithm repeats this process until the bandwidth of each ABR source is frozen.

One way to impose flow control on ABR sources that has been proposed is to use window flow control for each ABR connection at the hop level. That is, each ABR connection is subject to a window flow control algorithm for each hop along the path of the connection. Thus, each ABR connection is assigned a sequence of window sizes which specify the window size to be used at each hop along the path for the connection. Each node along the path of a connection reserves buffer space to hold a number of packets equal to the window size for the link incoming to that node. Packet loss is avoided by sending acknowledgements only after a packet has been forwarded to the next node along the path. This scheme results in backpressure that propagates toward the source in case of congestion. For example, if congestion occurs on a link, it inhibits acknowledgements from being sent back on the upstream link, causing the buffer space allocated to the connection at the node feeding the upstream link to fill. This will cause that node to delay sending acknowledgements to the node upstream to that node, and so forth. If congestion persists long enough, the source will eventually be inhibited from transmitting packets. It has been shown that if each node transmits buffered packets from ABR connections that pass through it on a round robin basis, that if window sizes are sufficiently large, and that if each source always has data to send, then the resulting bandwidths that the ABR sources receive are max–min fair [Hahne, 1991]. This scheme is also known as *credit-based flow control* [Kung and Morris, 1995], and algorithms have been proposed for adaptively changing the window sizes in response to demand.

Another way to achieve max–min fairness (or fairness with respect to any criterion for allocated bandwidths) is to explicitly compute the bandwidth allocations and enforce the allocations at the entry points of the network. Explicit enforcement of bandwidth allocations is known as *rate-based flow control*.

Bandwidth is defined in a time-average sense, and this needs to be specified precisely in order to enforce a bandwidth allocation. Suppose packet sizes are fixed. One way to precisely define conformance to a bandwidth allocation of ρ packets per unit time is as follows. In any interval of time of duration x, a source can transmit at most $\rho x + \sigma$ packets. The packet stream departing the source is then said to be (σ, ρ) *smooth*. The parameter σ is a constant positive integer and is a measure of the potential burstiness of the source. Over short intervals of time, the bandwidth of the source can be larger than ρ, but over sufficiently large (depending on the value σ) intervals of time, the bandwidth is essentially at most ρ. The source or network can insure conformance to this by using *leaky bucket flow control* [Turner, 1986], defined as follows.

A source is not allowed to transmit a packet unless it has a permit, and each transmitted packet by the source consumes one permit. As long as the number of permits is less than σ, the source receives new permits once every $(1/\rho)$ units of time. Thus, a source can accumulate up to σ permits. A packet that arrives when there are no permits can either be buffered by the source or discarded. Each ABR source could be assigned the parameters (σ, ρ), where ρ is the allocated bandwidth of the source, and σ represents the allocated burstiness of the source, in some sense.

Non-ABR sources require a certain minimum bandwidth from the network for satisfactory operation, and may also require that latency be bounded. Examples are multimedia sources, interactive database sources, distributed computing applications, and real-time applications. Leaky bucket flow control can also be applied to non-ABR sources. The parameter ρ specifies the maximum average bandwidth of the source, and the parameter σ measures the potential burstiness of the source, in some sense. If leaky bucket rate control is applied to all of the sources in the network, delay and buffering requirements can be controlled, and packet loss can be reduced or eliminated. Some recent analyses of this issue are presented in Cruz [1995, 1991a, 1991b]. We now briefly describe one important idea in these analyses.

A packet stream may become more bursty as it passes through network nodes. For example, suppose a packet stream arriving to a node is (σ, ρ) smooth, and that each packet has a delay in the node of at most d seconds. Consider the packet stream as it departs the node. Over any interval of length x, say, $[t, t+x]$, the packets that depart must have arrived to the switch in the interval $[t-d, t+x]$, since the delay is bounded by d. The number of these packets is bounded by $\rho(x+d) + \sigma = \rho x + (\sigma + \rho d)$, since the packet stream is (σ, ρ) smooth as it enters the node. Hence, the packet stream as it departs the node is $(\sigma + \rho d, \rho)$ smooth, and is potentially more bursty than the packet stream as it entered the node. Thus, as a packet stream passes through network nodes, it can get more bursty at each hop, and the buffering requirements will increase at each hop.

We note that buffer requirements at intermediate nodes can be reduced by using rate control (with buffering) to govern the transmission of packets *within* the network [Cruz, 1991b; Golestani, 1991]. This limits the burstiness of packet flow within the network, at the expense of some increase in average delay. It is also possible to reduce delay jitter by using this approach.

Defining Terms

ARQ protocol: An automatic repeat request (ARQ) protocol ensures reliable transport of packets between two points in a network. It involves labeling packets with sequence numbers and management of retransmission of lost or errored packets. It can be exercised on a link, or on an end-to-end basis.

Datagram network: A packet switched network that treats each packet as a separate entity. Each packet, or datagram, is forwarded to its destination independently. The data sources do not coordinate with the network or the destinations prior to sending a datagram. Datagrams may be lost in the network, without notification of the source.

Header field: A packet is generally divided into a header and a payload. The payload carries the data being transported, whereas the header contains control information for routing, sequence numbers for ARQ, flow control, error detection, etc. The header is generally divided into fields, and a header field contains information specific to a type of control function.

Packet switched network: A network where the fundamental unit of information is a packet. To achieve economy, the communication links of a packet switched network are statistically shared. Buffering is required at network nodes to resolve packet contentions for communication links.

Trunk reservation: A technique used in telephone networks whereby some portion of a communications link is reserved for circuits (trunks) that make most efficient use of the link. A link may reject a circuit request because of trunk reservation, so that other circuits which make better use of a the link may later be accepted.

Virtual circuit switched network: A packet switched network that organizes data packets according to the virtual circuit that they belong to. Packets from a given virtual circuit are routed along a fixed path, and each node along the path may reserve resources for the virtual circuit. A virtual circuit must be established before data can be forwarded along it.

References

Bertsekas, D. and Gallager, R. 1992. *Data Networks*, 2nd ed., Prentice–Hall, Englewood Cliffs, NJ.

Comer, D. 1991. *Internetworking with TCP/IP*, Vol. 1: *Principles, Protocols, and Architecture*, 2nd ed., Prentice–Hall, Englewood Cliffs, NJ.

Cruz, R.L. 1995. Quality of service guarantees in virtual circuit switched networks. *IEEE J. Selec. Areas Commum.*, 13(6):1048–1056.

Cruz, R.L. 1991a. A calculus for network delay, Part I: Network elements in isolation. *IEEE Trans. Inf. Th.*, 37(1):114–131.

Cruz, R.L. 1991b. A calculus for network delay, Part II: Network analysis. *IEEE Trans. Inf. Th.*, 37(1):132–141.

Cruz, R.L. and Tsai, J.T. 1996. COD: Alternative architectures for high speed packet switching. *IEEE/ACM Trans. Netwk.*, 4(1):11–21.

Golestani, S.J. 1991. Congestion-free communication in high speed packet networks. *IEEE Trans. Commun.*, 39(12):1802–1812.

Hahne, E. 1991. Round-robin scheduling for max-min fairness in data networks. *IEEE J. Selec. Areas Commun.*, 9(7):1024–1039.

Jacobson, V. 1988. Congestion avoidance and control. In *Proc. ACM SIGCOMM '88*, Stanford, CA, pp. 314–329.

Jain, R. 1986. A timeout-based congestion control scheme for window flow-controlled networks. *IEEE J. Selec. Areas Commun.*, 4(7):1162–1167.

Kung, H.T. and Morris, R. 1995. Credit-based flow control for ATM networks. *IEEE Netwk. Mag.*, 9(2):40–48.

Maxemchuk, N. 1987. Routing in the Manhattan street network. *IEEE Trans. Commun.*, 35(5):503–512.

Steenstrup, M., ed. 1995. *Routing in Communication Networks*. Prentice–Hall, Englewood Cliffs, NJ.

Turner, J.S. 1986. New directions in communications. *IEEE Commun. Mag.*, 24(10):8–15.

Further Information

Routing in Communication Networks, edited by Martha E. Steenstrup is a recent survey of the field of routing, including contributed chapters on routing in circuit switched networks, packet switched networks, high-speed networks, and mobile networks.

Internetworking with *TCP*, Vol. 1 by Douglas E. Comer provides an introduction to protocols used on the Internet, and provides pointers to information available on-line through the Internet.

Proceedings of the INFOCOM Conference are published annually by the Computer Society of the IEEE. These proceedings document some of the latest developments in communication networks.

The journal *IEEE/ACM Transactions on Networking* reports advances in the field of networks. For subscription information contact: IEEE Service Center, 445 Hoes Lane, P.O. Box 1331, Piscataway NJ, 088855-1331. Phone (800) 678-IEEE.

48
Transport Layer

48.1 Introduction 661
48.2 Transport Service 662
48.3 Data-Transfer Protocol 663
48.4 Connection-Management Protocol 665
48.5 Transport Protocols 667
48.6 Conclusions 670

A. Udaya Shankar
*University of Maryland,
College Park*

48.1 Introduction

The transport layer of a computer network is situated above the network layer and provides reliable communication service between any two users on the network. The users of the **transport service** are e-mail, remote login, file transfer, web browsers, etc. Within the transport layer are **transport entities**, one for each user. The entities execute distributed algorithms, referred to as **transport protocols**, that makes use of the message-passing service of the network layer to offer the transport service. The most well-known transport protocols today are the Internet Transmission Control Protocol (TCP), the original transport protocol, and the ISO TP4.

Historically, transport protocol design has been driven by the need to operate correctly inspite of unreliable network service and failure-prone networks and hosts. The channels provided by the network layer between any two entities can lose, duplicate, and reorder messages in transit. Entities and channels can fail and recover. An entity failure is fail-stop, that is, a failed entity performs no actions and retains no state information except for stable storage. A channel failure means that the probability of message delivery becomes negligible; that is, even with retransmissions a message is not delivered within a specified time. These fault-tolerance considerations are still valid in today's internetworks.

To a first approximation, we can equate transport service to reliable data transfer between any two users. But reliable data transfer requires resources at the entities, such as buffers and processes for retransmitting data, reassembling data, etc. These resources typically cannot be maintained across failures. Furthermore, maintaining the resources continuously for every pair of users would be prohibitively inefficient, because only a very small fraction of user pairs in a network exchange data with any regularity, especially in a large network such as the Internet.

Therefore, a transport service has two parts: **connection management** and **data transfer**. Data transfer provides for the reliable exchange of data between connected users. Connection management provides for the establishment and termination of connections between users. Users can open and close connections to other users, and can accept or reject incoming connection requests. Resources are acquired when a user enters a connection, and released when the user leaves the connection. An incoming connection request is rejected if the user has failed or its entity does not have adequate resources for new connections.

A key concern of transport protocols is to ensure that a connection is not infiltrated by old messages that may remain in the network from previous terminated connections. The standard techniques are to use the **3-way handshake** mechanism for connection management and the **sliding window** mechanism for data transfer within a connection. These mechanisms use **cyclic sequence numbers** to identify the connection attempts of a user and the data blocks within a connection. The protocol must ensure that received cyclic sequence numbers are correctly interpreted, and this invariably requires the network to enforce a **maximum message lifetime**.

In the following section, we describe the desired transport service. We then describe a simplified transport protocol that illustrates the core of TCP and ISO TP4. We conclude with a brief mention of performance issues and *minimum-latency* transport protocols.

Unless explicitly mentioned, we consider a **client–server** architecture. That is, the users of the transport layer are partitioned into clients and servers, and clients initiate connections to servers. Every user has a unique identity (id).

48.2 Transport Service

This section explains the correctness properties desired of the transport layer. We first define the notion of **incarnations**. An incarnation of a client is started whenever the client requests a connection to any server. An incarnation of a server is started whenever the server accepts a (potentially new) connection request from any client. Every incarnation is assigned an **incarnation number** when it starts; the incarnation is uniquely distinguished by its incarnation number and entity id.

Once an incarnation x of an entity c is started in an attempt to connect to an entity s, it has one of two possible futures. The first possibility is that at some point x becomes open and acquires an incarnation number y of some incarnation of s—we refer to this as "x becomes open to incarnation y of s"; at some later point x becomes closed. The second possibility is that x becomes closed without ever becoming open. This can happen to a client incarnation either because its connection request was rejected by the server or because of failure (in the server, the client, or the channels). It can happen to a server incarnation either because of failure or because it was started in response to a connection request that later turns out to be a duplicate request from some old, now closed, incarnation.

Because of failures, it is also possible that an incarnation x of c becomes open to incarnation y of s but y becomes closed without becoming open. This is referred to as a *half-open* connection.

A **connection** is an association between two open incarnations. Formally, a connection exists between incarnation x of entity c and incarnation y of entity s if y has become open to x and x has become open to y. The following properties are desired of connection management:

- Consistent connections: If an incarnation x of entity c becomes open to an incarnation y of entity s, then incarnation y is either open to x or will become open to x unless there are failures.
- Consistent data transfer: If an incarnation x of entity c becomes open to an incarnation y of entity s, then x accepts received data only if sent by y.
- Progress: If an incarnation x of a client requests a connection to a server, then a connection is established between x and an incarnation of the server within some specified time, provided the server does not reject x's request and neither client, server nor channels fail within that time.

Transport Layer 663

- Terminating handshakes: An entity cannot stay indefinitely in a state (or set of states) where it is repeatedly sending messages expecting a response that never arrives. (Such *infinite chatter* is worse than deadlock because in addition to not making progress, the protocol is consuming precious network resources.)

Given a connection between incarnations x and y, the following properties are desired of the data transfer between x and y:

- In-sequence delivery: Data blocks are received at entity $y(x)$ in the same order as they were sent by $x(y)$.
- Progress: A data block sent by $x(y)$ is received at $y(x)$ within some specified time, provided the connection is not terminated (either intentionally or due to failures) within that time.

48.3 Data-Transfer Protocol

This section describes the sliding-window method for achieving reliable flow controlled data transfer, *assuming that users are always connected and correctly initialized*. Later we add connection management to this protocol.

Consider two entities c and s connected by unreliable network channels. The user at c produces data blocks to be delivered to the user at s. Because the channels can lose messages, every data block has to be retransmitted until it is acknowledged. For throughput reasons, entity c should be able to have several data blocks outstanding, that is, sent but not acknowledged. Similarly, entity s should be able to buffer data blocks received out-of-sequence (due to retransmissions or network reordering).

Let us number the data blocks produced by user c with successively increasing sequence numbers, starting from 0. The sliding window mechanism maintains two windows of sequence numbers, one at each entity.

Entity c maintains the following variables:

- na: $\{0, 1, \ldots\}$; initially 0. This is the number of data blocks sent and acknowledged.
- ns: $\{0, 1, \ldots\}$; initially 0. This is the number of data blocks produced by the local user.
- sw: $\{0, 1, \ldots, SW\}$; initially 0. This is the maximum number of outstanding data blocks that the entity can buffer.

Data blocks na to $ns-1$ are outstanding and must be buffered. The entity can accept an additional $sw - ns + na$ data blocks from its user, that is, data blocks ns to $na + sw - 1$. The sequence numbers na to $na + sw - 1$ constitute the **send window**, and sw is its size.

Entity s maintains the following variables:

- nr: $\{0, 1, \ldots\}$; initially 0. This is the number of data blocks delivered to the user.
- rw: $\{0, 1, \ldots, RW\}$; initially 0. This is the maximum number of data blocks that the entity can buffer.

Data blocks 0 to $nr - 1$ have been delivered to the user in sequence. A subset, perhaps empty, of the data blocks nr to $nr + rw - 1$ have been received and are buffered. The sequence numbers nr to $nr + rw - 1$ constitute the **receive window**, and rw is its size.

The easiest way for an entity to identify a received data block or acknowledgement is for the message to include the sequence number of the concerned data block. But such a sequence number field would grow without bound. Instead the protocol uses cyclic sequence numbers; that is, j mod N, for some N, instead of the *unbounded* sequence number j.

To ensure that a received cyclic sequence number is correctly interpreted, it is necessary for the network to enforce a maximum message lifetime, that is, no message older than the lifetime remains in the network. It then suffices if N satisfies

$$N \geq SW + RW + \frac{L}{\delta}$$

where SW and RW are the maximum sizes of the send and receive windows, L is the maximum message lifetime, and δ is the minimum time between successive data block productions.

Flow control is another issue in data transfer, that is, entity c should not send data faster than the network or entity s can handle. By dynamically varying the send window size, the sliding window mechanism can also achieve flow control. In particular, consumer-directed flow control works as follows: entity s dynamically varies its receive window size to reflect local resource constraints, and regularly requests entity c to set its send window size to the receive window size. Note that in this case, the previous condition on N reduces to $N \geq 2RW + L/\delta$.

We finish this section with a specification of the sliding window protocol. The data messages of the protocol have the form (D, sid, rid, $data$, cj), where sid is the sender's id, rid is the intended receiver's id, $data$ is a data block, and cj is its cyclic sequence number.

The acknowledgment (ack) messages of the protocol have the form (ACK, sid, rid, cj, w), where sid and rid are as defined previously, cj is a cyclic sequence number, and w is a window size. When the message is sent, cj and w are set to the values of nr mod N and rw, respectively. Thus the message indicates the data block next expected in sequence. Because it acknowledges all earlier data blocks, it is referred to as a *cumulative ack*.

The events of the producer and consumer entities are shown in Fig. 48.1. There are two types of events. A *nonreceive* event has an enabling condition, denoted ec, and an action, denoted ac; the action can be executed whenever the event is enabled. A receive event for a message has only an action; it is executed whenever the message is received.

Entity c

Accept datablock from user (data)
ec: ns−na < sw
ac: buffer data as datablock ns;
    ns := ns+1

Send datablock (j)   // also resends
ec: 0 ≤ j ≤ ns−na−1
ac: Send (D, c, s, datablock j, j mod N)

Reception of (ACK, s, c, cj, w)
ac: tmp := (cj−na) mod N;
    if 1 ≤ tmp ≤ ns−na then
        delete datablocks na to na+tmp
            from send window;
        na := na+tmp;
        sw := w
    else if tmp = 0 then sw := max(sw, w)
    // else tmp > ns−na; do nothing

Entity s

Expand receive window
ec: rw < RW
ac: rw := rw+1

Deliver datablock to user (data)
ec: datablock nr in receive window
ac: data := datablock nr;
    nr := nr+1;
    rw := rw−1

Send acknowledgement
ec: true   // also resends
ac: Send (ACK, s, c, nr mod N, rw)

Reception of (D, c, s, data, cj)
ac: tmp := (cj−nr) mod N;
    if 0 ≤ tmp < rw then
        buffer data as datablock nr+tmp;
    // else tmp≥rw; do nothing

FIGURE 48.1 Events of sliding window protocol.

48.4 Connection-Management Protocol

This section describes a connection-management protocol. Traditional transport protocols, including TCP and TP4, identify successive incarnations by increasing, though not necessarily successively, incarnation numbers from some modulo-N space. Every entity uses a counter or a real-time clock to generate incarnation numbers for local incarnations.

Another feature of traditional transport protocols is that an entity stores a remote incarnation's number only while it is connected to the remote incarnation. This necessitates a 3-way handshake for connection establishment. A client that wants to connect to a server sends a connection request with its incarnation number, say, x. When the server receives this, it responds by sending a response containing x and a server incarnation number, say, y. When the client receives the response, it becomes open to y and responds by sending an ack containing x and y. The server becomes open when it receives the ack. The server could not become open when it received the connection request containing only x, because it may have been a duplicate from previous now terminated connection.

A 2-way handshake suffices for connection closing. An open entity sends a disconnect request that is acknowledged by the other entity.

A 2-way handshake also suffices for connection rejection. It is obvious that a server may have to reject a connection request of a client. What is not so obvious is that a client may have to reject a "connection request" of a server. Specifically, if a server receives an old connection request from a terminated incarnation of the client, the server attempts to complete the second stage of the 3-way handshake. In this case, the client has to reject the server.

The unreliable channels imply that a k-way handshake has the following structure: In every stage except the last, a message is sent repeatedly until the message of the next stage is received. The message of the last stage is sent only in response, otherwise the handshake would never terminate.

It is convenient to think of the protocol as a distributed system that is driven by user requests. Each user request causes the associated entity to initiate a 2- or 3-way handshake with the other entity. At each stage of the handshake, one entity learns something about the other entity and may issue an appropriate indication to its local user. At the end of the handshake, the protocol has served the user request. The protocol's behavior can be complex because two handshakes can be executing concurrently, with one of them conveying information that is relevant to the other.

There is an intricate relationship between the modulo-N space of the incarnation numbers and the handshaking algorithms, much more so than in the case of data transfer, since the latter assumes correctly initialized users. To achieve correct interpretation of received cyclic incarnation numbers, it is necessary to have bounds on message lifetime, incarnation lifetime, wait duration, and recovery duration. Under the reasonable assumption that the incarnation lifetime dominates the wait and recovery durations, it is sufficient and necessary to have

$$N \geq \frac{4L + I}{\alpha}$$

where L is the maximum message lifetime, I is the maximum incarnation lifetime, and α is the minimum time between successive incarnation creations at an entity. Most references in the literature incorrectly assume that $N \geq 2L/\alpha$ is sufficient.

This bound may not be satisfiable for exceedingly long-lived incarnations, say, of the order of days. In that case, if we assume that the probability of two successive connections having identical modulo-N client and server incarnation numbers is negligible (it is approximately $1/N^2$ under reasonable assumptions of incarnation lifetimes), then the following bound, which does not depend on I, suffices:

$$N \geq \frac{4L}{\alpha}$$

We now give a specification of the connection-management protocol.

A client entity maintains the following variables for each server s:

- *status*[s]: {CLOSED, OPENING, OPEN, CLOSING}; initially CLOSED. This is the status of the client's relationship with server s: CLOSED iff client has no incarnation involved with s. OPENING means client has an incarnation requesting a connection with s. OPEN means client has an incarnation open to s. CLOSING means client has an incarnation closing a connection with s.
- *lin*[s]: {NIL, 0, 1, ...}; initially NIL. This is the local incarnation number: NIL if *status*[s] equals CLOSED; otherwise identifies client incarnation involved with server s.
- *din*[s]: {NIL, 0, 1, ...}; initially NIL. This is the distant incarnation number: NIL if *status*[s] equals CLOSED or OPENING; otherwise identifies the incarnation of server s with which the client incarnation is involved.

A server entity maintains the following state variables for each client c:

- *status*[c]: {CLOSED, OPENING, OPEN}; initially CLOSED. This is status of server's relationship with client c: CLOSED iff server has no incarnation involved with c. OPENING means server has an incarnation accepting a connection request from c. OPEN means server has an incarnation open to c.
- *lin*[c]: {NIL, 0, 1, ...}; initially NIL. This is local incarnation number: NIL if *status*[c] equals CLOSED; otherwise identifies server incarnation involved with client c.
- *din*[c]: {NIL, 0, 1, ...}; initially NIL. This is the distant incarnation number. NIL if *status*[c] equals CLOSED; otherwise identifies the incarnation of client c with which the server incarnation is involved.

The messages of the protocol have the form (M, *sid*, *rid*, *sin*, *rin*), where M is the type of the message, *sid* is the sender's id, *rid* is the intended receiver's id, *sin* is the sender's incarnation number, and *rin* is the intended receiver's incarnation number. In some messages, *sin* or *rin* may be absent.

Each message is either a *primary* message or a *secondary* message. A primary message is sent repeatedly until a response is received or the maximum wait duration has elapsed. A secondary message is sent only in response to the reception of a primary message. Note that the response to a primary message may be another primary message, as in a 3-way handshake.

The messages sent by clients are as follows:

- (CR, *sid*, *rid*, *sin*): Connection request; sent when opening; primary message.
- (CRRACK, *sid*, *rid*, *sin*, *rin*): Acknowledgement to connection request reply (CRR); secondary message.
- (DR, *sid*, *rid*, *sin*, *rin*): Disconnect request; sent when closing; primary message.
- (REJ, *sid*, *rid*, *rin*): Reject response to a connection request reply that is received when closed. The *sin* of the received CRR is used as the value of *rin*; secondary message.

The messages sent by servers are as follows:

- (CRR, *sid*, *rid*, *sin*, *rin*): Reply to connection request in 3-way handshake; sent when opening; primary message.
- (DRACK, *sid*, *rid*, *sin*, *rin*): Response to disconnect request; secondary message.
- (REJ, *sid*, *rid*, *rin*): Reject response to a CR received when closed. The *sin* of the received message is used as the value of *rin*, secondary message.

The events of the client and server entities are shown in Figs. 48.2 and 48.3, assuming unbounded incarnation numbers.

Transport Layer

Client entity c: events concerning server s

ConnectRequest(s)
  ec: status[s] := CLOSED
  ac: status[s] := OPENING ; lin[s] := new incarnation number

DisconnectRequest(s)
  ec: status[s] = OPEN
  ac: status[s] := CLOSING

Abort(s)
  ec: status[s] ≠ CLOSED & "response timeout"
  ac: status[s] := CLOSED ; lin[s] := NIL ; din[s] := NIL

SendCR(s)
  ec: status[s] = OPENING
  ac: Send (CR, c, s, lin[s])

SendDR(s)
  ec: status[s] = CLOSING
  ac: Send (DR, c, s, lin[s], din[s])

Receive (CRR, s, c, sin, rin)
  ac: if status[s] = OPENING & rin = lin[s] then
     status[s] := OPEN ; din[s] := sin ;
     Send (CRRACK, c, s, lin[s], din[s])

    else if status[s] = OPEN & rin = lin[s] & sin = din[s] then
     // duplicate CRR
     Send (CRRACK, c, s, lin[s], din[s])

    else if status[s] = OPEN & rin = lin[s] & sin>din[s] then
     // server crashed, recovered, responding to old CR
     Send (REJ, c, s, sin) ; status[s] := CLOSED ;
     din[s] := NIL ; lin[s] := NIL

    else if status[s] is CLOSED or CLOSING then Send (REJ, c, s, sin)

Receive (REJ, s, c, rin)
  ac: if status[s] is OPENING or CLOSING & rin = lin[s] then
     status[s] := CLOSED ; din[s] := NIL ; lin[s] := NIL
    // else status[s] is OPEN or CLOSED; do nothing

Receive (DRACK, s, c, sin, rin)
  ac: if status[s] = CLOSING & rin = lin[s] & sin = din[s] then
     status[s] := CLOSED ; din[s] := NIL ; lin[s] := NIL
    // else status[s] is OPENING or OPEN or CLOSED; do nothing

FIGURE 48.2 Client events of connection management protocol.

48.5 Transport Protocols

A transport protocol between a client entity c and a server entity s consists of a connection-management protocol augmented with two data-transfer protocols, one for data transfer from c to s and another for data transfer from s to c. At each entity, the data-transfer protocol is initialized each time the entity becomes open and its events are executed only while open. The data-transfer messages are augmented by incarnation number fields, which are used by receiving entities to filter out data-transfer messages of old connections.

Server entity s: events concerning client c

Abort(c)
  ec: status[c] ≠ CLOSED & "response timeout"
  ac: status[c] := CLOSED ; lin[c] := NIL ; din[c] := NIL

SendCRR(c)
  ec: status[s] = OPENING
  ac: Send (CRR, s, c, lin[c], din[c])

Receive (CR, c, s, sin)
  ac: if status[c] = CLOSED & "rejecting connnections" then
        Send (REJ, s, c, sin) ;
      else if status[c] = CLOSED & "accepting connections" then
        lin[c] := new incarnation number ;
        status[c] := OPENING ; din[c] := sin
      else if status[c] = OPENING & sin>din[c] then
        // previous din[c] value was from some old CR
        din[c] := sin
      else if status[c] = OPEN & sin>din[c] then
        // client crashed, reconnecting
        if "willing to reopen" then
           lin[c] := new incarnation number ;
           din[c] := sin ; status[c] := OPENING
        else status[c] := CLOSED ; lin[c] := NIL ; din[c] := NIL
      // else status[c] = OPEN & sin ≤ din[c]; do nothing

Receive (CRRACK, c, s, sin,, rin,)
  ac: if status[c] = OPENING & sin = din[c] & rin = lin[c] then
        status[c] := OPEN
  // else status[c] is OPEN or CLOSED; do nothing

Receive (DR, c, s, sin, rin)
  ac: if status[c] = OPEN & sin = din[c] & rin = lin[c] then
        Send (DRACK, s, c, lin[c], din[c]) ;
        status[c] := CLOSED ; lin[c] := NIL ; din[c] := NIL
      else if status[c] = CLOSED then Send (DRACK, s, c, rin, sin) ;
  // else status[c] = OPENING; do nothing

Receive (REJ, c, s, rin)
  ac: if status[c] = OPENING & rin = lin[c] then
        status[c] := CLOSED ; lin[c] := NIL ; din[c] := NIL
  // else status[c] is OPEN or CLOSED; do nothing

FIGURE 48.3 Server events of connection management protocol.

We illustrate with the protocols of the previous sections. Start with the connection-management protocol of Sec. 48.4 between c and s. Add two sliding window protocols, one from c to s and one from s to c, as follows:

- At each entity, introduce variables *ns, na, sw*, and the send window buffers for the outgoing data transfer, and *nr, rw*, and the receive window buffers for the incoming data transfer. These data-transfer variables are initialized whenever the entity becomes open. Whenever the entity becomes closed, these variables are deallocated.

- Modify the client as follows. Add $status[s]$ = OPEN to the enabling condition of every data-transfer event. Add *sin* and *rin* fields to the sliding window protocol messages. When a data-transfer message is sent, *sin* is set to the local incarnation number *lin*[*s*] and *rin* is set to the remote incarnation number *din*[*s*]. When a data-transfer message is received, first test for *status*[*s*] = OPEN, *sin* = *din*[*s*] and *rin* = *lin*[*s*]. If the test fails, ignore the message, otherwise, process the message as in the sliding window protocol specification.

- Modify the server similarly.

There are various ways to extend and integrate the data-transfer protocols.

The messages of the two data-transfer protocols can be combined. For example, the data messages sent by an entity can have additional fields to *piggy-back* acknowledgement information for incoming data, that is, fields for *nr* and *rw*. This is done in TCP and ISO TP4.

The preceding protocol uses cumulative acknowledgments. We can also use *reject* messages to indicate a gap in the received data. Rejects allow the data source to retransmit missing data sooner than cumulative acks. The protocol can also use *selective* acknowledgements to indicate correctly received out-of-sequence data. This allows the data source to retransmit only what is needed, rather than everything outstanding. Selective acks and rejects are not used in TCP and ISO TP4, although there are studies indicating that they can improve performance significantly.

The preceding protocol uses fixed-size data blocks. An alternative is to send variable-sized data blocks. This can be done by augmenting the data messages with a field indicating the size of the data block. TCP does this with an octet, or byte, as the basic unit. A similar modification would be needed for selective acks.

The connection-management protocol can be extended and integrated in several ways.

The preceding protocol allows either user to close a connection at any point, without waiting for data transfer to be completed. An alternative is so-called *graceful closing*, where a user can close only its outgoing data transfer. The user must continue to handle incoming data until the remote user issues a close also. TCP has graceful closing, whereas ISO TP4 does not. It is a simple matter to add graceful closing to a protocol that does not have it.

It is possible to merge connection establishment, data transfer, and connection termination. The connection request can contain data, which would be delivered after the server becomes open. The connection request can also indicate that after the data is delivered the connection is to be closed. TCP allows this.

TCP uses a single 32-b cyclic sequence number space to identify both incarnations and data blocks. When an incarnation is created at an entity, an **initial sequence number** is chosen and assigned to the incarnation. Successive new message sent by the incarnation, whether of connection management or data transfer, occupy increasing sequence numbers starting from this initial sequence number.

TCP messages integrate both data transfer and connection management. Every TCP message has fields indicating the sequence number of the message, the next sequence number expected, the data segment (if any), the segment length, and receive window size. A connection-management message that requires an acknowledgement is considered to use up a sequence number, say, n. This means that the next new connection-management message sent by the entity, whether or not it requires an acknowledgement, would have the sequence number $n + 1$. The remote entity can acknowledge a connection-management message by sending a message of any type with its next expected sequence number field equal to $n + 1$.

The TCP messages SYN, SYN-ACK, ACK, FIN, FIN-ACK and RST correspond, respectively, to the messages CR, CRR, CRRACK, DR, DRACK, and REJ of our protocol.

TCP provides **balanced opening**, a service that is outside the client–server paradigm. Here, if two entities request connections to each other at the same time, a *single* connection is established. In the client–server architecture, each entity would be considered to have a client-half and a server-half, and two connections would be established. In fact, TCP's algorithm for balanced opening is flawed: it may result in valid connection requests being rejected and invalid connection requests leading to connections. Fortunately, no application seems to use TCP's balanced-opening service. ISO TP4 does not have balanced opening.

48.6 Conclusions

This chapter has described the service expected of a transport layer and a distributed algorithm, or protocol, that achieves this service. The protocol faithfully illustrates the inner workings of TCP and ISO TP4. We conclude by outlining two currently active research areas in transport protocols.

One area is that of minimum-latency transport protocols. The delay in connection establishment incurred by the 3-way handshake is unacceptable for many transaction-oriented applications such as remote procedure calls (RPCs). Note that although transaction data can be sent with a connection request, the server cannot process the transaction until it confirms that this is a new request. This has motivated the development of transport protocols where the server can determine the *newness* of a connection request as soon as it is received, thereby achieving connection establishment with a 2-way handshake.

To achieve this, the server has to retain information about clients even when it is not connected to them. Consider a 3-way handshake between client incarnation x and server incarnation y. If the server had remembered the incarnation number, say, z, that the client had previously used when it connected to the server, then the server could determine that the connection request with x was new (because x would be greater than z). In that case, the server could have become open at once, resulting in a 2-way handshake connection establishment. A server cannot be expected to indefinitely remember the last incarnation number of every client to which it was connected, due to the enormous number of clients in a typical internetwork. However, a caching scheme is feasible, and several have been proposed, culminating recently in a proposed modification to TCP.

An alternative to caching is to use timer-based mechanisms. Here also, a server is required to maintain information on each client it has served for a duration comparable to that in cache-based mechanisms (the major component in both is the message lifetime). In most timer-based protocols, if a client's entry is removed before the specified duration, for example, due to a crash or memory limitation, then the server can incorrectly accept old connection requests of that client. Simple connection management protocol (SCMP) is an exception: by assuming synchronized clocks, it maintains correctness but it may reject new connections for a period of time depending on clock skews and other parameters. In any case, timer-based approaches do not have a backup 3-way handshake.

Another area of current research concerns the performance of transport protocols. As mentioned earlier, transport protocol design has been historically driven by the need to operate correctly inspite of unreliable network service and failure-prone networks and hosts. This has resulted in the premise that transport protocols should have minimal knowledge of the network state. For example, message roundtrip times are the only knowledge that TCP has of the current network state.

In recent years, Internet traffic has grown both in quantity and variety, resulting in increasing utilization, delays, and often congestion. This has prompted the development of retransmission and windowing policies to reduce network congestion, mechanisms to reduce processing time for transport protocol messages, etc. However, this performance work is still governed by the premise, motivated by fault tolerance, that a transport protocol should have minimal knowledge of the network state.

It is not clear whether this premise would be valid for the high-speed networks of the near and far future. Current implementations of TCP use a very conservative flow control scheme: the source reduces its send window by a multiplicative factor whenever roundtrip times indicate network congestion, and increases the send window by an additive factor in the absence of such indication. Although this is very robust, it tends to underutilize the network resources. It should be noted that most protocol proposals for high-speed networks make use of some form of resource reservation to offer varying quality-of-service guarantees. It is not clear how this can be incorporated within the current TCP framework, or even whether it should be.

Defining Terms

3-way handshake: A handshake between two entities involving three messages, say $m1, m2, m3$. The initiating entity repeatedly sends $m1$. The other entity responds by repeatedly sending $m2$. The initiating entity responds by sending a single $m3$ for each $m2$ received.

Transport Layer 671

Balanced-opening: Where two entities requesting connections to each other at the same time establish a single connection.
Client–server architecture: A paradigm where the users of the transport layer consist of clients and servers, connections are always between clients and servers, and connections are always initiated by clients.
Connection: An association between two open incarnations.
Connection management: That portion of the transport service and protocol responsible for the establishment and termination of connections between users.
Cyclic sequence numbers: The modulo-N sequence numbers used for identifying incarnations and data blocks.
Data transfer: That portion of the transport service and protocol responsible for the reliable exchange of data between connected users.
Incarnation: An instance of an entity created for a connection attempt.
Incarnation numbers: Sequence numbers for identifying incarnations.
Initial sequence numbers: The incarnation numbers of TCP.
Maximum message lifetime: The maximum duration for which a message can reside in the network.
Receive window: The data blocks that can be buffered at the consuming entity.
Send window: The data blocks that can be outstanding at the producing entity.
Sliding-window mechanism: A method for sending data reliably over unreliable channels, using cyclic sequence numbers to identify data, a send window at the data producer, and a receive window at the data consumer.
Transport entities: The processes within the transport layer.
Transport protocols: The distributed algorithm executed by the transport entities communicating over the network service.
Transport service: The service expected of the transport layer.

Further Information

The specification of TCP can be found in:

Postel, J. 1981. *Transmission Control Protocol: DARPA Internet Program Protocol Specification*. Request for Comment RFC-793, STD-007, Network Information Center, Sept.

The specification of ISO TP4 can be found in:

International Standards Organization 1989. *Information Processing Systems – Open Systems Interconnection – Transport Protocol Specification Addendum 2: Class four operation over connectionless network service*. International Standard ISO 8073/DAD 2, ISO, April.

The full specification and analysis of the data-transfer protocol in Sec. 48.3, including selective acks and rejects, can be found in:

Shankar, A.U. 1989. Verified data transfer protocols with variable flow control, *ACM Trans. Comput. Syst.*, 7(3):281–316.

The full specification and analysis of the connection management protocol in Sec. 48.4, as well as caching-based protocols and references to timer-based protocols can be found in:

Shankar, A.U. and Lee, D. 1995. Minimum latency transport protocols with modulo-N incarnation numbers. *IEEE/ACM Trans. Networking*, 3(3):255–268.

Composing data transfer and connection management is explained in:

Shankar, A.U. 1991. Modular design principles for protocols with an application to the transport layer, *Proc. IEEE*, 79(12):1687–1709.

Balanced opening and the flaws of TCP are described in:

Murphy, S.L. and Shankar, A.U. 1991. Connection management for the transport layer: Service specification and protocol verification. *IEEE Trans. Commun.*, 39(12):1762–1775.

49
Gigabit Networks[1]

Jonathan M. Smith
University of Pennsylvania

49.1 Introduction 672
49.2 Asynchronous Transfer Mode Host Interfacing 674
49.3 Multimedia Architectures 675
49.4 Distributed Shared Memory Communications 677
49.5 Conclusions 678

49.1 Introduction

This chapter summarizes what we have learned in the past decade of research into extremely high throughput networks. Such networks are colloquially referred to as *gigabit networks* in reference to the billion bit per second throughput regime they now operate in. The engineering challenges are in the integration of fast transmission systems and high-performance engineering workstations.

High-throughput fiber optic networks, prototype high-speed packet switches, and high-performance workstations were all available in the late 1980s. A major engineering challenge was integrating these elements into a computer networking system capable of high application-to-application throughput. As a result of a proposal from D. Farber and R. Kahn (the Kahn/Farber Initiative) the U.S. Government [National Science Foundation (NSF) and DARPA] funded sets of collaborators in five gigabit testbeds. [Computer Staff, 1990]. These testbeds were responsible for investigating different issues, such as applications, metropolitan-area networks (MANs) vs local-area networks (LANs), and technologies such as high performance parallel interface (HIPPI), asynchronous transfer mode (ATM), and synchronous optical network (SONET) [Partridge, 1993]. The Aurora gigabit testbed linked University of Pennsylvania (Penn), Bellcore, IBM, and MIT, with gigabit transmission facilities provided by collaborators Bell Atlantic, MCI and Nynex, and was charged with exploring technologies for gigabit networking, whereas the other four testbeds were applications-focused and hence used off-the-shelf technologies. Results of Aurora work underpin today's high-speed network infrastructures.

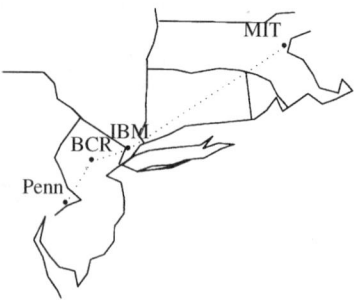

FIGURE 49.1   Aurora geography.

[1]Support for research reported by University of Pennsylvania came from Bellcore (through Project Dawn), IBM, Hewlett-Packard, National Science Foundation and Defense Advanced Research Projects Agency (DARPA) through Corporation for National Research Initiative (CNRI) under cooperative agreement NCR-89-19038, and the National Science Foundation under CDA-92-14924.

Gigabit Networks

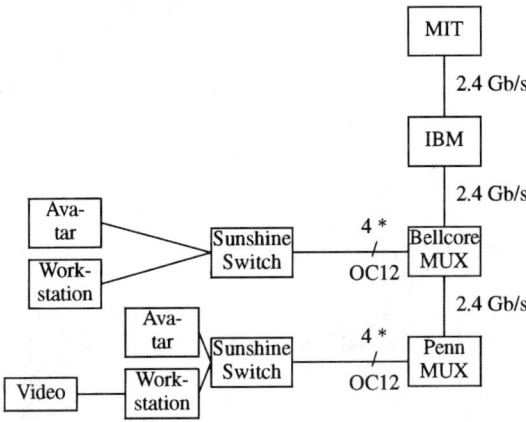

FIGURE 49.2 Partial Aurora logical topology.

Clark et al. [1993] set out the research goals and plans for the testbed and outline majors research directions addressed by the testbed. Aurora uniquely addressed the issues of providing switched infrastructure and high end-to-end throughput between workstation class machines. In contrast to supercomputers, these machines were, and are, the basis for most computing today.

Figures 49.1 and 49.2 provide illustrations of the Aurora geography and logical topologies, respectively.

Sections 49.2–49.4 describe ATM network host interface architectures that can operate in gigabit ranges, multimedia aspects of gigabit networks, and distributed shared memory as an applications programming interface for gigabit networks. Section 49.5 summarizes our state of knowledge. Table 49.1 provides a compact summary of some major Aurora milestones.

TABLE 49.1 Aurora Gigabit Testbed: Selected Milestones

Date	Milestone
5/6/93	2.4 Gb/s OC-48 SONET backbone operational Penn <=> Bellcore
5/7/93	End-to-end data between workstations at Penn and Bellcore, interoperating Penn and Bellcore ATM host interfaces
5/19/93	Sunshine switches ATM cells between IBM RS/6000 at Penn and IBM RS/6000 at Bellcore
6/7/93	Penn and Bellcore ATM interfaces interoperate through Sunshine
6/8/93	End-to-end video over ATM from Penn workstation w/Penn video card to Bellcore workstation display
10/26/93	2nd Sunshine operational, at Penn
11/12/93	Full-motion A/V teleconference over packet transfer mode (PTM)/SONET, Penn <=> IBM
12/31/93	25 Mb/s TCP/IP over Aurora switched loopback
2/25/94	Cheap Video ATM appliance running over Aurora
3/15/94	Telementoring interactive distance learning over Aurora Penn <=> Bellcore using Cheap Video NTSC/ATM
3/30/94	70 Mb/s TCP/IP over Aurora between RS/6000s
4/17/94	MNFS/AIX solving differential heat equations over Aurora
4/21/94	Avatar, w/audio VC operational Penn <=> Bellcore, and IBM PVS IBM <=> over PlaNET, and VuNet Penn <=> MIT
5/6/94	Avatar in operation Penn <=> MIT
12/31/94	Link to IBM and MIT taken out of operation
7/5/95	HP PA-RISC/Afterburner ATM Link Adapter achieves 144 Mb/s TCP/IP
8/22/95	ATM Link Adapter achieves 215 + Mb/s TCP/IP

49.2 Asynchronous Transfer Mode Host Interfacing

Aurora work showed that efficient, low-cost host/computer interfaces to ATM networks can be built and incorporated into a hardware/software architecture for workstation-class machines. This was believed problematic due to the nature of small, fixed-size ATM cells and their mismatch with workstation memory architectures. Penn designed [Traw and Smith, 1991] and implemented an ATM host interface for the IBM RISC System/6000 workstation, which connects to the machine's micro channel bus. It translates variable-size application data units into streams of fixed-size ATM cells using dedicated segmentation-and-reassembly logic. The novel application of a content-addressable memory, hardware-implemented linked-list manager and the reassembly pipeline structure allowed use of a low clock speed, and hence low-cost technologies. The cellification and decellification logic have measured performance which could support data rates of 600–700 Mb/s [Traw, 1995].

A major concern with advanced applications such as medical imaging and teleconferencing is privacy. Privacy transformations have traditionally been rather slow due to the *bit-complexity* of the substitution—and confusion—introducing operations. An augmentation of the network host interface with cryptographic hardware was designed [Smith, Traw, and Farber, 1992]. It was based on observations by Broscius and Smith [1991], which describes the use of parallelism to achieve high performance in an implementation of the National Bureau of Standards (NBS) data encryption standard. The board was implemented and achieved a measured performance of 100 Mb/s. Among the interesting features were the use of GaAs programmable logic arrays (PLAs) for the substitution boxes in the cipher and a scheme for unrolling the embedded loops using multiple instances of the hardware. The difficulty of getting data to and from the encrypting hardware through a bus remained. Smith, Traw, and Farber [1992] describe the history and motivation for the architecture, and explain how to insert a high-performance cryptographic chip (for example, the VLSI Technologies VM007 data encryption standard (DES) chip, which operates at 192 Mb/s) into the ATM host interface architecture. The resulting system is able to operate at full network speed while providing a per-cell (agile) per virtual channel identifier (VCI) rekeying; both the performance and the operation are transparent to the host computer, while providing much greater key control than possible with link encryption approaches.

Traw and Smith [1991] describe one of the two earliest workstation host interfaces for ATM networks, both done in Aurora. This interface chose an all-hardware solution, with careful separation of functions between hardware and software implementation. Traw and Smith [1993] report on the implementation of the ATM host interface and its support software. The architecture is presented in detail, and design decisions are evaluated. Later work [Smith and Traw, 1993] focused attention on the adaptor to application path through software, and some of the key design decisions embedded in the software are examined. Of particular interest are the system performance measures where the adaptor operates with a significant application workload present.

The initial software subsystem provided an application programmer interface roughly equivalent to a raw Internet protocol (IP) socket, and was able to achieve more than 90% of the hardware subsystem's performance, thus driving an optical carrier level 3 (OC-3) at its full 155 Mb/s rate. Key innovations were the reduction of data copying (through use of virtual memory (VM) support; this direction was later followed by others, including the University of Arizona team [Druschel, Peterson, and Davie, 1994] designing software for the Osiris [Davie, 1993] interface developed at Bellcore by Bruce Davie) and the partitioning of functions between hardware and software. As can be seen from Table 49.2 [Druschel et al., 1993] this reduction in data copying was necessitated by the memory bandwidth limitations of early-1990s workstations. The bottleneck on the IBM RS/6000 was initially the workstation's system bus to I/O bus interconnect [Traw and Smith, 1993] however, improvements to the I/O subsystem architecture moved the bottleneck to the physical link. For the Hewlett-Packard (HP) PA-RISC implementation [Traw, 1995] designed to demonstrate scaling of the host interface architecture to higher speeds, the bottleneck was the bus attachment (for this environment the SGC graphics bus served as the attachment point).

The HP PA-RISC/afterburner ATM link adapter held a record for highest reported transport control protocol (TCP)/IP/ATM performance of 215 + Mb/s for almost one year. This performance

TABLE 49.2 Workstation Memory Bandwidths

	Memory (Mb/s, peak)	CPU/Memory (Mb/s, sustained)		
		Copy	Read	Write
IBM RS/6000 340	2133	405(0.19)	605(0.30)	590(0.28)
Sun SS10/30	2300	220(0.10)	350(0.15)	330(0.14)
HP 9000/720	1600	160(0.10)	450(0.28)	315(0.20)
Dec 5000/200	800	100(0.13)	100(0.13)	570(0.71)

Source: Tabulation by Druschel, P., Abbott, M.B., Pagels, M.A., and Peterson, L.L., 1993. Network subsystem design. *IEEE Network* special issue, 7(4):8–17.

was measured between two HP PA-RISC 755s, connected by a 320 Mb/s SONET-compliant null modem, using the *netperf* test program. The best performance was achieved using a 32-kilobyte socket buffer size and 256-kilobyte packets.

Custom physical layer interfaces were implemented as daughter cards so that alternate physical layers (e.g., AMD TAXI and HP GLINK [Moore and Traw, 1995]) could be explored within the context of the Aurora testbed. The GLINK implementation allowed low-cost distribution of SONET-rate ATM over twisted pairs in networks that are about the diameter of a laboratory work area (50 ft.); coaxial cable extends the operational limitations of electrical GLINK to 300 ft.

Software for the IBM RS/6000 ATM interface was enhanced by the addition of TCP/IP support [Alexander, Traw, and Smith, 1994] implemented as a common I/O (CIO) loadable device driver, which allowed us to operate at 70 Mb/s sustained over the testbed. For the Aurora testbed, this was the first and fastest operational TCP/IP which carried traffic over the wide-area network (WAN). It has been used since to carry Mether-NFS (MNFS) distributed shared memory traffic over the testbed between Penn and Bellcore. When the IP is used as a component of the user datagram protocol (UDP)/IP protocol stack, over 90 Mb/s were obtained on an RS/6000 model 580 connected to an RS/6000 model 530.

Traw and Smith [1995] showed that host interfaces could be aggregated in a number of manners to support multiples of the bandwidth provided by a single adapter. The results of a simulation study showed that for hardware implementations, striping at the byte or ATM cell level might be appropriate; in this model the host adaptor would provide a protocol data unit (PDU) interface to the host and perform the striping transparently; Bellcore's Osiris interface performed cell striping and the IBM SIA performed byte striping. A software-implemented solution would stripe most effectively by using multiple interfaces to send multiple concurrent IP packets; then TCP/IP's facilities for in-order delivery of packets would compensate for the skew between links.

49.3 Multimedia Architectures

Multimedia architectures for gigabit endworking must be designed with scale, endpoint heterogeneity, and application requirements as the key driving elements. We devised an integrated multimedia architecture with which applications define which data are to be bundled together for transport and select which data are unbundled from received packages. This allows sources to choose the degree of resource allocation they wish to provide; receivers choose which elements of the package they wish to produce. Although potentially wasteful of bandwidth, the massive reduction in the multiplicity of customized channels allows sources to service a far greater number of endpoints and receivers to accommodate endpoint resources by reproducing what they are capable of. The scaling advantage of this approach is that much of the complexity of customization is moved closest to the point where customization is necessary, the endpoint.

Multimedia work included the development of custom hardware; for example, an early video capture board used for experiments between Penn and Bellcore was developed [Udani, 1992]. Experiments with this microchannel architecture adapter suggested that software-manipulated video would not operate with acceptable quality. This led to the all-hardware National Television Systems

Committee (NTSC)/ATM Avatar ATM appliance developed by Traw and Marcus for use in telementoring experiments linking Penn and Bellcore for purposes of undergraduate digital design projects focused on developing ATM hardware. The Avatar [Marcus and Traw, 1995] card, which supports NTSC video and CD-quality audio, is the first example of an ATM appliance, with a parts cost of under $300. Many of these cards were fabricated. They were used for distance teaching when connecting the Bellcore experimental Video Windows, for collaborative work between researchers at Penn and Bellcore, and for teleconferencing between Penn and Massachusetts Institute of Technology.

Much of the multimedia focus rested on the development of operating system abstractions that could support high-speed applications. These abstractions used the hardware and low-level operating system support developed for the IBM RISC System/6000 workstations equipped with the AIX operating system, an IBM implementation of UNIX. Key new ideas included a more general model of quality of service (QoS) requirements, and technical means for evaluating how any bandwidth allocation implementation requires support from the operating system scheduling mechanism for true *end-to-end* service delivery. Nahrstedt [1995] identified the software support services needed to provide QoS guarantees to advanced applications that control the characteristics of their networking system, and adapt within parameterized limits. These services form a kernel, or a least common subset of services, required to support advanced applications.

A logical relationship between applications-specified QoS [Nahrstedt and Smith, 1996], as well as operating system policy and mechanism, and network-provided QoS was developed. An example challenge is the kinematic data stream directing a robotic arm, which can tolerate neither packet drops nor packet delays—unlike video or audio, which can tolerate drops but not delays. The approach used of a bidirectional translator (like a compiler/uncompiler pair for a computer language) which resides between the network service interface and the application's service primitives. This can dynamically change QoS as application requirements or network capabilities change, allowing better use of network capacity, which can be mapped more closely to applications current needs than if a worst-case requirement is maintained. The implementation [Nahrstedt, 1995] outlined the complete requirements for such a strategy, including communication primitives and data necessary for translation between network and application. For example, an application request to zoom and refocus a camera on the busiest part of a scene will certainly require peer-to-peer communication between the application and the camera management entity. The network may need to understand the implications for interframe compression schemes and required bandwidth allocations. The translation method renegotiates QoS as necessary.

These ideas were described in Nahrstedt and Smith [1995], which describes a mechanism to provide bidirectional negotiation of quality-of-service parameters between applications and the other elements of a workstation participant in advanced networked applications. The scheme is modeled on a *broker*, a traditional mechanism for carrying on back-and-forth negotiations while filtering implementation details irrelevant to the negotiation. The QoS broker reflects both the dynamics of service demands for complex applications and the treatment of both applications and service kernels as first-class participants in the negotiation process. The QoS Broker was implemented in the context of a system for robotic teleoperation implemented over ATM in cooperation with Penn's General Robotics and Sensory Perception (GRASP) laboratory. The broker was implemented and evaluated as part of a complete end-to-end architecture presented by Nahrstedt [1995].

In the system, application requirements are determined by a negotiation protocol at startup. This turned out to be a major cost in the system, as the worst-case scheduler consumed considerable time in testing the feasibility of resource guarantees. Nonetheless, the system was capable of providing guaranteed services; a complete implementation, including a novel real-time protocol stack, is available in source-code form with anonymous FTP from `ftp.cis.upenn.edu`.

Gigabit multimedia is desired by the applications community; Bajcsy et al. [1994], describes the need for network support for a broader class of applications than audio/video. In particular, it has become clear that interaction with the physical world is among the most challenging applications for networking, as the QoS requirements for many systems will be sufficiently complex to cause interaction and competition between requirements. An example would be a tradeoff between throughput and reliability, which would tend one way for real-time video, while in the opposite direction for force-feedback data. The results could have considerable bearing on critical national challenges such as agile manufacturing, as software for a reliable gigabit network infrastructure providing end-to-end guarantees could be developed on the principles described in the thesis.

49.4 Distributed Shared Memory Communications

Farber [1995] proposed distributed shared memory (DSM) as a technology solution for integrating computation and communications more closely. This was one of the major investigations of the Aurora testbed, and the MNFS [Minnich, 1993] distributed shared memory has been used to support applications over the Aurora WAN such as a parallel heat equation solver. There were four major questions we sought to answer in the experimental evaluation of DSM in the Aurora project. Each of these were answered, as we outline subsequently; a more sweeping perspective was given by Farber in his 1995 ACM SIGCOMM Award Lecture. These four research questions were as follows:

1. Is DSM a reasonable abstraction for distributed programming? Yes, it is, as demonstrated by applications ported directly from shared-memory supercomputers. DSM is an abstraction for distributed applications programming. It has the ability to support programming with distributed control and shared data across a wide range of interconnected computer models. Distributed shared memory (sometimes also called distributed virtual memory) is an interesting communications paradigm for gigabit networks. DSM may provide the best path for optimizing the construction of distributed systems requiring high-speed networking, especially where the traditional balance between network speed and processing speed has been inverted. The rationale is that memory management is well understood, and that memory speeds represent the best case achievable for interprocess communication. A combination of cacheing and a cache policy known as prefetching can shape the traffic produced by the application.
2. Can DSM work over WANs? Yes, it can, and appears to work reasonably well, although many optimizations remain, such as better programming language interactions, cache management, techniques for preloading caches, and reductions in false sharing due to data layout.
3. What effect does increased bandwidth (e.g., gigabit per second) have on DSM performance delivered to applications? Shaffer's thesis [1996] showed that distributed shared memory was a viable technology for parallel applications even in a WAN environment. This speaks to the fundamental scientific questions about the relationships and tradeoffs between bandwidth and communications latency induced by propagation delay. A key insight was that for data-intensive applications, observed latency can be more a function of throughput than of physical propagation delay. This is due to the fact that as PDUs are used at higher levels of protocol architectures, the PDU does not arrive until its last bit has arrived. This means that throughput has a significant effect on latency observed at any layer other than physical, where the PDU can be considered to be a bit.
4. What effect does combining high bandwidth and high delay (in high *bandwidth * delay* product) networks have on the DSM performance delivered to applications? The key issue in the testbed specialization of the distributed shared memory paradigm was the effect of increased propagation delay on application performance. Shaffer demonstrates [1996] that application-measured latency is a function of both propagation delay and the *throughput delivered end-to-end*. Whereas the propagation delay is clearly a fundamental limit given speed-of-light limitations and the like, it may not be the dominant cost. The consequence of this is that high-bandwidth networks can reduce delay simply by reducing the latency component associated with throughput. This is especially true of data objects of a large enough size to be affected by throughput considerations; for example, virtual memory page

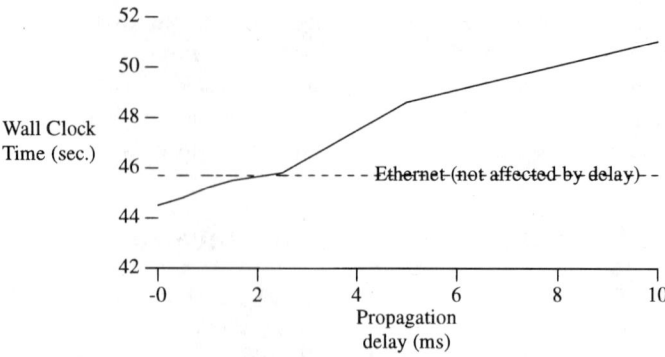

FIGURE 49.3 Performance of distributed heat equation solution, DSM/ATM.

sizes are typically 4096 or 8192 bytes. The DSM experiments on the testbed itself required the entire experimental ATM infrastructure built for Aurora. After the Bellcore to Penn span was terminated, an experimental output port controller (OPCv2) designed by Marcus [1996], was programmed to emulate selected delay characteristics. The experimental configuration studies the effect of a large *bandwidth* ∗ *delay* link by replacing an MNFS client machine connected via ethernet LAN to one connected through the ATM WAN, either directly (when the WAN operated) or by emulation with OPCv2. A parallel computation, a heat equation solver, uses a central difference method solution. This problem is extremely computation intensive, since the computational complexity of matrix solution is at least quadratic in terms of the problem size. Figure 49.3 plots completion time for a large problem instance (1200 × 1000 elements) against the delay induced by the OPCv2 for the ATM-connected host. The key observations to make are that the ATM solution outperforms the ethernet solution, with the same problem on the same software on the same machines, for a variety of emulated distances. Each millisecond is equivalent to 130 mi of fiber distance; thus a 1-ms delay, where the ATM configured system outperforms the ethernet LAN, represents the distance to Bell Communications Research from Penn. The delay measured to Bellcore and back through the Aurora WAN was 2.1 ms, or equivalent to 1.05 ms on this plot. The measured completion times for the computation are consistent between the real and emulated environments.

The experiments have shown that parallel applications running over the Aurora infrastructure execute as quickly as when run over a local ethernet LAN, giving one data point in the space of bandwidth ∗ delay tradeoffs. The OPCv2 allows the space to be explored more completely once the two measurements from the testbed configuration are used to anchor any LAN-based emulations to true WAN delay values.

49.5 Conclusions

> There is an old network saying: Bandwidth problems can be cured with money. Latency problems are harder because the speed of light is fixed—you can't bribe God.
> D. Clark [Patterson and Hennessy, 1996]

The Aurora gigabit testbed research had a fundamental influence on the design and development of gigabit network technology.

The ATM host interface work answered concerns about the viability of ATM segmentation and reassembly (SAR). It is now overwhelmingly clear that ATM SAR can operate at gigabit/second rates, and thus the performance concerns expressed when the testbeds were begun were largely nonissues.

ATM host interface hardware developed in Aurora has influenced all available commercial products, which resemble either the Penn ATM host interface with hardware SAR (though implemented with an application specific integrated circuit (ASIC) [Chu, 1996] rather than programmable logic arrays (PLAs) or the Bellcore Osiris (with software-directed SAR). Operating systems research at Penn and later work at the University of Arizona showed how to reduce copying between host interfaces and applications through careful management of DMA, buffer pools, and process data access semantics; these ideas are now appearing in software from vendors such as Sun Microsystems [Chu, 1996] and Silicon Graphics. It is thus also clear that operating systems can deliver gigabit range throughputs to applications with appropriate restructuring and rethinking of copying and protection boundary crossing. Issues we identified at Penn such as interprocess communication (IPC) latency, are now being studied by others such as Cornell University, using commercial ATM host interfaces; this work has had a considerable effect on the operating systems community: several ACM Symposium on Operating Systems Principles (SOSP) and USENIX papers have been stimulated by it. Among our other ATM contributions was a collaborative effort with Bellcore that produced an early ATM appliance, the Avatar NTSC/ATM video board developed at Penn and Bellcore for telementoring.

The still unanswered questions revolve around the discovery and evaluation of mechanisms that deliver a practical reduction in latency to applications. These include better cache management algorithms for distributed shared memory as well as techniques for lookahead referencing, or prefetching. Another important area is the reduction in operating system induced costs in network latency; the software overhead is equivalent to about 200 mi of fiber in the AIX implementation Penn did for Aurora. Of course, it should be noted that the primary target was high applications *throughput*.

The multimedia work in the gigabit networking research community has had impact on the operating systems and robotics communities. It has also pointed out some issues to be avoided in adapter design, such as head-of-line blocking observed in serial DMA of large ATM AAL3/4 CS-PDUs on the IBM RISC System/6000 system. The work influenced industry (particularly IBM Heidelberg) and has been covered in a reference work on multimedia [Steinmetz and Nahrstedt, 1995].

The DSM work remains controversial in the systems community as custom and inertia make new ideas slow to accept. The important observation we can draw from the work in Aurora is that if the mechanism becomes more widely accepted, there are now algorithms that can aid in the location and prefetching of data developed, and significant experimental evidence supporting the hypothesis that higher network bandwidths can aid distributed applications of any type to achieve higher performance.

References

Alexander, D.S., Traw, C.B.S., and Smith, J.M. 1994. Embedding high speed ATM in UNIX IP. In *USENIX High-Speed Networking Symposium*, Oakland, CA, pp. 119–121, Aug. 1–3.

Bajcsy, R., Farber, D.J., Paul, R.P., and Smith, J.M. 1994. Gigabit telerobotics: Applying advanced information infrastructure. In *1994 International Symposium on Robotics and Manufacturing*, Maui, HI, Aug.

Broscius, A.G. and Smith, J.M. 1991. Exploiting parallelism in hardware implementation of the DES. In *Proceedings, CRYPTO 1991 Conference*, ed. J. Feigenbaum, Santa Barbara, CA, pp. 367–376, Aug.

Chu, J. 1996. Zero-copy TCP in Solaris. In *Proceedings USENIX 1996 Annual Technical Conference*, San Diego, CA, pp. 253–264, Jan. 22–26.

Clark, D.D. Davie, B.S., Farber, D.J., Gopal, I.S., Kadaba, B.K., Sincoskie, W.D., Smith, J.M., and Tennenhouse, D.L. 1993. The AURORA gigabit testbed. *Computer Networks and ISDN Systems*, 25(6):599–621, North-Holland.

Computer Staff, 1990. Gigabit network testbeds. *IEEE Computer*, 23(9):77–80.

Davie, B.S. 1993. The architecture and implementation of a high speed host interface. *IEEE J. Selec. Areas Commun.*, Special issue on high-speed computer/network interfaces, 11(2):228–239.

Druschel, P., Abbott, M.B., Pagels, M.A., and Peterson, L.L. 1993. Network subsystem design. *IEEE Network*, Special issue: End-system support for high-speed networks (breaking through the network I/O bottleneck), 7(4):8–17.

Druschel, P., Peterson, L.L., and Davie, B.S. 1994. Experiences with a high-speed network adapter: A software perspective. In *Proceedings 1994 SIGCOMM Conference*, London, UK, pp. 2–13, Aug. 31–Sept. 2.

Farber, D.J. 1995. *The Convergence of Computers and Communications—Part 2*, ACM SIGCOMM Award Lecture, Aug. 30.

Marcus, W.S. 1996. An experimental device for multimedia experimentation. *IEEE/ACM Trans. Networking*, to appear.

Marcus W.S. and Traw, C.B.S. 1995. AVATAR: ATM video/audio transmit and receive. Distributed Systems Laboratory, University of Pennsylvania Tech. Rep. March.

Minnich, R.G. 1993. Mether-*NFS*: A modified NFS which supports virtual shared memory. In *Expepriences with Distributed and Mutltiprocessor Systems (SEDMS IV)*, pp. 89–107. USENIX Assoc., San Diego, CA.

Moore, A.M. and Traw, C.B.S. 1995. GLINK as a solution for local ATM distribution. In *Proceedings of the 1995 Design SuperCon Conference on Digital Communications Design*, San Jose, CA, pp. 5:1–5:20, Feb.

Nahrstedt, K. 1995. *An Architecture for Provision of End-to-End QoS Guarantees*, Ph.D. thesis, Tech. Rep. CIS Dept., Univ. Pennsylvania.

Nahrstedt, K. and Smith, J.M. 1995. The QoS broker. *IEEE Multimedia Mag.*, 2(1):53–67.

Nahrstedt, K. and Smith, J.M. 1996. Design, implementation and experiences of the OMEGA end-point architecture. *IEEE J. Selec. Areas Commun.*, Special issue on multimedia systems, to appear.

Partridge, C. 1993. *Gigabit Networking*, Addison-Wesley, Reading, MA.

Patterson, D.A. and Hennessy, J.L. 1996. *Computer Architecture: A Quantitative Approach*, 2nd ed., Morgan Kaufmann, San Francisco, CA.

Shaffer, J.H. 1996. *The Effects of High-Bandwidth Networks on Wide-Area Distributed Systems*, Ph.D. thesis, CIS Dept., Univ. Pennsylvania.

Smith, J.M. and Traw, C.B.S. 1993. Giving applications access of Gb/s networking. *IEEE Network*, Special issue: End-system support for high-speed networks (breaking through the network I/O bottleneck), 7(4):44–52.

Smith, J.M., Traw, C.B.S., and Farber, D.J. 1992. Cryptographic support for a gigabit network. In *Proceedings INET '92*, Kobe, Japan, pp. 229–237, Inaugural Conference of the Internet Society, June 15–18.

Steinmetz, R. and Nahrstedt, K. 1995. *Multimedia: Computing, Communications, and Applications*, Prentice-Hall, Englewood Cliffs, NJ.

Traw, C.B.S. 1995. *Applying Architectural Parallelism in High Performance Network Subsystems*, Ph.D. thesis, CIS Dept., Univ. Pennsylvania.

Traw, C.B.S. and Smith, J.M. 1991. A high-performance host interface for ATM networks. In *Proceedings SIGCOMM 1991*, Zurich, Switzerland, pp. 317–325, Sept. 4–6.

Traw, C.B.S. and Smith, J.M. 1993. Hardware/software organization of a high-performance ATM host interface. *IEEE J. Selec. Areas Commun.*, Special issue on high speed computer/network interfaces 11(2):240–253.

Traw, C.B.S. and Smith, J.M. 1995. Striping within the network subsystem. *IEEE Network* (July/Aug.):22–32.

Udani, S.K. 1992. *Architectural Considerations in the Design of Video Capture Hardware*, M.S.E thesis (EE), School of Engineering and Applied Sciences, Univ. Pennsylvania, April.

50
Local Area Networks

Thomas G. Robertazzi
State University of New York, Stony Brook

50.1 Introduction 681
50.2 Local Area Networks 681
    Carrier Sense Buses • Token Ring • Token Bus • Wireless LANs • Asyncronous Transfer Mode (ATM) LANs • Private Branch Exchange (PBX)
50.3 Metropolitan Area Networks: The LAN Interconnect 684
    Fiber Distributed Data Interface (FDDI) • Distributed Queue Dual Bus (DQDB)
50.4 The Future 684

50.1 Introduction

Local area networks (LANs) are computer networks operating over a small area such as a single department in a company. Typically, a single LAN will support a small number of users (say, less than 25). Most LANs installed to date have served to transport data only between terminals, personal computers, workstations, and computers. Future LANs may carry a variety of traffic, such as voice, video, and data.

In general, the elements of a computer or local area network must follow compatible rules of operation together to function effectively. These rules of operation are known as protocols. A variety of LANs, operating under different protocols, are available today. These are described subsequently.

From their inception in the early 1980s, LANs have been able to provide significantly higher data rates than the switched telephone network. Thus, interconnecting these high data rate LANs has been a problem. Two metropolitan area network (MAN) solutions for interconnection, fiber distributed data interface (FDDI), and distributed queue dual bus (DQDB), are also described at the end of this chapter.

50.2 Local Area Networks

There are six main types of local network architectures that have been commercially produced to date: carrier sense multiple access (CSMA) buses with collision detection (CSMA/CD), token rings, token buses, wireless LANs, asynchronous transfer mode (ATM) LANs and private branch exchanges (PBXs). The first four have been standardized in the IEEE 802 series standards.

Carrier Sense Buses

The IEEE 802.3 standard deals with a network architecture and protocol first constructed at Xerox in the 1970s and termed ethernet. All stations in an ethernet are connected, through interfaces, to a coaxial cable (or wire) that is run through the ceiling or floor near each user's computer equipment. In 1994 there were 50 million ethernet nodes with 15 million new nodes being added each year.

The coaxial cable essentially acts as a private radio channel for the users. An interesting protocol called carrier sense multiple access with collision detection is used in such a network. Each station constantly monitors the cable and can detect when it is idle (no user transmitting), when one user is transmitting (successfully), or when more than one user is simultaneously transmitting (resulting in an unsuccessful collision on the channel). The cable basically acts as a broadcast bus. Any station can transmit on the cable if the station detects it to be idle. Once a station transmits, other stations will not interrupt the transmission. As there is no central control in the network, occasionally two or more stations may attempt to transmit at about the same time. The transmissions will overlap and be unintelligible (collision). The transmitting stations will detect such a situation and and each will retransmit at a randomly chosen later time.

Ethernet and 802.3 networks have raw speeds of up to 10 million b/s. Idle time and collisions, however, can reduce the useful information throughput significantly. Coaxial cables, **twisted pair wire** or **fiber optics** can be used for these networks. The maximum length of these networks is limited by signal propagation delay. An 802.3 bus differs from an internal computer bus in size and in the lack of a bus controller.

Since the introduction of ethernet more than a dozen years ago, desktop computer capabilities and networking requirements have increased significantly. To meet these demands a 100-Mb/s (fast) ethernet has been developed by a consortium of companies. This will be standardized within IEEE 802.3. Media options for the 100-Mb/s ethernet include shielded or unshielded twisted pair as well as fiber optics. Wiring distances of up to 100 m between an end system and wiring closet can be supported. Adapters as well as repeaters that can operate at either 10 or 100 Mb/s will be available.

Token Ring

Token ring LANs were developed by IBM in the early 1980s. Topologically, stations are arranged in a circle with point-to-point links between neighbors. Transmissions flow in only one direction (clockwise or counter-clockwise). A message transmitted is relayed over the point-to-point links to the receiving station and then forwarded around the rest of the ring and back to the sender to serve as an acknowledgement.

Only a station possessing a special digital codeword known as a token may transmit. When a station is finished transmitting it passes the token to its downstream neighbor. Thus, there are no collisions in a token ring and utilization can approach 100% under heavy loads.

Because of the use of point-to-point links, token rings can use various transmission media such as twisted pair wire or fiber optic cables. The transmission speed of a token ring can range from 1 to 16 Mb/s depending on the type of point-to-point links used. Token rings are often wired in star configurations for ease of installation. Token rings are covered by the IEEE 802.5 standard. In 1994 there were 10 million token ring nodes with 3–4 million new nodes being added each year.

Token Bus

A token bus uses a coaxial cable along with the token concept to produce a LAN with improved thruput compared to the 802.3 protocol. That is, stations pass a token from one to another to determine which station currently has permission to transmit. Also, in a token bus (and in a token ring) response times can be deterministically bounded. This is important in factory automation where commands to machines must be received by set times. By way of comparison, response times in an ethernet-like network can only be probabilistically defined. For this reason, General Motor's Manufacturing Automation Protocol makes use of the token bus. Token buses can operate at 1, 5, and 10 Mb/s. Token bus operation is standardized in the IEEE 802.4 standard.

Wireless LANs

Wireless LANs use a common radio channel to provide LAN connectivity without any physical wiring. Protocols for wireless LANs are currently being standardized as the IEEE 802.11 standard. Although, one might consider the use of a CSMA/CD protocol in this environment, stations using radio technology are unable to listen to the same channel on which they are transmitting. Thus, it is not possible to implement the collision detection (CD) part of CSMA/CD in a radio environment. Therefore, a modified protocol known as carrier sense multiple access with collision avoidance (CSMA/CA) can be used.

In this variant on CSMA contention for the channel at the end of a successful transmission is mitigated by computing a stochastic idle time for each station during which a station puts off its transmission to see if the channel remains free. A second method of channel access in the form of polling from a master station can be provided for traffic with time delay constraints.

Wireless LANS have aggregate capacities from several hundred kilobit per second to the low megabit per second range. The future of wireless LANs is unclear since it is possible that this capacity will not be sufficient to meet new demands for services.

One possible way around the problem of the limited spectrum available for wireless LANs is to use infrared light as the transmission medium. There are both direct infrared systems (range: 1–3 mi) and nondirect systems (bounced off walls or ceilings). For small areas data rates are consistent with those of existing ethernet and token ring networks with 100-Mb/s systems on the horizon.

Asyncronous Transfer Mode (ATM) LANs

ATM LANs are relatively new to the LAN marketplace. This is a packet switching technology utilizing relatively short, fixed length packets to provide networking services. ATM was originally seen as a way to develop the next generation wide area telephone network using packet switching, rather than the more conventional circuit switching technology. It was envisioned as a way to transport video, voice, and data in an integrated network. A short packet size was chosen to meet several requirements including minimizing real-time queueing delay.

While progress on the original goal of using ATM technology in wide area telephone networks proceeded slowly because of the complexity of the challenge and the large investments involved, a number of smaller companies introduced ATM local area network products using much the same technology.

An ATM LAN consists of a switch into which are wired end users in a star-type topology. There are several possibilities for the internal architecture of the switch. A low-cost switch may essentially be a computer bus in a box. More sophisticated switches may use switching fabrics. These are very large-scale integrated (VLSI) implementations of patterned networks of simple switching elements, sometimes referred to as space division switching. A great deal of effort has gone into producing cost-effective ATM switches over the past 12 years. It should be pointed out that many of the issues that are not yet resolved for wide area network ATM (i.e., traffic policing, billing) are more tractable in the private ATM LAN environment.

ATM LANs can support a relatively small number of users at high data access rates (low megabit per second). Although ATM is good at handling mixed media traffic at high speeds, it remains to be seen if enough applications are developed requiring its high-bandwidth capability to make it a success. The cost effectiveness of ATM technology is another issue awaiting resolution.

Private Branch Exchange (PBX)

Historically, private branch exchanges were privately owned telephone switching computers that would be placed in the basement of a building and serve to interconnect phones in the building and provide access to outside lines provided by common carriers. However PBXs are now available that offer both telephone and data service. In a typical system a phone may have a data socket for terminals or workstations. PBXs are wired in a star topology with the PBX at the center of the star and each user wired directly to it.

50.3 Metropolitan Area Networks: The LAN Interconnect

Although several network architectures have been proposed for use as metropolitan area networks (MANs) the two that are closest to widespread commercial implementation are FDDI and DQDB. A key feature of a MAN is the ability to interconnect LANs. This is a problem because of the high data rates at which LANs operate.

Fiber Distributed Data Interface (FDDI)

The fiber distributed data interface is similar to a token ring LAN except that two rings, instead of one, may be used. Stations needing high reliability communication are connected to both rings. In the case of a break in the rings the network can be automatically reconfigured. FDDI rings operate at 100 Mb/s with a maximum of 500 nodes and a maximum fiber length of 200 km. In fact most actual FDDI installations have only a small number of nodes (such as routers). There is an American National Standards Institute (ANSI) standard for FDDI.

Distributed Queue Dual Bus (DQDB)

The distributed queue dual bus interface forms the basis of the IEEE 802.6 standard for MANs. DQDB is descended from the earlier queued packet and synchronous switch (QPSX), which was developed at the University of Western Australia and Telecom Australia. DQDB uses two unidirectional linear fiber optic buses. Stations are connected to both buses. Through the clever use of counters, the DQDB protocol provides approximate first-in first-out (FIFO) service to arriving packets. There are no collisions in DQDB so utilization can approach 100%. Bus speeds of 150 Mb/s are possible.

50.4 The Future

The future is likely to see an increase in data rates as fiber optic cables are widely deployed. This will spur the development of faster switching nodes through the use of parallel processing and VLSI implementation. Protocols will have to be simplified to increase processor throughput. New forms of traffic such as video and graphics will become more important. Computer networks in general, and LANs in particular, will proliferate throughout the world, making possible the ubiquitous transport of data between any two points on the globe.

Defining Terms

Area networks: LAN, within single building; MAN, metropolitan sized region; WAN, national/international region.
Coaxial cable: A shielded cable that conducts electrical signals and is used in bus type local area networks.
Fiber optic cable: A glass fiber cable that conducts light signals. Fiber optic cables can carry data at very high speeds and are immune to electrical interference.
Twisted pair wire: Two wires that are twisted together. Wires with different numbers of twists per unit length are available that provide different amounts of rejection to electromagnetic interference.

References

Black, U. 1989. *Data Networks: Concepts, Theory and Practice*, Prentice-Hall, Englewood Cliffs, NJ.

De Prycker, M. 1991. *Asynchronous Transfer Mode: Solution for Broadband ISDN*, Simon and Schuster, New York.

De Simone, A. and Nanda, S. 1995. Wireless Data: Systems, Standards, Services. *Wireless Networks*, 1(3):241–253.

Muller, N.J. 1995. *Wireless Data Networking*, Artech House, Boston, MA.

Peterson, L. and Davie, B. 1995. *Computer Networks: A Systems Approach*, Morgan Kaufman, San Francisco, CA.

Robertazzi, T.G. 1993. *Performance Evaluation of High Speed Switching Fabrics and Networks: ATM, Broadband ISDN and MAN Technology*, IEEE Press, Piscataway, NJ.

Sherer, P. 1994. The 100 Mbps Ethernet Standard. In Distinguished Lecture Series (IX) (videotape) Univ. Video Communications, Stanford, CA.

Spragins, J.D., Hammond, J.L., and Pawlikowski, K. 1991. *Telecommunications Protocols and Design*, Addison–Wesley, Reading, MA.

Walrand, J.N.D. 1991. *Communication Networks: A First Course*, Aksen Associates, Inc., and Richard Irwin Inc., Boston, MA.

Further Information

The following are the IEEE 802 series of standards related to local area networks.

IEEE 802.3: CSMA/CD bus protocol standard.
IEEE 802.4: Token bus standard.
IEEE 802.5: Token ring standard.
IEEE 802.6: DQDB metropolitan area network standard.
IEEE 802.11: CSMA/CA wireless LAN standard.

Tutorial articles on LANs appear in *IEEE Communications Magazine* and *IEEE Network* magazine. Technical articles on LANs appear in *IEEE Transactions on Networking, IEEE Transactions on Communications, IEEE Journal on Selected Areas in Communications,* and the journal *Wireless Networks*.

51
Asynchronous Time Division Switching[1]

Achille Pattavina
Politecnico di Milano, Italy

51.1 Introduction 686
51.2 The ATM Standard 687
51.3 Switch Model 688
51.4 ATM Switch with Blocking Multistage IN and Minimum Depth 690
51.5 ATM Switch with Blocking Multistage IN and Arbitrary Depth 692
51.6 ATM Switch with Nonblocking IN 696
51.7 Conclusions 699

51.1 Introduction

In the last decades separate communication networks have been deployed to support specific sets of services. For example, voice communication services (and, transparently, some low-speed data services) are supported by circuit switched networks, whereas packet switched networks have been specifically designed for low-to-medium speed data services. During the 1980s a worldwide research effort was undertaken to show the feasibility of packet switches capable of supporting narrowband services, such as voice and low-to-medium speed communications, together with broadband services, such as high-speed data communications and those services typically based on video applications.

The challenge of the forthcoming broadband-Integrated Services Digital Network (B-ISDN), as envisioned by the International Telecommunication Union–Telecommunication Standardization Sector (ITU-TSS), is to deploy a unique transport network based on *asynchronous time-division* (ATD) multiplexing and switching that provides a B-ISDN interface flexible enough to support all of today's services (voice and data) as well as future narrowband (for data applications) and broadband (typically for video applications) services to be defined. With ATD the bandwidth is not preallocated in the transmission and switching equipment of the broadband network so as to fully exploit by statistical multiplexing the available capacity of communication resources. ITU-TSS has defined a standard for the transfer mode based on ATD switching called the **asynchronous transfer mode (ATM)** [ITU, 1993].

---
[1]Work carried out at Politecnico di Milano/CEFRIEL, Milan (Italy), and supported by the Italian Ministry of University and Scientific Research (MURST), 60% funds.

This chapter is intended to present a brief review of the basic ATM concepts, as well as of the main architectural and performance issues of ATM switching. The basic principles of the asynchronous transfer mode are described in Sec. 51.2, whereas the model of an ATM switch is presented in Sec. 51.3 by discussing its main functions. Three main classes of ATM switching fabrics are identified therein based on the characteristics of their *interconnection network* (IN) and the basic characteristics and properties for each class are discussed in the following sections. ATM switches whose IN is blocking with minimum depth and arbitrary depth are dealt with in Sec. 51.4 and Sec. 51.5, respectively, whereas architectures with nonblocking IN are reported on in Sec. 51.6.

51.2 The ATM Standard

Time-division multiplexing (TDM) today is the most common technique used to transport on the same transmission medium tens or hundreds of channels, each carrying an independent digital flow of information, typically, digitized voice signals. A preventive per-channel allocation of the transmission capacity, in terms of a set of slots within a periodic frame, is done with TDM that time interleaves the supported channels onto the same transmission medium. Therefore, a rigid allocation of the transmission resource is obtained, in which the available bandwidth is fully used only if all of the channels are active at any time; that is, all of the channel slots carry user information. Therefore, TDM is well suited to support communication services with a high activity rate, as in the case of voice services, since other services whose information sources are active only for a small percentage of time, typically, data services, would waste transmission bandwidth.

The asynchronous time division (ATD) technique is intended to overcome this problem by enabling the sharing of transmission and switching resources by several channels without any preventive bandwidth allocation to the single users. Therefore, by means of a suitable storage capability, information from the single channels can be statistically multiplexed onto the same communication resource, thus avoiding resource wastage when the source activity is low. Asynchronous time-division multiplexing (ATDM) requires, however, each piece of information to be accompanied by the owner information, which is no longer given by the position of the information within a frame, as in the case of TDM. Switching ATDM channels requires the availability of ad-hoc packet switching fabrics designed to switch enormous amount of information when compared to the switching fabrics of current narrowband packet switched networks.

The B-ISDN envisioned by ITU-TSS is expected to support a heterogeneous set of narrowband and broadband services by sharing as much as possible the functionalities provided by a unique underlying transmission medium. ATM is the ATD-based transport mechanism to be adopted in the lower layers of the protocol architecture for this purpose [ITU, 1993]. Two distinctive features characterize an ATM network: (1) The user information is transferred through the network in small fixed-size units, called **ATM cells**, each 53 bytes long, divided into a *payload* (48 bytes) for the user information and a *header* (5 bytes) for control data. (2) It is a *connection-oriented* network, that is, cells are transferred onto previously setup virtual links identified by a label carried in the cell header. Both *virtual connections* (VC) and *virtual paths* (VP) are defined in the B-ISDN. A logical connection between two end users consists of a series of $n + 1$ virtual channels if n switching nodes are crossed. A virtual path is a bundle of virtual channels. Since a virtual channel is labeled by means of a hierarchical key virtual path identified/virtual channel identifier (VPI/VCI), a switching fabric can operate either a full VC switching or just a VP switching.

The B-ISDN protocol architecture includes three layers that, from the bottom up, are referred to as *physical layer, ATM layer* and *ATM adaptation layer* (AAL) [ITU, 1991]. The task of the physical layer is to provide a transmission capability for the transfer of ATM cells. Its functionalities include some basic tasks, such as timing, cell delineation, cell header verification, etc. Since the standard interface has the minimum rate of about 150 Mb/s, even ATM switching nodes of medium size with, say, 128 input and output links must be capable of carrying loads of tens of gigabits per second. The ATM layer includes the VCI/VPI translation, the header generation and extraction, and the multiplexing/demultiplexing of cells with different VCI/VPI onto the same physical layer connection. The purpose of the ATM adaptation layer is to add different sets of functionalities to the ATM layer, so as to differentiate the kind of services provided to the higher layers. Four service classes are defined to support connection-oriented and connectionless services. These services range from circuit emulation and variable bit rate video to low-speed data services. Even if the quality of service parameters to be guaranteed for each class are still to be defined, the B-ISDN network should be able to provide different traffic performance to each service class in terms of packet loss and delay figures.

51.3 Switch Model

Research in ATM switching has been developed worldwide for several years showing the feasibility of ATM switching fabrics. However, a unique taxonomy of ATM switching architectures is very hard to find, since different keys used in different orders can be used to classify ATM switches. Very briefly, we can say that most of the ATM switch proposals rely on the adoption for the interconnection network, which is the switch core, of multistage arrangements of very simple switching elements (SEs) each using the **cell self-routing** concept. This technique consists in allowing each SE to switch (route) autonomously the received cell(s) by only using a self-routing label preceding the cell. Even if other kinds of switching architectures can be used as well (e.g., shared memory or shared medium units), multistage INs provide the processing power required to carry the abovementioned load foreseen for small-to-large-size ATM switches. The interconnection network, which is the ATM switch core, operates by transferring in a time window, called the *slot*, the packets, which are received aligned by the IN from the switch inlets to the requested switch outlets.

Rather than using technological features of the switching architecture, the main key to classify the switch proposals can be found in the functional relationship setup between inlets and outlets by the switch. Multistage INs can be basically classified as **blocking** or **nonblocking**. In the former case different input/output (I/O) paths within the IN can share some interstage links; thus, the control of packet loss events requires the adoption of additional techniques, such as the packet storage capability in the SEs (**minimum-depth** INs), or deflection routing (**arbitrary-depth** INs). In the latter case different I/O paths are available, so that SEs do not need internal buffers and are much simpler to implement (a few tens of gates per SE). Nevertheless, these INs require more stages than blocking INs. A further distinctive feature of ATM switches is where the buffers holding cells are placed; three configurations are distinguished with reference to each single SE or to the whole IN, that is, **input queueing, output queueing**, and **shared queueing**.

We will refer here only to the cell switching of an ATM node, by discussing the operations related to the transfer of cells from the inputs to the outputs of the switch. Thus, all of the functionalities relevant to the setup and teardown of the virtual connections through the switch are intentionally disregarded. The general model of a $N \times N$ switch is shown in Fig. 51.1. The reference switch includes N *input port controllers* (IPC), N *output port controllers* (OPC), and an interconnection network (IN), where the IN is capable of switching up to K cells to the same OPC in one slot. The IN is said to have a speedup K if $K > 1$ and no speedup if $K = 1$, since an internal bit rate higher than the external rate (or an equivalent space-division technique) is required to allow the transfer of more than one cell to the same OPC. The IN is usually a multistage arrangement of very simple SEs, typically 2×2, which can be either provided with input/output/shared buffers (*SE queueing*), or unbuffered (*IN queueing*). In this last case input and output queueing, whenever adopted, take place at IPC and OPC, respectively, whereas shared queueing is accomplished by means of additional hardware associated to the IN. Depending on the type of interconnection network and the queueing

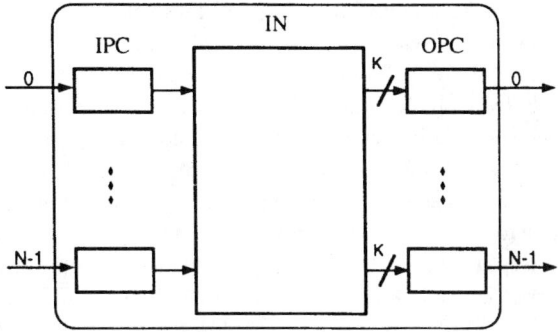

FIGURE 51.1 Model of ATM switch.

adopted, three classes of ATM switches will be distinguished:

- *Blocking multistage IN with minimum depth.* SE queueing and no speedup are adopted; the interconnection network is blocking and makes available only one path per I/O pair.
- *Blocking multistage IN with arbitrary depth.* IN queueing (typically, output queueing) and a speedup K are adopted; the interconnection network is blocking but makes available more than one path per I/O pair.
- *Nonblocking IN.* IN queueing and a speedup K are adopted.

In general, two types of conflicts characterize the switching operation in the interconnection network in each slot, the **internal conflicts** and the **external conflicts**. The former occur when two I/O paths compete for the same internal resource, that is, the same interstage link in a multistage arrangement, whereas the latter take place when more than K packets are switched in the same slot to the same OPC. An interconnection network $N \times N$ with speedup K, $K \leq N$, is said to be nonblocking if it guarantees absence of internal conflicts for any arbitrary switching configuration free from external conflicts for the given **network speedup** value K. That is, a nonblocking IN is able to transfer to the OPCs up to N packets per slot, in which at most K of them address the same switch output. Note that the adoption of output queues either in a SE or in the IN is strictly related to a full exploitation of the switch speedup; in fact, a structure with $K = 1$ does not require output queues, since the output port is able to transmit downstream one packet per slot. Whenever queues are placed in different elements of the ATM switch, it is assumed that there is no backpressure between queues, so that packets can be lost in any queue due to buffer saturation.

The main functions of the port controllers are:

- Rate matching between the input/output channel rate and the switching fabric rate
- Aligning cells for switching (IPC) and transmission (OPC) purposes (it requires a temporary buffer of one cell)
- Processing the cell received (IPC) according to the supported protocol functionalities at the ATM layer; a mandatory task is the routing (switching) function, that is, the allocation of a switch output and a new VPI/VCI to each cell, based on the VCI/VPI carried by the header of the received cell
- Attaching (IPC) and stripping (OPC) self-routing labels to each cell
- With IN queueing, storing (IPC) the packets to be transmitted and probing the availability of an I/O path through the IN to the addressed output if input queueing is adopted; queueing (OPC) the packets at the switch output if output queueing is adopted

The traffic performance of the nonblocking ATM switches will be described by referring to an offered uniform random traffic, with $p \, (0 < p \leq 1)$ denoting the probability that a packet is received at a switch input in a slot. All packet arrival events at IPCs are mutually independent, and each switch outlet is selected with the same probability by the cells. Typically, three parameters are used to describe the switching fabric performance, all referred to steady-state conditions: the *switch throughput* $\rho \, (0 < \rho \leq 1)$ (the traffic carried by the switch) expressed as a utilization factor of its output links; the *average packet delay* $T \, (T \geq 1)$, that is, average number of slots it takes for a packet received at a switch input to cross the network and thus be transmitted downstream by the addressed switch output, and the *packet loss probability* $\pi \, (0 < \pi \leq 1)$, defined as the probability that a packet received at a switch input is lost due to buffer overflow. The maximum throughput ρ_{max}, also referred to as *switch capacity*, indicates the load carried by the switch for an offered load $p = 1$.

Needless to say, our dream is a switching architecture with minimum complexity, capacity very close to 1, average packet delay less than a few slots, and a packet loss probability as low as desired, for example, less than 10^{-9}. The target loss performance is usually the most difficult task to achieve in an ATM switch.

51.4 ATM Switch with Blocking Multistage IN and Minimum Depth

The class of ATM switch characterized by a blocking multistage IN with minimum depth is based on the adoption of a *banyan network* as IN. Banyan networks are multistage arrangements of very simple 2×2 switching elements without internal speedup with interstage connection patterns such that only one path exists between any inlet and outlet of the network and different I/O paths can share one or more interstage links. SEs in a banyan network are organized in $n = \log_2 N$ stages, each comprising $N/2$ SEs. Figure 51.2 represents the *reverse baseline* network [Wu and Feng,

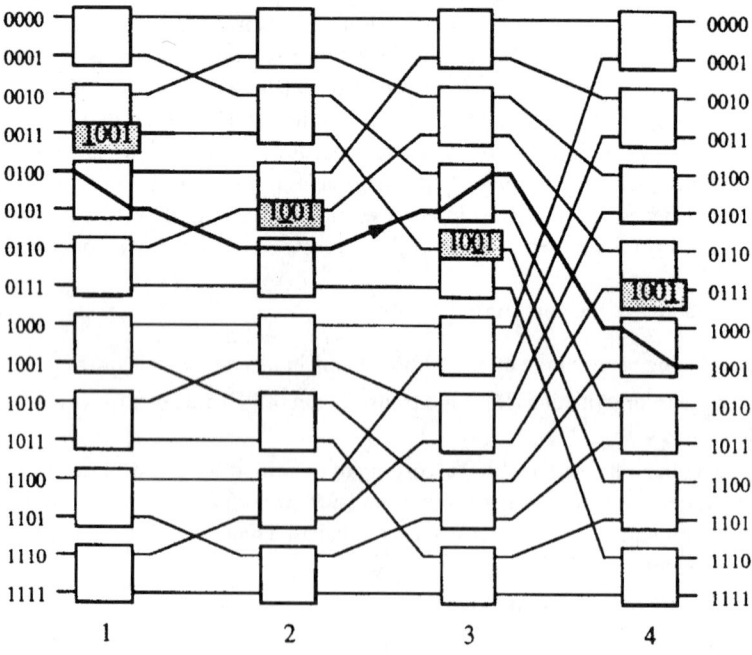

FIGURE 51.2 Reverse baseline network with self-routing example.

1980]. A SE with high and low inlets (and outlets) labeled 0 and 1, respectively, can assume only two states, *straight* giving the I/O paths 0-0 and 1-1 in the SE, and *cross* giving the I/O paths 0-1 and 1-0.

The availability of only one I/O path per inlet/outlet pair with interstage links shared by several I/O paths causes the banyan network blocking due to the internal conflicts. In spite of its high blocking probability, the key property of banyan networks that suggests their adoption in ATM switches is their cell self-routing capability: an ATM cell preceded by an address label, the *routing tag*, is given an I/O path through the network in a distributed fashion by the network itself. For a given topology this path is uniquely determined by the inlet address and by the routing tag, whose bits are used, one per stage, by the switching elements along the paths to route the cell to the requested outlet. The specific bit to be used, whose value 0 (1) means that the cell requires the high (low) SE outlet, depends on the banyan network topology. An example of self-routing in a reverse baseline network is shown in Fig. 51.2 where a cell received on inlet 4 addresses outlet 9.

As is clear from the preceding description, the operations of the SEs in the network are mutually independent, so that the processing capability of each stage in a $N \times N$ switch is $N/2$ times the processing capability of one SE. Thus, a very high parallelism is attained in packet processing within the IN of an ATM switch by relying on space-division techniques.

As already mentioned, interstage links are shared by several I/O paths that cause packet loss if two packets require the same link in a slot. Provision of a queueing capability in the SE (SE queueing) is therefore mandatory to be able to control the packet loss performance of the switch. In general, three configurations are considered: input queueing (IQ), output queueing (OQ) and shared queueing (SQ), where a buffer is associated with each SE inlet, each SE outlet or shared by all the SE inlets and outlets, respectively. Therefore the SE routing is operated after (before) the packet storage with IQ (OQ), both before and after the storage with SQ (see Fig. 51.3). Note that these three solutions have an increasing hardware complexity: each buffer is read and written at most once per slot with IQ, is read once and written twice per slot with OQ, is read twice and written twice per slot with SQ.

The maximum throughput of these networks increases with the total buffer size B (Fig. 51.4) from a value around $\rho_{max} = 0.4$–0.5 for all of the cases to a value close to $\rho_{max} = 0.75$ for input queueing and $\rho_{max} = 1.0$ for output and shared queueing. These values are the asymptotic throughputs expected, since a very large buffer in each SE implies that the throughput degradation is only due to internal conflicts. By definition these conflicts do not occur with OQ SEs and they give the asymptotic throughput of a single 2×2 SE with IQ and OQ SEs. Given a total SE buffer size, SQ gives the best cell loss performance; OQ is significantly better than IQ unless a very low offered load is considered, as shown in Fig. 51.5 for a total SE buffer of 16 ATM cells for a varying offered load p.

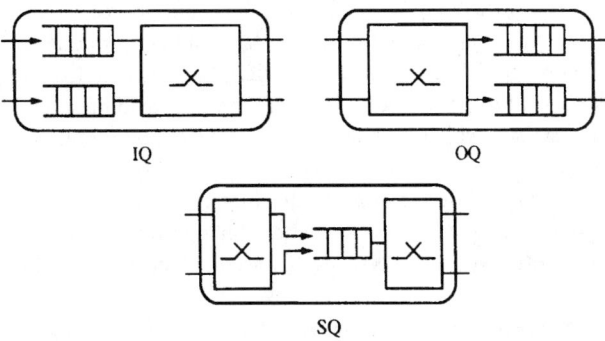

FIGURE 51.3 Structure of a 2×2 SE with IQ, OQ, and SQ.

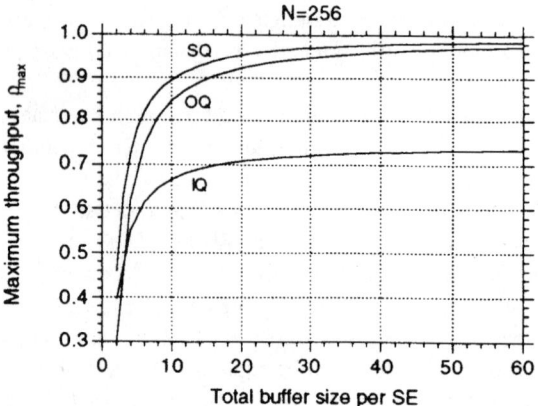

FIGURE 51.4 Throughput performance of banyan networks.

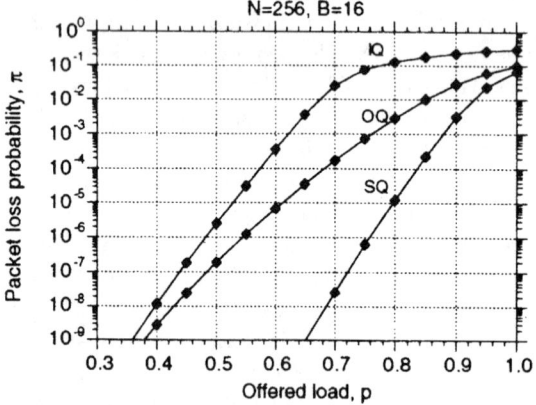

FIGURE 51.5 Loss performance of banyan networks.

51.5 ATM Switch with Blocking Multistage IN and Arbitrary Depth

Another class of ATM switches has been developed using the same concept of cell self-routing characterizing a banyan network but now based on the use of unbuffered SEs. The basic idea behind this type of switching fabric is that packet loss events that would occur because of multiple packets requiring the same interstage link are avoided by deflecting packets onto unrequested SE output links. Therefore, the packet loss performance is controlled by providing several paths between any inlet and outlet of the switch, which is generally accomplished by arranging a given number of self-routing stages cascaded one to another. Since each OPC of the switching fabric transmits at most one packet, queueing is still necessary when the interconnection network switches more than one packet in a slot to the same outlet interface (the IN has an internal speedup K). In this class of switch architectures, output queueing is obviously accomplished by providing each OPC of the switch with a queue.

The interconnection network of a $N \times N$ switch is a cascade of switching stages, each including $N/2$ unbuffered SEs, whose interstage link patterns depend on the specific ATM switch architecture. Unlike all other ATM switch classes, cells cross a variable number of stages before entering the addressed output queue, since direct connections to the output queues, through the SE local outlets, are also available from intermediate stages. Also the routing strategy operated by the switching block depends on the specific architecture. However, the general switching rule of this kind of architecture is to route the packets onto the local outlet addressed by the cell as early as possible. Apparently, those cells that do not reach this outlet at the last switching stage are lost.

The cell output address, as well as any other control information needed during the routing operation, is transmitted through the network in front of the cell itself as the cell routing tag. A feature common to all ATM switch architectures based on deflection routing is the packet self-routing principle, meaning that each packet carries all of the information needed in the switching element it is crossing to be properly routed toward the network outlet it is addressing. Compared to the simplest case of self-routing operated in a banyan network, where the routing is based on the analysis of a single bit of the output address, here the routing function can be a little more complicated and varies according to the specific architecture considered. Now, however, depending on the specific network architecture, the cell self-routing requires the processing of more than one single bit; in some cases the overall cell address must be processed in order to determine the path through the network for the cell.

Only two specific ATM switch architectures based on deflection routing will be described here to better clarify the principles of their internal operations, the *shuffleout* switch [Decina et al., 1991] and the *tandem banyan* switch [Tobagi et al., 1991], although other architectures have also been proposed.

Each of the K switching stages in the shuffleout switch includes $N/2$ switching elements of size 2×4, and the interstage connection pattern is a shuffle pattern. Thus, the network includes $N/2$ rows of SEs, numbered $0-N/2-1$, each having K SEs. A SE is connected to the previous stage by its two inlets and to the next stage by its two interstage outlets; all of the SEs in row i ($0 \leq i \leq N/2-1$) have access to the output queues interfacing the outlets $2i$ and $2i+1$ by means of the *local outlets*, thus accomplishing a speedup K. The shuffleout switch is shown in Fig. 51.6 for $N = 8$ and $K = 4$. The routing tag in the shuffleout switch is just the network output address.

The distributed routing algorithm adopted in the shuffleout interconnection *network is jointly* based on the *shortest path* and *deflection routing* principles. Therefore, a SE attempts to route the received cells along its outlet belonging to the minimum I/O path length to the required destination. The output distance d of a cell from the switching element it is crossing to the required outlet is defined as the minimum number of stages to be crossed by the cell in order to enter a SE interfacing

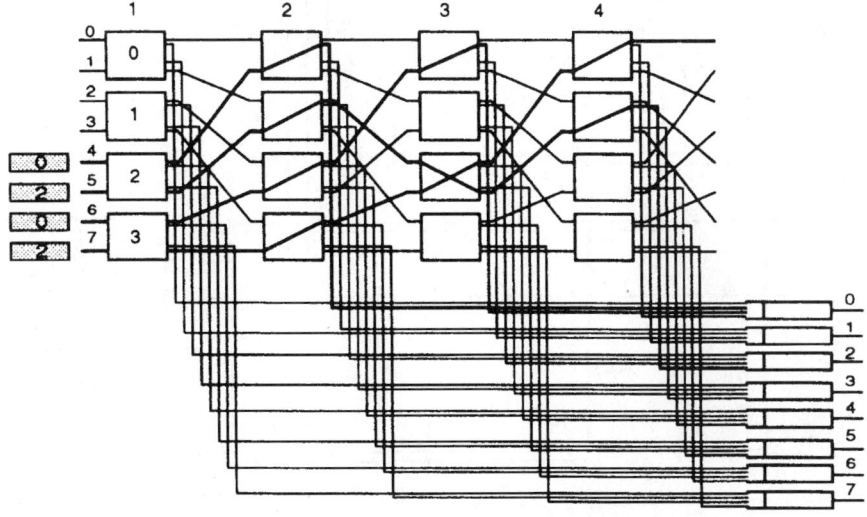

FIGURE 51.6 Routing example in shuffleout.

the addressed output queue. After reading the cell output address, the SE can easily compute the cell output distance, whose value ranges from 0 to $\log_2 N - 1$ because of the shuffle interstage pattern.

When two cells require the same SE outlet (either local or interstage), only one can be correctly switched, whereas the other must be transmitted to a nonrequested interstage outlet, due to the memoryless structure of the SE. Thus, conflicts are resolved by the SE applying the deflection routing principle: if the conflicting cells have different output distances, then the closest one is routed to its required outlet, whereas the other is deflected to the other interstage link. If the cells have the same output distance, a random choice is carried out. If the conflict occurs for a local outlet, the loser packet is deflected onto an interstage outlet that is randomly selected.

An example of packet routing is shown in Fig. 51.6, for $N = 8$. In the first stage both the SEs 2 and 3 receive two cells requiring the remote switch outlets 0 and 2, so that a conflict occurs for the their upper interstage link. In both cases the contending cells have the same distance ($d = 1$ in SE 2 and $d = 2$ in SE 3) and the random winner selection results in the deflection of the cells received on the low inlets. The two winner cells enter the output queue 0 at stages 2 and 3, whereas the other two cells addressing outlet 2 contend again at stage 3; the cell received at the low inlet now is the winner. This cell enters the output queue at the following stage, whereas the cell that has been deflected twice cannot reach the proper output queue within the fourth stage of the network and, thus, is lost.

Each output queue, which operates on a first-in first-out FIFO basis, is fed by K lines, one from each stage, so that up to K packets can be concurrently received in each slot. Since K can range up to several tens depending on the network parameter and performance target, it is necessary to limit the maximum number of packets entering the queue in the same slot. Therefore, a *concentrator* with size $K \times C$ is generally equipped in each output queue interface so that up to C packets can enter the queue concurrently. The number C of outputs from the concentrator and the output queue size B (cells) will be properly engineered so as to provide a given cell loss performance target.

In the $N \times N$ tandem banyan switch, shown in Fig. 51.7, K banyan networks are serially arranged so that the total number of stages is now nK ($n = \log_2 N$) each including 2×2 switching elements. Different topologies can be chosen for the basic banyan network of the tandem banyan switch, such as the omega [Lawrie, 1975], the baseline [Wu and Feng, 1980], and others. Each output queue is fed now by only K links, one from each banyan network. The first banyan network routes the received

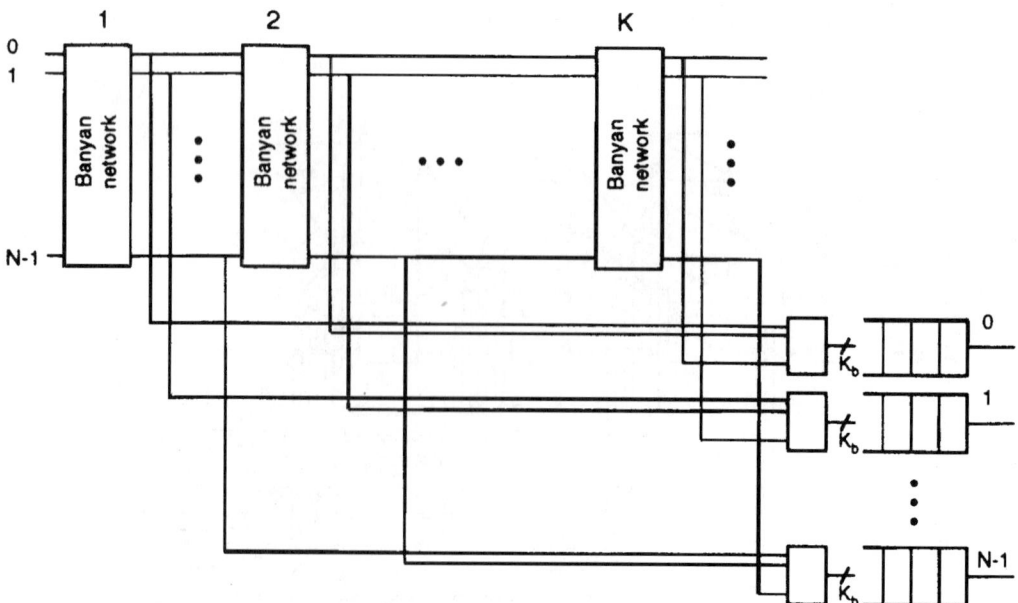

FIGURE 51.7 Architecture of the tandem banyan switch.

packets according to the very simple bit-by-bit self-routing (the specific bit to be used depends on the specific topology of the banyan network). In case of a conflict for the same SE outlet, the winner, which is chosen randomly, is routed correctly; the loser is deflected onto the other SE outlet, also by setting to 1 a proper field D of the routing tag, which is initially set to 0. Since each banyan network is a single I/O-path network, after the first deflection the packet cannot reach its addressed outlet in the same network. To prevent a deflected packet from causing the deflection of an undeflected packet at a later stage of the same network the SE always correctly routes the undeflected packet if two packets with different values of the field D are received.

The output queue i is fed by the outlet i of each of the K banyan networks, through proper packet filters that select for acceptance only those packets with $D = 0$, whose routing tag matches the output queue address. The kth banyan network ($k = 2, \ldots, K$) behaves accordingly in handling the packets received from the upstream network $k - 1$: it filters out all of the of undeflected packets and accepts only the deflected packets ($D = 1$). Analogously to all of the previous architectures, packets that emerge from the network K with $D = 1$ are lost.

Unlike shuffleout, here a much smaller number of links enter each output queue (one per banyan network), so that the output queue, in general, does not need to be equipped with a concentrator. On the other hand, in general cells cross a larger number of stages here since the routing of a deflected cell is not restarted just after a deflection, but rather, when the cell enters the next banyan network.

Unlike ATM switches with input queueing where the head-of-the-line blocking limits the throughput performance, the maximum throughput $\rho_{max} = 1$ can be achieved in an ATM switch with deflection routing. Therefore, our attention will be focused on the cell loss probability. Cells can be lost in different parts of the switch, that is, in the interconnection network (a cell reaches the last stage without entering the addressed local outlet), the concentrator, and the output queue. Loss events in the concentrator and in the output queue can be easily constrained to arbitrarily low values by properly designing the number of concentrator outlets and the output queue size. A much more critical parameter to achieve a low loss is the selection of the number of stages in the interconnection network that gives a particular low cell loss probability. Figure 51.8 shows that less than 20 (50) stages are needed in shuffleout to provide a cell loss probability lower than 10^{-7} for a switch with size $N = 32(1024)$ under maximum input load. Less stages are required for lower input loads. Similar comments apply to the tandem banyan switch referred to the number of networks rather than to the number of stages. Figure 51.9 also shows that the banyan network topology significantly affects the overall performance. Adopting an omega topology (O) requires less networks than with a baseline topology (B).

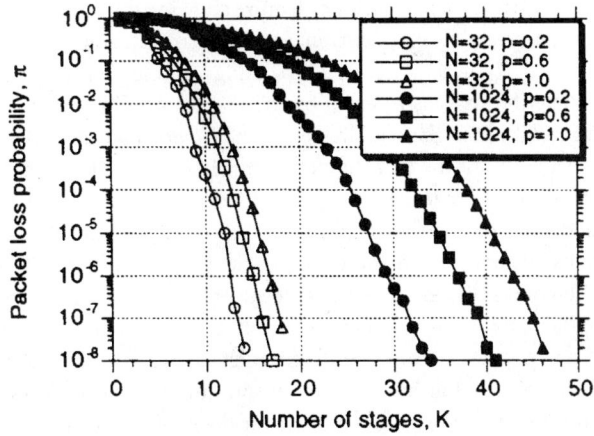

FIGURE 51.8 Loss performance in shuffleout.

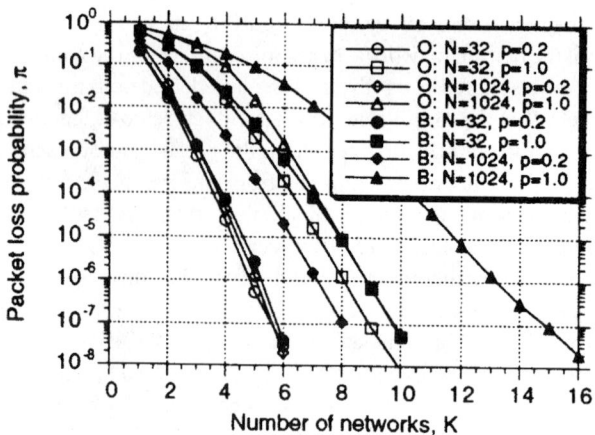

FIGURE 51.9 Loss performance in tandem banyan.

51.6 ATM Switch with Nonblocking IN

The third class of ATM switching fabrics is characterized by a nonblocking IN with packet queueing capability placed in the IPCs (input queueing), or in the OPCs (output queueing), or placed within the IN itself (shared queueing). Mixed queueing strategies have been proposed as well. Requiring a nonblocking IN means referring to a network free from internal conflicts, that is, in principle, to a crossbar network (switching in a $N \times N$ crossbar network is accomplished by N^2 crosspoints, each dedicated to a specific I/O pair). Nevertheless, the centralized control characterizing a crossbar network makes this solution infeasible for a very high-speed ATM environment. As in the previous ATM switch classes, the only viable solutions are based on the adoption of multistage arrangements of very simple SEs each capable of autonomously routing the cell to the required IN outlet.

The simplest ATM switch with nonblocking IN stores the cells in input queues located in the IPCs (IQ nonblocking IN). It is well known that a banyan network is nonblocking if the set of packets to be switched is *sorted* (by increasing or decreasing output addresses) and *compact* (packets received on adjacent network inlets). Therefore, the IN basically includes a sorting network and a routing banyan network, which are implemented as a multistage arrangement of very simple sorting and switching elements, respectively, whose size is typically 2×2. The number of stages of these networks is $n(n+1)/2$ for a Batcher sorting network [Batcher, 1968] and n for a banyan routing network, where $n = \log_2 N$, each stage including $N/2$ elements. The simplicity of these SEs, each requiring a gate count on the order of a few tens of gates, is the key feature allowing the implementation of these INs on a small number of chips by relying on the current very large size integrated (VLSI) complementary metallic oxide (CMOS) technology. An example of sorting-routing IN is given in Fig. 51.10 for an 8×8 ATM switch, where the six cells received by the IN are sorted by the Batcher network and offered as a compact set with increasing addresses to the n-cube banyan network. Note that the nonblocking feature of the IN means that the network is free from internal conflicts. Therefore, additional means must be available to guarantee the absence of external conflicts occurring when more than one cell addresses the same switch outlet. Input queues hold the packets that cannot be transmitted immediately, and specific hardware devices arbitrate among the head-of-the-line (HOL) cells in the different input queues to guarantee the absence of external conflicts.

An OQ $N \times N$ nonblocking ATM switch provides cell storage capability only at OPCs. Now the IN is able to transfer up to K packets from K different inlets to each output queue without blocking due to internal conflicts (it accomplishes an internal speedup K). Nevertheless, now there is no way of guaranteeing the absence of external conflicts for the speedup K, as N packets per slot can enter the IN without any possibility of them being stored to avoid external conflicts. Thus, here the packets in excess of K addressing a specific switch outlet in a slot are lost.

According to the first original proposal of an ATM switch with pure output queueing, known as

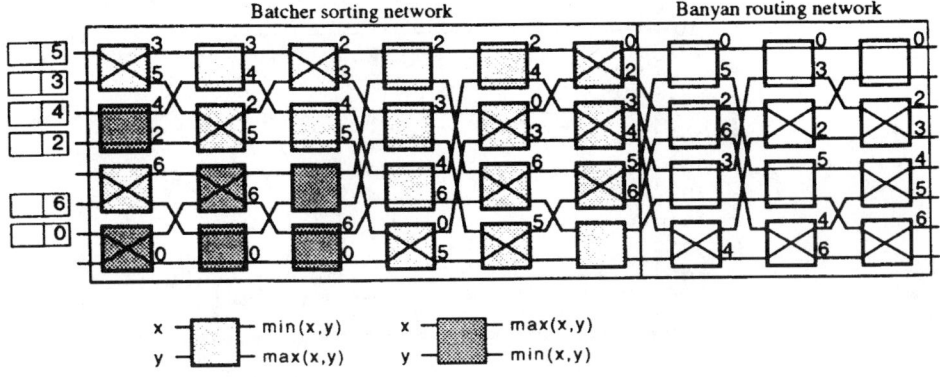

FIGURE 51.10 Example of cell switching in a Batcher-banyan network.

a *knockout* switch [Yeh et al., 1987], the IN includes a nonblocking $N \times N$ structure followed by as many output concentrators as the switch outlets. The $N \times N$ nonblocking structure is a set of N buses, each connecting one of the switch inlets to all of the N output concentrators. Each of these is a $N \times K$ network that is provided with N packet filters (one per attached bus) at its inputs, so as to drop all of the packets addressing different switch outlets.

A nonblocking $N \times N$ ATM switch with shared queueing behaves as a memory unit available in the IN and shared by all of the switch inlets to contain the packets destined for all of the switch outlets. The simplest implementation for this SQ switch model is given by a shared memory unit with N concurrent write accesses by the N inlets and up to N concurrent read accesses by the outlets. Clearly such structure shows implementation limits due to memory access speed, considering that the bit rate of the external ATM links is about 150 Mb/s. The current VLSI CMOS technology seems to allow the implementation of such $N \times N$ switches up to a moderate size, say, $N = 16$ or 32.

Multistage structures still provide a good answer for the implementation of larger size switches. The most important proposal is the *starlite* switch [Huang and Knauer, 1984], whose IN is a Batcher-banyan structure improved with additional hardware enabling feedback to the IN inlets those packets that cannot be switched to the addressed outlets because of external conflicts (the architecture does not have an internal speedup). Therefore, if P recirculation lines are available, then the size of the Batcher sorting network must be increased up to $(N + P) \times (N + P)$. These recirculation lines act as a shared buffer of size P.

The maximum throughput of an (unbuffered) crossbar network is $\rho_{\max} = 0.632$ because of the unavoidable external conflicts, whereas it drops to $\rho_{\max} = 0.586$ in an IQ nonblocking IN for an infinitely large switch ($N = \infty$). Thus, in spite of the queueing capability of this architecture, its capacity is even worse than in a crossbar switch. This phenomenon is explained by considering that holding in the HOL position those cells that cannot be transmitted because of external conflicts is a statistically worse condition than regenerating all of the cells slot by slot. Note that in a squared $N \times N$ switch with all active port controllers (PCs), a number h of blocked HOL cells imply h idle switch outputs, thus resulting in a carried load in the slot of only $(N-h)/N$. Nevertheless, behind a blocked HOL cell the queue can hold other cells addressing one of the idle switch outlets in the slot. Such an *HOL blocking* phenomenon is said to be responsible for the poor throughput performance of pure IQ switches.

At first glance a question could arise: What is the advantage of adding input queueing to a nonblocking unbuffered structure (the crossbar switch), since this results in a decrease of the switch capacity? The answer is provided in Fig. 51.11 showing the packet loss probability for a crossbar switch and an IQ switch with an input queue size B_i ranging from 1 to 32 cells. If the target loss probability is, say, 10^{-9}, we simply limit the offered load to $p = 0.3$ with $B_i = 8$, or to $p = 0.5$ for $B_i = 32$ with an IQ switch, whereas the packet loss probability of the crossbar switch is above 10^{-2} even for $p = 0.05$. Thus, input queueing does control the packet loss performance.

OQ nonblocking ATM switches show a much better traffic performance than IQ nonblocking

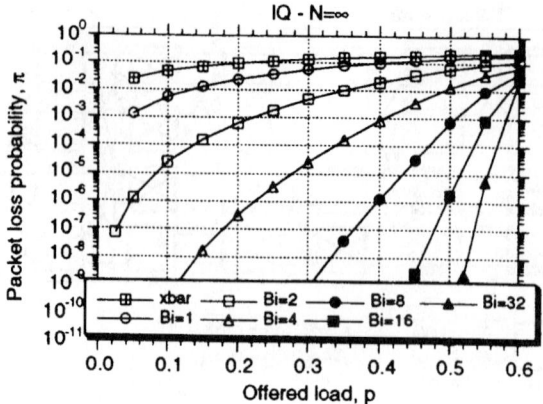

FIGURE 51.11 Loss performance with input queueing.

switches, as they do not have any HOL blocking phenomenon. In fact, the OQ switch capacity is $\rho_{max} = 1$ for $N = K$ and infinite output queue capacity. The loss performance is determined by the engineering of two parameters: the number of outlets per concentrator K, and the capacity of each output queue B_o. A speedup $K = 8$ is enough to guarantee a loss in the concentrator on the order of 10^{-8} with $p = 0.5$ and 10^{-6} with $p = 1.0$ for an arbitrary switch size. Moreover, the offered load p must be kept below a given threshold if we wish to keep the output queue capacity small by satisfying a given cell loss target (for example, $B_0 = 64$ guarantees loss figures below 10^{-7} for loads up to $p = 0.9$). The delay vs throughput performance in OQ switches is optimal since the packet delay is only determined by the congestion for the access to the same output link.

Nonblocking ATM switches using multiple queueing strategies have also been proposed so as to overcome the performance limits of structures with input or shared queueing. Architectures with either mixed input/output queueing (IOQ) [Lee, 1990] or mixed shared/output queueing (SOQ) [Giacopelli et al., 1991] have been proposed. These architectures are basically an expansion of the respective basic Batcher-banyan structures in which K banyan networks are now available to switch up to K cells to the same OPC (the switch has an internal speedup K) and OPCs are equipped with output queues, each with a capacity B_o. The main performance results attained by these mixed queueing architectures are that the HOL blocking characterizing IQ switches is significantly reduced in an IOQ switch and the number of recirculators P that give a certain loss performance decrease considerably in an SOQ switch compared to the basic SQ switch.

Table 51.1 provides the switch capacity of an IOQ switch for different speedups and output queue sizes for $B_i = \infty$ and $N = \infty$. Note that the first column only contains the numerical values of the independent variable for which the switch capacity is evaluated and its meaning is specified in each of the following columns. The switch capacity grows very fast with the speedup value K for an infinitely large output queue: $\rho_{max} = 0.996$ for $K = 4$, whereas it grows quite slowly with the output queue size given a switch speed-up K. In general, it is cheaper to provide larger output queues rather than increase the switch speedup, unless the switching modules are provided in VLSI chips that would prevent memory expansions. Thus, these results suggest that a prescribed switch capacity can be obtained acting on the output buffer size, given a speedup compatible with the selected architecture and the current technology.

TABLE 51.1 Switch Capacity with Mixed Input/Output Queueing

x	$B_o = \infty$ $\rho_m(K)$	$K = 2$ $\rho_m(B_o)$	$K = 4$ $\rho_m(B_o)$
1	0.586	0.623	0.633
2	0.885	0.754	0.785
4	0.996	0.830	0.883
8	1.000	0.869	0.938
16		0.882	0.967
32		0.885	0.982
64		0.885	0.990
128			0.994
256			0.996
∞	1.000	0.885	0.996

Analogously to OQ switches, pure shared queueing provides maximum switch capacity $\rho_{max} = 1$ for a shared queue with infinite capacity. The loss performance of SQ and SOQ architectures is

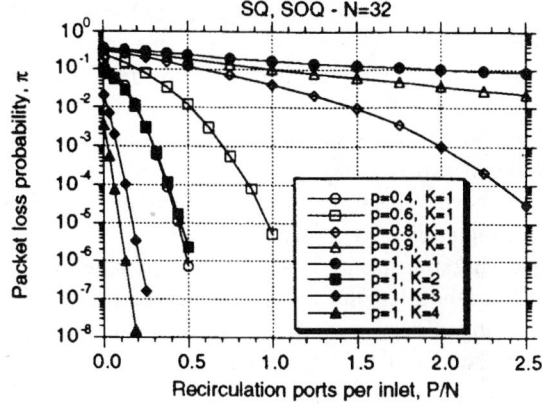

FIGURE 51.12 Loss performance with shared queueing.

given in Fig. 51.12 for $N = 32$ as a function of the normalized shared buffer size P/N for different offered loads. It is seen that even a small shared queue of 32 cells ($P/N = 1$) requires the offered load to be limited to $p = 0.6$ if loss figures below 10^{-5} are required. It is interesting to note how the SOQ switch capacity approaches 1 even with a small shared buffer capacity by adopting a very little speedup ($\rho_{max} = 1 - 10^{-6}$ for $K = 3$ and $P = 7$).

51.7 Conclusions

The main architectural and performance features of switching fabrics for broadband networks based on asynchronous time-division switching have been presented. It has been shown that multistage arrangements of very simple switching elements is the typical solution adopted in ATM switches, which enables the processing required by ATM cells to be compatible with the bit rate on the incoming links. All of the different switch classes presented are able to provide the required performance target in an ATM network, typically the cell loss probability, by means of a suitable limitation of the offered load. Choosing one architecture rather than another is just a matter of finding an acceptable tradeoff between implementation costs and traffic performance.

Defining Terms

Arbitrary depth network: Interconnection network providing more than one internal path per inlet/outlet pair.

Asynchronous transfer mode (ATM): Transport technique defined for the lower layers of the broadband communication network.

ATM cell: Basic fixed-size unit defined for the transport of user information in a broadband ATM network.

Blocking network: Network in which internal conflicts occur for an arbitrary permutation that is free from external conflicts for a given network speedup.

Cell self-routing: Property of a multistage interconnection network that consists of the autonomous routing of the packets in each switching element only based on a routing tag carried by the cell without the use of any central control.

Deflection routing: Routing technique that deviates a packet from its intended route through the network because of an internal conflict.

External conflict: Conflict between packets for the access to the same network outlet.

Input queueing: Queueing strategy in which the information units are stored in buffers associated with the inlets of the interconnection network (or a switching element).

Internal conflict: Conflict between packets for the access to the same link internal to the interconnection network.

Minimum depth network: Interconnection network providing only one internal path per inlet/outlet pair.

Network speed-up: Number of packets that can be concurrently received at each output interface of an interconnection network.

Nonblocking network: Network free from internal conflicts for an arbitrary permutation that is free from external conflicts for a given network speedup.

Output queueing: Queueing strategy in which the information units are stored in buffers associated with the outlets of the interconnection network (or a switching element).

Shared queueing: Queueing strategy in which the information units are stored in a buffer shared by all of the inlets and outlets of the interconnection network (or a switching element).

References

Batcher, K.E. 1968. Sorting networks and their applications. In *AFIPS Proceedings of Spring Joint Computer Conference*, pp. 307–314.

Decina, M., Giacomazzi, P., and Pattavina, A. 1991. Shuffle interconnection networks with deflection routing for ATM switching: The open-loop shuffleout. In *Proceedings of 13th International Teletraffic Congress*, Copenhagen, Denmark, pp. 27–34, June.

Giacopelli, J.N., Hickey, J.J., Marcus, W.S., Sincoskie, W.D., and Littlewood, M. 1991. Sunshine: A high performance self-routing broadband packet switch architecture. *IEEE J. Selec. Areas Commun.*, 9(Oct.):1289–1298.

Huang, A. and Knauer, S. 1984. Starlite: A wideband digital switch. In *Proceedings of GLOBECOM 84*, Atlanta, GA, pp. 121–125, Nov.

ITU. 1991. B-ISDN protocol reference model and its application. ITU-TI.321, International Telecommunications Union-Telecommunications Standardization Sector, Geneva, Switzerland.

ITU. 1993. B-ISDN asynchronous transfer mode functional characteristics. ITU-TI.150, International Telecommunications Union-Telecommunications Standardization Sector, Geneva, Switzerland.

Lawrie, D.H. 1975. Access and alignment of data in an array processor. *IEEE Trans. Comp.*, C-24(12):1145–1155.

Lee, T.T. 1990. A modular architecture for very large packet switches. *IEEE Trans. Commun.*, 38(7):1097–1106.

Tobagi, F.A., Kwok, T., and Chiussi, F.M. 1991. Architecture, performance and implementation of the tandem banyan fast packet switch. *IEEE J. Selec. Areas Commun.*, 9(8):1173–1193.

Wu, C.-L. and Feng, T.-Y. 1980. On a class of multistage interconnection networks. *IEEE Trans. Comp.*, C-29(Aug.):694–702.

Yeh, Y.S., Hluchyj, M.G., and Acampora, A.S. 1987. The knockout switch: A simple, modular architecture for high-performance packet switching. *IEEE J. Selec. Areas Commun.*, SAC-5(Oct.):1274–1283.

Further Information

A greater insight in the area of switching architectures for ATM networks can be obtained consulting the main Journals and Conference Proceedings in the field of Communications of the last 10 years or so. Two rather comprehensive selections of the most important papers in the area of ATM switching are collected in *Broadband Switching, IEEE Computer Society Press*, 1992 and *Performance Evaluation of High Speed Switching Fabrics and Networks, IEEE Press*, 1993.

52
Internetworking

52.1	Introduction... 701
52.2	Internetworking Protocols 702
	TCP/IP Internetworking • SNA Internetworking • SPX/IPX Internetworking
52.3	The Total Network Engineering Process 706
	Network Awareness • Network Design • Network Management
52.4	Internetwork Simulation 712
52.5	Internetwork Optimization 713
52.6	Summary 714

Harrell J. Van Norman
EG&G Mound Applied Technologies

52.1 Introduction

Engineering an internetwork is a lot like a farmer getting his apples delivered from an orchard to the supermarket. First, the farmer needs something to put the apples in: a bag, a box, a basket, or may be a crate; second, the farmer needs to drive on some roads over a specific route. Those same two problems confront an engineer trying to get data delivered from one local area network (LAN) to another.

Developing an internetwork involves finding something to put the apples in and providing a way to transport them from the orchard to the supermarket. The apples are the data and the container is the communications protocol that encapsulates or contains the data. Communications protocols envelope the application data and provide a container to ensure the goods are delivered safely and reliably. Internetworks require compatible protocols at both the transmitting and receiving sites. Various communications protocols are available, such as transmission control protocol/Internet protocol (TCP/IP), system network architecture (SNA), DECnet, IPX, AppleTalk, and XNS, to name a few.

Roads and routes the farmer travels on from the orchard to the supermarkets represent telecommunications circuits the communications protocols use to route the data from source to destination. Engineering an internetwork involves sizing circuits and designing routes for the efficient and optimal transfer of information between LANs. Providing cost-effective and properly performing internetworks involves designing the best type of circuits, either dedicated or switched, and sizing bandwidth requirements.

Internetworking is defined in the IBM dictionary of computing as "communication between two or more networks." Miller [1991] defined the term as "communication between data processing devices on one network and other, possibly dissimilar devices on another network." For the purpose of this chapter internetworking will take on the meaning of connecting two or more LANs to broaden the scope of communications.

52.2 Internetworking Protocols

Understanding, designing and optimizing an internetwork would be too difficult unless the problem were broken down into smaller subtasks. Communications architectures partition functionality into several layers, each providing some service to the layer above and using the services of the layer below. Figure 52.1 illustrates the seven-layer International Organization for Standardization (ISO) open systems interconnection (OSI) reference model. TCP/IP, SNA, and sequence packet exchange/Internet packet exchange (SPX/IPX) are mapped into the seven-layer model. Each of the communications protocols converge at the data link and physical layers; however, the architectures all diverge at the network layer. Each communications protocol is represented uniquely at the upper five layers of the OSI communications model.

The data link layer is relevant to internetworking because this is where bridges operate. The service provided by the data link layer is important to routers—the key internetworking device, operating at the network layer. The transport layer is the user of network services and is influenced by the routing decisions made at the network layer. These three layers are the primary focus of internetworking.

Two models represent how an internetwork appears to endnodes. In the **connectionless** or datagram model, an endnode transmits packets containing data and the destination address where the data is to be delivered. The network does the best to deliver the data, with some probability the data will get lost, duplicated, or damaged. Packets are individually routed and the network does not guarantee the packets will be delivered in the same order they were sent.

The connectionless model is similar to the postal system in which each letter enters the postal system with the address where it should be delivered. The postal system does the best to deliver the data, with some probability the letter will get lost of damaged. Letters are individually routed and a letter posted on Monday might get delivered after a letter posted on Tuesday, even if both were posted from the same location and addressed to the same location.

The other model is the **connection-oriented** or virtual circuit. Here an endnode first informs the network it wishes to start a conversation with some other endnode. The network then notifies the destination that a conversation is requested, and the destination accepts or refuses. This is similar to the telephone system, in which the caller informs the telephone system of the wish to start a

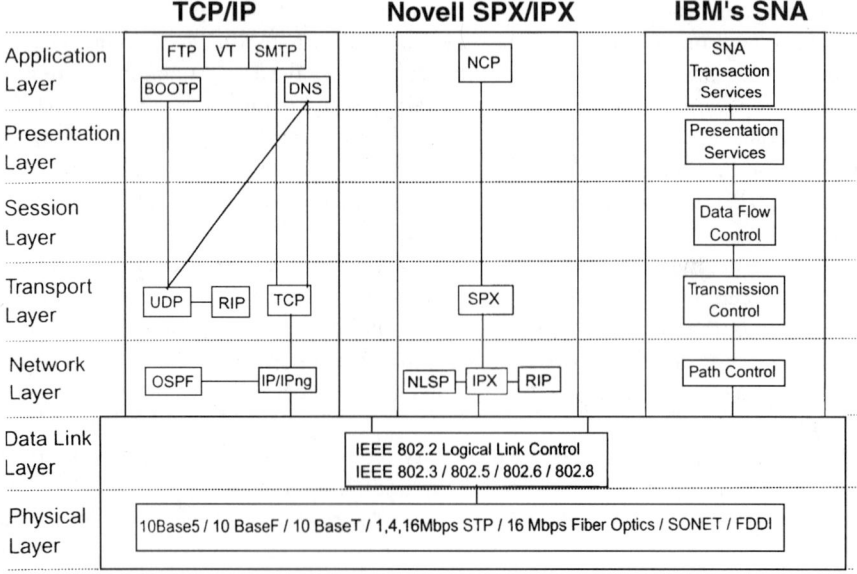

FIGURE 52.1 Internetworking communications protocol architectures.

Internetworking

conversation by dialing the telephone number of the destination. The telephone system establishes a path, reserves the necessary resources, contacts the destination by ringing its phone, and after the call is accepted, the conversation can take place. Often a connection-oriented service has the following characteristics [Perlman, 1992]:

1. The network guarantees packet delivery.
2. A single path is established for the call and all data follows that path.
3. The network guarantees a certain minimal amount of bandwidth.
4. If the network becomes overly utilized, future call requests are refused.

TCP/IP Internetworking

TCP/IP has emerged as the primary communications protocol for developing internetworks. Why? The answer is simple; interoperability, almost every conceivable combination of computer hardware and operating system has a driver available for TCP/IP. International travelers know English is the speaking world's most common second language. A native language may be preferred when speaking with local family and friends, but English is the world-wide second language and is predominately used to communicate internationally. Similarly, TCP/IP has emerged as the primary protocol for communicating between networks.

The suite of TCP/IP protocols, adopted in the 1970s by the U.S. Department of Defense, is the Internet backbone. Now, over 3.1 million users are connected to the Internet all speaking IP, up 81% from last year. The International Data Corporation (IDC), Framingham, MA estimates 4.5 million PCs are running TCP/IP and the cumulative installed base for TCP/IP is 8 million computers. TCP/IP is clearly the internetworking protocol of choice.

Endorsements of TCP/IP just keep coming, first Microsoft tucked a TCP/IP stack into Windows NT Version 3.1 and declared TCP/IP its strategic wide-area network protocol family. Most recently, a federally commissioned panel recommended that the government sanction TCP/IP as a standard and surrender its open systems interconnection-only stance in the Government OSI Profile (GOSIP). TCP/IP, the preferred protocol for internetworking, is also becoming the primary backbone protocol for local communications.

One of the major forces driving TCP/IP to the PC desktop is UNIX, which usually includes TCP/IP. UNIX has become the de facto standard enterprise server OS. As users downsize from mainframe platforms and adopt UNIX-based servers, they must outfit their PCs with access to TCP/IP. Downsizing is a major trend as processing power becomes more cost effective in the minicomputer and microcomputer platforms. An Intel 286-based PC introduced in 1983 provided 1 MIPS performance, in 1986 the i386 offered 5 MIPS, in 1990 the i486 came with 20 MIPS, and more than 100 MIPS was available in 1993 with the Pentium. Intel plans to routinely increase desktop processing power with the i686 in 1995, offering 175 MIPS, the i786, slated to provide 250 MIPS in 1997, and the i886 in 1999, capable of approximately 2000 MIPS.

TCP/IP is a connectionless protocol. TCP operates the transport layer and IP is the network layer protocol based on 4 octet addresses. A portion of the address indicates a link number and the remaining portion a system or host on the network. A 32-b **subnet mask** has 1s in the bits corresponding to the link portion of the IP address and 0s in the bits belonging to the host identifier field. If an IP address is X and its mask is Y, to determine whether a node with IP address Z is on the same link, AND the mask Y with X and AND the mask Y with Z, if the result is the same, then Z is on the same link.

Originally, IP addresses were grouped according to three distinct classes: Class A contained the network address field in the first octet and host identifier in the remaining 3 octets, class B contained the network address field in the first two octets and the host in the remaining two fields, and class C contained the network address field in the first three octets and the host identifier in the remaining octet. Today, IP allows a variable sized subnet mask and even permits noncontiguous 1s in the mask. This can result in incredibly confusing and computationally inefficient addressing structures. Regardless, each node on a link must know the subnet mask of its own IP address.

IP routers connect networks by routing packets based on destination link numbers, not destination host identifiers. If IP routers had to route packets to destination hosts, enormous memory would be necessary to contain information about every machine on the Internet. Consequently, IP routing does not route to the destination endnode, it routes only to the destination link.

TCP/IP does have a certain amount of overhead associated with address resolution protocol (ARP) queries and responses that can degrade network performance significantly. The ARP allows a host to find the physical address of a target host on the same link number, given only the target's IP address. For example, if host A wanted to resolve the IP address of host B, a broadcast query would be transmitted and received by all hosts, but only host B would respond with its physical address. When A receives the reply it uses the physical address to send packets directly to B. Considerable bandwidth and memory in routers are consumed by routers keeping track of endnodes rather than just keeping track of links [Comer, 1991].

Improvement of the TCP/IP protocol suite will come in new protocols and emerging classes of applications. The dynamic host configuration protocol (DHCP) is being standardized to automatically gather detail configuration data for a PC when it joins the network, creating a plug-and-play environment. Currently, network administrators must manually set parameters on each workstation, such as IP addresses and domains, but boot-P and DHCP will advance TCP/IP into an automated installation process.

Two classes of emerging applications for TCP/IP include: interenterprise and personal interenterprise. Interenterprise applications unite different organization's networks; for example, a retailer could order automatically from his suppliers or check credit-card information, all by TCP/IP. This brings electronic data interchange transactions, typically separate dial-up links, into the TCP/IP network. Personal interenterprise applications customize a PC view of the Internet, so that when connected specific information is automatically presented, such as stock portfolio data.

The Internet is the largest network in the world, a network of networks that unarguably demonstrates the usefulness of TCP/IP. But TCP/IP is suffering from its own success. By the year 2000, the Internet will likely have exhausted all available addressing space. The Internet's standards-setting body, the Internet Engineering Task Force, has adopted the next generation IP. Called **IPng**, this new version of the IP layer will solve the problem of dwindling address space by supporting 16-byte addresses that will allow billions more users to surf the Internet. The new protocol (to be assigned version number IP6) will not require any immediate changes to users' existing IP internetworks. However, users will need to upgrade their routers and eventually change applications if they want to take advantage of the advanced features of IPng, such as flow control, autoconfiguration, and encryption-based security.

SNA Internetworking

IBM's SNA traffic is difficult to integrate into a multiprotocol internetwork because it is a nonroutable protocol. NetBIOS and Digital's local area transport (LAT) are also nonroutable protocols. Although it is unfortunate, these nonroutable protocols are without network addresses. Without network layer functionality, only bridges can internetwork native SNA. There is no intrinsic reason why these protocols could not have been designed to run with network layer addresses, they just were not. Perhaps designers never believed LANs would be internetworked.

Routers, as opposed to bridges, are the primary internetworking devices to direct traffic from its source to its destination. Bridging is useful in simple point-to-point networks where alternate paths are not available and where protocol overhead is not a significant factor. Disadvantages of bridges relative to routers include:

- Bridges can use only a subnet of the topology (a spanning tree). Routers can use the best path that physically exists between source and destination.
- Reconfiguration after a topological change is an order of magnitude slower in bridges than routers.

- The total number of stations that can be interconnected through bridges is limited to the tens of thousands. With routers, the total size of the network is for all practical purposes unlimited (at least with ISO).
- Bridges limit the intermediate hop count to seven, significantly restricting the size of an internetwork.
- Bridges offer no firewall protection against broadcast storms.
- Bridges drop packets too large to forward, since the data link layer cannot fragment and reassemble packets.
- Bridges cannot give congestion feedback to the endnodes.

For the past several years, most corporations relied on token ring bridges implementing SNA's data link protocol, Logical Link Control 2 (LLC2) to give PCs on remote LANs a way to access IBM mainframe data. Frames are forwarded throughout the internetwork by token ring bridges using source-route bridging protocol to ensure data is delivered to the proper destination. The problem is that even a few seconds of congestion on the wide area network (WAN) link can drop the SNA/LLC2 sessions.

IBM's **data link switching (DLSw)** standard provides the reliability needed to integrate mission-critical SNA networks with IP backbones. DLSw products integrate SDLC with IP by stripping off the synchronous data link control (SDLC) header and replacing it with a TCP/IP header. Previous to DLSw routers, SNA data was encapsulated in TCP/IP, leaving the SDLC headers intact. Current testing of router vendor's DLSw implementations indicate they keep SNA sessions from being dropped over even the most congested internetwork links. Performance at full speed of the WAN link is another big improvement over token ring bridges. A successor to source-route bridging, DLSw routers deliver high-performance SNA internetworking even with significantly more complexity than source-route bridging.

The DLSw scheme allows routing IP and internetwork packet exchange (IPX) traffic alongside connection-oriented SNA. DLSw routers have the added complexity of blending connectionless LAN traffic with connection-oriented SNA traffic by sending messages to the stations on each end of the connections acknowledging that a path is still open between them. Preventing the SNA connection from timing out and dropping sessions involves considerable overhead traffic, such as frequent receiver ready, keep alive broadcasts, and timing signals for constant responses.

DLSw routers are criticized by some SNA users because of the inconsistent performance associated with contention-based internetworks. IP response time may be nearly instantaneous one time and may take a second the next, frustrating SNA users accustomed to consistent performance. DLSw can benefit large SNA networks suffering from time outs, hop limits, alternate routing disruptions, and the lack of security mechanisms. DLSw routing avails non-routable SNA traffic the robustness of nondisruptive alternate routing when links or nodes fail, the benefits from open nonproprietary protocols, and the cost effectiveness of standard TCP/IP-based solutions.

For IBM users, there is the option of remaining all "blue" with IBM's advanced peer-to-peer networking (APPN) family of protocols, the next generation of SNA. IBM continues to develop, position, and promote its own APPN as the only viable next-generation networking technology for its mainframe customers. The need for guaranteed response time in their mission-critical mainframe applications makes APPN an attractive option. Other advantages to the APPN approach include full class of service application priorities, enhanced fault management through NetView monitoring, the added efficiency from not encapsulating SNA into a routable protocols like TCP/IP and thereby doubling the addressing overhead, and improved accounting from SNA usage statistics.

SPX/IPX Internetworking

The number one LAN protocol is undoubtedly Novell's sequence packet exchange/Internet packet exchange (SPX/IPX), but most network administrators use TCP/IP for internetworking. As a result,

Novell has taken two significant steps to improve the performance of SPX/IPX for internetworking: **packet burst** and the network link services protocol (NLSP).

Before packet burst, the NetWare core protocol (NCP) was strictly a ping-pong discipline. The workstation and the server were limited to communicating by taking turns sending a single-packet request and receiving a single-packet response. Packet burst is an NCP call permitting a sliding window technique for multiple packet acknowledgment. By transmitting a number of packets in a burst, before waiting for an acknowledgment or request, packet burst eliminates the ping-pong effect and can provide up to a 60% reduction in the number of packets to read or write large files. Limited to only file reads and writes, packet burst cannot reduce the ping-pong method of all other NCP communications, such as opening directory handles, locating and opening files, and querying the bindery.

Rather than reducing the number of packets when the receiving station or internetwork becomes congested, as do most transport layer protocols, packet burst actually varies the interpacket gap (IPG). The IPG is the time between the end of one packet and beginning of the next packet. Varying the IPG for flow control is called packet metering.

The amount of data sent during a burst is called the message. Message size is based on the number of packets in the burst, called the window size, and the media access method of the link. The default window size has been optimized for most communications to 16 packets for a read and 10 for a write. For example, 16 reads on a token ring link with a 4096 packet size gives a 53-KB message. On an ethernet link, a read message is about 22-KB (16×1500 bytes). The number of packets in the burst window can be increased on high-speed links and where the degree of transmission success is high. If a packet or packets are lost in transmission for any reason, the client will send a request for retransmission of the lost packets. The server will retransmit only those packets that were lost. Once the client has all of the packets, it requests another message; however, since packets were lost, the IPG is increased for the next transmission. The IPG will continue to be increased until packets are no longer being lost.

NLSP is another internetworking improvement to NetWare. Novell developed the **link-state router** protocol NLSP, to replace the older IPX distance-vector based router information protocol (RIP). IPX RIP is similar to Cisco's IGRP, IP RIP, and RTMP in the AppleTalk suite. **Distance-vector routing** requires each node to maintain the distance from itself to each possible destination. Distance vectors are computed using the information in neighbors' distance vectors and store this information in routing tables that show only the next hop in the routing chain rather than an entire map of the network. As a result, RIP frequently broadcasts these summaries to adjacent routers.

Distance-vector routing is efficient for routers with limited memory and processing power, but newer routers typically have reduced instruction set computer (RISC) processors with power to spare. Link-state routers can provide full knowledge of the network including all disabled links with a single query. With distance-vector routers, this would involve querying most, if not all, of the routers. In addition, link state routers, like NSLP, converge more quickly than distance-vector routers, which cannot pass routing information on until distance vector has been recomputed. NSLP supports large, complex IPX internetworks without the overhead present in NetWare distance-vector routers.

NSLP routers compress service advertizing protocol (SAP) information and only broadcast SAP packets every two hours unless a change is made to network services, such as print servers or file servers. SAP broadcast overhead can be significant, especially with many servers or services. Novell has also reduced SAP traffic on the network by using NetWare Directory Services (NDS) to keep all network services information on a central server. NDS is available in NetWare 4.x and is similar to TCP/IP's domain name server (DNS). Novell also provides the capability to communicate with different systems through NetWare for SAA, NetWare NFS, and NetWare TCP/IP Gateway.

52.3 The Total Network Engineering Process

This guide to internetworking provides a step-by-step engineering approach. Migrating from a LAN or several LANs to an internetwork is no small undertaking. Key to the process of analyzing and designing internetworking strategies is employing analytic heuristic algorithms and simulation

Internetworking

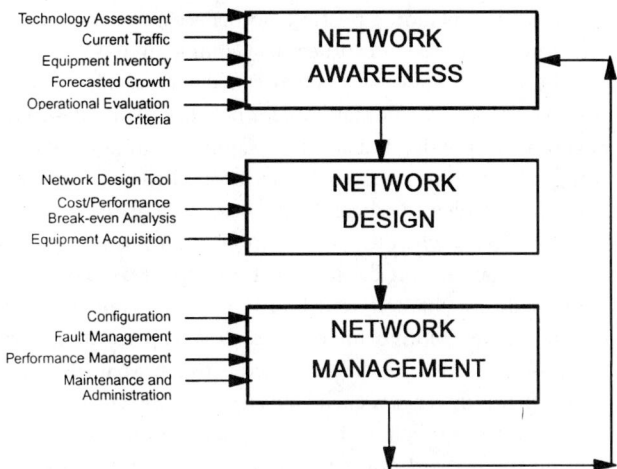

FIGURE 52.2 The total network engineering process.

techniques. Network design tools are essential to effectively evaluate alternative internetworking schemes while attempting to optimize communications resources. As new networking requirements surface, engineers design capacity and equipment to satisfy the demands. Facing requirements in an as-needed basis results in network evolution. Networks developed in an evolutionary manner provide incremental solutions to an array of individual requirements. With each new requirement the most cost-effective configuration of equipment and communications lines is determined, focusing on the specific need. Unfortunately, design decisions are often made without rigorous analysis of the entire network, addressing only the incremental expansion. This approach seldom provides the best design for the total communications environment.

The preferred method of addressing network requirements is through a structured-design approach. Here, alternative design techniques for the entire network are evaluated in terms of response time, cost, availability, and reliability. While avoiding excessive redesign and allowing for orderly capacity expansion, the process of network engineering evaluates current networking requirements and plans for future network expansion.

Figure 52.2 shows the **total network engineering process**, and its three phases: awareness, design, and management. **Network awareness** quantifies current traffic, equipment inventories, forecasted growth, and operational objectives. **Network design** uses design tools, performs a cost/performance breakeven analysis, and develops line configurations. **Network management** encompasses management of configuration, faults, performance, accounting, and security. Equipment and network software is acquired, verified, installed, maintained, and administered.

Network Awareness

The process of network awareness includes the following operations:

- Conducting a traffic inventory by collecting measurements of current traffic
- Mapping the physical inventory of networking equipment into the logical interconnections
- Forecasting requirements for network growth through surveying line management
- Qualifying through senior management the criteria used to evaluate operational tradeoffs

Many of these inputs are, of necessity, qualitative or conceptual in nature, whereas others can be quantified with varying degrees of accuracy. Using a network design process often requires a greater level of network awareness than most engineers generally possess. When an engineer has an accurate understanding of the network—including possible bottlenecks, response time spikes, and traffic patterns—the process of network design can begin. Gaining an accurate awareness of network traffic involves monitoring current loads via a network analysis tool, like the LANalyzer, Sniffer, or through the use of RMON probes. Traffic statistics for network planning should be collected, not only for all significant hours of each day, but also for a significant number of days in a year. CCITT recommends measuring carried traffic statistics for at least the 30 days (not necessarily consecutive) of the previous 12 months in which the mean busy hour traffic is the highest. A second-choice method recommended by CCITT, when the first method is not possible, consists in a measuring period of 10 consecutive normal working days during the busiest season of the year [CCITT, 1979] Both average and peak loads are characterized hourly over the sampling period to understand current bandwidth requirements. Monitoring new networked applications predicts their impacts to the internetwork. Traffic must also be differentiated between what stays local, remaining only within a LAN, and traffic that is internetworked between remote locations via WAN links.

Interviews with line management help to forecast growth requirements. Many managers have hidden agendas for using the network for a variety of business purposes. Without surveys or interviews of key management personnel, predicting future network requirements is impossible. Questions need to center around four key areas:

1. Future business communications requirements based on the use of advancing technologies
2. Planned facility changes and population shifts through developing new business partners
3. Forecasted needs for external connectivity with customers, vendors, and new markets
4. Changes in business practices based on new programs, applications, and remote systems access.

A little-publicized problem, but one that nevertheless causes big headaches, is developing an inventory of current network equipment. Obtaining an accurate image of the inventory database is helpful in the network engineering process to gain an improved network awareness. Based on network discovery techniques, like IP discovery and IPX SAP, routing nodes may be identified throughout the internetwork. These techniques only find network layer routing devices; however, internetworks are primarily based on routers rather than data link layer bridges. Once these devices are identified, MIB browsers are effective to understand what mix of protocol traffic is being routed throughout the internetwork. Finally, a topological map can be developed by translating the logical routing to physical locations.

Top level management must also be involved to quantify the operational evaluation criteria. Every business has established measures for effective operations. A network exists to meet the needs of those who use the network, not the preferences of those who maintain it. Therefore, objectives for the communications resources should be a reflection of those criteria held by the business as a whole. Five key operational evaluation criteria or network design goals characterize most networks: performance, availability, reliability, cost, and security. Manufacturing industries and startup companies often chose cost minimization, financial markets mandate high reliability, security is of utmost importance throughout classified governmental networks like the Department of Energy (DOE) nuclear weapons complex, and high performance is necessary to mass merchandise retail companies. Key operational evaluation criteria are described below.

Cost minimization is a weighty design factor that includes both monthly recurring and nonrecurring fees. To compare competing alternatives with different cost outlays, the time value of money must be considered and the net present value method is usually the best method for financial analysis. In the name of cost minimization, depending on the specific operational objective, one may be led to different solutions given the same problem. Variations of cost minimization goals include: build the cheapest network, get the most network for the dollar, maximize profit by allowing the firm to reach new markets, minimize risk of loss, or maximize growth opportunity.

All network designs include an inventory of lines, equipment, and software needed to make them run. Including all of the cost elements (installation costs, shipping costs, purchase price, monthly lease costs, maintenance costs, and service costs) into the optimization process prevents unpleasant surprises. Total network costs C_{TOT} includes the three major cost components in a network design: line costs, equipment costs, and software costs:

$$C_{TOT} = C_{AL} + C_{TL} + C_{IE} + C_{WAE} + C_S \qquad (52.1)$$

where:

C_{AL} = access line costs
C_{TL} = interoffice channel (IXC) line costs
C_{IE} = internetworking equipment costs (remote bridge, router, gateway, etc.)
C_{WAE} = WAN access equipment costs [modems, computer software unit/data software unit (CSU/DSU), channel banks, etc.]
C_S = software costs

Too often availability and reliability are mistakenly assumed to be synonymous. Reliability and serviceability are both components of availability. Network availability, a value between zero and one, reflects stability characteristics of communications systems connecting end users. Availability A is composed of two factors: Reliability, the probability a system will not fail during a given time, is expressed as mean time between failure (MTBF), and serviceability, including the time required to diagnosis and repair, is expressed as mean time to repair, (MTTR),

$$A = MTBF/(MTBF + MTTR) \qquad (52.2)$$

Performance is another key operational evaluation criteria. Represented by average transaction response times RT, it can be calculated by the following formula:

$$RT = I_T + H_T + O_T \qquad (52.3)$$

where

I_T = input message transmission time: number of bits in input message/link speed bits per second + propagation delay
H_T = host processing time
O_T = output message transmission time: number of bits in output message/link speed bits per second + propagation delay

At the completion of the network awareness phase the network planner must objectively evaluate whether or not this phase has been completed adequately for the purpose of the design process. Questions that might be addressed are:

- Do I have a reasonable assessment of current technology in this networking arena?
- Have I quantified the current network traffic?
- Is there a justifiable forecast of network growth?
- Are the operational objectives and goals quantified and agreed upon by upper level management?

Upon completion, quantifiable operational evaluation criteria are available as input into the network design phase.

Network Design

For too long networks merely evolved to their present condition by adding users and applications without analyzing performance impacts. The better way is for engineers to describe the topology and media via a network modeling tool. In many network design tools, models are constructed by connecting icons representing specific network devices or segments. Behind the model, powerful software simulates and analyzes the behavior of the network. Also, network conditions such as protocols, traffic patterns, and number of nodes are specified in the model. Traffic can be characterized from statistical traffic models or actual traces of network traffic taken from an analyzer. Simulation and performance evaluation predict network performance and explore future design options. Graphical user interfaces (GUI) simplify changing assumptions on number of users, type of media protocol stacks, application mix, traffic patterns, and so on. When the network model is in place, proposals for adding or deleting resources (such as controllers, terminals, multiplexers, protocol converters, remote front-end processors), redesigning application software, or changing routing strategies may be evaluated.

Network design consists of applying network design tools, evaluating alternative network configurations through a cost/performance breakeven analysis, and optimizing communications lines. Because there are many ways to configure a network and interconnect transmission components, it is not realistic to evaluate exhaustively every possible option. However, by using certain shortcuts, called heuristics (procedures that approximate an optimal condition using algorithms), the number of feasible designs can be scaled down to a manageable few.

Network design tools are developed using heuristic algorithms that abbreviate the optimization process greatly by making a few well-chosen approximations. Queuing theory is incorporated in the algorithms that calculate response times. A convenient notation used to label queues is:

$$A/B/n/K/p \quad (52.4)$$

where:

A = arrival process by which messages are generated
B = queue discipline, describing how messages are handled
n = number of servers available
K = number of buffering positions available
p = ultimate population size that may possibly request service

Some of the more typical queuing models pertaining to internetworking are as follows:

- Memoryless (exponential) interarrival distribution, memoryless service distribution, one server (M/M/1) is typical of situations involving a communications server or a file server of electronic mail.
- Memoryless (exponential) interarrival distribution, memoryless service distribution, c servers (M/M/c) is typical of situations involving dial-up ports for modem pools or private branch exchange (PBX) trunks.
- Memoryless (exponential) interarrival distribution, fixed (deterministic) service time, one server (M/D/1) is representative of a trunk servicing a packet switched network.

A number of internetworking situations require a more advanced model involving distributions other than exponential or deterministic. For the terminal nodes of communications system, the queue type that is most appropriate is M/G/1 distributions, referring to a generic service time represented by G. This means the process messages arrive are modeled as Markovian. The M/G/1 solution of queuing delay is given by the Pollaczek–Khinchin (PK) formula:

$$d = m + l * m^2 / [2(1-p)] \quad (52.5)$$

where

d = total delay
m = average time to transmit a message
l = number of arriving messages/second
p = utilization of the facility, expressed as a ratio

After a network model is produced, network engineers often use the tool to evaluate different scenarios of transmission components and traffic profiles. Through an iterative process of network refinement, "what if" questions may be addressed concerning various alternative configurations:

- What would be the impact on performance and utilization if more nodes are added to my network?
- How would response times change if my internetwork links were upgraded from voice grade private lines to 56-kb/s digital circuits; from fractional T-1 dedicated circuits to fast packet switched frame relay?
- How will downsizing to client/server technologies impact the network?
- Should I segment the network with high-speed switches or with traditional routers?
- How much additional capacity is available before having to reconfigure?
- What effect do priorities for applications and protocols have on user response times?

Each of the various alternative configurations will have associated costs and performance impacts. This produces a family of cost/performance curves that forms a solution set of alternative network designs. With the cost/performance curves complete, the final selection of an optimum configuration can be left up to the manager responsible for balancing budget constraints against user service levels.

At the completion of network design the network planner must objectively evaluate whether or not this phase has been completed adequately for the purpose of the design process. Questions that might be asked are:

- Is the model verified by accurately representing the inputs and generating output that satisfy the operational evaluation criteria?
- Is the cost/performance breakeven analysis selected from the family of feasible solutions the one best configuration based on the operational evaluation criteria?
- Is the equipment rated to function well within the design objectives?

When this phase is complete, an optimized implementable network design is available as input into the network management phase.

Network Management

Network management includes equipment acquisition, verification, installation, maintenance, and administration. This phase of the network engineering process feeds information into the network awareness component, which completes the cycle and begins the iterations of design and refinement.

Network design tools use statistical techniques to make certain assumptions about the behavior of a network and the profiles of transactions in the network. These assumptions in terms of lengths, distributions, and volumes of traffic make the mathematics easier, at the expense of accuracy. Operational networks often violate, to a certain extent, many design assumptions. The optimized network designs confront the real world of actual network performance during the network management phase. Detected variances between optimized network designs and actual network performance provide an improved awareness concerning host processing requirements, traffic profiles, and protocol characteristics.

In a nutshell, network management improves awareness. Design iterations and refinement based on improved network awareness progress toward the optimum. Network management refines operational evaluation criteria. Cost/performance ratios are evaluated against the operational objectives and goals an organization is employing. Network management avoids crises by reacting proactively to changing trends in utilization and performance.

At the completion of network management the network planner must objectively evaluate whether or not this phase has been completed adequately for the purpose of the design process. Questions that might be asked are:

- Is the network operating error free or are there periodic or isolated errors?
- Is the network performing properly by satisfying response time and availability goals?

When this phase is complete, an operational network is available.

Finally, the network engineer must verify and validate the model by asking: Is the model validated by the performance of the operational network? If this objective evaluation results in discrepancies between the model and the real world operational network, then there must be a return to the network awareness phase for model refinement and, possibly, improved optimization. Whenever new needs are placed on the network, the process must be reinitiated, with a revised needs analysis reflecting these new requirements. When the model is fully validated and verified, the network planner can exit the total network engineering decision process with an optimized operational network.

52.4 Internetwork Simulation

Today, with an overwhelming variety of alternative internetwork implementation options, engineers are often uncertain which technology is best suited for their applications. A few years ago there were relatively few systems from each of several vendors. Times have changed, and now a multitude of interconnection options exist. LAN media choices involve twisted pair, coaxial cable, fiber optics (single mode and multimode), and even wireless systems. LAN media access protocols include token ring (4 or 16 Mb/s) token bus, ethernet's carrier sense multiple access/collision delay (CSMA/CD), 100 Mb/s ethernet, 100 base VGAnyLAN, full duplex ethernet, switched ethernet, full duplex switched ethernet, fiber distributed data interface (FDDI), and distributed queue dual bus (DQDB). Cabling strategies include star-shaped intelligent hub wiring, distributed cabling interconnected through a high-speed backbone, and centralized tree-type cabling alternatives.

Discrete event **simulation** is the preferred design technique for internetworks involving varied LAN architectures and high-speed routers connected via WAN links. Here performance prediction is the primary objective. Engineers specify new or proposed LAN configurations with building blocks of common LAN components. LAN simulation tools predict measures of utilization, queuing buffer capacity, delays, and response times. Proposed configurations of departmental LANs or enterprise internetworks can be modeled, evaluating the capabilities of competing designs involving hardware components, protocols, and architectures. This can be achieved without learning complex programming languages or possessing a depth of queuing theory background. Several excellent design tools have been developed and marketed in recent years for simulation-based modeling and analysis of internetworks.

Can your present configuration handle an increased workload or are hardware upgrades needed? What part of the internetwork hardware should be changed: The backbone? The server? The router or bridge? How reliable is the internetwork, and how can it recover from component failure? Predictive methods preclude the unnecessary expense of modifying a network only to discover it still does not satisfy user requirements. What-if analysis applied to network models helps evaluate topologies and plot performance against workloads. Performance and reliability tradeoffs can be effectively identified with simulation. This is essential in service level planning.

Simulation is beneficial throughout the life cycle of a communications system. It can occur before significant development, during the planning stage, as an ongoing part of development, through installation, or as a part of ongoing operations. Alternatives of different configurations are compared and accurate performance is predicted before implementing new network systems. Presizing networks based on current workloads and anticipated growth demands and pretesting application performance impact are just a few of the benefits of simulation design.

Internetworking

Simulation tools expedite investigating internetwork capacities and defining growth. Knowing whether circuits, servers, node processing, or networked applications are the bottlenecks in performance is useful to any engineer. Managing incremental growth and predicting the impact of additional networking demands are other advantages. Justifying configuration designs to upper level management is another big asset of internetwork simulation.

52.5 Internetwork Optimization

Optimization tools balance performance and cost tradeoffs. Acceptable network performance is essential, but due to the non-linear nature of tariffs, cost calculations are critical. Tariff calculations are not trivial. Designs based on various routing, multiplexing, and bridging approaches can yield widely varying prices. Tariff data changes weekly, further complicating cost calculations. With new tariff filings there may be new optimal designs. The best topology of circuit routes and circuit speeds can significantly differ from previously optimal designs when re-evaluating with new tariff data. All optimization tools evaluate cost calculations against performance requirements.

Figure 52.3 graphically illustrates this cost/performance relationship. The ideal cost/performance point is where traffic matches capacity, with acceptable zones around this ideal. Acceptable oversizing is due to economies of scale or forecasted growth, and undersizing is due to excessive rebuilding costs. Outside this acceptable zone, networks need redesign. Where capacity significantly exceeds traffic, the network feels the effects of overbuilding (performance goals satisfied at unnecessarily high costs). Conversely, where traffic is greater than capacity, the network has been underbuilt (cost goals satisfied yet at the expense of performance objectives) and rebuilding is necessary.

Network optimization techniques have become even more complex due to recent introductions of new service offerings, such as frame relay, integrated services digital network (ISDN), synchronous digital network (SONET), SMDS and asynchronous transfer mode (ATM) and hubless digital services. These new services are offered by the major interexchange carriers, specialized fiber carriers, and regional Bell operating companies (RBOCs). In addition to reducing costs, new services when designed properly can also increase network reliability.

The enormous variability in circuit costs is illustrated in pricing a single 56-kb/s circuit. An interexchange circuit between Atlanta and Kansas City varies based on service types (DDS and fractional T-1) and service providers (AT&T, MCI, and WTG) between a high of $2751/month and a low of $795/month. Similarly, an intraLATA circuit between San Jose and Campbell, California

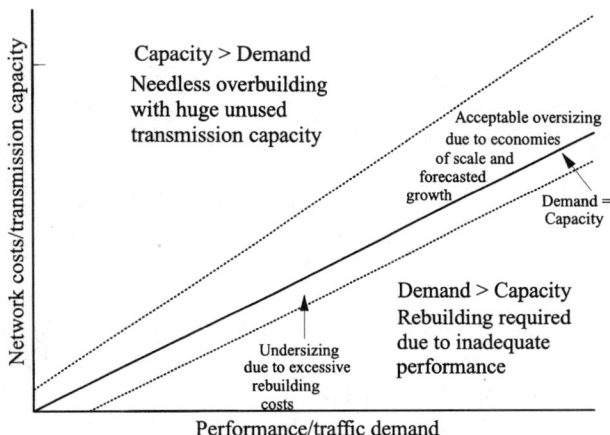

FIGURE 52.3 The cost/performance and capacity/demand relationship.

varies based on service types (DDS and new hubless digital) and service providers (AT&T, MCI, U.S. Sprint, and LEC-PacBell) between a high of $1339/month and a low of $136/month.

Furthermore, a variety of customer equipment-based multiplexing alternatives run the gamut from simple point-to-point multiplexing, to sophisticated dynamic bandwidth contention multiplexing, to multiplexer cascading, to colocating the CPE-based multiplexing equipment in the serving central office. Numerous networking options make the network engineer's job more complex, requiring optimization tools to evaluate these many new design techniques.

52.6 Summary

Information management today calls for effectively accommodating expansion of interconnected networks and accurately forecasting networking requirements. Only those who have an accurate awareness of their current communications requirements who can apply a structured design approach when evaluating competing networking techniques can confront this formidable challenge. James Jewett, co-founder of Telco Research, states, "Today, U.S. businesses frequently spend up to 40% more than is necessary on telecommunications services. Managers must be willing to scrutinize current communications networks and alternative networking techniques."

This is an age when communications budgets are being tightened, head counts are being trimmed, and major capital outlays are being delayed. Vendors selling LANs, internetworking gear, and long distance services feel cutbacks in flat revenues and intensely competitive bids. Network engineers feel the pinch as they try to meet increasingly demanding enduser performance objectives with dwindling budgets. Many network managers spend hundreds of thousands of dollars monthly on communications resources. These high costs reflect the importance of networks. Yet, with the high priority placed on LAN/WAN internetworking and the significant expenses devoted to maintain them, insufficient emphasis is placed on rigorous design and analysis. Network engineers tend to focus solely on operational capabilities. As long as the network is up and running, analysis or design is rarely conducted.

An engineering process of analysis and design for internetworks was presented. The process includes three interrelated phases: network awareness, network design, and network management. Each phase is part of a feedback control mechanism, progressing toward a network optimum. The industry emphasizes developing internetworking technologies; however, modeling techniques incorporating these technologies into operational networks has received little attention. Engineers applying this design and analysis process will be able to optimize their internetworks.

Defining Terms

Connectionless: A best-effort packet delivery model where the network does not guarantee delivery, and data may be lost, duplicated, damaged, or delivered out of order; also called datagram services.

Connection-oriented: A guaranteed packet delivery model where a virtual circuit with minimal bandwidth is established before any data transfer occurs and packets are delivered sequentially over a single path.

Data link switching (DLSw): A routing standard for integrating SNA data into IP-based internetworks by stripping off the SDLC header and replacing it with a TCP/IP header.

Distance-vector routing: An early routing information protocol implementation common to IPX RIP, IGRP, IP RIP, and AppleTalk's RTMP.

Internetwork optimization: This class of network design tools balance performance, cost, and reliability tradeoffs. Optimization algorithms apply heuristic procedures to greatly abbreviate the optimization process by making a few well-chosen approximations. Queuing theory is also incorporated in the algorithms to calculate response times.

Internetwork simulation: Discrete event simulation tools predict network performance and identify bottlenecks. The nondeterministic nature of internetworking communications is reflected in models by pseudorandom numbers that generate interarrival times of messages, lengths of message, time between failures, and transmission errors.

IPng: The new version of the IP layer that supports 16-byte addresses and offers advanced features such as flow control, autoconfiguration, and encryption-based security.

Link state routing: A new breed of efficient routing information protocols designed for large, complex internetworking environments and implemented in OSPF, IS-to-IS, and Novell's NSLP.

Network awareness: The first component in the network design process, in which existing technology is reviewed, future technology trends are evaluated, current traffic is determined, equipment is inventoried, and future growth is forecasted. An organization's unique operational evaluation criteria are also chosen during this phase.

Network design: The second component of the network design process, which includes with network design tools, performing a cost-performance breakeven analysis, and developing link configurations.

Network management: The third component of the network design process, which encompasses management of configuration, faults, performance, accounting, and security. Equipment and network software is acquired, verified, installed, maintained, and administered.

Packet burst: A sliding window implementation that allows transmitting multiple packets and receiving a multiple packet acknowledgment, greatly improving the efficiency of SPX/IPX internetworking.

Subnet mask: A four octet field used to identify the link portion of an IP address by setting bits to 1 where the network is to treat the corresponding bit in the IP address as part of the network address.

Total network engineering process: A process to engineer and optimize internetworks including the three phases of awareness, design, and management, with a feedback control mechanism, progressing toward a network optimum.

References

CCITT. 1979. Telephone operation: Quality of service and tariffs. *Orange Book*, Vol. II.2, International Telegraph and Telephone Consultative Committee. Geneva, Switzerland.

Comer, D.E. 1991. *Internetworking with TCP/IP*, Vol. I, Prentice–Hall, Englewood Cliffs, NJ.

Comer, D.E. and Stevens, D.L. 1991. *Internetworking with TCP/IP*, Vol. II, Prentice–Hall, Englewood Cliffs, NJ.

Ellis, R.L. 1986. *Designing Data Networks*, Prentice–Hall, Englewood Cliffs, NJ.

Marney-Petix, V.C. 1992. *Mastering Internetworking*, Numedia, Freemont, CA.

Marney-Petix, V.C. 1993. *Mastering Advanced Internetworking*, Numedia, Freemont, CA.

Miller, M.A. 1991. *Internetworking: A Guide to Network Communications*, M&T Books, San Mateo, CA.

Perlman, R. 1992. *Interconnections: Bridges and Routers*, Addison–Wesley, Reading, MA.

Shilling, G.D. and Miller, P.C. 1991. Performance modeling. *DataPro Netwk. Mgt.*, NM20(400):101–112.

Van Norman, H.J. 1992. *LAN/WAN Optimization Techniques*, Artech House, Boston, MA.

Further Information

The most complete reference is the three volume set, *Internetworking with TCP/IP* by Douglas E. Comer and David L. Stevens. Volume one covers principles, protocols and architectures design, and implementation; volume two covers design, implementation, and interfaces; volume three covers client/server programming and applications.

An excellent self-paced learning series covering the basics in a highly readable style with insights addressing an often misunderstood arena of communications technology is *Mastering Internetworking and Mastering Advanced Internetworking* by Victoria C. Marney-Petix.

The best technical treatment of the internal workings of bridges and routers is *Interconnections: Bridges and Routers* by Randia Perlman. It not only presents the algorithms and protocols but points out deficiencies, compares competing approaches and weighs engineering tradeoffs.

53
Architectural Framework for Asynchronous Transfer Mode Networks: Broadband Network Services

Gerald A. Marin
IBM

Raif O. Onvural
IBM

53.1 Introduction 717
53.2 Broadband Integrated Services Digital Network Framework .. 718
   Enabling Asynchronous Transfer Mode Networks
53.3 Architectural Drivers 720
   Simple Solutions: At a Cost
53.4 Review of Broadband Network Services 722
53.5 How Does It All Fit Together? 722
53.6 Broadband Network Services 724
   Access Services • Transport Services • Control Point Services
53.7 Conclusions 727

53.1 Introduction

The main feature of the integrated services digital network (ISDN) concept is the support of a wide range of audio, video, and data applications in the same network. Broadband ISDN (B-ISDN) is based on ISDN concepts, and has been evolving during the last decade by progressively incorporating directly into the network additional B-ISDN functions enabling new and advanced services. B-ISDN supports switched, semipermanent and permanent, and point-to-point and point-to-multipoint connections, and provides on-demand reserved and permanent services. B-ISDN connections support both circuit-mode and packet-mode services of a mono- and multimedia type with connectionless or connection-oriented nature, in bidirectional and unidirectional configurations.

Since the completion of the initial standardization work on B-ISDN in the late 1980s, there has been a tremendous amount of standards work toward defining B-ISDN interfaces, physical layers, signalling, and transfer mode. Asynchronous transfer mode (ATM) is the transfer mode of choice for B-ISDN. ATM is a connection-oriented, packet-switching technique. The ATM architecture and related standards define the basic framework on which different services with varying source characteristics and quality of service (QoS) requirements can be supported in an integrated manner. However, enabling B-ISDNs requires the design and development of a complete architectural framework that complements the ATM standards and includes network services such as congestion-control and path-selection frameworks, directory services, and group management protocols.

In this chapter, we first review the B-ISDN architecture, the architectural drivers such as high-bandwidth application requirements, and the changes in the networking infrastructure that need to be taken into account to enable ATM networks. We then present a high-level overview of a comprehensive, high-speed multimedia architecture developed by IBM to address the challenges of enabling ATM networks.

53.2 Broadband Integrated Services Digital Network Framework

B-ISDN is envisioned as providing universal communications based on a set of standard interfaces and scaleability in both distance and speed. B-ISDN physical layer interfaces, including the ones currently being defined, scale from 1.5 to 622 Mb/s. We have already started to see ATM being deployed in the local area, while a number of network providers have started to experiment with ATM in the wide area, providing seamless integration from local to wide area networks. Furthermore, B-ISDN services conceptually include every application that we can think of as we go forward into the future:

- Applications with or without timing relationship among the end users
- Variable and continuous bit rate services
- Connectionless and connection-oriented services

Voice and video are two applications that require an end-to-end timing relationship whereas there is no such requirement for data services. Most applications generate traffic at varying rates. Voice traffic alternates between talk spurts and silent periods. The amount of data generated per frame in a video service varies, depending on the compression algorithm used and how much information from the previous frame(s) can be used to reconstruct a new frame at the destination station while providing the required QoS. Variable bit rate (VBR) service refers to a source traffic behavior in which the rate of traffic submitted to the network varies over time. Early video and audio services in early ATM networks are expected to be supported using continuous bit rate (CBR) traffic, in which the traffic is submitted to the network at a constant rate. This is mainly because of the fact that the bandwidth scalability provided by ATM is a relatively new concept, and our previous experience in supporting such services is based on circuit switching, that is, PCM voice and H.222 video. As we understand traffic patterns and source characteristics better, VBR service is expected to replace CBR service for most real-time applications.

ATM is a connection-oriented service, that is, end-to-end connections must be established prior to data transmission. However, most network applications in current packet-switching networks are developed on top of a connectionless service. In order to protect customer investment in current applications, ATM should support connectionless service. This is provided mainly by providing an overlay on top of ATM.

Enabling Asynchronous Transfer Mode Networks

The main question in enabling ATM networks is how to make it all work as an integrated network. The answer to this challenge is the ATM standards and the development of network services such as congestion control, bandwidth management, path selection, and multicast and group-management mechanisms that complement the ATM standards while achieving high-resource utilization in the network.

The ATM standards have been developed by the International Telecommunications Union–Telecommunications Standardization Sector (ITU-TS) and various national standardization organizations such as American National Standards Institute (ANSI) and ETSI. The ATM Forum, on the other hand, is a consortium of more than 500 companies established to speed up the development and deployment of ATM products through interoperability specifications based on standards whenever available. Hereafter, we do not differentiate between the standards and the ATM Forum's specifications.

Figure 53.1 illustrates the current ATM architecture. A variety of physical layer interfaces are defined, including different transmission mediums such as multimode and single-mode fiber, shielded twisted pair, and coax. Transmission speeds now vary from 45 Mb/s (DS-3) to 622 Mb/s synchronous optical network (SONET).

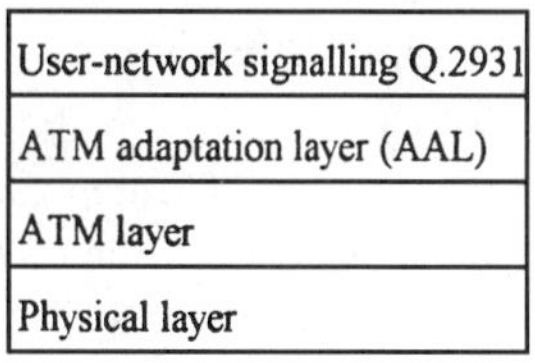

The ATM layer is based on the use of 53-byte fixed-size cells composed of a 5-byte header and a 48-byte payload. Of the 5-byte cell header, 28 b are used for routing. There is an 8-b header error check field to ensure the integrity of the cell header. This leaves only 4 b, of which 1 b is used for cell loss priority (high or low), and 3 b are used to identify the type of payload information and carry the explicit forward-congestion indicator. As discussed next, there is not really much time to do any processing in the network nodes as the transmission speeds increase. Some networking experts may have desired an extra byte or two to provide more features at the ATM layer. However, it would have been difficult and/or expensive to process this additional information as transmission speeds reach several gigabits per second.

| User-network signalling Q.2931 |
| ATM adaptation layer (AAL) |
| ATM layer |
| Physical layer |

FIGURE 53.1 Current ATM framework.

Nevertheless, more features than the ATM cell header supports are required to meet application requirements. Considering a wide variety of services with different requirements, the ITU-TS grouped functions that are common to most applications at the ATM adaptation layer (AAL). AAL type 1 is defined for CBR services that require an end-to-end timing relationship. The VBR counterpart of these services will be supported via AAL type 2, which is currently being defined. AAL type 3/4 is defined for the initial connection-oriented and connectionless services. AAL type 5 is a more efficient way of supporting data services. All AAL services are defined among end stations only; they are transparent to intermediate switches in the network.

ATM connections may be pre-established or may be established dynamically on request. Connections that are pre-established are defined by the network's management layer (not discussed here). Switched connections, on the other hand, require the definition of interfaces and signalling at these interfaces. Signalling at an ATM user–network interface (UNI) provides the capabilities needed to signal the network and to request a connection to a destination. Various capabilities included in UNI version 3.0 [ATM, 1994] can be summarized as follows:

- Establishment of point-to-point and point-to-multipoint virtual channel connections
- Three different private ATM address formats
- One public ATM address format
- Symmetric and asymmetric QoS connections with declared QoS class
- Symmetric and asymmetric bandwidth connections with declared bandwidth
- Transport of network transparent parameters among end nodes
- Support of error handling

An ATM user signals to request a connection from the network to a particular destination(s) with a specified QoS. Information elements such as peak cell rate, sustainable cell rate, cell delay tolerance, maximum burst size, AAL type, and address of destination end node(s) are included in the signalling messages exchanged across an UNI. Despite the tremendous effort that has gone into the defining these specifications, it will take a lot more work to enable high-bandwidth, multimedia applications in ATM networks. Various additional services are required:

- Locating destination end node(s): The addresses of end nodes are provided in a related signalling message. Determining the exact location of these end nodes, however, is not a part of the current standards.
- Determining how much network resources are needed to provide the required QoS: Guaranteeing the QoS needs of network applications requires, in general, that resources in the network be reserved. Although the traffic characteristics would be included in the connection request, a way of determining the commitment of resources needed to support different applications is not a part of today's ATM.
- Determining a path in the network that can support the connection: Path selection framework in ATM networks is quite different from path selection in current packet switching networks. While meeting the QoS requirements of applications, it is highly desirable to minimize resources used by the network, in order to achieve economic resource utilization. This requires the solution of a constrained optimization problem unlike the shortest path problem posed by traditional packet networks.
- Establishing and managing connections in the network: Routing in ATM networks remains largely an unresolved issue. Routing metrics and the decision place (i.e., source routing) are the two areas that are being studied currently at the ATM Forum. However, the efficient distribution of control information in a subnetwork and the quick dissemination of connection establishment and termination flows are two areas that are not addressed in the ATM standards.

53.3 Architectural Drivers

Before discussing how some of these issues may be addressed, let us look at various application requirements and the changes in the networking infrastructure imposed by advances in the transmission technology.

Perhaps, one of the main drivers for emerging ATM networks is the deployment of multimedia applications. Although initial ATM networks may be deployed as data-overlay networks that keep voice separate, it is unavoidable that voice, video, and data will come together as multimedia applications begin to emerge (i.e., videoconferencing, video distribution, distant learning). After all, multimedia networks supporting multimedia applications have been the main driving force behind the B-ISDN initiative.

Multimedia applications differ from monoapplications in various ways. They often take place among a group involving more than two users. This necessitates the existence of group-management and multicast services. Group-management functions would basically allow users to join and leave the group as they wish, and manage the connection among the group's members. Multicast trees provide a connection path among these users efficiently compared with point-to-point connections. Yet, the construction of a multicast tree that satisfies end-to-end delay constraints while achieving high-resource utilization is a hard problem with an exponentially growing time complexity. Addressing this problem in real time requires the development of efficient heuristics.

Multimedia applications also combine the different QoS requirements of individual applications into a single application. This requires the integration of the service demands of different applications and the definition of network services that can support such requirements. As an example, voice and lip motion need to be tightly synchronized in a video service, whereas the synchronization of subtitles (i.e., data) to video does not need to be as stringent.

ATM networks will support a wide variety of traffic with different traffic characteristics. Using the source parameters defined in the ATM Forum's UNI 3.0 specification, a user may pass its peak and sustainable cell rates to the network during the connection-establishment phase. The burstiness (defined in this context as the ratio of peak to average bit rates) of various applications envisioned in ATM networks varies from 1 (e.g., for CBR service) to 100 s (e.g., for distributed computing). As the burstiness of a source increases, the predictability of its traffic decreases. Accordingly, it becomes harder for the network to accommodate highly bursty sources while achieving high resource utilization. Another challenge is the integration of highly bursty sources in the network with less bursty sources, say CBR services, while meeting the mean-delay and jitter requirements. Let us consider two applications at the opposite ends of the spectrum with respect to their burstiness, voice and distributed computing. When a distributed computing source becomes active, it generates traffic at the rate of, say, 100 Mb/s. An active period is followed by a silent period during which the source processes data in its memory and does not generate any network traffic. The high burstiness of the source would mean that the periods of activity persist for considerably less time than the periods of inactivity. On the other hand, consider a PCM voice application generating frames at the constant rate of 64 Kb/s. The challenge is to provide network services such that when the two sources are active simultaneously on one or more physical links, the QoS provided to the voice source does not degrade due to a sudden spurt of cells belonging to the distributed computing application. The problem of dealing with applications with widely varying burstiness in an integrated manner cannot be avoided as we move toward VBR multimedia services; network services that meet this challenge are needed to deploy integrated services.

In addition to having to deal with varying source characteristics, ATM networks should support services with quite different QoS requirements. For example, most data applications require fully reliable service (i.e., a banking transaction). But, these services can use the functions provided by a higher layer protocol (i.e., a transport service) on top of an ATM connection that will recover lost data by retransmission. Voice applications can tolerate moderate cell losses without affecting the quality of the service. Interactive VBR video services have very stringent loss requirements, since a lost cell could cause the loss of frame synchronization, which in turn would cause the service to degrade for seconds until the stream were refreshed. All of these applications with different requirements are integrated in the network. Network services designed to support these applications should take into consideration their QoS requirements explicitly, if high resource utilization in the network is desired.

In addressing the challenges of supporting application requirements, it is necessary to consider the constraints inherent to high-speed networking. As transmission speed increases, two considerations start to play important roles in designing network services: in-flight data becomes huge, and available processing time at the intermediate nodes decreases.

With high-speed transmission links in the network, the amount of in-flight data increases dramatically as the link propagation delay increases. As an example, when the one-way propagation delay on a T1 link is 20 ms (i.e., across the U.S.), there can be up to 3750 bytes simultaneously on the link at any particular time. This number increases to 2.5 million bytes on a 1 Gb/s link. If hop-by-hop flow control were applied between the two ends of such a link, which works well in current packet networks with low-speed links, millions of bytes of data could be lost by the time a control packet traversed the network from one end to the other. This phenomenon dramatically changes the congestion-control framework in high-speed networks. Another aspect of this is the challenge of providing efficient and accurate distribution of control information in the network, that is, link failures, resource reservation levels, actual link utilization, etc.

Another key challenge high-speed links introduce is to keep the functions at intermediate network nodes simple while providing enough functionality for high-resource utilization. Consider a 9.6-kb link and a processor with one million instructions per second capability. Then, up to 44,166 instructions can be executed for each cell while keeping the link full. When the transmission rate reaches gigabits per second, however, this number goes below one. Even with processors that run at 30 MIPS, the functionality at an intermediate node should be minimized to keep up with high-speed links. Most required functions need to be simple enough to be implemented in hardware. This, perhaps, is one of the main reasons for keeping the ATM cell header so simple.

Simple Solutions: At a Cost

Most of these challenges could be addressed rather easily if the network resource utilization were not taken into consideration. For example, all applications can be treated in the network with the most stringent loss and delay requirements. Though simple, this approach might prove to be quite costly, in that an artificially large amount of network resources might be reserved in the network to support applications that could tolerate moderate loss and/or delay.

As another example, a very small amount of buffering can be provided at each network node with deterministic multiplexing (i.e., peak rate bandwidth allocation). Although this scheme minimizes queueing delays at each node and guarantees practically no loss, it causes resource utilization in the network to be unacceptably low. As an illustration, consider a source with peak and average bit rate of 10 Mb/s and 1 Mb/s, respectively. On a 100-Mb/s link, only 10 of these connections can be multiplexed, which results in an average link utilization of 10%.

Similarly, point-to-point connections can be used to provide communication among a group of users. This might, however, artificially increase the number of links used to provide a multipoint communication compared with a multicast tree, and thereby reduce the availability of extra links for use in point-to-point connections to support traffic from other sources.

In summary, simple things can be done to enable ATM networks; the challenge, however, is to provide network services that will meet the application requirements economically.

53.4 Review of Broadband Network Services

This section presents a high-level review of IBM's broadband network services (BBNS) architecture. BBNS is a comprehensive approach to providing network services for high-speed, multimedia networking based on existing ATM standards and ATM Forum specifications. Conceptually, the BBNS architecture is broken down into the following three functional groups:

- *Access services* provide the framework for supporting standards-based and industry-pervasive interfaces and protocols.
- *Transport services* provide the pipes and the trees needed to switch voice, video, and data across the network in an integrated structure with appropriate QoS guarantees.
- *Control point services* manage network resources by providing the distributed brain power that enables all network services.

Access services can be thought of as locators and translators. They interface with the foreign protocols outside the BBNS network and translate them into what the network understands inside.

Transport services are the transfer modes defined in the architecture and related services developed to support different types of applications with varying service requirements. In addition to ATM, the BBNS architecture supports variable length packets to provide a migration path from the current state of the art in networking today to an all-ATM network. Other services developed include the datagram and multicast services and mechanisms to support different delay priorities.

Control point services mainly provide congestion control, path selection, group management for multimedia applications, directory services, and distribution of network control information.

53.5 How Does It All Fit Together?

Before proceeding with the details of the BBNS architecture, we discuss how the various functions required to enable high-speed networking relate to each other. This discussion uses call setup as a vehicle.

A new connection request is first received by an access agent at the edge of the network. The agent runs an address resolution mechanism to find the destination end node. Then, using the source traffic characteristics and application requirements, the amount of bandwidth required by the application is determined. Bandwidth calculation takes into consideration the cell loss requirement of the application, uses the source traffic characteristics, and determines a guaranteed upper bound on the bandwidth required to support the application when this connection is multiplexed with the connections already established in the network. Then, the path-selection algorithm is run to determine a path from the origin node to the destination node(s). In determining this path, the end-to-end delay requirement of the application is taken into consideration explicitly. In particular, the algorithm finds a minimum-hop path in the network that can support the bandwidth end-to-end requirements of the application.

Once a path is found, an end-to-end connection is established using internal signalling. The traffic can start flowing after every node along the path accepts the connection establishment request. In this framework, source routing is used; paths are determined by the nodes the connections originate. This approach requires that each node that supports an access service knows the topology of the network (which is subject to change due to link and node failures) and the utilization and current reservation levels of network resources. Although the topology of the network does not change that often, utilization levels change frequently. This necessitates an efficient means of distributing network control information that maximizes the amount of information available to each node while minimizing the amount of network control traffic overhead. These two objectives are contradictory and the design and development of this feature is a complex task.

At this stage, let us consider the network operating with traffic flowing through a large number of connections. At the edge of the network, it is necessary to monitor the traffic generated by each source to assure that each source stays within the parameters negotiated at the call establishment phase. Otherwise, nonconforming sources would cause the QoS provided to conforming sources to degrade. A leaky bucket mechanism is used for this purpose. The main challenge in using a leaky bucket is to determine its parameters such that its operation would be transparent to conforming sources while assuring that nonconforming traffic is detected.

Let us now focus on the nonconforming traffic. There are various reasons that a source might not stay within the parameters negotiated at the call setup time: the source might not be able to characterize its traffic behavior accurately, there might be an equipment malfunction, or the source might simply be cheating. There are two mechanisms developed in the BBNS architecture to provide some amount of forgiveness to nonconforming sources. The first one is the adaptation function which filters the traffic and estimates the amount of bandwidth required by connections based on the actual traffic generated by each source. The second mechanism is the use of so called red/green marking. In this mechanism, conforming traffic is marked green before it is transmitted. Red marking allows (some) nonconforming traffic to enter the network to achieve higher utilization of network resources. Doing so, however, requires the development of mechanisms so that the services provided to conforming traffic in the traffic are not affected by the non-conforming traffic when a node becomes congested.

Another feature required in the network is to minimize the negative effects of resource failures to the service provided to applications. A closely associated function is that of providing support for different priority connections. Nondisruptive path switching is used to reroute connections established on a link (node) around failures in a way that the impact of the failure to the service provided to these affected connections is minimal. Similarly, different connections may have different priorities. Depending on the availability of resources in the network, it may be necessary to take down some low-priority connections in order to accommodate a high-priority connection in the network. The main challenge here is to minimize the cascading effect that may occur when a connection that was taken down tries to re-establish a new connection.

A multimedia network should support connections with different delay requirements. The architecture classifies the multimedia traffic into three categories: real-time, nonreal-time, and nonreserved. Real-time traffic has the stringent delay and loss requirements and requires bandwidth reservation to guarantee its QoS metrics. Nonreal-time traffic also requires bandwidth reservation but only for the loss guarantee, that is, this type of traffic can tolerate moderate delay. Nonreserved traffic, on the other hand, can tolerate both loss and delay. Accordingly, there is no bandwidth reserved for this type of traffic.

53.6 Broadband Network Services

Although the availability of high-speed links in the network infrastructure is important, what really matters is what customers are able to do with them. Two important steps need to be taken to realize fully the promise of ATM. They are:

- The development of high-bandwidth applications
- The development of networking services based on the evolving standards

So far, we have discussed various application requirements and types of services that are required to enable high-bandwidth applications in ATM networks. In this section, we will present a more detailed view of IBM's broadband network services.

Current predominant networking technologies are designed to emphasize service for one or two of these QoS classes. For example, telex (delay and loss tolerant); telephony (delay intolerant); TV and community antenna television (CATV) (delay intolerant); and packet-switched networks like X.25, frame relay, APPN, transport control protocol/Internet protocol (TCP/IP), SNA (delay tolerant). The promise of ATM is that networks can be built on a single technology to accommodate all QoS classes. BBNS is a comprehensive approach to providing network services for high-speed, multimedia networks based on the existing ATM standards and ATM Forum implementation agreements. Various services developed in this architecture are presented next.

Access Services

While providing users with a high-speed, multimedia networking infrastructure for emerging high-bandwidth applications, it is necessary to support existing user applications that have been developed based on multiple protocols. Broadband network services achieve this through access services that support multiple protocols including frame relay, ATM, voice, IP, and HDLC data, to name a few. External nodes connected to BBNS do not require any special knowledge of any BBNS features, that is, these stations communicate with BBNS access services using their native protocol.

Functions performed by access services include the following:

- Understanding and interpreting the external service or protocol
- Performing an address resolution function to locate the target resource
- Maintaining and taking down connections across the network in response (for example) to connection requests received at the ATM UNI
- Selecting the switching mode and network connection parameters that are optimal for the external protocol and the service requested
- Mapping the QoS parameters of the external protocol the parameters used by the architecture to assure the availability of sufficient network resources to guarantee that the QoS requirements are met
- Ensuring fairness among users by calling on BBNS bandwidth management services.

Transport Services

BBNS transport services provide a transport facility across the network for user traffic generated at the edges of the network and accommodated by access services. These include transmission scheduling, hardware-based switching with multicast capability, and provision of a switching mode that meets the requirements of external protocols efficiently. In particular, the architecture supports the following switching modes:

- ATM switching
- Automatic network routing, which provides source-routing capability
- Label swap (with tree support)
- Remote access to label-swap tree

The architecture supports ATM natively. Depending on product implementation and customer choice, a network built based on this architecture can support network trunks that carry either ATM cells only, or a mix of ATM cells and variable length packets, with the former being transported in standard ATM cell format. Just as ATM cells provide native transport for the ATM UNI, BBNS offers support for other interfaces such as HDLC, frame relay, and IP. This feature allows customers to introduce early ATM applications using ATM transport without any changes or performance penalty in supporting their existing applications. However, any customer committed to ATM-only networking is fully supported through the BBNS architecture.

To provide service appropriate to the QoS classes discussed previously BBNS supports two types of traffic priorities: delay and loss. Three different delay priorities are defined: real-time reserved bandwidth, nonreal-time reserved bandwidth, and best-effort service. Logically, each delay class is routed through a corresponding queue in each node. Similarly, four different loss priorities are defined for real-time traffic, and two different loss priorities are defined for nonreal-time traffic. Combined with the power of BBNS bandwidth management, these delay and loss priorities offer the flexibility to support a broad range of QoS requirements and, thus, to handle efficiently the emerging multimedia applications.

Control Point Services

BBNS control point services control, allocate, and manage network resources. Among other things, these services include bandwidth reservation, route computation, topology updates, address resolution and the group management support needed for multipoint connections.

Some of these services are briefly described next.

Set Management

Set management provides the mechanism to group user resources for the purpose of defining multiple logical (i.e., virtual private) networks. This is a critical requirement for service-provider networks, where resources belonging to one customer may have to be isolated from those of other customers. The same mechanism can also be used to define multiple virtual local area networks (LANs) on a network.

Set management also gives a network the ability to learn about new resources and users automatically, thereby reducing system definition and increasing productivity. In essence, set management allows a network function or a user port to be defined only once to make it available to the entire network. Furthermore, it provides full flexibility and mobility in naming/addressing. A user can be moved from one point to another without being forced to change his external address (phone number, for example).

Bandwidth Management

BBNS bandwidth management, congestion control, and path selection enable a network to provide quality of service guarantees on measures such as end-to-end delay, delay jitter, and packet (or cell) loss ratio, while significantly reducing recurring bandwidth cost. This is achieved through efficient bandwidth reservation, load balancing, and dynamic bandwidth allocation.

For each reserved bandwidth connection requiring QoS guarantees, the architecture reserves a sufficient amount of bandwidth on the links traversed by the connection. The BBNS approach, which depends on the connection's traffic characteristics and QoS requirements, computes sufficient bandwidth to satisfy the QoS requirements, taking into account the statistical multiplexing at the links. BBNS path selection finds a path from source to destination that satisfies the connection's QoS requirements using algorithms that minimize the number of hops in the path and balance the network traffic load.

The architectural approach supports both reserved and nonreserved bandwidth transport of user frames or cells. Note that no QoS support is provided for nonreserved traffic. When reserved bandwidth connections are used to support bursty connections (like traditional data traffic), the challenge of setting network connection parameters can be significant. When a user has a good understanding of his traffic characteristics, the user may be unable to signal those characteristics to the network until the workstation implements a new interface, such as the ATM UNI. Even when it is possible to set traffic parameters correctly initially, the bandwidth requirements may change dramatically on the order of seconds to minutes. To mitigate these problems the architecture offers bandwidth adaptation (dynamic bandwidth allocation) as an optional added value. First, BBNS bandwidth management will enforce an existing traffic contract on a cell time scale of milliseconds or lower. Then, at user option, bandwidth management will learn the actual bandwidth requirements of a connection and adapt by rereserving (and possibly rerouting) the connection using a longer time scale on the order the roundtrip delay (seconds).

Efficient Distribution of Control Information

BBNS hardware-based multicast switching capability allows fast and efficient distribution of network control messages. Messages such as topology and link-state updates are multicast on the control-point (CP) spanning tree (a multipoint connection, or tree, that extends to every node in the network) at close to propagation speed. Furthermore, since these update messages are carried only by links on the CP spanning tree ($n - 1$ links in an n-node network), the bandwidth needed by the control traffic is greatly reduced. The BBNS CP spanning tree algorithm maintains the integrity of the spanning tree. The algorithm builds the tree and automatically repairs it in case of partitioning caused by link or node failures. The topology algorithm provides reliable delivery of multicast messages. Control messages such as connection setup requests and connection maintenance messages also benefit significantly from the hardware multicast capability. These messages are multicast linearly to every intermediate node along a connection's path, thus allowing connection setups and takedowns much faster than through hop-by-hop signalling methods.

Nondisruptive Path Switching (NDPS)

Nondisruptive path switching (NDPS) makes it possible for a network to reroute network connections automatically in the face of node and/or link failures, while minimizing user disruption. Although such a feature is common to other networking architectures, the BBNS version is extremely efficient. First, changes to the network, planned or unplanned, are nearly instantly known throughout the network, because of the CP spanning tree. Second, all rerouting decisions are made in parallel in a distributed fashion, rather than sequentially at a central control point. This allows for rapid network stabilization.

Call Pre-emption

Call pre-emption enables a network to provide different levels of network availability to users with different relative priorities. When required, call pre-emption will reroute existing connections to accommodate higher priority connections. BBNS call pre-emption strategy intelligently selects connections to be pre-empted, minimizing the possible rippling effects. NDPS is used to reroute the pre-empted connections; thus, disruption is kept to a minimum. The architecture gives the network operator the flexibility of assigning a wide range of priorities to network users.

53.7 Conclusions

There are various issues that need to be satisfactorily addressed in order for ATM networks to become a reality. In this chapter, we reviewed some of these issues based on application requirements and changing networking infrastructure. Then, we presented a high-level review of IBM's architecture for high-speed, multimedia networking that addresses some of these challenges while achieving high resource utilization in the network. ATM, related standards, and IBM's broadband network services architecture are evolving over time. Despite the tremendous amount of work completed on the standards and various networking issues, and the fact that vendors have been working on the design and development of an architectural framework to enable ATM networks, there still remain several issues that need to be satisfactorily resolved. This includes the design and development of applications that can take advantage of high bandwidth in customer premises and public networks as well as network services that can meet the application requirements while maximizing the usage of network resources. ATM is positioned as the transport method with the greatest potential to provide unified voice, video, and data service for high-speed networks supporting multimedia applications. Much work has been done in defining the B-ISDN standards in which ATM transport is featured. Both international standards bodies and the ATM Forum continue the prolific work to enhance the standards and implementers' agreements that are fundamental to realizing customer expectations. Yet, various vendors are expected to provide value-added services that complement the ATM standards thereby distinguishing their products from others.

Broadband network services is IBM's architecture for value add to various ATM networking that include:

- A set of transport services that allow efficient integration of traffic generated by different sources with varying delay and loss requirements
- Bandwidth management framework that provides quality of service guarantees while achieving high-resource utilization in the network
- Efficient distribution of control information in the network minimizing the resources used for network control overhead while updating control information at the network nodes (at propagation delay speeds)
- Set management techniques that define and maintain logical groups and allow the network to learn about new resources and users automatically
- Nondisruptive path switching that minimizes disruption to users in case of network resource failures
- Call pre-emption that allows priority handling of connections and minimizes possible cascading of connection takedowns
- A path selection framework that addresses the requirements of different types of applications with varying delay and loss demands while achieving high utilization of network resources. This framework includes both point-to-point and point-to-multipoint path construction
- The flexibility to help users evolve their existing applications and networks to ATM transport by matching the switching mode provided by the network to the native transport mode of existing protocols and applications.

References

ATM. 1994. User-network specification version 3.0, ATM Forum.
ITU-TS Recommendations, International Telecommunications Union-Telecommunications Standards Sector.
Onvural, R.O. 1994. *ATM: Performance Issues*, Artech House, Norwood, MA.
Prycker, Q. *Asynchronous Transfer Mode*.
IBM. *Broadband Network Services*, IBM Blue Book, Research Triangle Park, NC.

IV

Optical

54 Fiber Optic Communications Systems *Joseph C. Palais* 731
Introduction • Optical Communications Systems Topologies • Signal Quality • System Design

55 Optical Fibers and Lightwave Propagation *Paul Diament* 740
Transmission Along Fibers • Total Internal Reflection • Modes of Propagation • Parameters of Fibers • Attenuation • Dispersion • Graded-Index Fibers • Mode Coupling • Summary

56 Optical Sources *N.K. Dutta* 751
Introduction • Laser Designs • Quantum Well Lasers • Distributed Feedback Lasers • Surface Emitting Lasers • Laser Reliability

57 Optical Transmitters *A.J. Price and K.D. Pedrotti* 774
Introduction • Directly Modulated Laser Transmitters • Externally Modulated Optical Transmitters

58 Optical Receivers *R.G. Smith and B.L. Kasper* 789
Introduction • The Receiver • Receiver Sensitivity: General

59 Fiber Optic Connectors and Splices *William C. Young* 803
Introduction • Optical Fiber Coupling Theory • Multibeam Interference (MBI) Theory • Connector Design Aspects • Splicing Design Aspects • Conclusions

60 Passive Optical Components *Joseph C. Palais* 824
Introduction • Losses in a Passive Optical Component • Grin Rod Lens • Isolator • Directional Couplers • Star Coupler • Optical Filter • Attenuator • Circulator • Mechanical Switch • Polarization Controller

61 Semiconductor Optical Amplifiers *Daniel J. Blumenthal and Nitin C. Kothari* 832
Introduction • Principle of Operation • Types of Semiconductor Optical Amplifiers • Design Considerations • Gain Characteristics • Pulse Amplification • Multichannel Amplification • Applications

62 Optical Amplifiers *Anders Bjarklev* 848
Introduction • General Amplifier Concepts • Alternative Optical Amplifiers for Lightwave System Applications • Summary

63 Coherent Systems *Shinji Yamashita* 862
Introduction • Fundamentals of Coherent Systems • Modulation Techniques • Detection and Demodulation Techniques • Receiver Sensitivity • Practical Constraints and Countermeasures • Summary and Conclusions

64 Fiber Optic Applications *Chung-Sheng Li* 872
Introduction • Optical Interconnects • Local Area Networks and Input/Output (I/O) Interconnections • Access Networks • Wavelength-Division Multiplexing-Based All Optical Networks • Fiber Sensors

65 Wavelength-Division Multiplexed Systems and Applications *Mari W. Maeda* 883
Introduction • Optical Components for Wavelength-Division Multiplexed Systems • Wavelength-Division Multiplexed System Design • Trunk Capacity Enhancement Applications • Wavelength-Division Multiplexed Networking and Reconfigurable Optical Transport Layer • Summary

54
Fiber Optic Communications Systems

Joseph C. Palais
Arizona State University

54.1 Introduction 731
54.2 Optical Communications Systems Topologies 733
      Fibers • Other Components
54.3 Signal Quality 738
54.4 System Design 738

54.1 Introduction

Transmission via beams of light traveling over thin glass fibers is a relative newcomer to communications technology, beginning in the 1970s, reaching full acceptance in the early 1980s, and continuing to evolve since then [Chaffee, 1988]. Fibers now form a major part of the infrastructure for national telecommunications information highways in the U.S. and elsewhere and serve as the transmission media of choice for numerous local area networks. In addition, short lengths of fiber serve as transmission paths for the control of manufacturing processes and for sensor applications.

This section presents the fundamentals of fiber communications and lays foundations for the more detailed descriptions to follow in this chapter. Many full-length books (e.g., Hoss [1990], Jeunhomme [1990], Palais [1992], and Keiser [1991]) exist for additional reference.

Earlier chapters in this handbook covered data rates and bandwidth requirements for telephone systems and local area networks. The steadily increasing demand for information capacity has driven the search for transmission media capable of delivering the required bandwidths. Optical carrier transmission has been able to meet the demand and should continue to do so for many years. Reasons for this are given in the following paragraphs.

Optical communications refers to the transmission of information signals over carrier waves that oscillate at optical frequencies. Optical fields oscillate at frequencies much higher than radio waves or microwaves as indicated on the abbreviated chart of the electromagnetic spectrum in Fig. 54.1. Frequencies and wavelengths are indicated on the figure.

For historical reasons, optical oscillations are usually described by their wavelengths rather than their frequencies. The two are related by

$$\lambda = c/f \qquad (54.1)$$

where f is the frequency in hertz, λ is the wavelength, and c is the velocity of light in empty space

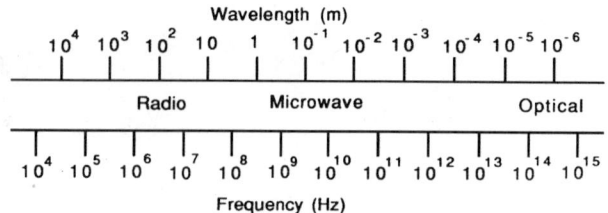

FIGURE 54.1 Portion of the electromagnetic spectrum.

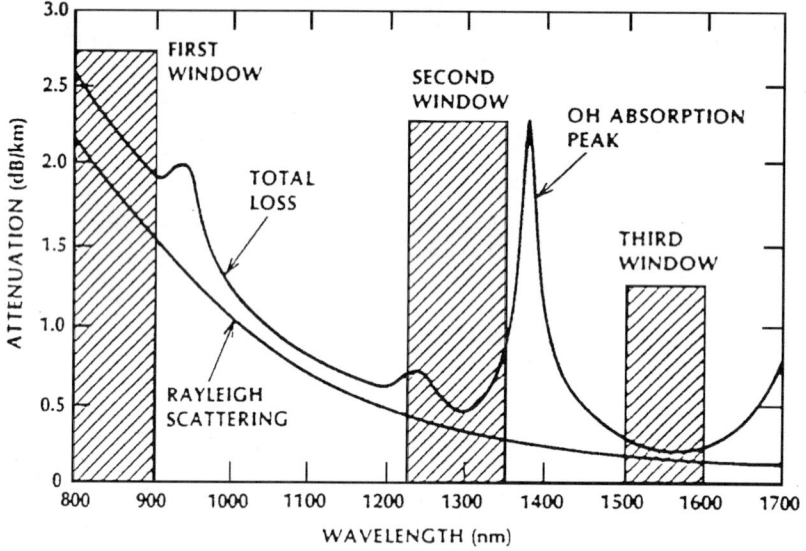

FIGURE 54.2 Attenuation of a silica glass fiber showing the three wavelength regions where most fiber systems operate. (*Source:* Palais, J.C. 1992. *Fiber Optic Communications*, 3rd ed., p. 112. Prentice-Hall, Englewood Cliffs, NJ. With permission.)

(3×10^8 m/s). A frequency of 3×10^{14} Hz corresponds to a wavelength of 10^{-6} m (a millionth of a meter is often called a *micrometer*). Wavelengths of interest for optical communications are on the order of a micrometer. Glass fibers have low loss in the three regions illustrated in Fig. 54.2, covering a range from 0.8 to 1.6 μm (800 to 1600 nm). This corresponds to a total bandwidth of almost 2×10^{14} Hz. The loss is specified in decibels, defined by

$$\text{dB} = 10 \log P_2/P_1 \tag{54.2}$$

where P_1 and P_2 are the input and output powers. Typically fiber transmission components are characterized by their loss or gain in decibels. The beauty of the decibel scale is that the total decibel value for a series of components is simply the sum of their individual decibel gains and losses.

Although the dB scale is relative, it can be made absolute by specifying a reference level. Two popular references are the milliwatt and microwatt. The respective scales are the dBm (decibels relative to a milliwatt) and the dBμ (decibels relative to a microwatt). That is,

$$P(\text{dBm}) = 10 \log P \text{ (mW)} \tag{54.3a}$$

$$P(\text{dB}\mu) = 10 \log P \text{ (}\mu\text{W)} \tag{54.3b}$$

Optical power measuring meters are often calibrated in dBm and dBμ. Light emitter power and receiver sensitivity (the amount of power required at the receiver) are often given in these units.

Losses in the fiber and in other components limit the length over which transmission can occur. Optical amplification and regeneration are needed to boost the power levels of weak signals for very long paths.

The characteristically high frequencies of optical waves (on the order of 2×10^{14} Hz) allow vast amounts of information to be carried. A single optical channel utilizing a bandwidth of just 1% of this center frequency would have an enormous bandwidth of 2×10^{12} Hz. As an example of this capacity, consider frequency-division multiplexing of commercial television programs. Since each TV channel occupies 6 MHz, over 300,000 television programs could be transmitted over a single optical channel. Although it would be difficult to actually modulate light beams at rates as high as those suggested in this example, tens of gigahertz rates have been achieved.

In addition to electronic multiplexing schemes, such as frequency-division multiplexing of analog signals and time-division multiplexing of digital signals, numerous optical multiplexing techniques exist for taking advantage of the large bandwidths available in the optical spectrum. They include **wavelength-division multiplexing (WDM)** and **optical frequency-division multiplexing (OFDM)**. These technologies allow the use of large portions of the optical spectrum. The total available bandwidth for fibers approaches 2×10^{14} Hz (corresponding to the 0.8–1.6 μm range). Such a huge resource is hard to ignore.

Although atmospheric propagation is possible, the vast majority of optical communications utilizes the waveguiding glass fiber.

A key element for optical communications, a coherent source of light, became available in 1960 with the demonstration of the first **laser**. This discovery was quickly followed by plans for numerous laser applications, including atmospheric optical communications. Developments on empty space optical systems in the 1960s laid the groundwork for fiber communications in the 1970s. The first low-loss optical waveguide, the glass fiber, was fabricated in 1970. Soon after, fiber transmission systems were being designed, tested, and installed. Fibers have proven to be practical for path lengths of under a meter to distances as long as needed on the Earth's surface and under its oceans (for example, almost 10,000 km for transpacific links).

54.2 Optical Communications Systems Topologies

Fiber communications are now common for telephone, local area, and cable television networks. Fibers are also found in short data links (such as required in manufacturing plants), closed-circuit video links, and sensor information generation and transmission.

A block diagram of a point-to-point fiber optical communications system appears in Fig. 54.3. This is the structure typical of the telephone network. The various components will be described later in this section and in succeeding sections of this chapter.

The fiber telephone network is digital, operating at data rates from a few megabit per second up to 2.5 Gb/s and beyond. At the 2.5-Gb/s rate, several thousand digitized voice channels (each operating at 64 kb/s) can be transmitted along a single fiber using time-division multiplexing (TDM).

Because cables may contain more than one fiber (in fact, some cables contain hundreds of fibers), a single cable may be carrying hundreds of thousands of voice channels. Rates in the tens of gigabit per second are attainable, further increasing the potential capacity of a single fiber.

Telephone applications may be broken down into several distinctly different areas: transmission between telephone exchanges, long-distance links, undersea links, and distribution in the local loop (that is, to subscribers). Although similarities exist among these systems, the requirements are somewhat different. Between telephone exchanges, large numbers of calls must be transferred over moderate distances. Because of the moderate path lengths, optical amplifiers or regenerators are not required. On the other hand, long-distance links (such as between major cities) require signal boosting of some sort (either regenerators or optical amplifiers). Undersea links (such as

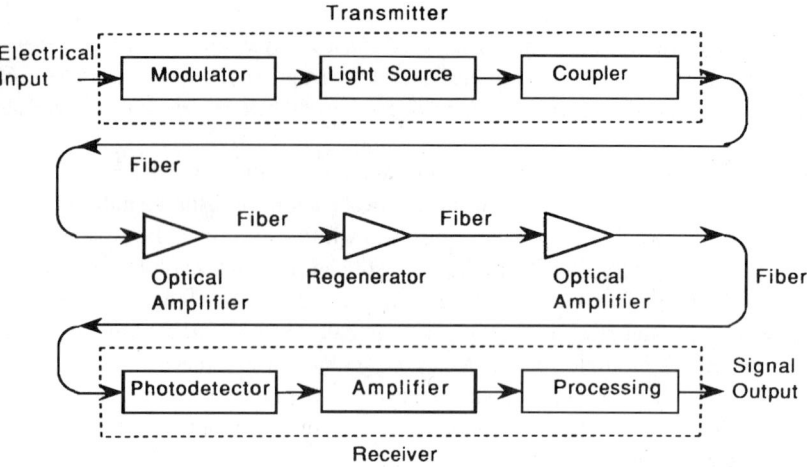

FIGURE 54.3 Major components in a point-to-point fiber transmission system.

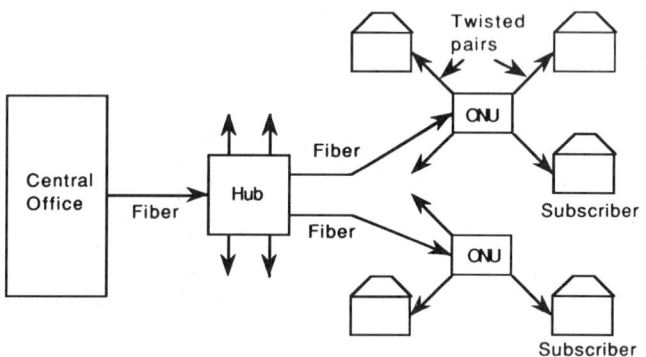

FIGURE 54.4 Fiber-to-the-curb network.

transatlantic or transpacific) require multiple boosts in the signal because of the long path lengths involved [Thiennot, Pirio, and Thomine, 1993].

The local loop does not involve long path lengths, but does include division of the optical power in order to share fiber transmission paths over all but the last few tens of meters or so into the subscriber's premises. One architechure for the subscriber distribution network, called fiber-to-the-curb (FTTC), is depicted in Fig. 54.4. Signals are transmitted over fibers through distribution hubs into the neighborhoods. The fibers terminate at optical network units (ONUs) located close to the subscriber. The ONU converts the optical signal into an electrical one for transmission over copper cables for the remaining short distance to the subscriber. Because of the power division at the hubs, optical amplifiers are needed to keep the signal levels high enough for proper signal reception.

Cable television distribution remained totally conducting for many years. This was due to the distortion produced by optical analog transmitters. Production of highly linear laser diodes [such as the distributed feedback (DFB) laser diode] in the late 1980s allowed the design of practical analog television fiber distribution links.

Conversion from analog to digital cable television transmission will be facilitated by the vast bandwidths that fiber make available and by signal compression techniques that reduce the required bandwidths for digital video signals.

Fiber Optic Communications Systems

Applications such as local area networks (LANs) require distribution of the signals over shared transmission fiber. Possible topologies include the passive star, the active star, and the ring network [Hoss, 1990]. These are illustrated in Figs. 54.5–54.7.

The major components found in optical communications systems (called out in Figs. 54.3–54.7) are: modulators, light sources, fibers, photodetectors, connectors, splices, directional couplers, star couplers, regenerators, and optical amplifiers. They are briefly described in the remainder of this section. More complete descriptions appear in the succeeding sections of this chapter.

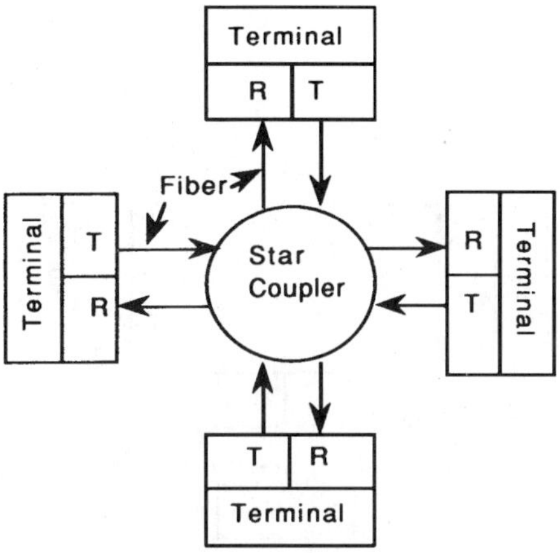

FIGURE 54.5 Passive star network: T represents an optical transmitter and R represents an optical receiver.

Fibers

Fiber links spanning more than a kilometer typically use silica glass fibers, as they have lower losses than either plastic or plastic cladded silica fibers. The loss properties of silica fibers were indicated in Fig. 54.2.

Material and waveguide **dispersion** cause pulse spreading, leading to intersymbol interference. This limits the fiber's bandwidth and, subsequently, its data-carrying capability. The amount of pulse spreading is given by

$$\Delta \tau = (M + M_g) L \Delta \lambda \qquad (54.4)$$

where M is the material dispersion factor and M_g is the waveguide dispersion factor, L is the fiber length, and $\Delta \lambda$ is the spectral width of the emitting light source. Because dispersion is wavelength

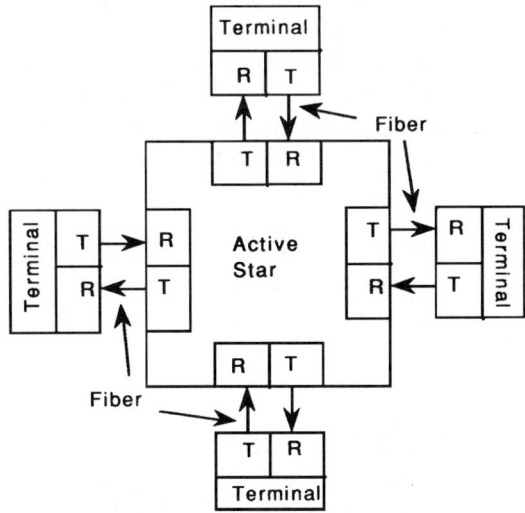

FIGURE 54.6 Active star network: T represents an optical transmitter and R represents an optical receiver. The active star consists of electronic circuits, which direct the messages to their intended destination terminals.

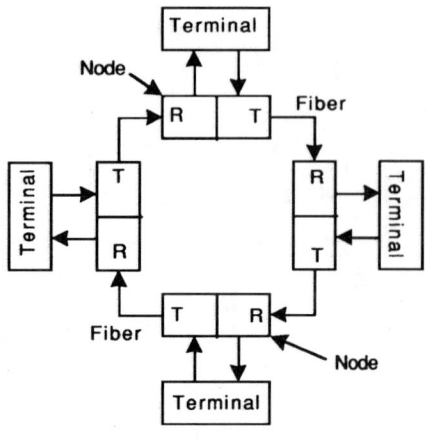

FIGURE 54.7 Ring network: T represents an optical transmitter and R represents an optical receiver. The nodes act as optical regenerators. Fibers connect the nodes together, while the terminals and nodes are connected electronically.

dependent, the spreading depends on the chosen wavelength and on the spectral width of the light source. The total dispersion $(M + M_g)$ has values near 120, 0, and 15 ps/(nm × km) at wavelengths 850, 1300, and 1550 nm, respectively.

The fiber's bandwidth and subsequent data rate is inversely proportional to the pulse spreading. As noted, dispersion is a minimum in the second window, actually passing through zero close to 1300 nm for conventional fibers. Special fibers are designed to compensate for the material dispersion by adjusting the waveguide dispersion to cancel it at the low-loss 1550-nm third window. Such fibers (called *dispersion shifted*) are ideal for long fiber links because of their low loss and low dispersion.

Multimode fibers allow more than one mode to simultaneously traverse the fiber. This produces distortion in the form of widened pulses because the energy in different modes travels at different velocities. Again, intersymbol interference occurs. For this reason multimode fibers are only used for applications where the bandwidth (or data rate) and path length are not large.

Single-mode fibers limit the propagation to a single mode, thus eliminating multimode spreading. Since they suffer only material and waveguide dispersive pulse spreading, these fibers (when operating close to the zero dispersion wavelength) have greater bandwidths than multimode fibers and are used for the longest and highest data rate systems.

Table 54.1 lists bandwidth limits for several types of fibers and Table 54.2 illustrates typical fiber sizes. Step index fibers (SI) have a core having one value of refractive index and a cladding of another value. Graded-index (GRIN) fibers have a core index whose refractive index decreases with distance from the axis and is constant in the cladding. As noted, single-mode fibers have the greatest bandwidths. To limit the number of modes to just one, the cores of single-mode fibers must be much smaller than those of multimode fibers.

Because of the relatively high loss and large dispersion in the 800-nm first window, applications there are restricted to moderately short path lengths (typically less than a kilometer). Because of the limited length, multimode fiber is practical in the first window. Light sources and photodetectors operating in this window tend to be cheaper than those operating at the longer wavelength second and third windows.

The 1300-nm second window, having moderately low losses and nearly zero dispersion, is utilized for moderate to long path lengths. Nonrepeatered paths up to 70 km or so are attainable in this window. In this window, both single-mode and multimode applications exist. Multimode is feasible for short lengths required by LANs (up to a few kilometer) and single-mode for longer point-to-point links.

TABLE 54.1 Typical Fiber Bandwidths

Fiber	Wavelength, nm	Source	Bandwidth, MHz × km
Multimode SI	850	LED	30
Multimode GRIN	850	LD	500
Multimode GRIN	1300	LD or LED	1000
Single mode	1300	LD	>10,000
Single mode	1550	LD	>10,000

Fiber systems operating in the 1550-nm third window cover the highest rates and longest unamplified, unrepeatered distances. Lengths on the order of 200 km are possible. Single-mode fibers are typically used in this window. Erbium-doped optical amplifiers operate in the third window, boosting the signal levels for very long systems (such as those traversing the oceans).

TABLE 54.2 Glass Fiber Sizes

Core, μm	Cladding, μm	Numerical Aperture	Operation
8	125	0.11	Single mode
50	125	0.2	Multimode
62.5	125	0.27	Multimode
85	125	0.26	Multimode
100	140	—	Multimode
200	280	—	Multimode

Other Components

Semiconductor **laser diodes (LD)** or **light-emitting diodes (LED)** serve as the light sources for most fiber systems. These sources are typically modulated by electronic driving circuits. The conversion from signal current i to optical power P is given by

$$P = a_0 + a_1 i \tag{54.5}$$

where a_0 and a_1 are constants. Thus, the optical power waveform is a replica of the modulation current.

For very high-rate modulation, external **integrated optic** devices are available to modulate the light beam after its generation by the source. Laser diodes are more coherent (they have smaller spectral widths) than LEDs and thus produce less dispersive pulse spreading, according to Eq. (54.4). In addition, laser diodes can be modulated at higher rates (tens of gigabit per second) than LEDs (which are limited to rates of just a few hundred megabit per second). LEDs have the advantage of lower cost and simpler driving electronics.

Photodetectors convert the optical beam back into an electrical current. Semiconductor **PIN photodiodes** and **avalanche photodiodes** (APD) are normally used. The conversion for the PIN diode is given by the linear equation

$$i = \rho P \tag{54.6}$$

where i is the detected current, P is the incident optical power, and ρ is the photodetector's **responsivity**. Typical values of the responsivity are on the order of 0.5 mA/mW.

Avalanche photodiodes follow the same equation but include an amplification factor that can be as high as several hundred. They improve the **sensitivity** of the receiver.

According to Eq. (54.6) the receiver current is a replica of the optical power waveform (which is itself a replica of the modulating current). Thus the receiver current is a replica of the original modulating signal current, as desired.

An optical regenerator (or repeater) consists of an optical receiver, electronic processor, and an optical transmitter. Regenerators detect (that is, convert to electrical signals) pulse streams that have weakened because of travel over long fiber paths, electronically determine the value of each binary pulse, and transmit a new optical pulse stream replicating the one originally transmitted. Using a series of regenerators spaced at distances of tens to hundreds of kilometers, total link lengths of thousands of kilometers are produced. Regenerators can only be used in digital systems.

Optical amplifiers simply boost the optical signal level without conversion to the electrical domain. This simplifies the system compared to the use of regenerators. In addition, optical amplifiers work with both analog and digital signals.

Splices and connectors are required in all fiber systems. Many types are available. Losses tend to be less than 0.1 dB for good splices and just a few tenths of a decibel for good connectors. Fibers are spliced either mechanically or by actually fusing the fibers together.

Directional couplers split an optical beam traveling along a single fiber into two parts, each traveling along a separate fiber. The splitting ratio is determined by the coupler design. In a star coupler (see Fig. 54.5), the beam entering the star is evenly divided among all of the output ports of the star. Typical stars operate as 8 × 8, 16 × 16 or 32 × 32 couplers. As an example, a 32 × 32 port star can accommodate 32 terminals on a LAN.

54.3 Signal Quality

Signal quality is measured by the **signal-to-noise ratio** (S/N) in analog systems and by the **bit error rate** (BER) in digital links.

The signal-to-noise ratio in a digital network determines the error rate and is given by

$$\frac{S}{N} = \frac{(M\rho P)^2 R_L}{M^n 2e R_L B(I_D + \rho P) + 4kTB} \tag{54.7}$$

where P is the received optical power, ρ is the detector's unamplified responsivity, M is the detector gain if an APD is used, n (usually between 2 and 3) accounts for the excess noise of the APD, B is the receiver's bandwidth, k is the Boltzmann constant ($k = 1.38 \times 10^{-23}$ J/K), e is the magnitude of the charge on an electron (1.6×10^{-19} C), T is the receiver's temperature in kelvin, I_D is the detector's dark current, and R_L is the resistance of the load resistor that follows the photodetector.

The first term in the denominator of Eq. (54.7) is caused by shot noise and the second term is attributed to thermal noise in the receiver. If the shot noise term dominates (and the APD excess loss and dark current are negligible), the system is *shot-noise limited*. If the second term dominates, the system is *thermal-noise limited*.

In a thermal-noise limited system, the probability of error P_e (which is the same as the bit error rate) is

$$P_e = 0.5 - 0.5\, \text{erf}\bigl(0.354\sqrt{S/N}\bigr) \tag{54.8}$$

where erf is the error function, tabulated in many references [Palais, 1992]. An error rate of 10^{-9} requires a signal-to-noise ratio of nearly 22 dB ($S/N = 158.5$).

54.4 System Design

System design involves assuring that the signal level at the receiver is sufficient to produce the desired signal quality. The difference between the power available from the transmitting light source (e.g., P_t in dBm) and the receiver's sensitivity (e.g., P_r in dBm) defines the system power budget L. Thus, the power budget is the allowed accumulated loss for all system components and is given (in decibels) by

$$L \text{ (dB)} = P_t \text{ (dBm)} - P_r \text{ (dBm)} \tag{54.9}$$

In addition to assuring sufficient available power, the system must meet the bandwidth requirements for the given information rate. This requires that the bandwidths of the transmitter, the fiber, and the receiver are sufficient for transmission of the message.

Defining Terms

Avalanche photodiode: Semiconductor photodetector that has internal gain caused by avalanche breakdown.

Bit error rate: Fractional rate at which errors occur in the detection of a digital pulse stream. It is equal to the probability of error.

Dispersion: Wavelength-dependent phase velocity commonly caused by the glass material and the structure of the fiber. It leads to pulse spreading because all available sources emit light covering a (small) range of wavelengths. That is, the emissions have a nonzero spectral width.

Integrated optics: Technology for constructing one or more optical devices on a common waveguiding substrate.

Laser: A source of coherent light. That is, a source of light having a small spectral width.

Laser diode: A semiconductor laser. Typical spectral widths are on the order of 1–5 nm.

Light-emitting diode: A semiconductor emitter whose radiation typically is not as coherent as that of a laser. Typical spectral widths are on the order of 20–100 nm.

Multimode fiber: A fiber that allows the propagation of many modes.

Optical frequency-division multiplexing: Multiplexing many closely spaced optical carriers onto a single fiber. Theoretically, hundreds (and even thousands) of channels can be simultaneously transmitted using this technology.

PIN photodiode: Semiconductor photodetector converting the optical radiation into an electrical current.

Receiver sensitivity: The optical power required at the receiver to obtain the desired performance (either the desired signal-to-noise ratio or bit error rate).

Responsivity: The current produced per unit of incident optical power by a photodetector.

Signal-to-noise ratio: Ratio of signal power to noise power.

Single-mode fiber: Fiber that restricts propagation to a single mode. This eliminates modal pulse spreading, increasing the fiber's bandwidth.

Wavelength-division multiplexing: Multiplexing several optical channels onto a single fiber. The channels tend to be widely spaced (e.g., a two-channel system operating at 1300 nm and 1550 nm).

References

Chaffee, C.D. 1988. *The Rewiring of America*, Academic, Orlando, FL.

Hoss, R.J. 1990. *Fiber Optic Communications*, Prentice–Hall, Englewood Cliffs, NJ.

Jeunhomme, L.B. 1990. *Single-Mode Fiber Optics*, 2nd ed., Marcel Dekker, New York.

Keiser, G. 1991. *Optical Fiber Communications*, McGraw–Hill, New York.

Palais, J.C. 1992. *Fiber Optic Communications*, 3rd ed., Prentice–Hall, Englewood Cliffs, NJ.

Thiennot, J., Pirio, F., and Thomine, J.-B. 1993. Optical undersea cable systems trends. *Proc. IEEE*, 81(11):1610–11.

Further Information

Information on optical communications is included in several professional society journals. These include the *IEEE Journal of Lightwave Technology* and the *IEEE Photonics Technology Letters*. Valuable information is also contained in several trade magazines such as *Lightwave* and *Laser Focus World*.

55
Optical Fibers and Lightwave Propagation

Paul Diament
Columbia University

55.1	Transmission Along Fibers	740
55.2	Total Internal Reflection	741
55.3	Modes of Propagation	742
55.4	Parameters of Fibers	743
55.5	Attenuation	744
55.6	Dispersion	744
55.7	Graded-Index Fibers	745
55.8	Mode Coupling	746
55.9	Summary	747

55.1 Transmission Along Fibers

Light that finds itself at one end of an **optical fiber** may or may not find its way to the other end. For a randomly chosen combination of the properties and configuration of the light, the optical power has only a small chance of propagating significant distances along the fiber. With proper design of the fiber and of the mechanism for launching the light, however, the signal carried by the light can be transmitted over many kilometers without severe leakage or loss [Diament, 1990; Palais, 1988; Jones, 1988; Cheo, 1985; Basch, 1987; Green, 1993; Hoss and Lacy, 1993; Buckman, 1992; Sterling, 1987].

In its simplest form, an optical fiber consists of a long cylindrical **core** of glass, surrounded by a **cladding** of slightly less **optically dense** glass. See Fig. 55.1. Since both the core and cladding are transparent, propagation depends on the ability of the combination to confine the light and thwart its escape across the side of the fiber. It is the interface between the core and cladding that acts as a reflector to confine any light that would otherwise escape the core and leak away from the cladding into the outside world. The mechanism for confinement to the core is that of **total internal reflection**, a process that is easy to understand for a planar interface between two dissimilar dielectric media [Diament, 1990; Born and Wolf, 1965]. See Fig. 55.2. That process applies as well to the cylindrical geometry of the interface between the core and cladding of the optical fiber.

Optical Fibers and Lightwave Propagation

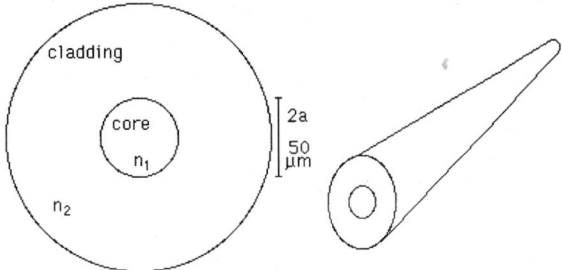

FIGURE 55.1 Structure of an optical fiber: core of radius a, refractive index n_1, surrounded by cladding of index n_2. Typical core diameter is 50 μm.

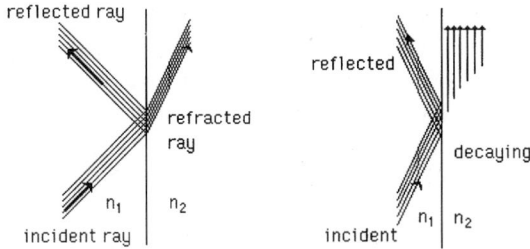

FIGURE 55.2 Ordinary refraction at a planar interface between two media (left), with angle of incidence less than the critical angle, compared with total internal reflection (right), with angle of incidence exceeding the critical angle. Light is incident from the denser medium ($n_1 > n_2$).

55.2 Total Internal Reflection

When light strikes the planar interface between two dissimilar transparent media, part of the light is transmitted and the rest is reflected. The proportions of each depend on the impedance mismatch between the two media; this in turn depends on the dissimilarity of the indices of refraction of the two materials. The **index of refraction** n measures how much more slowly light travels in the medium than it does in a vacuum.

When light strikes the interface at an angle to the normal, rather than head-on, the light that is transmitted is **refracted**, or bent away from its original direction, in accordance with **Snell's law**,

$$n_1 \sin \theta_1 = n_2 \sin \theta_2 \tag{55.1}$$

where n_1 and n_2 are the refractive indices of the two media and θ_1 and θ_2 are the angles of the ray to the normal to the interface in the two regions. See Fig. 55.3. If the light is incident onto the interface from the denser medium and if the ray's direction is close to grazing, so that the angle of incidence θ_1 is sufficiently large, then it may happen that $n_1 \sin \theta_1$ exceeds n_2 and there is then no solution for the angle of refraction θ_2 from Snell's law, since the quantity $\sin \theta_2$ can never exceed unity. Instead of partial reflection into the denser medium and partial refraction into the other region, there is then no transmission into the less dense medium. Rather, all of the light is reflected back into the denser medium. Total internal reflection occurs (internal to the denser medium in which the light originated) and the escape of the light into the less dense medium has been thwarted.

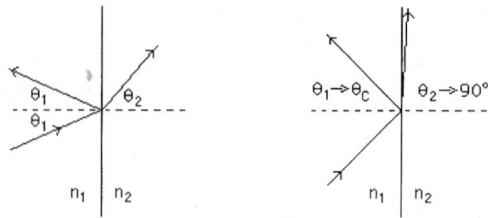

FIGURE 55.3 Snell's law determines the angle of refraction θ_2 at an interface; the angle of reflection θ_1 equals the angle of incidence. The case $n_1 > n_2$ is illustrated. When the angle of incidence θ_1 reaches the critical value $\theta_c = \sin^{-1}(n_2/n_1)$, refraction is at $\theta_2 = 90°$. Beyond the critical angle, there is total internal reflection.

There is more to the process of total internal reflection at a planar boundary than merely reflection from the interface. Although the light does not leak into the less dense medium and thereby get transmitted outward, it does seep across the interface and propagate, but only parallel to that interface surface [Diament, 1990]. In the direction normal to the interface, there is no propagation on the less dense side of the surface but the strength of the light that appears there decays exponentially. The spatial rate of decay increases with the angle of incidence, beyond the critical angle at which total internal reflection begins. If the decay is sufficiently strong, the light that seeps across is effectively constrained to a thin layer beyond the interface and propagates along it, while the light left behind is reflected back into the denser region with its original strength. The **reflection coefficient** is 100%. There is undiminished optical power flow along the reflected ray in the denser medium, but only storage of optical energy in the light that seeps across the interface into the less dense region.

Much the same process occurs along the curved, cylindrical interface between the core and cladding of the optical fiber. When the light strikes this interface at a sufficiently high grazing angle, total internal reflection occurs, keeping the optical power propagating at an angle within the core and along the axial direction in the cladding. The reflected light strikes the core again on the other side of the axis and is again totally internally reflected there. This confines the light to the core and to a thin layer beyond the interface in the cladding and keeps it propagating indefinitely along the axial direction. The overall result is that the light is guided by the fiber and not permitted to leak away across the side of the fiber, provided the light rays within the core encounter the interface at a sufficiently grazing angle. See Fig. 55.4.

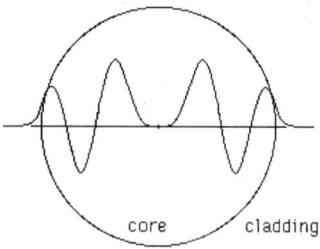

FIGURE 55.4 Standing wave across the core, formed by a repeatedly reflecting ray, with decaying field strength in the cladding, formed by the process of total internal reflection at the cylindrical boundary. The entire field pattern propagates along the axis of the fiber.

55.3 Modes of Propagation

For the cylindrical geometry of core and cladding in a fiber, the equivalent of a ray of light in an unbounded medium is any one of a set of **modes** of optical propagation. These refer to configurations of electromagnetic fields that satisfy **Maxwell's equations** and the boundary conditions imposed by the core-cladding interface and also conform to the total internal reflection requirement of grazing incidence to allow guided-wave propagation of the mode along the fiber [Diament, 1990; Born and Wolf, 1965].

Optical Fibers and Lightwave Propagation

These modes are nearly all **hybrid**, meaning that neither the electric nor the magnetic field of the mode is fully transverse to the axial direction of propagation, as can be the case for **waveguides** formed by hollow conducting pipes. Whereas the field configuration in a rectangular metallic waveguide varies as a sine function across the guide, that of an optical fiber follows a **Bessel function** variation. Bessel functions oscillate, much like trigonometric ones, but not with a constant amplitude. In the cladding, the fields decay in the radial direction away from the interface, but not exponentially as at a planar interface under total internal reflection. Rather, the fields in the cladding decay as **modified Bessel functions**, but the effect is the same: confinement to a thin layer just beyond the interface.

Just as the modes of hollow metal pipes are classified as **transverse electric** (TE) and **transverse magnetic** (TM), the hybrid modes of optical fibers also come in two distinguishable varieties [Schlesinger, Diament, and Vigants, 1961], labeled HE and EH, with pairs of subscripts appended to indicate the number of oscillations of the fields azimuthally (around the axis) and radially (across the core). For practical designs of optical fibers, dictated by the smallness of optical wavelengths, the indices of refraction of the core and cladding are very close to each other and there exist combinations of the hybrid modes that are very nearly of the more familiar transverse, **linearly polarized** types and of greatly simplified description; these are designated LP modes [Gloge, 1971]. The **dominant mode**, which has no constraint of a minimum frequency, is the HE_{11} mode, also designated the LP_{01} mode in the simplified version. All other modes have a **cutoff** frequency, a minimum frequency for axial propagation of a confined mode. Even the dominant mode, in practical cases, has fields that are virtually unconfined when the frequency is too low.

In frequency, then, optical fibers behave as **high-pass filters**. In angular distribution, fibers allow propagation of only those rays that are directed in a sufficiently narrow cone about the axis to satisfy the conditions of total internal reflection.

55.4 Parameters of Fibers

A few parameters are in common use to describe the ability of the fiber to transmit light. One is the **numerical aperture**, *NA*, defined for the fiber as

$$NA = \sqrt{(n_1^2 - n_2^2)} \tag{55.2}$$

and measuring the dissimilarity of the core and cladding indices, n_1 and n_2. This number corresponds more generally for any optical interface to

$$NA = n \sin \theta_{max} \tag{55.3}$$

a quantity that is conserved across any such interface, by Snell's law, and that measures how great the deviation of an incoming ray from the axial direction can get and still allow capture by, and propagation along, the optical system.

The numerical aperture enters into the formula for the key fiber parameter V, often referred to as the **normalized frequency** or sometimes as the **mode volume parameter**. This is defined for a **step-index fiber** of core radius a and for light of vacuum wavelength λ as

$$V = (2\pi a/\lambda) NA \tag{55.4}$$

The V parameter combines many physical attributes of the fiber and the light source; it is proportional to the frequency of the light and to the size of the core, as well as to the numerical aperture. It determines whether a particular mode of the fiber can propagate; in effect, it tells whether the light in that mode strikes the core–cladding interface at a sufficiently grazing angle to be captured by the fiber. Consequently, it also furnishes an estimate of how many modes can propagate along the fiber. For large values of V, the approximate number of distinguishable modes that are above cutoff and can therefore propagate is given by $V^2/2$.

For many purposes, including efficiency of launching and detection of the light and maintenance of the integrity of the signal carried by the light, it is undesirable to have a multiplicity of modes able to propagate along the fiber. For single-mode operation, the parameter V must be small, less than 2.4048 [which is the argument at which the Bessel function $J_0(x)$ first goes through zero]. This is not easy to achieve, because wavelengths of light are so tiny: for a core diameter of $2a = 50$ μm and an operating wavelength of $\lambda = 0.8$ μm, we have $V \approx 200(NA)$ and the core and cladding indices have to be extremely close, yet dissimilar, to attain **single-mode operation**. A much thinner core is needed for a single-mode fiber than for **multimode operation**. For the more readily achieved refractive index differences and core diameters, NA may be about 0.15 and several hundred modes can propagate along the fiber. A consequence of this multiplicity of modes is that each travels along the fiber at its own speed and the resulting confusion of arrival times for the various fractions of the total light energy carried by the many modes contributes to signal distortion ascribed to **dispersion**.

55.5 Attenuation

Dispersion limits the operation of the fiber at high data rates; losses limit it at low data rates. Light attenuation due to losses is ascribable to both **absorption** and **scattering**. Absorption occurs when impurities in the glass are encountered; the hydroxyl ion of water present in the glass as an impurity contributes heavily to absorption, along with certain metal ions and oxides. Careful control of impurity levels during fabrication of fibers has allowed light attenuation due to absorption to be kept below 1 dB/km.

Scattering is caused by imperfections in the structure of the fiber. A lower limit on losses is due to **Rayleigh scattering**, which arises from thermal density fluctuations in the glass, frozen into it during manufacture. The typical attenuation constant α from Rayleigh scattering varies as the inverse fourth power of the optical wavelength λ,

$$\alpha = \alpha_0 (\lambda_0/\lambda)^4; \quad \alpha_0 = 1.7 \text{ dB/km}, \lambda_0 = 0.85\,\mu\text{m}. \tag{55.5}$$

This makes operation at longer wavelengths less subject to losses, but molecular absorption bands at wavelengths beyond about 1.8 μm must be avoided. Rayleigh scattering is excessive below about 0.8 μm and, between these limits, the hydroxyl ion absorption peak near 1.39 μm must also be avoided. This has left three useful transmission *windows* for fiber optic systems, in the ranges 0.8–0.9 μm, 1.2–1.3 μm, and 1.55–1.6 μm.

55.6 Dispersion

Dispersion is a major concern in a communication system. It refers to processes that cause different components of the light signal to be affected differently. In particular, the speeds at which the different components are propagated along the fiber can vary enough to result in significantly different transmission delays or arrival times for the various portions of the signal. The outcome is distortion of the signal when it is reconstructed from all its components at the receiver or **repeater**. The signal components referred to may be the various frequencies in its spectrum, or they can comprise the multiplicity of modes that each carry their respective fraction of the total optical power to be transmitted.

For optical fibers, there are two aspects to dispersion effects. One is **material dispersion**, also called **intramodal** or **chromatic dispersion**; the other is termed **multimode dispersion**, or also **intermodal dispersion**. Material dispersion arises from the properties of the fiber material that cause the index of refraction to depend on frequency. Since the light source is never truly **monochromatic** but, rather, has a **linewidth**, the various frequencies in the light signal are subject to slightly different

Optical Fibers and Lightwave Propagation

refractive index values, which entails different speeds of transmission along the fiber even if there is only one mode of propagation. It is the spread in transmission delays that distorts the signal, and this spread depends on the second derivative of the refractive index with respect to wavelength. One way to minimize this source of dispersion is to operate at about 1.3 μm because silica glass has a second derivative that goes through zero at that wavelength.

Multimode dispersion arises from the fact that the optical energy is distributed among the many modes of propagation of the fiber and each mode has its own **group velocity**, the rate at which a signal carried by that mode is transmitted. The relevant measure of dispersion is then the difference between the propagation delays of the slowest and the fastest of the participating modes. For a step-index fiber, multimode dispersion is much more severe than material dispersion, although it is somewhat mitigated by the fact that different modes do not propagate independently in a real, imperfect fiber. The phenomena of **mode mixing** and of greater attenuation of high-order modes (those with many field oscillations within the core) tend to reduce the differences in arrival times of the various modes that carry the signal. Nevertheless, intermodal dispersion imposes a severe limitation on data transmission along a step-index fiber.

55.7 Graded-Index Fibers

The way to reduce the deleterious effects of multimode dispersion on the propagation of a signal along a fiber becomes clear when the mechanism of this sort of dispersion is understood. That different modes should exhibit significantly different propagation delays follows from consideration of where the bulk of the field energy resides in each mode. As mentioned, high-order modes have field profiles that oscillate more than those of low-order modes. This includes oscillations in both the radial and the azimuthal directions. Examination of the behavior of Bessel functions indicates that high-order modes have most of their energy nearer to the core–cladding boundary than do the low-order ones. The azimuthal variation of the fields corresponds to propagation at an angle to the fiber axis, so that there is progression azimuthally, around the axis, even as the mode propagates along the axis.

For the high-order modes, the azimuthal component of the propagation is more significant than it is for low-order modes, with the result that the overall progression of the high-order mode is in a spiral around the axis, whereas the low-order modes tend to progress more nearly axially than helically. The spiral trajectories of the energy also tend to concentrate far from the axis, nearer to the cladding, as compared to the straighter ones of the low-order modes, whose energy is massed near the axis and travels more nearly straight along it, as suggested in Fig. 55.5. The difference in propagation delay is then readily understood as arising from the longer, spiral path of the modal energy for high-order modes, as compared with the shorter, more direct path taken by the energy in the low-order modes.

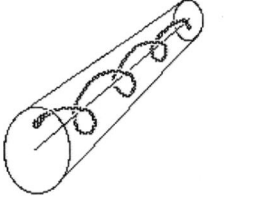 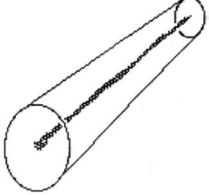

FIGURE 55.5 Spiral path of light energy in higher order modes, compared to more direct trajectory for lower order ones. High-order modes have their energy concentrated near the core–cladding interface; lower order ones are concentrated closer to the axis. The spiral path takes longer to traverse than the direct one.

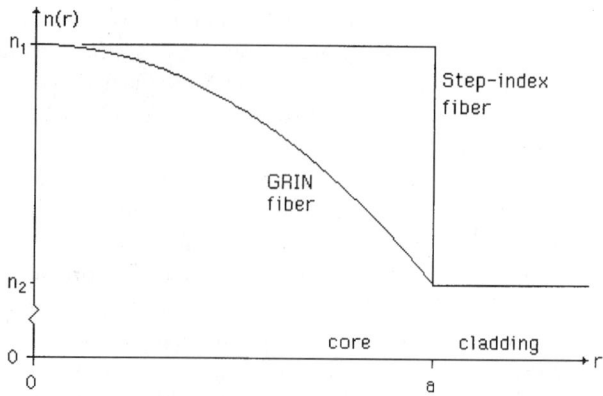

FIGURE 55.6 Radial index profiles of step-index and graded-index fibers. The broken vertical axis is a reminder that the maximum core index n_1 is quite close to the cladding index n_2 for practical fibers.

This picture of the underlying mechanism for multimode dispersion suggests an equalization scheme aimed at slowing down the low-order modes that are concentrated near the axis, while speeding up the high-order ones that spiral around as they progress and are to be found primarily near the core–cladding interface, farther from the axis. This is achieved by varying the index of refraction of the core gradually, from a high value (for slower speed) near the axis to a lower one (for higher speed) near the cladding. Fibers so fabricated are termed **graded-index** (or **GRIN**) fibers. It is found that a refractive index profile that varies approximately (but not precisely) quadratically with radius can be effective in reducing multimode dispersion dramatically, by equalizing the speeds of propagation of various modes of that fiber. Figure 55.6 illustrates the radial variation of the refractive index $n(r)$ for both a step-index fiber and a GRIN fiber.

55.8 Mode Coupling

One more consideration affecting propagation along an optical fiber is the possibility of **mode coupling**. Although the theoretical modes of propagation travel independently along fibers of perfect constitution and perfect geometry, real fibers inevitably have imperfections that can cause different modes to couple and exchange energy between them as they propagate.

A fiber may be imperfect in geometry, in that it may not be perfectly straight axially, or its core may not be perfectly round, or it may have been bent. The core and cladding may deviate from their nominal refractive indices. More importantly, there may be fluctuations in their material properties, or the graded-index profile may not be the ideal one. All such imperfections give rise to coupling between modes of propagation of the ideal version of the fiber, such that some of the power in one mode is drained and feeds another mode. When the other mode becomes the stronger one, energy is fed back to the first mode and thereafter sloshes back and forth repeatedly between the modes as the light travels along the fiber. How many modes participate in this periodic exchange of energy depends on the nature of the fluctuations of the imperfections.

A **perturbation analysis** of the effects of fluctuations in some parameter that would be constant in an ideal fiber reveals that the relevant property is the **spatial Fourier spectrum** of the fluctuations. The strength of the coupling between any two modes is proportional to the statistical average of the **power spectrum** of the fluctuations, evaluated at the difference between the **propagation constants** of the two modes in the ideal structure.

This implies that deviations on a large scale, such as gradual bending of the fiber, will affect only a certain group of modes. In this case, the spatial power spectrum exhibits a relatively narrow bandwidth, near zero **spatial frequency**. This allows only modes whose ideal propagation constants

Optical Fibers and Lightwave Propagation

are very close to each other to be coupled by the large-scale fluctuations. Such deviations therefore couple only modes that are nearly **degenerate** (have nearly equal propagation constants).

On the other hand, periodic fluctuations, such as striations in the material properties, exhibit a spatial spectrum that is peaked in a spectral region corresponding to the period of the striations. In that case, coupling will be significant for modes whose propagation constants differ by the **spatial wavenumber** associated with the periodicity of the fluctuations.

One effect of coupling among modes is to increase losses along the fiber, because energy can get transferred to modes that are not trapped within the core. Some of that energy can then leak out into the cladding and be lost before it can return to the propagating mode. Another effect, however, is to average the dispersion associated with many modes and thereby reduce the overall **pulse spreading**. For example, for closely coupled modes, propagation delays may vary as only the square root of the distance traveled, rather than with the more severe variation, proportional to distance, for uncoupled modes.

55.9 Summary

The step-index optical fiber comprises a dielectric cylindrical core surrounded by a cladding of slightly lower refractive index. This constitutes a dielectric cylinder, which is capable of acting as a waveguide. The mechanism involved in confining and guiding the optical power is that of total internal reflection at the cylindrical boundary between the core and the cladding. This phenomenon results in a **standing wave** pattern within the core and a decaying field in the cladding, with the entire field pattern propagating along the cylinder. The general modes of propagation are hybrid modes, having axial components of both the electric and the magnetic fields, and they advance along the fiber by spiraling around the axis. The exceptional transverse electric or transverse magnetic types are the azimuthally symmetric ones, which do not twist about the axis as they progress along the fiber. The dominant mode is hybrid, however.

Because the optical wavelengths are so short, the core is normally many wavelengths across, with the result that a large number of modes can propagate along the fiber, even when the refractive indices of the core and cladding are kept quite close to each other. Unless single-mode fiber operation is achieved, this entails multimode dispersion effects. Efforts to overcome the limitations this imposes on data transmission have led to the development of graded-index fibers, with an advantageous refractive index profile of the core that tends to equalize the propagation delays of different modes. With careful fabrication of the fibers and with selections of appropriate source wavelengths and bandwidths, attenuation along the fiber can be kept quite low and dispersive effects can be controlled, making optical fibers the medium of choice for propagation of optical signals.

Defining Terms

Absorption: Process by which light energy is drained from a wave and transferred to the material in which it propagates.

Bessel function: A type of mathematical function that oscillates and also decays in amplitude; descriptive of the radial dependence of the fields in a uniform-index core.

Chromatic dispersion: The variation of a material's index of refraction with frequency or wavelength.

Cladding: The outer, optically less dense portion of an optical fiber, surrounding the core.

Core: The inner, optically dense portion of an optical fiber, surrounded by the cladding.

Cutoff: The condition (usually too low an operating frequency) under which a mode is not capable of propagating; the cutoff frequency is the minimum frequency for propagation of a mode.

Degenerate modes: Modes whose propagation constants are equal, or nearly equal.

Dispersion: The property of a fiber (or any wave-propagating system) that causes different frequencies (or operating wavelengths) to propagate at different speeds; it is a cause of distortion of signals transmitted along the structure.

Dominant mode: The mode that can propagate under conditions (usually a specific frequency range) that do not allow any other modes to propagate.

Graded-index fiber: The type of optical fiber whose core has a nonuniform index of refraction.

GRIN fiber: Graded-index fiber.

Group velocity: The velocity of propagation of a signal whose spectrum occupies a narrow range of frequencies about some carrier frequency; the rate of change of frequency with respect to propagation constant.

High-pass filter: A circuit or structure that passes high-frequency signals but blocks low-frequency ones.

Hybrid: Describing a mode for which neither the electric nor the magnetic field vector is fully transverse to the direction of propagation.

Index of refraction: Ratio of the vacuum speed of light to its speed in the medium.

Intermodal dispersion: Dispersion effect that arises from a signal's being carried by many modes that travel at different speeds.

Intramodal dispersion: The variation of a material's index of refraction with frequency or wavelength, resulting in dispersive effects for a single mode.

Linearly polarized: Descriptive of a mode of propagation whose field vectors each maintain a fixed orientation in space as they oscillate in time.

Linewidth: The range of frequency or wavelength contained within a nominally monochromatic signal.

Material dispersion: The variation of a material's index of refraction with frequency or wavelength.

Maxwell's equations: The mathematical laws governing the propagation of light and other electromagnetic waves.

Mode: Field configuration of light with well-defined propagation characteristics.

Mode coupling: Process by which different modes of propagation can interchange their energy, instead of propagating independently of each other.

Mode mixing: Process of transfer or interchange of light energy among different modes of propagation.

Mode volume parameter: The V parameter, which determines how many modes can propagate along an optical fiber; it is proportional to the frequency of the light, to the size of the core, and to the numerical aperture.

Modified Bessel function: A type of mathematical function that either grows or decays in amplitude; descriptive of the radial dependence of the fields in a uniform-index cladding.

Monochromatic: Of a single frequency or wavelength.

Multimode dispersion: Dispersion effect that arises from a signal's being carried by many modes that travel at different speeds.

Multimode operation: Operation of an optical fiber under conditions that allow many modes to propagate along it.

Normalized frequency: The V parameter, which determines whether a mode can propagate along an optical fiber; it is proportional to the frequency of the light, to the size of the core, and to the numerical aperture.

Numerical aperture (NA): A measure of the ability of an optical structure or instrument to capture incident light; $NA = n \sin\theta$, where n is the refractive index and θ is the maximum angle of deviation from the optical axis of a ray that can still be captured by the aperture.

Optical fiber: A long, thin glass structure that can guide light.

Optically dense: Transmitting light at a relatively slow speed; having a high refractive index.

Perturbation analysis: Mathematical process of determining approximately the effects of a relatively weak disturbance of some system that is otherwise in equilibrium.

Power spectrum: Description of some quantity that varies in space or time in terms of the decomposition of its power into constituent sinusoidal components of different spatial or temporal frequencies or periodicities.

Propagation constant: The spatial rate of phase progression of a wave.

Pulse spreading: Process whereby the width of a pulse increases as it propagates.

Rayleigh scattering: A scattering process associated with thermal density fluctuations in the constitution of the glass.

Reflection coefficient: Ratio of strength of reflected light to that of the incident light.

Refraction: Process of bending a ray of light away from its original direction.

Repeater: Electronic circuitry that takes in a degraded signal and restores or reconstitutes it for further transmission.

Scattering: Process by which portions of light energy are redirected (by encounters with microscopic obstacles or inhomogeneities) to undesired directions and thereby lost from the propagating wave.

Single-mode operation: Operation of an optical fiber under conditions that allow only one mode to propagate along it.

Snell's law: Law governing reflection (reflection angle equals incidence angle) and refraction (component of propagation vector tangential to the interface is preserved) of a ray of light encountering the interface between media of different optical density.

Spatial Fourier spectrum: Description of some quantity that varies in space in terms of its decomposition into constituent sinusoidal components of different spatial frequencies or periodicities.

Spatial frequency: The number of cycles per unit length of a quantity that varies periodically in space.

Spatial wavenumber: Reciprocal of a quantity's interval of periodicity in space.

Standing wave: Resultant of counterpropagating waves, with a spatial distribution that does not move.

Step-index fiber: The type of optical fiber whose core and cladding each have a uniform index of refraction.

Total internal reflection: Process of reflecting light incident from a denser medium toward a less dense one, at an angle sufficiently grazing to avert transmission into the less dense medium. All of the incident light power is reflected, although optical fields do appear beyond the interface.

Transverse electric: Descriptive of a mode of propagation whose electric field vector is entirely transverse to the direction of propagation.

Transverse magnetic: Descriptive of a mode of propagation whose magnetic field vector is entirely transverse to the direction of propagation.

Waveguide: A structure that can guide the propagation of light or electromagnetic waves along itself.

References

Basch, E.E.B. ed. 1987. *Optical-Fiber Transmission*, Howard W. Sams, Indianapolis, IN.
Born, M. and Wolf, E. 1965. *Principles of Optics*, Pergamon Press, Oxford, England.
Buckman, A.B. 1992. *Guided-Wave Photonics*, Saunders College Publ., Fort Worth, TX.
Cheo, P.K. 1985. *Fiber Optics Devices and Systems*, Prentice–Hall, Englewood Cliffs, NJ.
Diament, P. 1990. *Wave Transmission and Fiber Optics*, Macmillan, New York.
Gloge, D. 1971. Weakly guiding fibers. *Applied Optics*, 10:2252–2258.
Green, P.E., Jr. 1993. *Fiber Optic Networks*, Prentice–Hall, Englewood Cliffs, NJ.

Hoss, R.J. and Lacy, E.A. 1993. *Fiber Optics*, Prentice–Hall, Englewood Cliffs, NJ.

Jones, W.B., Jr. 1988. *Introduction to Optical Fiber Communication Systems*, Holt, Rinehart and Winston, New York.

Palais, J.C. 1988. *Fiber Optic Communications*, Prentice–Hall, Englewood Cliffs, NJ.

Schlesinger, S.P., Diament, P., and Vigants, A. 1961. On higher-order hybrid modes of dielectric cylinders. *IRE Trans. Micro. Theory & Tech.*, MTT-8(March):252–253.

Sterling, D.J., Jr. 1987. *Technician's Guide to Fiber Optics*, Delmar, Albany, NY.

Further Information

Periodicals that can help one keep up to date with developments in fiber optics and related systems include the following.

Journal of Lightwave Technology, The Institute of Electrical and Electronics Engineers, Inc., New York.

IEEE Journal of Quantum Electronics, The Institute of Electrical and Electronics Engineers, Inc., New York.

Journal of the Optical Society of America, Optical Society of America, New York.

Optics Letters, Optical Society of America, New York.

Applied Optics, Optical Society of America, New York.

Bell System Technical Journal, AT&T, New York.

Optical Engineering, Society of Photo-Optical Instrumentation Engineers, Bellingham, WA.

Optics Communications, North-Holland, Amsterdam.

Photonics Spectra, Optical Publ., Pittsfield, MA.

Lightwave, the Journal of Fiber Optics, Howard Rausch Associates, Waltham, MA.

56
Optical Sources

	56.1	Introduction 751
	56.2	Laser Designs 752
	56.3	Quantum Well Lasers 756
		Strained Quantum-Well Lasers
	56.4	Distributed Feedback Lasers 760
		Tunable Lasers
N.K. Dutta	56.5	Surface Emitting Lasers 764
AT&T Bell Laboratories	56.6	Laser Reliability 768

56.1 Introduction

Phenomenal advances in research results, development, and application of optical sources have occurred over the last decade. The two primary optical sources used in telecommunications are the semiconductor laser and the light emitting diode (**LED**). The LEDs are used as sources for low data rate (< 200 Mb/s) and short-distance applications, and lasers are used for high data rate and long-distance applications. The fiber optic revolution in telecommunication which provided several orders of magnitude improvement in transmission capacity at low cost would not have been possible without the development of reliable semiconductor lasers. Today, semiconductor lasers are used not only for fiber optic transmission but also in optical reading and recording (e.g., CD players), printer, FAX machines, and in numerous applications as a high-power laser source. Semiconductor injection lasers and LEDs continue to be the source of choice for various system applications primarily because of their small size, simplicity of operation, and reliable performance.

This chapter describes the fabrication, performance characteristics, current state of the art, and research directions for semiconductor lasers. The focus of this chapter is lasers needed for fiber optic transmission systems. For early work and thorough discussion of semiconductor lasers and LEDs see Agrawal and Dutta [1986], Casey and Panish [1988], Kressel and Butler [1977], and Thompson [1980].

The semiconductor injection laser was invented in 1962 [Holonyuk and Bevacqua, 1962; Nathan et al., 1962; Quist et al., 1962]. With the development of **epitaxial** growth techniques and the subsequent fabrication of double heterojunction, laser technology advanced rapidly in the 1970s and 1980s [Agrawal and Dutta, 1986; Casey and Panish, 1988; Kressel and Butler, 1977; Thompson, 1980]. The demonstration of continuous wave (CW) operation of the semiconductor laser in the early 1970s [Hayashi et al., 1970] was followed by an increase in development activity in several industrial laboratories. This intense development activity in the 1970s was aimed at improving the performance characteristics and reliability of lasers fabricated using the AlGaAs material system [Kressel and Butler, 1977]. These lasers emit near 0.8 μm and were deployed in the early fiber optic transmission systems in the later 1970s and early 1980s.

The optical fiber has zero **dispersion** near the 1.3-μm wavelength and has lower loss near the 1.55-μm wavelength. Thus, semiconductor lasers emitting near 1.3 and 1.55 μm are of interest for fiber optic transmission application. Lasers emitting at these wavelengths are fabricated using the InGaAsP/InP materials system and were first fabricated in 1976 [Hsieh, 1976]. Many of the fiber optic transmission systems around the world that are in use or are currently being deployed utilize lasers emitting near 1.3 or 1.55 μm.

Initially these lasers were fabricated using a liquid phase epitaxy (LPE) growth technique. With the development of metal-organic chemical vapor deposition (MOCVD) and of the gas source molecular beam epitaxy (GSMBE) growth techniques in the 1980s, not only the reproducibility of the fabrication process has improved but these developments have also led to advances in laser designs such as **quantum well** lasers and very high-speed lasers using semi-insulating Fe doped InP current blocking layers [Agrawal and Dutta, 1986, Chap. 4].

56.2 Laser Designs

A schematic of a typical double heterostructure used for laser fabrication is shown in Fig. 56.1. It consists of n-InP, undoped $In_xGa_{1-x}As_{1-y}P_y$, p-InP, and p-InPGaAsP grown over (100) oriented n-InP substrate. The undoped $In_xGa_{1-x}As_{1-y}P_y$ layer is the light emitting layer (active layer). It is lattice matched to InP for $x \simeq 0.45y$. The **band gap** of the $In_xGa_{1-x}As_{1-y}P_y$ material (lattice matched to InP), which determines the laser wavelength, is given by [Nahory et al., 1978]

$$E_g \text{ (eV)} = 1.35 - 0.72y + 0.12y^2 \tag{56.1}$$

For lasers emitting near 1.3 μm, $y \simeq 0.6$. The double heterostructure material can be grown by LPE, GSMBE, or MOCVD growth technique.

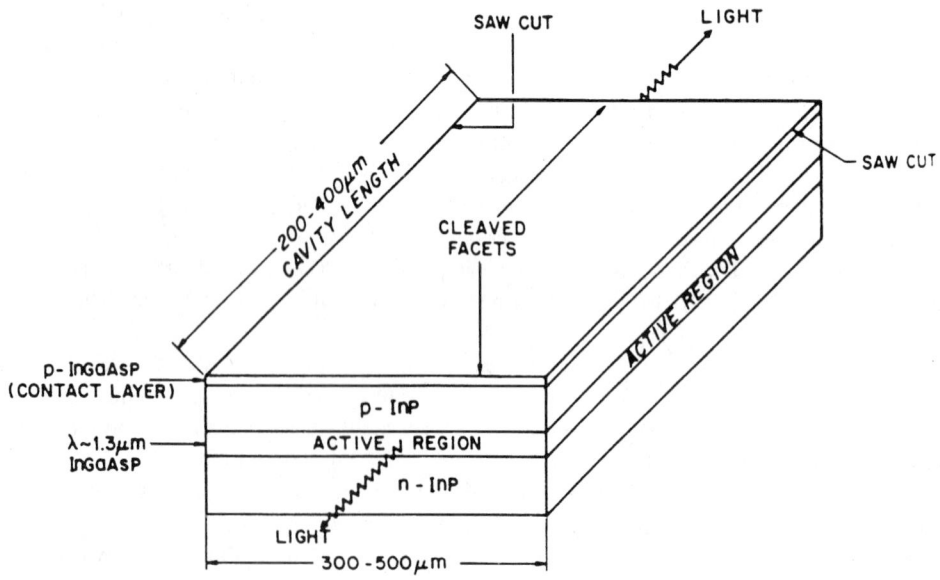

FIGURE 56.1 Schematic of a double heterostructure laser material.

Optical Sources

The double heterostructure material shown in Fig. 56.1 can be processed to produce lasers in several ways. Perhaps the simplest is the broad area laser which involves putting contacts on the p and n side and then cleaving. Such lasers do not have transverse mode confinement or current confinement, which leads to high threshold and nonlinearities in light vs current characteristics. Several laser designs has been developed to address these problems. Among them are the gain guided laser, the weak **index guided** laser, and the buried heterostructure (strong index guided) laser. A typical version of each of these laser structures is shown in Fig. 56.2. The gain guided structure uses a dielectric layer for current confinement. The current is injected in the opening in the dielectric (typically, 6–12 μm wide), which produces gain in that region and, hence, the lasing mode is confined to that region. The weak index guided structure has a ridge etched on the wafer; a dielectric layer surrounds the ridge. The current is injected in the region of the ridge, and the optical mode overlaps the dielectric (which has a low index) in the ridge. This results in weak index guiding.

The buried heterostructure design shown in Fig. 56.2 has the active region surrounded (buried) by lower index layers. The fabrication process of this laser involves growing a double heterostructure, etching a mesa using a dielectric mask, and then regrowing the layer surrounding the active region using a second epitaxial growth step. The second growth can be a single Fe doped InP semi-insulating layer or a combination of p-InP, n-InP, and Fe:InP layers. Generally, the MOCVD growth process is used for the growth of the regrown layer. Researchers have often given different names to be particular buried heterostructure laser design that they discovered. These are described in detail in Agrawal and Dutta [1986, Chap. 5]. For the structure of Fig. 56.2, the Fe doped InP layer provides both optical confinement to the lasing mode and current confinement to the active region. Buried heterostructure (BH) lasers are generally used in communication system applications because a properly designed strong index guided buried heterostructure design has superior mode stability, higher **bandwidth**, and superior linearity in light–current characteristics compared to the gain guided and weak index guided designs. Early recognition of these important requirements of communication grade lasers led to intensive research on InGaAs BH laser designs all over the world

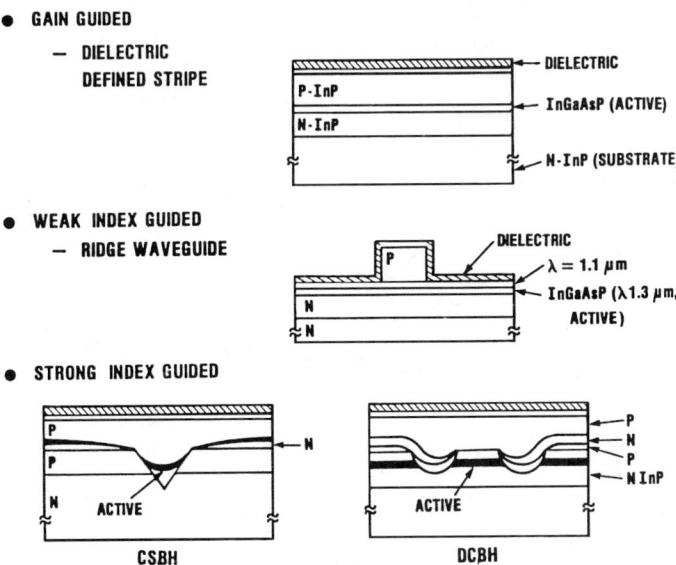

FIGURE 56.2 Schematic of a gain guided, weak index guided, and strong index guided buried heterostructure laser.

in the 1980s. It is worth mentioning that BH lasers are more complex and difficult to fabricate compared to the gain guided and weak index guided lasers.

The light vs current (L vs I) characteristics at different temperatures of an InGaAsP BH laser emitting at 1.3 μm are shown in Fig. 56.3. Typical **threshold current** of a BH laser at room temperature is in the 5–10 mA range. For gain guided and weak index guided lasers, typical room temperature threshold currents are in the 25–50 and 50–100 mA range, respectively. The external differential quantum efficiency defined as the derivative of the L vs I characteristics above threshold is $\sim$0.2–0.25 mW/mA/facet for a cleaved uncoated laser emitting near 1.3 μm.

An important characteristic of the semiconductor laser is that its output can be modulated easily simply by modulating the injection current. The relative magnitude of the modulated light output is plotted as a function of the modulation frequency of the current in Fig. 56.4 at different optical output powers. The laser is of the BH type (shown in Fig. 56.2) and the modulation current amplitude is 5 mA. Note that the 3-dB frequency to which the laser can

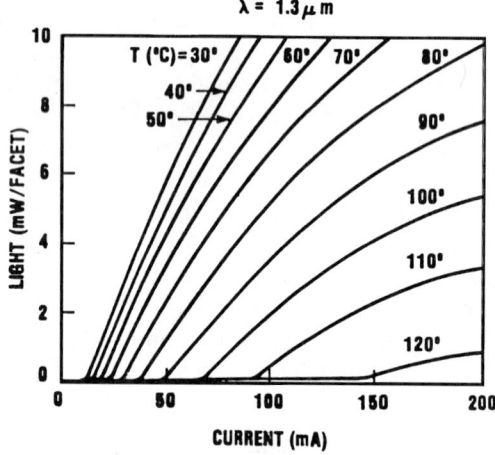

FIGURE 56.3 Light vs current characteristics of an InGaAsP buried heterostructure laser emitting at 1.3 μm.

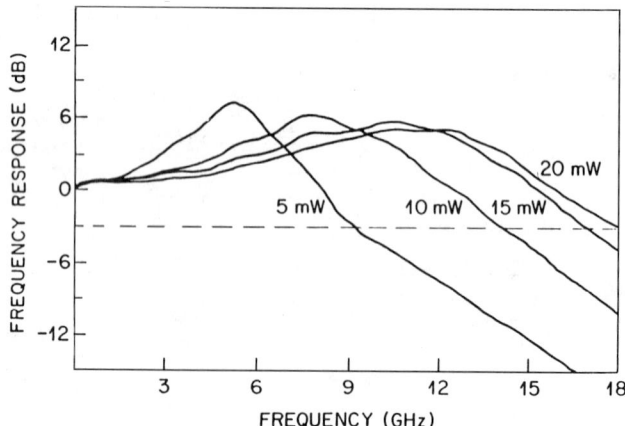

FIGURE 56.4 Modulation response of a laser at different optical output powers.

Optical Sources

be modulated increases with increasing output power, and the modulation response is maximum at a certain frequency (ω_r). The resonance frequency ω_r is proportional to the square root of the optical power. The modulation response determines the data transmission rate capability of the laser, for example, for 10-Gb/s data transmission, the 3-dB bandwidth of the laser must exceed 10 GHz. However, other system level considerations, such as allowable error rate penalty, often introduce much more stringent requirements on the exact modulation response of the laser.

A semiconductor laser with cleaved facets generally emits in a few longitudinal modes of the cavity. Typical spectrum of a laser with cleaved facets is shown in Fig. 56.5. The discrete emission wavelengths are separated by the longitudinal cavity mode spacing, which is ~ 10 Å for a laser ($\lambda \sim 1.3$ μm) with 250-μm cavity length. Lasers can be made to emit in a single frequency using frequency selective feedback, for example, using a grating internal to the laser cavity as described in Sec. 56.4.

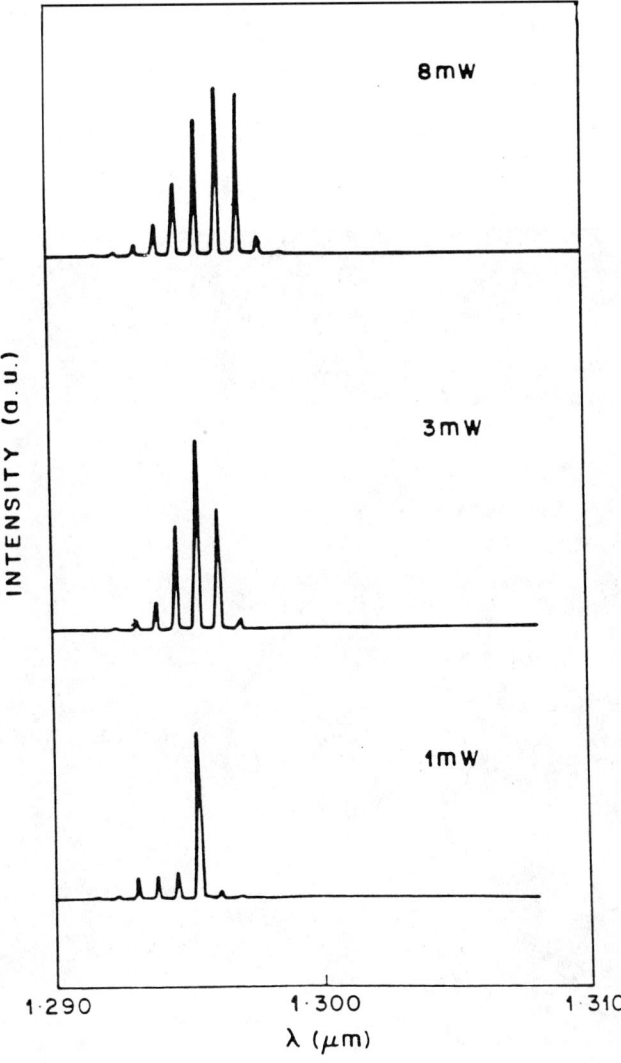

FIGURE 56.5 Emission spectrum of a laser with cleaved facets.

56.3 Quantum Well Lasers

Thus far we have described the fabrication and performance characteristics of a regular double heterostructure (DH) laser, which has an active region ~ 0.1 to 0.2 μm thick. Beginning with the 1980s, lasers with very thin active regions, quantum well lasers, were being developed in many research laboratories [Capasso and Magaritondo, 1987; Dutta et al., 1985a, 1985b, 1985c, 1985d; Hersee, DeCremoux, and Duchemin, 1987; Holonyak et al., 1980a, 1980b; Tsang, 1981, 1986]. Quantum well (QW) lasers have active regions of ~ 100 Å thick, which restricts the motion of the carriers (electrons and holes) in a direction normal to the well. This results in a set of discrete energy levels and the density of states is modified to a two-dimensional-like density of states. This modification of the density states results in several improvements in laser characteristics such as lower threshold current, higher efficiency, and higher modulation bandwidth and lower CW and dynamic spectral width. All of these improvements were first predicted theoretically and the demonstrated experimentally [Arakawa and Yariv, 1985, 1986; Chiu and Yariv, 1982; Dutta, 1983; Dutta and Nelson, 1982; Smith, Abram, and Burt, 1983; Sugimura 1983a, 1983b, 1984; Yariv, Lindsey, and Sivan, 1985].

The development of InGaAsP QW lasers was made possible by the development of MOCVD and GSMBE growth techniques. The transmission electron micrograph (TEM) of a multiple QW laser structure is shown in Fig. 56.6. Shown are five InGaAs quantum wells grown over n-InP substrate. The well thickness is 70 Å, and the wells are separated by barrier layers of InGaAsP $\lambda \sim 1.1$ μm. Multiquantum well (MQW) lasers with threshold current densities of 600 A/cm^2 have been fabricated [Tsang et al., 1990].

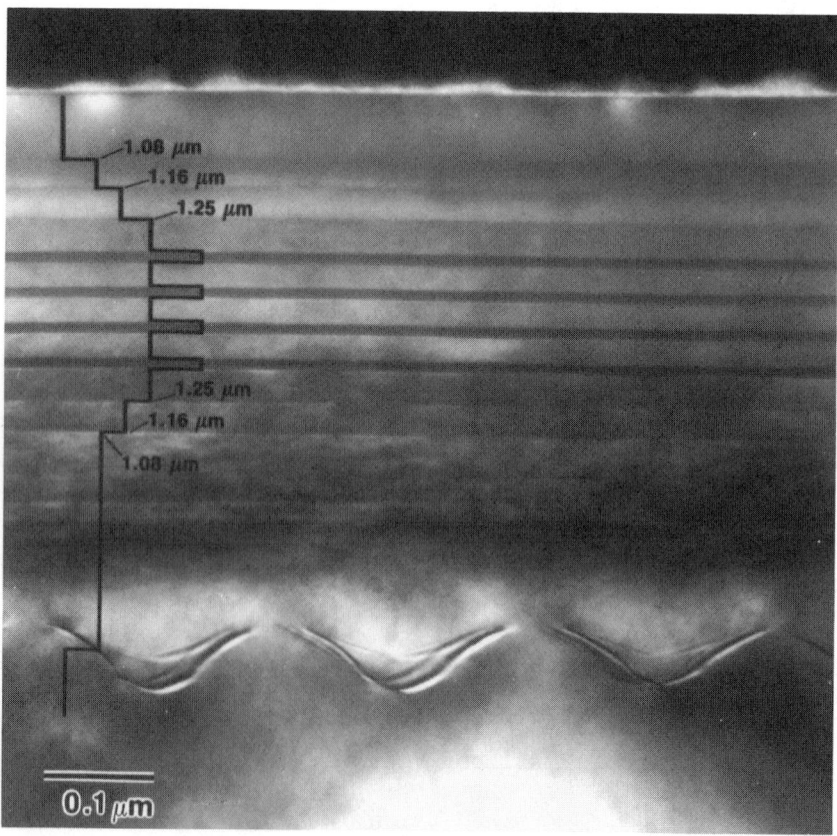

FIGURE 56.6 The transmission electron micrograph of a multiquantum well laser structure.

Optical Sources

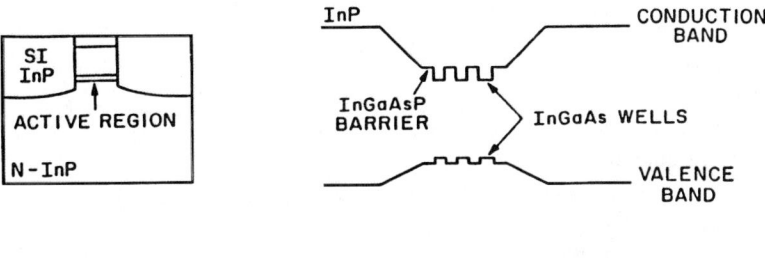

FIGURE 56.7 Schematic of multiquantum well buried heterostructure laser.

The schematic of a MQW BH laser is shown in Fig. 56.7. The laser has a MQW active region and it utilizes Fe doped InP semi-insulating layers for current confinement and optical confinement. The light vs current characteristics of a MQW BH laser are shown in Fig. 56.8. The laser emits near 1.55 μm. The MQW lasers have lower threshold currents than regular DH lasers. Also, the two-dimensional-like density of states of the QW lasers makes the transparency current density of these lasers significantly lower than that for regular DH lasers. This allows the fabrication of very low-threshold (I_{th} ~1 mA) lasers using high-reflectivity coatings.

The optical gain (g) of a laser at a current density J is given by

$$g = a(J - J_0) \tag{56.2}$$

where a is the gain constant and J_0 is the transparency current density. The cavity loss α is given by

$$\alpha = \alpha_c + \frac{1}{L} \ln\left(\frac{1}{R_1 R_2}\right) \tag{56.3}$$

where α_c is the free carrier loss, L is the length of the optical cavity and R_1 and R_2 are the reflectivity of the two facets. At threshold, gain equals loss; hence, it follows from Eqs. (56.2) and (56.3) that the threshold current density (J_{th}) is given by

$$J_{th} = J_0 + \frac{\alpha_c}{a} + \frac{1}{aL} \ln\left(\frac{1}{R_1 R_2}\right) \tag{56.4}$$

Thus, for a laser with high-reflectivity facet coatings ($R_1 \simeq R_2 \simeq 1$) and with low loss ($\alpha_c \sim 0$), $J_{th} \simeq J_0$. For a QW laser, $J_0 \sim 50$ A/cm^2 and for a DH laser, $J_0 \sim 700$ A/cm^2; hence, it is possible to get much lower threshold current using QW laser as the active region.

The light vs current characteristics of a QW laser with high-reflectivity coatings on both facets are shown in Fig. 56.9 [Temkin et al., 1990]. The threshold current at room temperature is ~1.1 mA. The laser is 170 μm long and has 90% and 70% reflective coating at the facets. Such low-threshold lasers are important for array applications.

Recently, QW lasers were fabricated that have higher modulation bandwidth than regular DH lasers. The current confinement and optical confinement in this laser are carried out using MOCVD grown Fe doped InP lasers similar to that shown in Fig. 56.2. The laser structure is then further modified by using a small contact pad and etching channels around the active region mesa. These modifications are designed to reduce the capacitance of the laser structure. A 3-dB bandwidth of 25 GHz is obtained [Morton et al., 1992].

Strained Quantum-Well Lasers

Quantum well lasers have also been fabricated using an active layer whose lattice constant differs slightly from that of the substrate and cladding layers. Such lasers are known as strained quantum-well lasers. Over the last few years, strained quantum well lasers have been extensively investigated all over the world [Beernik, York, and Coleman, 1989; Bour et al., 1988; Fisher et al., 1987; Koren et al., 1990; Laidig, Lin, and Caldwell 1985; Temkin, Tanbun-Ek, and Logan 1990b; Thijs et al., 1991; Thijs and van Dongen, 1989]. They show many desirable properties such as (1) a very low-threshold current density and (2) a lower linewidth than regular MQW lasers both under CW operation and under modulation. The origin of the improved device performance lies in the band-structure changes induced by the mismatch-induced strain [Corzine, Yan, and Coldren, in press; Loehr and Singh, 1991]. Strain splits the heavy-hole and the light-hole valence bands at the Γ point of the Brillouin zone where the bands gap is minimum in direct band-gap semiconductors.

Two material systems have been widely used for strained quantum well lasers: (1) InGaAs grown over InP by the MOCVD or the CBE growth technique [Koren et al., 1990; Laidig, Lin, and Caldwell, 1985; Temkin, Tanbun-Ek, and Logan 1990b; Thijs et al., 1991; Thijs and van Dogen, 1989] and (2) InGaAs grown over GaAs by the MOCVD or the MBE growth technique [Beernik, York, and Coleman, 1989; Bour et al., 1988; Fischer et al., 1987]. The former material system is of importance for low-chirp semiconductor laser for lightwave system applications, whereas the latter material system has been used to fabricate high-power lasers emitting near 0.98 μm, a wavelength of interest for pumping erbium-doped fiber amplifiers.

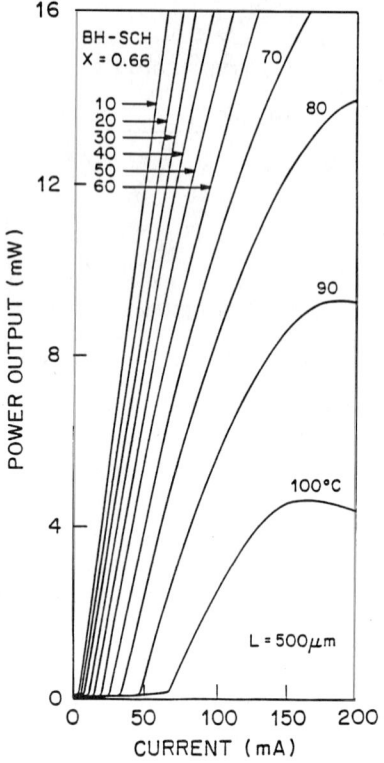

FIGURE 56.8 Light vs current characteristics of a multiquantum well buried heterostructure laser at different temperatures.

The alloy $In_{0.53}Ga_{0.47}As$ has the same lattice constant as InP. Semiconductor lasers with an $In_{0.53}Ga_{0.47}As$ active region have been grown on InP by the MOCVD growth technique. Excellent material quality is also obtained for $In_xGa_{1-x}As$ alloys grown over InP by MOCVD for nonlattice-matched compositions. In this case the laser structure generally consists of one or many $In_xGa_{1-x}As$ quantum well layers with InGaAsP barrier layers whose composition is lattice matched to that of InP. For $x < 0.53$, the active layer in these lasers is under tensile stress, whereas for $x > 0.53$, the active layer is under compressive stress.

Superlattice structures of InGaAs/InGaAsP with tensile and compressive stress have been grown by both MOCVD and CBE growth techniques over an n-type InP substrate. Figure 56.10 shows the broad-area threshold current density as a function of cavity length for strained MQW lasers with four $In_{0.65}Ga_{0.35}As$ quantum wells [Tsang et al., 1990]. The active region in this laser is under 0.8% compressive strain. Also shown for comparison is the threshold current density as a function of cavity length of MQW lattice-matched lasers with $In_{0.53}Ga_{0.47}As$ wells. The entire laser structure, apart from the quantum well composition, is identical for the two cases. The threshold current density is lower for the compressively strained MQW structure than that for the lattice-matched MQW structure.

Buried heterostructure (BH) lasers have been fabricated using compressive and tensile strained MQW lasers. Lasers with compressive strain have a lower threshold current than that for lasers with

Optical Sources

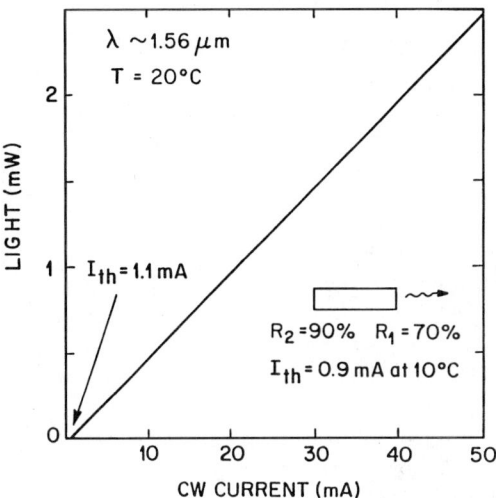

FIGURE 56.9 Light vs current of a quantum well laser with high reflectivity coatings on both facets.

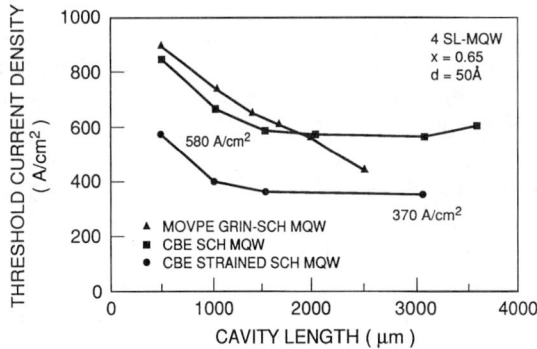

FIGURE 56.10 Broad-area threshold current density as a function of cavity length for strained and lattice-matched InGaAs/InP MQW lasers.

tensile strain. This can be explained by splitting of the light-hole and heavy-hole bands under stress [Adams 1986; Yablonovitch and Kane, 1986].

Strained quantum well lasers fabricated using $In_xGa_{1-x}As$ layers grown over a GaAs substrate have been extensively studied [Beernik, York, and Coleman, 1989; Bour et al., 1988; Chand et al., 1991; Choi and Wang, 1990; Dutta et al., 1991, 1990a, 1990b; Fisher et al., 1987; Kuo et al., 1991; Okayasu et al., 1990; Wu et al., 1991]. The lattice constant of InAs is 6.06 Å and that of GaAs in 5.654 Å. The $In_xGa_{1-x}As$ alloy has a lattice constant between these two values, and to a first approximation it can be assumed to vary linearly with x. Thus, an increase in the In mole fraction x increases the lattice mismatch relative to the GaAs substrate and, therefore, produces larger compressive strain on the active region.

A typical laser structure grown over the n-type GaAs substrate is shown in Fig. 56.11 for this material system. It consists of a MQW active region with one to four $In_xGa_{1-x}As$ wells separated by

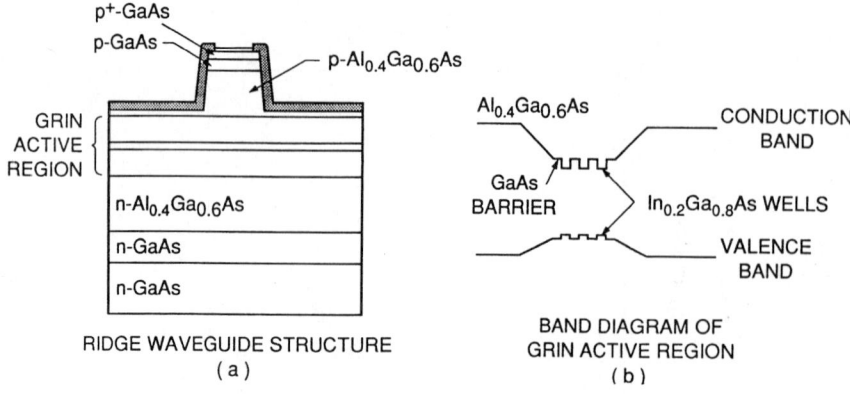

FIGURE 56.11 Typical $In_xGa_{1-x}As/GaAs$ MQW laser structure.

GaAs barrier layers. The entire MQW structure is sandwiched between n- and p-type $Al_{0.3}Ga_{0.7}As$ cladding layers, and the P-cladding layer is followed by a p-type GaAs contact layer. Variations of this structure with different cladding layers or large optical cavity designs have been reported. Emission wavelength depends on the In composition, x. As x increases, the emission wavelength increases and for x larger than a certain value (typically, ~ 0.25), the strain is too large to yield high-quality material.

For $x \simeq 0.2$, the emission wavelength is near 0.98 μm, a wavelength region of interest for pumping fiber amplifiers. Threshold current density as low as 47 A/cm^2 has been reported for $In_{0.2}Ga_{0.8}As/GaAs$ strained MQW lasers. High-power lasers have been fabricated using $In_{0.2}Ga_{0.8}As/GaAs$ MQW active region. Single-mode output powers of greater than 200 mW have been demonstrated using a ridge-waveguide-type laser structure. CW output powers greater than 1 W have been demonstrated using a laser-array geometry [Dutta et al., 1990b]. Good reliability results have been obtained for these strained quantum well lasers [Yano et al., 1991].

Frequency chirp of strained and unstrained QW lasers has been investigated. Strained QW lasers (InGaAs/GaAs) exhibit the lowest chirp (or dynamic linewidth) under modulation. The lower chirp of strained QW lasers is consistent with a small linewidth enhancement factor measured in such devices. A correlation between the measured chirp and linewidth enhancement factor for regular double-hetero-structure, strained and unstrained QW lasers is shown in Table 56.1. The high efficiency, high power, and low chirp of strained and unstrained QW lasers make these devices attractive candidates for lightwave transmission applications.

TABLE 56.1 Linewidth enhancement factor and full width at half-maximum chirp for various lasers

	Linewidth Enhancement Factor	FWHM[a] Chirp at 50 mA (1 Gb/s)
DH laser	5.5	1.2 A
MQW laser ($\lambda = 1.5\,\mu$m)	3.5	0.6 A
Strained MQW InGaAs/GaAs laser ($\lambda = 1\,\mu$m)	1.0	0.2 A
Strained MQW InGaAsP/InP laser laser ($\lambda = 1.5\,\mu$m)	2.0	0.4 A

[a] Full width at half-maximum.

56.4 Distributed Feedback Lasers

Semiconductor lasers fabricated using the InGaAsP material system are widely used as sources in many lightwave transmission systems. One measure of the transmission capacity of a system is the data rate. Thus the drive toward higher capacity pushes the systems to higher data rates where the chromatic dispersion of the fiber plays an important role in limiting the distance between

Optical Sources

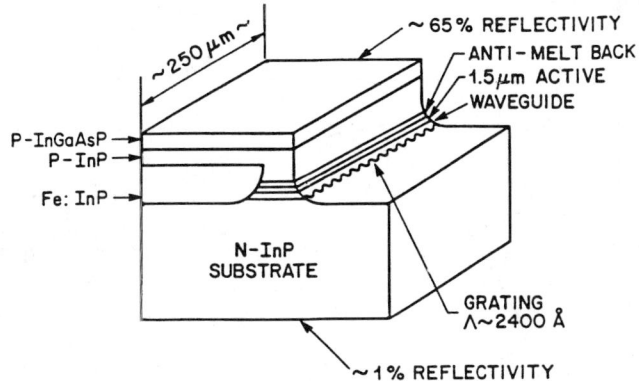

FIGURE 56.12 Schematic of a capped mesa buried heterostructure (CMBH) distributed feedback laser.

regenerators. Sources emitting in a single wavelength help reduce the effects of chromatic dispersion and are therefore used in most systems operating at high data rates (> 1.5 Gb/s).

The single-wavelength laser source used in most commercial transmission systems is the **distributed feedback (DFB)** laser where a diffraction grating etched on the substrate close to the active region provides frequency selective feedback, which makes the laser emit in a single wavelength. This section reports the fabrication, performance characteristics and reliability of DFB lasers [Agrawal and Dutta 1986, Chap. 7].

The schematic of a DFB laser structure is shown in Fig. 56.12. The fabrication of the device involves the following steps. First, a grating with a periodicity of 2400 Å is fabricated on a (100) oriented n-InP substrate using optical holography and wet chemical etching. Four layers are then grown over the structure. These layers are (1) n-InGaAsP ($\lambda \sim 1.3$ μm) waveguide layer, (2) undoped InGaAsP ($\lambda \sim 1.55$ μm) active layer, (3) p-InP cladding layer and (4) a p-InGaAsP ($\lambda \sim 1.3$ μm) contact layer. Mesas are then etched on the wafer using a SiO_2 mask and wet chemical etching. Fe doped InP semi-insulating layers are grown around the mesas using the MOCVD growth technique. The semi-insulating layers help confine the current to the active region and also provide index guiding to the optical mode. The SiO_2 stripe is then removed and the p-InP cladding layer and a p-InGaAsP contact layer are grown on the wafer using the vapor phase epitaxy growth technique. The wafer is then processed to produce 250-μm-long laser chips using standard metallization and cleaving procedures. The final laser chips have antireflection coating (<1%) at one facet and high-reflection coating ($\sim$65%) at the back facet. The asymmetric facet coatings help remove the degeneracy between the two modes in the stop band.

The CW light vs current characteristics of a laser are shown in Fig. 56.13. Also shown is the measured spectrum at different output powers. The threshold current of these lasers is in the 15–20 mA range.

For high-fiber-coupling efficiency, it is important that the laser emit in the fundamental transverse mode. The measured far-field pattern parallel and perpendicular to the junction plane at different output powers of a device is shown in Fig. 56.14. The figure shows that the laser operates in the fundamental transverse mode in the entire operating power range from threshold to 60 mW. The full width at half-maximum of the beam divergences parallel and perpendicular to the junction plane are 40° and 30°, respectively.

The dynamic spectrum (or chirp) of the laser under modulation is an important parameter when the laser is used as a source for transmission. The measured 20-dB full width is shown in Fig. 56.15 at two different data rates as a function of bias level. For a laser biased above threshold, the chirp width is nearly independent of the modulation rate.

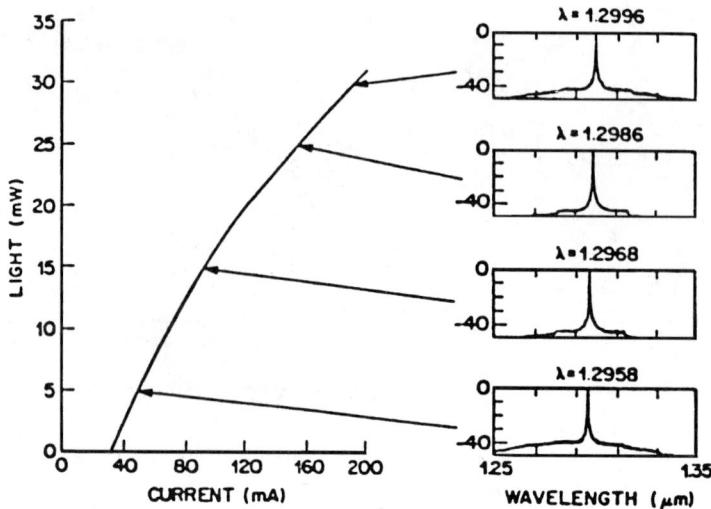

FIGURE 56.13 CW light vs current characteristics and measured spectrum at different output powers. Temperature = 30°C.

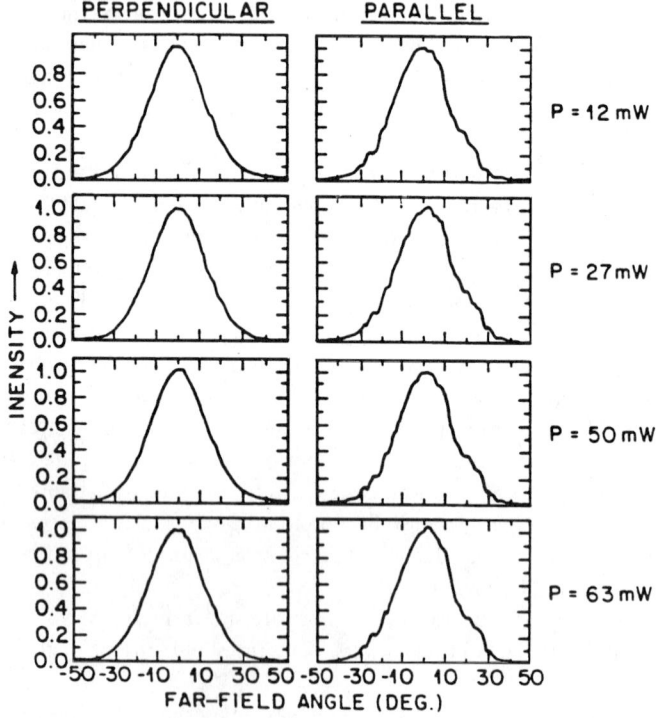

FIGURE 56.14 Measured far-field pattern parallel and perpendicular to the junction plane.

Tunable Lasers

Tunable semiconductor lasers are needed for many applications. Examples of application is lightwave transmission systems are (1) wavelength-division multiplexing where signals at many distinct wavelengths are simultaneously modulated and transmitted through a fiber and (2) coherent

Optical Sources

transmission systems where the wavelength of the transmitted signal must match that of the local oscillator. Several types of tunable laser structures have been reported in the literature [Dutta et al., 1986; Koch and Koren, 1991; Koch et al., 1988; Liou, Dutta, and Burrus, 1987; Suematsu, Arai, and Kishino, 1983; Tanbun-Ek et al., 1990]. Two principle schemes are (1) multisection DFB laser and (2) multisection distributed Bragg reflector (DBR) laser. The multisection DBR lasers generally exhibit higher tunability than that for the multisection DFB lasers.

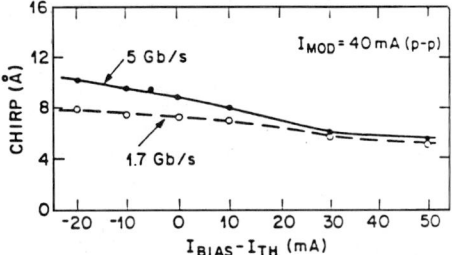

FIGURE 56.15 Measured 20-dB full width of the dynamic linewidth is plotted as a function of bias level for operation at 5 and 1.7 Gb/s.

The design of a multisection DBR laser is shown in Fig. 56.16 schematically [Koch et al., 1988]. The three sections of this device are (1) the active region section that provides the gain, (2) the grating section that provides the tunability, and (3) the phase-tuning section that is needed to access all wavelengths continuously. The current through each of these sections can be varied independently. The tuning mechanism can be understood by noting that the emission wavelength λ of a DBR laser is given by

$$\lambda = 2\bar{\mu}\Lambda \qquad (56.5)$$

where Λ is the grating period and $\bar{\mu}$ is the effective refractive index of the optical mode in the grating section. The latter can be changed simply by varying the current in the grating section.

The extent of wavelength tunability of a three-section DBR laser is shown in Fig. 56.17 [Koch and Koren, 1991]. Measured wavelengths are plotted as a function of phase-section current for different currents in the tuning section. A tuning range in excess of 6 nm can be obtained by controlling currents in the grating and phase-tuning sections.

An important characteristic of lasers for applications requiring a high degree of coherence is the spectral width (linewidth) under CW operation. The CW linewidth depends on the rate of spontaneous emission in the laser cavity. For coherent transmission applications, the CW linewidth must be quite small. The minimum linewidth allowed depends on the modulation format used. For differential phase-shift keying (DPSK) transmission, the minimum linewidth is approximately given by $B/300$ where B is the bit rate. Thus, for 1-Gb/s transmission rate, the minimum linewidth is $\sim$3 MHz.

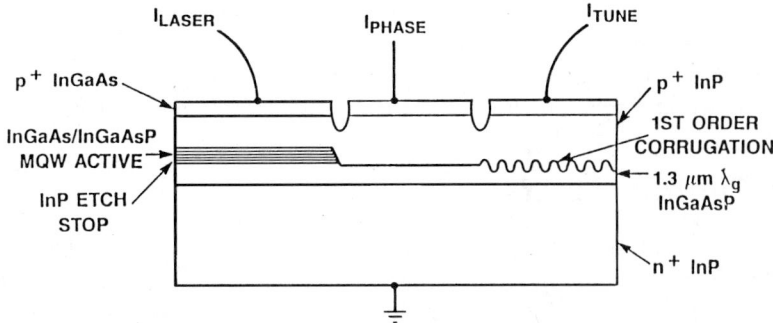

FIGURE 56.16 Schematic of a multisection DBR laser. The laser has a MQW active region. The three sections are optically coupled by the thick waveguide layer below the MQW active region.

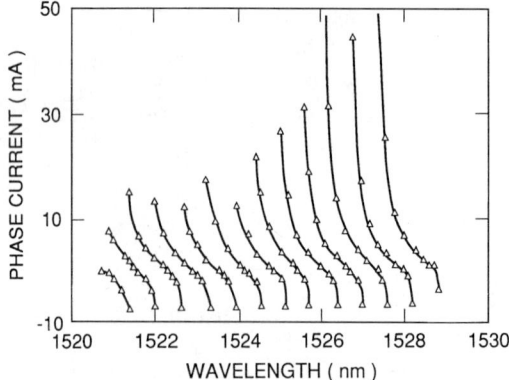

FIGURE 56.17 Frequency tuning characteristics of a three-section MQW DBR laser.

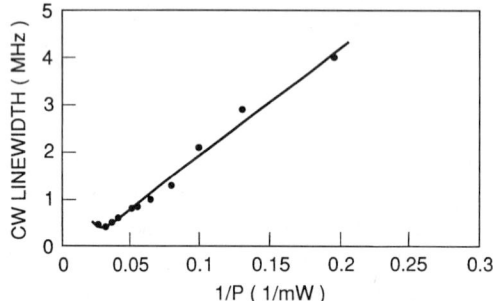

FIGURE 56.18 Measured CW linewidth plotted as a function of the inverse of the output power for a MQW DFB laser with a cavity length of 850 μm.

The CW linewidth of a laser decreases with increasing length and varies as $1+\alpha^2$, where α_c is the linewidth enhancement factor. Since α_c is smaller for a MQW laser, the linewidth of DFB or DBR lasers utilizing MQW active region is smaller than that for lasers with regular DH active region. The linewidth varies inversely with the output power at low powers (<10 mW) and shows saturation at high powers. The measured linewidth as a function of output power of a 850-μm long DFB laser with MQW active region is shown in Fig. 56.18. The minimum linewidth of 350 kHz was observed for this device at an operating power of 25 mW. For multisection DBR lasers of the type shown in Fig. 56.16, the linewidth varies with changes in currents in the phase-tuning and the grating sections.

56.5 Surface Emitting Lasers

Semiconductor lasers described in the previous sections have cleaved facets that form the optical cavity. The facets are perpendicular to the surface of the wafer and light is emitted parallel to the surface of the wafer. For many applications requiring a two-dimensional laser array or monolithic integration of lasers with electronic components (e.g., optical interconnects), it is desirable to have the laser output normal to the surface of the wafer. Such lasers are known as surface emitting lasers (SEL). A class of surface emitting lasers also has an optical cavity normal to the surface of the wafer [Chang-Hasnain et al., 1990; Geels and Coldren, 1990; Geels, Corzine, and Coldren, 1991; Iga,

Optical Sources

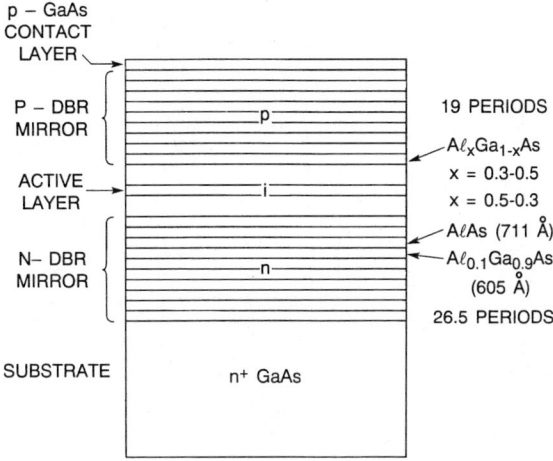

FIGURE 56.19 Schematic illustration of a generic SEL structure utilizing distributed Bragg mirrors formed by using multiple semiconductor layers. DBR pairs consist of AlAs (711 Å thick) and $Al_{0.1}Ga_{0.9}As$ (605 Å thick) alternate layers. Active layer could be either a quantum well or similar to a regular double heterostructure laser.

Koyama, and Kinoshita, 1988; Jewel et al., 1991; Soda et al., 1979; Tai et al., 1990a, 1990b]. These devices are known as vertical-cavity surface emitting lasers (VCSEL) in order to distinguish them from other surface emitters.

A generic SEL structure utilizing multiple semiconductor layers to form a Bragg reflector is shown in Fig. 56.19. The active region is sandwiched between n- and p-type cladding layers, which are themselves sandwiched between the two n- and p-type Bragg mirrors. The Bragg mirrors consist of alternating layers of low-index and high-index materials. The thicknesses of each layer is one quarter of the wavelength of light in the medium. Such periodic quarter-wave thick layers can have very high reflectivity [Born and Wolf, 1977].

For a SEL to have a threshold current density comparable to that of an edge emitting laser, the threshold gains must be comparable for the two devices. The threshold gain of an edge emitting laser is ~ 100 cm^{-1}. For a SEL with an active-layer thickness of 0.1 μm, this value corresponds to a single-pass gain of $\sim 1\%$. Thus, for the SEL device to lase with a threshold current density comparable to that of an edge emitter, the mirror reflectivities must be >99%.

The reflectivity spectrum of a SEL structure is shown in Fig. 56.20. The reflectivity is >99% over a 10-nm band. The drop in reflectivity in the middle of the band is due to the Fabry–Perot mode.

The number of pairs needs to fabricate a high-reflectivity mirror depends on the refractive index of layers of a pair. For large index differences fewer pairs are needed. For example, in the case of alternating layers of ZnS and CaF_2 for which the index differences is 0.9, only six pairs are needed for a reflectivity of 99%. By contrast, for an InP/InGaAsP ($\lambda \sim 1.3$ μm) layer pair for which the index difference is 0.3, more than 40 pairs are needed to achieve a reflectivity of 99%.

The energy-gap difference at the alternating layers of a Bragg mirror results in potential barriers at heterointerfaces. These potential barriers impede the flow of carriers through a Bragg-reflector mirror stack, resulting in high resistance of the device. It turns out that using graded heterobarrier interfaces rather than an abrupt interface reduces the series resistance significantly without compromising the reflectivity.

Five principal structures used for SEL fabrication are (1) etched mesa structure [Jewel et al., 1989], (2) ion-implanted structure [Lee et al., 1990], (3) dielectric isolated structure [Tai et al., 1989],

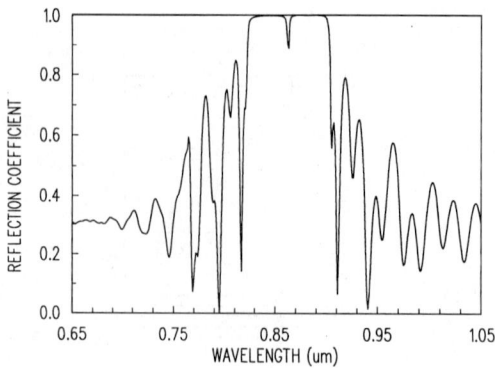

FIGURE 56.20 Reflectivity of a SEL wafer consisting of top and bottom DBR mirrors.

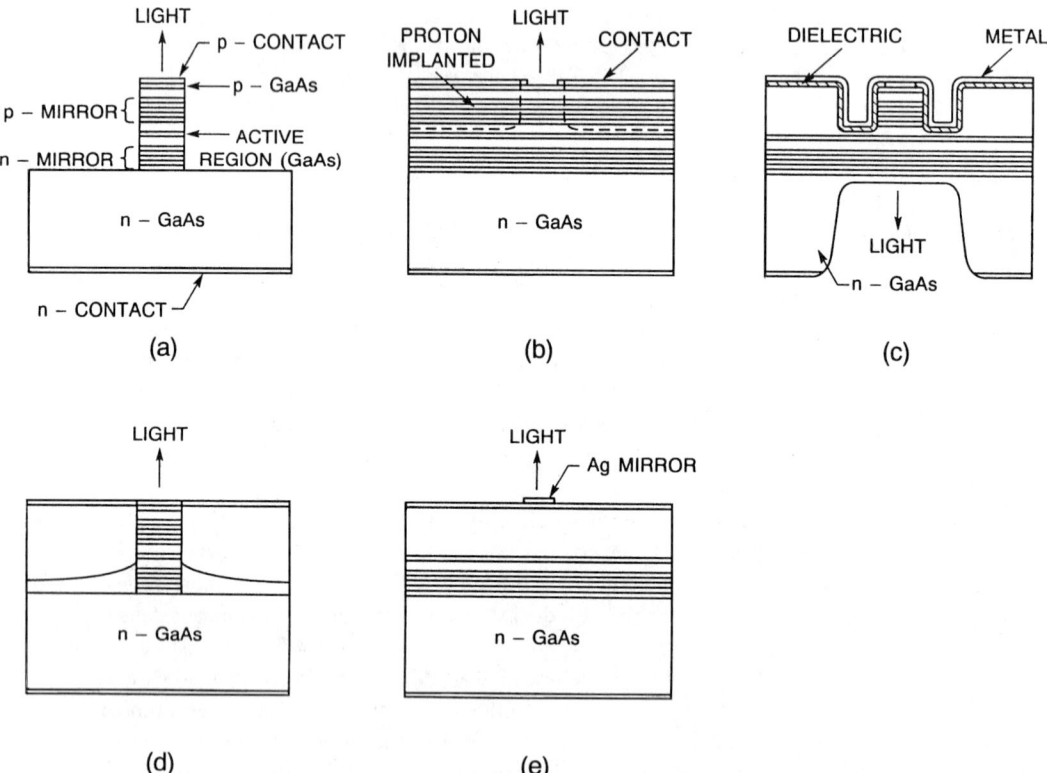

FIGURE 56.21 GaAs/AlGaAs SEL structure: (a) etched-mesa structure, (b) ion-implanted top emitting structure, (c) dielectric isolated structure, (d) buried heterostructure, and (e) metallic reflector structure.

(4) buried heterostructure [Ibaraki et al., 1989], and (5) metallic reflector structure [Schubert et al., 1990]. Schematics of these devices are shown in Fig. 56.21 for the case of GaAs active region.

In addition to very high-reflectivity mirrors, the key element necessary for the fabrication of good etched mesa SELs is the use of an etching process that produces surfaces with very little nonradiative recombination. Threshold current of 0.7 mA and 2 mA have been reported for

Optical Sources

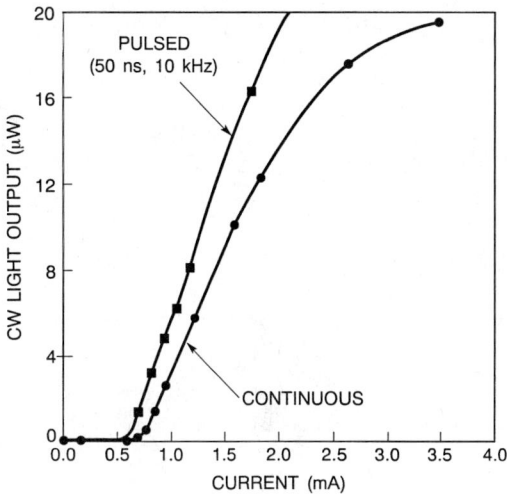

FIGURE 56.22 *L–I* characteristics of a low-threshold InGaAs/GaAs surface emitting laser. Calculated top and bottom mirror reflectivities of this structure were 0.99998 and 0.9998, respectively. The light is emitted from the bottom.

InGaAs/GaAs ($\lambda \sim 1~\mu$m) and GaAs/AlGaAs ($\lambda \sim 0.85~\mu$m) etched mesa lasers, respectively. The *L–I* characteristics of a low-threshold etched mesa type SEL are shown in Fig. 56.22 [Geels et al., 1990]. Modulation bandwidths of an ion implanted SEL structure are shown in Fig. 56.23. The ion implanted single-mode SELs typically have threshold currents of 3 mA [Dutta et al., 1991; Gourley et al., 1989; Hasnain et al., 1991a, 1991b].

In a version of the SEL structure, the quantum well gain regions are located at the maximum of the standing-wave pattern in the Fabry–Perot cavity. This structure is known as the resonant periodic-gain structure [Corzine et al., 1989; Raja et al., 1989]. It allows for the maximum coupling between the gain region and the optical field, resulting in a very low threshold for such devices.

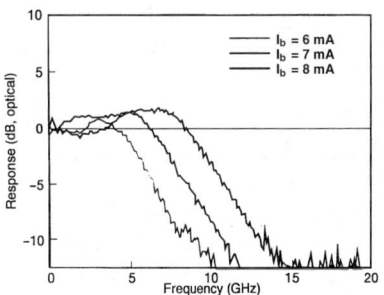

FIGURE 56.23 Small-signal modulation characteristics of a proton-implanted GaAs/AlGaAs surface emitting laser.

As mentioned previously, the refractive-index difference between InGaAsP ($\lambda \sim 1.3~\mu$m) and InP layers is smaller than that occurring in GaAs SELs. Therefore, a larger number of pairs (typically, 40–50) are needed to produce a high-reflectivity (>99%) mirror. Such mirror stacks have been grown by chemical-beam epitaxy (CBE), metal-organic chemical vapor deposition and molecular-beam epitaxy (MBE) growth techniques [Tai et al., 1990]. Such mirror stacks along with a dielectric Si/SiO$_2$ stack on the p-side have been used to fabricate InGaAsP SELs [Yang et al., 1990]. The threshold current of these lasers is generally higher than that for cleaved facet lasers. As the mirror fabrication technology develops further, a reduction in threshold current is likely to follow.

For many applications of SEL arrays, such as for optical interconnection systems, each laser in the array should be biased and controlled individually. Independently addressable laser arrays have been fabricated [Chang-Hasnain et al., 1991]. It is even possible to design multi-wavelength SEL arrays such that each laser operates at a slightly different wavelength using a growth technique where the Bragg wavelength of mirror stacks is varied across a wafer. Such laser arrays have been used in

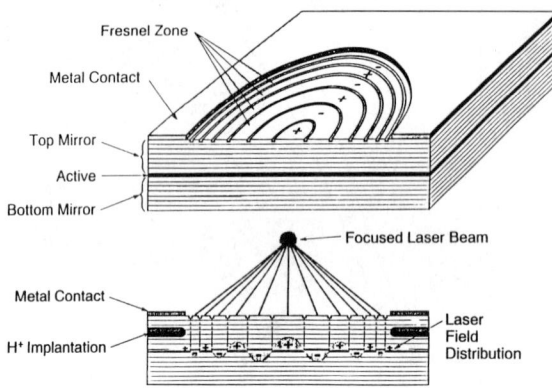

FIGURE 56.24 Schematic of a self-focusing surface emitting laser which has Fresnel zones internal to the optical cavity.

transmission experiment [Maeda et al., 1991]. Recently, a new SEL design has been demonstrated whose output is focused to a single spot [Vakhshoori et al., 1994]. The laser has a large area (~100 μm diameter) and it has Fresnel zonelike structure etched on the top mirror (Fig. 56.24). The lasing mode with the lowest loss has π phase shift in the near field as it traverses each zone. The laser emits 500 mW in a single mode.

56.6 Laser Reliability

The performance characteristics of injection lasers used in lightwave systems can degrade during their operation. The degradation is generally characterized by a change in the operational characteristics of the lasers and in some cases are associated with the formation and/or multiplication of defects in the active region. The degraded lasers usually exhibit an increase in the threshold current, which is often accompanies by a decrease in the external differential quantum efficiency. For single-wavelength lasers the degradation may be a change in the spectral characteristics, namely, the degraded device may no longer emit in a single wavelength although the threshold or the light output at a given current has changed very little. The dominant mechanism responsible for the degradation is determined by any or all of the several fabrication processes including epitaxial growth, wafer quality, and device processing and bonding [AT&T, 1985; DeLoach et al., 1973; Gordon, Nash, and Hartman, 1983; Hartman and Dixon, 1975; Johnston and Miller, 1973; Joyce et al., 1985; Matsui, Ishida, and Nannichi, 1975; Mizuishi et al., 1983; Nash et al., 1985; Petroff and Hartman, 1973; Petroff, Johston, and Hartman, 1974; Petroff and Lang, 1977; Uedo et al., 1982; Uedo et al., 1980; Yamakoshi et al., 1981].

Some lasers exhibit an initial rapid degradation after which the operating characteristics of the lasers are very stable. Given a population of lasers, it is possible to quickly identify the *stable* lasers by a high-stress test (also known as the purge test) [Gordon et al., 1983; Nash et al., 1985]. The stress test implies that operating the laser under a set of high-stress conditions (e.g., high current, high temperature, high power) would cause the weak lasers to fail and stabilize the possible winners. Observations on the operating current after stress aging have been reported by Nash et al. [1985]. Figure 56.25 shows the data on aging on InGaAsP ($\lambda \sim 1.3$ μm) lasers. Similar lasers were used in the first submarine fiber optic cable.

Optical Sources

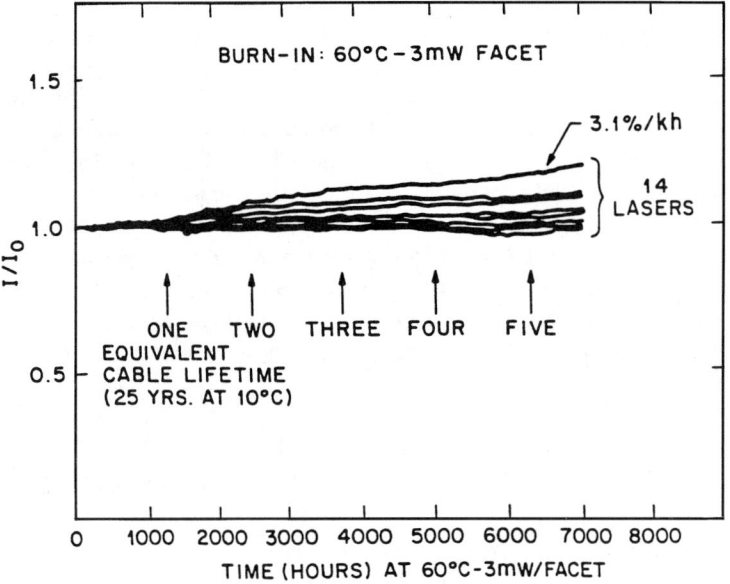

FIGURE 56.25 Operating current for 3-mW output at 60°C as a function of operating time. These data were generated for 1.3-μm InGaAsP lasers in the first submarine fiber optic cable.

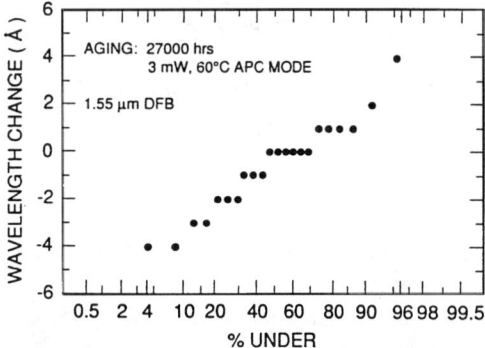

FIGURE 56.26 Change in emission wavelength after aging is plotted in the form of a normal probability distribution.

The expected operating lifetime of a semiconductor laser is generally determined by accelerated aging at high temperatures and using an activation energy. The lifetime (τ) at a temperature T is experimentally found to vary as [Hartman and Dixon, 1975; Joyce et al., 1985]

$$\tau = \tau_0 \exp(-E_a/kT) \tag{56.6}$$

where E_a is the activation energy and k is the Boltzmann constant. The measured activation energy for InGaAsP lasers is in the range of $\simeq$0.5–0.7 eV.

For some applications such as coherent transmission systems, the absolute wavelength stability of the laser is important. The measured change in emission wavelength at 100 mA of single-wavelength

DFB lasers before and after aging of several devices is shown in Fig. 56.25. Note that most of the devices do not exhibit any change in wavelength and the standard deviation of the change is less than 2 Å. This suggests that the absolute wavelength stability of the devices is adequate for coherent transmission applications.

Another parameter of interest in certifying the spectral stability of a DFB laser is the change in dynamic linewidth or chirp with aging. Since the chirp depends on data rate and the bias level, a certification of the stability of the value of the chirp is, in general, tied to the requirements of a transmission system application. A measure of the effect of chirp on the performance of a transmission system is the dispersion penalty. We have measured the dispersion penalty of several lasers at 600 Mb/s for a dispersion of 1700 ps/nm before and after aging. The medium change in dispersion penalty for 42 devices was less than 0.1 dB, and no single device showed a change larger than 0.3 dB. This suggests that the dynamic linewidth or the chirp under modulation is stable.

Defining Terms

Band gap: The energy difference between the lowest point in the conduction band and highest point in the valence band

Bandwidth: The modulation frequency up to which the device responds.

Dispersion: Variation of refractive index with wavelength.

Distributed feedback: Mechanism by which the optical feedback in a laser is spatially distributed.

Epitaxy: The process of growing one material over another with no change in lattice characteristics through the boundary.

Index guiding: Wave propagation in a medium surrounded by medium of higher index.

LED: Light emitting diode.

Quantum well: Very thin (~100-A semiconductor regions bounded by higher band gap semiconductor.

Threshold current: The current at which the light output of the laser rises sharply.

References

Adams, A.R. 1986. *Electron. Lett.*, 22:249.
Agrawal, G.P. and Dutta, N.K. 1986. *Long Wavelength Semiconductor Lasers*, 2nd ed. 1993, van Nostrand Reinhold, New York.
Arakawa, Y. and Yariv, A. 1985. *IEEE J. Quantum Electron.*, QE-21:1666.
Arakawa, Y. and Yariv, A. 1986. *IEEE J. Quantum Electron.*, QE-22:1887.
AT&T. 1985. Special issue *AT&T Tech. J.*, 64:3.
Beernik, K.J., York, P.K., and Coleman, J.J. 1989. *Appl. Phys. Lett.*, 25:2582.
Born, M. and Wolf, E. 1977. *Principles of Optics*, Sec. 1.6.5, p. 69. Pergamon Press, New York.
Bour, D.P., Gilbert, D.B., Elbaum, L., and Harvey, M.G. 1988. *Appl. Phys. Lett.*, 53:2371.
Capasso, F. and Margaritondo, G. ed. 1987. *Heterojunction Band Discontinuation: Physics and Applications*, North-Holland, Amsterdam.
Casey, H.C. Jr. and Panish, M.B. 1978. *Heterostructure Lasers*, Academic Press, New York.
Chand, N., Becker, E.E., van der Ziel, J.P., Chu, S.N.G., and Dutta, N.K. 1991. *Appl. Phys. Lett.*, 58:1704.
Chang-Hasnain, C.J., Maeda, M.W., Stoffel, N.G., Harbison, J.P., and Florez, L.T., 1990. *Electron. Lett.*, 26:940.
Chang-Hasnain, C.J., Harbison, J.P., Zah, C.E., Maeda, M.W., Florez, L.T., Stoffel, N.G., and Lee, T.P. 1991. *IEEE J. Quantum Electron.*, 27:1368.
Chiu, L.C. and Yariv, A. 1982. *IEEE J. Quantum Electron.*, QE-18:1406.
Choi, H.K. and Wang, C.A. 1990. *Appl. Phys. Lett.*, 57:321.

Corzine, S.W., Geels, R.S., Scott, J.W., Yan, R.H., and Coldren, L.A. 1989. *IEEE J. Quantum Electron.*, QE-25:1513.

Corzine, S.W., Yan, R., and Coldren, L.A., in press. Optical gain in III-V bulk and quantum well semiconductors, In *Quantum Well Lasers*, ed. P. Zory, Academic Press, New York.

DeLoach, B.C., Jr., Hakki, B.W., Hartman, R.L., and D'Asaro, L.A., 1973. *Proc. IEEE*, 61:1042.

Dutta, N.K. 1983. *J. Appl. Phys.*, 54:1236.

Dutta, N.K., Lopata, J., Berger, P.R., Sivco, D.L., and Cho, A.Y. 1991. *Electron. Lett.*, 27:680.

Dutta, N.K., Napholtz, S.G., Yen, R., Brown, R.L., Shen, T.M., Olsson, N.A., and Craft, D.C. 1985a. *Appl. Phys. Lett.*, 46:19.

Dutta, N.K., Napholtz, S.G., Yen, R., Wessel, T., and Olsson, N.A., 1985b. *Appl. Phys. Lett.*, 46:1036.

Dutta, N.K. and Nelson, R.J. 1982. *J. Appl. Phys.*, 53:74.

Dutta, N.K., Piccirilli, A. B., Cella, T., and Brown, R.L. 1986. *Apply Phys. Lett.*, 48:1501.

Dutta, N.K., Tu, L.W., Zydzik, G.J., Hasnain, G., Wang, Y.H., and Cho, A.Y. 1991. *Electron. Lett.*, 27:208

Dutta, N.K., Wessel, T., Olsson, N.A., Logan, R.A., and Yen, R. 1985c. *Appl. Phys. Lett.*, 46:525.

Dutta, N.K., Wessel, T., Olsson, N.A., Logan, R.A., Yen, R., and Anthony, P.J. 1985d. *Electron. Lett.*, 21:571.

Dutta, N.K., Wynn, J.D., Lopata, J., Sivco, D.L., and Cho, A.Y. 1990a. *Electron. Lett.*, 26:1816.

Dutta, N.K., Wynn, J.D., Sivco, D.L., and Cho, A.Y. 1990b. *Appl. Phys. Lett.*, 56:2293.

Fischer, S.E., Fekete, D., Feak, G.B., and Ballantyne, J.M. 1987. *Appl. Phys. Lett.*, 50:714.

Geels, R.S. and Coldren, L.A., 1990. *Appl. Phys. Lett.*, 57:1605.

Geels, R.S., Corzine, S.W., Scott, J.W., Young, D.B., and Coldren, L.A. 1990. *IEEE Photonic Tech. Lett.*, 2:234.

Geels, R.S., Corzine, S.W., and Coldren, L.A., 1991. *IEEE J. Quantum Electron.*, 27:1359.

Gordon, E.I., Nash, F.R., and Hartman, R.L. 1983. *IEEE Electron Dev. Lett.*, ELD-4:465.

Gourley, P.L., Lyo, S.K., Brennan, T.M., Hammons, B.E., Schaus, C.P., and Sun, S. 1989. *Appl. Phys. Lett.*, 55:2698.

Hall, R.N., Fenner, G.E., Kingsley, J.D., Soltys, T.J., and Carlson, R.O. 1962. *Phys. Ref. Lett.*, 9.

Hartman, R.L. and Dixon, R.W. 1975. *Appl. Phys. Lett.*, 26:239.

Hasnain, G., Tai, K., Dutta, N.K., Wang, Y.H., Wynn, J.D., Weir, B.E., and Cho, A. Y., 1991a. *Electron. Lett.*, 27:915.

Hasnain, G., Tai, K., Yang, L., Wang, Y.H., Fischer, R.J., Wynn, J.D., Weir, B.E., Dutta, N.K., and Cho, A.Y. 1991b. *IEEE J. Quantum Electron.*, QE-27:1377.

Hayashi, I., Panish, M.B., Foy, P.W., and Sumski, S. 1970. *Appl. Phys. Lett.*, 47:109.

Hersee, S.D., DeCremoux, B., and Duchemin, J.P. 1987. *Appl. Phys. Lett.*, 44:476.

Holonyuk, N., Jr. and Bevacqua, S.F. 1962. *Appl. Phys. Lett.*, 1:82.

Holonyak, N. Jr., Kolbas, R.M., Dupuis, R.D., and Dapkus, P.D. 1980a. *IEEE J. Quantum Electron.*, QE-16:170.

Holonyak, N. Jr., Kolbas, R.M., Laidig, W.D., Vojak, B.A., Hess, K., Dupuis, R.D., and Dapkus, P.D. 1980b. *J. Appl. Phys.*, 51:1328.

Hsieh, J.J. 1976. *Appl. Phys. Lett.*, 28:283.

Ibaraki, A., Kawashima, K., Furusawa, K., Ishikawa, T., Yamayachi, T., and Niina, T. 1989. *Japan J. Appl. Phys.*, 28:L667.

Iga, K., Koyama, F., and Kinoshita, S. 1988. *IEEE J. Quantum Electron.*, 24:1845.

Jewell, J.L., Harbison, J.P., Scherer, A., Lee, Y.H., and Florez, L.T. 1991. *IEEE J. Quantum Electron.*, 27, 1332.

Jewell, J.L., Scherer, A., McCall, S.L., Lee, Y.H., Walker, S.J., Harbison, J.P., and Florez, L.T. 1989. *Electron Lett.*, 25:1123.

Johnston, W.D. and Miller, B. I. 1973. *Appl. Phys. Lett.*, 23:1972.

Joyce, W.B., Liou, K.Y., Nash, F.R., Bossard, P.R., and Hartman, R.L. 1985. *AT&T Tech. J.*, 64:717.

Koch, T.L. and Koren, U. 1991. *IEEE J. Quantum Electron.*, QE-27:641.

Koch, T.L., Koren, U., Gnall, R.P., Burrus, C.A., and Miller, B.I. 1988. *Electron. Lett.*, 24:1431.
Koren, U., Oron, M., Young, M.G., Miller, B.I., DeMiguel, J.L., Raybon, G., and Chien, M. 1990. *Electron. Lett.*, 26:465.
Kressle, H. and Butler, J.K. 1977. *Semiconductor Lasers and Heterojection LEDs*, Academic Press, New York.
Kuo, J.M., Wu, M.C., Chen, Y.K., and Chin, M.A. 1991. *Appl. Phys. Lett.*, 59:2781.
Laidig, W.D., Lin, Y.F., and Caldwell, P.J. 1985. *J. Appl. Phys.*, 57:33.
Lee, Y.H., Tell, B., Brown-Goebeler, K.F., Jewell, J.L., Leibenguth, R.E., Asom, M.T., Livescu, G., Luther, L., and Mattera, V.D. 1990. *Electron. Lett.*, 26:1308.
Liou, K.Y., Dutta, N.K., and Burrus, C.A. 1987. *Appl. Phys. Lett.*, 50:489.
Loehr, J.P. and Singh, J. 1991. *IEEE J. Quantum Electron*, 27:708
Maeda, M.W., Chang-Hasnain, C., von Lehman, A., Izadpanah, H., Lin, C., Iqbal, M.Z., Florez, L., and Harbison, J. 1991. *Photonic Tech. Lett.*, 3:863.
Matsui, J., Ishida, R., and Nannichi, Y. 1975. *Japan J. Appl. Phys.*, 14:1555.
Mizuishi, K., Sawai, M., Todoroki, S., Tsuji, S., Hirao, M., and Nakamura, M. 1983. *IEEE J. Quantum Electron.*, QE-19:1294.
Morton, P., Logan, R.A., Tanbun-Ek, T., Sciortino, P.F., Sergent, A.M., Montgomery, R.K., and Lee, B.T. 1992. *Electron. Lett.*, 28:2156.
Nahory, R.E., Pollack, M.A., Johnston, W.D. Jr., and Barns, R.L. 1978. *Appl. Phys. Lett.*, 33:659.
Nash, F.R., Sundburg, W.J., Hartman, R.L., Pawlik, J.R., Ackerman, D.A., Dutta, N.K., and Dixon, R.W. 1985. *AT&T Tech. J.*, 64:809.
Nathan, M.I., Dumke, W.P., Burns, G., Dill, F.H. Jr., and Ziegler, H.J. 1962. *Appl. Phys. Lett.*, 1:63.
Okayasu, M., Fukuda, M., Takeshita, T., and Uehara, S. 1990. *IEEE Photonic Tech. Lett.*, 2:689.
Petroff, P.M. and Hartman, R.L. 1973. *Appl. Phys. Lett.*, 23:469.
Petroff, P. M., Johnston, W.D., Jr., and Hartman, R.L. 1974. *Appl. Phys. Lett.*, 25:226.
Petroff, P.M. and Lang, D.V. 1977. *Appl. Phys. Lett.*, 31:60.
Quist, T.M., Retiker, R.H., Keyes, R.J., Krag, W.E., Lax, B., McWhorter, A.L., and Ziegler, H.J. 1962. *Appl. Phys. Lett.*, 1:91.
Raja, M.Y.A., Brueck, S.R.J., Osinski, M., Schaus, C.F., McInerney, J.G., Brennan T.M., and Hammons, B.E., 1989. *IEEE J. Quantum Electron.*, QE-25:1500.
Schubert, E.F., Tu, L.W., Kopf, R.F., Zydzik, G.J., and Deppe, D.G. 1990. *Appl. Phys. Lett.*, 57:117.
Smith, C., Abram, R.A., and Burt, M.G. 1983. *J. Phys. C*, 16:L171.
Soda, H., Iga, K., Kitahara, C., Suematsu, Y. 1979. *Japan J. Appl. Phys.*, 18:2329.
Suematsu, Y., Arai, S., and Kishino, K., 1983. *J. Lightwave Tech.* LT-1:161.
Sugimura, A. 1983a. *Appl. Phys. Lett.*, 43:728.
Sugimura, A. 1983b. *IEEE J. Quantum Electron.*, QE-19:923.
Sugimura, A. 1984. *IEEE J. Quantum Electron.*, QE-20:336.
Tai, K., Choa, F.S., Tsang, W.T., Chu, S.N.G., Wynn, J.D., and Sergent, A.M. 1991. *Electron. Lett.*, 27:1514.
Tai, K., Fisher, R.J., Seabury, C.W., Olsson, N.A., Huo, D.T.C., Ota, Y., and Cho, A.Y., 1989. *Appl. Phys. Lett.*, 55:2473.
Tai, K., Hasnain, G., Wynn, J.D., Fischer, R.J., Wang, Y.H., Weir, B., Gamelin, J., and Cho, A.Y. 1990a. *Electron. Lett.*, 26:1628.
Tai, K., Yang, L., Wang, Y.H., Wynn, J.D., and Cho, A. Y., 1990b. *Appl. Phy. Lett.*, 56:2496.
Tanbun-Ek, T., Logan, R.A., Chu, S.N.G., and Sergent, A.M. 1990. *Appl. Phys. Lett.*, 57:2184.
Temkin, H., Dutta, N.K., Tanbun-Ek, T., Logan, R.A., and Sergent, A.M. 1990. *Appl. Phys. Lett.*, 57:1610.
Temkin, H., Tanbun-Ek, T., and Logan, R.A. 1990b. *Appl. Phys. Lett.*, 56:1210.
Temkin, H., Tanbun-Ek, T., Logan, R.A., Coblentz, D.A., and Sergent, A.M. 1991. *IEEE Photonic Tech. Lett.*, 3:100.

Thijs, P.J.A., Tiemeijer, L.F., Kuindersma, P.I., Binsma, J.J.M., and van Dongen, T. 1991. *IEEE J. Quantum Electron.*, 27:1426.

Thijs, P.J.A. and van Dongen, T. 1989. *Electron. Lett.*, 25:1735.

Thompson, G.H.B. 1980. *Physics of Semiconductor Lasers*, Wiley, New York.

Tsang, W.T. 1981. *Appl. Phys. Lett.*, 39:786.

Tsang, W.T. 1986. *IEEE J. Quantum Electron.*, QE-20:119.

Tsang, W.T., Yang, L., Wu, M.C., Chen, Y.K., and Sergent, A.M. 1990. *Electron. Lett.*, 2033.

Udeo, O., Umebu, I., Yamakoshi, S., and Kotani, T. 1982. *J. Appl. Phys.*, 53:2991.

Uedo, O., Yamakoshi, S., Komiya, S., Akita, K., and Yamaoka, T. 1980. *Appl. Phys. Lett.*, 36:300.

Vakhshoori, D., Wynn, J.D., and Leibenguth, R.E. 1994. *Appl. Phys. Lett.*, 65:144.

Wu, M.C., Chen, Y.K., Hong, M., Mannaerts, J.P., Chin, M.A., and Sergent A.M. 1991. *Appl. Phys. Lett.*, 59:1046.

Yablonovitch, E. and Kane, E.O. 1986. *J. Lightwave Tech.*, LT-4:50.

Yamakoshi, S., Abe, M., Wada, O., Komiya, S., and Sakurai, T. 1981. *IEEE J. Quantum Electron.*, QE-17:167.

Yang, L., Wu, M.C., Tai, K., Tanbun-Ek, T., and Logan, R.A. 1990. *Appl. Phys. Lett.*, 56:889.

Yano, F., Yoshikuni, Y., Fukuda, M., and Yoshida, J. 1991. *Photonic Tech. Lett.*, 3:877.

Yariv, A., Lindsey, C., and Sivan, V. 1985. *J. Appl. Phys.*, 58:3669.

Further Information

Agrawal, G.P. and Dutta, N.K. 1993. *Semiconductor Lasers*, van Nostrand Reinhold, New York.

Bhattacharya, P. 1994. *Semiconductor Optoelectronic Devices*, Prentice–Hall, Englewood Cliffs, NJ.

57
Optical Transmitters

A.J. Price
Rockwell International

K.D. Pedrotti
Rockwell International

57.1 Introduction..774
57.2 Directly Modulated Laser Transmitters............................775
   Laser Diode Characteristics • Electronic Driver Circuit Design • Optoelectronic Package Design • Transmitter Performance Evaluation
57.3 Externally Modulated Optical Transmitters 782
   Characteristics of External Modulators • Modulator Driver Circuit Design

57.1 Introduction

In the last 15 years optical fiber transmission systems have progressed enormously in terms of information handling capacity and link distance. Some major advances in transmitter technology, both electronic and optoelectronic, have made this possible. The single longitudinal mode infrared emission from semiconductor distributed feedback (DFB) laser diodes can be coupled efficiently into todays low-loss silica single-mode fibers. In addition, these lasers can be switched on and off with transition times on the order of 100 ps to provide data rates up to approximately 10 Gb/s. The outstanding performance of these laser diodes has in fact provoked considerable effort to improve the switching speed of digital integrated circuits in order to fully utilize the capability of the optoelectronic components.

The performance of directly modulated laser transmitters greatly surpasses that of light emitting diode (LED) transmitters both in terms of optical power coupled into the fiber and switching speed. Short-distance (<10 km), low-capacity (<1 Gb/s) links use LED transmitters. For these applications, such as for local area networks, for example, LED transmitters can provide a cost-effective alternative to the laser.

Where high bit rates and a large regenerator spacing are required, optical pulse distortion occurring in the fiber between the transmitter and receiver is very important. Because of the advent of the optical amplifier, these requirements are now commonplace in terrestrial as well as submarine communication systems because amplifiers can cost effectively eliminate the loss of very long fiber links. With loss out of the way, distortion becomes a major issue. Phase/frequency shifts accompany the switching of semiconductor lasers. Commonly referred to as **chirp**, these shifts cause rapid deterioration of the pulse shape in dispersive optical fibers, thereby limiting the reach of directly modulated transmitters. For example, at a bit rate of 2.5 Gb/s, a distance of 200 km represents the limit of what is currently achievable using conventional single-mode fiber (assuming an operating wavelength in the 1550 nm low-loss window). Higher performance transmitters use

Optical Transmitters

external modulators to exceed these limits. External modulators exist which produce little or no switching phase shift.

The following sections include some information about the characteristics of laser diodes used in directly modulated optical transmitters and the electronic circuits to drive them. A brief description of packaging technology is also provided. A section is included on the performance characterization of high-speed digital lightwave transmitters. The use of external modulators introduces a new set of design requirements. These are also discussed.

Whereas digital transmission dominates the field of optical fiber transmission systems, analog systems are also important for the transmission of cable TV signals and microwave signals for remote antenna location and phase array radar systems, for example. The requirements for analog transmitters depend on the type of signal to be transmitted and are generally quite different from those of digital systems. Properties which are not very important in digital systems, such as linearity and signal-to-noise ratio, are very important in analog systems because, once degraded, the signal quality cannot be restored. More information about analog transmission is contained in Kolner and Dalfi [1987], Atlas [1996], and Linder et al. [1995].

57.2 Directly Modulated Laser Transmitters

The design of a state-of-the-art laser transmitter requires the combining of several different technologies. The laser diode is, of course, the critical component whose properties govern the transmitter performance. Its electrical and optical properties are both very important. The electrical driver circuit and the link between it and the laser are also critical design areas. Finally, the package design is a specialized combination of a microwave and optical enclosure.

Laser Diode Characteristics

The laser diode must launch optical pulses into the optical fiber at the required wavelength, with high spectral purity, with a high-**extinction ratio** and with a low-intensity noise. These properties have been discussed in a previous chapter. The optical format is typically **non-return to zero (NRZ)**, but **return-to-zero (RZ)** format is also used. The ideal optical signal will reproduce the data stream faithfully without **jitter**, ringing, or chirp. These cannot be eliminated entirely, and each contributes to the overall system performance.

A look at the **light, current (L/I) characteristic** of a DFB laser reveals a lot about its basic character (see Fig. 57.1). The light output is relatively small until the current reaches a certain value, called the **threshold current**. Thereafter the optical intensity rises approximately linearly with increasing current. For digital modulation, the current to the laser switches between two levels, the 0 level current being near the threshold current and the 1 level current being higher. The problems associated with typical laser diodes are that the curve bends over at high current and tends to shift and bend to the right with increasing temperature and as the laser ages. In addition, the characteristic often exhibits small kinks, which are

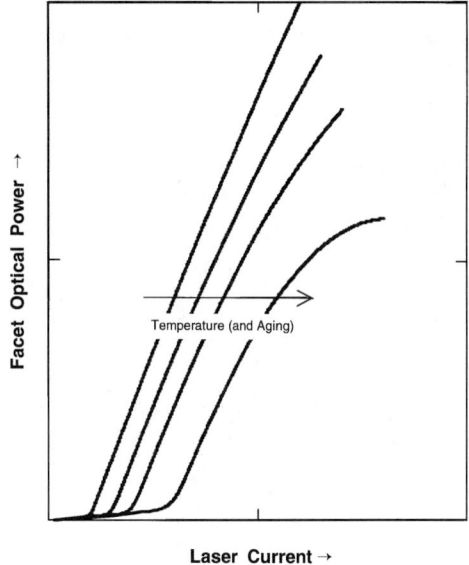

FIGURE 57.1 L/I (light vs current) characteristic of a DFB laser.

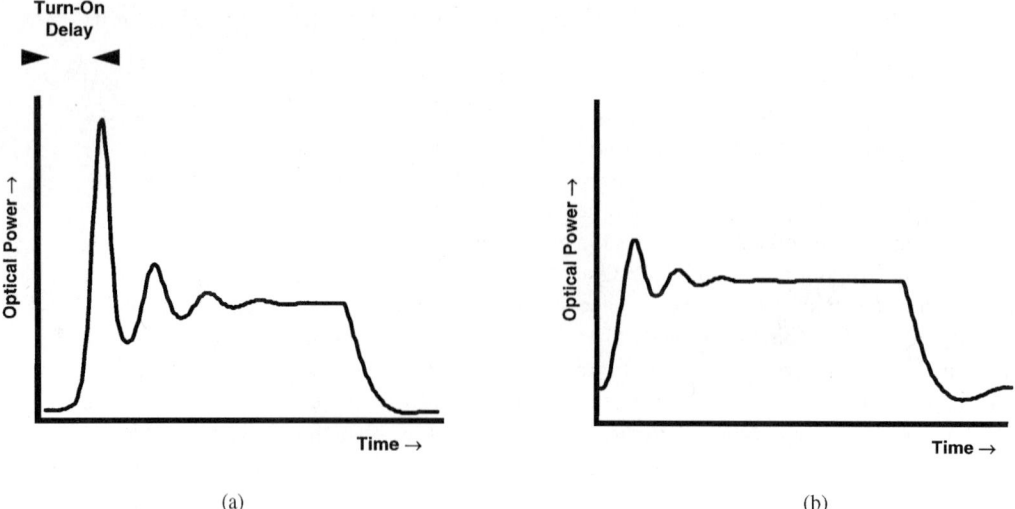

FIGURE 57.2 Switching waveforms of DFB lasers: (a) logic 0 bias below the threshold current and (b) logic 0 bias above the threshold current.

usually related to the level of optical power reflected back into the device [Twu et al., 1992]. Wavelength fluctuations accompany these kinks. In high bit rate systems an **optical isolator** prevents reflections from disturbing the laser. The changes due to temperature and time can, to a large extent, be countered using electronic feedback circuitry, as described in the next section. Often a temperature-controlled laser mount is also used. Keeping the temperature constant also reduces wavelength drift, and a low temperature extends the laser's lifetime.

Measurement of laser output while switching the input current between two levels indicates that the dynamic response of these diodes is not straightforward. When switching from a low optical power to a higher power, there is first a turn-on delay, which has a random component. A rapid increase in power follows, ringing at the **relaxation resonance** frequency. An unwanted frequency modulation accompanies the ringing. The turn-off is a smooth exponential decay, which is somewhat slower than the turn-on transition time. Typical waveforms for various current levels are shown in Fig. 57.2. The turn-on delay decreases as the off-state (bias) current is increased. For many digital applications, the bias is carefully set close to the laser threshold current to minimize the turn-on delay while at the same time providing a high-extinction ratio.

The dynamic behavior just described can be modeled using the **laser rate equations**. The single-mode rate equations are frequently presented as:

$$\frac{d}{dt}N = \frac{I_A}{q \cdot V_{act}} - \frac{N}{\tau_n} - g_0 \cdot (N - N_{0g}) \cdot (1 - \varepsilon \cdot S) \cdot S \qquad (57.1)$$

$$\frac{d}{dt}S = \left[\Gamma \cdot g_0 \cdot (N - N_{0g}) \cdot (1 - \varepsilon \cdot S) \frac{1}{\tau_p}\right] \cdot S + \Gamma \cdot \beta \cdot \frac{N}{\tau_n} \qquad (57.2)$$

Where:

N = electron density
q = electron charge
τ_n = electron lifetime
N_{0g} = electron density for zero gain

Optical Transmitters

S = photon density
τ_p = photon lifetime
I_A = current injected into active layer
V_{act} = volume of active layer
g_0 = differential gain constant
ε = gain compression factor
Γ = optical confinement factor
β = spontaneous emission factor

Computer programs intended for nonlinear circuit analysis, such as the Berkeley SPICE simulation program, can be used to solve these equations by defining a circuit to represent them [Tucker, 1985; Wedding, 1987]. The electron and photon densities are represented by circuit node voltages. The technique is powerful because it enables additional electronic circuitry to be included in the analysis. For example, the driver circuit, bond wires, and the laser electrical parasitics, as well as the rate equations, can all be modeled in the same analysis. This provides a versatile tool for analyzing the dynamic properties of laser transmitters.

Electronic Driver Circuit Design

A simple approach to driving a laser diode is to connect the output of a current driver circuit directly to the laser diode, as shown in Fig. 57.3. This approach is frequently adopted when the operating frequency is below 1 GHz. However, the inductance of the bond wires combined with the capacitance associated with the driver circuit and the laser can significantly degrade the risetime and cause ringing in the current pulses. This approach is difficult at higher frequencies because it might be impossible to reduce the inductance sufficiently. For narrow-bandwidth designs, sometimes a matching circuit can be used to cancel the effect of the inductance, but for very high-speed digital designs a different approach is required. Flip-chip laser mounting on top of the driver circuit [Pedrotti et al., 1992], and fully integrated lasers and drivers [Eriksson et al., 1996] are the subject of ongoing research, but typically some form of transmission line is needed between driver and laser. This leads to new requirements for the driver circuit. It is generally difficult to implement a characteristic transmission line impedance as low as the typical dynamic resistance of a laser (3–8 Ω) and so a technique frequently used is to increase its load resistance to a more manageable value using a closely mounted series resistor. The series resistor increases the power dissipation of the resultant load for a given modulation current and also the required voltage swing. The former has adverse consequences related to laser thermal effects, and the latter affects the driver circuit requirements. Low impedances are normally used (in the range 12–25 Ω), although, when the driver circuit is completely separated from the laser package, the standard 50-Ω transmission line impedance is frequently used for convenience, allowing the use of standard coaxial connectors (such as sub-miniature A (SMA) type) and commercial amplifiers to drive the laser. Testing is also simplified because most high-frequency test equipment uses 50-Ω interconnects as standard. An example of a current switch driver circuit with transmission line link to the laser is shown in Fig. 57.4. The modulation current is equal to the current in the tail of the differential output pair, which is controlled by the voltage at the *Vgmod* input. The laser bias current (the logic 0 current) is determined by the current in the separate bias source. One example of a high-speed GaAs hetero-junction bipolar transistor (HBT) laser driver circuit is that reported

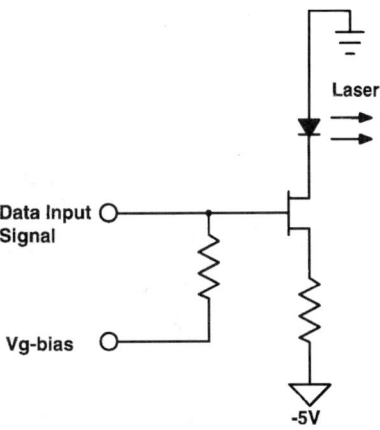

FIGURE 57.3 FET laser driver directly connected to laser diode.

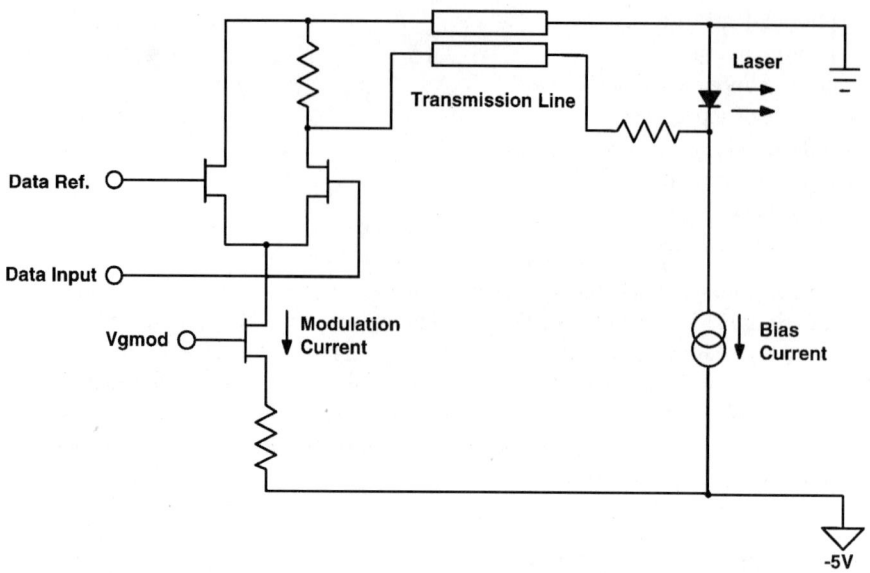

FIGURE 57.4 Current switch laser driver with transmission line coupling to laser diode.

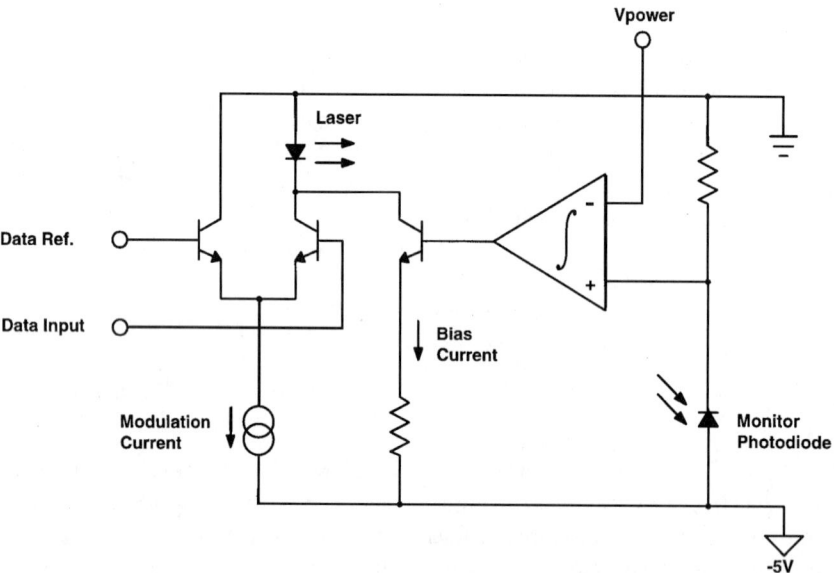

FIGURE 57.5 Average power control circuit.

in Runge et al. [1992], which is a monolithic design with several stages of current amplification for the laser drive. This circuit includes a Cherry–Hooper amplifier stage driving a cascaded differential pair output driver.

Typically, various feedback loops are used in practical laser transmitter designs to compensate for the behavior of laser diodes when influenced by variations in data rate, temperature, and aging effects. The simplest average power control circuit is shown in Fig. 57.5. The laser is connected in the collector lead of a typical differential pair driver stage. The data signal is applied to one side of the differential pair and a reference voltage is applied to the other. In addition, the drive transistor is connected in parallel with another transistor, which serves as a bias current source. This current

Optical Transmitters

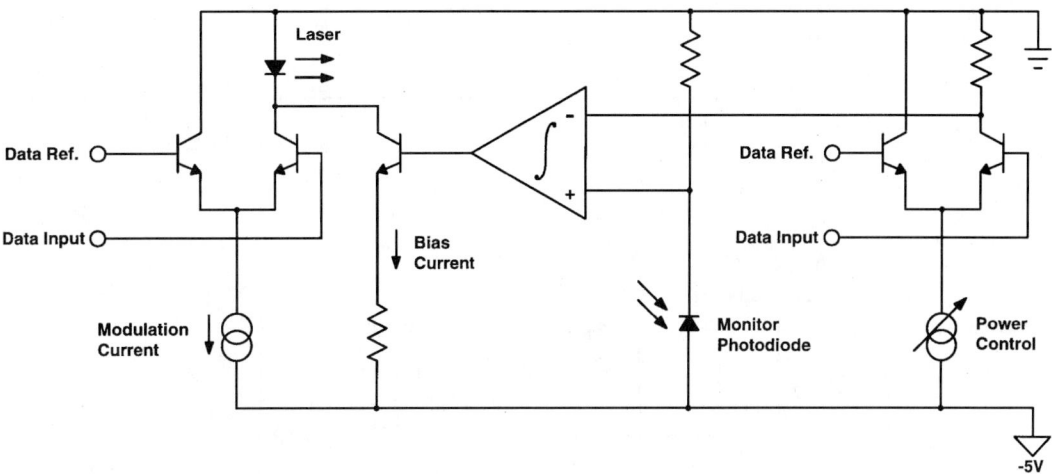

FIGURE 57.6 Average power control with mark density compensation.

source is controlled by the output of a monitor photodiode. A portion of the output power from the laser, typically taken from the laser's rear facet, is incident on this detector. The circuit keeps the 0 and 1 levels constant with variations in threshold current as long as the slope of laser's L/I characteristic does not change.

The average number of 1s relative to the number of 0s, the mark density, is linearly related to the average power. If the mark density of the data changes, then the laser bias point will shift using the circuit of Fig. 57.5. In the extreme case of a long string of zeros, the average power feedback circuit will continue to increase this zero level up to the setpoint for the average power. To alleviate this problem a circuit similar to that shown in Fig. 57.6 can be used. The V_{power} reference input of the average power feedback circuit is replaced with a signal proportional to the mark density. Now in the case of a long string of zeros, the reference voltage increases, causing the laser bias current to decrease, maintaining the correct bias current to the laser. This approach now compensates for threshold current shifts and mark density variation but not for changes in the slope of the L/I curve.

An approach that allows slope correction is given in Smith [1978]. To detect whether the zero level is near threshold, the amplitude of the current is modulated at a low frequency when the data input is zero. By detecting the slow ripple from the photodiode output at this modulation frequency, a signal whose amplitude is sensitive to the L/I curve slope at the logical zero level is obtained. Since the slope changes abruptly around the threshold current, the logical zero can be maintained close to the threshold current using a feedback circuit. When used in conjunction with an average power controller, a constant amplitude high-extinction ratio optical signal can be maintained for changes in both threshold current and slope.

For high bit rate transmitters, sometimes the 0 level current must be set significantly above the threshold current in order to reduce turn-on delay, ringing, and chirp. The resulting extinction ratio penalty is often a small price to pay for the improved system performance when chromatic dispersion is present. In such circumstances the circuit shown in Fig. 57.7 may be used [Chen, 1980]. Here, in addition to average power control, the AC component of the detected current is fed back to the modulation control of the laser driver. By keeping this constant, slope variations are compensated, and the extinction ratio is held approximately constant. Mark density compensation can be incorporated, as shown in Fig. 57.6. A disadvantage of the circuit shown is that a broadband monitor photodiode might be needed. This can be avoided by adding to the bias current a small low-frequency tone with an amplitude proportional to the modulation current. This can be detected accurately with low-frequency circuitry to obtain a measure of the optical modulation amplitude.

Various other methods have been investigated for use in the stabilization of laser output. Among these are encoding to ensure a suitable balance of 1s and 0s in the circuit, encoding both clock and

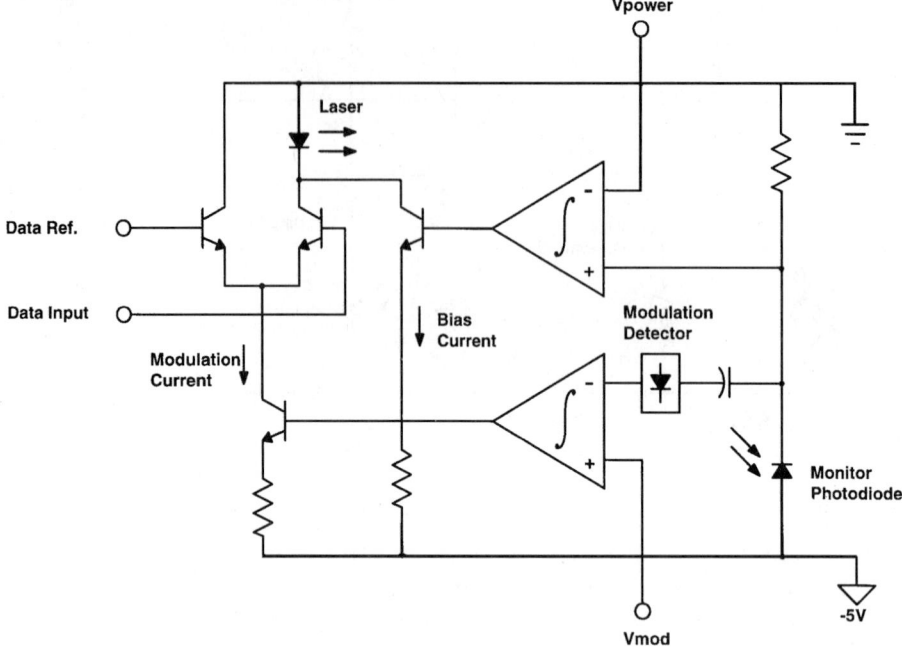

FIGURE 57.7 Average power and modulation amplitude control circuit.

data, and using the clock component to adjust the laser drive condition. Albanese [1978] developed a method to monitor the laser threshold without the use of a monitor photodiode by monitoring the change in laser electrical characteristics near threshold.

Optoelectronic Package Design

An illustrative example of a laser diode package is shown in Fig. 57.8. In this particular example, the laser diode is mounted on a diamond heat spreader, which is mounted on a ceramic carrier. Lasers such as these, which may be purchased from many vendors, allow ease of handling, testing, and replacement without handling the device itself. Laser and carrier are mounted on a **thermoelectric**

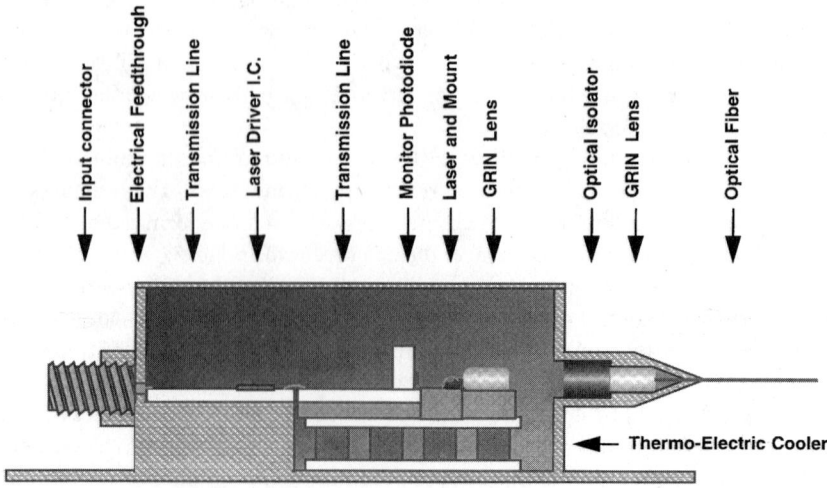

FIGURE 57.8 Laser package.

Optical Transmitters

cooler with a thermistor to monitor the temperature and a large-area optical detector, which is used to monitor the optical power by intercepting the backward wave emission of the laser. It is generally preferred to limit heat generating components mounted on the cooler to a minimum to keep down the cooling requirements. Frequently, the temperature controller will be set to keep the temperature near 25°C or a little higher in order to maximize the range over which stabilization can be maintained (the cooler can be used to heat as well as cool). The electrical link from the driver circuit is provided by a transmission line, as shown in the diagram. This critical link is designed carefully taking into account dimensional tolerances in order to repeatedly obtain good performance. To couple power into the optical fiber, a graded index (GRIN) or spherical lens is mounted close to the front facet of the laser to produce a collimated beam. The beam passes through a small optical isolator and is focused into the fiber using a second lens. An angled optical arrangement is used to reduce reflections. The alignment and stability of this arrangement is critical to achieve a low coupling loss of about 3 dB. For applications that do not require the use of an isolator to reduce reflections, adequate coupling efficiency may be obtained by directly coupling from the laser into the fiber. The end of the fiber is normally tapered, and the tip is formed into a convex lens to match the laser light into the fiber.

Transmitter Performance Evaluation

The methods of evaluating digital optical transmitters and systems have to a certain extent been standardized in recent years due to the introduction of synchronous optical network (SONET) and synchronous digital hierarchy (SDH) optical systems standards [Bell, 1994]. One of the purposes of these standards is to specify a set of requirements to allow interoperability of equipment from different manufacturers. An additional objective is that transmitters from one manufacturer operate with receivers made by another manufacturer. Clearly, to achieve this their characteristics must be sufficiently well specified.

Evaluating the performance of an optical transmitter requires certain measurements of the optical signal produced. Using a broadband optical detector, an electrical signal proportional to the optical intensity is obtained, allowing the intensity modulation to be examined using standard electrical measurement techniques. The signal is filtered using a low-pass filter with well-defined characteristics to provide a standard receiver response. For example, a fourth-order Bessel–Thompson filter with a 3-dB bandwidth equal to 0.75 of the bit rate is specified for testing SONET transmitters. Using an oscilloscope the signal can then be displayed in the form of an **eye pattern**. The signal shape must meet certain criteria defined by an eye pattern mask (see Fig. 57.9). Measurement

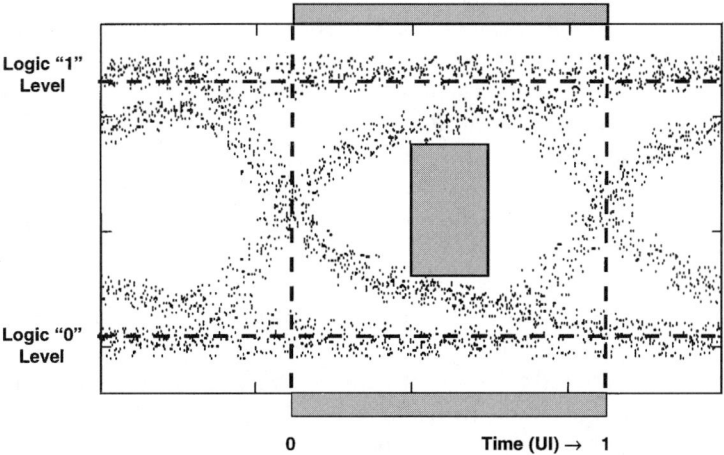

FIGURE 57.9 Eye pattern mask test.

software overlays these masks on the oscilloscope screen, measures violations of the limits specified by the masks, and also measures the other characteristics such as extinction ratio and jitter.

Eye pattern measurements of the transmitter output signal do not give any indication of phase or frequency modulation of the optical signal. These, however, are critically important for long-distance transmission through dispersive fiber. The spectral width is one indicator which is helpful here. A wide spectrum is very susceptible to the effects of chromatic dispersion in the transmission medium. For this reason, a specification is imposed on the spectral width of the transmitted signal. In addition, degradation of the signal must be tested by passing it through a length of optical fiber. Typically, the signal power required at a reference optical receiver input, for example, defined by a 10^{-10} error rate, when the signal is passed through a given length of optical fiber is compared with the power required when the transmitted signal is passed through a very short length of fiber. The optical power going into the receiver is adjusted using an optical attenuator. The measured power ratio is usually expressed in decibels of **dispersion power penalty** for a given chromatic dispersion. A small value is indicative of a high-performance, low-chirp transmitter.

57.3 Externally Modulated Optical Transmitters

Absorptive modulators and interferometric types are considered here. Because the former type can be made of the same materials used in the fabrication of lasers, the popularity of these lies in the possibility of integrating them with the laser diode, to save cost [Watanabe, Sato, and Soda, 1992]. The most commonly used interferometric type, the **Mach–Zehnder**, normally uses the electro-optic properties of lithium niobate to provide very precise control over the amplitude and phase of the signal, for high-performance applications.

Characteristics of External Modulators

Electroabsorption (EA) modulators are monolithic semiconductor devices, usually indium phosphide (InP) based. They rely on either the Franz–Keldysh effect (FKE) or the quantum-confined Stark effect (QCSE) and have optical waveguide p-i-n structures grown epitaxially typically using the InGaAsP-InP or InGaAs-InAlAs material systems [Chin, 1995; Kawano et al., 1992]. Compatibility with laser structures makes the integration of laser and modulator possible and very attractive. The advantages of using these modulators compared with direct laser modulation are that the chirp associated with the laser relaxation resonance is avoided (although there is some inherent chirp in the modulators [Koyama and Iga, 1988]), the input impedance is high, and these modulators are capable of very high-speed modulation.

A comprehensive comparison of the relative merits of different types of EA modulators is beyond the scope of this text. Some of the factors that affect their suitability in different applications are drive level requirements, extinction ratio, bandwidth, operating wavelength range, chirp, temperature stability, and, for analog systems, linearity.

Electrically, EA modulators are diodes that operate in reverse bias. The absorption is dependent on the applied voltage. For high-frequency applications, the capacitance of the device must be considered. A sufficiently low-driving impedance is used so that the low-pass pole caused by the capacitor does not limit the bandwidth too much. A transmission line link to the drive circuit will frequently be required. In that case the line could be terminated using a resistor connected in parallel with the modulator at the end of the transmission line to prevent reflections. For bandpass applications microwave matching techniques are used to improve the modulation efficiency. The bandwidth can also be improved by incorporating the device capacitance into a lumped transmission line structure. The electrical techniques appropriate to the use of these devices are those that also apply to many other high-impedance devices such as FETs and varactor diodes.

Mach–Zehnder modulators operate using interference by splitting the input into two paths, which are later recombined (see Fig. 57.10). Using lithium niobate or another electro-optic material for

Optical Transmitters

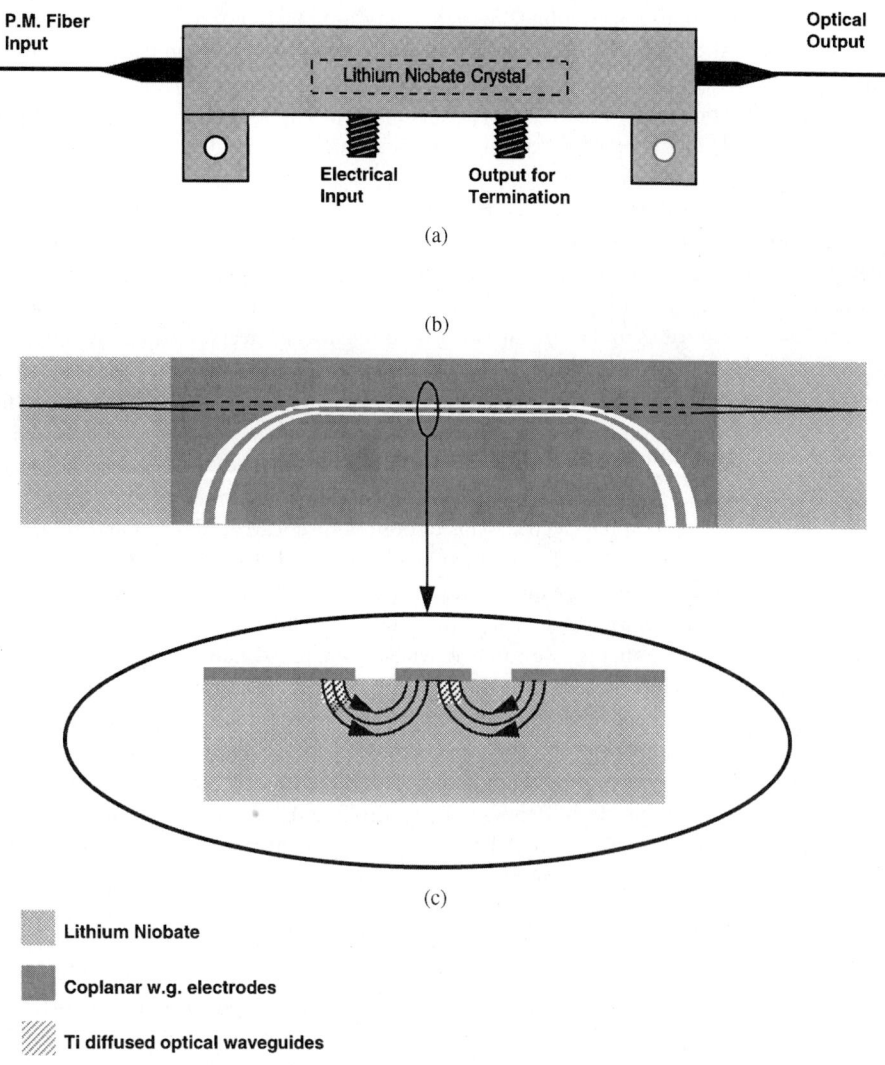

FIGURE 57.10 Mach–Zehnder modulators: (a) package outline, (b) lithium niobate crystal, and (c) detail.

the optical waveguides, the phase delay is dependent on the electric field applied to the waveguides. Phase modulators can therefore be incorporated in one or both of the split paths allowing electrical control of the phase of the two signals. Depending on their relative phases when they are recombined, constructive or destructive interference occurs, resulting in amplitude modulation. The phase modulator electrodes are usually formed into transmission lines following the path of the optical waveguides. In this way, over a certain path length, an applied electrical pulse will continue to modulate an optical signal as both waves progress along their respective waveguides. If the velocities are not well matched, which is usually the case, the optical and electrical waves will not track each other over a long distance and the high-frequency response is affected. For a given required bandwidth the length of interaction must be limited, which in turn limits the phase shift induced by a given applied voltage. Typically, therefore, high-bandwidth designs require relatively high-drive voltages compared with lower bandwidth designs. The drive voltage requirements are usually denoted by a value for $V\pi$, the change in voltage required to produce π-rad phase shift between the two paths. This corresponds to the voltage needed to go from maximum to minimum amplitude. Typical values are in the order of 10 V for a device with a 10-GHz modulation bandwidth, though this is continually being improved. The extinction ratio for these devices, the ratio of maximum to minimum value expressed in decibels, is dependent on the amplitude balance in the two paths, as well as the level of other unwanted transmission modes through the device. Typical values are >20 dB.

In modulators for optical transmitters the path lengths may be several centimeters, equivalent to many wavelengths of the light passing through them. Although great care is taken to maintain device symmetry, the path lengths are not generally identical, and so the required voltage for maximum amplitude transmission is usually some random value. In addition, temperature gradients, stresses, and the presence of impurities in the optical waveguides all have an effect on the relative phase. For these reasons, care is taken in the packaging and package mounting to avoid mechanical stresses. The terminating resistors for the electrical transmission lines are usually isolated from the lithium niobate crystal to prevent heating. Measures are taken to minimize the absorption and subsequent diffusion of impurities in order to reduce long term drifts in the phase difference.

Various transmission line electrode arrangements are possible. Commonly, a single 50-Ω coplanar waveguide is designed to apply electric fields to both optical waveguides simultaneously such that one optical path is retarded while the other is advanced. If the phase shifts are equal but opposite in the two paths, the resultant amplitude modulation is free of phase modulation (chirp); whereas if only one path is phase modulated, the resultant modulation contains both amplitude and phase modulation components. Researchers have discovered [Gnauch et al., 1991] that under certain conditions of chromatic dispersion, transmission distances are actually enhanced if a small amount of phase modulation accompanies the amplitude modulation by comparison with a completely chirp-free modulation, so now some commercial modulators provide this optimum amount of chirp. Another type of modulator simply provides two separate electrical inputs, each one controlling one arm of the Mach–Zehnder. By controlling the relative amplitude and sign of the two inputs, the amount and direction of the accompanying chirp can be precisely controlled. In addition, a differential output driver circuit need supply only half the voltage swing to each input compared with single-ended drive.

Chirp control is a very important issue when high bit rates are combined with long-distance transmission through dispersive fiber. Another approach to phase control is to add a phase modulator in cascade with the Mach–Zehnder. It can, of course, be integrated with the Mach–Zehnder by extending it and incorporating another electrode on the input or output waveguide. The drive waveforms must then be coordinated between the Mach–Zehnder and the phase modulator to obtain the required simultaneous phase and amplitude modulation. Allowance must be made for the optical delay between the two modulators, and delay must be provided in the drive signals to compensate.

Whichever electrode configuration is used, the output intensity vs input voltage is usually the same shape for all Mach–Zehnder modulators. It is shown in Fig. 57.11, and is sinusoidal in character. For digital modulation, the 1 level is aligned with a peak and the 0 level is aligned with an adjacent trough. Restated, the modulator is biased to a half-way point and a modulation swing equal to $V\pi$ (peak-to-peak) is applied. If the modulator has chirp, the sign of the chirp will depend on the sign of the slope, and so the correct side of the peak must be used. Some advantages of the sinusoidal characteristic are readily apparent. First, because the transfer function is flat at both the peaks and the troughs, noise or

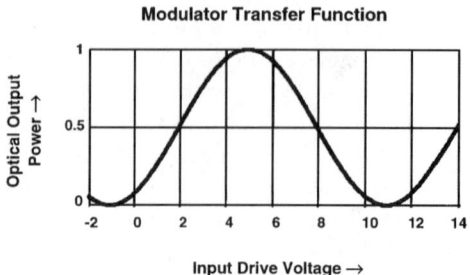

FIGURE 57.11 Electrical/optical (E/O) transfer function of Mach–Zehnder modulator.

Optical Transmitters 785

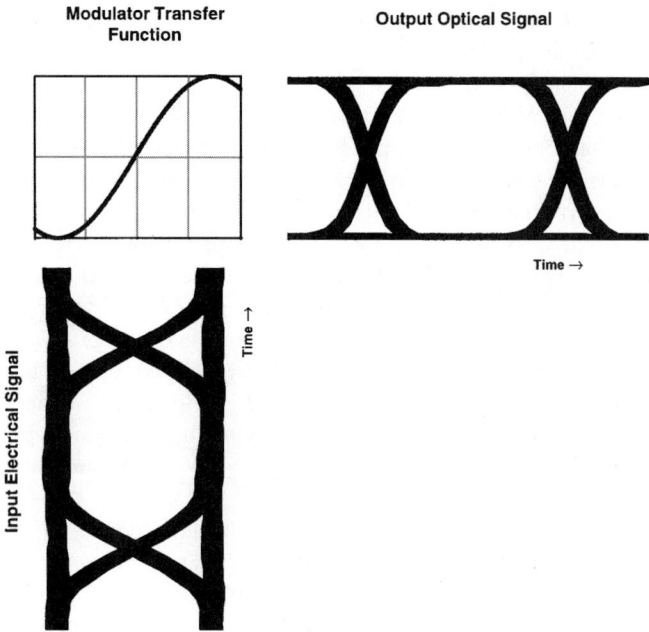

FIGURE 57.12 Effect of E/O transfer function on optical waveform.

spreading of the electrical logic levels will have little effect on the optical intensity levels. Second, the gradient of the transfer function is steepest at the bias point. Provided the bandwidth of the modulator is sufficiently high, this can result in a reduction in the output optical transition times compared with the electrical transition times of the drive waveform. The two factors together give rise to an optical output waveform, which is superior in many ways to the electrical waveform used to generate it. This is in marked contrast with the case of direct laser modulation, where the output optical waveform not only contains the imperfections of the drive waveform, but also includes some added aberrations caused by the laser itself. The waveform improvement is demonstrated in Fig. 57.12.

For most intensity modulation (IM) transmitters the bias point is at the half-maximum power point, corresponding to phase quadrature between the two paths. The intensity vs drive voltage characteristic is symmetrical, and the curve is most linear at this point. For most direct detection systems (those in which the received optical power is directly converted to a proportional electrical signal) this is the preferred operating point. However, for some applications, the modulator can be biased at the minimum amplitude point. The modulator then behaves very like the balanced mixers used, for example, as upconverters in microwave radio systems. An electrical signal applied to the modulator will be reproduced at optical frequencies in two sidebands on either side of the (suppressed) optical carrier. This means that many of the techniques used to generate radio signals can also be applied to the generation of optical signals. The upconversion is quite linear (sinusoidal in fact) so that the electrical signal properties, both time domain and frequency domain, are preserved in the conversion. If the electrical signal is filtered to reduce its electrical bandwidth, the optical sidebands will have the same bandwidth. Two references provide examples of such a use, Izutsu et al. [1981] and Price and Le Mercier [1995].

Modulator Driver Circuit Design

A common feature of EA modulators and Mach–Zehnder modulators is that electrically, by contrast with laser diodes, they present a high impedance to the driving circuit. A low-output impedance circuit, such as an emitter follower, is therefore appropriate for situations where the driver is connected directly to the modulator. However, as is true for laser driver circuits, the requirements change somewhat when the driver must be separated from the modulator, or when the bond wire inductance becomes a significant factor. Then resonances must be dampened using an appropriate driving resistance or load resistor. In the case of a transmission line link between driver and modulator, or when the modulator uses transmission line electrodes, it is essential to either resistively terminate the line, drive the line using a resistive source, or do both. Again, a 50-Ω transmission line is frequently chosen for convenience and standardization.

When transmission lines are used, the main driver amplifier requirements for both laser drivers and modulator drivers become very similar: they must generate electrical pulses of the required power in the transmission line. The current switch topology of Fig. 57.4 is appropriate for modulators as well as lasers. The main difference is that the required voltage swings are significantly greater for modulators, especially existing Mach–Zehnder designs, compared with laser drivers. For this reason, relatively high-power broadband 50-Ω amplifiers are frequently used to drive modulators. Power output requirements between 20 and 30 dB (1 mW) are typical. Hybrid GaAs metal-semiconductor field-effect transister (MESFET) designs are used. Integrated circuit solutions include distributed amplifiers.

Automatic control algorithms for Mach–Zehnder modulators contrast those of other optical transmitters because the transfer characteristic (see Fig. 57.11) is quite different and there exists the potential for multiple operating points and an unlimited lateral drift due to temperature and aging. Typically, some optical power is tapped from the output of the modulator and detected in a low-bandwidth optical detector. One method to automatically control the bias point of the modulator is to apply a low-frequency modulation to the gain of the driver amplifier. A signal will be generated in the detector if the bias point is not at the quadrature point as intended. By comparing the phase of the detected signal with the original low-frequency modulation, using a synchronous detector, an error signal can be obtained, which is used to force the bias point to the quadrature point. Inverting the sense of the feedback loop forces the bias point to move to the adjacent quadrature point of opposite slope, causing an inversion of the optical data and the chirp. Methods also exist to optimize the modulation amplitude in order to maintain a good optical output waveform. A problem with some Mach–Zehnder modulators is that over a period of time the phase shift might continue to drift over many cycles of phase. An automatic bias circuit is not able to continue to apply higher and higher voltages to track this drift, so a warning circuit is sometimes included to indicate that the voltage might soon reach a limit, after which it would fail to track. At a convenient time, the circuit is reset by remote control before failure occurs.

Defining Terms

Chirp: A frequency or phase shift associated with the switching on or off of a laser diode or modulator.

Dispersion power penalty: The power penalty in decibels associated with the increase in receiver optical power level required to achieve a specified error rate, for example, 10^{-10}, when the signal is subjected to a specified amount of chromatic dispersion.

Electroabsorptive (EA) modulator: Semiconductor modulators using the Franz–Keldysh effect (FKE) or the quantum-confined Stark effect (QCSE).

Extinction ratio: The ratio of maximum to minimum insertion loss for an amplitude modulator. The term also applies to the ratio (in decibels) of the average optical energy in a logic 1 level to the average optical energy in a logic 0 level of a digitally modulated optical signal.

Eye pattern: Picture built up by overlaying traces corresponding to all possible sequences of a digital waveform. (See Fig. 57.9.)

Jitter: Random and systematic deviations of the arrival times of pulses or transitions in digitally transmitted data. Observed as a horizontal spreading of an eye pattern displayed on an oscilloscope. (See Fig. 57.12.)

Laser rate equations: Set of equations representing the dynamic behavior of electron and photon population densities in a laser.

Laser threshold current: Current threshold at which the gain inside the laser cavity exceeds the losses. Below the threshold current light emission is spontaneous. Above the threshold current stimulated emission also occurs.

L/I characteristic: Graph of light output vs injected current, which exhibits a characteristic knee at the laser threshold current (see Fig. 57.1).

Mach–Zehnder modulator: A two-path optical interferometer that uses phase shifters in one or both paths to provide amplitude modulation.

Nonreturn to zero (NRZ): Digital format in which the logic 1 pulses occupy the whole bit period (cf. return to zero).

Optical isolator: A device that utilizes the Faraday effect to allow transmission through the device in one direction only. These devices are used to isolate lasers from the effects of reflections.

Relaxation resonance: Internal resonance between electrons and photons in a laser diode.

Return to zero (RZ): Digital format in which the logic 1 pulses occupy a fraction (usually 50%) of the bit period.

Thermoelectric (TE) cooler: Device utilizing the Peltier effect to pump heat from one surface to the other. Reversing the direction of current flow reverses the direction of heat flow, so that the same device may be used to heat or cool as necessary.

Vπ: When referred to a phase modulator, it is the voltage change required to give a phase shift of π rad (180°). For a Mach–Zehnder modulator this corresponds to the voltage required to change from maximum to minimum extinction.

References

Albanese, A. 1978. An automatic bias control (ABC) circuit for injection lasers. *Bell Syst. Tech. J.*, 57:1533–1544.

Atlas, D.A. 1996. On the overmodulation limit in externally modulated lightwave AM-VSB CATV systems. *IEEE Phot. Tech. Lett.*, 8(5):697–699.

Bell. 1994. Physical layer. In *Synchronous Optical Network (SONET) Transport Systems: Common Generic Criteria*, Chap. 4, Issue 1, GR-253-CORE, Bell Communications Research.

Chen, F.S. 1980. Simultaneous feedback control of bias and modulation currents for injection lasers. *Elect. Lett.*, 16(1):7–8.

Chin, M.K. 1995. Comparative analysis of the performance limits of Franz–Keldysh effect and quantum-confined Stark effect electroabsorption waveguide modulators. *IEE Proc. Optoelectron.*, 142(2):109–114.

Eriksson, U., Evaldsson, P., Donegan, J.F., Jordan, C., Hegarty, J., Hiei, F., and Ishibashi, A. 1996. Vertical integration of an InGaAs/InP HBT and a 1.55 um strained MQW p-substrate laser. *IEE Proc. Optoelectron.*, 143(1):107–109.

Gnauck, A.H., Korotky, S.K., Veselka, J.J., Nagel, J., Kemmerer, C.T., Minford, W.J., and Moser, D.T. 1991. Dispersion penalty reduction using an optical modulator with adjustable chirp. *Elec. Lett.*, 28:954–955.

Izutsu, M. et al. 1981. Integrated optical SSB modulator/frequency shifter. *IEEE J. Quant. Electron.*, QE-17(11):2225.

Kawano, K., Wakita, K., Mitomi, O., Kotaka, I., and Naganuma, M. 1992. Design of InGaAs-InAlAs multiple-quantum-well (MQW) optical modulators. *IEEE J. Quant. Electron.*, 28(1):224–230.

Koyama, F. and Iga, K. 1988. Frequency chirping in external modulators. *J. Lightwave Tech.*, 6:87–93.

Kolner, B.H. and Dolfi, D.W. 1987. Intermodulation distortion and compression in an integrated electrooptic modulator. *Appl. Optics*, 26(17):3676–3680.

Linder, N., Kiesel, P., Kneissl, M., Knuepfer, B., Quassowski, S., Doehler, G.H., and Traenkle, G. 1995. Linearity of double heterostructure electroabsorptive waveguide modulators. *IEEE J. Quant. Electron.*, 31(9):1674–1681.

Pedrotti, K.D., Seabury, C.W., Sheng, N.H., Lee, C.P., Agarwal, R., Chen, A.D.M., and Renner, D. 1992. 6-GHz operation of a flip-chip mounted 1.3-um laser diode on an AlGaAs/GaAs HBT laser driver circuit. In *OFC' 92*, ThJ7, p. 241.

Price, A.J. and Le Mercier, N. 1995. Reduced bandwidth optical digital intensity modulation with improved chromatic dispersion tolerance. *Elec. Lett.*, 31(1):58–59.

Runge, K., Detlef, D., Standley, R.D., Gimlett, J.L., Nubling, R.B., Peterson, R.L., Beccue, S.M., Wang, K.C., Sheng, N.H., Chang, M.F., Chen, D.M., and Asbeck, P.M. 1992. AlGaAs/GaAs HBT IC's for high-speed lightwave transmission systems. *EEE JSSC*, 27(10):1332–1341.

Smith, D.W. 1978. Laser level-control circuit for high-bit-rate systems using a slope detector. *Elec. Lett.*, 14(24):775–776.

Tucker, R.S. 1985. High-speed modulation of semiconductor lasers, *J. Lightwave Tech.*, LT-3(6):1180–1192.

Twu, Y., Parayanthal, P., Dean, B.A., and Hartman, R.L. 1992. Studies of reflection effects on device characteristics and system performances of 1.5 um semiconductor DFB lasers. *J. Lightwave Tech.*, 10(9):1267–1271.

Watanabe, T., Sato, K., and Soda, H. 1992. Low drive voltage and low chirp modulator integrated DFB laser light source for multiple-gigabit systems. *Fuj. Sci. Tech. J.*, 28(1):115–121.

Wedding, B. 1987. SPICE simulation of laser diode modules. *Elect. Lett.*, 23(8):383–384.

Further Information

Alferness, R.C. 1982. Waveguide electrooptic modulators. *IEEE Trans. Microwave Th. and Tech.*, MTT-30:1121–1137.

Gu, X. and Blank, L.C. 1993. 10 Gbit/s unrepeatered three-level optical transmission over 100 km of standard fiber. *Elec. Lett.*, 29(25):2209-2211.

Henry, P.S. 1985. Lightwave primer. *IEEE J. Quant. Electron.*, QE-21:1862–1879.

Ortel. 1992. *RF/Microwave Fiber Optic Link Design Guide*, Ortel Corp.

Patel, B.L. et al. 1992. Transmission at 10 Gb/s over 100 km using a high performance electroabsorption modulator and the direct prechirping technique. In *ECOC' 92*, postdeadline paper Th PD 1.3, Berlin, Germany.

Pedrotti, K.D., Zucca, F., Zampardi, P.J., Nary, S.M., Beccue, S.M. Runge, K., Meeker, D., Penny, J., and Wang, K.C. 1995. HBT transmitter and data regenerator arrays for WDM optical communications application. *IEEE J. Solid-State Circs.*, 30(10):1141–1144.

Shumate, P.W. 1988. Lightwave transmitters. In *Optical Fiber Telecommunications II*, ed. S.E. Miller and I.P. Kaminow, Chap. 19, 723–757. Academic Press.

Wedding, B., Franz, B., and Junginger, B. 1994. 10-Gb/s optical transmission up to 253 km via standard single-mode fiber using the method of dispersion-supported transmission. *J. Lightwave Tech.*, 12(10):1720–1727.

58
Optical Receivers

R.G. Smith
AT&T Bell Laboratories (retired)

B.L. Kasper
AT&T Bell Laboratories

58.1 Introduction 789
   Goal
58.2 The Receiver 789
   The Photodetector • Preamplifier • Postamplifier • Filtering • Timing Recovery • Decision Circuit
58.3 Receiver Sensitivity: General 793
   Signal Current • Sources of Noise

58.1 Introduction

The optical receiver is an integral part of an optical link. Its function is to detect the received optical power; amplify the signal; retime, regenerate, and reshape the signal where it may remain as an electrical signal (terminal); or be used to drive an optical transmitter which then feeds another fiber section.

Goal

The goal of the optical receiver is to achieve the desired system performance while requiring the minimum amount of received optical power. System performance criteria include bit error ratio (BER) and eye margin for digital systems, signal-to-noise ratio (SNR) for analog systems, and dynamic range (ratio of largest to smallest signal levels for which the BER or SNR can be achieved).

Over the past 20 years there has been a large amount of work on optical receivers. This work is well covered by review papers and book chapters [Kasper, 1988; Muoi, 1984; Smith and Personick, 1982]. Although the field is fairly mature there are still advances being made, such as the use of *optical preamplifiers* to improve receiver sensitivity [Olsson, 1989].

58.2 The Receiver

A block diagram of the optical receiver is shown in Fig. 58.1. The receiver consists of a photodetector, a preamplifier, postamplifier, filter, timing recovery, and decision circuit.

The Photodetector

The purpose of the photodetector is to convert the received optical signal into a photocurrent. It is essentially a photon-to-electron converter. The most widely used forms of photodetectors used in optical communications are the p-i-n photodiode and the avalanche photodiode (APD). In the case of the p-i-n a single electron-hole pair is generated for each photon absorbed by the detector.

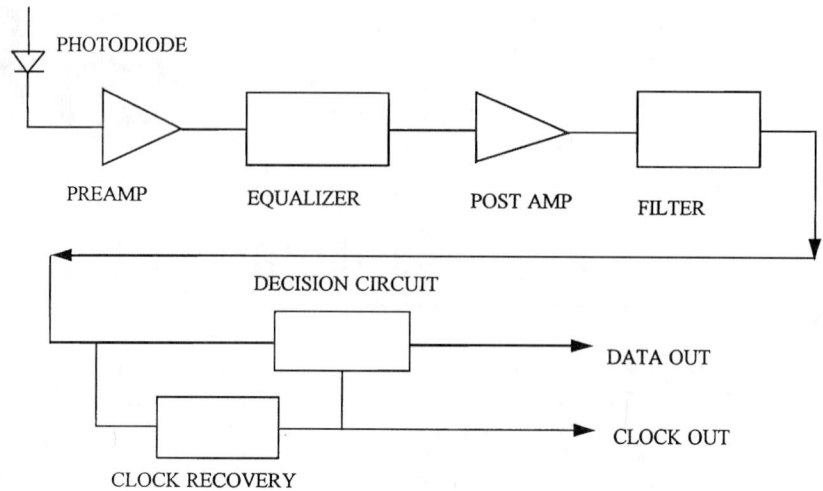

FIGURE 58.1 Schematic diagram of an optical receiver.

In the case of the APD, electronic gain within the detector itself generates secondary electron-hole pairs, multiplying the photocurrent, and generally improving the receiver sensitivity. (See chapter on Optical Detectors for a discussion of photodetectors.)

Preamplifier

The preamplifier or front end is the most important element of the receiver in determining the optical sensitivity. Its principal function is to amplify the generally small photocurrent with the least amount of added noise. Under most circumstances (especially with a p-i-n detector) the preamplifier noise is dominant in determining the sensitivity. The preamplifier must also have sufficient frequency response so as not to severely distort the received signal shape (*eye margin*). In addition to noise level and frequency response, the front end must be able to respond to a wide range of received optical power levels without severely distorting the signal (dynamic range). Often these objectives are in conflict and compromises must be made.

Optimized optical receivers differ in several ways from receivers used at RF or microwave frequencies. This difference arises from the fact that the photodetectors generally used (p-i-n, APD) are current sources, and therefore inherently high impedance. Optimal signal generation therefore requires the front end to have a high impedance, which differs from other types of receivers which are typically driven by low-impedance detectors.

A second major difference between optical systems derives from the relatively large energy of a photon. For a given received power, there are orders of magnitude fewer photons at optical frequencies than at microwave or lower frequencies. In the limit of low received optical power, and hence low number of received photons and photogenerated hole-electron pairs, the random arrival rate results in signal related shot noise, which is often important in determining receiver sensitivity.

Optical receiver front ends can be broadly classed into one of three types: high-impedance or *integrating* front end; *transimpedance* front end; and low-impedance front end, typically a 50-Ω impedance level. Although the distinction is not hard and fast, these categories characterize the various approaches to receiver design.

The high-impedance front end achieves the best sensitivity for a given detector. This design approach minimizes the thermal noise generated by resistors, given by $4kT/R$, by increasing their values such that the noise is reduced to a level where it is small compared to other noise sources. As

Optical Receivers

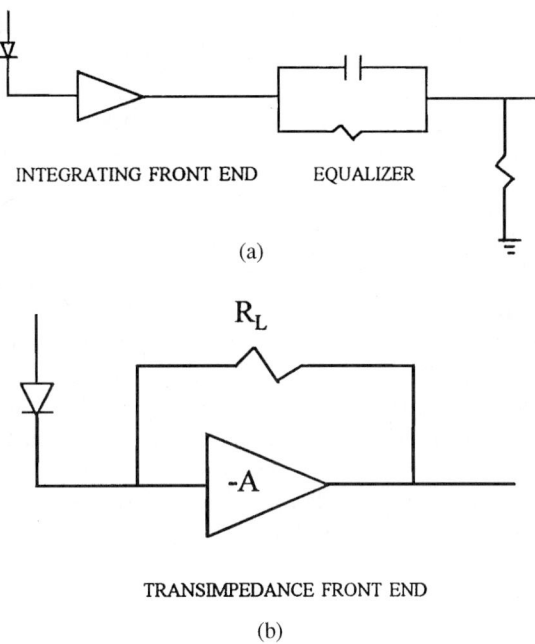

FIGURE 58.2 Schematic diagrams: (a) an integrating front end with equalization and (b) a transimpedance front end.

a result of the increased impedance level the frequency response, determined by the combination of the resistor and the capacitance at the input node of the receiver, is generally very much limited, typically being a factor of 10 or more smaller than that required to reproduce the signal shape. Such a receiver integrates the signal, hence the sometimes used description as an integrating front end. The accepted approach to the limited bandwidth is to equalize the signal after sufficient amplification. A simplified schematic of an equalized, high-impedance front end is shown in Fig. 58.2(a).

The high-impedance front end reduces the thermal noise associated with the input resistor, which has two major consequences. The first is related to the dynamic range of the receiver. Although the signal fidelity is restored by the equalizer, the low-frequency components of the received signal generate large voltages at the input node of the receiver. At received optical signals well in excess of the minimum this signal buildup will eventually cause the active device (input transistor) to overload, distorting the signal, or perhaps cutting off the receiver. For many applications, dynamic ranges of 20–30 dB of optical power (40–60 dB electrical) are required. This requirement and the limited dynamic range of the high-impedance type of front end limit the usefulness of this approach.

The second consequence of the equalization is to accentuate the noise generated at the output of the amplifying stage. Because of the effect of the equalization, the noise spectral density associated with the output stage is proportional to the square of the input capacitance. Optimization (reduction) of the equivalent noise of the output stage, and hence the receiver front end, requires the minimization of the input capacitance, a goal of all optical receiver designs. The details of such optimization and the relevant formulas giving the analytical expressions for the resulting noise can be found in Kasper [1988], Muoi [1984] and Smith and Personick [1982], and are summarized subsequently.

At the other extreme of receiver design is the low-impedance front end typified by a 50-Ω input impedance amplifier. The low-impedance front end is the least sensitive of the receivers, as its noise level is much larger than that of both the high-impedance and transimpedance approaches. Low-impedance receivers find use in laboratory experiments and are used in practical optical communication systems only in conjunction with optical preamplifiers.

The transimpedance receiver is schematically shown in Fig. 58.2(b). It is essentially a current-to-voltage converter. The bandwidth of the receiver is often dominated by the parasitic capacitance of the feedback resistor, which can be very small. Hence, for a given resistor value the bandwidth of the transimpedance receiver will be greater than the high-impedance receiver. Under practical conditions, the transimpedance receiver is usually designed with a bandwidth sufficient to reproduce the signal without recourse to equalization. Typically, this results in a higher noise level and hence a poorer sensitivity compared to a high-impedance receiver. It does, however have an improved dynamic range as the input node is effectively a low impedance due to the electrical feedback. The transimpedance receiver is generally used in optical systems as it offers a good compromise between the various receiver requirements.

Early receivers were realized with discrete transistors. At the present time many designs are realized using integrated circuits, especially Si bipolar and GaAs field-effect transistor (FET) designs. The GaAs receiver designs generally provide superior receiver sensitivity but are more expensive than their Si bipolar counterparts. Metal-oxide-semiconductor (MOS) integrated circuit (IC) technology has not found much application in optical receiver front ends, but is used in associated circuitry where the speed is sufficient. Generally speaking, the best sensitivities are still achieved with discrete FET front ends but for economic considerations integrated circuits are widely used.

Postamplifier

Front ends, as usually employed, typically generate relatively small output signal levels. The output from the front end is usually amplified by a postamplifier. Because the range of optical input signals can be as much as 30 dB (60-dB electrical) the postamplifier usually employs some form of output level control. This is typically achieved by an automatic gain control (AGC) function or by using a limiting amplifier. The output signal level is typically of the order of 1 V, which is sufficient to toggle the decision circuit.

Filtering

The main purpose of the filter shown in Fig. 58.1 is to improve the SNR by reducing the noise. Under ideal conditions, where the effective bandwidths of the components in the system prior to the filter are large, the filter function is such that the signal in a given time slot has zero signal level in all adjacent time slots. The effective bandwidth of such a filter is approximately 0.6–0.7 times the signalling rate, B. In a practical system, the pulse shape incident upon the filter is not ideal as a result of limited bandwidth in the transmitter, front end, postamplifier, and dispersion in the transmission fiber. The actual filter function required is one which transforms the waveform prior to the filter into one that produces an acceptable waveform at the decision circuit, while minimizing the noise. In many practical lightwave systems, there is little or no actual filtering done due to the limited bandwidth of components preceeding the filter. Often the filtering is accomplished by the preamplifier or front end.

Timing Recovery

In nearly all lightwave systems it is necessary to recover the clock in the process of regenerating the signal. Several commonly used methods of clock recovery make use of phase locked loops or surface acoustic wave filters, with the latter being the most widely used for high-frequency systems which have strict requirements on the phase transfer function. Because the commonly used nonreturn to zero (NRZ) format does not contain energy at the clock frequency it is necessary to use some form of nonlinear device, such as a rectifier or exclusive OR gate to produce frequency components at or near the clock frequency. This signal is then passed through a narrowband filter such as a surface acoustic wave (SAW) filter or dielectric resonator to extract the clock. The extracted clock, suitably phased, is used to clock the decision circuit, and may be used elsewhere in the terminal or regenerator as needed.

Decision Circuit

The decision circuit receives the signal (plus noise) from the post amplifier along with the recovered clock. The decision circuit produces an output which is reshaped and retimed.

58.3 Receiver Sensitivity: General

The sensitivity of a receiver is usually measured in terms of the average optical power required to produce the desired level of system performance. For digital systems the measure of performance is the BER or bit error ratio (often referred to as the bit error rate, even though incorrectly). The performance measure commonly used is for a $BER = 10^{-9}$, although many systems require significantly better performance. For analog applications the measure of performance is usually the carrier-to-noise ratio (CNR). Power is measured in terms of decibel referred to 1 mW of received power. Further, for digital systems, sensitivity is measured in terms of the *average* power received, taking into account the population of marks and spaces (i.e., 1s and 0s). The average power is thus approximately one-half of the peak power or 3 dB below the peak received power. In order to evaluate the receiver sensitivity it is necessary to evaluate both the photogenerated signal current and the sources of noise which corrupt the signal.

Signal Current

The photocurrent generated by a received optical power P is given by

$$I_s = \left[\frac{q\eta}{h\upsilon}\right] P \quad \text{(p-i-n)} \tag{58.1}$$

where h is Planck's constant ($6.623E\text{-}34$ J/s), υ is the optical frequency, q is the electronic charge, and η is the quantum efficiency of the detector (defined in terms of the number of photoelectrons generated per photon received). In well-designed detectors the quantum efficiency is usually between 0.8 and 1.0. When a p-i-n detector is used Eq. (58.1) becomes the signal current; for an avalanche photodiode the signal current is given by

$$I = MI_s = M\left[\frac{q\eta}{h\upsilon}\right] P \quad \text{(APD)} \tag{58.2}$$

where M is the expectation value of the avalanche gain.

Sources of Noise

Amplifier or Circuit Noise

As the received signal level (photocurrent generated in the photodetector) is small it must be amplified by an electronic amplifier, which contributes noise to the system. The principal sources of noise in the amplifier are thermal noise from resistors and noise generated in the amplifying transistors. It is common practice in the analysis of optical receivers to refer all noise sources to the input node where they can be compared to the signal current, Eq. (58.1) or (58.2).

In evaluating the contributions of the various sources of noise, account is taken of the frequency response of the various components of the receiver. The details of this computation are too detailed to be included in this summary and may be found in Personick [1973, 1979] and in Smith and

Personick [1982]. We will here summarize the essential results described in the aforementioned references.

The variance of the circuit noise for an FET front end is given by

$$\langle i^2 \rangle_{\text{ckt}} = \frac{4kT}{R_L} \mathfrak{I}_2 B + 2q I_g \mathfrak{I}_2 B + \frac{4kT\Gamma}{g_m}(2\pi C_T)^2 \mathfrak{I}_3 B^3 \quad \text{(FET)} \qquad (58.3)$$

where k is Boltzmann's constant, T the absolute temperature, R_L the load resistance at the input node, q the electronic charge, I_g the gate leakage current, g_m the transconductance of the input transistor, C_T the total capacitance at the input node, B the design bit rate, and Γ is a parameter characterizing the channel noise, which has a value ≈ 1. The parameters $\mathfrak{I}_2$ and $\mathfrak{I}_3$, referred to as the *Personick integrals*, are dimensionless numbers, which derive from the details of the transfer function of the system. Typical values of $\mathfrak{I}_2$ and $\mathfrak{I}_3$ are 0.6 and 0.1, respectively. (A detailed explanation of the Personick integrals may be found in Personick [1973] and Smith and Personick [1982].) The B^3 dependence of the last term arises from the referral of the output noise source to the input.

The corresponding equation for the variance of the noise for a bipolar front end, neglecting the effect of noise due to base resistance, is given by

$$\langle i^2 \rangle_{\text{ckt}} = \frac{4kT}{R_L} \mathfrak{I}_2 B + 2q I_b \mathfrak{I}_2 B + \frac{2q I_c}{g_m^2}(2\pi C_T)^2 \mathfrak{I}_3 B^3 \quad \text{(bipolar)} \qquad (58.4)$$

For a bipolar transistor $I_c = \beta I_b$, and the transconductance g_m is proportional to the collector current I_c; hence, the noise can be minimized by operating the input transistor at the optimum collector current. The minimum bipolar circuit noise given by

$$\langle i^2 \rangle_{\text{bipckt/min}} = \frac{4kT}{R_L} \mathfrak{I}_2 B + 8\pi kT C_T \sqrt{\frac{\mathfrak{I}_2 \mathfrak{I}_3}{\beta}} \qquad (58.5)$$

Signal Shot Noise

The random arrival rates of photons at the detector result in an additional source of noise, which is proportional to the received photocurrent, referred to as *signal-dependent shot noise*. The spectral density of this noise source is given by

$$\frac{d}{df}\langle i^2 \rangle_{\text{sig}} = 2qI \qquad (58.6)$$

where I is the average current. This signal-dependent noise constitutes a major difference in optical systems compared to radio frequency and microwave systems. The variance of this contribution to the noise is found by integrating the noise spectral density over the transfer function of the system giving

$$\langle i^2 \rangle_{\text{sig}} = 2q I \mathfrak{I}_1 B \qquad (58.7)$$

where $\mathfrak{I}_1$ is another Personick integral, which has a value ≈ 0.5, and B is the bit rate.

Shot Noise with an Avalanche Photodiode

In many receivers an avalanche photodiode is used in place of a p-i-n photodiode. As discussed in the chapter on Optical Detectors in this volume, the APD multiplies the received primary photocurrent

Optical Receivers

given by Eq. (58.1) by an average multiplication factor M,

$$I_{\text{APD}} = MI = M\left[\frac{q\eta}{h\upsilon}\right]P$$

For receivers employing an APD this becomes the signal current. [See Eq. (58.2).]

If the avalanche multiplication process were ideal (i.e., deterministic) the shot noise spectral density associated with the signal would be given by

$$\frac{d}{df}\langle i^2\rangle = 2qIM^2 \tag{58.8}$$

The avalanche process is, however, a random process and not deterministic. This randomness adds additional noise, which is accounted for by a multiplicative noise factor, referred to as the excess noise factor, and denoted $F(M)$. In this case the signal-dependent shot noise spectral density becomes

$$\frac{d}{df}\langle i^2\rangle_{\text{sig}} = 2qIM^2F(M) \tag{58.9a}$$

and the mean square noise after integration with respect to frequency is

$$\langle i^2\rangle_{\text{sig}} = 2qIM^2F(M)\mathfrak{I}_1 B \tag{58.9b}$$

Many approximations for $F(M)$ have been given. Among the most commonly used are

$$F(M) = M^x \tag{58.10a}$$

and

$$F(M) = M\left[1 - (1-k)\left(\frac{M-1}{M}\right)^2\right] \tag{58.10b}$$

In Eq. (58.10a), x is an empirical constant, usually between 0 and 1, and in Eq. (58.10b), k is the ratio of the ionization coefficients of holes and electrons (not to be confused with Boltzmann's constant). The value of k is between 0 and 1 if the avalanche is initiated by the most ionizing carrier, as it should be in a well-designed device. Equation (58.10b) provides a better approximation to the excess noise factor and is generally used.

Sensitivity of a Digital Receiver

In a digital system the objective is to correctly identify transmitted marks (1s) and spaces (0s). Incorrectly identifying the transmitted signal produces an error. The probability of making an error in turn depends on the signal level and the probability distribution of the system noise. Under many circumstances, but not always, the probability distribution of the noise is Gaussian, with an associated standard deviation σ.

Gaussian Approximation

The assumption of a Gaussian distribution permits simplified analytical approximations to sensitivity calculations, which are generally close to the more rigorous computational approaches. In optical systems the noise levels associated with the different signal states are not necessarily identical. The standard deviation of the noise associated with a transmitted 1 is denoted by σ_1 and for a transmitted 0 by σ_0. The ratios of the signal levels relative to the decision level D normalized to the associated root mean square (rms) noise levels are given by

$$Q_1 = \frac{S_1 - D}{\sigma_1} \qquad (58.11a)$$

and

$$Q_0 = \frac{D - S_0}{\sigma_0} \qquad (58.11b)$$

With these definitions, and using the Gaussian approximation, the probability of misinterpreting a transmitted 0 as a 1 is given by

$$P(0, 1) = \frac{1}{\sqrt{2\pi}\sigma_0} \int_D^\infty \exp\left[\frac{-(S-D)^2}{2\sigma_0^2}\right] dS = \frac{1}{\sqrt{2\pi}} \int_{Q_0}^\infty \exp\left[-\frac{Q^2}{2}\right] dQ$$

$$\approx \frac{1}{\sqrt{2\pi}Q_0} \exp\left(-\frac{Q_0^2}{2}\right) \qquad (58.12a)$$

and the probability of misinterpreting a transmitted 1 is given by

$$P(1, 0) = \frac{1}{\sqrt{2\pi}\sigma_1} \int_{-\infty}^D \exp\left[\frac{-(S-D)^2}{2\sigma_1^2}\right] dS$$

$$= \frac{1}{\sqrt{2\pi}} \int_{-\infty}^{Q_1} \exp\left[-\frac{Q^2}{2}\right] dQ \approx \frac{1}{\sqrt{2\pi}Q_1} \exp\left[-\frac{Q_1^2}{2}\right] \qquad (58.12b)$$

To a good approximation the minimum probability of error occurs when each probability of error is the same, that is, $P(0, 1) = P(1, 0)$, resulting in $Q_1 = Q_0 = Q$ with

$$Q = \frac{S_1 - S_0}{\sigma_1 + \sigma_0} \qquad (58.13)$$

The probability of error as a function of the Q factor is shown in Fig. 58.3. The desired error

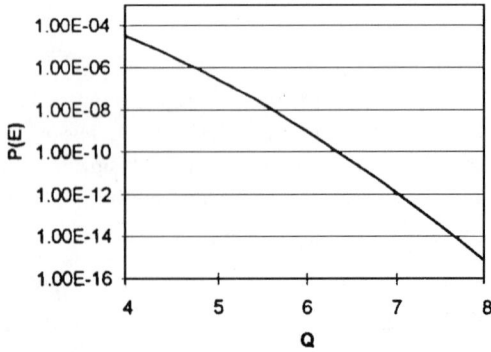

FIGURE 58.3 Probability of error vs Q value.

Optical Receivers

performance determines the value of Q. The average required signal level is given by

$$\overline{S} = \frac{S_1 + S_0}{2} \quad (58.14)$$

with

$$S_1 - S_0 = Q(\sigma_1 + \sigma_0) \quad (58.15)$$

When the decision level is not optimally placed ($Q_1 \neq Q_2$) the probability of error becomes

$$P(E) = 0.5[P(1,0) + P(0,1)] \quad (58.16)$$

Receiver Sensitivity: p-i-n Detector

We are now in a position to calculate the sensitivity of a receiver using a p-i-n detector. Combining Eqs. (58.1) and (58.15) we find that the required *peak* signal current is

$$I_s = Q(\sigma_1 + \sigma_0) \quad (58.17)$$

where we assume the signal in the 0 state is identically zero and $S_1 = I_s$. The variance of the noise σ^2 is given by the sum of the circuit noise and the signal shot noise, Eq. (58.7).

$$\sigma^2 = \langle i^2 \rangle_{\text{ckt}} + \langle i^2 \rangle_{\text{sig}} = \langle i^2 \rangle_{\text{ckt}} + 2qI_s\mathfrak{J}_1 B \quad (58.18)$$

The circuit noise for an FET front end is calculated from Eq. (58.3) and for an optimally biased bipolar front end by Eq. (58.5). Since the signal current equals I_s in the 1 state and 0 in the 0 state the variances become

$$\sigma_0^2 = \langle i^2 \rangle_{\text{ckt}} \quad (58.19\text{a})$$

and

$$\sigma_1^2 = \langle i^2 \rangle_{\text{ckt}} + 2qI_s\mathfrak{J}_1 B \quad (58.19\text{b})$$

Substituting Eqs. (58.19a) and (58.19b) into Eq. (58.17) gives

$$I_s = 2\sigma_0 Q + 2qQ^2\mathfrak{J}_1 B \quad (58.20)$$

The peak optical power is found from Eq. (58.1) and the sensitivity, measured in terms of the average power, is given by

$$\eta\overline{P}_s = \left(\frac{h\nu}{q}\right)[\sigma_0 Q + qQ^2\mathfrak{J}_1 B] \quad \text{(p-i-n)} \quad (58.21)$$

Under most circumstances, with a p-i-n detector, the second term in Eq. (58.21) is small compared to the first leading to the commonly used expression for the p-i-n sensitivity

$$\eta\overline{P}_s = \left(\frac{h\nu}{q}\right)Q\langle i^2 \rangle_{\text{ckt}}^{1/2} = \frac{1.24 Q}{\lambda}\langle i^2 \rangle_{\text{ckt}}^{1/2} \quad (58.22)$$

where λ is the wavelength expressed in micrometers.

Receiver Sensitivity: Avalanche Photodiode Detector

For an APD receiver the signal current is given by Eq. (58.2) and the variances of the noise in the two signal states become

$$\sigma_0^2 = \langle i^2 \rangle_{\text{ckt}} \tag{58.23a}$$

and

$$\sigma_1^2 = \langle i^2 \rangle_{\text{ckt}} + 2q I_s M^2 F(M) \Im_1 B \tag{58.23b}$$

These relations neglect the effect of photodetector dark current and also assume the received signal is identically zero when a 0 is received. A more general account may be found in Smith and Personick [1982]. Substituting these equations into Eq. (58.17) gives the result

$$\eta \overline{P}_s = \left(\frac{h\upsilon}{q}\right)\left[\frac{\sigma_0 Q}{M} + qQ^2 \Im_1 BF(M)\right] \tag{58.24}$$

A plot of this expression vs the gain of the APD, M, is shown in Fig. 58.4 for a receiver operating at 622 Mb/s [Tzeng, 1994]. For low gain the sensitivity is seen to improve, reaching an optimum and then to decline for further increases in the gain. The initial improvement is associated with the amplification of the signal current by the APD, whereas the decrease at high gain results from the excess signal shot noise associated with the random nature of the avalanche process.

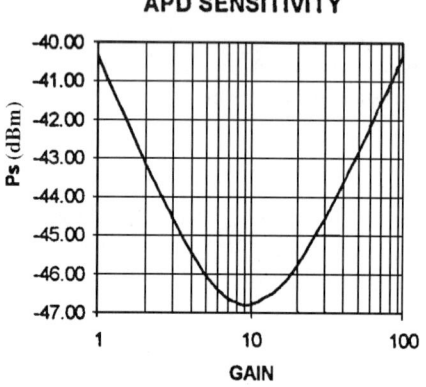

FIGURE 58.4 Sensitivity of an APD receiver at 622 Mb/s as a function of the APD gain.

Receiver Sensitivity: Optical Preamplifier

A recent advance in optical fiber technology is the development of optical amplifiers. The most widely used is the Er-doped fiber amplifier EDFA (see Chapter 62 in this volume for a detailed discussion of optical amplifiers). One application of optical amplifiers is their use as an optical preamplifier preceding the optical receiver previously described. The optical preamplifier amplifies the received optical signal prior to detection by the photodetector. As with other types of amplifiers, optical amplifiers introduce noise resulting from spontaneous emission in the amplifying region. This type of noise is referred to as *amplified spontaneous emission* (ASE).

Figure 58.5 is a schematic of an optical preamplifier showing the important parameters defined as follows:

- η_{in} = coupling efficiency of transmission fiber to the optical amplifier
- η_{out} = coupling of optical amplifier to photodetector input including transmission loss of the optical filter
- η = detector efficiency
- G = small signal gain of the optical amplifier
- F_m = noise figure of the optical amplifier
- B_o = bandwidth of the optical filter, Hz
- B_e = electrical bandwidth of the receiver, Hz

The noise figure of the amplifier is defined as $2n_{\text{sp}}$ where n_{sp} is the spontaneous emission factor. See Chapter 62 on Optical Amplifiers in this volume for more detail.

Optical Receivers

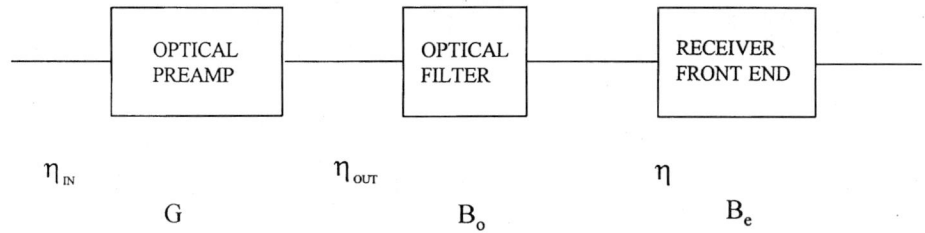

FIGURE 58.5 Schematic diagram of an optical preamplifier.

Photocurrents

In receivers incorporating optical preamplifiers, p-i-n detectors are usually employed. For this case the signal photocurrent is given by

$$I_s = \left(\frac{\eta q}{h\upsilon}\right)\eta_{\text{in}}\eta_{\text{out}}GP_s \qquad (58.25)$$

and the photocurrent generated by the ASE is given by

$$I_{\text{ASE}} = F_m q \eta_{\text{out}} \eta (G-1) B_o \qquad (58.26)$$

Noise Sources

The photodetector behaves as a square law detector. The generated photocurrent, which is proportional to the square of the incident optical field, contains components proportional to signal∗signal, ASE∗ASE, and signal∗ASE. Following the analysis of Olsson [1989] the components of the noise are given by

$$\sigma_{\text{shot}}^2 = 2q(I_s + I_{\text{ASE}})B_e \qquad (58.27a)$$

$$\sigma_{\text{spon-spon}}^2 = \frac{1}{2}F_m^2 q^2 \eta_{\text{out}}^2 \eta^2 (G-1)^2 (2B_o - B_e) B_e \qquad (58.27b)$$

$$\sigma_{\text{sig-spon}}^2 = [2q F_m \eta_{\text{out}} \eta (G-1)]\left[\left(\frac{\eta q}{h\upsilon}\right) P_s \eta_{\text{in}} \eta_{\text{out}} G\right] B_e$$

$$= 2q^2 F_m \eta_{\text{in}} \eta_{\text{out}}^2 \eta^2 G(G-1)\left(\frac{1}{h\upsilon}\right) P_s B_e \qquad (58.27c)$$

The first term, σ_{shot}^2, is the shot noise generated by the amplified signal and ASE and contains a term proportional to the received signal. The second term, $\sigma_{\text{spon-spon}}^2$, results from the ASE beating with itself in the detector. Since the bandwidth of the optical gain is large ($\approx 10^{12}$ Hz) compared to the signal bandwidth in common usage the spectral width of the ASE is normally reduced by the use of an optical filter of bandwidth B_o. B_o can be reduced in principle until it approaches B_e. The problems of achieving a narrow optical filter bandwidth and of aligning the optical filter with the signal wavelength set the practical limit on B_o. The noise term $\sigma_{\text{spon-spon}}^2$ is independent of the signal level and hence is present in both the 0 and 1 states of the signal.

The term $\sigma_{\text{sig-spon}}^2$ results from the mixing of the signal and ASE fields. This noise term is proportional to the received optical signal power, and in an ideal system when the received optical power is zero in the 0 state, $\sigma_{\text{sig-spon}}^2 = 0$. The total noise associated with the 0 state, assuming zero power received, is given by

$$\sigma_0^2 = \sigma_{ckt}^2 + \sigma_{shot}^2(ASE) + \sigma_{spon\text{-}spon}^2 \qquad (58.28)$$

and for the 1 state the noise is given by

$$\sigma_1^2 = \sigma_{ckt}^2 + \sigma_{shot}^2(ASE + \text{signal}) + \sigma_{spon\text{-}spon}^2 + \sigma_{sig\text{-}spon}^2 \qquad (58.29)$$

The receiver sensitivity can be calculated by combining Eqs. (58.13) and (58.25–58.29) yielding

$$I_s = 2Q^2\left[qB_e + qF_m\eta_{out}\eta(G-1)B_e + \frac{\sigma_0}{Q}\right] \qquad (58.30)$$

Typically, G is greater than or equal to 30 dB (1000) in which case the first term in brackets is negligible. Further, since $\sigma_{spon\text{-}spon}^2$ varies as the square of the gain while σ_{shot}^2 varies only as the first power of the gain and σ_{ckt}^2 is independent of the gain,

$$\sigma_0^2 \approx \sigma_{spon\text{-}spon}^2 \qquad (G \gg 1) \qquad (58.31)$$

Substituting this limit gives

$$I_s = 2Q^2\left\{qF_m\eta_{out}\eta(G-1)B_e\left[1 + \frac{1}{Q}\left(\frac{B_o}{B_e} - \frac{1}{2}\right)^{\frac{1}{2}}\right]\right\} \qquad (58.32)$$

and the sensitivity measured in terms of the average received power incident on the optical amplifier in the limit of large gain is

$$\eta_{in}\overline{P}_s = Q^2(h\upsilon)F_m B_e\left[1 + \frac{1}{Q}\left(\frac{B_o}{B_e} - \frac{1}{2}\right)^{\frac{1}{2}}\right] \qquad (58.33)$$

When the optical filter bandwidth is comparable to the electrical bandwidth (practical only at very high bit rates) and for very low BER ($Q \gg 1$), the limiting sensitivity becomes

$$\eta_{in}\overline{P}_s \approx Q^2(h\upsilon)F_m B_e \qquad (58.34)$$

When the bandwidth of the optical filter is large compared to the electrical bandwidth, Eq. (58.33) applies.

The value of the noise figure F_m depends on the detailed characteristics of the optical amplifier. Under ideal conditions of perfect inversion of the gain medium, $F_m = 2$. Typical values are somewhat higher. With an electrical bandwidth $B_e = B/2$, the limiting sensitivity for a receiver using an optical preamplifier is

$$\eta_{in}\overline{P}_s = Q^2(h\upsilon)B \qquad (58.35)$$

Sensitivity Limits

Quantum Limit. In the limit that the sensitivity is given by the statistics of the received photons which are characterized by a Poisson process the ultimate sensitivity for a $BER = 1E\text{-}9$ is 10 detected photons per bit received [Henry, 1985], which corresponds to

$$\eta\overline{P}_{limit} = 10(h\upsilon)B \qquad (58.36)$$

Optical Amplifier Limit. With an optimized optical amplifier the limiting sensitivity for $BER = 1E\text{-}9\ (Q = 6)$ is

$$\eta_{in}\overline{P}_{s\text{-opt.ampl}} = 36(h\upsilon)B \qquad (58.37)$$

which is approximately four times the quantum limit. Demonstrated sensitivities are within a factor of three to four of this value at bit rates greater than 1 GB/s. See Park and Grandlund [1994] for a discussion of experimental results.

Practical Avalanche Photodiode Performance. The ultimate receiver sensitivity achievable with an APD is a function of the circuit noise and the excess noise factor associated with the avalanche process. Because the excess noise factor increases as the avalanche gain is increased the optimum gain is in the range of 10–30 for existing APDs, and is significantly below the gains used in optically preamplified receivers. The achievable sensitivity is accordingly poorer than the limit for an optical preamp, being in the range of a few hundred to as much as several thousand photons per bit, and hence between a factor of 10–100 worse than the quantum limit.

Practical p-i-n Performance. In the absence of a gain mechanism, either avalanche or optical amplification, the receiver sensitivity is determined by the circuit noise, hence, depending on the design of the receiver front end. Typically, the sensitivity of a p-i-n-based receiver will be ≈ 10 dB poorer than an APD-based receiver and, hence, typically a few thousand photons per bit.

Defining Terms

Amplified spontaneous emission (ASE): The fundamental source of noise in optical amplifiers.
Avalanche photodiode (APD): A photodiode with internal current gain resulting from impact ionization.
Excess noise factor (ENF): A measure of the amount by which the short noise of an APD exceeds that of an ideal current amplifier.
Front end (FN): The input stage of the electronic amplifier of an optical receiver.
Integrating front end (IFE): A type of front end with limited bandwidth which integrates the received signal; requires subsequent equalization to avoid intersymbol interference.
Transimpedance front end (TFE): A type of front end which acts as a current to voltage converter; usually does not require equalization.

References

Henry, P.S. 1985. Lightwave primer, *IEEE J. Quantum Electr.*, QE-21:1862–1879.
Kasper, B.L. 1988. Receiver design. In *Optical Fiber Telecommunications II*, ed. S.E. Miller and I.P. Kaminov, pp. 689–722. Academic Press, San Diego, CA.
Muoi, T.V. 1984. Receiver design for high-speed optical-fiber systems. *IEEE J. Lightwave Tech.*, LT-2:243–267.
Olsson, N.A. 1989, Lightwave systems with optical amplifiers. *IEEE J. Lightwave Tech.*, LT-7:1071–1082.
Park, Y.K. and Grandlund 1994. Optical preamplifier receivers: Application to long-haul digital transmission, *Optical Fiber Tech.*, 1:59–71.
Personick, S.D. 1973. Receiver design for digital fiber optic communication systems I. *Bell Syst. Tech. J.*, 52:843–874.
Personick, S.D. 1979. Receiver design. In *Optical Fiber Telecommunications*, ed. S.E. Miller and A.G. Chenoweth, pp. 627–651, Academic Press, New York.

Smith, R.G. and Personick, S.D. 1982. Receiver design for optical fiber communication systems. In *Semiconductor Devices for Optical Communication, Topics in Applied Physics*, Vol. 39, ed. H. Kressel, pp. 89–160. Springer–Verlag, Berlin.

Tzeng, L.D. 1994. Design and analysis of a high-sensitivity optical receiver for SONET OC-12 systems, *J. Lightwave Tech.*, LT-12:1462–1470.

Further Information

This section has focused on direct detection of digital signals, which constitute the most widely used format in optical communication systems. Coherent optical systems have been studied for some time but have not yet found practical application. A discussion of this technology and the associated receiver sensitivities may be found in Chapter 63 on Coherent Systems in this volume. Another application of optical fiber technology is in the distribution of analog video, which is being employed by community antenna television (CATV) distribution companies. A discussion of the sensitivity of analog receivers may be found in Smith and Personick [1982].

59

Fiber Optic Connectors and Splices

59.1 Introduction.. 803
59.2 Optical Fiber Coupling Theory..................................... 804
  Multimode Fiber Joints • Single-Mode Fiber Joints • Reflectance Factors
59.3 Multibeam Interference (MBI) Theory........................... 811
  MBI Effects on Transmission • MBI Effects on Reflectance
59.4 Connector Design Aspects 815
  Introduction (Types of Connectors) • Factors Contributing to Insertion Loss • Factors Contributing to Reflectance (Physical Core-to-Core Contact)
59.5 Splicing Design Aspects.. 820
  Mechanical Splices • Fusion Splices
59.6 Conclusions 822

William C. Young
Bell Communications Research

59.1 Introduction

In recent years the state of the art of optical fiber technology has progressed to where the achievable attenuation levels for the fibers are very near the limitations due to Rayleigh scattering. As a result, optical fibers, and particularly the single-mode fibers that are used in today's communications systems, can be routinely fabricated with attenuation levels below 0.5 dB/km at a wavelength of 1300 nm and 0.25 dB/km at 1550 nm. Employing these fibers in optical communications systems requires precise jointing devices such as connectors and splices. Considering the small size of the fiber cores, less than 10 μm in diameter for single mode and 100 μm for multimode fibers, it is not surprising that these interconnecting devices can easily introduce significant optical losses. Furthermore, since single-mode fibers have practically unlimited bandwidth, it is also not surprising that they have become the choice for optical communications applications. To provide low-loss connectors and splices for these single-mode fibers, reliable and cost-effective alignment accuracies in the submicrometer range are required.

This chapter will review the fundamental technology that is presently used for optical connectors and splices. In particular, since single-mode fibers dominate optical communications systems and also require the greatest precision and performance, we will focus mainly on the jointing of these fibers. However, for completeness we will also briefly review multimode fiber jointing technology as well.

Before reviewing the technology we should define the implied differences between an optical connector and a splice. The term *connector* is commonly used when referring to the jointing of two optical fibers in a manner that not only permits but also anticipates the unjointing, or unconnecting, through the design intent. Optical connectors are commonly used for terminating components, system configuration and reconfiguration, testing, and maintenance. In contrast, the term *splice* is commonly used when referring to the jointing of two optical fibers in a manner that does not lend itself to unjointing. There are two types of splices, mechanical splices that join two butt-ended fibers and use index matching material between their endfaces and fusion splices where the two fibers are heated to just below their melting point and are pressed together to form a permanent joint. Splices are commonly used when the total optical fiber span length can be realized only by the concatenation of shorter sections of fiber, and also for repairing severed fiber/cables.

59.2 Optical Fiber Coupling Theory

The optical forward coupling efficiency of a connector or splice, that is, the insertion loss, and the optical backward coupling efficiency (reflectance) are the main optical performance criteria for optical fiber joints, and therefore, these two parameters are used to evaluate the joint's optical performance over various climatic and environmental conditions [Miller, Mettler, and White, 1986].

The factors affecting the optical coupling efficiency of optical fiber joints can be divided into two groups, extrinsic and intrinsic factors (Table 59.1).

Factors extrinsic to the optical fiber, such as lateral or transverse offset between the fiber cores, longitudinal offset (endface separation), and angular misalignment (tilt), are directly influenced by the techniques used to join the fibers. In contrast, intrinsic factors are directly related to the particular properties of the two optical fibers that are joined. In the case of single-mode fibers the important intrinsic factor is the fiber's mode field diameter. With multimode fibers the intrinsic factors are the core diameter and numerical aperture (NA). The fiber's core concentricity and outside diameter are also commonly referred to as intrinsic factors, but practical connectors and splices must be designed to accommodate variations in these two parameters.

TABLE 59.1 Factors Affecting Coupling Efficiency

Extrinsic factors:
Lateral fiber core offset
Longitudinal offset
Angular misalignment
Reflections

Intrinsic factors:
Mismatch in fiber core diameters (mode field diameters)
Mismatch in index profiles

The effect that these factors have on coupling efficiency may also depend on the characteristics of the optical source, and in the case of multimode fiber joints, on the relative location of the joint in the optical path. For example, in the case of an incoherent optical source such as a light emitting diode (LED), the mode power distribution (MPD) along a continuous fiber's path largely depends on the length and curvature of the fiber. With these sources and fibers the effect a particular offset will have on the coupling efficiency will decrease until a steady-state mode distribution is realized. Because of these MPD effects, a much practiced method of evaluating multimode fiber joints is to use an over-moded launch condition (uniform) followed by a mode filter that selectively filters the higher-ordered modes thereby establishing a steady-state MPD condition. Characterizing multimode fiber joints in this manner eliminates the uncertainty of different MPD effects. Of course, in use, the coupling efficiency of the connector or splice is dependent on the particular MPD (a function of the optical source and location along the optical path). In the case of single-fiber joints it is only necessary that the fiber is operated in the single-mode wavelength regime.

Multimode Fiber Joints

Various experiments [Chu and McCormick, 1978] and analytical models have been used to quantify the effects that extrinsic factors have on coupling efficiencies for butt-jointed multimode fibers. Based on these studies we have plotted (Fig. 59.1) the coupling loss as a function of normalized lateral, longitudinal, and angular offset for graded-index multimode fibers.

Fiber Optic Connectors and Splices

In Fig. 59.1, the coupling losses for various offsets are plotted for both a uniform MPD as well as a steady-state MPD. From these curves it can be seen that as the quality of the alignments improve, the effect of the presence of higher order modes diminishes. For example, today's connectors and splices have high control on angular (tilt) and longitudinal (gap) offsets, and as a result the coupling loss is mainly due to lateral offset. Therefore, as can be seen in Fig. 59.1, a connection of a typical graded-index multimode fiber having a 50-μm core diameter and having a lateral offset of about 10 μm has a coupling loss of 0.25 dB with a steady-state MPD, whereas with a uniform MPD the same

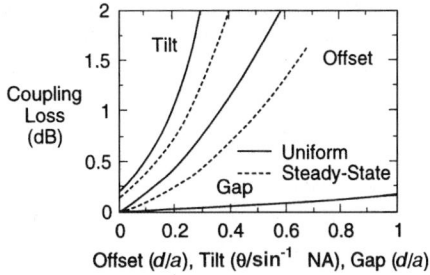

FIGURE 59.1 Coupling loss vs offsets for butt-jointed graded-index multimode fibers (where a is the fiber's core radius and d is the offset in micrometers).

joint has a loss of 0.5 dB. It is therefore very important, in the case of multimode fibers, that the particular MPD be known at the site of each joint for correct estimation of the loss incurred in the system. It should also be pointed out that due to cross-coupling effects the offsets at a joint can also change the MPD directly after the joint. Thus, in summary, because of the many factors that can change the MPD in a multimode fiber, such as optical launch conditions, offsets at joints, fiber length and deployment conditions, mode mixing and differential mode attenuation of the particular fibers, there is no unique coupling or insertion loss for multimode fiber joints that is based only on their physical fiber alignment. It therefore is very important for the system designer to analyze the system for these factors and then make the appropriate assignment for expected loss of the connector and splices.

Intrinsic factors can also have a large effect on multimode fiber joints. Therefore, when evaluating the joint loss of multimode fiber joints one also must consider the characteristics of the fibers on either side of the joint, as well as the direction of propagation of the optical power through the joint. Again, various experiments and analytical models have been used to quantify these effects. Based on these studies, the dependence of the coupling loss on mismatches in numerical apertures and core radii are summarized in Fig. 59.2.

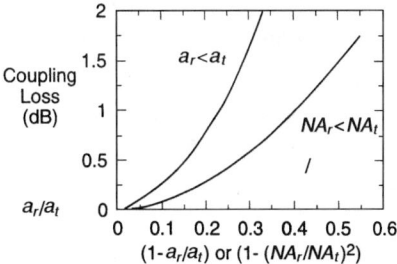

FIGURE 59.2 Coupling loss vs mismatches in numerical aperture NA and fiber core radius a for graded-index multimode fiber having a steady-state MPD. The subscripts r and t represent the receiving and transmitting fibers.

As can be seen in Fig. 59.2, we have plotted the coupling loss for the cases when the transmitting fiber has the larger core radius and larger NA. For the opposite cases there is no loss incurred at the joints. Referring to Fig. 59.2, for a mismatch in core diameters of 48 and 52 μm the expected loss is about 0.25 dB, and for a mismatch in numerical aperture of 0.20–0.23 it is about 0.35 dB. These values are significant when compared to expected loss due solely to extrinsic alignment inaccuracies, and are also propagation direction dependent. Therefore, the performance of multimode fiber connectors and splices are very deployment dependent requiring a complete understanding of the characteristics of the optical path before expected loss values can be calculated with confidence.

Single-Mode Fiber Joints

In the case of single-mode fiber joints, it has been shown that the fields of single-mode fibers being used today are nearly Gaussian, particularly those designed for use at a wavelength of 1310 nm. Therefore, the coupling losses for the joints can be calculated by evaluating the coupling between two misaligned Gaussian beams [Marcuse, 1977]. Based on this model, the following general formula has been developed [Nemota and Makimoto, 1979] for calculating the insertion loss (IL) between two single-mode fibers that have equal or unequal mode field diameters (an intrinsic factor) and lateral, longitudinal, and angular offsets, as well as reflections (extrinsic factors) as defined in Fig. 59.3

$$IL = -10 \log\left[\left(\frac{16 n_f^2 n_g^2}{(n_f + n_g)^4}\right) \frac{4\sigma}{q} \exp\left(-\frac{\rho\mu}{q}\right)\right] \text{ dB}$$

where

$$\rho = 0.5(K w_t)^2; \qquad q = G^2 + (\sigma + 1)^2$$
$$\mu = (\sigma + 1)F^2 + 2\sigma FG \sin q + \sigma(G^2 + \sigma + 1)\sin^2\theta$$
$$F = \frac{2x}{K w_t^2}; \qquad G = \frac{2z}{K w_t^2} \qquad (59.1)$$
$$\sigma = \left(\frac{w_r}{w_t}\right)^2; \qquad K = \frac{2\pi n_g}{\lambda}$$

Where:

n_f, n_g = refractive indices of the fiber and the medium between the fibers, respectively
w_t, w_r = mode field radii of the transmitting and receiving fibers, respectively
λ = operating wavelength
x, z = lateral and longitudinal offset, respectively
σ = is the angular misalignment

Although the fields of dispersion-shifted fibers (fibers having minimum dispersion near 1550 nm) are not truly Gaussian, this general formula is still applicable, particularly for the small offsets and misalignments that are required by connectors and splices designed for communication applications (<0.5 dB). In the preceding general formula, terms in the first parentheses, inside the square bracket, account for the loss due to reflection at two fiber/gap interfaces, which can be as high as 0.32 dB for air as a gap material. If fiber core-to-core contact is established in a connector or perfect index matching is used in a mechanical splice, this loss reduces to zero. The formula, however, does not account for the loss variation due to interference effects caused by multiple beam interference that may exist in joints with fiber endface separation [Wagner and Sandahl, 1982]. Even though high-performance connectors are designed for fiber core-to-core contact and therefore should not be subject to reflections caused by an air gap, the lack of workable standards and manufacturing tolerances may result in endface separation. Even if the end separation is less than a micrometer it can still have severe results on insertion loss. Although index matching material is used in mechanical splices, the refractive index of the material is temperature dependent and therefore the resulting index mismatches can also have an impact on insertion loss. The effects from multiple beam interference and its impact on connector and splice performance is discussed in a following section.

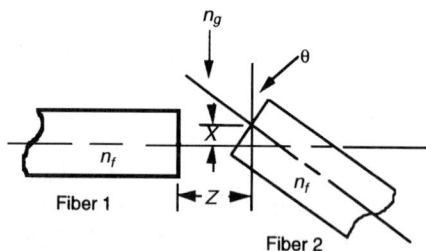

FIGURE 59.3 Schematic representation of a fiber butt joint having lateral and longitudinal offset, angular misalignment.

The preceding general formula can be reduced to determine the coupling loss due only to a mode field diameter mismatch (intrinsic factor). Assuming no extrinsic factors, and assuming

Fiber Optic Connectors and Splices

core-to-core contact, the formula reduces to

$$\text{coupling loss} = -10 \log \left[4 \left(\frac{w_t}{w_r} + \frac{w_r}{w_t} \right)^{-2} \right] \text{dB} \qquad (59.2)$$

Using this expression, the loss due to a 10% mismatch in mode field diameters, typical for today's commercial fibers and with no other factors present, is calculated to be 0.05 dB. Generally, to control the fiber's cutoff wavelength and zero dispersion wavelength, the mode field diameters of communications single-mode fibers are usually within 10%. A typical requirement range for the mode field diameter of **dispersion-unshifted single-mode fiber** is from 8.8 to 9.5 μm. It can also be seen in the preceding expression that contrary to multimode fiber joints, the loss of a single-mode fiber joint having mode field diameter mismatches is independent of the direction of propagation through the joint. In Fig. 59.4, the coupling loss between two fibers having different mode field radii is plotted as a function ratio of transmitting fiber's radii to the receiving fiber's radii.

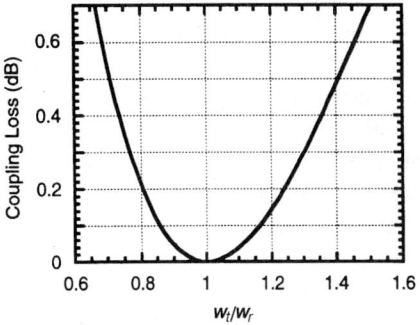

FIGURE 59.4 Plot of coupling loss vs ratio of mode field radii.

Although for a given fiber type and design the mode field diameters are usually held to within 10%, the loss in Fig. 59.4 is plotted for ratios from 0.6 to 1.54. This large range is needed to describe the intrinsic loss for fiber joints between dispersion-shifted and **dispersion-unshifted fibers**. This ensemble of fibers may be experienced when and if dispersion-shifted fibers become prevalent in future fiber deployments. It is common to see references that the mode field diameter of dispersion-shifted fiber is similar to that of today's fiber of choice, 1310-nm optimized dispersion-unshifted fiber. Although this is true, if one uses the intended wavelength of propagation, 1550-nm for dispersion-shifted and 1310-nm for unshifted fibers, it becomes important to look at the mode field diameters at similar wavelengths. This condition can exist when these different fiber types meet in a connector joint. Furthermore, both wavelengths, 1310 and 1550 nm, can be experienced. Table 59.2 summarizes some typical mode field diameters.

It is noteworthy to look at the intrinsic loss for 1310 nm propagation. The diameter ratios for this wavelength of propagation are about 0.70 or 1.43 depending on which is the receiving fiber. As Fig. 59.4 shows, the intrinsic loss due only to mode field diameter mismatch is 0.55 dB. For the case of 1550-nm propagation, the ratios are 0.77 and 1.30, and the loss due to mode field diameter mismatch is 0.30 dB. Of course, this loss must be added to any extrinsic loss due to existing misalignments.

Continuing, we can also use this general formula to analyze the types of alignments that are necessary for high-performance connectors and splices. As previously stated, angular misalignments and longitudinal offsets (endface separation) are more easily minimized, in connectors and splices, than lateral offsets. Therefore, the main objective in connector and splice design is controlling the lateral offset between the adjacent fibers. For this reason, the preceding general formula can be

TABLE 59.2 Mode Field Diameters of Different Fiber Types

Fiber Type	Mode Field Diameter (wavelength, 1310 nm)	Mode Field Diameter (wavelength, 1550 nm)
dispersion-unshifted, μm	9.3	10.5
dispersion-shifted, μm	6.5	8.1

reduced to the following expression that describes the dependence of insertion loss on lateral offset.

$$\text{insertion loss} = 4.343 \left(\frac{x}{w}\right)^2 \text{ dB} \tag{59.3}$$

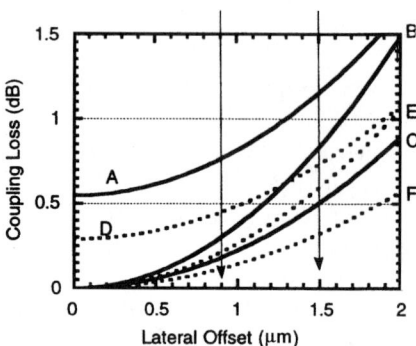

FIGURE 59.5 Plot of coupling loss vs lateral offset for various fiber type combinations.

This relationship assumes that identical fibers are in the joint, which is a valid assumption considering the small variation in mode field diameter in the manufacture of today's fibers and its effect on coupling loss. It also assumes no reflections, that is, zero end separation or perfect index matching. This expression has been used to plot insertion loss vs lateral offset in Fig. 59.5 for both dispersion-shifted (DS) and unshifted (US) fibers at both 1310- and 1550 nm wavelengths. Also plotted (using the preceding general formula) are losses for joints between these two different types of fibers. It should be noted that Eq. (59.3) has been verified in numerous experiments [Kummer and Fleming, 1982].

In Fig. 59.5, coupling loss (insertion loss) is plotted as a function of lateral offset between various combinations of fiber types. Curve A represents a joint between dispersion-shifted and dispersion-unshifted fibers, and curve B represents a joint between two dispersion-shifted fibers; both curves are for a propagation wavelength of 1310 nm. Also plotted is curve C, which represents a joint between two dispersion-unshifted fibers, and curve D a joint between a dispersion-shifted and dispersion-unshifted fibers; both of these curves are for a propagation wavelength of 1550 nm. Finally, joints between two dispersion-shifted and two dispersion-unshifted fibers, both for a wavelength of 1550 nm, are shown by curves E and F, respectively.

A typical specification for today's connectors is a maximum loss of 0.5 dB and an average loss of 0.25 dB. As can be seen, a lateral alignment of 1.5 μm is required for 0.5 dB and about 1 μm for 0.25 dB loss, using today's standard dispersion-unshifted fibers and an operating wavelength of 1310 nm (curve C). The information in Fig. 59.4 can then be used to predict the performance of today's connectors (connectors having 1.5- and 1-μm lateral offsets) with the various combinations of fiber types and operating wavelengths. Table 59.3 summarizes these losses for a 1.5-μm lateral offset.

TABLE 59.3 Insertion Loss for 1.5-μm Lateral Offset

	US/US Fibers, dB	DS/DS Fibers, dB	DS/US Fibers, dB
1310 nm (λ)	0.5	0.8	1.1
1550 nm (λ)	0.3	0.6	0.7

The preceding discussion is based on the fact that today's insertion loss specifications for high-performance connectors, dominated by lateral offsets, are usually a maximum of 0.5 dB and show averages of 0.25 dB that implies lateral offsets of 1.5 and 1 μm, respectively. In addition, to minimize reflections and insertion loss, the connectors must achieve fiber core-to-core contact throughout the operating range. How this submicrometer control is achieved in these butt-joint ferrule connectors is discussed in a following section.

Reflectance Factors

Reflectance is the ratio of reflected power to incident power at a fiber joint, expressed in decibels. Reflectance can therefore be defined by the following expression:

$$\text{reflectance} = 10 \log \left(\frac{\text{reflected power}}{\text{incident power}}\right) \text{ dB} \tag{59.4}$$

Fiber Optic Connectors and Splices

Historically, insertion loss has been the only performance criterion for connectors and splices. Recently, it has been recognized that a fiber joint's reflectance performance is also important because optical reflections from refractive index discontinuities in the fiber path can adversely affect the performance of direct-detection, coherent-detection, and analog lightwave communications systems, particularly in single-mode fiber applications [Mazurczyk, 1981]. In cases having multiple reflectors in the fiber path (such as multiple connectors and splices), high reflectance may also result in the interferometric conversion of laser phase noise to intensity noise, resulting in systems degradations and/or measurement limitations. Optical feedback may also cause detrimental changes in the performance characteristics of active components, resulting in further system degradations and measurement limitations. For example, optical feedback into a laser cavity can increase mode-partition noise and add noise caused by reflection-induced power fluctuations. In optical amplifiers, reflections can cause lasing or multiple reflection noise, thereby severely affecting amplifier characteristics, and drastically restricting the amounts of amplifier gain and/or reflectance that can be allowed in high-speed systems.

Thus, besides insertion loss, the reflectance of optical fiber connectors and splices has become an important optical performance criterion. In this regard, requirements for reflectance are not only routinely specified but are presently becoming more stringent, changing from the minimum of -30 to -40 dB.

Connectors

Since high-performance connectors are usually designed for **physical core-to-core contact** to reduce insertion loss, optically transparent fiber joints might be expected (effectively an optically continuous length of fiber). To understand why physical contact between fiber endfaces does not yet result in reflectances that are equivalent to a continuous fiber, we recall that when bulk silica glass is polished, a thin layer with a refractive index higher than the bulk glass is usually formed at the polished surface [Rayleigh, 1937]. This increase in refractive index is the result of compaction of the fiber endface material (increase in density) caused by the mechanics of the polishing process. Almost all connectors use polished endfaces. This polishing-induced increase in the refractive index of the fiber endface creates an interior index discontinuity parallel to the fiber endface. This index discontinuity occurs at a fraction of a micrometer from the fiber endface, and even when fiber endface contact is achieved these two index discontinuities within the fibers result in increased reflectance. Figure 59.6 shows a model of the endface for the **cleaved fiber** [Fig. 59.6(a)] and for the polished fiber [Fig. 59.6(b)].

The refractive index of the bulk fiber material (here the small difference between the core and cladding index is neglected), the gap material (generally air), and the index of the thin high-index layer are denoted by n, n_1, and n_2, respectively. The thickness of the high-index layer is denoted by h.

When physical contact between two polished fiber endfaces occurs, even though the interface between the fiber endface and the gap material disappears, two index discontinuities separated by a distance $2h$ still remain near the joint (as shown in Fig. 59.7).

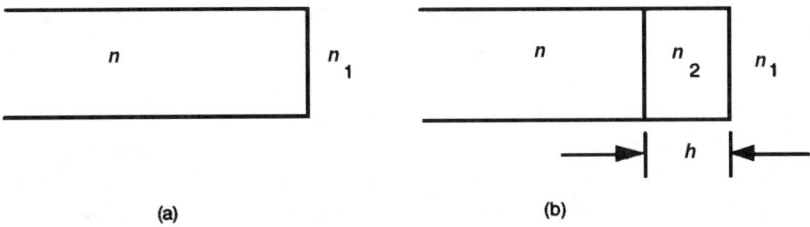

FIGURE 59.6 The model of (a) a cleaved and (b) a polished fiber endface depicting the high-index layer on the polished endface.

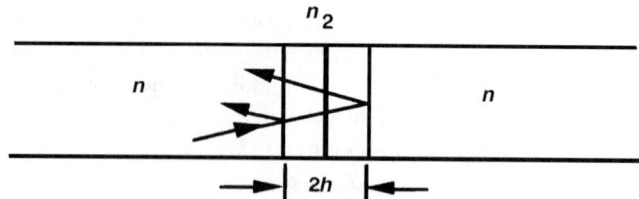

FIGURE 59.7 Schematic representation of two beam interference of a fiber joint having a high index-layer.

These two interfaces can cause a significant degradation in the connector's reflectance performance while only a negligible increase in insertion loss. This discussion assumes that the fiber endfaces are polished flat. Many of the commercially available ferrule-type physical-contact (PC) connectors use nonflat endfaces, as shown in Fig. 59.8.

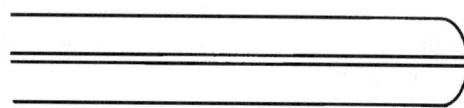

FIGURE 59.8 A nonflat, polished fiber endface.

Usually, the fiber endface is polished spherically convex, as shown in Fig. 59.8. The reason for using a nonflat fiber endface is to ensure fiber core-to-core contact. Because of point or line contact between the two contacting spherical surfaces, the axial compressive spring force in the connector body provides the necessary force to cause **controlled elastic deformation** of the fiber/ferrule endface, thereby achieving fiber endface contact. Thus, the model presented in Fig. 59.7 still holds for such *nonflat* fiber endfaces, and the reflectance performance is still limited by the two interfaces formed by the high-index layers previously described.

Furthermore, since the refractive index n_2 and the thickness h of the high-index layer at the polished fiber endface are dependent on various polishing conditions, such as the polishing material, polishing pressure, grit size of the polishing material, etc., the reflectance of PC connectors depends critically on the polishing technique. For example, polishing with diamond with grit size of 1 or 3 μm or with 1-μm grit size aluminum-oxide coated on a polyester sheet produces a smooth fiber endface having a high-index layer of almost 0.09 μm thick with the refractive index of about 1.54 compared 1.47 for bulk fiber glass. With such a layer present on a fiber endface, a theoretical value of approximately −30 dB is predicted for the reflectance of such a PC connector. Experimentally, −34 dB has been measured for similar connectors indicating good agreement with theoretical predictions.

When it was recently realized that it was possible to obtain better reflectance performance by using enhanced polishing techniques to significantly minimize the thickness of the compacted endface layer, the performance for some connectors changed from −30 to −40 dB. However, some of these techniques may also produce fiber endfaces that are recessed from the ceramic ferrule endface therefore making it more difficult to achieve fiber core contact in the connector. The reflectance performance also depends on various environmental conditions. Some conditions may produce an air gap in connectors designed for physical contact, resulting in multiple beam reflections.

Splices

Recall that there are two types of fiber splices, fusion splices and mechanical splices. Fusion of the two fibers together, with respect to reflectance, results in an optically transparent length of fiber. With these types of splices it is very difficult to measure any level of reflectance. On the other hand, because the index matching material used for mechanical splices is temperature dependent, mechanical splices can be exhibit wide ranges of reflectance. It is not uncommon for a mechanical splice, cleaved perpendicular fiber endfaces, to have reflectances between −25 and −45 dB over a temperature range of −40° and 80°F. In order to avoid this temperature dependence cleaved angled fiber endfaces have been recently employed. With endface angles between 5° and 10°, and index matching material, mechanical splices with reflectance performances equal to fusion splices can be easily achieved. Furthermore, it can be theoretically shown that the fibers having angled endfaces can be joined within the splice hardware without regard to ensuring that the endfaces are parallel.

59.3 Multibeam Interference (MBI) Theory

MBI Effects on Transmission

The previous sections discussed factors that cause insertion loss and reflectance in the butt coupling of two single-mode fibers. It was shown that as long as fiber endface contact is achieved with connectors or perfect index matching is achieved with splices the dominating factor that causes insertion loss in typical high-performance connectors and splices is lateral offset. Furthermore, as the connector's reflectance performance became an important performance criterion, it became a requirement that fiber core-to-core contact be achieved in order to minimize both insertion loss and reflectance. Consequently, it is necessary to ensure fiber endface contact over the expected environmental and climatic conditions. When contact is lost, an air gap is created between the fibers and two glass/air interfaces are formed. The Fresnel reflection at the two interfaces (here, for the sake of clarity, cleaved fiber endface fiber endfaces are assumed) results in an additional insertion loss of 0.32 dB, as shown by the following formula:

$$\text{insertion loss} = -10\log(1 - 2R) \text{ dB}$$

where

$$R = \text{Fresnel reflecion} = \left(\frac{n_f - n_g}{n_f + n_g}\right)^2 \tag{59.5}$$

Although an increase in insertion loss of 0.32 dB nearly doubles the typical performance of high-performance connectors, it is even more important that the two adjacent fibers and the air gap in the connection form a Fabry–Perot etalon. The resulting multiple beam interference, from this gap, can cause additional increases in insertion loss. In analyzing this effect, consider the two beams, as shown in Fig. 59.9.

Figure 59.9 represents an air gap between two fiber endfaces. The two beams shown in Fig. 59.9 will interfere with each other both constructively and destructively, depending on the physical path length difference $2x$. The effect that this two-beam interference can have on the insertion loss performance of the connector can be quite severe. The following expression can be used to evaluate the effect of this multiple beam interference for the small end separations that unfortunately might develop in some high-performance connectors. In cases having larger end separation (not applicable to connectors or splices), it is necessary to include the effects of the coherence length of the optical

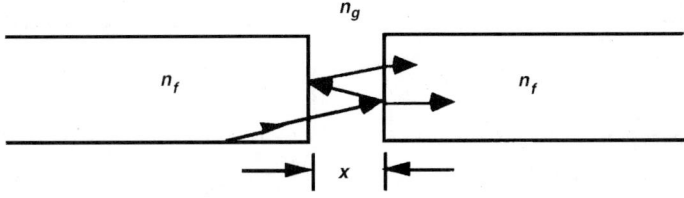

FIGURE 59.9 Multiple beam interference model for an air gap between two single-mode fibers.

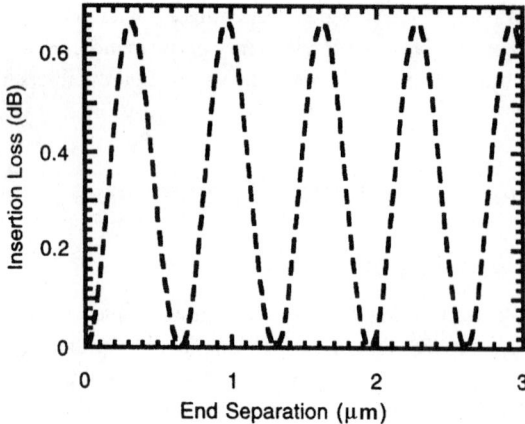

FIGURE 59.10 Plot of insertion loss vs endface separation of a butt-joint connector.

source, aperturing of the fiber core, and diffraction of the emitted beam.

$$IL = -10\log\left[1 - 2R\left(1 - \cos\left(\frac{4\pi n_g s}{\lambda}\right)\right)\right] \text{ dB}$$

where

$$R = \left(\frac{n_g - n_f}{n_g + n_f}\right)^2 \tag{59.6}$$

This expression was used to plot insertion loss vs end separation shown in Fig. 59.10. As shown, the insertion loss is highly dependent on the fiber endface separation x, and as a result, the insertion loss can increase by almost 0.7 dB for just a submicrometer air gap. When the realistic case of a polished connector is considered, the presence of a polishing-induced high-index layer on the fiber endface can increase the peak-to-peak variation in insertion loss from 0.7 dB to as much as 1.1 dB [Shah, Young, and Curtis, 1987]. For a performance level of 0.5 dB, such large loss variations are clearly not acceptable and, therefore, must be avoided.

MBI Effects on Reflectance

Next, we consider the effect of multiple beam interference on reflectance performance. In analyzing this effect, consider the two reflected beams shown in Fig. 59.11.

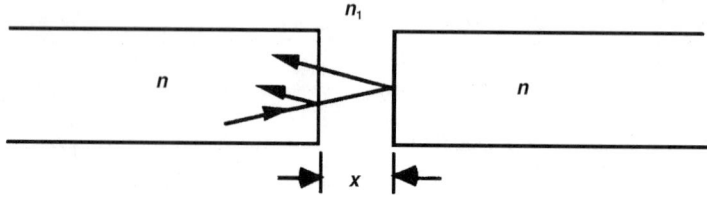

FIGURE 59.11 Schematic representation of two beam interference of a fiber joint having cleaved fiber endfaces.

Fiber Optic Connectors and Splices

We begin by analyzing the reflectance of a fiber butt joint that has fiber endface separation without index matching, an air gap. This type of fiber joint results in two fiber/air interfaces, and the reflectance is caused by the resulting Fresnel reflections. Again, in cases that have a large air gap, with reference to the coherence length of the optical source, multiple-beam interference effects can be neglected and the following expression can be used to calculate a reflectance of -11.4 dB:

$$\text{reflectance} = 10\log(2R) \text{ dB}$$

where

$$R = \left(\frac{n_g - n_f}{n_g + n_f}\right)^2 \tag{59.7}$$

Where
n_g = refractive index of gap
n_f = refractive index of fiber

Even though a reflectance level of -11.4 dB is much greater than today's typical performance requirement of <-40 dB, interference effects can occur with air-gap joints that are even more of a concern. Referring to Fig. 59.11, the two reflected beams shown can interfere with each other both constructively and destructively depending on the physical path length difference x, and the effect that this two-beam interference can have on the reflectance performance of the connector can be quite severe. Figure 59.12 illustrates the results of an often verified theoretical analysis.

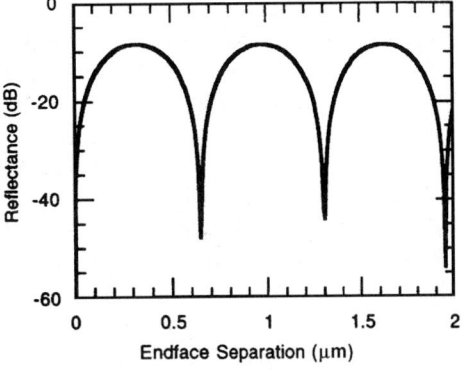

FIGURE 59.12 Theoretical plot of reflectance vs endface separation for two cleaved fiber endfaces.

As Fig. 59.12 shows, a large dependence exists on the fiber endface separation x, and as a result, the reflectance can vary from -8.5 dB to $-\infty$ within a quarter-wavelength change in separation. Since the reflectance at its lowest value is an extremely sensitive function of air gap (asymptotic), its average value tends to be very close to the largest reflectance value, -8.5 dB for the case shown in Fig. 59.11. As previously discussed, a large change in the insertion loss (around 0.7 dB for cleaved endfaces) will simultaneously occur. Thus, when any air gap exists, the connector reflectance performance usually will be severely degraded and approach -8.5 dB, as compared to the extremely low reflectance when fiber endface contact is achieved.

In the case of the polished fiber endfaces, Fig. 59.13 is a schematic of the two-beam reflection model when a high-index layer exists on both fiber endfaces and physical contact between fiber

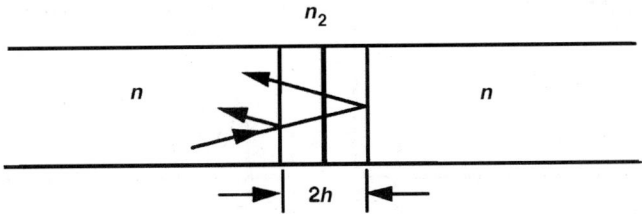

FIGURE 59.13 Schematic representation of two-beam interference of a fiber joint having a high-index layer.

endfaces is achieved. As Fig. 59.13 shows, the two-beam reflection model used in Fig. 59.11 can still be applied to determine the reflectance performance of the connection between two polished fiber endfaces under physical contact.

The amplitudes and relative phase-shifts of the two beams reflected from the polishing-induced index discontinuities (shown in Fig. 59.13) are directly determined by the refractive index of the polishing-induced layer n_2 and the total thickness of this layer $2h$, thereby determining the connector's reflectance performance. Assuming the refractive index of the layer is 1.54, the reflectance performance can be calculated as a function of the total thickness $2h$. As Fig. 59.14 shows, the reflectance increases from $-\infty$ to -40 dB as the layer thickness increases from 0 to 0.0075 μm (total thickness is 0.015 μm).

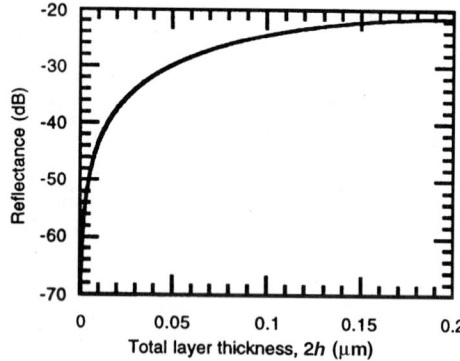

FIGURE 59.14 Reflectance vs high-index layer thickness for index of 1.54.

The reflectance attains its maximum value of about -21 dB for a layer thickness of 0.1 μm (total thickness is 0.2 μm), for a wavelength of 1310 nm, and then decreases with further increase in layer thickness. As Fig. 59.14 shows, reflectances of -30, -40, and -60 dB require polishing techniques that result in a layer thickness of 0.009, 0.0075, and 0.00075 μm, respectively.

Since physical contact may be lost due to environmental factors such as temperature changes and/or due to variations in fiber endface profiles resulting in an air gap at the joint, further analysis of the interference effects is required. Using the model in Fig. 59.15, three types of plugs, polished with different polishing procedures (different high-index layer thickness), are considered.

The hypothetical plugs are such that under physical contact condition, the connector reflectance values are -60, -40, and -30 dB, respectively. It is assumed that the index of refraction of the high index layer is 1.54 and that the plugs have different layer thicknesses that yield different reflectance performance.

The reflectances of each of these joints plotted as a function of air gap between the fiber endfaces at a wavelength of 1310 nm are shown in Fig. 59.16.

The plot is an expanded view of the region near zero endface separation showing a rapid asymptotic decrease in reflectance from its value at physical contact to $-\infty$ dB at a small nonzero separation. This effect is not consistent with the cleaved fiber endfaces, where the first reflectance minimum occurs at physical contact. The separations for minimum reflectance for the three plugs are 0.00038, 0.0056, and 0.0385 μm, respectively. Beyond these separations, the reflectance again increases. Thus, for polished fiber endfaces, a situation exists wherein the reflectance performance actually improves within a small range of nonzero endface separation.

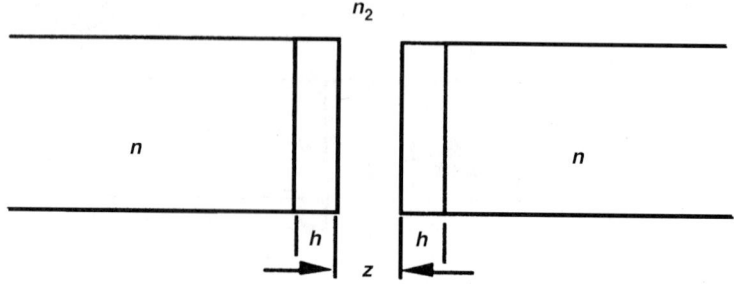

FIGURE 59.15 Air-gap joint with two polished fibers having high-index layers.

Fiber Optic Connectors and Splices

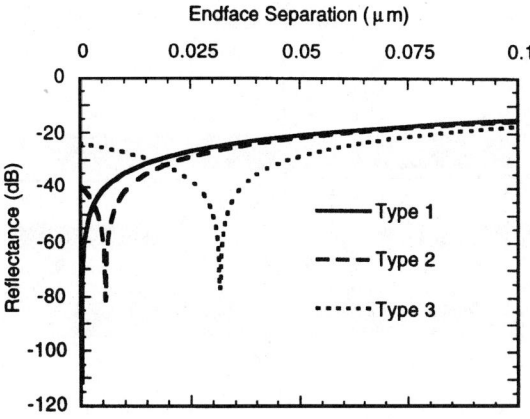

FIGURE 59.16 Reflectance vs endface separation for three different polishing procedures, −60, −40, and −30 dB; type 1, type 2, and type 3, respectively.

In summary, excellent reflectance performance can be achieved with single-mode fiber, physical-contact connectors when appropriate endface polishing techniques are used. In particular, with the enhanced polishing techniques that have been recently developed, reflectance performance of < −40 dB can be routinely obtained, provided fiber endface contact is maintained. However, if the fiber/ferrule endface polishing techniques are not fully under control, the performance of physical contact connectors may not be maintained over all environmental conditions or with random intermating of connectors using different endface polishing treatments.

59.4 Connector Design Aspects

Introduction (Types of Connectors)

There are two types of connector designs that have received attention, the expanded-beam type and the butt-joint connector. The expanded-beam connectors have not yet found acceptance in communications systems, and are unlikely to find a niche (based on past results); therefore, this chapter does not discuss them in detail. However, for the sake of completeness, we will briefly describe the philosophy of this type of connector design.

The goal of the expanded-beam connector is to make the connector more tolerant to lateral alignment inaccuracies. To accomplish this, the fibers are terminated with a collimating/focusing element (lenses, as shown in Fig. 59.17) that expands and collimates the beam, thereby facilitating connecting and disconnecting [Nica and Tholen, 1981].

Initially, it appears that the expanded-beam connector has a large advantage over the butt coupling of fibers. However, there are aspects of the design that require the same degree of precision alignment. Although it is a fact that the expanded and collimated beam is more tolerant to lateral offsets, angular misalignments of the collimated beam, even as small as a few arc-seconds, can result in large insertion losses. These angular misalignments of the expanded beam result in a lateral displacement of the focus point, resulting in high coupling losses between the fiber and lens. Consequently, the fiber and lens must be aligned to the same submicrometer tolerances that are required in butt coupling of the fibers. Furthermore, to avoid spherical aberrations, and the resulting wavelength dependency of insertion loss, aspheric lenses must be used. Because of these concerns, and the general complexity of the designs, expanded-beam connectors have yet to find general acceptance in communications systems.

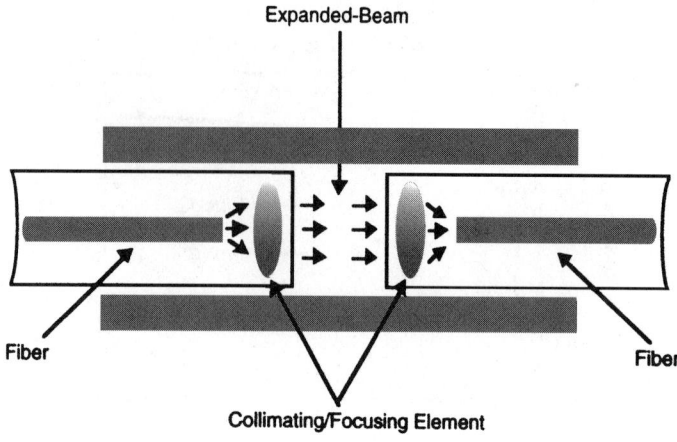

FIGURE 59.17 Schematic of an expanded-beam type optical fiber connector.

Communications systems, both multimode and single-mode fiber systems, have successfully used butt-joint connectors since the very first applications in the mid-1970s. Furthermore, despite the submicrometer alignment tolerances required, more than 30 million butt-joint connectors have been manufactured and deployed. This chapter focuses on the cylindrical-ferrule, physical-contact, butt-joint connectors that are the choice for optical communications systems. Two popular versions of these connectors, bayonet-lock fastened, and push/pull snap-lock fastened, are shown in Figs. 59.18 and 59.19, respectively.

Factors Contributing to Insertion Loss

Minimizing the lateral offsets between butt-coupled fibers and achieving fiber core-to-core contact are the two most challenging aspects of designing high-performance single-mode fiber connectors. The model in Fig. 59.20 shows the major inaccuracies encountered in minimizing the centering of the fiber's optical field within a single-mode fiber ferrule.

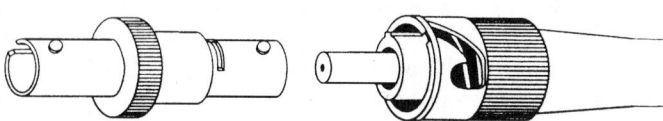

FIGURE 59.18 Bayonet-lock fastened cylindrical ferrule ST type connector.

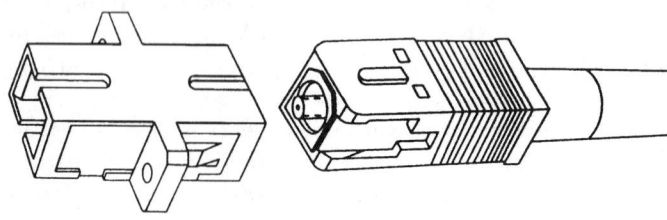

FIGURE 59.19 Push/pull snap-lock fastened cylindrical ferrule SC type connector.

Fiber Optic Connectors and Splices

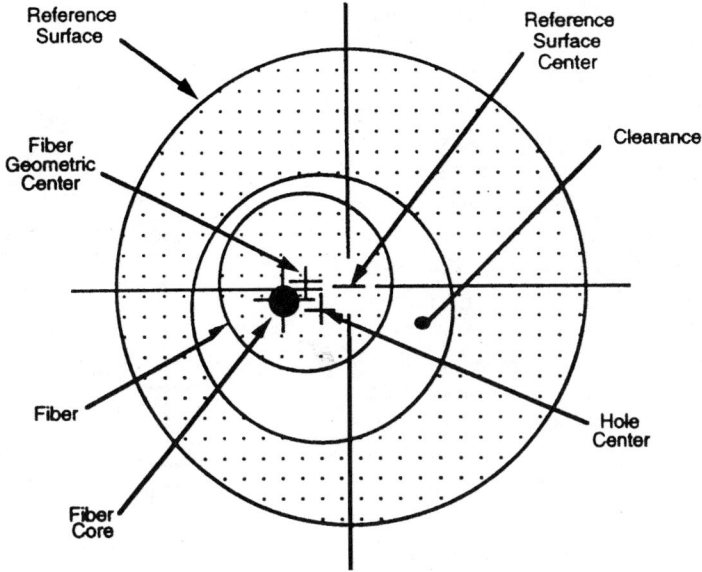

FIGURE 59.20 Model showing inaccuracies encountered in assembling a connector ferrule.

In analyzing the model, the first consideration is the quality of the fiber. As Fig. 59.20 shows, the axial offsets between the geometric center of the fiber and the fiber core center (optical center of the mode field diameter), and the clearance between the ferrule's hole and the outside diameter of the fiber, are major contributors to the final quality of the connector ferrule assembly (the lateral offset between the reference surface center of the ferrule and the fiber core).

The contributions from the physical dimensions of the ferrule must be also considered. Similarly to the fiber, fabrication of the connector ferrule results in additional inaccuracies that must be accommodated in the connector design. The outside diameter of the ferrule will vary, as will the hole diameter and the offset between the hole center and the reference surface center. As can be seen in Fig. 59.20, the offset that must be minimized is the offset between the fiber core center and the reference surface center. Recall that when two single-mode fibers are butt coupled, in order to achieve insertion losses <0.25 dB, the total lateral offset should be less than 1 μm.

Statistical analyses of the inaccuracies shown in Fig. 59.20, such as Monte Carlo simulations, have been carried out to quantify the tolerances required for a specific optical performance [Young, 1984]. These statistical analyses have also been experimentally verified. First, to achieve total lateral offsets of <1 μm, the diametral clearance between the hole in the ferrule and the fiber must also be <1 μm.

To more fully understand the difficulty in analyzing the required lateral offsets in the ferrule assembly, two popular ferrule assembly techniques are examined. To begin our analysis, recall that to achieve 1-μm lateral alignment between fiber cores within a connection, which is required for 0.25-dB insertion loss, we need on average 0.5-μm centering of the fiber core within each ferrule. This means that if the ferrules have the same outside diameters, and the ferrule axes are perfectly aligned by the alignment sleeve, then we need to consider only the overlap area that is established by the locus of the core centers in each ferrule assembly. This type of ferrule having the fiber cores centered to within less than 1 μm is shown in Fig. 59.21.

If we consider the popular ceramic cylindrical ferrule connectors, the manufacturing processes do not routinely yield the necessary hole centering. To overcome this aspect of these ferrules, **performance keying** or tuning is frequently used. Figure 59.22 demonstrates how performance keying can be used to achieve a similar size area for the locus of fiber core centers that is achieved in the directly centered ferrules. In performance keying, the location of the key is determined after the fibers are inserted into the ferrules and the connector endface is finished. Usually the keying is done by measuring the insertion loss as a function of ferrule rotation against a standard reference ferrule. The use of performance keying is widespread with single-mode fiber cylindrical ferrule connectors.

Finally, aligning the two ferrules and realizing a connection requires an alignment sleeve. This alignment sleeve is usually allowed to float in a housing that has the appropriate coupling features (screw threads, etc.). This component is referred to as either a coupling or an adaptor. A schematic representations of a cylindrical ferrule connector is shown in Fig. 59.23. At first observation it might appear that the cylindrical alignment sleeve is a trivial component. However, this is not the case. The materials of the sleeve have ranged from plastic, austenitic stainless steel, copper alloys, and glass, to ceramic. It appears there is a tendency today to use **single-slit ceramic sleeves.** The concern over the alignment sleeve designs for the cylindrical ferrule connector is driven by the fact that the sleeve must coaxially align ferrules that have unequal diameters. Even though the diameter of ferrules can be manufactured to a tolerance of 1 μm, an insertion loss performance of <0.25 dB requires lateral alignment of the cores of <1 μm. Additionally, this 1-μm tolerance has to accommodate all inaccuracies, including the fiber core offsets within the ferrules. For these reasons the alignment sleeve is indeed an important component of an optical fiber connector.

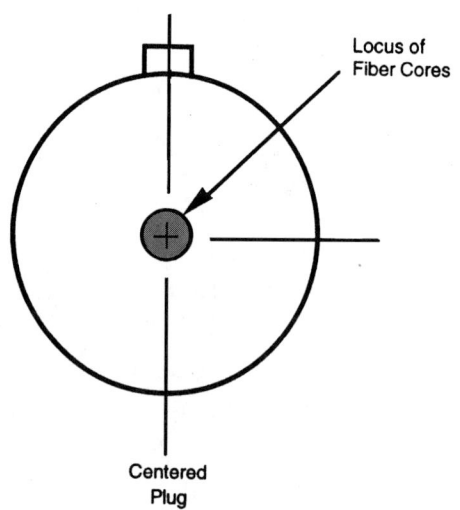

FIGURE 59.21 Ferrule having fiber cores centered to required tolerance.

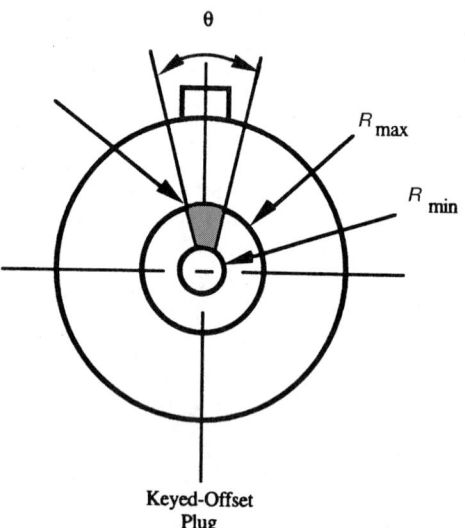

FIGURE 59.22 Model of a performance keyed ferrule.

Factors Contributing to Reflectance (Physical Core-to-Core Contact)

As mentioned previously, besides achieving submicrometer lateral alignments, in order to realize low insertion loss and low reflectance, high-performance connectors must achieve and maintain fiber core-to-core contact. Again referring to physical contact cylindrical ferrule connectors, at first glance (Fig. 59.23) it appears trivial to achieve physical core contact. However, one has to investigate the geometry in finer detail.

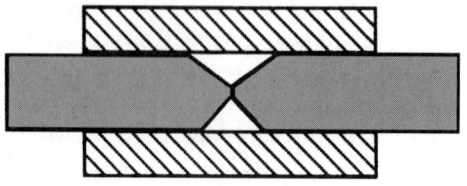

FIGURE 59.23 Schematic of a cylindrical ferrule alignment mechanism.

Fiber Optic Connectors and Splices

As stressed in this chapter, the lateral offset between adjacent single-mode fibers must be controlled to less than 1 μm in order to achieve 0.25 dB insertion loss performance. The cylindrical geometry therefore must maintain very small diametral clearances between the ferrules and the alignment sleeve to minimize lateral offsets. These small clearances require precise perpendicularity of the ferrule/fiber endfaces to achieve physical contact of the fiber cores. In practice, this required precision in perpendicularity has not been realized. As a result, the usual practice is to use spherically polished endfaces [Shintaku, Nagase, and Sugita, 1991]. In theory, the spherical endface should result in the fiber core being the highest point on the ferrule endface, thereby always achieving core contact. However, since the material of choice for manufacturing the submicrometer precision ferrule is a

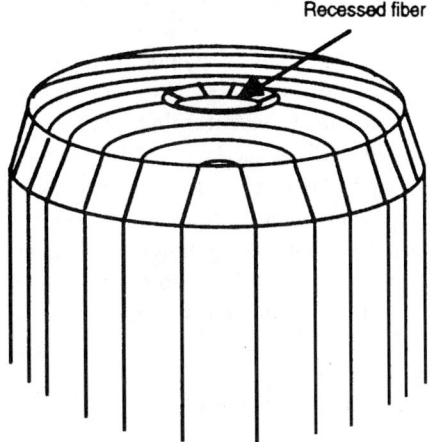

FIGURE 59.24 Ferrule endface with recessed fiber.

ceramic, spherical endface polishing tends to result in a slight recessing of the fiber endface within the ferrule tip (Fig. 59.24). This recessing is caused by the differential polishing rates of the ceramic ferrules and the glass fibers. This recessing of the fiber endfaces is overcome in the cylindrical ferrule connector by using **controlled elastic deformation** of the ferrules.

Therefore, when two ferrules are forced into contact (Fig. 59.25) by the compression springs within the fastening hardware, line contact is established and large hertz contact forces are exerted on the ferrules' endfaces. If one correctly selects the ferrule's material properties, specifies the radius of the sphere, and controls the amount of fiber recess, then a **controlled elastic deformation** of the ferrule can be used that results in fiber core-to-core contact. If these parameters are not controlled, an unstable physical contact connection can result.

To summarize this section on fiber core-to-core contact connectors, the **controlled elastic deformation** technique with type PC connectors is presently the overwhelming choice for low reflectance and low insertion loss connections. There are many details and features that must be controlled to achieve and maintain this core-to core contact to provide the appropriate elastic deformation within the connectors. Some of these are manufacturing know-how and must be fully developed and socialized within the connector community before appropriate intermateability standards can be realized.

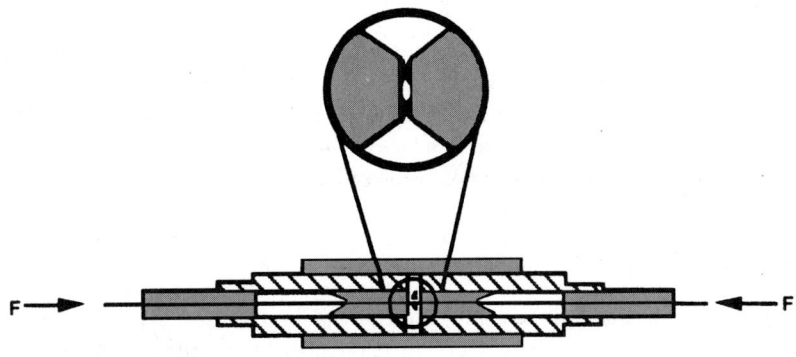

FIGURE 59.25 Ferrule endface contact of spherical ferrule endfaces with fiber recess.

59.5 Splicing Design Aspects

Fiber splices are used in applications where the goal is to achieve a permanent joint between two fibers. Since the splice will be permanent there are no intermateability concerns as with connectors. Without these concerns splices are easily designed to have very low insertion losses, <0.1 dB, and very low reflectance, <−55 dB. With the recent improvement in fiber geometry, multiple fiber splices are now becoming feasible as well. As previously mentioned, there are two types of fiber splices, mechanical splices and fusion splices. The recent use of angled endfaces in mechanical splices have made their optical performance equal to that of fusion. A large differentiator between them is the required level of initial capital investment. Fusion splicing is capital intensive where mechanical splicing is more or less a buy-as-needed process.

Mechanical Splices

Mechanical splicing techniques use butt-joint coupling similar to those used in connectors. However, the fact that the splice is permanent implies there does not need to be any intermateable parts. This fact results in their higher optical performance and lower cost. Essentially, the mechanical splice is a consumable part that mechanically aligns the fibers by referencing on their outside diameters [Patterson, 1988]. The splicing process is quite simple. The plastic coating on the fiber is removed and the fiber endface is cleaved, either perpendicular or with an angle. The two cleaved fibers are inserted or placed in the consumable part (the splice) and are usually held in place by a simple clamping method. The low loss is attributable to the use of a continuous alignment surface similar to a v-groove, and the use of an index matching material in the very small gap that exists between the fibers to eliminate reflections. A typical mechanical splice is depicted in Fig. 59.26.

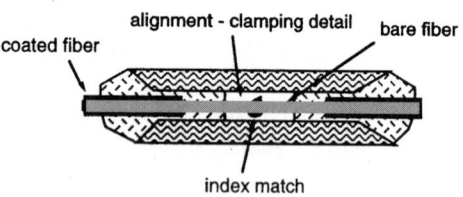

FIGURE 59.26 Schematic of a typical mechanical splice.

The first mechanical splices and most used today employ perpendicular endfaces. Since the cleaved endfaces are not exactly perpendicular, fiber core contact is not achieved within the splice. Consequently there exists a gap, which is filled with index matching material. Because the refractive index of the available materials varies with temperature, two index mismatches are created, one at each fiber end. These index mismatches result in multiple beam interference as in the case with connectors. These mismatches are not sufficient to cause significant increases in insertion loss (about 0.002 dB) but can severely effect the reflectance performance of the splice. As previously mentioned in this chapter, reflectances ranging between −25 and −40 dB over the expected temperature range can be realized. With the recent reflectance requirements of less than −40 dB, angled endfaces are starting to be used in mechanical splices to avoid this temperature dependence [Young, Shah, and Curtis, 1988]. Simple cleaving tools have been developed that provide endface angles ranging between 5° and 10°. These angled endfaces can be used in existing mechanical splices and thereby yield reflectance performance equal to that of fusion splices.

Fusion Splices

Much progress has been made in the performance of equipment used to perform arc fusion of optical fibers. Essentially the process is quite simple, the fibers' butt ends are heated to just below their melting point and pressed together to form a continuous fiber [Kato et al. 1984]. After the fusion process is complete a strength reinforcement member is applied to protect the bare fiber. Table 59.4 lists the typical steps used in the fusion process. In Fig. 59.27 a simple schematic of a holding and alignment mechanism that is part of most splicing machines is shown. In reality these

Fiber Optic Connectors and Splices

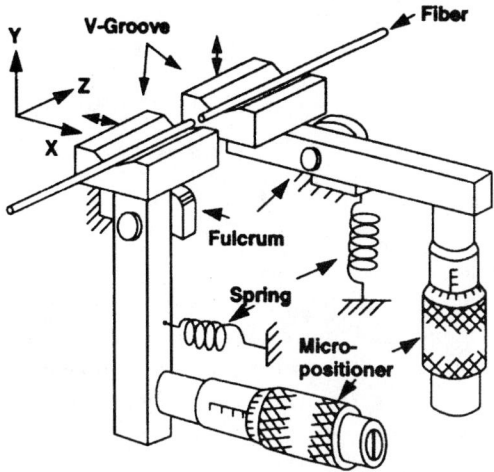

FIGURE 59.27 Schematic of holding and alignment mechanism.

mechanisms are quite sophisticated, involving computer controlled movements to achieve accurate alignment and positioning. The machines are fully automatic and also provide an estimation of the splice loss.

A example of a fusion splicing cycle is depicted in Fig. 59.28. The cycle is usually fully automatic. The operator loads the cleaved fibers in the v-grooves and grossly positions the ends. The cycle begins by moving the fibers into contact and then separating them a fixed amount. Then the arc is energized with a specific current and the fibers are moved toward each other. This step is used to clean and round the edges of the fiber ends. When the fiber ends reach the contact point the current is increased to soften the fibers, and the fibers are driven past the original contact point (pressed together). Finally the current is decreased and brought to zero. Although the process seems quite simple, the tolerances for these movements and currents can be quite demanding.

In the fusion splicing of multimode fibers, the surface tension between the fiber endfaces, while softened, can be used to self-align the fibers with reference to their physical geometric axes (outside diameters). This technique makes the fusion splicing of multimode fibers easy and reliable. With the small mode field diameters of single-mode fibers and the existence of fiber core eccentricities, this self-alignment technique has usually needed to be avoided. To avoid this self-alignment requires precise control of the temperature or softness of the end faces during fusion, the distance the fibers move beyond their original contact location (overlap distance), and the fiber cleave angles. Cleave angles of less than 1°, specific arc duration times within 0.1 s, and overlaps to within 1 μm are usually maintained. Also, since the insertion loss target is typically <0.1 dB it usually requires that the fibers are aligned with respect to their mode field diameters not their outside diameters. Various techniques have been used to achieve and monitor these alignments. For these reasons the fusion splicer has developed into a very sophisticated apparatus.

TABLE 59.4 Typical Steps in Fusion Splicing

- Cable structure removal
- Fiber coating removal
- Cleaning of bare fiber
- Fiber cleaving
- Fusion process
- Installation of strength reinforcement detail

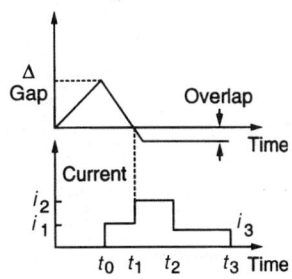

FIGURE 59.28 Example of a fusion splicing cycle.

Recently, there has been a marked improvement in the concentricity and diameters of single-mode fibers. These improvements have had a positive impact on fusion splicing of these fibers. The self-aligning of the fibers is becoming acceptable. Also the relaxed tolerances on fiber cleave angles and overlap distances have resulted in simpler and less costly splicers, as well as the multiple fiber fusion splices.

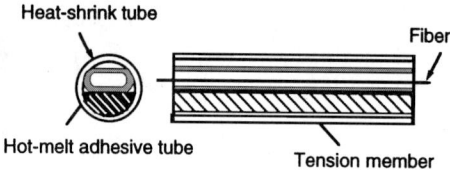

FIGURE 59.29 Fusion splice reinforcement detail.

Finally, to complete the splicing operation the splice must be reinforced. Typically, the tensile strength of an unprotected or reinforced splice is between 10 and 15% of the original coated fiber's strength. The strength of the splice can be improved to between 50 and 70% with the use of a reinforcement detail. A example of a reinforcement detail is shown in Fig. 59.29.

59.6 Conclusions

Major advances have been made in the past 20 years with regard to fiber optic connector and splicing technology. The performance of these components have dropped from 1.5 to 0.25 dB for connectors and to <0.1 dB for splices. Along with impressive improvement transmission performance, reflectance performances have also improved dramatically. Connectors and mechanical splices have been reported and marketed with reflectances <-55 dB. These performances have also been achieved with a drastic reduction in cost. However, in spite of a lot of work over the past 15 years, the intermateability standards for these connectors are not yet published. It hopeful that intermateability standards will soon be finalized at the international level.

With regard to fusion splices there has also been impressive improvement in recent years. Much of this improvement can be attributed to the improvement in the geometry of the optical fibers themselves. The increase in the precision of the fibers has resulted in simpler and less costly splicing machines, as well as the feasibility of multifiber splicing.

Defining Terms

Cleaved fiber: A fiber having an endface prepared by a combination of a prescribed stress distribution and an induced flaw that follows the Griffith fracture theory.

Controlled elastic deformation: A technique used to achieve core-to-core contact when the fiber endfaces have been recessed below the endface of the connector's ferrule.

Dispersion-unshifted fiber: A single-mode fiber designed to have a minimum dispersion at a wavelength of 1550 nm, the intended operating wavelength window.

Dispersion-unshifted single-mode fiber: A single-mode fiber designed to have minimum dispersion at a wavelength of 1310 nm, the intended operating wavelength window.

Performance keying: A method used to overcome the errors in coaxially aligning the fiber core center with the ferrule's reference surface axis resulting in low insertion loss connections.

Physical core-to-core contact: A connector designed to achieve physical and optical contact between the core areas of adjacent fibers, thereby minimizing insertion loss and reflectance caused by reflections.

Reflectance: Reflectance is analogous with the term return loss, the only difference being that one term is the negative of the other.

Single-slit alignment ceramic sleeves: A ceramic sleeve having a precise inside diameter and a logitudinal slit along it length that increases its compliance to varying ferrule diameters.

References

Chu, T.C. and McCormick, G.R. 1978. Measurement of loss due to offset, end separation and angular misalignment in graded-index fibers. *Bell Sys. Tech. J.*, 57(3):595–602.

Kato, Y. Tanifuji, T. Kashima, N., and Arioka, R. 1984. Arc-fusion splicing of single-mode fibers. *Appl. Opt.*, 23:2654–2658.

Kummer, R.B. and Fleming, S.R. 1982. Monomode optical fiber splice loss. In *Proceedings of Optical Fiber Communications Conference.*

Marcuse, D. 1977. Loss analysis of single-mode fiber splices. *Bell Syst. Tech. J.*, 56:703–117.

Mazurczyk, V.J. 1981. Sensitivity of single-mode buried heterostructure lasers to reflected power at 274 Mbit/s. *Electron. Lett.*, 17:143.

Miller, C.M., Mettler, S.C., and White, L. 1986, *Optical Fiber Splices and Connectors*, Marcel Dekker, New York.

Nemota, S. and Makimoto, T. 1979. Analysis of splice loss in single-mode fibers using a Gaussian field approximation. *Opt. Quantum Elec.*, 11:447.

Nica, A. and Tholen, A. 1981. High-efficient ball-lens connector and related functional devices for single-mode fibers. In *Proceedings of 7th European Conference on Optical Communications.*

Patterson, R.A. 1988. Mechanical optical fiber splice containing and articulated conformable metallic element. In *Proceedings of International Wire and Cable Symposium*, Reno, NV.

Rayleigh, Lord. 1937. The surface layer of polished silica and glass with further studies on optical contact. *Proc. R. Soc. London, Ser. A*, 160:507–526.

Shah, V., Young, W.C., and Curtis, L. 1987. Large fluctuations in transmitted power at fiber joints with polished endfaces. In *Proceedings of OFC'87*, TUF4, Reno, NV.

Shintaku, T., Nagase, R., and Sugita, E. 1991. Connection mechanism of physical-contact optical fiber connectors with spherical convex polished ends. *Appl. Opt.*, 30(36).

Wagner, R.E. and Sandahl, C.R. 1982. Interference effects in optical fiber connections. *J. Appl. Opt.*, 21(8):1381–1385.

Young, W.C. 1984. Single-mode fiber connectors—Design aspects. In *Proceedings 17th Annual Connector and Interconnection Tech. Symposium*, Electronic Connector Study Group. pp. 295–301.

Young, W.C., Shah, V., and Curtis, L. 1988. Optimization of return loss and insertion loss performance of single-mode fiber mechanical splices. In *Proceedings of International Wire and Cable Symposium*, Reno, NV.

Further Information

Several technical conferences provide the latest trends and results on fiber optic connectors and splices. Two of the major conferences to follow are the Optical Fiber Communications (OFC) Conference sponsored by the Optical Society of America (OSA) and IEEE (LEOS) and the International Wire and Cable Symposium sponsored by the U.S. Army Electronics Command, Fort Monmouth, NJ. Both of these conferences are held annually.

Three major journals that report latest work are the *Journal of Lightwave Technology* (OSA), *Applied Optics* (OSA), and *Electronic Letters* (IEE).

Generic requirements and future trends can be followed in the *Bellcore Digest of Technical Information*, Piscataway, NJ.

To follow the standardization of connectors and splices in the U.S. the Telecommunications Industries Association (TIA), Washington, D.C. is recommended. To follow international standardization refer to the International Electrotechnical Commission in Geneva.

60
Passive Optical Components

Joseph C. Palais
Arizona State University

60.1 Introduction 824
60.2 Losses in a Passive Optical Component 824
60.3 Grin Rod Lens 825
60.4 Isolator 825
60.5 Directional Couplers 826
60.6 Star Coupler 827
60.7 Optical Filter 828
60.8 Attenuator 829
60.9 Circulator 829
60.10 Mechanical Switch 830
60.11 Polarization Controller 830

60.1 Introduction

Passive optical components play a large role in making simple fiber systems (such as point-to-point links) practical and extending their applications to more complicated systems (such as local-area and subscriber-distribution networks). This chapter limits its coverage to a few of the more important bulk passive components not covered elsewhere in this book.

The components to be described are the following optical devices: GRIN rod lens, isolator, directional coupler, star coupler, optical filter, attenuator, circulator, mechanical switch, and polarization controller.

60.2 Losses in a Passive Optical Component

A generic passive optical component is shown in Fig. 60.1. It has several ports (as few as two and, in some cases, over 100), all of which can be input or output ports. The transmission loss generally refers to the coupling efficiency between any two ports. In terms of decibels, the transmission loss is then

$$L = -10\log(P_{out}/P_{in}) \tag{60.1}$$

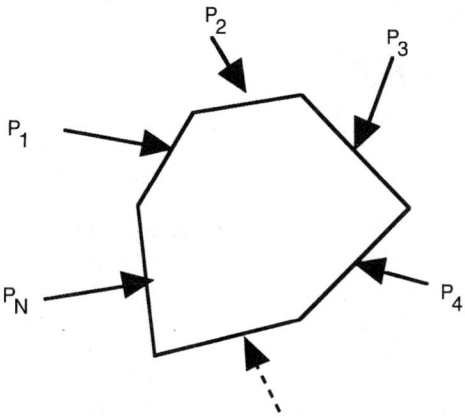

FIGURE 60.1 Passive optical component with N ports.

where P_{in} and P_{out} are the input and output powers at the two ports being considered. **Insertion loss** is commonly used to refer to the loss in a desired direction of signal travel and **isolation** or **directionality** is used for the transmission loss between two ports where coupling is undesirable. Typical losses for different devices will be presented in the paragraphs where the devices themselves are described.

60.3 Grin Rod Lens

The GRIN rod lens is a short cylindrical length of graded index (GRIN) material that functions as a lens. It is shown schematically in Fig. 60.2 being used to couple between two fibers. An advantage of this connector is the improved tolerance to lateral offsets because of the expanded beam. Another advantage is that significant gaps can exist between the lenses without large losses because of the collimation of the beam. This property also allows the GRIN rod lens to be part of functional devices where gaps are needed to accommodate additional components such as filters, beamsplitters, and gratings. Examples of useful devices employing the GRIN lens will be described later.

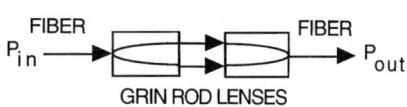

FIGURE 60.2 GRIN rod lens connector.

60.4 Isolator

An isolator, illustrated in Fig. 60.3, is a one-way transmission line. Its defining characteristics are insertion loss (the loss in the direction of allowed propagation) and isolation (the loss in the undesired direction of propagation). Typical insertion losses are less than 1 dB and isolations are 25 dB or more.

FIGURE 60.3 Isolator showing preferred direction of beam travel.

Isolators are used to protect laser diodes from back reflections, which can increase the laser noise level. Isolators are also needed in optical amplifiers to prevent feedback, which could initiate unwanted oscillations of a beam bouncing back and forth through the amplifier due to reflections.

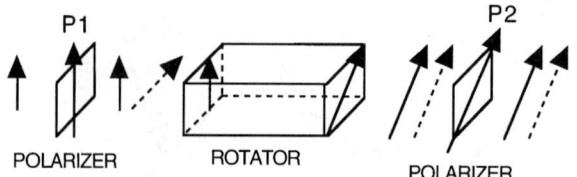

FIGURE 60.4 Principle of the optical isolator. P1 and P2 are polarizers that allow propagation of waves polarized in the directions indicated by the long arrows. The shorter solid arrows refer to waves traveling in the preferred direction (left to right). The dashed arrows refer to waves traveling in the reverse (undesired) direction. Ideally, no wave will get past polarizer P1 (traveling to the left) because the wave will enter P1 at 90° with respect to the allowed direction of wave polarization.

The principle of the isolator is illustrated in Fig. 60.4. A polarizer aligns the beam along one direction. The polarization is rotated 45° by a **Faraday rotator** to align it with an output polarizer. For a beam traveling in the opposite direction (to the left in the figure), the beam is rotated such that it aligns at 90° to the polarizer. It is thus blocked from further travel.

60.5 Directional Couplers

The four-port directional coupler is shown schematically in Fig. 60.5 [Palais, 1992]. Arrows in the diagram indicate the directions of desired power flow. Power entering port 1 is divided between ports 2 and 3 according to the desired splitting ratio. Any desired split can be obtained, but the most common one is a 50 : 50 split. Under ideal (lossless) conditions, the insertion loss to ports 2 and 3 would then be 3 dB and the isolation (coupling loss to port 4) would be (in decibels) infinite. In practice there are **excess losses** associated with the coupler. Excess losses, which can be as low as a few tenths of a decibel for a good coupler, add to the splitting loss. That is, the total insertion loss is the sum of the splitting loss and the excess loss. Typical insertion losses for the 3-dB coupler are, thus, around 3.3 dB. Typical isolations are 40 dB or more.

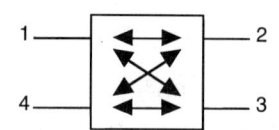

FIGURE 60.5 A four-port directional coupler.

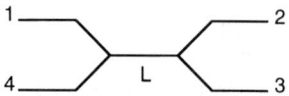

FIGURE 60.6 Fused biconically tapered directional coupler. Two fibers are fused over a length L.

Applications include local monitoring of the power levels of the transmitting light source, tapping power from a data bus in a local-area network, signal splitting in a subscriber distribution system, power splitting in an optical sensor, and combining signals in a heterodyne (coherent) receiver.

The fused biconically tapered coupler is one popular construction technique. Two fibers are fused together over a short length by heating. During the fusing process the fibers are stretched creating a tapered region in the fused section. The fused fiber coupler is shown schematically in Fig. 60.6. The amount of light coupled between the two fibers depends on the coupling length (L) and can be varied from 0 to 100%. A 50% split, creating a 3-dB coupler, is most common.

Directional couplers can also be constructed using GRIN rod lenses. Illustrated in Fig. 60.7 are a prism/beamsplitter and a reflective coating directional coupler.

Passive Optical Components

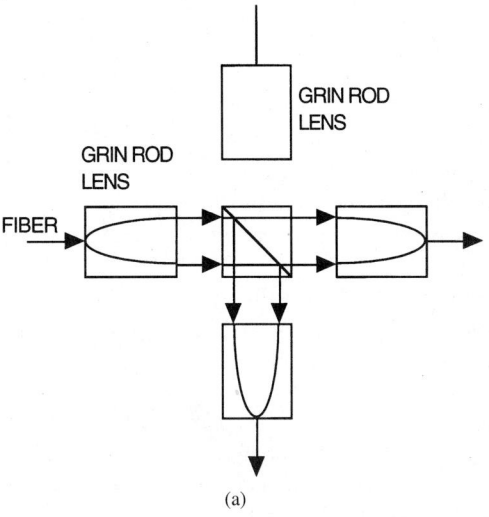

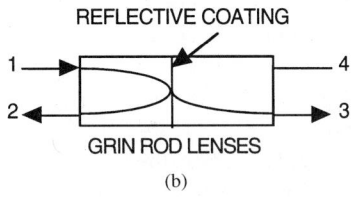

FIGURE 60.7 Directional couplers: (a) prism/beam-splitter coupler and (b) GRIN rod lens reflective coupler.

60.6 Star Coupler

The star coupler is a multiport device (more than four ports) which commonly distributes power entering one port equally among a number of output ports. Referring to Fig. 60.8, power entering any of the N input ports is equally distributed among all of the N output ports. As an example, an ideal 16×16 star coupler would deliver 1/16 of the input power to each output port, resulting in an insertion loss of 12.04 dB. If the coupler were not ideal, for example, having a 2-dB excess loss, the actual insertion loss would be 14.04 dB.

The star coupler is heavily used in the passive star local-area network architecture. Using the 16×16 coupler just described, 16 terminals (each with its own transmitter and receiver) can be interconnected.

The most common implementation is the fused tapered star coupler, as illustrated in Fig. 60.9 where 8 fibers are fused to produce an 8×8 star. The 16×16 star would require that 16 fibers be fused together.

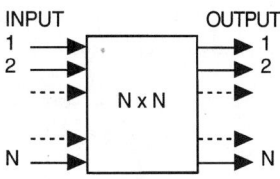

FIGURE 60.8 A star coupler.

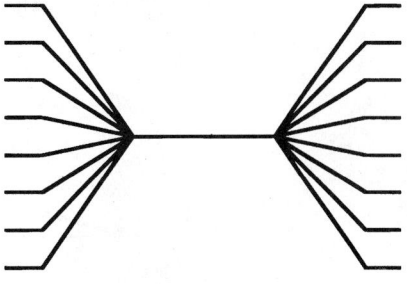

FIGURE 60.9 An 8×8 fused star coupler.

60.7 Optical Filter

Optical filters separate (or combine) the power carried by different wavelengths that travel through a fiber. Just like electronic filters, they pass or block different wavelengths. Unlike electronic filters, they often also spatially separate the various wavelengths present, as indicated in Fig. 60.10.

Optical filters are necessary in wavelength-division multiplexing (WDM) systems where multiple optical carriers traverse the same fiber to increase system capacity.

The fused directional coupler can be used for a two-channel system because the splitting is wavelength dependent. As an example, referring to the directional coupler in Fig. 60.6, if signals at wavelengths 1300 and 1550 nm are received along the same fiber, they can be separated by a coupler that transmits the 1300-nm beam from port 1 to 2 and the 1550-nm beam from port 1 to 3. At the transmitter, the same coupler can combine the two signals onto a fiber connected to port 1 if the 1300-nm signal enters port 2 and the 1550-nm signal enters port 3. As is commonly the case, the same device can serve as either a **multiplexer** or a **demultiplexer**.

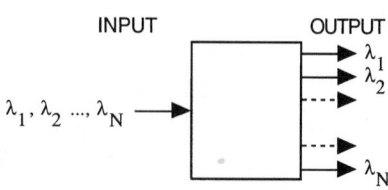

FIGURE 60.10 Optical filter used as a demultiplexer.

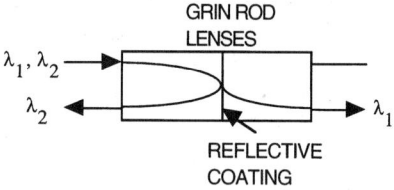

FIGURE 60.11 Two-wavelength demultiplexer. By reversing the direction of the arrows, the device acts as a multiplexer.

The reflective coupler drawn in Fig. 60.7(b) can also serve as a two-channel multiplexer/demultiplexer. In this case, as illustrated in Fig. 60.11, the coating is reflective at one of the system wavelengths and transmissive at the other.

WDM systems carrying more than two channels require more complex multiplexers/demultiplexers. Two such devices are sketched in Fig. 60.12. One uses a GRIN rod to collimate the beam diverging from the input fiber. The collimated beam strikes the blazed grating, which directs different wavelengths into different directions. The lens than refocuses the collimated (but redirected) beams

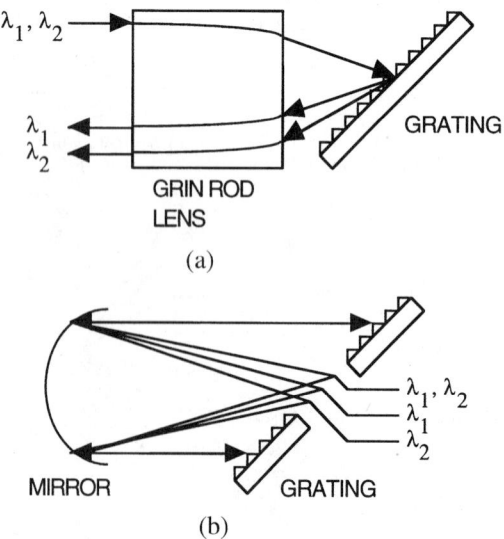

FIGURE 60.12 Two multiplexers/demultiplexers: (a) grating and GRIN rod lens; (b) grating and curved reflector.

Passive Optical Components

onto different output fibers leading to the receivers. In the second device, a curved reflector performs the same function as the GRIN rod lens. Both of these components can multiplex/demultiplex 10 or more channels (wavelengths). Insertion losses and isolations are on the order of 1 dB and 25 dB, respectively.

Another filter is based on Bragg reflection by a waveguide grating imbedded in the fiber, as illustrated in Fig. 60.13. This device has a periodic refractive index variation for a short length along the axis of the fiber. The variation is produced by exposing the fiber to a periodic variation in ultraviolet light, which permanently changes the refractive index of the glass fiber, thus producing a grating. The transmission characteristics are determined by the grating periodicity and the depth of the refractive index changes.

FIGURE 60.13 Bragg reflection filter.

60.8 Attenuator

Optical attenuators make it possible to change the light intensity in a fiber. Applications include testing receivers at various light levels (for example to determine the receiver's dynamic range) and adjusting the light intensity to prevent saturation of the receiver.

An absorbing type attenuator appears in Fig. 60.14. The absorbing wheel has a different amount of absorption at different points so that rotating it provides a variation in the attenuation. Misaligned fibers can also be used as attenuators, as indicated in Fig. 60.15. Varying the gap or the lateral offset provides variable attenuation.

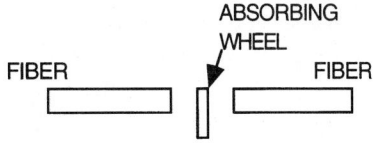

FIGURE 60.14 Fiber variable attenuator using an absorbing wheel.

FIGURE 60.15 Fiber variable attenuator using movable offset fibers.

60.9 Circulator

An optical circulator, whose operation is sketched in Fig. 60.16, directs the signal sequentially from one port to the next. That is, an input at port 1 exits at port 2, an input at port 2 exits at port 3, and an input at port 3 exits at port 1. Insertion losses are typically just under 1 dB and isolations are above 25 dB.

The circulator can be used to decouple transmitted and received signals traveling along the same fiber (that is, in a bidirectional communications system such as that sketched in Fig. 60.17).

A circulator uses a Faraday rotator, similar to that found in the optical isolator. A fuller description of the circulator appears in Kashima, 1995.

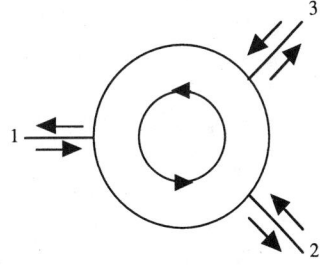

FIGURE 60.16 Fiber circulator.

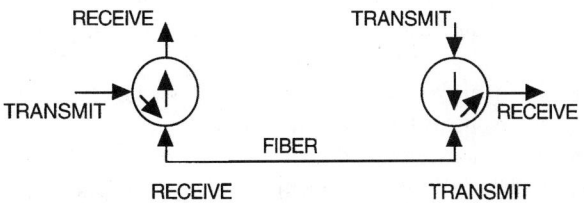

FIGURE 60.17 Full duplex fiber transmission system.

60.10 Mechanical Switch

An optical switch allows an incoming signal to be diverted to any one of two or more different output fibers as schematically shown in Fig. 60.18. A mechanical switch does this by physical movement of the fibers or some other element of the switch. The insertion loss (coupling to the desired port) should be no more than a few tenths of a decibel and the isolation (undesired coupling to any other port) should be more than 40 dB. Switching speeds are on the order of a few milliseconds, not fast by electronic and optical measures, but quick enough for many fiber applications.

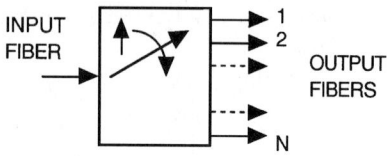

FIGURE 60.18 An $1 \times N$ mechanical switch.

On/off switches allow power to be supplied to a system under test only when needed. Multiple output port switches find applications in measurement systems (for example, to selectively measure the output power from a sequence of fibers). Switches can also redirect signals from a disabled fiber link onto an alternate fiber route. They are used to produce bypass paths directing signals around disabled equipment, such as a node in a local-area network.

In an on/off switch, a block is inserted into a small gap between fibers in the off position and removed in the on position. In a $1 \times N$ switch, the end of the transmitting fiber physically moves into alignment with the desired output fiber.

60.11 Polarization Controller

The polarization controller is a fiber device for producing a desired state of polarization. Although most fiber systems operate independently of the state of polarization of the optical beam, coherent systems and certain sensors are very much dependent on the state of polarization of the optical signal.

A simple polarization controller is constructed by winding the fiber around a disk whose diameter is on the order of a centimeter [Chen, 1996]. Usually, two or three of these fiber disks are connected in series (Fig. 60.19). Rotating the disks about the axis of the fiber produces a differential change in the refractive index of the two normal orthogonally polarized beams traversing the fiber. Any desired polarization state can be obtained with the appropriate rotation of the disks.

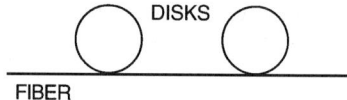

FIGURE 60.19 Fiber polarization controller.

Defining Terms

Demultiplexer: Separates different wavelengths entering from a single transmission fiber and directs them to separate receivers.

Directionality: The fractional change in power level (expressed in decibels) when a signal passes between two ports which are ideally meant to be uncoupled.

Excess losses: The fractional amount of input power (expressed in decibels) that reaches all the desired output ports of an optical component.

Faraday rotator: A component which rotates the polarization of the electric field passing through it. Often the rotation of an incident linearly polarized beam is 45°. The rotation is nonreciprocal.

Insertion loss: The fractional change in power level (expressed in decibels) when a signal passes between two ports which are designed to be coupled.

Isolation: Same as directionality.

Multiplexer: Combines two or more wavelengths from separate light sources onto a single fiber for transmission.

References

Chen, C.-L. 1996. *Elements of Optoelectronics and Fiber Optics*, Irwin, Chicago, IL.
Kashima, N. 1995. *Passive Optical Components for Optical Fiber Transmission*, Artech, Boston, MA.
Palais, J.C. 1992. *Fiber Optic Communications*, 3rd ed., Prentice–Hall, Englewood Cliffs, NJ.

Further Information

Information on components for optical fiber systems appears in several professional society journals. These include the *IEEE Journal of Lightwave Technology*, the *IEEE Photonics Technology Letters*, and the Optical Society of America's *Applied Optics*. Valuable information is also contained in several trade magazines such as *Lightwave* and *Laser Focus World*.

61
Semiconductor Optical Amplifiers

Daniel J. Blumenthal
Georgia Institute of Technology

Nitin C. Kothari
Georgia Institute of Technology

61.1 Introduction.. 832
61.2 Principle of Operation 833
61.3 Types of Semiconductor Optical Amplifiers 834
61.4 Design Considerations... 835
61.5 Gain Characteristics.. 836
   Small-Signal Gain and Bandwidth of Traveling Wave Amplifiers • Small-Signal Gain and Bandwidth of FP Amplifiers • Wavelength Dependence of Gain • Gain Saturation • Dynamic Range • Polarization Sensitivity • Amplifier Noise
61.6 Pulse Amplification 842
61.7 Multichannel Amplification.................................. 844
61.8 Applications 844

61.1 Introduction

The deployment of optical amplifiers in fiber optic communications systems has received widespread attention over the past several years. Signal repeaters are needed to overcome inherent transmission losses including fiber absorption and scattering, distribution losses, and connector and component losses. Traditional repeaters are based on optoelectronic conversion where the optical signal is periodically converted to an electronic signal, remodulated onto a new optical signal, and then transmitted onto the next fiber section. An alternative approach is to use repeaters based on optical amplification. In optical repeaters, the signal is amplified directly without conversion to electronics. This approach offers distinct advantages including longer repeater spacings, simultaneous multichannel amplification, and a bandwidth commensurate with the transmission window of the optical fiber. The deployment of optical amplifiers in operational networks has reduced maintainance costs and provided a path for upgrading the existing fiberplants since the amplifiers and fibers are transparent to signal bandwidth.

Two classes of optical amplifiers are used in fiber-based systems: active fiber amplifiers and semiconductor optical amplifiers (SOAs). In this chapter we concentrate on semiconductor amplifiers. Fiber-based amplifiers have advanced more rapidly into the deployment phase due to their high-output saturation power, high gain, polarization insensitivity, and long excited state lifetime that reduces crosstalk effects. Fiber amplifiers have been successfully used in the 1.55-μm fiber transmission window but are not ideally suited for the 1.31-μm fiber transmission window. Fiber amplifiers are covered in detail in the companion chapter in this Handbook. Recent developments in SOAs have led to dramatic improvements in gain, saturation power, polarization sensitivity, and crosstalk rejection. Semiconductor amplifiers also have certain characteristics that make their use in optical networks very desirable: (1) they have a flat gain bandwidth over a relatively wide wavelength range that allows them to simultaneously amplify signals of different wavelengths, (2) they are simple devices that can be integrated with other semiconductor based circuits, (3) their gain can be switched at high speeds to provide a modulation function, and (4) their current can be monitored to provide simultaneous amplification and detection. Additionally, interest in semiconductor amplifiers has been motivated by their ability to operate in the 1.3-μm fiber transmission window. For these reasons, semiconductor optical amplifiers continue to be studied and offer a complimentary optical amplification component to fiber based amplifiers.

In this chapter, we first discuss the principle of operation of SOAs followed by amplifier design considerations. The gain characteristics are described next with discussion on small-signal gain, wavelength dependence, gain saturation, dynamic range, polarization sensitivity, and noise. Amplification of optical pulses and multichannel amplification are treated followed by a brief discussion of applications.

61.2 Principle of Operation

A semiconductor optical amplifier operates on the principle of stimulated emission due to interaction between input photons and excited state electrons and is similar to a laser in its principle of operation. The semiconductor can be treated as a two energy level system with a ground state (valence band) and excited state (conduction band) as shown in Fig. 61.1. Current is injected into the semiconductor to provide an excess of electrons in the conduction band. When a photon is externally introduced into the amplifier, it can cause an electron in the conduction band to recombine with a hole in the valence band, resulting in emission of a second photon identical to the incident photon (**stimulated emission**). As these photons propagate in the semiconductor, the stimulated emission process occurs over and over again, resulting in **stimulated amplification** of the optical input. Since an SOA can amplify input photons, we can assign it a gain. As will be discussed in further detail, the **material gain** is only a function of the basic device composition and operating conditions, whereas the **amplifier gain** defines the relationship between the input and output optical power.

In addition to amplification of the input photons (signal), it is important to consider how noise is generated within the amplifier when an electron in the conduction band spontaneously recombines with a hole without the aid of a photon. This process, known as **spontaneous emission**, results in an emitted photon with energy equal to the energy difference between the electron and hole.

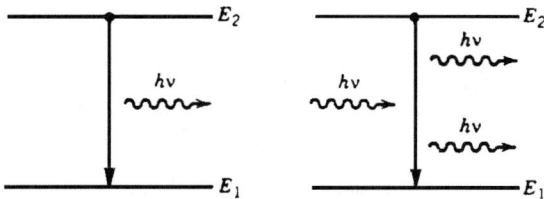

FIGURE 61.1 Spontaneous and stimulated emission occuring between the two energy states of an atom.

Photons created from spontaneous emission will propagate in the amplifier and experience gain through stimulated emission. This process leads to amplification of spontaneous emission and is called **amplified spontaneous emission (ASE)**. ASE is a noise mechanism as it is not related to the input signal and is a random process. Additionally, ASE takes away amplifier gain that would otherwise be available to the signal.

61.3 Types of Semiconductor Optical Amplifiers

SOAs can be classified as either **subthreshold** or **gain clamped**. Subthreshold amplifiers are lasers operated below threshold and gain clamped amplifiers are lasers operated above threshold but used as amplifiers. Subthreshold SOAs can be further classified according to whether optical feedback is used. If the SOA amplifies the optical signal in a single pass, it is referred to as a **traveling wave amplifier** (TWA) as shown in Fig. 61.2. The second type of subthreshold amplifier is a **resonant amplifier**, which contains a gain medium and some form of optical feedback. In this case the gain is resonantly enhanced at the expense of limiting the gain bandwidth to less than that of the TWA case for an equivalent material. An example of a resonant amplifier is the **Fabry–Perot (FP) amplifier** shown in Fig. 61.2. The FP amplifier has mirrors at the input and output ends, which form a resonant cavity. The resonant cavity creates an optical **comb filter** that filters the gain profile into uniformly spaced **longitudinal modes** (also see Fig. 61.2). The TWA configuration has a bandwidth limited by the material gain itself but is relatively flat which is desirable for an optical communications system application. Typical 3-dB bandwidths are on the order of 60–100 nm. Although the FP amplifier exhibits very large gain at wavelengths corresponding to the longitudinal modes, the gain rapidly decreases when the input wavelength is offset from the peak wavelengths. This makes the gain strongly dependent on the input wavelength and sensitive to variations that may occur in an optical communications system.

Gain clamped semiconductor optical amplifiers operate on the principle that the gain can be held constant by a primary lasing mode and signals can be amplified if their wavelength is located away from the main lasing mode. Laser structures suitable for this approach are **distributed feedback (DFB)** and **distributed Bragg reflector (DBR)** lasers, which lase into a single longitudinal mode. Since only a single mode oscillates, the remainder of the gain profile is available for amplification. Figure 61.3 illustrates the output optical spectra of a gain clamped amplifier with the main lasing mode and the amplified signal identified. A primary advantage of gain clamped amplifiers is a reduction in crosstalk in multichannel amplification applications discussed in Sec. 61.7.

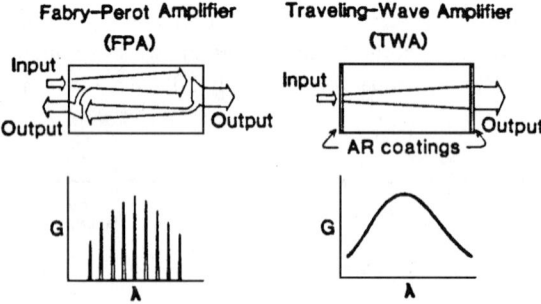

FIGURE 61.2 Classification of semiconductor optical amplifiers. *Source:* Saitoh, T. and Mukai, T. 1991. In *Coherence, Amplification, and Quantum Effects in Semiconductor Lasers*, ed. Y. Yamamoto, Wiley.

Semiconductor Optical Amplifiers

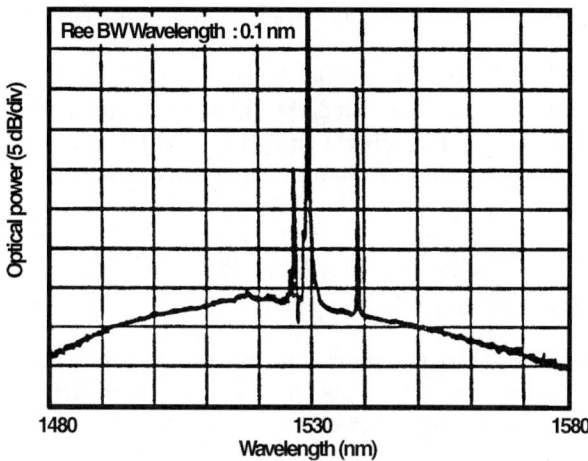

FIGURE 61.3 Output optical spectrum of a gain clamped amplifier with the main lasing mode and the identified amplified signal.

61.4 Design Considerations

For optical communications systems, traveling wave SOAs are desirable due to the wide gain bandwidth and relatively small variation in gain over a wide signal wavelength range. Most practical TWAs exhibit some small ripples in the gain spectrum that arise from residual facet reflections. A large effort has been devoted to fabricate amplifiers with low cavity resonances to reduce **gain ripple**. For an amplifier with facet reflectivities R_1 and R_2, the peak-to-valley ratio of the output intensity ripple is given by [Dutta and Simpson, 1993]

$$V = \left[\frac{1 + G_s \sqrt{R_1 R_2}}{1 - G_s \sqrt{R_1 R_2}} \right] \tag{61.1}$$

where G_s is the single-pass gain of the amplifier. For the ideal case $R_1, R_2 \to 0$, $V = 1$, that is, no ripple at cavity mode frequencies in the output spectrum. A practical value of V should be less than 1 dB. Thus, for an amplifier designed to provide gain $G_s = 25$ dB, the facet reflectivities should be such that $\sqrt{R_1 R_2}$ must be less than 3.6×10^{-4}.

Three principle schemes exist for achieving low facet reflectivities. They are: (1) antireflection dielectric coated amplifiers [Olsson, 1989], (2) buried facet amplifiers [Dutta et al., 1990], and (3) tilted facet amplifiers [Zah et al., 1987]. In practice, very low facet reflectivities are obtained by monitoring the amplifier performance during the coating process. The effective reflectivity can be estimated from the peak-to-peak ripple at the FP mode spacings caused by residual reflectivity. Reflectivities less than 10^{-4} over a small range of wavelengths are possible using antireflection coatings [Olsson, 1989]. In buried facet structures, a transparent window region is inserted between the active layer ends and the facets [Dutta et al., 1990]. The optical beam spreads in this window region before arriving at the semiconductor–air interface. The reflected beam spreads even farther and does not couple efficiently into the active layer. Such structures can provide reflectivities as small as 10^{-4} when used in combination with antireflection dielectric coatings. Another way to suppress the FP modes of the cavity is to slant the waveguide (gain region) from the cleaved facet so that the light incident on it does not couple back into the waveguide [Zah et al., 1987]. The process essentially decreases the effective reflectivity. A combination of antireflection coating and the tilted stripe can produce reflectivities less than 10^{-4}. A disadvantage of this structure is that the effective reflectivity of the higher order modes can increase causing the appearance of higher order modes at the output, which may reduce fiber-coupled power significantly.

Another important consideration in amplifier design is choice of semiconductor material composition. The appropriate semiconductor bandgap must be chosen so that light at the wavelength of interest is amplified. For amplification of light centered around 1.55 μm or 1.31 μm, the InGaAsP semiconductor material system must be used with optical gain centered around the wavelength region of interest.

61.5 Gain Characteristics

The fundamental characteristics of the optical amplifiers such as small-signal gain, gain bandwidth, gain saturation, polarization sensitivity of the gain, and noise are critical to amplifier design and use in systems. The component level behavior of the optical amplifier can be treated in a manner similar to electronic amplifiers.

Small-Signal Gain and Bandwidth of Traveling Wave Amplifiers

The optical power gain of an SOA is measured by injecting light into the amplifier at a particular wavelength and measuring the optical output power at that wavelength. The gain depends on many parameters including the input signal wavelength, the input signal power, material gains and losses, amplifier length, current injection level, etc. For a TWA operated with low-input optical power, the small-signal gain is given by the unsaturated single-pass gain of the amplifier

$$G_0(\nu) = \frac{P_{\text{out}}(\nu)}{P_{\text{in}}(\nu)} = \exp[(\Gamma g_0(\nu) - \alpha_m)L] \qquad (61.2)$$

where Γ is the **optical mode confinement factor**, g_0 is the **unsaturated material gain coefficient**, α_m is the **absorption coefficient**, and L is the amplifier length. The material gain coefficient g_0 has a Lorentzian profile if the amplifier is modeled as a two-level atomic system. An important distinction is made between the material gain bandwidth and the amplifier signal bandwidth. The 3-dB bandwidth (full width at half-maximum) of g_0 is $\Delta\nu_g = 1/\pi T_2$ and the 3-dB bandwidth of the TWA signal gain [Eq. (61.2)], is

$$\Delta\nu_A = \Delta\nu_g \left(\frac{\ln 2}{\Gamma g_0 L - \ln 2} \right) \qquad (61.3)$$

Figure 61.4 illustrates that the material gain bandwidth is greater than the amplifier small-signal gain bandwidth.

Small-Signal Gain and Bandwidth of FP Amplifiers

The small-signal power transmission of a FP amplifier shows enhancement of the gain at transmission peaks and is given by [Saitoh and Mukai, 1991]

$$G_0^{FP}(\nu) = \frac{(1 - R_1)(1 - R_2)G_0(\nu)}{(1 - \sqrt{R_1 R_2}G_0(\nu))^2 + 4\sqrt{R_1 R_2}G_0(\nu)\sin^2[\pi(\nu - \nu_0)/\Delta\nu]} \qquad (61.4)$$

where ν_0 is the cavity resonant frequency and $\Delta\nu$ is the **free spectral range** (also called the longitudinal mode spacing) of the SOA. The single-pass small-signal gain G_0 is given by Eq. (61.2). Note that for $R_1 = R_2 = 0$, G_0^{FP} reduces to that of a TWA. The 3-dB bandwidth B (full width at

Semiconductor Optical Amplifiers

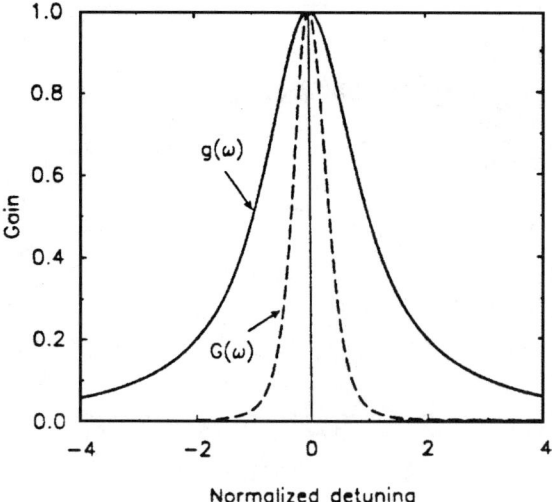

FIGURE 61.4 Lorentzian material gain profile and the corresponding amplifier gain spectrum for an amplifier modeled as a two-level atomic system. *Source:* Agrawal, G.P. 1995. In *Semiconductor Lasers: Past, Present, and Future*, ed. G.P. Agrawal, AIP Press.

half-maximum) of an FP amplifier is expressed as [Saitoh and Mukai, 1991]

$$B = \frac{2\Delta \nu}{\pi} \sin^{-1} \left[\frac{1 - \sqrt{R_1 R_2} G_0}{(4G_0 \sqrt{R_1 R_2})^{1/2}} \right] \tag{61.5}$$

whereas the 3-dB bandwidth of a TWA is three orders of magnitude larger than that of the FP amplifier since it is determined by the full gain width of the amplifier medium itself. Figure 61.5 shows the small-signal (unsaturated) gain spectra of a TWA within one free spectral range [Saitoh,

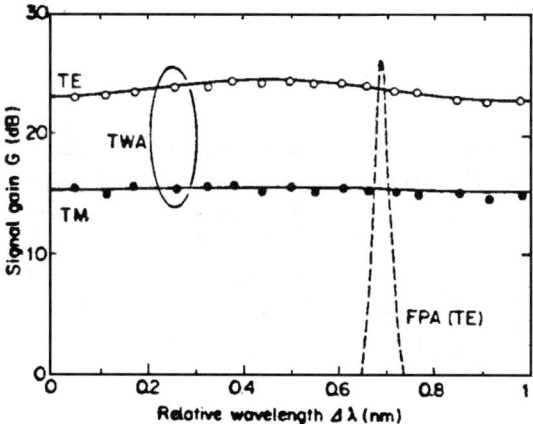

FIGURE 61.5 Unsaturated gain spectra of a TWA and an FP amplifier within one free spectral range. *Source:* Saitoh, T., Mukai, T., and Noguchi, Y. 1986. In *First Optoelectronics Conference Post-Deadline Papers Tech. Digest*, Paper B11-2, The Institute of Electronics and Communications Engineers of Japan, Tokyo.

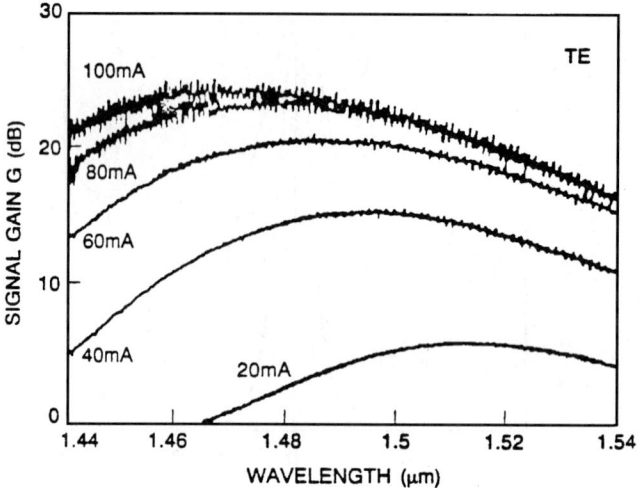

FIGURE 61.6 Gain spectrum of a TWA at several current levels. *Source:* Saitoh, T. and Mukai, T. 1987. In *Electron. Lett.*, 23:218.

Mukai, and Noguchi, 1986]. Solid curves represent theoretical FP curves fitted to the TWA experimental data. Experimental FP amplifier data is also shown (dashed curve). It can be seen that the signal gain fluctuates smoothly over the entire free spectral range of the TWA in contrast to the FP amplifier, where signal gain is obtained only in the vicinity of the resonant frequencies.

Wavelength Dependence of Gain

As seen previously, the amplifier gain varies with the input wavelength for both the TWA and FP cases. The peak and width of the gain can also vary as a function of injection current. The material gain vs wavelength for a typical amplifier is shown in Fig. 61.6 as a function of injection current. As the injection current increases, the peak gain increases and the location of the peak gain shifts toward shorter wavelengths.

Gain Saturation

The gain of an optical amplifier, much like its electronic counterpart, can be dependent on the input signal level. This condition, known as **gain saturation**, is caused by a reduction in the number of electrons in the conduction band available for stimulated emission and occurs when the rate of input photons is greater than the rate at which electrons used for stimulated emission can be replaced by current injection. The time it takes for the gain to recover is limited by the spontaneous lifetime and can cause intersymbol interference among other effects as will be described. The **saturated material gain coefficient** is given by

$$g(\nu) = \frac{g_0(\nu)}{1 + \dfrac{P}{P_s}} \qquad (61.6)$$

where g_0 is the unsaturated material gain coefficient, P is the total optical power in the active layer of the amplifier, and P_s is the saturation power defined as the light intensity that reduces the material gain g to half its value ($g_0/2$). In general, the optical saturation power is related to the carrier lifetime but can be reduced as the optical intensity in the amplifier increases. The single-pass

Semiconductor Optical Amplifiers

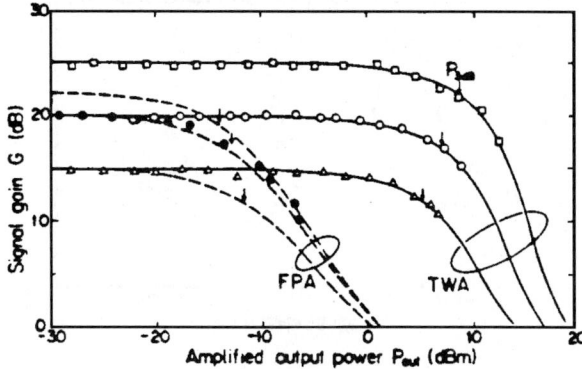

FIGURE 61.7 Theoretical and experimental signal gain of a TWA and FP amplifier as a function of amplified output power. *Source:* Saitoh, T. and Mukai, T. 1987. In *Electron. Lett.*, 23:218.

saturated signal gain $G(\nu)$ is given by [Saitoh and Mukai, 1991]

$$G(\nu) = G_0(\nu) \exp\left[-\frac{P_{\text{out}} - P_{\text{in}}}{P_s}\right] = G_0 \exp\left[-\frac{G-1}{G}\frac{P_{\text{out}}}{P_s}\right] \qquad (61.7)$$

where P_{in} and P_{out} are the input and the output optical powers and G_0 is the unsaturated gain given by Eq. (61.2). Figure 61.7 shows experimental gain saturation characteristics of both FP and TWAs [Saitoh and Mukai, 1987] along with theoretical curves.

Dynamic Range

The dynamic range is defined as the range of input power for which the amplifier gain will remain constant. The gain saturation curves shown in Fig. 61.8 [Adams et al., 1985] show the relationship

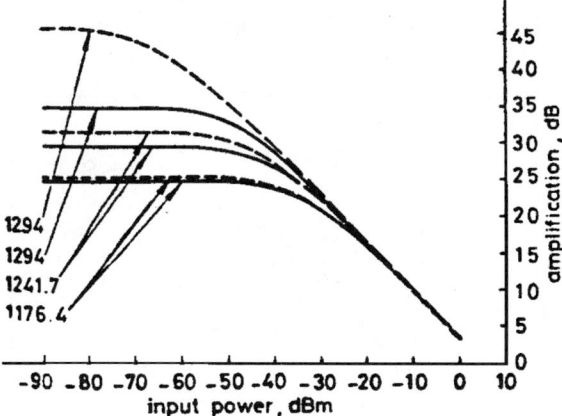

FIGURE 61.8 Effect of amplifier injection current on variation in gain as a function of optical input power. Flat regions are unsaturated gain. Range of unsaturated gain vs input power changes as a function of unsaturated gain. Values indicated on curves are for injected current density. Dashed curves are obtained without taking into account ASE. *Source:* Adams, M.J. et al. 1985. In *IEE Proceedings* Part J, 132:58.

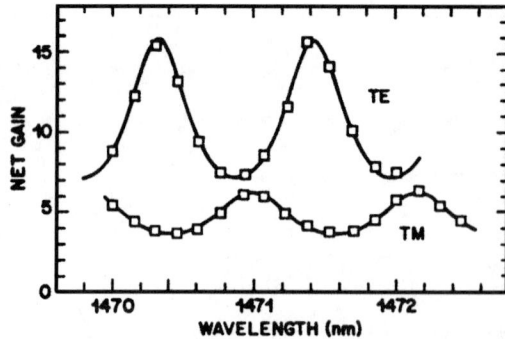

FIGURE 61.9 Theoretical and experimental gain spectra for TE- and TM-polarized input signals. *Source:* Jopson, R.M. et al. 1986. *Electron. Lett.*, 22:1105.

between input power and gain for various current injection levels. The flat regions are the unsaturated regions. As the gain increases, the 3-dB rolloff moves toward lower input power leading to a decrease in dynamic range.

Polarization Sensitivity

In general, the optical gain of an SOA is **polarization dependent**. It differs for the transverse-electric (TE) and transverse-magnetic (TM) polarizations. Figure 61.9 shows the gain spectra of a TWA for both TE and TM polarization states [Jopson et al., 1986]. The polarization-dependent gain feature of an amplifier is undesirable for lightwave system applications where the polarization state changes with propagation along the fiber.

Several methods of reducing or compensating for the gain difference between polarizations have been demonstrated [Saitoh and Mukai, 1991]. A successful technique that leads to polarization dependent gain of on the order of 1 dB involves the use of material strain [Dubovetsky et al., 1994]. The measured gain as a function of injection current for TE and TM polarized light for a tilted-facet amplifier is shown in Fig. 61.10 [Zah et al., 1987]. Figure 61.11 shows the measured optical gain

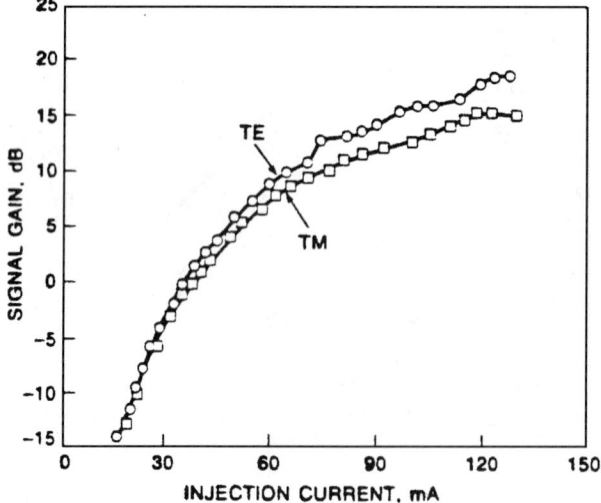

FIGURE 61.10 Measured gain as a function of injection current. *Source:* Zah, C.E. et al. 1987. *Electron. Lett.*, 23:990.

Semiconductor Optical Amplifiers

plotted as a function of output power for regular double heterostructure (DH) and multiquantum well (**MQW**) amplifiers [Dutta and Simpson, 1993]. The multiquantum well amplifier result is shown for the TE mode, whereas the gain difference between TE and TM modes for double heterostructure amplifiers was less than 1 dB.

Amplifier Noise

Semiconductor optical amplifiers add noise to the amplified optical signal in the form of amplified spontaneous emission. This noise can reduce the signal-to-noise ratio of the detected output and can reduce the overall gain available to the signal through gain saturation. ASE forms a noise floor with a wavelength dependence approximately that of the gain. Figure 61.12 shows a typical optical spectrum at the output with the amplified signal and the ASE. Semiconductor optical amplifiers can be characterized by a noise figure. Typical noise figures are in the range of 5–8 dB and can be calculated from the carrier density due to current injection N, the carrier density required to reach transparency N_0, the saturated gain g, and the internal amplifier losses α_{int} [Saitoh and Mukai, 1991]

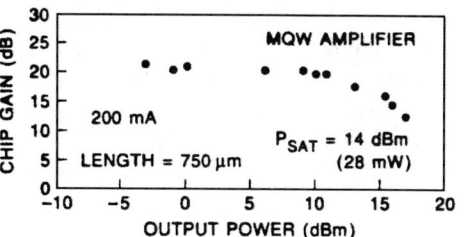

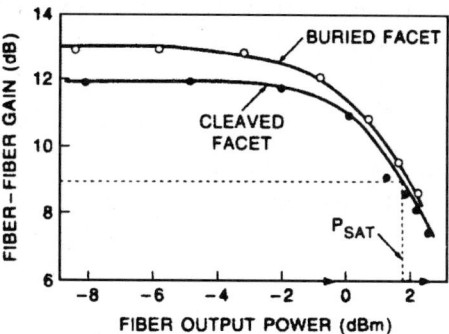

FIGURE 61.11 Measured gain as a function of output power for regular double heterostructure and multiquantum well amplifiers. *Source:* Dutta, N.K. and Simpson, J.R. 1993. In *Progress in Optics*, ed. E. Wolf, Vol. 31, North-Holland.

$$F_n = \frac{SNR_{\text{in}}}{SNR_{\text{out}}} = 2\left(\frac{N}{N-N_0}\right)\left(\frac{g}{g-\alpha_{\text{int}}}\right) \qquad (61.8)$$

In calculating signal-to-noise ratio, it is useful to consider the total output of an SOA after it has been converted to an electronic signal using a photodetector. Here we consider direct detection where a square law detector is used to convert the optical signal to an electrical signal. In addition to the amplified signal, several noise terms result from interaction between signal and noise in the

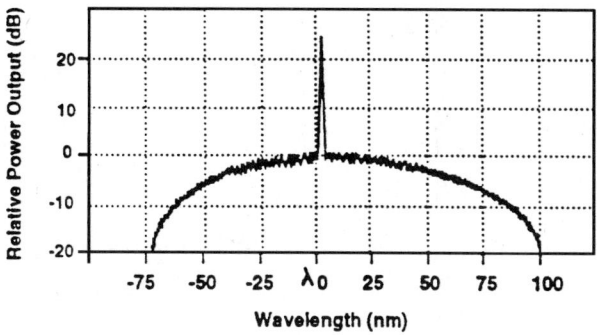

FIGURE 61.12 A typical optical spectrum at the output with the amplified signal and the amplified spontaneous emission. *Source:* BT&D Technologies Publication No. DS007, 1992.

square law detection process [Mukai, Yamamoto, and Kimura, 1982; Yamamoto, 1980]: amplified signal shot noise, spontaneous emission shot noise, signal-spontaneous beat noise, spontaneous-spontaneous beat noise, and signal excess noise. The first and third types of noise are proportional to the signal power, whereas the second and fourth types of noise are generated from the ASE that exists independently of the input signal. For a single amplifier, the two beat noise terms dominate as they are G times stronger [Saitoh and Mukai, 1991] than the shot noise contributions. The spontaneous-spontaneous beat noise can be reduced by loading a narrowband optical filter matched to the signal frequency, since this noise arises from the beat between the ASE components over a wide gain spectrum [Saitoh and Mukai, 1991]. For a direct detection system with a high-gain optical amplifier prior to the photodetector, the electrical SNR is given by [Jopson and Darcie, 1991]

$$SNR_{dd} = \frac{(P_{in}/h\nu)^2}{[FP_{in}/h\nu + (F/2)^2 \delta\nu_{opt}]B_e} \quad (61.9)$$

where the input power to the amplifier is P_{in}, the optical bandwidth of the output of the amplifier is $\delta\nu_{opt}$, the receiver electrical bandwidth is B_e, the amplifier noise figure is F, and the optical frequency is ν.

61.6 Pulse Amplification

The large bandwidth of a TWA suggests that they are capable of amplifying very fast optical pulses without significant pulse distortion. However, the output pulses can be distorted either when the optical power exceeds the saturation power or when the pulse width gets short compared to the carrier lifetime [Agrawal and Olsson, 1989].

The Gaussian pulse and its spectrum for various small-signal gains G_0 when the pulse width τ_p is much smaller compared to carrier lifetime τ_c is shown in Fig. 61.13 [Agrawal and Olsson, 1989]. The input pulse energy E_{in} is equal to 0.1 times the saturation energy E_{sat} of the SOA. The pulse gets distorted such that its leading edge becomes sharper compared with the trailing edge. When the pulse width is on the order of the carrier lifetime, saturation can lead to a closing of the eye of a pseudo-random bit stream as shown in Fig. 61.14 [Jopson and Darcie, 1991]. If a sequence of short pulses (bits) are injected, where the bit duration is less than the carrier lifetime, **bit patterning** can occur.

When the pulse width is comparable to τ_c, the saturated gain has time to recover during the pulse. Figure 61.15(a) shows [Agrawal and Olsson, 1989] the effect of gain recovery on the output

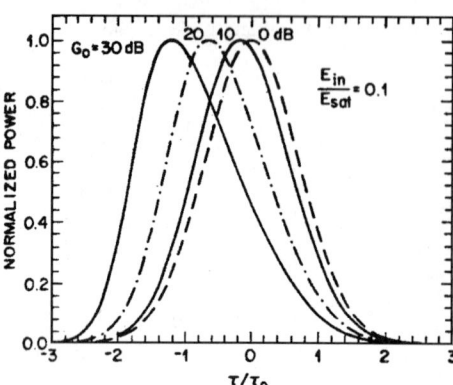
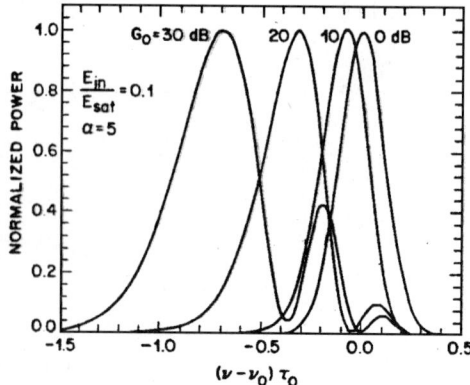

FIGURE 61.13 Output pulse shapes and spectra for several values of the unsaturated gain when the input pulse is Gaussian. The ratio of input pulse energy to the saturation energy is 0.1. *Source:* Agrawal, G.P. and Olsson, N.A. 1989. *IEEE J. Quant. Electron.*, 25:2297.

Semiconductor Optical Amplifiers

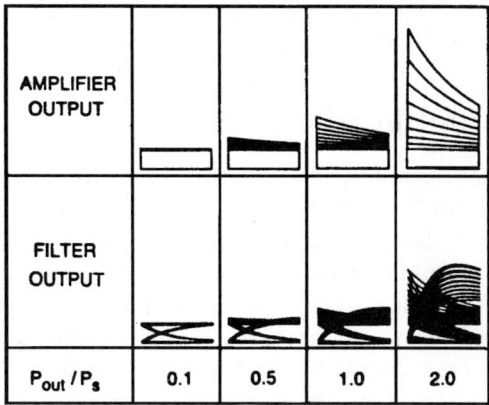

FIGURE 61.14 Effect of gain saturation on binary bit eye diagram. Upper figures show amplifier output as output power varies with saturation power. Decrease in bit height as a function of time is due to time constant of gain saturation. Lower figures show effect of electronic bandpass filtering prior to level discrimination in an optical receiver. Bit period is half the gain saturation time constant. *Source:* Jopson, R.M. and Darcie, T.E. 1991. In *Coherence, Amplification, and Quantum Effects in Semiconductor Lasers*, ed. Y. Yamamoto, Wiley.

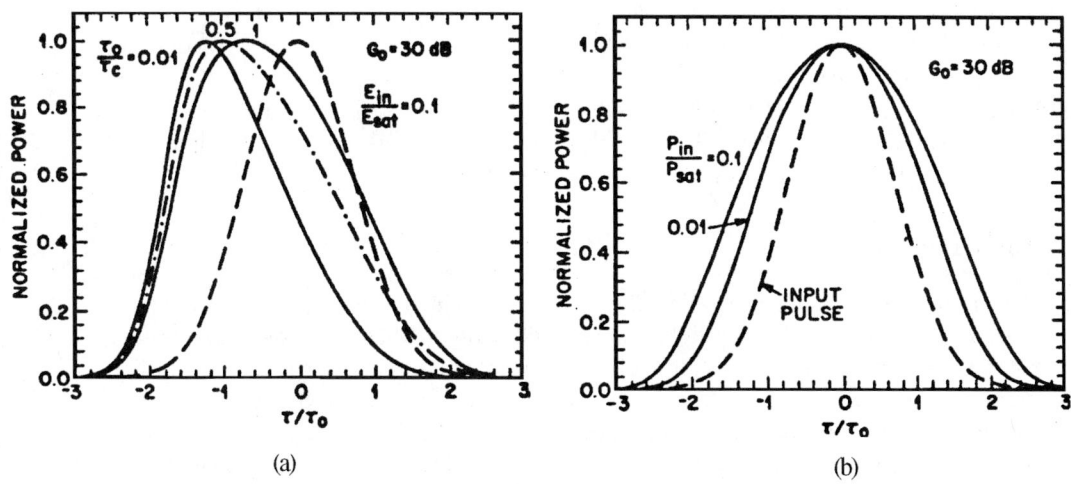

FIGURE 61.15 Output pulse shapes for (a) several values of pulse width in the partial gain recovery regime and (b) several values of input power when the pulse width is much larger than the carrier life time. *Source:* Agrawal, G.P. and Olsson, N.A. 1989. *IEEE J. Quant. Electron.*, 25:2297.

pulse shape when the input pulse is Gaussian. As the input pulse width is increased, the output pulse becomes less asymmetric and becomes broader than the input pulse. The gain recovery mainly affects the trailing edge of the pulse. In the presence of partial gain recovery, the output spectrum also becomes less asymmetric and the spectral shift becomes smaller. When $\tau_p \gg \tau_c$, the gain recovery is complete, and the output pulse as well as its spectrum become symmetric. Figure 61.15(b) shows

[Agrawal and Olsson, 1989] the output pulses when $\tau_p \gg \tau_c$. The pulse is also broadened because the peak experiences less amplification than the wings because of gain saturation.

61.7 Multichannel Amplification

A TWA can greatly simplify optical repeaters used in **wavelength-division-multiplexed** (WDM) systems, since its wide gain bandwidth allows a single amplifier to simultaneously amplify signals of different wavelengths. In practice, however, several phenomena in SOAs can induce **interchannel crosstalk: cross-gain modulation** (XGM) and **four-wave mixing** (FWM). To understand XGM, the optical power in Eq. (61.6) should be replaced by the total power in all channels. Thus, the gain of a specific channel is saturated not only by its own power but also by the power of other channels. If channel powers change depending on bit patterns as in amplitude-shift keying (ASK), then the signal gain of one channel changes from bit to bit, and the change also depends on the bit pattern of the other channels [Agrawal, 1995]. The amplified signal appears to fluctuate randomly, which degrades the signal-to-noise ratio at the receiver. This crosstalk can be largely avoided by operating SOAs in the unsaturated regime. It is absent for phase-shift keying (PSK) and frequency-shift keying (FSK) systems, since the power in each channel, and therefore the total power, remains constant with time [Agrawal, 1995].

The four-wave mixing causes the generation of new optical frequencies in closely spaced WDM systems. This phenomenon is similar to interharmonic distortion in electronic systems. The presence of multiple wavelengths in the amplifier results in nonlinear amplification. Two wavelengths can generate additional optical frequencies, as shown in Figure 61.16. If these frequencies coincide with existing channels, crosstalk results. This crosstalk can be incoherent [Darcie and Jopson, 1988] or coherent [Blumenthal and Kothari, 1996] in direct detection systems and limitations on the input power can be computed for each case [Darcie and Jopson, 1988; Blumenthal and Kothari, 1996] for a given number of channels, channel spacing, and amplifier gain. Unequal channel spacing can be used to minimize crosstalk effects due to FWM [Forghieri et al., 1994].

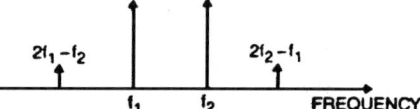

FIGURE 61.16 Generation of new frequencies due to FWM interaction between the signals at frequencies f_1 and f_2.

61.8 Applications

Semiconductor optical amplifiers have multiple potential uses in optical communication systems (see Fig. 61.17). Power boosting, optical preamplification prior to photodetection, compensation of distribution losses and in-line amplification are examples of transmission applications. In-line amplifiers eliminate the need for electronic regenerators and allow multichannel amplification. As a booster amplifier, the transmission distance can be increased by 10–100 km. SOAs can be integrated with a semiconductor laser to obtain a high-power, narrow-linewidth optical source useful for coherent communication systems [Koch and Koren, 1991; Glance et al., 1992]. Transmission distance can also be increased by putting an amplifier to operate the receiver in the shot noise limit.

Another potential application of SOAs is in the area of **dispersion compensation**. Midpoint dispersion compensation in a fiber link is achieved by using the SOA to perform frequency conversion through four-wave mixing [Tatham, Sherlock and Westbrook, 1993]. This process inverts the ordering of the optical spectrum within the pulse thereby reversing the effect of dispersion in the second section of fiber. An SOA can also be used as a booster amplifier to balance the effect of **frequency chirp** in direct drive semiconductor lasers [Yazaki et al., 1992].

A third class of application, **photonic switching**, may require switching of signals in space and/or wavelength. SOAs can be integrated in **space switches** as transmission gates [Gustavsson et al., 1992] with very high-extinction ratio in the off state. SOAs can also be used to translate signals between

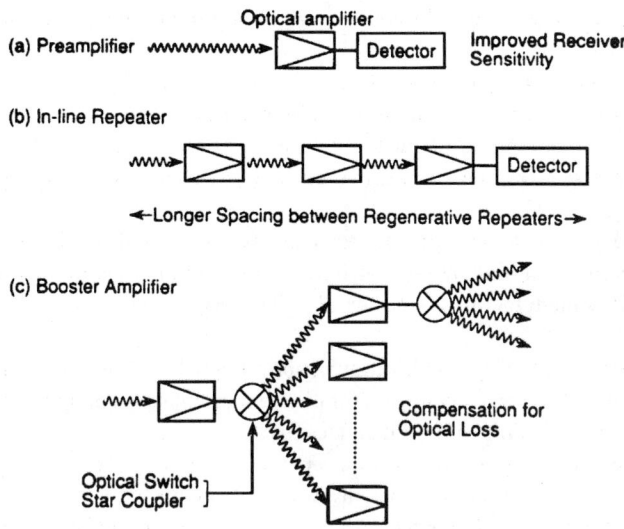

FIGURE 61.17 Configurations for SOA applications. *Source:* Yamamoto, Y. and Mukai, T. 1989. In *Optical and Quant. Electron.*, 21:S1.

wavelengths using wavelength conversion techniques such as cross-gain saturation [Valiente, Simon, and LeLigne, 1993] or four-wave mixing [Zhou et al., 1994].

Acknowledgments

This work was supported in part by the National Science Foundation under a National Young Investigator (NYI) award (contract number ECS-9457148) and the Air Force Office of Scientific Research (AFOSR) (contract number F49620-95-1-0024).

Defining Terms

Absorption coefficient: Amount of optical power absorbed in material per unit length.
Amplified spontaneous emission (ASE): Stimulated replication of photons that were originally generated spontaneous emission.
Amplifier gain: Ratio of optical power at input of amplifier to optical power at output of amplifier.
Bit patterning: Influence of a previous bit on subsequent bits in a bit stream due to gain saturation effects.
Comb filter: Transfer function of resonant structure that has repeating periodic passband.
Cross-gain modulation (XGM): Modulation of amplifier gain at one wavelength due to variation in signal at another wavelength.
Dispersion compensation: Realignment of phases for different spectral components in a pulse so that pulse broadening can be reduced.
Distributed Bragg reflector laser (DBR): Laser structure with Bragg reflectors at ends of cavity.
Distributed feedback laser (DFB): Laser structure with Bragg reflector in middle of cavity and without mirrors at ends.
Fabry–Perot (FP) amplifier: Resonant amplifier using mirrors at input and output for feedback.
Four-wave mixing (FWM): Generation of new frequencies, or intermodulation products, through gain nonlinearity.
Free spectral range: Spacing between longitudinal modes in resonant cavity.

Frequency chirp: Change in optical frequency as a function of time across a modulated optical signal.

Gain clamped: Operation of gain structure above condition where gain equals cavity losses.

Gain ripple: Periodic variation in gain over amplifier bandwidth.

Gain saturation: Reduction in gain due to increase in optical power.

Interchannel crosstalk: Transfer of information between channels due to a physical crosstalk mechanism.

Longitudinal modes: Single passband of a resonant cavity with comb-filter type transfer function.

Material gain: Number of photons generated per incident photon per unit length of material.

Optical mode confinement factor: Amount of optical signal that passes through amplifying medium.

Photonic switching: Switching or redirection of optical signals without conversion to electronics.

Polarization dependent: Variation in gain for different orientations of optical field.

Resonant amplifier: Amplifier with gain and feedback.

Saturated material gain: Optical signal-level dependent gain.

Space switches: Connection through spatially discrete elements.

Spontaneous emission: Emission of a photon by random transition from a high-energy state to a low-energy state.

Stimulated amplification: Stimulated replication of photons through stimulated emission.

Stimulated emission: Emission of a photon due to transition from a high-energy state to a low-energy state that was caused by in initial photon.

Subthreshold: Operation of gain structure below condition where gain equals cavity losses.

Traveling wave amplifier (TWA): Single-pass amplifier with optical gain.

Unsaturated material gain coefficient: Gain provided to optical signal that is not signal-level dependent.

Wavelength-division multiplexing (WDM): Transmission of simultaneous channels on parallel optical wavelengths.

References

Adams, M.J., Collins, J.V., and Henning, I.D. 1985. *IEE Proceedings* Part J, 132:58.

Agrawal, G.P. 1995. Semiconductor laser amplifiers. In *Semiconductor Lasers: Past, Present, and Future*, ed. G. P. Agrawal, American Institute of Physics Press.

Agrawal, G.P. and Olsson, N.A. 1989. *IEEE J. Quant. Electron.*, 25:2297.

Blumenthal, D.J. and Kothari, N.C. 1996. *IEEE Photon. Tech. Lett.*, 8:133.

Darcie, T.E. and Jopson, R.M. 1988. *Electron. Lett.*, 24:638.

Dubovitsky, S., Mathur, A., Steier, W.H., and Dapkus, P.D. 1994. *IEEE Photon. Tech. Lett.*, 6:176.

Dutta, N.K., Lin, M.S., Piccirilli, A.B., Brown, R.L., and Chakrabarti, U.K. 1990. *J. Appl. Phys.*, 67:3943.

Dutta, N.K. and Simpson, J.R. 1993. Optical amplifiers. In *Progress in Optics*, ed. E. Wolf, Vol. 31, North-Holland.

Forghieri, F., Tkach, R.W., Chraplyvy, A.R., and Marcuse, D. 1994. *IEEE Photon. Tech. Lett.*, 6:754.

Glance, B., Wiesenfeld, J.M., Koren, U., Gnauck, A.H., Presby, H.M., and Jourdan, A. 1992. *Electron. Lett.*, 28:1714.

Gustavsson, M., Lagerstrom, B., Thylen, L., Jansson, M., Lundgren, L., Morner, A.-C., Rask, M., and Stolz, B. 1992. *Electron. Lett.*, 28:2223.

Jopson, R.M. and Darcie, T.E. 1991. Semiconductor laser amplifiers in high-bit-rate and wavelength-division-multiplexed optical communications systems. In *Coherence, Amplification, and Quantum Effects in Semiconductor Lasers*, ed. Y. Yamamoto, Wiley.

Jopson, R.M., Eisenstein, G., Hall, K.L., Raybon, G., Burrus, C.A., and Koren, U. 1986. *Electron. Lett.*, 22:1105.

Koch, T.L. and Koren, U. 1991. *IEEE J. Quant. Electron.*, 27:641.
Mukai, T., Yamamoto, Y., and Kimura, T. 1982. *IEEE J. Quant. Electron.*, QE-18:1560.
Olsson, N.A. 1989. *IEEE J. Lightwave Tech.*, 7:1071.
Saitoh, T. and Mukai, T. 1987. *Electron. Lett.*, 23:218.
Saitoh, T., and Mukai, T. 1991. Traveling-wave semiconductor laser amplifiers. In *Coherence, Amplification, and Quantum Effects in Semiconductor Lasers*, ed. Y. Yamamoto, Wiley.
Saitoh, T., Mukai, T., and Noguchi, Y. 1986. Fabrication and gain characteristics of a 1.5 μm GaInAsP traveling-wave optical amplifier. In *First Optoelectronics Conference Post-Deadline Papers Tech. Digest*, The Institute of Electronics and Communications Engineers of Japan, Tokyo, Paper B11-2.
Tatham, M.C., Sherlock, G., and Westbrook, L.D. 1993. *Electron. Lett.*, 29:1851.
Valiente, I., Simon, J.C., and LeLigne, M. 1993. *Electron. Lett.*, 29:502.
Yamamoto, Y. 1980. *IEEE J. Quant. Electron.*, QE-16:1073.
Yazaki, P.A., Komori, K., Arai, S., Endo, A., and Suematsu, Y. 1992. *IEEE J. Lightwave Tech.*, 10:1247.
Zah, C.E., Osinski, J.S., Caneau, C., Menocal, S.G., Reith, L.A., Salzman, J., Shokoohi, F.K., and Lee, T.P. 1987. *Electron. Lett.*, 23:990.
Zhou, J., Park, N., Dawson, J.W., Vahala, K.J., Newkirk, M.A., and Miller, B.I. 1994. *IEEE Photon. Tech. Lett.*, 6:50.

Further Information

Coherence, Amplification, and Quantum Effects in Semiconductor Lasers, ed. Y. Yamamoto, Wiley, 1991.

Agrawal, G.P. and Dutta, N.K. 1993. *Semiconductor Lasers*, Van Nostrand Reinhold.

Optical Amplifiers

62.1 Introduction.. 848
62.2 General Amplifier Concepts 849
62.3 Alternative Optical Amplifiers for Lightwave System Applications 852
  Fiber-Raman Amplifiers • Brillouin Amplifiers • Semiconductor Optical Amplifiers • Rare-Earth-Doped Fiber Amplifiers • Rare-Earth-Doped Waveguide Amplifiers • Optical Parametric Amplifiers • Comparison of Amplifier Characteristics
62.4 Summary 858

Anders Bjarklev
Technical University of Denmark

62.1 Introduction

From the first demonstration of optical communication a primary drive in the research activity has been directed toward constant increase in the system capacity. Until very recently the ultimate capacity limits have been determined by the spectral bandwidth of the signal source and of the fundamental fiber parameters: attenuation and dispersion. On the basis of the limiting term of the transmission link, systems have been denoted as either *loss limited* or *dispersion limited*, and it can be seen that there have been alternations between these two extremes over the past 20 years. In the mid-1980s the international development had reached a state at which not only dispersion-shifted fibers were available but also spectrally pure signal sources emerged. The long-haul optical communication systems were, therefore, clearly loss limited and their problems had to be overcome by periodic regeneration of the optical signals at repeaters applying conversion to an intermediate electrical signal. Because of the complexity and lack of flexibility of such regenerators, the need for optical amplifiers was obvious.

The technological challenge was to develop a practical way of obtaining the needed gain, that is, to develop amplifiers with a complexity and application flexibility superior to the electrical regenerators. Several means of doing this had been suggested in the 1960s and 1970s, including direct use of the transmission fiber as gain medium through nonlinear effects [Agrawal, 1989], semiconductor amplifiers with common technical basis in the components used for signal sources [Simon, 1983], or doping optical waveguides with an active material (rare-earth ions) that could provide the gain [Koester and Snitzer, 1964]. First, however, with the pronounced technological need for optical amplifiers and the spectacular results on erbium-doped-fiber amplifiers [Mears et al., 1987], an intense worldwide research activity on optical amplifiers was initiated. This resulted in the appearance of commercially available packaged erbium-fiber amplifier modules as early as 1990, and the rapidly expanding technology and transfer of knowledge from research to practical

Optical Amplifiers

system implementation make it clear that optical amplifiers and, especially, the erbium-doped-fiber amplifier are bringing about a revolution in communication systems.

The following section provides a short review of the general properties and limitations of optical amplifiers. Thereafter, the specific technologies available for the realization of optical amplifiers are reviewed, and their key parameters are compared. Finally, the chapter includes a short review of some of most significant results on optical amplifiers in optical communication systems.

62.2 General Amplifier Concepts

In order to clarify and define the need for optical amplifiers, we discuss here some concepts general to all optical amplifiers. In an ideal optical amplifier the signal would be amplified by adding, in phase, a well-defined number of photons to each incident photon. This means that a bit sequence or an analog optical signal simply would increase its electromagnetic field strength, but not change its shape by passage through the optical amplifier, as illustrated in Fig. 62.1(a). In a perfect amplifier this process would take place independent of the wavelength, state of polarization, intensity, (bit) sequence, and optical bandwidth of the incident light signal, and no interaction would take place if more than one signal were amplified simultaneously. In practice, however, the optical gain depends not only on the wavelength (or frequency) of the incident signal, but also on the electromagnetic field intensity at any point inside the amplifier. Details of wavelength and intensity dependence of the optical gain depend on the amplifying medium.

Before we turn to describing the different types of amplifiers, however, it is informative to illustrate some general concepts of practical optical amplifiers. To do so, we consider a case in which the gain medium is modeled as a homogeneously broadened two-level system. In such a medium, the gain coefficient (i.e., the gain per unit length) can be written as [Agrawal, 1992]:

$$g(\omega) = \frac{g_0}{1 + (\omega - \omega_0)^2 T_2^2 + P_s/P_{\text{sat}}} \quad (62.1)$$

Here g_0 is the peak value of the gain coefficient determined by the pumping level of the amplifier, ω is the optical angular frequency of the incident signal, and ω_0 is the atomic transition angular frequency. P_s is the optical power of the signal and P_{sat} is the saturation power, which depends on the gain medium parameters such as fluorescence lifetime τ and the transition cross section. The parameter T_2 in Eq. (62.1) is normally denoted as the dipole relaxation time [Agrawal, 1992].

Two important amplifier characteristics are described in Eq. (62.1). First, if the signal power ratio obeys $P_s/P_{\text{sat}} \ll 1$ throughout the amplifier, the amplifier is said to be operated in the unsaturated region. The gain coefficient is then maximum when the incident angular frequency ω coincides with the atomic transition angular frequency ω_0. The gain reduction for angular frequencies different from ω_0 is generally given by a more complex function than the Lorentzian profile, but this simple example allows us to define the general property of the gain bandwidth. This is normally defined as the full-width half-maximum (FWHM) value of the gain coefficient spectrum $g(\omega)$. For the Lorentzian spectrum the gain bandwidth is given by $\Delta\omega_g = 2/T_2$.

From the communication system point of view, it is more natural to use the related concept of amplifier bandwidth instead of the gain bandwidth, which is related to a point within the gain medium. The difference becomes clear, when we consider the amplifier gain G defined as

$$G = P_s^{\text{out}}/P_s^{\text{in}} \quad (62.2)$$

where P_s^{in} is the input power and P_s^{out} the output power of a continuous wave (CW) signal being amplified. The amplifier gain G may be found by using the relation

$$\frac{dP}{dz} = gP \quad (62.3)$$

where $P(z)$ is the optical power at a distance z from the amplifier input end. If the gain coefficient

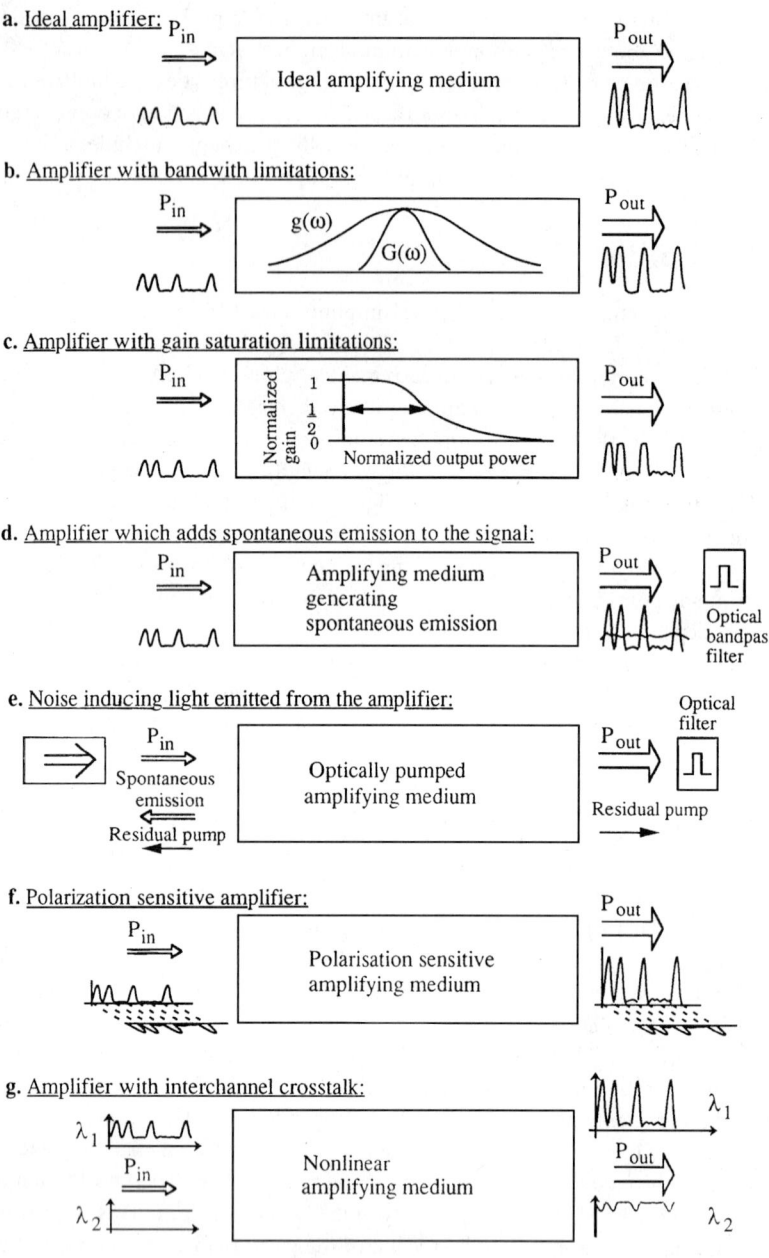

FIGURE 62.1 Optical amplifiers with different limiting properties: (a) ideal optical amplifier, (b) optical amplifier with a Gaussian bandwidth limitation, (c) optical amplifier with saturation limitation, (d) optical amplifier adding noise to the signal, (e) optical amplifier with isolator and filter to limit residual pump and ASE disturbance in the surrounding system, (f) polarization sensitive optical amplifier, and (g) optical amplifier inducing interchannel crosstalk.

Optical Amplifiers

$g(\omega)$, for simplicity, is considered constant along the amplifier length, the solution of Eq. (62.3) is an exponentially growing signal power $P(z) = P_s^{\text{in}} \exp(gz)$. For an amplifier length L, we then find that

$$G(\omega) = \exp[g(\omega)L] \qquad (62.4)$$

Equation (62.4) illustrates the frequency dependence of the amplifier gain G, and shows that $G(\omega)$ decreases much faster than $g(\omega)$ with the signal detuning $(\omega - \omega_0)$ because of the exponential dependence of G on g. As shown in Fig. 62.1(b), the amplifier bandwidth $\Delta\omega_A$, which is smaller than the gain bandwidth, results in signal distortion in the case where a broadband optical signal is transmitted through the amplifier.

Another important limitation of the nonideal amplifier is expressed in the power dependence of the gain coefficient. This property, which is known as gain saturation, is included in the example of Eq. (62.1), and it appears in all cases where the term P_s/P_{sat} is nonnegligible. Since the gain coefficient is reduced when the signal power P_s becomes comparable to the saturation power P_{sat}, the amplification factor (or amplifier gain) G will also decrease. This is illustrated in Fig. 62.1(c), where it is shown how G decreases from its unsaturated value G_0.

Gain saturation might be seen as a serious limitation for multiwavelength optical communication systems. However, gain saturated amplifiers have practical use as their self-regulating effect on the signal output power is going to be used in planned transoceanic communication links including many concatenated amplifiers. The principle is as follows: If the signal level in a chain of amplifiers is unexpectedly increased along the chain, due to saturation the following amplifiers will be operated at a lower gain and vice versa for a sudden signal power decrease.

Besides the bandwidth and gain saturation limitations of practical optical amplifiers another property is inherent for practical amplifiers today. Optical amplifiers, in general, will add spontaneously emitted or scattered photons to the signal during the amplification process, and this will consequently lead to a degradation of the signal-to-noise ratio (SNR). This property is illustrated in Fig. 62.1(d). The SNR degradation is quantified through a parameter F, normally denoted as the amplifier noise figure, which is defined as the SNR ratio between input and output [Yariv, 1990],

$$F = \frac{\text{SNR}_{\text{in}}}{\text{SNR}_{\text{out}}} \qquad (62.5)$$

It should be noted that it is common practice to refer the SNR to the electrical power generated when the optical signal is converted to electrical current by using a photodetector. The noise figure as defined in Eq. (62.5) therefore, in general, would depend on several detector parameters, which determine the shot noise and thermal noise associated with the practical detector. However, the influence of detector parameters will not help us to clarify the amplifier noise properties, and it is therefore advantageous to consider an ideal detector, whose performance is limited by shot noise only [Yariv, 1990].

In practice, the spontaneous emission is reduced by optical filtering of the received signal. Therefore, the SNR generally also will be dependent on the bandwidth of the optical filters and the spectral power distribution of the spontaneous emission from the amplifier. However, since this filtering is a process independent of the amplifier properties, it is also common practice to eliminate this ambiguity by considering an ideal filter [Agrawal, 1992]. Such an ideal filter is introduced only to allow the signal and the spontaneous emission within the signal bandwidth to pass to the detector. Therefore, it will only be the spontaneous emission spectral power density at the signal wavelength that enters the ideal detector, and the noise figure becomes independent of the spectral shape of the spontaneous emission. Note that since the amplifier deteriorates the signal-to-noise ratio, that is, $\text{SNR}_{\text{in}} > \text{SNR}_{\text{out}}$, the noise figure will always obey the relation $F > 1$.

For a practical communication system the amplifier spontaneous emission may have another influence besides that described through the noise figure (i.e., besides the effect that it adds fluctuations to the amplified optical signal power, which are converted to current fluctuations during the photodetection process). Namely, the spontaneous emission, which is emitted from the amplifier input end, may enter the signal source (a semiconductor laser), where it can result in performance disturbances. Therefore, it is often necessary to include isolation between amplifier and light source to avoid additional noise in the system.

Therefore, it must be considered that the optical communication system has to be protected against undesired emission from the amplifier, and it is necessary to include the special properties of optically pumped amplifiers. These may also transmit residual pump power onto the transmitter and/or detector. The actual effect of this will be strongly dependent on the spectral properties of the pump light and whether the signal source or detector is sensitive to such radiation. As shown in Fig. 62.1(e) this often implies the inclusion of separate isolators and filters (if not already included for direct reduction of noise-induced signal power fluctuations).

Another property that has to be evaluated for the optical amplifier is the polarization sensitivity. A high-polarization sensitivity means that the amplifier gain G differs for different polarization directions of the input signal. This is shown in Fig. 62.1(f) for two orthogonal linear polarization states. Since optical communication systems normally do not include polarization control, and because the polarization state is likely to vary due to external factors such as mechanical pressure, acoustic waves, and temperature variations, amplifier polarization sensitivity is naturally an undesired property in most cases. Therefore, for amplifiers that are inherently polarization sensitive, it has been a primary goal to reduce or even eliminate the amplifier output power variation due to changes in the signal input polarization state.

All of the limiting properties, which we have previously discussed, are not surprisingly distinctly different for different types of amplifiers. This is also the case for the crosstalk limitation that relates to multichannel applications of optical amplifiers. In contrast to the ideal case, where all signal channels (or wavelengths) are amplified by the same amount, undesired nonlinear phenomena may introduce interchannel crosstalk (i.e., the modulation of one channel is affected or modified by the presence of another signal channel). The effect of crosstalk is illustrated in Fig. 62.1(g), where one channel is considered to carry a modulated signal, whereas the other carries a continuous wave (CW) signal. By passage of the amplifier the CW has been modulated by the signal carried in the other channel. We will return to the more specific nature of these nonlinear phenomena in connection with the discussion of the different optical amplifiers.

The final limiting factors that should be mentioned here are closely related to the physical environment in which the amplifier is placed. These may include sensitivity toward vibrations, radioactive radiation, chemicals, pressure, and temperature. However, since at least amplifiers for optical communication systems generally are placed in controlled environments, we will not go further into a detailed discussion of such properties at this point.

The relative importance of the different limiting factors as just discussed depends on the actual amplifier application. One application is as replacement of electronic regenerators; in such cases the amplifiers are placed at a considerable distance from the transmitter and receiver and they are denoted as in-line amplifiers. The optical amplifier may also be used to increase the transmitter power by placing an amplifier just after the transmitter. Such amplifiers are called power amplifiers, or boosters, because their main purpose is to boost the transmitted power. Long-distance systems may also be improved by the inclusion of so-called preamplifiers, which are placed just before the receiver. Finally, optical amplifiers may find use in local area networks in which they can compensate for distribution losses. Thereby, the number of nodes in the networks may be increased.

62.3 Alternative Optical Amplifiers for Lightwave System Applications

The most important factors that limit the performance of practical optical amplifiers have now been identified. The next task then is to investigate the technologies available for obtaining realistic

Optical Amplifiers

amplifiers. Basically, six different ways of obtaining optical amplification have to be considered: **fiber-Raman** and **fiber-Brillouin amplifiers, semiconductor optical amplifiers (SOA)**, rare-earth-doped fiber amplifiers, **rare-earth-doped waveguide amplifiers**, and **parametric amplifiers**. The aim of the present description is to present the basic physical properties of these candidates as relates to their applicability as practical optical amplifiers.

Fiber-Raman Amplifiers

The fiber-Raman amplifier works through the nonlinear process of **stimulated Raman scattering (SRS)** occurring in the silica fiber itself. SRS is an interaction between light and the vibrational modes of silica molecules. The process appears because the SRS converts a small fraction of the incident power from one optical beam (the pump) to another optical beam (the signal) at a frequency down-shifted by an amount determined by the vibrational modes of the medium.

The light generated by SRS is called the Stokes wave, and Stokes waves are generated both copropagating and counterpropagating with the incident beam. As early as 1962 it was observed that for very intense pump waves a new phenomenon of SRS can occur in which the Stokes wave grows rapidly inside the medium such that most of the pump energy appears in it. If the signal angular frequency ω_s and pump angular frequency ω_p are chosen with a difference of $\Omega_R = \omega_p - \omega_s$ (the Stokes shift) corresponding to the value dictated by the vibrational energy of the molecules, the signal may experience Raman gain.

A significant feature of the Raman gain in silica fibers is that it extends over a large frequency range (up to 40 THz with a typical FWHM width of 6 THz) with a broad dominant peak near 13 THz. This is due to the amorphous nature of fused silica in which the molecular vibrational frequencies spread out into bands overlapping each other and creating a continuum (depending on the fiber core composition). The relatively large bandwidth of fiber Raman amplifiers and a very large saturation power (typically around 1 W) makes them attractive for optical communication systems. Also their noise properties are good, since a noise figure close to 3 dB is obtainable. On the other hand, one of the considerable drawbacks of this amplifier type is that to obtain a 20-dB gain in a 100-km-long standard fiber for optical communication more than 2.5 W of pump power at 1480 nm is needed [Agrawal, 1989]. Another important limitation of fiber-Raman amplifiers is their polarization sensitivity. Experiments show the necessity of matching the polarization directions of pump and signal, since SRS ceases to occur for the case of orthogonal polarizations. Because polarization control is not normally employed in optical communication systems, this property introduces an additional complexity, which together with the high pump power requirements make the Raman amplifier only slightly attractive from the point of view of the optical communication system designer. Table 62.1 lists key parameters for the fiber-Raman amplifier.

Brillouin Amplifiers

Another type of amplifier that makes use of the physical properties of the silica transmission fiber itself is the **stimulated Brillouin scattering (SBS)** amplifier [Agrawal, 1989]. The nonlinear SBS process appears through the generation of a backward propagating Stokes wave that carries most of the input energy, and is in this sense similar to the SRS process. There are, however, major differences between the SRS and SBS processes. One aspect is that the Stokes wave in the case of SBS propagates backwards and, in contrast to SRS, the input power levels needed to obtain SBS are much lower ($\approx$1-mW CW pump power is sufficient to obtain SBS). Note also that the SBS depends on the spectral width associated with the pump wave and it nearly ceases to occur for short pump pulses. Finally, the Stokes shift in SBS ($\approx$10 GHz) is smaller by three orders of magnitude compared with that occurring in SRS. The differences between SRS and SBS may be explained from a single fundamental difference; optical phonons participate in the Raman process, whereas acoustical phonons (forming acoustic waves) participate in the Brillouin process. In the ideal case Brillouin

TABLE 62.1 Comparison of Optical Amplifiers Around 1550 nm

Property	Raman Amplifier	Brillouin Amplifier	SOA	EDFA	RE-Doped Waveguide Amplifiers	Optical Parametric Amplifiers
Small-signal gain	>40 dB	>40 dB	>30 dB	>50 dB	20 dB	16 dB
Efficiency	0.08 dB/mW	5.5 dB/mW	28 dB/mA	11 dB/mW	0.1 dB/mW	10^{-4} dB/mW
Output power	1 W	1 mW	>0.1 W	>0.5 W	2 mW	
Distortion/Crosstalk	Negligible	Negligible	Significant (in saturation)	Negligible	Negligible	
Dynamic performance	>20 Gb/s	<100 MHz	>25 Gb/s	>100 Gb/s	>100 Gb/s	
Gain bandwidth	A few tens of nm	<100 MHz	60–70 nm	$\approx$30 nm	A few tens of nm	5000 GHz
Noise figure	$\approx$3 dB	>15 dB	5–7 dB	$\approx$3 dB	>3 dB	
Polarisation sensitivity	Significant	None	<A few dB	Insignificant (<0.1 dB)	<1 dB	<A few dB
Coupling loss (to fiber)	<1 dB	<1 dB	A few dB	<1 dB		

scattering should not occur in the forward direction in an optical fiber, but due to spontaneous or thermal Brillouin scattering small amounts of forward scattered light may be measured.

The obtainable SBS gain is nearly three orders of magnitude larger than the Raman gain [Agrawal, 1989]. This means that the fiber Brillouin amplifier can amplify a signal by 30–35 dB with a small pump power in the range of 1–10 mW. Also the saturation power of fiber-Brillouin amplifiers is different from that of the Raman amplifier; in the case of SBS it is in the milliwatt range. However, from the point of view of the optical communication system designer, who in the mid-1980s was seeking an efficient optical amplifier, the central problem of fiber Brillouin amplifiers was their narrow gain bandwidth (less than 100 MHz). This narrow bandwidth, which due to fiber inhomogeneities is broadened even compared to the bulk silica value of around 17 MHz [Agrawal, 1992], makes fiber Brillouin amplifiers less suitable as power amplifiers, preamplifiers, or in-line amplifiers in generally used intensity-modulated direct detection systems. Fiber-Brillouin amplifiers need to be pumped by narrow linewidth semiconductor lasers (<10 MHz), which obviously limit their potential use in optical communication systems, since in practice a wavelength tuneable narrow bandwidth pump is required. As a consequence of this, SBS has, since its first observation in optical fibers in 1972, primarily been studied as a nonlinear phenomenon limiting transmitter power in fiber optic communication systems. It is, on the other hand, important to note that the narrow bandwidth nature of SBS lately has attracted attention, for example, in connection with channel selection in multichannel lightwave systems.

Another property that adds to reservations about fiber Brillouin amplifiers is their disadvantageous noise properties. Noise appears due to a large population of acoustic phonons at room temperature and noise figures as large as 15 dB are likely to appear. This makes the Brillouin amplifier highly problematic both as preamplifier and for in-line applications.

Semiconductor Optical Amplifiers

The semiconductor optical amplifier, in contrast to both Raman and Brillouin fiber amplifiers, is fabricated in a material and structure differently than silica transmission fiber. Therefore, light has to be coupled from fiber to SOA and vice versa, which inevitably results in coupling losses. Unless special action is taken, the SOA itself also will experience relatively large feedback due to reflections occurring at the cleaved end facets (32% reflectivity due to refractive index differences between semiconductor and air), resulting in sharp and highly disadvantageous gain reduction between the cavity resonances of the Fabry–Perot resonator. Therefore, it is necessary to design traveling wave (TW) type SOAs by suppressing the reflections from the end facets; a common solution is the inclusion of antireflection coating of the end facets. It turns out that to avoid the amplifier bandwidth being determined by the cavity resonances rather than the gain spectrum itself, it becomes necessary to require that the facet reflectivities satisfy the condition [Agrawal, 1992]: $G\sqrt{R_1 R_2} < 0.17$. Here, G is the single-pass amplification factor, and R_1 and R_2 are the power reflection coefficients at the input and output facets, respectively. For an SOA designed to provide a 30-dB gain, this condition will mean that $\sqrt{R_1 R_2} < 0.17 \cdot 10^{-4}$, which is difficult to obtain in a predictable and reproducible manner by antireflection coating alone. Additional methods to reduce the reflection feedback include designs in which the active-region stripe is tilted from the facet normal and introduction of a transparent window region (nonguiding) between the active-layer ends and the antireflection coated facets. Thereby, reflectivities as small as 10^{-4} may be provided, and the SOA bandwidth will be determined by the gain medium. Typically, 3-dB amplifier bandwidths of 60–70 nm may be obtained.

The gain in an SOA depends on the carrier population, which changes with the signal power and the injection current. Note that in contrast to the alternative optical amplifiers the SOA is not optically pumped, and an important property is the very short effective lifetime of the injected carriers (i.e., of the order of 100 ps). This property becomes specifically relevant for multichannel applications of SOAs, where crosstalk limitations are in question. This crosstalk originates from two nonlinear phenomena: cross saturation and four-wave mixing. The former appears because the signal in one channel through stimulated recombinations affects the carrier population and, thereby, the gain of other channels. This significant problem may only be reduced by operating the amplifier well-below saturation, but this is not an easy task due to the relatively limited saturation output power of 5–10 mW. Four-wave mixing also appears because stimulated recombinations affect the carrier number. More specifically, the carrier population may be found to oscillate at the beat frequencies between the different channels, whereby both gain and refractive index are modulated. The multichannel signal therefore creates gain and index gratings, which will introduce interchannel crosstalk by scattering a part of the signal from one channel to another. Finally, it should here be mentioned that due to their large bandwidth, SOAs have the potential of amplifying ultra-short pulses (a few picoseconds). However, here the gain dynamics will play an important role and saturation induced self-phase modulation has to be considered. In other words, the SOA will introduce a frequency chirp, which may find positive use in terms of dispersion compensation.

The noise properties of the SOA are determined by two factors. One is the emission due to spontaneous decays and the other is the result of nonresonant internal losses α_{int} (e.g., free carrier absorption or scattering loss), which reduce the available gain from g to $(g - \alpha_{\text{int}})$. Also residual facet reflectivities (i.e., loss of signal input power) increase the noise figure. Typical values of the noise figure for SOAs are 5–7 dB.

An undesirable characteristic of the early SOAs is the polarization sensitivity, which appears because the amplifier gain differs for the transverse-electric (TE) and transverse-magnetic (TM) modes in the semiconductor waveguide structure. However, intense research over the last few years has reduced the problem, and an SOA with much reduced polarization sensitivity has been reported by many groups [Stubkjaer et al., 1993]. For such state-of-the-art SOAs a net gain of 25 dB may be achieved for a bias current of 100 mA. At this gain level the polarization sensitivity is ±1 dB, and it becomes insignificant for gain levels lower than ≈20 dB. Other methods of using serial or parallel coupled amplifiers or two passes through the same amplifier are also suggested, but such schemes increase complexity, cost, and stability requirements.

To conclude the discussion of the semiconductor optical amplifier, it is important to note that SOAs through the 1980s have shown potential for optical communication applications. However, drawbacks such as polarization sensitivity, interchannel crosstalk, and large coupling losses have to be overcome before their use becomes of practical interest. In favor of the SOA are the large amplifier bandwidth and the possibility of monolithical optoelectronic integration, especially within the receiver, where the input signal powers are weak enough to avoid undesired nonlinearities. Therefore, it is likely that SOAs will find application in future optical communication systems. It should be mentioned that the SOA itself can be used as an amplifier and a detector at the same time, because the voltage across the pn-junction depends on the carrier density, which again interacts with the optical input signal. Promising results for transparent channel drops, channel monitoring, and automatic gain control have been demonstrated [Stubkjaer et al., 1993]. Note that possible future applications of SOAs will not be in their ability as amplifiers alone, but rather as wavelength conversion elements and optical gates [Stubkjaer et al., 1993].

Rare-Earth-Doped Fiber Amplifiers

Parallel to the maturation of semiconductor optical amplifiers, another development has taken place, which in only three years has demonstrated a large impact on optical communication systems. With a point of reference in work on rare-earth- (RE-) doped glass lasers initiated as early as 1963 [Koester and Snitzer, 1964], the first rare-earth-doped fiber amplifiers (as possible useful devices for telecommunication applications) were demonstrated in 1987 [Mears et al., 1987]. Progress since then has multiplied to the extent that amplifiers today offer far-reaching new opportunities in telecommunication networks.

The possible operational wavelengths of rare-earth-doped fiber amplifiers are determined by the emission spectra of the rare-earth ions moderately modified by the host material in which they are embedded. Only a few rare-earth materials become relevant for optical communication purposes, primarily erbium, which may provide amplification in the 1550 nm wavelength band, and neodymium or praseodymium, which may be operated around the 1300 nm band. The absorption spectrum holds accurate information about the location of possible pump wavelengths and the ability to excite the rare-earth ions to higher energy levels. From this higher energy level the electron can relax, transferring its packet of energy to the ground state either radiatively or nonradiatively. The nonradiative de-excitation involves the creation of phonons (i.e., a quantized vibration of the surrounding medium), whereas the radiative decay to lower energy levels takes one of two forms. These are known as spontaneous and stimulated emission, and in both cases photons are emitted. Spontaneous emission always takes place when the electrons of a collection of atoms are in an excited state; therefore, spontaneously emitted light may not be avoided in a fiber amplifier. Stimulated emission is the process that allows signal amplification to take place and therefore is the desired property of the fiber amplifier. The process may be explained as follows. A photon incident on the medium, with an energy equal to the difference in energy of the ground state and an excited state, promotes de-excitation, with the creation of a photon that is in phase with the incident photon. When the ions are pumped to energy levels above the upper laser level, the population cascades down through the intervening levels by nonradiative transitions and eventually leads to a population of the upper laser level. This level is normally described as metastable, since transitions from this to lower manifolds only appear because the crystal field has broken the inversion symmetry of the ion's environment, permitting the electric dipole transitions to occur. Note that for this reason the involved lifetimes in the upper laser level (typically, 10–14 ms for erbium) are many orders of magnitude larger than the nonradiative lifetimes of the higher energy levels.

Focusing on the **erbium-doped-fiber amplifier (EDFA)**, which may be fabricated using a silica glass host, semiconductor pump sources have been successfully used to pump amplifiers in the 800-, 980-, and 1480-nm absorption bands. In addition, pumping at lower wavelengths is also applicable, but less interesting due to the lack of practical pump sources [Bjarklev, 1993]. Basic differences exist between the application of the three mentioned pump choices. First, it should be noted that amplification occurs according to a three-level scheme when 800- or 980-nm pumping is applied, but the erbium ion works as a two-level system when 1480-nm pumping is used. Furthermore, 800-nm pumping is much less efficient than 980-nm pumping due to a pronounced excited-state absorption (ESA) of pump photons [Bjarklev, 1993]. Because of these differences, today only

980- and 1480-nm pumping is considered in practical system applications. In cases where very low-noise figures are required, 980-nm pumping is the preferred choice, since the three-level nature of the system makes excitation of all erbium ions (full inversion) possible resulting in a noise figure very close to 3 dB. For 1480-nm pumping only 70% of the erbium ions may be excited to the upper laser level, and noise figures down to 5 dB are generally obtainable. However, 1480-nm pumping still provides highly efficient amplifiers, and more reliability testing of the semiconductor pump sources have been performed at this wavelength, making 1480-nm pumped amplifiers preferred for many present applications.

EDFAs readily provide small-signal gain in the order of 30–40 dB, but values as high as 54 dB have been demonstrated [Hansen, Dybdal, and Larsen, 1992]. The EDFA also provides high-saturation powers, and signal output powers of more than 500 mW have been demonstrated. In such power amplifier applications 80% or more of the pump photons may be converted to signal photons, indicating the high efficiency of the EDFA. A 30-nm bandwidth of the EDFA may easily be obtained, and although this is not quite as large as that of the SOA, it will be quite sufficient for most multiwavelength applications of optical amplifiers.

The long lifetime in the upper laser level makes the EDFA an almost ideal energy reservoir, and it is important to note that, with the power levels used in modern optical communication systems, one signal pulse will only interact with a very small fraction of the erbium ions within the EDFA. The following pulse is, therefore, very unlikely to interact with the same erbium ions, and it will remain unaffected by other pulses. For this reason, in practice no crosstalk will be seen in fiber amplifiers. The EDFA is also very insensitive to polarization variations of the signal, and only systems with cascaded amplifiers have lately demonstrated degradation due to polarization effects. A typical gain variation of 0.1 dB is found [Mazurczyk and Zyskind, 1993] when the polarization state of the signal is switched between orthogonal linearly polarized states. Finally, it should be mentioned that since the EDFA is an optical fiber itself, very low coupling losses (comparable to ordinary splice losses) are involved in the application of EDFAs. The key numbers describing the EDFA are shown in Table 62.1.

In contrast to the EDFA, the 1300-nm fiber amplifiers have to deal with a number of difficulties, and their applicability still remains an unanswered question. The neodymium-doped fiber amplifier operates at a wavelength that is too high (around 1340 nm) to provide a realistic gain and so the best candidate in the 1300-nm wavelength band is the **praseodymium-doped fiber amplifier** (PDFA). The primary difficulty is that for 1300-nm applications the rare-earth dopants have to be placed in a host material other than silica (otherwise no net gain will appear) [Bjarklev, 1993]. Until now the most promising host materials have been different compositions of fluoride glasses, which suffer from low strength and hygroscopy. However, a small-signal gain of 34 dB (fiber-to-fiber) and a saturated output power of 200 mW have been demonstrated for a praseodymium-doped fiber amplifier module. Another problem is that the 1300-nm amplifiers normally have a lower pump efficiency than EDFAs. However, the latest results present values as high as 4 dB/mW for praseodymium-doped amplifiers, which also show noise figures in the 5–8 dB range.

Rare-Earth-Doped Waveguide Amplifiers

Rare-earth-doped waveguide amplifiers are only different from fiber amplifiers in the respect that the rare-earth material is embedded in a planar optical waveguide and not a fiber. This means at first glance that only another waveguide geometry has to be considered. However, two important factors have to be considered in the design of rare-earth-doped waveguide amplifiers. First, the background loss is several orders of magnitude larger in a planar waveguide (around 0.1 dB/cm) than in a fiber, and second, a much higher rare-earth-dopant concentration is necessary in the relatively short planar waveguides. This raises the problem of energy transfer between the rare-earth ions and a resulting lower amplifier efficiency. Not surprisingly, the focus has been on the erbium-doped planar waveguide amplifiers in silica, and integration with a 980/1530-nm wavelength division multiplexer have been presented [Hattori et al., 1994]. In this component a maximum gain of 13.5 dB for a 600-mW pump power was shown, but gain values up to 20 dB have been demonstrated in other amplifiers. Of course, these early amplifiers are far from optimized, but the results indicate that the rare-earth-doped planar waveguides have realistic and attractive application possibilities in integrated-optical devices. We may consider other host materials such as lithiumniobat, and new applications beside optical communication. On the other hand, it is not realistic to expect that the rare-earth-doped waveguide amplifier will completely replace EDFAs as high-performance amplifiers in optical communication systems.

Optical Parametric Amplifiers

Finally, it should be mentioned that the possibility of using second-order nonlinearity in electro-optic materials for amplification exists. In this case, energy is transferred from a strong optical pump to the signal with the aid of an optical idler in a degenerate four-wave-mixing process. It is referred to as a parametric process as it originates from light-induced modulation of a medium parameter such as the refractive index, and its origin lies in the nonlinear response of bound electrons of the material to an applied optical field. The main difficulty is that phase matching of the three interacting waves has to be maintained. Furthermore, a high-pump-power requirement is a serious limitation, and parametric amplifiers are at present inferior to erbium-doped and semiconductor optical amplifiers. However, such phase sensitive amplifiers are suitable for squeezed-state amplification, and they have no lower limit for the noise figure in the limit of high amplification [Kumar, Kath, and Li, 1994].

Comparison of Amplifier Characteristics

Table 62.1 lists the key parameters of all of the different optical amplifiers discussed. In principle, the different types may be operated at any wavelength, except for the erbium-doped amplifiers. However, since the 1550-nm wavelength band is one of the most important for optical communication, Table 62.1 refers to this wavelength band, and as such it serves as a source of comparison between the different amplifiers.

62.4 Summary

Through the description of the alternative optical amplifiers for optical communication systems we learn that fiber-Raman and fiber-Brillouin amplifiers have not been able to completely fulfill the demands of optical communication systems, and it was with the demonstration of practical EDFAs that the development of optical amplifiers for communication systems really first started. The tremendous progress provided by the EDFA over the past five years has been complemented by the development of high-performance SOAs, and lately, rare-earth doped planar waveguide amplifiers and parametric amplifiers have been investigated. However, the last two still have to mature before a practical system application is demonstrated. For these reasons, we end this description by looking at the system performance of the EDFA and to a minor extent the SOA.

The main reason for the strong impact that especially the EDFAs have had and will have on data communication systems is to be found in the fact that they are unique in two respects: they are useful for amplifying input signals of different bit rates or formats, and they can be used to achieve simultaneous amplification of multiwavelength optical signals over a wide spectral region. These multiwavelength optical signals can carry different signal formats including digital data, digital video, analog video, etc., allowing flexible upgrading in a broadband distribution network by adding extra wavelength division multiplexed (WDM) channels as needed. With optical amplifiers, therefore, the bottleneck of the narrow and fixed bandwidth of electrical repeaters is avoided. The outstanding properties offered by the EDFAs can be summarized as follows: high gain, high-power conversion efficiency, low noise, low crosstalk, high-saturation power, polarization insensitivity,

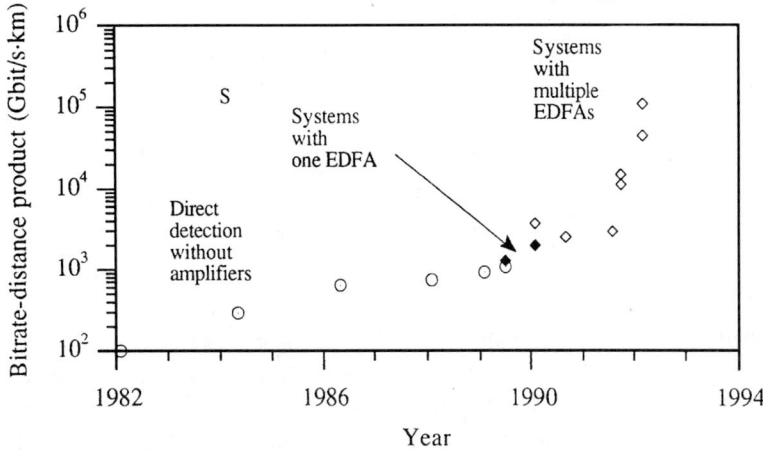

FIGURE 62.2 Capability of optical direct detection systems measured as reported bit rate distance product, against data of report.

broad spectral bandwidth, very low-coupling losses, and low cost. This short list of fiber amplifier properties underlines the huge potential of these components, which most likely will extend to a large variety of applications in the years to come.

Some of the most significant system capability results achieved over the last few years are plotted in Fig. 62.2 as a function of the publication date. For comparison, the highest performance results of direct detection systems without optical amplifiers are also included. Compared to these, it is obvious that only by including EDFAs at the ends of the transmission link (as either power boosters or preamplifiers) can a significant bit rate distance improvement be obtained. One of the highest capacities achieved with only one EDFA is 2000 Gb/s · km, which was demonstrated by transmitting 20 Gb/s over a distance of 100 km. This is equivalent to the transmission of 300,000 simultaneous telephone channels transmitted over 100 km.

Not surprisingly, most of the later system experiments include one or a multiple of in-line EDFAs, and transmission over transoceanic distances is obtained in this way. The breakthrough for capacity increase that happened with the availability of optical amplifiers and especially EDFAs in 1989 is therefore evident. It should also be mentioned that all reported high performance system experiments today use optical amplifiers. The present record is as much as 45 Tb/s · km, with which, according to Fig. 62.2, we have obtained a bit rate distance product improvement of a factor of 100 compared to systems without EDFAs.

However, limits also exist for direct detection systems with EDFAs. In contrast to electronic repeaters, the EDFA cannot regenerate the shape of the optical pulses. Thus, in dispersive fiber links, the transmission length will be dispersion limited, especially at high bit rates. However, methods to avoid this are studied intensively at present time, and optical amplifiers can also help to overcome the dispersion limitation by exploiting the nonlinear effects.

From the results in Fig. 62.2, it is evident that optical amplifiers have demonstrated their relevance in high-speed, long-distance repeaterless communication systems. The applications here are many, such as undersea and off-shore links, island jumping, and trunk lines between main exchanges. However, it is also relevant to consider that many medium-distance applications exist (i.e., interconnection of points spaced by 100–300 km). In such cases (and especially for submarine applications), it may be of interest to avoid the in-line amplifier, since the system maintenance and reliability problems would then be considerably reduced. Additionally, the EDFA will find many applications in optical signal distribution systems, and splits up to 39 million subpaths have been demonstrated for only two stages of amplification. In short, the EDFA has revolutionized thinking on future optical fiber networks.

Finally, note that the application of SOAs in a 10-Gb/s preamplified receiver today realizes system sensitivities of −28 dB (1 mW) with a fiber to fiber gain of the SOA of 18.5 dB. The polarization dependency of the receiver has been brought below 0.5 dB. For comparison, a fiber preamplifier used with the same front-end and the same transmitter resulted in a sensitivity of −34 dB(1 mW) [Stubkjaer et al., 1993].

Defining Terms

Erbium-doped fiber amplifier (EDFA): Fused silica fiber codoped with erbium as the active material providing gain in the 1.55-mm wavelength range. The erbium ions are most commonly pumped at either 0.98 or 1.48 mm using high-power semiconductor pump sources.

Fiber Brillouin amplifier: Optical fiber (typical of fused silica) providing signal gain through the process of SBS.

Fiber Raman amplifier: Optical fiber (typical of fused silica) providing signal gain through the process of SRS.

Parametric optical amplifier: Amplifier in which optical power is trnasferred from a strong optical pump to a signal through a degenerate four-wave mixing process. The term parametric amplifier refer to the property of using light-induced modulation of a medium parameter such as the refractive index.

Praseodymium-doped fiber amplifier (PDFA): Optical fiber (typically manufactured from a fluoride glass composition) codoped with praseodymium in order to obtain gain in the 1.3-mm wavelength range through stimulated emission.

Rare-earth-doped waveguide amplifier: Dielectric waveguide doped with a rare earth material. This material has the effect of providing gain through stimulated emission, and the necessary power for this process appears through absorption of pump photons (propagating within the waveguide structure at a different frequency than the signal).

Semiconductor optical amplifier (SOA): Electrically pumped active semiconductor component providing optical gain. SOAs have a large potential as amplifiers integrated with receivers, for example, and as wavelength conversion elements.

Stimulated Brillouin scattering (SBS): Interaction between light and acoustic phonons (forming acoustic waves) resulting in conversion of light at one frequency (the pump) to another (the signal). SBS couples energy from one mode to another propagating in the opposite direction, and is a quite narrowband phenomenon (bandwidth less than 100 MHz).

Stimulated Raman scattering (SRS): Interaction between light and vibrational modes (optical phonons) of the transmission medium that converts a fraction of the incident power from one optical beam (the pump) to another optical beam (the signal) at a frequency down shifted by an amount determined by the vibrational modes of the medium.

References

Agrawal, G.P. 1989. *Nonlinear Fiber Optics*, Academic Press, London.

Agrawal, G.P. 1992. *Fiber-Optic Communication Systems*, Wiley, New York.

Bjarklev, A. 1993. *Optical Fiber Amplifiers: Design and System Applications*, Artech House, Boston, MA.

Hansen, S.L., Dybdal, K., and Larsen, C.C. 1992. Upper gain limit in Er-doped fiber amplifiers due to internal Rayleigh backscattering. *Optical Fiber Communications, OFC'92*, San Jose, CA. 1:TuL4.

Hattori, K., Kitagawa, T., Oguma, M., Ohmori, Y., and Horiguchi, M. 1994. Erbium-doped silica-based planar-waveguide amplifier integrated with a 980/1530-nm WDM coupler. *Optical Fiber Communications, OFC'94*, San Jose, CA. 1:FB2.

Koester, C.J., and Snitzer, E. 1964. Amplification in a fiber laser. *Appl. Opt.*, 3(10):1182.

Kumar, P., Kath, W.L., and Li, R.-D. 1994. Phase-sensitive optical amplifiers. *Integrated Photonics Research IPR'94*, San Francisco CA. 1:SaB1.

Mazurczyk, V.J., and Zyskind, J.L. 1993. Polarization hole burning in erbium doped fiber amplifiers. *Conference on Lasers and Electro-Optics, CLEO'93*, Baltimore, MD, May 2–7. CPD26.

Mears, R.J., Reekie, L., Jauncey, I.M., and Payne, D.N. 1987. Low-noise erbium-doped fibre amplifier operating at 1.54 μm. *Electron. Lett.*, 23(19):1026.

Simon, J.C. 1983. Semiconductor laser amplifier for single mode optical fiber communications. *J. Opt. Commun.*, 4(2):51–62.

Stubkjaer, K.E., Mikkelsen, B., Durhuus, T., Joergensen, C.G., Joergensen, C., Nielsen, T.N., Fernier, B., Doussiere, P., Leclerc, D., and Benoit, J. 1993. Semiconductor optical amplifiers as linear amplifiers, gates and wavelength converters. *Proceeding of 19th European Conference on Optical Communication, ECOC'93*. 1:TuC5.

Yariv, A. 1990. Signal-to-noise considerations in fiber links with periodic or distributed optical amplification. *Opt. Lett.*, 15(19):1064–1066.

Further Information

A through treatment of fiber optical amplifiers and their applications in optical communication systems can be found in *Optical Fiber Amplifiers: Design and System Applications* by A. Bjarklev, 1993, Artech House.

Fiber lasers and amplifiers are treated in detail in *Rare Earth Doped Fiber Lasers and Amplifiers* by M.J.F. Digonnet, 1993, Marcel Dekker Inc., and in *Optical Fiber Lasers & Amplifiers* by P.W. France, 1991, Blackie and Son Ltd.

The journals *IEE Electronics Letters*, the *IEEE Photonics Technology Letters*, and *IEEE Journal of Lightwave Technology* reports advances in optical amplifiers and optical communication systems. Also Proceedings of the Annual European Conference on Optical Communication (ECOC) and the annual OSA Conference on Optical Fiber Communication (OFC) present state-of-the-art results on optical amplifiers. The annual SPIE conference OE/FIBERS also presents valuable information relevant to the subject.

63
Coherent Systems

63.1	Introduction	862
63.2	Fundamentals of Coherent Systems	863
63.3	Modulation Techniques	864
63.4	Detection and Demodulation Techniques	865
63.5	Receiver Sensitivity	865
63.6	Practical Constraints and Countermeasures	867
	Phase and Intensity Noise in the Laser Diode • Frequency Response and Bandwidth of Modulators and Receivers • Polarization Fluctuation in the Fiber • Dispersion and Nonlinearity in the Fiber	
63.7	Summary and Conclusions	870

Shinji Yamashita
The University of Tokyo

63.1 Introduction

Optical fiber communications have been expected to play a leading role in communication networks and have already been put into commercial operation in domestic and international trunk lines. The present systems employ intensity modulation/direct detection (IM/DD), where the transmitter laser diode (LD) is directly on–off modulated, and the signal is directly detected by an avalanche photodiode. Although advantageous in simplicity and low cost, these systems are as primitive as the radio communication systems prior to 1930, and do not fully utilize the potential ultra-high capacity of an optical fiber. On the other hand, in radio communications, the **heterodyne scheme** has become common, and sophisticated coherent modulations such as frequency or phase modulation are also widely used. Immediately after the invention of the laser, some heterodyne-type optical communication systems had been conceived, where they were assumed to use gas lasers and lens waveguides, and were far from practical. After the advent of LDs and optical fibers in 1970, the IM/DD was the only scheme; the heterodyne scheme had been given up because of the poor spectral purity and frequency stability of the LDs and because of the multimode nature of an optical fiber.

Interest in heterodyne-type optical fiber communications, hereafter called coherent optical fiber communications or coherent systems including the **homodyne scheme**, was revived around 1980 [Okoshi and Kikuchi, 1980], when the truly single-frequency LDs such as distributed-feedback (DFB) LDs and single-mode optical fibers began to be available. Since then, research on coherent systems has been expanded, along with the drastic improvement of LD's spectral purity and the spread of the single-mode optical fibers, because the coherent systems have two main advantages: (1) dramatic improvement of the receiver sensitivity and the resulting elongation of the repeaterless transmission distance and (2) drastic improvement of the frequency selectivity and the resulting possibility of dense frequency-division multiplexing (FDM) [Okoshi and Kikuchi, 1988; Smith, 1987; Linke and Gnauck, 1988; Ryu, 1995]. Many experimental demonstrations of long distance and FDM systems have been reported and recently field tests of a practical 2.5-Gb/s trunk system over 300 km [Hayashi, Ohkawa, and Yanai, 1993] and a 128-channel FDM distribution system [Oda et al., 1994] have been reported.

This chapter describes the fundamentals, practical implementation, and constraints and countermeasures of coherent optical fiber communication systems. The emphasis is on digital systems.

63.2 Fundamentals of Coherent Systems

Basic construction of the coherent optical fiber communication system is shown in Fig. 63.1. Amplitude, frequency, or phase of an optical carrier is modulated at a transmitter directly or by using a modulator. The electric field of the optical signal is expressed as

$$E_S(t) = E_S d(t) \exp[\omega_S t + \phi_S + \theta(t)] \qquad (63.1)$$

where E_S, ω_S, and ϕ_S are the field amplitude, frequency, and phase of the carrier; and $d(t)$ is the baseband information impressed on the amplitude and $\theta(t)$ is that on the frequency or phase. After passing through a single-mode optical fiber, the optical signal is mixed with a local oscillator (LO) light using a beam splitter (BS) or a fiber coupler, and detected by the pin photodiode (PD). When we denote the LO light field as

$$E_{LO}(t) = E_{LO} \exp(\omega_{LO} t + \phi_{LO}) \qquad (63.2)$$

the detected photocurrent becomes, assuming no mismatch of the **states of polarization (SOP)** between the signal and the LO light fields and ignoring the loss,

$$\begin{aligned} i_{photo}(t) &= K|E_S(t) + E_{LO}(t)|^2 \\ &= KE_S^2 d(t)^2 + KE_{LO}^2 \\ &\quad + 2KE_S E_{LO} d(t)\cos[\omega_{IF} t + \phi_{IF} + \theta(t)] \end{aligned} \qquad (63.3)$$

where $\omega_{IF} = \omega_S - \omega_{LO}$, $\phi_{IF} = \phi_S - \phi_{LO}$ and K is a constant. The third term in Eq. (63.3) is a beat component between the signal and the LO light. It is found to be a down converted signal from the optical carrier frequency into the microwave beat frequency. This beat frequency is called an intermediate frequency (IF), and this process is called optical heterodyne detection. The obtained

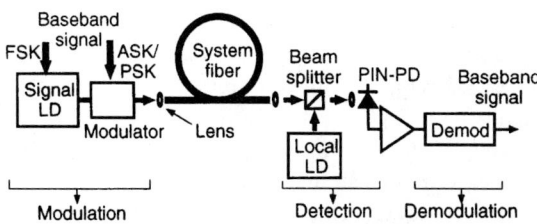

FIGURE 63.1 A basic construction of the coherent optical fiber communication system.

IF signal is fed into a heterodyne demodulator, which is basically identical to that used in radio communications, and then the baseband signal is regenerated.

In a special case where the IF is equal to zero, that is, where the optical carrier and LO light have the same frequency, it is called optical homodyne detection. In this case, the baseband signal is obtained directly; however, both the frequencies and phases of the optical carrier and LO light must be matched exactly.

The high receiver sensitivity near the shot-noise limit of a coherent receiver stems from the fact that the shot noise due to the high-power LO light predominates over the thermal noise that a DD receiver suffers from, and the high-frequency selectivity from an electrical filter with sharp cutoff characteristics can be used at the IF stage instead of a dull optical filter.

63.3 Modulation Techniques

In digital coherent systems, information is impressed on an optical carrier in one of three ways: amplitude-shift keying (ASK), frequency-shift keying (FSK), and phase-shift keying (PSK). These are basically identical to those of digital radio communication systems. In a binary ASK, $d(t)$ in Eq. (63.1) is equal to 1 for mark data and 0 for space data; similarly, $\theta(t) = +\Delta\omega t/2$ and $-\Delta\omega t/2$ in a binary FSK, and $\theta(t) = 0$ and π in a binary PSK, respectively. PSK signals can also be expressed as $d(t) = 1$ and -1, which means that a binary PSK signal is a carrier-suppressed binary ASK signal.

The FSK systems are classified by the magnitude of the frequency deviation $\Delta\omega$. Normalizing it by the bit rate R_b, the modulation index m is defined as $m = \Delta\omega/2\pi R_b$. When $m \geq 2$, it is referred to as wide-deviation FSK, and when $m \leq 2$, as narrow-deviation FSK. The minimum value of m is 0.5, which is referred to as minimum-shift keying (MSK). If the FSK signal has continuous phases between the mark and the space, which is the case of direct FSK modulation of an LD, it is called continuous-phase FSK (CPFSK).

In many cases, the ASK and PSK signals are produced by using LiNbO$_3$ external waveguide modulators, whereas the FSK signal is obtained by directly modulating the drive current of the transmitter DFB-LD. Direct ASK modulation of the LD, that is, the IM used in the conventional systems, is not available because of the accompanying large frequency shift (chirping). In the LiNbO$_3$ crystal, it is possible to change the refractive index by applying an electric field (electrooptic effect). Thus, the LiNbO$_3$ waveguide can be used directly as a phase modulator as shown in Fig. 63.2(c), or as an amplitude modulator by forming a Mach–Zehnder interferometer as shown in Fig. 63.2(a). High-speed modulation over 10 Gb/s has been realized by making these modulators travelling-wave types. A drawback of the use of these LiNbO$_3$ modulators is that high voltage around 10 V needs to be applied to get π phase shift; therefore, a high-power, high-speed, and high-cost driver is required. In contrast, the frequency can be modulated over 10 Gb/s by using the LD's chirping characteristics, that is, by modulating its drive current slightly through a bias circuit, as shown in Fig. 63.2(b), which is easy and simplifies the transmitter.

The data can be transmitted more efficiently by using multi-level keying formats, which are common in modern radio systems. In coherent optical fiber communication systems, however, their use has not yet been considered seriously, partly because the added complication in the system and technical difficulties, and partly because the available vast bandwidth in an optical fiber has not yet been fully utilized.

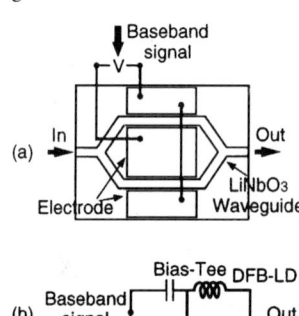

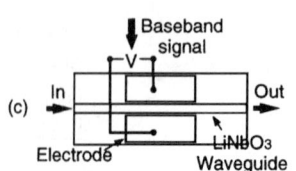

FIGURE 63.2 Modulation schemes: (a) ASK external modulator, (b) FSK direct modulation, and (c) PSK external modulator.

Coherent Systems

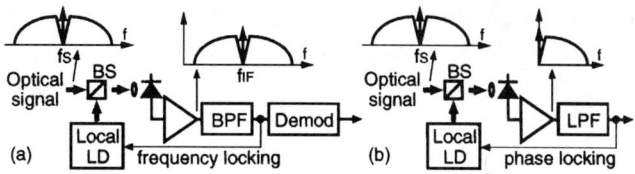

FIGURE 63.3 Detection schemes: (a) heterodyne and (b) homodyne.

63.4 Detection and Demodulation Techniques

Detection schemes are basically classified into heterodyne and homodyne types by the IF as previously described. A heterodyne-detected IF signal is a down converted signal from the optical carrier frequency into the microwave IF, that is, the heterodyne-scheme has spectral transparency. Demodulation must be performed to get the baseband signal after detection, whereas only the IF is to be locked, as shown in Fig. 63.3(a). Required bandwidth of the PD and IF amplifiers is usually 5–6 times larger than the baseband, which makes it difficult to detect a high bit rate signal.

On the other hand, in a homodyne receiver, detection and demodulation are performed simultaneously, and the baseband signal is obtained directly. Only the bandwidth identical to the baseband is required; therefore, it is suitable for detection of a high bit rate signal. However, an optical phase-locked loop (PLL) is indispensable, as shown in Fig. 63.3(b). Though there are some realizations such as the pilot carrier PLL and the Costas PLL [Smith, 1987], the optical PLL is extremely difficult due to the large **phase noise** of the LDs.

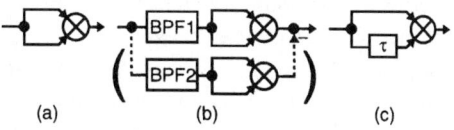

FIGURE 63.4 Demodulation schemes: (a) square-law demodulator for ASK, (b) single- or dual-filter demodulator for wide-deviation FSK, and (c) differential demodulator for narrow-deviation FSK and PSK.

Demodulators in heterodyne receivers are also basically identical to those used in digital radio receivers. In most cases, however, asynchronous (square law or differential) demodulators are used because synchronous demodulations are difficult because of the large phase noise of the LDs. An ASK signal is demodulated by a square-law demodulator [Fig. 63.4(a)]. For a wide-deviation FSK signal, a band pass filter is used to extract the mark or space spectrum and convert it to an ASK signal, which is fed into a square-law demodulator [Fig. 63.4(b)]. The signal-to-noise (SN) ratio is enhanced if both the mark and space spectra are demodulated using dual filters, and subtracted. A differential demodulator is used for a narrow-deviation FSK signal as a frequency discriminator [Fig. 63.4(c)]. For a CPFSK signal obtained by the direct modulation of an LD, the optimum delay τ becomes $1/2mR_b$ [Linke and Gnauck, 1988]. Likewise, a PSK signal is demodulated by a differential demodulator having the delay τ of 1-bit period $1/R_b$. In this case, however, the demodulator output is encoded according to $d(t) \cdot d(t - \tau)[d(t) = \pm 1]$, and a decoder is required to get the original signal $d(t)$. This implementation is called differential PSK (DPSK).

63.5 Receiver Sensitivity

In the detection of the light in the classical coherent state, a quantum mechanical noise exists, which is called shot noise. In a DD receiver, the shot-noise power is proportional to the signal power P_S. However, the thermal noise predominates over the shot noise when P_S is small, and it makes the DD receiver much less sensitive.

On the other hand, in a coherent receiver, the shot-noise power is proportional to the LO power P_{LO} and predominates over the thermal noise when $P_{LO} \gg P_S$; thus, highly sensitive receivers near the shot-noise limit are possible. Since the beat signal power is proportional to $P_S P_{LO}$, the shot-noise-limited SN ratio for the heterodyne scheme is given by

$$\gamma_{het} = \frac{\eta P_S}{2\hbar \omega B} \tag{63.4}$$

and for the homodyne scheme is

$$\gamma_{hom} = \frac{\eta P_S}{\hbar \omega B} \tag{63.5}$$

where $\hbar$ is the Planck's constant, η the quantum efficiency of the PD (ideally, $\eta = 1$), and B the receiver bandwidth. The SN ratio improvement by 3 dB in a homodyne scheme stems from the fact that the required receiver bandwidth becomes half [Okoshi and Kikuchi, 1988]. The expression of the SN ratio is simplified using the number of received photons per bit N. If we assume that the receiver bandwidth is half of the bit rate (Shannon's essential limit), $N = P_S/2\hbar\omega B$. In this case, $\gamma_{het} = \eta N$, and $\gamma_{hom} = 2\eta N$.

The bit error rate (BER) is approximately given as

$$\text{BER} = \frac{1}{2}\exp\left(-\frac{\gamma}{n}\right) \tag{63.6}$$

where n is an integer inherent in the modulation formats. For ASK and single-filter FSK, $n = 4$; for dual-filter FSK, $n = 2$; and for differential FSK and PSK, $n = 1$. Equation (63.6) holds for both heterodyne and homodyne schemes. Using Eq. (63.6), we can compute the required N for various schemes to achieve a prescribed BER. Table 63.1 shows the result of a computation assuming that $\eta = 1$ and BER $= 10^{-9}$. The required beat linewidth is also shown and will be described in the next section. The PSK homodyne scheme has the highest sensitivity ($N_{required} = 10$), the PSK and differential FSK heterodyne schemes are lower by 3 dB and the next best sensitivity ($N_{required} = 20$), the ASK homodyne and dual-filter FSK heterodyne schemes have a further 3-dB decrease (with $N_{required} = 40$), and the ASK and single-filter FSK heterodyne is the lowest ($N_{required} = 80$). Note that the optical power or the number of photons is defined to be the peak values.

For example, we will consider a 10-Gb/s PSK homodyne system having an ideal performance. If we assume that a signal at 1.55 μm of 10 mW is incident to the system fiber, the number of photons in one bit is calculated as $P_S/\hbar\omega R_b = 7.8 \times 10^6$. As $N_{required} = 10$ for an ideal PSK homodyne receiver, it has the power margin of $7.6 \times 10^5 = 58.9$ dB. The loss in a fiber at 1.55 μm is less than 0.2 dB/km; therefore, unrepeated transmission of about 300 km is possible in this case.

The differences in sensitivity among the schemes can be attributed to the degree of coherence of the light utilizes in modulation and demodulation; the PSK scheme is coherent modulation, the ASK, incoherent. The PSK homodyne is the most coherent system. As the degree of coherence of the system is enhanced, its realization becomes more and more difficult.

TABLE 63.1 Receiver sensitivity and linewidth requirement

Detection	Modulation, Demodulation	$N_{required}$	Required Beat Linewidth
Homodyne	PSK	10	$5 \times 10^{-4} R_b$
	ASK	40	$5 \times 10^{-4} R_b$
Heterodyne	DPSK	20	$3\text{-}7 \times 10^{-3} R_b$
	FSK(narrow-dev.), differential	20	$3\text{-}20 \times 10^{-3} R_b$
	FSK(wide-dev.), dual-filter	40	$0.1\text{-}0.2 R_b$
	FSK(wide-dev.), single-filter	80	$0.1\text{-}0.2 R_b$
	ASK, square-law	80	$0.1\text{-}0.2 R_b$

63.6 Practical Constraints and Countermeasures

The coherent systems promise excellent performance compared with the conventional IM/DD system as previously mentioned. However, practical systems have many difficulties, such as the phase and intensity noise of the LD, frequency response and bandwidth of the modulators and receivers, fluctuation of the SOP in a fiber, and the dispersion and nonlinearity in a fiber.

Phase and Intensity Noise in the Laser Diode

The prime concern in realizing coherent systems has been the the spectral purity of the signal and local LDs. Because the spectral shape of a single-mode laser under continuous wave (CW) operation becomes Lorentzian, the spectral purity is characterized by the linewidth of the full width at half-maximum (FWHM). The origin of the spectral broadening is phase noise generated by spontaneous emission events. Typical linewidth of a commonly used DFB-LD is around 10 MHz, which is in contrast to electrical or microwave oscillators having linewidth less than 1 Hz. The required beat linewidth to hold the sensitivity penalty below 1 dB is shown for various schemes in Table 63.1 [Linke and Gnauck, 1988]. It is proportional to the bit rate for all schemes, which can be understood by the fact that the phase correlation between adjacent bits becomes higher as the bit duration becomes shorter. As mentioned previously, the homodyne schemes require extremely narrow linewidth. It is tolerated in heterodyne schemes, but PSK and narrow-deviation FSK schemes require still narrower linewidth. In contrast, ASK and wide-deviation FSK can accept wider linewidth. As the bit rate has grown higher recently (typically, 2.5 Gb/s), the requirement has been relaxed and PSK or the narrow-deviation FSK schemes have become common. Nevertheless, homodyne schemes are still difficult.

To avoid the extremely severe linewidth requirement of the homodyne schemes, the **phase-diversity receiver** has been proposed, as shown in Fig. 63.5(a), in which two orthogonal in-phase (I) and quadrature (Q) homodyne-detected signals $i_I(t)$ and $i_Q(t)$ are produced using a so-called 90° optical hybrid as

$$i_I(t) = 2KE_S E_{LO} d(t) \cos[\phi_{IF} + \theta(t)] \qquad (63.7)$$

$$i_Q(t) = -2KE_S E_{LO} d(t) \sin[\phi_{IF} + \theta(t)] \qquad (63.8)$$

By feeding the signals to demodulators (square-law demodulators for ASK and differential demodulators for PSK) and combining, the baseband signal is obtained irrespective of the phase term ϕ_{IF} [Kazovky, 1989]. The 90° optical hybrid uses the properties of SOPs of the signal and LO light; I and Q signals are produced by combining the signal of a 45° linear SOP and the LO of a circular SOP, and by splitting them using a polarization beam splitter (PBS).

If the I and Q signals are up converted by two electrical LO waves having mutually orthogonal phases $\cos \omega_{IF2} t$ and $\sin \omega_{IF2} t$ as shown in Fig. 63.5(b) instead of demodulating them, the combined up converted signal becomes

$$i_{IF2}(t) = 2KE_S E_{LO} d(t) \cos[\omega_{IF2} t + \phi_{IF} + \theta(t)] \qquad (63.9)$$

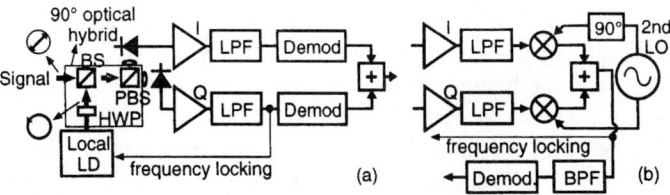

FIGURE 63.5 Phase-diversity receivers: (a) single-stage and (b) double-stage.

which is identical to a heterodyne-detected IF signal having the IF of ω_{IF2}. The baseband signal is obtained by feeding it to a heterodyne demodulator. It is called the double-stage phase-diversity (DSPD) receiver [Okoshi and Yamashita, 1990]. The receiver sensitivities and the linewidth requirement of the phase diversity and DSPD receivers are both the same with the heterodyne receivers.

An LD not only has phase noise but also **intensity noise**, which is also generated by spontaneous emission events. Especially, the intensity noise in a local LD produces additional large noise through the second term in Eq. (63.3). It can be suppressed by using a **balanced receiver** as shown in Fig. 63.6. When we use a 50 : 50 BS or fiber coupler, the two outputs have the same power but the relative phases between the signal and LO fields have a 180° difference, because the phase does not shift on transmission but shifts by 90° on reflection. The two outputs are detected independently and subtracted to be

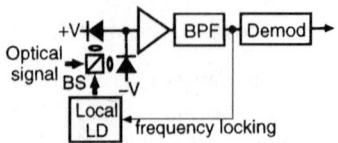

FIGURE 63.6 Balanced heterodyne receiver.

$$i_{\text{bal}}(t) = K|E_S(t) + E_{\text{LO}}(t)|^2 - K|E_S(t) - E_{\text{LO}}(t)|^2$$
$$= 4KE_S E_{\text{LO}} d(t)\cos[\omega_{\text{IF}} t + \phi_{\text{IF}} + \theta(t)] \qquad (63.10)$$

It is found that terms of the direct-detected signal and LO lights are suppressed. The balanced receiver has another merit in that it can effectively utilize the optical power that is unused in the single-detector receiver. The subtraction is easily performed by combining two PDs in tandem. Therefore, the balanced receiver is implemented in most systems. It can also be combined with phase-diversity receivers.

Frequency Response and Bandwidth of Modulators and Receivers

In most of the recent experiments, CPFSK modulation is employed because there is no need for an external modulator and direct modulation is much easier, as shown in Fig. 63.2(b). However, the normal DFB-LDs have different frequency responses for direct modulation between the low- and high-frequency regions, and cause distortion of the signal waveform. This is a result of the synergism of the heat effect and carrier effect on the modulation of the LDs. To avoid this, we can use an equalization circuit before the bias circuit for direct modulation, or use the more sophisticated LDs, having flat frequency response such as three-electrode distributed Bragg reflector (DBR) LDs. The LiNbO$_3$ modulators used for ASK/PSK modulations do not have this problem, although a high-power, high-speed driver is necessary and the transmitter becomes more complicated.

In a heterodyne receiver, the bandwidth of the PD and IF amplifiers must be over 5–6 times the baseband as shown in Fig. 63.3(a), which makes high-speed signal (over 5 Gb/s) detection difficult. In this point, a homodyne receiver is advantageous because the required bandwidth is only the baseband, as shown in Fig. 63.3(b). The homodyne receiver has another advantage in that it has no image interference from adjacent channels when it is used in an FDM system. These are the reasons why the phase-diversity techniques have been studied.

In principle, it is possible to build an FDM system with an arbitrarily large number of channels having channel spacing equal to the bit rate R_b. Actually, this is, of course, not exact since the modulated spectrum can be much wider and some guard band is necessary. The narrow-deviation CPFSK signal has a spectrum as narrow as R_b, and the required channel spacing has been found to be about $3R_b$ in a homodyne receiver [Linke and Gnauck, 1988]. It is increased to be $6R_b$ in a heterodyne receiver due to image interference. It should be noted that the balanced receiver is indispensable in the FDM system in order to reject the beat components between channels.

Polarization Fluctuation in the Fiber

In the derivation of Eq. (63.3), we assumed the SOPs of the signal and LO light fields are matched. In practice, however, the SOP of the signal fluctuates in a fiber due to the birefringence change caused by the external stress or the temperature change. There are four countermeasures: (1) use of a polarization-maintaining (PM) fiber such as a PANDA (Polarization-maintaining AND Absorption-reducing) or a bow-tie fiber as the system fiber, (2) use of an active polarization controller at the receiver, (3) use of a polarization-scrambling scheme in which the SOP of the signal is scrambled within one bit at the transmitter, and (4) use of a **polarization-diversity receiver**. Among these countermeasures, the fourth is considered to be superior and is implemented in many receivers. Figure 63.7 shows the polarization-diversity receiver. It uses an optical circuit similar to the 90° optical hybrid. The LO light is set to be a 45° linear SOP, and split by the PBS to be x and y linear SOPs having the same amplitude. The PBS splits the signal to be $k : 1-k$ according to the power ratio of x to y elements. Then the IF signals detected by the x and y branches have the amplitude ratio of $\sqrt{k} : \sqrt{1-k}$. Just like the phase-diversity receiver, they are fed to demodulators having square-law characteristics, and combined to be baseband signal independent of the value of k. The polarization diversity scheme can be combined with phase diversity and balanced receivers.

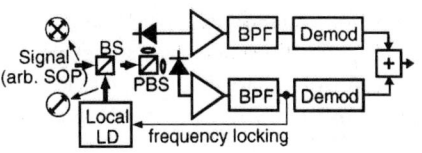

FIGURE 63.7 Polarization-diversity heterodyne receiver.

Dispersion and Nonlinearity in the Fiber

As bit rate, incident power, transmission distance, and number of channels are increased, optical fiber properties other than transmission loss become increasingly important. The loss of the single-mode optical fiber (0.2 dB/km at 1.55 μm and 0.35 dB/km at 1.3 μm) is the basic limitation on transmission distance. However, dispersions and various nonlinear effects can dominate under certain conditions [Linke and Gnauck, 1988].

The single-mode optical fiber has **chromatic dispersion** around 17 ps/km·nm in the lowest loss window at 1.55 μm. It is a phenomenon in which the speed of the light is different in frequency and acts on the spectrum spread by the modulation. As the different spectral components arrive at different times, it produces distortion in the demodulated waveform, resulting in intersymbol interference in the received signal and degradation of the receiver sensitivity. Its effect is proportional to $R_b^2 L$, where L is the total system length, and therefore becomes more significant as the bit rate becomes higher. A solution is the use of a dispersion-shifted fiber, which has zero dispersion around 1.55 μm, but it costs much more and is inconsistent with the normal fiber already installed for conventional systems. In a heterodyne receiver, we have another solution: the use of an equalizer at the IF stage [Iwashita and Takachio, 1990]. A microstrip line (typically 20 cm long for the compensation of 100-km fiber) has been found to be adequate as the equalizer because it has the reverse characteristics in dispersion. It is possible because the heterodyne scheme is spectrally transparent, that is, the IF spectrum is the down converted optical spectrum, as shown in Fig. 63.3(a). In a homodyne or a phase-diversity receiver, the detected spectrum is folded into baseband and thus equalization is not possible. The DSPD receiver, in contrast, can equalize at the second IF stage. Thus the DSPD receiver has both the advantages of the heterodyne and the homodyne schemes.

The birefringence in the fiber causes not only the fluctuation of the SOP but also **polarization mode dispersion (PMD)**. It is a phenomenon where two orthogonal polarization modes propagate at different speeds in a fiber, and the different spectral components arrive at different SOPs. The propagation delay is known to be proportional to $\sqrt{L}$. The PMD in a typical fiber is not very large (0.1–2.0 ps/$\sqrt{\text{km}}$), but it becomes important in a long-distance high-speed system.

Other problems in the optical fiber are the various nonlinear effects [Agrawal, 1989]. For coherent systems, the **stimulated Brillouin scattering (SBS)**, the **stimulated Raman scattering (SRS)**, and **nonlinear Kerr effect** are the main problems. The SBS and SRS are phenomena where the light incident to a fiber is back reflected due to the interaction between a photon and a phonon when the power exceeds the threshold. The threshold of the SBS is typically 3–4 mW, whereas the fiber amplifier can amplify the signal to 10–100 mW when used as the postamplifier after the transmitter LD. Fortunately, the bandwidth of the SBS is as narrow as 100 MHz. The SBS can be avoided if the high-speed carrier-suppressed modulations (PSK or CPFSK) or the optical spread-spectrum modulations are employed. The threshold of the SRS is much higher ($\sim$1 W) in a single-channel system, although the SRS brings crosstalk at lower power in an FDM system. The Raman induced crosstalk is known to be reduced by closely spacing the channels, which can be done easily in a coherent system.

The nonlinear Kerr effect is a third-order nonlinear effect. In the presence of the Kerr effect, the refractive index of the fiber becomes $n = n_0 + n_2|E|^2$, where n_0 and n_2 are the linear and nonlinear refractive indices. In the ASK system, the modulation in amplitude causes modulation in phase (AM-PM conversion). It is called self-phase modulation (SPM). It causes chirping in the signal spectrum, and produces waveform distortion in conjugation with the chromatic dispersion. In an FDM system, the phase of a channel may be modulated by the amplitude of the other channel. This is called cross-phase modulation (XPM). SPM and XPM can be avoided by using the modulations having constant envelope (PSK or FSK). The most severe restriction in FDM systems is four-wave mixing (FWM), in which the two lightwaves $E_1 \exp(j\omega_1 t)$ and $E_2 \exp(j\omega_2 t)$ generate another two lightwaves $E_3 \exp(j\omega_3 t)$ and $E_4 \exp(j\omega_4 t)$ through the Kerr effect, where $\omega_3 = 2\omega_1 - \omega_2$ and $\omega_4 = 2\omega_2 - \omega_1$. The generated lights may interfere the other channels. It has also been pointed out that, in long-distance single-channel systems using optical amplifiers, phase noise is enhanced through the nonlinear interactions between the signal and amplified spontaneous emission (ASE) noise, which makes the long-distance coherent system difficult [Ryu, 1992; Ryu, 1995].

63.7 Summary and Conclusions

Coherent techniques offer the potential of nearly ideal receiver sensitivity together with frequency selectivity improvements. Most of the technical obstacles described here have been overcome, and coherent systems are ready to be practical. Unfortunately, the advent and recent development of the erbium-doped fiber amplifiers has reduced the first advantage and seems to keep the coherent systems away from practical use. Nevertheless, the second advantage still remains very important because it enables us to exploit the vast bandwidth of an optical fiber. Moreover, coherent techniques are found to be effective to enhance the performance of the amplifier systems [Yamashita and Okoshi, 1994]. Coherent techniques will certainly play a leading role in future communication networks.

Defining Terms

Balanced receiver: A coherent receiver having two detectors combined in tandem, which can suppress intensity noise.

Chromatic dispersion: A phenomenon in which the speed of the light is different in optical frequency.

Heterodyne scheme: A coherent detection scheme in which the optical signal is down converted to the microwave intermediate frequency.

Homodyne scheme: A coherent detection scheme in which the optical signal is down converted directly to the baseband.

Intensity noise: Random intensity fluctuation of the light emitted from a laser, which is generated by spontaneous emission events.

Nonlinear Kerr effect: The third-order nonlinear effect in a fiber, which changes the refractive index in proportion to the intensity of the light.

Phase-diversity receiver: A homodyne receiver composed of a 90° optical hybrid and two sets of detectors and demodulators, in which the linewidth requirement is very relaxed.

Phase noise: Random phase fluctuation of the light emitted from a laser, which is generated by spontaneous emission events.

Polarization-diversity receiver: A heterodyne receiver comprising an optical hybrid and two sets of detectors and demodulators, which is insensitive to polarization fluctuation.

Polarization mode dispersion (PMD): A phenomenon in a fiber, in which two orthogonal polarization modes propagate at different speeds.

State of polarization (SOP): A spatial trace drawn by the electric field vector of the lightwave, which is characterized by amplitude and phase differences between the vertical and longitudinal wave components.

Stimulated Brillouin scattering (SBS), and stimulated Raman scattering (SRS): Phenomena in a fiber where the incident light is back reflected due to the interaction between a photon and a phonon when the power exceeds the threshold.

References

Agrawal, G.P. 1989. *Nonlinear Fiber Optics*, Academic Press, New York.

Hayashi, Y., Ohkawa, N., and Yanai, D. 1993. Estimated performance of 2.488 Gb/s CPFSK optical non-repeated transmission system employing high-output power EDFA boosters, *J. Lightwave Tech.*, 11(8):1369–1376.

Iwashita, K. and Takachio, N. 1990. Chromatic dispersion compensation in coherent optical communications, *J. Lightwave Tech.*, 8(3):367–375.

Kazovsky, L.G. 1989. Phase- and polarization-diversity coherent optical techniques, *J. Lightwave Tech.*, 7(2):279–292.

Linke, R.A. and Gnauck, A.H. 1988. High-capacity coherent lightwave systems, *J. Lightwave Tech.*, 6(11):1750–1769.

Oda, K., Toba, H., Nosu, K., Kato, K., and Hibino, Y. 1994. Long-term error-free operation of a fully-engineered 128 channel optical FDM distribution-system. *Elec. Lett.*, 30(1):62–63.

Okoshi, T. and Kikuchi, K. 1980. Frequency stabilization of semiconductor lasers for heterodyne communication systems. *Elec. Lett.*, 16(5):179–181.

Okoshi, T. and Kikuchi, K. 1988. *Coherent Optical Fiber Communications*, KTK/Kluwer, Tokyo.

Okoshi, T. and Yamashita, S. 1990. Double-stage phase-diversity (DSPD) optical receiver: Analysis and experimental confirmation of the principle, *J. Lightwave Tech.*, 8(3):376–386.

Ryu, S. 1992. Signal linewidth broadening due to nonlinear Kerr effect in long-haul coherent systems using cascaded optical amplifiers. *J. Lightwave Tech.*, 10(10):1450–1457.

Ryu, S. 1995. *Coherent Lightwave Communication Systems*, Artech House Publishers, Boston/London.

Smith, D.W. 1987. Techniques for multigigabit coherent optical transmission. *J. Lightwave Tech.*, 5(10):1466–1478.

Yamashita, S. and Okoshi, T. 1994. Suppression of beat noise from optical amplifiers using coherent receivers. *J. Lightwave Tech.*, 12(6):1029–1035.

Further Information

More details on the general aspects of coherent systems are described in Okoshi and Kikuchi [1988] and Ryu [1995]. Details on practical constraints can be found in Linke and Gnauck [1988] and Smith [1987]. Current research is reported in a number of journals, including IEEE/OSA *Journal of Lightwave Technology, IEEE Photonics Technology Letters,* and *IEE Electronics Letters.* Annual international conferences of interest include the Conference on Optical Fiber Communications (OFC), the European Conference on Optical Communication (ECOC), the International Conference on Integrated Optics and Optical Fiber Communication (IOOC), the Opto-Electronics Conference (OEC), and many more.

64
Fiber Optic Applications

64.1	Introduction	872
64.2	Optical Interconnects	873
	Interconnect Media • Backplane Optical Interconnects • Board and Multichip-Module Optical Interconnects	
64.3	Local Area Networks and Input/Output (I/O) Interconnections	876
64.4	Access Networks	878
64.5	Wavelength-Division Multiplexing-Based All Optical Networks	879
	Broadcast-and-Select Networks • Wavelength-Routed Networks	
64.6	Fiber Sensors	880

Chung-Sheng Li
IBM T.J. Watson Research Center

64.1 Introduction

The recent rapid advance of optical technologies include the following:

- Low-loss single-mode fiber
- Single longitudinal mode distributed feedback (DFB) laser diodes
- Low-threshold-current multiple-quantum-well (MQW) lasers
- Vertical cavity surface emitting lasers (VCSEL)
- High-speed p-i-n and MSM photodetectors and receivers
- Optical amplifiers
- Integrated optics

These enable the deployment of fiber optical components to both computing and communication systems. The rapid commercialization of these components already has had a profound impact in the following areas:

- Optical interconnect within a system or between systems
- Local area networks
- Access networks
- Backbone networks
- Fiber optic sensors

In this chapter, we will examine each individual application in detail.

64.2 Optical Interconnects

Current metal technology has difficulty in supporting bus and backplane bandwidth exceeding several hundred megahertz. Therefore, a serious bandwidth bottleneck exists between processor and memory or between processors. Thus, new technology is required to overcome these obstacles in order to create low-latency, high-bandwidth communication paths whose characteristics match the performance of ever-increasing processor speed. Different metal-based technology standards such as Rambus [Farmwald and Mooring, 1992], Ramlink [Gjessing et al., 1992], and scalable coherent interface (SCI) [IEEE, 1992] have been proposed for parallel metal interconnects to meet the requirements of systems with clock speeds up to 500 MHz and are in the process of being implemented. These systems show that with careful engineering such as the use of a controlled-impedance transmission line, fully differential signalling (SCI and Ramlink), and limiting the interconnection structure to point-to-point (SCI and Ramlink) can push the metal interconnection technology to higher bandwidths than could be supported with existing technology. Nevertheless, the cost of the new technology is initially high, and there is no guarantee that the solutions can be scaled to clock speeds higher than their present limits. If any of these interconnection technologies is going to succeed, it must be available at low cost in high volume, and must be able to evolve to support future clock speeds of 1 GHz and beyond.

Optical interconnections provide an alternative technology to solve the interconnection problem. The multigigabit bandwidth allowed by this technology is more than sufficient for applications, such as communications within a multimedia system, for the foreseeable future. Furthermore, it is easier to control reflections in this technology for both point-to-point and multidrop structures than in metal links [Stone and Cook, 1991]. Optical links generally exhibit less noise problems due to ground-loop noise because fibers do not carry currents as do metal links. For these reasons, optical interconnections may be an attractive alternate technology for building high-speed board and backplane interconnections in future multimedia systems.

An optical link can be designed to be an almost one-to-one replacement for metal point-to-point or multidrop connections. The conventional line driver is replaced by a laser driver and an edge-emitting or surface-emitting laser diode/LED, or a laser diode and an external modulator such as a Mach–Zehnder interferometer, directional coupler, total internal reflection (TIR) modulator spatial light modulator (SLM), self-electro-optic device (SEED), or vertical-to-surface transmission electrophotonic device (VSTEP) at the transmission end. The conventional line receiver is replaced by a light sensitive device such as a p-i-n diode or metal-semiconductor-metal (MSM) photodetector and an amplifier at the receiving end. The light can be guided from the transmission end to the receiving end through single-mode or multimode fiber ribbon cable, polyimide, or silica-on-silicon channel waveguides, or free-space microlenses and/or holograms.

Figure 64.1 shows the structure of a typical optical interconnect, which consists of a driver array, a laser diode or LED (LD/LED) array, a waveguide or fiber ribbon array, a photodetector array (p-i-n or MSM), and a receiver array. Using optical interconnects for high-bandwidth communication channels between boxes have been demonstrated, for example, in [Lockwood et al., 1995]. It is conceivable that optical interconnect can also be used within a box (both at the board and backplane levels), shown in Fig. 64.2.

Interconnect Media

Possible media that can be used for optical interconnects include:

- *Free-space interconnect.* Light travels fastest in free space. In addition, free-space interconnects also offer the highest density and the most sophisticated interconnection patterns. Unfortunately, bulk optical elements, such as lenses, holograms, beam splitters, etc., are usually unavoidable in free-space optical interconnects and thus make the alignment of optical beams very difficult and unstable with respect to environmental disturbances.

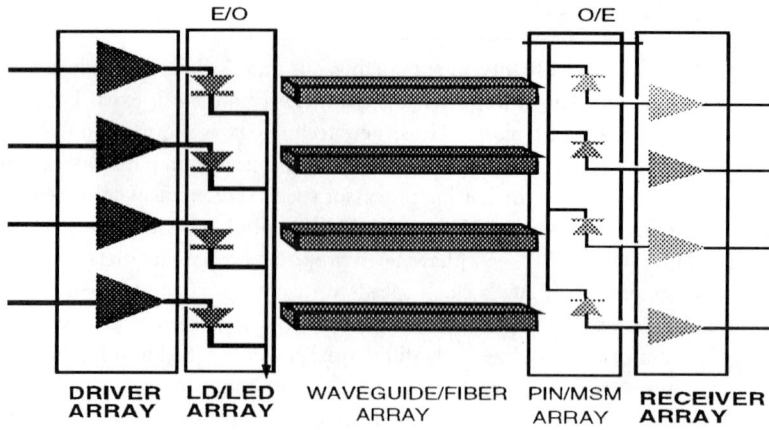

FIGURE 64.1 Structure of an optical interconnect system.

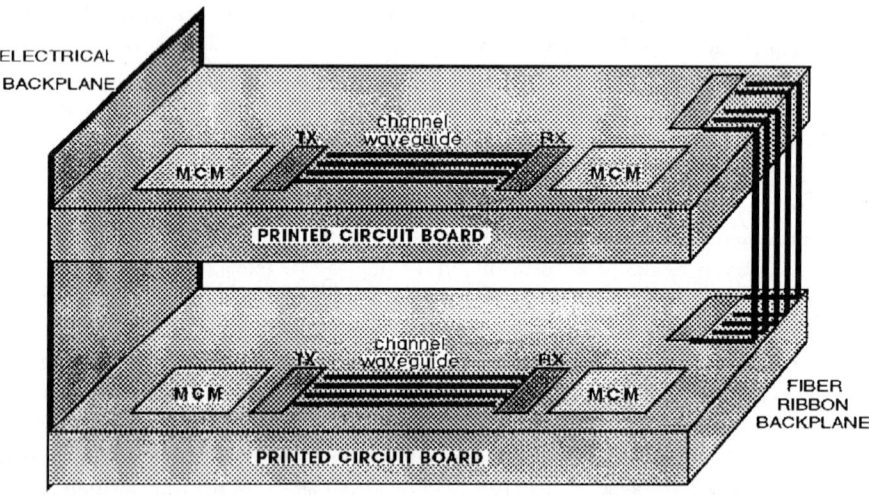

FIGURE 64.2 A packaging structure uses optical interconnect at board and backplane levels.

- *Optical fiber ribbon.* Optical fiber has the least loss compared with the other two media, and most of the technologies used in fabrication are already mature. Fiber ribbon cable also has the potential of providing reasonable interconnect density with regular interconnection pattern. However, fibers are incompatible with the existing packaging technology at the board or multichip module (MCM) level and they are not suitable for interconnects with very short distance or complicated patterns due to the possibly excessive volume occupied by the fiber cable. A lot of research effort has thus been devoted to develop compatible connector and packaging technologies to interconnect fiber ribbon and optical transceiver arrays.

- *Planar optic waveguide.* Passive planar optic waveguides are emerging as a viable alternative to optical fiber for very short distance interconnects. They have a higher propagation loss than optical fiber (0.01 ~ 0.5 dB/cm as compared to 0.2 dB/km) but use technologies that are compatible with existing PCB or MCM technology. Therefore, they are more suitable for short-distance dense interconnect applications. However, coupling of light into and from the waveguides is also difficult and careful alignment cannot be avoided.

Fiber Optic Applications 875

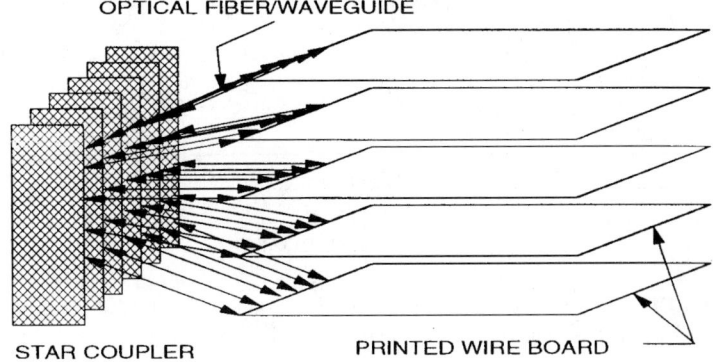

FIGURE 64.3 Optical backplane interconnects: star couplers are used to combine and redistribute the data signals.

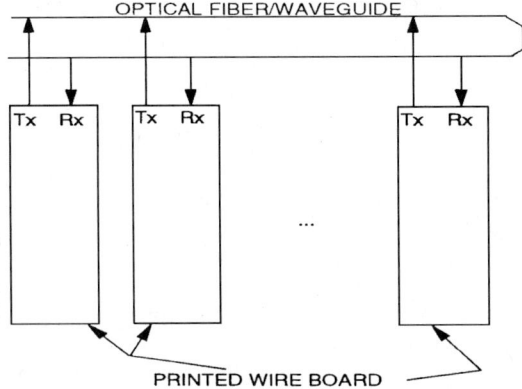

FIGURE 64.4 Optical backplane interconnects: a topological bus is used to provide a communication path between any two boards connected to the backplane.

Backplane Optical Interconnects

The function of a backplane is to provide a logical bus for all of the boards connected to it. Free space, fibers, and planar waveguides are all suitable for backplane interconnects. An optical backplane can be achieved through the use of star couplers, as shown in Fig. 64.3. Each board in the architecture occupies one input port and one output port from each of the star couplers so that signals input to any of the input ports will be broadcast to all of the output ports. The total number of star couplers required can be greatly reduced by multiplexing several channels into a single waveguide, with each channel using a different wavelength.

On the other hand, a topological bus can also be used to interconnect from one board to another, as shown in Fig. 64.4. The bus is either folded back at the end or two independent buses are used because a unidirectional optical bus structure is usually easier to implement.

Board and Multichip-Module Optical Interconnects

Board-level optical interconnects have to provide interconnects between different single chip modules (SCMs) or MCMs, whereas MCM-level optical interconnects have to provide interconnects

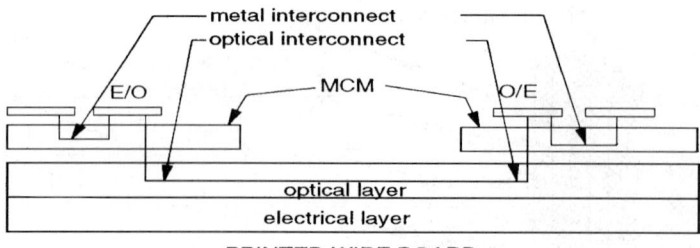

FIGURE 64.5 E/O and O/E conversion: conversion is performed at the same packaging level as the electrical signal is generated.

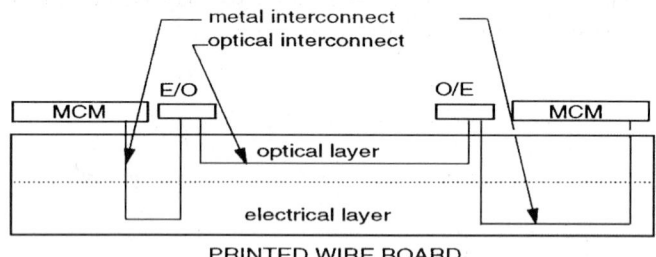

FIGURE 64.6 E/O and O/E conversion: conversion is performed at the next higher packaging level.

between unpackaged wire-bonded or solder-ball-bonded flipped chips. At the board level, the electrical to optical (E/O) and optical to electrical (O/E) conversion can be performed within an SCM/MCM, or through separate special-purpose E/O and O/E chips. Similarly, the E/O and O/E conversion at the MCM level can be performed within the chip where the logical signals are generated or via separate special-purpose E/O and O/E chips on an MCM.

If the E/O and O/E conversion is performed before the package is connected to the next higher level, as shown in Fig. 64.5, the electrical discontinuity can be minimized but the optical alignment is more difficult. On the other hand, more electrical discontinuity and, thus, more signal distortion are introduced if the E/O and O/E conversion is performed after the package is connected to the next level, as shown in Fig. 64.6. However, this is acceptable for applications that require only moderate data rates.

In both cases, there already exist multiple layers of metal interconnect to provide signal lines as well as power and ground plane. Optical interconnects can be developed on top of these metal interconnect layers to allow optical signals to propagate from one chip/module to another chip/module. In some cases, more than one optical layer may be necessary in order to provide a sufficient interconnect density (such as at the MCM level) just similar to its metal counterpart.

64.3 Local Area Networks and Input/Output (I/O) Interconnections

Examples of local area networks and input/output (I/O) channels that use fiber optics include:

- Fiber distributed data interface (FDDI), FDDI-II
- Fiber channel
- Gigabit ethernet (802.3z)

Fiber Optic Applications

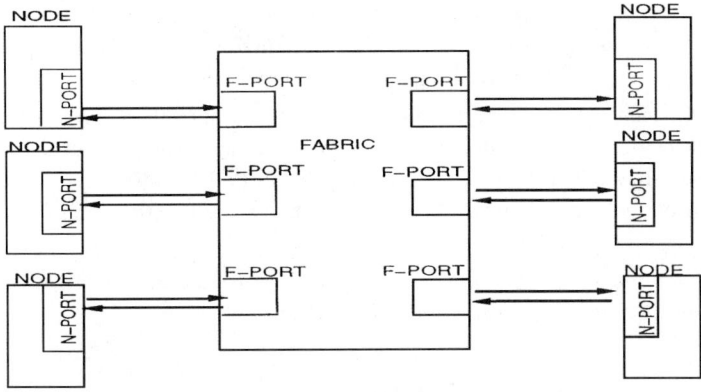

FIGURE 64.7 Several nodes interconnected by a fabric.

Since the protocol of FDDI and ethernet have been described elsewhere, this section will only focus on the fiber channel.

The major application area of fiber channel, which became an American National Standards Institute (ANSI) standard in 1994, is to provide high-bandwidth (up to 100 Megabyte/s) connectivity to cluster of supercomputers, mainframes, workstations, disk arrays, and other peripheral devices. Its origin has strong ties with high performance parall interface (HIPPI) and enterprise systems connection (ESCON), both of which provide circuit switching, and small computer systems interface (SCSI), which is the primary interconnection mechanism for peripheral devices. Fiber channel defines a point-to-point or switched point-to-point link, which includes definitions of the physical layer, transmission code, and higher level functions such as flow control and multiplexing. Figure 64.7 shows a collection of nodes (called *N_port*) interconnected by a fabric through the fabric ports (called *F_port*).

The fiber channel standard includes definitions of the following functions:

- *Physical layer.*   The physical layer specifies four data rates: 133, 266, 533, and 1062.5 Mb/s. Because of the adoption of 8B/10B line coding technique, there is a 25% increase in the transmission data rate. The transmission media defined for a fiber channel includes miniature coax, multimode and single-mode fibers. The physical layer also defines bit synchronization, serial-to-parallel conversion, and an open fiber control protocol. This protocol performs automatic shutdown of the laser diode when there is a fiber cut and automatically establishes a full duplex link when the fiber connection is restored.

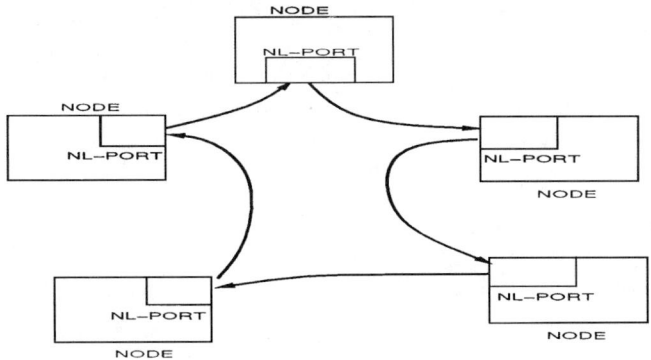

FIGURE 64.8 Several nodes interconnected by a fabric.

- *Data link layer.* This layer specifies a line coding technique, the 8B/10B code, to ensure sufficient transitions are present in the data stream. In addition, byte-to-word conversion and word synchronization are also performed in this layer.
- *Arbitrated loop layer.* The arbitrated loop layer specifies the loop control and bypass mechanisms for an end node when such an end node participates in the connection of a loop, as shown in Fig. 64.8. This configuration has the advantage of reduced cost at the cost of reduced performance, and is particularly attractive for disk storage applications.
- *Frame processing layer.* The processing and validation of the integrity of each received frame including CRC checking is performed in this layer.

64.4 Access Networks

Access networks provide interconnections between the customer premises equipment (CPE) and the central offices. The average distance of these types of networks is usually less than 3 mi. The components of access networks are extremely cost sensitive. Although the price of fiber has gone down dramatically, the price of a single-mode distributed feedback (DFB) laser has not. The design of the access network is thus to minimize the use of active devices such as the laser diodes or confine the use of such devices within central offices. The current fiber deployment strategies are divided into the following two categories: hybrid fiber coaxial (HFC) and fiber to the home (or fiber in the loop). Although fiber to the home is the ultimate answer to broadband services, HFC seems to be the most tangible interim solution.

The hybrid fiber coaxial cable structure deploys both the fiber and the coaxial cable. The basic fiber/coax network is already taking shape as a result of the cable TV delivery system. Signals in most of the existing HFC networks are transmitted using fiber with broadband RF spectrum to an headend. A typical architecture is shown in Fig. 64.9. The signals are then converted back into

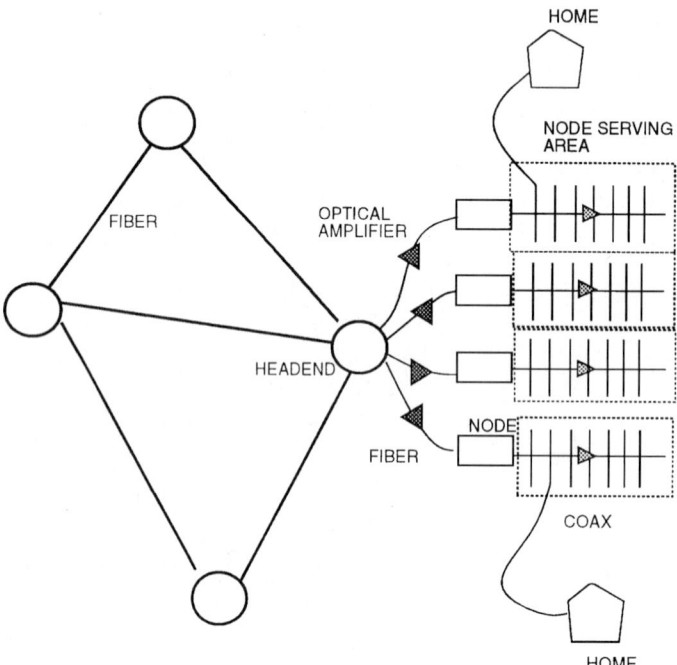

FIGURE 64.9 A broadcast-and-select network.

an electrical RF signal and distributed to the home. The major advantages of this star-configured fiber network with coax bus are as follows [Paff, 1995]:

- Better quality video
- More bandwidth possible with lower beginning investment
- Better reliability
- Dedicated feed into and out of a small area

Although the current HFC network is a broadcast and isolated architecture, a lot of research has been devoted into the next generation HFC network that can provide an interactive and broadband element within the public telecommunication fabric.

64.5 Wavelength-Division Multiplexing-Based All Optical Networks

The advantages of a wavelength-division multiplexing- (WDM-)based optical network can be grouped into two aspects. One aspect lies in its potential in utilizing the vast bandwidth offered by the single-mode optical fiber since there are no electronic components in the network. The other equally important aspect is its simple network control. To transmit a packet, the source only needs to identify the destination through a unique wavelength. There is no need to know about the exact network topology for route selection. There is no buffering within the network and, therefore, no need for flow control or congestion control at or below the network layer. Only the associated end nodes need to do handshaking between themselves.

Two types of WDM-based networks are commonly discussed:

- Broadcast-and-select networks
- Wavelength routed networks

Broadcast-and-Select Networks

A broadcast-and-select network consists of nodes interconnected to each other via a star coupler, as shown in Fig. 64.10. An optical fiber link, called the *uplink*, carries signals from each node to the star. The star combines the signals from all of the nodes and distributes the resulting optical signal equally among all its outputs. Another optical fiber link, called the *downlink*, carries the combined signal from an output of the star to each node. Examples of such networks are Lambdanet [Goodman et al., 1990] and Rainbow [Janniello, Ramaswami, and Steinberg, 1993].

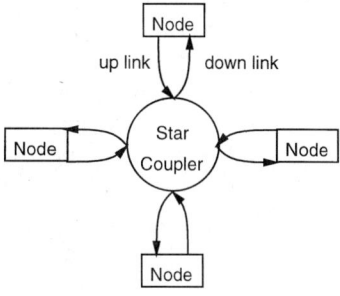

FIGURE 64.10 A broadcast-and-select network.

Wavelength-Routed Networks

A wavelength-routed network is shown in Fig. 64.11. The network consists of *static* or *reconfigurable* wavelength routers interconnected by fiber links. Static routers provide a fixed, nonreconfigurable routing pattern. A reconfigurable router, on the other hand, allows the routing pattern to be changed dynamically. These routers provide static or reconfigurable *lightpaths* between end nodes. A lightpath is a connection consisting of a path in the network between the two nodes and a wavelength assigned on the path. End nodes are attached to the wavelength routers. One or more controllers that perform the network management functions are attached to the end node(s).

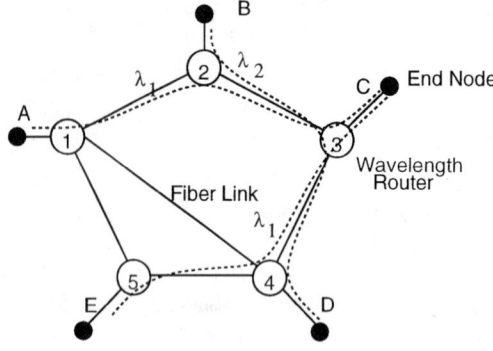

FIGURE 64.11 A wavelength-routed network.

64.6 Fiber Sensors

It is well known that optical techniques, interferometry in particular, offer extremely high precision. However, it is difficult to apply these high-precision techniques outside of a laboratory for the following two reasons:

- Environmental perturbations tend to cause rapid misalignment of the optical components.
- Optical components have to be within line of sight and thus remote sensing is difficult to achieve.

This situation has changed dramatically recently due to the incorporation of fiber optic waveguides into sensors where the fiber optic waveguides can act as the sensor or serve as a transceiver to transfer light signals to a remote sensor. Several examples of the application of optical techniques are as follows:

- *Biomedical transducer* [Mignani and Baldini, 1995]. The development of optic fiber sensors for medical applications dates back to the 1970s. The current-generation fiber optical sensors, which are mostly composed of silica and plastic fibers, are coupled to sensitive fiber sections or optrodes. These sensors are usually based on intensity modulation interrogation schemes, as shown in Fig. 64.12. Applications of these sensors include cardiovascular and intensive care, angiology, gastroenterology, ophthalmology, oncology, and neurology, to name just a few.
- *Chemical transducer.* Because of their immunity to electromagnetic interference (EMI), electric and intrinsic safety, possibility of miniaturization, and biocompatibility, optical fiber

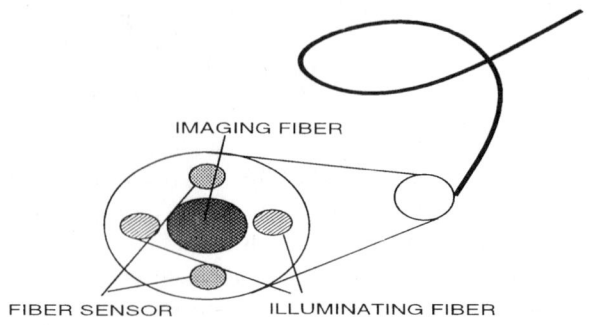

FIGURE 64.12 Configuration of a biosensor.

Fiber Optic Applications

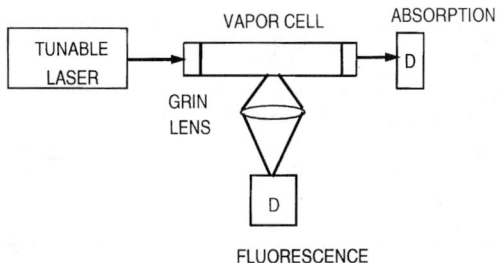

FIGURE 64.13 Configuration of a spectroscope setup for measuring absorption spectrum.

pH sensors are ideal for blood monitoring [Deboux et al., 1995]. These sensors are based on the observation of the changes in the optical properties of colorimetric or fluorimetric pH indicators. The changes are measured by reflection or through interaction with the evanescent field changes in the absorption due to changes in color of colorimetric indicators. Optical sensors based on optical time-domain reflectometry can also be used to detect the water ingress point.

- *Intelligent sensors.* Fiber optical sensors can be embedded in cement mortar bodies for nondestructive monitoring of structures and for deformation measurement [Habel and Polster, 1995].
- *Magnetic/electric field transducer* [Ezekiel et al., 1995]. Magnetic and electric fields can be measured through spectroscopic techniques, as shown in Fig. 64.13. These techniques usually involve the excitation of atoms and molecules in a cell using a tunable light source, such as a laser, and measuring the resulting absorption or fluorescence. In this way, it is possible to identify the type of molecule in the vapor, as well as the concentration. This method is generally referred to as a single-step or single-photon spectroscopy, since the excitation involves a single pair of energy levels. Recently, there are also new techniques that utilize two-step or two-photon spectroscopy, in which two radiation fields simultaneously interact with the atom or molecule in the vapor.
- *Moisture transducer.* Instead of using a large number of electrically passive sensors, moisture can be detected through the use of a single fiber based on optical time-domain reflectometry. This technique enables the presence of water to be detected as a linear function of the fiber length.

References

Deboux, B.J.-C., Lewis, E., Scully, P.J., and Edwards, R. 1995. A novel technique for optical fiber pH sensing based on methylene blue absorption. *J. Lightwave Tech.*, 13(7).

Ezekiel, S., Smith, S.P., Shahriar, M.S., and Hemmer, P.R. 1995. New opportunities in fiber-optic sensors. *J. Lightwave Tech.*, 13(7).

Farmwald, M.P. and Mooring, D. 1992. A fast path to one memory. *IEEE Spect.* 29(Oct.):50–51.

Gjessing, S., Gustavson, D.B., James, D.V., Stone, G., and Wiggers, H. 1992. A RAM link for high speed. *IEEE Spect.* 29(Oct.):52–53.

Goodman, M.S., Kobrinski, H., Vecchi, M., Bulley, R.M., and Gimlett, J.M. 1990. The LAMBDANET multiwavelength network: architecture, applications and demonstrations. *IEEE J. Selec. Areas Commun.*, 8(6):995–1004.

Habel, W.R. and Polster, H. 1995. The influence of cementitious building materials on polymeric surfaces of embedded optical fibers for sensors. *J. Lightwave Tech.*, 13(7).

IEEE. 1992. Scalable coherent interface: Logical, physical, and cache coherence specifications. IEEE P1594 Working Group, IEEE Standard Department.

Janniello, F.J., Ramaswami, R., and Steinberg, D.G. 1993. A prototype circuit-switched multi-wavelength optical metropolitan-area network. *J. Lightwave Tech.*, 11 (May/June): 777–782.

Lockwood, J.W., Duan, H., Morikuni, J.J., Kang, S.-M., Akkineni, S., and Cambell, R.H. 1995. Scalable optoelectronic ATM networks: The iPOINT fully functional testbed. *J. Lightwave Tech.*, 13(6):1093–1103.

Mignani, A.G. and Baldini, F. 1995. In-vivo biomedical monitoring by fiber optic systems. *J. Lightwave Tech.*, 13(7).

Paff, A. 1995. Hybrid fiber/coax in the public telecommunications infrastructure. *IEEE Commun. Mag.*, 33(4).

Stone, H.S. and Cocke, J. 1991. Computer architecture in the 1990s. *IEEE Comput. Mag.*, 24(9): 30–38.

65
Wavelength-Division Multiplexed Systems and Applications

Mari W. Maeda
Bellcore

65.1 Introduction...883
65.2 Optical Components for Wavelength-Division Multiplexed Systems...883
  Transmitter • Multiplexer and Demultiplexer • Optical Amplifier
65.3 Wavelength-Division Multiplexed System Design 885
65.4 Trunk Capacity Enhancement Applications 887
65.5 Wavelength-Division Multiplexed Networking and Reconfigurable Optical Transport Layer 888
65.6 Summary 889

65.1 Introduction

Research in wavelength-division multiplexing (WDM) technology is driven by the desire to exploit the immense bandwidth available for transmission in a single-mode optical fiber. The low-loss wavelength range around 1.55 μm alone can support over 10 THz of bandwidth. However, this bandwidth cannot be directly utilized by time-division multiplexed (TDM) systems due to the speed limitations posed by fiber dispersion. WDM offers a more practicable solution of multiplexing many high-speed channels at different optical carrier frequencies and transmitting them over the same fiber. It is expected that WDM technology will play a greater role in the coming years as the telecommunications networks evolve to meet the growing demand for bandwidth.

In this chapter, we describe the fundamental concepts of WDM systems, beginning with a review of key optical components and their performance requirements. This is followed by a discussion of system design issues and a review of several promising applications for telecommunications networks.

65.2 Optical Components for Wavelength-Division Multiplexed Systems

WDM systems are commonly classified into three general groups based on the choice of channel spacing and the aggregate transmission capacity supported. The simplest configuration, often

© 1997 by Bell Communications Research, Inc. With permission.

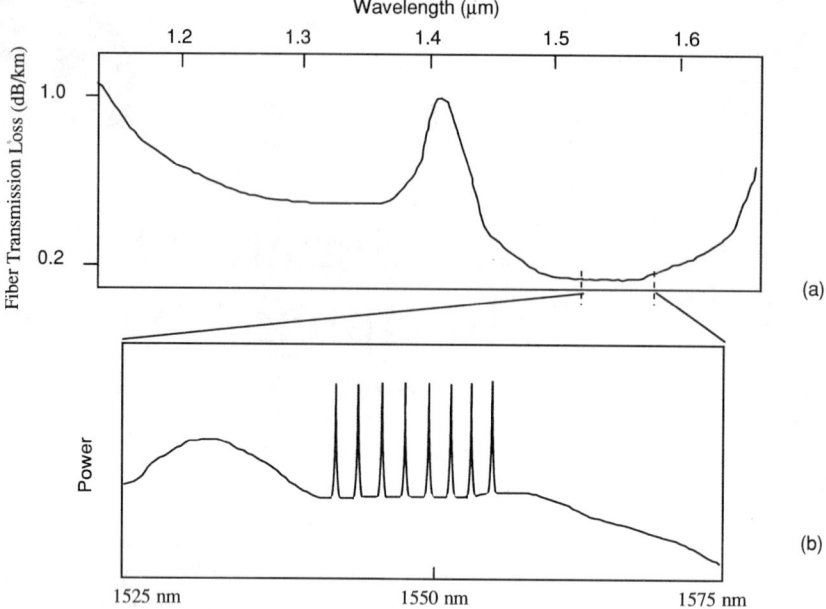

FIGURE 65.1 (a) Fiber transmission loss showing the low-loss windows at 1.3 and 1.5 μm. (b) Transmission spectrum for an eight-channel WDM system using erbium-doped fiber amplifier (EDFA). The background detail shows the spontaneous emission spectrum from EDFA.

known as the *coarse* WDM architecture, supports transmission of two channels in the broad low-loss windows of the fiber at 1.3 μm and at 1.55 μm. A greater capacity can be supported by *dense* WDM systems, where several wavelengths are transmitted in the 1.55-μm region with a typical channel spacing on the order of one to several nanometers, as shown in Fig. 65.1. Optical frequency-division multiplexed (OFDM) systems support yet greater density of channels with the spacing in the range of 0.1 nm (~10 GHz). Although both dense WDM and OFDM systems can support aggregate transmission capacity of over 100 Gb/s, dense WDM is becoming widely accepted as offering a more practical, near-term solution in various applications and will be the primary focus of this chapter.

Figure 65.2 shows a basic WDM system for point-to-point high-capacity transmission applications. The signals at different wavelengths are optically multiplexed so that N wavelengths are simultaneously carried on a common transmission fiber. Optical amplifiers may be used to compensate for fiber attenuation and component loss. At the receiving end, the signals are demultiplexed

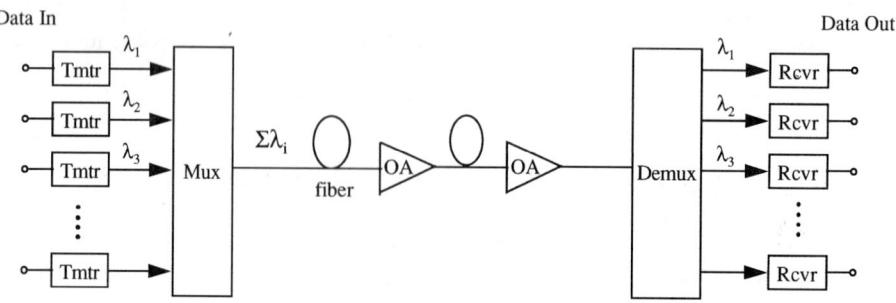

FIGURE 65.2 Basic WDM system configuration, optical amplifier (OA).

and detected by optoelectronic receivers. The key optical components for WDM systems are now reviewed.

Transmitter

Lasers for high-speed transmission systems must emit single wavelengths (i.e., a single longitudinal mode) and support high output power (>1-mW coupled into fiber pigtail) over a large modulation bandwidth with a specified spectral stability. Output spectral characteristic is one of the key considerations in designing high-speed systems where the transmission distances can be limited by fiber chromatic dispersion. Well-confined optical spectrum is especially important for WDM since the channel spacing must be selected to minimize interchannel crosstalk. Optical sources that are in widest use today are distributed feedback (DFB) lasers. DFB lasers can support multigigabit per second rate intensity modulation through direct injection current modulation and produce an output signal with a stable spectrum over the entire modulation bandwidth. For applications requiring ~10-Gb/s modulation rates, external modulation can be used to reduce the extent of wavelength chirping that normally accompanies direct modulation. It should be noted that incoherent sources such as light emitting diodes (LEDs) or super luminescent diodes together with wavelength filters can also be used to generate quasi-monochromatic signals [Iannone, 1995]. This so-called *spectral-slicing* technique is applicable for systems that emphasize simplicity and cost effectiveness of the transmitter design.

Multiplexer and Demultiplexer

Key considerations in selecting the multiplexing and demultiplexing technologies are the passband characteristics, insertion loss, polarization sensitivity, and thermal stability. For systems with wavelength spacing in the range of 1–10 nm, commonly used passive de/multiplexers are based on diffraction gratings and multilayer interference filters. For systems employing narrower channel spacing in the range of 0.1 nm, cascaded Mach–Zehnder or Fabry–Perot (FP) interferometer devices may be used for filtering the desired channel. A commonly used alternative to using a wavelength-selective coupling device is to use a passive N by N star coupler, which evenly splits each input signal between N outgoing ports. The intrinsic splitting loss of a star coupler may be compensated by the use of an optical amplifier. Star couplers are often employed in a *broadcast and select* configuration together with tunable filters on the receiving end. Examples of tunable filters include index-tuned liquid crystal FP filter, acousto-optic filter, and DFB filter amplifier. For further details on tunable filter technology and their applications in multiaccess networks, readers may consult Mestdagh [1995].

Optical Amplifier

Recent technological advances have shown optical amplification to be practical in the form of erbium-doped fiber amplifiers (EDFA). EDFAs possess the desirable features of high fiber-to-fiber gain (20–40 dB), high output power [>10 mW], and quantum limited noise figure, and offer approximately 30 nm of useful gain bandwidth in the 1535–1565 nm range [Desurvire, 1994]. For WDM systems employing cascaded EDFAs, cumulative gain nonuniformity over the gain spectrum can be significant and channel equalization is necessary (for example, by using filters or a signal pre-emphasis technique). The high signal powers made possible with EDFAs also raise the issue of signal degradations due to fiber nonlinearity, which must be carefully addressed especially in long-haul, high-speed system design.

65.3 Wavelength-Division Multiplexed System Design

Successful WDM system design entails a tradeoff of channel spacing, per channel bitrate, and feasible transmission distances based on various physical and practical constraints. Table 65.1

TABLE 65.1 Sources of limitations in WDM systems, their effects and possible solutions

Limitation Sources	Effect	Solutions
Fiber dispersion	waveform distortion at high bit rates	• reduce signal chirp (e.g., use external modulator) • use dispersion-shifted fiber • use dispersion compensation
SPM/XPM	waveform distortion at high bit rates and high launch powers	• use dispersion-shifted fiber • use dispersion compensation • increase channel spacing • reduce power
Four-wave mixing	nonlinear crosstalk and loss at high launch powers	• increase channel spacing • avoid dispersion-shifted fiber • reduce per-channel power
Amplifier gain nonuniformity	uneven gain for different λ	• use equalization or pre-emphasis • use narrow spacing to confine the channel spread
Wavelength drift	crosstalk and signal fading	• require active λ stabilization • increase channel spacing

lists key factors limiting system performance along with possible solutions for relaxing the constraints.

The maximum transmission distances that can be attained without optoelectronic regeneration is limited by three characteristics of the fiber: linear loss, chromatic dispersion, and nonlinearity. Linear loss in the fiber, due primarily to Rayleigh scattering and molecular absorption, can be largely compensated by the use of optical amplifiers. For systems employing multigigabit per second channel rates, viable transmission distances become limited by the fiber dispersion and optical spectral characteristics. For example, with a *conventional* (nondispersion shifted) single-mode fiber that is widely deployed in the field, the dispersion level of ~17 ps/km-nm at 1550 nm limits the transmission distances to a mere 11 km when launching a 10-Gb/s signal with a spectral width of 1.0 nm. However, when the spectral width is reduced to 0.1 nm, the transmission distance is extended to 65 km. Such reduction in the optical spectral width can be attained by replacing direct laser modulation with the use of an external modulator. The chromatic dispersion problem can be also alleviated with the use of dispersion-shifted fiber whose zero-dispersion wavelength is in the 1550-nm range instead of the 1300 nm for conventional fiber. However, when such redeployment of fiber is not an option, as the case may be when upgrading a network using already installed fibers, dispersion management techniques can be used to extend the transmission span. For example, a dispersion compensating device or a length of negative-dispersion fiber can be inserted periodically into the fiber span to correct the signal distortions [Lin et al., 1980; Onaka et al., 1996]. In laboratory experiments, several different dispersion management techniques have been successfully employed to extend the regeneratorless transmission spans into thousands of kilometers range.

When the transmitted signal powers are high, as they are in systems employing optical amplifiers, the effects of fiber nonlinearities must be taken into consideration during system design [Chraplyvy, 1990]. Fiber nonlinearities cause energy transfer between propagating wavelengths and lead to nonlinear loss and crosstalk if the system parameters such as launch powers and channel spacing are not carefully selected. One such effect is nonlinear phase modulation, which results from the power dependence of the refractive index. For single-wavelength systems, this nonlinearity takes the form of self-phase modulation (SPM), whereas in WDM systems cross-phase modulation (XPM) plays a greater role as the phase of one wavelength is influenced by the intensity of all other wavelengths [Marcuse, 1994]. SPM and XPM broadens the signal spectrum, causing pulse distortions due to fiber chromatic dispersion. Degradations can be alleviated by reducing the launched signal power, increasing the channel spacing, and by using a passive dispersion compensation technique mentioned earlier.

Another nonlinear effect that has a deleterious effect on WDM systems is four-wave mixing (FWM), where nonlinear interaction between three wavelengths generates a fourth wavelength, analogous to third-order intermodulation in radio systems. FWM interaction between different combinations of multiplexed wavelengths results in an appearance of interfering light at a signal channel, causing a crosstalk penalty. Because FWM requires optical phase matching between the interacting signals, the effect can be reduced by using unequal or large channel spacings and by locally avoiding dispersion of magnitude less than about 1–2 ps/nm-km in the transmission fiber. The latter can be accomplished by using the dispersion compensating technique whereby the dispersion level in an individual fiber segment is kept greater than the required 1–2 ps/nm-km but the overall dispersion is minimized.

There are other factors, in addition to fiber nonlinearity, that influence the choice of WDM channel spacing. When the wavelength spacing is very small, it is possible to pack in more channels into the wavelength regime where the EDFA gain is relatively uniform. However, when the spacing is reduced to a fraction of a nanometer, the design requirements for various optical components become quite stringent. For example, stabilization of optical sources and multiplexers becomes necessary so that wavelength drifts do not cause signal fading and crosstalk. Optical demultiplexers and filters also require more careful design to minimize interchannel crosstalk. Although there have been demonstrations of OFDM systems using wavelength spacing in the range of 0.1 nm (∼10 GHz) [Nosu et al., 1993], a typically selected spacing for most applications is in the range from 0.6 nm to several nanometers. Indeed, a preliminary agreement in the International Telecommunication Union (ITU) standards endorses the usage of a 100-GHz (∼1.1 nm) frequency grid referenced to an absolute frequency of 193.100 THz (1552.52 nm).

65.4 Trunk Capacity Enhancement Applications

Near-term applications for high-speed WDM systems can be found in fiber trunk capacity enhancements. Currently available upgrade options are: (1) use of higher speed TDM systems, (2) use of additional fibers, and (3) use of wavelength multiplexing. Although a mix of three approaches will most likely be used in different parts of the network depending on the need and the economics, a transmission technique of combining high-speed TDM with WDM is especially attractive for long-haul applications since cost saving is made possible through fiber and amplifier sharing. Also, in those parts of the network where existing capacity is nearing exhaustion, the use of WDM offers a lower cost alternative to installing new cables or constructing new ducts.

Figure 65.3 shows the growing transmission capacity demonstrated in laboratory experiments from the mid-1980s to 1996. A two to three orders of magnitude increase in per-fiber capacity has been achieved with improvements in device speeds and through the use of WDM. A record 1.1-Tb/s transmission over 150 km was demonstrated in early 1996 with the use of 55 wavelength sources each operating at 20 Gb/s [Onaka et al., 1996]. An equally remarkable trend has been witnessed in the transmission distances attained with WDM systems. In a recent experiment, eight 5-Gb/s channels were transmitted over 4500 km of fiber using 69 optical amplifiers [Suyama et al., 1996]. A separate experiment has demonstrated the feasibility of transmitting 20 5-Gb/s WDM channels over 9100 km of fiber in a circulating-loop configuration [Bergano et al., 1996].

Although it will be a while before terabit per second systems become a commercial reality, deployment of WDM systems has begun in terrestrial and undersea networks. Most terrestrial long distance networks in the U.S. are currently based on single-wavelength 2.5-Gb/s [optical carrier-OC-48] SONET systems. Next generation TDM systems operating at the speed of 10 Gb/s (OC-192) are nearing deployment while simultaneously WDM-based upgrade solutions are becoming commercially available. Network architects are also investigating the use of WDM in global undersea communication networks. For example, the Africa Optical Network, which will encircle the African continent with 40,000-km trunk and branch network in the year 1999, will utilize WDM technique [Marra et al., 1996]. Of eight wavelengths to be transported over the main ring fibers, four wavelengths will carry interregional traffic while the remaining four wavelengths will carry intraregional traffic. Over the trans-Atlantic and trans-Pacific spans, single-wavelength systems incorporating

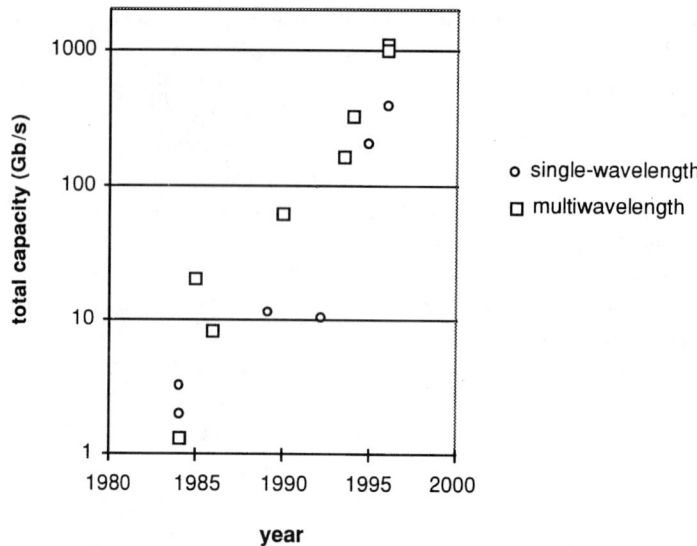

FIGURE 65.3 Trend in experimental demonstrations: total transmission capacity over one single-mode fiber for single-wave systems and for WDM systems.

optical amplifiers are currently deployed. However, the feasibility of upgrading the existing cable capacity using WDM has been experimentally verified [Feggeler et al., 1996] and WDM technology will no doubt play an important role in future generations of transoceanic networks.

65.5 Wavelength-Division Multiplexed Networking and Reconfigurable Optical Transport Layer

Whereas point-to-point WDM systems are becoming a commercial reality, the role of WDM in a reconfigurable optical transport network is being researched by a number of laboratories around the world [Hill et al., 1993; Sato, 1994; Brackett et al., 1993; Alexander et al., 1993]. Such WDM networks can be dynamically reconfigured to carry existing transport signals and are especially suited for supporting high-capacity demands that vary with time, a situation that may become likely as broadband services become widespread [Berthold, 1996]. WDM network may consist of nodes or network elements that can multiplex, switch, and terminate wavelengths. For example, an N by N wavelength cross connect routes wavelengths from input WDM ports to output WDM ports. A wavelength add–drop multiplexer (W-ADM) performs the function of assigning an input wavelength and multiplexing this signal to be routed over the network. A chain of W-ADMs can be deployed in a linear or ring configuration, similar to SONET ADMs.

In addition to reconfigurability and capacity enhancement, the WDM network can offer an advantage of builtin network survivability. For example, in an event of a fiber failure, the entire multiplexed signal traversing the failed link can be switched and routed on a protection fiber using a loopback mechanism similar to that used in SONET rings. Alternatively, a wavelength path can be protected on an end-to-end basis using a pre-configured protection path that becomes activated only after the primary path failure.

Figure 65.4 shows an example design for a W-ADM with single and multiwavelength ports. An incoming wavelength at a single-wavelength port is wavelength adapted through a receiver/laser pair or multiplexed directly through the optical cross connect fabric. Incoming multiplexed signals are demultiplexed and dropped or transparently routed through the optical cross connect fabric. In contrast to SONET ADMs, electronic signal processing and regeneration are kept to a minimum

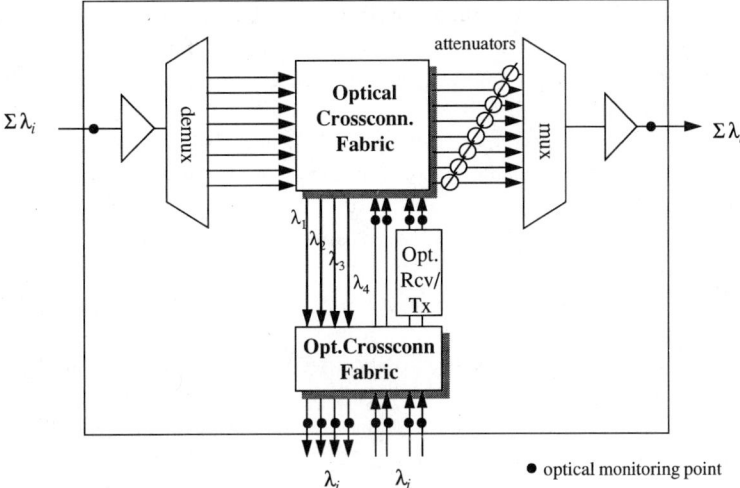

FIGURE 65.4 WDM add drop multiplexer. Part of the design supporting WDM signals in only one direction is shown. Signal levels, noise levels, and wavelength readings are tracked at various monitor points.

with most signal processing being done in the optical domain. The key active elements are optical cross connects for routing signals and optical amplifiers for compensating signal loss. The integrity of the signals is monitored at various points internal to the node, allowing dynamic adjustments to be made in the event of signal degradations.

Research is currently being pursued to define various aspects of the network architecture and the nodal design, including the performance requirements, protection switching mechanisms, and methods to support embedded management channels. Because the viability of WDM networks hinges strongly on the economics, both cost effectiveness and manageability of the network are being closely examined by a number of research consortia.

65.6 Summary

There have been rapid advances in the lightwave technology over the past decade, especially in the areas of single-frequency sources, optical amplifiers, wavelength multiplexers, demultiplexer and optical receivers. These advances have made possible a tremendous increase in the per-fiber transmission capacity, as exemplified by the terabit per second WDM transmission experiments demonstrated in early 1996. WDM systems are beginning to be deployed in the terrestrial networks, as well as in undersea global networks and are expected to play a greater role in the future as the demand for bandwidth continues to grow and the need for economical capacity upgrades becomes more compelling.

References

Alexander, S.B. et al. 1993. A precompetitive consortium on wide-band all-optical networks. *J. Lightwave Tech.*, 11(May/June):714–735.

Bergano, N. et al. 1996. 100 Gb/s Error Free Transmission over 9100 km using twenty 5 Gb/s WDM data channels, Proceedings OFC'96, Postdeadline Paper 23-1, Feb.

Berthold, J. 1996. Application of WDM in local exchange networks, Proceedings OFC'96, Paper ThN1, Feb.

Brackett, C.A. et al. 1993. A scaleable multiwavelength multihop optical network: A proposal for research on all-optical networks. *J. Lightwave Tech.*, 11(May/June):736–753.

Chraplyvy, A.R. 1990. Limitations on lightwave communications imposed by optical-fiber nonlinearities. *J. Lightwave Tech.*, 8:1548.

Desurvire, E. 1994. *Erbium-Doped Fiber Amplifiers*, Wiley, New York.

Feggeler, J. et al. 1996. WDM transmission measurements on installed optical amplifier undersea cable systems, Proceedings OFC'96, Paper TuN3, Feb.

Gnauck, A. et al. 1996. One terabit/s transmission experiment, Proceedings OFC'96, Postdeadline Paper 20-2, Feb.

Hill, G.R. et al. 1993. A transport network layer based on optical network elements. *J. Lightwave Tech.*, 11(May/June):667–679.

Iannone, P. 1995. A WDM PON architecture with bi-directional optical spectral slicing, Proceedings OFC'95, Paper TuK2, Feb.

Lin, C. et al. 1980. Optical-pulse equalization of low-dispersion transmission in single-mode fibers in the 1.3–1.7 μm spectral region. *Opt. Lett.*, 5(Nov.):476–478.

Marcuse, D. 1994. Dependence of cross-phase modulation on channel number in fiber WDM systems. *J. Lightwave Tech.*, 12(May):885–898.

Marra, W. et al. 1996. Africa ONE: The Africa optical network. *IEEE Commun. Mag.*, 34(Feb):50–57.

Mestdagh, D.J.G. 1995. *Fundamentals of Multi-Access Optical Fiber Networks*, Artech House, Norwood, MA.

Nosu, K. et al. 1993. 100 Channel optical FDM technology and its applications to optical FDM channel-based networks. *J. Lightwave Tech.*, 11(May/June):764–776.

Onaka, H. et al. 1996. 1.1 Tb/s WDM transmission over a 150 km 1.3 mm zero-dispersion single-mode fiber, Proceedings OFC'96, Postdeadline Paper 19-1, Feb.

Sato, K. 1994. Network reliability enhancement with optical path layer technologies. *IEEE J-SAC*, 12:159–170.

Suyama, M. et al. 1996. Improvement of WDM transmission performance by non-soliton RZ coding—A demonstration using 5 Gb/s 8-channel 4500 km straight line testbed, Proceedings OFC'96, Postdeadline Paper 26-2, Feb.

Further Information

For futher information on WDM technology and networking applications, see *Fiber-Optic Networks* by P.E. Green and *Special Issue on Broadband Optical Networks, IEEE Journal of Lightwave Technology, Vol. 11, May 1993*. Readers interested in access network applications may consult the article, "Technology and System Issues for a WDM-Based Fiber Loop Architecture," by S. Wagner and H. Lemberg, in the *IEEE Journal of Lightwave Technology*, Vol. 7, pp. 1759–1768, Nov. 1989. The latest developments in WDM components, systems, and applications are published each year in the *Technical Digest of Optical Fiber Communication Conference*.

V

Satellite

66 Geostationary Communications Satellites and Applications *Bruce R. Elbert* 893
Introduction • Satellite Network Fundamentals • Satellite Application Types

67 Satellite Systems *Robert L. Douglas* 912
Introduction • Satellite Systems • Transponder Systems • Launching Satellites • Station Keeping and Stabilization • Electrical Power Subsystem

68 The Earth Station *David R. Pritchard* 922
Introduction • Components of the Earth Station • Earth Station Site Selection • Power Distribution • Batteries • Antenna Foundation Requirements • Site Heating, Ventilation, and Air-Conditioning • Safety Considerations • Operation and Maintenance • Conclusion

69 Satellite Transmission Impairments *Louis J. Ippolito, Jr.* 935
Introduction • Attenuation on Slant Paths • Depolarization • Radio Noise • Scintillation • Summary

70 Satellite Link Design *Peter P. Nuspl and Jahangir A. Tehrani* 949
Introduction • Uplink and Downlink Equations • Interference Equations • Intermodulation Equations • Allowances, Losses and Margins • Sums of Link Equations • Designed Bit Error Ratio and Required C/N_0 • Numbers of Carriers, EIRP, and Power per Carrier • Other Issues • Summary • Appendix A: Some Calculations • Appendix B: Calculation of Antenna Discrimination

71 The Calculation of System Temperature for a Microwave Receiver *Wilbur Pritchard* 966
Noise, Antenna, and System Temperatures

72 Onboard Switching and Processing *A.K. Elhakeem* 976
Introduction • Onboard Switching Types • Summary

73 Path Diversity *Curt A. Levis* 996
Concepts • Site Diversity Processing • Site Diversity for Rain-Fade Alleviation • Optical Satellite Site Diversity • Site Diversity for Land–Mobile Satellite Communications • Site Diversity for Arctic Satellite Communications • Microscale Diversity for VSATs • Orbital Diversity

74 Mobile Satellite Systems *John Lodge and Michael Moher* 1015
Introduction • The Radio Frequency Environment and Its Implications • Satellite Orbits • Multiple Access • Modulation and Coding • Trends in Mobile Satellite Systems

75 Satellite Antennas *Yeongming Hwang and Youn Choung* 1032
Introduction • Horn Antennas and Reflector Feeds • Reflector Antennas • Phased Array Antennas • Tracking, Telemetry, and Command Antennas • Current and Planned LEO/MEO Mobile Satellite Antennas • Radiometer Antennas • Space Qualification • Future Trends and Further Study

76 Tracking and Data Relay Satellite System *Erwin C. Hudson* 1054
Introduction • TDRS System Overview • TDRSS Communications Design • TDRS Relay Satellite Design • TDRS Link Budget Examples • Summary

66
Geostationary Communications Satellites and Applications[1]

66.1	Introduction 893
66.2	Satellite Network Fundamentals .... 895 Satellite Orbits and Constellations • Frequency Spectrum • Network Topology • Satellite Versus Terrestrial Communication • Satellite and Earth Station Designs
66.3	Satellite Application Types .... 904 Television and Video Services • Trends in Satellite Voice Communications • Digital Information Services • Mobile and Personal Communications

Bruce R. Elbert
Hughes Space and Communications International, Inc. El Segundo, California

66.1 Introduction

Communication satellites, whether in geostationary Earth orbit (GEO) or low Earth orbit (LEO), provide an effective platform to relay radio signals between points on the ground. The users who employ these signals enjoy a broad spectrum of telecommunication services on the ground, the sea, and in the air. In recent years, such systems have become practical to the point where a typical household can have its own direct-to-home (DTH) satellite dish. Whether we reach the stage of broad acceptance depends on the competition with the more established broadcasting media, including over-the-air TV and cable TV. The success of British Sky Broadcasting in the U.K. market shows this is entirely possible. In any case, GEO and non-GEO satellites will continue to offer unique benefits for other services such as mobile communications and multimedia, but it is up to us to exploit them.

Technology and applications have progressed significantly since the late 1980s; however, the basic principles remain the same. Satellite communication applications extend throughout human activity, both occupational and recreational. Many large companies have built their foundations on TV, data communications, information distribution, maritime communications, and remote

---

[1] Adapted with permission from *The Satellite Communication Applications Handbook*, by Bruce Elbert, published by Artech House, Inc., Norwood, MA, 1997.

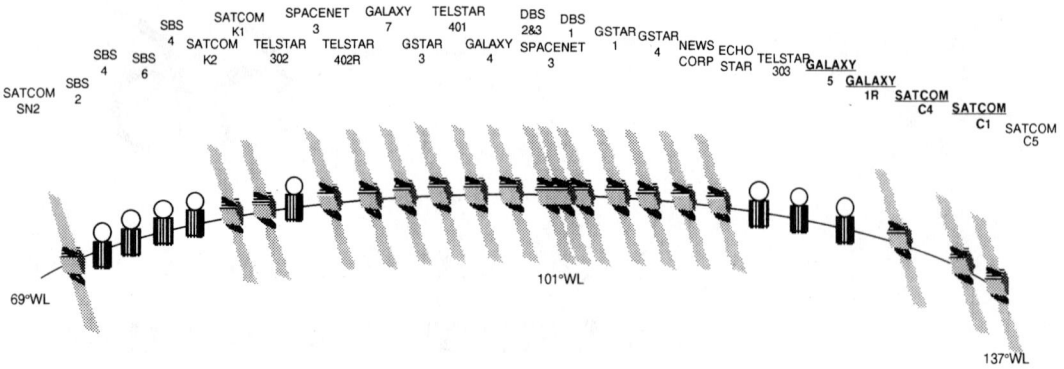

FIGURE 66.1 An illustration of the U.S. GEO orbital arc as of 1996, highlighting the most popular cable TV satellites.

monitoring. For many others, satellites have become a hidden asset. In the public sector, satellite applications are extremely effective in emergency situations where terrestrial lines and portable radios are not available or ineffective for a variety of reasons.

From this experience, those of us who have offered satellite services to large user communities know that the orbit position is paramount to achieving a solid business. Consider as an example the U.S. geostationary orbital service arc shown in Fig. 66.1. Highlighted are the most popular cable TV satellites: Galaxy I-R, Satcom C-1, Galaxy 5, and Satcom C-3. These four satellites collect revenue comparable to the remaining 25 satellites. Users of these *hot birds* pay a premium for access to the ground infrastructure of cable TV receiving antennas, much like tenants in a premium shopping mall pay to be in an outstanding location and in proximity to the most attractive department stores in the city. In the case of cable TV, access is everything because the ground antenna is, in turn, connected to households where cable services are consumed and paid for. For new satellite operators to get into such an established market often requires them to subsidize users by providing new antennas or paying for repointing. This then reduces the potential profitability as these switching costs must come out of the forthcoming revenues.

Satellite operators, who invest in the satellites and make capacity available to their customers, generally prefer that users own their own Earth stations. This is because installing antennas and associated indoor electronics is costly for satellite service providers. Once working, this investment must be maintained and upgraded to meet evolving needs. The owner/user benefits because: (1) there is complete control of the service facility and (2) advances in microcircuitry and satellite sophistication have reduced the cost and complexity of ownership and operation. A typical small Earth station is no more complex than a cellular telephone or video cassette recorder (VCR). Larger Earth stations such as TV uplinks and international telephone gateways are certainly not a consumer item, but the advantages of control typically motivate leading TV networks and communication companies to take these on.

Several U.S. corporations attempted to introduce DTH service in the mid-1980s at a time when cable TV was still establishing itself. The first entrants experienced great difficulties with the limited performance of low- and medium-power Ku-band satellites, hampering the capacity of the networks and affordability of the home receiving equipment. Europe and Japan had problems of their own creating viable DTH systems with the technologies at hand, choosing first to launch high-power Ku-band satellites with only a few operating channels. It was not until BSkyB and NHK were able to bring attractive programming to the public on their respective satellites that consumers adopted the services.

In the U.S., DTH first emerged in the 1980s through the backyard C-band satellite TV business that allowed consumers to pull existing cable TV programming from cable TV hot birds like Galaxy I and Satcom 3R. This clearly demonstrated the principle that people would vote with their money for a wide range of attractive programming, gaining access to services that were either not available or priced out of reach. Early adopters of these dishes paid a lofty price per installation because the programming was essentially free since the signals were unscrambled. Later, HBO and other cable networks scrambled their programming using the Videocipher 2 system, resulting in a halt to expansion of backyard dishes. This market settled back into the doldrums for several years.

In 1994, Hughes Electronics introduced its DIRECTV^{TM} service through a high-power BSS satellite (labeled DBS-1 in Fig. 66.1). With a combined total of over 150 channels of programming, DIRECTV, Inc., and another service provider, USSB, demonstrated that DTH could be both a consumer product and a viable alternative to cable. An older competing service, Primestar, was first introduced by TCI and other cable operators as a means to serve users who were beyond the reach of their cable systems. This subsequently provided the platform for a new and improved Primestar, rebuilt along lines very similar to DIRECTV. By converting from an analog to a digital system and expanding programming, the Primestar service has seen much greater demand than when it was first introduced in 1992.

Satellite communication applications can establish a solid business for companies that know how to work the details to satisfy customer needs. Another example is IDB Communications Group. From a very modest start from distributing radio programs by satellite, IDB branched out into every conceivable area of the telecommunication industry. Whether the requirement is to cover a media event in Antarctica or to connect Moscow to the rest of the world with voice and video links, IDB proved its worth. They capitalized on the versatility of satellite communication in the pursuit of a growing company and a demanding base of investors. Sadly, IDB ran into financial difficulties in 1994 and had to merge with LDDS, the fourth largest long distance carrier in the U.S. IDB's early success points up that needs can be met and money is to be made in the field of satellite communication.

66.2 Satellite Network Fundamentals

A commercial satellite communication system is composed of the space segment and the ground segment, illustrated in Fig. 66.2. In this simple representation of a GEO system, the space segment includes the satellites operating in orbit and a tracking, telemetry, and command (TT&C) facility that controls and manages their operation. This part of the system represents a substantial investment, where the satellite operating lifetime is typically of the order of 12 years based on fuel availability. Satellites tend to operate peacefully after they have survived launch and initiation of service. Failures of on-board equipment are generally transparent to users because the satellite operator can activate spare electronic equipment, particularly traveling wave tubes and solid state power amplifiers (the critical elements in the communications subsystem). From a user's perspective, the more important portion of the system is the ground segment, which is the focus of this chapter. It contains Earth stations of various types: large hubs or gateways and much smaller stations that render services directly to individual users.

Every satellite application achieves its effectiveness by building on the strengths of the satellite link. A satellite is capable of performing as a microwave repeater for Earth stations that are located within its coverage area, determined by the altitude of the satellite and the design of its antenna system [Elbert, 1987]. The arrangement of three basic orbit configurations is shown in Fig. 66.3. A GEO satellite can cover nearly one-third of the Earth's surface, with the exception of the polar regions. This includes more than 99% of the world's population and essentially all of its economic activity.

Satellite Orbits and Constellations

The LEO and medium Earth orbit (MEO) approaches require more satellites to achieve this level of coverage. Because of the fact that the satellites move in relation to the surface of the Earth,

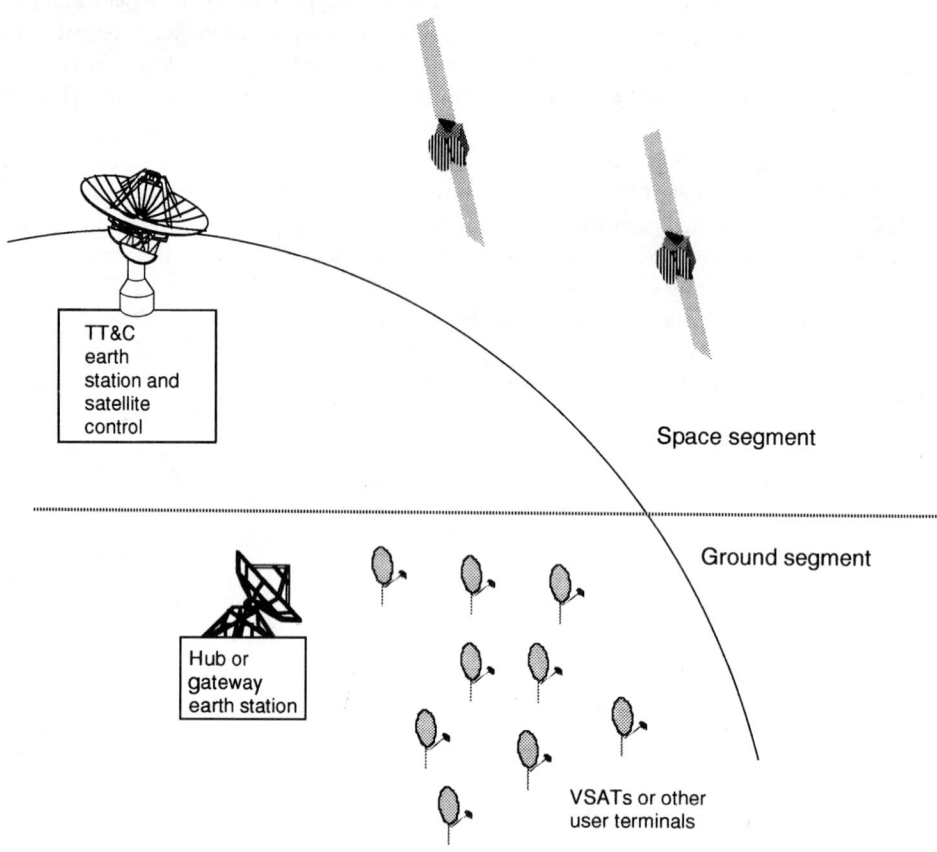

FIGURE 66.2 Elements of a satellite system, including the space and ground segments.

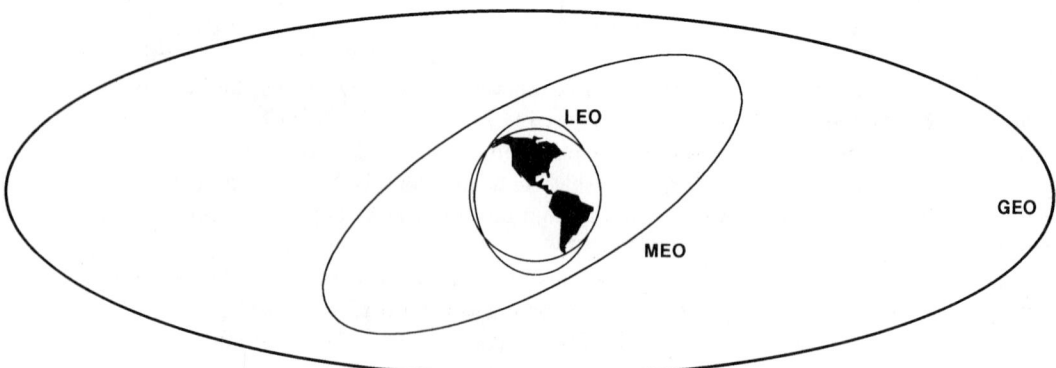

FIGURE 66.3 The three most popular orbits for communication satellites are low Earth orbit (LEO), medium Earth orbit (MEO), and geostationary Earth orbit (GEO). The respective altitude ranges are LEO (500–900 km), MEO (5,000–12,000 km), and GEO (36,000 km). Only one orbit per altitude is illustrated, even through there is a requirement for constellations of LEO and MEO satellites to provide continuous service.

a full complement of satellites (called a constellation) must be operating to provide continuous, unbroken service. The tradeoff here is that the GEO satellites, being more distant, incur a longer path length to Earth stations, whereas the LEO systems promise short paths not unlike those of terrestrial systems. The path length introduces a propagation delay since radio signals travel at the speed of light. Depending on the nature of the service, the increased delay of MEO and GEO orbits may impose some degradation on quality or throughput. The extent to which this materially affects the acceptability of the service depends on many factors.

At the time of this writing, the only one non-GEO system, Orbcom, has entered commercial service with its first two satellites. The applications of Orbcom include paging and low-speed burst data. Systems capable of providing telephone service, such as Iridium and I-CO Communications, will achieve operation around the year 2000. On the other hand, GEO satellite systems continue to be launched for literally all services contemplated for satellite, and the ones currently operating have lives (based on fuel reserves for stationkeeping) that will carry them through the year 2010.

Frequency Spectrum

Satellite communication services are provided through GEO satellites operating between 1 and 30 GHz. As microwave frequencies, transmissions to and from the satellite propagate along a line-of-sight path and experience free space loss that increases as the square of the distance. As indicated in Fig. 66.4, the spectrum is allocated in blocks which are designated L-, C-, X-, Ku-, and Ka-bands. The scale along the x axis is logarithmic in order to show all of the satellite bands; however, observe that the bandwidth available for applications increases in real terms as one moves toward the right (i.e., frequencies above 4 GHz). Also, the precise amount of spectrum that is available for services in a given region or country is usually less than Fig. 66.4 indicates. Precise definition of the services

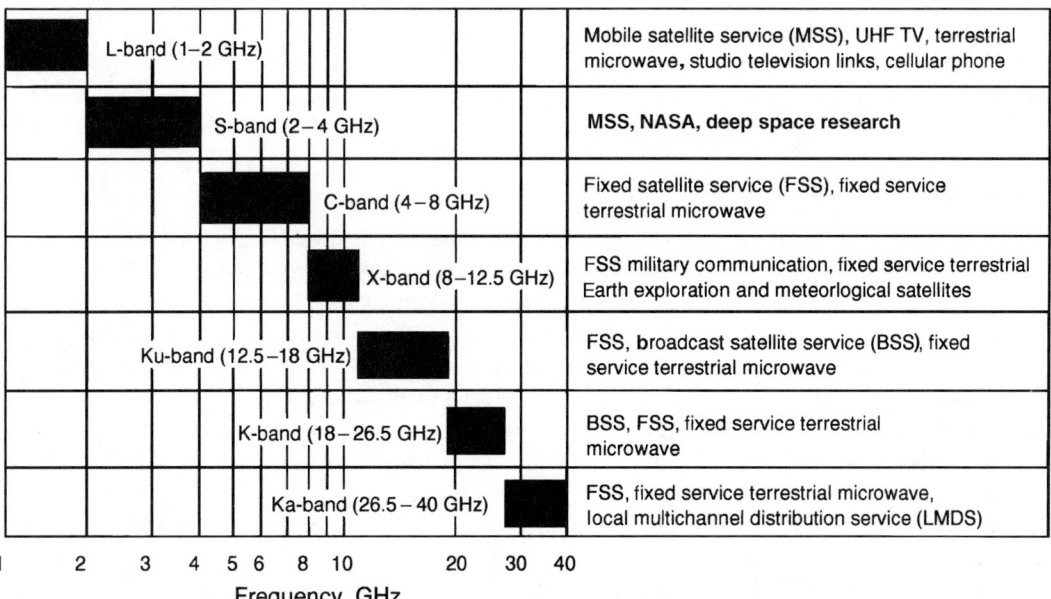

FIGURE 66.4 The general arrangement of the frequency spectrum that is applied to satellite communications and other radiocommunication services. Indicated are the short-hand letter designations along with an explanation of the typical applications. Note that the frequency ranges indicated are the general ranges and do not correspond exactly to the ITU frequency allocations and allotments.

and their allocated bandwidths can found in the current edition of the Radio Regulations, published by the International Telecommunication Union in Geneva, Switzerland.

As a general rule, the lower the band is in frequency, the better the propagation characteristics. This is countered by the second general principle, which is that the higher the band, the more bandwidth that is available. The mobile satellite service (MSS) is allocated to the L and S bands, where there is a greater degree of diffraction bending and penetration of physical media like foliage and non-metallic buildings. This tends to improve service in a mobile environment. Yet, the bandwidth available between 1 and 2.5 GHz, where MSS applications are authorized, must be shared not only among MSS applications, but with all kinds of radio broadcast and point-to-point services as well. Therefore, the competition is keen for this spectrum due to its excellent space and terrestrial propagation characteristics. The rapid rollout of wireless services such as cellular radiotelephone and personal communications services (PCS) will probably conflict with advancing GEO and non-GEO MSS systems.

Fixed satellite service (FSS) refers to applications between fixed Earth stations on the ground. These satellite deliver a wide range of services for video, telephone, and data communication. Wideband services such as DTH and broadband integrated services digital network (ISDN) using asynchronous transfer mode (ATM) can be accommodated at frequencies above 3 GHz, where there is more than 10 times the bandwidth available. Add to this the benefit of using directional ground antennas that effectively multiply the number of orbit positions. Some wideband services have begun their migration from the well-established world of C band to Ku band and Ka band. The X-band allocation for FSS has been reserved by governments for their internal uses, notably for military fixed communications.

Higher satellite effective isotropic radiated power (EIRP) used at Ku and Ka bands allows the use of relatively small Earth station antennas. An important factor in the use of these higher bands is the additional attenuation produced by heavy rain. Additional link margin is demanded in cases where the service must achieve an availability of greater than about 99%. The bands below Ku (e.g., C, S, and L) experience relatively minor amounts of rain attenuation and therefore can use their power to improve other aspects of service. C band, in particular, should maintain its strength for video distribution to cable systems and TV stations, particularly because of the extensive investment in uplink and downlink antennas and electronic equipment. One needs to identify and evaluate all of the critical factors before making a selection.

Network Topology

Applications are delivered through a network architecture that falls into one of three categories: point-to-point (mesh), point-to-multipoint (broadcast), and multipoint interactive [very small aperture terminal (VSAT)]. Mesh-type networks mirror the telephone network. They allow Earth stations to communicate directly with each other on a one-to-one basis. To make this possible, each Earth station in the network must have sufficient transmit and receive performance to exchange information with its least-effective partner. Generally, all such Earth stations have similar antennas and transmitter systems, so that their network is completely balanced. Links between pairs of stations can be operated on a full-time basis for the transfer of broadband information such as TV or multiplexed voice and data. Alternatively, links can be established only when needed to transfer information, either by user scheduling (reservation system) or on demand (demand assignment system).

By taking advantage of the broadcast capability of a GEO satellite, the point-to-multipoint network supports the distribution of information from a source (the hub Earth station) to a potentially very large number of users of that information (the remote Earth stations). A broadcast of information by the satellite is more efficient than terrestrial arrangements using copper wires, fiber optic cables, or multiple wireless stations. Therefore, any application that uses this basic feature will tend to provide value to the business.

TABLE 66.1 The Primary Strengths of Satellite Communications

Feature of Satellite Service	Application
Wide-area coverage	Domestic, regional, global
Wide bandwidth	Up to 1 GHz per coverage
Independent of land-based networks	Does not require connection to terrestrial infrastructure
Rapid installation	Individual sites can be installed and activated in one day for VSAT or two months for major hub
Low cost per added site	Depends on type of service; can be as low as $600 for DTH
Uniform service characteristics	Determined by coverage and type of transmission system
Total service from a single provider	By the satellite operator or a separate organization that leases transponder capacity
Mobile communication	Requires line-of-sight path over the coverage area

Many applications employ two-way links, which may or may not use the broadcast feature. The application of the VSAT to interactive data communication applications has proven very successful in many lines of business. A hub and spoke network using VSATs competes very effectively with most terrestrial network topologies that are designed to accomplish the same result. This is because the satellite provides the common point of connection for the network, eliminating the requirement for an separate physical link between the hub and each remote point. Other interactive applications can employ point-to-point links to mimic the telephone network, although this tends to be favored for rural and mobile services. Coming generation of Ka-band satellite networks that support very low-cost VSATs are expected to reduce barriers to entry into satellite communication.

Satellite Versus Terrestrial Communication

The degree to which satellite communication is superior to terrestrial alternatives depends on many interrelated factors, many of which are listed in Table 66.1. Although satellite communication will probably never overtake terrestrial telecommunications on a major scale, these strengths can produce very effective niches in the marketplace. Once the satellite operator has placed the satellite into service, a network can easily be installed and managed by a single organization. This is possible on a national or regional basis (including global in the case of non-GEO systems). The satellite delivers the same consistent set of services independently of the ground infrastructure, and the costs of doing this are potentially much lower as well. For the long term, the ability to serve mobile stations and to provide communications instantly are features that will gain strength in a changing world.

Satellite and Earth Station Designs

Originally, Earth stations were large, expensive, and located in rural areas so as not to interfere with terrestrial microwave systems that operate in the same frequency bands. These monsters had to use wideband terrestrial links to reach the closest city. Current emphasis is on customer premise Earth stations: simple, reliable, and low cost. An example of a Ku-band VSAT is illustrated in Fig. 66.5. Home receiving systems for DTH service are also low in cost and quite inconspicuous. Expectations are that the coming generation of Ka-band VSATs will reduce the cost and thereby increase their use in smaller businesses and, ultimately, the home. There is the prospect of satellites delivering improved Internet access ahead of the terrestrial networks, which have yet to deliver on the promise of ubiquitous wide digital bandwidth.

The other side of the coin is that as terminals have shrunk in size, satellites have grown in power and sophistication. There are three general classes of satellites used in commercial service, each designed for a particular mission and capital budget. Smaller satellites provide a basic number of

FIGURE 66.5 An example of a small customer premise VSAT used in data communication. (*Source*: Photograph courtesty of Hughes Network Systems.)

transponders, usually in a single-frequency band. One such satellite can be launched by the old and reliable Delta II rocket or dual-launched on the Ariane 4 or 5. Satellite operators in the U.S., Canada, Indonesia, and China have established themselves in business through this class of satellite. The Measat satellite, illustrated in Fig. 66.6, is an example of this class of vehicle which delivers a total of 18 transponders to Malaysia and neighboring countries in Asia. Introduction of mobile service in the LEO will involve satellites of this class as well.

Moving up to the middle range of spacecraft, we find designs capable of operating in two frequency bands simultaneously. Galaxy VII, shown in Fig. 66.7, provides 24 C-band and 24 Ku-band transponders to the U.S. market. A dual payload of this type increases capacity and decreases the cost per transponder. First generation DTH and MSS applications are also provided by these spacecraft. Because of the increased weight of this size of vehicle, launch service must be provided by larger rockets such as the Atlas 2, Ariane 4 or 5, the Russian Proton, or the Chinese Long March 2E or 3B (an improved Delta capable of launching such spacecraft was under development at the time of this writing).

Finally, the largest satellites serve specialized markets where the highest possible power or payload weight are required. The generation of GEO mobile satellites that are capable of serving handheld phones fall into this classification. An example of one of these satellites, APMT, is shown with its 12-m antenna deployed in Fig. 66.8. Also, the trends to use the smallest possible DTH home receiving antenna and to cover the largest service area combine to demand the largest possible spacecraft. The total payload power of such satellites reaches 8 kW, which is roughly eight times that of Measat. Designs are on drawing boards at the time of this writing for satellites that can support payload powers of up to 16 kW.

FIGURE 66.6 Measat, first launched in 1996, is a small satellite of the Hughes HS-376 class. With 1100 W of payload power, it provides both low-power C-band and medium-power Ku-band transponders to Malaysia and surrounding countries in the Asia–Pacific region. (*Source:* Photograph courtesy of Hughes Space and Communications Company.)

FIGURE 66.7 Galaxy VII, launched in 1993, is a medium-power Hughes HS-601 satellite that provides 24 low-power C-band and 24 medium-power Ku-band transponders to the U.S. market. It is used by cable TV and over-the-air broadcasting networks to distributed programming to affiliates. (*Source:* Photograph courtesy of Hughes Space and Communications Company.)

FIGURE 66.8 APMT is a regional GEO mobile communications satellite of the largest HS-601 HP class produced by Hughes. It serves the Asia Pacific region and employs a 12-m reflector to serve mobile and fixed users, many of which will use handheld user terminals. (*Source:* Photograph courtesy of Hughes Space and Communications Company.)

These general principles lead to a certain set of applications that serve telecommunication users. Satellite manufacture is a strong and viable business, as three American and a like number of European firms compete in the global market. Earth stations are produced by specialists in the U.S., Canada, Europe, Japan, and Korea. However, world-class consumer electronics and telecommunication equipment firms are lining up for the opportunity to address the megamarkets in DTH and MSS systems. In the next section, we review the most popular applications.

66.3 Satellite Application Types

Applications in satellite communication have evolved over the years to adapt to competitive markets. Evolutionary development, described in Elbert [1987], is a natural facet of the technology because satellite communication is extremely versatile. This is important to its extension to new applications in the coming century.

Television and Video Services

Commercial TV is the largest segment of the entertainment industry; it also represents the most financially rewarding user group to satellite operators. It may have taken 10 or more years for the leading networks in the U.S. and Europe to adopt satellites for distribution of their signals, but by 1985, it is the mainstay. Prior to this date, pioneering efforts in Indonesia and India allowed these countries to introduce nationwide TV distribution via satellite even before the U.S. made the conversion from terrestrial microwave. European TV providers pooled their resources through the European Broadcasting Union (EBU) and the EUTELSAT regional satellite system. Very quickly, the leading nations of Asia and Latin America adopted satellite TV delivery, rapidly expanding this popular medium to global levels.

There are four fundamental ways that GEO satellites are used to serve the TV industry:

- Point-to-point transmission of video information from a program source to the studio (also called *backhaul*)
- Point-to-multipoint distribution of TV network programming from the studio to the local broadcast station
- Point-to-multipoint distribution of cable TV programming from the studio to the local cable TV network
- Point-to-multipoint distribution of TV network and/or cable TV programming from the studio directly to the subscriber (i.e., DTH)

The first technique is a part of the three that follow, since programming material must be acquired from a variety of sources. A simplified view of the latter three is given in Fig. 66.9. In each case, the video channels are transmitted to the satellite by an uplink Earth station with the capability to obtain and assemble the programming. This can involve fixed antennas to backhaul programming from other satellites, tape play machines and video servers to play back prerecorded material, and fiber optic connections to distant studios. DTH systems with their smaller antennas require higher satellite power than cable TV and TV distribution. Hence, a common satellite for all three classes is rarely the case.

Over-the-Air TV Broadcasting

Video distribution by satellite is now standard for TV broadcasting in the very high-frequency (VHF) and ultra high-frequency (UHF) bands, which use local TV transmitters to cover a city or area. The satellite is used to carry the network signal from a central studio to multiple receive Earth stations, each connected to a local TV transmitter. This has been called TV distribution or

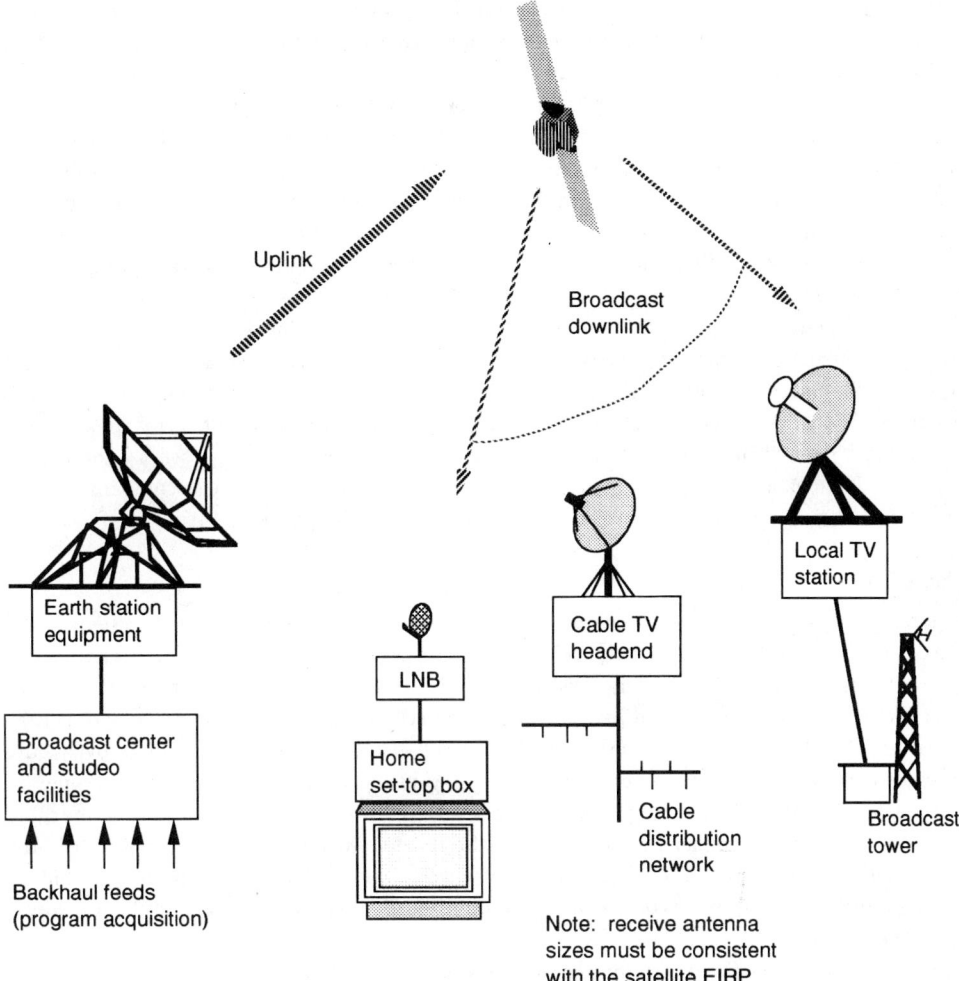

FIGURE 66.9 Framework for satellite communication applications in local broadcasting, cable TV, and direct-to-home satellite broadcast.

TV rebroadcast. When equipped with transmit uplink equipment, the remote Earth station can also transmit a signal back to the central studio to allow the station to originate programming for the entire network. U.S. TV networks such as CBS employ these reverse point-to-point links for on-location news reports. The remote TV uplink provides a transmission point for local sporting and entertainment events in the same city. This is popular in the U.S., for example, to allow baseball and football fans to see their home team play an away-from-home game in a remote city. More recently, TV networks employ fiber optic transmission services from a national fiber network, but satellites continue as the most flexible routing mechanism.

Revenue for local broadcast operations are available from two potential sources: advertisers and public taxes. From its beginnings in the U.S., TV provided an excellent medium to influence consumers on their purchase behavior. In exchange for watching commercials for soap, airlines, and automobiles, the consumer is entertained for virtually nothing. This has produced a large industry in the U.S. as stations address local advertisers and the networks promote nationwide advertising.

An alternative approach was taken in many European countries and Japan, where government-operated networks were the first to appear. In this case, the consumer is taxed on each TV set in operation. These revenues are then used to operate the network and to produce the programming. The BBC in the U.K. and NHK in Japan are powerhouses in terms of their programming efforts and broadcast resources. However, with the rapid introduction of truly commercial networks, cable TV, and DTH, these tax-supported networks are experiencing funding difficulties.

Public TV in the U.S. developed after commercial TV was well established. Originally called educational TV, this service existed in a fragmented way until a nonprofit organization called the Public Broadcasting Service (PBS) began serving the nation by satellite in 1978. The individual stations are supported by the local communities through various types of donations. Some are attached to universities; others get their support by donations from individuals and corporations. PBS itself acquires programming from the member stations and from outside sources such as the BBC. Programs are distributed to the members using satellite transponders purchased by the U.S. government. Therefore, it must compete with other government agencies for Congressional support. PBS programming is arguably of better quality than some of the popular shows on the commercial networks. Even though PBS addresses a relatively narrow segment, its markets are now under attack by even more targeted cable TV networks such as A&E, the Discovery Channel, the Learning Channel, and a more recent entry, the History Channel. All of these competitors built their businesses on satellite delivery to cable systems and DTH subscribers.

Cable TV

Begun as a way to improve local reception in the fringe areas and thinly populated sections of the U.S., cable TV has established itself as the dominant entry point of TV programming. Its popularity grew after pioneering cable TV networks such as HBO and Turner Broadcasting (which are not available over the air) were added by cable operators seeking more subscribers. These organizations, joined by Viacom, Disney, ABC, and NBC, used C-band satellite transmission to bypass the conventional TV networks to deliver programming to cable head ends and thence to the home. By 1980, 40% of urban-U.S. homes were using cable to receive the local TV stations (because the cable provided a more reliable signal); at the same time, the first nationwide cable networks were included as a sweetener and additional revenue source. During the 1980s, cable TV became an eight billion dollar industry and the prototype for this medium in Europe, Latin America, and the developed parts of Asia.

Cable TV networks offer programming as a subscriber service to be paid for on a monthly basis, or as an almost free service such as commercial TV broadcasting. HBO and the Disney Channel are examples of premium (pay) services, whereas the Discovery Channel and TBS are examples of commercial channels that receive most of their revenue from advertisers. The leading premium channels in North America and Europe are immensely successful in financial terms, but the business has yet to be firmly established in economies with low-income levels.

Cable TV became the first to offer a wide range of programming options that are under the direct control of the service provider. This assures that subscription fees and service charges can be collected from subscribers. If the fees are not paid, the service is terminated. Wireless cable, a contradiction in terms but nevertheless a viable alternative to wired cable, is gaining popularity in Latin America and Asia. The principle here is that the cable channels are broadcast locally in the microwave band to small dishes located at subscribers. Just as in the case of DTH, wireless cable depends on some form of conditional access control that allows the operator to electronically disconnect a nonpaying user.

Direct-to-Home Satellite Broadcasting

The last step in the evolution of the satellite TV network is DTH. After a number of ill-fated ventures during the early 1980s, DTH appears to be ready to take off. B Sky B in the U.K., NHK in Japan, DIRECTV in the U.S. and STAR TV in Asia are now established businesses, with other broadcasters following suit. Through its wide area broadcast capability, a GEO satellite is uniquely situated to deliver the same signal throughout a country or region at an attractive cost per user. The particular economics of this delivery depend on the following factors:

- The size of the receiving antennas: Smaller antennas are easier to install and maintain and are cheaper to purchase in the first place. They are also less noticeable (something which is desirable in some cultures but not necessarily in all others). However, the diameter of the receiving antenna is inversely proportional to the square root of the effective radiated power of the satellite.
- The design of the equipment: It is simple to install and operate (this author's RCA^{TM} brand Digital Satellite System (DSS) installation, needed to receive DIRECTV, took only 2 h: 105 min to run the cables and 15 min to install and point the dish).
- Several users can share the same antenna: This is sensible if the antenna is relatively expensive, say in excess of $1000; otherwise, users can afford their own. A separate receiver is needed for each independent TV watcher (i.e., for private channel surfing).
- The number of transponders that can be accessed through each antenna: The trend is now toward locating more than one satellite in the same orbit position, to increase the available number of transponders. The more channels that are available at the same slot, the more programming choices that the user will experience.
- The number of TV channels that can be carried by each transponder: Capacity is multiplied through digital compression techniques such as Moving Pictures Expert Group (MPEG) 2. This multiplies the quantity of simultaneous programs by five or more.

The ideal satellite video network delivers its programming to the smallest practical antenna on the ground, has a large number of channels available (greater than 100), and permits some means for users to interact with the source of programming. A simple connection to the public-switched telephone network (PSTN) allows services to be ordered directly by the subscriber. Thus, the system emulates a direct point-to-point connection between user and programming supplier, something that broadband fiber optic networks promise but have yet to deliver on a wide-scale basis.

Trends in Satellite Voice Communications

Voice communications are fundamentally based on the interaction between two people. It was recognized very early in the development of satellite networks that the propagation delay imposed by the GEO tends to degrade the quality of interactive voice communications, at least for some percentage of the population. However, voice communications represent a significant satellite application due to the other advantages of the medium. For example, many developing countries and lightly inhabited regions of developed countries continue to use satellite links as an integral part of the voice network infrastructure. Furthermore, an area where satellite links are essential for voice communications is the mobile field.

The public-switched telephone network within and between countries is primarily based on the requirements of voice communications, representing something in the range of 60–70% of all interactive traffic. The remainder consists of facsimile (fax) transmissions, low- and medium-speed data (both for private networks and access to public network services such as the Internet), and various systems for monitoring and controlling remote facilities. The principal benefit of the PSTN is that it is truly universal. If you can do your business within the limits of 3000 Hz of bandwidth and can tolerate the time needed to establish a connection through its dialup facility, the PSTN is your best bet.

Propagation delay has become an issue when competitively priced digital fiber optic networks are introduced. Prior to 1985 in the U.S., AT&T, MCI, and others were using a significant amount of analog telephone channels both on terrestrial and satellite links. An aggressive competitor in the form of U.S. Sprint invested in an all digital network that employed fiber optic transmission. Sprint expanded their network without microwave or satellite links, and introduced an all-digital service at a time when competition in long distance was heating up. Their advertising claimed that calls over their network were so quiet "you can hear a pin drop." This campaign was so successful that both MCI and AT&T quickly shifted their calls to fiber, resulting in rapid turn down of both satellite voice channels and analog microwave systems.

Satellite-based telephone service is acceptable in cases where it is the best alternative. For example, wired networks cannot provide mobile communications on the oceans and in very remote areas. Here, digital voice over satellite is widely accepted. A particular application, called mesh telephony, allows separate VSATs to operate as independent connection points for users or small villages. Gateways in major cities connect calls between these remote VSATs and subscribers on the PSTN. Major telephone Earth stations are part of the international telecommunication infrastructure, connecting to and between countries that are not yet served by modern fiber optic systems. The basic voice transmission quality is the same on satellite or terrestrial fiber because it is determined by the process of converting from analog to digital format. All long-distance transmission systems, satellite and cable, require echo cancellors to maintain quality. GEO satellite links experience a delay of about 260 ms and so are different from their terrestrial counterparts. However, they can be quite acceptable given that the application favors this medium for economic or practical reasons.

The economics of satellite voice communications are substantially different from those of the terrestrial PSTN, even given the use of digital technology on both approaches. With low-cost VSAT technology and high-powered satellites at Ku and Ka bands, satellite voice is the cheapest and quickest way to reach remote areas where terrestrial facilities are not available. It will be more attractive to install a VSAT than to extend a fiber optic cable over a distance greater than several hundred meters. A critical variable in this case is the cost of the VSAT, which is expected to drop from the $10,000 level to perhaps as low as $1500 before 2000.

Fiber optic transmission is not the only terrestrial technology that can address the voice communication needs of subscribers. There is now a new class of terrestrial wireless system called fixed cellular. Low-cost cordless phones or simple radio terminals are placed in homes or offices, providing access to the PSTN through a central base station. The base stations are concentrating points for traffic and can be connected to the PSTN by fiber or even satellite links. The cost of the base station and network control is kept low by not incorporating the automatic hand-off feature of cellular mobile radio. Instead, user terminals of different types make the connection through the closest base station, which remains in the same operating mode throughout the call. Fixed cellular can grow in cost beyond VSATs as more cells are added.

New classes of public network services are expected in coming years, under the general category of broadband communications. The underlying technology is asynchronous transfer mode, a flexible high-speed packet-switched architecture that integrates all forms of communications. ATM services can be delivered through fiber optic bandwidths and advanced digital switching systems, and include the following, among others:

- High-speed data on demand (384 kb/s to 155 Mb/s, and greater)
- Multichannel voice
- Video services on demand
- High resolution fax (300 dots per inch and better)
- Integrated voice/data/video for enhanced Internet services

Because of the high cost of upgrading the terrestrial telephone plant to provide ATM services on a widespread basis, many of these services will not appear in many places for some time. However, they represent the capability of the coming generation of public networks, being implemented around the globe.

Satellite communication currently addresses these services as well and, in fact, has been the only means available to provide them on a consistent basis. In private network applications, satellite communication has a big advantage for delivering broadband services to a large quantity of sites. Fiber optic networks are attractive for intra- and intercity public networks, and can offer broadband point-to-point transmission, which is potentially lower in cost per user. Yet, this is easier to do with a satellite because it provides a common traffic concentration point in the sky (a principle of traffic engineering). The bandwidth is used more effectively and, therefore, the network can carry more telephone conversations and generate more revenue.

Satellite networks are very expandable because all points are independent and local terrain does not influence the performance of link. Consider the example of the largest German bank, Deutsche Bank, which needed to offer banking services in the new states of the former East Germany. The telecom infrastructure in East Germany in 1990, although the best in eastern Europe, was very backward by western European standards. Deutsche Bank installed medium-sized Earth stations at new bank locations and was then able to offer banking services that were identical to those of their existing branches in the west. Another excellent example is Wal*Mart Department Stores, reviewed later in this chapter.

Digital Information Services

Satellite networks are able to meet a wide variety of data communication needs of businesses and noncommercial users such as government agencies. The wide-area coverage feature combined with the ability to deliver relatively wide bandwidths with a consistent level of service make satellite links attractive in the developing world as well as in the geographically larger developed countries and regions.

The data that is contained within the satellite transmission can take many forms over a wide range of digital capacities. The standard 36-MHz transponder, familiar to users of C-band worldwide, can easily transfer over 60 Mb/s, suitable for supercomputer applications and multimedia. Most applications do not need this type of bandwidth; therefore, this maximum is often divided up among multiple users who employ a multiple access system of some type. In fact, the multiple access techniques used on the satellite mirror the approaches used in local-area networks (LANs) and wide-area networks (WANs) over terrestrial links. As in any multiple access scheme, the usable capacity decreases as the number of independent users increases. Satellite data networks employing VSATs offer an alternative to terrestrial networks composed of fiber optics and microwave radio. There is even a synergy between VSATs and the various forms of terrestrial networks, as both can multiply the effectiveness of their counterpart.

Some of the must successful users of VSATs are familiar names in North American consumer markets. Wal*Mart, the largest U.S. department store chain, was an early adopter of the technology and has pushed its competitors to use VSATs in much the same manner that they pioneered. Chevron Oil also was first among the gasoline retailers to install VSATs at all of their company-owned filling stations to speed customer service and gain a better systemwide understanding of purchasing trends.

Whereas voice networks are very standardized, data networks cover an almost infinite range of needs, requirements, and implementations. In business, information technology (IT) functions are often an element of competitive strategy [Elbert, 1992]. In other words, a company that can use information and communications more effectively than its competitors could enjoy a stronger position in the ultimate market for its products or services. A data communication system is really only one element of an architecture that is intended to perform business automation functions.

A given IT application using client/server computing networks or broadband multimedia will demand efficient transfer of data among various users. Satellite communication introduces a relatively large propagation delay, but this is only one factor in the overall response time and throughput. These include: the input data rate (also in bits per second), the propagation delay through the link, the processing and queuing delay in data communication equipment, any contention for a terrestrial data line, and the final processing time in the end computer system (client or server). This is shown in Fig. 66.10. Each of these contributors must be considered carefully when comparing terrestrial and satellite links.

Propagation delay from a satellite link could reduce the throughput if there is significant interaction between the two ends of the link. The worst-case condition occurs where the end devices

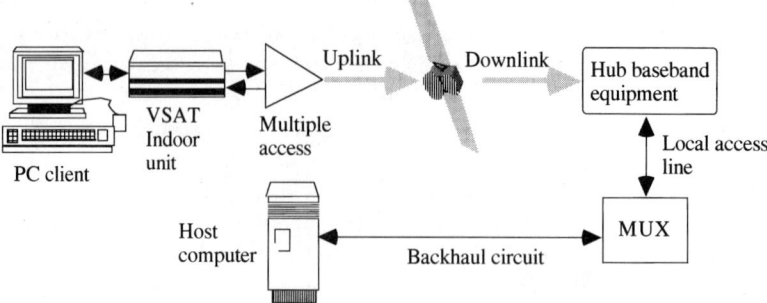

FIGURE 66.10 Contributors to the throughput and end-to-end delay of a data communication link involving a VSAT and a hub.

need to exchange control information to establish a connection or confirm receipt. Modern data communication standards, such as transmission control protocol/internet protocol (TCP/IP) and systems network architecture/synchronous data link control (SNA/SDLC), guard against the loss of throughput by only requesting retransmission of blocks of data which have errors detected in them by the receiver. However, to optimize throughput, there is a need to test and tune the circuit for the expected delay, type of information, and error rate.

The main deterrent to wider usage of digital satellite links is the cost of the terminal equipment, which today is much higher than that of a personal computer (PC). True low-cost VSATs have yet to appear at the time of this writing. Once the $2000 U.S. barrier is broken, many residential and small business users will be able to afford to purchase their own systems. This is not unlike the situation with backyard TVRO dishes, which during the 1980s were selling above $2000 U.S. Through innovations such as DIRECTV, Inc's digital satellite system, which sells for around $600 U.S., the home dish has become as attainable as the VCR. This also has benefits for data applications. If the bulk of the data transfer is outbound only (so that the user installation is receive only from the satellite), the DSS type of installation easily can deliver over 20 Mb/s of downlink capacity. A key example is the DirectPC^{TM} service from Hughes Network Systems that provides turbospeed Internet access service in the U.S.

Mobile and Personal Communications

The world has experienced an explosion of the demand for mobile telephone and data communications. The basis for this is the technology of cellular radio systems, which allow handheld mobile phones to connect to the public network as if they had access by wire. Availability of cellular radio at a reasonable cost has allowed a much larger subscriber base to develop than was possible during earlier generations of mobile phone technology. Now, we are prepared for the advance of cellular like systems that connect users directly to orbiting satellites, thus eliminating the terrestrial base station (a much smaller quantity of gateway Earth stations are still needed to connect mobile calls to PSTN subscribers).

Satellite-based mobile radio was first introduced to the commercial market in the late 1970s in the form of maritime mobile satellite service (MMSS) by COMSAT. Within a 10-year period, MMSS overtook the radio telegraph as the primary means of allowing ships to communicate. Ships ceased carrying conventional radio stations in the 1990s, which have been completely replaced by MMSS terminals. Today, the International Maritime Satellite (Inmarsat) system serves commercial ships and has been extended to commercial airlines as well.

Such as INTELSAT, Inmarsat serves the needs of its owners and their respective clients. Within the last five years, Inmarsat introduced land-based communications in a service called land mobile satellite service (LMSS). The first LMSS terminals are portable so that they can be set up quickly from a stationary location.

In 1994, Optus Communications of Australia began providing the Mobilesat LMSS service to its domestic market. Mobilesat follows the model of the second generation of LMSS phones, allowing subscribers to communicate from moving vehicles. Optus is the pioneer in this type of system, having directed the development of the ground infrastructure and mobile terminals. The Optus B series of satellites includes the necessary L-band payload for Mobilesat, and also carries Ku-band transponders to connect to gateways to the PSTN. The American Mobile Satellite Corp (AMSC) and Telesat Mobile Inc. extended this concept to the North American market in 1996. In this case, a satellite is dedicated to providing mobile telephone services.

A new generation of LMSS is expected before the year 2000, capable of supporting handheld phones similar to those used in cellular. In fact, the promoters of this advanced class of satellite network show models of modified cellular phones to promote their concept. The two basic approaches being taken are either the use of GEO, with satellites that serve specific countries or regions, and the LEO or MEO approach, where an orbiting constellation of satellites covers the entire globe. Figure 66.3 illustrates only one orbit with a single LEO satellite, where in the typical LEO constellation, the globe is covered by multiple orbits each containing several satellites. The number of required orbits and satellites decreases as the orbit altitude is increased. The minimum number of three satellites to provide global coverage (except the polar regions) is obtained at GEO. The advantage of the GEO approach is its lower initial cost and possibly lower risk, whereas the advantage of the LEO approach is the lower propagation delay and ability to connect users to each other anywhere in the world.

Many of these wireless networks have value to business as a vehicle for extensive data communications and paging services. Mobile data services have been around in the U.S. for over a decade, first introduced for private purposes. Federal Express equips their drivers with radio-based data communications terminals, and today a majority of competing delivery companies have also adopted the technology. An explosion may occur before the year 2000 due to the creation of several satellite mobile data networks in the U.S. and other countries. There is now the prospect of providing a public data network service to all users who wish to access information databases and other facilities while on the move. This is something that the cellular network has had difficulty demonstrating, due to technical issues of maintaining a solid circuit connection during the entire data call.

All together, the MSS is poised to become a substantial contributor to the future of the satellite industry. Spacecraft manufacturers are developing technical designs which even a few years ago looked such as science fiction (see Fig. 66.7). Once again, today's science fiction becomes tomorrow's science fact.

References

Adamson, S. et al. 1995. Advanced satellite applications—potential markets. Booz Allen & Hamilton, Noyes Data Corp., Park Ridge, NJ.

Elbert, B.R. 1987. *Introduction to Satellite Communication*, Artech House, Norwood, MA.

Elbert, B.R. 1992. *Networking Strategies for Information Technology*, Artech House, Norwood, MA.

Elbert, B.R. and Martyna, B. 1994. *Client/Server Computing—Architecture, Applications and Distributed Systems Management*, Artech House, Norwood, MA.

Elbert, B.R. 1997. *The Satellite Communication Applications Handbook*, Artech House, Norwood, MA.

Macario, R.C.V. 1991. On behalf of The Institution of Electrical Engineers. In *Personal and Mobile Radio Systems*, Peter Peregrims, Ltd., London, U.K.

Morgan, W.L. and Gordon, G.D. 1989. *Communications Satellite Handbook*, Wiley, New York.

Schwartz, M. 1987. *Telecommunication Networks*, Addison–Wesley, Reading, MA, p. 522.

67
Satellite Systems

67.1	Introduction	912
67.2	Satellite Systems	913
67.3	Transponder Systems	913
	Satellite Footprints • Transponder Analysis • Keeping the Pipeline Full • Television Transponder Systems • Antenna Subsystem	
67.4	Launching Satellites	917
67.5	Station Keeping and Stabilization	918
67.6	Electrical Power Subsystem	919

Robert L. Douglas
University of Memphis

67.1 Introduction

Communication satellites in **geostationary** orbits have had a tremendous impact on our lives by providing almost instantaneous worldwide electronic communications. The unique capabilities of communication systems that use geostationary satellites have revolutionized the design of telecommunication and television systems.

The purpose of this chapter is to provide an explanation of the specialized techniques that are used to communicate electronically using satellite repeaters or **transponders**. This chapter will be useful to anyone interested in learning about satellite communication techniques. Engineers and technicians, who are already familiar with communication concepts, can use this chapter to learn about the unique characteristics of communicating by satellite. Teachers and students will find this chapter to be an excellent reference in communication concepts and techniques. Managers and nontechnical people can use this chapter as a reference to obtain an understanding of the concepts and technical terms used by communication engineers. A glossary of satellite communication terms used in this chapter is provided in the Defining Terms section at the end of the chapter.

A communication satellite, often called a *COMSAT*, is a spacecraft that receives electrical signals from a transmitter on the Earth, amplifies these signals, changes the carrier frequencies, and then retransmits the amplified signals back to receivers on the Earth [Douglas, 1988]. Since communication satellites simply amplify and retransmit signals, they are often called repeaters or relays. Communication satellites are usually placed in geostationary Earth orbits, which allows the satellites to appear stationary to the transmitters and receivers on the Earth. Placing communication satellites in this special orbit prevents problems caused by the Doppler shift. One **Earth station** transmitter and one satellite can provide signals to receivers in an area that covers almost one-third of the Earth's surface. Satellites have revolutionized electronic communications since the first **commercial satellite** was launched in 1965. There are now more than 200 communication satellites in orbit around the world, and new satellites are launched regularly. Communication satellites have made it possible to routinely receive live television coverage of important news and sporting events from anywhere in the world. Satellites have also improved worldwide telephone communications, navigation, and weather information.

Arthur C. Clarke suggested the amazingly simple concept of communicating by satellites in stationary orbit in 1945. He called the satellite repeaters "extra-terrestrial relays" [Clarke, 1945]. At that time television was being developed commercially, and Clarke maintained that it would be much cheaper to develop and use communication satellites than to build a ground-based (**terrestrial**) network of transmitters and receivers with repeaters every 30–50 mi to amplify and retransmit the signals. He was correct, of course, but the technology needed to build the satellites and place them in geostationary orbit was not developed until the 1960s. By that time many countries had developed extensive terrestrial systems for the distribution of television and other wideband communication signals. Communication satellites are rapidly replacing these terrestrial systems.

Communication satellites have several advantages when compared to other forms of electronic communications. One of these advantages is the wide area covered by a single transmitter. The transmitting antennas on a communication satellite can be designed to provide signals to service areas as small as a city or as large as a country. The maximum area that can be covered by a geostationary satellite is approximately one-third of the Earth's surface. With satellites, it is just as easy to receive a signal in a remote location such as a farm or an island as it is to receive the signal in a city. One satellite can easily provide coverage for the entire U.S. Another advantage of a communication satellite is the low average operating cost. Manufacturing and launching a satellite is very expensive, but a typical satellite will operate continuously for 8–12 years. This extended operating lifetime makes satellite communication systems very competitive with terrestrial microwave systems and transoceanic cables.

67.2 Satellite Systems

A typical geostationary satellite, which is used to receive and retransmit signals from Earth, is launched from the Earth or a spacecraft and placed in orbit at a particular longitude. After a satellite has been placed in geostationary orbit, a *station keeping subsystem* in the satellite is activated from a control station on the Earth to maintain the satellite's position within plus or minus 0.1° of longitude and **equatorial inclination**. An *electrical power subsystem* in the satellite is activated to generate and regulate the electrical power needed by the transponders and other equipment in the satellite. An *antenna subsystem* provides the receiving and transmitting antennas needed for the transponder and other special purpose antennas such as an omnidirectional antenna that is used as a beacon. The antenna subsystem on each satellite provides the radiation pattern needed to transmit and receive signals to and from specific regions on the Earth. A *command and control telemetry subsystem* is provided on each satellite to transmit operating information about the satellite's systems to the Earth and to receive and process control signals from the Earth to the satellite. The *transponder subsystem* on each satellite provides the low-noise amplifiers (**LNAs**) the **downconverters**, and the **high-power amplifiers** (**HPAs**) needed to receive signals from the Earth station transmitters and to retransmit signals to the Earth station receivers. A typical example of each of these subsystems is further described later in this chapter. Communication satellites have a typical lifetime of 8–12 years. The amount of fuel (gas) in the station keeping subsystem is one of the factors that limits the useful lifetime of a satellite. The gradual deterioration of the solarcells that are used in the electrical power subsystem also limits the useful lifetime.

67.3 Transponder Systems

The primary purpose of communication satellites in geostationary orbit is to receive and retransmit signals from the Earth. Transponders provide the electronic equipment needed to accomplish

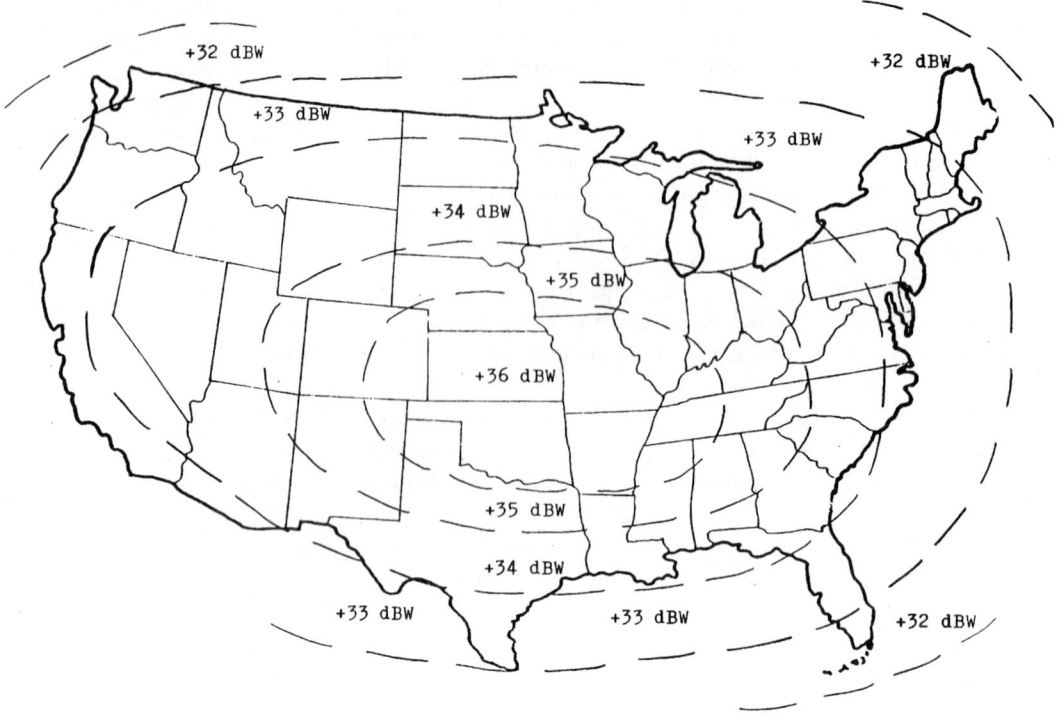

FIGURE 67.1 Satellite footprint.

this objective. A communication satellite is capable of receiving and retransmitting one or more independent signals. Each **uplink** signal is amplified and translated to a lower **downlink** carrier frequency by electronic circuits in the satellite. The electronic equipment used to amplify and shift the frequency of each signal is called a transponder. Typically, each satellite has 16–24 active transponders plus 6–8 spares.

Satellite Footprints

The radiation pattern of a satellite transmitting antenna determines the region on Earth where the downlink signal can be received. The radiation pattern for a transmitting antenna of any satellite can be superimposed onto a map of the region covered to display the **effective isotropic radiated power** (**EIRP**) radiated by the satellite to each area of the region. Maps that give the EIRP for a satellite are called satellite **footprints**. Vertical and horizontal radiation patterns for a satellite are usually combined on one footprint. Figure 67.1 shows a typical satellite footprint. The footprint given in Fig. 67.1 can be used to determine the EIRP for any location in the continental U.S. For example, the EIRP for Memphis, Tennessee, is +36 **dBW** (decibel referenced to 1 W) and the EIRP for Boston, Massachusetts, is +33 dBW. The actual power of a signal received from the satellite represented by this footprint is determined by subtracting the **space loss** to any Earth location from the EIRP.

Transponder Analysis

Figure 67.2 provides a functional block diagram of a simplified transponder, which has only one input signal. The uplink signal received from an Earth station transmitter is amplified by the combination of the receiving antenna of the satellite and a low-noise amplifier. Tunnel diode

Satellite Systems

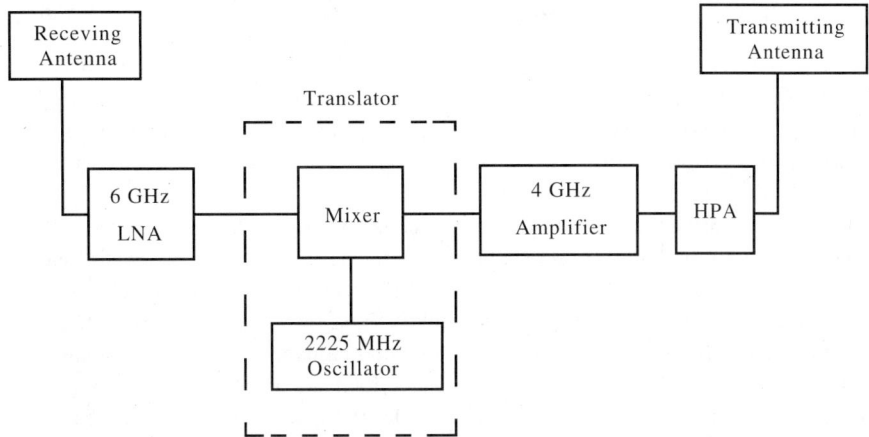

FIGURE 67.2 Transponder block diagram.

amplifiers, operated at reduced temperatures, are often used in the LNAs in satellite receivers. After amplification, the output signal from the LNA is mixed in a **translator** with a signal from a fixed frequency oscillator. The output signal from the mixer contains the sum and difference frequencies of the inputs. A filter passes the difference frequencies and blocks the sum frequencies. The output of the downconverter contains the same information as the input, but the center frequency of the carrier has been shifted down by 2225 MHz. The output of the downconverter is amplified in a high-power amplifier to increase the power of the downlink signal. Very few electronic devices provide the power output needed (10–50 W) and the lifetime required (up to 12 years without maintenance) for the 4–20 GHz signals amplified in the HPAs. Most of the HPAs used in satellite transponders are either **traveling wave tube amplifiers (TWTA)** or **gallium-arsenide field-effect transistor (GaAsFET)** amplifiers. GaAsFETs have replaced TWTAs at 4 GHz as new satellites have been launched to replace older satellites. GaAsFETs HPA are called **solid-state power amplifiers (SSPAs)**. TWTAs are still being used at 12 GHz and 20 GHz. The output of the HPA is amplified by the transmitting antenna and radiated back to Earth.

As an example of a simplified satellite transponder, assume that the uplink carrier signal has a frequency of 5945 MHz ± 18 MHz. This signal is amplified by the receiving antenna and the LNA and then mixed with a 2225-MHz signal. The output of the mixer is 3720 MHz ± 18 MHz (5945 − 2225) and 8170 MHz ± 18 MHz (5945 + 2225). The 3720 MHz ± 18 MHz signals pass through a filter to the HPA. The HPA and the transmitting antenna amplify and concentrate the output signal from the filter to provide the effective power needed to radiate this 3720 MHz ± 18 MHz downlink signal back to Earth. Notice that the uplink signal is not demodulated in the satellite.

Keeping the Pipeline Full

Satellite transponder systems provide a bandwidth of 500 MHz to over 2 GHz to the signals being relayed. **Frequency- and/or time-division multiplexing** is used to take full advantage of this available bandwidth. The design of a specific transponder system depends on the type of **multiplexing** used. For example, analog television video and audio signals can be frequency modulated onto transponder channels, which have a bandwidth of 40 MHZ. Telecommunication systems use different multiplexing techniques and therefore require different transponder designs to fully utilize the available bandwidth.

Television Transponder Systems

Most of the **C-band** communication satellites use two 12-channel transponders to receive, translate, and retransmit 24 independent channels simultaneously. One of these transponders receives and transmits signals using vertically polarized antennas while the other transponder receives and transmits signals using horizontally polarized antennas. Each LNA amplifies one of these uplink signals and then shifts the uplink frequencies in all 12 channels to corresponding downlink frequencies. The output of the translator is split into 12 separate signals and each signal is amplified by a separate HPA. The outputs of the HPAs are recombined and transmitted back to the Earth. Figure 67.3 provides a functional block diagram of a typical 4/6-GHz transponder system that has 24 channels. Two receiving antennas are used to separate the uplink signal into a vertically polarized and a horizontally polarized signal. The vertically polarized signal contains all of the even-numbered channels from 2 to 24, and the horizontally polarized signal contains all of the odd-numbered channels 1 through 23. Each channel has a bandwidth of 36 MHz, and there is a 4-MHz guardband between each channel. Two separate LNA-translator circuits are required because the horizontally polarized and vertically polarized signals **reuse** the same frequencies. After amplification in the HPAs, the even-numbered channels are recombined and radiated to the Earth using a vertically polarized

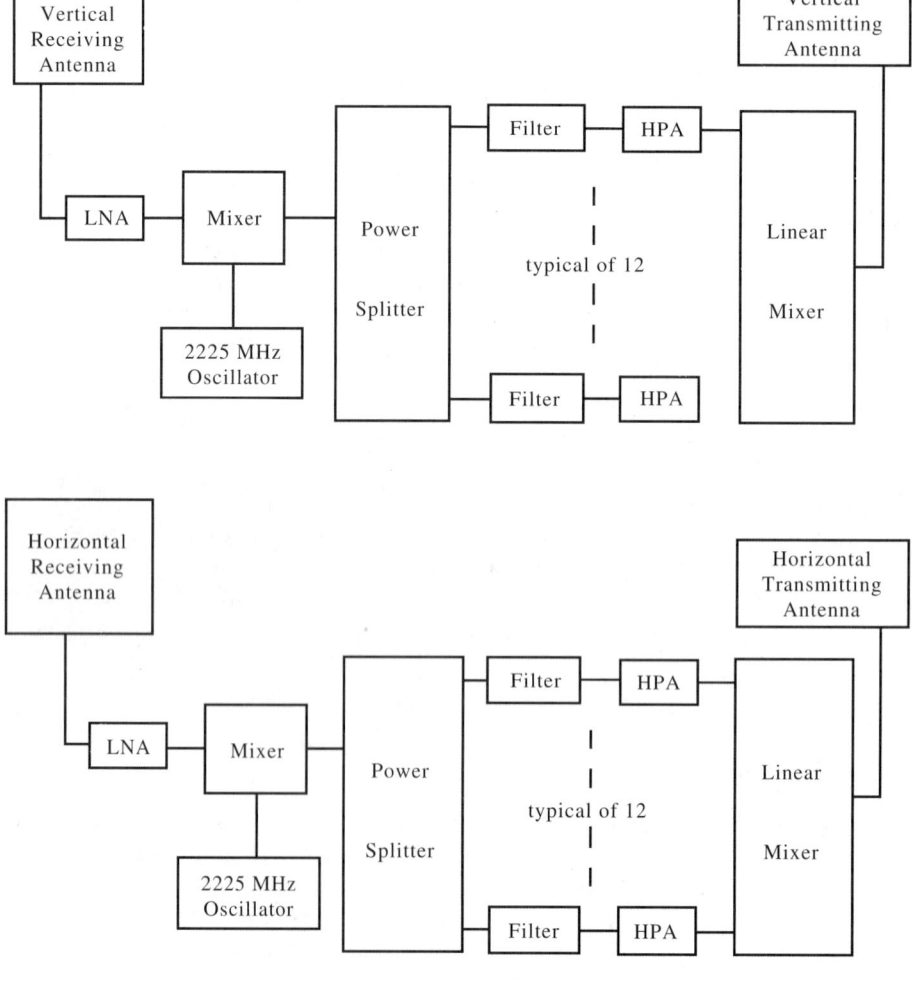

FIGURE 67.3 Simplified 4/6-GHz transponder system.

Satellite Systems

transmitting antenna and the odd-numbered channels are recombined and radiated to the Earth using a horizontally polarized transmitting antenna.

Redundancy is used in most transponder systems to improve reliability. For example, most satellites usually have 4–6 spare transponders and one or more spare LNA translators that are used to replace defective circuits. The command and telemetry subsystem is used by a controller on Earth to switch spare circuits in and out of the system. An input multiplexer uses bandpass filters to separate the uplink signal into individual channels between the down converter and the HPA. There is a separate HPA for each channel. An output multiplexer (diplexer) is used to recombine the outputs of the HPAs into a composite downlink signal.

Antenna Subsystem

A satellite usually has an omnidirectional antenna that is used to receive command signals from the Earth and to transmit telemetry signals to Earth. These command signals are used to control all of the equipment in the satellite. For example, a command signal is used to start the **apogee** motor to transfer a satellite from elliptical orbit to geostationary orbit. Command signals are also used to start or stop the spinning of a satellite and to position the attitude of a satellite so that the other satellite antennas are pointing toward the Earth. Telemetry signals are used to provide information about a satellite to the Earth control station. An omnidirectional antenna must be used for the telemetry and control subsystem during the launching and positioning of the satellite because the other antennas on the satellite may not be pointing toward the Earth. The directional antennas can be used for the command and telemetry signals after a satellite has been properly positioned and oriented.

Some satellites have Earth coverage antennas. Earth coverage antennas are horn (cone) shaped with radiation patterns that cover all of the Earth's surface visible from the geostationary orbit. Earth coverage antennas can be used to transmit or receive signals to or from any place on the Earth. A beamwidth of 17.2° is required for Earth coverage from the geostationary orbit.

Parabolic antennas are used on satellites to generate spot-beams or radiation patterns that cover only a portion of Earth's surface. In general, the diameter of parabolic antennas and the diameter of a radiation pattern on Earth are inversely related. Large parabolic antennas are used to transmit or to receive signals from a small specific region on Earth. Smaller parabolic antennas are used for larger areas.

The radiation pattern of a parabolic reflector can be tailored to cover a particular region by using multiple feed horns that are positioned to provide the coverage desired. For example, the radiation pattern of an antenna can be tailored to cover the continental U.S. (**CONUS**), Alaska, or portions of Canada. Tailoring a radiation pattern to cover a specific region on Earth is called beam shaping. One particular communication satellite uses two 60-in-diameter parabolic reflectors with multiple feed horns to obtain a beamwidth of 7° by 3.2° for CONUS coverage. Additional feed horns are used to provide a 4° by 2° beam for Alaska, a 3° beam for Hawaii, and a 3° beam for Puerto Rico [Gould and Lum]. Satellites that use vertical and horizontal **polarization** to reuse the available bandwidth often have two transmit parabolic antennas (one for vertical polarization and one for horizontal polarization) and two receive parabolic antennas (one for vertical polarization and one for horizontal polarization). Many satellites have two receiving and two transmitting antennas. One antenna is used to receive vertically polarized uplink signals, and the other receiving antenna is used for horizontally polarized uplink signals. Similarly, one transmitting antenna is used for vertically polarized downlink signals and the other for horizontally polarized downlink signals.

67.4 Launching Satellites

A communication satellite is launched into a geostationary orbit using either an unmanned launch vehicle that is expendable or a manned Space Shuttle that is reusable. The launch vehicle has one or more rocket stages that provide the acceleration needed to place a satellite into Earth orbit. The satellite is normally placed in a circular or elliptical Earth orbit by the launch vehicle and then transferred into a geostationary orbit.

An unmanned launch vehicle usually places the satellite directly into an elliptical Earth orbit that is called a transfer orbit. The highest point in this elliptical orbit is called the apogee and the lowest point is called the **perigee**. The launch is designed to provide an apogee that is the same height as the geostationary orbit. A special rocket in the satellite provides the acceleration needed to transfer the satellite from an elliptical orbit into a geostationary orbit. This special rocket in a satellite is called an apogee motor or an **apogee kick motor** (**AKM**). Elliptical orbits have an angle of inclination that corresponds to the latitude of the launch site. For example, satellites launched from Cape Canaveral, Florida, have an elliptical orbit with an inclination of approximately 27.5°. The apogee motor must also provide the acceleration to change the angle of inclination to zero as a satellite's orbit is changed from elliptical to geostationary.

Space Shuttles provide an alternate method to launch communication satellites. One or more satellites are stored in the payload bay of a Space Shuttle. The Space Shuttle, which is manned, is then launched into a circular orbit that is approximately 200 mi above Earth's surface. Astronauts unload the satellites from the payload bay, and place them into an elliptical Earth orbit. The satellites are then transferred from this elliptical orbit into a geostationary orbit. Launching satellites with a Space Shuttle is less expensive than using unmanned launch vehicles because Space Shuttles can be reused. Space Shuttles have also been used to retrieve satellites from space.

67.5 Station Keeping and Stabilization

Ideally, a satellite in geostationary orbit should remain in a fixed position because the gravitational and centrifugal forces on it are equal. Unfortunately, the gravitational forces on a geostationary satellite are not constant. The gravitational forces of the sun and moon on a geostationary satellite cause a change in the inclination angle of the satellite. The station keeping subsystem in a satellite provides small thruster rockets that are periodically used to move the satellite back to an inclination angle of zero degrees. Orbital calculations also assume that the mass of the Earth is uniformly distributed in a perfect sphere. Actually, the Earth is slightly egg shaped. This error causes a change in the longitude of geostationary satellites. The station keeping subsystem is also used to keep geostationary satellites in their assigned longitudes. The thruster rockets used for station keeping use gas that is stored in tanks in the satellites. Hydrazine gas is often used for the thruster rockets on satellites. The amount of gas stored in the tanks for the thruster rockets is one of the limits on the effective lifetime of a geostationary satellite.

The attitude of a geostationary satellite refers to the orientation of the satellite with respect to Earth. For example, the antennas on a satellite must be positioned so that their radiation pattern covers a specific region on Earth. Once the attitude of a satellite has been established, it is maintained by a stabilization subsystem. There are two different types of stabilization subsystems used in communication satellites.

Spinner-type satellites provide stabilization by causing the body of the satellite to rotate at several hundred revolutions per minute. The angular momentum of the spinning satellite acts like a gyroscope to stabilize the attitude of the satellite. Antennas on spinner satellites are **despun** because they must maintain a fixed pointing position and polarization. The bodies of spinner satellites are usually cylindrical in shape, and these satellites have a pendulumlike device called a **nutation damper** to damp out any periodic wobbling motion that may occur in the spin axis.

Another type of stabilization system uses three inertia wheels to create three-dimensional gyroscopic forces. Satellites that use **three-dimensional stabilization** have one inertia wheel rotating in each dimension. Each inertia wheel is a completely enclosed rotating sphere or cylinder that provides stabilization in one dimension. The three dimensions are referred to as pitch, roll, and yaw. The body of a three-dimensional satellite is usually square or rectangular in shape since it does not spin. The velocities of the inertia wheels can be changed to provide nutation damping.

67.6 Electrical Power Subsystem

Solar cells which convert sunlight into electrical energy are used to generate the electrical power used in the satellites. Satellites require a large number of solar cells to generate the electrical power needed. The solar cells are mounted on satellites using two different methods. Spinner satellites have their solar cells mounted on the rotating cylindrical body of the satellite so that a portion of the solar cell array is always exposed to sunlight as the body of the satellite rotates. This method is inefficient because few of the solar cells are continuously exposed to direct sunlight. The electrical power generated by the solar cells on a spinner satellite must be transferred by slip rings from the body of the satellite to the equipment that is despun. Three-dimensional stabilized satellites (that do not spin) use large flat panels or arrays of solar cells. These arrays are positioned so that all of the solar cells on the panel are exposed to direct sunlight simultaneously. These arrays of solar cells are often called **solar sails**. Solar sails are usually folded against the body of three- dimensional stabilized satellites during launch so that the satellite will fit inside of the launch vehicle. The solar sails are deployed after a satellite has been placed in geostationary orbit and stabilized. Solar sails must be rotated by stepper-motors once every 24 h to keep the solar cells pointing toward the sun. DC-to-DC converters are used to change the electrical output of the solar cells into the voltages needed throughout a satellite.

The relative positions of the sun, the Earth, and a geostationary satellite change continuously as the Earth rotates on its axis and revolves around the sun. A satellite eclipse occurs when the Earth moves between the sun and a satellite. Since the solar cells on a satellite cannot generate electrical power during a satellite eclipse, satellites must have storage batteries to provide electrical power during this time. These batteries are recharged by the solar cells between eclipses.

The use of solar cells limits the lifetime of a satellite. The electrical output of a solar cell gradually decreases with age; after 8–10 years, the electrical output from a solar-cell will decrease by about 20%. Eventually, there is not enough electrical power available to keep all of the transponders operating. However, the lifetime of a satellite can be extended by turning one or more of the transponders off. Communication satellites are usually replaced after about 10–12 years of continuous service.

Defining Terms

Apogee: The place in an elliptical orbit that is farthest from Earth.
Apogee kick motor (AKM): A special rocket in a satellite that is used to provide the acceleration needed to transfer the satellite from an elliptical orbit into a geostationary orbit.
C-band: A 4/6-GHz band of microwave frequencies used in many satellite communication systems.
Commercial satellites: Satellites owned and operated by private companies.
CONUS: The contiguous states in the U.S.
Demand assignment: A communication system that assigns channels to users upon demand.
Despun: The parabolic antennas on a spinner satellite do not spin because they must maintain a fixed pointing position and polarization.
Direct broadcast satellite (DBS): A communication system that broadcasts signals directly from a satellite into homes and businesses.
Domestic satellite (DOMSAT): A satellite that provides communication service within a country.
Downconverter: An electronic circuit that is used to translate a modulated signal to a lower band of frequencies.
Downlink: The path from a satellite to an Earth station.
Earth station: The site of the transmitting and receiving equipment used to communicate with satellites.
Effective isotropic radiated power (EIRP): The effective power radiated from a transmitting antenna when an **isotropic radiator** is used to determine the gain of the antenna.

Elevation: The angle at which a satellite is viewed from a site on Earth.
Equatorial inclination: The angle that the orbital plane of a satellite makes with the equator.
Equatorial orbit: A satellite is in an equatorial orbit when the orbital plane includes the equator. The angle of inclination for an equatorial orbit is zero degrees.
F (noise factor): A method used to specify noise in electronic circuits and systems.
Footprint: The radiation pattern of the transmitting antenna of a satellite superimposed onto a map of the region covered.
Frequency-division multiplexing: The combining of two or more input signals which are separated by frequency into a composite output signal.
Frequency reuse: The use of two different antenna polarizations to transmit and receive two independent signals that use the same frequency bands.
Gain/temperature figure of merit (G/T): The ratio of antenna gain to system noise temperature in decible per keluin. G/T is used to specify the quality of a **satcom** receiving system.
Gallium-arsenide field-effect transistor (GaAsFET): A special type of transistor that is used in low-noise amplifiers and high-power amplifiers.
Geostationary: A special Earth orbit that allows a satellite to remain in a fixed position above the equator.
Geosynchronous: Any Earth orbit in which the time required for one revolution of a satellite is an integral portion of a sidereal day.
High-power amplifier (HPA): A linear power amplifier that is used to amplify modulated signals at microwave frequencies.
Hydrazine: A gas that is used in the thruster rockets on satellites.
Inclinometers: A tool that is used to measure the angle of elevation of an antenna.
Isotropic radiator: A reference antenna that radiates energy in all directions from a point source.
Ku-band: A 12/14-GHz band of microwave frequencies used in many satellite communication systems.
LNA: Low-noise amplifier.
Look angles: The azimuth and elevation angles used to point or aim an Earth station antenna toward a satellite.
Multiple access: Several users have access to the same satellite channels.
Multiplexing: The combining of two or more input signals into a composite output signal.
Nutation damper: A pendulumlike device that is used to damp out any periodic wobbling motion that occurs in the spin axis of a satellite.
Perigee: The place on an elliptical orbit that is nearest to the Earth.
Polarization: The direction of the electric field radiated from an antenna.
Polar orbit: A satellite is in a polar orbit when the orbital plane includes the North and South Poles. The angle of inclination for a polar orbit is 90°.
Satcom: A satellite communication system.
SATCOM: A series of satellites owned by RCA.
Solar sails: A large flat array of solar cells.
SSPA: Solid-state power amplifier.
Space loss: The attenuation that occurs as an electromagnetic wave propagates through space.
Spinner: A satellite that spins at several hundred revolutions per minute to stabilize the attitude of the satellite.
Terrestrial: Satcom equipment on the Earth.
Three-dimensional stabilization: Three inertia wheels are used to provide roll, pitch, and yaw stabilization of a satellite.
Time-division multiplexing: The combining of two or more input signals that are separated by time into a composite output signal.
Translator: The shifting of an AM or FM signal up or down in frequency without changing the frequency spacing or relative amplitudes of the carrier and sideband components.

Transponder: The equipment in a satellite that is used to receive the uplink signals and transmit the downlink signals.

Traveling wave tube amplifier (TWTA): A microwave amplifier that is used as the high-power amplifiers in many satellites.

Uplink: The path from an Earth station to a satellite.

References

Clarke, A.C. 1945. Extraterrestrial Relays. *Wireless World*, 51(Oct.):305–308.

Douglas, R.L. 1988. *Satellite Communications Technology*, Prentice-Hall, Englewood Cliffs, NJ.

Gould, R.G. and Lum, Y.F. *Communication Satellite Systems*, IEEE Press.

Further Information

For further information on satellite systems and transponders, see:

- *Modern Electronic Communication Techniques*, Harold B. Killen, Macmillan, 1985.
- *Electronic Communication Techniques*, Paul H. Young, Macmillan, 1994.

68
The Earth Station

David R. Pritchard
*Norlight Telecommunications/
Teleport Chicago*

68.1 Introduction 922
68.2 Components of the Earth Station 923
68.3 Earth Station Site Selection 927
68.4 Power Distribution 929
68.5 Batteries 930
68.6 Antenna Foundation Requirements 930
68.7 Site Heating, Ventilation, and Air-Conditioning 931
68.8 Safety Considerations 932
68.9 Operation and Maintenance 933
68.10 Conclusion 933

68.1 Introduction

The term **Earth station** is used to refer to a ground-based communications facility made up of the transmitters, receivers, and antennas used to relay signals to or from a communications satellite. By definition, the Earth station also includes the equipment required to interconnect the satellite signals to the end user, whether the user is located at the Earth station facility or at a remote location. Information may be transmitted or received in the form of analog or digital carriers for voice, data or video transmission. The signal is transmitted, or uplinked, from the ground station to a satellite **transponder** (Fig. 68.1). The satellite transponder relays the signal back to Earth where it can be received, or downlinked, anywhere within the coverage area known as the **footprint**. The majority of commercial Earth stations are designed to operate with satellites in **geosynchronous** orbit operated by AT&T, GE, Hughes, and other domestic and international carriers. Geosynchronous satellites are positioned approximately 22,300 mi. above the equator and rotate at the same speed as the Earth. From a fixed point on the ground, a satellite in geosynchronous orbit appears stationary and can provide continuous service to the end user. Figure 68.2 is an aerial view of the Teleport Chicago Earth station showing multiple Earth station antennas used for video and audio transmission.

There are approximately 34 commercial satellites that can be accessed by Earth stations in North America, which provide domestic C- and Ku-band satellite service (not including the direct broadcast satellites).

The Earth Station

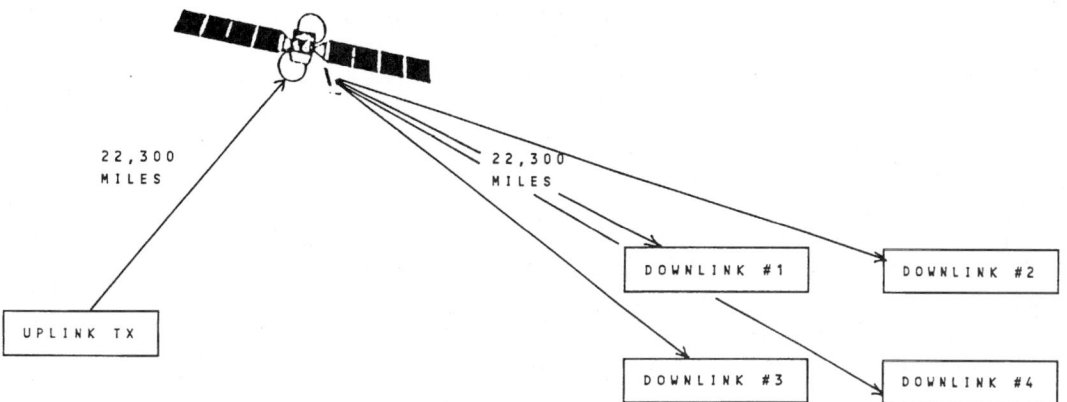

FIGURE 68.1 Typical satellite uplink/downlink system.

FIGURE 68.2 Teleport Chicago Earth station.

68.2 Components of the Earth Station

An Earth station will consist of one or more antennas, either stationary or motorized, equipped to receive and/or transmit signals to a communications satellite. Commercial satellites for voice, data, and video services operate at C- and Ku-band frequencies. C-band satellites operate with uplink frequencies from 5.9 to 6.4 GHz and downlink frequencies from 3.7 to 4.2 GHz. Ku-band satellites operate with uplink frequencies from 14.0 to 14.5 GHz and downlink frequencies from 11.7 to 12.2 GHz. The Ku-band satellites providing international service operate with slightly different frequency plans. Because of the shorter wavelength of the higher frequencies used, Ku-band satellite systems experience attenuation during periods of heavy rain. Overall link availability of Ku systems will be decreased depending on the annual rainfall amount in the geographic area where the Earth station is located. C-band transmissions are not appreciably affected by rainfall.

Some of the newer generation satellites have both C- and Ku-band capability. Some of their transponders can actually be configured by the satellite operator to receive a ground transmission on C-band and relay it back to Earth on Ku-band. Most C-band satellites operate with 24 channels or transponders each using a bandwidth of 40 MHz although only 36 MHz is actually used to provide for a guard band at each end of the transponder. The 36-MHz bandwidth will support one standard National Television Systems Committee (NTSC) television signal, including audio channels, using FM modulation. It will also support multiple analog or digital carriers for voice, data, or compressed video transmission. Ku-band satellites do not follow the same transponder plan as with C-band. Many Ku satellite transponders have wider bandwidths available. In many cases, two FM video carriers may operate side by side on one Ku-band transponder. Both C- and Ku-band satellites receive and transmit signals on alternate polarization to reduce the chance of interference from adjacent transponders in the frequency plan.

The antenna itself can vary in design but it is usually a parabolic dish with a feedhorn assembly. The feedhorn is the point at which satellite signals are collected from the antenna reflector surface to be fed to receive equipment. It is also the point where transmit signals are sent out to the reflector surface of the antenna to be focused into a beam aimed toward the satellite. The diameter of the reflective surface determines the gain and beamwidth of the antenna. The gain of the antenna becomes an important part of calculating a system link budget to determine the quality and availability of the signal.

Parabolic dish antennas can use many types of feedhorn assemblies. The most common are the prime-focus feed (Fig. 68.3) or the **Cassegrain** (Fig. 68.4) feed system. The prime focus system

FIGURE 68.3 Prime-focus feed system.

The Earth Station

FIGURE 68.4 Cassegrain feed system.

uses a feedhorn placed at the focal point of the reflector. Transmit or receive signals are directed from the reflector to the feed point. In the Cassegrain feed system, a subreflector is used to direct the transmit or receive signals to a feedhorn extending from the hub of the antenna. The actual feedhorn is hollow and the opening is covered with a protective window. The material used for the window is transparent to the satellite signal and keeps moisture or foreign objects from entering the assembly. Cassegrain feed systems are usually pressurized with nitrogen or dry air to help prevent moisture from degrading the signal. Pressurization systems should be monitored for any sign of leakage since damage to the feed window can occur from hail, birds, or vandalism. Even a small amount of moisture can affect transmit cross polarization, attenuate the transmit and receive signals, and generally degrade system performance. It is advisable to have an alarm tied into the pressurization system so that corrective action can quickly be taken if a feed window is damaged.

Feedhorn assemblies can be designed to accommodate up to four interface ports for transmitters and receivers. Four-port feedhorns are the most complex and are sometimes referred to as frequency reuse feeds. They contain a diplexing assembly at the base of the feedhorn to separate both transmit and receive frequencies. The diplexer also divides signals into two orthogonal ports for transmit and two for receive. This provides signals from both polarizations used by the satellite. Part of the diplexer assembly also contains a transmit reject filter to prevent signals that are being transmitted to the satellite from interfering with the signals being received within the feedhorn assembly. Some international and non-commercial satellite applications use circular or linear polarization for transmission, which requires a different diplexer design. The Federal Communications Commission (FCC) requires that satellite transmit antennas be able to maintain a radiation pattern that limits the amount of energy radiated in the sidelobes. This is to reduce the potential for interference to adjacent satellites in orbit. Antennas meeting this requirement are advertised as being 2° compliant and usually require adjustments of the subreflector (or feedhorn, in prime-focus systems) at the time of installation to meet the FCC requirement. To certify the antenna is 2° compliant, pattern testing is performed. The received signal level of an unmodulated carrier (transmitted from a different antenna) is observed as the antenna under test is moved in azimuth and elevation. The level of the sidelobes in relation to the main carrier is observed on the spectrum analyzer and a plot is made in both the azimuth and elevation axes to determine if 2° performance is being achieved. Transmit pattern testing may also be performed on the antenna in the same manner, but it is difficult to perform without causing interference to adjacent satellites as the antenna is moved in azimuth. Receive-only antennas are not required to be 2° compliant by law. However, they may suffer from some adjacent satellite interference if they are not compliant because of the closer spacing used by satellites in orbit today. Specific information with regard to FCC antenna performance requirements can be found in Part 25 of The Code of Federal Regulations.

Depending on the Earth station location, snow and ice can be a factor. The antenna reflector will collect ice and snow, reducing gain and affecting the transmit pattern and antenna gain. Earth station antennas can be equipped with deicing systems that keep the reflector clear during storm conditions. Deice systems use either electrical heating elements or gas-fired furnaces to supply heat to the rear of the reflector surface. Cassegrain antenna designs usually include electric heaters in the feedhorn and subreflector surfaces to keep those areas free of snow and ice. Automatic sensors can be mounted at the antenna to detect snow and ice and place the deice system into operation.

Received satellite signals are extremely low in level at the antenna feed point and require amplification before they can be fed to the satellite receiver or downconvertor. The amplifier must be capable of considerable gain with low noise, and it is usually mounted at the feedhorn to achieve the best results. This portion of the system is referred to as the **low-noise amplifier (LNA)** or **low-noise block downconverter (LNB)**. Two LNA or LNBs are required if both polarizations are used on the receive system. The LNA and LNB both use gallium-arsenide field effect transistors (GAsFET) to achieve the low-noise performance. The LNA amplifies the incoming satellite signal and passes it to the receiver at the original operating frequency. The LNB amplifies the received signal and then converts it to an L-band frequency of 950–1450 MHz. Noise performance of an LNA or LNB is measured in degrees kelvin. Today it is not uncommon to find LNAs or LNBs with noise temperatures of 30°K or better. A lower noise temperature increases overall system gain. The gain of the LNA or LNB along with its noise temperature are used when calculating total receive system gain.

Older system designs use an LNA as the heart of the receive system. Because the signal travels from the antenna to the receiver at the original operating frequency, a low-loss feedline has to be used.

Most newer satellite receivers are designed to operate with an L-band frequency input of 950–1450 MHz so an LNB can be used at the antenna. The L-band signal can be more easily fed from the LNB directly to the receiver using a conventional shielded coax. Both the LNA or LNB receive their DC power through the signal cable. The LNA/LNB power supply is located internally on some satellite receivers or it can be a separate component in the Earth station.

A newer style receiver with an LNB input can still be used with an LNA equipped antenna. A device called a block downconverter can be inserted between the LNA output and the receiver. The block downconverter will convert the LNA signal to the L-band frequencies required by the newer receiver.

For analog FM video transmission, a device called an **exciter** is used as the first stage in the uplink chain. The exciter performs baseband processing of the video signal and includes separate subcarrier modulators for the audio portion of the transmission. FCC regulations require that all video Earth stations include an **automatic transmitter identifier signal (ATIS)** in their carrier. This signal consists of an additional subcarrier modulated with Morse code, which contains the Earth station FCC callsign and telephone number. The ATIS signal is used to help protect against unintentional interference by a video uplink. It is used by the satellite controller to identify stray carriers which have inadvertently been left transmitting to the satellite. Newer video exciters include an onboard ATIS generator. There are outboard models available which may be added on to older units.

The processed video, audio, and ATIS subcarriers are applied to the modulator stage of the exciter. The intermediate frequency (IF) output of the modulator is then applied to the upconvertor, which is the last stage of the exciter. The upconvertor takes the IF signal and multiplies it to the desired uplink frequency, which is selected by the operator. The exciter output is then sent to the **high-power amplifier (HPA)**.

For data or compressed video signals, a modulator or **codec** is used instead of an exciter. The modulator or codec produces an IF output, which is fed to an outboard upconvertor. This **upconvertor** performs the same function as the upconvertor in a video exciter. The operator must tune the unit to the correct frequency for the desired satellite transponder. The output is then fed to the HPA. The HPA is the last device in the transmit chain prior to the antenna. This amplifier will develop the power necessary to send the signal from the Earth station antenna to the satellite. The exact amount of power required will vary depending on the type of modulation used, the transmission line loss, the gain of the antenna, and the characteristics of the satellite transponder. Most HPAs used for fixed Earth station applications use a **traveling-wave tube (TWT)** or **klystron** as the main amplifying component. TWTs are wide bandwidth devices and do not require any tuning before transmission. Klystrons are manufactured in single- or multiple-channel configurations. The multiple-channel klystrons have internal mechanical tuning cavities, which must be set to the desired transponder frequency before transmitting. Some Earth stations and some mobile uplink trucks will use a solid-state HPA mounted at the antenna feed point. This minimizes transmission line loss and reduces the output power requirement of the amplifier. Solid-state HPAs are not suited for all applications as they cannot develop the power levels that a TWT or Klystron amplifier can.

HPAs usually will have an input attenuator to control the level of drive from the exciter or upconvertor. This attenuator is used to set the power output level of the HPA. The attenuator may operate manually but some amplifiers use a motorized system for remote control of output power. The HPA output is connected to the antenna using a waveguide. Provision must be included in the system design for routing the HPA output to the appropriate antenna and polarization if the system is used to provide occasional service to multiple users. Provision should also be included for routing the HPA into a dummy test load. This helps facilitate testing of the HPA and reduces the chance that the system may be inadvertently left transmitting to the satellite. In redundant transmit systems, the backup HPA is usually connected to a dummy load unless a failure occurs on the primary unit.

68.3 Earth Station Site Selection

Earth stations in the U.S. are required to be licensed by the Federal Communications Commission before they can be placed into service. Part 25 of the Code of Federal Regulations should be referred to for the information needed to make an application for station authorization. Part of the procedure will include performing a frequency coordination analysis. This will determine the impact of a new Earth station to existing common carrier facilities. Local ordinances may also affect placement of transmit Earth stations for aesthetic, health, or safety reasons. Placement of a receive-only Earth station is limited by several factors. One is the potential for interference from ground-based communication systems. C-band downlink frequencies are also used by common carriers for point-to-point microwave communications. However, common carrier use of these frequencies has diminished in many areas as higher capacity fiber optic systems replace microwave radio as the means for transmission. This has made it less likely that that interference from these carriers will occur to Earth stations. Nonetheless, a spectrum analyzer should be used with a low-noise amplifier and handheld horn antenna to survey any potential Earth station site for external interference. There are companies that will perform this service for a fee during the initial site selection process. Ku-band downlink frequencies are also shared by common carriers for temporary local television transmission. Common carriers are secondary users of this band and are required by the FCC to coordinate transmissions with all satellite users in their service area. It is not uncommon, however, to experience interference from these terrestrial systems, especially in metropolitan areas where signals may be reflected into downlink antennas from nearby structures. Antennas can be equipped with mechanical shielding devices, which help reduce received interference. These devices may reduce overall system gain, but when used in combination with outboard filters in either the RF or IF stages of a receiver, interference can be reduced to a tolerable level.

In North America, the domestic satellite arc extends from 63° to 139° west longitude. A potential Earth station site should have a clear view of the entire satellite arc. Depending on geographic location, North American Earth stations can also see some of the international satellites. This will extend the arc clearance required.

Potential Earth station locations should be studied for adjacent trees or buildings that may cause obstructions. It is advisable to check with local zoning authorities to determine any future building plans near the proposed site location when making the initial visual inspection. An inclinometer or transit can be used to calculate elevation clearance. There are many shareware and commercially available computer programs that will calculate each satellites azimuth and elevation angle as viewed from the geographic coordinates of the proposed site.

Local authorities will also be able to provide information regarding zoning restrictions that may limit the placement of Earth station facilities. In some cases, it may only be a matter of constructing a suitable fence to comply with local ordinances but in others, a complete hazard assessment may need to be performed.

Earth stations may be colocated with the end user of the satellite service they provide. This is true for many video applications where programming originates or terminates at a television studio. In other cases, provision must be made for relaying the signal to the end user. This can be done through fiber optic cable, microwave radio, or a combination of both when route diversity is required. It is important to note that if the Earth station will use terrestrial microwave radio channels to receive or transmit satellite signals to the end user, the site selection process will need to include the feasibility of erecting a tower to support the microwave antennas. A microwave path study will then need to be performed to determine the tower height required. Path length and the frequency band used in a microwave interconnection will determine the potential impact of rain attenuation.

The frequency used for the interconnecting microwave facilities should be carefully planned. C-band Earth stations transmit in the range of 5.9–6.4 GHz, which is shared by telephone common carriers. Most Earth stations will not use microwave radio in this frequency band because it is likely to receive interference from the higher power signals transmitted from the Earth station to the satellite. Earth stations operated by common carriers usually use microwave channels in the 12-, 18-, or 23-GHz common carrier bands when fiber optic cable are not practical. Although this solves the problem of possible interference to the microwave channels from C-band transmissions, the 12-GHz common carrier band falls within some international satellite downlink frequency plans. Therefore, these microwave channels may interfere with some international satellite downlinks.

Local telephone and alternate access carriers should be contacted during the site selection process to determine the availability and cost for providing telephone and fiber optic interconnections to the Earth station. Plan ahead for future growth when contracting for these facilities to be brought into the site. At the time of installation, it is advisable to provide surge protection on all incoming telephone circuits to protect equipment from transients occurring elsewhere on the line due to lightning strikes, etc.

The local electric utility should be contacted to determine any special construction that will be required to bring sufficient power to the site. If the Earth station is to be manned, gas, water, and sewer utilities should be contacted to determine their installation costs into the site. If a hot-air deicing system is going to be used on the antenna, it is advisable to plan a route for the installation of the gas line prior to the beginning of construction.

68.4 Power Distribution

Reliability of the power system is another major concern for the Earth station. Depending on the application, different levels of primary and backup power systems may be required. To increase reliability, some Earth stations use two different sources of commercial power from the utility company that are fed from different grids. The electrical service entrance must be properly sized by calculating the load for all anticipated equipment. The calculated load should include ancillary systems such as antenna deicing and heating, ventilation, and air-conditioning (HVAC) systems. The electric service entrance should also be equipped with surge protectors to help guard against equipment damage from lightning strikes. The Earth station equipment building or shelter should be designed so that sufficient room is provided for the power distribution equipment required. It is sometimes helpful to provide separate distribution for critical and noncritical loads. When backup systems are used there are sometimes capacity limitations that require noncritical systems to remain off-line during power outages.

One or more outboard engine generators (Fig. 68.5) may be used to supply primary or standby power. If a generator is planned, the required capacity should be calculated by determining the initial total site load and the potential for growth. The maximum rate of fuel consumption per hour can then be calculated and a fuel storage tank can be sized accordingly. Consideration must be given to the placement of a fuel tank so that it remains accessible at all times for refueling. Local ordinances may stipulate whether the tank can be above or below ground and they will also dictate if special spill containment basins will be required. Newly installed underground tanks will require automatic leak detection systems and double-walled piping to comply with recent federal Environmental Protection Agency (EPA) rulings. As part of a standby generator system, an electrical or mechanical transfer switch will need to be installed. The transfer switch will sense a failure of the incoming utility power feed and automatically place the generator on-line. Because of the initial current demand placed on a generator at the moment of transfer, time-delay relays can be installed

FIGURE 68.5 Diesel engine generator.

to stagger the start-up time of high-current devices such as air-conditioner compressors. This arrangement can help keep the line frequency stable especially on generators that use a mechanical governor system. Electrical governors tend to provide better line regulation under varying loads.

Earth stations which utilize electric heaters on the antennas for deicing may wish to include an option for using either the commercial AC power or the standby generator to supply power. Using the generator for powering deicing systems is a good way to exercise the unit under an actual load condition.

It is a good idea to include remote alarms on the generator to monitor critical parameters such as oil pressure, water temperature, and fuel level. Most generators have built-in alarms that will initiate a shutdown if a dangerous situation occurs. However, few generators have factory installed prealarms to alert the user to a potential shutdown condition. It is also a good idea to have the generator vendor install prealarms so that the Earth station operator can take corrective action be taken before a shutdown occurs.

Critical Earth station components can be powered from an **uninterruptable power system (UPS)**. An UPS system will keep electrical service from being interrupted while backup generators come on-line during a power failure. As with a standby generator, the UPS system should have the capacity to accommodate all of the existing and anticipated site load. It may not be practical or economically feasible to run the entire Earth station from UPS power. In this case, the electrical distribution should be planned to route the UPS system to critical loads only.

Satellite transmission equipment is often designed with redundant power supplies. The UPS system can be used as one source of power to feed the equipment, and the utility power can be used as the other source. This will prevent an interruption to transmission if one power source fails.

An important part of any installation will be the grounding system. Grounding must conform to the National Electrical Code. A poor ground system becomes a safety hazard for personnel working at the site and also increases the risk of damage from a lightning strike. All antenna systems should be grounded and if a tower is used for microwave interconnection, it should also be grounded. Equipment racks in the Earth station shelter or building should be tied into the same ground system as all of the outdoor equipment.

68.5 Batteries

In some applications storage batteries are used to power equipment in the event of commercial AC failure. Storage batteries will also be part of an UPS system. Batteries must be properly housed and maintained to ensure safe, reliable operation. Instructions provided by the battery manufacturer should always be followed regarding the storage, installation, and maintenance of batteries.

The temperature of the batteries will affect their ability to deliver power under load and can also limit their overall life expectancy. In general, the batteries should be located in an area that will be kept between 72° and 78°F. Earth station design should include a separate room for batteries to provide a proper environment and safe operation. Local ordinances will address any special requirements for battery room designs such as spill containment or explosion-proof fixtures. Entrance to battery rooms should be restricted to maintenance personnel only and no one should be permitted to enter the room without proper safety clothing and eye protection. Special safety equipment including eye wash stations and clean-up equipment should be located in every battery room.

68.6 Antenna Foundation Requirements

The antenna manufacturer will provide specific information regarding the foundation requirements for each particular model. Because of the complexity of foundation design, it is recommended that a professional engineer be consulted. The information provided by the antenna manufacturer will include wind loading data and anchor bolt patterns. These should be provided to the foundation engineer before the design process begins.

FIGURE 68.6 Antenna pad construction.

For antennas requiring a reinforced concrete foundation or "pad," the engineer must take several items into consideration. Soil bearing capacity should be measured from samples obtained at the site. Once an excavation is made, additional samples may be taken from the base to determine if any special soil reinforcement is required prior to the construction of the foundation. The engineer will specify the minimum strength that the concrete must achieve after curing. This will be based on the analysis of wind load requirements and the soil condition. The design of the pad should also specify any steel reinforcements required.

The installation contractor should be familiar with local zoning and building ordinances and obtain all of the necessary permits before work begins. The contractor should also be provided with a survey of the site to ensure proper placement of the foundation. It is essential that the foundation is orientated correctly so that the desired portion of the satellite arc can be reached. Placement of each antenna foundation should be carefully planned so that any future antenna installations will not have a portion of their desired arc coverage obstructed by existing antenna installations.

As construction progresses, the foundation will be framed (Fig. 68.6) and a template can be laid out for the setting of anchor bolts before the concrete is poured. Most antenna manufacturers will provide a template to speed the layout of the anchor bolts. Test cylinders should be taken by the contractor while the concrete is being poured. These cylinders are normally tested by an independent laboratory after the concrete is cured to determine the strength it has achieved. The concrete strength needs to meet or exceed the figure specified by the design engineer before the antenna can be erected.

68.7 Site Heating, Ventilation, and Air-Conditioning

Adequate heating and cooling must be provided at the Earth station site to prevent equipment failures from occurring. In a small Earth station shelter, redundant window style air conditioners may be sufficient for cooling. If heating is required, a small wall-mounted electric furnace will usually work well. In larger facilities, designing the heating and cooling systems becomes more complicated. In many cases, it is wise to contract the services of an HVAC engineer to assist with the project. Some level of cooling redundancy is required in large facilities since the loss of air conditioning during the summer months will almost certainly cause equipment to overheat in a short amount of time. In many cases, heating a large facility in the winter months is not even required because the heat given off by the equipment may be utilized. The heat load may be such that the Earth station will require air conditioning all year long. When calculating the site heat load, be sure to include the heat given off by the HPA dummy loads.

Generally, the HPAs used in Earth stations are air-cooled devices. This requires that a minimum amount of air flow be maintained through the cabinet to adequately cool the TWT or klystron. Although many of the newer domestic satellites require less operating power than earlier designs, high-power amplifiers tend to produce the same amount of heat output whether they are operating at 30 or 3000 W of power. In large Earth station installations, multiple amplifier exhausts should be ducted directly outdoors and may require the use of an external exhaust fan. This type of system will most likely use an external air intake fan, as well.

HVAC contractors can also include ductwork in the system design to take a portion of the HPA exhaust air and heat other parts of the building. A concern with large air handler systems is that any failure of an intake or exhaust motor will result in rapid heat buildup in the amplifier and increases the risk of failure. Intake louvers should contain air filters, which must be serviced regularly. If filters become clogged, air flow can drop significantly, which also increases the risk of failure. An air flow alarm should be a part of the air handling system design to warn the operator of a problem.

68.8 Safety Considerations

There are many safety concerns when operating an Earth station. The Earth station operator should ensure that the facility is compliant with current Occupational Safety and Health Administration (OSHA) regulations. It is worth pointing out some special concerns that apply to Earth station safety. Fire suppression equipment must be provided. Local ordinances may dictate the type of fire suppressant to be used, and special care must be taken to select a suppressant that can safely be used on live electrical equipment. Unmanned sites should have remote alarms for smoke or fire detection which are monitored 24 h a day. The local fire department should visit the Earth station so that they can document any special access requirements and be trained on the special requirements if a fire occurs in the transmission equipment. Operating personnel should be trained on the use of fire extinguishers and any other suppressant systems on site.

Earth station facilities present an RF radiation exposure hazard to operation and maintenance personnel. HPAs are provided with interlocks to protect maintenance personnel from the risk of high voltage and exposure to RF radiation. These interlocks should be tested regularly. Warning lights can be mounted at antenna foundations, which automatically illuminate during transmission. This will help alert maintenance personnel working in the vicinity of the antenna that the system is active.

Klystron tubes contain beryllium oxide (BeO). If the klystron is physically damaged, a risk of exposure to toxic fumes exists. Maintenance personnel should always follow the manufacturers recommended handling procedures to minimize the risk of damage to the klystron.

The HPAs contain voltages which can be lethal upon contact. During maintenance, it is sometimes necessary to make adjustments to an HPA while the high voltage is present. Maintenance personnel should never perform work on an energized HPA alone.

Motorized antennas can be remotely controlled and may move without warning. Many antenna control systems provide for local overrides at the foundation to be used by maintenance personnel. The override prevents the antenna motors from engaging while work is being performed. These overrides should be used in addition to standard OSHA lockout/tagout procedures to fully protect maintenance personnel from possible injury.

The Earth Station

Finally, site security should be considered. Most facilities will have a fence to control access to the site and in use special lighting for added security. If the site is unmanned, it is wise to have remote intrusion alarms report to a 24-h monitoring center to help prevent theft or vandalism.

68.9 Operation and Maintenance

With limited exception, anytime a satellite transponder is accessed by an Earth station, the satellite controller must be notified by the operator performing the uplink. For Earth stations that carry occasional video services, the Earth station may need to contact the satellite controller many times per day. Each satellite carrier has specific transponder access procedures. In general, the following guidelines should be followed by Earth stations when accessing a video transponder:

1. Ensure the antenna is properly oriented on the correct satellite.
2. Ensure that the exciter and HPA are properly tuned and routed to a dummy load or disabled.
3. Set the HPA for a power level of approximately 10 W.
4. Verify the correct transmit polarization setting.
5. Set the exciter or upconvertor to produce an unmodulated carrier.
6. Contact the satellite controller and request permission to access the transponder. Verify the service order and time of service with the controller.
7. Begin transmission at low power when authorized by the controller. Be prepared to immediately cease transmission if commanded to do so by the controller.
8. The controller will verify proper frequency and polarization setting. Be prepared to rotate the antenna polarization, if asked to do so.
9. Gradually increase output power to operating level when permitted to by the controller.
10. Apply modulation and ATIS carrier when permitted to do so by the controller.
11. Provide the Earth station telephone number where you may be reached at anytime during the transmission.
12. Contact the satellite controller with the good-night time when service is terminated.

The majority of unintentional satellite interference occurs from operator error. The chance of interference is greatly reduced when Earth station operators are properly trained and follow the access procedures for every transmission.

Maintenance of the Earth station is largely a matter of following equipment manufacturers recommended service intervals. The major source of downtime results from equipment failure due to poor maintenance practices. Antenna drive system lubrication should be checked at regular intervals, especially when an antenna is not moved for an extended period of time. HVAC and equipment air filters should be checked weekly and changed as necessary. Many equipment failures have resulted from overheating due to poor air circulation. Equipment fans or blowers should be regularly checked for the same reason. Storage batteries need to be serviced at the manufacturers suggested interval for proper cell voltage and specific gravity.

68.10 Conclusion

New spacecraft are being launched and the demand for transmission service from Earth station facilities continues to increase. Proper planning during the design and construction of an Earth station facility is the key to starting a successful uplink or downlink operation. A well planned and maintained Earth station staffed by trained operators will continue to be a part of a vital link to global telecommunications.

Defining Terms

Automatic transmitter identification system (ATIS): This is a Morse code modulated subcarrier that is required by the FCC to be transmitted with all video carriers.

Cassegrain: This is a style of satellite antenna that uses two reflectors to direct energy to the feedhorn. One reflector is the main body of the antenna, which directs energy to a smaller reflector mounted at the focal point of the main reflector. The energy is then directed into the feedhorn itself.

Codec: A codec is an acronym for coder/decoder. The coder takes an analog voice, data, or video source and converts it to a digital output for transmission. Conversely, a decoder takes a received digital bit stream and converts it back into an analog signal.

Earth station: An Earth station is a ground-based communications facility that is made up of the transmitters, receivers, and antennas used to relay signals to or from a communications satellite.

Exciter: An exciter is a device that takes a voice, data or video signal and processes it for uplink. The exciter contains a modulator to convert the incoming signal to an intermediate frequency (IF). It also contains an upconvertor to take the processed IF and multiply it to the radio frequency (RF) desired. A video exciter usually contains audio subcarrier modulators to be combined with the video signal.

Footprint: The footprint is a satellite transponder's coverage area as seen from space. The size of the footprint and the level of the signal received in different areas of the footprint will vary depending on the design of the satellite and its orbital location.

Geosynchronous: An orbit at approximately 22,300 mi. above the Earth's equator. This allows satellites to rotate at the same speed as the Earth itself. The satellites then appear stationary from a point on the ground.

High-power amplifier (HPA): The last device before the antenna in most installations. This amplifier develops the power necessary to send a signal to the satellite.

Klystron: A transmitting tube used at ultra high frequency (UHF) and microwave frequencies to produce relatively high-power output over a wide bandwidth.

Low-noise amplifier (LNA): An high-gain amplifier usually of GAsFET design that amplifies the incoming satellite signal to a level suitable to be fed to the receiver. It is designed to add as little noise as possible to the signal during the amplification process.

Low-noise block downconverter (LNB): The same as an LNA except it converts the incoming satellite signal to an L-band frequency (950–1450 MHz) for use by newer style receivers.

Transponder: The device internal to the satellite that receives a ground-based uplink transmission and retransmits it back to Earth on the corresponding downlink frequency.

Traveling-wave tube (TWT): A medium-to-high power transmitting tube used for microwave and satellite transmission. It operates in a linear mode with wide bandwidth.

Upconvertor: A device used to multiply an IF signal to the desired satellite operating frequency. It can be internal to the exciter or a seperate outboard device. The upconvertor output is fed to the HPA.

Uninterruptable power system (UPS): This is a device which supplies continous AC power to the load during a failure of the normal power source. It utilizes a large standby battery bank and invertor system to rpoduce AC power when the normal source fails.

Further Information

Michael J., Downey, M.J., and Edington, B.G. 1993. *Engineers Handbook for Earth Station Design and Implementation*, M/R Communications.

Weinhouse, N. 1992. *Satellite Uplink Operator Training Course*, Norman Weinhouse Associates. Federal Communications Commission, Part 25, Code of the Federal Regulations, Office of the Federal Register, National Archives and Records Administration, U.S. Government Printing Office.

The first two publications cover the Earth station in greater detail and the third contains the rules and regulations for licensing and operating an Earth station.

69
Satellite Transmission Impairments

Louis J. Ippolito Jr.
Stanford Telecom

69.1	Introduction	935
69.2	Attenuation on Slant Paths	936
	Gaseous Attenuation • Rain Attenuation • ITU-R Rain Attenuation Model • Crane Global Rain Attenuation Prediction Model • Cloud and Fog Attenuation	
69.3	Depolarization	942
	Rain Depolarization • Ice Depolarization	
69.4	Radio Noise	944
69.5	Scintillation	945
	Amplitude Scintillation Prediction	
69.6	Summary	947

69.1 Introduction

The effects of the Earth's atmosphere on radiowaves propagating between Earth and space is a constant concern in the design and performance of space communications systems. The problems become more acute for systems operating in the bands above 10 GHz, where radiowave links can be adversely affected by atmospheric gases (primarily oxygen and water vapor), rain, clouds, fog, and scintillation. These conditions, when present alone or in combination on the Earth–space radiowave link, can cause uncontrolled variations in signal amplitude, phase, polarization, and angle of arrival, which result in a reduction in the quality of analog transmissions and an increase in the error rate of digital transmissions.

Propagation impairments on radiowaves above 10 GHz are primarily caused by constituents in the troposphere, which extends from the Earth's surface to about 20 km. Degradations induced in the ionosphere generally affect frequencies well below 10 GHz, and the ionosphere is essentially transparent to the radiowave at frequencies above 10 GHz.

Even apparent clear sky conditions can produce propagation effects, which can degrade or change characteristics of the transmitted radiowave. Gases present in the Earth's atmosphere, particularly oxygen and water vapor, interact with the radiowave and reduce the signal amplitude by an absorption process. Turbulence or rapid temperature variations in the transmission path can cause amplitude and phase scintillation or depolarize the wave. Clouds, fog, dirt, sand, and even severe air pollution can cause observable propagation effects. Background sky noise will always be present and contributes directly to the noise performance of the communications receiver system.

The degrading effect of precipitation in the transmission path is a major concern associated with satellite communications, particularly for those systems operating at frequencies above 10 GHz. At those frequencies, absorption and scattering caused by rain (and to a lesser extent hail, ice crystals, or wet snow) can cause a reduction in transmitted signal amplitude (attenuation), which can reduce the reliability and/or performance of the communications link. Other effects can be generated by precipitation in the Earth–space path. They include depolarization, rapid amplitude and phase fluctuations (scintillation), antenna gain degradation, and bandwidth coherence reduction.

The major factors affecting Earth–space paths in the bands above 10 GHz are as follows:

- Gaseous attenuation
- Hydrometeor attenuation
- Depolarization
- Radio noise
- Scintillation

Extensive experimental research has been performed on the direct measurement of propagation effects on Earth–space paths, beginning in the late 1960s, with the availability of propagation beacons on geostationary satellites [Ippolito, 1986]. Satellites involved included the NASA Applications Technology Satellites (ATS-5 and ATS-6); the Canadian/U.S. Communications Technology Satellite (CTS); the domestic U.S. COMSTAR communications satellites; the ETS-II, CS, and BSE satellites of Japan; SIRIO and ITALSAT of (Italy), and the OTS (European Space Agency) Olympus Satellite. The recently launched NASA Advanced Communications Technology Satellite (ACTS) is providing continued measurements in the 20 and 30 GHz bands.

Concise modeling and prediction procedures for the propagation factors, based on data and measurements from the satellite beacons, have evolved over several iterations and have been successfully applied in the design and performance of satellite telecommunications systems for several years. The following sections provide a review and assessment of the available propagation models and their application to satellite communications systems.

69.2 Attenuation on Slant Paths

A radiowave propagating through Earth's atmosphere can experience a reduction in signal level from one or more of several constituents that may be present in the path. The primary constituents that can cause radiowave attenuation on a slant path are atmospheric gases and hydrometeors.

Gaseous atmospheric attenuation is caused primarily by oxygen and water vapor, and is an absorption process. Hydrometeor attenuation experienced by a radiowave involves both absorption and scattering processes. Rain attenuation can produce major impairments in Earth–space communications, particularly in the frequency bands above 10 GHz. Cloud and fog attenuation are much less severe than rain attenuation; however, they must be considered in link calculations, particularly for frequencies above 15 GHz. Dry snow or ice particle attenuation is usually so low that it is unobservable on space communications links operating below 30 GHz.

Gaseous Attenuation

Gaseous attenuation is the reduction in signal amplitude caused by the gaseous constituents of the atmosphere that are present in the transmission path. Only oxygen and water vapor have observable resonance frequencies in the radiowave bands used for space communications. Oxygen has a series of very close absorption lines near 60 GHz and an isolated absorption line at 118.74 GHz. Water vapor has lines at 22.3, 183.3, and 323.8 GHz. Oxygen absorption involves magnetic dipole changes, whereas water vapor absorption consists of electric dipole transitions between rotational states.

Satellite Transmission Impairments

The former International Radio Consultative Committee (CCIR) [now the Radiocommunication Sector of the International Telecommunications Union, (ITU-R)], developed a prediction method to calculate the median gaseous attenuation expected on a slant path, for a given set of parameters [ITU-R, 1995a]. The ITU-R method is based on the specific attenuation and temperature dependence algorithms developed by Gibbons [1986]. The method is applicable up to 350 GHz, except for the high oxygen absorption bands around 60 GHz.

The input parameters required for the calculation are: frequency f (GHz); path elevation angle θ (°); height above mean sea level of Earth terminal, h_s (km); surface temperature T (°C); and water vapor density at the surface, for the location of interest, ρ_w (g/m³). If ρ_w is not available from local weather services, representative median values can be obtained from the ITU [ITU-R, 1995b], that presents global maps with average monthly values of ρ_w for regions of the world.

Four steps are required to calculate the total gaseous absorption with the ITU-R model:

1. Determine the *specific attenuation* values for oxygen and water vapor.
2. Determine the *equivalent heights* for oxygen and water vapor.
3. Adjust for *surface temperature*.
4. Determine the *total slant path gaseous attenuation* through the atmosphere.

The specific attenuation at the surface for dry air (oxygen), γ_0, (at a surface pressure of 1013 mbar) is determined from the following.

For $f < 57$ GHz,

$$\gamma_0 = \left[7.19 \times 10^{-3} + \frac{6.09}{f^2 + 0.227} + \frac{4.81}{(f^2 - 57)^2 + 1.50} \right] f^2 \times 10^{-4} \text{ dB/km} \quad (69.1)$$

For $63 < f < 350$ GHz,

$$\gamma_0 = \left[3.79 \times 10^{-7} f + \frac{0.265}{(f^2 - 63)^2 + 1.59} + \frac{0.028}{(f^2 - 118)^2 + 1.47} \right]$$
$$\times (f + 198)^2 \times 10^{-3} \text{ dB/km} \quad (69.2)$$

The specific attenuation for water vapor, γ_w, is found from

$$\gamma_w = \left[0.05 + 0.0021 \rho_w + \frac{3.6}{(f^2 - 22.2)^2 + 8.5} + \frac{10.6}{(f^2 - 183.3)^2 + 9.0} \right.$$
$$\left. + \frac{8.9}{(f - 325.4)^2 + 26.3} \right] f^2 \rho_w \times 10^{-4} \text{ dB/km} \quad (69.3)$$

The total slant path gaseous attenuation through the atmosphere, A_g, is found from

$$A_g = \frac{\gamma_0 h_0 e^{-\frac{h_s}{h_0}} + \gamma_w h_w}{\sin \theta} \text{ dB} \quad (69.4)$$

where

$h_0 = 6$ km   for $f < 57$ GHz

$h_0 = 6 + \dfrac{40}{(f - 118.7)^2 + 1}$ km   for $63 < f < 350$ GHz

$h_w = h_{w_0} \left[1 + \dfrac{3}{(f - 22.2)^2 + 5} + \dfrac{5}{(f - 183.3)^2 + 6} + \dfrac{2.5}{(f - 325.4)^2 + 4} \right]$ km

The factor h_{w_0} is 1.6 km for clear weather and 2.1 km during rain. The higher value equivalent height factor for rain accounts for the observed increase in the altitude supporting water vapor during the occurrence of rain.

Figure 69.1 shows the total gaseous attenuation calculated from the ITU-R Gaseous attenuation model for a range of frequencies and for elevation angles from 5° to 45°. A temperature of 20°C and water vapor density of 7.5 g/m³ were assumed. These values correspond to a moderate (42% relative humidity), clear day. The elevation above sea level was set at 0 km.

Rain Attenuation

The attenuation of a radiowave of wavelength λ propagating through a volume of rain of extent L is given by [Ippolito, 1986]

$$a \text{ (dB)} = 4.343 \int_0^L \left[N_0 \int Q_t(r, \lambda, m) e^{-\Lambda r} \, dr \right] d\ell \tag{69.5}$$

where Q_t is the total attenuation cross section for a water drop of radius r and refractive index m; N_0 and Λ are empirical constants from the rain drop size distribution, $n(r) = N_0 e^{-\Lambda r}$. The integration over l is taken over the extent of the rain volume in the direction of propagation. Both Q_t and the rain drop size distribution will vary along the path, and these variabilities must be included in the integration process. A determination of the variations along the propagation path is very difficult to obtain, particularly for slant paths to an orbiting satellite. These variations must be approximated or treated statistically for the development of useful rain attenuation prediction models.

Over the past several years extensive efforts have been undertaken to develop reliable techniques for the prediction of path rain attenuation for a given location and frequency, and the availability of satellite beacon measurements has provided a database for the validation and refinement of the prediction models. Virtually all of the prediction techniques use surface measured rain rate as the statistical variable and assume an aR^b approximation between rain rate R and rain induced attenuation. This approximation shows excellent agreement with direct Mie calculations of specific attenuation for both spherical and nonspherical rain drops. The a and b coefficients for nonspherical drops, that are more representative of a true rain drop, can be calculated by applying techniques similar to the Mie calculations for a spherical drop. An extensive tabulation of a and b coefficients for nonspherical (oblate spheroidal) drops is available from the ITU-R [1995d].

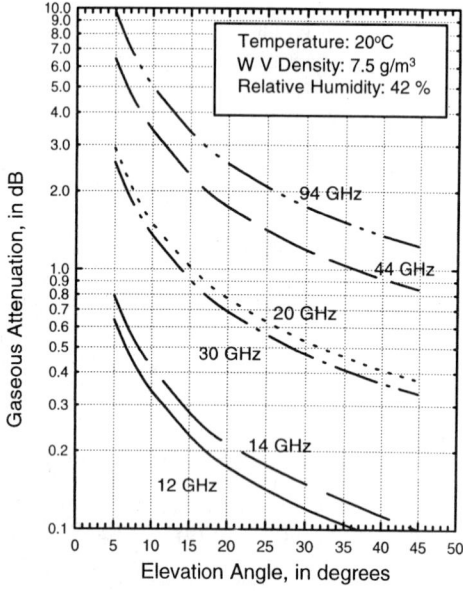

FIGURE 69.1 Total gaseous attenuation for a satellite link, under moderate humidity conditions.

The two major rain attenuation prediction models used for the determination of rain attenuation on space paths are the ITU-R Rain model [ITU-R, 1995a] and the Crane global model [Crane, 1980, 1982]. Both models require as inputs: operating frequency f, elevation angle to the satellite θ, polarization tilt angle δ, ground station elevation above sea level h_s, ground station latitude ϕ,

Satellite Transmission Impairments

and the annual point rain rate distribution $R(p)$. The ITU-R Rain Attenuation model develops the annual attenuation distribution from the 0.01% rain rate value, whereas the Crane model utilizes the full rain rate distribution, from 0.001–2%, to develop the attenuation distribution. Both models have demonstrated good prediction results when compared with measured data from the global database of the ITU-R, with the global model showing a slightly better performance for locations in the U.S.

ITU-R Rain Attenuation Model

The International Telecommunications Union (ITU) first adopted a procedure for the prediction of attenuation caused by rain at its XVth Plenary Assembly in Geneva in February 1982. This result was preceded by several years of intense deliberations by representatives of ITU-R Study Group V from several nations. The procedure provided the basis for rain attenuation calculations required for international planning and coordination meetings, and regional and world administrative radio conferences. The current ITU-R model is based on a global map of 14 rain climatic zones with associated rainfall intensity cumulative distributions for each of the zones [ITU-R, 1995a]. Average annual rain rates are given for exceedance times from 0.001 to 1.0%.

The ITU-R model requires the rain rate exceeded for 0.01% of an average year, $R_{0.01}$, for the ground location of interest. If this rain rate is not available from local weather data sources, it is determined from the ITU-R global climate zone map and the corresponding rain rate tables [ITU-R, 1995c]. The model assumes that the horizontal extent of the rain is coincident with the ambient 0°C isotherm height, that will vary with location, season of the year, time of day, etc. The average value of the 0°C isotherm height h_r is determined from [ITU-R, 1995e],

$$
\begin{aligned}
h_r &= 6.725 - 0.075\phi & &\text{for} & \phi &> 23° & &\text{Northern Hemisphere} \\
h_r &= 5 & & & 0 &\leq \phi \leq 23° & &\text{Northern Hemisphere} \\
h_r &= 5 & & & 0 &\leq \phi \leq -21° & &\text{Southern Hemisphere} \\
h_r &= 7.1 + 0.1\phi & & & -71° &\leq \phi \leq -21° & &\text{Southern Hemisphere} \\
h_r &= 0 & & & \phi &< -71° & &\text{Northern Hemisphere}
\end{aligned}
\quad (69.6)
$$

where ϕ is the latitude of the ground location, in degrees north (+) or south (−).

The slant path length L_s is found as

$$
L_s = \frac{2(h_r - h_s)}{\left(\sin^2\theta + \frac{(h_r - h_s)}{4250}\right)^{1/2} + \sin\theta}
\quad (69.7)
$$

A slant path length reduction factor is then applied to account for the horizontal nonuniformity of rain for the 0.01% of the time condition. The reduction factor r_p is given by

$$
r_p = \frac{1}{1 + \dfrac{L_s \cos\theta}{35 e^{-0.015 R_{0.01}}}}
\quad (69.8)
$$

The attenuation, in decibel, exceeded for 0.01% of an average year is then obtained from

$$
A_{0.01} = a R_{0.01}^b L_s r_p
\quad (69.9)
$$

where a and b are frequency dependent coefficients available from the ITU [1995d].

The attenuation exceeded for other percentages, p, of an average year, can be determined from

$$
A_p = 0.12 A_{0.01} p^{-(0.546 + 0.043 \log p)} \qquad 0.001\% \leq p \leq 1\%
\quad (69.10)
$$

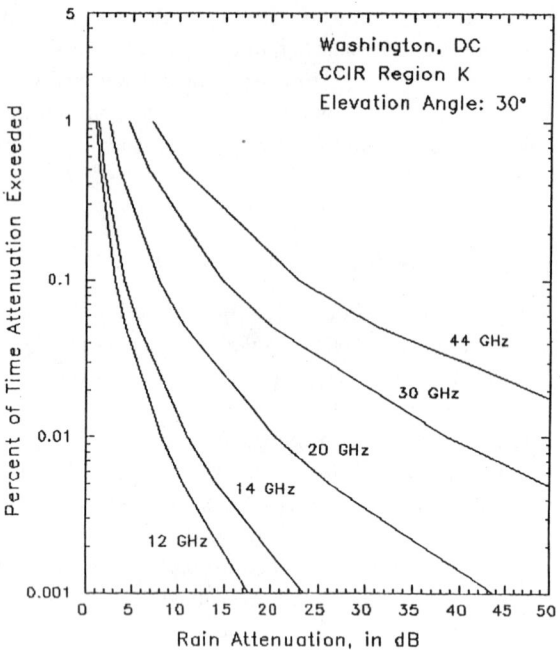

FIGURE 69.2 Rain attenuation distributions from the ITU-R rain attenuation prediction model.

Figure 69.2 shows rain attenuation distributions determined from the ITU-R prediction model for a ground terminal in Washington, D.C. (ITU-R rain climate zone K), at an elevation angle of 30°, for frequencies from 12 to 44 GHz. The distributions are representative of the prediction curves that can be developed from the ITU-R rain attenuation model.

Crane Global Rain Attenuation Prediction Model

The first published model, which provided a self-contained prediction procedure for global application, was developed by Crane [1980]. The Crane-based model, usually referred to as the global model, is based on the use of geophysical data to determine the surface point rain rate, point-to-path variations in rain rate, and the height dependency of attenuation, given the surface point rain rate or the percentage of the year the attenuation value is exceeded. The model also provides estimates of the expected year-to-year and station-to-station variations of the attenuation prediction for a given percent of the year.

The global model, like the ITU-R model, provides global climate regions, but the 12 climate regions for the global model differ from the ITU-R model regions previously discussed.

The mean slant-path rain attenuation at each probability of occurrence, p, is determined as follows.

For $0 < D \leq d$,

$$A(p) = \frac{aR(p)^b}{\cos\theta}\left[\frac{e^{UbD}-1}{Ub}\right] \quad (69.11)$$

For $d < D \leq 22.5$,

$$A(p) = \frac{aR(p)^b}{\cos\theta}\left[\frac{e^{Ubd}-1}{Ub} - \frac{X^b e^{Ybd}}{Yb} + \frac{X^b e^{YbD}}{Yb}\right] \quad (69.12)$$

Satellite Transmission Impairments

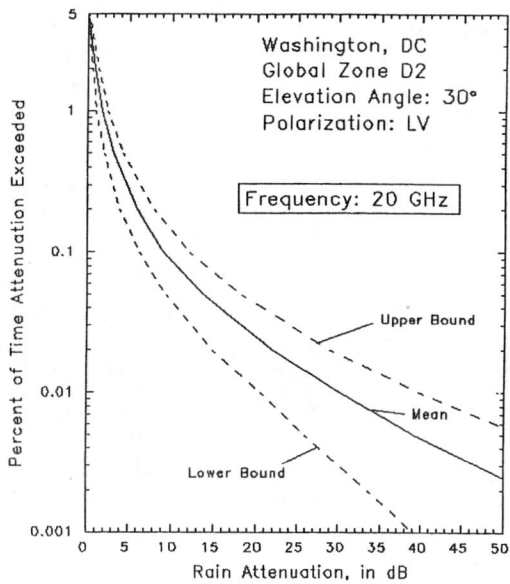

FIGURE 69.3 Rain attenuation distribution from the global rain attenuation prediction model.

For $D > 22.5$, calculate $A(p)$ with $D = 22.5$, and the rain rate $R'(p)$ at the probability value

$$p' = \left(\frac{22.5}{D}\right)p \tag{69.13}$$

Figure 69.3 shows the rain attenuation distributions determined from the global model for a ground terminal in Washington, D.C. (global rain climate zone D2), at an elevation angle of 30° and an operating frequency of 20 GHz. The mean distribution and the upper and lower bounds are shown on the figure.

The global model and the ITU-R model will often give different predictions. The global model tends to give slightly better correlation with measured data in the U.S., whereas the ITU-R model shows better results for Europe. Both models are very effective and give similar results, however, in accessing the effects of rain attenuation up to 5–8 dB, where most links are designed to operate.

Cloud and Fog Attenuation

Although rain is the most significant hydrometeor affecting radiowave propagation in the frequency bands above 10 GHz, the influence of clouds and fog can also be present on an Earth–space path. Clouds and fog generally consist of water droplets of less than 0.1 mm in diameter, whereas raindrops typically range from 0.1 to 10 mm in diameter. Clouds are water droplets, not water vapor; however, the relative humidity is usually near 100% within the cloud. High-level clouds, such as cirrus, are composed of ice crystals, which do not contribute substantially to radiowave attenuation but can cause depolarization effects. The average liquid water content of clouds varies widely, ranging from 0.05 to over 2 g/m^3. Peak values exceeding 5 g/m^3 have been observed in large cumulus clouds associated with thunderstorms; however, peak values for fair weather cumulus are generally less than 1 g/m^3.

A detailed study of the radiowave propagation effects of clouds at various locations in the U.S. by Slobin resulted in the development of a cloud model that provides estimates of cloud attenuation and noise temperature on radiowave links [Slobin, 1982]. Extensive data on cloud characteristics, such as type, thickness, and coverage, were gathered from twice-daily radiosonde measurements and hourly temperature and relative humidity profiles at stations in the contiguous U.S., Alaska, and Hawaii. Regions of statistically consistent clouds were determined, and resulted in the definition of 15 cloud regions. The Slobin model defines 12 cloud types based on liquid water content, cloud thickness, and base heights above the surface. Several of the more intense cloud types include two cloud layers, and the combined effects of both are included in the model. Several other cloud attenuation models have been published, and they provide results similar to the Slobin model for frequencies up to about 40 GHz [Gerace and Smith, 1990].

Fog results from the condensation of atmospheric water vapor into droplets that remain suspended in air. There are two types of fog, advection fog, which forms in coastal areas when warm, moist air moves over colder water, and radiation fog, which forms at night, usually in valleys, low areas, and along rivers. Typical fog layers are only 50–100 m in height, and hence the extent of fog on a slant path is very small, even for low elevation angles. Fog attenuations are much less than 1 dB for frequencies up to 100 GHz. A prediction model, developed by Altshuler, is available for the estimation of fog attenuation for frequencies above 30 GHz [Altshuler, 1984 rev. 1986].

69.3 Depolarization

Depolarization involves a change in the polarization characteristics of a radiowave caused by (1) hydrometeors, primarily rain or ice particles; or (2) multipath propagation. A depolarized radiowave will have its polarization state altered such that power is transferred from the desired polarization state to an undesired orthogonally polarized state, resulting in interference or crosstalk between the two orthogonally polarized channels. Rain and ice depolarization can be a problem in the frequency bands above about 12 GHz, particularly for frequency reuse communications links, which employ dual independent orthogonal polarized channels in the same frequency band to increase channel capacity. Multipath depolarization is generally limited to very low elevation angle space communications, and will be dependent on the polarization characteristics of the receiving antenna.

The cross-polarization discrimination *XPD* is defined for linearly polarized waves as

$$XPD = 20 \log \frac{|E_{11}|}{|E_{12}|} \qquad (69.14)$$

where E_{11} is the received electric field in the copolarized (desired) direction and E_{12} is the electric field converted to the orthogonal cross-polarized (undesired) direction.

Rain Depolarization

Rain induced depolarization is produced from a differential attenuation and phase shift caused by nonspherical raindrops. As the size of rain drops increase, their shape tends to change from spherical (the preferred shape because of surface tension forces) to oblate spheroids with an increasingly pronounced flat or concave base produced from aerodynamic forces acting upward on the drops [Pruppacher and Pitter, 1971]. Furthermore, raindrops may also be inclined to the horizontal (canted) because of vertical wind gradients. Therefore, the depolarization characteristics of a linearly polarized radiowave will depend significantly on the transmitted polarization angle. When measurements of depolarization observed on a radiowave path were compared with rain attenuation measurements coincidently observed on the same path, it was noted that the relationship between the measured statistics of *XPD* and copolarized attenuation *A* for a given probability *p* could be well approximated by the relationship

$$XPD_p(\text{dB}) = U - V \log[A_p(\text{dB})] \qquad (69.15)$$

where U and V are empirically determined coefficients, which depend on frequency, polarization angle, elevation angle, canting angle, and other link parameters. This discovery was similar to the aR^b relationship observed between rain attenuation and rain rate. For most practical applications semiempirical relations can be used for the U and V coefficients. The coefficients, as specified by the ITU-R [1995a], are valid for freqeuncies from 8 to 35 GHz, and are determined from

$$\begin{aligned} U &= 30 \log f - 10 \log[0.516 - 0.484 \cos(4\delta)] - 40 \log(\cos\theta) + 0.0052\,\sigma^2 \\ V &= 12.8 f^{0.19} \quad \text{for } 8 \le f \le 206 \text{ Hz} \\ &= 22.6 \quad \text{for } 20 \le f \le 356 \text{ Hz} \end{aligned} \qquad (69.16)$$

where f (GHz) is the frequency, δ (°) is the tilt angle of the polarization with respect to the horizontal (for circular polarization $\delta = 45°$), and θ (°) is the elevation angle of the path. The elevation angle must be $\le 60°$ for this relationship to be valid. Here, θ(°) is the standard deviation of the raindrop canting angle.

Ice Depolarization

Depolarization due to ice crystals has been observed on Earth–space paths from measurements with satellite beacons. This condition is characterized by a strong depolarization accompanied by very low copolarized attenuation. Also, abrupt changes in the XPD have been observed to coincide with lightning discharges in the area of the slant path, suggesting a change in the alignment of the ice crystals. Ice depolarization occurs when the ice particles are not randomly oriented but have a preferred alignment or direction. Ice crystals most often appear in the shape of needles or plates, and their major axes can abruptly be aligned by an event such as a lightning discharge. When this occurs, differential phase changes in the radiowave can cause an increase in the level of the cross polarized signal level, with little or no change to the copolarized signal. Ice crystals produce nearly pure differential phase shift, without any accompanying differential attenuation, which accounts for the depolarization effects observed in the absence of copolarized attenuation. Ice depolarization effects have been observed in the U.S., Europe, and Japan, at frequencies from 4 to 30 GHz. Ice depolarization often precedes the onset of severe thunderstorm conditions, and XPD degradations of 5–8 dB have been observed. Statistically, however, it occurs for only a very small percent of the time that XPD degradations are observed (1–3%), and so it is generally not a severe impediment to system performance.

The contribution of ice to the XPD distribution experienced on a link can be estimated by a procedure developed by the ITU [1995a]. The total XPD, caused by rain and ice, for p% of the time, XPD_p, is determined from

$$XPD_p = XPD_{\text{rain}} - \left[\frac{0.3 + 0.1 \log p}{2} XPD_{\text{rain}}\right], \text{dB} \qquad (69.17)$$

where XPD_{rain} is the XPD distribution caused by rain, in decibel, either measured or determined from the procedure in the previous section. The term in brackets is the contribution due to the presence of ice on the path.

Figure 69.4 shows the XPD distributions for rain XPD_{rain} and rain plus ice XPD_p calculated from the rain attenuation distribution of Fig. 69.3, for Washington D.C., at a frequency of 20 GHz. The contribution from ice is seen to be very small, except for for high percent values, that is, for 0.1% or

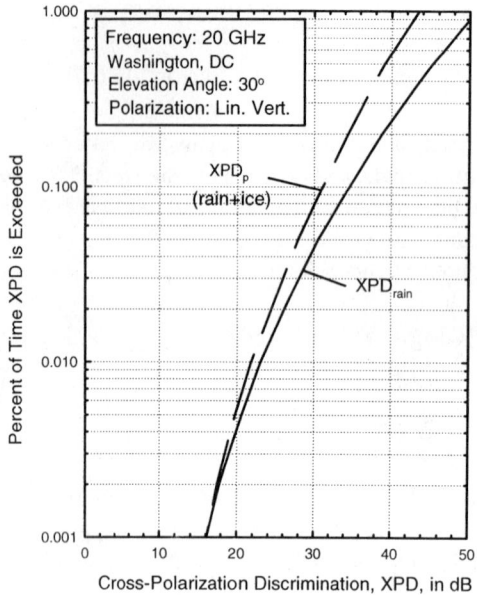

FIGURE 69.4 Effect of ice on cross-polarization discrimination distribution.

greater. At these percentage values the *XPD* is already high, that is, >30 dB, so system performance will not be significantly degraded by a 1 or 2 dB decrease in *XPD*.

69.4 Radio Noise

The same constituents of the Earth's atmosphere (oxygen, water vapor, rain, clouds, etc.) that cause absorptive attenuation on a radiowave will also produce thermal noise emissions, which are a function of the intensity of the attenuation. This effect follows directly from Kirchoff's law, which specifies that noise emission from a medium in thermal equilibrium must equal its absorption. This result is valid at any frequency. The noise temperature t_s produced by an attenuating medium can be quantitatively described by application of the equation of radiative transfer,

$$t_s = \int_0^\infty t(r)\gamma(r)e^{-\int_0^r \gamma(r')dr'}\,dr \qquad (69.18)$$

where $t(r)$ (K) is the physical temperature of the medium, $\gamma(r)$ (km^{-1}) is the absorption coefficient of the medium, and r (km) is the distance along the path from the antenna. If $t(r)$ is replaced by a mean path temperature t_m, the preceding integral equation simplifies to

$$t_s = t_m\left(1 - 10^{-\frac{A(\text{dB})}{10}}\right) \qquad (69.19)$$

where $A(\text{dB})$ is the attenuation of the absorbing medium. The mean path temperature t_m is difficult to measure directly, for a satellite slant path. Wulfsburg [1964], using a model atmosphere, developed an expression for estimating the mean path temperature from the surface temperature,

$$t_m = 1.12 t_g - 50 \qquad (69.20)$$

Satellite Transmission Impairments

where t_g (K) is the ground surface temperature. Using this result, t_m ranges from about 255 to 290 K for the surface temperature range of 0–30°C.

Simultaneous measurements of rain attenuation and noise temperature on a slant path using satellite beacons have shown that the best overall statistical correlation of the noise temperature and attenuation measurements occurs for t_m values between 270–280 K for the vast majority of the reported measurments [Ippolito, 1971; Strickland, 1974; Hogg and Chu, 1975]. Similar values for gaseous attenuation or cloud attenuation have also been observed. The noise temperature introduced by the attenuating medium will add directly to the receiver system noise temperature and will degrade the overall performance of the link. The noise power *increase* occurs coincident with the signal power *decrease* due to path attenuation; both effects are additive and contribute to a reduction in link carrier to noise ratio.

69.5 Scintillation

Scintillation describes the condition of rapid fluctuations of the signal parameters of a transmitted radiowave caused by time-dependent irregularities in the transmission path. Signal parameters affected include amplitude, phase, angle of arrival, and polarization. Scintillation can be induced in the ionosphere and in the troposphere. The mechanisms and characteristics of ionospheric and tropospheric scintillation differ, and they generally impact different ranges of frequencies. Electron density irregularities occurring in the ionosphere can affect frequencies up to about 6 GHz. Tropospheric scintillation is typically produced by refractive index fluctuations in the first few kilometers of altitude and is caused by high-humidity gradients and temperature inversion layers. The effects are seasonally dependent, vary day to day, and vary with the local climate. Tropospheric scintillation has been observed on line-of-sight links up through 10 GHz and on Earth–space paths at frequencies to above 30 GHz. To a first approximation, the refractive index structure in the troposphere can be considered horizontally stratified, and variations appear as thin layers, which change with altitude. Slant paths at low elevation angles, that is, highly oblique to the layer structure, thus tend to be effected most significantly by scintillation conditions.

Amplitude Scintillation Prediction

Quantitative estimates of the level of amplitude scintillation produced by a turbulent layer in the troposphere are determined by assuming small fluctuations on a thin turbulent layer and applying turbulence theory considerations [Tatarski, 1961]. Karasawa, Yamada, and Allnutt [1899] developed a model to calculate the amplitude fluctuations caused by tropospheric scintillation, based on monthly averages of temperature and relative humidity. The model utilizes low elevation angle measurements in Japan at 11 and 14 GHz. The ITU-R [1995a] extended and updated the procedure, which includes estimates of signal fading as a time percentage as well as standard deviation of signal fluctuations. The rms amplitude fluctuation, in dB, is determined from

$$\sigma_x = 0.0228\, \sigma_{\text{xREF}}\, f^{0.45} (\csc \theta)^{1.3} [G(D)]^{0.5}, \qquad \theta \geq 5° \qquad (69.21\text{a})$$

$$\sigma_x = 0.0228\, \sigma_{\text{xREF}}\, f^{0.45} \left(\frac{2}{\sqrt{\sin^2 \theta + (4.71 \times 10^{-4})} + \sin \theta} \right)^{1.3} [G(D)]^{0.5},$$
$$\theta < 50° \qquad (69.21\text{b})$$

with

$$\sigma_{\text{xREF}} = 0.15 + \frac{118.51}{(T+273)^2} U e^{\frac{19.7T}{(T+273)}} \qquad (69.22)$$

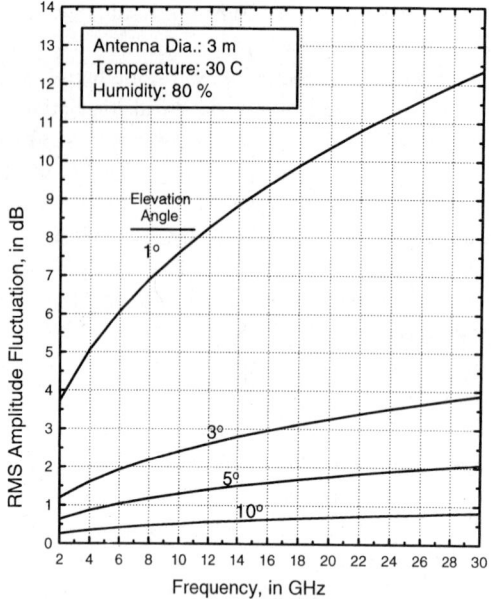

FIGURE 69.5 RMS amplitude fluctuation caused by scintillation, as a function of frequency and elevation angle.

where f (GHz) is the frequency, θ (°) is the elevation angle, D (m) is the antenna diameter, T (°C) is the average *monthly* temperature, and U (%) is the average monthly relative humidity. $G(D)$ is the antenna averaging factor of Tatarski [1961].

Figure 69.5 shows the rms amplitude fluctuation calculated from the extended CCIR prediction model as a function of frequency, for elevation angles from 1 to 10°. A 3-m antenna diameter was assumed, with an average monthly temperature of 30°C and an average monthly humidity of 80%. Significant amplitude fluctuations are observed for all frequencies at a 1° elevation angle. The amplitude fluctuations drop quickly as the elevation angle is increased, and at 10° the effect is very small, even at the higher frequencies. The model predicts a slight increase of σ_x with increasing average temperature, and a stronger increase in σ_x with increasing average humidity.

Estimates of the time statistics of signal fading due to scintillation can also be determined from the ITU-R prediction model. The signal fade level $x(p)$, corresponding to p percent of the cumulative time distribution, is given by

$$x(p) = [3.0 - 0.061 \log^3 p - 0.072 \log^2 p - 1.71 \log p]\sigma_x \qquad (69.23)$$

where σ_x (dB) is the rms amplitude fluctuation, determined from Eqs. (69.21a) and (69.21b), and $0.01\% \leq p \leq 50\%$. Figure 69.6 shows the cumulative distribution of signal fading due to scintillation for a 12-GHz link, as calculated from the model. A 3-m antenna diameter was assumed, with a T of 30°C and a U of 80%. The plots again show severe effects at low elevation angles, but at 10° or higher the fade level drops off to only 2–3 dB.

The ITU-R scintillation model supports the observation that scintillation is more pronounced under high-temperature and high-humidity conditions, such as in tropical regions or in temperate regions in summer. Significant fading, in excess of 10 dB at low elevation angles, can be expected. At low elevation angles, fading caused by scintillation could exceed attenuation caused by rain, particularly at percentages above about 1%, since the cumulative distributions for scintillation show high fading at the higher time percentages, that is, 1–10% and higher.

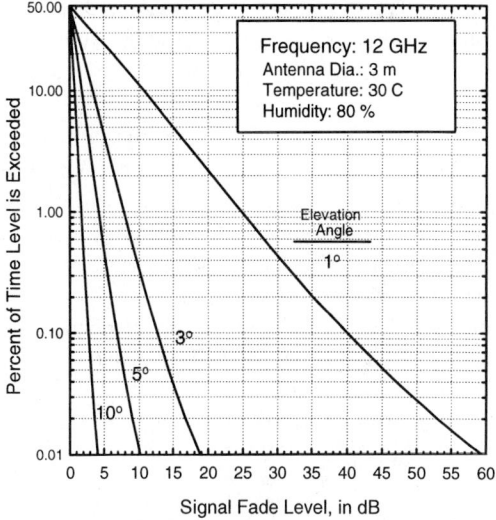

FIGURE 69.6 Cumulative distribution of signal fading due to scintillation on a 12-GHz link.

69.6 Summary

This chapter has reviewed the major propagation impairments affecting Earth–space radiowave links and has provided the latest validated modeling and prediction techniques available for the determination of these effects. Modeling and prediction procedures for gaseous attenuation, rain attenuation, cloud and fog attenuation, rain and ice depolarization, radio noise, and amplitude scintillation are included. The models are validated with extensive satellite beacon measurements obtained throughout the world, and are applicable to radiowave links operating up to 100 GHz and a wide range of elevation angles and link conditions.

References

Altshuler, E.A. 1984. A simple expression for estimating attenuation by fog at millimeter wavelengths. *IEEE Trans. Ant. Prop.*, AP-32(7); rev. 1986. AP-34(8).

Crane, R.K. 1980. Prediction of attenuation by rain. *IEEE Trans. Commun.*, COM-28(Sept.):1717–1733.

Crane, R.K. 1982. A two component rain model for the prediction of attenuation statistics. *Radio Sci.*, 17(6):1371–1388.

Gerace, G.C. and Smith, E.K. 1990. A comparison of cloud models. *IEEE Ant. Prop. Mag.*, 32(5):32–37.

Gibbins, C.J. 1986. Improved algorithms for the determination of specific attenuation at sea level by dry air and water vapor, in the frequency range 1–350 GHz. *Radio Sci.*, 21(Nov.–Dec.):949–954.

Hogg, D.C. and Chu, T.-S. 1975. The role of rain in satellite communications. *Proc. IEEE*, 63(9): 1308–1331.

Ippolito, L.J. 1971. Effects of precipitation on 15.3 and 31.65 GHz Earth-space transmissions with the ATS-V satellite. *Proc. IEEE*, 59(Feb.):189–205.

Ippolito, L.J. 1986. *Radiowave Propagation In Satellite Communications*, Van Nostrand Reinhold, New York.

ITU-R. 1995a. Recommendation ITU-R PN. 618-3. Propagation data and prediction methods required for the design of earth-space telecommunications systems. In *Propagation in Non-Ionized Media, 1994 PN Series Volume*, pp. 329–343, International Telecommunications Union, Geneva.

ITU-R. 1995b. Recommendation ITU-R PN. 676-1. Attenuation by atmospheric gases in the frequency range 1–350 GHz. In *Propagation in Non-Ionized Media, 1994 PN Series Volume*, pp. 227–237, International Telecommunications Union, Geneva.

ITU-R. 1995c. Recommendation ITU-R PN. 837-1. Characteristics of Precipitation for propagation modelling. In *Propagation in Non-Ionized Media, 1994 PN Series Volume*, pp. 238–241, International Telecommunications Union, Geneva.

ITU-R. 1995d. Recommendation ITU-R PN. 838. Specific attenuation model for rain for use in precipitation methods. In *Propagation in Non-Ionized Media, 1994 PN Series Volume*, pp. 242–243, International Telecommunications Union, Geneva.

ITU-R. 1995e. Recommendation ITU-R PN. 839. Rain height model for prediction methods. In *Propagation in Non-Ionized Media, 1994 PN Series Volume*, p. 244, International Telecommunications Union, Geneva.

Karasawa, Y., Yamada, M., and Allnutt, J.E. 1988. A new prediction method for tropospheric scintillation on Earth-space paths. *IEEE Trans. Ant. Prop.*, 36(11):1608–1614.

Pruppacher, H.R. and Pitter, R.L. 1971. A semi-empirical determination of the shape of cloud and rain drops. *J. Atmos. Sci.*, 28(Jan.):86–94.

Slobin, S.D. 1982. Microwave noise temperature and attenuation of clouds: Statistics of these effects at various sites in the United States, Alaska, and Hawaii. *Radio Sci.*, 17(6):1443–1454.

Strickland, J.I. 1974. The measurement of slant path attenuation using radar, radiometers and a satellite beacon. *J. Res. Atmos.*, 8:347–358.

Tatarski, V.I. 1961. *Wave Propagation in a Turbulent Medium*, McGraw Hill, New York.

Wulfsberg, K.N. 1964. Sky noise measurements at millimeter wavelengths. *Proc. IEEE*, 52(March): 321–322.

Further Information

For further information on propagation impairments on satellite communications links, see *Radiowave Propagation In Satellite Communications*, L.J. Ippolito, Van Nostrand Reinhold, 1986, and *Satellite-to-Ground Radiowave Propagation*, J.E. Allnutt, Peter Peregrinus, Ltd., London, 1989. For more details on modeling development and propagation databases, see the publications of Study Group 3 of the International Telecommunications Union Radiocommunications Sector (ITU-R) (formerly the CCIR), Geneva.

70
Satellite Link Design

Peter P. Nuspl[1]
*W.L. Pritchard & Co.,
Bethesda, MD*

Jahangir A. Tehrani[2]
INTELSAT

- 70.1 Introduction 949
- 70.2 Uplink and Downlink Equations 950
- 70.3 Interference Equations 951
- 70.4 Intermodulation Equations.. 954
- 70.5 Allowances, Losses and Margins 954
- 70.6 Sums of Link Equations 955
- 70.7 Designed Bit Error Ratio and Required C/N_0 957
- 70.8 Numbers of Carriers, EIRP, and Power per Carrier 959
- 70.9 Other Issues 962
- 70.10 Summary 962
- Appendix A: Some Calculations 963
- Appendix B: Calculation of Antenna Discrimination 963

70.1 Introduction

This chapter presents a unified approach to satellite link designs for geosynchronous satellites with wideband **transponders** for telecommunications services; it presents and compares **transparent** and **regenerative** link designs. A transponder, as the term is commonly used in telecommunications, is a wideband frequency-translating power amplifier; for example, a C-band transponder could have a useable bandwidth of 72 MHz with an uplink center frequency at 6.25 GHz and a downlink center frequency at about 4.0 GHz. Input and output multiplex filters in the satellite payload provide necessary channelization. As just defined, without other signal processing, this is a transparent transponder, often referred to as a *bent pipe*. A regenerative payload has, in addition, appropriate demultiplexing filters with selected transmission signals as outputs, demodulators (digital), and possibly forward error correcting (FEC) decoders for each of these signals, followed by suitable encoding and modulation. Such onboard processing adds complexity, power consumption, and mass to payloads, but there are many substantial benefits which are being quantified [Nuspl et al., 1987; Mathews, 1988]. In the following sections, general link equations are described and then specialized to digital links of transparent and regenerative types.

In the tables there are side-by-side comparisons of the transparent and regenerative forms; terminology and units are given in the central column, or in the notes to each table. With a few notable exceptions, the forms of the equations are identical, but transparent and regenerative values

---
[1] Formerly, Senior Scientific Advisor in R&D at INTELSAT.
[2] The views expressed herein are those of the authors and do not necessarily represent the views of INTELSAT.

must be used as appropriate. Where the notation is not explicit or for emphasis, regenerative terms are given in bold. A fundamental attribute of transparent links is that uplink noise, interference and intermodulation, add directly (*noise power addition*); a key feature of regenerative links is that uplink and downlink are separated, and the *errors are additive* but not the noise. As numerical examples, very small aperture terminal (VSAT) parameters are given in some tables and star VSAT designs for transparent and regenerative links are presented as examples.

The notations used subsequently and in the tables are as follows:

For A and B in units such as deciBel Hertz,

$$A \oplus B = -10 \log_{10}[10^{-A/10} + 10^{-B/10}] \quad \text{(noise) power addition}$$

$$A \ominus B = -10 \log_{10}[10^{-A/10} - 10^{-B/10}] \quad \text{(noise) power subtraction}$$

The operator $\oplus$ is used to denote multiple summations: $\oplus_1^n$. Also, the identities $[A \oplus B] \pm X \equiv [A \pm X] \oplus [B \pm X]$ are useful.

For A, B, and C with consistent units, including deciBel losses, the root sum square (RSS) operator & is defined as follows:

$$A \,\&\, B \,\&\, C = \text{RSS}(A, B, C) = \sqrt{A^2 + B^2 + C^2} \quad \text{root sum squared}$$

70.2 Uplink and Downlink Equations

It is useful to state the uplink and downlink equations separately, for **clear-sky conditions** and with thermal noise contributions only, as in Table 70.1. The saturating flux density ϕ_{sat} is a significant

TABLE 70.1 Uplink and Downlink Equations, Thermal Noise, Clear Sky

Transparent[a]	Name	Unit	Regenerative[b]
$\phi = \phi_{\text{sat}} + IBO - APA_u$	Operating flux density	dBW/m²	$\phi = \phi_{\text{sat}} + \boldsymbol{IBO_u} - APA_u$
$\left.\dfrac{C}{N_0}\right\|_u = \phi - G_{1m2} + \left.\dfrac{G}{T}\right\|_s - k$	Uplink carrier power-to-noise density	dBHz	$\left.\dfrac{C}{N_0}\right\|_u = \phi - G_{1m2} + \left.\dfrac{G}{T}\right\|_s - k$
$P_d = EIRP_{\text{sat}} + OBO$	Downlink EIRP	dBW	$P_d = EIRP_{\text{sat}} + \boldsymbol{OBO_d}$
$\left.\dfrac{C}{N_0}\right\|_d = P_d + APA_d + \left.\dfrac{G}{T}\right\|_e - k - Lsp_d - La_d$	Downlink carrier power-to-noise density	dBHz	$\left.\dfrac{C}{N_0}\right\|_d = P_d + APA_d + \left.\dfrac{G}{T}\right\|_e - k - Lsp_d - La_d$

APA = antenna pattern advantage with respect to edge of coverage
$EIRP$ = equivalent isotropically radiated power at the edge of coverage
IBO = input backoff (negative)
OBO = output backoff (negative)
u = uplink
d = downlink
ϕ_{sat} = saturating flux density at the edge of coverage, dBW/m²

G_{1m2} = gain of 1 square meter at f_u, dBi
G/T = gain-to-temperature, dB K^{-1} [dB/K]
s = satellite receiver
e = Earth station receiver
k = -228.6 Boltzmann's constant, dBW/K-Hz
Lsp_d = spreading loss at f_d, dB
La_d = atmospheric loss at f_d, dB

[a] In a transparent link, OBO is determined by IBO, through the I/O transfer characteristic of the transponder amplifier.

[b] In a regenerative link, $\boldsymbol{IBO_u}$ is a design parameter for the uplink and $\boldsymbol{OBO_d}$ is the key design parameter for the downlink; through the I/O transfer characteristic of the transponder amplifier, the corresponding $\boldsymbol{IBO_d}$ is found and set onboard.

Satellite Link Design

uplink parameter and is based on single-carrier measurements; it is usually selectable from a range of settings. Since the transponder amplifier [a traveling wave tube amplifier (TWTA), solid state power amplifier (SSPA), or a linearized TWTA] is nonlinear, for multicarrier operation it is necessary to operate it backed off from saturation; even then, intermodulation products are produced, however, at reduced levels. The G/T of a (satellite or Earth station) receiver is often used as a figure of merit. The effective system temperature T is in kelvin and is calculated in Chapter 76. The saturated equivalent isotropically radiated power (EIRP) of the transponder is the major downlink parameter, whereas the input backoff (IBO) is the only link variable amenable to design once the satellite has been launched.

70.3 Interference Equations

More and more, nonthermal contributions to links are becoming very significant. Table 70.2 is a concise form of the various carrier-to-interference densities C/I_0 due to **interference** and **intermodulation**; note that cochannel interference (CCI) and adjacent satellite interference (ASI) are treated differently.

Cochannel interference arises from reuse of frequency bands by spatial or polarization isolations of the signals; CCI is regarded as being under the designer's control. The carriers in the cochannel transponders can operate at the same frequency; however, they are assigned to the transponders that operate with beams that are spatially isolated. In cross polarization the signals have opposite polarization. The amount of cochannel interference depends on the modulation characteristics of

TABLE 70.2 Interference and Intermodulation Equations[a]

Transparent	Name	Unit	Regenerative
$\left.\dfrac{C}{I_{0\text{cci}}}\right\|_u = \oplus_1^n [\phi - \phi_i + \Psi_{i,up}] + B$	Carrier-to-CCI density up	dBHz	$\left.\dfrac{C}{I_{0\text{cci}}}\right\|_u = \oplus_1^n [\phi - \phi_i + \Psi_{i,up}] + B$
$\left.\dfrac{C}{I_{0\text{cci}}}\right\|_d = \oplus_1^m [P_d - P_j + \Psi_{j,dn}] + B$	Carrier-to-CCI density down	dBHz	$\left.\dfrac{C}{I_{0\text{cci}}}\right\|_d = \oplus_1^m [P_d - P_j + \Psi_{j,dn}] + B$
$EIRPt_u = \phi - G_{1m2} + Lsp_u + La_u$	Total EIRP up	dBW	$EIRPt_u = \phi - G_{1m2} + Lsp_u + La_u$
$\left.\dfrac{C}{I_{0\text{asi}}}\right\|_u = EIRPt_u - EIRP_{0,u}$	Carrier-to-ASI density up	dBHz	$\left.\dfrac{C}{I_{0\text{asi}}}\right\|_u = EIRPt_u - EIRP_{0,u}$
$\left.\dfrac{C}{I_{0\text{asi}}}\right\|_d = P_d + APA_d + \Gamma_{asi,d} - EIRP_{0,d}$	Carrier-to-ASI density down	dBHz	$\left.\dfrac{C}{I_{0\text{asi}}}\right\|_d = P_d + APA_d + \Gamma_{asi,d} - EIRP_{0,d}$
$C/IM_{0\text{hpa}} = EIRPt_u - HPA_{IM0L} + 10\log 4000$	Carrier-to-HPA density	dBHz	$C/IM_{0\text{hpa}} = EIRPt_u - HPA_{IM0L} + 10\log 4000$
$C/IM_0 = C/IM + B$	Carrier-to-IM density	dBHz	$C/IM_0 = C/IM + B$

CCI = cochannel interference
ASI = adjacent satellite interference
IM = Intermodulation noise power in the carrier bandwidth
ϕ_i = interfering flux density, dBW/m^2
P_j = interfering EIRP, dBW
B = useable bandwidth, dBHz
Ψ = isolation, dB

Lsp_u = spreading loss at f_u, dB
La_u = atmospheric loss at f_u, dB
$\Gamma_{asi,d}$ = E/S antenna discrimination, dB
$EIRP_{0,u}$ = EIRP density
        ASI toward wanted satellite, dBW/Hz
$EIRP_{0,d}$ = EIRP density
        ASI toward wanted E/S, dBW/Hz
HPA_{IM0L} = HPA IM density limit, dBW/4kHz

[a] In multicarrier operation, intermodulation products in the band B depend on the transponder operating point, and C/IM is obtained from simulations or measurements.

the carriers, the relative levels and frequency separations between the carriers, and the isolations among the beams. Assuming equal powers in the beams, the beam isolations contribution for the uplink and downlink can be calculated as follows:

1. Spatially isolated beams

    $\Psi_{up} = G_{nw} - G_{ni}$ uplink isolation at f_u, dB
    $\Psi_{dn} = G_{nw} - G_{mw}$ downlink isolation at f_d, dB
    Where
    w = wanted (desired)
    i = interfering
    G_{nw} = spacecraft antenna gain
    in the desired sense of polarization of beam n (or m) toward the wanted w Earth station, dBi[2]
    G_{ni} = spacecraft antenna gain
    in the desired sense of polarization of beam n (or m) toward interfering (i) Earth station, dBi

2. Oppositely polarized beams

    $\Psi_{up} = G_{nw} - 10 \log_{10}[10^{G_{nix}/10} + 10^{(G_{ni}-\rho)/10}]$ dB
    $\Psi_{dn} = G_{mw} - 10 \log_{10}[10^{G_{mwx}/10} + 10^{(G_{mw}-\rho)/10}]$ dB
    Where
    G_{nix} = spacecraft antenna gain
    in the opposite sense of polarization of beam n (or m) toward interfering (i) Earth station, dBi
    G_{mwx} = spacecraft antenna gain
    in the opposite sense of polarization of beam n (or m) toward wanted (w) Earth station
    ρ = Earth station polarization purity (Receive or transmit), dB

These equations for the oppositely polarized signals assume circular polarization with a relative tilt angle of 45° between the major axis of the polarization ellipses associated with the Earth station and spacecraft antennas. The tilt angle is unknown and can be anywhere between 0° and 90°. An average value of 45° results in factors of 10 (power addition) as shown in the preceding equations. However, the worst case corresponds to a relative tilt angle of 0°, which results in factors of 20 to be used in the preceding equations (voltage addition).

Assuming that the cochannel carriers (wanted and interfering) are similar (i.e., same type of modulation and comparable bandwidth, but not of equal powers), and ignoring antenna pattern advantages, then the ratios of the wanted (w) to the interfering (i up, j down) carrier powers can be calculated as follows.

For n uplink interferers, $i = 1, \ldots, n$,

$$(C/I)_{cc,u} = \oplus_1^n [\phi_w - \phi_i + \Psi_{i,up}] \text{ dB} \tag{70.1a}$$

and for the same ϕ_{sat} in all cochannel transponders,

$$= IBO_w + \oplus_1^n [-IBO_{cci} + \Psi_{i,up}] \text{ dB}$$

and for a constant IBO_I in the interfering channels,

$$= IBO_w - IBO_I + \Psi_{net,u} \text{ dB} \tag{70.1b}$$

---

[2] Here dBi means antenna gain with respect to isotropic, expressed in deciBel.

Satellite Link Design

and for m downlink interferers, $j = 1, \ldots, m$,

$$(C/I)_{cc,d} = \oplus_1^m [EIRP_{sat,w} + OBO_w - (EIRP_{sat,j} + OBO_j) + \Psi_{j,dn}] \text{ dB} \quad (70.2a)$$

and for the same $EIRP_{sat}$ in all cochannels and the corresponding constant output backoff OBO_J,

$$= OBO_w - OBO_J + \Psi_{net,d} \text{ dB} \quad (70.2b)$$

where ϕ, EIRP, and backoff terms are defined in the tables and

$$\Psi_{net,u} = \oplus_1^n [\Psi_{i,up}] \text{ dB}$$
$$\Psi_{net,d} = \oplus_1^m [\Psi_{j,dn}] \text{ dB}$$

Form a) of these C/I is reported in Table 70.2, whereas Table 70.7.2 applies form b).

In the case of cross-polarization interference, the effective isolation is calculated as power addition of the polarization isolations of the spacecraft and the Earth station antennas:

$$\Psi_{eff} = \Psi_{sc,x} \oplus \Psi_{es,x} \text{ dB}$$

When a number of cochannel transponders exists, the net **isolation** is the power addition of the isolation values for each cochannel transponder, as explained previously and detailed in the following example.

Example: Calculations of net isolations assuming similar co-channel carriers with equal input/output back offs and beam edge conditions.

Assumptions: INTELSAT VI satellite [Brown et al., 1990]

- Desired beam is West Hemi.
- Interfering beams are East Hemi, North West Zone, North East Zone, South West Zone, South East Zone.

Spacecraft beam isolation:

1. W. Hemi to E. Hemi isolation = 27.0 dB (spatial)
2. W. Hemi to N.E. zone isolation = 30.0 dB (spatial $\oplus$ cross polarization)
3. W. Hemi to S.E. zone isolation = 30.0 dB (spatial $\oplus$ cross polarization)
4. W. Hemi to N.W. zone isolation = 27.0 dB (cross polarization)
5. W. Hemi to S.W. zone isolation = 27.0 dB (cross polarization)
6. Spacecraft beam isolation = $\oplus_1^5$ = 21.0 dB (noise power add)

Uplink isolation:

7. Interfering Earth station-1 cross-polarization isolation = 27.3 dB
8. Interfering Earth station-2 cross-polarization isolation = 27.3 dB
9. Total Earth station cross-polarization isolation = 27.3 $\oplus$ 27.3 = 24.3 dB
10. Uplink net isolation, $\Psi_{net,u} = \Psi_{eff}$ = 21.0 $\oplus$ 24.3 = 19.3 dB

Downlink isolation:

11. Spacecraft beam isolation (same as in uplink) = 21.0 dB
12. Desired Earth station cross-polarization isolation = 27.3 dB
13. Downlink net isolation, $\Psi_{net,d}$ = 21.0 $\oplus$ 27.3 = 20.1 dB

Adjacent satellite interference is not under the designer's control (unless the neighboring satellite is in the same system and the links can be codesigned), but ASI is subject to coordination. Hence, the EIRP density from the adjacent satellite is a given external parameter; $I_{0,\text{asi}}$ does not change as signal power C is designed. The **discrimination** $\Gamma_{\text{asi},d}$ of the user's own antenna is the key parameter to choose in the design of the $(C/I_0)_{\text{asi},d}$.

As examples of typical ASI, for Intermediate Data Rate (IDR) [quaternary phase-shift keying (QPSK) modulation with rate-3/4 FEC] transmission and reception by Revised Standard A INTELSAT [1992] Earth stations, INTELSAT VII hemispheric transponders, the EIRP densities ($EIRP_0$) have a nominal value of -32.8 dBW/Hz in the uplink, and a corresponding -43.7 dBW/Hz in the downlink.[3] Table 70.2 shows the equations for $C/I_{0\text{asi}}$ in the links.

70.4 Intermodulation Equations

These brief link designs can hardly do justice to intermodulation issues. The fundamental point is that all amplifiers exhibit nonlinear behavior; this results in intermodulation products, some of which fall within the design band. The only major exceptions are single-carrier operations, as in full-transponder time-division multiple access (TDMA) or in multiplexed time-division multiplexing (TDM) downlinks: there are no intermodulation products, and C/IM is then assigned a very high value. C/IM can be modeled and simulated, or actually measured as input backoff is changed. Table 70.2 assumes C/IM characteristics are known for multicarrier operations. For designs with only a few carriers, C/IM_0 should be calculated for the specific carrier of interest.

70.5 Allowances, Losses and Margins

Table 70.3 is a summary example of applicable **allowances**, propagation (and other) **losses** and (equipment) **margins**.

In order to meet the yearly link **availability** requirement, the amounts of uplink and downlink rain margins need to be determined. The rain statistics of the uplink and the downlink sites need to be studied and the total yearly allocation for link availability needs to be apportioned between the uplink and the downlink portions of the link. In the uplink, the amount of rain attenuation corresponding to the uplink allocation (in percent time of the year) is determined from the statistics; this is called the *uplink attenuation loss* (i.e., the required uplink margin in transparent links). In the downlink, the amount of rain attenuation plus the increase in the receiving Earth station system noise associated with the downlink allocation (in percent time of the year) is determined; this is called the *downlink rain degradation* (i.e., the required downlink margin). A particular link needs to satisfy both the uplink and the downlink margins. This assumes the effect of rain depolarization is negligible.

For a transparent link, the uplink margin is satisfied by allotting the required uplink attenuation loss to the (up)link. Assuming that it does not simultaneously rain on both the uplink and the downlink, with the allotted uplink margin, the available downlink (rain) margin $M_{a,d}$ is calculated for the transparent link. If the available downlink margin is less than the required downlink margin, then the uplink allotment is increased until the downlink margin is satisfied. The system margin M_{sys} for transparent links is the maximum of these two uplink margins.

For a regenerative link, the allotments to each portion of the link are the required margins in that portion.

The numerical values in Table 70.3 are for 99.6% (up and down) link availability of C-band VSAT services. Modem margins in Table 70.3 apply at the design bit error ratio (BER) of 10^{-6} for the

---

[3]These are averages of maximum and minimum values contained in INTELSAT [1992].

Satellite Link Design

TABLE 70.3 Allowances, Losses and Margins[a] (Example for VSAT Networks)

Transparent		Name	Unit	Regenerative	Up/Down
%$TERN$ = 10% (of total, clear sky)	10	Terrestrial noise	%		10/10
$TERN = -10\log_{10}[1-\%TERN/100)]$	0.5	Allowance	dB	$TERN_u/TERN_d$	0.5/0.5
		Other losses			
HPA instabilities and uplink tracking loss	0.5		dB		0.5/—
Downlink tracking loss	0.8		dB		—/0.8
E/S equipment noise (5% total)	0.2		dB		0.1/0.1
OL	RSS 0.9	Allowance	dB	OL_u/OL_d	RSS 0.5/0.8
$AOL = TERN + OL$	1.4	$TERN$ and OL	dB	AOL_u/AOL_d	1.0/1.3
		Propagation[b]			
$L_{r,u}$	1.0	Rain att'n loss, up	dB	$L_{r,u}$	1.0/—
$L_{s,u}$	1.2	Tropo scint, up	dB	$L_{s,u}$	1.2/—
$L_{p,u} = L_{r,u}$ & $L_{s,u}$	1.6	Prop losses, up	dB	$L_{p,u} = L_{r,u}$ & $L_{s,u}$	1.6/—
$L_{D,d}$	1.0	Degraded, down	dB	$L_{D,d}$	—/1.0
$L_{s,d}$	1.2	Tropo scint, down	dB	$L_{s,d}$	—/1.2
$L_{p,d} = L_{D,d}$ & $L_{s,d}$	1.6	Prop losses, down	dB	$L_{p,d} = L_{D,d}$ & $L_{s,d}$	—/1.6
M_m	1.7	Coded modem margins	dB	$M_{m,u}/M_{m,d}$	1.5/1.7
$(C/N_0)_{\text{other};\Delta} = [(C/N_0) \ominus (C/N_0)_d] + \Delta + L_{p,u}$		Other than thermal down[c]	dBHz	n.a.	
$(C/N_0)_{d;D} = (C/N_0)_d + \Delta - L_{p,d} + L_{p,u}$ see Tables 70.1, 70.4		Degraded C/N_0, down	dBHz	n.a.	
$M_{d:D} = (C/N_0)_{ta} + \Delta + L_{p,u} - [(C/N_0)_{\text{other};\Delta} \oplus (C/N_0)_{d;D}]$		Degradation margin, down	dB	n.a.	
$M_{\text{sys}} = \max\{L_{p,u}, L_{p,u} + \Delta; M_{d:D} \leq L_{p,d}\}$		Link margins	dB	$M_u = L_{p,u}/M_d = L_{p,d}$	1.6/1.6

AOL = allowance for $TERN$ and OL
HPA = high-power amplifier at E/S
OL = other losses
$TERN$ = terrestrial noise
& = root sum squared
$\oplus, \ominus$ = noise power addition, subtraction

L_r, L_s = rain, scintillation losses, dB
$L_{D,d}$ = loss in Degraded downlink: rain attenuation plus degradation due to increased system noise temperature
L_p = propagation losses: up, down, dB
Δ = additional uplink margin, dB

[a] See discussion and sample calculation of $M_{a,d}$ and M_{sys} (system margin) for transparent links.
[b] VSAT link availability is typically 99.6%, equally apportioned to the uplink and downlink.
[c] $(C/N_0)_{\text{other}}$ is the portion of $(C/N_0)_{ta}$ which is not affected by downlink fading.
[d] Adjacent channel interference effects are assumed negligible.

degraded sky. Onboard decoding would require an *additional* implementation margin of 0.4 dB in the uplink.

70.6 Sums of Link Equations

Table 70.4 presents sums of the C/I_0 equations for the uplink, downlink, and transparent link, including interference (CCI and ASI), intermodulation, allowances, losses, and margins, so that these apply for the degraded sky (also called *threshold*, meaning below which performance is not acceptable for the specified percentage of time).

TABLE 70.4 Sums of Link Equations

Transparent, Clear Sky[a]	Name	Unit	**Regenerative**, Degraded Sky[a]
$(C/N_0)_{su} = (C/N_0)_u$ $\oplus (C/I_{0\,cci})_u$ $\oplus (C/I_{0\,asi})_u$ $\oplus C/IM_{0\,hpa}$[b]	Sum uplink C-to-N_0	dBHz	$\mathbf{(C/N_0)_{su} = [(C/N_0)_u}$ $\mathbf{\oplus (C/I_{0\,cci})_u}$ $\mathbf{\oplus (C/I_{0\,asi})_u}$ $\mathbf{\oplus C/IM_{0\,hpa}] - M_u}$
n.a.	Available, up C-to-N_0	dBHz	$(C/N_0)_{a,u} = (C/N_0)_{su} - AOL_u$
$(C/N_0)_{sd} = (C/N_0)_d$ $\oplus (C/I_{0\,cci})_d$ $\oplus (C/I_{0\,asi})_d$ $\oplus C/IM_0$	Sum downlink C-to-N_0	dBHz	$\mathbf{(C/N_0)_{sd} = [(C/N_0)_d - M_d]}$ $\mathbf{\oplus (C/I_{0\,cci})_d}$ $\mathbf{\oplus (C/I_{0\,asi})_d}$ $\mathbf{\oplus C/IM_0}$
$(C/N_0)ta = [(C/N_0)_{su} \oplus (C/N_0)_{sd}]$ $- AOL \text{ (excluding } M_{\text{sys}})$[c]	(total) available C-to-N_0	dBHz	$(C/N_0)_{a,d} = (C/N_0)_{sd} - AOL_d$

[a] Compare the different forms for the transparent and regenerative links. For regenerative links, the C/N_0 sums are shown in bold.

[b] $C/IM_{0\,hpa}$ accounts for intermodulation products from all other Earth stations (primarily from within the design network) radiating into the band B.

[c] For transparent links, $M_{d:D}$ is the degradation margin due to downlink; M_{sys} accounts for uplink and downlink rain losses, as in Table 70.5; see the discussion.

TABLE 70.5 Designed BER and Required C/N_0

Transparent	Name	Unit	**Regenerative**
BER[a]	Designed bit error ratio		$\mathbf{BER = BER_{up} + BER_{dn}}$
Designed BER determines E_b/N_0	Energy-per-bit to noise-density	dB	(Decoded) BER_{up} determines (E_b/N_0) up (Decoded) BER_{dn} determines (E_b/N_0) down.
n.a.	Required C/N_0 up	dBHz	$\left.\dfrac{C}{N_0}\right|_{r,u} = \left.\dfrac{E_b}{N_0}\right|_u + R_b + M_{m,u}$
$\left.\dfrac{C}{N_0}\right|_r = \left.\dfrac{E_b}{N_0}\right|_r + R_b + M_m + M_{\text{sys}}$	Required C/N_0 (total), down	dBHz	$\left.\dfrac{C}{N_0}\right|_{r,d} = \left.\dfrac{E_b}{N_0}\right|_d + R_b + M_{m,d}$

E_b = energy per *information* bit, W-s
M_m = modem margin, dB
$M_{m,u}, M_{m,d}$ = modem margins, dB
M_{sys} = system margin, dB (transparent)
R_b = information rate, $mr R_{\text{sym}}$, b/s
$m = \log_2 M$, M-ary alphabet of channel symbols ($M = 2, 4, 8$)
r = input bits/output bits, **code rate**
R_{sym} = channel symbol rate, symbol/s

[a] Curves of *BER* vs E_b/N_0 *for coded modems* are used; in principle, such curves are independent of the information rate.

For a transparent link, note the noise-power combining of uplink and downlink terms, but the exclusion of M_{sys}, which is taken into account in the required C/N_0 in Table 70.5.

Regeneration separates the available uplink and downlink C/N_0 sum equivalents; allowances could be equal or less than for the transparent case. For the regenerative case the losses and allowances are applied to the uplink and downlink separately as illustrated in the right column.

In some situations, links are dominated by a particular contribution, and there is not much scope for design. But in designing most multicarrier transparent links, the total available C/N_0

Satellite Link Design

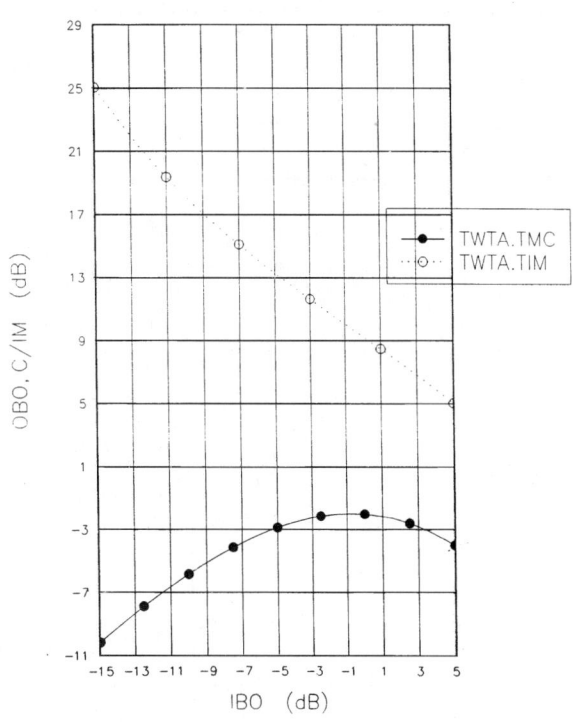

FIGURE 70.1 Output backoff and C/IM vs input backoff.

can be optimized; of course, the selection of IBO/OBO is the ultimate design for the desired links. Figure 70.1 illustrates typical TWTA transfer characteristics and intermodulation characteristics. Having thermal and interference components which increase with less backoff (more carrier power), and another component due to intermodulation products which decreases with less backoff (more nonlinear operation), a maximum value for available C/N_0 can be found.

Example: Optimum backoffs. Figure 70.2 illustrates a typical design, which selects the best input/output backoffs for transparent and regenerative links. In multicarrier mode, regenerative links can apply much more uplink backoff and use slightly more (negative) downlink backoff.

70.7 Designed Bit Error Ratio and Required C/N_0

For digital links, bit error ratio is the most common form of expressing performance. A service BER requirement is in conjunction with a needed availability, usually expressed in percentage of time annually; for VSAT networks operating in C-bands, Tables 70.3 and 70.7.3 assume 99.6% link availability. Given the design values of (uplink and downlink) E_b/N_0 for each carrier, the corresponding required C/N_0 are readily calculated for the three links, as in Table 70.5; Table 70.7.5 gives numerical examples. The use of system and modem margins should be noted.

A classic form of comparing transparent and regenerative links is the **iso-BER curve**, as uplink and downlink values are varied [Mathews, 1988], illustrated in Fig. 70.3. These are required values, so less costly links are lower and to the left on these curves. Point A (4.9, 24.8) off the transparent

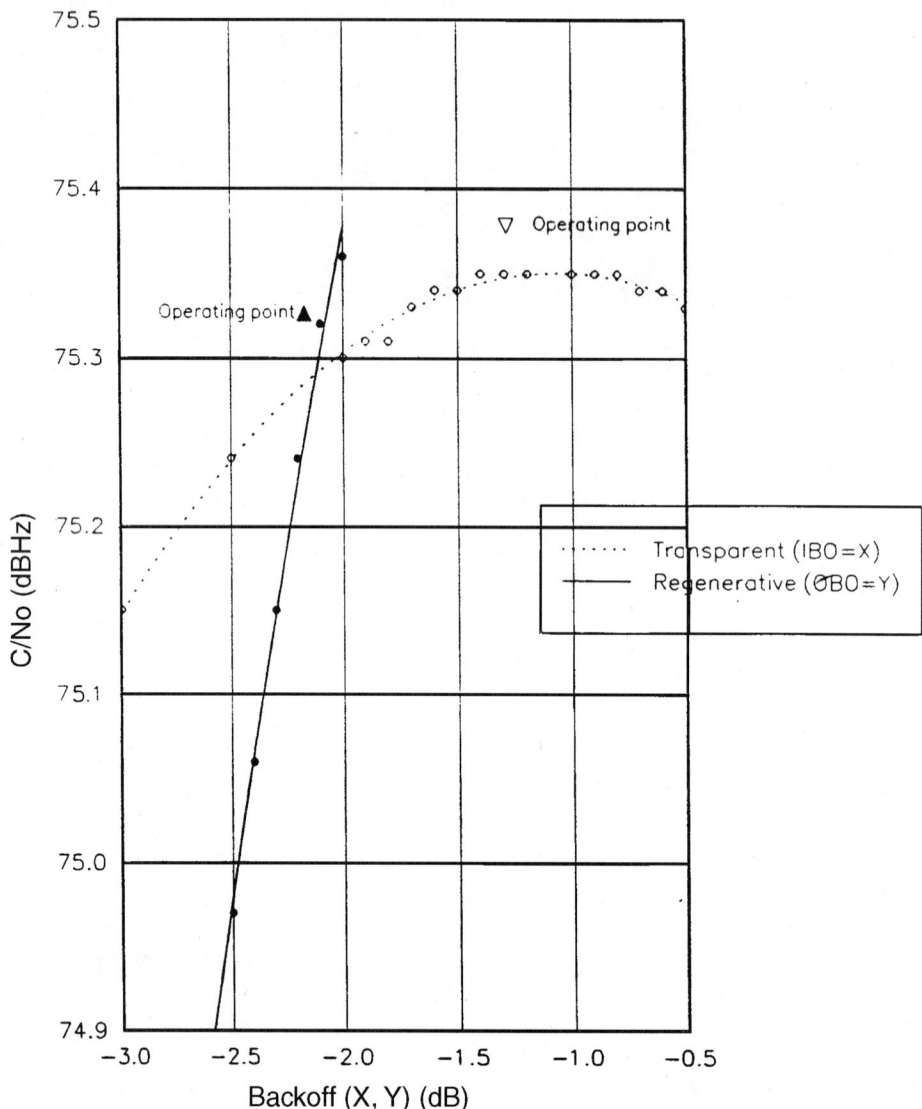

FIGURE 70.2 Selecting IBO for maximum C/N_0.

curve represents a split in noise contributions, with almost all in the downlink. As an example on the regenerative curve, for a downlink-conserving design such as at point B (4.8, 12.8), BER_{up} would be set at a negligible fraction (<1%) of the total BER, so that the highest BER_{dn} can be tolerated, to get the lowest required $(E_b/N_0)_{dn}$. Between transparent and regenerative points A and B, compare the uplink difference of 12 dB and the downlink difference of only 0.1 dB, in favor of using regeneration.

As another example, a link needing an uplink-conserving design is digital satellite news gathering (SNG), where transportability and low-power consumption are important in the uplinking Earth station, and the receive station is usually large. Another link design is required and would show a tradeoff between transparent and regenerative operation.

Satellite Link Design

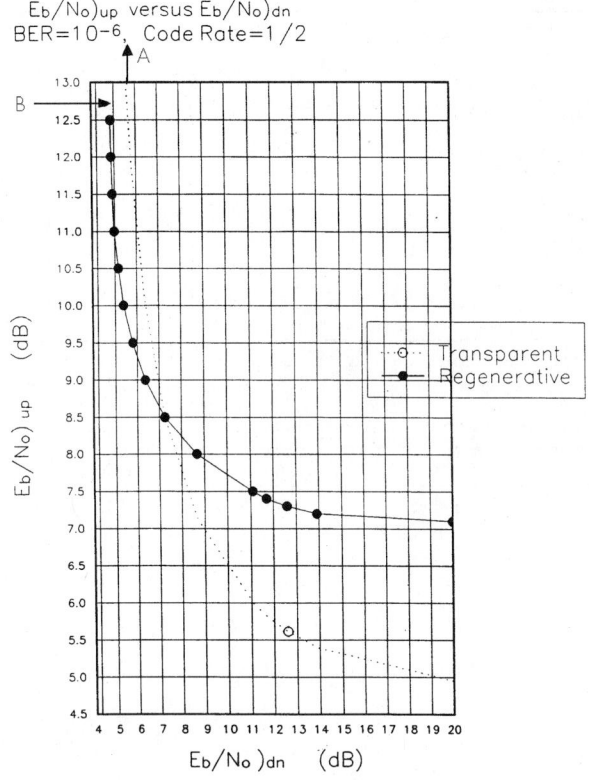

FIGURE 70.3 Iso-BER curves, comparing transparent and regenerative cases.

70.8 Numbers of Carriers, EIRP, and Power per Carrier

Table 70.6 shows the calculations of the numbers of carriers that can be supported by the links; of course the lesser of n_u and n_d is selected. The power per carrier at an Earth station can then be calculated; with regeneration the uplink generally requires much less power. The power per carrier in the downlink is also calculated for both types of links.

TABLE 70.6 Number of Carriers, EIRP and Power per Carrier

Transparent[a]	Name	Unit	Regenerative[b]
n.a.	Uplink carriers		$n_u = 10^{[(C/N_0)_{a,u}-(C/N_0)_{r,u}]/10}$
n.a.	Downlink carriers		$n_d = 10^{[(C/N_0)_{a,d}-(C/N_0)_{r,d}]/10}$
$n = 10^{[(C/N_0)_{ta}-(C/N_0)_r]/10}$	Number of carriers		$n = \min(n_u, n_d)$
$EIRPc_u = EIRPt_u - 10\log n$	EIRP per carrier, up	dBW	$EIRPc_u = EIRPt_u - 10\log n$
$Pc_u = EIRPc_u - G_u + Lf_u$	Power per carrier, up	dBW	$Pc_u = EIRPc_u - G_u + Lf_u$
$EIRPc_d = P_d - 10\log n$	EIRP per carrier, down	dBW	$EIRPc_d = P_d - 10\log n$

Lf_u = feed loss in earth station, dB
G_u = antenna gain at f_u, dBi

[a] In a transparent link, the number of carriers n is calculated from the total available (*including* M_{sys}) and required C/N_0 values.

[b] In a regenerative link, n_u is calculated from the total uplink EIRP controlled by **IBO**$_u$ and the *selected* uplink BER; n_u is usually higher than n_d (except under E/S power limitations); and n_d is calculated from the total available and required C/N_0 values for the downlink. Each calculated n, n_u, and n_d is checked against its bandlimited value.

TABLE 70.7.0 Numerical Examples, Hub to VSAT[a]

Transponder, INTELSAT VI	$B = 72.0$ MHz (78.6 dBHz)
	$\phi_{sat} = -77.6$ dBW/m², (high gain); $(G/T)_s = -9.2$ dB/K
	$EIRP_{sat} = 31.0$ dBW, (hemi beam)
Uplink, from several hubs	$D_u = 5.0$ m; $f_u = 6.130$ GHz; $APA_u = 0$ dB; $G_{1m2} = 37.2$ dBi/m²
	$IBO_{cci} = -3$ dB; $(HPA_{IM0L}) = 21.0$ dBW/4 kHz; $Lf_u = 0.7$ dB;
Downlink, to many VSATs	$D_d = 1.8$ m; $f_d = 3.905$ GHz; $APA_d = 1$ dB; $(G/T)_e = 15.0$ dB/K
	$OBO_{cci} = -2$ dB
Service parameters	$R_b = 64$ kbit/s (48.1 dBHz); $BER = 1 \times 10^{-6}$
	Availability = 99.6%: see Table 70.3.

[a] $X = IBO$, $Y(X) = OBO$, $Z(X) = C/IM$.

TABLE 70.7.1 Uplink and Downlink Thermal Equations, Clear Sky

Transparent		Name	Unit	Regenerative	
$\phi = (-77.6) + X - 0.0$ $= -77.6 + X$		Operating flux density	dBW/m²	$\phi = (-77.6) + X_u - 0.0$ $= -77.6 + X_u$	
$\left.\dfrac{C}{N_0}\right\|_u = \phi - 37.2 + (-9.2) - (-228.6)$ $= 104.6 + X$	#1t	Uplink carrier power-to-noise density	dBHz	$\left.\dfrac{C}{N_0}\right\|_u = \phi - 37.2 + (-9.2) - (-228.6)$ $= 104.6 + X_u$	#1r
$P_d = 31.0 + Y(X)$		Downlink EIRP	dBW	$P_d = 31.0 + Y(X_d)$	
$\left.\dfrac{C}{N_0}\right\|_d = P_d + 1.0 + 15.0 - (-228.6)$ $\quad -196.4 - 0.3$ $= 78.9 + Y(X)$	#2t	Downlink carrier power-to-noise density	dBHz	$\left.\dfrac{C}{N_0}\right\|_d = P_d + 1.0 + 15.0 - (-228.6)$ $\quad -196.4 - 0.3$ $= 78.9 + Y(X_d)$	#2r

TABLE 70.7.2 Interference and Intermodulation Equations

Transparent		Name	Unit	Regenerative	
$\left.\dfrac{C}{I_{0\,cci}}\right\|_u = X - (-3.0) + 19.3 + 78.6$ $= 100.9 + X$	#3t	Carrier-to-CCI density up	dBHz	$\left.\dfrac{C}{I_{0\,cci}}\right\|_u = X_u - (-3.0) + 19.3 + 78.6$ $= 100.9 + X_u$	#3r
$\left.\dfrac{C}{I_{0\,cci}}\right\|_d = Y(X) - (-2.0) + 20.1 + 78.6$ $= 100.7 + Y(X)$	#4t	Carrier-to-CCI density down	dBHz	$\left.\dfrac{C}{I_{0\,cci}}\right\|_d = Y(X_d) - (-2.0) + 20.1 + 78.6$ $= 100.7 + Y(X_d)$	#4r
$EIRPt_u = \phi - 37.2 + 200.4 + 0.3$ $= 85.9 + X$		Total EIRP up	dBW	$EIRPt_u = \phi - 37.2 + 200.4 + 0.3$ $= 85.9 + X_u$	
$\left.\dfrac{C}{I_{0\,asi}}\right\|_u = EIRPt_u - (-32.8)$ $= 118.7 + X$	#5t	Carrier-to-ASI density up	dBHz	$\left.\dfrac{C}{I_{0\,asi}}\right\|_u = EIRPt_u - (-32.8)$ $= 118.7 + X_u$	#5r
$\left.\dfrac{C}{I_{0\,asi}}\right\|_d = P_d + 1.0 + 12.4 - (-43.7)$ $= 88.1 + Y(X)$	#6t	Carrier-to-ASI density down	dBHz	$\left.\dfrac{C}{I_{0\,asi}}\right\|_d = P_d + 1.0 + 12.4 - (-43.7)$ $= 88.1 + Y(X_d)$	#6r
$C/IM_{0\,hpa} = EIRPt_u - 21.0 + 36.0$ $= 100.9 + X$	#7t	Carrier-to-IM density E/S HPA	dBHz	$C/IM_{0\,hpa} = EIRPt_u - 21.0 + 36.0$ $= 100.9 + X_u$	#7t
$C/IM_0 = Z(X) + 78.6$ $= 78.6 + Z(X)$	#8t	Carrier-to-IM density (satellite)	dBHz	$C/IM_0 = Z(X_d) + 78.6$ $= 78.6 + Z(X_d)$	#8r

Example: Hub to VSAT Stations. Tables 70.1–70.6 have documented some simplified link design equations; as another set of illustrative examples, the corresponding Tables 70.7.0–70.7.6 provide numerical values of a typical star VSAT link design, in the direction hub to VSAT. In Table 70.7.0 are the salient assumptions about the transponder, the Earth stations, and the service at 64-kb/s information rate. It is assumed that several large stations would act as hubs (gateways), and that multicarrier frequency-division multiple access (FDMA) is the access mode.

Satellite Link Design

TABLE 70.7.3 Allowance, Losses and Margins

Transparent	Name	Unit	Regenerative	
$AOL = 0.5 + 0.9 = 1.4$	TERN and OL	dB	AOL_u/AOL_d	1.0/1.3
	Propagation:			
$L_{p,u} = 1.0 \& 1.2 = 1.6$	Prop losses, up	dB	$L_{p,u} = L_{r,u} \& L_{s,u}$	1.6/—
$L_{p,d} = 1.0 \& 1.2 = 1.6$	Prop losses, down	dB	$L_{p,d} = L_{D,d} \& L_{s,d}$	—/1.6
$M_m = 1.7$	Coded modem margin	dB	$M_{m,u}/M_{m,d}$	1.5/1.7
$(C/N_0)_{\text{other};\Delta} = [74.6 \ominus 76.9]$ $+1.6 = 80.1$	opt Other than thermal	dBHz	n.a.	
$(C/N_0)_{d;D} = 76.9 + 1.6 - 1.6 = 76.9$	opt Degraded	dBHz	n.a.	
$M_{d;D} = 74.6 + 1.6 - (80.1 \oplus 76.9) = 1.0$	opt Available margin, down	dB	n.a.	
$M_{\text{sys}} = 1.6$	Up, downlink margins	dB	$M_u = 1.6/M_d = 1.6$	

TABLE 70.7.4 Sums of Link Equations

Transparent			Name	Unit Regenerative		
$(C/N_0)_{su} = [104.6 + X]$	{103.3	#1t	Sum uplink	dBHz $(C/N_0)_{su} = [104.6 + X_u]$	{103.3	#1r
$\oplus [100.9 + X]$	99.6	#3t	C-to-N_0	$\oplus [100.9 + X_u]$	99.6	#3r
$\oplus [118.7 + X]$	117.4	#5t	$X = -1.3$	$\oplus [118.7 + X_u]$	117.4	#5r
$\oplus [100.9 + X]$	99.6}	#7t	(Figure 70.2)	$\oplus [100.9 + X_u] - 1.6$	99.6} $- 1.6$	#7r
$= 97.0 + X$	$= 95.7$	opt	$Y = -2$ (Figure 70.1)	$= 97.0 + X_u] - 1.6$	$= 94.1$	
n.a.			Available, up C-to-N_0	dBHz $(C/N_0)_{a,u} = 94.0 - 1.0$ $= 93.0$ including M_u		
$(C/N_0)_{sd} = [78.9 + Y(X)]$	{76.9	#2t	Sum	dBHz $(C/N_0)_{sd} = [78.9 + Y(X_d) - 1.6]$	{75.2	#2r
$\oplus [100.7 + Y(X)]$	98.7	#4t	downlink	$\oplus 100.7 + Y(X_d)$	98.6	#4r
$\oplus [88.1 + Y(X)]$	86.1	#6t	C-to-N_0	$\oplus 88.1 + Y(X_d)$	86.0	#6r
$\oplus [78.6 + Z(X)]$	88.9}	#8t	$Z = 10.3$	$\oplus 78.6 + Z(X_d)$	89.0}	#8r
	$= 76.1$	opt			$= 74.7$	opt
$(C/N_0)_{ta} = \{95.7 \otimes 76.1\}$ -1.4 $= 74.6$ (excluding M_{sys})		opt	(Total) Available C-to-N_0	dBHz $(C/N_0)_{a,d} = 74.7 - 1.3$ $= 73.4$ including M_d		opt

Transparent: $X_{\text{opt}} = -1.3$, $Y_{\text{opt}} = -2.0$, $Z = 10.3$, Regenerative: $X_u = -14.3$, $Y_{\text{opt}} = -2.1$, $Z = 10.4$.

TABLE 70.7.5 Designed BER and Required C/N_0

Transparent	Name	Unit	Regenerative
$BER = 1.0 \times 10^{-6}$	Designed bit error ratio		$BER = 1.0 \times 10^{-12} + 1.0 \times 10^{-6}$
Designed BER determines E_b/N_0	Energy-per-bit to noise-density	dB	$(E_b/N_0)_u = 12.8$ $(E_b/N_0)_d = 4.8$
n.a.	Required C/N_0 up	dBHz	$\left.\dfrac{C}{N_0}\right|_{r,u} = 12.8 + 48.1 + 1.5$ $= 62.4$ excluding M_u
$\left.\dfrac{C}{N_0}\right|_r = 4.8 + 48.1 + 1.7 + 1.6$ $= 56.2$ (including M_{sys})	Required C/N_0 (total), down	dBHz	$\left.\dfrac{C}{N_0}\right|_{r,d} = 4.8 + 48.1 + 1.7$ $= 54.6$ excluding M_d

The importance of backoff is indicated with the variables X, $Y(X)$, and $Z(X)$; then the best IBO $= X$ and OBO $= X_d$ are selected as already exemplified in Fig. 70.2. Table 70.7.4 has numbered entries with X, $Y(X)$, and $Z(X)$ and also numerical values when the best operating point is utilized. In Table 70.7.3 M_{sys} is determined, from which the required C/N_0 for the transparent case is calculated.

TABLE 70.7.6 Number of Carriers, EIRP and Power per Carrier

Transparent	Name	Unit	Regenerative	Design
n.a.	Uplink carriers		$n_u = 10^{[93.1 \ominus 62.4]/10}$ $= 1174$	$n_u = 10^{(-81.1-62.4)/10}$ $= 75$
n.a.	Downlink carriers		$n_d = 10^{[73.4-54.6]/10}$ $= 75$	$n_d = 75$
$n = 10^{[(74.6-56.2]/10}$ $= 69$ opt	Number of carriers		$n = \min(n_u, n_d)$ $= 75$ opt	
$EIRPc_u = 85.9 + (-1.3)$ $-18.4 = 66.2$ opt	EIRP per carrier, up	dBW	$\mathbf{EIRPc_u} = 85.9 + (\mathbf{-1.3})$ $-18.8 = 65.8$	$X_u = -12.0$ $EIRPc_u = 53.8$
$Pc_u = 66.2 - 47.9 + 0.7$ $= 19.0 \{79 W\}$ opt	Power per carrier, up	dBW	$\mathbf{Pc_u} = 65.8 - 47.9 + 0.7$ $= 18.6 \{72 W\}$	$Pc_u = 6.6 \text{ (5 W)}$
$EIRPc_d = 31.0 + (-2.0)$ $-18.4 = 10.6$ opt	EIRP per carrier, down	dBW	$\mathbf{EIRPc_d} = 31.0 + (\mathbf{-2.1})$ $-18.8 = 10.1$ opt	

Table 70.7.6 shows that 69 carriers can be supported in the transparent link, whereas 75 carriers are possible with regeneration; and the downlink EIRPs are similar; the small differences are due to the downlink-limited condition with the small (1.8-m) antennas. However, Table 70.7.6 also shows a design on the right in which the regenerative uplink can be backed off an additional 12 dB, to yield an uplink power of only 6.6 dBW per carrier, which is less than 5 W. In the transparent link, the hub(s) needs 79 W per carrier, and so a large high power amplifier (HPA) is required.

There would be similar differences in the direction from VSAT to hub.

70.9 Other Issues

The link designs and examples are for star VSAT networks, which work with a few large hubs. Even more interesting would be a mesh VSAT designs where there is a need to conserve uplink and downlink power, and interference from similar systems is more severe due to lower isolations. Also, these link designs have excluded bandwidth improvement factors for intermodulation, which take into account sparse use of spectrum; likewise, bandwidth allocation factors and overheads have been excluded. In particular for small to medium size antennas, outer coding can be usefully applied. These designs have no demand assignment multiple access (DAMA) signalling, which would need to be robustly designed. As to other issues, data-only links can take advantage of automatic repeat request (ARQ) so that throughput and basic BER can be selected accordingly. Links with voice/video-only can be operated with relatively poor BER, and effective error concealment techniques can be applied.

70.10 Summary

This chapter has presented a unified but simplified set of link designs for transparent and regenerative links. Several numerical examples illustrate the methods and the differences in link designs.

All link designs must properly take into account thermal, interference and intermodulation contributions, as well as allowances, losses, and margins. Transparent links, which are most prevalent now, suffer noise power addition of up to ten entries; regenerative links, which are being developed in many forms and can be expected in next generation satellites, separate the uplink and downlink so that each has only up to four entries. Regenerative links require an uplink modem margin (for the payload demodulator) but other allowances, losses, and margins are generally lower than for transparent links.

Appendix A: Some Calculations

Boltzmann's constant:
$$k = -228.6 \quad \text{dBJ/K}$$

Propagation velocity in free space:
$$c = 2.998 \times 10^8 \quad \text{m/s}$$

Wavelength (Lambda)
$$\lambda = \frac{c}{f} \quad \text{m}$$

where f is the frequency of operation [up f_u, down f_d], in Hertz.

The maximum antenna gain relative to isotropic for a parabolic antenna, up and down:

$$G = 10 \log_{10}\left(\eta\left(\frac{\pi D}{\lambda}\right)^2\right) \quad \text{dBi}$$

where η is the antenna efficiency (nominal 0.6) and D is the aperture diameter in meters.

Gain of ideal 1 m²:
$$G_{1m2} = 21.455 + 20 \log F \quad \text{dBi/m}^2$$

where F is the (approximate) frequency in gigaHertz.

Typical atmospheric loss, up and down:

$$
\begin{aligned}
L_a = 0.25 \quad &\text{dB} \quad 2 < F < 5 \quad \text{GHz} \\
0.33 \quad & \quad\quad 5 < F < 10 \quad \text{GHz} \\
0.53 \quad & \quad\quad 10 < F < 13 \quad \text{GHz} \\
0.73 \quad & \quad\quad 13 < F \quad \text{GHz}
\end{aligned}
$$

Free space spreading loss up and down:
$$L_{sp} = 32.45 + 20 \log_{10}(FR) \quad \text{dB}$$

where R is the (slant-range) distance to the satellite in meters. [Typically, at 10° elevation angle, $R = 40{,}587{,}000$ m to geostationary Earth orbit (GEO).]

Appendix B: Calculation of Antenna Discrimination[4]

Gain mask:
$$
\begin{aligned}
G_m &= G - 0.0025[\theta D/\lambda]^2 \quad &\text{mainlobe} \quad &\text{dBi} \\
G_1 &= 2 + 15 \log_{10}(D/\lambda) \quad &\text{first sidelobe} & \\
G_s &= 32 - 25 \log_{10} \theta \quad &\text{other sidelobes} &
\end{aligned}
$$

[4] Based on International Telecommunications Union (ITU) Rec. 465-4 and Report 391-6.

where G is the maximum gain (as previously calculated) and θ is the satellite spacing in degrees (nominal 3°).

$$\delta_m = (20\lambda/D)\sqrt{(G - G_1)} \text{ degree}$$
$$\delta_1 = \max[1°, 100\lambda/D°]$$

Antenna discrimination for receive Earth station:

IF	$\theta < \delta_m$	$\Gamma_{asi,d} = G - G_m$	of mainlobe,	dB
IF	$\delta_m \leq \theta < \delta_1$	$\Gamma_{asi,d} = G - G_1$	of first sidelobe	dB
IF	$\delta_1 \leq \theta < 48$	$\Gamma_{asi,d} = G - G_s$	of other sidelobes	dB
IF	$48 \leq \theta$	$\Gamma_{asi,d} = G - (-10)$		dB

Defining Terms

Allowance: An allowance in deciBels is subtracted from a sum of carrier-to-noise densities which would have been available in a link; usually such an allowance accounts for factors external to a given system (and beyond the system designer's control).

Availability: Usually expressed as a percentage of a year, availability is the total time duration over which the specified BER is met or exceeded.

Clear-sky condition: Only atmospheric losses are added to spreading losses; this excludes weather effects.

Code rate: The code rate (actually a *ratio*) for forward error correcting codes is the ratio of input to output bits (or symbols), such as rate 1/2, rate 3/4 for binary codes, and rate 239/255 for a typical Reed–Solomon code.

Degraded-sky condition: In the uplink there is additional attenuation due to rain; in the downlink there is rain attenuation and also an increased effective receiver system temperature, due to the increased sky noise.

Discrimination: Antenna discrimination is the differential gain compared to maximum for an antenna in the specified direction; usually, masks are provided for the main lobe, the first sidelobe, and other sidelobes.

Interference: The generic term that includes all *unwanted* signals that affect a telecommunications link, excluding thermal noise and intermodulation products; interference can be man-made or natural; it can be internal or external to a system.

Intermodulation: A generic term; *intermodulation products* are produced in nonlinear devices (amplifiers) and they are dependent on the wanted signal levels, modulation formats, and frequency plan.

Iso-BER curve: A graph of constant (iso) bit error ratio with uplink and downlink parameters on the y- and x-axis respectively; C/N_0 or E_b/N_0 are usually used.

Isolation: An antenna beam's minimum discrimination to a cochannel or cross-polarization channel.

Margin: In deciBel, a margin is added to required values, for example to E_b/N_0 required for a target BER, and reflects uncertainties in equipment performance; this is sometimes called *implementation margin*.

Propagation loss or **downlink degradation:** In deciBel, propagation loss is subtracted from all carrier-to-noise (equivalent) densities where C (signal power) is affected but I_o is not; rain attenuation, increased sky noise, depolarization, and tropospheric scintillation are typical examples of such losses.

Regenerative: A link that, in comparison to transparent, has additional functions in the satellite payload, such as demultiplexing and demodulation of the communications signals in the uplink, and modulation for the downlink; this is sometimes called *onboard regeneration (OBR)*. OBR and additional functions such as possibly FEC decoding, re-encoding, baseband switching or IF routing, and TDM constitute what are usually called *onboard processing (OBP)*.

Transparent: A transparent link uses a transponder that has an output signal format that is exactly the same as the input format.

Transponder: Consists of a wideband receiver, frequency-translating equipment and a power amplifier; its associated input and output multiplex filters determine its center frequency and bandwidth; it would be more correctly called a *repeater*.

Acknowledgment

The authors wish to acknowledge William Salazar and Luis Gonzales for their dedicated work in preparing the link budget programs.

References

Brown, M.P., Duesing, R.W., Nguyen, L.N., Sandrin, W.A., and Tehrani, J.A. 1990. INTELSAT VI transmission design and computer system models for FDMA services. *COMSAT Tech. Rev.*, 20(2):373–399.

INTELSAT. 1992. *INTELSAT Intersystem Coordination Manual*, IICM-314, Rev. 1A, Dec.

Mathews, N. 1988. Performance evaluation of regenerative digital satellite links with FEC codecs. *Satellite Integrated Communications Networks*, Proc. Tirrenia Workshop, North-Holland, UK.

Nuspl, P., Peters, R., Abdel-Nabi, T., and Mathews, N., 1987. On-board processing for communications satellites: Systems and benefits. *Int. J. Satellite Commun.*, 5(2):65–76.

Further Information

For a general treatment of propagation topics, see the book by J.E. Allnutt, 1989, *Satellite-to-Ground Radiowave Propagation, Theory, Practice and System Impact at Frequencies Above* 1 GHz, Peter Peregrinus Ltd. UK.

For another treatment of (transparent) link budgets, see Chapter 6 of *Satellite Communications* by D. Roddy, 1989, Prentice Hall, Englewood Cliffs, NJ.

There was a special issue on on-board processing, edited by P. Nuspl, *International Journal of Satellite Communications*, Vol. 5, No. 2, Wiley-Interscience, London, UK, April-June, 1987. See B. Pontano, P. de Santis, L. White, and F. Faris, "On-Board Processing Architectures for International Communications Satellite Applications," *Proceedings*, 13th American Institute of Aeronautics and Astronautics (AIAA) International Communication Satellite Systems Conference and Exhibit, Los Angeles, CA, Paper 90-0853, March 1990. Also, see several sessions in Proceedings of International Conference on Digital Satellite Communications ICDSC-8, Guadeloupe, April 1989; ICDSC-9, Copenhagen, May 1992; 14th AIAA Conference, Washington, DC, March 1992; and 15th AIAA Conference, San Diego, March 1994.

Also, trends for future payloads are described in "Advanced Communications Technologies for Future INTELSAT Systems," P.P. Nuspl, 1992, Proceedings, Advanced Methods for Satellite and Deep Space Communications, 1992 DLR Seminar, Bonn, Germany, September. (DLR is the German Aerospace Research Center near Munich, Germany.)

71
The Calculation of System Temperature for a Microwave Receiver

Wilbur Pritchard
W.L. Pritchard & Co., Inc.
Bethesda, MD 20814

71.1 Noise, Antenna, and System Temperatures.................. 966
Thermal Noise Density • Noise Temperature • Noise Figure • Networks in Tandem • Antenna Temperature • The General Microwave Receiver

71.1 Noise, Antenna, and System Temperatures

Thermal Noise Density

The ultimate capacity of any communications channel is determined by the ratio C/N_o of carrier power to thermal **noise density** and the purpose of RF link analysis is to calculate this ratio. N_o is the composite noise spectral density and is given by $N_o = kT_s$, where T_s is the composite system temperature. It is the temperature to which a resistor at the system input, with the system having only its gain and generating no **noise**, would have to be heated to generate the same amount of random noise as is observed at the real system receiver output. This temperature characterizes the entire system noise performance and is intended to comprise the effects of **thermal noise**, receiver noise, and noise available at the antenna terminals. The latter noise itself comprises two components, the thermal noise of the surroundings with which the antenna interchanges energy and external sources of noise in the sky.

All noise considerations are based on the irreducible thermal noise due to the random motion of electrons in conductors. The full quantum expression for the spectral density of this noise is

$$N_o = \frac{hf}{e^{\frac{hf}{kT}} - 1} - hf \tag{71.1}$$

In the range of frequencies of usual interest in satellite communication $kT \gg hf$ and we have Nyquist's familiar formula for thermal noise density

$$N_o = kT \tag{71.2}$$

The Calculation of System Temperature for a Microwave Receiver

In a bandwidth B the resulting voltage across a resistance R is given by Nyquist's formula $\sqrt{4kTBR}$. If we consider a source of impedance R generating this voltage, then the power available into a matched load R is simply kTB. This concept of available power is used almost exclusively in considering the noise performance of composite systems.

Noise Temperature

Figure 71.1 is an arbitrary two-port network characterized by a power gain G and excess noise characterized by parameters to be defined.

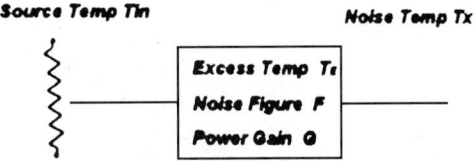

FIGURE 71.1 Noise parameters: basic network.

We define the **noise temperature** at a point as that temperature to which a matched resistor must be heated to generate the same available noise as is measured at that point. Thus, the noise at any point x in the chain is given by $N_x = kT_x B$. If the point in question is the input, then the noise temperature is often written as T_i and, more specifically, if an antenna is at the input then the temperature is written as Ta.

A different and altogether basic characterization is found in the **system temperature** T_s, which is the temperature to which a resistor at the input must be heated to generate a noise equal to the observed total noise output. Other sources of noise are neglected. The output noise can now be written as $N_x = GkT_s B$, and $GT_s = T_x$. The system temperature will clearly be higher than the input source temperature by virtue of the extra noise generated by the network. We use this idea to define an **excess temperature** T_E, characteristic of the network itself, and to write the noise temperature at any point T_x, by the equations

$$T_s = T_{\text{in}} + T_E$$
$$T_x = G(T_E + T_{\text{in}}) \tag{71.3}$$

Performance figures and noise temperatures quoted by receiver manufacturers are usually this excess temperature T_E, naturally enough, since it is a measure of the receiver quality, but it is a frequently seen error to use that temperature in the calculation of system performance. That computation must use the system temperature, which requires the addition of the **antenna temperature**, and consideration of other sources of noise.

Noise Figure

An older, and still much used, figure of merit for network noise performance is **noise figure** F. It is based on the idea that the reduction in available **carrier-to-noise ratio** (CNR) from the input to the output of a network is a measure of the noise added to the system by the network and thus a measure of its noise quality. The difficulty with that concept in its naive form is that the same network will seem better if the input CNR is poorer and vice versa. To avoid this anomaly the definition is based on the assumption that the available noise power at the input is always taken at room temperature, arbitrarily taken in radio engineering work to be 290 K.

$$F = \frac{C_{\text{in}}/N_{\text{in}}}{C_{\text{out}}/N_{\text{out}}} \tag{71.4}$$

This can be rewritten several ways. Note that the input and output carrier levels are related by the available **power gain** and, in accordance with our definition, the input noise level is simply $kT_o B$.

One form, occasionally taken as a definition, is that the output noise is given by

$$N_{\text{out}} = FGkT_oB \tag{71.5}$$

Note well that, although the definition of noise figure assumes the input at room temperature, it can be measured and used at any input temperature if the calculations are made correctly. We can understand this better by rewriting the preceding equation as

$$N_{\text{out}} = GkT_oB + (F-1)GkT_oB \tag{71.6}$$

$$N_{\text{out}} = GkT_{\text{in}}B + (F-1)GkT_oB \tag{71.7}$$

In the form of Eq. (71.6) we can identify the first term as attributable to the amplified input noise and the second term as the excess noise generated by the network. In fact we can change the term T_o to T_{in} where T_{in} is an arbitrary input source temperature, for instance, that of a receiver antenna. By comparing the second term of Eq. (71.7) to that of Eq. (71.3) we see that they are both terms for the excess noise and thus write

$$T_E = (F-1)T_o \tag{71.8}$$

This is a particularly useful relation inasmuch as the need to go back and forth between the two characterizations of excess noise is frequent.

The Hot Pad

A very important special case is the noise characterization of a plain attenuator, at either room temperature 290 K or some arbitrary temperature T. Transmission line loss between an antenna and the first stage of RF amplification, typically a **low-noise amplifier** (LNA), the loss in the atmosphere due to atmospheric gases such as oxygen and water vapor, and the loss due to rain can all be considered as **hot pads**. It can be shown [Mumford and Scheibe, 1968] that the noise temperature of such an attenuator is

$$T_N = \frac{T_{\text{in}} + (L-1)T}{L} \tag{71.9}$$

Networks in Tandem

To be able to apply our set of definitions and relations to a real microwave receiver, we need a relation between the characteristics of two individual networks and their composite characteristics when the networks are in tandem or cascade (see Fig. 71.2).

The calculation of the noise performance of two stages in tandem is done by equating the noise output as calculated from the combined parameters, either F or T_E, with the noise output as calculated by taking the noise output of the first stage as the noise input for the second stage. From

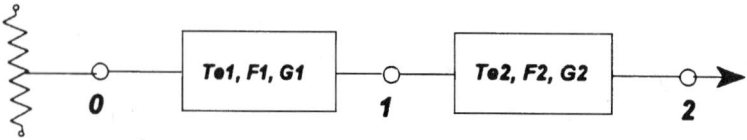

FIGURE 71.2 Two networks in tandem.

Eq. (71.3) we have, for the output noise in terms of the noise at x,

$$T_x = G_1(T_{\text{in}} + T_{e1})$$
$$T_{\text{out}} = G_2(T_x + T_{e2}) \quad (71.10)$$
$$T_{\text{out}} = G_1 G_2(T_{\text{in}} + T_{e1}) + G_2 T_{e2}$$

We can also write the output noise as $G_1 G_2(T_{\text{in}} + T_e)$, again by using Eq. (71.3). By equating the two expressions for output noise we arrive at the fundamental relation for the composite

$$T_e = T_{e1} + \frac{T_{e2}}{G_1} \quad (71.11)$$

The identical approach can be used with excess noise being characterized by noise figures, or Eq. (71.11) can be converted using the relation between excess noise temperature and the noise figure given by Eq. (71.7). In either case the result is

$$F = F_1 + \frac{(F_2 - 1)}{G_1} \quad (71.12)$$

Antenna Temperature

We have been using the term antenna temperature casually with the understanding that the input to a microwave receiver is usually connected to an antenna and that in some way the antenna can be considered as generating an equivalent input temperature. This is indeed the case, and we can make the idea more precise by defining the antenna temperature as the weighted average temperature of the surroundings with which the antenna exchanges energy. For example, if two narrow beam antennas communicated with each other with no loss to external radiation, and if one of these antennas were connected to a matched resistor at a temperature T_A, then T_A would be the antenna temperature of the other antenna.

An Earth station antenna typically exchanges energy with the sky in the vicinity of the satellite at which it is pointed, the ground via its **side lobes**, and the sun, usually in the antenna side lobe structure but occasionally in the main beam. A satellite antenna exchanges energy, principally with the ground, but also with the sky and occasionally with the moon and sun. Obviously the satellite orbit and antenna coverage determine the situation. Antenna temperature is much affected by atmospheric losses and especially by rain attenuation. By way of example, a narrow beam antenna pointed at the clear sky might have a temperature of 20 K but this will increase to more than 180 K with 4.0 dB of rain loss, as is easily calculated from the hot pad formula (71.9). With a **receiver excess temperature** of 50 K the system temperature will increase from 70 to 230 K or 5.0 dB. Note that this is a greater loss than that due to the signal attenuation by the rain. This is a frequently overlooked effect of rain attenuation. The implication for system design, at those frequencies where rain loss is anything other than negligible, is significant. The margin achieved using high-performance, low-noise receivers, is something of an illusion. The margin disappears quickly just when one needs it the most. The performance improvement achieved with a larger antenna is better than that achieved with a low-noise receiver because the deterioration of the former because of rain attenuation is only that due to the signal loss, whereas the latter suffers the signal loss and the increase in antenna temperature.

The antenna temperature can, in principle, be calculated from the definition by evaluating the integral of the temperature, weighted by the **antenna gain**, both as a function of solid angle. Due allowance must be made for reflections. The exact expression is

$$T_a = \frac{1}{4\pi}\int G_1 T_{sky} d\Omega + \frac{1}{4\pi}\int G_2[(1-\rho^2)T_g + \rho^2 T_{sky}]d\Omega_2 \quad (71.13)$$

Where:

G_1 = antenna gain in a sky direction
G_2 = antenna gain in a ground direction
Ω_1 = solid angle in sky directions
Ω_2 = solid angle in ground directions
T_g = ground temperature
T_{sky} = **sky temperature**
ρ = voltage reflection factor of the ground

The sky temperature, the starting point for most calculations of antenna temperature, is found by first determining the clear sky temperature, normally very low. The spectral absorption of atmospheric gases raises this temperature in accordance with the hot pad effect and yields a sky temperature that is a function of frequency and angle of elevation. Figure 71.3 is a plot of this sky

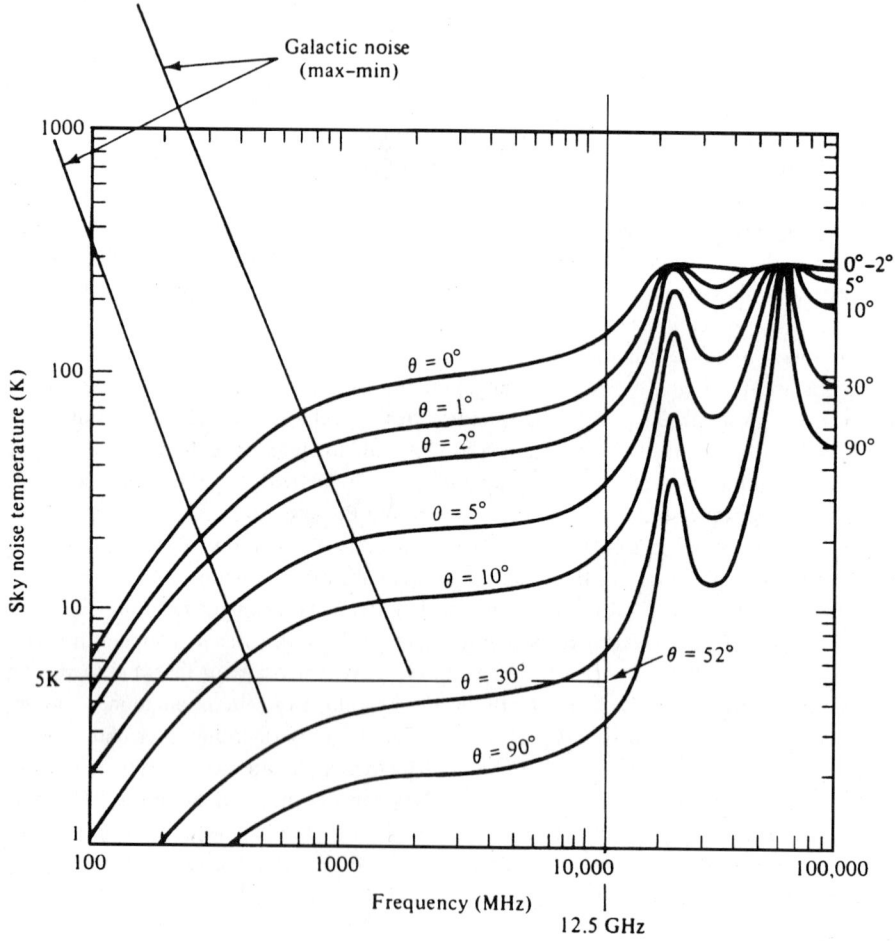

FIGURE 71.3 Sky noise temperature calculated for various elevation angles.

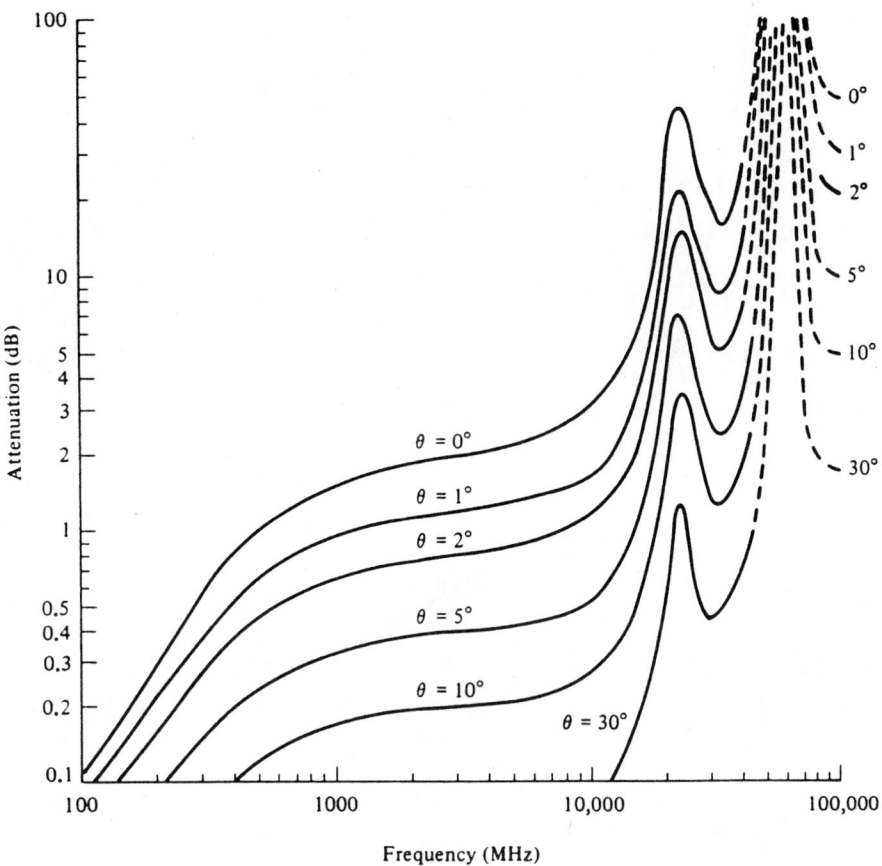

FIGURE 71.4 One way signal attenuation through troposphere at various elevation angles.

temperature vs. frequency with angle of elevation as parameter. Note the presence of a microwave window in the frequency region high enough for the galactic noise to have fallen to a low value, nominally the 3.5-K microwave background, a remnant of the big bang, and low enough to avoid the increases due to atmospheric gas absorption. The frequency sensitivity largely mirrors the absorption spectra of these gases (shown in Fig. 71.4) and the dependence on angle of elevation seen in Fig. 71.5 results from the increased path length through the atmosphere at low angles. Note that there are regions in the sky where there is radio noise at a higher level than shown and generated by various galactic sources. Some of these sources of radio noise are used to calibrate the noise performance of Earth stations.

Ground and rain temperatures are typically taken to be $T_o = 290$ K. The integration requires a knowledge of the antenna pattern in spherical coordinates and the location of ground masses, reflecting surfaces, reflection factors, and sky temperatures. It can be carried out numerically but normally it is complicated and not worth the trouble. A result sufficiently accurate for most practical purposes can be found using

$$T_a = a_1 T_{\text{sky}} + a_2 T_g + a_3 T_{\text{sun}} \tag{71.14}$$

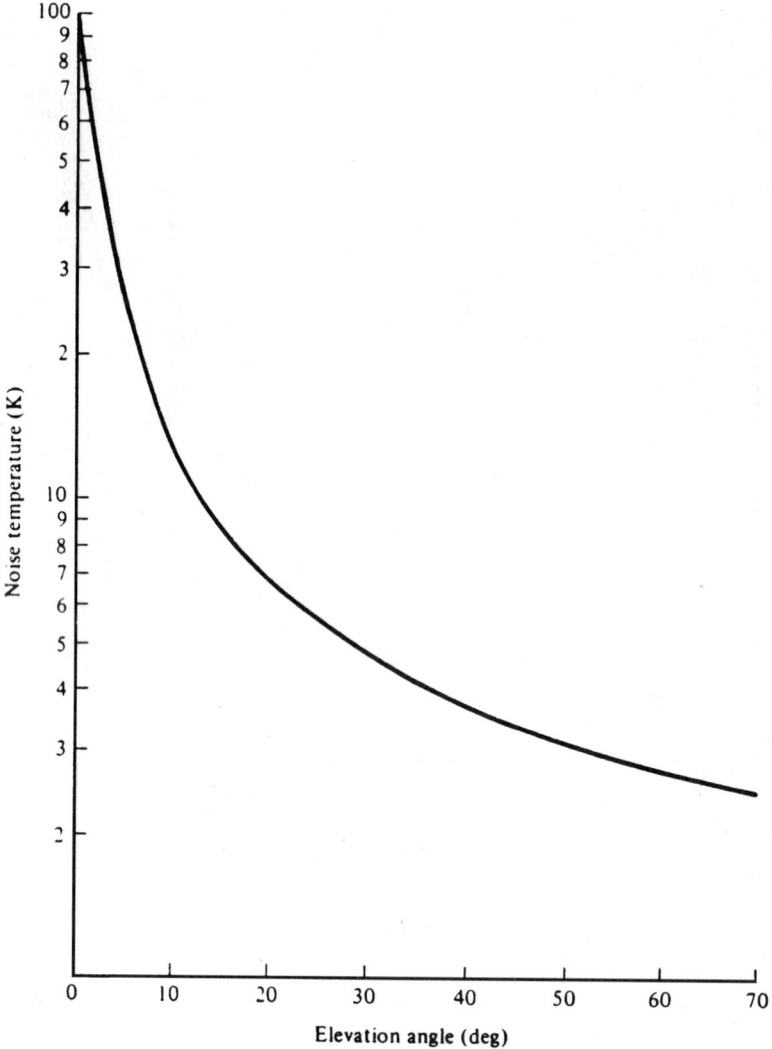

FIGURE 71.5 Sky noise temperature at 4 GHz versus elevation angle.

where the terms a_1, a_2, and a_3 are the relative solid angles covered by the main beam, side lobes looking at the ground, and side lobe looking at the sun, respectively. They can be approximated by

$$a_1 = \frac{1}{4\pi}(G_{sky}\Omega_{sky} + \rho^2 G_g \Omega_g)$$
$$a_2 = \frac{\Omega_g}{4\pi} G_g (1 - \rho^2) \tag{71.15}$$
$$a_3 = \rho \frac{\Omega_s}{4\pi} G_s / L_r$$

The idea is to obtain an average of the temperature weighted by the antenna gain. The Ω and G are the solid angles and gains, respectively, of the main beam, side lobes that see the ground, and side lobe that sees the sun. Here, ρ is the voltage reflection factor of the ground, and p is a factor to allow for the sun's radiation being randomly polarized while the antenna receives linear or

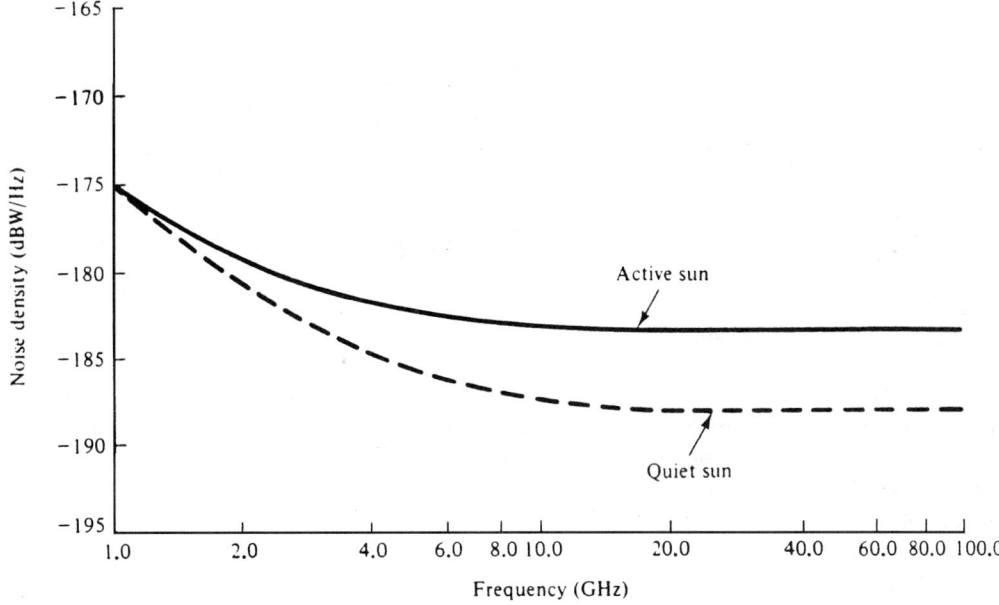

FIGURE 71.6 Noise spectral density of quiet and active sun. (From *NASA Propagation Effects Handbook for Satellite System Design*, ORI TR 1983.)

circular polarization only; p is typically $1/2$. The assumption is that the sun is not in the main beam but rather in a side-lobe where the gain is much lower, and this reduces the effect of the sun's high temperature. If the sun is in the main beam, and if the antenna beam is less than or equal to the sun's subtended angle (0.5°), then the increase in temperature is formidable and produces a **sun outage**. The sun's apparent temperature can be found from Fig. 71.6.

Note that rain loss L_r has the good effect of shielding the receiver from the sun. On a beam reasonably wide compared to the sun, the margin normally available to protect against rain loss, say 6.0 dB or so at Ku band, is sufficient to ameliorate the **sun interference** to a bearable level even with the sun in the main beam. An antenna of 50 cm in diameter for the direct reception of satellite TV has a beamwidth of 3.5–4.0°, and a solid angle 50 times that of the sun. In effect this reduces the sun's temperature by 50 times.

The General Microwave Receiver

Figure 71.7 shows the functional diagram of a general microwave receiver. Almost all satellite and Earth station receivers fit a diagram of this sort.

We can use Eqs. (71.11) and (71.12) pairwise to combine the stages and yield an overall result. The antenna noise, discussed in the last section, is taken as the input, and we assume that there are no significant noise contributions beyond the mixer **down converter**. An important and useful point concerns the noise figure of a matched passive attenuator. Such a device has the result of reducing the available carrier power by its loss in decibels, while not changing at all the available thermal noise. Thus, its noise figure, defined as the ratio of input to output carrier to noise ratios, is simply equal to its loss. The result, from combining the stages of the receiver two at a time is

$$T_s = T_a + (L-1)T_o + LT_r + L(F-1)\frac{T_o}{G_r} \qquad (71.16)$$

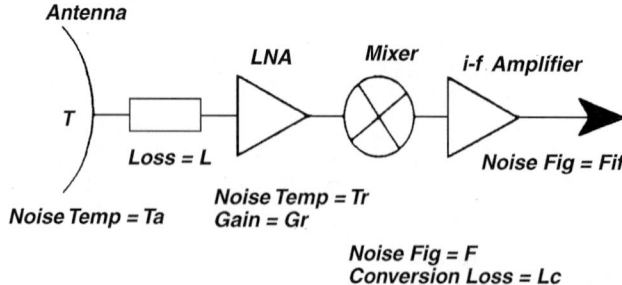

FIGURE 71.7 General microwave receiver.

This equation is informative. Besides showing the more or less obvious advantages of low-noise receivers, it highlights the unexpectedly devastating effect of the loss L in the line between the antenna and the LNA. Even 1.0 dB here produces a temperature increase of 75 K. Such a loss is by no means unrealizable in a long line such as, might be necessary between a prime focus feed and a convenient receiver location in a large antenna. The need to place the LNA directly at the horn feed, or to use a Cassegrainian type of antenna so that the LNA can be placed at the reflector vertex, is clear.

Defining Terms

Antenna gain: Directivity, ratio of power transmitted in a particular direction to the power that would be transmitted in that direction if the antenna were an isotropic radiator.
Antenna temperature: Noise temperature at the antenna terminals
Carrier to noise ratio: Ratio of carrier to available noise power.
Down converter: Mixer to reduce RF carrier frequency to more convenient and constant value for amplification.
Excess noise temperature: Apparent temperature of extra noise above input noise added by any black box.
Hot pad: Input attenuator at some temperature higher than room temperature (rain, for example, at room temperature compared to sky noise at a few Kelvins.
Low-noise amplifier (LNA): First amplifier in receiver, just after antenna, with very good excess noise characteristics.
Noise: Random electrical signals that interface with signal detection.
Noise density: Noise in Watts per unit bandwidth.
Noise figure: Ratio of input to output carrier to noise ratio.
Noise temperature: Apparent temperature of resistance producing the observed or calculated noise.
Power gain: Ratio of available power at output of a network to available power at the input.
Receiver temperature: Excess temperature of entire receiver including low-noise amplifiers and down converter.
Side lobe: Part of the antenna radiation pattern off the main been but still radiating measurable energy.
Sky temperature: Apparent noise temperature of the sky.
Sun interference: Increase in antenna noise because antenna is pointed at the sun.
Sun outage: Sun interference so great that receiver is blinded completely.
System temperature: Apparent temperature of resistance producing the observed noise of the entire system if connected to the input.
Thermal noise: Noise due to random electron motions dependent on temperature.

References

Bousquet, M., Maral, G., and Pares, J. 1982. *Les Systémes de Télécommunications par Satellites*, Masson et Cie, Éditeurs, Paris.

Ippolito, L.J., 1986. *Radio Wave Propagation and Satellite Communication Systems*, Van Nostrand, New York.

Ippolito, L.J., Kaul, R.D., and Wallace, R.G. 1986. *Propagation Effects Handbook for Satellite Systems Design*, NASA Ref. Pub. 1082(03), 3rd ed., Washington, DC, June.

Jasik, J. 1961. *Antenna Engineering Handbook*, McGraw-Hill, New York.

Kraus, J.D. 1950. *Antennas*, McGraw-Hill, New York.

Lin, S.H. 1976. Empirical rain attenuation model for Earth-satellite path. *IEEE Trans. Commun.* COM-27(5):812–817.

Mumford, W.W. and Scheibe E.H. 1968. *Noise Performance Factors in Communications Systems*, Horizon House-Microwave Inc., Dedham, Mass.

Pritchard, W.L., Suyderhoud, H.G., and Nelson, R.A. 1993. *Satellite Communication Systems Engineering*, 2nd ed., PTR Prentice Hall, Englewood Cliffs, NJ.

Onboard Switching and Processing

A.K. Elhakeem
Concordia University

72.1 Introduction.. 976
   Satellite Switching (SS) Advantages • Applications
72.2 Onboard Switching Types 978
   Reconfigurable Static IF Switching • Regenerative IF Switching • Onboard Baseband Switching • Baseband Fast Packet Switches • Photonic Baseband Switches • Asynchronous Transfer Mode- (ATM-) Oriented Switches
72.3 Summary 994

72.1 Introduction

The **onboard** (**OB**) combination of **satellite switching** (**SS**) and spot beam utilization is intended to enable a large number of small and inexpensive mobile as well as fixed site terminals to communicate and integrate with commercial broadcasting, integrated and various voice, data, and video applications. This interconnection should not only serve as a standby for terrestrial and optical networks, but also as a competing alternative. With the launching of low Earth orbit satellites (LEOs) [Louie, Rouffet, and Gilhausen, 1992] approaching, the roundtrip satellite propagation delays will become an order of magnitude less than those of geostationary Earth orbits (GEOs). Also, ground terminals will be cheaper, thus providing the edge for competing with ground-based networks. The concept of a bent pipe satellite will be long gone and users now operating in few zones around the globe will be divided into many zones with one satellite antenna supporting local communication of each zone and the switch onboard the satellite routing the traffic from zone to zone within the cluster (Fig. 72.1). If a constellation of LEOs or medium Earth orbit satellites (MEOs) is to cover the globe, then each satellite will handle one cluster, but intrasatellite links have to be provided by a ground network or by a space-based network for users belonging to different clusters to intracommunicate. OBSS systems apply to GEO, LEO, and MEO constellations. Operating in this environment while accommodating various **packet switching** (**PS**) and **circuit switching** (**CS**) types of traffic affects the selection of the OB switch type and technology. Before looking at the various OB switch types, details, performance, and fault tolerance, we state their main advantages and applications.

Satellite Switching (SS) Advantages

1. Terminal cost is reduced by lowering transmission power (smaller user antennas used). Also reducing the number of users per zone reduces user bit rate and hence cost.

Onboard Switching and Processing

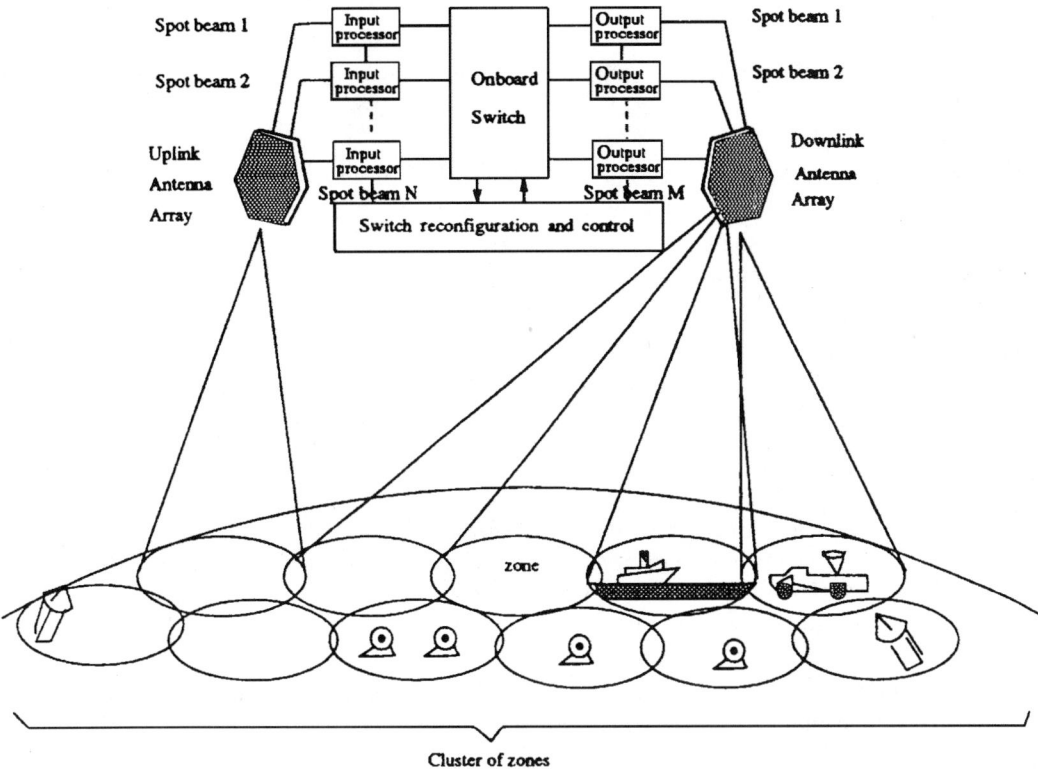

FIGURE 72.1 Onboard (OB) processing satellite using multiple-beam antennas and onboard switching.

2. Up- and downlinks are decoupled, thus allowing their separate optimization. For example, for stream (CS) and wideband Integrated Services Digital Network (WBISDN) users in a certain zone, time-division multiplexing (TDM) uplink and TDM downlink techniques are adequate. On the other hand, for a multiple of bursty Earth stations, in another zone, a **demand assignment multiple access–time-division multiple access (DAMA–TDMA)** uplink/TDM downlink technique allows the optimal sharing of the uplink capacity. A third zone in heavy rain or under shadowing fading conditions might prefer a switching satellite that has code-division multiple access (CDMA) in the uplink and/or downlink.
3. The number of possible virtual connections is reduced from an order of N^2 to the order of N where N is the number of users.
4. Flexibility and reconfigurability includes, among other things, reprogramming of onboard processing (OBP) control memories, reconfiguration of Earth stations (time/frequency plans for each connection setup), accommodation of both circuit and packet switched traffic, adopting different flow and congestion control measures for each zone depending on current traffic loads and types, and accommodation of multirate and/or dynamically changing bit rate and and bit rate conversion on board.
5. SS has natural amenability to DAMA and ISDN services.
6. There is enhanced fault tolerance; a single satellite failure handling a cluster of zones can be easily replaced by another within hours or minutes thus averting the disasters of losing customers for a few days or months. A single failing antenna onboard the satellite can be easily replaced by another hopping antenna.

These advantages come with a price, that is, the cost and complexity of the satellite itself.

7. SS increases the total capacity of the satellite system in terms of number of circuits and services provided within the same bandwidth.

Applications

- Internetworking, low-cost and low-power user terminals, fixed site and mobile communications
- Video conferencing
- Interconnection of business oriented private networks, local area networks (LANs), metropolitan area networks (MANs), wide area networks (WANs)
- Narrowband and wideband ISDN
- Interconnection of all of the preceding to and through the public telephone network (PTN)
- Supercomputers networking
- Multiprotocol interconnection
- Universal personal communication (UPC)

72.2 Onboard Switching Types

- Reconfigurable static IF switching, also called **microwave switch matrix** (**MSM**)
  This has fixed (nonhopping) beams. The access techniques could be **frequency-division multiple access** (**FDMA**), TDMA, or multifrequency-TDMA (MF-TDMA) giving rise to three kinds of switched satellites, that is, SS-FDMA, SS-TDMA, SS-MF-TDMA. The first, SS-FDMA, is the most common.
- Reconfigurable static intermediate frequency (IF) switching with regeneration and forward error correction but no baseband processing
- Baseband switching with fixed or hopping (scanning) spot beams
- Fast packet switching
- Photonic baseband switches
- Asynchronous transfer mode (ATM) oriented switches

Reconfigurable Static IF Switching

Both the uplinks and downlinks access techniques could be TDMA, FDMA, or MF-TDMA. However, the most commonly used are FDMA and TDMA on the uplinks and downlinks, respectively. This choice results in a reduced terminal ground station cost. FDMA is selected on the uplinks to eliminate the need for high-power amplifiers (HPA) at the user terminal [an FDMA user transmits all of the time but with a fraction of the bandwidth (BW)]. On the other hand, to utilize the full power of the OB (HPA), it should operate near saturation, thus increasing the downlinks signal and allowing very small aperture terminals (VSAT) operation. Simply a high-power high bit rate TDM operation in the downlinks is efficient. No demodulation or error correction takes place onboard. Fixed beams are used, and the number of such beams in the uplinks is typically less than that of the downlinks. Stations use frequency synthesizers to transmit the information using an RF carrier corresponding to destination. This form of FDMA is called single channel per carrier (SCPC) and is appropriate for low rate users (64 kb/s). This carrier is demand assigned to the user call via a signalling channel (which could be ground based or an out of band SCPC channel). Onboard, the various RF carriers pass through the low noise amplifier then become down converted to IF (Fig 72.2). Demultiplexing separates the various user signals that form the inputs to the static MSM. These are routed through the MSM to their destinations. The connection pattern of the switch (which input connected to which output) is configured by the master ground station or ground network control center (NCC) control channel. However, some of the connections may be permanent to allow broadcasting. Taking the example of the advanced communications technology satellite (ACTS) system of NASA [Cashman, 1992], the uplinks access technique is TDMA not FDMA, and the MSM is controlled by a digital controller and a dual storage bank of memories. The MSM switch is configured according to the 5-kb/s, telemetry, tracking, and command signal (TT&C)

Onboard Switching and Processing

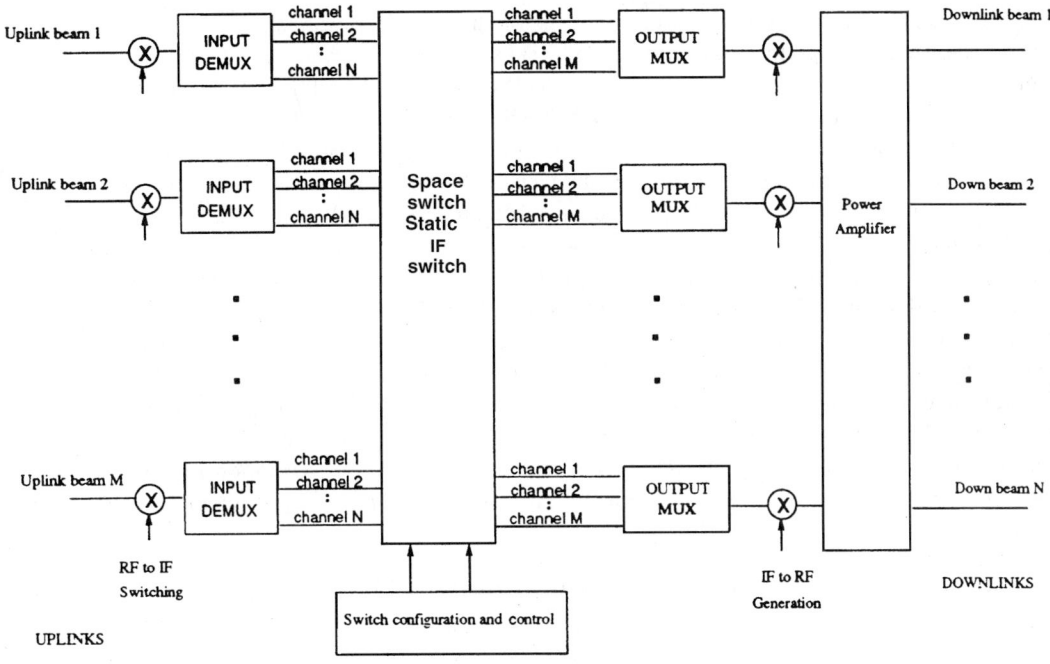

FIGURE 72.2 Nonregenerative onboard IF switching.

issued from the master ground station at the beginning of each superframe of 1-min duration. The TDMA frame is 1 or 32 ms and the switching time is approximately 100 ns. One bank of memory will be in active control of the switch while the other is waiting for TT&C new configurations. In the case of SCPC uplinks, the switch is configured on call basis (slower basis). Three active fixed or hopping beams exist on the MSM input and output. However, the MSM size is 4×4 to allow for redundancy. The output signals of the MSM switch going to same zone are grouped (multiplexed). The resultant is subsequently translated to RF frequency, amplified by solid-state power amplifiers (SSPA), and transmitted via output beams. An in-band or out-of-band signalling scheme is used by the user terminals to establish the call. The master control station finally assigns the right resources to the call, for example, one out of a pool on the FDMA uplinks and one time slot of a certain TDMA frame on the downlinks for that call. This reservation goes to the OBS, which becomes correspondingly configured for the call duration. We notice here the isolation between the up- and downlinks provided by OBS, for example, in the preceding case the uplink user transmits continuously using the assigned FDMA SCPC, i.e., disjoint from the TDM process taking place on the downlinks. Amplifying these FDMA signals OB creates intermodulation products (IMP). SCPC also suffers from its inability to meet variable bit rate (VBR) services. Multichannel per carrier (MCPC)/TDMA or MF/TDMA is typically employed to reduce these IMP. Here a number of users (terminals) are allocated channels (time slots) on a TDM frame, which is transmitted on a single carrier.

Other disadvantages of FDMA oriented MSM are: the difficulty of dynamic networks reconfiguration without data loss, the difficulty of operation with hopping beams, and broadcasting of multiple carrier transmissions to many destinations.

Regenerative IF Switching

This is similar to the previous static IF switching in the following respects: switching of IF signals, employing TDMA, FDMA, or MF-TDMA access techniques with the FDMA uplinks/MF-TDMA downlinks as the preferred technique, no baseband processing being involved, a station simultaneously transmitting/receiving from many stations having to use a different carrier for each source (destination), programming reconfiguration via fixed connections and/or under the TT&C commands, use of low noise amplifiers (LNAs) and SSPAs, and RF to IF and IF to RF translations.

However, it differs in the following respects. The IF signals (following RF to IF translation and demultiplexing) are demodulated, forward error corrected, then forward error encoded [using error detecting and correcting coding (EDAC)] and then modulated [continuous phase shift keying (CPSK) or minimum shift keying (MSK)] again before feeding them to the MSM switch. If both uplinks and downlinks are of the MF-TDMA kind, then the various uplinks user TDM channels (slots) of each beam form the IF inputs to the switch. At the switch output, output IF slots are multiplexed again, and the resultant modulates the RF carrier of the downlinks beam. The regeneration involved in the demodulation–modulation processes allows the separation of uplinks and downlinks noises, fading, interference effects, and so forth. Also the decoding–encoding could be done at different rates thus providing for more protection against rain in the downlinks, for example, by decreasing forward error correction (FEC) encoding rate. These regenerations lead to better links performances and better amenability to MF-TDMA generation that needs higher transmission rates. Nevertheless, regenerative IF switching suffers the same aforementioned disadvantages of static IF switching.

Implementation of Regenerative and Static IF Switches via MSM Technology

Basically, this is a solid-state programmable crossbar switch. Signal switching at the matrix crosspoints is performed using dual-gate field-effect transistor (FET) active switches (employing GaAs technology for switches with IF bandwidths equal to or exceeding 2.5 GHz). In recent implementations, such as the NASA ACTS OBS [Cashman, 1992], input signal distribution and output signal combining are performed using passive recursive couplers (Fig. 72.3).

Among many alternatives, such as single pole, rearrangeable switching, and so forth, the coupler crossbar implementation was selected by both ACTS and other NASA systems for its simple control, low insertion loss, and amenability to broadcasting. In this architecture, a switching element [Fig. 72.3(b)] is situated between two directional couplers. If the data is to flow from input i to output j, the switch will be ON and the input and output couplers pass the data; otherwise, the switch is OFF and the data is reflected and absorbed by the coupler loads. To equalize the output power at all N input couplers connected to the N output lines, the coupler coefficients increase as the output number increases, that is, $\alpha_j = 1/(N - j + 2)$. The same logic α_j applies to the output coupler coefficient β_i as in Fig. 72.3(b). Switches at the various cross points are independently controlled to provide broadcasting as well as point-to-point connections. Moreover, the switch design could be modular both electrically and mechanically (as in NASA designs); for example, different switch points or parts of the switch can be repaired and tested independently, thus improving cost and reliability, or larger switches can be built from smaller modules.

The switch elements could be built using ferrite switches, p-i-n diodes, or single gate FETs. The first option is too slow, the second needs larger amounts of DC power, the third yields no gain during ON states; dual gate FETs are typically preferred. These use Schottky barrier diodes with very low junction capacitance, thus allowing the switching from gain to high insertion loss in about 100 ps. Two FETs are used, with the RF input signal applied to the first and the switch control applied to the second. This provides for excellent isolation between the DC control (which ranges from 30-dB loss to 10-dB gain) and the RF input. One could envision an IF switch with constant coupling coefficients (α_j, β_i) and variable switch amplifier gain. Typically, however, the earlier variable coupling gains implementation is preferred.

The switch states (configuration) and latency of each cross point are decided upon by ground control signals. This control unit is typically called control distribution unit (CDU). One bank of the dual control memory will be in active control of the switch states while the other bank receives the fresh ground control commands. These ground control signals also cause the exchange of the two memory bank identities to take account, for example, termination and/or new call establishments.

Onboard Switching and Processing

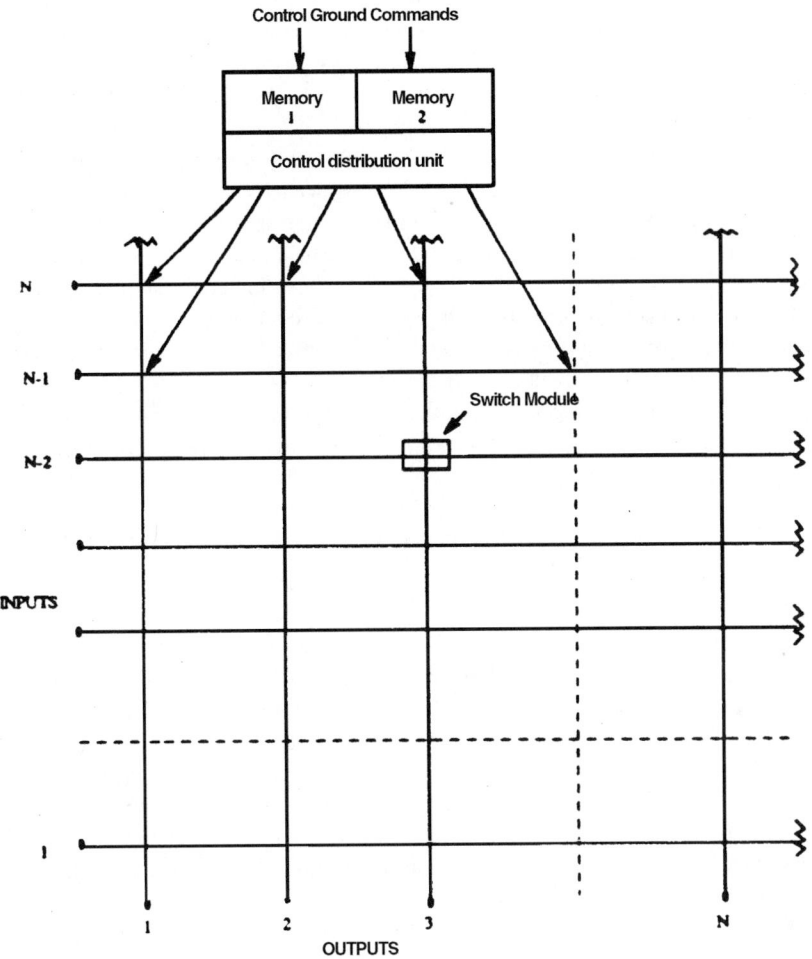

FIGURE 72.3(a) $N \times N$ IF switch matrix using crossbar configuration.

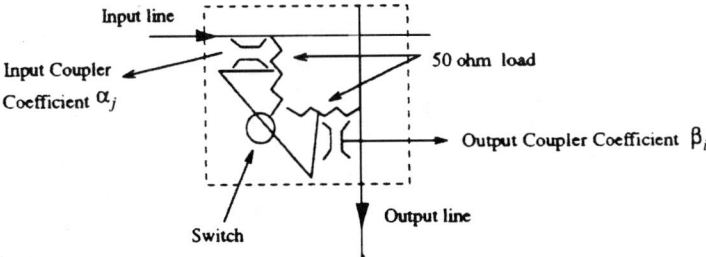

FIGURE 72.3(b) Switch module units: $\alpha_j =$ couplering coefficient of jth output cross point on the same input line; $\beta_i =$ couplering coefficient of ith cross point on the same output line.

The switch sequence (logical signals) that controls the switching state of all cross points is periodically dumped from the control memory till fresh control commands are loaded to the other memory bank from the ground control station, in which case the switch control signals are dumped from the other memory bank. The reconfiguration rate is a few microseconds and the switching time is in the 2–10 ns range.

The reliability of the IF switch matrix is highly dependent on the failure rate of the GaAs FET, which is roughly equal to 23×10^{-9} (including the associated passive elements). This reliability can be improved by adding extra rows and columns so as to provide alternate routes from inputs to outputs that do not pass by the failed cross points [the dashes of Fig. 72.3(a)]. Additional reliability could be achieved by using multiple ($N \times N$) switching planes in parallel at the expense of cost and weight. For the switch control unit, using the memory decoder yields a reliability of 0.8331 for 10 years operation. This can be improved further by using simple forward error correction techniques to correct for possible memory read or write errors.

For an $N \times N$ cross point IF switch and if the time frame has M data units, the control memory size is given by $M(N \log_2 N)$, since N cross points are controlled by $\log_2 N$ bits of memory (if decoding is used).

Implementation of Regenerative and Static IF Switches Using Optical Wave Guides

Future low orbit satellite clusters will probably communicate through intersatellite optical links at date rates of the order of few gigabits per second. This will make the instant and immediate switching of these signals in the optical domain a great asset compared to the classic tedious and slow approach of optical to electric conversion, switching of IF or baseband signals, then electrical to optical conversion at the switch outputs. In both IF and baseband implementations of optical switches, a directional coupler is the basic switching element.

The directional coupler (Fig. 72.4) consists of two waveguides situated close enough to allow the propagating modes (carrying the information signals to be switched) to interact (couple) and transfer power from one waveguide to another. Electrical control is still applied to allow the ON or OFF switching of the input signal to one or both outputs. This controls the index of refraction and, therefore, the coupling coefficients.

In the OFF state, a 3-dB optical isolation is achievable between outputs. If the switching voltage is 25 V, coupling lengths lie in the range of 1–2 cm, the loss associated with single mode fiber termination is of the order of a few decibels, and switching times lie in the 1–100 ns range. It is possible, however, to lower these to 50 ps by using traveling wave electrodes or by using free space optical switching techniques.

To reduce the N^2 complexity of MSM cross point switches using optical waveguides, different configurations are used. The one shown in Fig. 72.4(b) uses only $N(N-1)/2$, where $N = 6$ couplers, thus saving, for example, $(36 - 15) = 21$ couplers for a (6×6) switch.

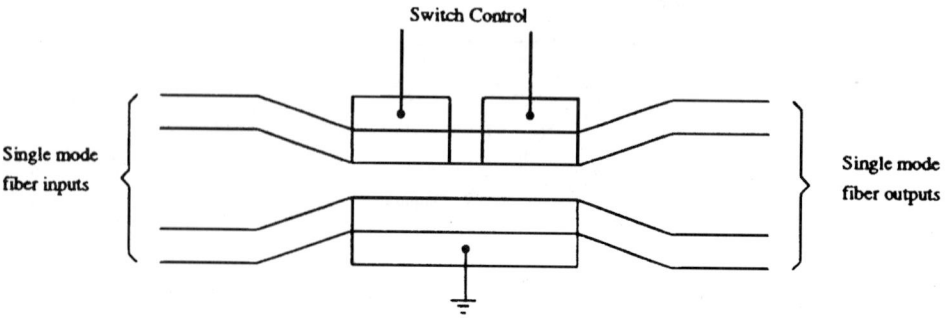

FIGURE 72.4(a) A 2×2 all optical switch module using directional coupler.

Onboard Switching and Processing

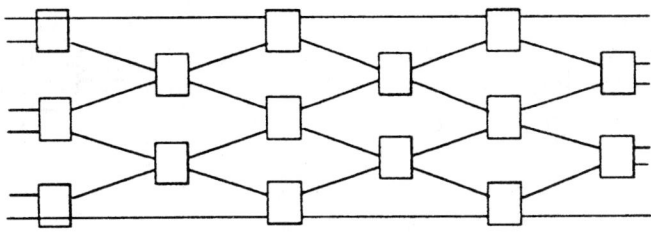

FIGURE 72.4(b) Multistage implementation of an optical $N \times N$ switch using $N(N-1)/2$.

Optical waveguides are not without disadvantages; for example, switch size is limited by the area of available substrate, high insertion loss (due to fan in losses of waveguides), high level of crosstalk, internal blocking in Benes, Banyan, or general multistage types, which eventually lead to difficulties in operation in broadcasting modes.

To enhance the reliability of optical wave guide IF switching architectures, replication in the case of $N \times N$ cross point structure or addition of a few input and output lines, and dilation (addition of more parallel lines between stages) in the case of Benes and multistage structures are generally recommended.

Onboard Baseband Switching

The uplink signals are downconverted, demodulated (possibly FEC decoded), switched, and then modulated, FEC encoded, upconverted, and transmitted downlink (Fig. 72.5). Switching takes place at baseband level, thus allowing for a much needed flexibility. For example, a station with only one carrier at hand can transmit to many destinations via different TDMA slots. These transmissions could reach the destinations on the different carriers, the different bit and FEC rates,

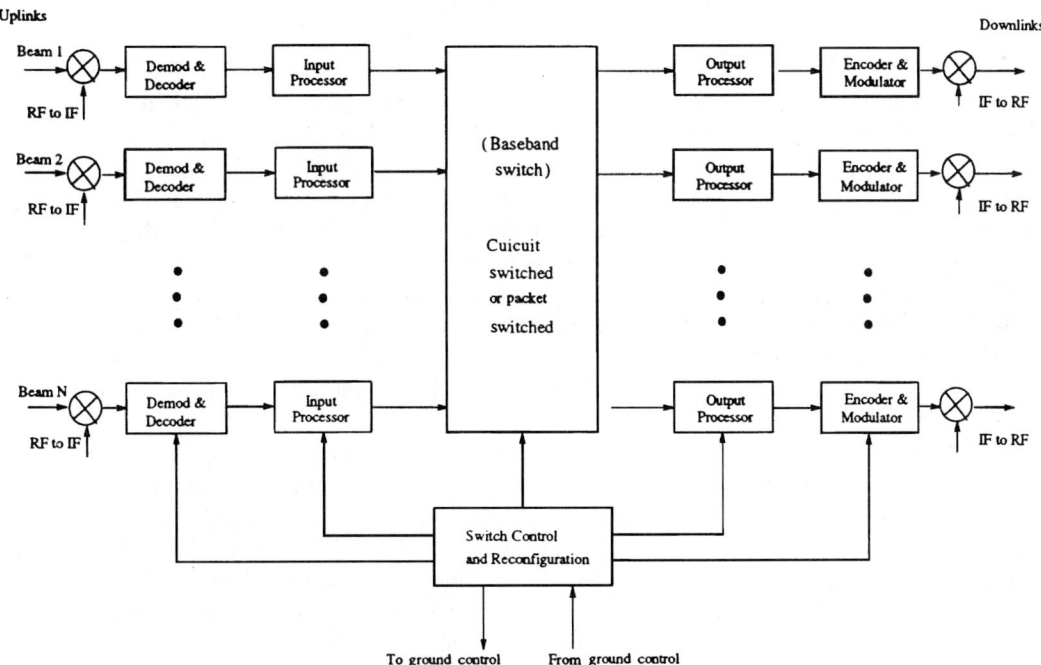

FIGURE 72.5 Building blocks of onboard baseband switching.

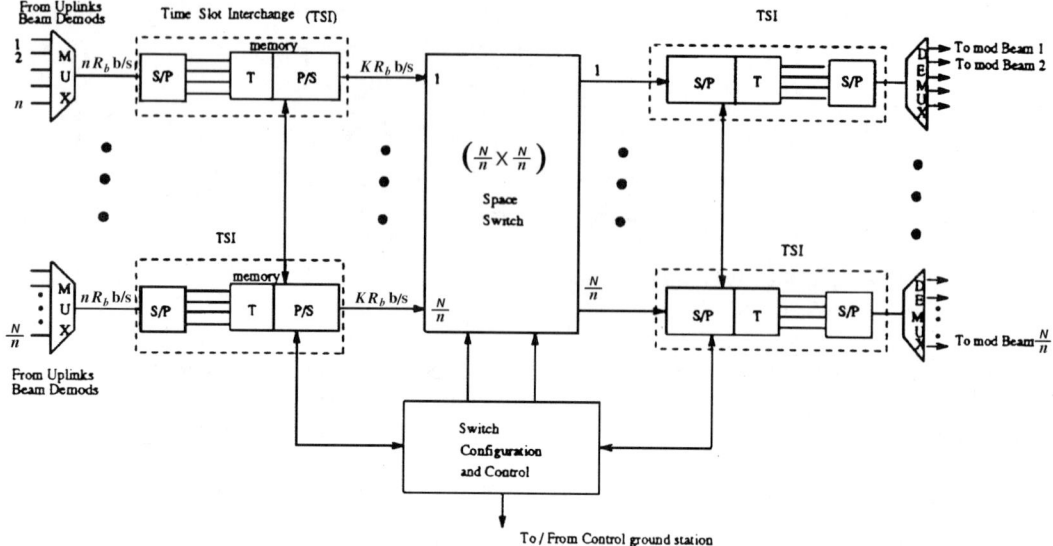

FIGURE 72.6(a)   T–S–T baseband switching.

the different modulation formats, and so forth, that best suit the fading environment and other conditions in the downlink. Multicasting and accommodation of a mixture of high and low bit rates without needing multicarriers as in IF switching are some of the added benefits.

Baseband switching can be oriented toward circuit or packet oriented traffic. When large portions of the network traffic and/or most of the users are of the circuit switched type, and/or switch reconfiguration is infrequent and onboard, baseband circuit switching is more efficient. For a satellite network carrying mostly packet switched traffic, and/or circuit switched traffic with frequent configuration, and/or a combination thereof, a packet switch is more efficient. Hybrid packet/circuit baseband switching is also a viable alternative that has gained attention recently.

Baseband Circuit Switching Categories

Time–space–time (T–S–T), common memory (T-stage), distributed memory, and high-speed fiber optic ring are the main types of this category.

(T–S–T) Baseband Switching.   This is used for multigigabit per second operation and has been adopted in the NASA ACTS system and for European Space Agency (ESA) prototypes. It has also been in use for some time in terrestrial networks (AT&T no. 4ESS switch). The basic idea is to reduce the space switch complexity by premultiplexing n of input data units (e.g., bytes), thus reducing the number of inputs and outputs of the space switch. However, this reduced switch has to switch data much faster, as will follow.

Following RF to IF conversion, filtering, and data demodulation and forward error decoding in each of the (N/n) uplink banks [Fig. 72.6(a)], the n baseband data units of each multiplexer (with a combined rate of nR_b b/s) are further multiplexed via the time slot interchange (TSI) box [Fig. 72.6(b)] to yield a bit rate at the TSI equal to kR_b b/s, where $k \geq 2n$ to allow for a strictly internally nonblocking switch. Internal nonblocking means that there are enough virtual paths inside the switch to connect any of the (N/n) inputs to any of the (N/n) outputs. The TSI is basically a serial to parallel converter followed by a T-stage (memory), which stores the data units of the uplink frames, followed by a parallel to serial converter. At the space switch output, another TSI executes a reversed sequence of events that finally feeds all data units of the downlink frame to the data forward error encoder and modulator, after which IF to RF conversion and subsequent transmission in the downlink takes place. To see the reduced complexity of this approach compared

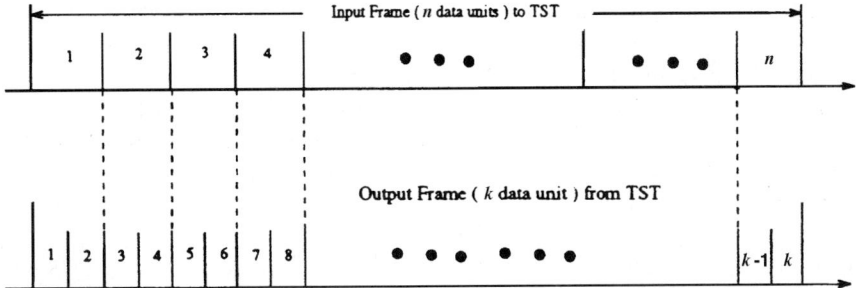

FIGURE 72.6(b)   Time slot interchange (TSI) reading from T-stage is at least twice as fast as writing into this stage ($k \geq 2n - 1$).

to a conventional cross point switch, we take $N = 1024$, multiplexing of $n = 128$ slots (each can serve one unit of data), $k = 2n = 256$. The size of the space switch is $(N/n \times N/n)$, that is, (8×8) a great simplification compared to the corresponding conventional cross point switch of (1024×1024) complexity. The relation $k \geq 2n$ was derived in the 1950s for a class of three-stage Clos space switch, which resembles the T–S–T switch, and it effectively says that the speed of the space stage inputs (output of TSI unit) should be at least twice as fast as that of the TSI input [Fig. 72.6(b)]. Memory read and write from the T stages limit the speed of the external lines (from data demodulator and FEC decoders). If we take the frame time $T = 4$ ms, 1 byte as the data unit, input data rate of $R_b = 4$ Mb/s per input beam, the T-stage total memory size should be $2n \cdot T_F \cdot R_b \cong 4$ Mb. The total memory requirements for all input and output T-stages become $4(8+8) = 64$ Mb/s. The total switch throughput $= NR_b = 1024 \cdot 4$ Mb/s ≈ 4 Gb/s, which can be increased by increasing (N/n), that is, the number of pins of the space switch, and the memory access time per word of the input or output T-stages (assuming a 16-b-wide memory bus) is given by $1/(KR_b/16) = 16$ ns.

If current technology does not permit, one has to reduce n and/or the input data rate R_b. Dual memories (T-stages) of size $2n \cdot T_F \cdot R_b$ bytes are used on both input and output TSIs such that simultaneous read and write to the same memory can be achieved. Also, pipelined operation is assumed such that while bytes of the new frame are shifted at the input speed and stored in T-stages, bytes of the previous frames are shifted out to the output stage at the output speed.

To accommodate a multiplexity of user terminals of high and low speeds, additional multiplexing of lower rate users [not shown in Fig. 72.6(a)] is implemented at the input lines (prior to input TSI stage). This makes for a square $N/n \times N/n$ switch structure.

A sophisticated control processor (consisting mainly of memories and TSI timers) reconfigures the switch in a conflict free manner during the call establishment period for CS traffic and each frame for PS traffic. This consumes time and hardware since all three stages (input TSI, space switch, output TSI) have to be correspondingly adjusted. For example, if a user in input beam 3 has a call in slot 5 of the input frame to a user in output beam 6 who listens in slot 9 of the output frame, then the space switch control part makes sure that input beam (3) is connected to output beam (6) in each frame for one slot time. The TSI control at the switch inputs and outputs make sure that the input byte is written in the fifth location in input memory and the output byte read from the ninth location.

To improve T–S–T reliability, (N/n) active T-stages and r redundant ones are used in both input and output sides of the space switch, which is by itself replicated. Also error detection and correction can be used to protect the memories from permanent and transient errors. For example, 16 out of 20 (N/n out of $N/n + r$) redundancy and 1 out of 2 redundancy for space switch yields 99% reliability for 10 years.

It is also noted that the T–S–T architecture could in principle be used for routing packet switched traffic as well. However, switch control will be derived in this case from the individual input packet headers, a process that slows the switch compared to the case of CS traffic. It is also generally assumed that all corresponding time slots of all inputs to TSIs are bit synchronized. This requirement will add costly hardware, especially in the case of hybrid PS/CS traffic; for example, some preambles will be stripped at the switch input, while some central headers will be routed through the switch. Furthermore, T–S–T switching has the advantage of expandability to higher speeds, and the disadvantage of encountering two switching TDM frame delays (corresponding to the input and output TSI stages).

Baseband Memory Switch (MS). This is the simplest and least costly architecture [Evans et al., 1986]. During a certain TDM frame, demodulated data bytes from all uplink beams are stored in a single double (dual) data memory (Fig. 72.7) in a sequential manner under the control of a set of counters or control memory. Parallel in time, the stored bytes from the previous switching TDM frame are served to the appropriate switch output under the supervision of the control memory. For one double dual memory to replace the many T-stages of the previous T–S–T architecture, this memory has to be large enough and fast enough to store the data bytes of the synchronized M input TDM frames in one frame time. A high-speed bus and a multiplexer (MUX) are thus used, and a demultiplexer (DEMUX) distributes the data memory contents into the various output channels under the control of simple counters and the configuration memory. Memory switches could be asymmetric (number of inputs $\neq$ number of outputs) and are internally nonblocking. The data memory size is $(2NS)$, where S is the number of TDM slots per frame and N is the number of output beams. This is half the memory requirements of the T-stages of T–S–T switches and is typically organized into 4 or 8 smaller memory blocks.

The contents of control memory serve as the addresses to the data memory and together with a simple counter associated with each output they read the right location of data memory in the right TDM slot of each output frame. The control memory is reconfigured with each new call. However, this is less intensive in time and in control hardware compared to the T–S–T case.

This architecture could possibly route packet switched traffic, in which case the switch control is derived from the destination in the packet header, which is more involved and time critical.

Bus speed and memory access time upperbounds the capacity of this architecture compared to the T–S–T that grows independently of these. Also, the memory is a reliability hazard avoided by

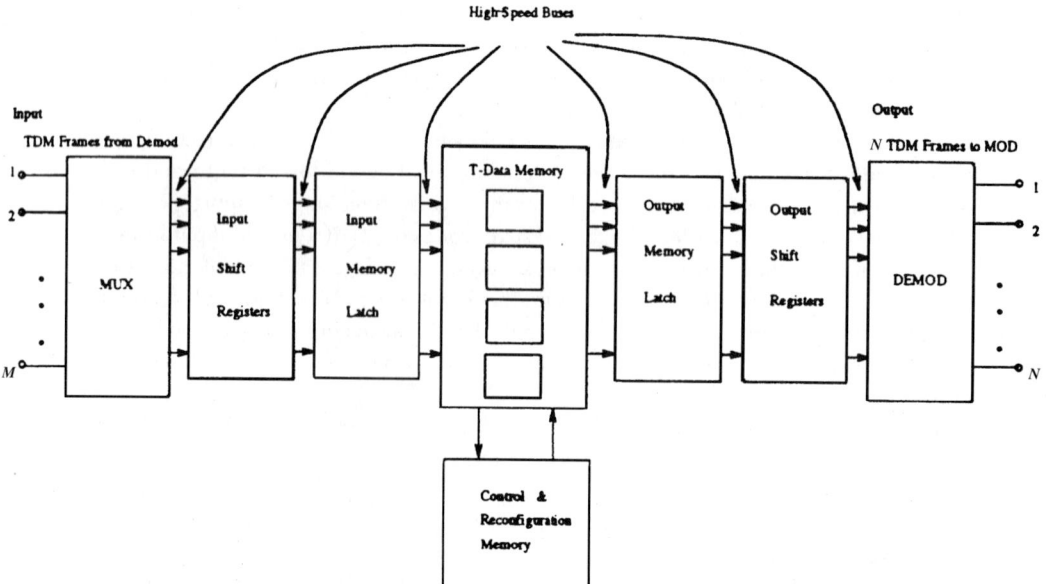

FIGURE 72.7 A baseband memory switch.

Onboard Switching and Processing

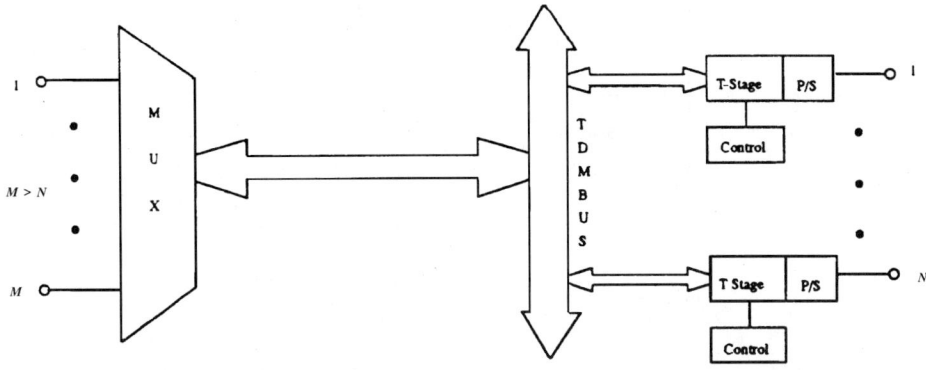

FIGURE 72.8 Distributed memory switch (DMS).

memory replication, and immunity to soft (intermittent) errors is provided by error detection and correction between data memory banks and read and write latches.

Power and weight are important considerations, and in recent designs (for example, the PSDE-SAT2 of the ESA), memory switches were approximately four times lighter than T–S–T switches and they took only 1/10th of the chips.

Distributed Memory Switch (DMS). To improve the reliability of the **memory switch** (**MS**) architecture, the memory could be split into N T-stages at the output side (Fig. 72.8). This way, failure of one or more of the T-stages means that only few (not all) beams are out of service.

Multiplexing of the uplink baseband TDM frames and sequential writing to the TDM bus takes place as in MS. All T-stages hear the broadcast TDM data; however, under distributed control and timers, only the appropriate data is sequentially written at the right time slot into the applicable T-stage. Reading data from T-stage to output beams is done via random access in different time slots of the output TDM frame. The memory size is still the same as in MS, and the distributed memory switch (DMS) is a strictly nonblocking switch. Redundancy is achieved by duplicating each T-stage; the MUX is also duplicated. Control units are also distributed and duplicated for fault tolerance. Reconfiguration of the T-stages is simple compared to T–S–T switches.

However, T–S–T architecture has better expansion capabilities than MS or DMS structures. Though bus speed is not a problem in the DMS, the bus is overloaded by the presence of the many T-stages, thus increasing the bus driver power consumption. Error detecting and correcting codes are also applied to the individual T-stages to protect against reading and writing errors.

Baseband Fast Packet Switches

Circuit switches are more or less amenable to handling packet switched traffic [Inukai and Pontano, 1993]. For example, the MS architecture becomes self-routing (and thus able to handle packet switched traffic) by stripping the headers of the packets that contain the destination and storing them in the control memory, which then selects the appropriate data unit from the data memory to the appropriate output register. In the case of the DMS architecture, these stripped packet headers are routed (using an address bus) to the distributed control units at the various output beams. A fiber optical token access scheme ring such as that of the fiber distributed data interface (FDDI) standard is directly adaptable to both CS and PS traffic.

MS, DMS, and T–S–T switch throughputs for PS traffic are reduced by the described header stripping, bus contention, memory size, and so forth. A self-routing multistage switch is generally preferred for handling PS traffic if throughputs higher than a few gigabits per second are required. A control memory is unnecessary in this case since part of the packet header is used to route the

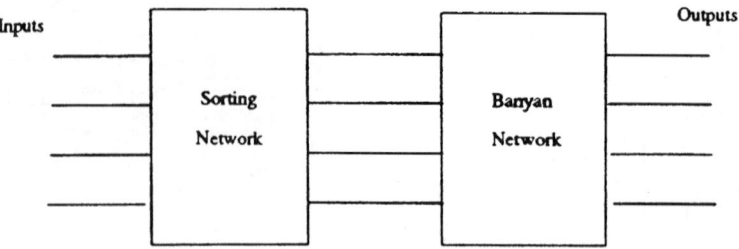

FIGURE 72.9(a) A sorter-Banyan baseband switch.

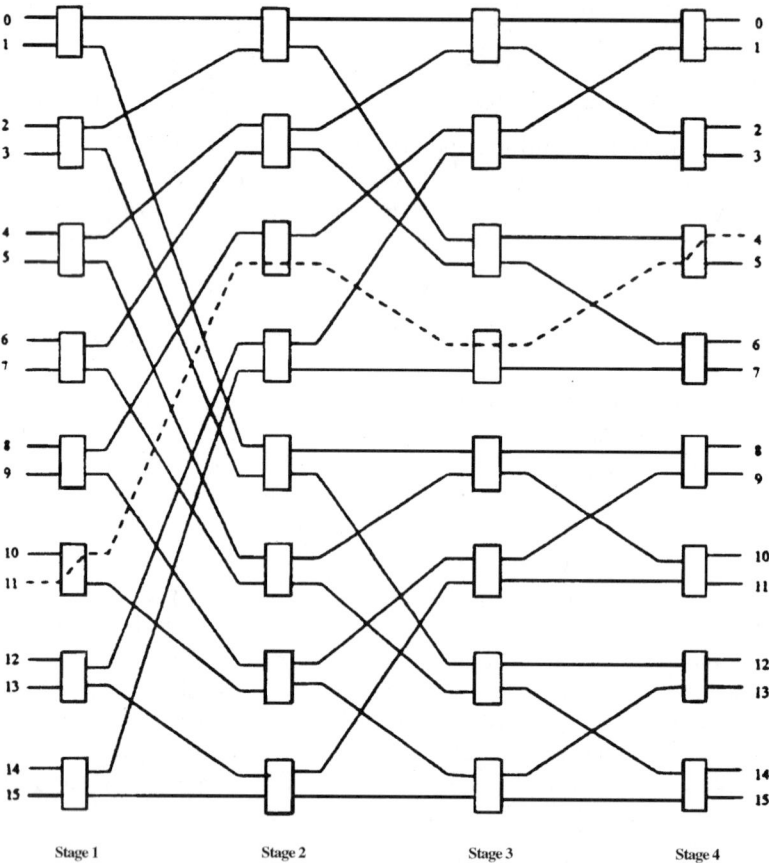

FIGURE 72.9(b) Banyan network with four stages and 16 input terminals. The dotted path corresponds to a packet header of 0100.

switch through each stage. Error detection and correction are used to protect against misrouting the packet due to header errors. However, complex data and control memories are not used; rather a set of cascaded (2 × 2) module switches is used. A typical multistage switch of the Banyan switch (BS) type is shown in Fig. 72.9(a) [the sorted **Banyan network** (**BN**) or Batcher Banyan network] [Hui and Arthurs, 1987].

The heart of this is the Banyan network (also called delta network), which is detailed in Fig. 72.9(b) for a 16 × 16 BN. The packet finds its path automatically by self-routing within the various stages in the $N \times N$ rectangular Banyan network, which is composed of ($\log_2 N = 4$) stages. Each stage

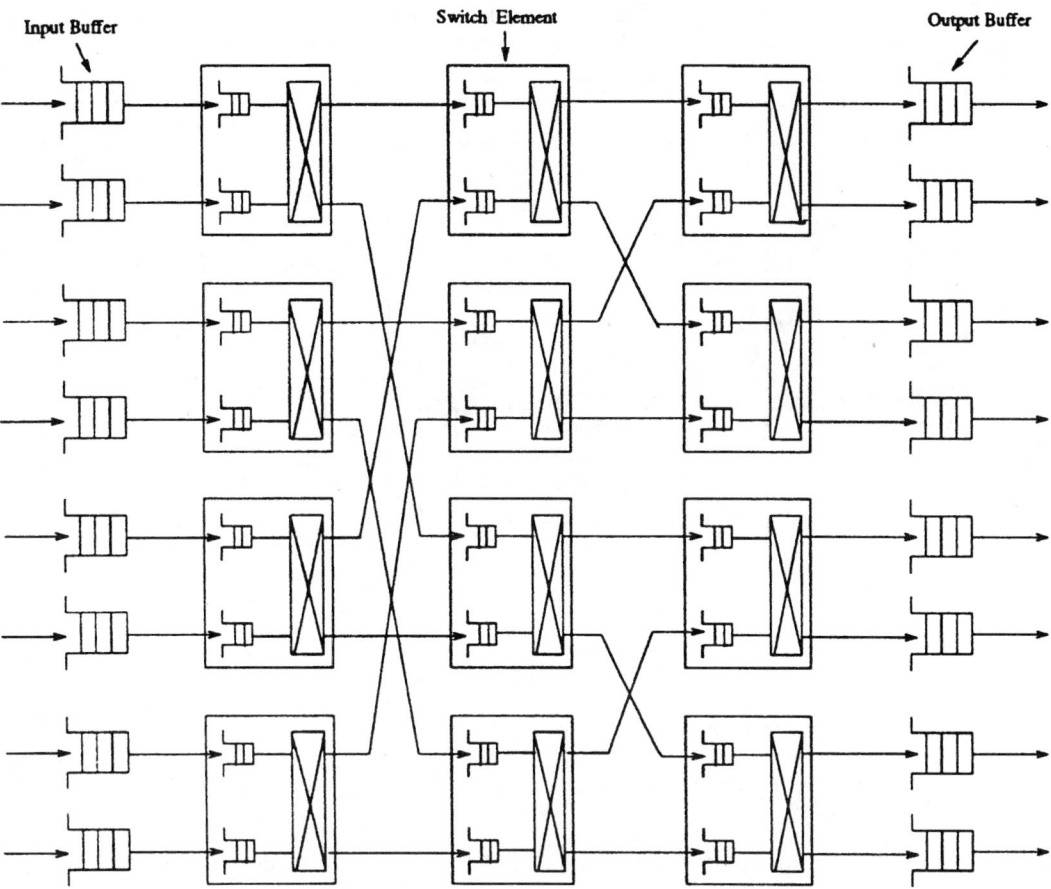

FIGURE 72.10 A buffered Banyan network.

consists of $(N/2)$ smaller switching elements. Each element is a small (2×2) binary switch. A control bit of 0 or 1 moves the input packet to the upper or lower element output, respectively.

A BN has the property that there is exactly one path from any input to any output but has the disadvantages of exhibiting internal blocking, that is, two packets from two inputs colliding at the same intermediate switching element output, even if they were destined to different final outputs. All of the previous CS described earlier have the advantage of being strictly nonblocking in this sense. However, both types (internally blocking and nonblocking switches) may have final output blocking, which is due to two packets from different inputs destined to same output. To solve the two problems of internal and output blocking, the sorting network of Fig. 72.9(a) should be used or one has to replace the whole figure by an alternate buffered Banyan network (Fig.72.10). In the latter, a limited buffering capacity exists at each (2×2) module input. If two input packets are destined to the same output, one packet is routed to the output and the other is temporarily stored for subsequent routing in the following time slot.

Among recent buffered Banyan implementations is the Caso–Banyan baseband switch of Toshiba [Shobtake and Kodama, 1990]. In that implementation, the FIFO property of the 2×2 switch element was violated such that when two head of line data units of the two input buffers contended for the same output, only one is forwarded to this output and another packet (not the head of line) of the other input is forwarded to the other output. It has been found that the unit data (packet, cell, byte) blocking rate is 1% of that given by conventional buffered Banyan networks at normalized input utilization factor $\rho = 0.85$. Now the sorting network approach of Fig. 72.9(a) [Hui and Arthurs, 1987] is used to reorder packets at the N input requests so they appear ordered according

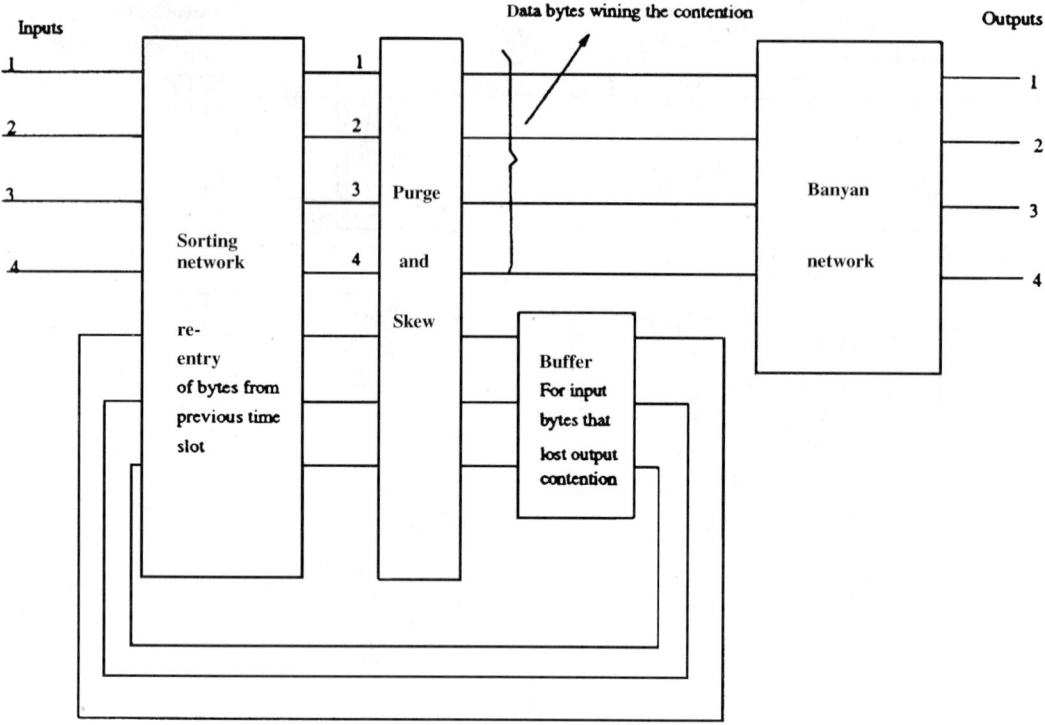

FIGURE 72.11 A Starlite network.

to their packet destination at the sorter N outputs (on the first phase). Also, in this phase only short connection requests are made by various inputs. This packet rearrangement guarantees no internal blocking when feeding these to the subsequent BN.

Output blocking is avoided by purging contending packets and acknowledging only the input requests that have priority (in the second phase), and forwarding only one of the sorter output packets to the BN (in the third phase).

Unacknowledged inputs will try retransmission of their output blocked packets in subsequent time slots. This three-phase algorithm slows the switching speed and lowers the achieved throughputs. Other implementations, such as the Starlite switch (Fig. 72.11), still have BN and use buffers at the output to store contending packets that did not win the output contention for subsequent routing in the following time slots, but do not have a separate request (phase 1) of the three-phase algorithm underlying Fig. 72.9(a). The Starlite approach [Huang and Knauer, 1984], however, has more hardware (buffers and control), and may produce packets out of sequence at the switch outputs.

Having realized that increasing the speed of Banyan type networks (reducing the blocking) was achieved only at the expense of increasing the hardware, a straightforward solution of *replicated-Banyan switches* provided modularity as well. Here, the number of internal paths (space diversity) is increased by distributing input traffic among many Banyan switches in different planes, thus reducing internal blockings. Output blocking may still be decreased by incorporating buffers at the various outputs of each switching plane. *Internal speed up* could be used to speed the data passage through the switch elements in all of the previous architectures (sorted Banyan, replicated Banyan, buffered Banyan, etc.) in a manner similar to that used in baseband CS (T–S–T, MS, DM). This leads to an almost strictly internal nonblocking switch as was previously outlined and a *space* switch component with less inputs and outputs.

Similar to baseband CS, the Banyan switches suffer from both *stuck at* and intermittent faults. To increase their reliability, replication of switching planes, *dilation* (addition of parallel links and more switching stages within the same plane), and using EDAC codes to protect the packet headers before and after the baseband switches are the state-of-the-art techniques of fault tolerance in the baseband PS case [Kumar, 1989]. Also, self-testing hardware and testing sequences have been designed and added onboard to detect the presence of stuck at faults.

Gamma switching networks [sometimes called inverse augmented data manipulator (IADM)] have a multistage structure, which uses a 3 × 3 switching element and allows broadcasting, a property not readily available in the Banyan switching networks presented.

Omega and *close* switching networks also belong to the multistage group but could be made strictly nonblocking under certain conditions. Lack of amenability to broadcasting is the common thread in most of the Banyan-based and multistage networks. To have this capability, replicated or dilated Banyan networks are essential and/or *copy* networks could be used.

Photonic Baseband Switches

These are ultra fast and efficient switches that can handle a mix of CS and PS traffic and have an inherent broadcasting capability. TDM-based optical rings, optical token rings, **wave division multiplexing** (WDM) optical rings, and knockout switches are the basic types in this category. A single fiber-based ring can support a multitude of switched traffic to a total sum of a few gigabits per second. Optical technology (fibers, waveguides, free space optical transmission) provides a switching media whose bandwidths lie in the terahertz region, hence are bounded only by the processing speed of input and output interfaces, and stringent bit snchronization requirements around the ring, etc. Optical switch throughput exceeding 80 Gb/s can be accommodated with present day technology.

A TDM Optical Ring Switch

Each beam will have one input and one output processor on the optical ring see Fig. 72.12, which mainly shows the baseband stages). The input processor includes not only the optical transmitter,

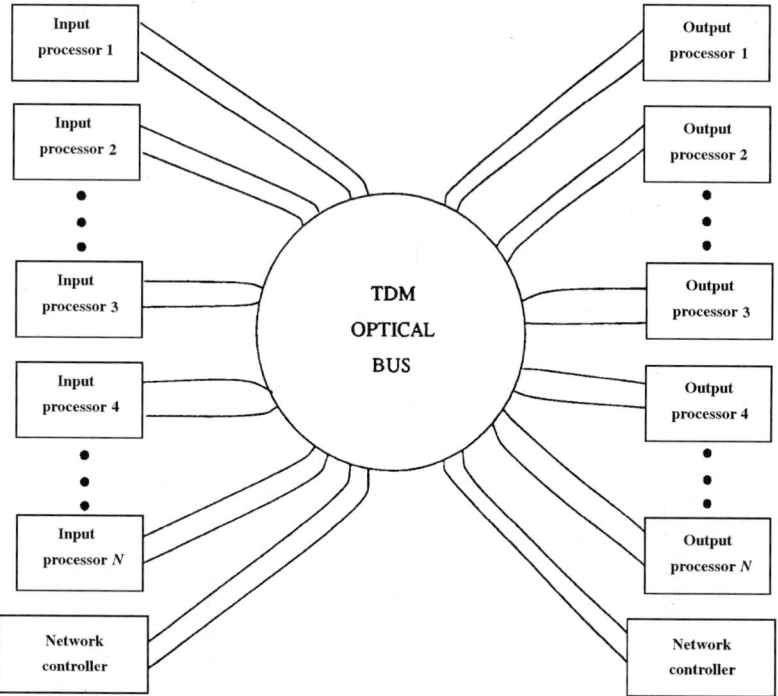

FIGURE 72.12 Onboard TDM optical ring switch.

but also EDAC decoding, packet assembly, descrambling, deinterleaving, burst and/or frame synchronization, input traffic demultiplexing, and possibly bit stuffing for CS traffic. The time frame on the optical ring is divided into slots (that serve, for example, one byte for CS traffic or one packet for PS traffic) and each input processor of an uplink beam is allocated a certain part of this ring switching frame (say, 8 slots) for its traffic to all output processors of the downlink. The optical ring traffic controller, which has a dedicated control packet at the ring TDM frame start, keeps the ring in sync, issues the frame start and ring management commands, and reconfigures (allocates slots) the TDM ring frame among the users (input processors).

For PS traffic the output processor of each beam receives all packets of the ring frame but processes only the ones destined to this beam by a filtering mechanism to recognize the destination address in each packet. For CS traffic, the output processor selects the appropriate slots of the frame based on the controller commands that are updated by the arrival of each CS call. Alternately, the input processor may insert dummy destination bits (addresses) in each CS packet only for routing purposes within the switch (TDM ring). These addresses will be filtered by the intended output processor. Note that dummy bits could be used even in the first approach to have the same packet size for both CS and PS traffic flowing through the ring so as to simplify synchronization and control hardware. Broadcasting is easily achieved by having all output processors identifying certain broadcast bits in the packet header and processing the packet accordingly.

The output processor of each beam will have the optical receiver, modulator, scrambler, interleaver, EDAC encoding circuits, and packet reassembly and multiplexing of the various packets on the TDM frame to the downlink. The reliability of the ring can be increased by using dual rings as in the FDDI standard, where failure of some parts or a whole ring does not affect the ring operation. Also, a bypass switch allows the isolation of faulty input and/or output processors. It is to be noted, however, that the access technique in FDDI is token ring and not TDM.

Wave Division Multiplexed (WDM) Baseband Switches

Figure 72.13 shows a WDM switch that uses a star coupler [Hinton, 1990]. This is a device with M optical inputs and N optical outputs that combines the M optical input signals, then divides them among the N optical outputs. Packetized baseband information is buffered at the various inputs. The input laser has a fixed wavelength. The output receiver should be tuned to the same wavelength as that of the intended input. Once both the transmitter and receiver lasers are triggered by the star controller, the input buffer serially shifts the data bits into the laser transmitter. This controller also

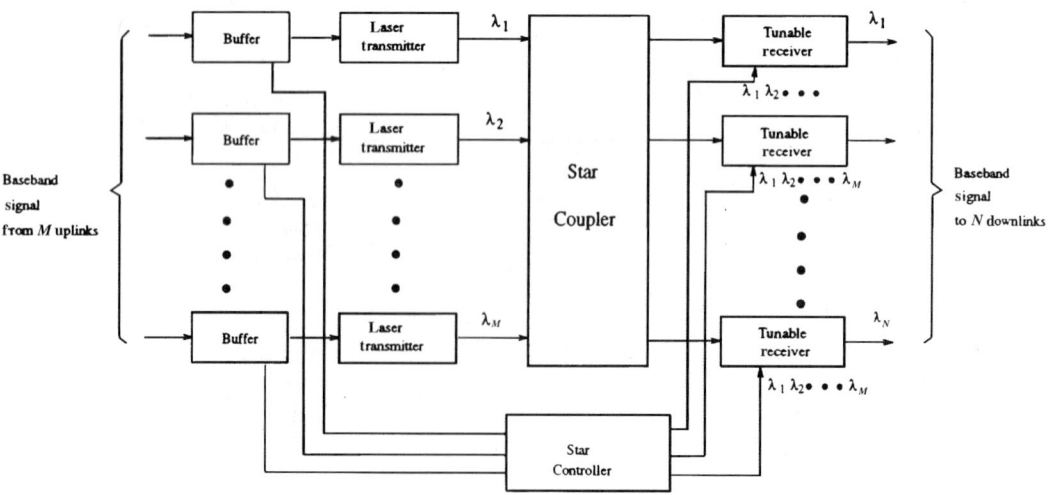

FIGURE 72.13 A WDM-based onboard baseband switch.

resolves the contention for switch outputs. This way, only one of the input packets is shifted, while the other contending input packet is left in the input buffer for the next time slot (packet time).

This represents the PS mode of operation and the amenability to broadcasting is clear, if a certain input packet wavelength is to be broadcast to a set of outputs, all involved output receivers are tuned to the same input wavelength. For CS operation the control commands are received from the ground station via out-of-band signalling channels. These reconfigure the star controller of the WDM switch by the arrival or termination of each new call. These commands may also be easily derived from a signalling TDM subframe slots (in-band signalling).

A mix of CS and PS traffic is easily accommodated, however, dummy bits would be inserted into CS traffic or the packet header stripped from PS traffic before entering the switch so as to have the same packet size passing through the WDM switch. Also, this switch becomes very efficient for directly switching CS intersatellite links in the optical domain.

For a cluster of low orbit satellites (LEO) (64 in the Global Star proposal [Louie, Rouffet, and Gilhausen, 1992]), optical intersatellite communications may render ground gateways (hubs) unnecessary, and will hence reduce the end-to-end user delay. However, with an efficient all optical switch such as WDM, we save the expensive and delay prone optical/electronic and electronic/optical conversions. Reconfiguration control of CS traffic is still done electronically by the arrival of each new call and does not have to be ultrafast. This could be achieved by an out-of-band signal emanating from the ground control station. This signal carries the time slot plan to each involved LEO, to enable it to communicate with its neighboring satellites. WDM technology has reached its maturity in terrestrial applications. Among WDM experiments is the HYPASS, where receiver wavelengths are fixed and the input laser transmitters are correspondingly tuned to the destination address. There is nothing in the weight, size, or cost that will forbid the use of these WDM switches in intersatellite switching applications so as to provide a total switch throughput over 50 Gb/s.

Asynchronous Transfer Mode-(ATM-) Oriented Switches

Onboard switches handling PS and CS traffic could become more involved with the passing of wideband ISDN and multimedia multirate services through the multibeam satellite. With the advent of ATM-based WBISDN, signals of different kinds (video, voice, data, facsimile, etc.) are all physically transported on a standard size packet called a *cell*.

Processing a standard size cell through the OB self-routing fast packet switch (multistage or Banyan types) makes the dummy bits addition and header stripping circuits mentioned earlier unnecessary. All services (PS, CS, variable bit rate, etc.) will have the same cell format. The cell has a header part and a data part. The cell size is 53 bytes, and the header mainly consists of two parts (Fig. 72.14). The first is the 12-b virtual path identifier (VPI) and the second is a 16-b virtual circuit identifier (VCI), which is followed by 12 b of error detection. The VPI is used to route the cell from satellite to next satellite or to ground, etc., enroute to the destination across the internetwork. The second cell field is a VCI between the two communicating entities. Cells find their way onboard the

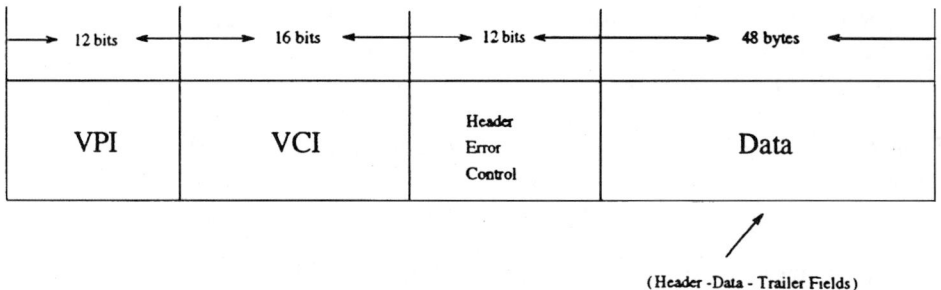

FIGURE 72.14 ATM cell user network interface (UNI).

satellite fast packet switch by a route identifier. This route identifier is obtained by feeding the VPI to a translation lookup table.

However, this translation process may become unnecessary if the VPI is configured such that its first few bits are the routing tag through the satellite packet type switch. This is one of many possible scenarios. If the VPIs are not permanently assigned to ground user stations, the satellite should also have OB translation tables (to obtain switching routing tag) for both the VPI and the VCI of each incoming cell, otherwise different cells from different source–destination pairs may have the same value in their VPI and VCI fields. This implies that the satellite will have the baseband routing switch (ABN type, for example) as well as an ATM switch. This complicates the OB processing, slows down the operation, and is not recommended. Thus far, we have assumed an ATM type of user station. For stations launching packets, not cells, to the satellite, the satellite should still have an ATM switch onboard and a packet processing facility to divide the packet into cells, to add VPIs and VCIs to cells, and so forth. For ground user stations launching CS traffic, the ATM switch onboard will assign a VPI, and a VCI to the incoming traffic and then route the traffic through the onboard baseband switch (BN, photonic, etc.). It is easily seen that unless the ATM users of the world have the ATM interface, the traffic arriving to the satellite during the early migration years of ATM technology may have cell, packet, or CS format thus complicating onboard processing. However, this might be inevitable if satellite vendors are to provide the same service to ground-based optical network users on standby or even on a competitive basis. Once ATM supersedes other switching modes OB processing will be less cumbersome except for the VPI and VCI processors referred to earlier.

72.3 Summary

Numerous switching techniques are in use or being developed for ground-based networks. We have tried to present the most relevant to the satellite environment. Knockout switches, token ring switches, and back to back hybrid packet/circuit switching, such as that of the ACTS system, are some of the techniques that could find application OB.

To compare the different switching architectures, one has to look at their throughput blocking, queueing delay, and reliability performance (see the Further Information section).

Cost of hardware, weight, power consumption, and size of the OB switch are crucial considerations. Comparing the different architectures based on these could be a less quantifiable process. Comparing them on a basis such as amenability to migration to ATM environments, and to different uplinks and downlinks access techniques could be both art and science.

Defining Terms

Banyan network (BN): A self-routing switching technique for baseband signals.
Circuit switching (CS): A channel sharing technique where users are assigned physical or virtual circuits throughout their communications.
Demand assignment multiple access (DAMA): A multiaccess technique where each user is assigned a varying size time and/or frequency slot depending on his call needs.
Frequency-division multiple access (FDMA): A multiaccess techniques where each user is assigned a certain frequency band for his call.
Memory switch (MS): A baseband switching technique for OB processing.
Microwave switch matrix (MSM): A switching technique for microwave signals.
Onboard (OB): Part of the satellite payload.
Packet switching (PS): A transmission and networking technique where the data units are called packets.
Satellite switching (SS): A Satellite that radirects signals in the time, and/or frequency and/or space domains.

Time-division multiple access (TDMA): A multiaccess technique where each user is assigned one or more slots in a certain time frame for his call.

Time space time (T-S-T): A baseband switching technique based on multiplexing in the time and space domains.

Wave-division multiplexing (WDM): A multiplexing technique for optical fiber networks also proposed for OB switching.

References

Cashman, W.F. 1992. ACTS multibeam communications package, description and performance characterization. In *Proc. AIAA 92 Conference*, AIAA, Washington, DC, pp. 1151–1161, March.

Evans, B.G. et al. 1986. Baseband switches and transmultiplexers for use in an on-board processing mobile/business satellite system. *IEE Proc.*, Pt. F, 133(4):336–363.

Hinton, S.H. 1990. Photonic switching fabrics. *IEEE Commun. Mag.*, 28(4):71–89.

Huang, A. and Knauer. 1984. Starlite—A wideband digital switch. In *Proc. IEEE GlobeCom-84*, pp. 5.3.1–5.3.5, Nov.

Hui, J.Y. and Arthurs, E. 1987. A broadband packet switch for integrated transport. *IEEE J. Selec. Areas Commun.*, SAC-5 (8):1264–1273.

Inukai, T., Faris F., and Shyy, D.J. 1992. On-board processing satellite network architectures for broadband ISDN. In *Proc. AIAA 92 Conference*, AIAA, Washington, DC, pp. 1471–1483, March.

Inukai, T. and Pontano, A.B. 1993. Satellite onboard baseband switching architectures. *European Trans. Telecomm.*, 4(1):53–61.

Karol, M.J., Hluchyj, M.G., and Morgan, S.P. 1987. Input versus output queueing on a space-division packet switch. *IEEE Trans. Com. on Comm.*, COM-35(Dec.):1347–1356.

Kumar, V.P. and Reibman, A.L. 1989. Failure dependent performance analysis of a fault-tolerant multistage interconnection network. *IEEE Trans. Comp.*, 38(12):1703–1713.

Kwan, R.K. et al. 1992. Technology requirements for mesh VSAT applications. In *Proc. AIAA 92 Conference*, AIAA, Washington, DC, pp. 1304–1314, March.

Louie, M., Rouffet, D., and Gilhausen, K.S. 1992. Multiple access techniques and spectrum utilization of the GlobalStar mobile satellite system. In *Proc. AIAA 92 Conference*, AIAA, Washington, DC, pp. 903–911, March.

Maronicchio, F. et al. 1988. Italsat—The first preoperational SS-TDMA system. In *Proc. IEEE Globe-Com 88 Conference*, pp. 53.6.1–53.6.5, March.

Oie, Y., Suda, S., Murata, M., and Miyahara, H. 1990. Survey of the performance of nonblocking switches with FIFO input buffers. In *Proc. IEEE ICC 90 Conference*, pp. 316.1.1–316.1.5, April.

Saadawi, T., Ammar, M., and Elhakeem, A.K. 1994. *Fundamentals of Telecommunications Networks*, Chap. 12, Wiley Interscience, New York.

Shabtake, Y. and Kodama, T. 1990. A cell switching algorithm for the buffored Banyan network. In *Proc. IEEE ICC 90 Conference*, pp. 316.4.1–316.4.7, April.

Further Information

Kwan et al. [1992] and Cashman [1992] provide a good treatment of the architectures and applications of microwave IF switches. Maronicchio et al. [1988], Evans et al. [1986] and Inukai and Pontano [1993] cover on-board baseband switches. The architectures and analysis of Banyan and other terrestrial baseband classes are investigated in Hui and Arthurs [1987], Huang and Knauer [1984], Carol, Hluchyj, and Morgan [1987], Oie et al. [1990], and Saadawi, Ammar, and Elhakeem [1994]. Photonic switching and LEO satellites are covered in Hinton [1990] and Louie, Rouffet, and Gilhausen [1992]. Fault tolerant switches are discussed in Kumar and Reibman [1989], which contains an exhaustive list of references.

73
Path Diversity

73.1	Concepts	996
73.2	Site Diversity Processing	997
73.3	Site Diversity for Rain-Fade Alleviation	997
	Measurements • Data Presentation • Diversity Improvement Factor and Diversity Gain • Isolation Diversity Gain • Prediction Models • Empirical models • Analytical Models	
73.4	Optical Satellite Site Diversity	1008
73.5	Site Diversity for Land–Mobile Satellite Communications	1008
73.6	Site Diversity for Arctic Satellite Communications	1008
73.7	Microscale Diversity for VSATs	1010
73.8	Orbital Diversity	1010

Curt A. Levis
The Ohio State University

73.1 Concepts

Propagation impairments along the path between satellite and Earth station are often spatially inhomogeneous. For example, cells of severe rain are generally a few kilometers in horizontal and vertical extent; the regions of destructive interference due to scintillations or to multipath generally have dimensions of a few meters, and so do the field variations due to roadside trees in a satellite communications system for land-mobile use. One way to ameliorate such impairments is to vary the path. When an additional Earth station is used for this purpose, this is called site diversity [Fig. 73.1(a)]; when an alternate satellite is used, it is termed orbital diversity [Fig. 73.1(b)].

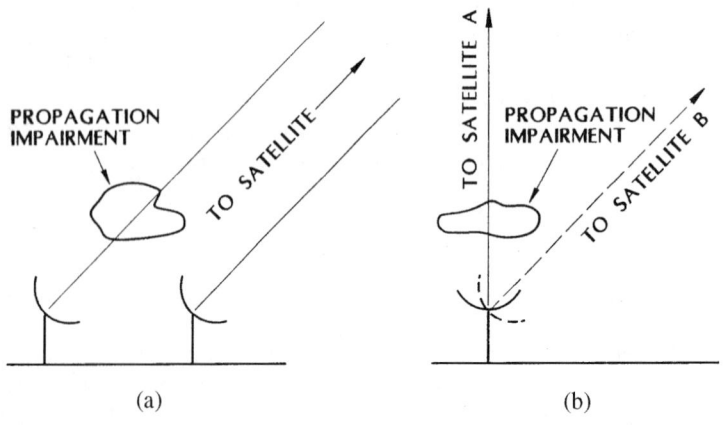

FIGURE 73.1 Path diversity concepts: (a) site diversity and (b) orbital diversity.

Site and orbital diversity are not mature disciplines; there are very few data on the operational performance of complete systems. Research has focused on the propagation aspects and, to a lesser extent, on processor design and performance.

In principle, site and orbital diversity may be used on both uplinks and downlinks. In practice, the downlink problem is usually the more critical because of the limited power available aboard satellites, whereas the uplink problem is the more difficult when the processing is performed on the ground, where the synchronization of signals at the satellite cannot be observed directly. For these reasons, site diversity processing has centered primarily on downlinks. Of course, attenuation data are applicable to both downlink and uplink design.

73.2 Site Diversity Processing

For site diversity reception, the signals from the various terminals must be linked to a common point, synchronized, and combined or selected according to some control algorithm. In principle, this can be done at intermediate frequency (IF), but baseband seems to be the preferred implementation. The links must have high reliability, e.g., coaxial lines or fiber optics. For wide-area service using very small aperture terminals (VSATs) accessed through a metropolitan area network (MAN), the MAN itself has been proposed as the link [Spracklen, Hodson, and Heron, 1993].

Synchronization generally involves both a fixed and a variable delay. For low data rates (e.g., 2 Mb/s), the fixed delay may be a digital first-in first-out (FIFO) memory; for higher rates (e.g., 34 Mb/s) a surface acoustic wave (SAW) delay line at IF has been proposed. The variable delay may be an elastic digital memory, which may be implemented as an addressable latch coupled to a multiplexer [Di Zenobio et al., 1988].

One method of combining signals is to switch to the best of the diversity branches, with carrier-to-noise ratio (C/N) or bit error rate (BER) as criterion. To avoid excessive switching when signal qualities are approximately equal, hysteresis in the switching algorithm is helpful. One algorithm causes switchover when

$$(E_0 \geq N) \wedge (E_s \leq N/H) = \text{TRUE} \tag{73.1}$$

where E_0 denotes the number of errors in a given time interval in the channel currently on line, E_s the same for the current standby channel, and H is a hysteresis parameter appropriate for the current signal quality. The time interval is chosen to give N errors for a desired BER threshold [Russo, 1993].

A second approach is a weighted linear combination of the signals, with the weight for each branch determined on the basis of either C/N or BER. Considerable theory is available for designing optimized weight controls [Hara and Morinaga, 1990a, 1990b]. Such systems should outperform switching; however, their application has been considered only recently. Linear combining techniques developed for base-station site diversity in cellular systems may be adaptable to satellite systems.

73.3 Site Diversity for Rain-Fade Alleviation[1]

The concept of path diversity for rain-fade alleviation was first utilized on line-of-sight microwave paths; recognition of its applicability to satellites seems to date to the mid-1960s. It is most effective with systems having an appreciable rain margin, since cells of heavy rain generally do not extend far horizontally (Fig. 73.2). It is less effective at low elevation angles, for which substantial attenuation may result from stratiform rain.

---

[1]This section includes material from Lin, K.-T. and Levis, C.A. 1993. Site diversity for satellite Earth terminals and measurements at 28 GHz. *Proc. IEEE*, 81(6):897–904. © 1993 IEEE. With permission.

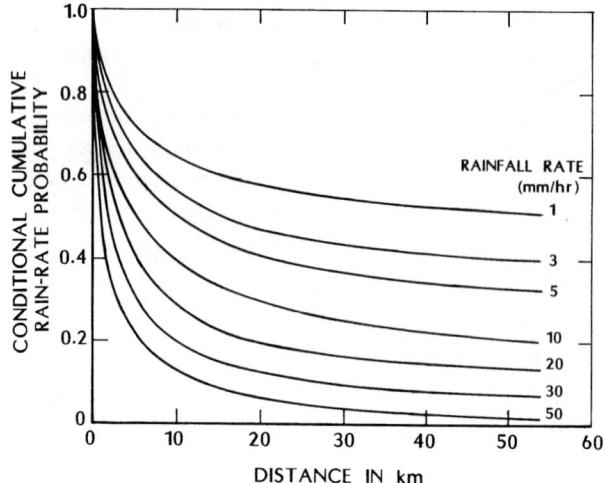

FIGURE 73.2 Probability that a rain-rate will be exceeded simultaneously at two stations, divided by the probability that it will be exceeded at a specific one of the stations. [*Source:* Adapted from Yokoi, H., Yamada, M., and Ogawa, A. 1974. Measurement of precipitation attenuation for satellite communications at low elevation angles. *J. Rech. Atmosph.* (France), 8:329–338. With permission.]

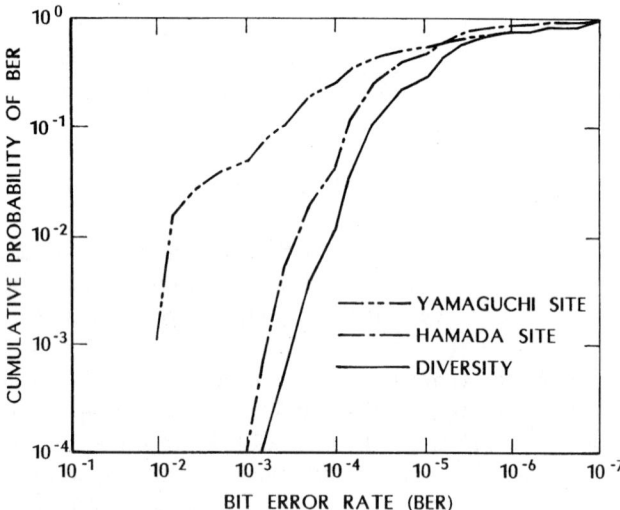

FIGURE 73.3 Bit error rate probability for an operational experiment in Japan. (*Source:* Watanabe et al. 1983. Site diversity and up-path power control experiments for TDMA satellite link in 4/11 GHz bands. In *Sixth International Conference on Digital Satellite Communications*, pp. IX-21–IX-28. IEEE, New York. © 1983 IEEE. With permission.)

Measurements

An experiment with up- and downlink site diversity using switching based on BER determined from the forward error correcting (FEC) code was conducted in Japan with Intelsat V at approximately 6° elevation at 14 GHz (up) and 11 GHz (down) with a site separation of 97 km. Up- and downlinks were switched simultaneously. The data rate was 120 Mb/s. Results during a rain event appear in Fig. 73.3.

TABLE 73.1 Some Site-Diversity Propagation Experiments

Reference	Location	Frequency, GHz	Baseline, km
Bostian et al. [1990] (B, R)[a]	Blacksburg, VA	11.4	7.3
Färber et al. [1991] (B)[a]	Netherlands	12.5, 30	10
Fionda, Falls, and Westwater [1991] (S)[a]	Denver, CO	20.6, 31.6	50
Goldhirsh [1984] (R)[a]	Wallops Island, VA	19.0, 28.6	≤ 35
Goldhirsh [1982] (R)[a]	Wallops Island, VA	28.6	≤ 35
Goddard and Cherry [1984] (R)[a]	Chilbolton, UK	11.6	≤ 40
Lin, Bergmann, and Pursley [1980] (S)[a]	NJ, GA, IL, CO	13–18	11–33[b]
Lin and Levis [1993] (S)[a]	Columbus, OH	28.6	9.0
Pratt, Bostian, and Stutzman [1989] (R, B)[a]	Blacksburg, VA	11.5	7.3
Rogers and Allnutt [1990] (S)[a]	Atlanta, GA	12.0	37.5
Witternigg et al. [1993] (S)[a]	Graz, Austria	12.0	9–26[b]

[a] (B) = satellite beacon, (S) = sky noise, (R) = radar.
[b] Includes multiple (3- or 4-site) diversity.

Diversity rain-fade measurements have been obtained by four different methods. The most direct measures the signal from a satellite beacon at two or more Earth stations; unfortunately, opportunities for such measurements have been quite limited. A second approach utilizes the thermodynamic law that absorbing materials at a finite temperature emit noise. By pointing radiometers in the proposed satellite direction and measuring the excess sky noise, signal attenuation along the path can be inferred. It is difficult to measure attenuations in excess of about 15 dB by this method, and there are some uncertainties in converting the noise to equivalent attenuation; nevertheless, much useful data has been obtained from sky noise measurements. An efficient third means of data acquisition is to measure the radar reflectivity along the proposed paths. Through careful calibration and data processing, useful data may be obtained, although some assumptions must be made about the raindrop size distributions in converting reflectivity data to attenuation. An advantage of this method is that several diversity site configurations can be investigated in one experiment, provided the attenuations on all paths occur within the scanned radar volume. Finally, fast-response measurements with rain-gauge networks were the earliest means to infer propagation data, and rain-gauge measurements are still often correlated with propagation measurements.

A tabulation of site-diversity experiments prior to 1983 appears in the literature [Ippolito, 1989]. Table 73.1 lists additional experiments, with no pretension to completeness.

Data Presentation

The natural form of the data is a time series of attenuation samples for each path over one or more years. Statistical means are used to reduce the data to more convenient forms. A common graphical form of data presentation is in the form of fade-level statistics, where the ordinate represents the percentage of the time during which the attenuation specified by the abscissa is exceeded. Figure 73.4 shows the fade-level statistics for the two individual sites of a specific site diversity experiment and the corresponding statistics for ideal switching and maximal-ratio combining, both based on C/N. The effectiveness of the use of diversity is shown by the displacement, downward and to the left, of the combined-signal curves relative to those of the individual sites.

Propagation experiments and calculations have been interpreted traditionally in terms of switching to the best signal, with hysteresis neglected. Unless otherwise indicated, all diversity data presented here and in the literature should be interpreted accordingly.

For the system designer, statistics of the duration of fades and of the time between fades are also of interest. Figure 73.5 shows the fade-duration statistics for single sites and the corresponding diversity statistics; the **interfade interval** statistics for the same diversity experiment are shown in Fig. 73.6. The benefit of diversity is evident from the general reduction of fade durations and increase in the time between fades.

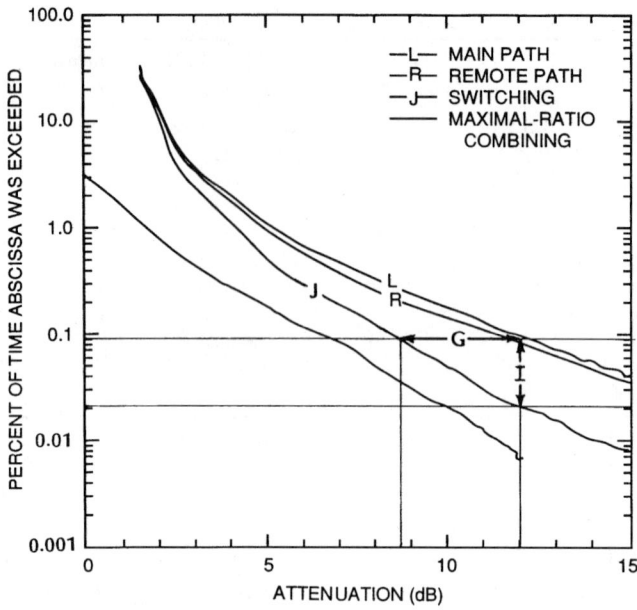

FIGURE 73.4 Fade-level statistics for single sites and two methods of processing at 28.6 GHz in Columbus, OH, with 9-km site separation and 25.6° elevation. [*Source:* Adapted from Lin, K.-T. and Levis, C.A. 1993. Site diversity for satellite Earth terminals and measurements at 28 GHz. *Proc. IEEE*, 81(6):897–904. © 1993 IEEE. With permission.]

Fade-level statistics for a 3-site diversity configuration are shown in Fig. 73.7. It has been shown that average fade durations may be shortened by adding additional diversity sites, even if the fade-level statistics do not show a corresponding improvement [Witternig et al., 1993].

Diversity Improvement Factor and Diversity Gain

Two criteria of diversity effectiveness have been defined in terms of fade-level distributions. One is the **diversity improvement factor**, defined [CCIR, 1990] as the ratio p_1/p_2, where p_1 denotes the percent of the time a given attenuation will be exceeded for the single-site distribution, and p_2 is the corresponding percentage for the combined-signal distribution. The diversity improvement factor (many authors use the term diversity advantage) may be defined with respect to any of the individual sites. If the system is balanced, that is, if the statistics for all of the participating sites are essentially the same, it is common practice to average the single-site attenuations (in decibel) at each probability level for use as the single-site statistics. The diversity improvement factor I is a measure of the vertical distance between the single-site and diversity curves, and it is proportional to that distance if the time-percentage scale is logarithmic. In Fig. 73.4, the diversity improvement factor at the 12-dB single-site attenuation level for switching-combining is shown to be about $(0.09/0.021) = 4.3$ for that experiment. The diversity improvement factor specifies the improvement in system reliability due to use of diversity. For the example of Fig. 73.4, if the system has a 12-dB rain margin, the probability of rain outage would be reduced by a factor of 4.3.

The most commonly used criterion for diversity effectiveness is the **diversity gain**, defined as the attenuation difference (in decibel) between the single-site attenuation and the diversity system attenuation at a given exceedence level. It is a measure of the horizontal distance between the single-site and diversity fade statistics curves, and it is proportional to that distance if the attenuation scale is linear in decibel. It represents the decrease, due to diversity, in the required rain fade margin. In Fig. 73.4, the distance G shows the diversity gain for that experiment for switching-combining at the

Path Diversity

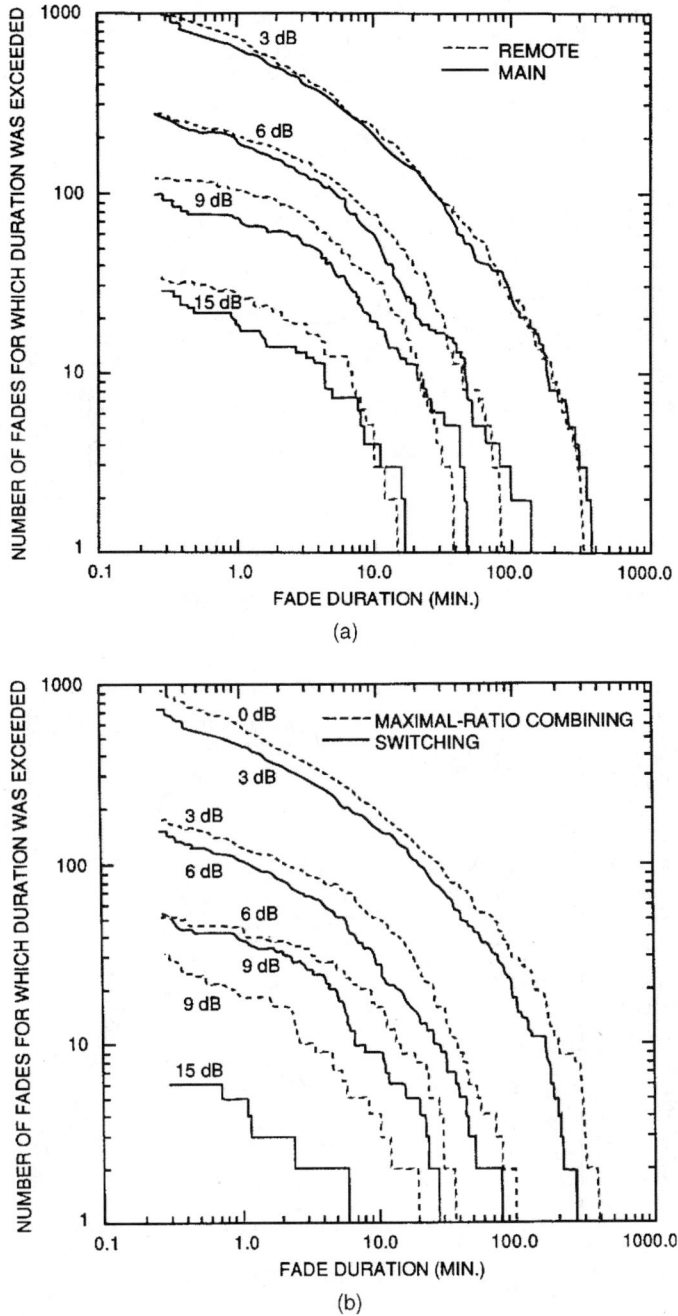

FIGURE 73.5 Fade-duration statistics for one year at 28.6 GHz in Columbus, OH, with 9-km site separation and 25.6° elevation: (a) single sites and (b) diversity for two methods of processing. [*Source:* Adapted from Lin, K.-T. and Levis, C.A. 1993. Site diversity for satellite Earth terminals and measurements at 28 GHz. *Proc. IEEE*, 81(6):897–904. © 1993 IEEE. With permission.]

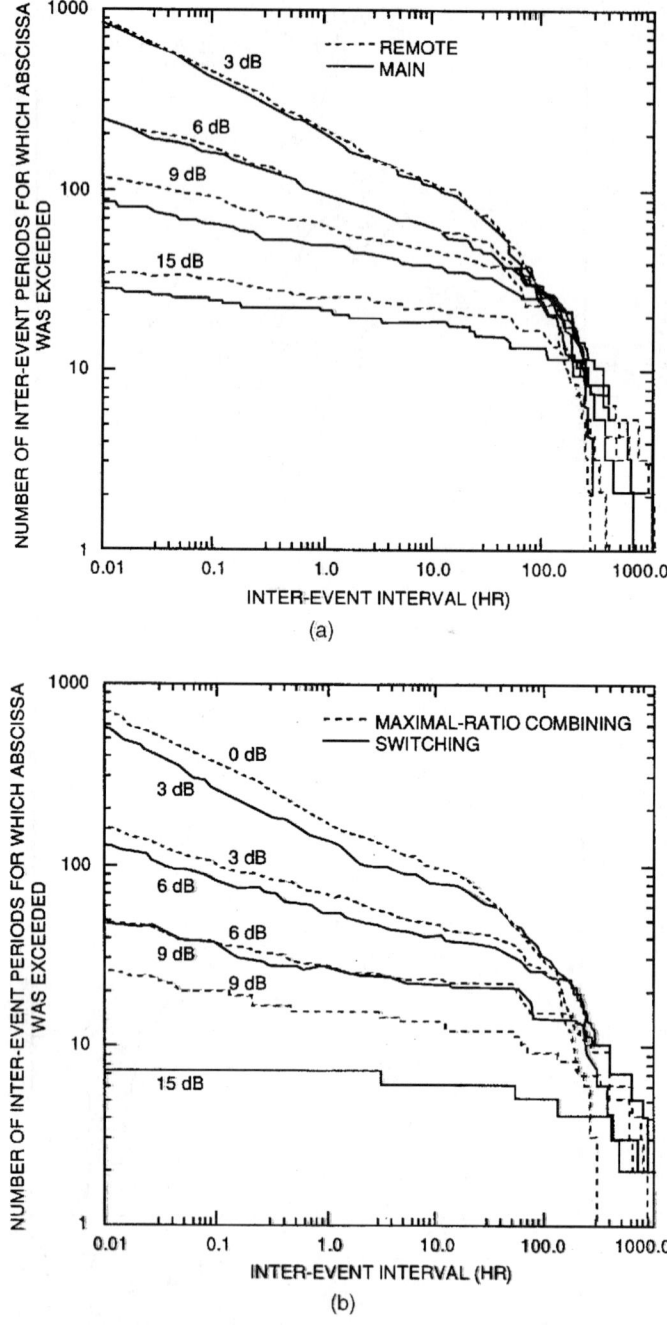

FIGURE 73.6 Interfade interval statistics for one year at 28.6 GHz in Columbus, OH, with 9-km site separation and 25.6° elevation: (a) single sites and (b) diversity for two methods of processing. [*Source:* Adapted from Lin, K.-T. and Levis, C.A. 1993. Site diversity for satellite Earth terminals and measurements at 28 GHz. *Proc. IEEE*, 81(6):897–904. © 1993 IEEE. With permission.]

Path Diversity

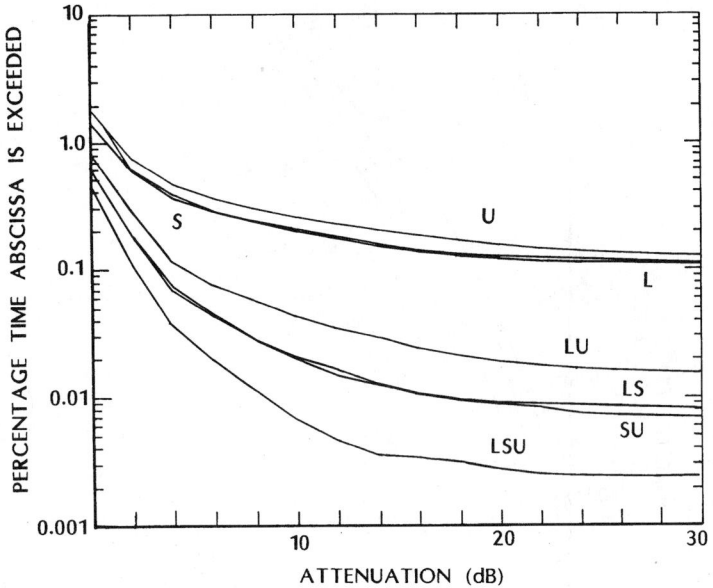

FIGURE 73.7 Single-site, two-site diversity, and triple diversity fade-level statistics at 19 GHz for three sites near Tampa, FL, with separations of 11, 16, and 20 km and 55° elevation. [*Source:* Tang, D.D., Davidson, D., and Bloch, S.C. 1982. Diversity reception of COMSTAR satellite 19/29-GHz beacons with the Tampa Triad, 1978–1981. *Radio Sci.*, 17(6):1477–1488. © 1982 *Radio Science*. With permission.]

0.09% exceedence level as $(12.0 - 8.7) = 3.3$ dB. For an allowed rain failure rate of 0.09%, the fade margin could be reduced from 12 dB for single-site operation to 8.7 dB for switching-diversity operation.

Although diversity gain is defined as the attenuation difference between single-site and diversity systems at a given exceedence level p, it should be stated and graphed as a function of the corresponding single-site attenuation level $A_s(p)$; for example, in Fig. 73.4, $G(12 \text{ dB}) = 3.3$ dB. This point is trivial when dealing with a single set of fade-level statistics, but it is significant when different sets of statistics are compared. The dependence of G on A_s has been found to be relatively independent of single-site statistics; this is not true of the dependence of G on p. Conversely, the model adopted by the CCIR for diversity improvement factor implies that the diversity improvement factor I is relatively independent of single-site statistics when it is specified as a function of p, but not when specified as a function of A_s.

Diversity gain and improvement are clearly not independent. Given the single-site fade-level statistics and the diversity improvement factor at all attenuation levels, the diversity fade-level statistics curve can be determined and, from it, diversity gain. Similarly, the diversity improvement factor can be determined if the single-site fade-level statistics and the diversity gain are given. An exception occurs in the practical case where reception is limited (e.g., by noise considerations) to a certain attenuation level. In this case, diversity data may extend to lower time percentages than the single-site data, and diversity gain will be undefined for such time percentages (e.g., in Fig. 73.4 for $p < 0.04\%$). Diversity gain and improvement factor can be defined for all methods of diversity signal processing (e.g., both switching and maximal-ratio combining in Fig. 73.4), but current literature references ordinarily imply switching to the best available signal.

A **relative diversity gain** can be defined [Goldhirsh, 1982] as the ratio of the diversity gain for a given site separation to the diversity gain for a very large site separation, say, 35 km, with other system parameters held constant. The relative diversity gain is determined by the rain-rate profile along the Earth–satellite path and should be essentially independent of frequency and of rain drop-size distribution.

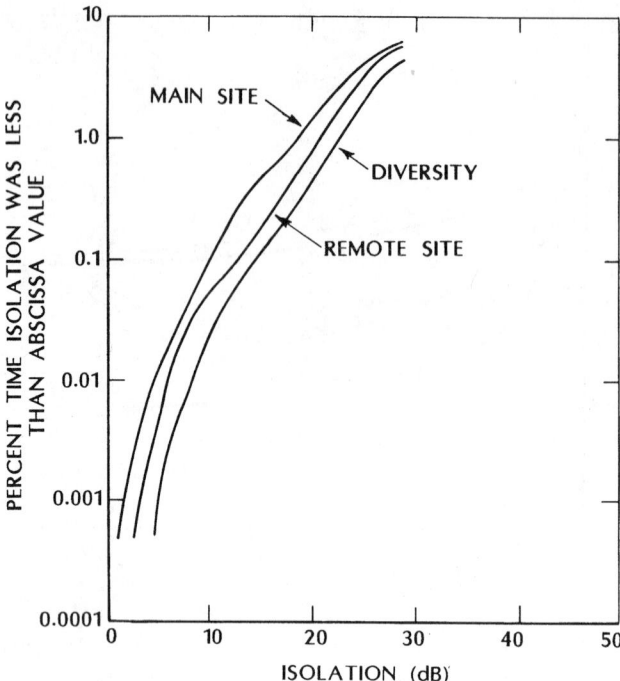

FIGURE 73.8 Polarization isolation statistics at 11.6 GHz in Blacksburg, VA, for 7.3 km site separation and 10.7° elevation. [*Source:* Towner et al. 1982. Initial results from the VPI & SU SIRIO diversity experiment. *Radio Sci.*, 17(6):1489–1494. © 1982 Radio Science. With permission.]

Isolation Diversity Gain

For systems utilizing polarization diversity to allow frequency reuse, isolation-level statistics, analogous to fade-level statistics, may be generated (Fig. 73.8). **Isolation diversity gain** may then be defined as

$$G_{\text{ID}}(p) = I_j(p) - I_s(p) \qquad (73.2)$$

where I_j and I_s denote, respectively, the diversity and single-site isolation, both at a selected time percentage p for which these isolations were not exceeded, for example, the horizontal displacement of the diversity curve with respect to a single-site curve in Fig. 73.8.

Prediction Models

Empirical models have been generated, generally by proposing a mathematical relationship on theoretical or observational grounds and adjusting the coefficients by regression analysis. Analytical models are based on theoretical models of the structure of rain or of the attenuation statistics of signals in the presence of rain. In practice, empirical models are much easier to apply, but they give much less information, being limited to a specific parameter, such as diversity gain.

Empirical models

An empirical model for diversity gain in decibels [Hodge, 1982] has been adopted by the CCIR [1990],

$$G_D = G_d G_f G_\theta G_\phi \qquad (73.3)$$

where

$$G_d = a(1 - e^{-bd}) \tag{73.4}$$

$$a = 0.78A - 1.94(1 - e^{-0.11A}) \tag{73.5}$$

$$b = 0.59(1 - e^{-0.1A}) \tag{73.6}$$

$$G_f = e^{-0.025f} \tag{73.7}$$

$$G_\theta = 1 + 0.006\,\theta \tag{73.8}$$

$$G_\phi = 1 + 0.002\,\phi \tag{73.9}$$

where d is the site separation in kilometers, A is the single-site attenuation in decibel, θ is the path elevation angle in degrees, f is the frequency in gigahertz, and $\phi \leq 90°$ is the baseline orientation angle with respect to the azimuth direction of the propagation path. When tested against the CCIR data set, the arithmetic mean and standard deviation were found to be 0.14 dB and 0.96 dB, respectively, with an rms error of 0.97 dB. Many comparisons of this model with experiment are found in the literature, mostly with good agreement. Two examples are given in Fig. 73.9.

Another empirical model [Rogers and Allnutt, 1984] is based on the typical behavior of diversity gain as a function of single-site attenuation (see Fig. 73.10). In this model, the offset between the ideal and observed diversity gain is attributed to correlated impairment on both paths due to stratified rain. For modeling purposes, a time percentage of 0.3% is assumed for the correlation, and the offset is computed as the single-site path attenuation for the rain-rate, which is exceeded 0.3% of the time. A straight line parallel to the ideal diversity-gain curve ($G_D = A_s$) is thus determined. The *knee* is determined by the rain rate at which the two paths become uncorrelated, that is, the onset of the convective regime, which is taken as 25 mm/h. Thus the knee attenuation is calculated as the single-path attenuation for a rain-rate of 25 mm/h. The model is completed by a straight line from the knee to the origin.

An empirical expression for the relative diversity gain is

$$G_{D,\text{rel}} = 1 - 1.206 e^{-0.531\sqrt{d}} \qquad 1 \leq d \leq 30 \tag{73.10}$$

where d is the site spacing in kilometer [Goldhirsh, 1982].

An empirical model for the diversity improvement factor, which has been adopted by the CCIR [1990] but is not as widely supported in the literature as the Hodge diversity gain model, is

$$I = \frac{1}{1+\beta^2} \times \left(1 + \frac{100\beta^2}{P_1}\right) \tag{73.11}$$

where

$$\beta^2 = 10^{-4} d^{1.33} \tag{73.12}$$

the site separation in kilometer again is d, and P_1 represents the single-site time percentage of exceedence of a given attenuation, that is, the ordinate of a fade-level statistics diagram such as Fig. 73.4. For the usual case of small β^2, this reduces to

$$I = 1 + \frac{100\beta^2}{P_1} \tag{73.13}$$

In Fig. 73.11, experimental data is compared with this model.

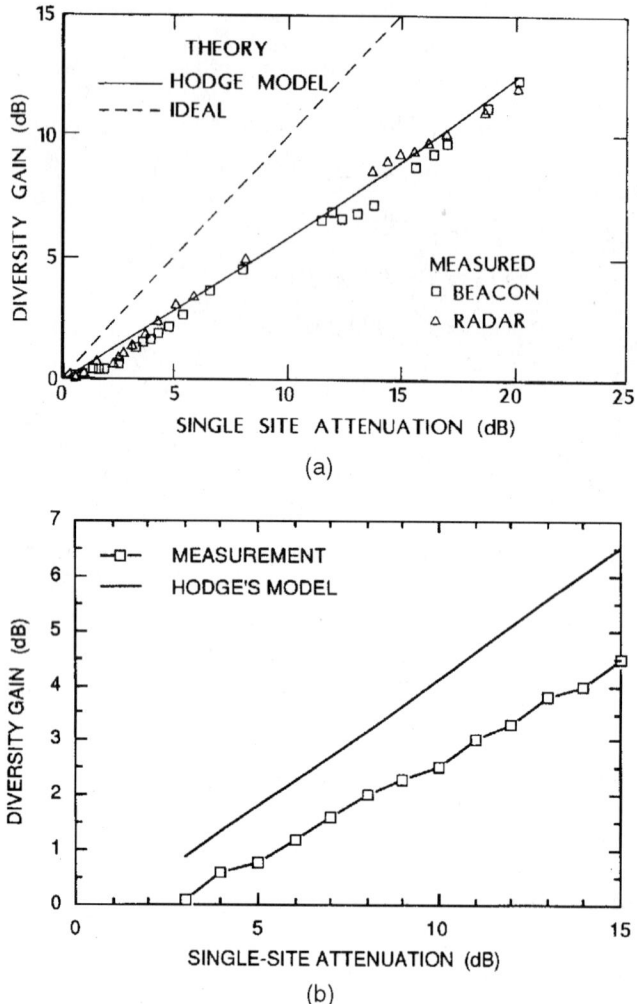

FIGURE 73.9 Comparison of Hodge's model with experiment: (a) 11.4-GHz, Blacksburg, VA, 7.3-km baseline, 18° elevation, [*Source:* Bostian et al. 1990. Satellite path diversity reception at 11.4 GHz: Direct measurements, radar observations, and model predictions. *IEEE Trans. Ant. Prop.*, AP-38(7):1035–1038. © 1990 IEEE. With permission.] (b) 28.6-GHz, Columbus, OH, 25.6° elevation, 9-km baseline. [*Source:* Adapted from Lin, K.-T. and Levis, C.A. 1993. Site diversity for satellite Earth terminals and measurements at 28 GHz. *Proc. IEEE*, 81(6):897–904. © 1993 IEEE. With permission.]

Analytical Models

Analytical models are usually more cumbersome and computationally intensive than empirical models, but they are potentially applicable to a wider class of problems, for example, the determination of diversity gain for three or more sites.

Predictions based on a rain model consisting of storm cells and debris which are mutually uncorrelated, with distributions and parameters based on quite detailed meteorological considerations, show rms deviations of 59% in attenuation and 222% in probability when compared to 48 diversity pairs in the CCIR data base [Crane and Shieh, 1989].

Path Diversity

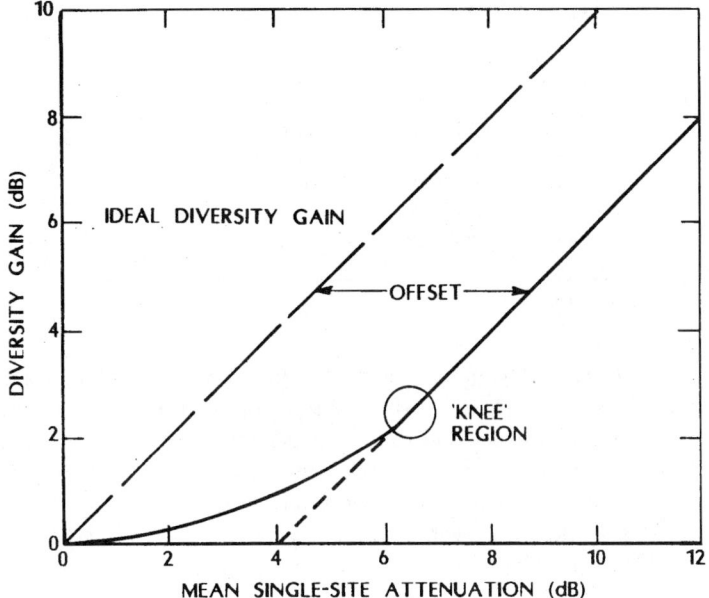

FIGURE 73.10 Site diversity model. [*Source:* Rogers, D.V. and Allnutt, J.E. 1984. Evaluation of a site diversity model for satellite communications systems. *IEE Proc. F.*, 131(5):501–506. © 1984 IEE. With permission.]

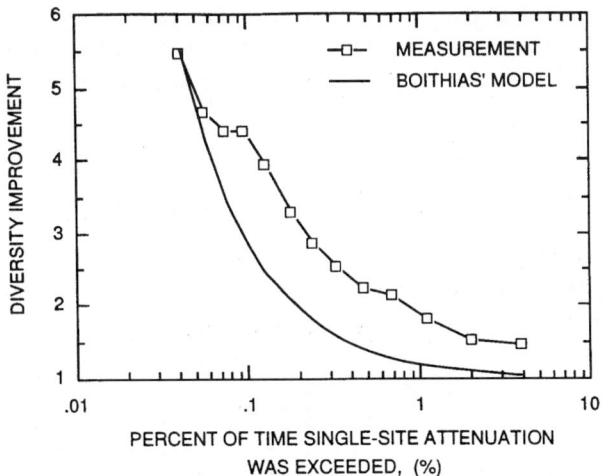

FIGURE 73.11 Comparison of CCIR diversity improvement factor model, due to Boithias, with experiment at 28.6 GHz in Columbus, OH, for 9-km site separation, 25.6° elevation. [*Source:* Adapted from Lin, K.-T. and Levis, C.A. 1993. Site diversity for satellite Earth terminals and measurements at 28 GHz. *Proc. IEEE*, 81(6):897–904. © 1993 IEEE. With permission.]

A model based on circularly cylindrical rain cells of constant rain rate and uniform horizontal distribution was used in Mass [1987]. The cell diameters are lognormally distributed, with the median dependent on rain rate, and the cylinder heights are given by four equally probable values, which are also dependent on rain rate. Calculated attenuation differences are compared graphically with experimental values for some 20 paths, principally at yearly exceedence levels of 0.1, 0.01, and 0.001%. A bias of 0.2 dB and rms deviation of 0.9 dB can be calculated from these data.

TABLE 73.2 Mathematically Based Models

Reference	Probability Distribution	Spatial Correlation
Matricciani [1983]	lognormal	$e^{-\beta d}$
Kanellopoulos and Koukoulas [1987]	lognormal	$G/(G^2 + d^2)^{1/2}$
Kanellopoulos and Koukoulas[a] [1990]	lognormal	$G/(G^2 + d^2)^{1/2}$
Kanellopoulos and Ventouras [1990a]	lognormal	$e^{-\alpha\sqrt{d}}$
Kanellopoulos and Ventouras [1990b]	lognormal	$G/(G^2 + d^2)^{1/2}$
Kanellopoulos and Ventouras[a] [1992]	gamma	$G/(G^2 + d^2)^{1/2}$
Koukoulas and Kanellopoulos [1990]	gamma, lognormal	$e^{-\alpha\sqrt{d}}$

[a]Three-site diversity.

When a suitable n-dimensional joint rain-rate probability distribution is assumed for n diversity paths, and a suitable spatial correlation coefficient is assumed for the rain rate, more detailed meteorological considerations can be avoided. Models using this approach are listed in Table 73.2. In most cases, the model results are compared with a limited set of experimental data.

A comparison of seven models for diversity gain with numerous measurements [Bosisio, Capsoni, and Matricciani, 1993] should be used with caution because of the weight it accords differences at low attenuations, which generally have little influence on system performance.

73.4 Optical Satellite Site Diversity

For optical Earth–space paths, cloud cover is the most serious source of obscuration, with correlation distances extending, in some cases, to hundreds of miles. The link availability using a number of NASA sites singly and in three- and four-site diversity configurations has been calculated from meteorological data; the best single-site availability was 75%, although several four-site configurations were predicted to have 98% availability [Chapman and Fitzmaurice, 1991].

73.5 Site Diversity for Land–Mobile Satellite Communications

In the UHF and 1.5-GHz frequency bands considered for the land-mobile service, the principal impairment mechanisms are shadowing (for example, by the woody parts of trees) and multipath due to terrain. Site separations of 1–10 m have been found useful for country roads, and an expression

$$I(A, d) = 1 + [0.2 \log_e(d) + 0.23]A \qquad (73.14)$$

for the diversity improvement factor has been proposed, where d denotes distance in meters and A the single-site fade depth in decibel [Vogel, Goldhirsh, and Hase, 1992].

Because fade margins are likely to be low in these applications, linear signal combining is attractive. An optimization of weights has been proposed, based on modeling the signal as having a coherent (direct) and an incoherent (scattered) component in the presence of fast Rician noise. Land, sea, and aircraft applications of this model differ in the representation of the incoherent component [Hara and Morinaga, 1990a, 1990b]. A Bayes-test algorithm for weight optimization has also been proposed [Sandström, 1991].

73.6 Site Diversity for Arctic Satellite Communications

Propagation measurements at 6 GHz in the Canadian arctic at 80°N latitude have shown clear-air fading and enhancements in the summer under conditions of sharply defined temperature inversions at the low elevation angles required for geostationary satellites [Strickland, 1981]. These variations

Path Diversity

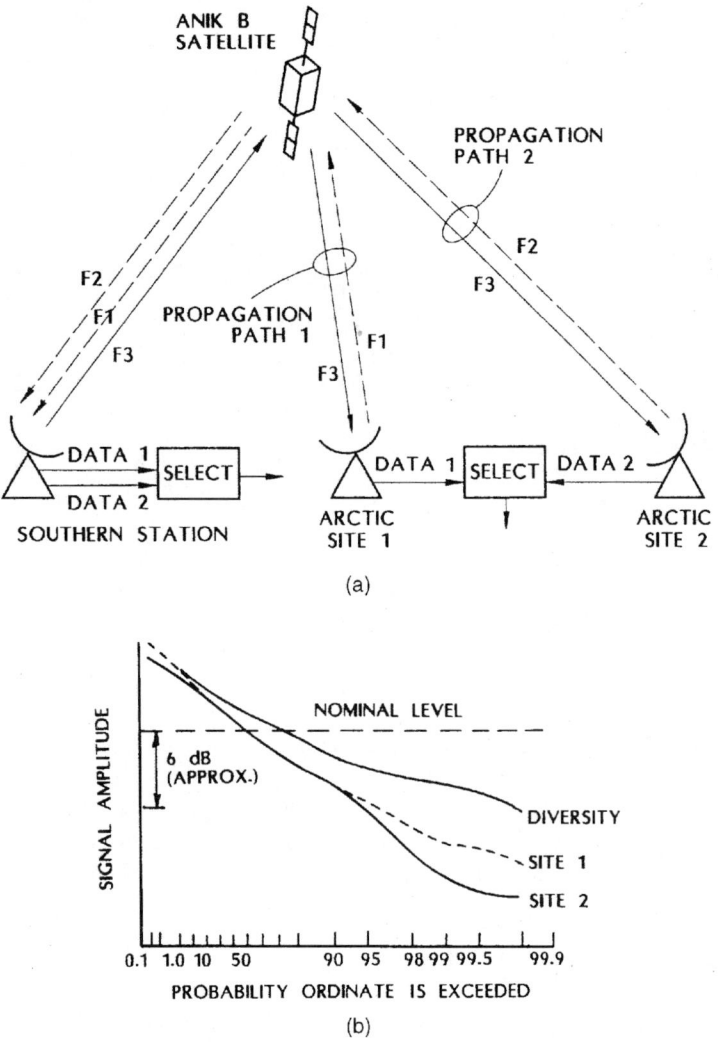

FIGURE 73.12 Signal-level statistics for a diversity experiment at 4/6 GHz at Eureka, Canada (80° N) with 150-m vertical antenna separation: (a) experiment configuration and (b) signal statistics. [*Source:* Adapted from Mimis, V. and Smalley, A. 1982. Low elevation angle site diversity satellite communications for the Canadian arctic. *IEEE International Conference on Communications*, pp. 4A.4.1–4.A.4.5 (IEEE Cat. 0536-1486/82). © IEEE. With permission.]

were equally well correlated for horizontal site separations of 480 m and vertical separations of 20 m, but were qualitatively uncorrelated at vertical separations of 180 m. These results appear to be predicted well by modeling the interface as sinusoidally corrugated, resulting in vertical coverage gaps of size

$$g = 1.039 \times 10^{-3} \left[\frac{\lambda^2}{A} \left(\frac{\Delta N}{\phi} \right)^2 h^2 \right]^{1/3} \qquad (73.15)$$

where λ is the corrugation wavelength, A its amplitude, h its height above the observer, and ϕ the elevation angle of the path, with g, λ, h, and A in meters, ΔN in N-units, and ϕ in degrees.

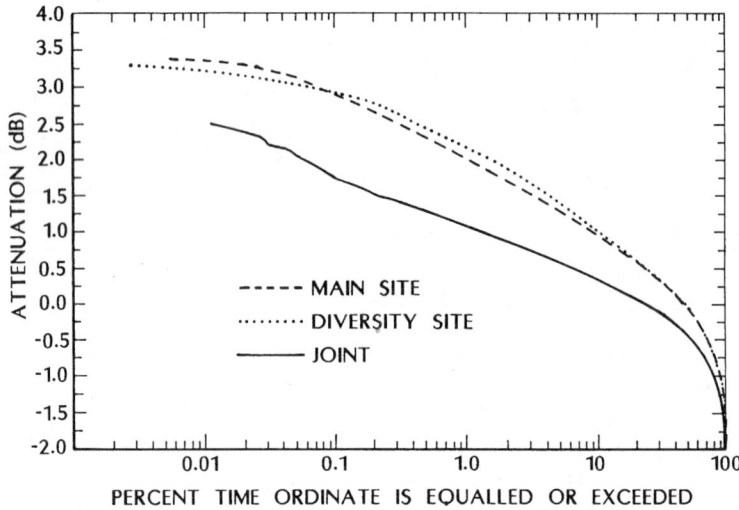

FIGURE 73.13 Microscale attenuation statistics at 20 GHz in Blacksburg, VA, with a 48-m baseline, 14° elevation, from a 1-h sample. [*Source:* Cardoso, J.C., Safaai-Jazi, A., and Stutzman W.L. 1993. Microscale diversity in satellite communications. *IEEE Trans. Ant. Prop.*, 41(6):801–805. © 1993 IEEE. With permission.]

In a systems experiment, a diversity system at this location communicated with a single southern station via the Anik B satellite in the 6/4-GHz frequency bands. Both C/N and BER were used, separately, as switching criteria, with data rates up to 1.544 Mb/s. On the S–N path, a single frequency was transmitted, on the N–S path each diversity site transmitted on its own carrier to allow signal selection at the southern station. Figure 73.12 shows the experiment configuration and results.

73.7 Microscale Diversity for VSATs

The large separations required for site-diversity protection from rain fades may be too expensive for low-cost systems utilizing very small aperture terminals, but smaller separations can protect against fading due to scintillations [Cardoso, Safaai-Jazi, and Stutzman, 1993]. The improvement may be small in cases where rain fades predominate. Fade-level statistics for a separation of 48 m are shown in Fig. 73.13. The correlation of the scintillations is examined experimentally and theoretically in Haidara [1993].

73.8 Orbital Diversity

Orbital diversity is likely to be less effective than site diversity because the propagation impairment always is near the Earth terminals, so that propagation effects tend to be more correlated for orbital diversity. The calculated tradeoff between the required satellite separation angle, as viewed from the Earth station, for orbital diversity and the separation of sites for site diversity, which will give identical diversity gain, at 19 GHz at Fucino, Italy, is shown in Fig. 73.14. Economic factors favor putting the redundancy on the ground unless the system is usable by a great many Earth stations. A system utilizing spare channels on the Italsat and Olympus satellites and 100 Earth stations has been analyzed ignoring the economic aspects [Matricciani, 1987]. Although significant power margin reductions are attainable, it was concluded that many problems remain to be solved before orbital diversity can be considered practical.

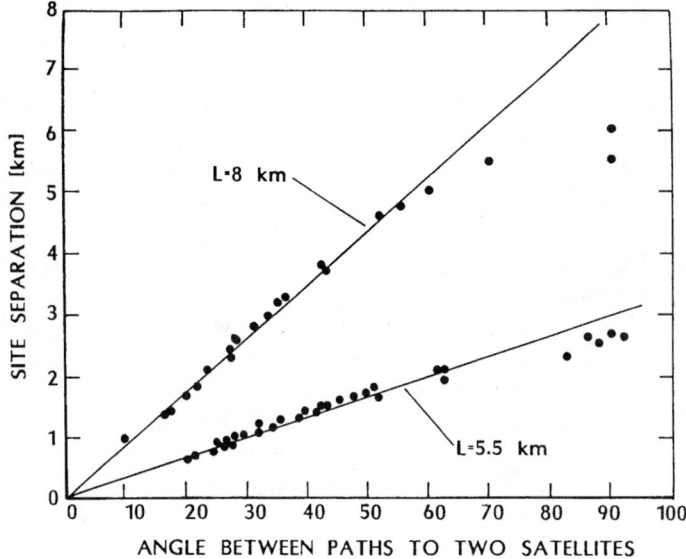

FIGURE 73.14 Relationship between site separation (for site diversity) and angular separation (for orbital diversity) for the same gain, calculated from radar data for the Fucino plain, Italy, at 19 GHz. L is the slant-path length through the rain for the site-diversity systems. The orbital-diversity station is located midway between the site-diversity stations. [*Source:* Matricciani, E. 1983. An orbital diversity model for earth to space links under rain and comparisons with site diversity. *Radio Sci.* 18(4):583–588. © 1983 *Radio Science.* With permission.]

Defining Terms

Diversity gain: The difference in the attenuation, measured in decibel, which is exceeded a given percentage of the time at a single reference site, and the attenuation which is exceeded the same percentage of the time for the combined diversity signal.

Diversity improvement factor: The ratio of the percent of the time that a given attenuation will be exceeded at a single reference site to the percent of the time that the same attenuation will be exceeded for the combined diversity signal.

Interfade interval: A time interval during which the attenuation remains less than a given threshold, while it is exceeded immediately before and after the interval.

Isolation diversity gain: The difference between a polarization isolation value in decibel of a diversity system, such that the observed isolation is less than this value a given percentage of time, and the corresponding value for a single reference site.

Relative diversity gain: The ratio of the diversity gain for a given site separation to that for a very large site separation, typically 35 km, with other system parameters held constant.

References

Bosisio, A.V., Capsoni, C., and Matricciani, E. 1993. Comparison among prediction methods of site diversity system performances. In *8th International Conference on Antennas and Propagation (ICAP 93)*, Vol. 1, pp. 60–63. (IEE Conf. Publ. No. 370.)

Bostian, C.W., Pratt, T., Stutzman, W.L., and Porter, R.E. 1990. Satellite path diversity reception at 11.4 GHz: Direct measurements, radar observations, and model predictions. *IEEE Trans. Ant. Prop.*, AP-38(7):1035–1038.

CCIR 1990. Propagation data and prediction methods required for Earth-space telecommunication systems. In *Reports of the CCIR, 1990*, Rept. 564-4, Annex to Vol. V, Propagation in non-ionized media, pp. 458–463, 486–488. International Telecommunication Union, Geneva, Switzerland.

Cardoso, J.C., Safaai-Jazi, A., and Stutzman, W.L. 1993. Microscale diversity in satellite communications. *IEEE Trans. Ant. Prop.*, 41(6):801–805.

Chapman, W. and Fitzmaurice, M. 1991. Optical space-to-ground link availability assessment and diversity requirements. In *Free-Space Laser Communications Technology III (Proc. SPIE 1417)*, pp. 63–74.

Crane, R.K. and Shieh, H.-C. 1989. A two-component rain model for the prediction of site diversity performance. *Radio Sci.*, 24(6):641–665.

Di Zenobio, D., Lombardi, P., Migliorni, P., and Russo, E. 1988. A switching circuit scheme for a satellite site diversity system. In *Proceedings of 1988 IEEE International Symposium on Circuits and Systems*, Vol. 1, pp. 119–122. (IEEE Cat. 88CH2458-8.)

Färber, K., Mawira, A., Quist, J., and Verhoef, G.J.C. 1991. 12 & 30 GHz Olympus propagation beacon experiment of PTT-Netherlands: Space diversity and frequency dependence of the co-polar phase. In *7th International Conference on Antennas and Propagation (ICAP 91)*, Part 1, pp. 476–479. (IEE Conf. Publ. 333.)

Fionda, E., Falls, M.J., and Westwater, E.R. 1991. Attenuation statistics at 20.6, 31.65, and 52.86 GHz derived from emission measurements by ground-based microwave radiometers. *IEE Proc., H* 138(1):46–50.

Goddard, J.W.F. and Cherry, S.M. 1984. Site diversity advantage as a function of spacing and satellite elevation angle, derived from dual-polarization radar data. *Radio Sci.*, 19(1):231–237.

Goldhirsh, J. 1982. Space diversity performance prediction for Earth-satellite paths using radar modeling techniques. *Radio Sci.*, 17(6):1400–1410.

Goldhirsh, J. 1984. Slant path rain attenuation and path diversity statistics obtained through radar modeling of rain structure. *IEEE Trans. Ant. Prop.*, AP-32(1):54–60.

Haidara, F.M. 1993. Characterization of tropospheric scintillations on Earth-space paths in the Ku and Ka frequency bands using the results from the Virginia Tech OLYMPUS Experiment. Ph.D. Dissertation, Bradley Dept. of Elect. Eng., Virginia Polytech. Inst. & State Univ., Blacksburg, VA.

Hara, S. and Morinaga, N. 1990a. Optimum post-detection diversity of binary DPSK system in fast Rician fading channel. *Trans. Inst. Electron. Inf. Commun. Eng.* (Japan), E-73(2):220–228.

Hara, S. and Morinaga, N. 1990b. Post-detection combining diversity improvement of four-phase DPSK system in mobile satellite communications. *Electron. Commun. Japan 1*, 73(7):68-75. [Translated from *Denshi Joho Tsushin Gakkai Ronbunshi*, 1989. 72-BII(7):304–309.]

Hodge, D.B. 1982. An improved model for diversity gain on earth-space propagation paths. *Radio Sci.*, 17(6):1393–1399.

Ippolito, L.J. 1989. *Propagation Effects Handbook for Satellite Systems Design*, 4th ed., NASA Ref. Publ. 1082(04):7-78–7-107.

Kanellopoulos, J.D. and Koukoulas, S.G. 1987. Analysis of the rain outage performance of route diversity systems. *Radio Sci.*, 22(4):549–565.

Kanellopoulos, J.D. and Koukoulas, S.G. 1990. Prediction of triple-site diversity performance in Earth-space communication. *J. Electromagn. Waves Appl.* (Netherlands), 4(4):341–358.

Kanellopoulos, J. and Ventouras, S. 1990a. A modification of the predictive analysis for the multiple site diversity performance taking into account the stratified rain. *Eur. Trans. Telecommun. Relat. Tech.*, 1(1):49–57.

Kanellopoulos, J. and Ventouras, S. 1990b. A unified analysis for the multiple-site diversity outage performance of single/dual-polarized communication systems. *Eur. Trans. Telecommun. Relat. Tech.*, 1(6):625–632.

Kanellopoulos, J.D. and Ventouras, S. 1992. A model for the prediction of the triple-site diversity performance based on the gamma distribution. *IEICE Trans. Commun.* (Japan), E75-B(4):291–297.

Koukoulas, S.G. and Kanellopoulos, J.D. 1990. A model for the prediction of the site diversity performance based on the two-dimensional gamma distribution. *Trans. Inst. Elec. Inf. Commun. Eng.* (Japan), E-73(2):229–236.

Lin, K.-T. and Levis, C.A. 1993. Site diversity for satellite Earth terminals and measurements at 28 GHz. *Proc. IEEE*, 81(6):897–904.

Lin, S.H., Bergmann, H.J., and Pursley, M.V. 1980. Rain attenuation on Earth-satellite paths—Summary of 10-year experiments and studies. *Bell Syst. Tech. J.*, 59(2):183–228.

Mass, J. 1987. A simulation study of rain attenuation and diversity effects on satellite links. *COMSAT Tech. Rev.*, 17(1):159–188.

Matricciani, E. 1983. An orbital diversity model for earth to space links under rain and comparisons with site diversity. *Radio Sci.*, 18(4):583–588.

Matricciani, E. 1987. Orbital diversity in resource-shared satellite communication systems above 10 GHz. *IEEE J. Sel. Areas Commun.*, SAC-5(4):714–723.

Mimis, V. and Smalley, A. 1982. Low elevation angle site diversity satellite communications for the Canadian arctic. In *IEEE International Conference on Communications*, pp. 4A.4.1–4A.4.5 (IEEE Cat.0536-1486/82).

Pratt, T., Bostian, C.W., and Stutzman, W.L. 1989. Diversity gain and rain height statistics for slant paths from radar measurements. In *6th International Conference on Antennas and Propagation (ICAP 89)*, pp. 340–344. (IEE Conf. Publ. 301.)

Rogers, D.V. and Allnutt, J.E. 1984. Evaluation of a site diversity model for satellite communications systems. *IEE Proc. F*, 131(5):501–506.

Rogers, D.V. and Allnutt, J.E. 1990. Results of a 12-GHz radiometric site diversity experiment at Atlanta, Georgia. *COMSAT Tech. Rev.*, 20(1):97–103.

Russo, E. 1993. Implementation of a space diversity system for K_a-band satellite communications. In *Proceedings of IEEE International Conference on Communications '93*, pp. 1468–1474. (IEEE Cat. 0-7803-0950-2/93.)

Sandström, H., 1991. Optimum processing of QPSK signals for site diversity. *Int. J. Satel. Commun.*, 9:93–97.

Spracklen, C.T., Hodson, K., and Heron, R. 1993. The application of wide area diversity techniques to Ka band VSATs. In *Electron. Div. Colloq. Future of Ka Band for Satell. Commun.*, pp. 5/1–5/8. (IEE Colloq. Dig. 1993/215.)

Strickland, J.I., 1981. Site diversity measurements of low-angle fading and comparison with a theoretical model. *Ann. Télécommun.* (France), 36(7–8):457–463.

Tang, D.D., Davidson, D., and Bloch, S.C. 1982. Diversity reception of COMSTAR satellite 19/29-GHz beacons with the Tampa Triad, 1978–1981. *Radio Sci.*, 17(6):1477–1488.

Towner, G.C., Marshall, R.E., Stutzman, W.L., Bostian, C.W., Pratt, T., Manus, E.A., and Wiley, P.H. 1982. Initial results from the VPI&SU SIRIO diversity experiment. *Radio Sci.*, 17(6):1489–1494.

Vogel, W.J., Goldhirsh, J., and Hase, Y. 1992. Land-mobile-satellite fade measurements in Australia. *J. Spacecr. Rockets*, 29(1):123–128.

Watanabe, T., Satoh, G., Sakurai, K., Mizuike, T., and Shinonaga, H. 1983. Site diversity and up-path power control experiments for TDMA satellite link in 14/11 GHz bands. In *Sixth International Conference on Digital Satellite Communications*, pp. IX-21–IX-28. IEEE, New York.

Witternig, N., Kubista, E., Randeu, W.L., Riedler, W., Arbesser-Rastburg, B., and Allnutt, J.E. 1993. Quadruple-site diversity experiment in Austria using 12 GHz radiometers. *IEE Proc. H*, 140(5):354–360.

Yokoi, H., Yamada, M., and Ogawa, A. 1974. Measurement of precipitation attenuation for satellite communications at low elevation angles. *J. Rech. Atmosph.* (France), 8:329–338.

Further Information

L.J. Ippolito, 1989, gives a very comprehensive treatment of propagation impairments and their amelioration, including path diversity systems.

The fine treatment of the general propagation problem in *Satellite-to-Ground Radiowave Propagation*, 1989, by J.E. Allnutt includes material on path diversity in Chapters 4 and 10.

The excellent text *Radiowave Propagation in Satellite Communications* by Louis J. Ippolito Jr., 1986, includes a good overview of space diversity in Chapter 10, including the application of Hodge's model to the performance prediction of switching diversity systems.

Although Lin et al., 1980, is limited to the research of one particular organization, it is so broad as to furnish much insight into the general problem of impairments and of space diversity as a means of alleviation.

A good review of early work with copious references appears in J.E. Allnutt, 1978, "Nature of space diversity in microwave communications via geostationary satellites: a review," *Proc. IEE*, 125(5):369–376.

A discussion of rain storms useful for path diversity applications is found in R.R. Rogers, 1976, "Statistical rainstorms models; their theoretical and physical foundation," *IEEE Trans. Antennas Propag.*, AP-24(4):547–566.

74
Mobile Satellite Systems

John Lodge
Communications Research Centre, Ottawa, Canada

Michael Moher
Communications Research Centre, Ottawa, Canada

74.1 Introduction 1015
74.2 The Radio Frequency Environment and Its Implications 1016
74.3 Satellite Orbits 1020
74.4 Multiple Access 1024
74.5 Modulation and Coding 1025
74.6 Trends in Mobile Satellite Systems 1028

74.1 Introduction

Mobile satellite systems are capable of delivering a range of services to a wide variety of terminal types. Examples of mobile satellite terminal platforms include land vehicles, aircraft, marine vessels, and remote data collection and control sites. Services can also be provided to portable terminals, which can range in size from that of a briefcase to that of a handheld telephone. Many services include position determination, using some combination of the inherent system signal processing and inputs from a global positioning system (GPS) receiver, with transmission of that information via the mobile satellite communications system. The types of the communication channels available to the user can be subdivided into three categories; store-and-forward packet data channels, interactive packet data channels, and circuit-switched channels. Store-and-forward packet channels, which are the easiest to implement, allow for the transmission of small quantities of user data with delivery times that can be several minutes or more. This channel type is usually acceptable for services such as vehicle position reports (for example, when tracking truck trailers, rail cars, or special cargoes), paging, vehicle routing messaging, telemetry, telexes, and for some emergency and distress signalling. For many emergency and distress applications, however, as well as for interactive messaging (such as inquiry-based services), a delay of up to several minutes is unacceptable. For these applications the second channel type, interactive packet data channels, is required. Finally, for applications involving real-time voice communications or the transmission of large amounts of data (for example, facsimile and file transfers) circuit-switched channels are needed. In many cases, voice services provide a mobile telephone capability similar to those offered by cellular telephone, but offering much broader geographical coverage. In these cases a gateway station, which may provide an interface between the satellite network and the public switched telephone network (PSTN), communicates with the mobile via the satellite. In other cases mobile radio voice services can be offered to a closed user group (such as a government or large company), with the satellite communications being provided between the mobiles and a base station.

Mobile satellite systems share many aspects with other satellite systems as well as with terrestrial mobile radio systems (see Sec. 6.1). With respect to other satellite systems, the main differences are due to the small size and mobility of the user's terminal. The largest antennas used by mobile satellite terminals are 1-m parabolic dish antennas for large shipboard installations. At the other extreme are antennas small enough to be part of a handheld terminal. Also, the transmit power level of the terminal is limited due to small antenna size and often due to restrictions on available battery power. These factors have the consequence that mobile satellite systems operate at a significantly lower power margin than most mobile radio systems. Therefore, mobile satellite services are very limited unless there is line-of-sight transmission to the satellite.

74.2 The Radio Frequency Environment and Its Implications

When studying any radio system, it is generally instructive to begin by considering the basic link equation. This is particularly true for mobile satellite systems because the combination of limited electrical power on the satellite and at the mobile terminals and relatively small antennas on the mobile terminals results in a situation where transmitted power is a critical resource that should be conserved subject to the requirement of providing a sufficiently high signal-to-noise ratio to support the desired grade of service. Here, we only consider a simplified link equation with the objective of providing some context for subsequent issues. More detailed treatments of topics affecting link design are provided in Secs. 5.4 (satellite transmission impairments), 5.5 (satellite link design), and 5.6 (noise, antenna, and system temperature).

Consider the following simplified link equation:

$$C/N_o = P_t + G_t - L_p + G_r - T_r - k - L \text{ dB-Hz} \tag{74.1}$$

Where:

C/N_o = ratio of the signal power to the noise power spectral density after being received and amplified, dB-Hz
P_t = RF power delivered to the transmitting antenna, dBW
G_t = gain of the transmitting antenna relative to isotropic radiation, dBi
L_p = free-space path loss, dB
G_r = gain of the receiving antenna, dBi
T_r = composite noise temperature of the receiving system, dBK
k = Boltzmann's constant expressed in decibels, −228.6 dBW/K-Hz
L = composite margin accommodating losses due to a variety of transmission impairments, dB

A link equation of this type can be applied to either the **forward direction** or the **return direction**, as well as to the corresponding links between the satellite and the gateway or base station. Also, many satellites merely amplify and frequency translate the received signal from the uplink prior to transmission over the downlink. Therefore the end-to-end signal-to-noise ratio must incorporate both uplink and downlink noise. Further examination of some of the constituents of the link equation can provide some interesting insights. For instance, the transmitting antenna gain is a measure of an antenna's ability to convert the electrical radio signal into an electromagnetic waveform and then to focus that energy toward the receiving antenna. Similarly, the receiving antenna gain is a measure of an antenna's ability to collect the transmitted energy and then to convert the energy into an electrical signal that can be amplified and processed by the receiver. In general, the greater the gain of an antenna, the narrower is its beam. As an example, consider a circular parabolic antenna. Its antenna gain is given by the following expression:

$$G = 10\log_{10}(\Omega \pi^2 D^2/\lambda^2) \text{ dBi} \tag{74.2}$$

Where:

Ω = efficiency of the antenna
D = diameter of the antenna, m
λ = wavelength of the RF signal, m

Although the circular parabolic antenna is only one example of many possible antennas, this equation provides an order-of-magnitude indication of the size of antenna needed to provide a desired amount of gain.

The free space path loss is given by

$$L_p = 10 \log_{10}[(4\pi r)^2/\lambda^2] \text{ dB} \tag{74.3}$$

where r is the distance in meters between the terminal and the satellite.

Note that the path loss and both the transmitting and receiving antenna gains increase with the square of the radio frequency. From Eq. (74.1) this would imply that given constraints on the dimensions of the antennas, there is a net benefit in using relatively high radio frequencies. Although there is some truth to this, the benefits are not as great as one might think because the composite propagation losses also tend to increase with the radio frequency, as is described in Sec. 5.4 on satellite transmission impairments. Frequencies used for delivering mobile satellite services range from about 130 MHz up to 30 GHz! Use of the lowest frequency bands is dominated by low data rate store-and-forward services provided with low Earth-orbit (LEO) satellites. Most mobile satellite telephone services, with typical transmission rates ranging from several kilobit per second (kbps) up to 64 kbps, are provided in frequency bands located between 1500 and 2500 MHz. Interactive packet data and circuit-switched services, at transmission rates up to several megabit per second, are planned for the higher frequency bands. This correlation between transmission rates and carrier frequencies is a result of both the preceding link considerations as well as the fact that larger spectrum allocations are available in the higher frequency bands than in the lower ones.

Efficient use of the satellite's two critical resources, spectrum and power, can be achieved if the satellite's antenna system covers the appropriate area of the Earth's surface with multiple beams instead of one large beam. The total allocated system bandwidth is divided into a number of distinct subbands, which need not be of equal bandwidths, and each beam is then assigned a subband in such a way that some desired minimum isolation between beams with the same subband is maintained. This situation is illustrated in Fig. 74.1. Although not necessarily the case in an operational system, this example illustrates the beams as being circular and equal in area. Here, there are 4 distinct subbands and 16 beams. Note that in this example each frequency subband is used four times. In general, the frequency reuse factor is defined to be the ratio of the number of beams to the number of distinct frequency subbands. Improved power efficiency is a direct result of the higher satellite antenna gain implicit in the smaller beam area. Ideally, the satellite should be able to apportion dynamically its total transmit power among the individual beams depending upon the loading in each beam.

The satellite mobile channel is subject to multipath fading and shadowing just as the terrestrial mobile channel. The physical mechanisms for these propagation phenomena are the same. Shadowing is due to foliage, buildings, or terrain blocking the line-of-sight signal from the satellite. Many propagation campaigns have been performed to investigate the statistics of shadowing in mobile satellite environments. For example, Fig. 74.2 shows some results of data gathered near Ottawa, Canada, in June of 1983 [Butterworth, 1984]. Note that these curves exhibit a *knee* between 50 and 90%. For the portion of the curves to the left of the knee, the characteristics are dominated by multipath fading, whereas shadowing is the dominating factor to the right of the knee. With adequate margins, it is reasonable to expect mobile satellite systems to maintain communications links in environments characterized by fading and light shadowing, but it is not practical to overcome heavy shadowing for most services. For this reason, availability targets between 90 and 95%, corresponding to margins of less than 10 dB, are common for land mobile satellite systems

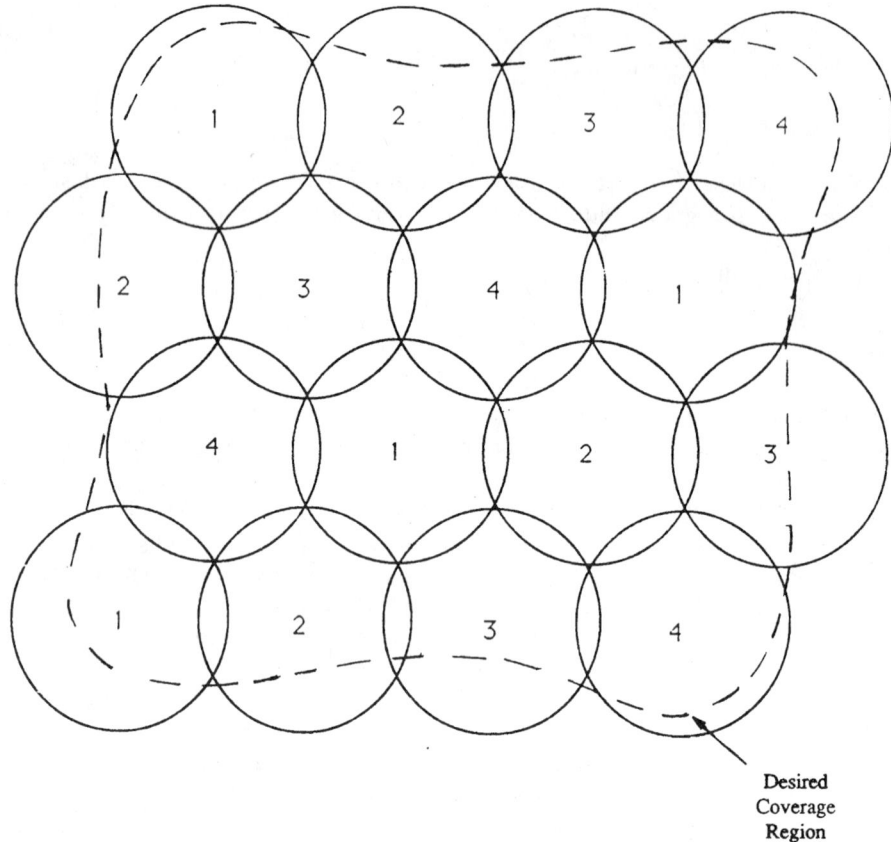

FIGURE 74.1 An example illustrating frequency reuse with a multiple-beam satellite. Here, the total allocated system frequency band is divided into four subbands. The assignment of the subbands to the beams is denoted by the number shown in each beam.

delivering services to users in rural, highway, and suburban environments. Availability figures for urban environments tend to be much lower. Exceptions include very low-capacity services, such as paging, which may require margins as high as 30 dB to allow for building penetration. Of course, the margin that is required to meet a given level of availability in a particular set of user environments is dependent on many factors including the elevation angle to the satellite, the type of terrain, the density and height of the vegetation, and the number of man-made obstacles.

The shadowing is often modeled as a log-normal process [Butterworth, 1984; Loo, 1984] where the direct path power (relative to the median power) has a distribution given by

$$P[S < x_{\mathrm{dB}}] = 0.5 + 0.5 \, \mathrm{erf}\left(\frac{x_{\mathrm{dB}} - m_s}{\sqrt{2}\sigma_s}\right) \qquad (74.4)$$

where S is the received signal strength in decibel, m_s is the average decibel shadowing loss, and σ_s is the decibel variance of the shadowing loss. The parameters of this distribution depend not only on environment but on the elevation angle to the satellite and on frequency. For a variety of situations, estimates of these parameters can be derived from the empirical results, of propagation campaigns, provided in ITU recommendations [1992, 1995] and in Stutzman [1988].

Multipath fading results from interference caused by reflections of the radio signal from the immediate environment, with the time-varying nature being caused by motion of the terminal

Mobile Satellite Systems

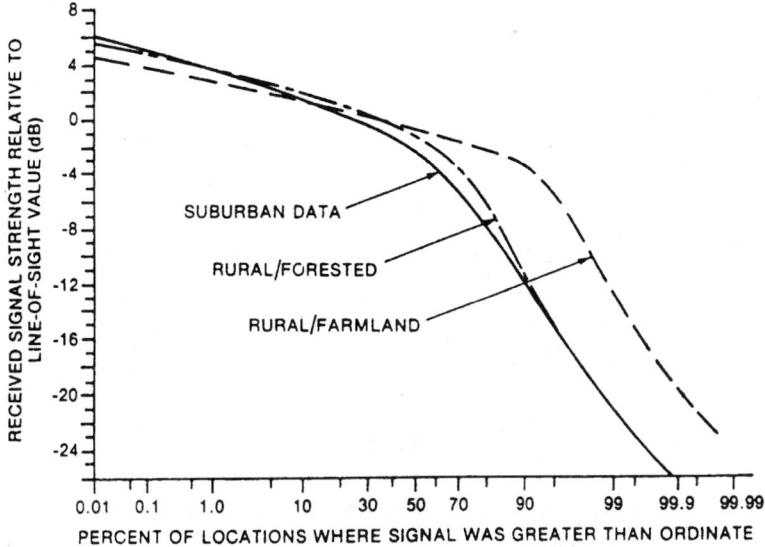

FIGURE 74.2 Distribution functions for data recorded near Ottawa, Canada, in June 1983. The measurements were made with an omnidirectional antenna at a frequency of 1542 MHz. The elevation angle (the angle between a plane parallel to the Earth's surface and the line-of-sight path to the satellite) was approximately 20°. (*Source:* Butterworth, 1984.)

through the environment. In satellite applications, the reflections are generally too weak to be useful by themselves but not so weak as to be ignored. Consequently, the satellite mobile channel is usually modeled as a Rician fading channel consisting of an unshadowed direct path and a diffuse component, which has a complex Gaussian distribution (Rayleigh distributed amplitude and uniformly distributed phase). The amplitude distribution of the resulting channel is the noncentral chi-squared distribution with two degrees of freedom (assuming a unit amplitude direct path)

$$P[R > r] = 1 - e^{-(a^2+b^2)/2} \sum_{k=0}^{\infty} \left(\frac{a}{b}\right)^k I_k(ab) \qquad (74.5)$$

where $a = \sqrt{K}$, $b = r/\sigma$, and $I_k(\cdot)$ is the kth-order modified Bessel function. The quantity K is the ratio of the direct path power to the average power of the multipath $2\sigma^2$. Rician channels are characterized by their K factor. The probability of a fade of a given depth is shown in Fig. 74.3 for various K factors. Note that there is a greater than 50% probability that the diffuse component will increase the signal strength. K factors with vehicle-mounted antennas (automobile, aircraft, or ship) are typically 10 dB and higher, with the worst-case value of 10 dB occurring at low elevation angles. For handheld terminals, where the antenna has little directivity, lower K factors can be expected.

The time-varying nature of the fading, in particular, the duration of the fades, is characterized by its spectrum. For land–mobile applications the Clarke model, which assumes uniformly distributed scatterers in a plane about the moving terminal, is often used. The maximum Doppler frequency observed is related to the transmission frequency by the expression $f_d = v/\lambda$, where v is the velocity of the terminal, and λ is the transmission wavelength. The Clarke model for the diffuse spectrum is given by

$$S(f) = \frac{1}{\sqrt{1 - (f/f_d)^2}} \qquad (74.6)$$

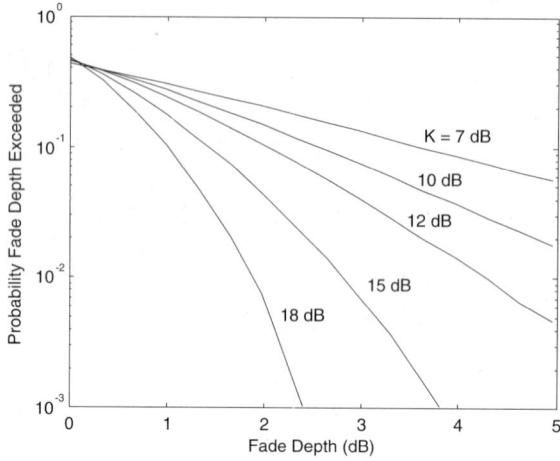

FIGURE 74.3 Cumulative fade distribution for the Rician channel as a function of K factor.

and the corresponding time correlation of the diffuse component is given by

$$R(\tau) = J_0(2\pi f_d \tau) \tag{74.7}$$

where $J_0(\cdot)$ is the zeroth-order Bessel function.

For the land–mobile channel, some effort has been devoted to providing an appropriate model combining the effects of shadowing and multipath fading, which is appropriate for satellite applications. An example of these is the Lutz model [Lutz et al., 1991], which consists of two states, a good state and a bad state. In the good state, the channel is Rician. In the bad state, the line-of-sight signal is essentially blocked due to significant shadowing. The parameters of the model are the distribution of the dwell times in each state, the K factor, and the distribution of the shadowing, which are selected to model the environment of interest.

In aeronautical–mobile, the fading spectrum is generally modeled as Gaussian. However, measurements in this case indicate that the fading bandwidth is not strongly dependent on the aircraft velocity. The fading tends to be slow and is due to reflections from a limited portion of the Earth's surface and from the aircraft's body. At L-band frequencies (1.6 GHz) the fading bandwidths for aeronautical–mobile channels range from 20 to 100 Hz with 20 Hz being more prevalent [Neul et al., 1987].

The physics involved in determining an appropriate model for the maritime channel is quite similar to that for the aeronautical case [Moreland, 1987]. Consequently, the fading spectrum is generally modeled as Gaussian. However, the fading bandwidth is typically less than 1 Hz [Schweikert and Hagenauer, 1983].

In general, multipath in satellite–mobile channels is not time dispersive. Measurements indicate that for practical bandwidths (<2 MHz), this is true even for wideband channels [Jahn et al., 1995]. A possible exception to this is the high-altitude aeronautical channel where one may occasionally receive distant reflections from the sea surface or ice. For this case, the strength of the reflected path is strong enough to be problematic only at elevation angles below 20°.

74.3 Satellite Orbits

Initial mobile satellite systems used geostationary orbits and were generally aimed at mobile terminals with significant antenna gain (>8 dB). These include a number of regional mobile satellite services as well as the Inmarsat international service. The geostationary orbits provide good coverage in the low- to mid-latitude regions. Service availability to mobile units in the mid- to higher latitude regions can be compromised due to blockage at the lower elevation angles. To achieve a higher service availability to a greater portion of the Earth's surface necessitates the use of nongeostationary orbits.

With nongeostationary orbits, coverage from one satellite is time-varying, although the orbits are often chosen to be a submultiple of either the solar day (24 h) or the sidereal day (23 h 56 m 4 s). With the latter, the satellite ground tracks are periodic, whereas with the former the ground tracks drift approximately 1° longitudinally each day. As a result of this time-varying behavior, a constellation of satellites must be employed to provide continuous coverage.

A variety of nongeostationary orbits and constellations have been proposed. These range from the LEO systems at altitudes for 500 to 2000 km, through the medium Earth-orbit (MEO) systems at altitudes from 9,000–14,000 km to the highly elliptical orbit (HEO) systems. One constraint on orbits are the Van Allen radiation belts, which encircle the planet [Hess, 1968]. The primary belt from 2,000–9,000 km and the secondary belt from 14,000–19,000 km can restrict operation at these altitudes because of the greater radiation hardness requirements on the satellite components.

Constellations are generally configured as a number of orbital planes with a number of satellites in each plane. The number of satellites in the constellation generally decreases with the altitude. Most examples fall into the category of inclined circular orbits [Ballard, 1980] where the inclination refers to the angle the orbital plane makes with the equator. The extreme case of this are the polar-orbiting constellations [Adams and Rider, 1987]. The coverage provided by all of these constellations is global in nature but the minimum elevation angle and number of satellites visible simultaneously at any location depends on the constellation and its altitude. In general, however, simple constellations can provide good coverage of the Earth. As an example, the cumulative distribution of the elevation angles (averaged over time and location) for a 12 satellite constellation is shown in Fig. 74.4. This figure indicates that the elevation to the primary visible satellite is always greater than 23°; 90% of the time/locations it is always greater than 30°. Figure 74.4 also indicates that a second satellite is always visible with an elevation of at least 10°.

An alternative to the inclined circular orbits are the elliptical orbits of which the Molniya [Ruddy, 1981] and Loopus [Dondl, 1984] are well-known examples. A constellation of satellites in highly

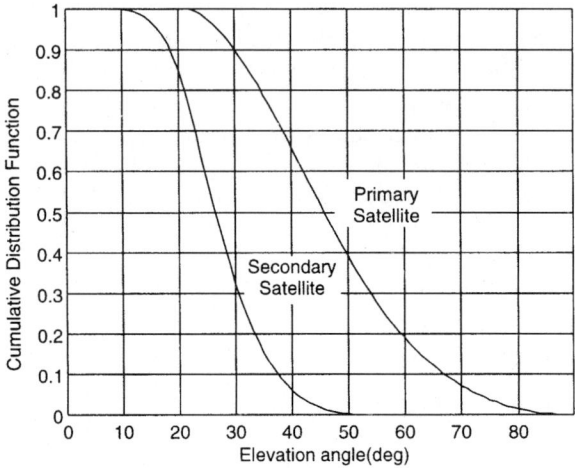

FIGURE 74.4 Cumulative distribution of elevation angle to primary satellite with a constellation of four inclined planes of three satellites at an altitude of 10,355 km.

elliptical orbits can approximate the effect of an overhead geosynchronous satellite for high-latitude regions. However, the useful coverage with elliptical orbits is restricted (generally they are only active near the orbit apogee), and they are generally intended for regional services.

Whereas nongeostationary orbits ameliorate a number of the propagation phenomena affecting communications with a mobile terminal, they also introduce a number of challenges. In particular, with nongeostationary satellites, a number of quantities such as path delay and Doppler shifts become not only position dependent but also time varying, depending on the satellite orbital speed. The period of a satellite in an elliptic or circular orbit around the Earth is

$$T = \frac{2\pi a^{3/2}}{\sqrt{GM}} \quad (74.8)$$

where a is the semimajor axis of the orbit (the radius in the case of a circular orbit) and GM is the Earth's gravitational parameter ($GM = 3.99 \times 10^{14}$ m^3/s^2).

As with a geostationary satellite, the propagation delay and path loss vary with user position (range), and the general expression for satellite range as a function of elevation angle θ is given by

$$r(\theta) = R_e(\sqrt{(1 + h/R_e)^2 - \cos^2\theta} - \sin\theta) \quad (74.9)$$

where R_e is the Earth's radius (6378 km) and h is the satellite altitude. The corresponding delay is $\tau = r/c$. Unlike geostationary satellite systems, there can be significant Doppler and propagation delay variations due to the relative velocity of the user and the satellite with nongeostationary mobile satellite systems. Both of these quantities are proportional to the range rate, with the delay variation given by

$$\frac{d\tau}{dt} = \frac{1}{c}\frac{dr}{dt} \quad (74.10)$$

where τ is the propagation delay, r is the range, and the Doppler shift is given by

$$f_s = \frac{1}{\lambda}\frac{dr}{dt} \quad (74.11)$$

Note that this Doppler frequency is distinct from the maximum Doppler rate used to characterize multipath fading. The latter corresponds to the mobile terminal velocity relative to its immediate environment. There are a number of contributors to the range rate. They include the satellite motion, the Earth's rotation, and the velocity of the mobile terminal relative to the Earth. In a number of simple cases involving circular orbits, the contribution of these various components can be calculated explicitly. When the user lies in the orbital plane, such that the geometry is as shown in Fig. 74.5, then the range rate due to satellite motion alone is given by

$$\left(\frac{dr}{dt}\right)_{sat} = -\frac{2\pi R_e}{T}\cos\theta \quad (74.12)$$

In addition, the velocity of the mobile terminal due to the Earth's rotation is given by

$$v_r = \frac{2\pi R_e}{T_e}\cos\gamma \quad (74.13)$$

in a direction of constant latitude, where γ is the latitude of the mobile terminal, and T_e is the period of the Earth's rotation (23 h 56 m 4 s). In the general case, the range rate is the vector sum of the velocities of these different contributors projected along the range vector. This depends on

Mobile Satellite Systems

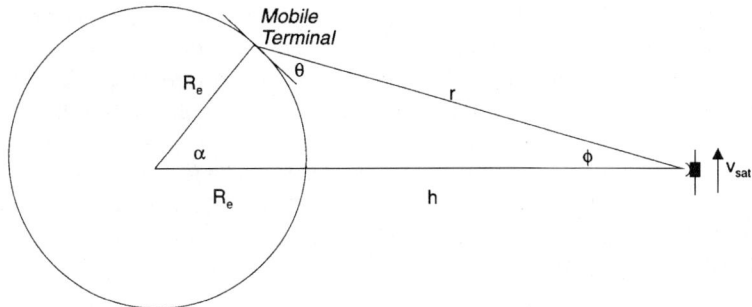

FIGURE 74.5 Illustration of satellite geometry for calculating range and range rates.

the mobile terminal position, as well as velocity, and does not have a simple explicit formula [Vilar and Austin, 1991]. However, the general circular-orbit case can be upperbounded by Eqs. (74.12) and (74.13) to give

$$\left|\frac{dr}{dt}\right| \leq 2\pi R_e \cos\theta \left(\frac{1}{T} + \frac{\cos\gamma}{T_e}\right) \qquad (74.14)$$

This shows that both the Doppler and delay rate are zero at the subsatellite point and increase toward the satellite horizon, although for Doppler the increase depends on the relative direction. The Doppler can range from zero for a geostationary orbits to greater than 42 kHz for a LEO satellite (1000 km) at 2.0 GHz. The corresponding delay rate can range from zero to 21 μs/s.

Another parameter of interest is the rate of change of Doppler. There is no simple expression for the Doppler rate but, in a similar manner, it can be upper bounded by

$$\left|\frac{df_s}{dt}\right| \leq \frac{(2\pi)^2 R_e}{\lambda} \left(\frac{R_e + h}{r}\frac{1}{T^2} + \frac{\cos\gamma}{T_e^2}\right) \qquad (74.15)$$

Expressions (74.14) and (74.15) do not include the effect of mobile terminal motion relative to the Earth. The Doppler rate can range from zero for a geostationary satellite to 300 Hz/s for a LEO satellite at 1000 km at 2.0 GHz depending on the positions of the user and the satellite.

In addition to these potential influences of nongeostationary satellites on modem design, there can also be a significant impact on the multiple access strategy due to differential delay and differential Doppler across a spotbeam. Assuming the differential effects of the Earth's rotation across a spotbeam can be neglected, the differential delay can be upper bounded by

$$\Delta\tau \leq \frac{r(\theta_{\min}) - r(\theta_{\max})}{c} \qquad (74.16)$$

where $\theta_{\min}$ and $\theta_{\max}$ are the minimum and maximum elevations for the spotbeam of interest. The greatest differential delay occurs at the edge of coverage. For many scenarios, this differential delay can be on the order of milliseconds, which can have an impact on multiple access strategies that require the time-synchronization of users.

The differential Doppler across a spotbeam depends on the location of the spotbeam relative to the orbital plane of the satellite. If one neglects the differential effects of the Earth's rotation across a spotbeam, one can bound the differential Doppler in spotbeams along the orbital plane by

$$\Delta f_s \leq \frac{2\pi R_e}{T}(\cos\theta_{\min} - \cos\theta_{\max}) \qquad (74.17)$$

This bound assumes that the whole spotbeam lies completely in either the fore or aft coverage of the satellite, but indicates that the maximum differential Doppler occurs across a spotbeam covering the subsatellite point. For many scenarios, the differential Doppler across a spotbeam can be on the order of kilohertz, and thus can also influence the design of a multiple access strategy.

It should be noted that lower altitude satellites must provide the same RF power flux density as higher altitude satellites. Typically, however, higher altitude satellites are used to cover larger areas, and to provide an equivalent return link service must have larger antennas and more spotbeams. Consequently, although fewer of the higher altitude satellites are required to provide global coverage, they are generally larger and more complex.

74.4 Multiple Access

Satellites are frequently designed with antennas large enough to provide multiple spotbeams for mobile satellite services. This not only reduces the power requirements but also allows frequency reuse between the beams. The result is a system that is similar to terrestrial cellular systems with the following differences.

1. Isolation between co-channel users is controlled by the spacecraft antenna characteristics and not by propagation losses.
2. The near–far problem is less critical and so power-control is less of a factor. Because of the delay constraints in satellite systems, power control is generally limited to counteracting the effects of variation in antenna gain and average propagation losses. This, however, means that cobeam users in the forward direction will generally be received at different power levels.
3. Because of the distortion caused by the curvature of the Earth's surface, the spotbeams do not cover the same surface area. In particular, the outer beams can be significantly extended in the radial direction as illustrated in Fig. 74.6; this figure is a two-dimensional projection of the coverage pattern on the Earth that preserves radial proportions but results in some angular distortion. The consequence of this distortion is that the outer beams may be required to serve significantly more users; furthermore, the outer beams are subject to more adjacent beam interference for similar reasons. A complicating factor is that the outer beams are also the most likely to be in the field of view of other satellites in the constellation. The satellite antenna can be designed to provide equal size spotbeams on the Earth, but this policy penalizes the communications performance of a large number of users for the benefit of none.

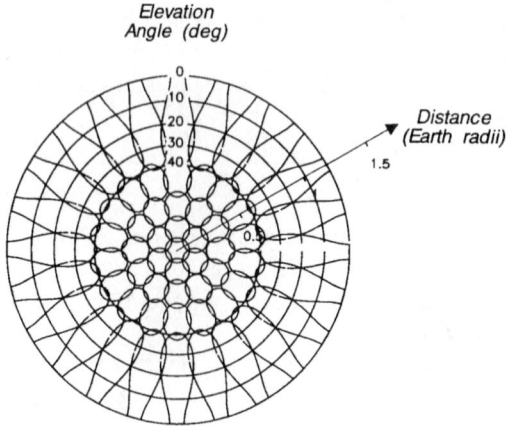

FIGURE 74.6 Spotbeam pattern projected onto the Earth's surface: 4-dB contours with edge of coverage at 10° elevation.

Mobile Satellite Systems

4. With many constellations there can be significant overlapping coverage from multiple satellites, and this allows the possibility of diversity transmissions. A whole range of diversity techniques are possible for all multiple access strategies depending on the power, bandwidth, and complexity constraints. The simplest approach is selection diversity during the call setup. More complex approaches include dynamic techniques such as **switched diversity** in the forward direction based on return-link signal strength, and **maximal ratio combining** in the return direction.
5. This brings up a final difference with respect to terrestrial systems in that, with a fixed antenna cluster, the satellite cells and base station are moving with respect to the mobile rather than vice versa. This raises issues regarding handover procedures and, in the light of multiple satellite coverage, the orthogonality of user accesses.

In general, multiple access techniques for mobile satellite applications can be divided into two groups: wideband techniques, such as code-division multiple access (CDMA) [Gilhousen et al., 1990], and narrowband techniques, such as frequency-division multiple access (FDMA) and **narrowband time-division multiple access** (TDMA). Each technique has its strengths and weaknesses, which depend on the application and implementation, but there are a number of common aspects. The power efficiency of all techniques can be improved by adding forward error correction coding but with a bandwidth penalty. The exception to this is asynchronous CDMA, which does not suffer a bandwidth penalty but, generally, must tolerate greater co-user interference. There are synchronous versions of all of these techniques (synchronization at the chip level for CDMA, and at the bit and frequency level for FDMA and TDMA) that can be used to improve spectral efficiency on the forward link where one can easily synchronize transmissions. For applications in the 1- and 2-GHz range, the propagation characteristics of wideband techniques (<2 MHz) and narrowband techniques are similar.

Interference tolerance, in particular multiple access interference, is one aspect that differs between wideband and narrowband systems. With CDMA, interference due to co-users is averaged over all co-users and affects all users. With narrowband techniques, interference generally results in greater degradation but only of individual carriers. It is this interference tolerance that determines the frequency reuse capability and the bandwidth-limited capacity. When comparing the bandwidth-limited capacities of various approaches, care must be taken to be equitable. Interference has a probability distribution due to the random location (and isolation) of cochannel users relative to the desired user. In both wideband and narrowband techniques, it is the tails of this distribution that determine the effect of interference on the service availability.

An additional issue is the design of the ancillary channels for signalling and control for a multi-beam satellite system and their integration with the traffic services with a minimum of power and bandwidth requirements.

74.5 Modulation and Coding

Along with the choice of multiple access strategy, modulation and coding schemes must be selected to meet the stringent bandwidth and power efficiency requirements of the mobile satellite communications system. With respect to the choice of modulation scheme, binary phase-shift keying (BPSK) has been popular for low data rate store-and-forward systems due to its simplicity and robustness. However, BPSK is very spectrally inefficient when transmitted from a terminal with a nonlinear power amplifier (e.g., a power-efficient class-C amplifier). Aeronautical satellite communications systems have improved on the spectral efficiency, without sacrificing robustness or compatibility with nonlinear amplifiers, by using $\pi/2$-**BPSK** coupled with 40% square-root raised-cosine pulse shaping.

The most popular choice of modulation technique for higher rate services has been quadrature phase-shift keying (QPSK). QPSK can be considered as two BPSK waveforms orthogonally

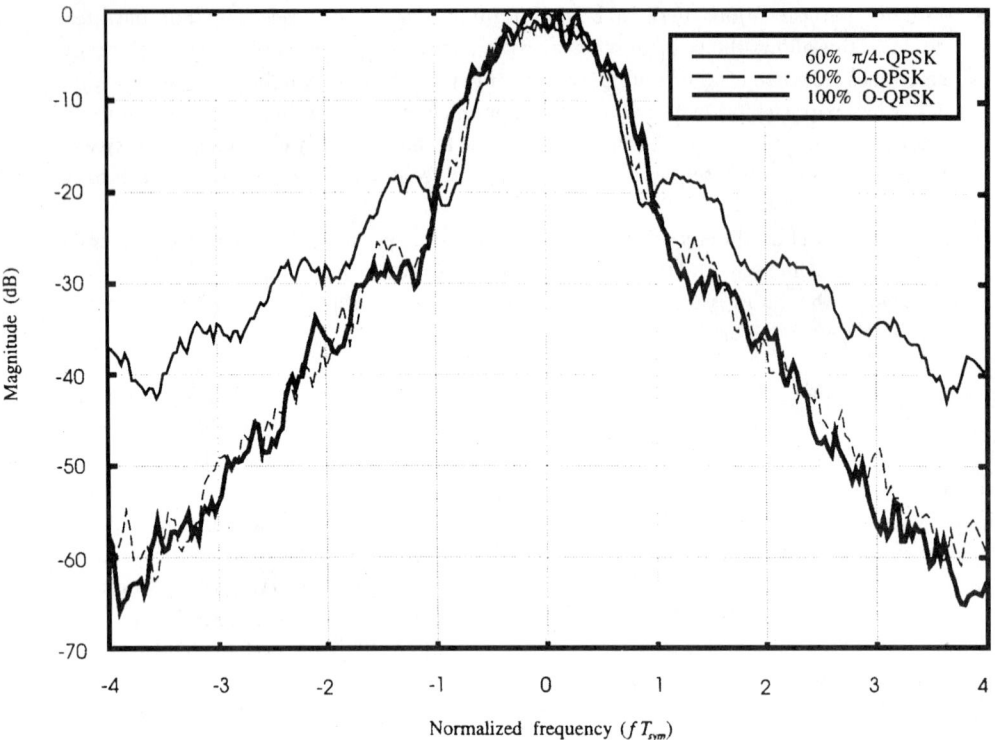

FIGURE 74.7 The transmitted power spectrum of $\pi/4$-QPSK (60% square-root raised-cosine pulse shaping) and offset-QPSK (60% and 100% square-root raised-cosine pulse shaping) following hard-limiting amplification. (*Source:* Patenaude and Lodge, 1992.)

modulated: one by a sine wave at the carrier frequency and the other by the corresponding cosine wave. Consequently, for a given choice of pulse shaping filter and for a linear channel the spectral efficiency (in terms of the ratio of the number of bit per second transmitted to the occupied bandwidth in Hertz) for QPSK is twice that for BPSK. With standard QPSK the symbol timing is identical in the sine and cosine channels. A fairly linear transmitting power amplifier is required in the mobile terminal if good power efficiency is to be maintained with standard QPSK. As was the case for BPSK, variations of QPSK can be used that are reasonably tolerant to amplifier nonlinearity. Figure 74.7 [Patenaude and Lodge, 1992] illustrates the transmitted power spectrum of $\pi/4$-**QPSK** (60% square-root raised-cosine pulse shaping) and **offset-QPSK** (60% and 100% square-root raised-cosine pulse shaping) after being amplified by a hard-limiting amplifier (which is the extreme example of a nonlinear power amplifier). Note that even in this severe case, offset-QPSK with 60% pulse shaping is quite spectrally efficient. In fact, this is the choice of modulation scheme for a number of mobile satellite communications systems including Inmarsat's M, B, and Aeronautical systems.

The viability of mobile satellite systems is dependent on excellent power efficiency. Consequently, detection is often based on phase coherent approaches rather than the less power efficient options of differentially coherent and noncoherent detection that are commonly used for terrestrial systems. Increasingly, traditional carrier tracking schemes, such as phase-locked loop-based techniques, are being replaced by digital schemes such as the one described in A.J. Viterbi and A.M. Viterbi [1983]. The fading and shadowing nature of the channel can result in frequent cycle slips of the carrier recovery algorithm, resulting in a phase ambiguity (which is inherent in the PSK signal) that must be resolved. Differential phase encoding of the transmitted symbols is a fairly traditional approach for eliminating the detrimental effects of the phase ambiguity. However, differential encoding tends to increase the average bit error rate for a given symbol error rate and is consequently avoided in many mobile satellite system signal designs. A more common approach is to periodically insert a known bit or symbol (referred to as a *reference symbol*) or group of bits (referred to as a *unique word*) into the transmitted signal, which can be used to detect the occurrence of cycle slips and then to resolve the resulting carrier phase ambiguity. In the extreme case, for which the rate of insertion is sufficiently greater than the required tracking bandwidths, carrier tracking can be based solely on the reference symbols [Moher and Lodge, 1989].

Forward error control (FEC) coding is an important tool for achieving good power and bandwidth efficiency. The choice of FEC coding technique is dependent on the number of bits that are to be coded as an identifiable group. First, consider the case for which only several tens of bits are to be coded into a codeword. This situation naturally arises in a number of circumstances including short data packets containing signalling information or user data, and for digital voice transmission for which additional delays (due to coding) of greater than a couple of voice frames is unacceptable and only the most sensitive bits in the frame are protected with FEC coding. In these cases, block codes such as the Golay code and Bose-Chaudhuri Hocquenghem (BCH) codes are common choices. Another possibility is the use of convolutional coding with puncturing [Yasuda, Kashiki, and Hirata, 1984] used to achieve the desired code rate. Convolutional codes with a constraint-length of 7 have been popular choices for early systems, with constraint-length 9 codes being proposed for later ones. Usually prior to convolutional encoding, a field of *flush bits* is appended to the end of the field of data bits so that the last data bits to be decoded have a similar level of integrity as the rest of the data bits. The field of flush bits can represent a significant (and undesirable) overhead for the transmission of very short data blocks. Fortunately, the requirement for such a field can be obviated if tail biting [Ma and Wolf, 1986] is used. Usually, convolutional coding is preferable to block coding because it is more amenable to the use of soft decisions.

The second scenario to consider is the transmission of a continuous data stream (e.g., digital voice) or data in large fields. In this case, convolutional coding is a reasonable choice for services for which high average bit error rates (i.e., higher than 10^{-4}) are acceptable. For applications requiring higher quality data, reasonable choices include concatenated coding (with a convolutional code as the inner code and a Reed–Solomon code as the outer code) and turbo coding [Berrou, Glavieux, and Thitimajshima, 1993]. As an example of the relative power efficiency, Fig. 74.8 shows the performance of several of these alternatives where the only channel impairment is additive white Gaussian noise. Curves A, B, C, and D are all for rate-1/2 coding. Curve D is a constraint length-7 convolutional code. Curve C is the same convolutional code, punctured to rate-3/5, concatenated with a (255, 213) Reed–Solomon code. Curves A and B represent the performance of turbo coding for block lengths of 10,000 and 300 data bits, respectively. Curves E and F provide an indication of the tradeoff between bandwidth and power efficiency. All use the convolutional codec of curve D, so the level of code complexity is fixed. Curve E is punctured to a rate of 3/4, whereas curve F uses pragmatic coding [Viterbi et al., 1989] to achieve a rate of 1 b per dimension. For the range of spectral efficiencies considered here and for a fixed level of coding complexity, the rule of thumb is that a 1-dB improvement in bandwidth efficiency results in approximately a 1-dB degradation in power efficiency. Note that trading bandwidth against power efficiency, using puncturing or pragmatic coding, is equally applicable for concatenated and turbo coding schemes.

Fading and shadowing can cause reductions in the level of the received signal to a point where reliable transmission is no longer possible. If these signal outages are of short duration, they can successfully be compensated for by interleaving the order of the coded data prior to modulation in the transmitter, and then deinterleaving the data prior to decoding [Clark and Cain, 1981]. This effectively shuffles the received bits having low received signal power with the received bits having higher signal power. Of course, for the combination of coding and interleaving to eliminate errors in the decoded data bits, the interleaving must be performed over a signal duration that is several times greater than the duration of the outage. Interleaving adds to the total transmission delay and,

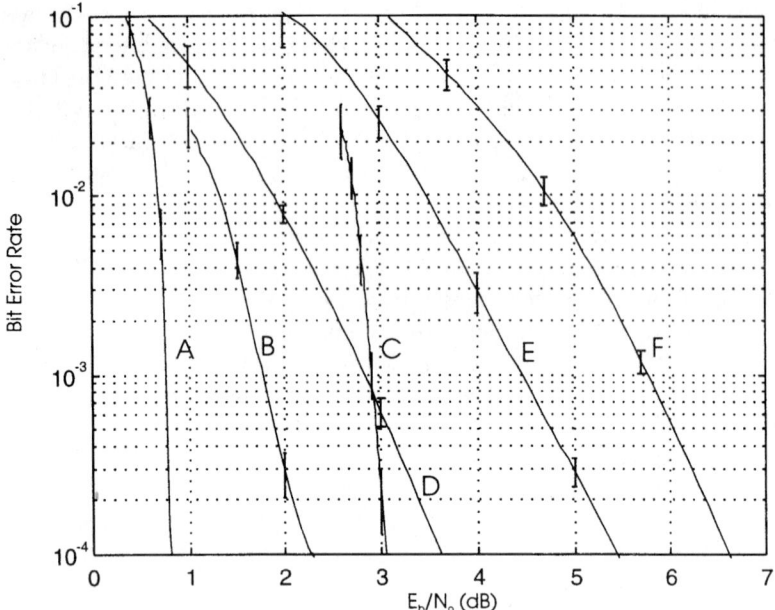

FIGURE 74.8 The performance of various coding schemes in an additive white Gaussian noise environment:
A: rate-1/2 turbo code with a block size of 10,000 b,
B: rate-1/2 turbo code with a block size of 300 b,
C: rate-1/2 concatenated convolution/Reed-Solution code,
D: rate-1/2 constraint length 7 convolution code,
E: rate-3/4 (punctured) constraint length 7 convolutional code,
F: rate-1 (pragmatic) constraint length 7 convolutional code.
(*Source:* Provided by P. Guinand, Communications Research Centre, Ottawa, Ontario, Canada.)

consequently, the duration of outages that can be dealt with is dependent on the delay constraint of the given mobile satellite service.

Inevitably some of the outages will have a duration too great to be handled by interleaving and coding. In such cases, many services will simply be unavailable. However, some source codecs (e.g., certain vocoders) have the capability to gracefully handle several sequential missing or erroneous frames. Also, applications that can tolerate long delays can overcome fairly long outages by employing automatic repeat request (ARQ) schemes. ARQ schemes are treated in detail in Chapter 15 of this handbook.

74.6 Trends in Mobile Satellite Systems

An increasing variety of services are being offered to travelers via mobile satellite systems, with the data rates for the high-end services increasing steadily. In parallel with this, there is continuing pressure to reduce the size of the mobile terminals. The result is that the size of the system's space segment is growing through some combination of an increased number of satellites and larger, more complex satellites. For a given orbit altitude, the size of the satellite antenna is growing dramatically. A corresponding increase can be seen in system complexity, with the apportioning of this complexity between the space segment and the ground infrastructure varying from one system to another. System designs that place much of this complexity in the space segment are characterized by onboard switching and processing (see Chapter 72) and, in some cases, intersatellite links. System designs that place much of the complexity in the ground infrastructure are characterized by a large number of coordinated and very sophisticated gateway stations.

The number of customers for mobile satellite services has begun to grow significantly, and the forecasts are for truly explosive growth in the future. Many organizations are proposing new systems to meet this anticipated demand, with new frequency bands being allocated to accommodate them. Even with these new bands, increasing frequency congestion will be experienced if the forecasted traffic volumes materialize. The first generation mobile satellite systems were predominantly power limited. However, subsequent generations will be increasingly bandwidth limited.

The traffic volumes for many early generation mobile satellite systems are fairly symmetric (in the sense that the volume of traffic in the forward direction is approximately the same as that for the return direction) because voice telephone applications are dominant. However, it is anticipated that an increasing portion of the traffic will support services such as database access, message retrieval (e.g., e-mail and voice mail), and broadcast information. In general, these services result in much greater volumes of data being transmitted in the forward direction than in the return direction. Fortunately, this asymmetric traffic requirement is a good match with the technological situation, in a couple of respects. First, in the forward direction, improved bandwidth efficiency can be achieved by taking advantage of the fact that the system can be designed to accurately coordinate the carrier frequency, carrier phase, and symbol timing of each transmission within a beam. Consequently, techniques such as synchronous CDMA, orthogonal frequency-division multiplexing, and wavelet-packet multiple access can be employed. Second, unlike the return direction for which the mobile terminal's power amplifier places fairly severe constraints on the dynamic range of the transmitted waveform, the forward link channel is incrementally linear. By *incrementally linear* we mean that each carrier appears to be transmitted through a linear channel, with intermodulation due to the nonlinearity of the satellite transponder's power amplifier resulting in an additive noiselike process. Given that both the traffic requirements and the technological situation are asymmetric, the transmission schemes selected for these mobile satellite services will differ significantly between the forward and the return direction.

Defining Terms

Forward direction: Direction of transmission that is from the satellite to the mobile.

Maximal ratio combining: A combining technique for which signals from two or more communication paths are aligned in phase, weighted proportional to their voltage-to-noise power ratios, and then summed [Proakis, 1989].

Narrowband TDMA: A multiple access strategy where a small number of users, typically less than 10, share a single frequency channel in time. Larger numbers of users are accommodated by using a number of narrowband TDMA carriers of different frequencies.

Offset-QPSK: A variation of QPSK for which the symbol timing in the sine channel is offset by half of a symbol period relative to that in the cosine channel.

Return direction: Direction of transmission that is from the mobile to the satellite.

Switched diversity: The process of switching to the best of two or more communication paths [Jakes, 1974].

$\pi/2$-BPSK: A variation of BPSK for which subsequent symbols experience a relative phase shift of $\pi/2$ rad.

$\pi/4$-QPSK: A variation of QPSK for which subsequent symbols experience a relative phase shift of $\pi/4$ rad.

References

Adams, W.S. and Rider, L. 1987. Circular polar constellations providing continuous single or multiple coverage above a specified latitude. *J. Astron. Sci.*, 35(April–June):155–192.

Ballard, A.H. 1980. Rosette constellations of earth satellites. *IEEE Trans. Aero. Elec. Sys.*, AES-16(5):656–673.

Berrou, C., Glavieux, A., and Thitimajshima, P. 1993. Near Shannon limit error-correcting coding and decoding: Turbo-codes. In *Proceedings of the IEEE International Conference on Communications* (ICC'93), pp. 1064–1071. May.

Butterworth, J.S. 1984. Propagation measurements for land-mobile satellite systems at 1542 MHz. Communications Research Centre Tech. Note No. 723, Dept. of Communications, Ottawa, Ontario, Canada, Aug.

Clark, G.C., Jr. and Cain, J.B. 1981. Interleaver structures for coded systems. In *Error-Correction Coding for Digital Communications*, Sec. 8.3, Plenum Press, New York.

Dondl, P. 1984. Loopus opens a new dimension in satellite communications. *Int. J. Sat. Comm.*, 2(4):241–250.

Gilhousen, K.S., Jacobs, I.M., Padovani, R., and Weaver, L.A. Jr. 1990. Increased capacity using CDMA for mobile satellite communications *IEEE J. Selec. Areas Commun.*, 8(4):503–514.

Hess, W.N. 1968. *The Radiation Belt and Magnetosphere*, Blaisdell, Waltham, MA.

ITU. 1992, 1995. Propagation data required for the design of earth-space land mobile telecommunication systems. Recommendation ITU-R PN.681 (1992) and ITU-R PN.681-1 (1995), International Telecommunications Union, Geneva, Switzerland.

Jahn A., Lutz, E., Sforza, M., and Buonomo, S. 1995. A wideband channel model for land mobile satellite systems. In *Proceedings of the International Mobile Satellite Conference*, pp. 122–127. Ottawa, June.

Jakes, W.C. ed. 1993. *Microwave Mobile Communications*, IEEE Press, New York (originally published 1974).

Loo, C. 1984. A statistical model for land-mobile satellite link. In *Proceedings of the International Conference on Communications*, pp. 588–594.

Lutz, E., Cygan, D., Dippold, M., Dolainsky, M., and Papke, W. 1991. The land mobile satellite communication channel—Recording, statistics and channel model. *IEEE Trans. Veh. Tech.*, 40(2):375–386.

Ma, H.H. and Wolf, J.K. 1986. On tail biting convolutional codes. *IEEE Trans. Commun.*, COM-34(2):104–111.

Moher, M.L. and Lodge, J.H. 1989. TCMP—A modulation and coding strategy for Rician fading channels. *IEEE J. on Selec. Areas Commun.*, 7(9):1347–1355.

Moreland, K.W. 1987. Ocean scatter propagation models for aeronautical and maritime satellite communication applications. M.Sc. Thesis, Dept. of Systems and Computer Engineering, Carleton Univ., Ottawa, Canada, Feb.

Neul, A., Hagenauer, J., Papke, W., Dolainsky, F., and Edbauer, F. 1987. Propagation measurements for the aeronautical satellite channel. *Proc. IEEE. Veh. Tech. Conf.*, Tampa, FL, pp. 90–97, June.

Patenaude, F. and Lodge, J. 1992. Non-coherent detection of hardlimited O-QPSK. In *Proceedings of the 1992 International Conference on Selected Topics in Wireless Communications*, Vancouver, Canada, pp. 93–97, June.

Proakis, J.G. 1989. *Digital Communications*, 2nd ed., McGraw–Hill, New York.

Ruddy, J.M. 1981. A novel non-geostationary satellite communications system. In *Proceedings of the International Conference on Communications*, Vol. 3, Denver, CO, June.

Schweikert, R. and Hagenauer, J. 1983. Channel modelling and multipath compensation with forward error correction for small satellite ship earth stations. In *Proceedings of the 6th International Conference on Digital Satellite Communications*, Phoenix, AZ, pp. XII-32–38.

Stutzman, W.L. ed. 1988. Mobile satellite propagation measurements and modeling: A review of results for systems engineers. In *Proceedings of the International Mobile Satellite Conference*, Pasadena, CA, pp. 107–117, June.

Vilar, E. and Austin, J. 1991. Analysis and correction techniques of Doppler shift for non-geosynchronous communication satellites. *Int. J. Sat. Comm.*, 9(2):123–136.

Viterbi, A.J. and Viterbi, A.M. 1983. Nonlinear estimation of PSK-modulated carrier phase with applications to burst digital transmission. *IEEE Trans. Info. Theory*, IT-29(4):543–551.

Viterbi, A.J., Wolf, J.K., Zehavi, E., and Padovani, R. 1989. A pragmatic approach to trellis-coded modulation. *IEEE Commun. Mag.*, 27(7):11–19.

Yasuda, Y., Kashiki, K., and Hirata, Y. 1984. High rate punctured convolutional codes for soft decision Viterbi decoding. *IEEE Trans. Commun.*, COM-32(3):315–319.

Further Information

A number of tutorial papers on mobile satellite systems have appeared in recent years. These include:

Kato, S. 1995. Personal communication systems and low earth orbit satellites. *Proceedings of the Space and Radio Science Symposium*, U.R.S.I., Brussels, Belgium, pp. 30–42, April.

Lodge, J. 1991. Mobile satellite communications systems: Toward global personal communications. *IEEE Commun. Mag.*, 29(11):24–30.

Wu, W.W., Miller, E.F., Pritchard, W.L., and Pickholtz, R.L. 1994. Mobile satellite communications. *Proc. IEEE*, 82(9):1431–1448.

Special issues of journals, dedicated to mobile satellite communications, are:

Kucar, A.D., Kato, S., Hirata, Y., and Lundberg, O., Ed. 1992. Special Issue on Satellite Systems and Services for Travelers, *IEEE J. on Select. Areas Commun.*, 10(8), Oct. 1992.

Maral, E., ed. 1994. Special issue on personal communication via satellite. *Int. J. of Sat. Commun.*, 12(1), Jan.–Feb. 1994.

A comprehensive book that contains a number of sections that are highly relevant to the discussion of multiple access techniques is:

Abramson, N. ed. 1993. *Multiple Access Communications: Foundations for Emerging Technologies*, IEEE Press, New York.

75
Satellite Antennas

75.1	Introduction	1032
75.2	Horn Antennas and Reflector Feeds	1034
	Horn Antennas • Cup-Dipole Antennas • Frequency Reuse Feed and Polarization Diversity • Monopulse Feed	
75.3	Reflector Antennas	1038
75.4	Phased Array Antennas	1044
75.5	Tracking, Telemetry, and Command Antennas	1046
75.6	Current and Planned LEO/MEO Mobile Satellite Antennas	1048
75.7	Radiometer Antennas	1050
75.8	Space Qualification	1050
75.9	Future Trends and Further Study	1051

Yeongming Hwang
City University of Hong Kong

Youn Choung
TRW Space and Electronics Group

75.1 Introduction

Satellite antennas provide not only for telecommunications, but also communications for controlling and commanding the satellite in a prescribed **orbit**: an invisible tower in space [Hwang, 1992]. Its capabilities have evolved from being only a radio relay station to being an onboard processing base station. Most communications satellites are boosted into geosynchronous Earth orbit (GEO). At a particular altitude, with an angular speed exactly matching the Earth's rotational rate, the satellite is motionless as viewed from the Earth. Three such satellites in synchronous orbit can cover over 90% of all inhabited regions of the Earth. Signals sent up from the ground and transmitted down from the satellite travel 19,300 nmi and suffer large propagation loss. A high-gain antenna is required to compensate for the path loss. The antenna can be a simple one for generating a circular pencil **beam** for Earth coverage or a much more complicated antenna generating multiple shaped beams and/or reconfigurable beams for **frequency reuse** with spatial and polarization diversities. As mobile communications services evolve from a regional to a global scale, the satellites play an important role in providing the low-power transmitters needed to broadcast from relatively unobtrusive base stations distributed throughout space.

In order to provide low-power, personal hand-held telephone services, the coverage area would normally be divided into cells, each cell covered by a separate antenna beam. For a GEO satellite this beam would require a very large antenna which could not be implemented even with deployable antenna technology. In addition, the time delay of the half-hop propagation for a GEO satellite varies between 119 and 139 μs, becoming a problem in voice communication. This time delay can be reduced by the use of low Earth orbit (LEO) or medium Earth orbit (MEO) satellites. LEO and MEO orbits are particularly suitable for cellular mobile communication systems since the cell size remains constant throughout the orbit. Several LEO and MEO systems have been proposed,

Satellite Antennas

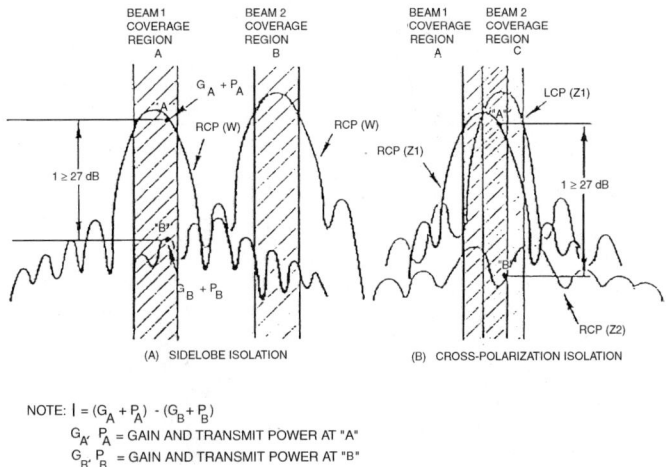

FIGURE 75.1 Transmit isolation requirements.

among them Iridium, Globalstar, and Odyssey. The range of possible communication distance, propagation loss, propagation delay, and satellite view time vs orbital attitude are key parameters in the system tradeoff and design.

To meet these requirements, the antenna radiation pattern varies from omnidirectional to highly directional. It can be fixed or reconfigured to meet the needs for each system configuration. A shaped beam is required in order to constrain the energy to conform to the coverage area to improve equivalent isotropically radiated power (EIRP) and reduce interference between beams. It can be achieved either by a multiple beam antenna system or a shaped reflector system. Beam reconfiguration is required to accommodate the change of radiation pattern for a time-domain multiple access (TDMA) system.

Conventional **antenna parameters** such as half-power beamwidth (HPBW), antenna radiation efficiency, and aperture efficiency are insufficient to describe the performance of the system. In addition to the gain ripple and gain slope over the coverage and frequency band, isolation or interference among these beams is important in the system. They are assessed in terms of spatial and polarization isolation, and both must be considered. Figure 75.1 depicts a 27-dB isolation between two transmit beams. Similarly, the receive isolation is defined as the isolation of two received signals of equal incident flux density. Beam shaping efficiency is used to evaluate the gain over a highly shaped coverage. An ideally shaped beam is one that has maximum antenna gain without sidelobes and achieves constant antenna radiation intensity over the area. Therefore, the theoretically achieved maximum gain is

$$G = 41253/A_{\text{cov}}$$

where A_{cov} is the coverage area in degrees square. The performance is evaluated in terms of beam shaping efficiency η defined as

$$\eta = G_{\min} A_{\text{cov}}/41253$$

where $G_{\min}$ is the minimum gain over A_{cov}. Practically, the gain-area product is between 10,000 and 18,000, depending on the size of antenna aperture and the shape of coverage.

The tracking, telemetry, and command (TT&C) antenna provides ranging, telemetry, and command operation throughout all mission phases after launch vehicle separation. An autotracking antenna is required for two reasons: to improve the pointing accuracy of the highly shaped beam

TABLE 75.1 Satellite Antennas

Area Coverage	Function	Antenna Types	Antenna Gain	Design Challenge	Name of Satellites
Earth coverage	±9° FOV	• Corrugated horn • Multiflare horn	Medium	Single horn with diplexer to cover both up- and down-links with good axial ratio	Most of all communication satellites
Omni coverage Spot coverage	TT&C with 2 operation modes: 1. transfer orbit 2. on-orbit	• Conspiral antenna • Biconical antenna • Horn antenna	Low Low Medium/low	• Nearly omni-directional • Spacecraft body blockage • Reliability	Most of all communication satellites
Pencil beam coverage	Limited portion of FOV	Reflector antenna • Single feed • Multiple feeds	High	• Gimbal Steering • Beam forming network (BFN)	INTELSAT-VIIs NASA ACTS MILSTAR
Shaped beam coverage	Multiple beams for frequency reuse	• Multibeam antennas • Shaped reflector antennas	High	• Isolation between beams • Switching BFN	INTELSAT-V INTELSAT-VI INTELSAT-VII
Nulling coverage	Maximize signal-to-interference power ratio	Reflector antennas • Multiple feeds • Adaptive BFN Phased array antennas • Adaptive BFN	High	• Adaptive nulling algorithm • Adaptive BFN • Calibration and characterization of BFN	MILSTAR
Finite coverage	Passive microwave radiometer	Reflector antennas	High	• Size of reflector • Surface tolerence	

where the performance degradation is not accepted in the system due to satellite pointing error, and to communicate with a moving target or user.

Successful observation of weather and climate from satellites has led to other new applications such as the measurement of atmospheric parameters by seeing through cloud canopies. A microwave radiometer is a good candidate for this application. However, at least a 4.4-m antenna would be required to provide 30-km nadir spatial resolution. This type of antenna is still under development. The most common types of antennas being used to meet these requirements are summarized in Table 75.1.

Satellite antennas differ from other antennas in several respects. A satellite must be designed to withstand the dynamic mechanical and thermal stress during the launch and in the space environment. The materials used become a key factor. Other design constraints are size and weight. The antenna has to fit the shroud of the launching vehicle. As a result, the reflector used for the high-gain antenna must be stowed using either a solid reflector or a furlarable meshed reflector. Weight cannot exceed that weight budget allocated in the system. The challenge of satellite antenna design for the future will be to make them lighter, smaller, and cheaper.

75.2 Horn Antennas and Reflector Feeds

Currently the reflector antenna is the most widely used satellite antenna, but future satellites may feature phased array antennas capable of rapid and low-loss beam scan, achievable with the design maturity of solid-state microwave amplifiers and receivers. The reflector antenna may have either a single feed or multiple feeds, depending on the coverage requirement; the phased array antenna in contrast has many radiating elements. The best candidates for feeds and radiating elements are horns and cup dipoles. Above 6 GHz horns are usually used since the waveguide is low loss and

Satellite Antennas

TABLE 75.2 Comparison Among Various Horns

	Open-End Waveguide	Conical Horn	Dual-Mode Horn	Corrugated Horn	Multiflare Horn
Bandwidth:	Wide	Wide	Narrow	Wide	Moderate
Pattern symmetry:	No	No	Yes	Yes	Yes
Phase center location:	On the aperture	Inside aperture	Inside aperture	Inside aperture	On/Inside the aperture
VSWR for large aperture:	Good	Excellent	Frequency sensitive	Poor without matching section	Excellent
Fabrication:	Easy	Easy	Easy	Difficult	Easy
Applications:	Array element	Array element	High-efficiency feed for reflector	High-efficiency feed for reflector	High-efficiency feed for reflector

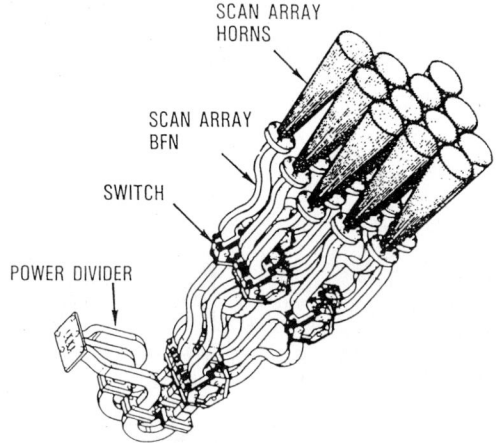

FIGURE 75.2 Feed array for NASA ACTS MBA.

small in size. Cup dipoles are typically used below 4 GHz since coaxial cables are smaller and lighter than waveguides.

Horn Antennas

There are five types of horn antennas: open-ended waveguide, conical horn, dual-mode horn, corrugated horn, and multiflare horn. Table 75.2 compares pros and cons among these five horns. The open-ended waveguide or the conical horn are most frequently used for multiple beam antennas (MBA) with phased array elements since their apertures can be small. Figure 75.2 shows a 30-GHz feed array for the NASA Advanced Communication Technology Satellite (ACTS). The advantage of this type of horn is its simplicity and ease of fabrication. The disadvantage is that the radiation pattern is not circularly symmetrical, and so antenna efficiency is low and a high axial ratio is obtained for the circular polarization.

The dual-mode horn antenna, also called the Potter horn, generates a circularly symmetric pattern by adding transverse electric (TE_{11}) and transverse magnetic (TM_{11}) modes in proper amplitude and phase [Potter, 1963]. The TM_{11} mode is excited by a simple waveguide step. However, due to the phase velocity difference, proper phasing of the two modes occurs only in a limited-frequency band with less than 5% bandwidth. Therefore the dual-mode horn is only used in narrow bandwidth applications.

The corrugated horn carries a hybrid mixture of TE_{11} and TM_{11} modes through the grooved horn walls; this fundamental hybrid mode is called the hybrid electric (HE_{11}) mode. The TE and TM

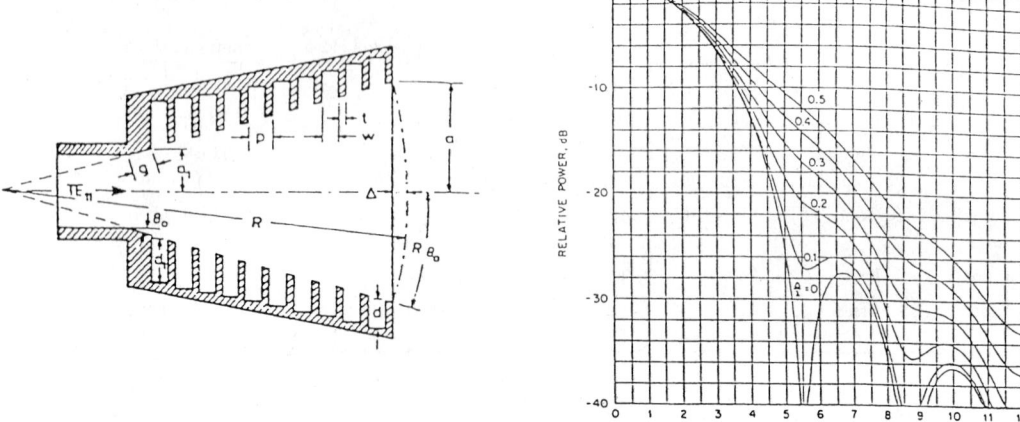

FIGURE 75.3 Narrow-flare corrugated horn.

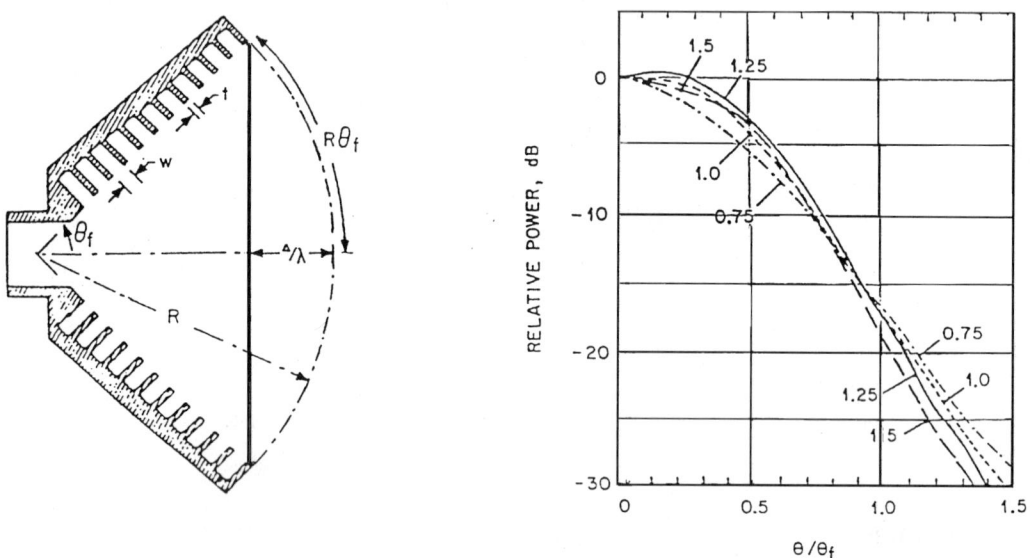

FIGURE 75.4 Wide-flare corrugated horn.

components in the HE_{11} mode have the same cutoff and phase velocity; therefore, the beamwidth limitation of the dual-mode (or multiflare) horn is removed. A detailed theoretical analysis has been given by Clarricoats and Saha [1971]. One design problem of the corrugated horn is that without a matching section at the horn throat region it has poor voltage standing wave ratio (VSWR), especially over the lower portion of the operating frequency band. Two kinds of matching sections can be used to improve VSWR: varying groove depths and ring-loaded grooves [James, 1982]. The corrugated horn can easily be designed up to 60% bandwidth using a varying grooves matching section, and using a ring-loaded matching section, can even be designed to a 2 : 1 bandwidth. One drawback of the corrugated horn is that the fabrication of the grooved walls is difficult, resulting in high manufacturing cost. Yet the corrugated horn is the most used reflector feed and is the best candidate for the Earth coverage antenna. A corrugated horn is classified either as a narrow- or as a wide-flare horn. Figure 75.3 shows the antenna geometry and the normalized radiation patterns for the narrow-flare corrugated horn. Figure 75.4 shows those for the wide-flare corrugated horn.

Satellite Antennas

A multiflare horn is formed of several flared waveguide sections. This mechanically simple structure leads to very low fabrication cost. The smooth wall, as well as the flare structure, provides good RF match to the waveguide. A low VSWR is always achieved. A conical multiflare horn generates higher order TE_{1n} and TM_{1n} modes at the flare change. The circularly symmetric pattern is obtained by adding these higher modes (mainly TM_{11}) to the fundamental TE_{11} in proper ratio. The analysis is based on mode matching at the flare change to compensate for phase front curvature change. The differential phase shift among modes in a multiflare horn is less than that in a dual-mode horn. Thus, the multiflare horn is capable of achieving about 15% bandwidth. A three-section multiflare horn is comparable (or even smaller) in size to a corrugated horn with the same beamwidth [Chen, Choung, and Palma, 1986]. The multiflare horn has advantages over the dual-mode horn in terms of bandwidth and over the corrugated horn in terms of manufacturing simplicity. Multiflare horns are currently utilized as spot beam feeds for NASA ACTS.

Cup-Dipole Antennas

A cup-dipole antenna has crossed dipoles in a cylinder and has good axial ratio for circular polarization. The antenna is compact and easy to manufacture. There are two types, as shown in Fig. 75.5: unequal arms with single coaxial cable and equal arms using two coaxial cables and a 90° hybrid. The cup dipole with unequal arms is mechanically simple but can only generate single circular polarization with less than 5% frequency bandwidth. The cup dipole with equal arms has dual circular polarization with about 20% bandwidth, but it is mechanically complex. Several design variations of the cup dipole antenna such as the short backfire and cavity-back helix are described in Chap. 4 of Kumar and Hristov [1989]. The dipole is about 0.45λ and the cup diameter is from 0.7λ to 3λ depending on the beamwidth and/or the array spacing.

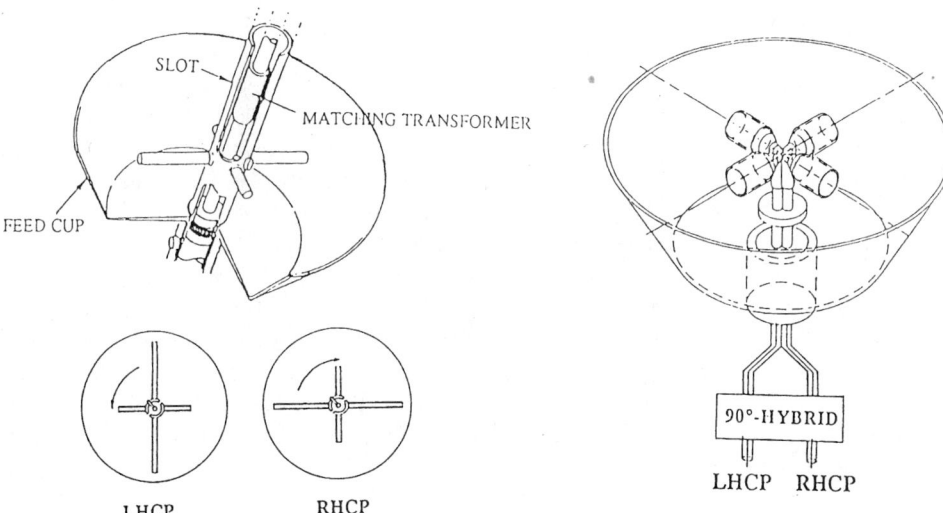

FIGURE 75.5 Two types of CP cup dipole.

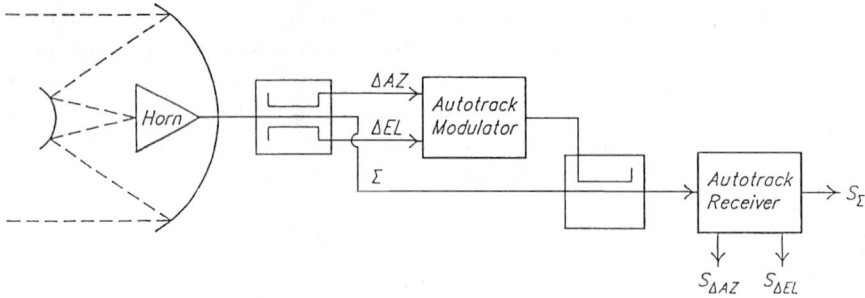

FIGURE 75.6 Block diagram of monopulse tracking.

Frequency Reuse Feed and Polarization Diversity

A given frequency band can be used more than once at the satellite to increase the communications capability of a given link. Since different communications channels will exist at the same carrier frequencies, each reuse channel must be isolated.

There are two techniques to achieve this required isolation. One way is to use two orthogonal polarizations. One channel carries a vertical polarization [or right-hand circular polarization (RHCP)], and the other channel carries a horizontal polarization [or left-hand circular polarization (LHCP)] with more than 30-dB isolation. This can be done using a polarizer (for the CP case) and an orthomode transducer placed right after the feed horn. Another way is to use two spot beams pointing at different locations with high (about 30 dB) spatial isolation. This can be achieved using two feeds (or feed clusters) of the reflector to generate low sidelobes. Low sidelobes can be achieved with minimum spillover energy by the use of a high taper feed on the reflector.

Monopulse Feed

Point-to-point satellite communications require a narrow beamwidth to obtain high antenna gain. But the satellite jitters, causing a pointing error. In order to maintain reliable communications, a satellite antenna must be accurately pointed toward a signal source. To achieve accurate pointing, the satellite employs an autotracking system, providing tracking signals related to pointing errors in elevation and azimuth. The tracking signals control a feedback servoloop of the satellite to orient the satellite as required, positioning the antenna accurately with respect to the signal source. A block diagram of the monopulse tracking is shown in Fig. 75.6. It is achieved using a single corrugated horn and a tracking coupler. The coupler uses a longitudinal traveling wave mechanism and utilizes the circular TE_{11} and TE_{21} modes that provide the sum and the difference channels, respectively, to achieve wideband monopulse tracking. The main beam is obtained on the boresight axis of the antenna when only the dominant mode (TE_{11}) is excited at the aperture of the feed. When the beam is off the boresight axis, higher order modes (TE_{21}) are excited in the feed. The amplitude of the TE_{21} mode is proportional to the angle off boresight, and a null in the higher order mode is obtained at the boresight.

75.3 Reflector Antennas

The reflector antenna, a high-gain antenna used to support high data rates with low transmitter power, is the most desirable type for communication antennas because of its light weight, structural simplicity, and design maturity. Its disadvantage is that the reflector has to be offset to avoid the

Satellite Antennas

TABLE 75.3 Reflector Antenna Configurations Used in Communication Antenna Systems

Type	Features and Application	Shape
1. Center-fed parabolic reflector	• Simple, symmetrical light weight structure • Aperture blockage causing sidelobe level and reducing peak gain • High feed line loss • Pencil beam	
2. Cassegrain reflector system	• Short, low feed line loss • Shaped reflector may be used to increase efficiency • Aperture blockage causing sidelobe level and reducing efficiency • Pencil beam	
3. Offset parabolic reflector	• No aperture blockage • Convenience for spacecraft mounting with feed embedded below the platform • Shaped beam by using multi-horns or shaping reflector surface	
4. Offset Cassegrain (or Gregorian) reflector	• No or minimum aperture blockage • Worse scan loss • Fixed or scanning beam	
5. Overlapping dual-gridded parabolic reflector	• Offset configuration • Polarization reuse with 33-dB isolation • Shaped beam • Thermal distortion of the gridded surface	
6. Multiband dual FSS reflector	• FSS subreflector or overlapping subreflectors in which one is an FSS • Same main reflector aperture for multifrequency band operation • Compact • Thermal distortion of FSS	
7. Deployable reflector	• Meshed reflector that can be folded • Allow to use larger reflector aperture size • Tradeoff betwen weight and surface accuracy • Passive intermodulation may be generated • Use mostly for lower frequency band system	

aperture blockage. The offset destroys the rational symmetry of the optical aperture and limits the achievable performance as compared to that of a lens antenna.

A reflector antenna system consists of one or more **reflector surfaces**. The surface can be solid, gridded, printed substrate, or meshed. It may be paraboloid, hyperboloid, spheroid, ellipsoid, or general in shape. Seven antenna types that have been most commomly used for satellite communication systems are summarized in Table 75.3. They include the parabolic reflector and the

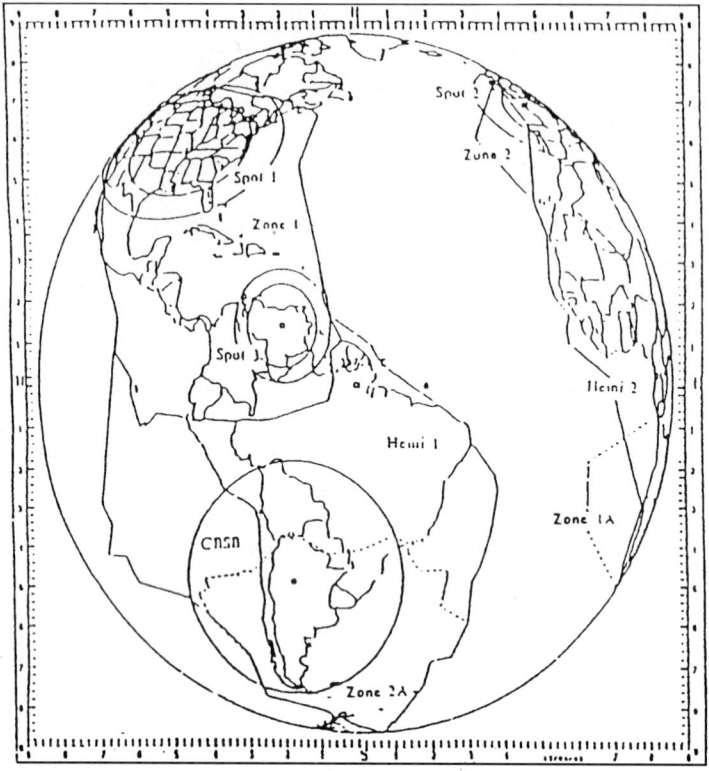

FIGURE 75.7 INTELSAT VII hemi/zone coverage.

Cassegrain reflector, both either center fed or offset fed; the overlapped dual-gridded reflector; a multireflector system with a frequency selective surface (FSS) subreflector, and the deployable reflector antenna.

The center-fed parabolic and Cassegrain reflectors are used for a high-gain antenna where the radiation pattern is of circular shape. When the radiation pattern must be shaped to form a contour beam, a multiple beam reflector antenna system or shaped reflector antenna system can be utilized. One of the most complicated multiple beam antenna systems in use was designed for the INTELSAT V–VII series. The INTELSAT VII satellite used a 2.44-m **graphite fiber reinforced plastic (GFRP)** reflector for the 4-GHz frequency band over the Atlantic, Pacific, and Indian Oceans. The antenna provides shaped hemispheric and zone coverage as shown in Fig. 75.7. The coverage was determined by the traffic demand in each region. The isolation between zone and hemispheric beams is accomplished by polarization diversification, whereas the isolation between the east and west hemispheric beams is accomplished by spatial diversification. There are 110 feeds used in the feed system. Each feed is a 1.13 wavelength GFRP circular horn element arranged in triangular grid, as depicted in Fig. 75.8. The beam forming network consists of air supported striplines and power dividers sandwiched between ground planes. The gain of the hemispheric beam is about 15 dBi, and the gain of the zone beam is 22 dBi. (Decibels over an isotropic radiator.) The isolation achieved for both polarization and spatial diversification is about 27 dB.

Another complex reflector antenna in flight is NASA ACTS [Choung and Wong, 1968] operating at Ka-band (30 GHz for receive and 20 GHz for transmit) (See Fig. 75.9). The MBA is designed to cover the continental U.S. (CONUS). The coverage and polarization plan are shown in Fig. 75.10. The MBA provides three fixed spot beams (Cleveland, Atlanta, and Tampa) for high-burst-rate (HBR) operations and two scanning beams for low-burst-rate (LBR) operations. The scanning beams

Satellite Antennas

FIGURE 75.8 INTELSAT VII transmit feed array.

provide coverage for two scan sectors (east and west) and 13 scan spots (Seattle, San Francisco, Los Angeles, Phoenix, Denver, White Sands, Dallas, Kansas City, New Orleans, Memphis, Huntsville, Houston, and Miami). ACTS has two electrically identical MBAs (one for transmit and one for receive) with half-power beamwidth (HPBW) on the order of $0.35°$. Each MBA shown in Fig. 75.10 has one main reflector, a dual polarized subreflector, and two orthogonal feed assemblies. Both the low sidelobe fixed beams and the provision for overlapping fixed and scanning beams can be achieved by using an offset Cassegrain antenna with a dual-gridded subreflector consisting of two offset hyperbolas arranged in a piggyback configuration. The front subreflector is gridded to pass one sense of polarization and reflect the orthogonal polarization. The back subreflector is solid and reflects the polarization transmitted by the front subreflector. The focal axes of the two subreflectors are symmetrically displaced from the symmetry plane of the main reflector by an angle of $10°$ so that two orthogonally polarized feed assemblies can be placed in their respective focal regions without mechanical interference.

A multireflector antenna system with a FSS subreflector is used for multiband operation. The FSS is designed to be transparent for one frequency band and to be reflective for another band. The feeds are placed at two foci of the reflector system. The typical FSS antenna has a main reflector, an FSS subreflector using etched Jerusalem cross elements, a cluster of receive feeds, and a transmit feed, as shown in Fig. 75.11. The receive feed cluster may allow adaptive nulling and the main reflector may be gimballed to allow beam scan within the Earth coverage. This design concept is applicable to a MILSTAR MBA.

Successful deployment and application of the 30-ft-diam reflector on the ATS-6 satellite demonstrated the feasibility of large deployable reflectors. The deployable reflector antenna became a space proven technology that has been used for satellite applications. The use of a deployable reflector antenna provides an allowable larger aperture that can be stowed for launch, then deployed on

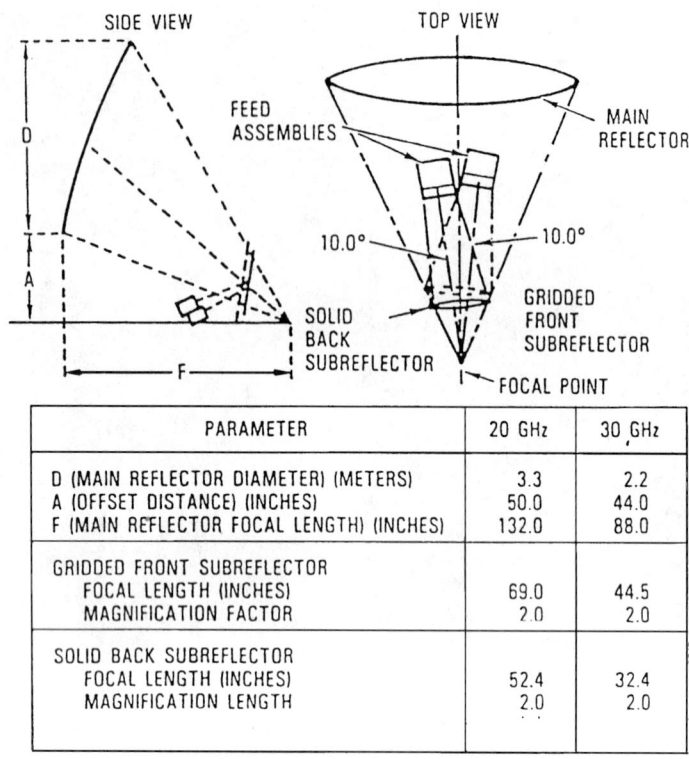

FIGURE 75.9 NASA ACTS MBA geometry.

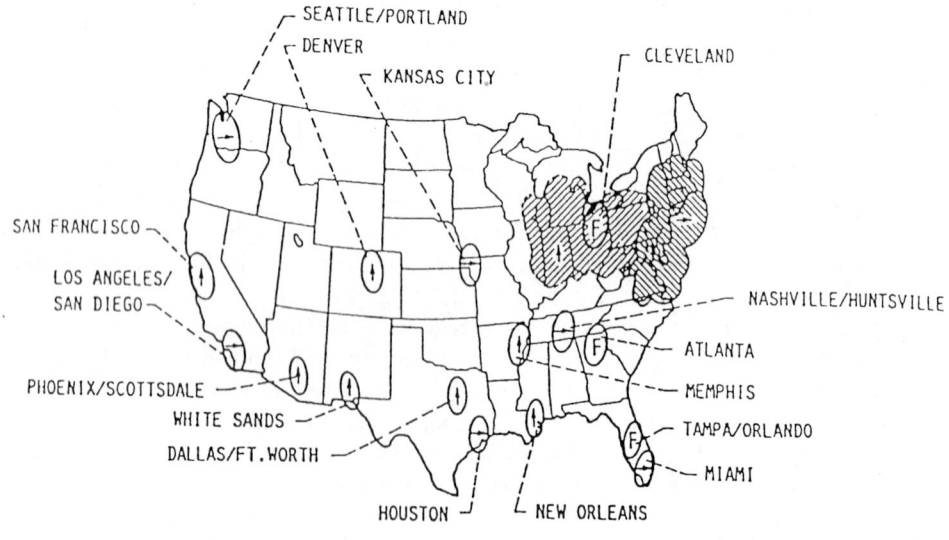

FIGURE 75.10 NASA ACTS coverage and polarization plan.

Satellite Antennas

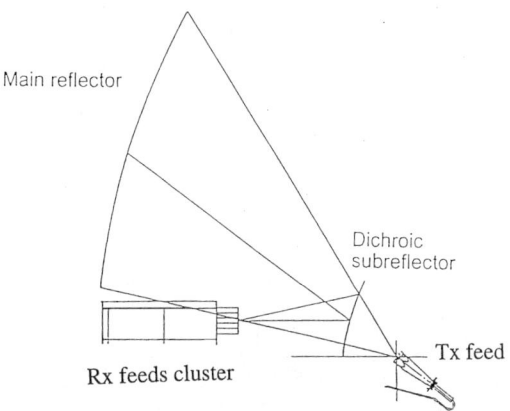

FIGURE 75.11 A typical FSS reflector antenna.

FIGURE 75.12 TDRSS mesh antenna.

orbit. The reflector utilizes GFRP ribs to shape and support the reflective mesh surface, as shown in Fig. 75.12. The number of ribs is based on a tradeoff considering surface tolerance and weight. As the number of ribs increases, the surface error decreases, while weight increases. This type of antenna has also been used for NASA's Tracking and Data Relay Satellite system (TDRSS) single access (SA) antenna as shown in Fig. 75.13. Each TDRSS spacecraft has two SA antennas that are used for communication with user satellites in low Earth orbit.

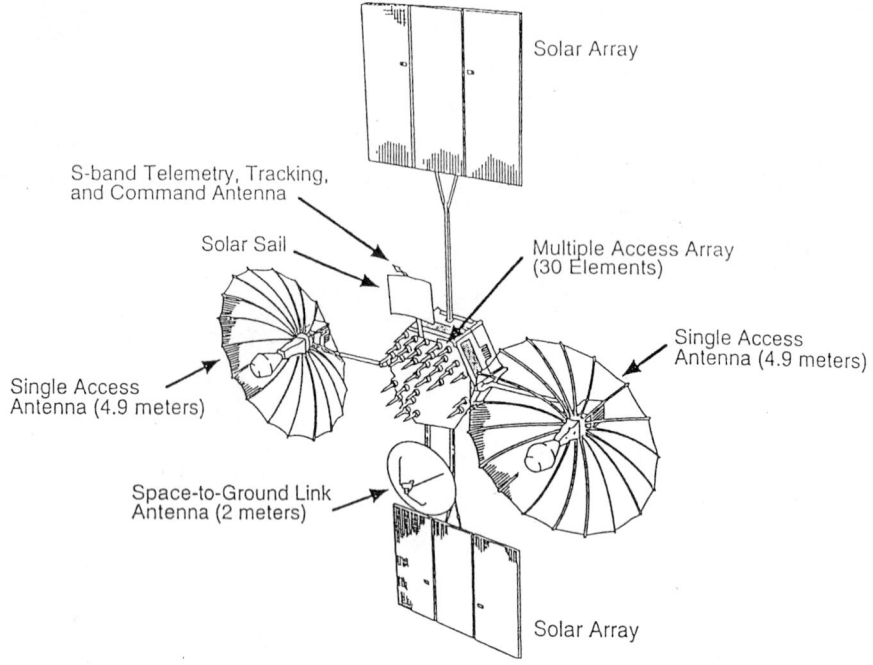

FIGURE 75.13   Tracking and data relay satellite.

75.4 Phased Array Antennas

The complexity of a phased array is highly dependent on the grating lobe requirements. A phased array antenna for a communication satellite, especially for geosynchronous orbit, has a limited field of view (FOV) and thus does not require grating lobe free antenna design. It only requires that the grating lobe be out of the FOV. Thus, the antenna design can be simpler, using high-gain elements with larger spacing.

The array gain is approximately expressed as

$$G = 10 \, \text{Log}_{10} N + G_e$$

where N is the number of elements and G_e is the element gain in decibels. A larger element has higher gain, and thus the number of elements is minimized for the given gain. Also mutual coupling between elements is insignificant if the element spacing is larger than two wavelengths. If the gain requirement is stringent, a large number of elements is required and the mechanical complexity increases. In this case, the trade should be performed with comparison to a multibeam antenna.

The most widely used radiating elements in a satellite phased array antenna system are the horn, the dipole, the helix, and the microstrip patch, as summarized in Table 75.4. The horn is most frequently used. The helix is used for a high-gain antenna system used in mobile communication or in a multiple access (MA) system like TDRSS. The waveguide slot array antenna is also used for a high-gain and high-power communications system. Most of waveguide slot array antennas are linearly polarized. If circular polarization is required, it can be obtained by putting a layer of meanderline polarizer in front of the array to transform a linearly polarized wave to a circularly polarized one. The microstrip patch is used where a low profile and compact size are required.

The TDRSS MA antenna [Imbriale and Wong, 1977] is an S-band, 30-element phased array antenna for both transmit and receive. Between transmit and receive 12 elements are diplexed, and the remaining 18 elements are receive only. The MA antenna is spacecraft body mounted with a

Satellite Antennas

TABLE 75.4 Types of Satellite Phased Array Antenna System

Type	Features and Applications
Horn array antenna	• Good electrical performance • Used as a feed array in a multibeam antenna system
Dipole array antenna	• Low profile and simple
Helix array antenna	• High gain is required • Mobile antenna system
Waveguide slot array antenna	• Low profile • High power • Meanderline polarizer is required for circular polarization
Active MMIC phased array	• High gain as the BFN loss is compensated for by active device

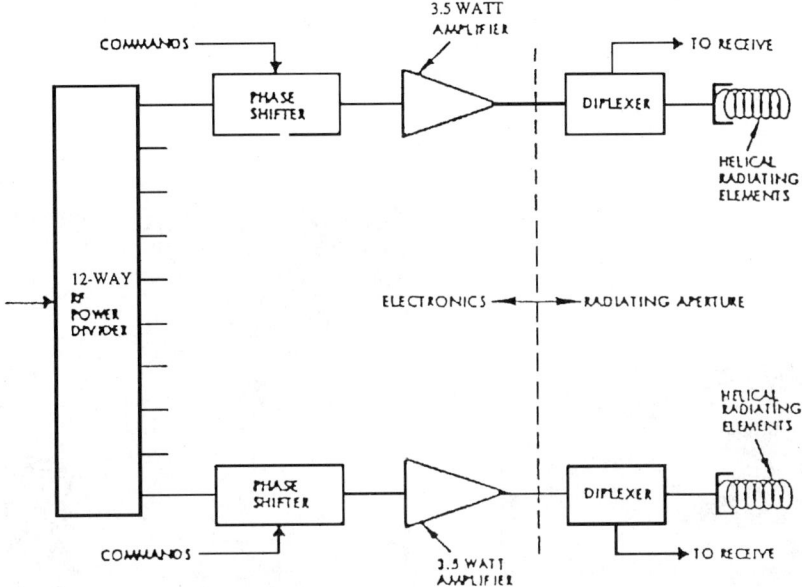

FIGURE 75.14 TDRSS MA transmit phased array antenna.

fixed ground plane and points to the spacecraft nadir as shown in Fig. 75.13. The radiating elements are helices arranged in a triangular grid. The circularly polarized helical element is lightweight and provides 13 dB of gain over ±13.5° FOV. Figure 75.14 shows a block diagram of the transmit array. The transmitter splits the input signal (2.1064 GHz) with a 12-way power divider. Each element is individually phase shifted and amplified to form a transmit beam. Figure 75.15 shows a block diagram of the receive array. The receive signals (2.2875 GHz) from each of the MA users are received and preamplified at each of the 30 receive elements. The preamplified signals are then multiplexed. The beam forming is performed at IF level. Recent progress made in the areas of GaAs monolithic microwave integrated circuit (MMIC) design and processing in conjunction with advanced microwave packaging techniques has made MMIC phased array antennas attractive for the satellite that provides one or more pencil beams capable of being steered at electronic speed. The beam forming network's high loss is compensated for by the active devices. COMSAT Laboratories has built a 64-element prototype Ku-band active phased array [Zaghloul et al, 1990] for application

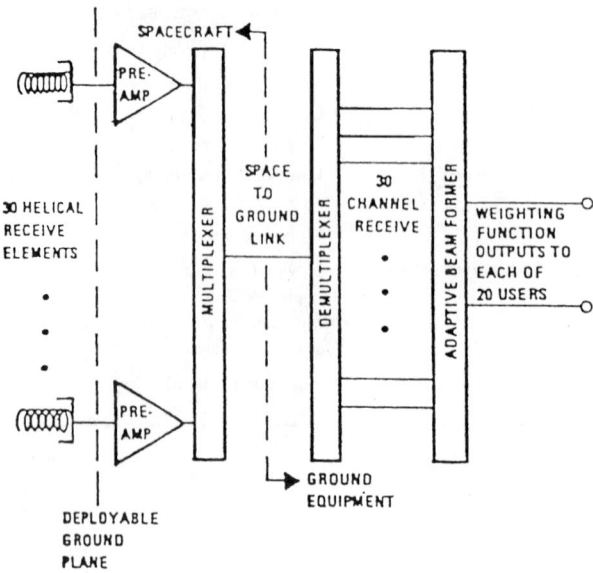

FIGURE 75.15 TDRSS MA receive phased array antenna.

(a)

64-Element Ku-band active phased array system.
(b)

FIGURE 75.16 COMSAT Ku-band MMIC phased array antenna: (a) element and (b) array.

to INTELSAT coverage of the Atlantic Ocean region with four dual-polarized reconfigurable shaped beams. A 3.4λ square horn arranged in a triangular lattice is used. Figure 75.16 shows a single active element and the 64-element active phased array system.

75.5 Tracking, Telemetry, and Command Antennas

The TT&C antenna must provide broad coverage throughout the spacecraft mission, including initial acquisition, transfer orbit, and normal on-orbit operations. It must also provide nearly omnidirectional coverage for the unlikely event of altitude failure. If the satellite starts to tumble, satellite control commands from the ground must be received for any arbitrary orientation of the

Satellite Antennas

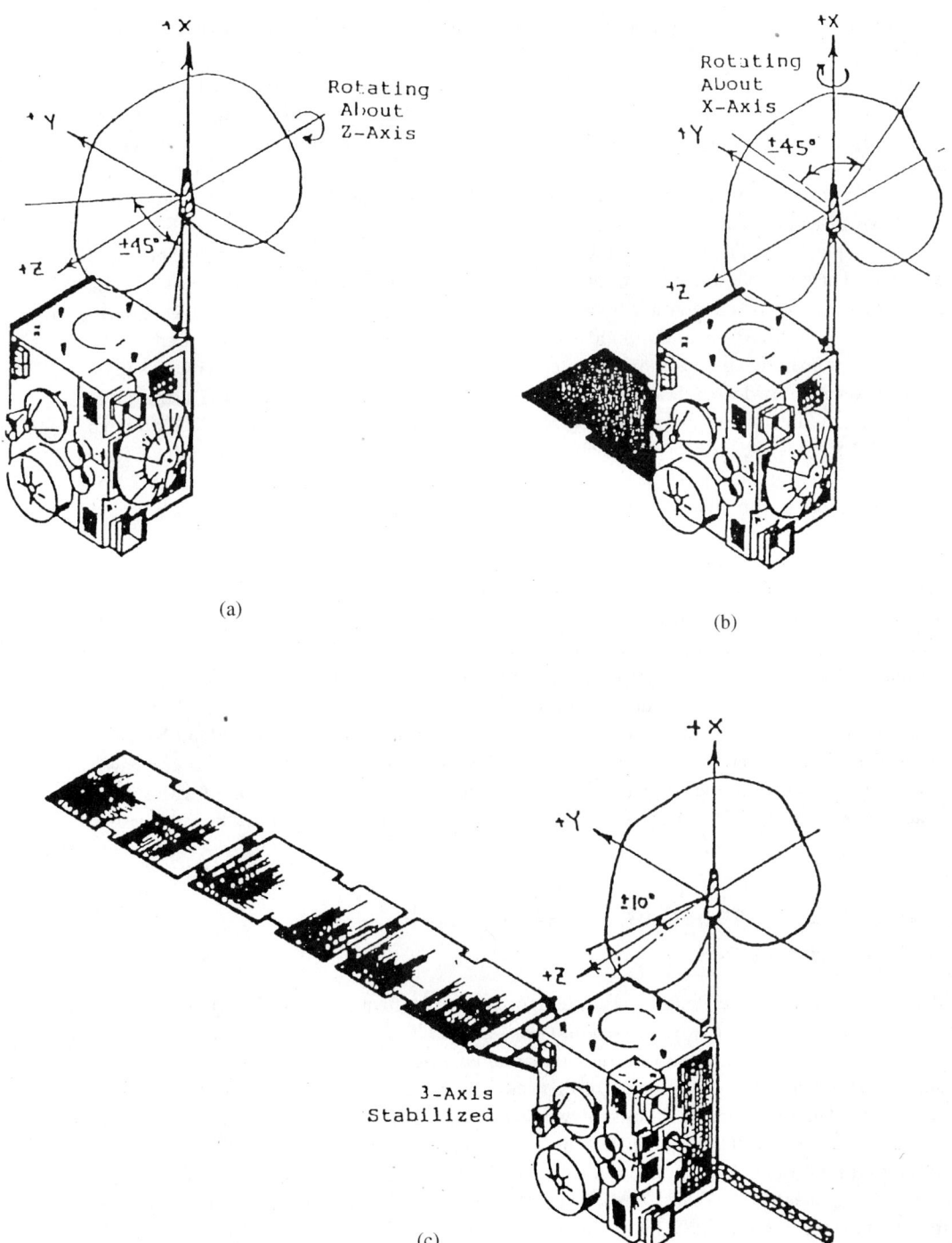

FIGURE 75.17 TT&C antenna and its operation modes: (a) initial acquisition, (b) transfer orbit, and (c) on-orbit operation.

satellite. During on-orbit operation, the antenna provides Earth coverage for the lifetime of the satellite.

To achieve omnidirectional coverage, two hemispheric antennas can be mounted on opposite sides of the satellite and the appropriate antenna selected. For on-orbit Earth coverage, either one of the hemispheric antennas or the separate horn antenna can be used. In this case, high system margin is achieved, but switching complexity between antennas remains an issue. The design challenge for the TT&C antenna is to use only a single antenna having a broad beam cardioid radiation pattern with lower system margin.

Most frequently used TT&C antennas are conical spiral antennas [Yeh and Mei, 1967] or biconical antennas [Barrow, Chu, and Jansen, 1939]. A loosely wound four-arm conical spiral antenna mounted at the end of a deployable boom is ideal for the single antenna approach. Figure 75.17 illustrates various TT&C operation modes using a single antenna. This approach is used for the geostationary operational environmental satellite (GOES) weather satellite. This antenna is simple and lightweight, provides nearly identical patterns, and has good polarization purity. During initial acquisition, the antenna is deployed, but no solar panel deployment has taken place. The coverage region of interest is a 45° conical region about the z-axis. In transfer orbit operation, a single solar panel is deployed, and the coverage region of interest is a 45° conical region about the x-axis. For on-orbit operation with full deployment of the solar panels and the solar sail, the coverage region is only a 9° cone about the z-axis.

The conical spiral antenna can provide satisfactory coverage under free-space conditions, but the interference and shadowing caused by the spacecraft could cause serious pattern degradations that result in inadequate coverage in some regions of space. To calculate the antenna pattern in the presence of the spacecraft, the geometrical theory of diffraction (GTD) [Lee, Schenuum, and Hwang, 1991] is used as the best method of analysis. The GTD is a high-frequency solution. Using this approach, a ray optics technique can be applied to determine the fields incident on the various scatterers such as edges, corners, and curved surfaces. The fields diffracted are found by using the GTD solutions in terms of rays summed with the geometrical optics components and edge-diffracted components in the far field. The basic approach is to model the spacecraft with simple structural forms such as flat plates and cylinders; this has been used very successfully in the past.

75.6 Current and Planned LEO/MEO Mobile Satellite Antennas

Most of the LEO/MEO satellite antennas are designed to provide frequency reuse for cellular systems. As a result, the coverage areas are divided into cells. Each cell is covered by an antenna beam. The LEO/MEO satellites orbit in the Van Allen radiation belts, which lie between the magnetosphere and the ionosphere. The environment of these two orbital satellites is somewhat different from that of GEO. The Van Allen belts consist of two regions containing energetic protons and electrons traveling around the Earth's magnetic field lines. LEO satellites are orbiting under the most intense and damaging portions of the lower Van Allen belts. The energetic particles can damage the exposed antenna components. The eclipse portions and the eclipse durations of the LEO/MEO satellites are also different from those of GEO. They may spend 35% or more of their time in the Earth's shadow.

The logical choice for the LEO/MEO satellite antennas is the multiple beam antenna system. The types of antennas selected can be phased arrays or reflector multiple beam antenna systems. Variable beamwidth, sidelobe control, and/or power level in each beam can also be incorporated in the system. The phased array can be an active MMIC phased array with distributed output power amplifiers.

Even though many mobile communication satellite systems have been proposed, at this writing communication antennas are still in the design and development phase. The antennas can range

Satellite Antennas

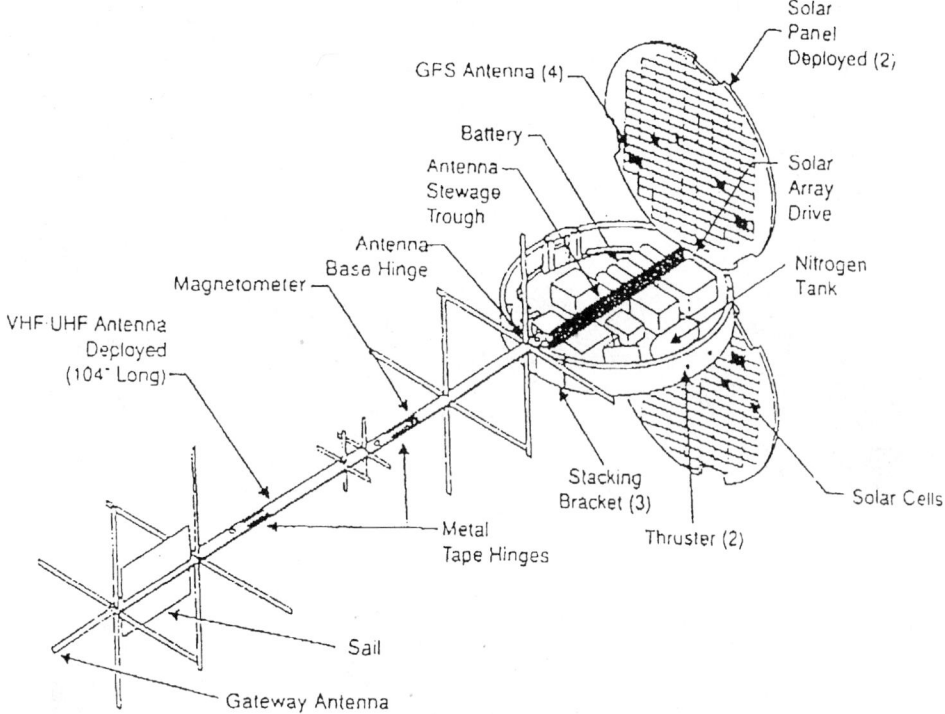

FIGURE 75.18 Orbcomm's satellite configuration.

from the simplest system, such as the VHF/UHF cross-dipole array antenna in Orbcomm's Microstar satellites, to the very complicated Ka-band MBA system in Teledesic satellites. Not all of the proposed antenna systems are described in the open literature. Figure 75.18 depicts the Orbcomm's spacecraft [Schoen and Locke, 1995] in which the antenna is designed for uplink frequencies from 148 to 150.05 MHz and downlink frequencies from 137 to 138 Mhz.

The currently proposed antennas in the Globalstar mobile satellite system [Schindall, 1995] are L- and S-band active MMIC phased array antennas. The antenna design may be changed in the final development stage. Each antenna will generate 16 beams that cover the 5760-km-diam circle on Earth. The antenna beam is shaped to yield higher gain at the edge of coverage to compensate for the higher range losses, thus providing nearly constant power throghout the operating region. Figure 75.19 depicts the 91-element transmit array. The receive antenna is separated from the transmit antenna and is almost identical to the transmit antenna except that high-power amplifiers are replaced by low-noise amplifiers. The element consists of microstrip patch, filters, interfacing substrates, and hybrids used for generating circular polarization. Heat sink is used for heat dissipation. The beam forming network is implemented in stripline and comprises 32 layer Teflon dielectric substrates with required printed power dividers or combiners. The transmit antenna operates at 2.4835–2.5 GHz with minimum 50-dB gain, whereas the receive antenna operates at 1.61–1.6265 GHz with minimum 40-dB gain.

The counterpart in the Odyssey communication satellite system [Speltz, 1996] is a phased array antenna system. The antenna design may be changed in the final development stage. Two direct radiating arrays have been developed: an L-band downlink and an S-band uplink. Each antenna generates 61 spot beams. Hybrid networks driven by amplifiers provide real-time power reconfiguration to meet varying demands in each region, and the satellite antennas are designed to provide coverage to a portion of the total area visible to the satellite.

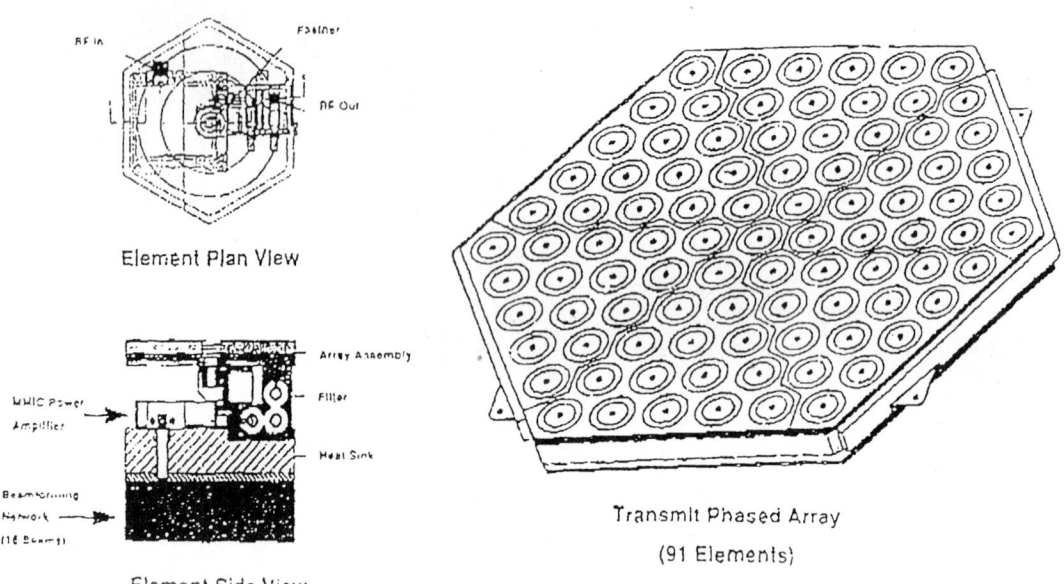

FIGURE 75.19 Globalstar active MMIC phased array antenna.

75.7 Radiometer Antennas

A satellite passive radiometer has unique properties that are not found in other measurement systems: it provides consistent viewing angles to a given location. The capability to observe the same location for a long time prolongs the sensing dwell time to achieve the required signal-to-noise ratio of the measurement within the spectral resolution limits of the instrument. Such an ability to provide a temperature profile at frequent intervals has proven significantly beneficial to meteorological analysis and forecasting based on modeling simulation.

The radiometer is a multichannel microwave (90–230 GHz) system. Temperature profiles are made near the center and on the wings of the 118-GHz oxygen absorption region. Moisture profiles can be made using the 183-GHz water vapor absorption band. The determination of precipitation coverage and rate can be done at several frequencies at the atmospheric windows at 92, 104, 150, and 230 GHz. All channels can operate simultaneously from the same resolution area using coincidental beams. However, beamwidths of different channels need not be the same.

Key design parameters are beam efficiency, antenna losses, and antenna physical temperature, in addition to conventional requirements such as directional gain, low sidelobes, mismatch, and polarization purity. The beam efficiency is the fraction of total power contained within the main beam of the correct polarization. It should be greater than 85%.

An antenna of at least 4.4-m aperture is required to provide 30-km nadir spatial resolution. The 4.4 m is also the largest antenna that fits the Space Shuttle cargo bay without folding. However, it is difficult to build a light-weight antenna with dimensional stability and surface accuracy of 2 mils in a space thermal environment. The scanning of a 4.4-m antenna generates angular momentum, which must be compensated for by the spacecraft attitude control subsystem.

75.8 Space Qualification

Qualification tests of all flight hardware are required to show suitability of the design for the space environment. Although the space or operating environment cannot be exactly duplicated, the tests are conducted in a controlled manner so that the hardware that passes the tests will function successfully.

The tests can be grouped into two main categories: external environment during and after launch and operating environment. The external environment tests include zero gravity, high vacuum, solar radiation, γ-radiation, neutron radiation, electromagnetic pulse, and mechanical launch environment.

Four main components of the launch environment are steady-state acceleration, low-frequency vibration (5–100 Hz), vibro/acoustic (20–2000 Hz), and pyrotechnic shock. Space is a vacuum with diurnal and seasonal temperature fluctuation; the thermal vacuum test is for survivability in such environment. If the component must survive nuclear weapons effects or laser attack, it may also undergo γ-ray testing and electromagnetic compatibility testing.

The operating environment refers to the circumstances in which the components or communication system must operate. Two factors are considered most in the operating environment: **multipactor and passive intermodulation (PIM)**. Multipactor discharge can cause noise in a sensitive system, heating of the surface, and gas discharge into the local atmosphere. The temperature can rise to the point that it discolors or burns the components. PIM takes place in high-power applications, especially where the antenna is designed for both transmit and receive functions. PIM may interfere with the receive signals. Careful attention should be paid to the design and fabrication of any junction between microwave components. Materials play an important role in the design's ability to meet these stringent requirements. Graphite epoxy is widely used for the reflector and other composite materials.

Space qualification is a lengthy and expensive process. Any new hardware component or system has to go through a dedicated qualification phase. Components of lesser importance can be qualified by similarity, that is, the component and the environment are identical to previously qualified hardware.

75.9 Future Trends and Further Study

Communications satellite antenna systems have undergone drastic changes since the advent of satellites. The changes can be divided into five major phases. In the first phase, antenna systems produced only simple circular beams used in the first generation satellites in the early 1960s. The second phase started in 1970s. In this phase, a shaped beam directed the energy to the coverage area to improve effective isotropic radiation power, reducing the output of the power amplifiers. A major change occurred in the third phase (1980s), when the demand for increased communication satellite capacity directed frequency reuse. Typical technology during this period used a multibeam antenna system to provide highly shaped patterns and several simultaneous beams, such as that used for INTELSAT V. Limited beam reconfiguration was required to accommodate the change of radiation pattern from one satellite orbital location to another. The fourth phase (1990s) was characterized by full exploitation of the beam reconfiguration. Beam reconfiguration was also extended from space to channel. Typical examples are the direct broadcast satellite and NASA ACTS, which can provide coverage of either half CONUS or full CONUS. The fifth phase (2000s and beyond) will represent rapid antenna pattern control capability, required to provide adaptive nulling and coverage for mobile communications satellites. Table 75.5 summarizes technology development and risks at each of the five phases.

The design of a satellite antenna system is driven by the system requirements, technology advancement, and maturity of developed components and composite materials. Recent research and development in MMIC technology, photonic array antennas, and superconductive materials in the communications system will be incorporated into future satellite antenna systems. Present satellite antenna systems are usually designed separately, then integrated into other subsystems, that is, filters, multiplexers, and power and receive amplifiers. Future communications systems will be designed and developed as integrated systems.

TABLE 75.5 Evolution of Satellite Antenna System Technology

Year	Antenna Radiation Patterns	Key Components	Required Technology Development
1960s	Simple beam	• Antenna connectors	• Lightweight • Space qualified materials
1970s	Simple shaped beam	• Reflector • Deployable mesh reflector • Lens	• Lightweight • Space qualified materials
1980s	Reconfigurable over space	• Reflector • Lens • Gridded reflector • Shaped reflector • FSS reflector • Variable power divider (VPD)/ Variable phase shifter (VPS) • Beam forming network (BFN) components	• Wider bandwidth • Computer-aided design (CAD) tools • Switches
1990s	Reconfigurable over space and channel	• Ditto • MMIC LNAs • MMIC VPD/VPS • BFN controller	• CAD tools • Higher frequency BFN • Switches • MMIC solid-state power amplifier (SSPA)
2000s and beyond	Reconfigurable over space, channel and time	• Large optics • Photonic phased array • Active MMIC phased array	• MMIC phased array • FSS • Multiband BFN components

Communications antenna systems design involves a tradeoff between the space segment and the ground segment. Higher EIRP in the space segment enables a lower cost and a smaller ground antenna system. A lower EIRP satellite antenna system reduces the complexity and cost of the space segment; however, a larger ground antenna system is required to compensate for the space loss. This incurs higher cost for the ground segment.

Defining Terms

Antenna performance parameters: Include beam shaping efficiency that is used to indicate how well a beam is shaped as compared to an ideally shaped beam; circular polarizations, right-hand circular polarization (RHCP), and left-hand circular polarization; copolarization and cross-polarization isolations/interferences; equivalent isotropically radiated power (EIRP); field of view (FOV); half-power beamwidth (HPBW); and voltage standing wave ratio (VSWR).

Beam: A ray of electromagnetic field, which may be omnidirectional, pencil, shaped, and/or reconfigurable, the beam that will change as a function of time and space.

Frequency reuse: A frequency band can be used more than once via space and/or polarization diversifications.

Graphite fiber reinforced plastic surface (GFRP): A perfectly conducting material that is light and has a very low-thermal-expansion coefficient.

Multipactor and passive intermodulation (PIM): Interference or noises due the heating of the surface, gas discharge, or higher order modes generated in the system.

Orbits: Geosynchronous Earth orbit (GEO), an orbit that is 19,3000 nmi away from the Earth; low Earth orbit (LEO); medium Earth orbit (MEO).

Reflector surfaces: Include solid, gridded for polarization purification, frequency selective surface (FSS) that will pass one frequency band and reject another band(s), and meshed.

Space qualification: Qualification test of flight hardware.

Types of antennas: Include cup-dipole antennas, deployable reflector antennas that are stowed for launch and deployed on orbit, horn antennas, monolithic microwave integrated circuit (MMIC) phased array antennas, radiometer antennas, shaped beam antennas whose radiation pattern is shaped to conform the contour of the coverage area, and tracking, telemetry, and command (TT&C) antennas that provide communications during initial acquisition, transfer orbit, and on-orbit operations.

References

Barrow, W., Chu, L., and Jansen, J. 1939. Biconical electromagnetic horns. *Proc. IRE*, (Dec.):769–779.

Chen, C., Choung, Y., and Palma, D. 1986. A compact multiflare horn design for spacecraft reflector antennas. *1986 IEEE AP-S Symp. Digest*, pp. 907–910, June.

Choung, Y. and Wong, W. 1986. Multibeam antenna design and development for NASA advanced communications technology satellite. In *IEEE/GLOBECOM Conference*, Houston, TX, pp. 16.2.1–16.2.6, Dec.

Clarricoats, P. and Saha, P. 1971. Propagation and radiation behavior of corrugated feeds. *Proc. IEE*, 118(Sept.):1176–1186.

Hwang, Y. 1992. Satellite antennas. *IEEE Proc.*, 80(1):183–193.

Imbriale, W. and Wong, G. 1977. An S-band phased array for multiple access communications. *NTC '77*, pp. 19.3.1–19.3.7.

James, G. 1982. TE_{11} to HE_{11} mode converters for small angle corrugated horns. *IEEE Trans. AP*, AP-30(6):1057–1062.

Kumar, A. and Hristov, H. 1989. *Microwave Cavity Antennas*, pp. 101–212, Artech House, Boston, MA.

Lee, E., Schenuum, G., and Hwang, Y. 1991. Scattering measurements for I–VII TC&R antennas on a spacecraft mockup. In *1991 IEEE AP-S Symposium*, London, Canada, pp. 1690–1693, June.

Potter, P. 1963. A new horn antenna with suppressed sidelobes and equal beamwidths. *Microwave J.*, VI(June):71–78.

Schindall, J. 1995. Concept and implementation of the globalstar mobile satellite system. *1995 International Mobile Satellite Conference*, Ottawa, Canada, pp. A-11–16.

Schoen, D. and Locke, P. 1995. Orbcomm—Initial operations. In *1995 International Mobile Satellite Conference*, Ottawa Canada, pp. 397–400.

Speltz, L. 1996. Personal communications satellite systems. 18th Annual Pacific Telecommunications Conf., Honolulu, HI, Jan. 14–18.

Yeh, Y. and Mei, K. 1967. Theory of conical equiangular-swpiral antennas. *IEEE Trans. AP-S.*, Pt. 1, (Sept.):634–639.

Yeh, Y. and Mei, K. 1968. Theory of conical equiangular-spiral antennas. *IEEE Trans. AP-S.*, Pt. 2, (Jan.):14–21.

Zaghloul, A., Hwang, Y., Sobello, R., and Assal, F. 1990. Advances in multibeam communications satellite antennas. *IEEE Proc.*, 78(7):1214–1232.

76
Tracking and Data Relay Satellite System

Erwin C. Hudson
TRW Space & Electronics Group

76.1 Introduction 1054
76.2 TDRS System Overview 1055
76.3 TDRSS Communications Design 1056
76.4 TDRS Relay Satellite Design 1058
   White Sands Complex • User Satellite Transponder
76.5 TDRS Link Budget Examples 1064
76.6 Summary 1065

76.1 Introduction

The tracking and data relay satellite system (TDRSS) was developed in the late 1970s and early 1980s by NASA Goddard Space Flight Center (GSFC) with an industry team comprising Western Union Space Communication Company (now GTE Contel), TRW Space & Electronics Group and Harris Government Communications Division. TDRSS is operated for NASA by GTE Contel and Allied Technical Services Corporation.

The TDRSS mission is to provide communications and tracking services to low-Earth orbiting user satellites. User satellites include the Landsat 4 and 5 multispectral imaging satellites, the Hubble Space Telescope, and the Compton Gamma Ray Observatory (GRO). TDRSS is also the primary communication and tracking system for the Space Transportation System (STS) providing data, voice, and video links for the Space Shuttle missions.

The TDRS system consists of three segments: the space segment encompassing the tracking and data relay satellite (TDRS) vehicles in **geosynchronous orbit**; the ground segment, which includes the White Sands complex (WSC) in New Mexico and the GRO remote tracking station (GRTS) in Australia; and the user segment, which includes the TDRSS transponders on user satellites orbiting the Earth. As shown in Fig. 76.1, forward links to user satellites are established through the WSC up to a TDRS and then down to the user satellites in low Earth orbits. Return links from the user satellites are relayed up through a TDRS and then down to the WSC.

Prior to TDRSS, NASA's worldwide network of tracking stations provided communications in 5–15 min contacts as user satellites occasionally passed over a station's coverage area. The TDRS relay satellites, from their vantage point 35,786 km above the Earth, allow near-continuous coverage of low-Earth orbiting satellites. TDRSS eliminates the necessity of operating a network of ground-based tracking stations, while providing lower cost and higher performance communications for NASA.

Tracking and Data Relay Satellite System

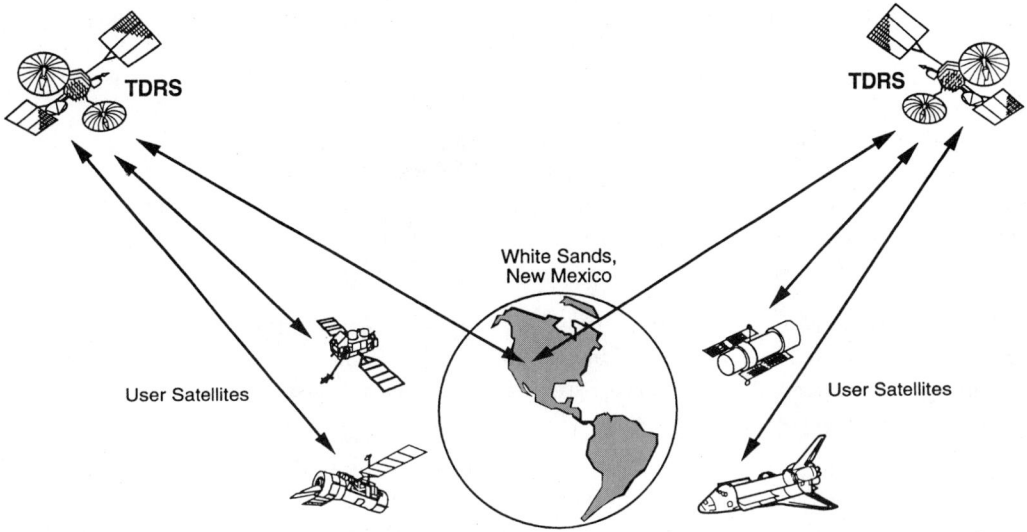

FIGURE 76.1 Tracking and data relay satellite system concept.

TDRS-A, the first TDRS satellite, was carried into orbit on Shuttle mission STS-6 in April 1983. TDRS-B was lost in the Space Shuttle Challenger accident in January 1986. With Shuttle mission STS-26 in September 1988, the second relay satellite, TDRS-C, was deployed and the TDRS system became fully operational. As of mid-1996, TDRSS is a mature system with a constellation of six relay satellites on orbit.

76.2 TDRS System Overview

TDRSS provides command, telemetry, range tracking, Doppler tracking, and wideband data relay services to user satellites operating in low Earth orbits. User satellite payload operations control centers (POCCs) interface with TDRSS through the NASA communications (NASCOM) network. The user satellite operator sends a schedule request to the network control center (NCC), which determines whether TDRSS has the space and ground resources to honor the request. If the schedule request can be honored, the NCC sends a schedule order message through NASCOM to initiate service. At the scheduled time, the requested communication services are established between the user POCC and the user satellite, all automatically and without manual intervention.

TDRSS provides four user services: command, tracking, telemetry, and mission data communications. Command links are forward services, telemetry and mission data links are return services, and range tracking is both a forward and return (turn-around) service. TDRSS can also perform Doppler tracking using only a return service.

Command service. User satellite operators send instructions and other data to their space vehicles through command channels. Command links may be used, for example, to operate sensors, fire thrusters, or to reconfigure onboard equipment to recover from failures.

Telemetry service. Satellites are designed to telemeter state-of-health and operational status data to their controllers. Satellite telemetry typically includes vehicle attitude, temperatures of sensitive equipment, power system voltages and currents, and other critical parameters.

Tracking service. Space missions such as Earth sensing require accurate knowledge of the orbital position of the sensor. In addition to Doppler frequency measurements, which provide an estimate of range rate, TDRSS performs user satellite ranging by sending a **pseudonoise (PN) coded** ranging signal on a round-trip path from the TDRSS ground station, to the user satellite, and back to the TDRSS ground station. The round-trip time delay is determined by correlating the transmitted ranging code with the returned version of the code and adjusting time delays until the two copies match. With proper time delay calibration, the round-trip distance can be determined to within a few meters.

TDRSS uses ground-based transponders at known locations to perform rapid and highly accurate orbit determination for each TDRS. The NASA Flight Dynamics Facility precisely determines the orbital elements of each user satellite using the TDRS orbital data and the ranging and Doppler measurements to the user transponders.

Mission data relay service. Mission data is the experiment data, scientific measurements, sensor data, or other digital information generated by a satellite in orbit. While many scientific missions have data rate requirements of 1 Mb/s or less, Earth sensing and imaging missions may require tens to hundreds of megabit per second.

A typical TDRSS contact involves a forward link command service to a user satellite and a return link data relay service back through the WSC to the user POCC. Based on the schedule order parameters, the required TDRS and WSC equipment is assigned and configured prior to the contact. At the specified time, the TDRS points the assigned antenna toward the user satellite and activates the forward signal. The forward signal frequency is initially offset and then continuously adjusted to compensate for Doppler shift so that the user receives the signal at a constant frequency. The user satellite acquires the forward signal and establishes simultaneous two-way communications by transmitting a return link signal back to the TDRS. At the end of the contact, the TDRS and WSC equipment assignments are released, and the equipment is made available to support other users.

76.3 TDRSS Communications Design

TDRSS provides a complete set of communication capabilities for existing, planned, and future user satellites. The system supports a variety of frequencies, data rates, modulation types, and signal formats [Poza, 1979]. The relay satellites themselves are relatively simple analog repeaters; the communications processing is done in the ground facilities where the equipment can be maintained, upgraded, and modified as required.

The primary TDRSS constellation consists of two satellites over the Atlantic Ocean, at 41° West and 46° West longitude, and two satellites over the Pacific Ocean, at 171° West and 174° West longitude. These orbital locations were selected to maximize worldwide coverage while providing line-of-sight visibility to White Sands. The dry desert climate in southern New Mexico results in low rain losses and excellent link availability, even at low elevation angles; this is one of the principal reasons NASA selected White Sands for the TDRSS ground site.

The TDRSS constellation has an oval-shaped zone-of-exclusion centered over the Indian Ocean where coverage is not available below 1200 km due to blockage by the Earth. TDRS-A, now beyond the end of its 10 year design life, has been placed over the Indian Ocean at 85° East longitude to provide extended service to the Gamma Ray Observatory. Since TDRS-A is not visible from White Sands, NASA has established the GRO remote tracking station in Canberra, Australia for command and control of TDRS-A.

Each TDRS communicates with its ground control site over **Ku-band** space-to-ground links. User satellites communicate over space-to-space links with the TDRS relay satellites at **S-band** and/or Ku-band. The TDRSS frequency plan is shown in Fig. 76.2.

TDRSS services are referred to as either single access or multiple access. Single access service is established using one of two large reflector antennas on the TDRS. Multiple access service is provided with an S-band array antenna on each TDRS. Multiple access service supports up to 20 simultaneous **code-division multiple access (CDMA)** users in the return direction while providing a single beam in the forward direction to one user satellite at a time.

Tracking and Data Relay Satellite System

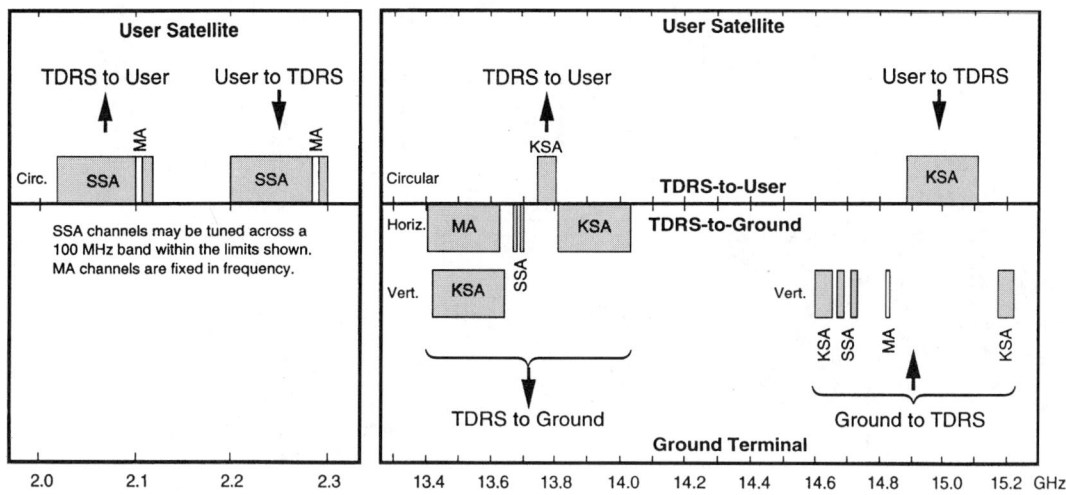

FIGURE 76.2 TDRSS mission communications frequency plan.

TABLE 76.1 TDRSS Forward Service Capabilities

Forward Service	Frequency F_{fwd}, GHz	Bandwidth, MHz	Forward Link Data Rate	Ranging Rate ($\approx$3 Mchips/sec)
MA	2.1064	6	100 b/s–10 kb/s	$F_{fwd} \times 31/221/96$
SSA	2.0304–2.1133	20	100 b/s–300 kb/s	$F_{fwd} \times 31/221/96$
KSA	13.7750	50	1 kb/s–25 Mb/s	$F_{fwd} \times 31/1469/96$

TABLE 76.2 TDRSS Return Service Capabilities

Return Service	Frequency F_{rtn}, GHz	Bandwidth, MHz	Return Link Data Rate	Ranging Rate ($\approx$3 Mchips/sec)[a]
MA	2.2875	5	100 b/s–50 kb/s	$F_{rtn} \times 31/240/96$
SSA	2.2050–2.2950	10	100 b/s–6 Mb/s	$F_{rtn} \times 31/240/96$
KSA	15.0034	225	1 kb/s–300 Mb/s	$F_{rtn} \times 31/1600/96$

[a] Return link ranging is not available when the SSA data rate exceeds 3.15 Mb/s, or when the KSA data rate exceeds 150.15 Mb/s.

Table 76.1 describes the three forward link services available through TDRSS. Multiple access (MA) forward service uses a 6-MHz bandwidth channel centered at a fixed frequency of 2.1064 GHz. S-band single access (SSA) forward channels are 20 MHz wide and are tunable over the 2.0304–2.1133 GHz band. Ku-band single access (KSA) forward channels are centered at a fixed 13.7750 GHz and have 50-MHz bandwidths. The forward link ranging code rates, nominally 3 Mchips/sec, are coherently related to the forward link carrier frequency by the ratios in Table 76.1. All TDRSS frequency conversions are performed coherently to maintain this relationship between the code rate and the carrier frequency [Gagliardi, 1979].

Table 76.2 describes the three TDRSS return link services. MA return service provides users with a 5-MHz bandwidth channel centered at a fixed 2.2875 GHz. SSA return channels are 10 MHz wide and tunable over the 2.2050–2.2950 GHz band. KSA return channels, centered at a fixed 15.0034 GHz, are 225 MHz wide and support data rates up to 300 Mb/s. The MA and SSA return carrier frequencies are generated by multiplying the forward link frequencies by the ratio 240/221. The KSA return carrier is derived in the same manner using the turn-around ratio 1600/1469. The

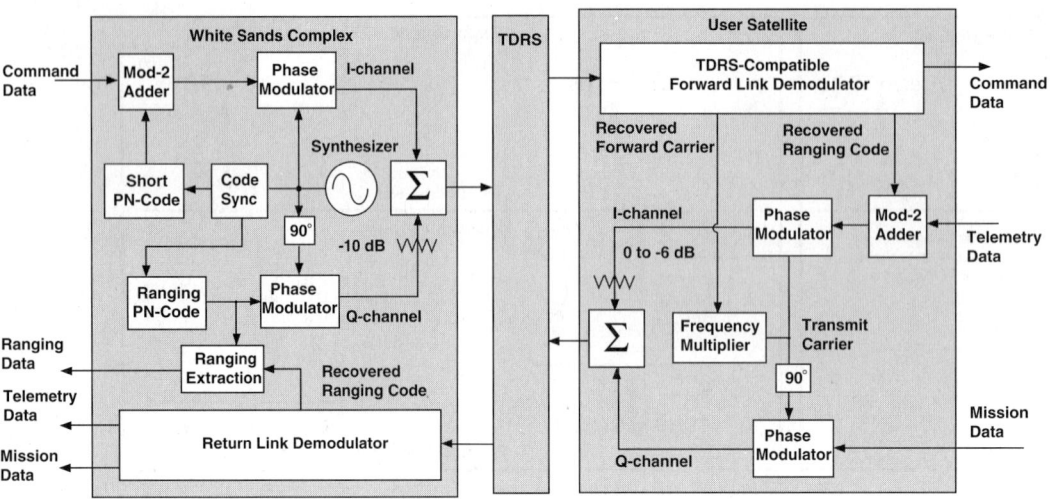

FIGURE 76.3 TDRSS forward and return link signal design.

return link ranging code rates are the same as the forward link rates and are tied to the carrier frequency by the ratios given in Table 76.2.

Figure 76.3 shows a typical TDRSS forward signal transmitted up from the WSC to a user satellite, and the return signal from the user satellite back down to the WSC. Unlike ground-based satellite tracking systems, the TDRS transmits high-power signals down from above with the Earth in the background. For data rates below 300 kb/s, TDRSS must use **spread-spectrum** signals rather than narrowband signals to avoid violating international limitations on microwave flux density [NASA 1988, App. G].

A typical TDRSS forward signal is **quadrature phaseshift keying (QPSK)** modulated with the quadrature (Q) channel attenuated 10 dB below the in-phase (I) channel. A short 1023-bit PN-code running at a chip rate of approximately 3 MHz (per Table 76.1) is modulo-2 added to the command data and modulated onto the I-channel. The Q-channel is modulated with a long 1023×256 bit PN-code running at the same chip rate as the short code. The starting points or epochs of the short and long codes are synchronized to simplify user acquisition. The short code is used by the user satellite transponder for initial acquisition and the long code is used for turn-around ranging.

A typical TDRSS return signal is QPSK modulated with telemetry data on the I-channel and mission data on the Q-channel. The I-channel may be attenuated up to 6 dB relative to the Q-channel. The return link carrier is generated by frequency multiplying the recovered forward link carrier by the appropriate turn-around ratio. The user satellite telemetry data on the I-channel is modulo-2 added to the ranging code recovered from the forward signal. When the mission data rate on the Q-channel is less than 300 kb/s, a PN-code may be modulo-2 added to the data to improve interference rejection.

Although PN-coded QPSK is the most common TDRSS waveform, additional signal types, waveform designs, and communication modes are available. The *Space Network Users Guide* (STDN 101.2) [NASA, 1988] is a complete reference for the entire set of frequency bands, waveforms, and data rates available through TDRSS.

76.4 TDRS Relay Satellite Design

Each TDRS relay satellite consists of a communications payload module containing the antennas and electronics to perform the TDRSS mission and a spacecraft bus module that provides a three-axis

Tracking and Data Relay Satellite System

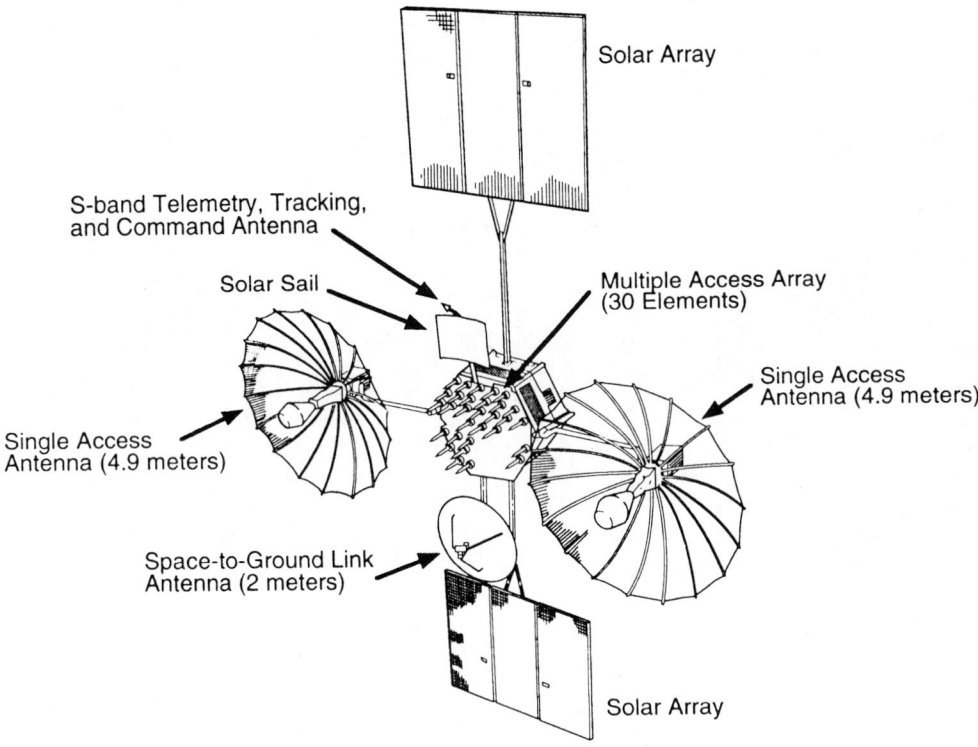

FIGURE 76.4 Tracking and data relay satellite (TDRS-G shown).

stabilized platform and support functions for the payload [Harper and Woodson, 1981]. A sketch of the TDRS vehicle is shown in Fig. 76.4.

The TDRS frame is constructed primarily of aluminum and consists of a six-sided outer structure built around a cylindrical inner section. The cylindrical section supports the antenna platform on the top, attaches to the launch vehicle adapter on the bottom, and contains two pressurized fuel tanks in its interior. The six-sided enclosure houses the communications payload and spacecraft bus electronics on the interior facing sides of 12 rectangular panels. The weight of the fully assembled and fueled satellite, including the launch vehicle adapter, is 2230 kg.

Two rectangular solar arrays, each approximately 3.8 × 3.8 m, are attached to the north and south sides of the spacecraft. The two arrays contain 28,080 silicon solar cells and produce greater than 1700 W at end-of-life. A solar array drive motor slowly rotates each array once per 24 h to track the sun as the TDRS orbits the Earth.

The TDRS antenna suite consists of two 4.9-m single access antennas, a 30-element multiple access array, and a 2-m space-to-ground link antenna. The single access antennas are mechanically gimbaled and steerable over ±22.5° East–West and ±31.0° North–South. The multiple access array is electronically steerable over a ±13.0° conical field-of-view.

Attached directly to the base of each single access antenna is an electronics compartment, which houses the high-power transmitters and low noise receivers. Placing this equipment at the antenna minimizes signal losses between the antenna and the electronics, improving receiver sensitivity and radiated power.

The block diagram in Fig. 76.5 shows the signal flow through the TDRS communication payload. Backup equipment, frequency generators, TDRS command and telemetry processors, and other support equipment are not shown.

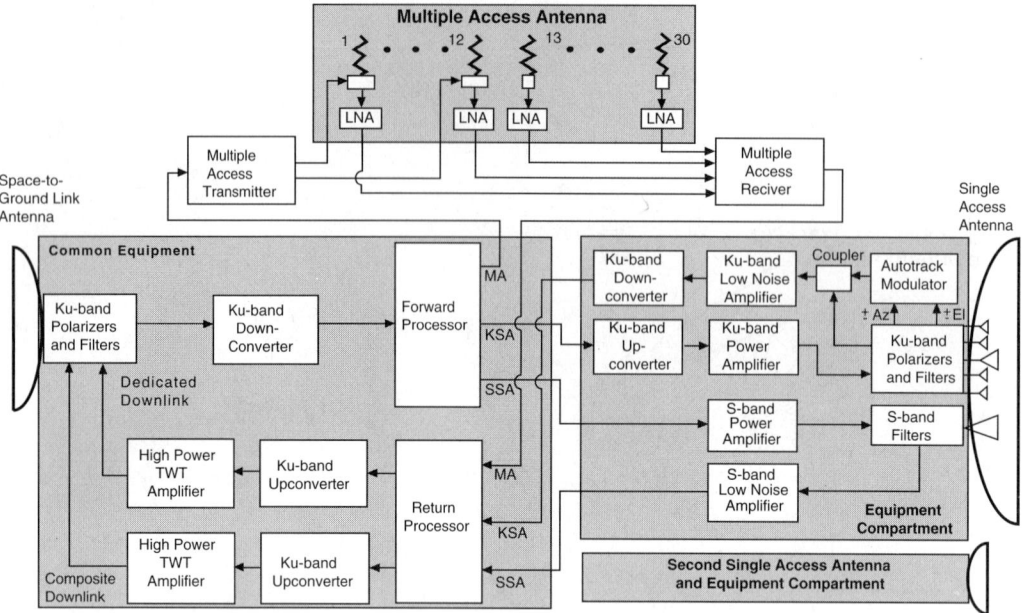

FIGURE 76.5 Tracking and data relay satellite payload block diagram.

The TDRS interface with the WSC is at Ku-band through the space-to-ground link antenna. The frequency-division multiplexed forward signals from the WSC are downconverted and then demultiplexed in the forward processor. The forward processor performs automatic level control on the MA, SSA, and KSA forward signals and tunes the SSA signals to the commanded center frequencies. The forward processor then sends the SSA and KSA signals to the single access antennas and the MA signal to the MA transmitter.

The KSA forward signals are upconverted to 13.7750 GHz in the antenna compartment and transmitted through a 1.5-W power amplifier. The SSA forward signals are sent to the antenna centered at the commanded frequency between 2.0304 and 2.1133 GHz, and are transmitted through a 26-Watt power amplifier. Both KSA and SSA transmitters may be used simultaneously through one single access antenna for dual-band user service.

The MA phased array has 30 elements, 12 of which can simultaneously transmit and receive; the remaining 18 elements are receive only. The MA transmitter splits the 2.1064-GHz MA forward signal using a 12-way divider and adjusts each of these components in phase to steer the forward beam. Eight signals are used for normal operations and the remaining four are spares. Each transmit signal is routed through a 3.6-W power amplifier in the MA transmitter and a composite beam is formed in the direction of the user satellite.

All 30 MA array antenna elements have low noise amplifiers, providing 30-MA return signals centered at 2.2875 GHz to the MA receiver. The MA receiver frequency-division multiplexes these 30 signals on 7.5-MHz centers into a 225-MHz wide signal. The MA receiver sends the multiplexed MA return signal to the return processor for downlink transmission. MA return array processing is performed on the ground rather than onboard to minimize the complexity of the satellite.

The single access antennas have both S-band and Ku-band low noise amplifiers for SSA and KSA return services. The KSA return signal at 15.0034 GHz is downconverted in the antenna equipment compartment, and both the SSA and KSA return signals are sent to the return processor for downlink transmission. The return processor tunes the center frequencies of the SSA return channels over the 2.2050–2.2950 GHz receive band.

The single access antennas have narrow 0.27° half-power beamwidths at Ku-band and autotracking is available to ensure accurate beam pointing. TDRS performs KSA autotracking using four error-sensing horns around the main Ku-band receive feed. The autotrack modulator scans these error signals, and amplitude modulates azimuth and elevation errors onto the KSA return signal. The autotrack modulator has a low modulation index so that the autotrack modulation has essentially no effect on the phase modulated KSA user data. Autotrack error signals are recovered at the TDRS ground site and used to generate antenna pointing commands that are sent up to the TDRS. Autotrack improves pointing from a 99% circular error probability of 0.22° using open-loop pointing alone to better than 0.06° with closed-loop autotrack, reducing potential KSA pointing loss from 9.6 dB to 0.6 dB.

The return processor frequency division multiplexes the 30-MA return element signals, the SSA return signals and one of the two KSA return signals into a composite downlink. The composite downlink is amplified through a 30-W traveling-wave tube amplifier (TWTA) and transmitted on horizontal polarization. The return processor performs automatic level control on the return signals to keep the TWTA in its highly linear region, at an output power of approximately 3 W, to avoid intermodulation products between the many composite carriers. The remaining KSA return signal is transmitted on a dedicated downlink. The dedicated downlink uses an identical 30-W TWTA in its quasilinear region, at approximately 12-W output power, to prevent suppression of the autotrack modulation on the KSA return carrier. The dedicated downlink band overlaps the MA portion of the composite downlink band and the dedicated downlink is transmitted down on vertical polarization to avoid interference.

The TDRSS satellites are launched on the Space Shuttle and use Boeing's inertial upper stage (IUS) to fly to geosynchronous orbit. The TDRS/IUS vehicle, in the configuration shown in Fig. 76.6, is released from the cargo bay; the IUS performs an initial burn to place the TDRS into an elliptical transfer orbit; the first stage of the IUS is discarded. Several hours later at the peak or apogee of the transfer orbit, the IUS second stage does an injection burn to insert the TDRS into geosynchronous orbit. The TDRS begins deployment of its solar arrays and antennas while still attached to the IUS to take advantage of the combined TDRS/IUS moment of inertia. After the IUS separates, the TDRS completes its deployments and initiates normal spacecraft operations. After approximately a month of on-orbit calibration and evaluation, the new TDRS is ready to begin providing relay services to user satellites.

White Sands Complex

The White Sands complex consists of two facilities, approximately 5 km apart, located on the western edge of the White Sands Missile Base in southern New Mexico. To the south is the original White Sands ground terminal (WSGT), which became operational in 1983. A new facility, called the second TDRSS ground terminal (STGT), was completed in 1995. WSGT has been temporarily closed and is being upgraded to the STGT hardware and software configuration.

A diagram of the White Sands ground equipment is shown in Fig. 76.7. Three 18-m Ku-band antennas provide the microwave interface with the TDRS relay satellites. The Ku-band low noise receivers are colocated with the antennas to minimize losses. Each White Sands ground terminal consists of satellite control equipment to operate the TDRS relay satellites and communications/tracking equipment to provide user services. Racks of command transmission, telemetry reception, and ranging/Doppler equipment perform the control and monitoring functions required to operate the relay satellites. Forward link user communications equipment consist of encoders, modulators, equalizers, transmitters, and multiplexers. Return link user communications equipment consist of demultiplexers, demodulators, equalizers, and decoders.

FIGURE 76.6 TDRS-A launched on Shuttle mission STS-6, April 1983.

User Satellite Transponder

User satellites carry a TDRSS-compatible transponder to interface with the TDRS system over the space-to-space links. User transponders may be S-band, Ku-band, or have a combination of S-band and Ku-band capabilities. Users with mission data rate requirements below 6 Mb/s may be S-band only; users above 6 Mb/s usually have both S-band and Ku-band capabilities.

Figure 76.8 has a diagram of a typical user transponder subsystem capable of both S-band and Ku-band operations. The subsystem consists of a two-axis gimbaled antenna module and a signal processing module. The antenna module may be mounted on a boom away from the body of the user satellite to allow an unobstructed line of sight to TDRS; the signal processing module is mounted directly to the user vehicle [Landon and Raymond, 1982].

Tracking and Data Relay Satellite System

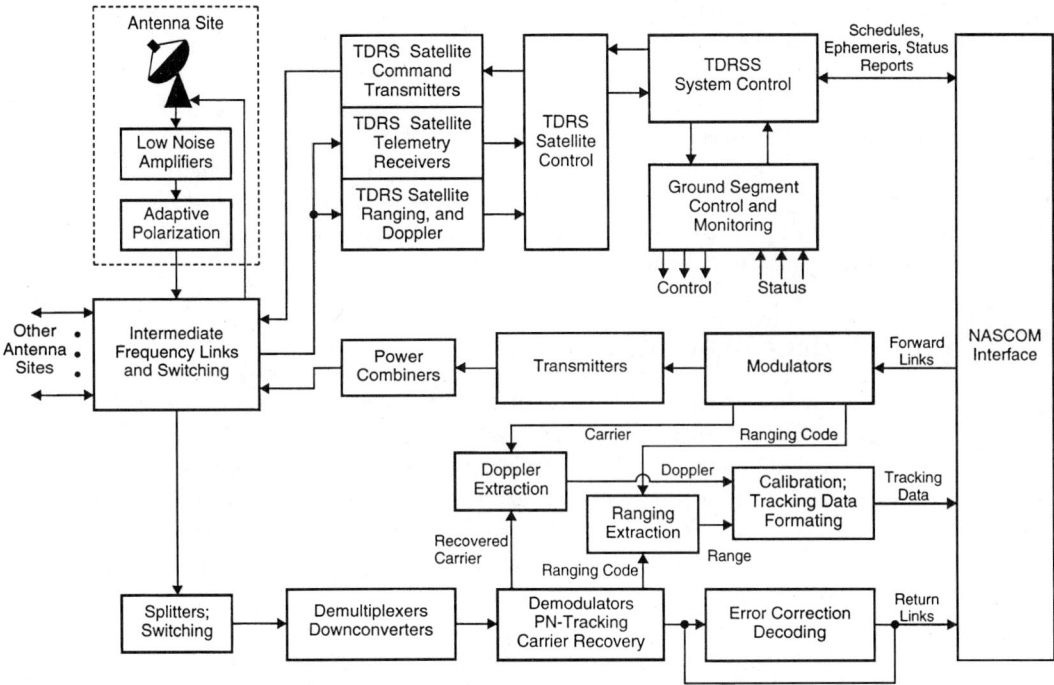

FIGURE 76.7 White Sands ground equipment block diagram.

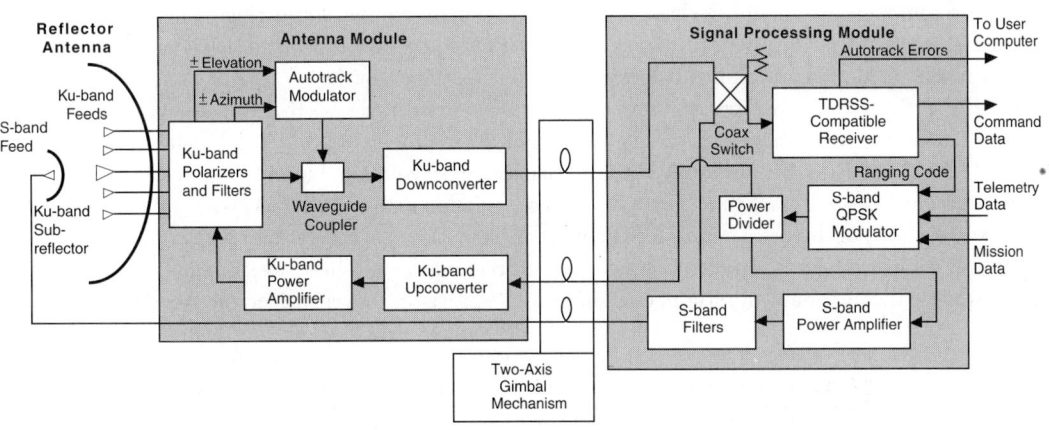

FIGURE 76.8 TDRSS user satellite transponder block diagram.

The antenna module provides the mounting structure for a dual-frequency S/Ku-band reflector antenna, typically 1–2 m in diameter. The antenna compartment houses a Ku-band low-noise preamplifier/downconverter for receiving along with an upconverter and a 10–25 W Ku-band power amplifier for transmitting. Cable losses at S-band are low enough that amplification and frequency conversion at the antenna are not required. Most high data rate users also have Ku-band tracking feeds and autotrack electronics.

The antenna module and the signal processing modules are connected with a number of wires and coax cables. Service loops in the gimbal drive assembly allow the antenna module to be articulated for greater than hemispherical coverage.

The signal processing module contains a TDRSS-compatible spread-spectrum receiver that demodulates either the S-band forward signal or the downconverted Ku-band forward signal. The receiver recovers the command data, extracts the ranging code, and derives frequency references by phase tracking the forward link carrier. The spread-spectrum receiver also processes the autotrack error signals required for onboard autotracking.

A QPSK modulator in the signal processing module modulates the return signal with the mission data and turn-around ranging code. The return link signal goes directly to the S-band power amplifier for S-band return service or to the Ku-band upconverter and power amplifier in the antenna module for Ku-band return service.

76.5 TDRS Link Budget Examples

Appendix A of the *Space Network Users Guide* (STDN 101.2) [NASA, 1988] describes the TDRS link budget methodology and contains example budgets for each user service.

On the forward links, TDRS guarantees a minimum **effective isotropic radiated power (EIRP)** for SSA, MA, and KSA service. The user satellite designer is responsible for having adequate receive sensitivity to close the link, allowing for maximum distance, polarization mismatch, and the bit error rate (BER) performance of the user satellite demodulator.

In the return direction, a minimum received power (P_{rec}) at each TDRS antenna aperture is specified in STDN 101.2 as a function of data rate to provide an end-to-end BER of 10^{-5} or better. The user satellite designer is responsible for transmitting enough power to close the link, accounting for maximum distance, polarization mismatch, and any signal distortions that exceed those allowed in STDN 101.2.

Table 76.3 summarizes the Gamma Ray Observatory link budget for SSA forward service at 2.1064 GHz. The waveform is 1:10 unbalanced QPSK with a PN ranging code on the I-channel and PN-spread command data on the Q-channel. A robust command channel is critical to ensuring the health and safety of a satellite in orbit, and as a result, GRO acquires the command signal with 19-dB margin and demodulates command data with almost 15-dB margin.

Table 76.4 shows a GRO link budget for one of several SSA return services that GRO is designed to use. The return center frequency, generated by multiplying the forward carrier frequency by 240/221, is 2.2875 GHz. The waveform is 1 : 4 unbalanced QPSK with ranging and 32 kb/s telemetry data modulated on the I-channel and with 512 kb/s mission data on the Q-channel. Like the command channel, telemetry is critical to spacecraft operations and is designed for greater than a 10 dB link margin. The mission data channel has 4.2-dB margin, which assures excellent performance while allowing for any uncertainties and unexpected variations over temperature and life.

TABLE 76.3 Gamma Ray Observatory SSA Forward Link Budget

Parameter	SSA Forward Service
TDRS SSA forward EIRP per STDN 101.2	43.5 dBW
Space loss from TDRS to user (44,000 km)	−191.8 dB
Polarization mismatch loss	−0.2 dB
GRO received power	−148.5 dBWi
GRO antenna gain (1.5 m, 48% efficiency)	27.6 dBi
Pointing loss (3σ error of $1.1°$)	−0.4 dB
Effective received power at GRO	−121.3 dBW
Effective thermal noise density (81,800 K)	−179.8 dBW-Hz
GRO power received-to-noise ratio (P_{rec}/No)	58.5 dB-Hz
Minimum Prec/No for acquisition and ranging	39.5 dB-Hz
Acquisition and ranging link margin	19.0 dB
Minimum P_{rec}/No for 1-kb/s command data	43.7 dB-Hz
Command link margin (10^{-5} BER)	14.8 dB

TABLE 76.4 Gamma Ray Observatory SSA Return Link Budget

Parameter	SSA Return Service
GRO power amplifier (5.2 W)	7.2 dBW
Power losses from amplifier to antenna	−5.9 dB
Antenna gain (1.5 m, 52% efficiency)	27.9 dBi
GRO pointing loss (3σ error of 1.1°)	−0.4 dB
GRO transmitter EIRP	28.8 dBW
Polarization mismatch loss	−0.2 dB
Space loss from GRO to TDRS (44,000 km)	−192.5 dB
TDRS received power	−163.9 dBWi
I-channel power sharing (1/5)	−7.0 dB
I-channel distortion and interference losses	−0.7 dB
I-channel received power at TDRS	−171.6 dBWi
TDRS STDN 101.2 requirement (10^{-5} BER)	−176.8 dBWi
Link margin (Ranging, 32-Kb/s telemetry)	10.2 dB
Q-channel power sharing (4/5)	−1.0 dB
Q-channel distortion and interference losses	−1.7 dB
Q-channel received power at TDRS	−166.6 dBWi
TDRS STDN 101.2 requirement (10^{-5} BER)	−170.8 dBWi
Link margin (512 Kb/s mission data)	4.2 dB

76.6 Summary

TDRSS establishes a complete space communications infrastructure, providing command, telemetry, tracking, and wideband data relay services to NASA satellites in low Earth orbit. Using relatively simple relay satellites in geosynchronous orbits and a sophisticated ground complex at White Sands, TDRSS user satellites have access to a full complement of forward and return communication capabilities at S-band and Ku-band.

NASA has placed an order with Hughes Space and Communications Company for three new TDRS satellites to allow replacement of the existing vehicles as they complete their useful lives on orbit. Launches of these next generation relay satellites are expected to begin by the late 1990s, allowing NASA to continue to provide TDRSS communications services to the Space Shuttle and other user satellites well beyond the turn of the century.

Defining Terms

Code-division multiple access: Use of families of near-orthogonal codes to allow several data streams to occupy the same frequency band and pass through a single channel.

Effective isotropic radiated power: Transmitted flux density expressed in terms of the equivalent input power to an ideal unity gain transmit antenna.

Geosynchronous orbit: A 24-h orbit, approximately 35,800 km above the Earth, in which a satellite appears from the ground to be almost stationary.

Ku-band: The frequency band between 12.5 and 18.0 GHz, in which several subbands are allocated for space communications and space research.

Pseudonoise code: Binary codes, generated using linear shift registers and sequential feedback, that appear random and have good correlation properties.

Quadrature phase shift keying: A modulation technique in which pairs of data bits are encoded onto a carrier by rotating the phase to one of four biorthogonal positions.

S-band: The frequency band between 1.55 and 5.20 GHz, in which several subbands are allocated for space communications and space research.

Spread-spectrum: Techniques, such as modulo-2 adding a PN-code to binary communications data, that increase bandwidth and reduce the power density of the signal.

References

Agrawal, B.N. 1986. *Design of Geosynchronous Spacecraft*, Prentice–Hall, Englewood Cliffs, NJ.
Gagliardi, R.M. 1979. Phase coherency in tracking and data relay satellites. *IEEE Trans. Comm.*, COM-27(10):1527–1537.
Gagliardi, R.M. 1984. *Satellite Communications*, Lifetime Learning Pub., Belmont, CA.
Harper, W.R. and Woodson, W.L. 1981. The tracking and data relay satellite. *ISA Trans.*, 20(2):49–62.
Landon, R.B. and Raymond, H.G. 1982. Ku-band satellite communications via TDRSS, *9th AIAA Communication Satellite Systems Conference*, 82-0457.
NASA. 1988. *Space Network (SN) Users Guide*, Goddard Space Flight Center, Greenbelt, MD. STDN 101.2 Rev. 6.
Poza, H.B. 1979. TDRSS telecommunications payload: An overview. *IEEE Trans. Aero. Elec. Syst.*, AES-15(3):414–429.

Further Information

TDRSS was a special topic at the 1977 National Telecommunications Conference (NTC'77) and the collection of TDRSS-related papers in the NTC'77 Conference Record (IEEE Publication 77CH1292-2 CSCB) provides an excellent overview of the TDRS system. TDRSS user services and the TDRS user interfaces are fully specified in the *Space Network Users Guide* [NASA, 1988]. For a general overview of spacecraft design, including orbital, launch vehicle, and payload electronics considerations, see *Design of Geostationary Spacecraft* [Agrawal, 1986]. *Satellite Communications* [Gagliardi, 1984] is a good introductory text covering digital communications through satellite channels.

VI

Wireless

77 **Mobile Radio: An Overview** *Andy D. Kucar* 1069
   Introduction • Prologue • A Glimpse of History • Πάντα ρει (Panta Rhei) • Repertoire of Systems and Services • The Airwaves Management • Operating Environment • Service Quality • Network Issues and Cell Size • Coding and Modulation • Speech Coding • Macrodiversity and Microdiversity • Multiplex and Multiple Access • System Capacity • Conclusion

78 **Base Station Subsystems** *Chong Kwan Un and Chong Ho Yoon* 1090
   Introduction • System Architectures • Analysis of Call Handling Schemes • Summary

79 **Access Methods** *Bernd-Peter Paris* 1104
   Introduction • Relevant Wireless Communication System Characteristics • Frequency Division Multiple Access • Time Division Multiple Access • Code Division Multiple Access • Comparison and Outlook

80 **Location Strategies for Personal Communications Services** *Ravi Jain, Yi-Bing Lin, and Seshadri Mohan* 1116
   Introduction • An Overview of PCS • IS-41 Preliminaries • Global System for Mobile Communications • Analysis of Database Traffic Rate for IS-41 and GSM • Reducing Signalling During Call Delivery • Per-User Location Caching • Caching Threshold Analysis • Techniques for Estimating Users' LCMR • Discussion • Conclusions

81 **Cell Design Principles** *Michel Daoud Yacoub* 1146
   Introduction • Cellular Principles • Performance Measures and System Requirements • System Expansion Techniques • Basic Design Steps • Traffic Engineering • Cell Coverage • Interference • Conclusions

82 **Microcellular Radio Communications** *Raymond Steele* 1160
   Introducing Microcells • Highway Microcells • City Street Microcells • Indoor Microcells • Microcellular Infrastructure • Multiple Access Issues • Discussion

83 **Fixed and Dynamic Channel Assignment** *Bijan Jabbari* 1175
   Introduction • The Resource Assignment Problem • Fixed Channel Assignment • Enhanced Fixed Channel Assignment • Dynamic Channel Assignment • Conclusion

84 **Propagation Models** *Theodore S. Rappaport, Rias Muhamed, and Varun Kapoor* 1182
   Introduction • Free-Space Propagation Model • The Three Basic Propagation Mechanisms • Reflection • Diffraction • Scattering • Practical Propagation Models • Small-Scale Fading

85 **Power Control** *Roman Pichna and Qiang Wang* 1197
   Introduction • Cellular Systems and Power Control • Power Control Examples • Summary

86 **Second Generation Systems** *Marc Delprat and Vinod Kumar* 1208
   Introduction • Basic Features and System Architecture in Cellular, Cordless, and PMR • Systems Description • Interworking and Compatibility • Further Improvements of Radio Interface Performances • Conclusion

87 **The Pan-European Cellular System** *Lajos Hanzo* 1226
   Introduction • Overview • Logical and Physical Channels • Speech and Data Transmission •

Transmission of Control Signals • Synchronization Issues • Gaussian Minimum Shift Keying Modulation • Wideband Channel Models • Adaptive Link Control • Discontinuous Transmission • Summary

88 The IS-54 Digital Cellular Standard *Paul Mermelstein* 1246
Introduction • Modulation of Digital Voice and Data Signals • Speech Coding Fundamentals • Channel Coding Considerations • VSELP Encoder • Linear Prediction Analysis and Quantization • Bandwidth Expansion • Quantizing and Encoding the Reflection Coefficients • VSELP Codebook Search • Long-Term Filter Search • Orthogonalization of the Codebooks • Quantizing the Excitation and Signal Gains • Channel Coding and Interleaving • Bad Frame Masking • Conclusions

89 CDMA Technology and the IS-95 North American Standard *Arthur H.M. Ross and Klein S. Gilhousen* 1257
General Overview • A Short History • Overview of CDMA • The CDMA System • Multiple Forms of Diversity • Mobile Power Control for CDMA Digital Cellular • Low Transmit Power • Privacy • Mobile Station Assisted Soft Handoff • Capacity • Soft Capacity • Transition to CDMA • Overview of the IS-95 CDMA Standard

90 Japanese Cellular Standard *K. Kinoshita and Masao Nakagawa* 1276
Cellular Market in Japan • Standardization Process • General • System Overview

91 The British Cordless Telephone Standard: CT-2 *Lajos Hanzo* 1289
History and Background • The CT-2 Standard • The Radio Interface • Burst Formats • Signalling Layer Two (L2) • CPP-Initiated Link Setup Procedures • CFP-Initiated Link Setup Procedures • Handshaking • Main Features of the CT-2 System

92 Digital European Cordless Telephone *Saf Asghar* 1305
Introduction • Application Areas • DECT/ISDN Interworking • DECT/GSM Interworking • DECT Data Access • How DECT functions • Architectural Overview

93 The RACE Program *Stanley Chia* 1327
Introduction • RACE I Mobile Project • RACE II Mobile Project Line • Beyond the RACE Program

94 Half-Rate Standards *Wai-Yip Chan, Ira Gerson, and Toshio Miki* 1338
Introduction • Speech Coding for Cellular Mobile Radio Communications • Codec Selection and Performance Requirements • Speech Coding Techniques in the Half-Rate Standards • Channel Coding Techniques in the Half-Rate Standards • The Japanese Half-Rate Standard • The European GSM Half-Rate Standard • Conclusions

95 Modulation Methods *Gordon L. Stüber* 1353
Introduction • Basic Description of Modulated Signals • Analog Frequency Modulation • Phase Shift Keying (PSK) and $\pi/4$-QPSK • Continuous Phase Modulation (CPM) and MSK • Gaussian Minimum Shift Keying • Orthogonal Frequency Division Multiplexing (OFDM) • Conclusions

96 Wireless LANs *Suresh Singh* 1367
Introduction • Physical Layer Design • MAC Layer Protocols • Network Layer Issues • Transport Layer Design • Conclusions

97 Wireless Data *Allen H. Levesque and Kaveh Pahlavan* 1380
Introduction • Characteristics of Wireless Data Networks • Market Issues • Modem Services Over Cellular Networks • Private Data Networks • Cellular Data Networks and Services • Other Planned Systems • Conclusions

77

Mobile Radio: An Overview

77.1	Introduction	1069
77.2	Prologue	1070
77.3	A Glimpse of History	1070
77.4	Πάντα ρει (Panta Rhei)	1072
77.5	Repertoire of Systems and Services	1073
77.6	The Airwaves Management	1078
77.7	Operating Environment	1078
77.8	Service Quality	1079
77.9	Network Issues and Cell Size	1080
77.10	Coding and Modulation	1081
77.11	Speech Coding	1083
77.12	Macrodiversity and Microdiversity	1084
77.13	Multiplex and Multiple Access	1085
77.14	System Capacity	1086
77.15	Conclusion	1087

Andy D. Kucar
4U Communications Research Inc.

77.1 Introduction

The focus of this section is on terrestrial and satellite mobile radio communications. This includes: *cellular radio systems* such as existing North American **advanced mobile phone service (AMPS)**, Japanese **mobile communication systems (MCS)**, Scandinavian **nordic mobile telephone (system) (NMT)**, British **total access communication system (TACS)**, **groupe spécial mobile (GSM)**, **digital AMPS**, and spread-spectrum **code division multiple access (CDMA)**; *cordless telephony systems* such as existing *CT1* and *CT2* and the proposed *CT2Plus*, *CT3*, and *digital European cordless telecommunications* (*DECT*); *mobile radio data systems* such as *ARDIS* and *RAM*; projects known as **personal communications networks (PCN)**, **personal communications systems (PCS)**, and **future public land mobile communications systems (FPLMTS)**; *satellite mobile radio systems*, such as existing *INMARSAT* and *OmniTRACS* and the proposed *INMARSAT*, *MSAT*, *Iridium*, *Globalstar*, and *ORBCOMM*. After brief prologue and historical overview, technical issues, such as the repertoire of systems and services, the airwaves management, the operating environment, service quality, network issues and cell size, channel coding and modulation, speech coding, diversity, multiplex and multiple access (**frequency division multiple accesss, FDMA**; **time division multiple accesss**,

TDMA; CDMA) are discussed. Potential economical and sociological impacts of the mobile radio communications in the wake of the redistribution of airwaves at World Administrative Radio Conferences are also addressed.

Most existing mobile radio communications systems collect the information on network behavior, users' positions, etc., with the purpose of enhancing the performance of communications, improving handover procedures, and increasing the system capacity. Coarse positioning is usually achieved inherently, whereas more precise *navigation* can be achieved by employing *LORAN-C* and/or **global positioning system (GPS)** signals, or some other means, at the marginal expense in cost and complexity.

77.2 Prologue

Mobile radio systems provide their users with opportunities to travel freely within the service area being able to communicate with any telephone, fax, data modem, and electronic mail subscriber anywhere in the world; to determine their own positions; to track precious cargo; to improve the management of fleets of vehicles and the distribution of goods; to improve traffic safety; and to provide vital communication links during emergencies, search and rescue operations, etc. These *tieless (wireless, cordless)* communications, exchanges of information, determination of position, course, and distance traveled are made possible by the unique property of the radio to employ an *aerial (antenna)* for radiating and receiving electromagnetic waves. When the user's radio antenna is stationary over a prolonged period of time, the term *fixed radio* is used; a radio transceiver capable of being carried or moved around, but stationary when in operation, is called a *portable radio*; a radio transceiver capable of being carried and used, by a vehicle or by a person on the move, is called *mobile radio*. Individual radio users may communicate directly or via one or more intermediaries, which may be *passive radio repeater(s), base station(s),* or *switch(es)*. When all intermediaries are located on the Earth, the terms *terrestrial radio system* and *radio system* have been used; when at least one intermediary is satellite borne, the terms *satellite radio system* and *satellite system* have been used. According to the location of a user, the terms *land, maritime, aeronautical, space*, and *deep-space radio systems* have been used. The second unique property of all terrestrial and satellite radio systems is that they all share the same natural resource—the *airwaves (frequency bands and the space)*.

Recent developments in **microwave monolithic integrated circuit (MMIC), application specific integrated circuit (ASIC)**, analog/digital signal processing (A/DSP), and battery technology, supported by **computer aided design (CAD)** and robotics manufacturing allow a viable implementation of miniature radio transceivers. The continuous flux of market forces (excited by the possibilities of a myriad of new services and great profits), international and domestic standard forces (who manage the common natural resource—the airwaves), and technology forces (capable of creating viable products), acted harmoniously and created a broad choice of communications (voice and data), information, and navigation systems, which propelled an explosive growth of mobile radio services for travelers.

77.3 A Glimpse of History

> *Many things have an epoch, in which they are found at the same time in several places, just as the violets appear on every side in spring.*
>
> Farkas Wolfgang Bolyai, in 1823

Late in the 19th century Heinrich Rudolf Hertz, Nikola Tesla, Alexander Popov, Edouard Branly, Oliver Lodge, Guglielmo Marconi, Adolphus Slaby, and some other engineers and scientists experimented with the transmission and reception of electromagnetic waves. In 1898 Tesla demonstrated in Madison Square Garden a radio remotely controlled boat; the same year Marconi established the first wireless ship-to-shore telegraph link with the royal yacht Osborne; these events are now

Mobile Radio: An Overview

accepted as the birth of the mobile radio. Since that time, mobile radio communications have provided safe navigation for ships and airplanes, saved many lives, dispatched diverse fleets of vehicles, won many battles, generated many new businesses, etc. A summary of some of the key historical developments related to the commercial mobile radio communications is provided in Table 77.1.

Satellite mobile radio systems launched in the 1970s and early 1980s use ultra high frequency (UHF) bands around 400 MHz and around 1.5 GHz for communications and navigation services.

In the 1950s and 1960s, numerous private mobile radio networks, **citizen band (CB) mobile radio**, ham operator mobile radio, and portable home radio telephones used diverse types and brands of radio equipment and chunks of airwaves located anywhere in the frequency band from

TABLE 77.1 A Summary of Events Related to Mobile Radio Communications

Year	Event
1898	Nikola Tesla demonstrated a radio remotely controlled boat in New York. Guglielmo Marconi established the wireless ship-to-shore telegraph link in England.
1903	First International Radiotelegraphic Conference held in Berlin.
1908	Public radio telephone between ships and land in Japan was established.
1921	Police car radio dispatch service was introduced in Detroit, MH police department.
1945	During WW II, significant progress in design and widespread use of mobile radio.
1958	LORAN-C commercial operation started. The initial development was started during WW II.
1964	Railway telephone service on Japanese Tokaido bullet train was introduced.
1968	Carterphone decision. FCC allows non-Bell equipment to be connected to (Bell) network.
1971	Fully automatic radiotelephone system, the B network, was introduced in West Germany. Later extended to the corresponding networks in Austria, Luxemburg, and the Netherlands.
1974	U.S. FCC allocated 40-MHz frequency band, paving the way for establishing what is now known as advanced mobile phone service.
1976	MARISAT consortium initiated commercial service for mobile maritime users, providing full duplex voice, data, and teleprinter services worldwide.
1979	Mobile communications system MCS-L1 introduced by NTT Japan.
1982	The Conference of European Postal and Telecommunications Administrations established Groupe Spécial Mobile with the mandate to define future Pan-European cellular radio standard.
1982	INMARSAT began providing similar services as MARISAT.
1982	Cospas—1 inclined orbit satellite was launched, with a search and rescue package compatible with future global maritime distress and safety system (FGMDSS) onboard.
1983	SARSAT search and rescue instrument package was placed onboard of U.S. National Oceanic and Atmospheric Administration satellite NOAA-8 and launched.
1984	First interagency tests of global positioning system receivers conducted in California.
1985	Total access communications system was introduced in U.K.
1985	CD900 cellular mobile radio system was introduced in West Germany.
1987	Japan launched its own experimental satellite ETS-V.
1988	Geostar introduced its *link one* radio-determination services. The radio-determination information is obtained from a LORAN-C receiver and sent over an L-band satellite payload toward Earth.
1988	Qualcomm/Omninet started its two-way data communication and radio determination (using a LORAN-C receiver) OmniTRACS services.
1988	Second high-capacity land mobile communications system (MCS-L2) was introduced in Japan.
1990	Pegasus rocket launched from the wing of a B-52; the rocket injected its 423-lb payload into a 273×370 nmi $94°$ inclined orbit.
1993	GSM (now global system for mobile communications) in commercial use. After almost two decades of studies and experiments, sponsored by Canadian and U.S. tax payers, North American mobile satellite system MSAT is entering its realization stage, Field trials of CTx, DCT, CDMA, TDMA, FDMA mobile radio communications systems in progress worldwide.

near 30 MHz to 3 GHz. Then, in the 1970s, Ericsson introduced the NMT system, and AT&T Bell Laboratories introduced AMPS. The impact of these two public land mobile telecommunication systems on the standardization and prospects of mobile radio communications may be compared with the impact of Apple and IBM on the personal computer industry. In Europe, systems such as AMPS competed with NMT systems; in the rest of the world, AMPS, backed by Bell Laboratories' reputation for technical excellence and the clout of AT&T, became de facto and de jure the technical standard (British TACS and Japanese MCS-L1 are based on AMPS). In 1982, the **Conference of European Postal and Telecommunications Administrations (CEPT)** established GSM with the mandate to define future Pan-European cellular radio standards. On January 1, 1984, during the phase of explosive growth of AMPS and similar cellular mobile radio communications systems and services, came the divestiture (breakup) of AT&T.

77.4 Πάντα ρει (Panta Rhei)

Based on the solid foundation established in the 1970s, the buildup of mobile radio systems and services in 1990s is continuing at a 20–50% rate per year, worldwide. Terrestrial mobile radio systems offer analog voice and low-to-medium rate data services compatible with existing public switching telephone networks in scope but with poorer voice quality and lower data throughput. Satellite mobile radio systems currently offer analog voice, low-to-medium rate data, radio determination, and global distress safety services for travelers. By the end of 1988 (1994) there were about 2 (8) million cellular telephones in North America, and additional 2 (8) million in the rest of the world. There are about 40 million cordless phones and about 9 million pagers in North America and about the same number in the rest of the world. Considerable progress has been made in recent years [Davis et al. eds., 1984; Cox, Hirade, and Mahmoud eds., 1987; Mahmoud, Rappaport, and Öhrvik eds., 1989; Kucar ed., 1991; Rhee and Lee eds., 1991; Steele ed., 1992; Chuang et al. eds., 1993; Cox and Greenspan eds., 1995].

Equipment miniaturization and price are important constraints on the systems providing these services. In the early 1950s, mobile radio equipment used a considerable amount of a car's trunk space and challenged the capacity of a car's alternator/battery source, while in transmit mode; today, the pocket-size (7.7 oz ≈ 218 g) handheld cellular radio telephone provides 45 min of talk capacity. The average cost of the least expensive models of battery powered cellular mobile radio telephones has dropped proportionally and has broken the $500 U.S. barrier.

There is a rapidly expanding market of *portable* communications, primarily devoted to the *indoor* (in building, around building) environment. Today, these cordless (wireless, fiberless) radio systems offer telepoint services similar in scope to those provided by the public telephone booths; their objectives are to provide a broad range of services similar to ones currently offered by the **public switched telephone network (PSTN)** and the **integrated service digital network (ISDN)**.

Mobile satellite systems are expanding in many directions: large and powerful single unit geostationary systems; medium-sized, low-orbit multisatellite systems; and small-sized, low-orbit multisatellite systems, launched from a plane, [Kucar et al. eds., 1992; Del Re et al. eds., 1995].

The growth and profit potentials of the mobile radio communications market attracted the *big league* players (network, systems, and switching). This caused profound changes in research and development, standardization, and the decision-making processes in the mobile radio communications industry. In the search for El Dorado, the mobile radio communications industry is following two main paths: terrestrial and satellite. The terrestrial mobile radio pioneers, now accompanied by large marketing teams, favor existing cellular radio systems concepts; the newcomers with telephony, switching, and software backgrounds promote **cordless telephony (CT)**, PCN, and PCS; those with backgrounds in administration promote FPLMTS concepts. The satellite mobile radio pioneers build on existing and new *geostationary* satellite systems, whereas the newcomers promote *inclined orbit* concepts. The promoters of each concept may further be subdivided into analog and digital, FDMA, TDMA, and spread spectrum CDMA, etc.

77.5 Repertoire of Systems and Services

The variety of services offered to travelers essentially consists of information in analog and/or digital form. Although most of today's traffic consists of analog voice transmitted by analog frequency modulation FM (or phase modulation PM), digital signalling and a combination of analog and digital traffic might provide superior frequency reuse capacity, processing, and network interconnectivity. By using a powerful and affordable microprocessor and digital signal processing chips, a myriad of different services particularly well suited to the needs of people on the move could be realized economically. A brief description of a few elementary systems/services currently available to travelers will follow. Some of these elementary services can be combined within the mobile radio units for a marginal increase in the cost and complexity with the respect to the cost of a single service system; for example, a mobile radio communications system can include a positioning receiver, digital map, etc.

Terrestrial Systems

In a terrestrial mobile radio network, a repeater was usually located at the nearest summit, offering maximum service area coverage. As the number of users increased, the available frequency spectrum became unable to handle the increased traffic, and a need for frequency reuse arose. The service area was split into many small subareas called cells, and the term cellular radio was born. The frequency reuse offers an increased system capacity, whereas the smaller cell size can offer an increased service quality but at the expense of increased complexity of the user's terminal and network infrastructure. The tradeoffs between real estate availability (base stations) and cost, the price of equipment (base and mobile), network complexity, and implementation dynamics dictate the shape and the size of a cellular network.

Satellite Systems

These employ one or more satellites to serve as base station(s) and/or repeater(s) in a mobile radio network. The position of satellites relative to the service area is of crucial importance for the coverage, service quality, price, and complexity of the overall network. When a satellite encompasses the Earth in 24-h periods, the term *geosynchronous orbit* has been used. An orbit that is inclined with the respect to the equatorial plane is called an inclined orbit; an orbit with a 90° inclination is called a *polar orbit*. A circular geosynchronous 24-h orbit over the equatorial plane (0° inclination) is known as *geostationary orbit*, since from any point at the surface of the Earth the satellite appears to be stationary; this orbit is particularly suitable for the land mobile services at low latitudes and for maritime and aeronautical services at latitudes of <80°. Systems that use geostationary satellites include INMARSAT, MSAT, and AUSSAT. An elliptical geosynchronous orbit with the inclination angle of 63.4° is known as *tundra orbit*. An elliptical 12-h orbit with the inclination angle of 63.4° is known as *Molniya orbit*. Both tundra and Molniya orbits have been selected for the coverage of the northern latitudes and the area around the North Pole; for users at those latitudes, the satellites appear to wander around the zenith for a prolonged period of time. The coverage of a particular region (*regional coverage*) and the whole globe (*global coverage*) can be provided by different constellations of satellites including those in inclined and polar orbits. For example, inclined circular orbit constellations have been proposed for GPS (18–24 satellites, 55–63° inclination), Globalstar (48 satellites, 47° inclination), and Iridium (66 satellites, 90° inclination, polar orbits) systems; all three systems will provide global coverage. ORBCOM system employs Pegasus launchable low-orbit satellites to provide uninterrupted coverage of the Earth below ±60° latitudes and an intermittent but frequent coverage over the polar regions.

Satellite antenna systems can have one (*single-beam global system*) or more beams (*multibeam spot system*). The multibeam satellite systems similar to the terrestrial cellular system, employs antenna directivity to achieve better frequency reuse, at the expense of system complexity.

Radio Paging

This is a nonspeech, one-way (from base station toward travelers), personal selective calling system with alert and without message or with defined message, such as numeric or alphanumeric. A person wishing to send a message contacts a system operator by PSTN and delivers his message. After an acceptable time (queuing delay), a system operator forwards the message to the traveler by radio repeater (FM broadcasting transmitter, VHF or UHF dedicated transmitter, satellite, cellular radio system). After receiving the message, a traveler's small (roughly the size of a cigarette pack) receiver (pager) stores the message into its memory and on demand either emits alerting tones or displays the message.

Examples. The Swedish system uses a 57-kHz subcarrier on FM broadcasting transmitters. The United States systems employ 150-, 450-, and 800-MHz mobile radio frequencies. The RPC1 system used in the United Kingdom, United States, Australia, New Zealand, the People's Republic of China, and Finland employs 150-MHz mobile radio frequencies. The Japanese system operates around 250 MHz, etc.

Global Distress Safety System (GDSS)

Here geostationary and inclined orbit satellites transfer emergency calls sent by vehicles to the central Earth station. Examples are COSPAS, **search and rescue satellite aided tracking system (SARSAT)**, **geostationary operational environmental satellites (GOES)**, and **search and rescue satellite (SERES)**. The recommended frequency for this transmission is 406.025 MHz.

Global Positioning System (GPS)

The United States Department of Defense Navstar GPS 18–24 planned satellites in inclined orbits emit L band ($L1 = 1575.42$ MHz, $L2 = 1227.6$ MHz) spread spectrum signals from which an intelligent microprocessor-based receiver will be able to extract extremely precise time and frequency information and accurately determine its own three-dimensional position, velocity, and acceleration worldwide. The coarse accuracy of <100 m available to commercial users has been demonstrated by using a handheld receiver. An accuracy of meters or centimeters is possible by using the precise (military) codes and/or differential GPS (additional reference) principals. [ION, 1980, 1984, 1986, 1993]

Glonass

This is the Russian's counterpart of the United States's GPS. It uses frequencies between 1602.56 MHz and 1615.50 MHz to achieve goals similar to GPS. Other systems have been studied by the European Space Agency (Navsat) and by West Germany (Granas, Popsat, and Navcom).

LORAN-C

This is the 100-kHz frequency navigation system that provides a positional accuracy between 10–150 m. A user's receiver measures the time difference between the master station transmitter and secondary stations signals and defines his hyperbolic line of position. North American LORAN-C coverage includes the Great Lakes, Atlantic, and Pacific Coast, with decreasing signal strength and accuracy as the user approaches the Rocky Mountains from the east. Similar radionavigation systems are the 100-kHz Decca and 10-kHz Omega.

Inmarsat

This communications system consists of three operational geostationary payloads located at 26° W (Atlantic Ocean), 63° E (Indian Ocean), and 180° W (Pacific Ocean). The standard-A L band system, by employing a 0.79–1.95-m-diam pointing antenna and about 200 kg of above/below deck equipment, can provide analog voice telephony, telex, facsimile, up to 56 kb/s data, group call broadcasting, and emergency calls to maritime users. The Standard-B system will provide digital voice (about 9.6 kb/s), data, and telex services, by employing smaller equipment than standard A.

Mobile Radio: An Overview

The standard-C system, which employs a small antenna (about the size of a half-liter can) and a small transceiver (roughly the size of a telephone book directory) can offer up to 600 b/s data. Standard-M system is planned for land mobile and maritime mobile users, while aeronautical systems will provide data and voice services to the air travelers.

Volna

This is a Soviet system of satellites, which in conjunction with satellites and with L-band transponders on Raduga and Gorizont satellites, will provide voice and data services to a fleet of ships and aircrafts, worldwide.

Airphone

This is a public, fully automatic, air-to-ground telephone system that operates in the 900-MHz band using 6-kHz single-sideband (SSB) transmission. Each ground transceiver, by emitting an effective isotropic radiated power of 3 dBW, serves a cell with a radius of about 400 km. An aircraft uses two blade antennas, four transceivers each radiating 7 dBW, a telephone set, and an airborne computer that directs all call logistics.

Dispatch

This two-way radio land mobile or satellite system, with or without connection to the PSTN, consists of an operating center controlling the operation of a fleet of vehicles such as aircrafts, taxis, police cars, tracks, rail cars, etc. A summary of some of existing and planned terrestrial systems, including MOBITEX RAM and ARDIS, is given in Table 77.2. The OmniTRACS dispatch system employs Ku-band geostationary satellite located at 103° W to provide two-way digital message and position reporting (derived from incorporated satellite-aided LORAN-C receiver), throughout the contiguous U.S. (CONUS).

Cellular Radio or Public Land Mobile Telephone System

This offers a full range of services to the traveler that are equivalent to those provided by PSTN. Some of the operating cellular radio systems are: the North American AMPS, the Japanese land MCS-L1 and MCS-L2, the Nordic NMT-450 and NMT-900, the German C450, the Italian public land mobile

TABLE 77.2 Comparison of Dispatch Systems

Parameter	US	Sweden	Japan	Australia
TX freq. band, MHz				
Base	935–940	76.0–77.5	850–860	865.00–870.00
	851–866			415.55–418.05
Mobile	896–901	81.0–82.5	905–915	820.00–825.00
	806–821			406.10–408.60
Duplexing method	FDD/semi, full	FDD/semi	FDD/semi	FDD/semi, full
RF channel bw, kHz	12.5	25.0	12.5	25.0
	25.0			12.5
RF channel rate, kb/s	≤ 9.6	1.2	1.2	≤ 9.6
Number of traffic ch.	400	60 ?	799	200
	600			
Modulation type				
Voice	FM	FM	FM	FM
Data	FSK	MSK-FM	MSK-FM	FSK

Similar systems are used in the Netherlands, U.K., former U.S.S.R. and France. ARDIS is a commercial system compatible with U.S. specifications. MOBITEX/RAM is a commercial system compatible with U.S. specifications. *Source*: 4U Communications Research Inc., 1995.02.23–22:39, updated: 1994.10.31.

TABLE 77.3 Comparison of Cellular Mobile Radio Systems

Parameter	AMPS	MCS–L1 MCS–L2	NMT	C450	TACS	GSM	PCN	IS–54
TX freq., MHz								
Base	869–894	870–885	935–960	461–466	935–960	890–915	1710–1785	869–894
Mobile	824–849	925–940	890–915	451–456	890–915	935–960	1805–1880	824–849
Multiple access	FDMA	FDMA	FDMA	FDMA	FDMA	TDMA	TDMA	TDMA
Duplex method	FDD	FDD	FDD	FDD	FDD	FDD	FDD	FDD
Channel bw, kHz	30.0	25.0 / 12.5	12.5	20.0 / 10.0	25.0	200.0	200.0	30.0
Traffic channels per RF channel	1	1	1	1	1	8	16	3
Total traffic ch.	832	600 / 1200	1999	222 / 444	1000	125 × 8	375 × 16	832 × 3
Voice	analog	analog	analog	analog	analog	RELP	RELP	VSELP
Sylabic comp.	2:1	2:1	2:1	2:1	2:1	—	—	—
Speech rate, kb/s	—	—	—	—	—	13.0	6.7	8.0
Modulation	PM	PM	PM	PM	PM	GMSK	GMSK	$\pi/4$[1]
Peak dev., kHz	±12	±5	±5	±4	±9.5	—	—	—
Ch. rate, kb/s	—	—	—	—	—	270.8	270.8	48.6
Control	digital	digital	digital	digital	digital	digital	digital	digital
Modulation	FSK	FSK	FFSK	FSK	FSK	GMSK	GMSK	$\pi/4$
BB waveform	Manch.	Manch.	NRZ	NRZ	Manch.	NRZ	NRZ	NRZ
Peak dev., kHz	±8	±4.5	±3.5	±2.5	±6.4	—	—	—
Ch. rate, kb/s	10.0	0.3	1.2	5.3	8.0	270.8	270.8	48.6
Channel coding	BCH	BCH	B1	BCH	BCH	RS	RS	Conv.
Base→mobile	(40, 28)	(43, 31)	burst	(15, 7)	(40, 28)	(12, 8)	(12, 8)	1/2
Mobile→base	(48, 36)	a.(43, 31) p.(11, 07)	burst	(15, 7)	(48, 36)	(12, 8)	(12, 8)	1/2

[1] $\pi/4$ corresponds to the $\pi/4$ shifted differentialy encoded QPSK with $\alpha = 0.35$ square root raised-cosine filter.
Source: 4U Communications Research Inc., 1995.02.23–22:39, updated: 1994.10.31.

radio communication system at 450 MHz, the French radiotelephone multiservice network at 200, 400-MHz RADIOCOM 2000, and the United Kingdom's TACS. The technical characteristics of some of existing and planned systems are summarized in Table 77.3.

Cordless Telephony

The first generation of the U.K.'s cordless telephones (coded CT1) was developed as the answer to the large quantities of imported, technically superior, yet unlicensed mobile radio equipment. The simplicity and cost effectiveness of CT1 analog radio and base station products using eight RF channels and FDMA scheme stem from their applications limited to incoming calls from a limited number of mobile users to the isolated telepoints. As the number of users grew, so did the cochannel interference levels, while the quality of the service deteriorated. Anticipating this situation, the second generation digital cordless telecommunications radio equipment and *common air interface* standards (CT2/*CAI*), incompatible with the CT1 equipment, have been developed. CT2 schemes employ digital voice but the same FDMA principles as the CT1 schemes. Network and frequency reuse issues necessary to accommodate anticipated residential, business, and telepoint traffic growth have not been addressed adequately. Recognizing these limitations and anticipating the market requirements, different FDMA, TDMA, CDMA, and hybrid schemes aimed at cellular mobile and digital cordless telecommunications (DCT) services have been developed. The technical characteristics of some schemes are given in Table 77.4.

Mobile Radio: An Overview

TABLE 77.4 Comparison of Digital Cordless Telephone Systems

Parameter	CT2Plus	CT3	DECT	CDMA
Multiple access method	(F/T)DMA	TDMA	TDMA	CDMA
Duplexing method	TDD	TDD	TDD	FDD
RF channel bw, MHz	0.10	1.00	1.73	2×1.25
RF channel rate, kb/s	72	640	1152	1228.80
Number of traffic ch. per one RF channel	1	8	12	32
Burst/frame length, ms	1/2	1/16	1/10	n/a
Modulation type	GFSK	GMSK	GMSK	BPSK/QPSK
Coding	Cyclic, RS	CRC 16	CRC 16	Conv 1/2, 1/3
Transmit power, mW	≤ 10	≤ 80	≤ 100	≤ 10
Transmit power steps	2	1	1	many
TX power range, dB	16	0	0	≥ 80
Vocoder type	ADPCM	ADPCM	ADPCM	CELP
Vocoder rate, kb/s	fixed 32	fixed 32	fixed 32	up to 8
Max data rate, kb/s	32	ISDN 144	ISDN 144	9.6
Processing delay, ms	2	16	16	80
Reuse efficiency[3]				
Minimum	1/25	1/15	1/15	1/4
Average	1/15	1/07	1/07	2/3
Maximum	1/02[1]	1/02[1]	1/02[1]	3/4
Theor. number of vc. per cell and 10 MHz	100×1	10×8	6×12	4×32
Practical per 10 MHz				
Minimum	4	5–6	5–6	32 (08)[2]
Average	7	11–12	11–12	85 (21)
Maximum	50[1]	40[1]	40[1]	96 (24)

[1] The capacity (in the number of voice channels) for a single isolated cell.
[2] The capacity in parentheses may correspond to a 32 kb/s vocoder.
[3] Reuse efficiency and associate capacities reflect our own estimates.
Source: 4U Communications Research Inc., 1995.02.23–22:39 updated: 1994.10.31.

Future Public Land Mobile Telecommunications Systems

This is a huge international administrative project, for which tasks and objectives are presented in ITU-R Document 8–1/292. It discusses different terrestrial and satellite mobile radio communications and broadcasting systems, the transmission of data, voice, and images, at rates between 8–1920 kb/s, and a very broad range of services and technical and administrative issues.

Amateur Satellite Services

These started in 1965 when the OSCAR 3 satellite was launched; successive OSCAR/AMSAT satellites use 144-, 432-, 1270-, and 2400-MHz carrier frequencies. The Russia's Iskra satellites use 21/29 MHz and RS-3 satellites use 145/29-MHz carrier frequencies.

Vehicle Information System

This is a synonym for the variety of systems and services aimed toward traffic safety and location. This includes traffic management, vehicle identification, digitized map information and navigation, radio navigation, speed sensing and adaptive cruise control, collision warning and prevention, etc.

Some of the vehicle information systems can easily be incorporated in mobile radio communications transceivers to enhance the service quality and capacity of respective communications systems.

77.6 The Airwaves Management

The airwaves (frequency spectrum and the space surrounding us) are a limited natural resource shared among several different radio users (military, government, commercial, public, and amateur). Its sharing (among different users and services described in the preceding section; TV and sound broadcasting, etc.), coordination, and administration is an ongoing process exercised on national as well as on international levels. National administrations (**Federal Communications Commission, FCC**, in the U.S., **Department of Communications, DOC**, in Canada, etc.), in cooperation with users and industry, set the rules and procedures for planning and utilization of scarce frequency bands. These plans and utilizations have to be further coordinated internationally.

The **International Telecommunications Union (ITU)** is a specialized agency of the United Nations, stationed in Geneva, Switzerland, with more than 150 government members, responsible for all policies related to radio, telegraph, and telephone. According to the ITU, the world is divided into three regions: region 1—Europe including the former Soviet Union, Mongolia, Africa, and the Middle East west of Iran; region 2—the Americas, and Greenland; and region 3—Asia (excluding parts west of Iran and Mongolia), Australia, and Oceania. Historically, these three regions have developed, more or less independently, their own frequency plans, which best suit local purposes. With the advent of satellite services and globalization trends, the coordination between different regions becomes more urgent. Frequency spectrum planning and coordination is performed through ITU's bodies, such as Comité Consultatif de International Radio (CCIR), now ITU-R; International Frequency Registration Board (IFRB), now ITU-R; World Administrative Radio Conference (WARC); and Regional Administrative Radio Conference (RARC).

ITU-R, through its study groups, deals with technical and operational aspects of radio communications. Results of these activities have been summarized in the form of reports and recommendations published every four years or more [ITU, 1990]. The IFRB serves as a custodian of the common and scarce natural resource, the airwaves; in its capacity, the IFRB records radio frequencies, advises the members on technical issues, and contributes on other technical matters. Based on the work of ITU-R and the national administrations, ITU members convene at appropriate RARC and WARC meetings, where documents on frequency planning and utilization, the *radio regulations*, are updated. Actions on a national level follow; see ITU, 1986 and 1992.

The far-reaching impact of the mobile radio communications on economies and the well being of the three main trading blocks, other developing and third world countries, potential manufacturers and users makes the airways (frequency spectrum) even more important.

77.7 Operating Environment

While traveling, a customer i.e., user of cellular mobile radio system, may experience sudden changes in signal quality caused by his movements relative to the corresponding base station and surroundings, multipath propagation, and unintentional jamming, such as man–made noise, adjacent channel interference, and cochannel interference inherent to the cellular systems. Such an environment belongs to the class of nonstationary random fields, where experimental data is difficult to obtain and their behavior is hard to predict and model satisfactorily. When reflected signal components become comparable in level to the attenuated direct component and their delays comparable to the inverse of the channel bandwidth, *frequency selective fading* occurs. The reception is further degraded due to movements of a user, relative to reflection points and relay station, causing the Doppler frequency shifts. The simplified model of this environment is known as the *Doppler multipath Rayleigh channel*.

Mobile Radio: An Overview

The existing and planned cellular mobile radio systems employ sophisticated narrowband and wideband filtering, interleaving, coding, modulation, equalization, decoding, carrier and timing recovery, and multiple access schemes. The cellular mobile radio channel involves a *dynamic interaction* of signals arriving via different paths, adjacent and cochannel interference, and noise. Most channels exhibit some degree of memory; the description of which requires higher order statistics of (*spatial* and *temporal*) multidimensional random vectors (amplitude, phase, multipath delay, Doppler frequency, etc.) to be employed. This may require the evaluation of the usefulness of existing radio channel models and the eventual development of more accurate ones.

Cell engineering, prediction of service area and service quality, in an ever changing mobile radio channel environment, is a very difficult task. The average path loss depends on terrain microstructure within a cell, with considerable variation between different types of cells (i.e., urban, suburban, and rural environments). A variety of models based on experimental and theoretical work have been developed to predict path radio propagation losses in a mobile channel. Unfortunately, none of them are universally applicable. In almost all cases, an excessive transmitting power is necessary to provide an adequate system performance.

The first generation mobile satellite systems employ geostationary satellites (or payloads piggy backed on a host satellite) with small 18-dBi antennas covering the whole globe. When the satellite is positioned directly above the traveler (at zenith), a near constant signal environment, known as the *Gaussian channel*, is experienced. The traveler's movement relative to the satellite is negligible (i.e., Doppler frequency is practically equal zero). As the traveler moves—north or south, east or west—the satellite appears lower on the horizon. In addition to the direct path, many significant strength reflected components are present, resulting in a degraded performance. Frequencies of these components fluctuate due to movement of the traveler relative to the reflection points and the satellite. This environment is known as the *Doppler Ricean channel*. An inclined orbit satellite located for a prolonged period of time above 45° latitude north and 106° longitude west could provide travelers all over the U.S. and Canada, including the far North, a service quality unsurpassed by either geostationary satellite or terrestrial cellular radio. Similarly, a satellite located at 45° latitude north and 15° longitude east could provide travelers in Europe with improved service quality.

Inclined orbit satellite systems can offer a low startup cost, a near Gaussian channel environment, and improved service quality. Low-orbit satellites, positioned closer to the service area, can provide high-signal levels and short (a few milliseconds long) delays, and offer compatibility with the cellular terrestrial systems. These advantages need to be weighted against network complexity, intersatellite links, tracking facilities, etc.

77.8 Service Quality

The primary and the most important measure of service quality should be customer satisfaction. The customer's needs, both current and future, should provide guidance to a service offerer and an equipment manufacturer for both the system concept and product design stages. Acknowledging the importance of every single step of the complex service process and architecture, attention is limited here to a few technical merits of quality.

1. *Guaranteed quality* level is usually related to a percentage of the service area coverage for an adequate percentage of time.
2. *Data service quality* can be described by the average bit error rate (e.g., **BER** $< 10^{-5}$), packet BER (PBER $< 10^{-2}$), signal processing delay (1–10 ms), multiple access collision probability ($<20\%$), the probability of a false call (false alarm), the probability of a missed call (miss), the probability of a lost call (synchronization loss), etc.
3. *Voice quality* is usually expressed in terms of the mean opinion score (MOS) of subjective evaluations by service users. **MOS** marks are: bad = 0, poor = 1, fair = 2, good = 3,

and excellent = 4. MOS for PSTN voice service, pooled by leading service providers, relates the poor MOS mark to a signal–to–noise ratio (S/N) in a voice channel of $S/N \approx 35$ dB, whereas an excellent score corresponds to $S/N > 45$ dB. Currently, the users of the mobile radio services are giving poor marks to the voice quality associated with a $S/N \approx 15$ dB and an excellent mark for $S/N > 25$ dB. It is evident that there is a significant difference (20 dB) between the PSTN and mobile services. If digital speech is employed, both the speech and the speaker recognition have to be assessed. For more objective evaluation of speech quality under real conditions (with no impairments, in the presence of burst errors during fading, in the presence of random bit errors at BER = 10^{-2}, in the presence of Doppler frequency offsets, in the presence of truck acoustic background noise, in the presence of ignition noise, etc.), additional tests, such as the diagnostic acceptability measure (DAM), diagnostic rhyme test (DRT), Youden square rank ordering, Sino–Graeco–Latin square tests, etc., can be performed.

77.9 Network Issues and Cell Size

To understand ideas and technical solutions offered in existing schemes, and in proposals such as *cordless telephony (CT)*, DCT, PCS, PCN, etc., one need also to analyze the reasons for their introduction and success. Cellular mobile services are flourishing at an annual rate of 20–40%, worldwide. These systems (such as AMPS, NMT, TACS, MCS, etc.), use FDMA and digital modulation schemes for access, and command and control purposes and analog phase/frequency modulation schemes for the transmission of an analog voice. Most of the network intelligence is concentrated at fixed elements of the network including base stations, which seem to be well suited to the networks with a modest number of medium- to large-sized cells. To satisfy the growing number of potential customers, more cells and base stations were created by the cell splitting and frequency reuse process. Technically, the shape and size of a particular cell is dictated by the base station antenna pattern and the topography of the service area. Current terrestrial cellular radio systems employ cells with 0.5–50 km radius. The maximum cell size is usually dictated by the link budget, in particular, the gain of a mobile antenna and available output power. This situation arises in a rural environment, where the demand on capacity is very low and cell splitting is not economical. The minimum cell size is usually dictated by the need for an increase in capacity, in particular, in downtown cores. Practical constraints, such as real estate availability and price, and construction dynamics limit the minimum cell size to 0.5–2 km. In such types of networks, however, the complexity of the network and the cost of service grow exponentially with the number of base stations, whereas the efficiency of present handover procedures becomes inadequate.

Antennas with an omnidirectional pattern in a horizontal direction but with about 10 dBi gain in the vertical direction provide the frequency reuse efficiency of $N_{FDMA} = 1/12$. Base station antennas with similar directivity in the vertical direction and 60° directivity in the horizontal direction (a cell is divided into six sectors) can provide the reuse efficiency $N_{FDMA} = 1/4$, this results in a threefold increase in the system capacity; if CDMA is employed instead of FDMA, an increase in reuse efficiency $N_{FDMA} = 1/4 \rightarrow N_{CDMA} = 2/3$ may be expected.

Recognizing some of the limitations of existing schemes and anticipating the market requirements, the research in TDMA schemes aimed at cellular mobile and DCT services, and in CDMA schemes aimed towards mobile satellite system, cellular and personal mobile applications, has been initiated. Although employing different access schemes, TDMA (CDMA) network concepts rely on a smart mobile/portable unit that scans time slots (codes) to gain information on network behavior, free slots (codes), etc., improving frequency reuse and handover efficiency while hopefully keeping the complexity and cost of the overall network at reasonable levels. Some of the proposed system concepts depend on low-gain (0-dBi) base station antennas deployed in a license-free, uncoordinated fashion; small size cells (10–1000 m in radius) and an emitted isotropic radiated power of about 10 mW [+10 dB(1 mW)] per 100 kHz have been anticipated. A frequency reuse efficiency of

$N = 1/9$ to $N = 1/36$ has been projected for DCT systems. $N = 1/9$ corresponds to the highest user capacity with the lowest transmission quality, whereas $N = 1/36$ has the lowest user capacity with the highest transmission quality. This significantly reduced frequency reuse capability of proposed system concepts will result in significantly reduced system capacity, which need to be compensated by other means, including new spectra.

In practical networks, the need for a capacity (and frequency spectrum) is distributed unevenly in space and time. In such an environment, the capacity and frequency reuse efficiency of the network may be improved by dynamic channel allocation, where an increase in the capacity at a particular hot spot may be traded for the decrease in the capacity in cells surrounding the hot spot, the quality of the transmission, and network instability.

To cover the same area (space) with smaller and smaller cells, one needs to employ more and more base stations. A linear increase in the number of base stations in a network usually requires an exponential increase in the number of connections between base stations, switches, and network centers. These connections can be realized by fixed radio systems (providing more frequency spectra will be available for this purpose), or, more likely, by a cord (wire, cable, fiber, etc.).

The first generation geostationary satellite system antenna beam covers the entire Earth (i.e., the cell radius equals $\approx$6500 km). The second generation geostationary satellites will use larger multibeam antennas providing 10–20 beams (cells) with 800–1600 km radius. Low-orbit satellites such as Iridium will use up to 37 beams (cells) with 670 km radius. The third generation geostationary satellite systems will be able to use very large reflector antennas (roughly the size of a baseball stadium) and provide 80–100 beams (cells) with a cell radius of $\approx$200 km. If such a satellite is tethered to a position 400 km above the Earth, the cell size will decrease to $\approx$2 km in radius, which is comparable in size with today's small-size cell in terrestrial systems. Yet, such a satellite system may have the potential to offer an improved service quality due to its near optimal location with respect to the service area. Similarly to the terrestrial concepts, an increase in the number of satellites in a network will require an increase in the number of connections between satellites and/or Earth network management and satellite tracking centers, etc. Additional factors that need to be taken into consideration include price, availability, reliability, and timeliness of the launch procedures, a few large vs many small satellites, tracking stations, etc.

77.10 Coding and Modulation

The conceptual transmitter and receiver of a mobile system may be described as follows. The transmitter signal processor accepts analog voice and/or data and transforms (by analog and/or digital means) these signals into a form suitable for a double-sided suppressed carrier amplitude modulator (also called quadrature amplitude modulator, QAM). Both analog and digital input signals may be supported, and either analog or digital modulation may result at the transmitter output. Coding and interleaving can also be included. Very often, the processes of coding and modulation are performed jointly; we will call this joint process *codulation*. A list of typical modulation schemes suitable for transmission of voice and/or data over a Doppler affected Ricean channel, which can be generated by this transmitter, is given in Table 77.5.

Existing cellular radio systems such as AMPS, TACS, MCS, and NMT employ hybrid (analog and digital) schemes. For example, in access mode AMPS uses a digital codulation scheme (BCH coding and frequency-shift keying, FSK, modulation). While in the information exchange mode, the frequency modulated analog voice is merged with discrete SAT and/or ST signals and occasionally blanked to send a digital message. These hybrid codulation schemes exhibit a constant envelope and as such allow the use of dc power efficient nonlinear amplifiers. On the receiver side, these schemes can be demodulated by an inexpensive but efficient limiter/discriminator device. They require modest to high $C/N (= 10$–$20)$ dB, are very robust in adjacent (a spectrum is concentrated near the carrier) and cochannel interference (up to $C/I = 0$ dB, due to capture effect) cellular radio environment, and react quickly to the signal fade outages (no carrier, code, or frame

TABLE 77.5 Modulation Schemes, Glossary of Terms

Abbreviation	Description	Remarks/Use
ACSSB	Amplitude companded single sideband	Satellite transmission
AM	Amplitude modulation	Broadcasting
APK	Amplitude phase keying modulation	
BLQAM	Blackman quadrature amplitude modulation	
BPSK	Binary phase shift keying	Spread spectrum systems
CPFSK	Continuous phase frequency shift keying	
CPM	Continuous phase modulation	
DEPSK	Differentially encoded PSK (with carrier recovery)	
DPM	Digital phase modulation	
DPSK	Differential phase shift keying (no carrier recovery)	
DSB-AM	Double sideband amplitude modulation	
DSB-SC-AM	Double sideband suppressed carrier AM	Includes digital schemes
FFSK	Fast frequency shift keying (MSK)	
FM	Frequency modulation	Broadcasting, AMPS voice
FSK	Frequency shift keying	AMPS data and control
FSOQ	Frequency shift offset quadrature modulation	
GMSK	Gaussian minimum shift keying	GSM voice, data and control
GTFM	Generalized tamed frequency modulation	
HMQAM	Hamming quadrature amplitude modulation	
IJF	Intersymbol jitter free (SQORC)	
LPAM	L-ary pulse amplitude modulation	
LRC	LT symbols long raised cosine pulse shape	
LREC	LT symbols long rectangularly encoded pulse shape	
LSRC	LT symbols long spectrally raised cosine scheme	
MMSK	Modified minimum shift keying	
MPSK	M-ary phase shift keying	
MQAM	M-ary quadrature amplitude modulation	Subclass of DSB-SC-AM
MQPR	M-ary quadrature partial response	Radio-relay transmission
MQPRS	M-ary quadrature partial response system	$\equiv$ MQPR
MSK	Minimum shift keying	
m-h	Multi-h CPM	
OQPSK	Offset (staggered) quadrature phase shift keying	
PM	Phase modulation	Low-capacity radio
PSK	Phase shift keying	
QAM	Quadrature amplitude modulation	
QAPSK	Quadrature amplitude phase shift keying	
QPSK	Quadrature phase shift keying	$\equiv$ 4 QAM, low-capacity radio
QORC	Quadrature overlapped raised cosine	
SQAM	Staggered quadrature amplitude modulation	
SQPSK	Staggered quadrature phase shift keying	
SQORC	Staggered quadrature overlapped raised cosine	
SSB	Single sideband	Low- and high-capacity radio
S3MQAM	Staggered class-3 quadrature amplitude modulation	
TFM	Tamed frequency modulation	
TSI QPSK	Two-symbol-interval QPSK	
VSB	Vestigial sideband	TV
WQAM	Weighted quadrature amplitude modulation	Includes most digital schemes
XPSK	Crosscorrelated PSK	
$\pi/4$ QPSK	$\pi/4$ shift QPSK	IS-54 TDMA voice and data
3MQAM	Class-3 quadrature amplitude modulation	
4MQAM	Class-4 quadrature amplitude modulation	
12PM3	12-state PM with 3 b correlation	

Source: 4U Communications Research Inc., 1995.02.23–22:39, updated: 1994.10.31.

synchronization). Frequency selective and Doppler affected mobile radio channels will cause modest to significant degradations known as the random phase/frequency modulation.

Tightly filtered codulation schemes, such as $\pi/4$ QPSK additionally filtered by a square root raised-cosine filter, exhibit a nonconstant envelope, which demands (quasi) linear, less dc power efficient amplifiers to be employed. On the receiver side, these schemes require complex demodulation receivers, a linear path for signal detection, and a nonlinear one for reference detection—differential detection or carrier recovery. When such a transceiver operates in a selective fading multipath channel environment, additional countermeasures (inherently sluggish equalizers, etc.) are necessary to improve the performance i.e., reduce the bit-error-rate floor. These codulation schemes require modest $C/N (= 8$–$16)$ dB and perform modestly in adjacent and/or cochannel (up to $C/I = 8$ dB) interference environment.

Codulation schemes employed in spread spectrum systems use low-rate-coding schemes and mildly filtered modulation schemes. When equipped with sophisticated amplitude gain control on the transmit and receive side and robust rake receiver, these schemes can provide superior $C/N (= 4$–10 dB) and $C/I (<0$ dB) performance.

77.11 Speech Coding

Human vocal tract and voice receptors, in conjunction with language redundancy (coding), are well suited for face to face conversation. As the channel changes (e.g., from telephone channel to mobile radio channel), different coding strategies are necessary to protect the loss of information.

In (analog) companded phase modulation/frequency modulation (PM/FM) mobile radio systems, speech is limited to 4 kHz, compressed in amplitude (2:1), pre-emphasized, and phase/frequency modulated. At a receiver, inverse operations are performed. Degradation caused by these conversions and channel impairments results in lower voice quality. Finally, the human ear and brain have to perform the estimation and decision processes on the received signal.

In digital schemes sampling and digitizing of an analog speech (source) are performed first. Then, by using knowledge of the properties of the human vocal tract and of the language itself, a spectrally efficient source coding is performed. A high rate 64-kb/s, 56-kb/s, and adaptive differential pulse code modulation (ADPCM) 32-kb/s digitized voice complies with ITU-T recommendations for toll quality but may be less practical for the mobile environment. One is primarily interested in 8–16-kb/s rate speech coders, which might offer satisfactory quality, spectral efficiency, robustness, and acceptable processing delays in a mobile radio environment. A summary of the major speech coding schemes is provided in Table 77.6.

At this point, a partial comparison between analog and digital voice should be made. The quality of 64-kb/s digital voice, transmitted over a telephone line, is essentially the same as the original analog voice (they receive nearly equal MOS). What does this near equal MOS mean in a radio environment? A mobile radio conversation consists of one (mobile to home) or a maximum of two (mobile to mobile) mobile radio paths, which dictate the quality of the overall connection. The results of a comparison between analog and digital voice schemes in different artificial mobile radio environments have been widely published. Generally, systems that employ digital voice and digital codulation schemes seem to perform well under modest conditions, whereas analog voice and analog codulation systems outperform their digital counterparts in fair and difficult (near threshold, in the presence of strong cochannel interference) conditions. Fortunately, present technology can offer a viable implementation of both analog and digital systems within the same mobile/portable radio telephone unit. This would give every individual a choice of either an analog or digital scheme, better service quality, and higher customer satisfaction. Tradeoffs between the quality of digital speech, the complexity of speech and channel coding, as well as dc power consumption have to be assessed carefully and compared with analog voice systems.

TABLE 77.6 Digitized Voice, Glossary of Terms

ADM	Adaptive delta modulation
ADPCM	Adaptive differential pulse code modulation
ACIT	Adaptive code subband excited transform (GTE)
APC	Adaptive predictive coding
APC-AB	APC with adaptive bit allocation
APC-HQ	APC with hybrid quantization
APC-MQL	APC with maximum likelihood quantization
AQ	Adaptive quantization
ATC	Adaptive transform coding
BAR	Backward adaptive re-encoding
CELP	Code excited linear prediction
CVSDM	Continuous variable slope delta modulation
DAM	Diagnostic acceptability measure
DM	Delta modulation
DPCM	Differential pulse code modulation
DRT	Diagnostic rhyme test
DSI	Digital speech interpolation
DSP	Digital signal processing
HCDM	Hybrid companding delta modulation
LDM	Linear delta modulation
LPC	Linear predictive coding
MPLPC	Multi pulse LPC
MSQ	Multipath search coding
NIC	Nearly instantanous companding
PVXC	Pulse vector excitation coding
PWA	Predicted wordlength assignment
QMF	Quadrature mirror filter
RELP	Residual excited linear prediction
RPE	Regular pulse excitation
SBC	Subband coding
TASI	Time assigned speech interpolation
TDHS	Time domain harmonic scaling
VAPC	Vector adaptive predictive coding
VCELP	Vector code excited linear prediction
VEPC	Voice excited predictive coding
VQ	Vector quantization
VQL	Variable quantum level coding
VSELP	Vector–sum excited linear prediction
VXC	Vector excitation coding

Source: 4U Communications Research Inc., 1995.02.23–22:39, updated: 1994.10.31.

77.12 Macrodiversity and Microdiversity

Macrodiversity

In a cellular system, the base station is usually located in the barocenter of the service area (center of the cell). Typically, the base antenna is omnidirectional in azimuth but with about 6–10 dBi gain in elevation and serves most of the cell area (e.g., > 95%). Some parts within the cell may experience a lower quality of service because the direct path signal may be attenuated due to obstruction losses caused by buildings, hills, trees, etc. The closest neighboring (the first tier) base stations serve corresponding neighboring areas cells by using different sets of frequencies, eventually causing adjacent channel interference. The second closest neighboring (the second tier) base stations might use the same frequencies (frequency reuse) causing cochannel interference. If the same real estate (base stations) is used in conjunction with 120° directional (in azimuth) antennas, the designated area may be served by three base stations. In this configuration one base station serves three cells by using three 120° directional antennas. Therefore, the same number of existing base stations equipped with new directional antennas and additional combining circuitry is required to serve the same number of cells, yet in a different fashion. The mode of operation in which two or more base stations serve the same area is called *macrodiversity*. Statistically, three base stations are able to provide a better coverage of an area similar in size to the system with a centrally located base station. The directivity of a base station antenna (120° or even 60°) provides additional discrimination against signals from neighboring cells, therefore reducing adjacent and cochannel interference (i.e., improving reuse efficiency and capacity). Effective improvement depends on the terrain configuration and the combining strategy and efficiency. However, it requires more complex antenna systems and combining devices.

Microdiversity

Microdiversity is when two or more signals are received at one site (base or mobile).

Space diversity systems employ two or more antennas spaced a certain distance apart from one another. A separation of only $\lambda/2 = 15$ cm at $f = 1$ GHz, which is suitable for implementation on the mobile side, can provide a notable improvement in some mobile radio channel environments. Microspace diversity is routinely used on cellular base sites. Macrodiversity is also a form of space diversity.

Field-component diversity systems employ different types of antennas receiving either the electric or the magnetic component of an electromagnetic signal.

Frequency diversity systems employ two or more different carrier frequencies to transmit the same information. Statistically, the same information signal may or may not fade at the same time at different carrier frequencies. Frequency hopping and very wideband signalling can be viewed as frequency diversity techniques.

Time diversity systems are primarily used for the transmission of data. The same data is sent through the channel as many times as necessary, until the required quality of transmission is achieved (automatic repeat request, ARQ). *Would you please repeat your last sentence* is a form of time diversity used in a speech transmission.

The improvement of any diversity scheme is strongly dependent on the combining techniques employed, i.e., the selective (switched) combining, the maximal ratio combining, the equal gain combining, the feedforward combining, the feedback (Granlund) combining, majority vote, etc.

77.13 Multiplex and Multiple Access

Communications networks for travelers have two distinct directions: the *forward link* i.e., from the base station (via satellite) to the traveler, and the *return link* i.e., from a traveler (via satellite) to the base station. In the forward direction a base station distributes information to travelers according to the previously established protocol, i.e., no multiple access is involved. In the reverse direction many travelers make attempts to access one of the base stations. This occurs in so-called *control channels*, in a particular time slot, at a particular frequency, or by using a particular code. If collisions occur, customers have to wait in a queue and try again until success is achieved. If successful (i.e., no collision occurred), a particular customer will exchange (automatically) the necessary information for call setup. The network management center (NMC) will verify the customer's status, his credit rating, etc. Then, the NMC may assign a channel frequency, time slot, or code on which the customer will be able to exchange information with his correspondent. The optimization of the forward and reverse links may require different coding and modulation schemes and different bandwidths in each direction.

In forward link, there are three basic distribution (multiplex) schemes: one that uses discrimination in frequency between different users and is called *frequency division multiplex (FDM)*, another that discriminates in time and is called *time division multiplex (TDM)*, and the last having different codes based on spread spectrum signalling, that is known as *code division multiplex (CDM)*. It should be noted that hybrid schemes using a combination of basic schemes can also be developed.

In the reverse link, there are three basic access schemes: one that uses discrimination in frequency between different users and is called FDMA, another that discriminates in time and is called TDMA, and the last having different codes based on spread spectrum signalling that is known as CDMA. It should be noted that hybrid schemes using a combination of basic schemes can also be developed.

A performance comparison of multiple access schemes is a very difficult task. The strengths of FDMA schemes seem to be fully exploited in narrowband channel environments. To avoid the use of equalizers, channel bandwidths as narrow as possible should be employed; yet, in such narrowband channels the quality of service is limited by the maximal expected Doppler frequency and practical stability of frequency sources. Current practical limits are about 5 kHz.

The strengths of both TDMA and CDMA schemes seem to be fully exploited in wideband channel environments. TDMA schemes need many slots (and bandwidth) to collect information on network behavior. Once the equalization is necessary (at bandwidths >20 kHz), the data rate should be made as high as possible to increase frame efficiency and freeze the frame to ease equalization; yet, high-data rates require high-RF peak powers and a lot of signal processing power, which may be difficult to achieve in handheld units. Current practical bandwidths are about 0.1–1.0 MHz.

CDMA schemes need large spreading (processing) gains (and bandwidth) to realize spread spectrum potentials; yet, high-data rates require a lot of signal processing power, which may be difficult to achieve in handheld units. Current practical bandwidths are about 1.2 MHz. Narrow frequency bands seem to favor FDMA schemes, since both TDMA and CDMA schemes require more spectrum to fully develop their potentials. Once the adequate power spectrum is available, however, the later two schemes may be better suited for a complex (micro)cellular network environment. Multiple access schemes are also message sensitive. The length and type of message and the kind of service will influence the choice of multiple access, ARQ, frame and coding, among others.

77.14 System Capacity

The recent surge in the popularity of cellular radio and mobile service in general has resulted in an overall increase in traffic and a shortage of available system capacity in large metropolitan areas. Current cellular systems exhibit a wide range of traffic densities, from low in rural areas to overloading in downtown areas, with large daily variations between peak hours and quiet night hours. It is a great system engineering challenge to design a system that will make optimal use of the available frequency spectrum, offering a maximal traffic throughput (e.g., erlangs/megahertz/service area) at an acceptable service quality, constrained by the price and size of the mobile equipment. In a cellular environment, the overall system capacity in a given service area is a product of many (complexly interrelated) factors including the available frequency spectra, service quality, traffic statistics, type of traffic, type of protocol, shape and size of service area, selected antennas, diversity, frequency reuse capability, spectral efficiency of coding and modulation schemes, efficiency of multiple access, etc.

In the 1970s, so-called analog cellular systems employed omnidirectional antennas and simple or no diversity schemes offering modest capacity, which satisfied a relatively low number of customers. Analog cellular systems of the 1990s employ up to 60° sectorial antennas and improved diversity schemes; this combination results in a three- to fivefold increase in capacity. A further (twofold) increase in capacity can be expected from narrowband analog systems (25 kHz → 12.5 kHz); however, slight degradation in service quality might be expected. These improvements spurred the current growth in capacity, the overall success and prolonged life of analog cellular radio.

There are also numerous marketing results, where a 10- to 20-fold increase in capacity has been claimed. In this kind of campaign new (our) digital systems of the 21st century, operating under nice conditions, are usually compared with the old (their) systems of 1970s, operating under the worst conditions. There are plenty of ways of increasing the capacity of cellular radio, acquiring new frequency spectra is perhaps the easiest but not necessary the most cost effective way.

77.15 Conclusion

In this section, a broad repertoire of terrestrial and satellite systems and services for travelers is briefly described. The technical characteristics of the dispatch, cellular, and cordless telephony systems are tabulated for ease of comparison. Issues such as operating environment, service quality, network complexity, cell size, channel coding and modulation (codulation), speech coding, macro- and microdiversity, multiplex and multiple access, and the mobile radio communications system capacity are discussed.

Presented data reveals significant differences between existing and planned terrestrial cellular mobile radio communications systems and between terrestrial and satellite systems. These systems use different frequency bands, different bandwidths, different codulation schemes, different protocols, etc. (i.e., they are not compatible).

What are the technical reasons for this incompatibility? In this section, performance dependence on multipath delay (related to the cell size and terrain configuration), Doppler frequency (related to the carrier frequency, data rate, and the speed of vehicles), and message length (may dictate the choice of multiple access) are briefly discussed. A system optimized to serve the travelers in the Great Plains may not perform very well in mountainous Switzerland; a system optimized for downtown cores may not be well suited to a rural environment; a system employing geostationary (above equator) satellites may not be able to serve travelers at high latitudes very well; a system appropriate for slow moving vehicles may fail to function properly in a high-Doppler shift environment; a system optimized for voice transmission may not be very good for data transmission, etc. A system designed to provide a broad range of services to everyone, everywhere may not be as good as a system designed to provide a particular service in a particular local environment, as a decathlete world champion may not be as successful in competitions with specialists in particular disciplines.

There are plenty of opportunities, however, where compatibility between systems, their integration, and frequency sharing may offer improvements in service quality, efficiency, cost, and capacity (and therefore availability). Terrestrial systems offer a low startup cost and a modest cost per user in densely populated areas. Satellite systems may offer a high quality of the service and may be the most viable solution to serve travelers in scarcely populated areas, on oceans, and in the air. Terrestrial systems are confined to two dimensions, and radio propagation occurs in the near horizontal sectors. Barostationary satellite systems use the narrow sectors in the user's zenith nearly perpendicular to the Earth's surface having the potential for frequency reuse and an increase in the capacity in downtown areas during peak hours. A call setup in a forward direction (from the PSTN via base station to the traveler) may be a very cumbersome process in a terrestrial system when a traveler to whom a call is intended is roaming within an unknown cell; however, this is very easily realized in a global beam satellite system.

Defining Terms[1]

AMPS: Advanced mobile phone service
ASIC: Application specific integrated circuits
BER: Bit error rate

---

[1] *Source:* 4U, Communications Research Inc., 1995.02.23–22:39, updated: 1995.02.18.

CAD: Computer aided design
CB: Citizen band (mobile radio)
CDMA: Spread spectrum code division multiple access
CEPT: Conference of European Postal and Telecommunications (Administrations)
CT: Cordless telephony
DOC: Department of Communications (Canada)
DSP: Digital signal processing
FCC: Federal Communications Commission (U.S.)
FDMA: Frequency division multiple access
FPLMTS: Future public land mobile telecommunications systems
GDSS: Global distress safety system
GOES: Geostationary operational environmental satellites
GPS: Global positioning system
GSM: Groupe Spécial Mobile (now global system for mobile communications)
ISDN: Integrated service digital network
ITU: International Telecommunications Union
MMIC: Microwave monolitic integrated circuits
MOS: Mean opinion score
NMC: Network management center
NMT: Nordic mobile telephone (system)
PCN: Personal communications networks
PCS: Personal communications systems
PSTN: Public switched telephone network
SARSAT: Search and rescue satellite aided tracking system
SERES: Search and rescue satellite
TACS: Total access communication system
TDMA: Time division multiple access
WARC: World administrative radio conference

References

Chuang, J.C.-I., Anderson, J.B., Hattori, T., and Nettleton, R.W. eds. 1993. Special Issue on wireless personal communications. *IEEE J. Selected Areas in Comm.* Pt. I. 11(6).

Chuang, J.C.-I., Anderson, J.B., Hattori, T., and Nettleton, R.W. eds. 1993. Special Issue on wireless personal communications. *IEEE J. Selected Areas in Comm.* Pt. II. 11(7).

Cox, D.C., Hirade, K., and Mahmoud, S.A. eds. 1987. Special Issue on Portable and mobile communications. *IEEE J. Selected Areas in Comm.* 5(4).

Cox, D.C. and Greenstein, L.J. eds. 1995. Special Issue on Wireless personal communications. *IEEE Comm. Mag.* 33(1).

Davis, J.H., Mikulski, J.J., Porter, P.T., and King, B.L. eds. 1984. Special Issue on Mobile radio communications. *IEEE J. Selected Areas in Comm.* 2(4).

Del Re, E., Devieux, C.L., Jr., Kato, S., Raghavan, S., Taylor, D., and Ziemer, R. eds. 1995. Special Issue on Mobile satellite communications for seamless PCS. *IEEE J. on Selected Areas in Comm.* 13(2).

International Telecommunication Union. 1986. *Radio Regulations,* 1982 ed. rev. 1985 and 1986.

International Telecommunications Union, 1990. Mobile, Radio determination, Amateur and Related Satellite Services. Recommendations of the CCIR, (also Resolutions and Opinions). Vols. VIII, XVIIth Plenary Assembly, Düsseldorf, 1990.

International Telecommunication Union. 1992. Final Acts of the World Administrative Radio Conference for Dealing with Frequency Allocations in Certain Parts of the Spectrum (WARC-92) at Málaga-Torremolinos, Geneva.

ION Global Positioning System. 1980–1993. Reprinted by Inst. of Navigation. Vol. I. Washington, D.C, 1980; Vol. II. Alexandria, VA, 1984; Vol. III. Alexandria, VA, 1986; Vol. IV. Alexandria, VA, 1993.

Kucar, A.D. ed. 1991. Special Issue on Satellite and terrestrial systems and services for travelers. *IEEE Comm. Mag.* 29(11).

Kucar, A.D., Kato, S., Hirata, Y., and Lundberg, O. eds. 1992. Special Issue on Satellite systems and services for travelers. *IEEE J. on Selected Areas in Comm.* 10(8).

Mahmoud, S.A., Rappaport, S.S., and Öhrvik, S.O. eds. 1989. Special Issue on Portable and mobile communications. *IEEE J. on Selected Areas in Comm.* 7(1).

Reports of the CCIR, 1990 (also Decisions). Land Mobile Service, Amateur Service, Amateur Satellite Service, Annex 1 to Volumes VIII, XVIIth Plenary Assembly, Düsseldorf, 1990.

Reports of the CCIR, 1990 (also Decisions). Maritime Mobile Service. Annex 2 to Vol. VIII, XVIIth Plenary Assembly, Düsseldorf, 1990.

Rhee, S.B. and Lee, W.C.Y. eds. 1991. Special Issue on Digital cellular technologies. *IEEE Trans. on Vehicular Tech.* 40(2).

Steele, R. ed. 1992. Special Issue on PCS: The second generation. *IEEE Comm. Mag.* 30(12).

Further Information

This trilogy written by participants in AT&T Bell Labs projects on research and development in mobile radio is the Bible of diverse cellular mobile radio topics.

Jakes, W.C., Jr. ed. 1974. *Microwave Mobile Communications*, John Wiley & Sons, Inc. New York, 1974.

AT&T Bell Labs Technical Personnel, 1979. Advanced mobile phone service (AMPS). *Bell System Technical Journal*, 58(1).

Lee, W.C.Y. 1982. *Mobile Communications Engineering*, McGraw-Hill Book Co. New York, 1982.

An in-depth understanding of design, engineering, and use of cellular mobile radio networks, including PCS and PCN, requires knowledge of diverse subjects, such as three-dimensional cartography, electromagnetic propagation and scattering, antennas, analog and digital communications, project engineering, etc. The following is a list of books relating to these topics.

Balanis, C.A. 1982. *Antenna Theory Analysis and Design*, Harper & Row, New York, 1982.

Bowman, J.J., Senior, T.B.A., and Uslenghi, P.L.E. 1987. *Electromagnetic and Acoustic Scattering by Simple Shapes*, revised printing. Hemisphere Pub. Corp. New York, 1987.

Kucar, A. D. 1995. *Satellite and Terrestrial Radio Systems: Fixed, Mobile, PCS and PCN, Radio vs. Cable. A Practical Approach*, Stridon Press Inc.

Proakis, J.G., 1983. *Digital Communications*, McGraw-Hill Book Co. New York.

Sklar, B. 1988. *Digital Communications, Fundamentals and Applications*, Prentice-Hall Inc., Englewood Cliffs, NJ.

Snyder, J.P. 1987. *Map Projection—A Working Manual*, U.S. Geological Survey Professional Paper 1395, United States Government Printing Office, Washington DC, 2nd printing. 1989.

Spilker, J.J., Jr. 1977. *Digital Communications by Satellite*, Prentice-Hall Inc., Englewood Cliffs, NJ.

Van Trees, H. L., 1968–1971. *Detection, Estimation, and Modulation Theory*, Pt. I 1968, Pt. II and III, 1971, John Wiley & Sons, Inc., New York.

78
Base Station Subsystems

Chong Kwan Un
Korea Advanced Institute of Science and Technology

Chong Ho Yoon
Hankook Aviation University

78.1 Introduction 1090
78.2 System Architectures 1090
   Base Transceiver Subsystems (BTS) • Base Station Controller (BSC)
78.3 Analysis of Call Handling Schemes 1094
   System Modeling • Analysis of Call Handling Schemes [Yoon and Un, 1993] • Numerical Examples
78.4 Summary 1102

78.1 Introduction

In this chapter, a **base station subsystem** (**BSS**) providing the interface between a mobile switching center (MSC) and a mobile is described. To investigate the functions of a BSS, three system configurations of the BSS are presented according to the channel access mechanism, the most widely known being frequency division multiple access (FDMA), time division multiple access (TDMA), and code division multiple access (CDMA) [Ehrlich, Fisher, and Wingard, 1979; Uebayashi, Ohno, and Nojima, 1993; Qualcomm, 1992]. Also, a teletraffic performance model of BSS and numerical results are presented.

A subscriber in a cell originates a call via an idle channel available among a given number of radio channels to the BSS in a cell. When the subscriber enters an adjacent cell, the BSS tries to acquire an idle channel in the new cell. If there are one or more idle channels, the call is successfully handed off without a service breakdown. Otherwise, the call is forced to terminate before completion. Since the quality of the telephone service is enhanced if the rate of breakdown during a conversation is lower than the blocking of **originating calls** (OCs), a **handoff** call (HC) must have a priority over an OC. To reduce this probability, some schemes based on cutoff priority with a fixed number of **guard channels** (GCs) have been introduced by several researchers [Yoon and Un, 1993; Hong and Rappaport, 1986; El-Dolil, Wong, and Steele, 1989; Guerin, 1988]. In this chapter we also present three call handling schemes with and without guard channels.

78.2 System Architectures

A BSS consists of several **base transceiver subsystems** (**BTS**) and a **base station controller** (**BSC**). Several configurations of the BSS might be possible, as shown in Fig. 78.1. For suburb sites Figs. 78.1(a)–78.1(d) are preferred, and for downtown sites Fig. 78.1(c) is preferred.

To investigate the functions of BSS, the functional diagrams of three BSSs are shown in Fig. 78.2, according to the channel access mechanism. Figure 78.2(a) is ATT's BSS for advance mobile phone

Base Station Subsystems

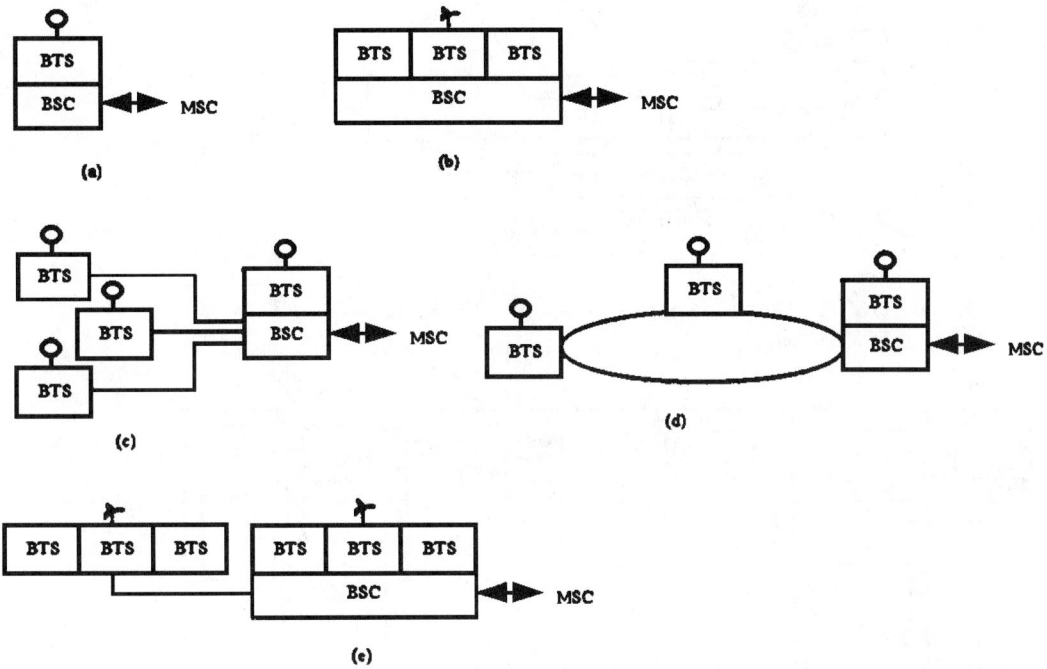

FIGURE 78.1 Configurations of BSS: (a) Combined omni, (b) combined star, (c) star, (d) ring, and (e) urban star configurations.

service (AMPS) using FDMA [Ehrlich, Fisher, and Wingard, 1979], Fig. 78.2(b) is NTT's BSS using TDMA [Uebayashi, Ohno, and Nojma, 1993], and Fig. 78.2(c) is Qualcomm's BSS using CDMA [1992].

Base Transceiver Subsystems (BTS)

A BTS provides RF radiation and reception with an appropriate channel access mechanism (e.g., FDMA, TDMA, or CDMA), and voice and data transmission interfaces between itself and the BSC. Typically, a BTS consists of several receive and transmit antennas, RF distributor, modulators and demodulators, and T1/E1 trunk line interfaces for voice and data traffic. For CDMA systems, a global positioning system (GPS) receive antenna is additionally included. Functions of each block are as follows.

1. *Antennas:* When each BTS functions in the omnidirectional mode, an omnidirectional transmit antenna and two-branch space-diversity receive antennas are used. If the BTS is configured in the directional mode with three 120° sectors, a directional transmit antenna and two receive antennas per each sector are used. The RF power is typically below 45 W. For the CDMA system, a GPS antenna is added which receives ticks of 1 pulse per second with a 10-MHz reference clock to generate system clocks.
2. *RF distributor:* It combines several carriers from amplifiers to transmit antennas with power amplifiers boosting the modulated signals to high-power signals. The transmit frequencies of FDMA, TDMA, and CDMA systems are in the range of 870–890 MHz, 940–956 MHz, and 869–894 MHz, respectively. The receive frequencies of FDMA, TDMA, and CDMA systems are in the range of 825–845 MHz, 810–826 MHz, and 824–849 MHz, respectively.
3. *Modulators and demodulators:* A modulator generates carrier signals of voice, supervisory audio tone (SAT), pilot, synch, and paging data. Each demodulator receives a two-diversity

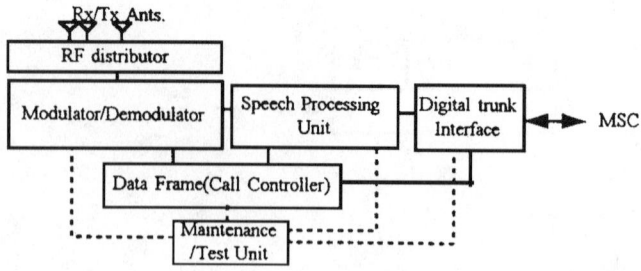

FIGURE 78.2(a) BSS structures with different channel access schemes: FDMA system.

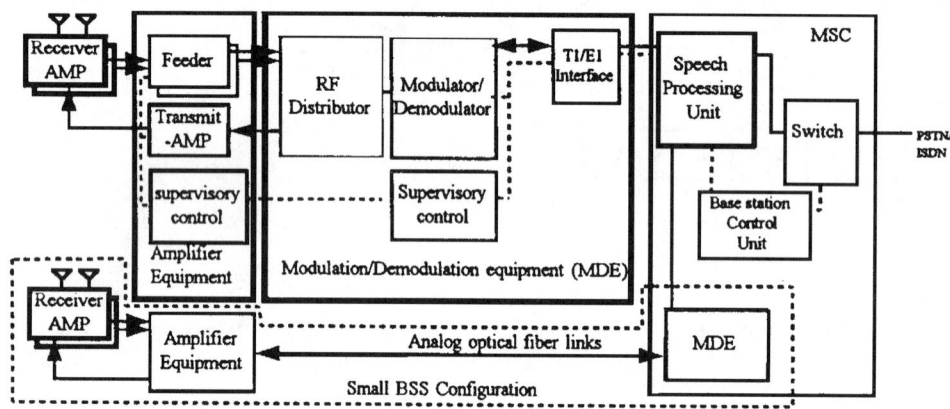

FIGURE 78.2(b) BSS structures with different channel access schemes: TDMA system.

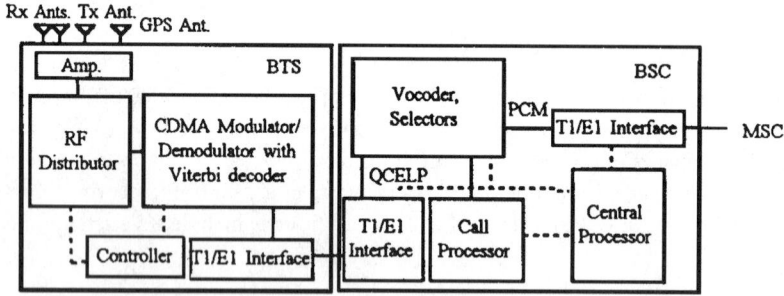

FIGURE 78.2(c) BSS structures with different channel access schemes: CDMA system.

input derived from the two receiving antennas. With these inputs and a local oscillator signal, it demodulates a baseband voice/SAT or data signals.

For the FDMA system, voice and data are modulated as phase modulation (PM) and frequency modulation (FM), respectively. The carrier spacing of the FDMA is 30 kHz, and the data transmission rate is 10 kb/s. The TDMA system uses a three-channel TDMA per carrier scheme with 42 kb/s $\pi/4$-shift quadrature differential phase shift keying (QDPSK) with a Nyquist filter whose rolloff factor is 0.5. The carrier spacing of the TDMA is 25 kHz. Thus, the bandwidth per channel is 8.3 kHz (=25 kHz/3). The CDMA system generates 19.2 Ksymbols per second and uses a convolutional encoder for either 9.6, 4.8, 2.4, or 1.2 kb/s

voice and data, and a Viterbi decoder for providing a forward error correction mechanism over the multipath fading wireless channel.
4. *Trunk interface:* It performs voice and data communication between BTS and BSC over digital links, operating at T1 or E1 rate. Typically, each slot per T1/E1 frame is allocated for delivering a single voice channel traffic. A 64 kb/s slot of the TDMA system, however, can carry three 11.2 kb/s voice channels, and the CDMA system uses a compressed voice packet transmission scheme with an HDLC frame format to increase the channel efficiency.

Base Station Controller (BSC)

A BSC locates mobiles to the cell with highest signal strength (handoff), and performs call setup, call supervision, and call termination. Also, it performs remotely ordered equipment testing, updates the location information of mobile stations, and provides data transmission interfaces between itself and the MSC. The BSC consists of speech processing units, a call controller, a central processor, a maintenance and test unit, and digital trunk interfaces. In particular, selectors providing soft handoffs are included in the CDMA BSC. Functions of each block are as follows.

1. *Speech processing unit/vocoder:* It provides the per channel audio-level speech and data paths interface between MSC and RF. Certain orders and request data signals can be added on the transmitter path before the modulator, and these data signals must be removed from the receiver path after the demodulator.

    The speech processing unit employs one of two signalling methods for sending the orders and requests signalling over a voice channel without interfering with voice conversation. The two methods are the blank-and-burst and dim-and-burst modes. With the blank-and-burst mode, the signals are sent in the form of a binary data message over the voice channel by momentarily muting the voice and inserting a binary data sequence. The data sequence requires approximately one-tenth of a second. However, the blank-and-burst signal is sent and replaces voice traffic temporarily. For the CDMA system, the dim-and-burst as well as the blank-and-burst mode can be used. When a vocoder desires to transmit at its maximum rate under the dim-and-burst mode, it is permitted to supply data at half of this rate. The remaining rate is used for signalling and overhead.

    For the digital cellular system, both pulse code modulation (PCM) digitized voice and data suffer from the coding used to increase the whole system capacity. The TDMA system uses a 11.2 kb/s vector sum excited linear prediction (VSELP) transmission code, which consists of 6.7 kb/s source traffic and additional preamble bits, sync bits, and control bits. In the CDMA system, the vocoder handles variable rate vocoding [transforms the Qualcomm developed code excited linear prediction code (QCELP) to PCM, or vice versa]. The vocoder has a variable rate that supports 8, 4, 2, and 1 kb/s operation and corresponds to channel rates of 9.6, 4.8, 2.4, and 1.2 kb/s. For example, a low-rate vocoder, running at 4 kb/s would increase the system capacity by a factor of 1.7 times with some degraded voice quality.

    For the TDMA and CDMA systems, echo cancellors are added, which eliminate echo due to a 2-W/4-W hybrid transformer in the public switch telephone network.

2. *Data frame/call controller:* It maintains an independent setup channel for the shared use of the BTS in communicating with all mobiles within its zone. Only data traffic is transmitted on the setup channel. Any mobile wishing to initiate a call monitors the forward setup channel (land to mobile). If the channel is idle, the mobile can transmit a call request or a page response. Otherwise, the mobile must wait a short time interval and monitor the channel again, until the forward channel is idle.

    The call processor of the FDMA system initiates the hard handoff procedure with the MSC. To determine when and if a handoff is necessary, a signal-strength measurement or a phase range measurement in the speech processing unit is made once every few seconds on each active voice channel. The soft handoffs between BTSs of the CDMA system are handled in selectors.

The call processors must also detect and control the signalling tone, off and on hooks, and transmitter power control. In addition, the processor provides a paging procedure to find a mobile station when a land-originating call is initiated.

3. *Central controller:* It allocates or de-allocates voice and data channels, communicates with the MSC and BTS, and controls the maintenance and test unit.
4. *Digital trunk interface:* It performs voice and data communication between BTS and BSC or between BSC and MSC over digital links, operating at T1 or E1 rate.
5. *Selector for the CDMA system:* The selector handles soft handoffs between BTSs. The soft handoff allows both the original cell and a new cell to serve the call temporarily during the handoff transition. The transition is from the originating cell to both cells and then to the new cell. Thus, it provides the make-before-break switching function. After a cell is initiated, the mobile continues to scan neighboring cells to determine if the signal from another cell becomes comparable to that of the original cell. Then, it sends a control message to the MSC, which states that the new cell is now strong and identifies the new cell. The MSC indicates the handoff by establishing a link to the mobile through the new cell while maintaining the old link. While the mobile is located in the transition region between the two cells, the call is supported by both cells until the mobile notifies the MSC that one of the links is no longer useful. This decision is performed by the selector, which selects an appropriate BTS among the current three engaged BTSs.

78.3 Analysis of Call Handling Schemes

System Modeling

Here, three call handling schemes for a BSS are presented and analyzed. It is assumed that the BSS can handle two types of calls (originating and prioritized handoff calls) and store these calls in a finite storage buffer. One scheme is a prioritized handoff scheme with GCs to reduce the blocking probability of HCs with a penalty on OCs. Under this scheme, handoff calls can be stored exclusively and access all free channels without restriction, whereas OCs have access to idle channels, except for a fixed number of guard channels. The other two schemes are prioritized handoff schemes without GCs to increase the total grade of service by reducing the blocking probability of OCs without a severe penalty on the grade of service for HCs. Under these two schemes without guard channels, both types of calls are allowed to be stored, and prioritized handoff calls push out originating calls if the buffer is full.

Figure 78.3 shows a BSS which has a set of C duplex channels with a finite call storage buffer of size $K - C, C \leq K$. Prioritized HCs and ordinary OCs are handled in the BSS. We assume that these calls have the same exponential service time distribution with rate of μ, but different arrival rates of the Poisson process, λ_1 and λ_2, respectively. Here, λ_1 is assumed to be proportional to λ_2 such that $\lambda_1 = P_h \lambda_2$, where P_h is the probability of having an HC. Then, the total and arrival rate λ_T is $\lambda_1 + \lambda_2$, and the total traffic load per channel ρ_T is $\lambda_T/C\mu$. A hexagonal cell shape is assumed for the system. The cell radius R for a hexagonal cell is defined as the maximum distance from the BSS to the cell boundary. With the cell radius, the average new call origination rate per cell is given by $\lambda_2 \equiv 3\sqrt{3/2}\lambda_0 R^2$, where λ_0 denotes the arrival rate of originating calls per unit area [Hong and Rapport, 1986]. Under these assumptions, we can treat the BSS as an $M/M/C/K$ priority queueing

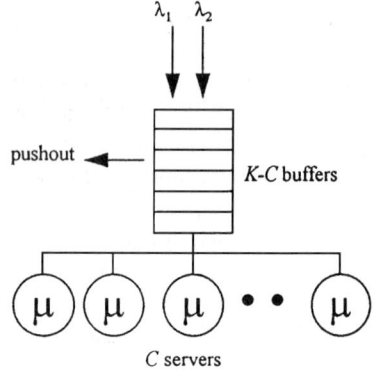

FIGURE 78.3 Base station model.

system which is formed by a set of C channels with a finite system (buffer plus channels) of size $K, C \leq K$.

Analysis of Call Handling Schemes [Yoon and Un, 1993]

Call Handling Scheme with Guard Channels (Scheme 1)

Under scheme 1, HCs can access C_g guard channels, $C_g \leq C$, exclusively among C channels. The remaining $(C - C_g)$ channels are shared by both types of calls. An OC is blocked if the number of available channels is less than or equal to C_g. If C channels have already been in use, $(K - C)$ HCs are buffered, but no buffering of OCs takes place. Hong and Rappaport [1986] analyzed this scheme with an infinite storage buffer for HCs. Here, we slightly modify their result to analyze our finite storage buffer model.

In Fig. 78.4, we show the state-transition-rate diagram for this scheme. Let P_j be the steady-state probability of a total of j calls being in the system. Using the state-transition-rate diagram, we obtain the probability P_j as

$$P_j = \begin{cases} \dfrac{(\lambda_1 + \lambda_2)^j}{j! \mu^j} P_0 & \text{for } 1 \leq j \leq C - C_g \\ \dfrac{(\lambda_1 + \lambda_2)^{(C-C_g)} \lambda_1^{(j-C+C_g)}}{j! \mu^j} P_0 & \text{for } C - C_g + 1 \leq j \leq C \\ \dfrac{(\lambda_1 + \lambda_2)^{(C-C_g)} \lambda_1^{(j-C+C_g)}}{C! \mu^C (C\mu)^{j-C}} P_0 & \text{for } C + 1 \leq j \leq K \end{cases} \quad (78.1)$$

where

$$P_0 = \left[\sum_{j=0}^{C-C_g} \dfrac{(\lambda_1 + \lambda_2)^j}{j! \mu^j} + \sum_{j=C-C_g+1}^{C} \dfrac{(\lambda_1 + \lambda_2)^{(C-C_g)} \lambda_1^{(j-C+C_g)}}{j! \mu^j} \right.$$

$$\left. + \sum_{j=C+1}^{K} \dfrac{(\lambda_1 + \lambda_2)^{(C-C_g)} \lambda_1^{(j-C+C_g)}}{C! \mu^C (C\mu)^{j-C}} \right]^{-1}$$

Since an HC is blocked only when the buffer is full, its blocking probability P_{B1} is given by $P_{B1} = P_K$. Also, the blocking probability of an OC, P_{B2}, is given by the sum of the probabilities that the number of calls in the system is larger than or equal to $(C - C_g)$, that is

$$P_{B2} = \sum_{j=C-C_g}^{K} P_j.$$

Using Little's formula for a steady-state queueing system [Kleinrock, 1975], we can obtain the

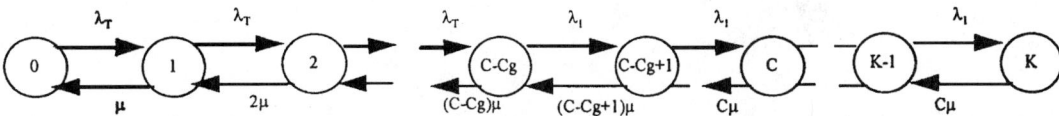

FIGURE 78.4 State-transition-rate diagram for scheme 1.

average waiting time of an HC, which is successfully served as

$$E[W_1] = \sum_{j=C}^{K} \frac{(j-C)P_j}{\lambda_1(1-P_{B1})}$$

Call Handling Schemes without Guard Channels

Here, we consider two prioritized call handling schemes without GCs as efficient call handling schemes for increasing the total grade of service (GOS).

Scheme 2. Under scheme 2, any type of calls has access to all channels as long as there is more than one free channel. When all channels are busy, these calls will be queued next to the newest HC in a buffer or will be queued at the first position of the buffer if there is no HC in the buffer. If an OC is in the position K and an arrival occurs, then it gets pushed out and the new arrival is queued next to the newest HC. When the system is full and the buffer has no more OCs, any type of arriving call is blocked. Accordingly, the HCs will be served by the first-in, first-out (FIFO) rule with the head-of-line priority basis, whereas the OCs will be served by the last-in, first-out (LIFO) rule with the pushout basis.

Let P_j, where $0 \leq j \leq K$, be the probability of having j calls in a basic $M/M/C/K$ system, and let $P_{j,k}$ be the probability of having j calls ($j = 0, 1, \ldots, K$) in the system and k OCs ($k = 0, 1, \ldots, j - C$) in the buffer. In Fig. 78.5, we show the state-transition-rate diagram for this

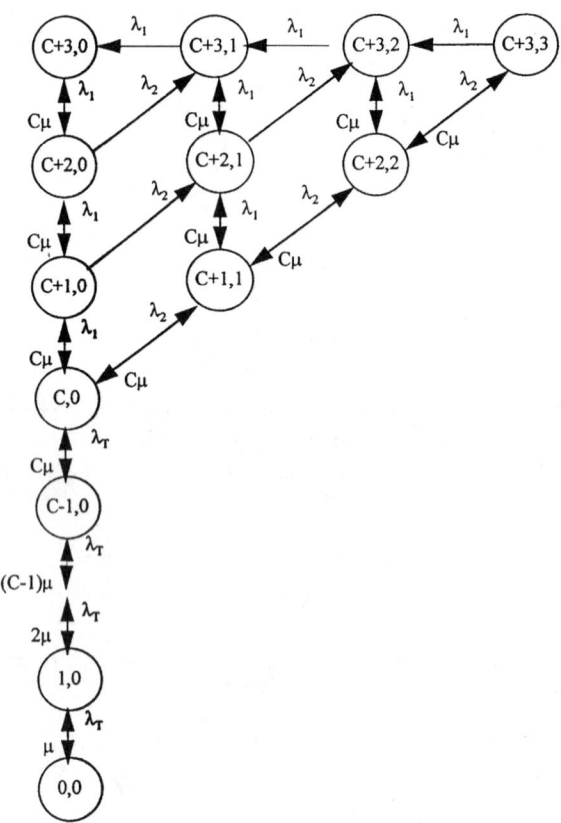

FIGURE 78.5 State-transition-rate diagrams for scheme 2 and scheme 3.

Base Station Subsystems

scheme. The states may be expressed by the following balanced equations:

$$(\lambda_T + C\mu)P_{j,k} = \lambda_2 P_{j-1,k-1} + \lambda_1 P_{j-1,k} + C\mu P_{j+1,k}$$
$$+ C\mu P_{j+1,k+1} \cdot \delta(j-k-C), \quad \text{for } C \leq j \leq K, k \leq j - C$$
$$C\mu P_{K,0} = \lambda_1 P_{K-1,0} + \lambda_2 P_{K,1}$$
$$(\lambda_1 + C\mu)P_{K,k} = \lambda_2 P_{K-1,k-1} + \lambda_1 P_{K-1,k} + \lambda_1 P_{K,k+1}, \quad \text{for } 1 \leq k \leq K - C$$

(78.2)

where $\delta(\cdot)$ is a Dirac delta function. The solutions for the part $j - k < 1$ or $k < 0$ are all zero, and those for the part $0 \leq j \leq C$ and $k = 0$ are $P_{j,0} = \rho_T^j/j! \cdot P_0$. Since an HC is blocked only when the system is full and the buffer has no more OCs, its blocking probability is given by $P_{B1} = P_{K,0}$.

Also, the blocking probability of an OC, P_{B2}, is given by the sum of the probability of blocking at its arrival and the probability of being pushed out caused by the arrival next to it. Since $\lambda_T P_K = \lambda_1 P_{B1} + \lambda_2 P_{B2}$, we can obtain P_{B2} as $P_{B2} = (\lambda_T P_K - \lambda_1 P_{B1})/\lambda_2$.

Next, we find the conditional waiting time probability distribution function (PDF) of each call getting served eventually. Since an HC gets served by the pure death process with rate of $C\mu$, its conditional waiting time PDF is given by

$$W_1(t) = W_1(0) + \left[\sum_{j=C}^{K-1} \sum_{k=0}^{j-C} P_{j,k} \int_{0+}^{t} \frac{C\mu(C\mu x)^{j-C-k}}{(j-C-k)!} e^{-C\mu x} dx \right.$$
$$\left. + \sum_{k=1}^{K-C} P_{K,k} \int_{0+}^{t} \frac{C\mu(C\mu x)^{K-C-k}}{(K-C-k)!} e^{-C\mu x} dx \right] \bigg/ (1 - P_{B1}) \quad t > 0$$

(78.3)

where

$$W_1(0) = \sum_{j=0}^{C-1} P_j/(1 - P_{B1})$$

The second term of Eq. (78.3) corresponds to the case when an HC arrives at the system that is not full, and it joins at the position $j - C - k + 1, C \leq j < K$. As a result, k OCs change their positions after the HC. The third term corresponds to the case when an HC arrives at the full system, and it joins at the position $K - C - k + 1, k > 0$. Thus, $(k - 1)$ OCs change their position after the HC and the oldest OC in the buffer gets pushed out. Since

$$\int_t^\infty \frac{m(mx)^k}{k!} e^{-mx} dx = \sum_{i=0}^{k} \frac{(mt)^i e^{-mt}}{i!}$$

we can rewrite Eq. (78.3) as

$$W_1(t) = 1 - \left[\sum_{j=C}^{K-1} \sum_{k=0}^{j-C} P_{j,k} \sum_{i=0}^{j-C-k} \frac{(C\mu t)^i}{i!} e^{-C\mu t} \right.$$
$$\left. + \sum_{k=1}^{K-C} P_{K,k} \sum_{i=0}^{K-C-k} \frac{(C\mu t)^i}{i!} e^{-C\mu t} \right] \bigg/ (1 - P_{B1}), \quad t > 0 \quad (78.4)$$

To find the conditional waiting time PDF of an OC that is served successfully, we consider an OC arriving at the position i in the system. The OC may move into the new position $i-1$ with probability $C\mu/(\lambda_T + C\mu)$ or the position $i+1$ with probability $\lambda_T/(\lambda_T + C\mu)$. After a number of transitions in the buffer, the OC will be served eventually if it enters one of C servers, or it will get pushed out if it moves to the position $K+1$. Let $u_{i,n}$ be the probability that an OC arrives at position i and gets served after n transitions. It can be obtained using the method of random walk with absorbing barriers at C and $K+1$ as [Feller, 1968]

$$u_{i,n} = \sum_{k=0}^{\infty} \left(\frac{q}{r}\right)^{ka} \cdot w_{2ka+(i-C),n} - \sum_{k=0}^{\infty} \left(\frac{q}{r}\right)^{ka-(i-C)} \cdot w_{2ka-(i-C),n} \qquad (78.5)$$

where

$$q = \lambda_T/(\lambda_T + C\mu), \qquad r = 1-q, \qquad a = K+1-C,$$

$$w_{i,n} = \frac{i}{n} \cdot \binom{n}{(n+i)/2} \cdot q^{(n-i)/2} \cdot r^{(n+i)/2}$$

and where $n \geq i$ and n of $w_{i,n}$ should have the same parity.

Also, let $g_{i,t}$ be the probability that an OC arrives at the position i and gets served after t. Since the time during which it completes after n transitions in the buffer has a gamma distribution of order $(n-1)$, it is obtained as

$$g_{i,t} = \sum_{n=1}^{\infty} u_{i,n} \cdot \frac{(\lambda_T + C\mu)^n t^{n-1}}{(n-1)!} \cdot e^{-(\lambda_T + C\mu)t}, \qquad C < i \leq K \qquad (78.6)$$

When the system size is j ($j < K$) with k ($k \geq 0$) OCs in the buffer, an OC arrives at the position $j-k+1$. If the system size is K with k ($k > 0$) OCs in the buffer, the OC arrives at the position $K-k+1$ by pushing out the oldest OC in the buffer. Thus, the conditional waiting time PDF of the OC which gets served is obtained with

$$W_2(0) = \sum_{j=0}^{C-1} P_{j,0}/(1-P_{B2})$$

as

$$W_2(t) = W_2(0)$$

$$+ \frac{\sum_{j=C}^{K-1}\sum_{k=0}^{j-C} P_{j,k} \int_{0+}^{t} g_{j-k+1,x} dx + \sum_{k=1}^{K-C} P_{K,k} \int_{0+}^{t} g_{K-k+1,x} dx}{1 - P_{B2}}, \qquad t > 0$$

$$= 1 - \left[\sum_{j=C}^{K-1}\sum_{k=0}^{j-C} P_{j,k} \sum_{n=1}^{\infty} u_{j-k+1,n} \sum_{i=0}^{n-1} \frac{[(\lambda_T + C\mu)t]^i}{i!} \cdot e^{-(\lambda_T + C\mu)t} \right.$$

$$\left. + \sum_{k=1}^{K-C} P_{K,k} \sum_{n=1}^{\infty} u_{K-k+1,n} \sum_{i=0}^{n-1} \frac{[(\lambda_T + C\mu)t]^i}{i!} \cdot e^{-(\lambda_T + C\mu)t} \right] \bigg/ (1 - P_{B2}),$$

$$t > 0 \qquad (78.7)$$

Base Station Subsystems

Scheme 3. Under scheme 3, any type of call has access to all channels as long as there is more than one free channel. When all channels are busy, HCs are served as under scheme 2. However, OCs will be queued next to the newest OC in a buffer or will be queued at the first position of the buffer, if the buffer is empty. When the system is full, an arriving OC is blocked. In addition, if an OC is in the position K and an HC arrives, then the OC gets pushed out and the new arrival is queued next to the newest HC. Thus, the HCs will be served by the FIFO rule with the head-of-line priority basis as done under scheme 2, whereas the OCs will be served by the FIFO rule with the pushout basis.

The state-transition-rate diagram for scheme 3 is the same as scheme 2. Thus, the blocking probabilities of HCs and OCs for scheme 3 are the same as scheme 2. Since the waiting time distribution of an HC is independent of the service rule of OCs, it is also obtained from Eq. (78.4). However, the waiting time distribution of an OC is different from the result of scheme 2 as follows.

Noting that an HC arriving only forces a move of the position of OC in i to the new position $i + 1$, we can obtain the probability of $u_{i,n}$ that an OC arrives at position i in the system and gets served after n transitions by using the result of Eq. (78.5) with $q = \lambda_1/(\lambda_1 + C\mu)$. Also, since the number of transitions in the buffer depends on the arrival rate of HCs and the service rate, the probability that an OC is in position i and the OC gets served after t is obtained from Eq. (78.6) as

$$g_{i,t} = \sum_{n=1}^{\infty} u_{i,n} \cdot \frac{(\lambda_1 + C\mu)^n t^{n-1}}{(n-1)!} \cdot e^{-(\lambda_1 + C\mu)t}, \quad C < i \leq K$$

When the system size is j, $j < K$, an OC arriving can join at position $j+1$ without considering the number of OCs in the buffer. It is blocked, however, when the system is full. Thus, the conditional waiting time PDF of the OC which gets served is obtained with

$$W_2(0) = \sum_{j=0}^{C-1} P_{j,0}$$

as

$$W_2(t) = W_2(0) + \left[\sum_{j=C}^{K-1} \sum_{k=0}^{j-C} P_{j,k} \int_{0+}^{t} g_{j+1,x} dx \right] \Big/ (1 - P_{B2})$$

$$= 1 - \frac{\sum_{j=C}^{K-1} \sum_{k=0}^{j-C} P_{j,k} \sum_{n=1}^{\infty} u_{j+1,n} \sum_{i=0}^{n-1} \frac{((\lambda_1 + C\mu)t)^i}{i!} \cdot e^{-(\lambda_1 + C\mu)t}}{(1 - P_{B2})}$$

$$t > 0 \quad (78.8)$$

Numerical Examples

In this section, some numerical examples are given to show the performance characteristics of the three call handling schemes for a personal portable radio telephone system. In these numerical examples, parameters are set as follows: $K = 30$, $C = 20$, $R = 0.8$ km, and time is normalized by the average holding time $(1/\mu)$.

In Fig. 78.6, we show the blocking probabilities of OCs and HCs vs the rate of OCs per unit area λ_0/km^2 and the total traffic load per channel ρ_T for the three schemes with the handoff probability $P_h = 0.1$. For P_{B2}, it is observed that the call handling schemes without GCs (schemes 2 and 3) are superior to the one with GCs (scheme 1), but the opposite holds for P_{B1}. This tradeoff becomes manifest as the number of guard channels C_g increases. The higher blocking probability of HCs under schemes without GCs, however, may be sufficiently small to be acceptable in the case where a typical requirement of P_{B2} is 0.5.

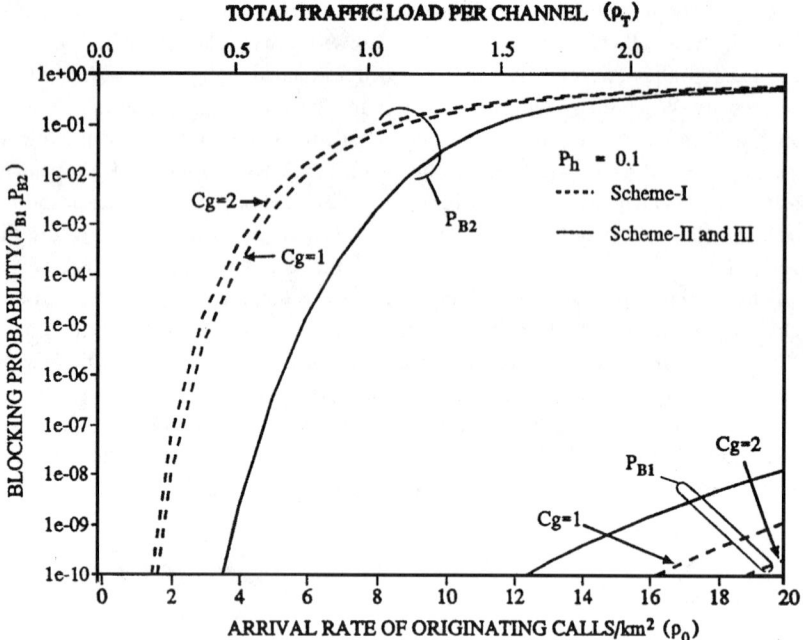

FIGURE 78.6 Comparison of blocking probabilities of three call handling schemes for $P_h = 0.1$.

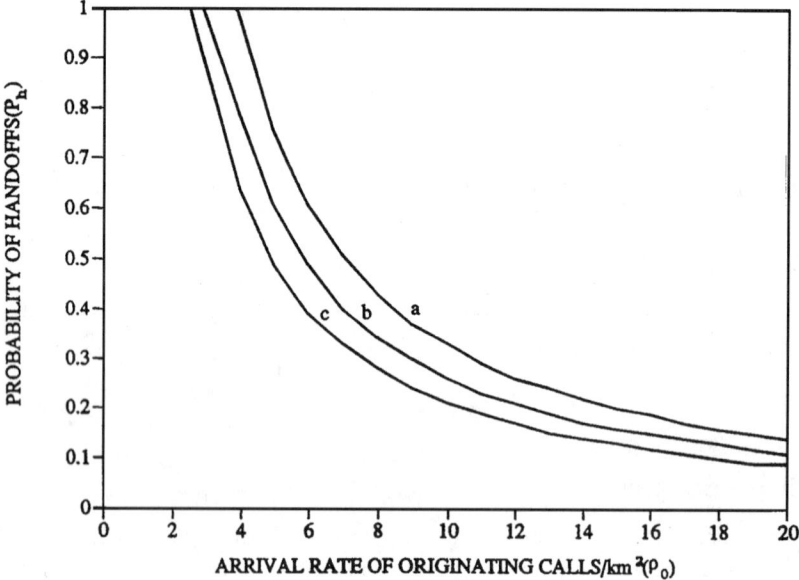

FIGURE 78.7 Boundaries between scheme 1 with a single GC and a scheme without GCs (scheme 2 or 3) for $P_h = 0.1$ and the value of $(1 - \alpha)$ equal to (a) 10^{-5}, (b) 10^{-6}, (c) 10^{-7}.

To make the choice of call handling schemes from the system provider's perspective by neglecting the waiting time of calls, we here define a cost function of P_{B1} and P_{B2}, which indicates the relative importance of blocking for HCs and OCs, as $CF = \alpha \cdot P_{B1} + (1 - \alpha) \cdot P_{B2}$, where α is in the interval $[0, 1]$. In Fig. 78.7, we show boundaries for the choice between the call handling schemes without GCs (schemes 2 and 3) and the one with single GC (scheme 1), which are obtained by

Base Station Subsystems

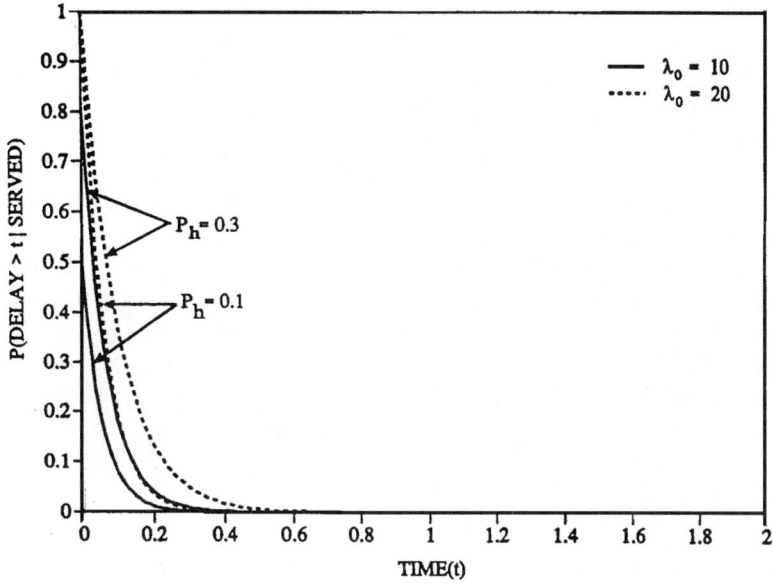

FIGURE 78.8 Comparison of waiting time distributions of HCs under schemes without guard channels, schemes 2 and 3.

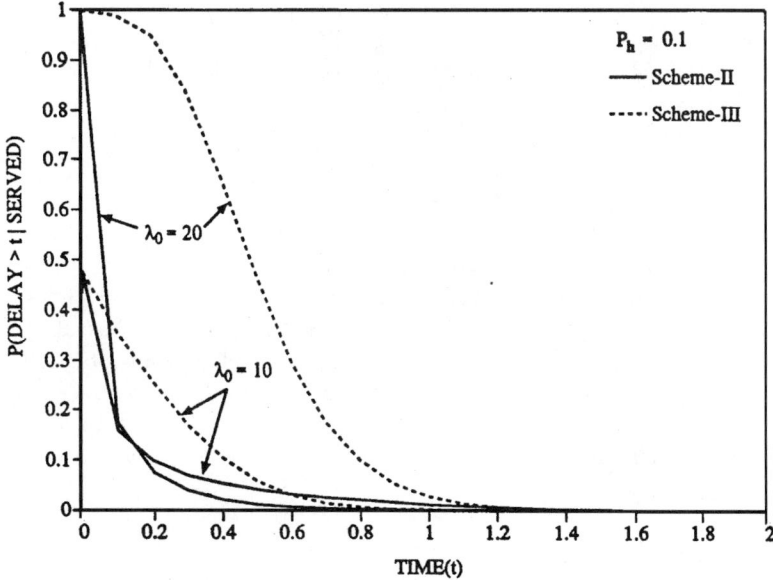

FIGURE 78.9 Comparison of waiting time distributions of OCs for $P_h = 0.1$ under schemes without guard channels, schemes 2 and 3.

comparing each CF. For given P_h, λ_0, and α, the region to the left of each curve is preferred to the call handling schemes without GCs, whereas that to the right of the curve is preferred to the one with a single GC. From the result, the call handling schemes without GCs (schemes 2 and 3) have an advantage over the one with GCs for a typical cell in the residential area, where the handoff probability is low.

We now show the waiting time distributions of OCs and HCs. Since the waiting time of OCs can be neglected in a typically lightly loaded cell, we here consider a heavy loaded environment (the total traffic load per channel is larger than 1). In Fig. 78.8, the waiting time distributions of HCs under schemes without guard channels (schemes 2 and 3) is shown. Since HCs are served with priority over OCs, the results of schemes 2 and 3 are the same. Although an HC suffers a longer delay as P_h or λ_0 increases, the delay is sufficiently short to be acceptable.

In Fig. 78.9, the waiting time distributions of OCs under schemes 2 and 3 with $P_h = 0.1$ are shown. From these results, one can find that scheme 2 shows a shorter average waiting time than scheme 3, although it has a longer tail of waiting time distribution. We also note that the average delay for $\lambda_0 = 20$ is shorter, relative to $\lambda_0 = 10$, under scheme 2. This effect is explained by the fact that as λ_0 increases, OCs that do not get served quickly are more likely to get pushed out. Considering the results of schemes without guard channels and the one with guard channels, one can find a tradeoff between the waiting time of an OC and its blocking probability. This tradeoff can be managed to satisfy both the system provider and subscribers by controlling the finite buffer length.

78.4 Summary

In this chapter we have described three system configurations of the base station subsystems for cellular mobile communication networks. We have observed that the prioritized pushout call handling schemes without guard channels reduce the call blocking probability of ordinary calls without a severe penalty on prioritized handoff calls. Also, with a cost function from the system provider's perspective, we have shown that two prioritized pushout call handling schemes without guard channels are better than the scheme with guard channels. By examining the waiting time distribution and average waiting time of originating calls under a heavy loaded environment, we have also shown that the scheme 2 performs better than the scheme 3. Thus, the scheme 2 can be a good candidate as a call handling scheme for a typical cell in a residential area where the handoff probability is low.

Defining Terms

Base station controller: A system for managing BTSs.
Base station subsystem: A system that consists of BTSs and a BSC.
Base transceiver subsystem: A system with RF transceivers and interfaces between BSC and itself.
Guard channels: Several reserved channels used by higher priority calls exclusively.
Handoff: The procedure initiated by a base station when the mobile unit moves out of the coverage area of the base station without interrupting the call. Typically, the handoff calls have a priority over originating calls.
Originating calls: Calls initiated in a cell.

References

Ehrlich, N., Fisher, R., and Wingard, T. 1979. Cell-site hardware, *Bell Sys. Tech. J.* 58(1):153–159.
El-Dolil, S.A., Wong, W., and Steele, R. 1989. Teletraffic performance of highway microcells with overlay macrocell. *IEEE J. Select. Areas Commun.* 7(1):71–78.
Feller, W. 1968. *An Introduction to Probability Theory and its Application*, Vol. 1, John Wiley and Sons, New York, pp. 349–370.
Guerin, R. 1988. Queueing-blocking system with two arrival streams and guard channels. *IEEE Trans. Commun.* COM-36(2):153–163.
Hong, D.H. and Rappaport, S.S. 1986. Traffic model and performance analysis for cellular mobile radio telephone systems with prioritized and nonprioritized handoff procedures. *IEEE Trans. Vech. Tech.* VT-28 (3):77–92.

Kleinrock, L. 1975. *Queueing System*, Vol. I, John Wiley and Sons, New York.
Qualcomm. 1992. *The CDMA Network Engineering Handbook*, Nov.
Uebayashi, S., Ohno, K. and Nojima, T. 1993. Development of TDMA cellular base station equipment, *43rd Vehicular Tech. Conf.* Secaucus, NJ, pp. 566–569.
Yoon, C.H. and Un, C.K. 1993. Performance of personal radio telephone system with and without guard channels. *IEEE J. Select. Areas Commun.* 11(6):911–917.

Further Information

For a more general description on microcell systems, see the paper by Sanrnecki, Vinodral, Javed, O'Kelly, and Dick in *IEEE Communications Magazine*, 1993, Vol. 31, No. 4.

The paper by Pavlidou presents the numerical method for finding the delay and blocking probabilities of the voice and data integrated cellular mobile system, in *IEEE Transactions on Communications*, 1994, Vol. 42, No. 2/3/4.

79

Access Methods

Bernd-Peter Paris
George Mason University

79.1 Introduction 1104
79.2 Relevant Wireless Communication System Characteristics ... 1106
79.3 Frequency Division Multiple Access 1107
   Channel Considerations • Influence of Antenna Height • Example 79.1: CT2 • Further Remarks
79.4 Time Division Multiple Access 1109
   Propagation Considerations • Initial Channel Assignment • Example 79.2: GSM • Further Remarks
79.5 Code Division Multiple Access 1112
   Propagation Considerations • Multiple-Access Interference • Further Remarks
79.6 Comparison and Outlook 1113

79.1 Introduction

The radio channel is fundamentally a broadcast communication medium. Therefore, signals transmitted by one user can potentially be received by all other users within range of the transmitter. Although this high connectivity is very useful in some applications, like broadcast radio or television, it requires stringent access control in wireless communication systems to avoid, or at least to limit, interference between transmissions. Throughout, the term wireless communication systems is taken to mean communication systems that facilitate two-way communication between a portable radio communication terminal and the fixed network infrastructure. Such systems range from mobile cellular systems through personal communication systems (PCS) to cordless telephones.

The objective of wireless communication systems is to provide communication channels on demand between a portable radio station and a radio port or base station that connects the user to the fixed network infrastructure. Design criteria for such systems include **capacity**, cost of implementation, and quality of service. All of these measures are influenced by the method used for providing multiple-access capabilities. However, the opposite is also true: the access method should be chosen carefully in light of the relative importance of design criteria as well as the system characteristics.

Multiple access in wireless radio systems is based on insulating signals used in different connections from each other. The support of parallel transmissions on the uplink and downlink, respectively, is called multiple access, whereas the exchange of information in both directions of a connection is referred to as **duplexing**. Hence, multiple access and duplexing are methods that facilitate the sharing of the broadcast communication medium. The necessary insulation is achieved by assigning to each transmission different components of the domains that contain the signals. The signal domains commonly used to provide multiple access capabilities include the following.

Access Methods

Spatial domain: All wireless communication systems exploit the fact that radio signals experience rapid attenuation during propagation. The propagation exponent ρ on typical radio channels lies between $\rho = 2$ and $\rho = 6$ with $\rho = 4$ a typical value. As signal strength decays inversely proportional to the ρth power of the distance, far away transmitters introduce interference that is negligible compared to the strength of the desired signal. The cellular design principle is based on the ability to reuse signals safely if a minimum reuse distance is maintained. Directional antennas can be used to enhance the insulation between signals. We will not focus further on the spatial domain in this treatment of access methods.

Frequency domain: Signals which occupy nonoverlapping frequency bands can be easily separated using appropriate bandpass filters. Hence, signals can be transmitted simultaneously without interfering with each other. This method of providing multiple access capabilities is called **frequency-division multiple access (FDMA)**.

Time domain: Signals can be transmitted in nonoverlapping time slots in a round-robin fashion. Thus, signals occupy the same frequency band but are easily separated based on their time of arrival. This multiple access method is called **time-division multiple access (TDMA)**.

Code domain: In **code-division multiple access (CDMA)** different users employ signals that have very small cross-correlation. Thus, correlators can be used to extract individual signals from a mixture of signals even though they are transmitted simultaneously and in the same frequency band. The term code-division multiple-access is used to denote this form of channel sharing. Two forms of CDMA are most widely employed and will be described in detail subsequently, frequency hopping (FH) and direct sequence (DS).

System designers have to decide in favor of one, or a combination, of the latter three domains to facilitate multiple access. The three access methods are illustrated in Fig. 79.1. The principal idea in all three of these access methods is to employ signals that are orthogonal or nearly orthogonal.

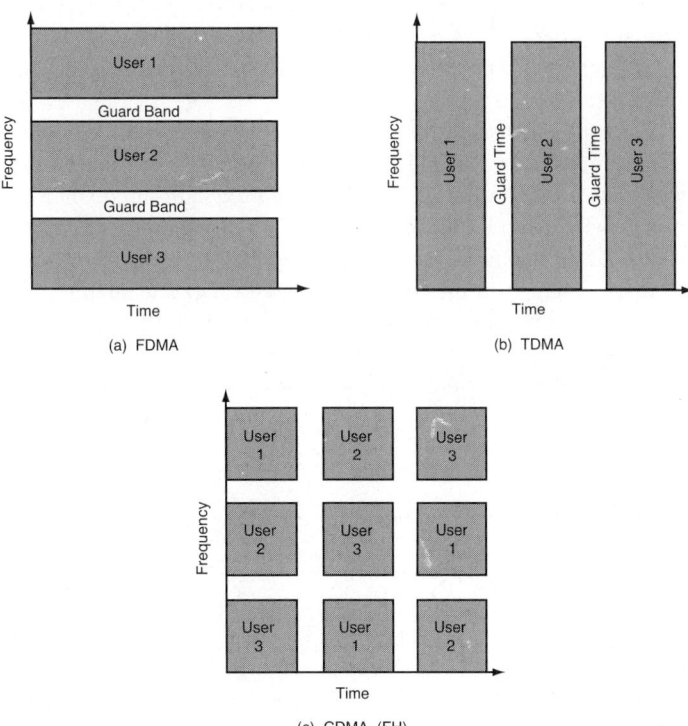

FIGURE 79.1 Multiple-access methods for wireless communication systems.

Then, correlators that project the received signal into the subspace of the desired signal can be employed to extract a signal without interference from other transmissions.

Preference for one access method over another depends largely on overall system characteristics, as we will see in the sequel. No single access method is universally preferable, and system considerations should be carefully weighed before the design decision is made. Before going into the detailed description of the different access methods, we will discuss briefly the salient features of some wireless communication systems. This will allow us later to assess the relative merits of the access methods in different scenarios.

79.2 Relevant Wireless Communication System Characteristics

Modern wireless radio systems range from relatively simple cordless telephones to mobile cellular systems and the emerging personal communication systems (PCS). It is useful to consider such diverse systems as cordless telephone and mobile cellular radio to illustrate some of the fundamental characteristics of wireless communication systems [Cox, 1992].

A summary of the relevant parameters and characteristics for cordless telephone and cellular radio is given in Table 79.1. As evident from that table, the fundamental differences between the two systems are speech quality and the area covered by a base station. The high speech quality requirement in the cordless application is the consequence of the availability of tethered access in the home and office and the resulting direct competition with wire-line telephone services. In the mobile cellular application, the user has no alternative to the wireless access and may be satisfied with lower, but still acceptable, quality of service.

In cordless telephone applications the transmission range is short because the base station can simply be moved to a conveniently located wire-line access point (wall jack) to provide wireless network access where desired. In contrast, the mobile cellular base station must provide access for users throughout a large geographical area of up to approximately 30 km (20 mi) around the base station. This large coverage area is necessary to economically meet the promise of of uninterrupted service to roaming users.

The different range requirements directly affect the transmit power and antenna height for the two systems. High-power transmitters used in mobile cellular user sets consume far more power than even complex signal processing hardware. Hence, sophisticated signal processing, including speech compression, voice activity detection, error correction and detection, and adaptive equalization, can be employed without substantial impact on the battery life in portable hand sets. Furthermore, such techniques are consistent with the goals of increased range and support of large numbers of users with a single, expensive base station. On the other hand, the high mobile cellular base station antennas introduce delay spreads that are one or two orders of magnitude larger than those commonly observed in cordless telephone applications.

TABLE 79.1 Summary of Relevant Characteristics of Cordless Telephone and Cellular Mobile Radio

Characteristic or Parameter	Cordless Telephone	Cellular Radio
Speech quality	Toll quality	Varying with channel quality; possibly decreased by speech pause exploitation
Transmission range	<100 m	100 m–30 km
Transmit power	Milliwatts	Approx. 1 W
Base station antenna height	Approx. 1 m	Tens of meters
Delay spread	Approx. 1 μs	Approx. 10 μs
Complexity of base station	Low	High
Complexity of user set	Low	High

Clearly, the two systems just considered are at extreme ends of the spectrum of wireless communications systems. Most notably, the emerging PCS systems fall somewhere between the two. However, the comparison above highlights some of the system characteristics that should be considered when discussing access methods for wireless communication systems.

79.3 Frequency Division Multiple Access

As mentioned in Sec. 79.1, in FDMA nonoverlapping frequency bands are allocated to different users on a continuous time basis. Hence, signals assigned to different users are clearly orthogonal, at least ideally. In practice, out-of-band spectral components can not be completely suppressed leaving signals not quite orthogonal. This necessitates the introduction of guard bands between frequency bands to reduce adjacent channel interference, i.e., inference from signals transmitted in adjacent frequency bands; see also Fig. 79.1(a).

It is advantageous to combine FDMA with time-division duplexing (TDD) to avoid simultaneous reception and transmission that would require insulation between receive and transmit antennas. In this scenario, the base station and portable take turns using the same frequency band for transmission. Nevertheless, combining FDMA and frequency division duplex is possible in principle, as is evident from the analog FM-based systems deployed throughout the world since the early 1980s.

Channel Considerations

In principle there exists the well-known duality between TDMA and FDMA; see Bertsekas and Gallagher, 1987, p. 113 ff. In the wireless environment, however, propagation related factors have a strong influence on the comparison between FDMA and TDMA. Specifically, the duration of a transmitted symbol is much longer in FDMA than in TDMA. As an immediate consequence, an equalizer is typically not required in an FDMA-based system because the delay spread is small compared to the symbol duration.

To illustrate this point, consider a hypothetical system that transmits information at a constant rate of 50 kb/s. This rate would be sufficient to support 32-kb/s adaptive differential pulse code modulation (ADPCM) speech encoding, some coding for error protection, and control overhead. If we assume further that some form of QPSK modulation is employed, the resulting symbol duration is 40 μs. In relation to delay spreads of approximately 1 μs in the cordless application and 10 μs in cellular systems, this duration is large enough that only little intersymbol interference is introduced. In other words, the channel is frequency nonselective, i.e., all spectral components of the signal are affected equally by the channel. In the cordless application an equalizer is certainly not required; cellular receivers may require equalizers capable of removing intersymbol interference between adjacent bits. Furthermore, it is well known that intersymbol interference between adjacent bits can be removed without loss in SNR by using maximum-likelihood sequence estimation; e.g., Proakis, 1989, p. 622.

Hence, rather simple receivers can be employed in FDMA systems at these data rates. However, there is a flip side to the argument. Recall that the Doppler spread, which characterizes the rate at which the channel impulse response changes, is given approximately by $B_d = v/cf_c$, where v denotes the speed of the mobile user, c is the propagation speed of the electromagnetic waves carrying the signal, and f_c is the carrier frequency. Thus, for systems operating in the vicinity of 1 GHz, B_d will be less than 1 Hz in the cordless application and typically about 100 Hz for a mobile traveling on a highway. In either case, the signal bandwidth is much larger than the Doppler spread B_d, and the channel can be characterized as slowly fading. Whereas this allows tracking of the carrier phase and the use of coherent receivers, it also means that fade durations are long in comparison to the symbol duration and can cause long sequences of bits to be subject to poor channel conditions. The problem is compounded by the fact that the channel is frequency nonselective because it implies that the entire signal is affected by a fade.

To overcome these problems either time diversity, frequency diversity, or spatial diversity could be employed. Time diversity can be accomplished by a combination of coding and interleaving if the fading rate is sufficiently large. For very slowly fading channels, such as the cordless application, the necessary interleaving depth would introduce too much delay to be practical. Frequency diversity can be introduced simply by slow frequency hopping, a technique that prescribes users to change the carrier frequency periodically. Frequency hopping is a form of spectrum spreading because the bandwidth occupied by the resulting signal is much larger than the symbol rate. In contrast to direct sequence spread spectrum discussed subsequently, however, the instantaneous bandwidth is not increased. The jumps between different frequency bands effectively emulate the movement of the portable and, thus, should be combined with the just described time-diversity methods. Spatial diversity is provided by the use of several receive or transmit antennas. At carrier frequencies exceeding 1 GHz, antennas are small and two or more antennas can be accommodated even in the hand set. Furthermore, if FDMA is combined with time-division duplexing, multiple antennas at the base station can provide diversity on both uplink and downlink. This is possible because the channels for the two links are virtually identical, and the base station, using channel information gained from observing the portable's signal, can transmit signals at each antenna such that they combine coherently at the portable's antenna. Thus, signal processing complexity is moved to the base station extending the portable's battery life.

Influence of Antenna Height

In the cellular mobile environment base station antennas are raised considerably to increase the coverage area. Antennas mounted on towers and rooftops are a common sight, and antenna heights of 50 m above ground are no exceptions. Besides increasing the coverage area, this has the additional effect that frequently there exists a better propagation path between two base station antennas than between a mobile and the base station; see Fig. 79.2.

Assuming that FDMA is used in conjunction with TDD as specified at the beginning of this section, then base stations and mobiles transmit on the same frequency. Now, unless there is tight synchronization between all base stations, signals from other base stations will interfere with the reception of signals from portables at the base station. To keep the interference at acceptable levels, it is necessary to increase the reuse distance, i.e., the distance between cells using the same frequencies. In other words, sufficient insulation in the spatial domain must be provided to facilitate the separation of signals. Note that these comments apply equally to cochannel and adjacent channel interference.

This problem does not arise in cordless applications. Base station antennas are generally of the same height as user sets. Hence, interference created by base stations is subject to the same

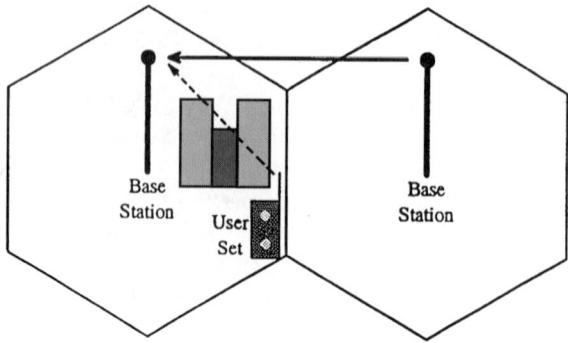

FIGURE 79.2 High base station antennas lead to stronger propagation paths between base stations than between a user set and its base stations.

propagation conditions as signals from user sets. Furthermore, in cordless telephone applications there are frequently attenuating obstacles, such as walls, between base stations that reduce intracell interference further. Note that this reduction is vital for the proper functioning of cordless telephones since there is typically no network planning associated with installing a cordless telephone. As a safety feature, to overcome intercell interference, adaptive channel management strategies based on sensing interference levels can be employed.

Example 79.1: CT2

The CT2 standard was originally adopted in 1987 in Great Britain and improved with a common air interface (CAI) in 1989. The CAI facilitates interoperability between equipment from different vendors whereas the original standard only guarantees noninterference. The CT2 standard is used in home and office cordless telephone equipment and has been used for telepoint applications [Goodman, 1991b].

CT2 operates in the frequency band 864–868 MHz and uses carriers spaced at 100 kHz. FDMA with time division duplexing is employed. The combined gross bit rate is 72 kb/s, transmitted in frames of 2-ms duration of which the first-half carries downlink and the second-half carries uplink information. This setup supports a net bit rate of 32 kb/s of user data (32-kb/s ADPCM encoded speech) and 2-kb/s control information in each direction. The CT2 modulation technique is binary frequency shift keying.

Further Remarks

From the preceding discussion it is obvious that FDMA is a good candidate for applications like cordless telephone. In particular, the simple signal processing makes it a good choice for inexpensive implementation in the benign cordless environment. The possibility of concentration of signal processing functions in the base station strengthens this aspect.

In the cellular application, on the other hand, FDMA is inappropriate because of the lack of built-in diversity and the potential for severe intercell interference between base stations. A further complication arises from the difficulty of performing handovers if base-stations are not tightly synchronized.

For PCS the decision is not as obvious. Depending on whether the envisioned PCS application resembles more a cordless private branch exchange (PBX) than a cellular system, FDMA may be an appropriate choice. We will see later that it is probably better to opt for a combined TDMA/FDMA or a CDMA-based system to avoid the pitfalls of pure FDMA systems and still achieve moderate equipment complexities.

Finally, there is the problem of channel assignment. Clearly, it is not reasonable to assign a unique frequency to each user as there are not sufficient frequencies and the spectral resource would be unused whenever the user is idle. Instead, methods that allocate channels on demand can make much more efficient use of the spectrum. Such methods will be discussed further during the description of TDMA systems.

79.4 Time Division Multiple Access

In TDMA systems users share the same frequency band by accessing the channel in non-overlapping time intervals in a round-robin fashion [Falconer, Adachi, and Gudmundson, 1995]. Since the signals do not overlap, they are clearly orthogonal, and the signal of interest is easily extracted by switching the receiver on only during the transmission of the desired signal. Hence, the receiver filters are simply windows instead of the bandpass filters required in FDMA. As a consequence, the guard time between transmissions can be made as small as the synchronization of the network permits. Guard times of 30–50 μs between time slots are commonly used in TDMA-based systems. As a consequence, all users must be synchronized with the base station to within a fraction of the guard time. This is achievable by distributing a master clock signal on one of the base station's broadcast channels.

TDMA can be combined with TDD or frequency-division duplexing (FDD). The former duplexing scheme is used, for example, in the Digital European Cordless Telephone (DECT) standard and is well suited for systems in which base-to-base and mobile-to-base propagation paths are similar, i.e., systems without extremely high base station antennas. Since both the portable and the base station transmit on the same frequency, some signal processing functions for the downlink can be implemented in the base station, as discussed earlier for FDMA/TDD systems.

In the cellular application, the high base station antennas make FDD the more appropriate choice. In these systems, separate frequency bands are provided for uplink and downlink communication. Note that it is still possible and advisable to stagger the uplink and downlink transmission intervals such that they do not overlap, to avoid the situation that the portable must transmit and receive at the same time. With FDD the uplink and downlink channel are not identical and, hence, signal processing functions can not be implemented in the base-station; antenna diversity and equalization have to be realized in the portable.

Propagation Considerations

In comparison to a FDMA system supporting the same user data rate, the transmitted data rate in a TDMA system is larger by a factor equal to the number of users sharing the frequency band. This factor is eight in the pan-European global system for mobile communications (GSM) and three in the advanced mobile phone service (D-AMPS) system. Thus, the symbol duration is reduced by the same factor and severe intersymbol interference results, at least in the cellular environment.

To illustrate, consider the earlier example where each user transmits 25 K symbols per second. Assuming eight users per frequency band leads to a symbol duration of 5 μs. Even in the cordless application with delay spreads of up to 1 μs, an equalizer may be useful to combat the resulting interference between adjacent symbols. In cellular systems, however, the delay spread of up to 20 μs introduces severe intersymbol interference spanning up to 5 symbol periods. As the delay spread often exceeds the symbol duration, the channel can be classified as frequency selective, emphasizing the observation that the channel affects different spectral components differently.

The intersymbol interference in cellular TDMA systems can be so severe that linear equalizers are insufficient to overcome its negative effects. Instead, more powerful, nonlinear decision feedback or maximum-likelihood sequence estimation equalizers must be employed [Proakis, 1991]. Furthermore, all of these equalizers require some information about the channel impulse response that must be estimated from the received signal by means of an embedded training sequence. Clearly, the training sequence carries no user data and, thus, wastes valuable bandwidth.

In general, receivers for cellular TDMA systems will be fairly complex. On the positive side of the argument, however, the frequency selective nature of the channel provides some built-in diversity that makes transmission more robust to channel fading. The diversity stems from the fact that the multipath components of the received signal can be resolved at a resolution roughly equal to the symbol duration, and the different multipath components can be combined by the equalizer during the demodulation of the signal. To further improve robustness to channel fading, coding and interleaving, slow frequency hopping and antenna diversity can be employed as discussed in connection with FDMA.

Initial Channel Assignment

In both FDMA and TDMA systems, channels should not be assigned to a mobile on a permanent basis. A fixed assignment strategy would either be extremely wasteful of precious bandwidth or highly susceptible to cochannel interference. Instead, channels must be assigned on demand. Clearly, this implies the existence of a separate uplink channel on which mobiles can notify the base

Access Methods

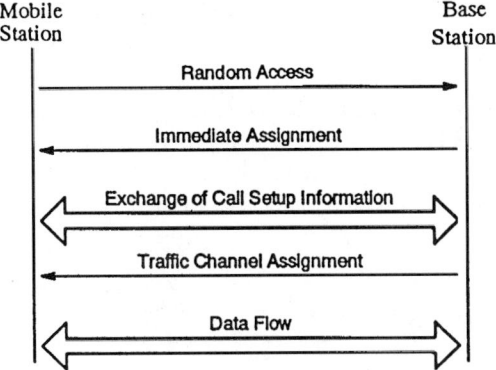

FIGURE 79.3 Mobile-originating call establishment.

station of their need for a traffic channel. This uplink channel is referred to as the **random-access channel** because of the type of strategy used to regulate access to it.

The successful procedure for establishing a call that originates from the mobile station is outlined in Fig. 79.3. The mobile initiates the procedure by transmitting a request on the random-access channel. Since this channel is shared by all users in range of the base station, a random access protocol, like the ALOHA protocol, has to be employed to resolve possible collisions. Once the base station has received the mobile's request, it responds with an immediate assignment message that directs the mobile to tune to a dedicated control channel for the ensuing call setup. Upon completion of the call setup negotiation, a traffic channel, i.e., a frequency in FDMA systems or a time slot in TDMA systems, is assigned by the base station and all future communication takes place on that channel. In the case of a mobile-terminating call request, the sequence of events is preceded by a paging message alerting the base station of the call request.

Example 79.2: GSM

Named after the organization that created the system standards (Groupe Speciale Mobile) this pan-European digital cellular system has been deployed in Europe since the early 1990s [Hodges, 1990]. GSM uses combined TDMA and FDMA with frequency-division duplex for access. Carriers are spaced at 200 kHz and support eight TDMA time slots each. For the uplink the frequency band 890–915 MHz is allocated, whereas the downlink uses the band 935–960 MHz. Each time slot is of duration 577 μs, which corresponds to 156.26-b periods, including a guard time of 8.25-b periods. Eight consecutive time slots form a GSM frame of duration 4.62 ms.

The GSM modulation is Gaussian minimum shift keying with time-bandwidth product of 0.3, i.e., the modulator bandpass has a cutoff frequency of 0.3 times the bit rate. At the bit rate of 270.8 kb/s, severe intersymbol interference arises in the cellular environment. To facilitate coherent detection, a 26-b training sequence is embedded into every time slot. Time diversity is achieved by interleaving over 8 frames for speech signals and 20 frames for data communication. Sophisticated error-correction coding with varying levels of protection for different outputs of the speech coder is provided. Note that the round-trip delay introduced by the interleaver is on the order of 80 ms for speech signals. GSM provides slow frequency hopping as a further mechanism to improve the efficiency of the interleaver.

Further Remarks

In cellular systems, such as GSM or the North-American D-AMPS, TDMA is combined with FDMA. Different frequencies are used in neighboring cells to provide orthogonal signalling without the need for tight synchronization of base stations. Furthermore, channel assignment can then be performed in each cell individually. Within a cell, one or more frequencies are shared by users in the time domain.

From an implementation standpoint TDMA systems have the advantage that common radio and signal processing equipment at the base station can be shared by users communicating on the same frequency. A somewhat more subtle advantage of TDMA systems arises from the possibility of monitoring surrounding base stations and frequencies for signal quality to support mobile assisted handovers.

79.5 Code Division Multiple Access

CDMA systems employ wideband signals with good cross-correlation properties [Kohno, Meidan, and Milstein, 1995]. That means the output of a filter matched to one user's signal is small when a different user's signal is input. A large body of work exists on spreading sequences that lead to signal sets with small cross correlations [Sarwate and Pursley, 1980]. Because of their noise-like appearance such sequences are often referred to as pseudonoise (PN) sequences, and because of their wideband nature CDMA systems are often called spread-spectrum systems.

Spectrum spreading can be achieved mainly in two ways: through frequency hopping as explained earlier or through direct sequence spreading. In direct sequence spread spectrum, a high-rate, antipodal pseudorandom spreading sequence modulates the transmitted signal such that the bandwidth of the resulting signal is roughly equal to the rate of the spreading sequence. The cross correlation of the signals is then largely determined by the cross-correlation properties of the spreading signals. Clearly, CDMA signals overlap in both time and frequency domains but are separable based on their spreading waveforms.

An immediate consequence of this observation is that CDMA systems do not require tight synchronization between users as do TDMA systems. By the same token, frequency planning and management are not required as frequencies are reused throughout the coverage area.

Propagation Considerations

Spread spectrum is well suited for wireless communication systems because of its built-in frequency diversity. As discussed before, in cellular systems the delay spread measures several microseconds and, hence, the coherence bandwidth of the channel is smaller than 1 MHz. Spreading rates can be chosen to exceed the coherence bandwidth such that the channel becomes frequency selective, i.e., different spectral components are affected unequally by the channel and only parts of the signal are affected by fades. Expressing the same observation in time domain terms, multipath components are resolvable at a resolution equal to the chip period and can be combined coherently, for example, by means of a RAKE receiver [Proakis, 1989]. An estimate of the channel impulse response is required for the coherent combination of multipath components. This estimate can be gained from a training sequence or by means of a so-called pilot signal.

Even for cordless telephone systems, operating in environments with submicrosecond delay spread and corresponding coherence bandwidths of a few megahertz, the spreading rate can be chosen large enough to facilitate multipath diversity. If the combination of multipath components already described is deemed too complex, a simpler, but less powerful, form of diversity can be used that decorrelates only the strongest received multipath component and relies on the suppression of other path components by the matched filter.

Multiple-Access Interference

If it is possible to control the relative timing of the transmitted signals, such as on the downlink, the transmitted signals can be made perfectly orthogonal, and if the channel only adds white

Gaussian noise, matched filter receivers are optimal for extracting a signal from the superposition of waveforms. If the channel is dispersive because of multipath, the signals arriving at the receiver will no longer be orthogonal and will introduce some multiple-access interference, i.e., signal components from other signals that are not rejected by the matched filter.

On the uplink, extremely tight synchronization between users to within a fraction of a chip period, which is defined as the inverse of the spreading rate, is generally not possible, and measures to control the impact of multiple-access interference must be taken. Otherwise, the near–far problem, i.e., the problem of very strong undesired users' signals overwhelming the weaker signal of the desired user, can severely decrease performance. Two approaches are proposed to overcome the near–far problem: power control with soft handovers and multiuser detection.

Power control attempts to ensure that signals from all mobiles in a cell arrive at the base station with approximately equal power levels. To be effective, power control must be accurate to within about 1 dB and fast enough to compensate for channel fading. For a mobile moving at 55 mph and transmitting at 1 GHz, the Doppler bandwidth is approximately 100 Hz. Hence, the channel changes its characteristic drastically about 100 times per second and on the order of 1000 b/s must be sent from base station to mobile for power control purposes. As different mobiles may be subject to vastly different fading and shadowing conditions, a large dynamic range of about 80 dB must be covered by power control. Notice, that power control on the downlink is really only necessary for mobiles that are about equidistant from two base stations, and even then neither the update rate nor the dynamic range of the uplink is required.

The interference problem that arises at the cell boundaries where mobiles are within range of two or more base stations can be turned into an advantage through the idea of soft handover. On the downlink, all base stations within range can transmit to the mobile, which in turn can combine the received signals to achieve some gain from the antenna diversity. On the uplink, a similar effect can be obtained by selecting the strongest received signal from all base stations that received a user's signal. The base station that receives the strongest signal will also issue power control commands to minimize the transmit power of the mobile. Note, however, that soft handover requires fairly tight synchronization between base stations, and one of the advantages of CDMA over TDMA is lost.

Multiuser detection is still an emerging technique. It is probably best used in conjunction with power control. The fundamental idea behind this technique is to model multiple-access interference explicitly and devise receivers that reject or cancel the undesired signals. A variety of techniques have been proposed ranging from optimum maximum-likelihood sequence estimation via multistage schemes, reminiscent of decision feedback algorithms, to linear decorrelating receivers. An excellent survey of the theory and practice of multiuser detection is given by Verdu, 1992.

Further Remarks

CDMA systems work well in conjunction with frequency division duplexing. This arrangement decouples the power control problem on the uplink and downlink, respectively.

Signal quality enhancing methods, such as time diversity through coding and interleaving, can be applied just as with the other access methods. In spread spectrum systems, however, coding can be built into the spreading process, avoiding the loss of bandwidth associated with error protection. Additionally, CDMA lends itself naturally to the exploitation of speech pauses that make up more than half the time of a connection. If no signals are transmitted during such pauses, then the instantaneous interference level is reduced and the total number of users supportable by the system can be approximately doubled.

79.6 Comparison and Outlook

The question of which of the access methods is best does not have a single answer. Based on the preceding discussion FDMA is only suited for applications such as cordless telephone with very small cells and submicrosecond delay spreads. In cellular systems and for most versions of personal communication systems, the choice reduces to TDMA vs CDMA.

In terms of complexity, TDMA receivers require adaptive, nonlinear equalizers when operating in environments with large delay spreads. CDMA systems, in turn, need RAKE receivers and sophisticated power control algorithms. In the future, some form of multiple-access interference rejection is likely to be implemented as well. Time synchronization is required in both systems, albeit for different reasons. The additional complexity for coding and interleaving is comparable for both access methods.

An often quoted advantage of CDMA systems is the fact that the performance will degrade gracefully as the load increases. In TDMA systems, in turn, requests will have to be blocked once all channels in a cell are in use. Hence, there is a hard limit on the number of channels per cell. There are proposals for extended TDMA systems, however, that incorporate reassignment of channels during speech pauses. Not only would such extended TDMA systems match the advantage of the exploitation of speech pauses of CDMA systems, they would also lead to a soft limit on the system capacity. The extended TDMA proposals would implement the statistical multiplexing of the user data, e.g., by means of the packet reservation multiple access protocol [Goodman, 1991a]. The increase in capacity depends on the acceptable packet loss rate; in other words, small increases in the load lead to small increases in the packet loss probability.

Many comparisons in terms of capacity between TDMA and CDMA can be found in the recent literature. Such comparisons, however, are often invalidated by making assumptions that favor one access method over the other. An important exception constitutes the recent paper by Wyner, 1994. Under a simplified model that nevertheless captures the essence of cellular systems, he computes the Shannon capacity. Highlights of his results include the following.

- TDMA is distinctly suboptimal in cellular systems.
- When the signal-to-noise-ratio is large, CDMA appears to achieve twice the capacity of TDMA.
- Multiuser detectors are essential to realize near-optimum performance in CDMA systems.
- Intercell interference in CDMA systems has a detrimental effect when the signal-to-noise ratio is large, but it can be exploited via diversity combining to increase capacity when the signal-to-noise ratio is small.

More research along this avenue is necessary to confirm the validity of the results. In particular, incorporation of realistic channel models into the analysis is required. However, this work represents a substantial step towards quantifying capacity increases achievable with CDMA.

Defining Terms

Capacity: Shannon originally defined capacity as the maximum data rate which permits error-free communication in a given environment. A looser interpretation is normally employed in wireless communication systems. Here capacity denotes the traffic density supported by the system under consideration normalized with respect to bandwidth and coverage area.

Code-division multiple access (CDMA): Systems use signals with very small cross-correlations to facilitate sharing of the broadcast radio channel. Correlators are used to extract the desired user's signal while simultaneously suppressing interfering, parallel transmissions.

Duplexing: Refers to the exchange of messages in both directions of a connection.

Frequency-division multiple access (FDMA): Simultaneous access to the radio channel is facilitated by assigning nonoverlapping frequency bands to different users.

Multiple access: Denotes the support of simultaneous transmissions over a shared communication channel.

Random-access channel: This uplink control channel is used by mobiles to request assignment of a traffic channel. A random access protocol is employed to arbitrate access to this channel.

Time-division multiple access (TDMA): Systems assign nonoverlapping time slots to different users in a round-robin fashion.

References

Bertsekas, D. and Gallager, R. 1987. *Data Networks*, Prentice Hall, Englewood Cliffs, NJ.

Cox, D.C. 1992. Wireless network access for personal communications, *IEEE Comm. Mag.*, pp. 96–115.

Falconer, D.D., Adachi, F., and Gudmundson, B. 1995. Time division multiple access methods for wireless personal communications. *IEEE Comm. Mag.* 33(1):50–57.

Goodman, D. 1991a. Trends in cellular and cordless communications. *IEEE Comm. Mag.* pp 31–40.

Goodman, D.J. 1991b. Second generation wireless information networks. *IEEE Trans. on Vehicular Tech.* 40(2):366–374.

Hodges, M.R.L. 1990. The GSM radio interface. *Br. Telecom Tech. J.* 8(1):31–43.

Kohno, R., Meidan, R., and Milstein, L.B. 1995. Spread spectrum access methods for wireless communications. *IEEE Comm. Mag.* 33(1):58.

Proakis, J.G. 1989. *Digital Communications*. 2nd ed. McGraw-Hill, New York.

Proakis, J.G . 1991. Adaptive equalization for TDMA digital mobile radio. *IEEE Trans. on Vehicular Tech.* 40(2):333–341.

Sarwate, D.V. and Pursley, M.B. 1980. Crosscorrelation properties of pseudorandom and related sequences. *Proceedings of the IEEE*, 68(5):593–619.

Verdu, S. 1992. Multi-user detection. In *Advances in Statistical Signal Processing—Vol. 2: Signal Detection*, JAI Press, Greenwich, CT.

Wyner, A.D. 1994. Shannon-theoretic approach to a Gaussian cellular multiple-access channel. *IEEE Trans. on Information Theory*, 40(6):1713–1727.

Further Information

Several of the IEEE publications, including the *Transactions on Communications, Journal on Selected Areas in Communications, Transactions on Vehicular Technology, Communications Magazine,* and *Personal Communications Magazine* contain articles the on subject of access methods on a regular basis.

80
Location Strategies for Personal Communications Services

Ravi Jain
Bell Communications Research

Yi-Bing Lin
Bell Communications Research

Seshadri Mohan[1]
Bell Communications Research

80.1	Introduction	1116
80.2	An Overview of PCS	1117
	Aspects of Mobility—Example 80.1 • A Model for PCS	
80.3	IS-41 Preliminaries	1121
	Terminal/Location Registration • Call Delivery	
80.4	Global System for Mobile Communications	1123
	Architecture • User Location Strategy	
80.5	Analysis of Database Traffic Rate for IS-41 and GSM	1126
	The Mobility Model for PCS Users • Additional Assumptions • Analysis of IS-41 • Analysis of GSM	
80.6	Reducing Signalling During Call Delivery	1129
80.7	Per-User Location Caching	1129
80.8	Caching Threshold Analysis	1131
80.9	Techniques for Estimating Users' LCMR	1134
	The Running Average Algorithm • The Reset-K Algorithm • Comparison of the LCMR Estimation Algorithms	
80.10	Discussion	1141
	Conditions When Caching Is Beneficial • Alternative Network Architectures • LCMR Estimation and Caching Policy	
80.11	Conclusions	1143

80.1 Introduction

The vision of nomadic personal communications is the ubiquitous availability of services to facilitate exchange of information (voice, data, video, image, etc.) between nomadic end users independent of time, location, or access arrangements. To realize this vision, it is necessary to locate

---

[1]Address correspondence to: Seshadri Mohan, MCC-1A216B, Bellcore, 445 South St, Morristown, NJ 07960; Phone: 201-829-5160, Fax: 201-829-5888, E-mail: **smohan@thumper.bellcore.com**.

© 1996 by Bell Communications Research, Inc. Used with permission. The material in this chapter appeared originally in the following IEEE publications: S. Mohan and R. Jain. 1994. Two user location strategies for personal communications services, *IEEE Personal Communications: The Magazine of Nomadic Communications and Computing*, pp. 42–50, Feb., and R. Jain, C.N. Lo, and S. Mohan. 1994. A caching strategy to reduce network impacts of PCS, J-SAC Special Issue on Wireless and Mobile Networks, Aug.

users that move from place to place. The strategies commonly proposed are two-level hierarchical strategies, which maintain a system of mobility databases, home location registers (HLR) and visitor location resisters (VLR), to keep track of user locations. Two standards exist for carrying out two-level hierarchical strategies using HLRs and VLRs. The standard commonly used in North America is the EIA/TIA Interim Standard 41 (IS 41) [EIA/TIA, 1991] and in Europe the Global System for Mobile Communications (GSM) [Mouly and Pautet, 1992; Lycksell, 1991]. In this chapter, we refer to these two strategies as *basic* location strategies.

We introduce these two strategies for locating users and provide a tutorial on their usage. We then analyze and compare these basic location strategies with respect to load on mobility databases and signalling network. Next we propose an auxiliary strategy, called the *per-user caching* or, simply, the *caching* strategy, that augments the basic location strategies to reduce the signalling and database loads.

The outline of this chapter is as follows. In Sec. 80.2 we discuss different forms of mobility in the context of personal communications services (PCS) and describe a reference model for a PCS architecture. In Secs. 80.3 and 80.4, we describe the user location strategies specified in the IS-41 and GSM standards, respectively, and in Sec. 80.5, using a simple example, we present a simplified analysis of the database loads generated by each strategy. In Sec. 80.6, we briefly discuss possible modifications to these protocols that are likely to result in significant benefits by either reducing query and update rate to databases or reducing the signalling traffic or both. Section 80.7 introduces the caching strategy followed by an analysis in the next two sections. This idea attempts to exploit the spatial and temporal locality in calls received by users, similar to the idea of exploiting locality of file access in computer systems [Silberschatz and Peterson, 1988]. A feature of the caching location strategy is that it is useful only for certain classes of PCS users, those meeting certain call and mobility criteria. We encapsulate this notion in the definition of the user's call-to-mobility ratio (CMR), and local CMR (LCMR), in Sec. 80.8. We then use this definition and our PCS network reference architecture to quantify the costs and benefits of caching and the threshold LCMR for which caching is beneficial, thus characterizing the classes of users for which caching should be applied. In Sec. 80.9 we describe two methods for estimating users' LCMR and compare their effectiveness when call and mobility patterns are fairly stable, as well as when they may be variable. In Sec. 80.10, we briefly discuss alternative architectures and implementation issues of the strategy proposed and mention other auxiliary strategies that can be designed. Section 80.11 provides some conclusions and discussion of future work.

The choice of platforms on which to realize the two location strategies (IS-41 and GSM) may vary from one service provider to another. In this paper, we describe a possible realization of these protocols based on the advanced intelligent network (AIN) architecture (see Bellcore, 1991, and Berman and Brewster, 1992), and signalling system 7 (SS7). It is also worthwhile to point out that several strategies have been proposed in the literature for locating users, many of which attempt to reduce the signalling traffic and database loads imposed by the need to locate users in PCS.

80.2 An Overview of PCS

This section explains different aspects of mobility in PCS using an example of two nomadic users who wish to communicate with each other. It also describes a reference model for PCS.

Aspects of Mobility—Example 80.1

PCS can involve two possible types of mobility, terminal mobility and personal mobility, that are explained next.

Terminal Mobility: This type of mobility allows a terminal to be identified by a unique terminal identifier independent of the point of attachment to the network. Calls intended for that terminal can therefore be delivered to that terminal regardless of its network point of attachment. To facilitate terminal mobility, a network must provide several functions, which include those that locate, identify, and validate a terminal and provide services (e.g., deliver calls) to the terminal based on the location information. This implies that the network must store and maintain the location information of the terminal based on a unique identifier assigned to that terminal. An example of a terminal identifier is the IS-41 EIA/TIA cellular industry term mobile identification number (MIN), which is a North American Numbering Plan (NANP) number that is stored in the terminal at the time of manufacture and cannot be changed. A similar notion exists in GSM (see Sec. 80.4).

Personal Mobility: This type of mobility allows a PCS user to make and receive calls independent of both the network point of attachment and a specific PCS terminal. This implies that the services that a user has subscribed to (stored in that user's service profile) are available to the user even if the user moves or changes terminal equipment. Functions needed to provide personal mobility include those that identify (authenticate) the end user and provide services to an end user independent of both the terminal and the location of the user. An example of a functionality needed to provide personal mobility for voice calls is the need to maintain a user's location information based on a unique number, called the universal personal telecommunications (UPT) number, assigned to that user. UPT numbers are also NANP numbers. Another example is one that allows end users to define and manage their service profiles to enable users to tailor services to suit their needs. In Sec. 80.4, we describe how GSM caters to personal mobility via smart cards.

For the purposes of the example that follows, the terminal identifiers (TID) and UPT numbers are NANP numbers, the distinction being TIDs address terminal mobility and UPT numbers address personal mobility. Though we have assigned two different numbers to address personal and terminal mobility concerns, the same effect could be achieved by a single identifier assigned to the terminal that varies depending on the user that is currently utilizing the terminal. For simplicity we assume that two different numbers are assigned.

Figure 80.1 illustrates the terminal and personal mobility aspects of PCS, which will be explained via an example. Let us assume that users Kate and Al have, respectively, subscribed to PCS services from PCS service provider (PSP) A and PSP B. Kate receives the UPT number, say, 500 111 4711, from PSP A. She also owns a PCS terminal with TID 200 777 9760. Al too receives his UPT number 500 222 4712 from PSP B, and he owns a PCS terminal with TID 200 888 5760. Each has been provided a personal identification number (PIN) by their respective PSP when subscription began. We assume that the two PSPs have subscribed to PCS access services from a certain network provider such as, for example, a local exchange carrier (LEC). (Depending on the capabilities of the PSPs, the access services provided may vary. Examples of access services include translation of UPT number to a routing number, terminal and personal registration, and call delivery. Refer to Bellcore, 1993a, for further details). When Kate plugs in her terminal to the network, or when she activates it, the terminal registers itself with the network by providing its TID to the network. The network creates an entry for the terminal in an appropriate database, which, in this example, is entered in the terminal mobility database (TMDB) A. The entry provides a mapping of her terminal's TID, 200 777 9760, to a routing number (RN), RN1. All of these activities happen without Kate being aware of them. After activating her terminal, Kate registers herself at that terminal by entering her UPT number (500 111 4711) to inform the network that all calls to her UPT number are to be delivered to her at the terminal. For security reasons, the network may want to authenticate her and she may be prompted to enter her PIN number into her terminal. (Alternatively, if the terminal is equipped with a smart card reader, she may enter her smart card into the reader. Other techniques, such as, for example, voice recognition, may be employed). Assuming that she is authenticated, Kate has now registered herself. As a result of personal registration by Kate, the network creates an entry for her in the personal mobility database (PMDB) A that maps her UPT number to the TID of the terminal at which she registered. Similarly, when Al activates his terminal and registers

Location Strategies for Personal Communications Services 1119

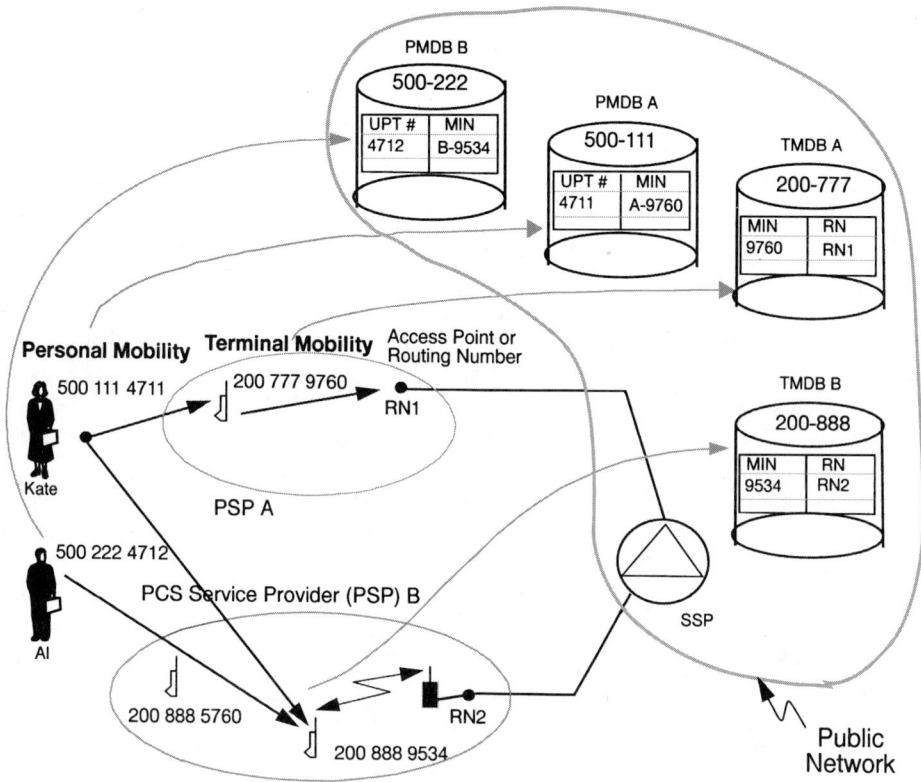

FIGURE 80.1 Illustrating terminal and personal mobility.

himself, appropriate entries are created in TMDB B and PMDB B. Now Al wishes to call Kate and, hence, he dials Kate's UPT number (500 111 4711). The network carries out the following tasks.

1. The switch analyzes the dialed digits and recognizes the need for AIN service, determines that the dialed UPT number needs to be translated to a RN by querying PMDB A and, hence, it queries PMDB A.
2. PMDB A searches its database and determines that the person with UPT number 500 111 4711 is currently registered at terminal with TID 200 777 9760.
3. PMDB A then queries TMDB A for the RN of the terminal with TID 200 777 9760. TMDB A returns the RN (RN1).
4. PMDB A returns the RN (RN1) to the originating switch.
5. The originating switch directs the call to the switch RN1, which then alerts Kate's terminal. The call is completed when Kate picks up her terminal.

Kate may take her terminal wherever she goes and perform registration at her new location. From then on, the network will deliver all calls for her UPT number to her terminal at the new location. In fact, she may actually register on someone else's terminal too. For example, suppose that Kate and Al agree to meet at Al's place to discuss a school project they are working on together. Kate may register herself on Al's terminal (TID 200 888 9534). The network will now modify the entry corresponding to 4711 in PMDB A to point to B 9534. Subsequent calls to Kate will be delivered to Al's terminal.

The scenario given here is used only to illustrate the key aspects of terminal and personal mobility; an actual deployment of these services may be implemented in ways different from those suggested here. We will not discuss personal registration further. The analyses that follow consider only terminal mobility but may easily be modified to include personal mobility.

A Model for PCS

Figure 80.2 illustrates the reference model used for the comparative analysis. The model assumes that the HLR resides in a service control point (SCP) connected to a regional signal transfer point (RSTP). The SCP is a storehouse of the AIN service logic, i.e., functionality used to perform the processing required to provide advanced services, such as speed calling, outgoing call screening, etc., in the AIN architecture (see Bellcore, 1991 and Berman and Brewster, 1992). The RSTP and the local STP (LSTP) are packet switches, connected together by various links such A links or D links, that perform the signalling functions of the SS7 network. Such functions include, for example, global title translation for routing messages between the AIN switching system, which is also referred to as the service switching point (SSP), and SCP and IS-41 messages [EIA/TIA, 1991]. Several SSPs may be connected to an LSTP.

The reference model in Fig. 80.2 introduces several terms which are explained next. We have tried to keep the terms and discussions fairly general. Wherever possible, however, we point to equivalent cellular terms from IS-41 or GSM.

For our purposes, the geographical area served by a PCS system is partitioned into a number of radio port coverage areas (or cells, in cellular terms) each of which is served by a radio port (or, equivalently, base station) that communicates with PCS terminals in that cell. A registration area (also known in the cellular world as location area) is composed of a number of cells. The base stations of all cells in a registration area are connected by wireline links to a mobile switching center (MSC). We assume that each registration area is served by a single VLR. The MSC of a registration area is responsible for maintaining and accessing the VLR and for switching between radio ports. The VLR associated with a registration area is responsible for maintaining a subset of the user information contained in the HLR.

Terminal registration process is initiated by terminals whenever they move into a new registration area. The base stations of a registration area periodically broadcast an identifier associated with that area. The terminals periodically compare an identifier they have stored with the identifier to the registration area being broadcast. If the two identifiers differ, the terminal recognizes that it has

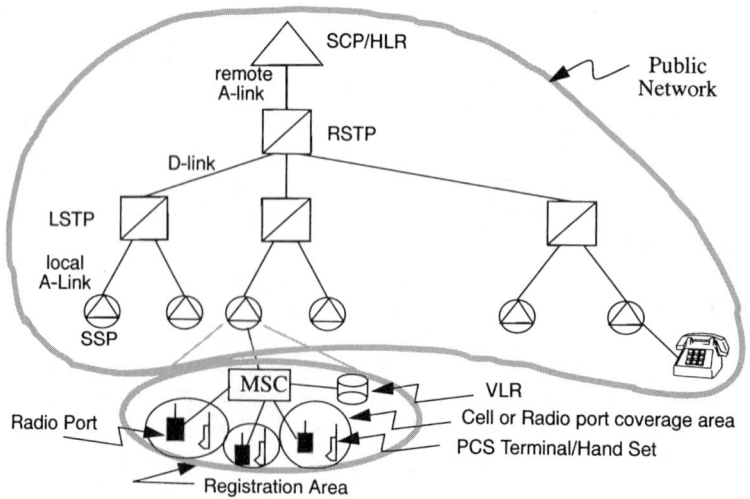

FIGURE 80.2 Example of a reference model for a PCS.

moved from one registration area to another and will, therefore, generate a registration message. It also replaces the previous registration area identifier with that of the new one. Movement of a terminal within the same registration area will not generate registration messages. Registration messages may also be generated when the terminals are switched on. Similarly, messages are generated to deregister them when they are switched off.

PCS services may be provided by different types of commercial service vendors. Bellcore, 1993a describes three different types of PSPs and the different access services that a public network may provide to them. For example, a PSP may have full network capabilities with its own switching, radio management, and radio port capabilities. Certain others may not have switching capabilities, and others may have only radio port capabilities. The model in Fig. 80.2 assumes full PSP capabilities. The analysis in Sec. 80.5 is based on this model and modifications may be necessary for other types of PSPs.

It is also quite possible that one or more registration areas may be served by a single PSP. The PSP may have one or more HLRs for serving its service area. In such a situation users that move within the PSP's serving area may generate traffic to the PSP's HLR (not shown in Fig. 80.2) but not to the network's HLR (shown in Fig. 80.2). In the interest of keeping the discussions simple, we have assumed that there is one-to-one correspondence between SSPs and MSCs and also between MSCs, registration areas, and VLRs. One impact of locating the SSP, MSC, and VLR in separate physical sites connected by SS7 signalling links would be to increase the required signalling message volume on the SS7 network. Our model assumes that the messages between the SSP and the associated MSC and VLR do not add to signalling load on the public network. Other configurations and assumptions could be studied for which the analysis may need to be suitably modified. The underlying analysis techniques will not, however, differ significantly.

80.3 IS-41 Preliminaries

We now describe the message flow for call origination, call delivery, and terminal registration, sometimes called location registration, based on the IS-41 protocol. This protocol is described in detail in EIA/TIA, 1991. Only an outline is provided here.

Terminal/Location Registration

During IS-41 registration, signalling is performed between the following pairs of network elements:

- New serving MSC and the associated database (or VLR)
- New database (VLR) in the visited area and the HLR in the public network
- HLR and the VLR in former visited registration area or the old MSC serving area.

Figure 80.3 shows the signalling message flow diagram for IS-41 registration activity, focusing only on the essential elements of the message flow relating to registration; for details of variations from the basic registration procedure, see Bellcore, 1993a.

The following steps describe the activities that take place during registration.

1. Once a terminal enters a new registration area, the terminal sends a registration request to the MSC of that area.
2. The MSC sends an authentication request (AUTHRQST) message to its VLR to authenticate the terminal, which in turn sends the request to the HLR. The HLR sends its response in the authrqst message.
3. Assuming the terminal is authenticated, the MSC sends a registration notification (REGNOT) message to its VLR.
4. The VLR in turn sends a REGNOT message to the HLR serving the terminal. The HLR updates the location entry corresponding to the terminal to point to the new serving MSC/VLR.

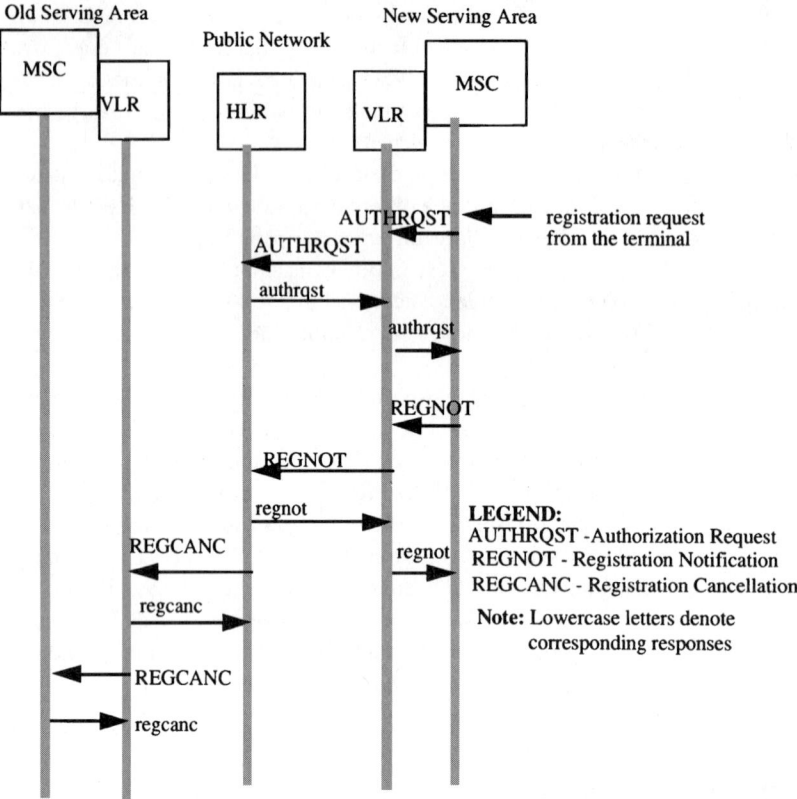

FIGURE 80.3 Signalling flow diagram for registration in IS-41.

The HLR sends a response back to the VLR, which may contain relevant parts of the user's service profile. The VLR stores the service profile in its database and also responds to the serving MSC.

5. If the user/terminal was registered previously in a different registration area, the HLR sends a registration cancellation (REGCANC) message to the previously visited VLR. On receiving this message, the VLR erases all entries for the terminal from the record and sends a REGCANC message to the previously visited MSC, which then erases all entries for the terminal from its memory.

The protocol shows authentication request and registration notification as separate messages. If the two messages can be packaged into one message, then the rate of queries to HLR may be cut in half. This does not necessarily mean that the total number of messages are cut in half.

Call Delivery

The signalling message flow diagram for IS-41 call delivery is shown in Fig. 80.4. The following steps describe the activities that take place during call delivery.

1. A call origination is detected and the number of the called terminal (for example, MIN) is received by the serving MSC. Observe that the call could have originated from within the public network from a wireline phone or from a wireless terminal in an MSC/VLR serving area. (If the call originated within the public network, the AIN SSP analyzes the dialed digits and sends a query to the SCP.)

Location Strategies for Personal Communications Services

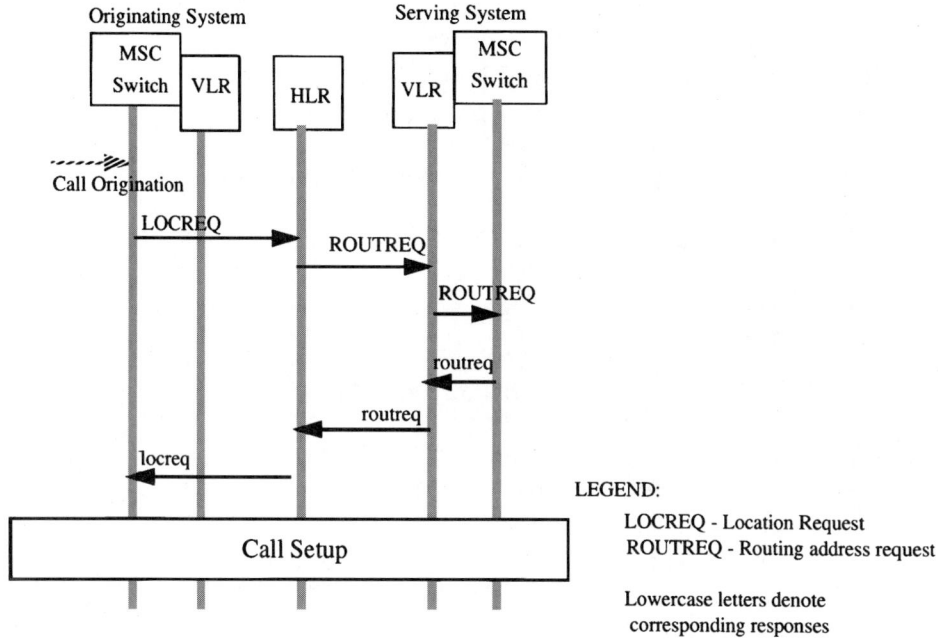

FIGURE 80.4 Signalling flow diagram for call delivery in IS-41.

2. The MSC determines the associated HLR serving the called terminal and sends a location request (LOCREQ) message to the HLR.
3. The HLR determines the serving VLR for that called terminal and sends a routing address request (ROUTEREQ) to the VLR, which forwards it to the MSC currently serving the terminal.
4. Assuming that the terminal is idle, the serving MSC allocates a temporary identifier, called a temporary local directory number (TLDN), to the terminal and returns a response to the HLR containing this information. The HLR forwards this information to the originating SSP/MSC in response to its LOCREQ message.
5. The originating SSP requests call setup to the serving MSC of the called terminal via the SS7 signalling network using the usual call setup protocols.

Similar to the considerations for reducing signalling traffic for location registration, the VLR and HLR functions could be united in a single logical database for a given serving area and collocated; further, the database and switch can be integrated into the same piece of physical equipment or be collocated. In this manner, a significant portion of the messages exchanged between the switch, HLR and VLR as shown in Fig. 80.4 will not contribute to signalling traffic.

80.4 Global System for Mobile Communications

In this section we describe the user location strategy proposed in the European Global System for Mobile Communications (GSM) standard and its offshoot, digital cellular system 1800 (DCS1800). There has recently been increased interest in GSM in North America, since it is possible that early deployment of PCS will be facilitated by using the communication equipment already available from European manufacturers who use the GSM standard. Since the GSM standard is relatively unfamiliar to North American readers, we first give some background and introduce the various abbreviations. The reader will find additional details in Mouley and Pautet, 1992. For an overview on GSM, refer to Lycksell, 1991.

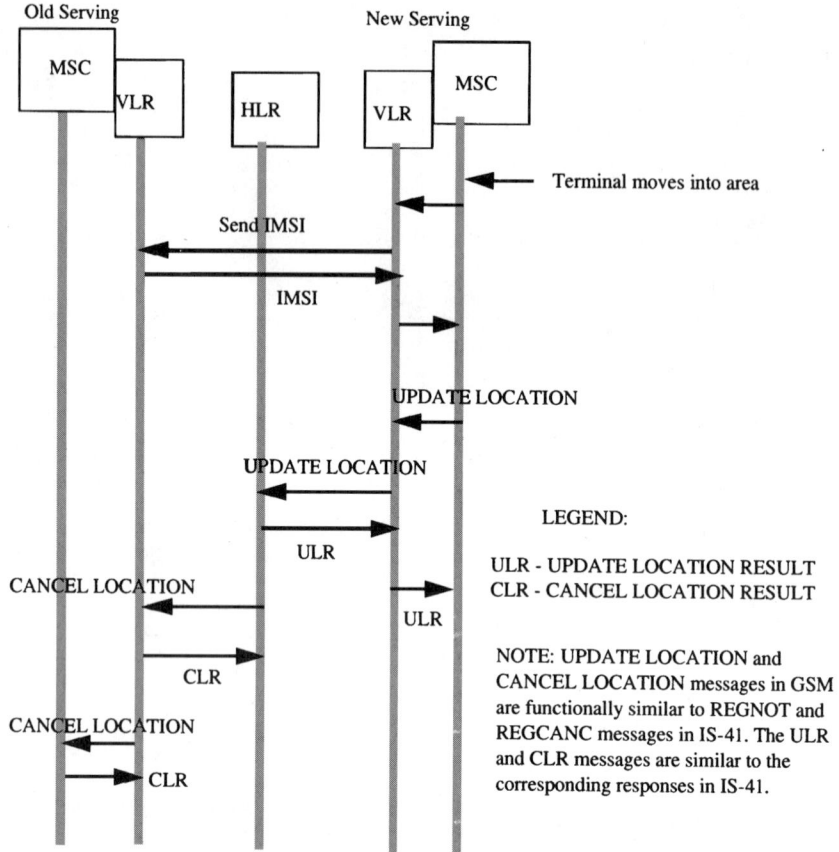

FIGURE 80.5 Flow diagram for registration in GSM.

The abbreviation GSM originally stood for Groupe Special Mobile, a committee created within the pan-European standardization body Conference Europeenne des Posts et Telecommunications (CEPT) in 1982. There were numerous national cellular communication systems and standards in Europe at the time, and the aim of GSM was to specify a uniform standard around the newly reserved 900-MHz frequency band with a bandwidth of twice 25 MHz. The phase 1 specifications of this standard were frozen in 1990. Also in 1990, at the request of the United Kingdom, specification of a version of GSM adapted to the 1800-MHz frequency, with bandwidth of twice 75 MHz, was begun. This variant is referred to as DCS1800; the abbreviation GSM900 is sometimes used to distinguish between the two variations, with the abbreviation GSM being used to encompass both GSM900 and DSC1800. The motivation for DCS1800 is to provide higher capacities in densely populated urban areas, particularly for PCS. The DCS1800 specifications were frozen in 1991, and by 1992 all major GSM900 European operators began operation.

At the end of 1991, activities concerning the post-GSM generation of mobile communications were begun by the standardization committee, using the name universal mobile telecommunications system (UMTS) for this effort. In 1992, the name of the standardization committee was changed from GSM to special mobile group (SMG) to distinguish it from the 900-MHz system itself, and the term GSM was chosen as the commercial trademark of the European 900-MHz system, where GSM now stands for global system for mobile communications.

The GSM standard has now been widely adopted in Europe and is under consideration in several other non-European countries, including the United Arab Emirates, Hong Kong, and New Zealand. In 1992, Australian operators officially adopted GSM.

Architecture

In this section we describe the GSM architecture, focusing on those aspects that differ from the architecture assumed in the IS-41 standard.

A major goal of the GSM standard was to enable users to move across national boundaries and still be able to communicate. It was considered desirable, however, that the operational network within each country be operated independently. Each of the operational networks is called a public land mobile network (PLMN) and its commercial coverage area is confined to the borders of one country (although some radio coverage overlap at national boundaries may occur), and each country may have several competing PLMNs.

A GSM customer subscribes to a single PLMN called the home PLMN, and subscription information includes the services the customer subscribes to. During normal operation, a user may elect to choose other PLMNs as their service becomes available (either as the user moves or as new operators enter the marketplace). The user's terminal [GSM calls the terminal a mobile station (MS)] assists the user in choosing a PLMN in this case, either presenting a list of possible PLMNs to the user using explicit names (e.g., DK Sonofon for the Danish PLMN) or choosing automatically based on a list of preferred PLMNs stored in the terminal's memory. This PLMN selection process allows users to choose between the services and tariffs of several competing PLMNs. Note that the PLMN selection process differs from the cell selection and handoff process that a terminal carries out automatically without any possibility of user intervention, typically based on received radio signal strengths and, thus, requires additional intelligence and functionality in the terminal.

The geographical area covered by a PLMN is partitioned into MSC serving areas, and a registration area is constrained to be a subset of a single MSC serving area. The PLMN operator has complete freedom to allocate cells to registration areas. Each PLMN has, logically speaking, a single HLR, although this may be implemented as several physically distributed databases, as for IS-41. Each MSC also has a VLR, and a VLR may serve one or several MSCs. As for IS-41, it is interesting to consider how the VLR should be viewed in this context. The VLR can be viewed as simply a database off loading the query and signalling load on the HLR and, hence, logically tightly coupled to the HLR or as an ancillary processor to the MSC. This distinction is not academic; in the first view, it would be natural to implement a VLR as serving several MSCs, whereas in the second each VLR would serve one MSC and be physically closely coupled to it. For GSM, the MSC implements most of the signalling protocols, and at present all switch manufacturers implement a combined MSC and VLR, with one VLR per MSC [Mouly and Pautet, 1992].

A GSM mobile station is split in two parts, one containing the hardware and software for the radio interface and the other containing subscribers-specific and location information, called the subscriber identity module (SIM), which can be removed from the terminal and is the size of a credit card or smaller. The SIM is assigned a unique identity within the GSM system, called the international mobile subscriber identity (IMSI), which is used by the user location strategy as described the next subsection. The SIM also stores authentication information, services lists, PLMN selection lists, etc., and can itself be protected by password or PIN.

The SIM can be used to implement a form of large-scale mobility called SIM roaming. The GSM specifications standardize the interface between the SIM and the terminal, so that a user carrying his or her SIM can move between different terminals and use the SIM to personalize the terminal. This capability is particularly useful for users who move between PLMNs which have different radio interfaces. The user can use the appropriate terminal for each PLMN coverage area while obtaining the personalized facilities specified in his or her SIM. Thus, SIMs address personal mobility. In the European context, the usage of two closely related standards at different frequencies, namely, GSM900 and DCS1800, makes this capability an especially important one and facilitates interworking between the two systems.

User Location Strategy

We present a synopsis of the user location strategy in GSM using call flow diagrams similar to those used to describe the strategy in IS-41.

In order to describe the registration procedure, it is first useful to clarify the different identifiers used in this procedure. The SIM of the terminal is assigned a unique identity, called the IMSI, as already mentioned. To increase confidentiality and make more efficient use of the radio bandwidth, however, the IMSI is not normally transmitted over the radio link. Instead, the terminal is assigned a temporary mobile subscriber identity (TMSI) by the VLR when it enters a new registration area. The TMSI is valid only within a given registration area and is shorter than the IMSI. The IMSI and TMSI are identifiers that are internal to the system and assigned to a terminal or SIM and should not be confused with the user's number that would be dialed by a calling party; the latter is a separate number called the mobile subscriber integrated service digital network (ISDN) number (MSISDN), and is similar to the usual telephone number in a fixed network.

We now describe the procedure during registration. The terminal can detect when it has moved into the cell of a new registration area from the system information broadcast by the base station in the new cell. The terminal initiates a registration update request to the new base station; this request includes the identity of the old registration area and the TMSI of the terminal in the old area. The request is forwarded to the MSC, which, in turn, forwards it to the new VLR. Since the new VLR cannot translate the TMSI to the IMSI of the terminal, it sends a request to the old VLR to send the IMSI of the terminal corresponding to that TMSI. In its response, the old VLR also provides the required authentication information. The new VLR then initiates procedures to authenticate the terminal. If the authentication succeeds, the VLR uses the IMSI to determine the address of the terminal's HLR.

The ensuing protocol is then very similar to that in IS-41, except for the following differences. When the new VLR receives the registration affirmation (similar to regnot in IS-41) from the HLR, it assigns a new TMSI to the terminal for the new registration area. The HLR also provides the new VLR with all relevant subscriber profile information required for call handling (e.g., call screening lists, etc.) as part of the affirmation message. Thus, in contrast with IS-41, authentication and subscriber profile information are obtained from both the HLR and old VLR and not just the HLR.

The procedure for delivering calls to mobile users in GSM is very similar to that in IS-41. The sequence of messages between the caller and called party's MSC/VLRs and the HLR is identical to that shown in the call flow diagrams for IS-41, although the names, contents and lengths of messages may be different and, hence, the details are left out. The interested reader is referred to Mouly and Pautet, 1992, or Lycksell, 1991, for further details.

80.5 Analysis of Database Traffic Rate for IS-41 and GSM

In the two subsections that follow, we state the common set of assumptions on which we base our comparison of the two strategies.

The Mobility Model for PCS Users

In the analysis that follows in the IS-41 analysis subsection, we assume a simple mobility model for the PCS users. The model, which is described in Thomas, Gilbert, and Mazziotto, 1988, assumes that PCS users carrying terminals are moving at an average velocity of v and their direction of movement is uniformly distributed over $[0, 2\pi]$. Assuming that the PCS users are uniformly populated with a density of ρ and the registration area boundary is of length L, it has been shown that the rate of registration area crossing R is given by

$$R = \frac{\rho v L}{\pi} \qquad (80.1)$$

Using Eq. (80.1), we can calculate the signalling traffic due to registration, call origination, and delivery. We new need a set of assumptions so that we may proceed to derive the traffic rate to the databases using the model in Fig. 80.2.

Additional Assumptions

The following assumptions are made in performing the analysis.

- 128 total registration areas
- Square registration area size: $(7.575 \text{ km})^2 = 57.5 \text{ km}^2$, with border length $L = 30.3$ km
- Average call origination rate = average call termination (delivery) rate = 1.4/h/terminal
- Mean density of mobile terminals = $\rho = 390/\text{km}^2$
- Total number of mobile terminals = $128 \times 57.4 \times 390 = 2.87 \times 10^6$
- Average call origination rate = average call termination (delivery) rate = 1.4/h/terminal
- Average speed of a mobile, $v = 5.6$ km/h
- Fluid flow mobility model

The assumptions regarding the total number of terminals may also be obtained by assuming that a certain public network provider serves 19.15×10^6 users and that 15% (or 2.87×10^6) of the users also subscribe to PCS services from various PSPs.

Note that we have adopted a simplified model that ignores situations where PCS users may turn their handsets on and off that will generate additional registration and deregistration traffic. The model also ignores wireline registrations. These activities will increase the total number of queries and updates to HLR and VLRs.

Analysis of IS-41

Using Eq. (80.1) and the parameter values assumed in the preceding subsection, we can compute the traffic due to registration. The registration traffic is generated by mobile terminals moving into a new registration area, and this must equal the mobile terminals moving out of the registration area, which per second is

$$R_{\text{reg, VLR}} = \frac{390 \times 30.3 \times 5.6}{3600\pi} = 5.85$$

This must also be equal to the number of deregistrations (registration cancellations),

$$R_{\text{dereg, VLR}} = 5.85$$

The total number of registration messages per second arriving at the HLR will be

$$R_{\text{reg, HLR}} = R_{\text{reg, VLR}} \times \text{total No. of registration areas} = 749$$

The HLR should, therefore, be able to handle, roughly, 750 updates per second. We observe from Fig. 80.3 that authenticating terminals generate as many queries to VLR and HLR as the respective number of updates generated due to registration notification messages.

The number of queries that the HLR must handle during call origination and delivery can be similarly calculated. Queries to HLR are generated when a call is made to a PCS user. The SSP that receives the request for a call, generates a location request (LOCREQ) query to the SCP controlling the HLR. The rate per second of such queries must be equal to the rate of calls made to PCS users.

TABLE 80.1 IS-41 Query and Update Rates to HLR and VLR

Activity	HLR Updates/s	VLR Updates/s	HLR Queries/s	VLR queries/s
Mobility-related activities at registration	749	5.85	749	5.85
Mobility-related activities at deregistration		5.85		
Call origination				8.7
Call delivery			1116	8.7
Total (per RA)	5.85	11.7	14.57	23.25
Total (Network)	749	1497.6	1865	2976

This is calculated as

$$R_{\text{CallDeliv, HLR}} = \text{call rate per user} \times \text{total number of users}$$
$$= \frac{1.4 \times 2.87 \times 10^5}{3600}$$
$$= 1116$$

For calls originated from a mobile terminal by PCS users, the switch authenticates the terminal by querying the VLR. The rate per second of such queries is determined by the rate of calls originating in an SSP serving area, which is also a registration area (RA). This is given by

$$R_{\text{CallOrig, VLR}} = \frac{1116}{128} = 8.7$$

This is also the number of queries per second needed to authenticate terminals of PCS users to which calls are delivered:

$$R_{\text{CallDeliv, VLR}} = 8.7$$

Table 80.1 summarizes the calculations.

Analysis of GSM

Calculations for query and update rates for GSM may be performed in the same manner as for IS-41, and they are summarized in Table 80.2. The difference between this table and Table 80.1 is that in GSM the new serving VLR does not query the HLR separately in order to authenticate the terminal during registration and, hence, there are no HLR queries during registration. Instead, the entry (749 queries) under HLR queries in Table 80.1, corresponding to mobility-related authentication activity at registration, gets equally divided between the 128 VLRs. Observe that with either protocol the total database traffic rates are conserved, where the total database traffic for the entire network is given by the sum of all of the entries in the last row total (Network), i.e.,

$$\text{HLR updates} + \text{VLR updates} + \text{HLR queries} + \text{VLR queries}$$

From Tables 80.1 and 80.2 we see that this quantity equals 7087.

The conclusion is independent of any variations we may provide to the assumptions in earlier in the section. For example, if the PCS penetration (the percentage of the total users subscribing to PCS services) were to increase from 15 to 30%, all of the entries in the two tables will double and, hence, the total database traffic generated by the two protocols will still be equal.

TABLE 80.2 GSM Query and Update Rates to HLR and VLR

Activity	HLR Updates/s	VLR Updates/s	HLR Queries/s	VLR Queries/s
Mobility-related activities at registration	749	5.85		11.7
Mobility-related activities at deregistration		5.85		
Call origination				8.7
Call delivery			1116	8.7
Total (per VLR)	749	11.7	1116	29.1
Total (Network)	749	1497.6	1116	3724.8

80.6 Reducing Signalling During Call Delivery

In the preceding section, we provided a simplified analysis of some scenarios associated with user location strategies and the associated database queries and updates required. Previous studies [Meier-Hellstern and Alonso, 1992; Lo, Wolff, and Bernhardt, 1992] indicate that the signalling traffic and database queries associated with PCS due to user mobility are likely to grow to levels well in excess of that associated with a conventional call. It is, therefore, desirable to study modifications to the two protocols that would result in reduced signalling and database traffic. We now provide some suggestions.

For both GSM and IS-41, delivery of calls to a mobile user involves four messages: from the caller's VLR to the called party's HLR, from the HLR to the called party's VLR, from the called party's VLR to the HLR, and from the HLR to the caller's VLR. The last two of these messages involve the HLR, whose role is to simply relay the routing information provided by the called party's VLR to the caller's VLR. An obvious modification to the protocol would be to have the called VLR directly send the routing information to the calling VLR. This would reduce the total load on the HLR and on signalling network links substantially. Such a modification to the protocol may not be easy, of course, due to administrative, billing, legal, or security concerns. Besides, this would violate the query/response model adopted in IS-41, requiring further analysis.

A related question which arises is whether the routing information obtained from the called party's VLR could instead be stored in the HLR. This routing information could be provided to the HLR, for example, whenever a terminal registers in a new registration area. If this were possible, two of the four messages involved in call delivery could be eliminated. This point was discussed at length by the GSM standards body, and the present strategy was arrived at. The reason for this decision was to reduce the number of temporary routing numbers allocated by VLRs to terminals in their registration area. If a temporary routing number (TLDN in IS-41 or MSRN in GSM) is allocated to a terminal for the whole duration of its stay in a registration area, the quantity of numbers required is much greater than if a number is assigned on a per-call basis. Other strategies may be employed to reduce signalling and database traffic via intelligent paging or by storing user's mobility behavior in user profiles (see, for example, Tabbane, 1993). A discussion of these techniques is beyond the scope of the paper.

80.7 Per-User Location Caching

The basic idea behind per-user location caching is that the volume of SS7 message traffic and database accesses required in locating a called subscriber can be reduced by maintaining local storage, or cache, of user location information at a switch. At any switch, location caching for a given user should be employed only if a large number of calls originate for that user from that switch, relative to the user's mobility. Note that the cached information is kept at the switch from which calls originate, which may or may not be the switch where the user is currently registered.

Location caching involves the storage of location pointers at the originating switch; these point to the VLR (and the associated switch) where the user is currently registered. We refer to the procedure of locating a PCS user a *FIND* operation, borrowing the terminology from Awerbuch and Peleg, 1991. We define a basic *FIND*, or *BasicFIND*( ), as one where the following sequence of steps takes place.

1. The incoming call to a PCS user is directed to the nearest switch.
2. Assuming that the called party is not located within the immediate RA, the switch queries the HLR for routing information.
3. The HLR contains a pointer to the VLR in whose associated RA the subscriber is currently situated and launches a query to that VLR.
4. The VLR, in turn, queries the MSC to determine whether the user terminal is capable of receiving the call (i.e., is idle) and, if so, the MSC returns a routable address (TLDN in IS-41) to the VLR.
5. The VLR relays the routing address back to the originating switch via the HLR.

At this point, the originating switch can route the call to the destination switch. Alternately, *BasicFIND*( ) can be described by pseudocode as follows. (We observe that a more formal method of specifying PCS protocols may be desirable).

> *BasicFIND*( ){
>   Call to PCS user is detected at local switch;
>   *if* called party is in same RA *then* return;
>   Switch queries called party's HLR;
>   Called party's HLR queries called party's current VLR, *V*;
>   *V* returns called party's location to HLR;
>   HLR returns location to calling switch;
> }

In the *FIND* procedure involving the use of location caching, or *CacheFIND*( ), each switch contains a local memory (cache) that stores location information for subscribers. When the switch receives a call origination (from either a wire-line or wireless caller) directed to a PCS subscriber, it first checks its cache to see if location information for the called party is maintained. If so, a query is launched to the pointed VLR; if not, *BasicFIND*( ), as just described, is followed. If a cache entry exists and the pointed VLR is queried, two situations are possible. If the user is still registered at the RA of the pointed VLR (i.e., we have a *cache hit*), the pointed VLR returns the user's routing address. Otherwise, the pointed VLR returns a *cache miss*.

> *CacheFIND*( ){
>   Call to PCS user is detected at local switch;
>   *if* called is in same RA *then* return;
>   *if* there is no cache entry for called user
>   *then* invoke *BasicFIND*( ) and return;
>   Switch queries the VLR, *V*, specified in the cache entry;
>   *if* called is at *V*, *then*
>   *V* returns called party's location to calling switch;
>   *else* {
>   *V* returns "miss" to calling switch;
>   Calling switch invokes *BasicFIND*( );
>   }
> }

When a cache hit occurs we save one query to the HLR [a VLR query is involved in both *CacheFIND( )* and *BasicFIND( )*], and we also save traffic along some of the signalling links; instead of four message transmissions, as in *BasicFIND( )*, only two are needed. In steady-state operation, the cached pointer for any given user is updated only upon a miss.

Note that the *BasicFIND( )* procedure differs from that specified for roaming subscribers in the IS-41 standard EIA/TIA, 1991. In the IS-41 standard, the second line in the *BasicFIND( )* procedure is omitted, i.e., every call results in a query of the called user's HLR. Thus, in fact, the procedure specified in the standard will result in an even higher network load than the *BasicFIND( )* procedure specified here. To make a fair assessment of the benefits of *CacheFIND( )*, however, we have compared it against *BasicFIND( )*. Thus, the benefits of *CacheFIND( )* investigated here depend specifically on the use of caching and not simply on the availability of user location information at the local VLR.

80.8 Caching Threshold Analysis

In this section we investigate the classes of users for which the caching strategy yields net reductions in signalling traffic and database loads. We characterize classes of users by their CMR. The CMR of a user is the average number of calls to a user per unit time, divided by the average number of times the user changes registration areas per unit time. We also define a LCMR, which is the average number of calls to a user from a given originating switch per unit time, divided by the average number of times the user changes registration areas per unit time.

For each user, the amount of savings due to caching is a function of the probability that the cached pointer correctly points to the user's location and increases with the user's LCMR. In this section we quantify the minimum value of LCMR for caching to be worthwhile. This caching threshold is parameterized with respect to costs of traversing signalling network elements and network databases and can be used as a guide to select the subset of users to whom caching should be applied. The analysis in this section shows that estimating user's LCMRs, preferably dynamically, is very important in order to apply the caching strategy. The next section will discuss methods for obtaining this estimate.

From the pseudocode for *BasicFIND( )*, the signalling network cost incurred in locating a PCS user in the event of an incoming call is the sum of the cost of querying the HLR (and receiving the response), and the cost of querying the VLR which the HLR points to (and receiving the response). Let

α = cost of querying the HLR and receiving a response

β = cost of querying the pointed VLR and receiving a response

Then, the cost of *BasicFIND( )* operation is

$$C_B = \alpha + \beta \tag{80.2}$$

To quantify this further, assume costs for traversing various network elements as follows.

A_l = cost of transmitting a location request or response message on A link between SSP and LSTP

D = cost of transmitting a location request on response message or D link

A_r = cost of transmitting a location request or response message on A link between RSTP and SCP

L = cost of processing and routing a location request or response message by LSTP

R = cost of processing and routing a location request or response message by RSTP

H_Q = cost of a query to the HLR to obtain the current VLR location

V_Q = cost of a query to the VLR to obtain the routing address

Then, using the PCS reference network architecture (Fig. 80.2),

$$\alpha = 2(A_l + D + A_r + L + R) + H_Q \tag{80.3}$$
$$\beta = 2(A_l + D + A_r + L + R) + V_Q \tag{80.4}$$

From Eqs. (80.2)–(80.4)

$$C_B = 4(A_l + D + A_r + L + R) + H_Q + V_Q \tag{80.5}$$

We now calculate the cost of *CacheFIND*( ). We define the *hit ratio* as the relative frequency with which the cached pointer correctly points to the user's location when it is consulted. Let

p = cache hit ratio
C_H = cost of the *CacheFIND*( ) procedure when there is a hit
C_M = cost of the *CacheFIND*( ) procedure when there is a miss

Then the cost of *CacheFIND*( ) is

$$C_C = p\,C_H + (1-p)C_M \tag{80.6}$$

For *CacheFIND*( ), the signalling network costs incurred in locating a user in the event of an incoming call depend on the hit ratio as well as the cost of querying the VLR, which is stored in the cache; this VLR query may or may not involve traversing the RSTP. In the following, we say a VLR is a *local* VLR if it is served by the same LSTP as the originating switch, and a *remote* VLR otherwise. Let

q = Prob (VLR in originating switch's cache is a local VLR)
δ = cost of querying a local VLR
ϵ = cost of querying a remote VLR
η = cost of updating the cache upon a miss

Then,

$$\delta = 4A_l + 2L + V_Q \tag{80.7}$$
$$\epsilon = 4(A_l + D + L) + 2R + V_Q \tag{80.8}$$
$$C_H = q\delta + (1-q)\epsilon \tag{80.9}$$

Since updating the cache involves an operation to a fast local memory rather than a database operation, we shall assume in the following that $\eta = 0$. Then,

$$C_M = C_H + C_B = q\delta + (1-q)\epsilon + \alpha + \beta \tag{80.10}$$

From Eqs. (80.6), (80.9) and (80.10) we have

$$C_C = \alpha + \beta + \epsilon - p(\alpha + \beta) + q(\delta - \epsilon) \tag{80.11}$$

For net cost savings we require $C_C < C_B$, or that the hit ratio exceeds a *hit ratio threshold* p_T, derived using Eqs. (80.6), (80.9), and (80.2),

$$p > p_T = \frac{C_H}{C_B} = \frac{\epsilon + q(\delta - \epsilon)}{\alpha + \beta} \tag{80.12}$$

$$= \frac{4A_l + 4D + 4L + 2R + V_Q - q(4D + 2L + 2R)}{4A_l + 4D + 4A_r + 4L + 4R + H_Q + V_Q} \tag{80.13}$$

Equation (80.13) specifies the hit ratio threshold for a user, evaluated at a given switch, for which local maintenance of a cached location entry produces cost savings. As pointed out earlier, a given user's hit ratio may be location dependent, since the rates of calls destined for that user may vary widely across switches.

The hit ratio threshold in Eq. (80.13) is comprised of heterogeneous cost terms, i.e., transmission link utilization, packet switch processing, and database access costs. Therefore, numerical evaluation of the hit ratio threshold requires either detailed knowledge of these individual quantities or some form of simplifying assumptions. Based on the latter approach, the following two possible methods of evaluation may be employed.

1. Assume one or more cost terms dominate, and simplify Eq. (80.13) by setting the remaining terms to zero.
2. Establish a common unit of measure for all cost terms, for example, *time delay*. In this case, A_l, A_r, and D may represent transmission delays of fixed transmission speed (e.g., 56 kb/s) signalling links, L and R may constitute the sum of queueing and service delays of packet switches (i.e., STPs), and H_Q and V_Q the transaction delays for database queries.

In this section we adopt the first method and evaluate Eq. (80.13) assuming a single term dominates. (In Sec. 80.9 we present results using the second method). Table 80.3 shows the hit ratio threshold required to obtain net cost savings, for each case in which one of the cost terms is dominant.

In Table 80.3 we see that if the cost of querying a VLR or of traversing a local A link is the dominant cost, caching for users who may move is never worthwhile, regardless of users' call reception and mobility patterns. This is because the caching strategy essentially distributes the functionality of the HLR to the VLRs. Thus, the load on the VLR and the local A link is always increased, since any move by a user results in a cache miss. On the other hand, for a fixed user (or telephone), caching is always worthwhile. We also observe that if the remote A links or HLR querying are the bottlenecks, caching is worthwhile even for users with very low hit ratios.

As a simple average-case calculation, consider the net network benefit of caching when HLR access and update is the performance bottleneck. Consider a scenario where $u = 50\%$ of PCS users receive $c = 80\%$ of their calls from $s = 5$ RAs where their hit ratio $p > 0$, and $s' = 4$ of the SSPs at those RAs contain sufficiently large caches. Assume that caching is applied only to this subset of users and to no other users. Suppose that the average hit ratio for these users is $p = 80\%$, so that 80% of the HLR accesses for calls to these users from these RA are avoided. Then the net saving in the accesses to the system's HLR is $H = (u\,c\,s'\,p)/s = 25\%$.

We discuss other quantities in Table 80.3 next. It is first useful to relate the cache hit ratio to users' calling and mobility patterns directly via the LCMR. Doing so requires making assumptions about the distribution of the user's calls and moves. We consider the steady state where the incoming call stream from an SSP to a user is a Poisson process with arrival rate λ, and the time that the user

TABLE 80.3 Minimum Hit Ratios and LCMRs for Various Individual Dominant Signalling Network Cost Terms

Dominant Cost Term	Hit ratio Threshold, p_T	LCMR Threshold, $LCMR_T$	LCMR Threshold ($q = 0.043$)	LCMR Threshold ($q = 0.25$)
A_l	1	∞	∞	∞
A_r	0	0	0	0
D	$1 - q$	$1/q - 1$	22	3
L	$1 - q/2$	$2/q - 1$	45	7
R	$1 - q/2$	$2/q - 1$	45	7
H_Q	0	0	0	0
V_Q	1	∞	∞	∞

resides in an RA has a general distribution $F(t)$ with mean $1/\mu$. Thus,

$$LCMR = \frac{\lambda}{\mu} \tag{80.14}$$

Let t be the time interval between two consecutive calls from the SSP to the user and t_1 be the time interval between the first call and the time when the user moves to a new RA. From the random observer property of the arrival call stream [Feller, 1966], the hit ratio is

$$p = \Pr[t < t_1] = \int_{t=0}^{\infty} \lambda e^{-\lambda t} \int_{t_1=t}^{\infty} \mu[1 - F(t_1)] \, dt_1 \, dt$$

If $F(t)$ is an exponential distribution, then

$$p = \frac{\lambda}{\lambda + \mu} \tag{80.15}$$

and we can derive the *LCMR threshold*, the minimum LCMR required for caching to be beneficial assuming incoming calls are a Poisson process and intermove times are exponentially distributed,

$$LCMR_T = \frac{p_T}{1 - p_T} \tag{80.16}$$

Equation (80.16) is used to derive LCMR thresholds assuming various dominant costs terms, as shown in Table 80.3.

Several values for $LCMR_T$ in Table 80.3 involve the term q, i.e., the probability that the pointed VLR is a local VLR. These values may be numerically evaluated by simplifying assumptions. For example, assume that all of the SSPs in the network are uniformly distributed amongst l LSTPs. Also, assume that all of the PCS subscribers are uniformly distributed in location across all SSPs and that each subscriber exhibits the same incoming call rate at every SSP. Under those conditions, q is simply $1/l$. Consider the case of the public switched telephone network. Given that there are a total of 160 local access transport area (LATA) across the 7 Regional Bell Operating Company (RBOC) regions [Bellcore, 1992c], the average number of LATAs, or l, is 160/7 or 23. Table 80.3 shows the results with $q = 1/l$ in this case.

We observe that the assumption that all users receive calls uniformly from all switches in the network is extremely conservative. In practice, we expect that user call reception patterns would display significantly more locality, so that q would be larger and the LCMR thresholds required to make caching worthwhile would be smaller. It is also worthwhile to consider the case of a RBOC region with PCS deployed in a few LATA only, a likely initial scenario, say, 4 LATAs. In either case the value of q would be significantly higher; Table 80.3 shows the LCMR threshold when $q = 0.25$.

It is possible to quantify the net costs and benefits of caching in terms of signalling network impacts in this way and to determine the hit ratio and LCMR threshold above which users should have the caching strategy applied. Applying caching to users whose hit ratio and LCMR is below this threshold results in net increases in network impacts. It is, thus, important to estimate users' LCMRs accurately. The next section discusses how to do so.

80.9 Techniques for Estimating Users' LCMR

Here we sketch some methods of estimating users' LCMR. A simple and attractive policy is to not estimate these quantities on a per-user basis at all. For instance, if the average LCMR over all users in a PCS system is high enough (and from Table 80.3, it need not be high depending on which network elements are the dominant costs), then caching could be used at every SSP to yield net

Location Strategies for Personal Communications Services 1135

system-wide benefits. Alternatively, if it is known that at any given SSP the average LCMR over all users is high enough, a cache can be installed at that SSP. Other variations can be designed.

One possibility for deciding about caching on a per-user basis is to maintain information about a user's calling and mobility pattern at the HLR and to download it periodically to selected SSPs during off-peak hours. It is easy to envision numerous variations on this idea.

In this section we investigate two possible techniques for estimating LCMR on a per-user basis when caching is to be deployed. The first algorithm, called the *running average* algorithm, simply maintains a running average of the hit ratio for each user. The second algorithm, called the *reset-K* algorithm, attempts to obtain a measure of the hit ratio over the recent history of the user's movements. We describe the two algorithms next and evaluate their effectiveness using a stochastic analysis taking into account user calling and mobility patterns.

The Running Average Algorithm

The running average algorithm maintains, for every user that has a cache entry, the running average of the hit ratio. A running count is kept of the number of calls to a given user, and, regardless of the *FIND* procedure used to locate the user, a running count of the number of times that the user was at the same location for any two consecutive calls; the ratio of these numbers provides the measured running average of the hit ratio. We denote the measured running average of the hit ratio by p_M; in steady state, we expect that $p_M = p$. The user's previous location as stored in the cache entry is used only if the running average of the hit ratio p_M is greater than the cache hit threshold p_T. Recall that the cache scheme outperforms the basic scheme if $p > p_T = C_H/C_B$. Thus, in steady state, the running average algorithm will outperform the basic scheme when $p_M > p_T$.

We consider, as before, the steady state where the incoming call stream from an SSP to a user is a Poisson process with arrival rate λ, and the time that the user resides in an RA has an exponential distribution with mean $1/\mu$. Thus $LCMR = \lambda/\mu$ [Eq. (80.14)] and the location tracking cost at steady state is

$$C_C = \begin{cases} p_M C_H + (1 - p_M) C_B, & p_M > p_T \\ C_B, & \text{otherwise} \end{cases} \quad (80.17)$$

Figure 80.6 plots the cost ratio C_C/C_B from Eq. (80.17) against *LCMR*. (This corresponds to assigning uniform units to all cost terms in Eq. (80.13), i.e., the second evaluation method as discussed in Sec. 80.8. Thus, the ratio C_C/C_B may represent the percentage reduction in user location time with the caching strategy compared to the basic strategy.) The figure indicates that in the steady state, the caching strategy with the running average algorithm for estimating LCMR can significantly outperform the basic scheme if *LCMR* is sufficiently large. For instance with $LCMR \sim 5$, caching can lead to cost savings of 20–60% over the basic strategy.

Equation (80.17) (cf., solid curves in Fig. 80.6) is validated against a simple Monte Carlo simulation (cf., dashed curves in Fig. 80.6). In the simulation, the confidence interval for the 95% confidence level of the output measure C_C/C_B is within 3% of the mean value. This simulation model will later be used to study the running average algorithm when the mean of the movement distribution changes from time to time [which cannot be modeled by using Eq. (80.17)].

One problem with the running average algorithm is that the parameter p is measured from the entire past history of the user's movement, and the algorithm may not be sufficiently dynamic to adequately reflect the recent history of the user's mobility patterns.

The Reset-*K* Algorithm

We may modify the running average algorithm such that p is measured from the recent history. Define every K incoming calls as a *cycle*. The modified algorithm, which is referred to as the

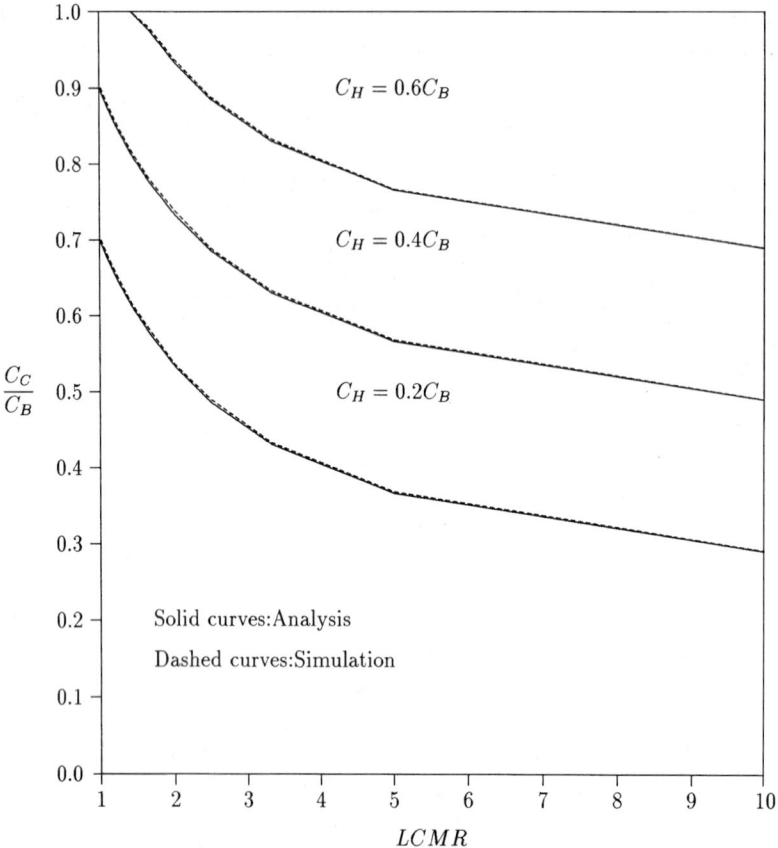

FIGURE 80.6 The location tracking cost for the running average algorithm.

reset-K algorithm, counts the number of cache hits n in a cycle. If the measured hit ratio for a user, $p_M = n/K \geq p_T$, then the cache is enabled for that user, and the cached information is always used to locate the user in the next cycle. Otherwise, the cache is disabled for that user, and the basic scheme is used. At the beginning of a cycle, the cache hit count is reset, and a new p_M value is measured during the cycle.

To study the performance of the reset-K algorithm, we model the number of cache misses in a cycle by a Markov process. Assume as before that the call arrivals are a Poisson process with arrival rate λ and the time period the user resides in an RA has an exponential distribution withe mean $1/\mu$. A pair (i, j), where $i > j$, represents the state that there are j cache misses before the first i incoming phone calls in a cycle. A pair $(i, j)^*$, where $i \geq j \geq 1$, represents the state that there are $j - 1$ cache misses before the first i incoming phone calls in a cycle, and the user moves between the ith and the $i + 1$ phone calls. The difference between (i, j) and $(i, j)^*$ is that if the Markov process is in the state (i, j) and the user moves, then the process moves into the state $(i, j + 1)^*$. On the other hand, if the process is in state $(i, j)^*$ when the user moves, the process remains in $(i, j)^*$ because at most one cache miss occurs between two consecutive phone calls.

Figure 80.7(a) illustrates the transitions for state $(i, 0)$ where $2 < i < K + 1$. The Markov process moves from $(i - 1, 0)$ to $(i, 0)$ if a phone call arrives before the user moves out. The rate is λ. The process moves from $(i, 0)$ to $(i, 1)^*$ if the user moves to another RA before the $i + 1$ call arrival. Let $\pi(i, j)$ denote the probability of the process being in state (i, j). Then the transition

Location Strategies for Personal Communications Services

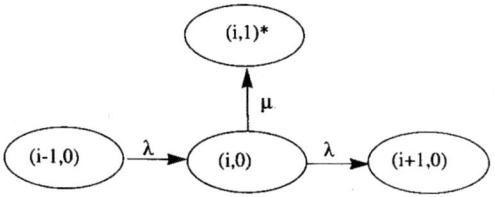

(a) Transitions for state (i,0) (2 < i < K+1)

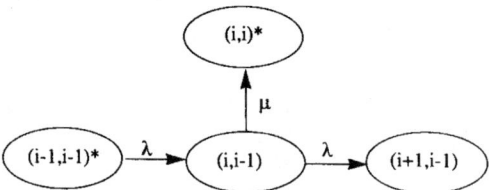

(b) Transitions for state (i,i-1)(1 < i < K+1)

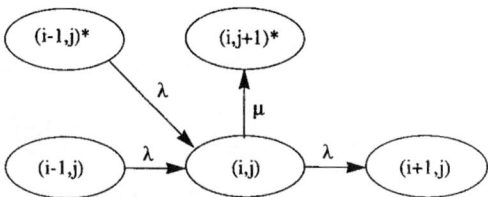

(c) Transitions for state (i,j) (2< i < K+1, 0 < j < i-1)

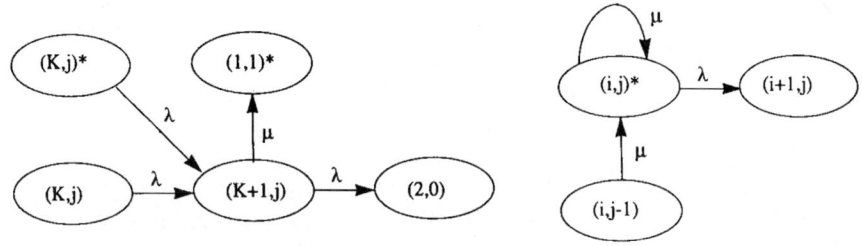

(d) Transitions for state (K+1,j) (0 < j < K+1)

(e) Transitions for state (i,j)* (0 < j ≤ i, 1 < i < K

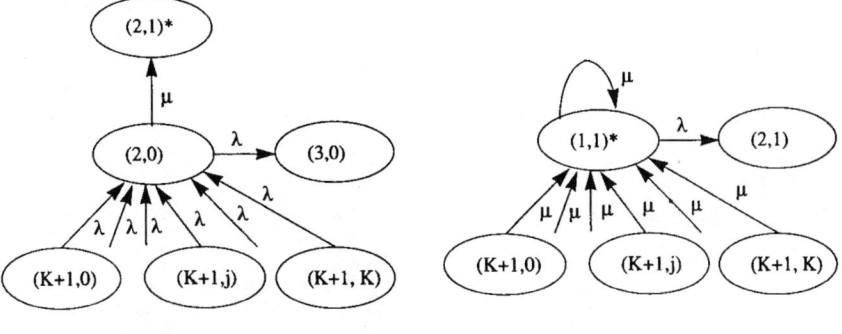

(f) Transitions for state (2,0)

(g) Transitions for state (1,1)*

FIGURE 80.7 State transitions.

equation is

$$\pi(i, 0) = \frac{\lambda}{\lambda + \mu} \pi(i - 1, 0), \qquad 2 < i < K + 1 \qquad (80.18)$$

Figure 80.7(b) illustrates the transitions for state $(i, i-1)$ where $1 < i < K+1$. The only transition into the state $(i, i-1)$ is from $(i-1, i-1)^*$, which means that the user always moves to another RA after a phone call. [Note that there can be no state $(i-1, i-1)$ by definition and, hence, no transition from such a state.] The transition rate is λ. The process moves from $(i, i-1)$ to $(i, i)^*$ with rate μ, and moves to $(i+1, i-1)$ with rate λ. Let $\pi^*(i, j)$ denote the probability of the process being in state $(i, j)^*$. Then the transition equation is

$$\pi(i, i-1) = \frac{\lambda}{\lambda + \mu} \pi^*(i - 1, i - 1), \qquad 1 < i < K + 1 \qquad (80.19)$$

Figure 80.7(c) illustrates the transitions for state (i, j) where $2 < i < K+1, 0 < j < i-1$. The process may move into state (i, j) from two states $(i-1, j)$ and $(i-1, j)^*$ with rate λ, respectively. The process moves from (i, j) to $(i, j+1)^*$ or $(i+1, j)$ with rates μ and λ, respectively. The transition equation is

$$\pi(i, j) = \frac{\lambda}{\lambda + \mu} [\pi(i-1, j) + \pi^*(i-1, j)],$$
$$2 < i < K + 1, \qquad 0 < j < i - 1 \qquad (80.20)$$

Figure 80.7(d) illustrates the transitions for state $(K+1, j)$ where $0 < j < K+1$. Note that if a phone call arrives when the process is in (K, j) or $(K, j)^*$, the system enters a new cycle (with rate λ), and we could represent the new state as $(1, 0)$. In our model, we introduce the state $(K+1, j)$ instead of $(1, 0)$, where

$$\sum_{0 \leq j \leq K} \pi(K + 1, j) = \pi(1, 0)$$

so that the hit ratio, and thus the location tracking cost, can be derived [see Eq. (80.25)]. The process moves from $(K+1, j)$ [i.e., $(1, 0)$] to $(1, 1)^*$ with rate μ if the user moves before the next call arrives. Otherwise, the process moves to $(2, 0)$ with rate λ. The transition equation is

$$\pi(K + 1, j) = \frac{\lambda}{\lambda + \mu} [\pi(K, j) + \pi^*(K, j)], \qquad 0 < j < K + 1 \qquad (80.21)$$

For $j = 0$, the transition from $(K, j)^*$ to $(K+1, 0)$ should be removed in Fig. 80.7(d) because the state $(K, 0)^*$ does not exist. The transition equation for $(K+1, 0)$ is given in Eq. (80.18). Figure 80.7(e) illustrates the transitions for state $(i, j)^*$ where $0 < j < i, 1 < i < K+1$. The process can only move to $(i, j)^*$ from $(i, j-1)$ (with rate μ). From the definition of $(i, j)^*$, if the user moves when the process is in $(i, j)^*$, the process remains in $(i, j)^*$ (with rate μ). Otherwise, the process moves to $(i+1, j)$ with rate λ. The transition equation is

$$\pi^*(i, j) = \frac{\mu}{\lambda} \pi(i, j - 1), \qquad 0 < j \leq i, \qquad 1 < i < K + 1, \qquad i \geq 2 \quad (80.22)$$

The transitions for $(2, 0)$ are similar to the transitions for $(i, 0)$ except that the transition from $(1, 0)$

is replaced by $(K+1, 0), \ldots, (K+1, K)$ [cf., Fig. 80.7(f)]. The transition equation is

$$\pi(2, 0) = \frac{\lambda}{\lambda + \mu} \left[\sum_{0 \leq j \leq K} \pi(K+1, j) \right] \tag{80.23}$$

Finally, the transitions for $(1, 1)^*$ is similar to the transitions for $(i, j)^*$ except that the transition from $(1, 0)$ is replaced by $(K+1, 0), \ldots, (K+1, K)$ [cf., Fig. 80.7(g)]. The transition equation is

$$\pi^*(1, 1) = \frac{\mu}{\lambda} \left[\sum_{0 \leq j \leq K} \pi(K+1, j) \right] \tag{80.24}$$

Suppose that at the beginning of a cycle, the process is in state $(K+1, j)$, then it implies that there are j cache misses in the previous cycle. The cache is enabled if and only if

$$p_M \geq p_T = \frac{C_H}{C_B} \Rightarrow 1 - \frac{j}{K} \geq \frac{C_H}{C_B} \Rightarrow 0 \leq j \leq \left\lceil K\left(1 - \frac{C_H}{C_B}\right) \right\rceil$$

Thus, the probability that the measured hit ratio $p_M < p_T$ in the previous cycle is

$$Pr[p_M < p_T] = \frac{\sum_{\lceil k[1-(C_H/C_B)]\rceil < j \leq K} \pi(K+1, j)}{\sum_{0 \leq j \leq K} \pi(K+1, j)}$$

and the location tracking cost for the reset-K algorithm is

$$C_C = C_B Pr[p_M < p_T] + (1 - Pr[p_M < p_T])$$

$$\times \left\{ \sum_{0 \leq j \leq K} \left(\frac{(K-j)C_H}{K} + \frac{j(C_H + C_B)}{K} \right) \left[\frac{\pi(K+1, j)}{\sum_{0 \leq i \leq K} \pi(K+1, i)} \right] \right\} \tag{80.25}$$

The first term Eq. (80.25) represents the cost incurred when caching is disabled because the hit ratio threshold exceeds the hit ratio measured in the previous cycle. The second term is the cost when the cache is enabled and consists of two parts, corresponding to calls during which hits occur and calls during which misses occur. The ratio in square brackets is the conditional probability of being in state $\pi(K+1, j)$ during the current cycle.

The numerical computation of $\pi(K+1, j)$ can be done as follows. First, compute $a_{i,j}$ and $b_{i,j}$ where $\pi(i, j) = a_{i,j}\pi^*(1, 1)$ and $\pi^*(i, j) = b_{i,j}\pi^*(1, 1)$. Note that $a_{i,j} = 0(b_{i,j} = 0)$ if $\pi(i, j)[\pi^*(i, j)]$ is not defined in Eqs. (80.18)–(80.24). Since

$$\sum_{i,j} [\pi(i, j) + \pi^*(i, j)] = 1$$

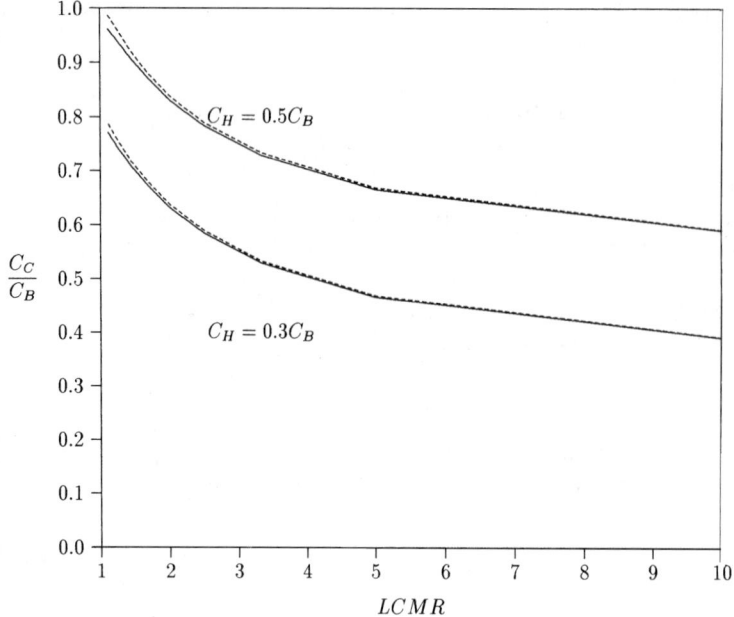

FIGURE 80.8 The location tracking costs for the reset-K algorithm; $K = 20$.

we have

$$\pi^*(1, 1) = \frac{1}{\sum_{i,j}(a_{i,j} + b_{i,j})}$$

and $\pi(K + 1, j)$ can be computed and the location tracking cost for the reset-K algorithm is obtained using Eq. (80.25).

The analysis is validated by a Monte Carlo simulation. In the simulation, the confidence interval for the 98% confidence level of the output measure C_C/C_B is within 3% of the mean value. Figure 80.8 plots curves for Eq. (80.25) (the solid curves) against the simulation experiments (the dashed curves) for $K = 20$ and $C_H = 0.5C_B$ and $0.3C_B$, respectively. The figure indicates that the analysis is consistent with the simulation model.

Comparison of the LCMR Estimation Algorithms

If the distributions for the incoming call process and the user movement process never change, then we would expect the running average algorithm to outperform the reset-K algorithm (especially when K is small) because the measured hit ratio p_M in the running average algorithm approaches the true hit ratio value p in the steady state. Surprisingly, the performance for the reset-K algorithm is roughly the same as the running average algorithm even if K is as small as 10. Figure 80.9 plots the location tracking costs for the running average algorithm and the reset-K algorithm with different K values.

The figure indicates that in steady state, when the distributions for the incoming call process and the user movement process never change, the running average algorithm outperforms reset K, and a large value of K outperforms a small K but the differences are insignificant.

If the distributions for the incoming call process or the user movement process change from time to time, we expect that the reset-K algorithm outperforms the running average algorithm. We have

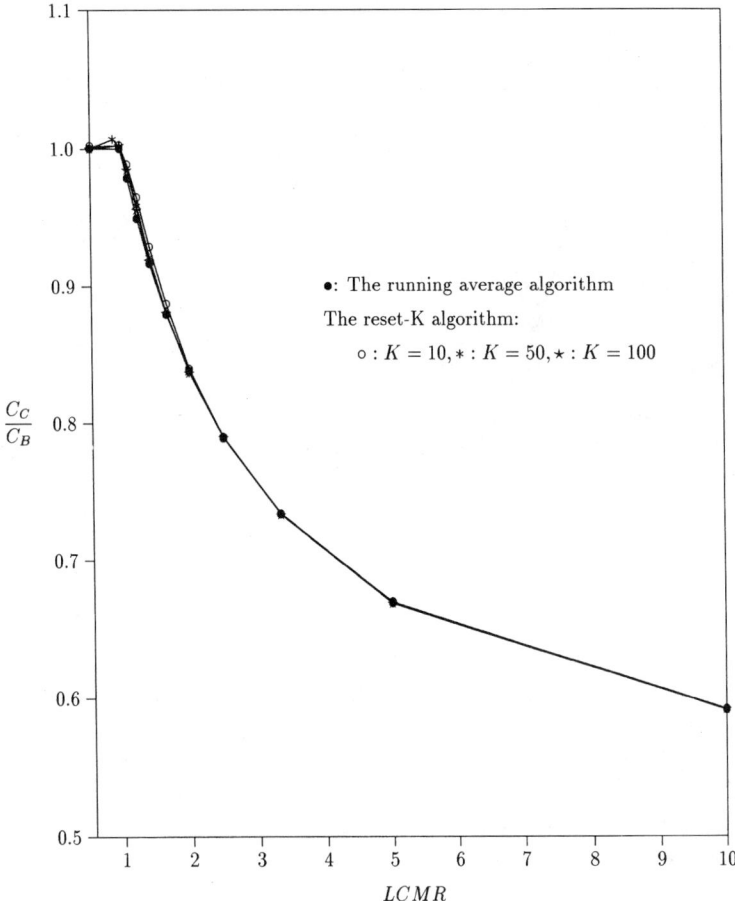

FIGURE 80.9 The location tracking costs for the running average algorithm and the reset-K algorithm; $C_H = 0.5C_B$.

examined this proposition experimentally. In the experiments, 4000 incoming calls are simulated. The call arrival rate changes from 0.1 to 1.0, 0.3, and then 5.0 for every 1000 calls (other sequences have been tested and similar results are observed). For every data point, the simulation is repeated 1000 times to ensure that the confidence interval for the 98% confidence level of the output measure C_C/C_B is within 3% of the mean value. Figure 80.10 plots the location tracking costs for the two algorithms for these experiments. By changing the distributions of the incoming call process, we observe that the reset-K algorithm is better than the running average algorithm for all C_H/C_B values.

80.10 Discussion

In this section we discuss aspects of the caching strategy presented here. Caching in PCS systems raises a number of issues not encountered in traditional computer systems, particularly with respect to architecture and locality in user call and mobility patterns. In addition, several variations in our reference assumptions are possible for investigating the implementation of the caching strategies. Here we sketch some of the issues involved.

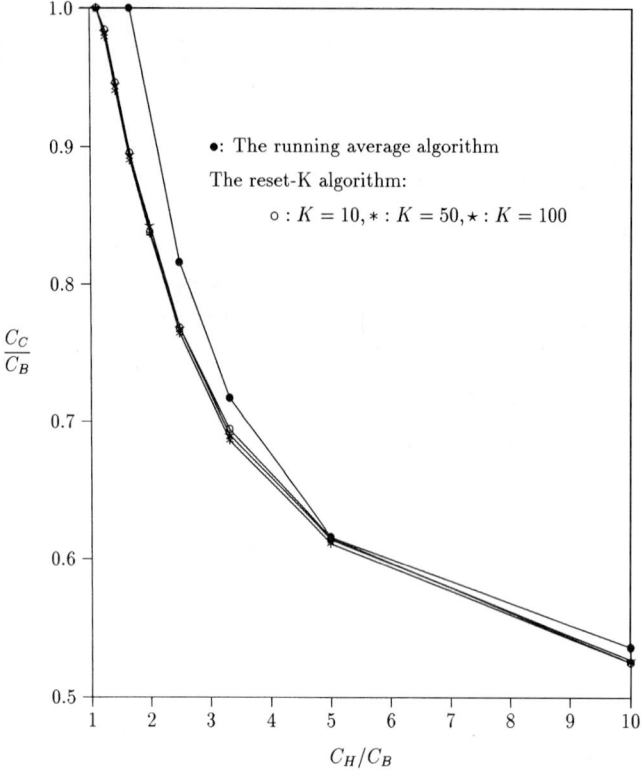

FIGURE 80.10 Comparing the running average algorithm and the reset-K algorithm under unstable call traffic.

Conditions When Caching Is Beneficial

We summarize the conditions for which the auxiliary strategies are worthwhile, under the assumptions of our analysis.

The caching strategy is very promising when the HLR update (or query load) or the remote A link is the performance bottleneck, since a low $LCMR$ ($LCMR > 0$) is required. For caching, the total database load and signalling network traffic is reduced whenever there is a cache hit. In addition, load and traffic is redistributed from the HLR and higher level SS7 network elements (RSTP, D links) to the VLRs and lower levels where excess network capacity may be more likely to exist. If the VLR is the performance bottleneck, the caching strategy is not promising, unless the VLR capacity is upgraded.

The benefits of the caching strategy depend on user call and mobility patterns when the D link, RSTP, and LSTP are the performance bottlenecks. We have used a Poisson call arrival model and exponential intermove time to estimate this dependence. Under very conservative assumptions, for caching to be beneficial requires relatively high $LCMR$ (25–50); we expect that in practice this threshold could be lowered significantly (say, $LCMR > 7$). Further experimental study is required to estimate the amount of locality in user movements for different user populations to investigate this issue further. It is possible that for some classes of users data obtained from active badge location system studies (e.g., Fishman and Mazer, 1992) could be useful. In general, it appears that caching could also potentially provide benefits to some classes of users even when the D link, the RSTP, or the LSTP are the bottlenecks.

We observe that more accurate models of user calling and mobility patterns are required to help resolve the issues raised in this section. We are currently engaged in developing theoretical models for user mobility and estimating their effect on studies of various aspects of PCS performance [Lin and Jain, 1993].

Alternative Network Architectures

The reference architecture we have assumed (Fig. 80.2) is only one of several possible architectures. It is possible to consider variations in the placement of the HLR and VLR functionality, (e.g., placing the VLR at a local SCP associated with the LSTP instead of at the SSP), the number of SSPs served by an LSTP, the number of HLRs deployed, etc. It is quite conceivable that different regional PCS service providers and telecommunications companies will deploy different signalling network architectures, as well as placement of databases for supporting PCS within their serving regions [Russo et al., 1993]. It is also possible that the number and placement of databases in a network will change over time as the number of PCS users increases.

Rather than consider many possible variations of the architecture, we have selected a reference architecture to illustrate the new auxiliary strategies and our method of calculating their costs and benefits. Changes in the architecture may result in minor variations in our analysis but may not significantly affect our qualitative conclusions.

LCMR Estimation and Caching Policy

It is possible that for some user populations estimating the LCMR may not be necessary, since they display a relatively high-average LCMR. For some populations, as we have shown in Sec. 80.9, obtaining accurate estimates of user LCMR in order to decide whether or not to use caching can be important in determining the net benefits of caching.

In general, schemes for estimating the LCMR range from static to dynamic and from distributed to centralized. We have presented two simple distributed algorithms for estimating LCMR, based on a long-range and short-range running calculation; the former is preferable if the call and mobility pattern of users is fairly static, whereas the latter is preferable if it is variable. Tuning the amount of history that is used to determine whether caching should be employed for a particular user is an obvious area for further study but is outside the scope of this chapter.

An alternative approach is to utilize some user-supplied information, by requesting profiles of user movements (e.g., see Tabbane, 1993 and Jain, 1995) and to integrate this with the caching strategy. A variation of this approach is to use some domain knowledge about user populations and their characteristics.

A related issue is that of cache size and management. In practice it is likely that the monetary cost of deploying a cache may limit its size. In that case, cache entries may not be maintained for some users; selecting these users carefully is important to maximize the benefits of caching. Note that the cache hit ratio threshold cannot necessarily be used to determine which users have cache entries, since it may be useful to maintain cache entries for some users even though their hit ratios have temporarily fallen below the threshold. A simple policy that has been found to be effective in computer systems in the least recently used (LRU) policy [Silbershatz and Peterson, 1988] in which cache entries that have been least recently used are discarded; LRU may offer some guidance in this context.

80.11 Conclusions

We began this chapter with an overview of the nuances of PCS, such as personal and terminal mobility, registration, deregistration, call delivery, etc. A tutorial was then provided on the two most common strategies for locating users in PCS, in North American interim standard IS-41 and the Pan-European standard GSM. A simplified analysis of the two standards was then provided to show the reader the extent to which database and signalling traffic is likely to be generated by PCS services. Suggestions were then made that are likely to result in reduced traffic.

Previous studies [Lo, Wolff, and Bernhardt, 1992; Meier-Hellstern and Alonso, 1992; Lo, Mohan, and Wolff, 1993; and Lo and Wolff, 1993] of PCS-related network signalling and data management functionalities suggest a high level of utilization of the signalling network in supporting call and mobility management activities for PCS systems. Motivated by the need to evolve location strategies to reduce signalling and database loads, we then presented an auxiliary strategy, called per-user caching, to augment the basic user location strategy proposed in standards [EIA/TIA, 1991; Mouly and Pautet, 1992].

Using a reference system architecture for PCS, we quantified the criteria under which the caching strategy produces reductions in the network signalling and database loads in terms of users' LCMRs. We have shown that, if the HLR or the remote A link in an SS7 architecture is the performance bottleneck, caching is useful regardless of user call and mobility patterns. If the D link or STPs are the performance bottlenecks, caching is potentially beneficial for large classes of users, particularly if they display a degree of locality in their call reception patterns. Depending on the numbers of PCS users who meet these criteria, the system-wide impacts of these strategies could be significant. For instance, for users with $LCMR \sim 5$ and stable call and move patterns, caching can result in cost reduction of 20–60% over the basic user location strategy *BasicFIND( )* under our analysis. Our results are conservative in that the *BasicFIND( )* procedure we have used for comparison purposes already reduces the network impacts compared to the user location strategy specified in PCS standards such as IS-41.

We have also investigated in detail two simple on-line algorithms for estimating users' LCMRs and examined the call and mobility patterns for which each would be useful. The algorithms allow a system designer to tune the amount of history used to estimate a users' LCMR and, hence, to attempt to optimize the benefits due to caching. The particular values of cache hit ratios and LCMR thresholds will change with variations in the way the PCS architecture and the caching strategy is implemented, but our general approach can still be applied. There are several issues deserving further study with respect to deployment of the caching strategy, such as the effect of alternative PCS architectures, integration with other auxiliary strategies such as the use of user profiles, and effective cache management policies.

Recently, we have augmented the work reported in this paper by a simulation study in which we have compared the caching and basic user location strategies [Harjono, Jain, and Mohan, 1994]. The effect of using a time-based criterion for enabling use of the cache has also been considered [Lin, 1994]. We have proposed elsewhere, for users with low CMRs, an auxiliary strategy involving a system of forwarding pointers to reduce the signalling traffic and database loads [Jain and Lin, 1995], a description of which is beyond the scope of this chapter.

Acknowledgment

We acknowledge a number of our colleagues in Bellcore who have reviewed several previous papers by the authors and contributed to improving the clarity and readability of this work.

References

Awerbuch, B. and Peleg, D. 1991. Concurrent online tracking of mobile users. In *Proc. SIGCOMM Symp. Comm. Arch. Prot.* Oct.

Bellcore. 1991. Advanced intelligent network release 1 network and operations plan, Issue 1. Tech. Rept. SR-NPL-001623. June. Bell Communications Research, Morristown, NJ.

Bellcore. 1993a. Personal communications services (PCS) network access services to PCS providers, Special Report SR-TSV-002459, Oct. Bell Communications Research, Morristown, NJ.

Bellcore. 1992c. Switching system requirements for interexchange carrier interconnection using the integrated services digital network user part (ISDNUP). Tech. Ref. TR-NWT-000394. Dec. Bell Communications Research. Morristown, NJ.

Berman, R.K. and Brewster, J.H. 1992. Perspectives on the AIN architecture. *IEEE Comm. Mag.* 1(2):27–32.

Electronic Industries Association/Telecommunications Industry Association. 1991. Cellular radio telecommunications intersystem operations. Tech. Rept. IS-41. Rev. B. July.

Feller, W. 1966. *An Introduction to Probability Theory and Its Applications.* John Wiley, New York.

Fishman, N. and Mazer, M. 1992. Experience in deploying an active badge system. In *Proc. Globecom Workshop on Networking for Pers. Comm. Appl.* Dec.

Harjono, H., Jain, R. and Mohan, S. 1994. Analysis and simulation of a cache-based auxiliary location strategy for PCS. In *Proc. IEEE Conf. Networks Pers. Comm.*

Jain, R. 1995. A classification scheme for user location strategies in personal communications services systems. Aug. Submitted for publication.

Jain, R. and Lin Y.-B. 1995. An auxiliary user location strategy employing forwarding pointers to reduce network impacts of PCS. *ACM Journal on Wireless Info. Networks (WINET)*, 1(2).

Lin, Y.-B. 1994. Determining the user locations for personal communications networks. *IEEE Trans. Vehic. Tech.* (Aug):466–473.

Lo, C., Mohan, S., and Wolff, R. 1993. A comparison of data management alternatives for personal communications applications. Second Bellcore Symposium on Performance Modeling, SR-TSV-002424, Nov. Bell Communications Research, Morristown, NJ.

Lo, C.N., Wolff, R.S., and Bernhardt, R.C. 1992. An estimate of network database transaction volume to support personal communications services. In *Proc. Intl. Conf. Univ. Pers. Comm.*

Lo, C. and Wolff, R. 1993. Estimated network database transaction volume to support wireless personal data communications applications. In *Proc. Intl. Conf. Comm.* May.

Lycksell, E. 1991. GSM system overview. Tech. Rept. Swedish Telecom. Admin. Jan.

Meier-Hellstern, K. and Alonso, E. 1992. The use of SS7 and GSM to support high density personal communications. In *Proc. Intl. Conf. Comm.*

Mohan, S. and Jain, R. 1994. Two user location strategies for PCS. *IEEE Pers. Comm. Mag.* Premiere issue. (Feb):42–50.

Mouly, M. and Pautet, M.B. 1992. *The GSM System for Mobile Communications.* M. Mouly, 49 rue Louise Bruneau, Palaiseau, France.

Russo, P., Bechard, K., Brooks, E., Corn, R.L., Honig, W.L., Gove, R., and Young, J. 1993. IN rollout in the United States. *IEEE Comm. Mag.* (March):56–63.

Silberschatz, A. and Peterson, J. 1988. *Operating Systems Concepts.* Addison–Wesley, Reading, MA.

Tabbane, S. 1992. Comparison between the alternative location strategy (AS) and the classical location strategy (CS). Tech. Rept. Rutgers Univ. WINLAB. July. Rutgers, NJ.

Tabbane, S. 1993. Evaluation of an alternative location strategy for future high density wireless communications systems. Tech. Rept. WINALAB-TR-51, Rutgers Univ. WINLAB. Jan. Rutgers, NJ.

Thomas, R., Gilbert, H., and Mazziotto, G. 1988. Influence of the mobile station on the performance of a radio mobile cellular network. In *Proc. 3rd Nordic Seminar.* Sept.

81
Cell Design Principles

81.1	Introduction	1146
81.2	Cellular Principles	1147
81.3	Performance Measures and System Requirements	1148
81.4	System Expansion Techniques	1148
81.5	Basic Design Steps	1149
81.6	Traffic Engineering	1150
81.7	Cell Coverage	1151
	Propagation Model • Base Station Coverage • Application Examples	
81.8	Interference	1155
	Adjacent Channel Interference • Cochannel Interference	
81.9	Conclusions	1158

Michel Daoud Yacoub
University of Campinas

81.1 Introduction

Designing a cellular network is a challenging task that invites engineers to exercise all of their knowledge in telecommunications. Although it may not be necessary to work as an expert in all of the fields, the interrelationship among the areas involved impels the designer to naturally search for a deeper understanding of the main phenomena. In other words, the time for segregation, when radio engineers and traffic engineers would not talk to each other, at least through a common vocabulary, is probably gone.

A great many aspects must be considered in a cellular network planning. The main ones include the following.

Radio Propagation: Here the topography and the morphology of the terrain, the urbanization factor and the clutter factor of the city, and some other aspects of the target geographical region under investigation will constitute the input data for the radio coverage design.

Frequency Regulation and Planning: In most countries there is a centralized organization, usually performed by a government entity, regulating the assignment and use of the radio spectrum. The frequency planning within the assigned spectrum should then be made so that interferences are minimized and the traffic demand is satisfied.

Modulation: As far as analog systems are concerned, the narrowband FM is widely used due to its remarkable performance in the presence of fading. The North American Digital Cellular

Standard IS-54 proposes the $\pi/4$ differential quadrature phase-shift keying ($\pi/4$ DQPSK) modulation, whereas the Global Standard for Mobile Communications (GSM) establishes the use of the Gaussian minimum-shift keying (GMSK).

Antenna Design: To cover large areas and for low-traffic applications omnidirectional antennas are recommended. Some systems at their inception may have these characteristics, and the utilization of omnidirectional antennas certainly keeps the initial investment low. As the traffic demand increases, the use of some sort of capacity enhancement technique to meet the demand, such as replacing the omnidirectional by directional antennas, is mandatory.

Transmission Planning: The structure of the channels, both for signalling and voice, is one of the aspects to be considered in this topic. Other aspects include the performance of the transmission components (power capacity, noise, bandwidth, stability, etc.) and the design or specification of transmitters and receivers.

Switching Exchange: In most cases this consists of adapting the existing switching network for mobile radio communications purposes.

Teletraffic: For a given grade of service and number of channels available, how many subscribers can be accommodated into the system? What is the proportion of voice and signalling channels?

Software Design: With the use of microprocessors throughout the system there are software applications in the mobile unit, in the base station, and in the switching exchange.

Other aspects, such as human factors, economics, etc., will also influence the design.

This chapter outlines the aspects involving the basic design steps in cellular network planning. Topics, such as traffic engineering, cell coverage, and interference, will be covered, and application examples will be given throughout the section so as to illustrate the main ideas. We start by recalling the basic concepts including *cellular principles, performance measures and system requirements*, and *system expansion techniques*.

81.2 Cellular Principles

The basic idea of the cellular concept is *frequency reuse* in which the same set of channels can be reused in different geographical locations sufficiently apart from each other so that *cochannel interference* be within tolerable limits. The set of channels available in the system is assigned to a group of *cells* constituting the *cluster*. Cells are assumed to have a *regular hexagonal* shape and the number of cells per cluster determines the *repeat pattern*. Because of the hexagonal geometry only certain repeat patterns can tessellate. The number N of cells per cluster is given by

$$N = i^2 + ij + j^2 \qquad (81.1)$$

where i and j are integers. From Eq. (81.1) we note that the clusters can accommodate only certain numbers of cells such as 1, 3, 4, 7, 9, 12, 13, 16, 19, 21, ..., the most common being 4 and 7. The number of cells per cluster is intuitively related with system capacity as well as with transmission quality. The fewer cells per cluster, the larger the number of channels per cell (higher traffic carrying capacity) and the closer the cocells (potentially more cochannel interference). An important parameter of a cellular layout relating these entities is the D/R ratio, where D is the distance between cocells and R is the cell radius. In a hexagonal geometry it is found that

$$D/R = \sqrt{3N} \qquad (81.2)$$

81.3 Performance Measures and System Requirements

Two parameters are intimately related with the grade of service of the cellular systems: carrier-to-cochannel interference ratio and blocking probability.

A high carrier-to-cochannel interference ratio in connection with a low-blocking probability is the desirable situation. This can be accomplished, for instance, in a large cluster with a low-traffic condition. In such a case the required grade of service can be achieved, although the resources may not be efficiently utilized. Therefore, a measure of efficiency is of interest. The **spectrum efficiency** η_s expressed in erlang per square meter per hertz, yields a measure of how efficiently space, frequency, and time are used, and it is given by

$$\eta_s = \frac{\text{number of reuses}}{\text{coverage area}} \times \frac{\text{number of channels}}{\text{bandwidth available}} \times \frac{\text{time the channel is busy}}{\text{total time of the channel}}$$

Another measure of interest is the **trunking efficiency** in which the number of subscribers per channel is obtained as a function of the number of channels per cell for different values of blocking probability. As an example, assume that a cell operates with 40 channels and that the mean blocking probability is required to be 5%. Using the erlang-B formula (refer to the Traffic Engineering section of this chapter), the traffic offered is calculated as 34.6 erlang. If the traffic per subscriber is assumed to be 0.02 erl, a total of 34.6/0.02 = 1730 subscribers in the cell is found. In other words, the trunking efficiency is 1730/40 = 43.25 subscribers per channel in a 40-channel cell. Simple calculations show that the trunking efficiency decreases rapidly when the number of channels per cell falls below 20.

The basic specifications require cellular services to be offered with a fixed telephone network quality. Blocking probability should be kept below 2%. As for the transmission aspect, the aim is to provide good quality service for 90% of the time. Transmission quality concerns the following parameters:

- Signal-to-cochannel interference (S/I_c) ratio
- Carrier-to-cochannel interference ratio (C/I_c)
- Signal plus noise plus distortion-to-noise plus distortion ($SINAD$) ratio
- Signal-to-noise (S/N) ratio
- Adjacent channel interference selectivity (ACS)

The S/I_c is a subjective measure, usually taken to be around 17 dB. The corresponding C/I_c depends on the modulation scheme. For instance, this is around 8 dB for 25-kHz FM, 12 dB for 12.5-kHz FM, and 7 dB for GMSK, but the requirements may vary from system to system. A common figure for $SINAD$ is 12 dB for 25-kHz FM. The minimum S/N requirement is 18 dB, whereas ACS is specified to be no less than 70 dB.

81.4 System Expansion Techniques

The obvious and most common way of permitting more subscribers into the network is by allowing a system performance degradation but within acceptable levels. The question is how to objectively define what is acceptable. In general, the subscribers are more likely to tolerate a poor quality service rather than not having the service at all. Some alternative expansion techniques, however, do exist that can be applied to increase the system capacity. The most widely known are as follows.

Adding New Channels: In general, when the system is set up not all of the channels need be used, and growth and expansion can be planned in an orderly manner by utilizing the channels that are still available.

Cell Design Principles

Frequency Borrowing: If some cells become more overloaded than others, it may be possible to reallocate channels by transferring frequencies so that the traffic demand can be accommodated.

Change of Cell Pattern: Smaller clusters can be used to allow more channels to attend a bigger traffic demand at the expense of a degradation of the transmission quality.

Cell Splitting: By reducing the size of the cells, more cells per area, and consequently more channels per area, are used with a consequent increase in traffic capacity. A radius reduction by a factor of f reduces the coverage area and increases the number of base stations by a factor of f^2. Cell splitting usually takes place at the midpoint of the congested areas and is so planned in order that the old base stations are kept.

Sectorization: A cell is divided into a number of sectors, three and six being the most common arrangements, each of which is served by a different set of channels and illuminated by a directional antenna. The sector, therefore, can be considered as a new cell. The base stations can be located either at the center or at the corner of the cell. The cells in the first case are referred to as center-excited cells and in the second as corner-excited cells. Directional antennas cut down the cochannel interference, allowing the cocells to be more closely spaced. Closer cell spacing implies smaller D/R ratio, corresponding to smaller clusters, i.e., higher capacity.

Channel Allocation Algorithms: The efficient use of channels determines the good performance of the system and can be obtained by different channel assignment techniques. The most widely used algorithm is based on fixed allocation. Dynamic allocation strategies may give better performance but are very dependent on the traffic profile and are usually difficult to implement.

81.5 Basic Design Steps

Engineering a cellular system to meet the required objectives is not a straightforward task. It demands a great deal of information, such as market demographics, area to be served, traffic offered, and other data not usually available in the earlier stages of system design. As the network evolves, additional statistics will help the system performance assessment and replanning. The main steps in a cellular system design are as follows.

Definition of the Service Area: In general, the responsibility for this step of the project lies on the operating companies and constitutes a tricky task, because it depends on the market demographics and, consequently, on how much the company is willing to invest.

Definition of the Traffic Profile: As before, this step depends on the market demographics and is estimated by taking into account the number of potential subscribers within the service area.

Choice of Reuse Pattern: Given the traffic distribution and the interference requirements a choice of the reuse pattern is carried out.

Location of the Base Stations: The location of the first base station constitutes an important step. A significant parameter to be taken into account in this is the relevance of the region to be served. The base station location is chosen so as to be at the center of or as close as possible to the target region. Data, such as available infrastructure and land, as well as local regulations are taken into consideration in this step. The cell radius is defined as a function of the traffic distribution. In urban areas, where the traffic is more heavily concentrated, smaller cells are chosen so as to attend the demand with the available channels. In suburban and in rural areas, the radius is chosen to be large because the traffic demand tends to be small. Once the placement of the first base station has been defined, the others will be accommodated in accordance with the repeat pattern chosen.

Radio Coverage Prediction: Given the topography and the morphology of the terrain, a radio prediction algorithm, implemented in the computer, can be used to predict the signal strength in the geographic region. An alternative to this relies on field measurements with the use of appropriate equipment. The first option is usually less costly and is widely used.

Design Checkup: At this point it is necessary to check whether or not the parameters with which the system has been designed satisfy the requirements. For instance, it may be necessary to re-evaluate the base station location, the antenna height, etc., so that better performance can be attained.

Field Measurements: For a better tuning of the parameters involved, field measurements (radio survey) should be included in the design. This can be carried out with transmitters and towers provisionally set up at the locations initially defined for the base station.

The cost assessment may require that a redesign of the system should be carried out.

81.6 Traffic Engineering

The starting point for engineering the traffic is the knowledge of the required grade of service. This is usually specified to be around 2% during the busy hour. The question lies on defining the busy hour. There are usually three possible definitions: 1) busy hour at the busiest cell, 2) system busy hour, and 3) system average over all hours.

The estimate of the subscriber usage rate is usually made on a demographic basis from which the traffic distribution can be worked out and the cell areas identified. Given the repeat pattern (cluster size), the cluster with the highest traffic is chosen for the initial design. The traffic A in each cell is estimated and, with the desired blocking probability $E(A, M)$, the erlang-B formula as given by Eq. (81.3) is used to determine the number of channels per cell, M

$$E(M, A) = \frac{A^M/M!}{\sum_{i=0}^{M} A^i/i!} \qquad (81.3)$$

In case the total number of available channels is not large enough to provide the required grade of service, the area covered by the cluster should be reduced in order to reduce the traffic per cell. In such a case, a new study on the interference problems must be carried out. The other clusters can reuse the same channels according to the reuse pattern. Not all channels need be provided by the base stations of those cells where the traffic is supposedly smaller than that of the heaviest loaded cluster. They will eventually be used as the system grows.

The traffic distribution varies in time and space, but it is commonly bell shaped. High concentrations are found in the city center during the rush hour, decreasing toward the outskirts. After the busy hour and toward the end of the day, this concentration changes as the users move from the town center to their homes. Note that because of the mobility of the users handoffs and roaming are always occurring, reducing the channel holding times in the cell where the calls are generated and increasing the traffic in the cell where the mobiles travel. Accordingly, the erlang-B formula is, in fact, a rough approximation used to model the traffic process in this ever-changing environment. A full investigation of the traffic performance in such a dynamic system requires all of the phenomena to be taken into account, making any traffic model intricate. Software simulation packages can be used so as to facilitate the understanding of the main phenomena as well as to help system planning. This is a useful alternative to the complex modeling, typically present in the analysis of cellular networks, where closed-form solutions are not usually available

On the other hand, conventional traffic theory, in particular, the erlang-B formula, is a handy tool widely used in cellular planning. At the inception of the system the calculations are carried

Cell Design Principles

out based on the best available traffic estimates, and the system capacity is obtained by grossly exaggerating the calculated figures. With the system in operation some adjustments must be made so that the requirements are met.

The approach just mentioned assumes the simplest channel assignment algorithm: the fixed allocation. It has the maximum spatial efficiency in channel reuse, since the channels are always assigned at the minimum reuse distance. Moreover, because each cell has a fixed set of channels, the channel assignment control for the calls can be distributed among the base stations.

The main problem of fixed allocation is its inability to deal with the alteration of the traffic pattern. Because of the mobility of the subscribers, some cells may experience a sudden growth in the traffic offered, with a consequent deterioration of the grade of service, whereas other cells may have free channels that cannot be used by the congested cells.

A possible solution for this is the use of dynamic channel allocation algorithms in which the channels are allocated on a demand basis There is an infinitude of strategies using the dynamic assignment principles, but they are usually complex to implement. An interim solution can be exercised if the change of the traffic pattern is predictable. For instance, if a region is likely to have an increase of the traffic on a given day (say, a football stadium on a match day), a mobile base station can be moved toward such a region in order to alleviate the local base.

Another specific solution uses the traffic available at the boundary between cells that may well communicate with more than one base station. In this case, a call that is blocked in its own cell can be directed to the neighboring cell to be served by its base station. This strategy, called *directed retry*, is known to substantially improve the traffic capacity. On the other hand, because channels with marginally acceptable transmission quality may be used, an increase in the interference levels, both for adjacent channel and cochannel, can be expected. Moreover, subscribers with radio access only to their own base will experience an increase in blocking probability.

81.7 Cell Coverage

The propagation of energy in a mobile radio environment is strongly influenced by several factors, including the natural and artificial relief, propagation frequency, antenna heights, and others. A precise characterization of the signal variability in this environment constitutes a hard task. Deterministic methods, such as those described by the *free space, plane earth,* and *knife-edge diffraction* propagation models, are restricted to very simple situations. They are useful, however, in providing the basic mechanisms of propagation. Empirical methods, such as those proposed by many researchers (e.g., Egli, 1957; Okumura et al., 1968; Lee, 1986; Ibrahim and Parsons, 1983; and others), use curves and/or formulas based on field measurements, some of them including deterministic solutions with various correction factors to account for the propagation frequency, antenna height, polarization, type of terrain, etc. Because of the random characteristics of the mobile radio signal, however, a single deterministic treatment of this signal will certainly lead the problem to a simplistic solution. Therefore, we may treat the signal on a statistical basis and interpret the results as random events occurring with a given probability. The cell coverage area is then determined as the proportion of locations where the received signal is greater than a certain threshold considered to be satisfactory.

Suppose that at a specified distance from the base station the *mean signal strength* is considered to be known. Given this we want to determine the cell radius such that the mobiles experience a received signal above a certain threshold with a stipulated probability. The mean signal strength can be determined either by any of the prediction models or by field measurements. As for the statistics of the mobile radio signal, five distributions are widely accepted today: lognormal, Rayleigh, Suzuki [Suzuki, 1977], Rice, and Nakagami. The lognormal distribution describes the variation of the mean signal level (large-scale variations) for points having the same transmitter–receiver antennas separation, whereas the other distributions characterize the instantaneous variations (small-scale variations) of the signal. In the calculations that follow we assume a lognormal environment. The other environments can be analyzed in a like manner; although this may not be of interest if some sort of diversity is implemented, because then the effects of the small-scale variations are minimized.

Propagation Model

Define m_w and k as the mean powers at distances x and x_0, respectively, such that

$$m_w = k\left(\frac{x}{x_0}\right)^{-\alpha} \tag{81.4}$$

where α is the path loss coefficient. Expressed in decibels, $M_w = 10\log m_w$, $K = 10\log k$ and

$$M_w = K - 10\alpha\log\left(\frac{x}{x_0}\right) \tag{81.5}$$

Define the received power as $w = v^2/2$, where v is the received envelope. Let $p(W)$ be the probability density function of the received power W, where $W = 10\log w$. In a lognormal environment, v has a lognormal distribution and

$$p(W) = \frac{1}{\sqrt{2\pi}\,\sigma_w}\exp\left(-\frac{(W-M_w)^2}{2\sigma_w^2}\right) \tag{81.6}$$

where M_W is the mean and σ_w is the standard deviation, all given in decibels. Define w_T and $W_T = 10\log w_T$ as the threshold above which the received signal is considered to be satisfactory. The probability that the received signal is below this threshold is its *probability distribution function* $P(W_T)$, such that

$$P(W_T) = \int_{-\infty}^{W_T} p(W)\,dW = \frac{1}{2} + \frac{1}{2}\mathrm{erf}\left[\frac{(W_T-M_W)^2}{2\sigma_w^2}\right] \tag{81.7}$$

where erf( ) is the error function defined as

$$\mathrm{erf}(y) = \frac{2}{\sqrt{\pi}}\int_0^y \exp(-t^2)\,dt \tag{81.8}$$

Base Station Coverage

The problem of estimating the cell area can be approached in two different ways. In the first approach, we may wish to determine the proportion β of locations at x_0 where the received signal power w is above the threshold power w_T. In the second approach, we may estimate the proportion μ of the circular area defined by x_0 where the signal is above this threshold. In the first case, this proportion is averaged over the perimeter of the circumference (cell border); whereas in the second approach, the average is over the circular area (cell area).

The proportion β equals the probability that the signal at x_0 is greater than this threshold. Hence,

$$\beta = \mathrm{prob}(W \geq W_T) = 1 - P(W_T) \tag{81.9}$$

Using Eqs. (81.5) and (81.7) in Eq. (81.9) we obtain

$$\beta = \frac{1}{2} - \frac{1}{2}\mathrm{erf}\left[\frac{W_T - K + 10\alpha\log(x/x_0)}{\sqrt{2}\,\sigma_w}\right] \tag{81.10}$$

This probability is plotted in Fig. 81.1, for $x = x_0$ (cell border).

Cell Design Principles

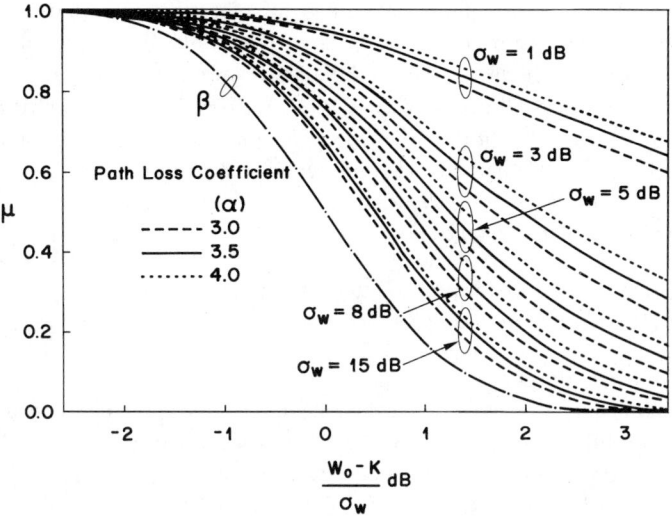

FIGURE 81.1 Proportion of locations where the received signal is above a given threshold; the dashdot line corresponds to the β approach and the other lines to the μ approach.

Let prob$(W \geq W_T)$ be the probability of the received power W being above the threshold W_T within an infinitesimal area dS. Accordingly, the proportion μ of locations within the circular area S experiencing such a condition is

$$\mu = \frac{1}{S} \int_S [1 - P(W_T)] \, dS \tag{81.11}$$

where $S = \pi r^2$ and $dS = x \, dx \, d\theta$. Note that $0 \leq x \leq x_0$ and $0 \leq \theta \leq 2\pi$. Therefore, solving for $d\theta$, we obtain

$$\mu = 2 \int_0^1 u\beta \, du \tag{81.12}$$

where $u = x/x_0$ is the normalized distance.

Inserting Eq. (81.10) in Eq. (81.12) results in

$$\mu = 0.5 \left\{ 1 + \text{erf}(a) + \exp\left(\frac{2ab+1}{b^2}\right) \left[1 - \text{erf}\left(\frac{ab+1}{b}\right) \right] \right\} \tag{81.13}$$

where $a = (K - W_T)/\sqrt{2}\sigma_w$ and $b = 10\alpha \log(e)/\sqrt{2}\sigma_w$.

These probabilities are plotted in Fig. 81.1 for different values of standard deviation and path loss coefficients.

Application Examples

From the theory that has been developed it can be seen that the parameters affecting the probabilities β and μ for cell coverage are the path loss coefficient α, the standard deviation σ_w, the required threshold W_T, and a certain power level K, measured or estimated at a given distance from the base station.

The applications that follow are illustrated for two different standard deviations: $\sigma_w = 5$ dB and $\sigma_w = 8$ dB. We assume the path loss coefficient to be $\alpha = 4$ (40 dB/decade), the mobile station receiver sensitivity to be -116 dB (1 mW), and the power level estimated at a given distance from the base station as being that at the cell border, $K = -102$ dB (1 mW). The receiver is considered to operate with a *SINAD* of 12 dB for the specified sensitivity. Assuming that cochannel interference levels are negligible and given that a signal-to-noise ratio S/N of 18 dB is required, the threshold W_T will be -116 dB (1 mW) $+ (18 - 12)$ dB (1 mW) $= -110$ dB (1 mW).

Three cases will be explored as follows.

Case 81.1: We want to estimate the probabilities β and μ that the received signal exceeds the given threshold 1) at the border of the cell, probability β and 2) within the area delimited by the cell radius, probability μ.

Case 81.2: It may be interesting to estimate the cell radius x_0 such that the received signal be above the given threshold with a given probability (say 90%) 1) at the perimeter of the cell and 2) within the cell area. This problem implies the calculation of the mean signal strength K at the distance x_0 (the new cell border) of the base station. Given K and given that at a distance x_0 (the former cell radius) the mean signal strength M_w is known [note that in this case $M_w = -102$ dB (1 mW)], the ratio x_0/x can be estimated.

Case 81.3: To fulfill the coverage requirement, rather than calculating the new cell radius, as in case 81.2, a signal strength at a given distance can be estimated such that a proportion of the locations at this distance, proportion β, or within the area delimited by this distance, proportion μ, will experience a received signal above the required threshold. This corresponds to calculating the value of the parameter K already carried out in case 81.2 for the various situations.

The calculation procedures are now detailed for $\sigma_w = 5$ dB. Results are also shown for $\sigma_w = 8$ dB.

Case 81.1: Using the given parameters we obtain $(W_T - K)/\sigma_w = -1.6$. With this value in Fig. 81.1, we obtain the probability that the received signal exceeds -116 dB (1 mW) for $S/N = 18$ dB given that at the cell border the mean signal power is -102 dB (1 mW) given in Table 81.1.

Note, from Table 81.1, that the signal at the cell border exceeds the receiver sensitivity with 97% probability for $\sigma_w = 5$ dB and with 84% probability for $\sigma_w = 8$ dB. If, on the other hand, we are interested in the area coverage rather than in the border coverage, then these figures change to 100% and 95%, respectively.

Case 81.2: From Fig. 81.1, with $\beta = 90\%$ we find $(W_T - K)/\sigma_w = -1.26$. Therefore, $K = -103.7$ dB (1 mW). Because $M_w - K = -10\alpha \log(x/x_0)$, then $x_0/x = 1.10$. Again, from Fig. 81.1, with $\mu = 90\%$ we find $(W_T - K)/\sigma_w = -0.48$, yielding $K = -107.6$ dB (1 mW). Because $M_w - K = -10\alpha \log(x/x_0)$, then $x_0/x = 1.38$. These results are summarized in Table 81.2, which shows the normalized radius of a cell where the received signal power is above -116 dB (1 mW) with 90% probability for $S/N = 18$ dB, given that at a reference distance from the base station (the cell border) the received mean signal power is -102 dB (1 mW).

Note, from Table 81.2, that in order to satisfy the 90% requirement at the cell border the cell radius can be increased by 10% for $\sigma_w = 5$ dB. If, on the other hand, for the same standard deviation the 90% requirement is to be satisfied within the cell area, rather than at the cell border, a substantial gain in power is achieved. In this case, the cell radius can be increased by a factor of 1.38. For

TABLE 81.1 Case 81.1 Coverage Probability

Standard Deviation, dB	β Approach (Border Coverage), %	μ Approach (Area Coverage), %
5	97	100
8	84	95

Cell Design Principles

TABLE 81.2 Case 81.2 Normalized Radius

Standard Deviation, dB	β Approach (Border Coverage)	μ Approach (Area Coverage)
5	1.10	1.38
8	0.88	1.27

TABLE 81.3 Case 81.3 Signal Power

Standard Deviation dB	β Approach (Border Coverage), dB (1 mW)	μ Approach (Area Coverage), dB (1 mW)
5	−103.7	−107.6
8	−99.8	−106.2

$\sigma_w = 8$ dB and 90% coverage at the cell border, the cell radius should be reduced to 88% of the original radius. For area coverage, an increase of 27% of the cell radius is still possible.

Case 81.3: The values of the mean signal power K are taken from case 81.2 and shown in Table 81.3, which shows the signal power at the cell border such that 90% of the locations will experience a received signal above -116 dB for $S/N = 18$ dB.

81.8 Interference

Radio-frequency interference is one of the most important issues to be addressed in the design, operation, and maintenance of mobile communication systems. Although both intermodulation and intersymbol interferences also constitute problems to account for in system planning, a mobile radio system designer is mainly concerned about adjacent-channel and cochannel interferences.

Adjacent Channel Interference

Adjacent-channel interference occurs due to equipment limitations, such as frequency instability, receiver bandwidth, filtering, etc. Moreover, because channels are kept very close to each other for maximum spectrum efficiency, the random fluctuation of the signal, due to fading and near–far effect, aggravates this problem.

Some simple, but efficient, strategies are used to alleviate the effects of adjacent channel interference. In narrowband systems, the total frequency spectrum is split into two halves so that the reverse channels, composing the uplink (mobile to base station) and the forward channels, composing the downlink (base station to mobile), can be separated by half of the spectrum. If other services can be inserted between the two halves, then a greater frequency separation, with a consequent improvement in the interference levels, is accomplished. Adjacent channel interference can also be minimized by avoiding the use of adjacent channels within the same cell. In the same way, by preventing the use of adjacent channels in adjacent cells a better performance is achieved. This strategy, however, is dependent on the cellular pattern. For instance, if a seven-cell cluster is chosen, adjacent channels are inevitably assigned to adjacent cells.

Cochannel Interference

Undoubtedly the most critical of all interferences that can be engineered by the designer in cellular planning is cochannel interference. It arises in mobile radio systems using cellular architecture because of the frequency reuse philosophy.

A parameter of interest to assess the system performance in this case is the carrier-to-cochannel interference ratio C/I_c. The ultimate objective of estimating this ratio is to determine the reuse distance and, consequently, the repeat pattern. The C/I_c ratio is a random variable, affected by random phenomena such as 1) location of the mobile, 2) fading, 3) cell site location, 4) traffic distribution, and others. In this subsection we shall investigate the **outage probability**, i.e., the probability of failing to achieve adequate reception of the signal due to cochannel interference. This parameter will be indicated by $p(CI)$. As can be inferred, this is intrinsically related to the repeat pattern.

Cochannel interference will occur whenever the wanted signal does not simultaneously exceed the minimum required signal level s_0 and the n interfering signals, $i_1, i_2, \ldots, i_n$, by some protection ratio r. Consequently, the conditional outage probability, given n interferers, is

$$p(CI \mid n) = 1 - \int_{s_0}^{\infty} p(s) \int_{0}^{s/r} p(i_1) \int_{0}^{(s/r)-i_1} p(i_2) \cdots \\ \times \int_{0}^{(s/r)-i_1-\cdots-i_{n-1}} p(i_n) \, di_n \cdots di_2 \, di_1 \, ds \qquad (81.14)$$

The total outage probability can then be evaluated by

$$p(CI) = \sum_{n} p(CI \mid n) p(n) \qquad (81.15)$$

where $p(n)$ is the distribution of the number of active interferers.

In the calculations that follow we shall assume an interference-only environment, i.e., $s_0 = 0$, and the signals to be Rayleigh faded. In such a fading environment the probability density function of the signal-to-noise ratio x is given by

$$p(x) = \frac{1}{x_m} \exp\left(-\frac{x}{x_m}\right) \qquad (81.16)$$

where x_m is the mean signal-to-noise ratio. Note that $x = s$ and $x_m = s_m$ for the wanted signal, and $x = i_j$ and $x_m = i_{mj}$ for the interfering signal j, with s_m and i_{mj} being the mean of s and i_j, respectively.

By using the density of Eq. (81.16) in Eq. (81.14) we obtain

$$p(CI \mid n) = \sum_{j=1}^{n} \prod_{k=1}^{j} \frac{z_k}{1+z_k} \qquad (81.17)$$

where $z_k = r s_m / i_{mk}$

If the interferers are assumed to be equal, i.e., $z_k = z$ for $k = 1, 2, \ldots, n$, then

$$p(CI \mid n) = 1 - \left(\frac{z}{1+z}\right)^n \qquad (81.18)$$

Define $Z = 10 \log z$, $S_m = 10 \log s_m$, $I_m = 10 \log i_m$, and $R_r = 10 \log r$. Then, $Z = S_m - (I_m + R_r)$. Equation (81.18) is plotted in Fig. 81.2 as a function of Z for $n = 1$ and $n = 6$, for the situation in which the interferers are equal.

If the probability of finding an interferer active is p, the distribution of active interferers is given by the binomial distribution. Considering the closest surrounding cochannels to be the most relevant interferers we then have six interferers. Thus

$$p(n) = \binom{6}{n} p^n (1-p)^{6-n} \qquad (81.19)$$

Cell Design Principles

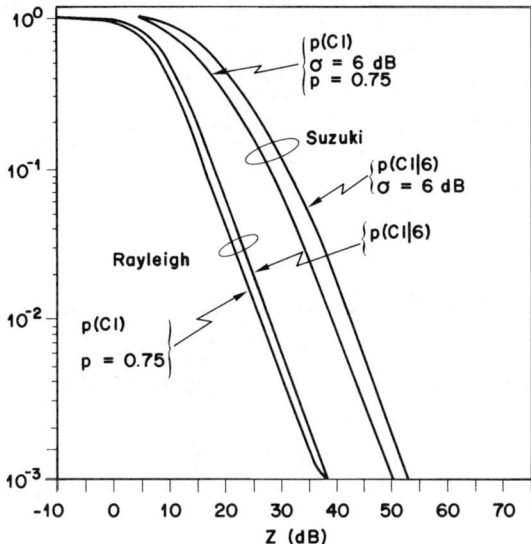

FIGURE 81.2 Conditional and unconditional outage probability for $n = 6$ interferes in a Rayleigh environment and in a Suzuki environment with $\sigma = 6$ dB.

For equal capacity cells and an evenly traffic distribution system, the probability p is approximately given by

$$p = \sqrt[M]{B} \tag{81.20}$$

where B is the blocking probability and M is the number of channels in the cell.

Now Eqs. (81.20), (81.19), and (81.18) can be combined into Eq. (81.15) and the outage probability is estimated as a function of the parameter Z and the channel occupancy p. This is shown in Fig. 81.2 for $p = 75\%$ and $p = 100\%$.

A similar, but much more intricate, analysis can be carried out for the other fading environments. Note that in our calculations we have considered only the situation in which both the wanted signal and the interfering signals experience Rayleigh fading. For a more complete analysis we may assume the wanted signal to fade differently from the interfering signals, leading to a great number of possible combinations. A case of interest is the investigation of the influence of the standard deviation in the outage probability analysis. This is illustrated in Fig. 81.2 for the Suzuki (lognormal plus Rayleigh) environment with $\sigma = 6$ dB.

Note that by definition the parameter z is a function of the carrier-to-cochannel interference ratio, which, in turn, is a function of the reuse distance. Therefore, the outage probability can be obtained as a function of the cluster size, for a given protection ratio.

The ratio between the mean signal power s_m and the mean interfering power i_m equals the ratio between their respective distances d_s and d_i such that

$$\frac{s_m}{i_m} = \left(\frac{d_s}{d_i}\right)^{-\alpha} \tag{81.21}$$

where α is the path loss coefficient. Now, 1) let D be the distance between the wanted and interfering base stations, and 2) let R be the cell radius. The cochannel interference worst case occurs when

TABLE 81.4 Probability of Cochannel Interference in Different Cell Clusters

		\multicolumn{2}{c}{Rayleigh}	\multicolumn{2}{c}{Suzuki $\sigma = 6$ dB}
N	$Z+R$, dB	$p=75\%$	$p=100\%$	$p=75\%$	$p=100\%$
1	−4.74	100	100	100	100
3	10.54	31	40	70	86
4	13.71	19	26	58	74
7	19.40	4.7	7	29	42
12	24.46	1	2.1	11	24
13	25.19	0.9	1.9	9	22

Outage Probability, %

the mobile is positioned at the boundary of the cell, i.e., $d_s = R$ and $d_i = D - R$. Then,

$$\frac{i_m}{s_m} = \left(\frac{D}{R} - 1\right)^{-\alpha} \tag{81.22a}$$

or, equivalently,

$$S_m - I_m = 10\alpha \log\left(\frac{D}{R} - 1\right) \tag{81.22b}$$

In fact, $S_m - I_m = Z + R_r$. Therefore,

$$Z + R_r = 10\alpha\log(\sqrt{3N} - 1) \tag{81.23}$$

With Eq. (81.23) and the curves of Fig. 81.2, we can compare some outage probabilities for different cluster sizes. The results are shown in Table 81.4 where we have assumed a protection ratio $R_r = 0$ dB. The protection ratio depends on the modulation scheme and varies typically from 8 dB (25-kHz FM) to 20 dB [single sideband (SSB) modulation].

Note, from Table 81.4, that the standard deviation has a great influence in the calculations of the outage probability.

81.9 Conclusions

The interrelationship among the areas involved in a cellular network planning is substantial. Vocabularies belonging to topics, such as radio propagation, frequency planning and regulation, modulation schemes, antenna design, transmission, teletraffic, and others, are common to all cellular engineers.

Designing a cellular network to meet system requirements is a challenging task which can only be partially and roughly accomplished at the design desk. Field measurements play an important role in the whole process and constitute an essential step used to tune the parameters involved.

Defining Terms

Outage probability: The probability of failing to achieve adequate reception of the signal due to, for instance, cochannel interference.

Spectrum efficiency: A measure of how efficiently space, frequency, and time are used. It is expressed in erlang per square meter per hertz.

Trunking efficiency: A function relating the number of subscribers per channel and the number of channels per cell for different values of blocking probability.

References

Egli, J. 1957. Radio above 40 Mc over irregular terrain. *Proc. IRE.* 45 (10):1383–1391.

Hata, M. 1980. Empirical formula for propagation loss in land-mobile radio services. *IEEE Trans. Vehicular Tech.* VT-29:317–325.

Ho, M.J. and Stüber, G.L. 1993. Co-channel interference of microcellular systems on shadowed Nakagami fading channels. *Proc. IEEE Vehicular Tech. Conf.* pp. 568–571.

Ibrahim, M.F. and Parsons, J.D. 1983. Signal strength prediction in built-up areas, Part I: median signal strength. *Proc. IEE* Pt. F. (130):377–384.

Lee, W.C.Y. *Mobile Communications Design Fundamentals*, Howard W. Sams, Indianapolis, IN.

Leonardo, E.J. and Yacoub, M.D. 1993a. A statistical approach for cell coverage area in land mobile radio systems. Proceedings of the 7th IEE Conf. on Mobile and Personal Comm., Brighton, UK, Dec. pp. 16–20.

Leonardo, E.J. and Yacoub, M.D. 1993b. (Micro) Cell coverage area using statistical methods. Proceedings of the IEEE Global Telecom. Conf. GLOBECOM'93, Houston, TX, Dec. pp. 1227–1231.

Okumura, Y., Ohmori, E., Kawano, T., and Fukuda, K. 1968. Field strength and its variability in VHF and UHF land mobile service. *Rev. Elec. Comm. Lab.* 16 (Sept.–Oct.):825–873.

Reudink, D.O. 1974. Large-scale variations of the average signal. In *Microwave Mobile Communications*, pp. 79–131, John Wiley, New York.

Sowerby, K.W. and Williamson, A.G. 1988. Outage probability calculations for multiple cochannel interferers in cellular mobile radio systems. *IEE Proc.* Pt. F. 135(3):208–215.

Suzuki, H. 1977. A statistical model for urban radio propagation. *IEEE Trans. Comm.* 25(7):673–680.

Further Information

The fundamentals of mobile radio engineering in connection with many practical examples and applications as well as an overview of the main topics involved can be found in Yacoub, M.D., *Foundations of Mobile Radio Engineering*, CRC Press, Inc. Boca Raton, Fl, 1993.

82
Microcellular Radio Communications

82.1	Introducing Microcells	1160
82.2	Highway Microcells	1161
	Spectral Efficiency of Highway Microcellular Network	
82.3	City Street Microcells	1164
	Teletraffic Issues	
82.4	Indoor Microcells	1169
82.5	Microcellular Infrastructure	1170
	Radio over Fiber • Miniaturized Microcellular BSs	
82.6	Multiple Access Issues	1172
82.7	Discussion	1173

Raymond Steele
*Southampton University
and
Multiple Access
Communications Ltd.*

82.1 Introducing Microcells

In mobile radio communications an operator will be assigned a specific bandwidth W in which to operate a service. The operator will, in general, not design the mobile equipment, but purchase equipment that has been designed and standardized by others. The performance of this equipment will have a profound effect on the number of subscribers the network can support, as we will show later. Suppose the equipment requires a radio channel of bandwidth B. The operator can therefore fit $N_T = W/B$ channels into the allocated spectrum W.

Communications with mobiles are made from fixed sites, known as base stations (BSs). Clearly, if a mobile travels too far from its BS, the quality of the communications link becomes unacceptable. The perimeter around the BS where acceptable communications occur is called a cell and, hence, the term cellular radio. BSs are arranged so that their radio coverage areas, or cells, overlap, and each BS may be given $N = N_T/M$ channels. This implies that there are M BS and each BS uses a different set of channels.

The number N_T is relatively low, perhaps only 1000. As radio channels cannot operate with 100% utilization, the cluster of BSs or cells has fewer than 1000 simultaneous calls. In order to make the business viable, more users must be supported by the network. This is achieved by repeatedly reusing the channels. Clusters of BSs are tessellated with each cluster using the same N_T channels. This means that there are users in each cluster using the same frequency band at the same time, and inevitably there will be interference. This interference is known as cochannel interference. Cochannel cells, i.e., cells using the same channels, must be spaced sufficiently far apart for the interference levels to be acceptable. A mobile will therefore receive the wanted signal of power S and a total interference power of I, and the signal-to-interference ratio (SIR) is a key system design parameter.

Suppose we have large cells, a condition that occurs during the initial stages of deploying a network when coverage is important. For a given geographical area G_A we may have only one cluster of seven cells, and this may support some 800 simultaneous calls in our example. As the subscriber base grows, the number of clusters increases to, say, 100 with the area of each cluster being appropriately decreased. The network can now support some 80,000 simultaneous calls in the area G_A. As the number of subscribers continues to expand, we increase the number of clusters. The geographical area occupied by each cluster is now designed to match the number of potential users residing in that area. Consequently, the smallest clusters and, hence, the highest density of channels per area is found in the center of cities. As each cluster has the same number of channels, the smaller the clusters and, therefore, the smaller the cells, the greater the **spectral efficiency** measured in erlang per hertz per square meter. Achieving this higher spectral efficiency requires a concomitant increase in the infrastructure that connects the small cell BSs to their base station controller (BSC). The BSCs are part of the nonradio part of the mobile network that is interfaced with the public switched telephone network (PSTN) or the integrated service digital network (ISDN).

As we make the cells smaller, we change from locating the BS antennas on top of tall buildings or hills, where they produce large cells or macrocells, to the tops of small buildings or the sides of large buildings, where they form minicells, to lamp post elevations, where they form **street microcells**. Each decrease in cell size is accompanied by a reduction in the radiated power levels from the BSs and from the mobiles. As the BS antenna height is lowered, the neighboring buildings and streets increasingly control the radio propagation. This chapter is concerned with microcells and microcellular networks. We commence with the simplest type of microcells, namely, those used for highways.

82.2 Highway Microcells

Since their conception by Steele and Prabhu, 1985, many papers have been published on **highway microcells**, ranging from propagation measurements to teletraffic issues [Chia et al., 1987; El-Dolil, Wong, and Steele, 1989; Steele and Nofal, 1992; Green, 1990; Keenan and Motley, 1990; Saleh and Valenzula, 1987; Steele, 1992; Merrett, Cooper, and Symington; Steele and Williams, 1993]. Figure 82.1 shows the basic concepts for a highway microcellular system having two cells per cluster. The highway is partitioned into contiguous cigar-shaped segments formed by directional antennas. Omnidirectional antennas can be used at junctions, roundabouts, cloverleaf, and other road intersections. The BS antennas are mounted on poles at elevations of some 6–12 m. Figure 82.2 shows received signal levels as a function of the distance d between BS and MS for different roads

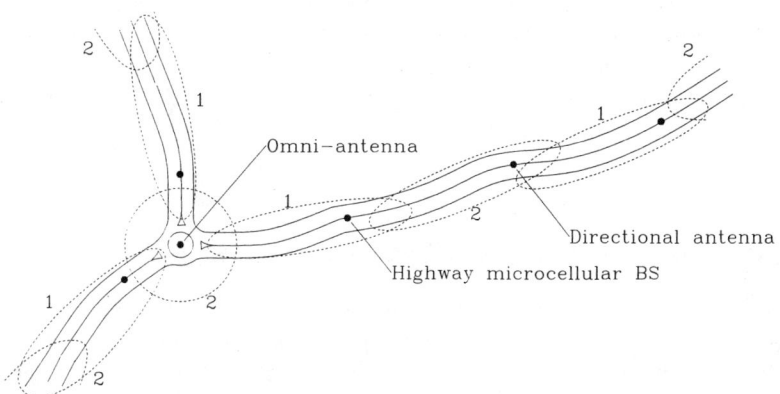

FIGURE 82.1 Highway microcellular clusters. Microcells with the same number use the same frequency set.

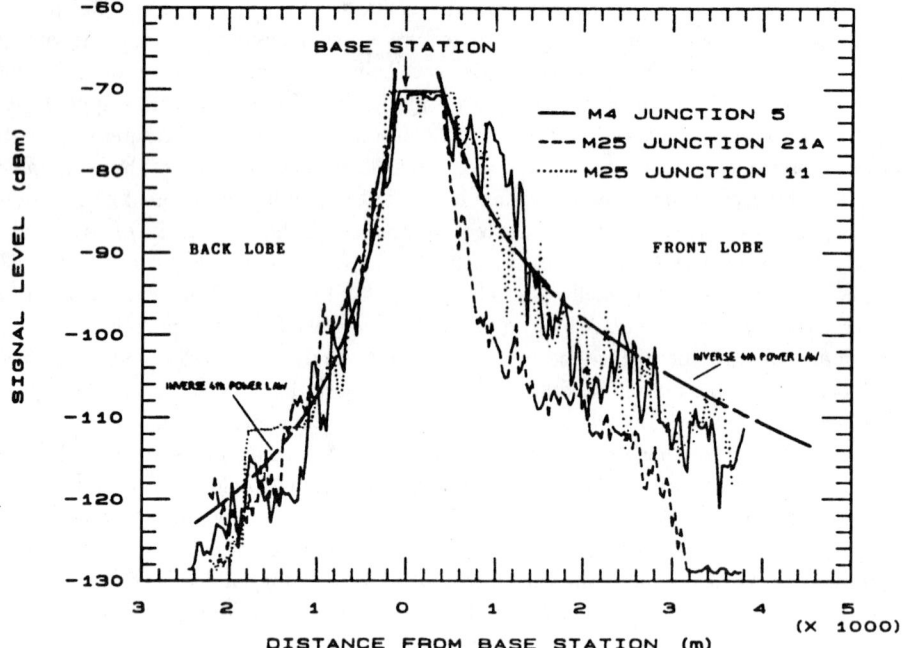

FIGURE 82.2 Overlayed received signal strength profiles of various highway cells including the inverse fourth power law curve for both the front and back antenna lobes. *Source*: Chia et al. 1987. Propagation and bit error ratio measurements for a microcellular system. *JIERE*. 57(6):5255–5266. With permission.

[Chia et al. 1987]. The average loss in received signal level, or path loss, is approximately inversely proportional to d^4. The path loss is associated with a slow fading component that is due to the variations in the terrain, the road curvature and cuttings, and the presence of other vehicles.

The curves in the figure are plotted for an 18-element yagi BS antenna having a gain of 15 dB and a front-to-back ratio of 25 dB. In Fig. 82.2 reference is made to junctions on different motorways, e.g., junction 5 on motorway M4. This is because the BS antennas are mounted at these road junctions with the yagi antenna pointing along the highway in order to create a cigar-shaped cell. The flat part of the curve near the BS is due to the MS receiver being saturated by high-signal levels. Notice that the curve related to M25, junction 11, decreases rapidly with distance when the MS leaves the immediate vicinity of the BS. This is due to the motorway making a sharp turn into a cutting, causing the MS to lose line of sight (LOS) with the BS. Later the path loss exponent is approximately 4. Experiments have shown that using the arrangement just described, with each BS transmitting 16 mW at 900 MHz, 16 kb/s noncoherent frequency shift keying, two-cell clusters could be formed where each cell has a length along the highway ranging from 1 to 2 km. For MSs traveling at 110 km/h the average handover rate is 1.2 per minute [Chia et al. 1987].

Spectral Efficiency of Highway Microcellular Network

Spectral efficiency is a key system parameter. The higher the efficiency, the greater will be the teletraffic carried by the network for the frequency band assigned by the regulating authorities per unit geographical area. We define the spectral efficiency in mobile radio communications in erlang per hertz per square meter as

$$\eta \stackrel{\Delta}{=} A_{CT}/S_T W \qquad (82.1)$$

although erlang per megahertz per square kilometer is often used. In this equation, A_{CT} is the total traffic carried by the microcellular network,

$$A_{CT} = C A_C \tag{82.2}$$

where C is the number of microcells in the network and A_C the carried traffic by each microcellular BS. The total area covered of the tessellated microcells is

$$S_T = C S \tag{82.3}$$

where S is the average area of a microcell, whereas the total bandwidth available is

$$W = MNB \tag{82.4}$$

whose terms M, N, and B were defined in Sec. 82.1. Substituting Eqs. 82.2–82.4 into Eq. (82.1), yields

$$\eta = \frac{\rho}{SMB} \tag{82.5}$$

where

$$\rho = A_C/N \tag{82.6}$$

is the utilization of each BS channel.

If the length of each microcell is L, there are n up lanes and n down lanes, and each vehicle occupies an effective lane length V, which is speed dependent, then the total number of vehicles in a cell is

$$K = 2nL/V \tag{82.7}$$

Given that all vehicles have a mobile terminal, the maximum number of mobiles in a cell is K. In a highway microcell we are not interested in the actual area $S = 2nL$ but in how many vehicles can occupy this area, namely, the effective area K. Notice that K is largest in a traffic jam when all vehicles are stationary and V only marginally exceeds the vehicle length. Given that N is sufficiently large, η is increased when the traffic flow is decreased.

Using fixed channel assignment (FCA) with frequency division multiple access (FDMA) or with time division multiple access (TDMA), the cluster size M can be two. Using dynamic channel assignment (DCA) with TDMA, or code division multiple access (CDMA), causes the spectral efficiency η to be very high because for a given traffic utilization ρ and channel bandwidth B, the S is small (as we are considering microcells), and M may be thought of as 1, or less, due to sectorization. The total traffic A_{CT}, given by Eq. (82.2), is also very high because by making L relatively short, C is accordingly high.

The traffic carried by a microcellular BS is

$$A_C = [\lambda_N(1 - P_{bn}) + \lambda_H(1 - P_{fhm})]\overline{T}_H \tag{82.8}$$

where P_{bn} is the probability of a new call being blocked, P_{fhm} is the probability of handover failure when mobiles enter the microcell while making a call and concurrently no channel is available, λ_N and λ_H are the new call and handover rates, respectively, and $\overline{T}_H$ is the mean channel holding time of all calls. For the simple case where no channels are exclusively reserved for handovers, $P_{bn} = P_{fhm}$, and

$$A_C = \lambda_T \overline{T}_H (1 - P_{bn}) = A(1 - P_{bn}) \qquad (82.9)$$

where

$$\lambda_T = \lambda_N + \lambda_H \qquad (82.10)$$

and A is the total offered traffic. The mathematical complexity resides in calculating A and P_{bn}, and the reader is advised to consult El-Dolil, Wong, and Steele, 1989, and Steele and Nofal, 1992.

Priority schemes have been proposed whereby N channels are available for handover, but only $N - N_h$ for new calls. Thus N_h channels are exclusively reserved for handover [El-Dolil, Wong, and Steele, 1989]. While P_{bn} marginally increases, P_{fhm} decreases by orders of magnitude for the same average number of new calls per sec per microcell. This is important as people prefer to be blocked while attempting to make a call compared to having a call in progress terminated due to no channel being available on handover. An important enhancement is to use an oversailing macrocellular cluster, where each macrocell supports a microcellular cluster. The role of the macrocell is to provide channels to support microcells that are overloaded and to provide communications to users who are in areas not adequately covered by the microcells [El-Dolil, Wong, and Steele, 1989]. When vehicles are in a solid traffic jam, there are no handovers and so N_h should be zero. When traffic is flowing fast, N_h should be high. Accordingly a useful strategy is to make N_h adaptive to the new call and handover rates [Steele, Nofal, and El-Dolil, 1990].

82.3 City Street Microcells

We will define a city street microcell as one where the BS antenna is located below the lowest building. As a consequence, the diffraction over the buildings can be ignored, and the heights of the buildings are of no consequence. Roads and their attendant buildings form trenches or canyons through which the mobiles travel. If there is a direct line-of-sight path between the BS and a MS and a ground-reflected path, the received signal level vs BS–MS distance is as shown in Fig. 82.3. Should there be two additional paths from rays reflected from the buildings, then the profile for this four-ray situation is also shown in Fig. 82.3. These theoretical curves show that as the MS travels from the

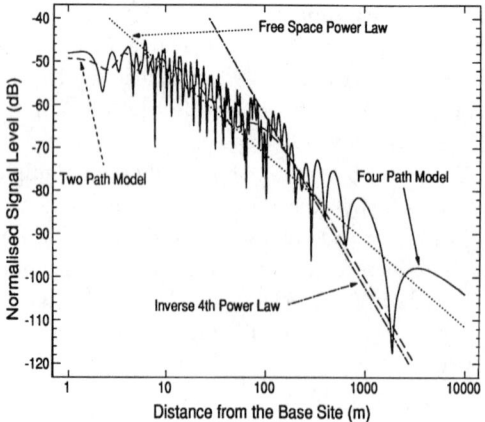

FIGURE 82.3 Signal level profiles for the two- and four-path models. Also shown are the free space and inverse fourth power laws. *Source:* Green. 1990. Radio link design for microcellular system, *British Telecom. Tech. J.* 8(1):85–96. With permission.

BS the average received signal level is relatively constant and then decreases relatively rapidly. This is a good characteristic as it offers a good signal level within the microcell, and the interference into adjacent microcells falls off rapidly with distance.

In practice there are many paths, but there is often a dominant one. As a consequence the fading is Rician [Stelle, 1992]. The Rician distribution approximates to a Gaussian one when the received signal is from a dominant path with the power in the scattered paths being negligible, to a Rayleigh one when there is no dominant path. Macrocells usually have Rayleigh fading, whereas in microcells the fading only occasionally becomes Rayleigh and is more likely to be closer to Gaussian. This means that the depth of the fades in microcells are usually significantly smaller than in macrocells enabling microcellular communications to operate closer to the receiver noise floor without experiencing error bursts and to accommodate higher cochannel interference levels. Because of the small dimensions of the microcells, the delays between the first and last significant paths is relatively small compared to the corresponding delays in macrocells. Consequently, the impulse response is generally shorter in microcells and, therefore, the transmitted bit rate can be significantly higher before intersymbol interference is experienced compared to the situation in macrocells. Microcells are, therefore, more spectrally efficient with an enhanced propagation environment.

There are two types of these city street microcells, one for pedestrians and the other for vehicles. In general, there will be more portables carried by pedestrians than mobile stations in cars. Also, as cars travel more quickly than people, their microcells are accordingly larger than for pedestrians. The handover rates for portables and vehicular MS may be similar, and networks must be capable of handling the many handovers per call that may occur. In addition, the time available to complete a handover may be very short compared to those in macrocells.

City street microcells are irregular when the streets are irregular as demonstrated by the MIDAS[1] plot of a BS in Southampton city area displayed in Fig. 82.4. To achieve a contiguous coverage we site the BSs one at a time. Having sited the first BS and located the microcellular boundary along the streets, we locate adjacent BSs such that their boundaries butt with each other along the main streets. Unless many microcellular BSs are deployed, there will be some secondary streets where there will be insufficient signal levels. Those areas that are not covered by the microcellular BS will be accommodated by an oversailing macrocellular BS that services the complete cluster of microcellular BSs. Figure 82.5 shows a cluster of microcells; the oversailing macrocell could be sited outside the area of this figure. We emphasize that total coverage by microcells in a typical city center is difficult to achieve, and it is vital that oversailing macrocells are used to cover these dead spots. The macrocell also facilitates handovers and efficient microcellular channel utilization.

There are important differences between highway microcells and city microcells, which relate to their one- and two-dimensional characteristics. A similar comment applies to street microcells and hexagonal cells. Basically, the buildings have a profound effect on cochannel interference. The buildings shield much of the cochannel interference, and the double regression path loss law of microcells [Green, 1990] also decreases interference if the break-distance constitutes the notional microcell boundary. City microcellular clusters may have as few as two microcells, but four is more typical, and in some topologies six or more may be required. The irregularity of city streets means that some signals can find paths through building complexes to give cochannel interference where it is least expected.

Teletraffic Issues

Consider the arrangement where each microcellular cluster is overlaid by a macrocell. The macrocells are also clustered. The arrangement is shown in Fig. 82.6. The total traffic carried is

$$A_{CT} = C_m A_{cm} + C_M A_{CM} \qquad (82.11)$$

[1] MIDAS is a propriety software outdoor planning tool developed by Multiple Access Communications Ltd.

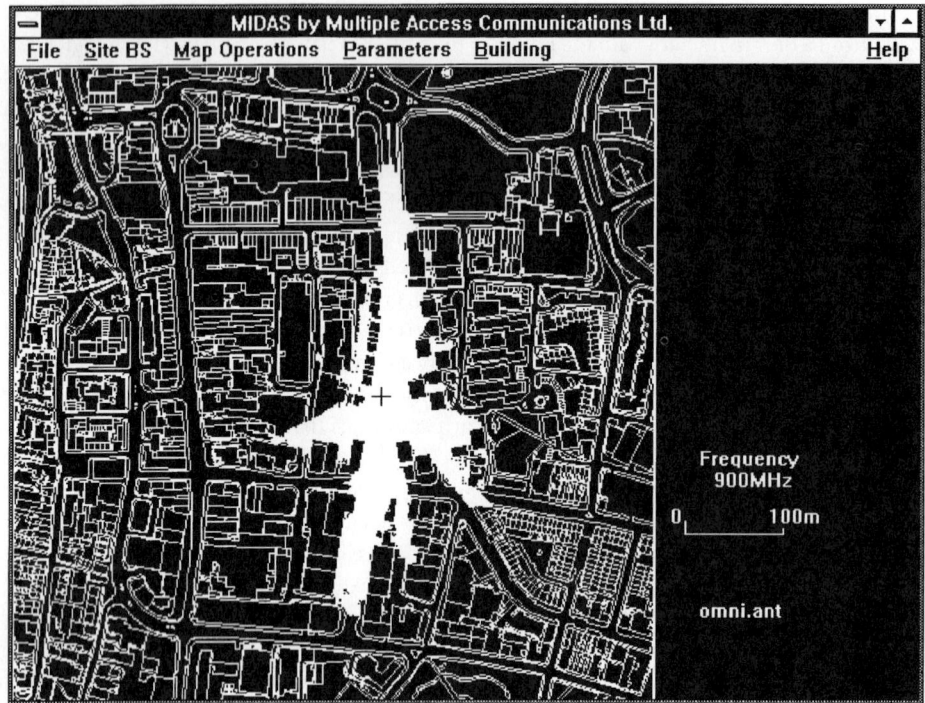

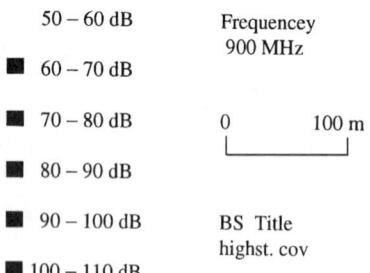

FIGURE 82.4 MIDAS plot of a city street microcell in Southampton. The plot is normally in colour, with each colour signifying a range of path loss in dB. In this plot the white area represents a path loss of up to 80 dB, and the cross shows the base station location.

Microcellular Radio Communications 1167

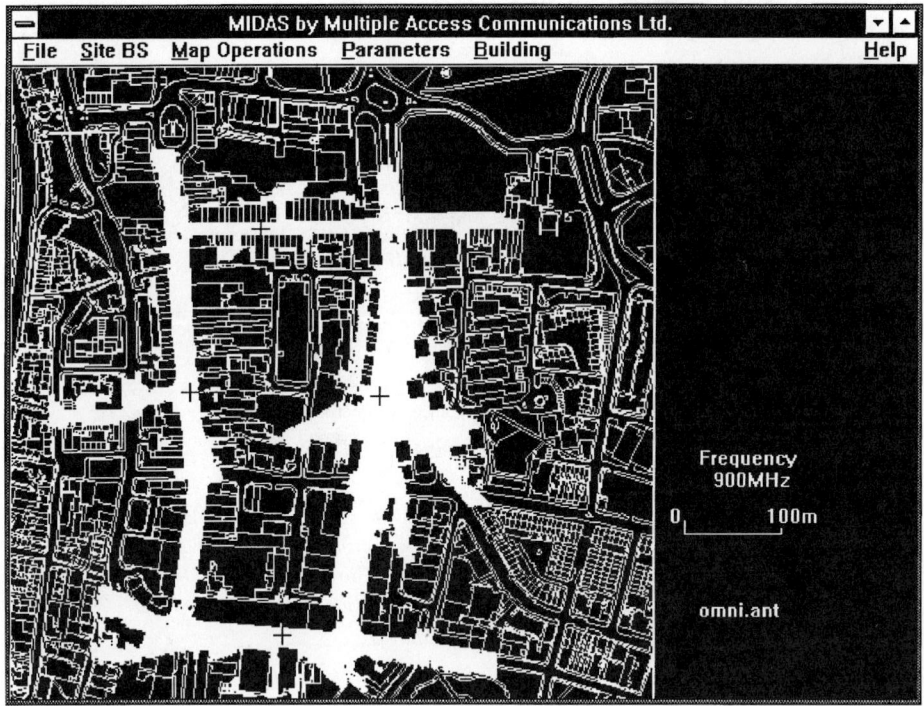

ⓒ Crown copyright

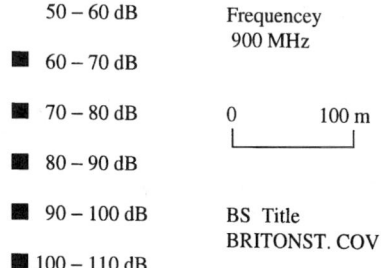

FIGURE 82.5 Cluster of microcells in Southampton city centre. The plot is normally in colour, with each colour signifying a range of path loss in dB. In this plot the white area represents a path loss of up to 80 dB, and the crosses show the microcellular base station locations.

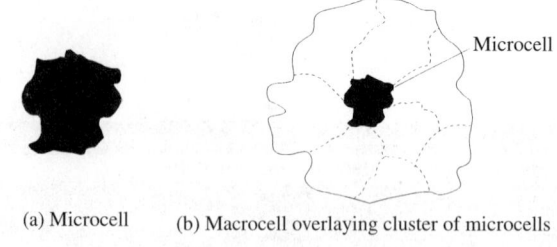
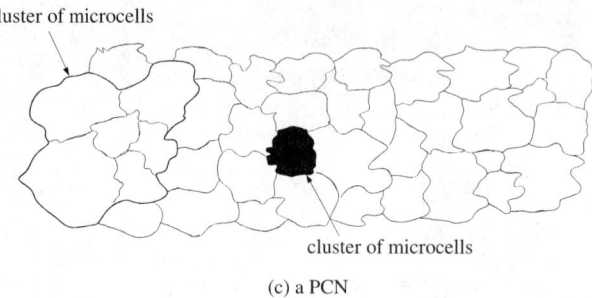

(a) Microcell   (b) Macrocell overlaying cluster of microcells

(c) a PCN

FIGURE 82.6 Microcellular clusters with oversailing macrocellular clusters. Each macrocell is associated with a particular microcellular cluster. *Source*: Steele and Williams. 1993. Third generation PCN and the intelligent multimode mobile portable. *IEE Elect. and Comm. Eng. J.* 5(3):147–156. With permission.

where C_m and C_M are number of microcells and macrocells in the network, respectively. Each microcellular BS has N channels and carries A_{cm} erlang. The corresponding values for the macrocellular BSs are N_0 and A_{CM}. The channel utilization for the network is

$$\rho_2 = \frac{C_m A_{cm} + C_M A_{CM}}{C_m N + C_M N_0} = \frac{M A_{cm} + A_{CM}}{MN + N_0} \tag{82.12}$$

where M is the number of microcells per cluster.

The spectral efficiency is found by noting that the total bandwidth is

$$B_T = B_c(MN + M_0 N_0) \tag{82.13}$$

where B_c is the effective channel bandwidth, and M_0 is the number of macrocells per macrocellular cluster. The traffic carried by a macrocellular cluster and its cluster of microcells is

$$A_M = A_{CM} M_0 + M_0(A_{cm} M) \tag{82.14}$$

over an area of

$$S_M = (S_m M) M_0 \tag{82.15}$$

where S_m is the area of each microcell.

The spectral efficiency is, therefore,

$$\eta = \frac{A_{CM} + A_{cm}M}{B_c(MN + M_0N_0)S_mM}$$
$$= \frac{\rho_2}{B_cS_mM}\left\{\frac{MN + N_0}{MN + M_0N_0}\right\} \qquad (82.16)$$

We note that by using oversailing macrocells to assist microcells experiencing overloading we are able to operate the microcells at high levels of channel utilization. However, the channel utilization of the macrocells must not be high if the probability of calls forced to terminate due to handover failure is to be minuscule.

82.4 Indoor Microcells

Microcellular BSs may be located within buildings to produce **indoor microcells** whose dimensions may extend from a small office, to part of larger offices, to a complete floor, or to a number of floors. The microcells are box-like for a single office microcell; or may contain many boxes, e.g., when a microcell contains contiguous offices. Furniture, such as bookcases, filing cabinets, and desks may represent large obstacles that may introduce shadowing effects. The signal attenuation through walls, floors, and ceilings may vary dramatically depending on the construction of the building. There is electromagnetic leakage down stairwells and through service ducting, and signals may leave the building and re-enter it after reflection and diffractions from other buildings.

Predicting the path loss in office microcells is, therefore, fraught with difficulties. At the outset it may not be easy to find out the relevant details of the building construction. Even if these are known, an estimation of the attenuation factors for walls, ceilings, and floors from the building construction is far from simple. Then there is a need to predict the effect of the furniture, effect of doors, the presence of people, and so on. Simple equations have been proposed. For example, the one by Keenam and Motley, 1990, who represent the path loss in decibels by

$$PL = L(V) + 20\log_{10}d + n_fa_f + n_wa_w \qquad (82.17)$$

where d is the straight line distance between the BS and the MS, a_f and a_w are the attenuation of a floor and a wall, respectively, n_f and n_w are the number of floors and walls along the line d, respectively, and $L(V)$ is a so-called clutter loss, which is frequency dependent. This equation should be used with caution. Researchers have made many measurements and found that even when they use computer ray tracing techniques the results can be considerably disparate. Errors having a standard deviation of 8–12 dB are not unusual at the time of writing. Given the wide range of path loss in mobile communications, however, and the expense of making measurements, particularly when many BS locations are examined, means that there is nevertheless an important role for planning tools to play, albeit their poorer accuracy compared to street microcellular planning tools.

As might be expected, the excess path delays in buildings is relatively small. The maximum delay spread within rooms and corridors may be <200 and 300 ns, respectively [Saleh and Valenzula, 1987]. The digital European cordless telecommunication (DECT) indoor system operates at 1152 kb/s without either channel coding or equalization [Steele, 1992]. This means that the median rms values are relatively low, and 25 ns has been measured [saleh and Valenzula, 1987]. When the delay spread becomes too high resulting in bit errors, the DECT system hops the user to a better channel using DCA.

82.5 Microcellular Infrastructure

An important requirement of first and second generation mobile networks is to restrict the number of base stations to achieve sufficient network capacity with an acceptably low probability of blocking. This approach is wise given the cost of base stations and their associated equipment, plus the cost and difficulties in renting sites. It is somewhat reminiscent of the situation faced by early electronic circuit designers who needed to minimize the number of tubes and later the number of discrete transistors in their equipment. It was the introduction of microelectronics that freed the circuit designer. We are now in an analogous situation where we need to free the network designers of the third generation communication networks, allowing the design to have microcellular BSs in the position where they are required, without being concerned if they are rarely used, and knowing that the cost of the microcellular network is a minor one compared to the overall network cost. This approach is equivalent to installing electric lighting where we are not unduly concerned if not all the lights are switched on at any particular time, preferring to be able to provide illumination where and when it is needed.

To realise high-capacity mobile communications we need to design microcellular BSs of negligible costs, of coffee mug dimensions, and with the ability to connect them at the cost of, say, electrical wiring in streets and buildings. Cordless telecommunication (CT) BSs are already of shoe-box size, and companies are designing coffee mug-size versions. The cost of these BSs will be low in mass production, and many BSs will be equivalent in cost to one first generation analog cellular BS. Microcellular BSs could be miniaturized, fully functional BSs achieved by using microelectronic techniques and by exploiting the low-radiated power levels (<10 mW) required. At the other extreme, the microcells could be formed using distribution points (DPs) that only have optical-to-microwave converters, microwave-to-optical converters, and linear amplifiers, with the remainder of the BS at another location. In between the miniaturized, fully functional BSs and the DPs there is a range of options that depends on how much complexity is built into the microcellular BS and how the intelligence of the network is distributed.

Radio over Fiber

The method of using DPs to form microcells is often referred to as radio over fiber (ROF) [Merrett, Cooper, and Symington]. Figure 82.7(a) shows a microcellular BS transmitting to a MS. When the DP concept is evoked, the microcellular BS contains electrical-to-optical (E/O) and optical-to-electrical (O/E) converters as shown in Fig. 82.7(b). The microwave signal that would have been radiated to the mobile is now applied, after suitable attenuation, to a laser transmitter. Essentially, the microwave signal amplitude modulates the laser, and the modulated signal is conveyed over a single-mode optical fiber to the distribution point. O/E conversion ensues followed by power amplification, and the resulting signal is transmitted to the MS. Signals from the MS are low-noise amplified and applied to the laser transmitter in the DP. Optical signals are sent from the DP to the BS where O/E conversion is performed followed by radio reception.

In general, the BS transceiver will be handling multicarrier signals for many mobiles, and the DP will accordingly be transceiving signals with many mobiles whose power levels may be significantly different, even when power control is used. Care must be exercised to avoid serious intermodulation products arising in the optical components.

Figure 82.7(c) shows the cositing of n microcellular BSs for use with DPs. This cositing may be conveniently done at a mobile switching center (MSC). Shown in the figure are the DPs and their irregular shaped overlapping microcells. The DPs can be attached to lamp posts in city streets, using electrical power from the electric light supply and the same ducting as used by the electrical wiring, or local telephone ducting. The DPs can also be attached to the outside of buildings. DPs within buildings may be conveniently mounted on ceilings.

The DP concept allows small, lightweight equipment in the form of DPs to be geographically distributed to form microcells; however, there are problems. The N radio carriers cause intermod-

Microcellular Radio Communications

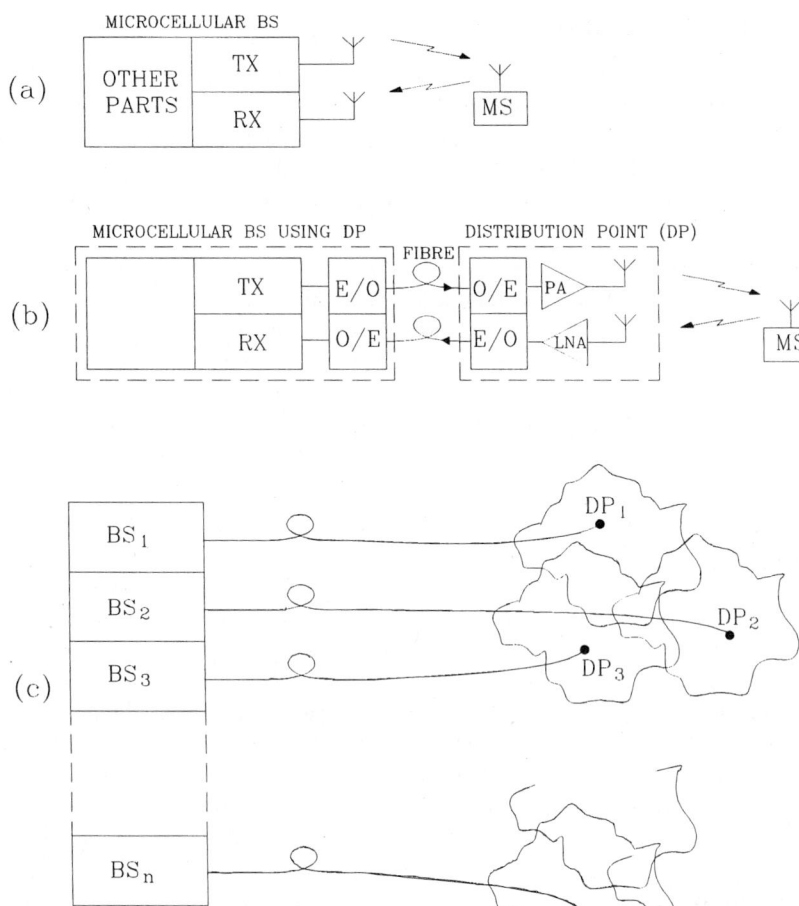

FIGURE 82.7 Creating microcells using DPs: (a) microcellular BS communicating with MS, (b) radio over fiber to distribution point, and (c) microcellular clusters using DPs.

ulation products (IMPs), which may be reduced by decreasing the depth of amplitude modulation for each radio channel. Unfortunately this also decreases the carrier-to-noise ratio (CNR) and the dynamic range of the link. With TDMA having many channels per carrier, we can decrease the number of radio carriers and make the IMPs more controllable. CDMA is particularly adept at coping with IMPs. The signals arriving from the MSs may have different power levels, in spite of power control. Because of the small size cells, the dynamic range of the signals arriving from MSs having power control should be <20 dB. If not, the power levels of the signals arriving at the DP from the MSs may need to be made approximately similar by individual amplification at the DP. We also must be careful to limit the length of the fiber as optical signals propagate along fibers much more slowly than radio signals propagate in free space. This should not be a problem in microcells, unless the fiber makes many detours before arriving at its DP.

The current cost of lasers is not sufficiently low for the ROF DP technique to be deployed. However, there is research into lasers, which are inherently simple, low cost, robust, and provide narrow line widths. There are also the developments in optoelectronic integrated circuits that may ultimately bring costs down. In addition, wavelength division multiplexing will bring benefits.

Miniaturized Microcellular BSs

The low-radiated power levels used by BSs and MSs in microcells have important ramifications on BS equipment design. Small fork combiners can be used along with linear amplifiers. Even FDMA BSs become simple when the high-radiated power levels are abandoned. It is the changes to the size of the RF components that enables the size of the microcellular BSs to be small.

The interesting question that next arises is, how much baseband signal processing complexity should the microcellular BS have? If the microcellular BSs are connected to nodes in an optical LAN, we can convey the final baseband signals to the BS, leaving the BS with the IF and front-end RF components. This means that processing of the baseband signals will be done at the group station (GS), which may be a MSC connected to the LAN. The GS will, therefore, transcode the signals from the ISDN into a suitable format. Using an optical LAN, however, and with powerful microelectronics, the transcoding and full BS operations could be done at each microcellular BS. Indeed, the microcellular BS may eventually execute many of the operations currently handled by the MSC.

82.6 Multiple Access Issues

There are three basic multiple access methods. Time division multiple access, frequency division multiple access, and spread spectrum multiple access (SSMA). SSMA comes in two versions; frequency-hopping SSMA and discrete-sequence SSMA. The latter is usually referred to as CDMA. There are also many hybrids of these systems. The principles of multiple access are described elsewhere in this book and will not be repeated here. Instead, we will comment on key factors that effect the choice of the multiple access method in microcellular environments.

As a preamble, if we observe the equations for spectral efficiency η, we see that η is inversely proportional to the number of microcells per cluster M. The smallest value of M is unity, where every microcell uses the same frequencies. Under these conditions the SIR will be low. Thus to achieve high η, we need a low value of M, and for an acceptable bit error rate (BER), we require the radio link to be able to operate with low values of SIRs. Because cellular radio operates in an intentional jamming environment, whereas CDMA was conceived to operate in a military environment where jamming by the enemy is expected, CDMA is a most appropriate multiple access method for cellular radio. The CDMA system, IS-95, will operate efficiently in single cell ($M = 1$) clusters where each cell is sectorized.

In highway microcells two-cell clusters can be used with TDMA and FDMA. Street microcells have complex shapes, see Figs. 82.4 and 82.5, and when FCA is used with TDMA and FDMA, there is a danger that high levels of interference will be ducted through streets and cause high-interference levels in a small segment of a microcell. To accommodate this phenomenon, the system must have a rapid handover (HO) capability, with HO to either a different channel at the same BS, to the interfering cell, or to an oversailing macrocell. CDMA is much more adept at handling this situation. The irregularity of street microcells, except in regularly shaped cities, such as midtown Manhattan, suggests that FCA should not be used. If it is, it requires $M \geq 4$. Instead, DCA should be employed. For example, when DCA is used with TDMA we abandon the notion of clusters of microcells. We may arrange for all microcells to have the same frequency set and design the system with accurate power control to contain the cochannel interference, and to encourage the MS to switch to another channel at the current BS or switch to a new BS directly when the SIR becomes below a threshold at either end of the duplex link. The application of DCA increases the capacity and can also contend with the situation where a MS suddenly experiences a rapid peak of cochannel interference during its travels.

The interference levels in CDMA in street microcells is mainly from users within its microcell, rather than from users in other microcells due to the shielding of the buildings. For CDMA to operate efficiently in street microcells, it should increase its chip rate to ensure it can exploit path diversity in its RAKE receiver. By increasing the chip rate, higher data rates can be accommodated and, hence, a greater variety of services. CDMA should be used in a similar way in office microcells. If the chip rate cannot be increased, however, the equipment installer can deploy a distributed antenna system where between each antenna a delay element is introduced. By this means path diversity gains are realized.

TDMA/DCA is appropriate for indoor microcells, where the complexity of the DCA is easier to implement compared to street microcells. FDMA should be considered for indoor microcells where it is well suited to provide high-bit-rate services since the transmitted rate is the same as the source rate. It also benefits from the natural shielding that exists within buildings to contain the cochannel interference and the low-power levels that simplify equipment design.

82.7 Discussion

At the time of writing, microcells are used in cordless telecommunications, where indoor microcells and outdoor telepoint microcells are used. There are very few microcells in cellular systems because there are no commercially available microcellular BSs. Nevertheless, operators have formed microcells using existing macrocellular BSs. Microcellular BSs, however, do exist in manufacturer's laboratories, and their entrance into the market is imminent. When microcells are deployed in large numbers, the vast increase in teletraffic will call for new network topologies and protocols.

In our deliberations we focused on highway microcells, city-street microcells, and indoor microcells. Minicells, where the BS antenna is below most of the buildings but above others, are currently being deployed. We may anticipate the fusion of the types of minicells and microcells. We will have microcells of strange shapes, like city street microcells but in three dimensions. Street microcells may serve the lower floors of buildings and vice versa. Microcells, located in minicell environments, may cover the streets as well as floors in neighboring buildings. We may also anticipate very small microcells, the so-called picocells. Indeed, we will have multicellular networks with multimode radio interfaces. This means that an intelligent multimode terminal with its supporting network will be required [Steele and Williams, 1993]. The role of microcells is to carry the high-bit-rate traffic and, hence, support a wide range of services. Our teletraffic equations tell us that microcellular personal communication networks will support orders more teletraffic than current conventional systems. Technology advancements will produce coffee cup size microcellular BSs and facilitate new network architectures that will eventually lead to the widespread concentration of intelligence at the BSs.

Defining Terms

Highway microcells: Segments of a highway having a base station and supporting mobile communications.
Indoor microcells: Small volumes of a building, e.g., an office, having a base station and supporting mobile communications.
Spectral efficiency: Has a special meaning in cellular radio. It is the traffic carried in erlang per hertz (or kilohertz) per area in square meters (or square kilometeres).
Street microcells: Small cells whose shape are determined by the street topology and their buildings. The base station antennas are below the urban skyline.

References

Chia, S.T.S., Steele, R., Green, E., and Baran, A. 1987. Propagation and bit error ratio measurements for a microcellular system. *JIERE*, Supplement 57(6):5255–5266.
El-Dolil, S.A., Wong, W.C., and Steele, R. 1989. Teletraffic performance of highway microcells with overlay macrocell. *IEEE JSAC*, 7(1):71–78.

Green, E. 1990. Radio link design for microcellular systems. *British Telecom Tech. J.* 8(1):85–96.

Keenan, J.M. and Motley, A.J. 1990. Radio coverage in buildings. *British Telecom. Tech. J.* 8(1):19–24.

Merrett, R.P., Cooper, A.J., and Symington, I.C. A cordless access system using radio-over-fiber techniques. IEEE VT-91:921–924.

Saleh, A.A.M. and Valenzula, R.A. 1987. A statistical model for indoor multipath propagation. *IEEE JSAC*, (Feb.):128–137.

Steele, R. 1992. *Mobile Radio Communications*, Pentech Press, London.

Steele, R. and Nofal, M. 1992. Teletraffic performance of microcellular personal communication networks. *IEE Proc-I*, 139(4):448–461.

Steele, R., Nofal, M., and El-Dolil, S. 1990. An adaptive algorithm for variable teletraffic demand in highway microcells. *Electronic Letters*, 26(14):988–990.

Steele R. and Prabhu, V.K. 1985. Mobile radio cellular structures for high user density and large data rates. In *Proc of the IEE*, Pt. F. (5):396–404.

Steele, R. and Williams, J.E.B. 1990. Third generation PCN and the intelligent multimode mobile portable, *IEE Elec. and Comm. Eng. J.* 5(3):147–156.

Further Information

The *IEEE Communications Magazine Special Issue* on an update on personal communications, Vol. 30, No. 12, Dec. 1992 provides a good introduction to microcells, particularly the paper by L.J. Greenstein, et al.

83
Fixed and Dynamic Channel Assignment

Bijan Jabbari
George Mason University

83.1 Introduction 1175
83.2 The Resource Assignment Problem 1175
83.3 Fixed Channel Assignment 1176
83.4 Enhanced Fixed Channel Assignment 1177
83.5 Dynamic Channel Assignment 1178
83.6 Conclusion 1180

83.1 Introduction

One of the important aspects of frequency reuse-based cellular radio as compared to early land mobile telephone systems is the potential for dynamic allocation of channels to traffic demand. This fact had been recognized from the early days of research (e.g., see chapter 7 in Jakes, 1994; Cox and Reudnik, 1972 and 1973) in this field. With the emergence of wireless personal communications and use of microcell with nonuniform traffic, radio resource assignment becomes essential to network operation and largely determines the available spectrum efficiency. The primary reason for this lies in the use of microcell in dense urban areas where distinct differences exist as compared to large cell systems due to radio propagation and fading effects that affect the interference conditions.

In this chapter, we will first review the channel reuse constraint and then describe methods to accomplish the assignment. Subsequently, we will consider variations of fixed channel assignment and discuss dynamic resource assignment. Finally, we will briefly discuss the traffic modeling aspect.

83.2 The Resource Assignment Problem

The resources in a wireless cellular network are derived either in frequency division multiple access (FDMA), time division multiple access (TDMA), or joint frequency–time (MC-TDMA) [Abramson, 1994]. In these channel derivation techniques, the frequency reuse concept is used throughout the service areas comprised of cells and microcells. The same channel is used by distinct terminals in different cells, with the only constraint of meeting a given interference threshold. In spread spectrum multiple access (SSMA) such as the implemented code division multiple access (CDMA) system (IS-95) [TIA, 1993] each subscriber spreads its transmitted signal over the same frequency band by using a pseudorandom sequence simultaneously. As any active channel is influenced by the others, a new channel can be set up only if the overall interference is below a given threshold. Thus, the problem of resource assignment in CDMA relates to transmission power control in forward (base station to mobile terminal) and reverse (mobile terminal to base station) channels. Of course, the problem of power control applies to TDMA and FDMA as well, but not to the extent that it impacts the capacity of CDMA. Here, we will focus on time- and frequency-based access methods. For preliminary results in CDMA the readers are referred to Everitt, 1994.

Fixed channel assignment (FCA) and dynamic channel assignment (DCA) techniques are the two extremes of allocating radio channels to mobile subscribers. For a specific grade of service and quality of transmission, the assignment scheme provides a tradeoff between spectrum utilization and implementation complexity. The performance parameters from a radio resource assignment point of view are interference constraints (quality of transmission link), probability of call blocking (grade of service), and the system capacity (spectrum utilization) described by busy hour erlang traffic that can be carried by the network. In a cellular system, however, there exist other functions, such as handoff and its execution or radio access control. These functions may be facilitated by the use of specific assignment schemes and, therefore, they should be considered in such a tradeoff [Everitt, 1994].

The problem of channel assignment can be described as the following: Given a set of channels derived from the specified spectrum, assign the channels and their transmission power such that for every set of assigned channels to cell i, $(C/I)_i > (C/I)_0$. Here, $(C/I)_0$ represents the minimum allowed carrier to interference and $(C/I)_i$ represents carrier to interference at cell i.

83.3 Fixed Channel Assignment

In fixed channel assignment the interference constraints are ensured by a frequency plan independent of the number and location of active mobiles. Each cell is then assigned a fixed number of carriers, dependent on the traffic density and cell size. The corresponding frequency plan remains fixed on a long-term basis. In reconfigurable FCA (sometimes referred to as flexible FCA), however, it is possible to reconfigure the frequency plan periodically in response to near/medium term changes in predicted traffic demand.

In FCA, for a given set of communications system parameters, $(C/I)_0$ relates to a specific quality of transmission link (e.g., probability of bit error or voice quality). This parameter in turn relates to the number of channel sets [Jakes, 1994] (or cluster size) given by $K = 1/3 \, (D/R)^2$. Thus, the ratio D/R is determined by $(C/I)_0$. Here D is the cochannel reuse distance and R is the cell radius. For example, in the North American cellular system advanced mobile phone service (AMPS), $(C/I)_0 = 18$ dB, which results in $K = 7$ or $D = 4.6R$. Here, we have used a propagation attenuation proportional to the fourth power of the distance. The radius of the cell is determined mainly by the projected traffic density. In Fig. 83.1 a seven cell cluster with frequency sets F1 through F7 (in cells designated A–G) has been illustrated. It is seen that the same set of frequencies is repeated two cells away.

The number of channels for each cell can be determined through the erlang-B formula (for example, see Cooper, 1990) by knowing the busy hour traffic and the desired probability of blocking (grade of service). Probability of blocking P_B is related to offered traffic A, and the number of channels per cell N by

$$P_B = \frac{A^N/N!}{\sum_{i=0}^{N} A^i/i!}$$

Fixed and Dynamic Channel Assignment

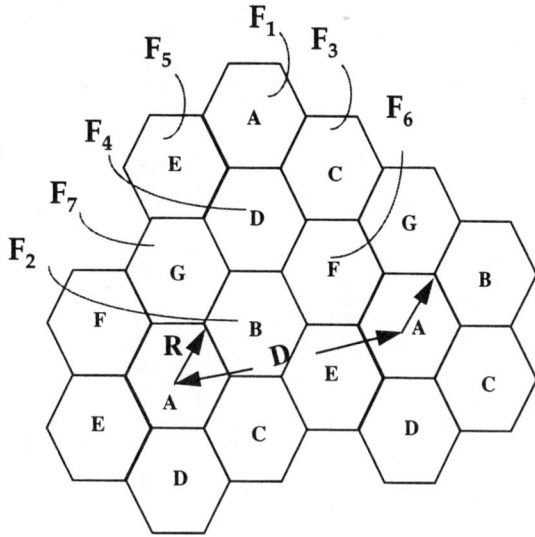

FIGURE 83.1 Fixed channel assignment.

This applies to the case of blocked calls cleared. If calls are delayed, the grade of service becomes the probability of calls being delayed P_Q and is given by the erlang-C formula [Cooper, 1990],

$$P_Q = \frac{\dfrac{A^N}{N!(1 - A/N)}}{\displaystyle\sum_{i=0}^{N-1} A^i/i! + \dfrac{A^N}{N!(1 - A/N)}}$$

FCA is used in almost all existing cellular mobile networks employing FDMA or TDMA. To illustrate resource assignment in FCA, we describe the FCA schemes used in gobal system for mobile communications (GSM or DCS) [Jabbari et al., 1995]. Here, the mobile terminal continuously monitors the received signal strength and quality of a broadcast channel along with the identity of the transmitting base station. This mechanism allows both the mobile and the network to keep track of the subscriber movements throughout the service area. When the mobile terminal tries and eventually succeeds in accessing the network (see Fig. 83.2), a two way control channel is assigned that allows for both authentication and ciphering mode establishment. On completion of these two phases, the setup phase initiates and a traffic channel is eventually assigned to the mobile terminal. Here, the terminal only knows the identity of the serving cell; the (control and traffic) radio channel assignment is a responsibility of the network and fulfills the original carrier to cell association, while simplifying the control functions during normal system operation.

83.4 Enhanced Fixed Channel Assignment

FCA has the advantage of having simple realization. Since the frequency sets are preassigned to cells based on long-term traffic demand, however, it cannot adapt to traffic variation across cells and, therefore, FCA will result in poor bandwidth utilization. To improve the utilization while maintaining the implementation simplicity, various strategies have been proposed as enhancements to FCA and deployed in existing networks. Two often used methods are *channel borrowing* and *directed retry*, which are described here.

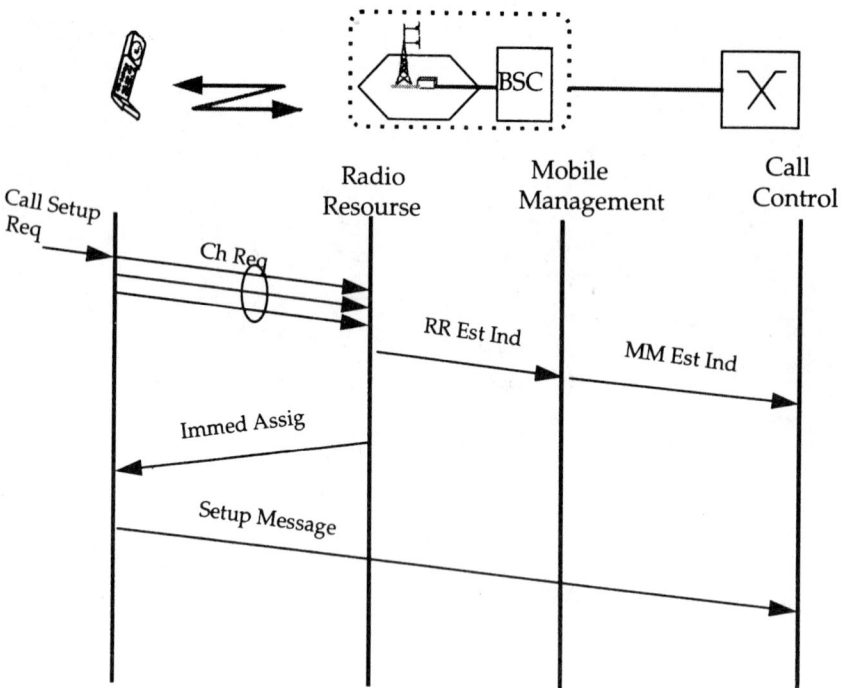

FIGURE 83.2 Radio resource assignment in GSM/DCS. *Source*: Jabbari, B. et al. 1995. Network issues for wireless personal communications. *IEEE Commun. Mag.* 33(1).

In the channel borrowing strategy [Jakes, 1994; Engel and Peritsky, 1973; Elnoubi, Singh, and Gupta, 1982; Zhang and Yum, 1989], channels that are not in use in their cells may be borrowed by adjacent cells with high offered traffic on a call-by-call basis. Borrowing of channels allows the arriving calls to be served in their own cell. This implies that there will be further restrictions in using the borrowed channels in other cells. Various forms of borrowing have been surveyed in Tekinay and Jabbari, 1991.

In directed retry [Eklundh, 1986; Everitt, 1990; 1994], a call to or from a mobile subscriber may try other cells with channels with sufficient signal strength meeting the C/I constraint if there are no channels available in its own cell to be served. In some cases it may be necessary to direct some of the calls in progress in a given congested cell to adjacent lightly loaded cells (those calls that can be served by adjacent cells) in order to accommodate the new calls in that given cell. This is referred to as directed handoff [Karlson and Eklundh, 1989; Everitt, 1990]. The combination of these two capabilities provides a significant increase in bandwidth utilization.

83.5 Dynamic Channel Assignment

In dynamic channel assignment [Cox and Reudnik, 1972; 1973; Engel and Peritsky, 1993; Anderson, 1973; Beck and Panzer, 1989; Panzer and Beck, 1990; Chuang, 1993; Dimitrijevic and Vucetic, 1993], the assignment of channels to cells occurs based on the traffic demand in the cells. In other words, channels are pooled together and assignments are made and modified in real time. Therefore, this assignment scheme has the potential to achieve a significantly improved bandwidth utilization when there are temporal or spatial traffic variations.

In DCA, the interference constraints are ensured by a real-time evaluation of the most suitable (less interfered) channels that can be activated in a given cell (reassignment may be needed). That is, the system behaves as if the frequency plan was dynamically changing to meet the actual radio link quality and traffic loads, realizing an implicit sharing of the frequency band under interference constraints.

The implementation of DCA generally is more complex due to the requirement for system-wide state information where the state refers to which channel in which cell is being used. Obviously, it is impractical to update the system state in a large cellular network at any time, especially those based on microcells, as the controller will be overloaded or call set delay will be unacceptable. Therefore, methods have been devised based on a limited state space centralized control [Everitt and Macfadyen, 1983] or based on a distributed control to perform the necessary updating. A method referred to as maximum packing suggested in Everitt and Macfadyen, 1983, records only the number of channels in use. Distributed control schemes, however, where channel assignments are made at mobile stations or base stations may be attractive. These schemes are particularly suitable for a mobile controlled resource assignment where each mobile measures the actually perceived interference and decides to utilize a radio resource in a completely decentralized way.

Design of DCA algorithms is critical to achieve the potential advantages in efficiency and robustness to traffic heterogeneity throughout cells as compared to FCA and enhanced FCA. Poor DCA algorithms, however, might lead to an uncontrolled global situation, i.e., the locally selected channel might be very good for the specific mobile terminal and at the same time very poor for the interference level induced to other traffic sources.

Two classes of *regulated DCA* and *segregation DCA* are discussed here due to their importance. In a regulated DCA, appropriate thresholds tend to maintain the current channel, avoiding useless handoffs that can tie up channels; in a segregation DCA, channel acquisition obeys priorities assigned to each channel. In the latter, the channels successfully activated in a cell have a higher assignment probability than those found to have a high interference in the cell; of course, priorities change in time depending on changes in the system status. These types of assignment lead to a (frequency) plan still changing dynamically but more slowly and, in general, only because of substantial load imbalance. In steady-state conditions, the plan either tends to be confirmed or fluctuates around a basic configuration, proven to be suboptimal in terms of bandwidth efficiency.

Both digital European cordless telecommunications (DECT) and cordless technology known as CT2 are employing the DCA technique [Tuttlebee, 1992], and the next generation cellular systems are foreseen to deploy it. In CT2 the handset and the base station jointly determine a free suitable channel to serve the call. In DECT, the mobile terminal (portable handset) not only recognizes the visited base station (radio fixed part) through a pilot signal but continuously scans all of the system channels and holds a list of the less interfered ones. The potentially available channels are ordered by the mobile terminal with respect to the measured radio parameters, namely, the radio signal strength indicator, measuring and combining cochannel, adjacent-channel, and intermodulation interference. When a connection has to be established, the best channel is used to communicate on the radio interface.

It is possible to have a hybrid of DCA and FCA in a cellular network in which a fraction of channels are fixed assigned and the remainder are allocated based on FCA. This scheme has less system implementation complexity than the DCA scheme but provides performance improvement (lower probability of call blocking) depending on the DCA–FCA channel partitioning [Kahwa and Georganas, 1978].

In general, DCA schemes cannot be considered independently of the adopted power control mechanism because the transmitted power from both mobile terminal and base station substantially affects the interference within the network. For a detailed discussion of DCA and power control, the readers are referred to [Chuang, Sollenberger, and Cox, 1994]. Despite the availability of several power control algorithms, much work is needed to identify realizable power and channel control algorithms that maximize bandwidth efficiency. Nevertheless, realization of power control mechanisms would require distribution of the system status information and information exchange between mobile terminal and network entities. This in turn will involve overhead and terminal power usage.

The performance of DCA depends on the algorithm implementing this capability [Everitt, 1990; Everitt and Manfield, 1989]. In general, due to interactions between different cells the performance of the system will involve modeling the system as a whole, as opposed to in FCA where cells are treated independently. Therefore, mathematical modeling and performance evaluation of DCA becomes quite complex. Simplifying assumptions may, therefore, be necessary to obtain approximate results [Prabhu and Rappaport, 1974; Everitt and Macfadyen, 1983]. Simulation techniques have been widely used in evaluation of DCA performance. For a representative performance characteristics of DCA and a comparison to enhanced FCA schemes the readers are referred to [Everitt, 1994].

83.6 Conclusion

In this chapter we have classified and reviewed channel assignment techniques. We have emphasized the advantages of DCA schemes over FCA in terms of bandwidth utilization in a heterogeneous traffic environment at the cost of implementation complexity. The DCA schemes are expected to play an essential role in future cellular and microcellular networks.

References

Abramson, N. 1994. Multiple access techniques for wireless networks. *Proc. of IEEE.* 82(9).

Anderson, L.G. 1973. A simulation study of some dynamic channel assignment algorithms in a high capacity mobile telecommunications system. *IEEE Trans. on Comm.* COM-21(11).

Beck, R. and Panzer, H. 1989. Strategies for handover and dynamic channel allocation in microcellular mobile radio systems. *Proceedings of the IEEE Vehicular Technology Conference.*

Chuang, J.C.-I. 1993. Performance issues and algorithms for dynamic channel assignment. *IEEE J. on Selected Areas in Comm.* 11(6).

Chuang, J.C.-I., Sollenberger, N.R., and Cox, D.C. 1994. A pilot-based dynamic channel assignment schemes for wireless access TDMA/FDMA systems. *Internat. J. of Wireless Inform. Networks,* Jan.

Cooper, R.B. 1990. *Introduction to Queueing Theory,* 3rd ed. CEEPress Books.

Cox, D.C. and Reudnik, D.O. 1972. A comparison of some channel assignment strategies in large-scale mobile communications systems. *IEEE Trans. on Comm.* COM-20(2).

Cox, D.C. and Reudnik, D.O. 1973. Increasing channel occupancy in large-scale mobile radio environments: dynamic channel reassignment. *IEEE Trans. on Comm.* COM-21(11).

Dimitrijevic, D. and Vucetic, J.F. 1993. Design and performance analysis of algorithms for channel allocation in cellular networks. *IEEE Trans. on Vehicular Tech.* 42(4).

Eklundh, B. 1986. Channel utilization and blocking probability in a cellular mobile telephone system with directed retry. *IEEE Trans. on Comm.* COM 34(4).

Elnoubi, S.M., Singh, R., and Gupta, S.C. 1982. A new frequency channel assignment algorithm in high capacity mobile communications. *IEEE Trans. on Vehicular Techno.* 31(3).

Engel, J.S. and Peritsky, M.M. 1973. Statistically-optimum dynamic server assignment in systems with interfering servers. *IEEE Trans. on Comm.* COM-21(11).

Everitt, D. 1990. Traffic capacity of cellular mobile communications systems. *Computer Networks and ISDN Systems,* ITC Specialist Seminar, Sept. 25–29, 1989.

Everitt, D. 1994. Traffic engineering of the radio interface for cellular mobile networks. *Proc. of IEEE.* 82(9).

Everitt, D.E. and Macfadyen, N.W. 1983. Analysis of multicellular mobile radiotelephone systems with loss. *BT Tech. J.* 2.

Everitt, D. and Manfield, D. 1989. Performance analysis of cellular mobile communication systems with dynamic channel assignment. *IEEE J. on Selected Areas in Comm.* 7(8).

Jabbari, B., Colombo, G., Nakajima, A., and Kulkarni, J. 1995. Network issues for wireless personal communications. *IEEE Comm. Mag.* 33(1).

Jakes, W.C. ed. 1994. *Microwave Mobile Communications*, Wiley, New York, 1974, reissued by IEEE Press.

Kahwa, T.J. and Georganas, N.D. 1978. A hybrid channel assignment scheme in Large scale, cellular-structured mobile communications systems. *IEEE Trans. on Comm.* COM-26(4).

Karlsson, J. and Eklundh, B. 1989. A cellular mobile telephone system with load sharing—an enhancement of directed retry. *IEEE Trans. on Comm.* COM 37(5).

Panzer, H. and Beck, R. 1990. Adaptive resource allocation in metropolitan area cellular mobile radio systems. *Proceedings of the IEEE Vehicular Technology Conference.*

Prabhu, V. and Rappaport, S.S. 1974. Approximate analysis for dynamic channel assignment in large systems with cellular structure. *IEEE Trans. on Comm.* COM-22(10).

Tekinay, S. and Jabbari, B. 1991. Handover and channel assignment in mobile cellular networks. *IEEE Comm. Mag.* 29(11).

Telecommunications Industry Association. 1993. TIA Interim Standard IS-95, CDMA Specifications.

Tuttlebee, W.H.W. 1992. Cordless personal communications. *IEEE Comm. Mag.* (Dec.).

Zhang, M. and Yum, T.-S.P. 1989. Comparison of channel-assignment strategies in cellular mobile telephone systems. *IEEE Trans. on Vehicular Tech.* 38(4).

84
Propagation Models

84.1	Introduction	1182
84.2	Free-Space Propagation Model	1183
84.3	The Three Basic Propagation Mechanisms	1183
84.4	Reflection	1183
	Ground Reflection (Two-Ray) Model	
84.5	Diffraction	1186
	Knife-Edge Diffraction Model	
84.6	Scattering	1187
84.7	Practical Propagation Models	1188
	Log-Distance Path Loss Model • Outdoor Propagation Models • Indoor Propagation Models	
84.8	Small-Scale Fading	1193
	Flat Fading • Frequency Selective Fading	

Theodore S. Rappaport
Virginia Polytechnic Institute and State University

Rias Muhamed
Virginia Polytechnic Institute and State University

Varun Kapoor
Virginia Polytechnic Institute and State University

84.1 Introduction

The radio channel places fundamental limitations on the performance of mobile communication systems. The transmission path between the transmitter and the receiver can vary from simple direct line of sight to one that is severely obstructed by buildings and foliage. The speed of motion impacts how rapidly the signal level fades as a mobile terminal moves in space. Unlike wired channels that are stationary and predictable, radio channels are extremely random and do not offer easy analysis. In fact, modeling the radio channel has historically been one of the challenging parts of any radio system design and is typically done in a statistical fashion, based on measurements made specifically for an intended communication system.

The mechanisms behind electromagnetic wave propagation are diverse but can generally be attributed to reflection, diffraction, and scattering. Most cellular radio systems operate in urban areas where there is no direct line-of sight path between the transmitter and the receiver, and the presence of high-rise buildings causes severe diffraction loss. Because of the multiple reflections from various objects, the electromagnetic waves travel along different paths of varying lengths. The interaction between these waves causes multipath fading at a specific location, and the strengths of the waves decrease as the distance between the transmitter and receiver increases.

Propagation models have traditionally focused on predicting the average received signal strength at a given distance from the transmitter, as well as the variability of the signal strength in spatial proximity to a particular location. A statistical representation of the mean signal strength for an arbitrary transmitter-receiver (T-R) separation distance is useful in estimating the radio coverage area of a transmitter, whereas radio modem design issues, such as antenna diversity and coding, require models that predict the signal variability over a very small distance. Propagation models that characterize signal strength over large T-R separation distances (several hundreds or thousands of wavelengths) are called **large-scale models**, whereas models that characterize the rapid fluctuations of the received signal strength over short distances (a few wavelengths) or short time durations (on the order of seconds) are called **small-scale models**.

84.2 Free-Space Propagation Model

Free space is an ideal propagation model that can be accurately applied only to satellite communication systems and short line-of-sight radio links. It demonstrates, however, how received power decays as a logarithmic function of the T-R separation, which is a fundamental characteristic of large-scale modeling. The power received by a receiver antenna, which is separated from a radiating transmitter antenna by a distance d, assuming a free space path between the antennas, is given by the Friis free-space equation

$$P_r = \frac{P_t G_t G_r \lambda^2}{(4\pi)^2 d^2 L} \tag{84.1}$$

where P_t the transmitted power, G_t the transmitter antenna gain, G_r the receiver antenna gain, d the T-R separation distance in meters, L the system losses ($L > 1$), and λ the wavelength in meters.

The Friis transmission formula gives an inverse square relationship between the received power and the T-R separation distance. This implies the received power decays at a rate of 20 dB/decade with distance.

The path loss is defined as the difference between the effective transmitted power and the received power and may include the effect of the antenna gains. The path loss PL in decibels for the free space model is given by

$$PL = -10 \log_{10} \frac{P_r}{P_t} = -10 \log_{10} \left[\frac{G_t G_r \lambda^2}{(4\pi)^2 d^2 L} \right] \tag{84.2}$$

84.3 The Three Basic Propagation Mechanisms

Reflection occurs when a propagating electromagnetic wave impinges upon an obstruction whose dimensions are very large when compared to the wavelength of the radio wave. Reflections from the surface of the earth and from buildings or walls produce reflected waves, which may interfere constructively or destructively at the receiver.

Diffraction occurs when the radio path between the transmitter and receiver is obstructed by a surface that has sharp irregularities (edges). The secondary waves resulting from the obstructing surface are present throughout the space and even behind the irregularity, giving rise to bending of waves about the irregularity, even when a line of sight does not exist. At high frequencies, diffraction, like reflection, depends on the geometry of the object, and the amplitude, phase, and polarization of the incident wave at the point of diffraction.

Scattering occurs when the medium through which the wave travels consists of objects with dimensions that are small compared to the wavelength and the number of obstacles per unit volume is quite large.

84.4 Reflection

When a radio wave propagating in one medium impinges upon another medium having a different dielectric constant, permeability, or conductivity, the wave is partially reflected and partially

TABLE 84.1 Ground Parameters at 100 MHz

Material	Permittivity, ε_r	Conductivity σ (s/m)
Poor ground	4	0.001
Typical ground	15	0.005
Good ground	25	0.02
Sea water	81	5.0
Fresh water	81	0.01

transmitted. For the case of a plane wave in air incident normally on a perfect conductor, the wave is completely reflected without loss of energy. If the plane wave is incident on a perfect dielectric, part of the energy is transmitted and part of the energy is reflected, and there is no loss of energy in absorption. The electric field intensity of the reflected wave and the incident wave are related through the Fresnel reflection coefficient Γ. The reflection coefficient is a function of the permittivity of the ground, wave polarization, angle of incidence, and the frequency of the propagating wave. Table 84.1 shows typical values of permittivity and conductivity for a few common types of ground when operating in the VHF frequency range.

Figure 84.1 shows the geometry for calculating the reflection coefficients. The subscripts v and h refer to the vertical and horizontal polarization, respectively, of the E-field, and the subscripts i and r refer to the incident and reflected E-field, respectively. The permittivity, permeability, and conductance of the two media are ε_1, μ_1, σ_1 and ε_2, μ_2, σ_2, respectively. For the case when the first medium is free space, the reflection coefficients for the two cases of vertical (parallel) and horizontal (perpendicular) polarization are given as (for vertical E-field polarization)

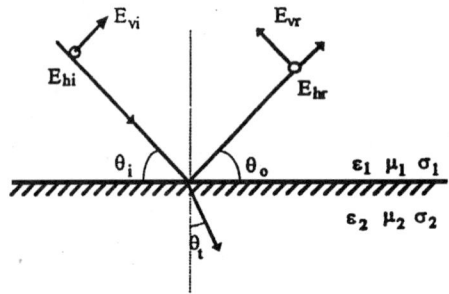

FIGURE 84.1 Geometry for reflection coefficient.

$$\Gamma_v = \frac{E_{vr}}{E_{vi}} = \frac{-\varepsilon_r \sin\theta_i - \sqrt{\varepsilon_r - \cos^2\theta_i}}{\varepsilon_r \sin\theta_i + \sqrt{\varepsilon_r - \cos^2\theta_i}} \quad (84.3)$$

and for horizontal E-field polarization

$$\Gamma_h = \frac{E_{hr}}{E_{hi}} = \frac{\sin\theta_i - \sqrt{\varepsilon_r - \cos^2\theta_i}}{\sin\theta_i + \sqrt{\varepsilon_r - \cos^2\theta_i}} \quad (84.4)$$

where ε_r is the relative permittivity of the second medium. The incident angle θ_i at which the reflection coefficient Γ_v is equal to zero is called the *Brewster angle* and is given by the value of θ_i that satisfies the equation

$$\sin\theta_i = \frac{\sqrt{\varepsilon_r - 1}}{\sqrt{\varepsilon_r^2 - 1}} \quad (84.5)$$

Ground Reflection (Two-Ray) Model

In a mobile radio channel a single direct path between the base station and the mobile seldom exists and, hence, the free-space propagation model of Eq. (84.2) is of limited use. The two-ray ground reflection model shown in Fig. 84.2 is a useful propagation model that is based on geometric optics

Propagation Models

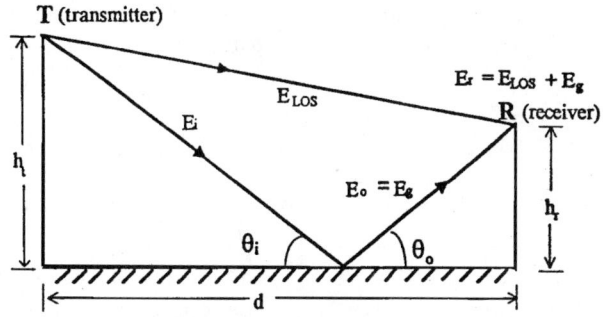

FIGURE 84.2 Two-ray ground reflection model.

and considers both the direct and ground reflected propagation path. This model assumes that the wavelength λ is much smaller than anything else encountered in the channel.

In most mobile communication systems where the separation distance is only a few tens of kilometers, the flat terrain assumption is a valid approximation. The total received E-field E_r is the resultant of the direct line-of-sight component E_{LOS} and a ground reflected component E_g, and is referenced to a close-in E-field measured at a small distance d_0.

Referring to Fig. 84.2, h_t is the height of the transmitter and h_r the height of the receiver. According to laws of reflection,

$$\theta_i = \theta_0 \quad \text{and} \quad E_0 = \Gamma E_i \tag{84.6}$$

where Γ is the reflection coefficient for ground. As θ_i approaches 0° (i.e., grazing incidence) the reflected wave is equal in magnitude and 180° out of phase with the incident wave. The resultant E-field, which is the sum of E_{LOS} and E_g, can be shown to be

$$|E_r(d)| = \frac{2E_{LOS}d_0}{d} \sin\frac{\theta_\Delta}{2} \tag{84.7}$$

where the phase difference (θ_Δ) is related to the path difference (Δ) between the direct and ground reflected paths by

$$\theta_\Delta = \frac{2\pi \Delta}{\lambda} \tag{84.8}$$

At large values of d,

$$\sin\frac{\theta_\Delta}{2} \approx \frac{\theta_\Delta}{2} = \frac{2\pi h_t h_r}{\lambda d} \tag{84.9}$$

and the received E-field in volts per meter is given by

$$E_R(d) \approx 2E_{LOS}\frac{2\pi h_t h_r d_0}{\lambda d^2} \approx \frac{k}{d^2} \tag{84.10}$$

where k is a constant related to E_{d_0}, the electric field measured at distance d_0 from the transmitter. The power received at d is related to the square of the electric field and can be expressed approximately as

$$P_r = P_t G_t G_r \frac{h_t^2 h_r^2}{d^4} \tag{84.11}$$

At large distances, the received power falls off at a rate of 40 dB/decade, and the received power and path loss become independent of frequency. The path loss in decibels for the two-ray model is approximated as

$$PL = -10\log_{10} G_t - 10\log_{10} G_r - 20\log_{10} h_t - 20\log_{10} h_r + 40\log_{10} d \quad (84.12)$$

84.5 Diffraction

Diffraction around the Earth's curvature makes it possible to transmit radio signals beyond the line of sight and allows an electromagnetic wave to propagate beyond obstructions. Although the field strength decreases rapidly as one moves deeper into the shadowed region, the field is still finite and has sufficient strength to produce a useful signal.

The phenomenon of diffraction can be easily explained using Huygen's principle, which states that all points on a wavefront can be considered as point sources for the production of secondary wavelets, and that these wavelets combine to produce a new wavefront in the direction of propagation. Diffraction is caused by the propagation of secondary wavelets into the shadowed region. The field strength of a diffracted wave in the shadowed region is the vector sum of all of the secondary wavelets.

Knife-Edge Diffraction Model

When shadowing is caused by a single object, such as a mountain or building, the attenuation caused by diffraction over such an object can be estimated by treating the obstruction as a diffracting knife edge. This is the simplest of diffraction models, and the diffraction loss in this case can be readily estimated using the classical Fresnel solution for the field behind a knife edge or half-plane. Figure 84.3 illustrates this approach.

The field strength at point R in the shadowed region (also called the **diffraction zone**) is a vector sum of the fields due to all of the secondary Huygen sources in the plane above the knife edge. The field strength E_d of a knife-edge diffracted wave is given by

$$E_d = E_0 F(v) = E_0 \frac{1+j}{2} \int_v^\infty \exp(-j\pi t^2/2)\, dt \quad (84.13)$$

where E_0 is the free space field strength in the absence of the knife edge and $F(v)$ is the complex Fresnel integral. The Fresnel integral is a function of the Fresnel–Kirchoff diffraction parameter v

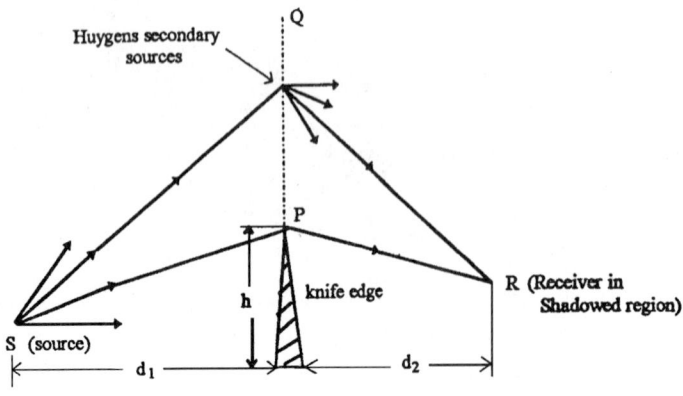

FIGURE 84.3 Knife-edge diffraction.

Propagation Models

which is defined as

$$v = h\sqrt{\frac{2(d_1 + d_2)}{\lambda d_1 d_2}} \qquad (84.14)$$

where h is the height of the knife edge and d_1 and d_2 are the distances of the knife edge from the transmitter and the receiver, respectively. If the knife edge protrudes above a line drawn from S to R, h and v are positive. If the obstruction lies below the line of sight, then both h and v are negative. The Fresnel integrals are available in tabular form for given values of the diffraction parameter v. The diffraction gain in decibels due to the presence of a knife edge is given by

$$G_d = 20 \log_{10} |F(v)| \qquad (84.15)$$

Since the expression for diffraction gain involves the Fresnel integral, which is difficult to compute, approximate numerical solutions are often used. An approximate solution, in decibels, provided by Lee, 1985, is given as

$$G_d = 0 \qquad 1 \leq v \qquad (84.16a)$$
$$G_d = 20 \log_{10}(0.5 + 0.62v) \qquad 0 \leq v \leq 1 \qquad (84.16b)$$
$$G_d = 20 \log_{10}(0.5 \exp(0.95v)) \qquad -1 \leq v \leq 0 \qquad (84.16c)$$
$$G_d = 20 \log_{10}(0.4 - \sqrt{0.1184 - (0.1v + 0.38)^2}) \qquad -2.4 \leq v \leq -1 \qquad (84.16d)$$
$$G_d = 20 \log_{10}(-0.225/v) \qquad v \leq -2.4 \qquad (84.16e)$$

Note that when $v = 0$, the diffraction gain is -6 dB. In many practical situations, especially in hilly terrain, the propagation path may consist of more than one obstruction, and it may be required to calculate the total diffraction losses due to all of the obstacles. Bullington, 1947, suggested that the series of obstacles be replaced by a single equivalent obstacle so that the path loss can be obtained using single knife-edge diffraction models. This method oversimplifies the calculations and often provides very optimistic estimates of the received signal strength. In a more rigorous treatment, Millington, Hewitt, and Immirzi, 1962, gave a wave-theory solution for the field behind two knife edges in series. This solution is very useful and can be applied easily for predicting diffraction losses due to two knife edges. Extending this to more than two knife edges, however, becomes a formidable mathematical problem. Many models that are mathematically less complicated have been developed to estimate the diffraction losses due to multiple obstructions [Epstein and Peterson, 1953; Deygout, 1966].

84.6 Scattering

The measured path loss in a mobile radio environment is often less than what is predicted by reflection and diffraction models alone. This is because when a radio wave impinges on a rough surface, the reflected energy is spread out (diffused) in all directions due to scattering. Objects such as trees, lamp posts, and rough surfaces tend to scatter energy towards a receiver. The roughness of a surface is often tested using the Rayleigh criterion, which defines a critical height h_c of surface protuberances for a given angle of incidence θ_i,

$$h_c = \frac{\lambda}{8 \cos \theta_i} \qquad (84.17)$$

A surface is considered smooth if its minimum to maximum protuberance h is less than h_c and is considered rough if the protuberance is greater than h_c. For rough surfaces, the reflection coefficient needs to be modified by a scattering loss factor ρ_s to account for the diminished specularly reflected field. Ament, 1953, derived this as

$$\rho_s = \exp\left[-8\left(\frac{\pi \sigma_h \cos \theta_i}{\lambda}\right)^2\right] \qquad (84.18)$$

where σ_h is the standard deviation of the surface height about the mean surface height. The scattering loss factor derived by Ament was modified by Boithias, 1987, to give better agreement with measured results,

$$\rho_s = \exp\left[-8\left(\frac{\pi \sigma_h \cos \theta_i}{\lambda}\right)^2\right] I_0\left[8\left(\frac{\pi \sigma_h \cos \theta_i}{\lambda}\right)^2\right] \qquad (84.19)$$

where I_0 is the Bessel function of the first kind and zeroth order.

Analysis based on the geometric theory of diffraction and physical optics can be used to determine the scattered field strength. For urban mobile radio systems, models based on the bistatic radar equation may be used to compute the scattering losses in the far field. The radar cross section (RCS) of a scattering object is defined as the ratio of the power density of the signal scattered in the direction of the receiver to the power density of the radio wave incident upon the scattering object and has units of square meters. The bistatic radar equation given by Eq. (84.20) describes the propagation of a wave traveling in free space and intercepted by a scattering object, and then reradiated in the direction of the receiver,

$$P_R(\text{dBm}) = P_T(\text{dBm}) + G_T(\text{dBi}) + 20 \log_{10} \lambda + \text{RCS}[\text{dBm}^2]$$
$$- 30 \log_{10}(4\pi) - 20 \log_{10} d_T - 20 \log_{10} d_R \qquad (84.20)$$

where d_T and d_R are the distance from the scattering object to the transmitter and receiver, respectively. This equation can only be applied to scatterers in the far field of both the transmitter and receiver.

84.7 Practical Propagation Models

Most radio propagation models are derived using a combination of analytical and empirical methods. The empirical approach is based on fitting curves or analytical expressions that recreate a set of measured data. This has the advantage of implicitly taking into account all propagation factors, both known and unknown. The validity of empirical models at transmission frequencies and environments other than those used to originally derive the model can only be established by verifying them with data taken from the specific area at the required transmission frequency.

Log-Distance Path Loss Model

Both theoretical and measurement-based propagation models show that the average received signal power decreases with distance raised to some exponent. Here the average path loss for an arbitrary T-R separation is expressed as a function of distance by using a path loss exponent.

$$\overline{PL}(d) \propto \left(\frac{d}{d_0}\right)^n \qquad (84.21)$$

TABLE 84.2 Path Loss Exponents for Different Environments

Environment	Path Loss Exponent, n
Free-space	2
Urban area cellular radio	2.7–4
Shadowed urban cellular radio	5–6
In building line of sight	1.6–1.8
Obstructed in building	4–6
Obstructed in factories	2–3

or in decibels

$$\overline{PL} = \overline{PL}(d_0) + 10n \log_{10}\left(\frac{d}{d_0}\right) \qquad (84.22)$$

where n is the path loss exponent that indicates the rate at which the path loss increases with distance, d_0 is the free space close-in reference distance, and d is the T-R separation distance. When plotted on a log-log scale, the path loss is a straight line with a slope equal to n. The value of n depends on the specific propagation environment. For example, in free space, n is equal to 2, and when obstructions are present it will have a larger value. Table 84.2 gives a listing of typical path loss exponents obtained in various mobile radio environments.

It is important to select a free space reference distance that is appropriate for the propagation environment. In large cellular systems, 1 km and 1 mile reference distances are commonly used [Lee, 1985], whereas in microcellular systems, much smaller distances are used. The reference distance should always be in the far field of the antenna ($d_0 > 2D^2/\lambda$, where D is the largest antenna dimension) so that near-field effects do not alter the reference path loss. The reference path loss is calculated using the free space path loss formula given by Eq. (84.2).

Measurements have shown that at any value of d, the path loss $PL(d)$ at a particular location is distributed log-normally (normal in decibels) about the mean distance-dependent value. That is, $PL(d) = \overline{PL}(d) + X_\sigma$, where X_σ is a zero-mean lognormally distributed random variable with standard deviation σ in decibels. The log-normal distribution describes the random shadowing effects, which occur over a large number of measurement locations, which have the same T-R separation but have different levels of clutter on the propagation paths. A path loss exponent n and standard deviation σ will therefore statistically describe the path loss for an arbitrary location having a specific T-R separation. In practice, based on measurement data, the value of n and σ are computed using linear regression such that the difference between the measured and estimated path losses is minimized in a mean square error sense.

Outdoor Propagation Models

Radio transmission in a mobile communications system often takes place over irregular terrain. The terrain profile of a particular area needs to be taken into account while estimating the path loss. The terrain profile may vary from a simple curved earth profile to a highly mountainous profile. The presence of trees, buildings, and other obstacles also need to be taken into account.

A number of propagation models are available to predict path loss over irregular terrain. Although all of these models aim to predict signal strength at a particular receiving point or a specific small area (called a sector), the methods vary widely in their approach, complexity, and accuracy. Most of these models are based on a systematic interpretation of measurement data obtained from the service area. One of the more popular outdoor propagation models is the Longley–Rice [Rice et al., 1967] model. This model is applicable to point-to-point communication systems in the frequency range from 40 MHz to 100 GHz, over different kinds of terrain. The median transmission loss is predicted using the path geometry of the terrain profile and the refractivity of the troposphere.

Geometric-optics techniques (two-ray ground reflection model) are used to predict signal strengths within the radio horizon. Diffraction losses over isolated obstacles are estimated using the Fresnel–Kirchoff knife-edge models. Forward scatter theory is used to make troposcatter predictions over long distances and far-field diffraction losses in double horizon paths are predicted using a modified Van der Pol–Bremmer method. The Longley–Rice propagation prediction model is also referred to as the *ITS irregular terrain model*.

The Longley–Rice model is also available as a computer program [Longley and Rice, 1968] to calculate long-term median transmission loss relative to free space loss over irregular terrain for frequencies between 20 MHz and 10 GHz. For a given transmission path, the program takes as its input the transmission frequency, path length, polarization, antenna heights, surface refractivity, effective radius of Earth, ground conductivity, ground dielectric constant, and climate. The program also operates on path-specific parameters such as horizon distance of the antennas, horizon elevation angle, angular transhorizon distance, terrain irregularity, and other specific inputs.

The Longley–Rice method operates in two modes. When a detailed terrain path profile is available, the path specific parameters mentioned in the preceding paragraph can be easily determined and the prediction is called a *point-to-point mode* prediction. On the other hand, if the terrain path profile is not available, the Longley–Rice method provides techniques to estimate the path specific parameters, and such a prediction is called as an *area mode* prediction.

There have been many modifications and corrections to the Longley–Rice model since its original publication. One important modification [Longley, 1978] deals with radio propagation in urban areas, and this is particularly relevant to mobile radio. This modification introduces an excess term as an allowance for the additional attenuation due to urban clutter near the receiving antenna. This extra term called the urban factor (UF) has been derived by comparing the predictions by the original Longley–Rice model with those obtained by Okumura, 1968.

The major shortcoming of the Longley–Rice model is that it does not detail any method for determining corrections due to environmental factors in the immediate vicinity of the mobile or consider any correction factors to be included to account for the effects of buildings and foliage. Further, multipath is not considered.

The Okumura Model

Okumura's model is one of the most widely used models for signal prediction in urban areas. This model is applicable for frequencies in the range 150–2000 MHz and distances of 1–100 km. It can be used for base station effective antenna heights ranging from 30 to 1000 m.

Okumura developed a set of curves giving the median attenuation relative to free space, A_{mu}, in an urban area over a quasismooth terrain with a base station effective antenna height h_{te} of 200 m and a mobile antenna height h_{re} of 3 m. These curves are plotted as a function of frequency in the range 100–3000 MHz and as a function of distance from the base station in the range 1–100 km. To determine the path loss using Okumura's model, the free space path loss between the points of interest is first determined, and then the value of $A_{mu}(f, d)$ as read from the curves is added to it along with correction factors to account for antennas not at the reference heights. The model can be expressed as

$$L_{50} = L_F + A_{mu}(f, d) + G(h_{te}) + G(h_{re}) \qquad (84.23)$$

where L_{50} is the median propagation loss in decibels, L_F the free space propagation loss, A_{mu} the median attenuation relative to free space, $G(h_{te})$ the base station antenna height gain factor, and $G(h_{re})$ the mobile antenna height gain factor.

Plots of $G(h_{te})$ and $G(h_{re})$ vs the respective effective antenna heights are available. These curves show that $G(h_{te})$ varies at a rate of 20 dB/decade and $G(h_{re})$ for effective antenna heights less than 3 m varies at a rate of 10 dB/decade. Based on the terrain-related parameters, various corrections can be applied to Okumura's model. Some of the important terrain-related parameters are the effective

Propagation Models

base station antenna height h_{te}, terrain undulation height Δh, isolated ridge height, average slope of the terrain, and the mixed land–sea parameter. Once the terrain-related parameters are calculated, the necessary correction factors can be added or subtracted as required. All of these correction factors are also available as Okumura curves.

Okumura's model is wholly based on measured data and does not provide any analytical explanations. For many situations, extrapolations of the derived curves can be made to obtain values outside the measurement range, and the validity of such extrapolations depends on the circumstances and the smoothness of the curve in question.

Okumura's model is considered to be among the best in terms of accuracy in path loss prediction for mature cellular radio systems, and offers standard deviations within 12 dB of measured data. It is also a very practical model that has become a standard for system planning in today's land mobile radio systems in Japan. The only major disadvantage with the model is its complexity and slow response to rapid changes in radio path profile. Generally, it is found that the model is particularly good in urban and suburban areas but not as good in rural areas over irregular terrain.

The Hata Model

The Hata model [Hata, 1980] is an empirical formulation of the graphical path loss information provided by Okumura. Hata presented the urban area propagation loss as the standard formula and supplied correction equations to the standard formula for application to other situations. The standard formula for median path loss in urban areas is given by

$$L_{50} = 69.55 + 26.16 \log_{10} f_c - 13.82 \log_{10} h_{te} - a(h_{re})$$
$$+ (44.9 - 6.55 \log_{10} h_{te}) \log_{10} d \qquad (84.24)$$

where loss is in decibels, f_c the frequency from 150 to 1500 MHz, h_{te} the effective transmitter (base station) antenna height ranging from 30 to 200 m, h_{re} is the effective receiver (mobile) antenna height ranging from 1 to 10 m, d is the T-R separation distance in kilometers, and $a(h_{re})$ is the correction factor for effective mobile antenna height, which is a function of the size of the service area (city). For a small to medium sized city, the correction factor is given by

$$a(h_{re}) = (1.1 \log_{10} f_c - 0.7) h_{re} - (1.56 \log_{10} f_c - 0.8) \text{ dB} \qquad (84.25)$$

and for a large city, it is given by

$$a(h_{re}) = 8.29 (\log_{10} 1.54 h_{re})^2 - 1.1 \text{ dB} \qquad \text{for } f_c \leq 200 \text{ MHz} \qquad (84.26a)$$
$$= 3.2 (\log 11.75 h_{re})^2 - 4.97 \text{ dB} \qquad \text{for } f_c \geq 400 \text{ MHz} \qquad (84.26b)$$

To obtain the path loss in decibels in a suburban area the formula is modified as

$$L_{50} = L_{50} \text{ (urban)} - 2[\log_{10}(f_c/28)]^2 - 5.4 \qquad (84.27)$$

and for the path loss in decibels in open areas the formula is modified as

$$L_{50} = L_{50} \text{ (urban)} - 4.78(\log_{10} f_c)^2 - 18.33 \log_{10} f_c - 40.98 \qquad (84.28)$$

Although Hata's formulation does not have any of the path-specific corrections available in the Okumura's model, the preceding expressions have significant practical value. The predictions of the Hata model compare very closely with that of the original Okumura's model as the difference is generally within 1 dB, so long as d exceeds 1 km.

Walfish and Bertoni Model

A recent model developed by Walfish and Bertoni, 1988, considers the impact of rooftops and building height by using diffraction to predict average signal strength at street level. The model considers the path loss S to be a product of three factors.

$$S = P_0 Q^2 P_1 \tag{84.29}$$

where P_0 represents free space path loss, which is the ratio of received to radiated power for isotropic antennas in free space, and is given by

$$P_0 = \left(\frac{\lambda}{4\pi d}\right)^2 \tag{84.30}$$

The factor Q^2 gives the reduction in the roof top signal at the row of buildings, which immediately shadows the receiver at the street level. The P_1 term is based on wedge diffraction and determines the signal loss from the rooftop to the street.

Indoor Propagation Models

With the advent of Personal Communication Systems (PCS), there is a great deal of interest in characterizing the radio communication channel inside a building. The indoor channel differs from the traditional mobile radio channel in two aspects, the interference effects and the fading rate. The interference caused by the presence of electronic equipment can be unpredictable. The dynamic range of fading experienced by a hand-held unit inside the building is often smaller than that experienced by a moving vehicle in an urban setting. Because of secondary effects, such as motion of people and doors being opened and closed, the channel characteristics change with time, although slowly. It also has been observed that propagation within buildings is strongly influenced by local features, such as the layout of the building, the construction materials, and the building type. This section outlines models and measurement results for propagation within buildings.

Saleh Model

Saleh and Valenzula, 1987, reported the results of indoor propagation measurements between two vertically polarized omnidirectional antennas located on the same floor of a medium sized office building. Measurements were made using 10-ns, 1.5-GHz, radar-like pulses. The method involved averaging the square law detected received pulse response while sweeping the frequency of the transmitted pulse. Using this method, multipath components within 5 ns were resolvable.

The results obtained by Saleh and Valenzula show that 1) the indoor channel is quasistatic or very slowly time varying and 2) the statistics of the channel's impulse response are independent of transmitting and receiving antenna polarization, if there is no line-of-sight path between them. They reported a maximum multipath delay spread of 100–200 ns within the rooms of a building and 300 ns in hallways. The measured rms delay spread within rooms had a median of 25 ns and a maximum of 50 ns. The signal attenuation with no line-of-sight path was found to vary over 60-dB range and obey a log-distance power law [Eqs. (84.21) and (84.22)] with an exponent between 3 and 4.

Saleh and Valenzuela developed a simple multipath model for indoor channels based on measurement results. The model assumed that the rays arrive in clusters. The amplitudes of the received ray are independent Rayleigh random variables with variances that decay exponentially with cluster delay as well as ray delay within a cluster. The corresponding phase angles are independent uniform random variables over $[0, 2\pi]$. The clusters and rays within a cluster form a Poisson arrival process with different rates. The clusters and rays have exponentially distributed interarrival times. The

Propagation Models

formation of the clusters is related to the building structure, whereas the rays within the cluster are formed by multiple reflections from objects in the vicinity of the transmitter and the receiver.

SIRCIM Model

Rappaport et al., 1989, reported results of measurements at 1300 MHz in five factory buildings and carried out subsequent measurements in other types of buildings. Multipath delays ranged from 40 to 800 ns. Mean multipath delay and rms delay spread values ranged from 30 to 300 ns, with median values of 96 ns in LOS paths and 105 ns in obstructed paths. Delay spreads were found to be uncorrelated with T-R separation but were affected by factory inventory, building construction materials, building age, wall locations, and ceiling heights. Measurements in a food processing factory that manufactures dry-goods and has considerably less metal inventory than other factories had an rms delay spread that was half of those observed in factories producing metal products. Newer factories, which incorporate steel beams and steel reinforced concrete in the building structure, have stronger multipath signals than older factories, which used wood and brick for perimeter walls. The data suggested that radio propagation in buildings may be described by a hybrid geometric/statistical model that accounts for both specular reflections from walls and ceilings and random scattering from inventory and equipment. The authors developed an elaborate empirically derived statistical model and computer code called Simulation of Indoor Radio Channel Impulse-response Models (SIRCIM) that recreates realistic samples of indoor channel measurements [Rappaport et al., 1991], and subsequent work developed models for predicting coverage between floors using building blue prints [Seidel and Rappaport, 1992].

84.8 Small-Scale Fading

In a mobile radio environment, the short-term fluctuations caused by multipath propagation is called small-scale fading, to distinguish it from the large-scale variation in mean signal level, which is dependent on T-R separation. Small-scale fading is caused by wave interference between two or more multipath components that arrive at the receiver while the mobile travels a short distance (a few wavelengths) or over a short period of time. These waves combine vectorally at the receiver antenna to give the resultant signal, which can vary widely in amplitude, depending on the distribution of phases of the waves and the bandwidth of the transmitted signal.

Different channel conditions produce different types of small-scale fading. The type of fading experienced by the mobile depends on the following factors.

- *Speed of the mobile unit:* The relative motion between the base transmitter and the mobile results in random frequency modulation due to different Doppler shifts on different multipath components.
- *The transmission bandwidth of the signal:* If the transmitted signal bandwidth is greater than the flat-fading bandwidth of the multipath channel, the received signal is distorted.
- *The time delay spread of the received signal:* The presence of reflecting objects and scatterers in the channel constitutes a constantly changing environment that dissipates the signal energy. These effects result in multiple signals that arrive at the receiving antenna displaced with respect to one another in time and space, resulting in different times of arrival and time delays.
- *Random phase and amplitude:* The random phase and amplitudes of the different multipath components arriving at the receiving antenna cause rapid fluctuations in signal strength.
- *Rate of change of the channel:* The temporal variations of the channel impulse response caused by the changing multipath geometry results in signal fading. In a mobile environment, channel variations may arise due to movements of the transmitter, receiver or objects in their vicinity.

Flat Fading

Small-scale fading is generally classified as being either *flat* or *frequency selective*. If the mobile radio channel has a constant gain and a linear phase response over a bandwidth that is greater than the bandwidth of the transmitted signal, then the received signal will undergo *flat fading*. In this type of fading, the spectral characteristics of the transmitted signal are preserved. The strength of the received signal, however, will change with time, due to fluctuations in the gain of the channel caused by multipath. Flat fading channels are also known as **amplitude varying channels** and are sometimes referred to as *narrowband channels* since the bandwidth of the applied signal is narrow when compared to the channel bandwidth.

Rayleigh fading: In a flat fading mobile radio channel, where either the transmitter or the receiver is immersed in cluttered surroundings, the envelope of the received signal will typically have a Rayleigh distribution. The Rayleigh distribution has a probability density function given as

$$p(r) = \frac{r e^{-(\frac{r^2}{2\sigma^2})}}{\sigma^2} \qquad 0 \leq r \leq \infty \qquad (84.31)$$

where σ^2 is the variance of the of the received signal r.

The *level crossing rate* and the *average fade duration* of a Rayleigh fading signal are two important statistics for determining error control codes and diversity schemes to be used in a communication system. The *level crossing rate* (LCR) is defined as the expected rate at which the received signal envelope, normalized to the local mean signal, crosses a specified level R, in a positive going direction. The number of level crossings per second is given by

$$N_R = \sqrt{2\pi} f_m \rho e^{-\rho^2} \qquad (84.32)$$

where $f_m = (v/\lambda)\cos\theta$ is the maximum Doppler frequency and $\rho = R/R_{\text{rms}}$ is the value of the specified signal level R normalized to the local rms amplitude of the fading envelope. The *average fade duration* is defined as the average period of time for which the received signal is below a specified level R. It depends primarily on the speed of the mobile and is given by

$$\bar{\tau} = \frac{e^{\rho^2} - 1}{\rho f_m \sqrt{2\pi}} \qquad (84.33)$$

The duration of a signal fade below a specified value of ρ determines the most likely number of signalling bits that will be lost during a fadeoutage event.

Rician fading: When there is a dominant signal component, such as a line-of-sight propagation path, the small-scale fading distribution is Rician. In such a situation, random components arriving at different angles are superimposed on a stationary signal. At the output of an envelope detector, this has the effect of adding a dc component to the random multipath. The Rician distribution is given by

$$p(r) = \frac{r}{\sigma^2} \exp - \left(\frac{r^2 + A^2}{2\sigma^2}\right) I_0\left(\frac{A r}{\sigma^2}\right), \quad \text{for } \{A \geq 0, \quad r \geq 0\} \qquad (84.34a)$$

$$= 0 \qquad \text{for } r < 0 \qquad (84.34b)$$

The Rician distribution is often described in terms of a parameter K, which is defined as the ratio between the deterministic signal power and the variance of the multipath,

$$K(\text{dB}) = 10 \log_{10} \frac{A^2}{2\sigma^2} \qquad (84.35)$$

The parameter A denotes the peak to peak amplitude of the dominant sine wave about the carrier in decibels. The parameter K completely specifies the Rician distribution. As $A \to 0$ and $K \to 0$, the dominant path diminishes and the Rician distribution degenerates to a Rayleigh distribution.

Frequency Selective Fading

If the channel has a constant gain and linear phase over a bandwidth that is much **smaller** than the bandwidth of the transmitted signal, then the channel induces **frequency selective fading**. Under such conditions, the channel impulse response has a multipath delay spread that is greater than the time duration of the transmitted message waveform. When this occurs, the received signal includes multiple versions of the transmitted waveform, each of which are delayed in time, and hence the received signal is distorted. Viewed in the frequency domain, this amounts to certain frequency components in the received signal spectrum having greater gains than others. Frequency selective fading channels are more difficult to model than flat fading channels since each multipath signal must be modeled, and the channel must be considered as a linear filter. When analyzing mobile communication systems, statistical impulse response models such as SIRCIM (for indoor channels) and SURP [Turin et al., 1972] or SMRCIM [Rappaport et al., 1993] (for outdoor channels) are generally used to study the effects of frequency selective small-scale fading.

Defining Terms

Diffraction: Bending of electromagnetic waves around obstructions with sharp irregularities.
Large-scale models: Radio propagation models that characterize signal strength variations over large transmitter–receiver separation distances.
Rayleigh fading: Random variations in received signal envelope which follow a Rayleigh probability distribution.
Reflection: Phenomenon by which part of an electromagnetic wave traveling from one medium to another is turned back at the surface of the boundary.
Rician fading: Random variations in received signal envelope which follow a Rician probability distribution.
Scattering: Diffusion of reflected electromagnetic energy in all directions.
Small-scale models: Radio propagation models that characterize fluctuations of the received signal strength over small distances or short time durations.

References

Ament, W.S. 1953. Toward a theory of reflection by a rough surface. *Proceedings of the IRE.* 41(1):142–146.
Boithias, L. 1987. *Radio Wave Propagation*, McGraw-Hill, New York.
Bullington, K. 1947. Radio propagation at frequencies above 30 megacycles. *Proceedings of the IEEE.* 35:1122–1136.
Deygout J. 1966. Multiple knife-edge diffraction of microwaves. *IEEE Trans. on Antennas and Propagation.* AP-14(4):480–489.
Epstein J. and Peterson, D.W. 1953. An exerimental study of wave propagation at 840 MC. *Proceedings of the IRE.* 41(5):595–611.
Hashemi, H. 1993. Indoor radio propagation channel. *Proceedings of the IEEE.* 81(7):941–968.
Hata, M. 1980. Empirical formula for propagation loss in land mobile radio services. *IEEE Trans. on Vehicular Tech.* VT-29(3):317–325.
Lee, W.C.Y. 1985. *Mobile Communications Engineering.* McGraw–Hill New York.
Longley, A.G. and Rice, P.L. 1968. Prediction of tropospheric radio transmission loss over irregular terrain; A computer method. ESSA Tech. Rept. ERL 79-ITS 67.

Longley, A.G. 1978. Radio propagation in urban areas. OT Rept. 78–144, April.

Millington, G., Hewitt, R., and Immirzi, F.S. 1962. Double knife edge diffraction in field strength predictions. *Proceedings of the IEE.* 109C:419–429.

Molkdar, D. 1991. Review on radio propagation into and within buildings. *IEE Proceedings,* 138(1).

Okumura, Y., Ohmori, E., Kawano, T., and Fakuda, K. 1968. Field strength and its variability in VHF and UHF land mobile radio service. *Rev. Elec. Communication Lab.* 16:825–873.

Rappaport, T.S., et al. 1991. Statistical channel impulse response models for factory and open plan building radio communication system design. *IEEE Trans. on Comm.* COM-39(5):794–806.

Rappaport, T.S., et al. 1993. Performance of decision feedback equalizers is simulated urban and indoor channels. Special issue on land/mobile/portable propagation, *IEICE Transactions on Communications,* Vol. E76-B, No. 2, Feb. 1993, Japan.

Rice, P.L., Longley, A.G., Norton, K.A., and Barsis, A.P. 1967. Transmission loss predictions for tropospheric communication circuits, NBS Tech Note 101, 2 vol. iss. May 7, 1965, rev. May 1, 1966, rev. Jan.

Saleh, A.M. and Valenzula, R.A. 1987. A statistical model for indoor multipath propagation. *IEEE J. Selected Areas in Comm.* SAC-5(2):128–137.

Seidel, S.Y. and Rappaport, T.S. 1992. 914 Mhz path loss prediction models for indoor wireless communications in multifloored buildings. *IEEE Trans. on Antenna and Propagation.* AP-40(2):1–11.

Turin, L.G. et al. 1972. A statistical model of urban multipath propagation. *IEEE Trans. on Vehicular Tech.* VT-21(1).

Walfish, J. and Bertoni, H.L. 1988. A theoretical model of UHF propagation in urban environments. *IEEE Trans. on Antennas and Propagation.* AP-36(Oct.):1788–1796.

Further Information

Recently, several books and survey papers on mobile radio propagation have been published. Two particularly well-written texts that cover the topic are *The Mobile Radio Propagation Channel,* by Parsons (Wiley 1992) and *Radio Wave Propagation,* by Griffiths (Wiley 1989). Two recent survey papers that cover the subject of indoor radio propagation are those by Molkdar, 1991 and Hashemi, 1993. The textbook *Wireless Communications* by Rappaport (Prentice-Hall) treats many propagation issues for wireless communications.

85
Power Control

Roman Pichna
University of Victoria

Qiang Wang
University of Victoria

85.1 Introduction 1197
85.2 Cellular Systems and Power Control 1198
85.3 Power Control Examples 1202
85.4 Summary 1205

85.1 Introduction

The growing demand for mobile communications is pushing the technological barriers of wireless communications. The available spectrum is becoming crowded and the old analog frequency division multiple access (FDMA) cellular systems no longer meet the growing demand for new services, higher quality, and spectral efficiency. A second generation of digital cellular mobile communication systems are on the horizon. The second generation systems are represented by two standards, the IS-54 and **global system for mobility** (**GSM**). Both are time division multiple access (TDMA) based digital cellular systems and offer a significant increase in spectral efficiency and quality of service. Concurrently, another digital cellular standard has been proposed and accepted, known as IS-95, which is based on direct sequence code division multiple access (DS/CDMA) technology and promises further increase in spectral efficiency.

The channel capacity of a cellular system is significantly influenced by the cochannel interference. To minimize the cochannel interference, several techniques are proposed: frequency reuse patterns, which ensure that the same frequencies are not used in adjacent cells; efficient power control, which minimizes the transmitted power; cochannel interference cancellation techniques; and orthogonal signalling (time, frequency, or code). All of these are being intensively researched, and some have already been implemented.

This chapter provides a short overview of power control. Since power control is a very broad topic, it is not possible to exhaustively cover all facets associated with power control. The interested reader can find additional information in the recommended reading found at the end of this chapter.

The following section (Sec. 85.2) provides a brief introduction into cellular networks and demonstrates the necessity of power control. The various types of power control are presented. The next section (Sec. 85.3) illustrates some applications of power control employed in various systems such as analog **advanced mobile phone service** (**AMPS**), GSM, DS/CDMA cellular standard IS-95, and digital cordless telephone standard **cordless telephone second generation** (**CT2**). A list of defining terms is provided at the end of the chapter.

85.2 Cellular Systems and Power Control

In cellular communication systems, the service area is divided into cells, each covered by a single base station. If, in the forward link (base station to mobile), all users served by all base stations share the same frequency, each communication between a base station and a particular user would also reach all other users in the form of cochannel interference. The greater the distance between the mobile and the interfering transmitter, however, the weaker the interference becomes due to the propagation loss. To ensure a good quality of service throughout the cell, the received signal in the fringe area of the cell must be strong. Once the signal has crossed the boundary of a cell, however, it becomes interference and is required to be as weak as possible. Since this is difficult, the channel frequency is usually not reused in adjacent cells in most of the cellular systems. If the frequency is reused, the cochannel interference damages the signal reception in the adjacent cell, and the quality of service severely degrades unless other measures are taken to mitigate the interference. Therefore, a typical reuse pattern reuses the frequency in every seventh cell (frequency reuse factor = 1/7). The only exception is for CDMA-based systems where the users are separated by codes, and the allocated frequency may be shared by all users in all cells.

Even if the frequency is reused in every seventh cell, there is still some cochannel interference arriving at the receiver. It is, therefore, very important to maintain a minimal transmitted level at the base station to keep the cochannel interference low, frequency reuse factor high, and therefore the capacity of the system and quality of service high.

The same principle applies in the reverse link (mobile to base station); the power control maintains the minimum necessary transmitted power for reliable communication. Several additional benefits can be gained from this strategy. The lower transmitted power conserves the battery energy allowing the mobile terminal (the portable) to be lighter and stay on the air longer. Furthermore, recent concerns about health hazards caused by the portable's electromagnetic emissions are also alleviated.

In the reverse link, the power control also serves to alleviate the near–far effect. If all mobiles transmitted at the same power level, the signal from a near mobile would be received as the strongest. The difference between the received signal strength from the nearest and the farthest mobile can be in the range of 100 dB, which would cause saturation of the weaker signals' receivers or an excessive amount of adjacent channel interference. To avoid this, the transmitted power at the mobile must be adjusted inversely proportional to the effective distance from the base station. The term effective distance is used since a closely located user in a propagation shadow or in a deep fade may have a weaker signal than a more distant user having excellent propagation conditions.

In a DS/CDMA system, power control is a vital necessity for system operation. The capacity of a DS/CDMA cellular system is interference limited since the channels are neither separated in frequency nor separated in time, and the cochannel interference is inherently strong. A single user exceeding the limit on transmitted power could inhibit the communication of all other users.

The power control systems have to compensate not only for signal strength variations due to the varying distance between base station and mobile but must also attempt to compensate for signal strength fluctuations typical of a wireless channel. These fluctuations are due to the changing propagation environment between the base station and the user as the user moves across the cell or as some elements in the cell move. There are two main groups of channel fluctuations: slow (i.e., **shadowing**) and fast **fading**.

As the user moves away from the base station, the received signal becomes weaker because of the growing propagation attenuation with the distance. As the mobile moves in uneven terrain, it often travels into a propagation shadow behind a building or a hill or other obstacle much larger than the wavelength of the frequency of the wireless channel. This phenomenon is called shadowing. Shadowing in a land-mobile channel is usually described as a stochastic process having log-normal distributed amplitude. For other types of channels other distributions are used, e.g., Nakagami.

Electromagnetic waves transmitted from the transmitter may follow multiple paths on the way from the transmitter to the receiver. The different paths have different delays and interfere at the antenna of the receiver. If two paths have the same propagation attenuation and their delay differs in an odd number of half-wavelengths (half-periods), the two waves may cancel each other at the antenna completely. If the delay is an even multiple of the half-wavelengths (half-periods), the two waves may constructively add, resulting in a signal of double amplitude. In all other cases (nonequal gains, delays not a multiple of half-wavelength), the resultant signal at the antenna of the receiver is between the two mentioned limiting cases. This fluctuation of the channel gain is called fading. Since the scattering and reflecting surfaces in the service area are randomly distributed (houses, trees, furniture, walls, etc.), the amplitude of the resulting signal is also a random variable. The amplitude of fading is usually described by a Rayleigh, Rice, or Nakagami distributed random variable.

Since the mobile terminal may move at the velocity of a moving car or even of a fast train, the rate of channel fluctuations may be quite high and the power control has to react very quickly in order to compensate for it. The rate of fading is usually expressed in terms of Doppler frequency.

The performance of the **reverse link** of DS/CDMA systems is most affected by the near–far effect and, therefore, very sophisticated power control systems in the reverse link that attempt to alleviate the effects of channel fluctuations must be used. Together with other techniques, such as micro- and macrodiversity, interleaving, and coding, the DS/CDMA cellular system is able to cope with the wireless channel extremely well.

The effective use of the **power control** in DS/CDMA cellular system enables the frequency to be reused in every cell, which in turn enables features such as the soft hand-off and base station diversity. All together, these help enhance the capacity of the system.

In the **forward link** of a DS/CDMA system, power control may also be used. It may vary the transmitted power to the mobile, but the dynamic range is smaller due to the shared spectrum and, thus, shared interference.

We can distinguish between two kinds of power control, the open-loop power control and the closed-loop power control. The open-loop power control estimates the channel and adjusts the transmitted power accordingly but does not attempt to obtain feedback information on its effectiveness. Obviously, the open-loop power control is not very accurate, but since it does not have to wait for the feedback information it may be relatively fast. This can be advantageous, in the case of a sudden channel fluctuation, such as a mobile driving from behind a big building. This fast action is required, for instance, in the reverse link of a DS/CDMA system where the sudden increase of received strength at the base station may suppress all other signals.

The principle operation of open-loop power control is shown in Fig. 85.1. The open-loop power control must base its action on the estimation of the channel state. In the reverse link it estimates the channel by measuring the received power level of the pilot from the base station in the forward link and sets the transmitted power level inversely proportional to it. Ideally, this ensures that the average power level received from the mobile at the base station remains constant irrespective of the channel variations. This approach, however, assumes that the forward and the reverse link signal strengths are closely correlated. Although forward and reverse link may not share the same frequency and, therefore, the fading is significantly different, the long-term channel fluctuations due to shadowing and propagation loss are basically the same.

The closed-loop power control system [Fig. 85.2(a)] may base its decision on an actual communication link performance metric, e.g., received signal power level, received signal-to-noise ratio, received bit-error rate, or received frame-error rate. In the case of the reverse link power control, this metric may be forwarded to the mobile as a base for an autonomous power control decision, or the metric may be evaluated at the base station and only a power control adjustment command is transmitted to the mobile. If the reverse link power control decision is made at the base station, it

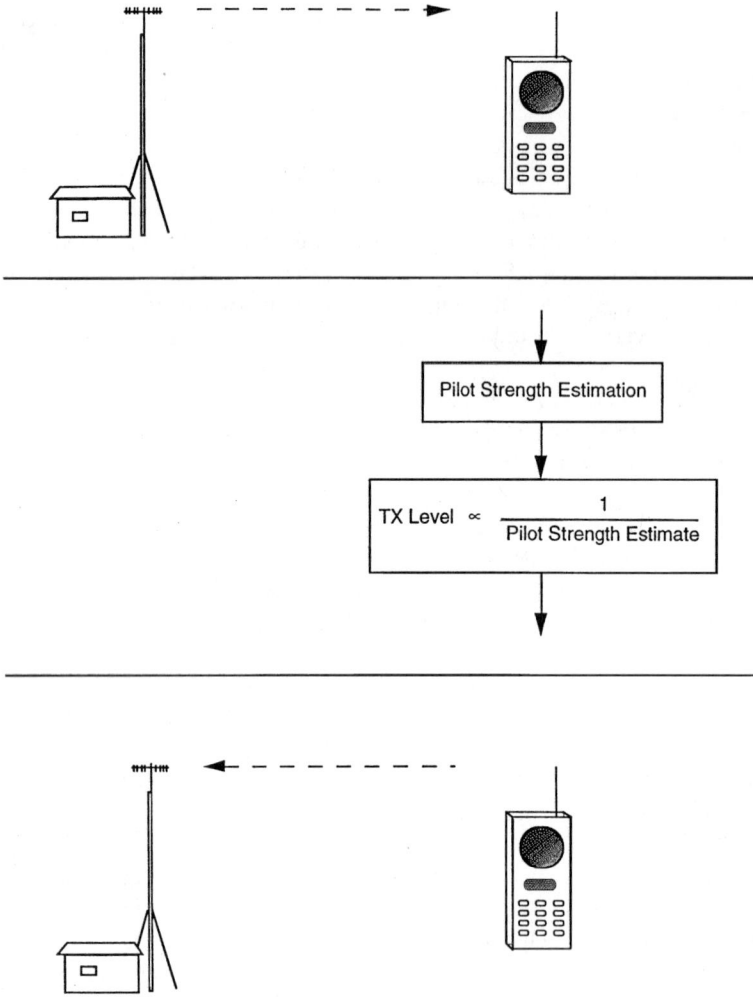

FIGURE 85.1 Reverse link open-loop power control.

may be based on the additional knowledge of the particular mobile's performance and/or a group of mobiles' performance (such as mobiles in a sector, cell, or even in a cluster of cells). If the power control decision for a particular mobile is made at the base station or at the switching office for all mobiles and is based on the knowledge of all other mobile's performance, it is called a centralized power control system. A centralized power control system may be more accurate than a distributed power control system, but it is much more complex in design, more costly, and technologically challenging.

In principle, the same categorization may be used for the power control in the forward link [Fig. 85.2(b)] except that in the forward link pilots from the mobiles are usually unavailable and only closed-loop power control is applied.

In the ideal case, power control compensates for the propagation loss, shadowing, and fast fading. There are many effects, however, that prevent the power control from becoming ideal. Fast fading rate, finite delays of the power control system, nonideal channel estimation, error in the power control command transmission, limited dynamic range, etc., all contribute to degrading the performance of the power control system. It is very important to examine the performance of power

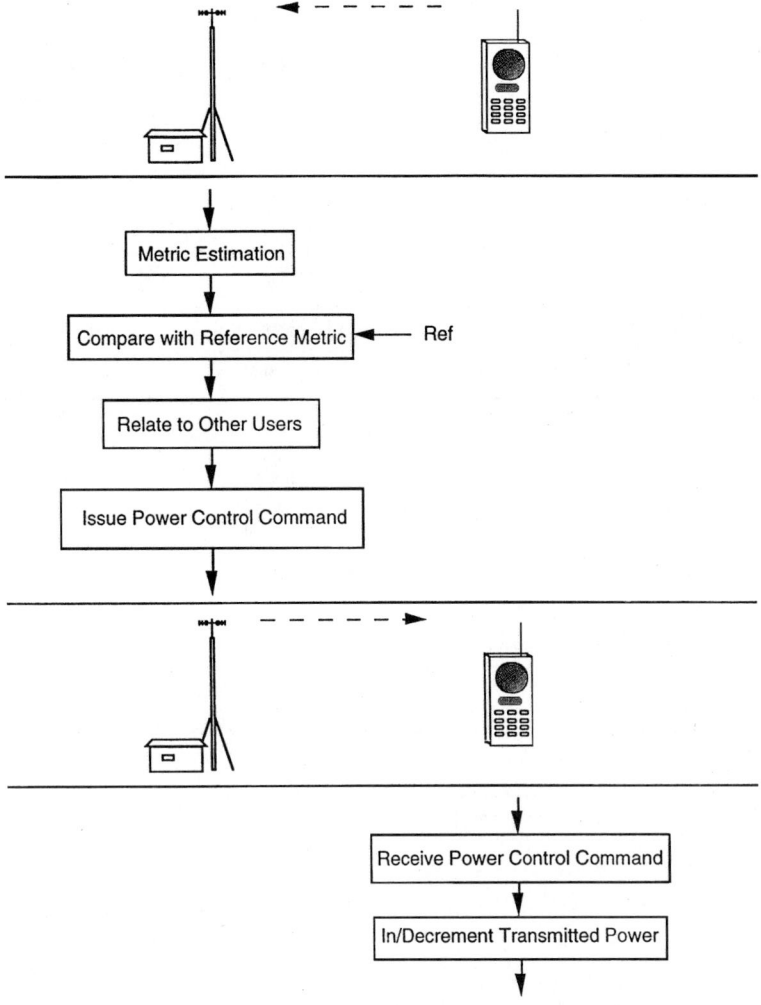

FIGURE 85.2(a) Reverse link closed-loop power control.

control under nonideal conditions since the research done has shown that the power control system is quite sensitive to some of these conditions [Viterbi and Zehavi, 1993]. Kudoh, 1993, simulated a nonideal closed-loop power control system. Errors in the system were represented by a log-normal distributed control error with standard deviation σ_E in decibels. Some results on capacity reduction are presented in Table 85.1.

The effects of Doppler and delay and feedback errors in power control loop on power control have also been studied. [Pichna et al., 1993].

TABLE 85.1 Capacity Reduction Versus Power Control Error

	$\sigma_E = 0.5$ dB, %	$\sigma_E = 1$ dB, %	$\sigma_E = 2$ dB, %	$\sigma_E = 3$ dB, %
Forward link	10	29	64	83
Reverse link	10	31	61	81

Source: Kudoh, E. 1993. On the capacity of DS/CDMA cellular mobile radios under imperfect transmitter power control. *IEICE Trans. Commun*, E76-B(April):886–893.

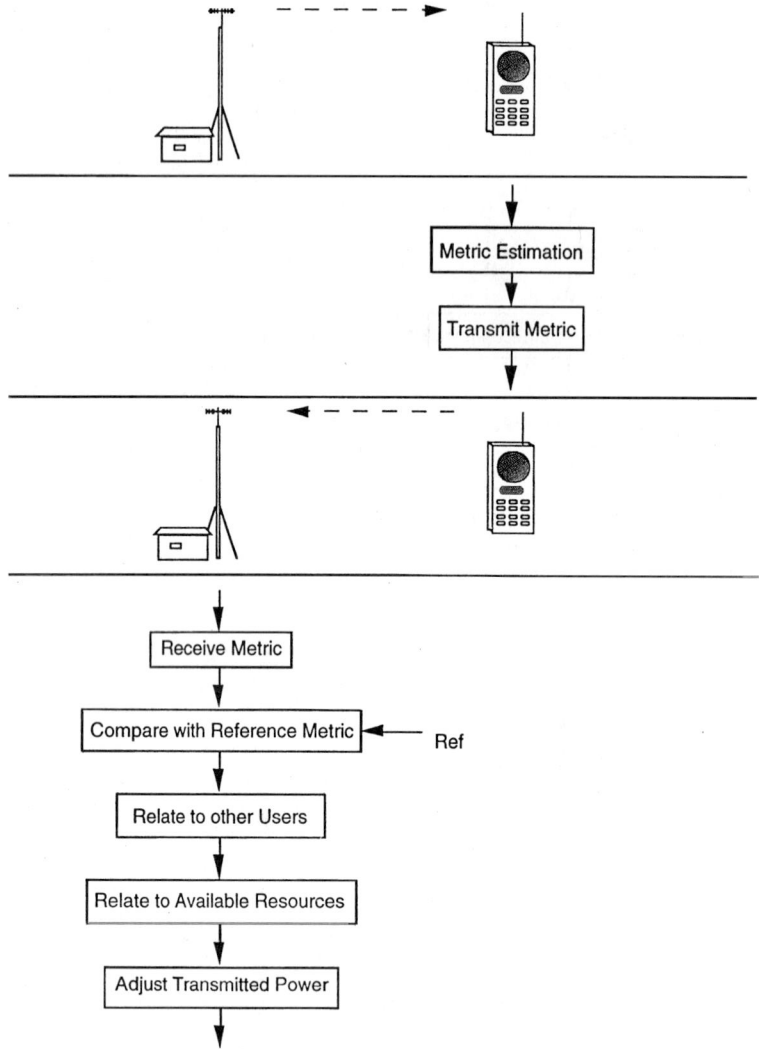

FIGURE 85.2(b) Forward link closed-loop power control.

85.3 Power Control Examples

In the following section, several applications of power control of analog and digital cellular systems are presented.

In the analog networks we may see power control implemented in both the reverse link and forward link [Lee, 1989]. Power control in the reverse link 1) reduces the chance of receiver saturation by a closely located mobile, 2) reduces the co-channel interference and thus increases the frequency-reuse factor and capacity, and 3) reduces the average transmitted power at the mobile thus conserving battery energy at the mobile.

The power control in the forward link 1) reduces co-channel interference and thus increases the frequency reuse factor and capacity and 2) reduces adjacent-channel interference and improves the quality of service.

One example of a power control system shown by Lee, 1993, was of an air-to-ground communication system. The relevant airspace is divided into six zones based on the aircraft altitude. The transmitted power at the aircraft is then varied in six steps based on the zone in which the aircraft is located. The power control system exhibits a total of approximately 28 dB of dynamic range. This reduces the co-channel interference and, due to the excellent propagation conditions in the free air, has a significant effect on the capacity of the system.

Another example of a power control system in an analog wireless network is in the analog part of the TIA standard IS-95 [TIA, 1993]. IS-95 standardizes a dual-mode FDMA/CDMA cellular system compatible with the present day AMPS analog FDMA cellular system.

The analog part of IS-95 divides the mobiles into three classes according to nominal effective radiated power (ERP) with respect to half-wave dipole at the mobile. For each class, the standard specifies eight power levels. Based on the propagation conditions, the mobile station may receive a power control command that specifies at what power level the mobile should transmit. The maximum change is 4 dB per step. See Table 85.2.

IS-95 supports further discontinuous transmission. This feature allows the mobile to vary its transmitted power between two states: low and high. These two states must be at least 8 dB apart.

As for the power control in a digital wireless system, three examples will be shown: GSM [Balston and Macario, 1993], CT2/CT2PLUS standard [DOC, 1993] for digital cordless telephones of second generation, and the IS-95 standard for digital cellular DS/CDMA system [TIA, 1993].

TABLE 85.2 Nominal ERP of the Mobile

Power level	Nominal ERP (dBW) of mobile		
	I	II	III
0	6	2	−2
1	2	2	−2
2	−2	−2	−2
3	−6	−6	−6
4	−10	−10	−10
5	−14	−14	−14
6	−18	−18	−18
7	−22	−22	−22

Source: Telecommunications Industry Association/Electronic Industries Association. 1993. Mobile Station-Base Station Compatibility Standard for Dual-Mode Wideband Spread Spectrum Cellular System, *TIA/EIA/IS-95 Interim Standard*.

TABLE 85.3 GSM Transmitter Classes

Power class	Base station power, W	Mobile station power, W
1	320	20
2	160	8
3	80	5
4	40	2
5	20	0.8
6	10	
7	5	
8	2.5	

Source: Balston, D.M. and Macario, R.C.V. 1993. *Cellular Radio Systems*, Artech House, Norwood, MA.

GSM is a Pan-European digital cellular system that was introduced in many countries during the 1992–1993 period. GSM is a digital TDMA system with a frequency hopping feature. The power control in GSM ensures that the mobile station uses only the minimum power level necessary for reliable communication with the base station. GSM defines eight classes of base stations and five classes of mobiles according to their power output, as shown on Table 85.3.

The transmitted power at the base station is controlled, nominally in 2-dB steps. The adjustment of the transmitted power reduces the intercell interference and, thus, increases the frequency reuse factor and capacity. The transmitted power at the base station may be decremented to a minimum of 13 dBm.

The power control of the mobile station is a closed-loop system controlled from the base station. The power control at the mobile sets the transmitted power to one of 15 transmission power levels spaced by 2 dB. Any change can be made only in steps of 2-dB during each time slot. Another task for the power control in GSM is to control graceful rampon and rampoff of the TDMA bursts since too steep slopes would cause spurious frequency emissions.

The dynamic range of the received signal at the base station may be up to 116 dB [Balston and Macario, 1993] and, thus, the near–far problem may also by experienced, especially if the problem occurs in adjacent time slots. In addition to power control, a careful assignment of adjacent slots can also alleviate the near–far effect.

The CT2PLUS standard [DOC, 1993] is a Canadian enhancement of the ETI CT2 standard. Both these standards allow power control in the forward and in the reverse link. Because of the expected small cell radius and relatively slow signal level fluctuation rate, given by the fact that the user of the portable is a pedestrian, the power control specifications are relatively simple. The transmission at the portable can have two levels: normal (full) and low. The low–normal difference is up to 20 dB.

The IS-95 standard represents a second generation digital wireless cellular using system using DS/CDMA. Since in a DS/CDMA system all users have the same frequency allocation, the cochannel interference is crucial for the performance of the system [Gilhousen et al., 1991]. The near–far effect may cause the received signal level to change up to 100 dB [Viterbi, 1994]. This considerable dynamic range is disastrous for a DS/CDMA where the channels are separated by a finite correlation between spreading sequences. This is further aggravated by the shadowing and the fading. The fading may have a relatively high rate since the mobile terminal is expected to move at the speed of a car. Therefore, the power control system must be very sophisticated. Power control is employed in both the reverse link and in the forward link. The reverse link power control serves the following two functions.

1. It equalizes the received power level from all mobiles at the base station. This function is vital for system operation. The better the power control performs, the more it reduces the cochannel interference and, thus, increases the capacity. The power control compensates for the near–far effect, shadowing, and partially for slow fading.
2. It minimizes the necessary transmission power level to achieve good quality of service. This reduces the cochannel interference, which increases the system capacity and alleviates health concerns. In addition, it saves the battery power. Viterbi, 1994, has shown up to 20–30-dB average power reduction compared to the AMPS mobile user as measured in field trials.

The forward link power control serves the following three functions.

1. It equalizes the system performance over the service area (good quality signal coverage of the worst-case areas).
2. It provides load shedding between unequally loaded cells in the service areas (e.g., along a busy highway) by controlling the intercell interference to the heavy loaded cells.
3. It minimizes the necessary transmission power level to achieve good quality of service. This reduces the cochannel interference in other cells, which increases the system capacity and alleviates health concerns in the area around the base station.

The reverse link power control system is composed of two subsystems: the closed-loop and the open-loop. The system operates as follows. Prior to the application to access, closed-loop power control is inactive. The mobile estimates the mean received power of the received pilot from the base station and the open-loop power control estimates the mean output power at the access channel [TIA, 1993]. The system then sets the closed-loop probing and estimates the mean output power,

$$\begin{aligned}\text{mean output power (dBm)} = &\; -\text{ mean input power (dBm)} \\ &\; - 73 \\ &\; + \text{NOM\_PWR (dB)} \\ &\; + \text{INIT\_PWR (dB)}\end{aligned} \quad (85.1)$$

where NOM_PWR and INIT_PWR are parameters obtained by the mobile prior to transmission. Subsequent probes are sent at increased power levels in steps until a response is obtained. The initial transmission on the reverse traffic channel is estimated as

$$\begin{aligned}\text{mean output power (dBm)} = &\; -\text{ mean input power (dBm)} \\ &\; - 73 \\ &\; + \text{NOM\_PWR (dB)} \\ &\; + \text{INIT\_PWR (dB)} \\ &\; + \text{the sum of all access probe} \\ &\quad \text{corrections (dB)}\end{aligned} \quad (85.2)$$

Once the first closed-loop power control bit is received the mean output power is estimated as

$$\begin{aligned}\text{mean output power (dBm)} = {} & -\text{ mean input power (dBm)} \\ & - 73 \\ & + \text{NOM\_PWR (dB)} \\ & + \text{INIT\_PWR (dB)} \\ & + \text{the sum of all access probe corrections (dB)} \\ & + \text{the sum of all closed-loop power control} \\ & \quad\text{corrections (dB)} \end{aligned} \qquad (85.3)$$

The ranges of the parameters NOM_PWR and INIT_PWR are shown in Table 85.4.

The closed-loop power control command arrives at the mobile every 1.25 ms (i.e., 800 b/s). Therefore, the base station estimates the received power level for approximately 1.25 ms. A closed-loop power control command can have only two values: 0 to increase the power level and 1 to decrease the power level. The mobile must respond to the power control command by setting the required transmitted power level within 500 μs. The total range of the closed-loop power control system is ±24 dB. The total supported range of power control (closed loop and open loop) must be at least ±32 dB.

TABLE 85.4 NOM_PWR and INIT_PWR Parameters

	Nominal value, dB	Range, dB
NOM_PWR	0	−8–7
INIT_PWR	0	−16–15

Source: Telecommunications Industry Association/Electronic Industries Association. 1993. Mobile station-base station compatibility standard for dual-mode wideband spread spectrum cellular system. *TIA/EIA/IS-95 Interim Standard.*

The behavior of the closed-loop power control system while the mobile receives base station diversity transmissions is straightforward. If all diversity transmitting base stations request the mobile to increase the transmitted power (all power control commands are 0), the mobile increases the power level. If at least one base station requests the mobile to decrease its power, the mobile decreases its power level.

The system also offers a feature of gated transmitted power for variable rate transmission mode. The gate-off state reduces the output power by at least 20 dB within 6 μs. This reduces the interference to the other users at the expense of transmitted bit rate. This feature may be used together with variable rate voice encoder or voice activated keying of the transmission.

The forward link power control works as follows. The mobile monitors the errors in the frames arriving from the base station. It reports the frame-error rate to the base station periodically. (Another mode of operation may report the error rate only if the error rate exceeds a preset threshold.) The base station evaluates the received frame-error rate reports and slightly adjusts its transmitting power. In this way, the base station may equalize the performance of the forward links in the cell or sector.

A system conforming with the standard has been field tested, and the results show that the power control is able to combat the channel fluctuation (together with other techniques such as RAKE reception) and achieve the bit energy to interference power density (E_b/I_0) necessary for a reliable service [Viterbi, 1994]. The histogram of the achieved E_b/I_0 is shown in Fig. 85.3.

Power control together with soft handoff determines the feasibility of the DS/CDMA cellular system and is crucial to its performance. QUALCOMM, Inc. has shown on field trials that their system conforms with the theoretical predictions and surpasses the capacity of other currently proposed cellular systems [Viterbi, 1994].

85.4 Summary

We have shown the basic principles of power control in wireless cellular networks as well as some examples of power control systems employed in some networks.

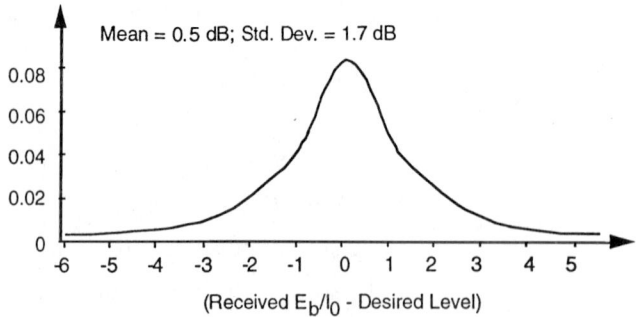

FIGURE 85.3 Differential E_b/I_0 Histogram. *Source:* Viterbi, A.J. 1994. The orthogonal-random waveform dichotomy for digital mobile personal communication. *IEEE Personal Comm.* 1(1st Quarter): 18–24.

In a wireless channel the channel transmission or channel gain is a random variable. If all transmitters in the system transmitted at equal and constant power levels, the received powers would be random.

In the reverse link (mobile to base station) each user has its own wireless channel, generally uncorrelated with all other users. The received signals at the base station are independent and random. Furthermore, since the users are randomly distributed over the cell, the distance between the mobiles and the base station may vary and so does the propagation loss. The differences between the strongest and the weakest received signal level may approach the order of 100 dB. This power level difference may cause saturation of the receivers at the base station even if they are allocated a different frequency or time slot. This phenomenon is called the near–far effect.

The near–far effect is especially detrimental for a DS/CDMA system where the frequency band is shared by all users and, for any given user, all other users' transmissions form the cochannel interference. Therefore, for the DS/CDMA system it is vitally important to efficiently mitigate the near–far effect.

The most natural way to mitigate the near–far effect is to power control the transmission in such a way that the transmitted power counterfollows the channel fluctuations and compensates for them. Then the received signal at the base station arrives at a constant amplitude.

The use of power control is not limited to the reverse link but is also employed in the forward link. The controlled transmission maintaining the transmitted level at the minimum acceptable level reduces the cochannel interference, which translates into an increased capacity of the system.

Since the DS/CDMA systems are most vulnerable to the near–far effect they have a very sophisticated power control system. In giving examples, we have concentrated on the DS/CDMA cellular system. We have also shown the power control used in other systems such as GSM, AMPS, and CT2.

Although there are more techniques available for mitigation of the near–far effect, power control is the most efficacious. As such, power control forms the core in the effort in combatting the near–far effect and channel fluctuations in general [Viterbi, 1994].

Defining Terms

Advanced mobile phone service (AMPS): Analog cellular system in North America.
Cordless telephone, second generation (CT2): A digital FDMA/TDD system.
Fading: Fast varying fluctuations of the wireless channel mainly due to the interference of time-delayed multipaths.
Forward link: Link from the base (fixed) station to the mobile (user, portable).
Groupe Spéciale Mobile (GSM): Recently referred to as the **global system for mobility**. An ETSI standard for digital cellular and microcellular systems.

Power control: Control system for controlling the transmission power. Used to reduce the co-channel interference and mitigate the near–far effect in the reverse link.

Reverse link: Link from the mobile (user, portable) to the base (fixed) station.

Shadowing: Slowly varying fluctuations of the wireless channel due mainly to the shades in propagation of electromagnetic waves. Often described by log-normal probability density function.

References

Balston, D.M. and Macario, R.C.V. 1993. *Cellular Radio Systems*, Artech House, Norwood, MA.

Department of Communications. 1993. ETI Interim Standard # I-ETS 300 131, Annex 1, Issue 2, Attachment 1. In *CT2PLUS Class 2: Specification for the Canadian Common Air Interface for Digital Cordless Telephony, Including Public Access Services, RS-130*, Communications Canada, Ottawa, ON.

Gilhousen, K.S., Jacobs, I.S., Padovani, R, Viterbi, A.J., Weaver, L.A., and Wheatley C.E., III. 1991. On the capacity of cellular CDMA system. *IEEE Trans. Veh. Tech.* 40(May):303–312.

Kudoh, E. 1993. On the capacity of DS/CDMA cellular mobile radios under imperfect transmitter power control. *IEICE Trans. Commun.* E76-B(April):886–893.

Lee, W.C.Y. 1989. *Mobile Cellular Telecommunications Systems*, McGraw-Hill, New York.

Lee, W.C.Y. 1993. *Mobile Communications Design Fundamentals*, 2nd ed., John Willey & Sons, New York.

Pichna, R., Kerr, R., Wang, Q., Bhargava, V.K., and Blake, I.F. 1993. CDMA cellular network analysis software. Final Rep. Ref. No. 36-001-2-3560/01-ST, prepared for Department of Communications, Communications Research Centre, Ottawa, ON. March.

Simon, M.K., Omura, J.K., Scholtz, R.A, and Levitt, B.K., 1994. *Spread Spectrum Communication Handbook*, McGraw-Hill, New York.

Telecommunications Industry Association/Electronic Industries Association. 1993. Mobile station-base station compatibility standard for dual-mode wideband spread spectrum cellular system. *TIA/EIA/IS-95 Interim Standard*, Arlington, VA. July.

Viterbi, A.J. and Zehavi, E. 1993. Performance of power-controlled wideband terrestrial digital communication. *IEEE Trans. Comm.* 41(April):559–569.

Viterbi, A.J. 1994. The orthogonal-random waveform dichotomy for digital mobile personal communication. *IEEE Personal Comm.* 1(1st Quarter):18–24.

Further Information

For general information see the following overview books.

Balston, D.M. and Macario, R.C.V. 1993. *Cellular Radio Systems*, Artech House, Norwood, MA.

Simon, M.K., Omura, J.K., Scholtz, R.A., Levitt, B.K., 1994. *Spread Spectrum Communication Handbook*, McGraw-Hill, New York.

For more details on power control in DS/CDMA systems consult the following.

Gilhousen, K.S., Jacobs, I.S., Padovani, R., Viterbi, A.J., Weaver, L.A., and Wheatley, C.E., III. 1991. On the capacity of cellular CDMA system. *IEEE Trans. Veh. Tech.* 40(May):303–312.

Viterbi, A.J. and Zehavi, E. 1993. Performance of power-controlled wideband terrestrial digital communication. *IEEE Trans. Comm.* 41(April):559–569.

Readers deeply interested in power control are recommended to see *IEEE Transactions on Communications*, *IEEE Transactions on Vehicular Technology*, and relevant issues of *IEEE Journal on Selected Areas in Communications*.

86
Second Generation Systems

Marc Delprat
Alcatel Mobile Communication

Vinod Kumar
Alcatel Mobile Communication

86.1	Introduction..1208
86.2	Basic Features and System Architecture in Cellular, Cordless, and PMR..1209
	Essential Features • Architecture
86.3	Systems Description..1211
	Global System for Mobile Communications/Digital Cellular System1800 (GSM/DCS1800) • Interim Standard 54 (IS-54) • Interim Standard 95 (IS-95) • Personal Digital Cellular (PDC) • Cordless Telecommunications 2 (CT2) • Digital European Cordless Telecommunications (DECT) • Personal Handy Phone System (PHPS) • Trans European Trunked Radio (TETRA) • Associated Public Safety Communications Officers Project 25 (APCO25)
86.4	Interworking and Compatibility 1220
86.5	Further Improvements of Radio Interface Performances......1222
	Variety of Speech Codecs • Packet Mode Services • Air Interface Adaptivity for Better Spectrum Efficiency • Enhanced Quality of Service Through Improved Radio Coverage
86.6	Conclusion 1223

86.1 Introduction

Designed during the 1980s, all of the so-called second generation mobile communication systems are digital. For voice calls, digitally encoded speech is transmitted on the radio interface using one of the many available digital modulation schemes. In view of the processing complexity required for these digital systems, two offered advantages are 1) the possibility of using spectrally efficient radio transmission schemes (e.g., time division multiple access [TDMA] or code division multiple access [CDMA]) in comparison to the analog frequency division multiple access (FDMA) schemes previously employed and 2) the facility of implementation of a wide variety of (integrated) speech and data services and security features (e.g., encryption).

Standardization has played an essential role in the development of second generation systems. In Europe, the need of a pan-European system to replace a large variety of disparate analog **cellular** systems was the major motivating factor behind the creation of the Global System for Mobile communications (GSM). In North America and Japan, where unique analog systems existed, the need to standardize, respectively IS-54, IS-95, and Personal Digital Cellular (PDC) for digital cellular applications arose from the lack of spectrum to serve the high traffic density areas [Cox,

Second Generation Systems

1992]. Additionally, some of the second generation systems, such as Digital European **Cordless** Telecommunications (DECT) and Personal Handy Phone System (PHPS) are the result of a need to offer wireless services in residential and office environments with low-cost subscriber equipment [Tuttlebee, 1992]. Both Trans European Trunked Radio systems (TETRA) and Associated Public Safety Communications Officers (APCO) Project 25 aim at providing some sort of unified systems for professional applications, especially security networks. Although second generation systems have also been designed for satellite and paging applications, this chapter concentrates on bidirectional land mobile radio systems.

The physical layer characteristics of all of these systems offer robust radio links paired with good spectral efficiency. The network related functionalities have been designed to offer secure communication to authenticated users even when roaming between various networks based on the same system.

After a brief description of the basic characteristics and generic system architecture of second generation wireless systems, this chapter presents the salient features of cellular, cordless, and **professional mobile radio** (PMR) systems from Europe, North America, and Japan. Interworking scenarios between some systems, problems related to coexistence of multiple radio interfaces, and future possible enhancements of wireless systems are finally presented.

86.2 Basic Features and System Architecture in Cellular, Cordless, and PMR

Essential Features

Services: Cellular and cordless systems are telephony-oriented and provide Integrated Services Digital Network (ISDN) type services, whereas the typical PMR operation mode is half-duplex group calls within fleets of users. A large variety of data services and supplementary services, tailored to system type, are offered by second generation systems.

Security: Enhanced security features (authentication, encryption) are currently offered by the cellular and cordless systems. PMR systems, for public-safety applications, are required to exhibit a high level of security.

Capacity: Cellular systems are designed for medium to high traffic density and cordless systems must provide very high capacity, up to 10,000 Erlang per square kilometer (E/km^2). PMR systems generally encounter low traffic density (<1 E/km^2). For optimized spectral efficiency, most cellular and cordless systems use medium- to wide-band TDMA or CDMA access, whereas PMR systems use narrow-band TDMA or FDMA access.

Range: In cellular systems the cell dimensions may vary from 0.3 km (in microcellular urban systems) to 30 km and above in low-density rural areas. In cordless systems it is typically limited to a few tens of meters (<200 m). In PMR systems it may vary from 2 km (urban portable coverage) to 30 km and above (rural mobile coverage).

Radio Resource Management: In cellular and in PMR, predefined channel allocation and frequency reuse patterns are employed. In high-capacity, interference limited cordless systems dynamic channel selection is preferred. This is based on radio measurements performed regularly by the mobile station (MS) and the base station (BS). Moreover, dynamic channel selection helps to alleviate some frequency planning issues in very small cell environment.

Mobility: Sophisticated mobility management (location updating, roaming, handover) is essential for cellular, highly desirable for some cordless applications (large business or Telepoint), and less important in most PMR applications (large cells and limited mobility users). In addition, the already standardized cordless systems have been optimized for slowly moving mobiles (<10 km/h).

Type of Traffic: Cellular and cordless systems are characterized mainly by individual calls between mobile and fixed users. The average call duration can be several minutes. PMR systems have mainly local group calls with short duration (<1 min) between mobile users and involving line stations of the PMR network (e.g., dispatcher). Here short call setup time is a major requirement (<0.5 s). Specific operational modes of PMR are open channel (permanent allocation of a channel with late entry facility) and direct mode (mobile-to-mobile direct communication).

Speech Quality and Delay: Since cordless systems are considered an extension of the Public Switched Telephone Network (PSTN), they require an equivalent speech quality and, therefore, use toll quality codecs such as the 32-kb/s adaptive differential pulse code modulation (ADPCM) scheme of the International Telecommunication Union (ITU-T) G.721 recommendation. Such low transmission delay codecs are helpful to avoid the need for additional echo control devices in the network. A relatively lower quality could be acceptable for cellular and PMR systems, which, therefore, use medium to low bit rate vocoders (4–16 kb/s).

Terminal Type: Cordless systems aim at providing small, simple, and low-cost portable stations by limiting the technical characteristics (low power, no equalization, simple speech codec). Cellular and PMR systems support both vehicular and portable stations. A clear tendency towards size reduction can be observed. Because of their professional use, PMR terminals are generally designed with tighter mechanical and environment constraints.

Level of Standardization: In addition to the air interface, cordless standards must specify at least PSTN access, and cellular standards generally specify the interfaces between network elements. PMR networks are often delivered as a complete system so that it is more important to standardize the intersystem interfaces than the internal network architecture.

Architecture

Typical cellular, cordless, and PMR network architectures are represented in Fig. 86.1. These are reference models which may not always fully apply to all systems described hereafter.

In cellular systems, the base station subsystem (BSS) comprises a controller (BSC) and radio transceivers (BS or BTS) which provide radio communication with MS in the covered area. The network subsystem (NSS) includes dedicated mobile switching equipment (MSC) linking all system elements through leased lines to PSTN, ISDN, and Packet Switched Public Data Network (PSPDN). The home and visitor location registers (HLR/VLR) are databases containing mobile subscriber data and used for subscriber registration and mobility management. Copies of the subscribers' secret keys are stored in the authentication center (AuC) and the mobile equipment serial numbers are stored in the equipment identity register (EIR). ITU-T signaling system no. 7 (SS#7) and related application protocols are often used in the mobile network. All system elements are operated, controlled, and maintained by the operation and maintenance center (OMC).

The cordless network architecture depends on application. For residential use, the portable station (PS) behaves like a regular telephone and has direct access to the PSTN through the private BS. In a public-access system, BS are connected to a local exchange (LE) containing a local database (DB) used for subscriber registration and mobility management in the covered area. The LEs are connected to the PSTN/ISDN (for the purpose of traffic routing) and to centralized elements of the cordless system through the PSPDN (for signaling exchange). These centralized elements perform control functions (user identification, charging, network management) and may contain a centralized database that stores location updates of the cordless subscribers and, therefore, enables routing of incoming calls. For business applications, the same private automatic branch exchange (PABX)

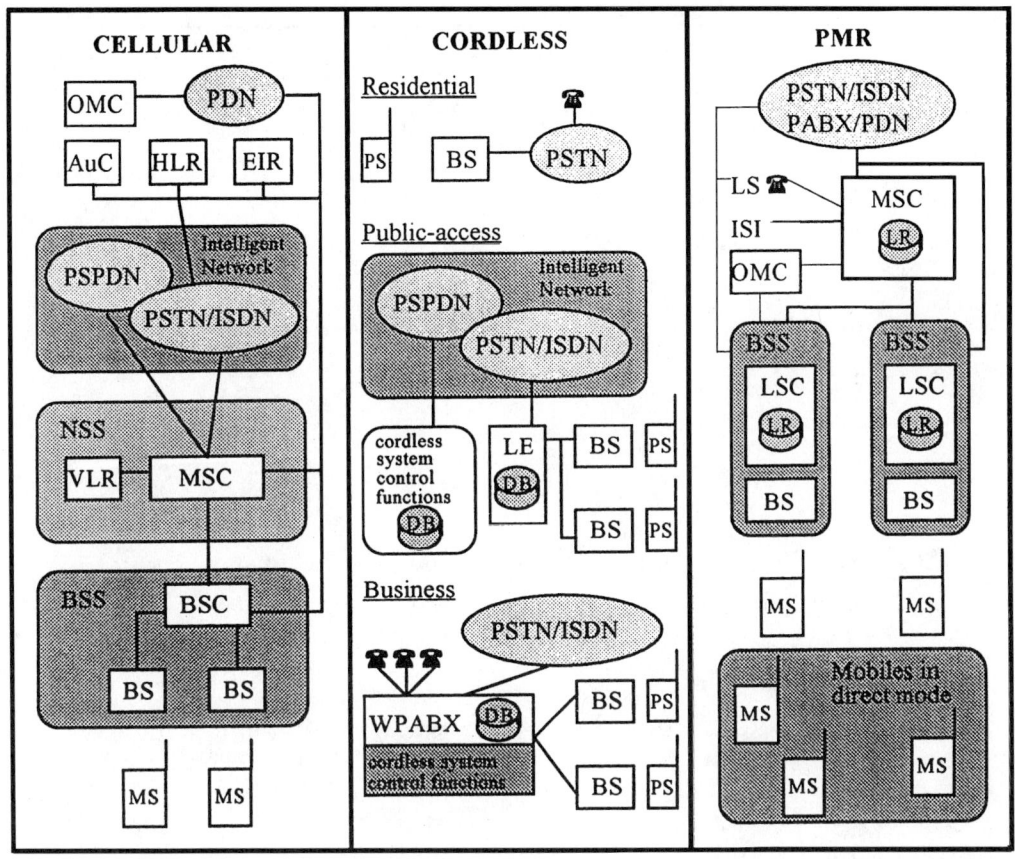

FIGURE 86.1 Network reference models.

may be used for both wire and wireless access. The wireless PABX (WPABX) interconnects the BSs of the private network. Cordless subscribers can therefore access other private wired subscribers or the PSTN/ISDN. The WPABX generally incorporates the subscriber DB and control functions of the cordless system.

PMR networks' architecture is somewhat similar to that of cellular networks, but the BSS is generally constituted of a single piece of equipment incorporating the BS and the local station controller (LSC). Since the LSC contains a copy of the subscribers location register (LR), local calls (which represent a significant part of the traffic) can be established and routed locally in the BSS, thus achieving a short call setup time and maintaining local operation even in fallback mode if the BSS-MSC link is interrupted. Intersite calls are routed through the MSC and access to other networks (PSTN/ISDN/PDN), or devices (PABX) may be provided either at MSC or BSS level. Line stations ([LS], e.g., dispatcher) may be connected directly to BSS or MSC, or through an intervening network (e.g., ISDN). Interworking of different PMR networks of the same standard is provided through the intersystem interface (ISI). Mobiles may also use direct simplex communications (direct mode), either autonomously or while keeping contact with the network (dual watch).

86.3 Systems Description

In this section, the major second generation systems are briefly described, namely, cellular: GSM/DCS1800, IS-54, IS-95, and PDC; cordless: CT2, DECT, and PHPS; and PMR: TETRA and APCO25. Table 86.1 provides a comparative summary of system characteristics.

TABLE 86.1 Air Interface Characteristics of Second Generation Systems

	Cellular				Cordless			PMR	
Standard	GSM (DCS 1800)	IS-54	IS-95	PDC	CT2	DECT	PHPS	TETRA	APCO Project 25
Frequency band (MHz)	Europe,	USA	USA	Japan	Europe & Asia	Europe	Japan	Europe	USA
Uplink	890–915 (1710–1785)	824–849	824–849	940–956 (1429–1441, 1453–1465)	864–868	1880–1900	1895–1907	380–400?	Various bands, e.g. 150–170 ~800
Downlink	935–960 (1805–1880)	869–894	869–894	810–826 (1477–1489, 1501–1513)					
Duplex spacing (MHz)	45 (95)	45	45	130 (48)	—	—	—	10?	?
Carrier spacing (kHz)	200	30	1250	25	100	1728	300	25	12.5 (6.25)
No. of radio channels in the frequency band	124 (374)	832	20	640	40	10	77	?	several hundreds of channel pairs
Multiple access	TDMA	TDMA	CDMA	TDMA	FDMA	TDMA	TDMA	TDMA	FDMA
Duplex mode	FDD	FDD	FDD	FDD	TDD	TDD	TDD	FDD	FDD
Number of channels per carrier	8 (half rate: 16)	3 (half rate: 6)	Max 64	3 (half rate: 6)	1	12	4	4	1
Modulation	GMSK	Π/4 DQPSK	QPSK BPSK	Π/4 DQPSK	GFSK	GFSK	Π/4 DQPSK	Π/4 DQPSK	C4FM or CQPSK
Carrier bit rate (kb/s)	270.8	48.6	1288	42	72	1152	384	36	9.6
Speech coder Net bit rate (kb/s)	RPE-LTP 13	VSELP 7.95	QCELP (var. rate: 8, 4, 2, 1)	VSELP 6.7	ADPCM 32	ADPCM 32	ADPCM 32	ACELP 4.5	IMBE 4.4
Channel coder for speech channels	1/2 rate convol. + CRC	1/2 rate convol. + CRC	1/2 (down) 1/3 (up) convol. + CRC	1/2 rate convol. + CRC	no	no	no	2/3 & 4/9 rates convol. + CRC	Golay & Hamming codes
Gross bit rate speech+channel coding (kb/s)	22.8	13	var. rate 19.2, 9.6, 4.8, 2.4	11.2	—	—	—	7.2	7.2
Frame size (ms)	4.6	40	20	20	2	10	5	57	20
MS transmission power (W)	Peak Aver. 20 2.5 8 1 5 0.625 2 0.25 DCS1800 1 0.125 0.25 0.031	Peak Aver. 9 3 4.8 1.6 1.8 1.6	0.6	Peak Aver. 2 0.66	Peak Aver. 0.01 0.005	Peak Aver. 0.25 0.01	Peak Aver. 0.08 0.01	Peak Aver. 10 2.5 3 0.75 1 0.25	?
Power control									
MS	Y	Y	Y	Y	low power mode in MS	N	Y	Y	N
BS	Y	Y	Y	Y		N	Y	N	N
Operational C/I (dB)	9	16	6	17	20	12	26	19	?
Equalizer	needed	needed	Rake receiver	option	no	option	no	option	no
Handover	Y	Y	Soft handoff	Y	N	Y	Y	option	option

Global System for Mobile Communications/Digital Cellular System1800 (GSM/DCS1800)

The GSM standard was specified by the European Telecommunications Standards Institute (ETSI) for pan-European digital cellular mobile radio services [European Telecommunications Standards Institute, 1990]. It was designed to answer the need for a common mobile communications standard throughout Europe, where a variety of incompatible analog cellular systems like the Nordic Mobile Telephone (NMT) and the Total Access Communications Systems (TACS) existed. After some years of intensive research efforts in the early to mid-1980s, GSM standard specification started. This was preceded with a comprehensive evaluation of several system implementations. The allocation of a dedicated pan-European frequency band around 900 MHz and the signature of a memorandum of understanding (MoU) between countries, which committed to launch nation-wide GSM networks with internetworking capabilities, were the other important steps leading to an European network.

Phase 1 of the GSM standard was completed by ETSI in 1990 and is the basis of currently implemented networks. It provides a variety of speech and data services. These services being offered progressively to the users include telephony, emergency calls, conference calls, fax transmission, short messages, and data transmission at various rates up to 9600 bit/s. Supplementary services like call forwarding, call barring and connected line identification are also defined.

The GSM network architecture closely follows the general principles introduced in the preceding section. All interfaces between network elements have been standardized, including MSC–BSC (A) and BSC–BTS (Abis) interfaces. Interfaces to AuC/HLR/EIR and to PSTN/ISDN use SS#7 with the mobile application part (MAP) protocol for noncircuit related signaling. This architecture, with a clear breakdown between the different machines and different functional parts (e.g., radio resource management in the BSC), facilitates evolutions.

The subscriber identity module (SIM) is the key to personal mobility, whereby a user can use any GSM terminal equipment just by inserting his SIM card. This smart card contains all subscriber data and is also used for basic security functions, such as subscriber identity authentication and key generation for traffic encryption on the air interface. This prevents fraudulous use of the system and ensures call privacy.

The GSM air interface is characterized by an eight-order TDMA scheme with frequency division duplex (FDD). The available frequency band in Europe is 2×25 MHz, with a radio channel spacing of 200 kHz. Data are modulated at 270 kb/s using Gaussian minimum shift keying (GMSK) modulation and transmitted in bursts of 577 μs. Each TDMA frame is constituted of eight time slots corresponding to eight separate physical channels. Each of these physical channels supports a combination of logical channels that are used in turn to carry signaling or traffic data. Slow frequency hopping is used to combat adverse propagation conditions, and most infrastructure implementations also include BTS receiver antenna diversity. The relatively low carrier to interference ratio (C/I) operational value (9 dB) is achieved through powerful channel coding, interleaving, and equalization techniques. Speech transmission is based on a linear prediction coder called regular pulse excited–long term prediction (RPE–LTP), which yields a net bit rate of 13 kb/s and a gross bit rate of 22.8 kb/s after channel coding. The air interface protocol follows a classical layered structure and includes a number of advanced features that are specific to mobile radio applications, such as mobile assisted handover (MAHO), power control (in both up and down links), and discontinuous transmission (DTX) based on voice activity detection (VAD).

GSM standardization in ETSI is still an ongoing process, and a number of additional services and features (multiparty calls, half-rate coder, general packet data service, etc.) will be available in phase 2. Standardization of a GSM adaptation tailored to railways application has also recently started. This will include specific features such as group calls or support of high-speed mobiles.

One important GSM extension is the Digital Cellular System-1800 (DCS1800) standard designed for Personal Communication Networks ([PCN], optimized for urban and suburban use) and for which various licences have already been granted in Europe. The main differences with GSM are the frequency band (around 1800 MHz), national roaming capabilities, and a reduced transmission power (hence a reduced cell size). It is also a candidate standard for the U.S. Personal Communication Services (PCS) in the 1900-MHz band.

Originally focused on the European market, the GSM/DCS1800 standard has now achieved world-wide credibility. More than 65 countries have already adopted the GSM standard, and at least 40 GSM/DCS1800 networks are currently in service around the world. These figures are constantly growing.

Interim Standard 54 (IS-54)

The main driving force for the definition of a second generation digital standard in North America was the rapidly growing demand for cellular services during the 1980s. This would have easily exceeded the capacity of the analog networks based on the Advanced Mobile Phone System ([AMPS] developed in the 1970s by Bell Laboratories). Accordingly, the new digital standard was specified by the Telecommunications Industry Association (TIA) upon request of the Cellular Telecommunications Industry Association (CTIA), with the major constraints to provide a significant increase in system capacity while maintaining upward compatibility with widespread AMPS (e.g., through dual mode BS and MS). The Federal Communications Commission (FCC), however, decided to open the existing cellular band (2×25 MHz in the 800-MHz range) to any suitable technology.

The IS-54 standard was finally selected from several proposals and published in January 1991 [Electronic Industries Association/Telecommunications Industry Association, 1991]. Both dual mode (AMPS/IS-54) mobile station and base station are specified in this standard, thus enabling the design of equipment capable of analog or digital operation. Other related standards have also been defined: IS-55 and IS-56 were defined for the performance specifications and measurement methods for mobile station and base station, respectively. Concerning network aspects, several standards have been developed by TIA since 1988 independently from the air interface design and are, hence, applicable to analog AMPS as well as IS-54 and other systems. The network reference model is as represented in Fig. 86.1 and the interfaces between network elements are specified in IS-41, except for the MSC-BSS interface, which is under-going evolution. IS-41 deals with automatic roaming (including validation and authentication of roamers), intersystem handover, and operation, administration, and maintenance, and others. IS-41 procedures are mapped on the transaction capabilities application part (TCAP) protocol (for transaction handling and message packaging) and at lower layers the transmission of network messages may use either X.25 or SS#7 formats. Other network related standards are IS-52, numbering plan; IS-53, supplementary services; IS-93, interfaces to other systems, e.g., PSTN; and IS-124, on-line exchange of call records.

Though less ISDN oriented than GSM, the IS-54 standard also supports several services, e.g., telephone service, short message service, and data services with a maximum transmission rate of 9.6 kb/s. Supplementary services include call forwarding, three-party call, and call barring. Security features include a personal identification number (PIN), subscriber authentication upon connection to the system, and voice as well as subscriber data encryption.

The IS-54 air interface uses TDMA/FDD technology with three channels per 30-kHz AMPS carrier. The modulation bit rate is 48.6 kb/s. $\Pi/4$ shifted differential quadrature phase shift keying (DQPSK) modulation is employed. For each (full rate) channel, the gross bit rate is 13 kb/s, and speech is encoded at 7.95 kb/s using a vector sum excited linear prediction (VSELP) algorithm. Advanced radio-link control based on power control and DTX improves spectrum efficiency. The air interface protocol, compatible with the AMPS protocol, includes an optional extended mode to allow for the addition of new system features and operational capabilities.

The traffic capacity achievable with IS-54 is expected to be three to four times that of existing AMPS systems. This capacity will be doubled with the introduction of a half-rate codec, currently under standardization. Using the original frequency band of the AMPS system, digital channels compliant to the IS-54 standard are now progressively replacing analog channels, thereby alleviating the shortage of spectrum while enabling a smooth transition from analog to digital. Network and

Second Generation Systems 1215

terminal equipment are already available from several manufacturers and commercial service is being offered in the largest U.S. cities.

Based on the IS-54 standard, Hughes has developed a new technology called E-TDMA, with operational networks in the U.S. and recently adopted for several regional networks in Russia and China. It is claimed to achieve a significantly higher capacity, taking advantage of advanced features such as half-rate coding, digital speech interpolation (DSI) and channel pools.

Interim Standard 95 (IS-95)

In 1991, Qualcomm demonstrated a CDMA digital cellular validation system, compliant with CTIA requirements for second generation cellular technology. The results of the field trials, conducted publicly with the support of a number of manufacturers and carriers, incited CTIA to request that TIA start the development of a wide-band (i.e., spread spectrum technology) digital cellular standard. IS-95 was then specified by TIA based on the Qualcomm proposal. Companion standards IS-96 and IS-97 have also been designed for the performance specifications and measurement methods for mobile station and base station, respectively.

IS-95 [Electronic Industries Association/Telecommunications Industry Association; 1993] is an air interface specification, meeting requirements similar to IS-54 (e.g., significant increase over analog system capacity, ease of transition, and compatibility with existing analog system) but with completely different technical choices, based on direct sequence (DS) CDMA. In DS-CDMA, a wide-band frequency channel is shared between several overlapping signals, each characterized by a specific pseudorandom binary sequence that spreads the initial spectrum of the data to be transmitted.

The waveform design uses a pseudorandom noise (PN) spreading sequence with a chip rate of 1.23 MHz. The transmitted signal bandwidth is about 1.25 MHz, i.e., one-tenth of the total bandwidth allocated to one cellular service carrier. Smooth transition from analog to digital can be achieved by removing initially only one or a small number of 1.25-MHz channels from the present analog service to provide digital service.

A combination of open- and closed-loop power control enables the mobile to operate at minimum required transmission (Tx) power level. A less sophisticated power control is also available in downlink. Path diversity is achieved with "soft handoff," a technique used during transitions between cells. This permits the instantaneous selection (on a frame-by-frame basis) of the best paths between a mobile and two or more cell sites. It requires, however, precise time synchronization (~ 1 μs), among all cell sites and extra network resources.

Wide-band transmission allows the use of powerful forward error correction (convolutional encoding with a constraint length $K = 9$) and modulation (PN modulation, quadriphase in downlink and biphase in uplink, based on orthogonal Walsh sequences in 64 dimensions). Rake receivers are implemented to process and combine signal (multipath) components.

The vocoder is an 8-kb/s specific code excited linear prediction (CELP) scheme called QCELP that incorporates variable bit rate capabilities. Selection of one of four rates (8-, 4-, 2-, or 1-kb/s) is based on adaptive energy thresholds on the input signal. For speech calls (voice activity factor <50%), a significant decrease in radio interference and in MS power consumption is thus achievable.

IS-95 employs a specific layered protocol with extension capabilities. The security functions are similar to those of IS-54 (e.g., same authentication mechanisms). Concerning the network aspects, IS-41 Revision C shall support intersystem procedures for IS-95 terminals. Several field trials using the Qualcomm system (origin of IS-95) have been successfully performed. Moreover, Korea has selected this technology for its second generation cellular system, and a full-scale commercial service is expected to start by end 1995.

Personal Digital Cellular (PDC)

As in North America and in Europe, the design of the PDC standard in Japan was motivated in the late 1980s by the saturation of analog cellular networks and by the need for new and enhanced services. After a study phase initiated by the Japanese Ministry of Posts and Telecommunications in April 1989, the PDC air interface standard was issued in April 1991 by the Research and Development Center for Radio Systems (RCR) under the name STD27 [RCR-STD 27, 1991]. It was complemented by network interface specifications providing the basis for a unified digital cellular system in Japan and enabling connectivity with fixed ISDN.

RCR STD27 is a common air interface specification. Though there are some similarities with the American IS-54 standard in terms of technical features, no compatibility with existing analog systems was required here. As a matter of fact, the new digital systems in Japan will benefit from a specific spectrum allocation, initially in the 800-MHz band and later in the 1.5-GHz band.

The carrier spacing is 25 kHz and the multiple access scheme is a TDMA/FDD of order 3 with the current full-rate codec and of order 6 with the future half-rate codec. The carrier bit rate is 42 kb/s. $\Pi/4$ DQPSK modulation is used. The full-rate speech codec employs a VSELP algorithm with a gross bit rate of 11.2 kb/s and a net bit rate of 6.7 kb/s, with forward error correction based on a convolutional code (rate $\frac{1}{2}$) and a cyclic redundancy check (CRC). A specific channel assignment procedure (flexible channel reuse from one BS to another) enables an increase of the system capacity. The standard also specifies power control and MAHO type handover procedures.

PDC systems will provide numerous services, including speech transmission, data transmission (G3-facsimile, modem, videotex), and short message service. Supplementary services such as calling line identification, call forwarding, or three-party call are also foreseen. The air interface protocol is ISDN oriented with a layered structure following the open systems interconnection (OSI) principles, including a link access protocol called LAPDM at layer 2 and a layer 3 divided into radio transmission management (RT), mobility management (MM), and call control (CC, based on ITU-T I.451). Security features include authentication and encryption.

Interfaces between network elements of a PDC system have been defined by cellular operators in Japan, except the A interface (BSS–MSC) which is left open for implementation. The network architecture follows the reference model of Fig. 86.1, ITU-T SS#7 is used between network elements and on the interface to other networks. The application protocols are an enhanced version of ISDN user part (ISUP) for circuit related signaling and MAP, developed as an application service element on TCAP, for noncircuit related signaling.

Commercial service with a PDC network was initiated by NTT in 1993 for the 800-MHz band and in 1994 for the 1.5-GHz band. Two other operators have launched digital cellular services in the 800-MHz band in 1994, and the government has recently decided to allow two new operators to offer digital service in the 1.5-GHz band. Enhancements of the PDC standard are also foreseen (half-rate codec, packet data, etc.).

Cordless Telecommunications 2 (CT2)

European research in digital cordless telephony started in the U.K. and Sweden in the early 1980s, mainly for WPABX applications. The first digital cordless standards were published in the U.K in 1987. Standards designed as a limited set of coexistence specifications (frequency band, transmitter power, interface to PSTN) resulted in several proprietary products with different and incompatible air interface characteristics. In 1989, the U.K. government issued four operator licences for public-access cordless systems (i.e., Telepoint), requesting the design of a common air interface (CAI) to allow interworking between systems. The CT2 CAI standard, developed in cooperation between U.K. manufacturers and various operators, was published in May 1989. Despite the initial ETSI choice for a different technology (DECT, described below), some manufacturers considered CT2 suitable for short term products. Thus, the CT2 CAI standard was finally endorsed by ETSI in November 1991 as an Interim European Telecommunication Standard (I-ETS) [European Telecommunications Standard Institute, 1991].

The CT2 CAI specifies incoming and outgoing calls for business and residential applications, whereas the Telepoint application was initially limited to outgoing calls only. Speech services and, recently, data services have also been specified, including asynchronous and synchronous data protocols with user rates up to 19.2 and 32 kb/s, respectively.

The air interface is designed for operation in the 800-MHz band, with a carrier spacing of 100 kHz. It employs a FDMA/time division duplex (TDD) access scheme where each carrier supports a single bidirectional communication using data blocks of 2-ms duration. Each block contains reserved fields for signaling information, with different multiplex formats for in-call signaling, normal call setup, or Telepoint access. The modulation is Gaussian frequency shift keying (GFSK) with a carrier bit rate of 72 kb/s. The speech codec is a 32-kb/s ADPCM scheme following ITU-T G.721. A 16-bit CRC enables detection of errors on signaling messages. Traffic channels are allocated using a dynamic channel assignment (DCA) technique. Simplified power control (normal/low power modes) can be implemented.

The protocol has an OSI-layered structure and includes a simplified and recoded version of Q.931 at layer 3, useful for expansion to new services and facilities. Security features include mobile authentication (mandatory for a public-access system) and network authentication (optional) using the UKF1 algorithm. The standard was subsequently enhanced (Revision 1) with additional features such as link re-establishment and location tracking (thus enabling the routing of incoming calls in a public-access system).

After several years of operation, the situation of CT2 is moderate, with some commercial success for Telepoint applications in Asia and more recently in Europe, contrasted to the failure of initial U.K. networks. An enhanced version, called CT2+, has also been introduced in Canada. Residential CT2 cordless products are essentially found as part of Telepoint packages. WPABX applications are emerging and several products have already been launched.

Digital European Cordless Telecommunications (DECT)

The standardization of a second generation cordless system covering a wide range of existing or emerging cordless applications and providing enhanced services with increased spectrum efficiency was initiated in Europe in 1985. A decision in 1988 favored the TDMA/TDD access scheme (based on CT3 designed in Sweden). The DECT standard, finally published in March 1992, benefits from a pan-European frequency allocation just below the 2-GHz frequency [European Telecommunications Standard Institute, 1992].

Basic capabilities (standard telephony features) and enhancements (ISDN interface, data transmission, privacy) common to all applications are specified by the DECT standard. Additional features are available depending on the cordless application. Residential applications can be enhanced with an intercom function. For public use, outgoing calls and authentication are basic whereas incoming call (log-on) and handover are enhancement features. The standard defines data transmission bearer services at basic bit rate multiples of 32 kb/s and up to 320 kb/s. This makes it particularly suitable for the implementation of ISDN connection-based services as well as X.25 or IEEE.802 services. The DECT protocol allows activation of ISDN supplementary services (through stimulus procedures), plus a few specific supplementary services (queue management, cost information). As security features, it includes user and network authentication, possibly using a smart card and encryption of user and/or signaling information.

System operation is based on ten carriers, spaced 1.728 MHz apart, in the 1880–1900-MHz range. GFSK modulation with a bit rate of 1152 kb/s is employed. The frame structure defines 24 time slots per 10-ms frame, therefore allowing 12 bidirectional speech calls per carrier with the 32-kb/s ADPCM codec (ITU-T G.721). Higher data rate links can be obtained in a multiple time slots configuration and/or using both uplink and downlink time slots (for asymmetric data transmission). Various logical channels are defined for in-call signaling, the control information being embedded with traffic data in each time slot. Error detection is provided for signaling information using a 16-bit CRC. DECT uses dynamic channel selection, which is equivalent to DCA in CT2, but here the TDMA mode offers more monitoring capabilities and, therefore, facilitates channel reallocation for intra- or intercell handover. The standard specifies a "make before break" seamless handover when BSs are synchronized. The protocol is OSI layered and draws extensively from ISDN and GSM protocols for layers 2 and 3.

Various DECT products for WPABX exist, and successful field trials for wireless access applications have been reported. ETSI is still in the process of enhancing the DECT standard with additional network access protocols, namely, the generic access profile ([GAP], aiming to provide a unique air interface protocol for all telephony applications), the data interoperability profile (for radio LAN applications), the ISDN/DECT interworking profile, and the GSM interoperability profile (for DECT extension to GSM networks).

Personal Handy Phone System (PHPS)

In 1989, the Japanese Ministry of Posts and Telecommunications (MPT) set up a group to define the requirements for the introduction of PHPS for digital cordless telephony applications. The definition of a common air interface (CAI) started in 1990, and after some field tests to validate the transmission method, the Telecommunication Technology Committee (TTC) of MPT reported on technical specifications in June 1992. The PHPS standard was finally published in March 1993 by the RCR under the name STD28 [Research and Development Center for Radio Systems, 1993]. Voice and data services have been defined, including circuit and packet data transmission with a maximum bit rate of 128 kb/s. Around 20 supplementary services are also specified (e.g., automatic call back, call forwarding, calling number identification) and the protocol incorporates expansion capabilities for additional features. For residential use, both incoming and outgoing calls are possible and, in addition, two terminals may communicate directly (without going through the base station) if they are close enough.

For public access, outgoing calls are supported and incoming calls may be offered through location registration. The capabilities of a public-access system also depend on the network architecture, an efficient configuration should benefit from intelligent network (IN) facilities of the public ISDN. BSs may also be connected to a local switch, performing call control and mobility management (both handover and internetwork roaming). Control functions of the cordless network (user identification, charging, network management) are performed in centralized equipment connected to the public ISDN, possibly through a PSPDN. SS#7 protocol is used between elements of the cordless network. A user PIN, validation and authentication procedures, as well as speech and data encryption ensure security in network operation.

The PHPS standard benefits from a frequency allocation of 77 carriers in the 1900-MHz band. The air interface is based on a TDMA/TDD access method with eight time slots per frame and allows four bidirectional communications per 300 kHz wide carrier. The modulation is $\Pi/4$ DQPSK with a carrier bit rate of 384 kb/s. Speech coding uses the G.721 ADPCM technique at 32 kb/s. Traffic data are transmitted in bursts containing 5 ms of coded speech or data embedded with associated signaling. The frame structure allows future introduction of half-rate (16-kb/s) and fourth-rate (8-kb/s) codecs. Error detection is performed on signaling using a 16-bit CRC.

Improved spectral efficiency is achievable through antenna diversity in the BS with postdetection selection for uplink and transmitter antenna selection for downlink. Power control is also available in both uplink and downlink. The BS starts its operation by quasistatic autonomous frequency assignment for the carrier supporting the main control channel. Dynamic channel selection is then performed per call for traffic channels. Synchronization of the BSs is desirable for fast handover and interference reduction. The air interface protocol architecture relies on a three-phase link setup process: radio channel access (with a simplified signaling structure), link connection, and communication (using a fully OSI-layered architecture).

In June 1993, MPT launched a vast system validation campaign involving three coordinating agencies, 40 manufacturers, and six operator groups. PHPS terminals are now available for residential use and several operators are expected to open commercial service for public-access systems in spring 1995.

Trans European Trunked Radio (TETRA)

Considering the growing demand for PMR systems and the need to cover the wide range of user requirements, the European Commission and ETSI decided, in 1988, to launch the standardization of a new digital trunked PMR system called TETRA. In fact, TETRA will be a set of standards including three air interface specifications for voice plus data (V+D), packet data optimized (PDO), and direct mode operation (DMO), all based on similar technical characteristics [European Telecommunications Standard Institute, 1994].

TETRA offers a wide range of services and facilities, allowing both typical PMR operation (i.e., group calls within fleets of users) and more advanced applications (secure speech and data, facsimile, file transfer, or fleet management). Circuit mode and packet mode data (connection oriented and connectionless) services have been defined with a maximum user rate of 19.2 kb/s, as well as a short message service. Around 30 supplementary services are being specified, including typical fixed network services (line identification, call forwarding) and more PMR-oriented services (various types of priority, late entry, discreet listening). The internal network architecture is not for standardization, but a number of interfaces will be specified (air interface, intersystem interface, terminal equipment interface, and PSTN/ISDN/PABX/PDN interfaces).

TETRA is capable of operating in frequency bands between 150 and 900 MHz, and its design is tailored to the 400-MHz band. A pan-European harmonized frequency allocation in the 380–400 MHz band will progressively become available for public safety applications. The V+D air interface uses a TDMA/FDD access scheme with four channels per carrier and a carrier spacing of 25 kHz. The modulation is $\Pi/4$ DQPSK and the carrier bit rate is 36 kb/s. The speech codec uses the algebraic CELP (ACELP) technique at 4.5 kb/s, and forward error correction is achieved through convolutional coding at various rates and a CRC, yielding a gross bit rate of 7.2-kb/s per channel. The protocol is OSI layered and exhibits some similarities with GSM and DECT protocols but is optimized to support the specific TETRA features. The V+D air interface will also include a number of advanced features, such as call re-establishment (handover), transmission trunking (whereby radio resource is allocated only for the duration of a transaction and then released), and uplink power control.

The PDO air interface is very similar to the V+D one, except that traffic data transmission is not based on a fixed time slot allocation, but on a random access technique coupled with a contention protocol (either Reservation-ALOHA or data-sense multiple access, depending on system load). The DMO air interface, currently under specification, will enable simplex communication between mobiles (without any relaying in the infrastructure). It will be derived from the V+D air interface, though allowing only one channel per 25-kHz carrier.

The TETRA standard will be suitable for small and large systems and for open trunks (with a commercial service) as well as for closed networks (dedicated to particular private users). It should also contribute to the development of the emerging mobile data market through its PDO version. Since European police and safety organizations are expected to be early TETRA customers, the standard fully takes into account the specific requirements of security networks (e.g., open channel, end-to-end encryption of speech and data, signaling encryption, user and network authentication).

The core of an approved TETRA standard (V+D and PDO air interfaces) should be published in 1995. Other parts of the standard are still under specification (DMO air interface, intersystem interface, supplementary services protocol). First pilot systems are expected in 1996, and commercial products should be available in 1998.

Associated Public Safety Communications Officers Project 25 (APCO25)

The U.S. Project 25 is a joint project between APCO, the National Association of State Telecommunications Directors (NASTD), and a number of federal agencies. It is the first ever standard setting effort involving public safety agencies at the local, state, and federal level. APCO Canada and the British Home Office also contribute to this effort.

The purpose of Project 25 is to develop standards for digital land mobile radio, tailored to public safety requirements. Unlike APCO16, the current requirement specification that covers existing analog systems, Project 25 will establish open system standards enabling interoperability in a multivendor environment. The objectives are to provide enhanced functionalities, to maximize spectrum efficiency, and to allow efficient and reliable intra-agency and inter-agency communications. The work began in 1989, following a FCC notice on advanced technology for the public safety radio services. Project 25 received support from many telecommunications companies and a MoU was developed with the TIA, allowing an industry advisory group (TIA–25) to carry out the technical specification work [Electronic Industries Association/Telecommunications Industry Association/Associated Public Safety Communications Officers, 1994].

The services defined are digital voice (individual, group, and broadcast calls), circuit data (protected or unprotected), packet data (acknowledged or unacknowledged) and a set of nine supplementary services including encryption. The standard allows traffic data encryption (four different types), signaling encryption and electronic serial number check (but no authentication). The air interface has been defined for two different operational modes: the conventional mode and the trunked mode. A talk-around functionality, allowing direct communication between mobiles has also been specified as part of the conventional mode.

The air interface relies on an FDMA/FDD access scheme with initially 12.5-kHz channel spacing and future migration to 6.25-kHz spacing. The standard will be suitable for operation in various frequency bands from below 100 to 1000 MHz. The modulation is a four-state continuous frequency (C4FM) scheme with a carrier bit rate of 9.6 kb/s; continuous quadrature PSK modulation is considered for the second phase with 6.25-kHz spacing. The frame structure is based on a 20-ms speech frame, associated signaling information being embedded with traffic data in each frame. The speech codec uses the improved multiband excitation technique at 4.4 kb/s, and forward error correction is provided through Hamming and Golay codes, yielding a gross bit rate of 7.2 kb/s. Handover mechanisms have also been defined.

The 12.5-kHz air interface standard is now published, and other interfaces have been specified (data port, data host, network management). The 6.25-kHz air interface and other specifications (telephone interconnect and intersystem interfaces) are still under consideration. Project 25 has defined a migration strategy that not only has backward compatibility to today's analog systems (APCO16) but also forward compatibility between receivers in the first and second generation of future radios.

The potential market includes U.S. public safety agencies and similar agencies in other countries. In the U.S., bids for public safety systems in the 800-MHz band asking for compliance with Project 25 have already been issued, and several manufacturers are developing Project 25 products and have recently demonstrated equipment for the conventional mode.

86.4 Interworking and Compatibility

A large variety of standardized second generation systems exists today. The spectrum in the 800–900-MHz and 1.5–1.9-GHz ranges has started to be pretty congested. Diversified service applications can be implemented at an optimized equipment/radio spectrum cost using products designed to one or the other standard.

Network level interworking between different systems coupled with the use of multistandard subscriber equipment can offer two major advantages, namely, suitable service quality in every environment at reasonable cost and completely seamless radio coverage and subscriber mobility from one application environment to the other.

In North America, IS-54 and IS-95 have been designed as compatibility standards with analog AMPS; at network level, interworking between the various cellular systems will be possible with future versions of IS-41. At another level, in Europe, applications of interconnection between DECT-based WPABX and MSC of a GSM/DCS1800 network have been already standardized. Figure 86.2 is a schematic representation of one such setup, wherein WPABXs

Second Generation Systems

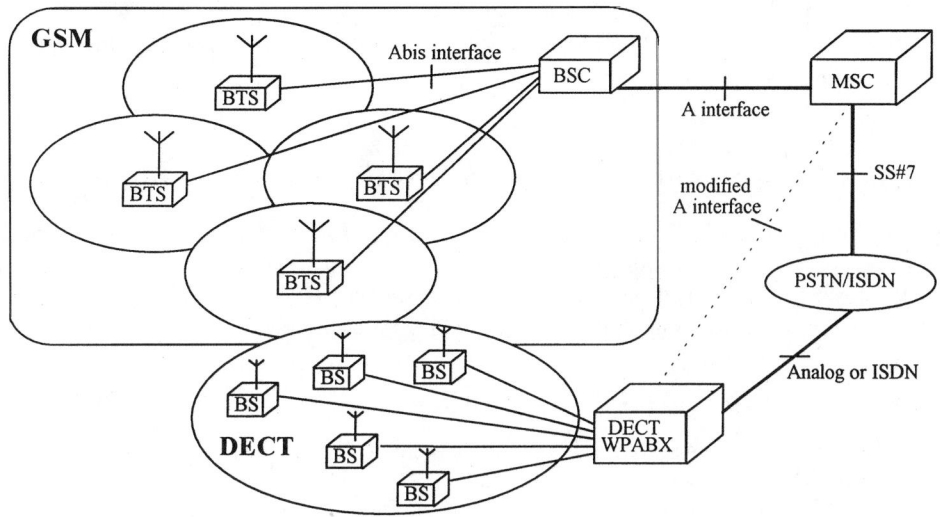

FIGURE 86.2 DECT/GSM interworking at network level.

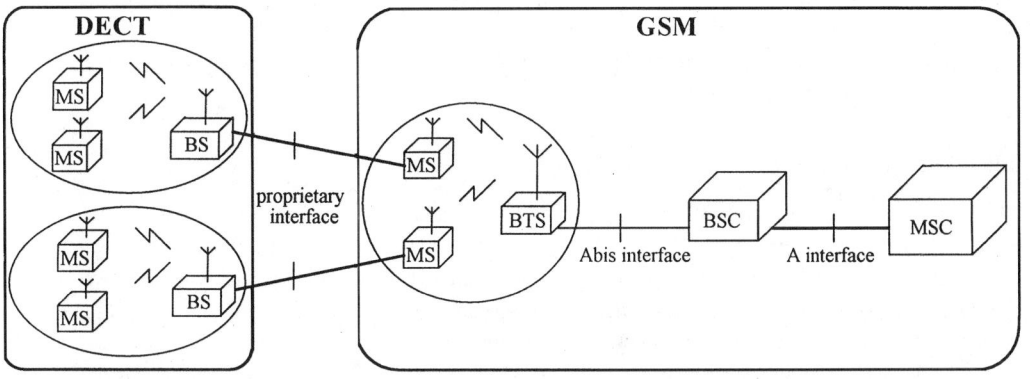

FIGURE 86.3 DECT/GSM interworking through interconnection of a GSM MS with a DECT BS.

controlling an island of DECT base stations for indoors or dense office coverage are shown to be interworking with MSC of a large cell GSM network. Figure 86.3 shows another such application where a GSM MS is connected, through a proprietary interface, to a DECT base station, which can handle multiple DECT subscribers simultaneously. This type of "moving cell" is of interest for handling multiple subscribers using low-cost handsets in fast moving vehicles like trains.

However this interworking may eventually lead to the coexistence of networks based on different standards in overlapping or geographically adjacent areas. Such radio coverage can create some intersystem, near–far interference problems either if two networks operate on the extremities of adjacent bands (e.g., DCS1800 and DECT) or if harmonic or other spurious emissions from a system are uncontrolled (e.g., disturbance of 1.8–1.9-GHz systems by 800–900-MHz ones).

Various scenarios of the type where the MS of a high-power cellular system can saturate the BS receiver of a lower power small-cell system have been analyzed. European standards have designed specifications on spurious emissions, Tx power classes, and minimum transmission/reception coupling loss in order to avoid any degradation in service quality. Similarly, in the U.S., all of the spectrum allocations for new services/systems by the FCC are accompanied with requirements on limited spurious emissions.

86.5 Further Improvements of Radio Interface Performances

Many ideas for the evolution of second generation systems have been put forward, and some of those are already under standardization and/or implementation. Enhanced service quality and further improvements in spectrum efficiency are the two most sought after goals. Most systems, especially cellular ones, have already embraced one or more of the following ideas

Variety of Speech Codecs

With the recent advances in speech coding techniques, low-delay toll quality at 8 kb/s and near-toll quality below 6 kb/s are now available. This enables the introduction of half-rate codecs in cellular systems, though with an increased complexity. In GSM/DCS1800, IS-54, and PDC this evolution was anticipated at an early stage, so the need for a half-rate speech channel was taken into account in the design of the TDMA frame structure. Standardization of half-rate codecs in the case of these three systems is now well advanced or completed. This will double the systems capacity (in terms of number of voice circuits per cell site) while maintaining a speech quality comparable to that available from related full-rate codecs. Alternatively, requirements for an enhanced speech quality have also been identified, especially for some emerging cellular applications. Standardization of an enhanced full-rate GSM codec has recently started, with the aim of achieving toll quality in good propagation environments.

Packet Mode Services

To benefit from better spectral efficiency properties of packet mode transmission, certain new services are being implemented. Cellular digital packet data (CDPD) and the implementation of relatively high-speed (>9.6 kb/s) data transmission in IS-95 are already well known. For GSM/DCS1800, the standardization of general packet radio service (GPRS) for packet mode data transmission is progressing well. New information multiplexing schemes (based on GSM/DCS1800 air interface), useful for high-reliability packet data transmission in transport applications are under development.

Air Interface Adaptivity for Better Spectrum Efficiency

TDMA burst and frame structure, channel coding, and interleaving are designed for worst-case channel distortions, i.e., large and very large cells, low-C/I ratio, and fast moving mobile. Handheld terminals are common in areas of high traffic density and usually operate in a small cell coverage. Good receiver performance can be ensured with shorter interburst guard time and shortened receiver training sequences usually included in the middle of each burst, and so the user information bit rate per hertz can be increased. A similar achievement is possible by decreasing the channel coding overhead wherever permissible. As an extreme case, various consecutive bursts of optimized size could be concatenated with a well-designed receiver training sequence, and the increase in MS to BS information rate can be exploited to offer high bit rate data services. "Allocation of bandwidth on demand" based on constant observation of quality of each communication is an adaptivity measure helpful to increase the spectral efficiency of the system. Also, variable rate speech coding techniques, either source or channel dependent, seem promising to improve quality and/or spectral efficiency. In particular, flexible allocation of a given gross bit rate between speech and channel coding on the basis of channel quality has been proposed. A set of parameters from the above list and some others like Tx power can be adapted dynamically. Both fast adaptation during a call and slow adaptation during call establishment on the basis of required and available quality of service are being considered for future enhancements of second generation systems.

Enhanced Quality of Service Through Improved Radio Coverage

The quality of service in digital wireless systems is based on the end-to-end bit error rate and on the continuity of radio links between the two ends. Interference free radio coverage with sufficient desired signal strength needs to be provided to achieve the above. Moreover, communication continuity has to be ensured between coverage areas with high traffic density (microcells) and low/medium traffic density (macrocells). Various second generation systems provide the possibility of implementing mechanisms such as slow frequency hopping, antenna diversity (also called microdiversity), and macrodiversity or multisite transmission. In addition, slow and fast and sometimes seamless handover methods are available through soft handover based on antenna diversity or frequency diversity, dynamic channel selection (as in DECT), fast mobile decided handovers between synchronized base stations using multisite operation, and overlaid cell structures whereby macrocells are superimposed on islands of microcells mainly to handle fast moving mobile subscribers. Combinations of these techniques can be implemented to further optimize overall radio coverage quality provided that the complexity of radio resource management is not unduly increased.

86.6 Conclusion

The availability of cost competitive digital and radio technology and constantly increasing demand for wireless services formed the two major factors for the success of second generation mobile communication systems described in this chapter. The fast world-wide acceptance and deployment of GSM is illustrative of this situation. Today all of the indicators point towards a constant growth of this demand in terms of quantity, quality, and variety of wireless services. Various international organizations are involved in the preparation of telecommunications for the 21st century.

Standardization of a universal mobile telecommunication system (UMTS) is one of the major working items in ETSI. Moreover, various research projects coordinated by the European Union are busy evaluating various technological possibilities for the implementation of UMTS. Support of voice, data, and multimedia services in the mobile environments is one of the major objectives of UMTS. Several industry partners believe that network interworking based on the enhanced second generation cellular and cordless systems such as GSM and DECT will offer a valid and economically viable platform for achieving the above stated objective. Integration of satellite systems in UMTS is being considered as well.

International Telecommunication Union (ITU) and, particularly, the Task Group 8/1 (TG 8/1) of ITU-R (radio) is working towards the standardization of a world-wide third generation mobile communication system called the Future Public Land Mobile System (FPLMTS). The more recently adopted acronym IMT-2000 for International Mobile Telecommunications-2000 better illustrates the purpose of this world-wide standardization effort. Like UMTS, the FPLMTS could be largely based on the interworking between various existing and newly created terrestrial or satellite mobile communication systems. The presently existing regular two-way flow of information between ETSI and ITU-R should be helpful in achieving technically coordinated standards for UMTS and FPLMTS.

The World Administration Radio Conference (WARC 92) earmarked some 220 MHz of spectrum for FPLMTS after the year 2000. Some portions of this spectrum are already being allocated on regional basis for immediate use (e.g., PCS in the U.S.). The second generation systems starting to be popular today may thus have many long years of prosperous life, especially if a transition to so-called feature rich third generation systems is possible through their evolutions.

Defining Terms

Cellular: Refers to public land mobile radio networks for generally wide area, e.g., national, coverage, to be used with medium- or high-power vehicular mobiles or portable stations and for providing mobile access to the PSTN. The network implementation exhibits a cellular architecture which enables frequency reuse in nonadjacent cells.

Cordless: These are systems to be used with simple low-power portable stations operating within a short range of a base station and providing access to fixed public or private networks. There are three main applications, namely, residential (at home, for plain old telephone service [POTS]), public access (in public places and crowded areas, also called Telepoint), and WPABX (providing cordless access in the office environment), plus emerging applications like radio access for local loop.

Professional (or private) mobile radio (PMR): Covers a large variety of land mobile radio systems designed for professional users. This includes small local systems as well as regional or national networks, and open systems (with a commercial service) or closed systems (dedicated to particular private users). In "conventional systems," each radio channel is permanently allocated to a given fleet of users (generally for low-density systems). In "trunked systems," radio resources are shared among users and allocated on a per-call or per-transaction basis.

References

Cox, D.C. 1992. Wireless Network Access for Personal Communications. *IEEE Communications Magazine*, (12):96–115.

Electronic Industries Association/Telecommunications Industry Association. 1991. Cellular System, Dual–Mode Mobile Station-Base Station Compatibility Standard, EIA/TIA Interim Standard-54, Electronic Industries Association, Washington, D.C.

Electronic Industries Association/Telecommunications Industry Association. 1993. Mobile Station-Base Station Compatibility Standard for Dual-Mode Wideband Spread Spectrum Cellular System, EIA/TIA Interim Standard-95, Electronic Industries Association, Washington, D.C.

Electronic Industries Association/Telecommunications Industry Associations and Associated Public Safety Communications Officers. 1994. APCO Project 25, IS-102 and related Telecommunications Systems Bulletin TSB-102. Electronic Industries Association, Washington, D.C.

European Telecommunications Standards Institute. 1990. GSM Recommendations Series 01–12. ETSI, Sophia Antipolis, France.

European Telecommunications Standards Institute. 1991. Common Air Interface Specification to be Used for the Interworking Between Cordless Telephone Apparatus in the Frequency Band 864.1 MHz to 868.1 MHz, Including Public Access Services, I-ETS 300 131, ETSI, Sophia Antipolis, France.

European Telecommunications Standards Institute. 1992. Digital European Cordless Telecommunications Common Interface, ETS 300 175, ETSI, Sophia Antipolis, France.

European Telecommunications Standards Institute. 1994. Trans European Trunked Radio System, draft prETS 300 392, draft prETS 300 393, draft prETS 300 394. ETSI, Sophia Antipolis, France.

Research and Development Center for Radio Systems. 1991. Personal Digital Cellular System Common Air Interface, RCR-STD27B. Research and Development Center for Radio Systems, Tokyo, Japan.

Research and Development Center for Radio Systems. 1993. Personal Handy Phone System: Second Generaration Cordless Telephone System Standard, RCR-STD28. Research and Development Center for Radio Systems, Tokyo, Japan.

Tuttlebee, W.H.W. 1992. Cordless Personal Communications. *IEEE Communications Magazine*, (12):42–53.

Further Information

European standards (GSM, CT2, DECT, TETRA) are published by ETSI Secretariat, 06921 Sophia Antipolis, France.

US standards (IS-54, IS-95, APCO) are published by Electronic Industries Association, Engineering Department, 2001 Eye Street, N.W. Washington D.C. 20006, USA.

Japanese standards (PDC, PHPS) are published by RCR (Research and Development Center for Radio Systems) 1-5-16, Toranomon, Minato-ku, Tokyo 105, Japan.

87
The Pan-European Cellular System

Lajos Hanzo
University of Southampton

87.1 Introduction 1226
87.2 Overview 1227
87.3 Logical and Physical Channels 1228
87.4 Speech and Data Transmission 1229
87.5 Transmission of Control Signals 1232
87.6 Synchronization Issues 1236
87.7 Gaussian Minimum Shift Keying Modulation 1237
87.8 Wideband Channel Models 1237
87.9 Adaptive Link Control 1239
87.10 Discontinuous Transmission 1241
87.11 Summary 1241

87.1 Introduction

Following the standardization and launch of the Pan-European digital mobile cellular radio system known as GSM, it is of practical merit to provide a rudimentary introduction to the system's main features for the communications practitioner. Since GSM operating licences have been allocated to 126 service providers in 75 countries, it is justifiable that the GSM system is often referred to as the Global System of Mobile communications.

The GSM specifications were released as 13 sets of recommendations [ETSI, 1988], which are summarized in Table 87.1, covering various aspects of the system [Hanzo and Stefanov, 1992].

After a brief system overview in Sec. 87.2 and the introduction of physical and logical channels in Sec. 87.3, we embark upon describing aspects of mapping logical channels onto physical resources for speech and control channels in Secs. 87.4 and 87.5, respectively. These details can be found in recommendations R.05.02 and R.05.03. These recommendations and all subsequently enumerated ones are to be found in ETSI, 1988. Synchronization issues are considered in Sec. 87.6. Modulation (R.05.04), transmission via the standardized wideband GSM channel models (R.05.05), as well as adaptive radio link control (R.05.06 and R.05.08), discontinuous transmission (DTX) (R.06.31), and voice activity detection (VAD) (R.06.32) are highlighted in Secs. 87.7–87.10, whereas a summary of the fundamental GSM features is offered in Sec. 87.11.

TABLE 87.1 GSM recomendations [R.01.01]

R.00	*Preamble* to the GSM recommendations
R.01	*General structure* of the recommendations, description of a GSM network, associated recommendations, vocabulary, etc.
R.02	*Service aspects*: bearer-, tele- and supplementary services, use of services, types and features of mobile stations (MS), licensing and subscription, as well as transferred and international accounting, etc.
R.03	*Network aspects*, including network functions and architecture, call routing to the MS, technical performance, availability and reliability objectives, handover and location registration procedures, as well as discontinuous reception and cryptological algorithms, etc.
R.04	*Mobile/base station (BS) interface and protocols*, including specifications for layer 1 and 3 aspects of the open systems interconnection (OSI) seven-layer structure.
R.05	*Physical layer on the radio path*, incorporating issues of multiplexing and multiple access, channel coding and modulation, transmission and reception, power control, frequency allocation and synchronization aspects, etc.
R.06	*Speech coding specifications*, such as functional, computational and verification procedures for the speech codec and its associated voice activity detector (VAD) and other optional features.
R.07	*Terminal adaptors for MSs*, including circuit and packet mode as well as voiceband data services.
R.08	*Base station and mobile switching center* (MSC) *interface*, and transcoder functions.
R.09	*Network interworking* with the public switched telephone network (PSTN), integrated services digital network (ISDN) and, packet data networks.
R.10	*Service interworking, short message service.*
R.11	*Equipment specification and type approval specification* as regards to MSs, BSs, MSCs, home (HLR) and visited location register (VLR), as well as system simulator.
R.12	*Operation and maintenance*, including subscriber, routing tariff and traffic administration, as well as BS, MSC, HLR and VLR maintenance issues.

87.2 Overview

The system elements of a GSM public land mobile network (PLMN) are portrayed in Fig. 87.1, where their interconnections via the standardized interfaces A and Um are indicated as well. The mobile station (MS) communicates with the serving and adjacent base stations (BS) via the radio interface Um, whereas the BSs are connected to the mobile switching center (MSC) through the network interface A. As seen in Fig. 87.1, the MS includes a mobile termination (MT) and a terminal equipment (TE). The TE may be constituted, for example, by a telephone set and fax machine. The MT performs functions needed to support the physical channel between the MS and the base station, such as radio transmissions, radio channel management, channel coding/decoding, speech encoding/decoding, and so forth.

The BS is divided functionally into a number of base transceiver stations (BTS) and a base station controller (BSC). The BS is responsible for channel allocation (R.05.09), link quality and power budget control (R.05.06 and R.05.08), signalling and broadcast traffic control, frequency hopping (FH) (R.05.02), handover (HO) initiation (R.03.09 and R.05.08), etc. The MSC represents the gateway to other networks, such as the public switched telephone network (PSTN), integrated services digital network (ISDN) and packet data networks using the interworking functions standardized in recommendation R.09. The MSC's further functions include paging, MS location updating (R.03.12), HO control (R.03.09), etc. The MS's mobility management is assisted by the home location register (HLR) (R.03.12), storing part of the MS's location information and routing incoming calls to the visitor location register (VLR) (R.03.12) in charge of the area, where the paged MS roams. Location update is asked for by the MS, whenever it detects from the received and decoded broadcast control channel (BCCH) messages that it entered a new location area. The HLR contains, amongst a number of other parameters, the international mobile subscriber identity (IMSI), which is used for the authentication (R.03.20) of the subscriber by his authentication center (AUC). This enables the system to confirm that the subscriber is allowed to access it. Every subscriber belongs to a home network and the specific services that the subscriber is allowed to use are entered into his HLR. The equipment identity register (EIR) allows for stolen, fraudulent, or faulty mobile stations

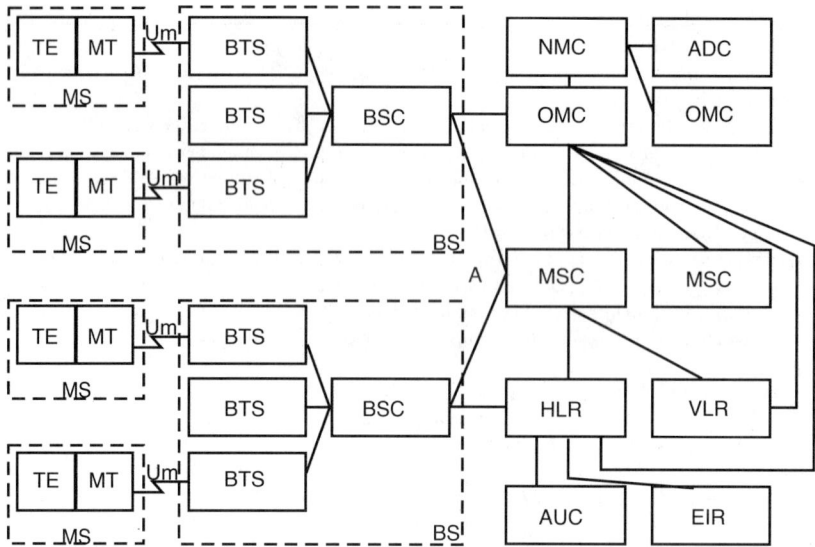

FIGURE 87.1 Simplified structure of GSM PLMN © ETT [Hanzo and Steele, 1994].

to be identified by the network operators. The VLR is the functional unit that attends to a MS operating outside the area of its HLR. The visiting MS is automatically registered at the nearest MSC, and the VLR is informed of the MSs arrival. A roaming number is then assigned to the MS, and this enables calls to be routed to it. The operations and maintenance center (OMC), network management center (NMC) and administration center (ADC) are the functional entities through which the system is monitored, controlled, maintained and managed (R.12).

The MS initiates a call by searching for a BS with a sufficiently high received signal level on the BCCH carrier; it will await and recognize a frequency correction burst and synchronize to it (R.05.08). Now the BS allocates a bidirectional signalling channel and also sets up a link with the MSC via the network. How the control frame structure assists in this process will be highlighted in Sec. 87.5. The MSC uses the IMSI received from the MS to interrogate its HLR and sends the data obtained to the serving VLR. After authentication (R.03.20) the MS provides the destination number, the BS allocates a traffic channel, and the MSC routes the call to its destination. If the MS moves to another cell, it is reassigned to another BS, and a handover occurs. If both BSs in the handover process are controlled by the same BSC, the handover takes place under the control of the BSC, otherwise it is performed by the MSC. In case of incoming calls the MS must be paged by the BSC. A paging signal is transmitted on a paging channel (PCH) monitored continuously by all MSs, and which covers the location area in which the MS roams. In response to the paging signal, the MS performs an access procedure identical to that employed when the MS initiates a call.

87.3 Logical and Physical Channels

The GSM logical traffic and control channels are standardized in recommendation R.05.02, whereas their mapping onto physical channels is the subject of recommendations R.05.02 and R.05.03. The GSM system's prime objective is to transmit the logical traffic channel's (TCH) speech or data information. Their transmission via the network requires a variety of logical control channels. The set of logical traffic and control channels defined in the GSM system is summarized in Table 87.2. There are two general forms of speech and data traffic channels: the full-rate traffic channels (TCH/F), which carry information at a gross rate of 22.8 kb/s, and the half-rate traffic channels (TCH/H), which communicate at a gross rate of 11.4 kb/s. A physical channel carries either a

The Pan-European Cellular System

TABLE 87.2 GSM Logical Channels ©ETT [Hanzo and Steele, 1994]

		Logical Channels			
Duplex BS ↔ MS Traffic Channels: TCH		Control Channels: CCH			
FEC-coded Speech	FEC-coded Data	Broadcast CCH BCCH BS → MS	Common CCH CCCH	Stand-alone Dedicated CCH SDCCH BS ↔ MS	Associated CCH ACCH BS ↔ MS
TCH/F 22.8 kb/s	TCH/F9.6 TCH/F4.8 TCH/F2.4 22.8 kb/s	Freq. Corr. Ch: FCCH	Paging Ch: PCH BS → MS	SDCCH/4	Fast ACCH: FACCH/F FACCH/H
TCH/H 11.4 kb/s	TCH/H4.8 TCH/H2.4 11.4 kb/s	Synchron. Ch: SCH	Random Access Ch: RACH MS → BS	SDCCH/8	Slow ACCH: SACCH/TF SACCH/TH SACCH/C4 SACCH/C8
		General Inf.	Access Grant Ch: AGCH BS → MS		

full-rate traffic channel, or two half-rate traffic channels. In the former, the traffic channel occupies one timeslot, whereas in the latter the two half-rate traffic channels are mapped onto the same timeslot, but in alternate frames.

For a summary of the logical control channels carrying signalling or synchronisation data, see Table 87.2. There are four categories of logical control channels, known as the BCCH, the common control channel (CCCH), the stand-alone dedicated control channel (SDCCH), and the associated control channel (ACCH). The purpose and way of deployment of the logical traffic and control channels will be explained by highlighting how they are mapped onto physical channels in assisting high-integrity communications.

A physical channel in a time division multiple access (TDMA) system is defined as a timeslot with a timeslot number (TN) in a sequence of TDMA frames. The GSM system, however, deploys TDMA combined with frequency hopping (FH) and, hence, the physical channel is partitioned in both time and frequency. Frequency hopping (R.05.02) combined with interleaving is known to be very efficient in combatting channel fading, and it results in near-Gaussian performance even over hostile Rayleigh-fading channels. The principle of FH is that each TDMA burst is transmitted via a different RF channel (RFCH). If the present TDMA burst happened to be in a deep fade, then the next burst most probably will not be. Consequently, the physical channel is defined as a sequence of radio frequency channels and timeslots. Each carrier frequency supports eight physical channels mapped onto eight timeslots within a TDMA frame. A given physical channel always uses the same TN in every TDMA frame. Therefore, a timeslot sequence is defined by a TN and a TDMA frame number FN sequence.

87.4 Speech and Data Transmission

The speech coding standard is recommendation R.06.10, whereas issues of mapping the logical speech traffic channel's information onto the physical channel constituted by a timeslot of a certain carrier are specified in recommendation R.05.02. Since the error correction coding represents part of this mapping process, recommendation R.05.03 is also relevant to these discussions. The example of the full-rate speech traffic channel (TCH/FS) is used here to highlight how this logical channel is mapped onto the physical channel constituted by a so-called normal burst (NB) of the TDMA frame structure. This mapping is explained by referring to Figs. 87.2 and 87.3. Then this example will

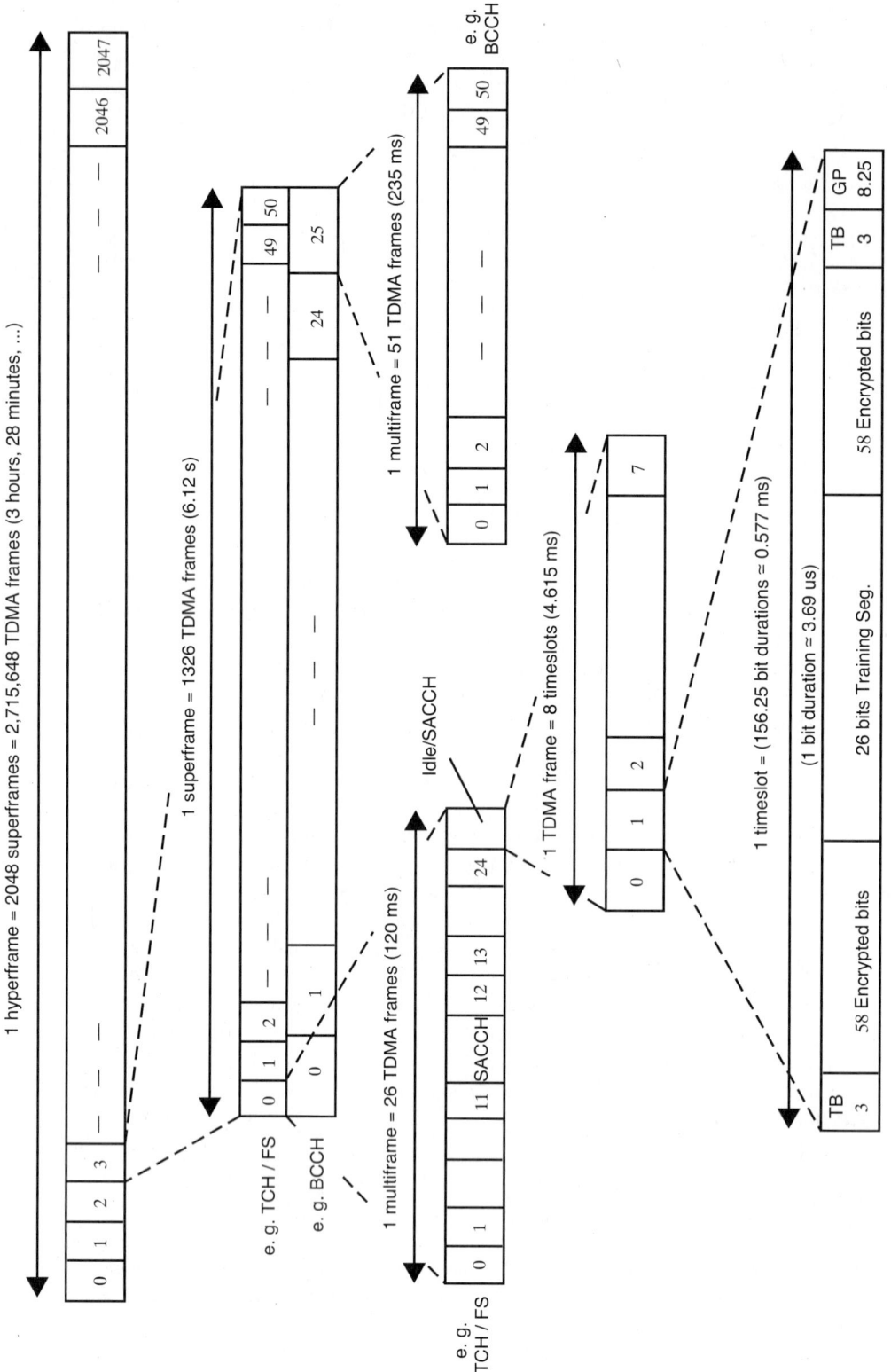

FIGURE 87.2 The GSM TDMA frame structure ©ETT [Hanzo and Steele, 1994].

The Pan-European Cellular System

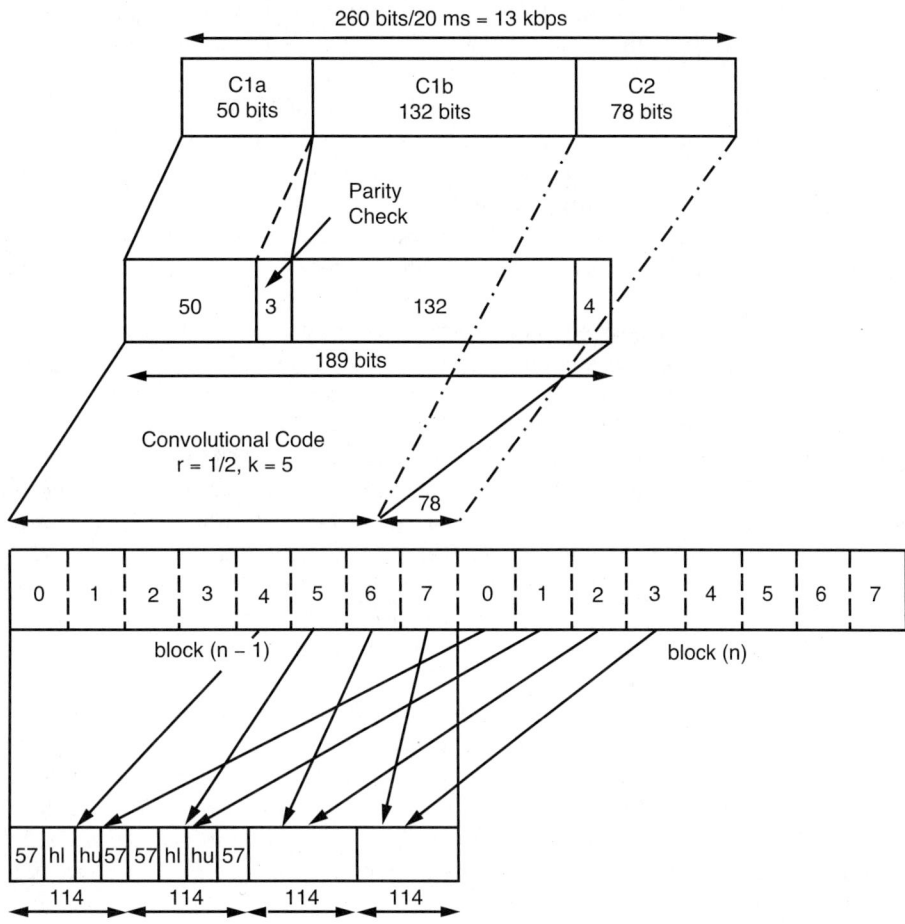

FIGURE 87.3 Mapping the TCH/FS logical channel onto a physical channel, ©ETT [Hanzo and Steele, 1994].

be extended to other physical bursts such as the frequency correction (FCB), synchronization (SB), access (AB), and dummy burst (DB) carrying logical control channels, as well as to their TDMA frame structures, as seen in Figs. 87.2 and 87.6.

The regular pulse excited (RPE) speech encoder is fully characterized in the following references: Vary and Sluyter, 1986; Salami et al., 1992; and Hanzo and Stefanov, 1992. Because of its complexity, its description is beyond the scope of this chapter. Suffice to say that, as it can be seen in Fig. 87.3, it delivers 260 b/20 ms at a bit rate of 13 kb/s, which are divided into three significance classes: class 1a (50 b), class 1b (132 b) and class 2 (78 b). The class-1a bits are encoded by a systematic (53, 50) cyclic error detection code by adding three parity bits. Then the bits are reordered and four zero tailing bits are added to perodically reset the memory of the subsequent half-rate, constraint length five convolutional codec (CC) CC(2, 1, 5), as portrayed in Fig. 87.3. Now the unprotected 78 class-2 bits are concatenated to yield a block of 456 b/20 ms, which implies an encoded bit rate of 22.8 kb/s. This frame is partitioned into eight 57-b subblocks that are block diagonally interleaved before undergoing intraburst interleaving. At this stage each 57-b subblock is combined with a similar subblock of the previous 456-b frame to construct a 116-b burst, where the flag bits hl and hu are included to classify whether the current burst is really a TCH/FS burst or it has been stolen by an urgent fast associated (FACCH) control channel message. Now the bits are encrypted and positioned in a NB, as depicted at the bottom of Fig. 87.2, where three tailing bits (TB) are added at both ends of the burst to reset the memory of the Viterbi channel equalizer (VE), which is responsible for removing both the channel-induced and the intentional controlled intersymbol interference [Steele ed., 1992].

The 8.25-b interval duration guard period (GP) at the bottom of Fig. 87.2 is provided to prevent burst overlapping due to propagation delay fluctuations. Finally, a 26-b equalizer training segment is included in the center of the normal traffic burst. This segment is constructed by a 16-b Viterbi channel equalizer training pattern surrounded by five quasiperiodically repeated bits on both sides. Since the MS has to be informed about which BS it communicates with, for neighboring BSs one of eight different training patterns is used, associated with the so-called BS color codes, which assist in identifying the BSs.

This 156.25-b duration TCH/FS NB constitutes the basic timeslot of the TDMA frame structure, which is input to the Gaussian minimum shift keying (GMSK) modulator to be highlighted in Sec. 87.7, at a bit rate of approximately 271 kb/s. Since the bit interval is $1/(271 \text{ kb/s}) = 3.69 \ \mu s$, the timeslot duration is $156.25 \cdot 3.69 \approx 0.577$ ms. Eight such normal bursts of eight appropriately staggered TDMA users are multiplexed onto one (RF) carrier giving, a TDMA frame of $8 \cdot 0.577 \approx 4.615$-ms duration, as shown in Fig. 87.2. The physical channel as characterized earlier provides a physical timeslot with a throughput of $114 \text{ b}/4.615 \text{ ms} = 24.7$ kb/s, which is sufficiently high to transmit the 22.8 kb/s TCH/FS information. It even has a reserved capacity of $24.7 - 22.8 = 1.9$ kb/s, which can be exploited to transmit slow control information associated with this specific traffic channel, i.e., to construct a so-called slow associated control channel (SACCH), constituted by the SACCH TDMA frames, interspersed with traffic frames at multiframe level of the hierarchy, as seen in Fig. 87.2.

Mapping logical data traffic channels onto a physical channel is essentially carried out by the channel codecs [Wong and Hanzo, 1992], as specified in recommendation R.05.03. The full- and half-rate data traffic channels standardized in the GSM system are: TCH/F9.6, TCH/F4.8, TCH/F2.4, as well as TCH/H4.8, TCH/H2.4, as was shown earlier in Table 87.2. Note that the numbers in these acronyms represent the data transmission rate in kilobits per second. Without considering the details of these mapping processes we now focus our attention on control signal transmission issues.

87.5 Transmission of Control Signals

The exact derivation, forward error correcting (FEC) coding and mapping of logical control channel information is beyond the scope of this chapter, and the interested reader is referred to ETSI, 1988 (R.05.02 and R.05.03) and Hanzo and Stefanov, 1992, for a detailed discussion. As an example, the mapping of the 184-b SACCH, FACCH, BCCH, SDCCH, PCH, and access grant control channel (AGCH) messages onto a 456-b block, i.e., onto four 114-b bursts is demonstrated in Fig. 87.4. A double-layer concatenated FIRE-code/convolutional code scheme generates 456 bits, using an overall coding rate of $R = 184/456$, which gives a stronger protection for control channels than the error protection of traffic channels.

Returning to Fig. 87.2 we will now show how the SACCH is accommodated by the TDMA frame structure. The TCH/FS TDMA frames of the eight users are multiplexed into multiframes of 24 TDMA frames, but the 13th frame will carry a SACCH message, rather than the 13th TCH/FS frame, whereas the 26th frame will be an idle or dummy frame, as seen at the left-hand side of Fig. 87.2 at the multiframe level of the traffic channel hierarchy. The general control channel frame structure shown at the right of Fig. 87.2 is discussed later. This way 24-TCH/FS frames are sent in a 26-frame multiframe during $26 \cdot 4.615 = 120$ ms. This reduces the traffic throughput to $(24/26) \cdot 24.7 = 22.8$ kb/s required by TCH/FS, allocates $(1/26) \cdot 24.7 = 950$ b/s to the SACCH and wastes 950 b/s in the idle frame. Observe that the SACCH frame has eight timeslots to transmit the eight 950-b/s SACCHs of the eight users on the same carrier. The 950-b/s idle capacity will be used in case of half-rate channels, where 16 users will be multiplexed onto alternate frames of the

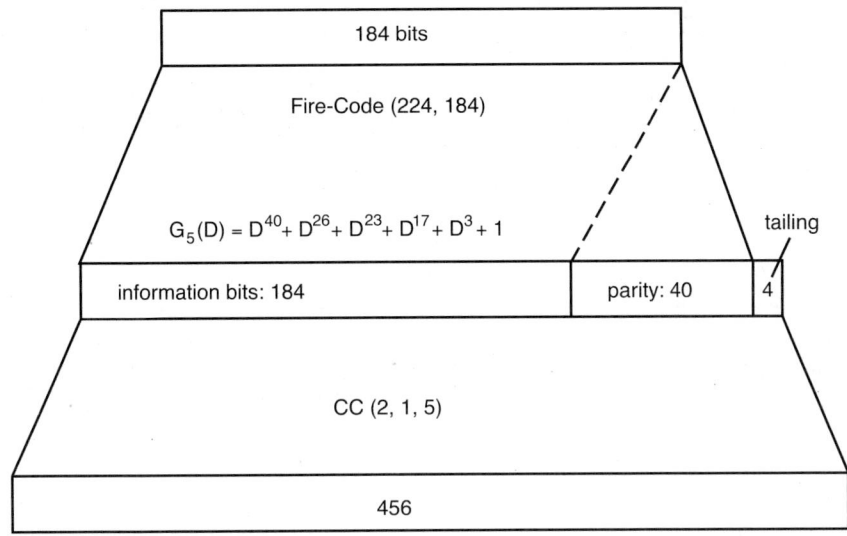

FIGURE 87.4 FEC in SACCH, FACCH, BCCH, SDCCH, PCH and AGCH, ©ETT [Hanzo and Steele, 1994].

TDMA structure to increase system capacity. Then 16, 11.4-kb/s encoded half-rate speech TCHs will be transmitted in a 120-ms multiframe, where also 16 SACCHs are available.

The FACCH messages are transmitted via the physical channels provided by bits stolen from their own host traffic channels. The construction of the FACCH bursts from 184 control bits is identical to that of the SACCH, as also shown in Fig. 87.4, but its 456-b frame is mapped onto eight consecutive 114-b TDMA traffic bursts, exactly as specified for TCH/FS. This is carried out by stealing the even bits of the first four and the odd bits of the last four bursts, which is signalled by setting $hu = 1$, $hl = 0$ and $hu = 0$, $hl = 1$ in the first and last bursts, respectively. The unprotected FACCH information rate is 184 b/20 ms = 9.2 kb/s, which is transmitted after concatenated error protection at a rate of 22.8 kb/s. The repetition delay is 20 ms, and the interleaving delay is $8 \cdot 4.615 = 37$ ms, resulting in a total of 57-ms delay.

In Fig. 87.2 at the next hierarchical level, 51-TCH/FS multiframes are multiplexed into one superframe lasting $51 \cdot 120$ ms = 6.12 s, which contains $26 \cdot 51 = 1326$-TDMA frames. In the case of 1326-TDMA frames, however, the frame number would be limited to $0 \leq FN \leq 1326$ and the encryption rule relying on such a limited range of FN values would not be sufficiently secure. Then 2048 superframes were amalgamated to form a hyperframe of $1326 \cdot 2048 = 2{,}715{,}648$-TDMA frames lasting $2048 \cdot 6.12$ s $\approx$ 3 h 28 min, allowing a sufficiently high FN value to be used in the encryption algorithm. The uplink and downlink traffic-frame structures are identical with a shift of three timeslots between them, which relieves the MS from having to transmit and receive simultaneously, preventing high-level transmitted power leakage back to the sensitive receiver. The received power of adjacent BSs can be monitored during unallocated timeslots.

In contrast to duplex traffic and associated control channels, the simplex BCCH and CCCH logical channels of all MSs roaming in a specific cell share the physical channel provided by timeslot zero of the so-called BCCH carriers available in the cell. Furthermore, as demonstrated by the right-hand side section of Fig. 87.2, 51 BCCH and CCCH TDMA frames are mapped onto a $51 \cdot 4.615 = 235$-ms duration multiframe, rather than on a 26-frame, 120-ms duration multiframe. In order to compensate for the extended multiframe length of 235 ms, 26 multiframes constitute a 1326-frame superframe of 6.12-s duration. Note in Fig. 87.5 that the allocation of the uplink and downlink frames is different, since these control channels exist only in one direction.

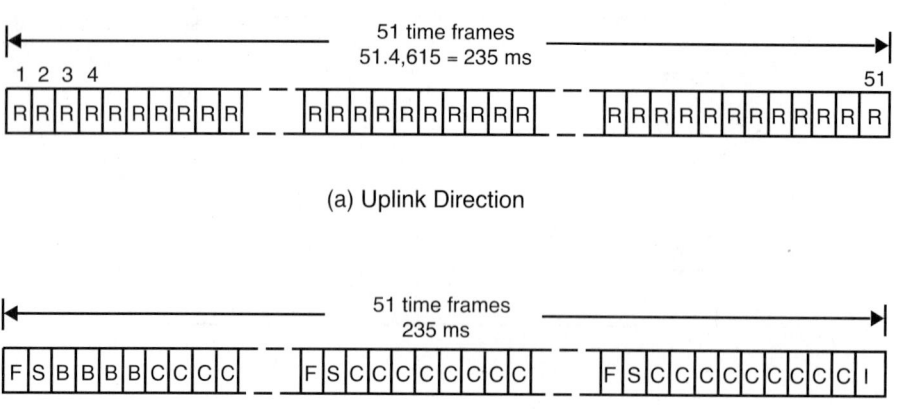

FIGURE 87.5 The control multiframe, ©ETT [Hanzo and Steele, 1994].

Specifically, the random access channel (RACH) is only used by the MSs in the uplink direction if they request, for example, a bidirectional SDCCH to be mapped onto an RF channel to register with the network and set up a call. The uplink RACH has a low capacity, carrying messages of 8-b/235-ms multiframe, which is equivalent to an unprotected control information rate of 34 b/s. These messages are concatenated FEC coded to a rate of 36 b/235 ms = 153 b/s. They are not transmitted by the NB derived for TCH/FS, SACCH, or FACCH logical channels, but by the AB, depicted in Fig. 87.6 in comparison to a NB and other types of bursts to be described later. The FEC coded, encrypted 36-b AB messages of Fig. 87.6, contain among other parameters, the encoded 6-b BS identifier code (BSIC) constituted by the 3-b PLMN color code and 3-b BS color code for unique BS identification. These 36 b are positioned after the 41-b synchronization sequence, which has a high wordlength in order to ensure reliable access burst recognition and a low probability of being emulated by interfering stray data. These messages have no interleaving delay, while they are transmitted with a repetition delay of one control multiframe length, i.e., 235 ms.

Adaptive time frame alignment is a technique designed to equalize propagation delay differences between MSs at different distances. The GSM system is designed to allow for cell sizes up to 35 km radius. The time a radio signal takes to travel the 70 km from the base station to the mobile station and back again is 233.3 μs. As signals from all the mobiles in the cell must reach the base station without overlapping each other, a long guard period of 68.25 b (252 μs) is provided in the access burst, which exceeds the maximum possible propagation delay of 233.3 μs. This long guard period in the access burst is needed when the mobile station attempts its first access to the base station or after a handover has occurred. When the base station detects a 41-b random access synchronization sequence with a long guard period, it measures the received signal delay relative to the expected signal from a mobile station of zero range. This delay, called the timing advance, is signalled using a 6-b number to the mobile station, which advances its timebase over the range of 0–63 b, i.e., in units

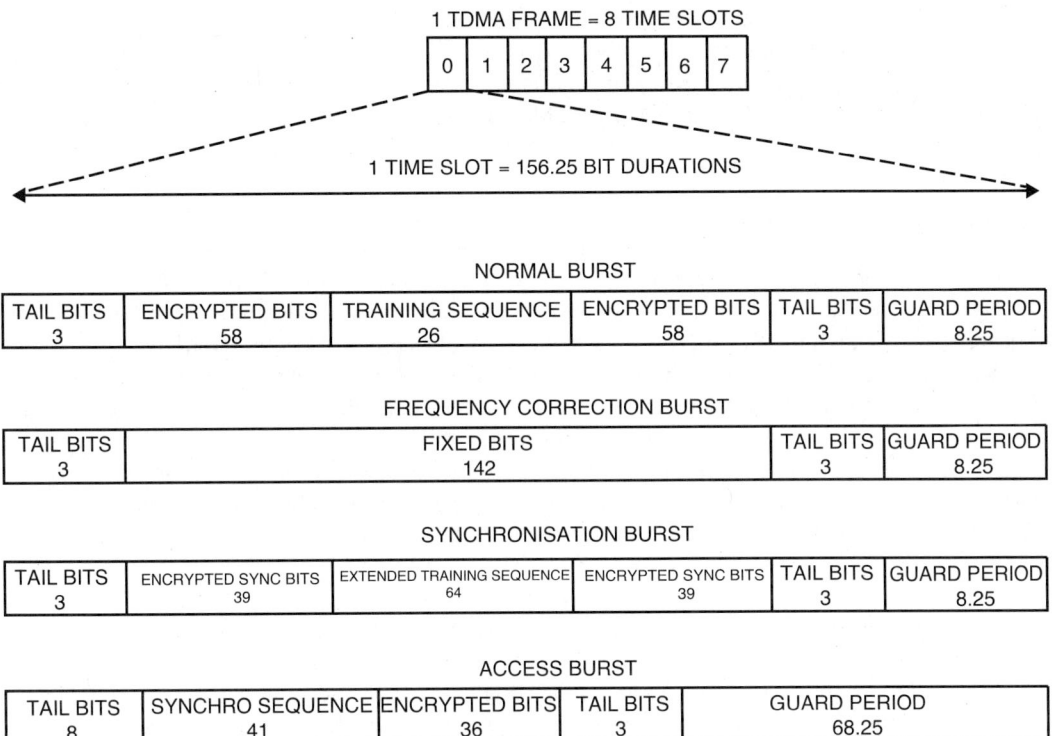

FIGURE 87.6 GSM burst structures, ©ETT, [Hanzo and Steele, 1994].

of 3.69 μs. By this process the TDMA bursts arrive at the BS in their correct timeslots and do not overlap with adjacent ones. This process allows the guard period in all other bursts to be reduced to $8.25 \cdot 3.69\ \mu s \approx 30.46\ \mu s$ (8.25 b) only. During normal operation, the BS continously monitors the signal delay from the MS and, if necessary, it will instruct the MS to update its time advance parameter. In very large traffic cells there is an option to actively utilize every second timeslot only to cope with higher propagation delays, which is spectrally inefficient, but in these large, low-traffic rural cells it is admissible.

As demonstrated by Fig. 87.2, the downlink multiframe transmitted by the BS is shared amongst a number of BCCH and CCCH logical channels. In particular, the last frame is an idle frame (I), whereas the remaining 50 frames are divided in five blocks of ten frames, where each block starts with a frequency correction channel (FCCH) followed by a synchronization channel (SCH). In the first block of ten frames the FCCH and SCH frames are followed by four BCCH frames and by either four AGCH or four PCH. In the remaining four blocks of ten frames, the last eight frames are devoted to either PCHs or AGCHs, which are mutually exclusive for a specific MS being either paged or granted a control channel.

The FCCH, SCH, and RACH require special transmission bursts, tailored to their missions, as depicted in Fig. 87.6. The FCCH uses frequency correction bursts (FCB) hosting a specific 142-b pattern. In partial response GMSK it is possible to design a modulating data sequence, which results in a near-sinusoidal modulated signal imitating an unmodulated carrier exhibiting a fixed frequency offset from the RF carrier utilized. The synchronization channel transmits SB hosting a $16 \cdot 4 = 64$-b extended sequence exhibiting a high-correlation peak in order to allow frame alignment with a quarter-bit accuracy. Furthermore, the SB contains $2 \cdot 39 = 78$ encrypted FEC-coded synchronization bits, hosting the BS and PLMN color codes, each representing one of eight legitimate identifiers. Lastly, the AB contain an extended 41-b synchronization sequence, and they are invoked to facilitate initial access to the system. Their long guard space of 68.25-b duration prevents frame overlap, before the MS's distance, i.e., the propagation delay becomes known to the BS and could be compensated for by adjusting the MS's timing advance.

87.6 Synchronization Issues

Although some synchronization issues are standardized in recommendations R.05.02 and R.05.03, the GSM recommendations do not specify the exact BS-MS synchronization algorithms to be used, these are left to the equipment manufacturers. A unique set of timebase counters, however, is defined in order to ensure perfect BS-MS synchronism. The BS sends FCB and SB on specific timeslots of the BCCH carrier to the MS to ensure that the MS's frequency standard is perfectly aligned with that of the BS, as well as to inform the MS about the required initial state of its internal counters. The MS transmits its uniquely numbered traffic and control bursts staggered by three timeslots with respect to those of the BS to prevent simultaneous MS transmission and reception, and also takes into account the required timing advance (TA) to cater for different BS-MS-BS round-trip delays.

The timebase counters used to uniquely describe the internal timing states of BSs and MSs are the quarter-bit number ($QN = 0$–624) counting the quarter-bit intervals in bursts, bit number ($BN = 0$–156), timeslot number ($TN = 0$–7) and TDMA Frame Number ($FN = 0$–$26 \cdot 51 \cdot 2048$), given in the order of increasing interval duration. The MS sets up its timebase counters after receiving a SB by determining QN from the 64-b extended training sequence in the center of the SB, setting $TN = 0$ and decoding the 78-encrypted, protected bits carrying the 25-SCH control bits.

The SCH carries frame synchronization information as well as BS identification information to the MS, as seen in Fig. 87.7, and it is provided solely to support the operation of the radio subsystem. The first 6 b of the 25-b segment consist of three PLMN color code bits and three BS color code bits supplying a unique BS identifier code (BSIC) to inform the MS which BS it is communicating with. The second 19-bit segment is the so-called reduced TDMA frame number RFN derived from the full TDMA frame number FN, constrained to the range of $[0$–$(26 \cdot 51 \cdot 2048) - 1] = (0$–$2{,}715{,}647)$ in terms of three subsegments $T1$, $T2$, and $T3$. These subsegments are computed as follows: $T1(11\,b) = [FN \text{ div } (26 \cdot 51)]$, $T2(5\,b) = (FN \text{ mod } 26)$ and $T3'(3b) = [(T3 - 1) \text{ div } 10]$, where $T3 = (FN \text{ mod } 5)$, whereas div and mod represent the integer division and modulo operations, respectively. Explicitly, in Fig. 87.7 $T1$ determines the superframe index in a hyperframe, $T2$ the multiframe index in a superframe, $T3$ the frame index in a multiframe, whereas $T3'$ is the so-called signalling block index [1–5] of a frame in a specific 51-frame control multiframe, and their roles are best understood by referring to Fig 87.2. Once the MS has received the SB, it readily computes the FN required in various control algorithms, such as encryption, handover, etc., as

$$FN = 51[(T3 - T2) \text{ mod } 26] + T3 + 51 \cdot 26 \cdot T1, \qquad \text{where } T3 = 10 \cdot T3' + 1$$

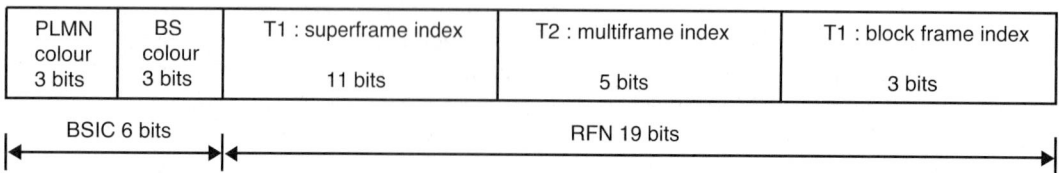

FIGURE 87.7 Synchronization channel (SCH) message format, ©ETT [Hanzo and Steele, 1994].

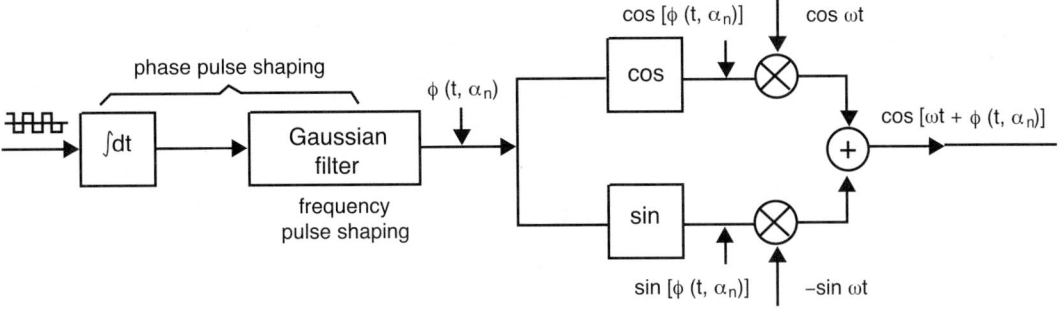

FIGURE 87.8 GMSK modulator schematic diagram, ©ETT [Hanzo and Steele, 1994].

87.7 Gaussian Minimum Shift Keying Modulation

The GSM system uses constant envelope partial response GMSK modulation [Steele ed., 1992] specified in recommendation R.05.04. Constant envelope, continuous-phase modulation schemes are robust against signal fading as well as interference and have good spectral efficiency. The slower and smoother are the phase changes, the better is the spectral efficiency, since the signal is allowed to change less abruptly, requiring lower frequency components. The effect of an input bit, however, is spread over several bit periods, leading to a so-called partial response system, which requires a channel equalizer in order to remove this controlled, intentional intersymbol interference (ISI) even in the absence of uncontrolled channel dispersion.

The widely employed partial response GMSK scheme is derived from the full response minimum shift keying (MSK) scheme. In MSK the phase changes between adjacent bit periods are piecewise linear, which results in discontinuous-phase derivative, i.e., instantaneous frequency at the signalling instants, and hence widens the spectrum. Smoothing these phase changes, however, by a filter having a Gaussian impulse response [Steele ed., 1992], which is known to have the lowest possible bandwidth, this problem is circumvented using the schematic of Fig. 87.8, where the GMSK signal is generated by modulating and adding two quadrature carriers. The key parameter of GMSK in controlling both bandwidth and interference resistance is the 3-dB down filter-bandwidth × bit interval product $(B \cdot T)$, referred to as normalized bandwidth. It was found that as the $B \cdot T$ product is increased from 0.2 to 0.5, the interference resistance is improved by approximately 2 dB at the cost of increased bandwidth occupancy, and best compromise was achieved for $B \cdot T = 0.3$. This corresponds to spreading the effect of 1 b over approximately 3-b intervals. The spectral efficiency gain due to higher interference tolerance and, hence, more dense frequency reuse was found to be more significant than the spectral loss caused by wider GMSK spectral lobes.

The channel separation at the TDMA burst rate of 271 kb/s is 200 kHz, and the modulated spectrum must be 40 dB down at both adjacent carrier frequencies. When TDMA bursts are transmitted in an on-off keyed mode, further spectral spillage arises, which is mitigated by a smooth power ramp up and down envelope at the leading and trailing edges of the transmission bursts, attenuating the signal by 70 dB during a 28- and 18-μs interval, respectively.

87.8 Wideband Channel Models

The set of 6-tap GSM impulse responses [Greenwood and Hanzo, 1992] specified in recommendation R.05.05 is depicted in Fig. 87.9, where the individual propagation paths are independent Rayleigh fading paths, weighted by the appropriate coefficients h_i corresponding to their relative powers portrayed in the figure. In simple terms the wideband channel's impulse response is measured by transmitting an impulse and detecting the received echoes at the channel's output in every D-spaced so-called delay bin. In some bins no delayed and attenuated multipath component

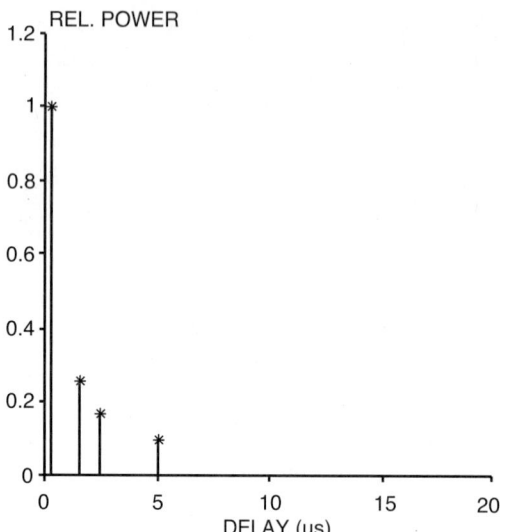
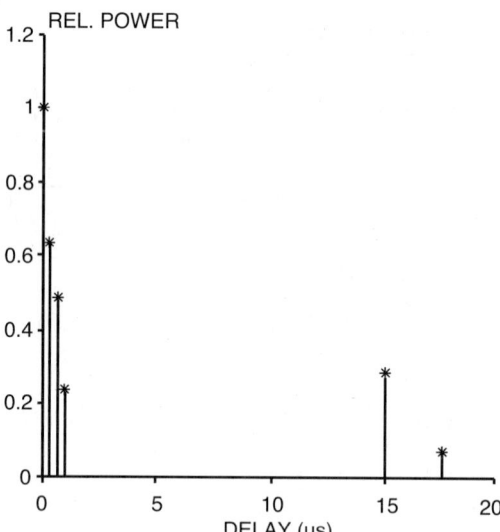
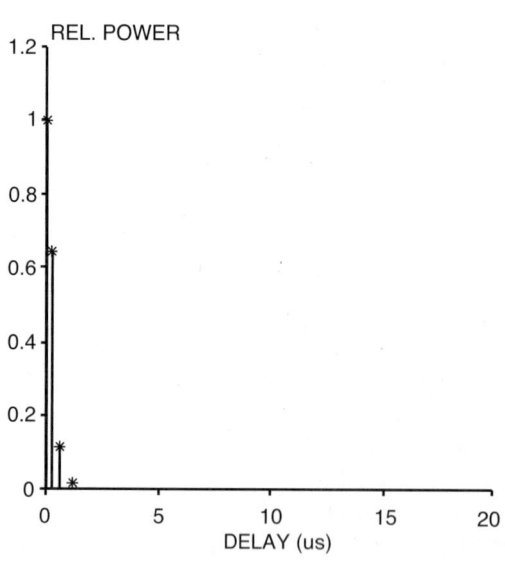
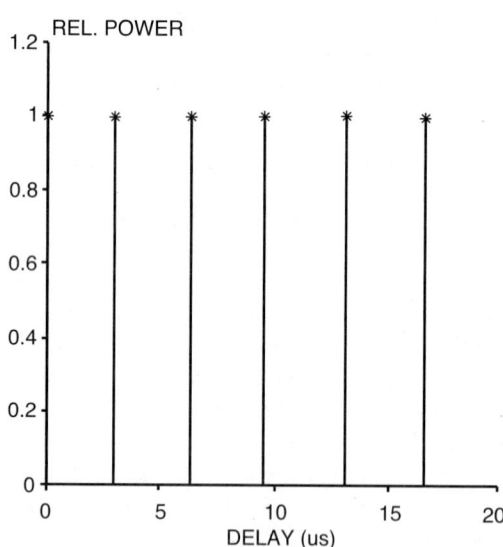

FIGURE 87.9 Typical GSM channel impulse responses, ©ETT [Hanzo and Steele, 1994].

is received, whereas in others significant energy is detected, depending on the typical reflecting objects and their distance from the receiver. The path delay can be easily related to the distance of the reflecting objects, since radio waves are travelling at the speed of light. For example, at a speed of 300, 000 km/s, a reflecting object situated at a distance of 0.15 km yields a multipath component at a round-trip delay of 1 μs.

The typical urban (TU) impulse response spreads over a delay interval of 5 μs, which is almost two 3.69-μs bit-intervals duration and, therefore, results in serious ISI. In simple terms, it can be treated as a two-path model, where the reflected path has a length of 0.75 km, corresponding to a reflector located at a distance of about 375 m. The hilly terrain (HT) model has a sharply decaying short-delay section due to local reflections and a long-delay path around 15 μs due to distant reflections. Therefore, in practical terms it can be considered a two- or three-path model having reflections from a distance of about 2 km. The rural area (RA) response seems the least hostile amongst all standardized responses, decaying rapidly inside 1-b interval and, therefore, is expected to be easily combated by the channel equalizer. Although the type of the equalizer is not standardized, partial response systems typically use VEs. Since the RA channel effectively behaves as a single-path nondispersive channel, it would not require an equalizer. The fourth standardized impulse response is artificially contrived in order to test the equalizer's performance and is constituted by six equidistant unit-amplitude impulses representing six equal-powered independent Rayleigh-fading paths with a delay spread over 16 μs. With these impulse responses in mind, the required channel is simulated by summing the appropriately delayed and weighted received signal components. In all but one case the individual components are assumed to have Rayleigh amplitude distribution, whereas in the RA model the main tap at zero delay is supposed to have a Rician distribution with the presence of a dominant line-of-sight path.

87.9 Adaptive Link Control

The adaptive link control algorithm portrayed in Fig. 87.10 and specified in recommendation R.05.08 allows for the MS to favor that specific traffic cell which provides the highest probability of reliable communications associated with the lowest possible path loss. It also decreases interference with other cochannel users and, through dense frequency reuse, improves spectral efficiency, whilst maintaining an adequate communications quality, and facilitates a reduction in power consumption, which is particularly important in hand-held MSs. The handover process maintains a call in progress as the MS moves between cells, or when there is an unacceptable transmission quality degradation caused by interference, in which case an intracell handover to another carrier in the same cell is performed. A radio-link failure occurs when a call with an unacceptable voice or data quality cannot be improved either by RF power control or by handover. The reasons for the link failure may be loss of radio coverage or very high-interference levels. The link control procedures rely on measurements of the received RF signal strength (RXLEV), the received signal quality (RXQUAL), and the absolute distance between base and mobile stations (DISTANCE).

RXLEV is evaluated by measuring the received level of the BCCH carrier which is continuously transmitted by the BS on all time slots of the B frames in Fig. 87.5 and without variations of the RF level. A MS measures the received signal level from the serving cell and from the BSs in all adjacent cells by tuning and listening to their BCCH carriers. The root mean squared level of the received signal is measured over a dynamic range from -103 to -41 dBm for intervals of one SACCH multiframe (480 ms). The received signal level is averaged over at least 32 SACCH frames ($\approx$15 s) and mapped to give RXLEV values between 0 and 63 to cover the range from -103 to -41 dBm in steps of 1 dB. The RXLEV parameters are then coded into 6-b words for transmission to the serving BS via the SACCH.

RXQUAL is estimated by measuring the bit error ratio (BER) before channel decoding, using the Viterbi channel equalizer's metrics [Steele ed., 1992] and/or those of the Viterbi convolutional decoder [Wong and Hanzo, 1992]. Eight values of RXQUAL span the logarithmically scaled BER range of 0.2–12.8% before channel decoding.

The absolute DISTANCE between base and mobile stations is measured using the timing advance parameter. The timing advance is coded as a 6-b number corresponding to a propagation delay from 0 to $63 \cdot 3.69\ \mu\text{s} = 232.6\ \mu\text{s}$, characteristic of a cell radius of 35 km.

While roaming, the MS needs to identify which potential target BS it is measuring, and the BCCH carrier frequency may not be sufficient for this purpose, since in small cluster sizes the same BCCH frequency may be used in more than one surrounding cell. To avoid ambiguity a 6-b BSIC is transmitted on each BCCH carrier in the SB of Fig. 87.6. Two other parameters transmitted in the BCCH data provide additional information about the BS. The binary flag called

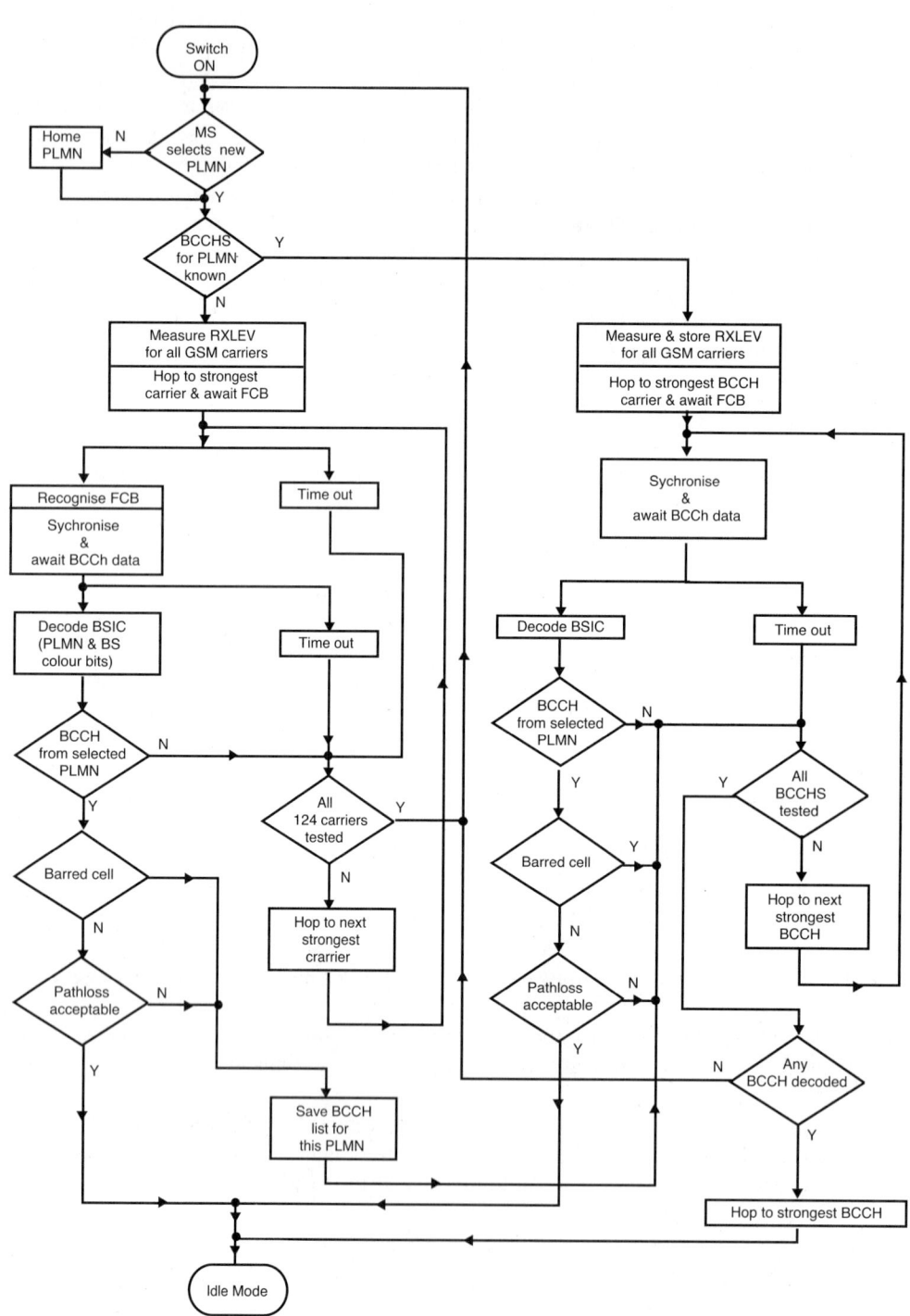

FIGURE 87.10 Initial cell selection by the MS, ©ETT [Hanzo and Steele, 1994].

PLMN_PERMITTED indicates whether the measured BCCH carrier belongs to a PLMN that the MS is permitted to access. The second Boolean flag, CELL_BAR_ACCESS, indicates whether the cell is barred for access by the MS, although it belongs to a permitted PLMN. A MS in idle mode, i.e., after it has just been switched on or after it has lost contact with the network, searches all 125 RF channels and takes readings of RXLEV on each of them. Then it tunes to the carrier with the highest RXLEV and searches for FCB in order to determine whether or not the carrier is a BCCH carrier. If it is not, then the MS tunes to the next highest carrier, and so on, until it finds a BCCH carrier, synchronizes to it and decodes the parameters BSIC, PLMN_PERMITTED and CELL_BAR_ACCESS in order to decide whether to continue the search. The MS may store the BCCH carrier frequencies used in the network accessed, in which case the search time would be reduced. Again, the process described is summarized in the flowchart of Fig. 87.10.

The adaptive power control is based on RXLEV measurements. In every SACCH multiframe the BS compares the RXLEV readings reported by the MS or obtained by the base station with a set of thresholds. The exact strategy for RF power control is determined by the network operator with the aim of providing an adequate quality of service for speech and data transmissions while keeping interferences low. Clearly, adequate quality must be achieved at the lowest possible transmitted power to keep cochannel interferences low, which implies contradictory requirements in terms of transmitted power. The criteria for reporting radio link failure are based on the measurements of RXLEV and RXQUAL performed by both the mobile and base stations, and the procedures for handling link failures result in the re-establishment or the release of the call, depending on the network operator's strategy.

The handover process involves the most complex set of procedures in the radio-link control. Handover decisions are based on results of measurements performed both by the base and mobile stations. The base station measures RXLEV, RXQUAL, DISTANCE, and also the interference level in unallocated time slots, whereas the MS measures and reports to the BS the values of RXLEV and RXQUAL for the serving cell and RXLEV for the adjacent cells. When the MS moves away from the BS, the RXLEV and RXQUAL parameters for the serving station become lower, whereas RXLEV for one of the adjacent cells increases.

87.10 Discontinuous Transmission

Discontinuous transmission (DTX) issues are standardized in recommendation R.06.31, whereas the associated problems of voice activity detection VAD are specified by R.06.32. Assuming an average speech activity of 50% and a high number of interferers combined with frequency hopping to randomize the interference load, significant spectral efficiency gains can be achieved when deploying discontinuous transmissions due to decreasing interferences, while reducing power dissipation as well. Because of the reduction in power consumption, full DTX operation is mandatory for MSs, but in BSs, only receiver DTX functions are compulsory.

The fundamental problem in voice activity detection is how to differentiate between speech and noise, while keeping false noise triggering and speech spurt clipping as low as possible. In vehicle-mounted MSs the severity of the speech/noise recognition problem is aggravated by the excessive vehicle background noise. This problem is resolved by deploying a combination of threshold comparisons and spectral domain techniques [ETSI, 1988; Hanzo and Stefanov, 1992]. Another important associated problem is the introduction of noiseless inactive segments, which is mitigated by comfort noise insertion (CNI) in these segments at the receiver.

87.11 Summary

Following the standardization and launch of the GSM system its salient features were summarized in this brief review. Time division multiple access (TDMA) with eight users per carrier is used at a multiuser rate of 271 kb/s, demanding a channel equalizer to combat dispersion in large cell environments. The error protected chip rate of the full-rate traffic channels is 22.8 kb/s, whereas in half-rate channels it is 11.4 kb/s. Apart from the full- and half-rate speech traffic channels, there are 5 different rate data traffic channels and 14 various control and signalling channels to support the system's operation. A moderately complex, 13 kb/s regular pulse excited speech codec with long term predictor (LTP) is used, combined with an embedded three-class error correction codec and multilayer interleaving to provide sensitivity-matched unequal error protection for the speech bits. An overall speech delay of 57.5 ms is maintained. Slow frequency hopping at 217 hops/s yields substantial performance gains for slowly moving pedestrians.

TABLE 87.3 Summary of GSM features

System feature	Specification
Up-link bandwidth, MHz	890–915 = 25
Down-link bandwidth, MHz	935–960 = 25
Total GSM bandwidth, MHz	50
Carrier spacing, KHz	200
No. of RF carriers	125
Multiple access	TDMA
No. of users/carrier	8
Total No. of channels	1000
TDMA burst rate, kb/s	271
Modulation	GMSK with BT = 0.3
Bandwidth efficiency, b/s/Hz	1.35
Channel equalizer	yes
Speech coding rate, kb/s	13
FEC coded speech rate, kb/s	22.8
FEC coding	Embedded block/convolutional
Frequency hopping, hop/s	217
DTX and VAD	yes
Maximum cell radius, km	35

Constant envelope partial response GMSK with a channel spacing of 200 kHz is deployed to support 125 duplex channels in the 890–915-MHz up-link and 935–960-MHz down-link bands, respectively. At a transmission rate of 271 kb/s a spectral efficiency of 1.35-bit/s/Hz is achieved. The controlled GMSK-induced and uncontrolled channel-induced intersymbol interferences are removed by the channel equalizer. The set of standardized wideband GSM channels was introduced in order to provide bench markers for performance comparisons. Efficient power budgeting and minimum cochannel interferences are ensured by the combination of adaptive power and handover control based on weighted averaging of up to eight up-link and down-link system parameters. Discontinuous transmissions assisted by reliable spectral-domain voice activity detection and comfort-noise insertion further reduce interferences and power consumption. Because of ciphering, no unprotected information is sent via the radio link. As a result, spectrally efficient, high-quality mobile communications with a variety of services and international roaming is possible in cells of up to 35 km radius for signal-to-noise and interference ratios in excess of 10–12 dBs. The key system features are summarized in Table 87.3.

Defining Terms

A3: Authentication algorithm
A5: Cyphering algorithm
A8: Confidential algorithm to compute the cyphering key
AB: Access burst
ACCH: Associated control channel
ADC: Administration center
AGCH: Access grant control channel
AUC: Authentication center
AWGN: Additive Gaussian noise
BCCH: Broadcast control channel
BER: Bit error ratio
BFI: Bad frame indicator flag
BN: Bit number
BS: Base station

BS-PBGT: BS powerbudget: to be evaluated for power budget motivated handovers
BSIC: Base station identifier code
CC: Convolutional codec
CCCH: Common control channel
CELL_BAR_ACCESS: Boolean flag to indicate, whether the MS is permitted to access the specific traffic cell
CNC: Comfort noise computation
CNI: Comfort noise insertion
CNU: Comfort noise update state in the DTX handler
DB: Dummy burst
DL: Down link
DSI: Digital speech interpolation to improve link efficiency
DTX: Discontinuous transmission for power consumption and interference reduction
EIR: Equipment identity register
EOS: End of speech flag in the DTX handler
FACCH: Fast associated control channel
FCB: Frequency correction burst
FCCH: Frequency correction channel
FEC: Forward error correction
FH: Frequency hopping
FN: TDMA frame number
GMSK: Gaussian minimum shift keying
GP: Guard space
HGO: Handover in the VAD
HLR: Home location register
HO: Handover
HOCT: Handover counter in the VAD
HO_MARGIN: Handover margin to facilitate hysteresis
HSN: Hopping sequence number: frequency hopping algorithm's input variable
IMSI: International mobile subscriber identity
ISDN: Integrated services digital network
LAI: Location area identifier
LAR: Logarithmic area ratio
LTP: Long term predictor
MA: Mobile allocation: set of legitimate RF channels, input variable in the frequency hopping algorithm
MAI: Mobile allocation index: output variable of the FH algorithm
MAIO: Mobile allocation index offset: intial RF channel offset, input variable of the FH algorithm
MS: Mobile station
MSC: Mobile switching center
MSRN: Mobile station roaming number
MS_TXPWR_MAX: Maximum permitted MS transmitted power on a specific traffic channel in a specific traffic cell
MS_TXPWR_MAX(n): Maximum permitted MS transmitted power on a specific traffic channel in the nth adjacent traffic cell
NB: Normal burst
NMC: Network management center
NUFR: Receiver noise update flag
NUFT: Noise update flag to ask for SID frame transmission
OMC: Operation and maintenance center
PARCOR: Partial correlation
PCH: Paging channel

PCM: Pulse code modulation
PIN: Personal identity number for MSs
PLMN: Public land mobile network
PLMN_PERMITTED: Boolean flag to indicate whether the MS is permitted to access the specific PLMN
PSTN: Public switched telephone network
QN: Quarter bit number
R: Random number in the authentication process
RA: Rural area channel inpulse response
RACH: Random access channel
RF: Radio frequency
RFCH: Radio frequency channel
RFN: Reduced TDMA frame number: equivalent representation of the TDMA frame number that is used in the synchronization channel
RNTABLE: Random number table utilized in the frequency hopping algorithm
RPE: Regular pulse excited
RPE-LTP: Regular pulse excited codec with long term predictor
RS-232: Serial data transmission standard equivalent to CCITT V24. interface
RXLEV: Received signal level: parameter used in handovers
RXQUAL: Received signal quality: parameter used in handovers
S: Signed response in the authentication process
SACCH: Slow associated control channel
SB: Synchronization burst
SCH: Synchronization channel
SCPC: Single channel per carrier
SDCCH: Stand-alone dedicated control channel
SE: Speech extrapolation
SID: Silence identifier
SIM: Subscriber identity module in MSs
SPRX: Speech received flag
SPTX: Speech transmit flag in the DTX handler
STP: Short term predictor
TA: Timing advance
TB: Tailing bits
TCH: Traffic channel
TCH/F: Full-rate traffic channel
TCH/F2.4: Full-rate 2.4-kb/s data traffic channel
TCH/F4.8: Full-rate 4.8-kb/s data traffic channel
TCH/F9.6: Full-rate 9.6-kb/s data traffic channel
TCH/FS: Full-rate speech traffic channel
TCH/H: Half-rate traffic channel
TCH/H2.4: Half-rate 2.4-kb/s data traffic channel
TCH/H4.8: Half-rate 4.8-kb/s data traffic channel
TDMA: Time division multiple access
TMSI: Temporary mobile subscriber identifier
TN: Time slot number
TU: Typical urban channel inpulse response
TXFL: Transmit flag in the DTX handler
UL: Up link
VAD: Voice activity detection
VE: Viterbi equalizer
VLR: Visiting location register

References

European Telecommunications Standardization Institute. 1988. Group Speciale Mobile or Global System of Mobile Communication (GSM) Recommendation, ETSI Secretariat, Sophia Antipolis Cedex, France.

Greenwood, D. and Hanzo, L. 1992. Characterisation of mobile radio channels, In *Mobile Radio Communications*. ed. R. Steele, Chap. 2, pp. 92–185. IEEE Press–Pentech Press, London.

Hanzo, L. and Stefanov, J. 1992. The Pan-European digital cellular mobile radio system—known as GSM. In *Mobile Radio Communications*, ed. R. Steele, Chap. 8, pp. 677–773, IEEE Press–Pentech Press, London.

Hanzo, L. and Steele, R. 1994. The Pan-European mobile radio system, Pts. 1 and 2, *European Trans. on Telecomm.*, 5(2):245–276.

Salami, R.A., Hanzo, L. et al. 1992. Speech coding. In *Mobile Radio Communications*, ed. R. Steele, Chap. 3, pp. 186–346. IEEE Press–Pentech Press, London.

Steele, R. ed., 1992. *Mobile Radio Communications*, IEEE Press–Pentech Press, London.

Vary, P. and Sluyter, R.J. 1986. MATS-D speech codec: Regular-pulse excitation LPC, *Proceedings of Nordic Conference on Mobile Radio Communications*. pp. 257–261.

Wong, K.H.H. and Hanzo, L. 1992. Channel coding. In *Mobile Radio Communications*, ed. R. Steele, Chap. 4, pp. 347–488. IEEE Press–Pentech Press, London.

88
The IS-54 Digital Cellular Standard

Paul Mermelstein
INRS-Télécommunications
University of Québec

88.1	Introduction	1246
88.2	Modulation of Digital Voice and Data Signals	1247
88.3	Speech Coding Fundamentals	1248
88.4	Channel Coding Considerations	1249
88.5	VSELP Encoder	1249
88.6	Linear Prediction Analysis and Quantization	1249
88.7	Bandwidth Expansion	1252
88.8	Quantizing and Encoding the Reflection Coefficients	1252
88.9	VSELP Codebook Search	1252
88.10	Long-Term Filter Search	1253
88.11	Orthogonalization of the Codebooks	1253
88.12	Quantizing the Excitation and Signal Gains	1253
88.13	Channel Coding and Interleaving	1254
88.14	Bad Frame Masking	1255
88.15	Conclusions	1255

88.1 Introduction

The goals of this chapter are to give the reader a tutorial introduction and high-level understanding of the techniques employed for speech transmission by the IS-54 digital cellular standard. It builds on the information provided in the standards document but is not meant to be a replacement for it. Separate standards cover the control channel used for the setup of calls and their handoff to neighboring cells, as well as the encoding of data signals for transmission. For detailed implementation information the reader should consult the most recent standards document [TIA, 1992].

IS-54 provides for encoding bidirectional speech signals digitally and transmitting them over cellular and microcellular mobile radio systems. It retains the 30-kHz channel spacing of the earlier advanced mobile telephone service (AMPS), which uses analog frequency modulation for speech transmission and frequency shift keying for signalling. The two directions of transmission use frequencies some 45 MHz apart in the band between 824 and 894 MHz. AMPS employs one channel per conversation in each direction, a technique known as frequency division multiple access (FDMA). IS-54 employs time division multiple access (TDMA) by allowing three, and in the future six, simultaneous transmissions to share each frequency band. Because the overall 30-kHz

The IS-54 Digital Cellular Standard

channelization of the allocated 25 MHz of spectrum in each direction is retained, it is also known as a FDMA-TDMA system. In contrast, the later IS-95 standard employs code division multiple access (CDMA) over bands of 1.23 MHz by combining several 30-kHz frequency channels.

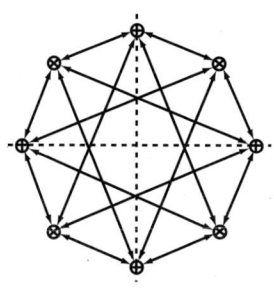

FIGURE 88.1 Constellation for $\pi/4$ shifted QPSK modulation. *Source*: TIA, 1992. Cellular System Dual-mode Mobile Station–Base Station Compatibility Standard TIA/EIA IS-54. With permission.

Each frequency channel provides for transmission at a digital bit rate of 48.6 kb/s through use of differential quadrature-phase shift key (DQPSK) modulation at a 24.3-kBd channel rate. The channel is divided into six time slots every 40 ms. The full-rate voice coder employs every third time slot and utilizes 13 kb/s for combined speech and channel coding. The six slots provide for an eventual half-rate channel occupying one slot per 40 ms frame and utilizing only about 6.5 kb/s for each call. Thus, the simultaneous call carrying capacity with IS-54 is increased by a factor 3(factor 6 in the future) above that of AMPS. All digital transmission is expected to result in a reduction in transmitted power. The resulting reduction in intercell interference may allow more frequent reuse of the same frequency channels than the reuse pattern of seven cells for AMPS. Additional increases in erlang capacity (the total call-carrying capacity at a given blocking rate) may be available from the increased trunking efficiency achieved by the larger number of simultaneously available channels. The first systems employing dual-mode AMPS and TDMA service were put into operation in 1993.

88.2 Modulation of Digital Voice and Data Signals

The modulation method used in IS-54 is $\pi/4$ shifted differentially encoded quadrature phase-shift keying (DPSK). Symbols are transmitted as changes in phase rather than their absolute values. The binary data stream is converted to two binary streams X_k and Y_k formed from the odd- and even-numbered bits, respectively. The quadrature streams I_k and Q_k are formed according to

$$I_k = I_{k-1} \cos[\Delta\phi(X_k, Y_k)] - Q_{k-1} \sin[\Delta\phi(X_k, Y_k)]$$
$$Q_k = I_{k-1} \sin[\Delta\phi(X_k, Y_k)] + Q_{k-1} \cos[\Delta\phi(X_k, Y_k)]$$

where I_{k-1} and Q_{k-1} are the amplitudes at the previous pulse time. The phase change $\Delta\phi$ takes the values $\pi/4$, $3\pi/4$, $-\pi/4$, and $-3\pi/4$ for the dibit (X_k, Y_k) symbols (0,0), (0,1), (1,0) and (1,1), respectively. This results in a rotation by $\pi/4$ between the constellations for odd and even symbols. The differential encoding avoids the problem of 180° phase ambiguity that may otherwise result in estimation of the carrier phase.

The signals I_k and Q_k at the output of the differential phase encoder can take one of five values, 0, ± 1, $\pm 1/\sqrt{2}$ as indicated in the constellation of Fig. 88.1. The corresponding impulses are applied to the inputs of the I and Q baseband filters, which have linear phase and square root raised cosine frequency responses. The generic modulator circuit is shown in Fig. 88.2. The rolloff factor α determines the width of the transition band and its value is 0.35,

$$|H(f)| = \begin{cases} 1, & 0 \leq f \leq (1-\alpha)/2T \\ \sqrt{1/2\{1 - \sin[\pi(2fT-1)/2\alpha]\}}, & (1-\alpha)/2T \leq f \leq (1+\alpha)/2T \\ 0, & f > (1+\alpha)/2T \end{cases}$$

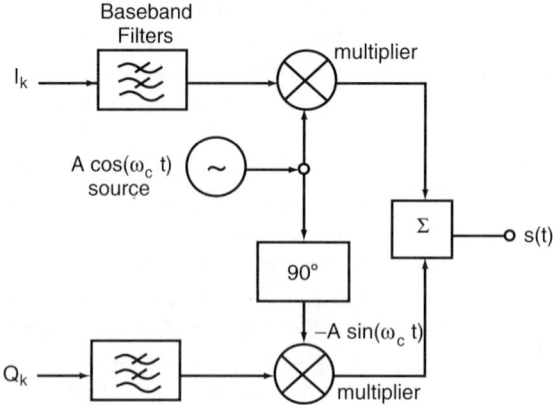

FIGURE 88.2 Generic modulation circuit for digital voice and data signals. *Source:* TIA, 1992. Cellular System Dual-mode Mobile Station–Base Station Compatibility Standard TIA/EIA IS-54.

88.3 Speech Coding Fundamentals

The IS-54 standard employs a vector-sum excited linear prediction (VSELP) coding technique. It represents a specific formulation of the much larger class of code-excited linear preduction (CELP) coders [Atal and Schroeder, 1984] that have proved effective in recent years for the coding of speech at moderate rates in the range 4–16 kb/s. VSELP provides reconstructed speech with a quality that is comparable to that available with frequency modulation and analog transmission over the AMPS system. The coding rate employed is 7.95 kb/s. Each of the six slots per frame carry 260 b of speech and channel coding information for a gross information rate of 13 kb/s. The 260 b correspond to 20 ms of real time speech, transmitted as a single burst.

For an excellent recent review of speech coding techniques for transmission, the reader is referred to Gersho, 1994. Most modern speech coders use a form of analysis by synthesis coding where the encoder determines the coded signal one segment at a time by feeding candidate excitation segments into a replica of a synthesis filter and selecting the segment that minimizes the distortion between the original and reproduced signals. Linear prediction coding (LPC) techniques [Atal and Hanauer, 1971] encode the speech signal by first finding an optimum linear filter to remove the short-time correlation, passing the signal through that LPC filter to obtain a residual signal, and encoding this residual using much fewer bits than would have been required to code the original signal with the same fidelity. In most cases the coding of the residual is divided into two steps. First, the long-time correlation due to the periodic pitch excitation is removed by means of an optimum one-tap filter with adjustable gain and lag. Next, the remaining residual signal, which now closely resembles a white-noise signal, is encoded. Code-excited linear predictors use one or more **codebooks** from which they select replicas of the residual of the input signal by means of a closed-loop error-minimization technique. The index of the codebook entry as well as the parameters of all the filters are transmitted to allow the speech signal to be reconstructed at the receiver. Most code-excited coders use trained codebooks. Starting with a codebook containing Guassian signal segments, entries that are found to be used rarely in coding a large body of speech data are iteratively eliminated to result in a smaller codebook that is considered more effective.

The speech signal can be considered quasistationary or stationary for the duration of the speech frame, of the order of 20 ms. The parameters of the short-term filter, the LPC coefficients, are determined by analysis of the autocorrelation function of a suitably windowed segment of the input signal. To allow accurate determination of the time-varying pitch lag as well as simplify the computations, each speech frame is divided into four 5-ms subframes. Independent pitch filter computations and residual coding operations are carried out for each subframe.

The speech decoder attempts to reconstruct the speech signal from the received information as best possible. It employs a codebook identical to that of the encoder for excitation generation and, in the absence of transmission errors, would produce an exact replica of the signal that produced the minimized error at the encoder. Transmission errors do occur, however, due, to signal fading and excessive interference. Since any attempt at retransmission would incur unacceptable signal delays, sufficient error protection is provided to allow correction of most transmission errors.

88.4 Channel Coding Considerations

The sharp limitations on available bandwidth for error protection argue for careful consideration of the sensitivity of the speech coding parameters to transmission errors. Pairwise interleaving of coded blocks and convolutional coding of a subset of the parameters permit correction of a limited number of transmission errors. In addition, a cyclic redundancy check (CRC) is used to determine whether the error correction was successful. The coded information is divided into three blocks of varying sensitivity to errors. Group 1 contains the most sensitive bits, mainly the parameters of the LPC filter and frame energy, and is protected by both error detection and correction bits. Group 2 is provided with error correction only. The third group, comprising mostly the fixed codebook indices, is not protected at all.

The speech signal contains significant temporal redundancy. Thus, speech frames within which errors have been detected may be reconstructed with the aid of previously correctly received information. A bad-frame masking procedure attempts to hide the effects of short fades by extrapolating the previously received parameters. Of course, if the errors persist, the decoded signal must be muted while an attempt is made to hand off the connection to a base station to/from which the mobile may experience better reception.

88.5 VSELP Encoder

A block diagram of the VSELP speech encoder [Gerson and Jasiuk, 1990] is shown in Fig. 88.3. The excitation signal is generated from three components, the output of a long term or pitch filter, as well as entries from two codebooks. A weighted synthesis filter generates a synthesized approximation to the frequency-weighted input signal. The weighted mean square error between these two signals is used to drive the error minimization process. This weighted error is considered to be a better approximation to the percpetually important noise components than the unweighted mean square error. The total weighted square error is minimized by adjusting the pitch lag and the codebook indices as well as their gains. The decoder follows the encoder closely and generates the excitation signal identically to the encoder but uses an unweighted linear-prediction synthesis filter to generate the decoded signal. A spectral postfilter is added after the synthesis filter to enhance the quality of the reconstructed speech.

The precise data rate of the speech coder is 7950 b/s or 159 b per time slot, each corresponding to 20 ms of signal in real time. These 159 b are allocated as follows: 1) short-term filter coefficients, 38 bits; 2) frame energy, 5 bits; 3) pitch lag, 28 bits; 4) codewords, 56 bits; and 5) gain values, 32 bits.

88.6 Linear Prediction Analysis and Quantization

The purpose of the LPC analysis filter is to whiten the spectrum of the input signal so that it can be better matched by the codebook outputs. The corresponding LPC synthesis filter $A(z)$ restores the short-time speech spectrum characteristics to the output signal. The transfer function of the

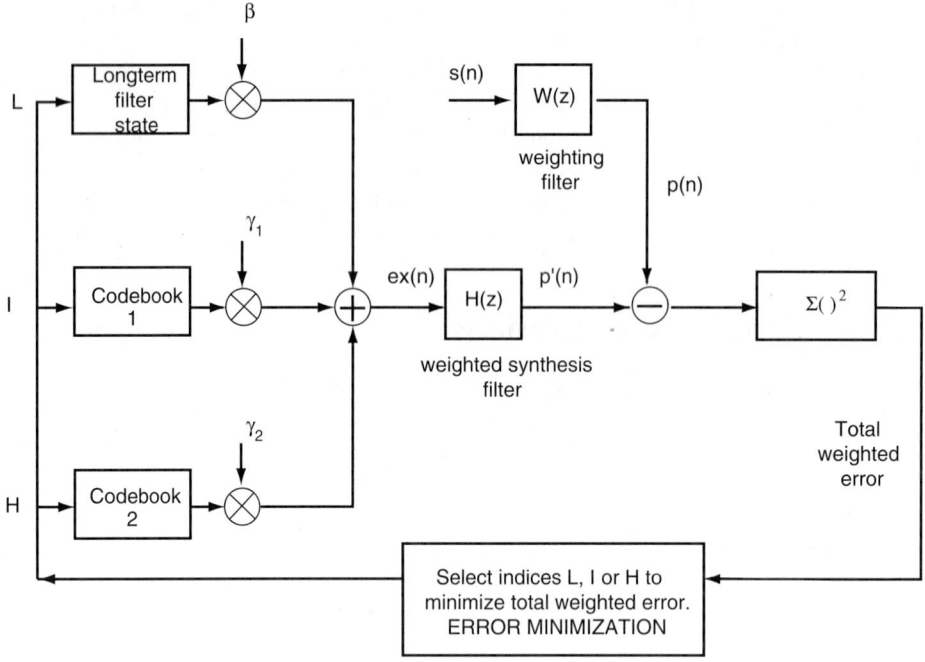

FIGURE 88.3 Black diagram of the speech encoder in VSELP. TIA. 1992. Cellular system Dual-mode Mobile Station–Base Station Compatibility Standard. TIA/EIA IS-54.

tenth-order synthesis filter is given by

$$A(z) = \frac{1}{1 - \sum_{i=1}^{N_p} \alpha_i z^{-i}}$$

The filter predictor parameters $\alpha_1, \ldots, \alpha_{N_p}$ are not transmitted directly. Instead, a set of reflection coefficients $r_1, \ldots, r_{N_p}$ are computed and quantized. The predictor parameters are determined from the reflection coefficients using a well-known backward recursion algorithm [Makhoul, 1975].

A variety of algorithms are known that determine a set of reflection coefficients from a windowed input signal. One such algorithm is the fixed point **covariance lattice**, FLAT, which builds an optimum inverse lattice stage by stage. At each stage j, the sum of the mean-squared forward and backward residuals is minimized by selection of the best reflection coefficient r_j. The analysis window used is 170 samples long, centered with respect to the middle of the fourth 5-ms subframe of the 20-ms frame. Since this centerpoint is 20 samples from the end of the frame, 65 samples from the next frame to be coded are used in computing the reflection coefficient of the current frame. This introduces a lookahead delay of 8.125 ms.

The FLAT algorithm first computes the covariance matrix of the input speech for $N_A = 170$ and $N_p = 10$,

$$\phi(i, k) = \sum_{n=N_p}^{N_A - 1} s(n - i)s(n - k), \qquad 0 \leq i, \quad k \leq N_p,$$

Define the forward residual out of stage j as $f_j(n)$ and the backward residual as $b_j(n)$. Then the autocorrelation of the initial forward residual $F_0(i, k)$ is given by $\phi(i, k)$. The autocorrelation of

The IS-54 Digital Cellular Standard

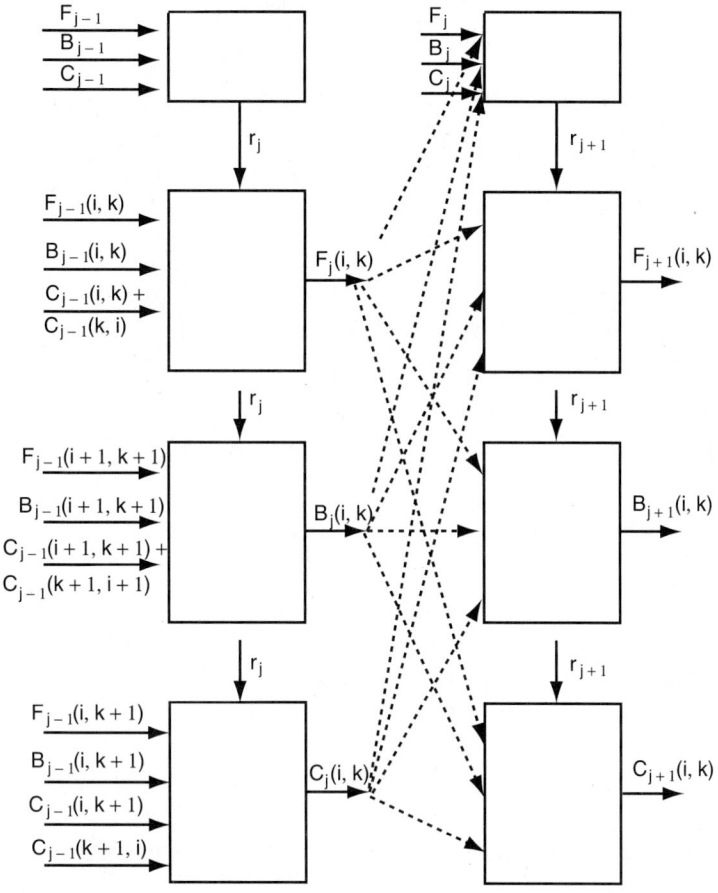

FIGURE 88.4 Block diagram for lattice covariance computations.

the initial backward residual $B_0(i, k)$ is given by $\phi(i+1, k+1)$ and the initial cross correlation of the two residuals is given by $C_0(i, k) = \phi(i, k+1)$ for $0 \leq i, k \leq N_{p-1}$. Initially j is set to 1. The **reflection coefficient** at each stage is determined as the ratio of the cross correlation to the mean of the autocorrelations. A block diagram of the computations is shown in Fig. 88.4. By quantizing the reflection coefficients within the computation loops, reflection coefficients at subsequent stages are computed taking into account the quantization errors of the previous stages. Specifically,

$$C'_{j-1} = C_{j-1}(0, 0) + C_{j-1}(N_p - j, N_p - j)$$
$$F'_{j-1} = F_{j-1}(0, 0) + F_{j-1}(N_p - j, N_p - j)$$
$$B'_{j-1} = B_{j-1}(0, 0) + B_{j-1}(N_p - j, N_p - j)$$

and

$$r_j = \frac{-2C'_{j-1}}{F'_{j-1} + B'_{j-1}}$$

Use of two sets of correlation values separated by $N_p - j$ samples provides additional stability to the computed reflection coefficients in case the input signal changes form rapidly.

Once a quantized reflection coefficient r_j has been determined, the resulting auto- and cross correlations can be determined iteratively as

$$F_j(i,k) = F_{j-1}(i,k) + r_j[C_{j-1}(i,k) + C_{j-1}(k,i)] + r_j^2 B_{j-1}(i,k)$$
$$B_j(i,k) = B_{j-1}(i+1,k+1) + r_j[C_{j-1}(i+1,k+1) + C_{j-1}(k+1,i+1)]$$
$$+ r_j^2 F_{j-1}(i+1,k+1)$$

and

$$C_j(i,k) = C_{j-1}(i,k+1) + r_j[B_{j-1}(i,k+1) + F_{j-1}(i,k+1)]$$
$$+ r_j^2 C_{j-1}(k+1,i)$$

These computations are carried out iteratively for r_j, $j = 1, \ldots, N_p$.

88.7 Bandwidth Expansion

Poles with very narrow bandwidths may introduce undesirable distortions into the synthesized signal. Use of a binomial window with effective bandwidth of 80 Hz suffices to limit the ringing of the LPC filter and reduce the effect of the LPC filter selected for one frame on the signal reconstructed for subsequent frames. To achieve this, prior to searching for the reflection coefficients, the $\phi(i,k)$ is modified by use of a window function $w(j)$, $j = 1, \ldots, 10$, as follows:

$$\phi'(i,k) = \phi(i,k)w(|i-k|)$$

88.8 Quantizing and Encoding the Reflection Coefficients

The distortion introduced into the overall spectrum by quantizing the reflection coefficients diminishes as we move to higher orders in the reflection coefficients. Accordingly, more bits are assigned to the lower order coefficients. Specifically, 6, 5, 5, 4, 4, 3, 3, 3, 3, and 2 b are assigned to $r_1, \ldots, r_{10}$, respectively. Scalar quantization of the reflection coefficients is used in IS-54 because it is particularly simple. **Vector quantization** achieves additional quantizing efficiencies at the cost of significant added complexity.

It is important to preserve the smooth time evolution of the linear prediction filter. Both the encoder and decoder linearly interpolate the coefficients α_i for the first, second and third subframes of each frame using the coefficients determined for the previous and current frames. The fourth subframe uses the values computed for that frame.

88.9 VSELP Codebook Search

The codebook search operation selects indices for the long-term filter (pitch lag L) and the two codebooks I and H so as to minimize the total weighted error. This closed-loop search is the most computationally complex part of the encoding operation, and significant effort has been invested to minimize the complexity of these operations without degrading performance. To reduce complexity, simultaneous optimization of the codebook selections is replaced by a sequential optimization procedure, which considers the long-term filter search as the most significant and therefore executes it first. The two vector-sum codebooks are considered to contribute less and less to the minimization of the error, and their search follows in sequence. Subdivision of the total codebook into two vector sums simplifies the processing and makes the result less sensitive to errors in decoding the individual bits arising from transmission errors.

Entries from each of the two vector-sum codebooks can be expressed as the sum of basis vectors. By orthogonalizing these basis vectors to the previously selected codebook component(s), one ensures that the newly introduced components reduce the remaining errors. The subframes over which the codebook search is carried out are 5 ms or 40 samples long. An optimal search would need exploration of a 40-dimensional space. The vector-sum approximation limits the search to 14 dimensions after the optimal pitch lag has been selected. The search is further divided into two stages of 7 dimensions each. The two codebooks are specified in terms of the fourteen, 40-dimensional basis vectors stored at the encoder and decoder. The two 7-b indices indicate the required weights on the basic vectors to arrive at the two optimum codewords.

The codebook search can be viewed as selecting the three best directions in 40-dimensional space, which when summed result in the best approximation to the weighted input signal. The gains of the three components are determined through a separate error minimization process.

88.10 Long-Term Filter Search

The long-term filter is optimized by selection of a lag value that minimizes the error between the weighted input signal $p(n)$ and the past excitation signal filtered by the current weighted synthesis filter $H(z)$. There are 127 possible coded lag values provided corresponding to lags of 20–146 samples. One value is reserved for the case when all correlations between the input and the lagged residuals are negative and use of no long term filter output would be best. To simplify the convolution operation between the impulse response of the weighted synthesis filter and the past excitation, the impulse response is truncated to 21 samples or 2.5 ms. Once the lag is determined, the untruncated impulse response is used to compute the weighted long-term lag vector.

88.11 Orthogonalization of the Codebooks

Prior to the search of the first codebook, each filtered basis vector may be made orthogonal to the long-term filter output, the zero-state response of the weighted synthesis filter $H(z)$ to the long-term prediction vector. Each orthogonalized filtered basis vector is computed by subtracting its projection onto the long-term filter output from itself.

Similarly, the basis vectors of the second codebook can be orthogonalized with respect to both the long-term filter output and the first codebook output, the zero-state response of $H(z)$ to the previously selected summation of first-codebook basis vectors. In each case the codebook excitation can be reconstituted as

$$u_{k,i}(n) = \sum_{m=1}^{M} \theta_{im} v_{k,m}(n)$$

where $k = 1, 2$ for the two codebooks, $i = I$ or H the 7-b code vector received, $v_{k,m}$ are the two sets of basis vectors, and $\theta_{im} = +1$ if bit m of codeword $i = 1$ and -1 if bit m of codeword $i = 0$. Orthogonalization is not required at the decoder since the gains of the codebooks outputs are determined with respect to the weighted nonorthogonalized code vectors.

88.12 Quantizing the Excitation and Signal Gains

The three codebook gain values β, γ_1, and γ_2 are transformed to three new parameters GS, $P0$ and $P1$ for quantization purposes. GS is an energy offset parameter that equalizes the input and output signal energies. It adjusts the energy of the output of the LPC synthesis filter to equal the energy computed for the same subframe at the encoder input. $P0$ is the energy contribution of the long-term prediction vector as a fraction of the total excitation energy within the subframe.

Similarly, $P1$ is the energy contribution of the code vector selected from the first codebook as a fraction of the total excitation energy of the subframe. The transformation reduces the dynamic range of the parameters to be encoded. An 8-b vector quantizer efficiently encodes the appropriate $(GS, P0, P1)$ vectors by selecting the vector which minimizes the weighted error. The received and decoded values β, γ_1, and γ_2 are computed from the received $(GS, P0, P1)$ vector and applied to reconstitute the decoded signal.

88.13 Channel Coding and Interleaving

The goals of channel coding are to reduce the impairments in the reconstructed speech due to transmission errors. The 159 b characterizing each 20-ms block of speech are divided into two classes, 77 in class 1 and 82 in class 2. Class 1 includes the bits in which errors result in a more significant impairment, whereas the speech quality is considered less sensitive to the class-2 bits. Class 1 generally includes the gain, pitch lag, and more significant reflection coefficient bits. In addition, a 7-b cyclic redundancy check is applied to the 12 most perceptually significant bits of class 1 to indicate whether the error correction was successful. Failure of the CRC check at the receiver suggests that the received information is so erroneous that it would be better to discard it than use it. The error correction coding is illustrated in Fig. 88.5.

The error correction technique used is rate 1/2 convolutional coding with a constraint length of 5 [Lin and Costello, 1983]. A tail of 5 b is appended to the 84 b to be convolutionally encoded to result in a 178-b output. Inclusion of the tail bits ensures independent decoding of successive time slots and no propagation of errors between slots.

Interleaving the bits to be transmitted over two time slots is introduced to diminish the effects of short deep fades and to improve the error-correction capabilities of the channel coding technique. Two speech frames, the previous and the present, are interleaved so that the bits from each speech block span two transmission time slots separated by 20 ms. The interleaving attempts to separate

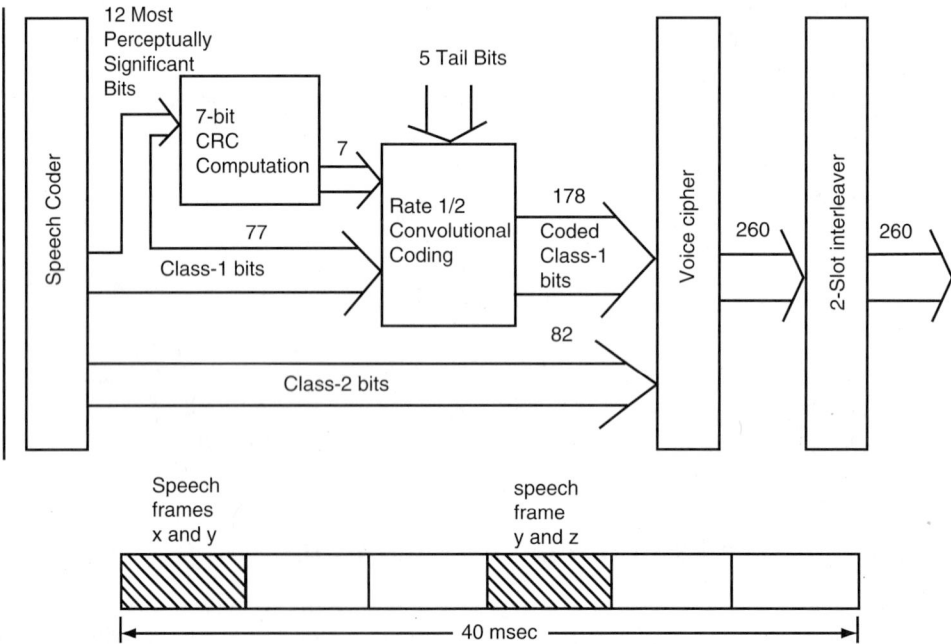

FIGURE 88.5 Error correction insertion for speech coder. Source TIA, 1992. Cellular Systems Dual-Mode Mobile Station–Base Station Compatibility Standards. TIA/EIA IS-54. With permission.

the convolutionally coded class-1 bits from one frame as much as possible in time by inserting noncoded class-2 bits between them.

88.14 Bad Frame Masking

A CRC failure indicates that the received data is unusable, either due to transmission errors resulting from a fade, or from pre-emption of the time slot by a control message (fast associated control channel, FACCH). To mask the effects that may result from leaving a gap in the speech signal, a masking operation based on the temporal redundancy between adjacent speech blocks has been proposed. Such masking can at best bridge over short gaps but cannot recover loss of signal of longer duration. The bad frame masking operation may follow a finite state machine where each state indicates an operation appropriate to the elapsed duration of the fade to which it corresponds. The masking operation consists of copying the previous LPC information and attenuating the gain of the signal. State 6 corresponds to error sequences exceeding 100 ms, for which the output signal is muted. The result of such a masking operation is generation of an extrapolation in the gap to the previously received signal, significantly reducing the perceptual effects of short fades. No additional delay is introduced in the reconstructed signal. At the same time, the receiver will report a high frequency of bad frames leading the system to explore handoff possibilities immediately. A quick successful handoff will result in rapid signal recovery.

88.15 Conclusions

The IS-54 digital cellular standard specifies modulation and speech coding techniques for mobile cellular systems that allow the interoperation of terminals built by a variety of manufacturers and systems operated across the country by a number of different service providers. It permits speech communication with good quality in a transmission environment characterized by frequent multipath fading and significant intercell interference. Generally, the quality of the IS-54 decoded speech is better at the edges of a cell than the corresponding AMPS transmission due to the error mitigation resulting from channel coding. Near a base station or in the absence of significant fading and interference, the IS-54 speech quality is reported to be somewhat worse than AMPS due to the inherent limitations of the analysis–synthesis model in reconstructing arbitrary speech signals with limited bits. At this time the TIA is considering introducing an alternative higher quality 8-kb/s speech coding technique for applications where higher speech quality is desirable. The International Telecommunications Union (ITU) is in the process of selecting such a coding algorithm, which promises speech quality equivalent to wireline telephony links in the absence of transmission errors [Salami et al., 1994]. The selection of the standard placed significant emphasis on the ability to implement the transceiver and encoder/decoder in a variety of mobile and portable terminals at low cost and with low-power dissipation. The standard selected reflects a practically reasonable compromise between high performance and low complexity. The low-weight terminals permitting hours of talk time in mobile environments attest to the benefits of the compromises arrived at in the definition of the standard.

Defining Terms

Codebook: A set of signal vectors available to both the encoder and decoder.

Covariance lattice algorithm: An algorithm for reduction of the covariance matrix of the signal consisting of several lattice stages, each stage implementing an optimal first-order filter with a single coefficient.

Reflection coefficient: A parameter of each stage of the lattice linear prediction filter that determines 1) a forward residual signal at the output of the filter-stage by subtracting from the forward residual at the input a linear function of the backward residual, also 2) a backward residual at the output of the filter stage by subtracting a linear function of the forward residual from the backward residual at the input.

Vector quantizer: A quantizer that assigns quantized vectors to a vector of parameters based on their current values by minimizing some error criterion.

References

Atal, B.S. and Hanauer, S.L. 1971. Speech analysis and synthesis by linear prediction of the speech wave. *J. Acoust. Soc. Am.* 50:637–655.

Atal, B.S. and Schroeder, M. 1984. Stochastic coding of speech signals at very low bit rates. *Proc. Int. Conf. Comm.* pp. 1610–1613.

Gersho, A. 1994. Advances in speech and audio compression. *Proc. IEEE.* 82:900–918.

Gerson, I.A. and Jasiuk, M.A. 1990. Vector sum excited linear prediction (VSELP) speech coding at 8 kbps. *Int. Conf. Acoust. Speech and Sig. Proc.* ICASSP90. pp. 461–464.

Lin S. and Costello, D. 1983. *Error Control Coding: Fundamentals and Application*, Prentice Hall, Englewood Cliffs NJ.

Makhoul, J. 1975. Linear prediction, a tutorial review. *Proc. IEEE.* 63:561–580.

Salami, R., Laflamme, C., Adoul, J.P., and Massaloux, D. 1994. A toll quality 8 kb/s speech codec for the personal communication system (PCS). *IEEE Trans. Vehic. Tech.* 43:808–816.

Telecommunications Industry Association, 1992. EIA/TIA Interim Standard, Cellular System Dual-mode Mobile Station–Base Station Compatibility Standard IS-54B, TIA/EIA, Washington, D.C.

Further Information

For a general treatment of speech coding for telecommunications, see N.S. Jayant and P. Noll, *Digital Coding of Waveforms*, Prentice Hall, Englewood, NJ, 1984. For a more detailed treatment of linear prediction techniques, see J. Markel and A. Gray, *Linear Prediction of Speech*, Springer–Verlag, NY, 1976.

# 89

CDMA Technology and the IS-95 North American Standard

Arthur H.M. Ross
QUALCOMM Incorporated

Klein S. Gilhousen
QUALCOMM Incorporated

89.1	General Overview	1257
89.2	A Short History	1258
89.3	Overview of CDMA	1259
89.4	The CDMA System	1262
89.5	Multiple Forms of Diversity	1263
89.6	Mobile Power Control for CDMA Digital Cellular	1264
	The Mobile Fading Environment • Open-Loop Control • Closed-Loop Control • Other Benefits and Features	
89.7	Low Transmit Power	1267
89.8	Privacy	1268
89.9	Mobile Station Assisted Soft Handoff	1268
89.10	Capacity	1269
	Voice Activity Detection • Frequency Reuse • Sectorization • Low E_b/N_0 (or C/I) and Error Protection • Effect of Thermal Noise • Net Capacity Calculation	
89.11	Soft Capacity	1272
89.12	Transition to CDMA	1272
89.13	Overview of the IS-95 CDMA Standard	1272

89.1 General Overview

Code division multiple access (CDMA) provides a superior, spectrally efficient, digital solution for the second generation of cellular, wireless telephony, and personal communications systems (PCS) services. The CDMA air interface is near optimum in its use of the subscriber station transmitter power, enabling the widespread commercial use of low-cost, lightweight, hand-held portable units that have vastly superior battery life. The technology is also near optimum in its link budgets, minimizing the number of base stations required for an excellent grade of service coverage. As customer penetration in a given service area increases, the system is called upon to deliver more capacity. CDMA has demonstrated a capacity increase over advanced mobile phone service (AMPS) of at least a factor of ten, which means that up to ten times fewer base stations will be required when customer demand for service increases. The use of soft handoff nearly eliminates the annoyance of dropped calls, fading, and poor voice quality. In short, no other air interface technology comes close to its performance.

CDMA, in this context, means not just generic code division multiple access, but the specific implementation of it described in the air interface standard: *TIA/EIA/IS-95-A: Mobile Station–Base Station Compatibility Standard for Dual-Mode Wideband Spread Spectrum Cellular System*. Compliance with the requirements in this document assures subscribers and service providers that equipment of all types will interoperate satisfactorily.

IS-95-A originated with a system design pioneered by QUALCOMM, beginning in April 1989. The early requirements were defined by the cellular carriers with input from manufacturers and carriers and by public testing. After the initial proposal and presentation of a draft air interface by QUALCOMM, a standards committee composed of system operators, subscriber equipment vendors, infrastructure equipment vendors, and test equipment vendors reviewed, revised, and formalized the air interface. The formal adoption of what is now the IS-95-A air interface took place in December of 1993.

This overview provides an overall understanding of the basic principles of CDMA system design and facilitates comprehension of the IS-95-A standard.

89.2 A Short History

In September 1988 the Cellular Telecommunications Industry Association (CTIA), out of concern that the burgeoning subscriber population would soon overwhelm the capacity of the already stressed AMPS facilities, even with cell subdivision, adopted a resolution they called the user performance requirements (UPR). The UPR expressed a need for a new all-digital air interface design that would achieve the following key goals.

- Tenfold increase over analog capacity
- Long life and adequate growth of second-generation technology
- Ability to introduce new features
- Quality improvements
- Privacy
- Ease of transition and compatibility with existing analog system
- Early availability and reasonable costs for dual mode radios and cells
- Cellular open network architecture (CONA)

An engineering subcommittee of the Telecommunications Industry Association was tasked with the development of such a standard. The result, after about two years of development, was the IS-54 air interface, sometimes known as digital AMPS (D-AMPS). D-AMPS uses time division multiple access techniques, in conjunction with low-bit-rate speech coding, to support not one but three conversations on each 30-kHz radio channel. The three-way time division of each channel, although an improvement over analog AMPS, is not responsive to the ten times capacity goal.

Meanwhile, QUALCOMM engineers, skeptical of even the modest capacity claims made on behalf of D-AMPS, began investigating an alternative technology, based on spread-spectrum techniques. The properties of spread spectrum are well understood, but it had at the time seen little, if any, commercial application. The primary applications had been military, usually motivated either by antijam or low-probability-of-intercept requirements.

For D-AMPS to achieve even the modest capacity gain claimed for it, the carrier-to-interference (C/I) ratio needed must be no worse than that required for AMPS. QUALCOMM believed that the C/I was no better than AMPS, and might actually be worse, leading to capacity gain less than three times. As little public information is available about the few D-AMPS systems in operation, the real performance achieved by this technology cannot be easily determined.

Recognizing that management of interference is the key to achieving the dramatic capacity improvement goal of the UPR, QUALCOMM took the radical step of dropping the 30-kHz channelization. As discussed later in this chapter, if a large number of conversations can be somehow overlapped in a common spectral band, then averaging of the interference takes place and the frequency reuse rules are radically altered. In addition to the use of common spectrum, it would be necessary to be near optimum in all other aspects of signal design, modulation, and coding, and to use a high-quality variable rate speech coder with full error protection of the coded speech.

Initial estimates of the potential CDMA capacity gain were very encouraging. They ranged up to, perhaps, 20 times AMPS, well in excess of the UPR goal, and almost seven times better than the purported three-times capacity increase of D-AMPS.

With extensive cellular industry support, QUALCOMM developed a demonstration CDMA system compliant with the CTIA requirements as defined by the UPR. This system's field trials were conducted publicly with the support and participation of infrastructure equipment manufacturers including Northern Telecom, AT&T, and Motorola; subscriber equipment manufacturers including Motorola, OKI Telecom, Clarion, Sony, Alps Electric, Nokia and Matsushita-Panasonic; and major cellular carries including PacTel Cellular (now AirTouch), Ameritech Mobile, NYNEX Mobile, GTE Mobile Communications, Bell Atlantic Mobile Systems, US West New Vector, Group, and Bell Cellular of Canada.

In July, 1990 a draft of the proposed new standard was reviewed by many major carriers and equipment manufactures. The document was revised as a result of their comments, and on October 1, 1990, Revision 1.0 of the standard was released.

The field trial results were formally presented to the CTIA on Dec. 5, 1991. Subsequently, on Jan. 6, 1992 the CTIA Board of Directors unanimously adopted a resolution that stated "CTIA further requests that the Telecommunications Industry Association (TIA) prepare 'structurally' to accept contributions regarding wideband (cellular) systems. This should be a separate effort not diluting the IS-54 revision process."

On Feb. 11, 1992 the TIA's cellular and common carrier radio section recommended, by unanimous vote, that the TR45 Committee address standardization activities regarding wideband spread-spectrum digital technologies. Accordingly, and responding to the desires of the user and service provider communities, the TIA TR45 Committee in March, 1992 created a new engineering subcommittee, TR45.5, to develop spread-spectrum digital cellular standards.

The original QUALCOMM common air interface document was formally submitted to TR45.5 in April of 1992. After extensive discussion, detailed critical review, and revision, the CDMA common air interface, was formally adopted as North American digital cellular standard IS-95 on July 16, 1993. Subsequent minor improvements resulted in publication of Revision A in May of 1995.

89.3 Overview of CDMA

CDMA is a modulation and multiple access scheme based on spread-spectrum communication, a well-established technology that QUALCOMM has been applying to digital cellular radio communications and advanced wireless technologies. The approach will solve the near-term capacity concerns of major markets and the industry's long-term need for an economic, efficient, and truly portable communications.

Ever since the second pair of wireless telegraphs came into existence, we have been confronted with the problem of multiple access to the frequency spectrum without mutual interference. In the early days of wireless telegraphy, both frequency division in the form of resonant antennas and time division in the form of schedules, as well as netted operations were employed. As the number of wireless radios in operation increased and as the technology allowed, it became necessary to impose some discipline on the process in the form of frequency allocations. This has grown over the years to the complex process we have today for world-wide frequency allocations and licensing by service type.

The multiple access problem can be thought of as a filtering problem. There are many simultaneous users that want to use the same electromagnetic spectrum, and there is a choice of an array of filtering and processing techniques that allow the different signals to be separately received and demodulated without excessive mutual interference. The techniques that have long been used include: propagation mode selection, spatial filtering with directive antennas, frequency filtering, and time sharing. Over the last 40 years, techniques involving spread spectrum modulation have evolved in which more complex waveforms and filtering processes are employed.

Propagation mode selection involves a proper choice of operating frequency and antenna so that signals propagate between the intended communicators but not between (very many) other communicators. Frequency reuse in cellular mobile telephone systems is an example of this technique carried to a great degree of sophistication.

Spatial filtering uses the properties of directive antenna arrays to maximize response in the direction of desired signals and to minimize response in the direction of interfering signals. The current analog cellular system uses sectorization to a good advantage to reduce interference from cochannel users in nearby cells.

With frequency division multiple access (FDMA), a traffic channel is a relatively narrow band in the frequency domain into which a signal's transmission power is concentrated. Different signals are assigned different frequency channels. Interference to and from adjacent channels is limited by the use of bandpass filters that pass signal energy within the specified narrow frequency band while rejecting signals at other frequencies. The analog FM cellular system uses FDMA.

FDMA spectral efficiency in a cellular system is determined by the modulation spectral efficiency (the information bit rate per hertz of bandwidth) and the frequency reuse factor. The U.S. analog cellular system, divides the allocated spectrum into 30-kHz bandwidth channels; narrowband FM modulation is employed, resulting in a modulation efficiency of 1 call per 30 kHz of spectrum. Because of interference, the same frequency cannot be used in every cell. The frequency reuse factor is a number representing how often the same frequency can be reused. To provide acceptable call quality, a carrier-to-interference ratio (C/I) of 18 dB or greater is needed. Empirical results have shown that in most cases this level of C/I requires a reuse factor of seven. The resulting capacity is one call per 210 kHz of spectrum in each cell. Note that by increasing the number of cells, an arbitrarily high capacity can be obtained but with increased equipment costs. In addition, there is also a cost of increasing handoff rates as mobile stations move through smaller coverage areas.

With time division multiple access (TDMA), a traffic channel consists of a time slot in a periodic train of time intervals making up a frame. A given signal's energy is confined to one of these time slots. Adjacent channel interference is limited by the use of a time gate that only passes signal energy that is received at the proper time. Some systems use a combination of FDMA and TDMA. The TIA Digital Cellular Standard, IS-54-B, now IS-136, uses 30-kHz FDMA channels that are subdivided into three time slots for TDMA transmissions. One time slot is required for each call when employing 8 kb/s vocoders.

TDMA spectral efficiency is determined in a manner similar to that used for FDMA. The IS-54-B TDMA standard provides a basic modulation efficiency of three voice calls per 30 kHz of bandwidth. The currently accepted frequency reuse criteria is similar to the analog design. The resulting capacity is one call per 70 kHz of spectrum or three times that of the analog FM system.

With CDMA, (see Fig. 89.1) each signal consists of a different pseudorandom binary sequence that modulates a carrier, spreading the spectrum of the waveform. A large number of CDMA signals share the same frequency spectrum. If CDMA is viewed in either the frequency or time domain, the multiple access signals appear to be on top of each other. The signals are separated in the receivers by using a correlator that accepts only signal energy from the selected binary sequence and despreads its spectrum. The other users' signals, whose codes do not match, are not despread in bandwidth and, as a result, contribute only to the noise and represent a self-interference generated by the

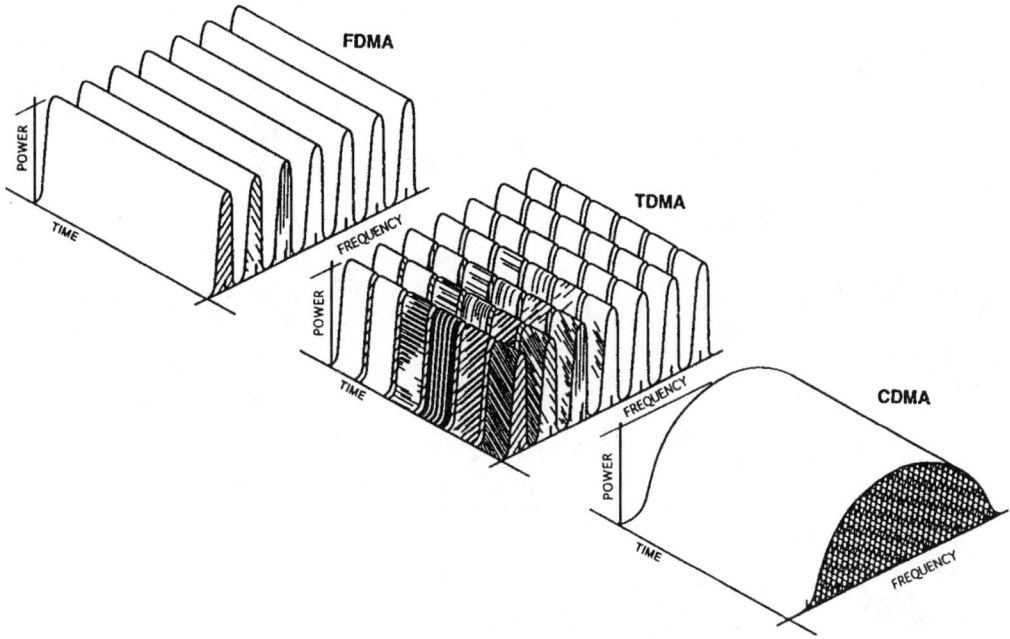

FIGURE 89.1 *Frequency and time domain representations of FDMA, TDMA, and CDMA.* Unlike FDMA or TDMA, CDMA has multiple users simultaneously sharing the same wideband channel. Individual users are selected by correlation processing of the pseudonoise waveform.

system. All of the desired signal's energy will pass through a narrow-bandwidth filter following the correlator, while the interfering signals energy is reduced by the ratio of the bandwidth before the correlator to the bandwidth after the correlator, greatly improving the signal-to-noise ratio for the desired signal. This improvement ratio is known as the processing gain.

The increased signal-to-noise ratio for the desired signal is shown in Fig. 89.2. The signal-to-interference ratio is determined by the ratio of desired signal power to the sum of the power of all of the other signals and is enhanced by the system processing gain or the ratio of spread bandwidth to baseband data rate. The major parameters that determine the CDMA digital cellular system capacity are processing gain, required E_b/N_0, voice duty cycle, frequency reuse efficiency, and the number of sectors in the cell.[1] The CDMA cellular telephone system achieves a spectral efficiency of up to 20 times the analog FM system efficiency when serving the same area with the same antenna system when the antenna system has three sectors per cell. This is a capacity of up to one call per 10 kHz of spectrum.

In the cellular radio frequency reuse concept, interference is accepted but controlled with the goal of increasing system capacity. CDMA does this effectively because it is inherently an excellent anti-interference waveform. Since all calls use the same frequencies, CDMA frequency reuse efficiency is determined by a small reduction in the signal-to-noise ratio caused by system users in neighboring cells. CDMA frequency reuse efficiency is approximately 2/3 compared to 1/7 for narrowband FDMA systems. The CDMA system can also be a hybrid of FDMA and CDMA techniques where the total system bandwidth is divided into a set of wideband channels, each of which contains a large number of CDMA signals.

---

[1] E_b/N_0 is defined as the bit energy to noise power spectral density: comparable to C/I.

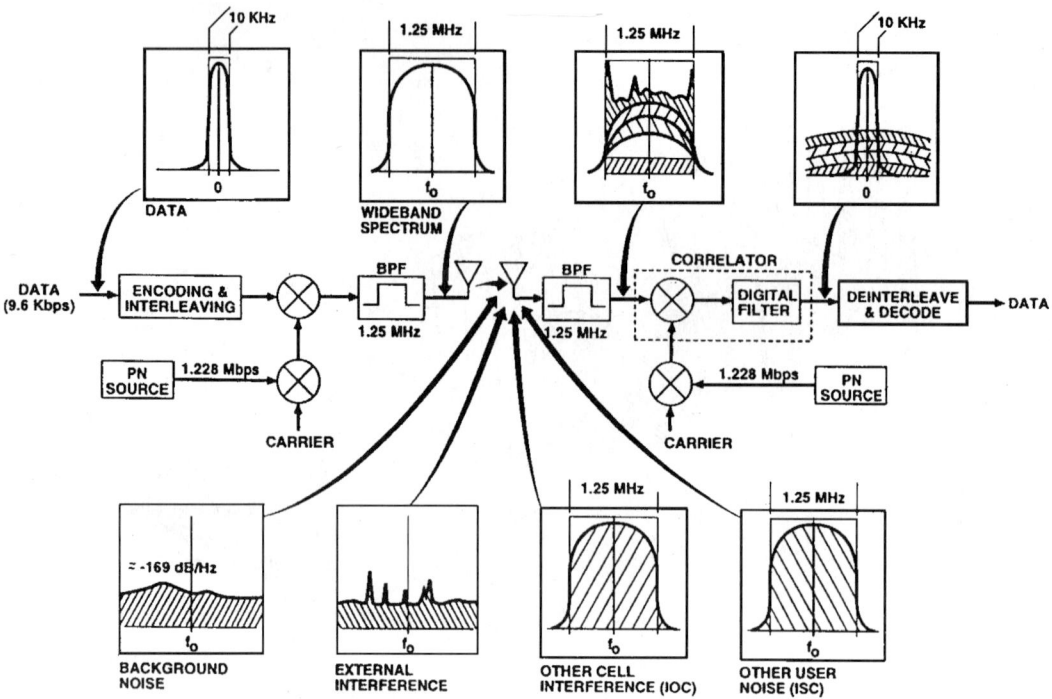

FIGURE 89.2 View of the CDMA concept. The desired signal is selected from four different sources of interference. The dominant source is system self-interference produced by other users of the same cell. This source is controlled by closed-loop power control.

89.4 The CDMA System

The multiple access scheme exploits isolation provided by the antenna system, geometric spacing, power gating of transmissions by voice activity, power control, a very efficient modem, and a signal design that uses very powerful error correction coding.

A combination of open-loop and closed-loop power control (through measurements of the received power at the mobile station and the base station) commands the mobile station to make power adjustments in order to maintain only the power level required for adequate performance. This minimizes interference to other users, helps to overcome fading, and conserves battery power in the mobile station.

The CDMA digital cellular waveform design uses a pseudorandom noise (PN) spread spectrum carrier. The chip rate of the PN spreading sequence was chosen so that the resulting bandwidth is about 1.25 MHz after filtering or approximately one-tenth of the total bandwidth allocated to one cellular service carrier.

The Federal Communications Commission (FCC) has allocated a total of 25 MHz for mobile station to cell site and 25 MHz for cell site to mobile station for the provision of cellular services. The FCC has divided this allocation equally between two service providers, the A and the B carriers, in each service area. Because of the time sequence of the FCC's actions in allocating the cellular spectrum, the 12.5 MHz allocated to each carrier for each direction of the link is further subdivided into two subbands. For the B carriers, the subbands are 10 MHz and 2.5 MHz each. For the A carriers, the subbands are 11 MHz and 1.5 MHz each. A signal bandwidth of less than 1.5 MHz fits into any of the subbands, whereas a bandwidth of less than 2.5 MHz fits into all but one subband.

A set of ten 1.25-MHz-bandwidth CDMA channels can be used by each operator if the entire allocation is converted to CDMA. Initially, only one or a small number of 1.25-MHz channels needs to be removed from the present FM analog service to provide digital service. This facilitates the deployment by introducing a gradual reduction in analog capacity. Each 1.25-MHz CDMA segment can provide about twice the capacity of the entire 12.5-MHz allocation using the present FM system. Some frequency guard band is necessary if there are adjacent high-power cellular (or other) frequencies in use, and the maximum capacity of the CDMA cell is required. Capacity can be sacrificed for decreased guard band if desired. Adjacent CDMA channels need not employ a guard band.

89.5 Multiple Forms of Diversity

In relatively narrowband modulation systems, such as analog FM modulation employed by the first-generation cellular phone system, the existence of multiple paths causes severe fading. With wideband CDMA modulations, however, the different paths may be independently received greatly reducing the severity of the multipath fading. Multipath fading is not completely eliminated because multipaths that cannot be independently processed by the demodulator occasionally occur. This will result in some fading behavior.

Diversity is the favored approach to mitigate fading. There are three major types of diversity: time, frequency, and space. Time diversity can best be obtained by the use of interleaving and error correction coding. Wideband CDMA offers a form of frequency diversity by spreading the signal energy over a wide bandwidth; frequency selective fading usually affects only a 200–300 kHz portion of the signal bandwidth. Space or path diversity is obtained three different ways by providing the following.

- Multiple signal paths through simultaneous links from the mobile station to two or more cell sites (soft handoff).
- Exploitation of the multipath environment through spread-spectrum processing (rake receiver), allowing signals arriving with different propagation delays to be received separately and combined.
- Multiple antennas at the cell site.

The following are different types of diversity employed in the CDMA system to greatly improve performance.

- Time diversity: symbol interleaving, error detection, and correction coding
- Frequency diversity: 1.25-MHz wideband signal
- Space (path) diversity: dual cell site receive antennas, multipath rake receivers, and multiple cell sites (soft handoff)

Antenna diversity can easily be provided in FDMA and TDMA systems. Time diversity can be provided in all digital systems that can tolerate the required higher transmitted symbol rate needed to make the required error correction process effective. The remaining methods, however, can only be provided easily with CDMA. A unique feature of direct sequence CDMA is the ability to provide extensive path diversity; the greater the order of diversity in a system, the better the performance in this difficult propagation environment. Additional diversity of a different mode is much more powerful than additional numbers of the same type of diversity because the fading processes are more likely to be independent with different diversity modes.

Multipath processing takes the form of parallel correlators for the PN waveform. The mobile and cell receivers employ three and four parallel correlators, respectively. Receivers using parallel correlators (sometimes called rake receivers) allow individual path arrivals to be tracked independently, and the sum of their received signal strengths is then used to demodulate the signal. Although there is fading on each arrival, the fades are independent. Demodulation based on the sum of the signals is then much more reliable. The multiplicity of correlators is also the basis for the simultaneous tracking of signals from two different cells and allows the subscriber unit to control the soft handoff.

89.6 Mobile Power Control for CDMA Digital Cellular

Spread-spectrum techniques, long established for antijam and multipath rejection applications, have also been proposed for CDMA to support simultaneous digital communication among a large community of relatively uncoordinated users. It has frequently been pointed out that the mobile to base link in such a system is subject to a near far problem in which a mobile station close to the base has a much lower path loss to the station than far away mobile stations. If all of the mobiles were to use the same transmitter power, then the close-by mobile would apparently jam the far-away mobile stations. Thus, the need for a mobile power control system is postulated in order mitigate this problem.

Before proceeding further, it is interesting to determine the performance of a CDMA system with no power control at all in order to have a standard with which to compare various power control techniques. We will use as an example system, the system used by IS-95. This system uses a 1.25-MHz-bandwidth (W) direct sequence spread-spectrum modulation, a maximum data rate of 9600 b/s (R_b) and requires an bit energy to noise density ratio, E_b/N_0, of 6 dB in nominal conditions, depending on the fading environment, for good quality reception.

The near–far problem can be reduced to a classical jamming problem for which the J/S equation[2] applies,

$$\frac{J}{S} = \frac{W/R_b}{E_b/N_0} \qquad (89.1)$$

With the given values, the J/S is 15 dB. This means that a jamming or interfering signal can be 15 dB stronger than a desired signal before the desired signal quality is affected (15 dB is a factor of 30 in power). This is a result of CDMA's unique ability to discriminate against undesired signals whose spectrum spreading codes do not match the desired signal's code.

Consider a case with only two mobiles in one sector of a cell, an adequate link will be obtained for both mobiles as long as one mobile is not more than 15 dB closer (in path loss) to the base station than the other mobile. In an environment in which the path loss is proportional to the fourth power of distance (the usual case in cellular), this corresponds to a factor of 2.4 in distance. For mobiles randomly and uniformly distributed in the sector area of a cell, the probability that both will achieve an acceptable quality link is 83%. Note that a narrowband cellular system, such as the analog FM cellular system, will achieve an acceptable quality with about the same probability when operating under the same propagation assumptions (fourth power, 8-dB σ, frequency reuse = 1/7, three sectored cell). Since both provide two active calls per 1.25 MHz of spectrum within the sectored cell, one could conclude that CDMA with no power control at all has about the same capacity as the existing analog FM system.

The heart of the CDMA capacity advantage is that the jamming can be the aggregate effect of the other users of the system. If each station optimally controls its transmitter power so that all signals arrive at the cell with the equal power, then Eq. (89.1) defines a limit on system capacity.

The Mobile Fading Environment

First, consider the dynamic range requirement. A cellular mobile may be anywhere in the cell from right under the base station antenna tower to perhaps 5–10 miles distant. In an environment where

---

[2] J/S is the jamming margin or the factor by which a jammer can exceed the signal strength of a desired signal before interference results.

CDMA Technology and the IS-95 North American Standard　　　　　　　　　　　　　　　　1265

propagation goes as the fourth power of distance, as in most cellular service areas, it has been found that the total dynamic range of path loss is on the order of 80 dB. This means that the mobile transmitter must vary its power from a few tens of nanowatts up to the order of 1 W.

Another problem with power control is that the path loss can vary rapidly because of multipath induced Rayleigh fading. In multipath fading, the signal arrives at the receiver after traveling directly between antennas and also after being reflected from building, hills, etc., in the physical environment. When such multiple signals arrive at the receive antenna, the RF phase difference between the signals can be such that the signals may cancel at one moment and be enhanced the next. Such fading commonly causes fluctuations of 20–30 dB while the mobile travels a distance of only 1 ft. If the mobile is traveling on the freeway at speeds of 100 ft/s, clearly the fade rate can exceed 100 Hz. Actually, such deep and fast fades, although relatively common with narrowband waveforms is relatively uncommon with CDMA because of the mitigating effects of the various diversity modes in the system.

An additional complicating factor is that the multipath fading of signals to the mobile is not necessarily the same as fading from the mobile. This is caused primarily by the fact that the signals to and from the mobile are separated by 45 MHz in the frequency domain. This is usually great enough to decouple any dependency between fading in the two directions.

Open-Loop Control

The wide dynamic range portion of the problem is best dealt with using an open-loop power control technique in which the mobile determines an estimate of the path loss between the cell and the mobile. This is done by measuring the received signal level at the mobile. This is accomplished by utilizing the automatic gain control (AGC) circuitry of the receiver. The AGC circuits operate on the receiver's IF frequency amplifiers so that the input to the receiver's A/D converters is held constant. The mobile transmitter circuits use IF amplifier circuits identical to those used in the receiver. The AGC control voltage, offset by another control signal, is used to control the gain of the transmitter IF amplifiers exactly in step with the receiver's IF gain. Thus, if the mobile moves closer to the cell increasing the received signal level, the receiver AGC will reduce the receiver IF amplifier gain, and the transmitter IF gain. This will result in a proportionally lower mobile transmitter power. This is as it should be since the mobile is now closer to the cell and must reduce transmitter power in order to maintain a constant receive power at the cell base station.

The equation which governs the operation of the open-loop control (with power in decibels referred to 1.mW) is

$$\text{transmit power in dBm} = -73 - \text{receive power in dBm} + \text{parameters}$$

Thus, when the mobile received power is −90 dBm, the mobile transmitter power will be +17 dB (1 mW) in a nominal sized cell. The parameters are used to adjust the open-loop power control for different sized cells and different cell ERP (Effective radiated power) and receiver sensitivities.

The circuitry for mechanizing the open-loop power control has proven to be quite simple and reliable. The desired dynamic range has been achieved with an accuracy of on the order of ±6 dB. As stated earlier however, even with perfect open-loop control, because of the lack of reciprocity of path loss due to multipath fading, the signals arriving at the cell may still be significantly different from the desired power level.

Closed-Loop Control

The closed-loop control has the function of controlling the mobile transmit power so that the desired SNR is received at the cell. Note that this is SNR control not just power control. The difference will become clear further on.

Each cell receiver (there is one for every call in progress being received by this cell) forms an estimate of the received SNR of its mobile's signal. This estimate is made every 1.25 ms. The SNR measurement is compared with the set point SNR. If the received SNR is too high, a decrease power command is sent. If the SNR is too low, an increase power command is sent. A power command is sent every 1.25 ms providing an 800 b/s control stream rate.

The power commands are inserted into the data stream being transmitted to the mobile being controlled. The command bits are simply written over data channel symbols. (The data channel error detection and correction can easily fill in the missing symbols.) The power control commands themselves are not protected by error detection and correction, primarily because the necessary delay involved in error correction is intolerable to the closed-loop control.

The mobile receiver picks off the power control commands as they emerge from the demodulator. The bits are accumulated in a digital word that represents the total accumulated correction to the mobile's power. The value represented by the accumulated correction is converted to an analog voltage and then added to the open-loop control voltage derived from the receiver AGC signal and applied to the transmitter gain control circuits. A total dynamic range of closed-loop control is limited to ±24 dB. This is adequate to correct the combined errors already described. The loop is also fast enough with the 800 b/s control stream rate to keep up with most of the fast changing multipath induced Rayleigh fading.

The power control system has been provided with a number of setable parameters to adapt its use to different conditions. The open-loop control parameter is used primarily to adjust for different cell sizes and ERP. Primary closed-loop parameters include the power control step size and the dynamic range. Nominal step size is 1.0 dB. Note that every cell can have a different parameter set. The parameter values are communicated to each mobile by a broadcast message transmitted on the sync channel.

Other Benefits and Features

Besides controlling the mobile power received by each cell and, therefore, the self-interference environment, the power control system also provides an automatically adaptive anti-interference function. For example, suppose that a jammer transmitter suddenly starts transmission near a cell site. The immediate effect is that all of the mobiles' SNRs are degraded. The closed-loop power control system will respond by transmitting a series of power-up commands until the desired SNR is restored. Possibly, not all mobiles will be able to comply with the power increase commands because of already being at extreme range. These calls will be lost unless they can be handed off to another cell. This is facilitated by reducing the cell's transmit power when interference is present so as to shrink the cell and move the handoff boundary closer, thus accomplishing the desired objective. Usually, however, all mobiles will be able to continue without interruption of their calls. When the interference ceases, the mobiles will automatically be commanded to reduce power to the normal level.

A similar automatic adaption occurs when a cell has an unusually heavy demand compared to its neighbors. The power control system in the heavily loaded cell will command its mobiles to increase power. This will result in de-emphasizing the interference received from neighbor cells, allowing temporarily higher capacity in this cell. This will, of course, increase interference into the neighbor cells, but the premise was that a heavily loaded cell was surrounded by more lightly loaded cells and, thus, the added interference could be well tolerated. This flexible borrowing of neighboring cells' capacity has no counterpart in analog FM or in digital TDMA systems.

Another advantage of the power control system for handheld portable units is that battery drain is minimized by the power control system. Clearly, if high power need not be transmitted most of the time, then the corresponding battery drain need not be incurred. Note that CDMA phones average only 2 mW transmit power compared to analog FM of 600 mW.

Yet another capability of the power control system is in the use of adaptive SNR thresholds. Rather than fixing the SNR threshold at a constant value, the system allows the SNR to vary as a

CDMA Technology and the IS-95 North American Standard

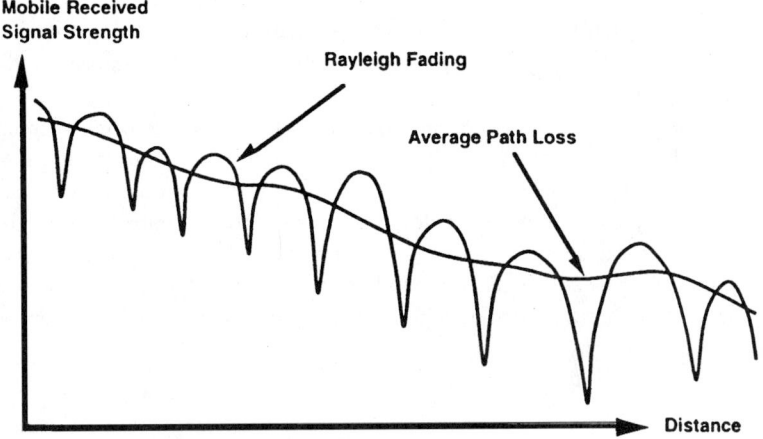

FIGURE 89.3 Mobile received signal strength in log-normal shadowing and Rayleigh fading.

function of system loading and of link quality. Consider that if the system load drops to 50% of maximum then up to 3 dB additional SNR could be allowed to every mobile. This much of an increase in a coded digital system such as CDMA will produce virtually perfect links.

It has also been noted that less SNR is required by mobiles that are motionless, or nearly so. This can be used to advantage by reducing SNR to those mobiles that are achieving a low-channel frame error rate, while increasing the SNR to those mobiles that are moving and, thus, achieving a higher frame error rate. This better balances the grade of service seen by the subscribers, while maximizing system capacity. Figure 89.3 shows a typical situation as a mobile drives away from the cell-site. Depicted are both log-normal shadowing and Rayleigh fading.

89.7 Low Transmit Power

Besides directly improving capacity, one of the more important results of reducing the required E_b/N_0 (signal-to-interference level) is the reduction of transmitter power required to overcome noise and interference. This reduction means that mobile stations also have reduced transmitter

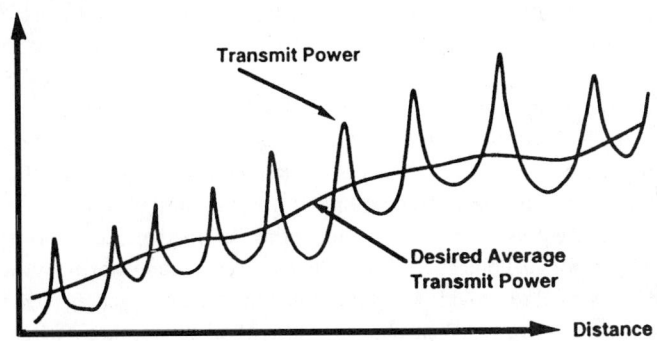

FIGURE 89.4 Transmit power, open-loop control only.

output requirements that reduce cost and allow lower power units to operate at larger ranges than the similarly powered analog or TDMA units. Furthermore, a reduced transmitter output requirement increases coverage and penetration and may also allow a reduction in the number of cells required for coverage. (See Figs. 89.4 and 89.5.)

An even greater gain is the reduction of average (rather than peak) transmitted power that is realized because of the power control used in CDMA. Most of the time, propagation conditions are benign. But because of occasional severe fading, narrowband systems must always transmit with enough power to override the occasional fades. CDMA uses power control to provide only the power required at the time it is actually needed, and thus reduces the average power by transmitting at high levels only during fades. In effect, the link margin in CDMA is kept in reserve ready for use when needed but when not needed does not contribute interference to other users or increase battery drain.

89.8 Privacy

The scrambled form of CDMA signals provides for a very high degree of privacy and makes this digital cellular system inherently more immune to cross talk, inexpensive scanning receivers, and air-time fraud. The standard includes the authentication and voice privacy features specified in EIA/TIA/IS-54-B even though the CDMA architecture inherently provides voice privacy and provisions for extended protection. The digital voice channel is, of course, amenable to direct encryption using Data encryption standards (DES) or other standard encryption techniques.

89.9 Mobile Station Assisted Soft Handoff

Soft handoff allows both the original cell and a new cell to temporarily serve the call during the handoff transition. The transition is from the original cell to both cells and then to the new cell. Not only does this greatly minimize the probability of a dropped call, but it also makes the handoff virtually undetectable by the user. In this regard, the analog system (and the digital TDMA-based systems) provides a break-before-make switching function, whereas the CDMA-based soft handoff system provides a make-before-break switching function.

After a call is initiated, the mobile station continues to scan the neighboring cells to determine if the signal from another cell becomes comparable to that of the original cell. When this happens, it indicates to the mobile station that the call has entered a new cell's coverage area and that a handoff can be initiated. The mobile station transmits a control message to the mobile switching center (MSC), which states that the new cell site is now strong and identifies the new cell site. The MSC initiates the handoff by establishing a link to the mobile station through the new cell while maintaining the old link. While the mobile station is located in the transition region between the two cell sites, the call is supported by communication through both cells; thereby eliminating the ping-ponging effect, or repeated requests to hand the call back and forth between two cell sites. The original cell site will only discontinue handling the call when the mobile station is firmly established in the new cell.

A soft handoff may persist for the entire duration of a call if the mobile remains in the overlapping coverage area of two cells. In this sense, it should be thought of more as a mode of operation than as a discrete event. Besides providing undetectable handoff, it also greatly improves the effectiveness of coverage in the difficult areas between base stations. In coverage design, sufficient coverage margin must be provided to cope with shadow fading. This is slow fading that occurs primarily because of signal path blockage by physical objects. It is usually modeled as a log-normal process with sigma of 8 dB. Coverage is designed to provide a certain probability of coverage, say 90%. To obtain 90% probability with a single log-normal, 8-dB sigma process, 10.25 dB of margin is required. But with soft handoff, there are two paths available, which means that if either path is adequate then the coverage will be satisfactory. This means that the margin can be reduced to about 4 dB, a very

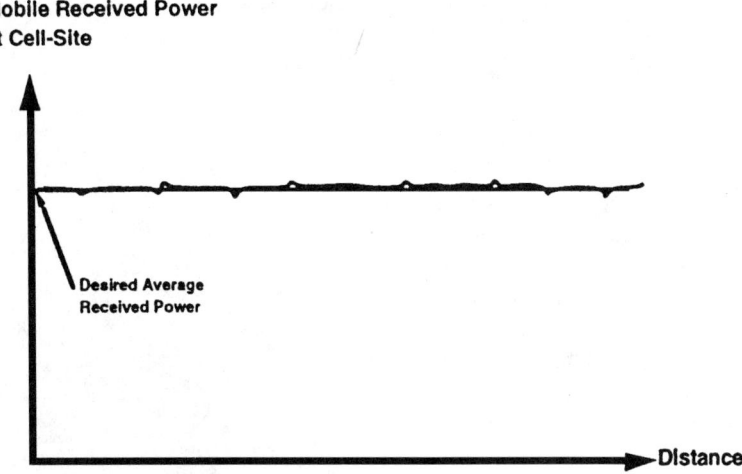

FIGURE 89.5 Mobile power received at cell site.

significant gain in coverage. This allows cells to be much larger in radius and greatly reduces the number of cells necessary to provide a certain quality of coverage.

89.10 Capacity

The primary parameters that determine CDMA digital cellular system capacity are processing gain, E_b/N_0 (with the required margin for fading), voice duty cycle, frequency reuse efficiency, and the number of sectors in the cell site antenna. Additionally, for a given blocking probability, the larger number of voice circuits provided by CDMA results in a significant increase in trunking efficiency, which serves a larger number of subscribers per voice circuit.

For example, if a spread spectrum bandwidth of 1.25 MHz is utilized by mobile stations transmitting continuously at 9600 b/s and if the modulation and coding technique utilized requires an E_b/N_0 of 6 dB, then up to 32 mobile stations could transmit simultaneously, as long as they are each power controlled to provide equal received power at the receiving location. In a CDMA cellular system, this capacity is reduced by interference received from neighboring cells and increased by other factors.

In CDMA, frequency reuse efficiency is determined by the signal-to-interference ratio that results from all of the system users within range, instead of the users in any given cell. Since the total capacity becomes quite large, the statistics of all of the users are more important than those of a single user. The law of large numbers can be said to apply. This means that the net interference to any given signal is the average of all of the users' received power times the number of users. As long as the ratio of received signal power to the average noise power density is greater than a threshold value, the channel will provide an acceptable signal quality. With TDMA and FDMA, interference is governed by a law of small numbers in which worst-case situations determine the percentage of time in which the desired signal quality will not be achieved.

Voice Activity Detection

In a typical full duplex two-way voice conversation, the duty cycle of each voice is less than 35%. It is difficult to exploit the voice activity factor in either FDMA or TDMA systems because of the time delay associated with reassigning the channel resource during the speech pauses. With CDMA, it is possible to reduce the transmission rate when there is no speech, and thereby substantially reduce interference to other users. Since the level of other user interference directly determines capacity, the capacity is increased by approximately a factor of two. This also reduces average mobile station transmit power requirements by approximately a factor of two.

Frequency Reuse

In CDMA, the wideband channel is reused in every cell. The total interference at the cell site to a given inbound mobile station signal is comprised of interference from other mobile stations in the same cell plus interference from mobile stations in neighboring cells. The frequency reuse efficiency of omnidirectional cells, defined as the ratio of interference from mobile stations within a cell to the total interference from all cells, is about 65%. Figure 89.6 shows, in an idealized geometry, the distribution of this interference from neighboring cells. Each cell in the first tier contributes about 6% of the total interference and so the entire first tier contributes an average of six times 6% or 36%; cells in the second and greater tiers contribute less than 4%.

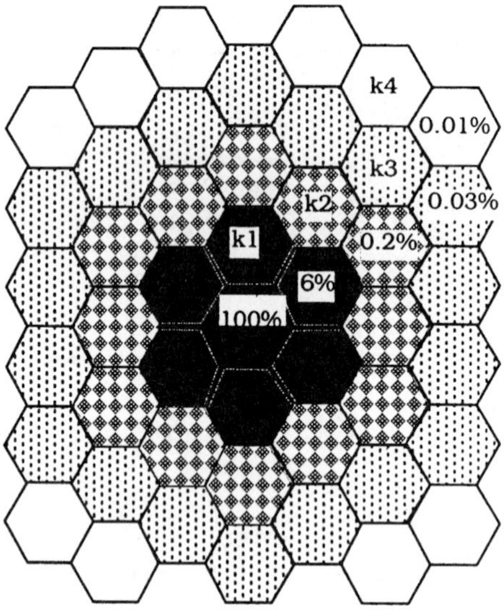

FIGURE 89.6 Interference contributions from neighboring cells. Significant contribution to interference seen by a cell as a result of activity in other cells is limited to the first tier of cells surrounding the cell.

Sectorization

Sectorization of CDMA cells is far more effective in increasing capacity than it is for narrowband FDMA and TDMA systems. Cells are typically sectored three ways, with 120°-beamwidth antennas. Because of the broad antenna patterns and frequent propagation anomalies, the coverage areas of these antennas overlap considerably. The isolation from one sector to another cannot be counted on to ensure good separation of narrowband signals. Frequency planning must take into account worst-case interference, which is only slightly reduced by the sectorization. The cellwise frequency reuse obtained in narrowband systems, thus, does not change when the cells are sectorized. A typical seven way reuse pattern remains seven way over all three sectors. That is, viewed sector-wise, it becomes 21-way reuse: each AMPS channel is reused in every 21st sector. The six channels that an AMPS omnidirectional cell has available in 1.25 MHz become two channels per sector in a three-way sectored cell.

Sector overlap in CDMA also has a deleterious effect due to interference. But the interference affects planning through its average value, not its worst-case, single station value. Therein lies a substantial advantage for CDMA. The increase in average interference due to sector overlap is small. Moreover it is partially mitigated by gain due to soft handoff between sectors. The result is that N-way sectorization of CDMA cells increases capacity nearly N-fold. Each sector has approximately the same capacity regardless of the degree of sectorization. Sectorization in AMPS, by contrast, entails little or no increase in capacity, only an increase in range.

Low E_b/N_0 (or C/I) and Error Protection

E_b/N_0 is the ratio of energy per bit to the noise power spectral density and is the standard figure of merit by which digital modulation and coding schemes are compared. It is directly analogous to the carrier-to-noise ratio C/N for analog FM modulation. Because of the wide channel bandwidth employed in the CDMA system, it is possible to use extremely powerful, high-redundancy error correction coding techniques. With narrowband digital modulation techniques, a much higher E_b/N_0 is required compared to CDMA because less powerful, low-redundancy error correction codes must be used to conserve channel bandwidth. The CDMA system employs the most powerful combination of forward error correction coding ever used in a mobile radio system together with an extremely efficient digital demodulator in its implementation of the CDMA digital cellular system. The lower E_b/N_0 increases capacity, increases cell radius of coverage, and decreases transmitter output power requirements.

Coding is provided for all transmitted bits not just for the most important bits as in other systems. Additionally, each frame is protected by a cyclic redundancy check code (CRCC) that allows data frames in error to be identified so that the vocoder can mitigate the effects of errored frames. The result is that isolated errored frames are essentially not noticeable by the user.

Effect of Thermal Noise

At each base station receiver, noise comes from two primary sources, thermal noise and mutual interference. The base station receiver power control system will adjust each mobile's transmitted power so that the desire SNR is obtained. As users are added, the interference to all users increases, and to maintain the desired SNR, the mobiles' power must be increased slightly. In the limit, one more mobile cannot be added without causing all mobiles' power to increase without bound.

In practice, the system is operated at a maximum capacity loading such that about half of the total receiver noise is from mutual interference and half is from thermal noise. Another way of looking at it is that we dedicate half of the theoretic capacity to the maintenance of stability in the power control system and to prevent excess mobile power from being used.

Net Capacity Calculation

The net capacity of the system can be calculated from all of the discussed considerations and Eq. (89.1). Taking F to be the fraction of same-sector interference and v to be the voice activity factor, the total interference disregarding thermal noise is

$$J = (N-1)SvF^{-1} \tag{89.2}$$

From Eq. (89.1) the system capacity limit per sector is approximately

$$N_p \approx \frac{W/R_b}{E_b/N_0} \frac{F}{v} \tag{89.3}$$

For each 1.25-MHz bandwidth CDMA channel, and 9600 b/s voice coding, the processing gain W/R_b is a factor of 128. This is reduced by the required E_b/N_0, a factor of 4 (6 dB), and by the neighbor cell interference, an additional factor of $F^{-1} \approx 3/2$. The voice activity gain contributes a factor of $v^{-1} \approx 2$, yielding $N_p \approx 128/3$ calls per sector. Unlike narrowband systems, antenna sectorization increases cell capacity in approximate proportion to the number of sectors, usually taken to be 3. The theoretical maximum capacity is, thus, about 128 calls per 1.25-MHz channel per cell.

For a practical system the power rise over thermal must be limited. From Eq. (89.1) it is readily shown that the relationship between rise over thermal is related to loading by

$$\frac{S_{\text{total}} + N_0 W}{N_0 W} \approx \frac{1}{1 - N/N_p} \tag{89.4}$$

A reasonable practical limit is a twofold increase, which by Eq. (89.4) reduces the capacity to $0.5N_p \approx 64/3$ calls per 1.25 MHz channel per sector, or 64 calls per cell. This is greater than ten times the capacity of the analog AMPS system, which provides only 6 calls per 1.25 MHz per cell. When blocking effects are considered, the capacity advantage widens significantly.

The calculated capacity has been verified in many, many field trials conducted all over the world. Never has the capacity been less than the desired minimum factor of ten improvement over AMPS.

89.11 Soft Capacity

In the present U.S. cellular environment, the FCC has allocated 25 MHz of spectrum that is equally split between two system operators in each service area. The spectrum is further divided between the cells, with a maximum of 57 analog FM channels in a three sector cell site. When demand for service is at a peak, the 58th caller in a given cell must be given a system busy signal. There is no way to add even one more signal to a fully occupied system. This call blocking behavior results in about a 35% loss of capacity. With the CDMA system, however, there is a much softer relationship between the number of users and the grade of service. For example, the system operator could decide to allow a small degradation in the bit error rate and increase the number of available channels during peak hours.

This capability is especially important for avoiding dropped calls at handoff because of a lack of channels. In the analog system and in digital TDMA, if a channel is not available, the call must be reassigned to a second candidate or it will be dropped at the handoff. With CDMA, however, the call can be accommodated if it is acceptable to slightly raise the users bit error rates until another call is completed.

It is also possible to offer a higher grade of service (at a higher cost to the user) where the high-grade user would obtain a larger fraction of the available power (capacity) than the low-grade user. Handoffs for high-grade users can be given priority over those for other users.

89.12 Transition to CDMA

In the initial introduction of CDMA service, a band segment of approximately 1.25 MHz is occupied by the CDMA operation because the spread-spectrum modulation requires this minimum bandwidth. An additional guard band of about 600 kHz must be provided for the first CDMA channel. The total represents only 15% of the present FDMA/FM system capacity, or about two analog FM channels per sector in a three sector cell. In return, however, the introductory single channel CDMA system allows up to 20 calls per sector.

Initially, a set of cells capable of covering the entire geographic area will be identified and equipped with omnidirectional or multisectored CDMA cell site equipment. This should be far fewer cells than required by the existing FM system. Although only the selected cells are equipped with CDMA cell site equipment, the 1.25-MHz segment is cleared out in all cells in the local area (i.e., the area of coverage), to prevent mutual interference between the FM and CDMA segments of the system.

As demand for CDMA service grows, additional omnidirectional cell sites are added and existing omnidirectional cells are converted to multisectored cells to increase capacity or improve coverage in the more difficult areas. Frequency planning is not necessary to support the change as additional cells are added or converted by sectorization; as demand for CDMA service grows beyond the capacity provided by the initial service, an additional 10% of the band segments can be removed from analog service and dedicated to the CDMA service. Each 1.25-MHz band segment requires an additional RF chain and power amplifier per sector. Additional modems are required to support the new channels.

89.13 Overview of the IS-95 CDMA Standard

The common air interface standard, IS-95-A, prescribes, in considerable detail, the behavior of dual-mode CDMA/AMPS subscriber stations, and, to a lesser extent, the base stations. Use of compliant subscriber stations assures base station manufacturers of predictable behavior on which

TABLE 89.1 IS-95 CDMA Standard Documentation

Section	Title	Topic
1	General	Defines the terms and numeric indications used in the document, including the time reference used in CDMA systems and the tolerances used throughout.
2	Requirements for mobile station analog operation	Describes the operation of the CDMA-analog dual-mode mobile stations operating in analog mode.
3	Requirements for base station analog operation	Describes the requirements for analog base stations.
4	Requirements for mobile station analog options	Describes the requirements for CDMA-analog dual-mode mobile stations which use the 32-digit dialing option on the reverse analog control channel and also describes mobile station requirements for use of the optional extended protocol.
5	Requirements for base station analog options	Describes the base station requirements for CDMA-analog dual-mode mobile stations operating the CDMA mode.
6	Requirements for mobile station CDMA operation	Describes the requirements for CDMA-analog dual-mode mobile stations operating in the CDMA mode.
7	Requirements for base station CDMA operation	Describes the requirements for CDMA base stations.

TABLE 89.2 Appendices to IS-95 CDMA Standard

Appendix	Title	Topic
A	Message encryption and voice privacy	Describes the requirements for message encryption and voice privacy. Note: This appendix is governed by U.S. International Traffic and Arms Regulation (TIARA) and the Export Administration Regulations.
B	CDMA call flow examples	Provides examples of simple call flow in the CDMA system.
C	CDMA system layering	Describes the layers of the CDMA system: physical layer, link layer, multiplex sublayer and the control process layer.
D	CDMA constants	Values for the constant identifiers found in Secs. 6 and 7.
E	CDMA retrievable and setable parameters	Describes the mobile station parameters that the base station can set and retrieve.
F	Mobile station database	Database model that can be used for dual-mode mobile stations complying with this document.
G	Bibliography	This appendix is not considered part of the standard but lists documents that may be useful in implementing the standard.

to base system designs. An IS-95-A compliant subscriber station can obtain service by communicating with either an AMPS (analog FM) base station or with a CDMA base station. System selection depends on the availability of either system in the geographic area of the station, as well as its programmed preference. Minimum performance requirements for dual- mode base stations, IS-97, and subscriber stations, IS-98, supplement the basic air interface.

IS-95-A emphasizes subscriber station requirements because the subscriber side reflects all call processing features, and because the specification is simpler in the context of a single user. In contrast to the very detailed subscriber station requirements, base station requirements are sketchy and incomplete. Generally the standard prescribes only those base station requirements that are important for the design of subscriber stations, leaving unspecified behavior to the discretion of the vendors. Base stations are fielded in much smaller quantities, but at much greater cost per unit. Marketplace considerations therefore incentivize good designs. Minimal requirements facilitate innovation in those designs.

The standard specifies that mobile stations operating with analog base stations meet the analog compatibility provisions for mobile stations as specified in EIA/TIA/IS-54-B, *Dual-Mode Mobile Station–Base Station Compatibility Specification*, January 1992. The incorporation of the analog portions of EIA/TIA/IS-54-B instead of EIA/TIA-553 (*Mobile Station–Land Station Compatibility Specification*, September 1989) accommodates all the changes to analog operation imposed by the EIA/TIA/IS-54-B dual-mode standard. (See Tables 89.1 and 89.2.)

Defining Terms

Access channel: A reverse CDMA channel used by subscriber stations for communicating to the base station. The access channel is used for short signalling message exchanges, such as call originations, responses to pages, and registrations.

CDMA channel: The set of channels transmitted between the base station and the subscriber stations within a given CDMA frequency assignment. See also forward CDMA channel and reverse CDMA channel.

Code channel: A subchannel of a forward CDMA channel. A forward CDMA channel contains 64 code channels. Code channel zero is assigned to the pilot channel. Code channels 1–7 may be assigned to the either paging channels or the traffic channels. Code channel 32 may be assigned to either a sync channel or a traffic channel. The remaining code channels may be assigned to traffic channels.

Code division multiple access (CDMA): A technique for spread-spectrum multiple-access digital communications that creates channels through the use of unique code sequences.

Forward CDMA channel: A CDMA channel from a base station to subscriber stations. The forward CDMA channel contains one or more code channels that are transmitted on a CDMA frequency assignment using a particular pilot time offset. The code channels are associated with the pilot channel, sync channel, paging channels, and traffic channels.

Forward traffic channel: A code channel used to transport user and signalling traffic from the base station to the subscriber station.

Handoff: The act of transferring communication with a subscriber station from one base station to another. Hard handoff is characterized by temporary disconnection of the traffic channel. Soft handoff is characterized by simultaneous communication with a subscriber by more than one base station.

Paging: The act of seeking a subscriber station when a call has been placed to that subscriber station.

Paging channel: A code channel in a forward CDMA channel used for transmission of control information and pages from a base station to a subscriber station.

Pilot channel: An unmodulated, direct-sequence spread-spectrum signal transmitted continuously by each CDMA base station. The pilot channel allows a subscriber station to acquire the timing of the forward CDMA channel, provides a phase reference for coherent demodulation, and provides a means for signal strength comparisons between base stations for determining when to handoff.

Reverse CDMA channel: The CDMA channel from the subscriber station to the base station. From the base stations perspective, the reverse CDMA channel is the sum of all subscriber station transmissions on a CDMA frequency assignment.

Reverse link power control: Process ensuring that all subscriber signals arrive at a base station at their setpoint powers.

Reverse traffic channel: A reverse CDMA channel used to transport user and signalling traffic from a single subscriber station to one or more base stations.

Sync channel: Code channel 32 in the forward CDMA channel that transports the synchronization message to the subscriber station.

Traffic channel: A communication path between a subscriber station and a base station used for user and signalling traffic. The term traffic channel implies a forward traffic channel and reverse traffic channel pair. See also forward traffic channel and reverse traffic channel.

References

Bello, P.L. 1963. Characterization of randomly time-variant linear channels. *IEEE Trans. Comm. Syst.* CS-11:360–393.

Gilhousen, K.S., Jacobs, I.M., Padovani, R., Viterbi, A.J., Weaver, L.A., and Wheatley, C. A. 1991. On the capacity of a cellular CDMA system. *IEEE Trans. Veh. Tech.* VT-40(2):303–312.

Jakes, W.C., Jr. ed. 1974. *Microwave Mobile Communications*, John Wiley & Sons, New York.

Lee, W.C.Y. 1989. *Mobile Cellular Telecommunications Systems*, McGraw–Hill, New York.

Parsons, D. 1992. *The Mobile Radio Propagation Channel*, Wiley, New York.

Peterson, R.L., Ziemer, R.E., and Borth, D.E., *Introduction to Spread Spectrum Communications*, Prentice Hall, Englewood Cliffs, NJ.

Shannon, C.E. 1949. Communication in the presence of noise. *Proc. IEEE.* 37:10–21.

Simon, M.K., Omura, J.K., Scholtz, R.A., and Levitt, B.A. 1985. *Spread Spectrum Communications*, Vol. I, II, III. Computer Science Press, Rockville, MD.

Telecommunications Industry Association. 1995. Mobile Station-Base Station Compatibility Standard for Dual-Mode Wideband Spread Spectrum Cellular System. TIA/EIA/IS-95-A.

Turin, G.L. 1980. Introduction to spread spectrum antimultipath techniques and their application to urban digital radio. *Proc. IEEE.* 68:328–354.

Viterbi, A.J. 1995. *CDMA Principles of Spread Spectrum Communication*, Addison–Wesley, Reading, MA.

Viterbi, A.M. and Viterbi, A.J. 1993. Erlang capacity of a power controlled CDMA system. *IEEE J. on Selected Areas in Comm.* 11(6):892–900.

Viterbi, A.J., Viterbi, A.M., Gilhousen, K.S., and Zehavi, E. 1994. Soft handoff extends CDMA cell coverage and increases reverse link capacity. *IEEE J. on Selected Areas in Comm.* 12(8):1281–1288.

Viterbi, A.J., Viterbi, A.M., and Zehavi, E. 1993. Performance of power-controlled wideband terrestrial digital communication. *IEEE Trans. on Comm.* COM-41(4):559–569.

Further Information

Gilhousen, K.S., Jacobs, I.M., Padovani, R., Viterbi, A.J., Weaver, L.A., and Wheatley, C. A. 1991. On the capacity of a cellular CDMA system. *IEEE Trans. Veh. Tech.* VT-40(2):303–312.

Telecommunications Industry Association. 1995. Mobile Station-Base Station Compatibility Standard for Dual-Mode Wideband Spread Spectrum Cellular System. TIA/EIA/IS-95-A.

Turin, G.L. 1980. Introduction to spread spectrum antimultipath techniques and their application to urban digital radio. *Proc. IEEE.* 68:328–354.

Viterbi, A.J. 1995. *CDMA Principles of Spread Spectrum Communication*, Addison–Wesley, Reading, MA.

90
Japanese Cellular Standard

K. Kinoshita
NTT Do Co Mo

M. Nakagawa
Keio University

90.1 Cellular Market in Japan..1276
90.2 Standardization Process 1276
90.3 General...1277
   Scope of Application • Principle of Standardization
90.4 System Overview...1277
   Definition of the Interface • Services Provided by the System • Access method • Signalling System

90.1 Cellular Market in Japan

Recently amazing growth has been experienced in the analog market in Japan. Since 1988 this growth has been spurred by the competition provided by the **new common carrier (NCC)** cellular networks. As a result, the total number of cellular subscribers reached 1.3 million in 1992. It is presumed that growth will continue to accelerate, and the total is expected to reach 10 million by the year 2000. Therefore, the current analog system will soon meet saturation with the given frequency bands.

On the other hand, Japanese analog cellular telephone service is provided by two different systems, NTT systems and Japanese TACS systems. It is very important, however, that the terminals work anywhere and in any cellular network in Japan in order to make cellular telephones more popular. To realize roaming ability nationwide between several different cellular networks, standardizing air-interfaces between base stations and terminals is necessary.

90.2 Standardization Process

As just mentioned it is very important to standardize the air interface in developing digital systems. Therefore, in April 1989 the Ministry of Posts and Telecommunications (MPT) organized a research and study committee on digital cellular systems in Japan, and the committee started to study the technical requirements of the system. The Japanese digital cellular radio system committee, which was formed in July 1989 under the Telecommunications Technology Council organized by the consultative body of the minister of MPT, further studied the requirements after receiving the report from the research and study committee. The Japanese digital cellular radio system committee submitted the technical requirements for the Japanese digital cellular radio system to the MPT in June 1990. Under these requirements, radio equipment regulations provided by the MPT should be changed.

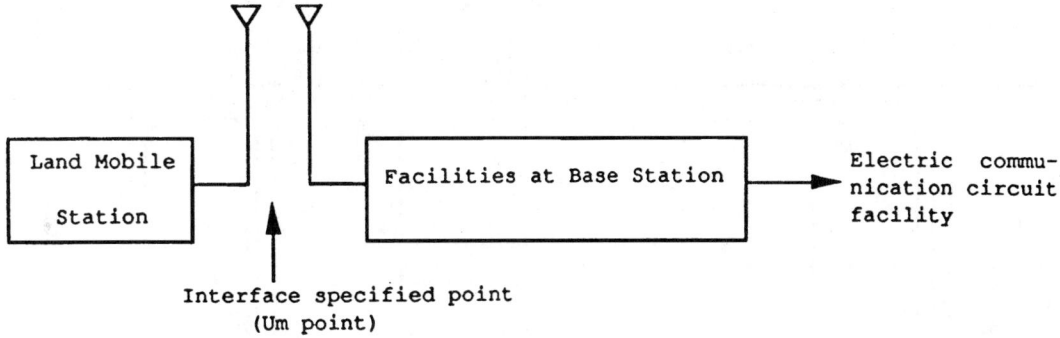

FIGURE 90.1 Structure of the digital cellular telecommunications system.

On the other hand, the Japanese Research and Development Center for Radio Systems (RCR) Standards Committee organized subcommittees in May 1990 for the standardization of Japanese digital cellular systems and for the treatment of intellectual proprietary rights (IPR) issues under the standards committee. The RCR standards committee is a private organization for standardization of Japanese radio systems comprising people from universities, users, telecommunication and consumer electronics manufacturers (including overseas companies), network operators, etc. This RCR standards committee draws up the detailed standard specifications for the connections between base stations and terminals in realizing the standard **Japanese digital cellular (PDC) system** based on the technical requirements.

90.3 General

Scope of Application

The digital cellular telecommunication system consists of land mobile stations (MS) and base station facilities as shown in Fig. 90.1. The Japanese standard regulates the radio interfaces and services for the digital cellular telecommunication system.

Principle of Standardization

In terms of mutual connectivity and compatibility, the Japanese standard defines the minimum level of specifications required for basic connections and services as the essential requirement and the minimum level of specifications required for whatever free choice is permitted, such as protocols, as optional standard to provide options and future expansion. Further, in order to provide options and future expansion capabilities as much as possible, care has been taken not to place restrictions on nonstandardized specifications.

Figure 90.2 outlines the relationship between standardized services and optional protocols used.

90.4 System Overview

Definition of the Interface

The points where the interface occurs in digital cellular telecommunication systems exist at the four locations, Um, R, S, and C, shown in Fig. 90.3. The definition of each interface point in Fig. 90.3 is as follows.

1. Interface Um: Interface between the MS and base station system. This interface shall conform to this standard.

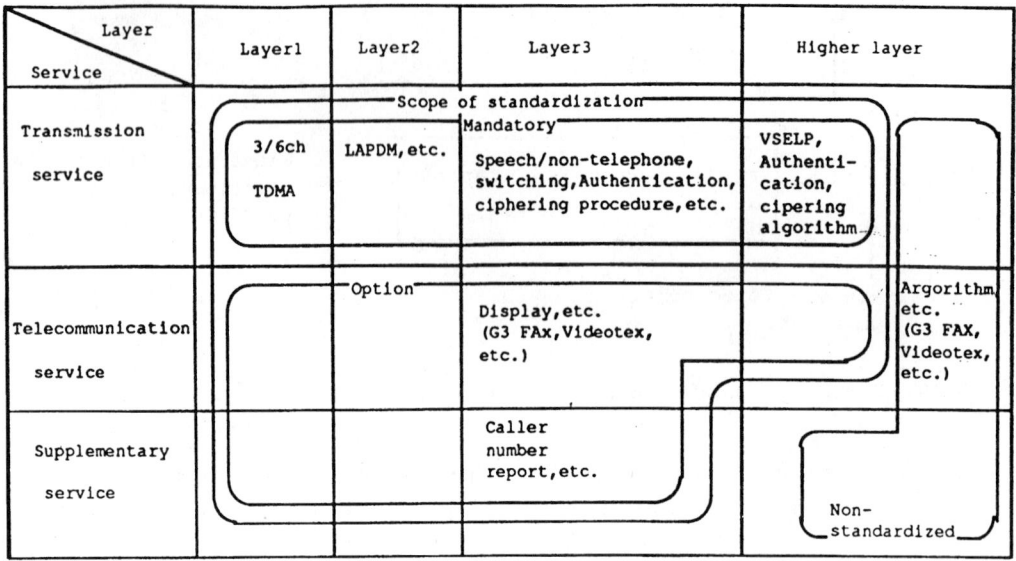

FIGURE 90.2 The relationship between services and protocols.

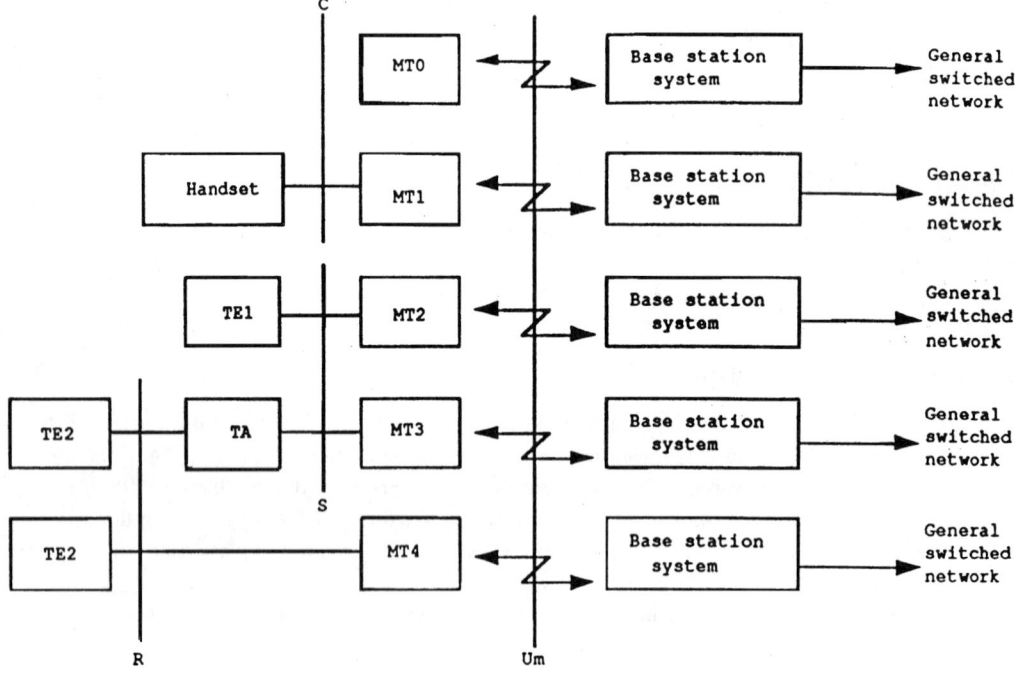

FIGURE 90.3 Interface points.

2. Interface R: Interface between non-I interface (I interface is conformed to integrated services digital network, ISDN) terminal equipment and either mobile terminal equipment or terminal adapters. This interface is not specified in this standard.
3. Interface S: Interface between either I interface terminal equipment or terminal adapters and the mobile terminal equipment. Conditions of this interface shall conform to the I interface standards.

4. Interface C: Interface between the mobile terminal equipment and handset within the MS. This interface is not specified in this standard.

Elements in Fig. 90.3 are defined as follows.

1. MT0: Mobile terminal equipment that contain man–machine interface facilities, such as terminals, as an integral part of it.
2. MT1: Mobile terminal equipment that can be connected with an external handset via interface C.
3. MT2, MT3: Mobile terminal equipment that offer a connection point with external terminal equipment by means of I interface standards.
4. MT4: Mobile terminal equipment that offer a connection point with external terminal equipment by means of non-I interface standards.
5. TE1: Terminal equipment conforming to I interface standards.
6. TE2: Terminal equipment not conforming to I Interface standards.
7. TA: Terminal adapters that perform conversation between the non-I interface condition and the I interface condition.

Services Provided by the System

Service Features

The services provided by Japanese digital cellular telecommunication system consist of several attributes listed in Table 90.1. For the service attribute consisting of more than one service item, one of those items can be selected.

Service Types

Bearer Services. The bearer services provided through the information transfer channel are listed in Table 90.2.

Teleservices. The teleservices provided via the information transfer channel are listed in Table 90.3.

Supplementary Services (Circuit Switched). Supplementary services that are provided for circuit switched service are listed in Tables 90.4–90.12. The numbers in the item column of Table 90.4–90.12 are the referenced CCITT recommendation numbers.

TABLE 90.1 Service Attributes

Service Attribute	Service Item
Information transfer capability	Speech Alternative speech or nontelephone Unrestricted digital information 3.1-KHz audio
Information transfer mode	Circuit Packet
Information transfer rate, kb/s	11.2 5.6
Frame structure, Hz	50 25
Communication configuration	Point to point Point to multipoint
Establishment of communication	Demand Reserved

TABLE 90.2 Bearer services

Item	Service
11.2-kb/s speech*	Transmission function suitable for speech communications. 11.2 kb/s VSELP codec is used. Bit transparency is not guaranteed.
5.6-kb/s speech*	Transmission function suitable for speech communications. 5.6 kb/s PSI-CELP codec is used. Bit transparency is not guaranteed.
8-kb/s unrestricted digital	Enables communications between a user terminal at MS with transfer rate of 8 kb/s and ISDN terminal using the 8 kb/s subrate.
64-kb/s unrestricted digital	Enables communications between a user terminal at MS with transfer rate of 64 kb/s and ISDN terminal.
Alternate 11.2-kb/s* speech/data	Supports voice and data. After a call is established, the transmission function of the network becomes switchable between 11.2-kb/s speech and 11.2-kb/s data according to requests by the user.
Alternate 5.6-kb/s* speech/data	Supports voice and data. After a call is established, the transmission function of the network becomes switchable between 5.6-kb/s speech and 11.2-kb/s data according to requests by the user.
Alternate 11.2-kb/s speech/8-kb/s unrestricted digital	Supports voice and 8-kb/s unrestricted digital information. After a call is established, the network's transmission function can be switched between 11.2-kb/s speech and 8-kb/s unrestricted digital information by the user.
Alternate 5.6-kb/s speech/8-kb/s unrestricted digital	Supports voice and 8-kb/s unrestricted digital information. After a call is established, the network's transmission function can be switched between 5.6-kb/s speech and 8-kb/s unrestricted digital information by the user.
Packet	ISDN packet service(X.31) and X.25 packet services are available.

*Items originally stipulated by Japanese standard (PDC system).

TABLE 90.3 Teleservices

Item	Service
G3 facsimile*	Enables communications between G3 fax terminals according to CCITT.T. 30 procedure.
G4 facsimile	Enables communications between G4 fax terminals.
Videotex	Video information transmission service using the Captain.
JUST-PC	Enables data communications between personal computers using the MPT recommended method.
JUST-MHS	Message handling service using a higher layer of MPT recommended method for PC data communications.
Modem (V.42 ANNEX)*	Enables data communications between personal computers using a modem, conforms to V.42 ANNEX.
Short message	Sends a message to a single user or broadcasts a message to multiple users and subsequently reports a reception acknowledgment to the party who transmitted the message.
Automobile location information service	Reports location of the specified MS to the user requesting such information.

*Items originally stipulated by Japanese standard (PDC system).

TABLE 90.4 Supplementary Services: Number Identification

Item	Service
Direct dial in (DDI) (I.251.1)	Enables a user to dial another user directly on an ISPBX or a private network.
Multiple subscriber number (MSN) (I.251.2)	Enables assigning multiple ISDN numbers to one interface.
Calling line identification presentation* (CLIP) (I.251.3)	Reports the calling user's number (including the subaddress if one exists) to the caller user.
Calling line identification restriction* (CLIR) (I.251.4)	Inhibits reporting of the calling user's number (including the subaddress if one exists) to the called user.

(*continues*)

Japanese Cellular Standard

TABLE 90.4 Supplementary Services: Number Identification (*continued*)

Item	Service
Connected line identification presentation (COLP) (I.251.5)	Reports the called user's number (including the subaddress if one exists) to the calling user.
Connected line identification restriction (COLR) (I.251.6)	Inhibits reporting of the called user's number (including the subaddress if one exists) to the calling user.
Malicious call identification (MCI) (I.251.7)	Allows the user to request the network to identify and memorize the information of the originator of the calls that are terminated by the user.
Subaddressing (SUB) (I.251.8)	Enables network to transmit the subaddress between users transparently.

*Items originally stipulated by Japanese standard (PDC system).

TABLE 90.5 Supplementary Services: Call Offering

Item	Services
Call transfer* (CT) (I.252.1)	Allows the user to transfer an active call to a third party. This service applies to both originating calls and terminating calls. It also differs from the call forwarding service that transfers a call from the called party before call establishment.
Call forwarding busy (CFB) (I.252.2)	Whereby a call is forwarded to another user when the called user is busy. The served user's originating service is unaffected.
Call forwarding no reply (CFNR) (I.252.3)	Whereby an unanswered call to a user is forwarded to another user. The served user's originating service is unaffected.
Call forwarding unconditional (CFU) (I.252.4)	Whereby the network forwards the call of a registered user to another user, regardless of the condition of the termination.
Call deflection* (CD) (I.252.5)	Upon receiving a call allows the user to choose if the call should be forwarded to another user or not.
Call forwarding no page response	Mobile communication specific service that forwards all incoming calls or incoming calls of specified basic service to another user when a paging response is not received.
Call forwarding not registered	Mobile communication specific service that forwards all incoming calls or incoming calls or specified basic service to another user when the location registration of the MS is not registered.
Call forwarding no radio resource	Mobile communication specific service that forwards all incoming calls or incoming calls of specified basic service to another user when the radio channel is congested.
Voice messaging function*	Function that on alerting, transfers the call to a voice messaging equipment to record a message instead of answering the call.
Line hunting (LH) (I.252.6)	Enables reception of a call by using a specific number of an interface featuring multiple channels and numbers.

*Items originally stipulated by this standard.

TABLE 90.6 Supplementary Services: Call Completion

Item	Service
Call waiting* (CW) (I.253.1)	Notifies the user of an incoming call on a call basic when no traffic channel is available.
Call hold (HOLD) (I.253.2)	Interrupts the existing call by setting it to the hold state. The held call may be reactivated if desired. After the call is interrupted, the traffic channel used for the call may be set on hold for use by newly incoming call.
Completion of calls to busy subscribers (CCBS) (I.253.3)	When the called user is busy, the network realerts the called user after it becomes idle. When the called user is available for the call, the network reports this to the originating user and may subsequently set up a call if necessary.

*Items originally stipulated by this standard (PDC system).

TABLE 90.7 Supplementary Services: Multiparty

Item	Service
Conference calling (CONF) (I.254.1)	Enables the user to communicate with several other users simultaneously.
Three-party service* (3PTY) (I.245.2)	Allows the user to hold the active call and make an additional call to the third user. It subsequently allows switching between the two calls, and/or the release of one call while maintaining the other. Optionally, this service enables the conference calling so that the three parties can talk simultaneously.

*Items originally stipulated by this standard (PDC system).

TABLE 90.8 Supplementary Services: Community of Interest

Item	Service
Closed user group (CUG) (I.255.1)	Enables users to form a group to/from which user access is restricted. One user can be a member of one or more CUGs. Generally, a member user of a CUG can only communicate with other users in the same CUG and cannot communicate with users outside the group. Specific CUG members are additionally allowed to originate calls outside the group or terminate calls from outside the group.
Private numbering plan (PNP) (I.255	Allows users to originate or terminate calls using user defined private numbers.

TABLE 90.9 Supplementary Services: Charging

Item	Service
Credit card calling (CRED) (I.256.1)	Puts call charges on a credit card account.
Advice of charge* (AOC) (I.256.2)	Advises the user of charging information on a call-by-call basis.
Reverse charging (REV) (I.256.3)	Puts call charges on the called party upon request by the originating party and at the consent of the called party.
Free phone	Puts call charge on the called party throughout the nation or a specified region when free phone number is dialed.

*Items originally stipulated by this standard (PDC system).

TABLE 90.10 Supplementary Services: Additional Information Transfer

Item	Service
User-to-user signalling (UUS) (I.257.1)	Allows the user to transfer the user-to-user information through the signalling channel in association with the call.

TABLE 90.11 Supplementary Services: Origination and Termination Restriction

Item	Service
Outgoing call barring*	Restricts outgoing calls based on the called party number or the location of the called terminal. This service can be set for all or for specified basic services; however, it does not restrict termination of a call and origination of an emergency call.
Incoming call barring	Restricts incoming calls. This service is set for all or for specified basic services. Outgoing calls from a terminal are not restricted.

*Items originally stipulated by this standard (PDC system).

Access method

Core parameters
See Table 90.13.

Time Division Multiple Access (TDMA) system
The three-channel multiplex TDMA system is used as the access method for the radio channel in the digital cellular telecommunication system. The six-channel multiplex TDMA system is an optional function for mobile stations whereas the three-channel multiplex TDMA system is a mandatory function (see Table 90.14).

Functional Structure of Radio Channel
The functional structure of radio channel is shown in Fig. 90.4.

Broadcast Channel (BCCH). BCCH is a unidirectional channel used by the base station system to broadcast the system control information related to location registration, channel structure, system state, etc., to land mobile stations.

TABLE 90.12 Supplementary Services: Other

Item	Service
Priority connection and channel hold (CH)	Allows the following operation by setting priority classes to an MS. 1. If an MS or a terminal of higher priority class originates a call when all radio CHs are busy, a radio channel used for an MS or a terminal of lower priority class is disconnected for subsequent use for higher priority class. 2. The radio channel used for the higher priority or important communication is held after such communications have been completed.

TABLE 90.13 Core Parameters

Frequency bands, MHz	Base station transmit frequency	810–826 1477–1501
	Mobile station transmit frequency	940–956 1429–1453
Send/receive distance, MHz		130/48
Error correction mode	CCH TCH	BCH Convolutional code
Zone composition		3 sector 4 iteration
Others		Equalizer (option)* Diversity (option)

*According to propagation experiments, the average delay spread in Tokyo area is 1 μs. Even in the Kofu area, which is supposed to get more delay spread, only about 10% of the area has a result of over 5-μs delay spread. It has been concluded that it is not always necessary to use an equalizer with diversity or adequate assignment of base stations.

TABLE 90.14 TDMA system parameters

Item	At full-rate codec	At half-rate codec
Multiplexed No. of channels	3	6
Carrier frequency separation, KHz	50 (25 interleave)	
Modulation system	$\pi/4$ shift QPSK (Rolloff factor = 0.5)	
Transmission rate, kb/s	42	
Info. bit rate, kb/s	11.2	5.6

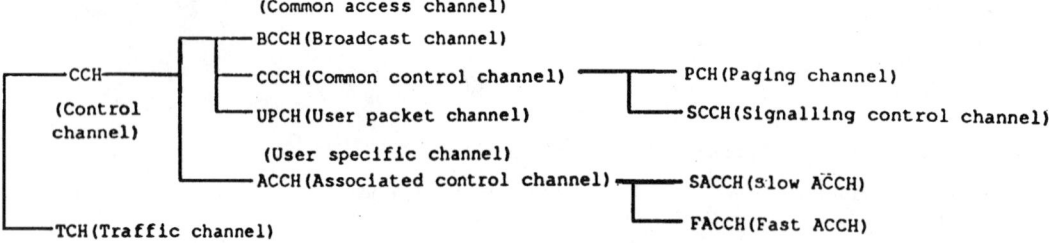

FIGURE 90.4 Functional structure of radio channels.

Common Control Channel (CCCH). CCCH is a bidirectional channel for transmitting signalling information. The following two types of CCCH exist.

1. Paging channel (PCH) is a point-to-multipoint unidirectional channel used for transmitting the common information from the base station system to mobile stations within a wide area, i.e., paging area, which is composed of multiple calls. It is used for paging and grouping control for intermittent reception by the MS.
2. Signalling control channel (SCCH) SCCH is a point-to-multipoint bidirectional channel used for transmitting information from/to the base station system to/from mobile stations when a cell area within which the mobile station is located is known to the base station system. A SCCH is prepared for transfer of the cell specific information by using different frequencies on a cell-by-cell basis. The uplink channel (from MS to base station system) is operating in the random access mode.

User Packet Channel (UPCH). UPCH is a point-to-multipoint bidirectional channel that transfers the control signal and user packet data. The uplink channel is operating in the random access mode.

Associated Control Channel (ACCH). ACCH is a point-to-point bidirectional channel associated with the TCH and is used for transferring signalling information and user packet data. The normal ACCH is called a slow ACCH (SACCH). In addition to the SACCH, there is a fast ACCH (FACCH), which is established by temporarily stealing the TCH to perform high-speed data transfer.

Traffic Channel (TCH). TCH is a point-to-point bidirectional channel that transfers the user information and its control signal. The TCH carries voice and facsimile signals.

Radio Circuit Control

Control Procedure. The control procedure is specified to enable the MS originating and terminating call connections, the location registration by mobile stations, the channel handover during a call, the service identification, etc. These controls shall be exactly performed by using commonly and independently assigned slots.

Slot Configuration. The slot configuration shall be in accordance with Fig. 90.5. The configuration is designed to meet the following requirements.

1. Uplink reception processing and downlink transmission processing can be carried out in a time sequential manner at a base station.
2. The base station, which enables collision control of the random access channel, can transmit a downlink signal after having confirmed the reception of uplink signal.
3. Duplexer at the MS can be simplified.
4. Antenna switching diversity can easily be implemented at the MS.

Japanese Cellular Standard

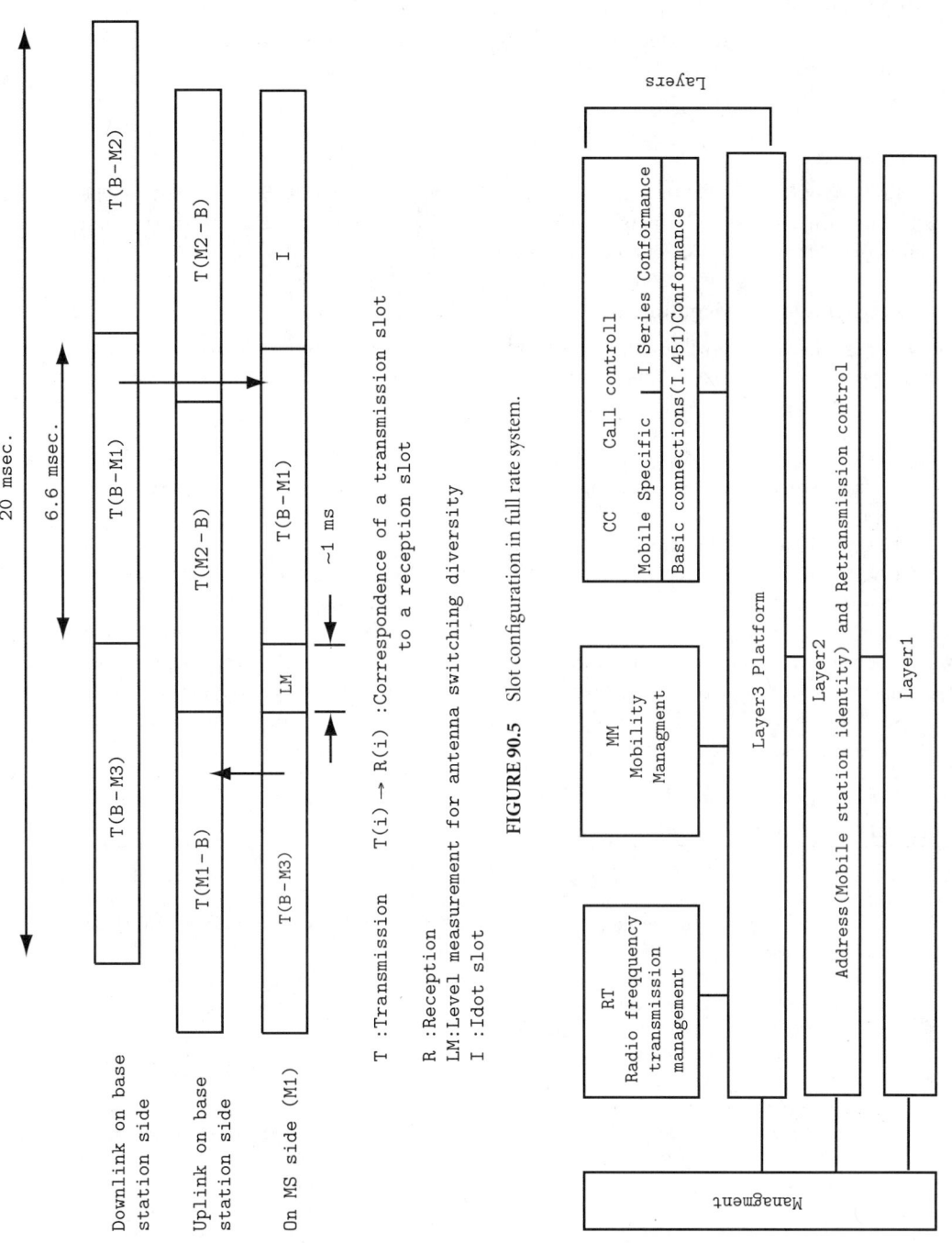

FIGURE 90.5 Slot configuration in full rate system.

FIGURE 90.6 Signalling system structure.

Signalling System

Signal System Structure

The signalling system structure of the digital cellular telecommunication system is shown in Fig. 90.6. The signalling system will have a layered structure made up of layers 1–3, which conforms to the OSI reference model.

Layered Structure

The definitions of layers 1–3 of the signalling system are shown in Table 90.15. The layer-3 will be classified into call control (CC), mobility management (MM), and radio frequency transmission (RT) functions according to CCIR and CCITT recommendations.

Characteristics of the Signalling System

The structure of layers 1–3 will feature an expandable design, while ensuring high serviceability, such as connection quality, etc., and system economy, i.e., signalling efficiency. The signalling format for each layer is depicted in Fig. 90.7.

1. Layer 1 assembles and disassembles layers 2 and 3 using the error correction and bit interleaving. Signals of layers 2–3, along with the preamble, synchronization (sync.) word, and supplementary information, form a slot.
2. Layer 2 will consist of address part and control part.
3. Layer 3 will feature a common platform used by CC, MM, and RT functions. This platform makes efficiency of signal transmission high, as well as shortening the time required for the service. For example, call origination by an MS, RT and CC of layer 3 will report radio frequency condition information and setup information, respectively, to the base station. The layer-3 common platform allows layers 1 and 2 to deal with information such as a signal so as to increase the efficiency of signal transmission and to shorten the service time.
4. Layer-3 messages will be configured as: (message) + (supplementary information)

TABLE 90.15 Definitions of Layers 1–3

Layer	Functional definitions		Examples
Layer 1	Ensures transmission of bit sequence using a communication circuit consisting of physical entities.		Radio signal transmission, frequency assignment, radio channel packet random access control, etc.
Layer 2	Located on layer 1 and offers highly reliable and transparent data transfer using the bit transmission function provided by layer 1.		Frame structure, procedural elements, data field, procedural specifications, etc.
Layer 3	Performs end-to-end data transfers between end system entities using the data transfer function provided by layer 2	RT	Specifies items related to radio frequency transmission control and performs establishment, maintenance, handover, etc., of the radio channel
		MM	Specifies items related to mobility management control, and performs location registration and authentication
		CC	Specifies items related to call connection control. Basic call connections will be in accordance with the CCITT I-recommendations. For supplementary services which require a large number of signals, mobile specific sequences of I-series recommendations, leading to reduction of time required for services.

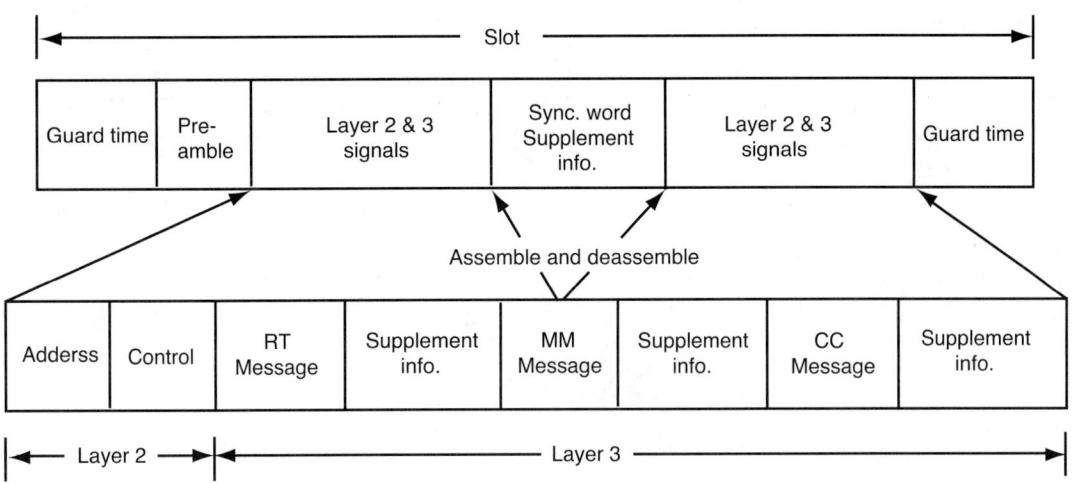

FIGURE 90.7 Signal format outline for layers 1–3.

5. MM and RT messages will have a fixed length, taking into account that these functions are rather limited.
6. CC will conform to the ISDN user-network interface layer 3 specifications (I.451), with emphasis on the harmonization with ISDN. The CC messages will be based on the I.451 format. The mobile specific format, however, will be used for coding of bearer capabilities, because the information transfer capability of a digital cellular telecommunication system differs from that of fixed networks. In addition, the caller number reporting is shortened from a full octet to 4 b.

Defining Terms

Common carrier: A company who possesses a big public telecommunication network, such as NTT.
Future public land mobile telecommunication system (FPLMTS): Japan will propose some systems including time division multiple access (TDMA) and code division multiple access (CDMA) for the future world standard microcellular system.
Japanese digital cellular system (JDC): Operations began in 1993 under the standard explained in this chapter.
New common carrier (NCC): A company of common carriers founded after the privatization in 1980s such as IDO.
Personal digital cellular system (PDC): Formerly called JDC, named changed in 1994.
Personal handy phone system (PHS): Personal mobile telecommunications system using microcells starts its operation in 1995. This type of microcellular system will be the first one in the world.

References

Kuramoto, M., Kinosita, K., Nakajima, A., Utano, T., and Murase, A. 1993. Overall system performance in a digital cellular system based on Japanese standard. *43rd IEEE Vehicular Technology Conference*, pp. 168–171.

Research & Development Center for Radio Systems (RCR). Personal Digital Cellular Telecommunication System RCR Standard-27C.

Further Information

See "Overview of Wireless Personal Communications," by J.E. Padgett, C.G. Gunther, and T. Hattori, *IEEE Communications Magazine*, Jan. 1995, or K. Kinoshita et al., "Development of a TDMA Digital Cellular System Based on Japanese Standard," *Proc. 41st. IEEE Vehicular Technology Conf.* pp. 642–645, 1991.

91
The British Cordless Telephone Standard: CT-2

Lajos Hanzo
University of Southampton

91.1 History and Background 1289
91.2 The CT-2 Standard 1290
91.3 The Radio Interface 1291
Transmission Issues • Multiple Access and Burst Structure • Power Ramping, Guard Period, and Propagation Delay • Power Control
91.4 Burst Formats 1293
91.5 Signalling Layer Two (L2) 1295
General Message Format • Fixed Format Packet
91.6 CPP-Initiated Link Setup Procedures 1299
91.7 CFP-Initiated Link Setup Procedures 1299
91.8 Handshaking 1300
91.9 Main Features of the CT-2 System 1302

91.1 History and Background

Following a decade of world-wide research and development (R&D), cordless telephones (CT) are now becoming widespread consumer products, and they are paving the way towards ubiquitous, low-cost personal communications networks (PCN) [Tuttlebee ed., 1990; Steele ed., 1992]. The two most well-known European representatives of CTs are the digital European cordless telecommunications (DECT) system [Ochsner, 1990; Asghar, 1995] and the CT-2 system [Steedman, 1990; Gardiner, 1990]. Three potential application areas have been identified, namely, domestic, business, and public access, which is also often referred to as telepoint (TP).

In addition to conventional voice communications, CTs have been conceived with additional data services and local area network (LAN) applications in mind. The fundamental difference between conventional mobile radio systems and CT systems is that CTs have been designed for small to very small cells, where typically benign low-dispersion, dominant line-of-sight (LOS) propagation conditions prevail. Therefore, CTs can usually dispense with channel equalizers and complex low-rate speech codecs, since the low-signal dispersion allows for the employment of higher bit rates before the effect of channel dispersion becomes a limiting factor. On the same note, the LOS propagation scenario is associated with mild fading or near-constant received signal level, and when combined with appropriate small-cell power-budget design, it ensures a high average signal-to-noise ratio (SNR). These prerequisites facilitate the employment of high-rate, low-complexity speech codecs, which maintain a low battery drain. Furthermore, the deployment of forward error correction codecs can often also be avoided, which reduces both the bandwidth requirement and the power consumption of the portable station (PS).

A further difference between public land mobile radio (PLMR) systems [Hanzo, 1995] and CTs is that whereas the former endeavor to standardize virtually all system features, the latter seek to offer a so-called access technology, specifying the common air interface (CAI), access and signalling protocols, and some network architecture features, but leaving many other characteristics unspecified. By the same token, whereas PLMR systems typically have a rigid frequency allocation scheme and fixed cell structure, CTs use dynamic channel allocation (DCA) [Jabbari, 1995]. The DCA principle allows for a more intelligent and judicious channel assignment, where the base station (BS) and PS select an appropriate traffic channel on the basis of the prevailing traffic and channel quality conditions, thus minimizing, for example, the effect of cochannel interference or channel blocking probability.

In contrast to PLMR schemes, such as the Pan-European global system of mobile communications (GSM) system [Hanzo, 1995], CT systems typically dispense with sophisticated mobility management, which accounts for the bulk of the cost of PLMR call charges, although they may facilitate limited hand-over capabilities. Whereas in residential applications CTs are the extension of the public switched telephone network (PSTN), the concept of omitting mobility management functions, such as location update, etc., leads to telepoint CT applications where users are able to initiate but not to receive calls. This fact drastically reduces the network operating costs and, ultimately, the call charge at a concommittant reduction of the services rendered.

Having considered some of the fundamental differences between PLMR and CT systems let us now review the basic features of the the CT-2 system.

91.2 The CT-2 Standard

The European CT-2 recommendation has evolved from the British standard MPT-1375 with the aim of ensuring the compatibility of various manufacturers' systems as well as setting performance requirements, which would encourage the development of cost-efficient implementations. Further standardization objectives were to enable future evolution of the system, for example, by reserving signalling messages for future applications and to maintain a low PS complexity even at the expense of higher BS costs. The CT-2 or MPT 1375 CAI recommendation is constituted by the four following parts.

1. *Radio interface:* Standardizes the radio frequency (RF) parameters, such as legitimate channel frequencies, the modulation method, the transmitter power control, and the required receiver sensitivity as well as the carrier-to-interference ratio (CIR) and the time division duplex (TDD) multiple access scheme. Furthermore, the transmission burst and master/slave timing structures to be used are also laid down, along with the scrambling procedures to be applied.
2. *Signalling layers one and two:* Defines how the bandwidth is divided among signalling, traffic data, and synchronization information. The description of the first signalling layer includes the dynamic channel allocation strategy, calling channel detection, as well as link setup and establishment algorithms. The second layer is concerned with issues of various signalling message formats, as well as link establishment and re-establishment procedures.
3. *Signalling layer three:* The third signalling layer description includes a range of message sequence diagrams as regards to call setup to telepoint BSs, private BSs, as well as the call clear down procedures.

4. *Speech coding and transmission:* The last part of the standard is concerned with the algorithmic and performance features of the audio path, including frequency responses, clipping, distortion, noise, and delay characteristics.

Having briefly reviewed the structure of the CT-2 recommendations let us now turn our attention to its main constituent parts and consider specific issues of the system's operation.

91.3 The Radio Interface

Transmission Issues

In our description of the system we will adopt the terminology used in the recommendation, where the PS is called cordless portable part (CPP), whereas the BS is referred to as cordless fixed part (CFP). The channel bandwidth and the channel spacing are 100 kHz, and the allocated system bandwidth is 40 MHz, which is hosted in the range of 864.15–868.15 MHz. Accordingly, a total of 40 RF channels can be utilized by the system.

The accuracy of the radio frequency must be maintained within ±10 kHz of its nominal value for both the CFP and CPP over the entire specified supply voltage and ambient temperature range. To counteract the maximum possible frequency drift of 20 kHz, automatic frequency correction (AFC) may be used in both the CFP and CPP receivers. The AFC may be allowed to control the transmission frequency of only the CPP, however, in order to prevent the misalignment of both transmission frequencies.

Binary frequency shift keying (FSK) is proposed, and the signal must be shaped by an approximately Gaussian filter in order to maintain the lowest possible frequency occupancy. The resulting scheme is referred to as Gaussian frequency shift keying (GFSK), which is closely related to Gaussian minimum shift keying (GMSK) [Steele ed., 1992] used in the DECT [Asghar, 1995] and GSM [Hanzo, 1995] systems.

Suffice to say that in M-arry FSK modems the carrier's frequency is modulated in accordance with the information to be transmitted, where the modulated signal is given by

$$S_i(t) = \sqrt{\frac{2E}{T}} \cos[\omega_i t + \Phi] \qquad i = 1, \ldots, M$$

and E represents the bit energy, T the signalling interval length, ω_i has M discrete values, whereas the phase Φ is constant.

Multiple Access and Burst Structure

The so-called TDD multiple access scheme is used, which is demonstrated in Fig. 91.1. The simple principle is to use the same radio frequency for both uplink and downlink transmissions between

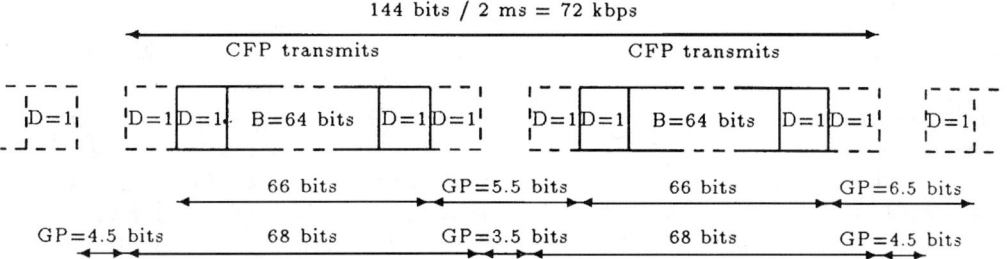

FIGURE 91.1 M1 burst and TDD frame structure.

the CPP and the CFP, respectively, but with a certain staggering in time. This figure reveals further details of the burst structure, indicating that 66 or 68 b per TDD frame are transmitted in both directions.

There is a 3.5- or 5.5-b duration guard period (GP) between the uplink and downlink transmissions, and half of the time the CPP (the other half of the time the CFP) is transmitting with the other part listening, accordingly. Although the guard period wastes some channel capacity, it allows a finite time for both the CPP and CFP for switching from tranmission to reception and vice versa. The burst structure of Fig. 91.1 is used during normal operation across an established link for the transmission of adaptive differential pulse code modulated (ADPCM) speech at 32 kb/s according to the CCITT G721 standard in a so-called B channel or bearer channel. The D channel, or signalling channel, is used for the transmission of link control signals. This specific burst structure is referred to as a multiplex one (M1) frame.

Since the speech signal is encoded according to the CCITT G721 recommendation at 32 kb/s the TDD bit rate must be in excess of 64 kb/s in order to be able to provide the idle guard space of 3.5- or 5.5-b interval duration plus some signalling capacity. This is how channel capacity is sacrificed to provide the GP. Therefore, the transmission bit rate is stipulated to be 72 kb/s and the transmission burst length is 2 ms, during which 144-b intervals can be accommodated. As it was demonstrated in Fig. 91.1, 66 or 68 b are transmitted in both the uplink and downlink burst, and taking into account the guard spaces, the total transmission frame is constituted by $(2 \cdot 68) + 3.5 + 4.5 = 144$ b or equivalently, by $(2 \cdot 66) + 5.5 + 4.5 = 144$ b. The 66-b transmission format is compulsory, whereas the 68-b format is optional. In the 66-b burst there is one D bit dedicated to signalling at both ends of the burst, whereas in the 68-b burst the two additional bits are also assigned to signalling. Accordingly, the signalling rate becomes 2 b/2 ms or 4 b/2 ms, corresponding to 1 kb/s or 2 kb/s signalling rates.

Power Ramping, Guard Period, and Propagation Delay

As mentioned before and suggested by Fig. 91.1, there is a 3.5- or 5.5-b interval duration GP between transmitted and received bursts. Since the signalling rate is 72 kb/s, the bit interval becomes about $1/(72 \text{ kb/s}) \approx 13.9$ μs and, hence, the GP duration is about 49 μs or 76 μs. This GP serves a number of purposes. Primarily, the GP allows the transmitter to ramp up and ramp down the transmitted signal level smoothly over a finite time interval at the beginning and end of the transmitted burst. This is necessary, because if the transmitted signal is toggled instantaneously, that is equivalent to multiplying the transmitted signal by a rectangular time-domain window function, which corresponds in the frequency domain to convolving the transmitted spectrum with a sinc function. This convolution would result in spectral side-lobes over a very wide frequency range, which would interfere with adjacent channels. Furthermore, due to the introduction of the guard period, both the CFP and CPP can tolerate a limited propagation delay, but the entire transmitted burst must arrive within the receivers' window, otherwise the last transmitted bits cannot be decoded.

Power Control

In order to minimize the battery drain and the cochannel interference load imposed upon cochannel users, the CT-2 system provides a power control option. The CPPs must be able to transmit at two different power levels, namely, either between 1 and 10 mW or at a level between 12 and 20 dB lower. The mechanism for invoking the lower CPP transmission level is based on the received signal level at the CFP. If the CFP detects a received signal strength more than 90 dB relative to 1 μV/m, it may instruct the CPP to drop its transmitted level by the specified 12–20 dB. Since the 90-dB gain factor corresponds to about a ratio of 31,623, this received signal strength would be equivalent for a 10-cm antenna length to an antenna output voltage of about 3.16 mV. A further beneficial ramification of using power control is that by powering down CPPs that are in the vicinity of a telepoint-type multiple-transceiver CFP, the CFP's receiver will not be so prone to being desensitised by the high-powered close-in CPPs, which would severely degrade the reception quality of more distant CPPs.

91.4 Burst Formats

As already mentioned in the previous section on the radio interface, there are three different subchannels assisting the operation of the CT-2 system, namely, the *voice/data channel* or *B channel*, the *signalling channel* or *D channel*, and the *burst synchronization channel* or *SYN channel*. According to the momentary system requirements, a variable fraction of the overall channel capacity or, equivalently, a variable fraction of the bandwidth can be allocated to any of these channels. Each different channel capacity or bandwidth allocation mode is associated with a different burst structure and accordingly bears a different name. The corresponding burst structures are termed as multiplex one (M1), multiplex two (M2), and multiplex three (M3), of which multiplex one used during the normal operation of established links has already been described in the previous section. Multiplex two and three will be extensively used during link setup and establishment in subsequent sections, as further details of the system's operation are unravelled.

Signalling layer one (L1) defines the burst formats multiplex one–three just mentioned, outlines the calling channel detection procedures, as well as link setup and establishment techniques. *Layer two (L2)* deals with issues of acknowledged and unacknowledged information transfer over the radio link, error detection and correction by retransmission, correct ordering of messages, and link maintenance aspects.

The burst structure multiplex two is shown in Fig. 91.2. It is constituted by two 16-b D-channel segments at both sides of the 10-b *preamble (P)* and the 24-b frame synchronization pattern (SYN), and its signalling capacity is 32 b/2 ms = 16 kb/s. Note that the M2 burst does not carry any B-channel information, it is dedicated to synchronization purposes. The 32-b D-channel message is split in two 16-b segments in order to prevent that any 24-b fraction of the 32-b word emulates the 24-b SYN segment, which would result in synchronization misalignment.

Since the CFP plays the role of the master in a telepoint scenario communicating with many CPPs, all of the CPP's actions must be synchronized to those of the CFP. Therefore, if the CPP attempts to initiate a call, the CFP will reinitiate it using the M2 burst, while imposing its own timing structure. The 10-b preamble consists of an alternate zero/one sequence and assists in the operation of the clock recovery circuitry, which has to be able to recover the clock frequency before the arrival of the SYN sequence, in order to be able to detect it. The SYN sequence is a unique word determined by computer search, which has a sharp autocorrelation peak, and its function is discussed later. The way the M2 and M3 burst formats are used for signalling purposes will be made explicit in our further discussions when considering the link setup procedures.

The specific SYN sequences used by the CFP and the CPP are shown in Table 91.1 along with the so-called *channel marker (CHM)* sequences used for synchronization purposes by the M3 burst format. Their differences will be made explicit during our further discourse. Observe from the table that the sequences used by the CFP and CPP, namely, SYNF, CHMF and SYNP, CHMP, respectively,

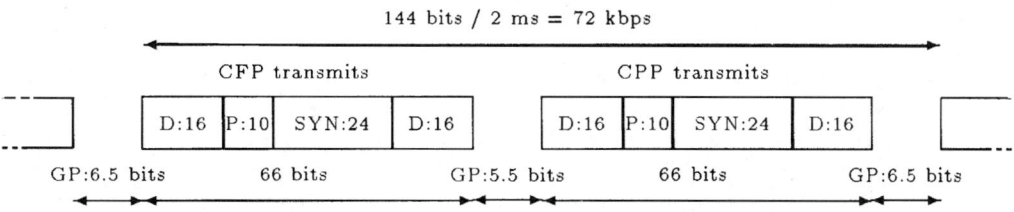

FIGURE 91.2 CT2 multiplex two burst structure.

TABLE 91.1 CT-2 Synchronization Patterns

	MSB (sent last)					LSB (sent first)
CHMF	1011	1110	0100	1110	0101	0000
CHMP	0100	0001	1011	0001	1010	1111
SYNCF	1110	1011	0001	1011	0000	0101
SYNCP	0001	0100	1110	0100	1111	1010

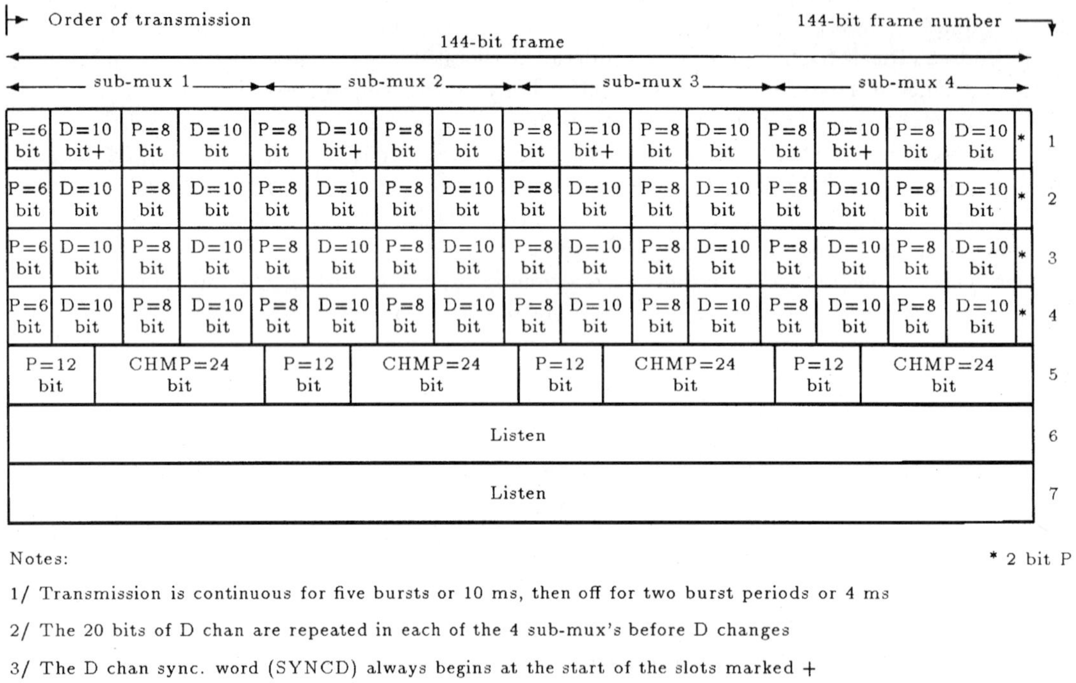

FIGURE 91.3 CT2 multiplex three burst structure.

are each other's bit-wise inverses. This was introduced in order to prevent CPPs and CFPs from calling each other directly. The CHM sequences are used, for instance, in residential applications, where the CFP can issue an M2 burst containig a 24-b CHMF sequence and a so-called poll message mapped on to the D-channel bits in order to wake up the specific CPP called. When the called CPP responds, the CFP changes the CHMF to SYNF in order to prevent waking up further CPPs unnecessarily.

Since the CT-2 system does not entail mobility functions, such as registration of visiting CPPs in other than their own home cells, in telepoint applications all calls must be initiated by the CPPs. Hence, in this scenario when the CPP attempts to set up a link, it uses the so-called multiplex three burst format displayed in Fig. 91.3. The design of the M3 burst reflects that the CPP initiating the call is oblivious of the timing structure of the potentially suitable target CFP, which can detect access attempts only during its receive window, but not while the CFP is transmitting. Therefore, the M3 format is rather complex at first sight, but it is well structured, as we will show in our further discussions. Observe in the figure that in the M3 format there are five consecutive 2-ms long 144-b transmitted bursts, followed by two idle frames, during which the CPP listens in order to determine whether its 24-b CHMP sequence has been detected and acknowledged by the CFP. This process can be followed by consulting Fig. 91.6, which will be described in depth after considering the detailed construction of the M3 burst.

The first four of the five 2-ms bursts are identical D-channel bursts, whereas the fifth one serves as a synchronization message and has a different construction. Observe, furthermore, that both the first four 144-b bursts as well as the fifth one contain four so-called submultiplex segments, each of which hosts a total of $(6 + 10 + 8 + 10 + 2) = 36$ b. In the first four 144-b bursts there are $(6 + 8 + 2) = 16$ one/zero clock-synchronizing P bits and $(10 + 10) = 20$ D bits or signalling bits. Since the D-channel message is constituted by two 10-b half-messages, the first half of the D-message is marked by the + sign in the figure. As mentioned in the context of M2, the D-channel bits are split in two halves and interspersed with the preamble segments in order to ensure that these bits do not emulate valid CHM sequences. Without splitting the D bits this could happen upon concatenating the one/zero P bits with the D bits, since the tail of the SYNF and SYNP sequences is also a one/zero segment. In the fifth 144-b M3 burst, each of the four submultiplex segments is constituted by 12 preamble bits and 24 CPP channel marker (CHMP) bits.

The four-fold submultiplex M3 structure ensures that irrespective of how the CFP's receive window is aligned with the CPP's transmission window, the CFP will be able to capture one of the four submultiplex segments of the fifth M3 burst, establish clock synchronization during the preamble, and lock on to the CHMP sequence. Once the CFP has successfully locked on to one of the CHMP words, the corresponding D-channel messages comprising the CPP identifier can be decoded. If the CPP identifier has been recognized, the CFP can attempt to reinitialize the link using its own master synchronization.

91.5 Signalling Layer Two (L2)

General Message Format

The signalling L2 is responsible for acknowledged and un-acknowledged information transfer over the air interface, error detection and correction by retransmission, as well as for the correct ordering of messages in the acknowledged mode. Its further functions are the link end-point identification and link maintenance for both CPP and CFP, as well as the definition of the L2 and L3 interface.

Compliance with the L2 specifications will ensure the adequate transport of messages between the terminals of an established link. The L2 recommendations, however, do not define the meaning of messages, this is specified by L3 messages, albeit some of the messages are undefined in order to accommodate future system improvements.

The L3 messages are broken down to a number of standard packets, each constituted by one or more codewords (CW), as shown in Fig. 91.4. The codewords have a standard length of eight octets, and each packet contains up to six codewords. The first codeword in a packet is the so-called address codeword (ACW) and the subsequent ones, if present, are data codewords (DCW). The first octet of the ACW of each packet contains a variety of parameters, of which the binary flag **L3_END** is indicated in Fig. 91.4, and it is set to zero in the last packet. If the L3 message transmitted is mapped onto more than one packet, the packets must be numbered up to N. The address codeword is always preceded by a 16-b D-channel frame synchronization word **SYNCD**. Furthermore, each eight-octet CW is protected by a 16-b parity-check word occupying its last two octets. The binary Bose–Chaudhuri–Hocquenghem BCH(63,48) code is used to encode the first six octets or 48 b by adding 15 parity b to yield 63 b. Then bit 7 of octet 8 is inverted and bit 8 of octet 8 added such that the 64-b codeword has an even parity. If there are no D-channel packets to send, a 3-octet idle message **IDLE_D** constituted by zero/one reversals is transmitted. The 8-octet format of the ACWs and DCWs is made explicit in Fig. 91.5, where the two parity check octets occupy octets 7 and 8. The first octet hosts a number of control bits. Specifically, bit 1 is set to logical one for an ACW and to zero for a DCW, whereas bit 2 represents the so-called format type FT bit. $FT = 1$ indicates that variable length packet format is used for the transfer of L3 messages, whereas $FT = 0$ implies that a fixed length link setup is used for link end point addressing end service requests. FT is only relevant to ACWs, and in DCWs it has to be set to one.

ACW
DCW1
- - - - -
DCW5
a/ L2 packet: Contains at least an ACW and 0...5 DCWs
Packet 1
Packet 2
- - - - -
Packet N
ACW
L3_END=1
ACW
L3_END=1
ACW
L3_END=0
b/ L3 message: N packets

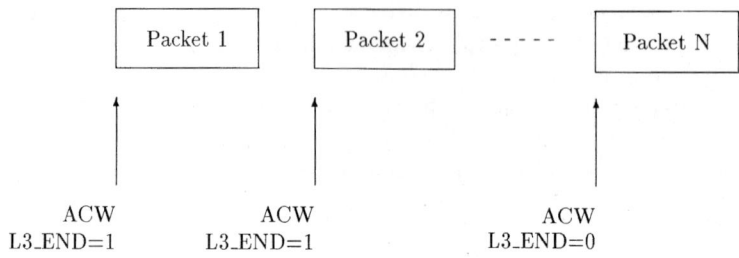

FIGURE 91.4 General L2 and L3 message format.

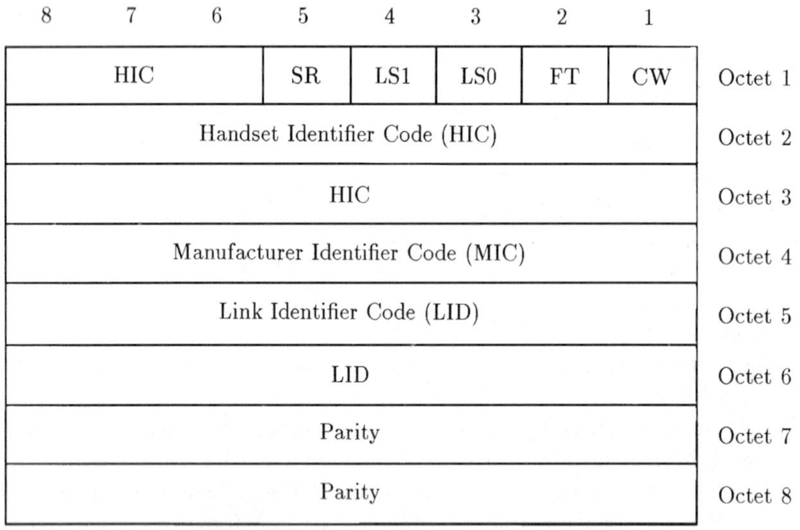

8
7
6
5
4
3
2
1
HIC
SR
LS1
LS0
FT
CW
Octet 1
Handset Identifier Code (HIC)
Octet 2
HIC
Octet 3
Manufacturer Identifier Code (MIC)
Octet 4
Link Identifier Code (LID)
Octet 5
LID
Octet 6
Parity
Octet 7
Parity
Octet 8

FIGURE 91.5 Fixed format packets mapped on M1, M2, and M3 during link initialization and on M1 and M2 during handshake.

Fixed Format Packet

As an example, let us focus our attention on the fixed format scenario associated with $FT = 0$. The corresponding codeword format defined for use in M1, M2, and M3 for link initiation and in M1 and M2 for handshaking is displayed in Fig. 91.5. Bits 1 and 2 have already been discussed, whereas the 2-bit link status (LS) field is used during call setup and handshaking. The encoding of the four possible LS messages is given in Table 91.2. The aim of these LS messages will

TABLE 91.2 Encoding of Link Status Messages

LS1	LS0	Message
0	0	Link_request
0	1	Link_grant
1	0	ID_OK
1	1	ID_lost

The British Cordless Telephone Standard: CT-2

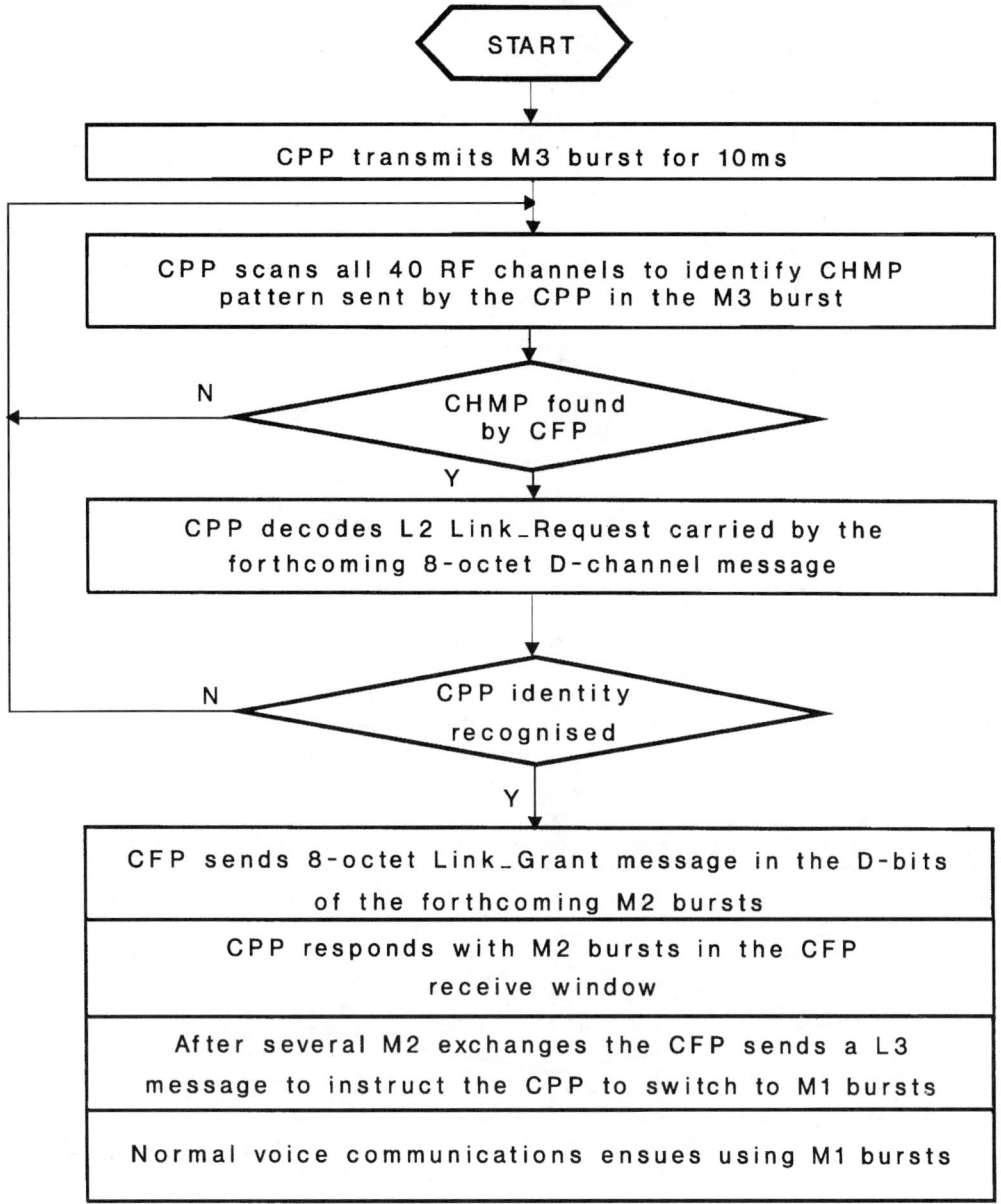

FIGURE 91.6 Flowchart of the CT-2 link initialization by the CPP.

become more explicit during our further discussions with reference to Fig. 91.6 and Fig. 91.7. Specifically, **link_request** is transmitted from the CPP to the CFP either in an M3 burst as the first packet during CPP-initiated call setup and link re-establishment, or returned as a poll response in an M2 burst from the CPP to the CFP, when the CPP is responding to a call. **Link_grant** is sent by the CFP in response to a link_request originating from the CPP. In octets 5 and 6 it hosts the so-called link identification (LID) code, which is used by the CPP, for example, to address a specific CFP or a requested service. The LID is also used to maintain link reference during handshake exchanges and link re-establishment. The two remaining link status handshake messages, namely, ID_OK and ID_lost, are used to report to the far end whether a positive confirmation of adequate link quality has been received within the required time-out period. These issues will be revisited during our further

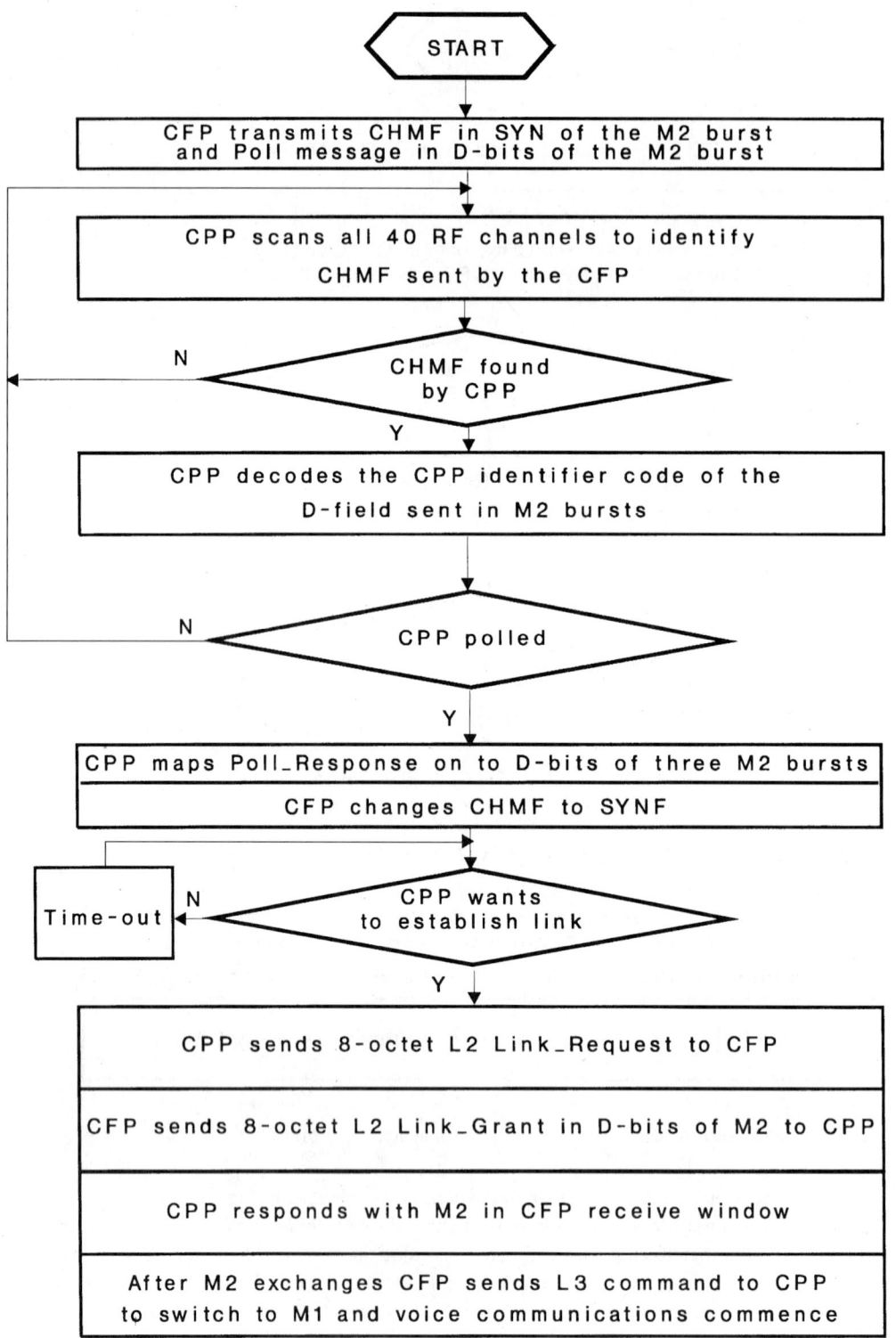

FIGURE 91.7 Flowchart of the CT-2 link initialization by the CFP.

elaborations. Returning to Fig. 91.5, we note that the fixed packet format ($FT = 0$) also contains a 19-b handset identification code (HIC) and an 8-b manufacturer identification code (MIC). The concatenated HIC and MIC fields jointly from the unique 27-b portable identity code (PIC), serving as a link end-point identifier. Lastly, we have to note that bit 5 of octet 1 represents the signalling rate (SR) request/response bit, which is used by the calling party to specify the choice of the 66- or 68-b M1 format. Specifically, $SR = 1$ represents the four bit/burst M1 signalling format. The first 6 octets are then protected by the parity check information contained in octets 7 and 8.

91.6 CPP-Initiated Link Setup Procedures

Calls can be initiated at both the CPP and CFP, and the call initation and detection procedures invoked depend on which party initiated the call. Let us first consider calling channel detection at the CFP, which ensues as follows. Under the instruction of the CFP control scheme, the RF synthesizer tunes to a legitimate RF channel and after a certain settling time commences reception. Upon receiving the M3 bursts from the CPP, the automatic gain control (AGC) circuitry adjusts its gain factor, and during the 12-b preamble in the fifth M3 burst, bit synchronization is established. This specific 144-b M3 burst, is transmitted every 14 ms, corresponding to every seventh 144-b burst. Now the CFP is ready to bit-synchronously correlate the received sequences with its locally stored CHMP word in order to identify any CHMP word arriving from the CPP. If no valid CHMP word is detected, the CFP may retune itself to the next legitimate RF channel, etc.

As mentioned, the call identification and link initialization process is shown in the flowchart of Fig. 91.6. If a valid 24-b CHMP word is identified, D-channel frame synchronization can take place using the 16-b SYNCD sequence and the next 8-octet L2 D-channel message delivering the link_request handshake portrayed earlier in Fig. 91.5 and Table 91.2 is decoded by the CFP. The required $16 + 64 = 80$ D bits are accommodated in this scenario by the $4 \cdot 20 = 80$ D bits of the next four 144-b bursts of the M3 structure, where the 20 D bits of the four submultiplex segments are transmitted four times within the same burst before the D message changes. If the decoded LID code of Fig. 91.5 is recognized by the CFP, the link may be reinitialized based on the master's timing information using the M2 burst associated with SYNF and containing the link_grant message addressed to the specific CPP identified by its PID.

Otherwise the CFP returns to its scanning mode and attempts to detect the next CHMP message. The reception of the CFP's 24-b SYNF segment embedded in the M2 message shown previously in Fig. 91.2 allows the CPP to identify the position of the CFP's transmit and receive windows and, hence, the CPP now can respond with another M2 burst within the receive window of the CFP. Following a number of M2 message exchanges, the CFP then sends a L3 message to instruct the CPP to switch to M1 bursts, which marks the commencement of normal voice communications and the end of the link setup session.

91.7 CFP-Initiated Link Setup Procedures

Similar procedures are followed when the CPP is being polled. The CFP transmits the 24-b CHMF words hosted by the 24-b SYN segment of the M2 burst shown in Fig. 91.2 in order to indicate that one or more CPPs are being paged. This process is displayed in the flowchart of Fig. 91.7, as well as in the timing diagram displayed in Fig. 91.8. The M2 D-channel messages convey the identifiers of the polled CPPs.

The CPPs keep scanning all 40 legitimate RF channels in order to pinpoint any 24-b CHMF words. Explicitly, the CPP control scheme notifies the RF synthesizer to retune to the next legitimate RF channel if no CHMF words have been found on the current one. The synthesizer needs a finite time to settle on the new center frequency and then starts receiving again. Observe in Fig. 91.8 that at this stage only the CFP is transmitting the M2 bursts; hence, the uplink-half of the 2-ms TDD frame is unused.

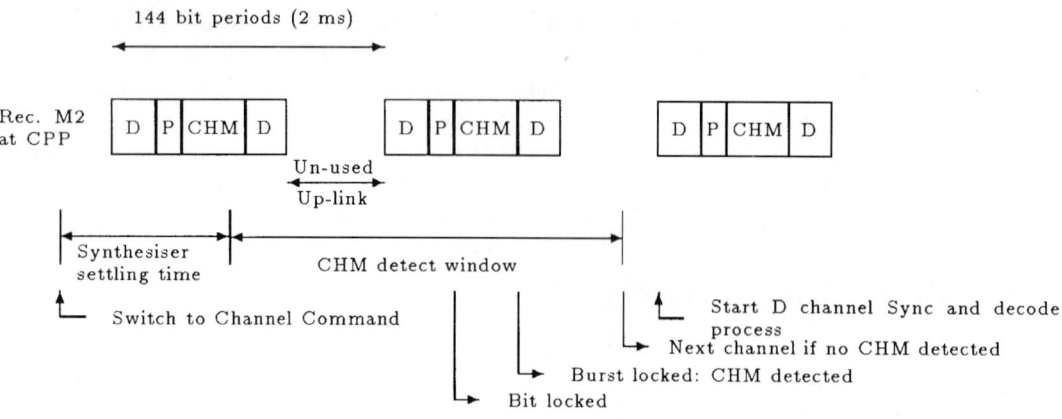

FIGURE 91.8 CT-2 call detection by the CPP.

Since the M2 burst commences with the D-channel bits arriving from the CFP, the CPP receiver's AGC will have to settle during this 16-b interval, which corresponds to about $16 \cdot 1/[72 \text{ kb/s}] \approx 0.22$ ms. Upon the arrival of the 10 alternating one–zero preamble bits, bit synchronization is established. Now the CPP is ready to detect the CHMF word using a simple correlator circuitry, which establishes the appropriate frame synchronization. If, however, no CHMF word is detected within the receive window, the synthesizer will be retuned to the next RF channel, and the same procedure is repeated, until a CHMF word is detected.

When a CHMF word is correctly decoded by the CPP, the CPP is now capable of frame and bit synchronously decoding the D-channel bits. Upon decoding the D-channel message of the M2 burst, the CPP identifier (ID) constituted by the LID and PID segments of Fig. 91.5 is detected and compared to the CPP's own ID in order to decide as to whether the call is for this specific CPP. If so, the CPP ID is reflected back to the CFP along with a SYNP word, which is included in the SYN segment of an uplink M2 burst. This channel scanning and retuning process continues until a legitimate incoming call is detected or the CPP intends to initiate a call.

More precisely, if the specific CPP in question is polled and its own ID is recognized, the CPP sends its poll_response message in three consecutive M2 bursts, since the capacity of a single M2 burst is 32 D bits only, while the handshake messages of Fig. 91.5 and Table 91.2 require 8 octets preceded by a 16-b SYNCD segment. If by this time all paged CPPs have responded, the CFP changes the CHMF word to a SYNF word, in order to prevent activating dormant CPPs who are not being paged. If any of the paged CPPs intends to set up the link, then it will change its poll_response to a L2 link_request message, in response to which the CFP will issue an M2 link_grant message, as seen in Fig. 91.7, and from now on the procedure is identical to that of the CPP-initiated link setup portrayed in Fig. 91.6.

91.8 Handshaking

Having established the link, voice communications is maintained using M1 bursts, and the link quality is monitored by sending handshaking (HS) signalling messages using the D-channel bits. The required frequency of the handshaking messages must be between once every 400 ms and 1000 ms. The CT-2 codewords ID_OK, ID_lost, link_request and link_grant of Table 91.2 all represent valid handshakes. When using M1 bursts, however, the transmission of these 8-octet messages using the 2- or 4-b/2ms D-channel segment must be spread over 16 or 32 M1 bursts, corresponding to 32 or 64 ms.

Let us now focus our attention on the *handshake protocol* shown in Fig. 91.9. Suppose that the CPP's handshake interval of $T_{htx\_p} = 0.4$ s since the start of the last transmitted handshake has

The British Cordless Telephone Standard: CT-2

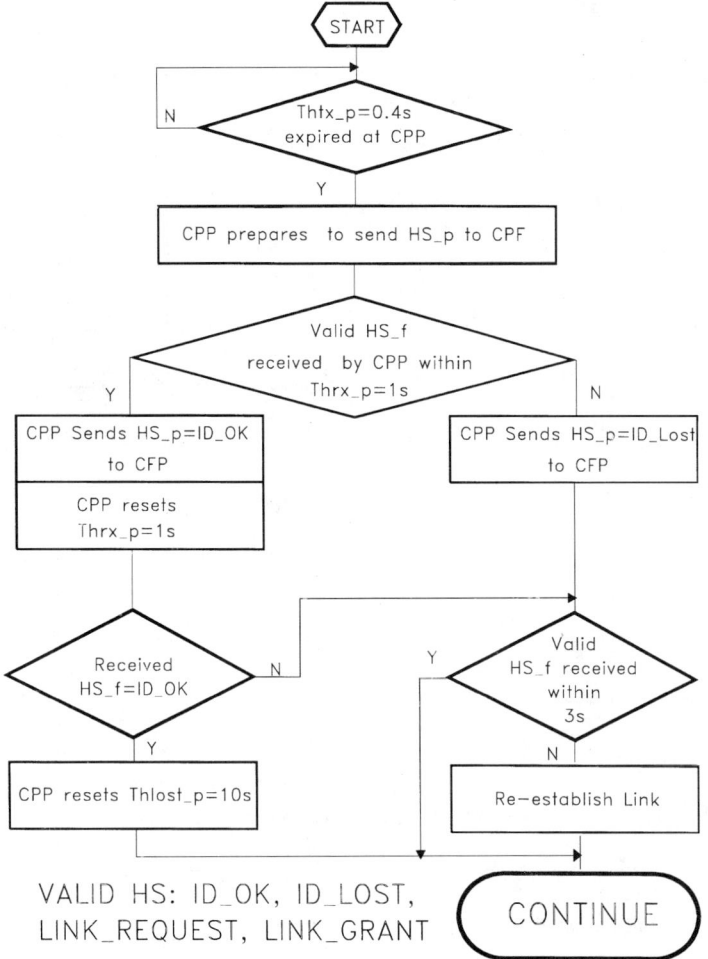

FIGURE 91.9 CT-2 handshake algorithms.

expired, and hence the CPP prepares to send a handshake message HS_p. If the CPP has received a valid HS_f message from the CFP within the last Thrx_p = 1s, the CPP sends an HS_p = ID_OK message to the CFP, otherwise an ID_Lost HS_p. Furthermore, if the valid handshake was HS_f = ID_OK, the CPP will reset its HS_f lost timer Thlost_p to 10 s. The CFP will maintain a 1-s timer referred to as Thrx_f, which is reset to its initial value upon the reception of a valid HS_p from the CPP.

The CFP's actions also follow the structure of Fig. 91.9 upon simply interchanging CPP with CFP and the descriptor _p with _f. If the Thrx_f = 1 s timer expires without the reception of a valid HS_p from the CPP, then the CFP will send its ID_Lost HS_f message to the CPP instead of the ID_OK message and will not reset the Thlost_f = 10 s timer. If, however, the CFP happens to detect a valid HS_p, which can be any of the ID_OK, ID_Lost, link_request and link_grant messages of Table 91.2, arriving from the CPP, the CFP will reset its Thrx_f = 1 s timer and resumes transmitting the ID_OK HS_f message instead of the ID_Lost. Should any of the HS messages go astray for more than 3 s, the CPP or the CFP may try and re-establish the link on the current or another RF channel. Again, although any of the ID_OK, ID_Lost, link_request and link_grant represent valid handshakes, only the reception of the ID_OK HS message is allowed to reset the Thlost = 10 s timer at both the CPP and CFP. If this timer expires, the link will be relinquished and the call dropped.

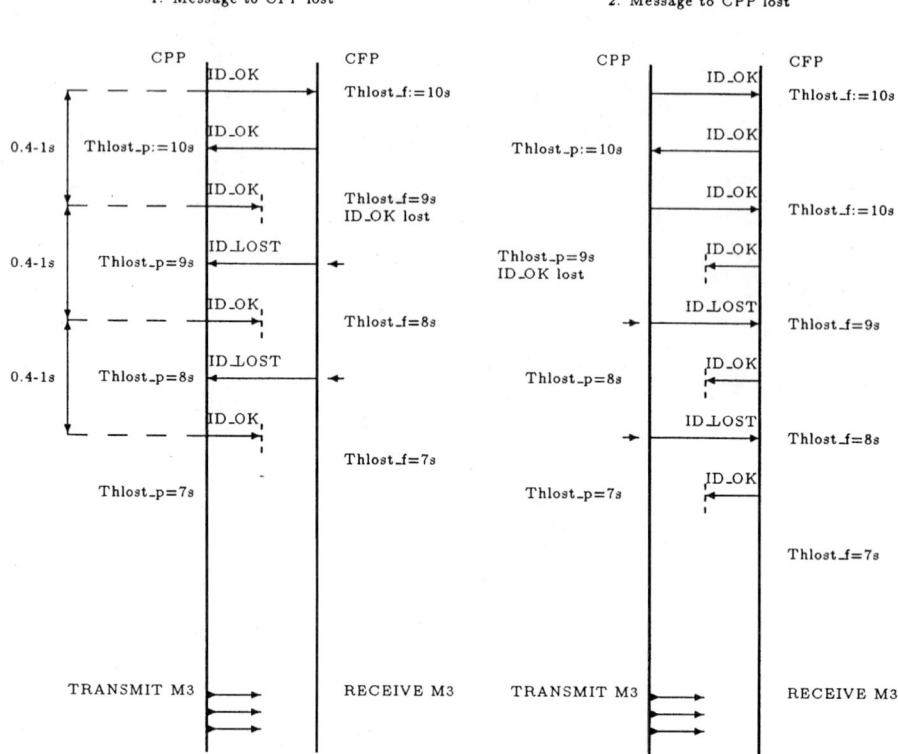

FIGURE 91.10 Handshake loss scenarios.

The handshake mechanism is further augmented by referring to Fig. 91.10, where two different scenarios are examplified, portraying the situation when the HS message sent by the CPP to the CFP is lost or, conversely, that transmitted by the CFP is corrupted.

Considering the first scenario, during error-free communications the CPP sends HS_p = ID_OK, and upon receiving it the CFP resets its Thlost_f timer to 10 s. In due course it sends an HS_f = ID_OK acknowledgement, which also arrives free from errors. The CPP resets the Thlost_f timer to 10 s and, after the elapse of the 0.4–1 s handshake interval, issues an HS_p = ID_OK message, which does not reach the CFP. Hence the Thlost_f timer is now reduced to 9 s and an HS_f = ID_Lost message is sent to the CPP. Upon reception of this, the CPP now cannot reset its Thlost_p timer to 10 s but can respond with an HS_p = ID_OK message, which again goes astray, forcing the CFP to further reduce its Thlost_f timer to 8 s. The CFP issues the valid handshake HS_f = ID_Lost, which arrives at the CPP, where the lack of HS_f = ID_OK reduces Thlost_p to 8 s. Now the corruption of the issued HS_p = ID_OK reduces Thlost_f to 7 s, in which event the link may be reinitialized using the M3 burst. The portrayed second example of Fig. 91.10 can be easily followed in case of the scenario when the HS_f message is corrupted.

91.9 Main Features of the CT-2 System

In our previous discourse we have given an insight in the algorithmic procedures of the CT-2 MPT 1375 recommendation. We have briefly highlighted the four-part structure of the standard dealing with the radio interface, signalling layers 1 and 2, signalling layer 3, and the speech coding issues, respectively. There are forty 100-kHz wide RF channels in the band 864.15–868.15 MHz, and the 72 kb/s bit stream modulates a Gaussian filtered FSK modem. The multiple access technique is

TDD, transmitting 2-ms duration, 144-b M1 bursts during normal voice communications, which deliver the 32-kb/s ADPCM-coded speech signal. During link establishment the M2 and M3 bursts are used, which were also portrayed in this treatise, along with a range of handshaking messages and scenarios.

Defining Terms

AFC: Automatic frequency correction
CAI: Common air interface
CFP: Cordless fixed part
CHM: Channel marker sequence
CHMF: CFP channel marker
CHMP: CPP channel marker
CPP: Cordless portable part
CT: Cordless telephone
DCA: Dynamic channel allocation
DCW: Data code word
DECT: Digital European cordless telecommunications system
FT: Frame format type bit
GFSK: Gaussian frequency shift keying
GP: Guard period
HIC: Handset identification code
HS: Handshaking
ID: Identifier
L2: Signalling layer 2
L3: Signalling layer 3
LAN: Local area network
LID: Link identification
LOS: Line of sight
LS: Link status
M1: Multiplex one burst format
M2: Multiplex two burst format
M3: Multiplex three burst format
MIC: Manufacturer identification code
MPT-1375: British CT2 standard
PCN: Personal communications network
PIC: Portable identification code
PLMR: Public land mobile radio
SNR: Signal-to-noise ratio
SR: Signalling rate bit
SYN: Synchronization sequence
SYNCD: 16-b D-channel frame synchronization word
TDD: Time division duplex multiple access scheme
TP: Telepoint

References

Asghar, S. 1995. Digital European cordless telephone (DECT), In *The Mobile Communications Handbook*, Chap. 30, CRC Press, Inc. Boca Raton, FL.

Gardiner, J.G. 1990. Second generation cordless (CT-2) telephony in the UK: telepoint services and the common air-interface, Elec. & Comm. Eng. J. (April):71–78.

Hanzo, L. 1995. The Pan-European mobile radio system, In *The Mobile Communications Handbook*, Chap. 25, CRC Press, Inc. Boca Raton, FL.

Jabbari, B. 1995. Dynamic channel assignment, In *The Mobile Communications Handbook*, Chap. 21, CRC Press, Inc. Boca Raton, FL.

Ochsner, H. 1990. The digital European cordless telecommunications specification DECT. In *Cordless telecommunication in Europe*, ed. W.H.W. Tuttlebee, pp. 273–285. Springer–Verlag.

Steedman, R.A.J. 1990. The Common Air Interface MPT 1375. In *Cordless Telecommunication in Europe*, ed. W.H.W. Tuttlebee, pp. 261–272, Springer–Verlag.

Steele, R. ed. 1992. *Mobile Radio Communications*, Pentech Press, London.

Tuttlebee, W.H.W. ed. 1990. *Cordless Telecommunication in Europe*, Springer–Verlag.

92
Digital European Cordless Telephone

Saf Asghar
Advanced Micro Devices, Inc.

92.1 Introduction 1305
92.2 Application Areas 1306
92.3 DECT/ISDN Interworking 1307
92.4 DECT/GSM Interworking 1307
92.5 DECT Data Access 1307
92.6 How DECT functions 1308
92.7 Architectural Overview 1308
Baseband Architecture • Voice Coding and Telephony Requirements • Telephony Requirements • DECT Protocol Model • Physical Layer • MAC Layer • Data Link Control (DLC) Layer • Network Layer • Lower Layer Management Entity (LLME) • Physical Layer • MAC Interface: Digital-Service Access Point (D-SAP) • LLME Interface: PM-SAP • Physical Layer Procedures • Medium Access Layer • MAC Primitives • Connection Oriented Primitives • Connectionless and Broadcast Primitives • Low-Level Management Entity • MAC Procedures • Data Link Layer • Lower Layer Management Entity • Data Link Control Primitives • Network Layer • LLME • Modulation Method • Radio Frequency Architecture

92.1 Introduction

Cordless technology, in contrast to cellular radio, primarily offers access technology rather than fully specified networks. The digital European cordless telecommunications (DECT) standard, however, offers a proposed network architecture in addition to the air interface physical specification and protocols but without specifying all of the necessary procedures and facilities. During the early 1980s a few proprietary digital cordless standards were designed in Europe purely as coexistence standards. The U.K. government in 1989 issued a few operator licenses to allow public-access cordless known as telepoint. Interoperability was a mandatory requirement leading to a common air interface (CAI) specification to allow roaming between systems. This particular standard (CT2/CAI), has been described in the previous chapter. The European Telecommunications Standards Institute (ETSI) in 1988 took over the responsibility for DECT. After formal approval of the specifications by the ETSI technical assembly in March 1992, DECT became a European telecommunications standard, ETS300-175 in August 1992. DECT has a guaranteed pan-European frequency allocation, supported and enforced by European Commission Directive 91/297. The CT2 specification has been adopted by ETSI alongside DECT as an interim standard I-ETSI 300 131 under review.

92.2 Application Areas

Initially, DECT was intended mainly to be a private system, to be connected to a private automatic branch exchange (PABX) to give users mobility, within PABX coverage, or to be used as a single cell at a small company or in a home. As the idea with telepoint was adopted and generalized to public access, DECT became part of the public network. DECT should not be regarded as a replacement of an existing network but as created to interface seamlessly to existing and future fixed networks such as public switched telephone network (PSTN), integrated services digital network (ISDN), global system for mobile communications (GSM), and PABX. Although telepoint is mainly associated with CT2, implying public access, the main drawback in CT2 is the ability to only make a call from a telepoint access point. Recently there have been modifications made to the CT2 specification to provide a structure that enables users to make and receive calls. The DECT standard makes it possible for users to receive and make calls at various places, such as airport/railroad terminals, and shopping malls. Public access extends beyond telepoint to at least two other applications: replacement of the wired local loop, often called cordless local loop (CLL), (Fig. 92.1) and neighborhood access, Fig. 92.2. The CLL is a tool for the operator of the public network. Essentially, the operator will install a multiuser base station in a suitable campus location for access to the public network at a subscriber's telephone hooked up to a unit coupled to a directional antenna. The advantages of CLL are high flexibility, fast installation, and possibly lower investments. CLL does not provide mobility. Neighborhood access is quite different from CLL. Firstly, it offers mobility to the users and, secondly, the antennas are not generally directional, thus requiring higher field strength (higher output power or more densely packed base stations). It is not difficult to visualize that CLL systems could be merged with neighborhood access systems in the context of establishments, such as supermarkets, gas stations, shops etc., where it might be desirable to set up a DECT system for their own use and at the same time also provide access to customers. The DECT standard already includes signalling for authentication, billing, etc. DECT opens possibilities for a new operator structure, with many diversified architectures connected to a global network operator (Fig. 92.3). DECT is designed to have extremely high capacity. A small size is used, which may seem an expensive approach for covering large areas. Repeaters placed at strategic locations overcome this problem.

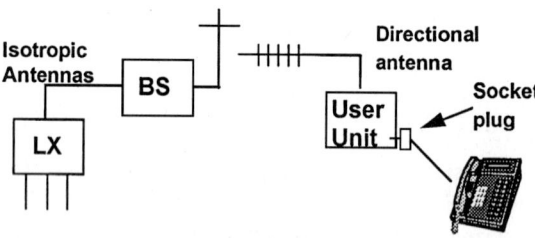

FIGURE 92.1

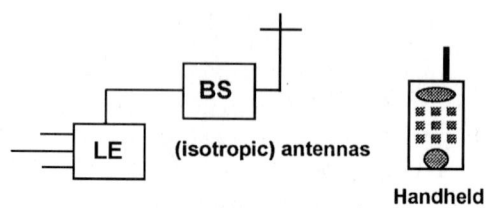

FIGURE 92.2

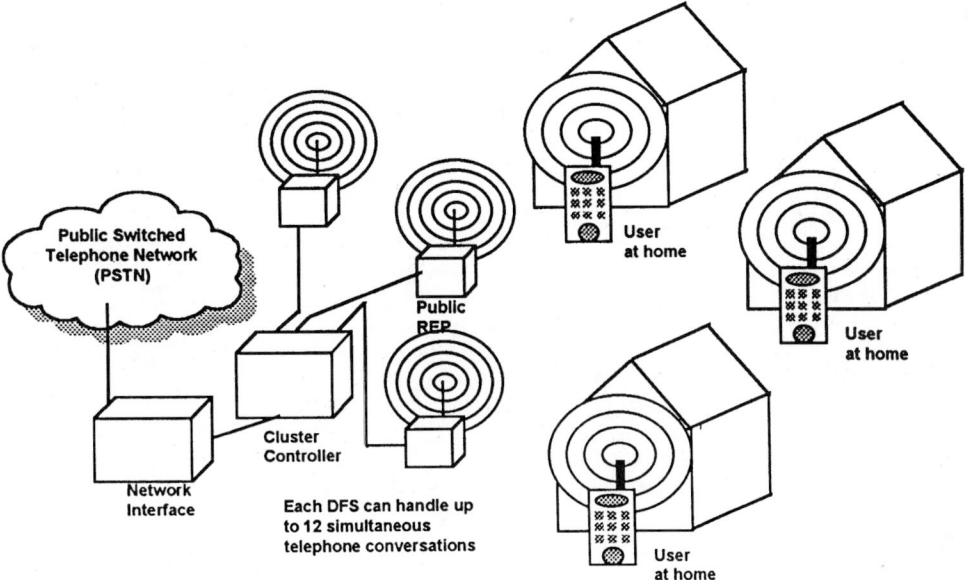

FIGURE 92.3

92.3 DECT/ISDN Interworking

From the outset, a major objective of the DECT specification was to ensure that ISDN services were provided through the DECT network. Within the interworking profile two configurations have been defined: DECT end system and DECT intermediate system. In the end system the ISDN is terminated in the DECT fixed system (DFS). The DFS and the DECT portable system (DPS) may be seen as a ISDN terminal equipment (TE1). The DFS can be connected to an S, S/T, or a P interface. The intermediate system is fully transparent to the ISDN. The S interface is regenerated even in the DPS. Both configurations have the following services specified: 3.1-kHz telephony, i.e, standard telephony; 7-kHz telephony; i.e, high-quality audio; video telephony; group III fax, modems, X.25 over the ISDN; and telematic services, such as group IV fax, telex, and videotex.

92.4 DECT/GSM Interworking

Groupe Speciale Mobile (GSM) is a pan-European standard for digital cellular radio operational throughout the European community. ETSI has the charter to define an interworking profile for GSM and DECT. The profile describes how DECT can be connected to the fixed network of GSM and the necessary air interface functions. The users obviously benefit from the mobility functions of GSM giving DECT a wide area mobility. The operators will gain access to another class of customer. The two systems when linked together will form the bridge between cordless and cellular technologies. Through the generic access profile ETSI will specify a well-defined level of interoperability between DECT and GSM. The voice coding aspect in both of these standards is different; therefore, this subject will be revisited to provide a sensible compromise.

92.5 DECT Data Access

The DECT standard is specified for both voice and data applications. It is not surprising that ETSI confirmed a role for DECT to support cordless local area network (LAN) applications. A new technical committee, ETSI RES10 has been established to specify the high performance European radio LAN similar to IEEE 802.11 standard in the U.S. (Table 92.1).

TABLE 92.1 DECT Characteristics

Parameters	DECT
Operating frequency, MHz	1880–1990 (Europe)
Radio carrier spacing, MHz	1.728
Transmitted data rate, Mb/s	1.152
Channel assignment method	DCA
Speech data rate, kb/s	32
Speech coding technique	ADPCM G.721
Control channels	In-call-embedded (various logical channels C, P, Q, N)
In-call control channel data rate, kb/s	4.8 (plus 1.6 CRC)
Total channel data rate, kb/s	41.6
Duplexing technique	TDD
Multiple access-TDMA	12 TDD timeslots
Carrier usage-FDMA/MC	10 carriers
Bits per TDMA timeslot, b	420 (424 including the 2 field)
Timeslot duration (including guard time), μs	417
TDMA frame period, ms	10
Modulation technique	Gaussian filtered FSK
Modulation index	0.45–0.55
Peak output power, mW	250
Mean output power, mW	10

92.6 How DECT functions

DECT employs frequency division multiple access (FDMA), time division multiple access (TDMA), and time division duplex (TDD) technologies for transmission. Ten carrier frequencies in the 1.88- and 1.90-GHz band are employed in conjunction with 12 time slots per carrier TDMA and 10 carriers per 20 MHz of spectrum FDMA. Transmission is through TDD. Each channel has 24 time slots, 12 for transmission and 12 for receiving. A transmission channel is formed by the combination of a time slot and a frequency. DECT can, therefore, handle a maximum of 12 simultaneous conversations. TDMA allows the same frequency to use different time slots. Transmission takes place for 10 ms, and during the rest of the time the telephone is free to perform other tasks, such as channel selection. By monitoring check bits in the signalling part of each burst, both ends of the link can tell if reception quality is satisfactory. The telephone is constantly searching for a channel for better signal quality, and this channel is accessed in parallel with the original channel to ensure a seamless changeover. Call handover is also seamless, each cell can handle up to 12 calls simultaneously, and users can roam around the infrastructure without the risk of losing a call. Dynamic channel assignment (DCA) allows the telephone and base station to automatically select a channel that will support a new traffic situation, particularly suited to a high-density office environment.

92.7 Architectural Overview

Baseband Architecture

A typical DECT portable or fixed unit consists of two sections: a baseband section and a radio frequency section. The baseband partitioning includes voice coding and protocol handling (Fig. 92.4).

Voice Coding and Telephony Requirements

This section addresses the audio aspects of the DECT specification. The CT2 system as described in the previous chapter requires adaptive differential pulse code modulation (ADPCM) for voice coding. The DECT standard also specifies 32-kb/s ADPCM as a requirement. In a mobile environment it is debatable whether the CCITT G.721 recommendation has to be mandatory. In the handset or the mobile it would be quite acceptable in most cases to implement a compatible or a less

Digital European Cordless Telephone

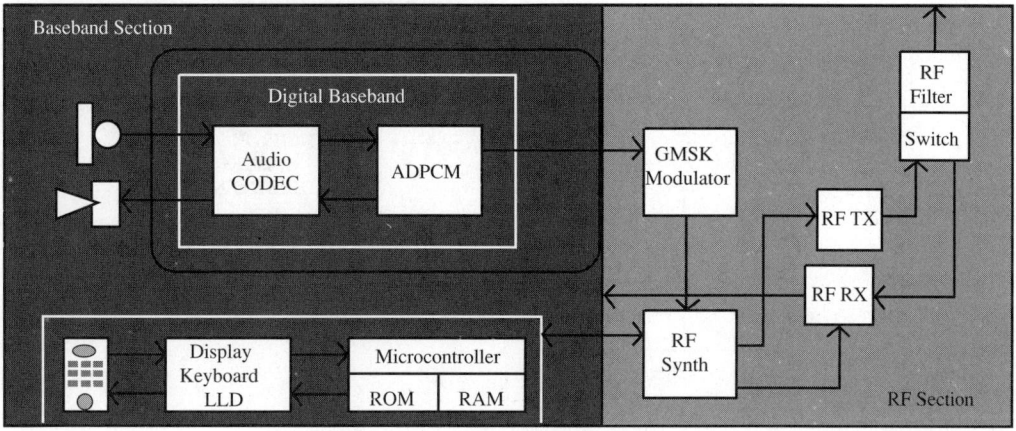

FIGURE 92.4

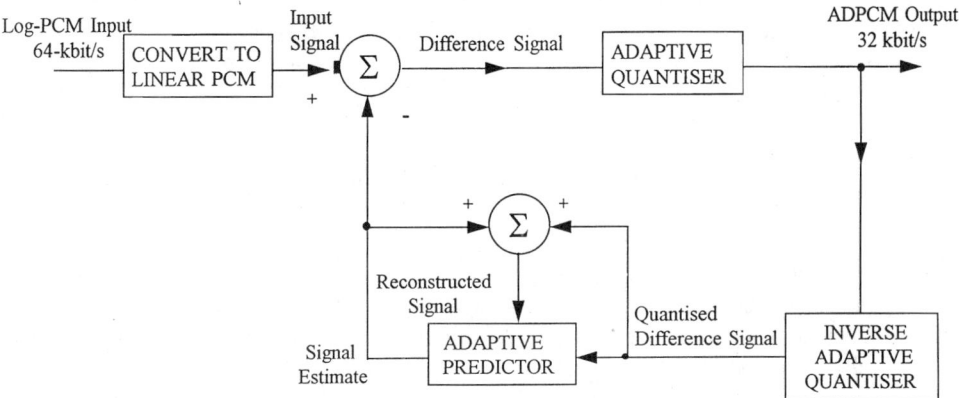

FIGURE 92.5 ADPCM encoder.

complex version of the recommendation. We are dealing with an air interface and communicating with a base station that in the residential situation terminates with the standard POTS line, hence compliance is not an issue. The situation changes in the PBX, however, where the termination is a digital line network. DECT is designed for this case, hence compliancy to the voice coding recommendation becomes important. Adhering to this strategy for the base station and the handset has some marketing advantages.

G.721 32-kb/s ADPCM from its inception was adopted to coexist with G.711 64-kb/s pulse code modulation (PCM) or work in tandem, the primary reason being an increase in channel capacity. For modem type signalling, the algorithm is suboptimal in handling medium-to-high data rates, which is probably one of the reasons why there really has not been a proliferation of this technology in the PSTN infrastructure. The theory of ADPCM transcoding is available in books on speech coding techniques, e.g., O'Shaughnessy, 1987.

The ADPCM transcoder consists of an encoder and a decoder. From Figs. 92.5 and 92.6 it is apparent that the decoder exists in the encoder structure. A benefit derived from this structure allows for efficient implementation of the transcoder.

The encoding process takes a linear speech input signal (the CCITT specification relates to a nonwireless medium such as a POTS infrastructure), and subtracts its estimate derived from earlier input signals to obtain a difference signal. This difference signal is 4-b coded with a 16-level adaptive

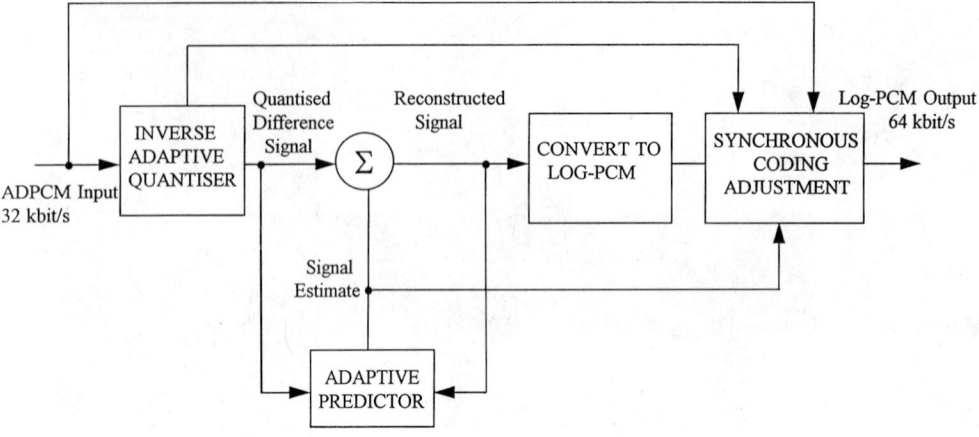

FIGURE 92.6 ADPCM decoder.

quantizer every 125 μs, resulting in a 32-kb/s bit stream. The signal estimate is constructed with the aid of the inverse adaptive quantizer that forms a quantized difference signal that added to the signal estimate is also used to update the adaptive predictor. The adaptive predictor is essentially a second-order recursive filter and a sixth-order nonrecursive filter,

$$S_0(k) = \sum_{i=1}^{2} a_i(k-1)\varepsilon_r(k-i) + \sum_{i=1}^{6} b_i(k-1)d_q(k-i) \qquad (92.1)$$

where coefficients a and b are updated using gradient algorithms.

As suggested, the decoder is really a part of the encoder, that is, the inverse adaptive quantizer reconstructs the quantized difference signal, and the adaptive predictor forms a signal estimate based on the quantized difference signal and earlier samples of the reconstructed signal, which is also the sum of the current estimate and the quantized difference signal as shown in Fig. 92.6. Synchronous coding adjustment tries to correct for errors accumulating in ADPCM from tandem connections of ADPCM transcoders.

ADPCM is basically developed from PCM. It has good speech reproduction quality, comparable to PSTN quality, which therefore led to its adoption in CT2 and DECT.

Telephony Requirements

A general cordless telephone system would include an acoustic interface, i.e., microphone and speaker at the handset coupled to a digitizing compressor/decompressor analog to uniform PCM to ADPCM at 32 kb/s enabling a 2:1 increase in channel capacity as a bonus. This digital stream is processed as described in earlier sections to be transmitted over the air interface to the base station where the reverse happens, resulting in a linear or a digital stream to be transported over the land-based network. The transmission plans for specific systems have been described in detail in Tuttlebee ed., 1995.

An important subject in telephony is the effect of network echoes [Weinstein, 1977]. Short delays are manageable even if an additional delay of, say, less than 15 μs is introduced by a cordless handset. Delays of a larger magnitude, in excess of 250 μs (such as satellite links [Madsen and Fague, 1993]), coupled to cordless systems can cause severe degradation in speech quality and transmission; a small delay introduced by the cordless link in the presence of strong network echoes is undesirable. The DECT standard actually specifies the requirement for network echo control. Additional material can be obtained from the relevant CCITT documents [CCITT, 1984–1985].

Digital European Cordless Telephone

DECT Protocol Model

This section provides an overview of the software layer entities and the message interfaces between the software layers for the DECT common interface software package (Fig. 92.7).

The functionality of the DECT protocol is described in the ETSI specifications ETS 300 175-1–ETS 300 175-5.

The DECT protocol model is based on the International Standards Organization (ISO) open systems interconnection (OSI) seven-layer model. The complete DECT air interface corresponds to the first three ISO OSI layers; however, DECT defines four layers of protocol: physical, medium access control, data link, and network (Fig. 92.8).

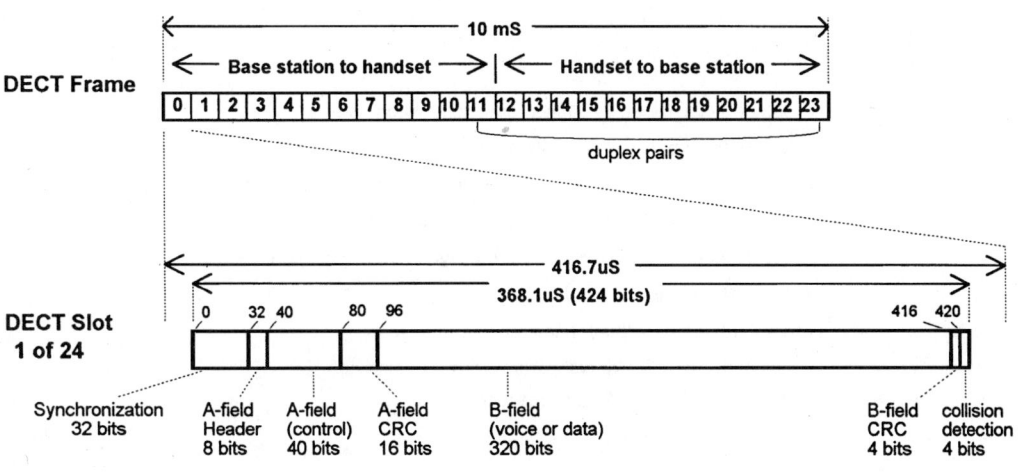

DECT Hardware Model

Multiframe Structure

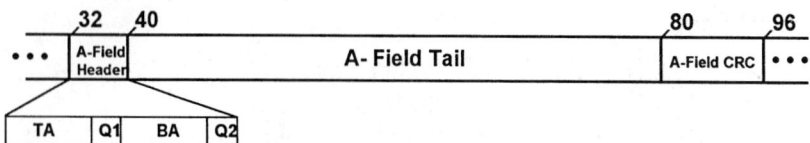

- TA - 3 bits used to define the type of message in the A-field tail.
- BA - 3 bits used to define the type of message in the B-field.
- Q1,Q2 - 2 bits providing information on signal quality at the other device

A-Field Header

- TA=000: Ct packet 0 - Higher layer control information
- TA=001: Ct packet 1 - Higher layer control information
- TA=010: Nt on connectionless bearer - Base station identification
- TA=011: Nt - Identification message for traffic bearers
- TA=100: Qt - Slot, frame, multiframe synchronization information
- TA=101: Escape
- TA=110: Mt - MAC layer control
- TA=111(PP): Mt for first portable part transmission
  TA=111(RFP): Pt - Paging message

A-Field Tail Formats

DECT Multiframe - RFP to PP

0	1	2	3	4	5	6	7	8	9	10	11	12	13	14	15
Pt	Mt	Pt	Mt	Pt	Mt	Pt	Mt	Qt	Mt	Pt	Mt	Pt	Mt		Mt
	Ct		Ct		Ct		Ct		Ct		Ct		Ct		Ct
Nt	Nt	Nt	Nt	Nt	Nt	Nt	Nt		Nt	Nt	Nt	Nt	Nt	Nt	Nt

160mS

DECT Multiframe - PP to RFP

0	1	2	3	4	5	6	7	8	9	10	11	12	13	14	15
Mt		Mt		Mt		Mt		Mt		Mt		Mt		Mt	
Ct		Ct		Ct		Ct		Ct		Ct		Ct		Ct	
Nt	Nt	Nt	Nt	Nt	Nt	Nt	Nt	Nt	Nt	Nt	Nt	Nt	Nt	Nt	Nt

160mS

T-MUX Algorithm

Digital European Cordless Telephone

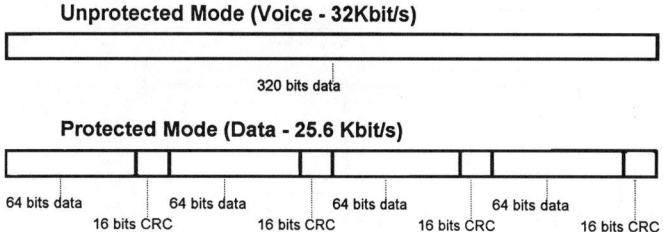

Protected/Unprotected B-Feild

- Unprotected mode used for voice information
- Protected mode used for transferring higher layer control information.

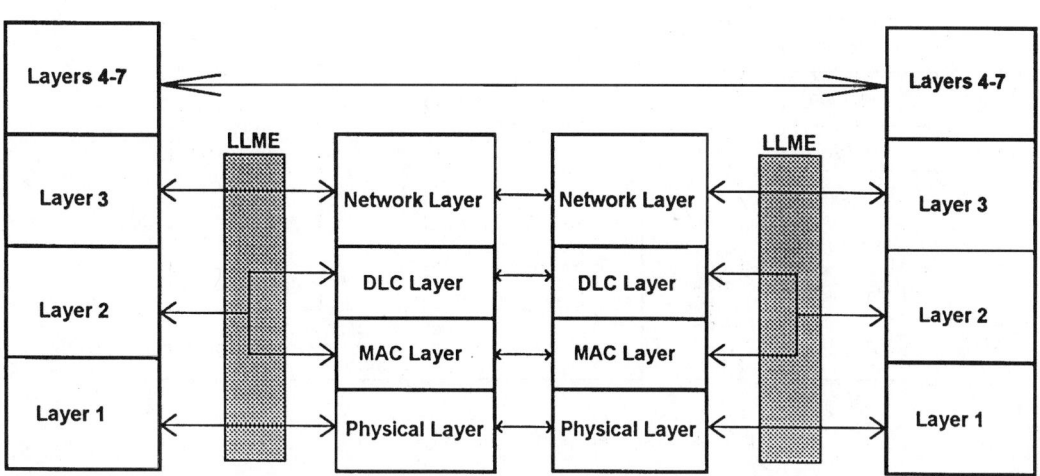

FIGURE 92.7 DECT protocol structure.

The complete DECT air interface corresponds to the first three ISO OSI layers; however, since the OSI Layer does not adequately consider multiple access to one transmission medium, the DECT structure uses four layers for the node-to-node communication: physical, medium access control, data link, and network layers (Fig. 92.9).

These first protocol layers serve to support the creation of a functional data link through the cordless network whereas layers 4, 5, 6, and 7 are concerned with supporting communications between the end users/networks.

Physical Layer

Divides the radio spectrum into physical channels using TDMA operation on ten RF carriers.

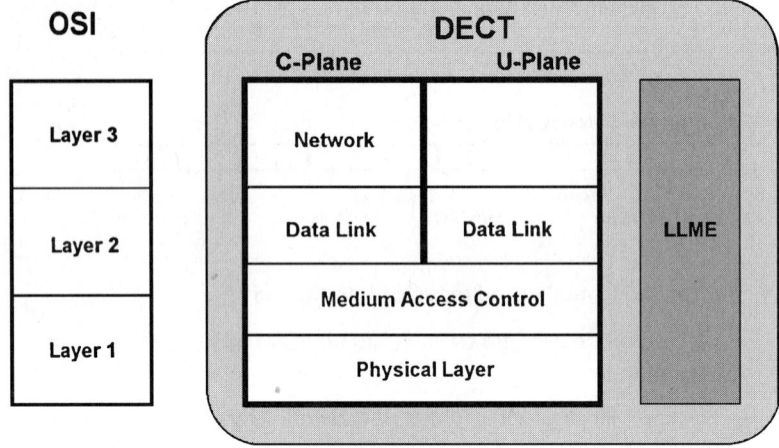

FIGURE 92.8

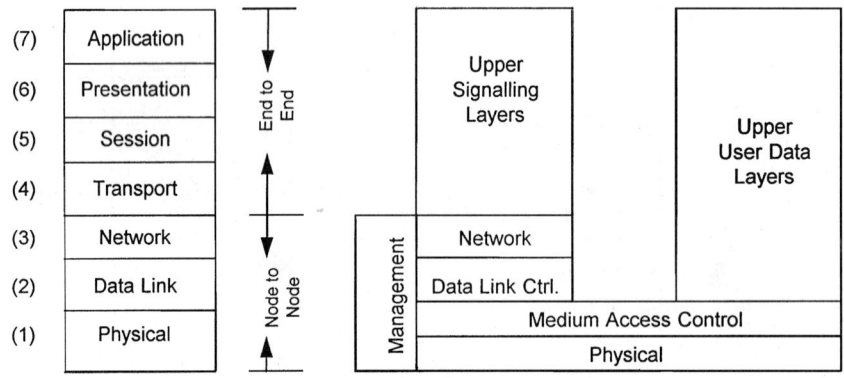

FIGURE 92.9   DECT Protocol Layers.

MAC Layer

The MAC layer selects the physical channels and then establishes or releases connections on those channels. It also multiplexes/demultiplexes control information in slot-sized packets.

These functions provide three services: broadcast service, connection oriented service, and a connectionless service. The broadcast service multiplexes broadcast information into the A-field, and this field appears as part of all active transmissions. In the absence of user traffic, at least one physical channel broadcasts.

Data Link Control (DLC) Layer

The DLC layer provides a reliable data link to the network layer. The DECT DLC layer separates the operation into two planes: the C plane and the U plane. The C plane is common to all applications and provides reliable links for the transmission of internal control signalling and limited user information traffic. It uses LAPC for full error control. The U plane provides a family of alternative services optimized to the specific application. i.e., transparent unprotected service for voice, circuit mode, and packet mode.

Network Layer

This is the main signalling layer of the protocol and is similar to the ISDN layer 3 protocol. It supports the establishment, maintenance, and release of the connections. Some network layer services are: call control, supplementary services, connection oriented message service, Connection less message service, and mobility management.

Lower Layer Management Entity (LLME)

The LLME interfaces with all of the previous layers to provide procedures that concern more than one layer. Most of these procedures have only local significance.

Physical Layer

The physical layer is responsible for the segmentation of the transmission media into physical channels using TDMA on ten carriers between 1880 and 1900 MHz. Each carrier contains a TDMA structure defined as 24 timeslots in a 10-ms frame. These timeslots can be used to transmit data packets.

Each burst contains a 32-b synchronization field and a data field. The synchronization field is used for clock and packet synchronization. The data field is received by the MAC layer.

MAC Interface: Digital-Service Access Point (D-SAP)

The physical layer communicates to the MAC layer via the D-SAP digital-service access point (D-SAP). The D-SAP is mainly used to exchange D fields. D-field segments may be passed in either direction. The following primitives are exchanged though the D-SAP:

    PL-TX
    PL-RX
    PL-FREQ-ADJ

LLME Interface: PM-SAP

The primitives passed through PM-SAP are mainly used to invoke and control physical layer processes. The following primitives are exchanged through PM-SAP:

    PL-ME-SYNC
    PL-ME-SIG_STR
    PL-ME-TIME_ADJ

Prior to the use of a DECT physical channel, the receiver measures the strength of signals on that physical channel using the PL-ME_STR primitive. Using these signal measurements, the LLME produces two ordered lists: least interfered channels and channels with greatest field strength (PP only).

Physical Layer Procedures

- Addition of the synchronization field and transmission
- Packet reception and removal of synchronization field
- Signal strength measurement
- Synchronization pulse detection
- Timing adjustment
- Frequency adjustment

- Transmission and reception of the Z field
- Sliding collision detection

Medium Access Layer

The medium access layer specifies three groups of MAC services: the broadcast message control service (BMC), the connectionless message control service (CMC), and the multibearer control service (MBC). It also specifies the logical channels that are used by the MAC services and how they are multiplexed and mapped into the service data units that are interchanged with the physical layer.

The MAC layer controls the reception of data for short, half, and full slots, and the transmission of full slots by issuing primitives to the physical layer. Full slots are numbered $K = 0–23$, slot numbers (frame timing) are only defined in a special Q-channel message that is transmitted at a low rate by all FPs.

The MAC layer superimposes a multiframe structure on the TDMA frame. This is a time division multiplex of 16 frames. The multiframe numbering is defined the same way for FPs and PPs. Frame numbers (multiframe timing) are not included in a transmission and must be interpolated from a multiframe marker included in frame 8 in all FP transmissions. The MAC layer software should support MAC messages.

MAC Primitives

The MAC layer interfaces with the physical layer, the data link layer, and the LLME using four types of primitives: request (Req), indicate (Ind), response (Res), and confirm (Cfm). A Cfm primitive only occurs at a confirmation of an action initiated by a Req primitive. A Res primitive can only follow a Ind primitive.

Connection Oriented Primitives

MAC-CON: Connection setup
MAC-MOD: Connection modification
MAC-CO_DTR: CO data transmit ready
MAC-CO_DATA: CO data transfer
MAC-RES_DLC: Restart DLC
MAC-DIS: Connection release
MAC-BW: MAC bandwidth
MAC-ENC_KEY: Load encryption key
MAC-ENC_EKS: Enable/disable encryption

Connectionless and Broadcast Primitives

MAC-PAGE: Paging
MAC-DOWN_CON: Downlink connection
MAC-UP_CON: Uplink connection

Low-Level Management Entity

MAC-ME-CON: Connection setup
MAC-ME-CON_ALL: Connection setup allowed
MAC-ME-REL: Bearer release
MAC-ME-REL_REP: MBC release support
MAC-ME-RFP_PRELOAD: FP information preloading

MAC-ME-PT_PRELOAD: PT information preloading
MAC-ME-INFO: Systems information output
MAC-ME-EXT: Extended system info
MAC-ME-CHANMAP: Channel map
MAC-ME-STATUS: Status report
MAC-ME-ERROR: Error reports

MAC Procedures

Broadcast and Connectionless Procedures
- Downlink broadcast and connectionless procedures
- Uplink Connectionless procedures

Connection/Oriented Service Procedures
- C/O connection setup
- C/O connection modification
- C/O connection release
- C/O bearer setup
- C/O bearer handover
- C/O bearer release
- C/O data transfer
- MOD-2 protected I-channel operation
- Higher layer unprotected information (IN) and MAC error detection services (IP).

LLME Procedures
- Broadcasting
- Extended system information
- PP states and transitions
- Physical channel selection
- In-connection quality control
- Maximum allowed system load at RFPs
- PMID and FMID definitions
- RFP idle receiver scan sequence
- PT fast setup receiver scan sequence

Data Link Layer

As indicated previously, the data link control layer contains two independent planes of protocol: the C plane and the U plane. The C plane is the control plane of the DECT protocol stacks, and the U plane is the user plane of the DECT protocol stacks. This plane contains most of the end-to-end and (external) user information and control.

C Plane

The C plane is common to all applications and provides reliable links for the transmission of internal control signalling and limited user information traffic. It uses LAPC for full error control. The DLC C plane provides two independent services: the data link service (LAPC + Lc) and the broadcast service (Lb). Each of these services is completely independent and is accessed through independent SAPs.

C Plane Data Link Services

These services are provided by two protocol entities called LAPC and Lc. These entities separate the link access protocol functions from the lower link control functions. Each independent data link has an associated instance of these entities. The data link service is accessed via S-SAP.

C Plane Broadcast Service

This service contains only one instance of the Lb lower entity. This entity provides a restricted broadcast service in the down link direction and uses the dedicated MAC broadcast service. The broadcast service is accessed via B-SAP.

U-Plane Services

These services are application dependent. Each U plane is divided into an upper (LUx) entity and a lower (FBx) entity. The upper entity contains all of the procedures, and the lower entity buffers and fragments the U-plane frames from the MAC layer.

The following LUx members are defined by the protocol:

- LU1: Transparent Unprotected service (TRUP)
- U2: Frame relay service (FREL)
- LU3: Frame Switching service (FSW)
- LU4: Forward error correction service (FEC)
- LU5: Basic rate adaption service (BRAT)
- LU6: Secondary rate adaption service (SRAT)
- LU7-LU15: Reserved
- LU16: Escape for nonstandard family (ESC)

Lower Layer Management Entity

The LLME Provides coordination and control for the C-plane and U-plane processes. The LLME controls the routing of the C-plane and U-plane frames from the available MAC connections and controls the opening and closing (handover) of the MAC connections in response to service demands.

Data Link Control Primitives

These primitives describe the DLC interactions with other layers.

Primitives to the Network Layer via S-SAP

DL-ESTABLISH
DL-RELEASE
DL-DATA
DL-UNIT-DATA
DL-SUSPEND
DL-RESUME
DL-ENC_KEY
DL-ENCRYPT

Primitives to the Network Layer via B-SAP

DL-BROADCAST
DL-EXPEDITED

Network Layer

The DECT protocol specifies the C plane of the Network Layer. The C-plane contains all of the internal signalling information.

Entities

The network layer protocols are grouped as:

- Call control (CC) entity
- Supplementary services: call independent supplementary services (CISS) entity
- Connection oriented message service (COMS) entity
- Connectionless message service (CLMS) entity
- Mobility management (MM) entity
- Link control entity (LCE)

Call control (CC). This is the main service instance, and provides a set of procedures to establish, maintain, and release circuit switched services, as well as support for all call related signalling.

Supplementary services: CRSS and CISS. Supplementary services provide additional capabilities to be used with bearer services and teleservices. Two types of supplementary services are defined: call related supplementary services (CRSS) and call independent supplementary services (CISS). CRSS are explicitly associated with a single instance of a CC; CISS may refer to all CC instances.

Connection oriented message service (COMS). COMS offers point-to-point connection oriented packet service. This service only supports packet mode calls and offers a faster and simpler call establishment than the CC entity.

Connectionless message service (CLMS). CLMS offers a connectionless point-to-point or point-to-multipoint service.

Mobility manager (MM). This entity handles functions necessary for the secure provision of DECT services and supports, in particular, incoming calls. The MM procedures are described in seven groups:

- Identity procedures
- Authentication procedure
- Location procedure
- Access rights procedure
- Key allocation procedure
- Parameter retrieval procedure
- Ciphering related procedure

Link control entity (LCE). This is the lowest entity in the network layer. It performs the following tasks:

- Supervises the lower layer link states for every data link endpoint in the C plane
- Downlink routing
- Uplink routing
- Queuing of messages to all C-plane data link endpoints
- Creates and manages the LCD-REQUEST-PAGING messages (B-SAP)
- Queues and submits other messages to B-SAP
- Assigns new data link endpoint identifiers (DLEI)
- Assigns layer 2 instances to existing data link endpoints
- Reports data link failures to all layer 3 instances that are using the link

LLME

All of the network layer entities interface to the LLME which provides coordination of the operation between different network layer entities and also between the network and lower layers. The LLME interfaces with the physical, MAC, DLC, and NWK layers to provide procedures that concern more than one layer. Most of these procedures have only local significance.

LLME Physical Layer: PM-SAP

The primitives passed through PM-SAP are mainly used to invoke and control physical layer processes. The following primitives are exchanged through PM-SAP: PL-ME-SYNC, PL-ME-SIG_STR, and PL-ME-TIME_ADJ. Using these primitives the LLME implements procedures to produce: the list of quietest channels, the list of physical channels with greatest field strength (PP only), and timing information.

Prior to the use of a DECT physical channel, the receiver measures the strength of signals on that physical channel using the PL-ME_STR primitive. Using these signal measurements, the LLME produces two ordered lists: a list of least interfered channels and a list channels with greatest field strength (PP only). Using the PL-ME-SYNC primitives and interworking with higher layer detection of slot numbers, the LLME extracts timing information to establish the slot and frame timing.

MAC Layer

- Creation, maintenance and release of bearers, by activating and deactivating pairs of physical channels.
- Physical channels management, including the choice of free physical channels and the assessment of quality of received signals.

Connection and control.

    MAC-ME-CON: Connection setup
    MAC-ME-CON_ALL: Connection setup allowed
    MAC-ME-REL: Bearer release
    MAC-ME-REL_REP: MBC release support

System information and identities.

    MAC-ME-RFP_PRELOAD: FP information preloading
    MAC-ME-PT_PRELOAD: PT information preloading
    MAC-ME-INFO: Systems information output
    MAC-ME-EXT: Extended system information
    MAC-ME-CHANMAP: Channel map
    MAC-ME-STATUS: Status report
    MAC-ME-ERROR: Error reports

Procedures.

- Broadcasting
- Extended system information
- PP states and transitions
- Physical channel selection
- In-connection quality control
- Maximum allowed system load at RFPs

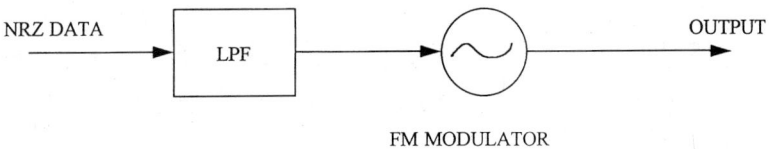

FIGURE 92.10 Premodulation baseband-filtered MSK.

- PMID and FMID definitions
- RFP idle receiver scan sequence
- PT fast setup receiver scan sequence

Data Link Layer (DLC)

DLC covers routing of C-plane and U-plane data to suitable connections and connection management, which includes the establishment and release of connections in response to network layer demands. It also provides coordination and control for the C-plane and U-plane processes. The LLME controls the routing of the C-plane and U-plane frames from the available MAC connections and controls the opening and closing (handover) of the MAC connections in response to service demands.

Network Layer

The network layer provides service negotiation and mapping. All of the network layer entities interface to the LLME, which provides coordination of the operation between different network layer entities and also between the network and lower layers.

Modulation Method

The modulation method for DECT is Gaussian filtered frequency shift keying (GFSK) with a nominal deviation of 288 kHz [Madsen and Fague, 1993]. The BT, i.e., Gaussian filter bandwidth to bit ratio, is 0.5 and the bit rate is 1.152 Mb/s. Specification details can be obtained from the relevant ETSI documents listed in the reference section.

Digital transmission channels in the radio frequency bands, including the DECT systems, present serious problems of spectral congestion and introduce severe adjacent/cochannel interference problems. There were several schemes employed to alleviate these problems: new allocations at high frequencies, use of frequency-reuse techniques, efficient source encoding, and spectrally efficient modulation techniques.

Any communication system is governed by mainly two criteria, transmitted power and channel bandwidth. These two variables have to be exploited in an optimum manner in order to achieve maximum bandwidth efficiency, defined as the ratio of data rate to channel bandwidth (units of bits/Hz/s) [Pasupathy, 1979]. GMSK/GFSK has the properties of constant envelope, relatively narrow bandwidth, and coherent detection capability. Minimum shift keying (MSK) can be generated directly from FM, i.e., the output power spectrum of MSK can be created by using a premodulation low-pass filter. To ensure that the output power spectrum is constant, the low-pass filter should have a narrow bandwidth and sharp cutoff, low overshoot, and the filter output should have a phase shift $\pi/2$, which is useful for coherent detection of MSK; see Fig. 92.10.

Properties of GMSK satisfy all of these characteristics. We replace the low-pass filter with a premodulation Gaussian low-pass filter [Murota and Hirade, 1981]. As shown in Fig. 92.11, it is relatively simple to modulate the frequency of the VCO directly by the baseband Gaussian pulse stream, however, the difficulty lies in keeping the center frequency within the allowable value. This becomes more apparent when analog techniques are employed for generating such signals. A possible solution to this problem in the analog domain would be to use a phase-lock loop (PLL)

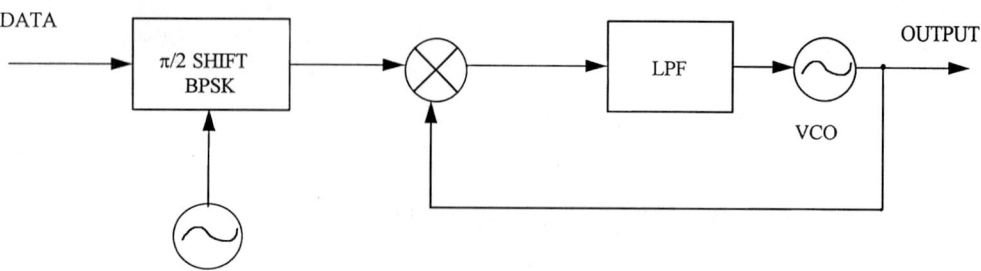

FIGURE 92.11 PLL-type GMSK modulator.

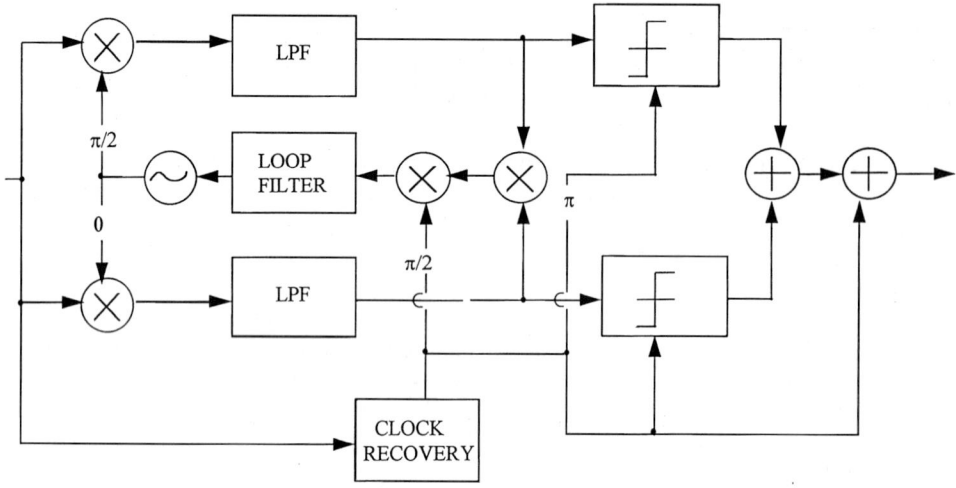

FIGURE 92.12 Costas Loop.

modulator with a precise transfer function. It is desirable these days to employ digital techniques, which are far more robust in meeting the requirements talked about earlier. This would suggest an orthogonal modulator with digital waveform generators [de Jager and Dekker, 1978].

The demodulator structure in a GMSK/GFSK system is centered around orthogonal coherent detection, the main issue being recovery of the reference carrier and timing. A typical method, is described in de Buda, 1972, where the reference carrier is recovered by dividing by four the sum of the two discrete frequencies contained in the frequency doubler output, and the timing is recovered directly from their difference. This method can also be considered to be equivalent to the Costas loop structure as shown in Fig. 92.12.

In the following are some theoretical and experimental representations of the modulation technique just described. Considerable literature is available on the subject of data and modulation schemes and the reader is advised to refer to Pasupathy, 1979, and Murota and Hirade, 1981, for further access to relevant study material.

Radio Frequency Architecture

We have discussed the need for low power consumption and low cost in designing cordless telephones. These days digital transmitter/single conversion receiver techniques are employed to provide highly accurate quadrature modulation formats and quadrature down conversion schemes that allow a great deal of flexibility to the baseband section. Generally, one would have used digital

Digital European Cordless Telephone

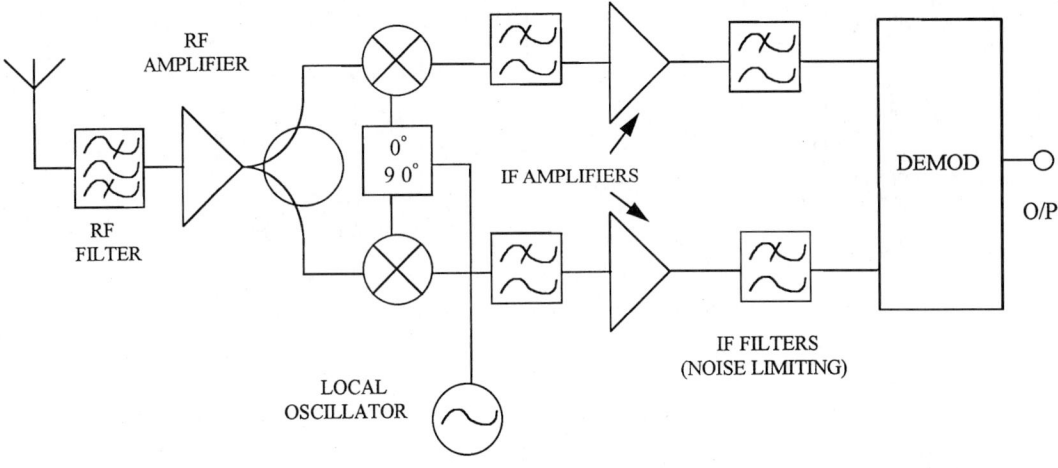

FIGURE 92.13 Direct conversion receiver architecture.

signal processors to perform most of the demodulation functions at the cost of high current consumption. With the advent of application specific signal processing, solutions with these techniques have become more attractive.

From a system perspective, range, multipath, and voice quality influence the design of a DECT phone. A high bit rate coupled with multipath reflections in an indoor environment makes DECT design a challenging task. The delay spread (multipath) can be anywhere in the 100–200 ns range, and a DECT bit time is 880 ns. Therefore, a potential delay spread due to multipath reflections is 1–20% of a bit time. Typically, antenna diversity is used to overcome such effects.

DECT employs a TDMA/TDD method for transmission, which simplifies the complexity of the radio frequency end. The transmitter is on for 380 ms or so. The receiver is also only on for a similar length of time.

A single conversion radio architecture requires fast synthesizer switching speed in order to transmit and receive on as many as 24 timeslots per frame. In this single conversion transmitter structure, the synthesizer has to make a large jump in frequency between transmitting and receiving, typically in the order of 110 MHz. For a DECT transceiver, the PLL synthesizer must have a wide tuning bandwidth at a high-frequency reference in addition to good noise performance and fast switching speed. The prescaler and PLL must consume as low a current as possible to preserve battery life.

In the receive mode the RF signal at the antenna is filtered with a low-loss antenna filter to reduce out-of-band interfering signals. This filter is also used on the transmit side to attenuate harmonics and reduce wideband noise. The signal is further filtered, shaped, and down converted as shown in the Fig. 92.13. The signal path really is no different from most receiver structures. The challenges lie in the implementation, and this area has become quite a competitive segment, especially in the semiconductor world.

The direct conversion receiver usually has an intermediate frequency nominally at zero frequency, hence the term zero IF. The effect of this is to fold the spectrum about zero frequency, which results in the signal occupying only one-half the bandwidth. The zero IF architecture possesses several advantages over the normal superheterodyne approach. First, selectivity requirements for the RF filter are greatly reduced due to the fact that the IF is at zero frequency and the image response is coincident with the wanted signal frequency. Second, the choice of zero frequency means that the

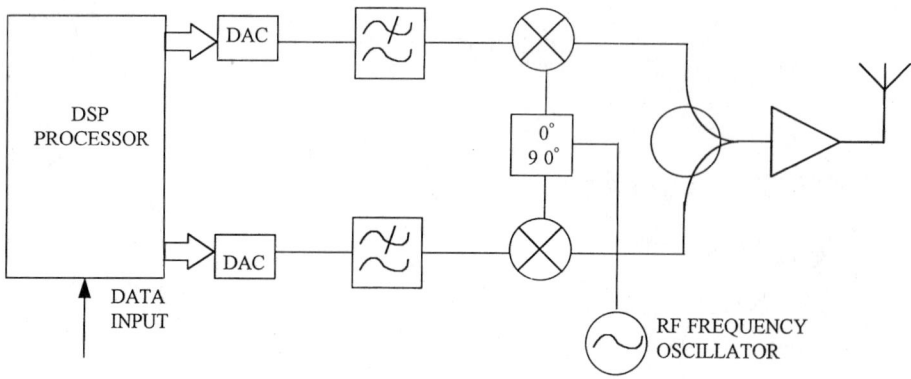

FIGURE 92.14 Transmit section.

bandwidth for the IF paths is only half the wanted signal bandwidth. Third, channel selectivity can be performed simply by a pair of low-bandwidth low-pass filters.

For the twin IF chains of a direct conversion receiver, automatic gain control (AGC) is always required due the fact that each IF channel can vary between zero and the envelope peak at much lower rates than the highest signal bandwidth frequency. An additional requirement in newer systems is received signal strength indication (RSSI) to measure the signal or interference level on any given channel.

Moving on to the transmitter architecture (shown in Fig. 92.14 is a typical I-Q system), it is safe to say that the task of generating an RF signal is much simpler than receiving it. A transmitter consists of three main components: a final frequency generator, a modulator, and the power amplifier. These components can all be combined in common circuits, i.e., frequency synthesizer with inbuilt modulator. The problem of generating a carrier at a high frequency is largely one of frequency control. The main approach for accurately generating an output frequency from a crystal reference today is the PLL, and there is considerable literature available on the subject [Gardner, 1979]. In the modulation stage, depending upon the tightness of the phase accuracy specification of a cordless system, it may be necessary to apply tight control on the modulation index to ensure that the phase path of the signal jumps exactly in 90° increments. Description of an I-Q modulator is also given in a previous chapter.

Defining Terms

AGC: Automatic gain control
ARQ: Automatic repeat request
AWGN: Additive white Gaussian noise
BABT: British approvals board for telecommunications
Base Station: The fixed radio component of a cordless link. This may be single-channel (for domestic) or multi-channel (for Telepoint and business)
BER: Bit error rate (or ratio)
CCITT: Comitè Consultatif International des Tèlègraphes et Tèlèphones, part of the ITU
CEPT: Conference of European Posts and Telecommunications Administrations
CPFSK: Continuous phase frequency shift keying
CPP: Cordless portable part, the cordless telephone handset carried by the user
CRC: Cyclic redundancy check
CT2: Second generation cordless telephone-digital
D Channel: Control and information data channel (16 kb/s in ISDN)

DCT: Digital cordless telephone
DECT: Digital European cordless telecommunications
DLC: Data link control layer, protocol layer in DECT
DSP: Digital signal processing
DTMF: Dual tone multiple frequency (audio tone signalling system)
ETSI: European Telecommunications Standards Institute
FDMA: Frequency division multiple access
FSK: Frequency shift keying
GMSK: Gaussian filtered minimum shift keying
ISDN: Integrated services digital network
ITU: International Telecommunications Union
MPT 1375: U.K. standard for common air interface (CAI) digital cordless telephones
MSK: Minimum shift keying
PSK: Phase shift keying
RES 3: Technical subcommittee, radio equipment and systems 3 of ETSI, responsible for the specification of DECT
RSSI: Received signal strength indication
SAW: Surface acoustic wave
TDD: Time division duplex
TDMA: Time division multiple access

References

Cheer, A.P. 1985. Architectures for digitally implemented radios, IEE Colloquium on Digitally Implemented Radios, London.

Comité Consultatif International des Télégraphes et Téléphones. 1984, "32 kbits/sec Adaptive Differential Pulse Code Modulation (ADPCM)," CCITT Red Book, Fascicle III.3, Rec G721.

Comité Consultatif International des Télégraphes et Téléphones. 1984–1985. *General Characteristics of International Telephone Connections and Circuits*, CCITT Red Book, Vol. 3, Facicle III.1, Rec. G101–G181.

de Buda, R. 1972. Coherent demodulation of frequency shifting with low deviation ratio, *IEEE Trans.* COM-20 (June):466–470.

de Jager, F. and Dekker, C.B. 1978. Tamed frequency modulation. A novel method to achieve spectrum ecomony in digital transmission, *IEEE Trans. in Comm.* COM-20 (May):534–542.

Dijkstra, S. and Owen, F. 1994. The case for DECT, *Mobile Comms Int.* :60–65.

European Telecommunications Standards Inst. 1992. RES-3 DECT Ref. Doc. ETS 300 175-1 (Overview). Oct. ETSI Secretariat, Sophia Antipolis Ceder, France.

European Telecommunications Standards Inst. 1992. RES-3 DECT Ref. Doc. ETS 300 175-2 (Physical Layer) Oct. ETSI Secretariat, Sophia Antipolis Ceder, France.

European Telecommunications Standards Inst. 1992. RES-3 DECT Ref. Doc. ETS 300 175-3 (MAC Layer) Oct. ETSI Secretariat, Sophia Antipolis Ceder, France.

European Telecommunications Standards Inst. 1992. RES-3 DECT Ref. Doc. ETS 300 175-4 (Data Link Control Layer) Oct. ETSI Secretariat, Sophia Antipolis Ceder, France.

European Telecommunications Standards Inst. 1992. RES-3 DECT Ref. Doc. ETS 300 175-5 (Network Layer) Oct. ETSI Secretariat, Sophia Antipolis Ceder, France.

European Telecommunications Standards Inst. 1992. RES-3 DECT Ref. Doc. ETS 300 175-6 (Identities and Addressing) Oct. ETSI Secretariat, Sophia Antipolis Ceder, France.

European Telecommunications Standards Inst. 1992. RES-3 DECT Ref. Doc. ETS 300 175-7 (Security Features) Oct. ETSI Secretariat, Sophia Antipolis Ceder, France.

European Telecommunications Standards Inst. 1992. RES-3 DECT Ref. Doc. ETS 300 175-8 (Speech Coding & Transmission) Oct. ETSI Secretariat, Sophia Antipolis Ceder, France.

European Telecommunications Standards Inst. 1992. RES-3 DECT Ref. Doc: ETS 300 175-9 (Public Access Profile) Oct. ETSI Secretariat, Sophia Antipolis Ceder, France.

European Telecommunications Standards Inst. 1992. RES-3 DECT Ref. Doc. ETS 300 176 (Approval Test Spec) Oct. ETSI Secretariat, Sophia Antipolis Ceder, France.

Gardner, F.M. 1979. *Phase Lock Techniques*, Wiley-Interscience, New York.

Madsen, B. and Fague, D. 1993. Radios for the future: designing for DECT. *RF Design.* (April):48–54.

Murota, K. and Hirade, K. 1981,GMSK modulation for digital mobile telephony, *IEEE Trans.*, COM-29(7), 1044–1050.

Olander, P. 1994. DECT a powerful standard for multiple applications, *Mobile Comms Int.*. 14–16.

O'Shaughnessy, D. 1987. *Speech Communication*, Addison-Wesley. Reading, MA.

Pasupathy, S. 1979. Minimum shift keying: spectrally efficient modulation. *IEEE Comm. Soc. Mag.* 17(4):14–22.

Tuttlebee, W.H.W., ed., 1995. *Cordless Telecommunications Worldwide*, Springer–Verlag.

Weinstein, S.B. 1977. Echo cancellation in the telephone network, *IEEE Comm. Soc. Mag.* 15(1):9–15.

93
The RACE Program

Stanley Chia
BT Laboratories

93.1 Introduction 1327
93.2 RACE I Mobile Project 1328
93.3 RACE II Mobile Project Line 1329
  Code Division Multiple Access • Advanced Time Division Multiple Access (ATDMA) • Mobile Network • Mobile Audio-Visual Terminal • Mobile Broadband System • Smart Antenna Technology for Advanced Mobile Infrastructure • Integration of Satellite in Future Mobile Network • Radio Front End
93.4 Beyond the RACE Program 1336

93.1 Introduction

The objective of the European Union's program for research and development in advanced communications technologies in Europe (RACE) is to enable the development of advanced techniques and technologies that will facilitate the introduction of innovative telecommunications services and improve quality and cost effectiveness of traditional ones. They must meet the significant demand for fast data transmission between computers and contend with the rapid growth in visual communications, thus offering the capabilities of telecommunications to a wider community with better adaptation to human communication needs. With the collaborative research activities being prenormative and precompetitive, the aim is to create a series of common architectures, open interfaces, and standardized protocols in order to ensure system interoperability.

The RACE program started in 1987 as a ten-year program to be carried out in two phases, RACE I and RACE II, preceded by a definition phase. The main goal of the program is to contribute to the introduction of **integrated broadband communications (IBC)** into community-wide services, taking into account the evolving integrated service digital network and national introduction strategies. During the progress of the program, it becomes obvious that telecommunications and advanced information services will play a key role in world socio-economic development as we approach the 21st century. This offers tremendous market opportunities for new services and infrastructure implementation, and, thus, accelerates the maturity of the integrated broadband communication concept and technologies. It is also noted that an underlying objective of RACE is to "promote the competitiveness of the Community's telecommunications industry, operators and service providers in order to make available to final users the services which will sustain the competitiveness of the European economy and contribute to maintaining and creating employment in Europe." The program has, indeed, created large-scale awareness of the market opportunities within the industry that will accompany the implementation of the next generation of telecommunications services and systems in Europe. In addition, it is contributing to the creation of awareness among users of the advantage that application of advanced communications will deliver.

Up to now mobile services are divided essentially into four groups, cellular, paging, cordless, and private mobile radio, each providing services using a number of different networks and incompatible systems. The standardization of systems, interfaces to users and service providers, and the decreasing cost of terminals and networks have created economies of scale and high-usage levels. This rapid market penetration, however, will lead to capacity shortage by the turn of the century. A third generation mobile system is envisaged by the European Union to accommodate the requirements of higher capacity, service integration, and universal personal telecommunications. This third generation mobile system will provide seamless connections between mobile and fixed network environments. A prerequisite for this is a flexible air interface that can adapt to meet varying service and operational requirements as may be found within domestic and business customer premises networks and public networks. In the mid-1980s work began internationally, in the International Telecommunication Union-Radio (ITU-R, previously CCIR) Interim Working Party 8/13 (now Task Group 8/1), on systems that would supersede existing cellular and cordless systems. In Europe, the work of the RACE mobile definition phase project and then the full mobile project in the RACE I program laid conceptual foundations for a **universal mobile telecommunications system** (UMTS). Subsequently, this evolved into a major project line in the RACE II program and has since begun to be standardized within the European Telecommunication Standard Institute (ETSI).

93.2 RACE I Mobile Project

The RACE I program is designed to evaluate the options and clear the path for standardization. Specifically, the RACE I mobile project has contributed to the development of third generation mobile telecommunication systems, conceived as the universal mobile telecommunication system and mobile broadband system (MBS), intended to realize true personal mobile radio communications, at the least, from anywhere within Europe, and allow people to communicate freely with each other in domestic or business environments, urban or rural settings, and fixed locations or roving vehicles; see Fig 93.1. The project has paved the way for the creation of the ETSI subtechnical committee SMG-5 charged to provide standardization for the UMTS. The project has also made significant

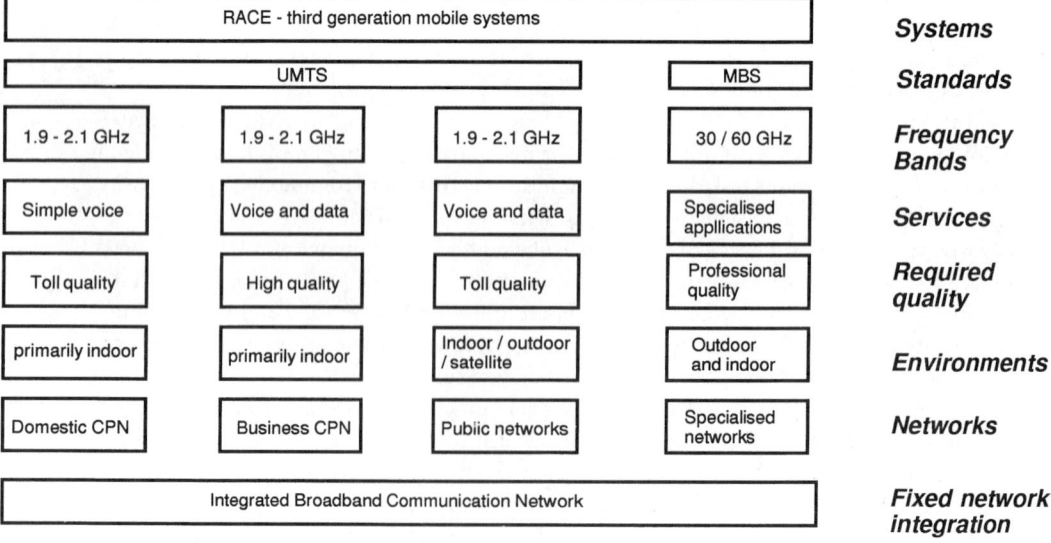

FIGURE 93.1 Schematic overview of UMTS and MBS.

contributions towards the allocation of the radio spectrum requirements for future public land mobile telecommunications systems. It was on the basis of the RACE work that the **WARC'92** decided to allocate 230 MHz in the 2-GHz band that will replace the current pan-European Global System for Mobile Communication (GSM) system by the early part of the 21st century. In addition, the RACE I project also provided the technologies and service foundations for mobile broadband systems operating in the vicinity of 60 GHz.

The UMTS is designed to be a universal multiservice, multifunction digital system, evolving from currently operational or planned second generation systems. The provision of user data rate of up to 2 Mb/s will be achieved through the UMTS in the 2-GHz frequency band and will allow seamless interface between the mobile and the fixed network. This will also allow integrated service digital network (ISDN) services in the fixed network to be extended into the mobile domain. For user data rates in excess of 2 Mb/s, the use of the 60-GHz band will be required to provide sufficient bandwidth and to share the broadband transport infrastructure and functionality directly. In brief, the RACE I mobile project has made significant contributions to the creation of a system description document for the third generation mobile system, the definition of a flexible air interface, and the key functions of the fixed network, supporting the allocation of suitable frequencies and establishing common functional specifications for the UMTS as the basis for the contribution to standards. Indeed, the project has provided a firm foundation for the second phase of the project: RACE II.

93.3 RACE II Mobile Project Line

The RACE II program is structured into project lines containing groups of related projects in specific subject areas. For mobile applications, the mobile and personal communications project line is directly related to the implementation of third generation wireless communications. The European Commission recognized that "recent market developments and initiatives taking place in Japan and North America, indicate that in order to meet the challenges of globalization and international competition, very significant efforts should be devoted in particular to the development of mobile/personal communications networks, systems, products and services, as a means of capturing a meaningful percentage of world market." This provided a strong impetus to the European Union to direct the development not only of a Pan-European market but also of the world markets. The primary objective of the project line, particularly for UMTS, is the contribution to standardization activities. This emphasis is justified by the fact that mobile standards are becoming increasingly complex and can only be achieved efficiently through close collaboration of operators and manufacturers. Activities in support of standards development are as follows:

1. The study of network architecture, signalling protocol, network management, mobility management, radio and fixed network resource management, security aspects, and billing and accounting.
2. The evaluation of two different multiple access techniques for the radio interface based on novel spread spectrum techniques and the enhancement of the relatively mature time division multiple access technology through computer simulation and real-time testbeds with a set of common assumptions and performance assessment scenarios.
3. The study of the integration of UMTS satellite components into the fixed infrastructure, including radio interface compatibility as well as signalling and interworking issues.
4. The research and development of technologies, models, and tools in areas including adaptive antenna design and applications, modulation, source and channel coding, terminal design, and radio-wave propagation characteristics.
5. The study of the marketing aspects, service creation, quality of service, service requirements, evolution and implementation strategies, operational and functional requirements in anticipation of the increasing number and types of services currently envisaged for UMTS, and the reduction of transmission rates due to more efficient source coding and data compression techniques.

The time frame for the UMTS activities is designed to be compatible with the expected availability of a standard around 1998, which will allow the commercial exploitation to take place by the turn of the century. By contrast, the standardization of mobile broadband systems is likely to take place over a longer period of time and the current activities are focused on the development of the enabling technologies and the evaluation of the economic feasibility of mobile broadband services.

The mobile and personal communication project line is divided into a number of projects in order to evaluate the market and technical requirements, as well as the physical implementation aspects, of the UMTS. Specifically fixed network issues are studied within the MONET project, whereas multiple access techniques are addressed by the CODIT and ATDMA projects. The latter two projects investigate the relative merits of spread spectrum multiple access and time division multiple access (TDMA), respectively. The issue of integrating satellite mobile components of the UMTS into the terrestrial network is addressed by SIANT. This project will interact with MONET on the fixed network aspect and CODIT and ATDMA on air interface definition issues. The MAVT project is aimed at developing a mobile audio-visual terminal demonstrator to support bandwidth efficient video services over mobile radio channels. It evaluates different source and channel coding techniques for low-bit-rate image and speech transmission for future mobile terminals. The MBS project, which studies the requirement of supporting high-bit-rate services using millimeter-wave technology, will have strong interaction with the MONET project on network architecture and evolution, as well as CODIT and ATDMA projects on air interface and cellular coverage definition. The project TSUNAMI, which researches into smart antenna technologies for universal advanced mobile infrastructure, is considering the benefits that adaptive antenna technology can bring to both air interface techniques, as well as developing appropriate component technologies. Finally, the GIRAFE project is expected to have close interaction with MAVT, CODIT, ATDMA, and SIANT on terminal technology issues. In the following the key project areas are described in more details.

Code Division Multiple Access

The overall objective of the CODIT project is to explore the potential of code division multiple access for a future high-capacity UMTS. A system concept based on CDMA is established, together with advanced radio frequency technologies and cellular architecture and techniques. A system demonstrator comprising test mobile stations, radio base stations, and a radio network controller is to be designed and built in order to validate the system concept through laboratory bench test, field trials, and computer simulation.

The CODIT system is proposed to be able to handle a large number of users in pico-, micro-, and macro-cells with simple spectrum management, which would accommodate multiple operators as well as private networks. Operating environments will be for both indoor and outdoor with a high grade of service. Low-power terminals will be supported for high-quality voice services. It will also be able to support variable data rates for advanced data service and will be compatible with existing data networks.

It is noted that UMTS will need to accommodate different bit rates with instantaneous data rates as high as 2 Mb/s. Following the evaluation of different spread spectrum multiple access schemes, a direct-sequence CDMA (DS-CDMA) scheme with three different chip rates is adopted. The CODIT system will, therefore, be able to support three different bandwidths of approximately 1, 5, and 20 MHz. For each bandwidth, there is an associated chip rate. The radio access scheme, which is based on direct sequence code division multiple access, allows flexibility when choosing the appropriate coding of information at any given time instant. During the course of a frame, which has a duration of 10 ms, the coding is not changed. The selected parameters, however, can change from frame to frame, which is utilized by the speech codec for adaptation. The speech codec, which is based on a codebook excited linear prediction algorithm, can adaptively vary its rate from 16 kb/s down to 4 kb/s. In order to minimize the requirements of synchronization and avoid code planning, user information is spread with pseudonoise (PN) sequences. Only one long PN sequence

The RACE Program

is adopted, and each link is assigned a unique phase shift. In addition, the system is unsynchronized in the sense that mobile stations connected to a base station are not synchronized, and base stations are not synchronized among themselves. An exception is during soft handover when synchronous reception at the mobile station is required to combine two signals from two different base stations arriving within the time window of the Rake receiver of a mobile station. With measurements from the mobile station, the time of transmission from the base stations is adjusted on a call-by-call basis in order to achieve synchronous reception at the former.

A computer simulation of the system in accordance with the proposed air interface has been set up and is used to verify the specification and produce radio interface performance. Simulation results have shown that system capacity is dependent on the environment in which the system operates and is typically lower in picocellular and microcellular than in macrocellular environments. It is, therefore, important that capacity is to be boosted in microcell and picocell environments in order to ensure that DS-CDMA is to be an all round solution for UMTS service provision. This can be achieved by the use of antenna diversity, macrodiversity techniques, and the use of special spreading codes. Furthermore, the positioning of base stations for a DS-CDMA system could affect capacity. The reduction in cellular system capacity from hexagonally distributed base stations to randomly distributed base stations is about 15%. Macrodiversity, however, can be used to improve both uplink and downlink capacity, improve signal quality, and enable a make-before-break soft handover routine at the expense of more modems in the base station and more infrastructure.

Advanced Time Division Multiple Access (ATDMA)

The main objective of the ATDMA project is to develop and quantify the potential of a time division multiple access (TDMA) system for UMTS. Given that the maximum capabilities of TDMA are far from being attained by present systems (e.g., GSM and DECT), the target of the project is to develop a set of techniques that can improve the overall system capacity by an order of magnitude, while at the same time improving the quality of service, grade of service, and range of services. The goals of the ATDMA project are to compare the performance of an improved TDMA system with other radio access techniques and to evaluate the system architecture developed and implemented in the testbed. The real-time testbed is intended to provide a signal processing platform for the implementation of transport techniques and a protocol test platform for the control techniques, where the latter is aimed at assessing the effectiveness of the link adaptation concept.

The project is structured into four technical areas: dealing with techniques studies, channel characterization, testbed implementation, and system performance evaluation. System evaluation will be performed using both real-time testbed and computer simulations. The real-time system demonstrator comprising two mobile stations, two radio base stations, a network emulator, and a radio channel simulator will be implemented to evaluate the performance of the radio frequency subsystem, transport and low-level control subsystem. This testbed will support a range of service demonstrations, including advanced video and voice coding technologies. The services will operate under different environments taking into account the needs of multiple operators and low-cost terminals. To achieve this, the radio control issues, such as channel allocation strategies, resource reservation techniques, signalling channel assignments, and response time of resource management mechanisms, are to be optimized in accordance with allowable interference and noise levels, transmission bandwidth, modulation schemes, forward error correction codes, robustness of the source, and channel coding, etc.

The impact of duplexing on the division of the UMTS frequency band and the system complexity and flexibility were examined resulting in the conclusion that both time and frequency division duplex must be employed. The transport interface will support two carrier symbol rates of 450 kb/s for macrocells and 1.8 Mb/s for both microcells and picocells. In addition, it will support the following: a common modulation scheme for all environments, a single demodulator-channel estimator, a prefilter equalizer, flexible burst and frame structures, a speech codec with gross rate of 13 kb/s, including error protection and net speech rate 6.4 and 9.6 kb/s, and adaptive error protection schemes.

The key to the capacity gain is by the dynamic utilization of radio resource so as to optimize the system performance from a global perspective. This will be realised via the combined use of packet access and a generic air interface that supports three cell types, picocell, microcell, and macrocell, and incorporates static and dynamic adaptation strategies. Specifically, static adaptation will enable system parameter changes to occur on a medium-to-long term basis. This form of adaptation will be used at call setup and during interoperator or intercell type handover to set the air interface's initial conditions. Dynamic adaptation will be performed during each call in order to cope with the time-varying radio-wave propagation characteristics of an environment and the traffic loading on the terminal and the base station. From the transport and control plane point of view, two adaptation mechanisms can be differentiated. The transport plane consists of transport techniques (including modulation, equalization, burst and frame structure, etc.) and error protection schemes (including automatic repeat request, forward error correction, etc.) which could be modified dynamically. Evidently, the forward error correction code rate, slot allocation, and interleaving depth could be mutually interchanged to minimize the required average resource allocation while still maintaining a rated quality of service and the need to minimize the overall spectrum demand. The dynamic adaptation can also be realised in the control plane. This includes the use of adaptive power control, handover algorithm optimization, packet transmission with bandwidth allocation on demand for mixed voice and data services, and link adaptation, which dynamically adapts the parameters of the air interface during a call in order to cope with varying propagation, interference, quality of service, and traffic loading conditions. Thus, if the rated quality of service objectives can be relaxed during overloading conditions, then this mechanism can offer a form of soft capacity.

Mobile Network

The main aim of the MONET project is to develop network standards for UMTS that will integrate the infrastructure for mobile and fixed communications, as well as offer the same range of services as provided by the fixed networks. The underlying challenge is to define a fixed infrastructure capable of supporting a huge volume of mobile connected traffic over a wide geographical area. To achieve this, it is essential that new concepts for handover, call handling, location management, security, telecommunication management, and database and base station interconnection are developed in order to minimize the signalling load due to the mobility of the user. Furthermore, the UMTS network architecture has to be aligned with the integrated broadband communication network so that intelligent network (IN) and universal personal telecommunications features can be exploited.

The design of a network to support a distributed system like UMTS is a complex undertaking. The design methodology is based on the ITU-T (previously CCITT) three-stage methodology. As the starting pointing for the construction of the functional architecture and network architecture, functional models of the most prominent mobility procedures (e.g., location management, handover, and call handling) are analyzed. A functional architecture will be defined covering the domestic, business, and public environments. The mapping of functional entities to the network elements may be different for different types of environment. A broad range of possibilities for environments exists ranging from a single location area to more advanced ones comprising several location areas and paging areas. Other possible arrangements may depend on the size of cells and the expected traffic volumes. A network architecture will be defined for the domestic, business and public environments. The network architecture identifies the physical entities of the system and the physical interfaces between them; only interfaces that are relevant for standardization processes are identified. The functional model will be used for the functional descriptions of the mobility procedures. These detailed descriptions will allow the precise definition of the requirements on the functional entities of the model in terms of capabilities.

The UMTS protocol framework defines the overall structure of the application layer for UMTS signalling protocols. It contains a list of all application layer signalling protocols to be used and the application entities used to host the protocols. The design of UMTS application layer signalling protocols follows a number of new paths. Two new features are protocol reuse and protocol flexibility. Current signalling protocols have been designed for specific systems with little or no consideration given to the reuse of the resulting protocols in other systems. In addition, current signalling protocols are very tightly coupled to a single physical network topology. In UMTS, however, potentially very different topologies could be used in different environments. Thus, protocols must not be optimized for a single topology nor for a single allocation of functions. To facilitate the flexible reuse of protocols between different systems, a clear separation between the system functionality and the supporting protocol must be made. Thus, a protocol should only reflect the communication aspects of the functionality it supports. As a consequence, system specification should only indicate how relevant protocols are to be used whereas the protocol specification itself is provided separately.

A major goal with respect to the implementation of UMTS is to allow its integration into networks for fixed communication. It is because UMTS and B-ISDN are likely to mature in the same time frame that integration of UMTS into B-ISDN is logical. The B-ISDN is expected to be present in major parts of the world during the introduction of UMTS to the market. This integration scenario, thus, underlines the idea of UMTS to be a world standard: UMTS will be the wireless access for the B-ISDN. For techno-economical reasons, it is also considered important to allow integration of UMTS into networks for fixed communication and to reuse existing and forthcoming infrastructure as much as possible. It also enables UMTS to progress rapidly such that it can exploit the advanced signalling possibilities and services of the B-ISDN. In addition the B-ISDN protocols and functions will be used where possible complete with UMTS specific protocols and functions. Regarding the available bandwidth, B-ISDN will not impose limitations on the service provisioning by UMTS. As far as the integration process is concerned, promising developments in the intelligent network arena could facilitate a convenient route. As a word of caution, although there is a strong emphasis on integration aspects in the UMTS requirements, it still might require a fall-back option to implement UMTS as a stand-alone system (especially in areas where no B-ISDN is available). This scenario requires the UMTS to be able to operate optionally with no constraints from existing or forthcoming systems. The result is that both the radio access part and the fixed part of the network are UMTS specific, together with new UMTS specific network components, such as base stations, switches, and control equipment.

The intelligent network concept plays an important role in existing and future telecommunication networks, including mobile systems, in Europe. Originally, the IN was conceived as a means for service providers to provide rapid creation and introduction of new services upon existing fixed networks. For each mobility procedure, a part of the functions can be implemented in IN whereas the remaining part is implemented in the backbone network. In ETSI IN capability set one, service control and data functions are centralized. The need for future development of IN towards wider service coverage with interconnection of different IN networks has already been recognized and this requires decentralized service control. The additional signalling efficiency requirements of handover, and the structure of distributed databases associated with UMTS also reinforced the need for distribution of service control and data functions. Up to now IN has only addressed services that are applied to calls that are in the process of setting up, in-progress, or clearing down. Hence there is great emphasis on the basic call state model in the switch. Mobility procedures, however, like location updating can occur while there is no active call. Developments are necessary in order to manage these noncall related functions. In the current IN, only one service at a time can be activated from the basic call. In a mobile system, however, several services and mobility procedures can be active in parallel. For example, a location updating or a handover might occur during invocation of another IN service. To support this, a more advanced mechanism for interaction between basic call control and service control is required.

Security is an indispensable feature of any wireless communication system. The development of a security architecture for UMTS has taken into account two main factors: The first is that security requirements, which have an equivalent in existing systems, can be fulfilled in a novel and more efficient way due to the advances in the field of data and network security and the supporting technology. It is now possible to use a wider range of security mechanisms in order to realize the basic security services and the associated key management. The second factor is that UMTS offers new features, which distinguish it from existing systems. These new features also lead to new security requirements and demand new security solutions. It is generally assumed that a UMTS user is represented by a smart card in security procedures. This smart card plays a role in most of the user related features and is called a subscription identification device. These user related features are 1) multiuser terminals, 2) direct support to UPT, 3) new forms of payment, 4) user anonymity, 5) incontestable charging, 6) detection of modification, 7) authentication of a service provider by the user, and 8) authentication of a network operator by the user. Network related features are 1) authentication during handover, 2) incontestable charging between network operator and service provider, 3) restricted availability of authentication information to visited network operator, 4) security in the fixed network, and 5) mobile customer premises network (CPN).

Subscriber accounting can be divided into three subprocesses: usage metering, charging, and billing. Metering of the use of public UMTS resources is done in the originating and terminating public UMTS. The charges for a UMTS call are based on an origin–destination relationship, not on the actual route used. All public UMTS users associated with a subscriber have a home operator. Since the originating network knows the terminating network, the charges for the first part of the connection can be deduced from a so-called mobile originating usage record. As the called party might be roaming, however, the charges for the last part of the connection are to be deduced from a so-called mobile terminating usage record. Furthermore, as part of the charging process is responsible for collecting usage records, all usage records of a particular call that are registered by visited public UMTS operators must eventually be transferred to the home UMTS operator of the involved users. This is done only in the case of billing the subscriber. A UMTS subscriber shall be billed by his home operator only. It is preferred that all UMTS services used are specified in one bill. Call charges are billed either to the home operator of a subscriber or directly to the user.

Interoperator accounting again is proposed to follow the three stages of usage metering, charging, and billing. Unlike subscriber accounting, usage metering information is to be based on the actual route of a call. Interoperator accounting must be verifiable. This implies that when traffic over a given connection or link between two operators is to be charged, usage metering needs to be done on both outgoing and incoming traffic. Since part of the charging process is responsible for collecting usage records, all usage records that are registered by visited UMTS operators must eventually be sent to the home UMTS operators or to a clearinghouse. The billing process compiles the usage records and clears with other operators. This can be done directly among operators or through a clearinghouse.

Mobile Audio-Visual Terminal

The main objective of the MAVT project is to develop video and audio coding schemes for the transmission of multimedia services in a mobile environment taking into account user and service requirements, network and channel characteristics, as well as terminal architecture. A demonstrator will be implemented with low-bit-rate video and audio coding algorithms in accordance with the ISO MPEG4 coding standard. The demonstrator will be interfaced with the CODIT and ATDMA testbed for performance evaluation.

Following the development of the $p \times 8$ kb/s source coding scheme and the low-bit-rate audio algorithm, the possibility of transmission of video data via mobile radio channels has been confirmed. New algorithms based on current standards [ITU-T Rec H.61—Codec for audio visual services at $n \times 384$ kbits, the International Standards Organization, Moving Picture Experts Group (MPEG),

International Standards Organization, Joint Photographic Experts Group (JPEG)] are developed by optimizing existing low-bit-rate coding schemes in terms of video and audio coding delay. Rate compatible punctured convolutional codes, together with schemes that enable flexible exchange between source and channel data rate, as well as combined source/channel coding methods are evaluated.

Mobile Broadband System

The MBS project addresses the system concepts, techniques, and technologies required for the realization of a mobile broadband system. It also identifies the potential market and economical issues relating to the widespread introduction of the corresponding systems and services. The project also aims at demonstrating the industrial capability to produce the subsystems required by future high-data rate (154 Mb/s) mobile communication systems in a cost-effective manner. In addition, the system aspects, radio access schemes, network management issues, integration with Integrated Broadband Communications Networks (IBCN), broadband wireless local area networks (LANs) and multimedia applications are to be studied. It is clear that MBS will be a mobile extension to the B-ISDN and that it will encompass services to be provided by other systems such as HIPERLAN (High performance radio local area network). Even though MBS will be able to support low-bit-rate services, it will not be a replacement of UMTS but an enhancement, providing its applications mainly to the professional users.

From the study, it was identified that forward mobile controlled handover is the most appropriate scheme to support continuous mobility for mobile broadband services. Studies on antenna configuration suitable for elongated street cells have indicated the use of a single radiator solution based on a dielectric antenna structure as the best solution. The use of adaptive beam forming antennas in order to improve spectrum efficiency and transmission quality is potentially feasible.

Smart Antenna Technology for Advanced Mobile Infrastructure

The TSUNAMI project addresses the development of adaptive antenna component technologies at frequencies and bandwidths appropriate for UMTS. Components of interest include array antennas, radio frequency, intermediate frequency and digital beam forming networks, digital signal processor, and efficient adaptive array control algorithms. The improvement of mobile system performance through the increase in antenna directivity will also be addressed.

Adaptive antenna component specifications and prototypes are generated using a top down approach. The primary focus will be the use of adaptive antennas at the base stations. This will consist of a balanced combination of requirement definition, traffic analysis, propagation analysis, and measurement and performance evaluation using field trials and computer simulations. Radio planning and traffic engineering models will be developed based on the coverage characteristics of adaptive base station antennas.

Integration of Satellite in Future Mobile Network

The SIANT project is designed for the evaluation and identification of the requirements for the integration of satellites into the UMTS. The project is charged to optimize the whole mobile network taking into account the terrestrial backbone network defined by the MONET project. The project is further responsible for providing a set of recommendations on service aspects, and radio aspects, network aspects, and security aspects for the satellite integration of the UMTS.

The project sets out to define mobile services to be offered by considering terrestrial UMTS and MBS services. The result is a set of operational and functional requirements satisfying user needs, as well as network interoperability constraints. The definition of the air interface will take into account the features derived either from ATDMA or CODIT projects. Key issues to be addressed are the requirements and limitations for the integration of satellite systems, optimization of radio resource management, orbit selection and satellite architecture, technology aspects for the terminals, and the air interface definition.

Radio Front End

The GIRAFE project is to investigate the application of microelectronics integration and packaging techniques to the radio front end for mobile and wireless telephones of different standards at 1.5–2.3 GHz by creating a library of basic radio frequency building blocks, including transmit mixer, receive mixer, phase shifter, voltage control oscillator, and fully integrated phase lock loop in a high-frequency silicon bipolar process. It will also investigate and develop novel techniques that will have a major impact on low-cost, high-volume packaging and external passive inductive components for radio frequency integrated front end applications.

Key issues to be addressed are 1) multisystem requirement specification, 2) low-voltage, low-power radio frequency designs, 3) low-cost external planar passive inductive components for matching and filtering, and 4) low-cost, high-volume radio frequency packaging techniques. The goals are to compile a set of common functional specifications for several mobile and wireless standards and to develop a library of low-voltage, low-power, highly integrable basic radio frequency building blocks. In addition it is also expected to develop external radio frequency planar inductive passive components and low-cost, high-volume radio frequency packaging techniques.

93.4 Beyond the RACE Program

It is recognized by the European Union that there is a continuing need for communication research and development in Europe. Future work will build on the achievements of the RACE program and contribute further to the success of European activities in the area. Based on the view of the RACE management committee and other expressed opinions of leading experts in the field, a new program is considered by the European Union to follow on from RACE, taking into account the changed situation in view of the contribution from RACE. This new program is entitled research and development in advanced communications technologies and services (ACTS) [Dasilva and Fernandes, 1995]. This will benefit the well-established practice of collaboration on a European level. An important distinction between ACTS and RACE is that the former focuses on operational trials and the rationale for performing all of the research and development, whereas the latter focuses on technology and service development. The ACTS program is budgeted for resource of in excess of 100 million ECU, thereby giving a new impetus to the further development of mobile and personal communication systems.

Defining Terms

Integrated broadband communications: A global concept that covers all kinds of communications and technical and operational means to offer services. Its related infrastructure and services will be offered by network operators and service providers using mobile, terrestrial, broadcast, and satellite transmissions, with a range of equipment from different manufacturers adapted for business and civilian application.

Universal mobile telecommunications system: The European version of the third generation mobile system that is being standardized by the European Telecommunications Standard Institution.

WARC'92: The World Administration Radio Congress, which identified the radio spectrum for future public land mobile telecommunications systems internationally in 1992.

References

Chia, S. 1992. The universal mobile telecommunication system. *IEEE Comm. Mag.* Dec., 30(2):54–62.

Chia, S.T.S. and Grillo, D. 1992. UMTS—mobile communications beyond the year 2000: requirements, architecture and system options. *IEE Electronics and Comm. Eng. J.* Oct., 4(5):331–340.

Commission of the European Communities. 1993. *Proceeding of RACE Mobile Telecommunications Workshop*, Metz, France. Directorate General XIII. Brussels. June.

Commission of the European Communities. 1994. Research and technology development in advanced communications technologies in Europe-RACE 1994. Directorate General XIII. Annual Rept. Feb. Brussels, Belgium.

Commission of the European Communities. 1994. Research and technology development in advanced communications technologies in Europe-RACE 1994. Sec. 2.1, p. 3. The RACE Programme-Objectives.

Commission of the European Communities. 1994. Research and technology development in advanced communications in Europe-RACE 1994. Sec. 6.3.2.1, p. 43. "Project Line 3-Mobile and Personal Communications Overview of Project Line, Introduction."

Commission of the European Communities. 1994. Research and technology development in advanced communications technologies in Europe-RACE 1994. Sec.2.1, p. 3. The RACE Programme–Objectives.

Commission of the European Communities. 1994. *Proceeding of RACE Mobile Telecommunications Workshop*, Amsterdam. Directorate General XIII. Brussels. Vol. 1 and 2, May.

Dasilva, J. and Fernandes, B. E. 1995. The European research programme for advanced mobile systems. *IEEE Personal Comm. Mag.* Feb., 2(1):14–19.

Hsing, T.R., Chen, C.T., and Bellisio, J.A. 1995. Video communications and services in the copper loop. *IEEE Com. Mag.* Jan., 31(1):62–68.

1993. Special issue on the European path toward UMTS, *IEEE Personal Comm. Mag.* Feb., 2(1).

Further Information

The *IEE Electronics Communication Engineering Journal* frequently publishes review papers submitted by the RACE projects. These papers are normally identified by RACE section in each issue. Another active forum for detailed information of the RACE projects is the Subtechnical Committee, Special Mobile Group 5 of the European Telecommunications Standards Institute (ETSI), which is charged to study universal mobile telecommunications system matters. This body meets quarterly with a significant amount of contributions submitted by RACE participants as temporary documents to support standardization. These temporary documents can be obtained by applying directly to the ETSI. Finally, the European Union Directorate General XIII RACE Office in Brussels publishes annual reports and workshop digest on all RACE projects.

94
Half-Rate Standards

Wai-Yip Chan
Illinois Institute of Technology

Ira Gerson
Motorola Corporate Systems Research Laboratories

Toshio Miki
NTT Mobile Communication Network, Inc.

94.1 Introduction..1338
94.2 Speech Coding for Cellular Mobile Radio Communications..1339
94.3 Codec Selection and Performance Requirements...............1340
94.4 Speech Coding Techniques in the Half-Rate Standards........1341
94.5 Channel Coding Techniques in the Half-Rate Standards......1342
94.6 The Japanese Half-Rate Standard 1344
94.7 The European GSM Half-Rate Standard 1346
94.8 Conclusions 1350

94.1 Introduction

A half-rate speech coding standard specifies a procedure for digital transmission of speech signals in a digital cellular radio system. The speech processing functions that are specified by a half-rate standard are depicted in Fig. 94.1. An input speech signal is processed by a *speech encoder* to generate a digital representation at a *net bit rate* of R_s bits per second. The encoded bit stream representing the input speech signal is processed by a *channel encoder* to generate another bit stream at a *gross bit rate* of R_c bits per second, where $R_c > R_s$. The channel encoded bit stream is organized into data frames, and each frame is transmitted as payload data by a radio-link access controller and modulator. The net bit rate R_s counts the number of bits used to describe the speech signal, and the difference between the gross and net bit rates ($R_c - R_s$) counts the number of error protection bits needed by the *channel decoder* to correct and detect transmission errors. The output of the channel decoder is given to the *speech decoder* to generate a *quantized* version of the speech encoder's input signal. In current digital cellular radio systems that use time-division multiple access (TDMA), a voice connection is allocated a fixed transmission rate (i.e., R_c is a constant). The operations performed by the speech and channel encoders and decoders and their input and output data formats are governed by the half-rate standards.

Globally, three major TDMA cellular radio systems have been developed and deployed. The initial digital speech services offered by these cellular systems were governed by *full-rate standards*. Because of the rapid growth in demand for cellular services, the available transmission capacity in some areas is frequently saturated, eroding customer satisfaction. By providing essentially the same voice quality but at half the gross bit rates of the full-rate standards, half-rate standards can readily double the number of callers that can be serviced by the cellular systems. The gross bit rates of the full-rate and half-rate standards for the European Groupe Speciale Mobile (GSM), Japanese Personal Digital Cellular[1] (PDC), and North American cellular (IS-54) systems are listed

---

[1] Personal Digital Cellular was formerly Japanese Digital Cellular (JDC).

Half-Rate Standards

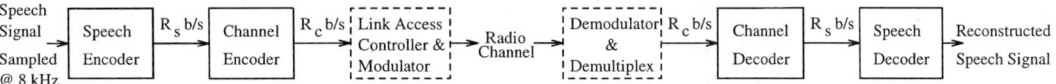

FIGURE 94.1 Digital speech transmission for digital cellular radio. Boxes with solid outlines represent processing modules that are specified by the half-rate standards.

TABLE 94.1 Gross Bit Rates Used for Digital Speech Transmission in Three TDMA Cellular Radio Systems

Standard Organization and Digital Cellular System	Gross Bit Rate, b/s Full Rate	Half Rate
European Telecommunications Standards Institute (ETSI), GSM	22,800	11,400
Research & Development Center for Radio Systems (RCR), PDC	11,200	5,600
Telecommunication Industries Association (TIA), IS-54	13,000	6,500

in Table 94.1. The three systems were developed and deployed under different time tables. Their disparate full- and half-bit rates partly reflect this difference. At the time of writing (January, 1995), the European and the Japanese systems have each selected an algorithm for their respective half-rate **codec**. Standardization of the North American half-rate codec has not reached a conclusion as none of the candidate algorithms has fully satisfied the standard's requirements. Thus, we focus here on the Japanese and European half-rate standards and will only touch upon the requirements of the North American standard.

94.2 Speech Coding for Cellular Mobile Radio Communications

Unlike the relatively benign transmission media commonly used in the public-switched telephone network (PSTN) for analog and digital transmission of speech signals, mobile radio channels are impaired by various forms of fading and interference effects. Whereas proper engineering of the radio link elements (modulation, power control, diversity, equalization, frequency allocation, etc.) ameliorates fading effects, burst and isolated bit errors still occur frequently. The net effect is such that speech communication may be required to be operational even for bit-error rates greater than 1%. In order to furnish reliable voice communication, typically half of the transmitted payload bits are devoted to error correction and detection.

It is common for low-bit-rate speech codecs to process samples of the input speech signal one frame at a time, e.g., 160 samples processed once every 20 ms. Thus, a certain amount of time is required to gather a block of speech samples, encode them, perform channel encoding, transport the encoded data over the radio channel, and perform channel decoding and speech synthesis. These processing steps of the speech codec add to the overall end-to-end transmission delay. Long transmission delay hampers conversational interaction. Moreover, if the cellular system is interconnected with the PSTN and a four-wire to two-wire (analog) circuit conversion is performed in the network, feedbacks called *echoes* may be generated across the conversion circuit. The echoes can be heard by the originating talker as a delayed and distorted version of his/her speech and can be quite annoying. The annoyance level increases with the transmission delay and may necessitate (at additional costs) the deployment of **echo cancellers**.

A consequence of user mobility is that the level and other characteristics of the acoustic background noise can be highly variable. Though acoustic noise can be minimized through suitable acoustic transduction design and the use of adaptive filtering/cancellation techniques [Ohya, Suda, and Miki, 1994; Suda, Ikeda, and Ikedo, 1994; Gibson, Koo, and Gray, 1991], the speech encoding algorithm still needs to be robust against background noise of various levels and kinds (e.g., babble, music, noise bursts, and colored noise).

Processing complexity directly impacts the viability of achieving a circuit realization that is compact and has low-power consumption, two key enabling factors of equipment portability for the end user. Factors that tend to result in low complexity are fixed-point instead of floating-point computation, lack of complicated arithmetic operations (division, square roots, transcendental functions), regular algorithm structure, small data memory, and small program memory. Since, in general, better speech quality can be achieved with increasing speech and channel coding delay and complexity, the digital cellular mobile-radio environment imposes conflicting and challenging requirements on the speech codec.

94.3 Codec Selection and Performance Requirements

The half-rate speech coding standards are drawn up through competitive testing and selection. From a set of candidate codec algorithms submitted by contending organizations, the one algorithm that meets basic selection criteria and offers the best performance is selected to form the standard. The codec performance measures and codec testing and selection procedures are set out in a test plan under the auspices of the organization (Table 94.1) responsible for the standardization process [see, e.g., Telecommunication Industries Association, 1993]. Major codec characteristics evaluated are speech quality, delay, and complexity. The full-rate codec is also evaluated as a *reference codec*, and its evaluation scores form part of the selection criteria for the codec candidates.

The speech quality of each candidate codec is evaluated through listening tests. To conduct the tests, each candidate codec is required to process speech signals and/or encoded bit streams that have been preprocessed to simulate a range of operating conditions: variations in speaker voice and level, acoustic background noise type and level, channel error rate, and stages of **tandem coding**. During the tests, subjects listen to processed speech signals and judge their quality levels or annoyance levels on a five-point opinion scale. The opinion scores collected from the tests are suitably averaged over all trials and subjects for each test condition [see Jayant and Noll, 1984, for mean opinion score (MOS) and degradation mean opinion score]. The categorical opinion scales of the subjects are also calibrated using *modulated noise reference units* (*MNRUs*) [Dimolitsas, Corcoran, and Baraniecki, 1994]. Modulated noise better resembles the distortions created by speech codecs than noise that is uncorrelated with the speech signal. Modulated noise is generated by multiplying the speech signal with a noise signal. The resultant modulated noise is scaled to a desired power level and then added to the uncoded (clean) speech signal. The ratio between the power level of the speech signal and that of the modulated noise is expressed in decibels and given the notation *dBQ*. Under each test condition, subjects are presented with speech signals processed by the codecs as well as speech signals corrupted by modulated noise. Through presenting a range of modulated-noise levels, the subjects' opinions are calibrated on the dBQ scale. Thereafter, the mean opinion scores obtained for the codecs can also be expressed on that scale.

For each codec candidate, a profile of scores is compiled, consisting of speech quality scores, delay measurements, and complexity estimates. Each candidate's score profile is compared with that of the reference codec, ensuring that basic requirements are satisfied [see, e.g., Masui and Oguchi, 1993]. An overall figure of merit for each candidate is also computed from the profile. The candidates, if any, that meet the basic requirements then compete on the basis of maximizing the figure of merit.

Basic performance requirements for each of the three half-rate standards are summarized in Table 94.2. In terms of speech quality, the GSM and PDC half-rate codecs are permitted to underperform their respective full-rate codecs by no more than 1 dBQ averaging over all test conditions and no more than 3 dBQ within each test condition. More stringently, the North American half-rate codec is required to furnish a speech-quality profile that is statistically equivalent to that of the North American full-rate codec as determined by a specific statistical procedure for multiple comparisons [TIA, 1993]. Since various requirements on the half-rate standards are set relative to their full-rate counterparts, an indication of the *relative* speech quality between the three half-rate standards can be deduced from the test results of De Martino [1993] comparing the three full-rate

Half-Rate Standards

TABLE 94.2 Basic Performance Requirements for the Three Half-Rate Standards

Digital Cellular Systems	Min. Speech Quality, dBQ Rel. to Full Rate	Max. Delay, ms	Max. Complexity Rel. to Full Rate
Japanese (PDC)	−1 average, −3 maximum	94.8	3×
European (GSM)	−1 average, −3 maximum	90	4×
North American (IS-54)	Statistically equivalent	100	4×

codecs. The maximum delays in Table 94.2 apply to the total of the delays through the speech and channel encoders and decoders (Fig. 94.1). Codec complexity is computed using a formula that counts the computational operations and memory usage of the codec algorithm. The complexity of the half-rate codecs is limited to 3 or 4 times that of their full-rate counterparts.

94.4 Speech Coding Techniques in the Half-Rate Standards

Existing half-rate and full-rate standard coders can be characterized as *linear-prediction based analysis-by-synthesis* (LPAS) speech coders [Gersho, 1994]. LPAS coding entails using a time-varying all-pole filter in the decoder to synthesize the quantized speech signal. A short segment of the signal is synthesized by driving the filter with an *excitation* signal that is either *quasiperiodic* (for *voiced* speech) or *random* (for *unvoiced* speech). In either case, the excitation signal has a *spectral envelope* that is relatively flat. The synthesis filter serves to shape the spectrum of the excitation input so that the spectral envelope of the synthesized output resembles the filter's magnitude frequency response. The magnitude response often has prominent peaks; they render the *formants* that give a speech signal its phonetic character. The synthesis filter has to be adapted to the current frame of input speech signal. This is accomplished with the encoder performing a linear prediction (LP) analysis of the frame: the inverse of the all-pole synthesis filter is applied as an LP *error filter* to the frame, and the values of the filter parameters are computed to minimize the energy of the filter's output error signal. The resultant filter parameters are quantized and conveyed to the decoder for it to update the synthesis filter.

Having executed an LP analysis and quantized the synthesis filter parameters, the LPAS encoder performs analysis-by-synthesis (ABS) on the input signal to find a suitable excitation signal. An ABS encoder maintains a *copy* of the decoder. The encoder examines the possible outputs that can be produced by the decoder copy in order to determine how best to instruct (using transmitted information) the actual decoder so that it would output (synthesize) a good approximation of the input speech signal. The decoder copy tracks the state of the actual decoder, since the latter evolves (under ideal channel conditions) according to information received from the encoder. The details of the ABS procedure vary with the particular excitation model employed in a specific coding scheme. One of the earliest seminal LPAS schemes is *code excited linear prediction (CELP)* [Gersho, 1994]. In CELP, the excitation signal is obtained from a **codebook** of *code vectors*, each of which is a candidate for the excitation signal. The encoder searches the codebook to find the one code vector that would result in a best match between the resultant synthesis output signal and the encoder's input speech signal. The matching is considered best when the energy of the difference between the two signals being matched is minimized. A *perceptual weighting filter* is usually applied to the difference signal (prior to energy integration) to make the minimization more relevant to human perception of speech fidelity. Regions in the frequency spectrum where human listeners are more sensitive to distortions are given relatively stronger weighting by the filter and vice versa. For instance, the concentration of spectral energy around the formant frequencies gives rise to stronger *masking* of coder noise (i.e. rendering the noise less audible) and, therefore, weaker weighting can be applied to the formant frequency regions. For masking to be effective, the weighting filter has to be adapted to the time-varying speech spectrum. Adaptation is achieved usually by basing the weighting filter parameters on the synthesis filter parameters.

The CELP framework has evolved to form the basis of a great variety of speech coding algorithms, including all existing full- and half-rate standard algorithms for digital cellular systems. We outline next the basic CELP encoder-processing steps, in a form suited to our subsequent detailed descriptions of the PDC and GSM half-rate coders. These steps have accounted for various computational efficiency considerations and may, therefore, deviate from a conceptual functional description of the encoder constituents.

1. LP analysis on the current frame of input speech to determine the coefficients of the all-pole synthesis filter;
2. quantization of the LP filter parameters;
3. determination of the open-loop **pitch period** or lag;
4. adapting the perceptual weighting filter to the current LP information (and also pitch information when appropriate) and applying the adapted filter to the input speech signal;
5. formation of a filter cascade (which we shall refer to as *perceptually weighted synthesis filter*) consisting of the LP synthesis filter, as specified by the quantized parameters in step 2, followed by the perceptual weighting filter;
6. subtraction of the *zero-input response* of the perceptually weighted synthesis filter (the filter's decaying response due to past input) from the perceptually weighted input speech signal obtained in step 4;
7. an *adaptive codebook* is searched to find the most suitable periodic excitation, i.e., when the perceptually weighted synthesis filter is driven by the best code vector from the adaptive codebook, the output of the filter cascade should best match the difference signal obtained in step 6;
8. one or more nonadaptive excitation codebooks are searched to find the most suitable random excitation vectors that, when added to the best periodic excitation as determined in step 7 and with the resultant sum signal driving the filter cascade, would result in an output signal best matching the difference signal obtained in step 6.

Steps 1–6 are executed once per frame. Steps 7 and 8 are executed once for each of the *subframes* that together constitute a frame. Step 7 may be skipped depending on the pitch information from step 3, or if step 7 were always executed, a *nonperiodic excitation* decision would be one of the possible outcomes of the search process in step 7. Integral to steps 7 and 8 is the determination of gain (scaling) parameters for the excitation vectors. For each frame of input speech, the filter and excitation and gain parameters determined as outlined are conveyed as encoded bits to the speech decoder.

In a properly designed system, the data conveyed by the channel decoder to the speech decoder should be free of errors most of the time, and the speech signal synthesized by the speech decoder would be identical to that as determined in the speech encoder's ABS operation. It is common to enhance the quality of the synthesized speech by using an adaptive *postfilter* to attenuate coder noise in the perceptually sensitive regions of the spectrum. The postfilter of the decoder and the perceptual weighting filter of the encoder may seem to be functionally identical. The weighting filter, however, influences the selection of the best excitation among available choices, whereas the postfilter actually shapes the spectrum of the synthesized signal. Since postfiltering introduces its own distortion, its advantage may be diminished if tandem coding occurs along the end-to-end communication path. Nevertheless, proper design can ensure that the net effect of postfiltering is a reduction in the amount of audible codec noise [Chen and Gersho, 1995]. Excepting postfiltering, all other speech synthesis operations of an LPAS decoder are (effectively) duplicated in the encoder (though the converse is not true). Using this fact, we shall illustrate each coder in the sequel by exhibiting only a block diagram of its encoder or decoder but not both.

94.5 Channel Coding Techniques in the Half-Rate Standards

Crucial to the maintainence of quality speech communication is the ability to transport coded speech data across the radio channel with minimal errors. Low-bit-rate LPAS coders are particularly sensitive to channel errors; errors in the bits representing the LP parameters in one frame, for instance, could result in the synthesis of nonsensical sounds for longer than a frame duration. The error rate of a digital cellular radio channel with no channel coding can be catastropically high for LPAS coders. The amount of tolerable transmission delay is limited by the requirement of interactive communication and, consequently, *forward error control* is used to remedy transmission errors. "Forward" means that channel errors are remedied in the receiver, with no additional information from the transmitter and, hence, no additional transmission delay. To enable the channel decoder to correct channel errors, the channel encoder conveys more bits than the amount generated by the speech encoder. The additional bits are for error *protection*, as errors may or may not occur in any particular transmission epoch. The ratio of the number of encoder input (information) bits to the number of encoder output (code) bits is called the (channel) *coding rate*. This is a number no more than one and generally decreases as the error protection power increases. Though a lower channel coding rate gives more error protection, fewer bits will be available for speech coding. When the channel is in good condition and, hence, less error protection is needed, the received speech quality could be better if bits devoted to channel coding were used for speech coding. On the other hand, if a high channel coding rate were used, there would be uncorrected errors under poor channel conditions and speech quality would suffer. Thus, when nonadaptive forward error protection is used over channels with nonstationary statistics, there is an inevitable tradeoff between quality degradation due to uncorrected errors and that due to expending bits on error protection (instead of on speech encoding).

Both the GSM and PDC half-rate coders use *convolutional coding* [Proakis, 1995] for error correction. Convolutional codes are sliding or sequential codes. The encoder of a rate m/n, $m < n$ convolutional code can be realized using m shift registers. For every m information bits input to the encoder (one bit to each of the m shift registers), n code bits are output to the channel. Each code bit is computed as a modulo-2 sum of a subset of the bits in the shift registers. Error protection overhead can be reduced by exploiting the unequal sensitivity of speech quality to errors in different positions of the encoded bit stream. A family of *rate-compatible punctured convolutional codes* (RCPCCs) [Hagenauer, 1988] is a collection of related convolutional codes; all of the codes in the collection except the one with the lowest rate are derived by *puncturing* (dropping) code bits from the convolutional code with the lowest rate. With an RCPCC, the channel coding rate can be varied on the fly (i.e., variable-rate coding) while a sequence of information bits is being encoded through the shift registers, thereby imparting on different segments in the sequence different degrees of error protection.

For decoding a convolutional coded bit stream, the *Viterbi algorithm* [Proakis, 1995] is a computationally efficient procedure. Given the output of the demodulator, the algorithm determines the most likely sequence of data bits sent by the channel encoder. To fully utilize the error correction power of the convolutional code, the amplitude of the demodulated *channel symbol* can be quantized to more bits than the minimum number required, i.e., for subsequent *soft decision decoding*. The minimum number of bits is given by the number of channel-coded bits mapped by the modulator onto each channel symbol; decoding based on the minimum-rate bit stream is called *hard decision* decoding. Although soft decoding gives better error protection, decoding complexity is also increased.

Whereas convolutional codes are most effective against randomly scattered bit errors, errors on cellular radio channels often occur in bursts of bits. These bursts can be broken up if the bits put into the channel are rearranged after demodulation. Thus, in *block interleaving*, encoded bits are read into a matrix by row and then read out of the matrix by column (or vice versa) and then passed on to the modulator; the reverse operation is performed by a *deinterleaver* following demodulation. Interleaving increases the transmission delay to the extent that enough bits need to be collected in order to fill up the matrix.

Owing to the severe nature of the cellular radio channel and limited available transmission capacity, uncorrected errors often remain in the decoded data. A common countermeasure is to append an error detection code to the speech data stream prior to channel coding. When residual channel errors are detected, the speech decoder can take various remedial measures to minimize the

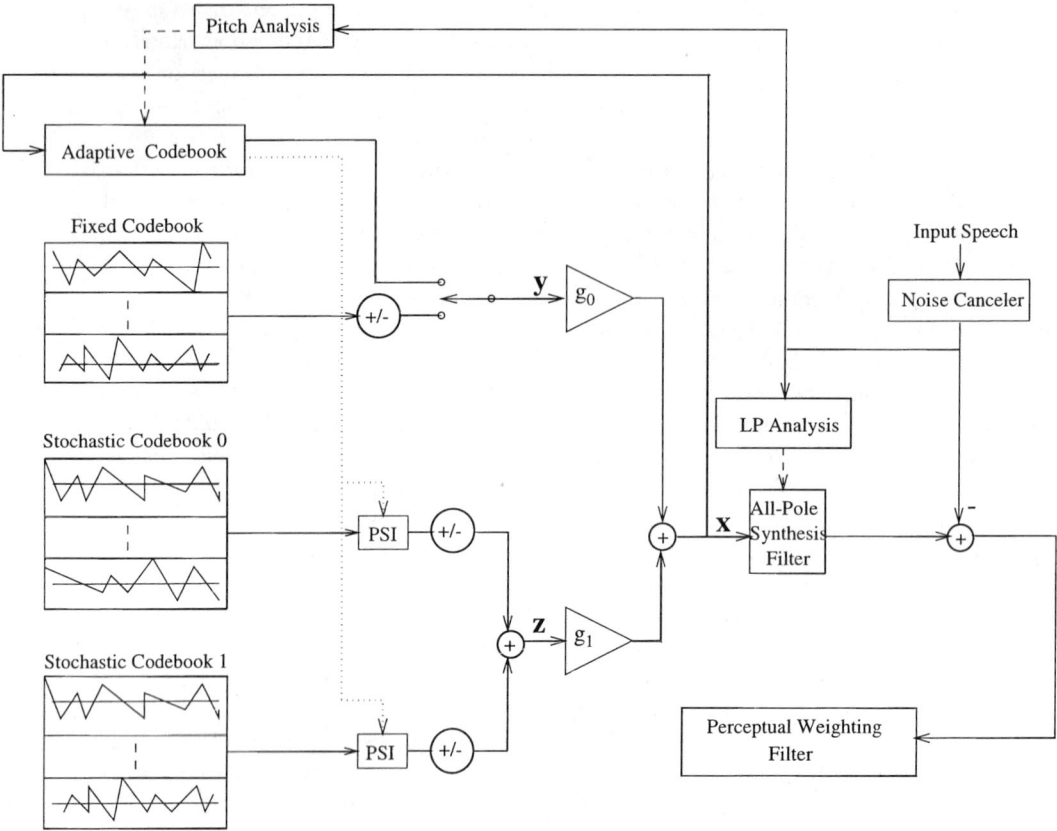

FIGURE 94.2 Basic structure of the PSI-CELP encoder.

negative impact on speech quality. Common measures are repetition of speech parameters from the most recent good frames and gradual muting of the possibly corrupted synthesized speech.

The PDC and GSM half-rate standard algorithms together embody some of the latest advances in speech coding techniques, including: *multimodal coding* where the coder configuration and bit allocation change with the type of speech input; *vector quantization* (*VQ*) [Gersho and Gray, 1991] of the LP filter parameters; higher precision and improved coding efficiency for pitch-periodic excitation; and postfiltering with improved tandeming performance. We next explore the more distinctive features of the PDC and GSM speech coders.

94.6 The Japanese Half-Rate Standard

An algorithm was selected for the Japanese half-rate standard in April 1993, following the evaluation of 12 submissions in a first round, and four final candidates in a second round [Masui and Oguchi, 1993]. The selected algorithm, called pitch synchronous innovation CELP[2] (PSI-CELP), met all of the basic selection criteria and scored the highest among all candidates evaluated. A block diagram of the PSI-CELP encoder is shown in Fig. 94.2, and bit allocations are summarized in Table 94.3. The complexity of the coder is estimated to be approximately 2.4 times that of the PDC full-rate

---

[2]There were two candidate algorithms named PSI-CELP in the PDC half-rate competition. The algorithm described here was contributed by NTT Mobile Communications Network, Inc. (NTT DoCoMo).

Half-Rate Standards

TABLE 94.3 Bit Allocations for the PSI-CELP Half-Rate PDC Speech Coder

Parameter	Bits	Error Protected Bits
LP synthesis filter	31	15
Frame energy	7	7
Periodic excitation	8×4	8×4
Stochastic excitation	10×4	0
Gain	7×4	3×4
Total	138	66

coder. The frame size of the coder is 40 ms, and its subframe size is 10 ms. These sizes are longer than those used in most existing CELP-type standard coders. However, LP analysis is performed twice per frame in the PSI-CELP coder.

A distinctive feature of the PSI-CELP coder is the use of an adaptive noise canceller [Ohya, Suda, and Miki, 1994; Suda, Ikeda, and Ikedo, 1994] to suppress noise in the input signal prior to coding. The input signal is classified into various modes, depending on the presence or absence of background noise and speech and their relative power levels. The current active mode determines whether *Kalman filtering* [Gibson, Koo, and Gray, 1991] is applied to the input signal and whether the parameters of the Kalman filter are adapted. Kalman filtering is applied when a significant amount of background noise is present or when both background noise and speech are strongly present. The filter parameters are adapted to the statistics of the speech and noise signals in accordance with whether they are both present or only noise is present.

The LP filter parameters in the PSI-CELP coder are encoded using VQ. A tenth-order LP analysis is performed every 20 ms. The resultant filter parameters are converted to 10 *line spectral frequencies* (LSFs).[3] The LSF parameters have a naturally increasing order, and together are treated as the ordered components of a vector. Since the speech spectral envelope tends to evolve slowly with time, there is intervector dependency between adjacent LSF vectors that can be exploited. Thus, the two LSF vectors for each 40-ms frame are paired together and jointly encoded. Each LSF vector in the pair is split into three subvectors. The pair of subvectors that cover the same vector component indexes are combined into one composite vector and vector quantized. Altogether, 31 b are used to encode a pair of LSF vectors. This three-way *split VQ*[4] scheme embodies a compromise between the prohibitively high complexity of using a large vector dimension and the performance gain from exploiting intra- and intervector dependency.

The PSI-CELP encoder uses a perceptual weighting filter consisting of a cascade of two filter sections. The sections exploit the pitch-harmonic structure and the LP spectral-envelope structure of the speech signal, respectively. The pitch-harmonic section has four parameters, a pitch lag and three coefficients, whose values are determined from an analysis of the periodic structure of the input speech signal. Pitch-harmonic weighting reduces the amount of noise in between the pitch harmonics by aggregating coder noise to be closer to the harmonic frequencies of the speech signal. In high-pitched voice, the harmonics are spaced relatively farther apart, and pitch-harmonic weighting becomes correspondingly more important.

The excitation vector x (Fig. 94.2) is updated once every subframe interval (10 ms) and is constructed as a *linear combination* of two vectors

$$x = g_0 y + g_1 z \tag{94.1}$$

where g_0 and g_1 are scalar gains, y is labeled as the *periodic* component of the excitation and z as the *stochastic* or *random* component. When the input speech is voiced, the ABS operation would find a value for y from the *adaptive codebook* (Fig. 94.2). The codebook is constructed out of past samples of the excitation signal x; hence, there is a feedback path into the adaptive codebook in Fig. 94.2. Each code vector in the adaptive codebook corresponds to one of the 192 possible pitch lag L values available for encoding; the code vector is populated with samples of x beginning with the Lth sample backward in time. L is not restricted to be an integer, i.e., *fractional pitch period* is permitted.

---

[3] Also known as line spectrum pairs (LSPs).
[4] Matrix quantization is another possible description.

Successive values of L are more closely spaced for smaller values of L; short, medium, and long lags are quantized to one-quarter, one-half, and one sampling-period resolution, respectively. As a result, the *relative* quantization error in the encoded pitch frequency (which is the reciprocal of the encoded pitch lag) remains roughly constant with increasing pitch frequency. When the input speech is unvoiced, y would be obtained from the fixed codebook (Fig. 94.2). To find the best value for y, the encoder searches through the aggregate of 256 code vectors from both the adaptive and fixed codebooks. The code vector that results in a synthesis output most resembling the input speech is selected. The best code vector thus chosen also implicitly determines the voicing condition (voiced/unvoiced) and the pitch lag value L^* most appropriate to the current subframe of input speech. These parameters are said to be determined in a *closed-loop* search.

The stochastic excitation z is formed as a sum of two code vectors, each selected from a *conjugate codebook* (Fig. 94.2) [Ohya, Suda, and Miki, 1994]. Using a pair of conjugate codebooks each of size 16 code vectors (4 b) has been found to improve robustness against channel errors, in comparison with using one single codebook of size 256 code vectors (8 b). The synthesis output due to z can be decomposed into a sum of two orthorgonal components, one of which points in the same direction as the synthesis output due to the periodic excitation y and the other component points in a direction orthogonal to the synthesis output due to y. The latter synthesis output component of z is kept, whereas the former component is discarded. Such decomposition enables the two gain factors g_0 and g_1 to be separately quantized. For voiced speech, the conjugate code vectors are preprocessed to produce a set of *pitch synchronous innovation* (PSI) vectors. The first L^* samples of each code vector are treated as a fundamental period of samples. The fundamental period is replicated until there are enough samples to populate a subframe. If L^* is not an integer, interpolated samples of the code vectors are used (upsampled versions of the code vectors can be precomputed). PSI has been found to reinforce the periodicity and substantially improve the quality of synthesized voiced speech.

The postfilter in the PSI-CELP decoder has three sections, for enhancing the formants, the pitch harmonics, and the high frequencies of the synthesized speech, respectively. Pitch-harmonic enhancement is applied only when the adaptive codebook has been used. Formant enhancement makes use of the decoded LP synthesis filter parameters, whereas a refined pitch analysis is performed on the synthesized speech to obtain the values for the parameters of the pitch-harmonic section of the postfilter. A first-order high-pass filter section compensates for the low-pass spectral tilt [Chen and Gersho, 1995] of the formant enhancement section.

Of the 138 speech data bits generated by the speech encoder every 40-ms frame, 66 b (Table 94.3) receive error protection and the remaining 72 speech data bits of the frame are not error protected. An error detection code of 9 *cyclic redundancy check* (*CRC*) bits is appended to the 66 b and then submitted to a rate 1/2, punctured convolutional encoder to generate a sequence of 152 channel coded bits. Of the unprotected 72 b, the 40 b that index the excitation codebooks (Table 94.3) are remapped or *pseudo-Gray coded* [Zeger and Gersho, 1990] so as to equalize their channel error sensitivity. As a result, a bit error occuring in an index word is likely to cause about the same amount of degradation regardless of the bit error position in the index word. For each speech frame, the channel encoder emits 224 b of payload data. The payload data from two adjacent frames are interleaved before transmission over the radio link.

Uncorrected errors in the most critical 66 b are detected with high probability as a CRC error. A finite state machine keeps track of the recent history of CRC errors. When a sequence of CRC errors is encountered, the power level of the synthesized speech is progressively suppressed, so that muting is reached after four consecutive CRC errors. Conversely, following the cessation of a sequence of CRC errors, the power level of the synthesized speech is ramped up gradually.

94.7 The European GSM Half-Rate Standard

A *vector sum excited linear prediction* (VSELP) coder, contributed by Motorola, Inc., was selected in January 1994 by the main GSM technical committee as a basis for the GSM half-rate standard. The

Half-Rate Standards

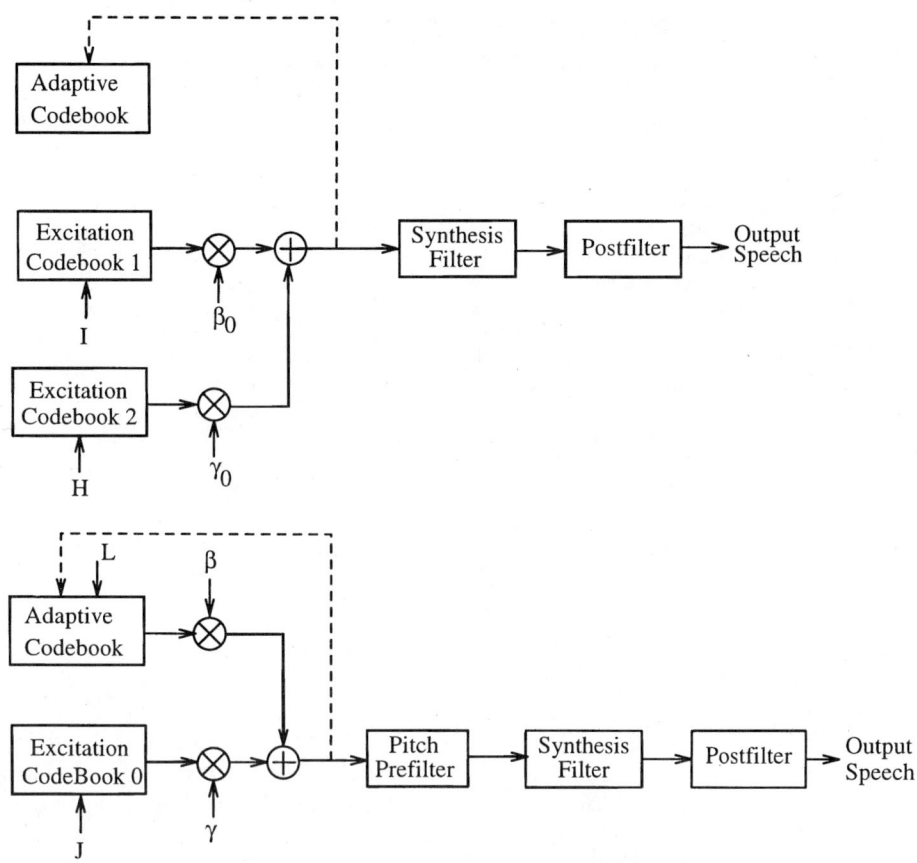

FIGURE 94.3 Basic structure of the GSM VSELP decoder. Top is for mode 0 and bottom is for modes 1, 2, and 3.

standard was finally approved in January 1995. VSELP is a generic name for a family of algorithms from Motorola; the North American full-rate and the Japanese full-rate standards are also based on VSELP. All VSELP coders make use of the basic idea of representing the excitation signal by a linear combination of *basis vectors* [Gerson and Jasiuk, 1990]. This representation renders the excitation codebook search procedure very computationally efficient. A block diagram of the GSM half-rate decoder is depicted in Fig. 94.3 and bit allocations are tabulated in Table 94.4. The coder's frame size is 20 ms, and each frame comprises four subframes of 5 ms each. The coder has been optimized for execution on a processor with 16-b word length and 32-b accumulator. The GSM standard is a *bit exact* specification: in addition to specifying the codec's processing steps, the numerical formats and precisions of the codec's variables are also specified.

The synthesis filter coefficients in GSM VSELP are encoded using the *fixed point lattice technique* (FLAT) [Gerson et al., 1994] and vector quantization. FLAT is based on the *lattice filter* representation of the linear prediction error filter. The tenth-order lattice filter has 10 stages, with the ith stage, $i \in \{1, \ldots, 10\}$, containing a a *reflection coefficient* parameter r_i. The lattice filter has an *order-recursion* property such that the best prediction error filters of all orders less than ten are all embedded in the best tenth-order lattice filter. This means that once the values of the lower order reflection coefficients have been optimized, they do not have to be reoptimized when a higher order predictor is desired; in other words, the coefficients can be optimized sequentially from low to high orders. On the other hand, if the lower order coefficients were suboptimal (as in the case when the coefficients

TABLE 94.4 Bit Allocations for the VSELP Half-Rate GSM Coder

Parameter	Bits/subframe	Bits/frame
LP synthesis filter		28
Soft interpolation		1
Frame energy		5
Mode selection		2
Mode 0		
Excitation code I	7	28
Excitation code H	7	28
Gain code G_s, P_0	5	20
Mode 1, 2, and 3		
Pitch lag L (first subframe)		8
Difference lag (subframes 2, 3, 4)	4	12
Excitation code J	9	36
Gain code G_s, P_0	5	20
Total		112

are quantized), the higher order coefficients could still be selected to minimize the prediction *residual* (or error) energy at the output of the higher order stages; in effect, the higher order stages can compensate for the suboptimality of lower order stages.

In the GSM VSELP coder, the ten reflection coefficients $\{r_1, \ldots, r_{10}\}$ that have to be encoded for each frame are grouped into three coefficient vectors $\mathbf{v}_1 = [r_1 r_2 r_3]$, $\mathbf{v}_2 = [r_4 r_5 r_6]$, $\mathbf{v}_3 = [r_7 r_8 r_9 r_{10}]$. The vectors are quantized sequentially, from $\mathbf{v}_1$ to $\mathbf{v}_3$, using a b_i-bit VQ codebook C_i for $\mathbf{v}_i$, where b_i, $i = 1, 2, 3$ are 11, 9, and 8 b, respectively. The vector $\mathbf{v}_i$ is quantized to minimize the prediction error at the energy output of the jth stage of the lattice filter where r_j is the highest order coefficient in the vector $\mathbf{v}_i$. The computational complexity associated with quantizing $\mathbf{v}_i$ is reduced by searching only a small subset of the code vectors in C_i. The subset is determined by first searching a *prequantizer* codebook of size c_i bits, where c_i, $i = 1, \ldots, 3$ are 6, 5, and 4 b, respectively. Each code vector in the prequantizer codebook is associated with $2^{b_i - c_i}$ code vectors in the target codebook. The subset is obtained by pooling together all of the code vectors in C_i that are associated with the top few best matching prequantizer code vectors. In this way, a factor of reduction in computational complexity of nearly $2^{b_i - c_i}$ is obtained for the quantization of $\mathbf{v}_i$.

The half-rate GSM coder changes its configuration of excitation generation (Fig. 94.3) in accordance with a *voicing mode* [Gerson and Jasiuk, 1992]. For each frame, the coder selects one of four possible voicing modes depending on the values of the *open-loop* pitch-prediction gains computed for the frame and its four subframes. Open loop refers to determining the pitch lag and the pitch-predictor coefficient(s) via a direct analysis of the input speech signal or, in the case of the half-rate GSM coder, the perceptually weighted (LP-weighting only) input signal. Open-loop analysis can be regarded as the opposite of closed-loop analysis, which in our context is synonymous with ABS. When the pitch-prediction gain for the frame is weak, the input speech signal is deemed to be unvoiced and mode 0 is used. In this mode, two 7-b *trained* codebooks (excitation codebooks 1 and 2 in Fig. 94.3) are used, and the excitation signal for each subframe is formed as a linear combination of two code vectors, one from each of the codebooks. A trained codebook is one designed by applying the coder to a representative set of speech signals while optimizing the codebook to suit the set. Mode 1, 2, or 3 is chosen depending on the strength of the pitch-prediction gains for the frame and its subframes. In these modes, the excitation signal is formed as a linear combination of a code vector from an 8-b adaptive codebook and a code vector from a 9-b trained codebook (Fig. 94.3). The code vectors that are summed together to form the excitation signal for a subframe are each scaled by a gain factor (β and γ in Fig. 94.3). Each mode uses a gain VQ codebook specific to that mode.

As depicted in Fig. 94.3, the decoder contains an adaptive pitch prefilter for the voiced modes and an adaptive postfilter for all modes. The filters enhance the perceptual quality of the decoded speech and are not present in the encoder. It is more conventional to locate the pitch prefilter as a section of the postfilter; the distinctive placement of the pitch prefilter in VSELP was chosen to reduce artifacts caused by the time-varying nature of the filter. In mode 0, the encoder uses an LP spectral weighting filter in its ABS search of the two excitation codebooks. In the other modes, the encoder uses a pitch-harmonic weighting filter in cascade with an LP spectral weighting filter for searching excitation codebook 0, whereas only LP spectral weighting is used for searching the adaptive codebook. The pitch-harmonic weighting filter has two parameters, a pitch lag and a coefficient, whose values are determined in the aforementioned open-loop pitch analysis.

A code vector in the 8-b adaptive codebook has a dimension of 40 (the duration of a subframe) and is populated with past samples of the excitation signal beginning with the Lth sample back from the present time. L can take on one of 256 different integer and fractional values. The best adaptive code vector for each subframe can be selected via a complete ABS; the required exhaustive search of the adaptive codebook is, however, computationally expensive. To reduce computation, the GSM VSELP coder makes use of the aforementioned open-loop pitch analysis to produce a list of *candidate lag values*. The open-loop pitch-prediction gains are ranked in decreasing order, and only the lags corresponding to top-ranked gains are kept as candidates. The final decisions for the four L values of the four subframes in a frame are made jointly. By assuming that the four L values can not vary over the entire range of all possible 256 values in the short duration of a frame, the L of the first subframe is coded using 8 b, and the L of each of the other three subframes is coded *differentially* using 4 b. The 4 b represent 16 possible values of deviation relative to the lag of the previous subframe. The four lags in a frame trace out a *trajectory* where the change from one time point to the next is restricted; consequently, only 20 b are needed instead of 32 b for encoding the four lags. Candidate trajectories are constructed by linking top ranked lags that are commensurate with differential encoding. The best trajectory among the candidates is then selected via ABS.

The trained excitation codebooks of VSELP have a special vector sum structure that facilitates fast searching [Gerson and Jasiuk, 1990]. Each of the 2^b code vectors in a b-bit trained codebook is formed as a linear combination of b *basis vectors*. Each of the b scalar weights in the linear combination is restricted to have a binary value of either 1 or -1. The 2^b code vectors in the codebook are obtained by taking all 2^b possible combinations of values of the weights. A substantial storage saving is incurred by storing only b basis vectors instead of 2^b code vectors. Computational saving is another advantage of the vector-sum structure. Since filtering is a linear operation, the synthesis output due to each code vector is a linear combination of the synthesis outputs due to the individual basis vectors, where the same weight values are used in the output linear combination as in forming the code vector. A vector sum codebook can be searched by first performing synthesis filtering on its b basis vectors. If, for the present subframe, another trained codebook (mode 0) or an adaptive codebook (mode 1, 2, 3) had been searched, the filtered basis vectors are further orthogonalized with respect to the signal synthesized from that codebook, i.e., each filtered basis vector is replaced by its own component that is orthogonal to the synthesized signal. Further complexity reduction is obtained by examining the code vectors in a sequence such that two successive code vectors differ in only one of the b scalar weight values; that is, the entire set of 2^b code vectors is searched in a *Gray coded* sequence. With successive code vectors differing in only one term in the linear combination, it is only necessary in the codebook search computation to progressively track the difference [Gerson and Jasiuk, 1990].

The total energy of a speech frame is encoded with 5 b (Table 94.4). The two gain factors (β and γ in Fig. 94.3) for each subframe are computed after the excitation codebooks have been searched and are then transformed to parameters G_s and P_0 to be vector quantized. Each mode has its own 5-b gain VQ codebook. G_s represents the energy of the subframe relative to the total frame energy, and P_0 represents the fraction of the subframe energy due to the first excitation source (excitation codebook 1 in mode 0, or the adaptive codebook in the other modes).

An *interpolation bit* (Table 94.4) transmitted for each frame specifies to the decoder whether the LP synthesis filter parameters for each subframe should be obtained from interpolating between the decoded filter parameters for the current and the previous frames. The encoder determines the value of this bit according to whether interpolation or no interpolation results in a lower prediction residual energy for the frame. The postfilter in the decoder operates in concordance with the actual LP parameters used for synthesis.

The speech encoder generates 112 b of encoded data (Table 94.4) for every 20-ms frame of the speech signal. These bits are processed by the channel encoder to improve, after channel decoding at the receiver, the uncoded bit-error rate and the detectability of uncorrected errors. Error detection coding in the form of 3 CRC bits is applied to the most critical 22 data bits. The combined 25 b plus an additional 73 speech data bits and 6 *tail bits* are input to an RCPCC encoder (the tail bits serve to bring the channel encoder and decoder to a fixed terminal state at the end of the payload data stream). The 3 CRC bits are encoded at rate 1/3 and the other 101 b are encoded at rate 1/2, generating a total of 211 channel coded bits. These are finally combined with the remaining 17 (uncoded) speech data bits to form a total of 228 b for the payload data of a speech frame. The payload data from two speech frames are interleaved for transmission over four timeslots of the GSM TDMA channel.

With the Viterbi algorithm, the channel decoder performs soft decision decoding on the demodulated and deinterleaved channel data. Uncorrected channel errors may still be present in the decoded speech data after Viterbi decoding. Thus, the channel decoder classifies each frame into three integrity categories: bad, unreliable, and reliable, in order to assist the speech decoder in undertaking error concealment measures. A frame is considered bad if the CRC check fails or if the received channel data is close to more than one candidate sequence. The latter evaluation is based on applying an adaptive threshold to the metric values produced by the Viterbi algorithm over the course of decoding the most critical 22 speech data bits and their 3 CRC bits. Frames that are not bad may be classified as unreliable, depending on the metric values produced by the Viterbi algorithm and on channel reliability information supplied by the demodulator.

Depending on the recent history of decoded data integrity, the speech decoder can take various error concealment measures. The onset of bad frames is concealed by repetition of parameters from previous reliable frames, whereas the persistence of bad frames results in power attenuation and ultimately muting of the synthesized speech. Unreliable frames are decoded with normality constraints applied to the energy of the synthesized speech.

94.8 Conclusions

The half-rate standards employ some of the latest techniques in speech and channel coding to meet the challenges posed by the severe transmission environment of digital cellular radio systems. By halving the bit rate, the voice transmission capacity of existing full-rate digital cellular systems can be doubled. Although advances are still being made that can address the needs of quarter-rate speech transmission, much effort is currently devoted to enhancing the speech quality and robustness of full-rate (GSM and IS-54) systems, aiming to be closer to *toll quality*. On the other hand, the imminent introduction of competing wireless systems that use different modulation schemes [e.g., coded division multiple access (CDMA)] and/or different radio frequencies [e.g., personal communications systems (PCS)] is poised to alleviate congestion in high-user-density areas.

Defining Terms

Codebook: An ordered collection of all possible values that can be assigned to a scalar or vector variable. Each unique scalar or vector value in a codebook is called a *codeword*, or *code vector* where appropriate.

Codec: A contraction of *(en)coder–decoder*, used synonymously with the word *coder*. The encoder and decoder are often designed and deployed as a pair. A half-rate standard codec performs speech as well as channel coding.

Echo canceller: A signal processing device that, given the source signal causing the echo signal, generates an estimate of the echo signal and subtracts the estimate from the signal being interfered with by the echo signal. The device is usually based on a discrete-time adaptive filter.

Pitch period: The fundamental period of a voiced speech waveform that can be regarded as periodic over a short-time interval (quasiperiodic). The reciprocal of pitch period is *pitch frequency* or simply, *pitch*.

Tandem coding: Having more than one encoder–decoder pair in an end-to-end transmission path. In cellular radio communications, having a radio link at each end of the communication path could subject the speech signal to two passes of speech encoding–decoding. In general, repeated encoding and decoding increases the distortion.

Acknowledgment

The authors would like to thank Erdal Paksoy and Mark A. Jasiuk for their valuable comments.

References

Chen, J.-H. and Gersho, A. 1995. Adaptive postfiltering for quality enhancement of coded speech. *IEEE Trans. Speech & Audio Proc.* 3(1) 59–71.

De Martino, E. 1993. Speech quality evaluation of the European, North-American and Japanese speech codec standards for digital cellular systems. In *Speech and Audio Coding for Wireless and Network Applications*, ed. B.S. Atal, V. Cuperman, and A. Gersho, pp. 55–58, Kluwer Academic Publishers, Norwell, MA.

Dimolitsas, S., Corcoran, F.L., and Baraniecki, M.R. 1994. Transmission quality of North American cellular, personal communications, and public switched telephone networks. *IEEE Trans. Veh. Tech.* 43(2):245–251.

Gersho, A. 1994. Advances in speech and audio compression. *Proc. IEEE.* 82(6) 900–918.

Gersho, A. and Gray, R.M. 1991. *Vector Quantization and Signal Compression*, Kluwer Academic Publishers, Norwell, MA.

Gerson, I.A. and Jasiuk, M.A. 1990. Vector sum excited linear prediction (VSELP) speech coding at 8 kbps. In *Proceedings, IEEE Intl. Conf. Acoustics, Speech, & Sig. Proc.* pp. 461–464, April.

Gerson, I.A. and Jasiuk, M.A. 1992. Techniques for improving the performance of CELP—type speech coders. *IEEE J. Sel. Areas Comm.* 10(5):858–865.

Gerson, I.A., Jasiuk, M.A., Nowack, J.M., Winter, E.H., and Müller, J.-M. 1994. Speech and channel coding for the half-rate GSM channel. In *Proceedings, ITG-Report 130 on Source and Channel Coding*, pp. 225–232. Munich, Germany, Oct.

Gibson, J.D., Koo, B., and Gray, S.D. 1991. Filtering of colored noise for speech enhancement and coding. *IEEE Trans. Sig. Proc.* 39(8):1732–1742.

Hagenauer, J. 1988. Rate-compatible punctured convolutional codes (RCPC codes) and their applications. *IEEE Trans. Comm.* 36(4):389–400.

Jayant, N.S. and Noll, P. 1984. *Digital Coding of Waveforms*, Prentice-Hall, Englewood Cliffs, NJ.

Masui, F. and Oguchi, M. 1993. Activity of the half rate speech codec algorithm selection for the personal digital cellular system. *Tech. Rept. of IEICE.* RCS93-77(11):55–62 (in Japanese).

Ohya, T., Suda, H., and Miki, T. 1994. 5.6 kbits/s PSI-CELP of the half-rate PDC speech coding standard. In *Proceedings, IEEE Veh. Tech. Conf.* pp. 1680–1684, June.

Proakis, J.G. 1995. *Digital Communications*, 3rd ed. McGraw-Hill, New York.

Suda, H., Ikeda, K., and Ikedo, J. 1994. Error protection and speech enhancement schemes of PSI-CELP, *NTT R&D*. (Special issue on PSI-CELP speech coding system for mobile communications), 43(4):373–380 (in Japanese).

Telecommunication Industries Association (TIA). 1993. Half-rate speech codec test plan V6.0. TR45.3.5/93.05.19.01.

Zeger, K. and Gersho, A. 1990. Pseudo-Gray coding. *IEEE Trans. Comm.* 38(12):2147–2158.

Further Information

Additional technical information on speech coding can be found in the books, periodicals, and conference proceedings that appear in the list of references. Other relevant publications not represented in the list are *Speech Communication*, Elsevier Science Publishers; *Advances in Speech Coding*, B. S. Atal, V. Cuperman, and A, Gersho, eds., Kluwer Academic Publishers; and *Proceedings of the IEEE Workshop on Speech Coding*.

95
Modulation Methods

95.1 Introduction..1353
95.2 Basic Description of Modulated Signals..........................1354
95.3 Analog Frequency Modulation 1355
95.4 Phase Shift Keying (PSK) and $\pi/4$-QPSK 1356
95.5 Continuous Phase Modulation (CPM) and MSK..............1359
95.6 Gaussian Minimum Shift Keying................................1361
95.7 Orthogonal Frequency Division Multiplexing (OFDM).......1363
95.8 Conclusions 1365

Gordon L. Stüber
Georgia Institute of Technology

95.1 Introduction

Modulation is the process where the message information is added to the radio carrier. Most first generation cellular systems such as the advanced mobile telephone system (AMPS) use analog **frequency modulation** (**FM**), because analog technology was very mature when these systems were first introduced. Digital modulation schemes, however, are the obvious choice for future wireless systems, especially if data services such as wireless multimedia are to be supported. Digital modulation can also improve spectral efficiency, because digital signals are more robust against channel impairments. Spectral efficiency is a key attribute of wireless systems that must operate in a crowded radio frequency spectrum.

To achieve high spectral efficiency, modulation schemes must be selected that have a high **bandwidth efficiency** as measured in units of bits per second per Hertz of bandwidth. Many wireless communication systems, such as cellular telephones, operate on the principle of frequency reuse, where the carrier frequencies are reused at geographically separated locations. The link quality in these systems is limited by cochannel interference. Hence, modulation schemes must be identified that are both bandwidth efficient and capable of tolerating high levels of cochannel interference. More specifically, digital modulation techniques are chosen for wireless systems that satisfy the following properties.

Compact Power Density Spectrum: To minimize the effect of adjacent channel interference, it is desirable that the power radiated into the adjacent channel be 60–80 dB below that in the desired channel. Hence, modulation techniques with a narrow main lobe and fast rolloff of sidelobes are desirable.

Good Bit-Error-Rate Performance: A low-bit-error probability should be achieved in the presence of cochannel interference, adjacent channel interference, thermal noise, and other channel impairments, such as fading and intersymbol interference.

Envelope Properties: Portable and mobile applications typically employ nonlinear (class C) power amplifiers to minimize battery drain. Nonlinear amplification may degrade the bit-error-rate performance of modulation schemes that transmit information in the amplitude of the carrier. Also, spectral shaping is usually performed prior to up-conversion and nonlinear amplification. To prevent the regrowth of spectral sidelobes during nonlinear amplification, the input signal must have a relatively constant envelope.

A variety of digital modulation techniques are currently being used in wireless communication systems. Two of the more widely used digital modulation techniques for cellular mobile radio are $\pi/4$ phase-shifted quadrature **phase shift keying** ($\pi/4$-QPSK) and **Gaussian minimum shift keying** (GMSK). The former is used in the North American IS-54 digital cellular system and Japanese Personal Digital Cellular (PDC), whereas the latter is used in the global system for mobile communications (GSM system). This chapter provides a discussion of these and other modulation techniques that are employed in wireless communication systems.

95.2 Basic Description of Modulated Signals

With any modulation technique, the bandpass signal can be expressed in the form

$$s(t) = \text{Re}\{v(t)e^{j2\pi f_c t}\} \tag{95.1}$$

where $v(t)$ is the complex envelope, f_c is the carrier frequency, and Re$\{z\}$ denotes the real part of z. For digital modulation schemes $v(t)$ has the general form

$$v(t) = A \sum_k b(t - kT, \mathbf{x}_k) \tag{95.2}$$

where A is the amplitude of the carrier $\mathbf{x}_k = (x_k, x_{k-1}, \ldots, x_{k-K})$ is the data sequence, T is the symbol or baud duration, and $b(t, \mathbf{x}_i)$ is an equivalent shaping function usually of duration T. The precise form of $b(t, \mathbf{x}_i)$ and the memory length K depends on the type of modulation that is employed. Several examples are provided in this chapter where information is transmitted in the amplitude, phase, or frequency of the bandpass signal.

The **power spectral density** of the bandpass signal $S_{ss}(f)$ is related to the power spectral density of the complex envelope $S_{vv}(f)$ by

$$S_{ss}(f) = \frac{1}{2}[S_{vv}(f - f_c) + S_{vv}(f + f_c)] \tag{95.3}$$

The power density spectrum of the complex envelope for a digital modulation scheme has the general form

$$S_{vv}(f) = \frac{A^2}{T} \sum_m S_{b,m}(f) e^{-j2\pi f mT} \tag{95.4}$$

where

$$S_{b,m}(f) = \frac{1}{2} E[B(f, \mathbf{x}_m) B^*(f, \mathbf{x}_0)] \tag{95.5}$$

$B(f, \mathbf{x}_m)$ is the Fourier transform of $b(t, \mathbf{x}_m)$, and $E[\cdot]$ denotes the expectation operator. Usually symmetric signal sets are chosen so that the complex envelope has zero mean, i.e., $E[b(t, \mathbf{x}_0)] = 0$. This implies that the power density spectrum has no discrete components. If, in addition, $\mathbf{x}_m$ and

Modulation Methods

x_0 are independent for $|m| > K$, then

$$S_{vv}(f) = \frac{A^2}{T} \sum_{|m|<K} S_{b,m}(f) e^{-j2\pi fmT} \tag{95.6}$$

95.3 Analog Frequency Modulation

With analog frequency modulation the complex envelope is

$$v(t) = A \exp\left[j2\pi k_f \int_0^t m(\tau)\,d\tau \right] \tag{95.7}$$

where $m(t)$ is the modulating waveform and k_f in Hz/v is the frequency sensitivity of the FM modulator. The bandpass signal is

$$s(t) = A \cos\left[2\pi f_c t + 2\pi k_f \int_0^t m(t)\,dt \right]. \tag{95.8}$$

The instantaneous frequency of the carrier $f_i(t) = f_c + k_f m(t)$ varies linearly with the waveform $m(t)$, hence the name frequency modulation. Notice that FM has a constant envelope making it suitable for nonlinear amplification. However, the complex envelope is a nonlinear function of the modulating waveform $m(t)$ and, therefore, the spectral characteristics of $v(t)$ cannot be obtained directly from the spectral characteristics of $m(t)$.

With the sinusoidal modulating waveform $m(t) = A_m \cos(2\pi f_m t)$ the instantaneous carrier frequency is

$$f_i(t) = f_c + \Delta_f \cos(2\pi f_m t) \tag{95.9}$$

where $\Delta_f = k_f A_m$ is the peak frequency deviation. The complex envelope becomes

$$v(t) = \exp\left[2\pi \int_0^t f_i(t)\,dt \right]$$
$$= \exp[2\pi f_c t + \beta \sin(2\pi f_m t)] \tag{95.10}$$

where $\beta = \Delta_f / f_m$ is called the modulation index. The bandwidth of $v(t)$ depends on the value of β. If $\beta < 1$, then narrowband FM is generated, where the spectral widths of $v(t)$ and $m(t)$ are about the same, i.e., $2f_m$. If $\beta \gg 1$, then wideband FM is generated, where the spectral occupancy of $v(t)$ is slightly greater than $2\Delta_f$. In general, the approximate bandwidth of an FM signal is

$$W \approx 2\Delta_f + 2f_m = 2\Delta_f \left(1 + \frac{1}{\beta}\right) \tag{95.11}$$

which is a relation known as Carson's rule. Unfortunately, typical analog cellular radio systems use a modulation index in the range $1 \lesssim \beta \lesssim 3$ where Carson's rule is not accurate. Furthermore, the message waveform $m(t)$ is not a pure sinusoid so that Carson's rule does not directly apply.

In analog cellular systems the waveform $m(t)$ is obtained by first companding the speech waveform and then hard limiting the resulting signal. The purpose of the limiter is to control the peak frequency deviation Δ_f. The limiter introduces high-frequency components that must be removed with a low-pass filter prior to modulation. To estimate the bandwidth occupancy, we first determine the ratio of the frequency deviation Δ_f corresponding to the maximum amplitude of $m(t)$, and the highest frequency component B that is present in $m(t)$. These two conditions are the most extreme cases, and the resulting ratio, $D = \Delta_f/B$, is called the *deviation ratio*. Then replace β by D and f_m by B in Carson's rule, giving

$$W \approx 2\Delta_f + 2B = 2\Delta_f\left(1 + \frac{1}{D}\right) \tag{95.12}$$

This approximation will overestimate the bandwidth requirements. A more accurate estimate of the bandwidth requirements must be obtained from simulation or measurements.

95.4 Phase Shift Keying (PSK) and $\pi/4$-QPSK

With **phase shift keying** (PSK), the equivalent shaping function in Eq. (95.2) has the form

$$b(t, \mathbf{x}_k) = \psi_T(t)\exp\left[j\frac{\pi}{M}x_k h_s(t)\right], \qquad \mathbf{x}_k = x_k \tag{95.13}$$

where $h_s(t)$ is a phase shaping pulse, $\psi_T(t)$ an amplitude shaping pulse, and M the size of the modulation alphabet. Notice that the phase varies linearly with the symbol sequence $\{x_k\}$, hence the name phase shift keying. For a modulation alphabet size of M, $x_k \in \{\pm 1, \pm 3, \ldots, \pm(M-1)\}$. Each symbol x_k is mapped onto $\log_2 M$ source bits. A QPSK signal is obtained by using $M = 4$, resulting in a transmission rate of 2 b/symbol.

Usually, the phase shaping pulse is chosen to be the rectangular pulse $h_s(t) = u_T(t) \triangleq u(t) - u(t-T)$, where $u(t)$ is the unit step function. The amplitude shaping pulse is very often chosen to be a square root raised cosine pulse, where the Fourier transform of $\psi_T(t)$ is

$$\Psi_T(f) = \begin{cases} \sqrt{T} & 0 \leq |f| \leq (1-\beta)/2T \\ \sqrt{\dfrac{T}{2}\left[1 - \sin\dfrac{\pi T}{\beta}\left(f - \dfrac{1}{2T}\right)\right]} & (1-\beta)/2T \leq |f| \leq (1+\beta)/2T \end{cases} \tag{95.14}$$

The receiver implements the same filter $\Psi_R(f) = \Psi_T(f)$ so that the overall pulse has the raised cosine spectrum $\Psi(f) = \Psi_R(f)\Psi_T(f) = |\Psi_T(f)|^2$. If the channel is affected by flat fading and additive white Gaussian noise, then this partitioning of the filtering operations between the transmitter and receiver will optimize the signal to noise ratio at the output of the receiver filter at the sampling instants. The rolloff factor β usually lies between 0 and 1 and defines the **excess bandwidth** $100\beta\%$. Using a smaller β results in a more compact power density spectrum, but the link performance becomes more sensitive to errors in the symbol timing. The IS-54 system uses $\beta = 0.35$, while PDC uses $\beta = 0.5$.

The time domain pulse corresponding to Eq. (95.14) can be obtained by taking the inverse Fourier transform, resulting in

$$\psi_T(t) = 4\beta \frac{\cos[(1+\beta)\pi t/T] + \sin[(1-\beta)\pi t/T](4\beta t/T)^{-1}}{\pi\sqrt{T}[1 - 16\beta^2 t^2/T^2]} \tag{95.15}$$

A typical square root raised cosine pulse with a rolloff factor of $\beta = 0.5$ is shown in Fig. 95.1. Strictly speaking the pulse $\psi_T(t)$ is noncausal, but in practice a truncated time domain pulse is used. For example, in Fig. 95.1 the pulse is truncated to $6T$ and time shifted by $3T$ to yield a causal pulse.

Unlike conventional QPSK that has four possible transmitted phases, $\pi/4$-QPSK has eight possible transmitted phases. Let $\theta(n)$ be the transmitted carrier phase for the nth epoch, and let

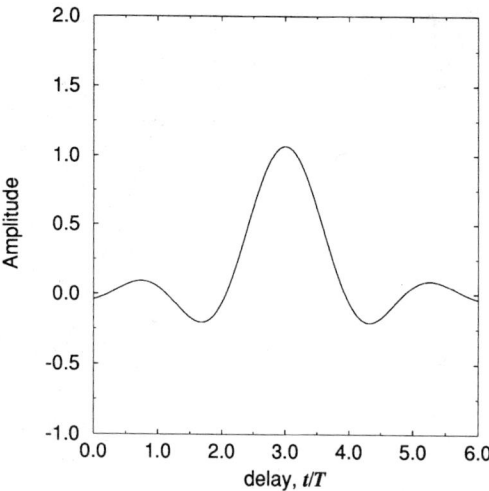

FIGURE 95.1 Square root raised cosine pulse with rolloff factor $\beta = 0.5$.

$\Delta\theta(n) = \theta(n) - \theta(n-1)$ be the differential carrier phase between epochs n and $n-1$. With $\pi/4$-QPSK, the transmission rate is 2 b/symbol and the differential phase is related to the symbol sequence $\{x_n\}$ through the mapping

$$\Delta\theta(n) = \begin{cases} -3\pi/4, & x_n = -3 \\ -\pi/4, & x_n = -1 \\ \pi/4, & x_n = +1 \\ 3\pi/4, & x_n = +3 \end{cases} \quad (95.16)$$

Since the symbol sequence $\{x_n\}$ is random, the mapping in Eq. (95.16) is arbitrary, except that the phase differences must be $\pm\pi/4$ and $\pm 3\pi/4$. The phase difference with the given mapping can be written in the convenient algebraic form

$$\Delta\theta(n) = x_n \frac{\pi}{4} \quad (95.17)$$

which allows us to write the equivalent shaping function of the $\pi/4$-QPSK signal as

$$b(t, \underline{x}_k) = \psi(t)\exp\left\{j\left[\theta(k-1) + x_k\frac{\pi}{4}\right]\right\}$$

$$= \psi_T(t)\exp\left[j\frac{\pi}{4}\left(\sum_{n=-\infty}^{k-1} x_n + x_k\right)\right] \quad (95.18)$$

The summation in the exponent represents the accumulated carrier phase, whereas the last term is the phase change due to the kth symbol. Observe that the phase shaping function is the rectangular pulse $u_T(t)$. The amplitude shaping function $\psi_T(t)$ is usually the square root raised cosine pulse in Eq. (95.15).

The phase states of QPSK and $\pi/4$-QPSK signals can be summarized by the signal space diagram in Fig. 95.2 that shows the phase states and allowable transitions between the phase states. However, it does not describe the actual phase trajectories. A typical diagram showing phase trajectories with

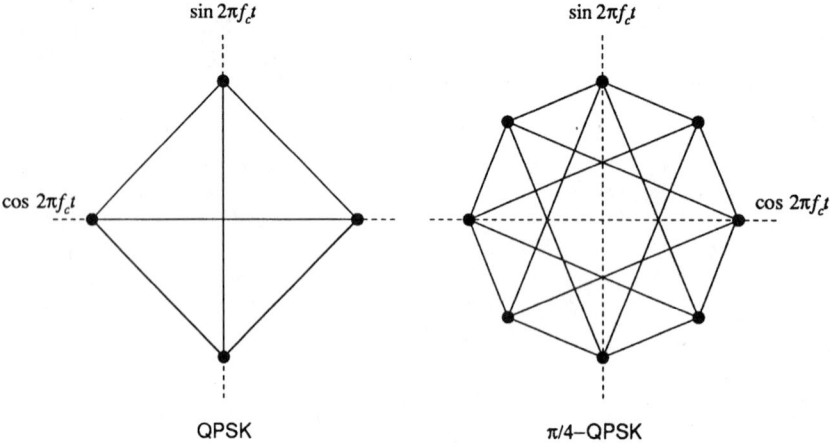

FIGURE 95.2 Signal-space constellations for QPSK and $\pi/4$-DQPSK.

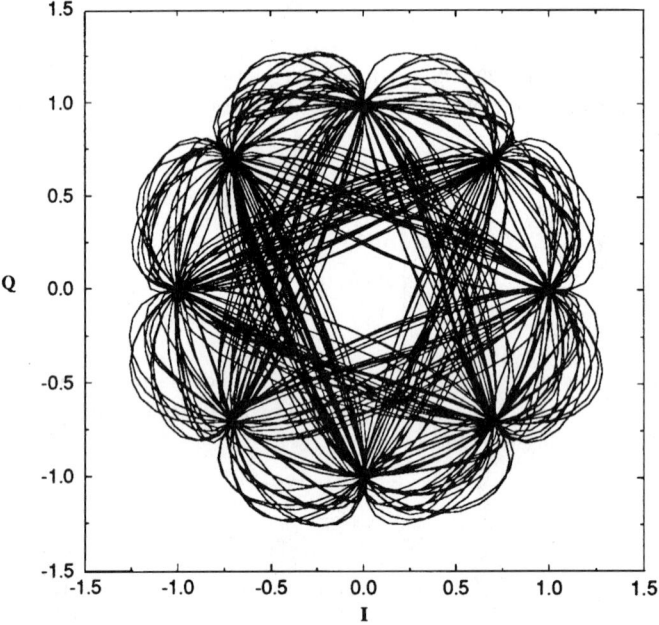

FIGURE 95.3 Phase diagram of $\pi/4$-QPSK with square root raised cosine pulse; $\beta = 0.5$.

square root raised cosine pulse shaping is shown in Fig. 95.3. Note that the phase trajectories do not pass through the origin. This reduces the envelope fluctuations of the signal making it less susceptible to amplifier nonlinearities and reduces the dynamic range required of the power amplifier.

The power density spectrum of QPSK and $\pi/4$-QPSK depends on both the amplitude and phase shaping pulses. For the rectangular phase shaping pulse $h_s(t) = u_T(t)$, the power density spectrum of the complex envelope is

$$S_{vv}(f) = \frac{A^2}{T} |\Psi_T(f)|^2 \qquad (95.19)$$

Modulation Methods

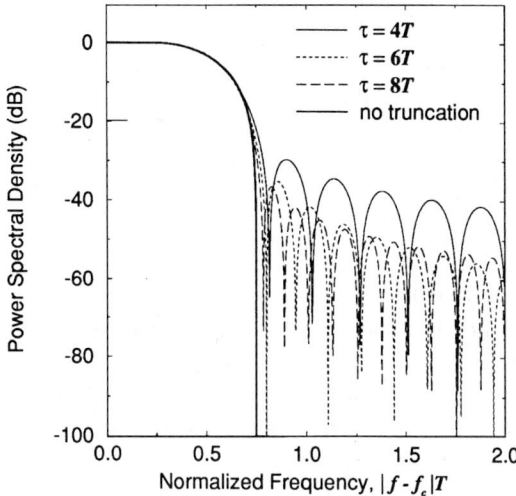

FIGURE 95.4 Power density spectrum of truncated square root raised cosine pulse with various truncation lengths; $\beta = 0.5$.

With square root raised cosine pulse shaping, $\Psi_T(f)$ has the form defined in Eq. (95.14). The power density spectrum of a pulse $\tilde{\psi}_T(t)$ that is obtained by truncating $\psi_T(t)$ to length τ can be obtained by writing $\tilde{\psi}_T(t) = \psi_T(t)\text{rect}(t/\tau)$. Then $\tilde{\Psi}_T(f) = \Psi_T(f) * \tau\text{sinc}(f\tau)$, where $*$ denotes the operation of convolution, and the power density spectrum is again obtained by applying Eq. (95.19). Truncation of the pulse will regenerate some side lobes, thus causing adjacent channel interference. Figure 95.4 illustrates the power density spectrum of a truncated square root raised cosine pulse for various truncation lengths τ.

95.5 Continuous Phase Modulation (CPM) and MSK

Continuous phase modulation (CPM) refers to a broad class of frequency modulation techniques where the carrier phase varies in a continuous manner. A comprehensive treatment of CPM is provided in [Anderson, Aulin, and Sundberg, 1986]. CPM schemes are attractive because they have constant envelope and excellent spectral characteristics. The complex envelope of any CPM signal is

$$v(t) = A \exp\left[j 2\pi k_f \int_{-\infty}^{t} \sum_n x_n h_s(\tau - nT)\, d\tau\right] \quad (95.20)$$

The instantaneous frequency deviation from the carrier is

$$f_{\text{dev}}(t) = k_f \sum_n x_n h_s(t - nT) \quad (95.21)$$

where k_f is the peak frequency deviation. If the frequency shaping pulse $h_s(t)$ has duration T, then the equivalent shaping function in Eq. (95.2) has the form

$$b(t, \mathbf{x}_k) = \exp\left\{j\left[\beta(T)\sum_{n=-\infty}^{k-1} x_n + x_k \beta(t)\right]\right\} u_T(t) \quad (95.22)$$

where

$$\beta(t) = \begin{cases} 0, & t < 0 \\ \dfrac{\pi h}{\int_0^T h_s(\tau)\,d\tau} \int_0^t h_s(\tau)\,d\tau, & 0 \le t \le T \\ \pi h, & t \ge T \end{cases} \quad (95.23)$$

is the phase shaping pulse, and $h = \beta(T)/\pi$ is called the modulation index.

Minimum shift keying (MSK) is a special form of binary CPM ($x_k \in \{-1, +1\}$) that is defined by a rectangular frequency shaping pulse $h_s(t) = u_T(t)$, and a modulation index $h = 1/2$ so that

$$\beta(t) = \begin{cases} 0, & t < 0 \\ \pi t/2T, & 0 \le t \le T \\ \pi/2, & t \ge T \end{cases} \quad (95.24)$$

Therefore, the complex envelope is

$$v(t) = A \exp\left(j \frac{\pi}{2} \sum_{n=-\infty}^{k-1} x_n + \frac{\pi}{2} x_k \frac{t - kT}{T}\right) \quad (95.25)$$

A MSK signal can be described by the phase trellis diagram shown in Fig. 95.5 which plots the time behavior of the phase

$$\theta(t) = \frac{\pi}{2} \sum_{n=-\infty}^{k-1} x_n + \frac{\pi}{2} x_k \frac{t - kT}{T} \quad (95.26)$$

The MSK bandpass signal is

$$\begin{aligned} s(t) &= A \cos\left(2\pi f_c t + \frac{\pi}{2} \sum_{n=-\infty}^{k-1} x_n + \frac{\pi}{2} x_k \frac{t - kT}{T}\right) \\ &= A \cos\left[2\pi \left(f_c + \frac{x_k}{4T}\right)t - \frac{k\pi}{2} x_k + \frac{\pi}{2} \sum_{n=-\infty}^{k-1} x_n\right] \quad kT \le t \le (k+1)T \end{aligned} \quad (95.27)$$

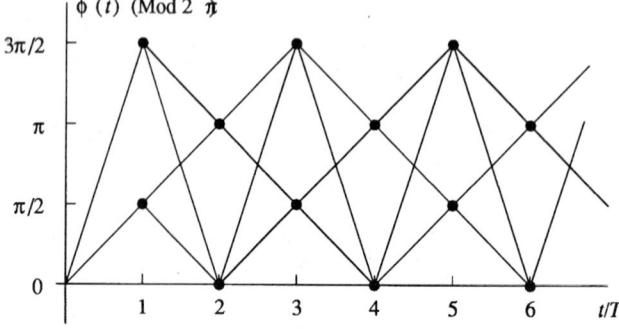

FIGURE 95.5 Phase-trellis diagram for MSK.

Modulation Methods

From Eq. (95.27) we observe that the MSK signal has one of two possible frequencies $f_L = f_c - 1/4T$ or $f_U = f_c + 1/4T$ during each symbol interval. The difference between these frequencies is $f_U - f_L = 1/2T$. This is the minimum frequency difference between two sinusoids of duration T that will ensure orthogonality with coherent demodulation [Proakis, 1995], hence the name minimum shift keying. By applying various trigonometric identities to Eq. (95.27) we can write

$$s(t) = A[x_k^I \psi(t - k2T) \cos(2\pi f_c t) - x_k^Q \psi(t - k2T - T) \sin(2\pi f_c t)],$$
$$kT \le t \le (k+1)T \quad (95.28)$$

where

$$x_k^I = -x_{k-1}^Q x_{2k-1}$$
$$x_k^Q = x_k^I x_{2k}$$
$$\psi(t) = \cos\left(\frac{\pi t}{2T}\right), \quad -T \le t \le T$$

Note that the x_k^I and x_k^Q are independent binary symbols that take on elements from the set $\{-1, +1\}$, and the half-sinusoid amplitude shaping pulse $\psi(t)$ has duration $2T$ and $\psi(t-T) = \sin(\pi t/2T)$, $0 \le t \le 2T$. Therefore, MSK is equivalent to offset quadrature amplitude shift keying (OQASK) with a half-sinusoid amplitude shaping pulse.

To obtain the power density spectrum of MSK, we observe from Eq. (95.28) that the equivalent shaping function of MSK has the form

$$b(t, \mathbf{x}_k) = x_k^I \psi(t) + j x_k^Q \psi(t - T) \quad (95.29)$$

The Fourier transform of Eq. (95.29) is

$$B(f, \mathbf{x}_k) = \left(x_k^I + j x_k^Q e^{-j2\pi fT}\right) \Psi(f) \quad (95.30)$$

Since the symbols x_k^I and x_k^Q are independent and zero mean, it follows from Eqs. (95.5) and (95.6) that

$$S_{vv}(f) = \frac{A^2 |\Psi(f)|^2}{2T} \quad (95.31)$$

Therefore, the power density spectrum of MSK is determined solely by the Fourier transform of the half-sinusoid amplitude shaping pulse $\psi(t)$, resulting in

$$S_{vv}(f) = \frac{16 A^2 T}{\pi^2} \left[\frac{\cos 2\pi fT}{1 - 16 f^2 T^2}\right]^2 \quad (95.32)$$

The power spectral density of MSK is plotted in Fig. 95.8. Observe that an MSK signal has fairly large sidelobes compared to $\pi/4$-QPSK with a truncated square root raised cosine pulse (c.f., Fig. 95.4).

95.6 Gaussian Minimum Shift Keying

MSK signals have all of the desirable attributes for mobile radio, except for a compact power density spectrum. This can be alleviated by filtering the modulating signal $x(t) = \sum_n x_n u_T(t - nT)$ with a low-pass filter prior to frequency modulation, as shown in Fig. 95.6. Such filtering removes the

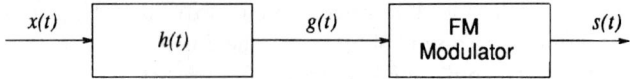

FIGURE 95.6 Premodulation filtered MSK.

higher frequency components in $x(t)$ and, therefore, yields a more compact spectrum. The low-pass filter is chosen to have 1) narrow bandwidth and a sharp transition band, 2) low-overshoot impulse response, and 3) preservation of the output pulse area to ensure a phase shift of $\pi/2$.

GMSK uses a low-pass filter with the following transfer function:

$$H(f) = A \exp\left\{-\left(\frac{f}{B}\right)^2 \frac{\ln 2}{2}\right\} \qquad (95.33)$$

where B is the 3-dB bandwidth of the filter and A a constant. It is apparent that $H(f)$ is bell shaped about $f = 0$, hence the name Gaussian MSK. A rectangular pulse $\text{rect}(t/T) = u_T(t + T/2)$ transmitted through this filter yields the frequency shaping pulse

$$h_s(t) = A\sqrt{\frac{2\pi}{\ln 2}}(BT) \int_{t/T-1/2}^{t/T+1/2} \exp\left\{-\frac{2\pi^2(BT)^2 x^2}{\ln 2}\right\} dx \qquad (95.34)$$

The phase change over the time interval from $-T/2 \leq t \leq T/2$ is

$$\theta\left(\frac{T}{2}\right) - \theta\left(\frac{-T}{2}\right) = x_0 \beta_0(T) + \sum_{\substack{n=-\infty \\ n \neq 0}}^{\infty} x_n \beta_n(T) \qquad (95.35)$$

where

$$\beta_n(T) = \frac{\pi h}{\int_{-\infty}^{\infty} h_s(v) \, dv} \int_{-T/2-nT}^{T/2-nT} h_s(v) \, dv \qquad (95.36)$$

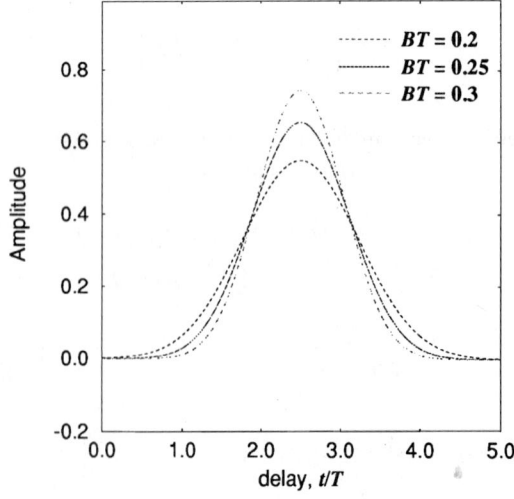

FIGURE 95.7 GMSK frequency shaping pulse for various normalized filter bandwidths BT.

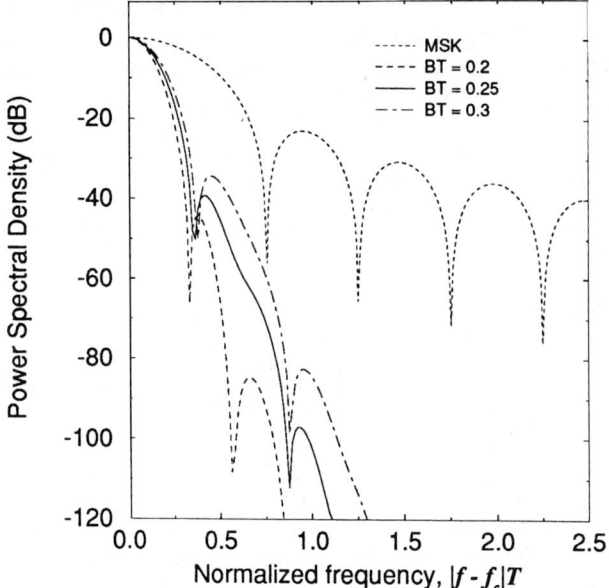

FIGURE 95.8 Power density spectrum of MSK and GMSK.

The first term in Eq. (95.35) is the desired term, and the second term is the intersymbol interference (ISI) introduced by the premodulation filter. Once again, with GMSK $h = 1/2$ so that a total phase shift of $\pi/2$ is maintained.

Notice that the pulse $h_s(t)$ is noncausal so that a truncated pulse must be used in practice. Figure 95.7 plots a GMSK frequency shaping pulse that is truncated to $\tau = 5T$ and time shifted by $2.5T$, for various normalized filter bandwidths BT. Notice that the frequency shaping pulse has a duration greater than T so that ISI is introduced. As BT decreases, the induced ISI is increased. Thus, whereas a smaller value of BT results in a more compact power density spectrum, the induced ISI will degrade the bit-error-rate performance. Hence, there is a tradeoff in the choice of BT. Some studies have indicated that $BT = 0.25$ is a good choice for cellular radio systems [Murota, Kinoshita, and Hirade, 1981].

The power density spectrum of GMSK is quite difficult to obtain, but can be computed by using published methods [Garrison, 1975]. Fig. 95.8 plots the power density spectrum for $BT = 0.2$, 0.25, and 0.3, obtained from Wesolowski, 1994. Observe that the spectral sidelobes are greatly reduced by the Gaussian low-pass filter.

95.7 Orthogonal Frequency Division Multiplexing (OFDM)

Orthogonal frequency division multiplexing (OFDM) is a modulation technique that has been recently suggested for use in cellular radio [Birchler and Jasper, 1992], digital audio broadcasting [Le Floch, Halbert-Lassalle, and Castelain, 1989], and digital video broadcasting. The basic idea of OFDM is to transmit blocks of symbols in parallel by employing a (large) number of orthogonal subcarriers. With block transmission, N serial source symbols each with period T_s are converted into a block of N parallel modulated symbols each with period $T = NT_s$. The block length N is chosen so that $NT_s \gg \sigma_\tau$, where σ_τ is the rms delay spread of the channel. Since the symbol rate on each subcarrier is much less than the serial source rate, the effects of delay spread are greatly reduced. This has practical advantages because it may reduce or even eliminate the need for equalization. Although the block length N is chosen so that $NT_s \gg \sigma_\tau$, the channel dispersion will still cause consecutive blocks to overlap. This results in some residual ISI that will degrade the performance. This residual ISI can be eliminated at the expense of channel capacity by using guard intervals between the blocks that are at least as long as the effective channel impulse response.

The complex envelope of an OFDM signal is described by

$$v(t) = A \sum_{k} \sum_{n=0}^{N-1} x_{k,n} \phi_n(t - kT) \qquad (95.37)$$

where

$$\phi_n(t) = \exp\left\{ j \frac{2\pi \left(n - \frac{N-1}{2}\right) t}{T} \right\} U_T(t), \qquad n = 0, 1, \ldots, N-1 \qquad (95.38)$$

are orthogonal waveforms and $U_T(t)$ is a rectangular shaping function. The frequency separation of the subcarriers, $1/T$, ensures that the subcarriers are orthogonal and phase continuity is maintained from one symbol to the next, but is twice the minimum required for orthogonality with coherent detection. At epoch k, N-data symbols are transmitted by using the N distinct pulses. The data symbols $x_{k,n}$ are often chosen from an M-ary **quadrature amplitude modulation** (M-QAM) constellation, where $x_{k,n} = x_{k,n}^I + j x_{k,n}^Q$ with $x_{k,n}^I, x_{k,n}^Q \in \{\pm 1, \pm 3, \ldots, \pm(N-1)\}$ and $N = \sqrt{M}$.

A key advantage of using OFDM is that the modulation can be achieved in the discrete domain by using either an inverse discrete Fourier transform (IDFT) or the more computationally efficient inverse fast Fourier transform (IFFT). Considering the data block at epoch $k = 0$ and ignoring the frequency offset $\exp\{-j[2\pi(N-1)t/2T]\}$, the complex low-pass OFDM signal has the form

$$v(t) = \sum_{n=0}^{N-1} x_{0,n} \exp\left\{ \frac{j2\pi nt}{NT_s} \right\}, \qquad 0 \leq t \leq T \qquad (95.39)$$

If this signal is sampled at epochs $t = kT_s$, then

$$v^k = v(kT_s) = \sum_{n=0}^{N-1} x_{0,n} \exp\left\{ \frac{j2\pi nk}{N} \right\}, \qquad k = 0, 1, \ldots, N-1 \qquad (95.40)$$

Observe that the sampled OFDM signal has duration N and the samples $v^0, v^1, \ldots, v^{N-1}$ are just the IDFT of the data block $x_{0,0}, x_{0,1}, \ldots, x_{0,N-1}$. A block diagram of an OFDM transmitter is shown in Fig. 95.9.

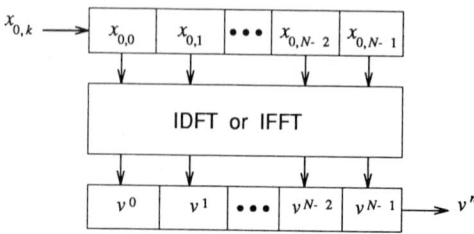

FIGURE 95.9 Block diagram of OFDM transmitter using IDFT or IFFT.

Modulation Methods

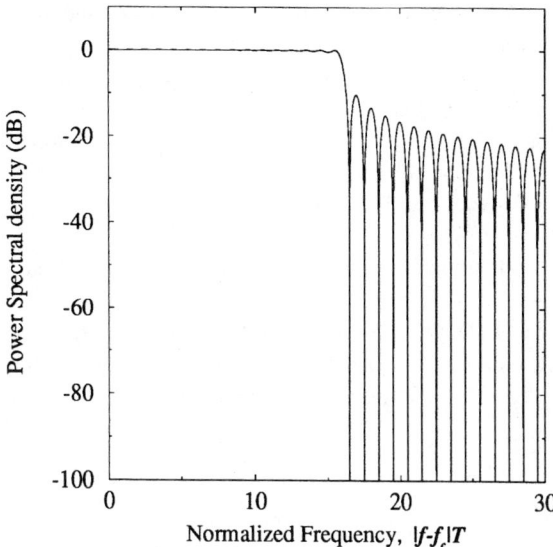

FIGURE 95.10 Power density spectrum of OFDM with $N = 32$.

The power spectral density of an OFDM signal can be obtained by treating OFDM as independent modulation on subcarriers that are separated in frequency by $1/T$. Because the subcarriers are only separated by $1/T$, significant spectral overlap results. Because the subcarriers are orthogonal, however, the overlap improves the spectral efficiency of the scheme. For a signal constellation with zero mean and the waveforms in Eq. (95.38), the power density spectrum of the complex envelope is

$$S_{vv}(f) = \frac{A^2}{T}\sigma_x^2 \sum_{n=0}^{N-1} \left|\text{sinc}\left[fT - \left(n - \frac{N-1}{2}\right)\right]\right|^2 \tag{95.41}$$

where $\sigma_x^2 = \frac{1}{2}E[|x_{k,n}|^2]$ is the variance of the signal constellation. For example, the complex envelope power spectrum of OFDM with $N = 32$ subcarriers is shown in Fig. 95.10.

95.8 Conclusions

A variety of modulation schemes are employed in wireless communication systems. Wireless modulation schemes must have a compact power density spectrum, while at the same time providing a good bit-error-rate performance in the presence of channel impairments such as cochannel interference and fading. The most popular digital modulation techniques employed in wireless systems are GMSK in the European GSM system, $\pi/4$-QPSK in the North American IS-54 and Japanese PDC systems, and OFDM in digital audio broadcasting systems.

Defining Terms

Bandwidth efficiency: Transmission efficiency of a digital modulation scheme measured in units of bits per second per Hertz of bandwidth.
Continuous phase modulation: Frequency modulation where the phase varies in a continuous manner.
Excess bandwidth: Percentage of bandwidth that is in excess of the minimum of $1/2T$ (T is the baud or symbol duration) required for data communication.

Frequency modulation: Modulation where the instantaneous frequency of the carrier varies linearly with the data signal.

Gaussian minimum shift keying: MSK where the data signal is prefiltered with a Gaussian filter prior to frequency modulation.

Minimum shift keying: A special form of continuous phase modulation having linear phase trajectories and a modulation index of 1/2.

Orthogonal frequency division multiplexing: Modulation by using a collection of low-bit-rate orthogonal subcarriers.

Phase shift keying: Modulation where the instantaneous phase of the carrier varies linearly with the data signal.

Power spectral density: Relative power in a modulated signal as a function of frequency.

Quadrature amplitude modulation: Modulation where information is transmitted in the amplitude of the cosine and sine components of the carrier.

References

Anderson, J.B., Aulin, T., and Sundberg, C.-E. 1986. *Digital Phase Modulation*, Plenum Press, New York.

Birchler, M.A., and Jasper, S.C, 1992. A 64 kbps digital land mobile radio system employing M-16QAM. *Proc. 5th Nordic Sem. Dig. Mobile Radio Commun.* Dec.:237–241.

Garrison, G.J. 1975. A power spectral density analysis for digital FM, *IEEE Trans. Commun.*, COM-23(Nov.):1228–1243.

Le Floch, B., Halbert-Lassalle, R., and Castelain, D. 1989. Digital sound broadcasting to mobile receivers, *IEEE Trans. Consum. Elec.* 35:(Aug.).

Murota, K. and Hirade, K. 1981. GMSK modulation for digital mobile radio telephony, *IEEE Trans. Commun.*, COM-29(July):1044–1050.

Murota, K., Kinoshita, K., and Hirade, K. 1981. Spectral efficiency of GMSK land mobile radio. *Proc. ICC'81.* (June):23.8.1.

Proakis, J.G. 1989. *Digital Communications*, 2nd ed. McGraw-Hill, New York.

Wesolowski, K. 1994. Private Communication.

Further Information

A good discussion of digital modem techniques is presented in *Advanced Digital Communications*, edited by K. Feher, Prentice-Hall, 1987.

Proceedings of various IEEE conferences such as the Vehicular Technology Conference, International Conference on Communications, and Global Telecommunications Conference, document the lastest development in the field of wireless communications each year.

Journals such as the *IEEE Transactions on Communications* and *IEEE Transactions on Vehicular Technology* report advances in wireless modulation.

96
Wireless LANs

96.1	Introduction 1367
96.2	Physical Layer Design 1368
96.3	MAC Layer Protocols 1369
	Reservation-TDMA (R-TDMA) • Distributed Foundation Wireless MAC (DFWMAC) • Randomly Addressed Polling (RAP)
96.4	Network Layer Issues 1373
	Alternative View of Mobile Networks • A Proposed Architecture • Networking Issues
96.5	Transport Layer Design 1376
96.6	Conclusions 1378

Suresh Singh
University of South Carolina

96.1 Introduction

A proliferation of high-performance portable computers combined with end-user need for communication is fueling a dramatic growth in wireless **local area network** (LAN) technology. Users expect to have the ability to operate their portable computer globally while remaining connected to communications networks and service providers. Wireless LANs and cellular networks, connected to high-speed networks, are being developed to provide this functionality.

Before delving deeper into issues relating to the design of wireless LANs, it is instructive to consider some scenarios of user mobility.

1. A simple model of user mobility is one where a computer is physically moved while retaining network connectivity at either end. For example, a move from one room to another as in a hospital where the computer is a hand-held device displaying patient charts and the nurse using the computer moves between wards or floors while accessing patient information.
2. Another model situation is where a group of people (at a conference, for instance) set up an ad-hoc LAN to share information as in Fig. 96.1.
3. A more complex model is one where several computers in constant communication are in motion and continue to be networked. For example, consider the problem of having robots in space collaborating to retrieve a satellite.

A great deal of research has focused on dealing with physical layer and **medium access control** (**MAC**) layer protocols. In this chapter we first summarize standardization efforts in these areas. The remainder of the chapter is then devoted to a discussion of networking issues involved in wireless LAN design. Some of the issues discussed include routing in wireless LANs (i.e., how does data find its destination when the destination is mobile?) and the problem of providing service guarantees

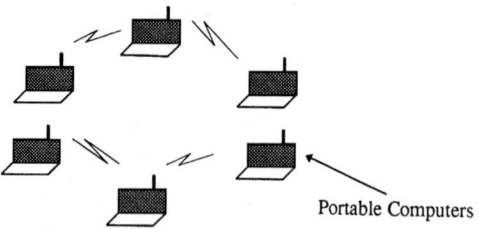

FIGURE 96.1 Ad-hoc wireless LAN.

to end users (e.g., error-free data transmission or bounded delay and bounded bandwidth service, etc.).

96.2 Physical Layer Design

Two media are used for transmission over wireless LANs, infrared and radio frequency. RF LANs are typically implemented in the industrial, scientific, and medical (ISM) frequency bands 902–928 MHz, 2400–2483.5 MHz and 5725–5850 MHz. These frequencies do not require a license allowing the LAN product to be portable, i.e., a LAN can be moved without having to worry about licensing.

IR and RF technologies have different design constraints. IR receiver design is simple (and thus inexpensive) in comparison to RF receiver design because IR receivers only detect the amplitude of the signal not the frequency or phase. Thus, a minimal of filtering is required to reject interference. Unfortunately, however, IR shares the electromagnetic spectrum with the sun and incandescent or fluorescent light. These sources of modulated infrared energy reduce the signal-to-noise ratio of IR signals and, if present in extreme intensity, can make the IR LANs inoperable. There are two approaches to building IR LANs.

1. The transmitted signal can be focused and aimed. In this case the IR system can be used outdoors and has an area of coverage of a few kilometers.
2. The transmitted signal can be bounced off the ceiling or radiated omnidirectionally. In either case, the range of the IR source is 10–20 m (i.e., the size of one medium-sized room).

RF systems face harsher design constraints in comparison to IR systems for several reasons. The increased demand for RF products has resulted in tight regulatory constraints on the allocation and use of allocated bands. In the U.S., for example, it is necessary to implement spectrum spreading for operation in the ISM bands. Another design constraint is the requirement to confine the emitted spectrum to a band, necessitating amplification at higher carrier frequencies, frequency conversion using precision local oscillators, and selective components. RF systems must also cope with environmental noise that is either naturally occurring, for example, atmospheric noise or man made, for example, microwave ovens, copiers, laser printers, or other heavy electrical machinery. RF LANs operating in the ISM frequency ranges also suffer interference from amateur radio operators.

Operating LANs indoors introduces additional problems caused by multipath propagation, Rayleigh fading, and absorption. Many materials used in building construction are opaque to IR radiation resulting in incomplete coverage within rooms (the coverage depends on obstacles within the room that block IR) and almost no coverage outside closed rooms. Some materials, such as white plasterboard, can also cause reflection of IR signals. RF is relatively immune to absorption and reflection problems. Multipath propagation affects both IR and RF signals. The technique to alleviate the effects of multipath propagation in both types of systems is the same use of aimed (directional) systems for transmission enabling the receiver to reject signals based on their angle of incidence. Another technique that may be used in RF systems is to use multiple antennas. The phase difference between different paths can be used to discriminate between them.

Wireless LANs

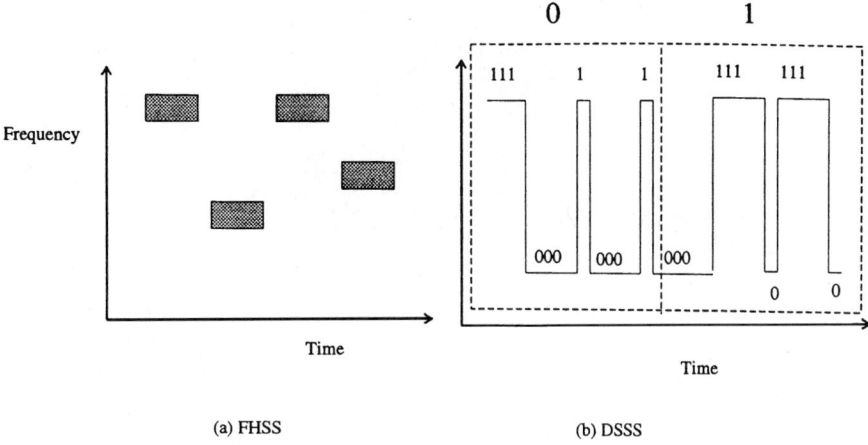

FIGURE 96.2 Spread spectrum.

Rayleigh fading is a problem in RF systems. Recall that Rayleigh fading occurs when the difference in path length of the same signal arriving along different paths is a multiple of half a wavelength. This causes the signal to be almost completely canceled out at the receiver. Because the wavelengths used in IR are so small, the effect of Rayleigh fading is not noticeable in those systems. RF systems, on the other hand, use wavelengths of the order of the dimension of a laptop. Thus, moving the computer a small distance could increase/decrease the fade significantly.

Spread spectrum transmission technology is used for RF-based LANs and it comes in two varieties: direct-sequence spread spectrum (DSSS) and frequency-hopping spread spectrum (FHSS). In a FHSS system, the available band is split into several channels. The transmitter transmits on one channel for a fixed time and then hops to another channel. The receiver is synchronized with the transmitter and hops in the same sequence; see Fig. 96.2(a). In DSSS systems, a random binary string is used to modulate the transmitted signal. The relative rate between this sequence and user data is typically between 10 and 100; see Fig. 96.2(b).

The key requirements of any transmission technology is its robustness to noise. In this respect DSSS and FHSS show some differences. There are two possible sources of interference for wireless LANs: the presence of other wireless LANs in the same geographical area (i.e., in the same building, etc.) and interference due to other users of the ISM frequencies. In the latter case, FHSS systems have a greater ability to avoid interference because the hopping sequence could be designed to prevent potential interference. DSSS systems, on the other hand, do exhibit an ability to recover from interference because of the use of the spreading factor [Fig. 96.2(b)].

It is likely that in many situations several wireless LANs may be collocated. Since all wireless LANs use the same ISM frequencies, there is a potential for a great deal of interference. To avoid interference in FHSS systems, it is necessary to ensure that the hopping sequences are orthogonal. To avoid interference in DSSS systems, on the other hand, it is necessary to allocate different channels to each wireless LAN. The ability to avoid interference in DSSS systems is, thus, more limited in comparison to FHSS systems because FHSS systems use very narrow subchannels (1 MHz) in comparison to DSSS systems that use wider subchannels (for example, 25 MHz), thus, limiting the number of wireless LANs that can be collocated. A summary of design issues can be found in Bantz and Bauchot, 1994.

96.3 MAC Layer Protocols

MAC protocol design for wireless LANs poses new challenges because of the in-building operating environment for these systems. Unlike wired LANs (such as the ethernet or token ring), wireless LANs operate in strong multipath fading channels where channel characteristics can change in very

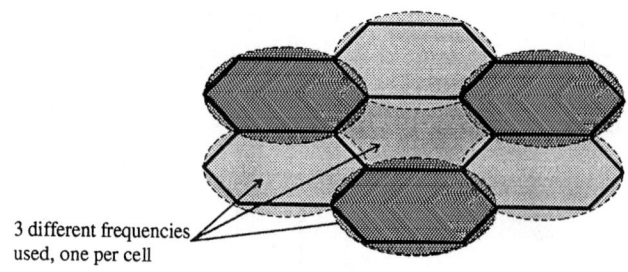

FIGURE 96.3 Cellular structure for wireless LANs (note frequency reuse).

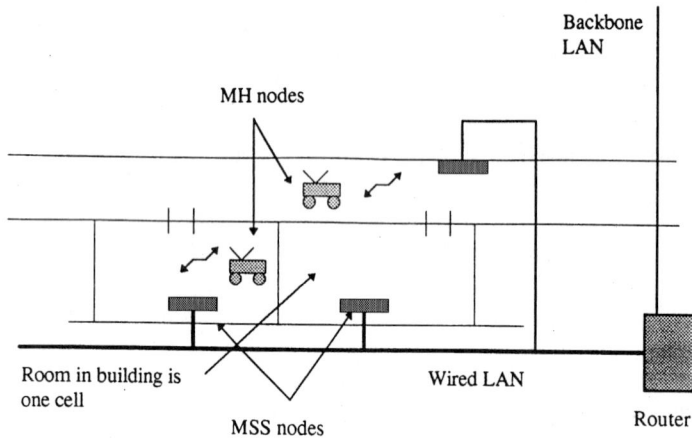

FIGURE 96.4 In-building LAN (made up of several wireless LANs).

short distances resulting in unreliable communication and unfair channel access due to capture. Another feature of the wireless LAN environment is that carrier sensing takes a long time in comparison to wired LANs; it typically takes between 30 and 50 μs (see Chen and Lee, 1994), which is a significant portion of the packet transmission time. This results in inefficiencies if the CSMA family of protocols is used without any modifications.

Other differences arise because of the mobility of users in wireless LAN environments. To provide a building (or any other region) with wireless LAN coverage, the region to be covered is divided into cells as shown in Fig. 96.3. Each cell is one wireless LAN, and adjacent cells use different frequencies to minimize interference. Within each cell there is an access point called a **mobile support station** (MSS) or base station that is connected to some wired network. The mobile users are called **mobile hosts (MH)**. The MSS performs the functions of channel allocation and providing connectivity to existing wired networks; see Fig. 96.4. Two problems arise in this type of an architecture that are not present in wired LANs.

1. The number of nodes within a cell changes dynamically as users move between cells. How can the channel access protocol dynamically adapt to such changes efficiently?
2. When a user moves between cells, the user has to make its presence known to the other nodes in the cell. How can this be done without using up too much bandwidth? The protocol used to solve this problem is called a handoff protocol and works along the following lines: A switching station (or the MSS nodes working together, in concert) collects signal strength information for each mobile host within each cell. Note that if a mobile host is near a

cell boundary, the MSS node in its current cell as well as in the neighboring cell can hear its transmissions and determine signal strengths. If the mobile host is currently under the coverage of MSS M1 but its signal strength at MSS M2 becomes larger, the switching station initiates a handoff whereby the MH is considered as part of M2's cell (or network).

The mode of communication in wireless LANs can be broken in two: communication from the mobile to the MSS (called *uplink* communication) and communication in the reverse direction (called *downlink* communication). It is estimated that downlink communication accounts for about 70–80% of the total consumed bandwidth. This is easy to see because most of the time users request files or data in other forms (image data, etc.) that consume much more transmission bandwidth than the requests themselves. In order to make efficient use of bandwidth (and, in addition, guarantee service requirements for real-time data), most researchers have proposed that the downlink channel be controlled entirely by the MSS nodes. These nodes allocate the channel to different mobile users based on their current requirements using a protocol such as **time division multiple access** (TDMA). What about uplink traffic? This is a more complicated problem because the set of users within a cell is dynamic, thus making it infeasible to have a static channel allocation for the uplink. This problem is the main focus of MAC protocol design.

What are some of the design requirements of an appropriate MAC protocol? The IEEE 802.11 recommended standard for wireless LANs has identified almost 20 such requirements, some of which are discussed here (the reader is referred to Chen, 1994, for further details). Clearly any protocol must maximize throughput while minimizing delays and providing fair access to all users. In addition to these requirements, however, mobility introduces several new requirements.

1. The MAC protocol must be independent of the underlying physical layer transmission technology adopted (be it DSSS, FHSS or IR).
2. The maximum number of users can be as high as a few hundred in a wireless LAN. The MAC protocol must be able to handle many users without exhibiting catastrophic degradation of service.
3. The MAC protocols must provide secure transmissions because the wireless medium is easy to tap.
4. The MAC protocol needs to work correctly in the presence of collocated networks.
5. It must have the ability to support ad-hoc networking (as in Fig. 96.1).
6. Other requirements include the need to support priority traffic, preservation of packet order, and an ability to support multicast.

Several contention-based protocols currently exist that could be adapted for use in wireless LANs. The protocols currently being looked by IEEE 802.11 include protocols based on **carrier sense multiple access (CSMA)**, polling, and TDMA. Protocols based on **code division multiple access** (CDMA) and **frequency division multiple access** (FDMA) are not considered because the processing gains obtained using these protocols are minimal while, simultaneously, resulting in a loss of flexibility for wireless LANs.

It is important to highlight an important difference between networking requirements of ad-hoc networks (as in Fig. 96.1) and networks based on cellular structure. In cellular networks, all communication occurs between the mobile hosts and the MSS (or base station) within that cell. Thus, the MSS can allocate channel bandwidth according to requirements of different nodes, i.e., we can use centralized channel scheduling for efficient use of bandwidth. In ad-hoc networks there is no such central scheduler available. Thus, any multiaccess protocol will be contention based with little explicit scheduling. In the remainder of this section we focus on protocols for cell-based wireless LANs only.

All multiaccess protocols for cell-based wireless LANs have a similar structure; see Chen, 1994.

1. The MSS announces (explicitly or implicitly) that nodes with data to send may contend for the channel.

2. Nodes interested in sending data contend for the channel using protocols such as CSMA.
3. The MSS allocates the channel to successful nodes.
4. Nodes transmit packets (contention-free transmission).
5. MSS sends an explicit acknowledgment (ACK) for packets received.

Based on this model we present three MAC protocols.

Reservation-TDMA (R-TDMA)

This approach is a combination of TDMA and some contention protocol (see PRMA in Goodman, 1990). The MSS divides the channel into slots (as in TDMA), which are grouped into frames. When a node wants to transmit it needs to reserve a slot that it can use in every consecutive frame as long as it has data to transmit. When it has completed transmission, other nodes with data to transmit may contend for that free slot. There are four steps to the functioning of this protocol.

 a. At the end of each frame the MSS transmits a feedback packet that informs nodes of the current reservation of slots (and also which slots are free). This corresponds to steps 1 and 3 from the preceding list.
 b. During a frame, all nodes wishing to acquire a slot transmit with a probability ρ during a free slot. If a node is successful it is so informed by the next feedback packet. If more than one node transmits during a free slot, there is a collision and the nodes try again during the next frame. This corresponds to step 2.
 c. A node with a reserved slot transmits data during its slot. This is the contention-free transmission (step 4).
 d. The MSS sends ACKs for all data packets received correctly. This is step 5.

The R-TDMA protocol exhibits several nice properties. First and foremost, it makes very efficient use of the bandwidth, and average latency is half the frame size. Another big benefit is the ability to implement power conserving measures in the portable computer. Since each node knows when to transmit (nodes transmit during their reserved slot only) it can move into a power-saving mode for a fixed amount of time, thus increasing battery life. This feature is generally not available in CSMA-based protocols. Furthermore, it is easy to implement priorities because of the centralized control of scheduling. One significant drawback of this protocol is that it is expensive to implement (see Barke and Badrinath, 1994).

Distributed Foundation Wireless MAC (DFWMAC)

The CSMA/CD protocol has been used with great success in the ethernet. Unfortunately, the same protocol is not very efficient in a wireless domain because of the problems associated with cell interference (i.e., interference from neighboring cells), the relatively large amount of time taken to sense the channel (see Glisic, 1991) and the hidden terminal problem (see Tobagi and Kleinrock, 1975a and 1975b). The current proposal is based on a CSMA/collision avoidance (CA) protocol with a four-way handshake; see Fig. 96.5.

The basic operation of the protocol is simple. All MH nodes that have packets to transmit compete for the channel by sending ready to transmit (RTS) messages using nonpersistent CSMA. After a station succeeds in transmitting a RTS, the MSS sends a clear to transmit (CTS) to the MH. The MH transmits its data and then receives an ACK. The only possibility of collision that exists is in the RTS phase of the protocol and inefficiencies occur in the protocol, because of the RTS and CTS stages. Note that unlike R-TDMA it is harder to implement power saving functions. Furthermore, latency is dependent on system load making it harder to implement real-time guarantees. Priorities are also not implemented. On the positive side, the hardware for this protocol is very inexpensive.

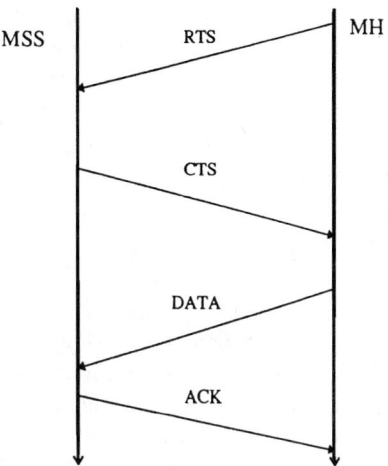

FIGURE 96.5 CSMA/CA and four-way handshaking protocol.

Randomly Addressed Polling (RAP)

In this scheme, when a MSS is ready to collect uplink packets it transmits a READY message. At this point all nodes with packets to send attempt to grab the channel as follows.

a. Each MH with a packet to transmit generates a random number between 0 and P.
b. All active MH nodes simultaneously and orthogonally transmit their random numbers (using CDMA or FDMA). We assume that all of these numbers are received correctly by the MSS. Remember that more than one MH node may have selected the same random number.
c. Steps a and b are repeated L times.
d. At the end of L stages, the MSS determines a stage (say, k) where the total number of distinct random numbers was the largest. The MSS polls each distinct each random number in this stage in increasing order. All nodes that had generated the polled random number transmit packets to the MSS.
e. Since more than one node may have generated the same random number, collisions are possible. The MSS sends a ACK or NACK after each such transmission. Unsuccessful nodes try again during the next iteration of the protocol.

The protocol is discussed in detail in Chen and Lee, 1993 and a modified protocol called GRAP (for group RAP) is discussed in Chen, 1994. The authors propose that GRAP can also be used in the contention stage (step 2) for TDMA- and CSMA-based protocols.

96.4 Network Layer Issues

An important goal of wireless LANs is to allow users to move about freely while still maintaining all of their connections (network resources permitting). This means that the network must route all packets destined for the mobile user to the MSS of its current cell in a transparent manner. Two issues need to be addressed in this context.

- How can users be addressed?
- How can active connections for these mobile users be maintained?

Ioanidis, Duchamp, and Maguire, 1991 propose a solution called the IPIP (IP-within-IP) protocol. Here each MH has a unique **internet protocol** (IP) address called its home address. To deliver a packet to a remote MH, the source MSS first broadcasts an address resolution protocol (ARP) request to all other MSS nodes to locate the MH. Eventually some MSS responds. The source MSS then encapsulates each packet from the source MH within another packet containing the IP address of the MSS in whose cell the MH is located. The destination MSS extracts the packet and delivers it to the MH. If the MH has moved away in the interim, the new MSS locates the new location of the MH and performs the same operation. This approach suffers from several problems as discussed in Teraoka and Tokoro, 1993. Specifically, the method is not scaleable to a network spanning areas larger than a campus for the following reasons.

1. IP addresses have a prefix identifying the campus subnetwork where the node lives; when the MH moves out of the campus, its IP address no longer represents this information.
2. The MSS nodes serve the function of routers in the mobile network and, therefore, have the responsibility of tracking all of the MH nodes globally causing a lot of overhead in terms of message passing and packet forwarding; see Ghai and Singh, 1994.

Teraoka and Tokoro, 1993, have proposed a much more flexible solution to the problem called virtual IP (VIP). Here every mobile host has a virtual IP address that is unchanging regardless of the location of the MH. In addition, hosts have physical network addresses (traditional IP addresses) that may change as the host moves about. At the transport layer, the target node is always specified by its VIP address only. The address resolution from the VIP address to the current IP address takes place either at the network layer of the same machine or at a gateway. Both the host machines and the gateways maintain a cache of VIP to IP mappings with associated timestamps. This information is in the form of a table called *address mapping table* (AMT). Every MH has an associated *home gateway*. When a MH moves into a new subnetwork, it is assigned a new IP address. It sends this new IP address and its VIP address to its home gateway via a *VipConn* control message. All intermediate gateways that relay this message update their AMT tables as well. During this process of updating the AMT tables, all packets destined to the MH continue to be sent to the old location. These packets are returned to the sender, who then sends them to the home gateway of the MH. It is easy to see that this approach is easily scaleable to large networks, unlike the IPIP approach.

Alternative View of Mobile Networks

The approaches just described are based on the belief that mobile networks are merely an extension of wired networks. Other authors [Singh, 1995] disagree with this assumption because there are fundamental differences between the mobile domain and the fixed wired network domain. Two examples follow.

1. The available bandwidth at the wireless link is small; thus, end-to-end packet retransmission for transmission control protocol (TCP)-like protocols (implemented over datagram networks) is a bad idea. This leads to the conclusion that transmission within the mobile network must be connection oriented. Such a solution, using virtual circuits (VC), is proposed in Ghai and Singh, 1994.
2. The bandwidth available for a MH with open connections changes dynamically since the number of other users present in each cell varies randomly. This is a feature not present in fixed high-speed networks where, once a connection is set up, its bandwidth does not vary much. Since bandwidth changes are an artifact of mobility and are dynamic, it is necessary to deal with the consequences (e.g., buffer overflow, large delays, etc.) locally to both, i.e., shield fixed network hosts from the idiosyncrasies of mobility as well as to respond to changing bandwidth quickly (without having to rely on end-to-end control). Some other differences are discussed in Singh, 1995.

A Proposed Architecture

Keeping these issues in mind, a more appropriate architecture has been proposed in Ghai and Singh, 1994, and Singh, 1995. Mobile networks are considered to be different and separate from wired networks. Within a mobile network is a three-layer hierarchy; see Fig. 96.6. At the bottom layer are the MHs. At the next level are the MSS nodes (one per cell). Finally, several MSS nodes are controlled by a **supervisor host (SH)** node (there may be one SH node per small building). The SH nodes are responsible for flow control for all MH connections within their domain; they are also responsible for tracking MH nodes and forwarding packets as MH nodes roam. In addition, the SH nodes serve as a *gateway* to the wired networks. Thus, any connection setup from a MH to a fixed host is broken in two, one from the MH to the SH and another from the SH to the fixed host. The MSS nodes in this design are simply connection endpoints for MH nodes. Thus, they are simple devices that implement the MAC protocols and little else. Some of the benefits of this design are as follows.

1. Because of the large coverage of the SH (i.e., a SH controls many cells) the MH remains in the domain of one SH much longer. This makes it easy to handle the consequences of dynamic bandwidth changes locally. For instance, when a MH moves into a crowded cell, the bandwidth available to it is reduced. If it had an open ftp connection, the SH simply buffers undelivered packets until they can be delivered. There is no need to inform the other endpoint of this connection of the reduced bandwidth.
2. When a MH node sets up a connection with a service provider in the fixed network, it negotiates some quality of service (QOS) parameters such as bandwidth, delay bounds, etc. When the MH roams into a crowded cell, these QOS parameters can no longer be met because the available bandwidth is smaller. If the traditional view is adopted (i.e., the mobile networks are extensions of fixed networks) then these QOS parameters will have to be renegotiated each time the bandwidth changes (due to roaming). This is a very expensive proposition because of the large number of control messages that will have to be exchanged. In the approach of Singh, 1995, the service provider will never know about the bandwidth changes since it deals only with the SH that is accessed via the wired network. The SH bears the responsibility of handling bandwidth changes by either buffering packets until the bandwidth available to the MH increases (as in the case of the ftp example) or it could discard a fraction of real-time packets (e.g., a voice connection) to ensure delivery of most of the packets within their deadlines. The SH could also instruct the MSS to allocate a larger amount of bandwidth to the MH when the number of buffered packets becomes large. Thus, the service provider in the fixed network is shielded from the mobility of the user.

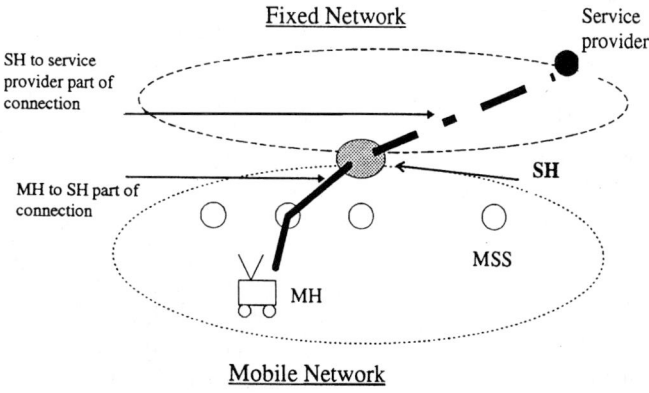

FIGURE 96.6 Proposed architecture for wireless networks.

Networking Issues

It is important for the network to provide connection-oriented service in the mobile environment (as opposed to connectionless service as in the internet) because bandwidth is at a premium in wireless networks, and it is, therefore, inadvisable to have end-to-end retransmission of packets (as in TCP). The proposed architecture is well suited to providing connection-oriented service by using VCs.

In the remainder of this section we look at how virtual circuits are used within the mobile network and how routing is performed for connections to mobile hosts. Every connection set up with one or more MH nodes as a connection endpoint is routed through the SH nodes and each connection is given a unique VC number. The SH node keeps track of all MH nodes that lie within its domain. When a packet needs to be delivered to a MH node, the SH first buffers the packet and then sends it to the MSS at the current location of the MH or to the predicted location if the MH is currently between cells. The MSS buffers all of these packets for the MH and transmits them to the MH if it is in its cell. The MSS discards packets after transmission or if the SH asks it to discard the packets. Packets are delivered in the correct order to the MH (without duplicates) by having the MH transmit the expected sequence number (for each VC) during the initial handshake (i.e., when the MH first enters the cell). The MH sends ACKs to the SH for packets received. The SH discards all packets that have been acknowledged. When a MH moves from the domain of SH1 into the domain of SH2 while having open connections, SH1 continues to forward packets to SH2 until either the connections are closed or until SH2 sets up its own connections with the other endpoints for each of MH's open connections (it also gives new identifiers to all these open connections). The detailed protocol is presented in Ghai and Singh, 1994.

The SH nodes are all connected over the fixed (wired) network. Therefore, it is necessary to route packets between SH nodes using the protocol provided over the fixed networks. The VIP protocol appears to be best suited to this purpose. Let us assume that every MH has a globally unique VIP address. The SHs have both a VIP as well as a fixed IP address. When a MH moves into the domain of a SH, the IP address affixed to this MH is the IP address of the SH. This ensures that all packets sent to the MH are routed through the correct SH node. The SH keeps a list of all VIP addresses of MH nodes within its domain and a list of open VCs for each MH. It uses this information to route the arriving packets along the appropriate VC to the MH.

96.5 Transport Layer Design

The transport layer provides services to higher layers (including the application layer), which include connectionless services like UDP or connection-oriented services like TCP. A wide variety of new services will be made available in the high-speed networks, such as continuous media service for real-time data applications such as voice and video. These services will provide bounds on delay and loss while guaranteeing some minimum bandwidth.

Recently variations of the TCP protocol have been proposed that work well in the wireless domain. These proposals are based on the traditional view that wireless networks are merely extensions of fixed networks. One such proposal is called I-TCP [Barke and Badrinath, 1994] for indirect TCP. The motivation behind this work stems from the following observation. In TCP the sender times out and begins retransmission after a timeout period of several hundred milliseconds. If the other endpoint of the connection is a mobile host, it is possible that the MH is disconnected for a period of several seconds (while it moves between cells and performs the initial greeting). This results in the TCP sender timing out and transmitting the same data several times over, causing the effective throughput of the connection to degrade rapidly. To alleviate this problem, the implementation of I-TCP separates a TCP connection into two pieces—one from the fixed host to another fixed host that is near the MH and another from this host to the MH (note the similarity of this approach with the approach in Fig. 96.6). The host closer to the MH is aware of mobility and has a larger timeout period. It serves as a type of gateway for the TCP connection because it sends ACKs back to the sender before receiving ACKs from the MH. The performance of I-TCP is far superior to traditional TCP for the mobile networks studied.

In the architecture proposed in Fig. 96.6, a TCP connection from a fixed host to a mobile host would terminate at the SH. The SH would set up another connection to the MH and would have the responsibility of transmitting all packets correctly. In a sense this is a similar idea to I-TCP except that in the wireless network VCs are used rather than datagrams. Therefore, the implementation of TCP service is made much easier.

A problem that is unique to the mobile domain occurs because of the unpredictable movement of MH nodes (i.e., a MH may roam between cells resulting in a large variation of available bandwidth in each cell). Consider the following example. Say nine MH nodes have opened 11-kb/s connections in a cell where the available bandwidth is 100 kb/s. Let us say that a tenth mobile host M10, also with an open 11-kb/s connection, wanders in. The total requested bandwidth is now 110 kb/s while the available bandwidth is only 100 kb/s. What is to be done? One approach would be to deny service to M10. However, this seems an unfair policy. A different approach is to penalize all connections equally so that each connection has 10-kb/s bandwidth allocated.

To reduce the bandwidth for each connection from 11 kb/s to 10 kb/s, two approaches may be adopted:

1. Throttle back the sender for each connection by sending control messages.
2. Discard 1-kb/s data for each connection at the SH. This approach is only feasible for applications that are tolerant of data loss (e.g., real-time video or audio).

The first approach encounters a high overhead in terms of control messages and requires the sender to be capable of changing the data rate dynamically. This may not always be possible; for instance, consider a teleconference consisting of several participants where each mobile participant is subject to dynamically changing bandwidth. In order to implement this approach, the data (video or audio or both) will have to be compressed at different ratios for each participant, and this compression ratio may have to be changed dynamically as each participant roams. This is clearly an unreasonable solution to the problem. The second approach requires the SH to discard 1-kb/s of data for each connection. The question is, how should this data be discarded? That is, should the 1 kb of discarded data be consecutive (or clustered) or uniformly spread out over the data stream every 1 s? The way in which the data is discarded has an effect on the final perception of the service by the mobile user. If the service is audio, for example, a random uniform loss is preferred to a clustered loss (where several consecutive words are lost). If the data is compressed video, the problem is even more serious because most random losses will cause the encoded stream to become unreadable resulting in almost a 100% loss of video at the user.

A solution to this problem is proposed in Seal and Singh, 1995, where a new sublayer is added to the transport layer called the *Loss profile transport sublayer* (*LPTSL*). This layer determines how data is to be discarded based on special transport layer markers put by application calls at the sender and based on negotiated loss functions that are part of the QOS negotiations between the SH and service provider. Figure 96.7 illustrates the functioning of this layer at the service provider, the SH, and the MH. The original data stream is broken into *logical segments* that are separated by markers (or flags). When this stream arrives at the SH, the SH discards entire logical segments (in the case of compressed video, one logical segment may represent one frame) depending on the bandwidth available to the MH. The purpose of discarding entire logical segments is that discarding a part of such a segment of data makes the rest of the data within that segment useless—so we might as well discard the entire segment. Observe also that the flags (to identify logical segments) are inserted by the LPTSL via calls made by the application layer. Thus, the transport layer or the LPTSL does not need to know encoding details of the data stream. This scheme is currently being implemented at the University of South Carolina by the author and his research group.

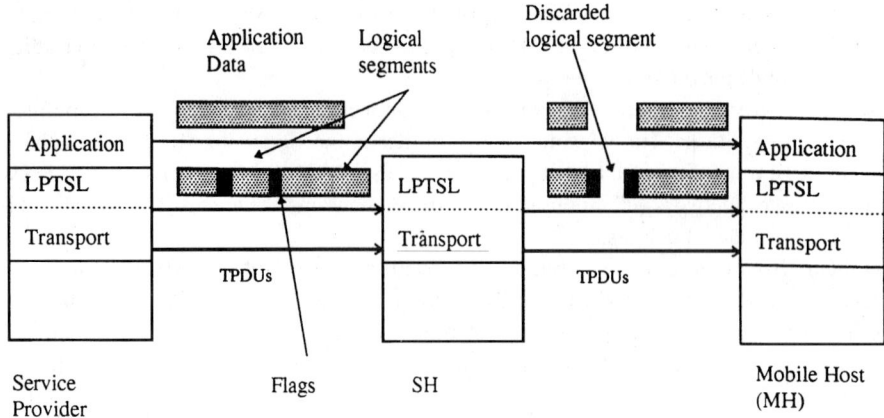

FIGURE 96.7 LPTSL, an approach to handle dynamic bandwidth variations.

96.6 Conclusions

The need for wireless LANs is driving rapid development in this area. The IEEE has proposed standards (802.11) for the physical layer and MAC layer protocols. A great deal of work, however, remains to be done at the network and transport layers. There does not appear to be a consensus regarding subnet design for wireless LANs. Our work has indicated a need for treating wireless LAN subnetworks as being fundamentally different from fixed networks, thus resulting in a different subnetwork and transport layer designs. Current efforts are underway to validate these claims.

Defining Terms

Carrier-sense multiple access (CSMA): Protocols such as those used over the ethernet.
Medium access control (MAC): Protocols arbitrate channel access between all nodes on a wireless LAN.
Mobile host (MH) nodes: The nodes of wireless LAN.
Supervisor host (SH): The node that takes care of flow-control and other protocol processing for all connections.

References

Bantz, D.F. and Bauchot, F.J. 1994. Wireless LAN design alternatives. *IEEE Network*. 8(2):43–53.
Barke, A. and Badrinath, B.R. 1994. I-TCP: indirect TCP for mobile hosts. Tech. Rept. DCS-TR-314, Dept. Computer Science, Rutgers Univ. Piscataway, NJ.
Chen, K.-C. 1994. Medium access control of wireless LANs for mobile computing. *IEEE Network*, 8(5):50–63.
Chen, K.-C. and Lee, C.H. 1993. RAP: a novel medium access control protocol for wireless data networks. *Proc. IEEE GLOBECOM'93*, IEEE Press, Piscataway, NJ 08854. 1713–1717.
Ghai, R. and Singh, S. 1994. An architecture and communication protocol for picocellular networks. *IEEE Personal Comm. Mag.* 1(3):36–46.
Glisic, S.G. 1991. 1-Persistent carrier sense multiple access in radio channel with imperfect carrier sensing. *IEEE Trans. on Comm.* 39(3):458–464.
Goodman, D.J. 1990. Cellular packet communications. *IEEE Trans. on Comm.* 38(8):1272–1280.
Ioanidis, J., Duchamp, D., and Maguire, G.Q. 1991. IP-based protocols for mobile internetworking. *Proc. of ACM SIGCOMM'91*, ACM Press, New York, NY 10036 (Sept.):235–245.

Seal, K. and Singh, S. 1995. Loss profiles: a quality of service measure in mobile computing. *J. Wireless Networks*, submitted for publication.

Singh, S. 1995. Quality of service guarantees in mobile computing. *J. of Computer Comm.*, to appear.

Teraoka, F. and Tokoro, M. 1993. Host migration transparency in IP networks: the VIP approach. *Proc. of ACM SIGCOMM*, ACM Press, New York, NY 10036 (Jan.):45–65.

Tobagi, F. and Kleinrock, L. 1975a. Packet switching in radio channels: Part I carrier sense multiple access models and their throughput delay characteristic. *IEEE Trans. on Comm.*, 23(12):1400–1416.

Tobagi, F. and Kleinrock, L. 1975b. Packet switching in radio channels: Part II the hidden terminal problem in CSMA and busy-one solution. *IEEE Trans. on Comm.*, 23(12):1417–1433.

Further Information

A good introduction to physical layer issues is presented in Bantz, 1994 and MAC layer issues are discussed in Chen, 1994. For a discussion of network and transport layer issues, see Singh, 1995 and Ghai and Singh, 1994.

97
Wireless Data

97.1	Introduction..1380
97.2	Characteristics of Wireless Data Networks.......................1381
	Radio Propagation Characteristics
97.3	Market Issues..1383
97.4	Modem Services Over Cellular Networks 1383
97.5	Private Data Networks 1384
	ARDIS • MOBITEX
97.6	Cellular Data Networks and Services 1387
	Cellular Digital Packet Data (CDPD) • Digital Cellular Data Services
97.7	Other Planned Systems 1391
	Trans-European Trunked Radio (TETRA)
97.8	Conclusions 1393

Allen H. Levesque
GTE Laboratories, Inc.

Kaveh Pahlavan
Worcester Polytechnic Institute

97.1 Introduction

Wireless data services and systems represent a rapidly growing and increasingly important segment of the communications industry. Whereas the wireless data industry is becoming increasingly diverse, one can identify two mainstreams that relate directly to users' requirement for data services. On one hand, there are requirements for relatively low-speed data services provided to mobile users over wide geographical areas, as provided by private mobile data networks and by data services carried on common-carrier cellular telephone networks. On the other hand, there are requirements for high-speed data services in local areas, as provided by cordless private branch exchange (PBX) systems and wireless local area networks (LANs), as well as by the emerging personal communications services (PCS). Personal communications services are treated in Chapter 18 and wireless LANs are treated in Chapter 34. In this chapter we mainly address wide-area wireless data systems, commonly called *mobile data systems*, and briefly touch upon data services to be incorporated into the emerging digital cellular systems.

Mobile data systems provide a wide variety of services for both business users and public safety organizations. Basic services supporting most businesses include electronic mail, enhanced paging, modem and facsimile transmission, remote access to host computers and office LANs, and information broadcast services. Public safety organizations, particularly law-enforcement agencies, are making increasing use of wireless data communications over traditional VHF and UHF radio dispatch networks and over public cellular telephone networks. In addition, there are wireless services supporting vertical applications that are more or less tailored to the needs of specific companies or industries, such as transaction processing, computer-aided delivery dispatch, customer service, fleet management, and emergency medical services. Work currently in progress to develop the national Intelligent vehicle highway system (IVHS) includes the definition of a wide array of new traveler services, many of which will be supported by standardized mobile data networks.

Much of the growth in use of wireless data services has been spurred by the rapid growth of the paging service industry and increasing customer demand for more advanced paging services, as well as the desire to increase work productivity by extending to the mobile environment the suite of digital communications services readily available in the office environment. There is also a desire to make more cost-efficient use of the mobile radio and cellular networks already in common use for mobile voice communications by incorporating efficient data transmission services into these networks. The services and networks that have evolved to date represent a variety of specialized solutions and, in general, they are not interoperable with each other. As the wireless data industry expands, there is an increasing demand for an array of attractively priced standardized services and equipment accessible to mobile users over wide geographic areas. Thus, we see the growth of nationwide privately operated service networks as well as new data services built upon the first and second generation cellular telephone networks. The establishment of new PCS systems in the 2-GHz bands will further extend this evolution.

In this chapter we describe the principal existing and evolving wireless data networks and the related standards activities now in progress. We begin with a discussion of the technical characteristics of wireless data networks.

97.2 Characteristics of Wireless Data Networks

From the perspective of the data user, the basic requirement for wireless data service is convenient, reliable, low-speed access to data services over a geographical area appropriate to the user's pattern of daily business operation. By low speed we mean data rates comparable to those provided by standard data modems operating over the public switched telephone network (PSTN). This form of service will support a wide variety of short-message applications, such as notice of electronic mail or voice mail, as well as short file transfers or even facsimile transmissions that are not overly lengthy. The user's requirements and expectations for these types of services are different in several ways from the requirements placed on voice communication over wireless networks. In a wireless voice service, the user usually understands the general characteristics and limitations of radio transmission and is tolerant of occasional *signal fades* and brief dropouts. An overall level of acceptable voice quality is what the user expects. In a data service, the user is instead concerned with the accuracy of delivered messages and data, the time-delay characteristics of the service network, the ability to maintain service while traveling about, and, of course, the cost of the service. All of these factors are dependent on the technical characteristics of wireless data networks, which we discuss next.

Radio Propagation Characteristics

The chief factor affecting the design and performance of wireless data networks is the nature of radio propagation over wide geographic areas. The most important mobile data systems operate in various land–mobile radio bands from roughly 100 to 200 MHz, the specialized mobile radio (SMR) band around 900 MHz, and the cellular telephone bands at 824–894 MHz. In these frequency bands, radio transmission is characterized by distance-dependent field strength, as well as the well-known effects of *multipath fading*, signal shadowing, and signal blockage. The signal coverage provided by a radio transmitter, which in turn determines the area over which a mobile data receiving terminal can receive a usable signal, is governed primarily by the *power–distance relationship*, which gives signal power as a function of distance between transmitter and receiver. For the ideal case of single-path transmission in free space, the relationship between transmitted power P_t and received power P_r is given by

$$P_r/P_t = G_t G_r (\lambda/4\pi d)^2 \qquad (97.1)$$

where G_t and G_r are the transmitter and receiver antenna gains, respectively, d is the distance between the transmitter and the receiver, and λ is the wavelength of the transmitted signal. In the mobile radio environment, the power-distance relationship is in general different from the free-space case just given. For propagation over an Earth plane at distances much greater than either the signal wavelength or the antenna heights, the relationship between P_t and P_r is given by

$$P_r/P_t = G_t G_r \left(h_1^2 h_2^2/d^4\right) \qquad (97.2)$$

where h_1 and h_2 are the transmitting and receiving antenna heights. Note here that the received power decreases as the fourth power of the distance rather than the square of distance seen in the ideal free-space case. This relationship comes from a propagation model in which there is a single signal reflection with phase reversal at the Earth's surface, and the resulting received signal is the vector sum of the direct line-of-sight signal and the reflected signal. When user terminals are used in mobile situations, the received signal is generally characterized by rapid fading of the signal strength, caused by the vector summation of reflected signal components, the vector summation changing constantly as the mobile terminal moves from one place to another in the service area. Measurements made by many researchers show that when the fast fading is averaged out, the signal strength is described by a Rayleigh distribution having a log-normal mean. In general, the power-distance relationship for mobile radio systems is a more complicated relationship that depends on the nature of the terrain between transmitter and receiver.

Various propagation models are used in the mobile radio industry for network planning purposes, and a number of these models are described in Bodson, McClure, and McConoughey eds., 1984. Propagation models for mobile communications networks must take account of the terrain irregularities existing over the intended service area. Most of the models used in the industry have been developed from measurement data collected over various geographic areas. A very popular model is the *Longley–Rice model* [Rice et al., 1967; Longley and Rice, 1968]. Many wireless networks are concentrated in urban areas. A widely used model for propagation prediction in urban areas is one usually referred to as the *Okumura–Hata model* [Okumura et al., 1968; Hata, 1980]. The Longley–Rice and Okumura–Hata propagation models are discussed in Chapter 22 of this handbook.

By using appropriate propagation prediction models, one can determine the range of signal coverage for a base station of given transmitted power. In a wireless data system, if one knows the level of received signal needed for satisfactory performance, the area of acceptable performance can, in turn, be determined. Cellular telephone networks utilize base stations that are typically spaced 1–5 mi apart, though in some mid-town areas, spacings of 1/2 mi or less are now being used. In packet-switched data networks, higher power transmitters are used, spaced about 5–15 mi apart.

An important additional factor that must be considered in planning a wireless data system is the in-building penetration of signals. Many applications for wireless data services involve the use of mobile data terminals inside buildings, for example, for trouble-shooting and servicing computers on customers' premises. Another example is wireless communications into hospital buildings in support of emergency medical services. It is usually estimated that in-building signal penetration losses will be in the range of 15–30 dB. Clearly, received signal strengths can be satisfactory in the outside areas around a building but totally unusable inside the building. This becomes an important issue when a service provider intends to support customers using mobile terminals inside buildings.

One important consequence of the rapid fading experienced on mobile channels is that errors tend to occur in bursts, causing the transmission to be very unreliable for short intervals of time. Another problem is signal dropouts that occur, for example, when a data call is handed over from one base station to another, or when the mobile user moves into a location that severely blocks the signal. Because of this, mobile data systems employ various error-correction and error-recovery techniques to insure accurate and reliable delivery of data messages.

97.3 Market Issues

Although the market for personal computers (PCs) is not growing as it has in past years, the market for portable computers such as laptops, pen-pads, and notebook computers is growing rapidly. Of greater importance to the wireless data communication industry, the market for networked portables is growing much faster than the market for portable computing. Wireless is the communication method of choice for portable terminals. Mobile data communication services discussed here provide a low-speed solution for wide area coverage. For high-speed and local communications, a portable terminal with wireless access can bring the processing and database capabilities of a large computer directly to specific locations for short periods of time, thus opening a horizon for new applications. For example, one can take portable terminals into classrooms for instructional purposes, or to hospital beds or accident sites for medical diagnosis.

97.4 Modem Services Over Cellular Networks

A simple form of wireless data communication now in common use is data transmission using modems or facsimile terminals over analog cellular telephone links. In this form of communication, the mobile user simply accesses a cellular channel just as he would in making a standard voice call over the cellular network. The user then operates the modem or facsimile terminal just as would be done from office to office over the PSTN. A typical connection is shown in Fig. 97.1, where the mobile user has a lap-top computer and portable modem in the vehicle, communicating with another modem and computer in the office. Typical users of this mode of communication include service technicians, real estate agents, and traveling sales people. In this form of communication, the network is not actually providing a data service but simply a voice link over which the data modem or fax terminal can interoperate with a corresponding data modem or fax terminal in the office or service center. The connection from the mobile telephone switching office (MTSO) is a standard landline connection, exactly the same as is provided for an ordinary cellular telephone call. Many portable modems and fax devices are now available in the market and are sold as elements of the so-called mobile office for the traveling business person. Law enforcement personnel are also making increasing use of data communication over cellular telephone and dispatch radio networks to gain rapid access to databases for verification of automobile registrations and drivers' licenses. Portable devices are currently available that operate at transmission rates up to 9.6 or 14.4 kb/s. Error-correction modem protocols such as MNP-10, V.34, and V.42 are used to provide reliable delivery of data in the error-prone wireless transmission environment.

In another form of mobile data service, the mobile subscriber uses a portable modem or fax terminal as already described but now accesses a modem provided by the cellular service operator

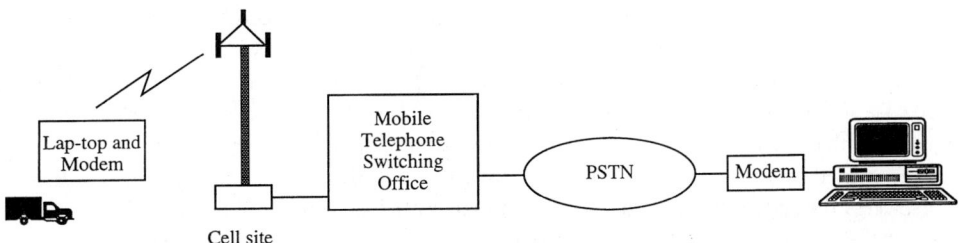

FIGURE 97.1 Modem operation over an analog cellular voice connection.

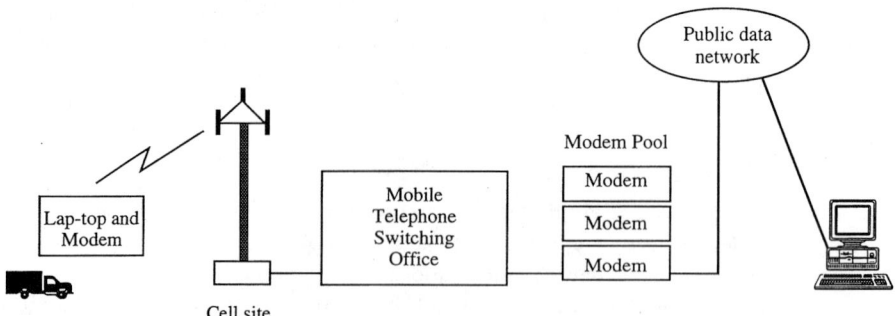

FIGURE 97.2 Cellular data service supported by modem pools in the network.

as part of a *modem pool*, which is connected to the MTSO. This form of service is shown in Fig. 97.2. The modem pool might provide the user with a choice of several standard modem types. The call connection from the modem pool to the office is a digital data connection, which might be supported by any of a number of public packet data networks, such as those providing X.25 service. Here, the cellular operator is providing a special service in the form of modem pool access, and this service in general carries a higher tariff than does standard cellular telephone service, due to the operator's added investment in the modem pools. In this form of service, however, the user in the office or service center does not require a modem but instead has a direct digital data connection to the desk-top or host computer.

Each of the types of wireless data transmission just described is in effect an appliqué onto an underlying cellular telephone service and, therefore, has limitations imposed by the characteristics of the underlying voice connection. That is, the cellular segment of the call connection is a circuit-mode service, which might be cost effective if the user needs to send long file transfers or fax transmissions but might be relatively costly if only short messages are to be transmitted and received. This is because the subscriber is being charged for a circuit-mode connection, which stays in place throughout the duration of the communication session, even if only intermittent short messages exchanges are needed. The need for systems capable of providing cost-effective communication of relatively short message exchanges led to the development of wireless packet data networks, which we describe next.

97.5 Private Data Networks

Here we describe two packet data networks that provide mobile data services to users in major metropolitan areas throughout the United States.

ARDIS

ARDIS is a two-way radio service developed as a joint venture between IBM and Motorola and first implemented in 1983. In mid-1994, IBM sold its interest in ARDIS to Motorola. The ARDIS network consists of four network control centers with 32 network controllers distributed through 1250 base station in 400 cities in the U.S. The service is suitable for two-way transfers of data files of size less than 10 kilobytes, and much of its use is in support of computer-aided dispatching, such as is used by field service personnel, often while they are on customers' premises. Remote users access the system from laptop radio terminals, which communicate with the base stations. Each of the ARDIS base stations is tied to one of the 32 radio network controllers, as shown in Fig. 97.3. The backbone of the network is implemented with leased telephone lines. The four ARDIS hosts, located in Chicago, New York, Los Angeles, and Lexington, KY, serve as access points

Wireless Data

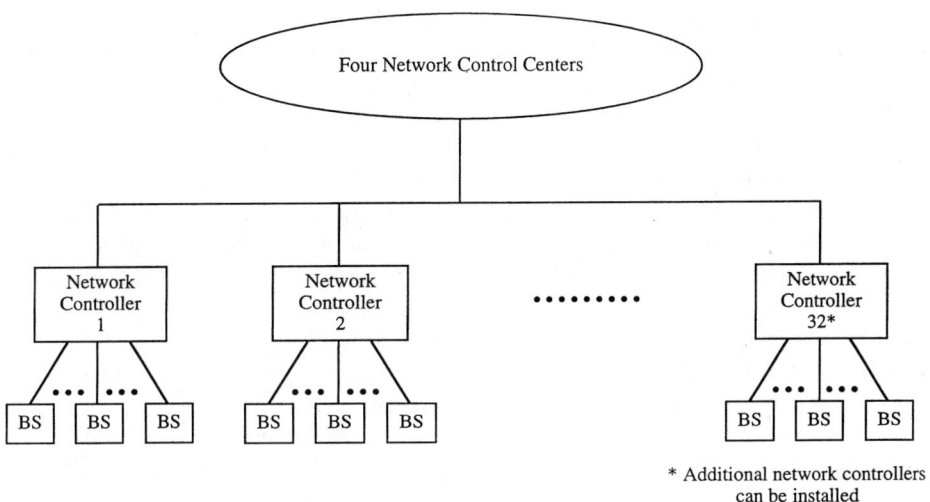

FIGURE 97.3 ARDIS network architecture.

for a customer's mainframe computer, which can be linked to an ARDIS host using async, bisync, SNA, or X.25 dedicated circuits.

The operating frequency band is 800 MHz, and the RF links use separate transmit and receive frequencies spaced by 45 MHz. The system was initially implemented with an RF channel data rate 4800 b/s per 25-kHz channel, using the MDC-4800 protocol. This has been upgraded to 19.2 kb/s, using the RD-LAP protocol, which provides a user data rate of about 8000 b/s. The system architecture is cellular, with cells overlapped to increase the probability that the signal transmission from a portable transmitter will reach at least one base station. The base station power is 40 W, which provides line-of-sight coverage up to a radius of 10–15 miles. The portable units operate with 4 W of radiated power. The overlapping coverage, combined with designed power levels, and error-correction coding in the transmission format, insures that the ARDIS can support portable communications from inside buildings, as well as on the street. This capability for in-building coverage is an important characteristic of the ARDIS service. The modulation technique is frequency-shift keying (FSK), the access method is frequency division multiple access (FDMA), and the transmission packet length is 256 bytes.

Although the use of overlapping coverage, almost always on the same frequency, provides reliable radio connectivity, it poses the problem of interference when signals are transmitted simultaneously from two adjacent base stations. The ARDIS network deals with this by turning off neighboring transmitters, for 0.5–1 s, when an outbound transmission occurs. This scheme has the effect of constraining overall network capacity.

The laptop portable terminals access the network using a random access method called data sense multiple access (DSMA) [Pahlavan and Levesque, 1995]. A remote terminal listens to the base station transmitter to determine if a busy bit is on or off. When the busy bit is off, the remote terminal is allowed to transmit. If two remote terminals begin to transmit at the same time, however, the signal packets may collide, and retransmission will be attempted, as in other contention-based multiple access protocols. The busy bit lets a remote user know when other terminals are transmitting and, thus, reduces the probability of packet collision.

MOBITEX

The MOBITEX system is a nationwide, interconnected trunked radio network developed by Ericsson and Swedish Telecom. The first MOBITEX network went into operation in Sweden in 1986, and networks have either been implemented or are being deployed in 13 countries. A MOBITEX operations association oversees the open technical specifications and coordinates software and hardware developments [Khan and Kilpatrick, 1995]. In the U.S., MOBITEX service was introduced by RAM Mobile Data in 1991 and now covers 7500 cities and towns, with automatic roaming across all service areas. By locating its base stations close to major business centers, the RAM Mobile system provides a degree of in-building signal coverage. Although the MOBITEX system was designed to carry both voice and data service, the U.S. and Canadian networks are used to provide data service only. MOBITEX is an intelligent network with an open architecture that allows establishing virtual networks. This feature facilitates the mobility and expandability of the network [Kilpatrick, 1992; Parsa, 1992].

The MOBITEX network architecture is hierarchical, as shown in Fig. 97.4. At the top of the hierarchy is the network control center (NCC), from which the entire network is managed. The top level of switching is a national switch (MHX1) that routes traffic between service regions. The next level comprises regional switches (MHX2s), and below that are local switches (MOXs), each of which handles traffic within a given service area. At the lowest level in the network, multichannel trunked-radio base stations communicate with the mobile and portable data sets. MOBITEX uses packet-switching techniques, as does ARDIS, to allow multiple users to access the same channel at the same time. Message packets are switched at the lowest possible network level. If two mobile users in the same service area need to communicate with each other, their messages are relayed through the local base station, and only billing information is sent up to the network control center.

The base stations are laid out in a grid pattern using the same frequency reuse rules as are used for cellular telephone networks. In fact, the MOBITEX system operates in much the same way as a cellular telephone system, except that handoffs are not managed by the network. That is, when a radio connection is to be changed from one base station to another, the decision is made by the mobile terminal, not by a network computer as in cellular telephone systems.

To access the network, a mobile terminal finds the base station with the strongest signal and then registers with that base station. When the mobile terminal enters an adjacent service area, it automatically re-registers with a new base station, and the user's whereabouts are relayed to the

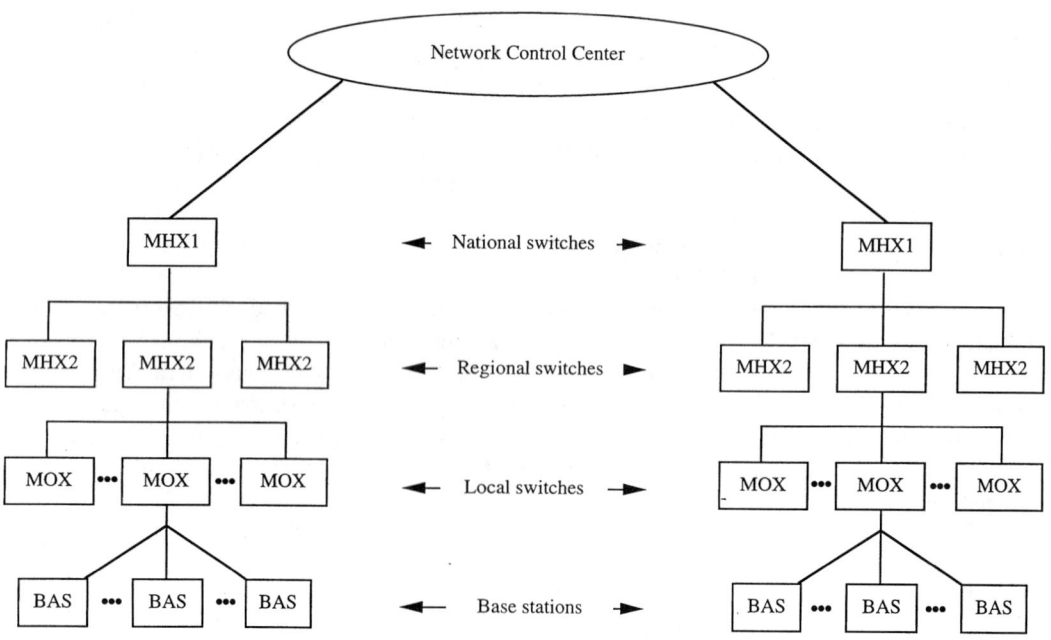

FIGURE 97.4 MOBITEX network architecture.

Wireless Data

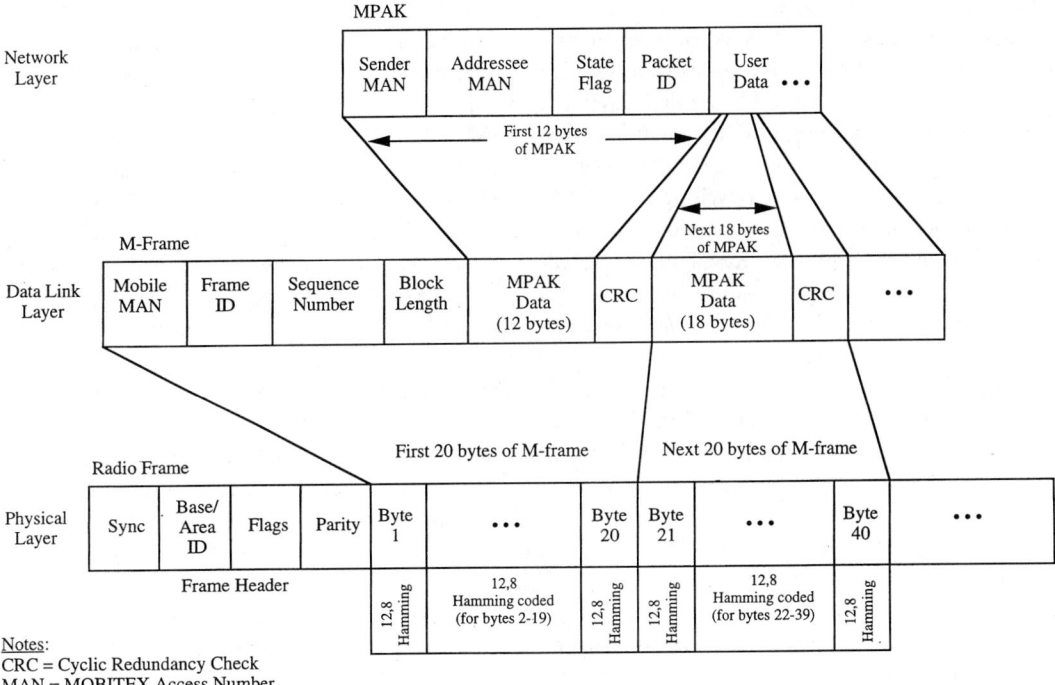

FIGURE 97.5 MOBITEX packet and frame structure at three layers of the protocol stack.

higher level network nodes. This provides automatic routing of messages bound for the mobile user, a capability known as *roaming*. The MOBITEX network also has a store-and-forward capability.

The mobile units transmit at 896–901 MHz and the base stations at 935 to 940 MHz. The system uses dynamic power setting, in the range of 100 mW–10 W for mobile units and 100 mW–4 W for portable units. The Gaussian minimum shift keying (GMSK) modulation technique is used, with $BT = 0.3$ and noncoherent demodulation. The transmission rate is 8000 b/s half-duplex in 12.5-kHz channels, and the service is suitable for file transfers up to 20 kilobytes. The MOBITEX system uses a proprietary network-layer protocol called MPAK, which provides a maximum packet size of 512 bytes and a 24-b address field. Forward-error correction, as well as retransmissions, are used to ensure the bit-error-rate quality of delivered data packets. Fig. 97.5 shows the packet structure at various layers of the MOBITEX protocol stack. The system uses the reservation-slotted ALOHA (R-S-ALOHA) random access method.

97.6 Cellular Data Networks and Services

Cellular Digital Packet Data (CDPD)

The cellular digital packet data (CDPD) system was designed to provide packet data services as an overlay onto the existing analog cellular telephone network, which is called advanced mobile phone service (AMPS). CDPD was developed by IBM in collaboration with the major cellular carriers. These companies will cover 95% of the U.S., including all major urban areas. Any cellular carrier owning a license for AMPS service is free to offer its customers CDPD service without any need for further licensing. A basic goal of the CDPD system is to provide data services on a noninterfering basis with the existing analog cellular telephone services using the same 30-kHz channels. This is accomplished in either of two ways. First, one or a few AMPS channels in each cell site can be devoted to CDPD service. Second, CDPD is designed to make use of a cellular channel that is temporarily not being used for voice traffic and to move to another channel when the current channel is allocated to voice service. The compatibility of CDPD with the existing cellular telephone system allows it to be installed in any AMPS cellular system in North America, providing data services that are not dependent on support of a digital cellular standard in the service area. The participating companies issued release 1.0 of the CDPD specification in July 1993, and release 1.1 was issued in late 1994 [CDPD, 1994]. At this writing (mid-1995), CDPD service is implemented in more than a dozen major market areas, and deployment in an additional 55 markets is planned by the end of 1995. Intended applications for CDPD service include: electronic mail, field support servicing, package delivery tracking, inventory control, credit card verification, security reporting, vehicle theft recovery, traffic and weather advisory services, and a potentially wide range of information retrieval services.

Although CDPD cannot increase the number of channels usable in a cell, it can provide an overall increase in user capacity if data users use CDPD instead of voice channels. This capacity increase would result from the inherently greater efficiency of a connectionless packet data service relative to a connection-oriented service, given bursty data traffic. That is, a packet data service does not require the overhead associated with setup of a voice traffic channel in order to send one or a few data packets. In the following paragraphs we briefly describe the CDPD network architecture and the principles of operation of the system. Our discussion follows Quick and Balachandran, 1993, closely.

The basic structure of a CDPD network (Fig. 97.6) is similar to that of the cellular network with which it shares transmission channels. Each mobile end system (M-ES) communicates with a mobile data base station (MDBS) using the protocols defined by the air-interface specification,

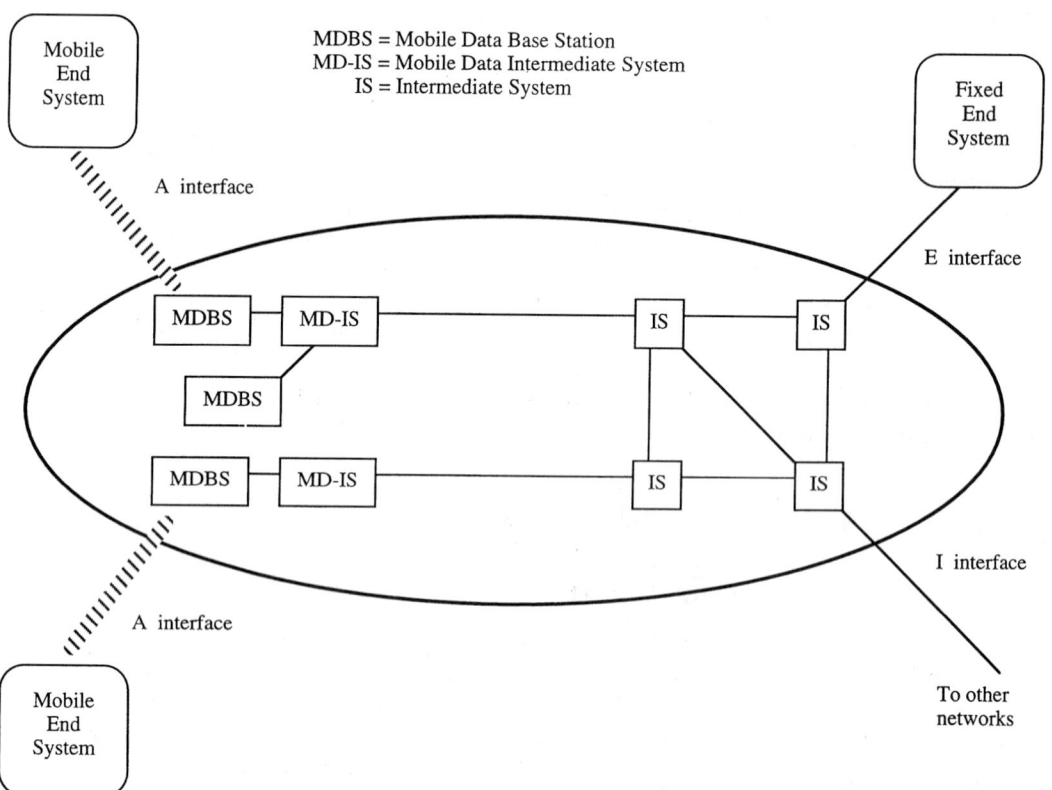

FIGURE 97.6 Cellular digital packet data network architecture.

Wireless Data

to be described subsequently. The MDBSs are expected to be collocated with the cell equipment providing cellular telephone service to facilitate the channel-sharing procedures. All of the MDBSs in a service area will be linked to a mobile data intermediate system (MD-IS) by microwave or wireline links. The MD-IS provides a function analogous to that of the mobile switching center (MSC) in a cellular telephone system. The MD-IS may be linked to other MD-ISs and to various services provided by end systems outside the CDPD network. The MD-IS also provides a connection to a network management system and supports protocols for network management access to the MDBSs and M-ESs in the network.

Service endpoints can be local to the MD-IS or remote, connected through external networks. A MD-IS can be connected to any external network supporting standard routing and data exchange protocols. A MD-IS can also provide connections to standard modems in the PSTN by way of appropriate modem interworking functions (modem emulators). Connections between MD-ISs allow routing of data to and from M-ESs that are roaming, that is, operating in areas outside their home service areas. These connections also allow MD-ISs to exchange information required for mobile terminal authentication, service authorization, and billing.

CDPD employs the same 30-kHz channelization as used in existing AMPS cellular systems throughout North America. Each 30-kHz CDPD channel will support channel transmission rates up to 19.2 kb/s. Degraded radio channel conditions, however, will limit the actual information payload throughput rate to lower levels, typically 5–10 kb/s, and will introduce additional time delay due to the error-detection and retransmission protocols.

The CDPD radio link physical layer uses GMSK modulation at the standard cellular carrier frequencies, on both forward and reverse links. The Gaussian pulse shaping filter is specified to have bandwidth-time product $B_b T = 0.5$. The specified $B_b T$ product assures a transmitted waveform with bandwidth narrow enough to meet adjacent-channel interference requirements, while keeping the intersymbol interference small enough to allow simple demodulation techniques. The choice of 19.2 kb/s as the channel bit rate yields an average power spectrum that satisfies the emission requirements for analog cellular systems and for dual-mode digital cellular systems.

The forward channel carries data packets transmitted by the MDBS, whereas the reverse channel carries packets transmitted by the M-ESs. In the forward channel, the MDBS forms data frames by adding standard high level data link control (HDLC) terminating flags and inserted zero bits, and then segments each frame into blocks of 274 b. These 274 b, together with an 8-b *color code* for MDBS and MD-IS identification, are encoded into a 378-b coded block using a (63, 47) Reed–Solomon code over a 64-ary alphabet. A 6-b synchronization and flag word is inserted after every 9 code symbols. The flag words are used for reverse link access control. The forward link block structure is shown in Fig. 97.7.

In the reverse channel, when an M-ES has data frames to send, it formats the data with flags and inserted zeros in the same manner as in the forward link. That is, the reverse link frames are segmented and encoded into 378-b blocks using the same Reed–Solomon code as in the forward

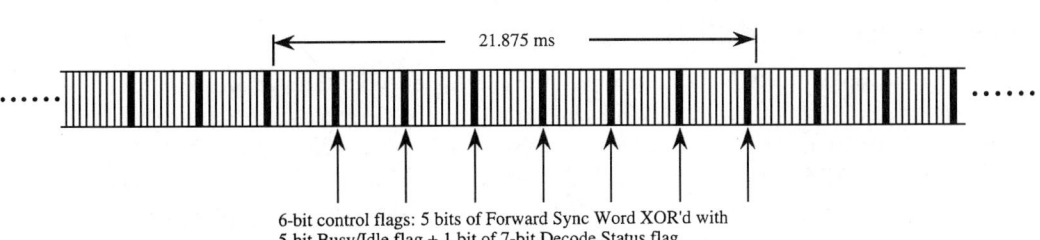

FIGURE 97.7 Cellular digital packet data forward link block structure.

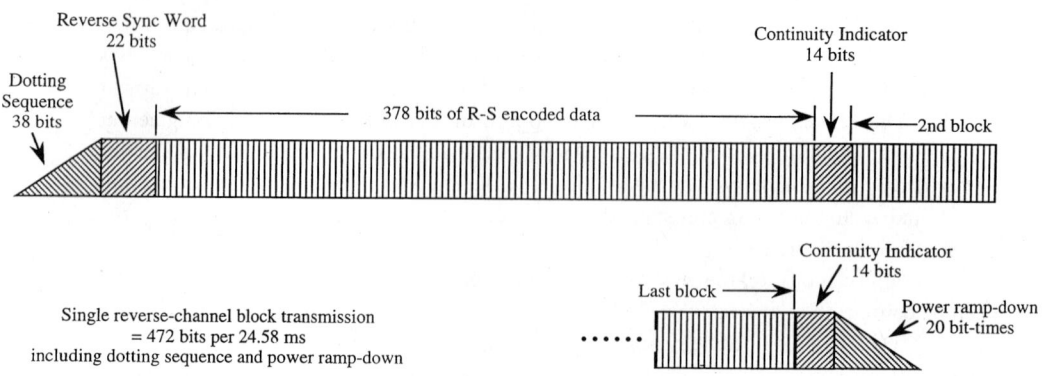

FIGURE 97.8 Cellular digital packet data reverse link block structure.

channel. The M-ES may form up to 64 encoded blocks for transmission in a single reverse channel transmission burst. During the transmission, a 7-b transmit continuity indicator is interleaved into each coded block and is set to all ones to indicate that more blocks follow, or all zeros to indicate that this is the last block of the burst. The reverse channel block structure is shown in Fig. 97.8.

The media access control (MAC) layer in the forward channel is relatively simple. The receiving M-ES removes the inserted zeros and HDLC flags and reassembles data frames that were segmented into multiple blocks. Frames are discarded if any of their constituent blocks are received with uncorrectable errors.

On the reverse channel (M-ES to MDBS), access control is more complex, since multiple M-ESs must share the channel. CDPD uses a multiple access technique called digital sense multiple access (DSMA), which is closely related to the carrier sense multiple access/collision detection (CSMA/CD) access technique.

The network layer and higher layers of the CDPD protocol stack are based on standard ISO and internet protocols. It is expected that the earliest CDPD products will use the internet protocols.

The selection of a channel for CDPD service is accomplished by the radio resource management entity in the MDBS. Through the network management system, the MDBS is informed of the channels in its cell or sector that are available either as dedicated data channels or as potential CDPD channels when they are not being used for analog cellular service, depending on which channel allocation method is implemented. For the implementation in which CDPD service is to use channels of opportunity, there are two ways in which the MDBS can determine whether the channels are in use. If a communication link is provided between the analog system and the CDPD system, the analog system can inform the CDPD system directly about channel usage. If such a link is not available, the CDPD system can use a forward power monitor (sniffer antenna) to detect channel usage on the analog system. Circuitry to implement this function can be built into the cell sector interface.

Digital Cellular Data Services

In response to the rapid growth in demand for cellular telephone service throughout the U.S and Canada, the Cellular Telecommunications Industry Association (CTIA) and the Telecommunications Industry Association (TIA) have been developing standards for new digital cellular systems to replace the existing AMPS cellular system. Two air-interface standards have now been published. The IS-54 standard specifies a three-slot TDMA system, and the I-95 standard specifies a CDMA spread spectrum system. In both systems, a variety of data services are being planned.

The general approach taken in the definition of IS-95 data services has been to base the services on standard data protocols, to the greatest extent possible [Tiedemann, 1993]. The previously-specified physical layer of the IS-95 protocol stack was adopted for the physical layer of the data services, with an appropriate radio link protocol (RLP) overlaid. The current standardization effort is directed to defining three primary services: 1) asynchronous data, 2) group-3 facsimile, and 3) packet data service carried over a circuit-mode connection. Later, attention will be given to other services, including synchronous data and contention-based packet data service.

IS-95 asynchronous data will be structured as a circuit-switched service. For circuit-switched connections, a dedicated path is established between the data devices for the duration of the call. It is used for connectivity through the PSTN requiring point-to-point communications to the common PC or fax user. There are several applications that fall into this category. For file transfer involving PC-to-PC communications the asynchronous data service is the desired cellular service mode. The service will employ an RLP to protect data from transmission errors caused by radio channel degradations at the air interface. The RLP employs automatic repeat request (ARQ), forward error correction (FEC), and flow control. Flow control and retransmission of data blocks with errors are used to provide an improved error performance in the mobile segment of the data connection at the expense of variations in throughput and delay. Typical raw channel rates for digital cellular transmission are measured at approximately a 10^{-2} bit-error rate. Acceptable data transmission, however, usually requires a bit-error rate of approximately 10^{-6} and achieving this requires the design of efficient ARQ and error-correction codes to deal with error characteristics in the mobile environment.

At this writing, the TIA CDMA data services task group has developed a service description for asynchronous data service standards. Work on defining a synchronous data service is underway, whereas work on contention-based packet data service standards has not yet begun. In parallel with the CDMA data services effort, anther TIA task group, TR45.3.2.5, has been defining standards for digital data services for the TDMA digital cellular standard IS-54 [Sacuta, 1992; Weissman, Levesque, and Dean, 1993]. As with the IS-95 data services effort, initial priority is being given to standardizing circuit-mode asynchronous data and group-3 facsimile services. As of this writing, a standard for asynchronous data and group 3 facsimile service has been completed [TIA, 1994].

97.7 Other Planned Systems

Trans-European Trunked Radio (TETRA)

As is the case in the U.S. and Canada, there is interest in Europe in establishing fixed wide-area standards for mobile data communications. Whereas the Pan-European standard for digital cellular, termed Global Systems for Mobile communications (GSM), will provide an array of data services, data will be handled as a circuit-switched service, consistent with the primary purpose of GSM as a voice service system. Therefore, the European Telecommunications Standards Institute (ETSI) has begun developing a public standard for trunked radio and mobile data systems. The standards, which are known generically as Trans-European Trunked Radio (TETRA), are the responsibility of the ETSI RES 6 subtechnical committee [Haine, Martin, and Goodings, 1992].

TETRA is being developed as a family of standards. One branch of the family is a set of radio and network interface standards for trunked voice (and data) services. The other branch is an air-interface standard optimized for wide-area packet data services for both fixed and mobile subscribers and supporting standard network access protocols. Both versions of the standard will use a common physical layer, based on $\pi/4$ dfferential quadrature phase shift keying ($\pi/4$-DQPSK) modulation operating at a channel rate of 36 kb/s in each 25-kHz channel.

Figure 97.9 is a simplified model of the TETRA network, showing the three interfaces at which services will be defined. The U_m interface is the radio link between the base station (BS) and the mobile station (MS). At the other side of the network is the fixed network access point (FNAP) through which mobile users gain access to fixed users. Fixed host computers and fixed data networks typically use standardized interfaces and protocols, and it is intended that the mobile segments of the TETRA network will utilize these same standards. Finally, there is the interface to the mobile data

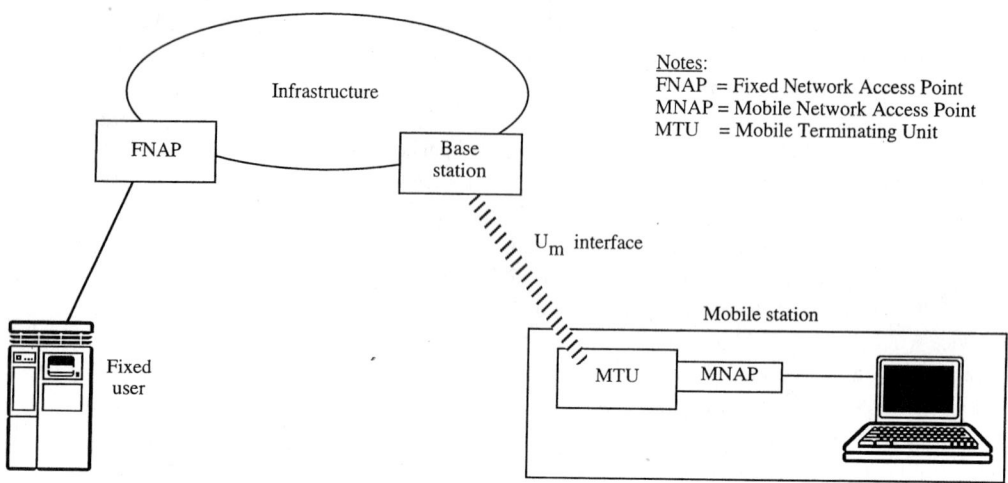

FIGURE 97.9 TETRA network interfaces.

user, the mobile network access point (MNAP). This is shown in the figure as a physical interface between a mobile terminating unit (MTU) and a data terminal. The MNAP, however, may be only a logical interface and not a physical interface, if the MS is an integrated device with no external data port. It is envisioned that all three interfaces, FNAP, MNAP, and the air interface U_m, will support data services with packet-mode protocols.

It is planned that TETRA will provide both connection-oriented and connectionless data services. Although work is still in progress on defining protocols for the three interfaces in the TETRA network model, some broad decisions have been made. It has been decided that the physical data port at the mobile station will be a true X.25 interface, which means that any attached data terminal

TABLE 97.1 Characteristics and Parameters of Five Mobile Data Services

System:	ARDIS	MOBITEX	CDPD	IS-95[b]	TETRA[b]
Frequency band					
Base to mobile, (MHz).	(800 band,	935–940[a]	869–894	869–894	(400 and
Mobile to base, (MHz).	45-kHz sep.)	896–901	824–849	824–849	900 Bands)
RF channel spacing	25 kHz (U.S.)	12.5 kHz	30 kHz	1.25 MHz	25 kHz
Channel access/ multiuser access	FDMA/ DSMA	FDMA/ dynamic- R-S-ALOHA	FDMA/ DSMA	FDMA/ CDMA-SS	FDMA/ DSMA & SAPR[c]
Modulation method	FSK, 4-FSK	GMSK	GMSK	4-PSK/DSSS	$\pi/4$-QDPSK
Channel bit rate, kb/s	19.2	8.0	19.2	9.6	36
Packet length	Up to 256 bytes (HDLC)	Up to 512 bytes	24–928 b	(Packet service-TBD)	192 b (short) 384 b (long)
Open architecture	No	Yes	Yes	Yes	Yes
Private or Public Carrier	Private	Private	Public	Public	Public
Service Coverage	Major metro. areas in U.S.	Major metro. areas in U.S.	All AMPS areas	All CDMA cellular areas	European trunked radio
Type of coverage	In-building & mobile	In-building & mobile	Mobile	Mobile	Mobile

[a] Frequency allocation in the U.S. in the U.K., 380–450 MHz band is used.
[b] IS-95 and TETRA data services standardization in progress.
[c] Slotted-ALOHA packet reservation

must implement the X.25 protocol. This will provide true peer-to-peer communication between the mobile and fixed ends of the call connection. For connectionless data services, the protocol definition is still incomplete, but it is envisioned that the protocols will be based on ISO standards for connectionless-mode network service [ISO, 1987].

The protocol for the radio link has not yet been defined, but it will certainly employ a combination of forward-error correction (FEC) coding, CRC error-detection coding, and an ARQ scheme. It has been reported that the TETRA standard is being designed to accommodate two popular forms of multiuser access, slotted Aloha (with and without packet reservation), and data-sense multiple access.

Table 97.1 compares the chief characteristics and parameters of the five wireless data services described.

97.8 Conclusions

Mobile data radio systems have grown out of the success of the paging-service industry and the increasing customer demand for more advanced services. Today 100,000 customers are using mobile data services and the industry expects 13 million users by the year 2000. This could be equivalent to 10–30% of the revenue of the cellular radio industry. Today, mobile data services provide length-limited wireless connections with in-building penetration to portable users in metropolitan areas. The future direction is toward wider coverage, higher data rates, and capability for transmitting longer data files.

References

Bodson, D., McClure, G.F., and McConoughey, S.R. eds. 1984. *Land-Mobile Communications Engineering*, Selected Reprint Ser., IEEE Press, New York.

CDPD Industry Coordinator. 1994. Cellular Digital Packet Data Specification, Release 1.1, November 1994, Pub. CDPD, Kirkland, WA.

Haine, J.L., Martin, P.M., and Goodings, R.L.A. 1992. A European standard for packet-mode mobile data, *Proceedings of Personal, Indoor, and Mobile Radio Conference (PIMRC'92)*, Boston, MA. Pub. IEEE, New York.

Hata, M. 1980. Empirical formula for propagation loss in land-mobile radio services. *IEEE Trans. on Vehicular Tech.* 29(3):317–325.

International Standards Organization (ISO), 1987. Protocol for providing the connectionless-mode network service. Pub. ISO 8473.

Khan, M. and Kilpatrick, J. 1995. MOBITEX and mobile data standards. *IEEE Comm. Maga.*, 33(3):96–101.

Kilpatrick, J.A., 1992. Update of RAM Mobile Data's packet data radio service. *Proceedings of the 42nd IEEE Vehicular Technology Conference (VTC'92)*, Denver, CO: 898–901, Pub. IEEE, New York.

Longley, A.G. and Rice, P.L. 1968. Prediction of tropospheric radio transmission over irregular terrain. A. computer method—1968, Environmental Sciences and Services Administration Tech. Rep. ERL 79-ITS 67, U.S. Government Printing Office, Washington, DC.

Okumura, Y., Ohmori, E., Kawano, T., and Fukuda, K., 1968. Field strength and its variability in VHF and UHF land-mobile service. *Review of the Electronic Communication Laboratory*, 16:825–873.

Pahlavan, K. and Levesque, A.H. 1994. Wireless data communications. *Proceedings of the IEEE*, 82(9):1398–1430.

Pahlavan, K. and Levesque, A.H. 1995. *Wireless Information Networks*, J. Wiley & Sons, New York.

Parsa, K. 1992. The MOBITEX packet-switched radio data system. *Proceedings of the Personal, Indoor and Mobile Radio Conference (PIMRC'92)*, Boston, MA: 534–538. Pub. IEEE, New York.

Quick, R.R., Jr. and Balachandran, K. 1993. Overview of the cellular packet data (CDPD) system. *Proceedings of the Personal, Indoor and Mobile Radio Conference (PIMRC'93)*, Yokoham, Japan: 338–343. Pub. IEEE, New York.

Rice, P.L., Longley, A.G., Norton, K.A., and Barsis, A.P. 1967. Transmission loss predictions for tropospheric communication circuits. National Bureau of Standards, Tech. Note 101, Boulder, CO.

Sacuta, A. 1992. Data standards for cellular telecommunications—a service-based approach. *Proceedings of the 42nd IEEE Vehicular Technology Conference*, Denver CO: 263–266. Pub. IEEE, New York.

Telecommunications Industry Association. 1994. Async data and fax. Project No. PN-3123, and Radio link protocol 1. Project No. PN-3306, Nov. 14. Issued by TIA, Washington, DC.

Tiedemann, E. 1993. Data services for the IS-95 CDMA standard. presented at Personal, Indoor and Mobile Radio Conf. PIMRC'93. Yokohama, Japan.

Weissman, D., Levesque, A.H., and Dean, R.A. 1993. Interoperable wireless data. *IEEE Comm. Mag.* 31(2):68–77.

Further Information

Pahlavan and Levesque, 1994 provides a comprehensive survey of the wireless data field as of mid-1994. The monthly journals *IEEE Communications Magazine* and *IEEE Personal Communications Magazine*, and the bimonthly journal *IEEE Transactions on Vehicular Technology* report advances in many areas of mobile communications, including wireless data. For subscription information contact: IEEE Service Center, 445 Hoes Lane, P. O. Box 1331, Piscataway, NJ, 08855-1131. Phone (800)678-IEEE.

VII

Source Compression

98 **Lossless Compression** *Khalid Sayood and Nasir D. Memon* 1397
Introduction • Entropy Coders • Universal Codes • Text Compression • Image Compression • Compression Packages • Lossless Compression Standards

99 **Facsimile** *Nasir Memon and Khalid Sayood* 1411
Introduction • Facsimile Compression Techniques • International Standards • Future Trends

100 **Speech** *Boneung Koo* 1424
Introduction • Properties of Speech Signals • Types of Speech Coding Algorithms • Quantization • Predictive Coders • Frequency-Domain Coders • Analysis-by-Synthesis Coders • Vocoders • Variable Bit Rate (VBR) Coding • Performance Evaluation • Speech Coding Standards • Concluding Remarks

101 **Still Image Compression and Its Optimization for Limited Bit-Depth Rendering Devices** *Ping Wah Wong* 1437
Introduction • Still Image Compression Techniques • The JPEG Compression Standard • Halftoning Techniques • Optimization of JPEG for Halftoning • Summary and Conclusions

102 **Video** *Eric Dubois* 1449
Introduction • Source Characteristics and Viewer Requirements • Coding Algorithms • Standards • Perspectives

103 **The High-Definition Television Grand Alliance System** *Eric Petajan* 1462
Introduction • System Requirements for Terrestrial Broadcast in the U.S. • Grand Alliance System Overview • HDTV System Descriptions • Scanning Format Applications • Conclusions

104 **Audio Coding** *Peter Noll* 1475
Introduction • Auditory Masking • Noise Shaping and Perception-Based Coding • Mean Opinion Score • Low Bit Rate Coding • Subband Coding: ITU-R G.722 Coder • Transform Coding • Subband/Transform Coding • ISO/MPEG Audio Coding • Multichannel Stereophony • Conclusion

105 **Cable** *Jeff Hamilton, Mark Kolber, Charles Schell, and Len Taupier* 1488
Introduction • Cable System Architecture • Source Origination and Head End • Transmission Channel • Consumer Premises Equipment • Access Control and Security

106 **Video Servers** *A.L. Narasimha Reddy and Roger Haskin* 1502
Introduction • Data Server • Video Networks • Network Multimedia

107 **Desktop Videoconferencing** *Madhukar Budagavi* 1516
Introduction • Overview • Videoconferencing over ISDN • Videoconferencing over Internet • Videoconferencing over General Switched Telephone Network • Concluding Remarks

98
Lossless Compression

98.1	Introduction	1397
98.2	Entropy Coders	1398
	Huffman Codes • Arithmetic Codes • Adaptive Huffman Coding • Adaptive Arithmetic Coding	
98.3	Universal Codes	1401
	Lynch–Davisson–Schalkwijk–Cover Codes • Syndrome Source Coding • Golomb Codes • Rice Codes • Ziv–Lempel Codes	
98.4	Text Compression	1403
	Context Models • State Models • Dictionary-Based Coding	
98.5	Image Compression	1406
	Prediction Models • Modeling Prediction Errors	
98.6	Compression Packages	1408
98.7	Lossless Compression Standards	1408
	Joint Bilevel Image Experts Group • Joint Photographic Experts Group	

Khalid Sayood
University of Nebraska-Lincoln

Nasir D. Memon
Northern Illinois University

98.1 Introduction

Lossless **compression** or reversible compression, as the name implies, denotes compression approaches in which the decompressed or reconstructed data exactly match the original. Theoretically the lossless compression rate in terms of bits per symbol is bounded below by the entropy of the source. For a general source $\mathcal{S}$ with alphabet $\mathcal{A} = \{1, 2, \ldots, m\}$, which generates a sequence $\{X_1, X_2, \ldots\}$, the entropy is given by

$$H(\mathcal{S}) = \lim_{n \to \infty} \frac{1}{n} G_n \qquad (98.1)$$

where

$$G_n = -\sum_{i_1=1}^{i_1=m} \sum_{i_2=1}^{i_2=m} \cdots \sum_{i_n=1}^{i_n=m} P(X_1 = i_1, X_2 = i_2, \ldots, X_n = i_n)$$
$$\times \log P(X_1 = i_1, X_2 = i_2, \ldots, X_n = i_n) \qquad (98.2)$$

and $\{X_1, X_2, \ldots, X_n\}$ is a sequence of length n from the source. The reason for the limit is to capture any structure that may exist in the source output. In the case where the source puts out independent, identically distributed (iid) symbols, Eq. (98.2) collapses to

$$G_n = n \sum_{i_1=1}^{i_1=m} P(X_1 = i_1) \log P(X_1 = i_1) \qquad (98.3)$$

and the entropy of the source is given by

$$H(\mathcal{S}) = \sum_{i=1}^{m} P(X = i) \log P(X = i) \qquad (98.4)$$

Although taking longer and longer blocks of symbols is certainly a valid method for extracting the structure in the data, it is generally not a feasible approach. Consider an alphabet of size m. If we encoded the output of this source in blocks of n, we would effectively be dealing with an alphabet of size m^n! Besides being impractical, if we assume the source generates strings of finite length, then it has been shown that extending the alphabet by taking longer and longer blocks of symbols turns out to be inherently inefficient.

A more effective strategy is to use a model to capture the inherent structure in the source. This model can be used in a number of different ways. One approach is to use the model to generate a *residual* sequence which is the difference between the actual source output and the model **predictions**. If the model accurately reflects the structure in the source output the residual sequence can be considered to be iid. Often a second stage model is used to further extract any structure that may remain in the residual sequence. The second stage modeling is often referred to as **error modeling**. Once we get (or assume that we have) an iid sequence, we can use entropy coding to obtain a coding rate close to the entropy as defined by Eq. (98.4).

Another approach is to use the model to provide a **context** for the encoding of the source output, and encode sequences by using the statistics provided by the model. The sequence is encoded symbol by symbol. At each step the model provides a probability distribution for the next symbol to the encoder, based on which the encoding of the next symbol is performed.

These approaches separate the task of lossless compression into a modeling task and a coding task. As we shall see in the next section, encoding schemes for iid sequences are known that perform optimally and, hence, the critical task in lossless compression is that of modeling. The model that is imposed on the source determines the rate at which we would be able to encode a sequence emitted by the source. Naturally, the model is highly dependent on the type of data being compressed. Later in this chapter we describe some popular modeling schemes for text and image data.

This chapter is organized as follows. We first describe two popular techniques for encoding the residuals, which assume knowledge of the probabilities of the symbols, followed by adaptive versions of these techniques. We then describe **universal coding** techniques, which do not require any a priori knowledge of the statistics of the source. Finally, we look at two of the three most popular areas for lossless compression, text and images. The third area, compression of facsimile is covered in the next chapter.

98.2 Entropy Coders

The idea behind **entropy coding** is very simple; use shorter codes for more frequently occurring symbols (or sets of symbols). This idea has been around for a long time and was used by Samuel Morse in the development of the Morse code. As the codes generated are variable length it is essential that a sequence of codewords be decoded to a unique sequence of symbols. One way of guaranteeing this is to make sure that no codeword is a prefix of another code. This is called the prefix condition and codes that satisfy this condition are called prefix codes. The prefix condition, while sufficient, is not necessary for unique decoding. However, it can be shown that given any uniquely decodable code that is not a prefix code, we can always find a prefix code that performs at least as well in terms

Huffman Codes

The **Huffman coding** algorithm was developed by David A. Huffman as part of a class assignment [Huffman, 1952]. The algorithm is based on two observations about optimum prefix codes:

1. In an optimum code, symbols that occur more frequently will have shorter codewords than symbols that occur less frequently.
2. In an optimum code, the two symbols that occur least frequently will have the same length.

The Huffman procedure is obtained by adding the simple requirement that the codewords corresponding to the two lowest probability symbols differ only in the last bit. That is, if γ and δ are the two least probable symbols in an alphabet, then, if the codeword for γ was $m * 1$, then the codeword for δ would be $m * 0$. Here m is a string of 1s and 0s, and $*$ denotes concatenation.

First the letters of the alphabet are sorted in decreasing probability order. A new alphabet is generated by combining the two lowest probability letters into a single letter whose probability is the sum of the two probabilities. When this composite letter is decomposed, its constituent letters will have codewords, which are identical except in the final bit. The process of sorting, then combining the two letters with the lowest probabilities to generate a new alphabet is continued until the new alphabet contains only two letters. We assign a codeword of 0 to one and 1 to the other, and now proceed to decompose the letters. At each step of the decomposition we will get a bit each of two codewords. At the end of the process we will have a prefix code for the entire alphabet. An example of this process is shown in Tables 98.1 and 98.2, where $C(a_k)$ is the codeword for a_k.

The rate for the Huffman code in bits per symbol can be shown to lie in the interval $[H(\mathcal{S}), H(\mathcal{S}) + p_{\max} + 0.086]$, where $p_{\max}$ is the probability of the most probable symbol [Gallagher, 1978]. The lower bound is achieved when the probabilities of the source symbols are all powers of two. If, instead of coding one symbol at a time, we block n symbols together the bounds on the coding rate are

$$H(\mathcal{S}) \leq R_H \leq H(\mathcal{S}) + \frac{p_{\max}^n + 0.086}{n} \qquad (98.5)$$

Thus, we can get the coding rate arbitrarily close to the **entropy**. Unfortunately this approach also means an exponential growth in the size of the source alphabet. A technique that avoids this

TABLE 98.1 Composition Process for Huffman Coding

Alphabet (Prob.)	Sorted (Prob.)	Composite (Prob.)	Composite (Prob.)	Sorted (Prob.)	Composite (Prob.)	Sorted (Prob.)
$a_1(0.2)$	$a_2(0.4)$	$a_2(0.4)$	$a_2(0.4)$	$a_2(0.4)$	$a_2(0.4)$	$a_3''(0.6)$
$a_2(0.4)$	$a_1(0.2)$	$a_1(0.2)$	$a_1(0.2)$	$a_3'(0.4)$	$a_3''(0.6)$	$a_2(0.4)$
					$a_3'' \Leftarrow a_3', a_1$	
$a_3(0.2)$	$a_3(0.2)$	$a_3(0.2)$	$a_3'(0.4)$	$a_1(0.2)$		
			$a_3' \Leftarrow a_3, a_4'$			
$a_4(0.1)$	$a_4(0.1)$	$a_4'(0.2)$				
		$a_4' \Leftarrow a_4, a_5$				
$a_5(0.1)$	$a_5(0.1)$					

TABLE 98.2 Code Assignment in Huffman Coding

	$a_3'' \Rightarrow a_3', a_1$	$a_3' \Rightarrow a_3, a_4'$	$a_4' \Rightarrow a_4, a_5$
$C(a_3'') = 0$	$C(a_3') = C(a_3'') * 0 = 00$	$C(a_3) = C(a_3') * 0 = \mathbf{000}$	$C(a_4) = C(a_4') * 0 = \mathbf{0010}$
$C(a_2) = 1$	$C(a_1) = C(a_3'') * 1 = \mathbf{01}$	$C(a_4') = C(a_3') * 1 = 001$	$C(a_5) = C(a_4') * 1 = \mathbf{0011}$

problem of exponential growth with block length is **arithmetic coding**, which is described in the next section.

Arithmetic Codes

Encoding sequences of symbols is more efficient than encoding individual symbols in terms of coding rate. However, Huffman encoding of sequences becomes impractical because to Huffman code a particular sequence of length m we need codewords for all possible sequences of length m. This latter fact causes an exponential growth in the size of the codebook. What we need is a way of assigning codewords to *particular* sequences without having to generate codes for all sequences of that length.

If we assume that all letters in a given alphabet occur with nonzero probability, the value of the cumulative density function *cdf* $F_X(X)$ of a sequence X is distinct from the value of the *cdf* for any other sequence of symbols. If we could impose an ordering on the sequences such that $X_i < X_j$ if $i < j$, then the half-open sets $[F_X(X_{i-1}), F_X(X_i))$ are disjoint, and any element in this set can be used as a unique tag for the sequence X_i. It can be shown that the binary representation of this number truncated to $\lceil \log(1/(P(X))) \rceil + 1$ bits is a unique code for this sequence [Cover and Thomas, 1991].

Arithmetic coding is a procedure that generates this unique code in an incremental fashion. That is, the code is developed and transmitted (or stored) as the sequence of symbols develops. The decoding procedure is also incremental in nature. Elements of the sequence can be identified as the code is received, without having to wait for the entire code to be received. The coding and decoding procedure requires the knowledge of the *cdf* $F_X(x)$ of the source (note that we need the *cdf* of the source *not* of the *sequences*).

It can be shown [Cover and Thomas, 1991] that the bounds on the coding rate for arithmetic coding R_A for a sequence of length n are

$$H(\mathcal{S}) \leq R_A \leq H(\mathcal{S}) + \frac{2}{n} \tag{98.6}$$

Comparing the upper bound to Eq. (98.5), it seems that Huffman coding will always have an advantage over arithmetic coding. However, recall that Huffman coding is not a realistic alternative for sequences of any reasonable length.

In terms of implementation, Huffman encoding is easier to implement, although the decoding complexity for the two is comparable. On the other hand, arithmetic coding can generally provide a coding rate closer to the entropy than Huffman coding. The exception being when the symbol probabilities are powers of 2, in which case the Huffman code exactly achieves entropy. Other cases where arithmetic coding does not provide much advantage over Huffman coding are when the alphabet size is relatively large. In these cases $p_{\max}$ is generally small, and Huffman coding will generate a rate close to the entropy. However, when there is a substantial imbalance in probabilities, especially in small alphabets, arithmetic coding can provide significant advantage. This is especially true in the coding of facsimile information where the base alphabet size is two with highly uneven probabilities. Arithmetic coding is also easier to use when multiple codes are to be used for the same source, as the setup cost is simply the generation of multiple *cdf*s. This situation occurs often in text compression where *context modeling* may require the use of a different arithmetic code for different contexts.

Adaptive Huffman Coding

The Huffman code relies on knowledge of source statistics for its efficiency. In many applications the source statistics are not known a priori and the Huffman code is implemented as a two-pass procedure. The statistics of the source are collected in the first pass. These statistics are used to generate the Huffman code, which is then used to encode the source in the second pass. In order to convert this algorithm into a one-pass procedure, the probabilities need to be estimated adaptively, and the code altered to reflect these estimates. Theoretically, if we wanted to encode the $(k + 1)$st symbol using the statistics of the first k symbols, we could recompute the code using the Huffman coding procedure each time a symbol is transmitted. However, this would not be a very practical approach due to the large amount of computation involved. The adaptive Huffman coding procedure is a computationally efficient procedure for estimating the probabilities and updating the Huffman code.

In the adaptive Huffman coding procedure, at the start of transmission neither transmitter nor receiver knows anything about the statistics of the source sequence. The tree at both the transmitter and the receiver consists of a single node, which corresponds to all symbols not yet transmitted, and has a weight of zero. As transmission progresses, nodes corresponding to symbols transmitted will be added to the tree, and the tree is reconfigured using a computationally efficient update procedure [Gallagher, 1978]. Prior to the beginning of transmission, a fixed code for each symbol is agreed upon between transmitter and receiver, which is used upon the first occurrence of the symbol.

The adaptive Huffman coding procedure provides an excellent alternative to Huffman coding in cases where the source statistics are unknown. The drawbacks are increased complexity and substantially increased vulnerability to channel errors. As both transmitter and receiver are building the code as the transmission proceeds, a single error can cause the building of different codes at transmitter and receiver effectively stopping any transfer of information. There is some overhead involved in transmitting the fixed codes for the first occurrence of each symbol; however, if the transmission is sufficiently long the effect of this overhead on the average transmission rate is minimal. Of course, if we try to combat the effect of channel errors with frequent resynchronization, the overhead can become a major factor.

Adaptive Arithmetic Coding

Adapting the arithmetic coder to changing statistics is relatively easy. The simplest way to do this is to use the number of times a symbol is encountered as an estimate of the probability and, hence, the cumulative distribution function. This approach can be refined to provide efficient implementation algorithms. More sophisticated algorithms have been developed for the particular case of binary alphabets. One of the more well-known adaptive arithmetic coders is the *Q-coder*, which is part of the Joint Photographic Expert Group (JPEG) and Joint Bilevel Image Experts Group (JBIG) standards.

98.3 Universal Codes

The coding schemes we have described previously depend on the source statistics being stationary, or at worst, slowly varying. In situations where this assumption does not hold, these coding schemes might provide undesirable results. Coding schemes, which optimally code sources with unknown parameters, are generally known as universal coding schemes [Davisson, 1973]. Although proofs of optimality for these schemes generally rely on asymptotic results, several universal coding schemes have also been shown to perform well in practical situations. We describe some of these in the following.

Lynch–Davisson–Schalkwijk–Cover Codes

In this approach the input is divided into blocks of n bits where $n = 2^m$ [Cover and Thomas, 1991]. For each block the Hamming weight w (the number of 1s in the n-bit long sequence) of the block is transmitted using m bits followed by $\lceil \log_2((n/w) - 1) \rceil$ bits to indicate a specific sequence out of all of the possible n bit sequences with w ones.

Syndrome Source Coding

This approach entails using a block error correcting decoder as a source encoder [Ancheta, 1977]. The source output is represented in binary form and divided into blocks of length n. Each block is then viewed as an error pattern and represented by the syndrome vector of an error correcting code. As different error correcting codes will correct a different number of errors, the syndrome vector is preceded by an index pointing to the code being used.

Golomb Codes

Golomb [1966] codes are unary codes, which are optimal for certain distributions. Given the parameter g for the Golomb code any integer n is represented by a prefix and a suffix. The prefix is the unary representation of $\lfloor n/g \rfloor$, and the suffix is the binary representation of n mod m. Although these codes are optimal for certain exponential distributions, they are not universal in the sense that sources can be found for which the average code length diverges.

Rice Codes

The Rice code can be seen as a universal version of the Golomb codes [Rice, Yeh, and Miller, 1991]. The Rice algorithm first converts the input into a sequence of nonnegative integers. This sequence is then coded in blocks of length J (a suggested choice for J is 16). Each block is coded using a variety of binary and unary codes depending on the properties of the particular block. The Rice code has been shown to be optimal over a range of entropies.

Ziv–Lempel Codes

The **Ziv–Lempel (LZ) codes** are a family of dictionary-based coding schemes that encode strings of symbols by sending information about their location in a dictionary. The basic idea behind these dictionary-based schemes is that for certain sources, certain patterns reoccur very frequently. These patterns can be made into entries in a dictionary. Then, all future occurrence of these patterns can be encoded via a pointer to the relevant entry in the dictionary. The dictionary can be static or adaptive. Most of the adaptive schemes have been inspired by two papers by Ziv and Lempel in 1977 and 1978 [Bell, Cleary, and Witten, 1990].

The 1977 algorithm and its derivatives use a portion of the already encoded string as the dictionary. For example, consider the encoding of the string *abracadabra*, where the underlined portion of the string has already been encoded. The string *abra* could then be encoded by simply sending the pair (7, 4), where the first number is the location of the previous occurrence of the string relative to the current position and the second number is the length of the match. How far back we search for a match depends on the size of a prespecified window, and may include all of the past history.

The 1978 algorithm actually builds a dictionary of all strings encountered. Each new entry in the dictionary is a previous entry followed by a letter from the source alphabet. The dictionary is seeded with the letters of the source alphabet. As the coding progresses the entries in the dictionary will consist of longer and longer strings. The most popular derivative of the 1978 algorithm is the LZW algorithm, a variant of which is used in the UNIX *compress* command, as well as the *GIF* image compression format, and the V.42-bis compression standard. We now describe the LZW algorithm.
LZW Algorithm. The LZW algorithm starts with a dictionary containing all of the letters of the alphabet [Bell, Cleary, and Witten, 1990]. Accumulate the output of the source in a string s as long as the string s is in the dictionary. If the addition of another letter α from the source output creates a string $s * \alpha$ that is not in the dictionary, send the index in the dictionary for s, add the string $s * \alpha$ to the dictionary, and start a new string that begins with the letter α. The easiest way to describe the LZW algorithm is through an example. Suppose we have a source that transmits symbols from the alphabet $A = \{a, b, c\}$, and we wish to encode the sequence *aabaabc*.... Assuming we are

using 4-b to represent the entries in the codebook our initial dictionary may look like this

a	0000
b	0001
c	0010

We will first look at what happens at the encoder. As we begin to encode the given sequence we have the string $s = a$, which is in the dictionary. The next letter results in the string $s * a = aa$, which is not in the dictionary. Therefore, we send 0000, which is the code for a, and add aa to our dictionary, which is now

a	0000
b	0001
c	0010
aa	0011

and we begin a new string $s = a$. This string is in our dictionary and so we read the next letter and update our string to $s * b = ab$, which is not in our dictionary. Thus, we add ab to our dictionary, send the code for a to the receiver and start a new string $s = b$. Again, reading in the next letter we have the string $s * a = ba$. This is not in our dictionary and so we send 0001, which is the code for b, add ba to our dictionary, and start a new string $s = a$. This time when we read a new letter, the new string $s * a = aa$ is in our dictionary and so we continue with the next letter to obtain the string $s * b = aab$. The string aab is not in the dictionary and so we send the 0011, which is the code for aa, and add the string aab into the dictionary. The dictionary at this stage looks like this.

a	0000
b	0001
c	0010
aa	0011
ab	0100
ba	0101
aab	0110

Notice that the entries in the dictionary are getting longer. The decoder starts with the same initial dictionary. When it receives the first codeword 0000, it decodes it as a and sets its string equal to a. Upon receipt of the next 4 b (which would again be 0000) it adds the string aa to the dictionary and initializes the string s to a. The update procedure at the decoder is the same as the update procedure at the encoder and so the decoder dictionary tracks the encoder, albeit with some delay. The only time this delay causes a problem is when the encoder encounters a sequence of the form $\alpha * r * \alpha * r * \alpha * \beta$ is to be encoded, where α and β are symbols r is a string, and $\alpha * r$ is already in the dictionary. The encoder sends the code for $\alpha * r$, and then adds $\alpha * r\alpha$ to the encoder dictionary. The next code transmitted by the encoder is the code for $\alpha * r\alpha$, which the decoder does not yet have. This problem can easily be handled as an exception.

We can see that after a while we will have exhausted all the 4-b combinations. At this point we can do one of several things. We can flush the dictionary and restart, treat the dictionary as a static dictionary, or add 1 b to our code and continue adding entries into the dictionary.

98.4 Text Compression

A text source has several very distinctive characteristics, which can be taken advantage of when designing text compression schemes. The most important characteristic from the point of view of compression is the fact that a text source generally contains a large number of recurring patterns (words). Because of this, directly encoding the output of a text source with an entropy coder would be highly inefficient. Consider the letter u. The probability of this letter occurring in a fragment of English text is 0.02. Therefore, if we designed an entropy coder for this source we would assign a relatively long codeword to this letter. However, if we knew that the preceding letter was q, the probability of the current letter being u is close to one. Therefore, if we designed a number of entropy coders, each being indexed by the previous character encoded, the number of bits used to represent u would be considerably less. Thus, to be efficient, it is important that we treat each symbol to be encoded in the context of the past history, and *context modeling* has become an important approach to text compression.

Another technique for modeling recurring patterns is the use of finite state machines. Although this approach has been less fruitful in terms of generating compression algorithms, it holds promise for the future.

Finally, an obvious approach to text compression is by the use of dictionaries. We have described some adaptive dictionary-based techniques in the section on Universal coding. We now briefly describe some of the more popular methods. More complete descriptions can be found in Bell, Cleary, and Witten [1990].

Context Models

In lossless compression both encoder and decoder are aware of exactly the same past history. In other words, they are aware of the context in which the current symbol occurs. This information can therefore be used to increase the efficiency of the coding process. The context is used to determine the probability distribution for the next symbol. This probability distribution can then be used by an entropy coder to encode the symbol. In practice this might mean that the context generates an pointer to a code table designed for that particular distribution. Thus, if the letter u follows a space it will be encoded using considerably more bits than if it followed the letter q.

Context-based schemes are generally adaptive as different text fragments can vary considerably in terms of repeating patterns. The probabilities for different symbols in the different contexts is updated as they are encountered. This means that one will often encounter symbols that have not been encountered before for any of the given contexts (this is known as the *zero frequency* problem). In adaptive Huffman coding this problem was resolved by sending a code to indicate that the following symbol was being encountered for the first time followed by a prearranged code for that symbol. There was a certain amount of overhead associated with it, but for a sufficiently long symbol string the additional rate due to the overhead was negligible. Unfortunately, in context-based encoding the zero-frequency problem is encountered often enough for overhead to be a problem. This is especially true for longer contexts. Consider a context model of order four (the context is determined by the last four symbols). If we take an alphabet size of 95, the possible number of contexts is 95^4, which is more than 81 million! Of course, most of these contexts will never occur, but still the zero-frequency problem will occur often enough for the effect of the overhead to be substantial. One way to at least reduce the size of the problem would be to reduce the order of the context. However, longer contexts are more likely to generate accurate predictions. This problem can be resolved by a process called *exclusion*, which approximates a blending strategy for combining the probabilities of a symbol with respect to contexts of different orders. The exclusion approach works by first attempting to find if the symbol to be encoded has a nonzero probability with respect to the maximum context length. If this is so, the symbol is encoded and transmitted. If not, an escape symbol is transmitted and the context size is reduced by one and the process is repeated. This procedure is repeated until a context is found with respect to which the symbol has a nonzero probability. To guarantee that this process converges, a null context is always included, with respect to which all symbols have equal probability. The probability of the escape symbol can be computed in a number of different ways leading to different implementations.

Context modeling-based text compression schemes include: *PPMC* [prediction by partial match (PPM), PPMC is a descendant of PPMA and PPMB], *DAFC*, which uses a maximum context order

Lossless Compression

TABLE 98.3 Results of Compression Experiments (bits per character)

Text	Size	Digram	LZB	LZFG	Huffman	DAFC	PPMC	WORD	DMC
bib(technical bibliography)	111261	6.42	3.17	2.90	5.24	3.84	2.11	2.19	2.28
book(Far from the Madding Crowd)	768771	5.52	3.86	3.62	4.58	3.68	2.48	2.70	2.51
geo(seismic data 32-b numbers)	102400	7.84	6.17	5.70	5.70	4.64	4.78	5.06	4.77
news(USENET batch file)	377109	6.03	3.55	3.44	5.23	4.35	2.65	3.08	2.89
obj(executable Macintosh file)	246814	6.41	3.14	2.96	6.80	5.77	2.69	4.34	3.08
paper(technical, troff format)	82199	5.60	3.43	3.16	4.65	3.85	2.45	2.39	2.68
progc(source code in C)	39611	6.25	3.08	2.89	5.26	4.43	2.49	2.71	2.98

Source: Reprinted from Bell, T.C., Witten, I.H., and Cleary, J.C. 1989. Modeling for text compression. *ACM Comput. Surveys*, Dec.

of one; and WORD, which uses two alphabets, one which only contains alphanumeric characters, with the other containing the rest. A performance comparison is shown in Table 98.3.

State Models

Context models discard sequential information beyond the size of the context. This information can be preserved and used by finite-state models. However, there is relatively less work available in this area. An adaptive technique for finite-state modeling named *dynamic Markov modeling* (DMC) was given by Cormack and Horspool [Bell, Cleary, and Witten, 1990]. In DMC one starts with some fixed initial model and rapidly adapts it to the input data. Adaptation is done by maintaining counts on the transitions from each state and cloning a state when the count exceeds a certain threshold. The size of the model grows rapidly and, hence, this technique has been found to be feasible only for binary input data, in which case each state in the model has only two output transitions. Although at first sight, DMC seems to offer the full generality of Markov models, it has been proven that they only generate finite-context models and, hence, are no more powerful then the techniques described earlier. Despite this fact, DMCs provide an attractive and efficient way of implementing finite-context models.

Dictionary-Based Coding

Dictionary-based coding techniques have always been popular for text compression. The dictionary can be static or it can adapt to the source characteristics. Most current adaptive dictionary-based coders are variations of the 1977 and 1978 algorithms due to Ziv and Lempel (LZ77 and LZ78). The LZ77 and LZ78 algorithms are described in the section on universal coding.

The most popular static dictionary-based coding scheme is *digram coding*. The dictionary consists of the letters of the alphabet followed by the most popular pairs of letters. For example, an 8-b dictionary for printable ascii characters could be constructed by following the 95 printable ascii characters by 161 most frequently occurring pairs of characters. The algorithm would parse the input a pair of characters at a time. If the pair existed in the dictionary, it would put out the code for the pair and advance the pointer by two characters. If not, it would encode the first element of the pair and then advance the pointer by one character. This is a very simple compression scheme; however, compared to the other schemes described here, the compression performance is not very good.

Table 98.3 shows a comparison of the various schemes described here. The LZB algorithm is a particular implementation of the LZ77 approach, whereas the LZFG algorithm is an implementation of the LZ78 approach.

98.5 Image Compression

Algorithms for text compression do not work well when applied to images. This is because such algorithms exploit the frequent reoccurrence of certain exact patterns that are very typical of textual data. In image data, there are no such exact patterns common to all images. Also, in image data, correlation among pixels exists along both the horizontal and vertical dimensions. Text compression algorithms, when applied to images, invariably fail to exploit some of these correlations. Hence, specific algorithms need to be designed for lossless compression of images.

As mentioned before, there are two basic components in a lossless compression technique, modeling and coding. In lossless image compression, the task of modeling is usually split into two stages. In the first step, a *prediction model* is used to predict pixel values and replace them by the error in prediction. If prediction is based on previously transmitted values, then knowing the prediction error and the prediction scheme, the receiver can recover the value of the original pixel.

In the second step, a *prediction error model* is constructed, which is used to drive a variable length coder for encoding prediction errors. As we have already mentioned, coding schemes are known that can code at rates very close to the model entropy. Hence, it is important to build accurate prediction error models. In the next two sections, we describe various schemes that have been proposed in literature for each of these two steps.

Prediction Models

Linear Predictive Models

Linear predictive techniques (also known as lossless DPCM) usually scan the image in raster order, predicting each pixel value by taking a linear combination of pixel values in a casual neighborhood. For example, we could use $(P[i-1, j] + P[i, j-1])/2$ as the prediction value for $P[i, j]$, where $P[i, j]$ is the pixel value in row i, column j. Schemes that predict the current pixel based on a two-dimensional neighborhood are known as *two-dimensional predictive schemes*. Otherwise, they are called *one-dimensional schemes*.

Despite their apparent simplicity linear predictive techniques are quite effective and give performance surprisingly close to more state-of-the-art techniques. If we assume that the image is being generated by an *autoregressive* (AR) *model*, coefficients that best fit given data in the sense of the L_2 norm can be computed. One problem with such a technique is that the implicit assumption of the data being generated by a stationary source is, in practice, seldom true for images. Hence, such schemes yield very little improvement over the simpler linear predictive techniques such as the example in the previous paragraph [Rabbani and Jones, 1991].

Significant improvements can be obtained for some images by adaptive schemes that compute optimal coefficients on a block-by-block basis or by adapting coefficients to local changes in image statistics. Improvements in performance can also be obtained by adaptively selecting from a set of predictors. A sample of such a scheme is given by Sayood and Anderson [1992] that involves switching between two simple predictors, depending on the value of the prediction error. Another simple strategy to adaptively select between predictors is to take the median from a set of predictors. Significant improvements can usually be obtained by using the median of the following three predictors: (1) $P[i, j-1]$, (2) $P[i-1, j]$, and (3) $P[i, j-1] + P[i-1, j] - P[i-1, j-1]$.

Context Models

In context-based models, data is partitioned into contexts, and statistics are collected for each context. These statistics are then used to perform prediction. One problem with such schemes when compressing gray-scale images is that the number of different contexts grows exponentially with the size of the context. Given the large alphabet size of image data, such schemes quickly become infeasible even for small context sizes. Some clever schemes have been devised for partitioning the large number of contexts to a more manageable size [Todd, Langdon, and Rissanen, 1985]. Such schemes, however, have all been for modeling the prediction errors after using a simple linear predictive scheme. We shall describe them in more detail in the next section. An exception to this rule is binary images, for which context-based prediction schemes are among the most efficient.

Multiresolution Models

Multiresolution models generate representations of an image with varying spatial resolution. This usually results in a pyramidlike representation of the image with each layer of the pyramid serving as a prediction model for the layer immediately below. The pyramid can be generated either in a top-down or a bottom-up manner. Transmission is generally done in a top-down manner. One advantage of such a scheme is that the receiver recovers the image level by level. After reconstructing each level the receiver can get an approximation of the entire image by interpolating unknown pixels. This leads to *progressive transmission*, a technique that is found useful in many applications.

One of the more popular such techniques is known as hierarchical interpolation (HINT), and was proposed by Endoh and Yamakazi [Rabbani and Jones, 1991]. The specific steps involved in HINT are as follows: First, the pixels labeled Δ in Fig. 98.1 are decorrelated using DPCM. In the second step, the intermediate pixels (o) are estimated by linear interpolation and replaced by the error in estimation. Then, the pixels X are estimated from Δ and o and replaced by the error. Finally, the pixels labeled $\star$ and then • are estimated from known neighbors and replaced. The reconstruction process proceeds in a similar manner.

FIGURE 98.1 The HINT scheme for hierarchical prediction.

Lossy Plus Lossless Models

Given the success of lossy compression schemes for efficiently constructing a low-bit-rate approximation, one way of obtaining a prediction model for an image is to use a lossy representation of an image as the prediction model. Such schemes are called *lossy plus lossless schemes* (LPL). Here, a low-rate lossy representation of the image is first transmitted. Subsequently, the difference between the lossy and the original is transmitted to yield lossless reconstruction. Although the lossy image and its residual are usually encoded independently, making use of the lossy image to encode the residual can lead to significant savings in bit rates. Generally, LPL schemes do not give as good a compression as other standard methods. However, they are found to be very useful in certain applications. One application is when a user is browsing through a database of images, looking for a specific image of interest.

Modeling Prediction Errors

If the residual image that comprises prediction errors can be treated as an iid source, then it can be efficiently coded using any of the standard variable length techniques, such as Huffman coding or arithmetic coding. Unfortunately, even after applying the most sophisticated prediction techniques, generally the residual image has ample structure that violates the iid assumption. Therefore, in order to encode the residual image efficiently we need to use better models that capture the structure that remains after prediction.

Schemes that do not use the iid assumption typically consist of the following two components:

1. A *family of probability mass functions*. This family could be static, in the sense that a single family is used globally over all images and, further, it remains unchanged during the entire process of coding. It could also be adaptive, in the sense that the distributions are updated

regularly during the coding process to better reflect the statistics of the specific image being encoded.
2. An *activity function*, which indicates which particular distribution is to be used to encode a particular pixel (or block of pixels). Such a function could be backward adaptive (on-line) or forward adaptive (off-line). In the former case, the currently active distribution is estimated from the neighboring pixels that have already been encoded. In the latter case, the model that best suits the current block of pixels is determined and its index is transmitted to the receiver.

Perhaps the best known and most simple technique for modeling and coding the prediction residual in lossless image compression is the Rice encoder described previously [Rice, Yeh, and Miller, 1991].

Another popular way to model prediction errors is by using contexts. As mentioned in the previous section, the number of parameters required to specify a context-dependent model can get very large due to the large alphabet size for typical images. One way to alleviate this problem is to partition the residual image alphabet, representing prediction errors, into equivalence classes called *error buckets*. In order to reduce contexts, error buckets of neighboring pixels can be fused into a single context. Error values within a bucket are assumed to be independent of context. Todd, Langdon, and Rissanen [1985] have obtained excellent results with 11 buckets. However, these results are obtained with a two-pass technique, which is generally not desirable in practical applications. Various one-pass implementations of context-based techniques have been reported in literature [Howard and Vitter, 1991].

98.6 Compression Packages

The last year has seen a rapid proliferation of compression packages for use in reducing storage requirements. Most of these are based on the LZ algorithms with a secondary variable length encoder, which is generally a Huffman encoder. These include the following:

- LZ77-based schemes: *arj, lha, Squeeze, UC2, pkzip, zip, zoo*
- LZ78 based schemes: *gif, compress, pak, Stacker, arc, pkarc*
- PPMC based schemes: *ha*

98.7 Lossless Compression Standards

Joint Bilevel Image Experts Group

The JBIG algorithm proposed by the Joint Bilevel Image Experts Group defines a standard for lossless image compression. The algorithm is mainly designed for bilevel or binary data, such as facsimile, but can also be used for image data with up to 256 b/pixel by encoding individual bit planes. The algorithm performs best for data up to 6 b/pixel, and beyond 6 b, other algorithms such as lossless JPEG give better performance.

JBIG is essentially a progressive scheme that encodes an image by building representations at lowering resolutions in a bottom-up manner. Resolution is halved at every stage. The image is then transmitted in a top-down manner, that is, lower resolutions first. Resolution reduction is done by using a block of pixels in the higher resolution layer and pixels already encoded in the lower resolution layer. These values are used as an index into a predefined lookup table that can be also specified by the user. The standard does not impose any restrictions on D, the number of resolution layers that are constructed. Indeed, D can be set to zero if progressive coding is of no utility. In this case, coding is said to be *single-progression sequential*, or just *sequential*.

The algorithm allows some degree of compatibility between the progressive and sequential modes. Images that have been encoded in a progressive manner can be decoded sequentially, that is, as just one layer. Images that have been encoded sequentially, however, cannot be decoded progressively.

TABLE 98.4 JPEG predictors for Lossless Coding

Mode	Prediction for $P[i, j]$
0	0 (No Prediction)
1	$P[i-1, j]$
2	$P[i, j-1]$
3	$P[i-1, j-1]$
4	$P[i, j-1] + P[i-1, j] - P[i-1, j-1]$
5	$P[i, j-1] + (P[i-1, j] - P[i-1, j-1])/2$
6	$P[i-1, j] + (P[i, j-1] - P[i-1, j-1])/2$
7	$(P[i, j-1] + P[i-1, j])/2$

This compatibility between progressive and sequential modes is achieved by partitioning an image into *stripes*, with each stripe representing a sequence of image rows with user defined height. If the image has multiple bit planes, then stripes from each bit plane can be interleaved. Each stripe is separately encoded, with the user defining the order in which these stripes are concatenated into the output data stream.

Joint Photographic Experts Group

The recently proposed JPEG still compression standard [Wallace, 1991] uses linear predictive techniques in its lossless mode. It provides eight different predictive schemes from which the user can select. Table 98.4 lists the eight predictors. The first scheme makes no prediction. The next three are one-dimensional predictors, and the last four are two-dimensional prediction schemes.

Defining Terms

Arithemetic coding: A coding techique that maps source sequences to intervals on the real line.
Compression ratio: Size of original data/size of compressed data.
Context model: A modeling technique often used for text and image compression in which the encoding of the current symbol is conditioned on its context.
Entropy: The average information content (usually measured in bits per symbol) of a data source.
Entropy coding: A fixed to variable length coding of statistically independent source symbols that achieves or approaches optimal average code length.
Error model: A model used to capture the structure in the residual sequence which represents the difference between the source sequence and source model predictions.
Huffman coding: Entropy coding using Huffman's algorithm, which maps source symbols to binary code words of integral length such that no code word is a prefix of a longer code word.
Predictive coding: A form of coding where a prediction is made for the current event based on previous events and the error in prediction is transmitted.
Universal code: A source coding scheme that is designed without knowledge of source statistics but that converges to an optimal code as the source sequence length approaches infinity.
Ziv–Lempel coding: A family of dictionary coding-based schemes that encode strings of symbols by sending information about their location in a dictionary. The LZ77 family uses a portion of the already encoded string as dictionary, and the LZ78 family actually builds a dictionary of strings encountered.

References

Ancheta, T.C., Jr. 1977. Joint source channel coding. Ph.D. Thesis, Univ. of Notre Dame, Aug.
Bell, T.C., Cleary, J.C., and Witten, I.H. 1990. *Text Compression*. Advanced reference series. Prentice–Hall, Englewood Cliffs, N.J.

Cover, T.M. and Thomas, J.A. 1991. *Elements of Information Theory*, Wiley series in telecommunications. Wiley.

Davisson, L.D. 1973. Universal noiseless coding. *IEEE Trans. Inf. Th.*, 19:783–795.

Gallagher, R.G. 1978. Variations on a theme by Huffman. *IEEE Trans. Inf. Th.*, IT-24(6):668–674.

Golomb, S.W. 1966. Run-length encodings. *IEEE Trans. Inf. Th.*, IT-12(July):399–401.

Howard, P.G. and Vitter, J.S. 1991. New methods for lossless image compression using arithmetic coding. In *Proceedings of the Data Compression Conference*, eds. J.H. Reif and J.A. Storer, pp. 257–266. IEEE Computer Society Press.

Huffman, D.A. 1952. A method for the construction of minimum redundancy codes. *Proc. IRE*, 40:1098–1101.

Rabbani, M. and Jones, P.W. 1991. *Digital Image Compression Techniques*, Vol. TT7, Tutorial texts series. SPIE Optical Engineering Press.

Rice, R.F., Yeh, P.S., and Miller, W. 1991. Algorithms for a very high speed universal noiseless coding module. Tech. Rept. 91-1, Jet Propulsion Lab., California Institute of Technology, Pasadena, Feb.

Sayood, K. and Anderson, K.S. 1992. A differential lossless compression scheme. *IEEE Trans. Acoustics, Speech, Signal Proc.*, Jan.

Todd, S., Langdon, G.G., and Rissanen, J.J. 1985. Parameter reduction and context selection for compression of gray scale images. *IBM J. Res. Dev.*, 29(March):88–193.

Wallace, G.K. 1991. The JPEG still picture compression standard. *Commun. ACM*, 34(April):31–44.

Further Information

An excellent source for further information on arithmetic coding, Ziv–Lempel algorithms, and text compression is the book *Text Compression*, by T.C. Bell, J.C. Cleary, and I.H. Whitten.

A good place to obtain information about compression packages as well as information about recent commercial developments is the frequently asked questions (*faq*) for the group comp.compression in the netnews hierarchy. This *faq* is maintained by Jean-Loup Gailly and is periodically posted in the comp.compression newsgroup. The *faq* can also be obtained by ftp from rtfm.mit.edu in pub/usenet/news.answers/compression-faq/part[1–3]. The information about compression packages was obtained from the *faq*.

99
Facsimile

99.1 Introduction..1411
99.2 Facsimile Compression Techniques...............................1412
   One-Dimensional Coding • Two-Dimensional Coding Schemes • Multilevel Facsimile Coding • Lossy Techniques • Pattern Matching Techniques
99.3 International Standards..1417
   CCITT Group 3 and 4—Recommendations T.4 and T.6 • The Joint Bi-Level Image Processing Group (JBIG) Standard • Comparison of MH, MR, MMR, and JBIG
99.4 Future Trends 1421

Nasir Memon
Northern Illinois University

Khalid Sayood
University of Nebraska-Lincoln

99.1 Introduction

A **facsimile** (**fax**) image is formed when a document is raster scanned by a light sensitive electronic device, which generates an electrical signal with a strong pulse corresponding to a dark dot on the scan line and a weak pulse for a white dot. In digital fax machines, the electrical signal is subsequently digitized to two levels and processed, before transmission over a telephone line. Modern digital fax machines partition a page into 2376 scan lines, with each scan line comprising 1728 dots. A fax document can, therefore, be viewed as a two-level image of size 2376 × 1728, which corresponds to 4,105,728 b of data. The time required to transmit this raw data over a 4800-b/s telephone channel would be more than 14 mins! Transmitting a 12 page document would require almost 3 h. Clearly this is unacceptable. To reduce the bit rates, some form of compression technique is required. Imposing the more realistic constraint of 1 min of transmission time per page leads us to the requirement of encoding a fax image at 0.07 b per pixel, for a **compression ratio** of almost 15 : 1. Fortunately, fax images contain sufficient redundancies and even higher than 15 : 1 compression can be achieved by state-of-the-art compression techniques.

Facsimile image compression provides one of the finest examples of the importance of the development of efficient compression technology in modern day communication. The field of facsimile image transmission has seen explosive growth in the last decade. One of the key factors behind this proliferation of fax machines has been the development and standardization of effective compression techniques. In the rest of this chapter we describe the different approaches that have been developed for the compression of fax data. For the purpose of discussion we classify the compression techniques into five different categories and give one or two representative schemes for each. We then describe international standards for facsimile encoding, the development of which has played a key role in the establishment of facsimile transmission as we know it today. Finally, we conclude with recent progress and anticipated future developments.

Source Compression

99.2 Facsimile Compression Techniques

Over the last three decades numerous different techniques have been developed for the compression of facsimile image data. For the purpose of discussion we classify such compression techniques into five different categories: (1) one-dimensional coding, (2) two-dimensional techniques, (3) multilevel techniques, (4) lossy techniques, and (5) pattern matching techniques. We discuss each approach in a separate subsection and describe one or two representative schemes.

One-Dimensional Coding

In Fig. 99.1 we show as examples two documents that are typically transmitted by a fax machine. One property that clearly stands out is the clustered nature of black (b) and white (w) pixels. The b and w pixels occur in bursts. It is precisely this property that is exploited by most facsimile compression techniques. A natural way to exploit this property is by *run length coding*, a technique used in some form or the other by a majority of the earlier schemes for facsimile image coding.

In runlength coding, instead of coding individual pixels, the lengths of the runs of pixels of the same color are encoded, following an encoding of the color itself. With a two-level image just encoding alternating runs is sufficient. To efficiently encode the runlengths we need an appropriate model. A simple way to obtain a model for the black and white runs is by regarding each scan line as being generated by the first-order Markov process, shown in Fig. 99.2, known as the *Capon model* for binary images [Capon, 1959]. The two states S_w and S_b shown in the figure represent the events that the current pixel is a white pixel or a black pixel, respectively. $P(w/b)$ and $P(b/w)$ represent *transition probabilities*. $P(w/b)$ is the probability of the next pixel being a white pixel when the current pixel is black and $P(b/w)$ is vice versa. If we denote the probabilities $P(w/b)$ and $P(b/w)$

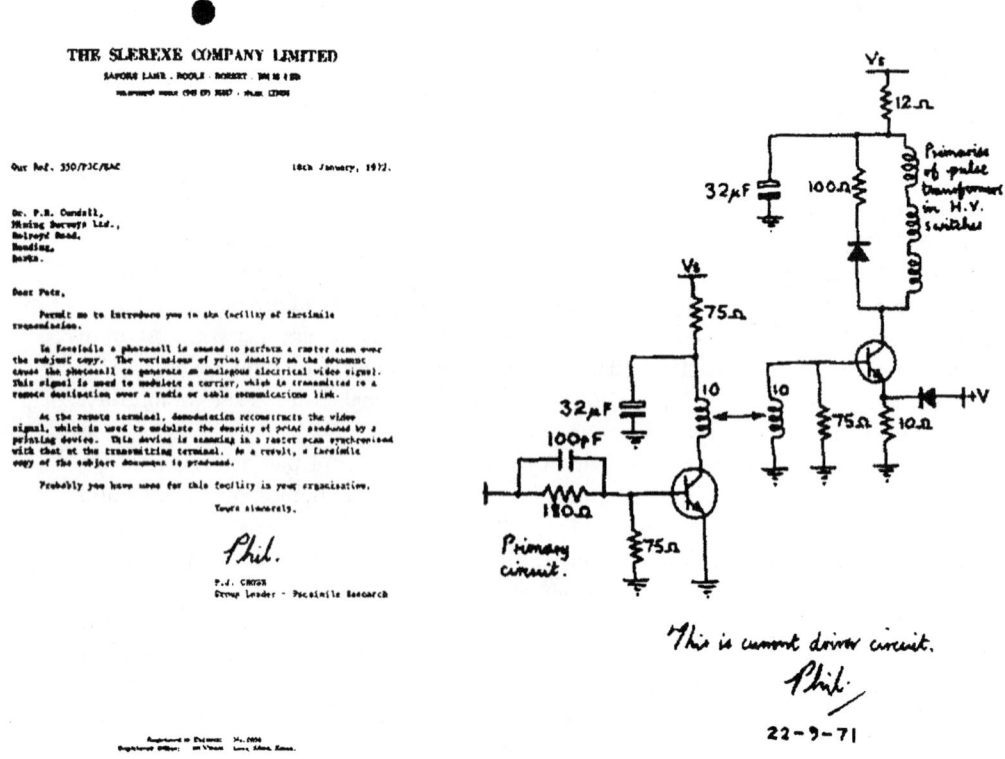

FIGURE 99.1 Example documents from the CCITT group 3 test images.

Facsimile

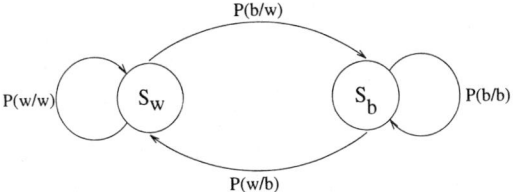

FIGURE 99.2 The Capon model for binary images.

by t_w and t_b, respectively, then the probability of a run of length r_k in a state s is given by

$$P(r_k \mid s) = t_s(1 - t_s)^{r_k - 1} \qquad s \in \{S_w, S_b\}$$

which gives us a *geometric distribution* for the runlengths. The expected runlength of black and white runs then turns out to be $1/t_b$ and $1/t_w$, respectively. The geometric distribution has been found to be an appropriate model for the runlengths encountered in special classes of facsimile images such as weather maps [Kunt and Johnsen, 1980]. However, for more structured documents such as letters that contain printed text, it turns out to be inadequate. Getting analytical models for runlengths of structured documents is difficult. In practice, models are obtained empirically by analyzing a set of typical images, and optimal variable length codes are then constructed based on the statistics of runlengths in this set. Usually two distinct sets of codewords are constructed for the black and white runs as the statistics for the two are found to be significantly different. The extra cost involved in maintaining two separate code tables is worth the improvement in compression obtained.

Two-Dimensional Coding Schemes

The amount of compression obtained by one-dimensional coding schemes described in the previous subsection is usually quite limited. This is because such schemes do not take into account vertical correlations, that is, the correlation between adjacent scan lines, typically found in image data. Vertical correlations are especially prominent in high resolution images that contain twice the number of scan lines per page. There have been many schemes proposed for taking vertical correlations into account. We will discuss a few that are representative.

One way to take vertical correlations into account is by encoding pixels belonging to k successive lines simultaneously. Many different techniques of this nature have been proposed in the literature, including *block coding, cascade division coding, quad-tree encoding*, etc. (for a review see Kunt and Johnsen [1980] and Yasuda [1980]). However, such techniques invariably fail to utilize correlations that occur across the boundaries of the blocks or bundles of lines that are being encoded simultaneously. A better way to exploit vertical correlations is to process pixels line by line as in one-dimensional coding, and make use of the information encountered in previous scan lines in order to encode the current pixel or sequence of pixels. Next we list three such techniques that have proven to be very successful.

Relative Element Address Designate (READ) Coding

Since two adjacent scan lines of a fax image are highly correlated, so are their corresponding runs of white and black pixels. Hence, the runlengths of one scan line can be encoded with respect to the runlengths of the previous scan line. A number of schemes based on this approach were developed in the late 1970s. Perhaps the best known among them is the *relative element address designate* (READ) coding technique that was a part of Japan's response to a call for proposals for an international standard [Yasuda, 1980]. In READ coding, prior to encoding a run length, we locate five *reference*

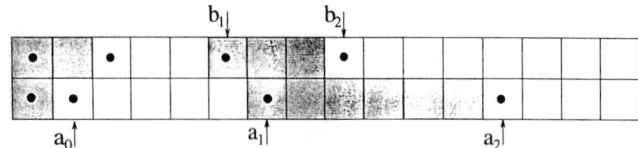

FIGURE 99.3 Two rows of an image, the transition pixels are marked with a dot.

pixels on the current and previous scan line. These pixels are denoted by $a_0, a_1, a_2, b_1,$ and b_2 respectively, and are identified as follows:

- a_0: This is the last pixel whose value is known to both encoder and decoder. At the beginning of encoding each line a_0 refers to an imaginary white pixel to the left of the first actual pixel. Although it often is a *transition pixel*, it does not have to be one.
- a_1: This is the first transition pixel to the right of a_0.
- a_2: This is the second transition pixel to the right of a_0.
- b_1: This is the first transition pixel on the line above the line currently being encoded to the right of a_0 whose color is the opposite of the color of a_0.
- b_2: This is the first transition pixel to the right of b_1 and on the same line as b_1.

For example, if the second row is the one being currently encoded, and we have encoded the pixels up to the second pixel, then the assignment of the different pixels is shown in Fig. 99.3. Note that while both the transmitter (encoder) and receiver (decoder) know the positions a_0, b_1, and b_2, the positions a_1 and a_2 are known only to the encoder. Coding is done in one of three modes depending on the relative positions of these pixels. If the run lengths on the current and previous line are similar, then the distance between a_1 and b_1 would typically be much smaller than the distance between a_0 and a_1. Hence, the current length can be specified by encoding the distance (a_1, b_1). This is called *vertical mode* coding. However, when the distance between a_1 and b_1 is large, that is, if there is no similar run on the previous line, then it is better to encode the runs (a_0, a_1) and (a_1, a_2) using one-dimensional runlength coding. This type of encoding is known as *horizontal mode* coding. A third type of coding, known as *pass mode*, is performed when the condition $a_0 \leq b - 1 < b_2 < a_1$ occurs. That is, we go through two runs in the previous line before completing the current run on the current line. In this case we simply advance the next pixel to be encoded to a_0', which is the pixel on the current line that is exactly under b_2. Before sending any runlengths, a codeword specifying the mode being used is transmitted. Additional details, including the specific codewords to be used, are given in Yasuda [1980].

Two-Dimensional Predictive Coding

In **predictive coding**, the image is scanned in some fixed order and a prediction is made of the current pixel based on the values of previously transmitted pixels. If the neighborhood employed to perform prediction contains pixels from both the previous and current scan lines, then the technique is referred to as two-dimensional prediction. Since prediction is being made on the basis of pixels known to the receiver, only the prediction error needs to be transmitted. With binary images, the prediction error sequence is again binary, with a 0 indicating no error and a 1 indicating that an error in prediction was made. If the prediction scheme is effective, then the prediction error sequence will contain many more zeroes than ones and hence can be coded more efficiently. If we fix the neighborhood used for prediction, then given a specific image the optimum prediction function that minimizes the probability of prediction error can be computed. However, such an optimum function varies from image to image. This fact limits the practical utility of predictive schemes. Prediction when used as a preprocessing step, however, can often enhance the performance of other facsimile compression techniques such as runlength coding [Yasuda, 1980].

Model-Based Coding

If we impose an n'th-order Markov model on a binary source, then its entropy is given by

$$\sum_{k=1}^{n} P(s_k)(P(x=0 \mid s_k) \cdot \log_2 P(x=0 \mid s_k) + P(x=1 \mid s_k) \cdot \log_2 P(x=1 \mid s_k))$$

where $s_1, \ldots, s_n$ are the states and x is the current pixel. When coding binary images, the states s_i are simply taken to be the different bit patterns that can occur in a particular neighborhood of n pixels that occur prior to x. Given the conditional probabilities mentioned, the source can be optimally encoded by using *arithmetic coding* [Rissanen and Langdon, 1979]. Note that since *Huffman coding* uses an integral number of bits to encode each source symbol, it is of little utility for encoding a binary source unless some form of alphabet extension is performed that blocks individual bits to build an extended alphabet set. Hence, model-based coding was not used for binary images until the early 1980s when the development of sophisticated arithmetic coding techniques enabled the encoding of sources at rates arbitrarily close to then entropies. In fact, it has been proven that model-based arithmetic coding is essentially superior to any other scheme that may encode more than one bit at a time [Langdon and Rissanen, 1981]. In practice, however, we do not have the exact conditional probabilities needed by the model. An estimate of these can be adaptively maintained by keeping track of the counts of black and white pixels encountered so far corresponding to every state. The recently finalized **Joint Bi-Level Image Processing Group (JBIG)** standard [Hampel and Arps, 1992] uses model-based arithmetic coding and significantly outperforms the previous standards for facsimile image compression for a wide variety of test images. The compression ratio obtained is especially superior when encoding half-tone images or mixed documents that contain graphics and text [Arps and Truong, 1994].

Multilevel Facsimile Coding

The techniques we have discussed so far can also be applied to facsimile images that have been digitized using more than two amplitude levels. An image containing 2^n gray levels, with $n \geq 2$, can be decomposed into n different bit planes, each of which can then be compressed by any two-level compression technique. Better compression can be obtained if pixel intensities are expressed by using a **Gray code** representation as compared to the standard binary number representation. This is because the Gray code representation guarantees that two numbers that differ in magnitude by one will differ in their representations in only a single bit position.

The bit-plane approach for coding multilevel images can be taken to its extreme by constructing a two-level bit plane for each of the 2^n gray levels in the image. The 2^n resulting *level planes* can then be compressed by some two-level compression technique. Among the 2^n different level planes, a single arbitrary one need not be encoded as it can be completely determined by the remaining $2^n - 1$ level planes. A comparison of level plane and bit plane coding has been made, and it appears that level plane coding performs better than bit plane coding for images that contain a relatively small number of gray levels (typically, 2–4 b per pixel) [Yasuda et al., 1985].

Another approach to coding multilevel facsimile images is to use one of the many techniques that have been developed for encoding gray scale video images. These techniques have been described in the previous section under the topic of lossless image compression. Such techniques typically perform better than bit-plane encoding and level-plane encoding when the number of gray levels present is relatively large (more than 6 b per pixel).

Compression ratios achieved by lossless techniques are usually very modest. Typical state-of-the-art lossless compression techniques can only achieve between 2–1 and 3–1 compression for images that have been acquired by a camera or some similar sensory device. Hence, it is quite common to use *lossy* or noninformation preserving compression techniques for multilevel images. State-of-the-art lossy techniques can easily achieve more than 15–1 compression while preserving excellent visual fidelity. A description of lossy techniques for multilevel images is given in a later section of this chapter.

Lossy Techniques

Besides multilevel facsimile images, lossy techniques can also be used for two-level images. Two types of lossy techniques have been used on two-level images. The first type consists of a large number of pre- and postprocessing techniques that are primarily used for enhancing subsequent lossless compression of two-level images. The scanning and spatial sampling process inherent in digital facsimile systems invariably leads to a high degree of jaggedness in the boundaries between black and white pixels. This jaggedness, besides reducing the visual quality of the reconstructed document also severely effects the compression ratios that can be obtained by breaking up long runs of uniform color. Hence, preprocessing techniques that filter out noise would not only improve picture quality but also reduce transmission time. Various such preprocessing techniques have been developed, a survey of which is given in [Yasuda, 1980].

A simple preprocessing technique is to remove isolated black points and bridge small gaps of white pixels between a sequence of black pixels. More sophisticated techniques employ morphological operators to modify local patterns such that subsequent compression is increased. Such techniques, however, introduce significant degradations in the image and hence require postprocessing of the reconstructed image at the receiving end. This fact limits their utility in commercial systems as they require the facsimile equipment at the receiving end to be equipped with circuitry to perform postprocessing.

An alternative approach to reduce jaggedness in a facsimile image is by modifying the **quantizer** that is used to obtain a two-level image from electrical impulses generated while scanning a document. One such quantizer, called the *notchless bilevel quantizer*, has been proposed [Yasuda, 1980], which adaptively adjusts the quantization level on the basis of preceding pixels. It has been shown that images obtained by using the notchless quantizer have considerably lower entropy and better visual quality.

The second class of lossy compression techniques for facsimile image data attempts to approximate the input image by replacing patterns extracted from the image with appropriate patterns from a library. Such schemes form an important special class of facsimile image compression techniques and are discussed in the next subsection.

Pattern Matching Techniques

Since digitized images used in facsimile transmission often contain mostly text, one way of compressing such images is to perform optical character recognition (OCR) and encode characters by their American Standard Code for Information Interchange (ASCII) code along with an encoding of their position. Unfortunately, the large variety of fonts that may be encountered, not to mention handwritten documents, makes character recognition very unreliable. Furthermore, such an approach limits documents that can be transmitted to specific languages making international communication difficult. However, an adaptive scheme that develops a library of patterns as the document is being scanned circumvents the problems mentioned. Given the potentially high compression that could be obtained with such a technique, many different algorithms based on this approach have been proposed and continue to be investigated [Pratt et al., 1980; Johnsen, Segen, and Cash, 1983; Witten, Moffat, and Bell, 1994].

Techniques based on pattern matching usually contain a *pattern isolater* that extracts patterns from the document while scanning it in raster order. A pattern is defined to be a connected group of black pixels. This pattern is then matched with the a library of patterns that has been accumulated thus far. If no close match is formed, then an encoding of the pattern is transmitted and the pattern is added to the library. The library is empty at the beginning of coding and gradually builds up as encoding progresses.

If a close match for the current pattern is found in the library, then the index of the library symbol is transmitted followed by an encoding of an offset with respect to the previous pattern that is needed to spatially locate the current pattern in the document. Since the match need not be exact, the *residue*, which represents the difference between the current pattern and its matching library symbol, also needs to be transmitted if lossless compression is required. However, if the transmission need not be information preserving, then the residue can be discarded. Most practical schemes discard at least part of the residue in order to obtain high compression ratios.

Although the steps just outlined represent the basic approach, there are a number of details that need to be taken care of for any specific implementation. Such details include the algorithm used for isolating and matching patterns, the encoding technique used for the patterns that do not find a close match in the library, algorithms for fast identification of the closest pattern in the library, distortion measures for closeness of match between patterns, heuristics for organizing and limiting the size of the library, etc. The different techniques reported in the literature differ in the way they tackle the issues listed. For a good survey of such techniques, the reader is referred to Witten, Moffat, and Bell [1994].

A real-time coder based on pattern matching was proposed by AT&T to the Consultative Committee on International Telephony and Telegraphy (CCITT) for incorporation into the international standard [Johnsen, Segen, and Cash, 1983]. The coder gave three times the compression given by the existing standard. The higher compression though came at the cost of loss in quality as the scheme proposed was not information preserving.

99.3 International Standards

Several standards for facsimile transmission have been developed over the past few decades. These include specific standards for compression. The requirements on how fast the facsimile of an A4 document (210 × 297 mm) is transmitted has changed over the last two decades, and the CCITT, a committee of the International Telecommunications Union (ITU) of the United Nations, has issued a number of recommendations based on the speed requirements at a given time. The CCITT classifies the apparatus for facsimile transmission into four groups. Although several considerations are used in this classification, if we only consider the time to transmit an A4 size document over the phone lines, the four groups are described as follows:

- **Group 1.** This apparatus is capable of transmitting an A4 size document in about 6 min over the phone lines using an analog scheme. The apparatus is standardized in Recommendation T.2.
- **Group 2.** This apparatus is capable of transmitting an A4 document over the phone lines in about 3 min. Group 2 apparatus also use an analog scheme and, therefore, do not use data compression. The apparatus is standardized in Recommendation T.3.
- **Group 3.** This apparatus uses a digitized binary representation of the facsimile. As it is a digital scheme it can, and does, use data compression and is capable of transmitting an A4 size document in about a minute. The apparatus is standardized in Recommendation T.4.
- **Group 4.** The speed requirement is the same as group 3. The apparatus is standardized in Recommendations T.6, T.503, T.521, and T.563.

CCITT Group 3 and 4—Recommendations T.4 and T.6

The recommendations for group 3 facsimile include two coding schemes: a one-dimensional scheme and a two-dimensional scheme. In the one-dimensional coding mode a runlength coding scheme is used to encode alternating white and black runs on each scan line. The first run is always a white run. If the first pixel is a black pixel, then we assume that we have a white run of length zero. A special end-of-line (EOL) code is transmitted at the end of every line. Separate Huffman codes are used for the black and white runs. Since the the number of runlengths is high, instead of generating a Huffman code for each runlength r_l, the runlength is expressed in the form

$$r_l = 64 * m + t \qquad \text{for } t = 0, 1 \ldots, 63, \text{ and } m = 1, 2, \ldots, 27 \qquad (99.1)$$

A runlength r_l is then represented by the codes for m and t. The codes for t are called the *terminating codes*, and the codes for m are called the *make-up codes*. If $r_l < 63$, then only a terminating code needs to be used. Otherwise, both a make-up code and a terminating code are used. This coding scheme is generally referred to as a **modified Huffman (MH)** scheme. The specific codewords to be used are prescribed by the standard and can be found in a variety of sources including Hunter and Robinson [1980]. One special property of the codewords is that a sequence of six zeroes cannot result no matter how they are concatenated. Hence the codeword 0000001 is used to indicate end-of-line.

For the range of m and t given, lengths of up to 1728 can be represented, which is the number of pixels per scan line in an A4 size document. However, if the document is wider, the recommendations provide for those with an optional set of 13 codes. The optional codes are the same for both black and white runs.

The two-dimensional encoding scheme specified in the group 3 standard is known as the **modified READ (MR) coding**. It is essentially a simplification of the READ scheme described earlier. In modified READ the decision to use the horizontal mode or the vertical mode is made based on the distance $a_1 b_1$. If $|a_1 b_1| \leq 3$, then the vertical mode is used, otherwise the horizontal mode is used. The codec also specifies a k factor that no more than $k - 1$ successive lines are two dimensionally encoded; k is 2 for documents scanned at low resolution and 4 for high resolution documents. This prevents vertical propagation of bit errors to no more than k lines.

The group 4 encoding algorithm, as standardized in CCITT recommendation T.6, is identical to the two-dimensional encoding algorithm in recommendation T.4. The main difference between T.6 and T.4 from the compression point of view is that T.6 does not have a one-dimensional coding algorithm, which means that the restriction specified by the k factor as described in the previous paragraph is also not present. This slight modification of the modified READ algorithm has earned it the name **modified modified READ (MMR)**. The group 4 encoding algorithm also does away with the end-of-line code, which was intended to be a form of redundancy to avoid image degradation due to bit errors. Another difference in the group 4 algorithm is the ability to encode lines having more than 2623 pixels. Such runlengths are encoded by using a mark-up code(s) of length 2560 and a terminating code of length less than 2560. The terminating code itself may consist of mark-up and terminating codes as specified by the group 3 technique.

Handling Transmission Errors

If facsimile images are transmitted over the existing switched telephone network, techniques for handling transmission errors are needed. This is because an erroneous bit causes the receiver to interpret the remaining bits in a different manner. With the one-dimensional modified Huffman coding scheme, resynchronization can quickly occur. Extensive studies of the resynchronization period for the group 3 one-dimensional coding schemes have been made. It was shown that in most cases the Huffman code specified resynchronizes quickly, with the number of lost pixels typically less than 50. For a document scanned at high resolution, this corresponds to a length of 6.2 mm on a scan line.

To handle transmission errors, CCITT has defined an optional error limiting mode and an error correcting mode. In the error limiting mode, which is used only with MH coding, each line of 1728 pixels is divided into 12 groups of 144 pixels each. A 12-b header is then constructed for the line indicating an all white group with a 0 and a nonwhite group with a one. The all white groups are not encoded and the nonwhite groups are encoded separately by using MH. This technique limits the effect of bit errors from propagating through an entire scan line.

The error correction mode breaks up the coded data stream into packets and attaches an error detecting code to each packet. Packets received in error are retransmitted as requested by the receiver but only after the entire page has first been transmitted. The number of retransmissions for any packet is restricted to not exceed four.

The Joint Bi-Level Image Processing Group (JBIG) Standard

The JBIG is a joint experts group of the International Standards Organization (ISO), International Electrotechnical Commission (IEC), and the CCITT. This experts group was jointly formed in 1988 to establish a standard for the progressive encoding of bilevel images. The JBIG standard can be viewed as a combination of two algorithms: a **progressive transmission** algorithm and a lossless compression algorithm. Each of these can be understood independently of the other.

Lossless Compression

The lossless compression algorithm uses a simple *context model* to capture the structure in the data. A particular arithmetic coder is then selected for each pixel based on its *context*. The context is made up of neighboring pixels. For example, in Fig. 99.4 the pixel to be coded is marked **X**, whereas the pixels to be used as the context are marked **O** or **A**. The **A** and **O** pixels are previously encoded pixels and are available to both encoder and decoder. The **A** pixel can be moved around in order to better capture any structure that might exist in the image. This is especially useful in half-toned images in which the **A** pixels are used to capture the periodic structure. The location and movement of the **A** pixel is transmitted to the decoder as side information.

The arithmetic coders specified in the JBIG standard is a special binary adaptive arithmetic coder known as the QM coder. The QM coder is a modification of an adaptive binary arithmetic coder called the Q coder [Pennebaker and Mitchell, 1988], which in turn is an extension of another binary adaptive arithmetic coder called the *skew* coder [Langdon and Rissanen, 1981]. Instead of dealing directly with the 0s and 1s put out by the source, the QM coder maps them into a more probable symbol (MPS) and less probable symbol (LPS). If 1 represents black pixels and 0 represents white pixels, then in a mostly black image 1 will be the MPS, whereas in an image with mostly white regions 0 will be the MPS. To make the implementation simple, the JBIG committee recommended several deviations from the standard arithmetic coding algorithm. The update equations in arithmetic coding that keep track of the subinterval to be used for representing the current string of symbols involve multiplications, which are expensive in both hardware and software. In the QM coder expensive multiplications are avoided and rescalings of the interval take the form of repeated doubling, which corresponds to a left shift in the binary representation. The probability q_c of the LPS for context C is updated each time a rescaling takes place and the context C is active. An ordered list of values for q_c is kept in a table. Every time a rescaling occurs, the value of q_c is changed to the next lower or next higher value in the table, depending on whether the rescaling was caused by the occurrence of an LPS or MPS. In a nonstationary situation, it may happen that the symbol assigned to LPS actually occurs more often than the symbol assigned to MPS. In this situation, the assignments are reversed; the symbol assigned the LPS label is assigned the MPS label and vice versa.

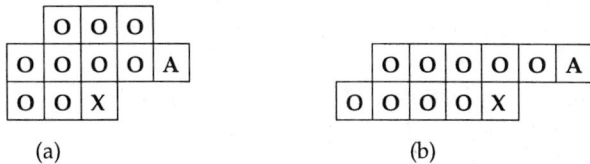

FIGURE 99.4 (a) Three line and (b) two line model template for lowest resolution layer.

The test is conducted every time a rescaling takes place. The decoder for the QM coder operates in much the same way as the encoder, by mimicking the encoder operation.

Progressive Transmission

In progressive transmission of an image a low resolution representation of the image is first sent. This low resolution representation requires very few bits to encode. The image is then updated, or refined, to the desired fidelity by transmitting more and more information. To encode an image for progressive transmission, we need to create a sequence of progressively lower resolution images from the original higher resolution image. The JBIG specification recommends generating one lower resolution pixel for each two by two block in the higher resolution image. The number of lower resolution images (called layers) is not specified by JBIG. However, there is a suggestion that the lowest resolution image is roughly 10–25 dot per inch (dpi). There are a variety of ways in which the lower resolution image can be obtained from a higher resolution image, including sampling and filtering. The JBIG specification contains a recommendation against the use of sampling. The specification provides a table-based method for resolution reduction. The table is indexed by the neighboring pixels shown in Fig. 99.5 in which the circles represent the lower resolution layer pixels and the squares represent the higher resolution layer pixels.

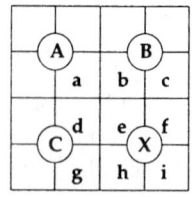

FIGURE 99.5 Pixels used to determine value of lower level pixel.

Each pixel contributes a bit to the index. The table was formed by computing the expression

$$4e + 2(b + d + f + h) + (a + c + g + i) - 3(B + C) - A$$

If the value of this expression is greater than 4.5 the pixel X is tentatively declared to be 1. The table has certain exceptions to this rule to reduce the amount of edge smearing, generally encountered in a filtering operation. There are also exceptions that preserve periodic patterns and dither patterns.

When the progressive mode is used for transmission, information from lower resolution layers can be used to improve compression. This is done by including pixels from lower resolution layers in the context used to encode a pixel in the current layer. The contexts used for coding the lowest resolution layer are those shown in Fig. 99.4. The contexts used in coding the higher resolution layer are shown in Fig. 99.6. In each context 10 pixels are used. If we include the 2 b required to indicate which context template is being used, 12 b will be used to indicate the context. This means that we can have 4096 different contexts.

The standard does not impose any restrictions on D, the number of resolution layers that are constructed. Indeed, D can be set to zero if progressive coding is of no utility. In this case, coding is said to be *single-progression sequential*, or just *sequential*. The algorithm allows some degree of compatibility between the progressive and sequential modes. Images that have been encoded in a progressive manner can be decoded sequentially, that is, as just one layer. Images that have been encoded sequentially, however, cannot be decoded progressively. This compatibility between progressive and sequential modes is achieved by partitioning an image into *stripes*, with each stripe representing a sequence of image rows with

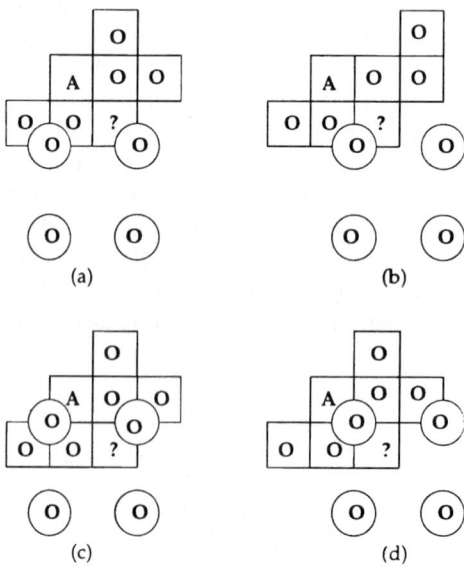

FIGURE 99.6 Contexts used in the coding of higher resolution layers.

TABLE 99.1 Comparison of Binary Image Coding Schemes [Arps and Truong, 1994]

Source Description	Original Size, pixels	MH, bytes	MR, bytes	MMR, bytes	JBIG, bytes
Letter	4352 × 3072	20,605	14,290	8,531	6,682
Sparse Text	4352 × 3072	26,155	16,676	9,956	7,696
Dense Text	4352 × 3072	135,705	105,684	92,100	70,703

user defined height. If the image has multiple bit planes, then stripes from each bit plane can be interleaved. Each stripe is separately encoded, with the user defining the order in which these stripes are concatenated into the output data stream.

Comparison of MH, MR, MMR, and JBIG

In the previous subsection we have seen four different facsimile coding algorithms that are part of different international standards. As we might expect the JBIG algorithm performs better than the MMR algorithm, which performs better than the MR algorithm, which in turn performs better than the MH algorithm. The level of complexity also follows the same trend, though one could argue that MMR is actually less complex than MR. A comparison of the schemes for some facsimile sources is shown in Table 99.1.

The modified READ algorithm was used with $K = 4$, whereas the JBIG algorithm was used with an adaptive 3 line template and adaptive arithmetic coder to obtain the results in this table. As we go from the one-dimensional MH coder to the two-dimensional MMR coder we get a factor of two reduction in file size for the sparse text sources. We get even further reduction when we use an adaptive coder and an adaptive model, as is true for the JBIG coder. When we come to the dense text, the advantage of the two-dimensional MMR over the one-dimensional MH is not as significant, as the amount of two-dimensional correlation becomes substantially less.

The compression schemes specified in T.4 and T.6 break down when we try to use them to encode half-tone images. This is to be expected as the model that was used to develop these coding schemes is not valid for half-tone images. The JBIG algorithm, with its adaptive model and coder suffers from no such drawbacks, and performs well for half-tone images as well [Arps and Truong, 1994].

99.4 Future Trends

The next decade will see continued progress in the development of facsimile technology. Future developments anticipated include proliferation of color facsimile, integration of facsimile equipment with personal computers, penetration of fax machines into the home market, telepublishing, and even distribution of newspapers through fax machines. Compression technology is expected to play a key role in these developments as new techniques have to be designed and incorporated into international standards.

For example, an immediate challenge that stands before the international community is the establishment of a compression standard for color facsimile. Although few color fax machines exist today, technological developments in printing technology are expected to bring their price down to reasonable levels.

Defining Terms

Compression ratio: Size of original data/size of compressed data.
Facsimile: The process by which a document is optically scanned and converted to electrical signals.

Facsimile image: The quantized digital image corresponding to the document that has been input to a facsimile machine.

Fax: Abbreviation for facsimile.

Gray code: A binary code for integers in which two integers that differ in magnitude by one differ in only one bit position.

Group 3: Facsimile apparatus capable of transmitting an A4 size document in about a minute. The apparatus is standardized in Recommendation T.4.

Group 4: Facsimile apparatus for sending a document over public data networks with virtually error-free reception. Standardized in Recommendations T.6, T.503, T.521, and T.563.

Joint Bi-Level Image Processing Group (JBIG): A group from International Standards Organization (ISO), International Electrotechnical Commission (IEC), and the CCITT. This experts group was jointly formed in 1988 to establish a standard for the progressive encoding of bilevel images. The term JBIG is also used to refer to the coding algorithm proposed by this committee.

Modified Huffman code (MH): One-dimensional coding scheme used by Group 3 equipment.

Modified modified READ code (MMR): Two-dimensional coding scheme used by Group 4 equipment.

Modified READ code (MR): Two-dimensional coding scheme used by Group 3 equipment.

Predictive coding: A form of coding where a prediction is made for the current event based on previous events and the error in prediction is transmitted.

Progressive transmission: A form of transmission in which a low resolution representation of the image is first sent. The image is then updated, or refined, to the desired fidelity by transmitting more and more information.

Quantizer: The process of converting analog data to digital form.

References

Arps, R. and Troung, T. 1994. Comparison of international standards for lossless still image compression. *Proc. IEEE*, 82(6):889–899.

Capon, J. 1959. A probablistic model for run-length coding of pictures. *IRE Trans. Inf. Th.* (Dec.): 157–163.

Hampel, H. Arps, R.B., et al. 1992. Technical features of the JBIG standard for progressive bi-level image compression. *Sig. Proc.*, 4(2):103–111.

Hunter, R. and Robinson, A.H. 1980. International digital facsimile standards. *Proc. IEEE*, 68(7):855–865.

Johnsen, O., Segen, J., and Cash, G.L. 1983. Coding of two-level pictures by pattern matching and substitution. *Bell Sys. Tech. J.*, 62(8):2513–2545.

Kunt, M. and Johnsen, O. 1980. Block coding of graphics: A tutorial review. *Proc. IEEE*, 68(7):770–786.

Langdon, G.G., Jr. and Rissanen, J. 1981. Compression of black-white images with arithmetic coding. *IEEE Trans. Commun.*, COM-29(6):858–867.

Pennebaker, W.B. and Mitchell, J.L. 1993. *JPEG Still Image Compression Standard*. Van Nostrand Reinhold.

Pratt, W., Capitant, P., Chen, W., Hamilton, E., and Wallis, R. 1980. Combined symbol matching facsimile data compression system. *Proc. IEEE*, 68(7):786–796.

Rissanen, J.J. and Langdon, G.G. 1979. Arithmetic coding. *IBM J. Res. Dev.*, 23(2):149–162.

Witten, I., Moffat, A., and Bell, T.C. 1994. *Managing Gigabytes: Compressing and Indexing Documents and Images*. Van Nostrand Reinhold.

Yasuda, Y. 1980. Overview of digital facsimile coding techniques in Japan. *Proc. IEEE*, 68(7):830–845.

Yasuda, Y., Yamakazi, Y., Kamae, T., and Kobayashi, K. 1985. Advances in fax. *Proc. IEEE*, 73(4):707–731.

Further Information

FAX—Facsimile Technology and Applications Handbook, 2nd edition by K. McConnell, D. Bodson, and R. Schaphorst published by Artech House, 685 Canton Street, Norwood, Massachusetts 02062 is an excellent single source on various aspects of facsimile technology, including compression.

Two comprehensive surveys by Yasuhiko Yasuda et al. and Yasuhiko Yasuda on coding techniques for facsimile have appeared in the *Proceedings of the IEEE* in 1980 and 1985. These surveys summarize most of the research that has been conducted on facsimile coding and contain an extensive list of references. In addition, the two issues that they appear in, July 1980 and April 1985, are both special issues on facsimile coding.

For a description of the CCITT standards, the best sources are the original documents containing the recommendations:

Standardization of Group 3 Facsimile Apparatus for Document Transmission, Recommendation T.4, 1980.

Facsimile Coding Schemes and Coding Control Functions for Group 4 Facsimile Apparatus, Recommendation T.6, 1984.

Progressive Bi-level Image Compression, Recommendation T.81, 1992. Also appears as ISO/IEC International Standard 11544: 1993.

These documents can be ordered from the International Telecommunication Union, Place Des Nations 1211, Geneva 20, Switzerland. They are also available from Omnicom, Phillips Business Information, 1201 Seven Locks Road, Suite 300, Potomac, Maryland 20854, fax: 1-800-666-4266.

A more recent survey by Arps and Troung (see reference list) compares the performance of different standards.

100
Speech

100.1	Introduction	1424
100.2	Properties of Speech Signals	1425
100.3	Types of Speech Coding Algorithms	1426
100.4	Quantization	1427
100.5	Predictive Coders	1427
100.6	Frequency-Domain Coders	1429
100.7	Analysis-by-Synthesis Coders	1430
100.8	Vocoders	1432
100.9	Variable Bit Rate (VBR) Coding	1433
100.10	Performance Evaluation	1434
100.11	Speech Coding Standards	1434
100.12	Concluding Remarks	1435

Boneung Koo
Kyonggi University

100.1 Introduction

Speech compression refers to the compact representation of speech signals, and speech coding refers to the digital representation of speech signals. Since the primary goal of speech coding is to compress the signal, that is, to reduce the number of bits required to represent it, the two terms, speech compression and speech coding, can be used interchangeably.

Coded speech will be transmitted or stored for a specific application. As the number of bits used in the representation of a signal is reduced, the effective bandwidth of the transmission channel will be increased and the amount of space required to store the signal will be decreased. Coded speech must be reconstructed, or decoded, back to analog form at the receiver. In this coding/decoding process, some amount of distortion results inevitably in the reconstructed speech. Hence, the ultimate goal of speech coding is to compress the signal while maintaining a prescribed level of reconstructed speech quality. Various coding algorithms differ in how to select signals or parameters that represent the speech efficiently. Those selected signals and/or parameters are then quantized and transmitted to the receiver for decoding.

Typical applications of speech coding include the conventional telephone network, digital cellular mobile radio, and secure military telecommunications. Recently, speech over computer and data networks has attracted attention with typical applications in multimedia and videoconferencing. Also, speech coding is an essential part of the future wireless personal communications system. Even though prices of processors, memories, and transmission media have become cheaper and processing speed higher than in the past, the importance of speech coding or compression has not diminished because demands for the more efficient use of hardware are ever increasing, which is especially true when speech is the dominant objective of a system.

Since the introduction of 64 kb/s log-pulse-code modulation (PCM) in the AT&T long-distance telephone network in the early 1970s, speech coding has been a major research area in telecommunications. In 1984, adaptive differential PCM (ADPCM) was adopted by the Consultative Committee on International Telephony and Telegraphy (CCITT) [now, the International Telecommunications Union–Telecommunications (ITU-T)] as a 32-kb/s international standard. Since the mid-1980s, there has been tremendous activity in the application and standardization of medium-to-low rate speech coding. An important contribution was made by code-excited linear prediction (CELP) introduced in 1985. Advances in computers and very large-scale integrated circuit (VLSI) technology have made it possible to implement complicated coding and signal processing algorithms in real time. Coders in the range of 8–13 kb/s have been adopted in several different digital cellular networks around the world. Versions of CELP are now the 16-kb/s international standard and the 4.8-kb/s U.S. governmental standard.

Principles and applications of typical speech coding algorithms are reviewed in this article. Properties of the speech signal are briefly reviewed, followed by a classification of coder types. Principles of quantization and coding algorithms are then presented. This presentation is not exhaustive; emphasis is given to coders employed in standards. Criteria to be considered for speech coder evaluation and subjective and objective measures for speech quality evaluation are presented. A summary of speech coding standards is given, followed by concluding remarks.

100.2 Properties of Speech Signals

Speech is nonstationary but its characteristics are slowly time varying. Thus, in many cases, a speech signal can be processed in short segments, typically of 5–30 ms in time. Each segment, or frame, can be classified as voiced, unvoiced, or silence. Voiced sounds are generated by passing quasiperiodic air pulses through the glottis, causing vibrations in the vocal cords. This results in a quasiperiodic nature in the time domain and harmonic structure in the frequency domain. An unvoiced sound is generated by constricting the vocal tract and forcing air through the constriction to create turbulence. This leads to a near-white noise source to excite the vocal tract such that the unvoiced segment looks like random noise in the time domain and broadband in the frequency domain.

In its simplest form, the speech production system can be modeled as a source-system model, that is, a linear system driven by an excitation source. In the conventional two-state approach, the excitation is a train of pulses for voiced segments, and white noise for unvoiced segments. In the more recent mixed excitation approach, a segment can be a consequence of both voiced and unvoiced excitation.

In general, energy of the voiced segment is higher than that of the unvoiced one. A speech segment that is not a consequence of speech activity is classified as silence or nonspeech. In telephone speech, approximately 50% of the talk time is known to be silence. This fact is utilized in some wireless cellular systems to increase the effective channel bandwidth via voice activity detection.

The period of the quasiperiodicity in the voiced segment is referred to as the pitch period in the time domain or the pitch or fundamental frequency in the frequency domain. The pitch of the voiced segments is an important parameter in many speech coding algorithms. It can be identified as the periodicity of the peak amplitudes in the time waveform and the fine structure of the spectrum. The pitch frequency of men and women usually lies in the range 50–250 Hz (4–20 ms), and 120–500 Hz (2–8.3 ms), respectively.

The bandwidth of speech rarely extends beyond 8 kHz. In wideband speech coding, bandwidth is limited to 7 kHz, and speech is sampled at 16 kHz. In telephony, bandwidth is limited to below 4 kHz (0.2–3.4 kHz, typically) and speech is sampled, usually, at 8 kHz. It is assumed throughout this chapter that input speech is bandlimited to below 4 kHz and sampled at 8 kHz unless otherwise specified.

The speech quality can be roughly classified into the following four categories. *Broadcast* quality refers to wideband speech, *toll* or *network* to narrowband (telephone) analog speech, *communications* to degraded but natural and highly intelligible quality, and, finally, *synthetic* to unnatural but intelligible quality typically represented by the linear predictives coding (LPC) **vocoder**. Speech quality produced by coders to be discussed in this chapter belongs to one of the last three categories.

100.3 Types of Speech Coding Algorithms

Speech coders studied so far can be classified broadly into three categories: **waveform coders**, vocoders, and **hybrid coders**. Waveform coders are intended directly at approximating the original waveform. The reconstructed sound may or may not be close to the original. On the other hand, vocoders are primarily aimed at approximating the sound and, consequently, the reconstructed waveform may or may not be close to the original. Coders that employ features of both waveform coders and vocoders are called hybrid coders. A classification of coders is shown in Table 100.1.

In waveform coders, all or parts of the original waveform are quantized for transmission. For example, in PCM, input speech sample itself is quantized, and in differential PCM (DPCM) and ADPCM, the prediction residual is quantized. Also, adaptive delta modulation (ADM), continuously variable-slope delta modulation (CVSD), adaptive predictive coding (APC), residual-excited linear prediction (RELP), subband coding (SBC), and adaptive transform coding (ATC) are waveform coders. Coders that employ predictors, such as ADPCM, ADM, APC, and RELP, are called predictive coders. SBC and ATC are frequency-domain coders in that coding operations can best be described in the frequency domain. Speech quality produced by waveform coders is generally high, however, at higher bit rates than vocoders.

In vocoders, the speech production model or other acoustic feature parameters that represent perceptually important elements of speech are estimated, quantized, and transmitted to the receiver, where speech is reconstructed based on these parameters. For this reason, vocoders are also called parametric coders and are speech specific in many cases. Vocoder types include channel, formant, phase, cepstral or hormomorphic vocoders, and LPC, sinusoidal transform coding (STC), and multiband excitation (MBE) vocoders. Vocoders can generally achieve higher compression ratios than waveform coders; however, they are known for artificial or unnatural speech quality, except for the recent improvement in STC and MBE. Among these vocoders, only LPC and MBE are discussed in this chapter for their historical and practical importance.

In hybrid coders, the high compression efficiency of vocoders and high-quality speech reproduction capability of waveform coders are combined to produce good quality speech at medium-to-low bit rates. The so called **analysis-by-synthesis** coders such as MPLP, RPE, VSELP, and CELP are hybrid coders.

TABLE 100.1 A Classification of Speech Coders

Type	Coding Algorithm
Waveform Coders:	PCM (pulse-code modulation), APCM (adaptive PCM)
	DPCM (differential PCM), ADPCM (adaptive DPCM)
	DM (delta modulation), ADM (adaptive DM)
	CVSD (continuously variable-slope DM)
	APC (adaptive predictive coding)
	RELP (residual-excited linear prediction)
	SBC (subband coding)
	ATC (adaptive transform coding)
Hybrid coders:	MPLP (multipulse-excited linear prediction)
	RPE (regular pulse-excited linear prediction)
	VSELP (vector-sum excited linear prediction)
	CELP (code-excited linear prediction)
Vocoders:	Channel, Formant, Phase, Cepstral, or Homomorphic
	LPC (linear predictive coding)
	STC (sinusoidal transform coding)
	MBE (multiband excitation), IMBE (improved MBE)

100.4 Quantization

The input to the quantizer is generally modeled as a random process with a continuous amplitude. The function of the quantizer is to produce an output with a discrete amplitude such that it can be encoded with a finite number of bits. Depending on the coding algorithm, the quantizer input can be raw speech samples, the prediction residual, or some parameters estimated at the transmitter and required at the receiver to reconstruct the speech. There are two types of quantizers: scalar and vector.

The scalar quantizer operates on a sample-by-sample basis. Parameters of a scalar quantizer are dynamic range, stepsize, input/output step values, and the number of output levels or the number of bits required to represent each output level. The difference between the input and output is called quantization noise. The quantizer should be designed such that the quantization noise power is minimized for the specified class of input signals. Usually, quantizers are designed according to the probability density function (pdf) of the signal to be quantized. The uniform quantizer works best for a uniform input. Otherwise, a logarithmic or pdf-optimized nonuniform quantizer should be used. Quantizer adaptations to changing input characteristics can take several forms, however, the stepsize adaptation to account for the changing dynamic range or variance is the most common. Detailed information can be found in Jayant and Noll [1984].

According to rate distortion theory, coding vectors is more efficient than coding scalars, and the efficiency is increased as the dimension of the vector is increased at the expense of computational complexity. The input to the vector quantizer (VQ) is a vector formed from consecutive samples of speech or prediction residual or from model parameters. The incoming vector is compared to each codeword in the codebook, and the address of the closest codeword is selected for transmission. The distance is measured with respect to a distortion measure. The simplest and the most common distortion measure is the sum of squared errors. Entries of the codebook are selected by dividing the vector space into nonoverlapping and exhaustive cells. Each cell, called the centroid, is assigned a unique address in the codebook. The quantization process is to find the address of the centroid of the cell that the input vector falls in.

Two important issues associated with vector quantization are the codebook design and the code search procedure. An iterative codebook design procedure of importance is called the Linde, Buzo and Gray (LBG) algorithm, which involves an initial guess and iterative improvement by using training vectors. The complexity of VQ increases exponentially with the vector dimension and codebook size, however, the code search time can be significantly reduced by using a structured codebook. See Makhoul et al. [1985] for more information.

100.5 Predictive Coders

Speech samples have redundancy due to correlation among samples. In predictive coders, redundancy is removed by using predictors. The difference between the input sample and the predictor output is called the prediction residual. There are two types of predictors widely used in speech coders for removing the redundancy: the short-term predictor (STP) and the long-term predictor (LTP). The STP and LTP model the short-term correlation and the long-term correlation (due to pitch), respectively. The synthesis filter associated with the STP reproduces the spectral coarse or formant structure (spectral envelope). The synthesis filter associated with the LTP is used to reconstruct the spectral fine structure of the voiced segment.

The STP can be used to obtain the prediction residual $e(n)$ shown in Fig. 100.1(a). The difference equation and the transfer function of the loop in Fig. 100.1(a) are $e(n) = s(n) - s_p(n)$ and $1 - A(z)$, respectively. The filter $A(z)$ produces a linear combination of past samples according to the relation

$$s_p(n) = \sum_{i=1}^{p} a_i s(n-i)$$

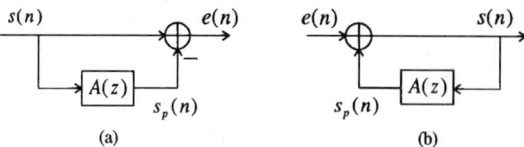

FIGURE 100.1 Linear predictor applications: (a) analysis and (b) synthetics.

and its transfer function is

$$A(z) = \sum_{i=1}^{p} a_i z^{-i} \tag{100.1}$$

where p is the prediction order and the a_i are linear prediction (LP) coefficients. Figure 100.1(b) is the STP synthesis loop, where the original signal is reconstructed by inserting back the redundancy. The transfer function of the synthesis loop in Fig. 100.1(b) is $\{1 - A(z)\}^{-1}$. A variety of algorithms are available for estimating the LP coefficients and details can be found in Rabiner and Schafer [1978].

Other types of predictors are the all-zero predictor and the pole-zero predictor. The output of an all-zero predictor is a linear combination of past prediction residuals. The pole-zero predictor is a combination of the all-pole predictor and the all-zero predictor. The pole-zero predictor, employed in ADPCM adopted in the CCITT standard, has been known to best model the speech signal. In practice, however, the all-pole predictor has been more widely used because of its adequate performance and computational advantage.

Redundancy due to long-term correlations can be reduced by using the LTP. A general form of the LTP transfer function is given by

$$A_L(z) = \sum_{i=-j}^{j} b_i z^{-i-\tau} \tag{100.2}$$

where τ is the pitch period in samples and the b_i are parameters called the pitch gain. Here j is usually a small integer of 0 or 1. Several algorithms are available for open-loop estimation of the LTP parameters. In analysis-by-synthesis coding, close-loop estimation can be used to produce significantly better results than open-loop estimation [Spanias, 1994].

In DPCM coding, the prediction residual instead of the original speech is computed, quantized, and transmitted. The STP is used to compute prediction residual. Because the variance of the prediction residual should be substantially smaller than that of the original speech, a quantizer with fewer bits can be used with little drop in reconstructed speech quality. The transmission rate or the data rate is thus determined by the number of bits required to represent the number of quantizer output levels. The decoder synthesizes the output by adding the quantized prediction residual and the predicted value.

Adaptation of the quantizer stepsize and predictor parameters improves the reconstructed speech quality, which is called ADPCM. The predictor can be forward adaptive or backward adaptive. In forward adaptation, predictor parameters are estimated at the encoder and transmitted to the receiver as side information. In backward adaptation, predictor parameters are estimated from the reconstructed data, which are also available at the receiver, and thus are not required to be transmitted. A variety of adaptive prediction algorithms can be found in Gibson [1980, 1984].

ADPCM is employed in the 32-kb/s CCITT standard, G.721. The coder consists of a 4-b adaptive quantizer and a backward-adaptive predictor with 2 poles and 6 zeros. The coder provides for toll

Speech

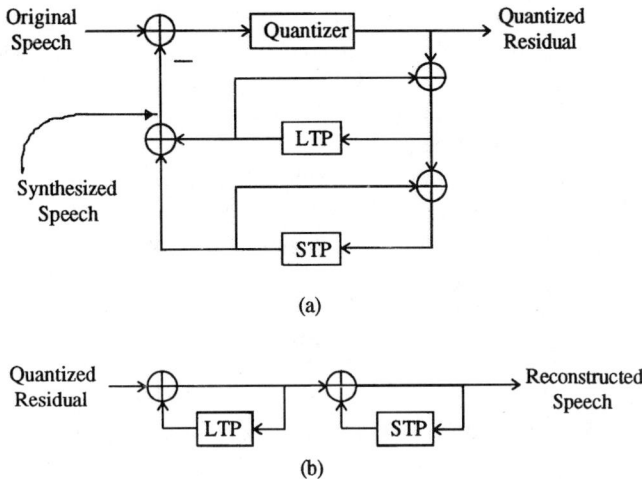

FIGURE 100.2 APC coder: (a) encoder and (b) decoder.

quality speech comparable to that of 64-kb/s log PCM. The G.721 coder was later modified for the 24- and 40-kb/s standard.

DM can be viewed as DPCM with 1-b quantizer and a first-order predictor. ADM employs an adaptive stepsize to track varying statistics in the input signal. Because of its simple structure in quantization and prediction, ADM generally requires a higher sampling rate than ADPCM. ADM generates very high-quality speech in the range of 32–48 kb/s using a very simple algorithm. A 48-kb/s ADM can produce toll-quality speech with a mean opinion score. (MOS) rating of 4.3. A well-known version of ADM is CVSD. More information on ADM and CVSD can be found in Jayant and Noll [1984].

In APC, the LTP or pitch predictor is used in addition to the STP in DPCM, reducing the variance of the prediction residual and allowing fewer bits to be used in the quantizer. Except for the LTP loop, the APC encoder in Fig. 100.2(a) is the same as the ADPCM encoder with an all-pole predictor. Transfer functions of the STP and LTP are shown in Eqs. (106.1) and (106.2), respectively. The prediction residual computed by using the predictors is encoded by a scalar quantizer on a sample-by-sample basis and transmitted to the receiver. Predictor parameters are estimated, quantized, and also transmitted to the receiver as side information. At the receiver shown in Fig. 100.2(b), speech is reconstructed by passing the quantized prediction residual through the LTP synthesizer and the STP synthesizer. APC provides toll quality speech at 16 kb/s and communications quality at 9.6 kb/s. The International Maritime Satellite Organization (Inmarsat)-B standard employs a 16-kb/s APC coder.

RELP has been studied to reduce the number of bits required for encoding the residual in APC. RELP is basically the same as APC except that only the low-frequency part of the residual is transmitted to the receiver with a decimation factor of 3 or 4. This is based on the observation that perceptually important pitch information is in the low-frequency band. At the receiver, the residual is recovered by nonlinear interpolation and used as an excitation to synthesize speech. The estimated acceptable subjective performance of RELP is limited to 9.6 kb/s or higher.

100.6 Frequency-Domain Coders

SBC and ATC are classified as frequency-domain coders in that coding operations rely on a frequency-domain representation of signals. The basic idea is to divide the speech spectrum into several frequency bands and encode each band separately.

In SBC, the signal is divided into several subbands by filter banks. Each band is then low-pass translated, decimated to the Nyquist sampling rate, encoded, and multiplexed for transmission. The number of bits allocated to each subband is determined by perceptual importance. Usually, more bits are assigned to lower bands to preserve pitch and formant information. Any coding scheme such as APCM, ADPCM, or VQ can be used to encode each subband. At the receiver, subband signals are demultiplexed, decoded, interpolated, translated back to their original spectral bands, and then summed up to reproduce the speech. The filter bank is a very important consideration for the implementation and speech quality of the SBC. An important class of filters is quadrature mirror filters (QMF), which deliberately allows aliasing between subbands in the analysis stage and eliminates it in the reconstruction stage, thus resulting in perfect reconstruction except for quantization noise.

An SBC/APCM coder operating at 16 and 24 kb/s was employed in the AT&T voice store-and-forward system, where a five-band nonuniform QMF is used as a filter bank and APCM coders are used as encoders.

More complicated frequency analysis than in SBC is involved in ATC. In ATC, a windowed frame of speech samples is unitary transformed into the frequency domain, and the transform coefficients are quantized and transmitted. At the receiver, the coefficients are inverse transformed and joined together to reproduce the synthesized speech. The unitary transform produces decorrelated spectral samples, the variances of which are slowly time varying so that redundancy can be removed. The Karhunen–Loeve transform (KLT) is the optimal unitary transform that maximally decorrelates the transform components, however, the discrete cosine transform (DCT) is the most popular in practice because of its near-optimality for first-order autoregressive (AR) processes and computational advantage. The DCT can be computed efficiently by using the fast fourier transform (FFT) algorithm.

The transform block size is typically in the range of 128–256, which is much greater than the number of subbands of the SBC, which is usually 4–16. Bit allocation to each band or transform coefficient can be fixed or adaptive. More bits are allocated to the coefficient that has more variance. Adaptive bit allocation is incorporated in ATC. ATC provides toll quality speech at 16 kb/s and communications quality at 9.6 kb/s.

100.7 Analysis-by-Synthesis Coders

A common feature of coders described so far is that they operate in open-loop fashion. Hence, there is no look back and control over the distortion in the reconstructed speech. Such coders are called **analysis-and-synthesis** coders.

In the **analysis-by-synthesis** class of coders, an excitation sequence is selected in a closed-loop fashion by minimizing the perceptually weighted error energy between the original speech and the reconstructed speech. The conceptual block diagram of an analysis-by-synthesis coder is shown in Fig. 100.3. The coder consists of the excitation generator, pitch synthesis filter, linear prediction synthesis filter, and perceptual weighting filter. The pitch and the LP synthesis filters are, respectively, the same as the LTP and the STP synthesis loops shown in Fig. 100.2(b). The weighting filter is used to perceptually shape the error spectrum by de-emphasizing noise in the formant nulls. The

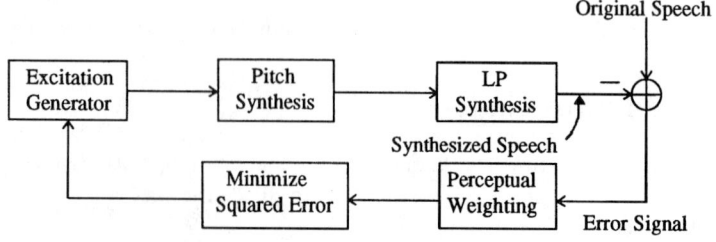

FIGURE 100.3 Analysis-by-synthesis coder.

transfer function of the weighting filter often used is given by

$$W(z) = \frac{1 - A(z/\gamma_1)}{1 - A(z/\gamma_2)}, \qquad 0 < \gamma_2 < \gamma_1 \leq 1$$

where $A(z)$ is the transfer function of the STP in Eq. (106.1). The value of (γ_1, γ_2) is typically $(1, 0.8)$ in conventional CELP coders, and $(0.9, 0.4)$ in the low-delay CELP coder to be described later.

The excitation generator forms or selects the excitation sequence that minimizes the perceptually weighted error criterion, which is usually a weighted sum of squares. The STP and LTP parameters are updated, typically every 10–30 ms and 5–15 ms, respectively. The STP is usually all-pole with a typical order of 10, and parameters are usually converted to another format (line spectrum pairs, for example) and vector quantized for coding efficiency. The pitch lag is usually searched over the range of 20–147 in samples (for 8-kHz sampling). There are several versions of coders depending on the type of excitation model, namely, MPLP, RPE, and CELP coders.

In MPLP coding, an excitation sequence is obtained by forming multiple nonuniformly spaced pulses of different amplitudes. The location and amplitude of each pulse is determined sequentially such that the weighted mean square error is minimized. Typically, 4–6 pulses in each 5-ms frame are used. The pulse amplitudes and locations are transmitted to the receiver. A 9.6-kb/s MPLP coder is used in the Skyphone, an airline communication service.

In RPE coding, an excitation sequence is formed by combining multiple pulses of different amplitudes as in MPLP coding; however, pulses are uniformly spaced. Hence, pulse positions are determined by specifying the location of the first pulse. The amplitude of each pulse is determined by solving a set of linear equations such that the weighted mean square error is minimized. Typically, 10–13 pulses in each 5-ms frame are used, and the pulse spacing is 3 or 4. A 13-kb/s RPE-LTP coder, an association of RPE with LTP, is used in the GSM Pan-European digital cellular standard.

In CELP coding, probably the most prevailing speech coding approach in recent years, the optimum excitation sequence is selected from a codebook, perhaps of random Gaussian sequences which minimizes the perceptually weighted mean squared error. The first CELP coder [Schroeder and Atal, 1985] used 1024 codes (vectors) each of length 40 samples (5 ms in 8-kHz sampling). Early CELP algorithms required heavy computations for the exhaustive codebook search and speech synthesis for each candidate code. A great deal of work followed to reduce computational complexity. Notable developments have been made in structured codebooks, such as sparse, overlapping and vector-sum excitation, and in the associated fast search algorithms. Also, ternary $(-1, 0, +1)$ valued codes can be used to save memory and for fast convolution without sacrificing speech quality.

CELP can provide good quality speech at low rates down to 4.8 kb/s. With the recent developments of high-speed and high-capacity DSP processors, CELP algorithms have become practical and actually adopted in several standards.

A typical example of CELP coding is the U.S. 4.8-kb/s Federal Standard (FS-1016) coder, also called the DoD-CELP due to its development by the Department of Defense. The 10th-order STP parameters are computed every 30 ms over the Hamming windowed frame of incoming speech. Excitation parameters are updated every 7.5 ms. The synthesis part of DoD-CELP is shown in Fig. 100.4, where the excitation is a combination of codes from a stochastic codebook and an adaptive codebook. The adaptive codebook, which consists of past excitation codes, replaces the

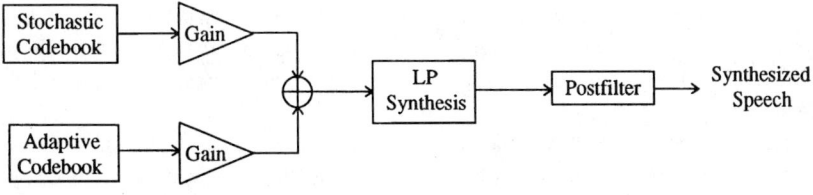

FIGURE 100.4 Speech synthesis in the FS-1016 CELP.

TABLE 100.2 Bit Allocations in the FS-1016 CELP

Parameters	Bits	Update, ms	Bit Rate, b/s
STP (10-LSPs)	34	30	1133.33
Stochastic codebook		7.5	1866.67
Index	9		
Gain	5		
Adaptive Codebook		7.5	1600
Index	7		
Gain	5		
Miscellaneous			200
Total			4800

Source: Atal, B.S., Cuperman, V., and Gersho, A., eds. 1991. *Advances in Speech Coding*, Kluwer Academic Publishers, Dordrecht, The Netherlands.

LTP synthesis, and the search for the optimum adaptive code is equivalent to search for the optimal pitch lag. The LP synthesis in Fig. 100.4 is same as the STP synthesis loop as shown in Fig. 100.1(b). In DoD-CELP and in many modern coders as well, the line spectrum pairs (LSPs), also called the line spectrum frequencies (LSFs), are used as the STP parameters for stability and quantization efficiency. The additive white-type noise contained in the CELP-coded speech can be reduced by using the postfilter of the form $(1 + cz^{-1})W(z)$, which essentially de-emphasizes spectral valleys. The range $0.2 \le c \le 0.4$ has been found to give reasonable subjective results.

The bit allocation scheme of the 4.8-kb/s DoD-CELP coder is shown in Table 100.2. It can be seen that more than two-thirds of the total bits are allocated to the codebook indices and gains that specify the excitation signal.

VSELP and low-delay CELP (LD-CELP) are versions of CELP coding. In the 8-kb/s VSELP algorithm, adopted in the time-domain multiple access (TDMA) North American digital cellular standard, IS-54, by the Cellular Telecommunications Industry Association (CTIA), the excitation is a combination of vectors (codes) from a pitch adaptive codebook and two highly structured codebooks. A 6.4 kb/s VSELP algorithm was adopted in the Japanese digital cellular standard. The LD-CELP algorithm is employed as a low-delay 16-kb/s international standard, G.728. The LD-CELP has a unique feature in that LTP is not used, and instead, the order of STP is increased to 50 to compensate for the lack of pitch information. The speech quality of LD-CELP is comparable to that of G.721, 32-kb/s ADPCM, and coding delay is less than 5 ms. More information on DoD-CELP, VSELP and LD-CELP coders can be found in Atal, Cuperman, and Gersho [1991].

100.8 Vocoders

LPC is the most widely studied of the very low rate vocoding methods. A classic example is the 2.4-kb/s LPC-10 algorithm adopted as the U.S. FS-1015 standard in 1977, which has been used for secure voice communications. The LPC-10 algorithm uses a 10th-order all-pole predictor and relies on the two-state (voiced and unvoiced) source excitation model. The speech synthesis model of LPC is shown in Fig. 100.5. The linear prediction parameters, gain, pitch period, and voicing/unvoicing (V/U) decision parameters are estimated, quantized, and transmitted to the receiver. At the receiver, speech is synthesized by using the impulse train for the voiced frame and white noise for the unvoiced frame. The impulse train, defined in the standard, is a periodic sequence that has a pitch period similar to a glottal pulse.

Reconstructed speech of an LPC vocoder operating at very low rates such as 2.4 kb/s is of synthetic quality due to the inaccuracy of the source excitation. The quality of the speech produced by LPC-10 is also synthetic in that it sounds artificial, unnatural and buzzy. The two-state excitation model was improved later by the **mixed excitation** source model, which reduced the buzziness and breathiness

Speech

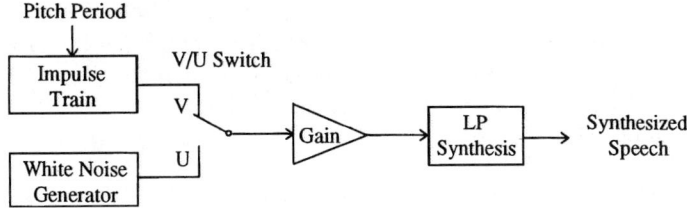

FIGURE 100.5 LPC synthesis model.

in the synthesized speech. A 2.4-kb/s mixed excitation LPC vocoder implemented in real time was reported in McCree and Barnwell [1993]. A 400 b/s LPC vocoder based on LPC-10 was also reported [Atal, Cuperman, and Gersho, 1991].

Other types of vocoding are STC and MBE, which belong to the class of sinusoidal coding. In sinusoidal coding, speech is analyzed and synthesized by using sinusoidal components. These coders have received attention recently because of their potential for high-quality speech coding at low bit rates.

The MBE takes a mixed excitation approach in that the short-time spectrum of a speech frame is treated as a collection of multiple bands, each of which is either voiced or unvoiced. The voiced bands are centered around the pitch harmonics and the unvoiced bands contribute to the random nature of the spectrum. The parameters of the MBE speech model consists of fundamental (pitch) frequency, the V/U decision for each band, and spectral magnitudes (envelope). An analysis-by-synthesis technique is used to estimate the parameters that minimize weighted error between the original spectrum and the synthetic spectrum. At the receiver, the voiced bands are synthesized in the time domain using a bank of harmonic sinusoids,

$$v(n) = \sum_{k=1}^{L} A_k \cos(k\omega_o n + \phi_k), \qquad n = 0, 1, \ldots, N$$

where A_k, $k\omega_o$, and ϕ_k are magnitudes, frequencies, and phases of L pitch harmonics respectively, ω_o is the pitch frequency, and N is the frame size. The unvoiced bands are synthesized in the frequency domain by scaling the normalized white noise spectrum by the spectral magnitude. Synthesized speech is obtained by adding the voiced bands and inverse Fourier transform of the unvoiced bands spectra. The phase information of the voiced harmonics is determined at the receiver by using partial information transmitted from the encoder and a phase prediction algorithm.

The 4.15 kb/s IBME, an improved MBE, coder was adopted as a 6.4-kb/s Inmarsat-M standard, which includes an additional 2.25 kb/s for forward error correction. In this coder, most of the bits are used for scalar quantization of spectral envelope parameters. Bit rate reduction was made possible by representing and vector quantizing the spectral envelope with LSPs that correspond to 10-pole LPC parameters. A 2.4-kb/s MBE coder and an even more compressed 1.2-kb/s MBE coder reportedly produced good quality speech. Refer to Atal, Cuperman, and Gersho [1991] for details on MBE coders. Information on the STC can be found in Gersho [1994] and Spanias [1994].

100.9 Variable Bit Rate (VBR) Coding

In variable bit rate (VBR) coding different speech segments are encoded at different bit rates. Variable rate coding can be effective for storage, packet voice, and multiple access channel applications. The two typical approaches for the VBR coding are in the variable update rates and the variable bit allocations for parameters from different speech segments. The VBR coding algorithms are usually based on conventional coders such as ADPCM, SBC, and CELP. A key element of VBR coding is the voice activity detection (VAD) used for voicing decisions and speech/nonspeech classifications.

An example of VBR coding is a variation of the CELP coder called Qualcomm CELP (QCELP), which is used in the CDMA North American digital cellular standard, IS-95, adopted by the CTIA.

QCELP operates at one of four predetermined rates, 8, 4, 2, or 1 kb/s, for each 20-ms analysis frame. The average bit rate is speech dependent with typical rate of 5.5 kb/s and provides near-toll quality speech output. Another example is variable rate phonetic segmentation (VRPS), where one of the several different CELP coder configurations including bit allocations is selected according to the phonetic classification (silence, unvoiced, voiced, onset). See Gersho [1994] for more information.

100.10 Performance Evaluation

Performance of a coding algorithm is primarily specified by the speech quality at the prescribed coding rate. When a particular speech coder is to be designed or selected for a particular application, other factors should also be considered. Those are coding delay (due to algorithmic and computational factors), complexity [of algorithm and implementation, often measured in million instructions per second (MIPS) of a computing processor], cost (of system and terminal), robustness (to background noise or channel impairments), tandeming (with itself or other coding system), transcoding (into another format, usually PCM), and voiceband data compatibility (e.g. modem and facsimile). The priority of consideration will depend on the target application. For example, in the public switched telephone network (PSTN) coding delay and tandeming can be more important than robustness, whereas coding rate and robustness can be more important than coding delay in mobile cellular applications.

Speech quality can be measured subjectively and objectively. Subjective measurements are obtained as consequences of listening tests, whereas objective measurements are computed directly from the coded speech parameters.

Typical subjective measures often quoted are MOS, diagnostic acceptibility measure (DAM), and diagnostic rhyme test (DRT). The MOS, probably the most widely used, is obtained by averaging test results rated by a group of listeners who are asked to quantify their perceptual impressions on a five-point scale (bad = 1, poor = 2, fair = 3, good = 4, excellent = 5). The DAM is a more systematic method of assessing speech quality, where the listeners are asked to grade 19 separate scales grouped in three categories on a 100% scale. This test is directed at medium-to-high quality speech and requires trained or experienced listeners. The DRT is directed at testing the intelligibility of the medium-to-low quality speech. In the DRT, the listener is asked to distinguish between word pairs of single syllable, such as meat–beat and sheet–cheat. Such subjective measures, however, can at best be a rough indicator of what to expect from a particular coder, and should be interpreted accordingly. More information and references on subjective measures can be found in Jayant and Noll [1984] and Deller, Proakis, and Hansen [1993].

A good objective measure should be able to consistently predict the subjective quality of the speech, and could be used in the design stage in controlling the amount of distortion that can be possibly introduced during the coding process. Widely used objective measures include mean squared error (MSE)-based measures, probably the most popular, such as signal-to-noise ratio (SNR) and segmental SNR (SEGSNR), and LPC-based measures, such as spectral distance and cepstral distance. Other time- or frequency-domain measures have been introduced and studied; however, a good objective measure that is consistent and provides strong correlation with the subjective quality has yet to be found. More discussion on objective measures can be found in Quackenbush, Barnwell, and Clements [1988].

100.11 Speech Coding Standards

A summary of typical speech coding standards adopted by industrial, national or international organizations is shown in Table 100.3. It can be seen in the table that most standards were established in the early 1990s, which can be regarded as a consequence of worldwide demand for bandwidth-efficient digital mobile voice communications and the technological maturity of speech coding algorithms in various applications. The bit rates of some coders such as 16 kb/s APC and 6.4 kb/s IMBE include error control codes, and thus the actual source coding rates are smaller than the listed.

TABLE 100.3 Speech Coding Standards

Bit Rate, kb/s	Algorithm	Standard (Year)	MOS
64	log PCM	CCITT G.711 (1972)	4.3
32	ADPCM	CCITT G.721 (1984)	4.1
16	LD-CELP	CCITT G.728 (1992)	4.0
16	APC	Inmarsat-B (1985)	—[a]
13	RPE-LTP	Pan-European DMR, GSM (1991)	3.5
9.6	MPLP	BTI Skyphone (1990)	3.4
8	VSELP	CTIA, IS-54 (1993)	3.5
6.7	VSELP	Japanese DMR (1993)	3.4
6.4	IMBE	Inmarsat-M (1993)	3.4
8/4/2/1 (var.)	QCELP	CTIA, IS-95 (1993)	3.4
4.8	CELP	US, FS-1016 (1991)	3.2
2.4	LPC-10	US, FS-1015 (1977)	2.3

[a] MOS data unavailable, however, reportedly near toll quality.

Sources: Spanias, A.S. 1994. Speech coding: A tutorial review. *Proc. IEEE*, 82(10):1541–1582; and Kondoz, A.M. 1994. *Digital Speech: Coding for Low Bit Rate Communications Systems*, Wiley, Chichester, West Sussex, England.

The MOS values vary among different tests, so it may not be appropriate as an absolute measure of comparison. Roughly speaking, coders with MOS rating of more than 4.0 in Table 100.3 can be considered to provide toll quality, whereas the rest provide near-toll or communications quality except for LPC-10.

100.12 Concluding Remarks

Principles and applications of typical coding algorithms for telephone speech have been reviewed in this section. Major research efforts in the last decade have concentrated on the analysis-by-synthesis technique in the bit range of 4–16 kb/s, and some of the results have turned into standards in cellular and mobile voice communications in the range of 8–13 kb/s. Important advances can also be typified by the standardization of CELP coders at 4.8 kb/s and the low-delay CELP coder at 16 kb/s, and the emergence of the sinusoidal vocoding approach (STC, MBE) in the 2.4–4.8 kb/s range.

Major research efforts in the future are expected to be concentrated on low bit range coding, say 4 kb/s or below, where perceptual factors of the human auditory system become more important [Jayant, 1992]. Continued activities for enhanced coder performance are also expected in the range of 4–8 kb/s for specific applications. In a practical aspect, half-rate ($\approx$4 kb/s) coding standards for cellular radio will soon be established, and speech coding algorithms to be included in the wireless personal communications systems will be a major issue for the next few years.

Defining Terms

Analysis-and-synthesis: Refers to open-loop operation where the analysis cannot be affected or changed by the synthesis.
Analysis-by-synthesis: Refers to closed-loop operation where the analysis can be affected or changed by the synthesis.
Hybrid coders: Coders that employ features of both waveform coders and vocoders.
Mixed excitation: A speech segment is modeled as a consequence of both voiced and unvoiced excitation.
Vocoders: Coders that are primarily aimed at reproducing the original sound.
Waveform coders: Coders that are primarily aimed at reproducing the original time-domain waveform.

References

Atal, B.S., Cuperman, V., and Gersho, A., eds. 1991. *Advances in Speech Coding*, Kluwer, Dordrecht, The Netherlands.

Deller, J.R. Jr., Proakis, J.G., and Hansen, J.H.L. 1993. *Discrete-Time Processing of Speech Signals*, Macmillan, New York.

Gersho, A. 1994. Advances in speech and audio compression. *Proc. IEEE*, 82(6):900–918.

Gibson, J.D. 1980. Adaptive prediction in speech differential encoding systems. *Proc. IEEE*, (April):488–525.

Gibson, J.D. 1984. Adaptive prediction for speech encoding. *ASSP Mag.*, 1:12–26.

Jayant, N.S. 1992. Signal compression: Technology targets and research directions. *IEEE Trans. Selec. Areas Commun.*, 10(5):796–818.

Jayant, N.S. and Noll, P. 1984. *Digital Coding of Waveforms: Principles and Applications to Speech and Video*, Prentice–Hall, Englewood Cliffs, NJ.

Kondoz, A.M. 1994. *Digital Speech: Coding for Low Bit Rate Communications Systems*, Wiley, Chichester, West Sussex, England.

Makhoul, J., Roucos, S., and Gish, H. 1985. Vector quantization in speech coding. *Proc. IEEE*, 73(11):1551–1588.

McCree, A.V. and Barnwell, T.P. 1993. Implementation and evaluation of 2400 bps mixed excitation LPC vocoder. In *Proc. ICASSP*, Minneapolis, MN, Vol. 2, pp. 159–162, April.

Quackenbush, S.R., Barnwell, T.P., and Clements, M.A. 1988. *Objective Measures of Speech Quality*, Prentice–Hall, Englewood Cliffs, NJ.

Rabiner, L.R. and Schafer, R.W. 1978. *Digital Processing of Speech Signals*, Prentice–Hall, Englewood Cliffs, NJ.

Schroeder, M.R. and Atal, B.S. 1985. Code-excited linear prediction (CELP): High quality speech at very low bit rates. In *Proc. ICASSP*, Tampa, FL, pp. 937–940, April.

Spanias, A.S. 1994. Speech coding: A tutorial review. *Proc. IEEE*, 82(10):1541–1582.

Further Information

A comprehensive tutorial and other references can be found in Spanias [1994]. A recent view on the state of the art can be found in Gersho [1994]. Journals specialized in speech coding are *IEEE Transactions on Speech and Audio Processing* (since 1993), *IEEE Transactions on Signal Processing* (1991–1993), *IEEE Transactions on Acoustics, Speech and Signal Processing* (1974–1990), and *IEEE Transactions on Communications*. The most recent activities can be found in the annual Proceedings of the International Conference on Acoustics, Speech and Signal Processing (ICASSP).

101

Still Image Compression and Halftoning

101.1	Introduction	1437
101.2	Still Image Compression Techniques	1438
	Differential Pulse Code Modulation • Transform Coding • Subband Coding • Vector Quantization	
101.3	The JPEG Compression Standard	1441
101.4	Halftoning Techniques	1443
	Ordered Dither • Error Diffusion • Minimization Approaches	
101.5	Optimization of JPEG for Halftoning	1445
101.6	Summary and Conclusions	1447

Ping Wah Wong
Hewlett Packard Laboratories

101.1 Introduction

The proliferation of digital imaging in recent years in commercial, industrial, and home applications such as image database, electronic photography, and desktop publishing has resulted in an explosion of image data. As a result interest in image compression has grown tremendously. Popular techniques in lossy still image compression include differential pulse code modulation (DPCM), transform coding, subband coding, vector quantization, and some hybrid techniques. All of these techniques manipulate, in one way or another, the image data into a suitable form to be quantized using either scalar or vector quantization, where the compression of data occurs. The quantization step also causes distortion between the reconstructed and the original image. There are some quantization strategies that take into account the characteristics of the human visual system, and they often lead to favorable results compared to nonperceptual coders. The main objective of image compression is to encode the raw image data into fewer number of bits while maintaining good image quality.

The work of the Joint Photographic Experts Group (JPEG) has resulted in a transform coder that uses the discrete cosine transform, along with uniform scalar quantization and entropy coding. The JPEG coder is of particular interest because it has been adopted by International Telecommunication Union-Telecommunication Standardization Sector (ITU-T) and International Standards Organization (ISO) as an international standard [ITU, 1993]. As a result of its standardization, many industrial applications have started to support JPEG as the desired method for still image compression. Since the JPEG encoding algorithm works on independent image blocks, it is amenable to simple and efficient hardware and software implementations. Specialized JPEG encoder and decoder chips that facilitate system integration are widely available from many semiconductor manufacturers. The performance of JPEG is very competitive with many other techniques over a large variety of natural images. Although it is possible to design, for example, a subband coder that outperforms JPEG, the other advantages of JPEG have often led the system designer to prefer JPEG in many industrial applications.

In a typical communications situation, compressed images are transmitted over telephone lines or a network, and then the decoded images are displayed or printed at the receiver. Many output devices such as liquid crystal display (LCD) or printers only support a limited number of output (color) graylevels. Hence, it is often necessary to apply some **halftoning** technique to convert the images into a limited bit-depth form for rendering purposes. This scenario offers an opportunity for optimizing the compression algorithm according to the characteristics of the output device so that improved compression performance can be realized.

In the following, the basic techniques for still image compression are described, with special emphasis on the JPEG compression standard. This is followed by a description of popular halftoning techniques, and the optimization of JPEG for the characteristics of halftoning and scaling in a rendering pipeline.

101.2 Still Image Compression Techniques

Among the popular techniques for still image compression are diffferential pulse code modulation, transform coding, subband coding, and vector quantization (VQ). Excellent reviews of these techniques can be found in Netravali and Limb [1980], Jain [1981], Wintz [1972], Woods [1991], Gersho and Gray [1992], and Jayant [1992]. We briefly describe here the basic principles of each of these techniques.

Differential Pulse Code Modulation

DPCM, as its name suggests, is a differential form of pulse code modulation. Rather than encoding the samples directly, it encodes

$$d_{m,n} = x_{m,n} - \hat{x}_{m,n}$$

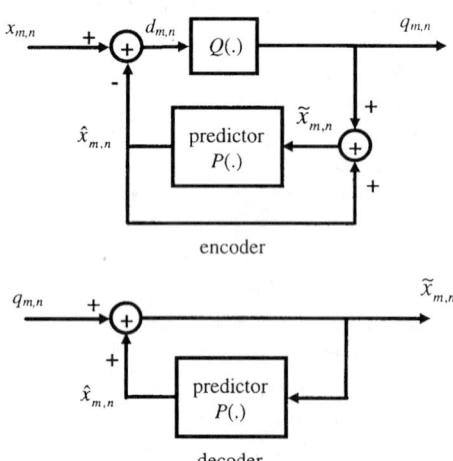

FIGURE 101.1 Encoder and decoder of a DPCM encoding system.

the difference between the sample $x_{m,n}$ and its prediction $\hat{x}_{m,n}$. It takes advantage of the fact that neighboring pixels in most images are highly correlated. In other words, the variations between neighboring samples tend to be much smaller than the full range of intensity values in the image. It is, therefore, advantageous to encode an error signal, i.e., the error between the present sample and its prediction, rather than directly encoding the image samples. In practice, a closed-loop DPCM system, as shown in the block diagram of Fig. 101.1 is often used. It is evident that the decoder contains an identical circuit as the feedback path of the encoder, and the prediction in the closed-loop system is based on a neighborhood of the decoded signal $\tilde{x}_{m,n}$. Note that in the special case where the predictor is a simple delay, and the quantizer is a 1-b hard limiter, the DPCM system becomes the well-known delta modulator.

It is evident that the complexity, as well as the performance, of a DPCM coder depends directly on the predictor. To design a good predictor, one normally requires some knowledge of the characteristics of the signal. In many practical systems, a simple linear predictor of the form

$$\hat{x}_{m,n} = \sum_{\mathcal{R}} h_{i,j} \tilde{x}_{m-i,n-j}$$

is used. The region $\mathcal{R}$ is usually formed by a straightly causal neighborhood of the current pixel location (m, n), where causality here is defined with respect to the scanning strategy. An optimum design of the linear predictor in the least square sense can be performed using the second-order characteristics (mean and autocorrelation) of the signal. An excellent review of linear prediction can be found in Makhoul [1975].

Transform Coding

Transform coding [Wintz, 1972] is a technique where one first converts the image into a collection of coefficients using an invertible transform. The coefficients are then encoded (quantized), and finally, the quantized coefficients are inverse transformed to form a reconstruction of the input image. This technique takes advantage of the fact that the information content of images is usually nonuniformly distributed, for example, in the frequency domain; hence, it is often more efficient to encode the transformed coefficients than to encode the image samples. Because of the energy packing property of the transform, a bit allocation is usually performed to determine the bit rates assigned to each coefficient subject to an overall rate constraint so that the overall distortion is minimized [Shoham and Gersho, 1988; Riskin, 1991]. Furthermore, the transformed coefficients tend to be less correlated than the original image samples, and hence, the encoder design usually becomes easier. Experimental results have indicated that transform coding can provide very good performance at low rates [Wintz, 1972].

Many transforms have been used in transform coding. Among the popular ones are discrete Fourier transform (DFT), discrete cosine transform (DCT), discrete sine transform, Hadamard transform, and Karhunen–Loève transform (KLT). In practice, the transforms are usually normalized so that they are unitary. In this case, the energy in the original signal is equal to the energy in the transform coefficients. The KLT is optimum for Gaussian signals in that it produces the minimum mean square error when scalar Lloyd–Max quantizers are used in the coder. It has the maximal energy packing property, and the transform coefficients are uncorrelated (independent in the Gaussian case). Unfortunately, the KLT is signal dependent, and a fast algorithm for its calculation does not exist (in the sense that a transform of a signal of length N cannot be computed at the order of $N \log N$ as in, for example, the DFT or DCT). Consequently, it is not very popular in practice. It has been shown that for first-order autoregressive signals with large correlation coefficients, the performance of DCT-based coders is close to optimum [Jain, 1981]. One particular DCT-based transform coder, the JPEG coder, is extremely important in practice, as it has been adopted by the ITU and ISO as an international standard for still image encoding [ITU, 1993]. The JPEG standard coding algorithm will be considered in detail in a later section.

Subband Coding

Subband coding is highly related to transform coding in that both techniques convert the image data into a collection of coefficients before they are encoded. The domain of the tranform can often be directly or indirectly interpreted as frequency. Subband coding takes advantage of the fact that the energy in images is typically distributed nonuniformly in the frequency

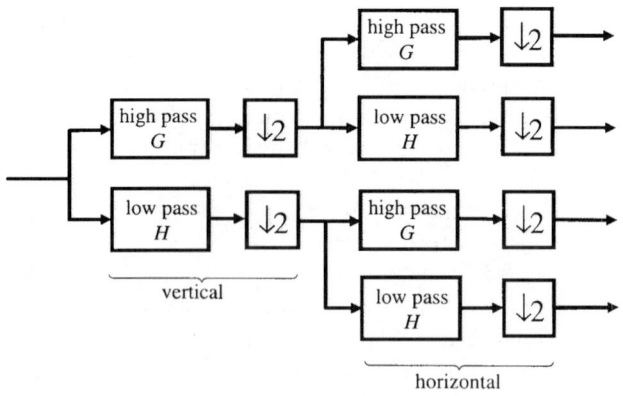

FIGURE 101.2 A separable filter bank for subband coding.

domain, and hence a decomposition into narrow frequency bands is an effective way of packing energy.

In subband coding of images, one typically uses a separable **quadrature mirror filter (QMF) bank** to decompose the signal into its subband signals [Woods and O'Neil, 1986]. A basic building block of the system consists of a four-band decomposition, as shown in Fig. 101.2. The input signal is filtered and down sampled into four subsignals. Each one of these subsignals can be further decomposed recursively into subband signals of narrower bandwidths. In a decomposition induced by a discrete orthogonal or biorthogonal wavelet decomposition [Mallat, 1989; Vetterli and Herley, 1992], one recursively decomposes only the low-pass subband at each stage, resulting in logarithmically spaced subbands. This is often referred to as constant Q analysis in that low-frequency subbands have small bandwidth but coarse spatial resolution, whereas high-frequency subbands have large bandwidth but fine spatial resolution. It is possible to incorporate the characteristics of the human visual response in subband coding, where bit rates are assigned to the subbands according to the sensitivity of the human visual system. The **perceptual coder** often gives output images that are judged to be more favorable by the human observer compared to images encoded by nonperceptual coders.

Coding gains of subband coding with ideal subband filters have been analyzed [Woods and O'Neil, 1986; Pearlman, 1991]. The rate distortion performance of subband coding using QMFs has also been reported [Fischer, 1992]. A good review of subband coding can be found in Woods [1991].

Vector Quantization

Vector quantization is a generalization of scalar quantization in that a vector of symbols is encoded as a unit [Gersho and Gray, 1992]. It is well known from information theory that VQ is an optimal approach in the sense that one can approach the performance indicated by the rate distortion function as the block size grows to infinity.

The basic form of vector quantization, full search vector quantization, consists of a codebook, an encoding procedure, and a decoding procedure. Without loss of generality, we assume that the codebook consists of K vectors each of length N. To encode an image, we first partition the image into blocks (vectors) of size N, that is, each vector is comprised of N elements. Our goal is to replace each image vector by a vector in the codebook that is the best match according to a distortion criterion. To do so, each image vector is compared with all of the vectors in the codebook, and then the one in the codebook that resulted in the smallest distortion is selected as a representation of the image block. That is, we encode a vector $x_{m,n}$ into $\hat{x}_{m,n}$ so that

$$\hat{x}_{m,n} = \arg\min_{v_j \in C} d(x_{m,n}, v_j)$$

where C is the codebook. Since the decoder also has a copy of the codebook, we only need to transmit the index j to the receiver to identify the selected code word. Hence, the bit rate is

$$R = \frac{\log_2 K}{N} \text{ bits/symbol}$$

At the decoder, we only need to look up the entry in the codebook corresponding to the index j, and insert the code vector v_j in the decoded image.

It is evident that the performance of a vector quantization system depends on the elements in the codebook. To construct a good codebook for image coding, one often uses a collection of typical images as training data, and designs the codebook so that it is optimal in the sense of minimizing the average distortion. An extremely important algorithm for designing codebooks for vector quantizers is the generalized Lloyd algorithm (also known as the LBG algorithm) [Linde, Buzo, and Gray, 1980], which is an extension of the Lloyd algorithm in scalar quantization to the vector case. It is well known that the generalized Lloyd algorithm generates locally optimum codebooks from a set of training data. Global optimality, however, cannot be guaranteed. The design of globally optimal vector quantizers is still an unsolved problem.

The complexity of the full search vector quantization algorithm grows exponentially with the coding rate and the vector size N. Over the years, there have been many variations that attempt to reduce the complexity. One attractive approach is tree-structured vector quantization, where the codebook is arranged in the form of a tree and, hence, can be searched very efficiently. Recently it has been shown that pruned tree-structured vector quantization can provide very similar performance to full search vector quantization, while the complexity is low due to the tree structure. Descriptions and references to many interesting variations of vector quantization can be found in Gersho and Gray [1992].

101.3 The JPEG Compression Standard

As described in the previous section, there are many competitive methods for the lossy compression of still images. Among these techniques, the lossy JPEG compression algorithm is of particular importance because it has been jointly adopted by ITU-T and ISO [ITU, 1993] as an international standard for coding still images. Many semiconductor manufacturers have implemented the lossy JPEG algorithm in hardware to facilitate its application in practical systems. As a result, JPEG has increasingly been used in many commercial applications.

In the JPEG image compression standard, there are actually two different algorithms; one is based on DPCM for loseless coding, whereas another one is based on transform coding for lossy coding. The lossless JPEG coder has thus far not been very popular. Therefore, we focus on the lossy mode of JPEG in the rest of this chapter. When we simply mention JPEG, we are referring to the lossy JPEG algorithm.

The JPEG algorithm is based on transform coding, in particular using the DCT. The coding algorithm can be represented by the block diagram of Fig. 101.3. It consists of several steps: discrete cosine transform, uniform scalar quantization, and **entropy coding**. A block diagram of the decoder is also shown in Fig. 101.3. As one would expect, the decoder invert the operations performed at the encoder.

Consider an image to be encoded using the JPEG coding algorithm. The image is first partitioned into 8×8 blocks. For each 8×8 block of pixels $x_{m,n}$ ($m = 0, 1, \ldots, 7; n = 0, 1, \ldots, 7$),

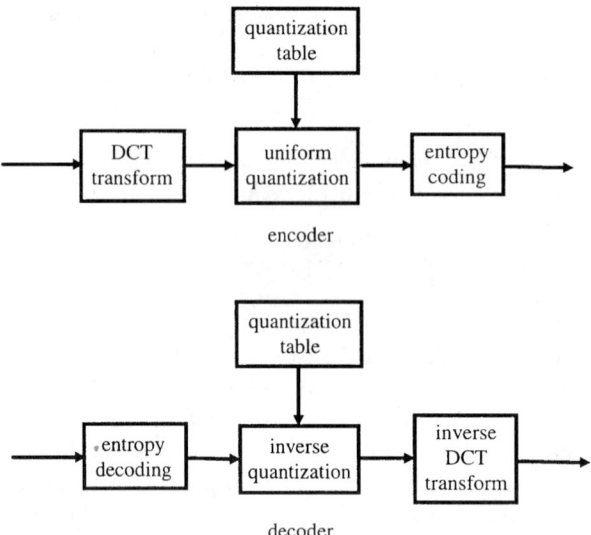

FIGURE 101.3 Block diagram of the encoder and decoder for lossy JPEG.

we perform a two-dimensional DCT to obtain the transformed coefficients $y_{m,n}$ ($m = 0, 1, \ldots, 7$; $n = 0, 1, \ldots, 7$). It can be verified that for original image data of 8-b resolution, the transformed coefficients require 11 b of precision. These coefficients are quantized using uniform scalar quantization, where the step sizes of the quantizers are obtained from a quantization table. The quantization table $q_{m,n}$ ($m = 0, 1, \ldots, 7; n = 0, 1, \ldots, 7$) is simply an 8×8 array of integers; each takes an integer value from 1 through 255. In the JPEG standard, two recommended quantization tables are provided: one for the luminance component and one for the chrominance components of the image. The user can also use a custom designed quantization table, which can be specified in the encoded data stream. Although it is possible to use an optimum scalar quantizer determined by the probability distribution of the DCT coefficients [Reininger and Gibson, 1983], uniform quantization is used in JPEG for its simplicity. The quantization is actually performed by dividing the DCT coefficient $y_{m,n}$ by the corresponding quantization table entry $q_{m,n}$, and then the result is rounded off to an integer, that is, $a_{m,n} = I(y_{m,n}/q_{m,n})$ where $I(\cdot)$ represents the rounding operator. As will be described shortly, an approximate reconstruction of $y_{m,n}$ is performed at the decoder by simply multiplying $a_{m,n}$ by $q_{m,n}$, that is, $\hat{y}_{m,n} = a_{m,n} q_{m,n}$. It is easy to see that the resolution (precision) of $y_{m,n}$ has been reduced according to the value of the quantization table entry $q_{m,n}$. In other words, the reconstruction $\hat{y}_{m,n}$ is not identical to $y_{m,n}$; it can only be specified to within a tolerance determined by the size of $q_{m,n}$. This is precisely the step in JPEG in which certain information in the original image is lost, that is, distortion is introduced. At the same time, this step also results in compression of the image. Following the quantization step, the indices $a_{m,n}$ are encoded using an entropy coder to further reduce the number of bits required to describe the image. To enable the coder to take maximum advantage of the structure of the coefficients (most energy in images are located at the low-frequency components, whereas high-frequency components are mostly zero), the quantized AC coefficients within a block are scanned and packed in a zig-zag fashion before entropy coding. The DC coefficients (one for each block) of the entire image, on the other hand, are coded using DPCM.

The JPEG decoder essentially performs the inverse function of the encoder. More specifically, the JPEG bit stream is first decoded using the appropriate entropy decoder, which produces another bit stream to be processed. The quantized coefficients of each block are then retrieved by unpacking

Still Image Compression and Halftoning

the decoded bit stream. Following this step, each index is multiplied by the suitable entry in the quantization table to produce a reconstructed DCT coefficient. Each 8 × 8 block of reconstructed DCT coefficients is then inverse DCT transformed, and the result is a decoded image block. All of the image blocks in an image are put together in the appropriate locations to give the decoded image.

As described before, the only place in the JPEG algorithm that causes distortion in the decoded image is the quantization step. Since the specific quantization step performed at each DCT coefficient is controlled by the corresponding quantization table entry, it is evident that the output (decoded) image depends strongly on the quantization table. We will consider a case where it is known beforehand that the decoded images will be displayed as halftones on a limited bit-depth device. An approach for optimizing the quantization table for specific halftoning and other output rendering algorithms will be described.

A well-known characteristic of JPEG coding, particular at low rates, is that the decoded image often exhibits blocking artifacts due to the inherent block processing approach. To this end, postprocessing methods for smoothing the decoded images can give better reconstructed images.

101.4 Halftoning Techniques

Given a continuous tone image of, for example, 256 levels, the goal of halftoning is to generate an image at two output levels, so that both images appear similar when observed from a distance. The corresponding technique for generating images with multiple, for example, 16, levels is called multitoning. The main application of halftoning and multitoning is for rendering or displaying images at limited bit-depth devices such as printers (typically, 2 output levels) or liquid crystal displays (typically, 16 output levels). Note that halftoning can also be viewed as a form of compression, where, for example, images of 8 b/pixel are converted to bilevel images (1 b/pixel) for rendering purposes.

There are three general techniques for halftoning: ordered dither, error diffusion, and minimization approaches. They are all different, not only in the halftoning procedures themselves, but also in the quality and characteristics of the halftones generated. In the following, each one of the popular halftoning techniques is described.

Ordered Dither

Among the three major classes of halftoning algorithms, ordered dithering is the simplest in terms of complexity. Consequently, it is very popular in many practical systems such as low-cost printers. In this algorithm, we use a dither matrix, which is an array of threshold values $a_{m,n}$. The dither matrix is tiled onto the continuous tone image $x_{m,n}$, where element by element thresholding is performed to generate a binary halftone $b_{m,n}$. The entire procedure can be represented by the block diagram of Fig. 101.4.

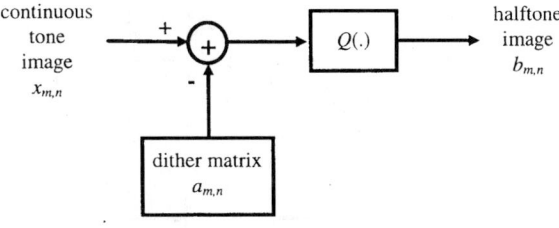

FIGURE 101.4 Halftoning by ordered dithering.

It is obvious that choices of the dither matrix will have a great influence on the characteristics and quality of the output halftones. The following 8 × 8 matrix:

$$\frac{1}{31}\begin{bmatrix} 18 & 20 & 19 & 16 & 13 & 11 & 12 & 15 \\ 27 & 28 & 29 & 22 & 4 & 3 & 2 & 9 \\ 26 & 31 & 30 & 21 & 5 & 0 & 1 & 10 \\ 23 & 25 & 24 & 17 & 8 & 6 & 7 & 14 \\ 13 & 11 & 12 & 15 & 18 & 20 & 19 & 16 \\ 4 & 3 & 2 & 9 & 27 & 28 & 29 & 22 \\ 5 & 0 & 1 & 10 & 26 & 31 & 20 & 21 \\ 8 & 6 & 7 & 14 & 23 & 25 & 24 & 17 \end{bmatrix}$$

generates halftones where the dots tend to form clusters, and the resulting halftone is called clustered dot dither. The halftones generated using the following matrix:

$$\frac{1}{63}\begin{bmatrix} 1 & 59 & 15 & 55 & 2 & 56 & 12 & 52 \\ 33 & 17 & 47 & 31 & 34 & 18 & 44 & 28 \\ 9 & 49 & 5 & 63 & 10 & 50 & 6 & 60 \\ 41 & 25 & 37 & 21 & 42 & 26 & 38 & 22 \\ 3 & 57 & 13 & 53 & 0 & 58 & 14 & 54 \\ 35 & 19 & 45 & 29 & 32 & 16 & 46 & 30 \\ 11 & 51 & 7 & 61 & 8 & 48 & 4 & 62 \\ 43 & 27 & 39 & 23 & 40 & 24 & 36 & 20 \end{bmatrix}$$

is called Bayer dither or dispersed dot dither due to the dispersed nature of the dots in the output halftones. More recently, a new class of ordered dither, called blue noise dither, has become popular. Typically, it requires dither matrices that are of substantially larger size (e.g., 128 × 128) than the traditional dither matrices, and it generates halftones that exhibit blue noise characteristics, that is, the noise energy in the halftone is primarily located at high frequencies.

Error Diffusion

Error diffusion is a relatively simple algorithm for generating very high-quality halftones. The generic error diffusion system can be represented by the block diagram of Fig. 101.5. It consists of a binary quantizer inside a feedback loop. The astute reader will readily recognize that it is very similar to the structure of DPCM coders. One intuitive view of error diffusion is that at each step the quantization error resulting from thresholding of the state variable $u_{m,n}$ is distributed to the future input pixels so that the overall quantization error can be reduced. The implementation of

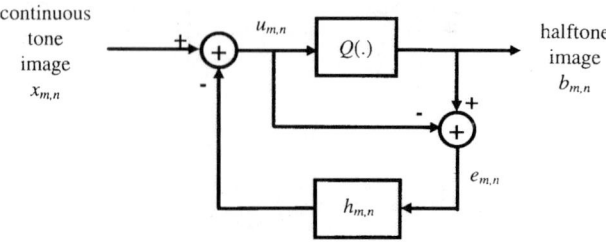

FIGURE 101.5 An error diffusion system for halftoning.

error diffusion is very easy, as one can simply iterate the nonlinear difference equations

$$u_{m,n} = x_{m,n} - \sum_{k,l} h_{k,l} e_{m-k,n-l}$$

$$e_{m,n} = b_{m,n} - u_{m,n} = Q(u_{m,n}) - u_{m,n}$$

$$Q(\alpha) = \begin{cases} 1 & \text{if } \alpha \geq 0.5 \\ 0 & \text{otherwise} \end{cases}$$

The filter coefficients $h_{m,n}$ are usually called the error diffusion kernel. A very popular error diffusion kernel was suggested by Floyd and Steinberg [1975], where

$$h_{0,1} = 7/16, \quad h_{1,-1} = 3/16, \quad h_{1,0} = 5/16, \quad h_{1,1} = 1/16$$

Because error diffusion generally produces very high-quality halftones at a relatively low complexity, it is a very popular algorithm in practical usage.

Minimization Approaches

In the minimization approach of halftoning, one first defines a distortion criterion between the continuous tone image $x_{m,n}$ and a halftone $b_{m,n}$. A popular distortion used in the literature is the weighted mean square error given by

$$d_{m,n} = \left(x_{m,n} - \sum_{i,j} v_{i,j} b_{m-i,n-j} \right)^2$$

where $v_{i,j}$ is the impulse response of a low-pass filter. Frequently, $v_{i,j}$ is chosen to mimic the frequency response of the human visual system. The halftoning algorithm is a minimization procedure that chooses a binary image to minimize $d_{m,n}$ on the average. Although the minimization approach for halftoning generally produces excellent image quality, it is also quite computationally intensive. Consequently, it has not been extensively used in practical systems.

101.5 Optimization of JPEG for Halftoning

In situations where we want to transmit an image to a remote location to be displayed or printed on a limited bit-depth device, we have the two options, as shown in Fig. 101.6:

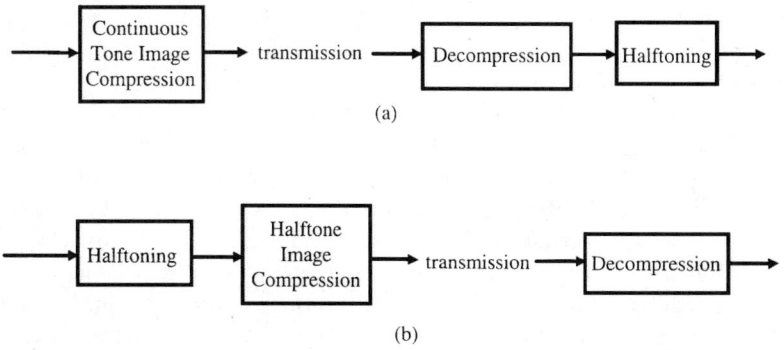

FIGURE 101.6 Two scenarios of image communication systems.

1. Compress the continuous tone image at the source. Decompress at the receiver and then halftone when the image is rendered.
2. Halftone the image at the source, compress the halftoned image, and then transmit the bit stream. Decompress the halftone at the receiver and then display the result.

We consider in this section the scenrio of Fig. 101.6(a), where a continuous tone image is to be compressed and then transmitted through a noiseless communications channel. In practice, we can use error correcting codes to protect the data to be transmitted, which implies that the assumption of a noiseless channel is not restrictive. At the receiver, the compressed data is decoded and then the image is rendered at the output device. In particular, we assume that we use JPEG as the compression method. Note that many output devices such as printers or liquid crystal displays are capable of rendering only a limited number of output levels. To be specific, we assume here that the output device is only capable of reproducing two levels. As a result, we need to halftone the output image in the rendering process. Recall that JPEG was designed for compressing continuous tone images that are to be displayed on continuous tone devices. Hence, one can expect that the default JPEG quantization table would not be optimum in general if we were to display the image as a halftone. Here we consider the optimization of the JPEG quantization table for the output halftone characteristics so that we can obtain an improved performance over the default JPEG [Vander Kam, Wong, and Gray, 1995]. We point out that this optimization approach can be generalized to other compression and halftoning techniques. For example, Neuhoff and Pappas [1994] have considered a subband coder that takes the characteristics of halftoning and scaling into account.

The rationale behind the improvement in performance is that the halftoning procedure can be viewed as a process that injects noise into the continuous tone image so that the output satisfies the limited bit-depth constraint. Each halftoning algorithm has a specific signature in the frequency domain, that is, each halftoning algorithm introduces noise to the original image with a majority of the energy concentrated in a few components in the frequency domain. If a JPEG compressed image is to be rendered using halftoning, it would be advantageous to encode more coarsely the frequency components that would eventually be corrupted by the halftoning procedure, whereas the frequency components that are not disturbed by halftoning should be quantized more finely. As a result, we can obtain an improvement in the quality of the output image over the default JPEG when coding both at the same rate.

The second factor affecting the optimization is the effect of scaling. To enable the viewer to render the output image at various sizes, we typically need to scale the output image before we halftone it. Scaling up can be viewed as an up sampling procedure followed by low-pass filtering, whereas scaling down can be accomplished as low-pass filtering followed by down sampling. In both cases, the scaling procedure shapes the spectrum of an image in a predictable way. Following the argument as before, we want to assign a higher bit rate to the frequency components corresponding to a large gain in the filter response, and a lower bit rate to the components where the attenuation caused by the scaling procedure is high.

The third factor is the response of the human visual system, which can be well approximated by a low-pass filter. We can also assign the bit rates according to the response of this system. The interpretation here is that we encode more precisely the frequency components that are more important to our visual system, and leave other components at lower rates.

A desired quantization error profile can be generated by combining the halftoning error spectrum, the response of the scaling procedure, and the human visual system characteristics. Using this profile, we can use a closed-loop design procedure to design a quantization table that will be optimum for the particular halftoning and scaling methods chosen [Vander Kam, Wong, and Gray, 1995]. It can be shown that the design procedure always converges, and hence the design will always succeed. For optimum compression performance, it is necessary to optimize the Huffman table as well, primarily because of the fact that the default JPEG Huffman table has not been optimized for low rate (<0.5 b/pixel) coding. Using both the optimized quantization and Huffman tables, the compression performance can typically be improved by more than 20% compared to the default JPEG, that is, one can compress typical images using 20% fewer bits than the default JPEG, while giving the same image quality.

101.6 Summary and Conclusions

Still image compression has become increasingly popular and important as the utilization of images has become more prevalent. We have considered a variety of popular still image compression techniques, including DPCM, transform coding, subband coding, and VQ. The emergence of the JPEG international image coding standard in recent years has made a profound impact in the utilization of images in many commercial, industrial, and home applications. It is extremely important in many situations where the various imaging devices that require compression must "speak a common language" so that communications or sharing data becomes possible. The JPEG compression standard enables the common link and hence plays an important role in the proliferation of imaging applications. Because of its importance, we have considered in this chapter the JPEG compression algorithm in depth.

Since an image must eventually be printed or displayed for it to be appreciated by human observers, it is important to consider image compression in conjunction with output devices. An important characteristic of output devices is that many of them, such as image printers or liquid crystal displays, are typically capable of rendering only a small number of output levels. In these cases, images will have to be halftoned and scaled before they are rendered. In view of this, we have described several popular halftoning algorithms, and also considered the optimization of JPEG for halftoning and scaling. Experimental results have consistently shown that an optimized JPEG algorithm for a given chain of output rendering algorithms can outperform the default JPEG by 20% or more.

Defining Terms

Entropy coding: A lossless coding method for compressing data to a level near the theoretical limit given by information theory—entropy.

Halftoning: A technique for generating a bilevel image from a continuous tone image so that they appear similar when observed from a distance. The corresponding techique for generating multilevel output is call multitoning.

Perceptual coder: A coder that encodes so that the quantization errors are distributed according to the characteristics of the human perception system.

Quadrature mirror filter bank: A bank of filters that decomposes a signal into subsignals, each of narrower bandwidth than the original, so that the original signal can be reconstructed perfectly from the subsignals.

References

Fischer, T.R. 1992. On the rate distortion efficiency of subband coding. *IEEE Trans. Inform. Theory*, 38(March):426–428.

Floyd, R. and Steinberg L. 1975. An adaptive algorithm for spatial gray scale. In *SID Int. Symp., Digest of Tech. Papers*, pp. 36–37.

Gersho, A. and Gray, R.M. 1992. *Vector Quantization and Signal Compression*. Kluwer Academic Publishers, Norwell, MA.

ITU. 1993. Information technology—digital compression and coding of continuous-tone still images, Part I: Requirements and guidelines. ITU-T Rec. T.81 | ISO/IEC 10918-1, International Telecommunications Union–Telecommunications Standardization Sector, Geneva, Switzerland.

Jain, A.K. 1981. Image data compression: A review. *Proc. IEEE*, 69(March):349–389.

Jayant, N. 1992. Signal compression: Technology targets and research directions. *IEEE J. Selec. Areas Commun.*, 10(June):796–818.

Linde, Y., Buzo, A., and Gray, R.M. 1980. An algorithm for vector quantizer design. *IEEE Trans. Commun.*, 28(Jan.):84–95.

Makhoul, J. 1975. Linear prediction: A tutorial review. *Proc. IEEE*, 63(April):561–579.

Mallat, S.G. 1989. A theory for multiresolution signal decomposition: The wavelet representation. *IEEE Trans. Pattern Anal. Machine Intell.*, 11(July):674–693.

Netravali, A.N. and Limb, J.O. 1980. Picture coding: A review. *Proc. IEEE*, 68(March):366–406.

Neuhoff, D.L. and Pappas, T.N. 1994. Perceptual coding of images for halftone display. *IEEE Trans. Image Proc.*, 3(Jan.):1–13.

Pearlman, W.A. 1991. Performance bounds for subband coding. In *Subband Image Coding*, ed. J.W. Woods, pp. 1–41, Kluwer Academic Publishers, Norwell, MA.

Reininger, R.C. and Gibson, J.D. 1983. Distributions of the two-dimensional DCT coefficients for images. *IEEE Trans. Commun.*, 31(June):835–839.

Riskin, E.A. 1991. Optimal bit allocation via the generalized BFOS algorithm. *IEEE Trans. Inform. Theory*, 37(March):400–402.

Shoham, Y. and Gersho, A. 1988. Efficient bit allocation for an arbitrary set of quantizers. *IEEE Trans. Acoust. Speech Signal Proc.*, 36(Sept.):1445–1453.

Vander Kam, R.A., Wong, P.W., and Gray, R.M. 1995. JPEG compression for a grayscale printing pipeline. In *Proc. SPIE*, Still Image Compression, 2418:229–240.

Vetterli, M. and Herley, C. 1992. Wavelets and filter banks: Theory and design. *IEEE Trans. Signal Proc.*, 40(Sept.):2207–2232.

Wintz, P.A. 1972. Transform picture coding. *Proc. IEEE*, 60(July):809–820.

Woods, J.W. ed. 1991. *Subband Image Coding*, Kluwer Academic Publishers, Norwell, MA.

Woods, J.W. and O'Neil, S.D. 1986. Subband coding of images. *IEEE Trans. Acoust. Speech Signal Proc.*, 34(Oct.):1278–1288.

Further Information

For further information on quantization and coding, see:

Jayant, N.S. and Noll, P. 1984. *Digital Coding of Waveforms*. Prentice–Hall, Englewood Cliffs, NJ.

Gersho, A. and Gray, R.M. 1992. *Vector Quantization and Signal Compression*. Kluwer Academic, Norwell, MA.

Details of algorithms and architectures on still image coding standards can be found in:

Pennebaker, W.J. and Mitchell, J.L. 1993. *JPEG Still Image Data Compression Standard*, Van Nostrand Reinhold, New York.

Bhaskaran, V. and Konstantinides, K. 1995. *Image and Video Compression Standards: Algorithms and Architectures*, Kluwer Academic Publishers, New York.

More information on halftoning techniques can be found in:

Ulichney, R.A. 1987. *Digital Halftoning*, MIT Press, Cambridge, MA.

102
Video

Eric Dubois
INRS-Télécommunications

102.1 Introduction 1449
102.2 Source Characteristics and Viewer Requirements 1450
102.3 Coding Algorithms 1453
    Motion Compensation • Predictive Coding • Interpolative Coding • Residual Coding • Buffer Control • Pyramidal Coding
102.4 Standards 1459
102.5 Perspectives 1460

102.1 Introduction

Time-varying imagery is a highly effective medium in a variety of applications including entertainment, interpersonal communication, information presentation and retrieval, and many others. We refer to a time-varying image signal in electronic form as a **video signal**. Until recently, video signals were stored and transmitted exclusively in analog form, and this continues to be the most common situation. The principal storage medium is analog video tape, and the main transmission channels are the over-the-air broadcast channel, cable, and satellite. However, due to recent advances in digital hardware and in image compression technology, systems for digital storage and transmission of video signals are becoming more common. It can be expected that digital representations of video will be predominant in the future. Video will be stored in digital form on disks and tape, and will be transmitted digitally over broadband networks, as well as wireless networks and broadcast channels.

Video signals are notorious for consuming huge amounts of bandwidth. For example, an analog National Television System Committee (NTSC) video signal uses a 6-MHz bandwidth, whereas an uncompressed digital NTSC video signal requires a bit rate of about 100 Mb/s. To allow economical storage and/or transmission, more efficient representations are required. This process is often viewed as a compression operation, wherein quality is sacrificed; an alternative view is that we seek a representation that will deliver maximum video quality for a given available data rate. At rates of 20–30 Mb/s, this would yield large images of high quality [high-definition television (HDTV)], whereas at low rates such as 64 kb/s and below, this would result in a system providing small pictures of rather low quality and with limited capability to render motion (videophone).

The techniques of source coding seek to exploit both redundancy in the source and the fidelity requirements of the receiver of the information in order to minimize distortion for a given data rate. If a stochastic model of the source and a fidelity criterion for the receiver are known, the optimal performance can be found using rate-distortion theory. However, an implementable coder with acceptable delay giving this optimal performance may not be available. In the case of time-varying imagery, neither general purpose stochastic models of the source nor mathematical fidelity criteria that capture all of the known properties of source and receiver are known at this time. Thus, optimal source coding for time-varying imagery does not currently rest on a firm theoretical basis. Rather, we have available an arsenal of tools that work well and that address, at least qualitatively, the known properties of the source and receiver.

The purpose of this chapter is to identify the properties of source and receiver that are exploited by video source coding algorithms, and to describe the tools that are available for source coding. The emphasis is on those aspects that are specific to time-varying imagery, since a previous section deals with still imagery. Thus, the use of motion information in the processing and compression of time-varying imagery is a pivotal element for this contribution. An overview of video compression standards is also given, followed by some perspectives for the future. An extensive treatment of video coding can be found in Netravali and Haskell [1995].

102.2 Source Characteristics and Viewer Requirements

A number of factors determine the source characteristics and the level of redundancy that can be exploited. These include the scanning format, the signal-to-noise ratio, the amount of detail, and the characteristics of the motion. Note that the source signal to be coded may not necessarily be the direct output of a video camera; it may be the result of some scan conversion or other preprocessing operations.

The scanning format encompasses the picture size and the sampling structure. We assume that the input to the source coder is fully digital, that is an ordered sequence of fixed length binary codewords representing the image color at a discrete set of points Ψ. This is denoted

$$u_i(x, y, t), \quad (x, y, t) \in \Psi_i, \quad i = 1, 2, 3 \qquad (102.1)$$

where u_1, u_2, u_3 represent coordinates in a suitable color space. These could be red-green-blue or luminance-chrominance components. Clearly black and white images would be represented by a single component. The sampling structure Ψ is a regular array of points such as a lattice, cropped spatially to the image frame size. Picture size may range from small windows on a videophone or workstation, say, of size 64 by 64, to HDTV pictures of size 1920 by 1080 and beyond. Two vertical-temporal scanning structures are widely used for video: **progressive scanning** and **interlaced scanning**. These scanning structures are shown in vertical-temporal profile in Fig. 102.1. In analog systems, where the capture, transmission, and display formats are the same, interlace is a method to achieve higher vertical resolution for a given vertical scan rate and system bandwidth. The vertical

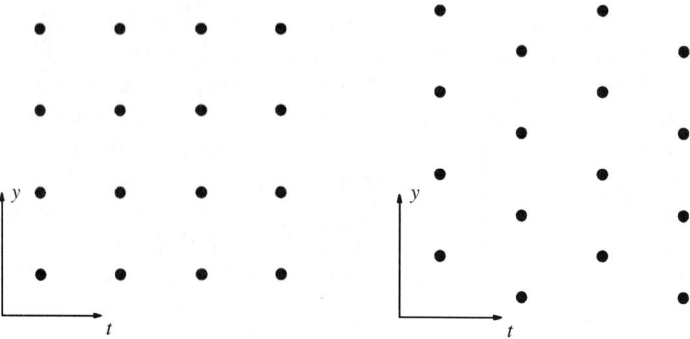

FIGURE 102.1 Profile view of scanning structures: (a) progressive and (b) interlace.

TABLE 102.1 Typical Scanning Formats Used for Digital Video Communications

Identifier	Luminance Size Hor.	Vert.	Chrominance Size Hor.	Vert.	Scanning Structure Interlace	Pict. Rate
QCIF	176	144	88	72	1:1	29.97
CIF	352	288	176	144	1:1	29.97
ITU-R 601	720	480	360	480	2:1	59.94
HDTV-P	1280	720	640	360	1:1	60.0
HDTV-I	1920	1080	960	540	2:1	60.0

scan rate of 50 or 60 Hz used in major worldwide systems is largely determined by the requirement to minimize the visibility of flicker at the display. Although interlace is a good technique in the context of analog systems, it usually leads to spatiotemporal **aliasing** in the presence of vertical motion, which makes subsequent processing, and especially source coding, much more difficult. (See Dubois [1992] for a discussion.) For this reason, there has been increasing focus on progressive scanning in recent system proposals. A detailed discussion of sampling structures and video signal sampling is given in Dubois [1985].

Although some standards allow considerable flexibility in the choice of picture format, a number of specific formats are in common use or have been proposed. Table 102.1 shows the parameters of some of these formats. The final coded bit rate is very directly related to the source sampling rate, although not linearly.

The signal-to-noise ratio of the source has a significant impact on compression efficiency. In general, noise has little redundancy and cannot be compressed very much; thus, it should be removed prior to coding if possible. There is, of course, a tradeoff between coding efficiency and the introduction of artifacts due to noise reduction. Valid assumptions must be made, or the processing may remove desired picture content such as snowstorms, fireworks, etc. Motion compensated processing is required for the most effective noise reduction [Dubois, 1992]. However, good image and noise models are necessary. The principle is to filter out those deviations from an assumed image model (e.g., that image intensity is constant along trajectories of motion) that match the characteristics of an assumed noise model (e.g., white noise with a specified variance).

The characteristics of the motion in the scene affect the amount of compression that is possible. Apparent motion in the image can be due to either camera motion (including zooming), object motion, or both. In a videophone application, most apparent motion may be due to movement of the person in front of the camera. In a surveillance application, most of it may be due to the scanning motion of the camera (except during an incident), whereas in a sports scene both camera motion and scene motion may be significant. Motion may in general be low, as in a videophone application, or very active, as in a sports application; it may be highly structured (rigid-body motion) or pseudorandom (rustling leaves). Knowledge of the motion characteristics can affect the coding algorithm; if there is unrestricted motion, a worst-case assumption must be made.

Redundancy in a source is related to the concept of predictability (although they are not the same). Given that a certain portion of the source signal has already been transmitted, the coder may attempt to predict what comes next. If the prediction is at all reliable, it is better to transmit information concerning the deviation from the prediction than the raw information. Predictability implies a certain regularity in the data that can be captured in statistical models. In the spatial dimension, if a certain pattern appears over a certain area, one may assume that the pattern will continue, and make the prediction on this basis. Of course, at an object boundary, the prediction will be wrong. However, if it is right most of the time, an overall coding gain can be achieved. In a moving picture, the assumption is that if an object is moving with a certain velocity and acceleration, it will continue to do so. Thus, if the velocity and acceleration are known, the coder can predict what the next frame will look like, and only transmit the error between this prediction and reality. This is the essence of motion-compensated coding. Again, at a scene change, the prediction will be completely wrong, and more information must be transmitted. However, on average there will be a substantial compression. Temporal change is not due solely to motion. The illumination may change, and the luminance of scene points may change as the angles between object surface, illumination source, and camera change. If we can model these effects as well, we can make better predictions and thereby improve coding performance.

In addition to the structural or statistical redundancy in the source, there is perceptual redundancy (also called irrelevancy). Specifically, information that cannot be perceived by the viewer need not be transmitted. The main phenomena related to this are spatial and temporal **masking**. Spatial masking (discussed in the chapter on still image coding) is the effect whereby certain types of distortion are less visible in the vicinity of high-contrast image detail than in flat low-contrast areas. Temporal masking is the effect where distortions are less visible in areas of the image that are newly exposed or contain fast motion. An important case here is the scene change. The viewer can tolerate a significant reduction in resolution just after a scene change. This is fortuitous, because this is exactly where the temporal motion-compensated prediction breaks down.

The goal of an efficient source coder is to maximize quality (or minimize distortion) subject to a number of constraints. In most applications of video compression, the quality or distortion are subjective measures determined by the visual properties of the viewer. However, if further signal processing or manipulation of the image sequence is to be performed, a more stringent objective criterion may be more appropriate, since impairments invisible to a viewer could have a negative effect on a signal processing operation such as chroma key. If subjective quality is the main concern, it would be very desirable to have a mathematical function that measures the subjective distortion. This distortion function could then be minimized during the optimization of the coding algorithm. Unfortunately, such a general purpose and widely accepted perceptual distortion measure is not available. Thus, properties of the human visual system are usually exploited based on qualitative or empirical considerations that are obtained under restricted conditions. However, this approach may still be quite effective [Schreiber, 1993; Jayant, Johnson, and Safranck, 1993]. The coding algorithm can then be evaluated using formal subjective testing under controlled conditions [Allnutt, 1983].

Other requirements may affect the type of coding algorithm that can be used. These include delay, random access, playback control, and scalability. Both the coding algorithm and the transmission channel introduce delay. Depending on the application, there may be tight bounds on this delay, especially in interactive applications. The most stringent application is in face-to-face communication, where a maximum end-to-end delay of about 200 ms can be tolerated. If the delay is longer than this, spontaneous interaction becomes difficult. In a database access application, somewhat longer delays of up to a few seconds may be acceptable. Finally, in an application such as video on demand, the delay constraint may be of no practical consequence, that is, any reasonable encoding system would meet the delay constraint.

Random access and playback control are requirements that may exist in services such as database access and video on demand. Some coding schemes are recursive in nature, and require that the sequence be decoded continuously from the beginning in order to view a particular frame. This may be acceptable in an application such as video conferencing, but it precludes random access and channel switching. Thus, the encoding algorithm must have *finite memory* in some sense to allow such a random access feature. The playback controls include fast-forward, reverse, and fast-reverse playback as well as pause functions. Again, general coding algorithms may not permit such features, and they must be incorporated as constraints if they are desired.

The coding algorithm must account for the properties of the channel or storage medium. The channel may allow only constant bit rate (CBR) operation, or it may permit variable bit rate (VBR) operation. In the former case, the coder must be designed to produce a fixed number of bits over a specified unit of time. In the latter case, the bit rate may vary over time, although some constraints negotiated at setup time must be respected (e.g., ratio of peak to average bit rate). Another requirement of a coding algorithm is acceptable performance in the presence of typical channel impairments. These channel impairments may vary considerably for different types of

Video

channels. For example, a dedicated digital link may have random errors with a very low probability of occurrence, whereas a wireless link may have bursty errors with a high probability of occurrence. Also, depending on the transport mechanism, the errors may occur in isolated bits, or they may involve an entire packet, consisting of a large number of bits. In general, when the coding efficiency for redundancy reduction is higher, the coded data is more vulnerable to transmission errors. Thus, overall system performance must be evaluated with the channel impairments taken into account in order to be realistic. If possible, joint source/channel coding should be considered.

In some situations, the same video signal must be delivered to a variety of receivers having different capabilities. Some examples include the broadcast of entertainment television to a population of receivers ranging from small tabletop units to large high-definition displays. Similarly, a database containing video data might be accessed by terminals ranging from relatively inexpensive home computers to powerful workstations, perhaps with three-dimensional display. The capability to transmit only that portion of the data required by the terminal, or to extract a subset of the data, is known as **scalability**. The most straightforward approach to this is simulcasting or multiple storage of the coded data for each type of receiver. This is complex and wasteful of storage or transmission capacity. Thus a more efficient approach is the concept of embedded coding, whereby a subset of the encoded data can be used to decode a lower version of the sequence suitable for display on the lower capability display.

Most of the preceding requirements can reduce the coding efficiency achievable, as compared to an unconstrained situation. In each case, the system designer must decide whether the benefits of the feature justify the additional complexity or performance loss.

102.3 Coding Algorithms

The goal of a source coding algorithm is usually to provide the best image quality for a given channel or data rate, subject to constraints on complexity, delay, or other requirements as discussed. In some circumstances, a certain quality may be prescribed, and a coding algorithm is sought that provides this quality at the lowest possible rate (again, subject to constraints). In time-varying imagery, most incremental information over time is carried by the motion occurring in the scene. Thus, the most highly efficient coding algorithms are based on motion estimation and compensation. This is the main feature distinguishing video coding from still image coding, and it represents the focus of this section. An image sequence is usually encoded by both intraframe coding techniques and interframe coding techniques. The intraframe coding algorithms are very similar to the still picture coding algorithms of the previous section. They are used for several reasons, including: to permit random access into the sequence, to allow fast forward and reverse playback, to provide robustness to channel errors, and as fallback when temporal coding fails (for example, at a scene change).

Motion Compensation

The three-dimensional motion of objects in a scene induces two-dimensional motion in the image plane. The relationship between them depends on the perspective transformation of the camera, and may be complex. The image of a given scene point traces out over time a trajectory in the space–time coordinates of the time-varying image (Fig. 102.2). The basic principle of motion-compensated coding is that the redundancy is very high along this trajectory. This is exploited using the coder structure of Fig. 102.3. The input signal is applied to a motion estimation module that determines a **motion field**. This motion field establishes the correspondences between image samples lying on the same trajectory in successive image pictures. Although the motion field specifies estimated displacements for each pixel in the image sequence, it may be specified parametrically over regions, as will be described subsequently. The motion field is then encoded with a lossless code for transmission to the receiver at rate R_M. Any quantization of the motion field is assumed to take place within the motion estimation module. The motion field is also supplied to the

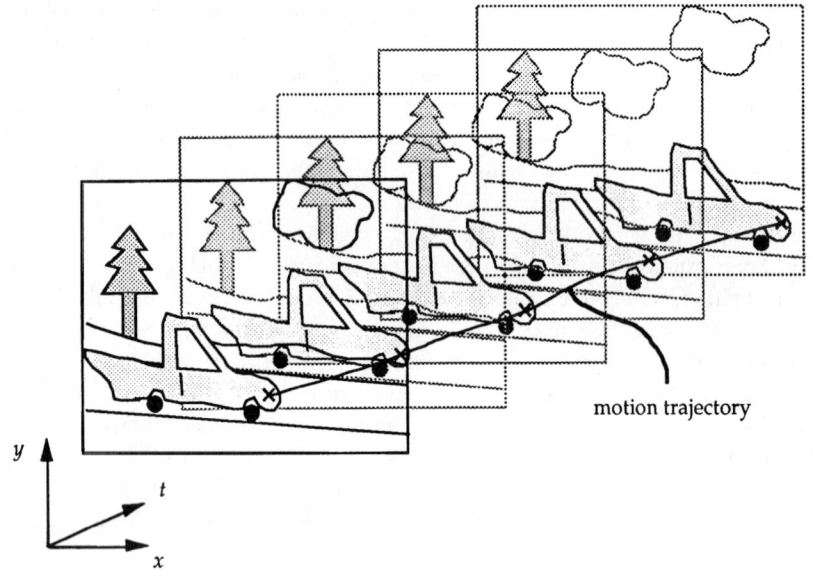

FIGURE 102.2 Trajectory in space time of a scene point.

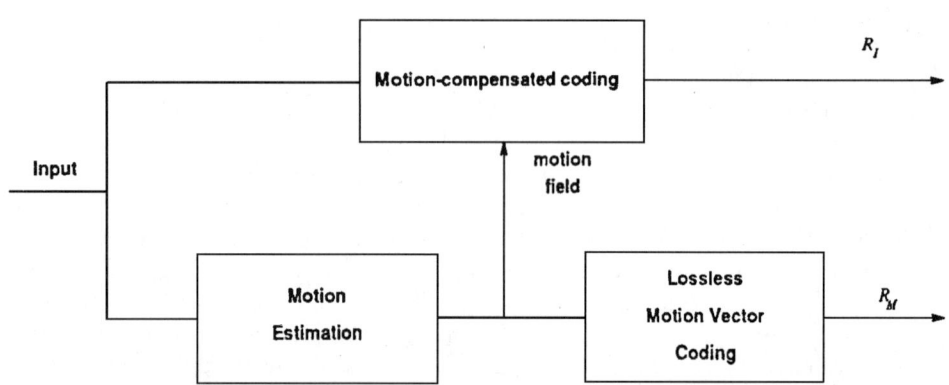

FIGURE 102.3 General structure of a motion-compensated video coder.

motion-compensated coding module, which performs a conditional coding of the image sequence given the estimated and quantized motion field. The output of this module (which includes any lossless coding) is transmitted to the receiver at rate R_I. It is possible for the motion estimator to also make use of previously coded/decoded data. This can be accommodated (conceptually) by the structure of Fig. 102.3 by assuming that the motion estimation module contains a copy of the motion-compensated coder. Of course, the motion-compensated coder would not be duplicated in a practical implementation; common elements would be shared between the motion estimation and motion-compensated coding modules.

The total rate R for coding the video sequence is the sum of R_M, the rate to transmit the motion information, and R_I, the rate to transmit the conditionally coded video information. The best way to partition the overall rate between these two components is an open problem. A greater precision in the motion field may lead to a lower rate R_I for the motion-compensated video but a higher rate R_M for the motion information. The optimal point will depend on the motion coding scheme and

Video

the motion-compensated video coding scheme. An extreme situation is when $R_M = 0$, where either no motion information is used, or motion is only computed from previously encoded video data. At the other extreme, we can imagine a situation where the motion field is coded very precisely, so that $R_I \ll R_M$.

There are two basic approaches for estimating the motion field: the pixel-based approach and the parametric approach. In the pixel-based approach, a displacement vector is estimated for each pixel in the picture using some robust algorithm. The resulting motion field is then encoded efficiently to achieve the desired rate. Significant compression is possible since the motion field is in general highly redundant (it is very smooth except at object boundaries). The decoded motion field is used in the motion-compensated coder. The parametric approach is more common in video coding, where the motion field is represented parametrically over image regions. The most common technique is to assume a constant displacement over a rectangular block of pixels. Higher order models such as the affine model have also been used. A more general approach is to assume that the motion is represented parametrically over nonrectangular regions. In this case, the specification of region boundaries is part of the motion representation. There are many algorithms to actually estimate the motion, including region matching, gradient-based optimization, and transform domain methods. See Mitiche and Bouthemy [1996] for a survey.

The two main techniques to perform conditional coding given motion, **predictive coding**, and **interpolative coding** are discussed in the next two sections.

Predictive Coding

The principle of predictive coding is that, given the motion information and previous decoded pictures, it is possible to form a good estimate of the next picture to be coded. Then, only the deviation from this estimate needs to be encoded and transmitted. A structure well suited to this is differential pulse code modulation (DPCM), shown in Fig. 102.4. In this system, the prediction is formed on the basis of previously reconstructed pictures using a local decoder at the transmitter. The prediction error is then spatially quantized and losslessly encoded at variable rate for transmission. The types of spatial quantization used are discussed in the section on residual coding.

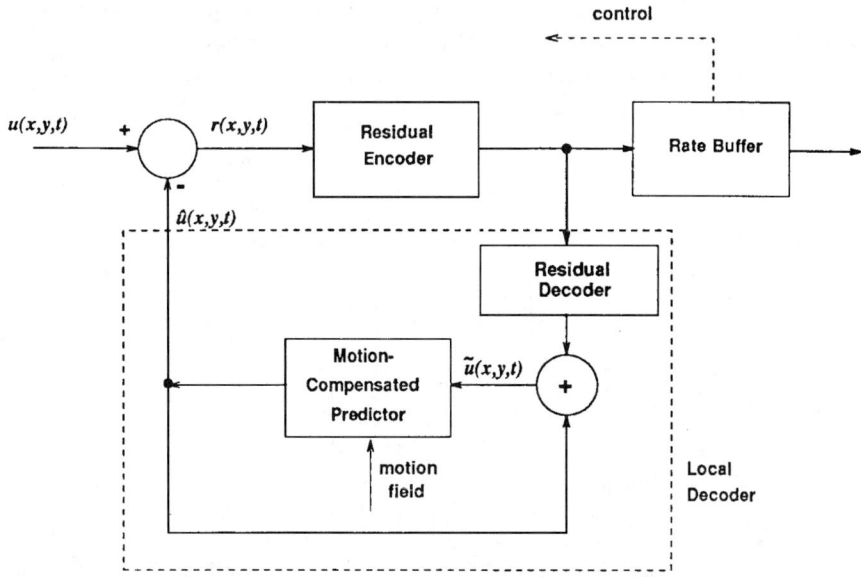

FIGURE 102.4 Motion-compensated DPCM encoder.

In the basic mode of operation, the image information at time t is predicted from the suitably displaced information in the previous vertical scan at time $t - T$, where T is the vertical scan time. The prediction of the sampled input image $u(x, y, t)$ is given by

$$\hat{u}(x, y, t) = \tilde{u}(x - d_x(x, y, t), y - d_y(x, y, t), t - T) \tag{102.2}$$

where $(d_x(x, y, t), d_y(x, y, t))$ is the estimated displacement at (x, y, t) and $\tilde{u}$ is the locally reconstructed output of the codec. If $(x - d_x, y - d_y, t - T)$ does not lie on the sampling grid, spatial interpolation must be used to compute $\hat{u}$ from existing samples of $\tilde{u}$ at time $t - T$. If the video signal is in interlaced format, each vertical scan of the image contains only part of the spatial information. Thus, a more elaborate prediction using at least the two previous pictures is very advantageous, as used in Moving Pictures Experts Group-2 (MPEG-2) standards [ISO/IEC, 1994].

A disadvantage of predictive coding is that the quantization error of the residual encoder is fed back to the predictor. This limits the compression efficiency, even if the source is very highly correlated along motion trajectories. One way to counteract this effect is to use interpolative coding for some pictures.

Interpolative Coding

Interpolative coding is sometimes called bidirectional predictive coding. In this method, it is assumed that some pictures spaced M pictures apart, called *reference pictures*, have already been encoded by some other method, say intraframe or predictive coding. Then, the intervening pictures are estimated based on the previous and subsequent reference pictures, using motion compensation. The estimation error is spatially quantized and losslessly coded for transmission. This coding sequence is illustrated with an example in Fig. 102.5. The pixels in a B picture can be estimated from the previous reference picture, the subsequent reference picture, or both, depending on which gives the best results. In this way, occlusion effects can be handled. However, the transmitted motion field must identify which mode is being used and incorporate the necessary motion vectors.

In general, the interpolatively coded pictures have lower error than predictively coded pictures at the same rate. However, the efficiency of DPCM decreases as the spacing between pictures is increased. It has been found that encoding two pictures out of every three by interpolative coding is a good compromise for a wide selection of video sequences. Interpolative coding is an open-loop method; there is no feedback of the quantization error as in DPCM.

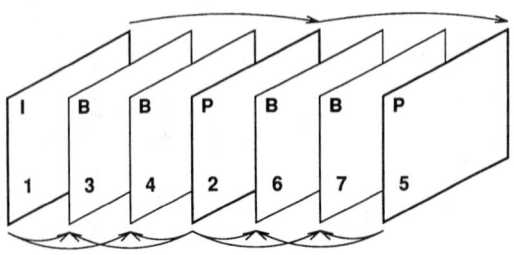

FIGURE 102.5 Example of picture structure for interpolative coding. Every third picture is encoded predictively. Two pictures out of every three are encoded interpolatively. The digit in the lower left corner of each picture specifies the order in which the pictures are coded.

Video

Residual Coding

The estimation error in DPCM or interpolative coding must be spatially quantized and losslessly encoded for transmission. This is a key element of the coding algorithm, because it is at this point that compression is achieved. The coding scheme should have good rate-distortion performance for the class of residual signals. There should be several versions of the residual coding algorithm available at different points on the rate-distortion curve to allow adaptive coding and buffer control, as discussed in the next subsection. Although scalar quantization can be used for residual coding, it is more effective to use some form of quantization with memory. The two most popular methods for this purpose are transform coding and subband coding (although the first can be viewed as a special case of the second). The reasons to use quantization with memory are twofold: (1) there is remaining correlation in the residual signal and (2) even for a memoryless source, quantization with memory can achieve greater compression, especially at low rates.

The methods used for residual coding are the same general methods that are applied to spatial coding. Since these are presented in the detail in the previous section, they are not discussed here. The methods that can be used include block transform coding, and in particular the discrete cosine transform (DCT), subband coding and wavelet coding, and vector quantization. The DCT method has been selected as the residual coding algorithm for all of the main compression standards developed to date. Although the basic algorithm is the same as for still pictures, the exact nature of the quantizers for the transform coefficients may be slightly different. The quantization operation is followed by a reversible code-assignment process. In general, the probability distribution of the quantized residual is highly nonuniform, so that variable-rate codeword assignment can yield a significant reduction in average bit rate, compared with fixed codeword length assignment. The methods that have been found effective include Huffman coding, runlength coding, and arithmetic coding. These techniques are discussed in detail in previous chapters. Because of the additional variability in residual statistics due to changing motion, the bit rate process can have significantly more variability in interframe coding than in spatial coding.

Buffer Control

The video coding algorithms described in preceding sections are characterized by a number of parameters, such as the quantizer scale factors, the input sampling structure, prefilter characteristics, etc. These parameters establish the rate and quality at which the coder operates. However, because of the time-varying nature of the video signal characteristics and the use of variable length codes, the short-term rate and quality change with time for a fixed set of parameter values. The rate can also vary as the coding mode changes, for example, the rate in intrapicture mode will be considerably higher than the rate in interpolative mode.

The transmission channel can be either a CBR link or a VBR link. In the former case, the variable rate data emitted by the coder must be smoothed out by a **rate buffer** (as in Fig. 102.4) to deliver data to the channel at a fixed constant rate. Because of constraints on delay and complexity, the buffer size is limited. As a result, there is a significant probability of buffer overflow or underflow. A control mechanism is required to adapt the coder parameters in order avoid buffer overflow while maintaining the highest and most uniform image quality possible. Even for a VBR channel, there will be constraints on the coder output bit rate process that must be satisfied, so that a rate buffer and control algorithm are still required.

The control algorithm periodically monitors the rate buffer occupancy, and changes the coder parameters to avoid potential overflow or underflow in such a way as to give the best overall image quality. Large or overly frequent parameter changes causing uneven quality should be avoided. Most techniques for buffer control have been heuristically derived. The occupancy of the rate buffer is used to determine the coding parameters (often only one, a scale factor for the DCT coefficient quantizers). At prespecified time intervals, the buffer fullness is evaluated and if it crosses certain thresholds in a given direction, the coding parameters are changed in a way to counteract the change in buffer occupancy. For example, if the buffer level is increasing, the quantization in the coder can be made coarser in order to reduce the number of bits being put into the buffer. The choices of coding modes and buffer switch points are made heuristically. More sophisticated techniques attempt to formulate a cost function that measures performance of the buffer control algorithm, and to use the history of the buffer occupancy and the statistics of the bit rate process in different coding modes to choose an optimal parameter adaptation strategy. See Zdepski, Raychaudhuri, and Joseph [1992] and Leduc [1994] for recent work.

Pyramidal Coding

Pyramidal coding is a method that can be used to perform embedded coding for scalable video applications [Girod, 1993]. It is capable of producing coded versions of a given video sequence at different levels of spatial, temporal, and amplitude resolution. The general structure of a pyramidal coder is shown in Fig. 102.6. The input is first coded at the lowest supported rate R_1 by the level 1 coder. The resulting bit stream can be decoded by any decoder supported by the service. This level 1 sequence would be at the lowest spatial, temporal, and amplitude resolution, whereas the input would have full spatial, temporal, and amplitude resolution. Thus, the level 1 coder would need to incorporate appropriate spatial and/or temporal downsampling. To obtain the next highest level in the hierarchy, the level 1 signal is locally decoded and used as a priori information in the conditional coding of the input. The incremental information is coded at a rate R_2, and together with the level 1 bit stream specifies the level 2 output. This process is repeated for as many levels as are desired. Conditional coding is usually achieved by interpolating a lower level signal to the same spatial and temporal resolution as the next level. This is then used to form a prediction of this higher level signal; the residual is then further coded for transmission.

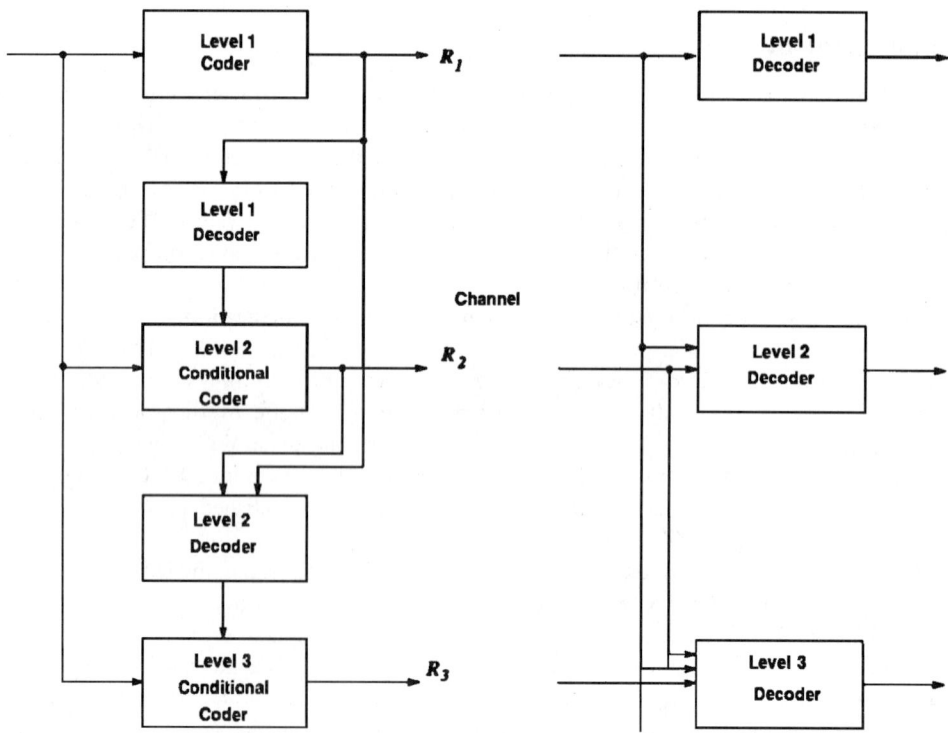

FIGURE 102.6 Pyramidal coder for embedded scalable video coding.

102.4 Standards

Standards play a major role in assuring the wide availability of image communication products that permit interoperability between equipment provided by different manufacturers. The first important standard for compressed video was the International Telecommunications Union–Telecommunications Standardization Sector (ITU-T) [formerly the Consultative Committee on International Telephony and Telegraphy (CCITT)] H.261 standard for videoconferencing. This was followed by the International Standards Organization (ISO)/IEC MPEG 1 and 2 standards. These standards are summarized in Table 102.2. A brief description of each follows. A more detailed overview of these standards is given in Aravind et al. [1993].

The ITU-T Recommendation H.261 specifies the video codec system as part of a group of standards for audiovisual telecommunication services [ITU-T, 1990]. In 1984, CCITT Study Group XV established a specialist group on coding for visual telephony, and the recommendation was completed and approved in December 1990. This transmission standard was developed for use in videophone and videoconferencing applications. The standard specifies bit rates of the form $p \times 64$ kb/s, where p is an integer between 1 and 30. The standard defines only the functionality of the decoder; the encoder is not explicitly defined, but it must produce a bit stream compatible with the standard decoder. The input signals are in QCIF or CIF format (Table 102.1). The coding algorithm uses motion-compensated temporal DPCM, with DCT coding of the residual. The motion compensation is block oriented, with a single motion vector for each 16×16 block of luminance pixels (referred to as a macroblock). The DCT is applied to 8×8 blocks of the residual signal. To meet the strict delay requirements, the prediction is always based on the immediately preceding picture. It is possible to set the temporal prediction to zero, so that the residual is equal to the input signal (intraframe coding mode), or to force the motion vector to be zero (interframe coding without motion compensation). The bit stream has a layered structure, with layers for the picture, group of blocks (defined as 3×11 macroblocks), macroblock, and block. The top three layers consist of a header containing synchronization information, parameter values, and user defined bits, followed by a number of units of the next lower layer. The lowest layer (block) simply contains encoded transform coefficients and an end-of-block (EOB) code. The coder designer must decide how to regulate the parameters to optimize quality, avoid buffer overflow or underflow, and meet certain constraints imposed by a hypothetical reference decoder (HRD).

MPEG under the ISO-IEC/JTC1/SC29/WG11 undertook in 1988 to develop a standard for video coding adapted to rates of about 1–1.5 Mb/s and suitable for digital storage media. The strict delay requirement of H.261 was relaxed, whereas additional features such as random access, fast forward/reverse, etc., were desired. This standard, referred to as MPEG-1, has now been approved as an international standard [ISO/IEC, 1993]. It builds considerably on the H.261 standard and has much commonality. It supports a wider range of input signal formats, although the input sequences are assumed to be progressively scanned. The basic techniques of motion-compensated temporal DPCM and DCT coding of the residual are maintained from H.261. However, the technique of interpolative coding has been added. The concepts of layers used in H.261 has been augmented with the sequence layer and the group of pictures layer. The group of pictures layer is the unit allowing random access into the sequence. The group of blocks layer is replaced with a more flexible slice layer. The macroblock and block layers are similar to H.261. The bit stream is characterized at the sequence layer by a bit rate and buffer size. The MPEG standard establishes constraints on the bit

TABLE 102.2 Standards for Compressed Video

Name	Identifier	Application	Typical Bit Rates
$p \times 64$	ITU-T Rec. H.261	teleconferencing	64 kbit/s–1.92 Mbit/s
MPEG-1	ISO/IEC 11172-2	digital storage media	1.0–1.5 Mbit/s
MPEG-2	ISO/IEC 13818-2	broad range	1.5–20 Mbit/s

stream to ensure that decoding will occur without buffer overflow for the given buffer size. See Le Gall [1992] for a more detailed overview of the MPEG algorithm.

In the second phase of MPEG, the algorithms have been extended to work with interlaced sequences and to integrate various forms of scalability [ISO/IEC, 1994]. The MPEG-2 standard development is near completion. It is considered for application to interlaced video such as the studio standard of ITU-R Recommendation 601 or HDTV. It extends the bit rate range of MPEG-1 up to about 20 Mb/s in currently studied implementations. MPEG-2 has introduced the concept of profiles and levels. A profile is a collection of signal processing tools (i.e., subset of the global syntax), whereas a level is related to the processing complexity and defines constraints on parameter values.

102.5 Perspectives

The video coding techniques described in the previous sections are well understood and will form the basis for many video services in the coming years. However, they do not approach fundamental limits of compression performance. Standards activity is currently addressing high-compression video coding under the phase 4 of MPEG. Techniques such as analysis-by-synthesis, object-oriented coding, and possibly even fractal coding will be brought to bear on the problem.

Defining Terms

Aliasing: An artifact caused by sampling a continuous signal with too low a sampling rate.
Interlaced scanning: A video scanning structure where the vertical position of horizontal scan lines in a given pass through the image is midway between the position of adjacent lines in the previous pass through the image.
Interpolative coding: A compression technique in which the coded signal is given by the quantized difference between the input signal and an estimation based on previously transmitted information occurring before and after the current input in time.
Masking: Reduction of the visibility of distortion by the image content.
Motion field: A vector field that determines the correspondences between image samples lying on the same trajectory in successive image pictures.
Predictive coding: A compression technique in which the coded signal is given by the quantized error between the input signal and a prediction based on previously transmitted information that occurs earlier in time.
Progressive scanning: A video scanning structure consisting of horizontal scan lines that lie in the same vertical position on each pass through the image.
Rate buffer: A first-in–first-out buffer memory that accepts the coded data from the codec at variable rate and emits it to the channel at fixed rate (or possibly at variable rate, but with smaller variability).
Scalability: The capability to easily derive multiple versions of the input signal at different spatial, temporal, and amplitude resolutions from the same coded data.
Video signal: A time-varying image signal in electronic form.

References

Allnutt, J. 1983. *Transmitted-Picture Assessment*, Wiley, Chichester, U.K.
Aravind, R., Cash, G.L., Duttweiler, D.L., Hang, H.-M., Haskell, B.G., and Puri, A. 1993. Image and video coding standards. *AT&T Tech. J.*, 72(1):67–89.
Dubois, E. 1985. The sampling and reconstruction of time-varying imagery with application in video systems. *Proc. IEEE*, 73(4):502–522.

Dubois, E. 1992. Motion-compensated filtering of time-varying images, *Multidimensional Systems and Signal Processing*, 3:211–239.

Girod, B. 1993. Scalable video for multimedia workstations, *Comp. & Graphics*, 17(3):269–276.

ISO/IEC. 1993. Information technology—Coding of moving pictures and associated audio for digital storage media at up to about 1.5 Mbit/s. ISO/IEC 11172, International Standards Organization.

ISO/IEC. 1994. Information technology—Generic coding of moving pictures and associated audio. JTC1/SC29, ISO/IEC 13818-2 Committee Draft, International Standards Organization.

ITU-T. 1990. Recommendation H.261—Video codec for audiovisual services at $p \times 64$ kbit/s. International Telecommunications Union–Telecommunications Standardization Sector, Geneva, Switzerland.

Jayant, N., Johnston, J., and Safranek, R. 1993. Signal compression based on models of human perception. *Proc. IEEE*, 81(10):1385–1422.

Leduc, J.-P. 1994. Bit-rate control for digital TV and HDTV codecs. *Sig. Proc.: Image Commun.*, 6(1):25–45.

Le Gall, D.J. 1992. The MPEG video compression algorithm, *Sig. Proc.: Image Commun.*, 4(2):129–140.

Mitiche, A. and Bouthemy, P. 1996. Computation and analysis of visual motion: A synopsis of current problems and methods. *Int. J. Comp. Vision*, to be published.

Netravali, A.N. and Haskell, B.G. 1995. *Digital Pictures: Representation, Compression, and Standards*, 2nd ed. Plenum, New York.

Schreiber, W.F. 1993. *Fundamentals of Electronic Imaging Systems: Some Aspects of Image Processing*, 3rd ed. Springer–Verlag, Berlin.

Zdepski, J., Raychaudhuri, D., and Joseph, K. 1991. Statistically based buffer control policies for constant rate transmission of compressed digital video. *IEEE Trans. Commun.*, 39(6):947–957.

Further Information

More details on the various aspects of video compression can be found in Netravali and Haskell [1995] and Schreiber [1993]. Current research is reported in a number of journals, including *IEEE Transactions on Image Processing, IEEE Transactions on Circuits and Systems for Video Technology, and Signal Processing: Image Communication*. Annual conferences of interest include the SPIE International Conference on Visual Communication and Image Processing, the IEEE International Conference on Image Processing, the European Conference on Signal Processing, and many more.

103
The High-Definition Television Grand Alliance System

Eric Petajan
Bell Laboratories

103.1 Introduction 1462
103.2 System Requirements for Terrestrial Broadcast in the U.S. ... 1463
103.3 Grand Alliance System Overview 1463
103.4 HDTV System Descriptions 1464
Scanning Formats • Motion Compensation • Motion Estimation • Predictive Motion Compensation Loop • Decoder Loop • Refreshing • Adaptive Selection and Quantization of Coefficients • Forward Analyzer • Variable Length Coding • Buffer Control • Compressed Video Formats • GA System Layer • GA Transmission System
103.5 Scanning Format Applications 1471
HDTV Interoperability with Computers
Migration to 1080 Progressive with 60 Frames Per Second • Compressed Video Storage and Manipulation
103.6 Conclusions 1473

103.1 Introduction

In 1987, the FCC chartered an advisory committee to recommend an advanced television system for the U.S. Many proposals were submitted and subsequently withdrawn from consideration or rejected by the Advisory Committee on Advanced Television Service (ACATS). From 1990 to 1992, the Advanced Television Test Center (ATTC) tested four all-digital systems, one analog high-definition television (HDTV) system, and one analog National Television Systems Committee (NTSC) enhancement system. The formation of the Grand Alliance (GA) resulted from the withdrawal by NHK of the only analog HDTV system from the competition and a stalemate between the other four all-digital systems. The members of the GA are AT&T, General Instrument Corporation, Massachusetts Institute of Technology (MIT), Philips Electronics North America Corporation, David Sarnoff Research Center, Thomson Consumer Electronics, and Zenith Electronics Corporation.

Digicipher^{TM} [GI, 1991] and channel compatible Digicipher^{TM} (CCDC) [MIT, 1992] HDTV systems were developed by a consortium between the MIT and General Instrument (GI); the digital spectrum compatible (DSC) HDTV system was developed by Zenith and AT&T; and the advanced digital television (AD-HDTV) [ATRC, 1992] HDTV system was developed by a consortium with Philips, Thomson, Sarnoff, NBC, and Compression Labs Inc. (CLI). The video coder in the

AD-HDTV system was based on the Moving Pictures Expert Group-1 (MPEG-1) [ISO, 1991; Le Gall, 1991] standard. The GA system [ATSC, 1995] is based on the MPEG-2 [ISO/IEC, 1994a, 1994b] standard and also contains features from all of the original systems. These features are described with particular emphasis on the video coding system.

103.2 System Requirements for Terrestrial Broadcast in the U.S.

The Federal Communications Commission (FCC) has declared that 6-MHz **terrestrial broadcast channels** will be allocated for HDTV[Service, 1993]. Most of these channels are NTSC **taboo channels** which contain interference from distant NTSC transmitters in the same band or adjacent channel interference from nearby transmitters. The FCC has required that the selected HDTV system provide acceptable viewer coverage without causing excessive interference with frequency-collocated or adjacent NTSC channels.

The selected HDTV system should provide greatly improved picture quality compared to NTSC. The spatial resolution should be at least twice that of NTSC horizontally (H) and vertically (V) without exhibiting **interlace artifacts**, and the HDTV system standard must avoid the **chrominance** artifacts and poor chrominance fidelity associated with NTSC. In addition, the audio must be delivered digitally using multichannel compression. Finally, HDTV will have a 16×9 aspect ratio for compatibility with film (compared to 4×3 for NTSC). These bandwidth and picture quality requirements demand the use of video compression techniques with the highest possible performance.

Receiver complexity should be minimized to ensure that manufacturers can eventually provide consumers with receivers at a cost that will not inhibit the proliferation of HDTV. This constraint implies that video coding and transmission techniques that place most of the complexity in the encoder rather than the decoder are preferred.

103.3 Grand Alliance System Overview

The GA system contains algorithms and components from each of the original digital systems. For example, perceptual modeling, bidirectional motion compensation, hierarchical motion estimation, and adaptive preprocessing, shown in Fig. 103.1, were used in one or more or the first systems. In addition, the Dolby AC-3 [ATSC, 1995] audio compression system was adopted by the GA and was previously used by General Instrument and Zenith/AT&T. The transmission system uses vestigial sideband modulation (VSB) with trellis coding and Reed–Solomon coding for error detection/correction. Testing of the prototype system is complete and the ACATS has issued its recommendation in favor of the Grand Alliance proposal. A U.S. HDTV standard is anticipated sometime in 1996.

One of the biggest challenges before the GA is to provide high picture quality given the bit rate constraints of terrestrial broadcasting in a 6-MHz channel. The presence of significant levels of noise and NTSC interference, and FCC service requirements limits the bit rate to about 20 Mb/s. Two of the key features of the GA system that help meet this challenge are the adaptive use of multiple formats (adaptive preprocessing and format conversion) and the use of forward analysis. The GA supports the transmission and display of $1280\,\text{H} \times 720\,\text{V}$ [SMPTE, 1995a, 1995b] active pixels at 60, 30, and 24 progressive frames/s $1920\,\text{H} \times 1080\,\text{V}$ [SMPTE, 1995a, 1995b] at 30 and 24 progressive frames/s, and $1920\,\text{H} \times 1080\,\text{V}$ at 30 interlaced frames/s. Except for the interlaced format, all high definition formats have square pixels and progressive scan for interoperability with computer display systems and graphics generation systems. In addition, a degree of compatibility with NTSC is retained with support of 59.94, 29.97, and 23.98 frames/s. A minor consequence of MPEG-2 compatibility is that the encoder and decoder must code 1088 lines while the receiver only displays 1080 lines. This avoids subdivision of 16×16 macroblocks with an insignificant impact on picture quality since the extra 8 lines are normally blank.

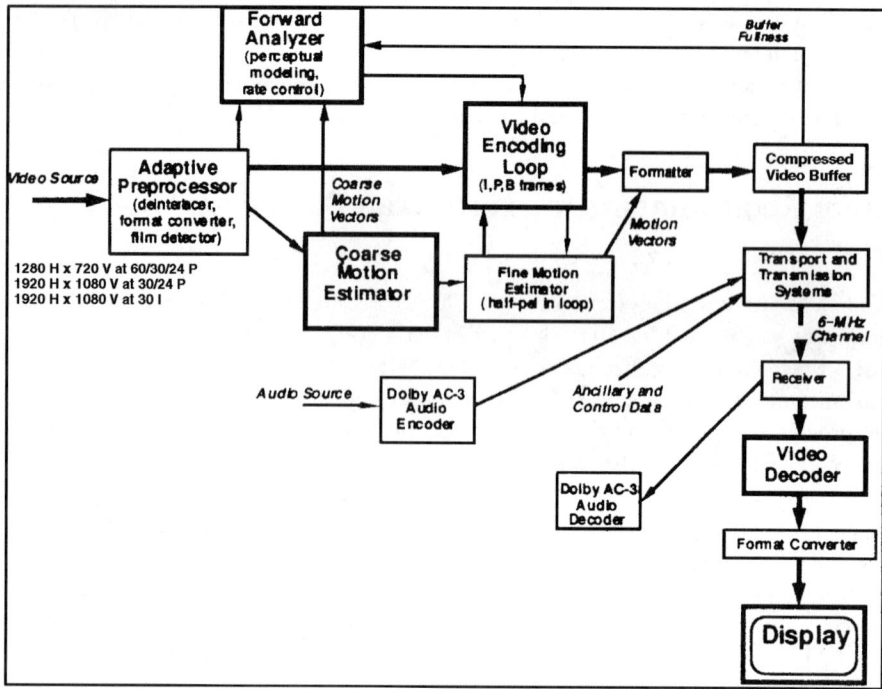

FIGURE 103.1 Grand Alliance system overview.

TABLE 103.1 Standard Resolution Formats Supported by the GA

Image Size	Aspect Ratio	Frame Rate[a] and Int vs Prog, frames/s
1920 H × 1080 V	16:9 square pixels	30 Int; 30 and 24 Prog
1280 H × 720 V	16:9 square pixels	60, 30 and 24 Prog
704 H × 480 V	4:3 or 16:9	30 Int; 60, 30 and 24 Prog
640 H × 480 V	4:3 square pixels	30 Int; 60, 30 and 24 Prog

[a]Note: All frame rates are optionally reduced by 1000/1001 : ~59.94, ~29.97, ~23.976.

More recently, the GA has included a complete set of standard definition formats which are compatible with VGA computer displays and digital video standards. These formats are included in Table 103.1.

One consequence of incorporating multiple format capability is the need for a scanning format transconversion subsystem before encoding and after decoding. The ability to decouple the transmission or coding format from the source format can result in higher decoded picture quality. For example, if interlaced (Int) video is presented to the system, then the user has the option of deinterlacing followed by transconversion to one of the progressive (Prog) formats (typically, 1280 × 720 at 60 frames/s). In addition, the output of the decoder must be converted to drive a given display format.

103.4 HDTV System Descriptions

Scanning Formats

Psychovisual research [Netravali, 1988] also indicates that loss of spatial resolution in moving areas of a picture is less visible than in stationary areas. Interlaced scanning is an attempt to take advantage of this for compression by spatiotemporal subsampling. Interlaced scanning was attractive for early TV systems (NTSC, PAL, and SECAM) because higher static spatial resolution could be achieved for still picture areas, given the bandwidth limitations of early cameras and displays. However, spatial and temporal aliasing will cause annoying artifacts for certain types of vertical detail (e.g., computer generated images, venetian blinds, and text). In this case, the effective vertical resolution must be greatly lowered by filtering to minimize the appearance of aliasing artifacts. In addition, motion compensated compression efficiency using interlaced scenes is much lower than that for the equivalent progressive scene. Two of the previously proposed HDTV systems (DSC and CCDC) use progressive scanning with 720 active lines per frame (1/59.94 s). The other two systems (AD-HDTV and Digicipher^{TM}) use interlaced scanning with 480 lines per field (1/59.94 s) or 960 lines per frame (1/29.97 s).

The choice of video scanning format for an advanced television (ATV) system depends on several competing factors. These factors are not only the subject of technical debate but economic and political debate as well. The GA approach to this issue is to support a reasonable variety of source, transmission, and display formats without adding excessive cost to the decoder. The FCC HDTV standard will primarily specify the transmission formats, whereas the choice of source and display formats will be more subject to forces within the production equipment and display equipment markets respectively.

All of the progressive GA formats have square pixels. One interlaced transmission format is supported mainly because of concerns about the near-term availability of progressive scan HDTV cameras. However, the GA has agreed to work toward the elimination of interlace in the transmission path in the future.

The progressive formats allow a trade between spatial and temporal resolution, and also between resolution and coding artifacts. In the case of 24- or 30-Hz film, the spatial resolution can be 1280×720 or 1920×1080 depending on the scene complexity. If the scene complexity is very high, then a choice of 1280×720 will reduce coding artifacts.

The display format is independent of the transmission formats since the display is not expected to switch between formats instantaneously or at all. A GA compliant decoder will provide frame buffering for all formats, and transconversion will be used to derive the display format. If interlace continues to be used for transmission (as opposed to source video storage or capture), then a deinterlacer must be provided in every receiver with a progressive display.

Motion Compensation

Temporal redundancy is removed by motion estimation and motion compensation [Netravali, 1988]. Motion vectors are computed between frames in the encoder and transmitted to the decoder. The GA system does not perform motion compensation on all frames. The I-frames are coded in isolation without any temporal prediction. The P-frames are coded using forward motion compensation starting with the last I-frame. The B-frames are coded using the surrounding I- and/or P-frames to perform forward and backward motion compensation. In addition, a given block in a P- or B-frame may be coded with or without motion compensation. One sequence of I-, B-, and P-frames is known in MPEG-1 and 2 as a group of pictures (GOP).

Motion Estimation

The motion estimator operates only on the luminance images and computes motion vectors using block matching. A motion vector is the spatial displacement between a block of pixels in the current frame and the most similar block in the previous frame (except for B-frames). The similarity or distance measure is the mean of the absolute value of the differences between corresponding pixels in the two blocks. Other distance measures (e.g., mean square error) may provide improved performance but are much more costly to implement than the mean absolute error.

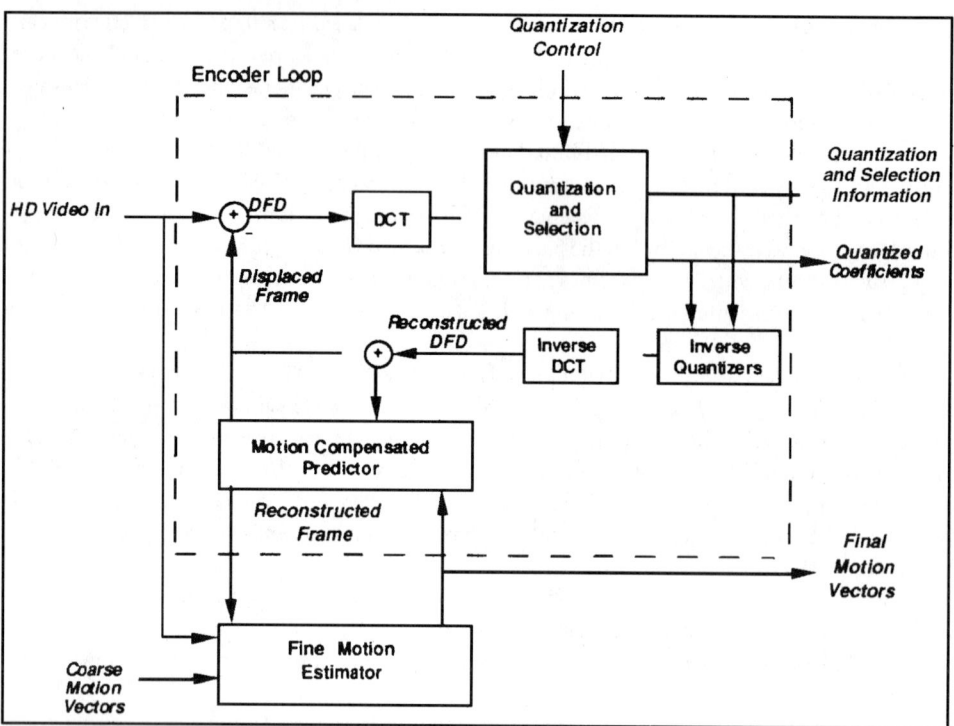

FIGURE 103.2 Video encoder loop.

The number of block matches (and therefore the motion estimator complexity) is proportional to the search area. The GA system uses hierarchical motion estimation to reduce implementation complexity. As shown in Fig. 103.2, the coarse motion vectors, which were computed on the original frames, are used as a reference for the fine and final motion vectors. The reconstructed frame is searched for a best match in the fine motion estimator but for greater hardware economy the search area is much smaller than the coarse stage. In the GA system a total search range of ± 64 H and ± 32 V is provided by the two stages for B-frames and ± 128 H for P-frames. This exceeds the motion estimator range of the predecessor systems.

Predictive Motion Compensation Loop

In the motion compensated prediction loop shown in Figs. 103.1 and 103.2, the motion vectors are applied to the reconstructed frame buffer, which contains a prediction of the decoded picture. The motion compensated predicted frame or displaced frame is subtracted from the original frame to produce the displaced frame difference (DFD). The DFD will be near zero intensity for areas with predictable motion, and nonzero where detail was uncovered or motion was not just a translation in the image plane.

The DFD is the signal which results from the removal of temporal redundancy. An 8×8 **discrete cosine transform (DCT)** is applied to the DFD in preparation for removal of spatial redundancy by coefficient quantization. A reconstructed form of the DFD is then produced by inverse quantization and inverse DCT. The reconstructed or predicted DFD is added to the motion compensated frame to produce the next reconstructed frame, which completes the coding cycle. **Coding artifacts** contained in the reconstructed frame from the current and previous loop cycles can be then be corrected in the DFD coding process.

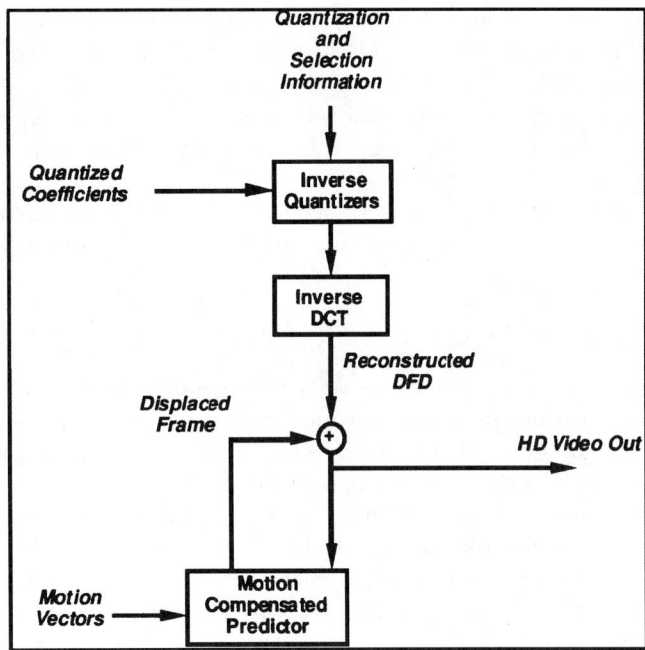

FIGURE 103.3 Video decoder loop.

Decoder Loop

The decoder loop is shown in Fig. 103.3. After deformatting the coded video bit stream, inverse quantization of DCT coefficients and inverse DCT, the reconstructed DFD enters the decoding loop. The decoded motion vectors are applied to the reconstructed frame, resulting in the reconstructed displaced frame. The reconstructed DFD is then added to the reconstructed displaced frame to produce the reconstructed frame.

A GA-compliant decoder must be able to decode bit streams which switch between GA scanning formats. The resulting video must be displayed without loss of picture at the transitions.

Refreshing

The video encoding process exploits temporal redundancy by temporal or inter-frame prediction. This prediction takes the form of motion vectors and an error signal resulting from applying the motion vectors. However, a pure motion compensated prediction loop is not practical without some form of prediction imperfection (i.e., forced *intraframe* coding). If a sequence of input images were perfectly predictable, the decoder would not receive any error signal and therefore could not reconstruct the picture after initialization. Decoder initialization is required after the channel is changed, or the signal is lost and then recovered.

Consider the coding of a still image. The motion vectors and the perfect DFD for a coded still will be zero. The reconstructed DFD in the decoder will also be zero. If the decoder were started with a sequence of zero DFD frames, then a blank or zero reconstructed frame would result. If the viewer changed to a channel transmitting a coded still, the still would therefore never appear. One way to prevent this occurrence is for a portion of the original picture to be mixed with the DFD in the encoder to allow the decoder to synchronize to the encoder after decoder initialization. This also improves recovery from channel impairments. The speed of recovery from channel changes and channel impairments is directly proportional to the amount of original in the mix. Both intraframe coded blocks (I-blocks) and frames (I-frames) are used to provide a mixture of predicted and original video in the transmitted bit stream. The use of I-blocks in a regular pattern that covers the entire image over multiple frames is known as progressive refreshing.

The space between I-frames and the number of consecutive B-frames is variable in the MPEG and the GA systems. Intraframe coding can be done using either I-frames or periodic I-blocks. Since intraframe coding produces a much higher bit rate than interframe coding, I-block refreshing provides easier control of the buffer fullness. However, if one desires to insert a commercial or other compressed video directly into a compressed program, I-frames provide clean insertion points.

Adaptive Selection and Quantization of Coefficients

Sufficient compression cannot be achieved unless a large fraction of the DCT coefficients are dropped and therefore not selected for **quantization** and transmission. The coefficients that are not selected are assumed to have zero value in the decoder. Several different schemes are used in the digital HDTV systems to code the coefficient selection information.

The GA system encodes the selections and runs of zeroes following a zigzag pattern through the array of frequency-ordered coefficients. The DC coefficients are coded differentially to take advantage of high-spatial correlation. The GA system uses one uniform quantizer for coefficient quantization. This method of coding the coefficient selection information and DC coefficients is used in MPEG-1 and 2.

Forward Analyzer

The forward analyzer manages the compressed video buffer and the distribution of quantization error in each coded picture. This is accomplished by monitoring the fullness of the buffer and controlling the compressed video data rate by adapting the amount of quantization error in a visibly uniform manner. The ability of the forward analyzer to anticipate changes in scene complexity is enhanced by the use of the coarse motion vectors which are computed before the picture reaches the encoding loop.

The coefficient quantization process reduces the number of levels used to represent a particular coefficient. This process is controlled by the forward analyzer, which inputs the fullness of the compressed video buffer, coarse motion vectors, and the original pictures, as shown in Fig. 103.1. The coarse motion vectors are used to form an ideal DFD, which provides a prediction of scene complexity ahead of the encoder loop. This prediction reduces the chance of buffer overflow or visible distortion by sudden increases in coarse quantization.

Variable Length Coding

The data representing information such as motion vectors, quantizer selection patterns, and transform coefficients are seldom statistically uniform. Usually, the data are statistically clustered, and the probability distributions can be estimated from analysis of real scenes. The use of variable length codes (VLCs) [Huffman, 1952] takes advantage of this statistical nonuniformity by assigning short codewords to the most frequent values, and assigning longer words to less frequent values. The GA system also provides separate VLC tables for inter- and intracoded coefficients and intra-DC coefficients as specified in the MPEG-2 standard.

The VLC decoder detects a unique start code, which indicates the beginning of a series of VLCs. The 8×8 blocks of pixels are grouped into 16×16 macroblocks. Horizontal runs of macroblocks are grouped into slices where each slice starts with a start code. Short slices reduce error propagation but generate more coding overhead. The GA constrains slices to start at the beginning of a macroblock row.

Buffer Control

Motion compensation, adaptive quantization, and variable length coding produce highly variable amounts of compressed video data as a function of time. For example, the compressed bit rate after a scene change can be several times greater than the bit rate in the channel. Therefore, compressed video data buffering in the encoder and decoder is required for efficient channel utilization.

Buffer size is constrained by the maximum tolerable delay through the system and by cost. The fullness of the buffer is controlled by adjusting the amount of distortion or quantization error in each image. In the encoder, the buffer will fill more quickly if the distortion is low. A feedback control system is required to regulate the distortion level, which controls the buffer fullness to prevent overflow. The design of this control system is complicated by the following:

- Delay between a change in the distortion level and the subsequent change in buffer fullness
- Perceptual constraints on the instantaneous and average distortion levels
- Difficulties in modeling the bit rate as a function of distortion level

The visibility of distortion due to an increase in scene complexity is minimized by smoothly increasing the distortion level. This is facilitated by accurately modeling the rate vs distortion level, because an accurate model allows the desired buffer fullness to be achieved for each frame. MPEG-2 specifies a standard rate control model which is parameterized. The parameter bit rate value must not exceed 48,500 for terrestrial broadcast and 97,000 for cable service. The maximum allowable bit rate for a GA decoder must be less than 80 Mb/s. The channel buffer size equals 8 Mb and the buffer size value (vbv) must not exceed 488.

Compressed Video Formats

The compressed video data must be organized into a format that can be reliably transmitted through an imperfect channel and can allow the decoder to recover quickly from the loss of data in the channel. In addition, storage and manipulation of the compressed video and transmission over alternate media (e.g., packet networks) motivates the inclusion of block-, frame-, and program-type descriptors in the format.

Robust transmission is primarily achieved by dividing the compressed video bit stream into fixed length blocks and applying error correction processing to each block. Error correction overhead or parity bits are added to each block before transmission. The receiver is then able to determine if a given block was corrupted in transmission. If a block was corrupted, error concealment is performed in the decoder to minimize the visual impact of the error.

The amount of picture that is degraded by the loss of a block of compressed video data depends on the format of the compressed video bit stream. Since most of the data consists of concatenated variable length codewords, VLC restart pointers must occur at regular intervals in the bit stream. More frequent placement of these pointers will reduce the amount of picture degradation but will also consume more of the bit rate.

GA System Layer

The GA transport is compatible with the MPEG-2 systems layer [ISO, 1994a, 1994b]. As shown in Fig. 103.4, the GA system uses a 188-byte transport packet, which is transmitted to the decoder after trellis coding and Reed–Solomon coding using 8-level vestigial sideband modulation (VSB) [Feher, 1987]. The link header provides simple packet synchronization and identification. A bit also indicates the presence of an adaptation header, which provides clock synchronization, and is also used for video, audio, and data presentation. The choice of a 188-byte packet length is a compromise between the desire to minimize header overhead with longer packets and the need for frequent headers for quick error recovery. Conveniently, each GA transport packet fits into 4 asynchronous transfer mode (ATM) cells with or without the link header.

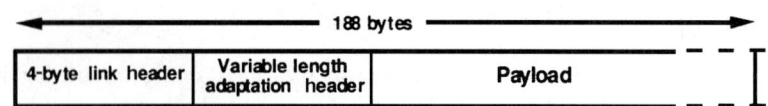

FIGURE 103.4 Grand Alliance transport packet.

Each transport packet header contains a packet identifier (PID) which indicates that audio, video, data, or program map table (PMT) is contained in the packet. The PMT contains the definition of the programs and must be updated at least every 400 ms. The program numbers are associated with the corresponding PMT PIDs via the program association table (PAT). The Advanced Television System Committee (ATSC) has specified additional constraints not detailed here.

GA Transmission System

As required by the FCC, a key feature of the GA system is its ability to reliably deliver ~20 Mb/s through the terrestrial TV broadcasting environment using one 6-MHz channel. The FCC recommends that each TV broadcaster be provided with an HDTV channel with equivalent service area, and that the current TV service is not significantly impacted by the HDTV service. These requirements are met by a combination of eight-level digital amplitude modulation (3 b/sample), Reed–Solomon forward error correction (R-S FEC) with 20 R-S parity bytes, a pilot tone, 1/16 field interleaving for burst immunity, 2/3 rate trellis coding, an NTSC rejection filter, and adaptive channel equalization. The suppressed carrier vestigial sideband amplitude modulation of the 8-level samples or symbols (8 VSB) is optimal for the noise- and interference-laden terrestrial broadcast environment, and the VSB signal appears as low-level random noise when interfering with NTSC. The 8 VSB is necessarily coupled with trellis coding for robust transmission. The 8 VSB alone provides 32.3 Mb/s (10.8 megasymbols/s) and the addition of trellis coding drops the payload to 21.5 Mb/s. RS-FEC further reduces the payload to 19.3 Mb/s. Figure 103.5 shows a VSB data frame. The GA also provides an optimal solution for cable TV with 16 VSB. Since trellis coding and its associated bit rate overhead are not needed for cable TV, the GA system can deliver 43 Mb/s over one 6-MHz cable channel. At the present time, the GA system does not support the type of switching between modulation modes (4, 8, or 16 VSB) required to avoid the cliff effect. The ACATS did not strongly support this feature after the first round of tests.

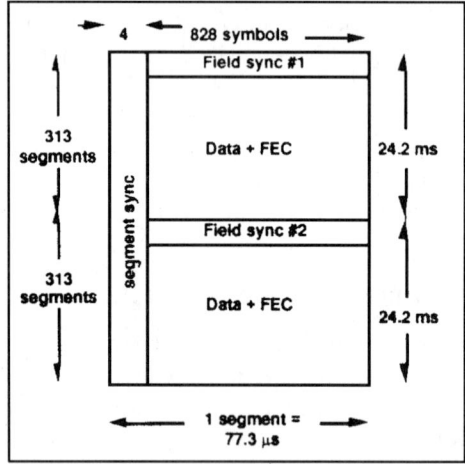

FIGURE 103.5 VSB data frame.

The small pilot tone and NTSC rejection filter specifically target interference from adjacent and frequency colocated NTSC channels. The pilot tone slightly increases the ratio of peak power to average power of the broadcast signal, which slightly increases visibility of the HDTV signal in some NTSC receptions. Its position in the band avoids NTSC carrier peaks primarily from sync signals. NTSC rejection filtering is automatically switched in when needed and slightly reduces the noise immunity of the HDTV reception. Additional synchronization symbols are embedded periodically in the modulated signal, which are not interleaved, RS, or trellis coded.

The performance of the GA-transmission system was tested by ACATS in terms of relative service area compared to the companion NTSC transmitter. The results of laboratory tests using terrestrial RF transmission simulators were input to a computer program which optimized the power, frequency assignment, and location of each GA transmitter. The program output both service areas and viewer populations. The number of stations with less HDTV service area than NTSC is estimated to be about 107 out of 1657, and 187 of the HDTV stations would have at least 20% more service area than NTSC.

103.5 Scanning Format Applications

The GA supports a range of scanning formats for coding and transmission. A given encoder input format does not necessarily indicate the use of that scanning format for encoding and transmission. Table 103.2 shows the advantages of each GA transmission format and indicates how a given encoder input format affects the choice of transmission format. Table 103.3 shows the relationship between transmission formats and display formats.

TABLE 103.2 Applications of Encoder Input Format/Transmission Format Combinations

Encoder Input Formats	Transmission Formats			
	720 × 1280P at 60 frames/s	720 × 1280P at 24/30 frames/s	1080 × 1920I 30 frames/s	1080 × 1920P at 24/30 frames/s
720 × 1280P at 60 frames/s	Sports, concerts animation, upconverted 480I, commercials, graphics	Adaptively reduce frame rate for lower bit rate or better picture when motion is slow	Not desirable	Adaptively increase spatial resolution and reduce frame rate for high detail/slow motion scenes
720 × 1280P at 24/30 frames/s	Adaptively increase frame rate by frame interpolation	Complex film scenes, complex graphics, complex animation	Not desirable	Adaptively increase spatial resolution
1080 × 1920I 30 frames/s	Deinterlaced scenes shot with current HDTV cameras	Deinterlaced complex scenes with reduced frame rate for lower bit rate or better picture when motion is slow	Interlaced camera generated scenes/no deinterlacer available	Deinterlaced scenes with reduced frame rate for lower bit rate or better picture when motion is slow
1080 × 1920P at 24/30 frames/s	Adaptively increase frame rate by frame interpolation while reducing spatial resolution for fast motion scenes	Reduced spatial resolution for lower bit rate or better picture	Not desirable	Film

TABLE 103.3 Applications of Transmission Format/Display Format Combinations

Transmission Formats	Display Formats		
	720 × 1280P at 60 frames/s	1080 × 1920I at 30 frames/s	1080 × 1920P at 60 frames/s
720 × 1280P at 60 frames/s	No conversion required	Have 1080 × 1920I display	Have 1080 × 1920P display
720 × 1280P at 24/30 frames/s	Use 3 : 2 or 2 : 2 pulldown	Have 1080 × 1920I display Need 3 : 2 or 2 : 2 pulldown	Have 1080 × 1920P display Need 3 : 2 or 2 : 2 pulldown
1080 × 1920I at 30 frames/s	Deinterlace in decoder to drive 720 × 1280P display	No conversion required	Deinterlace in decoder to drive 1080 × 1920P display
1080 × 1920P at 24/30 frames/s	Reduce spatial resolution and use 3 : 2 or 2 : 2 pulldown to drive 720 × 1280P display	Use 3 : 2 or 2 : 2 pulldown	Use 3 : 2 or 2 : 2 pulldown

HDTV Interoperability with Computers

Personal computers (PCs) today are commonly equipped with high-resolution progressive scan color monitors. Since the most costly part of an HDTV receiver is the display, the early consumers of HDTV may well be multimedia PC owners with HD video decoder cards. For example, the 1280 H × 720 V GA format fits neatly into the widely used 1280 H × 1024 V display with spare screen space for additional windows. High-resolution video on a PC is especially suited for interactive entertainment, which typically requires readable text, and the variety of input devices found on todays PCs. Of course, HDTV on a PC is also ideally suited for education, training, industrial imaging, video and film production, especially when postproduction processing is used. When widescreen HDTV displays become consumer products, those with progressive scanning will also be usable as computer displays.

The compressed HDTV bit stream could be received from terrestrial or satellite broadcast, via coaxial cable (cable TV), or high speed computer network. The bit rate for 24-frame/s film is easily limited to 10 Mb/s (for the 1280 H × 720 V format) for movie services over a local area network. Local storage of compressed HDTV is out of range of current computer disc-read-only memory (CD-ROM) drives but reliable storage and playback using consumer grade video tape technology has been demonstrated.

Migration to 1080 Progressive with 60 Frames Per Second

Enthusiasm has emerged to achieve compression of 1920 × 1080 at 60 frames/s within the 6-MHz terrestrial broadcast channel. This may not be possible for all scenes but may be achievable some of the time. In fact, the deinterlacing process produces an intermediate format of 1920 × 1080 at 60 frames/s which could be used as an input to an augmentation scheme, as shown in Fig. 103.6. A 30 frame/s subset would be coded in the standard way while the remaining frames would be predicted by an interpolator. The difference between the predicted interpolation and the actual frame would be coded and sent to the decoder. Appropriately computed motion vectors could be used for both motion compensation and interpolation. A similar scheme would start with 24- or 30-frame/s film and generate interpolated frames. This scheme would have the advantage of

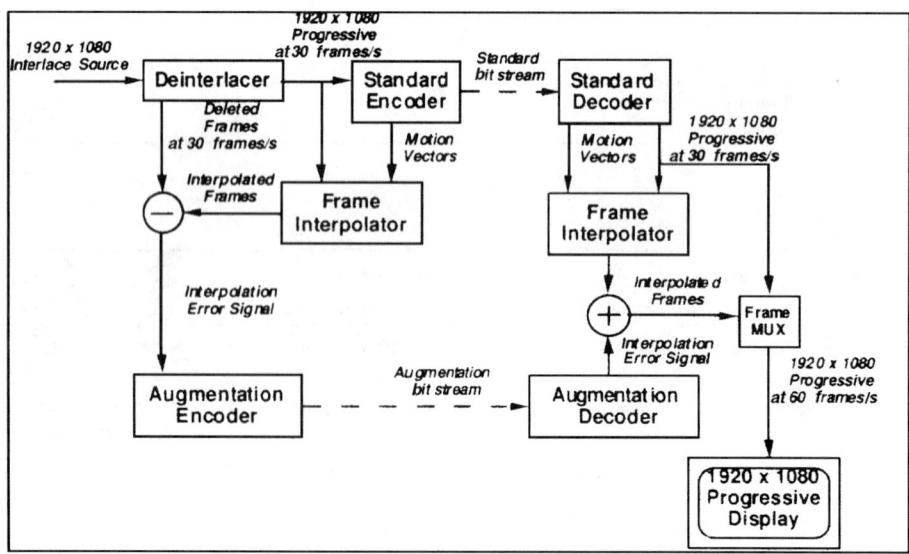

FIGURE 103.6 Possible migration scenario from interlaced source to 1080 progressive display.

providing improved temporal performance for existing film material when properly applied. A variety of migration schemes are currently under study for selection by the GA.

Compressed Video Storage and Manipulation

Techniques for efficient storage and manipulation of the compressed video (and audio) data will be necessary to provide consumers with video cassette recorder (VCR) capabilities such as speed search and freeze frame. Since all of the decoders considered here contain reconstructed frame storage, freeze frame is simply provided. Speed search can also be provided by all of these systems with varying degrees of efficiency and picture quality. This function is provided by subsampling the blocks of compressed video data from tapes that are moving at higher than normal speed across the head. The block descriptors are then used to partially reconstruct the video and the picture quality is directly proportional to the number of intraframe coded blocks in the compressed video bit stream. The ability to decode randomly selected portions of compressed video data also facilitates multimedia computing applications, where multiple received and stored video data streams could be manipulated quickly within a window system.

103.6 Conclusions

The GA provides a video compression solution for HDTV which is flexible enough to meet the needs of a wide variety of industries. A rational set of formats with simple conversions between them is provided to accommodate broadcasting, the computer/multimedia industry, computer graphics, industrial imaging, and the national information infrastructure. For example, 1280 × 720 is a convenient subset of a 1280 × 1024 graphics monitor, and the lower frame rate of film allows the compression of 1920 × 1080 for entertainment applications. The use of MPEG-2 syntax provides interoperability with other standards, provides flexible data services at the transport level, and facilitates industry compliance, acceptance, and cost effectiveness for the consumer. Since every GA decoder will accept the variety of GA formats, the future migration of the system to higher quality with evolving encoding methods is ensured.

Defining Terms

Chrominance: Color difference signals formed by weighted sums and differences of red, green, and blue channels.
Coding artifacts: Mostly block boundaries that come and go in detailed or noisy areas of the coded picture.
Discrete cosine transform (DCT): Converts 8 × 8 block of pixels into 8 × 8 block of frequency coefficients, which indicate energy in combined horizontal and vertical frequency spectrum.
Interlace artifacts: Primarily 30-Hz flicker from display of high vertical detail on an interlaced monitor.
Quantization: Mapping a set of intensity values to a smaller set of intensity values resulting in some loss of intensity information and possible coding artifacts.
Taboo channels: The 6-MHz channels, which are unused because of excessive interference from other NTSC channels.
Terrestrial broadcast: Television transmission from a ground station to homes.

References

ATRC. 1992. *Advanced Digital Television System Description,* Jan., Advanced Television Research Consortium.

ATSC. 1995. *ATSC Digital Television Standard*, Document A/53, Sept. 16., Advanced Television Systems Committee.

Feher, K. 1987. *Advanced Digital Communications*, Prentice–Hall, Englewood Cliffs, NJ.

GI. 1991. *DigiCipher^{TM} HDTV System Description*, Aug., General Instrument Corporation.

Huffman, D.A. 1952. A method for the construction of minimum redundancy codes. *Proc. IRE* 40, 1089.

ISO. 1991. ISO CD 11172-2: Coding of moving pictures and associated audio for digital storage media at up to about 1.5 Mbit/s, Nov., International Standards Organization.

ISO/IEC. 1994a. MPEG-2 systems. ISO/IEC IS 13818-1, International Standards. Organization/ International Electrochemical Commission.

ISO/IEC. 1994b. MPEG-2 video. ISO/IEC IS 13818-2, International Standards. Organization/International Electrochemical Commission.

LeGall, D. 1991. MPEG: A video compression standard for multimedia applications. *Trans. ACM*.

MIT. 1992. *Channel Compatible Digicipher HDTV System.* American Television Alliance, April, Massachusetts Institute of Technology, Cambridge, MA.

Netravali, H. 1988. *Digital Pictures: Representation and Compression.* Plenum Press, New York.

Service, F.A.C.o.A.T. 1993. Federal Communications Commission advanced television system recommendation. *IEEE Trans. Broadcasting*, 39(1):3–245.

SMPTE. 1995a. Proposed Standard for television, 1280 × 720 scanning and interface. SMPTE S17.392. Society of Motion Picture and Television Engineers.

SMPTE. 1995b. Standard for television, 1920 × 1080 scanning and interface. SMPTE 274M. Society of Motion Picture and Television Engineers.

Further Information

An excellent review of picture processing is found in Netravali [1988]. The ATSC Digital Television Standard (see references) is a well written and concise description of the proposed FCC HDTV standard. For a description of the ATTC test results see the Final Technical Report, FCC ACATS, October 31, 1995. A guide for using the Grand Alliance system can be found in the Guide to the use of the ATSC Digital Television Standard, Doc. A/54, October 4, 1995. A description of the Dolby AC-3 audio compression system is found in Digital Audio Compression Standard (AC-3), Doc. A/52 May 24, 1995.

104
Audio Coding

104.1	Introduction	1475
104.2	Auditory Masking	1477
104.3	Noise Shaping and Perception-Based Coding	1478
104.4	Mean Opinion Score	1479
104.5	Low Bit Rate Coding	1479
104.6	Subband Coding: ITU-R G.722 Coder	1480
104.7	Transform Coding	1481
104.8	Subband/Transform Coding	1482
104.9	ISO/MPEG Audio Coding	1482
104.10	Multichannel Stereophony	1484
104.11	Conclusion	1485

Peter Noll
Technische Universität Berlin

104.1 Introduction

High-quality digital compression of telephone speech, wideband speech, and wideband audio signals is of increasing interest in the fields of communication-based and storage-based applications. Higher bandwidths than the 300–3400 Hz telephone bandwidth result in major subjective improvements in represented audio quality. A high bandwidth improves the intelligibility and naturalness of speech, adds also a feeling of transparent communication, and eases speaker recognition. In the case of *wideband speech signals* with a typical 50–7000 Hz bandwidth, applications of high relevance are loudspeaker telephony, **Integrated Services Digital Network (ISDN)**, and video conferencing systems, and the use of commentary channels for broadcasting. Even higher bandwidths are necessary for high-quality audio. On a compact disc (CD), today's quasistandard of digital audio, or on digital audio tape (DAT), signals with a 10–20,000 Hz bandwidth and 44.1-kHz sampling rate are stored with a resolution of 16-b per sample.[1] The resulting net bit rate is $44.1 \times 16 = 705.6$ kb/s per mono channel. The CD, which had tremendous impact on the audio market, needs a significant overhead for a runlength-limited line code[2] that maps 8 information bits into 14 b, and for synchronization and error correction, resulting in a 49-b representation of each 16-b audio sample. Hence the total stereo bit rate is $1.41 \times 49/16 = 4.32$ Mb/s. Table 104.1 compares bit rates of the compact disc and the digital audio tape.

The compact disc with its 16-b pulse-code modulation (PCM) format has made digital audio popular and serves as an accepted *audio representation* standard. In audio production higher

---
[1]Note that DAT systems support also sampling rates of 32 and 48 kHz.
[2]The CD laser system cannot read less than 3 pits in a row and more than 11 consecutive zeros or ones. The line code, which is called eight-to-fourteen modulation (EFM), provides the appropriate runlength constraints.

TABLE 104.1 Bit Rates of CD and DAT (Stereophonic Signal at 44.1-kHz Sampling Frequency; Uniform 16-b PCM)

Storage Device	Audio Rate, Mb/s	Overhead, Mb/s	Total Bit Rate, Mb/s
CD	1.41	2.91	4.32
DAT	1.41	1.67	3.08

resolutions up to 24 b per sample are in use. Lower bit rates than those given by the 16-b PCM format are mandatory if audio signals are to be transmitted over channels of limited capacity or are to be stored on storage media of limited capacity. *Basic requirements* in the design of low bit rate audio coders are: (1) high quality of the reconstructed signals, (2) robustness to random and bursty channel bit errors and to packet (cell) losses, (3) graceful degradation of quality with increasing bit error rates in mobile radio and broadcast applications, and (4) low complexity and power consumption.

Typical application areas for *digital audio* with bandwidths of not less than 20 kHz are in the fields of audio production, distribution and program exchange, digital sound broadcasting (DSB), digital storage (archives, studios, consumer electronics), digital audio on personal computers (PCs), videoconferencing, multipoint interactive audiovisual communications, and in the field of enhanced quality TV systems.

In the case of *digital storage* of compressed audio a number of consumer products are already being offered. The coded bit streams are stored on magnetic tapes or discs or in erasable, programmable read-only memory (EPROM) memory cards. One example is Philips' Digital Compact Cassette (DCC), which stores 44.1-kHz sampled stereophonic data at a rate of 384 kb/s and uses a data compression algorithm called precision audio subband coding (PASC). PASC is essentially the same as layer I of the International Standards Organization/**Moving Pictures Expert Group (ISO/MPEG)** coding algorithm MPEG 1 except that it uses a simpler spectral analysis (see section on ISO/MPEG Audio Coding). Layer I or layer II of the MPEG 1 audio coding algorithm are used in the full-motion video option of interactive compact discs (CD-i). Such discs can store 70 min of full-motion video and audio, whereby the video part is coded using the MPEG 1 video coding algorithm.[3] A second example of digitally stored compressed audio is Sony's magneto-optical MiniDisc (MD) which makes use of the *adaptive transform acoustic coding* (ATRAC) compression algorithm (see section on Subband/Transform Coding). A third example is playback systems (e.g., for announcements) with the compressed data stored on flash EPROM memory cards. Future applications will also include interactive multimedia and interchange media services based on memory cards.

Delivery of digital audio signals must be possible over terrestrial and satellite-based digital broadcast and transmission systems including cellular mobile radio and cable-TV networks. Of great interest is a high-quality transmission of audio bit streams through narrowband ISDN networks operating at rates of 64 or 128 kb/s. ISDN and its backbone, broadband **asynchronous transfer mode (ATM)** networks, offer useful channels for a practical distribution of stereophonic and multichannel audio signals.

In the case of *broadcasting audio signals* major problems have to be solved to provide high quality for listeners using mobile and portable receivers since multipath interference and selective fading can otherwise cause main impairments. Broadcast network chains will include sections with different quality requirements ranging from production quality where editing, cutting, postprocessing, etc., has to be taken into account, down to emission quality. The **International Telecommunications Union-Radiocommunications (ITU-R)** has recommended that the ISO/MPEG layer II audio coding algorithm should be used for contribution, primary distribution to transmitters, and emission of digital audio signals (with a stereo bit rate of 256 kb/s for emission). In addition, the ISO/MPEG

---

[3] An ISO/MPEG bit stream supports both audio-only signals and multiplexed audio–visual signals.

Audio Coding

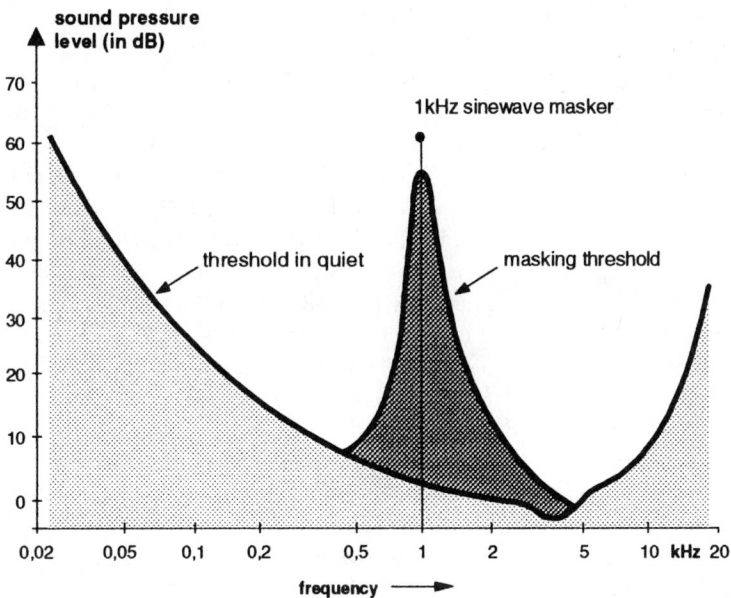

FIGURE 104.1 Threshold in quiet (lower curve) and masking threshold (here, pure tone masker). (*Source:* Noll, P. 1993. Wideband speech and audio coding. *IEEE Commun. Mag.*, 31(11):34–45. © 1993 IEEE. With permission.)

layer III coder is recommended for commentary links at a stereo bit rate of 128 kb/s. (A small fraction of the mentioned bit rates is reserved for data transmission.)

104.2 Auditory Masking

Digital representations of analog waveforms introduce some kind of distortions. A basic problem in the design of source coders is to achieve a given acceptable level of distortion with the smallest possible encoding bit rate. To reach this goal the encoding algorithm must be adapted both to the changing statistics of the source and to auditory perception. Auditory perception is based on critical band analyses in the inner ear where frequency-to-place transformations occur along the basilar membrane. The power spectra are not represented on a linear frequency scale but on frequency bands, called *critical bands*, with bandwidths on the order of 100 Hz below 500 Hz and with increasing bandwidths (up to 5000 Hz) at high signal frequencies [Zwicker, 1982]. Within critical bands the intensities of individual tones are summed by the ear. Up to 20,000-Hz bandwidth 26 critical bands have to be taken into account.[4] Audio coders that exploit auditory perception must be based on critical-band structured signal processing.

Auditory masking plays a major role in the design of bit rate efficient coders. It describes the effect that a low-level audio signal (the *maskee*) can become inaudible when a louder signal (the masker) occurs simultaneously. The effect of simultaneous masking and temporal masking (to be described) can be exploited in audio coding by transmitting only those details of the signal which are perceptible by ear. Such coders provide high coding quality without providing high signal-to-noise ratios. Without a masker, a signal is inaudible if its sound pressure level (SPL)[5] is below the absolute hearing threshold, called *threshold in quiet*, which depends on frequency and covers a dynamic range of more than 60 dB, as shown in the lower curve of Fig. 104.1. In the presence of a

---

[4]Telephone bandwidth (up to 3400 Hz) can be described by 16 critical bands.
[5]SPL = 0 dB relates to a sound pressure of 0.02 mN/m^2.

masker a *masking threshold* can be measured below which any signal will not be audible. This masking threshold, in the context of source coding also known as *threshold of just noticeable distortion*, depends on the SPL, the frequency of the masker, and on the characteristics of masker and maskee. If many maskers are present simultaneously in different frequency bands a *global masking threshold* can be derived from the individual masking thresholds.

Consider the example of the masking threshold for the SPL = 60-dB pure tone masker in Fig. 104.1: around 1 kHz the SPL of any maskee can be surprisingly high, it will be masked as long as its SPL is within the dark-hatched area. The distance between the level of the masker and the masking threshold is called *signal-to-mask ratio (SMR)*. SMR values of around 6 dB are needed to mask a tone by noise, much higher differences of 15 dB and more are needed to have noise being masked by a tone. Coding noise and low-level signals will not be audible as long as the SNR is higher than the SMR. The difference SMR − SNR is the *noise-to-mask ratio (NMR)*, it describes the safety margin for nonaudible noise and should be of negative sign.

We note in passing that, in addition to simultaneous masking of one sound by another one occurring at the same time, *temporal masking* plays also an important role in human auditory perception. It occurs when two sounds appear within a small interval of time; the stronger one masks the weaker one, regardless of whether the latter one occurs before or after it. Temporal masking can be used to mask pre-echoes caused by the spreading of a sudden large quantization error over the actual coding block [Noll, 1993].

104.3 Noise Shaping and Perception-Based Coding

The masking phenomenon depends on the spectral distribution of masker and maskee and on their variation with time. The masked signal may either consist of quantization (coding) noise and/or of low-level signal components. As long as these contributions are below the masking threshold they are irrelevant to the ear and, hence, inaudible.

The dependence of human auditory perception *on frequency* and the accompanying perceptual tolerance to errors can (and should) directly influence encoder designs; *dynamic noise-shaping* techniques can shift coding noise to frequency bands where it is not of perceptual importance. The noise shifting must be related to the actual short-term input spectrum in accordance with the signal-to-mask ratio and can be done in different ways. However, frequency weightings based on linear filtering cannot make full use of results from psychoacoustics, in *audio coding* noise-shaping parameters are, therefore, dynamically controlled in a more efficient way to exploit simultaneous masking and temporal masking in more detail. Note that in all examples the noise shaping is located in the encoder only, and so it does not contribute to the bit rate.

As one example of dynamic noise-shaping, *quantization noise feedback* can be used in predictive schemes. However, frequency domain coders with *dynamic allocations of bits* (and hence of quantization noise contributions) to subbands or transform coefficients offer an easier and more accurate way to control the quantization noise [Jayant and Noll, 1984].

Figure 104.2 depicts the structure of a perception-based coder that exploits auditory masking; buffering is not included. The encoding process is controlled by the SMR vs frequency curve from which the needed resolution (and hence the bit rate) in each critical band is derived. The SMR is determined from a high-resolution fast Fourier transform-(FFT-) based spectral analysis of the audio block to be coded. Principally, any coding scheme can be used that can be dynamically controlled by such perceptual information. If the necessary bit rate for a complete masking of distortions is available the coding scheme will be *perceptually transparent*, that is, the decoded signal is then *subjectively indistinguishable* from the source signal, for example, the compact disc reference [Jayant, 1992]. In practical designs we cannot go to the limits of masking or just noticeable distortion, since postprocessing of the acoustic signal (e.g., filtering in equalizers) by the end user and multiple encoding/decoding processes have to be considered. In addition, our current knowledge about auditory masking is very limited. Generalizations of masking results, derived for simple and

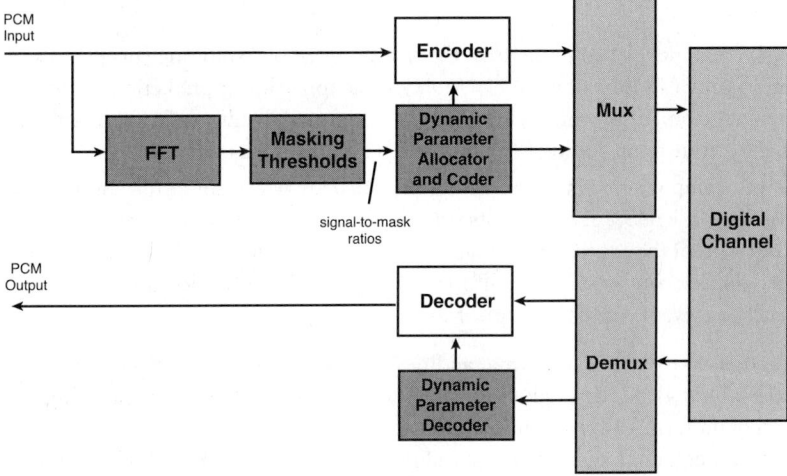

FIGURE 104.2 Block diagram of perception-based coders. (*Source:* Noll, P. 1993. Wideband speech and audio coding. *IEEE Commun. Mag.*, 31(11):34–45. © 1993 IEEE. With permission.)

stationary maskers and for limited bandwidths, may be appropriate for most source signals, but may fail for others. Therefore, as an additional requirement, we need a sufficient safety margin in practical designs of such perception-based coders.

104.4 Mean Opinion Score

Perceptual coding techniques ask for a distortion measure that specifies *perceptual similarity* between the uncoded source signal and its reconstructed (digitized) version. In audio coding the **mean opinion score (MOS)** has usually been applied to measure this similarity. Listeners classify the quality of coders on an N-point quality or impairment scale.

Table 104.2 shows the *signal impairment scale* and its associated numbering (ITU-R Rec. 562); see also Jayant and Noll [1984]. The impairment scale is extremely useful if coders with only small impairments have to be graded. The final result of such tests is an averaged judgement called the mean opinion score. In assessing the value of a given MOS rating, the standard deviation of this rating is also important.

TABLE 104.2 Five-Point MOS Impairment Scale

MOS Score	Impairment Scale
5	Imperceptible
4	Perceptible, but not annoying
3	Slightly annoying
2	Annoying
1	Very annoying

Recent results have shown that perception-based *objective* distortion measures that make use of results from psychoacoustics are quite promising measures of quality. For example, the *perceptual audio quality measure* [Beerends and Stemerdink, 1992] has been applied to audio signals in the ITU-R digital audio broadcasting tests and has given a correlation of 0.98 with a standard deviation of 0.17 (ITU-R Task Group 10/2, 1992).

104.5 Low Bit Rate Coding

Digital coding at high bit rates is dominantly waveform preserving, that is, the amplitude-vs-time waveform of the decoded signal approximates that of the input signal. Waveform coding principles have been covered in detail in Jayant and Noll [1984]. At lower bit rates the error criterion has to be in favor of an output signal that is useful to the human receiver rather than favoring a decoded signal that follows and preserves the input waveform. Basically any efficient source coding algorithm will do the following:

- Remove redundant components of the source signal by exploiting correlations between its samples either in time domain by employing adaptive linear prediction coding (LPC) or in frequency domain by employing high-resolution bandwidth splitting in **subband coding (SBC)** or **transform coding (TC)** schemes. Combinations thereof are also used.
- Remove components of the source signal, which are irrelevant to the auditory system, that is, to the ear. Irrelevancy manifests itself as unnecessary amplitude or frequency resolution; all portions of the source signal that are masked are irrelevant. Recently obtained results in audio bit rate compressions result from an efficient identification and subsequent removal of such irrelevant components prior to transmission.

First proposals to reduce PCM-format audio coding rates have followed those for speech coding [Jayant and Noll, 1984]. Modest reductions have been obtained by instantaneous companding (e.g., a conversion of uniform 14-b PCM into a 11-b nonuniform PCM presentation), forward-adaptive PCM (block companding) such as 16/14 scaling in digital satellite broadcasting systems (ITU-R Rec. J41, J42) and various forms of near instantaneously companded audio multiplex (NICAM) coding (ITU-R, Rec. 660). For example, the British Broadcasting Corporation (BBC) has used the NICAM 728 coding format for the transmission of sound in broadcast television networks; it uses 32-kHz sampling with 14-b initial quantization followed by a scaling to a 10-b format on the basis of 1-ms blocks [Hathaway, 1992]. To code the scalefactor, the information about the range of the quantizer, 3 b of side information are used. The net bit rate per audio channel is $10 \times 32 + 3 = 323$ kb/s. An overhead for error detection and synchronization increases the rate to 364 kb/s for a mono channel and to 728 kb/s for the stereo mode (hence the name NICAM 728).

Adaptive PCM schemes can solve the problem of providing a sufficient dynamic range for audio coding but they are not efficient compression schemes since they neither exploit statistical dependencies between samples nor remove sufficiently signal irrelevancies.

Redundancy reduction can easily be achieved by using frequency-domain coding, such as transform coding, subband coding, and hybrid forms thereof. In addition, and more importantly, frequency-domain coding offers an easy and (at least conceptually) direct way for perceptual coding, that is, for noise shaping and suppression of frequency components that need not to be transmitted. In the following we offer brief summaries of a number of audio coding algorithms, which are based on such frequency mappings as a first stage of audio coding. The differences are in the number of spectral components, the type of mapping (subband splitting vs discrete transform), and in the strategies for an efficient quantization of spectral components and masking of the resulting coding errors.

104.6 Subband Coding: ITU-R G.722 Coder

This 64-kb/s wideband speech coder can code signals with a 3-dB bandwidth of 50–7000 Hz [Mermelstein, 1988]. The ITU-R standard employs a two-band subband technique and a backward-adaptive **adaptive differential pulse-code modulation (ADPCM)** coding in each of the two subbands. *The subband splitting* is based on two identical quadrature mirror filters (QMF) to divide the wideband signal, sampled at 16 kHz, into two 8-kHz sampled components, called low subband and high subband. The filters overlap, and aliasing will occur because of subsampling of each of the components; the synthesis QMF filterbank at the receiver ensures that aliasing products are canceled. However, because of quantization error components on the two subbands aliasing will not be eliminated completely. Therefore 24-tap QMF filters with a stop-band rejection of 60 dB are employed.

The *coding* of the subband signals is based on ADPCM with backward-adaptive predictors and quantizers; their design is borrowed from ITU-R Rec.G.721. The quantizers, however, do not support the slowly adapting mode. High-quality coding is provided by a fixed bit allocation, where the low and high subband ADPCM coders use a 6-b per sample and a 2-b per sample quantizer, respectively. In the low subband the signal resembles the narrowband speech signal in most of its properties. A reduction of the quantizer resolution to 5 or 4 b per sample is possible to support transmission of auxiliary data at rates of 8 and 16 kb/s. Embedded encoding is used in the low subband ADPCM coding, that is, the adaptations of predictor and quantizer are always based only on the four most significant bits of each ADPCM codeword. Hence a stripping of one ot two least significant bits from the ADPCM codewords does not affect the adaptation processes and cannot lead to mistracking effects caused by different decoding processes in the receiver and transmitter.

Subjective tests have shown that the 64-kb/s ADPCM-coded version has the same MOS rating as the uncoded source, implying no measurable difference in subjective quality. In addition, the coder shows a graceful degradation of subjective quality with decreasing bit rate and increasing bit error rate. The coder also provides a good-to-fair quality for coding of audio signals. Table 104.3 compares MOS values for speech and audio signals at two bit rates. The resulting subjective quality is partly limited by the restriction on the bandwidth. Current activities in wideband speech coding concentrate on coding at 32 kb/s and below, with the 64-kb/s ITU-R standard serving as reference.

TABLE 104.3 MOS Performance of ITU-R G.722 Coder (7-kHz Bandwidth) [Mermelstein 1988]

Bit rate, kb/s	MOS Value for Speech	MOS Value for Audio
48	3.8	3.2
64	4.2	3.9

The ITU-R G.722 coder has recently been modified to support a 32-kHz sampling rate for audio signals of 15-kHz bandwidth [Iwadare et al., 1992]. At the resulting bit rate of 128 kb/s per mono channel MOS scores of 4.2 and 4.0 for the uncoded and the coded version, respectively, have been obtained on the basis of loudspeaker listening tests.

MPEG layer I and II coders are examples of subband coders with a high number of subbands. They are more suitable for perceptual coding; details will be given in a later section.

104.7 Transform Coding

The *conventional adaptive transform coding* (ATC) as proposed in the mid-1970s [Zelinski and Noll, 1977][6] employs a discrete cosine transform (DCT), which maps blocks of N samples into an equivalent number of transform coefficients. The logarithm of the absolute values of these coefficients form the DCT-based short-term log-spectral envelope; its encoded version is used to calculate, for each audio block, the optimum number of bits for each of the N quantizers for the transform coefficients (dynamic bit allocation). The encoder transmits the quantized transform coefficients and the encoded spectral envelope.

The conventional ATC has a number of shortcomings; it produces aliasing effects due to spectral overlap, and it produces pre-echoes because the inverse transform in the decoder spreads large quantization errors—possibly resulting from a sudden *attack* in the source signal—over the reconstructed block. In addition, inappropriate bit allocations occur if these allocations are based on an unweighted mean-squared coding error criterion. However, it was already stated in the first publication that the ATC performance could be improved if emphasis would be laid on irrelevancy reduction, and that the available bits had to be allocated dynamically such that the *perceived* distortion is minimum.

The problem of aliasing can be partly solved by introducing a modified discrete cosine transform (MDCT) with an 50% overlap between adjacent audio blocks. Pre-echoes can be reduced or avoided by using blocks of short lengths. However, shorter blocks cause an insufficient frequency resolution, and a larger percentage of the total bit rate is required for the transmission of side information.

---

[6]For the first time, the proposed ATC used DCT as the appropiate linear transform for speech, it introduced the technique of adaptive bit allocation, and it showed that a log-spectral envelope should be computed and parameterized to derive the appropriate bit allocation. The authors also proposed a bit allocation based on auditory perception.

A solution to this problem is to switch between block sizes of different lengths of, say, $N = 64$ to $N = 1024$ (*window switching*). The small blocks are only used to control pre-echo artifacts during nonstationary periods of the signal, otherwise the coder switches back to long blocks. It is clear that the block size selection has to be based on an analysis of the characteristics of the actual audio coding block. In addition, window switching has to be carried out such that the aliasing cancellation property of the transform is maintained.

Dolby's AC-2 coder uses a time-domain aliasing cancellation based on an alternate application of 512-point modified discrete sine and cosine transforms (MDST and MDCT) by means of 50% overlap/add windows. Adjacent transform coefficients are then grouped into subband ranges, which approximate the critical bands of human hearing [Davidson, Fielder, and Antill, 1990]. From the transform coefficients within such subbands a log-spectral envelope of the current block is estimated and used for a dynamic bit allocation routine. The coder uses no perceptual model except the critical band division. The AC-2 coder has been evaluated in the ITU-R process of digital audio broadcast standardization and has shown to be close in performance to the ISO/MPEG audio coding algorithm at its bit rate of 256 kb/s.

104.8 Subband/Transform Coding

Sony's *adaptive transform acoustic coding* (ATRAC) uses a hybrid frequency mapping: the signal is first split into three subbands (0–5.5, 5.5–11.0, and 11.0–22.0 kHz) reflecting auditory perception. A MDCT with 50% overlap is then applied to all subbands with dynamic window switching: the regular block size is $N = 512$, to provide optimum frequency resolution, but in the case of possible pre-echoes N is switched to 64 in the high-frequency bands and to $N = 128$ otherwise. The coder with its 44.1-kHz sampling rate and its 256-kb/s stereo bit rate has been developed for portable digital audio, specifically for the 64-mm optical or magneto-optical MiniDisc (MD), which has approximately one-fifth of the byte-capacity of a standard CD [Tsutsui et al., 1992].

104.9 ISO/MPEG Audio Coding

The Moving Pictures Expert Group within the International Organization of Standardization (ISO/MPEG) has provided a generic source coding method (MPEG 1) for storing audio-visual information on digital storage media at bit rates up to about 1.5 Mb/s [ISO/MPEG 1993,1994; Brandenburg and Stoll, 1992; Noll, 1993]. The audio coding part of this MPEG 1 standard is the first international standard in the field of digital audio compression and can be used both for audio–visual and audio-only applications. It is about to become a universal standard in many application areas with totally different requirements in the fields of consumer electronics, professional audio processing, telecommunications, and broadcasting. The coder provides a quality that is subjectively equivalent to CD quality (16-b PCM) at a stereo rate of 256 kb/s, and that is close to CD quality at a stereo rate of 128 kb/s. Because of its high dynamic range MPEG 1 audio quality exceeds that of a CD when referred to a, say, 18-b PCM original.

The MPEG 1 standard consists of three layers, I, II, and III, of increasing complexity and subjective performance. It supports sampling rates of 32, 44.1, and 48 kHz and bit rates between 32 kb/s (per monophonic channel) and 384 kb/s (per stereo channel) (layer II). An additional free format mode allows for any bit rate below 384 kb/s (layer II). The MPEG 2 standard includes an extension that adds sampling frequencies of 24, 22.05, and 16 kHz. It is expected that these sampling frequencies will be useful for the transmission of wideband speech and medium-quality audio signals. Indeed, a reduction in sampling frequency by a factor of two doubles the frequency resolution of the filter banks and transforms and will therefore improve the performance of the coders, in particular those with a low frequency resolution (MPEG 1, layers I and II). Only small modifications of ISO/MPEG phase 1 are needed for this extension, the bit stream syntax still complies with the standard.

Audio Coding

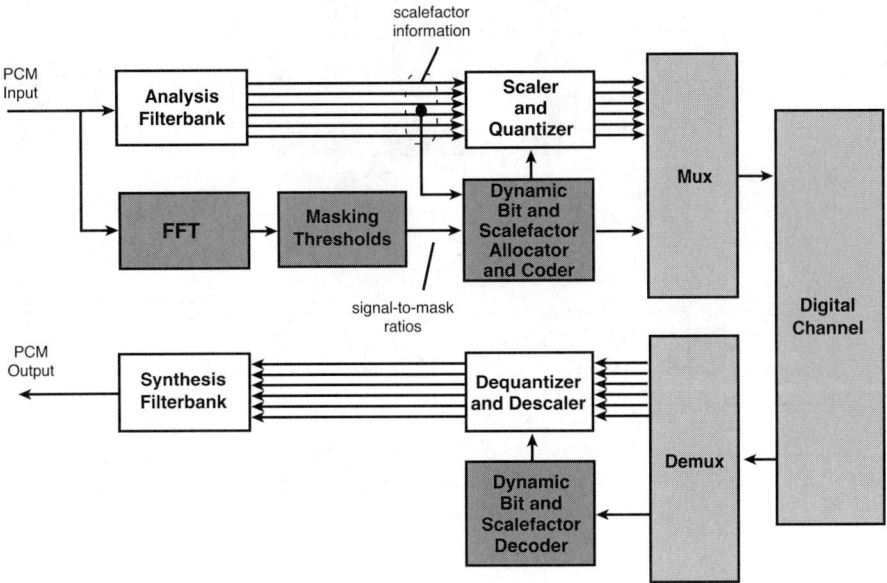

FIGURE 104.3 Structure of MPEG 1 audio encoder and decoder, layers I and II. (*Source:* Noll, P. 1993. Wideband speech and audio coding. *IEEE Commun. Mag.*, 31(11):34–45. ©1993 IEEE. With permission.)

It is worthwhile to note that the normative part of the standard describes the decoder and the meaning of the encoded bitstream, but that *the encoder is not defined*, thus leaving room for an evolutionary improvement of the encoder. In particular, different psychoacoustic models can be used ranging from very simple (or none at all) to very complex ones based on quality and implementability requirements (the standard gives two *examples* of such models). In addition, an encoder can be optimized for a certain application.

The MPEG 1 coder uses perception-based frequency-domain coding. In the first stage the audio signal is converted into 32 spectral subband components via an analysis filterbank. Each spectral component is quantized whereby the number of quantizer levels for each component is obtained from an dynamic bit allocation rule that is controlled by a psychoacoustic model that computes signal-to-mask ratios and obtains its information about the short-term spectrum by running an FFT in parallel to the encoder (see Fig. 104.3). The model is only needed in the encoder, which makes the decoder less complex, a desirable feature for audio playback and audio broadcasting applications.

MPEG 1 layer I contains the basic mapping of the digital audio input into 32 subbands via equally spaced bandpass filters (Fig. 104.3), fixed segmentation to format the data into blocks (8 ms at 48-kHz sampling rate) and quantization with block companding. At 48-kHz sampling rate each band has a width of 750 Hz. A polyphase filter structure is used for the frequency mapping; its filters are of order 511, which implies an impulse response of 5.33 ms length (at 48 kHz). In each subband blocks of 12 decimated samples are formed and for each block signal-to-mask ratios are calculated via an 512-point FFT.[7] For each subband the bit allocation selects a quantizer (out of a set of 15) such that both the bit rate requirement and the masking requirement are being met.

The *decoding* is straightforward: the subband sequences are reconstructed on the basis of 12-sample subband blocks taking into account the decoded scalefactor and bit allocation information. If a subband has no bits allocated to it, the samples in that subband are set to zero. Each time the

---

[7]A simpler short-term spectral analysis can be based on the powers of the individual 12-sample subband blocks. This approach is used in the digital compact disc (DCC) system.

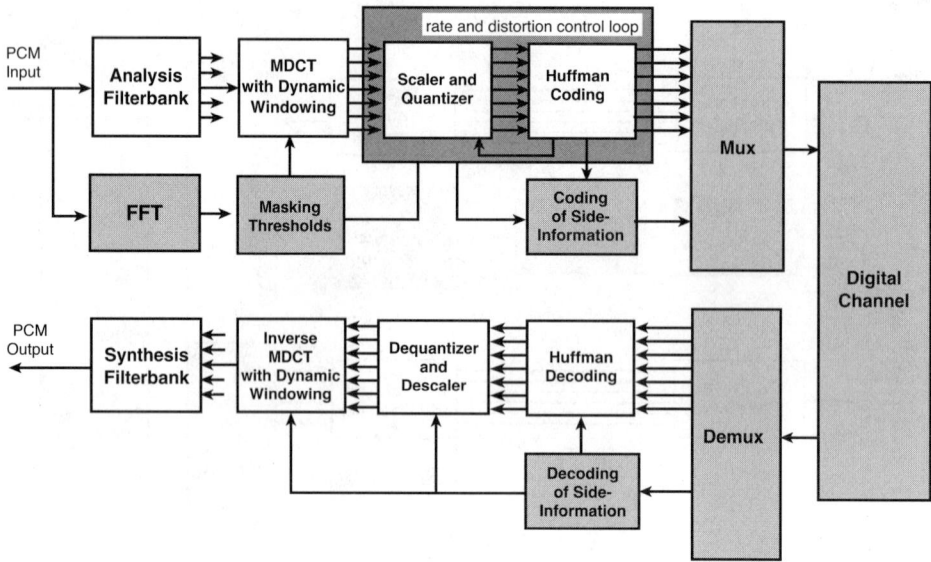

FIGURE 104.4 Structure of MPEG 1 audio encoder and decoder, layer III. (*Source:* Noll, P. 1993. Wideband speech and audio coding. *IEEE Commun. Mag.*, 31(11):34–45. © 1993 IEEE. With permission.)

subband samples of all 32 subbands have been calculated, they are applied to the synthesis filterbank, and 32 consecutive 16-b PCM format audio samples are calculated.

The MPEG 1 layer II coder is basically similar to the layer I coder but achieves a better performance through three modifications. First, the input to the psychoacoustic model is typically a 1024-point FFT leading to a finer frequency resolution for the calculation of the global masking threshold. Second, the overall scale factor side information is reduced by exploiting redundancies between the scale factors of three adjacent 12-sample blocks. Third, a finer quantization with a maximum of 16-b amplitude resolution is provided (which reduces the coding noise).

MPEG 1 layer III introduces many new features (see Fig. 104.4). It achieves its improvement mainly by an improved frequency mapping, an analysis-by-synthesis approach for the noise allocation, an advanced pre-echo control, and finally by nonuniform quantization with entropy (Huffman) coding. To achieve a higher frequency resolution closer to critical band partitions the 32-subband signals are subdivided further in frequency content by applying a switched 6-point or 18-point MDCT with 50% overlap to each of the subbands. The maximum number of frequency components is $32 \times 18 = 576$, each representing a bandwidth of only $24{,}000/576 = 41.67$ Hz. The term *hybrid filterbank* is used to describe such a cascade of polyphase filter bank and MDCT transform. The decoding follows that of the encoding process.

104.10 Multichannel Stereophony

A logical further step in digital audio is the definition of a universal loudspeaker reproduction standard to provide an improved stereophonic image not only for audio-only applications including teleconferencing, but also for enhanced television systems, in particular for high definition television (HDTV) systems and for digital storage media. Loudspeaker arrangements, referred to as 3/2-stereo, with a left and a right channel (L and R), an additional center channel C, and two side/rear surround channels (Ls and Rs), offer an improved acoustical representation. In particular, the three front loudspeakers lead to a stable frontal sound image and an enlarged listening area.

MPEG 2/Audio will provide a multichannel digital audio system, which supports stereophonic presentation with various combinations of channels including:

- 3 channels: 3/0-stereo   left, right, center
- 4 channels: 3/1-stereo   left, right, center, surround
- 5 channels: 3/2-stereo   left, right, center, surround left, surround right

In addition, MPEG 2 supports up to seven commentary (multilingual) channels and an optional low-frequency enhancement (subwoofer) channel.[8]

MPEG 2 will include two standardized coders, one of which will be forward and backwards compatible with MPEG 1; the second one will not be backwards compatible. Forward compatibility means that a future multichannel decoder should be able to decode properly MPEG 1 2/0-stereo signals, and backwards compatibility means that the existing MPEG 1 decoder can decode the basic 2/0-stereo signal from the five-channel bit stream. Because of its constraints caused by matrixing and dematrixing procedures the backwards compatible coder will need a higher bit rate for a given quality of the decoded bit stream.

Redundancies and irrelevancies, such as intra- and interchannel correlations and masking effects, respectively, are exploited to reduce the overall bit rate. In addition, stereo-irrelevant components of the multichannel signal may be identified and reproduced in a monophonic format to bring the bit rates further down. Although a bit rate of 384 kb/s, suitable for H0 channels of the ISDN hierarchy, may not offer transparent quality, it will be sufficient for many applications. Higher bit rates will be offered by the standards to support transparent reproductions.

A second example of multichannel coding is Dolby's AC-3 system that uses many features of its AC-2 system; for multichannel coding it makes use of perceptual phenomena, occurring if the human ear detects multiple signals, by combining a number of individual transform coefficients into a common coefficient. The AC-3 scheme has been chosen as the multichannel audio standard for the U.S. grand alliance HDTV system [ACATS, 1994].

104.11 Conclusion

Low bit rate digital audio is on its way to be applied in many different fields, such as consumer electronics, professional audio processing, telecommunications, and broadcasting. Perceptual coding in the frequency domain has paved the way to high compression rates. The ISO/MPEG 1 audio coding algorithm with its three layers has already been widely accepted as international standard. The MPEG 2 standard(s) for multichannel stereo will improve stereophonic images and will be of importance both for audio-only and multimedia applications. Other coding schemes serving different goals and meeting other requirements will evolve, and application of audio coding to broadcasting, mobile radio, and delivery over ISDN channels will become important issues.

In the area of broadcasting, mobile radio services are moving to portable and handheld devices, and new (third generation mobile communications) networks are evolving. Coders for these networks must not only operate at low bit rates but must be error robust in burst-error and packet-loss environments. Error concealment techniques will play a significant role, since due to the lack of available bandwidth, traditional channel coding techniques cannot sufficiently improve the reliability of the channel.

Emerging activities of the ISO/MPEG expert group aim at proposals for audio coding at very low bit rates (MPEG 4). Transmission of audio signals from and to remote databases could benefit significantly from these activities. The basic audio quality will be more important than compatibility with existing ISO/MPEG standards. This opens the door for completely new solutions. To meet the goals the introduction of enhanced perceptual models, an adaptation of time/frequency resolution, vector quantization, and analysis-by-synthesis coding strategies have to be considered.

---

[8] The subwoofer system reproduces frequencies below around 120 Hz with one or more loudspeakers, which can be positioned freely in the listening room.

Defining Terms

Adaptive differential pulse code modulation (ADPCM): A predictive coding technique with adaptation of quantizer step sizes and predictor coefficients based on the short-term statistics of the source signal.

Asynchronous transfer mode (ATM): Networks are based on packet (cell) switching, a random access technology. ATM provides efficient service for bursty data and for mixed traffic (data, audio, voice, video). In the ATM specification the cell size is fixed (53 bytes).

Auditory masking: Simultaneous masking describing the effect that a low-level audio signal (the maskee) can become inaudible when a louder signal (the masker) occurs simultaneously. Temporal masking describes the effect that a low-level audio signal can become inaudible before and after a louder signal has been switched on and off.

Integrated Services Digital Network (ISDN): Offers end-to-end digital connectivity to support a wide range of communication applications. Customers have physical access to one or two 64kb/s bearer (B-) channels and one data (D-) channel (the latter one supports signaling at 16 kb/s but can also carry user information). Higher rates than 64 or 128 kb/s are achievable using multiple calls with channel aggregation and may be more generally available in the future as multiple-rate services.

International Telecommunication Union (ITU): A worldwide standardization organization. ITU-R is its Radiocommunication section, formerly called the Consultative Committee on International Radio (CCIR), and ITU-T is its telecommunication section, formerly called the Consultative Committee on International Telephony and Telegraphy (CCITT).

Mean opinion score (MOS): A distortion measure that specifies *perceptual similarity* between an uncoded source signal and its coded version.

Moving Pictures Expert Group (MPEG): The Working Group 11 of the ISO/IEC/JTC1/SC2 standardization body. Its task is to propose algorithms for coding of moving pictures and associated audio.

Subband coding: Division of the source signal into bandpass signals (subbands) via a filterbank followed by individual coding of the subband signals.

Transform coding: Deriving transform (spectral) coefficients from blocks of the source signal via a linear transform followed by individual coding of the transform (spectral) coefficients.

References

ACATS. 1994. Grand alliance HDTV system specification, Ver. 1.0, Advisory Committee on Advanced Television Service.

Beerends, J.G. and Stemerdink, J.A. 1992. A perceptual audio quality measure. In *92nd AES Convention*, Preprint 3311, Vienna, Austria, March.

Brandenburg, K. and Stoll, G. 1992. The ISO/MPEG-audio codec: A generic standard for coding of high quality digital audio. In *92nd AES Convention*, Preprint 3336, Vienna, Austria, March.

Davidson, G., Fielder, L., and Antill, M. 1990. High-quality audio transform coding at 128 kbits/s. In *Proc. of the ICASSP*.

Hathaway, G.T. 1992. A NICAM digital stereophonic encoder. In *Audiovisual Telecommunications*, ed. N.D. Nigthingale, pp. 71–84. Chapman & Hall.

ISO/MPEG. 1992. MPEG 1: International Standard IS 11172-3 JTCl/SC29.

ISO/MPEG. 1994. MPEG 2: International Standard IS 13818-3.

Iwadare, M., Sugiyama, A., Hazu, F., Hirano, A., and Nishitani, T. 1992. A128 kb/s hi-fi audio CODEC based on adaptive transform coding with adaptive block size, *IEEE J. Selec. Areas Commun.*, 10(1):138–144.

Jayant, N.S. 1992. Signal compression: Technology targets and research directions, *IEEE J. Selec. Areas Commun.*, (June):796–818.

Jayant, N.S. and Noll, P. 1984. *Digital Coding of Waveforms: Principles and Applications to Speech and Video*, Prentice–Hall, Englewood Cliffs, NJ.

Mermelstein, P. 1988. G722, A new CCITT coding standard for digital transmission of wideband audio signals, *IEEE Commun. Mag.*, (Jan):8–15.

Noll, P. 1993. Wideband speech and audio coding, *IEEE Commun. Mag.*, 31(11):34–45.

Tsutsui, K., Suzuki, H., Shimoyoshi, O., Sonohara, Akagari, K., and Heddle, R.M. 1992. Adaptive transform acoustic coding for minidisc. In *93rd AES-Convention*, Preprint 3456, San Francisco, CA.

Zelinski, R. and Noll, P. 1977. Adaptive transform coding of speech signals. *IEEE Trans. Acoust., Speech Sig. Proc.*, (Aug.):299–309.

Zwicker, E. 1982. *Psychoacoustics*, Springer–Verlag, Berlin.

Further Information

There is a significant overlap between speech and audio coding. Therefore we recommend recent overview papers, such as "Speech Coding: A Tutorial Review," *Proc. IEEE*, Vol. 82, No. 10, Oct. 1994, pp. 1541–1582, by A. Spanias, and "Advances in Speech and Audio Compression," *Proc. IEEE*, Vol. 82, No. 6, June 1994, pp. 900–918, by A. Gersho. Both papers have a large number of references. The June edition has a special section on data compression with important papers on lossless coding and on vector quantization.

The technologies of Philips Digital Compact Cassette and Sony's MiniDisc storage systems are covered in detail in "Digital Compact Cassette" by A. Hoogendoorn and in "The Rewritable MiniDisc System" by T. Yoshida. Both papers were published in *Proc. IEEE*, Vol. 82, No. 10, Oct. 1994.

A more detailed version of this Handbook contribution was published under the title "Digital Audio Coding for Visual Communication" in the special issue on digital television of the *Proc. IEEE* (July 1995).

Network aspects are covered in "High Quality Networking of Audio-Visual Information" by N. Jayant, published in *IEEE Commun. Mag.*, Vol. 31, 1993, pp. 84–95.

Digital audio broadcasting is covered in detail in "Radio Broadcast Systems; Digital Audio Broadcasting (DAB) to mobile, portable and fixed receivers," Draft prETS 300401 of the European Telecommunications Standards Institute (ETSI), Jan. 1994.

Actual contributions to all aspects of digital audio processing and coding can be found at the biannual conventions of the Audio Engineering Society (AES) and the IEEE ASSP Workshops on Applications of Signal Processing to Audio and Acoustics.

The public C source MPEG 1 (layers I, II, and III) audio decoder software can be downloaded via anonymous ftp from fhginfo.fhg.de (153.96.1.4), directory /pub/layer3/public_c (Fraunhofer-Institut Erlangen, Germany). Layer III encoder *and* decoder software is available from the same source.

105
Cable

105.1	Introduction	1488
105.2	Cable System Architecture	1489
	Frequency Plans • Control Data Channel	
105.3	Source Origination and Head End	1491
	Off the Air Broadcast Signals • Direct Fiber Links • Analog and Digital Satellite Signals • AM and FM Microwave Signals • Commercial Insertion • Character Generators • Live Local Programming • Digital Satellite to Digital Cable Transcoding • Digital Video Servers	
105.4	Transmission Channel	1493
	Channel Characteristics • Modulation and Error Correction (Analog and Digital)	
105.5	Consumer Premises Equipment	1494
	The First Converters • The First Set-Top Converters • Set-Top Designs • Interdiction	
105.6	Access Control and Security	1497
	Access Control • Security Systems	

Jeff Hamilton
General Instrument Corporation

Mark Kolber
General Instrument Corporation

Charles Schell
General Instrument Corporation

Len Taupier
General Instrument Corporation

105.1 Introduction

Cable TV, or community antenna TV (CATV), began shortly after World War II. Television broadcast stations were established in most major cities. The frequencies used for transmission were and are primarily line-of-sight transmission and of a power sufficient to reach consumers in the city and close-in suburbs. As television caught on this gave rise to a large number of consumers who could not receive good quality pictures because they were in the shadow of a hill or too far from the transmitters.

The answer was to place a large antenna at a good receive site, on top of a mountain, for example, and distribute the signals to an entire community. The earliest such sites were in Astoria, OR and Williamsport, PA. The original cable was RG-11 military surplus from World War II.

Early CATV systems carried only 3 channels in the low-VHF band of 54–88 MHz. Over the years this evolved to 174, 220, 270, 300, 330, 450, and 550 as broadcast TV added the UHF band channels, pay TV arrived with such services as HBO, and distribution electronics improved to allow high-quality delivery of the ever larger number of channels. The typical new plant is designed for 750 MHz, with a capacity for over one hundred analog video channels.

The advent of digital video compression, based on the Moving Pictures Expert Group-2 (MPEG-2) standard, promises to increase this capacity by an order of magnitude. Initially digital services will be deployed on hybrid plants which carry a mix of analog and digital video signals. Typical plans are to offer 77 analog channels in the 50–550 MHz band which can be received by consumers with only analog equipment in the home. Added to this will be a band of digital audio and video services from 550 to 750 MHz. This 200-MHz digital cable band will provide the same capacity as the popular 32 transponder high-power DBS services.

New hybrid **set-top** converters will be deployed to subscribers of these services that receive both the analog and digital transmissions. Current compression technology provides typically, 4–10 channels in each 6-MHz band allowing well over 200 new video services in the initial 200-MHz digital band. As digital receiving equipment rolls out to more and more consumers, cable has the option of evolving to all digital transmission by progressively replacing analog channels with digital carriers.

From the early days of TV the consumer's choice has been to select from "what's on" a few channels. Cable added dozens more channels, and through its ability to charge on a per channel basis created the higher value subscription movie channels. Digital compression combined with the delivery of signals over fiber to neighborhoods of a few hundred homes now makes possible an efficient path to provide dedicated bandwidth to each subscriber. This capability amounts to *infinite bandwidth* in which everything a subscriber might want to view is on, by request. Instead of the consumers choosing among what programmers decide to send them, they will have the ability to tell the system what they want to see.

105.2 Cable System Architecture

The distribution system or cable plant that carries the signals from the **head end** to the subscribers can use various architectures. The common *tree and branch* architecture uses one inch or larger diameter solid aluminum outer conductor coaxial cable to carry the signal from the head end. This trunk cable runs through the system distributing the broadband signal to each neighborhood. Feeder lines that branch off the trunk typically use smaller diameter cable to carry the signal into the neighborhoods passing each subscriber's home. Drop cables lead from the curb side to each home and are typically 1/4 inch flexible coaxial cable. All CATV cables are 75-Ω characteristic impedance which is near optimum for low loss.

Devices called taps are used to extract a portion of the signal from a feeder to supply a drop cable. For example, a 20-dB tap extracts 1/100 of the power from the feeder causing a 20-dB loss at the tapped output but only a 0.04-dB loss to the main path.

CATV signals are measured in **decibel's relative to 1 millivolt (dBmV)**. For example, 0 dBmV corresponds to 1 mV, and +60 dBmV to 1 V. Signals are normally distributed on the trunk cable at a level from +30 to +50 dBmV per channel. The signal level delivered to the home is usually from 0 to +15 dBmV per channel. In order to maintain the signal level, trunk amplifiers and feeder amplifiers, also known as line extenders, are spaced along the lines about every 1/2 mi. This corresponds to 20–30 dB loss at the highest design frequency. A trunk line may have 40 or more amplifiers in cascade. Since the attenuation of the cables increases with increasing frequency, the amplifiers are equipped with equalizers to provide more gain to the higher frequencies. The ratio of cable loss in decibels between two frequencies is approximately equal to the square root of the ratio of the two frequencies. A cable with 7-dB loss at 55 MHz will have about 20-dB loss at 450 MHz. Some amplifiers include **automatic gain control (AGC)** to maintain the desired output level. The amplifier cases can also include diplexers and small upstream amplifiers to separate and amplify the typically 5–30 MHz return path signals. Amplifiers are usually powered by 60 or 90 V AC signals, which are carried along the same coaxial lines.

Many CATV systems now use a *hybrid fiber coax* (HFC) architecture. To minimize the number of trunk amplifiers through which the signal must pass, the signal is carried via fiber optic cables from the head end to multiple remote locations known as a nodes. The head-end optical transmitter normally uses a distributed feedback (DFB) laser diode generating a few milliwatts of optical power at 1310 nm. The entire RF spectrum is used to intensity modulate the optical output the laser. Since the fiber loss is only about 0.4 dB/km, the node can be located a relatively large distance from the head end, eliminating the need to cascade many amplifiers along a trunk. At the node, a photodetector receiver converts the optical signal back to RF for distribution on the coax. Return path signals are sent from the node back toward the head end through a separate fiber using less expensive Fabry–Perot laser sources.

The HFC architecture improves on the classic tree and branch in several areas. It provides many fewer active devices between the head end and home, which greatly improves system reliability. This reduction in the number of active devices also reduces maintenance costs, problem location, and recovery time. The fiber cables are not susceptible to noise ingress, which impairs TV pictures, nor do they radiate RF power, which can interfere with nearby radio communications.

The star configuration of HFC fiber plant design provides individual fiber runs to each neighborhood. This means the plant supports dedicated bandwidth from the head end to each node of as few as a few hundred homes. Currently, HFC plants carry the same broadcast TV signals to every node. As interactive and on-demand services are rolled out, the HFC plant will carry the unique signals demanded by the subscribers in each neighborhood on their node alone. This effectively multiplies bandwidth by as many times as there are nodes in a system. Telephony is a primary example of an interactive, on-demand service. A tree and branch cable plant could provide only a limited number of telephone channels before running out of shared bandwidth, particularly upstream. The HFC star architecture can provide dedicated bandwidth to and from each node with ample resources for full service.

Frequency Plans

The 5–30 MHz range, called the subband, is commonly used for return path or upstream (from subscriber to head end) transmissions. This is used for the relaying of remote broadcasts to the head end and for the return of billing and other information from the subscribers set-top. It can also be used for the upstream portion of telephony and interactive TV applications.

The range of frequencies used by CATV operators for downstream (sending video from the head end to the subscribers home) ranges from 50–200 MHz for older 23-channel systems to 50–550 MHz for 77-channel system. Newer systems are designed to pass up to 750 MHz or even 1000 MHz. Each analog **National Television Systems Committee (NTSC)** TV channel occupies 6 MHz. Using digital compression several video programs can be transmitted in each 6-MHz channel.

The bands for downstream transmission are:

- Low band: 54–88 MHz (off the air VHF channels 2–6)
- Mid band: 88–174 MHz
- High band: 174–216 MHz (off the air VHF channels 7–13)
- Super band: 216–300 MHz
- Hyper band: above 300 MHz

There are multiple plans for assigning channels within the available frequencies. The Electronics Industry Association (EIA) plan assigns channels to agree with the VHF off the air channel numbers where possible. Additional channels are spaced 6 MHz from these. The harmonically related carriers (HRC) plan places video carriers near multiples of 6 MHz. This helps reduces the visibility of interference from harmonic and intermodulation distortion created in the distribution equipment because the distortion products fall directly on and zero beat with other video carriers. The incrementally related carriers (IRC) plan is similar to the EIA plan except for channels 5 and 6. This plan reduces the visibility of intermodulation distortion products but not harmonic distortion products.

Since many of the frequencies used for CATV channels are assigned to various over the air services, it is important for the cable system to maintain shielding integrity, both to keep outside signals from interfering with the CATV system (ingress) and to keep CATV signals from radiating and causing interference to local radio traffic, such as the aircraft communication bands.

Cable

Control Data Channel

In some CATV systems, another signal, known as the out-of-band signal, is also sent from the head end to the subscriber. This signal is used to control addressable set-top **converters**. The converter contains a receiver that is always tuned to this signal regardless of which in-band channel is being viewed or power on/off. This signal is usually located in the FM radio band (88–108 MHz). The dedicated frequency and continuous reception by set-tops provides a reliable path for access control and user features such as program guides and downloaded graphical user interfaces.

105.3 Source Origination and Head End

The signals and programs that are to be delivered to CATV subscribers are all collected at a central location known as the head end. There can be many sources of signals including VHF and UHF off the air broadcast signals, analog and digital satellite signals, AM and FM microwave signals, prerecorded video tapes, character generators, digital video servers, and live locally originated programs.

Off the Air Broadcast Signals

The reception of distant off the air broadcast signals was the original need that created the CATV industry. Today as populated areas become more spread out, many people **subscribe** to CATV to improve the quality of their reception of broadcast signals.

At the head end, off the air signals are received using high-gain Yagi antennas mounted on a hill or tower to provide the maximum signal. The directional characteristics of the Yagi antenna also help to discriminate against reflections which cause ghosting. Advanced digital techniques are also available to cancel ghosting. A known reference signal is transmitted in the invisible vertical blanking interval section of the TV signal. The receiver compares the received version of the signal to the known reference signal and calculates a ghost profile. This profile is then subtracted from the entire video signal to cancel the ghost effect. Using these techniques, good off the air pictures can be received over 60–100 mi away.

Direct Fiber Links

The penetration of cable to the majority of TV viewers in the U.S. means more people watch network TV on cable than off air. This has made direct delivery of broadcast programming to the cable head end an attractive way to maximize the quality of these channels. Fiber is run from either the local broadcast studio or transmitter to the cable head end. There it is directly modulated onto a cable channel for distribution to subscribers.

Analog and Digital Satellite Signals

As the CATV industry grew, the concept of super stations and premium movie channels developed. These are distributed nationwide via geostationary communications satellites located 22,500 mi above the equator. Each programmer transmits their signal up to a satellite via an uplink, usually in the 5.9–6.4 GHz band. The satellite then transmits the signal back to Earth in the 3.7–4.2 GHz range or C band. Most analog CATV programming is concentrated on a few satellites known as cable birds. The transponders on these satellites generate 16–30 W of RF power, which usually requires a 10–30 ft diameter dish antenna at the cable head end. Each satellite contains 24 transponders and can carry 24 analog programs. Although, typically, a separate parabolic dish antenna is needed to receive each satellite, a specially shaped antenna can be used with multiple feeds to simultaneously receive multiple satellites.

Analog satellite signals are received in integrated receiver decoders (IRDs), at the cable head end. These devices tune the desired transponder frequency, demodulate the FM video signal, descramble the video signal, and decrypt and convert the digital audio back to analog form. They are authorized by the program providers access control and **security** system. The output analog video and audio go to a cable scrambler, modulator, and channel converter for transmission on the cable plant.

The enormous expense of satellite transponders drove the development of digital video compression for satellite transmission. Digital transmission of compressed video signals provides more efficient use of satellite transponders. Compression systems currently offer 4–10 video programs over a single transponder, greatly reducing the programmer's satellite link costs.

Digital compressed satellite signals are received in a digital IRD, which connects to the head-end systems in much the same way as an analog IRD. The digital IRD performs many more functions than an analog IRD but delivers similar outputs of analog video and audio signals. Differences include higher quality video signals due to digital transmission with no quality degradation of weak signals, high security due to digital encryption, and the availability of many more programs due to the efficiency of transmission.

AM and FM Microwave Signals

In some locations, it is impossible to receive off the air broadcast signals with the desired quality. In these cases, a dedicated point-to-point microwave link can be installed. The desired off the air signal is received at a location close to the transmitter or sometimes the microwave signal can be sent out from the TV station directly. The signal is beamed via very directional antennas directly to the head end. Both AM and FM modulation are used for microwave links.

Commercial Insertion

As special programming intended solely for CATV became available, the concept of local advertisement insertion developed. Local business can purchase advertising air time from the CATV operator. These commercials are inserted during defined commercial breaks in the programming. In the past advertisements were pre-recorded on video tape and automatically controlled to run at the correct time. Locally originated programs such as locally produced news, school board or local government meetings, and public access programming can also be recorded on video tape for transmission via CATV. New commercial insertion equipment uses compressed digital video sources. This provides greatly enhanced video quality, reliability, insertion tracking, scheduling, and methods for delivery of the program material to the insertion equipment. Commercial insertion into digital video channels on cable will require a new device called an add–drop multiplexer. This device will replace the compressed video bits of a satellite delivered advertisement with a local ad. This is all accomplished in the digital domain without any need to decompress the content.

Character Generators

Character generators are used to create bulletinboard style information channels. Programming, advertising, local government and weather information can be displayed. Usually, several screens worth of data are stored in the character generator and these are sequentially displayed. Computer graphics can also be included to enhance visual appeal. Expanded forms include characters generated over video stills or live video.

Live Local Programming

Many CATV operators carry local school or government meetings live. Often, a subband frequency (5–30 MHz) is used to carry the signal from the town hall, for example, via the cable plant return path back to the head end. At the head end the signal is converted to a normal forward path channel and sent to the subscribers via the distribution system.

Digital Satellite to Digital Cable Transcoding

Just as the integrated device called the IRD receives an analog or digital satellite service for use on cable a device called an integrated receiver transcoder (IRT) receives a compressed digital video multiplex for direct transmission to digital video subscribers without decompression at the head end. The IRT performs almost all of the functions required to pass a compressed multiplex of signals through from a digital satellite channel to a digital cable channel.

The IRT tunes a digital satellite signal, demodulates it, performs satellite link error correction, decrypts the multiple video and audio services in the multiplex, replaces the satellite access control information with new cable system control data, re-encrypts the services with cable security, inserts cable error control bits, and modulates the data onto a 64 quadrature amplitude modulation (QAM) carrier. Only a channel converter is required to prepare the signal for CATV transmission. This high density of signal processing means it takes much less rack space in the cable head end to add a channel of digital programming from satellite than an analog channel. At the same time the digital channel provides many simultaneous video programs to the subscriber.

Digital Video Servers

An expanded version of the video equipment used for digital commercial insertion can provide full length programs such as **pay-per-view** movies and movies with multiple start times. Video servers can be used to create programming for analog cable plants in place of video tape but their most efficient application is in digital cable plants. When digital set-tops are deployed a digital video server can output precompressed services to them for preprogrammed or for video on-demand services. The output is encrypted, multiplexed, and modulated in equipment that can also process streams from real-time video encoders.

105.4 Transmission Channel

Channel Characteristics

In order to deliver a high-quality signal, the CATV distribution plant must be designed to add a minimum of noise, distortion, and spurious signals. Various defects in the channel can cause visible impairments to the picture. The in-channel gain and group delay must be constant vs frequency to avoid ringing or loss of vertical detail. Excessive thermal noise causes snow or graininess in the picture. Spurious signals or intermodulation can cause herringbones or diagonal lines that move randomly through the picture. Table 105.1 is a list of some of the relevant parameters and values required for an excellent picture.

TABLE 105.1 Typical Maximum Impairments to Analog Video Signals

Gain flatness	±1 dB
Group delay variation	±170 ns
Hum modulation	±1% max
Signal to noise ratio	48 dB
Composite second order (CSO)	−50 dB
Composite triple beat (CTB)	−50 dB

Degradation from each part of the system is additive. The consumer's video signal has passed through a great number of active and passive components, each of which degrades the signal. Within the economic constraints on real products each system component must be designed to minimize its contribution to total impairment.

Digital transmission on cable will, in general, be much more robust, reliable, and easy to monitor than analog transmission. There are a few impairments that will effect digital transmission more than analog. The best example is burst noise. This shows up as a loss of signal for a fraction of one TV scan line in analog TV. This is fleeting and rarely objectionable. With digital compression all of the redundant information in the video signal has been removed. Any lost data will likely be noticeable. For this reason the QAM transmission for the cable channels has been designed to be highly robust even for bursts of noise over a full scan line long. Within the design limits digital video will contain no degradation from passing through the distribution system

Modulation and Error Correction (Analog and Digital)

Analog Modulation

Unscrambled analog video signals are carried in exactly the same modulation, vestigial sideband–amplitude modulation (AM-VSB) format as broadcast TV. This facilitates the use of TVs as the CATV reception device for simple services. The simplest forms of analog scrambling modify the modulation format by reducing signal power during the synchronizing periods of the video signal. Sync suppression has the added benefit of reducing peak load on the distribution system as the modulated TV signal is designed for peak power during these sync periods. More complex scrambling systems also perform various forms of video inversion. Both of these modified forms of modulation are suitable for CATV applications because of the robust and well-controlled reception conditions. High sync power, for example, is not required because there is no need to receive signals in the microvolt level from distant broadcast stations.

Digital Modulation

Just as analog satellite signals use frequency modulation (FM) digital satellite transmissions use quantenary phase-shift keying (QPSK). Both are wide bandwidth low carrier to noise modulation formats. Digital CATV systems use bandwidth-efficient high carrier to noise modulation called 64QAM and 256QAM. QAM is similar to QPSK in that two quadrature carriers are used but it adds amplitude modulation on each of the quadrature carriers. For example, 64QAM has eight states for each carrier instead of QPSK's two states. This results in 64 combinations and, therefore, each modulation symbol carries 6 bits of data. For 256QAM 8 bits are delivered for each symbol. These digital modulation formats deliver just over 30 and 40 million bits per second (Mb/s) in each 6-MHz CATV channel.

Digital Forward Error Correction (FEC)

A great advantage of digital transmission is its ability to build on the well-developed field of error correction. Sufficient redundancy is designed into the transmitted signal to allow correction of virtually all errors caused by transmission impairments. The International Telecommunications Union (ITU) standard J.83 annex B is the standard for digital cable transmission in the North America. It defines a concatenated coding system of Reed–Solomon (RS) block error correction with convolutional trellis coded modulation for highly robust performance in the cable channel. Interleaving is provided to correct bursts of errors of up to 95 μs.

After error correction the 64 and 256QAM modulation modes deliver 27 and 39 Mb/s of near error-free information, respectively, to the receiver.

105.5 Consumer Premises Equipment

The First Converters

The earliest cable systems had trouble carrying VHF channels 7–13 at their broadcast frequencies of 174–216 MHz. Because signal loss in coaxial cable increases at higher frequencies many more amplifiers were required, causing much higher distortion, than delivering channels 2–6 at 54–88 MHz. To solve this problem the early systems carried channels 7–13 in the frequency band *below* channel 2 (54 MHz). This frequency range is called the subband and is numbered T7–T13. The earliest converters *block converted* these subband frequencies back to the broadcast channels 7–13 near the subscribers location for reception on standards TVs.

The First Set-Top Converters

Early CATV set-top converters solved two technical problems: cochannel interference and the need for more channels. Without a set-top converter each channel was received by the TV on its cable channel frequency. Because TVs do not have effective RF shielding many subscribers received one or more broadcast channels strongly enough at the TVs input terminals to cause direct pick-up interference. This causes a ghost to be displayed on top of the signal received from the CATV system. Set-top converters allowed the CATV channel to be received in a well-shielded tuner and converted to a TV channel not broadcast in the local area. If the local stations transmitted on channels 2, 4, and 6, the output of the set-top would use channel 3 to the TV set. This eliminated the ghost effect and is still important today.

The second, and possibly the most important use of the set-top, was to extend channel capacity. The use of the UHF broadcast channel frequencies was not viable for early CATV systems due to their much higher transmission loss in coaxial cable. To go beyond the 12-VHF channels required the use of nonbroadcast channels. This required the use of a new tuner that could tune these cable-only channels. This also provided a rudimentary form of **access control**. Only those subscribers who had paid for the extended service were given a set-top that could tune the added channels.

Set-Top Designs

Modern CATV converters serve many additional functions. One of these is to provide tuners with good performance in the cable environment. To this day, almost all TV tuners are single conversion with one mixer and one oscillator. TV stations are allocated licenses on nonadjacent channels in each broadcast market. This means, for example, that if channel 3 is active in a certain market channels 2 and 4 will not be used. In addition, TV tuners are required to receive very low-signal levels, in the low-microvolt range to receive distant stations. For this reason, TV tuners must be high-gain, low-noise devices. When presented with strong signals on every channel as occurs in CATV transmission, standard TV tuners overload and generate distortion products, which range from imperceptible to very annoying in the TV picture. The signals in a properly maintained cable system are maintained at above 1 mV.

By around 1978, **descrambling** converters were given unique addresses to enable their control by computers located at the CATV system offices. At the same time, 400-MHz distribution plants were introduced; this boosted channel capacity to 52 channels and added the capability for two-way communications over the cable. At about the same time, wireless remote control of converter channel tuning was becoming available. By 1980, infrared (IR) control of converters was available to CATV operators. Microprocessors were added to converters that converted the remote control signals to tuning voltages and could interpret the addressable data sent from the cable companies' computers. This was really the start of the modern, descrambling converter as we know it today.

Addressable Converters

The addressable cable converter offers many advantages over the older preprogrammed converter. Initially, the most important advantage of was the ability to change services at the subscriber location, without requiring a technician to visit the home and replace the equipment. A service representative at the cable system office could enter the subscriber service change into the billing computer and the change was made immediately. It had an added benefit of upgrading the subscribers billing information at the same time. By about 1983, video demodulators were added to the converter. This allowed processing of baseband video and audio and added volume control as a converter feature. This was a major convenience for the many subscribers who did not have remote control TV sets. Baseband processing added the capability to descramble the higher security baseband scrambled video signals and provide audio privacy capability which removes the possibly offensive audio signals of channels not wanted by the subscriber. This was also the time many cable plants and converters increased bandwidth to 550 MHz with 77-channel capacity.

Impulse Pay-per-View Converters

The introduction of addressability simplified the process of ordering a one-time **event** over the cable. If a special event, such as a championship boxing match or a live concert, was offered on the cable a subscriber could order the event by calling the cable company, have it transmitted to home or place of business, and pay a one-time charge for the viewing. This system of pay-per-view was relatively successful but required planning by the subscriber well in advance of the event. It also required a large number of service representatives be available to staff the telephones.

Between 1983 and 1985, a new technology, **impulse pay-per-view**, was introduced using store-and-forward technology. This allowed a cable company to offer special events or movies 7 days a week, 24 hours a day. Subscribers could **purchase** the event shortly before or after the event started, using only their handheld remote control. The number of subscriber purchases increased dramatically. The converter became a two-way device transmitting the subscriber purchase information back to the cable system via a telephone modem or RF transmitter built into the converter. A record was made of every purchase made by the subscriber. Periodically the system polled all of its converters and uploaded purchases to the master billing computer at the cable office.

Over the years, many user friendly features were added to the converter. On-screen displays and downloadable electronic program guides are becoming common. Digital audio reception capability is being added to provide high-quality radio-type services as well as digital transmission of the audio signals for television channels. These added functions continue to make converters easier to use and improve the operators ability to provide enhanced services.

Digital Cable Converters

Digital cable converters add the ability to receive digital video transmissions on CATV plants. These sophisticated devices start with the most advanced form of analog converter and add QAM reception, MPEG-2 video decompression, Advanced Television System Committee (ATSC) or MPEG audio decompression, and secure digital decryption capabilities. One way to look at them is as a combination of an advanced analog converter and a digital DBS receiver.

Digital cable converters will finally remove the limits on channel capacity and picture quality to the home that have defined cable since its inception. The high security enabled by digital encryption technology will allow new services with very high value, while compression dramatically increases the channel capacity to carry them.

Cable Modems

Cable modems deliver high-bandwidth data service to consumers over the CATV plant. Typical rates for data delivery run to 27 Mb/s for downstream delivery to the consumer. Upstream rates for the various modem products vary widely from symmetrical service, where the upstream transmission matches the downstream rate, to products using telephone return channels limited to 28.8 kb/s. With as much as 1000 times the maximum rate available on dial-up telephone lines cable modems seek to provide enhanced Internet access and linked digital video and data services to personal computers.

Interdiction

Interdiction is an alternative to **scrambling** channels as a means to selectively deliver CATV services to paying subscribers. It has the advantages of using no set-top converter and delivering all paid services to any cable outlet in the home. This allows the unrestricted and direct connection of multiple VCRs and TVs and the ability to simultaneously watch and record two pay services. Modern interdiction systems are addressable and offer the same services as conventional cable systems. The common interdiction method injects a **jamming** oscillator onto the interdicted channel making it unwatchable. Interdictions disadvantages include the need for an interdiction module for every subscriber, not just those buying enhanced services as with scrambling, and the piracy threat of delivering clear services to every home up to the point of interdiction.

Each subscriber in an interdiction system will have a complete access control and jamming module in an off-premises enclosure. Enclosures typically service between 1 and 16 subscribers. Each subscriber module has a unique address, which can be addressed by the cable systems master computer. Each module can be authorized for any number of channels or disconnect service entirely.

The heart of the interdiction system is the jamming oscillator. Each oscillator consists of agile **voltage controlled oscillators (VCO's)** which sweep a band of channel frequencies. The number of **VCO's** is determined by the number of channel bands or the total bandwidth of the cable system to be processed. Typical VCO range is limited to approximately 50 MHz or 8 channels in an NTSC system. Lengthening the VCO range will reduce the on-channel time and will start degrading the effect of jamming on each channel. The VCO will inject a carrier between −250 and +750 kHz of the video carrier of the interdicted channel and skip those channels which the subscriber is authorized to view. Each subscriber module will consist of the jamming oscillator, an oscillator injector, **phase lock loop (PLL)** circuitry to control the oscillator, AGC control of the incoming channel signals, and a microprocessor.

105.6 Access Control and Security

Access control is applied to cable television systems to provide the means to offer services on a individually controlled basis. It allows consumers to choose, and pay for, only those services they desire. Various techniques enable the commonly encountered cable services such as subscriptions, events, pay-per-view movies, and **near video on-demand** movies. Access control has developed from simple physical schemes to sophisticated computer-controlled techniques.

Access control techniques themselves do not prevent unauthorized individuals from stealing services. That is the job of the security measures. When piracy becomes widespread the delivery of enhanced services can become either too expensive for the paying consumer and uneconomical for the operator. The inherent tradeoff of a security system is in how much to invest to hold off the pirates who would defeat it. Security measures protect revenues and make it worth deploying new services but increase the investment in set-tops and operating procedures.

Access Control

Access control is a combination of **access requirements** and **authorization rights**, thought of as lock and key, where access requirements are the lock and authorization rights are the key. Access control regulates access to system services, each identified by one or more **service codes**. These services, typically, are those for which cable customers must pay. Typically, the installation of a set-top in the home enables the systems access control capability.

Service codes are tokens used by the system to identify a particular service. The maximum number of service code values determines the maximum number of services supported by the system at any particular time. For example, if a typical system provides 256 service codes, then it supports up to 256 unique services. Since a single service might use more than one service code, however, the actual number of supported services is often less than 256. Access control systems for digital video offer many thousands of service codes to support the much larger number of digital channels and services.

Access Requirements

In order to access a service, that service must have a service code assigned and be authorized within the set-top box. Fundamentally, these are the only access requirements.

Authorization Rights

The system grants **authorization rights** to a service by turning on the service code assigned to that service in the set-top. Alternately, the system rescinds authorization rights to a service by turning off that service code.

Physical Access Control Techniques

The physical presence (or absence) of an electronic **trap** located between the cable itself and the converter is an early form of access control used to authorize subscription-type services. Typically, traps are on the utility pole where the residential drop originates. In this case a set-top converter may not be required. The cable installer provides access control by inserting or removing the trap device.

Computer-Controlled Access Control Techniques

In modern systems, a microcomputer in the set-top enforces access control by communicating electronically with head-end computers. The set-top only sends services to the television monitor if the service code is authorized. Depending on the type of service and on the capabilities of the set-top the control system can either directly or indirectly authorize services.

Head-End Authorization. Direct authorizations originate from the system head end. In this case, the head end sends a message to the set-top to authorize or to deauthorize a particular service code. The control computer **tags** the service by attaching this same service code to the signal. When the viewer ultimately selects the tagged service, the set-top reads the attached service code and checks to see if it is authorized. To make a subscription request, for example, the viewer calls an advertised telephone number and speaks to a service representative, who makes the authorization. To cancel the subscription (or event), the viewer requests the service representative to deauthorize the service. For subscriptions, the tag value never changes and an authorized viewer always has access. For events, the head end changes the tag value at the end of the event, forcing the set-top boxes to deauthorize and switch to a different screen. After the tag changes, head-end computers transmit messages to deauthorize the service code in all set-top boxes to allow the service code to be used again at a later time.

Self-Authorization. Self-authorizations schemes also exist for viewers to purchase services on impulse without any communication with the head-end. The set-top self-authorizes for a period of time and then either self-deauthorizes or waits for the head-end computer to deauthorize it. This technology requires the set-top to have some means to inform head-end computers of such impulse purchases by the viewer, such as a telephone modem or RF return transmitter. With this capability, the set-top self-authorizes services at any time, knowing that head-end computers will later read them and bill viewers appropriately.

Some form of deauthorization always follows self-authorization so the set-top terminates the service after it is completed, for example, when the movie ends. In a manner similar to authorization, there are two forms of deauthorization: deauthorization by the head end or self-deauthorization via a control timer.

The head-end deauthorization scheme enables services such as impulse pay-per-view. The viewer, on impulse, purchases a movie (or other offering) merely by requesting it via the handheld remote control unit or the set-top's front panel. The set-top tunes the service, reads the service code from the tag, and authorizes it. The box then delivers the service and stores the purchase in memory for later transfer to head-end computers.

At the end of the movie, the head-end changes the tag to a currently deauthorized value, forcing the set-top to stop delivering the service. As in the pay-per-view case previously discussed, head-end computers then deauthorize the service code in all set-tops.

The last case, where the set-top both authorizes and deauthorizes itself, enables a service type known as near video on-demand. In this service, the viewer purchases a movie that is actually running at staggered intervals on multiple cable channels. For example, a $2\frac{1}{2}$ hour movie might be running on five cable channels spaced out at 1/2 hour intervals. At the time of the purchase request, the set-top waits for the nearest movie start time (no more than 1/2 hour away) and automatically tunes the correct channel.

After tuning the movie's channel, the set-top reads the service code carried in the tag, self-authorizes, and stores purchase information for later transfer to the head end. The authorization time carried in the tag determines how long the authorization is valid. After the authorization time expires, the set-top automatically terminates that service by deauthorizing its service code.

While the timer is running and since all channels carrying the movie have the same tag, the viewer can, in effect, fast-forward or reverse the movie in 1/2 hour intervals by tuning to these different channels until the self-deauthorization occurs.

Security Systems

A security system can be defined as a series of techniques applied to the distribution of cable services that prohibit users from bypassing the system's access control facilities. Optimum security in cable systems is accomplished through two primary venues, legal actions (both litigation and legislation) and secure techniques (both physical and electronic schemes).

Legal Actions

The threat of legal action against the perpetrators is an important part of anti-piracy efforts. The federal law making cable theft a felony is a principle piece of legislation enabling law enforcement agencies to intervene in piracy cases. Many state statutes and some municipal codes also prohibit the sale, and a few even possession, of pirate devices. Cable equipment suppliers and cable operators support these laws by assisting law enforcement agencies at all levels of government wherever possible.

Secure Techniques

Secure techniques include both physical and electronic means. Physical schemes are used primarily within the head end and in the distribution paths for set-tops. They are primarily directed at keeping converters away from pirates who may modify them to provide service without payment. Electronic measures are based on coordination between installed set-top boxes and head-end equipment. They are performed in two primary areas:

- Security of the authorizations stored in specific set-top boxes
- Security of the signals transmitted downstream on the cable itself

The cable operator enters the set-top box's serial number into the access control computer. This number relates to a specific physical unit which is now known to the system. During a configuration process the control computer gives the set-top a unique address along with a system specifier. Using this unique address, the system then authorizes access to services ordered by the viewer.

To scramble an analog television signal, cable operators modify one or more of its key characteristics. Typical scrambled pictures appear faded or torn with the left-hand side of the picture on the right-hand side of the screen and vice versa. Techniques giving rise to these effects suppress the picture synchronization signals or invert the video information. Less common techniques change the order of the video lines in the picture, individually shift them right or left, or rotate them around a dynamically changing split point. Scrambling techniques may also modify the audio portion of the signal.

Electronic security for digital video signals is implemented as digital encryption of the compressed bit stream. The advanced art of encryption and key delivery can provide very high signal security on all of the digital services when well implemented.

Defining Terms

Access control: A scheme utilized in cable television systems to determine access rights as a function of predefined access requirements.

Access requirements: A set of conditions that must be satisfied prior to granting access to certain services carried on cable television systems. The system first assigns a service code to the service and then grants authorization rights for the service to the set-top box.

Authorization rights: A generic term describing the systems response to various authorization states of a service. If the state is ON, then the system delivers the service. If the state is OFF, then the system blocks the service.

Authorization state: Each service exists within the set-top box in one of two authorization states: on (or authorized) and off (or deauthorized).

Automatic gain control (AGC): An electronic circuit that detects the level of a signal and either increases or decreases its amplitude before passing the signal to the next stage.

Converter: A generic term referring to any device interposed between a cable system and a service rendering device (for example, a television monitor) which changes the cable signals into a form usable by the rendering device.

Decibel-millivolts (dBmV): A measure of power levels of a signal common in the cable TV field. It can be stated as the level at any point in a system expressed in decibels relative to 1 mV.

Descrambler: An element of some cable converters that is capable of descrambling secure cable signals.

Dither: The repeated sweep in frequency between two extremes.

Event: A type of service available on modern cable systems characterized by fixed start and stop times, which occurs, typically, once. Viewers purchase this service by calling a service agent who authorizes the service for that viewer and who records the purchase for later billing purposes.

Frequency shift keyed (FSK): A type of digital modulation in which a carrier is varied in frequency.

Head end: The point of origination of the cable TV signals before they are transmitted out on the system. Typically this equipment is located near the business office of the cable company.

Impulse pay-per-view: A type of service available on modern cable systems that enables viewers to purchase the service using their handheld remote units or the set-top front panel. The set-top box typically stores the purchase in memory and transfers it to the control computer at a later time.

Interdiction: The process of denying a particular service to a subscriber.

Jamming: Adding interference to a signal making it unusable.

National Television Systems Committee (NTSC): The U.S. broadcast standard for TV video transmission also used in Japan and many other countries.

Near video on demand: A type of service available on modern cable systems that enables viewers to purchase a service at nearly any time and that is always available on the system within a short interval. The set-top box stores the purchase in memory and transfers it to the control computer at a later time.

Pay-per-view: A type of service in which viewers pay to view a specific event, e.g., a movie.

Phase lock loop (PLL): A circuit used to stabilize oscillator circuits to some predetermined frequency by monitoring the actual oscillator frequency and sending a correction voltage to the oscillator tuning circuits.

Purchase: A viewer makes a purchase in one of two ways: either by calling a service agent, or on impulse by using the set-top remote unit or front panel controls.

Scrambling: Modifying standard video signals so they can only be received by authorized devices, typically set-top converters.

Security: A generic term referring to one of many techniques used on cable systems to prohibit viewers from bypassing the normal access control feature. Example techniques include legislation, litigation, and various electronic guises.

Service code: A symbol (or token) used by a cable system to identify and control access to a particular service. Service codes are authorized and deauthorized within the set-top box by the control computer.

Set-top: A specific type of converter that follows a cable systems access control constraints. Often set-tops include a descrambler and/or decryptor for changing scrambled or encrypted cable signals into a form usable by the subscriber.

Subscription: A type of service that after being authorized is available to the viewer 24 h per day throughout the billing cycle. Typically, the viewer places a request with a service agent who authorizes the service for the viewer and who records the purchase for later billing purposes. Early systems implemented subscriptions using traps.

Synthesizer: An oscillator capable of producing multiple frequencies from a single clock source.

Tag: Multiple symbols added to a service signal that are detected and used by a set-top box to deliver services. Signals are tagged with, at a minimum, the service code assigned to the service and are used in the set-top box to determine if delivery to the viewer is authorized.

Trap: A device used to control access to subscription services by blocking them at the point of delivery into the home, usually on the telephone pole where the cable drop originates.

Voltage controlled oscillator (VCO): An oscillator whose tank circuit contains a varacter diode, which changes capacity as its bias changes; the most common form of oscillator control in TV receivers and other communication equipment.

106
Video Servers

A.L. Narasimha Reddy
Texas A&M University

Roger Haskin
IBM Almaden Research Center

106.1 Introduction 1502
106.2 Data Server..1503
  Disk Scheduling • Multiprocessor Communication Scheduling • Other Issues
106.3 Video Networks 1508
  Interactive Television Server Architecture
106.4 Network Multimedia 1511
  Servers for Network Multimedia • Internet Video Servers

106.1 Introduction

The recent dramatic advances in communications technology has led to a demand to support multimedia data (digitized video and audio) across data networks. In part this demand comes from the personal computer marketplace, where the use of video and audio in PC applications and games is now pervasive, and in part the demand comes from the entertainment marketplace, where there is a move to use digital video to supplant the traditional analog methods of content creation and broadcast. This intersection of technologies and markets has led to the development of video servers.

Several telephone companies and cable companies are planning to install video servers that would serve video streams to customers over telephone lines or television cable. These projects aim to store movies in a compressed digital format and route the compressed video to the customer (through cable, telephone lines, or local area networks) where it can be uncompressed and displayed.

These projects aim to compete with the local video rental stores with better service: offering the ability to watch any movie at any time, avoiding the situation of all of the copies of the desired movie already being rented out, offering a wider selection of movies, offering other services such as tele-shopping and delivery of other video content such as interactive video games.

A number of business models have been proposed for such video servers. The business models include complete interactive video games and near-video-on-demand in hotels, among others. The systems' requirements vary considerably depending on the targeted markets. Near-video-on-demand envisions supplying popular/recent movies at regular intervals, say every 5 min, to requesting customers over some delivery medium. This form of service is expected to be popular in hotels, replacing the existing situation of making movies available at only few times during the day. Few tens of movies are expected to be available through such a server. **Video-on-demand** (**VOD**) envisions replacing conventional pay-per-view programming (as available on cable TV), which allows customers to view a small number of movies at widely spaced intervals, with a system that allows customers to choose from a large number of movies, and play them within a few minutes or seconds of when the customers request them. Some of these servers plan to provide the ability to supply VCR-like capabilities to pause/resume and fast forward/reverse a movie stream. Such servers are said to provide **interactive TV (ITV)** service. Interactive video servers could have wide applications. Some of the planned applications of interactive video servers include training, education, travel planning, etc. These interactive video servers play a predetermined set of video sequences based on the user's input. Interactive video games require a higher level of interaction with the video server since the playout of the video sequence varies more rapidly based on the user's input. Video teleconferencing requires two-way interactive response between the users. Which of these several business models will predominate remains to be seen.

Digital video broadcast (DVB) systems replace analog video channels with compressed digital video. DVB allows more channels to occupy the same frequency spectrum, and allows expensive and failure-prone electromechanical devices (e.g., video tape players) to be replaced with more reliable digital storage. In a DVB system, the video storage and transmission are identical to those in a VOD or ITV system. In DVB, however, instead of the video being under the control of a user, a piece of automation equipment controls the sequencing of video playback. Here, sequencing means creating a smooth, continuous video stream out of multiple independent pieces of stored video, for example, a series of commercials followed by a segment of a TV program, a station break, more commercials, etc. A DVB system is actually in this respect more comparable to an ITV system than to VOD, since the system is required to respond in timely fashion to commands from the automation equipment.

A number of pilot projects have been underway to study the feasibility of these approaches. Several major companies in the computer systems business (IBM, SGI, HPDEC), computer software (Oracle, Microsoft, IBM), telephone carriers (AT&T, US West, Bell Atlantic, Ameritech), cable companies (TCI, Time Warner, Cox cable) and equipment manufacturers (Scientific Atlanta, General Instrument) have been actively participating in these trials. The technical challenges in making these projects economic/business successes are daunting. In this chapter, we highlight some of the technical problems/approaches in building such video servers, with an emphasis on communication aspects of the problem.

Currently, with **Motion Pictures Expert Group-1 (MPEG-1)** compression at 1.5 Mb/s, a movie of roughly 90 min duration takes about 1 gigabyte of storage space. Storing such large data sets in memory is very expensive except for the most frequently requested movies. To limit the costs of storage, movies are typically stored on disks. It is possible to store infrequently accessed movies on tapes or other tertiary media. However, it has been shown that these media are not as cost-effective as magnetic disk drives even for infrequently accessed movies [Chervenak, 1994].

A video server storing about 1000 movies (a typical video rental store carries more) would have to spend about $100,000 just for storing the movies on disk at the current cost of $0.10/megabyte. If higher quality picture is desired and MPEG-2 compression is used at 6.0 Mb/s, then the cost of disk space goes up to $400,000. To this, the cost of computer systems required for storing this data needs to be added. Other nonrecurring costs include costs for enhancing the delivery medium and providing a **set top box** for controlling the video delivery. This requirement of large amounts of investment implies that the service providers need to centralize the resources and provide service to a large number of customers to amortize costs: However, the ever decreasing costs of magnetic storage (at the rate of 50% every 18 mo), and computing power (at the rate of 54% every year) may make these cost considerations less of a problem if the technical challenges can be overcome.

106.2 Data Server

A large video server highlights most of the technical issues/challenges in building a video server. A large video server may be organized as shown in Fig. 106.1. A number of nodes act as *storage nodes*. Storage nodes are responsible for storing video data either in memory, disk, tape or some other medium and delivering the required input/output (I/O) bandwidth to this data. The system

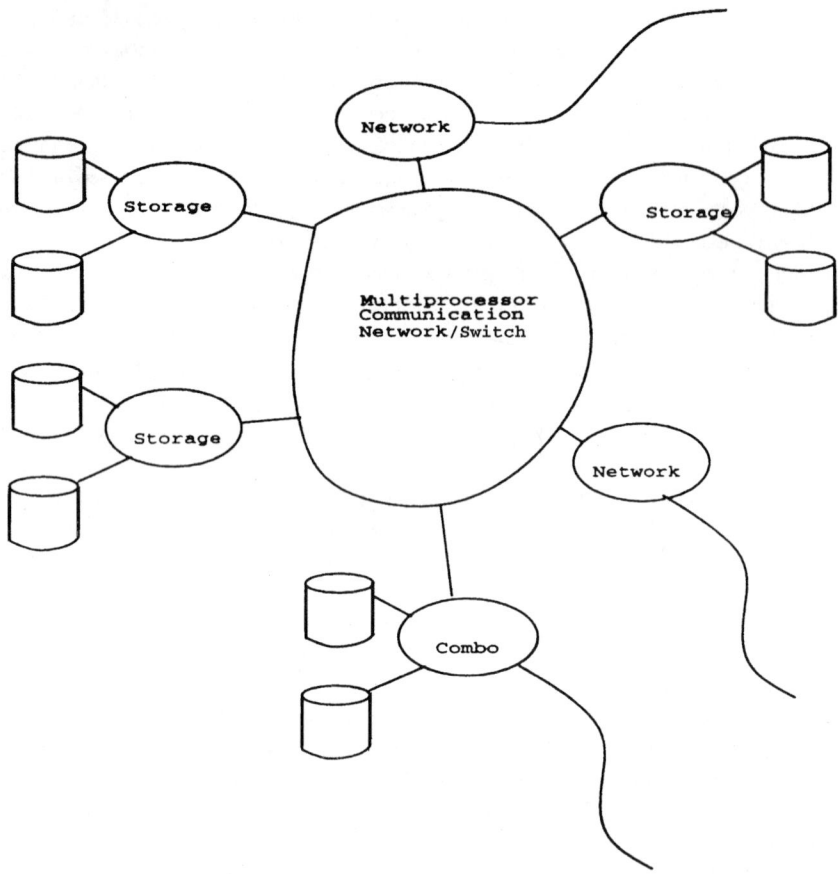

FIGURE 106.1 System model of a multiprocessor video server.

also has *network nodes*. These network nodes are responsible for requesting appropriate data blocks from storage nodes and routing them to the customers. Both these functions can reside on the same multiprocessor node, that is, a node can be a storage node, a network node, or both at the same time. Each request stream would originate at one of the several network nodes in the system, and this network node would be responsible for obtaining the required data for this stream from the various storage nodes in the system. The design of network nodes will change based on the delivery medium. The logical separation (whether or not they are physically separated) of nodes into network nodes and storage nodes makes it easier to adapt the video server organization to different delivery mediums.

The delivered data stream is fed to a set top box (STB) at the customer's site. The STB decompresses the digitally enoded data into a suitable input form for the TV monitor. The STB may provide additional features such as the ability to pause/resume and fast forward/reverse an incoming video stream. The design of the STB is a hotly discussed topic with proposals ranging from a dumb decoder to intelligent controller with most of the functions of a personal computer. The design of the STB depends on the functions provided, compression standard, bandwidth of the compressed stream, and input standard of the TV monitor, among other things. To enable future adaptation to evolving compression technology, some projects envision doing decompression completely in software by a processor in the set top box.

To obtain high I/O bandwidth, data has to be **striped** (distributed) across a number of storage nodes. If a movie is completely stored on a single disk, the number of streams requesting that movie will be limited by the disk bandwidth. To enable serving a larger number of streams of a single movie, each movie has to be striped across a number of nodes. As we increase the number of nodes for striping, we increase the bandwidth for a single movie. If all of the movies are striped across all of the nodes, we also improve the load balance across the system since every node in the system has to participate in providing access to each movie. A number of related issues such as fault-tolerance and incremental growth of the system are closely linked to striping characteristics, and we will discuss these issues later in the chapter. The unit of striping across the storage nodes is called a block. In earlier studies on disk scheduling [Reddy and Wyllie, 1992], 64–256 kilobytes is found to be a suitable disk block size for delivering high real-time bandwidth from the disk subsystem.

As a result of striping, a network node that is responsible for delivering a movie stream to the user may have to communicate with all of the storage nodes in the system during the playback of that movie. This results in a point-to-point communication from all of the storage nodes to the network node (possibly multiple times depending on the striping block size, the number of nodes in the system, and the length of the movie) during the playback of the movie. Each network node will be responsible for a number of movie streams. Hence, the resulting communication pattern is random point-to-point communication among the nodes of the system. It is possible to achieve some locality by striping the movies among a small set of nodes and restricting that the network nodes for a movie be among this smaller set of storage nodes.

The service for a video stream can be broken up into three components: (1) reading the requested block from the disk to a buffer in the storage node, (2) transferring this block from the storage node buffer to a buffer in the network node across the multiprocessor interconnection network, and (3) delivering the requested block to the user over the delivery medium. The critical issues in these three components of service are disk scheduling, interconnection network scheduling, and delivery guarantees over the delivery medium, respectively. We discuss each of these service components briefly here.

Disk Scheduling

Traditionally, disks have used seek optimization techniques such as **SCAN** or shortest seek time first (SSTF) for minimizing the arm movement in serving the requests. These techniques reduce the disk arm utilization by serving requests close to the disk arm. The request queue is ordered by the relative position of the requests on the disk surface to reduce the seek overheads. Even though these techniques utilize the disk arm efficiently, they may not be suitable for real-time environments since they do not have a notion of time or deadlines in making a scheduling decision.

In **real-time** systems, when requests have to be satisfied within deadlines, algorithms such as **earliest deadline first (EDF)** and least slack time first are used. The EDF algorithm is shown to be optimal [Liu and Layland, 1973] if the service times of the requests are known in advance. However, the disk service time for a request depends on the relative position of the request from the current position of the read–write head. Moreover, due to the overheads of seek time, strict real-time scheduling of the disk arm may result in excessive seek time cost and poor utilization of the disk.

New disk scheduling algorithms have been proposed recently that combine the real-time scheduling policies with the disk seek-optimizing techniques. Some of these algorithms include SCAN-EDF [Reddy and Wyllie, 1992] and grouped sweeping scheduling [Yu, Chen, and Kandlur, 1993]. These algorithms batch requests into rounds or groups based on request deadlines, serving requests within a round based on their disk track locations to optimize the seek time. These algorithms are shown to significantly improve real-time performance at smaller request blocks and improve nonreal-time performance at larger request blocks [Reddy and Wyllie, 1992]. Fig. 106.2 shows the impact of request block size on various scheduling algorithms. Other factors to be considered include multiple data stream rates, and variable bit rates.

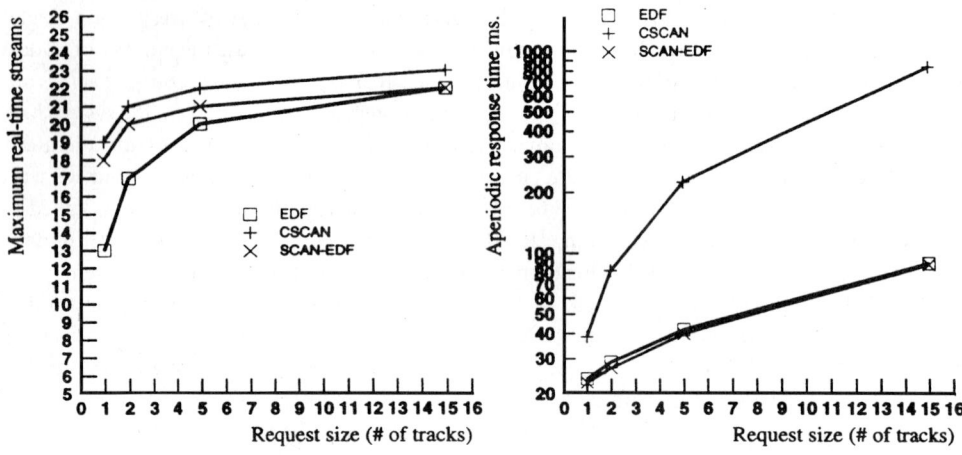

FIGURE 106.2 Performance impact of scheduling policies and request block size.

Multiprocessor Communication Scheduling

The communication scheduling problem deals with the issue of scheduling the network resources of the multiprocessor computer system for minimizing the communication delays. If the nodes in the multiprocessor system are interconnected by a complete crossbar network, there is no communication scheduling problem since any pair of nodes in the system can communicate without a conflict in the network. Regular distribution of video data over the storage networks enables us to guarantee that if source and destination pairs are scheduled without a conflict for the transmission of the first block of data then the entire playback of that video stream will not see any source and destination conflicts [Reddy, 1995]. Providing similar guarantees within the multiprocessor network depends on the network organization. Some interconnection networks such as Omega networks and hypercubes have been shown to exhibit properties that make it amenable to provide guarantees of no conflicts within the network. A simple solution such as a round-robin distribution of data blocks among the storage nodes suffices to provide such guarantees [Reddy, 1995] in these networks.

The basic approach to providing guarantees of conflict-free transfers relies on time-division multiplexing of the network resources. The network usage is divided into a number of slots. Each slot provides sufficient time to transfer a fixed block across the network under no conflicts. Each stream is assigned a fixed slot. The transfers are repeated after a fixed amount of time (a frame) to transfer the next block of each stream. The regular data distribution and the network properties ensure that if the first block of a stream is scheduled without any conflicts at the source or destination and within the network, the entire stream of blocks of that video stream can be transferred without any conflicts.

Other Issues

In this section, we briefly touch upon a number of issues that have significant impact on the design of a video server. The intent is to give the reader a feel of the problems and to provide starting pointers for further reading.

Buffering

Buffering can make a significant impact on the video server design. Once, the video stream is started on the consumer's monitor, the delivery of the video stream has a real-time nature, that is, the data has to be supplied at a constant rate. However, the system can control when the first block of the stream is delivered to the consumer's monitor (latency). Buffering can be used effectively to control the latency of delivery in video-on-demand applications. The more the data is buffered, the greater the stream startup latency, and the longer the time to serve a request block at the server. In more interactive applications, long latencies cannot be tolerated. Buffering can be used to increase the request block size or to provide extra time for serving a request of a video stream (deadline extension). Both these options, larger request blocks and deadline extensions are shown to improve the number of video streams that can be supported [Reddy and Wyllie, 1992].

Interactive Response

Providing a VCR-like capability of pause/resume and fast forward/reverse requires that sufficient resources be available at the server to absorb the variations in bandwidth demands due to such operations [Jayanta et al., 1994]. Pause operation may reduce the bandwidth demand at the server and the other three operations may increase the bandwidth requirements. To support such operations, the server has to determine that these bandwidth variations do not affect the delivery of other scheduled streams. It is also possible to keep the bandwidth requirements at the server constant by serving alternate copies of video streams that have less quality encoding at the higher frame rates required by the fast-forward and reverse operations [Chen, Kandlur and Yu, 1994]. Then the server can serve the alternate copies of video data during fast-forward and reverse operations without altering the schedule of the scheduled streams.

Fault Tolerance

Bandwidth considerations, load balancing, and other issues favor striping data across multiple disks and nodes in a multiprocessor video server. However, this increases the probability of losing service due to a failure of one of the disks or nodes in the system. In *line of business* applications such as ITV, VOD, and DVB, the video server must be highly reliable and tolerate hardware failures with minimal or no system downtime. To illustrate the importance of this, consider a large ITV system supporting 30,000 users (systems of this size are being built as of this writing [WSJ, 1995; Comput. Int. 1995; Bus. Wire, 1995a]). Assuming the primary application is pay-per-view movies at $5 apiece, the system is generating approximately $75,000/h! This is approximately a year's pay for the service person, who undoubtedly would not collect it for long if outages were frequent or prolonged.

Multilevel encoding has been proposed to tolerate against failures in the delivery medium. In these schemes, the video data is stored in the server in two layers, a base layer and an enhancement layer. When the available bandwidth is affected due to failures, the server can continue to provide a lower quality picture to the consumer by sending only the base layer and thereby putting less load on the network. A similar idea can be used for reducing the disk overhead by protecting against the loss of the base layer only and by not providing redundancy to protect against the loss of the enhancement layer.

As opposed to a conventional file or database system, loading from tape is not a desirable way to recover from storage device failures. At the 6-Mb/s video data rate common in most ITV systems, a 2-h movie requires 5.2 gigabyte of storage, which takes 1/2 h to restore even from a relatively high-performance 3 megabyte/s tape drive. Some ITV systems in use by customer trials have 1 tetrabyte online [WWW n.d.], which would require 97 h (or many tape drives!) to restore. This makes redundant storage [mirroring or redundant arrays of inexpensive disks (RAID)] a requirement.

Disk failures can be protected either by making duplicate copies of data (disk mirroring) or by parity protection (RAID techniques [Patterson, Gibson, and Katz, 1988]). These fault-tolerance techniques increase the availability of data and typically can tolerate the loss of a disk without losing access to data. However, these data protection techniques incur extra load under a failure and this extra workload needs to be factored into the design for providing guarantees after a failure. To protect against the failure of a SCSI bus (bus that interconnects multiple disks to a system's memory bus), it is possible to connect the disks to two SCSI buses such that the disks may be accessible through a second bus after a failure.

These techniques can guarantee availability of data but cannot guarantee availability of the necessary I/O bandwidth for timely delivery of data after a failure. Scheduling data delivery after a component failure is a difficult problem. Much work needs to be done in this area. Overdesigning the system is one of the possible options so that even after a failure the data can be delivered in a timely fashion. Dynamic resource allocation for tolerating failures is another possible option. In certain cases, only statistical guarantees may be provided as to the capability of providing the required real-time bandwidth after a failure.

Variable Bandwidth Devices

Some of the current disks utilize *zone-bit recording* which makes the track capacity a variable depending on the location of the track on the disk. This results in variable data rate depending on the track location. The variable data rate makes it harder to utilize the disk bandwidth efficiently. To guarantee the real-time requirements, the minimum supportable bandwidth can be assumed to be the average deliverable bandwidth. However, this results in inefficient use of the disk bandwidth. It is possible to use the remaining bandwidth for nonreal-time requests while utilizing the minimum disk bandwidth for real-time requests. Request spreading techniques [Birk, 1995] can be used to make sure that the requests are spread uniformly over the surface of the disk to utilize the disk at its average bandwidth.

106.3 Video Networks

The three most prominent types of digital networks used for ITV are **asynchronous digital subscriber loop (ADSL)**, **hybrid fiber-coax (HFC)**, and **asynchronous transfer mode (ATM)**. Although there are major differences between these networks, they all support both a high bandwidth video channel for sending video to the set top box, and a lower bandwidth control channel (usually bidirectional) for communicating user commands and other control information with the server. ADSL multiplexes the downstream video channel, a bidirectional 16-Kb/s control channel (which usually employs X.25 packet switching), and the existing analog telephone signal over the same copper twisted pair [Bell, 1993a, 1993b]. ADSL is designed to leverage the existing telephone company (telco) wiring infrastructure, and as such has been used in several VOD trials conducted by telcos [WNT, 1993; Plumb, 1993; Patterson, 1994].

ADSL is used for the connection between the customer premises and the telephone company central switching office. The server can either be located in the central office or at some centralized metropolitan or regional center, in which case the data travels between the server and the customer's central office over the telephone company's digital trunk lines. The video bandwidth ADSL is capable of supporting depends on the length of the subscriber loop (i.e., the distance between the central office and the customer premises). At present, ADSL will support 1.544 Mb/s (T1 bandwidth) over about 3 mi, which is the radius serviced by most urban central offices. MPEG compressed at T1 bandwidth can deliver VCR quality video, but is marginal for applications such as sports. The choice of T1 bandwidth is convenient because it is widely used in the telco trunk system and because T1 modems are readily available to connect the video server to the digital network.

As ADSL attempts to leverage the existing telephony infrastructure, hybrid fiber coax [Ciciora, 1995] leverages the existing cable TV plant, and as such is popular for ITV systems deployed by cable service providers. Standard cable plants deliver some number of analog TV channels (typically 70 channels using a total bandwidth of 450 MHz), which are fanned out to all subscribers. The cable system is structured as a large tree, with repeaters at each of the nodes and subscriber cable decoders at the leaves. However, 70 channels is barely a sufficient number for broadcast, and is not nearly sufficient for VOD or ITV. The basic idea behind HFC is to subdivide the tree into smaller subtrees (for example, one subtree per neighborhood), and feed each subtree independently by fiber from the cable company head end. Some number of channels in this subtree are still reserved for broadcast, and the remainder are available for interactive services. A segment of the cable's frequency range (typically 5–30 MHz) is reserved for upstream control signalling, and LAN-like packet switching is used on this upstream channel.

In an HFC system, each user of interactive services is assigned a channel for the period during which he is using the system. The server sends a command over the control channel to tell the STB which channel to select, and subsequently sends video to the STB over this channel. For security, STBs can prevent interactive channels from being selected manually.

In some early trials, a subset of the analog TV channels were reserved for ITV [UPT, 1993]. MPEG video from the server was converted to analog at the cable head end and multiplexed onto fiber. Since only a small number of channels (those not used for broadcast) on each subtree are available for ITV, the area covered by a subtree is small, and therefore a completely analog system is expensive.

Recently, new techniques [vestigial sideband (VSB), quadrature amplitude modulation (QAM)] have been developed that allow sending high-bandwidth digital data over analog TV channels. QAM-64, for example, allows 27 Mb/s of digital data to be carried by one 6-MHz analog TV channel. This payload can carry one high-bandwidth program [e.g. high-definition television (HDTV)], but is normally used to carry an MPEG-2 transport stream with a number of lower-bandwidth programs. For example, QAM-64 allows four 6-Mb/s MPEG-2 programs to be carried on a single 6-MHz analog TV channel, allowing a 70-channel cable TV plant to carry 280 channels. Higher bandwidth fiber-based systems are being deployed that support a downstream bandwidth of 750 MHz, which supports over 400 6 Mb/s MPEG programs. This is suitable for carrying a full complement of broadcast channels as well as a sufficient number of interactive channels for a neighborhood of 1000 households.

The downstream video path of a HFC video server is somewhat more complicated than its ADSL counterpart. The server sends individual MPEG programs to an MPEG multiplexor, which combines the independent programs into a single MPEG transport stream. The transport stream is then converted (e.g., by a QAM modulator) to a 6-MHz analog signal and then multiplexed onto one of the cable system's analog channels. The STB selects its MPEG program under control of the server from one of the MPEG transport streams.

Although ATM has not been deployed at nearly the rate anticipated a few years ago, it is still widely regarded as the future architecture for ITV. ATM has a number of strong advantages: it can support the high bandwidth required for video, it can multiplex video and control data over the same connection, and it allows bandwidth to be reserved over an end-to-end connection. A number of alternatives exist for transporting video over ATM. ATM defines a standard [ATM adaptive layer-1 (AAL1)] for continuous bit rate traffic, and another standard (AAL5) for packet-based data. Perhaps surprisingly, most ITV systems are using the packet-based approach. Even using AAL5, several alternatives exist: the ATM fabric can be used as a fast Internet protocol (IP) network, or a specialized video streaming protocol can be used [ATM, 1995]. Most ATM-based ITV systems implement the ATM Forum standard; examples of IP-based systems are discussed in a later section. Time–Warner's full-service network (FSN) ITV system, deployed in Orlando, Florida [Time, 1993], uses a combination of ATM and HFC. The server sends multiple MPEG program streams via ATM to a QAM modulator. From there, they are transmitted by fiber to the customer STB as previously discussed. Here, ATM is used primarily as a head-end switching fabric rather than as a distribution mechanism.

Interactive Television Server Architecture

Although the ITV server is often described as if it were a homogeneous black box, it is composed of a number of logically and, usually, physically independent pieces. Following is a description of an example ITV server built by IBM for the Hong Kong Telecom VOD trial [Haskin and Stein, 1995]. This system uses ADSL to distribute video, and is similar to ITV systems being deployed by some U.S. telcos. Figure 106.3 shows the trial setup used by IBM in Hong Kong. The ITV system as a

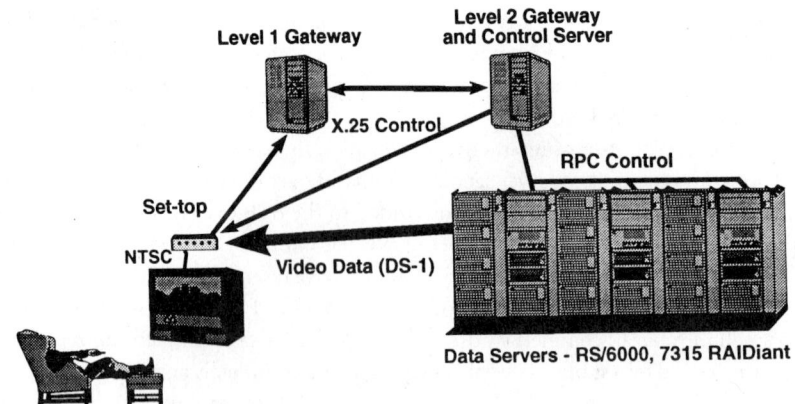

FIGURE 106.3 IBM's Hong Kong video server setup.

whole comprises an open system designed to allow an individual customer to connect to any of a variety of video servers. To this end, the network contains a component called the level 1 gateway, which serves as an intermediary between the customer and a corresponding level 2 gateway in each video server. This gateway architecture is the so-called video dial-tone system mandated by the FCC for ITV systems that use regulated common carriers. It is analogous to the standard telephony system, which allows a customer to connect to any desired long-distance carrier. The STB contains a control port, used to communicate with the video server and the level 1 gateway over the control channel using X.25, and a data port, over which the video server sends video and graphical images and downloads code to the STB at DS-1 data rate.

The video server contains an X.25 control port, used to communicate with the level 1 gateway and STB control ports. The video server's control interface to the level 1 gateway is called the level 2 gateway. This gateway architecture is what allows video servers from multiple service providers to connect to the system in an open manner. The video server also contains a number of DS-1 rate data ports sufficient to support the maximum expected number of active viewers.

Level 1 Gateway

The level 1 gateway (GWL1) provides the video dial tone for the set tops. The GWL1 processes requests for service from the set tops. If the set top is a valid subscriber, the GWL1 responds with a list of available video servers. Alternatively, and appropriate for many trials, the set top can be programmed to automatically select a default server in the initial request message. The GWL1 also establishes the high-speed data connection between the server and the STB. This is done in response to a request by the GWL2, specifying the server data port address and the STB identifier. The GWL1 maintains tables containing the data port address for each STB, and uses this table to issue commands to the data network switching equipment to connect the server port and the STB.

Level 2 Gateway

The level 2 gateway (GWL2) in the video server provides a standardized, open interface between the server and the network. As specified in [BA93b], the GWL2 performs the following functions:

- Connection to GWL1. Upon startup, the GWL2 initiates an X.25 connection to the GWL1 for the exchange of control information.
- Processing STB connection requests from GWL1. When the GWL1 receives a request from an STB to connect to a server, it sends a connection request to that server's GWL2. This request contains the STB's unique identification, its brand and version, and its network address.

- Initiating control and data connections to the STB. The GWL2 allocates an unused data output port, sends a request to the GWL1 to establish a connection between the data output port and the data input port of the STB, and finally establishes an X.25 control connection to the STB.

The GWL2 then downloads the version of set top enabling code (STEC) appropriate to that brand and version of STB. The STEC software provides the set top with the means to process the subsequent commands it will receive.

106.4 Network Multimedia

During the last several years the cost of computer power, storage devices, and storage media has dropped radically in the personal computer marketplace. This has led to the proliferation of interactive multimedia applications and games for the PC. To date, most multimedia applications run on stand-alone PCs, with digitized video and audio coming from local hard disks and compact disc read-only memorys (CD-ROMs). Increasingly, there has been a demand for file servers that support capture, storage, and playback of multimedia data. The reasons for this are identical to the ones that motivate the use of file servers for conventional data: sharing, security, data integrity, and centralized administration.

Servers for Network Multimedia

Although interactive TV is the glamor application for video servers, video playback over a local area network (LAN) is the bread and butter of the industry. A LAN multimedia system includes one or more servers, some number of multimedia client workstations, and a video-capable LAN. Applications for LAN multimedia include training videos, information kiosks, desktop news, and digital libraries.

The client workstation software architecture has much to do with the design of the multimedia LAN server. There are two basic client architectures, depending on whether or not the manufacturer-supplied driver software provided with the video codec is used.

Normally, the video codec (e.g., MPEG card and Indeo software codec) comes packaged with driver software that allows video to be played through the operating system multimedia support (e.g., the Windows Media Player or MCI programming interface). The loop to read a buffer of video data and present it on the display is implemented in this driver, which is a black box executable file supplied by the decoder manufacturer. Since this driver reads data through the file system interface, the server must make video available through that interface. Figure 106.4 shows a typical multimedia LAN server.

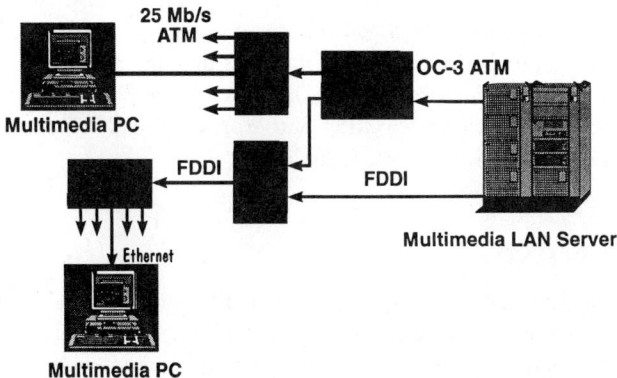

FIGURE 106.4 A multimedia LAN server.

Most client platforms (Windows, Macintosh, UNIX) support the notion of an installable file system (IFS), which the multimedia LAN server can use to provide video data. However, the performance characteristics of the multimedia IFS driver must be equivalent to local disk or CD-ROM. In particular, any readahead or extra buffering to smooth out variations in the network transfer rate must be implemented in the IFS driver, not in the video codec device driver (which is unaware that video is coming from the server as opposed to from local disk).

This file-based approach has the advantage of allowing the server to support almost any video codec. Furthermore, applications that use the operating system media player or multimedia programming interfaces (most multimedia applications fall into this class) need not be modified to use the server. For example, such servers allow copying a multimedia CD-ROM to the server, after which it can transparently be used by any number of clients simultaneously.

The server can either implement its own proprietary installable file system, or can use one designed to support a standard network operating system (NOS) protocol (e.g., Netware, LAN Manager, or NFS). A proprietary file system allows using more aggressive readahead and buffering or a network transport protocol tuned for video. However, a number of commercially available servers use standard NOS IFS drivers (or NOS clients). This has obvious advantages: it allows the server to support a large variety of platforms and makes it unnecessary for the user to install special software to use the server. Several commercial servers have been built using standard NOS clients: examples include IBM OS/2 LAN Server Ultimedia [IBM, 1993], Panasonic Video NFS [MEI, 1995], and IBM Multimedia LAN Server for AIX [IBM, 1996].

Contrary to popular belief, standard NOS network protocols work acceptably for video over a suitable LAN, that is, one with sufficiently high throughput, low latency, and low error rate. The need for throughput is obvious; low latency is required to prevent buffer starvation with the limited amount of readahead done by most standard NOS clients. For example, the NFS protocol supports a maximum 8 K read size. Application (e.g., codec driver) reads requests larger than 8 K are broken up by the NOS client and sent to the server as multiple 8-K requests. If the NOS client sends these requests synchronously (i.e., one at a time), the time to complete a request must average less than 54 ms to maintain a data rate of 1.2 Mb/s (the data rate for most CD-ROM MPEG). By supporting readahead (sending successive requests before previous ones complete), a NOS client can relax this latency requirement somewhat. Most NOS clients support readahead, but some older ones do not. The final requirement for low error rate stems from the fact that in most NOS clients error recovery is driven by timeouts. These timeouts are typically short enough to allow the occasional dropped packet without a visible interruption in video, but bursts of errors will cause video to degrade.

Modern technology allows configuring LANs to support video in a cost-effective manner. Switched ethernet, fiber distributed data interface (FDDI), and ATM LAN emulation are all suitable for video, and with high-performance routers can be scaled up to support a large building or campus.

A number of multimedia LAN server vendors have taken the approach of implementing proprietary streaming protocols to transport video over LANs. Typically, this forces the server vendor to supply special driver software for each video codec it wishes to support. This usually restricts such servers to a limited number of codecs and increases the time required to support new ones. However, a proprietary streaming protocol can, in principle, be designed to work better with a less-than-perfect network. Starlight Networks' Star Works and its MTP protocol is a successful example of this approach [Starlight, 1996a, 1996b].

Internet Video Servers

The recent explosion of interest in the Internet and the Worldwide Web have resulted in a number of attempts to support video over long distances using the Internet. These range from modest attempts to support low-resolution video over dial-up and Integrated Services Digital Network (ISDN) connections, to more ambitions attempts to transport high-resolution video over the MBONE wide-area network (WAN).

The earliest attempts to support video and audio over the Worldwide Web used store and forward. Using this technique, the video file is downloaded to the client workstation in its entirety, after which the operating system media player is launched to play the file. For all but the shortest videos, the delay to download before viewing is unacceptable.

Recently, a number of systems have become available that play video and audio in real time over the Internet. Most are implemented as a Web browser external viewer or plug in (the distinction is not important to this discussion), and most work more or less the same way. When the user clicks on a link to video (or audio) on a web page, the browser loads the contents of the file pointed to by the link and launches the viewer (or plug-in). This file, instead of containing the video data itself, contains a URL pointer to the data, often on a separate, dedicated video server. The viewer starts transferring data from the server, buffering it in random access memory (RAM) until it has a sufficient amount to begin playback. Video is decoded in software and displayed in a window; accompanying audio is played from the data buffer using the low-level audio subsystem programming interfaces (e.g., waveOutWrite on Windows).

In contrast to the file-based approach, this type of viewer usually does not take advantage of special hardware (e.g., MPEG cards) that might be on the system or that supports a limited number of cards using special drivers. Given the low video data rates possible over the Internet at present, this is usually not a serious consideration. On the other hand, rather than using a conventional network file system protocol (e.g., NFS) or even a conventional streaming protocol [e.g., transport control protocol (TCP)], these viewers usually implement special-purpose network transport protocols (layered on UDP) that are designed specifically for audio or video. Some additionally implement proprietary compression algorithms tailored to the Internet's low data rate and high error rate. Commercial products implementing proprietary transport protocols and/or compression algorithms include VDO Live [VDOnet, 1995], Real Audio, Xing StreamWorks [BusWire, 1995b], and InSoft [1996].

The research community and standards organizations have also been active in transporting multimedia over the internet. Real-time transport protocol (RTP) defines an encapsulation of any of a number of audio or video data formats into UDP datagrams that allows it to be streamed over the internet. Rather than retransmitting lost packets as does TCP, RTP timestamps packets to allow missing packets to be detected. RTP contains no flow-control mechanism, and leaves the task of recovering from lost packets to the video/audio codec. A number of commercial products are based on RTP (Precept FlashWare, Netscape LiveMedia), and several other products have announced intentions to support it (Xing, Real Audio, VDO Live).

At the high end, research is ongoing to stream video at high data rates over high-speed backbone networks such as the MBONE. The *vic* video conferencing system supports real-time video multicast using RTP [Schulzrinne et al., 1996]. Although the original version of vic supported only live video, Argonne National Laboratories has extended vic to support playback of stored video at high bandwidth (6 Mb/s) via RTP from IBM's Tiger Shark video server [Haskin, 1993] on an IBM SP-2 parallel computer [IBM, 1995]. Research is also underway to develop still better protocols. One such protocol, primary encoding transmission (PET) [Albanese et al., 1994], encodes MPEG data to add sufficient redundancy to tolerate high levels of packet loss while maintaining good picture quality. PET assigns priorities to each component of the MPEG data stream (headers, I-frames, and B-frames) and uses more redundancy for higher priority data. With only a modest space overhead, PET greatly increases MPEG's ability to tolerate data transmission errors. VTP [Chen et al. 1995] is an extension to RTP to provide a form of flow control by asking the server to slow down or speed up as network conditions dictate, and a form of error recovery by allowing the client to demand resend dropped video frames.

Defining Terms

Asynchronous digital subscriber loop (ADSL): Technology that allows video to be delivered over regular phone lines.

Data server: Computer system that stores and retrieves digital video.
Earliest deadline first (EDF): A real-time scheduling policy.
Hybrid fiber coax (HFC): Technology that allows bidirectional data transfer over cable.
Interactive TV (ITV): Where video delivery depends on the interactive commands from the customer.
MPEG: A video compression standard proposed by Motion Picture Experts Group.
Real-time/continuous media: Data that requires to be delivered within deadlines.
SCAN: A seek optimizing disk scheduling policy.
Set top box: A decoder used to decompress and convert digital video to appropriate input form suitable for customer's monitor.
Striping: Distribution of data over several storage nodes (or disks).
Video on demand (VOD): A service where a video stream is delivered on demand.

References

Albanese, A., Bloemer, J., Edmonds, J., and Luby, M. 1994. Priority encoding transmission. Int. Comp. Sci. Inst. Tech. Rep. No. TR-94-039, Aug.

ATM 1995. Doc. 95-0012r5. ATM Forum.

Bell, 1993a. ADSL video dial tone network interface requirements. TR-72535, Bell Atlantic, Inc., Dec.

Bell, 1993b. Signalling specification for video dial tone. TR-72540, Bell Atlantic Inc., Dec.

Birk, Y. 1995. Track pairing: A novel data layout for VOD servers with multi-zone recording disks. In. *Proc. of IEEE Conf. on Mult. Comput. and Systems.*

Bus. Wire 1995a. NEC to supply set-top boxes to Hong Kong telecom IMS. *Business Wire*, Nov. 14.

Bus. Wire 1995b. Xing streamworks squeezes video through 14.4 and 28.8 dial-up modems. *Business Wire*, Arroyo Grande, CA, Oct. 17.

Chen, M.S., Kandlur, D., and Yu, P.S. 1994. Support for fully interactive playout in a disk-array-based video server. *Proc. of ACM Multimedia Conf.*, pp. 391–398, Oct.

Chen, Z., Tan, S.M., Campbell, R.H., and Li. Y. 1995. Real time video and audio in the world wide web. In. *Proc. of the 4th Int. World Wide Web Conf.*, Dec. 11.

Chervenak, A. 1994. Tertiary storage: An evaluation of new applications. Ph.D Thesis, Univ. of Calif., Berkeley, CA.

Ciciora, W.S. 1995. Cable television in the United States—An overview. Cable Television Labs. Inc.

Comput. Int. 1995. Hong Kong to get world's first multimedia service. Computergram International, Nov. 8.

Furht, B. ed. 1995. *Multimedia Systems and Techniques*, Kluwer Academic Publishers.

Haskin, R. 1993. The shark continuous media file server. *Proc. of IEEE 1993 Spring COMPCON*, San Francisco, CA, pp. 12–17, Feb.

Haskin, R. and Stein, F. 1995. A system for the delivery of interactive television programming. In *Proc. of IEEE 1995 Spring COMPCON*, San Francisco, CA, pp. 209–216, March.

IBM 1993. IBM LAN server Ultimedia—multimedia for OS/2, DOS, and Windows clients. IBM Doc. 96F8520, Nov. 1993.

IBM 1995. The medium is the message: IBM, Argonne National Lab. host first bi-coastal "internews" conference. IBM Corp. press release IBM US 490, San Diego, CA, Dec. 6.

IBM 1996. IBM multimedia server for AIX, version 1.1. IBM Doc. 5967-213, Feb. 20.

Insoft 1996. Insoft targets internet real time multimedia users and developers with first interactive collaborative environment. Insoft Inc. press release, Mechanicsburg, PA, Jan. 8.

Jayanta, K.D., Salehi, J.D., Kurose, J.F., and Towsley, D. 1994. Providing VCR capabilities in large-scale video server. *Proc. of ACM Multimedia Conf.*, pp. 25–32, Oct.

Liu, C.L. and Layland, J.W. 1973. Scheduling algorithms for multiprogramming in a hard real-time environment. *J. ACM*, pp. 46–61.

MEI 1995. Matsushita develops "video server" technology for use on standard computer networks. Matsushita Electric Industrial Co., Ltd. press release, March 22.

Patterson, A. 1994. Hong Kong Telecom, IBM map video effort. *Electronic Engineering Times*, Aug. 1.

Patterson, D.A., Gibson, G., and Katz, R. H. 1988. A case for redundant arrays of inexpensive disks (RAID). *ACM SIGMOD Conf.*, June.

Plumb, L. 1993. Bell Atlantic demonstrates video on demand over existing telephone network. Bell Atlantic press release, June 14.

Reddy, A.L.N. 1995. Scheduling and data distribution in a multiprocessor video server. In *Proc. of Int. Conf. on Multimedia Computing and Systems*, pp. 256–263.

Reddy, A.L.N. and Wyllie, J. 1992. Disk scheduling in a multimedia I/O system. *Proc. of ACM Multimedia Conf.*, Aug.

Schulzrinne, H., Casner, S., Frederick, R., and Jacobson, V. 1996. A transport protocol for real-time applications. Internet RFC 1889, Internet Engineering Task Force, Jan.

Starlight 1996a. The challenges of networking video applications. Starlight Networks Tech. Rep., Starlight Networks Inc., Mountain View, CA, April 22.

Starlight 1996b. Effect of video on lan data traffic. Starlight Networks Tech. Rep., Starlight Networks Inc., Mountain View, CA, April 22.

Time 1993. Time Warner cable completes first phase of full service network. Time Warner press release, Dec. 16.

UPI 1993. Cox to launch interactive tests in Omaha. United Press International, Dec. 1.

VDOnet 1995. VDOlive will enable motion video on the internet. VDOnet Corp. press release (http://www.vdolive.com), Oct. 27.

WNT 1993. Bell Atlantic, IBM announce agreement for video on demand server. *World News Today*, Jan 8.

WSJ 1995. *Wall Street Journal*, Nov. 10.

WWW n.d. Basic plan for the multimedia experiments at the Tokyo metropolitan waterfront subcenter multimedia experiments. http://www.tokyo-teleport.co.jp/english/atms/atmsj201.htm.

Yu, P.S., Chen, M.S., and Kandlur, D.D. 1993. Grouped sweeping scheduling for DASD-based multimedia storage management. *Multimedia Sys.*, 1:99–109.

Further Information

The reader is referred to the book, *Multimedia Systems and Techniques*, edited by Borko Furht, published by Kluwer Academic Publishers.

107
Desktop Videoconferencing

107.1	Introduction	1516
107.2	Overview	1517
	Desktop Terminal • Network • Multipoint Control Unit/Multicast Router	
107.3	Videoconferencing over ISDN	1518
	Audio • Video • Multiplex • Control and Other Standards in H.320	
107.4	Videoconferencing over Internet	1521
	Audio and Video	
107.5	Videoconferencing over General Switched Telephone Network	1523
	Audio • Video • Multiplex • Control	
107.6	Concluding Remarks	1525

Madhukar Budagavi
Texas A&M University

107.1 Introduction

Videoconferencing allows groups of people and individuals in different locations to communicate with each other through the following media: (1) video, with which participants can see motion images of each other, (2) audio, with which participants can hear each other, and (3) **data**, with which participants can exchange images of documents and share files and applications. Traditional videoconferencing has been *room* based. Participants at a site have to go to conference rooms specially fitted with videoconferencing equipment in order to videoconference with others in similarly equipped rooms at remote sites. Desktop videoconferencing (DVC) takes videoconferencing to the familiar desktop personal computers of the users. DVC enables users to sit at their own desk and videoconference with others using their personal computers. It leads to a cheaper, more convenient, and flexible way of implementing a system for videoconferencing. Advances in video and audio compression technologies, network technology, and, more importantly, significant increases in the computing power of personal computers have played a part in making desktop videoconferencing feasible as well as popular.

The next section gives an overview of the various components that make up a DVC system. Desktop videoconferencing over the Integrated Services Digital Network (ISDN), the Internet, and the common analog general switched telephone network (GSTN) are described next. The emphasis of this chapter will be on discussing relevant standards and on highlighting some key technological advances that have made DVC possible.

107.2 Overview

A DVC system consists of three important components: the desktop terminal, the network, and the **multipoint** control unit/**multicast** router (MCU/MR). Figure 107.1 shows a typical DVC system configuration. The desktop terminal is the equipment (usually built around the PC or workstation, however, it could be stand alone) that resides at the user's end providing the necessary videoconferencing facilities to the user. It interfaces with input/output devices producing/consuming videoconferencing data for/from other users. The network is used for transporting videoconferencing data between participants, whereas the MCU/MR is responsible for coordinating and directing the flow of information between three or more videoconferencing users who wish to simultaneously communicate with each other.

Desktop Terminal

The desktop terminal consists of four processing blocks: (1) audio and video, (2) data, (3) control, and (4) multiplex–demultiplex blocks, as shown in Fig. 107.1.

- Audio and video processing block. This interfaces with audio and video (AV) input/output (I/O) devices. Video input is usually through a desktop video camera connected to the PC through special purpose video processing cards inside the PC. The video output is usually on the display monitor of the PC. Audio I/O equipment consists of microphones and headphones. If speakerphones are used, acoustic echo cancellation could be used to remove the echoes which result. The other important function of the AV processing block is compression/decompression of the AV signals. Compression is necessitated by the fact that current networks used for transporting videoconferencing data have limited bandwidth. Compression/decompression is done in software, using the local processor of the PC, in hardware by using AV codec chips, or by using special purpose digital signal processing chips.
- Data processing block. The term *data* here is used to specify information arising out of applications such as electronic whiteboards, still images/slides, shared files, and documents. The data processing block enables users to exchange such information and to work jointly on applications. Since applications usually reside on the PC of the user, data sharing is more convenient in desktop videoconferencing when compared to the room-based videoconferencing systems.
- Control block. As visible to the user, control is in the form of a graphical user interface (GUI) window on the PC screen with which a user can initiate conferences, set up preferred

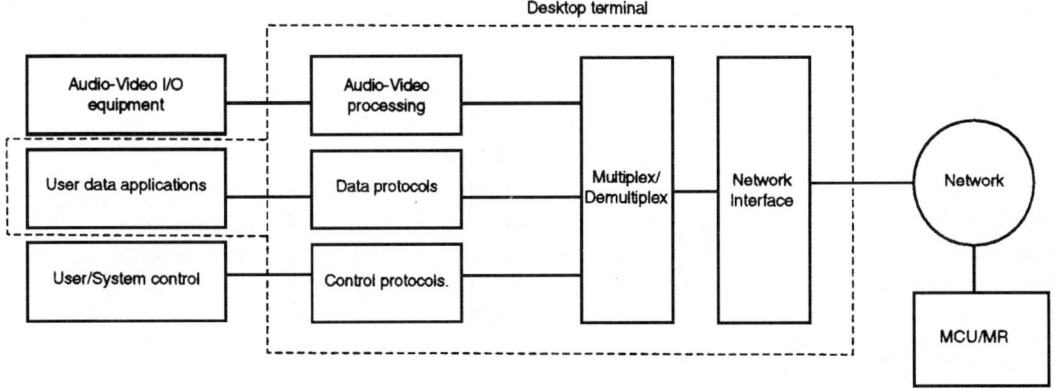

FIGURE 107.1 A general DVC system configuration.

data rates and compression methods for audio and video, and participate in interactive data applications. At the system level, the control unit is responsible for end-to-network signaling for accessing the network, and end-to-end signaling for proper operation of the desktop terminal. It may also be responsible for synchronization of the different media (audio/video/data) being transmitted and for receiver feedback for **quality of service** (QoS) monitoring.

- Multiplex–Demultiplex block. This multiplexes the resulting audio, video, data, and control signals into a single stream before transmission on the network. Similarly, the received bit stream is demultiplexed to obtain the audio, video, data, and control signals, which are then passed on to their respective processing blocks. The Multiplex–demultiplex block accesses the network through a suitable network interface.

Network

The network is the entity used for transporting desktop videoconferencing data between participants. For DVC applications it is desirable that the network provide a low and predictable delay connection; this is essential for transmission of audio and video data, which are real-time signals with strict delay requirements. There are two types of networks, circuit-switched networks and packet-switched networks, for which DVC applications/standards currently exist. Circuit-switched networks, which allocate a dedicated amount of bandwidth, provide a predictable delay connection, whereas packet-switched networks, which packetize data and transmit them over shared nondedicated bandwidth networks, cannot guarantee predictable timing of data delivery. Commonly used networks are the ISDN (a circuit-switched network), and local area networks (LANs) and the Internet (packet-switched networks). Standardization efforts are currently on for specifying videoconferencing standards for the GSTN and the asynchronous transfer mode (ATM) networks.

Multipoint Control Unit/Multicast Router

Videoconferencing can be classified as point-to-point or multipoint, based on the number of participants in a given session. A point-to-point videoconference has only two users who communicate with each other via audio, video, and shared data applications, whereas multipoint conferencing involves more than two participants and multimedia information from each participant has to be multicast to all others. Special network servers are required to coordinate the distribution of the videoconferencing data amongst the multiple participants in a multipoint videoconference. These servers are called multipoint control units in the terminology of circuit-switched networks. The equivalent function in packet-switched networks is carried out by multicast routers. Note that some packet-switched networks such as LANs inherently have multicasting capabilities.

107.3 Videoconferencing over ISDN

Integrated Services Digital Network is a digital communication channel. Current DVC products using ISDN utilize the basic rate interface (BRI), which provides two data channels of 64 kb/s (B-channel) and one signaling channel of 16 kb/s (D-channel). Most of the DVC systems incorporate H.320 standard [ITU-T, 1993], proposed by the International Telecommunications Union–Telecommunications Standardization Sector (ITU-T), in addition to any proprietary method, to transport their videoconferencing information over the ISDN.

H.320 is a suite of standards for videoconferencing over (primarily) the ISDN at rates from 64 kb/s to about 1920 kb/s. H.320 makes use of a number of other ITU-T standards, which can be integrated as shown in Fig. 107.2 (extracted from H.320), to form the complete DVC system. The standards for audio, video, control, and multiplex are summarized in Table 107.1.

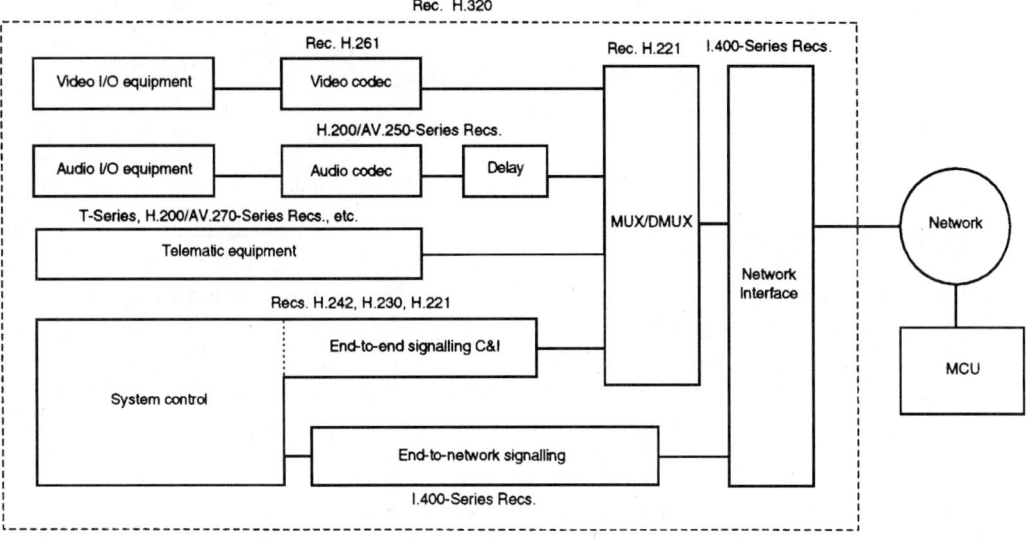

FIGURE 107.2 H.320 set of standards.

TABLE 107.1 H.320 Series of Recommendations

H.320	Narrowband visual telephone systems and terminal equipment
H.261	Video codec for audiovisual services at $p \times 64$ kb/s
G.711	Pulse coded modulation of voice frequencies
G.722	7-kHz audio coding within a 64-kb/s channel
G.728	Coding of speech at 16 kb/s using low-delay code excited linear prediction
H.221	Frame structure for a 64–1920 kb/s channel in audiovisual teleservices
H.230	Frame synchronous control and indication signals for audiovisual systems
H.242	System for establishing communication between audiovisual terminals using digital channels up to 2 Mb/s

Audio

H.320 supports both telephone quality (G.711 and G.728) as well as **wideband** quality (G.722) speech [Noll, 1995]. G.711 does the traditional A-law/μ-law quantization of speech whereas G.728 is based on the **analysis-by-synthesis** approach and makes use of the low-delay code excited linear prediction (LD-CELP) algorithm to compress speech to 16 kb/s. The wideband standard, G.722, splits the wideband speech into low and high band and does adaptive differential pulse code modulation (ADPCM) encoding of the two subband sequences. The type of audio coding used in a videoconferencing session is signaled by external means by using procedures defined in H.242 and H.230 control standards. For further details on audio coding standards, the reader is referred to Noll [1995].

Video

H.261 specifies the video coding standard for use in H.320. H.261 operates on video signals, which have to be in either of the two standard picture formats: the common intermediate format (CIF) or the quarter-CIF (QCIF). The **luminance** and **chrominance** resolutions of the CIF and QCIF formats are summarized in Table 107.2. Pictures in H.261 are coded either in intraframe or in interframe modes. In intraframe coding the image is encoded without any relation to the previous image, whereas in interframe coding, the current image is predicted from the previous image using **block motion compensation** and the difference between the current image, and the predicted image is encoded. A single **motion vector** is used for each *macroblock* (data corresponding to 16×16 blocks in an image). Depending on the mode of coding used, the macroblocks of either

TABLE 107.2 Picture Formats for H.261 and H.263

Picture Format	Recommendations	Luminance (pixels per line × lines)	Chrominance (pixels per line × lines)
Sub-QCIF	H.263	128 × 96	64 × 48
QCIF	H.261 and H.263	176 × 144	88 × 72
CIF	H.261 and H.263	352 × 288	176 × 144
4 CIF	H.263	704 × 576	352 × 288
16 CIF	H.263	1408 × 1152	704 × 576

the image or the residual image are split into blocks of size 8 × 8, which are then transformed using a two-dimensional discrete cosine transform (DCT). The resulting DCT coefficients are quantized, run-length encoded, and finally Huffman encoded to give a bit rate of $p \times 64$ kb/s, where p is an integer between 1 to 30. An overview of the H.261 standard is given in Schafer and Sikora [1995].

Multiplex

The H.221 standard determines the way in which audio, video, data, and control is multiplexed in H.320 before transmission on ISDN line. The H.221 multiplex breaks each 64 kb/s channel into 80 octet (bytes) frames. Multiplexing is done by allocating different parts of this 80 octet frame for carrying video, audio, data, and control information. A fixed part of the frame is used for signaling the service channel (SC). Figure 107.3 (extracted from H.221) illustrates a single H.221 frame. The SC in a frame consists of 8 b each of bit-rate allocation signal (BAS), frame alignment signal (FAS), and encryption control signal (ECS); the remaining bits (bits 25–80) of the SC can be used for transmitting videoconferencing data. The first 8 b of the SC form the FAS, which helps the receiver synchronize to the H.221 frames. The BAS channel consists of bits 9–16 of SC and is used to signal commands and **capabilities** to the other end. It is also used to transmit information describing the bit allocation between video, audio, and data channels in H.221 frames.

Control and Other Standards in H.320

Control messages, such as capabilities and commands, and other control information for use in H.320 are specified in ITU-T H.242 and H.230. ITU-T H.231 and H.243 cover procedures for

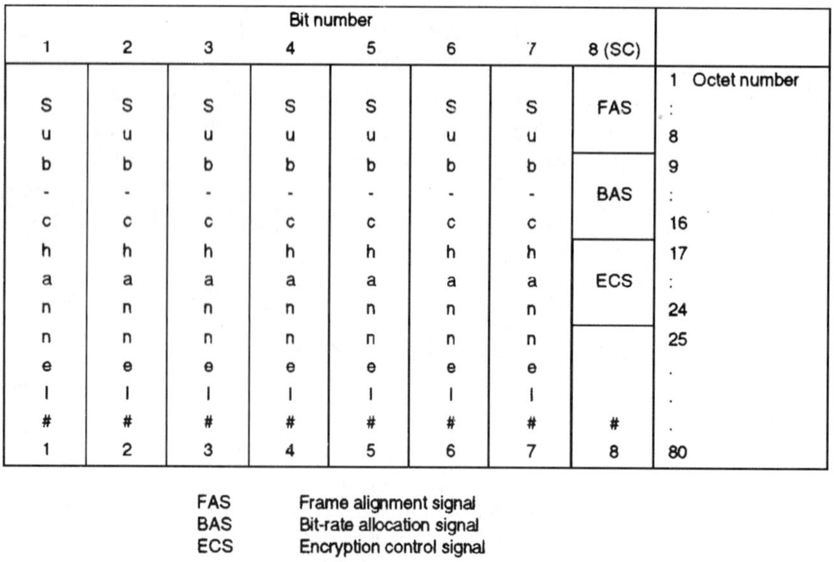

				Bit number						
1	2	3	4	5	6	7	8 (SC)			Octet number
S	S	S	S	S	S	S	FAS		1	
u	u	u	u	u	u	u			:	
b	b	b	b	b	b	b			8	
-	-	-	-	-	-	-	BAS		9	
c	c	c	c	c	c	c			:	
h	h	h	h	h	h	h			16	
a	a	a	a	a	a	a	ECS		17	
n	n	n	n	n	n	n			:	
n	n	n	n	n	n	n			24	
e	e	e	e	e	e	e			25	
l	l	l	l	l	l	l			.	
#	#	#	#	#	#	#	#		.	
1	2	3	4	5	6	7	8		80	

FAS Frame alignment signal
BAS Bit-rate allocation signal
ECS Encryption control signal

FIGURE 107.3 H.221 frame.

use by MCUs for multipoint conferences. ITU-T H.233 and H.234 specify encryption and key exchange for H.320. Encryption is to be used if secure videoconferencing, conferences that cannot be monitored as they pass through the network, is desired.

107.4 Videoconferencing over Internet

This section gives an overview of DVC over the Internet and describes some key technological advances that have made it possible on a worldwide scale. Unlike videoconferencing over ISDN discussed earlier, videoconferencing over the Internet is through packet-switched networks. Connectivity on the Internet is provided through the internet protocol (IP). The problem in using IP directly for DVC was that IP was a unicast (point-to-point communications) protocol and did not support efficient multicasting. Clearly, unicasting separate copies to each of the recipients is highly inefficient and leads to flooding of the network with redundant data. This problem is more acute in DVC applications where high rate video and audio data are involved. To overcome this problem, the IP multicast model was proposed in Deering and Cheriton [1990]. In IP multicast, the network does the replication as and when paths to different receivers diverge. Support for IP multicast is required both in network routers and switches and in the end desktop terminal. Since not all routers on the current Internet have multicasting capabilities, a (temporary) virtual network called the multicast backbone (MBONE) has been built on top of the existing Internet to provide IP multicast routing capabilities [Macedonia and Brutzman, 1994]. IP multicast addresses are officially categorized as IP Class D addresses. These addresses can be allocated dynamically, and they allow individual users to join or leave a multicast group at any time.

Two transport layer protocols were developed for use with IP: the transmission control protocol (TCP) and user datagram protocol (UDP). TCP was designed to provide a reliable point-to-point service for delivery of packet information in proper sequence, whereas the UDP was designed to simply provide a service for delivering packets to the destination without bothering about congestion control. TCP is found to be inappropriate for real-time transport of audio and video information as its error- and congestion-control mechanisms may result in indeterminate delays leading to discernible distortions and gaps in the real-time playout of the audio and video streams. Another drawback of TCP is that it does not support multicasting. The preceding problems motivated the development of the real-time transport protocol (RTP) [AVTWG-IETF, 1995] for transporting real-time audio and video data over the Internet. RTP was proposed by the Audio Video Transport Working Group (AVTWG) of the Internet Engineering Task Force (IETF). RTP is layered over UDP (note, however, that RTP is independent of the underlying network and can be used over other types of networks such as the ATM) and relies on the UDP to provide it services such as framing and multiplexing. RTP consists of two parts: a data part and a control part. The data part handles the transportation of the actual DVC data (audio and video) and takes care of timing and sequencing of data packets. The control part, real-time transport control protocol (RTCP), is responsible for managing control information such as sender identification, synchronization of different media (e.g., lip-syncing of audio and video), and receiver feedback. The receiver feedback is used to monitor QoS of the network. The QoS information may be used to adaptively change the source coder parameters so as to either decrease or increase the data rate as required.

Figure 107.4 shows the block diagram of a typical protocol stack for desktop videoconferencing over the Internet using RTP. Conferencing applications packetize audio, video, and data information and encapsulate them in RTP packets before transmitting them over the Internet.

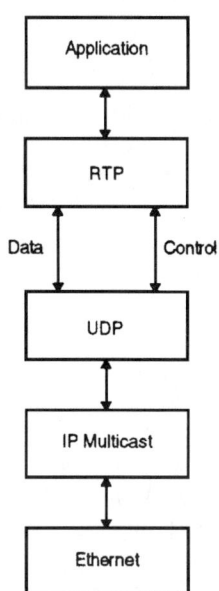

FIGURE 107.4 Protocol stack for DVC using RTP on Internet.

TABLE 107.3 MBONE Tools for the Internet

Network video (nv)	nv is a low-complexity software only video tool based on conditional replenishment and Haar transform.
Video conferencing (vic)	vic (a video tool) makes use of the Intra-H.261 scheme. It also supports the nv format, H.261, motion–Joint Photographic Experts Group format, and Sun's CellB format.
INRIA videoconferencing system (ivs)	ivs supports both audio and video coding. Audio coding is done using PCM and ADPCM, and video coding is done using H.261.
Visual audio tool (vat)	vat is an audioconferencing system that currently supports a variety of audio coding algorithms including PCM, ADPCM, and linear predictive coding.
Whiteboard (wb)	wb supports shared desktop whiteboarding and slide distribution.
Session directory (sd)	sd is a tool with which a user can easily join an available videoconferencing session or create and advertise new sessions.

RTP makes use of the underlying UDP and IP multicast to provide the Internet connection to the applications above it. Note that multiplexing of the information streams from the applications is implicitly achieved by the use of UDP. Some of the commonly used DVC applications on Internet are listed in Table 107.3. These applications are commonly known as MBONE tools, as their development was fostered by the deployment of IP multicast on the MBONE. Most of these tools are purely software based to reduce the costs involved in desktop videoconferencing. The audio and video coding algorithms used by these tools are now described.

Audio and Video

The audio algorithms used by the MBONE tools are the same as those used on circuit-switched networks, such as pulse code modulation (PCM), ADPCM, and linear predictive coding. However, for video coding, the error characteristics of the Internet and low-complexity requirements of the software only DVC systems have lead to the development of simpler coding algorithms which are better matched to the error characteristics of the Internet. In Internet, packet losses are often significant and compression schemes such as H.261, which are designed for low bit error rate channels, do not perform well. The performance degradation occurs when the interframe mode is used in H.261. Since macroblocks in the current frame are predicted from the previous frame and the error residual is transmitted, any packet loss leads to a mismatch in the state of the encoder and the decoder. This leads to rapid degradation in the video quality as subsequent interframe coded pictures are decoded. One solution is to use only intracoded pictures; however, this approach results in low compression efficiency because it does not make use of the redundancy that is typically present in the sequence of images arising in DVC applications. A solution to this problem is to use conditional replenishment. In this scheme, each video frame is partitioned into blocks and only the blocks that change beyond some threshold, when compared to their counterparts in the previous frame, are transmitted. The transmitted blocks are always intracoded to prevent errors in images due to lost packets from propagating. The MBONEtools, network video (nv) and video conferencing (vic) (see Table 107.3), make use of conditional replenishment. In nv, 8×8 image blocks from the conditional replenishment stage are Haar transformed, thresholded (coefficients with magnitude below the threshold are set to 0), and runlength encoded before transmission. The Haar transform, which involves only additions and subtractions for its computation, is used because of its low complexity. However, the compression achieved by Haar transform is inferior to that achieved by DCT. This lead to the adoption of DCT as the transform in the vic conditional replenishment-based video coding algorithm called the Intra-H261 algorithm [McCanne and Jacobson, 1995]. Also, vic incorporates a uniform quantizer (instead of the thresholder used in nv), a Huffman-runlength coder, and the syntax of H.261 to generate a H.261 compliant bit stream.

107.5 Videoconferencing over General Switched Telephone Network

General Switched Telephone Networks are the common analog telephone lines. ITU-T H.324 specifies the standard for videoconferencing over the GSTN by using V.34 data modems (H.324 is currently a draft recommendation) [ITU-T, 1995]. The bit rate targeted is very low at around 28.8 kb/s. A generic H.324 videoconferencing system is shown in Fig. 107.5. Like H.320, H.324 also invokes other ITU-T standards to provide complete videoconferencing functionality. Table 107.4 give a list of these standards. All H.324 terminals should support at least the V.34 modem, H.223 multiplex, and H.245 system control protocol, and depending on the application, could support audio and video codecs and data protocols.

Audio

G.723.1 is the recommendation for speech coding in the H.324 family of standards. G.723.1 specifies a dual rate speech coder with bit rates of 5.3 and 6.3 kb/s. The higher bit rate coder provides better quality than the lower bit rate coder. Both rates are a mandatory part of the encoder and decoder. The variable bitrate operation allows for dynamic adjustment of the speech bit rate depending on the requirements of other bit streams (such as video), which have to be simultaneously transmitted. G.723.1 is based on the principles of linear prediction (LP) and analysis-by-synthesis coding. The coder operates on speech frames of 30 ms duration (which in turn is split into four subframes of 7.5 ms each) and attempts to minimize a perceptually weighted error signal. The linear prediction

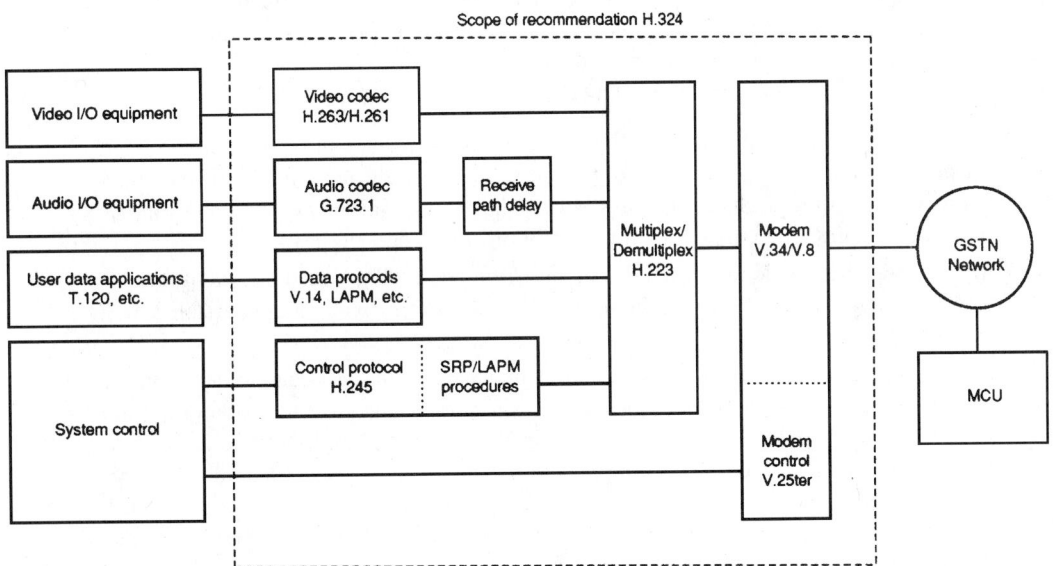

FIGURE 107.5 H.324 DVC system.

TABLE 107.4 H.324 Series of Recommendations

H.324	Terminal for low bit rate multimedia communication
H.263	Video coding for low bit rate communication
G.723.1	Dual rate speech coder for multimedia communication transmitting at 5.3 and 6.3 kb/s
H.223	Multiplexing protocol for low bit rate multimedia communication
H.245	Control protocol for multimedia communication

coefficients are vector quantized and transmitted once per frame. The excitation to the LP synthesis filter is modeled as a combination of two components: one from a fixed codebook and the other from an adaptive codebook. The adaptive codebook contribution corresponds to long-term prediction and reduces redundancy present in the speech signal due to the pitch periodicities. A fifth-order pitch predictor is used, and information concerning the pitch predictor (pitch lags and gains) are transmitted once every subframe. The main difference in coding for the two rates comes from the type of excitation used in the fixed codebook. The high rate coder uses multipulse maximum likelihood quantization (MP-MLQ) and the low rate coder uses algebraic codebook excited linear prediction (ACELP). Both schemes essentially model the excitation as a sequence of pulses, and information regarding the fixed codebook excitation is transmitted in the form of pulse position and signs once every subframe.

Video

H.263 is the video coding standard for low bit rate communication in H.324. H.263 is based on the H.261 recommendation and improves upon it to provide an equivalent subjective quality at less than half the bit rate. H.263 consists of a base level coder and four optional modes. H.263 specifies five standardized picture formats as listed in Table 107.2. The H.263 recommendation mandates that all decoders be able to operate using at least the sub-QCIF and QCIF formats and that the encoder be able to operate on at least either of the two formats, sub-QCIF or QCIF. Use of other picture formats is optional.

Some of the main differences between H.263 and H.261 are highlighted next. The reader is referred to the actual recommendation for the full description of the standard. Most of the improvements at the base level come from using motion vectors with half-pixel resolution and from using improved variable length coding for coding DCT coefficients. Also motion vectors and other overhead information are coded more efficiently than H.261.

H.263 specifies the following four optional modes:

- Unrestricted motion vector mode. In this mode motion vectors are allowed to point to pixels outside the picture. The values used for the nonexistent outside pixels are simply the values of the corresponding pixels on the nearest edge of the picture. This mode is found to improve the picture quality when there is significant motion at the edges of the image.
- Syntax-based arithmetic coding. In this mode arithmetic coding is used instead of variable length coding. This mode usually requires fewer bits than that required for variable length coding.
- Advanced prediction mode. This mode allows for allocation of separate motion vectors for each of the four 8×8 luminance blocks in a macroblock leading to a total of four motion vectors for a macroblock instead of one (as in the base H.263 coder). The use of four motion vectors is found to give a better prediction. The advanced prediction mode also uses overlapped block motion compensation (OBMC) where each 8×8 block in the macroblock is predicted as a weighted sum of three motion compensated blocks from the previous frame. OBMC results in less blocking artifacts and gives better subjective quality.
- PB frames. A PB frame consists of two image frames, P pictures and B pictures, coded as one unit. The name PB comes from the terminology used in Moving Pictures Expert Group (MPEG) coding. P pictures are interframe coded pictures and B pictures are pictures that are bidirectionally predicted, that is, predicted from both past and future pictures. In PB frames, the B picture is predicted from the last decoded picture and the P picture currently being decoded. This mode can be used to double the frame rate with only a modest increase in the coded video data rate.

A description of H.263 and performance comparisons of the four optional modes is given in Whybray and Ellis [1995].

Desktop Videoconferencing

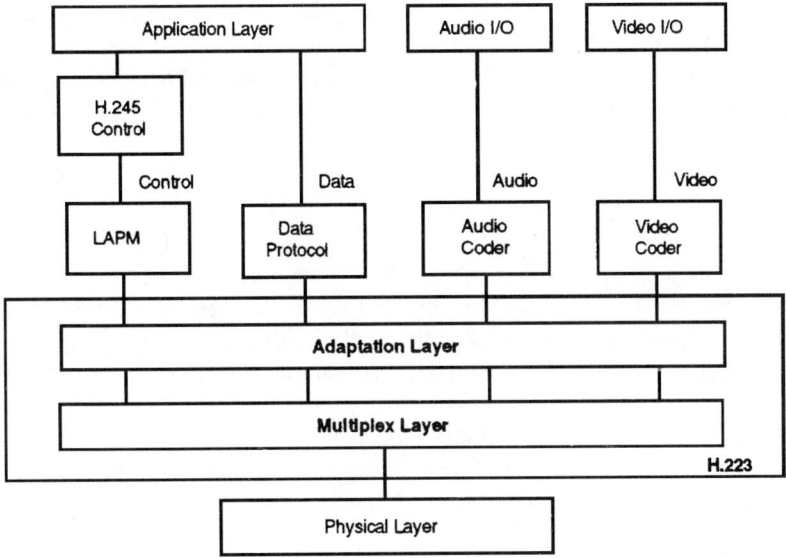

FIGURE 107.6 Protocol stack for H.223.

Multiplex

Video, audio, data, and control information is transmitted in H.324 on distinct logical channels. ITU-T H.223 determines the way in which the logical channels are mixed into a single bit stream before transmission over the GSTN. The H.223 multiplex consists of two layers, the multiplex layer and the adaptation layer, as shown in Fig. 107.6. The multiplex layer is responsible for multiplexing the various logical channels. It transmits the multiplexed stream in the form of packets. Each packet transmitted by the multiplex layer contains a one octet header and a variable number of information field octets carrying videoconferencing information. The header octet includes a multiplex code, which specifies, by indexing to a multiplex table, the logical channels to which each octet in the information field belongs. The adaptation layer adapts the information stream provided by the applications above it to the multiplex layer below it by adding, where appropriate, additional octets for the purposes of error control and sequence numbering. The type of error control used depends on the type of information (audio/video/data/control) being conveyed in the stream. The adaptation layer provides error control support in the form of cyclic redundancy check (CRC) bits and retransmission protocols.

Control

H.324 systems contain exactly one control channel in each direction. The messages and procedures used on these channels are specified in Recommendation H.245. The control channel is always carried on logical channel 0. The information carried on the control channel includes capabilities data, multiplex table entries, and mode preferences and flow control messages. The control channel also conveys messages for opening and closing logical channels.

107.6 Concluding Remarks

The preceding sections described some of the current standards and on-going efforts in desktop videoconferencing. Current research efforts in compression technology is targeted toward very low bit rate communications with significant efforts being put in that direction in phase 4 of MPEG standards activity. Two important application areas which might have significant bearing on desktop videoconferencing in the coming years are wireless personal communication systems and the Internet. In both these areas, efforts are on to provide multimedia services with high quality (comparable to that provided by ISDN) using improved compression algorithms and network technologies.

Defining Terms

Analysis-by-synthesis technique: In this technique the excitation to the LPC synthesis filter is chosen from a (predefined) set of possible excitation vectors such that it minimizes the perceptually weighted error between the input speech signal and its reconstructed version.

Block motion compensation: Motion compensated prediction is done on a block basis, that is, blocks of pixels are assumed to be displaced spatially in an uniform manner from one frame to another.

Capability: Describes the terminal's ability to process a particular kind of information stream.

Data: This term is also used to specify information arising in applications such as electronic whiteboards and still images/slides transmission.

Linear predictive coding: Models the speech as parameters of a linear filter.

Luminance and chrominance: Luminance is the brightness information in a video image, whereas chrominance is the corresponding color information. For H.261 and H.263, color is encoded as two color differences commonly called as Cb and Cr components.

Motion vector: Specifies the spatial displacement of a block of pixels from one frame to another.

Multicast: Transmission of the same information to multiple recipients in packet-switched networks such as the Internet.

Multipoint communication: Simultaneous communication between more than two participants. Terminology usually used in circuit-switched network applications.

Quality-of-service (QoS): Provided by Internet is measured in terms of parameters such as packet loss and packet delay variation.

Wideband speech: Traditional telephone quality speech is bandlimited to about 3 kHz. Wideband speech is bandlimited to 7 kHz and gives more natural sounding speech.

References

AVTWG-IETF. 1995. RTP profile for audio and video conferences with minimal control. RFC 1890, Audio Video Transport Working Group, Internet Engineering Task Force.

AVTWG-IETF. 1995. A transport protocol for real-time applications. RFC 1889, Audio Video Transport Working Group, Internet Engineering Task Force.

Deering, S.E. and Cheriton, D.R. 1990. Multicast routing in datagram internetworks and extended LANs. *ACM Trans. Comp. Sys.*, 8(May):85–110.

Deller, J.R., Jr., Proakis, J.G., and Hansen, J.H.L. 1993. *Discrete-Time Processing of Speech Signals*, Macmillian, New York.

Ebrahimi, T., Reusens, E., and Li, W. 1995. New trends in very low bitrate video coding. *Proc. IEEE*, 83(June):877–891.

Gersho, A. 1994. Advances in speech and audio compression. *Proc. IEEE*, 82(April):488–525.

ITU-T. 1993. Narrow-band visual telephone systems and terminal equipment. Recommendation H.320 International Telecommunications Union, Telecommunications Standardization Sector, Geneva.

ITU-T. 1995. Terminal for low bit rate multimedia communications. Draft Recommendation H.324 International Telecommunications Union, Telecommunications Standardization Sector, Geneva.

Macedonia, R.M. and Brutzman, D.P. 1994. MBONE provides audio and video across the internet. *Computer*, 27(April):30–36.

McCanne, S. and Jacobson, V. 1995. vic: A flexible framework for packet video. *In ACM Multimedia*, pp. 511–522, November, San Francisco.

Netravali, A.N. and Haskell, B.G. 1988. *Digital Pictures: Representation and Compression*, Plenum, New York.

Noll, P. 1995. Digital audio coding for visual communications. *Proc. IEEE*, 83(June):925–943.

Schafer, R. and Sikora, T. 1995. Digital video coding standards and their role in video communications. *Proc. IEEE*, 83(June):907–924.

Whybray, M.W. and Ellis, W. 1995. H.263–Video coding recommendation for PSTN videophone and multimedia. In IEE Colloquium Low Bit Image Coding, pp. 6/1-9. Digest No. 1995/154, June, London.

Further Information

A general overview of current audio and video coding standards is given in Noll [1995] and Schafer and Sikora [1995], respectively. Fundamental level books on video and speech processing include Netravali and Haskell [1988] and Deller, Proakis, and Hansen [1993]. Recent trends in speech coding research are given in Gersho [1994] and those in very low bit rate video coding are given in Ebrahimi, Reusens, and Li [1995]. Pointers to DVC on the Internet can be found on a number of world wide web sites including *http://www.lbl.gov/ctl/vconf-faq.html*, which lists the ftp sites from where most of the MBONE tools can be downloaded.

Current research relevant to desktop videoconferencing is reported in a number of journals including, *IEEE Transactions on Speech and Audio Processing, IEEE Transactions on Image Processing, IEEE Transactions on Circuit and Systems for Video Technology*, and *Signal Processing*. Conferences of interest include the International Conference on Image Processing (ICIP), International Conference on Acoustics Speech and Signal Processing (ICASSP), and ACM Multimedia.

VIII

Data Recording

108 Magnetic Storage *Jaekyun Moon* 1531
Introduction • Communication Channel Model of the Read/Write Process • Recent Advances in Coding and Detection • Fixed Delay Tree Search with Decision Feedback • Performance Comparisons • Future Considerations • Summary and Conclusions

109 Magneto-Optical Disk Data Storage *M. Mansuripur* 1546
Introduction • Preliminaries and Basic Definitions • The Optical Path • Automatic Focusing • Automatic Tracking • Thermomagnetic Recording Process • Magneto-Optical Readout • Materials of Magneto-Optical Data Storage

108
Magnetic Storage

108.1 Introduction 1531
108.2 Communication Channel Model of the Read/Write Process 1532
   Nonlinearity • Transition Noise
108.3 Recent Advances in Coding and Detection 1535
108.4 Fixed Delay Tree Search with Decision Feedback 1537
108.5 Performance Comparisons 1542
108.6 Future Considerations 1544
108.7 Summary and Conclusions 1544

Jaekyun Moon
University of Minnesota

108.1 Introduction

As the demand continues for increased speed and power in computer systems, there exists an equally important need for improved data storage capacity. The past three decades have seen a spectacular growth of data capacity in commercial magnetic storage devices. In 1996, disk drive products shipped by IBM Corporation boasts storage densities approaching 1 Gb in every square inch of disk surface. The data rates exceed 120 Mb/s while the bit error probabilities are kept at 10^{-14}.

Although the success of magnetic storage devices as secondary computer memory has been mainly due to continued improvements and innovations in heads and media technologies, advanced signal processing techniques have started to make significant contributions in improving reliability and capacity of magnetic storage devices. As such, they are quickly becoming a critical issue in designing an overall data storage system. Today's high-end disk drive systems, for example, employ the **partial response maximum likelihood (PRML)** technology, which has better immunity against random noise than the traditional peak detector. Many sophisticated coding and detection algorithms are now being actively investigated and seriously considered for application to commercial drives by all major companies in data storage industry. Also, an increasing number of universities are developing research programs in communications and signal processing for data storage channels, further spurring progress in this field.

There are several important factors one must consider in developing signal processing schemes suitable for magnetic storage systems. Among them are the cost constraint, power consumption issues, the data rate requirement.

In this chapter, we address magnetic storage system issues directly relevant to the design of data recovery schemes. In particular, we discuss the communication channel modeling of the **read/write process** in magnetic recording. Some of the unique characteristics and constraints of the magnetic storage channel are described. We also discuss different modulation coding and detection techniques that are considered suitable for magnetic storage. A particular equalization/detection technique called the **fixed delay tree search with decision feedback (FDTS/DF)** is described in detail. The bit error rate performances are also compared among different choices of modulation codes and detection schemes.

108.2 Communication Channel Model of the Read/Write Process

The received waveform in a digital pulse amplitude modulation (PAM) system is commonly expressed as

$$z(t) = \sum_k x_k s(t - kT) + n(t) \tag{108.1}$$

where x_k is the digital information sequence transmitted, $s(t)$ is the channel impulse response, $n(t)$ represents the additive noise observed at the receiver, and T is the symbol period. An analogous mathematical description exists for the read/write process of magnetic recording.

In magnetic recording, the binary data sequence x_k is first converted into a rectangular current waveform whose amplitude level swings between -1 and 1, reflecting the given binary information. This rectangular waveform is then applied to the write-head/medium/read-head assembly, which produces a differentiated and low-pass filtered version of the applied waveform. The reproduced waveform is also corrupted by noise. The overall write/read process can be modeled mathematically as

$$z(t) = \frac{dw(t)}{dt} * h(t) + n(t) \tag{108.2}$$

where $z(t)$ is the reproduced or readback waveform; $w(t)$ denotes the rectangular current waveform; $h(t)$ represents the low-pass filter type of response arising from the frequency-dependent signal loss terms due to the head-medium spacing in the read process, the read head gap effect, the medium thickness effect, and the imperfect writing of magnetic transitions; $n(t)$ is the additive white Gaussian noise due to the read head and electronics; and the asterisk denotes the convolution. Recording systems using thin-film media also exhibit a highly localized, nonadditive type of noise. We shall discuss this type of noise later in this section.

The input current waveform $w(t)$ is generated by multiplying each discrete-time input x_k by a rectangular pulse of duration T, that is,

$$w(t) = \sum_k x_k p(t - kT) \tag{108.3}$$

where

$$p(t) = u(t) - u(t - T) \tag{108.4}$$

with $u(t)$ representing the unit step response.

Combining Eqs. (108.2–108.4), the readback waveform can be written as

$$z(t) = \sum_k x_k [h(t - kT) - h(t - T - kT)] + n(t) \tag{108.5}$$

Magnetic Storage

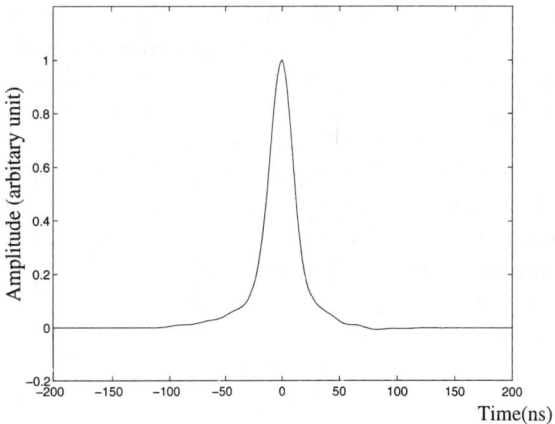

FIGURE 108.1 Step response measured from a real disk drive.

Comparing Eqs. (108.1) and (108.5), we see that the effective impulse response of the magnetic recording channel is $h(t) - h(t - T)$. This is usually called the pulse response since it is the head/medium response to a rectangular current pulse [$h(t)$ is called the step or transition response as it is the readback response to a single current transition]. The step response is often analytically modeled as the Lorentzian function,

$$h(t) = \frac{1/w}{1 + (t/w)^2} \qquad (108.6)$$

where w represents the width parameter, which depends on the physical width of the written transition as well as other head/medium properties. Figure 108.1 shows a step response acquired from an actual recording system employing a thin-film disk and a **magneto-resistive (MR) head**.

Equation (108.5) can also be written as

$$z(t) = \sum_k a_k h(t - kT) + n(t) \qquad (108.7)$$

where

$$a_k = x_k - x_{k-1}, \quad a_k \in \{-2, 0, 2\} \qquad (108.8)$$

Note that a_k is constrained to take on $+2$ and -2 in an alternating fashion.

Nonlinearity

Thus far, we have assumed that the channel is linear, that is, the signal portion of the received waveform can be constructed from linear superposition of isolated step responses. In practice, this is true only when the transitions are not very close to each other. As transitions approach each other closely, nonlinear distortion takes place, especially in high-speed disk recording systems [Palmer et al., 1987]. An important source of nonlinearity is the shift in transition positions, which occurs as the demagnetizing field of the previous transition influences the head field writing the current transition. Another type of nonlinearity is the broadening of the current transition as the head field gradient is reduced by the demagnetizing field from the previous transition. Thus, a transition is shifted earlier in time and tends to broaden if there is a previously written, nearby transition. Yet another form of nonlinearity, which becomes significant as transitions approach each other even closer, corresponds to partial erasure of adjacent transitions [Melas et al., 1987]. At the readback level, this appears as the sudden reduction of signal amplitude. Although the position shift can be eliminated to a large extent by precompensating the write current so that the actual written position of a transition coincides with the intended position, the transition broadening and partial erasure effects are difficult to avoid as linear density increases.

One way of overcoming this difficulty is to impose a minimum runlength code constraint on the input data sequence. A minimum runlength constraint, which is often called a d-*constraint*, forces runlengths of like symbols to be at least $(d + 1)$ and thereby separates magnetic transitions by at least $(d + 1)$ symbol intervals [Siegel and Wolf, 1991]. The d-constraint was originally introduced to magnetic recording in order to minimize the pulse overlapping to facilitate the operation of the simple peak detection circuitry, but as the linear density requirement becomes more stringent, there seems to be a more fundamental reason to employ the d-constraint.

In addition, the d-constraint actually overcomes the large rate loss associated with it and provides a coding gain at very high linear densities by eliminating certain error-prone data patterns [Immink, 1989; Moon and Carley, 1990]. The d-constraint can also provide robustness against channel parameter fluctuations [Immink, 1986].

In practice, transition spacing is also upper bounded by $(k + 1)$ symbol intervals to provide frequent information to the timing circuit. This is called the k-*constraint*. Runlength limited coding with both lower and upper limits on runlengths of like symbols is referred to as (d, k) coding.

Transition Noise

In magnetic recording systems employing thin-metallic media, the step response may change from one transition to the next due to random variations in the geometry of the magnetic transitions. This random deviation of the step pulse gives rise to what is known as transition noise. This type of noise depends on the written data pattern and cannot be modeled as additive noise. **Transition noise** can be a major problem in some high-density drives. One example is the experimental system used by IBM for its gigabit-per-square-inch demonstration held in 1989, where the transition noise was responsible for more than 90% of the total noise power [Howell et al., 1990].

A simple but general model for transition noise can be obtained using a simple argument. Transition noise arises because we do not have perfect control over the geometry of the written transitions. Figure 108.2 illustrates a typical zigzag transition geometry. Provided that the track width is considerably larger than the size of the individual tooth, which is typically the case, magnetization will be effectively averaged across the track in the read process. Thus, the detailed zigzag geometry is not important in determining the shape of the read pulse, as long as the correlation length of any statistical variations along the domain wall is small compared to the track width; more crucial factors are the average extent of the zigzags and the center position of the transition profile. The transition position corresponds to the position of the read pulse and the average extent of the zigzags is directly related to the width parameter of the read pulse. Let us denote $h(t, w)$ the readback response to a noise-free transition located at $t = 0$, where w is the width parameter. Let us also denote $h_k(t)$ as the transition response for the kth symbol interval. The subscript k emphasizes the underlying assumption that the step response is, in general, different from transition to transition. We assume that $h_k(t)$ is uniquely determined by two variables, the position and width parameters. The readback response to a noisy transition in the kth symbol interval can

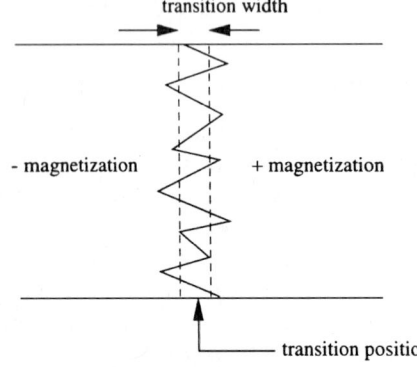

FIGURE 108.2 Typical transition geometry.

Magnetic Storage

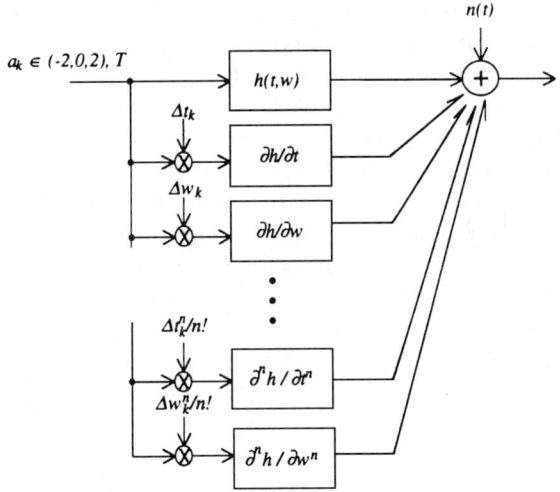

FIGURE 108.3 Channel model including transition noise sources.

then be written as

$$h_k(t) = h(t - kT - \Delta t_k, w + \Delta w_k) \tag{108.9}$$

where Δt_k and Δw_k are random parameters representing deviations in the position and width, respectively, from the nominal values. Taking an nth order Taylor series expansion, the preceding expression can be approximated as a linear sum of the noise-free response and residual responses due to deviations around the nominal position and width of the pulse,

$$\begin{aligned} h_k(t) \approx\ & h(t - kT, w) + \Delta t_k \frac{\partial h}{\partial t} + \Delta w_k \frac{\partial h}{\partial w} \\ & + \frac{(\Delta t_k)^2}{2!} \frac{\partial^2 h}{\partial t^2} + \frac{(\Delta w_k)^2}{2!} \frac{\partial^2 h}{\partial w^2} + \cdots \\ & + \frac{(\Delta t_k)^n}{n!} \frac{\partial^n h}{\partial t^n} + \frac{(\Delta w_k)^n}{n!} \frac{\partial^n h}{\partial w^n} \end{aligned} \tag{108.10}$$

The resulting channel model is depicted in Fig. 108.3, where a_k represents the transition sequence as defined in Eq. (108.8). The multiplicative nature of transition noise is clearly shown in Fig. 108.3. The partial derivatives weighted by the multiples of the random constants can be viewed as residual responses. The overall channel consists of the main path due to the nominal response and paths due to the residual responses.

108.3 Recent Advances in Coding and Detection

Advanced signal processing has been playing an increasingly important role in improving densities in magnetic storage devices. Until recently, simple analog peak detection has been the universal choice for data detection in magnetic data storage devices. A peak detector operates on the analog readback signal to determine the presence of a pulse within a predetermined window that is sliding in time. A readback pulse occurs where there is a transition in the input current waveform, and the written data pattern can be reconstructed by correctly identifying the pulse positions in the readback signal.

When the density requirement is moderate, combining a peak detector with an error-correcting code and a (d, k) **runlength-limited (RLL)** code provides an adequate means to recover data with a relatively low-implementation cost. However, as linear density increases, overlapping between neighboring pulses becomes severe, and the peak detector performance deteriorates rapidly due to large peak shifts and amplitude reductions.

Among notable recent advances in signal processing for magnetic recording is the partial response maximum likelihood (PRML) scheme, which combines **partial response (PR)** linear equalization with the Viterbi algorithm [Cideciyan et al., 1992]. The most popular choice of partial response is the class VI partial response (PR4) characterized by the $1 - D^2$ transfer function, where D represents a symbol delay. The PR4 target provides a reasonably good matching to the natural head/medium response of the magnetic storage channel, which is necessary to minimize equalization loss due to excessive noise enhancement. PR4 equalization also allows the interleaving of the readback signal into two dicode signals to which a simplified, **Viterbi-equivalent detection** algorithm can be applied [Ferguson, 1972; Wood, 1990]. The current PRML system developed by IBM operates on data sequences that are not constrained by the minimum runlength requirement (i.e., $d = 0$) and at current linear densities, this results in a significant code rate advantage over the conventional $(d = 1, k = 7)$ RLL code. Overall, the improved noise immunity and ability to combat intersymbol intersymbol interference (ISI) more effectively, coupled with the rate advantage of the $d = 0$ modulation code, makes the PRML technique considerably more attractive than the conventional peak detection scheme.

Another technique of recent interest is **decision feedback equalization (DFE)**. DFE does not force the readback response to a predetermined target and, thus, generally suffers from much smaller equalization loss than PRML techniques. When the feedback filter is implemented using a random access memory (RAM), DFE also can effectively counter the causal part of nonlinear distortion [Fisher, Cioffi and Melas, 1989]. Much effort is now being directed to efficient circuit implementation of DFE.

In the area of coding, an important development has taken place recently by the efforts of Karabed and Siegel [1991]. They have shown that a large coding gain is possible by imposing a code constraint that forces nulls in the power spectrum of written sequences that match with those in the channel transfer function. This is a remarkable result in the sense that code design, which has traditionally been pursued independently of the channel characteristics, can now take advantage of given spectral properties of the channel. A data recovery scheme, which combines a matched spectral null code with PRML, has been designed and implemented in very large-scale integrated (VLSI) chips [Thapar et al., 1993].

A general technique has also been developed for designing constrained codes with efficient encoders with sliding block decoders. This technique is based on constructing a constrained graph according to the given code constraint and performing the celebrated state splitting algorithm [Marcus, Siegel, and Wolf, 1992]. This technique can be applied to a very large class of codes, including all practical constrained codes encountered in data storage channels.

Recently, there has been considerable interest in understanding nonlinear distortion in high-density film disks and its impact on detection quality. As described earlier, nonlinear distortion occurs as the magnetic transitions are written very closely to one another at high linear densities. Whereas there have been efforts to deal directly with nonlinearities by modifying detection algorithms, there also has been considerable interest to avoid them by resorting to the minimum-runlength code constraint, which provides an increased spacing between adjacent transitions at the expense of a lower code rate. The rate 2/3 (1, 7) code runs at a clock rate 50% higher than that of the rate 8/9 (0, k) code; this puts the (1, 7) code at a considerable disadvantage for storage systems that are limited by the silicon speed. However, for systems that are not limited by the processing speed of the circuit and for which density is the most important concern, the (1, 7) code seems to be the better choice. This is especially true when the ratio of the width of the read pulse to the bit cell length is large (i.e., severe ISI), a situation in which the $d = 1$ code has been shown to exhibit a

Magnetic Storage

minimum-distance increasing property, resulting in a performance improvement upon the $d = 0$ codes despite a larger rate loss.

Among the approaches that employ the minimum runlength codes is the $(1, 7)$ maximum likelihood technique, which provides an efficient near-optimal sequence detection of $(1, 7)$ runlength-coded data [Patel, 1991]. The Viterbi detector can also be used in conjunction with a higher order partial response target, which offers a better matching to the unconditioned channel spectrum than the PR4 target [Thapar and Patel, 1987]. Another example of a detection scheme that provides efficient symbol recovery for minimum-runlength-limited channels is the fixed delay tree search with decision feedback [Moon and Carley, 1990]. The FDTS/DF is a hybrid scheme that combines a decision feedback equalizer structure and a depth-limited tree search processor that makes a decision based on computation and arbitration of path metrics. Recently, it has been shown that for the special case of the depth 1 tree and the $d = 1$ minimum runlength constraint, the need for explicit computation of path metrics can be avoided [Kenney and Carley, 1993]. This results in a considerable saving in hardware complexity and processing requirements.

In the next section, the FDTS/DF is discussed in detail. A technique to simplify the tree processor is described for a specific example of the $d = 1$ code constraint and a search depth of 1.

108.4 Fixed Delay Tree Search with Decision Feedback

The FDTS/DF algorithm combines a depth-limited tree search with a DFE. As in a standard DFE, the minimum-phase filtering is performed in the forward section to create a causal channel response. The feedback filter driven by past decisions then cancels some of the past intersymbol interference terms. This operation is similar to that of the DFE except that not all of the ISI terms are canceled by the feedback filter; in FDTS/DF operation τ most recent ISI terms are allowed to enter the decision element, which in turn utilizes the signal energy contained in the allowed ISI terms in making an improved decision. The decision is generated with a fixed delay of τ symbols.

The operation of the FDTS/DF is best illustrated with a simple example. Let us take $\tau = 1$. After the minimum-phase causal filtering by the forward filter and correct feedback cancellation, the effective channel impulse response seen by the decision element, the FDTS, would be $1 + f_1 D$, where the coefficient f_1 represents the amplitude of the single ISI term allowed to enter the decision element. The decision has to be made with a 1-symbol delay in order to cancel all ISI terms except the most recent one. Let r_k denote the observation sample, that is,

$$r_k = x_k + f_1 x_{k-1} + n_k \qquad (108.11)$$

where n_k represents the noise sample, which includes the residual ISI component in case the filters are designed based on the mean-squared-error (MSE) criterion. The task of the detector is to try to make a best decision on x_k given two successive observation samples r_k and r_{k+1} and all past decisions, $\hat{x}_{k-1}, \hat{x}_{k-2}, \hat{x}_{k-3}, \ldots$. A binary tree of look-ahead depth 1, shown in Fig. 108.4 can be used to describe the decision making process.

Following a typical tree or trellis representation of an ISI channel, the path metric can be calculated for each of the four look-ahead paths using r_k and r_{k+1} and the nominal signal levels associated with branches. The signal level associated with the two root branches can be calculated using the previous decision, $\hat{x}_{k-1}$. The root symbol associated with the most likely path

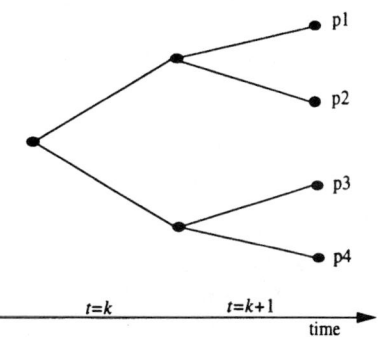

FIGURE 108.4 A binary tree with depth 1.

then becomes the decision symbol, and the two paths that diverge from the winning path at the root are discarded. In the next symbol interval, a new sample r_{k+2} is received and four new look-ahead paths are formed by branching out the two paths survived from the last interval. A decision on x_{k+1} is made in the same manner and the process continues. It can be shown that this algorithm is asymptotically equivalent to the delay-constrained optimum detector, provided that the previous decision is correct [Moon and Carley, 1994]. This algorithm is easily generalized into arbitrary depths. When τ is large, a variation to this algorithm exists that can simplify the path arbitration process using parallel exponential processes [Moon and Carley, 1994]. When a minimum runlength constraint is used, the single error event dominates in a maximum likelihood sequence detection (MLSD) system. It can be shown that when the single error event is the dominant source of error, the FDTS/DF with a small search depth can achieve nearly the MLSD performance. For the $d = 1$ constraint, for example, the FDTS/DF achieves a near MLSD performance with $\tau = 1$ or 2 at all reasonable densities [Moon and Carley, 1990].

In the following, we show how the tree search algorithm can be further simplified with essentially no additional loss of performance for some important practical case. Let us take a specific example of the $d = 1$ constraint and $\tau = 1$. For simplicity, we also assume that $f_1 = 1$, that is, the equalized channel impulse response seen by the decision device is given by $1 + D$. To begin, let us consider the $\tau = 1$ tree shown in Fig. 108.4. First assume that the previous input is -1 (i.e., $x_{k-1} = -1$). With this assumption and under the $d = 1$ constraint, the path $p2$ is not allowed as it contains two successive symbol transitions, and the decision rule can be described graphically as shown in Fig. 108.5. Each of the three paths $p1$, $p3$, and $p4$ is represented by a point, whose location is specified by the two successive noise-free signal levels y_k and y_{k+1} (for the $1 + D$ channel, $y_k = x_k + x_{k+1}$). The decision boundary, formed by two straight lines, is constructed using the nearest neighbor rule. That is, when the observation point specified by (r_k, r_{k+1}) is closer to $p1$ than either of $p3$ and $p4$ (i.e., when the observation falls in region I), the decision is made in favor of $+1$, the root symbol associated with $p1$; otherwise, the decision is made in favor of -1. Since $p1$ and $p3$ are the two closest points, the errors will be made mostly by confusion between these two paths. The probability of confusing $p1$ and $p4$ will become negligible at high signal-to-noise ratios (SNRs). This means that the decision boundary can be replaced, without significant performance

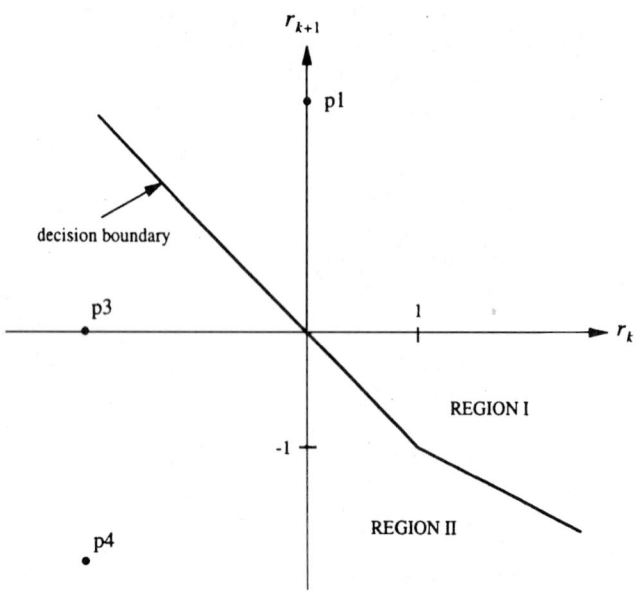

FIGURE 108.5 Optimal decision rule when $x_{k-1} = 1$.

Magnetic Storage

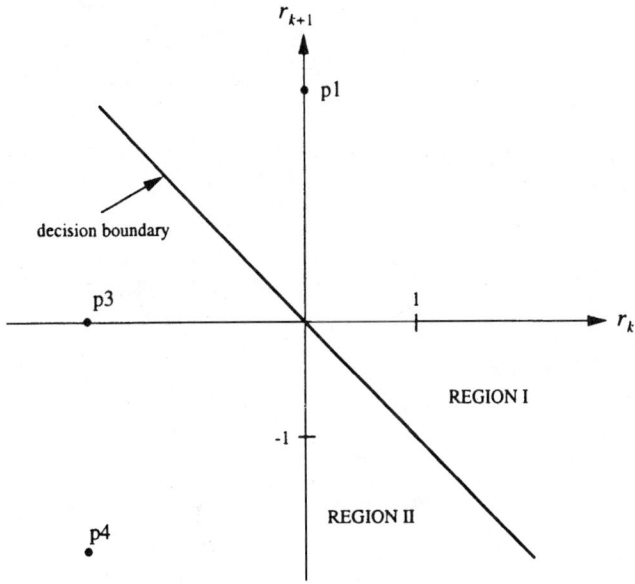

FIGURE 108.6 Simplified decision rule.

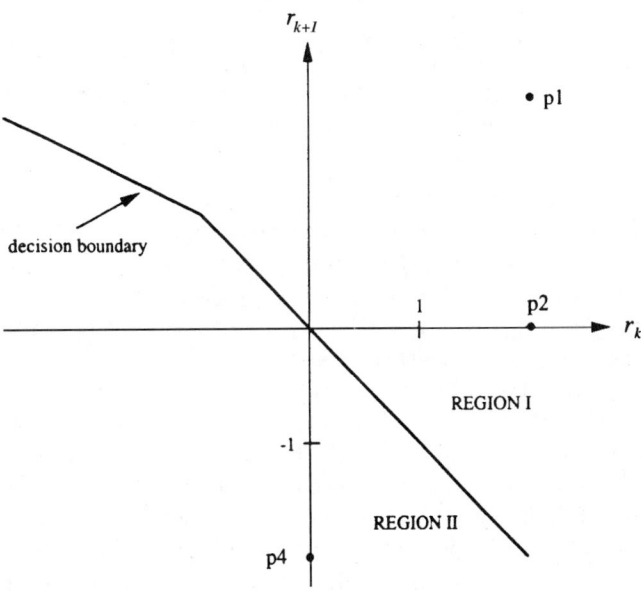

FIGURE 108.7 Optimal decision rule when $x_{k-1} = -1$.

loss, by a single straight line that divides $p1$ and $p3$ with equal distance, as shown in Fig. 108.6. This straight line boundary corresponds to a very simple decision rule: choose $\hat{x}_k = 1$ if $r_{k+1} \geq -r_k$ and choose $\hat{x}_k = -1$ if $r_{k+1} < -r_k$. Taking the similar steps for the case $x_{k-1} = 1$, we arrive at the decision rule depicted in Fig. 108.7. Using the same argument, the decision boundary can again be replaced by the straight line shown in Fig. 108.6. The simplified decision rule set by the decision regions in Fig. 108.6 applies to both $x_{k-1} = 1$ and $x_{k-1} = -1$ cases. This means that the simplified decision rule does not suffer from possible error propagation associated with the straightforward implementation of the FDTS, which uses two different decision rules depending

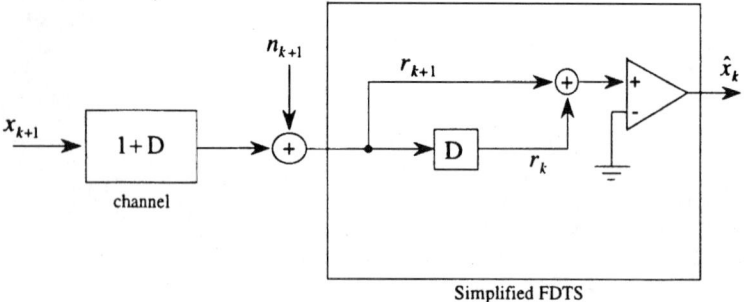

FIGURE 108.8 Simplified FDTS for the $(1 + D)$ channel with $d = 1$.

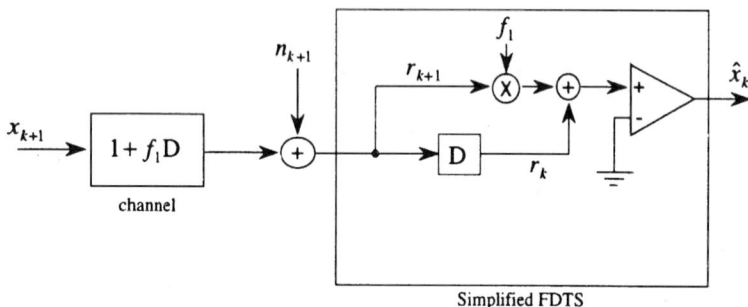

FIGURE 108.9 Simplified FDTS for the $(1 + f_1 D)$ channel with $d = 1$.

on the stored previous decision $\hat{x}_{k-1}$. Figure 108.8 shows the simple structure of this modified algorithm, which consists of a single delay, an adder, and a comparator. For channels of the form $1 + f_1 D$ with a positive f_1, a multiplier can be incorporated into the structure at the + input of the comparator as shown in Fig. 108.9. The asymptotic optimality of this simplified detector structure for any positive constant f_1 (f_1 is observed to be positive for all meaningful recording densities) can be proved using a simple worst-case margin analysis.

This simplified structure can be viewed also as a discrete-time matched filter for the truncated channel with transfer function $1 + f_1 D$, followed by a threshold detector. It should be pointed out, however, that the matched filter with a threshold detector does not normally provide an optimal detection quality when there is ISI. It is the $d = 1$ code constraint that makes this structure optimal even in the presence of significant ISI. The discrete-time matched filter, whose transfer function is given by $D + f_1$, can also be incorporated into the forward and feedback equalizers, as suggested by Kenney and Carley [1993]. This may provide an advantage in terms of the processing speed in VLSI implementation. The resulting structure then essentially becomes a DFE, except that the slicer makes a binary decision on a four-level input signal.

For the simplified structure, the worst-case signal separation for $x_k = 1$ and $x_k = -1$ at the slicing point is $2(1 + f_1^2)$. The performance advantage of the FDTS/DF over a standard DFE, assuming they are both used for the $d = 1$ constrained channel, can easily be understood. Assuming perfect cancellation of the tail of the channel response via the feedback equalizer, the signal separation for the DFE is 2. This means that the FDTS/DF improves upon the DFE by a factor of $1 + f_1^2$ in SNR.

The simplifying technique can be generalized to unconstrained input sequences and to arbitrary search depths. For an arbitrary depth τ, the observation space is $(\tau + 1)$ dimensional and hyperplanes can be used to form disjoint regions in this space. The first step in the simplification process is to identify a minimum number of hyperplanes that can be used to form a decision boundary to

Magnetic Storage

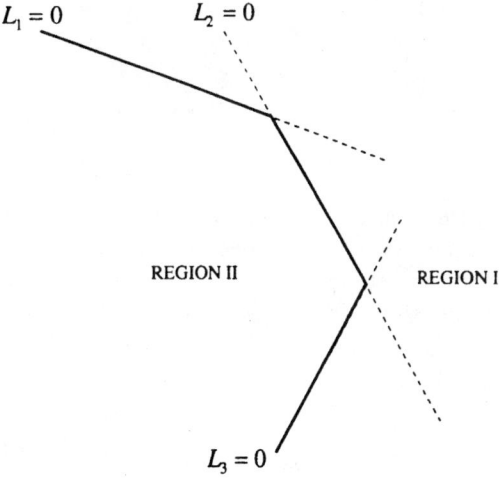

FIGURE 108.10 Hyperplanes used to form decision boundary.

separate the entire observation space into disjoint regions corresponding to two different binary decisions (see Fig. 108.10). To preserve the asymptotic optimality, the distance from any signal point (corresponding to a look-ahead path) to the decision boundary formed by hyperplanes should be greater than or equal to one-half of the minimum distance between any two diverging look-ahead paths. Generalizing the two-dimensional case, each hyperplane in a $(\tau + 1)$-dimensional space can be described as $L(r_k, r_{k+1}, \ldots, r_{k+\tau}, c) = 0$, where L is a linear function of the $(\tau + 1)$ observation variables $r_k, r_{k+1}, \ldots, r_{k+\tau}$, and some constant c. Consider Fig. 108.10, where three hyperplanes, expressed by $L_1 = 0$, $L_2 = 0$, and $L_3 = 0$, form the decision boundary in a multidimensional space. The right-hand side of a hyperplane L_i is specified by the inequality $L_i > 0$, and the left-hand side by $L_i < 0$. It is easy to see that a point given by $(r_k, r_{k+1}, \ldots, r_{k+\tau})$ will fall into region I if the following three inequalities are simultaneously satisfied:

$$L_i(r_k, r_{k+1}, \ldots, r_{k+\tau}, c_i) > 0, \qquad i = 1, 2, 3 \qquad (108.12)$$

The quantity $L_i(r_k, r_{k+1}, \ldots, r_{k+\tau}, c_i)$ can be obtained by adding the constant c_i to the output of a $(\tau + 1)$-tap FIR filter, driven by the observation sample. Therefore, the decision rule implied in Eq. (108.12) can be implemented by three parallel FIR filter and threshold device combinations followed by the logical AND operation, as shown in Fig. 108.11. The resulting structure is essentially

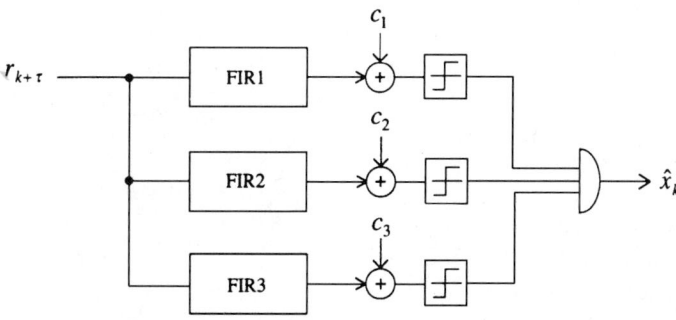

FIGURE 108.11 FIR filter implementation of the decision rule.

the adaptive linear neuron (Adaline) or the perceptron [Haykin, 1994], suggesting that for analytically intractable cases the tap weights of the FIR filters and the constants can be obtained using the training algorithms developed for neural networks.

108.5 Performance Comparisons

The bit error rate (BER) performances of different modulation code and detector combinations have been obtained using a channel simulation program that includes the transition noise sources described earlier. Before discussing the results, let us start with some definitions. The user density is defined to be the number of user bits that can be packed into PW50, the half-height width of the isolated transition response. The present state-of-the-art rigid disk drives operate at user density of approximately 2. The code rates of $(1, 7)$ and $(0, k)$ codes are typically 2/3 and 8/9, respectively.

In defining the signal-to-noise ratio of the channel, the signal power is assumed to be the square of the peak-to-zero amplitude of the readback waveform when the input current is a single-frequency square wave with transition spacing equal to PW50/2. The total noise consists of additive noise and transition noise. The additive noise power is the power of white noise in the Nyquist band corresponding to a sampling period of PW50/2. The transition noise power is that due to the first-order terms in the Taylor's series expansion of the channel transition response in Eq. (108.10). The total noise power is the sum of the additive noise power and transition noise power at the readback point.

BER simulation results are presented for different types of noise environments. An experimentally measured transition response shown in Fig. 108.1 is used for the simulation. Figure 108.12 shows the results obtained for a 90% additive noise dominant channel. The user density was at 2.5. The simple peak detector (PD) combined with a $(1, 7)$ code [referred to as PD$(1, 7)$] shows the worst performance, lagging about 10 dB in SNR behind the two best schemes, the FDTS/DF combined with a $(1, 7)$ code and the DFE with a $(0, k)$ code. The performances of the PR4 equalized maximum likelihood (ML) scheme operating on a $(0, k)$ code [referred to as VA/PR4$(0, k)$] and EPR4 [characterized by the transfer function $(1 - D)(1 + D)^2$] equalized ML scheme on a $(1, 7)$ code fall somewhere in between. It is noteworthy that the FDTS/DF scheme improves on the PR4 equalized ML detector despite the larger code rate loss. This is because at this relatively high density of 2.5, equalization loss of the PR4 system is quite severe as the mismatch between the natural channel and the PR4 target becomes pronounced. In general, for the additive noise dominant channels, equalization loss of fixed target equalizers becomes much larger than that of feedback equalizers as user density increases.

Figure 108.13 shows the BER simulations results for a 90% position jitter dominant channel. The noisy readback samples are generated assuming the random position jitter satisfies a

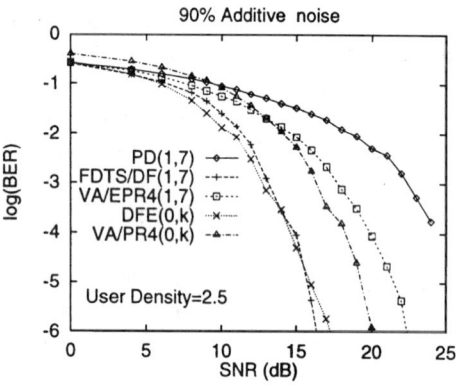

FIGURE 108.12 BER vs required channel SNR for additive noise dominant channel.

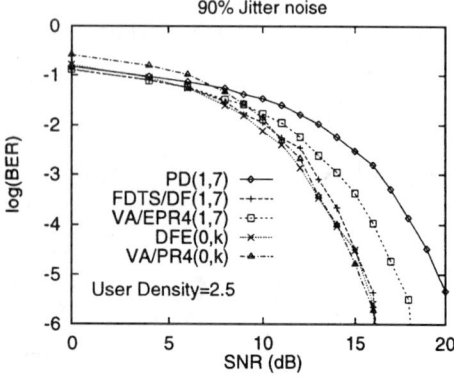

FIGURE 108.13 BER vs required channel SNR for position jitter dominant channel.

Magnetic Storage

truncated Gaussian distribution (transition position is confined within the symbol interval) and is independent from transition to transition. The results are markedly different from the additive noise channel. The power spectrum of the jitter noise component follows the signal spectrum very closely and has small high-frequency spectral contents. Therefore, equalization loss resulting from a high-frequency boost is relatively small. This is evident in the improved performance of the PR4 and EPR4 equalized detectors as well the peak detector, which all require large high-frequency boost in shaping the natural channel, compared to the additive noise dominant channel. All three detectors, the FDTS/DF on a $(1, 7)$ code, the DFE on a $(0, k)$ code, and the PR4 equalized ML on a $(0, k)$ code, perform comparably.

Figure 108.14 shows the similar plots for a 90% width variation dominant channel. All schemes except the peak detector fall in a narrow range of SNR for a given error rate. The peak detector again loses more than 6 dB in SNR compared to the FDTS/DF at the BER of 10^{-5}. The FDTS/DF with a $(1, 7)$ code has about 2-dB SNR advantage over the other three schemes.

An important aspect of transition noise that we have ignored in the present model is its dependence on the spacing between written transitions. In $(1, 7)$ encoded channels, the interaction between transitions is less severe and, thus, noise per transition is expected to be smaller than $(0, k)$ encoded channels. Therefore, in order to make a more accurate comparison in the transition noise channel, one needs to estimate the transition noise in the $(1, 7)$ and $(0, k)$ channels separately and incorporate the difference in the model.

The role of advanced detection schemes becomes more critical as the demand for higher user density continues. It is important to see how much improvement in detector error performance is achieved by using such advanced methods, as the density increases. Figure 108.15 shows a plot of the channel SNR required to obtain a fixed bit error rate of 10^{-4} vs user density. The noise is a mixture of 50% additive white Gaussian noise, 25% position jitter, and 25% width variation. At a low user density of 1.5, there is only a small performance difference among detectors.

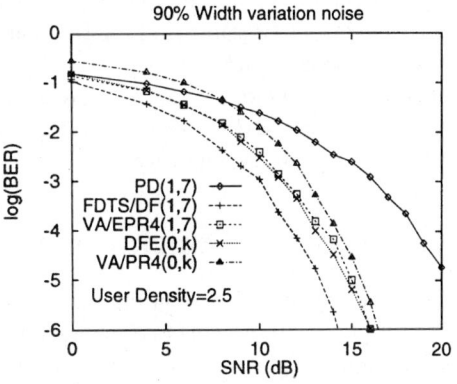

FIGURE 108.14 BER vs required channel SNR for width variation dominant channel.

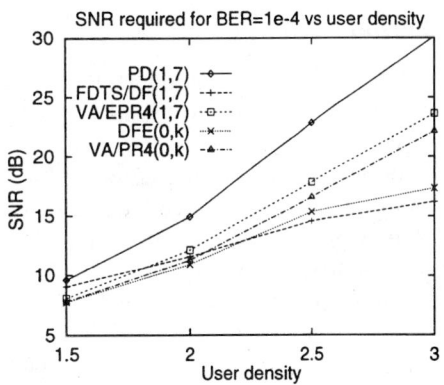

FIGURE 108.15 Detector performance with increasing density.

However, the peak detector shows degradation above user density 1.5, whereas the other detectors have comparable SNRs up to density 2. The difference between the Viterbi detectors and the decision feedback type detectors are observed above density 2. This is primarily caused by additive noise enhancement of the fixed-target equalizers such as PR4 or EPR4, which in turn is a result of the spectral mismatch between the target response and the natural channel response. This noise enhancement is relatively small in decision feedback type schemes, since the target response is effectively adjusted depending on the signal and noise spectra. As user density increases beyond 2.5, the FDTS/DF on $(1, 7)$ code and the DFE on $(0, k)$ code show substantial improvement over the equalized ML detectors. At user density 3, the FDTS has about 1-dB advantage over the DFE despite the code rate disadvantage.

108.6 Future Considerations

Although a brief description is given on common nonlinearities encountered in high-density magnetic channels, the performance study presented here does not include the nonlinear effects. The amount of nonlinearity present in the channel certainly depends on the minimum transition spacing allowed by the code. To investigate the impact of nonlinearities on a particular choice of code, understanding of nonlinearities on a quantitative level is required.

The transition noise model used in our simulation includes only the first-order effects of position jitter and width variation. Also, possible correlation between these two random parameters in a single transition, as well as neighboring transitions, has been ignored in our model. More accurate understanding of the statistics of transition noise on both the micromagnetic level and the system level is required to establish a highly reliable input/output model of recording for the purpose of predicting BERs. Once accurate noise and nonlinearity models are available, efforts can be directed to design coding and detection schemes that are optimal for the given characteristics of underlying transition noise and nonlinearity.

108.7 Summary and Conclusions

A brief description is given on the characteristics of the magnetic recording channel, including signal-dependent transition noise and nonlinearity. The potential of some advanced data detection schemes is investigated using a channel simulator program, which includes the effect of two different types of transition noise as well as additive Gaussian noise. Bit error rate simulation results indicate that transition noise has significantly different impact on detection performance from additive Gaussian noise.

Defining Terms

Decision feedback equalization (DFE): An equalization/detection method for intersymbol interference channels that utilizes data-driven feedback cancellation of intersymbol interference.

Fixed delay tree search with decision feedback (FDTS/DF): A suboptimal detection method that combines finite-depth tree search with partial cancellation of intersymbol interefence through feedback.

Magnetoresistive head: A type of read sensor used in today's high-end magnetic storage devices.

Partial response (PR): A data transmission method that allows a controlled amount of intersymbol interference.

Partial response maximum likelihood (PRML): A data detection technique that combines partial response equalization with Viterbi detection.

Read process: The process of retrieving written data from the storage medium.

Runlength-limited (RLL) codes: Codes that control the minimum and maximum separations of adjacent written magnetic transitions.

Transition noise: Media noise that arises from the written magnetic transitions.

Viterbi detection: A recursive implentation of maximum likelihood sequence esimation.

Write process: The process of recording data onto storage media.

References

Cideciyan, R.D., Dolivo, F., Hermann, R., Hirt, W., and Schott, W. 1992. A PRML system for digital magnetic recording. *IEEE JSAC*, 10(1):38–56.

Ferguson, M.J. 1972. Optimal reception for binary partial response channels. *Bell Syst. Tech. J.*, 51(2):493–505.

Fisher, K.D., Cioffi, J.M., and Melas, C.M. 1989. An adaptive DFE for storage channels suffering from nonlinear ISI. *Proceedings of the International Conference on Communication*, Boston, MA, June.

Haykin, S. 1994. *Neural Networks*, IEEE Press/Macmillan, New York.

Howell, T.D., McCown, D.P., Diola, T.D., Tang, Y.-S., Hense, K.R., and Gee, R.L. 1990. Error rate performance of experimental gigabit per square inch recording components. *IEEE Trans. Magnet.* (Sept.).

Immink, K.A.S. 1986. Coding methods for high-density optical recording. *Philips J. Res.*, 41:410–430.

Immink, K.A.S. 1989. Coding techniques for the noisy magnetic recording channel. *IEEE Trans. Commun.*, 37(5):413–419.

Karabed, R. and Siegel, P.H. 1991. Matched spectral-null codes for partial-response channels. *IEEE Trans. Inf.*, 37(2):818–855.

Kenney, J.G. and Carley, L.R. 1993. Multi-level decision feedback equalization for saturation recording. *IEEE Trans. Magnet.*, 29(3):2160–2171.

Marcus, B.H., Siegel, P.H., and Wolf, J.K. 1992. Finite-state modulation codes for data storage. *IEEE JSAC*, 10(1):5–37.

Melas, C.M., Arnett, P.C., Beardsley, I.A., and Palmer, D. 1987. Nonlinear superposition in saturation recording of disk media. *IEEE Trans. Magnet.*, MAG-23(5):2079–2081.

Moon, J. and Carley, L.R. 1990. Performance comparison of detection methods in magnetic recording. *IEEE Trans. Magnet.*, 26(6):3155–3172.

Moon, J. and Carley, L.R. 1994. Efficient sequence detection for intersymbol channels with runlength constraints. *IEEE Trans. Comm.* (Letter), (Sept.).

Palmer, D., Ziperovich, P., Wood, R., and Howell, T.D. 1987. Identification of nonlinear write effects using pseudorandom sequences. *IEEE Trans. Magnet.*, MAG-23(5):2377–2379.

Patel, A.M. 1991. A new digital signal processing channel for data storage products. *IEEE Trans. Magnet.*, 27(6).

Siegel, P.H. and Wolf, J.K. 1991. Modulation and coding for information storage. *IEEE Commun. Mag.*, 29(12):68–86.

Thapar, H. and Patel, A.M. 1987. A class of partial response systems for increasing storage density in magnetic recording. *IEEE Trans. Magnet.*, 23(5).

Thapar, H., Siegel, P., Shung, B., Rae, J., and Karabed, R. 1993. Trellis-coded partial response (TCPR): An improved recording method over PRML. *IEEE Trans. Magnet.* (Sept.).

Wood, R.W. 1990. Magnetic megabit. *IEEE Spectrum*, pp. 32–38.

Further Information

For an extensive treatment on constrained codes used for data storage channels, see *Coding Techniques for Digital Recorders*, K.A.S. Immink, Prentice Hall, 1991. For a comprehensive overview of magnetic recording technology, including computer tape and disk recording and consumer video and audio recording, consult *Magnetic Recording*, Mee and Daniel, McGraw Hill, 1989. For more detailed information on some coding and signal processing ideas for data storage, see the January 1992 issue of the *IEEE Journal on Selected Areas in Communications*. Also, each year, the Global Telecommunications Conference, the International Conference on Communications, and the Asilomar Conference on Signals, Systems and Computers run sessions dedicated to data storage. Finally, the September or the November issue of the *IEEE Transactions on Magnetics* includes papers presented in the annual International Conference on Magnetics, and contains a significant number of papers on communications and signal processing issues of data storage technology.

109
Magneto-Optical Disk Data Storage

109.1 Introduction 1546
109.2 Preliminaries and Basic Definitions 1547
   The Concept of Track • Disk Rotation Speed • Access Time
109.3 The Optical Path 1552
   Collimation and Beam-Shaping • Focusing by the Objective Lens
109.4 Automatic Focusing 1555
109.5 Automatic Tracking 1556
   Tracking on Grooved Regions • Sampled Tracking
109.6 Thermomagnetic Recording Process 1558
   Recording by Laser Power Modulation (LPM) • Recording by Magnetic Field Modulation
109.7 Magneto-Optical Readout 1561
   Differential Detection
109.8 Materials of Magneto-Optical Data Storage 1563

M. Mansuripur
*Optical Sciences Center,
University of Arizona, Tucson*

109.1 Introduction

Since the early 1940s, magnetic recording has been the mainstay of electronic information storage worldwide. Audio tapes provided the first major application for the storage of information on magnetic media. Magnetic tape has been used extensively in consumer products such as audio tapes and video cassette recorders (VCRs); it has also found application in backup/archival storage of computer files, satellite images, medical records, etc. Large volumetric capacity and low cost are the hallmarks of tape data storage, although sequential access to the recorded information is perhaps the main drawback of this technology. Magnetic hard disk drives have been used as mass storage devices in the computer industry ever since their inception in 1957. With an areal density that has doubled roughly every other year, hard disks have been and remain the medium of choice for secondary storage in computers.[1] Another magnetic data storage device, the floppy disk, has been successful in areas where compactness, removability, and fairly rapid access to the recorded information have been of prime concern. In addition to providing backup and safe storage, inexpensive floppies with their moderate capacities (2 megabyte on a 3.5-in-diam platter is typical

---

[1] At the time of this writing, achievable densities on hard disks are in the range of 10^7 b/cm^2. Random access to arbitrary blocks of data in these devices can take on the order of 10 ms, and individual read/write heads can transfer data at the rate of several megabits per second.

nowadays) and reasonable transfer rates have provided the crucial function of file/data transfer between isolated machines. All in all, it has been a great half-century of progress and market dominance for magnetic recording devices, which are only now beginning to face a potentially serious challenge from the technology of optical recording.

Like magnetic recording, a major application area for optical data storage systems is the secondary storage of information for computers and computerized systems. Like the high-end magnetic media, optical disks can provide recording densities in the range of 10^7 b/cm^2 and beyond. The added advantage of optical recording is that, like floppies, these disks can be removed from the drive and stored on the shelf. Thus the functions of the hard disk (i.e., high capacity, high data transfer rate, rapid access) may be combined with those of the floppy (i.e., backup storage, removable media) in a single optical disk drive. Applications of optical recording are not confined to computer data storage. The enormously successful audio **compact disk (CD)**, which was introduced in 1983 and has since become the de facto standard of the music industry, is but one example of the tremendous potentials of the optical technology.

A strength of optical recording is that, unlike its magnetic counterpart, it can support read-only, write-once, and erasable/rewritable modes of data storage. Consider, for example, the technology of optical audio/video disks. Here the information is recorded on a master disk, which is then used as a stamper to transfer the embossed patterns to a plastic substrate for rapid, accurate, and inexpensive reproduction. The same process is employed in the mass production of read-only files [CD read only memory (CD-ROM), and optical ROM (O-ROM)], which are now being used to distribute software, catalog, and other large data bases. Or consider the write-once read-many (WORM) technology, where one can permanently store massive amounts of information on a given medium and have rapid, random access to them afterwards. The optical drive can be designed to handle read only, WORM, and erasable media all in one unit, thus combining their useful features without sacrificing performance and ease of use, or occupying too much space. What is more, the media can contain regions with prerecorded information as well as regions for read/write/erase operations, both on the same platter. These possibilities open new vistas and offer opportunities for applications that have heretofore been unthinkable; the interactive video-disk is perhaps a good example of such applications.

In this chapter we lay out the conceptual basis for optical data storage systems; the emphasis will be on disk technology in general and magneto-optical disk in particular. The next section is devoted to a discussion of some elementary aspects of disk data storage including the concept of track and definition of the access time. Section 109.3 describes the basic elements of the optical path and its functions; included are the properties of the semiconductor laser diode, characteristics of the beam shaping optics, and certain features of the focusing objective lens. Because of the limited depth of focus of the objective and the eccentricity of tracks, optical disk systems must have a closed-loop feedback mechanism for maintaining the focused spot on the right track. These mechanisms are described in Secs. 109.4 and 109.5 for automatic focusing and automatic track-following, respectively. The physical process of thermomagnetic recording in magneto-optic (MO) media is described next, followed by a discussion of the MO readout process in Sec. 109.7. The final section describes the properties of the MO media.

109.2 Preliminaries and Basic Definitions

A disk, whether magnetic or optical, consists of a number of **tracks** along which the information is recorded. These tracks may be concentric rings of a certain width, W_t, as shown in Fig. 109.1. Neighboring tracks may be separated from each other by a guard band whose width we shall denote by W_g. In the least sophisticated recording scheme imaginable, marks of length Δ_0 are recorded along these tracks. Now, if each mark can be in either one of two states, present or absent, it may be associated with a binary digit, 0 or 1. When the entire disk surface of radius R is covered with such marks, its capacity C_0 will be

$$C_0 = \frac{\pi R^2}{(W_t + W_g)\Delta_0} \quad \text{bits per surface} \tag{109.1}$$

Consider the parameter values typical of current optical disk technology: $R = 67$ mm corresponding to 5.25-in-diam platters, $\Delta_0 = 0.5$ μm, which is roughly determined by the wavelength of the read/write laser diodes, and $W_t + W_g = 1$ μm for the track-pitch. The disk capacity will then be around 28×10^9 b, or 3.5 gigabytes. This is a reasonable estimate and one that is fairly close to reality, despite the many simplifying assumptions made in its derivation. In the following paragraphs we examine some of these assumptions in more detail.

The disk was assumed to be fully covered with information-carrying marks. This is generally not the case in practice. Consider a disk rotating at $\mathcal{N}$ revolutions per second (r/s). For reasons to be clarified later, this rotational speed should remain constant during the disk operation. Let the electronic circuitry have a fixed clock duration T_c. Then only pulses of length

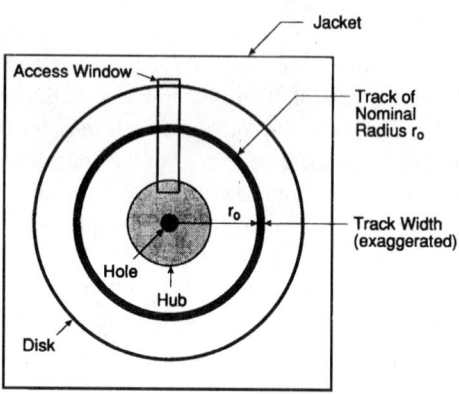

FIGURE 109.1 Physical appearance and general features of an optical disk. The read/write head gains access to the disk through a window in the jacket; the jacket itself is for protection purposes only. The hub is the mechanical interface with the drive for mounting and centering the disk on the spindle. The track shown at radius r_0 is of the concentric-ring type.

T_c (or an integer multiple thereof) may be used for writing. Now, a mark written along a track of radius r, with a pulse-width equal to T_c will have length ℓ, where

$$\ell = 2\pi \mathcal{N} r T_c \tag{109.2}$$

Thus for a given rotational speed $\mathcal{N}$ and a fixed clock cycle T_c, the minimum mark length ℓ is a linear function of track radius r, and ℓ decreases toward zero as r approaches zero. One must, therefore, pick a minimum usable track radius, $r_{\min}$, where the spatial extent of the recorded marks is always greater than the minimum allowed mark length, Δ_0. Equation (109.2) yields:

$$r_{\min} = \frac{\Delta_0}{2\pi \mathcal{N} T_c} \tag{109.3}$$

One may also define a maximum usable track radius $r_{\max}$, although for present purposes $r_{\max} = R$ is a perfectly good choice. The region of the disk used for data storage is thus confined to the area between $r_{\min}$ and $r_{\max}$. The total number N of tracks in this region is given by

$$N = \frac{r_{\max} - r_{\min}}{W_t + W_g} \tag{109.4}$$

The number of marks on any given track in this scheme is independent of the track radius; in fact, the number is the same for all tracks, since the period of revolution of the disk and the clock cycle uniquely determine the total number of marks on any individual track. Multiplying the number of usable tracks N with the capacity per track, we obtain for the usable disk capacity

$$C = \frac{N}{\mathcal{N} T_c} \tag{109.5}$$

Replacing for N from Eq. (109.4) and for $\mathcal{N}T_c$ from Eq. (109.3), we find

$$C = \frac{2\pi r_{\min}(r_{\max} - r_{\min})}{(W_t + W_g)\Delta_0} \tag{109.6}$$

If the capacity C in Eq. (109.6) is considered a function of $r_{\min}$ with the remaining parameters held constant, it is not difficult to show that maximum capacity is achieved when

$$r_{\min} = \frac{1}{2}r_{\max} \tag{109.7}$$

With this optimum $r_{\min}$, the value of C in Eq. (109.6) is only half that of C_0 in Eq. (109.1). In other words, the estimate of 3.5 gigabyte per side for 5.25-in disks seems to have been optimistic by a factor of two.

One scheme often proposed to enhance the capacity entails the use of multiple zones, where either the rotation speed $\mathcal{N}$ or the clock period T_c are allowed to vary from one zone to the next. In general, zoning schemes can reduce the minimum usable track radius below that given by Eq. (109.7). More importantly, however, they allow tracks with larger radii to store more data than tracks with smaller radii. The capacity of the zoned disk is somewhere between C of Eq. (109.6) and C_0 of Eq. (109.1), the exact value depending on the number of zones implemented.

A fraction of the disk surface area is usually reserved for **preformat** information and cannot be used for data storage. Also, prior to recording, additional bits are generally added to the data for **error-correction coding** and other house-keeping chores. These constitute a certain amount of overhead on the user data, and must be allowed for in determining the capacity. A good rule of thumb is that overhead consumes approximately 20% of the raw capacity of an optical disk, although the exact number may vary among the systems in use. Substrate defects and film contaminants during the deposition process can create bad **sectors** on the disk. These are typically identified during the certification process and are marked for elimination from the sector directory. Needless to say, bad sectors must be discounted when evaluating the capacity.

Modulation codes may be used to enhance the capcity beyond what has been described so far. Modulation coding does not modify the minimum mark length of Δ_0, but frees the longer marks from the constraint of being integer multiples of Δ_0. The use of this type of code results in more efficient data storage and an effective number of bits per Δ_0 that is greater than unity. For example, the popular (2, 7) modulation code has an effective bit density of 1.5 b per Δ_0. This or any other modulation code can increase the disk capacity beyond the estimate of Eq. (109.6).

The Concept of Track

The information on magnetic and optical disks is recorded along tracks. Typically, a track is a narrow annulus at some distance r from the disk center. The width of the annulus is denoted by W_t, whereas the width of the guard band, if any, between adjacent tracks is denoted by W_g. The track pitch is the center-to-center distance between neighboring tracks and is therefore equal to $W_t + W_g$. A major difference between the magnetic floppy disk, the magnetic hard disk, and the optical disk is that their respective track-pitches are presently of the order of 100, 10, and 1 μm. Tracks may be fictitious entities, in the sense that no independent existence outside the pattern of recorded marks may be ascribed to them. This is the case, for example, with the compact audio disk format where prerecorded marks simply define their own tracks and help guide the laser beam during readout. In the other extreme are tracks that are physically engraved on the disk surface before any data is ever recorded. Examples of this type of track are provided by pregrooved WORM and magneto-optical disks. Figure 109.2 shows micrographs from several recorded optical disk surfaces. The tracks along which the data are written are clearly visible in these pictures.

It is generally desired to keep the read/write head stationary while the disk spins and a given track is being read from or written onto. Thus, in an ideal situation, not only should the track

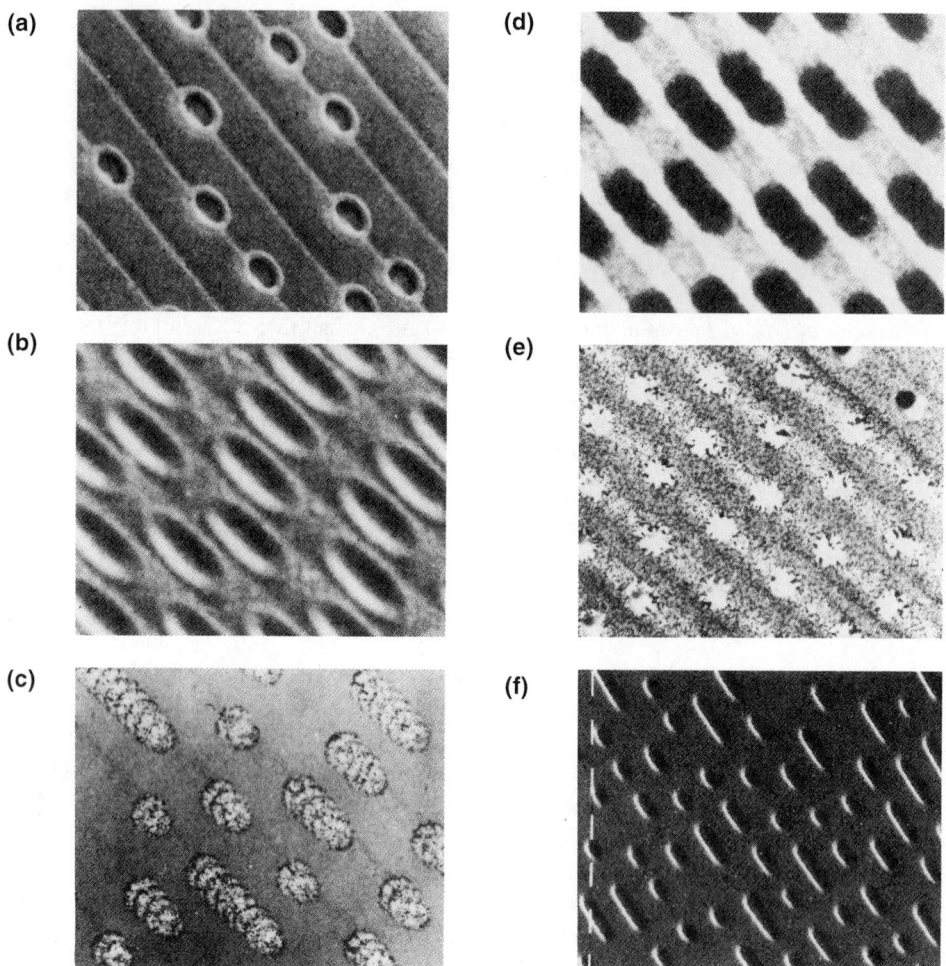

FIGURE 109.2 Micrographs of several types of optical storage media. The tracks are straight and narrow (track pitch = 1.6 μm), with an orientation angle of $\simeq -45°$: (a) ablative, write-once tellurium alloy; (b) ablative, write-once organic dye; (c) amorphous-to-crystalline, write-once phase-change alloy GaSb; (d) erasable, amorphous magneto-optic alloy GdTbFe; (e) erasable, crystalline-to-amorphous phase-change tellurium alloy; and (f) read-only CD-Audio, injection-molded from polycarbonate with a nickel stamper. (*Source:* 1989. *Ullmann's Encyclopedia of Industrial Chemistry*, 5th ed., Vol. A14, VCH, Weinheim, p. 196. With permission.)

be perfectly circular, but also the disk must be precisely centered on the spindle axis. In practical systems, however, tracks are neither precisely circular, nor are they concentric with the spindle axis. These eccentricity problems are solved in low-performance floppy drives by making tracks wide enough to provide tolerance for misregistrations and misalignments. Thus the head moves blindly to a radius where the track center is nominally expected to be, and stays put until the reading or writing is over. By making the head narrower than the track pitch, the track center is allowed to wobble around its nominal position without significantly degrading the performance during the read/write operation. This kind of wobble, however, is unacceptable in optical disk systems, which have a very narrow track, about the same size as the focused beam spot. In a typical situation arising in practice the eccentricity of a given track may be as much as ±50 μm while the track pitch is only about 1 μm, thus requiring active track-following procedures.

One method of defining tracks on an optical disk is by means of pregrooves, that are either etched, stamped, or molded onto the substrate. In grooved media of optical storage, the space between neighboring grooves is the so-called land [see Fig. 109.3(a)]. Data may be written in the grooves with the land acting as a guard band. Alternatively, the land regions may be used for recording while the grooves separate adjacent tracks. The groove-depth is optimized for generating an optical signal sensitive to the radial position of the read/write laser beam. For the push–pull method of track error detection the groove depth is in the neighborhood of $\lambda/8$, where λ is the wavelength of the laser beam.

In digital data storage applications, each track is divided into small segments or sectors intended for the storage of a single block of data (typically either 512 or 1024 bytes). The physical length of a sector is thus a few millimeters. Each sector is preceded by header information such as the identity of the sector, identity of the corresponding track, synchronization marks, etc. The header information may be preformatted onto the substrate, or it may be written on the storage layer prior to shipping the disk. Pregrooved tracks may be carved on the optical disk either as concentric rings or as a single continuous spiral. There are certain advantages to each format. A spiral track can contain a succession of sectors without interruption, whereas concentric rings may each end up with some empty space that is smaller than the required length for a sector. Also, large files may be written onto (and read from) spiral tracks without jumping to the next track which occurs when concentric tracks are used. On the other hand, multiple-path operations such as write-and-verify or erase-and-write which require two paths each for a given sector, or still-frame video are more conveniently handled on concentric-ring tracks.

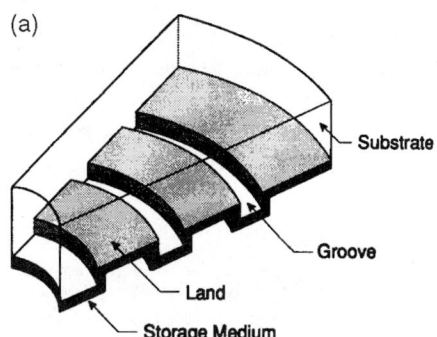

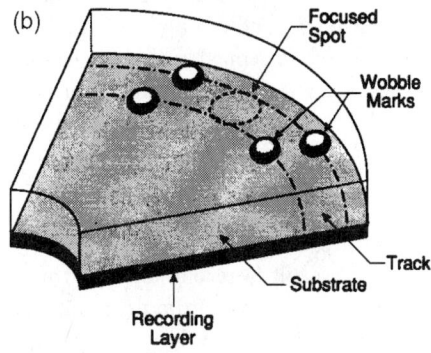

FIGURE 109.3 (a) Lands and grooves in an optical disk. The substrate is transparent, and the laser beam must pass through it before reaching the storage medium. (b) Sampled-servo marks in an optical disk. These marks, which are offset from the track-center, provide information regarding the position of focused spot.

Another track format used in practice is based on the sampled-servo concept. Here the tracks are identified by occasional marks placed permanently on the substrate at regular intervals, as shown in Fig. 109.3(b). Details of track-following by the sampled-servo scheme will follow shortly; suffice it to say at this point that servo marks help the system identify the position of the focused spot relative to the track-center. Once the position is determined it is fairly simple to steer the beam and adjust its position.

Disk Rotation Speed

When a disk rotates at a constant angular velocity ω, a track of radius r moves with the constant linear velocity $V = r\omega$. Ideally, one would like to have the same linear velocity for all of the tracks, but this is impractical except in a limited number of situations. For instance, when the desired mode of access to the various tracks is sequential, such as in audio and video disk applications, it is possible to place the head in the beginning at the inner radius, and move outward from the center thereafter while continuously decreasing the angular velocity. By keeping the product of r and ω constant, one can thus achieve constant linear velocity for all of the tracks.[2] Sequential access mode, however,

---

[2]In compact audio disk players the linear velocity is kept constant at 1.2 m/s. The starting position of the head is at the inner radius $r_{min} = 25$ mm, where the disk spins at 460 r/m. The spiral track ends at the outer radius $r_{max} = 58$ mm, where the disk's angular velocity is 200 r/m.

is the exception rather than the norm in data storage systems. In most applications, the tracks are accessed randomly with such rapidity that it becomes impossible to adjust the rotation speed for constant linear velocity. Under these circumstances the angular velocity is best kept constant during the normal operation of the disk. Typical rotation speeds are 1200 and 1800 r/m for slower drives, and 3600 r/m for the high-data-rate systems. Higher rotation rates (5000 r/m and beyond) are certainly feasible and will likely appear in future storage devices.

Access Time

The direct-access storage-device (DASD) used in computer systems for the mass storage of digital information is a disk drive capable of storing large quantities of data and accessing blocks of this data rapidly and in arbitrary order. In read/write operations it is often necessary to move the head to new locations in search of sectors containing specific data-items. Such relocations are usually time consuming and can become the factor that limits performance in certain applications. The access time τ_a is defined as the average time spent in going from one randomly selected spot on the disk to another. Also, τ_a can be considered the sum of a seek time τ_s, which is the average time needed to acquire the target track, and a latency τ_l, which is the average time spent on the target track while waiting for the desired sector. Thus,

$$\tau_a = \tau_s + \tau_l \qquad (109.8)$$

The latency is half the revolution period of the disk, since a randomly selected sector is, on the average, halfway along the track from the point where the head initially lands. Thus for a disk rotating at 1200 r/m $\tau_l = 25$ ms, whereas at 3600 r/m $\tau_l \simeq 8.3$ ms. The seek time, on the other hand, is independent of the rotation speed of the disk, but is determined by the traveling distance of the head during an average seek, as well as by the mechanism of head actuation. It can be shown that the average length of travel in a random seek is one-third of the full stroke. (In our notation the full stroke is $r_{max} - r_{min}$.) In magnetic disk drives where the head/actuator assembly is relatively light-weight (a typical Winchester head weighs about 5 g) the acceleration and deceleration periods are short, and seek times are typically around 10 ms in small format drives (i.e., 5.25 and 3.5 in). In optical disk systems, on the other hand, the head, being an assembly of discrete elements, is fairly large and heavy (typical weight $\simeq 100$ g), resulting in values of τ_s that are several times greater than those obtained in magnetic recording systems. The seek times reported for commercially available optical drives presently range from 20 ms in high performance 3.5-in drives to about 80 ms in larger (and slower) drives. We emphasize, however, that the optical disk technology is still in its infancy; with the passage of time, the integration and miniaturization of the elements within the optical head will surely produce light-weight devices capable of achieving seek times of the order of only a few milliseconds.

109.3 The Optical Path

The **optical path** begins at the light source which, in practically all laser disk systems in use today, is a semiconductor GaAs diode laser. Several unique features have made the laser diode indispensable in optical recording technology, not only for the readout of stored information but also for writing and erasure. The small size of this laser has made possible the construction of compact head assemblies, its coherence properties have enabled diffraction-limited focusing to extremely small spots, and its direct modulation capability has eliminated the need for external modulators. The laser beam is modulated by controlling the injection current; one applies pulses of variable duration to turn the laser on and off during the recording process. The pulse duration can be as short as a few nanoseconds, with rise and fall times typically less than 1 ns. Although readout can be accomplished at constant power level, that is, in continuous wave (CW) mode, it is customary for noise reduction purposes to modulate the laser at a high frequency (e.g., several hundred megahertz during readout).

Magneto-Optical Disk Data Storage

Collimation and Beam-Shaping

Since the cross-sectional area of the active region in a laser diode is only about 1 μm, diffraction effects cause the emerging beam to diverge rapidly. This phenomenon is depicted schematically in Fig. 109.4(a). In practical applications of the laser diode, the expansion of the emerging beam is arrested by a collimating lens, such as that shown in Fig. 109.4(b). If the beam happens to have aberrations (astigmatism is particularly severe in diode lasers) then the collimating lens must be designed to correct this defect as well.

In optical recording it is most desirable to have a beam with circular cross-section. The need for shaping the beam arises from the special geometry of the laser cavity with its rectangular cross-section. Since the emerging beam has different dimensions in the directions parallel and perpendicular to the junction, its cross section at the collimator becomes elliptical, with the initially narrow dimension expanding more rapidly to become the major axis of the ellipse. The collimating lens thus produces a beam with elliptical cross section. Circularization may be achieved by bending various rays of the beam at a prism, as shown in Fig. 109.4(c). The bending changes the beam's diameter in the plane of incidence, but leaves the diameter in the perpendicular direction intact.

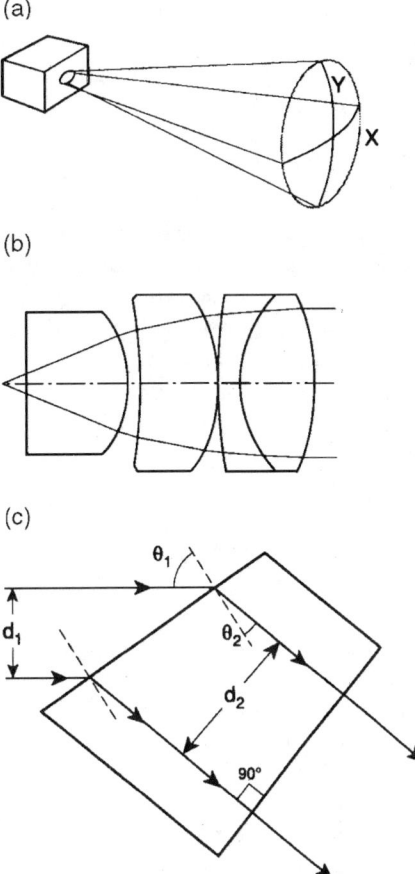

FIGURE 109.4 (a) Away from the facet, the output beam of a diode laser diverges rapidly. In general, the beam diameter along X is different from that along Y, which makes the cross section of the beam elliptical. Also, the radii of curvature R_x and R_y are not the same, thus creating a certain amount of astigmatism in the beam. (b) Multielement collimator lens for laser diode applications. Aside from collimating, this lens also corrects astigmatic aberrations of the beam. (c) Beam-shaping by deflection at a prism surface. θ_1 and θ_2 are related by the Snell's law, and the ratio d_2/d_1 is the same as $\cos\theta_2/\cos\theta_1$. Passage through the prism circularizes the elliptical cross section of the beam.

Focusing by the Objective Lens

The collimated and circularized beam of the diode laser is focused on the surface of the disk using an **objective lens**. The objective is designed to be aberration free, so that its focused spot size is limited only by the effects of diffraction. Figure 109.5(a) shows the design of a typical objective made from spherical optics. According to the classical theory of diffraction, the diameter of the beam, d, at the objective's focal plane is given by

$$d \simeq \frac{\lambda}{NA} \qquad (109.9)$$

where λ is the wavelength of light, and NA is the numerical aperture of the objective.[3]

---

[3] Numerical aperture is defined as $NA = n \sin\theta$, where n is the refractive index of the image space and θ is the half-angle subtended by the exit pupil at the focal point. In optical recording systems the image space is air whose index is very nearly unity; thus for all practical purposes $NA = \sin\theta$.

In optical recording it is desired to achieve the smallest possible spot, since the size of the spot is directly related to the size of marks recorded on the medium. Also, in readout, the spot size determines the resolution of the system. According to Eq. (109.9) there are two ways to achieve a small spot. First by reducing the wavelength and, second, by increasing the numerical aperture of the objective. The wavelengths currently available from GaAs lasers are in the range of 670–840 nm. It is possible to use a nonlinear optical device to double the frequency of these diode lasers, thus achieving blue light. Good efficiencies have been demonstrated by frequency doubling. Also recent developments in II–VI materials have improved the prospects for obtaining green and blue light directly from semiconductor lasers. Consequently, there is hope that in the near future optical storage systems will operate in the wavelength range of 400–500 nm. As for the numerical aperture, current practice is to use a lens with $NA \simeq 0.5$–0.6. Although this value might increase slightly in the coming years, much higher numerical apertures are unlikely, since they put strict constraints on the other characteristics of the system and limit the tolerances. For instance, the working distance at high numerical aperture is relatively short, making access to the recording layer through the substrate more difficult. The smaller depth of focus of a high numerical aperture lens will make attaining/maintaining proper focus more of a problem, whereas the limited field of view might restrict automatic track-following procedures. A small field of view also places constraints on the possibility of read/write/erase operations involving multiple beams.

The depth of focus of a lens, δ, is the distance away from the focal plane over which tight focus can be maintained [see Fig. 109.5(b)]. According to the classical diffraction theory

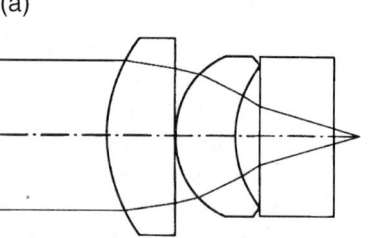

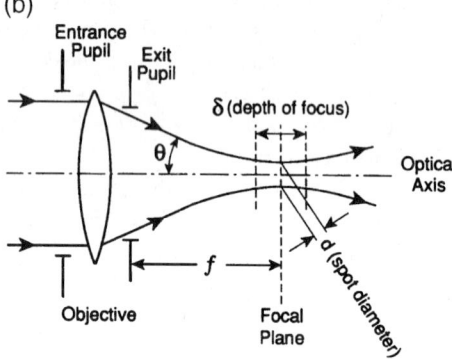

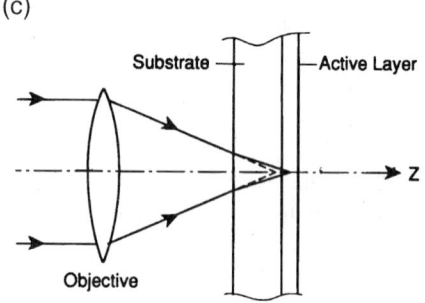

FIGURE 109.5 (a) Multi element lens design for a high numerical aperture video disc objective. (*Source: Kuntz, D. 1984. Specifying laser diode optics, Laser Focus, March. With permission.*) (b) Various parameters of the objective lens. The numerical aperture is $NA = \sin \theta$. The spot diameter d and the depth of focus δ are given by Eqs. (109.9) and (109.10), respectively. (c) Focusing through the substrate can cause spherical aberration at the active layer. The problem can be corrected if the substrate is taken into account in the design of the objective.

$$\delta \simeq \frac{\lambda}{NA^2} \qquad (109.10)$$

Thus for a wavelength of $\lambda = 700$ nm and $NA = 0.6$, the depth of focus is about ± 1 μm. As the disk spins under the optical head at the rate of several thousand revolutions per minute, the objective lens must stay within a distance of $f \pm \delta$ from the active layer if proper focus is to be maintained. Given the conditions under which drives usually operate, it is impossible to make rigid enough mechanical systems to yield the required positioning tolerances. On the other hand, it is fairly simple to mount the objective lens in an actuator capable of adjusting its position with the aid of closed-loop feedback control. We shall discuss the technique of automatic focusing in the next section. For now, let us emphasize that by going to shorter wavelengths and/or larger numerical apertures (as is required for attaining higher data densities) one will have to face a much stricter regime as far as automatic focusing is concerned. Increasing the numerical aperture is particularly worrisome, since δ drops with the square of NA.

A source of spherical aberrations in optical disk systems is the substrate through which the light must travel to reach the active layer of the disk. Figure 109.5(c) shows the bending of the rays at the disk surface that causes the aberration. This problem can be solved by taking into account the effects of the substrate in the design of the objective, so that the lens is corrected for all aberrations including those arising at the substrate. Recent developments in molding of aspheric glass lenses have gone a long way in simplifying the lens design problem. Figure 109.6 shows a pair of molded glass aspherics designed for optical disk system applications; both the collimator and the objective are single-element lenses, and are corrected for aberrations.

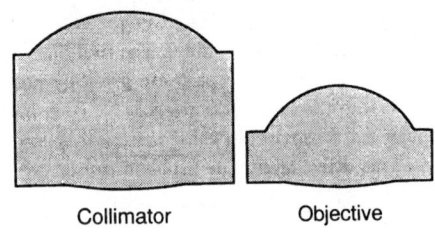

FIGURE 109.6 Molded glass aspheric lens pair for optical disk applications. These singlets can replace the multi element spherical lenses shown in Figs. 109.4(b) and 109.5(a).

109.4 Automatic Focusing

We mentioned in the preceding section that since the objective has a large numerical aperture ($NA \geq 0.5$), its depth of focus δ is rather shallow ($\delta \simeq \pm 1\,\mu\text{m}$ at $\lambda = 780$ nm). During all read/write/erase operations, therefore, the disk must remain within a fraction of a micrometer from the focal plane of the objective. In practice, however, the disks are not flat and they are not always mounted rigidly parallel to the focal plane, so that movements away from focus occur a few times during each revolution. The peak-to-peak movement in and out of focus may be as much as 100 μm. Without automatic focusing of the objective along the optical axis, this runout (or disk flutter) will be detrimental to the operation of the system. In practice, the objective is mounted on a small motor (usually a voice coil) and allowed to move back and forth in order to keep its distance within an acceptable range from the disk. The spindle turns at a few thousand revolutions per minute, which is a hundred or so revolutions per second. If the disk moves in and out of focus a few times during each revolution, then the voice coil must be fast enough to follow these movements in real time; in other words, its frequency response must extend to several kilohertz.

The signal that controls the voice coil is obtained from the light reflected from the disk. There are several techniques for deriving the focus error signal, one of which is depicted in Fig. 109.7(a). In this so-called obscuration method a secondary lens is placed in the path of the reflected light, one-half of its aperture is covered, and a split-detector is placed at its focal plane. When the disk is in focus, the returning beam is collimated and the secondary lens will focus the beam at the center of the split detector, giving a difference signal ΔS equal to zero. If the disk now moves away from the objective, the returning beam will become converging, as in Fig. 109.7(b), sending all the light to detector 1. In this case ΔS will be positive and the voice coil will push the lens towards the disk. On the other hand, when the disk moves close to the objective, the returning beam becomes diverging and detector 2 receives the light [see Fig. 109.7(c)]. This results in a negative ΔS that forces the voice coil to pull back in order to return ΔS to zero. A given focus error detection scheme is generally characterized by the shape of its focus error signal ΔS vs the amount of defocus Δz; one such curve is shown in Fig. 109.7(d). The slope of the focus error signal (FES) curve near the origin is of particular importance, since it determines the overall performance and stability of the servo loop.

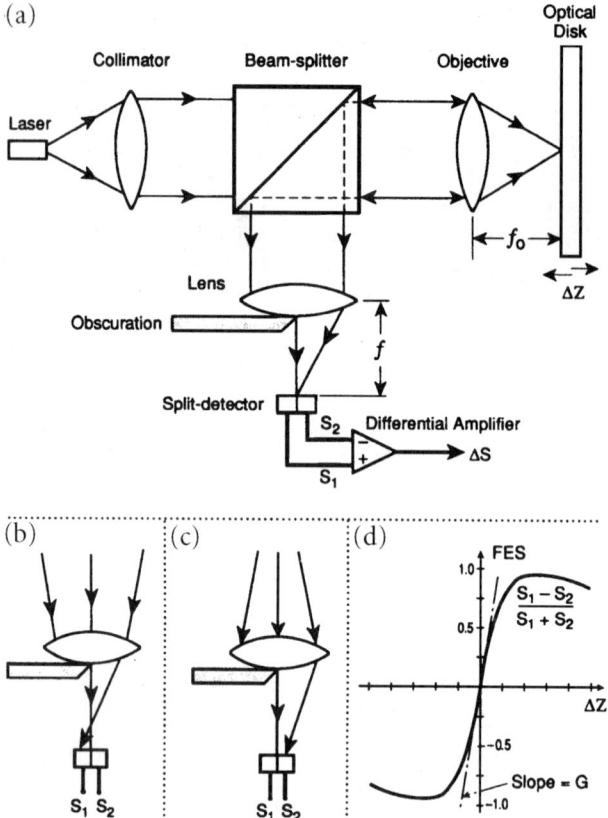

FIGURE 109.7 Focus error detection by the obscuration method. In (a) the disk is in focus, and the two halves of the split detector receive equal amounts of light. When the disk is too far from the objective (b), or too close to it (c), the balance of detector signals shifts to one side or the other. A plot of the focus error signal (FES) vs defocus is shown in (d), and its slope near the origin is identified as the FES gain G.

109.5 Automatic Tracking

Consider a track at a certain radial location, say r_0, and imagine viewing this track through the access window shown in Fig. 109.1. It is through this window that the head gains access to arbitrarily selected tracks. To a viewer looking through the window, a perfectly circular track centered on the spindle axis will look stationary, irrespective of the rotation rate. However, any eccentricity will cause an apparent radial motion of the track. The peak-to-peak distance travelled by a track (as seen through the window) depends on a number of factors including centering accuracy of the hub, deformability of the substrate, mechanical vibrations, manufacturing tolerances, etc. For a typical 3.5-in disk, for example, this peak-to-peak motion can be as much as 100 μm during one revolution. Assuming a revolution rate of 3600 r/m, the apparent velocity of the track in the radial direction will be several millimeter per second. Now, if the focused spot remains stationary while trying to from read or write to this track, it is clear that the beam will miss the track for a good fraction of every revolution cycle.

Practical solutions to the described problem are provided by **automatic tracking** techniques. Here the objective is placed in a fine actuator, typically a voice-coil, which is capable of moving the necessary radial distances and maintaining a lock on the desired track. The signal that controls the movement of this actuator is derived from the reflected light itself, which carries information about the position of the focused spot. There exist several mechanisms for extracting the track-error signal (TES); all of these methods require some sort of structure on the disk surface in order to identify the track. In the case of read-only disks (CD, CD-ROM, and video disk), the embossed pattern of data provides ample information for tracking purposes. In the case of write-once and erasable disks, tracking guides are carved on the substrate in the manufacturing process. As mentioned earlier, the two major formats for these tracking guides are pregrooves (for continuous tracking) and sampled-servo marks (for discrete tracking). A combination of the two schemes, known as continuous/composite format, is often used in practice. This scheme is depicted in Fig. 109.8, which shows a small section containing five tracks, each consisting of the tail-end of a groove, synchronization marks, a mirror area used for adjusting focus/track offsets, a pair of wobble marks for sampled tracking, and header information for sector identification.

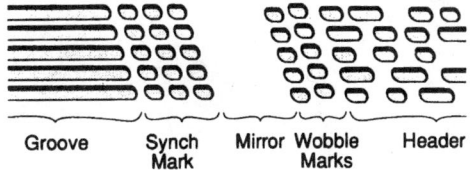

FIGURE 109.8 Servo fields in continuous/composite format contain a mirror area and offset marks for tracking. (*Source*: Marchant, A.B. 1990. *Optical Recording*, Addison–Wesley, Reading, MA., p. 264. With permission.)

Tracking on Grooved Regions

As shown in Fig. 109.3(a), grooves are continuous depressions that are either embossed or etched or molded onto the substrate prior to deposition of the storage medium. If the data is recorded on the grooves, then the lands are not used except for providing a guard band between neighboring grooves. Conversely, the land regions may be used to record the information, in which case grooves provide the guard band. Typical track-widths are about one wavelength of the light. The guard bands are somewhat narrower than the tracks; their exact shape and dimensions depending on the beam size, the required track-servo accuracy, and the acceptable levels of cross talk between adjacent tracks. The groove depth is usually around one-eighth of one wavelength ($\lambda/8$), since this depth can be shown to give the largest TES in the push–pull method. Cross sections of the grooves may be rectangular, trapezoidal, or triangular.

When the focused spot is centered on track, it is diffracted symmetrically from the two edges of the track, resulting in a balanced far-field pattern. As soon as the spot moves away from the center, the symmetry breaks down and the light distribution in the far field tends to shift to one side or the other. A split photodetector placed in the path of the reflected light can therefore sense the relative position of the spot and provide the appropriate feedback signal. This strategy is depicted schematically in Fig. 109.9(a); also shown in the figure are intensity plots at the detector plane for light reflected from various regions of the disk. Note how the intensity shifts to one side or the other depending on the direction of motion of the spot.

Sampled Tracking

Since dynamic track runout is usually a slow and gradual process, there is actually no need for continuous tracking as done on grooved media. A pair of embedded marks, offset from the track center as in Fig. 109.3(b), can provide the necessary information for correcting the relative position of the focused spot. The reflected intensity will indicate the positions of the two servo marks as two successive short pulses. If the beam happens to be on track, the two pulses will have equal magnitudes and there shall be no need for correction. If, on the other hand, the beam is off track, one of the pulses will be stronger than the other. Depending on which pulse is the strong one, the system will recognize the direction in which it has to move and will correct the error accordingly.

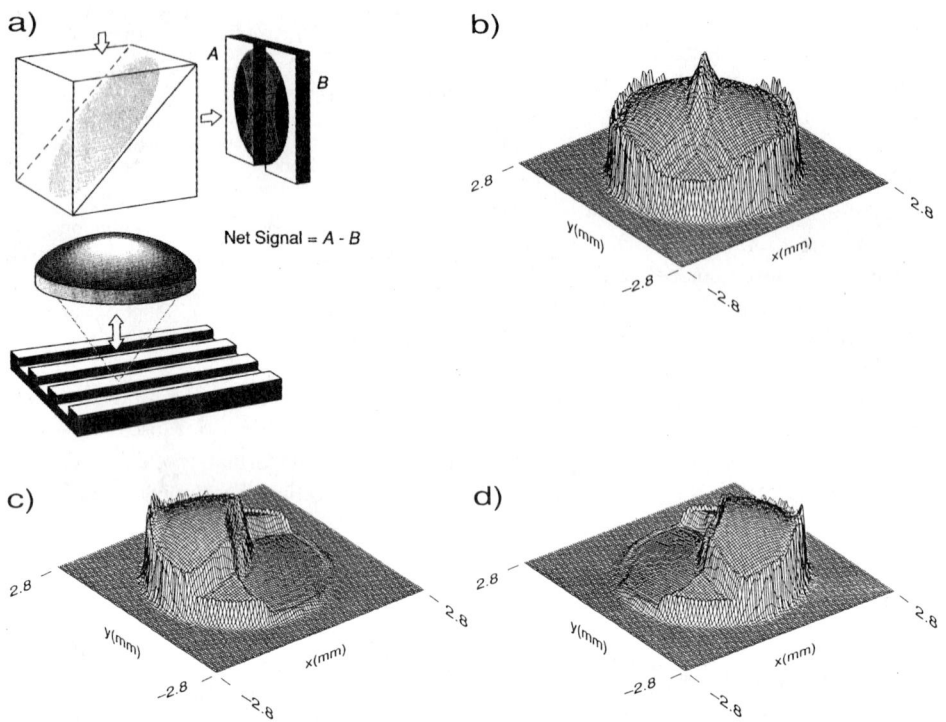

FIGURE 109.9 (a) Push–pull sensor for tracking on grooves. (*Source:* Marchant, A.B. 1990. *Optical Recording*, Addison-Wesley, Reading MA. 1990, p. 175. With permission.) (b) Calculated distribution of light intensity at the detector plane when the disk is in focus and the beam is centered on track. (c) Calculated intensity distribution at the detector plane with disk in focus but the beam centered on the groove edge. (d) Same as (c) except for the spot being focused on the opposite edge of the groove.

The servo marks must appear frequently enough along the track to ensure proper track following. In a typical application, the track might be divided into groups of 18 bytes, 2 bytes dedicated as servo offset areas and 16 bytes filled with other format information or left blank for user data.

109.6 Thermomagnetic Recording Process

Recording and erasure of information on a magneto-optical disk are both achieved by the **thermomagnetic process**. The essence of thermomagnetic recording is shown in Fig. 109.10. At the ambient temperature the film has a high magnetic coercivity[4] and therefore does not respond to the externally applied field. When a focused beam raises the local temperature of the film, the hot spot becomes magnetically soft (i.e., its coercivity drops). As the temperature rises, coercivity drops continuously until such time as the field of the electromagnet finally overcomes the material's resistance to reversal and switches its magnetization. Turning the laser off brings the temperatures back to normal, but the reverse-magnetized domain remains frozen in the film. In a typical situation in practice, the film thickness may be around 300 Å, laser power at the disk $\simeq$ 10 mW, diameter of the focused spot $\simeq$ 1 μm, laser pulse duration $\simeq$ 50 ns, linear velocity of the track $\simeq$ 10 m/s, and the magnetic field strength $\simeq$ 200 G. The temperature may reach a peak of 500 K at the center of

---

[4]Coercivity of a magnetic medium is a measure of its resistance to magnetization reversal. For example, consider a thin film with perpendicular magnetic moment saturated in the $+Z$ direction. A magnetic field applied along $-Z$ will succeed in reversing the direction of magnetization only if the field is stronger than the coercivity of the film.

Magneto-Optical Disk Data Storage

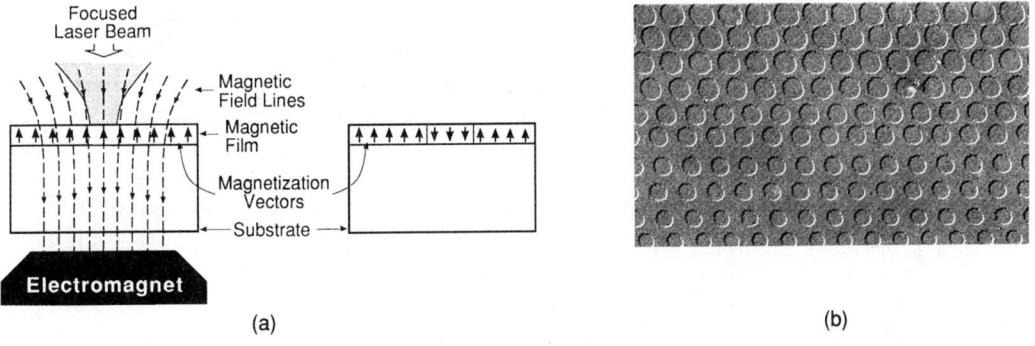

FIGURE 109.10 (a) Thermomagnetic recording process: the field of the electromagnet helps reverse the direction of magnetization in the area heated by the focused laser beam, (b) Lorentz micrograph of domains written thermomagnetically. The various tracks shown here were written at different laser powers, with power level decreasing from top to bottom. (*Source:* Greidanus, F. et al., 1989. Paper 26B-5, International Symposium on Optical Memory, Kobe, Japan, Sept. With permission.)

the spot, which is sufficient for magnetization reversal, but is not nearly high enough to melt or crystalize or in any other way modify the material's atomic structure.

The materials of magneto-optical recording have strong perpendicular magnetic anisotropy. This type of anisotropy favors the up and down directions of magnetization over all other orientations. The disk is initialized in one of these two directions, say, up, and the recording takes place when small regions are selectively reverse magnetized by the thermomagnetic process. The resulting magnetization distribution then represents the pattern of recorded information. For instance, binary sequences may be represented by a mapping of 0s to up-magnetized regains and 1s to down-magnetized regions [nonreturn to zero (NRZ)]. Alternatively, the NRZI scheme might be used, whereby transitions (up-to-down and down-to-up) are used to represent the 1s in the bit sequence.

Recording by Laser Power Modulation (LPM)

In this traditional approach to thermomagnetic recording, the electromagnet produces a constant field, whereas the information signal is used to modulate the power of the laser beam. As the disk rotates under the focused spot, the on/off laser pulses create a sequence of up/down domains along the track. The Lorentz electron micrograph in Fig. 109.10(b) shows a number of domains recorded by laser power modulation (LPM). The domains are highly stable and may be read over and over again without significant degradation. If, however, the user decides to discard a recorded block and to use the space for new data, the LPM scheme does not allow direct overwrite; the system must erase the old data during one disk revolution cycle and record the new data in a subsequent revolution cycle.

During erasure, the direction of the external field is reversed, so that up-magnetized domains in Fig. 109.10(a) now become the favored ones. Whereas writing is achieved with a modulated laser beam, in erasure the laser stays on for a relatively long period of time, erasing an entire sector. Selective erasure of individual domains is not practical, nor is it desired, since mass data storage systems generally deal with data at the level of blocks, which are recorded onto and read from individual sectors. Note that at least one revolution period elapses between the erasure of an old block and its replacement by a new block. The electromagnet therefore need not be capable of rapid switchings. (When the disk rotates at 3600 r/m, for example, there is a period of 16 ms or so between successive switchings.) This kind of slow reversal allows the magnet to be large enough to cover all of the tracks simultaneously, thereby eliminating the need for a moving magnet and an actuator. It also affords a relatively large gap between the disk and the magnet, which enables the use of double-sided disks and relaxes the mechanical tolerances of the system without over-burdening the magnet's driver.

The obvious disadvantage of LPM is its lack of direct overwrite capability. A more subtle concern is that it is perhaps unsuitable for the pulse width modulation (PWM) scheme of representing binary waveforms. Because of fluctuations in the laser power, spatial variations of material properties and lack of perfect focusing and track following, etc., the length of a recorded domain along the track may fluctuate in small but unpredictable ways. If the information is to be encoded in the distance between adjacent domain walls (i.e., PWM), then the LPM scheme of thermomagnetic writing may suffer from excessive domain-wall jitter. Laser power modulation works well, however, when the information is encoded in the position of domain centers [i.e., pulse position modulation (PPM)]. In general, PWM is superior to PPM in terms of the recording density, and, therefore, recording techniques that allow PWM are preferred.

Recording by Magnetic Field Modulation

Another method of thermomagnetic recording is based on magnetic field modulation (MFM) and is depicted schematically in Fig. 109.11(a). Here the laser power may be kept constant while the information signal is used to modulate the magnetic field. Photomicrographs of typical domain patterns recorded in the MFM scheme are shown in Fig. 109.11(b). Crescent-shaped domains are the hallmark of the field modulation technique. If one assumes (using a much simplified model) that the magnetization aligns itself with the applied field within a region whose temperature has

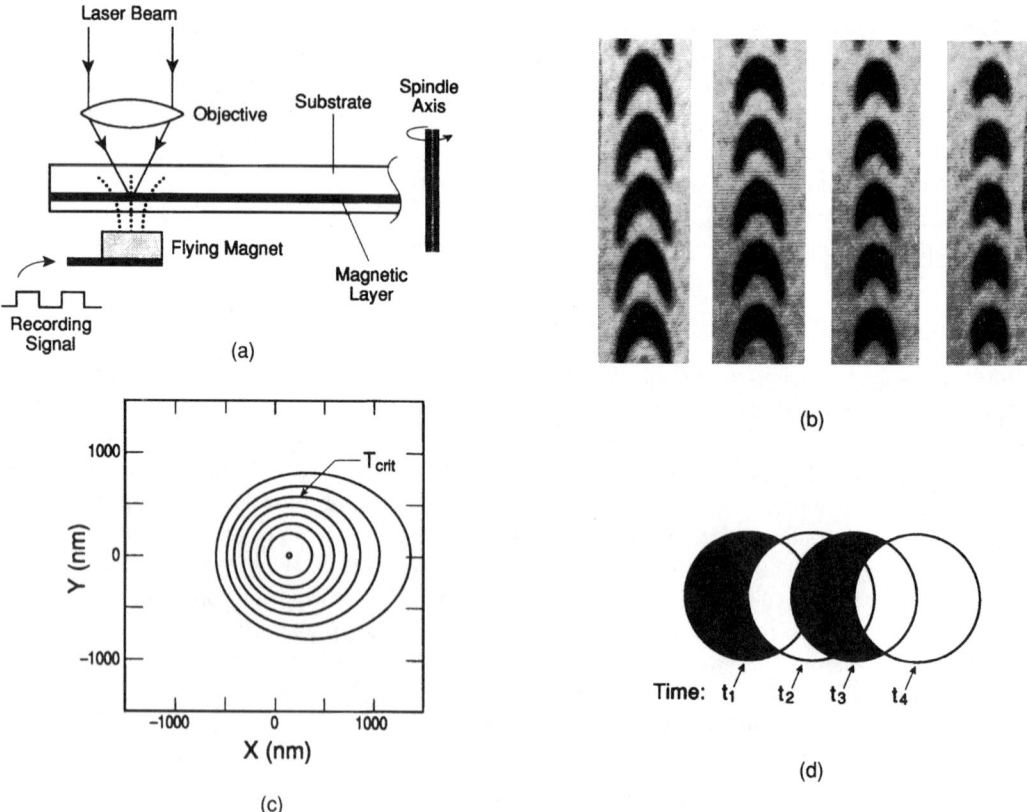

FIGURE 109.11 (a) Thermomagnetic recording by magnetic field modulation. The power of the beam is kept constant, while the magnetic field direction is switched by the data signal. (b) Polarized-light microphotograph of recorded domains. (c) Computed isotherms produced by a CW laser beam, focused on the magnetic layer of a disk. The disk moves with constant velocity under the beam. The region inside the isotherm marked as T_{crit} is above the critical temperature for writing, that is, its magnetization aligns with the direction of the applied field. (d) Magnetization within the heated region (above T_{crit}) follows the direction of the applied field, whose switchings occur at times t_n. The resulting domains are crescent shaped.

passed a certain critical value, T_{crit}, then one can explain the crescent shape of these domains in the following way: With the laser operating in the CW mode and the disk moving at constant velocity, temperature distribution in the magnetic medium assumes a steady-state profile, such as that shown in Fig. 109.11(c). Of course, relative to the laser beam, the temperature profile is stationary, but in the frame of reference of the disk the profile moves along the track with the linear track velocity. The isotherm corresponding to T_{crit} is identified as such in the figure; within this isotherm the magnetization aligns itself with the applied field. Figure 109.11(d) shows a succession of critical isotherms along the track, each obtained at the particular instant of time when the magnetic field switches direction. From this picture it is easy to infer how the crescent-shaped domains form, and also understand the relation between the waveform that controls the magnet and the resulting domain pattern.

The advantages of magnetic field modulation recording are that (1) direct overwriting is possible, and (2) domain wall positions along the track, being rather insensitive to defocus and laser power fluctuations, are fairly accurately controlled by the timing of the magnetic field switchings. On the negative side, the magnet must now be small and fly close to the disk surface if it is to produce rapidly switched fields with a magnitude of a hundred gauss or so. Systems that utilize magnetic field modulation often fly a small electromagnet on the opposite side of the disk from the optical stylus. Since mechanical tolerances are tight, this might compromise the removability of the disk. Moreover, the requirement of close proximity between the magnet and the storage medium dictates the use of single-sided disks in practice.

109.7 Magneto-Optical Readout

The information recorded on a perpendicularly magnetized medium may be read with the aid of the polar **magneto-optical Kerr effect**. When linearly-polarized light is normally incident on a perpendicular magnetic medium, its plane of polarization undergoes a slight rotation upon reflection. This rotation of the plane of polarization, whose sense depends on the direction of magnetization in the medium, is known as the polar Kerr effect. The schematic representation of this phenomenon in Fig. 109.12 shows that if the polarization vector suffers a counterclockwise rotation upon reflection from an up-magnetized region, then the same vector will rotate clockwise when the magnetization is down. A magneto-optical (MO) medium is characterized in terms of its reflectivity R and its Kerr rotation angle θ_k. R is a real number (between 0 and 1) that indicates the fraction of the incident power reflected back

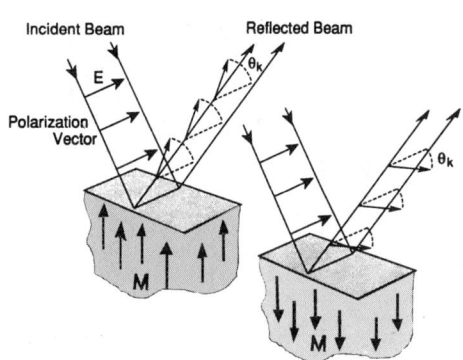

FIGURE 109.12 Schematic diagram describing the polar magneto-optical Kerr effect. Upon reflection from the surface of a perpendicularly magnetized medium, the polarization vector undergoes a rotation. The sense of rotation depends on the direction of magnetization **M**, and switches sign when **M** is reversed.

from the medium at normal incidence. Here θ_k is generally quoted as a positive number, but is understood to be positive or negative depending on the direction of magnetization; in MO readout, it is the sign of θ_k that carries the information about the state of magnetization, that is, the recorded bit pattern.

The laser used for readout is usually the same as that used for recording, but its output power level is substantially reduced in order to avoid erasing (or otherwise obliterating) the previously recorded information. For instance, if the power of the write/erase beam is 20 mW, then for the read operation the beam is attenuated to about 3 or 4 mW. The same objective lens that focuses the write beam is now used to focus the read beam, creating a diffraction-limited spot for resolving the recorded marks. Whereas in writing the laser was pulsed to selectively reverse magnetize small regions along the track, in readout it operates with constant power, that is, in CW mode. Both up- and down-magnetized regions are read as the track passes under the focused spot. The reflected beam, which is now polarization modulated, goes back through the objective and becomes collimated once again; its information content is subsequently decoded by polarization-sensitive optics, and the scanned pattern of magnetization is reproduced as an electronic signal.

Differential Detection

Figure 109.13 shows the differential detection system that is the basis of magneto-optical readout in practically all erasable optical storage systems in use today. The beam-splitter (BS) diverts half of the

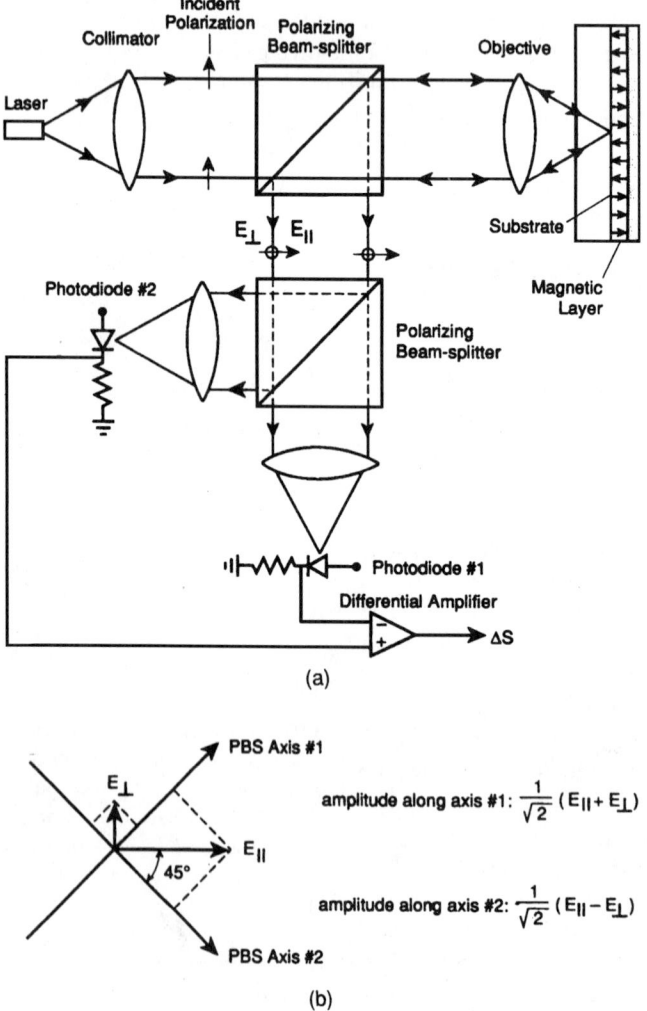

FIGURE 109.13 Differential detection scheme (a) utilizes a polarizing beam splitter and two photodetectors in order to convert the rotation of polarization to an electronic signal. $E_\parallel$ and $E_\perp$ are the reflected components of polarization; they are, respectively, parallel and perpendicular to the direction of incident polarization. The diagram in (b) shows the orientation of the PBS axes relative to the polarization vectors.

reflected beam away from the laser and into the detection module.[5] The polarizing beam-splitter (PBS) splits the beam into two parts, each carrying the projection of the incident polarization along one axis of the PBS, as shown in Fig. 109.13(b). The component of polarization along one of the axes goes straight through, whereas the component along the other axis splits off and branches to the side. The PBS is oriented such that in the absence of the Kerr effect its two branches will receive equal amounts of light. In other words, if the polarization, upon reflection from the disk, did not undergo any rotations whatsoever, then the beam entering the PBS would be polarized at 45° to the PBS axes, in which case it would split equally between the two branches. Under this condition, the two detectors generate identical signals and the differential signal ΔS will be zero. Now, if the beam returns from the disk with its polarization rotated clockwise (rotation angle = θ_k), then detector 1 will receive more light than detector 2, and the differential signal will be positive. Similarly, a counterclockwise rotation will generate a negative ΔS. Thus, as the disk rotates under the focused spot, the electronic signal ΔS reproduces the pattern of magnetization along the scanned track.

109.8 Materials of Magneto-Optical Data Storage

Amorphous rare earth transition metal alloys are presently the media of choice for erasable optical data storage applications. The general formula for the composition of the alloy may be written $(Tb_y Gd_{1-y})_x (Fe_z Co_{1-z})_{1-x}$ where terbium and gadolinium are the rare earth (RE) elements, whereas iron and cobalt are the transition metals (TM). In practice, the transition metals constitute roughly 80 atomic percent of the alloy (i.e., $x \simeq 20$). In the transition metal subnetwork, the fraction of cobalt is usually small, typically around 10%, and iron is the dominant element ($z \simeq 90$). Similarly, in the rare earth subnetwork Tb is the main element ($y \simeq 90$) whereas the gadolinium content is small or it may even be absent in some cases. Since the rare earth elements are highly reactive to oxygen, RE-TM films tend to have poor corrosion resistance and, therefore, require protective coatings. In multilayer disk structures the dielectric layers that enable optimization of the medium for the best optical/thermal behavior, also perform the crucial function of protecting the MO layer from the environment.

The amorphous nature of the material allows its composition to be continuously varied until a number of desirable properties are achieved. In other words, the fractions x, y, z of the various elements are not constrained by the rules of stoichiometry. Disks with very large areas can be coated uniformly with thin films of these media, and in contrast to polycrystalline films whose grains and grain boundaries scatter the beam and cause noise, amorphous films are continuous, smooth, and substantially free from noise. The films are deposited either by sputtering from an alloy target or by cosputtering from multiple elemental targets. In the latter case, the substrate moves under the various targets and the fraction of a given element in the alloy is determined by the time spent under each target, as well as the power applied to that target. During film deposition the substrate is kept at a low temperature (usually by chilled water) in order to reduce the mobility of deposited atoms and to inhibit crystal growth. The type of the sputtering gas (argon, krypton, xenon, etc.) and its pressure during sputtering, the bias voltage applied to the substrate, deposition rate, nature of the substrate and its pretreatment, temperature of the substrate all can have dramatic effects on the composition and short-range order of the deposited film. A comprehensive discussion of the factors that influence film properties will take us beyond the intended scope here; the interested reader may consult the vast literature of this field for further information.

---

[5]The use of an ordinary beam-splitter is an inefficient way of separating the incoming and outgoing beams, since half the light is lost in each pass through the splitter. One can do much better by using a so-called leaky beam-splitter.

Defining Terms

Automatic focusing: The process in which the distance of the disk from the objective's focal plane is continuously monitored and fed back to the system in order to keep the disk in focus at all times.

Automatic tracking: The process in which the distance of the focused spot from the track center is continuously monitored and the information fed back to the system in order to maintain the read/write beam on track at all times.

Compact disk (CD): A plastic substrate embossed with a pattern of pits that encode audio signals in digital format. The disk is coated with a metallic layer (to enhance its reflectivity) and read in a drive (CD player) that employs a focused laser beam and monitors fluctuations of the reflected intensity in order to detect the pits.

Error correction coding (ECC): Systematic addition of redundant bits to a block of binary data, as insurance against possible read/write errors. A given error correcting code can recover the original data from a contaminated block, provided that the number of erroneous bits is less than the maximum number allowed by that particular code.

Grooved media of optical storage: A disk embossed with grooves of either the concentric-ring type or the spiral type. If grooves are used as tracks, then the lands (i.e., regions between adjacent grooves) are the guard bands. Alternatively, lands may be used as tracks, in which case the grooves act as guard bands. In a typical grooved optical disk in use today the track width is 1.1 μm, the width of the guard band is 0.5 μm, and the groove depth is 70 nm.

Magneto-optical Kerr effect: The rotation of the plane of polarization of a linearly polarized beam of light upon reflection from the surface of a perpendicularly magnetized medium.

Objective lens: A well-corrected lens of high numerical aperture, similar to a microscope objective, used to focus the beam of light onto the surface of the storage medium. The objective also collects and recollimates the light reflected from the medium.

Optical path: Optical elements in the path of the laser beam in an optical drive. The path begins at the laser itself and contains a collimating lens, beam shaping optics, beam splitters, polarization-sensitive elements, photodetectors, and an objective lens.

Preformat: Information such as sector address, synchronization marks, servo marks, etc., embossed permanently on the optical disk substrate.

Sector: A small section of track with the capacity to store one block of user data (typical blocks are either 512 or 1024 bytes). The surface of the disk is covered with tracks, and tracks are divided into contiguous sectors.

Thermomagnetic process: The process of recording and erasure in magneto-optical media, involving local heating of the medium by a focused laser beam, followed by the formation or annihilation of a reverse-magnetized domain. The successful completion of the process usually requires an external magnetic field to assist the reversal of the magnetization.

Track: A narrow annulus or ringlike region on a disk surface, scanned by the read/write head during one revolution of the spindle; the data bits of magnetic and optical disks are stored sequentially along these tracks. The disk is covered either with concentric rings of densely packed circular tracks or with one continuous, fine-pitched spiral track.

References

Arimoto, A., Ojima, M., Chinone, N., Oishi, A., Gotoh, T., and Ohnuki, N. 1986. Optimum conditions for the high frequency noise reduction method in optical video disk players. *Appl. Opt.*, 25:1398.

Beck, J.W. 1970. Noise considerations of optical beam recording. *Appl. Opt.*, 9:2559.

Bouwhuis, G., Braat, J., Huijser, A., Pasman, J., Van Rosmalen, G., and Immink, K. S. 1985. *Principles of Optical Disk Systems*, Adam Hilger, Bristol. Chaps. 2 and 3.

Carslaw, H.S. and Jaeger, J.C. 1954. *Conduction of Heat in Solids*, Oxford Univ. Press, London.

Chikazumi, S. and Charap, S.H. 1964. *Physics of Magnetism*, Wiley, New York.

Egashira, K. and Yamada, R. 1974. Kerr effect enhancement and improvement of readout characteristics in MnBi film memory. *J. Appl. Phys.*, 45:3643–3648.

Hansen, P. and Heitman, H. 1989. Media for erasable magneto-optic recording. *IEEE Trans. Mag.*, 25:4390–4404.

Heemskerk, J. 1978. Noise in a video disk system: experiments with an (AlGa) As laser. *Appl. Opt.*, 17:2007.

Huth, B.G. 1974. Calculation of stable domain radii produced by thermomagnetic writing *IBM J. Res. Dev.*, pp. 100–109.

Immink, K.A.S. 1986. Coding methods for high-density optical recording. *Philips J. Res.*, 41:410–430.

Kivits, P., deBont, R., and Zalm, P. 1981. Superheating of thin films for optical recording. *Appl. Phys.*, 24:273–278.

Kryder, M.H. 1987. Data-storage technologies for advanced computing. *Sci. Am.* 257:116–125.

Maissel, L.I. and Glang, R., eds. 1970. *Handbook of Thin Film Technology*, McGraw–Hill, New York.

Malozemoff, A.P. and Slonczewski, J.C. 1979. *Magnetic Domain Walls in Bubble Materials*, Academic, New York.

Mansuripur, M. 1989. Certain computational aspects of vector diffraction problem. *J. Opt. Soc. Am. A*, 6:786–806.

Mansuripur, M., Connell, G.A.N., and Goodman, J.W. 1982a. Laser-induced local heating of multilayers. *Appl. Opt.*, 21:1106.

Mansuripur, M., Connell, G.A.N., and Goodman, J.W. 1982b. Signal and noise in magneto-optical read-out. *J. Appl. Phys.*, 53:4485.

Marchant, A.B. 1990. *Optical Recording*, Addison–Wesley, Reading, MA.

Patel, A.M. 1988. Signal and error-control coding. In *Magnetic Recording*, Vol. II, eds. C.D. Mee and E.D. Daniel, McGraw–Hill, New York.

Pershan, P.S. 1967. Magneto-optic effects. *J. Appl. Phys.*, 38:1482–1490.

Smith, D.O. 1965. Magneto-optical scattering multilayer magnetic and dielectric films. *Opt. Acta*, 12:13.

Special Issue. 1978. *Appl. Opt.*, July 1 (video disks).

Weissler, G.L. and Carlson, R.W., eds. 1979. *Vacuum Physics and Technology*, Vol. 14, Methods of Experimental Physics, Academic, New York.

Wolf, E. 1959. Electromagnetic diffraction in optical systems. I. An integral representation of the image field. *Proc. R. Soc. Ser. A*, 253:349–357.

Further Information

Proceedings of the Optical Data Storage Conference are published annually by SPIE, the International Society for Optical Engineering. These proceedings document the latest developments in the field of optical recording each year. Two other conferences in this field are the International Symposium on Optical Memory (ISOM), whose proceedings are published as a special issue of the Japanese Journal of Applied Physics, and the Magneto-Optical Recording International Symposium (MORIS), whose proceedings appear in a special issue of the Journal of the Magnetics Society of Japan.

Index

A

AALs (ATM adaptation layers), 534–536
AB (access burst), 1231, 1242
Absorption coefficient, 836
ACATS (Advisory Committee on Advanced Television Service), 1462
Access burst. *See* AB
Access channel, defined, **1274**
Access control, for TV, 1497–1498
Access grant control channel. *See* AGCH
Access methods for radio, 1104–1115
    CDMA, 1112–1113
    FDMA, 1107–1109
    methods compared, 1113–1114
    relevant characteristics, 1106–1107
    TDMA, 1109–1112
ACCH (associated control channel), 1229; defined, **1242**
ACD (automatic call distributor), 443–444
A/D (analog to digital) conversion, 107–117
Adaptive differential pulse code modulation coding. *See* ADPCM coding
Adaptive equalizers, 348, 352–356
Adaptive link control, for GSM, 1239–1241
Adaptive transform coding. *See* ATC
ADC (administration center), 1228; defined, **1242**
Additive white Gaussian noise. *See* AWGN
Adjacent channel interference, 1155
Adjacent satellite interference. *See* ASI
ADM (add/drop multiplexer), 551
Administration center. *See* ADC
ADPCM (adaptive differential pulse code modulation) coding
    defined, **1486**
    of speech, 1308–1310, 1425–1429, 1480–1481
    in videoconferencing, 1519
ADSL (asymmetric digital subscriber lines), 450–479
    coding, 467–468
    defined, **1513**
    DMT (discrete multitone), 454–471
    frame format, 468–470
    multiple applications, 466–467
    performance margin, 454
    range projections, 471–473
    system description, 451–454
    timing methods, 470–471
    VDSL (very high-speed digital subscriber lines), 473–4764
    as video network, 1508–1509
ADSL transmission unit-central office. *See* ATU-C
ADSL transmission unit-remote. *See* ATU-R
A/DSP (analog/digital signal processing), 1070
Advanced mobile phone service. *See* AMPS
Advanced time division multiple access. *See* ATDMA
Africa Optical Network, 887
AGC (automatic gain control), 1489; defined, **1500**
AGCH (access grant control channel), 1242
Aircraft communication system, 1202–1203
Air interfaces
    adaptivity, 1222
    comparison of types, 1257–1258
Airphone system, 1075
Airwaves management, 1078
AJ (antijam) communication systems, 199, 206–207
AKM (apogee kick motor), 918
A-law companding, 30–33, 369–371
Aliasing, 1451; defined, **1460**
Alliance for Telecommunications Industry Solutions. *See* ATIS
Allowances, and link design, 954–955, 961
ALOHA, 640
Alternate mark inversion. *See* AMI
AM (amplitude modulation)
    defined, **12**
    described, 3–9
    for TV, 1494
Amateur satellite services, 1077
American National Standards Institute. *See* ANSI
American Standard Code for Information Interchange. *See* ASCII
AMI (alternate mark inversion), 72, 393
Amplified spontaneous emission. *See* ASE
Amplifier gain, 833
Amplifier noise, 841–842
Amplifiers, optical. *See* Optical amplifiers
Amplitude distortion, 339
Amplitude modulation. *See* AM

Page on which term is defined is indicated in bold.

Amplitude scintillation, 945–947
Amplitude-shift keying. *See* ASK
AMPS (advanced mobile phone service)
   power control for, 1197, 1203
   defined, **1206**
   speech transmission techniques, 1246–1247, 1255
   see also IS-54
Analog/digital signal processing. *See* A/DSP
Analog hierarchy, 301
Analog linecard, 366
Analog modulation, 3–12
   amplitude modulation, 3–9
   angle modulation, 9–11
   quadrature multiplexing, 6
   single sideband, 6–8
   vestigial sideband, 8–9
   in wireless systems, 1353, 1355–1356
Analog telephone channels, 308–317
   attenuation distortion, 312
   circuit noise, 310–311
   crosstalk, 310
   envelope delay distortion, 312–313
   impulse noise, 312
   line conditioning, 313–316
   noise, 309–310
   telephone band, 308–309
Analog telephone instruments, 438–439
Analog-to-digital conversion. *See* A/D conversion
Analog voice, 1083
Analysis-and-synthesis coders, 1430; defined, **1435**
Analysis-by-synthesis coders, 1426, 1430–1432, 1519; defined, **1435, 1526**
Angle diversity, 216
Angle modulation, 9–11; defined, **12**
ANSI (American National Standards Institute)
   Committee T1, 161–162
   multiplexing standards, 378
   synchronization standards, 410
   test loops, 471–472
Antenna discrimination, 963–964
Antenna gain, 969–970
Antenna performance parameters, defined, **1052**
Antennas for satellites. *See* Satellite antennas
Antenna temperature, 967, 969–973
Antijam communication systems. *See* AJ systems
AP (application process), 567, 568, 569
APCO (Associated Public Safety Communications Officers) Project 25
   air interface characteristics, 1212
   described, 1219–1220
Aperiodic correlation, 101–103
APMT satellite, 900, 903
Apogee, 917
Apogee kick motor. *See* AKM
Application-oriented layer, 572
Application processes. *See* APs
Application-specific integrated circuit. *See* ASIC
ARAPNET, 574
Arbitrary-depth INs, 688

Arctic satellite communications, site diversity, 1008–1010
ARDIS, 1384–1387, 1392
Arithmetic coding, 1400, 1415
ARQ (automatic-repeat-request), 181–198
   basic principles, 182–183
   basic schemes, 183–188
   hybrid FEC/ARQ schemes, 194–195
   performances, 187–193
   protocol, and flow control, 657
   time diversity and, 217
   variant schemes, 193–1947
ASCII code, 1416
ASE (amplified spontaneous emission), 834
ASI (adjacent satellite interference), 951, 954
ASIC (application-specific integrated circuit), 1070
ASK (amplitude-shift keying), 864
Assignment of radio resources. *See* Channel assignment
Associated control channel. *See* ACCH
Associated Public Safety Communications Officers Project 25. *See* APCO Project 25
Asymmetric digital subscriber lines. *See* ADSL
Asynchronous digital hierarchy, 377
Asynchronous multiplexing, 90; defined, **92**
Asynchronous time division multiplexing. *See* ATDM
Asynchronous traffic, 612
Asynchronous transfer mode. *See* ATM
AT&T
   CCS6 system, 481–482
   DISCO, 526
   facsimile compression technique, 1417
   lithium niobate 16 × 16 switching demonstration, 526–527
   MMX (mastergroup multiplex), 305–306
   telephone service, 294–295
   transmission characteristics, 315, 316
ATC (adaptive transform coding), 1481–1482
ATDM (asynchronous time-division multiplexing) switching, 676, 687
ATDMA (advanced time division multiple access), 1331–1332
ATIS (Alliance for Telecommunications Industry Solutions), 378
ATIS (automatic transmitter identifier signal), 926
ATM (asynchronous transfer mode)
   adaptation layers, 534–536
   architectural framework for networks, 717–722
   architecture, 531–536
   for audio coding, 1476
   for Aurora testbed, 674–675, 678–679
   B-ISDN protocol layers, 587–589, 687
   BLANs, 615–617
   and broadband network services, 717–727
   cell mapping, 562
   cells, 687
   defined, **1486**
   digital cross-connects, 509–511
   interfaces, 536–541
   for ISDN, 585–589
   LANs, 681, 683

Page on which term is defined is indicated in bold.

Index

networking standards, 529–541
onboard switches, 993–994
SONET transmission, 550
standard, 687–688
switching, 432, 514, 686–700
UNI management information base, 537
viability of SAR, 678
in video networks, 1509
ATM architecture, 531–536
    AAL Type 1, 535–536
    AAL Type 2, 536
    AAL Type 3/4, 536
    AAL Type 5, 536
    ATM adaptation layer, 534–536, 687–688
    ATM layer, 532–534, 586–587, 687–688
    physical layer, 531–532, 687–688
ATM Forum, 529, 536–541
ATRAC coding, 1482
Attenuation distortion, 312–316
Attenuation on slant paths, 936–942
    cloud, 941–942
    fog, 941–942
    gaseous, 936–938
    rain, 938–941
Attenuator, optical, 829
ATU-C (ADSL transmission unit-central office), 452, 457–458, 466–467, 470
ATU-R (ADSL transmission unit-remote), 452, 459, 466–467, 470
AUC (authentication center), PLMN, 1227; defined, **1242**
Audio coding, 1475–1487
    auditory masking, 1477–1478
    for DECT, 1308–1310
    ISO/MPEG coding, 1482–1484
    low bit rate coding, 1479–1480
    multichannel stereophony, 1484–1485
    noise shaping, 1478–1479
    perception-based coding, 1478–1479
    subband coding, 1480–1482
    transform coding, 1481–1482
    and videoconferencing, 1519, 1522, 1523–1524
Auditory masking, 1477–1478; defined, **1486**
AUG-n (administrative unit group-n), 556
AU-n (administrative unit-n), 556
Aurora gigabit testbed, 672–679
Authentication center, PLMN. *See* AUC
Authorization rights, TV, 1497, 1498
Autocorrelation function, 62–63
    defined, **76**
    of pseudonoise sequences, 94–98
Automated attendant, 443
Automatic call distributor. *See* ACD
Automatic focusing, 1555–1556; defined, **1564**
Automatic gain control. *See* AGC
Automatic-repeat-request. *See* ARQ schemes
Automatic tracking, 1556–1558; defined, **1564**
Average transmitted power constraint, 334–335
AWGN (additive white Gaussian noise) channel, 141, 1242; defined, **153**

B

Backplane optical interconnects, 875
Balanced opening, 669
Balanced receiver, 868
Band gap, 752
Band-limited signals, 15
Bandpass digital data systems, 73–76
Bandpass signals
    defined, **16**
    and FDMA, 1260
    sampling of, 16–18
Bandwidth
    definitions, 274–276
    and digital system performance, 236–254
    efficiency, 1353
    efficiency plane, 238–239
    of lasers, 753, 755
Bandwidth-limited systems, 238, 240, 242–250
Banyan networks. *See* BNs
Banyan switches, 988–991
Barker sequences, 101–102, 270
Baseband fast packet switches, 987–991
Baseband pulses. *See* Baseband signalling
Baseband signaling, 318–338
    communication system model, 318–321
    constraints on, 334–336
    defined, **307**
    digital, 67–71
    echo cancellation, 415–421
    examples of, 319–321, 336–337
    eye diagrams, 327–329
    FDM, 302
    Nyquist criterion, 321–327
    PR (partial-response) 329–334
    and pulse shaping, 318–338
Base station controller. *See* BSC
Base stations. *See* BS
Base station subsystems. *See* BSSs
Base transceiver subsystems. *See* BTS
BASETs, 593–594
Basis function, 119; defined, **128**
Batcher-banyan network, 697–698
Bayes' theorem, 39–40
BBNS (broadband network services)
    access services, 724
    and ATM, 717–727
    control point services, 725–726
    reviewed, 722
    transport services, 724–725
BCCH (broadcast control channel), 1227–1228, 1242
BCH (Bose–Chaudhuri–Hocquenghem) codes, 245–246
B-channels, 578
Beam, for satellites, 1032; defined, **1052**
Beam shaping, 1553
Bearer services, 578
BEGIN message, 494–495
Bell, Alexander Graham, 292, 450
BELLCORE transmission characteristics, 314, 316
Bellman–Ford algorithm, 653

Page on which term is defined is indicated in bold.

Bellman's equation, 652
Bell System
　digital hierarchies, 377
　historical data, 293, 296–297
BER (bit error rate), 221–222, 276–277
　and link design, 956–961
　and service quality, 1079
Bessel's inequality, 124; defined, **128**
BH (buried heterostructure) lasers, 753–754, 757, 758
B-ICI (broadband interchange interface), 540–541
Binary codes, 233–234
Binary entropy function, 227
Binary N zero substitution. *See* BNZS
Binary signalling schemes, 146–147
Binary symmetric channel. *See* BSC
Biomedical transducer, 880
Biosensors, 880–881
Biphase code, 395–396
Bipolar coding, 72–73
Bipolar signalling, 387, 393
B-ISDN (broadband integrated services digital network), 584–589
　ATM specific layers, 587–589, 687–688
　ATM transfer mode, 717–727
　ATM transport mode, 529–541
　protocol reference model, 587
　signalling, 538, 589
　structure, 586
　see also ISDN
Bit-error probability, and channel coding, 248–249
Bit error rate. *See* BER
Bit interval, 142, 147; defined, **153**
Bit patterning, 842
Bit sensing photonic devices, 515–520
Bit sensing switching, 515–516
Bit stuffing, for digital hierarchies, 382
Bit transparent photonic devices, 525–526
Bit transparent switching, 515–516
BLANs (broadband local area networks), 611–620
　ATM, 615–617
　new applications, 618–619
　technologies, 613–615
　user requirements, 612–613
Block codes
　classes of, 170–171
　decoding of, 168–170
　simple binary, 233–234
　structure of, 168–170
　and system performance, 245–246
Blocking INs, 688
Blocking probability, 1148
Block motion compensation, 1519; defined, **1526**
BNs (Banyan networks), 690–698
　Batcher, 697–698
　switching, 988–991
　tandem, 693, 696
BNZS (binary N zero substitution) code, 396–398, 401
Bose–Chaudhuri–Hocquenghem codes. *See* BCH codes
Bounded distance decoding, 169; defined, **178**
Boyds, Charles Vernon, 296

BPSK modulation, 252, 1025
Bragg reflection filter, 829
Brillouin amplifiers, 853–854
British cordless telephone standard. *See* CT-2
Broadband DSC applications, 503–506
Broadband integrated services digital network. *See* B-ISDN
Broadband interchange interface. *See* B-ICI
Broadband local area networks. *See* BLANs
Broadband network services. *See* BBNS
Broadcast-and-select networks, 879
Broadcast topology, 624–627
BS (base station)
　coverage by, 1152–1155
　for GSM, 1227–1228
　location of, 1149
　microcellular, 1170–1172
　subsystems, 1090–1103
BSC (base station controller), 1090, 1093–1094
BSC (binary symmetric channel), 166, 232–233; defined, **178**
BSSs (base station subsystems), 1090–1103
　call handling schemes, 1094–1102
　defined, **1102**
　system architectures, 1090–1094
BTS (base transceiver subsystems), 1090–1093; defined, **1102**
Buffer control, 1457–1458, 1469
Buffering, for video, 1506–1507
Buffer requirements, 607–608
Buried heterostructure lasers. *See* BH lasers
Burst structure, of CT-2, 1291, 1293–1295
Bus topology, 624, 625–627

C

Cable TV, 1488–1501
　access control, 1497–1499
　consumer premises equipment, 1494–1497
　head end, 1491–1493
　satellites, 894, 904–906
　security, 1497–1499
　source origination, 1491–1493
　system architecture, 1489–1491
　transmission channel, 1493–1494
Caching. *See* Location caching
CAD (computer aided design), 1070
Call-by-call switching, 514, 515, 520–525
Call delivery
　in GSM, 1126
　in IS-41, 1122–1123
　reducing signalling during, 1129
Call distributors, 443–444
Called time switching, 297
Caller ID, 439–440
Call forwarding, 491
Call handling schemes, 1094–1102
Calling line identification, 491
Call pre-emption, 726
Call sequencer, 4435

Index

Campbell, G.A., 294
Capability, terminal, 1520; defined, **1526**
Capacity
   CDMA, 1269–1272
   defined, **1114**
   radio, 1104
Carrier sense multiple access. *See* CSMA
Carrier service area. *See* CSA
Carrier synchronization, 255–264
   modulated, 260–261
   unmodulated, 256–260
Carrier-to-cochannel interference, 1148, 1260
Carrier-to-interference ratio (C/I), 1260
Carrier-to-noise ratio. *See* CNR
Carty, John J., 295
Cassegrain feed system, 924–925
Catastrophic code, 174; defined, **178**
CATV. *See* Cable TV
Cauchy sequence, 120–122; defined, **128**
C-band satellites, 916, 923–924, 927–928
C-bit parity, 384
CB radio, 1071
CCI (co-channel interference), 210–211, 336, 951, 1147, 1155–1158
CCITT (Consultative Committee on International Telephony and Telegraphy)
   ADPCM standard, 1425
   digital hierarchies, 377
   facsimile transmission standards, 1417–1421
   FDM nomenclature, 303–307
   line code, 398
   open system standards, 574, 575
   speech coding standards, 1435
   synchronization standards, 410
   see also ITU
C conditioning, 313–316
CCS6 network, 481–482
CD (compact disk), 1547, 1564
CDMA (code division multiple access), 1257–1275
   base station subsystem for, 1090–1094
   capacity, 1269–1272
   channel assignment, 1175–1176
   commercial applications, 209–211
   compared with FDMA and TDMA, 1259–1261
   defined, **93**, **1065**, **1274**
   diversity types, 1263–1264
   and future UMTS, 1330–1331
   history of, 1258–1259
   IS-95 standard, 1272–1274
   MAC control schemes, 632
   for microcellular radio, 1172–1173
   for mobile radio, 1086
   multiple-access interference, 1112–1113
   overview, 1259–1262
   power control, 1264–1267
   propagation considerations, 1112
   as radio access method, 1112–1113
   RAKE receiver use, 213
   spread spectrum transmission, 91
   systems, 1262–1263

TDRSS service, 1056
   transmit power, 1267–1268
CDPD (cellular digital packet data), 1387–1390, 1392
Cell design, 1146–1159
   cell coverage, 1151–1158
   cellular principles, 1147
   design steps, 1149–1150
   expansion techniques, 1148–1149
   interference, 1155–1158
   performance measures, 1148
   traffic engineering, 1150–1151
Cell self-routing concept, 688
Cell splitting, 1149
Cellular data networks, 1387–1391
   CDPD, 1387–1390
   IS-95, 1390–1391
Cellular digital packet data. *See* CDPD
Cellular hub, DCS and, 503
Cellular mobile networks
   air interface characteristics, 1212
   architecture, 1210–1211
   base station subsystems, 1090–1103
   basic features, 1209–1210
   defined, **1224**
   power control, 1198–1201
Cellular network design. *See* Cell design
Cellular principles, 47
Cellular radio
   characteristics, 1106
   concept, 91
   defined, **93**
   mobile services, 1080
   satellite use, 910
   systems, 1075–1076
Cellular Telecommunications Industry Association. *See* CTIA
CELP (code-excited linear prediction) coding, 1341–1346, 1425–1426, 1431–1432, 1435
Centrex, 442–443
CEPT (Conference of European Posts and Telecommunications Administrations), 398, 1072
Channel assignment, 1175–1181
   dynamic channel, 1178–1180
   fixed channel, 1176–1178
   resource assignment problem, 1175–1176
Channel bank, 303
Channel bits, and coding, 248–249
Channel borrowing, 1177–1178
Channel coding
   half-rate, 1341–1344
   IS-54 standard, 1249, 1254–1255
   see also Coding
Channel distortion, 339–343
Channel equalization, 339–363
   channel distortion, 339–343
   decision-feedback equalizer, 356–359
   intersymbol interference, 243–248
   linear equalizers, 348–356
   maximum-likelihood sequence detection, 359–362

Page on which term is defined is indicated in bold.

Channel equalizers, 348–359
   adaptive, 348, 352–356
   decision-feedback, 356–359
   fractionally spaced, 349
   linear, 348–356
   preset, 348
   symbol-spaced, 349
   zero-forcing, 349
Channel fluctuations. *See* Fading; Shadowing
Channel identification, 456
Channel models, 131–140
   and baseband signals, 319
   block diagram of, 319
   digital, 138–139
   fading dispersive, 132–136
   line-of-sight, 136–138
Channel noise, 28–29
Channel translation equipment (CTE), 303
Chemical transducer, 880–881
Chip time, 201
Chirp (dynamic spectrum), 761, 774–775, 784–785
Chromatic dispersion, 869
Chrominance, 1463, 1519; defined, **1473**, **1526**
Circuit noise, 310–311; defined, **316**
Circuit switching. *See* CS
Circulator, optical, 829
Citizen band radio. *See* CB radio
City street microcells. *See* Street microcells
Clarke, Arthur C., 913
Clarke model, 1019
Clear-sky conditions, 950, 956, 960
Client-server architecture, 662
CLIP (calling line identity presentation), 491
CLIR (calling line identity restriction), 491
Closed systems, 568
Cloud attenuation, 941–942
CMI (code mark inversion), 396
CMR, in PCS, 1131–1134
CNR (carrier-to-noise ratio), 967, 1270
CO (central office) for DCSs, 499–500
Coarse alignment, 201
Coaxial cable, 682
Cochannel interference. *See* CCI
Codebook orthogonalization, 1253
Codebooks, 1248–1253, 1255, 1341
Codebook search, 1252–1253
Codec, 927
   defined, **1350**
   half-rate, 1339–1341
Code channel, defined, **1274**
Codec-filters. *See* PCM codec-filters
Code division multiple access. *See* CDMA
Code excited linear prediction coding. *See* CELP coding
Code mark inversion. *See* CMI
Code rate, 167; defined, **178**
Coding
   ADSL, 467–468
   audio, 1475–1487
   bipolar, 72–73
   block codes, 168–171
   channel coding, 231–233, 1249, 1254–1255
   communication system and, 224–226
   convolutional codes, 171–175
   for discrete-alphabet sources, 226–228
   facsimile image, 1412–1421
   FEC (forward error correction), 166–180
   information theory and, 224–235
   line, 71–73, 386–403
   lossless, 1398–1409
   for magnetic storage, 1535–1537
   for mobile radio, 1081–1083
   for mobile satellites, 1027–1028
   partial response, 72–73
   for power spectrum control, 71–73
   rate distortion theory, 229–231
   simple binary codes, 233–235
   source coding (video), 223–229
   speech coding, 1083–1084, 1222, 1338–1352, 1424–1436
   speech fundamentals, 1248–1249
   still image, 1427–1448
   system performance and, 245–253
   terminology, 178–179
   universal source, 228–229
   video, 1449–1461, 1519–1524
Coding artifacts, 1466; defined, **1473**
Coding gain, 167, 249–250; defined, **178**
Coding speech. *See* Speech coding
CODIT system, 1330–1331
Codulation, 1081–1083
Coherence bandwidth, 134; defined, **139**
Coherence time, 135
Coherent systems, 862–871
   basic construction, 863–864
   constraints, 867–870
   detection and demodulation techniques, 865
   modulation techniques, 864
   receiver sensibility, 865–866
Collimation, 1553, 1555
Collision avoidance schemes, 638–639
Collision detection, 681–682
Collision domains, 612
Column distance, in coding, 173; defined, **178**
Comb filter, 834
Common carrier, defined, **1287**
Common channel signal, 480–495
   connection control part, 487
   defined, **300**
   ISDN user part, 488–491
   message transfer part, 483–487
   signalling transport, 482–488
   signalling user parts, 488–495
   TCAP (transaction capabilities application part), 491–495
Communication link, 236–238
Communication scheduling, multiprocessor, 1506–1508
   buffering, 1506–1507
   fault tolerance, 1507–1508
   interactive response, 1507

Page on which term is defined is indicated in bold.

Index 1573

TDM, 1506
  variable bandwidth devices, 1508
Communication services
  bearer, 578
  conversational, 579
  definitions, 578–579
  distribution, 579
  interactive, 579
  supplementary, 579
  teleservices, 578–579
Communication systems
  block diagram of model, 318–321
  encoding and decoding for, 224–226
  information theory and, 224–226
  performance fundamentals, 236–254
Commutation
  defined, **93**
  and multiplexing, 90
Compact disk. *See* CD
Companding, 269–371
  A-law, 369–371
  defined, **369**
  mu-law, 369–371
  nonuniform quantizing and, 30–33
  phi-law, 30–33
Compandor, 31
Compression, lossless. *See* Lossless compression
Compression, source. *See* Source compression
Compression ratio, 1411; defined, **1421**
Computer aided design. *See* CAD
Computers
  closed systems, 568
  communication requirements, 567–568
  HDTV interoperability with, 1472
  OSI 7–layer model, 567–576
  OSI standards, 568–569, 574–575
Computer-telephony integration. *See* CTI
COMSAT, 912
Concatenated codes, 177; defined, **178**
Concatenation mechanism of SDH, 554
Conference of European Postal and Telecommunications Administrations. *See* CEPT
Connection, and incarnations, 662
Connectionless model, 702
Connectionless networks. *See* Datagram networks
Connectionless protocols, 651
Connection management, 661, 665–668
Connection-oriented model, 702–703
Connection-oriented protocol, 616, 651
Constellation of satellites, 897, 1021
Constraint length, 171–172
Consultative Committee on International Telephony and Telegraphy. *See* CCITT
Container-*n*, 555
Context-based compression models, 1406–1407
CONTINUE message, 495
Continuous-phase FSK. *See* CPFSK
Continuous-phase modulation. *See* CPM
Control signal transmission, 1232–1236
CONUS, 917

Conversion
  analog-to-digital (A/D), 107–117
  digital-to-analog (D/A), 107–117
  E/O and O/E, 876
Converters
  A/D and D/A, 107–117
  for TV, 1489, 1494–1497
Convolutional coding, 171–175, 1027
Cordless systems
  air interface characteristics, 1212
  architecture, 1210–1211
  basic features, 1209–1210
  defined, **1224**
Cordless telephone. *See* CT
Cordless telephone second generation. *See* CT2
Correlation between pseudonoise sequences, 94–106
Correlation receiver, 145
Countable set, 121
Couplers
  directional, 826–827
  star, 827
Covariance lattice algorithm, 1250–1251, 1255
CPE (customer premises equipment), 433–449
  classification by function, 438–440
  customer interface, 434–435
  defined, **433**
  PBXs (private branch exchanges), 440–449
  restricting regulations, 435–438
  standards, 435
  wiring and connectors, 447–448
CPFSK (continuous-phase frequency-shift keying), 284–286
  for coherent systems, 864, 868–870
CPI (Calling Party Identification), 439–440
CPM (continuous-phase modulation), 284–287
  defined, **287**
  for wireless systems, 1359–1360
Crane attenuation model, 940–941
CRC (cyclic redundancy check), 169, 1249, 1254–1255, 1271
  defined, **178**
CRC-6 (cyclic redundancy code with a 6-b remainder), 380
Cross-connect switching, 628
Cross-connect systems, digital. *See* DCSs
Cross correlation, 62–63, 98–101; defined, **76**
Cross-gain modulation. *See* XGM
Crosspoints, in switching, 426–427, 429
Crosstalk
  cancellation, 414
  categories of, 310
  defined, **316**
  far-end, 454
  and line coding, 386
  near-end, 453
Cryptographic chip, 674
CS (circuit switching), 425–426, 976
CSA (carrier service area)
  range, 471
  test loops, 473

Page on which term is defined is indicated in bold.

CSMA (carrier sense multiple access), 640–641
CSMA bus, 681–682
CSMA/CD LANs, 641–642
CSU (channel service unit), 445
CT (cordless telephony), 1072, 1076–1077, 1109
CT2 (cordless telephone second generation), 1289–1304
    air interface characteristics, 1212
    burst formats, 1293–1295
    defined, **1206**
    features summarized, 1216–1217, 1302–1303
    handshaking, 1300–1302
    link initialization, 1296–1300
    power control, 1203
    radio interface, 1290, 1291–1293
    signalling layers, 1290, 1295–1299
    standard, 1290–1291
    synchronization, 1293–1294
CT2PLUS, 1203
CTE (channel translation equipment), 303
CTI (computer-telephony integration), 448–449
CTIA (Cellular Telecommunications Industry Association), 1258–1259
Cumulative distribution function, 41
Cup-dipole antennas, 1037
Customer interface, for CPE, 434–435
Customer premises equipment. *See* CPE
Customers, in queuing, 78–86
Cycle frequency, 63; defined, **76**
Cyclic autocorrelation, 63, 67; defined, **76**
Cyclic code, 169; defined, **178**
Cyclic Hadamard difference sets, 97
Cyclic redundancy check. *See* CRC
Cyclic redundancy code with a 6-b remainder. *See* CRC-6
Cyclic sequence numbers, 662
Cyclostationary processes, 62–63; defined, **93**

D

D/A (digital-to-analog) conversion, 107–117
DA (demand assignment) control, 631, 633
DAMA (demand assignment multiple access), defined, **994**
DAMA-TDMA, 977
D-AMPS. *See* IS-54
DARPA (Defense Advanced Research Projects Agency), 574
DASD (direct-access storage device), 1552
Data, 1516
    defined, **1517**, **1526**
    striping, 1504–1505, 1514
    transfer, 661, 663–664
Database traffic rate, 1126–1129
Data communication network, 568
Data encryption standards, 674
Data exchange interface. *See* DXI
Datagram networks
    defined, **651**
    flow control in, 656–657
    routing in, 652–655
Data link. *See* DL
Data link switching. *See* DLSw
Data recording
    magnetic storage, 1531–1545
    magneto-optical disk storage, 1546–1565
Data servers, 1503–1508
    defined, **1514**
    see also Video servers
Data systems, wireless. *See* Wireless data systems
dBmV (decibel-millivolts), 1489; defined, **1500**
DBR (distributed Bragg reflector) lasers, 763–764, 834, 868
DCA (dynamic channel assignment), 1178–1180
DCS1800. *See* GSM
D-channels, 578
DCSs (digital cross-connect systems), 496–512
    ATM networks, 509–511
    functional description, 497–500
    hybrid networks, 507–508
    operations support, 508–509
    private networks, 506–507
    public networks, 500–506
    SONET equipment, 551
DCS3/1/0, 500–503
DCT (discrete cosine transform), 1466; defined, **1473**
Dc wander, 387
Decibel-millivolts. *See* dBmV
Decision boundary, 146; defined, **153**
Decision-directed mode, 354
Decision feedback equalization. *See* DFE
Decoding
    of block codes, 168–170
    of convolutional codes, 174–175
    see also Coding
DECT (digital European cordless telephone), 1305–1326
    air interface characteristics, 1212
    application areas, 1306–1307
    architectural overview, 1308–1324
    characteristics, 1308
    described, 1217–1218
    interworking with GMS, 1220–1221, 1307
    and ISDN, 1307
    modulation, 1321–1322
    protocol model, 1311–1313
    radio frequency, 1322–1324
Defense Advanced Research Projects Agency. *See* DARPA
Degraded-sky condition, 955–956
Dehopping, 203
Delay distortion, 339
Delay-locked lop, 201
Delay spread, 134
Delta modulation
    and A/D conversion, 111, 113
    coding, 395
    defined, **116**
Delta networks. *See* BNs (banyan networks)
Demand assignment control. *See* DA

Page on which term is defined is indicated in bold.

Index

Demand assignment multiple access. *See* DAMA
Demultiplexers
   defined, **830**
   optical filters as, 828
   for WDM systems, 885
Density functions, 41–45, 51–57
Department of Communications (Canada). *See* DOC
Depolarization, 942–944
Designed distance, in coding, 170; defined, **178**
Desktop terminal, 1517–1518, 1520, 1526
Desktop videoconferencing. *See* DVC
Despreading, 200
Despun antennas, 918
Detection, for magnetic storage, 1535–1537
Detection theory, 144–145
DFB (distributed feedback) lasers, 760–764, 775, 834
DFD (displaced frame difference), 1466–1468
DFE (decision feedback equalization), 356–360, 1536–1543; defined, **1544**
DFWMAC (distributed foundation wireless MAC), 1372
DH (double heterostructure) lasers, 752–755
Dialing, DTMF, 294
Dialogue, defined, **493**
Dictionary-based coding, 1405–1406
Differential detection, magneto-optical, 1562–1563
Differentially encoded quadrature PSK. *See* DPSK
Differential pulse code modulation coding. *See* DPCM coding
Diffraction
   defined, **1195**
   knife-edge, 1186–1187
   as propagation mechanism, 1183, 1186–1187
Digital advanced mobile phone service (D-AMPS). *See* IS-54
Digital biphase coding, 393–394, 395
Digital cellular capacity, 1261
Digital Cellular System-1800 (DCS1800). *See* GSM
Digital channel models, 138–139
Digital cross-connect systems. *See* DCSs
Digital European cordless telephone. *See* DECT
Digital hierarchies, 377–385
Digital information services, satellite use, 909–910
Digital modulation, 273–287
   defined, **287**
   lattices, 282–284
   with memory, 286–287
   multidimensional, 281–284
   one-dimensional, 278
   two-dimensional, 279–280
Digital networks, synchronous switching, 404–413
Digital signal levels. *See* DSs
Digital system performance, 236–254
   bandwidth considerations, 238–241
   bandwidth-limited systems, 242–250
   coded systems, 245–253
   communication link, 236–238
   direct-sequence spread-spectrum system, 250–253
   power considerations, 238–241
   power-limited system, 244–245
   uncoded systems, 242–245
Digital-to-analog conversion. *See* D/A conversion
Digital video broadcast systems. *See* DVB systems
Digital voice, glossary of terms, 1084
Digram coding, 1405
Dijkstra's algorithm, 653–654
Directed retry, 1177–1178
Directional couplers, 826–827
Directionality, and loss, 825; defined, **830**
Directly modulated laser transmitters, 775–782
Direct sequence. *See* DS
Direct-to-home satellites. *See* DTH satellites
Direct-to-line. *See* DTL
DISCO (distributed switching with centralized optics), 526–527
Discontinuous transmission. *See* DTX
Discrete-alphabet source coding, 226–228
Discrete cosine transform. *See* DCT
Discrete data channel, 166; defined, **178**
Discrete memoryless channel. *See* DMC
Discrete multitone. *See* DMT
Discrete probability theory, 36–41
Discrete-time Fourier transform. *See* DTFT
Discrimination, antenna, 954
Disk rotation speed, 1551–1552
Disk scheduling, for video, 1505–1506
Disk storage. *See* Magneto-optical disk storage
Disk technology, 1547–1552
Dispatch two-way radio system, 1075
Dispersion compensation, 844
Dispersion of optical fiber, 752, 782, 869–870
Dispersion power penalty, 782
Dispersive channels, 152–153; defined, **153**
Displaced frame difference. *See* DFD
Distance profile, 173; defined, **178**
Distance-vector routing, 706
Distortion measures, 229
Distortion, of channel, 339–343
Distributed Bragg reflector lasers. *See* DBR lasers
Distributed feedback lasers. *See* DFB lasers
Distributed queue dual bus. *See* DQDB
Distributed shared memory. *See* DSM
Distributed switching with centralized optics. *See* DISCO
Distribution networks, synchronous, 405–406
Distribution points. *See* DPs
Dither, 1443–1444
Diversity
   in CDMA, 1263–1264
   microscale, 1010
   optical, 1008
   orbital, 996, 1010–1011
   path, 996–1014
   site, 996–1010
   types of, 1263
Diversity gain, 1000–1008
Diversity improvement factor, 1000, 1007
Diversity techniques, 213–223
   BER (bit error rate), 221–222
   combining techniques, 217–221

Page on which term is defined is indicated in bold.

Diversity techniques, (Cont.)
  diversity branches, 215–217
  overview, 213–215
DL (data link), 380
DLSw (data link switching), 705
DMC (discrete memoryless channel), 231; defined, **235**
DMC (dynamic Markov modeling), 1405
DMH (digital multiplex hierarchy), 497
DMS (distributed memory switch), 987
DMT (discrete multitone), 450, 454–471
  described, 455–457
  implementation, 457–461
  performance analysis, 461–465
DOC (Department of Communications, Canada), 1078
Doppler rate, and mobile satellites, 1023
Doppler spread, 136; defined, **139**
Double heterostructure lasers. See DH lasers
Double-sideband carriers. See DSB carriers
Double-sideband-suppressed carrier. See DSB-SC
Downconverters, 913, 973
Downlink carrier, 914
Downlink degradation, 964
Downlink equations, 950–951, 955–962
DP (distribution point), 1170–1171
DPCM (differential pulse code modulation) coding, 260, 1426–1429, 1438–1439, 1455–1456
DPSK (differentially encoded quadrature PSK), 1247
DQDB (distributed queue dual bus), 684
DS (direct sequence) modulation, 200–202
DS (direct-sequence) systems, 206, 250–253
DSB (double-sideband) carriers, 4–6; defined, **12**
DS/CDMA, 1330–1331
  power control, 1198–1205
DSM (distributed shared memory) communications, 677–679
  Aurora research, 677–679
DS1 through 3, 378–384
DSS1 (digital subscriber signalling system), 481
DSU (data service unit), 445
DTFT (discrete-time Fourier transform), 64
DTH (direct-to-home) satellites, 893–895, 899, 904–907
DTL (direct-to-line), 306–307
DTMF (dual tone multifrequency) dialing, 294; defined, **300**
DTMF keypad, 438
DTX (discontinuous transmission), 1241
Duobinary coding, 401
Duobinary signalling, 330–332
Duplexing, 1104; defined, **1114**
DVB (digital video broadcast) systems, 1503
DVC (desktop videoconferencing), 1516–1527
  components, 1517–1518
  over GSTN, 1523–1525
  over Internet, 1521–1522
  over ISDN, 1518–1521
DXI (data exchange interface), 539–540
Dynamic channel assignment. See DCA
Dynamic Markov modeling. See DMC
Dynamic range, 839–840
Dynamic spectrum. See Chirp

E

EA (electroabsorption) modulators, 782
Earliest deadline first. See EDF
Early-late gate symbol synchronizer, 267
Earth station, 922–934
  antenna, 930–931
  components, 923–927
  defined, **922**
  designs, 899–900
  ownership of, 894
  power, 929–930
  safety considerations, 932–933
  site selection, 927–928
ECC (error-correction coding), 1549; defined, **1564**
Echo cancellation, 414–424
  ADSL, 460
  for baseband transmission, 415–421
  for passband transmission, 421–423
Echo cancelers, 296, 1339; defined, **300**
ECSA (Exchange Carrier Standards Association), 378
EDD (envelope delay distortion), 312–313, 314; defined, **316**
EDF (earliest deadline first) scheduling, 1505; defined, **1514**
EDFA (erbium-doped fiber amplifier), 856–860, 885
Edison, Thomas Alva, 294
EDSX (electronic digital signal cross connection), 504–506
Effective isotropic radiated power. See EIRP
EIA/TIA Interim Standard 41. See IS-41
EIA/TIA wiring standards, 447–448
EIRP (effective isotropic radiated power)
  defined, **237**
  link design and, 959–962
  satellite radiated, 914
  for TDRS, 1064
Electroabsorption modulators. See EA modulators
Electronic digital cross connection. See EDSX
Electronic telephones, 439
End-to-end service delivery, 676, 677
Entropy, 227; defined, **235**
Entropy coding, 1398–1401, 1441; defined, **1447**
Envelope delay, 339
Envelope delay distortion. See EDD
Epitaxial growth, 751
Equal gain combining, 220
Equalizers, channel. See Channel equalizers
Equatorial inclination, 913
Equipment for customer premises. See CPE (customer premises equipment)
Equivalent discrete-time transfer function, 322
Erasures, defined, **168**, **178**
Erbium-doped fiber amplifier. See EDFA
Error correction
  ARQ (automatic repeat request), 181–198
  defined, **197**
  FEC (forward error correction), 166–180
  procedures explained, 181
  in signalling link, 483–484

Index 1577

for TV, 1494
Error-correction coding. *See* ECC
Error detection, 181; defined, **197**
Error diffusion, and halftoning, 1444–1445
Error modeling, 1398
Error probability, 276–277; defined, **287**
Error rate monitor, 484
ESF (extended superframe format), 380
Ethernet, 591–596
   BLANs, 617
   802.3, 594, 611
   fiber optics use, 876–877
   history of, 592
   LANs, 682
   operation, 594–595
   overview, 591–592
   standards, 592–594
ETSI (European Telecommunications Standards Institute), 159–160, 1213, 1305, 1328
European cordless telephone. *See* DECT
European R&D. *See* RACE program
Excess bandwidth, 67, 324, 347, 1356; defined, **77**
Excess losses, 826; defined, **830**
Excess temperature, 967, 969
Exchange Carrier Standards Association. *See* ECSA
Exciter, 926–927
Expandor, 31
External conflicts in switching, 689
Externally modulated optical (Mach-Zehnder) transmitters, 782–786
Extinction ratio, 775
Eye diagrams, 327–329
Eye patterns, 345–346, 781–782

F

FA (fixed assignment) control, 631–632
Fabry-Perot amplifier. *See* FP amplifier
Facsimile, defined, **1411**, **1421**
Facsimile image, defined, **1411**, **1421**
Facsimile image compression, 1411–1423
   international standards, 1417–1421
   techniques, 1412–1417
   terms defined, **1447**
Facsimile machines, 444–445
   group types, 1417, 1422
   transmission errors, 1418–1419
Fade duration, 999–1003
Fade margins, 1008
Fading
   defined, **1206**
   diversity mitigation, 213–223, 1263
   flat, 1194
   frequency selective, 1195
   of mobile satellite, 1017–1020
   and power control, 1198–1199, 1204, 1264–1265
   Rayleigh, 1194, 1265, 1369
   Rician, 137, 1194–1195
   small-scale, 1193–1195
Fading dispersive channel model, 131–136

Faraday rotator, 826; defined, **830**
Far end alarm and control channel. *See* FEAC
Far end block error. *See* FEBE
Far-end crosstalk. *See* FEXT
FAS (frame alignment signal), 380
Fast frequency-hopped systems. *See* FFH systems
Fast packet switching, 613–615
Fault tolerance, in video server systems, 1507–1508
Fax. *See* Facsimile
FCA (fixed channel assignment), 1176–1178
FCC (Federal Communications Commission)
   airwaves management, 1078
   antenna requirements, 925–926
   CPE regulation, 433, 436–437
FDDI (fiber distributed data interface), 597–610
   access control, 635–637
   architecture, 598–600
   buffer requirements, 607–608
   fault management, 598–600
   fiber optics use, 876–877
   overview, 597–598
   protocol, 600–603
   for time-critical operations, 603–608
   for video, 1512
FDM (frequency-division multiplexing), 87–89, 301–307
   background on, 301–302
   defined, **93**
   DTL (direct to line), 306–307
   hierarchy, 303–306
   implementation, 302–303
   overview, 87–89
   pilots, 306
FDMA (frequency division multiple access)
   antenna height for, 1108–1109
   base station subsystem for, 1090–1094
   channel assignment in, 1175–1176, 1260
   channel considerations, 1107–1108
   defined, **1114**
   MAC control schemes, 632
   for microcellular radio, 1172–1173
   for mobile radio, 1086
   as radio access method, 1107–1109
   spectral efficiency, 1260
FDTS/DF (fixed delay tree search with decision feedback), 1532, 1537–1543; defined, **1544**
FEAC (far end alarm and control channel), 384
FEBE (far end block error), 547
FEC (forward error correction) coding, 166–180
   applications, 177–178
   block codes classes, 170–171
   block coding fundamentals, 168–170
   convolutional coding, 171–174
   decoding of block codes, 168–170
   decoding of convolutional codes, 174–175
   and frequency diversity, 216
   hybrid FEC/ARQ schemes, 194–195
   for mobile satellites, 1027
   terminology defined, **178–179**
   Trellis-coded modulation, 176–177
   for TV, 1494

Page on which term is defined is indicated in bold.

Federal Communications Commission. *See* FCC
Feedback channel, 183; defined, **197**
Feedback filter, 357
Feedforward filter, 357
FET-SEED (field effect transistor-self-electro-optic effect device), 516–524
FEXT (far-end crosstalk), 454
FFH (fast frequency-hopped) systems, 203
FH (frequency hopping), 91, 202–203, 1229; defined, **93**
Fiber channel
    fiber optics use, 877
    standard, 877–878
Fiber distributed data interface. *See* FDDI
Fiber optic applications, 872–882
    access networks, 878–879
    fiber sensors, 880–881
    input/output interconnections, 876–878
    LANs, 876–878
    optical interconnects, 873–876
    transmission systems, 751
    WDM-based networks, 879–880
Fiber-Raman amplifiers, 853
Fiber sensors, 880–881
Field effect transistor-self-electro-optic effect device. *See* FET-SEED
FIFO (first-in first-out), 684
Filters
    FIR (finite duration impulse response), 349–350, 357
    matched, 145, 325–327
    optical, 828–829
Fine alignment, 201
Finite duration impulse response filter. *See* FIR filter
Finite field, 168; defined, **178**
Finite power waveforms. *See* FPW
FIR (finite duration impulse response) filter, 349–350, 357
First-in first-out. *See* FIFO
Fixed assignment control. *See* FA
Fixed channel assignment. *See* FCA
Fixed delay tree search with decision feedback. *See* FDTS/DF
Fixed-reference D/A converters, 110; defined, 116
FKE (Franz-Keldysh effect), 782
Flash A/D conversion, 111; defined, **116**
FLAT algorithm, 1250
Flat fading, 1194–1195
Flat-top sampling, 20–21
Flow control, 650–660
    in datagram networks, 656–657
    and routing, 650–651
    in virtual circuit networks, 657–659
Flush bits, 1027
FM (frequency modulation), 9, 10–11, 1353
Focusing, disk
    automatic, 1555–1556, 1564
    by objective lens, 1553–1555

Fog attenuation, 941–942
Footprints, satellite, 914, 922
Foreign Attachment Rule, 435
Forward CDMA channel, defined, **1274**
Forward channel, 184; defined, **197**
Forward direction, 1016
Forward error correction coding. *See* FEC coding
Forward link power control, 1199–1202, 1204; defined, **1206**
Forward traffic channel, defined, **1274**
4B3T (4 binary 3 ternary) coding, 399, 400
Fourier sum, 124; defined, **129**
Fourier transform, 13–21, 64–65, 68, 76
Four-wave mixing. *See* FWM
FP (Fabry-Perot) amplifier, 834, 836–838
FPLMTS (future public land mobile telecommunications system), 1077, 1223; defined, **1287**
FPW (finite power waveforms), 46–51
    continuous, 46–47
    discrete, 47–48
    random, 48–50, 60–61
    time averages, 57–59
Fractionally spaced equalizer, 349
Frame alignment signal. *See* FAS3
Frame synchronization, 90, 255, 268–271; defined, **93**
Franz-Keldysh effect. *See* FKE
Free distance of code, 174; defined, **179**
Free-run accuracy, 410
Free-space interconnect, 873
Free-space propagation, 1183
Free spectral range, 836
Frequency chirp, 844
Frequency diversity, 216, 1263
Frequency division multiple access. *See* FDMA
Frequency division multiplexing. *See* FDM
Frequency-domain coders, 1429–1430
Frequency hopping. *See* FH
Frequency modulation. *See* FM
Frequency response, 312
Frequency reuse, 916, 1024
    cellular, 1147
    defined, **1052**
Frequency reuse efficiency, 1260–1261, 1269, 1270
Frequency reuse feed, beams for, 1032, 1038
Frequency-selective channel, 134–135
Frequency-selective fading, 134, 1195; defined, **139**
Frequency-shift keying. *See* FSK
Frequency spectrum, of satellites, 897–898
Friis free-space equation, 1183
FSK (frequency-shift keying)
    attenuation distortion, 312
    binary, 320
    and coherent systems, 864
    defined, **281–282**
FSS antenna, 1041
Full-duplex links, 184; defined, **198**
Future public land mobile telecommunications systems. *See* FPLMTS
FWM (four-wave mixing), 844

Page on which term is defined is indicated in bold.

Index   1583

defined, **386**, **404**
   with memory, 395–400
   for power spectrum control, 71–73
Line conditioning, 313–316; defined, **316**
Line-of-sight channel models, 136–138
Line-of-sight propagation. *See* LOS
Link budget, TDRS, 1064–1065
Link constraints, 242
Link design for satellites. *See* Satellite link design
Link equation, and satellite mobile channel, 1016–1017
Link grant, 1297–1298
Link initialization, for CT-2, 1296–1300
Link layer, 573
Link request, 1297–1298
Link-state router NLSP, 706
Lithium niobate 16 x 16 switching system, 526–527
Little's formula, 79–80
Lloyd algorithm, 1441
LMS algorithm, 354, 356
LMSS (land mobile satellite service), 910–911
LNAs (low-noise amplifiers), 913, 915, 926, 968
LNB (low-noise block downconverter), 926
Loading, in telephony, 456–457
Loading factor, 33
Local access and transport area. *See* LATA
Local area networks. *See* LANs
Location caching, 1129–1134
   estimating CMR and LCMR, 1134–1141
   per-user, 1129–1131
   strategies for, 1141–1143
   threshold analysis, 1131–1134
Location registration
   in GSM, 1123, 1126
   in IS-41, 1121–1122
Location strategies for PCSs, 1116–1145
   caching, 1129–1134
   GSM (Global System for Mobile Communications), 1123–1126, 1128–1129
   IS-41, 1121–1123, 1127–1128
   user LCMR estimates, 1134–1141
Log-distance path loss propagation, 1188–1189
Logical bus switching, 626, 628–629
Logical bus topology, 625–627
Loop noise equivalent bandwidth, 259
LORAN-C frequency navigation system, 1074
LOS (line-of-sight) propagation, 1289
Losses
   in passive optical devices, 824–827
   propagation, and link design, 954–955, 961
Lossless compression, 1397–1410
   comparison of schemes, 1405
   compression packages, 1408
   entropy coders, 1398–1401
   image compression, 1406–1408
   standards, 1408
   text compression, 1403–1406
   universal codes, 1401–1403
Lossy compression techniques, 1415–1416
Lossy plus lossless schemes. *See* LPL schemes
Low bit rate coding, 1479–1480

Low Earth orbit satellites. *See* LEO satellites
Low-noise amplifier. *See* LNA
Low probability of intercept. *See* LPI
LPC (linear prediction coding), 1248–1249, 1253, 1523–1524
   defined, **1526**
   see also CELP
LPI (low probability of intercept), 199, 207–209
LPL (lossy plus lossless) schemes, 1407
LPM (laser power modulation), 1559–1560
Lucent Technologies, 496
Luminance, 1519; defined, **1526**
Lynch-Davisson-Schalkwijk-Cover codes, 1401
LZ (Ziv-Lempel) codes, 228, 1402–1408; defined, **235**
LZW algorithm, 1402

M

MAC (medium access control) layer
   for DECT, 1314–1320
   distributed foundation wireless (DFWMAC), 1372
   for wireless LAN, 1367, 1369–1373
MACs (medium access control) systems, 623–634
   classification of schemes, 630–634
   commercial applications, 209–211
   control categories, 631–634
   control protocols, 592–594
   sublayer, 629–630
   topologies, 623–630
Mach-Zehnder modulators, 782–786
Macroblock, data, 1519
Macrocells, 1165–1169
Macrodiversity, 1084–1085
Magnetic/electric field transducer, 881
Magnetic field modulation. *See* MFM
Magnetic storage, 1531–1545
   coding advances, 1535–1537
   detection advances, 1535–1537
   fixed delay tree search with decision feedback, 1537–1542
   performance comparisons, 1542–1543
Magneto-optical disk storage, 1546–1565
   automatic focusing, 1555–1556
   automatic tracking, 1556–1558
   basic definitions, 1547–1552
   magneto-optical readout, 1561–1563
   materials, 1563
   optical path, 1552–1555
   thermomagnetic recording, 155801561
   track concept, 1547–1552
Magneto-optical Kerr effect, 1561; defined, **1564**
Magneto-optical readout, 1561–1563
Magneto-restive head. *See* MR
Management information base. *See* MIB
Manchester coding, 393–394, 401
MANs (metropolitan area networks), 684
MANs (multiple access networks), 622–649
   medium access controls, 630–634
   medium access systems, 623–634

Page on which term is defined is indicated in bold.

MANs (multiple access networks) (*Cont.*)
  polling-based networks, 634–639
  random-access protocols, 639–642
  spatial-reuse optical networks, 644–647
  topologies, 623–630
  wireless networks, 644–647, 683
Margins, equipment, 954–955, 961
Mark codes, 395–396
Marker (frame synchronization pattern), 268–270
Markov chain, 69, 80
Markov models, 1405
M-ary optimal receivers, 147–149
M-ary quadrature amplitude modulation. *See* M-QAM
M-ary signalling, 239–241
Masking
  auditory, 1477–1478
  defined, **1460**
  video, 1452
Mass functions, 41–45, 51–57
Mastergroups, 304
Master-slave hierarchy, 405
Matched filtering, 145, 325–327
Material gain, 833
MAU (media attachment unit), 591
MAVT (mobile audio-visual terminal) project, 1334–1335
Maximal ratio combining, 219–220, 1025
Maximum distance separable, in coding, 170; defined, **179**
Maximum likelihood sequence detection. *See* MLSD
Maximum message lifetime, 662
MBONE (multicast backbone), 1521–1522
MBS (mobile broadband system), 1335
MCU/MR (multipoint control unit/multicast router), 1517
Mean opinion score. *See* MOS
Mean square correlation, 104
Mean square equivalence, 143; defined, **153**
Mean square error. *See* MSE
Mean square estimate (MSE), 128
Measat satellite, 900, 901
Media-access control protocol, 611
Media attachment unit. *See* MAU
Medium access control systems. *See* MACs
Memory switches, 986–987
MEO (medium Earth orbit) satellites, 895–897
  antennas for, 1032, 1048–1049
Merit factor of sequences, 102
Mesh topology, 613–614, 624, 628–629
Metric, goodness measure, 174
Metric function, 119
Metric space, 119; defined, **129**
Metropolitan area networks. *See* MANs
MFM (magnetic field modulation), 1560–1561
MFSK (M-ary FSK) modulation, 240–242
MH (mobile host), 1370, 1373–1376
MH (modified Huffman) coding, 1418, 1421; defined, **1422**
MIB (management information base), 537, 613

Microcells, 1160–1174
  city street, 1164–1169
  highway, 1161–1164
  indoor, 1169
  overview, 1160–1161
Microcellular radio, 1160–1174
  infrastructure, 1170–1172
  microcells, 1160–1169
  and multiple access, 1172–1173
Microdiversity, 1010, 1085
Microwave receiver, system temperature, 966–975
Microwave signals, for TV, 1492
Microwave switch matrix. *See* MSM
Miller code, 395
MIN (multistage interconnection network), 613–614
Minimum-depth INs, 688
Minimum distance, in coding, 168, 170
Minimum shift keying. *See* MSK
Mixed excitation coding, 1432–1433; defined, **1435**
MLSD (maximum likelihood sequence detection), 359–362, 1538
M/M1 queuing system, 80–85
MMIC (monolithic microwave integrated circuit)
  antennas, 1045–1046
  and radio transceivers, 1070
MMR (modified modified READ) coding, 1418, 1421; defined, **1422**
MMSS (maritime mobile satellite service), 910
MMX (AT&T mastergroup multiplex), 305–306
Mobile audio-visual terminal project. *See* MAVT
Mobile broadband system. *See* MBS
Mobile communications systems
  second generation, 1208–1225
  third generation, 1223
Mobile data systems. *See* Wireless data systems
Mobile host. *See* MH
Mobile radio, 1069–1089
  airwaves management, 1078
  codulation, 1081–1083
  defining terms, 1087–1088
  historical summary, 1070–1072
  macrodiversity and microdiversity, 1084–1085
  modulation, 1081–1083
  multiplex and multiple access, 1085–1086
  network issues, 1080–1081
  operating environment, 1078–1079
  service quality, 1079–1080
  speech coding, 1083–1084
  system capacity, 1086–1087
  systems described, 1073–1078
Mobile satellite systems, 1015–1031
  coding, 1026–1028
  modulation, 1025–1026
  multiple access, 1024–1025
  for radio, 1073
  radio frequency environment, 1016–1020
  satellite orbits, 1020–1024
Mobile support station. *See* MSS
Mobility
  model for PCS users, 1126–1129

Page on which term is defined is indicated in bold.

Index

personal, 1118–1120
terminal, 1117–1120
of users, 1367, 1370
MOBITEX, 1385–1387, 1392
Model-based coding, 1415
MODEM (modulator/demodulator), 242–243, 247–250
Modems, 444, 446
  cable, 1496
  codec filters, 374–376
  services, 1383–1384
Modified Huffman coding. See MH coding
Modified modified READ coding. See MMR coding
Modified READ coding. See MR coding
Modulated signals, described, 1354–1355
Modulation
  analog, 3–12
  analog vs digital, 1353
  for coherent systems, 864–865
  for DECT, 1321–1322
  digital, 273–287
  glossary of terms, 1082
  for mobile radio, 1081–1083
  for mobile satellite systems, 1025–1026
  pulse code, 23–34
  signals described, 1354–1355
  for speech, 1308–1310, 1425–1429, 1480–1481
  for spread-spectrum communications, 200–206
  for TV, 1494
  for videoconferencing, 1519
  for wireless systems, 1353–1366
Modulation index, 10
Modulation methods for wireless systems, 1353–1366
  analog frequency, 1355–1356
  CPM (continuous phase), 1360–1361
  GMSK (Gaussian minimum shift keying), 1361–1363
  MSK (minimum shift keying), 1360–1361
  OFDM (orthogonal frequency division multiplexing), 1363–1365
  PSK (phase shift keying), 1356–1359
Modulator/demodulator. See MODEM
Moisture transducer, 881
MONET project, 1332–1334
Monolithic microwave integrated circuit antennas. See MMIC antennas
Monopulse feed, 1038
MOS (mean opinion score), 1479; defined, **1486**
Motion compensation
  HDTV, 1465, 1466
  video, 1453–1455
Motion estimation, for HDTV, 1465–1466
Motion field, 1453–1455; defined, **1460**
Motion vector, 1519, 1524; defined, **1526**
Moving Pictures Expert Group compression. See MPEG compression
MPEG compression, 1476, 1482–1484; defined, **1486**
MPEG-1 compression, 1503; defined, **1514**
MPEG2/Audio, 1484–1485
MPSK (M-ary phaseshift keying) modulation, 240–242
MPT-1375, 1290

M-QAM (M-ary quadrature amplitude modulation), 1364
MQW (multiquantum well) amplifiers, 841
MQW lasers, 756–760
MR (magneto-restive) head, 1533; defined, **1544**
MR (modified READ) coding, 1418, 1421; defined, **1422**
m sequences, 94, 95–97; defined, **105**
MS (memory switch), 986–987
MSE (mean square error), 352
MSE (mean square estimate), 128; defined, **129**
MSK (minimum shift keying), as modulation for wireless systems, 1360–1361
MSM (microwave switch matrix), 978, 980
MSOH (multiplex section overhead), 557–558
MSS (mobile satellite service), 898, 910–911
MSS (mobile support station), 1370–1374
MTIE (maximum time interval error), 409–411
MTP (message transfer part), 482–489
Mu-law companding, 369–371
Multicast backbone. See MBONE
Multicasting, 612; defined, **1526**
Multilevel signalling, 399–401
Multimedia, network, 676–677, 720–723, 1511–1513
Multipactor intermodulation, 1051
Multipath fading, 132
Multipath spread, 134; defined, **139**
Multiple access
  defined, **1114**
  filtering signals in, 1260
  for microcellular radio, 1172–1173
  for mobile radio, 1085–1086
  for mobile satellites, 1024–1025
  radio access methods, 1104–1115
Multiple access networks. See MANs
Multiplexers
  optical filters as, 828
  for SONET, 551
  for WDM systems, 885, 889
Multiplexing, 87–93
  analog, for telephony, 295
  asynchronous time division (ATDM), 686, 687
  defined, **87**, **301**, **622**
  digital, for telephony, 295
  digital hierarchical control, 497–498
  for DVC, 1518
  frequency, 87–89, 301–307
  methods of, 301
  for mobile radio, 1085–1086
  and multiple accessing, 622, 1085–1086
  schemes for digital hierarchies, 377–385
  space, 91
  spread-spectrum techniques, 91–92
  switching fabrics as analogous to, 427
  synchronous, 404
  time, 89–90
  time-division (TDM), 687
  for videoconferencing, 1520, 1525
Multiplying D/A converters, 110
Multipoint communication, defined, **1526**
Multipoint control unit/multicast router. See MCU/MR

Page on which term is defined is indicated in bold.

Multiquantum well amplifiers. *See* MQW amplifiers
Multiquantum well lasers. *See* MQW lasers
Multiresolution compression models, 1407
Multistage interconnection networks. *See* MINs
Multitone modulation, subbands for, 335
Mutual information, 230; defined, **235**

N

Nakagami distribution, 1198–1199
NASA
    ACTS, 1035, 1037, 1040–1042
    TDRS system, 1054–1066
National Bureau of Standards. *See* NBS
National standardization, 161–163
National Television Systems Committee. *See* NTSC
Natural sampling, 18–19
NBS (National Bureau of Standards) data encryption standard, 674
NCC (new common carrier) networks, 1276, 1287
NCP (NetWare core protocol), 706
NDPS (nondisruptive path switching), 726
Near-end crosstalk. *See* NEXT
Near-far effect, 1198–1206
Near video on-demand, 1497; defined, **1500**
NetWare core protocol. *See* NCP
Network awareness, 707–709
Network-dependent layers, 573
Network environment, 569
Network interface card. *See* NIC
Network level interworking, 1220–1222
Network link services protocol. *See* NLSP
Network management, 707, 711–712
Network Node Interface. *See* NNI
Network of networks, 293
Network operating system. *See* NOS
Network restoration, 505–506
Networks
    ATM, 717–727
    banyan, 690–698
    B-ISDN, 577–584
    BLAN (broadband local area), 611–621
    design, 707, 709–711
    ethernet, 591–596
    FDDI (fiber distributed data interface), 597–610
    gigabit, 672–680
    IN (interconnection), 688–689
    ISDN, 577–584
    LANs, 681–685
    mobile as distinct from wired, 1373–1376
    multimedia, 1511–1513
    satellite, 895–904
    servers for, 1511–1513
    telephone, 292–293
Networks in tandem (in cascade), 968–969
Network speedup, 689
Network synchronization, 404–413
    distribution networks, 405–406
    impairment characterization, 408–410

    impairment effects, 406–408
    standards, 410–412
Neuman-Hofman sequences, 270
New common carrier, defined, **1287**; *see also* NCC networks
NEXT (near-end crosstalk), 453
NIC (network interface card), 591
NLSP (network link services protocol), 706
NNI (Network Node Interface), 541
Noise
    amplifier, 841–842
    channel, 28–29
    circuit, 310–311
    defined, **309–310**
    diagram of, 309
    granular, 29
    idle channel, 29, 372–373
    impulse, 312
    and PCM system output, 28–30
    percent quantizing, 25–26
    sources of, 309–310
Noise density, thermal, 966–967
Noise figure, 967
Noise power addition, 950
Noise shaping, 1478–1479
Noise temperature, 967
Nonblocking INs, 688
Nongeostationary orbits, 1021–1023
Nonlinear Kerr effect, 870
Nonreturn to zero. *See* NRZ
Normed vector space, 120
North American asynchronous digital hierarchy, 377–385
NOS (network operating system), 1512
Notchless bilevel quantizer, 1416
NRZ (nonreturn to zero) codes, 387, 388–390, 396, 775
NSP (network service part), 482
NTSC (National Television Systems Committee), 1490, 1500
Nutation damper, 918
Nyquist A/D converters, 110
Nyquist band, 323, 327
Nyquist criterion
    baseband signalling and, 321–327
    defined, **322**
    intersymbol interference and, 321–325
    with matched filtering, 325–327
Nyquist frequency, 347
Nyquist interval, 15
Nyquist pulses, 323–324
Nyquist rate, 15–16; defined, **22**
Nyquist-Shannon interpolation formula, 15; defined, **22**

O

OB (onboard) switching, 976–995
    advantages of, 976–977
    ATM-oriented, 993–994

Page on which term is defined is indicated in bold.

Index

baseband, 983–993
reconfigurable static IF, 978–979
regenerative IF, 979–983
Objective lens focusing, 1553–1555; defined, **1564**
Observation space, 125
OC (originating call), 1090–1102
OC-*N* (optical carrier level-*N*), 543
OCR (optical character recognition), 1416
Odyssey satellite system, 1033, 1049
OFDM (orthogonal frequency division multiplexing), 1363–1365
Offset-QPSK, 1026
Okumura model, 1190–1191
Onboard switching. *See* OB switching
OOC (optical orthogonal codes), 104–105
Open systems interconnection. *See* OSI
Open systems interconnection environment. *See* OSIE
Operational environments, 569–570
Optical amplifiers, 848–861
    characteristics compared, 854, 858
    general concepts, 849–852
    SOAs (semiconductors), 832–847
    types, 852–858
    for WDM, 885
Optical carrier level-*N*. *See* OC-N
Optical character recognition. *See* OCR
Optical coherent systems. *See* Coherent systems
Optical components, passive, 824–831
Optical data storage. *See* Magneto-optical disk storage
Optical fiber communications. *See* Coherent systems
Optical fiber ribbon, 874
Optical filters, 828–829
Optical interconnects, 873–876
Optical isolator, 776
Optical mode confinement factor, 836
Optical orthogonal codes. *See* OOC
Optical parametric amplifiers, 858
Optical path
    defined, **1564**
    described, 1552–1555
Optical ring switches, 991–992
Optical sources, 751–773
    lasers, 751–773
    LEDs, 751, 774
Optical switching fabric, 520–525
Optical switching systems. *See* Photonic switching
Optical transmitters, 774–788
    directly modulated laser, 775–782
    externally modulated (Mach-Zehnder), 782–786
    performance evaluation, 781–782
Optimization, internetwork, 713–714
Optimum receivers, 141–154
    and detection theory, 144–145
    dispersive channels, 152–153
    Karhunen–Loäve expansion, 142–143
    M-ary signalling, 147–149
    Rayleigh channel, 150–152
    realistic channels, 149–152
    signal space, 146
    standard binary signalling schemes, 146–147

Optoelectronic package design, 780–781
Orbcom, 897, 1049, 1073
Orbital diversity, 996, 1010–1011
Orbits, satellite, 1020–1024
    geostationary, 1020–1021
    nongeostationary, 1021–1023
Ordered dither, 1443–1444
Originating calls. *See* OCs
Orthogonal frequency division multiplexing. *See* OFDM
Orthogonality principle, 128; defined, **129**
Orthogonalization of codebooks, 1253
Orthogonal vectors, 122
Orthonormal functions, 142; defined, **153**
Orthonormal vectors, 122; defined, **129**
OSI (open systems interconnection), 567–576
    computer communications requirements, 567
    ISO reference model, 569–573, 702
    7-layer model, 569–573
    signalling message control, 482–483
    standards, 568–569, 574–575
OSIE (open systems interconnection environment), 568–569, 574–575
Outage probability, 214, 1156–1158; *see also* BER
Outdoor propagation, 1189–1192
Output queueing, 688
Overhead bytes, 544
Overhead capacity
    defined, **385**
    in digital hierarchies, 378
Oversampling A/D converters, 110; defined, **116**

P

Packet burst, 706
Packet data services, 1387–1390
Packet delivery, for internetworking, 702–703
Packet mode services, 1222
Packet switched networks, 650–660
Packet switching. *See* PS
PACS (Personal Access Communications System), 337
Paging
    defined, **1274**
    radio, 1074
Paging channel, defined, **1274**
Pair selected ternary coding. *See* PST
PAM (pulse-amplitude modulation)
    baseband model, 320–321
    binary, 320
    defined, **33**, **278**
    intersymbol interference and, 343–348
    with matched filter, 326
    power spectral density, 68–71
    signal, 24–25
Pan-European cellular system. *See* GSM
Panta Rhei, 1072
Parameter estimation, and signal space representation, 126–128
Parametric optical amplifiers, 858
Parity-check bits, 182
Parity-check matrix, 169; defined, **179**

Page on which term is defined is indicated in bold.

Parseval's equality, 124; defined, **129**
Partial-period correlation, 103–104
Partial response coding, 72–73
Partial response equalization. *See* PR equalization
Partial response maximum likelihood. *See* PRML
Partial response signalling. *See* PR signalling
PAS (pulse amplitude switching), 430
Passband transmission, echo cancellation, 421–423
Passive intermodulation. *See* PIM
Passive optical components, 824–831
Pass mode coding, 1414
Path diversity, 996–1014
    for arctic communications, 1008–1010
    in CDMA, 1263
    for land-mobile communications, 1008
    microscale, 1010
    optical, 1008
    orbital, 996, 1010–1014
    and rain-fade, 997–1008
    site diversity, 996–1010
Path overhead bytes. *See* POH bytes
Pattern matching compression techniques, 1416
Payload capacity
    defined, **385**
    digital hierarchy use, 378
Payload type indicator. *See* PTI
Pay-per-view TV, 1493, 1496
PBS (Public Broadcasting System), 906
PBXs (private branch exchanges), 440–449
    ACD, 443–444
    Centrex, 442–443
    CTI, 448–449
    described, 440–442
    facsimile, 444–445
    ISDN arrangements, 445–447
    as LANs, 681–684
    modems, 444
    service units, 445
    voice processing, 443
PCM (pulse code modulation), 23–34
    and A/D conversion, 111
    advantages of, 23–24
    bandwidth, 27–28
    circuits, 26–27
    defined, **23**, **34**
    generation, 24–25
    noise effects, 28–30
    nonuniform quantizing, 30–33
    percent quantizing noise, 25–26
    and speech coding, 1425
    system design, 33
PCM codec-filters, 364–376
    described, 364
    linear for modems, 374–376
    uses in telephone networks, 365–367
    voice filters, 367–374
PCS (personal communications services)
    GSM protocol, 1123–1126
    IS-41 protocol, 1121–1123

    location caching, 1129–1134
    location strategies, 1116–1145
    mobility aspects, 1117–1120, 1126–1129
    overview, 1117–1121
    reference model, 1120–1121
PDC (Japanese personal digital cellular) systems, 1276–1288
    air interface characteristics, 1212
    cellular market in Japan, 1276
    defined, **1287**
    described, 1215–1216
    gross bit rates, 1338–1339
    half-rate standards, 1338–1341, 1344–1346
    standardization process, 1276–1277
    system overview, 1277–1287
PDFA (praseodymium-doped fiber amplifier), 857
PDH (plesiochronous digital hierarchy), 377, 542
Peak frequency deviation, 10
Peak-power constraint, 335
Percent quantizing noise, 25–26
Perception-based coding, 1478–1479
Perceptual coder, 1440; defined, **1447**
Perfect code, 169; defined, **179**
Perigee, 918
Personal access communications system. *See* PACS
Personal communications services. *See* PCS
Personal digital cellular standard. *See* PDC
Personal handy phone system. *See* PHPS
Personal mobility, 1118–1120
Per-user location caching, 1129–1131
Phased array antennas, 1044–1046
Phase-diversity receiver, 867
Phase-locked loop. *See* PLL
Phase modulation. *See* PM
Phase noise, 865
Phase-shift keying. *See* PSK
Phase transients, 410
Phase-tuning, 764
Phase velocity, 313
phi-law companding, 30–33
Photonic switching, 513–538
    baseband switches, 991
    bit-sensing devices, 515–520
    bit-sensing vs bit-transparent, 515
    bit-transparent devices, 525–526
    call-by-call switching, 514, 515, 520–525
    FET-SEED, 516–524
    optical amplifiers for, 844
    protection switching, 513–515
    role of, 515
    System 5, 520–525
PHPS (personal handy phone system)
    air interface characteristics, 1212
    defined, **1287**
    described, 1218
    Japanese, 337
Physical layer, 531–532, 573
Pilot channel, defined, **1274**
Pilots, FDM, 306
PIM (passive intermodulation), 1051

Page on which term is defined is indicated in bold.

Index

Pitch period, 1342
Plain old telephone service. *See* POTS
Planar optic waveguide, 874
Plesiochronous digital hierarchy. *See* PDH
Plesiochronous signals, 405
PLL (phase-locked loop), 257–260
PLMN (public land mobile network), 1075–1076, 1227–1228
PLMR (public land mobile radio), compared with CT systems, 1290
Plug-compatible systems, 568
PM (phase modulation), 9–11; defined, **12**
PMD (polarization mode dispersion), 869
PMR (professional or private mobile radio) systems
   air interface characteristics, 1212
   architecture, 1210–1211
   basic features, 1209–1210
   defined, **1224**
PN (pseudorandom noise) sequences, 94–106
   aperiodic correlation, 101–103
   correlation measures, 101–105
   definitions, 94, 105
   low crosscorrelation sequences, 98–101
   m sequences, 95–97
   overview, 95
   q-ary sequences, 97–98
PN spread spectrum carrier, 1262
PNNI (private network-node interface), 617
POH (path overhead) bytes, 544
Pointers, 408, 544–548, 556, 557
Point of presence. *See* POP
Point-to-multipoint connection, 539
Point-to-multipoint signalling messages, 539
Point-to-point connection, 538–539
Point-to-point signalling messages, 539
Poisson sum formula, 64–65, 76; defined, **77**
Polarization controller, 830
Polarization dependent gain, 840–841
Polarization diversity, 215–216, 869, 1038
Polarization mode dispersion. *See* PMD
Polarization multiplexing, 91; defined, **93**
Polar signalling, 387, 391–393, 401
Polling schemes, 633–639
POP (point of presence), 292; defined, **300**
Positional-priority schemes, 638
POTS (plain old telephone service), 291–300
   and ADSL, 450, 452
   future of, 299–300
   network, 292–293
   signalling, 297–298
   station apparatus, 293–295
   switching, 296–297
Power control, 1197–1207
   applications, 1201–1205
   for CDMA, 1264–1267
   for cellular systems, 1198–1201
   for CT-2, 1292–1293
   defined, **1207**
   forward link, 1199–1202, 1204
   reverse link, 1274
   types of, 1198–1202
Power-limited systems, 238, 240–241, 244–245
Power per carrier, 959–962
Power spectral density, 64, 68–73, 1353, 1354; defined, **77**
PR (partial response) equalization, 1536; defined, **1544**
PR (partial-response) signalling, 72–73, 324–334, 401
Praseodymium-doped fiber amplifier. *See* PDFA
Precoding
   adaptive, 335–336
   and PR signalling, 333–334
Prediction errors, 1407–1408
Predictive coding, 1414, 1427–1429, 1455–1456; defined, **1422**, **1460**
Preformat information, 1549; defined, **1564**
Premise wiring, 447–448
Presentation layer, 572–573
Preset equalizers, 348
Privacy issues, 674, 1268–1269
Private branch exchanges. *See* PBXs
Private data networks, 1384–1387
   ARDIS, 1384–1385
   and DCS, 506–507
   MOBITEX, 1385–1387
Private network-node interface. *See* PNNI
Private (or professional) mobile radio. *See* PMR
Private user-network interface. *See* PUNI
   PRM (protocol reference model)
   B-ISDN, 587
   ISDN, 581–582
PRML (partial response maximum likelihood), 1531; defined, **1544**
Probabilities, 35–61
   axiomatic formulas, 38–39
   Bayes' theorem, 39–41
   definitions, 60
   density functions, 41–45, 51–57
   discrete probability theory, 36–41
   mass functions, 41–45, 51–57
   theory of one random variable, 41–50
   theory of two random variables, 51–59
Probing schemes, 639
Processing gain, 200, 250–251
   in CDMA, 1261, 1269
Professional applications, unified systems for, 1209
Professional (or private) mobile radio. *See* PMR
Progressive scanning, 1450; defined, **1460**
Progressive transmission of facsimile images, 1420–1421; defined, **1422**
Projection theorem, 128
Propagation impairments, satellite. *See* Satellite transmission impairments
Propagation models, wireless, 1182–1196
   diffraction and, 1186–1187
   free-space models, 1183
   indoor, 1192–1193
   large-scale models, 1183
   outdoor, 1189–1190
   practical models, 1188–1193
   reflection and, 1183–1186

Page on which term is defined is indicated in bold.

Propagation models, wireless, 1182–1196
    scattering and, 1187–1188
    small-scale fading and, 1193–1195
    small-scale models, 1183
Propagation mode selection, 1260
Protection switching, 513–514, 515
Protocol layers, 570–573
    application, 572
    link, 573
    network, 573
    physical, 573
    presentation, 572–573
    session, 573
    summarized, 571
    transport, 573
Protocol reference model. *See* PRM
PRS (primary reference source), 405, 411
PS (packet switching), 425–426, 976
PSDN (public switched data network), 574
Pseudorandom noise (pseudonoise) sequences. *See* PN sequences
Pseudoternary coding, 393
PSI-CELP, 1344–1346
PSK (phase-shift keying)
    attenuation distortion, 312
    and coherent systems, 864
    defined, **287**
    intersymbol interference and, 343–348
    modulation techniques, 279
    modulation for wireless systems, 1356–1359
PST (pair selected ternary) coding, 399
PSTN (public switched telephone network), 308
    defined, **316**
    filters for modem applications, 374–376
    OSI standards, 574
    satellite use, 907–911
PTI (payload type indicator), 531
Public land mobile network. *See* PLMN
Public land mobile radio. *See* PLMR
Public switched data network. *See* PSDN
Public switched telephone network. *See* PSTN
Public TV, 906
Pull-in range, 410
Pulse amplification, 842–844
Pulse amplitude modulation. *See* PAM
Pulse amplitude switching. *See* PAS
Pulse code modulation. *See* PCM
Pulse shaping and baseband signalling, 318–338
    defined, **319**
    filter, 321
    orthogonal 320
    see also Baseband signalling
Puncturing, in coding, 174; defined, **179**
PUNI (private user-network interface), 617
Pyramidal coding, 1458

Q

QAM (quadrature amplitude modulation)
    carrier signals, 74–76
    constellations, 261–264
    defined, **77, 280**
    intersymbol interference and, 343–348
q-ary sequences, 97–98
QCSE (quantum-confined Stark effect), 782
QM (quadrature multiplexing), 6–7; defined, **12**
QMF (quadrature mirror filters), 1430, 1440; defined, **1447**
QoS (quality of service)
    defined, **1526**
    in DVC, 1518
    in LANs, 616
    requirements, 676
QPSK (quadrature phaseshift keying), 1025–1026, 1058
    modulation for TV, 1494
    modulation for wireless systems, 1356–1359
Quadrature amplitude modulation. *See* QAM
Quadrature mirror filters. *See* QMF
Quadrature multiplexing. *See* QM
Quadrature phaseshift keying. *See* QPSK
QUALCOMM, Inc., 1205, 1258–1259
Quality of service. *See* QoS
Quantization
    defined, **1473**
    for HDTV, 1468
Quantizer, 1416; defined, **1422**
Quantizing
    defined, **23, 34**
    nonuniform, 30–33
    percent noise, 25–26
    speech, 1427–1429
Quantizing distortion, and PCM codec-filters, 368–371
Quantum-confined Stark effect. *See* QCSE
Quantum well lasers. *See* QW lasers
Quaternary sequences, 100
Queuing, 78–86
    for ATM switches, 688–689
    averages for queue and server, 85–86
    Little's formula, 79–80
    M/M/1 system, 80–85
QW (quantum well) lasers, 756–760

R

RA (random access) control, 631, 632–633
RACE program, 1327–1337
    ATDMA project, 1331–1332
    CODIT system, 1330–1331
    GIRAFE project, 1336
    MAVT project, 1334–1335
    MBS project, 1335
    MONET project, 1330, 1332–1334
    network issues, 1332–1334
    SIANT project, 1335–1336
    TSUNAMI project, 1335
Radio, mobile. *See* Mobile radio
Radio access methods. *See* Access methods for radio
Radio coverage, needed improvements, 1223
Radio frequency, and mobile satellite, 1016–1020

Page on which term is defined is indicated in bold.

Index

Radiometer, 208, 1050
Radio noise, and satellite transmission, 944–945
Radio over fiber. *See* ROF
Radio paging, 1074
RAI (remote alarm indication), 382
Rain attenuation, 938–941
Rain depolarization, 942–943
Rain-fade alleviation, 997–1008
Raised cosine pulse, 324, 325
RAKE receiver, 213, 217
Random access channel, defined, **1114**
Random access control. *See* RA
Random access protocols, 639–642
Random phase channels, 150
Random phenomenon, 35
Random processes, 62–77
    definitions, 62–65
    properties, 65–67
Random variables
    density or mass functions, 44–45
    probability theory and, 35–36
    statistics of, 45
    theory of one variable, 41–50
    theory of two variables, 51–59
RAP (randomly addressed polling), 1373
Rare-earth-doped amplifiers, 856–858
Rate buffer, 1457; defined, **1460**
Rate distortion function, defined, **235**
Rate distortion theory, 229–231
Rayleigh channel, 150–152; defined, **153**
Rayleigh fading, 1194, 1265, 1266, 1369; defined, **1195**
READ (relative element address designate) coding, 1413–1414, 1418, 1422
Read/write process, 1531–1535
    channel model, 1532–1525
    defined, **1544**
Real systems environment, 569
Real-time (continuous time) systems, 1505; defined, **1514**
Real-time transport protocol. *See* RTP
Receiver excess temperature, 969
Receivers, microwave, 966–975
Receivers, optimum. *See* Optimum receivers
Receiver sensitivity, for coherent systems, 865–866
Receive window, 663
Reconfigurable static IF switching, 978–979
Recording of data. *See* Data recording
Redundancy, in video source, 1451–1452
Reference symbol, 1027
Reflection
    defined, **1195**
    ground, 1184–1186
    as propagation mechanism, 1183–1186
Reflection coefficient, 1251–1252, 1255–1256
Reflector antennas, 1034, 1038–1043
Reflector surfaces, 1039
Refreshing, for HDTV, 1467–1468
Regenerative IF switching, 979–983
Regenerative link designs, 949–962
Regional standardization, 159–160

Registration, terminal/location, 1121–1122, 1123, 1126
Relative diversity gain, 1003, 1005
Relative element address designate coding. *See* READ coding
Relaxation resonance, 776
Relay satellite, TDRS, 1058–1064
RELP (residual-excited linear prediction) coding, 1426, 1429
Remote alarm indication. *See* RAI
REN (ringer equivalence number), 437, 438
Repetition codes, 234
Research and development program, European. *See* RACE program
Reservation schemes, 633
Reset-K algorithm, 1135–1140
Residual coding, 1457
Residual-excited linear prediction coding. *See* RELP coding
Resonant amplifier, 834
Return direction, 1016
Return to zero. *See* RZ
Reverse CDMA channel, defined, **1274**
Reverse link power control, 1199–1201; defined, **1274**
Reverse traffic channel, defined, **1274**
Rice code, 1402
Rician fading, 137, 1194–1195
Riesz-Fischer theorem, 122–123; defined, **129**
Ring topology, 624
RIP (router information protocol), 706
RLL (runlength-limited) code, 1536; defined, **1544**
ROF (radio over fiber), 1170–1171
ROTORS, 470–471
Router information protocol. *See* RIP
Routing, 650–660
    in datagram networks, 652–655
    and flow control, 650–651
    hierarchical, 655–656
    in virtual circuit switched networks, 655
RSOH (regenerator section overhead), 557–558
R-TDMA (reservation TDMA), 1372
RTP (real-time transport protocol), 1513
Runlength coding, 1412
Runlength-limited code. *See* RLL code
Running average algorithm, 1135
RZ (return to zero) signalling, 387, 388, 393, 775

S

Saleh model, 1192–1193
Sampling, 13–22
    bandpass signals, 16–18
    flat-top, 20–21
    instantaneous, 13–15
    interval, 13
    natural, 18–19
    practical, 18–21
    rates, 13, 18
    of sinusoidal signals, 16
    theorem, 15–16, 21

Page on which term is defined is indicated in bold.

Index

SAR (segmentation and reassembly), 678
Satellite antennas, 1032–1053
   horn antennas, 1034–1038
   performance parameters, 1033, 1052
   phased array antennas, 1044–1046
   radiometer antennas, 1050
   reflector antennas, 1034, 1038–1043
   shaped-beam antennas, 1048–1049
   space qualification, 1050–1051
   TT&C antennas, 1046–1048
   types of, 1034, 1053
Satellite footprints, 914, 922
Satellite integration in future project. *See* SIANT
Satellite link design, 949–965
   allowances, 954–955
   designed BER, 957–958
   EIRP, 959–962
   interference equations, 951–954
   intermodulation equations, 954
   losses, 954–955
   margins, 954–955
   numbers of carriers, 959–962
   power per carrier, 959–962
   sums of link equations, 955–957
   transparent vs regenerative, 949–962
   uplink and downlink equations, 950–951
Satellite mobile systems. *See* Mobile satellite systems
Satellite orbits, 1020–1024
Satellites
   antennas, 1032–1053
   Earth station, 922–934
   electrical subsystem, 919
   GEO (geostationary Earth orbit), 893–911
   launching, 917
   link design, 949–965
   mobile systems, 1015–1031
   network fundamentals, 895–904
   propagation impairments, 935–948
   stabilization, 918
   station keeping, 918
   systems, 912–921
   TDRSS (tracking and data relay systems), 1054–1066
   transponder systems, 913–917
Satellite switching, 976–995
   advantages of, 976–977
   onboard switching, 976–995
Satellite transmission impairments, 935–948
   attenuation, 936–942
   depolarization, 942–944
   and path diversity, 996–1014
   radio noise, 944–945
   scintillation, 945–947
Saturated material gain coefficient, 838
S-band
   defined, **1065**
   TDRS use, 1056
SBC (subband coding), 1439–1440
SBS (stimulated Brillouin scattering), 853–854, 869–870
SC (suppressed carrier) modulation, 4–5; defined, **12**
Scalability, 1453; defined, **1460**

Scanning formats
   GA applications, 1471–1473
   for HDTV, 1464–1465
   video, 1450–1451
Scattering
   defined, **1195**
   as propagation mechanism, 1183, 1187–1188
Scattering function, 133
SCCP (signalling message control), 482–483, 487
Schwarz's inequality, 120; defined, **130**
Scintillation, and satellite transmission, 945–947
SDH (synchronous digital hierarchy) 554–564
   ATM mapping, 562
   frame structure, 554–562
   network block diagram, 561–562
   standards, 497–499, 562
   synchronization, 404, 408–411
SDMA (space-division multiple access), MAC control schemes, 632
SDS (sample description space), 40–41
SE (switching element), 688–696
Search algorithm, 201
Second generation mobile communication systems, 1208–1225
   architecture, 1210–1211
   basic features, 1209–1210
   interworking, 1220–1222
   performance improvements, 1222–1223
   systems description, 1211–1220
Section overheads, 557–558
Sectorization, in CDMA, 1270
Security networks, unified systems for, 1209
Security systems for TV, 1499
Segmentation and reassembly. *See* SAR
SEL (surface emitting lasers), 764–768
Selection combining, 217–219
Selective-repeat ARQ, 186–187, 190
Semiconductor lasers, 751–773
Semiconductor optical amplifiers. *See* SOAs
Send window, 663
Sensors, fiber optic, 880–881
SEP (signalling endpoint), 487
Sequence packet exchange/Internet packet exchange. *See* SPX/IPX
Serial-input-output chip. *See* SIO chip
Servers, data. *See* Data servers
Servers, in queuing, 79–86
Servers, video. *See* Video servers
Service codes, TV, 1497
Service outages, 505–506
Session layer, 573
Set partitioning, 177; defined, **179**
Set top box. *See* STB
Set top converters, 1489, 1495–1496; defined, **1501**
SF (superframe format), 379–380
SFH (slow frequency-hopped) systems, 203
SH (supervisor host), 1375
Shadowing
   defined, **1207**
   described, 1198

Page on which term is defined is indicated in bold.

Index

of mobile satellite, 1017–1020
and power control, 1198, 1204
Shannon, Claude, 224
Shannon bandwidth, 275–276
Shannon's channel coding theorem, 231–232
Shaped-beam antennas, 1048–1049
Shared-media LAN, 611
Shift-register sequence, 95; defined, **105**
SIANT (satellite integration in future) project, 1335–1336
Sigma-delta conversion, 111; defined, **116**
Sigma-delta PCM codec-filters, 374–376
Signal constellations, 283–284; defined, **287**
Signal detection, and signal space representation, 125–126
Signal fading
　diversity techniques to combat, 213–223
　see also Fading
Signalling
　baseband, 318–338
　binary, 146–147
　common channel, 480–495
　telephone network, 297–298
Signalling information, 514
Signalling link, 483
Signalling message handling, 485–486
Signalling network management, 486–487
Signalling network topology, 487–488
Signalling relation, 487–488
Signal space, 118–130
　decision boundary and, 146
　defined, **153**
　and parameter estimation, 126–128
　and signal detection, 125–126
　vector space fundamentals, 118–125
Signal-to-interference level, 1267–1268, 1269
Signal-to-noise ratio. See SNR
Signal vectors, 125; defined, **130**
Simulation, internetwork, 712–713
Single-parity-check codes, 234
Single sideband modulation. See SSB modulation
Sinusoidal signals, 16
SIO (serial-input-output) chip, 26–27
SIRCIM model, 1193
Site diversity, 996–1010
　for arctic communications, 1008–1010
　defined, **996**
　experiments, 999, 1009–1010
　for land-mobile communications, 1008
　optical satellite, 1008
　processing, 997
　for rain-fade, 997–1008
　16 × 16 photonic element, 526–527, 1519
Sky temperature, 970–972
Slant path attenuation. See Attenuation on slant path
Sliding window, 184–187, 190–193, 662–664
Slope tests, 309
Slow frequency-hopped (SFH) systems, 203
Small-scale fading, 1193–1195
SMARTNet, 645–646

SNC (subband coding), 1480–1481; defined, **1486**
SNR (signal-to-noise ratio), 276
　of amplifiers, 851
　defined, **287**
　gap, 462–463
　and power control systems, 1265–1267
SOAs (semiconductor optical amplifiers) 832–847
　design, 835–836
　gain characteristics, 836–842
　multichannel amplification, 844
　operation principles, 833–834
　overview, 854–856
　pulse amplification, 842–844
　types listed, 834
Soft capacity, of CDMA, 1272
Soft decision, 167; defined, **179**
Soft handoff, 1264; defined, **1274**
SOH (section overhead), 557–558
Solar cells, 919
Solar sails, 919
Solid-state power amplifiers. See SSPAs
SONET (synchronous optical network), 542–553
　and DCS systems, 497–500, 503–506
　equipment, 551
　frame, 543–550
　optical issues, 550–551
　standards, 551–552
SOP (states of polarization), 863
Source coding, 226–229, 1449–1461
　algorithms, 1453–1458
　for discrete-alphabet sources, 226–228
　source characteristics, 1450–1453
　standards, 1459–1460
　for universal sources, 228–229
Source compression
　facsimile image, 1411–1423
　lossless, 1397–1410
　still image, 1437–1448
　speech, 1424–1436
　see also Video
Space diversity, 215, 1263; see also Path diversity
Space-division multiple access. See SDMA
Space-division multiplexing, 91; defined, **93**
Space loss, 914
Space qualification, 1050–1051
Space switching, 297, 844; defined, **300**
Spanning set, 121; defined, **130**
Spatial filtering, 1260
Spatial reuse optical networks, 644–647
SPE (synchronous payload envelope), 543
Spectral constraints, 334–335
Spectral density, 64, 68–73
Spectral efficiency, 1161–1164
　of CDMA, FDMA, and TDMA, 1260
　defined, **1173, 1260**
Spectrum efficiency, 1148
　and air interface adaptivity, 1222–1223
　defined, **1158**
Speech blurring, 313
Speech codecs, 1222

Page on which term is defined is indicated in bold.

Speech coding, 1424–1436
  analysis-by-synthesis coders, 1430–1432
  in CDMA, 1259
  for cellular mobile radio, 1339–1342
  defined, **1424**
  for DVC, 1519, 1522, 1523–1524
  frequency-domain coders, 1429–1430
  fundamentals, 1248–1249
  half-rate standards, 1338–1352
  predictive coders, 1427–1429
  quantization, 1427
  standards, 1338–1352, 1434
  types of coders, 1426–1434
  VBR (variable bit rate), 1433–1434
Speech compression, 1424
Speech signals, 1425–1426
Speech transmission
  by GSM, 1229–1232
  by IS-54, 1246–1256
Spinner-type satellites, 918
Split phase coding, 393–394
Spontaneous emission, 833
Spotbeams, for mobile satellites, 1024
Spreading ratio, 200
Spread spectrum communications. *See* SS communications
Spread spectrum multiple access. *See* SSMA
SPX/IPX internetworking, 702, 705–706
Squared error distortion, 229
SRS (stimulated Raman scattering), 853, 869–870
SS (satellite switching), 976
SS (spread spectrum) communications, 199–212
  bandwidth, 200
  commercial applications, 209–211
  military applications, 206–209
  modulation techniques, 200–206
  multiplexing, 91–92
  overview, 199–200
  pulse shapes, 320–321
  terminology, 200
SSB (single sideband) modulation, 6–8; defined, **12**
SS-FDMA switching networks, 481–495
  signalling transport, 482–488
  signalling user parts, 488–495
  see also Common channel signal
SSMA (spread spectrum multiple access)
  and CDMA, 1264
  channel assignment, 1175–1176
  TDRSS use, 1058
SSPAs (solid-state power amplifiers), 915
Stabilization systems for satellites, 918–919
Standard array decoders, 169; defined, **179**
Standards
  for networking, 529–541
  for data encryption, 674
  for facsimile transmission, 1417–1421
  for Japanese cellular networks, 1276–1277
  for LAN architectures, 681–684
  for North American digital hierarchies, 378
  for open systems interconnection, 568–569, 574–575
  for speech coding, 1338–1352, 1434–1435
  for videoconferencing, 1518–1519
Standards setting bodies, 155–165
  coordination of, 163–164
  development cycle, 165
  global, 156–159
  national, 161–163
  regional, 151–160
  scientific, 164–165
Star coupler, 827
Star topology, 623, 645
States of polarization. *See* SOP
Statistical multiplexing, 90, 530
Stat-MUX, 90; defined, **93**
STB (set top box), 1503; defined, **1514**
STE (supergroup translation equipment), 304
Stereophony, multichannel, 1484–1485
Still image compression, 1437–1448
  compression techniques, 1438–1441
  halftoning, 1443–1447
  JPEG, 1441–1443, 1445–1447
Stimulated amplification, 833
Stimulated Brillouin scattering. *See* SBS
Stimulated emission, 833
Stimulated Raman scattering. *See* SRS
STM-1, 554–555, 558–561
STM-N, 555
Stochastic gradient algorithm 354
Stop-and-wait ARQ, 183–184, 188–189
Storage of data. *See* Magnetic storage; Magneto-optical disk storage
Store-and-forward switching, 628
STP (signal transfer point), 487–488
Strained quantum well lasers, 758–760
Stratum clocks, 405
Street microcells, 1164–1169, 1172; defined, **1173**
Striping data, 1504–1505; defined, **1514**
Strowger, Almon B., 296
STS-1 frame, 544–548
STS-3c, 550
STS-N, 543, 548
STS-Nc, 549
Stuffbits, 382
Subband coding. *See* SBC
Subnet mask, 703
Subscriber lines, asymmetric digital. *See* ADSL
Subscriber loops, 308–316; *see also* Analog telephone channels
Subthreshold SOAs, 834
Successive approximation, 111; defined, **116**
Sun interference, 973
Sun outage, 973
Superframe, 546
Superframe formats, 379–380
Supergroup bank, 304
Supergroup translation equipment. *See* STE
Supervisor host. *See* SH
Suppressed carrier modulation, 4; defined, **12**
Surface emitting lasers. *See* SEL
Switched diversity, 1025

Index

G

GA (Grand Alliance) system for HDTV, 1462–1474
    scanning format applications, 1471–1473
    system described, 1464–1471
GaAsFET (gallium-arsenide field-effect transistor), 915
Gain calibration, for codec-filters, 371–372
Gain characteristics of amplifiers, 836–842
Gain clamped SOAs, 834
Gain guided laser, 753
Gain ripple, 835
Gain saturated amplifiers, 851
Gain saturation, 838–839
Gain tracking, 373
Galaxy VII, 900, 902
Gallium-arsenide field-effect transistor. *See* GaAsFET
Gamma Ray Observatory link budget, 1064–1065
Gaseous attenuation, 936–938
Gateway, 503; defined, **511**
Gaussian filtered frequency shift keying. *See* GFSK
Gaussian minimum-shift keying. *See* GMSK
GDSS (global distress safety system), 1074
General switched telephone network. *See* GSTN
Generator matrix, 168; defined, **179**
Generator polynomial, 169; defined, **179**
GEO (geostationary Earth orbit) satellites, 893–921
    application types, 904–911
    network fundamentals, 895–904
    systems, 912–921
Geometric signal representation, 274; defined, **287**
Geostationary Earth orbit satellites. *See* GEO satellites
Geostationary orbits, 1020–1021
Geosynchronous orbit, 922, 1054
GFRP (graphite fiber reinforced plastic) reflector, 1040, 1043
GFSK (Gaussian filtered frequency shift keying), for DECT, 1321–1322
Gigabit networks, 672–680
    ATM host interfacing, 674–675
    Aurora testbed, 672–679
    DSM communications, 677–679
    multimedia architectures, 675–677
GIRAFE project, 1336
Global distress safety system. *See* GDSS
Global positioning system. *See* GPS
Global standardization, 156–159
Globalstar satellite, 1033, 1049–1050, 1073
Global system for mobile communications. *See* GSM
Glonass, 1074
GMSK (Gaussian minimum-shift keying)
    baseband pulse shapes, 336–337
    modulation for DECT, 1321–1322
    modulation for GSM, 1237
    modulation for wireless systems, 1361–1363
Go-back-N ARQ schemes, 185–186, 190–193
Gold sequences, 99, 104
Golomb codes, 1402
Gosset lattice, 283
GPS (global positioning system), 1070, 1074
Gram-Schmidt procedure, 123; defined, **129**

Grand Alliance system. *See* GA system
Granular noise, 29
Graphical user interface. *See* GUI
Graphite fiber reinforced plastic reflector. *See* GFRP reflector
Gray code, 1415; defined, **1422**
Grey, Elisha, 292
GRIN rod lens, 825
GRO. *See* Gamma Ray Observatory
Grooming, 497, 498, 500–501; defined, **512**
Ground reflection, 1184–1186
Group 3 facsimile machine, 1417–1419; defined, **1422**
Group 4 facsimile machine, 1417–1419; defined, **1422**
Group bank, 304
Group delay, 339
Group delay distortion, 312–313
Groupe Speciale Mobile. *See* GSM
Group translation equipment. *See* GTE
GSM (Global System for Mobile Communications or Groupe Speciale Mobile), 1123–1126, 1226–1245
    access methods, 1111
    adaptive link control, 1239–1241
    air interface characteristics, 1212
    architecture, 1125
    baseband pulse shapes, 336–337
    control signals transmission, 1232–1236
    database traffic rate, 1128–1129
    defined, **1206**
    described, 1123–1124, 1213–1214
    discontinuous transmission, 1241
    GMSK modulation, 336–337, 1237
    gross bit rates, 1338–1339
    half-rate standards, 1341, 1346–1350
    interworking with DECT, 1220–1221, 1307
    logical and physical channels, 1228–1229
    packet mode services, 1222
    power control, 1203
    speech and data transmission, 1229–1232
    summary of features, 1241–1242
    system elements, 1227–1228
    user location strategy, 1126
    VSELP encoder, 1346–1350
GSMPLMN, 1227–1228
GSTN (general switched telephone network), and videoconferencing, 1523–1525
GTE (group translation equipment), 304
Guard channels, 1095–1096
GUI (graphical user interface), 1517

H

H.261 standard for video coding, 1519–1520, 1522
H.263 standard for video coding, 1524
H.320 standard for videoconferencing, 1518–1521
H.324 standard for videoconferencing, 1523–1525
Half-duplex links, 183; defined, **198**
Half-rate speech coding, 1338–1352
    channel coding, 1342–1344

Page on which term is defined is indicated in bold.

Half-rate speech coding (*Cont.*)
   European GSM standard, 1346–1350
   Japanese PDC standard, 1344–1346
   techniques, 1339–1342
Halftoning, 1438, 1443–1447; defined, **1447**
Hamming codes, 234–235
Hamming correlation function, 104
Hamming distance, 167–168, 173; defined, **179**
Hamming distortion, 229
Handoff
   defined, **1274**
   soft, 1264, 1274
Handoff calls. *See* HCs
Handshaking, 662, 1296, 1300–1302, 1372–1373
Hard decisions, 166; defined, **179**
Hata model, 1191
HCs (handoff calls), 1090–1102; defined, **1102**
HDBN (high-density bipolar N) code, 398–399, 401
HDLC (high-level data link control), 380
HDSL (high-bit-rate digital subscriber lines), test loops, 471
HDTV (high-definition TV), 1462–1474
   GA (Grand Alliance) system overview, 1463–1464
   scanning format applications, 1471–1473
   system descriptions, 1464–1471
   terrestrial broadcast requirements, 1463
Head end, 1489, 1491–1493; defined, **1500**
Head-end authorization, 1498
Header field, 655
Heat loads, and switching, 519–520
Heterodyne detection schemes, 862, 865
Heterodyne-type optical fiber communications. *See* Coherent systems
HFC (hybrid fiber coax), 878–879, 1489–1490, 1509; defined, **1514**
Hierarchical interpolation. *See* HINT
High-bit-rate digital subscriber lines. *See* HDSL
High-definition TV. *See* HDTV
High-density bipolar N code. *See* HDBN
High-level data link control. *See* HDLC
High-power amplifiers. *See* HPAs
Highway microcells, 1161–1164, 1172; defined, **1173**
Hilbert spaces, 118–130
   defined, **130**
   separable, 121
Hilbert transform, 7
HINT (hierarchical interpolation), 1407
Holdover stability, 410
Homodyne detection schemes, 862, 865, 867
Homologation, 436
Horizontal mode coding, 1414
Horn antennas, 1034–1038
Hot pad, 968
HPAs (high-power amplifiers), 913, 927, 932
HTTP (hypertext transport protocol), 618–619
Hubbing networks, 501–502; defined, **512**
Huffman coding, 228, 1399–1408, 1415, 1418, 1422; defined, **235**
Hughes Electronics satellites, 895
Hummings, Henry, 294

Hybrid coders, 1426; defined, **1435**
Hybrid fiber coax. *See* HFC
Hybrid networks, DCS in, 507–508
Hybrid transformer, 294; defined, **300**
Hypertext transport protocol. *See* HTTP

I

IBC (integrated broadband communications), 1327
IBM's broadband network service. *See* BBNS
IC (integrated circuit), 107–114
Ice depolarization, 943–944
Ideal sampled signal, 14
Identity element, 119
Idle channel noise, 29, 372–373
IDLE-D, 1295
IEEE (Institute of Electrical and Electronic Engineers) standards board, 164–165, 574, 681
IF (intermediate frequency) switching, 978–983
IFS (installable file system), 1512
Implicit polling, 637–638
Impulse noise, 312; defined, **316**
Impulse pay-per-view, 1496; defined, **1500**
IMT-2000 (International Mobile Telecommunications-2000), 1223
IN (interconnection network), 688–689
Incarnations, 662
Independent vectors, 119
Index guided laser, 753
Indoor microcells, 1169; defined, **1173**
Indoor propagation models, 1192–1193
Information theory, 224–235
   channel capacity, 231
   channel coding, 231–233
   communication system, 224–226
   rate distortion theory, 229–231
   simple binary codes, 233–235
   source coding for discrete-alphabet sources, 226–228
   universal source coding, 223–224
Initial sequence number, 669
Inmarsat communications system, 1074–1075
Input/output interconnections. *See* I/O interconnections
Input queueing, 688
Insertion loss, 825; defined, **830**
Installable file system. *See* IFS
Institute of Electrical and Electronics Engineers. *See* IEEE
Integrated broadband communications. *See* IBC
Integrated circuits. *See* ICs
Integrated services digital network. *See* ISDN
Intelligent vehicle highway system. *See* IVHS
INTELSAT VII satellite, 1040–1041
Intensity noise, 868
Interactive TV. *See* ITV
Interactive video servers, 1502–1503; *see also* Video servers
Interchannel crosstalk, 844
Interconnection networks. *See* INs

Page on which term is defined is indicated in bold.

Index

Interconnections
    I/O (input/output), 876–878
    optical, 873–876
    telephony rules, 436–438
Interdiction for CATV, 1496–1497; defined, **1500**
Interexchange carriers, 292; defined, **300**
Interface card, 446
Interfaces, ATM, 536–541
Interfade interval, 999, 1002
Interference
    adjacent channel, 1155
    carrier-to-cochannel, 1148
    cochannel, 1147, 1155–1158
    defined, **964**
    radio-frequency, 1147, 1155–1158
    and satellite link design, 951–954, 960
Interim Standard 41. *See* IS-41
Interim Standard 54. *See* IS-54
Interim Standard 95. *See* IS-95
Interlace artifacts, 1463; defined, **1473**
Interlaced scanning, 1450; defined, **1460**
Interleaving code sequences, 177, 217; defined, 179
Intermediate frequency switching. *See* IF switching
Intermodulation
    defined, **964**
    and link design, 951–954, 960
International gateway, 503
International Mobile Telecommunications-2000. *See* IMT-2000
International Standards Organization. *See* ISO
International Telecommunications Union. *See* ITU
International Telegraph and Telephone Consultative Committee. *See* CCITT
Internet
    MBONE (multicast backbone), 1521–1522
    telephony, 449
    videoconferencing over, 1521–1522
    video servers on, 1512–1513
Internet protocol. *See* IP
Internetworking, 701–716
    defined, **701**
    DECT with GMS, 1220–1222
    optimization, 713–714
    protocols, 702–706
    simulation, 712–713
    total network engineering process, 706–712
Interpolative coding, 1456
Intersymbol interference. *See* ISI
Interworking, network. *See* Internetworking
I/O interconnections, 876–878
IP (internet protocol), 1373–1374, 1521–1522; *see also* TCP/IP
IPIP protocol, 1373
IPX RIP, 706
IRT (integrated receiver transcoder), 1493
IS-41
    call delivery, 1122–1123
    database traffic rate, 1127–1128
    interworking, 1220

    message flow, 1121–1123
    registration, 1121–1122
IS-54
    air interface characteristics, 1212
    baseband pulse, 337
    described, 1214–1215
    gross bit rates, 1338–1339
    half-rate standards, 1338–1341
    history of, 1258
    interworking, 1220
    speech transmission techniques, 1246–1256
IS-95
    air interface characteristics, 1212
    as CDMA standard, 1272–1274
    data services, 1390–1391, 1392
    described, 1215
    history of, 1258
    interworking, 1220
    power control, 1203–1204
    pulse shaping, 335, 337
ISDN (integrated services digital network), 577–589
    architecture, 579–581
    ATM transfer mode, 717–727
    basic principles, 577–579
    broadband (B-ISDN), 584–589
    communication services, 578–579
    CPEs, 445–447
    DECT interworking, 1307
    defined, **478**
    OSI standards, 574
    overview, 577–578
    protocol reference model, 581–582
    SS7 user part, 488–491
    test loops, 471
    user-network interface layers, 583–584
    videoconferencing over, 1518–1521
ISI (intersymbol interference)
    and channel distortion, 339, 343–348
    defined, **28**, **153**
    equalizers to combat, 348–361
    model of, 343–348
    Nyquist criterion and, 321–325
ISO (International Standards Organization) reference model, 569–573, 574
    application-oriented layers, 572
    internetworking architecture, 702
    logical structure, 570–571
    MPEG coding, 1482–1484
    network-dependent layers, 573
    presentation layer, 572–573
    vs TCP/IP protocol, 574
Iso-BER curve, 957–959
Isochronous traffic, 612
Isolation
    diversity gain, 1004
    and link design, 953
    and loss, 825
Isolator, 825–826
ISUP messages, 488–491

Page on which term is defined is indicated in bold.

ITU (International Telecommunications Union), 155–158, 410, 435
  described, 1078
  see also CCITT
ITU-R
  attenuation models, 937–940
  audio coding standards, 1476
  described, 1486
  FPLMTS standardization, 1223
ITU-T
  SDH standards, 562–563
  videoconferencing standards, 1518–1519, 1523–1525
  video standards, 1459
ITU-TS (International Telecommunications Union-Telecommunications Standards Sector), 529–530
ITV (interactive TV), 1503, 1507
  defined, **1514**
  server architecture, 1509–1511
IVHS (intelligent vehicle highway system), 1381
IVR (interactive voice response), 443

J

Japanese digital cellular system (JDC). *See* PDC
Japanese Telecommunication Technology Committee (TTC), 162–163
JBIG (Joint Bilevel Image Group), compression standards, 1408–1409, 1415, 1419–1421, 1422
JDC. *See* PDC
Jitter, 408–409, 775, 785
Joint Bilevel Image Group. *See* JBIG
Joint density and mass functions, 51–53
JPEG (Joint Photographic Experts Group)
  coder, 1437, 1445–1447
  compression standard, 1409, 1441–1443

K

Karhunen–Loäve expansion, 142–143; defined, **153**
Kasami sequences, 99, 104
Kerdock sequences, 100–101
Key telephones, 440
Kirchiff's law, 944
Klystron amplifiers, 927, 932
Knife-edge diffraction, 1186–1187
Kraft inequality, 226–227
KTS (key telephone system), 440
Ku-band
  defined, **1065**
  satellites, 923–924, 927–928
  TDRS use, 1056

L

LANs (local area networks), 681–685
  architectures, 681–684
  BLANs, 611–621
  CSMA/CD, 641–642
  defined, **681**
  devices, 446–447
  fiber optics use, 876–878
  internetworking, 701–716
  MAC sublayer, 629
  MAN interconnect, 684
  new applications, 618–619
  PBXs, 681, 683–684
  shared-media, 611
  token-bus, 681, 682
  token-ring, 634–635, 681, 682
  as video servers, 1511–1512
  virtual (VLANs), 617
  wireless, 681, 683, 1367–1379
Large-scale models, radio propagation, 1195
Lasers, 751–773
  designs, 752–755
  DFB (distributed feedback), 760–764
  QW (quantum well), 756–760
  reliability of, 768–770
  semiconductor, 751–773
  SEL (surface emitting), 764–768
  tunable, 762–764
Laser diodes, 774–781
  characteristics, 775–777
  for coherent systems, 862–868
  driving, 777–780
  package design, 780–781
  phase and intensity noise in, 867–868
Laser power modulation. *See* LPM
Laser rate equations, 776–777
Laser threshold current, 754, 775
Laser transmitters, 774–782
LATA (local access and transport area), 293; defined, **300**
Lattices, 282–284; defined, **287**
LCMR
  estimating, 1134–1141
  in PCS caching strategies, 1131–1134, 1142–1143
LEDs (light emitting diodes), 751, 774
Lempel-Ziv codes. *See* LZ codes
LEO (low Earth orbit) antennas, 1032, 1048–1049
LEO satellites, 893, 895–897
Level codes, 387
L/I (light vs current) characteristic, 775, 779
Light emitting diodes. *See* LEDs
Linear code, 168; defined, **179**
Linear combination, 119
Linear equalizers, 348–356
Linear manifold, 122
Linear PCM codec-filters, 374–376
Linear prediction coding. *See* LPC
Linear predictive compression, 1406
Linear space, 118
Linearly independent vectors, 119
Line coding, 386–403
  bandwidth comparison, 401–402
  common formats, 387–394
  considerations affecting choice of a code, 387

Page on which term is defined is indicated in bold.

Index

Switching
    ATD, 686, 687
    ATDM, 687
    ATM, 688–700
    banyan, 692–698
    BLAN/fast packet, 613–615
    conflicts in, 689
    data link, 705
    elements, 688
    for MACs, 623, 628–629
    mechanical optical, 830
    OB (onboard), 976–995
    optical techniques (photonics), 513–528
    queueing for, 688–699
    SS (satellite), 976–995
    in telephone system, 296–297
Switching fabrics, 425–432
    circuit switching, 425–426
    control of, 429–430
    crosspoints, 426–427
    division types, 427
    multistage, 439
    packet switching, 425–426
    time-division, 430–432
Switch models, 688–689
Symbol error probability, 276–277, 278, 281
Symbol-spaced equalizer, 349
Symbol synchronization, 255, 264–268
SYN (synchronization channel)
    for CT-2, 1293–1294, 1300
    defined, **1274**
Synchronization, 255–272
    carrier, 255, 256–264
    defined, **255**
    frame, 255, 268–271
    for GSM, 1236
    of networks, 404–413
    symbol, 255, 264–268
    see also Network synchronization
Synchronization channel. *See* SYN
Synchronous digital hierarchy. *See* SDH
Synchronous optical network. *See* SONET
Synchronous payload envelope. *See* SPE
Synchronous signals, definition, 542
Synchronous traffic, 612
Synchronous transport signal level-N. *See* STS-N
Synch symbol, 468
Syndrome source coding, 1402
Systematic code, 168; defined, **179**
System 5 switching, 520–525
System temperature, 966–975
    defined, **967**
    for microwave receiver, 966–975

T

Taboo channels, 1463; defined, **1473**
TACS systems, 1276
Tagging, TV, 1498; defined, **1501**

Tandem banyan, 693, 696
Tandem coding, 1340
Tap gain correlation function, 133
Target token rotation time. *See* TTRT
Tau-dither loop, 201
TC (transform coding), 1439, 1480, 1481–1482; defined, **1486**
TCAP (transaction capabilities application part), 491–495
TCM (trellis-coded modulation), 173, 176–177
TCP (transport control protocol), 651–652, 657
TCP/IP (transmission control protocol/internet protocol), 574–575, 674–675, 702–704
TDM (time-division multiplexing)
    basic principles, 89–90
    defined, **93**
    digital switching, 404
    for video transfer, 1506
TDM optical ring switch, 991–992
TDMA (time-division multiple access), 92
    advanced system ATDMA, 1331–1332
    base station subsystem for, 1090–1094
    channel assignment, 1110–1111, 1175–1176
    DAMA-TDMA, 977
    defined, **93**, **1115**
    for Japanese PDC, 1283
    MAC control schemes, 631–632
    for microcellular radio, 1172–1173
    for mobile radio, 1086
    narrowband, for mobile satellites, 1025
    propagation considerations, 1110
    as radio access method, 1109–1112
    R-TDMA, 1372
    spectral efficiency, 1260
TDRSS (Tracking and Data Relay Satellite System), 1054–1066
    antennas for, 1043–1046
    communications design, 1056–1058
    link budget, 1064–1065
    relay satellite, 1058–1064
    system overview, 1055–1056
    user services, 1055–1056
Telecommunications Industry Association. *See* TIA
Telephone networks
    analog, 308–317
    CT (cordless), 1072, 1076–1077, 1109
    PCM codec-filters for, 365–367
    PSTN diagram, 365–366
    satellite-based, 908–910
    services offered, 365–366
Telephones
    analog, 438–439
    CPE (customer premises equipment), 433–449
    electronic, 439
    ISDN, 446
    key, 440
Telephone service, plain old. *See* POTS
Teleport Chicago Earth Station, 922–923
Teleservices, 578
Television. *See* Video

Page on which term is defined is indicated in bold.

Temperature for microwave receiver, 966–975
10BROAD36, 593–594
Terminal equipment, FCC definitions, 433, 436
Terminal mobility, 1117–1120
Terminal registration
   in GSM, 1126
   in IS-41, 1121–1122
Ternary coding, 399–400
Terrestrial broadcast
   defined, **1473**
   system requirements, 1463
   vs satellite communications, 899
Terrestrial mobile radio, 1073
Terrestrial networks, 913, 1073
TETRA (trans-European trunked radio), 1391–1392
   air interface characteristics, 1212
   described, 1219
Text compression, 1403–1406
TH (time hopping) modulation, 203–206
Thermal noise, 950, 1271
Thermal noise density, 966–967
Thermoelectric cooler, 780–781
Thermomagnetic recording, 1558–1561; defined, **1564**
Threshold current, 754, 775
TIA (Telecommunications Industry Association), 162, 1259
TIA/EIA/IS-95-A. See IS-95
Time-ambiguity function, 68; defined, **77**
Time averages for waveforms, 57–61
Time diversity, 216–217, 1263
Time hopping modulation. See TH modulation
Time-division multiple access systems. See TDMA systems
Time-division multiplexing. See TDM
Time-division switching, 430–432
Time-frequency correlation function, 133
Time-limited signal, 21
Time-selective fading, 135; defined, **139**
Timeslot number. See TN
Time-space-time switching. See T-S-T switching
Time variance. See TVAR
Timing loops, 406
TLP (transmission level point), 371–372
TM (terminal multiplexer), 551
TMS (time multiplex switching), 431
TN (timeslot number), 1229
Token bus, 681, 682
Token passing schemes, 634–637
Token ring, 681, 682
Token ring networks, 597, 634–635
Token rotation time, 598
Tomlinson-Harashima precoding, 335
T-1 carrier, 90, 506–507; defined, **93**
Total network engineering process, 706–712
Total probability of error, 142; defined, **153**
Touchtone, 294
Tracking, 1547–1558
   automatic, 1556–1558
   concept of, 1549–1552
   defined, **1564**

   on grooves, 1557
   telemetry, 1046–1048
Tracking, telemetry, and command. See TT&C
Tracking algorithm, 201
Tracking and data relay satellite system. See TDRSS
Tracks, disk, 1547–1552
Traffic
   classes of, 612
   theory of, 1150–1151
Traffic channel, 1260; defined, **1274**
Traffic engineering, cellular, 1150–1151
Training mode, 354
Trans-European trunked radio. See TETRA
Transaction, defined, **493**
Transaction capabilities application part. See TCAP
Transducers, optical techniques, 880–881
Transform coding. See TC
Transition codes, 387
Transition noise, 1534–1535; defined, **1544**
Translator, satellite, 915
Transmission control protocol/internet protocol. See TCP/IP
Transmission impairments, satellite. See Satellite transmission impairments
Transmission level point. See TLP
Transmitters, optical. See Optical transmitters
Transoceanic networks, 887–888
Transparent link design, 949–962
Transponders, 912–917, 922–923
   defined, **949**
   TDRSS, 1062–1063
   transparent (bent pipe), 949
Transport control protocol. See TCP
Transport entities, 661
Transport layer, 661–671
   connection-management protocol, 665–667
   correctness properties, 662–663
   data-transfer protocol (sliding-window method), 663–664
   OSI, 573
   service, 661–663
   transport protocols, 667–669
Trap, TV, 1498; defined, **1501**
Traveler services. See Mobile radio
Traveling wave amplifier. See TWA
Traveling wave tube amplifiers. See TWT amplifiers
Trellis-coded modulation. See TCM
Tributaries, SDH, 556, 559–561
TRT (target rotation time), 636–637
Trunk capacity enhancement, for WDMs, 888–889
Trunking efficiency, 1148; defined, **1158**
Trunk reservation, 655
TSI (time-slot interchange), 431
T-S-T (time-space-time) switching, 984–986
TSUNAMI (smart antenna technology for advance mobile infrastructure), 1335
TT&C (tracking, telemetry, and command) antennas, 1046–1048
TTC (Japanese Telecommunications Technology Committee), 162–163

Page on which term is defined is indicated in bold.

Index

TTRT (target token rotation time), 599–609, 636–637
TU-3 pointer, 556–557
TU-*n* (tributary unit-*n*), 556
TUG-*n* (tributary unit group-*n*), 556
Tunable lasers, 762–764
TV. *See* Video
TVAR (time variance), 409–410
TWA (traveling wave amplifier), 834, 836–838
Twisted pair wire, 682
TWT (traveling wave tube) amplifiers, 915, 927, 932

U

UCD (uniform call distributor), 443–444
UMTS (universal mobile telecommunications system), 1223, 1328–1335
UNI (user-network interfaces) for ISDN, 537–538, 579–581, 583–584
Unipolar signalling, 387, 388–390, 401
Unique word, 1027
Universal codes, 1401–1403
Universal mobile telecommunications system. *See* UMTS
Universal source coding, 228–229
UNIX, 703
Unsaturated material gain coefficient, 836
Upconverter, 927
Uplink equations, 950–951, 955–962
Uplink signal, 914
UPR (user performance requirements), 1258–1259
UPS (uninterruptable power system), 930
User mobility, 1367, 1370
User-network interfaces. *See* UNI
User performance requirements. *See* UPR

V

VAD (voice activity detection), 1269–1270, 1433
VBR (variable bit rate), 1433–1434, 1452
VC (virtual channel), 586
VC-*n* (virtual container-*n*), 555
VCOs (voltage controlled oscillators), 1497; defined, **1501**
VDSL (very high-speed digital subscriber lines), 451, 473–476
Vector quantization, 1252, 1440–1441
Vector quantizer, 1256
Vector space, 118–130
  and block coding, 167
  defined, **118, 179**
  fundamentals, 118–125
  and signal representations, 125–128
Vector sum excited linear prediction. *See* VSELP
Vehicle information system, 1077
Vertical mode coding, 1414
Very large scale integration. *See* VLSI
Vestigial-sideband modulation. *See* VSB
Video

audio coding, 1475–1487
cable TV, 1488–1501
DVC (desktop videoconferencing), 1516–1527
HDTV (high-definition), 1462–1474
satellite use by, 894, 904–906
servers, 1502–1515
transponder systems, 916–917
Video coding
  H.261 standard, 1519–1520
  H.263 standard, 1524
  MBONE tools, 1522–1523
Video compression, 1449–1461
  coding algorithms, 1453–1458
  source characteristics, 1450–1453
  standards, 1459–1460
  viewer requirements, 1450–1453
Videoconferencing, desktop. *See* DVC
Video-on-demand. *See* VOD
Video servers, 1502–1515
  data server, 1503–1508
  network multimedia, 1511–1513
  system model, 1504
  video networks, 1508–1511
Video signal, 1449; defined, **1460**
Virtual channel. *See* VC
Virtual circuit, 702–703
Virtual circuit networks
  defined, **651**
  flow control in, 657–659
  routing in, 655
Virtual circuit switching, 426
Virtual containers, 555–557
Virtual path. *See* VP
Virtual tributary. *See* VT
Viterbi algorithm, 174–176; defined, **179**
Viterbi-equivalent detection, 1536; defined, **1544**
VLANs (virtual LANs), 618
VLC (variable length codes), 1468
VLSI (very large scale integration), 683
Vocoders, 1426, 1432–1433; defined, **1435**
VOD (video-on-demand), 1502–1503; defined, **1514**
Voice, digital, 1083–1084
Voice activity detection. *See* VAD
Voice band, 308–309; defined, **316**
Voice-channel bandwidth, 301–302
Voice coding. *See* Audio coding
Voice mail, 443
Voice PCM codec-filters, 367–374
  filtering, 367–368
  gain variations, 373–374
  idle channel noise, 372–373
  quantizing distortion, 368–371
  TLP (transmission level point), 371–372
Voice processing systems, 443
Volna satellite system, 1075
Voltage controlled oscillators. *See* VCOs
VP (virtual path), 586
VSAT (very small aperture terminal), 898–900, 908–910, 1010
VSB (vestigial-sideband) modulation, 8–9; defined, **12**

Page on which term is defined is indicated in bold.

VSELP (vector-sum excited linear prediction), 1248–1252, 1346–1350
VT (virtual tributaries), 544, 546–548

W

Walfish and Bertoni model, 1192
WAN (wide area network), 567, 677
Wander, 408–409
WARC'92 (World Administration Radio Conference), 1223, 1329
Watson, Thomas A., 292, 294, 295
Waveform coders, 1426; defined, **1435**
Waveforms, finite power, 46–51
Waveform time averages, 57–61
Waveguides, fiber optic, 880–881
Wavelength-division multiple access. *See* WDMA
Wavelength-routed networks, 879–880
Wavelet transforms, 127–128; defined, **130**
WDM (wave division multiplexing) systems, 883–890
    applications, 887–888
    baseband switches, 991–993
    design, 885–887
    limitations on, 885–887
    networking, 888–889
    optical components, 883–885
    optical networks, 879–880
    optical transport layer, 888–889
WDMA (wavelength-division multiple access)
    MAC control schemes, 632
    multichannel amplification, 844
Weighting, 310–311
Western Electric mastergroup, 304
White Sands complex, 1061, 1063
Wide area network. *See* WAN

Wideband DSC applications, 503–506
Wideband speech, 1519; defined, **1526**
Wireless data systems, 1380–1394
    CDPD (cellular networks and services), 1387–1391
    characteristics of, 1381–1383, 1392
    modem services, 1383–1384
    private networks (ARDIS and MOBITEX), 1384–1387
    radio propagation, 1381–1383
    TETRA, 1391–1392
Wireless LANS, 1367–1379
    MAC layer, 1369–1373
    network layer, 1373–1376
    physical layer, 1368–1369
    proposed architecture, 1373–1376
    transport layer, 1376–1377
Wireless networks
    channel assignment, 1175–1181
    LANs, 681–685
    multiple-access schemes, 642–644
    propagation models, 1182–1196
    proposed architecture, 1375–1376
World Administration Radio Congress. *See* WARC'92
WWW (Worldwide Web), and video support, 1512–1513

X

XGM (cross-gain modulation), 844

Z

Zero-forcing, 322
Zero-forcing equalizers, 349–350, 356
Ziv-Lempel codes. *See* LZ codes

Page on which term is defined is indicated in bold.